RSMeans

Building Construction Cost Data

57th Annual Edition

1999

Senior Editor

Phillip R. Waier, PE

Contributing Editors

Thomas J. Akins
Barbara Balboni
Howard M. Chandler
John H. Chiang, PE
Paul C. Crosscup
Jennifer L. Curran
Stephen E. Donnelly
J. Robert Lang
Robert C. McNichols
Robert W. Mewis
Melville J. Mossman, PE
John J. Moylan
Jeannene D. Murphy
Peter T. Nightingale
Jesse R. Page
Stephen C. Plotner
Michael J. Regan
Kornelis Smit
William R. Tennyson, II

Manager, Engineering Operations

John H. Ferguson, PE

President

Durwood S. Snead

Vice President and General Manager

Roger J. Grant

Vice President, Sales and Marketing

John M. Shea

Production Manager

Michael Kokernak

Production Coordinator

Marion E. Schofield

Technical Support

Michele S. Able
Wayne D. Anderson
Thomas J. Dion
Michael H. Donelan
Gary L. Hoitt
Paula Reale-Camelio
Kathryn S. Rodriguez
James N. Wills

Art Director

Helen A. Marcella

Book & Cover Design

Norman R. Forgit

Editorial Advisory Board

First Printing

Foreword

R.S. Means Co., Inc. is owned by CMD Group, a leading worldwide provider of proprietary construction information. CMD Group is comprised of three synergistic product groups crafted to be the complete resource for reliable, timely and actionable construction market data. In North America, CMD Group encompasses: Architects' First Source, an innovative product selection and specification solution in print and on the Internet; Construction Market Data (CMD), the source for construction activity information, as well as early planning reports for the design community; Associated Construction Publications, with 14 magazines, one of the largest editorial networks dedicated to U.S. highway and heavy construction coverage; Manufacturer's Survey Associates (MSA), a leading estimating and quantity survey firm in the U.S.; R.S. Means, the authority on construction cost data in North America; CMD Canada, the leading supplier of project information, industry news and forecasting data products for the Canadian construction industry; and BIMSA/Mexico, the dominant distributor of information on building projects and construction throughout Mexico. Worldwide, CMD Group includes Byggfakta Scandinavia, providing construction market data to Denmark, Estonia, Finland, Norway and Sweden; and Cordell Building Information Services, the market leader for construction and cost information in Australia.

Our Mission

Since 1942, R.S. Means Company, Inc. has been actively engaged in construction cost publishing and consulting throughout North America.

Today, over fifty years after the company began, our primary objective remains the same: to provide you, the construction and facilities professional, with the most current and comprehensive construction cost data possible.

Whether you are a contractor, an owner, an architect, an engineer, a facilities manager, or anyone else who needs a quick construction cost estimate, you'll find this publication to be a highly useful and necessary tool.

Today, with the constant flow of new construction methods and materials, it's difficult to find the time to look at and evaluate all the different construction cost possibilities. In addition, because labor and material costs keep changing, last year's cost information is not a reliable basis for today's estimate or budget.

That's why so many construction professionals turn to R.S. Means. We keep track of the costs for you, along with a wide range of other key information, from city cost indexes . . . to productivity rates . . . to crew composition . . . to contractor's overhead and profit rates.

R.S. Means performs these functions by collecting data from all facets of the industry, and organizing it in a format that is instantly accessible to you. From the preliminary budget to the detailed unit price estimate, you'll find the data in this book useful for all phases of construction cost determination.

The Staff, the Organization, and Our Services

When you purchase one of R.S. Means' publications, you are in effect hiring the services of a full-time staff of construction and engineering professionals.

Our thoroughly experienced and highly qualified staff works daily at collecting, analyzing, and disseminating comprehensive cost information for your needs. These staff members have years of practical construction experience and engineering training prior to joining the firm. As a result, you can count on them not only for the cost figures, but also for additional background reference information that will help you create a realistic estimate.

The Means organization is always prepared to help you solve construction problems through its five major divisions: Construction and Cost Data Publishing, Electronic Products and Services, Consulting Services, Insurance Division, and Educational Services.

Besides a full array of construction cost estimating books, Means also publishes a number of other reference works for the construction industry. Subjects include construction estimating and project and business management; special topics such as HVAC, roofing, plumbing, and hazardous waste remediation; and a library of facility management references.

In addition, you can access all of our construction cost data through your computer with Means CostWorks '99 CD-ROM, an electronic tool that offers over 50,000 lines of Means construction cost data.

What's more, you can increase your knowledge and improve your construction estimating and management performance with a Means Construction Seminar or In-House Training Program. These two-day seminar programs offer unparalleled opportunities for everyone in your organization to get updated on a wide variety of construction-related issues.

Means also is a worldwide provider of construction cost management and analysis services for commercial and government owners and of claims and valuation services for insurers.

In short, R.S. Means can provide you with the tools and expertise for constructing accurate and dependable construction estimates and budgets in a variety of ways.

Robert Snow Means Established a Tradition of Quality That Continues Today

Robert Snow Means spent years building his company, making certain he always delivered a quality product.

Today, at R.S. Means, we do more than talk about the quality of our data and the usefulness of our books. We stand behind all of our data, from historical cost indexes... to construction materials and techniques... to current costs.

If you have any questions about our products or services, please call us toll-free at 1-800-334-3509. Our customer service representatives will be happy to assist you.

Table of Contents

UNIT PRICES

GENERAL REQUIREMENTS	1
SITE WORK	2
CONCRETE	3
MASONRY	4
METALS	5
WOOD & PLASTICS	6
THERMAL & MOISTURE PROTECTION	7
DOORS & WINDOWS	8
FINISHES	9
SPECIALTIES	10
EQUIPMENT	11
FURNISHINGS	12
SPECIAL CONSTRUCTION	13
CONVEYING SYSTEMS	14
MECHANICAL	15
ELECTRICAL	16
SQUARE FOOT	17

REFERENCE INFORMATION

REFERENCE NUMBERS

CREWS

COST INDEXES

INDEX

How the Book Is Built: An Overview

A Powerful Construction Tool

You have in your hands one of the most powerful construction tools available today. A successful project is built on the foundation of an accurate and dependable estimate. This book will enable you to construct just such an estimate.

For the casual user the book is designed to be:

- quickly and easily understood so you can get right to your estimate
- filled with valuable information so you can understand the necessary factors that go into the cost estimate

For the regular user, the book is designed to be:

- a handy desk reference that can be quickly referred to for key costs
- a comprehensive, fully reliable source of current construction costs and productivity rates, so you'll be prepared to estimate any project
- a source book for preliminary project cost, product selections, and alternate materials and methods

To meet all of these requirements we have organized the book into the following clearly defined sections.

Quick Start

This one-page section (see following page) can quickly get you started on your estimate.

How To Use the Book: The Details

This section contains an in-depth explanation of how the book is arranged . . . and how you can use it to determine a reliable construction cost estimate. It includes information about how we develop our cost figures and how to completely prepare your estimate.

Unit Price Section

All cost data has been divided into the 16 divisions according to the MasterFormat system of classification and numbering as developed by the Construction Specifications Institute (CSI) and Construction Specifications Canada (CSC). For a listing of these divisions and an outline of their subdivisions, see the Unit Price Section Table of Contents.

Estimating tips are included at the beginning of each division.

Division 17: Quick Project Estimates:
In addition to the 16 Unit Price Divisions there is a S.F. (Square Foot) and C.F. (Cubic Foot) Cost Division, Division 17. It contains costs for 58 different building types that allow you to make a rough estimate for the overall cost of a project or its major components.

Reference Section

This section includes information on Reference Numbers, Change Orders, Crew Listings, Historical Cost Indexes, City Cost Indexes, Location Factors and a listing of Abbreviations. It is visually identified by a vertical gray bar on the edge of pages.

Reference Numbers: At the beginning of selected major classifications in the Unit Price Section are "reference numbers" shown in bold squares. These numbers refer you to related information in the Reference Section.

In this section, you'll find reference tables, explanations, and estimating information that support how we develop the unit price data. Also included are alternate pricing methods, technical data, and estimating procedures, along with information on design and economy in construction. You'll also find helpful tips on what to expect and what to avoid when estimating and constructing your project.

It is recommended that you refer to the Reference Section if a "reference number" appears within the major classification you are estimating.

Change Orders: This section includes information on the factors that influence the pricing of change orders.

Crew Listings: This section lists all the crews referenced in the book. For the purposes of this book, a crew is composed of more than one trade classification and/or the addition of power equipment to any trade classification. Power equipment is included in the cost of the crew. Costs are shown both with bare labor rates and with the installing contractor's overhead and profit added. For each, the total crew cost per eight-hour day and the composite cost per labor-hour are listed.

Historical Cost Indexes: These indexes provide you with data to adjust construction costs over time. If you know costs for a project completed in the past, you can use these indexes to calculate a rough estimate of what it would cost to construct the same project today.

City Cost Indexes: Obviously, costs vary depending on the regional economy. You can adjust the "national average" costs in this book to over 930 locations throughout the U.S. and Canada by using the data in this section. How to use information is included.

Location Factors, to quickly adjust the data to over 930 zip code areas, are included.

Abbreviations: A listing of abbreviations used throughout this book, along with the terms they represent, is included.

Index

A comprehensive listing of all terms and subjects in this book to help you find what you need quickly when you are not sure where it falls in MasterFormat.

The Scope of This Book

This book is designed to be as comprehensive and as easy to use as possible. To that end we have made certain assumptions and limited its scope in three key ways:

1. We have established material prices based on a "national average."
2. We have computed labor costs based on a 30-city "national average" of union wage rates.
3. We have targeted the data for projects of a certain size range.

For a more detailed explanation of how the cost data is developed, see "How To Use the Book: The Details."

Project Size

This book is aimed primarily at commercial and industrial projects costing $1,000,000 and up, or large multi-family housing projects. Costs are primarily for new construction or major renovation of buildings rather than repairs or minor alterations.

With reasonable exercise of judgment the figures can be used for any building work. *For civil engineering structures such as bridges, dams, highways, or the like, please refer to Means Heavy Construction Cost Data.*

Quick Start

If you feel you are ready to use this book and don't think you need the detailed instructions that begin on the following page, this Quick Start section is for you.

These steps will allow you to get started estimating in a matter of minutes.

1 Find each cost data section you need in the Unit Price Section Table of Contents.

The cost data has been divided into 16 divisions according to the CSI MasterFormat.

2 Turn to the indicated section and locate the line item you need for your estimate. Portions of a sample page layout appear here.

- If there is a reference number listed at the beginning of a section, for example, R031-050, it refers to additional information you may find useful. See the Reference Section for detailed information.
- Note the crew code designation. You'll find full descriptions of crews in the Crews Section, including labor-hour and equipment costs.

3 Determine the total number of units your job will require. Note that the unit of measure for the material you're using is listed under "UNIT."

- Bare Costs: These figures show unit costs for materials and installation. Labor and equipment costs are calculated according to crew costs and average daily output. Bare costs do not contain allowances for overhead, profit, or taxes.
- "Labor-hours" allows you to calculate the total labor-hours to complete that task. Just multiply the quantity of work by this figure for an estimate of activity duration.

4 Then multiply the total units by the "Total Incl. O&P," which stands for the total cost including the installing contractor's overhead and profit. (See the next pages for a complete explanation.)

- If the work is to be subcontracted, add the general contractor's markup, approximately 10%.

5 The price you calculate will be an estimate for a completed item of work.

6 Compile a list of all items included in the total project. Summarize cost information, and add project overhead.

Localize costs by using the City Cost Indexes or Location Factors found in the Reference Section.

For a more complete explanation of the way costs are derived, please see the following section.

Commonly Used Abbreviations

R.S. Means utilizes standard industry abbreviations. There is a complete glossary of abbreviations in the reference section. The following are a few of the most commonly used abbreviations you'll find in the book:

B.F.	Board Feet
C	Hundred; Centigrade
C.Y.	Cubic Yard (27 Cubic Feet)
Cwt	100 Pounds
Ea.	Each
Flr.	Floor
L.F.	Linear Foot
Lb.	Pound
MBF	Thousand Board Feet
Opng.	Opening
S.F.	Square Foot
SFCA	Square Foot Contact Area
S.Y.	Square Yard
Sq.	Square; 100 Square Feet
Sty.	Story
Surf.	Surface
V.L.F.	Vertical Linear Foot

Editors' Note: We urge you to spend time reading and understanding the supporting material. An accurate estimate requires experience, knowledge, and careful calculation. The more you know about how we at R.S. Means developed the data, the more accurate your estimate will be. In addition, it's important to take into consideration some of the reference material such as Crews Listing and the "reference numbers."

034 | Precast Concrete

034 100 | Structural Precast

			CREW	DAILY OUTPUT	LABOR-HOURS	UNIT	1999 BARE COSTS				TOTAL INCL O&P	
							MAT.	LABOR	EQUIP.	TOTAL		
132	0010	**LIFT SLAB** See division 033-130-5200	R033 -120									132
136	0010	**SLABS** Prestressed roof/floor members, solid, grouted, 4″ thick	R034 -030	C-11	3,600	.020	S.F.	4.65	.60	.43	5.68	6.65	136
	0050	6″ thick			4,500	.016		4.48	.48	.35	5.31	6.15	
	0100	8″ thick, hollow CN			5,600	.013		4.83	.39	.28	5.50	6.30	
	0150	10″ thick			8,800	.008		5.70	.25	.18	6.13	6.85	
	0200	12″ thick			8,000	.009		4.78	.27	.19	5.24	5.95	
140	0010	**TEES** Prestressed	R034 -030									140
	0020	Quad tee, short spans, roof		C-11	7,200	.010	S.F.	3.95	.30	.22	4.47	5.10	
	0050	Floor			7,200	.010		4	.30	.22	4.52	5.15	
	0200	Double tee, floor members, 60′ span			8,400	.009		6.10	.26	.18	6.54	7.35	

How to Use the Book: The Details

What's Behind the Numbers? The Development of Cost Data

The staff at R.S. Means continuously monitors developments in the construction industry in order to ensure reliable, thorough and up-to-date cost information.

While *overall* construction costs may vary relative to general economic conditions, price fluctuations within the industry are dependent upon many factors. Individual price variations may, in fact, be opposite to overall economic trends. Therefore, costs are continually monitored and complete updates are published yearly. Also, new items are frequently added in response to changes in materials and methods.

Costs—$ (U.S.)

All costs represent U.S. national averages and are given in U.S. dollars. The Means City Cost Indexes can be used to adjust costs to a particular location. The City Cost Indexes for Canada can be used to adjust U.S. national averages to local costs in Canadian dollars.

Material Costs

The R.S. Means staff contacts manufacturers, dealers, distributors, and contractors all across the U.S. and Canada to determine national average material costs. If you have access to current material costs for your specific location, you may wish to make adjustments to reflect differences from the national average. Included within material costs are fasteners for a normal installation. R.S. Means engineers use manufacturers' recommendations, written specifications and/ or standard construction practice for size and spacing of fasteners. Adjustments to material costs may be required for your specific application or location. Material costs do not include sales tax.

Labor Costs

Labor costs are based on the average of wage rates from 30 major U.S. cities. Rates are determined from labor union agreements or prevailing wages for construction trades for the current year. Rates along with overhead and profit markups are listed on the inside back cover of this book.

- If wage rates in your area vary from those used in this book, or if rate increases are expected within a given year, labor costs should be adjusted accordingly.

Labor costs reflect productivity based on actual working conditions. These figures include time spent during a normal workday on tasks other than actual installation, such as material receiving and handling, mobilization at site, site movement, breaks, and cleanup.

Productivity data is developed over an extended period so as not to be influenced by abnormal variations and reflects a typical average.

Equipment Costs

Equipment costs include not only rental, but also operating costs for equipment under normal use. The operating costs include parts and labor for routine servicing such as repair and replacement of pumps, filters and worn lines. Normal operating expendables such as fuel, lubricants, tires and electricity (where applicable) are also included. Extraordinary operating expendables with highly variable wear patterns such as diamond bits and blades are excluded. These costs are included under materials. Equipment rental rates are obtained from industry sources throughout North America—contractors, suppliers, dealers, manufacturers, and distributors.

Crew Equipment Cost/Day—The power equipment required for each crew is included in the crew cost. The daily cost for crew equipment is based on dividing the weekly bare rental rate by 5 (number of working days per week), and then adding the hourly operating cost times 8 (hours per day). This "Crew Equipment Cost/Day" is listed in Subdivision 016.

General Conditions

Cost data in this book is presented in two ways: Bare Costs and Total Cost including O&P (Overhead and Profit). General Conditions, when applicable, should also be added to the Total Cost including O&P. The costs for General Conditions are listed in Division 1 and the Reference Section of this book. General Conditions for the *Installing Contractor* may range from 0% to 10% of the Total Cost including O&P. For the *General* or *Prime Contractor*, costs for General Conditions may range from 5% to 15% of the Total Cost including O&P, with a figure of 10% as the most typical allowance.

Overhead and Profit

Total Cost including O&P for the *Installing Contractor* is shown in the last column on the Unit Price pages of this book. This figure is the sum of the bare material cost plus 10% for profit, the base labor cost plus total overhead and profit, and the bare equipment cost plus 10% for profit. Details for the calculation of Overhead and Profit on labor are shown on the inside back cover and in the Reference Section of this book. (See the "How To Use the Unit Price Pages" for an example of this calculation.)

Factors Affecting Costs

Costs can vary depending upon a number of variables. Here's how we have handled the main factors affecting costs.

Quality—The prices for materials and the workmanship upon which productivity is based represent sound construction work. They are also in line with U.S. government specifications.

Overtime—We have made no allowance for overtime. If you anticipate premium time or work beyond normal working hours, be sure to make an appropriate adjustment to your labor costs.

Productivity—The productivity, daily output, and labor-hour figures for each line item are based on working an eight-hour day in daylight hours in moderate temperatures. For work that extends beyond normal work hours or is performed under adverse conditions, productivity may decrease. (See the section in "How To Use the Unit Price Pages" for more on productivity.)

Size of Project—The size, scope of work, and type of construction project will have a significant impact on cost. Economies of scale can reduce costs for large projects. Unit costs can often run higher for small projects. Costs in this book are intended for the size and type of project as previously described in "How the Book Is Built: An Overview." Costs for projects of a significantly different size or type should be adjusted accordingly.

Location—Material prices in this book are for metropolitan areas. However, in dense urban areas, traffic and site storage limitations may increase costs. Beyond a 20-mile radius of large cities, extra trucking or transportation charges may also increase the material costs slightly. On the other hand, lower wage rates may be in effect. Be sure to consider both these factors when preparing an estimate, particularly if the job site is located in a central city or remote rural location.

In addition, highly specialized subcontract items may require travel and per diem expenses for mechanics.

Other factors—
- season of year
- contractor management
- weather conditions
- local union restrictions
- building code requirements
- availability of:
  - adequate energy
  - skilled labor
  - building materials
- owner's special requirements/restrictions
- safety requirements
- environmental considerations

Unpredictable Factors—General business conditions influence "in-place" costs of all items. Substitute materials and construction methods may have to be employed. These may affect the installed cost and/or life cycle costs. Such factors may be difficult to evaluate and cannot necessarily be predicted on the basis of the job's location in a particular section of the country. Thus, where these factors apply, you may find significant, but unavoidable cost variations for which you will have to apply a measure of judgment to your estimate.

Rounding of Costs

In general, all unit prices in excess of $5.00 have been rounded to make them easier to use and still maintain adequate precision of the results. The rounding rules we have chosen are in the following table.

Prices from . . .	Rounded to the nearest . . .
$.01 to $5.00	$.01
$5.01 to $20.00	$.05
$20.01 to $100.00	$.50
$100.01 to $300.00	$1.00
$300.01 to $1,000.00	$5.00
$1,000.01 to $10,000.00	$25.00
$10,000.01 to $50,000.00	$100.00
$50,000.01 and above	$500.00

Final Checklist

Estimating can be a straightforward process provided you remember the basics. Here's a checklist of some of the items you should remember to do before completing your estimate.

Did you remember to . . .
- factor in the City Cost Index for your locale
- take into consideration which items have been marked up and by how much
- mark up the entire estimate sufficiently for your purposes
- read the background information on techniques and technical matters that could impact your project time span and cost
- include all components of your project in the final estimate
- double check your figures to be sure of your accuracy
- call R.S. Means if you have any questions about your estimate or the data you've found in our publications

Remember, R.S. Means stands behind its publications. If you have any questions about your estimate . . . about the costs you've used from our books . . . or even about the technical aspects of the job that may affect your estimate, feel free to call the R.S. Means editors at 1-781-585-7898 or 1-800-448-8182.

Unit Price Section

Table of Contents

1

How to Use the Unit Price Pages

The following is a detailed explanation of a sample entry in the Unit Price Section. Next to each bold number below is the item being described with appropriate component of the sample entry following in parenthesis. Some prices are listed as bare costs, others as costs that include overhead and profit of the installing contractor. In most cases, if the work is to be subcontracted, the general contractor will need to add an additional markup (R.S. Means suggests using 10%) to the figures in the column "Total Incl. O&P."

1 Division Number/Title (034/Precast Concrete)

Use the Unit Price Section Table of Contents to locate specific items. The sections are classified according to the CSI MasterFormat.

2 Line Numbers (034 136 0100)

Each unit price line item has been assigned a unique 10-digit code based on the 5-digit CSI MasterFormat classification.

MasterFormat Mediumscope
MasterFormat Division

034 136
034 136 0100

Means Subdivision
Means Major Classification
Means Individual Line Number

3 Description (SLABS, etc.)

Each line item is described in detail. Sub-items and additional sizes are indented beneath the appropriate line items. The first line or two after the main item (in boldface) may contain descriptive information that pertains to all line items beneath this boldface listing.

Items which include the symbol **CN** are updated in the Key Material Price Section of the *Design/Build Intelligence* quarterly publication.

4 Reference Number Information

R034 -030 You'll see reference numbers shown in bold squares at the beginning of some major classifications. These refer to related items in the Reference Section, visually identified by a vertical gray bar on the edge of pages.

The relation may be: (1) an estimating procedure that should be read before estimating, (2) an alternate pricing method, or (3) technical information.

The "R" designates the Reference Section. The numbers refer to the MasterFormat classification system.

It is strongly recommended that you review all reference numbers that appear within the major classification you are estimating.

Example: The square number above is directing you to refer to the reference number R034-030. This particular reference number shows delivered and erected costs for standard prestressed, precast concrete shapes and other detail information.

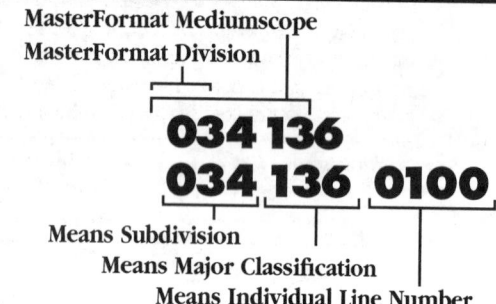

034 | Precast Concrete

		034 100 Structural Precast		CREW	DAILY OUTPUT	LABOR-HOURS	UNIT	1999 BARE COSTS				TOTAL INCL O&P	
								MAT.	LABOR	EQUIP.	TOTAL		
132	0010	**LIFT SLAB** See division 033-130-5200	R033 -120										132
136	0010	**SLABS** Prestressed roof/floor members, solid, grouted, 4" thick	R034 -030	C-11	3,600	.020	S.F.	.65	.60	.43	5.68	6.65	136
	0050	6" thick			4,500	.016		.48	.48	.35	5.31	6.15	
	0100	8" thick, hollow	CN		5,600	.013		.83	.39	.28	5.50	6.30	
	0150	10" thick			6,800	.008		.70	.25	.18	6.13	6.85	
	0200	12" thick			6,000	.009		.78	.27	.19	5.24	5.95	
140	0010	**TEES** Prestressed	R034 -030										140
	0020	Quad tee, short spans, roof		C-11	7,200	.010	S.F.	3.95	.30	.22	4.47	5.10	
	0050	Floor			7,200	.010		4	.30	.22	4.52	5.15	
	0200	Double tee, floor members, 60' span			8,400	.009		6.10	.26	.18	6.54	7.35	

Crew (C-11)

The "Crew" column designates the typical trade or crew used to install the item. If an installation can be accomplished by one trade and requires no power equipment, that trade and the number of workers are listed (for example, "2 Struc. Steel Workers"). If an installation requires a composite crew, a crew code designation is listed (for example, "C-11"). You'll find full details on all composite crews in the Crew Listings.

• For a complete list of all trades utilized in this book and their abbreviations, see the inside back cover.

Crews

Crew No.	Bare Costs		Incl. Subs O & P		Cost Per Labor-Hour	
Crew C-11	Hr.	Daily	Hr.	Daily	Bare Costs	Incl. O&P
1 Struc. Steel Foreman	$32.60	$260.80	$59.75	$478.00	$29.96	$53.07
6 Struc. Steel Workers	30.60	1468.80	56.10	2692.80		
1 Equip. Oper. (crane)	29.35	234.80	44.65	357.20		
1 Equip. Oper. Oiler	24.05	192.40	36.60	292.80		
1 Truck Crane, 150 Ton		1553.00		1708.30	21.57	23.73
72 L.H., Daily Totals		$3709.80		$5529.10	$51.53	$76.80

Productivity: Daily Output (5600)/Labor-Hours (.013)

The "Daily Output" represents the typical number of units the designated crew will install in a normal 8-hour day. To find out the number of days the given crew would require to complete the installation, divide your quantity by the daily output. For example:

Quantity	÷	Daily Output	=	Duration
15,000 S.F.	÷	5600 S.F./ Crew Day	=	2.68 Crew Days

The "Labor-Hours" figure represents the number of labor-hours required to install one unit of work. To find out the number of labor-hours required for your particular task, multiply the quantity of the item times the number of labor-hours shown. For example:

Quantity	x	Productivity Rate	=	Duration
15,000 S.F.	x	.013 Labor-Hours/ S.F.	=	195 Labor-Hours

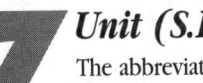

Unit (S.F.)

The abbreviated designation indicates the unit of measure upon which the price, production, and crew are based (S.F. = Square Foot). For a complete listing of abbreviations refer to the Abbreviations Listing in the Reference Section of this book.

Bare Costs:

Mat. (Bare Material Cost) (4.83)

The unit material cost is the "bare" material cost with no overhead and profit included. *Costs shown reflect national average material prices for January of the current year and include delivery to the job site. No sales taxes are included.*

Labor (.39)

The unit labor cost is derived by multiplying bare labor-hour costs for Crew C-11 by labor-hour units. The bare labor-hour cost is found in the Crew Section under C-11. (If a trade is listed, the hourly labor cost—the wage rate—is found on the inside back cover.)

Labor-Hour Cost Crew C-11	x	Labor-Hour Units	=	Labor
$29.96	x	.013	=	$.39

Equip. (Equipment) (.28)

Equipment costs for each crew are listed in the description of each crew. Tools or equipment whose value justifies purchase or ownership by a contractor are considered overhead as shown on the inside back cover. The unit equipment cost is derived by multiplying the bare equipment hourly cost by the labor-hour units.

Equipment Cost Crew C-11	x	Labor-Hour Units	=	Equip.
$21.57	x	.013	=	$.28

Total (5.50)

The total of the bare costs is the arithmetic total of the three previous columns: mat., labor, and equip.

Material	+	Labor	+	Equip.	=	Total
$4.83	+	$.39	+	$.28	=	$5.50

Total Costs Including O&P

This figure is the sum of the bare material cost plus 10% for profit; the bare labor cost plus total overhead and profit (per the inside back cover or, if a crew is listed, from the crew listings); and the bare equipment cost plus 10% for profit.

Material is Bare Material Cost + 10% = 4.83 + .48	=	$5.31
Labor for Crew C-11 = Labor-Hour Cost (53.07) x Labor-Hour Units (.013)	=	$ .68
Equip. is Bare Equip. Cost + 10% = .28 + .03	=	$ .31
Total	=	$6.30

3

Division 1
General Requirements

Estimating Tips

The General Requirements of any contract are very important to both the bidder and the owner. These lay the ground rules under which the contract will be executed and have a significant influence on the cost of operations. Therefore, it is extremely important to thoroughly read and understand the General Requirements both before preparing an estimate and when the estimate is complete, to ascertain that nothing in the contract is overlooked. Caution should be exercised when applying items listed in Division 1 to an estimate. Many of the items are included in the unit prices listed in the other divisions such as mark-ups on labor and company overhead.

010 Overhead & Miscellaneous Data

- Before determining a final cost estimate, it is a good practice to review all the items listed in subdivision 010 to make final adjustments for items that may need customizing to specific job conditions.

013 Submittals

- Requirements for initial and periodic submittals can represent a significant cost to the General Requirements of a job. Thoroughly check the submittal specifications when estimating a project to determine any costs that should be included.

014 Quality Control

- All projects will require some degree of Quality Control. This cost is not included in the unit cost of construction listed in each division. Depending upon the terms of the contract, the various costs of inspection and testing can be the responsibility of either the owner or the contractor. Be sure to include the required costs in your estimate.

015 Construction Facilities & Temporary Controls

- Barricades, access roads, safety nets, scaffolding, security and many more requirements for the execution of a safe project are elements of direct cost. These costs can easily be overlooked when preparing an estimate. When looking through the major classifications of this subdivision, determine which items apply to each division in your estimate.

016 Material & Equipment

- This subdivision contains transportation, handling, storage, protection and product options and substitutions. Listed in this cost manual are average equipment rental rates for all types of equipment. This is useful information when estimating the time and materials requirement of any particular operation in order to establish a unit or total cost.
- A good rule of thumb is that weekly rental is 3 times daily rental and that monthly rental is 3 times weekly rental.

- The figures in the column for Crew Equipment Cost represent the rental rate used in determining the daily cost of equipment in a crew. It is calculated by dividing the weekly rate by 5 days and adding the hourly operating cost times 8 hours.

017 Contract Closeout

- When preparing an estimate, read the specifications to determine the requirements for Contract Closeout thoroughly. Final cleaning, record documentation, operation and maintenance data, warranties and bonds, and spare parts and maintenance materials can all be elements of cost for the completion of a contract. Do not overlook these in your estimate.

018 Maintenance

- If maintenance and repair are included in your contract, they require special attention. To estimate the cost to remove and replace any unit usually requires a site visit to determine the accessibility and the specific difficulty at that location. Obstructions, dust control, safety, and often overtime hours must be considered when preparing your estimate.

Reference Numbers

Reference numbers are shown in bold squares at the beginning of some major classifications. These numbers refer to related items in the Reference Section. The reference information may be an estimating procedure, an alternate pricing method or technical information.

Note: Not all subdivisions listed here necessarily appear in this publication.

GENERAL REQUIREMENTS — **1**

010 000	Overhead		CREW	DAILY OUTPUT	LABOR-HOURS	UNIT	1999 BARE COSTS				TOTAL INCL O&P	
							MAT.	LABOR	EQUIP.	TOTAL		
004	0011	**ARCHITECTURAL FEES** R010 -010										**004**
	0020	For new construction										
	0060	Minimum				Project					4.90%	
	0090	Maximum									16%	
	0100	For alteration work, to $500,000, add to fee									50%	
	0150	Over $500,000, add to fee									25%	
012	0010	**CONSTRUCTION COST INDEX** (Reference) over 930 zip code locations in										**012**
	0020	The U.S. and Canada, total bldg cost, min. (Dalton, GA)				%					66.90%	
	0050	Average									100%	
	0100	Maximum (New York, NY)									133.90%	
014	0010	**CONSTRUCTION ECONOMIES** For bricklaying R042 -100										**014**
016	0010	**CONSTRUCTION MANAGEMENT FEES** $1,000,000 job, minimum				Project					4.50%	**016**
	0050	Maximum									7.50%	
	0300	$5,000,000 job, minimum									2.50%	
	0350	Maximum									4%	
018	0010	**CONSTRUCTION TIME** Requirements R010 -020										**018**
020	0010	**CONTINGENCIES** Allowance to add at conceptual stage				Project					15%	**020**
	0050	Schematic stage									10%	
	0100	Preliminary working drawing stage									7%	
	0150	Final working drawing stage									2%	
022	0010	**CONTRACTOR EQUIPMENT** See division 016 R016 -410										**022**
024	0010	**CREWS** For building construction, see How To Use This Book										**024**
028	0010	**ENGINEERING FEES** R010 -030										**028**
	0020	Educational planning consultant, minimum				Project					.50%	
	0100	Maximum				"					2.50%	
	0200	Electrical, minimum				Contrct					4.10%	
	0300	Maximum									10.10%	
	0400	Elevator & conveying systems, minimum									2.50%	
	0500	Maximum									5%	
	0600	Food service & kitchen equipment, minimum									8%	
	0700	Maximum									12%	
	0800	Landscaping & site development, minimum									2.50%	
	0900	Maximum									6%	
	1000	Mechanical (plumbing & HVAC), minimum									4.10%	
	1100	Maximum									10.10%	
	1200	Structural, minimum				Project					1%	
	1300	Maximum				"					2.50%	
034	0010	**FIELD OFFICE EXPENSE**										**034**
	0100	Field office expense, office equipment rental average				Month	133			133	146	
	0120	Office supplies, average				"	85			85	93.50	
	0125	Office trailer rental, see division 015-904										
	0140	Telephone bill; avg. bill/month incl. long dist.				Month	235			235	259	
	0160	Field office lights & HVAC				"	88			88	97	
036	0010	**FIELD PERSONNEL** Clerk average				Week		260		260	410	**036**
	0100	Field engineer, minimum						625		625	985	
	0120	Average						815		815	1,285	
	0140	Maximum						935		935	1,475	
	0160	General purpose laborer, average						860		860	1,355	
	0180	Project manager, minimum						1,180		1,180	1,860	
	0200	Average						1,320		1,320	2,085	
	0220	Maximum						1,490		1,490	2,350	

Important: See the Reference Section for critical supporting data - Reference Nos., Crews, & City Cost Indexes

010 000	Overhead		CREW	DAILY OUTPUT	LABOR-HOURS	UNIT	1999 BARE COSTS MAT.	LABOR	EQUIP.	TOTAL	TOTAL INCL O&P	
036	0240	Superintendent, minimum				Week		1,125		1,125	1,775	**036**
	0260	Average						1,245		1,245	1,965	
	0280	Maximum						1,405		1,405	2,215	
	0290	Timekeeper, average				↓		725		725	1,145	
038	0010	**HISTORICAL COST INDEXES** (Reference) Back to 1949										**038**
040	0010	**INSURANCE** Builders risk, standard, minimum	R010 -040			Job					.22%	**040**
	0050	Maximum									.59%	
	0200	All-risk type, minimum	R010 -050								.25%	
	0250	Maximum				↓					.62%	
	0400	Contractor's equipment floater, minimum	R010 -060			Value					.50%	
	0450	Maximum				"					1.50%	
	0600	Public liability, average				Job					1.55%	
	0800	Workers' compensation & employer's liability, average										
	0850	by trade, carpentry, general				Payroll		19.92%				
	0900	Clerical						.54%				
	0950	Concrete						19.12%				
	1000	Electrical						7.04%				
	1050	Excavation						11.59%				
	1100	Glazing						14.71%				
	1150	Insulation						18.86%				
	1200	Lathing						12.43%				
	1250	Masonry						18.01%				
	1300	Painting & decorating						15.62%				
	1350	Pile driving						30.30%				
	1400	Plastering						15.84%				
	1450	Plumbing						8.99%				
	1500	Roofing						34.61%				
	1550	Sheet metal work (HVAC)						12.77%				
	1600	Steel erection, structural						42.77%				
	1650	Tile work, interior ceramic						10.33%				
	1700	Waterproofing, brush or hand caulking						8.46%				
	1800	Wrecking						42.49%				
	2000	Range of 35 trades in 50 states, excl. wrecking, minimum						2%				
	2100	Average		↓		↓		18.30%				
	2200	Maximum						132.92%				
042	0010	**JOB CONDITIONS** Modifications to total										**042**
	0020	project cost summaries										
	0100	Economic conditions, favorable, deduct				Project					2%	
	0200	Unfavorable, add									5%	
	0300	Hoisting conditions, favorable, deduct									2%	
	0400	Unfavorable, add									5%	
	0500	General Contractor management, experienced, deduct									2%	
	0600	Inexperienced, add									10%	
	0700	Labor availability, surplus, deduct									1%	
	0800	Shortage, add									10%	
	0900	Material storage area, available, deduct									1%	
	1000	Not available, add									2%	
	1100	Subcontractor availability, surplus, deduct									5%	
	1200	Shortage, add									12%	
	1300	Work space, available, deduct									2%	
	1400	Not available, add				↓					5%	
046	0010	**LABOR INDEX** (Reference) For over 930 zip code locations in										**046**
	0020	the U.S. and Canada, minimum (Dalton, GA)				%		35%				
	0050	Average						100%				
	0100	Maximum (New York, NY)				↓		160.60%				

GENERAL REQUIREMENTS

1

		010 000	Overhead	CREW	DAILY OUTPUT	LABOR-HOURS	UNIT	MAT.	LABOR	EQUIP.	TOTAL	TOTAL INCL O&P	
									1999 BARE COSTS				
048	0010	**MAIN OFFICE EXPENSE** Average for General Contractors	R010 -050										048
	0020	As a percentage of their annual volume											
	0125	Annual volume under 1 million dollars					% Vol.				13.60%		
	0145	Up to 2.5 million dollars									8%		
	0150	Up to 4.0 million dollars									6.80%		
	0200	Up to 7.0 million dollars									5.60%		
	0250	Up to 10 million dollars									5.10%		
	0300	Over 10 million dollars									3.90%		
052	0010	**MARK-UP** For General Contractors for change	R010 -070										052
	0100	of scope of job as bid											
	0200	Extra work, by subcontractors, add					%					10%	
	0250	By General Contractor, add										15%	
	0400	Omitted work, by subcontractors, deduct										5%	
	0450	By General Contractor, deduct										7.50%	
	0600	Overtime work, by subcontractors, add										15%	
	0650	By General Contractor, add										10%	
	1000	Installing contractors, on his own labor, minimum							48.80%				
	1100	Maximum							102%				
054	0010	**MATERIAL INDEX** (Reference) For over 930 zip code locations in											054
	0020	the U.S. and Canada, minimum (Elizabethtown, KY)					%	91.80%					
	0040	Average						100%					
	0060	Maximum (Ketchikan, AK)						140.60%					
058	0010	**OVERHEAD** As percent of direct costs, minimum	R010 -050				%				5%		058
	0050	Average									12%		
	0100	Maximum	R010 -070								30%		
062	0010	**OVERHEAD & PROFIT** Allowance to add to items in this											062
	0020	book that do not include Subs O&P, average					%				25%		
	0100	Allowance to add to items in this book that											
	0110	do include Subs O&P, minimum					%					5%	
	0150	Average										10%	
	0200	Maximum										15%	
	0300	Typical, by size of project, under $100,000									30%		
	0350	$500,000 project									25%		
	0400	$2,000,000 project									20%		
	0450	Over $10,000,000 project									15%		
064	0010	**OVERTIME** For early completion of projects or where	R010 -110										064
	0020	labor shortages exist, add to usual labor, up to					Costs		100%				
068	0010	**PERFORMANCE BOND** For buildings, minimum	R010 -080				Job					.60%	068
	0100	Maximum					"					2.50%	
070	0010	**PERMITS** Rule of thumb, most cities, minimum					Job					.50%	070
	0100	Maximum					"					2%	
082	0010	**SMALL TOOLS** As % of contractor's work, minimum	R010 -050				Total					.50%	082
	0100	Maximum					"					2%	
086	0010	**TAXES** Sales tax, State, average	R010 -090				%	4.71%					086
	0050	Maximum						7.25%					
	0200	Social Security, on first $68,400 of wages	R010 -100						7.65%				
	0300	Unemployment, MA, combined Federal and State, minimum							2.60%				
	0350	Average							7%				
	0400	Maximum							8.50%				

Important: See the Reference Section for critical supporting data - Reference Nos., Crews, & City Cost Indexes

		013 100	**Progress Schedules**	CREW	DAILY OUTPUT	LABOR-HOURS	UNIT	MAT.	LABOR	EQUIP.	TOTAL	TOTAL INCL O&P	
												1999 BARE COSTS	
104	0010	**SCHEDULING** Critical path, as % of architectural fee, minimum					%					2%	**104**
	0100	Maximum					"					4%	
	0300	Computer-update, micro, no plots, minimum					Ea.					220	
	0400	Including plots, maximum					"					2,300	
	0600	Rule of thumb, CPM scheduling, small job					Job					.10%	
	0650	Large job					↓					.05%	
	0700	Cost control, small job										.15%	
	0750	Large job					↓					.04%	

		013 300	**Survey Data**										
306	0010	**SURVEYING** Conventional, topographical, minimum	A-7	3.30	7.273	Acre	16	175		191	287	**306**
	0100	Maximum	A-8	.60	53.333		48	1,250		1,298	1,975	
	0300	Lot location and lines, minimum, for large quantities	A-7	2	12		25	288		313	475	
	0320	Average	"	1.25	19.200		45	460		505	760	
	0400	Maximum, for small quantities	A-8	1	32	↓	72	745		817	1,225	
	0600	Monuments, 3' long	A-7	10	2.400	Ea.	19	57.50		76.50	110	
	0800	Property lines, perimeter, cleared land	"	1,000	.024	L.F.	.03	.58		.61	.92	
	0900	Wooded land	A-8	875	.037	"	.05	.85		.90	1.37	
	1100	Crew for building layout, 2 person crew	A-6	1	16	Day		410		410	625	
	1200	3 person crew	A-7	1	24			575		575	885	
	1300	4 person crew	A-8	1	32	↓		745		745	1,150	
	1500	Aerial surveying, including ground control, minimum fee, 10 acres				Total					5,000	
	1510	100 acres									8,300	
	1550	From existing photography, deduct				↓					1,220	
	1600	2' contours, 10 acres				Acre					400	
	1650	20 acres									270	
	1800	50 acres									85	
	1850	100 acres									72	
	2000	1000 acres									16.20	
	2050	10,000 acres				↓					10.35	
	2150	For 1' contours and										
	2160	dense urban areas, add to above				Acre					40%	
	3000	Inertial guidance system for										
	3010	locating coordinates, rent per day				Ea.					4,000	

		013 400	**Shop Drawings**										
406	0010	**MODELS** Cardboard & paper, 1 building, minimum				Ea.	550			550	605	**406**
	0050	Maximum				↓	1,200			1,200	1,325	
	0100	2 buildings, minimum					700			700	770	
	0150	Maximum				↓	1,600			1,600	1,750	
	0200	Plexiglass and metal, basic layout				SF Flr.	.06			.06	.07	
	0210	Including equipment and personnel				"	.23			.23	.25	
	0300	Site plan layout, minimum				Ea.	1,000			1,000	1,100	
	0350	Maximum				"	1,750			1,750	1,925	
408	0010	**RENDERINGS** Color, matted, 20" x 30", eye level,										**408**
	0020	1 building, minimum				Ea.	1,500			1,500	1,650	
	0050	Average					2,500			2,500	2,750	
	0100	Maximum					3,500			3,500	3,850	
	1000	5 buildings, minimum					3,000			3,000	3,300	
	1100	Maximum					6,000			6,000	6,600	
	2000	Aerial perspective, color, 1 building, minimum					2,500			2,500	2,750	
	2100	Maximum					6,000			6,000	6,600	
	3000	5 buildings, minimum					3,000			3,000	3,300	
	3100	Maximum				↓	10,000			10,000	11,000	

1

GENERAL REQUIREMENTS

013 | Submittals

013 800 | Construction Photos

		CREW	DAILY OUTPUT	LABOR-HOURS	UNIT	1999 BARE COSTS MAT.	LABOR	EQUIP.	TOTAL	TOTAL INCL O&P	
804	0010	**PHOTOGRAPHS** 8" x 10", 4 shots, 2 prints ea., std. mounting				Set	95			95	105	804
	0100	Hinged linen mounts					110			110	121	
	0200	8" x 10", 4 shots, 2 prints each, in color					190			190	209	
	0300	For I.D. slugs, add to all above					2.50			2.50	2.75	
	0500	Aerial photos, initial fly-over, 6 shots, 1 print ea., 8" x 10"					815			815	895	
	0550	11" x 14" prints					850			850	935	
	0600	16" x 20" prints					1,000			1,000	1,100	
	0700	For full color prints, add					40%				40%	
	0750	Add for traffic control area				▼	260			260	286	
	0900	For over 30 miles from airport, add per				Mile	4.80			4.80	5.30	
	1000	Vertical photography, 4 to 6 shots with										
	1010	different scales, 1 print each				Set	990			990	1,100	
	1500	Time lapse equipment, camera and projector, buy					3,500			3,500	3,850	
	1550	Rent per month				▼	520			520	570	
	1700	Cameraman and film, including processing, B.&W.				Day	580			580	640	
	1720	Color				"	650			650	715	

014 | Quality Control

014 100 | Testing Services

		CREW	DAILY OUTPUT	LABOR-HOURS	UNIT	1999 BARE COSTS MAT.	LABOR	EQUIP.	TOTAL	TOTAL INCL O&P	
108	0010	**FIELD TESTING**										108
	0015	For concrete building costing $1,000,000, minimum				Project					5,000	
	0020	Maximum									40,000	
	0050	Steel building, minimum									5,000	
	0070	Maximum									15,000	
	0100	For building costing, $10,000,000, minimum									32,000	
	0150	Maximum				▼					50,000	
	0200	Asphalt testing, compressive strength Marshall stability, set of 3				Ea.					165	
	0220	Density, set of 3									87	
	0250	Extraction, individual tests on sample									120	
	0300	Penetration									35	
	0350	Mix design, 5 specimens									200	
	0360	Additional specimen									40	
	0400	Specific gravity									45	
	0420	Swell test									70	
	0450	Water effect and cohesion, set of 6									200	
	0470	Water effect and plastic flow									70	
	0600	Concrete testing, aggregates, abrasion									120	
	0650	Absorption									46	
	0800	Petrographic analysis									850	
	0900	Specific gravity									45	
	1000	Sieve analysis, washed									65	
	1050	Unwashed									65	
	1200	Sulfate soundness									125	
	1300	Weight per cubic foot									40	
	1500	Cement, physical tests									300	
	1600	Chemical tests									270	
	1800	Compressive strength, cylinders, delivered to lab									13	
	1900	Picked up by lab, minimum									15	
	1950	Average				▼					20	

Important: See the Reference Section for critical supporting data - Reference Nos., Crews, & City Cost Indexes

1

GENERAL REQUIREMENTS

		CREW	DAILY OUTPUT	LABOR-HOURS	UNIT	1999 BARE COSTS				TOTAL INCL O&P	
014 100	**Testing Services**					MAT.	LABOR	EQUIP.	TOTAL		
108	2000	Maximum				Ea.					30	108
	2200	Compressive strength, cores (not incl. drilling)				↓					40	
	2250	Core drilling, 4" diameter (plus technician)				Inch					25	
	2260	Technician for core drilling				Hr.					50	
	2300	Patching core holes				Ea.					24	
	2400	Drying shrinkage at 28 days				↓					260	
	2500	Flexural test beams									65	
	2600	Mix design, one batch mix									285	
	2650	Added trial batches									132	
	2800	Modulus of elasticity									150	
	2900	Tensile test, cylinders									50	
	3000	Water-Cement ratio curve, 3 batches									155	
	3100	4 batches									205	
	3300	Masonry testing, absorption, per 5 brick									50	
	3350	Chemical resistance, per 2 brick									55	
	3400	Compressive strength, per 5 brick									70	
	3420	Efflorescence, per 5 brick									50	
	3440	Imperviousness, per 5 brick									96	
	3470	Modulus of rupture, per 5 brick									55	
	3500	Moisture, block only									35	
	3550	Mortar, compressive strength, set of 3									25	
	4100	Reinforcing steel, bend test									61	
	4200	Tensile test, up to #8 bar									35	
	4220	#9 to #11 bar									40	
	4240	#14 bar and larger									50	
	4400	Soil testing, Atterberg limits, liquid and plastic limits									61	
	4510	Hydrometer analysis									120	
	4530	Specific gravity									48	
	4600	Sieve analysis, washed									56.67	
	4700	Unwashed									65	
	4710	Consolidation test (ASTM D2435), minimum									320	
	4715	Maximum									475	
	4720	Density and classification of undisturbed sample									75	
	4735	Soil density, nuclear method, ASTM D2922-71									38.67	
	4740	Sand cone method ASTM D1556064									30.17	
	4750	Moisture content									5.33	
	4780	Permeability test, double ring infiltrometer									550	
	4800	Permeability, variable or constant head, undisturbed									205	
	4850	Recompacted									233.33	
	4900	Proctor compaction, 4" standard mold									125	
	4950	6" modified mold									75	
	5100	Shear tests, triaxial, minimum									450	
	5150	Maximum									600	
	5300	Direct shear, minimum									250	
	5350	Maximum									300	
	5550	Technician for inspection, per day, earthwork									215	
	5650	Bolting									270	
	5750	Roofing									250	
	5790	Welding				↓					260	
	5820	Non-destructive testing, dye penetrant				Day					320	
	5840	Magnetic particle									320	
	5860	Radiography									480	
	5880	Ultrasonic				↓					330	
	6000	Welding certification, minimum				Ea.					100	
	6100	Maximum				"					275	
	7000	Underground storage tank										

014 | Quality Control

014 100 | Testing Services

			DAILY OUTPUT	LABOR-HOURS	UNIT	1999 BARE COSTS				TOTAL INCL O&P	
						MAT.	LABOR	EQUIP.	TOTAL		
108	7500	Hydrostatic tank tightness test per tank, min.			Ea.					500	108
	7510	Maximum			"					1,000	
	7600	Vadose zone (soil gas) sampling, 10-40 samples, min.			Day					1,500	
	7610	Maximum			"					2,500	
	7700	Ground water monitoring incl. drilling 3 wells, min.			Total					5,000	
	7710	Maximum			"					7,000	
	8000	X-ray concrete slabs			Ea.					200	

015 | Construction Facilities & Temporary Controls

015 100 | Temporary Utilities

			CREW	DAILY OUTPUT	LABOR-HOURS	UNIT	1999 BARE COSTS				TOTAL INCL O&P	
							MAT.	LABOR	EQUIP.	TOTAL		
104	0010	**TEMPORARY UTILITIES**										104
	0100	Heat, incl. fuel and operation, per week, 12 hrs. per day	1 Skwk	100	.080	CSF Flr	5	2.24		7.24	9.10	
	0200	24 hrs. per day	"	60	.133		7.55	3.74		11.29	14.20	
	0350	Lighting, incl. service lamps, wiring & outlets, minimum	1 Elec	34	.235		2.03	7.50		9.53	13.45	
	0360	Maximum	"	17	.471		4.42	15		19.42	27.50	
	0400	Power for temp lighting only, per month, min/month 6.6 KWH								.75	1.18	
	0450	Maximum/month 23.6 KWH								2.85	2.85	
	0600	Power for job duration incl. elevator, etc., minimum								47	51.70	
	0650	Maximum								110	121	
	1000	Toilet, portable, see division 016-420-6410										

015 200 | Temporary Construction

			CREW	DAILY OUTPUT	LABOR-HOURS	UNIT	MAT.	LABOR	EQUIP.	TOTAL	TOTAL INCL O&P	
204	0010	**PROTECTION** Stair tread, 2" x 12" planks, 1 use	1 Carp	75	.107	Tread	3.87	2.91		6.78	8.85	204
	0100	Exterior plywood, 1/2" thick, 1 use		65	.123		1.23	3.36		4.59	6.65	
	0200	3/4" thick, 1 use		60	.133		1.75	3.64		5.39	7.65	
208	0010	**TEMPORARY CONSTRUCTION** See also division 015-300										208

015 250 | Construction Aids

			CREW	DAILY OUTPUT	LABOR-HOURS	UNIT	MAT.	LABOR	EQUIP.	TOTAL	TOTAL INCL O&P	
251	0010	**STAGING AIDS** and fall protection equipment										251
	0100	Sidewall staging bracket, tubular, buy				Ea.	30.50			30.50	33.50	
	0110	Cost each per day, based on 250 days use				Day	.12			.12	.13	
	0200	Guard post, buy				Ea.	15			15	16.50	
	0210	Cost each per day, based on 250 days use				Day	.06			.06	.07	
	0300	End guard chains, buy per set				Ea.	25			25	27.50	
	0310	Cost per set per day, based on 250 days use				Day	.12			.12	.13	
	1000	Roof shingling bracket, steel, buy				Ea.	6.40			6.40	7.05	
	1010	Cost each per day, based on 250 days use				Day	.03			.03	.03	
	1100	Wood bracket, buy				Ea.	11.50			11.50	12.60	
	1110	Cost each per day, based on 250 days use				Day	.05			.05	.05	
	2000	Ladder jack, aluminum, buy per pair				Pair	79			79	87	
	2010	Cost per pair per day, based on 250 days use				Day	.32			.32	.35	
	2100	Steel siderail jack, buy per pair				Pair	69			69	76	
	2110	Cost per pair per day, based on 250 days use				Day	.28			.28	.30	
	3000	Laminated wood plank, 2x10x16', buy				Ea.	42.50			42.50	47	
	3010	Cost each per day, based on 250 days use				Day	.17			.17	.19	
	3100	Aluminum scaffolding plank, 20" wide x 24' long, buy				Ea.	660			660	730	

Important: See the Reference Section for critical supporting data - Reference Nos., Crews, & City Cost Indexes

015 250		Construction Aids	CREW	DAILY OUTPUT	LABOR-HOURS	UNIT	1999 BARE COSTS				TOTAL INCL O&P	
							MAT.	LABOR	EQUIP.	TOTAL		
251	3110	Cost each per day, based on 250 days use				Day	2.65			2.65	2.91	251
	4000	Nylon full body harness, lanyard and rope grab				Ea.	233			233	257	
	4010	Cost each per day, based on 250 days use				Day	.93			.93	1.03	
	4100	Rope for safety line, 5/8" x 100' nylon, buy				Ea.	35			35	38.50	
	4110	Cost each per day, based on 250 days use				Day	.14			.14	.15	
	4200	Permanent U-Bolt roof anchor, buy				Ea.	30.50			30.50	33.50	
	4300	Temporary (one use) roof ridge anchor, buy				"	18.60			18.60	20.50	
	5000	Installation (setup and removal) of staging aids										
	5010	Sidewall staging bracket	2 Carp	64	.250	Ea.		6.80		6.80	10.75	
	5020	Guard post with 2 wood rails	"	64	.250			6.80		6.80	10.75	
	5030	End guard chains, set	1 Carp	64	.125			3.41		3.41	5.35	
	5100	Roof shingling bracket		96	.083			2.27		2.27	3.58	
	5200	Ladder jack	↓	64	.125			3.41		3.41	5.35	
	5300	Wood plank, 2x10x16'	2 Carp	80	.200			5.45		5.45	8.60	
	5310	Aluminum scaffold plank, 20" x 24'	"	40	.400			10.90		10.90	17.20	
	5410	Safety rope	1 Carp	40	.200			5.45		5.45	8.60	
	5420	Permanent U-Bolt roof anchor (install only)	2 Carp	40	.400			10.90		10.90	17.20	
	5430	Temporary roof ridge anchor (install only)	1 Carp	64	.125	↓		3.41		3.41	5.35	
252	0010	**SAFETY NETS** No supports, stock sizes, nylon, 4" mesh				S.F.	1.10			1.10	1.21	252
	0100	Polypropylene, 6" mesh					1.55			1.55	1.70	
	0200	Small mesh debris nets, 1/4" & 3/4" mesh, stock sizes					.72			.72	.79	
	0220	Combined 4" mesh and 1/4" mesh, stock sizes					1.80			1.80	1.98	
	0300	Monthly rental, 4" mesh, stock sizes, 1st month					.19			.19	.21	
	0320	2nd month rental					.15			.15	.17	
	0340	Subsequent months rental				↓	.15			.15	.17	
253	0010	**PERSONNEL PROTECTIVE EQUIPMENT**										253
	0015	Hazardous waste protection										
	0020	Respirator mask only, full face, silicone				Ea.	188			188	207	
	0030	Half face, silicone					21.50			21.50	23.50	
	0040	Respirator cartridges, 2 reg'd/mask, dust or asbestos					34.50			34.50	38	
	0050	Chemical vapor					28.50			28.50	31.50	
	0060	Combination vapor and dust					33.50			33.50	37	
	0100	Emergency escape breathing apparatus, 5 min					390			390	425	
	0110	10 min					435			435	475	
	0150	Self contained breathing apparatus with full face piece, 30 min					1,750			1,750	1,925	
	0160	60 min					2,850			2,850	3,150	
	0200	Encapsulating suits, limited use, level A					680			680	750	
	0210	Level B				↓	136			136	149	
	0300	Over boots, latex				Pr.	5.90			5.90	6.50	
	0310	PVC					14.80			14.80	16.30	
	0320	Neoprene					30			30	33	
	0400	Gloves, nitrile/PVC					3.49			3.49	3.84	
	0410	Neoprene coated				↓	19.20			19.20	21	
254	0014	**SCAFFOLDING, STEEL TUBULAR** Rent, 1 use per mo., no plank [R015 -100]										254
	0015	Set up and take down										
	0090	Building exterior, wall face, 1 to 5 stories	3 Carp	24	1	C.S.F.	24.50	27.50		52	69.50	
	0200	6 to 12 stories	4 Carp	21.20	1.509		24.50	41		65.50	91.50	
	0310	13 to 20 stories	5 Carp	20	2		24.50	54.50		79	113	
	0460	Building interior, wall face area, up to 16' high	3 Carp	25	.960		24.50	26		50.50	67.50	
	0560	16' to 40' high		23	1.043	↓	24.50	28.50		53	71.50	
	0800	Building interior floor area, up to 30' high		312	.077	C.C.F.	1.87	2.10		3.97	5.35	
	0900	Over 30' high	4 Carp	275	.116	"	1.87	3.18		5.05	7.05	
	0910	Steel tubular, heavy duty shoring, buy										
	0920	Frames 5' high 2' wide				Ea.	75			75	82.50	
	0925	5' high 4' wide				↓	85			85	93.50	

GENERAL REQUIREMENTS

1

015 250 | Construction Aids

		CREW	DAILY OUTPUT	LABOR-HOURS	UNIT	MAT.	LABOR	EQUIP.	TOTAL	TOTAL INCL O&P	
254						1999 BARE COSTS					**254**
0930	6' high 2' wide	R015 -100			Ea.	86			86	94.50	
0935	6' high 4' wide				↓	101			101	111	
0940	Accessories										
0945	Cross braces				Ea.	16			16	17.60	
0950	U-head, 8" x 8"					16			16	17.60	
0955	J-head, 4" x 8"					12			12	13.20	
0960	Base plate, 8" x 8"					13			13	14.30	
0965	Leveling jack				↓	30.50			30.50	33.50	
1000	Steel tubular, regular, buy										
1100	Frames 3' high 5' wide				Ea.	58			58	64	
1150	5' high 5' wide					67			67	73.50	
1200	6'-4" high 5' wide					84			84	92.50	
1350	7'-6" high 6' wide					145			145	160	
1500	Accessories cross braces					15			15	16.50	
1550	Guardrail post					15			15	16.50	
1600	Guardrail 7' section					7.25			7.25	8	
1650	Screw jacks & plates					24			24	26.50	
1700	Sidearm brackets					28			28	31	
1750	8" casters					33			33	36.50	
1800	Plank 2" x 10" x 16'-0"					42.50			42.50	47	
1900	Stairway section					235			235	259	
1910	Stairway starter bar					21			21	23.50	
1920	Stairway inside handrail					53			53	58.50	
1930	Stairway outside handrail					73			73	80.50	
1940	Walk-thru frame guardrail				↓	28			28	31	
2000	Steel tubular, regular, rent/mo.										
2100	Frames 3' high 5' wide				Ea.	3.75			3.75	4.13	
2150	5' high 5' wide					3.75			3.75	4.13	
2200	6'-4" high 5' wide					3.75			3.75	4.13	
2250	7'-6" high 6' wide					7			7	7.70	
2500	Accessories, cross braces					.60			.60	.66	
2550	Guardrail post					1			1	1.10	
2600	Guardrail 7' section					.75			.75	.83	
2650	Screw jacks & plates					1.50			1.50	1.65	
2700	Sidearm brackets					1.50			1.50	1.65	
2750	8" casters					6			6	6.60	
2800	Outrigger for rolling tower					3			3	3.30	
2850	Plank 2" x 10" x 16'-0"					5			5	5.50	
2900	Stairway section					10			10	11	
2910	Stairway starter bar					.10			.10	.11	
2920	Stairway inside handrail					5			5	5.50	
2930	Stairway outside handrail					5			5	5.50	
2940	Walk-thru frame guardrail				↓	2			2	2.20	
3000	Steel tubular, heavy duty shoring, rent/mo.										
3250	5' high 2' & 4' wide				Ea.	5			5	5.50	
3300	6' high 2' & 4' wide					5			5	5.50	
3500	Accessories, cross braces					1			1	1.10	
3600	U - head, 8" x 8"					1			1	1.10	
3650	J - head, 4" x 8"					1			1	1.10	
3700	Base plate, 8" x 8"					1			1	1.10	
3750	Leveling jack				↓	2			2	2.20	
6000	Heavy duty shoring for suspended slab forms to 8'-2" high, floor area										
6010	Set up and take down										
6100	1 use/month	4 Carp	36	.889	C.S.F.	29.50	24.50		54	70.50	
6150	2 uses/month	"	36	.889	"	14.80	24.50		39.30	54.50	
6500	To 14'-8" high										

Important: See the Reference Section for critical supporting data - Reference Nos., Crews, & City Cost Indexes

		015 250 │ Construction Aids	CREW	DAILY OUTPUT	LABOR-HOURS	UNIT	MAT.	LABOR	EQUIP.	TOTAL	TOTAL INCL O&P	
254	6600	1 use/month R015-100	4 Carp	18	1.778	C.S.F.	43	48.50		91.50	124	254
	6650	2 uses/month	"	18	1.778	"	21.50	48.50		70	101	
255	0011	**SCAFFOLDING SPECIALTIES**										255
	1200	Sidewalk bridge, heavy duty steel posts & beams, including										
	1210	parapet protection & waterproofing										
	1220	8' to 10' wide, 2 posts	3 Carp	15	1.600	L.F.	42	43.50		85.50	115	
	1230	3 posts	"	10	2.400	"	63	65.50		128.50	173	
	1500	Sidewalk bridge using tubular steel										
	1510	scaffold frames, including planking	3 Carp	45	.533	L.F.	4.72	14.55		19.27	28	
	1600	For 2 uses per month, deduct from all above					50%					
	1700	For 1 use every 2 months, add to all above					100%					
	1900	Catwalks, 32" wide, no guardrails, 6' span, buy				Ea.	120			120	132	
	2000	10' span, buy				"	190			190	209	
	2800	Hand winch-operated masons										
	2810	scaffolding, no plank moving required										
	2900	98' long, 10'-6" high, buy				Ea.	19,700			19,700	21,700	
	3000	Rent per month					790			790	865	
	3100	28'-6" high, buy					25,700			25,700	28,200	
	3200	Rent per month					1,025			1,025	1,125	
	3400	196' long, 28'-6" high, buy					49,700			49,700	54,500	
	3500	Rent per month					2,000			2,000	2,175	
	3600	64'-6" high, buy					73,000			73,000	80,000	
	3700	Rent per month					2,900			2,900	3,200	
	3720	Putlog, standard, 8' span, with hangers, buy					61			61	67	
	3730	Rent per month					10			10	11	
	3750	12' span, buy					92			92	101	
	3755	Rent per month					15			15	16.50	
	3760	Trussed type, 16' span, buy					210			210	231	
	3770	Rent per month					20			20	22	
	3790	22' span, buy					252			252	277	
	3795	Rent per month					30			30	33	
	3800	Rolling ladders with handrails, 30" wide, buy, 2 step					139			139	153	
	4000	7 step					405			405	450	
	4050	10 step					655			655	720	
	4100	Rolling towers, buy, 5' wide, 7' long, 10' high					1,150			1,150	1,250	
	4200	For 5' high added sections, to buy, add					188			188	207	
	4300	Complete incl. wheels, railings, outriggers,										
	4350	21' high, to buy				Ea.	1,925			1,925	2,125	
	4400	21' high, rent per month				"	138			138	152	
256	0010	**SWING STAGING**, 500 lb cap., 2' wide to 24' long, hand operated hoist										256
	0020	steel cable type, with 60' cables, buy				Ea.	3,650			3,650	4,000	
	0030	Rent per month				"	450			450	495	
	0600	Lightweight (not for masons) 24' long for 150' height.										
	0610	manual type, buy				Ea.	3,900			3,900	4,300	
	0620	Rent per month					520			520	575	
	0700	Powered, electric or air, to 150' high, buy					13,500			13,500	14,800	
	0710	Rent per month					840			840	925	
	0780	To 300' high, buy					14,300			14,300	15,700	
	0800	Rent per month					910			910	1,000	
	1000	Bosun's chair or work basket 3' x 3.5', to 300' high, electric, buy					7,250			7,250	7,975	
	1010	Rent per month					450			450	495	
	2200	Move swing staging (setup and remove)	E-4	2	16	Move		500	41	541	955	
257	0010	**PUMP STAGING**, aluminum R015-200										257
	0200	24' long pole section, buy				Ea.	335			335	365	

015 250 | Construction Aids

			CREW	DAILY OUTPUT	LABOR-HOURS	UNIT	MAT.	LABOR	EQUIP.	TOTAL	TOTAL INCL O&P	
							1999 BARE COSTS					
257	0300	18' long pole section, buy	R015 -200			Ea.	258			258	284	**257**
	0400	12' long pole section, buy					174			174	192	
	0500	6' long pole section, buy					92			92	101	
	0600	6' long splice joint section, buy					68			68	75	
	0700	Pump jack					111			111	123	
	0900	Foldable brace					48.50			48.50	53	
	1000	Workbench/back safety rail support					59			59	64.50	
	1100	Scaffolding planks/workbench, 14" wide x 24' long					545			545	600	
	1200	Plank end safety rail					183			183	202	
	1250	Safety net, 22' long					267			267	294	
	1300	System in place, 50' working height, per use based on 50 uses	2 Carp	84.80	.189	C.S.F.	4.95	5.15		10.10	13.55	
	1400	100 uses		84.80	.189		2.48	5.15		7.63	10.80	
	1500	150 uses		84.80	.189		1.66	5.15		6.81	9.90	
258	0010	**SURVEYOR STAKES** Hardwood, 1" x 1" x 48" long				C	41.50			41.50	46	**258**
	0100	2" x 2" x 18" long					50			50	55	
	0150	2" x 2" x 24" long					62			62	68	
	0200	2" x 2" x 30" long					71			71	78	
259	0010	**WEATHER STATION** Remote recording, minimum				Ea.	3,000			3,000	3,300	**259**
	0100	Maximum				"	7,100			7,100	7,800	

015 300 | Barriers & Enclosures

			CREW	DAILY OUTPUT	LABOR-HOURS	UNIT	MAT.	LABOR	EQUIP.	TOTAL	TOTAL INCL O&P	
302	0010	**BARRICADES** 5' high, 3 rail @ 2" x 8", fixed	2 Carp	30	.533	L.F.	10.80	14.55		25.35	35	**302**
	0150	Movable	"	20	.800	"	10.80	22		32.80	46.50	
	0300	Stock units, 6' high, 8' wide, plain, buy				Ea.	430			430	475	
	0350	With reflective tape, buy				"	525			525	580	
	0400	Break-a-way 3" PVC pipe barricade										
	0410	with 3 ea. 1' x 4' reflectorized panels, buy				Ea.	305			305	335	
	0500	Plywood with steel legs, 32" wide					70			70	77	
	0600	Telescoping Christmas tree, 9' high, 5 flags, buy					121			121	133	
	0800	Traffic cones, PVC, 18" high					8.70			8.70	9.55	
	0850	28" high					12.35			12.35	13.55	
	1000	Guardrail, wooden, 3' high, 1" x 6", on 2" x 4" posts	2 Carp	200	.080	L.F.	.99	2.18		3.17	4.53	
	1100	2" x 6", on 4" x 4" posts	"	165	.097		1.79	2.65		4.44	6.15	
	1200	Portable metal with base pads, buy					15			15	16.50	
	1250	Typical installation, assume 10 reuses	2 Carp	600	.027		1.58	.73		2.31	2.89	
	5000	Barricades, see also division 016-420-1600										
304	0010	**FENCING** Chain link, 11 ga, 5' high	2 Clab	100	.160	L.F.	4.04	3.43		7.47	9.85	**304**
	0100	6' high		75	.213		3.37	4.58		7.95	10.90	
	0200	Rented chain link, 6' high, to 500' (up to 12 mo.)		100	.160		2.19	3.43		5.62	7.80	
	0250	Over 1000' (up to 12 mo.)		110	.145		1.59	3.12		4.71	6.65	
	0350	Plywood, painted, 2" x 4" frame, 4' high	A-4	135	.178		4.06	4.71		8.77	11.80	
	0400	4" x 4" frame, 8' high	"	110	.218		7.05	5.80		12.85	16.75	
	0500	Wire mesh on 4" x 4" posts, 4' high	2 Carp	100	.160		5.55	4.37		9.92	13	
	0550	8' high	"	80	.200		8.35	5.45		13.80	17.80	
306	0010	**WINTER PROTECTION** Reinforced plastic on wood										**306**
	0100	framing to close openings	2 Clab	750	.021	S.F.	.35	.46		.81	1.11	
	0200	Tarpaulins hung over scaffolding, 8 uses, not incl. scaffolding		1,500	.011		.16	.23		.39	.54	
	0250	Tarpaulin polyester reinf. w/ integral fastening system 11 mils thick		1,600	.010		.73	.21		.94	1.14	
	0300	Prefab fiberglass panels, steel frame, 8 uses		1,200	.013		.70	.29		.99	1.22	

015 400 | Security

			CREW	DAILY OUTPUT	LABOR-HOURS	UNIT	MAT.	LABOR	EQUIP.	TOTAL	TOTAL INCL O&P	
480	0010	**WATCHMAN** Service, monthly basis, uniformed person, minimum				Hr.					7.80	**480**
	0100	Maximum									14.15	

015 400 | Security

		CREW	DAILY OUTPUT	LABOR-HOURS	UNIT	MAT.	LABOR	EQUIP.	TOTAL	TOTAL INCL O&P	
480	0200 Person and command dog, minimum				Hr.					10.30	480
	0300 Maximum									15.30	
	0500 Sentry dog, leased, with job patrol (yard dog), 1 dog				Week					195	
	0600 2 dogs				"					275	
	0800 Purchase, trained sentry dog, minimum				Ea.					800	
	0900 Maximum				"					2,000	

015 500 | Access Roads

		CREW	DAILY OUTPUT	LABOR-HOURS	UNIT	MAT.	LABOR	EQUIP.	TOTAL	TOTAL INCL O&P	
552	0010 ROADS AND SIDEWALKS Temporary										552
	0050 Roads, gravel fill, no surfacing, 4" gravel depth	B-14	715	.067	S.Y.	1.10	1.53	.31	2.94	3.94	
	0100 8" gravel depth	"	615	.078	"	2.20	1.77	.36	4.33	5.60	
	1000 Ramp, 3/4" plywood on 2" x 6" joists, 16" O.C.	2 Carp	300	.053	S.F.	1.45	1.46		2.91	3.89	
	1100 On 2" x 10" joists, 16" O.C.	"	275	.058		1.75	1.59		3.34	4.43	
	2200 Sidewalks, 2" x 12" planks, 2 uses	1 Carp	350	.023		.55	.62		1.17	1.59	
	2300 Exterior plywood, 2 uses, 1/2" thick		750	.011		.30	.29		.59	.79	
	2400 5/8" thick		650	.012		.35	.34		.69	.92	
	2500 3/4" thick		600	.013		.42	.36		.78	1.03	

015 600 | Temporary Controls

		CREW	DAILY OUTPUT	LABOR-HOURS	UNIT	MAT.	LABOR	EQUIP.	TOTAL	TOTAL INCL O&P	
602	0010 TARPAULINS Cotton duck, 10 oz. to 13.13 oz. per S.Y., minimum				S.F.	.38			.38	.42	602
	0050 Maximum					.55			.55	.61	
	0100 Polyvinyl coated nylon, 14 oz. to 18 oz., minimum					.45			.45	.50	
	0150 Maximum					.65			.65	.71	
	0200 Reinforced polyethylene 3 mils thick, white					.10			.10	.11	
	0300 4 mils thick, white, clear or black					.12			.12	.13	
	0400 5.5 mils thick, clear					.09			.09	.10	
	0500 White, fire retardant					.16			.16	.18	
	0600 7.5 mils, oil resistant, fire retardant					.17			.17	.19	
	0700 8.5 mils, black					.22			.22	.24	
	0710 Woven polyethylene, 6 mils thick					.45			.45	.50	
	0730 Polyester reinforced w/ integral fastening system 11 mils thick					1			1	1.10	
	0740 Mylar polyester, non-reinforced, 7 mils thick					1.10			1.10	1.21	

015 800 | Project Signs

		CREW	DAILY OUTPUT	LABOR-HOURS	UNIT	MAT.	LABOR	EQUIP.	TOTAL	TOTAL INCL O&P	
804	0010 SIGNS Hi-intensity reflectorized, no posts, buy				S.F.	12.05			12.05	13.25	804

015 900 | Field Offices & Sheds

		CREW	DAILY OUTPUT	LABOR-HOURS	UNIT	MAT.	LABOR	EQUIP.	TOTAL	TOTAL INCL O&P	
904	0010 OFFICE Trailer, furnished, no hookups, 20' x 8', buy	2 Skwk	1	16	Ea.	5,075	450		5,525	6,275	904
	0250 Rent per month					134			134	148	
	0300 32' x 8', buy	2 Skwk	.70	22.857		7,825	640		8,465	9,625	
	0350 Rent per month					158			158	174	
	0400 50' x 10', buy	2 Skwk	.60	26.667		13,400	750		14,150	15,900	
	0450 Rent per month					273			273	300	
	0500 50' x 12', buy	2 Skwk	.50	32		15,800	900		16,700	18,800	
	0550 Rent per month					315			315	345	
	0700 For air conditioning, rent per month, add					35.50			35.50	39	
	0800 For delivery, add per mile				Mile	1.50			1.50	1.65	
	1000 Portable buildings, prefab, on skids, economy, 8' x 8'	2 Carp	265	.060	S.F.	80	1.65		81.65	90.50	
	1100 Deluxe, 8' x 12'	"	150	.107	"	87	2.91		89.91	100	
	1200 Storage boxes, 20' x 8', buy	2 Skwk	1.80	8.889	Ea.	3,200	249		3,449	3,925	
	1250 Rent per month					66.50			66.50	73.50	
	1300 40' x 8', buy	2 Skwk	1.40	11.429		3,325	320		3,645	4,175	
	1350 Rent per month					85			85	93.50	

GENERAL REQUIREMENTS 1

016 400 | Equipment Rental

			UNIT	HOURLY OPER. COST	RENT PER DAY	RENT PER WEEK	RENT PER MONTH	CREW EQUIPMENT COST/DAY	
406	0010	**CONCRETE EQUIPMENT RENTAL**	R016 -410						**406**
	0100	without operators							
	0200	Bucket, concrete lightweight, 1/2 C.Y.	R033 -090	Ea.	.14	25	75	225	16.10	
	0300	1 C.Y.			.18	38.50	115	345	24.45	
	0400	1-1/2 C.Y.			.25	40	120	360	26	
	0500	2 C.Y.			.26	50	150	450	32.10	
	0600	Cart, concrete, self propelled, operator walking, 10 C.F.			1.24	66.50	200	600	49.90	
	0700	Operator riding, 18 C.F.			2.42	107	320	960	83.35	
	0800	Conveyer for concrete, portable, gas, 16" wide, 26' long			3.61	167	500	1,500	128.90	
	0900	46' long			3.71	227	680	2,050	165.70	
	1000	56' long			3.86	250	750	2,250	180.90	
	1100	Core drill, electric, 2-1/2 H.P., 1" to 8" bit diameter			.41	70	210	630	45.30	
	1150	11 H.P., 8" to 18" cores			.67	93.50	280	840	61.35	
	1200	Finisher, concrete floor, gas, riding trowel, 48" diameter			2.68	100	300	900	81.45	
	1300	Gas, manual, 3 blade, 36" trowel			2.27	56.50	170	510	52.15	
	1400	4 blade, 48" trowel			2.37	50	150	450	48.95	
	1500	Float, hand-operated (Bull float) 48" wide			.10	6.65	20	60	4.80	
	1570	Curb builder, 14 H.P., gas, single screw			1.60	58.50	175	525	47.80	
	1590	Double screw			2.32	61.50	185	555	55.55	
	1600	Grinder, concrete and terrazzo, electric, floor			1.34	58.50	175	525	45.70	
	1700	Wall grinder			.62	35	105	315	25.95	
	1800	Mixer, powered, mortar and concrete, gas, 6 C.F., 18 H.P.			.93	58.50	175	525	42.45	
	1900	10 C.F., 25 H.P.			1.34	50	150	450	40.70	
	2000	16 C.F.			1.60	66.50	200	600	52.80	
	2100	Concrete, stationary, tilt drum, 2 C.Y.			7.93	325	980	2,950	259.45	
	2120	Pump, concrete, truck mounted, 4" line, 80' boom			11.95	915	2,750	8,250	645.60	
	2140	5" line, 110' boom			13.29	1,100	3,275	9,825	761.30	
	2160	Mud jack, 50 C.F. per hr.			3.17	70	210	630	67.35	
	2180	225 C.F. per hr.			6.28	310	925	2,775	235.25	
	2600	Saw, concrete, manual, gas, 18 H.P.			2.83	76.50	230	690	68.65	
	2650	Self-propelled, gas, 30 H.P.			5.58	110	330	990	110.65	
	2700	Vibrators, concrete, electric, 60 cycle, 2 H.P.			.31	23.50	70	210	16.50	
	2800	3 H.P.			.36	33.50	100	300	22.90	
	2900	Gas engine, 5 H.P.			.67	33.50	100	300	25.35	
	3000	8 H.P.			.93	50	150	450	37.45	
	3100	Concrete transit mixer, hydraulic drive								
	3120	6 x 4, 250 H.P., 8 C.Y., rear discharge			30.08	925	2,775	8,325	795.65	
	3200	Front discharge			30.18	990	2,975	8,925	836.45	
	3300	6 x 6, 285 H.P., 12 C.Y., rear discharge			32.47	1,025	3,075	9,225	874.75	
	3400	Front discharge			32.65	1,075	3,250	9,750	911.20	
408	0010	**EARTHWORK EQUIPMENT RENTAL** Without operators	R016 -410							**408**
	0040	Aggregate spreader, push type 8' to 12' wide		Ea.	1.13	103	310	930	71.05	
	0050	Augers for truck or trailer mounting, vertical drilling	R022 -220							
	0055	Fence post auger, truck mounted		Ea.	10	465	1,400	4,200	360	
	0060	4" to 36" diam., 54 H.P., gas, 10' spindle travel	R022 -240		15.91	555	1,660	4,975	459.30	
	0070	14' spindle travel			16.38	650	1,955	5,875	522.05	
	0075	Auger, truck mounted, vertical drilling, to 25' depth	R022 -250		49.54	2,300	6,900	20,700	1,776	
	0080	Auger, horizontal boring machine, 12" to 36" diameter, 45 H.P.			7.31	325	980	2,950	254.50	
	0090	12" to 48" diameter, 65 H.P.	R023 -620		12.46	650	1,955	5,875	490.70	
	0100	Backhoe, diesel hydraulic, crawler mounted, 1/2 C.Y. cap.			11.43	410	1,225	3,675	336.45	
	0120	5/8 C.Y. capacity			15.53	450	1,350	4,050	394.25	
	0140	3/4 C.Y. capacity			17.39	515	1,550	4,650	449.10	
	0150	1 C.Y. capacity			22.20	615	1,850	5,550	547.60	
	0200	1-1/2 C.Y. capacity			26.59	835	2,500	7,500	712.70	
	0300	2 C.Y. capacity			42.85	1,125	3,400	10,200	1,023	
	0320	2-1/2 C.Y. capacity			59.38	2,025	6,100	18,300	1,695	
	0340	3-1/2 C.Y. capacity			79.44	2,525	7,600	22,800	2,156	
	0341	Attachments								

Important: See the Reference Section for critical supporting data - Reference Nos., Crews, & City Cost Indexes

GENERAL REQUIREMENTS

1

016 400 | Equipment Rental

			UNIT	HOURLY OPER. COST	RENT PER DAY	RENT PER WEEK	RENT PER MONTH	CREW EQUIPMENT COST/DAY	
408	0342	Bucket thumbs	R016 -410	Ea.	.46	293	880	2,650	179.70	408
	0345	Grapples			.31	300	900	2,700	182.50	
	0350	Gradall type, truck mounted, 3 ton @ 15' radius, 5/8 C.Y.	R022 -220		26.47	710	2,125	6,375	636.75	
	0370	1 C.Y. capacity			28.22	965	2,900	8,700	805.75	
	0400	Backhoe-loader, wheel type, 40 to 45 H.P., 5/8 C.Y. capacity	R022 -240		7.31	237	710	2,125	200.50	
	0450	45 H.P. to 60 H.P., 3/4 C.Y. capacity			8.60	250	750	2,250	218.80	
	0460	80 H.P., 1-1/4 C.Y. capacity	R022 -250		12.51	315	950	2,850	290.10	
	0470	112 H.P.,1-3/4 C.Y. loader, 1/2 C.Y. backhoe			15.55	385	1,150	3,450	354.40	
	0480	Attachments	R023 -620							
	0482	Compactor, 20,000 lb			1.29	188	565	1,700	123.30	
	0485	Hydraulic hammer, 750 ft-lbs			.41	305	915	2,750	186.30	
	0486	Hydraulic hammer, 1000 ft-lbs			.46	345	1,030	3,100	209.70	
	0500	Brush chipper, gas engine, 6" cutter head, 35 H.P.			4.12	137	410	1,225	114.95	
	0550	12" cutter head, 130 H.P.			12.51	190	570	1,700	214.10	
	0600	15" cutter head, 165 H.P.			16.17	255	765	2,300	282.35	
	0750	Bucket, clamshell, general purpose, 3/8 C.Y.			.80	60	180	540	42.40	
	0800	1/2 C.Y.			1.11	68.50	205	615	49.90	
	0850	3/4 C.Y.			1.12	76.50	230	690	54.95	
	0900	1 C.Y.			1.39	93.50	280	840	67.10	
	0950	1-1/2 C.Y.			1.65	128	385	1,150	90.20	
	1000	2 C.Y.			1.85	145	435	1,300	101.80	
	1010	Bucket, dragline, medium duty, 1/2 C.Y.			.36	38.50	115	345	25.90	
	1020	3/4 C.Y.			.57	48.50	145	435	33.55	
	1030	1 C.Y.			.60	51.50	155	465	35.80	
	1040	1-1/2 C.Y.			.77	70	210	630	48.15	
	1050	2 C.Y.			.84	86.50	260	780	58.70	
	1070	3 C.Y.			1.34	120	360	1,075	82.70	
	1200	Compactor, roller, 2 drum, 2000 lb., operator walking			1.88	130	390	1,175	93.05	
	1250	Rammer compactor, gas, 1000 lb. blow			.57	45	135	405	31.55	
	1300	Vibratory plate, gas, 13" plate, 1000 lb. blow			.71	41.50	125	375	30.70	
	1350	24" plate, 5000 lb. blow			1.75	63.50	190	570	52	
	1370	Curb builder/extruder, 14 H.P., gas, single screw			1.44	48.50	145	435	40.50	
	1390	Double screw			2.16	88.50	265	795	70.30	
	1750	Extractor, piling, see lines 2500 to 2750								
	1860	Grader, self-propelled, 25,000 lb.		Ea.	14.83	580	1,740	5,225	466.65	
	1910	30,000 lb.			17.82	720	2,165	6,500	575.55	
	1920	40,000 lb.			20.50	860	2,575	7,725	679	
	1930	55,000 lb.			33.48	1,300	3,865	11,600	1,041	
	1950	Hammer, pavement demo., hyd., gas, self-prop., 1000 to 1250 lb.			18.59	410	1,225	3,675	393.70	
	2000	Diesel 1300 to 1500 lb.			9.01	455	1,360	4,075	344.10	
	2050	Pile driving hammer, steam or air, 4150 ft.-lb. @ 225 BPM			1.80	288	865	2,600	187.40	
	2100	8750 ft.-lb. @ 145 BPM			3.24	410	1,235	3,700	272.90	
	2150	15,000 ft.-lb. @ 60 BPM			4.22	480	1,440	4,325	321.75	
	2200	24,450 ft.-lb. @ 111 BPM			4.65	585	1,760	5,275	389.20	
	2250	Leads, 15,000 ft.-lb. hammers		L.F.	.02	2.67	8	24	1.75	
	2300	24,450 ft.-lb. hammers and heavier		"	.04	3	9	27	2.10	
	2350	Diesel type hammer, 22,400 ft.-lb.		Ea.	6.17	480	1,440	4,325	337.35	
	2400	41,300 ft.-lb.			9.57	755	2,265	6,800	529.55	
	2450	141,000 ft.-lb.			19.68	1,450	4,325	13,000	1,022	
	2500	Vib. elec. hammer/extractor, 200 KW diesel generator, 34 H.P.			9.33	610	1,835	5,500	441.65	
	2550	80 H.P.			18.23	925	2,780	8,350	701.85	
	2600	150 H.P.			31.36	1,625	4,840	14,500	1,219	
	2700	Extractor, steam or air, 700 ft.-lb.			1.44	172	515	1,550	114.50	
	2750	1000 ft.-lb.			1.85	240	720	2,150	158.80	
	3000	Roller, tandem, gas, 3 to 5 ton			6.23	172	515	1,550	152.85	
	3050	Diesel, 8 to 12 ton			9.01	288	865	2,600	245.10	
	3100	Towed type, vibratory, gas 12.5 H.P., 2 ton			3.86	137	410	1,225	112.90	
	3150	Sheepsfoot, double 60" x 60"			4.74	145	435	1,300	124.90	

			UNIT	HOURLY OPER. COST	RENT PER DAY	RENT PER WEEK	RENT PER MONTH	CREW EQUIPMENT COST/DAY	
3200	Pneumatic tire diesel roller, 12 ton	R016 -410	Ea.	8.65	292	875	2,625	244.20	408
3250	21 to 25 ton			15.97	345	1,030	3,100	333.75	
3300	Sheepsfoot roller, self-propelled, 4 wheel, 130 H.P.	R022 -220		20.96	650	1,955	5,875	558.70	
3320	300 H.P.			26.68	765	2,300	6,900	673.45	
3350	Vibratory steel drum & pneumatic tire, diesel, 18,000 lb.	R022 -240		11.48	435	1,300	3,900	351.85	
3400	29,000 lb.			12.72	535	1,600	4,800	421.75	
3450	Scrapers, towed type, 9 to 12 C.Y. capacity	R022 -250		3.19	86.50	260	780	77.50	
3500	12 to 17 C.Y. capacity			5.93	258	775	2,325	202.45	
3550	Scrapers, self-propelled, 4 x 4 drive, 2 engine, 14 C.Y. capacity	R023 -620		60.87	1,875	5,650	17,000	1,617	
3600	2 engine, 24 C.Y. capacity			73.95	2,200	6,600	19,800	1,912	
3650	Self-loading, 11 C.Y. capacity			27.32	785	2,350	7,050	688.55	
3700	22 C.Y. capacity			33.07	1,200	3,600	10,800	984.55	
3710	Screening plant 110 hp. w / 5' x 10'screen			15.97	435	1,300	3,900	387.75	
3720	5' x 16' screen			17.61	515	1,550	4,650	450.90	
3850	Shovels, see Cranes division 016-460								
3860	Shovel/backhoe bucket, 1/2 C.Y.		Ea.	1.13	76.50	230	690	55.05	
3870	3/4 C.Y.			3.19	128	385	1,150	102.50	
3880	1 C.Y.			3.50	172	515	1,550	131	
3890	1-1/2 C.Y.			4.21	210	630	1,900	159.70	
3910	3 C.Y.			7.52	400	1,200	3,600	300.15	
4110	Tractor, crawler, with bulldozer, torque converter, diesel 75 H.P.			10.97	365	1,100	3,300	307.75	
4150	105 H.P.			14.68	500	1,500	4,500	417.45	
4200	140 H.P.			17	615	1,850	5,550	506	
4260	200 H.P.			29.87	1,000	3,000	9,000	838.95	
4310	300 H.P.			38.83	1,375	4,100	12,300	1,131	
4360	410 H.P.			47.48	1,675	5,000	15,000	1,380	
4380	700 H.P.			95.58	3,325	10,000	30,000	2,765	
4400	Loader, crawler, torque conv., diesel, 1-1/2 C.Y., 80 H.P.			12.51	415	1,250	3,750	350.10	
4450	1-1/2 to 1-3/4 C.Y., 95 H.P.			14.96	465	1,400	4,200	399.70	
4510	1-3/4 to 2-1/4 C.Y., 130 H.P.			18.80	600	1,800	5,400	510.40	
4530	2-1/2 to 3-1/4 C.Y., 190 H.P.			30.64	1,025	3,100	9,300	865.10	
4560	3-1/2 to 5 C.Y., 275 H.P.			42.81	1,325	4,000	12,000	1,142	
4610	Tractor loader, wheel, torque conv., 4 x 4, 1 to 1-1/4 C.Y., 65 H.P.			9.33	283	850	2,550	244.65	
4620	1-1/2 to 1-3/4 C.Y., 80 H.P.			11.89	375	1,125	3,375	320.10	
4650	1-3/4 to 2 C.Y., 100 H.P.			12.47	405	1,215	3,650	342.75	
4710	2-1/2 to 3-1/2 C.Y., 130 H.P.			18.81	500	1,500	4,500	450.50	
4730	3 to 4-1/2 C.Y., 170 H.P.			20.86	685	2,050	6,150	576.90	
4760	5-1/4 to 5-3/4 C.Y., 270 H.P.			38.31	1,025	3,100	9,300	926.50	
4810	7 to 8 C.Y., 375 H.P.			56.88	1,375	4,125	12,400	1,280	
4870	12-1/2 C.Y., 690 H.P.			115.60	2,400	7,200	21,600	2,365	
4880	Wheeled, skid steer, 10 C.F., 30 H.P. gas			4.53	125	375	1,125	111.25	
4890	1 C.Y., 78 H.P., diesel			6.28	300	900	2,700	230.25	
4891	Attachments for all skid steer loaders								
4892	Auger		Ea.	.12	83.50	250	750	50.95	
4893	Backhoe			.15	110	330	990	67.20	
4894	Broom			.16	107	320	960	65.30	
4895	Forks			.08	38.50	115	345	23.65	
4896	Grapple			.12	86.50	260	780	52.95	
4897	Concrete hammer			.25	180	540	1,625	110	
4898	Tree spade			.36	128	385	1,150	79.90	
4899	Trencher			.41	240	720	2,150	147.30	
4900	Trencher, chain, boom type, gas, operator walking, 12 H.P.			2.16	133	400	1,200	97.30	
4910	Operator riding, 40 H.P.			6.36	267	800	2,400	210.90	
5000	Wheel type, diesel, 4' deep, 12" wide			12.42	500	1,500	4,500	399.35	
5100	Diesel, 6' deep, 20" wide			15.65	735	2,200	6,600	565.20	
5150	Ladder type, diesel, 5' deep, 8" wide			9.41	375	1,125	3,375	300.30	
5200	Diesel, 8' deep, 16" wide			17.61	660	1,975	5,925	535.90	
5250	Truck, dump, tandem, 12 ton payload			17.41	375	1,125	3,375	364.30	

Important: See the Reference Section for critical supporting data - Reference Nos., Crews, & City Cost Indexes

016 400 | Equipment Rental

			UNIT	HOURLY OPER. COST	RENT PER DAY	RENT PER WEEK	RENT PER MONTH	CREW EQUIPMENT COST/DAY	
408	5300	Three axle dump, 16 ton payload	R016 -410	Ea.	20.28	465	1,400	4,200	442.25	408
	5350	Dump trailer only, rear dump, 16-1/2 C.Y.			3.45	175	525	1,575	132.60	
	5400	20 C.Y.	R022 -220		3.51	178	535	1,600	135.10	
	5450	Flatbed, single axle, 1-1/2 ton rating			10.87	142	425	1,275	171.95	
	5500	3 ton rating	R022 -240		11.43	147	440	1,325	179.45	
	5550	Off highway rear dump, 25 ton capacity			20.50	890	2,675	8,025	699	
	5600	35 ton capacity	R022 -250		32.27	1,300	3,900	11,700	1,038	
420	0010	**GENERAL EQUIPMENT RENTAL** Without operators	R021 -610							420
	0150	Aerial lift, scissor type, to 15' high, 1000 lb. cap., electric		Ea.	1.18	85	255	765	60.45	
	0160	To 25' high, 2000 lb. capacity	R021 -620		1.80	128	385	1,150	91.40	
	0170	Telescoping boom to 40' high, 750 lb. capacity, gas			6.42	385	1,150	3,450	281.35	
	0180	1000 lb. capacity	R022 -220		7.98	435	1,300	3,900	323.85	
	0190	To 60' high, 750 lb. capacity			8.47	525	1,575	4,725	382.75	
	0195	Air compressor, portable, 6.5 CFM, electric			.12	33.50	100	300	20.95	
	0196	gasoline			.13	41.50	125	375	26.05	
	0200	Air compressor, portable, gas engine, 60 C.F.M.			5.41	55	165	495	76.30	
	0300	160 C.F.M.			6.80	71.50	215	645	97.40	
	0400	Diesel engine, rotary screw, 250 C.F.M.			6.49	122	365	1,100	124.90	
	0500	365 C.F.M.			9.29	167	500	1,500	174.30	
	0600	600 C.F.M.			15.99	242	725	2,175	272.90	
	0700	750 C.F.M.			17.57	247	740	2,225	288.55	
	0800	For silenced models, small sizes, add			3%	5%	5%	5%		
	0900	Large sizes, add			5%	7%	7%	7%		
	0920	Air tools and accessories								
	0930	Breaker, pavement, 60 lb.		Ea.	.19	30	90	270	19.50	
	0940	80 lb.			.21	35	105	315	22.70	
	0950	Drills, hand (jackhammer) 65 lb.			.23	25	75	225	16.85	
	0960	Track or wagon, swing boom, 4" drifter			10.57	380	1,135	3,400	311.55	
	0970	5" drifter			11.38	650	1,955	5,875	482.05	
	0980	Dust control per drill			2.11	11.65	35	105	23.90	
	0990	Hammer, chipping, 12 lb.			.12	20	60	180	12.95	
	1000	Hose, air with couplings, 50' long, 3/4" diameter			.15	5	15	45	4.20	
	1100	1" diameter			.15	6.65	20	60	5.20	
	1200	1-1/2" diameter			.15	11.65	35	105	8.20	
	1300	2" diameter			.21	16.65	50	150	11.70	
	1400	2-1/2" diameter			.22	20	60	180	13.75	
	1410	3" diameter			.23	25	75	225	16.85	
	1450	Drill, steel, 7/8" x 2'				3.33	10	30	2	
	1460	7/8" x 6'				4	12	36	2.40	
	1520	Moil points			.82	2	6	18	7.75	
	1525	Pneumatic nailer w/accessories			.12	31.50	95	285	19.95	
	1530	Sheeting driver for 60 lb. breaker			.15	10	30	90	7.20	
	1540	For 90 lb. breaker			.15	16.65	50	150	11.20	
	1550	Spade, 25 lb.			.08	8.35	25	75	5.65	
	1560	Tamper, single, 35 lb.			.10	23.50	70	210	14.80	
	1570	Triple, 140 lb.			1.80	41.50	125	375	39.40	
	1580	Wrenches, impact, air powered, up to 3/4" bolt			.25	21.50	65	195	15	
	1590	Up to 1-1/4" bolt			.35	43.50	130	390	28.80	
	1600	Barricades, barrels, reflectorized, 1 to 50 barrels				2	6	18	1.20	
	1610	100 to 200 barrels				1.33	4	12	.80	
	1620	Barrels with flashers, 1 to 50 barrels				3	9	27	1.80	
	1630	100 to 200 barrels				2	6	18	1.20	
	1640	Barrels with steady burn type C lights				3.67	11	33	2.20	
	1650	Illuminated board, trailer mounted, with generator			.78	86.50	260	780	58.25	
	1670	Portable, stock, with flashers, 1 to 6 units				.77	2.30	6.90	.45	
	1680	25 to 50 units				.73	2.20	6.60	.45	
	1690	Butt fusion machine, electric			1.50	292	875	2,625	187	

GENERAL REQUIREMENTS 1

016 400 | Equipment Rental

			UNIT	HOURLY OPER. COST	RENT PER DAY	RENT PER WEEK	RENT PER MONTH	CREW EQUIPMENT COST/DAY	
420	1695	Electro fusion machine	R021 -610	Ea.	1.25	125	375	1,125	85	420
	1700	Carts, brick, hand powered, 1000 lb. capacity			1.11	21.50	65	195	21.90	
	1800	Gas engine, 1500 lb., 7-1/2' lift	R021 -620		1.65	95	285	855	70.20	
	1830	Distributor, asphalt, trailer mtd, 2000 gal., 38 H.P. diesel			7.54	465	1,400	4,200	340.30	
	1840	3000 gal., 38 H.P. diesel	R022 -220		8.12	500	1,500	4,500	364.95	
	1850	Drill, rotary hammer, electric, 1-1/2" diameter			.15	23.50	70	210	15.20	
	1860	Carbide bit for above				5.65	17	51	3.40	
	1870	Emulsion sprayer, 65 gal., 5 H.P. gas engine			.63	55	165	495	38.05	
	1880	200 gal., 5 H.P. engine			.67	58.50	175	525	40.35	
	1900	Fencing, see division 015-304 & 028-300								
	1920	Floodlight, mercury, vapor or quartz, on tripod								
	1930	1000 watt		Ea.	.12	16.65	50	150	10.95	
	1940	2000 watt			.12	40	120	360	24.95	
	1960	Floodlights, trailer mounted with generator, 2-1000 watt lights			1.80	96.50	290	870	72.40	
	2020	Forklift, wheeled, for brick, 18', 3000 lb., 2 wheel drive, gas			9.28	167	500	1,500	174.25	
	2040	28', 4000 lb., 4 wheel drive, diesel			6.39	200	600	1,800	171.10	
	2100	Generator, electric, gas engine, 1.5 KW to 3 KW			1.13	38.50	115	345	32.05	
	2200	5 KW			1.49	56.50	170	510	45.90	
	2300	10 KW			2.39	137	410	1,225	101.10	
	2400	25 KW			6.81	152	455	1,375	145.50	
	2500	Diesel engine, 20 KW			4.35	110	330	990	100.80	
	2600	50 KW			6.86	123	370	1,100	128.90	
	2700	100 KW			11.34	188	565	1,700	203.70	
	2800	250 KW			28.99	325	980	2,950	427.90	
	2850	Hammer, hydraulic, for mounting on boom, to 500 ft.-lb.			.98	128	385	1,150	84.85	
	2860	500 to 1200 ft.-lb.			2.52	267	800	2,400	180.15	
	2900	Heaters, space, oil or electric, 50 MBH			.12	26.50	80	240	16.95	
	3000	100 MBH			.12	31.50	95	285	19.95	
	3100	300 MBH			.12	50	150	450	30.95	
	3150	500 MBH			.15	66.50	200	600	41.20	
	3200	Hose, water, suction with coupling, 20' long, 2" diameter			.06	10	30	90	6.50	
	3210	3" diameter			.06	15	45	135	9.50	
	3220	4" diameter			.06	20	60	180	12.50	
	3230	6" diameter			.06	35	105	315	21.50	
	3240	8" diameter			.07	45	135	405	27.55	
	3250	Discharge hose with coupling, 50' long, 2" diameter			.05	6.65	20	60	4.40	
	3260	3" diameter			.05	8.35	25	75	5.40	
	3270	4" diameter			.06	13.35	40	120	8.50	
	3280	6" diameter			.06	31.50	95	285	19.50	
	3290	8" diameter			.08	33.50	100	300	20.65	
	3300	Ladders, extension type, 16' to 36' long				15	45	135	9	
	3400	40' to 60' long				31.50	95	285	19	
	3410	Level, laser type, for pipe laying, self leveling				102	305	915	61	
	3430	Manual leveling				78.50	235	705	47	
	3440	Rotary beacon with rod and sensor				103	310	930	62	
	3460	Builders level with tripod and rod				30	90	270	18	
	3500	Light towers, towable, with diesel generator, 2000 watt			1.55	117	350	1,050	82.40	
	3600	4000 watt			1.97	137	410	1,225	97.75	
	3700	Mixer, powered, plaster and mortar, 6 C.F., 7 H.P.			.82	53.50	160	480	38.55	
	3800	10 C.F., 9 H.P.			1.22	76.50	230	690	55.75	
	3850	Nailer, pneumatic			.08	23.50	70	210	14.65	
	3900	Paint sprayers complete, 8 CFM			.08	41.50	125	375	25.65	
	4000	17 CFM			.08	60	180	540	36.65	
	4020	Pavers, bituminous, rubber tires, 8' wide, 52 H.P., gas			14.94	575	1,725	5,175	464.50	
	4030	8' wide, 64 H.P., diesel			15.45	1,000	3,025	9,075	728.60	
	4050	Crawler, 10' wide, 78 H.P., gas			22.20	1,400	4,200	12,600	1,018	
	4060	10' wide, 87 H.P., diesel			22.80	1,075	3,200	9,600	822.40	
	4070	Concrete paver, 12' to 24' wide, 250 H.P.			24.27	1,525	4,550	13,700	1,104	

Important: See the Reference Section for critical supporting data - Reference Nos., Crews, & City Cost Indexes

GENERAL REQUIREMENTS

016 400 | Equipment Rental

			UNIT	HOURLY OPER. COST	RENT PER DAY	RENT PER WEEK	RENT PER MONTH	CREW EQUIPMENT COST/DAY	
420	4080	Placer-spreader-trimmer, 24' wide, 300 H.P.	R021 -610	Ea.	32.02	1,775	5,325	16,000	1,321	420
	4100	Pump, centrifugal gas pump, 1-1/2", 4 MGPH			.46	31.50	95	285	22.70	
	4200	2", 8 MGPH	R021 -620		.52	35	105	315	25.15	
	4300	3", 15 MGPH			1.24	46.50	140	420	37.90	
	4400	6", 90 MGPH	R022 -220		9.68	137	410	1,225	159.45	
	4500	Submersible electric pump, 1-1/4", 55 GPM			.36	36.50	110	330	24.90	
	4600	1-1/2", 83 GPM			.39	40	120	360	27.10	
	4700	2", 120 GPM			.41	45	135	405	30.30	
	4800	3", 300 GPM			.72	55	165	495	38.75	
	4900	4", 560 GPM			1.18	68.50	205	615	50.45	
	5000	6", 1590 GPM			5.36	207	620	1,850	166.90	
	5100	Diaphragm pump, gas, single, 1-1/2" diameter			.49	25	75	225	18.90	
	5200	2" diameter			.52	41.50	125	375	29.15	
	5300	3" diameter			.77	45	135	405	33.15	
	5400	Double, 4" diameter			1.55	86.50	260	780	64.40	
	5500	Trash pump, self-priming, gas, 2" diameter			1.24	40	120	360	33.90	
	5600	Diesel, 4" diameter			1.96	96.50	290	870	73.70	
	5650	Diesel, 6" diameter			5.03	162	485	1,450	137.25	
	5660	Rollers, see division 016-408								
	5700	Salamanders, L.P. gas fired, 100,000 B.T.U.		Ea.	.70	21.50	65	195	18.60	
	5705	14.7 kW			.31	13.35	40	120	10.50	
	5720	Sandblaster, portable, open top, 3 C.F. capacity			.19	50	150	450	31.50	
	5730	6 C.F. capacity			.33	60	180	540	38.65	
	5740	Accessories for above			.06	16.65	50	150	10.50	
	5750	Sander, floor			.12	38.50	115	345	23.95	
	5760	Edger			.10	26.50	80	240	16.80	
	5800	Saw, chain, gas engine, 18" long			.54	38.50	115	345	27.30	
	5900	36" long			1.13	73.50	220	660	53.05	
	5950	60" long			1.15	83.50	250	750	59.20	
	6000	Masonry, table mounted, 14" diameter, 5 H.P.			1.82	50	150	450	44.55	
	6050	Saw, portable cut-off, 8 H.P.			.82	58.50	175	525	41.55	
	6100	Circular, hand held, electric, 7-1/4" diameter			.16	20	60	180	13.30	
	6200	12" diameter			.27	36.50	110	330	24.15	
	6275	Shot blaster, walk behind, 20" wide			1.49	200	600	1,800	131.90	
	6300	Steam cleaner, 100 gallons per hour			.43	51.50	155	465	34.45	
	6310	200 gallons per hour			.70	58.50	175	525	40.60	
	6340	Tar Kettle/Pot, 400 gallon			.49	70	210	630	45.90	
	6350	Torch, cutting, acetylene-oxygen, 150' hose			7.53	25	75	225	75.25	
	6360	Hourly operating cost includes tips and gas			7.62				60.95	
	6410	Toilet, portable chemical				15	45	135	9	
	6420	Recycle flush type				18.35	55	165	11	
	6430	Toilet, fresh water flush, garden hose,				21.50	65	195	13	
	6440	Hoisted, non-flush, for high rise				18.35	55	165	11	
	6450	Toilet, trailers, minimum				33.50	100	300	20	
	6460	Maximum				100	300	900	60	
	6470	Trailer, office, see division 015-904								
	6500	Trailers, platform, flush deck, 2 axle, 25 ton capacity		Ea.	1.39	123	370	1,100	85.10	
	6600	40 ton capacity			1.71	228	685	2,050	150.70	
	6700	3 axle, 50 ton capacity			2.85	243	730	2,200	168.80	
	6800	75 ton capacity			3.66	325	975	2,925	224.30	
	6850	Trailer, storage, see division 015-904								
	6900	Water tank, engine driven discharge, 5000 gallons		Ea.	8.14	233	700	2,100	205.10	
	7000	10,000 gallons			11.07	335	1,000	3,000	288.55	
	7020	Transit with tripod				33.50	100	300	20	
	7030	Trench box, 3000 lbs. 6'x8'			.80	100	300	900	66.40	
	7040	7200 lbs. 6'x20'			1.46	172	515	1,550	114.70	
	7050	8000 lbs., 8' x 16'			1.61	182	545	1,625	121.90	
	7060	9500 lbs., 8'x20'			1.82	188	565	1,700	127.55	

GENERAL REQUIREMENTS 1

016 400 | Equipment Rental

			UNIT	HOURLY OPER. COST	RENT PER DAY	RENT PER WEEK	RENT PER MONTH	CREW EQUIPMENT COST/DAY	
420	7065	11,000 lbs., 8'x24'	R021 -610	Ea.	1.87	242	725	2,175	159.95	420
	7070	12,000 lbs., 10' x 20'			1.98	250	750	2,250	165.85	
	7100	Truck, pickup, 3/4 ton, 2 wheel drive	R021 -620		10.62	73.50	220	660	128.95	
	7200	4 wheel drive			12.06	81.50	245	735	145.50	
	7250	Crew carrier, 9 passenger	R022 -220		16.91	103	310	930	197.30	
	7290	Tool van, 24,000 G.V.W.			18.76	103	310	930	212.10	
	7300	Tractor, 4 x 2, 30 ton capacity, 195 H.P.			11.13	390	1,175	3,525	324.05	
	7410	250 H.P.			15	435	1,300	3,900	380	
	7500	6 x 2, 40 ton capacity, 240 H.P.			17.91	470	1,415	4,250	426.30	
	7600	6 x 4, 45 ton capacity, 240 H.P.			23.15	575	1,725	5,175	530.20	
	7620	Vacuum truck, hazardous material, 2500 gallon			13.89	300	900	2,700	291.10	
	7625	5,000 gallon			15.02	335	1,000	3,000	320.15	
	7650	Vacuum, H.E.P.A., 16 gal., wet/dry			.93	51.50	155	465	38.45	
	7655	55 gal, wet/dry			.80	51.50	155	465	37.40	
	7690	Large production vacuum loader, 3150 CFM			11.86	715	2,150	6,450	524.90	
	7700	Welder, electric, 200 amp			.81	31.50	95	285	25.50	
	7800	300 amp			1.09	68.50	205	615	49.70	
	7900	Gas engine, 200 amp			4.33	51.50	155	465	65.65	
	8000	300 amp			5.17	68.50	205	615	82.35	
	8100	Wheelbarrow, any size				8.35	25	75	5	
	8200	Wrecking ball, 4000 lb.			.41	61.50	185	555	40.30	
460	0010	**LIFTING AND HOISTING EQUIPMENT RENTAL**	R015 -010							460
	0100	without operators								
	0150	Crane, flatbed mntd, 3 ton cap.	R016 -410	Ea.	15.35	267	800	2,400	282.80	
	0200	Crane, climbing, 106' jib, 6000 lb. capacity, 410 FPM			26.58	1,225	3,650	11,000	942.65	
	0300	101' jib, 10,250 lb. capacity, 270 FPM	R022 -250		35.34	1,550	4,650	14,000	1,213	
	0400	Tower, static, 130' high, 106' jib,								
	0500	6200 lb. capacity at 400 FPM		Ea.	53.48	1,425	4,250	12,800	1,278	
	0600	Crawler, cable, 1/2 C.Y., 15 tons at 12' radius			19	535	1,600	4,800	472	
	0700	3/4 C.Y., 20 tons at 12' radius			20.20	550	1,650	4,950	491.60	
	0800	1 C.Y., 25 tons at 12' radius			20.85	565	1,700	5,100	506.80	
	0900	Crawler, cable, 1-1/2 C.Y., 40 tons at 12' radius			30.54	765	2,300	6,900	704.30	
	1000	2 C.Y., 50 tons at 12' radius			35.59	935	2,800	8,400	844.70	
	1100	3 C.Y., 75 tons at 12' radius			43.66	965	2,900	8,700	929.30	
	1200	100 ton capacity, standard boom			41.43	1,325	4,000	12,000	1,131	
	1300	165 ton capacity, standard boom			64.77	2,175	6,500	19,500	1,818	
	1400	200 ton capacity, 150' boom			120.73	2,325	7,000	21,000	2,366	
	1500	450' boom			135.50	3,000	9,000	27,000	2,884	
	1600	Truck mounted, cable operated, 6 x 4, 20 tons at 10' radius			14.07	665	2,000	6,000	512.55	
	1700	25 tons at 10' radius			20.80	1,075	3,200	9,600	806.40	
	1800	8 x 4, 30 tons at 10' radius			28.61	600	1,800	5,400	588.90	
	1900	40 tons at 12' radius			29.34	765	2,300	6,900	694.70	
	2000	8 x 4, 60 tons at 15' radius			44.87	900	2,700	8,100	898.95	
	2050	82 tons at 15' radius			45.56	1,625	4,900	14,700	1,344	
	2100	90 tons at 15' radius			48.95	1,025	3,100	9,300	1,012	
	2200	115 tons at 15' radius			51.15	1,875	5,600	16,800	1,529	
	2300	150 tons at 18' radius			76.61	1,575	4,700	14,100	1,553	
	2350	165 tons at 18' radius			77.65	2,200	6,600	19,800	1,941	
	2400	Truck mounted, hydraulic, 12 ton capacity			22.93	435	1,300	3,900	443.45	
	2500	25 ton capacity			23.67	600	1,800	5,400	549.35	
	2550	33 ton capacity			24.37	835	2,500	7,500	694.95	
	2600	55 ton capacity			34.14	865	2,600	7,800	793.10	
	2700	80 ton capacity			37.29	1,325	4,000	12,000	1,098	
	2800	Self-propelled, 4 x 4, with telescoping boom, 5 ton			10.18	315	950	2,850	271.45	
	2900	12-1/2 ton capacity			16.42	435	1,300	3,900	391.35	
	3000	15 ton capacity			18.20	485	1,450	4,350	435.60	
	3050	20 ton capacity			18.01	500	1,500	4,500	444.10	

Important: See the Reference Section for critical supporting data - Reference Nos., Crews, & City Cost Indexes

016 400 | Equipment Rental

			UNIT	HOURLY OPER. COST	RENT PER DAY	RENT PER WEEK	RENT PER MONTH	CREW EQUIPMENT COST/DAY	
460	3100	25 ton capacity	R015 -010	Ea.	20.99	635	1,900	5,700	547.90	**460**
	3200	Derricks, guy, 20 ton capacity, 60' boom, 75' mast			8.70	292	875	2,625	244.60	
	3300	100' boom, 115' mast	R016 -410		16.39	515	1,550	4,650	441.10	
	3400	Stiffleg, 20 ton capacity, 70' boom, 37' mast			11.56	385	1,150	3,450	322.50	
	3500	100' boom, 47' mast	R022 -250		18.06	625	1,875	5,625	519.50	
	3550	Helicopter, small, lift to 1250 lbs. maximum			257.50	2,425	7,250	21,800	3,510	
	3600	Hoists, chain type, overhead, manual, 3/4 ton			.06	6	18	54	4.10	
	3900	10 ton			.25	25	75	225	17	
	4000	Hoist and tower, 5000 lb. cap., portable electric, 40' high			4.03	172	515	1,550	135.25	
	4100	For each added 10' section, add				11.65	35	105	7	
	4200	Hoist and single tubular tower, 5000 lb. electric, 100' high			5.41	237	710	2,125	185.30	
	4300	For each added 6'-6" section, add			.73	20	60	180	17.85	
	4400	Hoist and double tubular tower, 5000 lb., 100' high			5.77	250	750	2,250	196.15	
	4500	For each added 6'-6" section, add			.06	16.65	50	150	10.50	
	4550	Hoist and tower, mast type, 6000 lb., 100' high			5.34	275	825	2,475	207.70	
	4570	For each added 10' section, add			.15	15	45	135	10.20	
	4600	Hoist and tower, personnel, electric, 2000 lb., 100' @ 125 FPM			10.04	685	2,050	6,150	490.30	
	4700	3000 lb., 100' @ 200 FPM			10.79	735	2,200	6,600	526.30	
	4800	3000 lb., 150' @ 300 FPM			11.47	790	2,370	7,100	565.75	
	4900	4000 lb., 100' @ 300 FPM			12.33	885	2,650	7,950	628.65	
	5000	6000 lb., 100' @ 275 FPM			12.93	925	2,775	8,325	658.45	
	5100	For added heights up to 500', add		L.F.		1.67	5	15	1	
	5200	Jacks, hydraulic, 20 ton		Ea.	.15	2.67	8	24	2.80	
	5500	100 ton		"	.19	20	60	180	13.50	
	6000	Jacks, hydraulic, climbing with 50' jackrods								
	6010	and control consoles, minimum 3 mo. rental								
	6100	30 ton capacity		Ea.	.06	100	300	900	60.50	
	6150	For each added 10' jackrod section, add				3	9	27	1.80	
	6300	50 ton capacity				158	475	1,425	95	
	6350	For each added 10' jackrod section, add				3.33	10	30	2	
	6500	125 ton capacity				415	1,250	3,750	250	
	6550	For each added 10' jackrod section, add				26.50	80	240	16	
	6600	Cable jack, 10 ton capacity with 200' cable				83.50	250	750	50	
	6650	For each added 50' of cable, add				8.35	25	75	5	
490	0010	**WELLPOINT EQUIPMENT RENTAL** See also division 021-444	R021 -440							**490**
	0020	Based on 2 months rental								
	0100	Combination jetting & wellpoint pump, 60 H.P. diesel		Ea.	3.48	242	725	2,175	172.85	
	0200	High pressure gas jet pump, 200 H.P., 300 psi		"	9.69	207	620	1,850	201.50	
	0300	Discharge pipe, 8" diameter		L.F.		.38	1.15	3.45	.25	
	0350	12" diameter				.57	1.70	5.10	.35	
	0400	Header pipe, flows up to 150 G.P.M., 4" diameter				.33	1	3	.20	
	0500	400 G.P.M., 6" diameter				.33	1	3	.20	
	0600	800 G.P.M., 8" diameter				.57	1.70	5.10	.35	
	0700	1500 G.P.M., 10" diameter				.60	1.80	5.40	.35	
	0800	2500 G.P.M., 12" diameter				1.17	3.50	10.50	.70	
	0900	4500 G.P.M., 16" diameter				1.50	4.50	13.50	.90	
	0950	For quick coupling aluminum and plastic pipe, add				1.53	4.60	13.80	.90	
	1100	Wellpoint, 25' long, with fittings & riser pipe, 1-1/2" or 2" diameter		Ea.		3.08	9.25	28	1.85	
	1200	Wellpoint pump, diesel powered, 4" diameter, 20 H.P.			3.49	140	420	1,250	111.90	
	1300	6" diameter, 30 H.P.			5.39	177	530	1,600	149.10	
	1400	8" suction, 40 H.P.			6.41	240	720	2,150	195.30	
	1500	10" suction, 75 H.P.			6.91	280	840	2,525	223.30	
	1600	12" suction, 100 H.P.			9.89	450	1,350	4,050	349.10	
	1700	12" suction, 175 H.P.			10.83	500	1,500	4,500	386.65	

			DAILY	LABOR-						TOTAL	
	017 100	**Final Cleaning**						1999 BARE COSTS			
			CREW	OUTPUT	HOURS	UNIT	MAT.	LABOR	EQUIP.	TOTAL	INCL O&P
104	0010	**CLEANING UP** After job completion, allow, minimum				Job					.30%	104
	0040	Maximum				"					1%
	0050	Cleanup of floor area, continuous, per day	A-5	24	.750	M.S.F.	1.65	16.10	1.79	19.54	29.50
	0100	Final	"	11.50	1.565	"	2.63	33.50	3.74	39.87	60
	0200	Rubbish removal, see division 020-620									

For information about Means Estimating Seminars, see yellow pages 11 and 12 in back of book

1

GENERAL REQUIREMENTS

Important: See the Reference Section for critical supporting data - Reference Nos., Crews, & City Cost Indexes

Division 2
Site Work

Estimating Tips

020 Subsurface Investigation & Demolition

In preparing estimates on structures involving earthwork or foundations, all information concerning soil characteristics should be obtained. Look particularly for hazardous waste, evidence of prior dumping of debris, and previous stream beds.

- The costs shown for selective demolition do not include rubbish handling or disposal. These items should be estimated separately using Means data or other sources.

021 Site Preparation & Excavation Support

- If possible visit the site and take an inventory of the type, quantity and size of the trees. Certain trees may have a landscape resale value or firewood value. Stump disposal can be very expensive, particularly if they cannot be buried at the site. Consider using a bulldozer in lieu of hand cutting trees.

- Estimators should visit the site to determine the need for haul road, access, storage of materials, and security considerations. When estimating for access roads on unstable soil, consider using a geotextile stabilization fabric. It can greatly reduce the quantity of crushed stone or gravel. Sites of limited size and access can cause cost overruns due to lost productivity. Theft and damage is another consideration if the location is isolated. A temporary fence or security guards may be required. Investigate the site thoroughly.

022 Earthwork

- Estimating the actual cost of performing earthwork requires careful consideration of the variables involved. This includes items such as type of soil, whether or not water will be encountered, dewatering, whether or not banks need bracing, disposal of excavated earth, length of haul to fill or spoil sites, etc. If the project has large quantities of cut or fill, consider raising or lowering the site to reduce costs while paying close attention to the effect on site drainage and utilities if doing this.

- If the project has large quantities of fill, creating a borrow pit on the site can significantly lower the costs. It is very important to consider what time of year the project is scheduled for completion. Bad weather can create large cost overruns from dewatering, site repair and lost productivity from cold weather.

025 Paving & Surfacing

- When estimating paving, keep in mind the project schedule. If an asphaltic paving project is in a colder climate and runs through to the spring, consider placing the base course in the autumn, then topping it in the spring just prior to completion. This could save considerable costs in spring repair. Keep in mind that prices for asphalt and concrete are generally higher in the cold seasons.

026 Piped Utilities
027 Sewerage & Drainage

- Never assume that the water, sewer and drainage lines will go in at the early stages of the project. Consider the site access needs before dividing the site in half with open trenches, loose pipe, and machinery obstructions. Always inspect the site to establish that the site drawings are complete. Check off all existing utilities on your drawing as you locate them. If you find any discrepancies, mark up the site plan for further research. Differing site conditions can be very costly if discovered later in the project.

029 Landscaping

- The timing of planting and guarantee specifications often dictate the costs for establishing tree and shrub growth and a stand of grass or ground cover. Establish the work performance schedule to coincide with the local planting season. Maintenance and growth guarantees can add from 20% to 100% to the total landscaping cost. The cost to replace trees and shrubs can be as high as 5% of the total cost depending on the planting zone, soil conditions and time of year.

Reference Numbers

Reference numbers are shown in bold squares at the beginning of some major classifications. These numbers refer to related items in the Reference Section. The reference information may be an estimating procedure, an alternate pricing method or technical information.

Note: Not all subdivisions listed here necessarily appear in this publication.

020 120	Std Penetration Tests	CREW	DAILY OUTPUT	LABOR-HOURS	UNIT	1999 BARE COSTS				TOTAL INCL O&P
						MAT.	LABOR	EQUIP.	TOTAL	
123 0010	**BORINGS** Initial field stake out and determination of elevations	A-6	1	16	Day		410		410	625	123
0100	Drawings showing boring details				Total		170		170	245
0200	Report and recommendations from P.E.						375		375	540
0300	Mobilization and demobilization, minimum	B-55	4	6	↓		129	160	289	380
0350	For over 100 miles, per added mile		450	.053	Mile		1.15	1.42	2.57	3.35
0600	Auger holes in earth, no samples, 2-1/2" diameter		78.60	.305	L.F.		6.60	8.15	14.75	19.20
0650	4" diameter		67.50	.356			7.65	9.45	17.10	22.50
0800	Cased borings in earth, with samples, 2-1/2" diameter		55.50	.432		12.50	9.30	11.50	33.30	41
0850	4" diameter	↓	32.60	.736		20	15.85	19.60	55.45	68
1000	Drilling in rock, "BX" core, no sampling	B-56	34.90	.458			11.15	17.25	28.40	36
1050	With casing & sampling		31.70	.505		12.50	12.30	18.95	43.75	53.50
1200	"NX" core, no sampling		25.92	.617			15	23	38	48.50
1250	With casing and sampling	↓	25	.640	↓	16	15.55	24	55.55	68
1400	Drill rig and crew with truck mounted auger	B-55	1	24	Day		515	640	1,155	1,500
1450	With crawler type drill	B-56	1	16	"		390	600	990	1,250
1500	For inner city borings add, minimum									10%
1510	Maximum									20%
125 0010	**DRILLING, CORE** Reinforced concrete slab, up to 6" thick slab										125
0020	Including bit, layout and set up									
0100	1" diameter core	B-89A	28	.571	Ea.	2.28	14.15	2.19	18.62	27.50
0150	Each added inch thick, add		300	.053		.40	1.32	.20	1.92	2.75
0300	3" diameter core		23	.696		5	17.20	2.66	24.86	35.50
0350	Each added inch thick, add		186	.086		.90	2.13	.33	3.36	4.70
0500	4" diameter core		19	.842		5	21	3.23	29.23	42
0550	Each added inch thick, add		170	.094		1.15	2.33	.36	3.84	5.35
0700	6" diameter core		14	1.143		8.30	28.50	4.38	41.18	58.50
0750	Each added inch thick, add		140	.114		1.40	2.83	.44	4.67	6.50
0900	8" diameter core		11	1.455		11.30	36	5.55	52.85	75
0950	Each added inch thick, add		95	.168		1.90	4.17	.65	6.72	9.35
1100	10" diameter core		10	1.600		15.10	39.50	6.15	60.75	86
1150	Each added inch thick, add		80	.200		2.50	4.95	.77	8.22	11.40
1300	12" diameter core		9	1.778		18.10	44	6.80	68.90	97
1350	Each added inch thick, add		68	.235		3	5.80	.90	9.70	13.50
1500	14" diameter core		7	2.286		22	56.50	8.75	87.25	123
1550	Each added inch thick, add		55	.291		3.80	7.20	1.11	12.11	16.75
1700	18" diameter core	↓	4	4		28.50	99	15.30	142.80	204
1750	Each added inch thick, add	↓	28	.571		5	14.15	2.19	21.34	30.50
1760	For horizontal holes, add to above				↓				30%	30%
1770	Prestressed hollow core plank, 6" thick									
1780	1" diameter core	B-89A	52	.308	Ea.	1.50	7.60	1.18	10.28	14.95
1790	Each added inch thick, add		350	.046		.25	1.13	.18	1.56	2.25
1800	3" diameter core		50	.320		3.30	7.90	1.23	12.43	17.50
1810	Each added inch thick, add		240	.067		.55	1.65	.26	2.46	3.49
1820	4" diameter core		48	.333		4.40	8.25	1.28	13.93	19.25
1830	Each added inch thick, add		216	.074		.75	1.83	.28	2.86	4.03
1840	6" diameter core		44	.364		5.45	9	1.39	15.84	21.50
1850	Each added inch thick, add		175	.091		.90	2.26	.35	3.51	4.95
1860	8" diameter core		32	.500		7.30	12.40	1.92	21.62	29.50
1870	Each added inch thick, add		118	.136		1.25	3.36	.52	5.13	7.25
1880	10" diameter core		28	.571		9.85	14.15	2.19	26.19	36
1890	Each added inch thick, add		99	.162		1.35	4	.62	5.97	8.45
1900	12" diameter core		22	.727		12	18	2.79	32.79	45
1910	Each added inch thick, add		85	.188	↓	2	4.66	.72	7.38	10.35
1950	Minimum charge for above, 3" diameter core		7	2.286	Total		56.50	8.75	65.25	98.50
2000	4" diameter core	↓	6.80	2.353	↓		58	9	67	102

Important: See the Reference Section for critical supporting data - Reference Nos., Crews, & City Cost Indexes

020 120 | Std Penetration Tests

			CREW	DAILY OUTPUT	LABOR-HOURS	UNIT	1999 BARE COSTS				TOTAL INCL O&P	
							MAT.	LABOR	EQUIP.	TOTAL		
125	2050	6" diameter core	B-89A	6	2.667	Total		66	10.20	76.20	115	125
	2100	8" diameter core		5.50	2.909			72	11.15	83.15	125	
	2150	10" diameter core		4.75	3.368			83.50	12.90	96.40	145	
	2200	12" diameter core		3.90	4.103			102	15.70	117.70	177	
	2250	14" diameter core		3.38	4.734			117	18.15	135.15	205	
	2300	18" diameter core		3.15	5.079			126	19.45	145.45	220	
	3010	Bits for core drill, diamond, premium, 1" diameter				Ea.	112			112	123	
	3020	3" diameter					274			274	300	
	3040	4" diameter					305			305	335	
	3050	6" diameter					490			490	535	
	3080	8" diameter					715			715	785	
	3120	12" diameter					1,100			1,100	1,200	
	3180	18" diameter					1,800			1,800	1,975	
	3240	24" diameter					2,400			2,400	2,650	
128	0010	TEST PITS Hand digging, light soil	1 Clab	4.50	1.778	C.Y.		38		38	60	128
	0100	Heavy soil	"	2.50	3.200			68.50		68.50	108	
	0120	Loader-backhoe, light soil	B-11M	28	.571			14.25	10.35	24.60	33.50	
	0130	Heavy soil	"	20	.800			19.95	14.50	34.45	47	
	1000	Subsurface exploration, mobilization				Mile				5	5.75	
	1010	Difficult access for rig, add				Hr.				100	115	
	1020	Auger borings, drill rig, incl. samples				L.F.				11.50	13.25	
	1030	Hand auger								17	19.55	
	1050	Drill and sample every 5', split spoon								15	17.25	
	1060	Extra samples				Ea.				20	23	

020 550 | Site Demolition

			CREW	DAILY OUTPUT	LABOR-HOURS	UNIT	MAT.	LABOR	EQUIP.	TOTAL	TOTAL INCL O&P	
554	0010	SITE DEMOLITION No hauling, abandon catch basin or manhole	B-6	7	3.429	Ea.		80	31.50	111.50	159	554
	0020	Remove existing catch basin or manhole, masonry		4	6			140	54.50	194.50	278	
	0030	Catch basin or manhole frames and covers, stored		13	1.846			43	16.85	59.85	85.50	
	0040	Remove and reset		7	3.429			80	31.50	111.50	159	
	0100	Roadside delineators, remove only	B-80	175	.183			4.29	3.08	7.37	10.05	
	0110	Remove and reset	"	100	.320			7.50	5.40	12.90	17.60	
	0600	Fencing, barbed wire, 3 strand	2 Clab	430	.037	L.F.		.80		.80	1.26	
	0650	5 strand	"	280	.057			1.23		1.23	1.93	
	0700	Chain link, posts & fabric, remove only, 8' to 10' high	B-6	445	.054			1.26	.49	1.75	2.50	
	0750	Remove and reset	"	70	.343			8	3.13	11.13	15.90	
	0800	Guiderail, corrugated steel, remove only	B-80A	100	.240			5.15	1.80	6.95	10.05	
	0850	Remove and reset	"	40	.600			12.85	4.49	17.34	25.50	
	0860	Guide posts, remove only	B-80B	120	.267	Ea.		6.10	2.36	8.46	12.10	
	0870	Remove and reset	B-55	50	.480			10.35	12.75	23.10	30	
	0900	Hydrants, fire, remove only	B-21A	5	8			213	78	291	410	
	0950	Remove and reset	"	2	20			530	196	726	1,025	
	1000	Masonry walls, block or tile, solid, remove	B-5	1,800	.031	C.F.		.74	.58	1.32	1.79	
	1100	Cavity wall		2,200	.025			.60	.48	1.08	1.46	
	1200	Brick, solid		900	.062			1.48	1.16	2.64	3.58	
	1300	With block back-up		1,130	.050			1.18	.93	2.11	2.85	
	1400	Stone, with mortar		900	.062			1.48	1.16	2.64	3.58	
	1500	Dry set		1,500	.037			.89	.70	1.59	2.15	
	1600	Median barrier, precast concrete, remove and store	B-3	430	.112	L.F.		2.59	4.07	6.66	8.50	
	1610	Remove and reset	"	390	.123	"		2.85	4.49	7.34	9.35	
	1710	Pavement removal, bituminous roads, 3" thick	B-38	690	.058	S.Y.		1.41	1.48	2.89	3.82	
	1750	4" to 6" thick		420	.095			2.32	2.43	4.75	6.25	
	1800	Bituminous driveways		640	.063			1.52	1.59	3.11	4.11	
	1900	Concrete to 6" thick, hydraulic hammer, mesh reinforced		255	.157			3.83	4	7.83	10.35	
	2000	Rod reinforced		200	.200			4.88	5.10	9.98	13.15	
	2100	Concrete, 7" to 24" thick, plain		33	1.212	C.Y.		29.50	31	60.50	80	

2 SITE WORK

020 550 | Site Demolition

		CREW	DAILY OUTPUT	LABOR-HOURS	UNIT	1999 BARE COSTS MAT.	LABOR	EQUIP.	TOTAL	TOTAL INCL O&P	
554	2200	Reinforced	B-38	24	1.667	C.Y.		40.50	42.50	83	110	554
	2300	With hand held air equipment, bituminous, to 6" thick	B-39	1,900	.025	S.F.		.57	.09	.66	1	
	2320	Concrete to 6" thick, no reinforcing		1,200	.040			.91	.15	1.06	1.59	
	2340	Mesh reinforced		1,400	.034			.78	.13	.91	1.36	
	2360	Rod reinforced	▼	765	.063	▼		1.43	.24	1.67	2.49	
	2400	Curbs, concrete, plain	B-6	360	.067	L.F.		1.56	.61	2.17	3.09	
	2500	Reinforced		275	.087			2.04	.80	2.84	4.05	
	2600	Granite		360	.067			1.56	.61	2.17	3.09	
	2700	Bituminous		528	.045			1.06	.41	1.47	2.11	
	2900	Pipe removal, sewer/water, no excavation, 12" diameter		175	.137			3.21	1.25	4.46	6.35	
	2930	15" diameter		150	.160			3.74	1.46	5.20	7.40	
	2960	24" diameter		120	.200			4.67	1.82	6.49	9.25	
	3000	36" diameter		90	.267			6.25	2.43	8.68	12.30	
	3200	Steel, welded connections, 4" diameter		160	.150			3.51	1.37	4.88	6.95	
	3300	10" diameter	▼	80	.300	▼		7	2.74	9.74	13.90	
	3500	Railroad track removal, ties and track	B-13	330	.170			3.94	1.66	5.60	8	
	3600	Ballast	B-14	500	.096	C.Y.		2.18	.44	2.62	3.89	
	3700	Remove and re-install, ties & track using new bolts & spikes		50	.960	L.F.		22	4.38	26.38	39	
	3800	Turnouts using new bolts and spikes	▼	1	48	Ea.		1,100	219	1,319	1,950	
	4000	Sidewalk removal, bituminous, 2-1/2" thick	B-6	325	.074	S.Y.		1.73	.67	2.40	3.42	
	4050	Brick, set in mortar		185	.130			3.03	1.18	4.21	6	
	4100	Concrete, plain, 4"		160	.150			3.51	1.37	4.88	6.95	
	4200	Mesh reinforced	▼	150	.160	▼		3.74	1.46	5.20	7.40	
	5000	Slab on grade removal, plain	B-5	45	1.244	C.Y.		29.50	23	52.50	71.50	
	5100	Mesh reinforced		33	1.697			40.50	31.50	72	97.50	
	5200	Rod reinforced	▼	25	2.240			53	42	95	129	
	5500	For congested sites or small quantities, add up to								200%	200%	
	5550	For disposal on site, add	B-11A	232	.069			1.72	3.62	5.34	6.65	
	5600	To 5 miles, add	B-34D	76	.105	▼		2.33	7.40	9.73	11.70	
558	0010	**HYDRODEMOLITION**, concrete pavement, 4000 PSI, 2" depth	B-5	500	.112	S.F.		2.66	2.09	4.75	6.45	558
	0120	4" depth		450	.124			2.95	2.32	5.27	7.15	
	0130	6" depth		400	.140			3.32	2.61	5.93	8	
	0410	6000 PSI, 2" depth		410	.137			3.24	2.55	5.79	7.85	
	0420	4" depth		350	.160			3.80	2.99	6.79	9.20	
	0430	6" depth		300	.187			4.43	3.49	7.92	10.75	
	0510	8000 PSI, 2" depth		330	.170			4.03	3.17	7.20	9.75	
	0520	4" depth		280	.200			4.74	3.73	8.47	11.50	
	0530	6" depth	▼	240	.233	▼		5.55	4.36	9.91	13.40	

020 600 | Building Demolition

		CREW	DAILY OUTPUT	LABOR-HOURS	UNIT	MAT.	LABOR	EQUIP.	TOTAL	TOTAL INCL O&P	
604	0010	**BUILDING DEMOLITION** Large urban projects, incl. 20 Mi. haul									604
	0012	No foundation or dump fees, C.F. is volume of building standing, steel	B-8	21,500	.003	C.F.	.07	.11	.18	.23	
	0050	Concrete		15,300	.004		.10	.15	.25	.32	
	0080	Masonry		20,100	.003		.08	.11	.19	.25	
	0100	Mixture of types, average	▼	20,100	.003		.08	.11	.19	.25	
	0500	Small bldgs, or single bldgs, no salvage included, steel	B-3	14,800	.003		.08	.12	.20	.25	
	0600	Concrete		11,300	.004		.10	.15	.25	.32	
	0650	Masonry		14,800	.003		.08	.12	.20	.25	
	0700	Wood	▼	14,800	.003	▼	.08	.12	.20	.25	
	1000	Single family, one story house, wood, minimum				Ea.			2,300	2,700	
	1020	Maximum							4,000	4,800	
	1200	Two family, two story house, wood, minimum							3,000	3,600	
	1220	Maximum							5,800	7,000	
	1300	Three family, three story house, wood, minimum				▼			4,000	4,800	

Important: See the Reference Section for critical supporting data - Reference Nos., Crews, & City Cost Indexes

		020 600 | **Building Demolition**	CREW	DAILY OUTPUT	LABOR-HOURS	UNIT	MAT.	1999 BARE COSTS LABOR	EQUIP.	TOTAL	TOTAL INCL O&P	
604	1320	Maximum				Ea.				7,000	8,400	604
	5000	For buildings with no interior walls, deduct				↓				50%		
608	0010	**DISPOSAL ONLY** Urban buildings with salvage value allowed										608
	0020	Including loading and 5 mile haul to dump										
	0200	Steel frame	B-3	430	.112	C.Y.		2.59	4.07	6.66	8.50	
	0300	Concrete frame		365	.132			3.05	4.79	7.84	9.95	
	0400	Masonry construction		445	.108			2.50	3.93	6.43	8.20	
	0500	Wood frame	↓	247	.194	↓		4.50	7.10	11.60	14.75	
612	0010	**DUMP CHARGES** Typical urban city, tipping fees only										612
	0100	Building construction materials				Ton					55	
	0200	Trees, brush, lumber									45	
	0300	Rubbish only									50	
	0500	Reclamation station, usual charge				↓					80	
616	0010	**EXPLOSIVE/IMPLOSIVE DEMOLITION** Large projects, no disposal/fee										616
	0020	based on building volume, steel building	B-5B	16,900	.003	C.F.		.07	.12	.19	.24	
	0100	Concrete building		16,900	.003			.07	.12	.19	.24	
	0200	Masonry building	↓	16,900	.003	↓		.07	.12	.19	.24	
	0400	Disposal of material, minimum	B-3	445	.108	C.Y.		2.50	3.93	6.43	8.20	
	0500	Maximum	"	365	.132	"		3.05	4.79	7.84	9.95	
620	0010	**RUBBISH HANDLING** The following are to be added to the										620
	0020	demolition prices										
	0400	Chute, circular, prefabricated steel, 18" diameter	B-1	40	.600	L.F.	19.35	13.25		32.60	42.50	
	0440	30" diameter	"	30	.800	"	25	17.70		42.70	55.50	
	0600	Dumpster, weekly rental, 1 dump/week, 6 C.Y. capacity (2 Tons)				Ea.					300	
	0700	10 C.Y. capacity (4 Tons)									385	
	0800	30 C.Y. capacity (10 Tons)									650	
	0840	40 C.Y. capacity (13 Tons)				↓					775	
	1000	Dust partition, 6 mil polyethylene, 4' x 8' panels, 1" x 3" frame	2 Carp	2,000	.008	S.F.	.40	.22		.62	.78	
	1080	2" x 4" frame	"	2,000	.008	"	.50	.22		.72	.89	
	2000	Load, haul to chute & dumping into chute, 50' haul	2 Clab	24	.667	C.Y.		14.30		14.30	22.50	
	2040	100' haul		16.50	.970			21		21	32.50	
	2080	Over 100' haul, add per 100 L.F.		35.50	.451			9.65		9.65	15.20	
	2120	In elevators, per 10 floors, add	↓	140	.114			2.45		2.45	3.86	
	3000	Loading & trucking, including 2 mile haul, chute loaded	B-16	45	.711			15.70	9.85	25.55	35.50	
	3040	Hand loading truck, 50' haul	"	48	.667			14.75	9.20	23.95	33	
	3080	Machine loading truck	B-17	120	.267			6.15	4.86	11.01	14.85	
	5000	Haul, per mile, up to 8 C.Y. truck	B-34B	1,165	.007			.15	.38	.53	.65	
	5100	Over 8 C.Y. truck	"	1,550	.005	↓		.11	.29	.40	.48	

		020 700 | **Selective Demolition**										
702	0010	**CEILING DEMOLITION**										702
	0200	Drywall, furred and nailed	2 Clab	800	.020	S.F.		.43		.43	.68	
	0220	On metal frame		760	.021			.45		.45	.71	
	0240	On suspension system, including system		720	.022			.48		.48	.75	
	1000	Plaster, lime and horse hair, on wood lath, incl. lath		700	.023			.49		.49	.77	
	1020	On metal lath		570	.028			.60		.60	.95	
	1100	Gypsum, on gypsum lath		720	.022			.48		.48	.75	
	1120	On metal lath		500	.032			.69		.69	1.08	
	1200	Suspended ceiling, mineral fiber, 2'x2' or 2'x4'		1,500	.011			.23		.23	.36	
	1250	On suspension system, incl. system		1,200	.013			.29		.29	.45	
	1500	Tile, wood fiber, 12" x 12", glued		900	.018			.38		.38	.60	
	1540	Stapled		1,500	.011			.23		.23	.36	
	1580	On suspension system, incl. system		760	.021			.45		.45	.71	
	2000	Wood, tongue and groove, 1" x 4"	↓	1,000	.016	↓		.34		.34	.54	

SITE WORK 2

SITE WORK **2**

			CREW	DAILY OUTPUT	LABOR-HOURS	UNIT	MAT.	LABOR	EQUIP.	TOTAL	TOTAL INCL O&P	
								1999 BARE COSTS				
702	2040	1" x 8"	2 Clab	1,100	.015	S.F.		.31		.31	.49	**702**
	2400	Plywood or wood fiberboard, 4' x 8' sheets	↓	1,200	.013	↓		.29		.29	.45	
704	0010	**CUTOUT DEMOLITION** Conc., elev. slab, light reinf., under 6 C.F.	B-9C	65	.615	C.F.		13.45	2.78	16.23	24	**704**
	0050	Light reinforcing, over 6 C.F.	"	75	.533	"		11.65	2.41	14.06	21	
	0200	Slab on grade to 6" thick, not reinforced, under 8 S.F.	B-9	85	.471	S.F.		10.30	2.12	12.42	18.55	
	0250	Not reinforced, over 8 S.F.		175	.229	"		4.99	1.03	6.02	9	
	0600	Walls, not reinforced, under 6 C.F.		60	.667	C.F.		14.55	3.01	17.56	26.50	
	0650	Not reinforced, over 6 C.F.	↓	65	.615			13.45	2.78	16.23	24	
	1000	Concrete, elevated slab, bar reinforced, under 6 C.F.	B-9C	45	.889			19.40	4.01	23.41	35	
	1050	Bar reinforced, over 6 C.F.	"	50	.800			17.50	3.61	21.11	31.50	
	1200	Slab on grade to 6" thick, bar reinforced, under 8 S.F.	B-9	75	.533	S.F.		11.65	2.41	14.06	21	
	1250	Bar reinforced, over 8 S.F.	"	105	.381	"		8.30	1.72	10.02	15	
	1400	Walls, bar reinforced, under 6 C.F.	B-9C	50	.800	C.F.		17.50	3.61	21.11	31.50	
	1450	Bar reinforced, over 6 C.F.	"	55	.727	"		15.90	3.28	19.18	28.50	
	2000	Brick, to 4 S.F. opening, not including toothing										
	2040	4" thick	B-9C	30	1.333	Ea.		29	6	35	52.50	
	2060	8" thick		18	2.222			48.50	10	58.50	87.50	
	2080	12" thick		10	4			87.50	18.05	105.55	158	
	2400	Concrete block, to 4 S.F. opening, 2" thick		35	1.143			25	5.15	30.15	45	
	2420	4" thick		30	1.333			29	6	35	52.50	
	2440	8" thick		27	1.481			32.50	6.70	39.20	58.50	
	2460	12" thick	↓	24	1.667			36.50	7.50	44	66	
	2600	Gypsum block, to 4 S.F. opening, 2" thick	B-9	80	.500			10.95	2.26	13.21	19.70	
	2620	4" thick		70	.571			12.50	2.58	15.08	22.50	
	2640	8" thick		55	.727			15.90	3.28	19.18	28.50	
	2800	Terra cotta, to 4 S.F. opening, 4" thick		70	.571			12.50	2.58	15.08	22.50	
	2840	8" thick		65	.615			13.45	2.78	16.23	24	
	2880	12" thick	↓	50	.800	↓		17.50	3.61	21.11	31.50	
	3000	Toothing masonry cutouts, brick, soft old mortar	1 Brhe	40	.200	V.L.F.		4.32		4.32	6.70	
	3100	Hard mortar		30	.267			5.75		5.75	8.95	
	3200	Block, soft old mortar		70	.114			2.47		2.47	3.84	
	3400	Hard mortar	↓	50	.160	↓		3.46		3.46	5.40	
	6000	Walls, interior, not including re-framing,										
	6010	openings to 5 S.F.										
	6100	Drywall to 5/8" thick	A-1	24	.333	Ea.		7.15	2.86	10.01	14.40	
	6200	Paneling to 3/4" thick		20	.400			8.60	3.43	12.03	17.30	
	6300	Plaster, on gypsum lath		20	.400			8.60	3.43	12.03	17.30	
	6340	On wire lath	↓	14	.571	↓		12.25	4.90	17.15	24.50	
	7000	Wood frame, not including re-framing, openings to 5 S.F.										
	7200	Floors, sheathing and flooring to 2" thick	A-1	5	1.600	Ea.		34.50	13.75	48.25	69	
	7310	Roofs, sheathing to 1" thick, not including roofing		6	1.333			28.50	11.45	39.95	57.50	
	7410	Walls, sheathing to 1" thick, not including siding	↓	7	1.143	↓		24.50	9.80	34.30	49.50	
706	0010	**DOOR DEMOLITION**										**706**
	0200	Doors, exterior, 1-3/4" thick, single, 3' x 7' high	1 Clab	16	.500	Ea.		10.75		10.75	16.90	
	0220	Double, 6' x 7' high		12	.667			14.30		14.30	22.50	
	0500	Interior, 1-3/8" thick, single, 3' x 7' high		20	.400			8.60		8.60	13.50	
	0520	Double, 6' x 7' high		16	.500			10.75		10.75	16.90	
	0700	Bi-folding, 3' x 6'-8" high		20	.400			8.60		8.60	13.50	
	0720	6' x 6'-8" high		18	.444			9.55		9.55	15	
	0900	Bi-passing, 3' x 6'-8" high		16	.500			10.75		10.75	16.90	
	0940	6' x 6'-8" high	↓	14	.571			12.25		12.25	19.30	
	1500	Remove and reset, minimum	1 Carp	8	1			27.50		27.50	43	
	1520	Maximum	"	6	1.333			36.50		36.50	57.50	
	2000	Frames, including trim, metal	A-1	8	1			21.50	8.60	30.10	43.50	
	2200	Wood	2 Carp	32	.500			13.65		13.65	21.50	
	3000	Special doors, counter doors	↓	6	2.667	↓		73		73	115	

Important: See the Reference Section for critical supporting data - Reference Nos., Crews, & City Cost Indexes

		020 700	Selective Demolition	CREW	DAILY OUTPUT	LABOR-HOURS	UNIT	1999 BARE COSTS MAT.	LABOR	EQUIP.	TOTAL	TOTAL INCL O&P	
706	3100		Double acting	2 Carp	10	1.600	Ea.		43.50		43.50	68.50	706
	3200		Floor door (trap type)		8	2			54.50		54.50	86	
	3300		Glass, sliding, including frames		12	1.333			36.50		36.50	57.50	
	3400		Overhead, commercial, 12' x 12' high		4	4			109		109	172	
	3440		20' x 16' high		3	5.333			146		146	229	
	3500		Residential, 9' x 7' high		8	2			54.50		54.50	86	
	3540		16' x 7' high		7	2.286			62.50		62.50	98	
	3600		Remove and reset, minimum		4	4			109		109	172	
	3620		Maximum		2.50	6.400			175		175	275	
	3700		Roll-up grille		5	3.200			87.50		87.50	137	
	3800		Revolving door		2	8			218		218	345	
	3900		Storefront swing door	▼	3	5.333	▼		146		146	229	
708	0010	**ELECTRICAL DEMOLITION**											708
	0020		Conduit to 15' high, including fittings & hangers										
	0100		Rigid galvanized steel, 1/2" to 1" diameter	1 Elec	242	.033	L.F.		1.05		1.05	1.58	
	0120		1-1/4" to 2"		200	.040			1.28		1.28	1.91	
	0140		2" to 4"		151	.053			1.69		1.69	2.53	
	0160		4" to 6"		80	.100			3.19		3.19	4.77	
	0200		Electric metallic tubing (EMT) 1/2" to 1"		394	.020			.65		.65	.97	
	0220		1-1/4" to 1-1/2"		326	.025			.78		.78	1.17	
	0260		3-1/2" to 4"	▼	95	.084	▼		2.69		2.69	4.02	
	0270		Armored cable, (BX) avg. 50' runs										
	0280		#14, 2 wire	1 Elec	690	.012	L.F.		.37		.37	.55	
	0290		#14, 3 wire		571	.014			.45		.45	.67	
	0300		#12, 2 wire		605	.013			.42		.42	.63	
	0310		#12, 3 wire		514	.016			.50		.50	.74	
	0320		#10, 2 wire		514	.016			.50		.50	.74	
	0330		#10, 3 wire		425	.019			.60		.60	.90	
	0340		#8, 3 wire	▼	342	.023	▼		.75		.75	1.12	
	0350		Non metallic sheathed cable (Romex)										
	0360		#14, 2 wire	1 Elec	720	.011	L.F.		.35		.35	.53	
	0370		#14, 3 wire		657	.012			.39		.39	.58	
	0380		#12, 2 wire		629	.013			.41		.41	.61	
	0390		#10, 3 wire	▼	450	.018	▼		.57		.57	.85	
	0400		Wiremold raceway, including fittings & hangers										
	0420		No. 3000	1 Elec	250	.032	L.F.		1.02		1.02	1.53	
	0440		No. 4000		217	.037			1.18		1.18	1.76	
	0460		No. 6000	▼	166	.048	▼		1.54		1.54	2.30	
	0500		Channels, steel, including fittings & hangers										
	0520		3/4" x 1-1/2"	1 Elec	308	.026	L.F.		.83		.83	1.24	
	0540		1-1/2" x 1-1/2"		269	.030			.95		.95	1.42	
	0560		1-1/2" x 1-7/8"	▼	229	.035			1.11		1.11	1.67	
	0600		Copper bus duct, indoor, 3 phase										
	0610		Including hangers & supports										
	0620		225 amp	2 Elec	135	.119	L.F.		3.78		3.78	5.65	
	0640		400 amp		106	.151			4.82		4.82	7.20	
	0660		600 amp		86	.186			5.95		5.95	8.85	
	0680		1000 amp		60	.267			8.50		8.50	12.70	
	0700		1600 amp		40	.400			12.75		12.75	19.10	
	0720		3000 amp	▼	10	1.600	▼		51		51	76.50	
	1300		Transformer, dry type, 1 ph, incl. removal of										
	1320		supports, wire & pipe terminations										
	1340		1 KVA	1 Elec	7.70	1.039	Ea.		33		33	49.50	
	1420		75 KVA	2 Elec	2.50	6.400	"		204		204	305	
	1440		3 Phase to 600V, primary										
	1460		3 KVA	1 Elec	3.85	2.078	Ea.		66.50		66.50	99	

SITE WORK **2**

33

SITE WORK

2

020 700	Selective Demolition	CREW	DAILY OUTPUT	LABOR-HOURS	UNIT	1999 BARE COSTS				TOTAL INCL O&P
						MAT.	LABOR	EQUIP.	TOTAL	
708										**708**
1520	75 KVA	2 Elec	2.70	5.926	Ea.		189		189	283
1550	300 KVA	R-3	1.80	11.111			350	75.50	425.50	610
1570	750 KVA	"	1.10	18.182	↓		575	123	698	995
1800	Wire, THW-THWN-THHN, removed from									
1810	in place conduit, to 15' high									
1830	#14	1 Elec	65	.123	C.L.F.		3.93		3.93	5.85
1840	#12		55	.145			4.64		4.64	6.95
1850	#10		45.50	.176			5.60		5.60	8.40
1860	#8		40.40	.198			6.30		6.30	9.45
1870	#6	↓	32.60	.245			7.85		7.85	11.70
1880	#4	2 Elec	53	.302			9.65		9.65	14.40
1890	#3		50	.320			10.20		10.20	15.25
1900	#2		44.60	.359			11.45		11.45	17.10
1910	1/0		33.20	.482			15.35		15.35	23
1920	2/0		29.20	.548			17.50		17.50	26
1930	3/0		25	.640			20.50		20.50	30.50
1940	4/0		22	.727			23		23	34.50
1950	250 kcmil		20	.800			25.50		25.50	38
1960	300 kcmil		19	.842			27		27	40
1970	350 kcmil		18	.889			28.50		28.50	42.50
1980	400 kcmil		17	.941			30		30	45
1990	500 kcmil	↓	16.20	.988	↓		31.50		31.50	47
2000	Interior fluorescent fixtures, incl. supports									
2010	& whips, to 15' high									
2100	Recessed drop-in 2' x 2', 2 lamp	2 Elec	35	.457	Ea.		14.60		14.60	22
2120	2' x 4', 2 lamp		33	.485			15.45		15.45	23
2140	2' x 4', 4 lamp		30	.533			17		17	25.50
2160	4' x 4', 4 lamp	↓	20	.800	↓		25.50		25.50	38
2180	Surface mount, acrylic lens & hinged frame									
2200	1' x 4', 2 lamp	2 Elec	44	.364	Ea.		11.60		11.60	17.35
2220	2' x 2', 2 lamp		44	.364			11.60		11.60	17.35
2260	2' x 4', 4 lamp		33	.485			15.45		15.45	23
2280	4' x 4', 4 lamp	↓	40	.400	↓		12.75		12.75	19.10
2300	Strip fixtures, surface mount									
2320	4' long, 1 lamp	2 Elec	53	.302	Ea.		9.65		9.65	14.40
2340	4' long, 2 lamp		50	.320			10.20		10.20	15.25
2360	8' long, 1 lamp		42	.381			12.15		12.15	18.15
2380	8' long, 2 lamp	↓	40	.400	↓		12.75		12.75	19.10
2400	Pendant mount, industrial, incl. removal									
2410	of chain or rod hangers, to 15' high									
2420	4' long, 2 lamp	2 Elec	35	.457	Ea.		14.60		14.60	22
2440	8' long, 2 lamp	"	27	.593	"		18.90		18.90	28.50
712	**0010 FLOORING DEMOLITION**									**712**
0200	Brick with mortar	2 Clab	475	.034	S.F.		.72		.72	1.14
0400	Carpet, bonded, including surface scraping		2,000	.008			.17		.17	.27
0440	Scrim applied		8,000	.002			.04		.04	.07
0480	Tackless		9,000	.002			.04		.04	.06
0600	Composition, acrylic or epoxy	↓	400	.040			.86		.86	1.35
0700	Concrete, scarify skin	A-1	225	.036			.76	.31	1.07	1.54
0800	Resilient, sheet goods	2 Clab	1,400	.011			.25		.25	.39
0820	For gym floors		900	.018			.38		.38	.60
0900	Vinyl composition tile, 12" x 12"		1,000	.016			.34		.34	.54
2000	Tile, ceramic, thin set		675	.024			.51		.51	.80
2020	Mud set		625	.026			.55		.55	.86
2200	Marble, slate, thin set		675	.024			.51		.51	.80
2220	Mud set	↓	625	.026	↓		.55		.55	.86

020 700	Selective Demolition	CREW	DAILY OUTPUT	LABOR-HOURS	UNIT	1999 BARE COSTS				TOTAL INCL O&P	
						MAT.	LABOR	EQUIP.	TOTAL		
712											**712**
2600	Terrazzo, thin set	2 Clab	450	.036	S.F.		.76		.76	1.20	
2620	Mud set		425	.038			.81		.81	1.27	
2640	Cast in place	↓	300	.053			1.14		1.14	1.80	
3000	Wood, block, on end	1 Carp	400	.020			.55		.55	.86	
3200	Parquet		450	.018			.49		.49	.76	
3400	Strip flooring, interior, 2-1/4" x 25/32" thick		325	.025			.67		.67	1.06	
3500	Exterior, porch flooring, 1" x 4"		220	.036			.99		.99	1.56	
3800	Subfloor, tongue and groove, 1" x 6"		325	.025			.67		.67	1.06	
3820	1" x 8"		430	.019			.51		.51	.80	
3840	1" x 10"		520	.015			.42		.42	.66	
4000	Plywood, nailed		600	.013			.36		.36	.57	
4100	Glued and nailed	↓	400	.020			.55		.55	.86	
8000	Remove flooring, bead blast, minimum	A-1A	1,000	.008			.22	.13	.35	.50	
8100	Maximum		400	.020			.56	.33	.89	1.24	
8150	Mastic only	↓	1,500	.005	↓		.15	.09	.24	.34	
714	**0010 FRAMING DEMOLITION**										**714**
1020	Concrete, average reinforcing, beams, 8" x 10"	B-9	120	.333	L.F.		7.30	1.50	8.80	13.10	
1040	10" x 12"		110	.364			7.95	1.64	9.59	14.30	
1060	12" x 14"		90	.444			9.70	2	11.70	17.50	
1200	Columns, 8" x 8"		120	.333			7.30	1.50	8.80	13.10	
1240	10" x 10"		120	.333			7.30	1.50	8.80	13.10	
1280	12" x 12"		110	.364			7.95	1.64	9.59	14.30	
1320	14" x 14"		100	.400			8.75	1.80	10.55	15.75	
1400	Girders, 14" x 16"		55	.727			15.90	3.28	19.18	28.50	
1440	16" x 18"		40	1	↓		22	4.51	26.51	39.50	
1600	Slabs, elevated, 6" thick		600	.067	S.F.		1.46	.30	1.76	2.62	
1640	8" thick		450	.089			1.94	.40	2.34	3.50	
1680	10" thick	↓	360	.111	↓		2.43	.50	2.93	4.37	
1900	Add for heavy reinforcement									25%	
2000	Steel framing, beams, 4" x 6"	B-13	500	.112	L.F.		2.60	1.10	3.70	5.25	
2020	4" x 8"		400	.140			3.25	1.37	4.62	6.55	
2080	8" x 12"		250	.224			5.20	2.20	7.40	10.50	
2200	Columns, 6" x 6"		400	.140			3.25	1.37	4.62	6.55	
2240	8" x 8"		350	.160			3.72	1.57	5.29	7.55	
2280	10" x 10"		320	.175			4.07	1.72	5.79	8.25	
2400	Girders, 10" x 12"		225	.249			5.80	2.44	8.24	11.70	
2440	10" x 14"		200	.280			6.50	2.75	9.25	13.10	
2480	10" x 16"		165	.339			7.90	3.33	11.23	15.90	
2520	10" x 24"	↓	125	.448			10.40	4.39	14.79	21	
3000	Wood framing, beams, 6" x 8"	B-2	275	.145			3.18		3.18	5	
3040	6" x 10"		220	.182			3.97		3.97	6.25	
3080	6" x 12"		185	.216			4.72		4.72	7.45	
3120	8" x 12"		140	.286			6.25		6.25	9.80	
3160	10" x 12"	↓	110	.364			7.95		7.95	12.50	
3400	Fascia boards, 1" x 6"	1 Clab	500	.016			.34		.34	.54	
3440	1" x 8"		450	.018			.38		.38	.60	
3480	1" x 10"		400	.020			.43		.43	.68	
3800	Headers over openings, 2 @ 2" x 6"		110	.073			1.56		1.56	2.45	
3840	2 @ 2" x 8"		100	.080			1.72		1.72	2.70	
3880	2 @ 2" x 10"	↓	90	.089			1.91		1.91	3	
4230	Joists, 2" x 6"	2 Clab	970	.016			.35		.35	.56	
4240	2" x 8"		940	.017			.37		.37	.57	
4250	2" x 10"		910	.018			.38		.38	.59	
4280	2" x 12"		880	.018			.39		.39	.61	
5400	Posts, 4" x 4"	↓	800	.020	↓		.43		.43	.68	

SITE WORK **2**

35

SITE WORK 2

020 700	Selective Demolition	CREW	DAILY OUTPUT	LABOR-HOURS	UNIT	MAT.	LABOR	EQUIP.	TOTAL	TOTAL INCL O&P	
714	5440	6" x 6"	2 Clab	400	.040	L.F.		.86		.86	1.35	714
	5480	8" x 8"		300	.053			1.14		1.14	1.80	
	5500	10" x 10"		240	.067			1.43		1.43	2.25	
	5800	Rafters, ordinary, 2" x 6"		850	.019			.40		.40	.64	
	5840	2" x 8"		837	.019			.41		.41	.65	
	6200	Stairs and stringers, minimum		40	.400	Riser		8.60		8.60	13.50	
	6240	Maximum		26	.615	"		13.20		13.20	21	
	6600	Studs, 2" x 4"		2,000	.008	L.F.		.17		.17	.27	
	6640	2" x 6"		1,600	.010	"		.21		.21	.34	
	9500	See Div. 020-620 for rubbish handling										
716	0010	**GUTTING** Building interior, including disposal, dumpster fees not included										716
	0500	Residential building										
	0560	Minimum	B-16	400	.080	SF Flr.		1.77	1.11	2.88	3.98	
	0580	Maximum	"	360	.089	"		1.97	1.23	3.20	4.42	
	0900	Commercial building										
	1000	Minimum	B-16	350	.091	SF Flr.		2.02	1.26	3.28	4.55	
	1020	Maximum	"	250	.128	"		2.83	1.77	4.60	6.35	
717	0010	**HAZARDOUS WASTE CLEANUP/PICKUP/DISPOSAL**										717
	0100	For contractor equipment, i.e. dozer,										
	0110	front end loader, dump truck, etc., see div. 016-408										
	1000	Solid pickup										
	1100	55 gal. drums				Ea.					200	
	1120	Bulk material, minimum				Ton					150	
	1130	Maximum				"					500	
	1200	Transportation to disposal site										
	1220	Truckload = 80 drums or 25 C.Y. or 18 tons										
	1260	Minimum				Mile					2.30	
	1270	Maximum				"					4	
	3000	Liquid pickup, vacuum truck, stainless steel tank										
	3100	Minimum charge, 4 hours										
	3110	1 compartment, 2200 gallon				Hr.					100	
	3120	2 compartment, 5000 gallon				"					100	
	3400	Transportation in 6900 gallon bulk truck				Mile					4.30	
	3410	In teflon lined truck				"					5	
	5000	Heavy sludge or dry vacuumable material				Hr.					100	
	6000	Dumpsite disposal charge, minimum				Ton					100	
	6020	Maximum				"					400	
718	0010	**HVAC DEMOLITION**										718
	0100	Air conditioner, split unit, 3 ton	Q-5	2	8	Ea.		236		236	355	
	0150	Package unit, 3 ton	Q-6	3	8			245		245	370	
	0200	Air conditioner, rooftop, gas heat, remove & reset, 10 ton		.30	80			2,450		2,450	3,700	
	0240	Remove & reset, 20 ton		.15	160			4,900		4,900	7,400	
	0300	Boiler, electric	Q-19	2	12			365		365	550	
	0340	Gas or oil, steel, under 150 MBH	Q-6	3	8			245		245	370	
	0380	Over 150 MBH	"	2	12			365		365	555	
	1000	Ductwork, 4" high, 8" wide	1 Clab	200	.040	L.F.		.86		.86	1.35	
	1100	6" high, 8" wide		165	.048			1.04		1.04	1.64	
	1200	10" high, 12" wide		125	.064			1.37		1.37	2.16	
	1300	12"-14" high, 16"-18" wide		85	.094			2.02		2.02	3.18	
	1400	18" high, 24" wide		67	.119			2.56		2.56	4.03	
	1500	30" high, 36" wide		56	.143			3.06		3.06	4.82	
	1540	72" wide		50	.160			3.43		3.43	5.40	
	3000	Mechanical equipment, light items. Unit is weight, not cooling.	Q-5	.90	17.778	Ton		525		525	795	
	3600	Heavy items	"	1.10	14.545	"		430		430	650	
	3700	Deduct for salvage (when applicable), minimum				Job					52	

Important: See the Reference Section for critical supporting data - Reference Nos., Crews, & City Cost Indexes

020 700 | Selective Demolition

		CREW	DAILY OUTPUT	LABOR-HOURS	UNIT	1999 BARE COSTS				TOTAL INCL O&P	
						MAT.	LABOR	EQUIP.	TOTAL		
718	3750	Maximum				Job					420	718
720	0010	**MILLWORK AND TRIM DEMOLITION**										720
	1000	Cabinets, wood, base cabinets	2 Clab	80	.200	L.F.		4.29		4.29	6.75	
	1020	Wall cabinets	"	80	.200	"		4.29		4.29	6.75	
	1060	Remove and reset, base cabinets	2 Carp	18	.889	Ea.		24.50		24.50	38	
	1070	Wall cabinets	"	20	.800	"		22		22	34.50	
	1100	Steel, painted, base cabinets	2 Clab	60	.267	L.F.		5.70		5.70	9	
	1120	Wall cabinets		60	.267	"		5.70		5.70	9	
	1200	Casework, large area		320	.050	S.F.		1.07		1.07	1.69	
	1220	Selective		200	.080	"		1.72		1.72	2.70	
	1500	Counter top, minimum		200	.080	L.F.		1.72		1.72	2.70	
	1510	Maximum	↓	120	.133			2.86		2.86	4.50	
	1550	Remove and reset, minimum	2 Carp	50	.320			8.75		8.75	13.75	
	1560	Maximum	"	40	.400	↓		10.90		10.90	17.20	
	2000	Paneling, 4' x 8' sheets, 1/4" thick	2 Clab	2,000	.008	S.F.		.17		.17	.27	
	2100	Boards, 1" x 4"		700	.023			.49		.49	.77	
	2120	1" x 6"		750	.021			.46		.46	.72	
	2140	1" x 8"		800	.020	↓		.43		.43	.68	
	3000	Trim, baseboard, to 6" wide		1,200	.013	L.F.		.29		.29	.45	
	3040	12" wide	↓	1,000	.016			.34		.34	.54	
	3080	Remove and reset, minimum	2 Carp	400	.040			1.09		1.09	1.72	
	3090	Maximum	"	300	.053			1.46		1.46	2.29	
	3100	Ceiling trim	2 Clab	1,000	.016			.34		.34	.54	
	3120	Chair rail		1,200	.013			.29		.29	.45	
	3140	Railings with balusters		240	.067	↓		1.43		1.43	2.25	
	3160	Wainscoting	↓	700	.023	S.F.		.49		.49	.77	
724	0010	**PLUMBING DEMOLITION**										724
	1020	Fixtures, including 10' piping										
	1100	Bath tubs, cast iron	1 Plum	4	2	Ea.		65		65	99	
	1120	Fiberglass		6	1.333			43.50		43.50	66	
	1140	Steel		5	1.600			52		52	79	
	1200	Lavatory, wall hung		10	.800			26		26	39.50	
	1220	Counter top		8	1			32.50		32.50	49.50	
	1300	Sink, steel or cast iron, single		8	1			32.50		32.50	49.50	
	1320	Double		7	1.143			37.50		37.50	56.50	
	1400	Water closet, floor mounted		8	1			32.50		32.50	49.50	
	1420	Wall mounted		7	1.143			37.50		37.50	56.50	
	1500	Urinal, floor mounted		4	2			65		65	99	
	1520	Wall mounted		7	1.143			37.50		37.50	56.50	
	1600	Water fountains, free standing		8	1			32.50		32.50	49.50	
	1620	Recessed		6	1.333	↓		43.50		43.50	66	
	2000	Piping, metal, to 2" diameter		200	.040	L.F.		1.30		1.30	1.98	
	2050	2" to 4" diameter	↓	150	.053			1.74		1.74	2.63	
	2100	4" to 8" diameter	2 Plum	100	.160			5.20		5.20	7.90	
	2150	8" to 16" diameter	"	60	.267	↓		8.70		8.70	13.15	
	2160	Deduct for salvage, aluminum scrap				Ton					525	
	2170	Brass scrap									420	
	2180	Copper scrap									1,025	
	2200	Lead scrap									235	
	2220	Steel scrap				↓					65	
	2250	Water heater, 40 gal.	1 Plum	6	1.333	Ea.		43.50		43.50	66	
726	0010	**ROOFING AND SIDING DEMOLITION**										726
	1000	Deck, roof, concrete plank	B-13	1,680	.033	S.F.		.77	.33	1.10	1.57	

020 700 | Selective Demolition

			CREW	DAILY OUTPUT	LABOR-HOURS	UNIT	1999 BARE COSTS MAT.	LABOR	EQUIP.	TOTAL	TOTAL INCL O&P	
726	1100	Gypsum plank	B-13	3,900	.014	S.F.		.33	.14	.47	.67	726
	1150	Metal decking	↓	3,500	.016			.37	.16	.53	.75	
	1200	Wood, boards, tongue and groove, 2" x 6"	2 Clab	960	.017			.36		.36	.56	
	1220	2" x 10"		1,040	.015			.33		.33	.52	
	1280	Standard planks, 1" x 6"		1,080	.015			.32		.32	.50	
	1320	1" x 8"		1,160	.014			.30		.30	.47	
	1340	1" x 12"		1,200	.013			.29		.29	.45	
	1350	Plywood, to 1" thick	↓	2,000	.008	↓		.17		.17	.27	
	2000	Gutters, aluminum or wood, edge hung	1 Clab	240	.033	L.F.		.72		.72	1.13	
	2100	Built-in		100	.080	"		1.72		1.72	2.70	
	2500	Roof accessories, plumbing vent flashing		14	.571	Ea.		12.25		12.25	19.30	
	2600	Adjustable metal chimney flashing		9	.889	"		19.05		19.05	30	
	2650	Coping, sheet metal, up to 12" wide	↓	240	.033	L.F.		.72		.72	1.13	
	2660	Concrete, up to 12" wide	2 Clab	160	.100	"		2.15		2.15	3.38	
	3000	Roofing, built-up, 5 ply roof, no gravel	B-2	1,600	.025	S.F.		.55		.55	.86	
	3001	Including gravel		890	.045			.98		.98	1.55	
	3100	Gravel removal, minimum		5,000	.008			.17		.17	.28	
	3120	Maximum		2,000	.020			.44		.44	.69	
	3400	Roof insulation board, up to 2" thick	↓	3,900	.010	↓		.22		.22	.35	
	3450	Roll roofing, cold adhesive	1 Clab	12	.667	Sq.		14.30		14.30	22.50	
	4000	Shingles, asphalt strip, 1 layer	B-2	3,500	.011	S.F.		.25		.25	.39	
	4100	Slate		2,500	.016			.35		.35	.55	
	4300	Wood	↓	2,200	.018	↓		.40		.40	.63	
	4500	Skylight to 10 S.F.	1 Clab	8	1	Ea.		21.50		21.50	34	
	5000	Siding, metal, horizontal		444	.018	S.F.		.39		.39	.61	
	5020	Vertical		400	.020			.43		.43	.68	
	5200	Wood, boards, vertical		400	.020			.43		.43	.68	
	5220	Clapboards, horizontal		380	.021			.45		.45	.71	
	5240	Shingles		350	.023			.49		.49	.77	
	5260	Textured plywood	↓	725	.011	↓		.24		.24	.37	
728	0010	**SAW CUTTING**, Asphalt, up to 3" deep	B-89	1,050	.015	L.F.	.24	.37	.29	.90	1.15	728
	0020	Each additional inch of depth		1,800	.009		.06	.22	.17	.45	.58	
	0400	Concrete slabs, mesh reinforcing, up to 3" deep		980	.016		.33	.40	.31	1.04	1.32	
	0420	Each additional inch of depth	↓	1,600	.010		.44	.24	.19	.87	1.06	
	0800	Concrete walls, hydraulic saw, plain, per inch of depth	B-89B	250	.064		.30	1.57	2.11	3.98	5.05	
	0820	Rod reinforcing, per inch of depth		150	.107		.42	2.61	3.52	6.55	8.30	
	1200	Masonry walls, hydraulic saw, brick, per inch of depth		300	.053		.30	1.31	1.76	3.37	4.26	
	1220	Block walls, solid, per inch of depth	↓	250	.064		.31	1.57	2.11	3.99	5.05	
	2000	Brick or masonry w/hand held saw, per inch of depth	A-1	125	.064		.24	1.37	.55	2.16	3.03	
	5000	Wood sheathing to 1" thick, on walls	1 Carp	200	.040			1.09		1.09	1.72	
	5020	On roof	"	250	.032	↓		.87		.87	1.37	
	9950	See also div. 020-125 core drilling										
730	0010	**TORCH CUTTING** Steel, 1" thick plate	1 Clab	32	.250	L.F.		5.35		5.35	8.45	730
	0040	1" diameter bar	"	210	.038	Ea.		.82		.82	1.29	
	1000	Oxygen lance cutting, reinforced concrete walls										
	1040	12" to 16" thick walls	1 Clab	10	.800	L.F.		17.15		17.15	27	
	1080	24" thick walls	"	6	1.333	"		28.50		28.50	45	
732	0010	**WALLS AND PARTITIONS DEMOLITION**										732
	0100	Brick, 4" to 12" thick	B-9C	220	.182	C.F.		3.97	.82	4.79	7.15	
	0200	Concrete block, 4" thick		1,000	.040	S.F.		.87	.18	1.05	1.58	
	0280	8" thick	↓	810	.049			1.08	.22	1.30	1.94	
	0300	Exterior stucco 1" thick over netting	B-9	3,200	.013			.27	.06	.33	.49	
	1000	Drywall, nailed	1 Clab	1,000	.008			.17		.17	.27	
	1020	Glued and nailed		900	.009			.19		.19	.30	
	1500	Fiberboard, nailed	↓	900	.009	↓		.19		.19	.30	

2 SITE WORK

Important: See the Reference Section for critical supporting data - Reference Nos., Crews, & City Cost Indexes

020 | Subsurface Investigation & Demolition

020 700 | Selective Demolition

			CREW	DAILY OUTPUT	LABOR-HOURS	UNIT	MAT.	LABOR	EQUIP.	TOTAL	TOTAL INCL O&P	
732	1520	Glued and nailed	1 Clab	800	.010	S.F.		.21		.21	.34	732
	2000	Movable walls, metal, 5' high		300	.027			.57		.57	.90	
	2020	8' high	↓	400	.020			.43		.43	.68	
	2200	Metal or wood studs, finish 2 sides, fiberboard	B-1	520	.046			1.02		1.02	1.61	
	2250	Lath and plaster		260	.092			2.04		2.04	3.21	
	2300	Plasterboard (drywall)		520	.046			1.02		1.02	1.61	
	2350	Plywood	↓	450	.053			1.18		1.18	1.86	
	3000	Plaster, lime and horsehair, on wood lath	1 Clab	400	.020			.43		.43	.68	
	3020	On metal lath		335	.024			.51		.51	.81	
	3400	Gypsum or perlite, on gypsum lath		410	.020			.42		.42	.66	
	3420	On metal lath	↓	300	.027			.57		.57	.90	
	3600	Plywood, one side	B-1	1,500	.016			.35		.35	.56	
	3750	Terra cotta block and plaster, to 6" thick	"	175	.137	↓		3.03		3.03	4.77	
	3800	Toilet partitions, slate or marble	1 Clab	5	1.600	Ea.		34.50		34.50	54	
	3820	Hollow metal	"	8	1	"		21.50		21.50	34	
	5000	Wallcovering, vinyl	1 Pape	700	.011	S.F.		.29		.29	.44	
	5040	Designer	"	480	.017	"		.42		.42	.64	
734	0010	**WINDOW DEMOLITION**										734
	0200	Aluminum, including trim, to 12 S.F.	1 Clab	16	.500	Ea.		10.75		10.75	16.90	
	0240	To 25 S.F.		11	.727			15.60		15.60	24.50	
	0280	To 50 S.F.		5	1.600			34.50		34.50	54	
	0320	Storm windows, to 12 S.F.		27	.296			6.35		6.35	10	
	0360	To 25 S.F.		21	.381			8.15		8.15	12.85	
	0400	To 50 S.F.		16	.500	↓		10.75		10.75	16.90	
	0600	Glass, minimum		200	.040	S.F.		.86		.86	1.35	
	0620	Maximum		150	.053	"		1.14		1.14	1.80	
	1000	Steel, including trim, to 12 S.F.		13	.615	Ea.		13.20		13.20	21	
	1020	To 25 S.F.		9	.889			19.05		19.05	30	
	1040	To 50 S.F.		4	2			43		43	67.50	
	2000	Wood, including trim, to 12 S.F.		22	.364			7.80		7.80	12.25	
	2020	To 25 S.F.		18	.444			9.55		9.55	15	
	2060	To 50 S.F.	↓	13	.615			13.20		13.20	21	
	5020	Remove and reset window, minimum	1 Carp	6	1.333			36.50		36.50	57.50	
	5040	Average		4	2			54.50		54.50	86	
	5080	Maximum	↓	2	4	↓		109		109	172	

020 750 | Concrete Removal

			CREW	DAILY OUTPUT	LABOR-HOURS	UNIT	MAT.	LABOR	EQUIP.	TOTAL	TOTAL INCL O&P	
754	0010	**FOOTINGS AND FOUNDATIONS DEMOLITION**										754
	0200	Floors, concrete slab on grade,										
	0240	4" thick, plain concrete	B-9C	500	.080	S.F.		1.75	.36	2.11	3.15	
	0280	Reinforced, wire mesh		470	.085			1.86	.38	2.24	3.35	
	0300	Rods		400	.100			2.18	.45	2.63	3.94	
	0400	6" thick, plain concrete		375	.107			2.33	.48	2.81	4.20	
	0420	Reinforced, wire mesh		340	.118			2.57	.53	3.10	4.62	
	0440	Rods	↓	300	.133	↓		2.91	.60	3.51	5.25	
	1000	Footings, concrete, 1' thick, 2' wide	B-5	300	.187	L.F.		4.43	3.49	7.92	10.75	
	1080	1'-6" thick, 2' wide		250	.224			5.30	4.18	9.48	12.85	
	1120	3' wide		200	.280			6.65	5.25	11.90	16.10	
	1140	2' thick, 3' wide	↓	175	.320			7.60	5.95	13.55	18.35	
	1200	Average reinforcing, add								10%	10%	
	1220	Heavy reinforcing, add				↓				20%	20%	
	2000	Walls, block, 4" thick	1 Clab	180	.044	S.F.		.95		.95	1.50	
	2040	6" thick		170	.047			1.01		1.01	1.59	
	2080	8" thick		150	.053			1.14		1.14	1.80	
	2100	12" thick	↓	150	.053	↓		1.14		1.14	1.80	

SITE WORK 2

39

020 750 | Concrete Removal

		CREW	DAILY OUTPUT	LABOR-HOURS	UNIT	1999 BARE COSTS				TOTAL INCL O&P	
						MAT.	LABOR	EQUIP.	TOTAL		
754	2200	For horizontal reinforcing, add				S.F.				10%	10%	**754**
	2220	For vertical reinforcing, add								20%	20%	
	2400	Concrete, plain concrete, 6" thick	B-9	160	.250			5.45	1.13	6.58	9.85	
	2420	8" thick		140	.286			6.25	1.29	7.54	11.20	
	2440	10" thick		120	.333			7.30	1.50	8.80	13.10	
	2500	12" thick	↓	100	.400			8.75	1.80	10.55	15.75	
	2600	For average reinforcing, add								10%	10%	
	2620	For heavy reinforcing, add								20%	20%	
	4000	For congested sites or small quantities, add up to				↓				200%	200%	
	4200	Add for disposal, on site	B-11A	232	.069	C.Y.		1.72	3.62	5.34	6.65	
	4250	To five miles	B-30	220	.109	"		2.64	7.25	9.89	12.05	
758	0010	**MASONRY DEMOLITION**										**758**
	1000	Chimney, 16" x 16", soft old mortar	A-1	24	.333	V.L.F.		7.15	2.86	10.01	14.40	
	1020	Hard mortar		18	.444			9.55	3.81	13.36	19.20	
	1080	20" x 20", soft old mortar		12	.667			14.30	5.70	20	29	
	1100	Hard mortar		10	.800			17.15	6.85	24	34.50	
	1140	20" x 32", soft old mortar		10	.800			17.15	6.85	24	34.50	
	1160	Hard mortar		8	1			21.50	8.60	30.10	43.50	
	1200	48" x 48", soft old mortar		5	1.600			34.50	13.75	48.25	69	
	1220	Hard mortar		4	2			43	17.15	60.15	86.50	
	2000	Columns, 8" x 8", soft old mortar		48	.167			3.58	1.43	5.01	7.20	
	2020	Hard mortar		40	.200			4.29	1.72	6.01	8.65	
	2060	16" x 16", soft old mortar		16	.500			10.75	4.29	15.04	21.50	
	2100	Hard mortar		14	.571			12.25	4.90	17.15	24.50	
	2140	24" x 24", soft old mortar		8	1			21.50	8.60	30.10	43.50	
	2160	Hard mortar		6	1.333			28.50	11.45	39.95	57.50	
	2200	36" x 36", soft old mortar		4	2			43	17.15	60.15	86.50	
	2220	Hard mortar	↓	3	2.667	↓		57	23	80	115	
	3000	Copings, precast or masonry, to 8" wide										
	3020	Soft old mortar	A-1	180	.044	L.F.		.95	.38	1.33	1.92	
	3040	Hard mortar	"	160	.050	"		1.07	.43	1.50	2.16	
	3100	To 12" wide										
	3120	Soft old mortar	A-1	160	.050	L.F.		1.07	.43	1.50	2.16	
	3140	Hard mortar	"	140	.057	"		1.23	.49	1.72	2.47	
	4000	Fireplace, brick, 30" x 24" opening										
	4020	Soft old mortar	A-1	2	4	Ea.		86	34.50	120.50	173	
	4040	Hard mortar		1.25	6.400			137	55	192	277	
	4100	Stone, soft old mortar		1.50	5.333			114	46	160	231	
	4120	Hard mortar		1	8	↓		172	68.50	240.50	345	
	5000	Veneers, brick, soft old mortar		140	.057	S.F.		1.23	.49	1.72	2.47	
	5020	Hard mortar		125	.064			1.37	.55	1.92	2.76	
	5100	Granite and marble, 2" thick		180	.044			.95	.38	1.33	1.92	
	5120	4" thick		170	.047			1.01	.40	1.41	2.03	
	5140	Stone, 4" thick		180	.044			.95	.38	1.33	1.92	
	5160	8" thick		175	.046	↓		.98	.39	1.37	1.97	
	5400	Alternate pricing method, stone, 4" thick		60	.133	C.F.		2.86	1.14	4	5.75	
	5420	8" thick	↓	85	.094	"		2.02	.81	2.83	4.07	

020 800 | Haz. Mat'l Abatement

		CREW	DAILY OUTPUT	LABOR-HOURS	UNIT	MAT.	LABOR	EQUIP.	TOTAL	TOTAL INCL O&P	
810	0010	**ASBESTOS ABATEMENT EQUIPMENT** and supplies, buy										**810**
	0200	Air filtration device, 2000 C.F.M.	R020 -820			Ea.	2,500			2,500	2,750	
	0250	Large volume air sampling pump, minimum					450			450	495	
	0260	Maximum					1,000			1,000	1,100	
	0300	Airless sprayer unit, 2 gun					2,000			2,000	2,200	
	0350	Light stand, 500 watt					250			250	275	

Important: See the Reference Section for critical supporting data - Reference Nos., Crews, & City Cost Indexes

020 800 | Haz. Mat'l Abatement

		CREW	DAILY OUTPUT	LABOR-HOURS	UNIT	1999 BARE COSTS				TOTAL INCL O&P
						MAT.	LABOR	EQUIP.	TOTAL	
810 0400	Personal respirators R020-820									**810**
0410	Negative pressure, 1/2 face, dual operation, min.				Ea.	22.50			22.50	25
0420	Maximum					24			24	26.50
0450	P.A.P.R., full face, minimum					400			400	440
0460	Maximum					700			700	770
0470	Supplied air, full face, incl. air line, minimum					450			450	495
0480	Maximum					600			600	660
0500	Personnel sampling pump, minimum					450			450	495
0510	Maximum					750			750	825
1500	Power panel, 20 unit, incl. G.F.I.					1,800			1,800	1,975
1600	Shower unit, including pump and filters					1,125			1,125	1,250
1700	Supplied air system (type C)					10,000			10,000	11,000
1750	Vacuum cleaner, HEPA, 16 gal., stainless steel, wet/dry					1,000			1,000	1,100
1760	55 gallon					2,200			2,200	2,425
1800	Vacuum loader, 9-18 ton/hr					90,000			90,000	99,000
1900	Water atomizer unit, including 55 gal. drum					230			230	253
2000	Worker protection, whole body, foot, head cover & gloves, plastic					30			30	33
2500	Respirator, single use					9			9	9.90
2550	Cartridge for respirator					36			36	39.50
2570	Glove bag, 7 mil, 50" x 64"					8.50			8.50	9.35
2580	10 mil, 44" x 60"					8.40			8.40	9.25
3000	HEPA vacuum for work area, minimun					1,000			1,000	1,100
3050	Maximum					4,500			4,500	4,950
6000	Disposable polyethelene bags, 6 mil, 3 C.F.					1.15			1.15	1.27
6300	Disposable fiber drums, 3 C.F.					6.50			6.50	7.15
6400	Pressure sensitive caution lables, 3" x 5"					.12			.12	.13
6450	11" x 17"					.25			.25	.28
6500	Negative air machine, 1800 C.F.M.					775			775	855
820 0010	**ASBESTOS ABATEMENT WORK AREA** Containment and preparation.									**820**
0100	Pre-cleaning, HEPA vacuum and wet wipe, flat surfaces	A-10	12,000	.005	S.F.		.16		.16	.26
0200	Protect carpeted area, 2 layers 6 mil poly on 3/4" plywood	"	1,000	.064		1.50	1.95		3.45	4.80
0300	Separation barrier, 2" x 4" @ 16", 1/2" plywood ea. side, 8' high	2 Carp	400	.040		1.25	1.09		2.34	3.10
0310	12' high		320	.050		1.40	1.36		2.76	3.69
0320	16' high		200	.080		1.50	2.18		3.68	5.10
0400	Personnel decontam. chamber, 2" x 4" @ 16", 3/4" ply ea. side		280	.057		2.50	1.56		4.06	5.20
0450	Waste decontam. chamber, 2" x 4" studs @ 16", 3/4" ply ea. side		360	.044		3	1.21		4.21	5.20
0500	Cover surfaces with polyethelene sheeting									
0501	Including glue and tape									
0550	Floors, each layer, 6 mil	A-10	8,000	.008	S.F.	.10	.24		.34	.50
0551	4 mil		9,000	.007		.06	.22		.28	.42
0560	Walls, each layer, 6 mil		6,000	.011		.09	.33		.42	.63
0561	4 mil		7,000	.009		.07	.28		.35	.52
0570	For heights above 12', add						20%			
0575	For heights above 20', add						30%			
0580	For fire retardant poly, add					100%				
0590	For large open areas, deduct					10%	20%			
0600	Seal floor penetrations with foam firestop to 36 Sq. In.	2 Carp	200	.080	Ea.	6.25	2.18		8.43	10.35
0610	36 Sq. In. to 72 Sq. In.		125	.128		12.50	3.49		15.99	19.25
0615	72 Sq. In. to 144 Sq. In.		80	.200		25	5.45		30.45	36
0620	Wall penetrations, to 36 square inches		180	.089		6.25	2.43		8.68	10.70
0630	36 Sq. In. to 72 Sq. In.		100	.160		12.50	4.37		16.87	20.50
0640	72 Sq. In. to 144 Sq. In.		60	.267		25	7.30		32.30	39
0800	Caulk seams with latex	1 Carp	230	.035	L.F.	.15	.95		1.10	1.66
0900	Set up neg. air machine, 1-2k C.F.M. /25 M.C.F. volume	1 Asbe	4.30	1.860	Ea.		56.50		56.50	91.50
830 0010	**DEMOLITION IN ASBESTOS CONTAMINATED AREA**									**830**
0200	Ceiling, including suspension system, plaster and lath	A-9	2,100	.030	S.F.		.93		.93	1.50

SITE WORK **2**

020 800 | Haz. Mat'l Abatement

		CREW	DAILY OUTPUT	LABOR-HOURS	UNIT	MAT.	LABOR	EQUIP.	TOTAL	TOTAL INCL O&P	
830	0210	Finished plaster, leaving wire lath	A-9	585	.109	S.F.		3.34		3.34	5.40	**830**
	0220	Suspended acoustical tile		3,500	.018			.56		.56	.90	
	0230	Concealed tile grid system		3,000	.021			.65		.65	1.05	
	0240	Metal pan grid system		1,500	.043			1.30		1.30	2.10	
	0250	Gypsum board		2,500	.026			.78		.78	1.26	
	0260	Lighting fixtures up to 2' x 4'		72	.889	Ea.		27		27	44	
	0400	Partitions, non load bearing										
	0410	Plaster, lath, and studs	A-9	690	.093	S.F.	.55	2.83		3.38	5.20	
	0450	Gypsum board and studs	"	1,390	.046	"		1.40		1.40	2.27	
	9000	For type C (supplied air) respirator equipment, add				%					10%	
840	0010	**BULK ASBESTOS REMOVAL**										**840**
	0020	Includes disposable tools and 1 suit and respirator/day/worker										
	0100	Beams, W 10 x 19	A-9	235	.272	L.F.		8.30		8.30	13.40	
	0110	W 12 x 22		210	.305			9.30		9.30	15	
	0120	W 14 x 26		180	.356			10.85		10.85	17.50	
	0130	W 16 x 31		160	.400			12.20		12.20	19.70	
	0140	W 18 x 40		140	.457			13.95		13.95	22.50	
	0150	W 24 x 55		110	.582			17.75		17.75	28.50	
	0160	W 30 x 108		85	.753			23		23	37	
	0170	W 36 x 150		72	.889			27		27	44	
	0200	Boiler insulation		480	.133	S.F.	.06	4.07		4.13	6.60	
	0210	With metal lath add				%				50%		
	0300	Boiler breeching or flue insulation	A-9	520	.123	S.F.		3.76		3.76	6.05	
	0310	For active boiler, add				%				100%		
	0400	Duct or AHU insulation	A-10B	440	.073	S.F.		2.22		2.22	3.59	
	0500	Duct vibration isolation joints, up to 24 Sq. In. duct	A-9	56	1.143	Ea.		35		35	56.50	
	0520	25 Sq. In. to 48 Sq. In. duct		48	1.333			40.50		40.50	65.50	
	0530	49 Sq. In. to 76 Sq. In. duct		40	1.600			49		49	79	
	0600	Pipe insulation, air cell type, up to 4" diameter pipe		900	.071	L.F.		2.17		2.17	3.50	
	0610	4" to 8" diameter pipe		800	.080			2.44		2.44	3.94	
	0620	10" to 12" diameter pipe		700	.091			2.79		2.79	4.50	
	0630	14" to 16" diameter pipe		550	.116			3.55		3.55	5.75	
	0650	Over 16" diameter pipe		650	.098	S.F.		3		3	4.85	
	0700	With glove bag up to 3" diameter pipe		100	.640	L.F.	3.15	19.55		22.70	35	
	1000	Pipe fitting insulation up to 4" diameter pipe		320	.200	Ea.		6.10		6.10	9.85	
	1100	6" to 8" diameter pipe		304	.211			6.40		6.40	10.35	
	1110	10" to 12" diameter pipe		192	.333			10.15		10.15	16.40	
	1120	14" to 16" diameter pipe		128	.500			15.25		15.25	24.50	
	1130	Over 16" diameter pipe		176	.364	S.F.		11.10		11.10	17.90	
	1200	With glove bag, up to 8" diameter pipe		40	1.600	Ea.	6.55	49		55.55	86	
	2000	Scrape foam fireproofing from flat surface		2,400	.027	S.F.		.81		.81	1.31	
	2100	Irregular surfaces		1,200	.053			1.63		1.63	2.63	
	3000	Remove cementitious material from flat surface		1,800	.036			1.08		1.08	1.75	
	3100	Irregular surface		1,400	.046			1.39		1.39	2.25	
	4000	Scrape acoustical coating/fireproofing, from ceiling		3,200	.020			.61		.61	.99	
	5000	Remove VAT from floor by hand		2,400	.027			.81		.81	1.31	
	5100	By machine	A-11	4,800	.013			.41	.01	.42	.67	
	5150	For 2 layers, add				%				50%		
	6000	Remove contaminated soil from crawl space by hand	A-9	400	.160	C.F.		4.88		4.88	7.90	
	6100	With large production vacuum loader	A-12	700	.091	"		2.79	.75	3.54	5.30	
	7000	Radiator backing, not including radiator removal	A-9	1,200	.053	S.F.		1.63		1.63	2.63	
	8000	Cement-asbestos transite board	2 Asbe	2,000	.008		.08	.24		.32	.48	
	8100	Transite shingle siding	A-10D	750	.043		.08	1.22	.93	2.23	3.02	
	8200	Shingle roofing	A-10B	2,000	.016		.07	.49		.56	.87	
	8250	Built-up, no gravel, non-friable	B-2	1,400	.029		.07	.62		.69	1.06	
	8300	Asbestos millboard	2 Asbe	1,000	.016		.08	.49		.57	.87	

Important: See the Reference Section for critical supporting data - Reference Nos., Crews, & City Cost Indexes

	020 800 | Haz. Mat'l Abatement	CREW	DAILY OUTPUT	LABOR-HOURS	UNIT	1999 BARE COSTS				TOTAL INCL O&P	
						MAT.	LABOR	EQUIP.	TOTAL		
840 9000	For type C (supplied air) respirator equipment, add				%					10%	**840**
850 0010	**WASTE PACKAGING, HANDLING, & DISPOSAL**										**850**
0100	Collect and bag bulk material, 3 C.F. bags, by hand	A-9	400	.160	Ea.	1.15	4.88		6.03	9.15	
0200	Large production vacuum loader	A-12	880	.073		.80	2.22	.60	3.62	5.10	
1000	Double bag and decontaminate	A-9	960	.067		2.30	2.03		4.33	5.80	
2000	Containerize bagged material in drums, per 3 C.F. drum	"	800	.080		6.50	2.44		8.94	11.10	
3000	Cart bags 50' to dumpster	2 Asbe	400	.040	↓		1.22		1.22	1.97	
5000	Disposal charges, not including haul, minimum				C.Y.					50	
5020	Maximum				"					175	
5100	Remove refrigerant from system	1 Plum	40	.200	Lb.		6.50		6.50	9.90	
9000	For type C (supplied air) respirator equipment, add				%					10%	
860 0010	**DECONTAMINATION CONTAINMENT AREA DEMOLITION** and clean-up										**860**
0100	Spray exposed substrate with surfactant (bridging)										
0200	Flat surfaces	A-9	6,000	.011	S.F.	.35	.33		.68	.92	
0250	Irregular surfaces		4,000	.016	"	.30	.49		.79	1.12	
0300	Pipes, beams, and columns		2,000	.032	L.F.	.55	.98		1.53	2.19	
1000	Spray encapsulate polyethelene sheeting		8,000	.008	S.F.	.30	.24		.54	.72	
1100	Roll down polyethelene sheeting		8,000	.008	"		.24		.24	.39	
1500	Bag polyethelene sheeting		400	.160	Ea.	.75	4.88		5.63	8.75	
2000	Fine clean exposed substrate, with nylon brush		2,400	.027	S.F.		.81		.81	1.31	
2500	Wet wipe substrate		4,800	.013			.41		.41	.66	
2600	Vacuum surfaces, fine brush	↓	6,400	.010	↓		.31		.31	.49	
3000	Structural demolition										
3100	Wood stud walls	A-9	2,800	.023	S.F.		.70		.70	1.13	
3500	Window manifolds, not incl. window replacement		4,200	.015			.46		.46	.75	
3600	Plywood carpet protection	↓	2,000	.032	↓		.98		.98	1.58	
4000	Remove custom decontamination facility	A-10A	8	3	Ea.	12.50	92		104.50	162	
4100	Remove portable decontamination facility	3 Asbe	12	2	"	12.50	61		73.50	112	
5000	HEPA vacuum, shampoo carpeting	A-9	4,800	.013	S.F.	.05	.41		.46	.72	
9000	Final cleaning of protected surfaces	A-10A	8,000	.003	"		.09		.09	.15	
870 0010	**ENCAPSULATION WITH SEALANTS**										**870**
0100	Ceilings and walls, minimum	A-9	21,000	.003	S.F.	.25	.09		.34	.43	
0110	Maximum		10,600	.006		.40	.18		.58	.74	
0200	Columns and beams, minimum		13,300	.005		.25	.15		.40	.52	
0210	Maximum		5,325	.012	↓	.45	.37		.82	1.09	
0300	Pipes to 12" diameter including minor repairs, minimum		800	.080	L.F.	.35	2.44		2.79	4.33	
0310	Maximum	↓	400	.160	"	1	4.88		5.88	9	
880 0010	**REMOVAL OF UNDERGROUND STORAGE TANKS** R020-880										**880**
0011	Petroleum storage tanks, non-leaking										
0100	Excavate & load onto trailer										
0110	3000 gal. to 5000 gal. tank	B-14	4	12	Ea.		273	54.50	327.50	485	
0120	6000 gal to 8000 gal tank	B-3A	3	13.333			305	238	543	735	
0130	9000 gal to 12000 gal tank	"	2	20	↓		455	355	810	1,100	
0190	Known leaking tank add				%				100%	100%	
0200	Remove sludge, water and remaining product from bottom										
0201	of tank with vacuum truck										
0300	3000 gal to 5000 gal tank	A-13	5	1.600	Ea.		43.50	105	148.50	181	
0310	6000 gal to 8000 gal tank		4	2			54.50	131	185.50	227	
0320	9000 gal to 12000 gal tank	↓	3	2.667	↓		72.50	175	247.50	300	
0390	Dispose of sludge off-site, average				Gal.					4	
0400	Insert solid carbon dioxide "dry ice" to produce inert gas										
0401	For cleaning/transporting tanks (1.5 lbs./100 gal. cap)	1 Clab	500	.016	Lb.	1.20	.34		1.54	1.86	
1020	Haul tank to certified salvage dump, 100 miles round trip	↓									

020 800 | Haz. Mat'l Abatement

			CREW	DAILY OUTPUT	LABOR-HOURS	UNIT	1999 BARE COSTS				TOTAL INCL O&P	
							MAT.	LABOR	EQUIP.	TOTAL		
880	1023	3000 gal. to 5000 gal. tank	R020 -880			Ea.				550	630	**880**
	1026	6000 gal. to 8000 gal. tank								650	750	
	1029	9,000 gal. to 12,000 gal. tank				↓				875	1,000	
	1100	Disposal of contaminated soil to landfill										
	1110	Minimum				C.Y.					100	
	1111	Maximum				"					300	
	1120	Disposal of contaminated soil to										
	1121	bituminous concrete batch plant										
	1130	Minimum				C.Y.					50	
	1131	Maximum				"					100	
	2010	Decontamination of soil on site incl poly tarp on top/bottom										
	2011	Soil containment berm, and chemical treatment										
	2020	Minimum	B-11C	100	.160	C.Y.	5.60	3.99	2.19	11.78	14.70	
	2021	Maximum	"	100	.160	↓	7.25	3.99	2.19	13.43	16.55	
	2050	Disposal of decontaminated soil, minimum									60	
	2055	Maximum	↓			↓					125	
890	0010	**OSHA TESTING**										**890**
	0100	Certified technician, minimum				Day					300	
	0110	Maximum				"					500	
	0200	Personal sampling, PCM analysis, minimum	1 Asbe	8	1	Ea.	2.75	30.50		33.25	52	
	0210	Maximum	"	4	2	"	3	61		64	102	
	0300	Industrial hygenist, minimum				Day					400	
	0310	Maximum				"					550	
	1000	Cleaned area samples	1 Asbe	8	1	Ea.	2.50	30.50		33	52	
	1100	PCM analysis		8	1		30	30.50		60.50	82	
	1110	Maximum	↓	4	2		3.09	61		64.09	102	
	1200	TEM analysis, minimum				↓					400	
	1210	Maximum									1,000	

021 100 | Site Clearing

			CREW	DAILY OUTPUT	LABOR-HOURS	UNIT	1999 BARE COSTS				TOTAL INCL O&P	
							MAT.	LABOR	EQUIP.	TOTAL		
104	0010	**CLEAR AND GRUB** Cut & chip light, trees to 6" diam.	B-7	1	48	Acre		1,100	1,175	2,275	3,025	**104**
	0150	Grub stumps and remove	B-30	2	12			290	800	1,090	1,325	
	0200	Cut & chip medium, trees to 12" diam.	B-7	.70	68.571			1,575	1,700	3,275	4,300	
	0250	Grub stumps and remove	B-30	1	24			580	1,600	2,180	2,625	
	0300	Cut & chip heavy, trees to 24" diam.	B-7	.30	160			3,675	3,950	7,625	10,100	
	0350	Grub stumps and remove	B-30	.50	48			1,150	3,200	4,350	5,300	
	0400	If burning is allowed, reduce cut & chip				↓					40%	
	3000	Chipping stumps, to 18" deep, 12" diam.	B-86	20	.400	Ea.		11.35	8.10	19.45	26.50	
	3040	18" diameter		16	.500			14.20	10.15	24.35	32.50	
	3080	24" diameter		14	.571			16.25	11.60	27.85	37.50	
	3100	30" diameter		12	.667			18.95	13.55	32.50	44	
	3120	36" diameter		10	.800			22.50	16.25	38.75	52.50	
	3160	48" diameter	↓	8	1	↓		28.50	20.50	49	65.50	
	5000	Tree thinning, feller buncher, conifer										
	5080	Up to 8" diameter	B-93	240	.033	Ea.		.95	1.62	2.57	3.23	
	5120	12" diameter		160	.050			1.42	2.44	3.86	4.84	
	5240	Hardwood, up to 4" diameter		240	.033			.95	1.62	2.57	3.23	
	5280	8" diameter	↓	180	.044			1.26	2.17	3.43	4.30	

Important: See the Reference Section for critical supporting data - Reference Nos., Crews, & City Cost Indexes

021 100 | Site Clearing

		CREW	DAILY OUTPUT	LABOR-HOURS	UNIT	1999 BARE COSTS MAT.	LABOR	EQUIP.	TOTAL	TOTAL INCL O&P	
104	5320	12" diameter	B-93	120	.067	Ea.		1.89	3.25	5.14	6.45	104
	7000	Tree removal, congested area, aerial lift truck										
	7040	8" diameter	B-85	7	5.714	Ea.		131	115	246	330	
	7080	12" diameter		6	6.667			153	135	288	385	
	7120	18" diameter		5	8			184	162	346	465	
	7160	24" diameter		4	10			230	202	432	575	
	7240	36" diameter		3	13.333			305	269	574	770	
	7280	48" diameter		2	20			460	405	865	1,150	
108	0010	**CLEARING** Brush with brush saw	A-1	.25	32	Acre		685	275	960	1,375	108
	0100	By hand	"	.12	66.667			1,425	570	1,995	2,875	
	0300	With dozer, ball and chain, light clearing	B-11A	2	8			199	420	619	770	
	0400	Medium clearing		1.50	10.667			266	560	826	1,025	
	0500	With dozer and brush rake, light		1	16			400	840	1,240	1,550	
	0550	Medium brush to 4" diameter		.60	26.667			665	1,400	2,065	2,575	
	0600	Heavy brush to 4" diameter		.40	40			995	2,100	3,095	3,850	
	1000	Brush mowing, tractor w/rotary mower, no removal										
	1020	Light density	B-84	2	4	Acre		114	104	218	288	
	1040	Medium density		1.50	5.333			151	139	290	385	
	1080	Heavy density		1	8			227	208	435	575	
116	0010	**FELLING TREES & PILING** With tractor, large tract, firm										116
	0020	level terrain, no boulders, less than 12" diam. trees										
	0300	300 HP dozer, up to 400 trees/acre, 0 to 25% hardwoods	B-10M	.75	16	Acre		415	1,500	1,915	2,300	
	0340	25% to 50% hardwoods		.60	20			520	1,875	2,395	2,875	
	0370	75% to 100% hardwoods		.45	26.667			695	2,525	3,220	3,850	
	0400	500 trees/acre, 0% to 25% hardwoods		.60	20			520	1,875	2,395	2,875	
	0440	25% to 50% hardwoods		.48	25			650	2,350	3,000	3,600	
	0470	75% to 100% hardwoods		.36	33.333			870	3,150	4,020	4,775	
	0500	More than 600 trees/acre, 0 to 25% hardwoods		.52	23.077			600	2,175	2,775	3,325	
	0540	25% to 50% hardwoods		.42	28.571			745	2,700	3,445	4,100	
	0570	75% to 100% hardwoods		.31	38.710			1,000	3,650	4,650	5,575	
	0900	Large tract clearing per tree										
	1500	300 HP dozer, to 12" diameter, softwood	B-10M	320	.038	Ea.		.98	3.53	4.51	5.40	
	1550	Hardwood		100	.120			3.13	11.30	14.43	17.25	
	1600	12" to 24" diameter, softwood		200	.060			1.56	5.65	7.21	8.60	
	1650	Hardwood		80	.150			3.91	14.15	18.06	21.50	
	1700	24" to 36" diameter, softwood		100	.120			3.13	11.30	14.43	17.25	
	1750	Hardwood		50	.240			6.25	22.50	28.75	34.50	
	1800	36" to 48" diameter, softwood		70	.171			4.47	16.15	20.62	24.50	
	1850	Hardwood		35	.343			8.95	32.50	41.45	49.50	
144	0010	**STRIPPING** Topsoil, and stockpiling, sandy loam										144
	0020	200 H.P. dozer, ideal conditions	B-10B	2,300	.005	C.Y.		.14	.36	.50	.61	
	0100	Adverse conditions	"	1,150	.010			.27	.73	1	1.22	
	0200	300 HP dozer, ideal conditions	B-10M	3,000	.004			.10	.38	.48	.57	
	0300	Adverse conditions	"	1,650	.007			.19	.69	.88	1.04	
	0400	400 HP dozer, ideal conditions	B-10X	3,900	.003			.08	.35	.43	.51	
	0500	Adverse conditions	"	2,000	.006			.16	.69	.85	1	
	0600	Clay, dry and soft, 200 HP dozer, ideal conditions	B-10B	1,600	.008			.20	.52	.72	.88	
	0601	Strip topsoil, clay, dry & soft, 200 HP dozer, ideal conditions		1,600	.008			.20	.52	.72	.88	
	0700	Adverse conditions		800	.015			.39	1.05	1.44	1.75	
	1000	Medium hard, 300 HP dozer, ideal conditions	B-10M	2,000	.006			.16	.57	.73	.86	
	1100	Adverse conditions	"	1,100	.011			.28	1.03	1.31	1.57	
	1200	Very hard, 400 HP dozer, ideal conditions	B-10X	2,600	.005			.12	.53	.65	.76	
	1300	Adverse conditions	"	1,340	.009			.23	1.03	1.26	1.49	

SITE WORK 2

021 150 | Selective Clearing

		CREW	DAILY OUTPUT	LABOR-HOURS	UNIT	MAT.	LABOR	EQUIP.	TOTAL	TOTAL INCL O&P	
154	0010	**SELECTIVE CLEARING**									154
	1000	Stump removal on site by hydraulic backhoe, 1-1/2 C.Y.									
	1040	4" to 6" diameter	B-17	60	.533	Ea.		12.30	9.70	22	29.50
	1050	8" to 12" diameter	B-30	33	.727			17.60	48.50	66.10	80
	1100	14" to 24" diameter		25	.960			23	64	87	106
	1150	26" to 36" diameter	↓	16	1.500	↓		36.50	100	136.50	166
	2000	Remove selective trees, on site using chain saws and chipper,									
	2050	not incl. stumps, up to 6" diameter	B-7	18	2.667	Ea.		61	66	127	168
	2100	8" to 12" diameter		12	4			92	99	191	252
	2150	14" to 24" diameter		10	4.800			110	119	229	300
	2200	26" to 36" diameter	↓	8	6	↓		138	148	286	380
	2300	Machine load, 2 mile haul to dump, 12" diam. tree, add								150	225

021 200 | Structure Moving

		CREW	DAILY OUTPUT	LABOR-HOURS	UNIT	MAT.	LABOR	EQUIP.	TOTAL	TOTAL INCL O&P	
204	0010	**MOVING BUILDINGS** One day move, up to 24' wide									204
	0020	Reset on new foundation, patch & hook-up, average move				Total					8,700
	0040	Wood or steel frame bldg., based on ground floor area	B-4	185	.259	S.F.		5.70	2.57	8.27	11.70
	0060	Masonry bldg., based on ground floor area	"	137	.350			7.65	3.47	11.12	15.80
	0200	For 24' to 42' wide, add				↓					15%
	0220	For each additional day on road, add	B-4	1	48	Day		1,050	475	1,525	2,175
	0240	Construct new basement, move building, 1 day									
	0300	move, patch & hook-up, based on ground floor area	B-3	155	.310	S.F.	5.75	7.15	11.30	24.20	30

021 400 | Dewatering

		CREW	DAILY OUTPUT	LABOR-HOURS	UNIT	MAT.	LABOR	EQUIP.	TOTAL	TOTAL INCL O&P	
404	0010	**DEWATERING** Excavate drainage trench, 2' wide, 2' deep	B-11C	90	.178	C.Y.		4.43	2.43	6.86	9.50	404
	0100	2' wide, 3' deep, with backhoe loader	"	135	.119			2.95	1.62	4.57	6.35
	0200	Excavate sump pits by hand, light soil	1 Clab	7.10	1.127			24		24	38
	0300	Heavy soil	"	3.50	2.286	↓		49		49	77
	0500	Pumping 8 hr., attended 2 hrs. per day, including 20 L.F.									
	0550	of suction hose & 100 L.F. discharge hose									
	0600	2" diaphragm pump used for 8 hours	B-10H	4	3	Day		78	11.10	89.10	132
	0650	4" diaphragm pump used for 8 hours	B-10I	4	3			78	23.50	101.50	146
	0800	8 hrs. attended, 2" diaphragm pump	B-10H	1	12			315	44.50	359.50	530
	0900	3" centrifugal pump	B-10J	1	12			315	58	373	545
	1000	4" diaphragm pump	B-10I	1	12			315	94	409	585
	1100	6" centrifugal pump	B-10K	1	12	↓		315	220	535	720
	1300	CMP, incl. excavation 3' deep, 12" diameter	B-6	115	.209	L.F.	8.95	4.88	1.90	15.73	19.45
	1400	18" diameter		100	.240	"	9.95	5.60	2.19	17.74	22
	1600	Sump hole construction, incl. excavation and gravel, pit		1,250	.019	C.F.	.56	.45	.18	1.19	1.51
	1700	With 12" gravel collar, 12" pipe, corrugated, 16 ga.		70	.343	L.F.	11.90	8	3.13	23.03	29
	1800	15" pipe, corrugated, 16 ga.		55	.436		14.65	10.20	3.98	28.83	36.50
	1900	18" pipe, corrugated, 16 ga.		50	.480		13.30	11.20	4.38	28.88	37
	2000	24" pipe, corrugated, 14 ga.		40	.600	↓	21.50	14	5.45	40.95	52
	2200	Wood lining, up to 4' x 4', add	↓	300	.080	SFCA	3	1.87	.73	5.60	7
	9950	See div. 021-444 for wellpoints									
	9960	See div. 021-484 for deep well systems									

021 440 | Wellpoints

		CREW	DAILY OUTPUT	LABOR-HOURS	UNIT	MAT.	LABOR	EQUIP.	TOTAL	TOTAL INCL O&P	
444	0010	**WELLPOINTS** For wellpoint equipment rental, see div. 016-490										444
	0100	Installation and removal of single stage system	R021 -440								
	0110	Labor only, .75 labor-hours per L.F., minimum	1 Clab	10.70	.748	LF Hdr		16.05		16.05	25
	0200	2.0 labor-hours per L.F., maximum	"	4	2	"		43		43	67.50
	0400	Pump operation, 4 @ 6 hr. shifts									
	0410	Per 24 hour day	4 Eqlt	1.27	25.197	Day		685		685	1,050
	0500	Per 168 hour week, 160 hr. straight, 8 hr. double time		.18	177	Week		4,825		4,825	7,350
	0550	Per 4.3 week month	↓	.04	800	Month		21,800		21,800	33,100

Important: See the Reference Section for critical supporting data - Reference Nos., Crews, & City Cost Indexes

021 440 | Wellpoints

			CREW	DAILY OUTPUT	LABOR-HOURS	UNIT	1999 BARE COSTS				TOTAL INCL O&P
							MAT.	LABOR	EQUIP.	TOTAL	
444	0600	Complete installation, operation, equipment rental, fuel &	R021 -440									444
	0610	removal of system with 2" wellpoints 5' O.C.									
	0700	100' long header, 6" diameter, first month	4 Eqlt	3.23	9.907	LF Hdr	100	269		369	520
	0800	Thereafter, per month		4.13	7.748		80	211		291	410
	1000	200' long header, 8" diameter, first month		6	5.333		100	145		245	330
	1100	Thereafter, per month		8.39	3.814		45	104		149	208
	1300	500' long header, 8" diameter, first month		10.63	3.010		35	82		117	163
	1400	Thereafter, per month		20.91	1.530		25	41.50		66.50	91
	1600	1,000' long header, 10" diameter, first month		11.62	2.754		30	75		105	147
	1700	Thereafter, per month		41.81	.765		15	21		36	48
	1900	Note: above figures include pumping 168 hrs. per week									
	1910	and include the pump operator and one stand-by pump.									

021 480 | Relief Wells

			CREW	DAILY OUTPUT	LABOR-HOURS	UNIT	MAT.	LABOR	EQUIP.	TOTAL	TOTAL INCL O&P
484	0010	**WELLS** For dewatering 10' to 20' deep, 2' diameter										484
	0020	with steel casing, minimum	B-6	165	.145	V.L.F.	2	3.40	1.33	6.73	8.95
	0050	Average		98	.245		4	5.70	2.23	11.93	15.75
	0100	Maximum		49	.490		10	11.45	4.47	25.92	33.50
	0300	For pumps for dewatering, see division 016-420-4100 to 4400									
	0500	For domestic water wells, see division 026-704									

021 520 | Shores

			CREW	DAILY OUTPUT	LABOR-HOURS	UNIT	MAT.	LABOR	EQUIP.	TOTAL	TOTAL INCL O&P
524	0010	**SHORING** Existing building, with timber, no salvage allowance	B-51	2.20	21.818	M.B.F.	645	475	78	1,198	1,550	524
	1000	With 35 ton screw jacks, per box and jack	"	3.60	13.333	Jack	40	291	47.50	378.50	550

021 560 | Underpinning

			CREW	DAILY OUTPUT	LABOR-HOURS	UNIT	MAT.	LABOR	EQUIP.	TOTAL	TOTAL INCL O&P
564	0010	**UNDERPINNING FOUNDATIONS** Including excavation,										564
	0020	forming, reinforcing, concrete and equipment									
	0100	5' to 16' below grade, 100 to 500 C.Y.	B-52	2.30	24.348	C.Y.	168	615	188	971	1,350
	0200	Over 500 C.Y.		2.50	22.400		152	565	173	890	1,250
	0400	16' to 25' below grade, 100 to 500 C.Y.		2	28		185	705	216	1,106	1,550
	0500	Over 500 C.Y.		2.10	26.667		175	670	206	1,051	1,475
	0700	26' to 40' below grade, 100 to 500 C.Y.		1.60	35		202	880	270	1,352	1,900
	0800	Over 500 C.Y.		1.80	31.111		185	785	240	1,210	1,700
	0900	For under 50 C.Y., add					10%	40%			
	1000	For 50 C.Y. to 100 C.Y., add					5%	20%			

021 610 | Sheet Piling

			CREW	DAILY OUTPUT	LABOR-HOURS	UNIT	MAT.	LABOR	EQUIP.	TOTAL	TOTAL INCL O&P
614	0010	**SHEET PILING** Steel, not incl. wales, 22 psf, 15' excav., left in place	B-40	10.81	5.920	Ton	795	163	178	1,136	1,350	614
	0100	Drive, extract & salvage		6	10.667		211	294	320	825	1,075
	0300	20' deep excavation, 27 psf, left in place	R021 -610	12.95	4.942		795	136	149	1,080	1,275
	0400	Drive, extract & salvage		6.55	9.771		211	270	294	775	1,000
	0600	25' deep excavation, 38 psf, left in place	R021 -620	19	3.368		795	93	101	989	1,150
	0700	Drive, extract & salvage		10.50	6.095		211	168	183	562	710
	0900	40' deep excavation, 38 psf, left in place **CN**		21.20	3.019		795	83.50	90.50	969	1,100
	1000	Drive, extract & salvage		12.25	5.224		211	144	157	512	645
	1200	15' deep excavation, 22 psf, left in place		983	.065	S.F.	9.25	1.80	1.96	13.01	15.30
	1300	Drive, extract & salvage		656	.098		2.37	2.69	2.93	7.99	10.30
	1500	20' deep excavation, 27 psf, left in place		960	.067		11.60	1.84	2	15.44	18.05
	1600	Drive, extract & salvage		640	.100		3.08	2.76	3.01	8.85	11.25
	1800	25' deep excavation, 38 psf, left in place		1,000	.064		17.10	1.77	1.92	20.79	24
	1900	Drive, extract & salvage		670	.096		4.22	2.64	2.87	9.73	12.15
	2100	Rent steel sheet piling and wales, first month				Ton	230			230	253
	2200	Per added month					23			23	25.50
	2300	Rental piling left in place, add to rental					450			450	495
	2500	Wales, connections & struts, 2/3 salvage					173			173	190

SITE WORK 2

021 610 | Sheet Piling

614			CREW	DAILY OUTPUT	LABOR-HOURS	UNIT	1999 BARE COSTS				TOTAL INCL O&P	614
							MAT.	LABOR	EQUIP.	TOTAL		
2700	High strength piling, 50,000 psi, add	R021 -610				Ton	60			60	66	
2800	55,000 psi, add						65			65	71.50	
3000	Tie rod, not upset, 1-1/2" to 4" diameter with turnbuckle	R021 -620					1,200			1,200	1,325	
3100	No turnbuckle						1,000			1,000	1,100	
3300	Upset, 1-3/4" to 4" diameter with turnbuckle						1,500			1,500	1,650	
3400	No turnbuckle						1,300			1,300	1,425	
3600	Lightweight, 18" to 28" wide, 7 ga., 9.22 psf, and											
3610	9 ga., 8.6 psf, minimum					Lb.	.50			.50	.55	
3700	Average						.55			.55	.61	
3750	Maximum						.62			.62	.68	
3900	Wood, solid sheeting, incl. wales, braces and spacers,											
3910	drive, extract & salvage, 8' deep excavation		B-31	330	.121	S.F.	1.52	2.79	.46	4.77	6.55	
4000	10' deep, 50 S.F./hr. in & 150 S.F./hr. out			300	.133		1.56	3.07	.51	5.14	7.10	
4100	12' deep, 45 S.F./hr. in & 135 S.F./hr. out			270	.148		1.61	3.41	.56	5.58	7.75	
4200	14' deep, 42 S.F./hr. in & 126 S.F./hr. out			250	.160		1.66	3.68	.61	5.95	8.30	
4300	16' deep, 40 S.F./hr. in & 120 S.F./hr. out			240	.167		1.71	3.84	.64	6.19	8.65	
4400	18' deep, 38 S.F./hr. in & 114 S.F./hr. out			230	.174		1.76	4	.66	6.42	8.95	
4500	20' deep, 35 S.F./hr. in & 105 S.F./hr. out			210	.190		1.82	4.38	.73	6.93	9.70	
4520	Left in place, 8' deep, 55 S.F./hr.			440	.091		2.73	2.09	.35	5.17	6.70	
4540	10' deep, 50 S.F./hr.			400	.100		2.88	2.30	.38	5.56	7.20	
4560	12' deep, 45 S.F./hr.			360	.111		3.04	2.56	.42	6.02	7.85	
4565	14' deep, 42 S.F./hr.			335	.119		3.22	2.75	.45	6.42	8.35	
4570	16' deep, 40 S.F./hr.			320	.125		3.42	2.88	.48	6.78	8.80	
4580	18' deep, 38 S.F./hr.			305	.131		3.65	3.02	.50	7.17	9.30	
4590	20' deep, 35 S.F./hr.			280	.143		3.91	3.29	.54	7.74	10.05	
4700	Alternate pricing, left in place, 8' deep			1.76	22.727	M.B.F.	615	525	86.50	1,226.50	1,600	
4800	Drive, extract and salvage, 8' deep			1.32	30.303	"	545	700	115	1,360	1,825	
5000	For treated lumber add cost of treatment to lumber											
5010	See division 063-102											

021 620 | Cribbing & Walers

624		CREW	DAILY OUTPUT	LABOR-HOURS	UNIT	MAT.	LABOR	EQUIP.	TOTAL	TOTAL INCL O&P	624
0010	**SOLDIER BEAMS & LAGGING** H piles with 3" wood sheeting										
0020	horizontal between piles, including removal of wales & braces										
0100	No hydrostatic head, 15' deep, 1 line of braces, minimum	B-50	545	.206	S.F.	6.45	5.40	2.74	14.59	19.05	
0200	Maximum		495	.226		7.15	5.95	3.01	16.11	21	
0400	15' to 22' deep with 2 lines of braces, 10" H, minimum		360	.311		7.60	8.20	4.14	19.94	26.50	
0500	Maximum		330	.339		8.60	8.95	4.52	22.07	29	
0700	23' to 35' deep with 3 lines of braces, 12" H, minimum		325	.345		9.90	9.10	4.59	23.59	31	
0800	Maximum		295	.380		10.75	10	5.05	25.80	34	
1000	36' to 45' deep with 4 lines of braces, 14" H, minimum		290	.386		11.10	10.15	5.15	26.40	35	
1100	Maximum		265	.423		11.75	11.15	5.65	28.55	37.50	
1300	No hydrostatic head, left in place, 15' dp., 1 line of braces, min.		635	.176		8.60	4.65	2.35	15.60	19.75	
1400	Maximum		575	.195		9.20	5.15	2.59	16.94	21.50	
1600	15' to 22' deep with 2 lines of braces, minimum		455	.246		12.90	6.50	3.28	22.68	28.50	
1700	Maximum		415	.270		14.35	7.10	3.59	25.04	31.50	
1900	23' to 35' deep with 3 lines of braces, minimum		420	.267		15.35	7	3.55	25.90	32.50	
2000	Maximum		380	.295		16.95	7.75	3.92	28.62	36	
2200	36' to 45' deep with 4 lines of braces, minimum		385	.291		18.45	7.65	3.87	29.97	37.50	
2300	Maximum		350	.320	"	21.50	8.45	4.26	34.21	42	
2350	Lagging only, 3" thick wood between piles 8' O.C., minimum	B-46	400	.120		1.43	2.96	.13	4.52	6.65	
2370	Maximum		250	.192		2.15	4.73	.21	7.09	10.45	
2400	Open sheeting no bracing, for trenches to 10' deep, min.		1,736	.028		.64	.68	.03	1.35	1.87	
2450	Maximum		1,510	.032		.72	.78	.04	1.54	2.13	
2500	Tie-back method, add to open sheeting, add, minimum								20%	20%	
2550	Maximum								60%	60%	

Important: See the Reference Section for critical supporting data - Reference Nos., Crews, & City Cost Indexes

021 620 | Cribbing & Walers

			CREW	DAILY OUTPUT	LABOR-HOURS	UNIT	1999 BARE COSTS MAT.	LABOR	EQUIP.	TOTAL	TOTAL INCL O&P	
624	2700	Tie-backs only, based on tie-backs total length, minimum	B-46	86.80	.553	L.F.	8.15	13.65	.61	22.41	32	624
	2750	Maximum		38.50	1.247	"	14.30	30.50	1.38	46.18	68.50	
	3500	Tie-backs only, typical average, 25' long		2	24	Ea.	360	590	26.50	976.50	1,400	
	3600	35' long		1.58	30.380	"	475	750	33.50	1,258.50	1,800	

021 680 | Slurry Wall

			CREW	DAILY OUTPUT	LABOR-HOURS	UNIT	1999 BARE COSTS MAT.	LABOR	EQUIP.	TOTAL	TOTAL INCL O&P	
684	0010	**SLURRY TRENCH** Excavated slurry trench in wet soils										684
	0020	backfilled with 3000 PSI concrete, no reinforcing steel										
	0050	Minimum	C-7	333	.216	C.F.	4.52	5.05	2.68	12.25	15.70	
	0100	Maximum		200	.360	"	7.55	8.35	4.46	20.36	26.50	
	0200	Alternate pricing method, minimum		150	.480	S.F.	9.05	11.15	5.95	26.15	34	
	0300	Maximum		120	.600		13.55	13.95	7.45	34.95	44.50	
	0500	Reinforced slurry trench, minimum	B-48	177	.316		6.80	7.60	11.15	25.55	31.50	
	0600	Maximum	"	69	.812		22.50	19.55	28.50	70.55	86.50	
	0800	Haul for disposal, 2 mile haul, excavated material, add	B-34B	99	.081	C.Y.		1.79	4.47	6.26	7.65	
	0900	Haul bentonite castings for disposal, add	"	40	.200	"		4.42	11.05	15.47	18.90	

022 | Earthwork

022 100 | Grading

			CREW	DAILY OUTPUT	LABOR-HOURS	UNIT	1999 BARE COSTS MAT.	LABOR	EQUIP.	TOTAL	TOTAL INCL O&P	
104	0010	**GRADING** Site excav. & fill, see div 022-200										104
	0020	Fine grading, see div 025-122										

022 200 | Excav./Backfill/Compact.

			CREW	DAILY OUTPUT	LABOR-HOURS	UNIT	1999 BARE COSTS MAT.	LABOR	EQUIP.	TOTAL	TOTAL INCL O&P	
204	0010	**BACKFILL** By hand, no compaction, light soil [R022 -220]	1 Clab	14	.571	C.Y.		12.25		12.25	19.30	204
	0100	Heavy soil		11	.727			15.60		15.60	24.50	
	0300	Compaction in 6" layers, hand tamp, add to above		20.60	.388			8.35		8.35	13.10	
	0400	Roller compaction operator walking, add	B-10A	100	.120			3.13	.93	4.06	5.85	
	0500	Air tamp, add	B-9C	190	.211			4.60	.95	5.55	8.30	
	0600	Vibrating plate, add	A-1	60	.133			2.86	1.14	4	5.75	
	0800	Compaction in 12" layers, hand tamp, add to above	1 Clab	34	.235			5.05		5.05	7.95	
	0900	Roller compaction operator walking, add	B-10A	150	.080			2.09	.62	2.71	3.88	
	1000	Air tamp, add	B-9	285	.140			3.07	.63	3.70	5.55	
	1100	Vibrating plate, add	A-1	90	.089			1.91	.76	2.67	3.84	
	1300	Dozer backfilling, bulk, up to 300' haul, no compaction	B-10B	1,200	.010			.26	.70	.96	1.17	
	1400	Air tamped	B-11B	240	.067			1.66	4.35	6.01	7.35	
	1600	Compacting backfill, 6" to 12" lifts, vibrating roller	B-10C	800	.015			.39	1.19	1.58	1.91	
	1700	Sheepsfoot roller	B-10D	750	.016			.42	1.29	1.71	2.05	
	1900	Dozer backfilling, trench, up to 300' haul, no compaction	B-10B	900	.013			.35	.93	1.28	1.56	
	2000	Air tamped	B-11B	235	.068			1.70	4.44	6.14	7.50	
	2200	Compacting backfill, 6" to 12" lifts, vibrating roller	B-10C	700	.017			.45	1.36	1.81	2.19	
	2300	Sheepsfoot roller	B-10D	650	.018			.48	1.48	1.96	2.37	
	2350	Spreading in 8" layers, small dozer	B-10B	1,060	.011			.30	.79	1.09	1.32	
208	0010	**BACKFILL, STRUCTURAL** Dozer or F.E. loader										208
	0020	From existing stockpile, no compaction										
	2000	75 H.P., 50' haul, sand & gravel	B-10L	1,100	.011	C.Y.		.28	.28	.56	.75	
	2020	Common earth		975	.012			.32	.32	.64	.84	
	2040	Clay		850	.014			.37	.36	.73	.97	
	2400	300' haul, sand & gravel		370	.032			.85	.83	1.68	2.21	

022 200 | Excav./Backfill/Compact.

			CREW	DAILY OUTPUT	LABOR-HOURS	UNIT	1999 BARE COSTS				TOTAL INCL O&P	
							MAT.	LABOR	EQUIP.	TOTAL		
208	2420	Common earth	B-10L	330	.036	C.Y.		.95	.93	1.88	2.49	**208**
	2440	Clay	↓	290	.041			1.08	1.06	2.14	2.83	
	3000	105 H.P., 50' haul, sand & gravel	B-10W	1,350	.009			.23	.31	.54	.70	
	3020	Common earth		1,225	.010			.26	.34	.60	.76	
	3040	Clay		1,100	.011			.28	.38	.66	.86	
	3300	300' haul, sand & gravel		465	.026			.67	.90	1.57	2.02	
	3320	Common earth		415	.029			.75	1.01	1.76	2.27	
	3340	Clay	↓	370	.032			.85	1.13	1.98	2.54	
	4000	200 H.P., 50' haul, sand & gravel	B-10B	2,500	.005			.13	.34	.47	.56	
	4020	Common earth		2,200	.005			.14	.38	.52	.64	
	4040	Clay		1,950	.006			.16	.43	.59	.72	
	4400	300' haul, sand & gravel		805	.015			.39	1.04	1.43	1.75	
	4420	Common earth		735	.016			.43	1.14	1.57	1.91	
	4440	Clay		660	.018			.47	1.27	1.74	2.13	
	5000	300 H.P., 50' haul, sand & gravel	B-10M	3,170	.004			.10	.36	.46	.54	
	5020	Common earth		2,900	.004			.11	.39	.50	.60	
	5040	Clay		2,700	.004			.12	.42	.54	.64	
	5400	300' haul, sand & gravel		1,500	.008			.21	.75	.96	1.15	
	5420	Common earth		1,350	.009			.23	.84	1.07	1.28	
	5440	Clay	↓	1,225	.010			.26	.92	1.18	1.41	
	6010	For trench backfill, see div. 022-254 & 258										
	6100	For compaction, see div. 022-222										
212	0010	**BORROW** Buy and load at pit, haul 2 miles round trip	R022 -240									**212**
	0020	and spread, with 200 H.P. dozer, no compaction										
	0100	Bank run gravel CN	B-15	600	.047	C.Y.	5.50	1.11	2.87	9.48	10.90	
	0200	Common borrow		600	.047		3.75	1.11	2.87	7.73	9	
	0300	Crushed stone, (1.40 tons per CY) 1-1/2"		600	.047		17.75	1.11	2.87	21.73	24.50	
	0320	3/4" CN		600	.047		17.75	1.11	2.87	21.73	24.50	
	0340	1/2"		600	.047		18.25	1.11	2.87	22.23	25	
	0360	3/8"		600	.047		19	1.11	2.87	22.98	26	
	0400	Sand, washed, concrete		600	.047		11.50	1.11	2.87	15.48	17.55	
	0500	Dead or bank sand		600	.047		3.75	1.11	2.87	7.73	9	
	0600	Select structural fill		600	.047		7.50	1.11	2.87	11.48	13.10	
	0700	Screened loam CN		600	.047		18.25	1.11	2.87	22.23	25	
	0800	Topsoil, weed free	↓	600	.047		14.15	1.11	2.87	18.13	20.50	
	0900	For 5 mile haul, add	B-34B	200	.040	↓		.88	2.21	3.09	3.78	
222	0010	**COMPACTION, STRUCTURAL** Steel wheel tandem roller, 5 tons	R022 -220	B-10E	8	1.500	Hr.		39	19.10	58.10	81	**222**
	0100	10 tons	B-10F	8	1.500	"		39	30.50	69.50	93.50	
	0300	Sheepsfoot or wobbly wheel roller, 8" lifts, common fill	B-10G	1,300	.009	C.Y.		.24	.43	.67	.84	
	0400	Select fill	"	1,500	.008			.21	.37	.58	.73	
	0600	Vibratory plate, 8" lifts, common fill	A-1	200	.040			.86	.34	1.20	1.73	
	0700	Select fill	"	216	.037	↓		.79	.32	1.11	1.60	
230	0010	**DRILLING ONLY** 2" hole for rock bolts, average	B-47	316	.076	L.F.		1.83	1.96	3.79	4.99	**230**
	0800	2-1/2" hole for pre-splitting, average	↓	600	.040			.96	1.03	1.99	2.62	
	1600	Quarry operations, 2-1/2" to 3-1/2" diameter	↓	715	.034	↓		.81	.86	1.67	2.20	
234	0010	**DRILLING AND BLASTING** Only, rock, open face, under 1500 C.Y.	B-47	225	.107	C.Y.	1.60	2.56	2.75	6.91	8.75	**234**
	0100	Over 1500 C.Y.		300	.080		1.60	1.92	2.06	5.58	7	
	0200	Areas where blasting mats are required, under 1500 C.Y.		175	.137		1.60	3.30	3.53	8.43	10.75	
	0250	Over 1500 C.Y.	↓	250	.096		1.60	2.31	2.47	6.38	8.05	
	0300	Bulk drilling and blasting, can vary greatly, average									5	
	0500	Pits, average									20	
	1300	Deep hole method, up to 1500 C.Y.	B-47	50	.480		1.60	11.55	12.35	25.50	33.50	
	1400	Over 1500 C.Y.		66	.364		1.60	8.75	9.35	19.70	25.50	
	1900	Restricted areas, up to 1500 C.Y.		13	1.846		1.60	44.50	47.50	93.60	123	
	2000	Over 1500 C.Y.	↓	20	1.200	↓	1.60	29	31	61.60	81	

Important: See the Reference Section for critical supporting data - Reference Nos., Crews, & City Cost Indexes

2 SITE WORK

022 200 | Excav./Backfill/Compact.

			CREW	DAILY OUTPUT	LABOR-HOURS	UNIT	MAT.	LABOR	EQUIP.	TOTAL	TOTAL INCL O&P	
234	2200	Trenches, up to 1500 C.Y.	B-47	22	1.091	C.Y.	4.64	26	28	58.64	76.50	234
	2300	Over 1500 C.Y.		26	.923		4.64	22	24	50.64	65.50	
	2500	Pier holes, up to 1500 C.Y.		22	1.091		1.60	26	28	55.60	73.50	
	2600	Over 1500 C.Y.		31	.774		1.60	18.60	19.95	40.15	53	
	2800	Boulders under 1/2 C.Y., loaded on truck, no hauling	B-100	80	.150			3.91	6.40	10.31	13	
	2900	Boulders, drilled, blasted	B-47	100	.240		1.60	5.75	6.20	13.55	17.50	
	3100	Jackhammer operators with foreman compressor, air tools	B-9	1	40	Day		875	180	1,055	1,575	
	3300	Track drill, compressor, operator and foreman	B-47	1	24	"		575	620	1,195	1,575	
	3500	Blasting caps				Ea.	3			3	3.30	
	3900	Blasting mats, rent, for first day					90			90	99	
	4000	Per added day					30			30	33	
	4200	Preblast survey for 6 room house, individual lot, minimum	A-6	2.40	6.667			171		171	260	
	4300	Maximum	"	1.35	11.852			305		305	460	
	4500	City block within zone of influence, minimum	A-8	25,200	.001	S.F.		.03		.03	.05	
	4600	Maximum	"	15,100	.002	"		.05		.05	.08	
238	0010	**EXCAVATING, BULK BANK MEASURE** Common earth piled [R022-240]										238
	0020	For loading onto trucks, add								15%	15%	
	0050	For mobilization and demobilization, see division 022-274 [R022-250]										
	0100	For hauling, see division 022-266										
	0200	Backhoe, hydraulic, crawler mtd., 1 C.Y. cap. = 75 C.Y./hr.	B-12A	600	.027	C.Y.		.71	.91	1.62	2.08	
	0250	1-1/2 C.Y. cap. = 100 C.Y./hr.	B-12B	800	.020			.53	.89	1.42	1.79	
	0260	2 C.Y. cap. = 130 C.Y./hr.	B-12C	1,040	.015			.41	.98	1.39	1.71	
	0300	3 C.Y. cap. = 160 C.Y./hr.	B-12D	1,280	.013			.33	1.68	2.01	2.36	
	0310	Wheel mounted, 1/2 C.Y. cap. = 30 C.Y./hr.	B-12E	240	.067			1.78	1.40	3.18	4.25	
	0360	3/4 C.Y. cap. = 45 C.Y./hr.	B-12F	360	.044			1.19	1.25	2.44	3.18	
	0500	Clamshell, 1/2 C.Y. cap. = 20 C.Y./hr.	B-12G	160	.100			2.67	3.26	5.93	7.65	
	0550	1 C.Y. cap. = 35 C.Y./hr.	B-12H	280	.057			1.53	2.05	3.58	4.57	
	0950	Dragline, 1/2 C.Y. cap. = 30 C.Y./hr.	B-12I	240	.067			1.78	2.19	3.97	5.10	
	1000	3/4 C.Y. cap. = 35 C.Y./hr.	"	280	.057			1.53	1.88	3.41	4.38	
	1050	1-1/2 C.Y. cap. = 65 C.Y./hr.	B-12P	520	.031			.82	1.45	2.27	2.84	
	1200	Front end loader, track mtd., 1-1/2 C.Y. cap. = 70 C.Y./hr.	B-10N	560	.021			.56	.63	1.19	1.55	
	1250	2-1/2 C.Y. cap. = 95 C.Y./hr.	B-100	760	.016			.41	.67	1.08	1.37	
	1300	3 C.Y. cap. = 130 C.Y./hr.	B-10P	1,040	.012			.30	.83	1.13	1.38	
	1350	5 C.Y. cap. = 160 C.Y./hr.	B-10Q	1,280	.009			.24	.89	1.13	1.36	
	1500	Wheel mounted, 3/4 C.Y. cap. = 45 C.Y./hr.	B-10R	360	.033			.87	.68	1.55	2.09	
	1550	1-1/2 C.Y. cap. = 80 C.Y./hr.	B-10S	640	.019			.49	.50	.99	1.30	
	1600	2-1/4 C.Y. cap. = 100 C.Y./hr.	B-10T	800	.015			.39	.56	.95	1.22	
	1650	5 C.Y. cap. = 185 C.Y./hr.	B-10U	1,480	.008			.21	.63	.84	1.01	
	1800	Hydraulic excavator, truck mtd, 1/2 C.Y. = 30 C.Y./hr.	B-12J	240	.067			1.78	2.65	4.43	5.65	
	1850	48 inch bucket, 1 C.Y. = 45 C.Y./hr.	B-12K	360	.044			1.19	2.24	3.43	4.27	
	3700	Shovel, 1/2 C.Y. capacity = 55 C.Y./hr.	B-12L	440	.036			.97	1.20	2.17	2.80	
	3750	3/4 C.Y. capacity = 85 C.Y./hr.	B-12M	680	.024			.63	.87	1.50	1.92	
	3800	1 C.Y. capacity = 120 C.Y./hr.	B-12N	960	.017			.45	.66	1.11	1.41	
	3850	1-1/2 C.Y. capacity = 160 C.Y./hr.	B-12O	1,280	.013			.33	.68	1.01	1.25	
	3900	3 C.Y. cap. = 250 C.Y./hr.	B-12T	2,000	.008			.21	.61	.82	1.01	
	4000	For soft soil or sand, deduct								15%	15%	
	4100	For heavy soil or stiff clay, add								60%	60%	
	4200	For wet excavation with clamshell or dragline, add								100%	100%	
	4250	All other equipment, add								50%	50%	
	4400	Clamshell in sheeting or cofferdam, minimum	B-12H	160	.100			2.67	3.59	6.26	8	
	4450	Maximum	"	60	.267			7.10	9.55	16.65	21.50	
	8000	For hauling excavated material, see div. 022-266										
242	0010	**EXCAVATING, BULK, DOZER** Open site										242
	2000	75 H.P., 50' haul, sand & gravel	B-10L	460	.026	C.Y.		.68	.67	1.35	1.78	

SITE WORK 2

	022 200 | Excav./Backfill/Compact.	CREW	DAILY OUTPUT	LABOR-HOURS	UNIT	MAT.	LABOR	EQUIP.	TOTAL	TOTAL INCL O&P	
242											**242**
2020	Common earth	B-10L	400	.030	C.Y.		.78	.77	1.55	2.05	
2040	Clay		250	.048			1.25	1.23	2.48	3.27	
2200	150' haul, sand & gravel		230	.052			1.36	1.34	2.70	3.56	
2220	Common earth		200	.060			1.56	1.54	3.10	4.09	
2240	Clay		125	.096			2.50	2.46	4.96	6.55	
2400	300' haul, sand & gravel		120	.100			2.61	2.57	5.18	6.85	
2420	Common earth		100	.120			3.13	3.08	6.21	8.20	
2440	Clay		65	.185			4.81	4.74	9.55	12.60	
3000	105 H.P., 50' haul, sand & gravel	B-10W	700	.017			.45	.60	1.05	1.35	
3020	Common earth		610	.020			.51	.68	1.19	1.54	
3040	Clay		385	.031			.81	1.08	1.89	2.44	
3200	150' haul, sand & gravel		310	.039			1.01	1.35	2.36	3.03	
3220	Common earth		270	.044			1.16	1.55	2.71	3.48	
3240	Clay		170	.071			1.84	2.46	4.30	5.55	
3300	300' haul, sand & gravel		140	.086			2.24	2.98	5.22	6.70	
3320	Common earth		120	.100			2.61	3.48	6.09	7.85	
3340	Clay		100	.120			3.13	4.17	7.30	9.40	
4000	200 H.P., 50' haul, sand & gravel	B-10B	1,400	.009			.22	.60	.82	1	
4020	Common earth		1,230	.010			.25	.68	.93	1.14	
4040	Clay		770	.016			.41	1.09	1.50	1.82	
4200	150' haul, sand & gravel		595	.020			.53	1.41	1.94	2.36	
4220	Common earth		516	.023			.61	1.63	2.24	2.72	
4240	Clay		325	.037			.96	2.58	3.54	4.32	
4400	300' haul, sand & gravel		310	.039			1.01	2.71	3.72	4.53	
4420	Common earth		270	.044			1.16	3.11	4.27	5.20	
4440	Clay		170	.071			1.84	4.93	6.77	8.30	
5000	300 H.P., 50' haul, sand & gravel	B-10M	1,900	.006			.16	.60	.76	.90	
5020	Common earth		1,650	.007			.19	.69	.88	1.04	
5040	Clay		1,025	.012			.31	1.10	1.41	1.68	
5200	150' haul, sand & gravel		920	.013			.34	1.23	1.57	1.87	
5220	Common earth		800	.015			.39	1.41	1.80	2.16	
5240	Clay		500	.024			.63	2.26	2.89	3.45	
5400	300' haul, sand & gravel		470	.026			.67	2.41	3.08	3.67	
5420	Common earth		410	.029			.76	2.76	3.52	4.20	
5440	Clay		250	.048			1.25	4.52	5.77	6.90	
5500	460 H.P., 50' haul, sand & gravel	B-10X	1,930	.006			.16	.72	.88	1.04	
5510	Common earth		1,680	.007			.19	.82	1.01	1.19	
5520	Clay		1,050	.011			.30	1.31	1.61	1.91	
5530	150' haul, sand & gravel		1,290	.009			.24	1.07	1.31	1.55	
5540	Common earth		1,120	.011			.28	1.23	1.51	1.79	
5550	Clay		700	.017			.45	1.97	2.42	2.86	
5560	300' haul, sand & gravel		660	.018			.47	2.09	2.56	3.03	
5570	Common earth		575	.021			.54	2.40	2.94	3.48	
5580	Clay		350	.034			.89	3.94	4.83	5.70	
6000	700 H.P., 50' haul, sand & gravel	B-10V	3,500	.003			.09	.79	.88	1.01	
6010	Common earth		3,035	.004			.10	.91	1.01	1.16	
6020	Clay		1,925	.006			.16	1.44	1.60	1.83	
6030	150' haul, sand & gravel		2,025	.006			.15	1.37	1.52	1.74	
6040	Common earth		1,750	.007			.18	1.58	1.76	2.01	
6050	Clay		1,100	.011			.28	2.51	2.79	3.21	
6060	300' haul, sand & gravel		1,030	.012			.30	2.68	2.98	3.42	
6070	Common earth		900	.013			.35	3.07	3.42	3.91	
6080	Clay		550	.022			.57	5.05	5.62	6.40	
246	**EXCAVATION, BULK, SCRAPERS**										**246**
0010											
0100	Elevating scraper 11 C.Y., sand & gravel 1500' haul R022 -240	B-33F	690	.020	C.Y.		.54	1.41	1.95	2.37	

Important: See the Reference Section for critical supporting data - Reference Nos., Crews, & City Cost Indexes

SITE WORK
2

SITE WORK | **2**

			CREW	DAILY OUTPUT	LABOR-HOURS	UNIT	1999 BARE COSTS				TOTAL INCL O&P	
022 200	**Excav./Backfill/Compact.**						MAT.	LABOR	EQUIP.	TOTAL		
246	0150	3000' haul	R022 -240	B-33F	610	.023	C.Y.		.61	1.59	2.20	2.68	**246**
	0200	5000' haul			505	.028			.73	1.92	2.65	3.24	
	0300	Common earth, 1500' haul			600	.023			.62	1.62	2.24	2.73	
	0350	3000' haul			530	.026			.70	1.83	2.53	3.09	
	0400	5000' haul			440	.032			.84	2.21	3.05	3.72	
	0500	Clay, 1500' haul			375	.037			.99	2.59	3.58	4.36	
	0550	3000' haul			330	.042			1.12	2.94	4.06	4.96	
	0600	5000' haul			275	.051			1.34	3.53	4.87	5.95	
	1000	Self propelled scraper, 14 C.Y. 1/4 push dozer, sand											
	1050	and gravel, 1500' haul		B-33D	920	.015	C.Y.		.40	2.07	2.47	2.89	
	1100	3000' haul			805	.017			.46	2.36	2.82	3.30	
	1200	5000' haul			645	.022			.57	2.95	3.52	4.12	
	1300	Common earth, 1500' haul			800	.017			.46	2.37	2.83	3.32	
	1350	3000' haul			700	.020			.53	2.71	3.24	3.80	
	1400	5000' haul			560	.025			.66	3.39	4.05	4.74	
	1500	Clay, 1500' haul			500	.028			.74	3.80	4.54	5.30	
	1550	3000' haul			440	.032			.84	4.32	5.16	6.05	
	1600	5000' haul			350	.040			1.06	5.45	6.51	7.55	
	2000	21 C.Y., 1/4 push dozer, sand & gravel, 1500' haul		B-33E	1,180	.012			.31	1.86	2.17	2.53	
	2100	3000' haul			910	.015			.41	2.41	2.82	3.27	
	2200	5000' haul			750	.019			.49	2.93	3.42	3.98	
	2300	Common earth, 1500' haul			1,030	.014			.36	2.13	2.49	2.89	
	2350	3000' haul			790	.018			.47	2.78	3.25	3.78	
	2400	5000' haul			650	.022			.57	3.38	3.95	4.58	
	2500	Clay, 1500' haul			645	.022			.57	3.40	3.97	4.62	
	2550	3000' haul			495	.028			.75	4.43	5.18	6.05	
	2600	5000' haul			405	.035			.91	5.40	6.31	7.35	
	2700	Towed, 10 C.Y., 1/4 push dozer, sand & gravel, 1500' haul		B-33B	560	.025			.66	2.89	3.55	4.18	
	2720	3000' haul			450	.031			.82	3.59	4.41	5.20	
	2730	5000' haul			365	.038			1.01	4.43	5.44	6.40	
	2750	Common earth, 1500' haul			420	.033			.88	3.85	4.73	5.60	
	2770	3000' haul			400	.035			.92	4.04	4.96	5.85	
	2780	5000' haul			310	.045			1.19	5.20	6.39	7.60	
	2800	Clay, 1500' haul			315	.044			1.17	5.15	6.32	7.45	
	2820	3000' haul			300	.047			1.23	5.40	6.63	7.85	
	2840	5000' haul			225	.062			1.64	7.20	8.84	10.40	
	2900	15 C.Y., 1/4 push dozer, sand & gravel, 1500' haul		B-33C	800	.017			.46	2.02	2.48	2.93	
	2920	3000' haul			640	.022			.58	2.53	3.11	3.67	
	2940	5000' haul			520	.027			.71	3.11	3.82	4.51	
	2960	Common earth, 1500' haul			600	.023			.62	2.69	3.31	3.91	
	2980	3000' haul			560	.025			.66	2.89	3.55	4.18	
	3000	5000' haul			440	.032			.84	3.67	4.51	5.35	
	3020	Clay, 1500' haul			450	.031			.82	3.59	4.41	5.20	
	3040	3000' haul			420	.033			.88	3.85	4.73	5.60	
	3060	5000' haul			320	.044			1.16	5.05	6.21	7.30	
250	0010	**EXCAVATING, STRUCTURAL** Hand, pits to 6' deep, sandy soil		1 Clab	8	1	C.Y.		21.50		21.50	34	**250**
	0100	Heavy soil or clay			4	2			43		43	67.50	
	0300	Pits 6' to 12' deep, sandy soil			5	1.600			34.50		34.50	54	
	0500	Heavy soil or clay			3	2.667			57		57	90	
	0700	Pits 12' to 18' deep, sandy soil			4	2			43		43	67.50	
	0900	Heavy soil or clay			2	4			86		86	135	
	1100	Hand loading trucks from stock pile, sandy soil			12	.667			14.30		14.30	22.50	
	1300	Heavy soil or clay			8	1			21.50		21.50	34	
	1500	For wet or muck hand excavation, add to above					%				50%	50%	
	2000	Machine excavation, for spread and mat footings, elevator pits,											

SITE WORK **2**

022 200	Excav./Backfill/Compact.	CREW	DAILY OUTPUT	LABOR-HOURS	UNIT	1999 BARE COSTS				TOTAL INCL O&P
						MAT.	LABOR	EQUIP.	TOTAL	
250	2001	and small building foundations									**250**
	2035	Common earth, hydraulic backhoe, 3/4 C.Y. bucket	B-12F	90	.178	C.Y.		4.75	4.99	9.74	12.70
	2040	1 C.Y. bucket	B-12A	108	.148			3.96	5.05	9.01	11.60
	2050	1-1/2 C.Y. bucket	B-12B	144	.111			2.97	4.95	7.92	9.95
	2060	2 C.Y. bucket	B-12C	200	.080			2.14	5.10	7.24	8.90
	2070	Sand and gravel, 3/4 C.Y. bucket	B-12F	100	.160			4.27	4.49	8.76	11.45
	2080	1 C.Y. bucket	B-12A	120	.133			3.56	4.56	8.12	10.40
	2090	1-1/2 C.Y. bucket	B-12B	160	.100			2.67	4.45	7.12	8.95
	3000	2 C.Y. bucket	B-12C	220	.073			1.94	4.65	6.59	8.05
	3010	Clay, till, or blasted rock, 3/4 C.Y. bucket	B-12F	80	.200			5.35	5.60	10.95	14.35
	3020	1 C.Y. bucket	B-12A	95	.168			4.50	5.75	10.25	13.20
	3030	1-1/2 C.Y. bucket	B-12B	130	.123			3.29	5.50	8.79	11.05
	3040	2 C.Y. bucket	B-12C	175	.091	↓		2.44	5.85	8.29	10.15
	9010	For mobilization or demobilization, see div. 022-274									
	9020	For dewatering, see div. 021-404									
	9022	For larger structures, see Bulk Excavation, div. 022-238									
	9024	For loading onto trucks, add								15%	
	9026	For hauling, see div. 022-266									
	9030	For sheeting or soldier beams & lagging, see div. 021-614 & 624									
	9040	For trench excavation of strip footings, see div. 022-254									
254	0010	**EXCAVATING, TRENCH** or continuous footing, common earth									**254**
	0020	No sheeting or dewatering included									
	0050	1' to 4' deep, 3/8 C.Y. tractor loader/backhoe	B-11C	150	.107	C.Y.		2.66	1.46	4.12	5.70
	0060	1/2 C.Y. tractor loader/backhoe	B-11M	200	.080			1.99	1.45	3.44	4.68
	0090	4' to 6' deep, 1/2 C.Y. tractor loader/backhoe	"	200	.080			1.99	1.45	3.44	4.68
	0100	5/8 C.Y. hydraulic backhoe	B-12Q	250	.064			1.71	1.58	3.29	4.33
	0110	3/4 C.Y. hydraulic backhoe	B-12F	300	.053			1.42	1.50	2.92	3.82
	0300	1/2 C.Y. hydraulic excavator, truck mounted	B-12J	200	.080			2.14	3.18	5.32	6.75
	0500	6' to 10' deep, 3/4 C.Y. hydraulic backhoe, 6' to 10' deep	B-12F	225	.071			1.90	2	3.90	5.10
	0510	1 C.Y. hydraulic backhoe	B-12A	400	.040			1.07	1.37	2.44	3.14
	0600	1 C.Y. hydraulic excavator, truck mounted	B-12K	400	.040			1.07	2.01	3.08	3.85
	0610	1-1/2 C.Y. hydraulic backhoe	B-12B	600	.027			.71	1.19	1.90	2.39
	0900	10' to 14' deep, 3/4 C.Y. hydraulic backhoe	B-12F	200	.080			2.14	2.25	4.39	5.70
	0910	1 C.Y. hydraulic backhoe	B-12A	360	.044			1.19	1.52	2.71	3.48
	1000	1-1/2 C.Y. hydraulic backhoe	B-12B	540	.030			.79	1.32	2.11	2.65
	1300	14' to 20' deep, 1 C.Y. hydraulic backhoe	B-12A	320	.050			1.34	1.71	3.05	3.91
	1310	1-1/2 C.Y. hydraulic backhoe	B-12B	480	.033			.89	1.48	2.37	2.98
	1320	2-1/2 C.Y. hydraulic backhoe	B-12S	850	.019			.50	1.99	2.49	2.95
	1400	By hand with pick and shovel 2' to 6' deep, light soil	1 Clab	8	1			21.50		21.50	34
	1500	Heavy soil	"	4	2			43		43	67.50
	1700	For tamping backfilled trenches, air tamp, add	A-1	100	.080			1.72	.69	2.41	3.46
	1900	Vibrating plate, add	B-18	230	.104	↓		2.31	.23	2.54	3.88
	2100	Trim sides and bottom for concrete pours, common earth	↓	1,500	.016	S.F.		.35	.03	.38	.60
	2300	Hardpan	↓	600	.040	"		.88	.09	.97	1.49
	2400	Pier and spread footing excavation, add to above				C.Y.				30%	30%
	3000	Backfill trench, F.E. loader, wheel mtd., 1 C.Y. bucket									
	3020	Minimal haul	B-10R	400	.030	C.Y.		.78	.61	1.39	1.87
	3040	100' haul	"	200	.060			1.56	1.22	2.78	3.75
	3080	2-1/4 C.Y. bucket, minimum haul	B-10T	600	.020			.52	.75	1.27	1.63
	3090	100' haul	"	300	.040	↓		1.04	1.50	2.54	3.25
258	0010	**EXCAVATING, UTILITY TRENCH** Common earth									**258**
	0050	Trenching with chain trencher, 12 H.P., operator walking									
	0100	4" wide trench, 12" deep	B-53	800	.010	L.F.		.27	.12	.39	.54
	0150	18" deep		750	.011			.29	.13	.42	.58
	0200	24" deep		700	.011			.31	.14	.45	.62
	0300	6" wide trench, 12" deep	↓	650	.012	↓		.33	.15	.48	.67

Important: See the Reference Section for critical supporting data - Reference Nos., Crews, & City Cost Indexes

			DAILY OUTPUT	LABOR-HOURS	UNIT	1999 BARE COSTS				TOTAL INCL O&P	
022 200 | Excav./Backfill/Compact.		CREW				MAT.	LABOR	EQUIP.	TOTAL		
258	0350	18" deep	B-53	600	.013	L.F.		.36	.16	.52	.73	**258**
	0400	24" deep		550	.015			.40	.18	.58	.79	
	0450	36" deep		450	.018			.48	.22	.70	.98	
	0600	8" wide trench, 12" deep		475	.017			.46	.20	.66	.93	
	0650	18" deep		400	.020			.54	.24	.78	1.10	
	0700	24" deep		350	.023			.62	.28	.90	1.26	
	0750	36" deep	▼	300	.027	▼		.73	.32	1.05	1.46	
	1000	Backfill by hand including compaction, add										
	1050	4" wide trench, 12" deep	A-1	800	.010	L.F.		.21	.09	.30	.43	
	1100	18" deep		530	.015			.32	.13	.45	.65	
	1150	24" deep		400	.020			.43	.17	.60	.87	
	1300	6" wide trench, 12" deep		540	.015			.32	.13	.45	.64	
	1350	18" deep		405	.020			.42	.17	.59	.86	
	1400	24" deep		270	.030			.64	.25	.89	1.28	
	1450	36" deep		180	.044			.95	.38	1.33	1.92	
	1600	8" wide trench, 12" deep		400	.020			.43	.17	.60	.87	
	1650	18" deep		265	.030			.65	.26	.91	1.30	
	1700	24" deep		200	.040			.86	.34	1.20	1.73	
	1750	36" deep	▼	135	.059	▼		1.27	.51	1.78	2.56	
	2000	Chain trencher, 40 H.P. operator riding										
	2050	6" wide trench and backfill, 12" deep	B-54	1,200	.007	L.F.		.18	.18	.36	.47	
	2100	18" deep		1,000	.008			.22	.21	.43	.56	
	2150	24" deep		975	.008			.22	.22	.44	.58	
	2200	36" deep		900	.009			.24	.23	.47	.63	
	2250	48" deep		750	.011			.29	.28	.57	.75	
	2300	60" deep		650	.012			.33	.32	.65	.87	
	2400	8" wide trench and backfill, 12" deep		1,000	.008			.22	.21	.43	.56	
	2450	18" deep		950	.008			.23	.22	.45	.59	
	2500	24" deep		900	.009			.24	.23	.47	.63	
	2550	36" deep		800	.010			.27	.26	.53	.70	
	2600	48" deep		650	.012			.33	.32	.65	.87	
	2700	12" wide trench and backfill, 12" deep		975	.008			.22	.22	.44	.58	
	2750	18" deep		860	.009			.25	.25	.50	.65	
	2800	24" deep		800	.010			.27	.26	.53	.70	
	2850	36" deep		725	.011			.30	.29	.59	.78	
	3000	16" wide trench and backfill, 12" deep		835	.010			.26	.25	.51	.68	
	3050	18" deep		750	.011			.29	.28	.57	.75	
	3100	24" deep	▼	700	.011	▼		.31	.30	.61	.80	
	3200	Compaction with vibratory plate, add								50%	50%	
262	0010	**FILL** Spread dumped material, by dozer, no compaction	B-10B	1,000	.012	C.Y.		.31	.84	1.15	1.40	**262**
	0100	By hand	1 Clab	12	.667	"		14.30		14.30	22.50	
	0500	Gravel fill, compacted, under floor slabs, 4" deep	B-37	10,000	.005	S.F.	.15	.11	.02	.28	.36	
	0600	6" deep		8,600	.006		.23	.13	.02	.38	.47	
	0700	9" deep		7,200	.007		.38	.15	.02	.55	.67	
	0800	12" deep		6,000	.008	▼	.52	.18	.03	.73	.89	
	1000	Alternate pricing method, 4" deep		120	.400	C.Y.	11.25	9.10	1.27	21.62	28	
	1100	6" deep		160	.300		11.25	6.80	.95	19	24	
	1200	9" deep		200	.240		11.25	5.45	.76	17.46	22	
	1300	12" deep	▼	220	.218	▼	11.25	4.96	.69	16.90	21	
	1500	For fill under exterior paving, see division 022-308										
266	0011	**HAULING** Excavated or borrow material, loose cubic yards										**266**
	0015	no loading included, highway haulers										
	0020	6 C.Y. dump truck, 1/4 mile round trip, 5.0 loads/hr.	B-34A	195	.041	C.Y.		.91	1.87	2.78	3.44	
	0030	1/2 mile round trip, 4.1 loads/hr.	▼	160	.050	▼		1.11	2.28	3.39	4.19	

SITE WORK **2**

			CREW	DAILY OUTPUT	LABOR-HOURS	UNIT	MAT.	LABOR	EQUIP.	TOTAL	TOTAL INCL O&P	
		022 200 | Excav./Backfill/Compact.						1999 BARE COSTS				
266	0040	1 mile round trip, 3.3 loads/hr.	B-34A	130	.062	C.Y.		1.36	2.80	4.16	5.15	266
	0100	2 mile round trip, 2.6 loads/hr.		100	.080			1.77	3.64	5.41	6.70	
	0150	3 mile round trip, 2.1 loads/hr.		80	.100			2.21	4.55	6.76	8.40	
	0200	4 mile round trip, 1.8 loads/hr.		70	.114			2.53	5.20	7.73	9.55	
	0310	12 C.Y. dump truck, 1/4 mile round trip 3.7 loads/hr.	B-34B	288	.028			.61	1.54	2.15	2.63	
	0320	1/2 mile round trip, 3.2 loads/hr.		250	.032			.71	1.77	2.48	3.03	
	0330	1 mile round trip 2.7, loads/hr.		210	.038			.84	2.11	2.95	3.61	
	0400	2 mile round trip, 2.2 loads/hr.		180	.044			.98	2.46	3.44	4.20	
	0450	3 mile round trip, 1.9 loads/hr.		170	.047			1.04	2.60	3.64	4.45	
	0500	4 mile round trip, 1.6 loads/hr.		125	.064			1.41	3.54	4.95	6.05	
	0540	5 mile round trip, 1 load/hr.		78	.103			2.27	5.65	7.92	9.70	
	0550	10 mile round trip, 0.60 load/hr.		58	.138			3.05	7.60	10.65	13.05	
	0560	20 mile round trip, 0.4 load/hr.		39	.205			4.53	11.35	15.88	19.40	
	0600	16.5 C.Y. dump trailer, 1 mile round trip, 2.6 loads/hr.	B-34C	280	.029			.63	2	2.63	3.17	
	0700	2 mile round trip, 2.1 loads/hr.		225	.036			.79	2.48	3.27	3.93	
	1000	3 mile round trip, 1.8 loads/hr.		193	.041			.92	2.90	3.82	4.59	
	1100	4 mile round trip, 1.6 loads/hr.		172	.047			1.03	3.25	4.28	5.15	
	1110	5 mile round trip, 1 load/hr.		108	.074			1.64	5.15	6.79	8.20	
	1120	10 mile round trip, .60 load/hr.		80	.100			2.21	7	9.21	11.10	
	1130	20 mile round trip, .4 load/hr.		54	.148			3.27	10.35	13.62	16.40	
	1150	20 C.Y. dump trailer, 1 mile round trip, 2.5 loads/hr.	B-34D	325	.025			.54	1.73	2.27	2.73	
	1200	2 mile round trip, 2 loads/hr.		260	.031			.68	2.16	2.84	3.42	
	1220	3 mile round trip, 1.7 loads/hr.		221	.036			.80	2.54	3.34	4.02	
	1240	4 mile round trip, 1.5 loads/hr.		195	.041			.91	2.88	3.79	4.56	
	1245	5 mile round trip, 1.1 load/hr.		143	.056			1.24	3.93	5.17	6.20	
	1250	10 mile round trip, .75 load/hr.		110	.073			1.61	5.10	6.71	8.05	
	1255	20 mile round trip, .5 load/hr.		78	.103			2.27	7.20	9.47	11.35	
	1300	Hauling in medium traffic, add								20%	20%	
	1400	Heavy traffic, add								30%	30%	
	1600	Grading at dump, or embankment if required, by dozer	B-10B	1,000	.012			.31	.84	1.15	1.40	
	1800	Spotter at fill or cut, if required	1 Clab	8	1	Hr.		21.50		21.50	34	
270	0010	**HORIZONTAL BORING** Casing only, 100' minimum,										270
	0020	not incl. jacking pits or dewatering										
	0100	Roadwork, 1/2" thick wall, 24" diameter casing	B-42	14	4.571	L.F.	45	110	80	235	315	
	0200	36" diameter		12	5.333		75	129	93.50	297.50	390	
	0300	48" diameter		12	5.333		105	129	93.50	327.50	425	
	0500	Railroad work, 24" diameter		10	6.400		45	155	112	312	420	
	0600	36" diameter		9.50	6.737		75	163	118	356	475	
	0700	48" diameter		9	7.111		105	172	125	402	530	
	0900	For ledge, add								145	175	
274	0010	**MOBILIZATION OR DEMOBILIZATION** Up to 50 miles										274
	0020	Dozer, loader, backhoe or excavator, 70 H.P.- 250 H.P.	B-34K	6	1.333	Ea.		29.50	152	181.50	213	
	0100	Above 250 H.P.		4	2			44	228	272	320	
	0300	Scraper, towed type (incl. tractor), 6 C.Y. capacity		3.75	2.133			47	244	291	340	
	0400	10 C.Y.		3.50	2.286			50.50	261	311.50	365	
	0600	Self-propelled scraper, 15 C.Y.		3.30	2.424			53.50	277	330.50	385	
	0700	24 C.Y.		3	2.667			59	305	364	425	
	0900	Shovel or dragline, 3/4 C.Y.		3.60	2.222			49	254	303	355	
	1000	1-1/2 C.Y.		3	2.667			59	305	364	425	
	1100	Delivery charge for small equipment on flatbed trailer, minimum									40	
	1150	Maximum									100	
	3000	For large pieces of equipment, allow for knockdown, assembly										
	3001	and lead and tail vehicles for highway transport										

R022-240 (reference, row 0040)
R016-410 (reference, row 0010 Mobilization)

Important: See the Reference Section for critical supporting data - Reference Nos., Crews, & City Cost Indexes

022 200 | Excav./Backfill/Compact.

		Description	CREW	DAILY OUTPUT	LABOR-HOURS	UNIT	1999 BARE COSTS				TOTAL INCL O&P	
							MAT.	LABOR	EQUIP.	TOTAL		
282	0010	**FILL** Borrow, load, 1 mile haul, spread with dozer										282
	0020	for embankments	B-15	1,200	.023	C.Y.	3.75	.56	1.44	5.75	6.55	
	0100	Select fill for shoulders & embankments	"	1,200	.023	"	7.50	.56	1.44	9.50	10.70	
	0201	For hauling over 1 mile, add to above per C.Y., div. 022-266				Mile				.62	.82	
286	0010	**LOAM OR TOPSOIL** Remove and stockpile on site										286
	0020	6" deep, 200' haul	B-10B	865	.014	C.Y.		.36	.97	1.33	1.63	
	0100	300' haul		520	.023			.60	1.61	2.21	2.69	
	0150	500' haul		225	.053			1.39	3.73	5.12	6.25	
	0200	Alternate method: 6" deep, 200' haul		5,090	.002	S.Y.		.06	.16	.22	.27	
	0250	500' haul		1,325	.009	"		.24	.63	.87	1.06	
	0400	Spread from pile to rough finish grade, F.E. loader, 1.5 C.Y.	B-10S	200	.060	C.Y.		1.56	1.60	3.16	4.16	
	0500	Up to 200' radius, by hand	1 Clab	14	.571			12.25		12.25	19.30	
	0600	Top dress by hand, 1 C.Y. for 600 S.F.	"	11.50	.696		18.25	14.90		33.15	43.50	
	0700	Furnish and place, truck dumped, screened, 4" deep	B-10S	1,300	.009	S.Y.	2.28	.24	.25	2.77	3.15	
	0800	6" deep	"	820	.015	"	2.92	.38	.39	3.69	4.23	

022 300 | Pavement Base

		Description	CREW	DAILY OUTPUT	LABOR-HOURS	UNIT	MAT.	LABOR	EQUIP.	TOTAL	TOTAL INCL O&P	
304	0010	**BASE** Prepare and roll sub-base, small areas to 2500 S.Y.	B-32A	1,500	.016	S.Y.		.42	.66	1.08	1.37	304
	0100	Large areas over 2500 S.Y.	B-32	3,700	.009	"		.23	.45	.68	.84	
308	0010	**BASE COURSE** For roadways and large paved areas										308
	0050	Crushed 3/4" stone base, compacted, 3" deep	B-36B	5,200	.012	S.Y.	3.07	.31	.63	4.01	4.55	
	0100	6" deep		5,000	.013		6.15	.32	.65	7.12	8	
	0200	9" deep		4,600	.014		9.25	.35	.71	10.31	11.45	
	0300	12" deep		4,200	.015		12.55	.38	.78	13.71	15.30	
	0301	Crushed 1-1/2" stone base, compacted to 4" deep		6,000	.011		3.78	.27	.54	4.59	5.15	
	0302	6" deep		5,400	.012		5.80	.30	.60	6.70	7.45	
	0303	8" deep		4,500	.014		7.55	.36	.73	8.64	9.70	
	0304	12" deep		3,800	.017		11.60	.43	.86	12.89	14.35	
	0350	Bank run gravel, spread and compacted										
	0370	6" deep	B-32	6,000	.005	S.Y.	2.34	.14	.28	2.76	3.10	
	0390	9" deep		4,900	.007		3.44	.17	.34	3.95	4.43	
	0400	12" deep		4,200	.008		4.68	.20	.40	5.28	5.90	
	0700	Liquid application to gravel base, asphalt emulsion	B-45	6,000	.003	Gal.	1.50	.07	.12	1.69	1.89	
	0800	Prime and seal, cut back asphalt		6,000	.003	"	1.77	.07	.12	1.96	2.19	
	1000	Macadam penetration crushed stone, 2 gal. per S.Y., 4" thick		6,000	.003	S.Y.	3	.07	.12	3.19	3.54	
	1100	6" thick, 3 gal. per S.Y.		4,000	.004		4.50	.10	.19	4.79	5.30	
	1200	8" thick, 4 gal. per S.Y.		3,000	.005		6	.13	.25	6.38	7.10	
	6000	Stabilization fabric, polypropylene, 6 oz./S.Y.	B-6	10,000	.002		1.20	.06	.02	1.28	1.43	
	8900	For small and irregular areas, add						50%	50%			

022 400 | Soil Stabilization

		Description	CREW	DAILY OUTPUT	LABOR-HOURS	UNIT	MAT.	LABOR	EQUIP.	TOTAL	TOTAL INCL O&P	
408	0010	**GROUTING, PRESSURE** Cement and sand, 1:1 mix, minimum	B-61	124	.323	Bag	7.90	7.40	2.88	18.18	23.50	408
	0100	Maximum		51	.784	"	7.90	18.05	7	32.95	44.50	
	0200	Cement and sand, 1:1 mix, minimum		250	.160	C.F.	15.85	3.68	1.43	20.96	24.50	
	0300	Maximum		100	.400		24	9.20	3.57	36.77	44.50	
	0400	Epoxy cement grout, minimum		137	.292		100	6.70	2.60	109.30	123	
	0500	Maximum		57	.702		100	16.15	6.25	122.40	142	
	0600	Structural epoxy grout				Gal.	45			45	49.50	
	0700	Alternate pricing method: (Add for materials)										
	0710	5 person crew and equipment	B-61	1	40	Day		920	355	1,275	1,825	

022 500 | Vibroflotation

		Description	CREW	DAILY OUTPUT	LABOR-HOURS	UNIT	MAT.	LABOR	EQUIP.	TOTAL	TOTAL INCL O&P	
504	0010	**VIBROFLOTATION** [R022 -510]										504
	0900	Vibroflotation compacted sand cylinder, minimum	B-60	750	.075	V.L.F.		1.86	1.34	3.20	4.33	

For expanded coverage of these items see *Means Heavy Construction Cost Data 1999*

SITE WORK 2

022 500 | Vibroflotation

		CREW	DAILY OUTPUT	LABOR-HOURS	UNIT	MAT.	LABOR	EQUIP.	TOTAL	TOTAL INCL O&P	
504	0950	Maximum	B-60	325	.172	V.L.F.		4.29	3.08	7.37	10	**504**
	1100	Vibro replacement compacted stone cylinder, minimum		500	.112			2.79	2	4.79	6.50	
	1150	Maximum		250	.224	↓		5.55	4.01	9.56	13	
	1300	Mobilization and demobilization, minimum		.47	119	Total		2,975	2,125	5,100	6,925	
	1400	Maximum	↓	.14	400	"		9,950	7,150	17,100	23,200	

R022 -510

022 700 | Slope/Erosion Control

		CREW	DAILY OUTPUT	LABOR-HOURS	UNIT	MAT.	LABOR	EQUIP.	TOTAL	TOTAL INCL O&P	
704	0010	**EROSION CONTROL** Jute mesh, 100 S.Y. per roll, 4' wide, stapled	B-80A	2,400	.010	S.Y.	.72	.21	.07	1	1.21	**704**
	0100	Plastic netting, stapled, 2" x 1" mesh, 20 mil	B-1	2,500	.010		.40	.21		.61	.77	
	0200	Polypropylene mesh, stapled, 6.5 oz./S.Y.		2,500	.010		1	.21		1.21	1.43	
	0300	Tobacco netting, or jute mesh #2, stapled	↓	2,500	.010	↓	.07	.21		.28	.41	
	1000	Silt fence, polypropylene, 3' high, ideal conditions	2 Clab	1,600	.010	L.F.	.35	.21		.56	.73	
	1100	Adverse conditions	"	950	.017	"	.28	.36		.64	.88	
	1200	Place and remove hay bales	A-2	3	8	Ton	50	172	57.50	279.50	385	
	1250	Hay bales, staked	"	1,800	.013	L.F.	6	.29	.10	6.39	7.15	
708	0010	**RETAINING WALLS** Aluminized steel bin, excavation										**708**
	0020	and backfill not included, 10' wide										
	0100	4' high, 5.5' deep	B-13	650	.086	S.F.	14.50	2	.85	17.35	20	
	0200	8' high, 5.5' deep		615	.091		16.70	2.12	.89	19.71	22.50	
	0300	10' high, 7.7' deep		580	.097		17.55	2.24	.95	20.74	24	
	0400	12' high, 7.7' deep		530	.106		18.95	2.46	1.04	22.45	26	
	0500	16' high, 7.7' deep		515	.109		20	2.53	1.07	23.60	27	
	0600	16' high, 9.9' deep		500	.112		21	2.60	1.10	24.70	29	
	0700	20' high, 9.9' deep		470	.119		24	2.77	1.17	27.94	31.50	
	0800	20' high, 12.1' deep		460	.122		25.50	2.83	1.19	29.52	33.50	
	0900	24' high, 12.1' deep		455	.123		27	2.86	1.21	31.07	36	
	1000	24' high, 14.3' deep		450	.124		28	2.89	1.22	32.11	37	
	1100	28' high, 14.3' deep	↓	440	.127	↓	29.50	2.96	1.25	33.71	38	
	1300	For plain galvanized bin type walls, deduct					10%					
	1800	Concrete gravity wall with vertical face including excavation & backfill										
	1850	No reinforcing										
	1900	6' high, level embankment	C-17C	36	2.306	L.F.	43	65.50	11.45	119.95	163	
	2000	33° slope embankment		32	2.594		40.50	74	12.85	127.35	175	
	2200	8' high, no surcharge		27	3.074		55.50	87.50	15.25	158.25	216	
	2300	33° slope embankment		24	3.458		67	98.50	17.15	182.65	248	
	2500	10' high, level embankment		19	4.368		79.50	124	21.50	225	305	
	2600	33° slope embankment		18	4.611	↓	110	131	23	264	355	
	2800	Reinforced concrete cantilever, incl. excavation, backfill & reinf.										
	2900	6' high, 33° slope embankment	C-17C	35	2.371	L.F.	40.50	67.50	11.75	119.75	163	
	3000	8' high, 33° slope embankment		29	2.862		47	81.50	14.20	142.70	195	
	3100	10' high, 33° slope embankment		20	4.150		61	118	20.50	199.50	276	
	3200	20' high, 500 lb. per L.F. surcharge	↓	7.50	11.067	↓	183	315	55	553	755	
	3500	Concrete cribbing, incl. excavation and backfill										
	3700	12' high, open face	B-13	210	.267	S.F.	20.50	6.20	2.62	29.32	35	
	3900	Closed face	"	210	.267	"	19.05	6.20	2.62	27.87	33.50	
	4100	Concrete filled slurry trench, see division 021-684										
	4300	Stone filled gabions, not incl. excavation,										
	4310	Stone, delivered, 3' wide										
	4350	Galvanized, 6' high, 33° slope embankment	B-13	49	1.143	L.F.	20	26.50	11.20	57.70	76	
	4500	Highway surcharge		27	2.074		40	48	20.50	108.50	142	
	4600	9' high, up to 33° slope embankment		24	2.333		45	54	23	122	159	
	4700	Highway surcharge		16	3.500		70	81.50	34.50	186	242	
	4900	12' high, up to 33° slope embankment		14	4		70	93	39	202	265	
	5000	Highway surcharge	↓	11	5.091		100	118	50	268	350	
	5950	For PVC coating, add					20%					

Important: See the Reference Section for critical supporting data - Reference Nos., Crews, & City Cost Indexes

022 700	Slope/Erosion Control	CREW	DAILY OUTPUT	LABOR-HOURS	UNIT	1999 BARE COSTS				TOTAL INCL O&P	
						MAT.	LABOR	EQUIP.	TOTAL		
708 7100	Segmental wall system, incl backfill, compaction, to 8'										**708**
7120	interlocking pins, no scaffolding, base, or tiebacks										
7140	8" x 18" x 21.5", 95 lbs	D-12	250	.128	S.F.	6.40	3.21		9.61	12.05	
7160	8" x 18" x 12.5", 85 lbs		330	.097		6.20	2.43		8.63	10.60	
7180	6" x 17.25" x 12", 72 lbs		375	.085		5.90	2.14		8.04	9.85	
7200	Beveled, 6" x 6" x 12", 68 lbs		350	.091		5.65	2.29		7.94	9.75	
7220	Step, 6" x 16" x 12", 85 lbs		300	.107		5.10	2.68		7.78	9.80	
7240	Caps, to 40 lbs		600	.053	L.F.	6.40	1.34		7.74	9.15	
7260	For reinforcing, add				S.F.				2	2.50	
8000	For higher walls, add components as necessary										
712 0010	**RIP-RAP** Random, broken stone										**712**
0100	Machine placed for slope protection	B-12G	62	.258	C.Y.	17.75	6.90	8.40	33.05	39.50	
0110	3/8 to 1/4 C.Y. pieces, grouted	B-13	80	.700	S.Y.	29	16.25	6.85	52.10	65	
0200	18" minimum thickness, not grouted	"	53	1.057	"	13.15	24.50	10.35	48	64	
0300	Dumped, 50 lb. average	B-11A	800	.020	Ton	10	.50	1.05	11.55	12.90	
0350	100 lb. average		700	.023		14.30	.57	1.20	16.07	17.95	
0370	300 lb. average		600	.027		16.65	.66	1.40	18.71	21	
0400	Gabions, galvanized steel mesh mats or boxes, stone filled, 6" deep	B-13	200	.280	S.Y.	11.85	6.50	2.75	21.10	26	
0500	9" deep		163	.344		13.50	8	3.37	24.87	31	
0600	12" deep		153	.366		15.65	8.50	3.59	27.74	34.50	
0700	18" deep		102	.549		25.50	12.75	5.40	43.65	54	
0800	36" deep		60	.933		42	21.50	9.15	72.65	90.50	
716 0010	**STONE WALL** Including excavation, concrete footing and										**716**
0020	stone 3' below grade. Price is exposed face area.										
0200	Decorative random stone, to 6' high, 1'-6" thick, dry set	D-1	35	.457	S.F.	7.60	11.25		18.85	26	
0300	Mortar set		40	.400		9.25	9.85		19.10	25.50	
0500	Cut stone, to 6' high, 1'-6" thick, dry set		35	.457		11.50	11.25		22.75	30	
0600	Mortar set		40	.400		13.50	9.85		23.35	30	
0800	Retaining wall, random stone, 6' to 10' high, 2' thick, dry set		45	.356		9.50	8.75		18.25	24	
0900	Mortar set		50	.320		11.50	7.85		19.35	25	
1100	Cut stone, 6' to 10' high, 2' thick, dry set		45	.356		14.75	8.75		23.50	30	
1200	Mortar set		50	.320		15.75	7.85		23.60	29.50	

022 800 | Soil Treatment

		CREW	DAILY OUTPUT	LABOR-HOURS	UNIT	MAT.	LABOR	EQUIP.	TOTAL	TOTAL INCL O&P	
804 0010	**TERMITE PRETREATMENT**										**804**
0020	Slab and walls, residential	1 Skwk	1,200	.007	SF Flr.	.19	.19		.38	.49	
0100	Commercial, minimum		2,496	.003		.16	.09		.25	.31	
0200	Maximum		1,645	.005		.13	.14		.27	.36	
0400	Insecticides for termite control, minimum		14.20	.563	Gal.	10	15.80		25.80	36	
0500	Maximum		11	.727	"	17.10	20.50		37.60	51	

023 | Tunneling, Piles & Caissons

023 100	Tunnel Construction	CREW	DAILY OUTPUT	LABOR-HOURS	UNIT	1999 BARE COSTS				TOTAL INCL O&P	
						MAT.	LABOR	EQUIP.	TOTAL		
150 0010	**MICROTUNNELING** Not including excavation, backfill, shoring,										**150**
0020	or dewatering, average 50'/day, slurry method										
0100	24" to 48" outside diameter, minimum				L.F.					600	
0110	Adverse conditions, add				%					50%	
1000	Rent microtunneling machine, average monthly lease				Month					80,000	
1010	Operating technician				Day					600	

For expanded coverage of these items see *Means Heavy Construction Cost Data 1999*

2 SITE WORK

023 100 | Tunnel Construction

			CREW	DAILY OUTPUT	LABOR-HOURS	UNIT	1999 BARE COSTS MAT.	LABOR	EQUIP.	TOTAL	TOTAL INCL O&P	
150	1100	Mobilization and demobilization, minimum				Job					40,000	150
	1110	Maximum				"					400,000	

023 400 | Tunnel Support Systems

			CREW	DAILY OUTPUT	LABOR-HOURS	UNIT	MAT.	LABOR	EQUIP.	TOTAL	TOTAL INCL O&P	
404	0011	**ROCK BOLTS**										404
	2020	Hollow core, prestressable anchor, 1" diameter, 5' long	2 Skwk	32	.500	Ea.	54	14		68	81.50	
	2025	10' long		24	.667		101	18.70		119.70	142	
	2060	2" diameter, 5' long		32	.500		203	14		217	245	
	2065	10' long		24	.667		390	18.70		408.70	460	
	2100	Super high-tensile, 3/4" diameter, 5' long		32	.500		12.80	14		26.80	36	
	2105	10' long		24	.667		23.50	18.70		42.20	55.50	
	2160	2" diameter, 5' long		32	.500		113	14		127	147	
	2165	10' long		24	.667		204	18.70		222.70	254	
	4400	Drill hole for rock bolt, 1-3/4" diam., 5' long (for 3/4" bolt)	B-56	17	.941			23	35.50	58.50	74.50	
	4405	10' long		9	1.778			43.50	67	110.50	141	
	4420	2" diameter, 5' long (for 1" bolt)		13	1.231			30	46.50	76.50	97	
	4425	10' long		7	2.286			55.50	86	141.50	181	
	4460	3-1/2" diameter, 5' long (for 2" bolt)		10	1.600			39	60	99	126	
	4465	10' long		5	3.200			78	120	198	252	

023 550 | Pile Driving

			CREW	DAILY OUTPUT	LABOR-HOURS	UNIT	MAT.	LABOR	EQUIP.	TOTAL	TOTAL INCL O&P	
554	0010	**MOBILIZATION** Set up & remove, air compressor, 600 C.F.M. [R023-620]	A-5	3.30	5.455	Ea.		117	13.05	130.05	198	554
	0100	1200 C.F.M.	"	2.20	8.182			176	19.55	195.55	298	
	0200	Crane, with pile leads and pile hammer, 75 ton	B-19	.60	106			2,950	2,400	5,350	7,475	
	0300	150 ton	"	.36	177			4,900	4,000	8,900	12,500	
	0500	Drill rig, for caissons, to 36", minimum	B-43	2	24			565	890	1,455	1,850	
	0600	Up to 84"	"	1	48			1,125	1,775	2,900	3,700	
	0800	Auxiliary boiler, for steam small	A-5	1.66	10.843			233	26	259	395	
	0900	Large	"	.83	21.687			465	52	517	785	
	1100	Rule of thumb: complete pile driving set up, small	B-19	.45	142			3,925	3,200	7,125	10,000	
	1200	Large	"	.27	237			6,550	5,325	11,875	16,700	
558	0011	**PILING SPECIAL COSTS** pile caps, see Division 033-130										558
	0500	Cutoffs, concrete piles, plain	1 Pile	5.50	1.455	Ea.		39.50		39.50	68.50	
	0600	With steel thin shell, add		38	.211			5.75		5.75	9.90	
	0700	Steel pile or "H" piles		19	.421			11.45		11.45	19.80	
	0800	Wood piles		38	.211			5.75		5.75	9.90	
	0900	Pre-augering up to 30' deep, average soil, 24" diameter	B-43	180	.267	L.F.		6.25	9.85	16.10	20.50	
	0920	36" diameter		115	.417			9.80	15.45	25.25	32.50	
	0960	48" diameter		70	.686			16.15	25.50	41.65	53	
	0980	60" diameter		50	.960			22.50	35.50	58	74	
	1000	Testing, any type piles, test load is twice the design load										
	1050	50 ton design load, 100 ton test				Ea.				14,000	15,000	
	1100	100 ton design load, 200 ton test								18,000	19,000	
	1150	150 ton design load, 300 ton test								22,500	24,000	
	1200	200 ton design load, 400 ton test								24,500	27,000	
	1250	400 ton design load, 800 ton test								28,350	31,500	
	1500	Wet conditions, soft damp ground										
	1600	Requiring mats for crane, add								40%	40%	
	1700	Barge mounted driving rig, add								30%	30%	

023 600 | Driven Piles

			CREW	DAILY OUTPUT	LABOR-HOURS	UNIT	MAT.	LABOR	EQUIP.	TOTAL	TOTAL INCL O&P	
604	0010	**PILES, CONCRETE** 200 piles, 60' long										604
	0020	unless specified otherwise, not incl. pile caps or mobilization										
	0050	Cast in place augered piles, no casing or reinforcing										
	0060	8" diameter	B-43	540	.089	V.L.F.	2.03	2.09	3.29	7.41	9.10	

SITE WORK 2

			DAILY OUTPUT	LABOR-HOURS	UNIT	1999 BARE COSTS				TOTAL INCL O&P
023 600	**Driven Piles**	CREW				MAT.	LABOR	EQUIP.	TOTAL	
604	0065	10" diameter	B-43	480	.100	V.L.F.	3.23	2.35	3.70	9.28	11.30
	0070	12" diameter		420	.114		4.55	2.69	4.23	11.47	13.85
	0075	14" diameter		360	.133		6.15	3.14	4.93	14.22	17.10
	0080	16" diameter		300	.160		8.25	3.76	5.90	17.91	21.50
	0085	18" diameter		240	.200		10.25	4.71	7.40	22.36	26.50
	0100	Cast in place, thin wall shell pile, straight sided,									
	0110	not incl. reinforcing, 8" diam., 16 ga., 5.8 lb./L.F.	B-19	700	.091	V.L.F.	3.81	2.52	2.05	8.38	10.60
	0200	10" diameter, 16 ga. corrugated, 7.3 lb./L.F.		650	.098		4.99	2.72	2.21	9.92	12.40
	0300	12" diameter, 16 ga. corrugated, 8.7 lb./L.F.		600	.107		6.50	2.94	2.40	11.84	14.60
	0400	14" diameter, 16 ga. corrugated, 10.0 lb./L.F.		550	.116		7.60	3.21	2.61	13.42	16.60
	0500	16" diameter, 16 ga. corrugated, 11.6 lb./L.F.		500	.128		9.35	3.53	2.87	15.75	19.25
	0800	Cast in place friction pile, 50' long, fluted,									
	0810	tapered steel, 4000 psi concrete, no reinforcing									
	0900	12" diameter, 7 ga.	B-19	600	.107	V.L.F.	11.65	2.94	2.40	16.99	20.50
	1000	14" diameter, 7 ga.		560	.114		12.70	3.15	2.57	18.42	22
	1100	16" diameter, 7 ga.		520	.123		14.95	3.40	2.76	21.11	25
	1200	18" diameter, 7 ga.		480	.133		17.50	3.68	2.99	24.17	28.50
	1300	End bearing, fluted, constant diameter,									
	1320	4000 psi concrete, no reinforcing									
	1340	12" diameter, 7 ga.	B-19	600	.107	V.L.F.	12.20	2.94	2.40	17.54	21
	1360	14" diameter, 7 ga.		560	.114		15.20	3.15	2.57	20.92	25
	1380	16" diameter, 7 ga.		520	.123		17.65	3.40	2.76	23.81	28
	1400	18" diameter, 7 ga.		480	.133		19.50	3.68	2.99	26.17	31
	1500	For reinforcing steel, add				Lb.	.52			.52	.57
	1700	For ball or pedestal end, add	B-19	11	5.818	C.Y.	76	161	131	368	495
	1900	For lengths above 60', concrete, add	"	11	5.818	"	79	161	131	371	495
	2000	For steel thin shell, pipe only				Lb.	.50			.50	.55
	2200	Precast, prestressed, 50' long, 12" diam., 2-3/8" wall	B-19	720	.089	V.L.F.	16	2.45	2	20.45	24
	2300	14" diameter, 2-1/2" wall		680	.094		21	2.60	2.11	25.71	29.50
	2500	16" diameter, 3" wall		640	.100		29	2.76	2.25	34.01	38.50
	2600	18" diameter, 3" wall	B-19A	600	.107		36.50	2.94	2.35	41.79	47.50
	2800	20" diameter, 3-1/2" wall		560	.114		43.50	3.15	2.52	49.17	56
	2900	24" diameter, 3-1/2" wall		520	.123		51.50	3.40	2.71	57.61	65
	3100	Precast, prestressed, 40' long, 10" thick, square	B-19	700	.091		7.70	2.52	2.05	12.27	14.90
	3200	12" thick, square		680	.094		8.55	2.60	2.11	13.26	16
	3400	14" thick, square		600	.107		9.80	2.94	2.40	15.14	18.25
	3500	Octagonal		640	.100		14	2.76	2.25	19.01	22.50
	3700	16" thick, square		560	.114		15.60	3.15	2.57	21.32	25
	3800	Octagonal		600	.107		16.80	2.94	2.40	22.14	26
	4000	18" thick, square	B-19A	520	.123		17.80	3.40	2.71	23.91	28
	4100	Octagonal	B-19	560	.114		19.90	3.15	2.57	25.62	30
	4300	20" thick, square	B-19A	480	.133		28	3.68	2.94	34.62	40.50
	4400	Octagonal	B-19	520	.123		22	3.40	2.76	28.16	32.50
	4600	24" thick, square	B-19A	440	.145		34.50	4.01	3.21	41.72	48
	4700	Octagonal	B-19	480	.133		31.50	3.68	2.99	38.17	44.50
	4750	Mobilization for 10,000 L.F. pile job, add		3,300	.019			.54	.44	.98	1.36
	4800	25,000 L.F. pile job, add		8,500	.008			.21	.17	.38	.53
608	0010	**PILES, STEEL** Not including mobilization or demobilization									
	0100	Step tapered, round, concrete filled									
	0110	8" tip, 60 ton capacity, 30' depth	B-19	760	.084	V.L.F.	5	2.32	1.89	9.21	11.40
	0120	60' depth		740	.086		5.65	2.39	1.94	9.98	12.30
	0130	80' depth		700	.091		5.85	2.52	2.05	10.42	12.85
	0150	10" tip, 90 ton capacity, 30' depth		700	.091		6.15	2.52	2.05	10.72	13.15
	0160	60' depth		690	.093		6.35	2.56	2.08	10.99	13.45
	0170	80' depth		670	.096		6.85	2.64	2.15	11.64	14.20

For expanded coverage of these items see Means Heavy Construction Cost Data 1999

023 600	Driven Piles	CREW	DAILY OUTPUT	LABOR-HOURS	UNIT	1999 BARE COSTS				TOTAL INCL O&P
						MAT.	LABOR	EQUIP.	TOTAL	
608 0190	12" tip, 120 ton capacity, 30' depth	B-19	660	.097	V.L.F.	8.75	2.68	2.18	13.61	16.40	608
0200	60' depth, 12" diameter		630	.102		7.75	2.80	2.28	12.83	15.70
0210	80' depth		590	.108		7.75	2.99	2.44	13.18	16.10
0250	"H" Sections, 50' long, HP8 x 36		640	.100		10.60	2.76	2.25	15.61	18.65
0400	HP10 X 42		610	.105		10.95	2.89	2.36	16.20	19.40
0500	HP10 X 57		610	.105		14.85	2.89	2.36	20.10	23.50
0700	HP12 X 53		590	.108		13.90	2.99	2.44	19.33	23
0800	HP12 X 74	B-19A	590	.108		19.25	2.99	2.39	24.63	28.50
1000	HP14 X 73		540	.119		19	3.27	2.61	24.88	29.50
1100	HP14 X 89		540	.119		23.50	3.27	2.61	29.38	34
1300	HP14 X 102		510	.125		26.50	3.46	2.77	32.73	37.50
1400	HP14 X 117		510	.125		30.50	3.46	2.77	36.73	42
1600	Splice on standard points, not in leads, 8" or 10"	1 Sswl	5	1.600	Ea.	55	49		104	151
1700	12" or 14"		4	2		80	61		141	200
1900	Heavy duty points, not in leads, 10" wide		4	2		85	61		146	206
2100	14" wide		3.50	2.286		110	70		180	249
2600	Pipe piles, 50' lg. 8" diam., 29 lb. per L.F., no concrete	B-19	500	.128	V.L.F.	9.45	3.53	2.87	15.85	19.40
2700	Concrete filled		460	.139		10.05	3.84	3.12	17.01	21
2900	10" diameter, 34 lb. per L.F., no concrete		500	.128		11.90	3.53	2.87	18.30	22
3000	Concrete filled		450	.142		13.15	3.92	3.19	20.26	24.50
3200	12" diameter, 44 lb. per L.F., no concrete		475	.135		15.15	3.72	3.03	21.90	26
3300	Concrete filled		415	.154		17.15	4.25	3.46	24.86	29.50
3500	14" diameter, 46 lb. per L.F., no concrete		430	.149		16.15	4.11	3.34	23.60	28
3600	Concrete filled		355	.180		19.05	4.97	4.05	28.07	33.50
3800	16" diameter, 52 lb. per L.F., no concrete		385	.166		17.95	4.59	3.73	26.27	31.50
3900	Concrete filled		335	.191		22	5.25	4.29	31.54	38
4100	18" diameter, 59 lb. per L.F., no concrete		355	.180		21.50	4.97	4.05	30.52	36
4200	Concrete filled		310	.206		28	5.70	4.64	38.34	45.50
4400	Splices for pipe piles, not in leads, 8" diameter	1 Sswl	4.67	1.713	Ea.	31.50	52.50		84	131
4500	14" diameter		3.79	2.111		41.50	64.50		106	164
4600	16" diameter		3.03	2.640		51	81		132	204
4800	Points, standard, 8" diameter		4.61	1.735		24.50	53		77.50	125
4900	14" diameter		4.05	1.975		34	60.50		94.50	148
5000	16" diameter		3.37	2.374		41	72.50		113.50	179
5200	Points, heavy duty, 10" diameter		2.89	2.768		33	84.50		117.50	191
5300	14" or 16" diameter		2.02	3.960		52.50	121		173.50	280
5500	For reinforcing steel, add		1,150	.007	Lb.	.40	.21		.61	.83
5700	For thick wall sections, add				"	.45			.45	.50
612 0010	**PILES, WOOD** Friction or end bearing, not including										612
0050	mobilization or demobilization	R023 -620								
0100	Untreated piles, up to 30' long, 12" butts, 8" points	B-19	625	.102	V.L.F.	5.25	2.83	2.30	10.38	13
0200	30' to 39' long, 12" butts, 8" points		700	.091		5.25	2.52	2.05	9.82	12.20
0300	40' to 49' long, 12" butts, 7" points		720	.089		5.50	2.45	2	9.95	12.30
0400	50' to 59' long, 13"butts, 7" points		800	.080		5	2.21	1.80	9.01	11.10
0500	60' to 69' long, 13" butts, 7" points		840	.076		5.65	2.10	1.71	9.46	11.55
0600	70' to 80' long, 13" butts, 6" points		840	.076		6.25	2.10	1.71	10.06	12.25
0800	Treated piles, 12 lb. per C.F.,									
0810	friction or end bearing, ASTM class B									
1000	Up to 30' long, 12" butts, 8" points	B-19	625	.102	V.L.F.	9.45	2.83	2.30	14.58	17.55
1100	30' to 39' long, 12" butts, 8" points		700	.091		9.35	2.52	2.05	13.92	16.70
1200	40' to 49' long, 12" butts, 7" points		720	.089		8.95	2.45	2	13.40	16.10
1300	50' to 59' long, 13" butts, 7" points		800	.080		10.90	2.21	1.80	14.91	17.60
1400	60' to 69' long, 13" butts, 6" points	B-19A	840	.076		13.50	2.10	1.68	17.28	20
1500	70' to 80' long, 13" butts, 6" points	"	840	.076		16.10	2.10	1.68	19.88	23
1600	Treated piles, C.C.A. 2.5# per C.F.									
1610	8" butts, 10' long	B-19	400	.160	V.L.F.	6	4.41	3.59	14	17.85

Important: See the Reference Section for critical supporting data - Reference Nos., Crews, & City Cost Indexes

023 600 | Driven Piles

			CREW	DAILY OUTPUT	LABOR-HOURS	UNIT	MAT.	LABOR	EQUIP.	TOTAL	TOTAL INCL O&P	
								1999 BARE COSTS				
612	1620	11' to 16' long	B-19	500	.128	V.L.F.	6	3.53	2.87	12.40	15.60	612
	1630	17' to 20' long		575	.111		6	3.07	2.50	11.57	14.40	
	1640	10" butts, 10' to 16' long		500	.128		6.50	3.53	2.87	12.90	16.15	
	1650	17' to 20' long		575	.111		6.50	3.07	2.50	12.07	14.95	
	1660	21' to 40' long		700	.091		6.50	2.52	2.05	11.07	13.55	
	1670	12" butts, 10' to 20' long		575	.111		7	3.07	2.50	12.57	15.50	
	1680	21' to 35' long		650	.098		7	2.72	2.21	11.93	14.60	
	1690	36' to 40' long		700	.091		7	2.52	2.05	11.57	14.10	
	1695	14" butts. to 40' long	▼	700	.091	▼	10	2.52	2.05	14.57	17.40	
	1700	Boot for pile tip, minimum	1 Pile	27	.296	Ea.	14	8.05		22.05	29.50	
	1800	Maximum		21	.381		42	10.35		52.35	64	
	2000	Point for pile tip, minimum		20	.400		16	10.90		26.90	36.50	
	2100	Maximum		15	.533		57.50	14.50		72	88.50	
	2300	Splice for piles over 50' long, minimum	B-46	35	1.371		40	34	1.52	75.52	102	
	2400	Maximum		20	2.400	▼	48	59	2.66	109.66	154	
	2600	Concrete encasement with wire mesh and tube		331	.145	V.L.F.	7.50	3.58	.16	11.24	14.40	
	2700	Mobilization for 10,000 L.F. pile job, add	B-19	3,300	.019			.54	.44	.98	1.36	
	2800	25,000 L.F. pile job, add	"	8,500	.008	▼		.21	.17	.38	.53	

023 700 | Bored/Augered Piles

			CREW	DAILY OUTPUT	LABOR-HOURS	UNIT	MAT.	LABOR	EQUIP.	TOTAL	TOTAL INCL O&P	
704	0010	**PRESSURE INJECTED FOOTINGS** or Displacement Caissons										704
	0100	incl. mobilization and demobilization, up to 50 miles										
	0200	Uncased shafts, 30 to 80 tons cap., 17" diam., 10' depth	B-44	88	.727	V.L.F.	12.20	19.85	8.90	40.95	56	
	0300	25' depth		165	.388		8.70	10.60	4.75	24.05	32.50	
	0400	80-150 ton capacity, 22" diameter, 10' depth		80	.800		15.25	22	9.80	47.05	63.50	
	0500	20' depth		130	.492		12.20	13.45	6.05	31.70	42.50	
	0700	Cased shafts, 10 to 30 ton capacity, 10-5/8" diam., 20' depth		175	.366		8.70	9.95	4.48	23.13	31	
	0800	30' depth		240	.267		8.15	7.25	3.26	18.66	24.50	
	0850	30 to 60 ton capacity, 12" diameter, 20' depth		160	.400		12.20	10.90	4.90	28	37	
	0900	40' depth		230	.278		9.40	7.60	3.41	20.41	26.50	
	1000	80 to 100 ton capacity, 16" diameter, 20' depth		160	.400		17.45	10.90	4.90	33.25	42.50	
	1100	40' depth		230	.278		16.25	7.60	3.41	27.26	34	
	1200	110 to 140 ton capacity, 17-5/8" diameter, 20' depth		160	.400		18.75	10.90	4.90	34.55	44	
	1300	40' depth		230	.278		17.45	7.60	3.41	28.46	35.50	
	1400	140 to 175 ton capacity, 19" diameter, 20' depth		130	.492		20.50	13.45	6.05	40	51.50	
	1500	40' depth	▼	210	.305	▼	18.75	8.30	3.73	30.78	38.50	
	1700	Over 30' long, L.F. cost tends to be lower										
	1900	Maximum depth is about 90'										

023 800 | Caissons

			CREW	DAILY OUTPUT	LABOR-HOURS	UNIT	MAT.	LABOR	EQUIP.	TOTAL	TOTAL INCL O&P	
804	0010	**CAISSONS** Incl. excav., concrete, 50 lbs. reinf. per C.Y., not										804
	0020	incl. mobilization, boulder removal, disposal										
	0100	Open style, machine drilled, to 50' deep, in stable ground, no										
	0110	casings or ground water, 18" diam., 0.065 C.Y./L.F.	B-43	200	.240	V.L.F.	4.82	5.65	8.90	19.37	24	
	0200	24" diameter, 0.116 C.Y./L.F.		190	.253		8.65	5.95	9.35	23.95	29	
	0300	30" diameter, 0.182 C.Y./L.F.		150	.320		13.50	7.55	11.85	32.90	39.50	
	0400	36" diameter, 0.262 C.Y./L.F.		125	.384		19.45	9.05	14.20	42.70	51	
	0500	48" diameter, 0.465 C.Y./L.F.		100	.480		34.50	11.30	17.75	63.55	75	
	0600	60" diameter, 0.727 C.Y./L.F.		90	.533		54	12.55	19.75	86.30	101	
	0700	72" diameter, 1.05 C.Y./L.F.		80	.600		78	14.10	22	114.10	133	
	0800	84" diameter, 1.43 C.Y./L.F.	▼	75	.640	▼	106	15.05	23.50	144.55	167	
	1000	For bell excavation and concrete, add										
	1020	4' bell diameter, 24" shaft, 0.444 C.Y.	B-43	20	2.400	Ea.	27	56.50	89	172.50	216	
	1040	6' bell diameter, 30" shaft, 1.57 C.Y.		5.70	8.421		96	198	310	604	760	
	1060	8' bell diameter, 36" shaft, 3.72 C.Y.		2.40	20		227	470	740	1,437	1,800	
	1080	9' bell diameter, 48" shaft, 4.48 C.Y.	▼	2	24	▼	273	565	890	1,728	2,150	

R023-620
R023-710
R023-810

SITE WORK **2**

023 800 | Caissons

		CREW	DAILY OUTPUT	LABOR-HOURS	UNIT	1999 BARE COSTS				TOTAL INCL O&P
						MAT.	LABOR	EQUIP.	TOTAL	
804	1100	10' bell diameter, 60" shaft, 5.24 C.Y.	B-43	1.70	28.235	Ea.	320	665	1,050	2,035	2,525	804
	1120	12' bell diameter, 72" shaft, 8.74 C.Y.		1	48		535	1,125	1,775	3,435	4,275
	1140	14' bell diameter, 84" shaft, 13.6 C.Y.	↓	.70	68.571	↓	830	1,625	2,525	4,980	6,225
	1200	Open style, machine drilled, to 50' deep, in wet ground, pulled									
	1300	casing and pumping, 18" diameter, 0.065 C.Y./L.F.	B-48	160	.350	V.L.F.	4.82	8.40	12.35	25.57	32
	1400	24" diameter, 0.116 C.Y./L.F.		125	.448		8.65	10.80	15.80	35.25	43.50
	1500	30" diameter, 0.182 C.Y./L.F.		85	.659		13.50	15.85	23.50	52.85	65
	1600	36" diameter, 0.262 C.Y./L.F.	↓	60	.933		19.45	22.50	33	74.95	92.50
	1700	48" diameter, 0.465 C.Y./L.F.	B-49	55	1.600		34.50	40	46	120.50	152
	1800	60" diameter, 0.727 C.Y./L.F.		35	2.514		54	63	72	189	239
	1900	72" diameter, 1.05 C.Y./L.F.		30	2.933		78	73.50	84	235.50	295
	2000	84" diameter, 1.43 C.Y./L.F.	↓	25	3.520	↓	106	88.50	101	295.50	370
	2100	For bell excavation and concrete, add									
	2120	4' bell diameter, 24" shaft, 0.444 C.Y.	B-48	19.80	2.828	Ea.	27	68	100	195	245
	2140	6' bell diameter, 30" shaft, 1.57 C.Y.		5.70	9.825		96	236	345	677	850
	2160	8' bell diameter, 36" shaft, 3.72 C.Y.	↓	2.40	23.333		227	560	825	1,612	2,025
	2180	9' bell diameter, 48" shaft, 4.48 C.Y.	B-49	3.30	26.667		273	670	765	1,708	2,200
	2200	10' bell diameter, 60" shaft, 5.24 C.Y.		2.80	31.429		320	790	900	2,010	2,600
	2220	12' bell diameter, 72" shaft, 8.74 C.Y.		1.60	55		535	1,375	1,575	3,485	4,475
	2240	14' bell diameter, 84" shaft, 13.6 C.Y.	↓	1	88	↓	830	2,200	2,525	5,555	7,200
	2300	Open style, machine drilled, to 50' deep, in soft rocks and									
	2400	medium hard shales, 18" diameter, 0.065 C.Y./L.F.	B-49	50	1.760	V.L.F.	4.82	44	50.50	99.32	131
	2500	24" diameter, 0.116 C.Y./L.F.		30	2.933		8.65	73.50	84	166.15	218
	2600	30" diameter, 0.182 C.Y./L.F.		20	4.400		13.50	110	126	249.50	330
	2700	36" diameter, 0.262 C.Y./L.F.		15	5.867		19.45	147	168	334.45	440
	2800	48" diameter, 0.465 C.Y./L.F.		10	8.800		34.50	221	253	508.50	665
	2900	60" diameter, 0.727 C.Y./L.F.		7	12.571		54	315	360	729	955
	3000	72" diameter, 1.05 C.Y./L.F.		6	14.667		78	370	420	868	1,125
	3100	84" diameter, 1.43 C.Y./L.F.	↓	5	17.600	↓	106	440	505	1,051	1,375
	3200	For bell excavation and concrete, add									
	3220	4' bell diameter, 24" shaft, 0.444 C.Y.	B-49	10.90	8.073	Ea.	27	203	232	462	605
	3240	6' bell diameter, 30" shaft, 1.57 C.Y.		3.10	28.387		96	715	815	1,626	2,125
	3260	8' bell diameter, 36" shaft, 3.72 C.Y.		1.30	67.692		227	1,700	1,950	3,877	5,050
	3280	9' bell diameter, 48" shaft, 4.48 C.Y.		1.10	80		273	2,000	2,300	4,573	6,000
	3300	10' bell diameter, 60" shaft, 5.24 C.Y.		.90	97.778		320	2,450	2,800	5,570	7,300
	3320	12' bell diameter, 72" shaft, 8.74 C.Y.		.60	146		535	3,675	4,200	8,410	11,000
	3340	14' bell diameter, 84" shaft, 13.6 C.Y.		.40	220	↓	830	5,525	6,325	12,680	16,600
	3600	For rock excavation, sockets, add, minimum		120	.733	C.F.		18.40	21	39.40	52
	3650	Average		95	.926			23.50	26.50	50	65.50
	3700	Maximum	↓	48	1.833	↓		46	52.50	98.50	131
	3900	For 50' to 100' deep, add				V.L.F.				7%	7%
	4000	For 100' to 150' deep, add								25%	25%
	4100	For 150' to 200' deep, add				↓				30%	30%
	4200	For casings left in place, add				Lb.	.50			.50	.55
	4300	For other than 50 lb. reinf. per C.Y., add or deduct				"	.50			.50	.55
	4400	For steel "I" beam cores, add	B-49	8.30	10.602	Ton	960	266	305	1,531	1,800
	4500	Load and haul excess excavation, 2 miles	B-34B	178	.045	C.Y.		.99	2.48	3.47	4.25
	4600	For mobilization, 50 mile radius, rig to 36"	B-43	2	24	Ea.		565	890	1,455	1,850
	4650	Rig to 84"	B-48	1.75	32			770	1,125	1,895	2,450
	4700	For low headroom, add									50%
	5000	Bottom inspection	↓ 1 Skwk	1.20	6.667	↓		187		187	295

R023 -810

2 SITE WORK

Important: See the Reference Section for critical supporting data - Reference Nos., Crews, & City Cost Indexes

024 520	Railroad Trackwork		CREW	DAILY OUTPUT	LABOR-HOURS	UNIT	1999 BARE COSTS				TOTAL INCL O&P	
							MAT.	LABOR	EQUIP.	TOTAL		
524	0010	**RAILROAD** Car bumpers, standard	R024 -520	B-14	2	24	Ea.	2,000	545	109	2,654	3,175	524
	0100	Heavy duty			2	24		3,775	545	109	4,429	5,150	
	0200	Derails hand throw (sliding)	R024 -530		10	4.800		700	109	22	831	965	
	0300	Hand throw with standard timbers, open stand & target			8	6		755	136	27.50	918.50	1,075	
	0400	Resurface and realign existing track			200	.240	L.F.		5.45	1.09	6.54	9.75	
	0600	For crushed stone ballast, add			500	.096	"	10.90	2.18	.44	13.52	15.90	
	0800	Siding, yard spur, level grade											
	0810	100 lb. rail, new material on wood ties		B-14	57	.842	L.F.	55	19.15	3.84	77.99	94.50	
	1000	Steel ties in concrete w/100# rail, fasteners & plates			22	2.182	"	98	49.50	9.95	157.45	196	
	1200	Switch timber, for a #8 switch, pressure treated			3.70	12.973	M.B.F.	580	295	59	934	1,175	
	1300	Complete set of timbers, 3.7 M.B.F. for #8 switch			1	48	Total	2,500	1,100	219	3,819	4,700	
	1400	Ties, concrete, 8'-6" long, 30" O.C.			80	.600	Ea.	89.50	13.65	2.74	105.89	123	
	1600	Wood, pressure treated, 6" x 8" x 8'-6", C.L. lots			90	.533		27	12.15	2.43	41.58	51	
	1700	L.C.L. lots			90	.533		28	12.15	2.43	42.58	52.50	
	1900	Heavy duty, 7" x 9" x 8'-6", C.L. lots			70	.686		29.50	15.60	3.13	48.23	60.50	
	2000	L.C.L. lots			70	.686		29	15.60	3.13	47.73	60	
	2200	Turnouts, #8, incl. 100 lb. rails, plates, bars, frog, switch pt.											
	2300	Timbers and ballast 6" below bottom of tie		B-14	.50	96	Ea.	17,400	2,175	440	20,015	23,000	
	2400	Wheel stops, fixed			18	2.667	Pr.	430	60.50	12.15	502.65	585	
	2450	Hinged			14	3.429	"	505	78	15.65	598.65	695	

024 820	Dredging		CREW	DAILY OUTPUT	LABOR-HOURS	UNIT	MAT.	LABOR	EQUIP.	TOTAL	TOTAL INCL O&P	
824	0010	**DREDGING** Mobilization and demobilization., add to below, minimum	B-8	.53	120	Total		2,900	4,350	7,250	9,225	824
	0100	Maximum	"	.10	640	"		15,300	23,000	38,300	48,900	
	0300	Barge mounted clamshell excavation into scows,										
	0310	Dumped 20 miles at sea, minimum	B-57	310	.155	C.Y.		3.79	3.69	7.48	9.90	
	0400	Maximum	"	213	.225	"		5.50	5.35	10.85	14.45	
	0500	Barge mounted dragline or clamshell, hopper dumped,										
	0510	pumped 1000' to shore dump, minimum	B-57	340	.141	C.Y.		3.46	3.37	6.83	9.05	
	0525	All pumping uses 2000 gallons of water per cubic yard										
	0600	Maximum	B-57	243	.198	C.Y.		4.84	4.71	9.55	12.65	
	1000	Hydraulic method, pumped 1000' to shore dump, minimum		460	.104			2.56	2.49	5.05	6.70	
	1100	Maximum		310	.155			3.79	3.69	7.48	9.90	
	1400	Into scows dumped 20 miles, minimum		425	.113			2.77	2.69	5.46	7.25	
	1500	Maximum		243	.198			4.84	4.71	9.55	12.65	
	1600	For inland rivers and canals in South, deduct								30%	30%	

024 880	Docks & Facilities		CREW	DAILY OUTPUT	LABOR-HOURS	UNIT	MAT.	LABOR	EQUIP.	TOTAL	TOTAL INCL O&P	
882	0010	**DOCKS** Floating, recreational, prefabricated galvanized steel with										882
	0020	polyethylene encased polystyrene, no pilings included	F-3	330	.121	S.F.	25	3.36	1.34	29.70	34	
	0200	Pile supported, shore constructed, bare, 3" decking		130	.308		15	8.55	3.41	26.96	33.50	
	0250	4" decking		120	.333		15	9.25	3.70	27.95	35	
	0400	Floating, small boat, prefab, no shore facilities, minimum		250	.160		6	4.43	1.77	12.20	15.50	
	0500	Maximum		150	.267		25	7.40	2.96	35.36	42.50	
	0700	Per slip, minimum (180 S.F. each)		1.59	25.157	Ea.	1,300	695	279	2,274	2,825	
	0800	Maximum		1.40	28.571	"	5,000	790	315	6,105	7,075	

SITE WORK 2

SITE WORK 2

025 100 | Walk/Rd/Parkng Paving

		CREW	DAILY OUTPUT	LABOR-HOURS	UNIT	MAT.	LABOR	EQUIP.	TOTAL	TOTAL INCL O&P	
104	0010	**ASPHALTIC CONCRETE PAVEMENT** for highways										104
	0020	and large paved areas R025-110									
	0080	Binder course, 1-1/2" thick	B-25	7,725	.011	S.Y.	2.03	.27	.23	2.53	2.91	
	0120	2" thick		6,345	.014		2.71	.33	.28	3.32	3.80	
	0160	3" thick		4,905	.018		4.02	.42	.36	4.80	5.50	
	0200	4" thick		4,140	.021		5.40	.50	.43	6.33	7.20	
	0300	Wearing course, 1" thick	B-25B	10,575	.009		1.56	.22	.19	1.97	2.27	
	0340	1-1/2" thick		7,725	.012		2.38	.30	.26	2.94	3.37	
	0380	2" thick		6,345	.015		3.20	.36	.32	3.88	4.43	
	0420	2-1/2" thick		5,480	.018		3.95	.42	.37	4.74	5.40	
	0460	3" thick		4,900	.020		4.70	.47	.41	5.58	6.35	
	0800	Alternate method of figuring paving costs										
	0810	Binder course, 1-1/2" thick	B-25	630	.140	Ton	26.50	3.29	2.84	32.63	37.50	
	0811	2" thick		690	.128		26.50	3	2.59	32.09	37	
	0812	3" thick		800	.110		26.50	2.59	2.24	31.33	36	
	0813	4" thick		850	.104		26.50	2.44	2.10	31.04	35.50	
	0850	Wearing course, 1" thick	B-25B	575	.167		29	4	3.54	36.54	42	
	0851	1-1/2" thick		630	.152		29	3.65	3.23	35.88	41	
	0852	2" thick		690	.139		29	3.33	2.95	35.28	40.50	
	0853	2-1/2" thick		745	.129		29	3.08	2.73	34.81	40	
	0854	3" thick		800	.120		29	2.87	2.54	34.41	39.50	
	1000	Pavement replacement over trench, 2" thick	B-37	90	.533	S.Y.	3.25	12.15	1.70	17.10	24.50	
	1050	4" thick		70	.686		6.45	15.60	2.18	24.23	34	
	1080	6" thick		55	.873		10.25	19.85	2.78	32.88	45.50	
108	0010	**ASPHALTIC CONCRETE** plant mix (145 lb. per C.F.) **CN** R025-110				Ton	26			26	28.50	108
	0200	All weather patching mix, hot					31			31	34	
	0250	Cold patch					36			36	39.50	
	0300	Berm mix					31			31	34	
112	0010	**CALCIUM CHLORIDE** Delivered, 100 lb. bags, truckload lots				Ton	400			400	440	112
	0200	Solution, 4 lb. flake per gallon, tank truck delivery				Gal.	.85			.85	.94	
120	0010	**CONCRETE PAVEMENT** Including joints, finishing, and curing										120
	0020	Fixed form, 12' pass, unreinforced, 6" thick	B-26	3,000	.029	S.Y.	14.55	.71	.63	15.89	17.80	
	0100	8" thick		2,750	.032		20.50	.77	.69	21.96	24.50	
	0200	9" thick		2,500	.035		23.50	.85	.76	25.11	28	
	0300	10" thick		2,100	.042		25.50	1.01	.90	27.41	30.50	
	0400	12" thick		1,800	.049		27	1.18	1.05	29.23	33	
	0500	15" thick		1,500	.059		30.50	1.42	1.26	33.18	37	
	0510	For small irregular areas, add						100%				
	0700	Finishing, broom finish small areas	2 Cefi	120	.133	S.Y.		3.49		3.49	5.20	
	1000	Curing, with sprayed membrane by hand	2 Clab	1,500	.011	"	.22	.23		.45	.60	
	1650	For integral coloring, see div. 033-126										
122	0010	**FINE GRADE** Area to be paved with grader, small area	B-11L	400	.040	S.Y.		1	1.44	2.44	3.12	122
	0100	Large area		2,000	.008			.20	.29	.49	.63	
	1100	Fine grade for slab on grade, machine		1,040	.015			.38	.55	.93	1.20	
	1150	Hand grading	B-18	700	.034			.76	.07	.83	1.27	
124	0011	**PAVING** Asphaltic concrete, parking lots & driveways										124
	0020	6" stone base, 2" binder course, 1" topping	B-25C	9,000	.005	S.F.	1.13	.13	.17	1.43	1.64	
	0300	Binder course, 1-1/2" thick		35,000	.001		.23	.03	.04	.30	.36	
	0400	2" thick		25,000	.002		.30	.05	.06	.41	.47	
	0500	3" thick		15,000	.003		.47	.08	.10	.65	.75	
	0600	4" thick		10,800	.004		.62	.11	.14	.87	1.01	
	0800	Sand finish course, 3/4" thick		41,000	.001		.15	.03	.04	.22	.25	
	0900	1" thick		34,000	.001		.19	.03	.05	.27	.31	
	1000	Fill pot holes, hot mix, 2" thick	B-16	4,200	.008		.38	.17	.11	.66	.80	
	1100	4" thick		3,500	.009		.56	.20	.13	.89	1.08	

Important: See the Reference Section for critical supporting data - Reference Nos., Crews, & City Cost Indexes

			CREW	DAILY OUTPUT	LABOR-HOURS	UNIT	1999 BARE COSTS				TOTAL INCL O&P	
		025 100 | Walk/Rd/Parkng Paving					MAT.	LABOR	EQUIP.	TOTAL		
124	1120	6" thick	B-16	3,100	.010	S.F.	.75	.23	.14	1.12	1.35	**124**
	1140	Cold patch, 2" thick	B-51	3,000	.016		.45	.35	.06	.86	1.11	
	1160	4" thick	↓	2,700	.018		.86	.39	.06	1.31	1.62	
	1180	6" thick	↓	1,900	.025	↓	1.33	.55	.09	1.97	2.43	
128	0010	**SIDEWALKS, DRIVEWAYS, & PATIOS** No base R025 -110										**128**
	0020	Asphaltic concrete, 2" thick	B-37	720	.067	S.Y.	2.74	1.52	.21	4.47	5.60	
	0100	2-1/2" thick	"	660	.073	"	3.47	1.65	.23	5.35	6.65	
	0300	Concrete, 3000 psi, CIP, 6 x 6 - W1.4 x W1.4 mesh,										
	0310	broomed finish, no base, 4" thick	B-24	600	.040	S.F.	1.02	1		2.02	2.66	
	0350	5" thick		545	.044		1.36	1.10		2.46	3.19	
	0400	6" thick	↓	510	.047		1.58	1.18		2.76	3.56	
	0450	For bank run gravel base, 4" thick, add	B-18	2,500	.010		.14	.21	.02	.37	.51	
	0520	8" thick, add	"	1,600	.015		.28	.33	.03	.64	.87	
	0550	Exposed aggregate finish, add to above, minimum	B-24	1,875	.013		.11	.32		.43	.61	
	0600	Maximum	"	455	.053		.35	1.32		1.67	2.42	
	1000	Crushed stone, 1" thick, white marble	2 Clab	1,700	.009		.18	.20		.38	.51	
	1050	Bluestone	"	1,700	.009		.18	.20		.38	.52	
	1700	Redwood, prefabricated, 4' x 4' sections	2 Carp	316	.051		4.53	1.38		5.91	7.15	
	1750	Redwood planks, 1" thick, on sleepers	"	240	.067	↓	3.17	1.82		4.99	6.35	

		025 150 | Unit Pavers										
154	0010	**ASPHALT BLOCKS,** 6"x12"x1-1/4", w/bed & neopr. adhesive	D-1	135	.119	S.F.	3	2.92		5.92	7.85	**154**
	0100	3" thick		130	.123		4.20	3.03		7.23	9.35	
	0300	Hexagonal tile, 8" wide, 1-1/4" thick		135	.119		3.20	2.92		6.12	8.05	
	0400	2" thick		130	.123		4.48	3.03		7.51	9.65	
	0500	Square, 8" x 8", 1-1/4" thick		135	.119		3	2.92		5.92	7.85	
	0600	2" thick	↓	130	.123		4.20	3.03		7.23	9.35	
	0900	For exposed aggregate (ground finish) add				↓	.50			.50	.55	
	0910	For colors, add					.35			.35	.39	
158	0010	**BRICK PAVING** 4" x 8" x 1-1/2", without joints (4.5 brick/S.F.)	D-1	110	.145	S.F.	2.04	3.58		5.62	7.80	**158**
	0100	Grouted, 3/8" joint (3.9 brick/S.F.)		90	.178		2.43	4.37		6.80	9.45	
	0200	4" x 8" x 2-1/4", without joints (4.5 bricks/S.F.)		110	.145		2.69	3.58		6.27	8.50	
	0300	Grouted, 3/8" joint (3.9 brick/S.F.)	↓	90	.178		2.48	4.37		6.85	9.55	
	0500	Bedding, asphalt, 3/4" thick	B-25	5,130	.017		.26	.40	.35	1.01	1.29	
	0540	Course washed sand bed, 1" thick	B-18	5,000	.005		.18	.11	.01	.30	.38	
	0580	Mortar, 1" thick	D-1	300	.053		.36	1.31		1.67	2.44	
	0620	2" thick		200	.080		.36	1.97		2.33	3.46	
	1500	Brick on 1" thick sand bed laid flat, 4.5 per S.F.		100	.160		2.96	3.94		6.90	9.35	
	2000	Laid on edge, 7.2 per S.F.		70	.229		2.06	5.60		7.66	11	
	2500	For 4" thick concrete bed and joints, add	↓	595	.027		.75	.66		1.41	1.86	
	2800	For steam cleaning, add	A-1	950	.008	↓	.05	.18	.07	.30	.42	
162	0010	**PRECAST CONCRETE PAVING SLABS**										**162**
	0710	Precast concrete patio blocks, 2-3/8" thick, colors, 8" x 16"	D-1	265	.060	S.F.	.95	1.49		2.44	3.36	
	0750	Exposed local aggregate, natural	2 Bric	250	.064		4.63	1.77		6.40	7.85	
	0800	Colors		250	.064		4.75	1.77		6.52	7.95	
	0850	Exposed granite or limestone aggregate		250	.064		5.50	1.77		7.27	8.80	
	0900	Exposed white tumblestone aggregate	↓	250	.064	↓	2.92	1.77		4.69	5.95	
166	0010	**STONE PAVERS**										**166**
	1100	Flagging, bluestone, irregular, 1" thick,	D-1	81	.198	S.F.	1.85	4.86		6.71	9.60	
	1150	Snapped random rectangular, 1" thick		92	.174		2.80	4.28		7.08	9.75	
	1200	1-1/2" thick		85	.188		3.37	4.63		8	10.90	
	1250	2" thick		83	.193		3.92	4.74		8.66	11.65	
	1300	Slate, natural cleft, irregular, 3/4" thick		92	.174		1.80	4.28		6.08	8.65	
	1350	Random rectangular, gauged, 1/2" thick		105	.152		3.90	3.75		7.65	10.15	
	1400	Random rectangular, butt joint, gauged, 1/4" thick	↓	150	.107	↓	4.19	2.62		6.81	8.70	

For expanded coverage of these items see _Means Heavy Construction Cost Data 1999_

2 SITE WORK

025 150	Unit Pavers	CREW	DAILY OUTPUT	LABOR-HOURS	UNIT	MAT.	LABOR	EQUIP.	TOTAL	TOTAL INCL O&P	
166	1500	For interior setting, add								25%	25%	166
	1550	Granite blocks, 3-1/2" x 3-1/2" x 3-1/2"	D-1	92	.174	S.F.	5.25	4.28		9.53	12.45	
	1600	4" to 12" long, 3" to 5" wide, 3" to 5" thick	↓	98	.163		4.38	4.02		8.40	11.05	
	1650	6" to 15" long, 3" to 6" wide, 3" to 5" thick		105	.152	↓	2.34	3.75		6.09	8.40	

025 250	Curbs	CREW	DAILY OUTPUT	LABOR-HOURS	UNIT	MAT.	LABOR	EQUIP.	TOTAL	TOTAL INCL O&P	
254	0010	**CURBS** Asphaltic, machine formed, 8" wide, 6" high, 40 L.F./ton	B-27	1,000	.032	L.F.	.58	.70	.07	1.35	1.83	254
	0100	8" wide, 8" high, 30 L.F. per ton		900	.036		.67	.78	.08	1.53	2.05	
	0150	Asphaltic berm, 12" W, 3"-6" H, 35 L.F./ton, before pavement	↓	700	.046		.80	1	.10	1.90	2.58	
	0200	12" W, 1-1/2" to 4" H, 60 L.F. per ton, laid with pavement	B-2	1,050	.038		.49	.83		1.32	1.85	
	0300	Concrete, wood forms, 6" x 18", straight	C-2A	500	.096		2.11	2.54		4.65	6.30	
	0400	6" x 18", radius		200	.240		2.22	6.35		8.57	12.35	
	0410	Steel forms, 6" x 18", straight		700	.069		2.57	1.82		4.39	5.65	
	0411	6" x 18", radius		400	.120	↓	3.46	3.18		6.64	8.75	
	0415	Machine formed, 6" x 18", straight	B-69A	2,000	.024		3.04	.57	.23	3.84	4.47	
	0416	6" x 18", radius	"	900	.053		3.17	1.27	.50	4.94	6	
	0421	Curb and gutter, straight										
	0422	with 6" high curb and 6" thick gutter, wood forms										
	0430	24" wide, .055 C.Y. per L.F.	C-2A	375	.128	L.F.	10.25	3.39		13.64	16.60	
	0435	30" wide, .066 C.Y. per L.F.		340	.141		11.15	3.74		14.89	18.10	
	0440	Steel forms, 24" wide, straight		700	.069		4.95	1.82		6.77	8.30	
	0441	Radius		300	.160		4.95	4.24		9.19	12.05	
	0442	30" wide, straight		700	.069		5.95	1.82		7.77	9.40	
	0443	Radius	↓	300	.160		5.95	4.24		10.19	13.15	
	0445	Machine formed, 24" wide, straight	B-69A	2,000	.024		4.95	.57	.23	5.75	6.60	
	0446	Radius		900	.053		4.95	1.27	.50	6.72	7.95	
	0447	30" wide, straight		2,000	.024		5.95	.57	.23	6.75	7.70	
	0448	Radius	↓	900	.053		5.95	1.27	.50	7.72	9.05	
	0550	Precast, 6" x 18", straight	B-29	700	.080		6.25	1.86	.91	9.02	10.80	
	0600	6" x 18", radius	"	325	.172		7.75	4	1.96	13.71	16.90	
	1000	Granite, split face, straight, 5" x 16"	D-13	500	.096		10	2.51	.78	13.29	15.75	
	1100	6" x 18"	"	450	.107		13.15	2.79	.87	16.81	19.75	
	1300	Radius curbing, 6" x 18", over 10' radius	B-29	260	.215	↓	16.10	5	2.45	23.55	28	
	1400	Corners, 2' radius		80	.700	Ea.	54	16.25	7.95	78.20	94	
	1600	Edging, 4-1/2" x 12", straight		300	.187	L.F.	5	4.34	2.12	11.46	14.60	
	1800	Curb inlets, (guttermouth) straight	↓	41	1.366	Ea.	120	31.50	15.55	167.05	199	
	2000	Indian granite (belgian block)										
	2100	Jumbo, 10-1/2"x7-1/2"x4", grey	D-1	150	.107	L.F.	1.75	2.62		4.37	6	
	2150	Pink		150	.107		2.15	2.62		4.77	6.45	
	2200	Regular, 9"x4-1/2"x4-1/2", grey		160	.100		1.70	2.46		4.16	5.70	
	2250	Pink		160	.100		2	2.46		4.46	6.05	
	2300	Cubes, 4"x4"x4", grey		175	.091		1.65	2.25		3.90	5.30	
	2350	Pink		175	.091		1.75	2.25		4	5.45	
	2400	6"x6"x6", pink	↓	155	.103	↓	3.60	2.54		6.14	7.90	
	2500	Alternate pricing method for indian granite										
	2550	Jumbo, 10-1/2"x7-1/2"x4" (30lb), grey				Ton	100			100	110	
	2600	Pink					125			125	138	
	2650	Regular, 9"x4-1/2"x4-1/2" (20lb), grey					120			120	132	
	2700	Pink					140			140	154	
	2750	Cubes, 4"x4"x4" (5lb), grey					200			200	220	
	2800	Pink					225			225	248	
	2850	6"x6"x6" (25lb), pink					140			140	154	
	2900	For pallets, add				↓	15			15	16.50	
258	0010	**EDGING**										258
	0050	Aluminum alloy, including stakes, 1/8" x 4", mill finish	B-1	390	.062	L.F.	1.75	1.36		3.11	4.07	

Important: See the Reference Section for critical supporting data - Reference Nos., Crews, & City Cost Indexes

025 250 | Curbs

			CREW	DAILY OUTPUT	LABOR-HOURS	UNIT	MAT.	LABOR	EQUIP.	TOTAL	TOTAL INCL O&P	
							\multicolumn{4}{c}{1999 BARE COSTS}		
258	0051	Black paint	B-1	390	.062	L.F.	2.03	1.36		3.39	4.37	258
	0052	Black anodized	↓	390	.062		2.34	1.36		3.70	4.72	
	0100	Brick, set horizontally, 1-1/2 bricks per L.F.	D-1	370	.043		.90	1.06		1.96	2.64	
	0150	Set vertically, 3 bricks per L.F.	"	135	.119		1.90	2.92		4.82	6.60	
	0200	Corrugated aluminum, roll, 4" wide	1 Carp	650	.012		.28	.34		.62	.84	
	0250	6" wide	"	550	.015		.35	.40		.75	1.01	
	0600	Railroad ties, 6" x 8"	2 Carp	170	.094		2.30	2.57		4.87	6.55	
	0650	7" x 9"		136	.118		2.55	3.21		5.76	7.85	
	0750	2" x 4"	↓	330	.048		2.67	1.32		3.99	5	
	0800	Steel edge strips, incl. stakes, 1/4" x 5"	B-1	390	.062		2.75	1.36		4.11	5.15	
	0850	3/16" x 4"	"	390	.062	↓	2.17	1.36		3.53	4.53	

025 300 | Athletic Pave/Surfacing

			CREW	DAILY OUTPUT	LABOR-HOURS	UNIT	MAT.	LABOR	EQUIP.	TOTAL	TOTAL INCL O&P	
304	0010	**RUNNING TRACK** Asphalt, incl base, 3" thick	B-37	300	.160	S.Y.	10.35	3.64	.51	14.50	17.60	304
	0100	Surface, latex rubber system, 3/8" thick, black	B-20	125	.192		4.65	4.67		9.32	12.45	
	0150	Colors		125	.192		8.25	4.67		12.92	16.45	
	0300	Urethane rubber system, 3/8" thick, black		120	.200		12.40	4.86		17.26	21.50	
	0400	Color coating	↓	115	.209	↓	13.50	5.10		18.60	23	
308	0010	**TENNIS COURT** Asphalt, incl. base, 2-1/2" thick, one court	B-37	450	.107	S.Y.	9.25	2.43	.34	12.02	14.30	308
	0200	Two courts		675	.071		7	1.62	.23	8.85	10.50	
	0300	Clay courts		360	.133		26.50	3.03	.42	29.95	34	
	0400	Pulverized natural greenstone with 4" base, fast dry		250	.192		26	4.37	.61	30.98	36	
	0800	Rubber-acrylic base resilient pavement	↓	600	.080		26	1.82	.25	28.07	32	
	1000	Colored sealer, acrylic emulsion, 3 coats	2 Clab	800	.020		3.73	.43		4.16	4.78	
	1100	3 coat, 2 colors	"	900	.018		4.60	.38		4.98	5.65	
	1200	For preparing old courts, add	1 Clab	825	.010	↓		.21		.21	.33	
	1400	Posts for nets, 3-1/2" diameter with eye bolts	B-1	3.40	7.059	Pr.	138	156		294	395	
	1500	With pulley & reel		3.40	7.059	"	189	156		345	455	
	1700	Net, 42' long, nylon thread with binder		50	.480	Ea.	170	10.60		180.60	204	
	1800	All metal	↓	6.50	3.692	"	287	81.50		368.50	445	
	2000	Paint markings on asphalt, 2 coats	1 Pord	1.78	4.494	Court	58.50	112		170.50	235	
	2200	Complete court with fence, etc., asphaltic conc., minimum	B-37	.20	240		11,700	5,450	765	17,915	22,200	
	2300	Maximum		.16	300		14,200	6,825	955	21,980	27,400	
	2800	Clay courts, minimum		.20	240		13,600	5,450	765	19,815	24,400	
	2900	Maximum	↓	.16	300	↓	16,700	6,825	955	24,480	30,200	

025 400 | Synthetic Surfacing

			CREW	DAILY OUTPUT	LABOR-HOURS	UNIT	MAT.	LABOR	EQUIP.	TOTAL	TOTAL INCL O&P	
404	0010	**TURF, ARTIFICIAL** Not including asphalt base or drainage, but										404
	0020	including cushion pad, over 50,000 S.F.										
	0200	1/2" pile and 5/16" cushion pad, standard	C-17	3,200	.025	S.F.	5.50	.71		6.21	7.15	
	0300	Deluxe		2,560	.031		6.50	.89		7.39	8.55	
	0500	1/2" pile and 5/8" cushion pad, standard		2,844	.028		8	.80		8.80	10.05	
	0600	Deluxe	↓	2,327	.034	↓	8.75	.98		9.73	11.20	
	0800	For asphaltic concrete base, 2-1/2" thick,										
	0900	with 6" crushed stone sub-base, add	B-25	12,000	.007	S.F.	1.07	.17	.15	1.39	1.61	

025 450 | Surfacing

			CREW	DAILY OUTPUT	LABOR-HOURS	UNIT	MAT.	LABOR	EQUIP.	TOTAL	TOTAL INCL O&P	
458	0010	**SEALCOATING** 2 coat coal tar pitch emulsion over 10,000 S.Y.	B-45	5,000	.003	S.Y.	.43	.08	.15	.66	.76	458
	0030	1000 to 10,000 S.Y.	"	3,000	.005		.43	.13	.25	.81	.96	
	0100	Under 1000 S.Y.	B-1	1,050	.023		.43	.51		.94	1.28	
	0300	Petroleum resistant, over 10,000 S.Y.	B-45	5,000	.003		.50	.08	.15	.73	.83	
	0320	1000 to 10,000 S.Y.	"	3,000	.005		.50	.13	.25	.88	1.03	
	0400	Under 1000 S.Y.	B-1	1,050	.023		.50	.51		1.01	1.35	

SITE WORK **2**

For expanded coverage of these items see *Means Heavy Construction Cost Data 1999*

025 450 | Surfacing

		CREW	DAILY OUTPUT	LABOR-HOURS	UNIT	1999 BARE COSTS				TOTAL INCL O&P
						MAT.	LABOR	EQUIP.	TOTAL	
458	0600	Non-skid pavement renewal, over 10,000 S.Y.	B-45	5,000	.003	S.Y.	.60	.08	.15	.83	.94	**458**
	0620	1000 to 10,000 S.Y.	"	3,000	.005		.60	.13	.25	.98	1.14
	0700	Under 1000 S.Y.	B-1	1,050	.023		.60	.51		1.11	1.46
	0800	Prepare and clean surface for above	A-2	8,545	.003			.06	.02	.08	.11
	1000	Hand seal asphalt curbing	B-1	4,420	.005	L.F.	.30	.12		.42	.52
	1900	Asphalt surface treatment, single course, small area									
	1901	0.30 gal/S.Y. asphalt material, 20#/S.Y. aggregate	B-91	5,000	.013	S.Y.	.70	.32	.30	1.32	1.60
	1910	Roadway or large area		10,000	.006		.64	.16	.15	.95	1.12
	1950	Asphalt surface treatment, dbl. course for small area		3,000	.021		1.30	.54	.50	2.34	2.81
	1960	Roadway or large area		6,000	.011		1.17	.27	.25	1.69	1.97
	1980	Asphalt surface treatment, single course, for shoulders		7,500	.009		.75	.22	.20	1.17	1.38
	2080	Sand sealing, sharp sand, asphalt emulsion, small area		10,000	.006		.50	.16	.15	.81	.96
	2120	Roadway or large area		18,000	.004		.43	.09	.08	.60	.70

025 500 | Bridges

		CREW	DAILY OUTPUT	LABOR-HOURS	UNIT	MAT.	LABOR	EQUIP.	TOTAL	TOTAL INCL O&P
504	0010	**BRIDGES** Pedestrian, spans over streams, roadways, etc.										**504**
	0020	including erection, not including foundations									
	0050	Precast concrete, complete in place, 8' wide, 60' span	E-2	215	.260	S.F.	25	7.75	4.71	37.46	46.50
	0100	100' span		185	.303		27.50	9	5.45	41.95	52
	0150	120' span		160	.350		30	10.40	6.30	46.70	57.50
	0200	150' span		145	.386		31	11.50	7	49.50	61.50
	0300	Steel, trussed or arch spans, compl. in place, 8' wide, 40' span		320	.175		35	5.20	3.16	43.36	51
	0400	50' span		395	.142		31.50	4.22	2.56	38.28	44.50
	0500	60' span		465	.120		31.50	3.59	2.18	37.27	43
	0600	80' span		570	.098		37.50	2.92	1.78	42.20	48
	0700	100' span		465	.120		52.50	3.59	2.18	58.27	66.50
	0800	120' span		365	.153		66.50	4.57	2.77	73.84	84
	0900	150' span		310	.181		70.50	5.40	3.26	79.16	90.50
	1000	160' span		255	.220		70.50	6.55	3.97	81.02	93.50
	1100	10' wide, 80' span		640	.087		36	2.60	1.58	40.18	46
	1200	120' span		415	.135		46.50	4.02	2.44	52.96	61
	1300	150' span		445	.126		52.50	3.75	2.27	58.52	66.50
	1400	200' span		205	.273		56	8.15	4.94	69.09	81
	1600	Wood, laminated type, complete in place, 80' span	C-12	203	.236		30	6.40	2.18	38.58	45.50
	1700	130' span	"	153	.314		31	8.45	2.90	42.35	51

025 800 | Pavement Marking

		CREW	DAILY OUTPUT	LABOR-HOURS	UNIT	MAT.	LABOR	EQUIP.	TOTAL	TOTAL INCL O&P
804	0010	**LINES ON PAV'T** Acrylic waterborne, white or yellow, 4" wide	B-78	20,000	.002	L.F.	.09	.05	.03	.17	.20	**804**
	0200	6" wide		11,000	.004		.09	.10	.05	.24	.30
	0500	8" wide		10,000	.005		.11	.10	.05	.26	.34
	0600	12" wide		4,000	.012		.20	.26	.13	.59	.77
	0620	Arrows or gore lines		2,300	.021	S.F.	.50	.46	.22	1.18	1.51
	0640	Temporary paint, white or yellow		15,000	.003	L.F.	.15	.07	.03	.25	.32
	0660	Removal	1 Clab	300	.027			.57		.57	.90
	0680	Temporary tape	2 Clab	1,500	.011		1.12	.23		1.35	1.59
	0710	Thermoplastic, white or yellow, 4" wide	B-79	15,000	.003		.56	.06	.05	.67	.76
	0730	6" wide		14,000	.003		.81	.06	.05	.92	1.04
	0740	8" wide		12,000	.003		1.09	.07	.06	1.22	1.37
	0750	12" wide		6,000	.007		1.62	.15	.11	1.88	2.14
	0760	Arrows		660	.061	S.F.	1.50	1.33	1.03	3.86	4.87
	0770	Gore lines		2,500	.016		1	.35	.27	1.62	1.95
	0780	Letters		660	.061		1.25	1.33	1.03	3.61	4.60
	0790	Layout of pavement marking	A-2	25,000	.001	L.F.		.02	.01	.03	.04
	0800	Parking stall, paint, white	B-78	440	.109	Stall	1.84	2.38	1.16	5.38	7.05
	1000	Street letters and numbers	"	1,600	.030	S.F.	.50	.65	.32	1.47	1.93

Important: See the Reference Section for critical supporting data - Reference Nos., Crews, & City Cost Indexes

SITE WORK 2

025 800 | Pavement Marking

		CREW	DAILY OUTPUT	LABOR-HOURS	UNIT	1999 BARE COSTS				TOTAL INCL O&P
						MAT.	LABOR	EQUIP.	TOTAL	
810	0010 **TRAFFIC CONTROL DEVICES**									**810**
	0100 Traffic channelizing pavement markers, layout only	A-7	2,000	.012	Ea.		.29		.29	.44
	0110 13" x 7-1/2" x 2-1/2" high, non-plowable install	2 Clab	96	.167		16.50	3.58		20.08	24
	0200 8" x 8"x 3-1/4" high, non-plowable, install		96	.167		18	3.58		21.58	25.50
	0230 4" x 4" x 3/4" high, non-plowable, install	↓	120	.133		2	2.86		4.86	6.70
	0240 9-1/4" x 5-7/8" x 1/4" high, plowable, concrete pav't	A-2A	70	.343		12	7.40	4.04	23.44	29
	0250 9-1/4" x 5-7/8" x 1/4" high, plowable, asphalt pav't	"	120	.200		2.30	4.31	2.36	8.97	11.80
	0300 Barrier and curb delineators,reflectorized, 2" x 4"	2 Clab	150	.107		1.30	2.29		3.59	5.05
	0310 3" x 5"	"	150	.107	↓	2.65	2.29		4.94	6.50
	0500 Rumble strip, polycarbonate									
	0510 24" x 3-1/2" x 1/2" high	2 Clab	50	.320	Ea.	5	6.85		11.85	16.30

026 | Piped Utilities

026 010 | Piped Utilities

		CREW	DAILY OUTPUT	LABOR-HOURS	UNIT	1999 BARE COSTS				TOTAL INCL O&P
						MAT.	LABOR	EQUIP.	TOTAL	
012	0010 **BEDDING** For pipe and conduit, not incl. compaction									**012**
	0050 Crushed or screened bank run gravel	B-6	150	.160	C.Y.	6.90	3.74	1.46	12.10	15
	0100 Crushed stone 3/4" to 1/2"		150	.160		17.75	3.74	1.46	22.95	27
	0200 Sand, dead or bank	↓	150	.160		3.75	3.74	1.46	8.95	11.55
	0500 Compacting bedding in trench	A-1	90	.089	↓		1.91	.76	2.67	3.84
014	0010 **EXCAVATION AND BACKFILL** See division 022-204 & 254									**014**
	0100 Hand excavate and trim for pipe bells after trench excavation									
	0200 8" pipe	1 Clab	155	.052	L.F.		1.11		1.11	1.74
	0300 18" pipe		130	.062	"		1.32		1.32	2.08
	0400 Underground tape, detectable aluminum, 2"		150	.053	C.L.F.	3.70	1.14		4.84	5.85
	0500 6"	↓	140	.057	"	9.25	1.23		10.48	12.15

026 050 | Manholes & Cleanouts

		CREW	DAILY OUTPUT	LABOR-HOURS	UNIT	1999 BARE COSTS				TOTAL INCL O&P
						MAT.	LABOR	EQUIP.	TOTAL	
054	0010 **UTILITY VAULTS** Precast concrete, 6" thick									**054**
	0050 5' x 10' x 6' high, I.D.	B-13	2	28	Ea.	1,250	650	275	2,175	2,700
	0100 6' x 10' x 6' high, I.D.		2	28		1,300	650	275	2,225	2,725
	0150 5' x 12' x 6' high, I.D.		2	28		1,375	650	275	2,300	2,825
	0200 6' x 12' x 6' high, I.D.		1.80	31.111		1,550	725	305	2,580	3,150
	0250 6' x 13' x 6' high, I.D.		1.50	37.333		2,025	870	365	3,260	4,000
	0300 8' x 14' x 7' high, I.D.	↓	1	56	↓	2,200	1,300	550	4,050	5,050
	0350 Hand hole, precast concrete, 1-1/2" thick									
	0400 1'-0" x 2'-0" x 1'-9", I.D., light duty	B-1	4	6	Ea.	330	133		463	575
	0450 4'-6" x 3'-2" x 2'-0", O.D., heavy duty	B-6	3	8	"	870	187	73	1,130	1,325

026 450 | Hydrants

		CREW	DAILY OUTPUT	LABOR-HOURS	UNIT	1999 BARE COSTS				TOTAL INCL O&P
						MAT.	LABOR	EQUIP.	TOTAL	
454	0010 **PIPING, WATER DISTRIBUTION** Mech. joints unless noted									**454**
	1000 Fire hydrants, two way; excavation and backfill not incl.									
	1100 4-1/2" valve size, depth 2'-0"	B-21	10	2.800	Ea.	730	70	13.60	813.60	925
	1120 2'-6"		10	2.800		770	70	13.60	853.60	970
	1140 3'-0"		10	2.800		815	70	13.60	898.60	1,025
	1300 7'-0"	↓	6	4.667		985	117	22.50	1,124.50	1,275
	2400 Lower barrel extensions with stems, 1'-0"	B-20	14	1.714		242	41.50		283.50	330
	2480 3'-0"	"	12	2		735	48.50		783.50	880

SITE WORK 2

For expanded coverage of these items see Means Heavy Construction Cost Data 1999

2 SITE WORK

026 650	Water Systems	CREW	DAILY OUTPUT	LABOR-HOURS	UNIT	1999 BARE COSTS				TOTAL INCL O&P
						MAT.	LABOR	EQUIP.	TOTAL	
652 0010	**CORROSION RESISTANCE** Wrap & coat, add to pipe, 4" dia.				L.F.	1.31			1.31	1.44	**652**
0040	6" diameter					1.38			1.38	1.52
0060	8" diameter					2.13			2.13	2.34
0100	12" diameter					3.17			3.17	3.49
0200	24" diameter					6			6	6.60
0500	Coating, bituminous, per diameter inch, 1 coat, add					.23			.23	.25
0540	3 coat					.37			.37	.41
0560	Coal tar epoxy, per diameter inch, 1 coat, add					.15			.15	.17
0600	3 coat					.29			.29	.32
686 0010	**PIPING, WATER DISTRIBUTION SYSTEMS** Pipe laid in trench,										**686**
0020	excavation and backfill not included									
1400	Ductile Iron, cement lined, class 50 water pipe, 18' lengths									
1410	Mechanical joint, 4" diameter	B-20	144	.167	L.F.	6.95	4.05		11	14.05
1420	6" diameter		126	.190		7.90	4.63		12.53	16
1430	8" diameter		108	.222		10.35	5.40		15.75	19.85
1440	10" diameter		90	.267		14	6.50		20.50	25.50
1450	12" diameter CN	B-21	72	.389		17.05	9.75	1.89	28.69	36
1460	14" diameter		54	.519		22	13	2.51	37.51	47.50
1470	16" diameter		46	.609		24	15.25	2.95	42.20	53.50
1480	18" diameter		42	.667		30.50	16.70	3.23	50.43	63
1490	24" diameter		35	.800		47	20	3.88	70.88	88
1550	Push on joint, 4" diameter	B-20	155	.155		6.35	3.77		10.12	12.90
1560	6" diameter		135	.178		7.25	4.32		11.57	14.80
1570	8" diameter		115	.209		9.95	5.10		15.05	18.95
1580	10" diameter		98	.245		15.60	5.95		21.55	26.50
1590	12" diameter		78	.308		16.45	7.50		23.95	30
1600	14" diameter	B-21	58	.483		18.10	12.10	2.34	32.54	41.50
1610	16" diameter		52	.538		25.50	13.50	2.61	41.61	52
1620	18" diameter		43	.651		28.50	16.30	3.16	47.96	60
1630	20" diameter		41	.683		31	17.10	3.31	51.41	64.50
1640	24" diameter		40	.700		40	17.55	3.39	60.94	75
1950	Butterfly valves with boxes, cast iron									
1970	4" diameter	B-20	6	4	Ea.	465	97.50		562.50	665
1990	6" diameter	"	5	4.800		490	117		607	725
2010	8" diameter	B-21	4	7		675	175	34	884	1,050
2030	10" diameter		3.50	8		980	200	39	1,219	1,425
2050	12" diameter		3	9.333		1,275	234	45.50	1,554.50	1,825
2070	14" diameter		2	14		2,325	350	68	2,743	3,175
2090	16" diameter		2	14		2,850	350	68	3,268	3,775
2650	Polyvinyl chloride pipe, class 160, S.D.R.-26, 1-1/2" diameter	B-20	300	.080	L.F.	.27	1.95		2.22	3.36
2700	2" diameter		250	.096		.38	2.33		2.71	4.10
2750	2-1/2" diameter		250	.096		.50	2.33		2.83	4.23
2800	3" diameter		200	.120		.75	2.92		3.67	5.45
2850	4" diameter		200	.120		1.18	2.92		4.10	5.90
2900	6" diameter		180	.133		2.50	3.24		5.74	7.85
2950	8" diameter	B-21	160	.175		4.25	4.38	.85	9.48	12.45
8000	Fittings, ductile iron, mechanical joint									
8010	90° bend 4" diameter	B-20	37	.649	Ea.	97.50	15.80		113.30	132
8020	6" diameter		25	.960		159	23.50		182.50	212
8040	8" diameter		21	1.143		230	28		258	297
8060	10" diameter	B-21	21	1.333		258	33.50	6.45	297.95	345
8080	12" diameter		18	1.556		390	39	7.55	436.55	495
8100	14" diameter		16	1.750		550	44	8.50	602.50	685
8120	16" diameter		14	2		625	50	9.70	684.70	775
8140	18" diameter		10	2.800		1,750	70	13.60	1,833.60	2,050

Important: See the Reference Section for critical supporting data - Reference Nos., Crews, & City Cost Indexes

026 650 | Water Systems

		CREW	DAILY OUTPUT	LABOR-HOURS	UNIT	1999 BARE COSTS				TOTAL INCL O&P
						MAT.	LABOR	EQUIP.	TOTAL	
8160	20" diameter	B-21	8	3.500	Ea.	1,950	87.50	17	2,054.50	2,300
8180	24" diameter		6	4.667		2,150	117	22.50	2,289.50	2,575
8200	Wye or tee, 4" diameter	B-20	25	.960		157	23.50		180.50	210
8220	6" diameter		17	1.412		169	34.50		203.50	240
8240	8" diameter		14	1.714		243	41.50		284.50	335
8260	10" diameter	B-21	14	2		510	50	9.70	569.70	650
8280	12" diameter		12	2.333		700	58.50	11.30	769.80	870
8300	14" diameter		10	2.800		670	70	13.60	753.60	860
8320	16" diameter		8	3.500		935	87.50	17	1,039.50	1,175
8340	18" diameter		6	4.667		1,825	117	22.50	1,964.50	2,200
8360	20" diameter		4	7		1,600	175	34	1,809	2,050
8380	24" diameter		3	9.333		2,800	234	45.50	3,079.50	3,525

686 | | | | | | | | | | | | 686

026 700 | Water Wells

		CREW	DAILY OUTPUT	LABOR-HOURS	UNIT	MAT.	LABOR	EQUIP.	TOTAL	TOTAL INCL O&P
0010	**WELLS** Domestic water									
0100	Drilled, 4" to 6" diameter	B-23	120	.333	L.F.		7.30	16.30	23.60	29.50
0200	8" diameter	"	95.20	.420	"		9.20	20.50	29.70	37
0400	Gravel pack well, 40' deep, incl. gravel & casing, complete									
0500	24" diameter casing x 18" diameter screen	B-23	.13	307	Total	20,000	6,725	15,000	41,725	49,100
0600	36" diameter casing x 18" diameter screen		.12	333	"	21,500	7,275	16,300	45,075	53,000
0800	Observation wells, 1-1/4" riser pipe		163	.245	V.L.F.	11	5.35	12	28.35	34
0900	For flush Buffalo roadway box, add	1 Skwk	16.60	.482	Ea.	30	13.50		43.50	54.50
1200	Test well, 2-1/2" diameter, up to 50' deep (15 to 50 GPM)	B-23	1.51	26.490	"	450	580	1,300	2,330	2,825
1300	Over 50' deep, add	"	121.80	.328	L.F.	12	7.20	16.05	35.25	42
1500	Pumps, installed in wells to 100' deep, 4" submersible									
1510	1/2 H.P.	Q-1	3.22	4.969	Ea.	425	146		571	690
1520	3/4 H.P.		2.66	6.015		475	177		652	795
1600	1 H.P.		2.29	6.987		525	205		730	890
1700	1-1/2 H.P.	Q-22	1.60	10		580	294	277	1,151	1,400
1800	2 H.P.		1.33	12.030		620	355	335	1,310	1,575
1900	3 H.P.		1.14	14.035		775	410	390	1,575	1,900
2000	5 H.P.		1.14	14.035		1,250	410	390	2,050	2,425
3000	Pump, 6" submersible, 25' to 150' deep, 25 H.P., 249 to 297 GPM		.89	17.978		3,775	530	500	4,805	5,500
3100	25' to 500' deep, 30 H.P., 100 to 300 GPM		.73	21.918		3,875	645	610	5,130	5,900
8000	Steel well casing	B-23A	3,020	.008	Lb.	.40	.19	.63	1.22	1.43
9950	See div. 021-444 for wellpoints									
9960	See div. 021-484 for drainage wells									

704 | | | | | | | | | | | | 704

026 800 | Fuel Distribution

		CREW	DAILY OUTPUT	LABOR-HOURS	UNIT	MAT.	LABOR	EQUIP.	TOTAL	TOTAL INCL O&P
0010	**GAS STATION PRODUCT LINE**									
0020	Primary containment pipe, fiberglass-reinforced									
0030	Plastic pipe 15' & 30' lengths									
0040	2" diameter	Q-6	425	.056	L.F.	2.97	1.73		4.70	5.90
0050	3" diameter		400	.060		4.40	1.83		6.23	7.60
0060	4" diameter		375	.064		5.75	1.96		7.71	9.30
0100	Fittings									
0110	Elbows, 90° & 45°, bell-ends, 2"	Q-6	24	1	Ea.	31	30.50		61.50	80.50
0120	3" diameter		22	1.091		35	33.50		68.50	89
0130	4" diameter		20	1.200		42	36.50		78.50	102
0200	Tees, bell ends, 2"		21	1.143		36.50	35		71.50	93.50
0210	3" diameter		18	1.333		40	41		81	106
0220	4" diameter		15	1.600		52	49		101	131
0230	Flanges bell ends, 2"		24	1		20	30.50		50.50	68.50
0240	3" diameter		22	1.091		22	33.50		55.50	74.50
0250	4" diameter		20	1.200		25	36.50		61.50	83

804 | | | | | | | | | | | | 804

026 800 | Fuel Distribution

		CREW	DAILY OUTPUT	LABOR-HOURS	UNIT	1999 BARE COSTS				TOTAL INCL O&P	
						MAT.	LABOR	EQUIP.	TOTAL		
804	0260	Sleeve couplings, 2"	Q-6	21	1.143	Ea.	7.75	35		42.75	61.50	804
	0270	3" diameter		18	1.333		10.75	41		51.75	73.50
	0280	4" diameter		15	1.600		15.50	49		64.50	91
	0290	Threaded adapters 2"		21	1.143		10.95	35		45.95	65
	0300	3" diameter		18	1.333		16.50	41		57.50	79.50
	0310	4" diameter		15	1.600		23	49		72	99.50
	0320	Reducers, 2"		27	.889		14.15	27		41.15	56.50
	0330	3" diameter		22	1.091		16.50	33.50		50	68.50
	0340	4" diameter		20	1.200		25	36.50		61.50	83
	1010	Gas station product line for secondary containment (double wall)									
	1100	Fiberglass reinforced plastic pipe 25' lengths									
	1120	Pipe, plain end, 3"	Q-6	375	.064	L.F.	3.98	1.96		5.94	7.35
	1130	4" diameter		350	.069		4.95	2.10		7.05	8.60
	1140	5" diameter		325	.074		5.50	2.26		7.76	9.45
	1150	6" diameter		300	.080		9.45	2.45		11.90	14.10
	1200	Fittings									
	1230	Elbows, 90° & 45°, 3"	Q-6	18	1.333	Ea.	37	41		78	102
	1240	4" diameter		16	1.500		70	46		116	147
	1250	5" diameter		14	1.714		145	52.50		197.50	240
	1260	6" diameter		12	2		150	61		211	258
	1270	Tees, 3"		15	1.600		53	49		102	133
	1280	4" diameter		12	2		85	61		146	186
	1290	5" diameter		9	2.667		158	81.50		239.50	297
	1300	6" diameter		6	4		165	122		287	365
	1310	Couplings, 3"		18	1.333		24	41		65	88
	1320	4" diameter		16	1.500		65	46		111	141
	1330	5" diameter		14	1.714		135	52.50		187.50	229
	1340	6" diameter		12	2		140	61		201	247
	1350	Cross-over nipples, 3"		18	1.333		5.75	41		46.75	68
	1360	4" diameter		16	1.500		6.75	46		52.75	77
	1370	5" diameter		14	1.714		10	52.50		62.50	90.50
	1380	6" diameter		12	2		10.50	61		71.50	104
	1400	Telescoping, reducers, concentric 4" x 3"		18	1.333		19.15	41		60.15	82.50
	1410	5" x 4"		17	1.412		50	43		93	121
	1420	6" x 5"		16	1.500		120	46		166	202

026 850 | Gas Distribution System

		CREW	DAILY OUTPUT	LABOR-HOURS	UNIT	MAT.	LABOR	EQUIP.	TOTAL	TOTAL INCL O&P	
854	0010	**PIPING, GAS SERVICE & DISTRIBUTION, POLYETHYLENE**									854
	0020	not including excavation or backfill									
	1000	60 psi coils, comp cplg @ 100', 1/2" diameter, SDR 9.3	B-20A	608	.053	L.F.	.35	1.36		1.71	2.49
	1040	1-1/4" diameter, SDR 11		544	.059		.57	1.52		2.09	2.98
	1100	2" diameter, SDR 11		488	.066		.72	1.70		2.42	3.41
	1160	3" diameter, SDR 11		408	.078		1.51	2.03		3.54	4.79
	1500	60 PSI 40' joints with coupling, 3" diameter, SDR 11	B-21A	408	.098		1.51	2.61	.96	5.08	6.70
	1540	4" diameter, SDR 11		352	.114		3.33	3.02	1.11	7.46	9.55
	1600	6" diameter, SDR 11		328	.122		10.90	3.24	1.19	15.33	18.30
	1640	8" diameter, SDR 11		272	.147		14.55	3.91	1.44	19.90	23.50

Important: See the Reference Section for critical supporting data - Reference Nos., Crews, & City Cost Indexes

2 SITE WORK

027 100 | Subdrainage Systems

		CREW	DAILY OUTPUT	LABOR-HOURS	UNIT	1999 BARE COSTS				TOTAL INCL O&P
						MAT.	LABOR	EQUIP.	TOTAL	
108	**0010**	**PIPING, SUBDRAINAGE, CONCRETE** R027-110								**108**
	0021	Not including excavation and backfill								
	3000	Porous wall concrete underdrain, std. strength, 4" diameter	B-20	335	.072	L.F.	1.78	1.74		3.52	4.70
	3020	6" diameter	"	315	.076		2.32	1.85		4.17	5.45
	3040	8" diameter	B-21	310	.090		2.86	2.26	.44	5.56	7.15
	3060	12" diameter		285	.098		6.05	2.46	.48	8.99	11
	3080	15" diameter		230	.122		6.95	3.05	.59	10.59	13.05
	3100	18" diameter	↓	165	.170		9.20	4.25	.82	14.27	17.65
	4000	Extra strength, 6" diameter	B-20	315	.076		2.29	1.85		4.14	5.45
	4020	8" diameter	B-21	310	.090		3.43	2.26	.44	6.13	7.80
	4040	10" diameter		285	.098		6.85	2.46	.48	9.79	11.90
	4060	12" diameter		230	.122		7.45	3.05	.59	11.09	13.60
	4080	15" diameter		200	.140		8.25	3.51	.68	12.44	15.30
	4100	18" diameter	↓	165	.170	↓	12	4.25	.82	17.07	21
110	**0010**	**PIPING, SUBDRAINAGE, CORRUGATED METAL**									**110**
	0021	Not including excavation and backfill								
	2010	Aluminum, perforated								
	2020	6" diameter, 18 ga.	B-14	380	.126	L.F.	2.57	2.87	.58	6.02	7.95
	2200	8" diameter, 16 ga.		370	.130		3.73	2.95	.59	7.27	9.35
	2220	10" diameter, 16 ga.		360	.133		4.67	3.03	.61	8.31	10.55
	2240	12" diameter, 16 ga.		285	.168		5.25	3.83	.77	9.85	12.60
	2260	18" diameter, 16 ga.	↓	205	.234	↓	7.85	5.30	1.07	14.22	18.05
	3000	Uncoated galvanized, perforated								
	3020	6" diameter, 18 ga.	B-20	380	.063	L.F.	4	1.54		5.54	6.80
	3200	8" diameter, 16 ga.	"	370	.065		5.50	1.58		7.08	8.55
	3220	10" diameter, 16 ga.	B-21	360	.078		8.25	1.95	.38	10.58	12.50
	3240	12" diameter, 16 ga.		285	.098		8.65	2.46	.48	11.59	13.85
	3260	18" diameter, 16 ga.	↓	205	.137	↓	13.20	3.42	.66	17.28	20.50
	4000	Steel, perforated, asphalt coated								
	4020	6" diameter 18 ga.	B-20	380	.063	L.F.	3.20	1.54		4.74	5.95
	4030	8" diameter 18 ga	"	370	.065		5	1.58		6.58	8
	4040	10" diameter 16 ga	B-21	360	.078		5.75	1.95	.38	8.08	9.80
	4050	12" diameter 16 ga		285	.098		6.60	2.46	.48	9.54	11.60
	4060	18" diameter 16 ga	↓	205	.137	↓	9	3.42	.66	13.08	16
112	**0010**	**PIPING, SUBDRAINAGE, VITRIFIED CLAY**									**112**
	0020	Not including excavation and backfill								
	3000	Perforated, 5' lengths, C700, 4" diameter	B-14	400	.120	L.F.	1.93	2.73	.55	5.21	7
	3020	6" diameter		315	.152		3	3.47	.69	7.16	9.45
	3040	8" diameter		290	.166		4.02	3.76	.75	8.53	11.15
	3060	12" diameter	↓	275	.175		8.15	3.97	.80	12.92	16
	4000	Channel pipe, 4" diameter	B-20	430	.056		2	1.36		3.36	4.34
	4020	6" diameter		335	.072		3	1.74		4.74	6.05
	4060	8" diameter	↓	295	.081		4.50	1.98		6.48	8.05
	4080	12" diameter	B-21	280	.100	↓	9.50	2.50	.49	12.49	14.90
114	**0010**	**PIPING, SUBDRAINAGE, POLYVINYL CHLORIDE**									**114**
	0020	Perforated, price as solid pipe, division 027-168								

027 150 | Sewage Systems

		CREW	DAILY OUTPUT	LABOR-HOURS	UNIT	1999 BARE COSTS				TOTAL INCL O&P
152	**0010**	**CATCH BASINS OR MANHOLES** not including footing, excavation,									**152**
	0020	backfill, frame and cover								
	0050	Brick, 4' inside diameter, 4' deep	D-1	1	16	Ea.	263	395		658	900
	0100	6' deep		.70	22.857		370	560		930	1,275
	0150	8' deep		.50	32	↓	475	785		1,260	1,750
	0200	For depths over 8', add		4	4	V.L.F.	100	98.50		198.50	263
	0400	Concrete blocks (radial), 4' I.D., 4' deep		1.50	10.667	Ea.	300	262		562	740
	0500	6' deep	↓	1	16	↓	390	395		785	1,050

027 150 | Sewage Systems

		CREW	DAILY OUTPUT	LABOR-HOURS	UNIT	1999 BARE COSTS				TOTAL INCL O&P
						MAT.	LABOR	EQUIP.	TOTAL	
152	0600	8' deep	D-1	.70	22.857	Ea.	510	560		1,070	1,425	152
	0700	For depths over 8', add		5.50	2.909	V.L.F.	76.50	71.50		148	195
	0800	Concrete, cast in place, 4' x 4', 8" thick, 4' deep	C-14H	2	24	Ea.	335	650	18.70	1,003.70	1,400
	0900	6' deep		1.50	32		485	865	25	1,375	1,925
	1000	8' deep		1	48		650	1,300	37.50	1,987.50	2,800
	1100	For depths over 8', add		8	6	V.L.F.	82	162	4.68	248.68	355
	1110	Precast, 4' I.D., 4' deep	B-22	4.10	7.317	Ea.	315	185	49.50	549.50	690
	1120	6' deep		3	10		425	253	68	746	935
	1130	8' deep		2	15		470	380	102	952	1,225
	1140	For depths over 8', add		16	1.875	V.L.F.	78.50	47.50	12.75	138.75	175
	1150	5' I.D., 4' deep	B-6	3	8	Ea.	445	187	73	705	860
	1160	6' deep		2	12		605	280	109	994	1,225
	1170	8' deep		1.50	16		760	375	146	1,281	1,575
	1180	For depths over 8', add		12	2	V.L.F.	99	46.50	18.25	163.75	202
	1190	6' I.D., 4' deep		2	12	Ea.	730	280	109	1,119	1,350
	1200	6' deep		1.50	16		950	375	146	1,471	1,800
	1210	8' deep		1	24		1,175	560	219	1,954	2,375
	1220	For depths over 8', add		8	3	V.L.F.	153	70	27.50	250.50	305
	1250	Slab tops, precast, 8" thick									
	1300	4' diameter manhole	B-6	8	3	Ea.	148	70	27.50	245.50	300
	1400	5' diameter manhole		7.50	3.200		247	75	29	351	420
	1500	6' diameter manhole		7	3.429		287	80	31.50	398.50	475
	1600	Frames & covers, C.I., 24" square, 500 lb.		7.80	3.077		213	72	28	313	380
	1700	26" D shape, 600 lb.		7	3.429		214	80	31.50	325.50	395
	1800	Light traffic, 18" diameter, 100 lb.		10	2.400		76.50	56	22	154.50	195
	1900	24" diameter, 300 lb.		8.70	2.759		138	64.50	25	227.50	279
	2000	36" diameter, 900 lb.		5.80	4.138		385	96.50	37.50	519	610
	2100	Heavy traffic, 24" diameter, 400 lb.		7.80	3.077		172	72	28	272	330
	2200	36" diameter, 1150 lb.		3	8		510	187	73	770	930
	2300	Mass. State standard, 26" diameter, 475 lb.		7	3.429		214	80	31.50	325.50	395
	2400	30" diameter, 620 lb.		7	3.429		243	80	31.50	354.50	425
	2500	Watertight, 24" diameter, 350 lb.		7.80	3.077		293	72	28	393	465
	2600	26" diameter, 500 lb.		7	3.429		330	80	31.50	441.50	525
	2700	32" diameter, 575 lb.		6	4		390	93.50	36.50	520	610
	2800	3 piece cover & frame, 10" deep,									
	2900	1200 lbs., for heavy equipment	B-6	3	8	Ea.	770	187	73	1,030	1,225
	3000	Raised for paving 1-1/4" to 2" high,									
	3100	4 piece expansion ring									
	3200	20" to 26" diameter	1 Clab	3	2.667	Ea.	102	57		159	202
	3300	30" to 36" diameter	"	3	2.667	"	143	57		200	247
	3320	Frames and covers, existing, raised for paving 2", including									
	3340	row of brick, concrete collar, up to 12" wide frame	B-6	18	1.333	Ea.	31	31	12.15	74.15	96
	3360	20" to 26" wide frame		11	2.182		41	51	19.90	111.90	146
	3380	30" to 36" wide frame		9	2.667		51	62.50	24.50	138	180
	3400	Inverts, single channel brick	D-1	3	5.333		57	131		188	267
	3500	Concrete		5	3.200		45	78.50		123.50	172
	3600	Triple channel, brick		2	8		86.50	197		283.50	400
	3700	Concrete		3	5.333		61	131		192	271
	3800	Steps, heavyweight cast iron, 7" x 9"	1 Bric	40	.200		8.35	5.50		13.85	17.75
	3900	8" x 9"		40	.200		12.50	5.50		18	22.50
	3928	12" x 10-1/2"		40	.200		13	5.50		18.50	23
	4000	Standard sizes, galvanized steel		40	.200		11.75	5.50		17.25	21.50
	4100	Aluminum		40	.200		13	5.50		18.50	23
162	0010	**PIPING, DRAINAGE & SEWAGE, CONCRETE**	R027 -110									162
	0020	Not including excavation or backfill									

027 150 | Sewage Systems

		CREW	DAILY OUTPUT	LABOR-HOURS	UNIT	1999 BARE COSTS				TOTAL INCL O&P
						MAT.	LABOR	EQUIP.	TOTAL	
1000	Non-reinforced pipe, extra strength, B&S or T&G joints									
1010	6" diameter	B-14	265.04	.181	L.F.	3.47	4.12	.83	8.42	11.15
1020	8" diameter		224	.214		3.81	4.87	.98	9.66	12.85
1030	10" diameter		216	.222		4.23	5.05	1.01	10.29	13.65
1040	12" diameter		200	.240		5.20	5.45	1.09	11.74	15.45
1050	15" diameter		180	.267		6.05	6.05	1.22	13.32	17.50
1060	18" diameter		144	.333		7.45	7.60	1.52	16.57	21.50
1070	21" diameter		112	.429		9.20	9.75	1.95	20.90	27.50
1080	24" diameter		100	.480		11.25	10.90	2.19	24.34	32
2000	Reinforced culvert, class 3, no gaskets									
2010	12" diameter	B-14	210	.229	L.F.	8.10	5.20	1.04	14.34	18.15
2020	15" diameter		175	.274		9.90	6.25	1.25	17.40	22
2030	18" diameter		130	.369		11.55	8.40	1.68	21.63	27.50
2035	21" diameter		120	.400		14.80	9.10	1.82	25.72	32.50
2040	24" diameter		100	.480		18.35	10.90	2.19	31.44	39.50
2045	27" diameter	B-13	92	.609		22	14.15	5.95	42.10	53
2050	30" diameter		88	.636		26.50	14.80	6.25	47.55	59
2060	36" diameter		72	.778		36	18.10	7.65	61.75	76
2070	42" diameter	B-13B	72	.778		44	18.10	11	73.10	88.50
2080	48" diameter		64	.875		57.50	20.50	12.40	90.40	108
2090	60" diameter		48	1.167		61.50	27	16.50	105	128
2100	72" diameter		40	1.400		95.50	32.50	19.80	147.80	178
2120	84" diameter		32	1.750		203	40.50	25	268.50	315
2140	96" diameter		24	2.333		243	54	33	330	390
2200	With gaskets, class 3, 12" diameter	B-21	168	.167		7.25	4.17	.81	12.23	15.45
2220	15" diameter		160	.175		8.70	4.38	.85	13.93	17.35
2230	18" diameter		152	.184		10.90	4.61	.89	16.40	20
2240	24" diameter		136	.206		16.30	5.15	1	22.45	27
2260	30" diameter	B-13	88	.636		22	14.80	6.25	43.05	54
2270	36" diameter	"	72	.778		32.50	18.10	7.65	58.25	72.50
2290	48" diameter	B-13B	64	.875		53.50	20.50	12.40	86.40	104
2310	72" diameter	"	40	1.400		138	32.50	19.80	190.30	225
2330	Flared ends, 6'-1" long, 12" diameter	B-21	190	.147		27.50	3.69	.71	31.90	36.50
2340	15" diameter		155	.181		30.50	4.52	.88	35.90	41.50
2400	6'-2" long, 18" diameter		122	.230		34	5.75	1.11	40.86	47
2420	24" diameter		88	.318		45	7.95	1.54	54.49	63.50
2440	36" diameter	B-13	60	.933		57	21.50	9.15	87.65	107
3040	Vitrified plate lined, add to above, 30" to 36" diameter				SFCA	3			3	3.30
3050	42" to 54" diameter, add					3.20			3.20	3.52
3060	60" to 72" diameter, add					3.75			3.75	4.13
3070	Over 72" diameter, add					4			4	4.40
3080	Radius pipe, add to pipe prices, 12" to 60" diameter				L.F.	50%				
3090	Over 60" diameter, add				"	20%				
3500	Reinforced elliptical, 8' lengths, C507 class 3									
3520	14" x 23" inside, round equivalent 18" diameter	B-21	82	.341	L.F.	40	8.55	1.66	50.21	59
3530	24" x 38" inside, round equivalent 30" diameter	B-13	58	.966		50	22.50	9.45	81.95	100
3540	29" x 45" inside, round equivalent 36" diameter		52	1.077		65	25	10.55	100.55	122
3550	38" x 60" inside, round equivalent 48" diameter		38	1.474		100	34.50	14.45	148.95	179
3560	48" x 76" inside, round equivalent 60" diameter		26	2.154		135	50	21	206	250
3570	58" x 91" inside, round equivalent 72" diameter		22	2.545		200	59	25	284	340
3780	Concrete slotted pipe, class 4 mortar joint									
3800	12" diameter	B-21	168	.167	L.F.	10.50	4.17	.81	15.48	19
3840	18" diameter	"	152	.184	"	16.25	4.61	.89	21.75	26
3900	Class 4 O-ring									
3940	12" diameter	B-21	168	.167	L.F.	12	4.17	.81	16.98	20.50
3960	18" diameter	"	152	.184	"	18	4.61	.89	23.50	28

162

R027 -110

CN

SITE WORK 2

SITE WORK 2

027 150 | Sewage Systems

		CREW	DAILY OUTPUT	LABOR-HOURS	UNIT	1999 BARE COSTS				TOTAL INCL O&P
						MAT.	LABOR	EQUIP.	TOTAL	
164	0010	**PIPING, STORM DRAINAGE, CORRUGATED METAL**									164
	0020	Not including excavation or backfill									
	2000	Corrugated metal pipe, galvanized and coated									
	2020	Bituminous coated with paved invert, 20' lengths									
	2040	8" diameter, 16 ga.	B-14	330	.145	L.F.	7.10	3.31	.66	11.07	13.75
	2060	10" diameter, 16 ga.		260	.185		8.90	4.20	.84	13.94	17.25
	2080	12" diameter, 16 ga.		210	.229		11.90	5.20	1.04	18.14	22.50
	2100	15" diameter, 16 ga.		200	.240		12.45	5.45	1.09	18.99	23.50
	2120	18" diameter, 16 ga.		190	.253		13.30	5.75	1.15	20.20	25
	2140	24" diameter, 14 ga. **CN**	↓	160	.300		26.50	6.80	1.37	34.67	41
	2160	30" diameter, 14 ga.	B-13	120	.467		28.50	10.85	4.58	43.93	53.50
	2180	36" diameter, 12 ga.		120	.467		50	10.85	4.58	65.43	77
	2200	48" diameter, 12 ga.	↓	100	.560		61.50	13	5.50	80	94.50
	2220	60" diameter, 10 ga.	B-13B	75	.747		101	17.35	10.55	128.90	150
	2240	72" diameter, 8 ga.	"	45	1.244	↓	135	29	17.60	181.60	213
	2500	Galvanized, 20' lengths									
	2520	8" diameter, 16 ga.	B-14	355	.135	L.F.	6.35	3.07	.62	10.04	12.45
	2540	10" diameter, 16 ga.		280	.171		6.70	3.90	.78	11.38	14.30
	2560	12" diameter, 16 ga.		220	.218		7.15	4.96	.99	13.10	16.75
	2580	15" diameter, 16 ga.		220	.218		9.50	4.96	.99	15.45	19.30
	2600	18" diameter, 16 ga.		205	.234		12.30	5.30	1.07	18.67	23
	2620	24" diameter, 14 ga.		175	.274		21	6.25	1.25	28.50	34
	2640	30" diameter, 14 ga.	↓	130	.369		26.50	8.40	1.68	36.58	44.50
	2660	36" diameter, 12 ga.	B-13	130	.431		34	10	4.23	48.23	58
	2680	48" diameter, 12 ga.		110	.509		53.50	11.85	4.99	70.34	82.50
	2700	60" diameter, 10 ga.	↓	78	.718	↓	86	16.70	7.05	109.75	128
	2780	End sections, 8" diameter	B-14	24	2	Ea.	52.50	45.50	9.10	107.10	139
	2785	10" diameter		22	2.182		54	49.50	9.95	113.45	148
	2790	12" diameter		35	1.371		65	31	6.25	102.25	127
	2800	18" diameter		30	1.600		72.50	36.50	7.30	116.30	145
	2810	24" diameter		25	1.920		105	43.50	8.75	157.25	193
	2820	30" diameter		25	1.920		209	43.50	8.75	261.25	310
	2825	36" diameter		20	2.400		299	54.50	10.95	364.45	430
	2830	48" diameter		10	4.800		670	109	22	801	930
	2835	60" diameter		5	9.600		1,175	218	44	1,437	1,700
	2840	72" diameter	↓	4	12	↓	1,375	273	54.50	1,702.50	2,000
	3000	Corrugated galvanized or alum. oval arch culverts, coated & paved									
	3020	17" x 13", 16 ga., 15" equivalent	B-14	200	.240	L.F.	11.95	5.45	1.09	18.49	23
	3040	21" x 15", 16 ga., 18" equivalent		150	.320		30.50	7.30	1.46	39.26	46.50
	3060	28" x 20", 14 ga., 24" equivalent		125	.384		24	8.75	1.75	34.50	42
	3080	35" x 24", 14 ga., 30" equivalent	↓	100	.480		50	10.90	2.19	63.09	74.50
	3100	42" x 29", 12 ga., 36" equivalent	B-13	100	.560		39.50	13	5.50	58	70
	3120	49" x 33", 12 ga., 42" equivalent		90	.622		45.50	14.45	6.10	66.05	79
	3140	57" x 38", 12 ga., 48" equivalent	↓	75	.747	↓	93	17.35	7.30	117.65	138
	3160	Steel, plain oval arch culverts, plain									
	3180	17" x 13", 16 ga., 15" equivalent	B-14	225	.213	L.F.	7.85	4.85	.97	13.67	17.30
	3200	21" x 15", 16 ga., 18" equivalent		175	.274		15.50	6.25	1.25	23	28
	3220	28" x 20", 14 ga., 24" equivalent	↓	150	.320		14.60	7.30	1.46	23.36	29
	3240	35" x 24", 14 ga., 30" equivalent	B-13	108	.519		30	12.05	5.10	47.15	57.50
	3260	42" x 29", 12 ga., 36" equivalent		108	.519		36	12.05	5.10	53.15	64
	3280	49" x 33", 12 ga., 42" equivalent		92	.609		42.50	14.15	5.95	62.60	75
	3300	57" x 38", 12 ga., 48" equivalent	↓	75	.747	↓	51	17.35	7.30	75.65	91
	3320	End sections, 17" x 13"		22	2.545	Ea.	34	59	25	118	157
	3340	42" x 29"	↓	17	3.294	"	172	76.50	32.50	281	345
	3360	Multi-plate arch, steel	B-20	1,690	.014	Lb.	.60	.35		.95	1.20

Important: See the Reference Section for critical supporting data - Reference Nos., Crews, & City Cost Indexes

027 150	Sewage Systems	CREW	DAILY OUTPUT	LABOR-HOURS	UNIT	1999 BARE COSTS				TOTAL INCL O&P	
						MAT.	LABOR	EQUIP.	TOTAL		
168	**0010** PIPING, DRAINAGE & SEWAGE, POLYVINYL CHLORIDE										**168**
	0020 Not including excavation or backfill										
	2000 10' lengths, S.D.R. 35, B&S, 4" diameter	B-20	375	.064	L.F.	2.17	1.56		3.73	4.84	
	2040 6" diameter		350	.069		3.95	1.67		5.62	7	
	2080 8" diameter	↓	335	.072		3.94	1.74		5.68	7.10	
	2120 10" diameter	B-21	330	.085		3.81	2.12	.41	6.34	7.95	
	2160 12" diameter		320	.087		5.05	2.19	.42	7.66	9.45	
	2200 15" diameter	↓	190	.147	↓	11	3.69	.71	15.40	18.70	
172	**0010** PIPING, DRAINAGE & SEWAGE, VITRIFIED CLAY C700										**172**
	0020 Not including excavation or backfill,										
	4030 Extra strength, compression joints, C425										
	5000 4" diameter x 4' long	B-20	265	.091	L.F.	1.77	2.20		3.97	5.40	
	5020 6" diameter x 5' long	"	200	.120		2.89	2.92		5.81	7.80	
	5040 8" diameter x 5' long	B-21	200	.140		4.08	3.51	.68	8.27	10.75	
	5060 10" diameter x 5' long		190	.147		6.70	3.69	.71	11.10	13.95	
	5080 12" diameter x 6' long		150	.187		8.80	4.67	.91	14.38	18	
	5100 15" diameter x 7' long		110	.255		16.05	6.35	1.23	23.63	29	
	5120 18" diameter x 7' long		88	.318		23.50	7.95	1.54	32.99	39.50	
	5140 24" diameter x 7' long	↓	45	.622		48	15.60	3.02	66.62	80.50	
	5160 30" diameter x 7' long	B-22	31	.968		83.50	24.50	6.55	114.55	138	
	5180 36" diameter x 7' long	"	20	1.500		123	38	10.20	171.20	207	
	6000 For 3' lengths, add					30%	30%				
	6020 For 2' lengths, add					40%	60%				
	6060 For plain joints, deduct					25%					
	7060 2' lengths, add to above				↓	40%					
174	**0010** SEWAGE PUMPING STATIONS Prefabricated steel, concrete										**174**
	0020 or fiberglass, 200 GPM	C-17D	.17	494	Total	27,000	14,100	3,225	44,325	55,500	
	0200 1,000 GPM		.07	1,200		40,000	34,200	7,850	82,050	106,500	
	0500 Add for generator unit, 200 GPM, steel		.34	247		20,500	7,050	1,625	29,175	35,500	
	0600 Concrete		.51	164		13,000	4,700	1,075	18,775	22,900	
	1000 Add for generator unit, 1,000 GPM, steel		.30	280		21,800	7,975	1,825	31,600	38,600	
	1200 Concrete	↓	.38	221		18,400	6,300	1,450	26,150	31,700	
	1500 For drilled water well, if required, add	B-23	.50	80	↓	5,300	1,750	3,900	10,950	12,900	
176	**0010** SEWAGE TREATMENT Plant, not incl. fencing or external piping										**176**
	0020 Steel packaged, blown air aeration plants										
	0100 1,000 GPD				Gal.				15	17.25	
	0200 5,000 GPD								10	11.50	
	0300 15,000 GPD								5.50	6.30	
	0400 30,000 GPD								5.20	6	
	0500 50,000 GPD								4	4.60	
	0600 100,000 GPD								3.50	4	
	0700 200,000 GPD								2.50	2.88	
	0800 500,000 GPD				↓				2.45	2.80	
	1000 Concrete, extended aeration, primary and secondary treatment										
	1010 10,000 GPD				Gal.				11	12.65	
	1100 30,000 GPD								5.50	6.35	
	1200 50,000 GPD								4.50	5.18	
	1400 100,000 GPD								3.50	4.05	
	1500 500,000 GPD				↓				2.50	2.90	
	1700 Municipal wastewater treatment facility										
	1720 1.0 MGD				Gal.				4.30	4.95	
	1740 1.5 MGD								4.25	4.90	
	1760 2.0 MGD								3.65	4.20	
	1780 3.0 MGD								2.85	3.30	
	1800 5.0 MGD				↓				2.60	3	

2 SITE WORK

027 150 | Sewage Systems

		CREW	DAILY OUTPUT	LABOR-HOURS	UNIT	1999 BARE COSTS				TOTAL INCL O&P	
						MAT.	LABOR	EQUIP.	TOTAL		
176	2000	Holding tank system, not incl. excavation or backfill									176
	2010	Recirculating chemical water closet	2 Plum	4	4	Ea.	660	130		790	925
	2100	For voltage converter, add	"	16	1		175	32.50		207.50	243
	2200	For high level alarm, add	1 Plum	7.80	1.026		100	33.50		133.50	161

027 350 | Wastewater System

		CREW	DAILY OUTPUT	LABOR-HOURS	UNIT	MAT.	LABOR	EQUIP.	TOTAL	TOTAL INCL O&P	
354	0010	WASTEWATER TREATMENT SYSTEM Fiberglass, 1,000 gallon	B-21	1.29	21.705	Ea.	2,650	545	105	3,300	3,900	354
	0100	1,500 gallon	"	1.03	27.184	"	5,800	680	132	6,612	7,600

027 400 | Septic Systems

		CREW	DAILY OUTPUT	LABOR-HOURS	UNIT	MAT.	LABOR	EQUIP.	TOTAL	TOTAL INCL O&P	
404	0010	SEPTIC TANKS Not incl. excav. or piping, precast, 1,000 gallon	B-21	8	3.500	Ea.	450	87.50	17	554.50	650	404
	0100	2,000 gallon	"	5	5.600		870	140	27	1,037	1,200
	0200	5,000 gallon	B-13	3.50	16		4,300	370	157	4,827	5,475
	0300	15,000 gallon, 4 piece	B-13B	1.70	32.941		11,000	765	465	12,230	13,800
	0400	25,000 gallon, 4 piece	↓	1.10	50.909		16,500	1,175	720	18,395	20,800
	0500	40,000 gallon, 4 piece		.80	70		27,000	1,625	990	29,615	33,300
	0520	50,000 gallon, 5 piece	B-13C	.60	93.333		31,100	2,175	1,875	35,150	39,700
	0540	75,000 gallon, cast in place	C-14C	.25	448		37,800	11,700	148	49,648	60,500
	0560	100,000 gallon	"	.15	746		46,800	19,500	246	66,546	83,000
	0600	High density polyethylene, 1,000 gallon	B-21	6	4.667		800	117	22.50	939.50	1,100
	0700	1,500 gallon	"	4	7		1,000	175	34	1,209	1,400
	1000	Distribution boxes, concrete, 7 outlets	2 Clab	16	1		75	21.50		96.50	117
	1100	9 outlets	"	8	2		225	43		268	315
	1150	Leaching field chambers, 13' x 3'-7" x 1'-4", standard	B-13	16	3.500		665	81.50	34.50	781	895
	1200	Heavy duty, 8' x 4' x 1'-6"		14	4		320	93	39	452	540
	1300	13' x 3'-9" x 1'-6"		12	4.667		910	108	46	1,064	1,225
	1350	20' x 4' x 1'-6"	↓	5	11.200		750	260	110	1,120	1,350
	1400	Leaching pit, precast concrete, 3' diameter, 3' deep	B-21	8	3.500		145	87.50	17	249.50	315
	1500	6' diameter, 3' section		4.70	5.957		375	149	29	553	680
	2000	Velocity reducing pit, precast conc., 6' diameter, 3' deep	↓	4.70	5.957	↓	225	149	29	403	515
	2200	Excavation for septic tank, 3/4 C.Y. backhoe	B-12F	145	.110	C.Y.		2.95	3.10	6.05	7.90
	2400	4' trench for disposal field, 3/4 C.Y. backhoe	"	335	.048	L.F.		1.28	1.34	2.62	3.41
	2600	Gravel fill, run of bank	B-6	150	.160	C.Y.	5.50	3.74	1.46	10.70	13.45
	2800	Crushed stone, 3/4"	"	150	.160	"	18.30	3.74	1.46	23.50	27.50

027 660 | Relining Exist. Pipelines

		CREW	DAILY OUTPUT	LABOR-HOURS	UNIT	MAT.	LABOR	EQUIP.	TOTAL	TOTAL INCL O&P	
664	0010	LINING PIPE with cement, incl. bypass and cleaning										664
	0020	Less than 10,000 L.F., urban, 6" to 10"	C-17E	130	.615	L.F.	5.80	17.50	.47	23.77	34.50
	0200	24" to 36"		90	.889		9.25	25.50	.68	35.43	51
	0300	48" to 72"	↓	80	1	↓	14.80	28.50	.76	44.06	62

028 | Site Improvements

028 100 | Irrigation Systems

		CREW	DAILY OUTPUT	LABOR-HOURS	UNIT	1999 BARE COSTS				TOTAL INCL O&P	
						MAT.	LABOR	EQUIP.	TOTAL		
104	0010	SPRINKLER IRRIGATION SYSTEM For lawns										104
	0100	Golf course with fully automatic system	C-17	.05	1,600	9 holes	75,000	45,500		120,500	154,500
	0200	24' diam. head at 15' O.C incl. piping, auto oper., minimum	B-20	70	.343	Head	16.50	8.35		24.85	31.50
	0300	Maximum		40	.600		38	14.60		52.60	65
	0500	60' diam. head at 40' O.C. incl. piping, auto oper., minimum		28	.857		50	21		71	88
	0600	Maximum	↓	23	1.043	↓	140	25.50		165.50	194

Important: See the Reference Section for critical supporting data - Reference Nos., Crews, & City Cost Indexes

	028 100 | **Irrigation Systems**	CREW	DAILY OUTPUT	LABOR-HOURS	UNIT	MAT.	LABOR	EQUIP.	TOTAL	TOTAL INCL O&P	
104	0800	Residential system, custom, 1" supply	B-20	2,000	.012	S.F.	.25	.29		.54	.74	104
	0900	1-1/2" supply	↓	1,800	.013	"	.28	.32		.60	.82	
	1020	Pop up spray head w/risers, hi-pop, full circle pattern, 4"	2 Skwk	76	.211	Ea.	3.17	5.90		9.07	12.80	
	1030	1/2 circle pattern		76	.211		3.17	5.90		9.07	12.80	
	1040	6", full circle pattern		76	.211		7.45	5.90		13.35	17.50	
	1050	1/2 circle pattern		76	.211		7.45	5.90		13.35	17.50	
	1060	12", full circle pattern		76	.211		9.25	5.90		15.15	19.50	
	1070	1/2 circle pattern		76	.211		9.30	5.90		15.20	19.55	
	1080	Pop up bubbler head w/risers, hi-pop bubbler head, 4"		76	.211		3.27	5.90		9.17	12.90	
	1090	6"		76	.211		7.45	5.90		13.35	17.50	
	1100	12"		76	.211		9.25	5.90		15.15	19.50	
	1110	Impact full/part circle sprinklers, 28'-54' 25-60 PSI		37	.432		8.80	12.15		20.95	29	
	1120	Spaced 37'-49' @ 25-50 PSI		37	.432		20.50	12.15		32.65	41.50	
	1130	Spaced 43'-61' @ 30-60 PSI		37	.432		42.50	12.15		54.65	65.50	
	1140	Spaced 54'-78' @ 40-80 PSI	↓	37	.432	↓	92.50	12.15		104.65	121	
	1145	Impact rotor pop-up full/part commercial circle sprinklers										
	1150	Spaced 42'-65' 35-80 PSI	2 Skwk	25	.640	Ea.	20.50	17.95		38.45	51	
	1160	Spaced 48'-76' 45-85 PSI	"	25	.640	"	6.20	17.95		24.15	35.50	
	1165	Impact rotor pop-up part. circle comm., 53'-75', 55-100 PSI, w/ acc										
	1170	Plastic case, metal cover	2 Skwk	25	.640	Ea.	134	17.95		151.95	176	
	1180	Rubber cover		25	.640		102	17.95		119.95	141	
	1190	Iron case, metal cover		22	.727		127	20.50		147.50	171	
	1200	Rubber cover		22	.727		133	20.50		153.50	179	
	1250	Plastic case, 2 nozzle, metal cover		25	.640		101	17.95		118.95	140	
	1260	Rubber cover		25	.640		107	17.95		124.95	146	
	1270	Iron case, 2 nozzle, metal cover		22	.727		143	20.50		163.50	190	
	1280	Rubber cover	↓	22	.727	↓	152	20.50		172.50	200	
	1282	Impact rotor pop-up full circle comm., 39'-99', 30-100 PSI										
	1284	Plastic case, metal cover	2 Skwk	25	.640	Ea.	101	17.95		118.95	140	
	1286	Rubber cover		25	.640		111	17.95		128.95	151	
	1288	Iron case, metal cover		22	.727		149	20.50		169.50	196	
	1290	Rubber cover		22	.727		154	20.50		174.50	202	
	1292	Plastic case, 2 nozzle, metal cover		22	.727		108	20.50		128.50	150	
	1294	Rubber cover		22	.727		113	20.50		133.50	156	
	1296	Iron case, 2 nozzle, metal cover		20	.800		144	22.50		166.50	194	
	1298	Rubber cover		20	.800		152	22.50		174.50	203	
	1305	Electric remote control valve, plastic, 3/4"		18	.889		15.55	25		40.55	56.50	
	1310	1"		18	.889		23	25		48	65	
	1320	1-1/2"	↓	18	.889	↓	49	25		74	93.50	
	1335	Quick coupling valves, brass, locking cover										
	1340	Inlet coupling valve, 3/4"	2 Skwk	18.75	.853	Ea.	38.50	24		62.50	80.50	
	1350	1"		18.75	.853		47.50	24		71.50	90	
	1360	Controller valve boxes, 6" round boxes		18.75	.853		3.27	24		27.27	41.50	
	1370	10" round boxes		14.25	1.123		7.20	31.50		38.70	57.50	
	1380	12" square box	↓	9.75	1.641	↓	13.90	46		59.90	88	
	1388	Electromech. control, 14 day 3-60 min, auto start to 23/day										
	1390	4 station	2 Skwk	1.04	15.385	Ea.	160	430		590	855	
	1400	7 station		.64	25		159	700		859	1,275	
	1410	12 station		.40	40		243	1,125		1,368	2,050	
	1420	Dual programs, 18 station		.24	66.667		1,200	1,875		3,075	4,275	
	1430	23 station	↓	.16	100	↓	1,575	2,800		4,375	6,150	
	1435	Backflow preventer, bronze, 0-175 PSI, w/valves, test cocks										
	1440	3/4"	2 Skwk	2	8	Ea.	97	224		321	460	
	1450	1"		2	8		99	224		323	465	
	1460	1-1/2"		2	8		200	224		424	575	
	1470	2"	↓	2	8	↓	214	224		438	590	

028 100 | Irrigation Systems

		CREW	DAILY OUTPUT	LABOR-HOURS	UNIT	MAT.	LABOR	EQUIP.	TOTAL	TOTAL INCL O&P	
										1999 BARE COSTS	
104	1475	Pressure vacuum breaker, brass, 15-150 PSI										**104**
	1480	3/4"	2 Skwk	2	8	Ea.	54	224		278	415
	1490	1"		2	8		81.50	224		305.50	445
	1500	1-1/2"		2	8		161	224		385	530
	1510	2"	▼	2	8	▼	149	224		373	520

028 200 | Fountains

		CREW	DAILY OUTPUT	LABOR-HOURS	UNIT	MAT.	LABOR	EQUIP.	TOTAL	TOTAL INCL O&P	
204	0010	**FOUNTAINS/AERATORS**										**204**
	0100	Pump w/controls									
	0200	Single phase, 100' chord, 1/2 H.P. pump	2 Skwk	4.40	3.636	Ea.	2,025	102		2,127	2,375
	0300	3/4 H.P. pump		4.30	3.721		2,275	104		2,379	2,675
	0400	1 H.P. pump		4.20	3.810		2,300	107		2,407	2,700
	0500	1-1/2 H.P. pump		4.10	3.902		2,425	109		2,534	2,825
	0600	2 H.P. pump		4	4		2,450	112		2,562	2,875
	0700	Three phase, 200' chord, 5 H.P. pump		3.90	4.103		6,500	115		6,615	7,325
	0800	7-1/2 H.P. pump		3.80	4.211		7,675	118		7,793	8,600
	0900	10 H.P. pump		3.70	4.324		7,725	121		7,846	8,700
	1000	15 H.P. pump		3.60	4.444		10,400	125		10,525	11,600
	1100	Nozzles, minimum		8	2		135	56		191	238
	1200	Maximum		8	2		246	56		302	360
	1300	Lights w/mounting kits, 200 watt		18	.889		295	25		320	365
	1400	300 watt		18	.889		330	25		355	400
	1500	500 watt		18	.889		360	25		385	435
	1600	Color blender	▼	12	1.333	▼	280	37.50		317.50	365

028 300 | Fences & Gates

		CREW	DAILY OUTPUT	LABOR-HOURS	UNIT	MAT.	LABOR	EQUIP.	TOTAL	TOTAL INCL O&P	
308	0010	**FENCE, CHAIN LINK INDUSTRIAL**, schedule 40										**308**
	0020	3 strands barb wire, 2" post @ 10' O.C., set in concrete, 6' H									
	0200	9 ga. wire, galv. steel	B-80	240	.133	L.F.	7	3.13	2.25	12.38	15
	0300	Aluminized steel		240	.133		9	3.13	2.25	14.38	17.20
	0500	6 ga. wire, galv. steel		240	.133		11.35	3.13	2.25	16.73	19.80
	0600	Aluminized steel		240	.133		13	3.13	2.25	18.38	21.50
	0800	6 ga. wire, 6' high but omit barbed wire, galv. steel		250	.128		11	3	2.16	16.16	19.10
	0900	Aluminized steel		250	.128		15.40	3	2.16	20.56	24
	0920	8' H, 6 ga. wire, 2-1/2" line post, galv. steel		180	.178		17.95	4.17	3	25.12	29.50
	0940	Aluminized steel		180	.178	▼	22	4.17	3	29.17	34.50
	1100	Add for corner posts, 3" diam., galv. steel		40	.800	Ea.	55	18.75	13.50	87.25	104
	1200	Aluminized steel		40	.800		66	18.75	13.50	98.25	116
	1300	Add for braces, galv. steel		80	.400		15	9.40	6.75	31.15	38.50
	1350	Aluminized steel		80	.400		20	9.40	6.75	36.15	44
	1400	Gate for 6' high fence, 1-5/8" frame, 3' wide, galv. steel		10	3.200		80	75	54	209	264
	1500	Aluminized steel	▼	10	3.200	▼	110	75	54	239	297
	2000	5'-0" high fence, 9 ga., no barbed wire, 2" line post,									
	2010	10' O.C., 1-5/8" top rail									
	2100	Galvanized steel	B-80	300	.107	L.F.	6	2.50	1.80	10.30	12.45
	2200	Aluminized steel		300	.107	"	7.25	2.50	1.80	11.55	13.85
	2400	Gate, 4' wide, 5' high, 2" frame, galv. steel		10	3.200	Ea.	100	75	54	229	286
	2500	Aluminized steel		10	3.200	"	110	75	54	239	297
	3100	Overhead slide gate, chain link, 6' high, to 18' wide		38	.842	L.F.	82.50	19.75	14.20	116.45	137
	3110	Cantilever type		48	.667		38	15.65	11.25	64.90	78.50
	3120	8' high		24	1.333		55	31.50	22.50	109	134
	3130	10' high	▼	18	1.778	▼	65	41.50	30	136.50	169
	5000	Double swing gates, incl. posts & hardware									
	5010	5' high, 12' opening	B-80	3.40	9.412	Opng.	263	221	159	643	805

Important: See the Reference Section for critical supporting data - Reference Nos., Crews, & City Cost Indexes

028 300 | Fences & Gates

		CREW	DAILY OUTPUT	LABOR-HOURS	UNIT	1999 BARE COSTS MAT.	LABOR	EQUIP.	TOTAL	TOTAL INCL O&P	
308	5020	20' opening	B-80	2.80	11.429	Opng.	340	268	193	801	1,000	308
	5060	6' high, 12' opening		3.20	10		475	235	169	879	1,075	
	5070	20' opening		2.60	12.308		655	289	208	1,152	1,400	
	5080	8' high, 12' opening		2.13	15.002		700	350	253	1,303	1,600	
	5090	20' opening		1.45	22.069		935	520	370	1,825	2,225	
	5100	10' high, 12' opening		1.31	24.427		800	575	410	1,785	2,225	
	5110	20' opening		1.03	31.068		1,200	730	525	2,455	3,025	
	5120	12' high, 12' opening		1.05	30.476		1,175	715	515	2,405	2,950	
	5130	20' opening	▼	.85	37.647	▼	1,500	885	635	3,020	3,725	
	5190	For aluminized steel add					20%					
	7795	Cantilever, manual, exp. roller, (pr), 40' wide x 8' high	B-22	1	30	Ea.	2,075	760	204	3,039	3,700	
	7800	30' wide x 8' high		1	30		1,750	760	204	2,714	3,325	
	7805	24' wide x 8' high	▼	1	30		1,300	760	204	2,264	2,825	
	7810	Motor operators for gates, (no elec wiring), 3' wide swing	2 Skwk	.50	32		765	900		1,665	2,275	
	7815	Up to 20' wide swing		.50	32		2,400	900		3,300	4,050	
	7820	Up to 45' sliding		.50	32	▼	2,425	900		3,325	4,100	
	7825	Overhead gate, 6' to 18' wide, sliding/cantilever		45	.356	L.F.	88.50	9.95		98.45	113	
	7830	Gate operators, digital receiver		7	2.286	Ea.	176	64		240	294	
	7835	Two button transmitter		24	.667		34.50	18.70		53.20	67.50	
	7840	3 button station		14	1.143		41	32		73	95.50	
	7845	Master slave system	▼	4	4	▼	150	112		262	345	
312	0010	**FENCE, CHAIN LINK RESIDENTIAL**, sch. 20, 11 ga. wire, 1-5/8" post										312
	0020	10' O.C., 1-3/8" top rail, 2" corner post, galv. stl. 3' high	B-1	500	.048	L.F.	3.17	1.06		4.23	5.15	
	0050	4' high		400	.060		3.79	1.33		5.12	6.25	
	0100	6' high		200	.120	▼	4.83	2.65		7.48	9.50	
	0150	Add for gate 3' wide, 1-3/8" frame, 3' high		12	2	Ea.	34	44		78	107	
	0170	4' high		10	2.400		42	53		95	130	
	0190	6' high		10	2.400		64.50	53		117.50	154	
	0200	Add for gate 4' wide, 1-3/8" frame, 3' high		9	2.667		42.50	59		101.50	140	
	0220	4' high		9	2.667		56	59		115	155	
	0240	6' high		8	3	▼	84.50	66.50		151	197	
	0350	Aluminized steel, 11 ga. wire, 3' high		500	.048	L.F.	3.79	1.06		4.85	5.85	
	0380	4' high		400	.060		4.48	1.33		5.81	7	
	0400	6' high		200	.120	▼	5.65	2.65		8.30	10.45	
	0450	Add for gate 3' wide, 1-3/8" frame, 3' high		12	2	Ea.	39	44		83	113	
	0470	4' high		10	2.400		52.50	53		105.50	142	
	0490	6' high		10	2.400		77	53		130	168	
	0500	Add for gate 4' wide, 1-3/8" frame, 3' high		10	2.400		52.50	53		105.50	141	
	0520	4' high		9	2.667		69.50	59		128.50	170	
	0540	6' high		8	3	▼	104	66.50		170.50	218	
	0620	Vinyl covered, 9 ga. wire, 3' high		500	.048	L.F.	3.98	1.06		5.04	6.05	
	0640	4' high		400	.060		4.92	1.33		6.25	7.50	
	0660	6' high		200	.120	▼	6.15	2.65		8.80	11	
	0720	Add for gate 3' wide, 1-3/8" frame, 3' high		12	2	Ea.	48	44		92	122	
	0740	4' high		10	2.400		63	53		116	153	
	0760	6' high		10	2.400		97.50	53		150.50	191	
	0780	Add for gate 4' wide, 1-3/8" frame, 3' high		10	2.400		63.50	53		116.50	154	
	0800	4' high		9	2.667		85.50	59		144.50	187	
	0820	6' high	▼	8	3	▼	132	66.50		198.50	249	
	0860	Tennis courts, 11 ga. wire, 2-1/2" post 10' O.C., 1-5/8" top rail										
	0900	10' high	B-1	155	.155	L.F.	8.30	3.43		11.73	14.50	
	0920	12' high		130	.185	"	9.85	4.08		13.93	17.20	
	1000	Add for gate 4' wide, 1-5/8" frame 7' high		10	2.400	Ea.	130	53		183	227	
	1040	Aluminized steel, 11 ga. wire 10' high		155	.155	L.F.	15	3.43		18.43	22	
	1100	12' high	▼	130	.185	"	15	4.08		19.08	23	

028 300 | Fences & Gates

		CREW	DAILY OUTPUT	LABOR-HOURS	UNIT	1999 BARE COSTS				TOTAL INCL O&P
						MAT.	LABOR	EQUIP.	TOTAL	
312	**1140** Add for gate 4' wide, 1-5/8" frame, 7' high	B-1	10	2.400	Ea.	180	53		233	282	**312**
	1250 Vinyl covered, 9 ga. wire, 10' high		155	.155	L.F.	15.90	3.43		19.33	23
	1300 12' high		130	.185	"	17.80	4.08		21.88	26
	1400 Add for gate 4' wide, 1-5/8" frame, 7' high		10	2.400	Ea.	155	53		208	255
320	**0010 FENCE, MISC. METAL** Chicken wire, posts @ 4', 1" mesh, 4' high	B-80	410	.078	L.F.	1.10	1.83	1.32	4.25	5.50	**320**
	0100 2" mesh, 6' high		350	.091		1	2.14	1.54	4.68	6.10
	0200 Galv. steel, 12 ga., 2" x 4" mesh, posts 5' O.C., 3' high		300	.107		1.50	2.50	1.80	5.80	7.50
	0300 5' high		300	.107		2	2.50	1.80	6.30	8.05
	0400 14 ga., 1" x 2" mesh, 3' high		300	.107		1.60	2.50	1.80	5.90	7.60
	0500 5' high		300	.107		2.20	2.50	1.80	6.50	8.30
	1000 Kennel fencing, 1-1/2" mesh, 6' long, 3'-6" wide, 6'-2" high	2 Clab	4	4	Ea.	250	86		336	410
	1050 12' long		4	4		300	86		386	465
	1200 Top covers, 1-1/2" mesh, 6' long		15	1.067		50	23		73	91
	1250 12' long		12	1.333		80	28.50		108.50	133
	1300 For kennel doors, see division 083-729									
	4500 Security fence, prison grade, set in concrete, 12' high	B-80	25	1.280	L.F.	20	30	21.50	71.50	92
	4600 16' high	"	20	1.600		24.50	37.50	27	89	114
	5000 Snow fence on steel posts 10' O.C., 4' high	B-1	500	.048		1.55	1.06		2.61	3.38
	5300 Tubular picket, steel, 6' sections, 1-9/16" posts, 4' high	B-80	300	.107		15.50	2.50	1.80	19.80	23
	5400 2" posts, 5' high		240	.133		21.50	3.13	2.25	26.88	31
	5600 2" posts, 6' high		200	.160		24.50	3.75	2.70	30.95	35.50
	5700 Staggered picket 1-9/16" posts, 4' high		300	.107		14	2.50	1.80	18.30	21.50
	5800 2" posts, 5' high		240	.133		23	3.13	2.25	28.38	33
	5900 2" posts, 6' high		200	.160		24	3.75	2.70	30.45	35
	6200 Gates, 4' high, 3' wide	B-1	10	2.400	Ea.	135	53		188	233
	6300 5' high, 3' wide		10	2.400		175	53		228	277
	6400 6' high, 3' wide		10	2.400		180	53		233	282
	6500 4' wide		10	2.400		210	53		263	315
324	**0010 FENCE, RAIL** Picket, No. 2 cedar, Gothic, 2 rail, 3' high	B-1	160	.150	L.F.	4.40	3.32		7.72	10.05	**324**
	0050 Gate, 3'-6" wide		9	2.667	Ea.	38	59		97	135
	0400 3 rail, 4' high		150	.160	L.F.	5.05	3.54		8.59	11.10
	0500 Gate, 3'-6" wide		9	2.667	Ea.	45.50	59		104.50	143
	0600 Open rail, rustic, No. 1 cedar, 2 rail, 3' high		160	.150	L.F.	3.95	3.32		7.27	9.55
	0650 Gate, 3' wide		9	2.667	Ea.	44	59		103	142
	0700 3 rail, 4' high		150	.160	L.F.	4.51	3.54		8.05	10.50
	0900 Gate, 3' wide		9	2.667	Ea.	56.50	59		115.50	155
	1200 Stockade, No. 2 cedar, treated wood rails, 6' high		160	.150	L.F.	5.50	3.32		8.82	11.25
	1250 Gate, 3' wide		9	2.667	Ea.	45	59		104	143
	1300 No. 1 cedar, 3-1/4" cedar rails, 6' high		160	.150	L.F.	13.50	3.32		16.82	20
	1500 Gate, 3' wide		9	2.667	Ea.	106	59		165	210
	1520 Open rail, split, No. 1 cedar, 2 rail, 3' high		160	.150	L.F.	4.01	3.32		7.33	9.60
	1540 3 rail, 4'-0" high		150	.160		5.20	3.54		8.74	11.30
	3300 Board, shadow box, 1" x 6", treated pine, 6' high		160	.150		8.10	3.32		11.42	14.15
	3400 No. 1 cedar, 6' high		150	.160		16	3.54		19.54	23
	3900 Basket weave, No. 1 cedar, 6' high		160	.150		15.90	3.32		19.22	22.50
	3950 Gate, 3'-6" wide		8	3	Ea.	101	66.50		167.50	216
	4000 Treated pine, 6' high		150	.160	L.F.	8.70	3.54		12.24	15.10
	4200 Gate, 3'-6" wide		9	2.667	Ea.	47.50	59		106.50	145
	5000 Fence rail, redwood, 2" x 4", merch grade 8'		2,400	.010	L.F.	.86	.22		1.08	1.30
	5050 Select grade, 8'		2,400	.010	"	2.66	.22		2.88	3.28
	6000 Fence post, select redwood, earthpacked & treated, 4" x 4" x 6'		96	.250	Ea.	8.55	5.55		14.10	18.10
	6010 4" x 4" x 8'		96	.250		10.15	5.55		15.70	19.85
	6020 Set in concrete, 4" x 4" x 6'		50	.480		11	10.60		21.60	29
	6030 4" x 4" x 8'		50	.480		13.10	10.60		23.70	31
	6040 Wood post, 4' high, set in concrete, incl. concrete		50	.480		6.40	10.60		17	24
	6050 Earth packed		96	.250		4.37	5.55		9.92	13.50

2 SITE WORK

Important: See the Reference Section for critical supporting data - Reference Nos., Crews, & City Cost Indexes

		028 300 | Fences & Gates	CREW	DAILY OUTPUT	LABOR-HOURS	UNIT	1999 BARE COSTS				TOTAL INCL O&P	
							MAT.	LABOR	EQUIP.	TOTAL		
324	6060	6' high, set in concrete, incl. concrete	B-1	50	.480	Ea.	8.20	10.60		18.80	26	324
	6070	Earth packed	↓	96	.250	↓	5.95	.55		11.50	15.25	

028 400 | Walk/Road/Parkg Appurt

			CREW	DAILY OUTPUT	LABOR-HOURS	UNIT	MAT.	LABOR	EQUIP.	TOTAL	INCL O&P	
404	0010	**GUIDE/GUARD RAIL** Corrugated steel, galv. steel posts, 6'-3" O.C.	B-80	850	.038	L.F.	10	.88	.63	11.51	13.05	404
	0200	End sections, galvanized, flared		50	.640	Ea.	40	15	10.80	65.80	79.50	
	0300	Wrap around end		50	.640	"	60	15	10.80	85.80	101	
	0400	Timber guide rail, 4" x 8" with 6" x 8" wood posts, treated		960	.033	L.F.	12	.78	.56	13.34	15.05	
	0600	Cable guide rail, 3 at 3/4" cables, steel posts, single face		900	.036		4.71	.83	.60	6.14	7.15	
	0700	Wood posts		950	.034		5.90	.79	.57	7.26	8.35	
	0900	Guide rail, steel box beam, 6" x 6"		120	.267		15.50	6.25	4.50	26.25	31.50	
	1100	Median barrier, steel box beam, 6" x 8"	↓	215	.149	↓	19.50	3.49	2.51	25.50	29.50	
	1200	Impact barrier, UTMCD, barrel type	B-16	30	1.067	Ea.	237	23.50	14.75	275.25	315	
	1400	Resilient guide fence and light shield, 6' high	B-2	130	.308	L.F.	17	6.70		23.70	29.50	
	1500	Concrete posts, individual, 6'-5", triangular	B-80	110	.291	Ea.	30	6.80	4.90	41.70	49	
	1550	Square	"	110	.291	"	32.50	6.80	4.90	44.20	52	
	2000	Median barrier, precast concrete, 3'-6" high, 2' wide, single face	B-29	380	.147	L.F.	28	3.42	1.68	33.10	37.50	
	2200	Double face	"	340	.165	"	23.50	3.83	1.87	29.20	34	
	2400	Speed bumps, thermoplastic, 10-1/2" x 2-1/4" x 48" long	B-2	120	.333	Ea.	80	7.30		87.30	99.50	
408	0010	**PARKING BARRIERS** Timber with saddles, treated type										408
	0100	4" x 4" for cars	B-2	520	.077	L.F.	2.58	1.68		4.26	5.50	
	0200	6" x 6" for trucks		520	.077	"	5.55	1.68		7.23	8.75	
	0400	Folding with individual padlocks		50	.800	Ea.	340	17.50		357.50	405	
	0600	Flexible fixed stanchion, 2' high, 3" diameter		100	.400		17.40	8.75		26.15	33	
	1000	Wheel stops, precast concrete incl. dowels, 6" x 10" x 6'-0"		120	.333		25.50	7.30		32.80	40	
	1100	8" x 13" x 6'-0"		120	.333		29.50	7.30		36.80	44	
	1200	Thermoplastic, 6" x 10" x 6'-0"	↓	120	.333		50	7.30		57.30	66.50	
	1300	Pipe bollards, conc filled/paint, 8' L x 4' D hole, 6" diam.	B-6	20	1.200		165	28	10.95	203.95	238	
	1400	8" diam.		15	1.600		250	37.50	14.60	302.10	350	
	1500	12" diam.	↓	12	2	↓	325	46.50	18.25	389.75	455	
	9000	Parking lot control, see division 111-501										
412	0010	**SIGNS** Stock, 24" x 24", no posts, .080" alum. reflectorized	B-80	70	.457	Ea.	25	10.70	7.70	43.40	52.50	412
	0100	High intensity		70	.457		45	10.70	7.70	63.40	74.50	
	0300	30" x 30", reflectorized		70	.457		48	10.70	7.70	66.40	78	
	0400	High intensity		70	.457		64	10.70	7.70	82.40	95.50	
	0600	Guide and directional signs, 12" x 18", reflectorized		70	.457		15.50	10.70	7.70	33.90	42	
	0700	High intensity		70	.457		15	10.70	7.70	33.40	41.50	
	0900	18" x 24", stock signs, reflectorized		70	.457		24	10.70	7.70	42.40	51.50	
	1000	High intensity		70	.457		38	10.70	7.70	56.40	67	
	1200	24" x 24", stock signs, reflectorized		70	.457		25	10.70	7.70	43.40	52.50	
	1300	High intensity		70	.457		45	10.70	7.70	63.40	74.50	
	1500	Add to above for steel posts, galvanized, 10'-0" upright, bolted		200	.160		28	3.75	2.70	34.45	40	
	1600	12'-0" upright, bolted		140	.229	↓	30	5.35	3.85	39.20	45.50	
	1800	Highway road signs, aluminum, over 20 S. F., reflectorized		350	.091	S.F.	12	2.14	1.54	15.68	18.20	
	2000	High intensity		350	.091		15	2.14	1.54	18.68	21.50	
	2200	Highway, suspended over road, 80 S.F. min., reflectorized		165	.194		15	4.55	3.27	22.82	27	
	2300	High intensity	↓	165	.194	↓	20	4.55	3.27	27.82	32.50	
416	0010	**STEPS** Incl. excav., borrow & concrete base, where applicable										416
	0100	Brick steps	B-24	35	.686	LF Riser	7.60	17.10		24.70	35	
	0200	Railroad ties	2 Clab	25	.640		2.75	13.75		16.50	24.50	
	0300	Bluestone treads, 12" x 2" or 12" x 1-1/2"	B-24	30	.800	↓	16.35	20		36.35	49	
	0500	Concrete, cast in place, see division 033-130										
	0600	Precast concrete, see division 034-804										
424	0010	**TRAFFIC SIGNALS** Mid block pedestrian crosswalk,										424
	0020	with pushbutton and mast arms	R-2	.30	186	Total	11,700	5,300	905	17,905	21,900	

2 SITE WORK

028 400	Walk/Road/Parkg Appurt	CREW	DAILY OUTPUT	LABOR-HOURS	UNIT	1999 BARE COSTS				TOTAL INCL O&P	
						MAT.	LABOR	EQUIP.	TOTAL		
424	0100	Intersection, 8 signals w/three sect. (2 each direction), programmed	R-2	.15	373	Total	24,500	10,600	1,800	36,900	45,100	424
	0120	For each additional traffic phase controller, add	L-9	1.20	30		1,500	745		2,245	2,875	
	0200	Semi-actuated, detectors in side street only, add		.81	44.444		2,500	1,100		3,600	4,550	
	0300	Fully-actuated, detectors in all streets, add		.49	73.469		4,500	1,825		6,325	7,925	
	0400	For pedestrian pushbutton, add		.70	51.429		3,500	1,275		4,775	5,925	
	0500	Optically programmed signal only, add per head		1.64	21.951		2,500	545		3,045	3,650	
	0600	School flashing system, programmed		.41	87.805	Signal	6,000	2,175		8,175	10,200	

028 600	Playfield Equipment										
604	0010	**BACKSTOPS** Baseball, prefabricated, 30' wide, 12' high & 1 overhang	B-1	1	24	Ea.	1,825	530		2,355	2,850	604
	0100	40' wide, 12' high & 2 overhangs	"	.75	32		1,800	710		2,510	3,100	
	0300	Basketball, steel, single goal	B-13	3.04	18.421		695	430	181	1,306	1,625	
	0400	Double goal	"	1.92	29.167		505	680	286	1,471	1,925	
	0600	Tennis, wire mesh with pair of ends	B-1	2.48	9.677	Set	1,000	214		1,214	1,425	
	0700	Enclosed court	"	1.30	18.462	Ea.	3,125	410		3,535	4,075	
	0900	Handball or squash court, outdoor, wood	2 Carp	.50	32		2,550	875		3,425	4,200	
	1000	Masonry handball/squash court	D-1	.30	53.333		19,600	1,300		20,900	23,600	
608	0010	**GOAL POSTS** Steel, football, double post	B-1	1.50	16	Pr.	1,350	355		1,705	2,025	608
	0100	Deluxe, single post		1.50	16		1,450	355		1,805	2,150	
	0300	Football, convertible to soccer		1.50	16		1,450	355		1,805	2,150	
	0500	Soccer, regulation		2	12		1,475	265		1,740	2,050	
610	0010	**MODULAR PLAYGROUND** Basic components										610
	0100	Deck, square, steel	B-1	1	24	Ea.	560	530		1,090	1,450	
	0110	Polyurethane		1	24		485	530		1,015	1,375	
	0120	Triangular, steel		1	24		320	530		850	1,200	
	0130	Post, steel, 5" square		18	1.333		17.20	29.50		46.70	65.50	
	0140	Aluminum, 2-3/8" square		20	1.200		14.45	26.50		40.95	58	
	0150	5" square		18	1.333	L.F.	19.20	29.50		48.70	67.50	
	0160	Roof, square poly, 54" side		18	1.333	Ea.	1,150	29.50		1,179.50	1,325	
	0170	Transfer module		3	8	"	1,725	177		1,902	2,175	
	0180	Guardrail, pipe		60	.400	L.F.	199	8.85		207.85	233	
	0190	Steps, deck-to-deck		8	3	Ea.	1,150	66.50		1,216.50	1,375	
	0200	Activity panel, minimum		2	12		262	265		527	710	
	0210	Maximum		2	12		600	265		865	1,075	
	0360	With guardrails		3	8		1,675	177		1,852	2,125	
	0370	Crawl tunnel, straight, 56" long		4	6		785	133		918	1,075	
	0380	90°		4	6		945	133		1,078	1,250	
	1200	Slide, tunnel		8	3		1,450	66.50		1,516.50	1,700	
	1210	Straight, poly		8	3		148	66.50		214.50	266	
	1220	Stainless steel, 54" high		6	4		181	88.50		269.50	340	
	1230	Curved, poly, 40" high		6	4		800	88.50		888.50	1,025	
	1240	Spyroslide, 56" - 72" high		5	4.800		2,475	106		2,581	2,900	
	1300	Ladder, vertical		5	4.800		405	106		511	615	
	1310	Horizontal, 8' long		5	4.800		630	106		736	860	
	1320	Corkscrew climber		3	8		430	177		607	750	
	1330	Fire pole		6	4		188	88.50		276.50	345	
	1340	Bridge, ring		4	6		805	133		938	1,100	
	1350	Clatter		4	6	L.F.	285	133		418	525	
612	0010	**PLAYGROUND EQUIPMENT** See also individual items										612
	0200	Bike rack, 10' long, permanent	B-1	12	2	Ea.	299	44		343	400	
	0400	Horizontal monkey ladder, 14' long		4	6		520	133		653	780	
	0590	Parallel bars, 10' long		4	6		174	133		307	400	
	0600	Posts, tether ball set, 2-3/8" O.D.		12	2		203	44		247	293	
	0800	Poles, multiple purpose, 10'-6" long		12	2	Pr.	110	44		154	191	

Important: See the Reference Section for critical supporting data - Reference Nos., Crews, & City Cost Indexes

028 600 | Playfield Equipment

			CREW	DAILY OUTPUT	LABOR-HOURS	UNIT	1999 BARE COSTS MAT.	LABOR	EQUIP.	TOTAL	TOTAL INCL O&P	
612	1000	Ground socket for movable posts, 2-3/8" post	B-1	10	2.400	Pr.	74	53		127	165	612
	1100	3-1/2" post		10	2.400	↓	77	53		130	169	
	1300	See-saw, steel, 2 units		6	4	Ea.	230	88.50		318.50	390	
	1400	4 units		4	6		800	133		933	1,100	
	1500	6 units		3	8		740	177		917	1,100	
	1700	Shelter, fiberglass golf tee, 3 person		4.60	5.217		2,150	115		2,265	2,550	
	1900	Slides, stainless steel bed, 12' long, 6' high		3	8		1,550	177		1,727	2,000	
	2000	20' long, 10' high		2	12		1,900	265		2,165	2,500	
	2200	Swings, plain seats, 8' high, 4 seats		2	12		815	265		1,080	1,325	
	2300	8 seats		1.30	18.462		1,500	410		1,910	2,300	
	2500	12' high, 4 seats		2	12		1,000	265		1,265	1,525	
	2600	8 seats		1.30	18.462		1,650	410		2,060	2,475	
	2800	Whirlers, 8' diameter		3	8		1,500	177		1,677	1,925	
	2900	10' diameter	↓	3	8	↓	2,175	177		2,352	2,675	
620	0010	**PLATFORM/PADDLE TENNIS COURT** Complete with lighting, etc.										620
	0100	Aluminum slat deck with aluminum frame	B-1	.08	300	Court	24,000	6,625		30,625	36,800	
	0500	Aluminum slat deck and wood frame	C-1	.12	266		24,400	6,900		31,300	37,600	
	0800	Aluminum deck heater, add	B-1	1.18	20.339		3,150	450		3,600	4,175	
	0900	Douglas fir planking and wood frame 2" x 6" x 30'	C-1	.12	266		22,500	6,900		29,400	35,600	
	1000	Plywood deck with steel frame		.12	266		23,500	6,900		30,400	36,700	
	1100	Steel slat deck with wood frame	↓	.12	266	↓	24,500	6,900		31,400	37,800	

028 700 | Site/Street Furnishings

			CREW	DAILY OUTPUT	LABOR-HOURS	UNIT	MAT.	LABOR	EQUIP.	TOTAL	TOTAL INCL O&P	
704	0010	**BENCHES** Park, precast concrete, w/backs, wood rails, 4' long	2 Clab	5	3.200	Ea.	310	68.50		378.50	450	704
	0100	8' long		4	4		635	86		721	830	
	0300	Fiberglass, without back, one piece, 4' long		10	1.600		355	34.50		389.50	445	
	0400	8' long		7	2.286		700	49		749	845	
	0500	Steel barstock pedestals w/backs, 2" x 3" wood rails, 4' long		10	1.600		630	34.50		664.50	750	
	0510	8' long		7	2.286		745	49		794	895	
	0520	3" x 8" wood plank, 4' long		10	1.600		635	34.50		669.50	755	
	0530	8' long		7	2.286		740	49		789	890	
	0540	Backless, 4" x 4" wood plank, 4' square		10	1.600		615	34.50		649.50	730	
	0550	8' long		7	2.286		585	49		634	720	
	0600	Aluminum pedestals, with backs, aluminum slats, 8' long		8	2		150	43		193	233	
	0610	15' long		5	3.200		250	68.50		318.50	385	
	0620	Portable, aluminum slats, 8' long		8	2		170	43		213	255	
	0630	15' long		5	3.200		275	68.50		343.50	415	
	0800	Cast iron pedestals, back & arms, wood slats, 4' long		8	2		460	43		503	575	
	0820	8' long		5	3.200		765	68.50		833.50	950	
	0840	Backless, wood slats, 4' long		8	2		400	43		443	510	
	0860	8' long		5	3.200		675	68.50		743.50	855	
	1700	Steel frame, fir seat, 10' long	↓	10	1.600	↓	150	34.50		184.50	219	
708	0010	**BLEACHERS** Outdoor, portable, 3 to 5 tiers, to 300' long, min.	2 Sswk	120	.133	Seat	29.50	4.08		33.58	40	708
	0100	Maximum, less than 15' long, prefabricated		80	.200		45	6.10		51.10	60.50	
	0200	6 to 20 tiers, minimum, up to 300' long		120	.133		29.50	4.08		33.58	39.50	
	0300	Max., under 15', (highly prefabricated, on wheels)	↓	80	.200	↓	61	6.10		67.10	78	
	0500	Permanent grandstands, wood seat, steel frame, 24" row										
	0600	3 to 15 tiers, minimum	2 Sswk	60	.267	Seat	57.50	8.15		65.65	78.50	
	0700	Maximum		48	.333		63.50	10.20		73.70	88.50	
	0900	16 to 30 tiers, minimum		60	.267		80	8.15		88.15	103	
	0950	Average		55	.291		106	8.90		114.90	132	
	1000	Maximum		48	.333		113	10.20		123.20	143	
	1200	Seat backs only, 30" row, fiberglass		160	.100		17	3.06		20.06	24.50	
	1300	Steel and wood	↓	160	.100	↓	21	3.06		24.06	28.50	

For expanded coverage of these items see *Means Site Work and Landscape Cost Data 1999*

028 700 | Site/Street Furnishings

			CREW	DAILY OUTPUT	LABOR-HOURS	UNIT	1999 BARE COSTS				TOTAL INCL O&P	
							MAT.	LABOR	EQUIP.	TOTAL		
708	1400	NOTE: average seating is 1.5' in width										708
712	0010	**PLANTER BLOCKS** Precast concrete, interlocking										712
	0020	"V" blocks for retaining soil	D-1	205	.078	S.F.	3.75	1.92		5.67	7.10	
716	0010	**PLANTERS** Concrete, sandblasted, precast, 48" diameter, 24" high	2 Clab	15	1.067	Ea.	425	23		448	505	716
	0100	Fluted, precast, 7' diameter, 36" high		10	1.600		1,150	34.50		1,184.50	1,300	
	0300	Fiberglass, circular, 36" diameter, 24" high		15	1.067		315	23		338	380	
	0400	60" diameter, 24" high		10	1.600		645	34.50		679.50	765	
	0600	Square, 24" side, 36" high		15	1.067		430	23		453	510	
	0700	48" side, 36" high		15	1.067		750	23		773	860	
	0900	Planter/bench, 72" square, 36" high		5	3.200		2,700	68.50		2,768.50	3,075	
	1000	96" square, 27" high		5	3.200		4,400	68.50		4,468.50	4,950	
	1200	Wood, square, 48" side, 24" high		15	1.067		755	23		778	865	
	1300	Circular, 48" diameter, 30" high		10	1.600		675	34.50		709.50	800	
	1500	72" diameter, 30" high		10	1.600		1,075	34.50		1,109.50	1,225	
	1600	Planter/bench, 72"	↓	5	3.200	↓	2,100	68.50		2,168.50	2,425	
720	0010	**TRASH CLOSURE** Steel with pullover cover										720
	0020	2'-3" wide, 4'-7" high, 6'-2" long	2 Clab	5	3.200	Ea.	680	68.50		748.50	860	
	0100	10'-1" long		4	4		900	86		986	1,125	
	0300	Wood, 10' wide, 6' high, 10' long	↓	1.20	13.333	↓	710	286		996	1,225	
724	0010	**TRASH RECEPTACLE** Fiberglass, 2' square, 18" high	2 Clab	30	.533	Ea.	200	11.45		211.45	238	724
	0100	2' square, 2'-6" high		30	.533		280	11.45		291.45	330	
	0300	Circular, 2' diameter, 18" high		30	.533		180	11.45		191.45	216	
	0400	2' diameter, 2'-6" high	↓	30	.533	↓	240	11.45		251.45	282	

029 100 | Shrub/Tree Transplanting

			CREW	DAILY OUTPUT	LABOR-HOURS	UNIT	1999 BARE COSTS				TOTAL INCL O&P	
							MAT.	LABOR	EQUIP.	TOTAL		
104	0010	**TREE GUYING** Including stakes, guy wire and wrap										104
	0100	Less than 3" caliper, 2 stakes	2 Clab	35	.457	Ea.	15	9.80		24.80	32	
	0200	3" to 4" caliper, 3 stakes	"	21	.762	"	17.60	16.35		33.95	45	
	1000	Including arrowhead anchor, cable, turnbuckles and wrap										
	1100	Less than 3" caliper, 3 anchors	2 Clab	20	.800	Ea.	45	17.15		62.15	76.50	
	1200	3" to 6" caliper, 4 anchors		15	1.067		65	23		88	108	
	1300	6" caliper, 6 anchors		12	1.333		80	28.50		108.50	133	
	1400	8" caliper, 8 anchors	↓	9	1.778	↓	80	38		118	148	

029 200 | Soil Preparation

			CREW	DAILY OUTPUT	LABOR-HOURS	UNIT	1999 BARE COSTS				TOTAL INCL O&P	
							MAT.	LABOR	EQUIP.	TOTAL		
208	0010	**PLANT BED PREPARATION**										208
	0100	Backfill planting pit, by hand, on site topsoil	2 Clab	18	.889	C.Y.		19.05		19.05	30	
	0200	Prepared planting mix	"	24	.667			14.30		14.30	22.50	
	0300	Skid steer loader, on site topsoil	B-62	340	.071			1.65	.33	1.98	2.92	
	0400	Prepared planting mix	"	410	.059			1.37	.27	1.64	2.42	
	1000	Excavate planting pit, by hand, sandy soil	2 Clab	16	1			21.50		21.50	34	
	1100	Heavy soil or clay	"	8	2			43		43	67.50	
	1200	1/2 C.Y. backhoe, sandy soil	B-11C	150	.107			2.66	1.46	4.12	5.70	
	1300	Heavy soil or clay	"	115	.139			3.47	1.90	5.37	7.45	
	2000	Mix planting soil, incl. loam, manure, peat, by hand	2 Clab	60	.267		25	5.70		30.70	36.50	
	2100	Skid steer loader	B-62	150	.160	↓	25	3.74	.74	29.48	34	
	3000	Pile sod, skid steer loader	"	2,800	.009	S.Y.		.20	.04	.24	.35	

Important: See the Reference Section for critical supporting data - Reference Nos., Crews, & City Cost Indexes

029 200 | Soil Preparation

		CREW	DAILY OUTPUT	LABOR-HOURS	UNIT	MAT.	LABOR	EQUIP.	TOTAL	TOTAL INCL O&P	
208	3100	By hand	2 Clab	400	.040	S.Y.		.86		.86	1.35	208
	4000	Remove sod, F.E. loader	B-10S	2,000	.006			.16	.16	.32	.42	
	4100	Sod cutter	B-12K	3,200	.005			.13	.25	.38	.48	
	4200	By hand	2 Clab	240	.067			1.43		1.43	2.25	

029 300 | Lawns & Grasses

		CREW	DAILY OUTPUT	LABOR-HOURS	UNIT	MAT.	LABOR	EQUIP.	TOTAL	TOTAL INCL O&P	
304	0010	SEEDING Mechanical seeding, 215 lb./acre (R029-310)	B-66	1.50	5.333	Acre	485	145	134	764	905	304
	0100	44 lb./M.S.Y.	"	2,500	.003	S.Y.	.15	.09	.08	.32	.39	
	0300	Fine grading and seeding incl. lime, fertilizer & seed,										
	0310	with equipment	B-14	1,000	.048	S.Y.	.16	1.09	.22	1.47	2.13	
	0600	Limestone hand push spreader, 50 lbs. per M.S.F.	1 Clab	180	.044	M.S.F.	3.25	.95		4.20	5.10	
	0800	Grass seed hand push spreader, 4.5 lbs. per M.S.F.	"	180	.044	"	10	.95		10.95	12.50	
	1000	Hydro or air seeding for large areas, incl. seed and fertilizer	B-81	8,900	.003	S.Y.	.15	.06	.07	.28	.34	
	1100	With wood fiber mulch added	"	8,900	.003	"	.25	.06	.07	.38	.45	
	1300	Seed only, over 100 lbs., field seed, minimum				Lb.	1			1	1.10	
	1400	Maximum					3.50			3.50	3.85	
	1500	Lawn seed, minimum					1.50			1.50	1.65	
	1600	Maximum					4.25			4.25	4.68	
	1800	Aerial operations, seeding only, field seed	B-58	50	.480	Acre	350	11.20	74.50	435.70	485	
	1900	Lawn seed		50	.480		510	11.20	74.50	595.70	660	
	2100	Seed and liquid fertilizer, field seed		50	.480		400	11.20	74.50	485.70	540	
	2200	Lawn seed		50	.480		550	11.20	74.50	635.70	705	
316	0010	SODDING 1" deep, bluegrass sod, on level ground, over 8 M.S.F.	B-63	22	1.818	M.S.F.	200	41	5.05	246.05	290	316
	0200	4 M.S.F.		17	2.353	"	220	53	6.55	279.55	330	
	0300	1000 S.F.		3.50	11.429	Ea.	250	258	32	540	715	
	0500	Sloped ground, over 8 M.S.F.		6	6.667	M.S.F.	200	151	18.55	369.55	475	
	0600	4 M.S.F.		5	8		220	181	22	423	550	
	0700	1000 S.F.		4	10		250	226	28	504	660	
	1000	Bent grass sod, on level ground, over 6 M.S.F.		20	2		460	45	5.55	510.55	580	
	1100	3 M.S.F.		18	2.222		500	50	6.20	556.20	635	
	1200	Sodding 1000 S.F. or less		14	2.857		550	64.50	7.95	622.45	715	
	1500	Sloped ground, over 6 M.S.F.		15	2.667		460	60.50	7.40	527.90	605	
	1600	3 M.S.F.		13.50	2.963		500	67	8.25	575.25	665	
	1700	1000 S.F.		12	3.333		300	75.50	9.25	384.75	460	

029 500 | Trees/Plants/Grnd Cover

		CREW	DAILY OUTPUT	LABOR-HOURS	UNIT	MAT.	LABOR	EQUIP.	TOTAL	TOTAL INCL O&P	
504	0010	GROUND COVER Plants, pachysandra, in prepared beds	B-1	15	1.600	C	34	35.50		69.50	93	504
	0200	Vinca minor, 1 yr, bare root		12	2	"	40	44		84	114	
	0600	Stone chips, in 50 lb. bags, Georgia marble		520	.046	Bag	3.36	1.02		4.38	5.30	
	0700	Onyx gemstone		260	.092		13.10	2.04		15.14	17.65	
	0800	Quartz		260	.092		5.50	2.04		7.54	9.25	
	0900	Pea gravel, truckload lots		28	.857	Ton	24.50	18.95		43.45	57	
516	0010	MULCH										516
	0100	Aged barks, 3" deep, hand spread	1 Clab	100	.080	S.Y.	2	1.72		3.72	4.90	
	0150	Skid steer loader	B-63	13.50	2.963	M.S.F.	220	67	8.25	295.25	355	
	0200	Hay, 1" deep, hand spread	1 Clab	475	.017	S.Y.	.25	.36		.61	.85	
	0250	Power mulcher, small	B-64	180	.089	M.S.F.	18.50	1.92	1.59	22.01	25	
	0350	Large	B-65	530	.030	"	18.50	.65	.86	20.01	22.50	
	0400	Humus peat, 1" deep, hand spread	1 Clab	700	.011	S.Y.	1.11	.25		1.36	1.61	
	0450	Push spreader	A-1	2,500	.003	"	1.39	.07	.03	1.49	1.67	
	0550	Tractor spreader	B-66	700	.011	M.S.F.	104	.31	.29	104.60	115	
	0600	Oat straw, 1" deep, hand spread	1 Clab	475	.017	S.Y.	.28	.36		.64	.88	
	0650	Power mulcher, small	B-64	180	.089	M.S.F.	25	1.92	1.59	28.51	32	
	0700	Large	B-65	530	.030	"	25	.65	.86	26.51	29.50	
	0750	Add for asphaltic emulsion	B-45	1,770	.009	Gal.	1.60	.23	.42	2.25	2.57	
	0800	Peat moss, 1" deep, hand spread	1 Clab	900	.009	S.Y.	1.55	.19		1.74	2	

029 500	Trees/Plants/Grnd Cover	CREW	DAILY OUTPUT	LABOR-HOURS	UNIT	1999 BARE COSTS				TOTAL INCL O&P	
						MAT.	LABOR	EQUIP.	TOTAL		
516 0850	Push spreader	A-1	2,500	.003	S.Y.	1.60	.07	.03	1.70	1.90	**516**
0950	Tractor spreader	B-66	700	.011	M.S.F.	175	.31	.29	175.60	194	
1000	Polyethylene film, 6 mil.	2 Clab	2,000	.008	S.Y.	.15	.17		.32	.44	
1100	Redwood nuggets, 3" deep, hand spread	1 Clab	150	.053	"	6	1.14		7.14	8.40	
1150	Skid steer loader	B-63	13.50	2.963	M.S.F.	600	67	8.25	675.25	775	
1200	Stone mulch, hand spread, ceramic chips, economy	1 Clab	125	.064	S.Y.	5.75	1.37		7.12	8.50	
1250	Deluxe	"	95	.084	"	8.60	1.81		10.41	12.30	
1300	Granite chips	B-1	10	2.400	C.Y.	28	53		81	115	
1400	Marble chips		10	2.400		105	53		158	200	
1500	Onyx gemstone		10	2.400		310	53		363	425	
1600	Pea gravel		28	.857		40.50	18.95		59.45	74.50	
1700	Quartz		10	2.400		135	53		188	233	
1800	Tar paper, 15 Lb. felt	1 Clab	800	.010	S.Y.	.40	.21		.61	.78	
1900	Wood chips, 2" deep, hand spread	"	220	.036	"	1.65	.78		2.43	3.05	
1950	Skid steer loader	B-63	20.30	1.970	M.S.F.	108	44.50	5.50	158	195	
520 0010	**PLANTING** Moving shrubs on site, 12" ball R029-545	B-62	28	.857	Ea.		20	3.98	23.98	35.50	**520**
0100	24" ball	"	22	1.091			25.50	5.05	30.55	45	
0300	Moving trees on site, 36" ball	B-6	3.75	6.400			150	58.50	208.50	296	
0400	60" ball	"	1	24			560	219	779	1,100	
524 0010	**SHRUBS** Broadleaf evergreen, planted in prepared beds										**524**
0100	Andromeda, 15"-18", container	B-1	96	.250	Ea.	14	5.55		19.55	24	
0200	Azalea, 15" - 18", container		96	.250		19	5.55		24.55	29.50	
0300	Barberry, 9"-12", container		130	.185		11.45	4.08		15.53	19	
0400	Boxwood, 15"-18", B & B		96	.250		16.25	5.55		21.80	26.50	
0500	Euonymus, emerald gaiety, 12" to 15", container		115	.209		12.25	4.62		16.87	21	
0600	Holly, 15"-18", B & B		96	.250		16.75	5.55		22.30	27	
0900	Mount laurel, 18" - 24", B & B		80	.300		48	6.65		54.65	63.50	
1000	Paxistema, 9 - 12" high		130	.185		9	4.08		13.08	16.30	
1100	Rhododendron, 18"-24", container		48	.500		25	11.05		36.05	45	
1200	Rosemary, 1 gal container		600	.040		35	.88		35.88	40	
2000	Deciduous, amelanchier, 2'-3', B & B		57	.421		55	9.30		64.30	75	
2100	Azalea, 15"-18", B & B		96	.250		18.25	5.55		23.80	28.50	
2300	Bayberry, 2'-3', B & B		57	.421		22.50	9.30		31.80	39	
2600	Cotoneaster, 15"-18", B & B		80	.300		14	6.65		20.65	26	
2800	Dogwood, 3'-4', B & B	B-17	40	.800		21	18.45	14.60	54.05	67.50	
2900	Euonymus, alatus compacta, 15" to 18", container	B-1	80	.300		20	6.65		26.65	32.50	
3200	Forsythia, 2'-3', container	"	60	.400		16	8.85		24.85	31.50	
3300	Hibiscus, 3'-4', B & B	B-17	75	.427		24	9.85	7.75	41.60	50.50	
3400	Honeysuckle, 3'-4', B & B	B-1	60	.400		17	8.85		25.85	32.50	
3500	Hydrangea, 2'-3', B & B	"	57	.421		20	9.30		29.30	36.50	
3600	Lilac, 3'-4', B & B	B-17	40	.800		38	18.45	14.60	71.05	86.50	
3900	Privet, bare root, 18"-24"	B-1	80	.300		5	6.65		11.65	15.95	
4100	Quince, 2'-3', B & B	"	57	.421		9	9.30		18.30	24.50	
4200	Russian olive, 3'-4', B & B	B-17	75	.427		17	9.85	7.75	34.60	42.50	
4400	Spirea, 3'-4', B & B	B-1	70	.343		14	7.60		21.60	27.50	
4500	Viburnum, 3'-4', B & B	B-17	40	.800		17.50	18.45	14.60	50.55	64	
528 0010	**SHRUBS AND TREES** Evergreen, in prepared beds, B & B										**528**
0100	Arborvitae pyramidal, 4'-5'	B-17	30	1.067	Ea.	35	24.50	19.45	78.95	98	
0150	Globe, 12"-15"	B-1	96	.250		10	5.55		15.55	19.70	
0300	Cedar, blue, 8'-10'	B-17	18	1.778		111	41	32.50	184.50	221	
0500	Hemlock, canadian, 2-1/2'-3'	B-1	36	.667		14.25	14.75		29	38.50	
0550	Holly, Savannah, 8' - 10' H		9.68	2.479		500	55		555	635	
0600	Juniper, andorra, 18"-24"		80	.300		14	6.65		20.65	26	
0620	Wiltoni, 15"-18"		80	.300		14.50	6.65		21.15	26.50	

Important: See the Reference Section for critical supporting data - Reference Nos., Crews, & City Cost Indexes

		029 500	Trees/Plants/Grnd Cover	CREW	DAILY OUTPUT	LABOR-HOURS	UNIT	1999 BARE COSTS				TOTAL INCL O&P	
								MAT.	LABOR	EQUIP.	TOTAL		
528	0640		Skyrocket, 4-1/2'-5'	B-17	55	.582	Ea.	38.50	13.40	10.60	62.50	75	528
	0660		Blue pfitzer, 2'-2-1/2'	B-1	44	.545		16	12.05		28.05	36.50	
	0680		Ketleerie, 2-1/2'-3'		50	.480		28	10.60		38.60	47.50	
	0700		Pine, black, 2-1/2'-3'		50	.480		29.50	10.60		40.10	49	
	0720		Mugo, 18"-24"	▼	60	.400		30	8.85		38.85	47	
	0740		White, 4'-5'	B-17	75	.427		44	9.85	7.75	61.60	72.50	
	0800		Spruce, blue, 18"-24"	B-1	60	.400		28	8.85		36.85	45	
	0840		Norway, 4'-5'	B-17	75	.427		56	9.85	7.75	73.60	85.50	
	0900		Yew, denisforma, 12"-15"	B-1	60	.400		22.50	8.85		31.35	39	
	1000		Capitata, 18"-24"		30	.800		21	17.70		38.70	51	
	1100		Hicksi, 2'-2-1/2'	▼	30	.800	▼	26	17.70		43.70	56.50	
532	0010	**TRAVEL** To all nursery items, for 10 to 20 miles, add					All					5%	532
	0100	30 to 50 miles, add					"					10%	
536	0010	**TREES** Deciduous, in prep. beds, balled & burlapped (B&B)											536
	0100	Ash, 2" caliper	B-17	8	4	Ea.	100	92	73	265	335	
	0200	Beech, 5'-6'		50	.640		200	14.75	11.65	226.40	256	
	0300	Birch, 6'-8', 3 stems	[R029 -545]	20	1.600		110	37	29	176	210	
	0500	Crabapple, 6'-8'		20	1.600		64	37	29	130	160	
	0600	Dogwood, 4'-5'		40	.800		61	18.45	14.60	94.05	112	
	0700	Eastern redbud 4'-5'		40	.800		115	18.45	14.60	148.05	172	
	0800	Elm, 8'-10'		20	1.600		70	37	29	136	166	
	0900	Ginkgo, 6'-7'		24	1.333		60	30.50	24.50	115	140	
	1000	Hawthorn, 8'-10', 1" caliper		20	1.600		80	37	29	146	177	
	1100	Honeylocust, 10'-12', 1-1/2" caliper		10	3.200		80	74	58.50	212.50	266	
	1300	Larch, 8'		32	1		50	23	18.20	91.20	111	
	1400	Linden, 8'-10', 1" caliper		20	1.600		95	37	29	161	194	
	1500	Magnolia, 4'-5'		20	1.600		55	37	29	121	149	
	1600	Maple, red, 8'-10', 1-1/2" caliper		10	3.200		125	74	58.50	257.50	315	
	1700	Mountain ash, 8'-10', 1" caliper		16	2		120	46	36.50	202.50	244	
	1800	Oak, 2-1/2"-3" caliper		3	10.667		159	246	194	599	770	
	2100	Planetree, 9'-11', 1-1/4" caliper		10	3.200		90	74	58.50	222.50	277	
	2200	Plum, 6'-8', 1" caliper		20	1.600		42	37	29	108	135	
	2300	Poplar, 9'-11', 1-1/4" caliper		10	3.200		51.50	74	58.50	184	235	
	2500	Sumac, 2'-3'		75	.427		24	9.85	7.75	41.60	50.50	
	2700	Tulip, 5'-6'		40	.800		48	18.45	14.60	81.05	97.50	
	2800	Willow, 6'-8', 1" caliper	▼	20	1.600	▼	63	37	29	129	158	

For information about Means Estimating Seminars, see yellow pages 11 and 12 in back of book

For expanded coverage of these items see *Means Site Work and Landscape Cost Data 1999*

Division Notes

	CREW	DAILY OUTPUT	LABOR-HOURS	UNIT	1999 BARE COSTS				TOTAL INCL O&P
					MAT.	LABOR	EQUIP.	TOTAL	

Division 3 Concrete

Estimating Tips

General

- Carefully check all the plans and specifications. Concrete often appears on drawings other than structural drawings, including mechanical and electrical drawings for equipment pads. The cost of cutting and patching is often difficult to estimate. See Division 020 for demolition costs.
- Always obtain concrete prices from suppliers near the job site. A volume discount can often be negotiated depending upon competition in the area. Remember to add for waste, particularly for slabs and footings on grade.

031 Concrete Formwork

- A primary cost for concrete construction is forming. Most jobs today are constructed with prefabricated forms. The selection of the forms best suited for the job and the total square feet of forms required for efficient concrete forming and placing are key elements in estimating concrete construction. Enough forms must be available for erection to make efficient use of the concrete placing equipment and crew.
- Concrete accessories for forming and placing depend upon the systems used. Study the plans and specifications to assure that all special accessory requirements have been included in the cost estimate such as anchor bolts, inserts and hangers.

032 Concrete Reinforcement

- Ascertain that the reinforcing steel supplier has included all accessories, cutting, bending and an allowance for lapping, splicing and waste. A good rule of thumb is 10% for lapping, splicing and waste. Also, 10% waste should be allowed for welded wire fabric.

033 Cast-in-Place Concrete

- When estimating structural concrete, pay particular attention to requirements for concrete additives, curing methods and surface treatments. Special consideration for climate, hot or cold, must be included in your estimate. Be sure to include requirements for concrete placing equipment and concrete finishing.

034 Precast Concrete
035 Cementitious Decks & Toppings

- The cost of hauling precast concrete structural members is often an important factor. For this reason, it is important to get a quote from the nearest supplier. It may become economically feasible to set up precasting beds on the site if the hauling costs are prohibitive.

Reference Numbers

Reference numbers are shown in bold squares at the beginning of some major classifications. These numbers refer to related items in the Reference Section. The reference information may be an estimating procedure, an alternate pricing method or technical information.

Note: Not all subdivisions listed here necessarily appear in this publication.

3 CONCRETE

031 100	Struct C.I.P. Formwork	CREW	DAILY OUTPUT	LABOR-HOURS	UNIT	MAT.	LABOR	EQUIP.	TOTAL	TOTAL INCL O&P	
							1999 BARE COSTS				
110	**0010** ACCESSORIES, ANCHOR BOLTS J-type, incl. nut and washer										**110**
0020	1/2" diameter, 6" long	1 Carp	90	.089	Ea.	1.11	2.43		3.54	5.05	
0050	10" long		85	.094		1.22	2.57		3.79	5.40	
0100	12" long		85	.094		1.27	2.57		3.84	5.45	
0200	5/8" diameter, 12" long		80	.100		1.40	2.73		4.13	5.85	
0250	18" long		70	.114		1.65	3.12		4.77	6.75	
0300	24" long		60	.133		1.90	3.64		5.54	7.85	
0350	3/4" diameter, 8" long		80	.100		1.70	2.73		4.43	6.20	
0400	12" long		70	.114		2.11	3.12		5.23	7.25	
0450	18" long		60	.133		2.58	3.64		6.22	8.60	
0500	24" long		50	.160		3.42	4.37		7.79	10.60	
0600	7/8" diameter, 12" long		60	.133		2.32	3.64		5.96	8.30	
0650	18" long		50	.160		2.96	4.37		7.33	10.10	
0700	24" long		40	.200		3.09	5.45		8.54	12	
0800	1" diameter, 12" long		55	.145		3.20	3.97		7.17	9.75	
0850	18" long		45	.178		3.61	4.85		8.46	11.60	
0900	24" long		35	.229		4.25	6.25		10.50	14.50	
0950	36" long		25	.320		5.60	8.75		14.35	19.90	
1200	1-1/2" diameter, 18" long		22	.364		8.95	9.95		18.90	25.50	
1250	24" long		18	.444		10.15	12.15		22.30	30.50	
1300	36" long		12	.667	▼	11.95	18.20		30.15	41.50	
1350	Larger sizes	▼	200	.040	Lb.	.88	1.09		1.97	2.69	
8000	Sleeves, see Division 031-126										
112	**0010** ACCESSORIES, CHAMFER STRIPS										**112**
2000	Polyvinyl chloride, 1/2" wide with leg	1 Carp	535	.015	L.F.	.48	.41		.89	1.17	
2200	3/4" wide with leg		525	.015		.55	.42		.97	1.26	
2400	1" radius with leg		515	.016		.80	.42		1.22	1.55	
2800	1-1/2" radius with leg		500	.016		1.75	.44		2.19	2.62	
5000	Wood, 1/2" wide		535	.015		.12	.41		.53	.77	
5200	3/4" wide		525	.015		.30	.42		.72	.98	
5400	1" wide	▼	515	.016	▼	.39	.42		.81	1.09	
114	**0010** ACCESSORIES, COLUMN FORM										**114**
1000	Column clamps, adjustable to 24" x 24", buy				Set	49.50			49.50	54.50	
1100	Rent per month					5.15			5.15	5.70	
1300	For sizes to 30" x 30", buy					62			62	68	
1400	Rent per month					6.50			6.50	7.15	
1600	For sizes to 36" x 36", buy					73.50			73.50	81	
1700	Rent per month					6.65			6.65	7.35	
2000	Bull winch (band iron) 36" x 36", buy					46			46	50.50	
2100	Rent per month					4			4	4.40	
2300	48" x 48", buy					48			48	53	
2400	Rent per month					5			5	5.50	
3000	Chain & wedge type 36" x 36", buy					67			67	73.50	
3100	Rent per month					7.50			7.50	8.25	
3300	60" x 60", buy					88			88	97	
3400	Rent per month					10			10	11	
4000	Friction collars 2'-6" dia., buy					650			650	715	
4100	Rent per month					64			64	70.50	
4300	4'-0" dia., buy					750			750	825	
4400	Rent per month				▼	82			82	90	
116	**0010** ACCESSORIES, DOVETAIL ANCHOR SYSTEM										**116**
0500	Anchor slot, galv., filled, 24 ga.	1 Carp	425	.019	L.F.	.56	.51		1.07	1.43	
0600	20 ga.		400	.020		.40	.55		.95	1.30	
0800	16 oz. copper, foam filled	▼	375	.021	▼	1.55	.58		2.13	2.63	

Important: See the Reference Section for critical supporting data - Reference Nos., Crews, & City Cost Indexes

		031 100	Struct C.I.P. Formwork	CREW	DAILY OUTPUT	LABOR-HOURS	UNIT	1999 BARE COSTS				TOTAL INCL O&P	
								MAT.	LABOR	EQUIP.	TOTAL		
116	0900		26 ga. stainless steel, foam filled	1 Carp	375	.021	L.F.	1.10	.58		1.68	2.13	116
	1200		Brick anchor, corr., galv., 3-1/2" long, 16 ga.	1 Bric	10.50	.762	C	22	21		43	56.50	
	1300		12 ga.		10.50	.762		32	21		53	67.50	
	1500		Flat, galv., 3-1/2" long, 16 ga.		10.50	.762		23.50	21		44.50	58.50	
	1600		12 ga.		10.50	.762		45.50	21		66.50	82.50	
	2000		Cavity wall, corr., galv., 5" long, 16 ga.		10.50	.762		25	21		46	60	
	2100		12 ga.		10.50	.762		36	21		57	72	
	3000		Furring anchors, corr., galv., 1-1/2" long, 16 ga.		10.50	.762		9.50	21		30.50	43	
	3100		12 ga.		10.50	.762		16	21		37	50	
	6000		Stone anchors, 3-1/2" long, galv., 1/8" x 1" wide		10.50	.762		73	21		94	113	
	6100		1/4" x 1" wide	▼	10.50	.762	▼	109	21		130	152	
120	0010	**ACCESSORIES, HANGERS**											120
	0020		Slab and beam form										
	0500		Banding iron, 3/4" x 22 ga, 14 L.F. per lb or										
	0550		1/2" x 14 ga, 7 L.F. per lb.				Lb.	.90			.90	.99	
	1000		Fascia ties, coil type add to frame ties below				C	55			55	60.50	
	1500		Frame ties to 8-1/8"					148			148	163	
	1550		8-1/8" to 10-1/8"					153			153	168	
	5000		Snap tie hanger, to 30" overall length, 4000 #					145			145	160	
	5050		30" to 36" overall length					155			155	171	
	5100		42" to 48" overall length				▼	165			165	182	
	5500		Steel beam hanger										
	5600		Flange to 8-1/8"				C	280			280	310	
	5650		8-1/8" to 10-1/8"					233			233	256	
	6000		Tie hangers to 24" overall length, 6000 #					320			320	350	
	6100		30" to 36" overall length					370			370	405	
	6500		Tie back hanger, up to 12-1/8" flange				▼	300			300	330	
	8500		Wire, black annealed, 9 ga				Cwt.	83			83	91.50	
	8600		16 ga				"	88			88	97	
122	0010	**ACCESSORIES, INSERTS**											122
	1000		All size nut insert, 5/8" & 3/4", incl. nut	1 Carp	84	.095	Ea.	3.05	2.60		5.65	7.45	
	2000		Continuous slotted, 1-5/8" x 1-3/8"										
	2100		3" long, 12 ga.	1 Carp	65	.123	Ea.	2.70	3.36		6.06	8.25	
	2150		6" long, 12 ga.		65	.123		3.55	3.36		6.91	9.20	
	2200		12" long, 8 ga.		65	.123		8.75	3.36		12.11	14.95	
	2250		24" long, 8 ga.		65	.123		14.10	3.36		17.46	21	
	2300		36" long, 8 ga.		60	.133		19	3.64		22.64	27	
	2350		60" long, 8 ga.	▼	55	.145	▼	32	3.97		35.97	41.50	
	7000		Threaded cast										
	7100		1/4" diameter bolt	1 Carp	84	.095	Ea.	1.40	2.60		4	5.65	
	7350		7/8" diameter bolt	"	84	.095	"	3.50	2.60		6.10	7.95	
	9000		Wedge										
	9050		For 5/8" diameter bolt	1 Carp	60	.133	Ea.	3.15	3.64		6.79	9.20	
	9100		For 3/4" diameter bolt	"	60	.133	"	7	3.64		10.64	13.45	
	9800		Cut washers, "Black"										
	9850		5/8" bolt				Ea.	.20			.20	.22	
	9950		For galvanized inserts, add				C	30%					
124	0010	**ACCESSORIES, REGLET** See division 076-210											124
125	0010	**SLAB TEXTURE STAMPING,** buy											125
	0020		Approx. 3 S.F.- 5 S.F. each, minimum				Ea.	40			40	44	
	0030		Average				"	44			44	48.50	
	0120		Per S.F. of tool, average				S.F.	48			48	53	
	0200		Commonly used chemicals for texture systems										
	0210		Hardeners w/colors average				S.F.	.40			.40	.44	

For expanded coverage of these items see *Means Concrete & Masonry Cost Data 1999*

CONCRETE

3

3 CONCRETE

		031 100	Struct C.I.P. Formwork	CREW	DAILY OUTPUT	LABOR-HOURS	UNIT	MAT.	LABOR	EQUIP.	TOTAL	TOTAL INCL O&P	
								1999 BARE COSTS					
125	0220		Release agents w/colors, average				S.F.	.15			.15	.17	125
	0225		Clear, average					.10			.10	.11	
	0230		Sealers, clear, average					.10			.10	.11	
	0240		Colors, average					.12			.12	.13	
126	0010	**ACCESSORIES, SLEEVES AND CHASES**											126
	0100		Plastic, 1 use, 9" long, 2" diameter	1 Carp	100	.080	Ea.	.55	2.18		2.73	4.05	
	0150		4" diameter		90	.089		1.62	2.43		4.05	5.60	
	0200		6" diameter		75	.107		2.84	2.91		5.75	7.70	
	0250		12" diameter		60	.133		6.65	3.64		10.29	13.05	
128	0010	**ACCESSORIES, SNAP TIES, FLAT WASHER**											128
	0100		3000 lb., to 8"				C	70			70	77	
	0200		11" & 12"					83			83	91	
	0250		16"					88			88	97	
	0300		18"					92			92	101	
	0500		With plastic cone, to 8"					62.50			62.50	68.50	
	0600		11" & 12"					73.50			73.50	81	
	0650		16"					78.50			78.50	86	
	0700		18"					81.50			81.50	90	
	1000		5000 lb., to 8"					100			100	110	
	1150		11" & 12"					114			114	125	
	1200		16"					126			126	138	
	1250		18"					131			131	144	
	1500		With plastic cone, to 8"					107			107	118	
	1600		11" & 12"					126			126	139	
	1650		16"					140			140	154	
	1700		18"					146			146	160	
130	0010	**ACCESSORIES, WALL AND FOUNDATION**											130
	2000		Footings, form braces, solid steel, adjustable				Ea.	6.60			6.60	7.30	
	2050		Spreaders for footer, adjustable				"	3.80			3.80	4.18	
	3000		Form oil, coverage varies greatly, minimum				Gal.	4.50			4.50	4.95	
	3050		Maximum				"	6.70			6.70	7.35	
	3500		Form patches, 1-3/4" diameter				C	8.80			8.80	9.70	
	3550		2-3/4" diameter				"	15.20			15.20	16.70	
	4000		Nail stakes, 3/4" diameter, 18" long				Ea.	.53			.53	.58	
	4050		24" long					.65			.65	.71	
	4200		30" long					.75			.75	.83	
	4250		36" long					.80			.80	.88	
132	0010	**EXPANSION JOINT** Keyed, cold, 24 ga, incl. stakes, 3-1/2" high	1 Carp	200	.040	L.F.	.80	1.09		1.89	2.60	132
	0050		4-1/2" high		200	.040		.90	1.09		1.99	2.71	
	0100		5-1/2" high		195	.041		1.12	1.12		2.24	2.99	
	0150		7-1/2" high		190	.042		1.35	1.15		2.50	3.30	
	0300		Poured asphalt, plain, 1/2" x 1"	1 Clab	450	.018		.31	.38		.69	.94	
	0350		1" x 2"		400	.020		1.13	.43		1.56	1.92	
	0500		Neoprene, liquid, cold applied, 1/2" x 1"		450	.018		1.50	.38		1.88	2.25	
	0550		1" x 2"		400	.020		5.70	.43		6.13	6.95	
	0700		Polyurethane, poured, 2 part, 1/2" x 1"		400	.020		1.75	.43		2.18	2.61	
	0750		1" x 2"		350	.023		6.90	.49		7.39	8.35	
	0900		Rubberized asphalt, hot or cold applied, 1/2" x 1"		450	.018		.58	.38		.96	1.24	
	0950		1" x 2"		400	.020		1.13	.43		1.56	1.92	
	1100		Hot applied, fuel resistant, 1/2" x 1"		450	.018		1	.38		1.38	1.70	
	1150		1" x 2"		400	.020		1.38	.43		1.81	2.20	
	2000		Premolded, bituminous fiber, 1/2" x 6"	1 Carp	375	.021		.36	.58		.94	1.31	
	2050		1" x 12"		300	.027		1.71	.73		2.44	3.03	

Important: See the Reference Section for critical supporting data - Reference Nos., Crews, & City Cost Indexes

			CREW	DAILY OUTPUT	LABOR-HOURS	UNIT	1999 BARE COSTS				TOTAL INCL O&P	
	031 100	**Struct C.I.P. Formwork**					MAT.	LABOR	EQUIP.	TOTAL		
132	2250	Cork with resin binder, 1/2" x 6"	1 Carp	375	.021	L.F.	1.75	.58		2.33	2.85	132
	2300	1" x 12"		300	.027		5.15	.73		5.88	6.80	
	2500	Neoprene sponge, closed cell, 1/2" x 6"		375	.021		1.73	.58		2.31	2.82	
	2550	1" x 12"		300	.027		6.40	.73		7.13	8.15	
	2750	Polyethylene foam, 1/2" x 6"		375	.021		.38	.58		.96	1.34	
	2800	1" x 12"		300	.027		2.10	.73		2.83	3.46	
	3000	Polyethylene backer rod, 3/8" diameter		460	.017		.02	.47		.49	.77	
	3050	3/4" diameter		460	.017		.07	.47		.54	.82	
	3100	1" diameter		460	.017		.13	.47		.60	.89	
	3500	Polyurethane foam, with polybutylene, 1/2" x 1/2"		475	.017		.80	.46		1.26	1.60	
	3550	1" x 1"		450	.018		1.40	.49		1.89	2.30	
	3750	Polyurethane foam, regular, closed cell, 1/2" x 6"		375	.021		.65	.58		1.23	1.63	
	3800	1" x 12"		300	.027		2.40	.73		3.13	3.79	
	4000	Polyvinyl chloride foam, closed cell, 1/2" x 6"		375	.021		1.62	.58		2.20	2.70	
	4050	1" x 12"		300	.027		5.05	.73		5.78	6.70	
	4250	Rubber, gray sponge, 1/2" x 6"		375	.021		2.58	.58		3.16	3.76	
	4300	1" x 12"	▼	300	.027	▼	10.35	.73		11.08	12.55	
	4500	Lead wool for joints, 1 ton lots				Lb.	2			2	2.20	
	4550	Retail				"	2.20			2.20	2.42	
	5000	For installation in walls, add						75%				
	5250	For installation in boxouts, add						25%				
138	0010	**FORMS IN PLACE, BEAMS AND GIRDERS** R031 -040										138
	0020	See also Elevated Slabs, division 031-150										
	0500	Beams & girders, exterior spandrel, plywood, 12" wide, 1 use R031 -050	C-2	225	.213	SFCA	2.29	5.70		7.99	11.45	
	0550	2 use		275	.175		1.19	4.65		5.84	8.60	
	0600	3 use R031 -060		295	.163		.91	4.34		5.25	7.80	
	0650	4 use		310	.155		.74	4.13		4.87	7.30	
	1000	Exterior spandrel, 18" wide, 1 use		250	.192		2.07	5.10		7.17	10.35	
	1050	2 use		275	.175		1.14	4.65		5.79	8.55	
	1100	3 use		305	.157		.83	4.20		5.03	7.50	
	1150	4 use		315	.152		.68	4.06		4.74	7.15	
	1500	Exterior spandrel, 24" wide, 1 use		265	.181		1.88	4.83		6.71	9.65	
	1550	2 use		290	.166		1.06	4.41		5.47	8.10	
	1600	3 use		315	.152		.75	4.06		4.81	7.25	
	1650	4 use		325	.148		.61	3.94		4.55	6.85	
	2000	Interior beam, 12" wide, 1 use		300	.160		2.23	4.27		6.50	9.15	
	2050	2 use		340	.141		1.27	3.76		5.03	7.30	
	2100	3 use		364	.132		.89	3.52		4.41	6.55	
	2150	4 use		377	.127		.73	3.39		4.12	6.15	
	2500	Interior beam, 24" wide, 1 use		320	.150		1.93	4		5.93	8.45	
	2550	2 use		365	.132		1.09	3.51		4.60	6.70	
	2600	3 use		385	.125		.77	3.32		4.09	6.10	
	2650	4 use		395	.122		.62	3.24		3.86	5.80	
	3000	Beam and Girder, encasing steel frame, hung, plywood, 1 use		325	.148		2.23	3.94		6.17	8.65	
	3050	2 use		390	.123		1.22	3.28		4.50	6.50	
	3100	3 use		415	.116		.89	3.08		3.97	5.85	
	3150	4 use		430	.112		.72	2.98		3.70	5.50	
	3500	Beam bottoms only, to 30" wide, plywood, 1 use		230	.209		3.32	5.55		8.87	12.40	
	3550	2 use		265	.181		1.86	4.83		6.69	9.65	
	3600	3 use		280	.171		1.33	4.57		5.90	8.65	
	3650	4 use		290	.166		1.08	4.41		5.49	8.15	
	4000	Beam sides only, vertical, 36" high, plywood, 1 use		335	.143		3.56	3.82		7.38	9.90	
	4050	2 use		405	.119		1.96	3.16		5.12	7.10	
	4100	3 use		430	.112		1.42	2.98		4.40	6.25	
	4150	4 use	▼	445	.108	▼	1.16	2.88		4.04	5.80	

For expanded coverage of these items see *Means Concrete & Masonry Cost Data 1999*

3 CONCRETE

			CREW	DAILY OUTPUT	LABOR-HOURS	UNIT	1999 BARE COSTS				TOTAL INCL O&P
031 100	**Struct C.I.P. Formwork**						MAT.	LABOR	EQUIP.	TOTAL	
138 4500	Sloped sides, 36" high, plywood, 1 use	R031 -040	C-2	305	.157	SFCA	3.50	4.20		7.70	10.45 **138**
4550	2 use			370	.130		1.95	3.46		5.41	7.60
4600	3 use	R031 -050		405	.119		1.40	3.16		4.56	6.50
4650	4 use			425	.113		1.14	3.01		4.15	6
5000	Upstanding beams, 36" high, plywood, 1 use	R031 -060		225	.213		4.21	5.70		9.91	13.60
5050	2 use			255	.188		2.35	5		7.35	10.50
5100	3 use			275	.175		1.70	4.65		6.35	9.15
5150	4 use			280	.171		1.38	4.57		5.95	8.70
142 0010	**FORMS IN PLACE, COLUMNS**	R031 -040									**142**
0500	Round fiberglass, 4 use per mo., rent, 12" diameter		C-1	160	.200	L.F.	2.10	5.15		7.25	10.45
0550	16" diameter	R031 -050		150	.213		2.25	5.50		7.75	11.15
0600	18" diameter			140	.229		2.50	5.90		8.40	12.05
0650	24" diameter	R031 -060		135	.237		3.20	6.15		9.35	13.15
0700	28" diameter			130	.246		3.50	6.35		9.85	13.85
0800	30" diameter			125	.256		4.10	6.60		10.70	14.90
0850	36" diameter			120	.267		4.70	6.90		11.60	16
1500	Round fiber tube, 1 use, 8" diameter			155	.206		1.84	5.35		7.19	10.45
1550	10" diameter			155	.206		2.57	5.35		7.92	11.25
1600	12" diameter			150	.213		3.10	5.50		8.60	12.05
1650	14" diameter			145	.221		3.97	5.70		9.67	13.30
1700	16" diameter			140	.229		5.40	5.90		11.30	15.25
1750	20" diameter			135	.237		8.05	6.15		14.20	18.50
1800	24" diameter			130	.246		9.80	6.35		16.15	21
1850	30" diameter			125	.256		14.25	6.60		20.85	26
1900	36" diameter			115	.278		19.05	7.20		26.25	32.50
1950	42" diameter			100	.320		39	8.25		47.25	55.50
2000	48" diameter			85	.376		48.50	9.75		58.25	69
2200	For seamless type, add						15%				
3000	Round, steel, 4 use per mo., rent, regular duty, 12" diameter		C-1	145	.221	L.F.	2.70	5.70		8.40	11.90
3050	16" diameter			125	.256		2.95	6.60		9.55	13.65
3100	Heavy duty, 20" diameter			105	.305		3.20	7.90		11.10	15.90
3150	24" diameter			85	.376		3.30	9.75		13.05	18.95
3200	30" diameter			70	.457		3.55	11.80		15.35	22.50
3250	36" diameter			60	.533		4.05	13.80		17.85	26
3300	48" diameter			50	.640		6.05	16.55		22.60	32.50
3350	60" diameter			45	.711		7.25	18.40		25.65	37
4000	Column capitals, steel, 4 uses/mo., 24" col, 4' cap diameter			12	2.667	Ea.	17.85	69		86.85	128
4050	5' cap diameter			11	2.909		19.40	75		94.40	140
4100	6' cap diameter			10	3.200		22	82.50		104.50	154
4150	7' cap diameter			9	3.556		24	92		116	172
4500	For second and succeeding months, deduct						50%				
5000	Plywood, 8" x 8" columns, 1 use		C-1	165	.194	SFCA	1.56	5		6.56	9.60
5050	2 use			195	.164		.90	4.24		5.14	7.65
5100	3 use			210	.152		.62	3.94		4.56	6.90
5150	4 use			215	.149		.51	3.85		4.36	6.60
5500	12" x 12" columns, 1 use			180	.178		1.58	4.59		6.17	9
5550	2 use			210	.152		.87	3.94		4.81	7.15
5600	3 use			220	.145		.63	3.76		4.39	6.60
5650	4 use			225	.142		.51	3.68		4.19	6.35
6000	16" x 16" columns, 1 use			185	.173		1.63	4.47		6.10	8.85
6050	2 use			215	.149		.87	3.85		4.72	7
6100	3 use			230	.139		.65	3.60		4.25	6.35
6150	4 use			235	.136		.53	3.52		4.05	6.15
6500	24" x 24" columns, 1 use			190	.168		1.75	4.35		6.10	8.80
6550	2 use			216	.148		.96	3.83		4.79	7.05
6600	3 use			230	.139		.66	3.60		4.26	6.40

Important: See the Reference Section for critical supporting data - Reference Nos., Crews, & City Cost Indexes

031 100 | Struct C.I.P. Formwork

			CREW	DAILY OUTPUT	LABOR-HOURS	UNIT	1999 BARE COSTS				TOTAL INCL O&P	
							MAT.	LABOR	EQUIP.	TOTAL		
142	6650	4 use	C-1	238	.134	SFCA	.57	3.47		4.04	6.10	142
	7000	36" x 36" columns, 1 use	R031-040	200	.160		1.73	4.13		5.86	8.40	
	7050	2 use	R031-050	230	.139		.97	3.60		4.57	6.70	
	7100	3 use		245	.131		.69	3.38		4.07	6.05	
	7150	4 use	R031-060	250	.128		.56	3.31		3.87	5.80	
	7500	Steel framed plywood, 4 use per mo., rent, 8" x 8"		340	.094		2.50	2.43		4.93	6.60	
	7550	10" x 10"		350	.091		2.25	2.36		4.61	6.20	
	7600	12" x 12"		370	.086		2.13	2.23		4.36	5.85	
	7650	16" x 16"		400	.080		2	2.07		4.07	5.45	
	7700	20" x 20"		420	.076		1.90	1.97		3.87	5.20	
	7750	24" x 24"		440	.073		1.80	1.88		3.68	4.94	
	7755	30" x 30"		440	.073	SFCA	1.80	1.88		3.68	4.94	
146	0010	**FORMS IN PLACE, CULVERT** 5' to 8' square or rectangular, 1 use	C-1	170	.188	SFCA	2.63	4.86		7.49	10.55	146
	0050	2 use		180	.178		1.39	4.59		5.98	8.80	
	0100	3 use	R031-050	190	.168		1.09	4.35		5.44	8.05	
	0150	4 use		200	.160		.94	4.13		5.07	7.55	
150	0010	**FORMS IN PLACE, ELEVATED SLABS**	R031-020									150
	0050	See also corrugated form deck, division 053-104										
	1000	Flat plate plywood to 15' high, 1 use	R031-050	470	.102	S.F.	2.49	2.72		5.21	7	
	1050	2 use		520	.092		1.37	2.46		3.83	5.40	
	1100	3 use	R031-060	545	.088		1	2.35		3.35	4.79	
	1150	4 use		560	.086		.81	2.29		3.10	4.48	
	1500	15' to 20' high ceilings, 4 use		495	.097		1.01	2.59		3.60	5.20	
	1600	21' to 35' high ceilings, 4 use		450	.107		1.20	2.84		4.04	5.80	
	2000	Flat slab with drop panels, to 15' high, 1 use		449	.107		2.57	2.85		5.42	7.30	
	2050	2 use		509	.094		1.58	2.51		4.09	5.70	
	2100	3 use		532	.090		1.29	2.41		3.70	5.20	
	2150	4 use		544	.088		1.13	2.35		3.48	4.95	
	2250	15' to 20' high ceilings, 4 use		480	.100		1.37	2.67		4.04	5.70	
	2350	20' to 35' high ceilings, 4 use		435	.110		1.51	2.94		4.45	6.30	
	3000	Floor slab hung from steel beams, 1 use		485	.099		1.58	2.64		4.22	5.90	
	3050	2 use		535	.090		1.13	2.39		3.52	5	
	3100	3 use		550	.087		.98	2.33		3.31	4.74	
	3150	4 use		565	.085		.91	2.26		3.17	4.56	
	3500	Floor slab, with 20" metal pans, 1 use		415	.116		3.45	3.08		6.53	8.65	
	3550	2 use		445	.108		1.73	2.88		4.61	6.40	
	3600	3 use		475	.101		1.15	2.69		3.84	5.50	
	3650	4 use		500	.096		1.20	2.56		3.76	5.35	
	3700	Floor slab with 30" pans, 1 use		418	.115		7.25	3.06		10.31	12.75	
	3720	2 use		455	.105		3.74	2.81		6.55	8.55	
	3740	3 use		470	.102		3.10	2.72		5.82	7.70	
	3760	4 use		480	.100		2.66	2.67		5.33	7.10	
	4000	Floor slab with 19" metal domes, 1 use		405	.119		3.45	3.16		6.61	8.75	
	4050	2 use		435	.110		2.35	2.94		5.29	7.20	
	4100	3 use		465	.103		1.95	2.75		4.70	6.50	
	4150	4 use		495	.097		1.20	2.59		3.79	5.40	
	4500	With 30" fiberglass domes, 1 use		405	.119		4.31	3.16		7.47	9.70	
	4520	2 use		450	.107		3	2.84		5.84	7.75	
	4530	3 use		460	.104		2.46	2.78		5.24	7.10	
	4550	4 use		470	.102		2.09	2.72		4.81	6.60	
	5000	Box out for slab openings, over 16" deep, 1 use		190	.253	SFCA	1.94	6.75		8.69	12.75	
	5050	2 use		240	.200	"	1.07	5.35		6.42	9.55	
	5500	Shallow slab box outs, to 10 S.F.		42	1.143	Ea.	9.75	30.50		40.25	59	
	5550	Over 10 S.F. (use perimeter)		600	.080	L.F.	1.30	2.13		3.43	4.79	
	6000	Bulkhead forms for slab, with keyway, 1 use, 2 piece		500	.096		1.56	2.56		4.12	5.75	
	6100	3 piece (see also edge forms)		460	.104		2.09	2.78		4.87	6.70	

031 100 | Struct C.I.P. Formwork

			CREW	DAILY OUTPUT	LABOR-HOURS	UNIT	1999 BARE COSTS MAT.	LABOR	EQUIP.	TOTAL	TOTAL INCL O&P	
150	6200	Bulkhead forms for slab, w/keyway expanded metal	R031 -020									150
	6210	In lieu of 2 piece form		C-1	1,100	.029	L.F.	1.12	.75		1.87	2.41
	6215	In lieu of 3 piece form	R031 -050		960	.033		1.12	.86		1.98	2.59
	6220	6" high, 4 uses			1,100	.029		1.35	.75		2.10	2.67
	6500	Curb forms, wood, 6" to 12" high, on elevated slabs, 1 use	R031 -060		180	.178	SFCA	1.44	4.59		6.03	8.85
	6550	2 use			205	.156		1.14	4.03		5.17	7.60
	6600	3 use			220	.145		.83	3.76		4.59	6.80
	6650	4 use			225	.142		.68	3.68		4.36	6.55
	7000	Edge forms to 6" high, on elevated slab, 4 use			500	.064	L.F.	.36	1.65		2.01	3
	7500	Depressed area forms to 12" high, 4 use			300	.107		.43	2.76		3.19	4.81
	7550	12" to 24" high, 4 use			175	.183		.58	4.73		5.31	8.10
	8000	Perimeter deck and rail for elevated slabs, straight			90	.356		9.10	9.20		18.30	24.50
	8050	Curved			65	.492		12.50	12.70		25.20	34
	8500	Void forms, round fiber, 3" diameter			450	.071		1.08	1.84		2.92	4.08
	8550	4" diameter			425	.075		1.40	1.95		3.35	4.61
	8600	6" diameter			400	.080		2.13	2.07		4.20	5.60
	8650	8" diameter			375	.085		3.46	2.21		5.67	7.25
	8700	10" diameter			350	.091		4.11	2.36		6.47	8.25
	8750	12" diameter			300	.107		4.95	2.76		7.71	9.80
	8800	Metal end closures, loose, minimum					C	30			30	33
	8850	Maximum					"	154			154	169
154	0010	**FORMS IN PLACE, EQUIPMENT FOUNDATIONS** 1 use		C-2	160	.300	SFCA	1.69	8		9.69	14.45	154
	0050	2 use			190	.253		.93	6.75		7.68	11.60
	0100	3 use			200	.240		.68	6.40		7.08	10.80
	0150	4 use			205	.234		.55	6.25		6.80	10.40
158	0010	**FORMS IN PLACE, FOOTINGS** Continuous wall, plywood, 1 use	R031 -050	C-1	375	.085	SFCA	1.93	2.21		4.14	5.60	158
	0050	2 use			440	.073		1.06	1.88		2.94	4.13
	0100	3 use	R031 -060		470	.068		.77	1.76		2.53	3.62
	0150	4 use			485	.066		.63	1.70		2.33	3.37
	0500	Dowel supports for footings or beams, 1 use			500	.064	L.F.	.64	1.65		2.29	3.30
	1000	Integral starter wall, to 4" high, 1 use			400	.080		1.20	2.07		3.27	4.57
	1500	Keyway, 4 use, tapered wood, 2" x 4"		1 Carp	530	.015		.19	.41		.60	.86
	1550	2" x 6"			500	.016		.27	.44		.71	.99
	2000	Tapered plastic, 2" x 3"			530	.015		.50	.41		.91	1.20
	2050	2" x 4"			500	.016		.65	.44		1.09	1.40
	2250	For keyway hung from supports, add			150	.053		.82	1.46		2.28	3.19
	3000	Pile cap, square or rectangular, plywood, 1 use		C-1	290	.110	SFCA	1.85	2.85		4.70	6.55
	3050	2 use			346	.092		1.02	2.39		3.41	4.88
	3100	3 use			371	.086		.74	2.23		2.97	4.33
	3150	4 use			383	.084		.60	2.16		2.76	4.06
	4000	Triangular or hexagonal caps, plywood, 1 use			225	.142		2.19	3.68		5.87	8.20
	4050	2 use			280	.114		1.20	2.95		4.15	5.95
	4100	3 use			305	.105		.87	2.71		3.58	5.20
	4150	4 use			315	.102		.71	2.63		3.34	4.91
	5000	Spread footings, plywood, 1 use			305	.105		1.49	2.71		4.20	5.90
	5050	2 use			371	.086		.82	2.23		3.05	4.41
	5100	3 use			401	.080		.60	2.06		2.66	3.90
	5150	4 use			414	.077		.49	2		2.49	3.68
	6000	Supports for dowels, plinths or templates, 2' x 2'			25	1.280	Ea.	3.10	33		36.10	55.50
	6050	4' x 4' footing			22	1.455		6.50	37.50		44	66
	6100	8' x 8' footing			20	1.600		13	41.50		54.50	79.50
	6150	12' x 12' footing			17	1.882		21	48.50		69.50	99.50
	7000	Plinths, 1 use			250	.128	SFCA	2.40	3.31		5.71	7.85
	7100	4 use			270	.119	"	.74	3.06		3.80	5.65

Important: See the Reference Section for critical supporting data - Reference Nos., Crews, & City Cost Indexes

031 100 | Struct C.I.P. Formwork

			CREW	DAILY OUTPUT	LABOR-HOURS	UNIT	1999 BARE COSTS MAT.	LABOR	EQUIP.	TOTAL	TOTAL INCL O&P	
162	0010	**FORMS IN PLACE, GRADE BEAM** Plywood, 1 use	R031 -050	C-2	530	.091	SFCA	1.36	2.41		3.77	5.30	162
	0050	2 use			580	.083		.75	2.21		2.96	4.29	
	0100	3 use	R031 -060		600	.080		.54	2.13		2.67	3.96	
	0150	4 use			605	.079		.44	2.12		2.56	3.82	
166	0010	**FORMS IN PLACE, MAT FOUNDATION** 1 use		C-2	290	.166	SFCA	1.49	4.41		5.90	8.60	166
	0050	2 use			310	.155		.82	4.13		4.95	7.40	
	0100	3 use			330	.145		.58	3.88		4.46	6.75	
	0120	4 use			350	.137		.48	3.66		4.14	6.30	
170	0010	**FORMS IN PLACE, SLAB ON GRADE**											170
	1000	Bulkhead forms with keyway, wood, 1 use, 2 piece		C-1	510	.063	L.F.	.65	1.62		2.27	3.26	
	1050	3 piece			400	.080		.80	2.07		2.87	4.13	
	1100	4 piece			350	.091		1.05	2.36		3.41	4.87	
	1400	Bulkhead forms w/keyway, 1 piece expanded metal, left in place											
	1410	In lieu of 2 piece form		C-1	1,375	.023	L.F.	1.12	.60		1.72	2.18	
	1420	In lieu of 3 piece form			1,200	.027		1.12	.69		1.81	2.31	
	1430	In lieu of 4 piece form			1,050	.030		1.12	.79		1.91	2.47	
	2000	Curb forms, wood, 6" to 12" high, on grade, 1 use			215	.149	SFCA	1.17	3.85		5.02	7.35	
	2050	2 use			250	.128		.65	3.31		3.96	5.90	
	2100	3 use			265	.121		.47	3.12		3.59	5.45	
	2150	4 use			275	.116		.38	3.01		3.39	5.15	
	3000	Edge forms, wood, 4 use, on grade, to 6" high			600	.053	L.F.	.24	1.38		1.62	2.43	
	3050	7" to 12" high			435	.074	SFCA	.63	1.90		2.53	3.68	
	3500	For depressed slabs, 4 use, to 12" high			300	.107	L.F.	.61	2.76		3.37	5	
	3550	To 24" high			175	.183		.79	4.73		5.52	8.30	
	4000	For slab blockouts, to 12" high, 1 use			200	.160		.62	4.13		4.75	7.20	
	4050	To 24" high, 1 use			120	.267		.78	6.90		7.68	11.70	
	4100	Plastic (extruded), to 6" high, multiple use, on grade			800	.040		.22	1.03		1.25	1.87	
	5000	Screed, 24 ga. metal key joint, see Div 031-132											
	5020	Wood, incl. wood stakes, 1" x 3"		C-1	900	.036	L.F.	.40	.92		1.32	1.89	
	5050	2" x 4"			900	.036	"	1.15	.92		2.07	2.72	
	6000	Trench forms in floor, wood, 1 use			160	.200	SFCA	1.90	5.15		7.05	10.25	
	6050	2 use			175	.183		1.13	4.73		5.86	8.70	
	6100	3 use			180	.178		.82	4.59		5.41	8.15	
	6150	4 use			185	.173		.67	4.47		5.14	7.80	
174	0010	**FORMS IN PLACE, STAIRS** (Slant length x width), 1 use	R031 -050	C-2	165	.291	S.F.	3.58	7.75		11.33	16.15	174
	0050	2 use			170	.282		1.97	7.55		9.52	14	
	0100	3 use	R031 -060		180	.267		1.43	7.10		8.53	12.75	
	0150	4 use			190	.253		1	6.75		7.75	11.70	
	1000	Alternate pricing method (0.7 L.F./S.F.), 1 use			100	.480	LF Rsr	4.40	12.80		17.20	25	
	1050	2 use			105	.457		2.60	12.20		14.80	22	
	1100	3 use			110	.436		2.05	11.65		13.70	20.50	
	1150	4 use			115	.417		1.75	11.15		12.90	19.45	
	2000	Stairs, cast on sloping ground (length x width), 1 use			220	.218	S.F.	2.25	5.80		8.05	11.65	
	2100	4 use			240	.200	"	1.25	5.35		6.60	9.80	
182	0010	**FORMS IN PLACE, WALLS**	R031 -010										182
	0100	Box out for wall openings, to 16" thick, to 10 S.F.		C-2	24	2	Ea.	19.60	53.50		73.10	106	
	0150	Over 10 S.F. (use perimeter)	R031 -050	"	280	.171	L.F.	1.76	4.57		6.33	9.15	
	0250	Brick shelf, 4" w, add to wall forms, use wall area abv shelf											
	0260	1 use	R031 -060	C-2	240	.200	SFCA	1.98	5.35		7.33	10.60	
	0300	2 use			275	.175		1.09	4.65		5.74	8.50	
	0350	4 use			300	.160		.79	4.27		5.06	7.55	
	0500	Bulkhead forms, with keyway, 1 use, 2 piece			265	.181	L.F.	2.50	4.83		7.33	10.35	
	0550	3 piece			175	.274	"	3.10	7.30		10.40	14.90	
	0600	Bulkhead forms w/keyway, 1 piece expanded metal, left in place											
	0610	In lieu of 2 piece form		C-1	800	.040	L.F.	1.05	1.03		2.08	2.78	
	0620	In lieu of 3 piece form		"	525	.061	"	1.05	1.58		2.63	3.63	

For expanded coverage of these items see *Means Concrete & Masonry Cost Data 1999*

031 100 | Struct C.I.P. Formwork

		CREW	DAILY OUTPUT	LABOR-HOURS	UNIT	MAT.	LABOR	EQUIP.	TOTAL	TOTAL INCL O&P
0700	Buttress forms, to 8' high, 1 use	C-2	350	.137	SFCA	2.97	3.66		6.63	9
0750	2 use		430	.112		1.63	2.98		4.61	6.50
0800	3 use		460	.104		1.19	2.78		3.97	5.70
0850	4 use		480	.100		.98	2.67		3.65	5.25
1000	Corbel (haunch) forms, to 12" wide, add to wall forms, 1 use		150	.320	L.F.	1.61	8.55		10.16	15.15
1050	2 use		170	.282		.88	7.55		8.43	12.80
1100	3 use		175	.274		.64	7.30		7.94	12.20
1150	4 use		180	.267		.52	7.10		7.62	11.75
2000	Job built plyform wall forms, to 8' high, 1 use, below grade		300	.160	SFCA	1.89	4.27		6.16	8.75
2050	2 use, below grade		365	.132		1.05	3.51		4.56	6.65
2100	3 use, below grade		425	.113		.75	3.01		3.76	5.55
2150	4 use, below grade		435	.110		.62	2.94		3.56	5.30
2400	Over 8' to 16' high, 1 use		280	.171		3.60	4.57		8.17	11.15
2450	2 use		345	.139		1.16	3.71		4.87	7.15
2500	3 use		375	.128		.83	3.41		4.24	6.25
2550	4 use		395	.122		.68	3.24		3.92	5.85
2700	Over 16' high, 1 use		235	.204		2.34	5.45		7.79	11.10
2750	2 use		290	.166		1.29	4.41		5.70	8.35
2800	3 use		315	.152		.93	4.06		4.99	7.45
2850	4 use		330	.145		.76	3.88		4.64	6.95
3000	For architectural finish, add		1,820	.026		.58	.70		1.28	1.75
3500	Polystyrene (expanded) wall forms									
3510	To 8' high, 1 use, left in place	1 Carp	295	.027	SFCA	1.55	.74		2.29	2.86
4000	Radial wall forms, smooth curved, 1 use	C-2	245	.196		2.30	5.20		7.50	10.75
4050	2 use		300	.160		1.26	4.27		5.53	8.10
4100	3 use		325	.148		.92	3.94		4.86	7.20
4150	4 use		335	.143		.75	3.82		4.57	6.80
4200	Wall forms, smooth curved, below grade, job built plyform, 1 use		225	.213		2.78	5.70		8.48	12
4210	2 use		225	.213		1.54	5.70		7.24	10.65
4220	3 use		225	.213		1.25	5.70		6.95	10.30
4230	4 use		225	.213		.90	5.70		6.60	9.95
4300	Curved, with 2' chords, 1 use		290	.166		1.89	4.41		6.30	9
4350	2 use		355	.135		1.04	3.60		4.64	6.80
4400	3 use		385	.125		.75	3.32		4.07	6.10
4450	4 use		400	.120		.61	3.20		3.81	5.70
4500	Over 8' high, 1 use		290	.166		.86	4.41		5.27	7.90
4525	2 use		355	.135		.48	3.60		4.08	6.15
4550	3 use		385	.125		.35	3.32		3.67	5.65
4575	4 use		400	.120		.29	3.20		3.49	5.35
4600	Retaining wall forms, battered, to 8' high, 1 use		300	.160		1.78	4.27		6.05	8.65
4650	2 use		355	.135		.98	3.60		4.58	6.75
4700	3 use		375	.128		.71	3.41		4.12	6.15
4750	4 use		390	.123		.54	3.28		3.82	5.75
4900	Over 8' to 16' high, 1 use		240	.200		1.94	5.35		7.29	10.55
4950	2 use		295	.163		1.07	4.34		5.41	7.95
5000	3 use		305	.157		.78	4.20		4.98	7.45
5050	4 use		320	.150		.63	4		4.63	7
5500	For gang wall forming, 192 S.F. sections, deduct					10%	10%			
5550	384 S.F. sections, deduct					20%	20%			
5750	Liners for forms (add to wall forms), A.B.S. plastic									
5800	Aged wood, 4" wide, 1 use	1 Carp	250	.032	SFCA	4.75	.87		5.62	6.55
5820	2 use		400	.020		2.63	.55		3.18	3.75
5840	4 use		750	.011		1.60	.29		1.89	2.22
5900	Fractured rope rib, 1 use		250	.032		7.40	.87		8.27	9.50
6000	4 use		750	.011		2.45	.29		2.74	3.16
6100	Ribbed look, 1/2" & 3/4" deep, 1 use		300	.027		5.15	.73		5.88	6.80

R031-010
R031-050
R031-060

3 CONCRETE

Important: See the Reference Section for critical supporting data - Reference Nos., Crews, & City Cost Indexes

				DAILY	LABOR-			1999 BARE COSTS				TOTAL	
031 100	**Struct C.I.P. Formwork**		CREW	OUTPUT	HOURS	UNIT	MAT.	LABOR	EQUIP.	TOTAL		INCL O&P	
182	6200	4 use	R031 -010	1 Carp	800	.010	SFCA	1.65	.27		1.92	2.25	**182**
	6300	Rustic brick pattern, 1 use			250	.032		4.80	.87		5.67	6.65	
	6400	4 use	R031 -050		750	.011		1.60	.29		1.89	2.22	
	6500	Striated, random, 3/8" x 3/8" deep, 1 use			300	.027		5.10	.73		5.83	6.75	
	6600	4 use	R031 -060	▼	800	.010	▼	1.65	.27		1.92	2.25	
	6800	Rustication strips, A.B.S. plastic, 2 piece snap-on											
	6850	1" deep x 1-3/8" wide, 1 use		C-2	400	.120	L.F.	3.75	3.20		6.95	9.20	
	6900	2 use			600	.080		2.10	2.13		4.23	5.65	
	6950	4 use			800	.060		1.25	1.60		2.85	3.90	
	7050	Wood, beveled edge, 3/4" deep, 1 use			600	.080		.22	2.13		2.35	3.60	
	7100	1" deep, 1 use			450	.107	▼	.35	2.84		3.19	4.86	
	7200	For solid board finish, uniform, 1 use, add to wall forms			300	.160	SFCA	.74	4.27		5.01	7.50	
	7300	Non-uniform finish		▼	250	.192		.65	5.10		5.75	8.75	
	7500	Lintel or sill forms, 1 use		1 Carp	30	.267		2.30	7.30		9.60	14	
	7520	2 use			34	.235		1.26	6.40		7.66	11.50	
	7540	3 use			36	.222		.92	6.05		6.97	10.55	
	7560	4 use		▼	37	.216		.75	5.90		6.65	10.10	
	7800	Modular prefabricated plywood, to 8' high, 1 use per month		C-2	910	.053		.97	1.41		2.38	3.27	
	7820	2 use per month			930	.052		.53	1.38		1.91	2.75	
	7840	3 use per month			950	.051		.39	1.35		1.74	2.55	
	7860	4 use per month			970	.049		.31	1.32		1.63	2.42	
	8000	To 16' high, 1 use per month			550	.087		1.23	2.33		3.56	5	
	8020	2 use per month			570	.084		.68	2.25		2.93	4.28	
	8040	3 use per month			590	.081		.49	2.17		2.66	3.95	
	8060	4 use per month			610	.079		.40	2.10		2.50	3.74	
	8100	Over 16' high, 1 use per month			550	.087		1.48	2.33		3.81	5.30	
	8120	2 use per month			570	.084		.82	2.25		3.07	4.43	
	8140	3 use per month			590	.081		.59	2.17		2.76	4.06	
	8160	4 use per month			610	.079		.48	2.10		2.58	3.83	
	8600	Pilasters, 1 use			270	.178		2.20	4.74		6.94	9.85	
	8620	2 use			330	.145		1.21	3.88		5.09	7.45	
	8640	3 use			370	.130		.88	3.46		4.34	6.40	
	8660	4 use			385	.125		.72	3.32		4.04	6.05	
	9000	Steel framed plywood, to 8' high, 1 use per month			600	.080		1.65	2.13		3.78	5.20	
	9020	2 use per month			640	.075		.91	2		2.91	4.15	
	9040	3 use per month			655	.073		.66	1.95		2.61	3.80	
	9060	4 use per month			665	.072		.54	1.92		2.46	3.62	
	9200	Over 8' to 16' high, 1 use per month			455	.105		1.80	2.81		4.61	6.40	
	9220	2 use per month			505	.095		.99	2.53		3.52	5.10	
	9240	3 use per month			525	.091		.72	2.44		3.16	4.62	
	9260	4 use per month			530	.091		.59	2.41		3	4.45	
	9400	Over 16' to 20' high, 1 use per month			425	.113		1.65	3.01		4.66	6.55	
	9420	2 use per month			435	.110		.91	2.94		3.85	5.65	
	9440	3 use per month			455	.105		.66	2.81		3.47	5.15	
	9460	4 use per month		▼	465	.103	▼	.54	2.75		3.29	4.92	
	9475	For elevated walls, add							10%				
	9480	For battered walls, 1 side battered, add						10%	10%				
	9485	For battered walls, 2 sides battered, add		▼				15%	15%				
186	0010	**GAS STATION FORMS** Curb fascia, with template,											**186**
	0050	12 ga. steel, left in place, 9" high		1 Carp	50	.160	L.F.	6.80	4.37		11.17	14.35	
	1000	Sign or light bases, 18" diameter, 9" high			9	.889	Ea.	38	24.50		62.50	80	
	1050	30" diameter, 13" high		▼	8	1		64.50	27.50		92	114	
	2000	Island forms, 10' long, 9" high, 3'- 6" wide		C-1	10	3.200		175	82.50		257.50	325	
	2050	4' wide			9	3.556		190	92		282	355	
	2500	20' long, 9" high, 4' wide			6	5.333		310	138		448	555	
	2550	5' wide		▼	5	6.400	▼	325	165		490	620	

031 100	Struct C.I.P. Formwork	CREW	DAILY OUTPUT	LABOR-HOURS	UNIT	1999 BARE COSTS				TOTAL INCL O&P	
						MAT.	LABOR	EQUIP.	TOTAL		
190	0010 SCAFFOLDING See division 015-254										190
192	0010 SHORES Erect and strip, by hand, horizontal members										192
	0500 Aluminum joists and stringers	2 Carp	60	.267	Ea.		7.30		7.30	11.45	
	0600 Steel, adjustable beams		45	.356			9.70		9.70	15.25	
	0700 Wood joists		50	.320			8.75		8.75	13.75	
	0800 Wood stringers		30	.533			14.55		14.55	23	
	1000 Vertical members to 10' high		55	.291			7.95		7.95	12.50	
	1050 To 13' high		50	.320			8.75		8.75	13.75	
	1100 To 16' high		45	.356			9.70		9.70	15.25	
	1500 Reshoring		1,400	.011	S.F.	.17	.31		.48	.68	
	1600 Flying truss system	C-17D	9,600	.009	SFCA		.25	.06	.31	.45	
	1760 Horizontal, aluminum joists, 6' to 30' spans, buy				L.F.	10.75			10.75	11.80	
	1770 Aluminum stringers, 12' & 16' spans				"	16.50			16.50	18.15	
	1810 Horizontal, steel beam, adjustable, 4' to 7' span				Ea.	110			110	121	
	1830 6' to 10' span					142			142	156	
	1920 9' to 15' span					260			260	286	
	1940 12' to 20' span					300			300	330	
	1970 Steel stringer, 6' to 15' span				L.F.	8.10			8.10	8.90	
	3000 Rent for job duration, aluminum, first month				SF Flr.	.25			.25	.28	
	3050 Steel				"	.20			.20	.22	
	3500 Vertical, adjustable steel, 5'-7" to 9'-6" high, 10,000# cap., buy				Ea.	59			59	65	
	3550 7'-3" to 12'-10" high, 7800# capacity					71			71	78	
	3600 8'-10" to 12'-4" high, 10,000# capacity					79			79	87	
	3650 8'-10" to 16'-1" high, 3800# capacity					86			86	94.50	
	4000 Frame shoring systems, aluminum, 10,000# per leg,										
	4050 6' wide, 5' & 6' high				Ea.	235			235	259	
	4100 5' to 7' post with base, jack screw & top plate				Set	52			52	57	
	5010 Steel, 10,000# per leg										
	5040 2' & 4' wide, 3', 4', 5' & 6' high				Ea.	97			97	107	
	5250 6' extension tube with adjusting collar					91			91	100	
	5550 Base plate					9.75			9.75	10.75	
	5600 12" adjustable leg					39			39	43	
	5650 Top plate					21			21	23	
	5750 Flying truss system				SFCA	9			9	9.90	
196	0010 SLIPFORMS Silos, minimum [R031 -030]	C-17E	3,885	.021	"	1	.59	.02	1.61	2.04	196
	0050 Maximum		1,095	.073	SFCA	1.40	2.08	.06	3.54	4.88	
	1000 Buildings, minimum		3,660	.022		1.20	.62	.02	1.84	2.32	
	1050 Maximum		875	.091		2.10	2.60	.07	4.77	6.50	
198	0010 WATERSTOP PVC, ribbed 3/16" thick, 4" wide	1 Carp	155	.052	L.F.	.74	1.41		2.15	3.03	198
	0050 6" wide		145	.055		1.28	1.51		2.79	3.78	
	0500 Ribbed, PVC, with center bulb, 9" wide, 3/16" thick		135	.059		1.76	1.62		3.38	4.49	
	0550 3/8" thick		130	.062		2.20	1.68		3.88	5.05	
	0800 Dumbbell type, PVC, 6" wide, 3/16" thick		150	.053		.92	1.46		2.38	3.30	
	0850 3/8" thick		145	.055		1.80	1.51		3.31	4.36	
	1000 9" wide, 3/8" thick, PVC, plain		130	.062		3.11	1.68		4.79	6.05	
	1050 Center bulb		130	.062		5.60	1.68		7.28	8.80	
	1250 Split PVC, 3/8" thick, 6" wide		145	.055		4.65	1.51		6.16	7.45	
	1300 9" wide		130	.062		6.85	1.68		8.53	10.15	
	2000 Rubber, flat dumbbell, 3/8" thick, 6" wide		145	.055		4.70	1.51		6.21	7.50	
	2050 9" wide		135	.059		7.75	1.62		9.37	11.10	
	2500 Flat dumbbell split, 3/8" thick, 6" wide		145	.055		6.75	1.51		8.26	9.80	
	2550 9" wide		135	.059		10	1.62		11.62	13.55	
	3000 Center bulb, 1/4" thick, 6" wide		145	.055		5.20	1.51		6.71	8.05	
	3050 9" wide		135	.059		9.50	1.62		11.12	13	

Important: See the Reference Section for critical supporting data - Reference Nos., Crews, & City Cost Indexes

031 | Concrete Formwork

031 100	Struct C.I.P. Formwork	CREW	DAILY OUTPUT	LABOR-HOURS	UNIT	1999 BARE COSTS				TOTAL INCL O&P	
						MAT.	LABOR	EQUIP.	TOTAL		
198	3500	Center bulb split, 3/8" thick, 6" wide	1 Carp	145	.055	L.F.	6.15	1.51		7.66	9.10	198
	3550	9" wide	↓	135	.059	↓	10.90	1.62		12.52	14.55	
	5000	Waterstop fittings, rubber, flat										
	5010	Dumbbell or center bulb, 3/8" thick,										
	5200	Field union, 6" wide	1 Carp	50	.160	Ea.	9.25	4.37		13.62	17.05	
	5250	9" wide		50	.160		13.25	4.37		17.62	21.50	
	5500	Flat cross, 6" wide		30	.267		34	7.30		41.30	49	
	5550	9" wide		30	.267		57	7.30		64.30	74	
	6000	Flat tee, 6" wide		30	.267		33.50	7.30		40.80	48.50	
	6050	9" wide		30	.267		47.50	7.30		54.80	64	
	6500	Flat ell, 6" wide		40	.200		32	5.45		37.45	43.50	
	6550	9" wide		40	.200		46.50	5.45		51.95	59.50	
	7000	Vertical tee, 6" wide		25	.320		28.50	8.75		37.25	45.50	
	7050	9" wide		25	.320		43.50	8.75		52.25	62	
	7500	Vertical ell, 6" wide		35	.229		25	6.25		31.25	37.50	
	7550	9" wide	↓	35	.229	↓	43	6.25		49.25	57.50	

032 | Concrete Reinforcement

032 100	Reinforcing Steel	CREW	DAILY OUTPUT	LABOR-HOURS	UNIT	1999 BARE COSTS				TOTAL INCL O&P	
						MAT.	LABOR	EQUIP.	TOTAL		
102	0010	**ACCESSORIES** Materials only										102
	0020	See also Form Accessories, division 031										
	0100	Beam bolsters, (BB) standard, lower, up to 1-1/2" high, plain				C.L.F.	33			33	36.50	
	0102	Galvanized					51			51	56	
	0104	Stainless					97			97	107	
	0106	Plastic					58			58	64	
	0108	Epoxy					80			80	88	
	0110	2-1/2" to 3" high, plain					144			144	158	
	0120	Galvanized					160			160	176	
	0140	Stainless					248			248	273	
	0160	Plastic					124			124	136	
	0162	Epoxy					211			211	232	
	0200	Upper, standard (BBU) to 1-1/2" high, plain					96			96	106	
	0210	2-1/2" to 3" high					173			173	190	
	0300	Beam bolster with plate (BBP) to 1-1/2" high, plain					144			144	158	
	0310	2-1/2" to 3" high					221			221	243	
	0500	Slab bolsters, continuous, plain (SB) 3/4" to 1" high, plain					29			29	32	
	0502	Galvanized					31			31	34	
	0504	Stainless					43			43	47.50	
	0506	Plastic					38			38	42	
	0510	1" to 2" high, plain					34			34	37.50	
	0515	Galvanized					400			400	440	
	0520	Stainless					760			760	835	
	0525	Plastic					450			450	495	
	0530	For bolsters with wire runners (SBR), add					400			400	440	
	0540	For bolsters with plates (SBP), add				↓	1,000			1,000	1,100	
	0700	Clip or bar ties, 16 ga., plain, 3" long				C	8.50			8.50	9.35	
	0710	4" long					9			9	9.90	
	0720	6" long					10.50			10.50	11.55	
	0730	8" long				↓	11			11	12.10	

032 100	Reinforcing Steel	CREW	DAILY OUTPUT	LABOR-HOURS	UNIT	MAT.	LABOR	EQUIP.	TOTAL	TOTAL INCL O&P	
						1999 BARE COSTS					
102	0900	Flange clips, expandable flanges, 10 ga., 12" O.C., continuous,										102
0910	galvanized, over 500 L.F., 4" to 8"				C.L.F.	31			31	34
0920	9" to 12"					42			42	46
0930	17" to 24"					44			44	48.50
1200	High chairs, individual, no plates (1 HC), to 3" high, plain				C	425			425	470
1202	Galvanized					57			57	62.50
1204	Stainless					118			118	130
1206	Plastic					55			55	60.50
1210	5" high, plain					62.50			62.50	69
1212	Galvanized					71			71	78
1214	Stainless					140			140	154
1216	Plastic					76			76	83.50
1220	8" high, plain					136			136	150
1222	Galvanized					174			174	191
1224	Stainless					288			288	315
1226	Plastic					171			171	188
1230	12" high, plain					284			284	310
1232	Galvanized					345			345	380
1234	Stainless					520			520	575
1236	Plastic					320			320	350
1240	15" high, plain					555			555	610
1242	Galvanized					590			590	650
1244	Stainless					875			875	960
1246	Plastic					575			575	635
1250	For each added 1" up to 24" high, plain, add					33			33	36.50
1252	Galvanized, add					38			38	42
1254	Stainless, add					45			45	49.50
1256	Plastic, add					38			38	42
1400	Individual high chairs, with plates, (HCP), to 5" high, add					110			110	121
1410	Over 5" high, add					121			121	133
1500	Bar chair (BC) for up to 1-3/4" high, plain					23			23	25.50
1520	Galvanized					26			26	28.50
1530	Stainless					69			69	76
1540	Plastic					45			45	49.50
1550	Joist chair (JC) for up to 1-1/2" high, plain					28			28	31
1580	Galvanized					34			34	37.50
1600	Stainless					49			49	54
1620	Plastic					49			49	54
1630	Epoxy					113			113	124
1700	Continuous high chairs, legs 8" O.C. (CHC) to 4" high, plain				C.L.F.	55			55	60.50
1705	Galvanized					64			64	70.50
1710	Stainless					136			136	150
1715	Plastic					69.50			69.50	76.50
1718	Epoxy					98			98	108
1720	6" high, plain					84			84	92.50
1725	Galvanized					108			108	119
1730	Stainless					159			159	175
1735	Plastic					108			108	119
1738	Epoxy					131			131	144
1740	8" high, plain					115			115	127
1745	Galvanized					151			151	166
1750	Stainless					189			189	208
1755	Plastic					150			150	165
1758	Epoxy					198			198	218
1760	12" high, plain					281			281	310
1765	Galvanized					325			325	355

Important: See the Reference Section for critical supporting data - Reference Nos., Crews, & City Cost Indexes

032 100	Reinforcing Steel	CREW	DAILY OUTPUT	LABOR-HOURS	UNIT	1999 BARE COSTS				TOTAL INCL O&P
						MAT.	LABOR	EQUIP.	TOTAL	
102	1770	Stainless				C.L.F.	450			450	495	102
1775	Plastic					320			320	355
1778	Epoxy					470			470	515
1780	15" high, plain					315			315	345
1785	Galvanized					365			365	400
1790	Stainless					360			360	395
1795	Plastic					360			360	395
1798	Epoxy					615			615	680
1800	For each added 1" up to 24" high, plain, add					26			26	28.50
1820	Galvanized, add					31			31	34
1840	Stainless, add					28			28	31
1860	Plastic, add					28			28	31
1900	For continuous bottom plate, (CHCP), add					123			123	135
1940	For upper continuous high chairs, (CHCU), add					123			123	135
1960	For galvanized wire runners, add					123			123	135
2100	Paper tubing, 4' lengths, for #2 & #3 bar					48			48	53
2120	For #6 bar				▼	66			66	72.50
2200	Screed base, 1/2" diameter, 2-1/2" high, plain				C	133			133	146
2210	Galvanized					138			138	152
2220	5-1/2" high, plain					160			160	176
2250	Galvanized					168			168	185
2300	3/4" diameter, 2-1/2" high, plain					167			167	184
2310	Galvanized					175			175	193
2320	5-1/2" high, plain					206			206	227
2350	Galvanized					220			220	242
2400	Screed holder, 1/2" diam. for 1" I.D. pipe, plain, 6" long					139			139	153
2420	12" long					230			230	253
2500	3/4" diameter, for 1-1/2" I.D. pipe, 6" long					250			250	275
2520	12" long					410			410	455
2700	Screw anchor for bolts, plain, 1/2" diameter					90			90	99
2720	1" diameter					271			271	298
2740	1-1/2" diameter					450			450	495
2800	Screw eye bolts, 1/2" x 5" long					1,100			1,100	1,200
2820	1" x 9" long					4,000			4,000	4,425
2840	1-1/2" x 14" long					10,200			10,200	11,200
2900	Screw anchor bolts, 1/2" x up to 7" long					420			420	460
2920	1" x up to 12" long					1,375			1,375	1,500
3000	Slab lifting inserts, single, 3/4" dia., galv., 4" high					280			280	310
3010	6" high					340			340	375
3030	7" high					390			390	430
3100	1" diameter, 5" high					440			440	485
3120	7" high					465			465	510
3200	Double lifting inserts, 1" diameter, 5" high					875			875	965
3220	7" high					925			925	1,025
3330	1-1/4" diameter, 5" high				▼	950			950	1,050
3500	Sleeper clips for wood sleepers, 20 ga., galv., 2" wide				M	330			330	365
3520	4" wide					410			410	450
3600	Spacers, plastic for 1" bar clearance, average					48			48	53
3620	For 2" bar clearance, average				▼	58			58	64
3800	Subgrade chairs, 1/2" diameter, 3-1/2" high				C	270			270	297
3850	12" high					770			770	845
3900	3/4" diameter, 3-1/2" high					350			350	385
3950	12" high					840			840	925
4200	Subgrade stakes, 3/4" diameter, 12" long					277			277	305
4250	24" long					375			375	415
4300	1" diameter, 12" long				▼	420			420	465

CONCRETE 3

032 100 | Reinforcing Steel

		CREW	DAILY OUTPUT	LABOR-HOURS	UNIT	1999 BARE COSTS				TOTAL INCL O&P
						MAT.	LABOR	EQUIP.	TOTAL	
102										**102**
4350	24" long				C	630			630	690
4500	Tie wire, 16 ga. annealed steel, under 500 lbs.				Cwt.	80			80	88
4520	2,000 to 4,000 lbs.				"	75			75	82.50
4550	Tie wire holder, plastic case				Ea.	31			31	34
4600	Aluminum case				"	36			36	39.50
104	0010	**COATED REINFORCING** Add to material								**104**
0100	Epoxy coated, A775				Cwt.	23.50			23.50	26
0150	Galvanized, #3					31.50			31.50	34.50
0200	#4					31.50			31.50	34.50
0250	#5					31			31	34
0300	#6 or over					31			31	34
1000	For over 20 tons, #6 or larger, minimum					28.50			28.50	31.50
1500	Maximum					34.50			34.50	38
107	0010	**REINFORCING IN PLACE** A615 Grade 60								**107**
0100	Beams & Girders, #3 to #7 **CN** R032-010	4 Rodm	1.60	20	Ton	535	610		1,145	1,650
0150	#8 to #18 R032-020		2.70	11.852		525	360		885	1,200
0200	Columns, #3 to #7		1.50	21.333		535	650		1,185	1,725
0250	#8 to #18 R032-040		2.30	13.913		525	425		950	1,300
0300	Spirals, hot rolled, 8" to 15" diameter		2.20	14.545		1,075	440		1,515	1,950
0320	15" to 24" diameter R032-050		2.20	14.545		1,025	440		1,465	1,900
0330	24" to 36" diameter		2.30	13.913		1,000	425		1,425	1,825
0340	36" to 48" diameter R032-060		2.40	13.333		1,025	405		1,430	1,825
0360	48" to 64" diameter		2.50	12.800		1,075	390		1,465	1,850
0380	64" to 84" diameter R032-080		2.60	12.308		1,125	375		1,500	1,900
0390	84" to 96" diameter		2.70	11.852		1,125	360		1,485	1,875
0400	Elevated slabs, #4 to #7		2.90	11.034		570	335		905	1,200
0500	Footings, #4 to #7		2.10	15.238		510	465		975	1,350
0550	#8 to #18		3.60	8.889		480	270		750	1,000
0600	Slab on grade, #3 to #7		2.30	13.913		510	425		935	1,300
0700	Walls, #3 to #7		3	10.667		510	325		835	1,125
0750	#8 to #18		4	8		510	243		753	980
1000	Typical in place, 10 ton lots, average		1.70	18.824		535	570		1,105	1,575
1100	Over 50 ton lots, average		2.30	13.913		515	425		940	1,300
1200	High strength steel, Grade 75, #14 bars only, add					60			60	66
2000	Unloading & sorting, add to above	C-5	100	.560			16.60	5.50	22.10	34
2200	Crane cost for handling, add to above, minimum		135	.415			12.30	4.07	16.37	25
2210	Average		92	.609			18.05	5.95	24	36.50
2220	Maximum		35	1.600			47.50	15.70	63.20	97
2400	Dowels, 2 feet long, deformed, #3	2 Rodm	520	.031	Ea.	.22	.94		1.16	1.86
2410	#4		480	.033		.40	1.01		1.41	2.18
2420	#5		435	.037		.62	1.12		1.74	2.61
2430	#6		360	.044		.89	1.35		2.24	3.32
2450	Longer and heavier dowels		725	.022	Lb.	.44	.67		1.11	1.64
2500	Smooth dowels, 12" long, 1/4" or 3/8" diameter		140	.114	Ea.	.67	3.47		4.14	6.75
2520	5/8" diameter		125	.128		1.18	3.89		5.07	8.05
2530	3/4" diameter		110	.145		1.46	4.42		5.88	9.25
2700	Dowel caps, 5" long, 1/2" to 3/4" diameter		800	.020		.21	.61		.82	1.28
2720	1-1/4" diameter		750	.021		.26	.65		.91	1.41
109	0010	**SPLICING REINFORCING BARS** Incl. holding bars in								**109**
0020	place while splicing R032-070									
0100	Butt weld columns #4 bars	C-5	190	.295	Ea.	1.01	8.75	2.89	12.65	18.95
0110	#6 bars		150	.373		1.50	11.05	3.66	16.21	24
0130	#10 bars		95	.589		2.03	17.45	5.80	25.28	38
0150	#14 bars		65	.862		2.54	25.50	8.45	36.49	55

Important: See the Reference Section for critical supporting data - Reference Nos., Crews, & City Cost Indexes

			DAILY	LABOR-			1999 BARE COSTS				TOTAL
032 100	**Reinforcing Steel**	CREW	OUTPUT	HOURS	UNIT	MAT.	LABOR	EQUIP.	TOTAL		INCL O&P
109	0280	Column splice clamps, sleeve & wedge, or end bearing	R032 -070									**109**
	0300	#7 or #8 bars		C-5	190	.295	Ea.	3.32	8.75	2.89	14.96	21.50
	0310	#9 or #10 bars			170	.329		3.42	9.75	3.23	16.40	23.50
	0320	#11 bars			160	.350		4.43	10.35	3.43	18.21	26
	0330	#14 bars			150	.373		5.35	11.05	3.66	20.06	28.50
	0340	#18 bars			140	.400		8.25	11.85	3.92	24.02	33.50
	0500	Reducer inserts for above, #14 to #18 bar						3.23			3.23	3.56
	0520	#14 to #11 bar						2.78			2.78	3.05
	0550	#10 to #9 bar						.84			.84	.92
	0560	#9 to #8 bar						.84			.84	.92
	0580	#8 to #7 bar						.83			.83	.91
	0600	For bolted speed sleeve type, deduct							15%			
	0800	Mechanical butt splice, sleeve type with filler metal, compression										
	0810	only, all grades, columns only #11 bars		C-5	68	.824	Ea.	13.30	24.50	8.10	45.90	64.50
	0900	#14 bars			62	.903		13.75	27	8.85	49.60	70
	0920	#18 bars			62	.903		16.80	27	8.85	52.65	73.50
	1000	125% yield point, grade 60, columns only, #6 bars			68	.824		18.05	24.50	8.10	50.65	70
	1020	#7 or #8 bars			68	.824		15.90	24.50	8.10	48.50	67.50
	1030	#9 bars			68	.824		15.95	24.50	8.10	48.55	67.50
	1040	#10 bars			68	.824		16.80	24.50	8.10	49.40	68.50
	1050	#11 bars			68	.824		21	24.50	8.10	53.60	73
	1060	#14 bars			62	.903		26.50	27	8.85	62.35	84.50
	1070	#18 bars			62	.903		39	27	8.85	74.85	97.50
	1200	Full tension, grade 60 steel, columns,										
	1220	slabs or beams, #6, #7, #8 bars		C-5	68	.824	Ea.	15.30	24.50	8.10	47.90	67
	1230	#9 bars			68	.824		17.10	24.50	8.10	49.70	68.50
	1240	#10 bars			68	.824		18.80	24.50	8.10	51.40	70.50
	1250	#11 bars			68	.824		22.50	24.50	8.10	55.10	74.50
	1260	#14 bars			62	.903		30	27	8.85	65.85	88
	1270	#18 bars			62	.903		49.50	27	8.85	85.35	109
	1400	If equipment handling not required, deduct							50%			
	1600	Mechanical threaded type, bar threading not included,										
	1700	Straight bars, #10 & #11		C-5	140	.400	Ea.	15.40	11.85	3.92	31.17	41
	1750	#14 bars			130	.431		18.15	12.75	4.23	35.13	46
	1800	#18 bars			75	.747		29.50	22	7.30	58.80	77
	2100	#11 to #18 & #14 to #18 transition			75	.747		31	22	7.30	60.30	79.50
	2400	Bent bars, #10 & #11			105	.533		27	15.80	5.25	48.05	62.50
	2500	#14			90	.622		36	18.45	6.10	60.55	77.50
	2600	#18			70	.800		52.50	23.50	7.85	83.85	106
	2800	#11 to #14 transition			75	.747		37.50	22	7.30	66.80	86
	2900	#11 to #18 & #14 to #18 transition			70	.800		52.50	23.50	7.85	83.85	106

032 200	**Welded Wire Fabric**										
207	0010	**WELDED WIRE FABRIC** ASTM A185	R032 -030									**207**
	0050	Sheets										
	0100	6 x 6 - W1.4 x W1.4 (10 x 10) 21 lb. per C.S.F.		2 Rodm	35	.457	C.S.F.	6.90	13.90		20.80	31.50
	0200	6 x 6 - W2.1 x W2.1 (8 x 8) 30 lb. per C.S.F.			31	.516		10.40	15.70		26.10	38.50
	0300	6 x 6 - W2.9 x W2.9 (6 x 6) 42 lb. per C.S.F.			29	.552		14.70	16.75		31.45	45
	0400	6 x 6 - W4 x W4 (4 x 4) 58 lb. per C.S.F.			27	.593		21.50	18		39.50	54.50
	0500	4 x 4 - W1.4 x W1.4 (10 x 10) 31 lb. per C.S.F.			31	.516		10.45	15.70		26.15	38.50
	0600	4 x 4 - W2.1 x W2.1 (8 x 8) 44 lb. per C.S.F.			29	.552		12.80	16.75		29.55	43
	0650	4 x 4 - W2.9 x W2.9 (6 x 6) 61 lb. per C.S.F.			27	.593		18.75	18		36.75	51.50
	0700	4 x 4 - W4 x W4 (4 x 4) 85 lb. per C.S.F.			25	.640		28.50	19.45		47.95	64.50
	0750	Rolls										
	0800	2 x 2 - #14 galv. @ 21 lb., beam & column wrap		2 Rodm	6.50	2.462	C.S.F.	14.80	75		89.80	145

032 200 | Welded Wire Fabric

		CREW	DAILY OUTPUT	LABOR-HOURS	UNIT	1999 BARE COSTS				TOTAL INCL O&P	
						MAT.	LABOR	EQUIP.	TOTAL		
207	0900 2 x 2 - #12 galv. for gunite reinforcing R032-030	2 Rodm	6.50	2.462	C.S.F.	21	75		96	152	207
	0950 Material prices for above include 10% lap										
	1000 Specially fabricated heavier gauges in sheets	4 Rodm	50	.640	C.S.F.		19.45		19.45	33.50	
	1010 Material only, minimum				Ton	520			520	570	
	1020 Average					725			725	800	
	1030 Maximum					930			930	1,025	
240	0010 **FIBROUS REINFORCING**										240
	0100 Synthetic fibers				Lb.	3.78			3.78	4.15	
	0110 1-1/2 lb. per C.Y., add to concrete				C.Y.	5.85			5.85	6.40	
	0150 Steel fibers				Lb.	.46			.46	.51	
	0155 25 lb. per C.Y., add to concrete				C.Y.	11.50			11.50	12.65	
	0160 50 lb. per C.Y., add to concrete					23			23	25.50	
	0170 75 lb. per C.Y., add to concrete					35.50			35.50	39	
	0180 100 lb. per C.Y., add to concrete					46			46	50.50	

032 300 | Stressing Tendons

		CREW	DAILY OUTPUT	LABOR-HOURS	UNIT	1999 BARE COSTS				TOTAL INCL O&P	
						MAT.	LABOR	EQUIP.	TOTAL		
307	0010 **PRESTRESSING STEEL** Post-tensioned in field R032-090										307
	0100 Grouted strand, 50' span, 100 kip	C-3	1,200	.053	Lb.	2.28	1.49	.13	3.90	5.15	
	0150 300 kip		2,700	.024		2.03	.66	.06	2.75	3.41	
	0300 100' span, 100 kip		1,700	.038		2.28	1.05	.09	3.42	4.38	
	0350 300 kip		3,200	.020		2.22	.56	.05	2.83	3.43	
	0500 200' span, 100 kip		2,700	.024		2.28	.66	.06	3	3.68	
	0550 300 kip		3,500	.018		2.22	.51	.05	2.78	3.35	
	0800 Grouted bars, 50' span, 42 kip		2,600	.025		.36	.69	.06	1.11	1.62	
	0850 143 kip		3,200	.020		.35	.56	.05	.96	1.37	
	1000 75' span, 42 kip		3,200	.020		.36	.56	.05	.97	1.39	
	1050 143 kip		4,200	.015		.33	.43	.04	.80	1.11	
	1200 Ungrouted strand, 50' span, 100 kip	C-4	1,275	.025		2	.78	.03	2.81	3.57	
	1250 300 kip		1,475	.022		2.10	.67	.03	2.80	3.50	
	1400 100' span, 100 kip		1,500	.021		1.95	.66	.03	2.64	3.32	
	1450 300 kip		1,650	.019		2.15	.60	.02	2.77	3.44	
	1600 200' span, 100 kip		1,500	.021		1.95	.66	.03	2.64	3.32	
	1650 300 kip		1,700	.019		2.15	.58	.02	2.75	3.41	
	1800 Ungrouted bars, 50' span, 42 kip		1,400	.023		.27	.71	.03	1.01	1.54	
	1850 143 kip		1,700	.019		.26	.58	.02	.86	1.33	
	2000 75' span, 42 kip		1,800	.018		.27	.55	.02	.84	1.26	
	2050 143 kip		2,200	.015		.26	.45	.02	.73	1.09	
	2220 Ungrouted single strand, 100' slab, 25 kip		1,200	.027		2.09	.82	.03	2.94	3.77	
	2250 35 kip		1,475	.022		1.95	.67	.03	2.65	3.34	

033 | Cast-In-Place Concrete

033 100 | Structural Concrete

		CREW	DAILY OUTPUT	LABOR-HOURS	UNIT	1999 BARE COSTS				TOTAL INCL O&P	
						MAT.	LABOR	EQUIP.	TOTAL		
102	0010 **AGGREGATE** Expanded shale, C.L. lots, 52 lb. per C.F., minimum R033-020				Ton	35			35	38.50	102
	0050 Maximum				"	45			45	49.50	
	0100 Lightweight vermiculite or perlite, 4 C.F. bag, C.L. lots R033-030				Bag	8.70			8.70	9.60	
	0150 L.C.L. lots				"	9.60			9.60	10.55	
	0250 Sand & stone, loaded at pit, crushed bank gravel R033-050				Ton	12.40			12.40	13.65	
	0350 Sand, washed, for concrete					8.25			8.25	9.05	

3 CONCRETE

033 100	Structural Concrete		CREW	DAILY OUTPUT	LABOR-HOURS	UNIT	1999 BARE COSTS				TOTAL INCL O&P	
							MAT.	LABOR	EQUIP.	TOTAL		
102	0400	For plaster or brick	R033 -020				Ton	16.15			16.15	17.80	**102**
	0450	Stone, 3/4" to 1-1/2"						13			13	14.30	
	0500	3/8" roofing stone & 1/2" pea stone	R033 -030					13			13	14.30	
	0550	For trucking 10 miles, add to the above						3.50			3.50	3.85	
	0600	30 miles, add to the above	R033 -050				↓	7.75			7.75	8.55	
	0850	Sand & stone, loaded at pit, crushed bank gravel					C.Y.	16.10			16.10	17.70	
	0950	Sand, washed, for concrete						11.15			11.15	12.25	
	1000	For plaster or brick						23			23	25.50	
	1050	Stone, 3/4" to 1-1/2"						18.40			18.40	20	
	1100	3/8" roofing stone & 1/2" pea stone						23			23	25.50	
	1150	For trucking 10 miles, add to the above						4.90			4.90	5.40	
	1200	30 miles, add to the above					↓	10.85			10.85	11.95	
	1310	Quartz chips, 50 lb. bags					Cwt.	15.25			15.25	16.80	
	1330	Silica chips, 50 lb. bags						8.35			8.35	9.20	
	1410	White marble, 3/8" to 1/2", 50 lb. bags					↓	12.60			12.60	13.85	
	1430	3/4" F.O.B. plant					Ton	25.50			25.50	28	
110	0010	**ANCHOR BOLTS** See division 031-110											**110**
114	0010	**CEMENT** Material only	R033 -020										**114**
	0050	Masonry, gray, T.L. or C.L. lots					Bag	5.80			5.80	6.40	
	0060	L.T.L. or L.C.L. lots	R033 -060					5.60			5.60	6.15	
	0100	Masonry, white, T.L. or C.L. lots						15.85			15.85	17.45	
	0120	L.T.L. or L.C.L. lots						16.40			16.40	18.05	
	0240	Portland, type I, plain/air entrained, TL lots, 94 lb bags					↓	7			7	7.70	
	0300	Trucked in bulk, per cwt **CN**					Cwt.	4.05			4.05	4.45	
	0400	Type III, high early strength, TL lots, 94 lb bags **CN**					Bag	7.90			7.90	8.70	
	0420	L.T.L. or L.C.L. lots						8.90			8.90	9.80	
	0500	White high early strength, T.L. or C.L. lots, bags						16.90			16.90	18.60	
	0520	L.T.L. or L.C.L. lots						21.50			21.50	23.50	
	0600	White, T.L. or C.L. lots, bags						16.50			16.50	18.15	
	0620	L.T.L. or L.C.L. lots					↓	20			20	22	
118	0010	**CONCRETE ADMIXTURES & SURFACE TREATMENTS**											**118**
	0040	Abrasives, aluminum oxide, over 20 tons					Lb.	.88			.88	.97	
	0070	Under 1 ton						.92			.92	1.01	
	0100	Silicon carbide, black, over 20 tons						1.20			1.20	1.32	
	0120	Under 1 ton					↓	1.25			1.25	1.38	
	0200	Air entraining agent, .7 to 1.5 oz. per bag, 55 gallon lots					Gal.	7.40			7.40	8.10	
	0220	5 gallon lots						8.45			8.45	9.25	
	0300	Bonding agent, acrylic latex (200-250 S.F. per gallon)						26			26	28.50	
	0320	Epoxy resin (70-80 S.F. per gallon)					↓	50			50	55	
	0400	Calcium chloride, 100 lb. bags, FOB plant, truckload lots					Ton	330			330	365	
	0420	Less than truckload lots					Bag	19.75			19.75	21.50	
	0500	Carbon black, liquid, 2 to 8 lbs. per bag of cement					Lb.	2.85			2.85	3.14	
	0600	Colors, integral, 2 to 10 lb. per bag of cement, minimum						1.30			1.30	1.43	
	0610	Average						2.07			2.07	2.28	
	0620	Maximum					↓	3.69			3.69	4.06	
	0700	Curing compound, (200 to 400 S.F. per gallon), 55 gal. lots					Gal.	9.90			9.90	10.90	
	0720	5 gallon lots						8.40			8.40	9.25	
	0800	Premium grade, (450 S.F. per gallon), 55 gallon lots						12.80			12.80	14.10	
	0820	5 gallon lots						14.25			14.25	15.70	
	0900	Dustproofing compound, (200-600 S.F./gal.), 55 gallon lots						7.75			7.75	8.55	
	0920	5 gallon lots					↓	8.90			8.90	9.80	
	1000	Epoxy dustproof coating, colors, (300-400 S.F. per coat),											
	1010	or transparent, (400-600 S.F. per coat)					Gal.	56			56	61.50	
	1100	Hardeners, metallic, 55 lb. bags, natural (grey)					Lb.	.76			.76	.84	

		033 100 | **Structural Concrete**		CREW	DAILY OUTPUT	LABOR-HOURS	UNIT	1999 BARE COSTS				TOTAL INCL O&P	
								MAT.	LABOR	EQUIP.	TOTAL		
118	1200	Colors, average					Lb.	.95			.95	1.05	**118**
	1300	Non-metallic, 55 lb. bags, natural (grey), minimum						.31			.31	.34	
	1310	Maximum						.35			.35	.39	
	1320	Non-metallic, colors, minimum						.43			.43	.47	
	1340	Maximum						.69			.69	.76	
	1400	Non-metallic, non-slip, 100 lb. bags, minimum						.43			.43	.47	
	1420	Maximum					↓	.85			.85	.94	
	1500	Solution type, (300 to 400 S.F. per gallon)					Gal.	5.80			5.80	6.40	
	1550	Release agent, for tilt slabs						6.40			6.40	7.05	
	1570	For forms, average						4.40			4.40	4.84	
	1600	Sealer, hardener and dustproofer, clear, 450 S.F., minimum						6.30			6.30	6.90	
	1620	Maximum						14			14	15.40	
	1700	Colors (300-400 S.F. per gallon)						36			36	39.50	
	1800	Set accelerator for below freezing, 1 to 1-1/2 gal. per C.Y.						5.25			5.25	5.80	
	1900	Set retarder, 2 to 4 fl. oz. per bag of cement					↓	10.75			10.75	11.80	
	2000	Waterproofing, integral 1 lb. per bag of cement					Lb.	.72			.72	.79	
	2100	Powdered metallic, 40 lbs. per 100 S.F., minimum						.95			.95	1.05	
	2120	Maximum					↓	1.95			1.95	2.15	
	2200	Water reducing admixture, average					Gal.	8.75			8.75	9.65	
122	0010	**CONCRETE, FIELD MIX** FOB forms 2250 psi	R033 -080				C.Y.	62.50			62.50	69	**122**
	0020	3000 psi					"	65			65	71.50	
126	0010	**CONCRETE, READY MIX** Regular weight	R033 -040										**126**
	0020	2000 psi					C.Y.	56.50			56.50	62	
	0100	2500 psi						57			57	63	
	0150	3000 psi **CN**	R033 -050					61			61	67	
	0200	3500 psi						61			61	67	
	0300	4000 psi	R033 -070					62.50			62.50	68.50	
	0350	4500 psi						63.50			63.50	69.50	
	0400	5000 psi **CN**						67.50			67.50	74	
	0411	6000 psi						77			77	84.50	
	0412	8000 psi						125			125	138	
	0413	10,000 psi						178			178	196	
	0414	12,000 psi						215			215	236	
	1000	For high early strength cement, add						10%					
	1010	For structural lightweight with regular sand, add						25%					
	2000	For all lightweight aggregate, add					↓	45%					
	3000	For integral colors, 2500 psi, 5 bag mix											
	3100	Red, yellow or brown, 1.8 lb. per bag, add					C.Y.	13.05			13.05	14.35	
	3200	9.4 lb. per bag, add						70.50			70.50	77.50	
	3400	Black, 1.8 lb. per bag, add						15.50			15.50	17.05	
	3500	7.5 lb. per bag, add						65			65	71.50	
	3700	Green, 1.8 lb. per bag, add						31			31	34	
	3800	7.5 lb. per bag, add					↓	149			149	163	
130	0010	**CONCRETE IN PLACE** Including forms (4 uses), reinforcing	R033 -010										**130**
	0050	steel, including finishing unless otherwise indicated											
	0300	Beams, 5 kip per L.F., 10' span	R033 -100	C-14A	15.62	12.804	C.Y.	207	350	44	601	835	
	0350	25' span		"	18.55	10.782		188	296	37	521	720	
	0500	Chimney foundations, industrial, minimum	R033 -130	C-14C	32.22	3.476		126	91	1.15	218.15	285	
	0510	Maximum		"	23.71	4.724		147	123	1.56	271.56	360	
	0700	Columns, square, 12" x 12", minimum reinforcing	R042 -055	C-14A	11.96	16.722		210	460	57	727	1,025	
	0720	Average reinforcing			10.13	19.743		310	540	67.50	917.50	1,275	
	0740	Maximum reinforcing			9.03	22.148		390	605	76	1,071	1,475	
	0800	16" x 16", minimum reinforcing			16.22	12.330		178	340	42	560	785	
	0820	Average reinforcing			12.57	15.911		293	435	54.50	782.50	1,075	
	0840	Maximum reinforcing			10.25	19.512		410	535	66.50	1,011.50	1,375	

Important: See the Reference Section for critical supporting data - Reference Nos., Crews, & City Cost Indexes

033 100 | Structural Concrete

		CREW	DAILY OUTPUT	LABOR-HOURS	UNIT	1999 BARE COSTS				TOTAL INCL O&P
						MAT.	LABOR	EQUIP.	TOTAL	
130	0900	24″ x 24″, minimum reinforcing	C-14A R033-010	23.66	8.453	C.Y.	149	232	29	410	565	130
0920	Average reinforcing		17.71	11.293		242	310	38.50	590.50	805
0940	Maximum reinforcing	R033-100	14.15	14.134		335	385	48.50	768.50	1,050
1000	36″ x 36″, minimum reinforcing		33.69	5.936		143	163	20.50	326.50	440
1020	Average reinforcing	R033-130	23.32	8.576		230	235	29.50	494.50	660
1040	Maximum reinforcing		17.82	11.223		315	310	38.50	663.50	880
1100	Columns, round, tied, 12″ diameter, minimum reinforcing	R042-055	20.97	9.537		217	261	32.50	510.50	690
1120	Average reinforcing		15.27	13.098		335	360	45	740	995
1140	Maximum reinforcing		12.11	16.515		450	455	56.50	961.50	1,275
1200	16″ diameter, minimum reinforcing		31.49	6.351		211	174	21.50	406.50	535
1220	Average reinforcing		19.12	10.460		340	287	36	663	875
1240	Maximum reinforcing		13.77	14.524		470	400	49.50	919.50	1,200
1300	20″ diameter, minimum reinforcing		41.04	4.873		200	134	16.65	350.65	450
1320	Average reinforcing		24.05	8.316		315	228	28.50	571.50	740
1340	Maximum reinforcing		17.01	11.758		430	320	40	790	1,025
1400	24″ diameter, minimum reinforcing		51.85	3.857		188	106	13.20	307.20	390
1420	Average reinforcing		27.06	7.391		305	203	25.50	533.50	690
1440	Maximum reinforcing		18.29	10.935		425	300	37.50	762.50	985
1500	36″ diameter, minimum reinforcing		75.04	2.665		176	73	9.10	258.10	320
1520	Average reinforcing		37.49	5.335		265	146	18.25	429.25	545
1540	Maximum reinforcing	▼	22.84	8.757		380	240	30	650	835
1900	Elevated slabs, flat slab, 125 psf Sup. Load, 20′ span	C-14B	38.45	5.410		138	148	17.75	303.75	405
1950	30′ span		50.99	4.079		126	112	13.40	251.40	330
2100	Flat plate, 125 psf Sup. Load, 15′ span		30.24	6.878		146	188	22.50	356.50	485
2150	25′ span		49.60	4.194		120	115	13.75	248.75	330
2300	Waffle const., 30″ domes, 125 psf Sup. Load, 20′ span		37.07	5.611		178	154	18.40	350.40	460
2350	30′ span		44.07	4.720		162	129	15.50	306.50	400
2500	One way joists, 30″ pans, 125 psf Sup. Load, 15′ span		27.38	7.597		209	208	25	442	590
2550	25′ span		31.15	6.677		196	183	22	401	530
2700	One way beam & slab, 125 psf Sup. Load, 15′ span		20.59	10.102		168	276	33	477	660
2750	25′ span		28.36	7.334		154	201	24	379	515
2900	Two way beam & slab, 125 psf Sup. Load, 15′ span		24.04	8.652		157	237	28.50	422.50	580
2950	25′ span	▼	35.87	5.799	▼	133	159	19	311	420
3100	Elevated slabs including finish, not									
3110	including forms or reinforcing									
3150	Regular concrete, 4″ slab	C-8	2,613	.021	S.F.	.77	.52	.25	1.54	1.91
3200	6″ slab		2,585	.022		1.21	.52	.25	1.98	2.40
3250	2-1/2″ thick floor fill		2,685	.021		.52	.50	.24	1.26	1.61
3300	Lightweight, 110# per C.F., 2-1/2″ thick floor fill		2,585	.022		.64	.52	.25	1.41	1.78
3400	Cellular concrete, 1-5/8″ fill, under 5000 S.F.		2,000	.028		.45	.67	.32	1.44	1.90
3450	Over 10,000 S.F.		2,200	.025		.35	.61	.29	1.25	1.65
3500	Add per floor for 3 to 6 stories high		31,800	.002			.04	.02	.06	.09
3520	For 7 to 20 stories high	▼	21,200	.003	▼		.06	.03	.09	.13
3800	Footings, spread under 1 C.Y.	C-14C	38.07	2.942	C.Y.	89.50	77	.97	167.47	223
3850	Over 5 C.Y.		81.04	1.382		82.50	36	.46	118.96	149
3900	Footings, strip, 18″ x 9″, plain		41.04	2.729		80.50	71.50	.90	152.90	204
3950	36″ x 12″, reinforced		61.55	1.820		83	47.50	.60	131.10	168
4000	Foundation mat, under 10 C.Y.		38.67	2.896		113	75.50	.96	189.46	247
4050	Over 20 C.Y.	▼	56.40	1.986		101	52	.66	153.66	194
4200	Grade walls, 8″ thick, 8′ high	C-14D	45.83	4.364		97.50	119	14.90	231.40	310
4250	14′ high		27.26	7.337		124	199	25	348	480
4260	12″ thick, 8′ high		64.32	3.109		89.50	84.50	10.65	184.65	244
4270	14′ high		40.01	4.999		99.50	136	17.10	252.60	345
4300	15″ thick, 8′ high		80.02	2.499		85.50	68	8.55	162.05	210
4350	12′ high		51.26	3.902		89	106	13.35	208.35	281
4500	18′ high	▼	48.85	4.094	▼	99	111	14	224	300

CONCRETE

3

033 100 | Structural Concrete

			CREW	DAILY OUTPUT	LABOR-HOURS	UNIT	1999 BARE COSTS				TOTAL INCL O&P	
							MAT.	LABOR	EQUIP.	TOTAL		
130	4520	Handicap access ramp, railing both sides, 3' wide	C-14H	14.58	3.292	L.F.	94.50	89	2.57	186.07	248	**130**
	4525	5' wide		12.22	3.928		108	106	3.06	217.06	290	
	4530	With cheek walls and rails both sides, 3' wide		8.55	5.614		96.50	151	4.38	251.88	350	
	4535	5' wide		7.31	6.566		97	177	5.10	279.10	395	
	4650	Slab on grade, not including finish, 4" thick	C-14E	60.75	1.449	C.Y.	73.50	39	.62	113.12	145	
	4700	6" thick	"	92	.957	"	70.50	25.50	.41	96.41	120	
	4751	Slab on grade, incl. troweled finish, not incl. forms										
	4760	or reinforcing, over 10,000 S.F., 4" thick slab	C-14F	3,425	.021	S.F.	.79	.52	.01	1.32	1.67	
	4820	6" thick slab		3,350	.021		1.16	.53	.01	1.70	2.09	
	4840	8" thick slab		3,184	.023		1.59	.56	.01	2.16	2.60	
	4900	12" thick slab		2,734	.026		2.38	.65	.01	3.04	3.63	
	4950	15" thick slab		2,505	.029		2.99	.71	.01	3.71	4.39	
	5000	Slab on grade, incl. textured finish, not incl. forms										
	5001	or reinforcing, 4" thick slab	C-14G	2,873	.019	S.F.	.79	.48	.01	1.28	1.61	
	5010	6" thick		2,590	.022		1.24	.53	.01	1.78	2.18	
	5020	8" thick		2,320	.024		1.62	.59	.02	2.23	2.70	
	5200	Lift slab in place above the foundation, incl. forms,										
	5210	reinforcing, concrete and columns, minimum	C-14B	2,113	.098	S.F.	4.74	2.69	.32	7.75	9.85	
	5250	Average		1,650	.126		5.20	3.45	.41	9.06	11.70	
	5300	Maximum		1,500	.139		5.80	3.79	.45	10.04	12.90	
	5500	Lightweight, ready mix, including screed finish only,										
	5510	not including forms or reinforcing										
	5550	1:4 for structural roof decks	C-14B	260	.800	C.Y.	91.50	22	2.62	116.12	139	
	5600	1:6 for ground slab with radiant heat	C-14F	92	.783		86.50	19.40	.41	106.31	125	
	5650	1:3:2 with sand aggregate, roof deck	C-14B	260	.800		91.50	22	2.62	116.12	139	
	5700	Ground slab	C-14F	107	.673		91.50	16.70	.35	108.55	127	
	5900	Pile caps, incl. forms and reinf., sq. or rect., under 5 C.Y.	C-14C	54.14	2.069		86	54	.68	140.68	182	
	5950	Over 10 C.Y.		75	1.493		83.50	39	.49	122.99	155	
	6000	Triangular or hexagonal, under 5 C.Y.		53	2.113		79	55	.70	134.70	176	
	6050	Over 10 C.Y.		85	1.318		83.50	34.50	.43	118.43	147	
	6200	Retaining walls, gravity, 4' high see division 022-708	C-14D	66.20	3.021		81	82	10.35	173.35	230	
	6250	10' high		125	1.600		72	43.50	5.45	120.95	154	
	6300	Cantilever, level backfill loading, 8' high		70	2.857		88	77.50	9.75	175.25	231	
	6350	16' high		91	2.198		85.50	59.50	7.50	152.50	197	
	6800	Stairs, not including safety treads, free standing, 3'-6" wide	C-14H	83	.578	LF Nose	5.75	15.60	.45	21.80	32	
	6850	Cast on ground		125	.384	"	4.05	10.35	.30	14.70	21	
	7000	Stair landings, free standing		200	.240	S.F.	2.25	6.50	.19	8.94	13	
	7050	Cast on ground		475	.101	"	1.30	2.73	.08	4.11	5.85	
134	0010	**CURING** Burlap, 4 uses assumed, 7.5 oz.	2 Clab	55	.291	C.S.F.	2.55	6.25		8.80	12.60	**134**
	0100	12 oz.		55	.291		3.74	6.25		9.99	13.90	
	0200	Waterproof curing paper, 2 ply, reinforced		70	.229		4.71	4.90		9.61	12.90	
	0300	Sprayed membrane curing compound		95	.168		2.42	3.61		6.03	8.35	
	0400	Curing blankets, 1" to 2" thick, buy, minimum				S.F.	1.10			1.10	1.21	
	0450	Maximum					2.75			2.75	3.03	
	0500	Electrically heated pads, 110 volts, 15 watts per S.F., buy					4.50			4.50	4.95	
	0600	20 watts per S.F., buy					6			6	6.60	
	0710	Electrically, heated pads, 15 watts/S.F., 20 uses, minimum					.16			.16	.18	
	0800	Maximum					.27			.27	.29	
138	0010	**CUTTING** Concrete see division 020-728										**138**
142	0010	**DAMPPROOFING** See division 071										**142**
146	0010	**EQUIPMENT** For placing conc. see div. 016-406										**146**
148	0010	**EXPANSION JOINT** See division 031-132										**148**

033 100 | Structural Concrete

			CREW	DAILY OUTPUT	LABOR-HOURS	UNIT	MAT.	LABOR	EQUIP.	TOTAL	TOTAL INCL O&P	
156	0010	**GROUT** Column & machine bases, non-shrink, metallic, 1" deep	1 Cefi	35	.229	S.F.	4.25	6		10.25	13.60	156
	0050	2" deep		25	.320		8.50	8.35		16.85	22	
	0300	Non-shrink, non-metallic, 1" deep		35	.229		7.50	6		13.50	17.15	
	0350	2" deep		25	.320		14.80	8.35		23.15	29	
160	0010	**GUNITE**										160
	0020	Applied in 1" layers, no mesh included	C-8	2,000	.028	S.F.	.92	.67	.32	1.91	2.41	
	0100	Mesh for gunite 2 x 2, #12, to 3" thick	2 Rodm	800	.020		.17	.61		.78	1.24	
	0150	Over 3" thick	"	500	.032		.21	.97		1.18	1.92	
	0300	Typical in place, including mesh, 2" thick, minimum	C-16	1,000	.072		1.68	1.83	.65	4.16	5.50	
	0350	Maximum		500	.144		2.60	3.67	1.29	7.56	10.15	
	0500	4" thick, minimum		750	.096		2.49	2.45	.86	5.80	7.60	
	0550	Maximum		350	.206		3.86	5.25	1.85	10.96	14.65	
	0900	Prepare old walls, no scaffolding, minimum	C-10	1,000	.024		.57	.59		1.16	1.52	
	0950	Maximum	"	275	.087		1.80	2.15		3.95	5.25	
	1100	For high finish requirement or close tolerance, add, minimum						50%				
	1150	Maximum						110%				
164	0010	**INSULATING CONCRETE** See division 033-130 and 035-212										164
168	0010	**PATCHING CONCRETE**										168
	0100	Floors, 1/4" thick, small areas, regular grout	1 Cefi	170	.047	S.F.	.07	1.23		1.30	1.91	
	0150	Epoxy grout	"	100	.080	"	.52	2.09		2.61	3.70	
	0300	Slab on Grade, cut outs, up to 50 C.F.	2 Cefi	50	.320	C.F.	4.52	8.35		12.87	17.45	
	2000	Walls, including chipping, cleaning and epoxy grout										
	2100	Minimum	1 Cefi	65	.123	S.F.	.12	3.22		3.34	4.94	
	2150	Average		50	.160		.19	4.18		4.37	6.45	
	2200	Maximum		40	.200		.35	5.25		5.60	8.20	
	2510	Underlayment, P.C based self-leveling, 4100 psi, pumped, 1/4"	C-8	20,000	.003		1.16	.07	.03	1.26	1.42	
	2520	1/2"		19,000	.003		2.09	.07	.03	2.19	2.45	
	2530	3/4"		18,000	.003		3.25	.07	.04	3.36	3.73	
	2540	1"		17,000	.003		4.41	.08	.04	4.53	5	
	2550	1-1/2"		15,000	.004		6.75	.09	.04	6.88	7.60	
	2560	Hand mix, 1/2"	C-18	4,000	.002		2.09	.05	.01	2.15	2.39	
	2610	Topping, P.C. based self-level/dry 6100 psi, pumped, 1/4"	C-8	20,000	.003		1.76	.07	.03	1.86	2.08	
	2620	1/2"		19,000	.003		3.17	.07	.03	3.27	3.63	
	2630	3/4"		18,000	.003		4.93	.07	.04	5.04	5.55	
	2660	1"		17,000	.003		6.70	.08	.04	6.82	7.50	
	2670	1-1/2"		15,000	.004		10.20	.09	.04	10.33	11.45	
	2680	Hand mix, 1/2"	C-18	4,000	.002		3.17	.05	.01	3.23	3.57	
172	0010	**PLACING CONCRETE** and vibrating, including labor & equipment [R033-090]										172
	0050	Beams, elevated, small beams, pumped	C-20	60	1.067	C.Y.		24.50	12	36.50	51.50	
	0100	With crane and bucket	C-7	45	1.600			37	19.85	56.85	79.50	
	0200	Large beams, pumped	C-20	90	.711			16.45	8	24.45	34.50	
	0250	With crane and bucket	C-7	65	1.108			26	13.75	39.75	55	
	0400	Columns, square or round, 12" thick, pumped	C-20	60	1.067			24.50	12	36.50	51.50	
	0450	With crane and bucket	C-7	40	1.800			42	22.50	64.50	89.50	
	0600	18" thick, pumped	C-20	90	.711			16.45	8	24.45	34.50	
	0650	With crane and bucket	C-7	55	1.309			30.50	16.25	46.75	65	
	0800	24" thick, pumped	C-20	92	.696			16.10	7.85	23.95	33.50	
	0850	With crane and bucket	C-7	70	1.029			24	12.75	36.75	51	
	1000	36" thick, pumped	C-20	140	.457			10.60	5.15	15.75	22	
	1050	With crane and bucket	C-7	100	.720			16.75	8.95	25.70	36	
	1400	Elevated slabs, less than 6" thick, pumped	C-20	140	.457			10.60	5.15	15.75	22	
	1450	With crane and bucket	C-7	95	.758			17.65	9.40	27.05	38	
	1500	6" to 10" thick, pumped	C-20	160	.400			9.25	4.50	13.75	19.35	
	1550	With crane and bucket	C-7	110	.655			15.20	8.10	23.30	32.50	
	1600	Slabs over 10" thick, pumped	C-20	180	.356			8.25	4	12.25	17.20	

For expanded coverage of these items see Means Concrete & Masonry Cost Data 1999

033 100 | Structural Concrete

		CREW	DAILY OUTPUT	LABOR-HOURS	UNIT	1999 BARE COSTS MAT.	LABOR	EQUIP.	TOTAL	TOTAL INCL O&P	
172	1650	With crane and bucket [R033-090]	C-7	130	.554	C.Y.		12.90	6.85	19.75	27.50	172
	1900	Footings, continuous, shallow, direct chute	C-6	120	.400			9.05	.62	9.67	14.75	
	1950	Pumped	C-20	150	.427			9.90	4.80	14.70	20.50	
	2000	With crane and bucket	C-7	90	.800			18.60	9.90	28.50	40	
	2100	Footings, continuous, deep, direct chute	C-6	140	.343			7.75	.53	8.28	12.65	
	2150	Pumped	C-20	160	.400			9.25	4.50	13.75	19.35	
	2200	With crane and bucket	C-7	110	.655			15.20	8.10	23.30	32.50	
	2400	Footings, spread, under 1 C.Y., direct chute	C-6	55	.873			19.70	1.36	21.06	32	
	2450	Pumped	C-20	65	.985			23	11.10	34.10	47.50	
	2500	With crane and bucket	C-7	45	1.600			37	19.85	56.85	79.50	
	2600	Footings, spread, over 5 C.Y., direct chute	C-6	120	.400			9.05	.62	9.67	14.75	
	2650	Pumped	C-20	150	.427			9.90	4.80	14.70	20.50	
	2700	With crane and bucket	C-7	100	.720			16.75	8.95	25.70	36	
	2900	Foundation mats, over 20 C.Y., direct chute	C-6	350	.137			3.10	.21	3.31	5.05	
	2950	Pumped	C-20	400	.160			3.71	1.80	5.51	7.75	
	3000	With crane and bucket	C-7	300	.240			5.60	2.98	8.58	11.90	
	3200	Grade beams, direct chute	C-6	150	.320			7.20	.50	7.70	11.80	
	3250	Pumped	C-20	180	.356			8.25	4	12.25	17.20	
	3300	With crane and bucket	C-7	120	.600			13.95	7.45	21.40	29.50	
	3500	High rise, for more than 5 stories, pumped, add per story	C-20	2,100	.030			.71	.34	1.05	1.48	
	3510	With crane and bucket, add per story	C-7	2,100	.034			.80	.43	1.23	1.71	
	3700	Pile caps, under 5 C.Y., direct chute	C-6	90	.533			12.05	.83	12.88	19.65	
	3750	Pumped	C-20	110	.582			13.45	6.55	20	28	
	3800	With crane and bucket	C-7	80	.900			21	11.15	32.15	45	
	3850	Pile cap, 5 C.Y. to 10 C.Y., direct chute	C-6	175	.274			6.20	.43	6.63	10.10	
	3900	Pumped	C-20	200	.320			7.40	3.60	11	15.45	
	3950	With crane and bucket	C-7	150	.480			11.15	5.95	17.10	24	
	4000	Over 10 C.Y., direct chute	C-6	215	.223			5.05	.35	5.40	8.25	
	4050	Pumped	C-20	240	.267			6.20	3	9.20	12.90	
	4100	With crane and bucket	C-7	185	.389			9.05	4.83	13.88	19.35	
	4300	Slab on grade, 4" thick, direct chute	C-6	110	.436			9.85	.68	10.53	16.10	
	4350	Pumped	C-20	130	.492			11.40	5.55	16.95	24	
	4400	With crane and bucket	C-7	110	.655			15.20	8.10	23.30	32.50	
	4600	Over 6" thick, direct chute	C-6	165	.291			6.55	.45	7	10.75	
	4650	Pumped	C-20	185	.346			8	3.90	11.90	16.75	
	4700	With crane and bucket	C-7	145	.497			11.55	6.15	17.70	24.50	
	4900	Walls, 8" thick, direct chute	C-6	90	.533			12.05	.83	12.88	19.65	
	4950	Pumped	C-20	100	.640			14.80	7.20	22	31	
	5000	With crane and bucket	C-7	80	.900			21	11.15	32.15	45	
	5050	12" thick, direct chute	C-6	100	.480			10.85	.75	11.60	17.70	
	5100	Pumped	C-20	110	.582			13.45	6.55	20	28	
	5200	With crane and bucket	C-7	90	.800			18.60	9.90	28.50	40	
	5300	15" thick, direct chute	C-6	105	.457			10.30	.71	11.01	16.85	
	5350	Pumped	C-20	120	.533			12.35	6	18.35	26	
	5400	With crane and bucket	C-7	95	.758	▼		17.65	9.40	27.05	38	
	5600	Wheeled concrete dumping, add to placing costs above										
	5610	Walking cart, 50' haul, add	C-18	32	.281	C.Y.		6.10	1.56	7.66	11.30	
	5620	150' haul, add		24	.375			8.15	2.08	10.23	15.10	
	5700	250' haul, add	▼	18	.500			10.85	2.77	13.62	20	
	5800	Riding cart, 50' haul, add	C-19	80	.112			2.44	1.04	3.48	4.99	
	5810	150' haul, add		60	.150			3.25	1.39	4.64	6.65	
	5900	250' haul, add	▼	45	.200	▼		4.33	1.85	6.18	8.85	
176	0010	**PRESTRESSED CONCRETE** pretensioned, see division 034 [R032-090]										176
	0020	See also division 032-307										
	0100	Post-tensioned in place, small job [R034-030]	C-17B	8.50	9.647	C.Y.	215	275	34	524	710	
	0200	Large job	"	10	8.200	"	161	233	28.50	422.50	580	

Important: See the Reference Section for critical supporting data - Reference Nos., Crews, & City Cost Indexes

		033 100	Structural Concrete	CREW	DAILY OUTPUT	LABOR-HOURS	UNIT	1999 BARE COSTS				TOTAL INCL O&P	
								MAT.	LABOR	EQUIP.	TOTAL		
180	0010	**SAWING CONCRETE** See division 020-728											180
184	0010	**STAIR TREAD INSERTS** Cast iron, abrasive, 3" wide	1 Carp	90	.089	L.F.	5.60	2.43		8.03	9.95	184
	0020	4" wide		80	.100		6.85	2.73		9.58	11.80	
	0040	6" wide		75	.107		8.40	2.91		11.31	13.85	
	0050	9" wide		70	.114		12.45	3.12		15.57	18.60	
	0100	12" wide	▼	65	.123		18.65	3.36		22.01	26	
	0300	Cast aluminum, compared to cast iron, deduct					10%					
	0500	Extruded aluminum safety tread, 3" wide	1 Carp	75	.107		4.75	2.91		7.66	9.80	
	0550	4" wide		75	.107		6.40	2.91		9.31	11.65	
	0600	6" wide		75	.107		10.75	2.91		13.66	16.40	
	0650	9" wide to resurface stairs	▼	70	.114	▼	15.15	3.12		18.27	21.50	
	1700	Cement filled pan type, plain	1 Cefi	115	.070	S.F.	2.15	1.82		3.97	5.10	
	1750	Non-slip	"	100	.080	"	3.23	2.09		5.32	6.65	
192	0010	**WATERPROOFING AND DAMPPROOFING** See division 071										192
	0050	Integral waterproofing, add to cost of regular concrete				C.Y.	4.20			4.20	4.62	
196	0010	**WINTER PROTECTION** For heated ready mix, add, minimum				▼	3.75			3.75	4.13	196
	0050	Maximum					4.75			4.75	5.20	
	0100	Protecting concrete and temporary heat, add, minimum	2 Clab	6,000	.003	S.F.	.15	.06		.21	.26	
	0150	Maximum, see also division 010-094	"	2,000	.008	"	.65	.17		.82	.98	
	0200	Temporary shelter for slab on grade, wood frame and polyethylene										
	0201	sheeting, minimum	2 Carp	10	1.600	M.S.F.	275	43.50		318.50	375	
	0210	Maximum	"	3	5.333	"	330	146		476	595	
	0300	See also Division 033-134										

		033 450	Concrete Finishing										
454	0010	**FINISHING FLOORS** Monolithic, screed finish	R033-130	1 Cefi	900	.009	S.F.		.23		.23	.35	454
	0050	Darby finish			750	.011			.28		.28	.42	
	0100	Screed and float finish	R033-140		725	.011			.29		.29	.43	
	0150	Screed, float, and broom finish			630	.013			.33		.33	.50	
	0200	Screed, float, and hand trowel			600	.013			.35		.35	.52	
	0250	Machine trowel		▼	550	.015			.38		.38	.57	
	0400	Integral topping and finish, using 1:1:2 mix, 3/16" thick		C-10	1,000	.024		.06	.59		.65	.95	
	0450	1/2" thick			950	.025		.15	.62		.77	1.10	
	0500	3/4" thick			850	.028		.23	.69		.92	1.30	
	0600	1" thick			750	.032		.30	.79		1.09	1.52	
	0800	Granolithic topping, laid after, 1:1:1-1/2 mix, 1/2" thick			590	.041		.16	1		1.16	1.70	
	0820	3/4" thick			580	.041		.24	1.02		1.26	1.81	
	0850	1" thick			575	.042		.32	1.03		1.35	1.91	
	0950	2" thick			500	.048		.65	1.18		1.83	2.50	
	1200	Heavy duty, 1:1:2, 3/4" thick, preshrunk, gray, 20 MSF			320	.075		.21	1.84		2.05	3.02	
	1300	100 MSF			380	.063		.21	1.55		1.76	2.58	
	1350	For colors, .50 psf, add, minimum			1,650	.015		.41	.36		.77	1	
	1400	Maximum		▼	1,500	.016		1.86	.39		2.25	2.65	
	1600	Exposed local aggregate finish, minimum		1 Cefi	625	.013		.36	.33		.69	.90	
	1650	Maximum			465	.017		1.14	.45		1.59	1.92	
	1800	Floor abrasives, .25 psf, add to above, aluminum oxide			850	.009		.15	.25		.40	.54	
	1850	Silicon carbide			850	.009		.18	.25		.43	.57	
	2000	Floor hardeners, metallic, light service, .50 psf, add			850	.009		.27	.25		.52	.67	
	2050	Medium service, .75 psf, add			750	.011		.39	.28		.67	.85	
	2100	Heavy service, 1.0 psf, add			650	.012		.54	.32		.86	1.07	
	2150	Extra heavy, 1.5 psf, add			575	.014		.62	.36		.98	1.22	
	2300	Non-metallic, light service, .50 psf, add			850	.009		.17	.25		.42	.56	
	2350	Medium service, .75 psf, add			750	.011		.24	.28		.52	.68	
	2400	Heavy service, 1.00 psf, add			650	.012		.31	.32		.63	.82	
	2450	Extra heavy, 1.50 psf, add		▼	575	.014	▼	.48	.36		.84	1.06	

3

CONCRETE

033 450 | Concrete Finishing

			CREW	DAILY OUTPUT	LABOR-HOURS	UNIT	1999 BARE COSTS				TOTAL INCL O&P	
							MAT.	LABOR	EQUIP.	TOTAL		
454	2600	Add for colored hardeners, metallic	R033 -130				50%					454
	2650	Non-metallic					25%					
	2800	Trap rock wearing surface for monolithic floors	R033 -140									
	2810	2.0 psf, add to above		C-10	1,250	.019	S.F.	.78	.47		1.25	1.58	
	3000	Floor coloring, dusted on, 0.5 psf per S.F., add to above, min.		1 Cefi	1,300	.006		.57	.16		.73	.87	
	3050	Maximum		"	625	.013	▼	2	.33		2.33	2.69	
	3100	Colors only, minimum					Lb.	.96			.96	1.06	
	3120	Maximum					"	3.36			3.36	3.70	
	3200	Integral colors, see division 033-126											
	3600	1/2" topping using 5 lb. per bag, regular colors		C-10	590	.041	S.F.	.18	1		1.18	1.72	
	3650	Blue or green		"	590	.041		.18	1		1.18	1.72	
	3800	Dustproofing, silicate liquids, 1 coat		1 Cefi	1,900	.004		.07	.11		.18	.24	
	3850	2 coats			1,300	.006		.13	.16		.29	.38	
	4000	Epoxy coating, 1 coat, clear			1,500	.005		.36	.14		.50	.61	
	4050	Colors			1,500	.005		.36	.14		.50	.61	
	4400	Stair finish, float			275	.029			.76		.76	1.13	
	4500	Steel trowel finish			200	.040			1.05		1.05	1.56	
	4600	Silicon carbide finish, .25 psf		▼	150	.053	▼	.50	1.39		1.89	2.63	
458	0010	**FINISHING WALLS** Break ties and patch voids		1 Cefi	540	.015	S.F.	.03	.39		.42	.61	458
	0050	Burlap rub with grout			450	.018		.03	.46		.49	.72	
	0100	Carborundum rub, dry			270	.030		.03	.77		.80	1.19	
	0150	Wet rub		▼	175	.046		.03	1.20		1.23	1.81	
	0300	Bush hammer, green concrete		B-39	1,000	.048		.03	1.09	.18	1.30	1.94	
	0350	Cured concrete		"	650	.074		.03	1.68	.28	1.99	2.95	
	0600	Float finish, 1/16" thick		1 Cefi	300	.027		.02	.70		.72	1.06	
	0700	Sandblast, light penetration		C-10	1,100	.022		.20	.54		.74	1.03	
	0750	Heavy penetration		"	375	.064	▼	.40	1.57		1.97	2.82	
	0800	Board finish, see division 031-182											
	0850	Rustication strips, see division 031-182											

034 100 | Structural Precast

			CREW	DAILY OUTPUT	LABOR-HOURS	UNIT	1999 BARE COSTS				TOTAL INCL O&P	
							MAT.	LABOR	EQUIP.	TOTAL		
104	0011	**BEAMS,** "L" shaped, 20' span, 12" x 20"	R034 -030	C-11	32	2.250	Ea.	1,350	67.50	48.50	1,466	1,675	104
	1000	Inverted tee beams, add to above, small beams					L.F.	15%					
	1050	Large beams					"	5.55			5.55	6.10	
	1200	Rectangular, 20' span, 12" x 20"		C-11	32	2.250	Ea.	925	67.50	48.50	1,041	1,200	
	1250	18" x 36"			24	3.		1,700	90	64.50	1,854.50	2,100	
	1300	24" x 44"			22	3.273		2,450	98	70.50	2,618.50	2,950	
	1400	30' span, 12" x 36"			24	3		2,175	90	64.50	2,329.50	2,600	
	1450	18" x 44"			20	3.600		3,050	108	77.50	3,235.50	3,625	
	1500	24" x 52"			16	4.500		4,325	135	97	4,557	5,100	
	1600	40' span, 12" x 52"			20	3.600		4,025	108	77.50	4,210.50	4,700	
	1650	18" x 52"			16	4.500		4,900	135	97	5,132	5,750	
	1700	24" x 52"			12	6		6,000	180	129	6,309	7,050	
	2000	"T" shaped, 20' span, 12" x 20"			32	2.250		1,600	67.50	48.50	1,716	1,950	
	2050	18" x 36"			24	3		2,550	90	64.50	2,704.50	3,050	
	2100	24" x 44"			22	3.273		3,600	98	70.50	3,768.50	4,200	
	2200	30' span, 12" x 36"		▼	24	3	▼	3,650	90	64.50	3,804.50	4,225	

Important: See the Reference Section for critical supporting data - Reference Nos., Crews, & City Cost Indexes

034 100 | Structural Precast

			CREW	DAILY OUTPUT	LABOR-HOURS	UNIT	1999 BARE COSTS				TOTAL INCL O&P	
							MAT.	LABOR	EQUIP.	TOTAL		
104	2250	18" x 44" R034 -030	C-11	20	3.600	Ea.	4,975	108	77.50	5,160.50	5,750	104
	2300	24" x 52"		16	4.500		5,150	135	97	5,382	6,000	
	2500	40' span, 12" x 52"		20	3.600		6,850	108	77.50	7,035.50	7,825	
	2550	18" x 52"		16	4.500		7,500	135	97	7,732	8,600	
	2600	24" x 52"		12	6		9,150	180	129	9,459	10,600	
112	0010	COLUMNS Rectangular to 12' high, small columns	C-11	120	.600	L.F.	66.50	18	12.95	97.45	119	112
	0050	Large columns		96	.750		105	22.50	16.20	143.70	174	
	0300	24' high, small columns		192	.375		99	11.25	8.10	118.35	138	
	0350	Large columns		144	.500		132	15	10.80	157.80	183	
120	0010	CRIBBING See under Retaining Walls, division 022-708										120
124	0010	CURBS Roadway type, see division 025-250										124
128	0010	JOISTS 40 psf L.L., 6" deep for 12' spans	C-12	600	.080	L.F.	5.70	2.16	.74	8.60	10.45	128
	0050	8" deep for 16' spans		575	.083		6.70	2.25	.77	9.72	11.75	
	0100	10" deep for 20' spans		550	.087		7.50	2.36	.81	10.67	12.85	
	0150	12" deep for 24' spans		525	.091		10.25	2.47	.84	13.56	16.10	
132	0010	LIFT SLAB See division 033-130-5200 R033 -120										132
136	0010	SLABS Prestressed roof/floor members, solid, grouted, 4" thick R034 -030	C-11	3,600	.020	S.F.	4.65	.60	.43	5.68	6.65	136
	0050	6" thick		4,500	.016		4.48	.48	.35	5.31	6.15	
	0100	8" thick, hollow CN		5,600	.013		4.83	.39	.28	5.50	6.30	
	0150	10" thick		8,800	.008		5.70	.25	.18	6.13	6.85	
	0200	12" thick		8,000	.009		4.78	.27	.19	5.24	5.95	
140	0010	TEES Prestressed R034 -030										140
	0020	Quad tee, short spans, roof	C-11	7,200	.010	S.F.	3.95	.30	.22	4.47	5.10	
	0050	Floor		7,200	.010		4	.30	.22	4.52	5.15	
	0200	Double tee, floor members, 60' span		8,400	.009		6.10	.26	.18	6.54	7.35	
	0250	80' span		8,000	.009		7.30	.27	.19	7.76	8.70	
	0300	Roof members, 30' span		4,800	.015		4.65	.45	.32	5.42	6.25	
	0350	50' span		6,400	.011		5.50	.34	.24	6.08	6.85	
	0400	Wall members, up to 55' high		3,600	.020		6.25	.60	.43	7.28	8.40	
	0500	Single tee roof members, 40' span		3,200	.023		5.70	.67	.49	6.86	8	
	0550	80' span		5,120	.014		7.15	.42	.30	7.87	8.95	
	0600	100' span		6,000	.012		11	.36	.26	11.62	13	
	0650	120' span		6,000	.012		11.50	.36	.26	12.12	13.55	
	1000	Double tees, floor members										
	1100	Lightweight, 20" x 8' wide, 45' span	C-11	20	3.600	Ea.	1,875	108	77.50	2,060.50	2,325	
	1150	24" x 8' wide, 50' span		18	4		2,025	120	86.50	2,231.50	2,525	
	1200	32" x 10' wide, 60' span		16	4.500		3,425	135	97	3,657	4,125	
	1250	Standard weight, 12" x 8' wide, 20' span		22	3.273		700	98	70.50	868.50	1,025	
	1300	16" x 8' wide, 25' span		20	3.600		930	108	77.50	1,115.50	1,300	
	1350	18" x 8' wide, 30' span CN		20	3.600		1,175	108	77.50	1,360.50	1,575	
	1400	20" x 8' wide, 45' span		18	4		1,325	120	86.50	1,531.50	1,775	
	1450	24" x 8' wide, 50' span		16	4.500		1,725	135	97	1,957	2,250	
	1500	32" x 10' wide, 60' span		14	5.143		3,150	154	111	3,415	3,875	
	2000	Roof members										
	2050	Lightweight, 20" x 8' wide, 40' span	C-11	20	3.600	Ea.	1,425	108	77.50	1,610.50	1,850	
	2100	24" x 8' wide, 50' span		18	4		1,900	120	86.50	2,106.50	2,400	
	2150	32" x 10' wide, 60' span		16	4.500		3,150	135	97	3,382	3,825	
	2200	Standard weight, 12" x 8' wide, 30' span		22	3.273		955	98	70.50	1,123.50	1,300	
	2250	16" x 8' wide, 30' span		20	3.600		1,000	108	77.50	1,185.50	1,375	
	2300	18" x 8' wide, 30' span		20	3.600		1,125	108	77.50	1,310.50	1,500	
	2350	20" x 8' wide, 40' span		18	4		1,150	120	86.50	1,356.50	1,550	

CONCRETE 3

034 100 | Structural Precast

			CREW	DAILY OUTPUT	LABOR-HOURS	UNIT	1999 BARE COSTS				TOTAL INCL O&P	
							MAT.	LABOR	EQUIP.	TOTAL		
140	2400	24" x 8' wide, 50' span	C-11	16	4.500	Ea.	1,525	135	97	1,757	2,025	**140**
	2450	32" x 10' wide, 60' span	↓	14	5.143	↓	2,675	154	111	2,940	3,350	

R034 -030

034 500 | Architectural Precast

			CREW	DAILY OUTPUT	LABOR-HOURS	UNIT	MAT.	LABOR	EQUIP.	TOTAL	TOTAL INCL O&P	
504	0011	**WALL PANELS** Material only										**504**
	0150	Low rise, 4' x 8'x 4" thick	C-11	320	.225	S.F.	8.70	6.75	4.85	20.30	27	
	0200	8'x 8' x 4" thick		576	.125		8.60	3.74	2.70	15.04	19.05	
	0250	8'x 16'x 4" thick		1,024	.070		8.55	2.11	1.52	12.18	14.80	
	0400	8'x 8', 4" thick, smooth gray		576	.125		8.60	3.74	2.70	15.04	19.05	
	0500	Exposed aggregate		576	.125		9.50	3.74	2.70	15.94	20	
	0600	High rise, 4' x 8' x 4" thick		288	.250		8.70	7.50	5.40	21.60	29	
	0650	8' x 8'x 4" thick		512	.141		8.60	4.21	3.03	15.84	20	
	0700	8' x 16' x 4" thick	↓	768	.094		8.55	2.81	2.02	13.38	16.60	
	0800	Insulated panel, 2" polystyrene, add					1.60			1.60	1.76	
	0850	2" urethane, add					2.10			2.10	2.31	
	1000	20' x 10', 6" thick, smooth gray CN	C-11	1,800	.040		11	1.20	.86	13.06	15.15	
	1100	Exposed aggregate	"	1,800	.040		12	1.20	.86	14.06	16.25	
	1200	Finishes, white, add					1.70			1.70	1.87	
	1250	Exposed aggregate, add					.80			.80	.88	
	1300	Granite faced, domestic, add				↓	27			27	29.50	
	2200	Fiberglass reinforced cement with urethane core										
	2210	R20, 8' x 8', minimum	E-2	750	.075	S.F.	8.30	2.22	1.35	11.87	14.55	
	2220	Maximum	"	600	.093	"	15.50	2.78	1.69	19.97	24	

R034 -010

034 700 | Tilt-Up Precast

			CREW	DAILY OUTPUT	LABOR-HOURS	UNIT	MAT.	LABOR	EQUIP.	TOTAL	TOTAL INCL O&P	
704	0010	**TILT-UP** Wall panel construction, walls only, 5-1/2" thick	C-14	1,600	.090	S.F.	2.44	2.39	.69	5.52	7.30	**704**
	0100	7-1/2" thick		1,550	.093		2.80	2.47	.71	5.98	7.80	
	0500	Walls and columns, 5-1/2" thick walls, 12" x 12" columns		1,565	.092		3	2.45	.70	6.15	8	
	0550	7-1/2" thick wall, 12" x 12" columns		1,370	.105	↓	3.90	2.80	.80	7.50	9.65	
	0800	Columns only, site precast, minimum		200	.720	L.F.	23	19.15	5.50	47.65	62	
	0850	Maximum	↓	105	1.371	"	42	36.50	10.45	88.95	116	

R034 -020

034 800 | Precast Specialties

			CREW	DAILY OUTPUT	LABOR-HOURS	UNIT	MAT.	LABOR	EQUIP.	TOTAL	TOTAL INCL O&P	
802	0010	**LINTELS**										**802**
	0800	Precast concrete, 4" wide, 8" high, to 5' long	D-1	175	.091	L.F.	5.60	2.25		7.85	9.65	
	0850	5'-12' long	D-4	190	.168		5.85	4.13	.58	10.56	13.50	
	1000	6" wide, 8" high, to 5' long		185	.173		8.35	4.24	.59	13.18	16.40	
	1050	5'-12' long		190	.168		6.95	4.13	.58	11.66	14.70	
	1200	8" wide, 8" high, to 5' long		185	.173		9.60	4.24	.59	14.43	17.75	
	1250	5'-12' long		190	.168		8.85	4.13	.58	13.56	16.80	
	1400	10" wide, 8" high, to 14' long		180	.178		31.50	4.36	.61	36.47	42.50	
	1450	12" wide, 8" high, to 19' long	↓	185	.173	↓	33.50	4.24	.59	38.33	44	
804	0010	**STAIRS**, Precast concrete treads on steel stringers, 3' wide	C-12	75	.640	Riser	53.50	17.30	5.90	76.70	92.50	**804**
	0300	Front entrance, 5' wide with 48" platform, 2 risers		16	3	Flight	276	81	27.50	384.50	465	
	0350	5 risers		12	4		315	108	37	460	555	
	0500	6' wide, 2 risers		15	3.200		315	86.50	29.50	431	515	
	0550	5 risers		11	4.364		335	118	40.50	493.50	600	
	0700	7' wide, 2 risers		14	3.429		380	92.50	31.50	504	600	
	0750	5 risers	↓	10	4.800		410	130	44.50	584.50	705	
	1200	Basement entrance stairs, steel bulkhead doors, minimum	B-51	22	2.182		460	47.50	7.80	515.30	590	
	1250	Maximum	"	11	4.364	↓	680	95.50	15.60	791.10	915	

035 200 | Lightweight Concrete

			CREW	DAILY OUTPUT	LABOR-HOURS	UNIT	MAT.	LABOR	EQUIP.	TOTAL	TOTAL INCL O&P	
204	0010	**CONCRETE CHANNEL SLABS** 2-3/4" or 3-1/2" thick, straight	C-12	1,575	.030	S.F.	3	.82	.28	4.10	4.90	**204**
	0050	Chopped up		785	.061		3	1.65	.56	5.21	6.50	
	0200	6" thick, span to 20'		1,300	.037		3.50	1	.34	4.84	5.80	
	0300	8" thick, span to 24'		1,100	.044		4	1.18	.40	5.58	6.70	
208	0010	**CONCRETE PLANK** Lightweight, nailable, T&G, 2" thick	C-12	1,800	.027	S.F.	2.53	.72	.25	3.50	4.18	**208**
	0050	2-3/4" thick		1,575	.030		3.10	.82	.28	4.20	5	
	0100	3-3/4" thick		1,375	.035		3.80	.94	.32	5.06	6	
	0150	For premium ceiling finish, add					.33			.33	.36	
	0200	For sloping roofs, slope over 4 in 12, add						25%				
	0250	Slope over 6 in 12, add						150%				
212	0010	**INSULATING** Lightweight cellular concrete roof fill	R035 -010									**212**
	0020	Portland cement and foaming agent	C-8	50	1.120	C.Y.	72	27	12.90	111.90	135	
	0100	Poured vermiculite or perlite, field mix,										
	0110	1:6 field mix	C-8	50	1.120	C.Y.	100	27	12.90	139.90	166	
	0200	Ready mix, 1:6 mix, roof fill, 2" thick		10,000	.006	S.F.	.56	.13	.06	.75	.89	
	0250	3" thick		7,700	.007		.84	.18	.08	1.10	1.28	
	0400	Expanded volcanic glass rock, with binder, minimum	2 Carp	1,500	.011		.25	.29		.54	.74	
	0450	Maximum	"	1,200	.013		.75	.36		1.11	1.40	
216	0010	**PRESTRESSED** Roof and floor members, see division 034	R034 -030									**216**

035 300 | Wood Fiber Systems

			CREW	DAILY OUTPUT	LABOR-HOURS	UNIT	MAT.	LABOR	EQUIP.	TOTAL	TOTAL INCL O&P	
304	0010	**FORMBOARD** Including sub-purlins	R051 -250									**304**
	0050	Non-asbestos fiber cement, 1/8" thick	C-13	2,950	.008	S.F.	2.10	.24	.03	2.37	2.76	
	0070	1/4" thick		2,950	.008		3.75	.24	.03	4.02	4.58	
	0100	Fiberglass, 1" thick, economy		2,700	.009		.95	.26	.03	1.24	1.54	
	1000	Poured gypsum, 2" thick, add to formboard above	C-8	6,000	.009		.75	.22	.11	1.08	1.30	
	1100	3" thick	"	4,800	.012		1.05	.28	.13	1.46	1.73	
306	0010	**WOOD FIBER** Lightweight cement system	R051 -250									**306**
	0050	Plank, beveled, 1" thick	2 Carp	1,000	.016	S.F.	1.85	.44		2.29	2.73	
	0100	Plank, T & G, 1-1/2" thick		975	.016		2.15	.45		2.60	3.07	
	0150	2" thick		950	.017		2.90	.46		3.36	3.91	
	0200	2-1/2" thick		925	.017		3	.47		3.47	4.04	
	0250	3" thick		900	.018		3.50	.49		3.99	4.61	
	0300	3-1/2" thick		875	.018		5	.50		5.50	6.30	
	0350	4" thick		850	.019		5.50	.51		6.01	6.85	
	1000	Bulb tee, sub-purlin and grout, 6' span, add	E-1	5,000	.005		1.62	.14	.02	1.78	2.05	
	1100	8' span	"	4,200	.006		1.67	.17	.02	1.86	2.16	
308	0010	**WOOD PLANK** Roof decks, see division 061-250 & 061-808										**308**

For information about Means Estimating Seminars, see yellow pages 11 and 12 in back of book

For expanded coverage of these items see *Means Concrete & Masonry Cost Data 1999*

	CREW	DAILY OUTPUT	LABOR-HOURS	UNIT	1999 BARE COSTS				TOTAL INCL O&P
					MAT.	LABOR	EQUIP.	TOTAL	

Important: See the Reference Section for critical supporting data - Reference Nos., Crews, & City Cost Indexes

Division 4
Masonry

Estimating Tips

041 Mortar & Masonry Accessories

- The terms *mortar* and *grout* are often used interchangeably, and incorrectly. Mortar is used to bed masonry units, seal the entry of air and moisture, provide architectural appearance, and allow for size variations in the units. Grout is used primarily in reinforced masonry construction and is used to bond the masonry to the reinforcing steel. Common mortar types are M(2500 psi), S(1800 psi), N(750 psi), and O(350 psi), and conform to ASTM C270. Grout is either fine or coarse, conforms to ASTM C476, and in-place strengths generally exceed 2500 psi. Mortar and grout are different components of masonry construction and are placed by entirely different methods. An estimator should be aware of their unique uses and costs.

042 Unit Masonry

- The most common types of unit masonry are brick and concrete masonry. The major classifications of brick are building brick (ASTM C62), facing brick (ASTM C216) and glazed brick, fire brick and pavers. Many varieties of texture and appearance can exist within these classifications, and the estimator would be wise to check local custom and availability within the project area. On repair and remodeling jobs, matching the existing brick may be the most important criteria.
- Brick and concrete block are priced by the piece and then converted into a price per square foot of wall. Openings less than two square feet are generally ignored by the estimator because any savings in units used is offset by the cutting and trimming required.

- All masonry walls, whether interior or exterior, require bracing. The cost of bracing walls during construction should be included by the estimator and this bracing must remain in place until permanent bracing is complete. Permanent bracing of masonry walls is accomplished by masonry itself, in the form of pilasters or abutting wall corners, or by anchoring the walls to the structural frame. Accessories in the form of anchors, anchor slots and ties are used, but their supply and installation can be by different trades. For instance, anchor slots on spandrel beams and columns are supplied and welded in place by the steel fabricator, but the ties from the slots into the masonry are installed by the bricklayer. Regardless of the installation method the estimator must be certain that these accessories are accounted for in pricing.

Reference Numbers

Reference numbers are shown in bold squares at the beginning of some major classifications. These numbers refer to related items in the Reference Section. The reference information may be an estimating procedure, an alternate pricing method or technical information.

Note: Not all subdivisions listed here necessarily appear in this publication.

		041 000 | Mortar	CREW	DAILY OUTPUT	LABOR-HOURS	UNIT	1999 BARE COSTS MAT.	LABOR	EQUIP.	TOTAL	TOTAL INCL O&P	
008	0010	**CEMENT** Gypsum 80 lb. bag, T.L. lots				Bag	11.25			11.25	12.40	008
	0050	L.T.L. lots					11.75			11.75	12.95	
	0100	Masonry, 70 lb. bag, T.L. lots **CN**					5.80			5.80	6.40	
	0150	L.T.L. lots					5.60			5.60	6.15	
	0200	White, 70 lb. bag, T.L. lots					15.85			15.85	17.45	
	0250	L.T.L. lots					16.40			16.40	18.05	
012	0010	**COLORS** 50 lb. bags (2 bags per M bricks), R041-100										012
	0020	range 2 to 10 lb. per bag of cement, minimum				Lb.	1.50			1.50	1.65	
	0050	Average					2.50			2.50	2.75	
	0100	Maximum					5			5	5.50	
016	0010	**GROUTING** Bond bms. & lintels, 8" dp., pumped, not incl. block R041-100										016
	0020	8" thick, 0.2 C.F. per L.F.	D-4	1,400	.023	L.F.	.66	.56	.08	1.30	1.69	
	0050	10" thick, 0.25 C.F. per L.F.		1,200	.027		.83	.65	.09	1.57	2.02	
	0060	12" thick, 0.3 C.F. per L.F.		1,040	.031		.99	.75	.11	1.85	2.37	
	0200	Concrete block cores, solid, 4" thk., by hand, 0.067 C.F./S.F.	D-8	1,100	.036	S.F.	.22	.92		1.14	1.66	
	0210	6" thick, pumped, 0.175 C.F. per S.F.	D-4	720	.044		.58	1.09	.15	1.82	2.49	
	0250	8" thick, pumped, 0.258 C.F. per S.F.		680	.047		.85	1.15	.16	2.16	2.90	
	0300	10" thick, pumped, 0.340 C.F. per S.F.		660	.048		1.13	1.19	.17	2.49	3.26	
	0350	12" thick, pumped, 0.422 C.F. per S.F.		640	.050		1.40	1.23	.17	2.80	3.62	
	0500	Cavity walls, 2" space, pumped, no shoring, 0.167 C.F./S.F.		1,700	.019		.55	.46	.06	1.07	1.39	
	0550	3" space, 0.250 C.F./S.F.		1,200	.027		.83	.65	.09	1.57	2.02	
	0600	4" space, 0.333 C.F. per S.F.		1,150	.028		1.10	.68	.10	1.88	2.36	
	0700	6" space, 0.500 C.F. per S.F.		800	.040		1.65	.98	.14	2.77	3.48	
	0800	Door frames, 3' x 7' opening, 2.5 C.F. per opening		60	.533	Opng.	8.25	13.05	1.82	23.12	31	
	0850	6' x 7' opening, 3.5 C.F. per opening		45	.711	"	11.60	17.40	2.43	31.43	42.50	
	2000	Grout, C476, for bond beams, lintels and CMU cores		350	.091	C.F.	3.31	2.24	.31	5.86	7.45	
020	0010	**LIME** Masons, hydrated, 50 lb. bag, T.L. lots **CN**				Bag	5.40			5.40	5.95	020
	0050	L.T.L. lots					5.65			5.65	6.20	
	0200	Finish, double hydrated, 50 lb. bag, T.L. lots					6.50			6.50	7.15	
	0250	L.T.L. lots					7.40			7.40	8.15	
024	0010	**MORTAR** R041-100										024
	0020	With masonry cement										
	0100	Type M, 1:1:6 mix	1 Brhe	143	.056	C.F.	3.17	1.21		4.38	5.35	
	0200	Type N, 1:3 mix R041-200	"	143	.056	"	2.96	1.21		4.17	5.15	
	2000	With portland cement and lime										
	2100	Type M, 1:1/4:3 mix	1 Brhe	143	.056	C.F.	3.82	1.21		5.03	6.10	
	2200	Type N, 1:1:6 mix, 750 psi		143	.056		3.11	1.21		4.32	5.30	
	2300	Type O, 1:2:9 mix (Pointing Mortar)		143	.056		3.02	1.21		4.23	5.20	
	2700	Mortar for glass block		143	.056		7.25	1.21		8.46	9.85	
	2800	Gypsum cement mortar					5.60			5.60	6.15	
	2900	Mortar for Fire Brick, 80 lb. bag, T.L. Lots				Bag	12.60			12.60	13.85	
028	0010	**POINTING MORTAR** See also Division 041-024				Lb.	.86			.86	.95	028
	0050	White				"	1.02			1.02	1.12	
032	0010	**SAND** For mortar, screened and washed, at the pit **CN**				Ton	14.45			14.45	15.90	032
	0050	With 10 mile haul					15.75			15.75	17.30	
	0100	With 30 mile haul					17			17	18.70	
	0200	Screened and washed, at the pit				C.Y.	20			20	22	
	0250	With 10 mile haul					22			22	24	
	0300	With 30 mile haul					23.50			23.50	26	
036	0010	**SURFACE BONDING** CMU walls with fiberglass mortar,										036
	0020	gray or white colors, not incl. block work	1 Bric	540	.015	S.F.	.50	.41		.91	1.19	
040	0010	**WATERPROOFING** Admixture, 1 qt. to 2 bags of masonry cement				Qt.	6			6	6.60	040

Important: See the Reference Section for critical supporting data - Reference Nos., Crews, & City Cost Indexes

041 | Mortar & Masonry Accessories

041 500 | Masonry Accessories

			CREW	DAILY OUTPUT	LABOR-HOURS	UNIT	MAT.	LABOR	EQUIP.	TOTAL	TOTAL INCL O&P	
								1999 BARE COSTS				
504	0010	ANCHOR BOLTS Hooked type with nut and washer, 1/2" diam., 8" long	1 Bric	200	.040	Ea.	.52	1.10		1.62	2.29	504
	0030	12" long		190	.042		1.22	1.16		2.38	3.15	
	0040	5/8" diameter, 8" long		180	.044		1.25	1.23		2.48	3.29	
	0050	12" long		170	.047		1.38	1.30		2.68	3.53	
	0060	3/4" diameter, 8" long		160	.050		1.70	1.38		3.08	4.03	
	0070	12" long		150	.053		2.11	1.47		3.58	4.61	
508	0010	CONTROL JOINT Rubber, 4" and wider wall	1 Bric	400	.020	L.F.	2.40	.55		2.95	3.50	508
	0050	PVC, 4" wall		400	.020		1.36	.55		1.91	2.36	
	0100	Rubber, 6" wall		320	.025		4.30	.69		4.99	5.80	
	0120	PVC, 6" wall		320	.025		1.62	.69		2.31	2.85	
	0140	Rubber, 8" and wider wall		280	.029		4.40	.79		5.19	6.05	
	0160	PVC, 8" wall		280	.029		1.67	.79		2.46	3.07	
	0180	12" wall		240	.033		2.79	.92		3.71	4.50	
512	0010	REINFORCING Steel bars A615, placed horiz., #3 & #4 bars R041 -500	1 Bric	450	.018	Lb.	.27	.49		.76	1.05	512
	0020	#5 & #6 bars		800	.010		.27	.28		.55	.72	
	0050	Placed vertical, #3 & #4 bars		350	.023		.27	.63		.90	1.27	
	0060	#5 & #6 bars		650	.012		.27	.34		.61	.82	
	0200	Joint reinforcing, regular truss, to 6" wide, mill std galvanized		30	.267	C.L.F.	10.20	7.35		17.55	22.50	
	0250	12" wide		20	.400		11.75	11.05		22.80	30	
	0400	Cavity truss with drip section, to 6" wide		30	.267		9.70	7.35		17.05	22	
	0450	12" wide		20	.400		11.05	11.05		22.10	29.50	
518	0010	WALL PLUGS For nailing to brickwork, 26 ga., galvanized, plain	1 Bric	10.50	.762	C	23.50	21		44.50	58.50	518
	0050	Wood filled		10.50	.762	"	80	21		101	121	
520	0010	WALL TIES To brick veneer, galv., corrugated, 7/8" x 7", 22 Ga.	1 Bric	10.50	.762	C	4.47	21		25.47	37.50	520
	0100	24 Ga.		10.50	.762		3.97	21		24.97	37	
	0150	16 Ga.		10.50	.762		13.45	21		34.45	47.50	
	0200	Buck anchors, galv., corrugated, 16 gauge, 2" bend. 8" x 2"		10.50	.762		93	21		114	135	
	0250	8" x 3"		10.50	.762		99	21		120	142	
	0600	Cavity wall, Z type, galvanized, 6" long, 1/4" diameter		10.50	.762		17.45	21		38.45	51.50	
	0650	3/16" diameter		10.50	.762		9.50	21		30.50	43	
	0800	8" long, 1/4" diameter		10.50	.762		23.50	21		44.50	58.50	
	0850	3/16" diameter		10.50	.762		10.45	21		31.45	44	
	1000	Rectangular type, galvanized, 1/4" diameter, 2" x 6"		10.50	.762		38.50	21		59.50	75	
	1050	2" x 8" or 4" x 6"		10.50	.762		29	21		50	64	
	1100	3/16" diameter, 2" x 6"		10.50	.762		17.45	21		38.45	51.50	
	1150	2" x 8" or 4" x 6"		10.50	.762		19.40	21		40.40	54	
	1500	Rigid partition anchors, plain, 8" long, 1" x 1/8"		10.50	.762		48	21		69	85.50	
	1550	1" x 1/4"		10.50	.762		94	21		115	136	
	1580	1-1/2" x 1/8"		10.50	.762		66.50	21		87.50	106	
	1600	1-1/2" x 1/4"		10.50	.762		158	21		179	206	
	1650	2" x 1/8"		10.50	.762		83	21		104	124	
	1700	2" x 1/4"		10.50	.762		225	21		246	281	
524	0010	VENT BOX Extruded aluminum, 4" deep, 2-3/8" x 8-1/8"	1 Bric	30	.267	Ea.	31.50	7.35		38.85	46	524
	0050	5" x 8-1/8"		25	.320		43.50	8.85		52.35	61.50	
	0100	2-1/4" x 25"		25	.320		60	8.85		68.85	80	
	0150	5" x 16-1/2"		22	.364		44	10.05		54.05	64	
	0200	6" x 16-1/2"		22	.364		65	10.05		75.05	87	
	0250	7-3/4" x 16-1/2"		20	.400		80	11.05		91.05	105	
	0400	For baked enamel finish, add					35%					
	0500	For cast aluminum, painted, add					60%					
	1000	Stainless steel ventilators, 6" x 6"	1 Bric	25	.320		90	8.85		98.85	113	
	1050	8" x 8"		24	.333		95	9.20		104.20	119	
	1100	12" x 12"		23	.348		110	9.60		119.60	136	
	1150	12" x 6"		24	.333		95	9.20		104.20	119	

MASONRY 4

For expanded coverage of these items see *Means Concrete & Masonry Cost Data 1999*

041 500	Masonry Accessories		CREW	DAILY OUTPUT	LABOR-HOURS	UNIT	1999 BARE COSTS				TOTAL INCL O&P	
							MAT.	LABOR	EQUIP.	TOTAL		
524	1200	Foundation block vent, galv., 1-1/4" thk, 8" high, 16" long, no damper	1 Bric	30	.267	Ea.	16	7.35		23.35	29	524
	1250	For damper, add	↓			↓	5			5	5.50	

042 | Unit Masonry

042 050	Chimneys		CREW	DAILY OUTPUT	LABOR-HOURS	UNIT	1999 BARE COSTS				TOTAL INCL O&P	
							MAT.	LABOR	EQUIP.	TOTAL		
054	0010	CHIMNEY See Div. 033-130 for foundation, add to prices below										054
	0100	Brick, 16" x 16", 8" flue, scaff. not incl.	D-1	18.20	.879	V.L.F.	13.95	21.50		35.45	49	
	0150	16" x 20" with one 8" x 12" flue	R042 -050	16	1		17.15	24.50		41.65	57.50	
	0200	16" x 24" with two 8" x 8" flues		14	1.143		22.50	28		50.50	68.50	
	0250	20" x 20" with one 12" x 12" flue		13.70	1.168		19.95	28.50		48.45	66.50	
	0300	20" x 24" with two 8" x 12" flues		12	1.333		26.50	33		59.50	80	
	0350	20" x 32" with two 12" x 12" flues	↓	10	1.600		33	39.50		72.50	97.50	
	1800	Metal, high temp. steel jacket, factory lining, 24" diam.	E-2	65	.862		174	25.50	15.55	215.05	253	
	1900	60" diameter	"	30	1.867		630	55.50	33.50	719	830	
	2100	Poured concrete, brick lining, 200' high x 10' diam.					5,150			5,150	5,675	
	2200	200' high x 18' diameter									4,995	
	2400	250' high x 9' diameter									3,920	
	2500	300' x 14' diameter									4,690	
	2700	400' high x 18' diameter									5,255	
	2800	500' x 20' diameter	↓			↓	9,200			9,200	10,100	

042 100	Brick Masonry		CREW	DAILY OUTPUT	LABOR-HOURS	UNIT	1999 BARE COSTS				TOTAL INCL O&P	
							MAT.	LABOR	EQUIP.	TOTAL		
104	0010	CAULKING See division 079-204										104
108	0010	COLUMNS Brick, scaffolding not included	R042 -100									108
	0050	8" x 8", 9 brick	D-1	56	.286	V.L.F.	3.32	7.05		10.37	14.60	
	0100	12" x 8", 13.5 brick		37	.432		4.99	10.65		15.64	22	
	0200	12" x 12", 20 brick		25	.640		7.40	15.75		23.15	32.50	
	0300	16" x 12", 27 brick		19	.842		9.95	20.50		30.45	43	
	0400	16" x 16", 36 brick		14	1.143		13.30	28		41.30	58	
	0500	20" x 16", 45 brick		11	1.455		16.60	36		52.60	74	
	0600	20" x 20", 56 brick		9	1.778		20.50	43.50		64	91	
	0700	24" x 20", 68 brick		7	2.286		25	56		81	115	
	0800	24" x 24", 81 brick		6	2.667		30	65.50		95.50	135	
	1000	36" x 36", 182 brick	↓	3	5.333	↓	67	131		198	278	
110	0010	COMMON BUILDING BRICK C62, TL lots, material only	R042 -120									110
	0020	Standard, minimum				M	250			250	275	
	0050	Average (select)	↓			"	310			310	340	
116	0010	COPING Stock units										116
	0050	Precast concrete, 10" wide, 4" tapers to 3-1/2", 8" wall	D-1	75	.213	L.F.	9.60	5.25		14.85	18.70	
	0100	12" wide, 3-1/2" tapers to 3", 10" wall		70	.229		10.10	5.60		15.70	19.85	
	0110	14" wide, 4" tapers to 3-1/2", 14" wall		65	.246		9.10	6.05		15.15	19.40	
	0150	16" wide, 4" tapers to 3-1/2", 14" wall		60	.267	↓	13.90	6.55		20.45	25.50	
	0250	Precast concrete corners		40	.400	Ea.	23.50	9.85		33.35	41.50	
	0300	Limestone for 12" wall, 4" thick		90	.178	L.F.	12.90	4.37		17.27	21	
	0350	6" thick	↓	80	.200	↓	15.15	4.92		20.07	24.50	

042 100 | Brick Masonry

			CREW	DAILY OUTPUT	LABOR-HOURS	UNIT	MAT.	LABOR	EQUIP.	TOTAL	TOTAL INCL O&P	
							1999 BARE COSTS					
116	0500	Marble, to 4" thick, no wash, 9" wide	D-1	90	.178	L.F.	18.70	4.37		23.07	27.50	116
	0550	12" wide		80	.200		27.50	4.92		32.42	38	
	0700	Terra cotta, 9" wide		90	.178		4.55	4.37		8.92	11.80	
	0750	12" wide		80	.200		7.60	4.92		12.52	16	
	0800	Aluminum, for 12" wall		80	.200		10.75	4.92		15.67	19.45	
124	0010	**FACE BRICK** C216, TL lots, material only R042 -120										124
	0300	Standard modular, 4" x 2-2/3" x 8", minimum				M	330			330	365	
	0350	Maximum **CN**					450			450	495	
	0450	Economy, 4" x 4" x 8", minimum					425			425	470	
	0500	Maximum					545			545	600	
	0510	Economy, 4" x 4" x 12", minimum					425			425	470	
	0520	Maximum					545			545	600	
	0550	Jumbo, 6" x 4" x 12", minimum					1,150			1,150	1,250	
	0600	Maximum					1,425			1,425	1,575	
	0610	Jumbo, 8" x 4" x 12", minimum					1,150			1,150	1,250	
	0620	Maximum					1,425			1,425	1,575	
	0650	Norwegian, 4" x 3-1/5" x 12", minimum					520			520	570	
	0700	Maximum					730			730	805	
	0710	Norwegian, 6" x 3-1/5" x 12", minimum					520			520	570	
	0720	Maximum					730			730	805	
	0850	Standard glazed, plain colors, 4" x 2-2/3" x 8", minimum					985			985	1,075	
	0900	Maximum					1,150			1,150	1,275	
	1000	Deep trim shades, 4" x 2-2/3" x 8", minimum					1,150			1,150	1,250	
	1050	Maximum					1,300			1,300	1,425	
	1080	Jumbo utility, 4" x 4" x 12"					750			750	825	
	1120	4" x 8" x 8"					1,100			1,100	1,200	
	1140	4" x 8" x 16"					2,500			2,500	2,750	
	1260	Engineer, 4" x 3-1/5" x 8", minimum					295			295	325	
	1270	Maximum					485			485	535	
	1350	King, 4" x 2-3/4" x 10", minimum					415			415	455	
	1360	Maximum					440			440	485	
	1400	Norman, 4" x 2-3/4" x 12"					415			415	455	
	1450	Roman, 4" x 2" x 12"					415			415	455	
	1500	SCR, 6" x 2-2/3" x 12"					415			415	455	
	1550	Double, 4" x 5-1/3" x 8"					415			415	455	
	1600	Triple, 4" x 5-1/3" x 12"					415			415	455	
	1770	Standard modular, double glazed, 4" x 2-2/3" x 8"					1,000			1,000	1,100	
	1850	Jumbo, colored glazed ceramic, 6" x 4" x 12"					1,500			1,500	1,650	
	2050	Jumbo utility, glazed, 4" x 4" x 12"					1,075			1,075	1,175	
	2100	4" x 8" x 8"					1,500			1,500	1,650	
	2150	4" x 16" x 8"					3,000			3,000	3,300	
	2160	Fire Brick, 2" x 2-2/3" x 9", minimum										
	2165	Maximum				M	910			910	1,000	
	2170	For less than truck load lots, add					10					
	2180	For buff or gray brick, add					15					
134	0010	**LINTELS** See division 051-232										134
154	0010	**SAWING** Brick or block by hand, per inch depth	D-5	300	.027	L.F.		.74		.74	1.14	154
158	0010	**SCAFFOLDING & SWING STAGING** Div. 015-254, 255 & 256 R015 -100										158
162	0010	**SIMULATED BRICK** Aluminum, baked on colors	1 Carp	200	.040	S.F.	2.15	1.09		3.24	4.09	162
	0050	Fiberglass panels		200	.040		2.25	1.09		3.34	4.20	
	0100	Urethane pieces cemented in mastic		150	.053		4.60	1.46		6.06	7.35	
	0150	Vinyl siding panels		200	.040		1.90	1.09		2.99	3.81	

MASONRY 4

042 100 | Brick Masonry

		CREW	DAILY OUTPUT	LABOR-HOURS	UNIT	1999 BARE COSTS				TOTAL INCL O&P	
						MAT.	LABOR	EQUIP.	TOTAL		
162	0160	Cement base, brick, incl. mastic	D-1	100	.160	S.F.	3.10	3.94		7.04	9.50	**162**
	0170	Corner		50	.320	V.L.F.	7.35	7.85		15.20	20.50	
	0180	Stone face, incl. mastic		100	.160	S.F.	6.95	3.94		10.89	13.75	
	0190	Corner		50	.320	V.L.F.	8	7.85		15.85	21	
166	0010	**STEAM CLEAN** Building, not incl. scaffolding, minimum	B-9	3,000	.013	S.F.		.29	.06	.35	.53	**166**
	0100	Maximum		1,500	.027	"		.58	.12	.70	1.05	
	0300	Common face brick		1,750	.023	S.F.		.50	.10	.60	.90	
	0400	Wire cut face brick		1,250	.032	"		.70	.14	.84	1.26	
170	0010	**STEPS** With select common	D-1	.30	53.333	M	310	1,300		1,610	2,400	**170**
182	0010	**VENT BOX** See division 041-524										**182**
184	0010	**WALLS** Building brick, including mortar										**184**
	0060	Includes 3% brick waste and 25% mortar waste	R042 -120									
	0140	4" thick, facing, 4" x 2-2/3" x 8"	R042 -180 D-8	1.45	27.586	M	290	695		985	1,400	
	0150	4" thick, as back-up, 6.75 bricks per S.F.		1.60	25		290	630		920	1,300	
	0204	8" thick, 13.50 bricks per S.F.	R042 -185	1.80	22.222		297	560		857	1,200	
	0250	12" thick, 20.25 bricks per S.F.		1.90	21.053		299	530		829	1,150	
	0304	16" thick, 27.00 bricks per S.F.		2	20		300	505		805	1,125	
	0500	Reinforced, 4" wall, 4" x 2-2/3" x 8"		1.40	28.571		300	720		1,020	1,450	
	0550	8" thick, 13.50 bricks per S.F.		1.75	22.857		370	575		945	1,300	
	0600	12" thick, 20.25 bricks per S.F.		1.85	21.622		375	545		920	1,250	
	0650	16" thick, 27.00 bricks per S.F.		1.95	20.513		375	515		890	1,225	
	0790	Alternate method of figuring by square foot										
	0800	4" wall, face, 4" x 2-2/3" x 8"	D-8	215	.186	S.F.	2.51	4.69		7.20	10.05	
	0850	4" thick, as back up, 6.75 bricks per S.F.		240	.167		1.96	4.20		6.16	8.70	
	0900	8" thick wall, 13.50 brick per S.F.		135	.296		4.01	7.45		11.46	16	
	1000	12" thick wall, 20.25 bricks per S.F.		95	.421		6.05	10.60		16.65	23	
	1050	16" thick wall, 27.00 bricks per S.F.		75	.533		8.15	13.45		21.60	30	
	1200	Reinforced, 4" x 2-2/3" x 8" , 4" wall		205	.195		1.97	4.92		6.89	9.80	
	1250	8" thick wall, 13.50 brick per S.F.		130	.308		4.02	7.75		11.77	16.50	
	1300	12" thick wall, 20.25 bricks per S.F.		90	.444		6.05	11.20		17.25	24	
	1350	16" thick wall, 27.00 bricks per S.F.		70	.571		8.15	14.40		22.55	31.50	
186	0010	**WALLS** Cavity, brick and CMU, no scaffolding										**186**
	0200	4" face brick, 4" block	D-8	165	.242	S.F.	3.40	6.10		9.50	13.25	
	0400	4" face brick, 6" block		145	.276		3.72	6.95		10.67	14.90	
	0600	4" face brick, 8" block		125	.320		3.80	8.05		11.85	16.75	
194	0010	**WINDOW SILL** Bluestone, thermal top, 10" wide, 1-1/2" thick	D-1	85	.188	S.F.	12.35	4.63		16.98	21	**194**
	0050	2" thick		75	.213	"	14.25	5.25		19.50	24	
	0100	Cut stone, 5" x 8" plain		48	.333	L.F.	10	8.20		18.20	24	
	0200	Face brick on edge, brick, 8" wide		80	.200		1.85	4.92		6.77	9.70	
	0400	Marble, 9" wide, 1" thick		85	.188		7	4.63		11.63	14.90	
	0600	Precast concrete, 4" tapers to 3", 9" wide		70	.229		9.05	5.60		14.65	18.70	
	0650	11" wide		60	.267		8.35	6.55		14.90	19.35	
	0700	13" wide, 3 1/2" tapers to 2 1/2", 12" wall		50	.320		12.15	7.85		20	25.50	
	0900	Slate, colored, unfading, honed, 12" wide, 1" thick		85	.188		14.75	4.63		19.38	23.50	
	0950	2" thick		70	.229		20.50	5.60		26.10	31.50	

042 200 | Concrete Unit Masonry

		CREW	DAILY OUTPUT	LABOR-HOURS	UNIT	MAT.	LABOR	EQUIP.	TOTAL	TOTAL INCL O&P	
216	0010	**CONCRETE BLOCK, BACK-UP** Scaffolding not included										**216**
	0020	Sand aggregate, 8" x 16" units, tooled joint 1 side	R042 -200									
	1000	Reinforced, alternate courses, 4" thick	D-8	435	.092	S.F.	.88	2.32		3.20	4.57	
	1100	6" thick		415	.096		1.24	2.43		3.67	5.15	
	1150	8" thick		395	.101		1.37	2.55		3.92	5.50	
	1200	10" thick		385	.104		1.95	2.62		4.57	6.20	

Important: See the Reference Section for critical supporting data - Reference Nos., Crews, & City Cost Indexes

042 200 | Concrete Unit Masonry

			DAILY	LABOR-			1999 BARE COSTS			TOTAL	
			CREW	OUTPUT	HOURS	UNIT	MAT.	LABOR	EQUIP.	TOTAL	INCL O&P
216	[1250	12" thick	D-9	365	.132	S.F.	2.12	3.24		5.36	7.40	216
220	0010	**CONCRETE BLOCK, DECORATIVE** Scaffolding not included										220
	0020	Embossed, simulated brick face									
	0100	8" x 16" units, 4" thick	D-8	400	.100	S.F.	2.11	2.52		4.63	6.25
	0200	8" thick		340	.118		2.90	2.96		5.86	7.80
	0250	12" thick		300	.133		3.73	3.36		7.09	9.30
	0400	Embossed both sides									
	0500	8" thick	D-8	300	.133	S.F.	3.22	3.36		6.58	8.75
	0550	12" thick	"	275	.145	"	4.27	3.67		7.94	10.40
	1000	Fluted high strength									
	1100	Flutes 1 side, 8" x 16" x 4" thick	D-8	345	.116	S.F.	2.59	2.92		5.51	7.40
	1150	Flutes 2 sides, 8" x 16" x 4" thick		335	.119		3.11	3.01		6.12	8.10
	1200	8" thick		300	.133		4.01	3.36		7.37	9.60
	1250	For special colors, add					.23			.23	.25
	1400	Deep grooved, smooth face									
	1450	8" x 16" x 4" thick	D-8	345	.116	S.F.	1.63	2.92		4.55	6.35
	1500	8" thick	"	300	.133	"	2.80	3.36		6.16	8.30
	2000	Formblock, incl. inserts & reinforcing									
	2100	8" x 16" x 8" thick	D-8	345	.116	S.F.	2.90	2.92		5.82	7.75
	2150	12" thick	"	310	.129	"	3.56	3.25		6.81	8.95
	2500	Ground face									
	2600	8" x 16" x 4" thick	D-8	345	.116	S.F.	3.89	2.92		6.81	8.80
	2650	6" thick		310	.129		4.32	3.25		7.57	9.80
	2700	8" thick		290	.138		4.99	3.48		8.47	10.90
	2750	12" thick	D-9	265	.181		5.85	4.46		10.31	13.40
	2900	For special colors, add, minimum					15%				
	2950	For special colors, add, maximum					45%				
	4000	Slump block									
	4100	4" face height x 16" x 4" thick	D-1	165	.097	S.F.	3.72	2.39		6.11	7.80
	4150	6" thick		160	.100		4.56	2.46		7.02	8.85
	4200	8" thick		155	.103		6.20	2.54		8.74	10.75
	4250	10" thick		140	.114		9.25	2.81		12.06	14.55
	4300	12" thick		130	.123		9.85	3.03		12.88	15.55
	4400	6" face height x 16" x 6" thick		155	.103		3.73	2.54		6.27	8.05
	4450	8" thick		150	.107		4.86	2.62		7.48	9.45
	4500	10" thick		130	.123		7.40	3.03		10.43	12.85
	4550	12" thick		120	.133		7.75	3.28		11.03	13.60
	5000	Split rib profile units, 1" deep ribs, 8 ribs									
	5100	8" x 16" x 4" thick	D-8	345	.116	S.F.	1.58	2.92		4.50	6.30
	5150	6" thick		325	.123		1.91	3.10		5.01	6.95
	5200	8" thick		305	.131		2.35	3.30		5.65	7.75
	5250	12" thick	D-9	275	.175		3.05	4.29		7.34	10.05
	5400	For special deeper colors, 4" thick, add					.17			.17	.19
	5450	12" thick, add					.32			.32	.35
	5600	For white, 4" thick, add					.69			.69	.76
	5650	6" thick, add					.92			.92	1.01
	5700	8" thick, add					1.13			1.13	1.24
	5750	12" thick, add					1.58			1.58	1.74
	6000	Split face or scored split face									
	6100	8" x 16" x 4" thick	D-8	350	.114	S.F.	1.87	2.88		4.75	6.55
	6150	6" thick		315	.127		2.29	3.20		5.49	7.50
	6200	8" thick		295	.136		3.61	3.42		7.03	9.25
	6250	12" thick	D-9	270	.178		4.84	4.37		9.21	12.15
	6400	For special deeper colors, 4" thick, add					.21			.21	.24
	6450	6" thick, add					.21			.21	.24

MASONRY 4

For expanded coverage of these items see *Means Concrete & Masonry Cost Data 1999*

4 MASONRY

042 200	Concrete Unit Masonry	CREW	DAILY OUTPUT	LABOR-HOURS	UNIT	MAT.	LABOR	EQUIP.	TOTAL	TOTAL INCL O&P	
						1999 BARE COSTS					
220											**220**
6500	8" thick, add				S.F.	.23			.23	.25	
6550	12" thick, add					.40			.40	.44	
6650	For white, 4" thick, add					.75			.75	.82	
6700	6" thick, add					.90			.90	.99	
6750	8" thick, add					1.15			1.15	1.27	
6800	12" thick, add					1.62			1.62	1.78	
7000	Scored ground face, 2 to 5 scores										
7100	8" x 16" x 4" thick	D-8	345	.116	S.F.	3.49	2.92		6.41	8.35	
7150	6" thick		310	.129		4.20	3.25		7.45	9.65	
7200	8" thick		290	.138		4.71	3.48		8.19	10.60	
7250	12" thick	D-9	265	.181		5.95	4.46		10.41	13.50	
8000	Hexagonal face profile units, 8" x 16" units										
8100	4" thick, hollow	D-8	345	.116	S.F.	1.97	2.92		4.89	6.70	
8200	Solid		345	.116		2.80	2.92		5.72	7.60	
8300	6" thick, hollow		310	.129		2.28	3.25		5.53	7.55	
8350	8" thick, hollow		290	.138		3.12	3.48		6.60	8.85	
8500	For stacked bond, add						26%				
8550	For high rise construction, add per story	D-8	67.80	.590	M.S.F.		14.85		14.85	23	
8600	For scored block, add					10%					
8650	For honed or ground face, per face, add				Ea.	.24			.24	.26	
8700	For honed or ground end, per end, add				"	2.43			2.43	2.67	
8750	For bullnose block, add					10%					
8800	For special color, add					13%					
224	0010	**CONCRETE BLOCK, INTERLOCKING** Scaffolding not incl., mortar incl.										**224**
0100	Not including grout or reinforcing										
0200	8" x 16" units, 2,000 psi, 8" thick	R042 -220	D-1	245	.065	S.F.	1.65	1.61		3.26	4.32	
0300	12" thick		220	.073		2.49	1.79		4.28	5.50	
0350	16" thick		185	.086		3.81	2.13		5.94	7.50	
0400	Including grout & reinforcing, 8" thick	D-4	245	.131		5.20	3.20	.45	8.85	11.15	
0450	12" thick		220	.145		6.15	3.56	.50	10.21	12.80	
0500	16" thick		185	.173		7.55	4.24	.59	12.38	15.55	
232	0010	**CONCRETE BLOCK, PARTITIONS** Scaffolding not included	R042 -200									**232**
0100	Acoustical slotted block										
0200	NRC .50, type A, 4" thick	D-8	315	.127	S.F.	3.11	3.20		6.31	8.40	
0210	NRC .65 type R, 4" thick		315	.127		4.88	3.20		8.08	10.35	
0250	NRC .50, type A, 6" thick		290	.138		4	3.48		7.48	9.80	
0260	NRC .65 type R, 6" thick		290	.138		5.25	3.48		8.73	11.20	
0400	NRC .45 type A-1, 8" thick		265	.151		4.59	3.80		8.39	10.95	
0410	NRC .55, type Q, 8" thick		265	.151		5.95	3.80		9.75	12.45	
0500	NRC .60 type R, 8" thick		265	.151		7.80	3.80		11.60	14.45	
0600	NRC .65 type RR, 8" thick		265	.151		10.50	3.80		14.30	17.50	
0700	NRC .65 type 4R-RF, 8" thick		265	.151		11.10	3.80		14.90	18.10	
0710	NRC .70 type R, 12" thick		245	.163		12.05	4.11		16.16	19.65	
1000	Lightweight block, tooled joints, 2 sides, hollow										
1100	Not reinforced, 8" x 16" x 4" thick	D-8	440	.091	S.F.	.86	2.29		3.15	4.50	
1150	6" thick		410	.098		1.22	2.46		3.68	5.15	
1200	8" thick		385	.104		3.01	2.62		5.63	7.40	
1250	10" thick		370	.108		2.61	2.72		5.33	7.10	
1300	12" thick	D-9	350	.137		3.28	3.37		6.65	8.85	
2000	Not reinforced, 8" x 24" x 4" thick, hollow		460	.104		.61	2.57		3.18	4.67	
2100	6" thick		440	.109		.86	2.68		3.54	5.10	
2150	8" thick		415	.116		1.01	2.85		3.86	5.55	
2200	10" thick		385	.125		1.29	3.07		4.36	6.20	
2250	12" thick		365	.132		1.54	3.24		4.78	6.75	
2800	Solid, not reinforced, 8" x 16" x 2" thick	D-8	440	.091		.68	2.29		2.97	4.31	

Important: See the Reference Section for critical supporting data - Reference Nos., Crews, & City Cost Indexes

042 200 | Concrete Unit Masonry

				DAILY OUTPUT	LABOR-HOURS	UNIT	1999 BARE COSTS				TOTAL INCL O&P	
			CREW				MAT.	LABOR	EQUIP.	TOTAL		
232	2900	4" thick	R042 -200	D-8	420	.095	S.F.	1.05	2.40		3.45	4.89	**232**
	2950	6" thick			390	.103		2.05	2.58		4.63	6.30	
	3000	8" thick			365	.110		3.01	2.76		5.77	7.60	
	3050	10" thick		↓	350	.114		2.78	2.88		5.66	7.55	
	3100	12" thick		D-9	330	.145	↓	3.28	3.58		6.86	9.15	
	4000	Regular block, tooled joints, 2 sides, hollow											
	4100	Not reinforced, 8" x 16" x 4" thick		D-8	430	.093	S.F.	.77	2.34		3.11	4.49	
	4150	6" thick			400	.100		1.13	2.52		3.65	5.15	
	4200	8" thick CN			375	.107		1.26	2.69		3.95	5.55	
	4250	10" thick		↓	360	.111		1.83	2.80		4.63	6.35	
	4300	12" thick		D-9	340	.141		2	3.47		5.47	7.60	
	4500	Reinforced alternate courses, 8" x 16" x 4" thick		D-8	425	.094		.84	2.37		3.21	4.62	
	4550	6" thick			395	.101		1.21	2.55		3.76	5.30	
	4600	8" thick			370	.108		1.34	2.72		4.06	5.70	
	4650	10" thick		↓	355	.113		2.03	2.84		4.87	6.65	
	4700	12" thick		D-9	335	.143		2.09	3.52		5.61	7.80	
	4900	Solid, not reinforced, 2" thick		D-8	435	.092		.64	2.32		2.96	4.30	
	5000	3" thick			430	.093		.85	2.34		3.19	4.58	
	5050	4" thick			415	.096		1.21	2.43		3.64	5.10	
	5100	6" thick			385	.104		1.32	2.62		3.94	5.50	
	5150	8" thick		↓	360	.111		1.89	2.80		4.69	6.45	
	5200	12" thick		D-9	325	.148		3.05	3.63		6.68	9	
	5500	Solid, reinforced alternate courses, 4" thick		D-8	420	.095		1.26	2.40		3.66	5.10	
	5550	6" thick			380	.105		1.37	2.65		4.02	5.65	
	5600	8" thick		↓	355	.113		1.95	2.84		4.79	6.55	
	5650	12" thick		D-9	320	.150	↓	2.78	3.69		6.47	8.80	
234	0010	**CONCRETE BRICK** C55, grade N, type I, scaffolding not included											**234**
	0100	Regular, 4 x 2-1/4 x 8		D-8	220	.182	Ea.	.26	4.58		4.84	7.40	
	0125	Rusticated, 4 x 2-1/4 x 8			220	.182		.29	4.58		4.87	7.40	
	0150	Frog, 4 x 2-1/4 x 8			220	.182		.28	4.58		4.86	7.40	
	0200	Double, 4 x 4-7/8 x 8		↓	180	.222	↓	.46	5.60		6.06	9.20	
236	0010	**CHIMNEY BLOCK** Scaffolding not included											**236**
	0220	1 piece, with 8" x 8" flue, 16" x 16"		D-1	28	.571	V.L.F.	9.95	14.05		24	33	
	0230	2 piece, 16" x 16"		↓	26	.615		11.50	15.15		26.65	36	
	0240	2 piece, with 8" x 12" flue, 16" x 20"		↓	24	.667	↓	14.70	16.40		31.10	41.50	
240	0010	**CONCRETE SCREEN BLOCK** Scaffolding not included											**240**
	0200	8" x 16", 4" thick		D-8	180	.222	S.F.	1.52	5.60		7.12	10.35	
	0300	8" thick			270	.148		2.24	3.73		5.97	8.25	
	0350	12" x 12", 4" thick			300	.133		1.85	3.36		5.21	7.25	
	0500	8" thick		↓	330	.121	↓	2.26	3.05		5.31	7.25	
248	0010	**GLAZED CONCRETE BLOCK** No scaffolding or reinforcing incl.											**248**
	0100	Single face, 8" x 16" units, 2" thick		D-8	360	.111	S.F.	5.70	2.80		8.50	10.60	
	0200	4" thick			345	.116		5.45	2.92		8.37	10.55	
	0250	6" thick			330	.121		5.80	3.05		8.85	11.10	
	0300	8" thick			310	.129		6.20	3.25		9.45	11.85	
	0350	10" thick		↓	295	.136		6.80	3.42		10.22	12.75	
	0400	12" thick		D-9	280	.171		7.10	4.22		11.32	14.35	
	0700	Double face, 8" x 16" units, 4" thick		D-8	310	.129		8.45	3.25		11.70	14.35	
	0750	6" thick			290	.138		9.25	3.48		12.73	15.60	
	0800	8" thick			270	.148	↓	9.65	3.73		13.38	16.40	
	1000	Jambs, bullnose or square, single face, 8" x 16", 2" thick			315	.127	Ea.	9.95	3.20		13.15	15.95	
	1050	4" thick			285	.140	"	9.75	3.54		13.29	16.25	
	1200	Caps, bullnose or square, 8" x 16", 2" thick			420	.095	L.F.	10.50	2.40		12.90	15.30	
	1250	4" thick		↓	380	.105	↓	11.35	2.65		14	16.55	

MASONRY 4

042 200 | Concrete Unit Masonry

			CREW	DAILY OUTPUT	LABOR-HOURS	UNIT	MAT.	LABOR	EQUIP.	TOTAL	TOTAL INCL O&P	
248	1500	Cove base, 8" x 16", 2" thick	D-8	315	.127	L.F.	6.10	3.20		9.30	11.70	248
	1550	4" thick		285	.140		6.20	3.54		9.74	12.35	
	1600	6" thick		265	.151		6.55	3.80		10.35	13.10	
	1650	8" thick		245	.163		6.95	4.11		11.06	14.05	
252	0010	INSULATION See also division 072-108										252
	0100	Inserts, styrofoam, plant installed, add to block prices										
	0200	8" x 16" units, 6" thick				S.F.	.74			.74	.81	
	0250	8" thick					.74			.74	.81	
	0300	10" thick					.87			.87	.96	
	0350	12" thick					.91			.91	1	
	0500	8" x 8" units, 8" thick					.60			.60	.66	
	0550	12" thick					.72			.72	.79	
256	0010	PARGETING Regular Portland cement, 1/2" thick	D-1	2.50	6.400	C.S.F.	12.85	157		169.85	259	256
	5100	Waterproof Portland cement	"	2.50	6.400	"	14.05	157		171.05	260	

042 300 | Reinforced Unit Masonry

			CREW	DAILY OUTPUT	LABOR-HOURS	UNIT	MAT.	LABOR	EQUIP.	TOTAL	TOTAL INCL O&P	
304	0010	CONCRETE BLOCK BOND BEAM Scaffolding not included										304
	0020	Not including grout or reinforcing										
	0100	Regular block, 8" high, 8" thick	D-8	565	.071	L.F.	1.77	1.78		3.55	4.71	
	0150	12" thick	D-9	510	.094		2.36	2.32		4.68	6.20	
	0500	Lightweight, 8" high, 8" thick	D-8	575	.070		1.57	1.75		3.32	4.46	
	0550	12" thick	D-9	520	.092		2.74	2.27		5.01	6.55	
	2000	Including grout and 2 #5 bars										
	2100	Regular block, 8" high, 8" thick	D-8	300	.133	L.F.	3.12	3.36		6.48	8.65	
	2150	12" thick	D-9	250	.192		4.14	4.72		8.86	11.90	
	2500	Lightweight, 8" high, 8" thick	D-8	305	.131		3.29	3.30		6.59	8.75	
	2550	12" thick	D-9	255	.188		4.52	4.63		9.15	12.20	
310	0010	CONCRETE BLOCK, EXTERIOR Not including scaffolding										310
	0020	Reinforced alt courses, tooled joints 2 sides, foam inserts										
	0100	Regular, 8" x 16" x 6" thick	D-8	390	.103	S.F.	1.27	2.58		3.85	5.40	
	0200	8" thick		365	.110		1.81	2.76		4.57	6.30	
	0250	10" thick		355	.113		2.27	2.84		5.11	6.90	
	0300	12" thick	D-9	330	.145		2.46	3.58		6.04	8.25	
	0500	Lightweight, 8" x 16" x 6" thick	D-8	410	.098		2.11	2.46		4.57	6.15	
	0600	8" thick		385	.104		3.07	2.62		5.69	7.45	
	0650	10" thick		370	.108		2.67	2.72		5.39	7.20	
	0700	12" thick	D-9	350	.137		3.34	3.37		6.71	8.95	
320	0010	CONCRETE BLOCK FOUNDATION WALL Scaffolding not included										320
	0050	Normal-weight, trowel cut joints, parged 1/2" thick, no reinforcing										
	0200	Hollow, 8" x 16" x 6" thick	D-8	450	.089	S.F.	1.36	2.24		3.60	4.98	
	0250	8" thick		430	.093		1.50	2.34		3.84	5.30	
	0300	10" thick		420	.095		2.08	2.40		4.48	6	
	0350	12" thick	D-9	395	.122		2.25	2.99		5.24	7.15	
	0500	Solid, 8" x 16" block, 6" thick	D-8	440	.091		1.55	2.29		3.84	5.25	
	0550	8" thick	"	415	.096		2.13	2.43		4.56	6.10	
	0600	12" thick	D-9	380	.126		3.30	3.11		6.41	8.45	
325	0010	CONCRETE BLOCK, HIGH STRENGTH Scaffolding not included										325
	0050	Hollow, reinforced alternate courses, 8" x 16" units										
	0200	3500 psi, 4" thick	D-8	430	.093	S.F.	1.21	2.34		3.55	4.97	
	0250	6" thick		400	.100		1.24	2.52		3.76	5.30	
	0300	8" thick		375	.107		1.78	2.69		4.47	6.15	
	0350	12" thick	D-9	340	.141		2.41	3.47		5.88	8.05	
	0500	5000 psi, 4" thick	D-8	430	.093		1.54	2.34		3.88	5.35	
	0550	6" thick		400	.100		1.99	2.52		4.51	6.10	

Important: See the Reference Section for critical supporting data - Reference Nos., Crews, & City Cost Indexes

MASONRY `4`

042 300 | Reinforced Unit Masonry

		CREW	DAILY OUTPUT	LABOR-HOURS	UNIT	MAT.	LABOR	EQUIP.	TOTAL	TOTAL INCL O&P	
							1999 BARE COSTS				
325	0600	8" thick	D-8	375	.107	S.F.	2.95	2.69		5.64	7.45	325
	0650	12" thick	D-9	340	.141	↓	3.23	3.47		6.70	8.95	
	1000	For 75% solid block, add					30%					
	1050	For 100% solid block, add					50%					
330	0010	**CONCRETE BLOCK, LINTELS** Scaffolding not included										330
	0100	Including grout and horizontal reinforcing										
	0200	8" x 8" x 8", 1 #4 bar	D-4	300	.107	L.F.	3.31	2.61	.36	6.28	8.10	
	0250	2 #4 bars		295	.108		3.43	2.66	.37	6.46	8.30	
	0400	8" x 16" x 8", 1 #4 bar		275	.116		5.80	2.85	.40	9.05	11.25	
	0450	2 #4 bars		270	.119		5.95	2.90	.41	9.26	11.45	
	1000	12" x 8" x 8", 1 #4 bar		275	.116		4.68	2.85	.40	7.93	10	
	1100	2 #4 bars		270	.119		4.79	2.90	.41	8.10	10.20	
	1150	2 #5 bars		270	.119		4.92	2.90	.41	8.23	10.35	
	1200	2 #6 bars		265	.121		5.10	2.96	.41	8.47	10.60	
	1500	12" x 16" x 8", 1 #4 bar		250	.128		7.70	3.14	.44	11.28	13.85	
	1600	2 #3 bars		245	.131		7.70	3.20	.45	11.35	13.95	
	1650	2 #4 bars		245	.131		7.80	3.20	.45	11.45	14.05	
	1700	2 #5 bars	↓	240	.133	↓	7.95	3.27	.46	11.68	14.30	
335	0010	**CONCRETE BLOCK COLUMN** or pilaster. Scaffolding not included.										335
	0050	Including vertical reinforcing (4-#4 bars) and grout										
	0160	1 piece unit, 16" x 16"	D-1	26	.615	V.L.F.	8.70	15.15		23.85	33	
	0170	2 piece units, 16" x 20"		24	.667		11.75	16.40		28.15	38.50	
	0180	20" x 20"		22	.727		16.85	17.90		34.75	46.50	
	0190	22" x 24"		18	.889		21	22		43	57	
	0200	20" x 32"	↓	14	1.143	↓	23	28		51	69	

042 350 | Masonry Panels

		CREW	DAILY OUTPUT	LABOR-HOURS	UNIT	MAT.	LABOR	EQUIP.	TOTAL	TOTAL INCL O&P	
354	0010	**WALL PANELS** Prefabricated, 4" thick, minimum	C-11	775	.093	S.F.	4.58	2.78	2	9.36	12.20	354
	0100	Maximum	"	500	.144	"	6.65	4.31	3.11	14.07	18.40	
	0200	4" brick & 2" concrete back-up, add					50%					
	0300	4" brick & 1" urethane & 3" concrete back-up, add					70%					

042 450 | Structural Facing Tile

		CREW	DAILY OUTPUT	LABOR-HOURS	UNIT	MAT.	LABOR	EQUIP.	TOTAL	TOTAL INCL O&P	
454	0010	**STRUCTURAL FACING TILE** Scaffolding not incl, functional colors										454
	0020	6T series, 5-1/3" x 12", 2.3 pieces per S.F., glazed 1 side, 2" thick	D-8	225	.178	S.F.	4.59	4.48		9.07	12	
	0100	4" thick		220	.182		5.65	4.58		10.23	13.30	
	0150	Glazed 2 sides		195	.205		9.10	5.15		14.25	18.05	
	0250	6" thick		210	.190		8.40	4.80		13.20	16.70	
	0300	Glazed 2 sides		185	.216		14.45	5.45		19.90	24.50	
	0400	8" thick		180	.222	↓	10.35	5.60		15.95	20	
	0500	Special shapes, group 1		400	.100	Ea.	4.47	2.52		6.99	8.85	
	0550	Group 2		375	.107		6.65	2.69		9.34	11.50	
	0600	Group 3		350	.114		8.30	2.88		11.18	13.65	
	0650	Group 4		325	.123		13.60	3.10		16.70	19.75	
	0700	Group 5		300	.133		19.55	3.36		22.91	26.50	
	0750	Group 6		275	.145	↓	28	3.67		31.67	36.50	
	1000	Fire rated, 4" thick, 1 hr. rating		210	.190	S.F.	11.05	4.80		15.85	19.60	
	1300	Acoustic, 4" thick	↓	210	.190	"	10.45	4.80		15.25	18.95	
	1400	For designer colors, add					25%					
	2000	8W series, 8" x 16", 1.125 pieces per S.F.										
	2050	2" thick, glazed 1 side	D-8	360	.111	S.F.	5.40	2.80		8.20	10.30	
	2100	4" thick, glazed 1 side		345	.116		5.85	2.92		8.77	11	
	2150	Glazed 2 sides		325	.123		9.05	3.10		12.15	14.75	
	2200	6" thick, glazed 1 side		330	.121		8.15	3.05		11.20	13.75	
	2250	8" thick, glazed 1 side	↓	310	.129	↓	9.35	3.25		12.60	15.30	

For expanded coverage of these items see *Means Concrete & Masonry Cost Data 1999*

042 450 | Structural Facing Tile

454			CREW	DAILY OUTPUT	LABOR-HOURS	UNIT	MAT.	LABOR	EQUIP.	TOTAL	TOTAL INCL O&P	
454	2500	Special shapes, group 1	D-8	300	.133	Ea.	7.80	3.36		11.16	13.80	454
	2550	Group 2		280	.143		10.10	3.60		13.70	16.70	
	2600	Group 3		260	.154		10.65	3.88		14.53	17.75	
	2650	Group 4		250	.160		24	4.03		28.03	32.50	
	2700	Group 5		240	.167		48	4.20		52.20	59.50	
	2750	Group 6		230	.174		52.50	4.38		56.88	64.50	
	3000	4" thick, glazed 1 side		345	.116	S.F.	9.85	2.92		12.77	15.35	
	3100	Acoustic, 4" thick		345	.116	"	6.95	2.92		9.87	12.20	
	3120	4W series, 8" x 8", 2.25 pieces per S.F.										
	3125	2" thick, glazed 1 side	D-8	360	.111	S.F.	2.60	2.80		5.40	7.20	
	3130	4" thick, glazed 1 side		345	.116		2.70	2.92		5.62	7.50	
	3135	Glazed 2 sides		325	.123		3.70	3.10		6.80	8.90	
	3140	6" thick, glazed 1 side		330	.121		3.60	3.05		6.65	8.70	
	3150	8" thick, glazed 1 side		310	.129		5.10	3.25		8.35	10.65	
	3155	Special shapes, group I		300	.133	Ea.	3.90	3.36		7.26	9.50	
	3160	Group II		280	.143	"	5.40	3.60		9	11.55	
	3200	For designer colors, add					25%					
	3300	For epoxy mortar joints, add				S.F.	.95			.95	1.05	

042 500 | Ceramic Veneer

508	0011	CAST CERAMIC FLOORING See division 096-354										508
510	0010	TERRA COTTA Coping, split type, not glazed, 9" wide	D-1	90	.178	L.F.	4.78	4.37		9.15	12.05	510
	0100	13" wide		80	.200		7.80	4.92		12.72	16.25	
	0200	Split type, glazed, 9" wide		90	.178		5.55	4.37		9.92	12.95	
	0250	13" wide		80	.200		9.05	4.92		13.97	17.60	
	0500	Partition or back-up blocks, scored, in C.L. lots										
	0700	Non-load bearing 12" x 12", 3" thick, Special Order Only	D-8	550	.073	S.F.	4.45	1.83		6.28	7.75	
	0750	4" thick, standard		500	.080		3.65	2.02		5.67	7.15	
	0800	6" thick		450	.089		3.70	2.24		5.94	7.55	
	0850	8" thick		400	.100		4.60	2.52		7.12	8.95	
	1000	Load bearing, 12" x 12", 4" thick, in walls		500	.080		3.70	2.02		5.72	7.20	
	1050	In floors		750	.053		3.70	1.34		5.04	6.15	
	1200	6" thick, in walls		450	.089		3.85	2.24		6.09	7.70	
	1250	In floors		675	.059		3.85	1.49		5.34	6.55	
	1400	8" thick, in walls		400	.100		4.70	2.52		7.22	9.05	
	1450	In floors		575	.070		4.70	1.75		6.45	7.90	
	1600	10" thick, in walls, special order		350	.114		5.45	2.88		8.33	10.50	
	1650	In floors, special order		500	.080		5.45	2.02		7.47	9.15	
	1800	12" thick, in walls, special order		300	.133		6.50	3.36		9.86	12.35	
	1850	In floors, special order		450	.089		6.50	2.24		8.74	10.65	
	2000	For reinforcing with steel rods, add to above					15%	5%				
	2100	For smooth tile instead of scored, add					.10			.10	.11	
	2200	For L.C.L. quantities, add					10%	10%				
512	0010	TERRA COTTA TILE On walls, dry set, 1/2" thick										512
	0100	Square, hexagonal or lattice shapes, unglazed	1 Tilf	135	.059	S.F.	4.25	1.58		5.83	7	
	0300	Glazed, plain colors		130	.062		5.70	1.64		7.34	8.65	
	0400	Intense colors		125	.064		8.80	1.71		10.51	12.20	

042 550 | Masonry Veneer

554	0010	BRICK VENEER Scaffolding not included, truck load lots	R042 -120									554
	0015	Material costs incl. 3% brick and 25% mortar waste										
	0020	Standard, select common, 4" x 2-2/3" x 8" (6.75/S.F.)	R042 -180	D-8	1.50	26.667	M	355	670		1,025	1,450	
	0050	Red, 4" x 2-2/3" x 8", running bond			1.50	26.667		380	670		1,050	1,475	
	0100	Full header every 6th course (7.88/S.F.)	R042 -500		1.45	27.586		375	695		1,070	1,500	
	0150	English, full header every 2nd course (10.13/S.F.)			1.40	28.571		375	720		1,095	1,550	

Important: See the Reference Section for critical supporting data - Reference Nos., Crews, & City Cost Indexes

042 550 | Masonry Veneer

		CREW	DAILY OUTPUT	LABOR-HOURS	UNIT	1999 BARE COSTS				TOTAL INCL O&P	
						MAT.	LABOR	EQUIP.	TOTAL		
554	0200	Flemish, alternate header every course (9.00/S.F.)	D-8	1.40	28.571	M	375	720		1,095	1,550	554
	0250	Flemish, alt. header every 6th course (7.13/S.F.)	R042-550	1.45	27.586		375	695		1,070	1,500	
	0300	Full headers throughout (13.50/S.F.)		1.40	28.571		375	720		1,095	1,550	
	0350	Rowlock course (13.50/S.F.)		1.35	29.630		375	745		1,120	1,575	
	0400	Rowlock stretcher (4.50/S.F.)		1.40	28.571		380	720		1,100	1,550	
	0450	Soldier course (6.75/S.F.)		1.40	28.571		380	720		1,100	1,550	
	0500	Sailor course (4.50/S.F.)		1.30	30.769		380	775		1,155	1,625	
	0601	Buff or gray face, running bond, (6.75/S.F.)		1.50	26.667		380	670		1,050	1,475	
	0700	Glazed face, 4" x 2-2/3" x 8", running bond		1.40	28.571		1,100	720		1,820	2,325	
	0750	Full header every 6th course (7.88/S.F.)		1.35	29.630		1,050	745		1,795	2,300	
	1000	Jumbo, 6" x 4" x 12",(3.00/S.F.)		1.30	30.769		1,300	775		2,075	2,625	
	1051	Norman, 4" x 2-2/3" x 12" (4.50/S.F.)		1.45	27.586		910	695		1,605	2,075	
	1100	Norwegian, 4" x 3-1/5" x 12" (3.75/S.F.)		1.40	28.571		590	720		1,310	1,775	
	1150	Economy, 4" x 4" x 8" (4.50 per S.F.)		1.40	28.571		480	720		1,200	1,650	
	1201	Engineer, 4" x 3-1/5" x 8", (5.63/S.F.)		1.45	27.586		345	695		1,040	1,450	
	1251	Roman, 4" x 2" x 12", (6.00/S.F.)		1.50	26.667		790	670		1,460	1,925	
	1300	S.C.R. 6" x 2-2/3" x 12" (4.50/S.F.)		1.40	28.571		910	720		1,630	2,125	
	1350	Utility, 4" x 4" x 12" (3.00/S.F.)		1.35	29.630		835	745		1,580	2,075	
	1360	For less than truck load lots, add					10					
	1400	For battered walls, add						30%				
	1450	For corbels, add						75%				
	1500	For curved walls, add						30%				
	1550	For pits and trenches, deduct						20%				
	1999	Alternate method of figuring by square foot										
	2000	Standard, sel. common, 4" x 2-2/3" x 8", (6.75/S.F.)	D-8	230	.174	S.F.	2.55	4.38		6.93	9.60	
	2020	Standard, red, 4" x 2-2/3" x 8", running bond (6.75/SF)		220	.182		2.55	4.58		7.13	9.90	
	2050	Full header every 6th course (7.88/S.F.)		185	.216		2.97	5.45		8.42	11.70	
	2100	English, full header every 2nd course (10.13/S.F.)		140	.286		3.81	7.20		11.01	15.40	
	2150	Flemish, alternate header every course (9.00/S.F.)		150	.267		3.39	6.70		10.09	14.20	
	2200	Flemish, alt. header every 6th course (7.13/S.F.)		205	.195		2.69	4.92		7.61	10.60	
	2250	Full headers throughout (13.50/S.F.)		105	.381		5.05	9.60		14.65	20.50	
	2300	Rowlock course (13.50/S.F.)		100	.400		5.05	10.10		15.15	21	
	2350	Rowlock stretcher (4.50/S.F.)		310	.129		1.71	3.25		4.96	6.95	
	2400	Soldier course (6.75/S.F.)		200	.200		2.55	5.05		7.60	10.65	
	2450	Sailor course (4.50/S.F.)		290	.138		1.71	3.48		5.19	7.30	
	2600	Buff or gray face, running bond, (6.75/S.F.)		220	.182		2.70	4.58		7.28	10.05	
	2700	Glazed face brick, running bond		210	.190		7.10	4.80		11.90	15.25	
	2750	Full header every 6th course (7.88/S.F.)		170	.235		8.30	5.95		14.25	18.30	
	3000	Jumbo, 6" x 4" x 12" running bond (3.00/S.F.)		435	.092		3.65	2.32		5.97	7.60	
	3050	Norman, 4" x 2-2/3" x 12" running bond, (4.5/S.F.)		320	.125		3.36	3.15		6.51	8.60	
	3100	Norwegian, 4" x 3-1/5" x 12" (3.75/S.F.)		375	.107		2.15	2.69		4.84	6.55	
	3150	Economy, 4" x 4" x 8" (4.50/S.F.)		310	.129		2.15	3.25		5.40	7.40	
	3200	Engineer, 4" x 3-1/5" x 8" (5.63/S.F.)		260	.154		1.93	3.88		5.81	8.15	
	3250	Roman, 4" x 2" x 12" (6.00/S.F.)		250	.160		4.68	4.03		8.71	11.40	
	3300	SCR, 6" x 2-2/3" x 12" (4.50/S.F.)		310	.129		4.19	3.25		7.44	9.65	
	3350	Utility, 4" x 4" x 12" (3.00/S.F.)		450	.089		2.45	2.24		4.69	6.15	
	3400	For cavity wall construction, add						15%				
	3450	For stacked bond, add						10%				
	3500	For interior veneer construction, add						15%				
	3550	For curved walls, add						30%				

042 700 | Glass Unit Masonry

			CREW	DAILY OUTPUT	LABOR-HOURS	UNIT	MAT.	LABOR	EQUIP.	TOTAL	INCL O&P	
704	0010	GLASS BLOCK Scaffolding not included	R042-700									704
	0100	Plain, 4" thick, under 1,000 S.F., 6" x 6" block	D-8	115	.348	S.F.	13.85	8.75		22.60	29	
	0150	8" x 8" block		160	.250		9.40	6.30		15.70	20	
	0200	12" x 12" block		175	.229		11.20	5.75		16.95	21.50	

For expanded coverage of these items see *Means Concrete & Masonry Cost Data 1999*

135

042 700 | Glass Unit Masonry

			CREW	DAILY OUTPUT	LABOR-HOURS	UNIT	1999 BARE COSTS				TOTAL INCL O&P	
							MAT.	LABOR	EQUIP.	TOTAL		
704	0300	1,000 to 5,000 S.F., 6" x 6" block	D-8	135	.296	S.F.	13.40	7.45		20.85	26.50	704
	0350	8" x 8" block		190	.211		9.05	5.30		14.35	18.25	
	0400	12" x 12" block		215	.186		10.85	4.69		15.54	19.25	
	0500	Over 5,000 S.F., 6" x 6" block		145	.276		12.80	6.95		19.75	25	
	0550	8" x 8" block		215	.186		8.65	4.69		13.34	16.85	
	0600	12" x 12" block		240	.167		10.35	4.20		14.55	17.95	
	0700	For solar reflective blocks, add					100%					
	0800	Under 1,000 S.F., 4" x 8" blocks	D-8	145	.276	S.F.	13.05	6.95		20	25	
	0850	Over 5,000 S.F.		170	.235		12.15	5.95		18.10	22.50	
	1000	Thinline, plain, 3-1/8" thick, under 1,000 S.F., 6" x 6" block		115	.348		11.60	8.75		20.35	26.50	
	1050	8" x 8" block		160	.250		6.65	6.30		12.95	17.10	
	1200	Over 5,000 S.F., 6" x 6" block		145	.276		10.80	6.95		17.75	22.50	
	1250	8" x 8" block		215	.186		6.20	4.69		10.89	14.10	
	1400	For cleaning block after installation (both sides), add		1,000	.040		.10	1.01		1.11	1.68	

R042 -700

042 900 | Adobe Masonry

			CREW	DAILY OUTPUT	LABOR-HOURS	UNIT	1999 BARE COSTS				TOTAL INCL O&P	
							MAT.	LABOR	EQUIP.	TOTAL		
904	0010	ADOBE BRICK Unstabilized, with adobe mortar (Southwestern States)										904
	0060	Brick, 4" x 3" x 8" (6.0 per S.F.)	D-8	1.60	25	M	300	630		930	1,300	
	0080	4" x 4" x 8" (4.5 per S.F.)		1.50	26.667		405	670		1,075	1,500	
	0100	4" x 4" x 14" (2.5 per S.F.)		1.35	29.630		450	745		1,195	1,650	
	0120	8" x 3" x 16" (3.0 per S.F.)		1.40	28.571		505	720		1,225	1,675	
	0140	4" x 3" x 12" (4.0 per S.F.)		1.45	27.586		465	695		1,160	1,575	
	0160	6" x 3" x 12" (4.0 per S.F.)		1.45	27.586		480	695		1,175	1,600	
	0180	4" x 5" x 16" (1.80 per S.F.)		1.35	29.630		495	745		1,240	1,700	
	0200	8" x 4" x 16" (2.25 per S.F.)		1.30	30.769		650	775		1,425	1,925	
	0220	10" x 4" x 14" (2.57 per S.F.)		1.25	32		600	805		1,405	1,900	
	0260	Brick, 4" x 3" x 8" (6.0 per S.F.)		266	.150	S.F.	1.80	3.79		5.59	7.90	
	0280	4" x 4" x 8" (4.5 per S.F.)		333	.120		1.83	3.03		4.86	6.70	
	0300	4" x 4" x 14" (2.5 per S.F.)		540	.074		1.13	1.87		3	4.14	
	0320	8" x 3" x 16" (3.0 per S.F.)		466	.086		1.52	2.16		3.68	5.05	
	0340	4" x 3" x 12" (4.0 per S.F.)		362	.110		1.86	2.78		4.64	6.40	
	0360	6" x 3" x 12" (4.0 per S.F.)		362	.110		1.92	2.78		4.70	6.45	
	0380	4" x 5" x 16" (1.80 per S.F.)		600	.067		.89	1.68		2.57	3.59	
	0400	8" x 4" x 16" (2.25 per S.F.)		577	.069		1.46	1.75		3.21	4.33	
	0420	10" x 4" x 14" (2.57 per S.F.)		555	.072		1.54	1.82		3.36	4.52	
	0440	Adobe, partially stabilized, add					10%					
	0480	Fully stabilized, add					25%					

044 | Stone

044 050 | Stone

			CREW	DAILY OUTPUT	LABOR-HOURS	UNIT	1999 BARE COSTS				TOTAL INCL O&P	
							MAT.	LABOR	EQUIP.	TOTAL		
054	0010	FACING PANELS Stone aggregate mounted on plywood,										054
	0020	see division 074-201										

044 100 | Rough Stone

			CREW	DAILY OUTPUT	LABOR-HOURS	UNIT	1999 BARE COSTS				TOTAL INCL O&P	
							MAT.	LABOR	EQUIP.	TOTAL		
104	0011	ROUGH STONE WALL, Dry										104
	0100	Random fieldstone, under 18" thick	D-12	60	.533	C.F.	9.75	13.40		23.15	32	
	0150	Over 18" thick	"	63	.508	"	12.50	12.75		25.25	33.50	

4
MASONRY

Important: See the Reference Section for critical supporting data - Reference Nos., Crews, & City Cost Indexes

044 | Stone

044 200 | Cut Stone

		CREW	DAILY OUTPUT	LABOR-HOURS	UNIT	MAT.	LABOR	EQUIP.	TOTAL	TOTAL INCL O&P	
204	**0010** **BLUESTONE** Cut to size										**204**
0500	Sills, natural cleft, 10" wide to 6' long, 1-1/2" thick	D-11	70	.343	L.F.	9.05	9		18.05	24	
0550	2" thick	"	63	.381		10.75	10		20.75	27.50	
1000	Stair treads, natural cleft, 12" wide, 6' long, 1-1/2" thick	D-10	115	.348		13.65	9.05	3.40	26.10	32.50	
1050	2" thick		105	.381		15.25	9.90	3.73	28.88	36	
1100	Smooth finish, 1-1/2" thick		115	.348		17.85	9.05	3.40	30.30	37.50	
1300	Thermal finish, 1-1/2" thick		115	.348		19.95	9.05	3.40	32.40	39.50	
1350	2" thick		105	.381		22.50	9.90	3.73	36.13	44.50	

044 300 | Simulated Stone

		CREW	DAILY OUTPUT	LABOR-HOURS	UNIT	MAT.	LABOR	EQUIP.	TOTAL	TOTAL INCL O&P	
304	**0010** **SIMULATED STONE**										**304**
0100	Insulated fiberglass panels, 5/8" ply backer	L-4	200	.120	S.F.	8.60	3.08		11.68	14.30	

044 500 | Stone Veneer

		CREW	DAILY OUTPUT	LABOR-HOURS	UNIT	MAT.	LABOR	EQUIP.	TOTAL	TOTAL INCL O&P	
504	**0011** **ASHLAR VENEER** 4" + or - thk, random or random rectangular R044-500										**504**
0150	Sawn face, split joints, low priced stone	D-8	140	.286	S.F.	4.71	7.20		11.91	16.40	
0200	Medium priced stone		130	.308		6.95	7.75		14.70	19.70	
0300	High priced stone		120	.333		9.10	8.40		17.50	23	
0600	Seam face, split joints, medium price stone		125	.320		6	8.05		14.05	19.15	
0700	High price stone		120	.333		10	8.40		18.40	24	
1000	Split or rock face, split joints, medium price stone		125	.320		8	8.05		16.05	21.50	
1100	High price stone		120	.333		10.50	8.40		18.90	24.50	
508	**0011** **LIGHTWEIGHT NATURAL STONE** Lava type										**508**
0100	Veneer, rubble face, sawed back, irregular shapes	D-10	130	.308	S.F.	5.10	8	3.01	16.11	21.50	
0200	Sawed face and back, irregular shapes	"	130	.308	"	5.25	8	3.01	16.26	21.50	

044 550 | Marble

		CREW	DAILY OUTPUT	LABOR-HOURS	UNIT	MAT.	LABOR	EQUIP.	TOTAL	TOTAL INCL O&P	
554	**0011** **MARBLE** Ashlar, split face, 4" + or - thick, random										**554**
0040	lengths 1' to 4' & heights 2" to 7-1/2", average	D-8	175	.229	S.F.	13.35	5.75		19.10	23.50	
0100	Base, polished, 3/4" or 7/8" thick, polished, 6" high	D-10	65	.615	L.F.	12.10	15.95	6	34.05	44.50	
0300	Carvings or bas relief, from templates, average		80	.500	S.F.	108	13	4.89	125.89	144	
0350	Maximum		80	.500	"	252	13	4.89	269.89	305	
0600	Columns, cornices, mouldings, etc.										
0650	Hand or special machine cut, average	D-10	35	1.143	C.F.	108	29.50	11.20	148.70	177	
0700	Maximum	"	35	1.143	"	232	29.50	11.20	272.70	315	
1000	Facing, polished finish, cut to size, 3/4" to 7/8" thick										
1050	Average	D-10	130	.308	S.F.	17.80	8	3.01	28.81	35	
1100	Maximum		130	.308		41	8	3.01	52.01	61	
1300	1-1/4" thick, average		125	.320		26	8.30	3.13	37.43	45	
1350	Maximum		125	.320		51.50	8.30	3.13	62.93	73	
1500	2" thick, average		120	.333		30	8.65	3.26	41.91	50	
1550	Maximum		120	.333		51.50	8.65	3.26	63.41	74	
2200	Window sills, 6" x 3/4" thick	D-1	85	.188	L.F.	6.70	4.63		11.33	14.55	
2500	Flooring, polished tiles, 12" x 12" x 3/8" thick										
2510	Thin set, average	D-11	90	.267	S.F.	8.25	7		15.25	19.95	
2600	Maximum		90	.267		30	7		37	44	
2700	Mortar bed, average		65	.369		8.40	9.70		18.10	24.50	
2740	Maximum		65	.369		27.50	9.70		37.20	45	
2780	Travertine, 3/8" thick, average	D-10	130	.308		11.40	8	3.01	22.41	28	
2790	Maximum	"	130	.308		28	8	3.01	39.01	46	
2800	Patio tile, non-slip, 1/2" thick, flame finish	D-11	75	.320		11.60	8.40		20	26	
2900	Shower or toilet partitions, 7/8" thick partitions										
3050	3/4" or 1-1/4" thick stiles, polished 2 sides, average	D-11	75	.320	S.F.	34.50	8.40		42.90	51	
3201	Soffits, add to above prices				"	20%	100%				
3210	Stairs, risers, 7/8" thick x 6" high	D-10	115	.348	L.F.	11.65	9.05	3.40	24.10	30.50	

For expanded coverage of these items see *Means Concrete & Masonry Cost Data 1999*

MASONRY 4

137

044 550 | Marble

			CREW	DAILY OUTPUT	LABOR-HOURS	UNIT	1999 BARE COSTS				TOTAL INCL O&P	
							MAT.	LABOR	EQUIP.	TOTAL		
554	3360	Treads, 12" wide x 1-1/4" thick	D-10	115	.348	L.F.	17	9.05	3.40	29.45	36.50	554
	3500	Thresholds, 3' long, 7/8" thick, 4" to 5" wide, plain	D-12	24	1.333	Ea.	12.85	33.50		46.35	66	
	3550	Beveled		24	1.333	"	14.95	33.50		48.45	68.50	
	3700	Window stools, polished, 7/8" thick, 5" wide	↓	85	.376	L.F.	12	9.45		21.45	28	

044 600 | Limestone

			CREW	DAILY OUTPUT	LABOR-HOURS	UNIT	MAT.	LABOR	EQUIP.	TOTAL	TOTAL INCL O&P	
604	0010	**LIMESTONE** See also Ashlar Veneer, division 044-504										604
	0020	Veneer facing panels										
	0500	Texture finish, light stick, 4-1/2" thick, 5'x 12'	D-4	300	.107	S.F.	18.70	2.61	.36	21.67	25	
	0750	5" thick, 5' x 14' panels	D-10	275	.145		22.50	3.77	1.42	27.69	32	
	1000	Sugarcube finish, 2" Thick, 3' x 5' panels		275	.145		11.10	3.77	1.42	16.29	19.60	
	1050	3" Thick, 4' x 9' panels		275	.145		11.55	3.77	1.42	16.74	20	
	1200	4" Thick, 5' x 11' panels		275	.145		15.40	3.77	1.42	20.59	24.50	
	1400	Sugarcube, textured finish, 4-1/2" thick, 5' x 12'		275	.145		19.90	3.77	1.42	25.09	29.50	
	1450	5" thick, 5' x 14' panels		275	.145	↓	23	3.77	1.42	28.19	33	
	2000	Coping, sugarcube finish, top & 2 sides		30	1.333	C.F.	51	34.50	13.05	98.55	124	
	2100	Sills, lintels, jambs, trim, stops, sugarcube finish, average		20	2		46	52	19.55	117.55	153	
	2150	Detailed		20	2	↓	61	52	19.55	132.55	169	
	2300	Steps, extra hard, 14" wide, 6" rise	↓	50	.800	L.F.	42	21	7.80	70.80	86.50	
	3000	Quoins, plain finish, 6"x12"x12"	D-12	25	1.280	Ea.	101	32		133	161	
	3050	6"x16"x24"	"	25	1.280	"	134	32		166	198	

044 650 | Granite

			CREW	DAILY OUTPUT	LABOR-HOURS	UNIT	MAT.	LABOR	EQUIP.	TOTAL	TOTAL INCL O&P	
651	0010	**GRANITE** Cut to size										651
	0050	Veneer, polished face, 3/4" to 1-1/2" thick										
	0150	Low price, gray, light gray, etc. **CN**	D-10	130	.308	S.F.	18.55	8	3.01	29.56	36	
	0220	High price, red, black, etc. **CN**	"	130	.308	"	31.50	8	3.01	42.51	50.50	
	0300	1-1/2" to 2-1/2" thick, veneer										
	0350	Low price, gray, light gray, etc.	D-10	130	.308	S.F.	20.50	8	3.01	31.51	38	
	0550	High price, red, black, etc.	"	130	.308	"	36	8	3.01	47.01	55	
	0700	2-1/2" to 4" thick, veneer										
	0750	Low price, gray, light gray, etc.	D-10	110	.364	S.F.	26	9.45	3.56	39.01	47	
	0950	High price, red, black, etc.	"	110	.364		41	9.45	3.56	54.01	64	
	1000	For bush hammered finish, deduct					5%					
	1050	Coarse rubbed finish, deduct					10%					
	1100	Honed finish, deduct					5%					
	1150	Thermal finish, deduct				↓	18%					
	2450	For radius under 5', add				L.F.	100%					
	2500	Steps, copings, etc., finished on more than one surface										
	2550	Minimum	D-10	50	.800	C.F.	72	21	7.80	100.80	120	
	2600	Maximum	"	50	.800	"	108	21	7.80	136.80	160	
	2800	Pavers, 4" x 4" x 4" blocks, split face and joints										
	2850	Minimum	D-11	80	.300	S.F.	10.30	7.90		18.20	23.50	
	2900	Maximum	"	80	.300	"	20	7.90		27.90	34.50	
	3500	Curbing, city street type, See Division 025-254										
	4000	Soffits, 2" thick, minimum	D-13	35	1.371	S.F.	31	36	11.20	78.20	102	
	4100	Maximum		35	1.371		62	36	11.20	109.20	136	
	4200	4" thick, minimum		35	1.371		42	36	11.20	89.20	114	
	4300	Maximum	↓	35	1.371	↓	79.50	36	11.20	126.70	155	

044 700 | Sandstone

			CREW	DAILY OUTPUT	LABOR-HOURS	UNIT	MAT.	LABOR	EQUIP.	TOTAL	TOTAL INCL O&P	
704	0011	**SANDSTONE OR BROWNSTONE**										704
	0100	Sawed face veneer, 2-1/2" thick, to 2' x 4' panels	D-10	130	.308	S.F.	12.45	8	3.01	23.46	29.50	
	0150	4" thick, to 3'-6" x 8' panels		100	.400		12.45	10.40	3.91	26.76	34	
	0300	Split face, random sizes	↓	100	.400	↓	8.95	10.40	3.91	23.26	30	

4

MASONRY

Important: See the Reference Section for critical supporting data - Reference Nos., Crews, & City Cost Indexes

044 700	Sandstone	CREW	DAILY OUTPUT	LABOR-HOURS	UNIT	1999 BARE COSTS				TOTAL INCL O&P	
						MAT.	LABOR	EQUIP.	TOTAL		
704	**0350**	Cut stone trim (limestone)									**704**
	0360	Ribbon stone, 4" thick, 5' pieces	D-8	120	.333	Ea.	110	8.40		118.40	134
	0370	Cove stone, 4" thick, 5' pieces		105	.381		110	9.60		119.60	136
	0380	Cornice stone, 10" to 12" wide		90	.444		136	11.20		147.20	166
	0390	Band stone, 4" thick, 5' pieces		145	.276		70	6.95		76.95	88
	0410	Window and door trim, 3" to 4" wide		160	.250		59.50	6.30		65.80	75.50
	0420	Key stone, 18" long	↓	60	.667	↓	63	16.80		79.80	95

044 750	Slate	CREW	DAILY OUTPUT	LABOR-HOURS	UNIT	1999 BARE COSTS				TOTAL INCL O&P	
						MAT.	LABOR	EQUIP.	TOTAL		
754	**0010**	**SLATE** Pennsylvania, blue gray to gray black; Vermont,									**754**
	0050	Unfading green, mottled green & purple, gray & purple									
	0100	Virginia, blue black									
	0200	Exterior paving, natural cleft, 1" thick									
	0500	24" x 24", Pennsylvania	D-12	120	.267	S.F.	8.65	6.70		15.35	19.90
	0550	Vermont		120	.267		11.30	6.70		18	23
	0600	Virginia	↓	120	.267	↓	12.65	6.70		19.35	24.50
	1000	Interior flooring, natural cleft, 1/2" thick									
	1300	24" x 24" Pennsylvania	D-12	120	.267	S.F.	5.95	6.70		12.65	16.95
	1350	Vermont		120	.267		8.90	6.70		15.60	20
	1400	Virginia	↓	120	.267	↓	9.20	6.70		15.90	20.50
	2000	Facing panels, 1-1/4" thick, to 4' x 4' panels									
	2100	Natural cleft finish, Pennsylvania	D-10	180	.222	S.F.	17.60	5.75	2.17	25.52	30.50
	2110	Vermont		180	.222		18.35	5.75	2.17	26.27	31.50
	2120	Virginia	↓	180	.222		20.50	5.75	2.17	28.42	34
	2150	Sand rubbed finish, surface, add					1.91			1.91	2.10
	2200	Honed finish, add					3.77			3.77	4.15
	2500	Ribbon, natural cleft finish, 1" thick, to 9 S.F.	D-10	80	.500		8.10	13	4.89	25.99	34.50
	2700	1-1/2" thick		78	.513		10.55	13.30	5	28.85	37.50
	2850	2" thick	↓	76	.526	↓	12.70	13.65	5.15	31.50	40.50
	3000	Roofing, see division 073-106									
	3500	Stair treads, sand finish, 1" thick x 12" wide									
	3600	3 L.F. to 6 L.F.	D-10	120	.333	L.F.	14.60	8.65	3.26	26.51	33
	3700	Ribbon, sand finish, 1" thick x 12" wide									
	3750	To 6 L.F.	D-10	120	.333	L.F.	9.75	8.65	3.26	21.66	27.50
	4000	Stools or sills, sand finish, 1" thick, 6" wide	D-12	160	.200		7.05	5		12.05	15.55
	4200	10" wide		90	.356		10.65	8.90		19.55	25.50
	4400	2" thick, 6" wide		140	.229		11.35	5.75		17.10	21.50
	4600	10" wide	↓	90	.356	↓	16.95	8.90		25.85	32.50
	4800	For lengths over 3', add					25%				

045 100	Masonry Cleaning	CREW	DAILY OUTPUT	LABOR-HOURS	UNIT	1999 BARE COSTS				TOTAL INCL O&P	
						MAT.	LABOR	EQUIP.	TOTAL		
102	**0010**	**CLEANING MASONRY** No staging included, unless noted									**102**
	0200	Chemical cleaning, brush and wash, minimum	D-1	800	.020	S.F.	.10	.49		.59	.87
	0220	Average		400	.040		.15	.98		1.13	1.70
	0240	Maximum	↓	330	.048		.20	1.19		1.39	2.07
	0400	High pressure water only, minimum	B-9	2,000	.020			.44	.09	.53	.79
	0420	Average	↓	1,500	.027	↓		.58	.12	.70	1.05

For expanded coverage of these items see *Means Concrete & Masonry Cost Data 1999*

MASONRY 4

4 MASONRY

045 100 | Masonry Cleaning

			CREW	DAILY OUTPUT	LABOR-HOURS	UNIT	1999 BARE COSTS				TOTAL INCL O&P	
							MAT.	LABOR	EQUIP.	TOTAL		
102	0440	Maximum	B-9	1,000	.040	S.F.		.87	.18	1.05	1.58	102
	2000	Steam cleaning, minimum		3,000	.013			.29	.06	.35	.53	
	2020	Average		2,500	.016			.35	.07	.42	.63	
	2040	Maximum	↓	1,500	.027			.58	.12	.70	1.05	
	4000	Add for masking doors and windows				↓					.80	
	4200	Add for pedestrian protection				Job					10%	
106	0010	**CLEAN AND POINT** Smooth brick	1 Bric	300	.027	S.F.	.22	.74		.96	1.39	106
	0100	Rough brick	"	265	.030	"	.23	.83		1.06	1.56	
108	0010	**WASHING BRICK** Acid wash, smooth brick [R045-100]	1 Bric	560	.014	S.F.	.22	.39		.61	.85	108
	0050	Rough brick		400	.020		.23	.55		.78	1.11	
	0060	Stone, acid wash	↓	600	.013	↓	.22	.37		.59	.81	
	1000	Muriatic acid, price per gallon in 5 gallon lots				Gal.	4			4	4.40	

045 200 | Masonry Restoration

			CREW	DAILY OUTPUT	LABOR-HOURS	UNIT	MAT.	LABOR	EQUIP.	TOTAL	TOTAL INCL O&P	
240	0010	**NEEDLE BEAM MASONRY** Incl. shoring 10' x 10' opening										240
	0400	Block, concrete, 8" thick	B-9	7.10	5.634	Ea.	28.50	123	25.50	177	253	
	0420	12" thick		6.70	5.970		40.50	130	27	197.50	280	
	0800	Brick, 4" thick with 8" backup block		5.70	7.018		40.50	153	31.50	225	320	
	1000	Brick, solid, 8" thick		6.20	6.452		28.50	141	29	198.50	285	
	1040	12" thick		4.90	8.163		40.50	178	37	255.50	365	
	1080	16" thick	↓	4.50	8.889		65.50	194	40	299.50	420	
	2000	Add for additional floors of shoring	B-1	6	4	↓	28.50	88.50		117	170	

045 550 | Flue Liners

			CREW	DAILY OUTPUT	LABOR-HOURS	UNIT	MAT.	LABOR	EQUIP.	TOTAL	TOTAL INCL O&P	
554	0010	**FLUE LINING** Including mortar joints, 8" x 8"	D-1	125	.128	V.L.F.	3.34	3.15		6.49	8.60	554
	0100	8" x 12"		103	.155		4.80	3.82		8.62	11.25	
	0200	12" x 12"		93	.172		5.70	4.23		9.93	12.85	
	0300	12" x 18"		84	.190		8.95	4.69		13.64	17.15	
	0400	18" x 18"		75	.213		12	5.25		17.25	21.50	
	0500	20" x 20"		66	.242		21	5.95		26.95	32.50	
	0600	24" x 24"		56	.286		24.50	7.05		31.55	38	
	1000	Round, 18" diameter		66	.242		25.50	5.95		31.45	37.50	
	1100	24" diameter	↓	47	.340	↓	32	8.35		40.35	48	

045 650 | Fire Brick

			CREW	DAILY OUTPUT	LABOR-HOURS	UNIT	MAT.	LABOR	EQUIP.	TOTAL	TOTAL INCL O&P	
652	0010	**FIRE BRICK** 9" x 2-1/2" x 4-1/2", low duty, 2000° F	D-1	.60	26.667	M	1,075	655		1,730	2,225	652
	0050	High duty, 3000° F	"	.60	26.667	"	1,550	655		2,205	2,725	
654	0010	**FIRE CLAY** Gray, high duty, 100 lb. bag				Bag	39.50			39.50	43.50	654
	0050	100 lb. drum, premixed (400 brick per drum)				Drum	50			50	55	
656	0010	**FIREPLACE** For prefabricated fireplace, see div. 103-054										656
	0100	Brick fireplace, not incl. foundations or chimneys										
	0110	30" x 29" opening, incl. chamber, plain brickwork	D-1	.40	40	Ea.	360	985		1,345	1,925	
	0200	Fireplace box only (110 brick)	"	2	8	"	116	197		313	435	
	0300	For elaborate brickwork and details, add					35%	35%				
	0400	For hearth, brick & stone, add	D-1	2	8	Ea.	132	197		329	450	
	0410	For steel angle, damper, cleanouts, add		4	4		92	98.50		190.50	254	
	0600	Plain brickwork, incl. metal circulator		.50	32		695	785		1,480	2,000	
	0800	Face brick only, standard size, 8" x 2-2/3" x 4"	↓	.30	53.333	M	360	1,300		1,660	2,450	

For information about Means Estimating Seminars, see yellow pages 11 and 12 in back of book

Division 5
Metals

Estimating Tips

050 Materials, Coatings & Fastenings

- Nuts, bolts, washers, connection angles and plates can add a significant amount to both the tonnage of a structural steel job as well as the estimated cost. As a rule of thumb add 10% to the total weight to account for these accessories.

- Type 2 steel construction, commonly referred to as "simple construction," consists generally of field bolted connections with lateral bracing supplied by other elements of the building, such as masonry walls or x-bracing. The estimator should be aware, however, that shop connections may be accomplished by welding or bolting. The method may be particular to the fabrication shop and may have an impact on the estimated cost.

052 Steel Joists

- In any given project the total weight of open web steel joists is determined by the loads to be supported and the design. However, economies can be realized in minimizing the amount of labor used to place the joists. This is done by maximizing the joist spacing and therefore minimizing the number of joists required to be installed on the job. Certain spacings and locations may be required by the design, but in other cases maximizing the spacing and keeping it as uniform as possible will keep the costs down.

053 Metal Decking

- The takeoff and estimating of metal deck involves more than simply the area of the floor or roof and the type of deck specified or shown on the drawings. Many different sizes and types of openings may exist. Small openings for individual pipes or conduits may be drilled after the floor/roof is installed, but larger openings may require special deck lengths as well as reinforcing or structural support. The estimator should determine who will be supplying this reinforcing. Additionally, some deck terminations are part of the deck package, such as screed angles and pour stops, and others will be part of the steel contract, such as angles attached to structural members and cast-in-place angles and plates. The estimator must ensure that all pieces are accounted for in the complete estimate.

055 Metal Fabrications

- The most economical steel stairs are those that use common materials, standard details and most importantly, a uniform and relatively simple method of field assembly. Commonly available A36 channels and plates are very good choices for the main stringers of the stairs, as are angles and tees for the carrier members. Risers and treads are usually made by specialty shops, and it is most economical to use a typical detail in as many places as possible. The stairs should be pre-assembled and shipped directly to the site. The field connections should be simple and straightforward to be accomplished efficiently and with a minimum of equipment and labor.

Reference Numbers

Reference numbers are shown in bold squares at the beginning of some major classifications. These numbers refer to related items in the Reference Section. The reference information may be an estimating procedure, an alternate pricing method or technical information.

Note: Not all subdivisions listed here necessarily appear in this publication.

		050 500	Metal Fastening	CREW	DAILY OUTPUT	LABOR-HOURS	UNIT	1999 BARE COSTS				TOTAL INCL O&P
								MAT.	LABOR	EQUIP.	TOTAL	
508	0010	**BOLTS & HEX NUTS** Steel, A307											508
	0100	1/4" diameter, 1/2" long					Ea.	.05			.05	.05
	0200	1" long						.06			.06	.06
	0300	2" long						.07			.07	.08
	0400	3" long						.10			.10	.11
	0500	4" long						.15			.15	.16
	0600	3/8" diameter, 1" long						.04			.04	.04
	0700	2" long						.07			.07	.08
	0800	3" long						.09			.09	.09
	0900	4" long						.22			.22	.24
	1000	5" long						.26			.26	.29
	1100	1/2" diameter, 1-1/2" long						.22			.22	.24
	1200	2" long						.24			.24	.27
	1300	4" long						.36			.36	.40
	1400	6" long						.49			.49	.54
	1500	8" long						.66			.66	.73
	1600	5/8" diameter, 1-1/2" long						.38			.38	.41
	1700	2" long						.42			.42	.46
	1800	4" long						.59			.59	.64
	1900	6" long						.80			.80	.88
	2000	8" long						1.06			1.06	1.16
	2100	10" long						1.26			1.26	1.38
	2200	3/4" diameter, 2" long						.66			.66	.73
	2300	4" long						.96			.96	1.06
	2400	6" long						1.23			1.23	1.35
	2500	8" long						1.54			1.54	1.69
	2600	10" long						2.07			2.07	2.28
	2700	12" long						2.35			2.35	2.59
	2800	1" diameter, 3" long						1.61			1.61	1.77
	2900	6" long						2.36			2.36	2.59
	3000	12" long						4.74			4.74	5.20
	3100	For galvanized, add						75%				
	3200	For stainless, add						350%				
510	0005	**CURB EDGING**											510
	0010	Steel angle w/anchors, on forms, 1" x 1", 0.8#/L.F.	E-4	350	.091	L.F.	1.36	2.84	.23	4.43	6.95
	0100	2" x 2" angles, 3.92#/L.F.		330	.097		4.19	3.02	.25	7.46	10.45
	0200	3" x 3" angles, 6.1#/L.F.		300	.107		6.70	3.32	.27	10.29	13.80
	0300	4" x 4" angles, 8.2#/L.F.		275	.116		8.15	3.62	.30	12.07	16
	1000	6" x 4" angles, 12.3#/L.F.		250	.128		11.65	3.98	.33	15.96	20.50
	1050	Steel channels with anchors, on forms, 3" channel, 5#/L.F.		290	.110		5.30	3.43	.28	9.01	12.45
	1100	4" channel, 5.4#/L.F.		270	.119		5.70	3.69	.30	9.69	13.35
	1200	6" channel, 8.2#/L.F.		255	.125		8.85	3.90	.32	13.07	17.25
	1300	8" channel, 11.5#/L.F.		225	.142		11.60	4.42	.37	16.39	21.50
	1400	10" channel, 15.3#/L.F.		180	.178		15.05	5.55	.46	21.06	27
	1500	12" channel, 20.7#/L.F.		140	.229		19.95	7.10	.59	27.64	35.50
	2000	For curved edging, add					35%	10%			
515	0010	**DRILLING** For anchors, up to 4" deep, incl. bit and layout										515
	0050	in concrete or brick walls and floors, no anchor										
	0100	Holes, 1/4" diameter	1 Carp	75	.107	Ea.	.08	2.91		2.99	4.67
	0150	For each additional inch of depth, add		430	.019		.02	.51		.53	.82
	0200	3/8" diameter		63	.127		.07	3.47		3.54	5.55
	0250	For each additional inch of depth, add		340	.024		.02	.64		.66	1.03
	0300	1/2" diameter		50	.160		.07	4.37		4.44	6.95
	0350	For each additional inch of depth, add		250	.032		.02	.87		.89	1.39

5

METALS

			DAILY	LABOR-		1999 BARE COSTS				TOTAL	
050 500 | Metal Fastening			CREW	OUTPUT	HOURS	UNIT	MAT.	LABOR	EQUIP.	TOTAL	INCL O&P
515	0400	5/8" diameter	1 Carp	48	.167	Ea.	.13	4.55		4.68	7.30	**515**
	0450	For each additional inch of depth, add		240	.033		.03	.91		.94	1.47
	0500	3/4" diameter		45	.178		.14	4.85		4.99	7.80
	0550	For each additional inch of depth, add		220	.036		.03	.99		1.02	1.60
	0600	7/8" diameter		43	.186		.17	5.10		5.27	8.20
	0650	For each additional inch of depth, add		210	.038		.04	1.04		1.08	1.69
	0700	1" diameter		40	.200		.20	5.45		5.65	8.80
	0750	For each additional inch of depth, add		190	.042		.05	1.15		1.20	1.87
	0800	1-1/4" diameter		38	.211		.29	5.75		6.04	9.35
	0850	For each additional inch of depth, add		180	.044		.07	1.21		1.28	1.99
	0900	1-1/2" diameter		35	.229		.47	6.25		6.72	10.30
	0950	For each additional inch of depth, add		165	.048		.12	1.32		1.44	2.21
	1000	For ceiling installations, add						40%			
	1100	Drilling & layout for drywall or plaster walls, no anchor									
	1200	Holes, 1/4" diameter	1 Carp	150	.053	Ea.	.01	1.46		1.47	2.30
	1300	3/8" diameter		140	.057		.01	1.56		1.57	2.46
	1400	1/2" diameter		130	.062		.01	1.68		1.69	2.65
	1500	3/4" diameter		120	.067		.02	1.82		1.84	2.88
	1600	1" diameter		110	.073		.03	1.99		2.02	3.15
	1700	1-1/4" diameter		100	.080		.04	2.18		2.22	3.48
	1800	1-1/2" diameter		90	.089		.06	2.43		2.49	3.88
	1900	For ceiling installations, add						40%			
520	0010	**EXPANSION ANCHORS** & shields										**520**
	0100	Bolt anchors for concrete, brick or stone, no layout and drilling									
	0200	Expansion shields, zinc, 1/4" diameter, 1" long, single	1 Carp	90	.089	Ea.	.85	2.43		3.28	4.76
	0300	1-3/8" long, double		85	.094		.97	2.57		3.54	5.10
	0400	3/8" diameter, 2" long, single		85	.094		1.45	2.57		4.02	5.65
	0500	2" long, double		80	.100		1.79	2.73		4.52	6.25
	0600	1/2" diameter, 2-1/2" long, single		80	.100		2.40	2.73		5.13	6.95
	0700	2-1/2" long, double		75	.107		2.32	2.91		5.23	7.15
	0800	5/8" diameter, 2-5/8" long, single		75	.107		3.43	2.91		6.34	8.35
	0900	3" long, double		70	.114		3.43	3.12		6.55	8.70
	1000	3/4" diameter, 2-3/4" long, single		70	.114		5.10	3.12		8.22	10.50
	1100	4" long, double		65	.123		6.80	3.36		10.16	12.80
	1410	Concrete anchor, w/rod & epoxy cartridge, 1-3/4" diameter x 15" long	E-22	20	1.200		72	34.50		106.50	134
	1415	18" long		17	1.412		86	40.50		126.50	159
	1420	2" diameter x 18" long		16	1.500		110	43		153	189
	1425	24" long		15	1.600		144	46		190	231
	1430	Chemical anchor, w/rod & epoxy cartridge, 3/4" diam. x 9-1/2" long		27	.889		10.45	25.50		35.95	52
	1435	1" diameter x 11-3/4" long		24	1		19.80	28.50		48.30	67.50
	1440	1-1/4" diameter x 14" long		21	1.143		38	33		71	93.50
	1500	Self drilling anchor, snap-off, for 1/4" diameter bolt	1 Carp	26	.308		.93	8.40		9.33	14.20
	1600	3/8" diameter bolt		23	.348		1.40	9.50		10.90	16.50
	1700	1/2" diameter bolt		20	.400		2.11	10.90		13.01	19.50
	1800	5/8" diameter bolt		18	.444		3.57	12.15		15.72	23
	1900	3/4" diameter bolt		16	.500		6.45	13.65		20.10	28.50
	2100	Hollow wall anchors for gypsum wall board, plaster or tile									
	2300	1/8" diameter, short				Ea.	.25			.25	.28
	2400	Long					.28			.28	.31
	2500	3/16" diameter, short					.59			.59	.65
	2600	Long					.63			.63	.69
	2700	1/4" diameter, short					.74			.74	.81
	2800	Long					.85			.85	.94
	3000	Toggle bolts, bright steel, 1/8" diameter, 2" long	1 Carp	85	.094		.29	2.57		2.86	4.36
	3100	4" long		80	.100		.38	2.73		3.11	4.72
	3200	3/16" diameter, 3" long		80	.100		.36	2.73		3.09	4.70

5

METALS

143

050 500	Metal Fastening	CREW	DAILY OUTPUT	LABOR-HOURS	UNIT	1999 BARE COSTS				TOTAL INCL O&P
						MAT.	LABOR	EQUIP.	TOTAL	
520 3300	6" long	1 Carp	75	.107	Ea.	.55	2.91		3.46	5.20	**520**
3400	1/4" diameter, 3" long		75	.107		.40	2.91		3.31	5
3500	6" long		70	.114		.60	3.12		3.72	5.55
3600	3/8" diameter, 3" long		70	.114		.88	3.12		4	5.90
3700	6" long		60	.133		1.30	3.64		4.94	7.20
3800	1/2" diameter, 4" long		60	.133		2.79	3.64		6.43	8.80
3900	6" long		50	.160		3.71	4.37		8.08	10.95
4000	Nailing anchors									
4100	Nylon nailing anchor, 1/4" diameter, 1" long				C	15.45			15.45	17
4200	1-1/2" long					19.90			19.90	22
4300	2" long					33			33	36.50
4400	Metal nailing anchor, 1/4" diameter, 1" long					21.50			21.50	24
4500	1-1/2" long					29.50			29.50	32.50
4600	2" long					38			38	41.50
5000	Screw anchors for concrete, masonry,									
5100	stone & tile, no layout or drilling included									
5200	Jute fiber, #6, #8, & #10, 1" long				Ea.	.18			.18	.20
5300	#12, 1-1/2" long					.26			.26	.29
5400	#14, 2" long					.42			.42	.46
5500	#16, 2" long					.43			.43	.47
5600	#20, 2" long					.69			.69	.76
5700	Lag screw shields, 1/4" diameter, short					.42			.42	.46
5800	Long					.50			.50	.55
5900	3/8" diameter, short					.73			.73	.80
6000	Long					.83			.83	.91
6100	1/2" diameter, short					1.09			1.09	1.20
6200	Long					1.29			1.29	1.42
6300	3/4" diameter, short					2.35			2.35	2.58
6400	Long					3.01			3.01	3.31
6600	Lead, #6 & #8, 3/4" long					.17			.17	.19
6700	#10 - #14, 1-1/2" long					.26			.26	.29
6800	#16 & #18, 1-1/2" long					.36			.36	.40
6900	Plastic, #6 & #8, 3/4" long					.04			.04	.04
7000	#8 & #10, 7/8" long					.04			.04	.04
7100	#10 & #12, 1" long					.05			.05	.06
7200	#14 & #16, 1-1/2" long					.06			.06	.07
8000	Wedge anchors, not including layout or drilling									
8050	Carbon steel, 1/4" diameter, 1-3/4" long	1 Carp	150	.053	Ea.	.33	1.46		1.79	2.66
8100	3 1/4" long		145	.055		.50	1.51		2.01	2.92
8150	3/8" diameter, 2-1/4" long		150	.053		.60	1.46		2.06	2.94
8200	5" long		145	.055		1.04	1.51		2.55	3.51
8250	1/2" diameter, 2-3/4" long		140	.057		.88	1.56		2.44	3.42
8300	7" long		130	.062		1.54	1.68		3.22	4.34
8350	5/8" diameter, 3-1/2" long		130	.062		1.66	1.68		3.34	4.47
8400	8-1/2" long		115	.070		3.24	1.90		5.14	6.55
8450	3/4" diameter, 4-1/4" long		115	.070		2.16	1.90		4.06	5.35
8500	10" long		100	.080		4.91	2.18		7.09	8.85
8550	1" diameter, 6" long		100	.080		7.40	2.18		9.58	11.60
8575	9" long		80	.100		9.60	2.73		12.33	14.85
8600	12" long		80	.100		10.40	2.73		13.13	15.70
8650	1-1/4" diameter, 9" long		70	.114		14.50	3.12		17.62	21
8700	12" long		60	.133		14.80	3.64		18.44	22
8750	For type 303 stainless steel, add					350%				
8800	For type 316 stainless steel, add					450%				
525 0010	**HIGH STRENGTH BOLTS**										**525**
0020	A325 Type 1, structural steel, bolt-nut-washer set									

(R050 -510)

Important: See the Reference Section for critical supporting data - Reference Nos., Crews, & City Cost Indexes

050 500 | Metal Fastening

				CREW	DAILY OUTPUT	LABOR-HOURS	UNIT	1999 BARE COSTS				TOTAL INCL O&P	
								MAT.	LABOR	EQUIP.	TOTAL		
525	0100	1/2" diameter x 1-1/2" long	R050 -510	1 Sswk	120	.067	Ea.	.24	2.04		2.28	4.01	525
	0120	2" long			120	.067		.27	2.04		2.31	4.04	
	0150	3" long			120	.067		.33	2.04		2.37	4.10	
	0170	5/8" diameter x 1-1/2" long			120	.067		.38	2.04		2.42	4.16	
	0180	2" long			120	.067		.42	2.04		2.46	4.20	
	0190	3" long			120	.067		.50	2.04		2.54	4.29	
	0200	3/4" diameter x 2" long			120	.067		.59	2.04		2.63	4.39	
	0220	3" long			115	.070		.69	2.13		2.82	4.66	
	0250	4" long			110	.073		.81	2.23		3.04	4.97	
	0300	6" long			105	.076		1.05	2.33		3.38	5.40	
	0350	8" long			95	.084		1.54	2.58		4.12	6.40	
	0360	7/8" diameter x 2" long			115	.070		.87	2.13		3	4.86	
	0365	3" long			110	.073		1.04	2.23		3.27	5.20	
	0370	4" long			105	.076		1.17	2.33		3.50	5.55	
	0380	6" long			100	.080		1.48	2.45		3.93	6.10	
	0390	8" long			90	.089		2.03	2.72		4.75	7.20	
	0400	1" diameter x 2" long			105	.076		1.24	2.33		3.57	5.65	
	0420	3" long			105	.076		1.43	2.33		3.76	5.85	
	0450	4" long			105	.076		1.57	2.33		3.90	6	
	0500	6" long			90	.089		1.92	2.72		4.64	7.10	
	0550	8" long			85	.094		2.76	2.88		5.64	8.35	
	0600	1-1/4" diameter x 3" long			85	.094		2.83	2.88		5.71	8.40	
	0650	4" long			80	.100		3.12	3.06		6.18	9.05	
	0700	6" long			75	.107		3.67	3.26		6.93	10.05	
	0750	8" long		▼	70	.114	▼	4.56	3.50		8.06	11.40	
	1020	A490, bolt-nut-washer set											
	1100	1/2" diameter x 1-1/2" long		1 Sswk	120	.067	Ea.	.30	2.04		2.34	4.07	
	1120	2" long			120	.067		.35	2.04		2.39	4.13	
	1150	3" long			120	.067		.41	2.04		2.45	4.19	
	1170	5/8" diameter x 1-1/2" long			120	.067		.42	2.04		2.46	4.21	
	1180	2" long			120	.067		.45	2.04		2.49	4.24	
	1190	3" long			120	.067		.56	2.04		2.60	4.35	
	1200	3/4" diameter x 2" long			120	.067		.68	2.04		2.72	4.49	
	1220	3" long			115	.070		1.04	2.13		3.17	5.05	
	1250	4" long			110	.073		.92	2.23		3.15	5.10	
	1300	6" long			105	.076		1.18	2.33		3.51	5.55	
	1350	8" long			95	.084		1.66	2.58		4.24	6.55	
	1360	7/8" diameter x 2" long			115	.070		.98	2.13		3.11	4.98	
	1365	3" long			110	.073		1.13	2.23		3.36	5.30	
	1370	4" long			105	.076		1.35	2.33		3.68	5.75	
	1380	6" long			100	.080		1.80	2.45		4.25	6.45	
	1390	8" long			90	.089		2.39	2.72		5.11	7.60	
	1400	1" diameter x 2" long			105	.076		1.40	2.33		3.73	5.80	
	1420	3" long			105	.076		1.54	2.33		3.87	5.95	
	1450	4" long			105	.076		1.96	2.33		4.29	6.45	
	1500	6" long			90	.089		2.22	2.72		4.94	7.45	
	1550	8" long			85	.094		3.11	2.88		5.99	8.75	
	1600	1-1/4" diameter x 3" long			85	.094		3.09	2.88		5.97	8.70	
	1650	4" long			80	.100		3.42	3.06		6.48	9.35	
	1700	6" long			75	.107		4.30	3.26		7.56	10.70	
	1750	8" long		▼ ▼	70	.114	▼	5.25	3.50		8.75	12.15	
530	0005	**LAG SCREWS**											530
	0010	Steel, 1/4" diameter, 2" long		1 Carp	200	.040	Ea.	.05	1.09		1.14	1.78	
	0100	3/8" diameter, 3" long			150	.053		.16	1.46		1.62	2.46	
	0200	1/2" diameter, 3" long		▼	130	.062	▼	.27	1.68		1.95	2.94	

050 500	Metal Fastening	CREW	DAILY OUTPUT	LABOR-HOURS	UNIT	MAT.	LABOR	EQUIP.	TOTAL	TOTAL INCL O&P	
							1999 BARE COSTS				
530 0300	5/8" diameter, 3" long	1 Carp	120	.067	Ea.	.53	1.82		2.35	3.45	**530**
535 0005	**MACHINE SCREWS**										**535**
0010	Steel, round head, #8 x 1" long				C	1.50			1.50	1.65	
0110	#8 x 2" long					3.39			3.39	3.73	
0200	#10 x 1" long					2.16			2.16	2.37	
0300	#10 x 2" long				↓	4.01			4.01	4.41	
540 0010	**MACHINERY ANCHORS** Standard, flush mounted,										**540**
0020	incl. stud w/fiber plug, connecting nut, washer & bolt										
0200	Material only, 1/2" diameter stud & bolt				Ea.	30.50			30.50	33.50	
0300	5/8" diameter					34.50			34.50	38	
0500	3/4" diameter					37.50			37.50	41.50	
0600	7/8" diameter					43.50			43.50	47.50	
0800	1" diameter					47.50			47.50	52.50	
0900	1-1/4" diameter				↓	60.50			60.50	66.50	
545 0010	**RIVETS**										**545**
0100	Aluminum rivet & mandrel, 1/2" grip length x 1/8" diameter				C	4.53			4.53	4.98	
0200	3/16" diameter					7			7	7.70	
0300	Aluminum rivet, steel mandrel, 1/8" diameter					4.23			4.23	4.65	
0400	3/16" diameter					6.35			6.35	7	
0500	Copper rivet, steel mandrel, 1/8" diameter					6.25			6.25	6.85	
0600	Monel rivet, steel mandrel, 1/8" diameter					19.30			19.30	21	
0700	3/16" diameter					40.50			40.50	44.50	
0800	Stainless rivet & mandrel, 1/8" diameter					12			12	13.20	
0900	3/16" diameter					18.20			18.20	20	
1000	Stainless rivet, steel mandrel, 1/8" diameter					7.60			7.60	8.35	
1100	3/16" diameter					13.75			13.75	15.10	
1200	Steel rivet and mandrel, 1/8" diameter					4.74			4.74	5.20	
1300	3/16" diameter				↓	7.15			7.15	7.85	
1400	Hand riveting tool, minimum				Ea.	76.50			76.50	84	
1500	Maximum					460			460	505	
1600	Power riveting tool, minimum					660			660	730	
1700	Maximum				↓	1,750			1,750	1,925	
550 0005	**POWDER ACTUATED** Tools & fasteners										**550**
0010	Stud driver, .22 caliber, buy, minimum				Ea.	287			287	315	
0100	Maximum				"	505			505	555	
0300	Powder charges for above, low velocity				C	15.05			15.05	16.55	
0400	Standard velocity					25			25	27.50	
0600	Drive pins & studs, 1/4" & 3/8" diam., to 3" long, minimum					22			22	24	
0700	Maximum				↓	57			57	63	
0800	Pneumatic stud driver for 1/8" diameter studs				Ea.	1,850			1,850	2,025	
0900	Drive pins for above, 1/2" to 3/4" long				M	400			400	440	
555 0005	**RAILROAD TRACK**	R024 -520									**555**
0010	Track bolts				Ea.	1.73			1.73	1.90	
0100	Joint bars				Pr.	26			26	28.50	
0200	Spikes				Ea.	.36			.36	.40	
0300	Tie plates				"	1.50			1.50	1.65	
560 0005	**WELDED SHEAR CONNECTORS**										**560**
0010	3/4" diameter, 3-3/16" long	E-10	1,030	.016	Ea.	.37	.49	.49	1.35	1.85	
0020	3-3/8" long		1,030	.016		.40	.49	.49	1.38	1.88	
0200	3-7/8" long		1,030	.016		.40	.49	.49	1.38	1.88	
0300	4-3/16" long		1,030	.016		.41	.49	.49	1.39	1.89	
0500	4-7/8" long	↓	1,030	.016	↓	.46	.49	.49	1.44	1.95	

Important: See the Reference Section for critical supporting data - Reference Nos., Crews, & City Cost Indexes

		050 500	Metal Fastening	CREW	DAILY OUTPUT	LABOR-HOURS	UNIT	1999 BARE COSTS				TOTAL INCL O&P	
								MAT.	LABOR	EQUIP.	TOTAL		
560	0600		5-3/16" long	E-10	1,030	.016	Ea.	.47	.49	.49	1.45	1.96	560
	0800		5-3/8" long		1,030	.016		.51	.49	.49	1.49	2	
	0900		6-3/16" long		1,000	.016		.52	.51	.51	1.54	2.06	
	1000		7-3/16" long		1,000	.016		.65	.51	.51	1.67	2.20	
	1100		8-3/16" long		1,000	.016		.67	.51	.51	1.69	2.23	
	1500		7/8" diameter, 3-11/16" long		1,030	.016		.58	.49	.49	1.56	2.08	
	1600		4-3/16" long		1,030	.016		.61	.49	.49	1.59	2.11	
	1700		5-3/16" long		1,030	.016		.63	.49	.49	1.61	2.13	
	1800		6-3/16" long		1,000	.016		.70	.51	.51	1.72	2.26	
	1900		7-3/16" long		1,000	.016		.84	.51	.51	1.86	2.41	
	2000		8-3/16" long		1,000	.016		.83	.51	.51	1.85	2.40	
565	0005	**WELDED STUDS**											565
	0010		1/4" diameter, 2-11/16" long	E-10	1,030	.016	Ea.	.16	.49	.49	1.14	1.62	
	0100		4-1/8" long		1,030	.016		.13	.49	.49	1.11	1.58	
	0200		3/8" diameter, 4-1/8" long		1,030	.016		.20	.49	.49	1.18	1.66	
	0300		6-1/8" long		1,030	.016		.25	.49	.49	1.23	1.72	
	0400		1/2" diameter, 2-1/8" long		1,030	.016		.19	.49	.49	1.17	1.65	
	0500		3-1/8" long		1,030	.016		.24	.49	.49	1.22	1.70	
	0600		4-1/8" long		1,030	.016		.29	.49	.49	1.27	1.76	
	0700		5-5/16" long		1,030	.016		.36	.49	.49	1.34	1.84	
	0800		6-1/8" long		1,000	.016		.39	.51	.51	1.41	1.92	
	0900		8-1/8" long		1,000	.016		.55	.51	.51	1.57	2.10	
	1000		5/8" diameter, 2-11/16" long		1,030	.016		.32	.49	.49	1.30	1.79	
	1010		4-3/16" long		1,030	.016		.48	.49	.49	1.46	1.97	
	1100		6-9/16" long		1,000	.016		.55	.51	.51	1.57	2.10	
	1200		8-3/16" long		1,000	.016		.79	.51	.51	1.81	2.36	
570	0005	**WELD ROD**											570
	0010		Steel, type E6011 (all purpose), 1/8" dia, less than 500#				Lb.	1.38			1.38	1.52	
	0100		500# to 2,000#					1.23			1.23	1.36	
	0200		2,000# to 5,000#					1.14			1.14	1.25	
	0300		5/32" diameter, less than 500#					1.32			1.32	1.45	
	0310		500# to 2,000#					1.20			1.20	1.31	
	0320		2,000# to 5,000#					1.09			1.09	1.20	
	0400		3/16" dia, less than 500#					1.32			1.32	1.45	
	0500		500# to 2,000#					1.19			1.19	1.30	
	0600		2,000# to 5,000#					1.08			1.08	1.19	
	0620		Steel, type E6010 (pipe), 1/8" dia, less than 500#					1.47			1.47	1.62	
	0630		500# to 2,000#					1.32			1.32	1.45	
	0640		2,000# to 5,000#					1.17			1.17	1.29	
	0650		Steel, type E7018 (low hydrogen), 1/8" dia, less than 500#					1.32			1.32	1.45	
	0660		500# to 2,000# **CN**					1.30			1.30	1.43	
	0670		2,000# to 5,000#					1.08			1.08	1.19	
	0700		Steel, type E7024 (jet weld), 1/8" dia, less than 500#					1.36			1.36	1.50	
	0710		500# to 2,000#					1.23			1.23	1.35	
	0720		2,000# to 5,000#					1.12			1.12	1.23	
	1550		Aluminum, type 4043, 1/8" dia, less than 10#					4.76			4.76	5.25	
	1560		10# to 60#					4.54			4.54	5	
	1570		Over 60#					4.30			4.30	4.72	
	1600		Aluminum, type 5356, 1/8" dia, less than 10#					4.77			4.77	5.25	
	1610		10# to 60#					4.54			4.54	4.99	
	1620		Over 60#					4.11			4.11	4.52	
	1900		Cast iron (cold welding), 1/8" dia, less than 500#					15.35			15.35	16.90	
	1910		500# to 1,000#					14.15			14.15	15.60	
	1920		Over 1,000#					13.05			13.05	14.35	

5

METALS

147

050 500 | Metal Fastening

		CREW	DAILY OUTPUT	LABOR-HOURS	UNIT	1999 BARE COSTS				TOTAL INCL O&P	
						MAT.	LABOR	EQUIP.	TOTAL		
570	2000	Stainless steel, type 308-16, 1/8" dia, less than 500#				Lb.	5.85			5.85	6.45	**570**
	2100	500# to 1000#					5.15			5.15	5.65	
	2220	Over 1000#				↓	4.78			4.78	5.25	
575	0005	**WELDING STRUCTURAL**										**575**
	0010	Field welding, 1/8" E6011, cost per welder, no operating engr. [R050-520]	E-14	8	1	Hr.	2.76	32.50	10.30	45.56	74.50	
	0200	With 1/2 operating engineer	E-13	8	1.500		2.76	46	10.30	59.06	95	
	0300	With 1 operating engineer	E-12	8	2	↓	2.76	60	10.30	73.06	115	
	0500	With no operating engineer, 2# weld rod per ton	E-14	8	1	Ton	2.76	32.50	10.30	45.56	74.50	
	0600	8# E6011 per ton	"	2	4		11	130	41	182	297	
	0800	With one operating engineer per welder, 2# E6011 per ton	E-12	8	2		2.76	60	10.30	73.06	115	
	0900	8# E6011 per ton	"	2	8	↓	11	239	41	291	465	
	1200	Continuous fillet, stick welding, incl. equipment										
	1300	Single pass, 1/8" thick, 0.1#/L.F.	E-14	240	.033	L.F.	.14	1.09	.34	1.57	2.52	
	1400	3/16" thick, 0.2#/L.F.		120	.067		.28	2.17	.69	3.14	5.05	
	1500	1/4" thick, 0.3#/L.F.		80	.100		.41	3.26	1.03	4.70	7.55	
	1610	5/16" thick, 0.4#/L.F.		60	.133		.55	4.35	1.37	6.27	10.05	
	1800	3 passes, 3/8" thick, 0.5#/L.F.		48	.167		.69	5.45	1.72	7.86	12.60	
	2010	4 passes, 1/2" thick, 0.7#/L.F.		34	.235		.96	7.65	2.42	11.03	17.75	
	2200	5 to 6 passes, 3/4" thick, 1.3#/L.F.		19	.421		1.79	13.75	4.33	19.87	31.50	
	2400	8 to 11 passes, 1" thick, 2.4#/L.F.	↓	10	.800		3.31	26	8.25	37.56	60.50	
	2600	For all position welding, add, minimum						20%				
	2700	Maximum						300%				
	2900	For semi-automatic welding, deduct, minimum						5%				
	3000	Maximum				↓		15%				
	4000	Cleaning and welding plates, bars, or rods										
	4010	to existing beams, columns, or trusses	E-14	12	.667	L.F.	.69	21.50	6.85	29.04	48.50	
580	0005	**WIRE**										**580**
	0010	Barbed wire, galvanized, domestic steel, hi-tensile 15-1/2 ga.				M.L.F.	23.50			23.50	26	
	0020	standard, 12-1/2 ga.					32			32	35.50	
	0030	14 ga.					27.50			27.50	30.50	
	0200	import, standard, 12-1/2 ga.					31.50			31.50	35	
	0205	Aluminum, 12-1/2 ga.					138			138	152	
	0210	Barbless wire, steel, galvanized 12-1/2 ga.				↓	29.50			29.50	32.50	
	0500	Helical fence topping, stainless steel				C.L.F.	144			144	159	
	0600	Hardware cloth galv., 1/4" mesh, 23 ga., 2' wide				C.S.F.	45			45	49.50	
	0700	3' wide					44			44	48	
	0900	1/2" mesh, 19 ga., 2' wide					38.50			38.50	42.50	
	1000	4' wide					38			38	41.50	
	1200	Chain link fabric, steel, 2" mesh, 6 ga, galvanized					113			113	124	
	1300	9 ga, galvanized					55			55	60.50	
	1350	Vinyl coated					45			45	49.50	
	1360	Aluminized					76			76	83.50	
	1400	2-1/4" mesh, 11.5 ga, galvanized					36			36	39.50	
	1600	1-3/4" mesh (tennis courts), 11.5 ga (core), vinyl coated					49			49	54	
	1700	9 ga, galvanized					64			64	70.50	
	2100	Welded wire fabric, 1" x 2", 14 ga.					27.50			27.50	30.50	
	2200	2" x 4", 12-1/2 ga.				↓	19.50			19.50	21.50	

Important: See the Reference Section for critical supporting data - Reference Nos., Crews, & City Cost Indexes

051 100	Bracing	CREW	DAILY OUTPUT	LABOR-HOURS	UNIT	MAT.	1999 BARE COSTS LABOR	EQUIP.	TOTAL	TOTAL INCL O&P	
108	0010	**BRACING**										108
	0300	Let-in, "T" shaped, 22 ga. galv. steel, studs at 16" O.C.	1 Carp	5.80	1.379	C.L.F.	37	37.50		74.50	99.50	
	0400	Studs at 24" O.C.		6	1.333		37	36.50		73.50	98	
	0500	16 ga. galv. steel straps, studs at 16" O.C.		6	1.333		48	36.50		84.50	110	
	0600	Studs at 24" O.C.	↓	6.20	1.290	↓	48	35		83	108	
110	0005	**PIPE SUPPORT FRAMING**										110
	0010	Under 10#/L.F.	E-4	3,900	.008	Lb.	.90	.26	.02	1.18	1.48	
	0200	10.1 to 15#/L.F.		4,300	.007		.78	.23	.02	1.03	1.30	
	0400	15.1 to 20#/L.F.		4,800	.007		.72	.21	.02	.95	1.19	
	0600	Over 20#/L.F.	↓	5,400	.006	↓	.66	.18	.02	.86	1.09	

051 200 | Structural Steel

			CREW	DAILY OUTPUT	LABOR-HOURS	UNIT	MAT.	1999 BARE COSTS LABOR	EQUIP.	TOTAL	TOTAL INCL O&P	
204	0010	**ANCHOR BOLTS**										204
	0020	See also divisions 031-110 and 041-504										
	0100	J-type, incl. nut, washer, 1/2" diameter x 6" long	2 Carp	70	.229	Ea.	1.11	6.25		7.36	11	
	0110	12" long		65	.246		1.27	6.70		7.97	11.95	
	0120	18" long		60	.267		1.73	7.30		9.03	13.35	
	0130	3/4" diameter x 8" long		50	.320		1.70	8.75		10.45	15.65	
	0140	12" long		45	.356		2.11	9.70		11.81	17.55	
	0150	18" long		40	.400		2.58	10.90		13.48	20	
	0160	1" diameter x 12" long		35	.457		3.20	12.50		15.70	23	
	0170	18" long		30	.533		3.61	14.55		18.16	27	
	0180	24" long		25	.640		4.25	17.45		21.70	32	
	0190	36" long		20	.800		5.60	22		27.60	40.50	
	0200	1-1/2" diameter x 18" long		22	.727		8.95	19.85		28.80	41	
	0210	24" long		16	1		10.15	27.50		37.65	54	
	0300	L-type, incl. hex nuts, 3/4" diameter x 12" long		45	.356		1.11	9.70		10.81	16.45	
	0310	18" long		40	.400		1.40	10.90		12.30	18.75	
	0320	24" long		35	.457		1.69	12.50		14.19	21.50	
	0330	30" long		30	.533		2.17	14.55		16.72	25.50	
	0340	36" long		25	.640		2.45	17.45		19.90	30	
	0350	1" diameter x 12" long		35	.457		2.23	12.50		14.73	22	
	0360	18" long		30	.533		2.74	14.55		17.29	26	
	0370	24" long		25	.640		3.29	17.45		20.74	31	
	0380	30" long		23	.696		3.82	19		22.82	34	
	0390	36" long		20	.800		4.34	22		26.34	39.50	
	0400	42" long		18	.889		5.50	24.50		30	44	
	0410	48" long		15	1.067		6.30	29		35.30	53	
	0420	1-1/4" diameter x 18" long		25	.640		5	17.45		22.45	33	
	0430	24" long		20	.800		5.95	22		27.95	41	
	0440	30" long		20	.800		6.75	22		28.75	42	
	0450	36" long		18	.889		7.65	24.50		32.15	46.50	
	0460	42" long		16	1		8.60	27.50		36.10	52.50	
	0470	48" long		14	1.143		9.65	31		40.65	59.50	
	0480	54" long		12	1.333		10.80	36.50		47.30	69.50	
	0490	60" long		10	1.600		11.65	43.50		55.15	81.50	
	0500	1-1/2" diameter x 18" long		22	.727		7.95	19.85		27.80	40	
	0510	24" long		19	.842		9	23		32	46	
	0520	30" long		17	.941		10.30	25.50		35.80	52	
	0530	36" long		16	1		11.95	27.50		39.45	56	
	0540	42" long		15	1.067		13.50	29		42.50	61	
	0550	48" long		13	1.231		15.20	33.50		48.70	69.50	
	0560	54" long		11	1.455		16.90	39.50		56.40	81	
	0570	60" long	↓	9	1.778	↓	18.55	48.50		67.05	97	

			DAILY	LABOR-		1999 BARE COSTS				TOTAL	
051 200 | Structural Steel		CREW	OUTPUT	HOURS	UNIT	MAT.	LABOR	EQUIP.	TOTAL	INCL O&P	
204	0580	1-3/4" diameter x 18" long	2 Carp	20	.800	Ea.	12.60	22		34.60	48.50	**204**
	0590	24" long		18	.889		14.45	24.50		38.95	54	
	0600	30" long		17	.941		16.35	25.50		41.85	58.50	
	0610	36" long		16	1		18.50	27.50		46	63.50	
	0620	42" long		14	1.143		20.50	31		51.50	71.50	
	0630	48" long		12	1.333		22.50	36.50		59	82.50	
	0640	54" long		10	1.600		25	43.50		68.50	96	
	0650	60" long		8	2		27	54.50		81.50	116	
	0660	2" diameter x 24" long		17	.941		18.85	25.50		44.35	61	
	0670	30" long		15	1.067		21	29		50	69	
	0680	36" long		13	1.231		23	33.50		56.50	78.50	
	0690	42" long		11	1.455		25.50	39.50		65	90.50	
	0700	48" long		10	1.600		28.50	43.50		72	99.50	
	0710	54" long		9	1.778		30	48.50		78.50	110	
	0720	60" long		8	2		32.50	54.50		87	122	
	0730	66" long		7	2.286		34.50	62.50		97	136	
	0740	72" long	▼	6	2.667	▼	37.50	73		110.50	157	
210	0005	**SUBPURLINS**										**210**
	0006	Bulb tees, painted, 32-5/8" O.C., 40 psf L.L.	R051 -250									
	0010	Up to 5'-6" span, 1.48 plf	E-1	5,900	.004	S.F.	.28	.12	.01	.41	.54	
	0050	6'-4" span, 1.68 plf		5,200	.005		.28	.14	.02	.44	.57	
	0060	7'-8" span, 1.87 plf		4,700	.005		.30	.15	.02	.47	.62	
	0100	8'-9" span, 2.15 plf		4,200	.006		.33	.17	.02	.52	.68	
	0200	10'-2" span, 3.19 plf		3,100	.008		.41	.23	.03	.67	.89	
	0300	12'-0" span, 3.87 plf	▼	2,700	.009	▼	.52	.27	.03	.82	1.08	
	0600	For galvanizing, add				Lb.	.21			.21	.23	
	1000	Truss tees, painted, 32-5/8" O.C., 40 psf L.L.										
	1100	2" high, up to 5'-9" span, 1.10 plf	E-1	7,000	.003	S.F.	.24	.10	.01	.35	.46	
	1110	6'-0" span, 1.27 plf		6,800	.004		.26	.11	.01	.38	.47	
	1120	6'-0" span, 1.33 plf		6,500	.004		.26	.11	.01	.38	.49	
	1130	7'-6" span, 1.78 plf		4,800	.005		.35	.15	.02	.52	.67	
	1200	2-1/2" high, up to 6'-9" span, 1.12 plf		7,000	.003		.24	.10	.01	.35	.46	
	1210	6'-9" span, 1.34 plf		6,500	.004		.26	.11	.01	.38	.49	
	1220	7'-0" span, 1.39 plf		6,200	.004		.28	.12	.01	.41	.51	
	1230	8'-9" span, 1.85 plf		4,700	.005		.38	.15	.02	.55	.70	
	1300	3" high, up to 7'-3" span, 1.14 plf		7,000	.003		.24	.10	.01	.35	.46	
	1310	9'-0" span, 1.88 plf		4,700	.005		.38	.15	.02	.55	.70	
	1400	3-1/2" high, up to 7'-9" span, 1.17 plf		6,900	.003		.25	.10	.01	.36	.46	
	1410	10'-9" span, 1.90 plf	▼	4,700	.005		.38	.15	.02	.55	.71	
	1420	For 24-5/8" spacing, add					33%	33%				
	1430	For 48-5/8" spacing, deduct	▼			▼	50%	50%				
212	0005	**CANOPY FRAMING**										**212**
	0010	6" and 8" members	E-4	3,000	.011	Lb.	.86	.33	.03	1.22	1.58	
215	0010	**CEILING SUPPORTS**										**215**
	1000	Entrance door/folding partition supports	E-4	60	.533	L.F.	13.20	16.60	1.37	31.17	46.50	
	1100	Linear accelerator door supports		14	2.286		60	71	5.85	136.85	202	
	1200	Lintels or shelf angles, hung, exterior hot dipped galv.		267	.120		9.75	3.73	.31	13.79	17.95	
	1250	Two coats primer paint instead of galv.		267	.120	▼	7.80	3.73	.31	11.84	15.80	
	1400	Monitor support, ceiling hung, expansion bolted		4	8	Ea.	209	249	20.50	478.50	710	
	1450	Hung from pre-set inserts		6	5.333		219	166	13.70	398.70	560	
	1600	Motor supports for overhead doors		4	8	▼	106	249	20.50	375.50	595	
	1700	Partition support for heavy folding partitions, without pocket		24	1.333	L.F.	30	41.50	3.43	74.93	113	
	1750	Supports at pocket only		12	2.667		60	83	6.85	149.85	226	
	2000	Rolling grilles & fire door supports		34	.941	▼	25.50	29.50	2.42	57.42	84.50	
	2100	Spider-leg light supports, expansion bolted to ceiling slab	▼	8	4	Ea.	86	124	10.30	220.30	335	

5

METALS

Important: See the Reference Section for critical supporting data - Reference Nos., Crews, & City Cost Indexes

051 200	Structural Steel		CREW	DAILY OUTPUT	LABOR-HOURS	UNIT	1999 BARE COSTS				TOTAL INCL O&P	
							MAT.	LABOR	EQUIP.	TOTAL		
215	2150	Hung from pre-set inserts	E-4	12	2.667	Ea.	90	83	6.85	179.85	259	215
	2400	Toilet partition support		36	.889	L.F.	30	27.50	2.28	59.78	86	
	2500	X-ray travel gantry support	▼	12	2.667	"	103	83	6.85	192.85	273	
220	0005	**COLUMNS** R051-210										220
	0010	Aluminum, extruded, stock units, 6" diameter	E-4	240	.133	L.F.	8.10	4.15	.34	12.59	16.90	
	0100	8" diameter		170	.188		10.80	5.85	.48	17.13	23	
	0200	10" diameter		150	.213		13.70	6.65	.55	20.90	28	
	0300	12" diameter		140	.229		23	7.10	.59	30.69	39	
	0400	15" diameter	▼	120	.267	▼	30	8.30	.69	38.99	49	
	0410	Caps and bases, plain, 6" diameter				Set	16.60			16.60	18.25	
	0420	8" diameter					19.40			19.40	21.50	
	0430	10" diameter					26			26	28.50	
	0440	12" diameter					39			39	43	
	0450	15" diameter					78.50			78.50	86	
	0460	Caps, ornamental, minimum					79.50			79.50	87.50	
	0470	Maximum				▼	910			910	1,000	
	0500	For square columns, add to column prices above				L.F.	50%					
	0600	Tubular aluminum	E-4	1,500	.021	Lb.	2.83	.66	.05	3.54	4.39	
	0700	Residential, flat, 8' high, plain		20	1.600	Ea.	54	50	4.11	108.11	155	
	0720	Fancy		20	1.600		94.50	50	4.11	148.61	200	
	0740	Corner type, plain		20	1.600		68.50	50	4.11	122.61	171	
	0760	Fancy	▼	20	1.600	▼	123	50	4.11	177.11	232	
	0800	Steel, concrete filled, extra strong pipe, 3-1/2" diameter	E-2	660	.085	L.F.	18.35	2.53	1.53	22.41	26	
	0930	6" diameter		1,200	.047		34.50	1.39	.84	36.73	41.50	
	1000	Lightweight units, 3-1/2" diameter		780	.072		2.17	2.14	1.30	5.61	7.55	
	1050	4" diameter	▼	900	.062	▼	3.26	1.85	1.12	6.23	8.10	
	1100	For galvanizing, add				Lb.	.35			.35	.39	
	1300	For web ties, angles, etc., add per added lb.	1 Sswk	945	.008		.81	.26		1.07	1.36	
	1500	Steel pipe, extra strong, no concrete, 3" to 5" diameter	E-2	16,000	.004		.62	.10	.06	.78	.93	
	1600	6" to 12" diameter		14,000	.004	▼	.61	.12	.07	.80	.96	
	1700	Steel pipe, extra strong, no concrete, 3" diameter x 12'-0"		60	.933	Ea.	83.50	28	16.85	128.35	159	
	1750	4" diameter x 12'-0"		58	.966		124	28.50	17.45	169.95	206	
	1800	6" diameter x 12'-0"		54	1.037		234	31	18.75	283.75	330	
	1850	8" diameter x 14'-0"		50	1.120		415	33.50	20	468.50	535	
	1900	10" diameter x 16'-0"		48	1.167		600	34.50	21	655.50	745	
	1950	12" diameter x 18'-0"		45	1.244	▼	815	37	22.50	874.50	985	
	3300	Structural tubing, square, A500GrB, 4" to 6" square, light section		11,270	.005	Lb.	.82	.15	.09	1.06	1.26	
	3600	Heavy section	▼	32,000	.002	"	.78	.05	.03	.86	.98	
	4000	Concrete filled, add				L.F.	2.81			2.81	3.09	
	4500	Structural tubing, sq, 4" x 4" x 1/4" x 12'-0"	E-2	58	.966	Ea.	134	28.50	17.45	179.95	218	
	4550	6" x 6" x 1/4" x 12'-0"		54	1.037		220	31	18.75	269.75	315	
	4600	8" x 8" x 3/8" x 14'-0"		50	1.120		455	33.50	20	508.50	580	
	4650	10" x 10" x 1/2" x 16'-0"		48	1.167	▼	845	34.50	21	900.50	1,025	
	5100	Structural tubing, rect, 5" to 6" wide, light section		9,500	.006	Lb.	.75	.18	.11	1.04	1.26	
	5200	Heavy section		31,200	.002		.75	.05	.03	.83	.96	
	5300	7" to 10" wide, light section		37,000	.002		.70	.05	.03	.78	.88	
	5400	Heavy section		68,000	.001	▼	.70	.02	.01	.73	.83	
	5500	Structural tubing, rect, 5" x 3" x 1/4" x 12'-0"		58	.966	Ea.	120	28.50	17.45	165.95	202	
	5550	6" x 4" x 5/16" x 12'-0"		54	1.037		188	31	18.75	237.75	281	
	5600	8" x 4" x 3/8" x 12'-0"		54	1.037		256	31	18.75	305.75	355	
	5650	10" x 6" x 3/8" x 14'-0"		50	1.120		410	33.50	20	463.50	530	
	5700	12" x 8" x 1/2" x 16'-0"	▼	48	1.167		755	34.50	21	810.50	915	
	5800	Adjustable jack post, 8' maximum height, 2-3/4" diameter					21			21	23	
	5850	4" diameter				▼	34			34	37.50	
	6000	Prefabricated, fireproof, with steel jackets and one coat	▼									

METALS 5

151

051 200 | Structural Steel

220			CREW	DAILY OUTPUT	LABOR-HOURS	UNIT	1999 BARE COSTS				TOTAL INCL O&P	220
							MAT.	LABOR	EQUIP.	TOTAL		
6100	shop paint, 2 to 4 hour rated, minimum	R051 -210	E-2	27,000	.002	Lb.	.43	.06	.04	.53	.62	
6200	Average			35,000	.002		.61	.05	.03	.69	.78	
6250	Maximum		▼	43,000	.001	▼	1.01	.04	.02	1.07	1.21	
6400	Mild steel, flat, 9" wide, stock units, painted, plain		E-4	160	.200	L.F.	4.61	6.20	.51	11.32	17	
6450	Fancy			160	.200		9.10	6.20	.51	15.81	22	
6500	Corner columns, painted, plain			160	.200		7.30	6.20	.51	14.01	20	
6550	Fancy		▼	160	.200		14.35	6.20	.51	21.06	27.50	
6800	W Shape, A36 steel, 2 tier, W8 x 24		E-2	1,080	.052		15.85	1.54	.94	18.33	21	
6850	W8 x 31			1,080	.052		20.50	1.54	.94	22.98	26	
6900	W8 x 48			1,032	.054		31.50	1.62	.98	34.10	39	
6950	W8 x 67			984	.057		44	1.69	1.03	46.72	52.50	
7000	W10 x 45			1,032	.054		29.50	1.62	.98	32.10	36.50	
7050	W10 x 68			984	.057		45	1.69	1.03	47.72	53.50	
7100	W10 x 112			960	.058		74	1.74	1.05	76.79	85.50	
7150	W12 x 50			1,032	.054		33	1.62	.98	35.60	40.50	
7200	W12 x 87			984	.057		57.50	1.69	1.03	60.22	67	
7250	W12 x 120			960	.058		79	1.74	1.05	81.79	91	
7300	W12 x 190			912	.061		125	1.83	1.11	127.94	142	
7350	W14 x 74			984	.057		49	1.69	1.03	51.72	57.50	
7400	W14 x 120			960	.058		79	1.74	1.05	81.79	91	
7450	W14 x 176		▼	912	.061	▼	116	1.83	1.11	118.94	132	

230												230
0010	**LIGHTWEIGHT FRAMING**	R051 -240										
0200	For steel studs see division 092-612											
0400	Angle framing, 4" and larger		E-4	3,000	.011	Lb.	.85	.33	.03	1.21	1.58	
0450	Less than 4" angles		"	1,800	.018	"	.91	.55	.05	1.51	2.06	
0460	1/2" x 1/2" x 1/8"		E-3	200	.120	L.F.	.18	3.75	.41	4.34	7.55	
0462	3/4" x 3/4" x 1/8"			160	.150		.51	4.69	.51	5.71	9.75	
0464	1" x 1" x 1/8"			135	.178		.72	5.55	.61	6.88	11.65	
0466	1-1/4" x 1-1/4" x 3/16"			115	.209		1.34	6.55	.72	8.61	14.20	
0468	1-1/2" x 1-1/2" x 3/16"			100	.240		1.63	7.50	.82	9.95	16.45	
0470	2" x 2" x 1/4"			90	.267		2.89	8.35	.91	12.15	19.50	
0472	2-1/2" x 2-1/2" x 1/4"			72	.333		3.71	10.40	1.14	15.25	24.50	
0474	3" x 2" x 3/8"			65	.369		5.35	11.55	1.27	18.17	28.50	
0476	3" x 3" x 3/8"		▼	57	.421	▼	6.50	13.15	1.44	21.09	33	
0600	Channel framing, 8" and larger		E-4	3,500	.009	Lb.	.93	.28	.02	1.23	1.57	
0650	Less than 8" channels		"	2,000	.016	"	.94	.50	.04	1.48	1.99	
0660	C2 x 1.78		E-3	115	.209	L.F.	1.67	6.55	.72	8.94	14.55	
0662	C3 x 4.1			80	.300		3.84	9.40	1.03	14.27	22.50	
0664	C4 x 5.4			66	.364		5.05	11.35	1.25	17.65	28	
0666	C5 x 6.7			57	.421		6.25	13.15	1.44	20.84	32.50	
0668	C6 x 8.2			55	.436		7.70	13.65	1.50	22.85	35	
0670	C7 x 9.8			40	.600		9.15	18.75	2.06	29.96	47	
0672	C8 x 11.5			36	.667		10.40	21	2.29	33.69	52	
0710	Structural bar tee, 3/4" x 3/4" x 1/8"			160	.150		1.09	4.69	.51	6.29	10.35	
0712	1" x 1" x 1/8"			135	.178		1.38	5.55	.61	7.54	12.40	
0714	1-1/2" x 1-1/2" x 1/4"			114	.211		2.18	6.60	.72	9.50	15.25	
0716	2" x 2" x 1/4"			89	.270		2.72	8.45	.92	12.09	19.45	
0718	2-1/2" x 2-1/2" x 3/8"			72	.333		4.90	10.40	1.14	16.44	26	
0720	3" x 3" x 3/8"			57	.421		6	13.15	1.44	20.59	32	
0730	Structural zee, 1-1/4" x 1-3/4" x 1-3/4"			114	.211		.69	6.60	.72	8.01	13.60	
0732	2-11/16" x 3" x 2-11/16"			114	.211		1.65	6.60	.72	8.97	14.65	
0734	3-1/16" x 4" x 3-1/16"			133	.180		2.28	5.65	.62	8.55	13.55	
0736	3-1/4" x 5" x 3-1/4"			133	.180		2.90	5.65	.62	9.17	14.20	
0738	3-1/2" x 6" x 3-1/2"			160	.150		4.01	4.69	.51	9.21	13.60	
0740	Junior beam, 3"		▼	80	.300	▼	4.62	9.40	1.03	15.05	23.50	

5 METALS

Important: See the Reference Section for critical supporting data - Reference Nos., Crews, & City Cost Indexes

				DAILY OUTPUT	LABOR-HOURS	UNIT	1999 BARE COSTS				TOTAL INCL O&P	
		051 200 | Structural Steel	CREW				MAT.	LABOR	EQUIP.	TOTAL		
230	0742	4″	E-3	72	.333	L.F.	6.25	10.40	1.14	17.79	27	230
	0744	5″		67	.358		8.10	11.20	1.23	20.53	31	
	0746	6″		62	.387		10.15	12.10	1.33	23.58	34.50	
	0748	7″		57	.421		12.40	13.15	1.44	26.99	39	
	0750	8″		53	.453		14.90	14.15	1.55	30.60	44	
	1000	Continuous slotted channel framing system, minimum	2 Sswk	2,400	.007	Lb.	1.20	.20		1.40	1.69	
	1200	Maximum	″	1,600	.010		2.16	.31		2.47	2.94	
	1300	Cross bracing, rods, 3/4″ diameter	E-3	700	.034		.96	1.07	.12	2.15	3.16	
	1310	7/8″ diameter		700	.034		.81	1.07	.12	2	2.99	
	1320	1″ diameter		700	.034		.81	1.07	.12	2	2.99	
	1330	Angle, 5″ x 5″ x 3/8″		2,800	.009		.84	.27	.03	1.14	1.44	
	1350	Hanging lintels, average		850	.028		.90	.88	.10	1.88	2.72	
	1380	Roof frames, 3′-0″ square, 5′ span	E-2	4,200	.013		.81	.40	.24	1.45	1.86	
	1400	Tie rod, not upset, 1-1/2″ to 4″ diameter, with turnbuckle	2 Sswk	800	.020		.90	.61		1.51	2.11	
	1420	No turnbuckle		700	.023		.81	.70		1.51	2.17	
	1500	Upset, 1-3/4″ to 4″ diameter, with turnbuckle		800	.020		1.02	.61		1.63	2.24	
	1520	No turnbuckle		700	.023		.90	.70		1.60	2.27	
	1600	Tubular aluminum framing for window wall, minimum		600	.027		4.03	.82		4.85	5.95	
	1800	Maximum		500	.032		11.90	.98		12.88	14.90	
232	0005	**LINTELS**										232
	0010	Plain steel angles, under 500 lb.	1 Bric	550	.015	Lb.	.60	.40		1	1.28	
	0100	500 to 1000 lb. CN		640	.013		.57	.35		.92	1.17	
	0200	1,000 to 2,000 lb.		640	.013		.53	.35		.88	1.12	
	0300	2,000 to 4,000 lb.		640	.013		.48	.35		.83	1.07	
	0500	For built-up angles and plates, add to above					.60			.60	.66	
	0700	For engineering, add to above					.15			.15	.16	
	0900	For galvanizing, add to above, under 500 lb.					.42			.42	.46	
	1000	Over 2,000 lb.					.30			.30	.33	
	2000	Steel angles, 3-1/2″ x 3″, 1/4″ thick, 2′-6″ long	1 Bric	50	.160	Ea.	8.10	4.42		12.52	15.75	
	2100	4′-6″ long		45	.178		14.60	4.91		19.51	23.50	
	2600	4″ x 3-1/2″, 1/4″ thick, 5′-0″ long		40	.200		18.60	5.50		24.10	29	
	2700	9′-0″ long		35	.229		33.50	6.30		39.80	47	
	3500	For precast concrete lintels, see div. 034-802										
234	0010	**PLATES** Struct. steel plates, 5-10 tons, 48″-60″ widths, 10′-60′ lengths										234
	0100	1/4″ thick				Cwt.	37.50			37.50	41.50	
	0200	5/16″ thick					37.50			37.50	41.50	
	0300	3/8″ thick					35.50			35.50	39	
	0400	1/2″ thick					35.50			35.50	39	
	0500	1″ thick					35.50			35.50	39	
	0600	2″ thick					36.50			36.50	40.50	
	0700	4″ thick					38.50			38.50	42.50	
	0800	6″ thick					40.50			40.50	44.50	
	0900	8″ thick					40.50			40.50	44.50	
238	0005	**SPACE FRAME**										238
	0010	Steel collars and channel members, 4′ modules, minimum	E-2	1,200	.047	S.F.	15	1.39	.84	17.23	19.85	
	0200	Maximum		900	.062		30	1.85	1.12	32.97	37.50	
	0400	5′ modules, minimum		1,300	.043		13.50	1.28	.78	15.56	17.95	
	0500	Maximum		1,000	.056		19	1.67	1.01	21.68	25	
	0600	Steel collars and square tubular members, 5′ modules, minimum		900	.062		11.50	1.85	1.12	14.47	17.15	
	0650	Maximum		600	.093		25.50	2.78	1.69	29.97	34.50	
	0700	Add to above for galvanizing, 4′ modular					1.90			1.90	2.09	
	0800	5′ modular					1.50			1.50	1.65	
	0900	Steel nodes & tubular members, 7′ to 10′ modules, minimum	E-2	1,650	.034		20.50	1.01	.61	22.12	25	
	0950	Maximum	″	825	.068		50	2.02	1.23	53.25	60	

5

METALS

		051 200 | **Structural Steel**		CREW	DAILY OUTPUT	LABOR-HOURS	UNIT	1999 BARE COSTS				TOTAL INCL O&P	
								MAT.	LABOR	EQUIP.	TOTAL		
240	0010	**STEEL CUTTING**											**240**
	0020	Hand burning, incl. preparation, torch cutting & grinding, no staging											
	0100	Steel to 1/2" thick		E-14	70	.114	L.F.		3.73	1.18	4.91	8.15	
	0150	3/4" thick		↓	50	.160			5.20	1.65	6.85	11.35	
	0200	1" thick		↓	45	.178	↓		5.80	1.83	7.63	12.60	
245	0005	**STRESSED SKIN** Roof & ceiling system											**245**
	0010	Spans to 100', minimum		E-2	1,150	.049	S.F.	4.73	1.45	.88	7.06	8.70	
	0100	Convex, spans to 200'			960	.058		7.70	1.74	1.05	10.49	12.65	
	0200	Double, spans to 300'		↓	760	.074	↓	11.70	2.19	1.33	15.22	18.15	
250	0010	**STRUCTURAL STEEL MEMBERS**											**250**
	0020	Common shapes, bolted connections, spans 10'-45', incl erection											
	0100	W 6 x 9	R051 -210	E-2	600	.093	L.F.	5.95	2.78	1.69	10.42	13.30	
	0120	x 16			600	.093		10.55	2.78	1.69	15.02	18.35	
	0140	x 20	R051 -240		600	.093		13.20	2.78	1.69	17.67	21	
	0300	W 8 x 10			600	.093		6.60	2.78	1.69	11.07	14	
	0320	x 15			600	.093		9.90	2.78	1.69	14.37	17.65	
	0350	x 21			600	.093		13.85	2.78	1.69	18.32	22	
	0360	x 24			550	.102		15.85	3.03	1.84	20.72	24.50	
	0370	x 28			550	.102		18.50	3.03	1.84	23.37	28	
	0500	x 31			550	.102		20.50	3.03	1.84	25.37	30	
	0520	x 35			550	.102		23	3.03	1.84	27.87	33	
	0540	x 48			550	.102		31.50	3.03	1.84	36.37	42.50	
	0600	W 10 x 12			600	.093		7.90	2.78	1.69	12.37	15.45	
	0620	x 15			600	.093		9.90	2.78	1.69	14.37	17.65	
	0700	x 22			600	.093		14.50	2.78	1.69	18.97	22.50	
	0720	x 26			600	.093		17.15	2.78	1.69	21.62	25.50	
	0740	x 33			550	.102		22	3.03	1.84	26.87	31.50	
	0900	x 49			550	.102		32.50	3.03	1.84	37.37	43	
	1100	W 12 x 14			880	.064		9.25	1.89	1.15	12.29	14.75	
	1300	x 22			880	.064		14.50	1.89	1.15	17.54	20.50	
	1500	x 26			880	.064		17.15	1.89	1.15	20.19	23.50	
	1520	x 35			810	.069		23	2.06	1.25	26.31	30.50	
	1560	x 50			750	.075		33	2.22	1.35	36.57	42	
	1580	x 58			750	.075		38.50	2.22	1.35	42.07	47.50	
	1700	x 72			640	.087		45.50	2.60	1.58	49.68	56.50	
	1740	x 87			640	.087		57.50	2.60	1.58	61.68	69.50	
	1900	W 14 x 26			990	.057		17.15	1.68	1.02	19.85	23	
	2100	x 30			900	.062		19.80	1.85	1.12	22.77	26.50	
	2300	x 34			810	.069		22.50	2.06	1.25	25.81	29.50	
	2320	x 43			810	.069		28.50	2.06	1.25	31.81	36	
	2340	x 53			800	.070		35	2.08	1.26	38.34	43.50	
	2360	x 74			760	.074		49	2.19	1.33	52.52	59	
	2380	x 90			740	.076		59.50	2.25	1.37	63.12	71	
	2500	x 120			720	.078		75.50	2.32	1.41	79.23	88.50	
	2700	W 16 x 26			1,000	.056		17.15	1.67	1.01	19.83	23	
	2900	x 31			900	.062		20.50	1.85	1.12	23.47	27	
	3100	x 40			800	.070		26.50	2.08	1.26	29.84	34	
	3120	x 50			800	.070		33	2.08	1.26	36.34	41.50	
	3140	x 67		↓	760	.074		44	2.19	1.33	47.52	54	
	3300	W 18 x 35		E-5	960	.083		23	2.52	1.14	26.66	31	
	3500	x 40			960	.083		26.50	2.52	1.14	30.16	34.50	
	3520	x 46			960	.083		30.50	2.52	1.14	34.16	39	
	3700	x 50			912	.088		31.50	2.65	1.20	35.35	40.50	
	3900	x 55			912	.088		34.50	2.65	1.20	38.35	44	
	3920	x 65		↓ ↓	900	.089	↓	43	2.69	1.22	46.91	53	

Important: See the Reference Section for critical supporting data - Reference Nos., Crews, & City Cost Indexes

051 200 | Structural Steel

			CREW	DAILY OUTPUT	LABOR-HOURS	UNIT	MAT.	LABOR	EQUIP.	TOTAL	TOTAL INCL O&P	
250	3940	x 76	R051 -210	E-5	900	.089	L.F.	50	2.69	1.22	53.91	61	**250**
	3960	x 86			900	.089		57	2.69	1.22	60.91	68.50	
	3980	x 106	R051 -240		900	.089		70	2.69	1.22	73.91	83	
	4100	W 21 x 44			1,064	.075		27.50	2.27	1.03	30.80	35.50	
	4300	x 50			1,064	.075		31.50	2.27	1.03	34.80	39.50	
	4500	x 62			1,036	.077		39	2.33	1.06	42.39	48.50	
	4700	x 68			1,036	.077		43	2.33	1.06	46.39	52.50	
	4720	x 83			1,000	.080		55	2.42	1.09	58.51	66	
	4740	x 93			1,000	.080		61.50	2.42	1.09	65.01	73	
	4760	x 101			1,000	.080		66.50	2.42	1.09	70.01	79	
	4780	x 122			1,000	.080		80.50	2.42	1.09	84.01	94	
	4900	W 24 x 55			1,110	.072		34.50	2.18	.99	37.67	43	
	5100	x 62			1,110	.072		39	2.18	.99	42.17	48	
	5300	x 68			1,110	.072		43	2.18	.99	46.17	52	
	5500	x 76			1,110	.072		48	2.18	.99	51.17	57.50	
	5700	x 84			1,080	.074		53	2.24	1.01	56.25	63	
	5720	x 94			1,080	.074		62	2.24	1.01	65.25	73	
	5740	x 104			1,050	.076		68.50	2.30	1.04	71.84	80.50	
	5760	x 117			1,050	.076		77	2.30	1.04	80.34	90	
	5780	x 146			1,050	.076		96.50	2.30	1.04	99.84	111	
	5800	W 27 x 84			1,190	.067		53	2.03	.92	55.95	62.50	
	5900	x 94			1,190	.067		59	2.03	.92	61.95	69.50	
	5920	x 114			1,150	.070		72	2.10	.95	75.05	84	
	5940	x 146			1,150	.070		92	2.10	.95	95.05	106	
	5960	x 161			1,150	.070		101	2.10	.95	104.05	117	
	6100	W 30 x 99			1,200	.067		62.50	2.01	.91	65.42	73	
	6300	x 108			1,200	.067		68	2.01	.91	70.92	79.50	
	6500	x 116			1,160	.069		71.50	2.08	.94	74.52	84	
	6520	x 132			1,160	.069		87	2.08	.94	90.02	101	
	6540	x 148			1,160	.069		97.50	2.08	.94	100.52	112	
	6560	x 173			1,120	.071		114	2.16	.98	117.14	131	
	6580	x 191			1,120	.071		126	2.16	.98	129.14	144	
	6700	W 33 x 118			1,176	.068		73	2.06	.93	75.99	84.50	
	6900	x 130			1,134	.071		80.50	2.13	.97	83.60	93.50	
	7100	x 141			1,134	.071		87	2.13	.97	90.10	101	
	7120	x 169			1,100	.073		112	2.20	.99	115.19	128	
	7140	x 201			1,100	.073		133	2.20	.99	136.19	151	
	7300	W 36 x 135			1,170	.068		82.50	2.07	.94	85.51	95.50	
	7500	x 150			1,170	.068		92	2.07	.94	95.01	106	
	7600	x 170			1,150	.070		112	2.10	.95	115.05	128	
	7700	x 194			1,125	.071		118	2.15	.97	121.12	134	
	7900	x 230			1,125	.071		139	2.15	.97	142.12	158	
	7920	x 260			1,035	.077		172	2.34	1.06	175.40	194	
	8100	x 300			1,035	.077		182	2.34	1.06	185.40	205	
255	0010	**STRUCTURAL STEEL PROJECTS** Bolted, unless noted otherwise	R050 -310										**255**
	0200	Apartments, nursing homes, etc., 1 to 2 stories		E-5	10.30	7.767	Ton	1,200	235	106	1,541	1,850	
	0300	3 to 6 stories	R050 -510	"	10.10	7.921		1,225	239	108	1,572	1,900	
	0400	7 to 15 stories		E-6	14.20	9.014		1,250	273	88	1,611	1,950	
	0500	Over 15 stories	R051 -210	"	13.90	9.209		1,300	279	90	1,669	2,025	
	0700	Offices, hospitals, etc., steel bearing, 1 to 2 stories **CN**		E-5	10.30	7.767		1,200	235	106	1,541	1,850	
	0800	3 to 6 stories	R051 -220	E-6	14.40	8.889		1,225	269	87	1,581	1,925	
	0900	7 to 15 stories			14.20	9.014		1,250	273	88	1,611	1,950	
	1000	Over 15 stories	R051 -230		13.90	9.209		1,300	279	90	1,669	2,025	
	1100	For multi-story masonry wall bearing construction, add							30%				
	1300	Industrial bldgs., 1 story, beams & girders, steel bearing	R051 -240	E-5	12.90	6.202		1,200	187	85	1,472	1,750	
	1400	Masonry bearing		"	10	8		1,200	242	109	1,551	1,875	

051 200 | Structural Steel		CREW	DAILY OUTPUT	LABOR-HOURS	UNIT	1999 BARE COSTS				TOTAL INCL O&P
						MAT.	LABOR	EQUIP.	TOTAL	
255 **1500**	Industrial bldgs., 1 story, under 10 tons,	R050 -310								**255**
1510	steel from warehouse, trucked		E-2	7.50	7.467	Ton	1,450	222	135	1,807	2,125
1600	1 story with roof trusses, steel bearing	R050 -510	E-5	10.60	7.547		1,425	228	103	1,756	2,075
1700	Masonry bearing		"	8.30	9.639		1,425	291	132	1,848	2,225
1900	Monumental structures, banks, stores, etc., minimum	R051 -210	E-6	13	9.846		1,200	298	96	1,594	1,950
2000	Maximum		"	9	14.222		2,000	430	139	2,569	3,125
2200	Churches, minimum	R051 -220	E-5	11.60	6.897		1,125	208	94.50	1,427.50	1,700
2300	Maximum		"	5.20	15.385		1,500	465	210	2,175	2,675
2800	Power stations, fossil fuels, minimum	R051 -230	E-6	11	11.636		1,200	350	114	1,664	2,075
2900	Maximum		"	5.70	22.456		1,800	680	219	2,699	3,425
2950	Nuclear fuels, non-safety steel, minimum	R051 -240		7	18.286		1,200	555	178	1,933	2,500
3000	Maximum			5.50	23.273		1,800	705	227	2,732	3,475
3040	Safety steel, minimum			2.50	51.200		1,750	1,550	500	3,800	5,225
3070	Maximum			1.50	85.333		2,300	2,575	835	5,710	8,050
3100	Roof trusses, minimum		E-5	13	6.154		1,475	186	84	1,745	2,050
3200	Maximum			8.30	9.639		2,500	291	132	2,923	3,425
3210	Schools, minimum			14.50	5.517		1,200	167	75.50	1,442.50	1,700
3220	Maximum			8.30	9.639		1,750	291	132	2,173	2,600
3400	Welded construction, simple commercial bldgs., 1 to 2 stories		E-7	7.60	10.526		1,225	320	155	1,700	2,075
3500	7 to 15 stories		E-9	8.30	15.422		1,425	465	171	2,061	2,575
3700	Welded rigid frame, 1 story, minimum		E-7	15.80	5.063		1,250	153	74.50	1,477.50	1,725
3800	Maximum		"	5.50	14.545		1,625	440	214	2,279	2,800
4000	High strength steels, add to A36 price, minimum						15			15	16.50
4100	Maximum						154			154	169
4300	Column base plates, light, up to 150 lb		2 Sswk	2,000	.008	Lb.	.66	.24		.90	1.18
4400	Heavy, over 150 lb		E-2	7,500	.007	"	.67	.22	.13	1.02	1.28
4600	Castellated beams, light sections, to 50#/L.F., minimum			10.70	5.234	Ton	1,375	156	94.50	1,625.50	1,900
4700	Maximum			7	8		1,750	238	145	2,133	2,500
4900	Heavy sections, over 50# per L.F., minimum			11.70	4.786		1,200	142	86.50	1,428.50	1,675
5000	Maximum			7.80	7.179		1,650	214	130	1,994	2,325
5500	Steel domes - see R131-310										
5700	Steel estimating weights per S.F. - see R051-220										
5900	Galvanizing structural steel in shop, under 1 ton, add to above					Ton	750			750	825
5950	1 ton to 20 tons						700			700	770
6000	Over 20 tons						650			650	715
6100	Cold galvanizing, brush		1 Psst	1,100	.007	S.F.	.06	.19		.25	.43
6125	Steel surface treatments										
6170	Wire brush, hand		1 Psst	240	.033	S.F.	.05	.87		.92	1.70
6180	Power tool		"	600	.013		.05	.35		.40	.72
6215	Pressure washing, 2800-6000 S.F./day		1 Pord	3,500	.002			.06		.06	.09
6220	Steam cleaning, 2800-4000 S.F./day			2,400	.003			.08		.08	.13
6225	Water blasting			3,200	.002			.06		.06	.10
6230	Brush-off blast		2 Pord	3,800	.004		.07	.10		.17	.23
6235	Blast (SSPC-6), loose scale, fine pwdr rust, 2.0 #/S.F.			2,500	.006		.13	.16		.29	.39
6240	Tight mill scale, little/no rust, 3.0 #/S.F.			1,920	.008		.20	.21		.41	.54
6245	Exist coat blistered/pitted, 4.0 #/S.F.			1,280	.013		.27	.31		.58	.78
6250	Exist coat badly pitted/nodules, 6.7 #/S.F.			770	.021		.45	.52		.97	1.28
6255	Near white blast (SSPC-10), loose scale, fine rust, 5.6 #/S.F.			1,025	.016		.38	.39		.77	1
6260	Tight mill scale, little/no rust, 6.9 #/S.F.			830	.019		.46	.48		.94	1.24
6265	Exist coat blistered/pitted, 9.0 #/S.F.			640	.025		.60	.62		1.22	1.61
6270	Exist coat badly pitted/nodules, 11.3 #/S.F.			510	.031		.76	.78		1.54	2.03
6510	Paints & protective coatings, sprayed										
6520	Alkyds, primer		2 Psst	3,600	.004	S.F.	.05	.12		.17	.27
6540	Gloss topcoats			3,200	.005		.04	.13		.17	.29
6560	Silicone alkyd			3,200	.005		.09	.13		.22	.35
6610	Epoxy, primer			3,000	.005		.15	.14		.29	.43

Important: See the Reference Section for critical supporting data - Reference Nos., Crews, & City Cost Indexes

			CREW	DAILY OUTPUT	LABOR-HOURS	UNIT	1999 BARE COSTS				TOTAL INCL O&P	
							MAT.	LABOR	EQUIP.	TOTAL		

051 200 | Structural Steel

255	6630	Intermediate or topcoat	R050 -310	2 Psst	2,800	.006	S.F.	.12	.15		.27	.42	255
	6650	Enamel coat			2,800	.006		.15	.15		.30	.45	
	6700	Epoxy ester, primer	R050 -510		2,800	.006		.20	.15		.35	.50	
	6720	Topcoats			2,800	.006		.10	.15		.25	.39	
	6810	Latex primer	R051 -210		3,600	.004		.05	.12		.17	.27	
	6830	Topcoats			3,200	.005		.05	.13		.18	.30	
	6910	Universal primers, one part, phenolic, modified alkyd	R051 -220		2,000	.008		.08	.21		.29	.47	
	6940	Two part, epoxy spray			2,000	.008		.13	.21		.34	.53	
	7000	Zinc rich primers, self cure, spray, inorganic	R051 -230		1,800	.009		.24	.23		.47	.70	
	7010	Epoxy, spray, organic			1,800	.009		.12	.23		.35	.58	
	7020	Above grade, simple structures, add	R051 -240						25%				
	7030	Intricate structures, add							50%				
275	0010	**VIBRATION PADS**											275
	0300	Laminated synthetic rubber impregnated cotton duck, 1/2" thick		2 Sswk	20	.800	S.F.	50	24.50		74.50	100	
	0400	1" thick			20	.800		101	24.50		125.50	156	
	0600	Neoprene bearing pads, 1/2" thick			24	.667		18.90	20.50		39.40	58.50	
	0700	1" thick			20	.800		39	24.50		63.50	87.50	
	0900	Fabric reinforced neoprene, 5000 psi, 1/2" thick			24	.667		8.50	20.50		29	47	
	1000	1" thick			20	.800		17	24.50		41.50	63.50	
	1200	Felt surfaced vinyl pads, cork and sisal, 5/8" thick			24	.667		22	20.50		42.50	62	
	1300	1" thick			20	.800		40	24.50		64.50	89.50	
	1500	Teflon bonded to 10 ga. carbon steel, 1/32" layer			24	.667		36	20.50		56.50	77	
	1600	3/32" layer			24	.667		54	20.50		74.50	97	
	1800	Bonded to 10 ga. stainless steel, 1/32" layer			24	.667		65	20.50		85.50	109	
	1900	3/32" layer			24	.667		83.50	20.50		104	130	
	2100	Circular, encased, rule of thumb					Kip	5.50			5.50	6.05	

051 400 | Structural Aluminum

404	0005	**ALUMINUM**											404
	0010	Structural shapes, 1" to 10" members, under 1 ton		E-2	1,050	.053	Lb.	1.69	1.59	.96	4.24	5.70	
	0050	1 to 5 tons			1,330	.042		1.57	1.25	.76	3.58	4.77	
	0100	Over 5 tons **CN**			1,330	.042		1.65	1.25	.76	3.66	4.86	
	0300	Extrusions, over 5 tons, stock shapes			1,330	.042		1.47	1.25	.76	3.48	4.66	
	0400	Custom shapes			1,330	.042		1.64	1.25	.76	3.65	4.84	

051 500 | Steel Wire Rope

508	0005	**WIRE ROPE**											508
	0010	6 x 19, bright, fiber core, 5000' rolls, 1/2" diameter					L.F.	.63			.63	.69	
	0050	Steel core						.72			.72	.80	
	0100	Fiber core, 1" diameter						2			2	2.20	
	0150	Steel core						2.20			2.20	2.42	
	0300	6 x 19, galvanized, fiber core, 1/2" diameter						1			1	1.10	
	0350	Steel core						1.12			1.12	1.24	
	0400	Fiber core, 1" diameter						3.03			3.03	3.34	
	0450	Steel core						3.22			3.22	3.54	
	0500	6 x 7, bright, IPS, fiber core, <500 L.F. w/acc., 1/4" diameter		E-17	6,400	.002		.65	.08		.73	.85	
	0510	1/2" diameter			2,100	.008		1.60	.24		1.84	2.20	
	0520	3/4" diameter			960	.017		2.89	.53		3.42	4.15	
	0550	6 x 19, bright, IPS, IWRC, <500 L.F. w/acc., 1/4" diameter			5,760	.003		.99	.09		1.08	1.25	
	0560	1/2" diameter			1,730	.009		1.61	.29		1.90	2.31	
	0570	3/4" diameter			770	.021		2.80	.66		3.46	4.28	
	0580	1" diameter			420	.038		4.75	1.20		5.95	7.40	
	0590	1-1/4" diameter			290	.055		7.90	1.74	.01	9.65	11.90	
	0600	1-1/2" diameter			192	.083		9.70	2.63	.01	12.34	15.50	
	0610	1-3/4" diameter		E-18	240	.167		15.45	5.10	2.14	22.69	28.50	
	0620	2" diameter			160	.250		19.80	7.65	3.20	30.65	39	

METALS **5**

157

508		051 500 | **Steel Wire Rope**	CREW	DAILY OUTPUT	LABOR-HOURS	UNIT	1999 BARE COSTS				TOTAL INCL O&P	508
							MAT.	LABOR	EQUIP.	TOTAL		
	0630	2-1/4" diameter	E-18	160	.250	L.F.	26.50	7.65	3.20	37.35	46	
	0650	6 x 37, bright, IPS, IWRC, <500 L.F. w/acc., 1/4" diameter	E-17	6,400	.002		1.17	.08		1.25	1.43	
	0660	1/2" diameter		1,730	.009		1.96	.29		2.25	2.69	
	0670	3/4" diameter		770	.021		3.14	.66		3.80	4.66	
	0680	1" diameter		430	.037		5.05	1.18		6.23	7.70	
	0690	1-1/4" diameter		290	.055		7.60	1.74	.01	9.35	11.55	
	0700	1-1/2" diameter		190	.084		10.85	2.66	.01	13.52	16.85	
	0710	1-3/4" diameter	E-18	260	.154		17.20	4.70	1.97	23.87	29.50	
	0720	2" diameter		200	.200		22.50	6.10	2.56	31.16	38	
	0730	2-1/4" diameter		160	.250		29.50	7.65	3.20	40.35	49.50	
	0800	6 x 19 & 6 x 37, swaged, 1/2" diameter	E-17	1,220	.013		1.18	.41		1.59	2.06	
	0810	9/16" diameter		1,120	.014		1.35	.45		1.80	2.32	
	0820	5/8" diameter		930	.017		1.63	.54		2.17	2.79	
	0830	3/4" diameter		640	.025		2.06	.79		2.85	3.72	
	0840	7/8" diameter		480	.033		2.59	1.05		3.64	4.78	
	0850	1" diameter		350	.046		3.17	1.44	.01	4.62	6.15	
	0860	1-1/8" diameter		288	.056		3.89	1.76	.01	5.66	7.50	
	0870	1-1/4" diameter		230	.070		4.73	2.20	.01	6.94	9.25	
	0880	1-3/8" diameter		192	.083		5.45	2.63	.01	8.09	10.85	
	0890	1-1/2" diameter	E-18	300	.133		6.60	4.07	1.71	12.38	16.45	

108		052 100 | **Steel Joists**	CREW	DAILY OUTPUT	LABOR-HOURS	UNIT	1999 BARE COSTS				TOTAL INCL O&P	108
							MAT.	LABOR	EQUIP.	TOTAL		
	0005	**COLD-FORMED JOISTS**										
	0010	Galvanized, 10" deep, 14 ga.	E-1	1,000	.024	L.F.	1.97	.72	.08	2.77	3.52	
	0700	12 gauge		1,000	.024		2.86	.72	.08	3.66	4.50	
	0900	12" deep, 14 gauge		880	.027		2.25	.82	.09	3.16	4	
	1000	12 gauge		880	.027		3.25	.82	.09	4.16	5.10	
110	0010	**OPEN WEB JOISTS**, Truckload lots										110
	0020	K series, horizontal bridging, spans up to 30', minimum	E-7	15	5.333	Ton	990	161	78.50	1,229.50	1,475	
	0050	Average		12	6.667		930	201	98	1,229	1,500	
	0080	Maximum		9	8.889		1,225	269	131	1,625	1,975	
	0130	8K1		1,200	.067	L.F.	2.37	2.01	.98	5.36	7.25	
	0140	10K1		1,200	.067		2.32	2.01	.98	5.31	7.20	
	0160	12K3		1,500	.053		2.65	1.61	.78	5.04	6.65	
	0180	14K3		1,500	.053		2.78	1.61	.78	5.17	6.80	
	0200	16K3		1,800	.044		2.92	1.34	.65	4.91	6.35	
	0220	16K6		1,800	.044		3.76	1.34	.65	5.75	7.25	
	0240	18K5		2,000	.040		3.57	1.21	.59	5.37	6.75	
	0260	18K9		2,000	.040		4.73	1.21	.59	6.53	8	
	0410	Span 30' to 50', minimum		17	4.706	Ton	875	142	69	1,086	1,300	
	0440	Average **CN**		17	4.706		900	142	69	1,111	1,325	
	0460	Maximum		10	8		915	242	118	1,275	1,550	
	0500	20K5		2,000	.040	L.F.	3.69	1.21	.59	5.49	6.85	
	0520	20K9		2,000	.040		4.86	1.21	.59	6.66	8.15	
	0540	22K5		2,000	.040		3.96	1.21	.59	5.76	7.15	
	0560	22K9		2,000	.040		5.10	1.21	.59	6.90	8.40	
	0580	24K6		2,200	.036		4.36	1.10	.53	5.99	7.35	

5
METALS

			DAILY	LABOR-		1999 BARE COSTS				TOTAL
052 100	**Steel Joists**	CREW	OUTPUT	HOURS	UNIT	MAT.	LABOR	EQUIP.	TOTAL	INCL O&P
110	0600	24K10	E-7	2,200	.036	L.F.	5.90	1.10	.53	7.53	9.05	110
	0620	26K6		2,200	.036		4.77	1.10	.53	6.40	7.80
	0640	26K10		2,200	.036		6.20	1.10	.53	7.83	9.40
	0660	28K8		2,400	.033		5.70	1.01	.49	7.20	8.65
	0680	28K12		2,400	.033		7.70	1.01	.49	9.20	10.80
	0700	30K8		2,400	.033		5.95	1.01	.49	7.45	8.90
	0720	30K12		2,400	.033		7.90	1.01	.49	9.40	11.05
	1010	CS series, horizontal bridging									
	1020	Spans to 30', minimum	E-7	15	5.333	Ton	815	161	78.50	1,054.50	1,275
	1040	Average		12	6.667		850	201	98	1,149	1,400
	1060	Maximum		9	8.889		945	269	131	1,345	1,675
	1100	10CS2		1,200	.067	L.F.	3.19	2.01	.98	6.18	8.15
	1120	12CS2		1,500	.053		3.41	1.61	.78	5.80	7.50
	1140	14CS2		1,500	.053		3.41	1.61	.78	5.80	7.50
	1160	16CS2		1,800	.044		3.62	1.34	.65	5.61	7.10
	1180	16CS4		1,800	.044		6.20	1.34	.65	8.19	9.90
	1200	18CS2		2,000	.040		3.83	1.21	.59	5.63	7
	1220	18CS4		2,000	.040		6.40	1.21	.59	8.20	9.85
	1240	20CS2		2,000	.040		4.05	1.21	.59	5.85	7.25
	1260	20CS4		2,000	.040		7.05	1.21	.59	8.85	10.55
	1280	22CS2		2,000	.040		4.26	1.21	.59	6.06	7.50
	1300	22CS4		2,000	.040		7.05	1.21	.59	8.85	10.55
	1320	24CS2		2,200	.036		4.26	1.10	.53	5.89	7.25
	1340	24CS4		2,200	.036		7.05	1.10	.53	8.68	10.30
	1360	26CS2		2,200	.036		4.26	1.10	.53	5.89	7.25
	1380	26CS4		2,200	.036		7.05	1.10	.53	8.68	10.30
	1400	28CS2		2,400	.033		4.47	1.01	.49	5.97	7.25
	1420	28CS4		2,400	.033		7.05	1.01	.49	8.55	10.10
	1440	30CS2		2,400	.033		4.69	1.01	.49	6.19	7.50
	1460	30CS4		2,400	.033		7.05	1.01	.49	8.55	10.10
	2000	LH series, bolted cross bridging									
	2020	Spans to 96', minimum	E-7	16	5	Ton	930	151	73.50	1,154.50	1,375
	2040	Average		13	6.154		1,025	186	90.50	1,301.50	1,550
	2080	Maximum		11	7.273		1,325	220	107	1,652	1,950
	2200	18LH04		1,400	.057	L.F.	6.15	1.73	.84	8.72	10.80
	2220	18LH08		1,400	.057		9.75	1.73	.84	12.32	14.75
	2240	20LH04		1,400	.057		6.15	1.73	.84	8.72	10.80
	2260	20LH08		1,400	.057		9.75	1.73	.84	12.32	14.75
	2280	24LH05		1,400	.057		6.70	1.73	.84	9.27	11.35
	2300	24LH10		1,400	.057		11.80	1.73	.84	14.37	17
	2320	28LH06		1,800	.044		8.20	1.34	.65	10.19	12.15
	2340	28LH11		1,800	.044		12.85	1.34	.65	14.84	17.25
	2360	32LH08		1,800	.044		8.75	1.34	.65	10.74	12.70
	2380	32LH13		1,800	.044		15.40	1.34	.65	17.39	20
	2400	36LH09		1,800	.044		10.80	1.34	.65	12.79	14.95
	2420	36LH14		1,800	.044		18.50	1.34	.65	20.49	23.50
	2440	40LH10		2,200	.036		10.80	1.10	.53	12.43	14.40
	2460	40LH15		2,200	.036		18.50	1.10	.53	20.13	23
	2480	44LH11		2,200	.036		11.30	1.10	.53	12.93	15
	2500	44LH16		2,200	.036		21.50	1.10	.53	23.13	26
	2520	48LH11		2,200	.036		11.30	1.10	.53	12.93	15
	2540	48LH16		2,200	.036		21.50	1.10	.53	23.13	26
	3010	DLH series, bolted cross bridging									
	3020	Spans to 144' (shipped in 2 pieces), minimum	E-7	16	5	Ton	995	151	73.50	1,219.50	1,450
	3040	Average		13	6.154		885	186	90.50	1,161.50	1,400
	3100	Maximum		11	7.273		1,050	220	107	1,377	1,650

5

METALS

159

052 100	Steel Joists	CREW	DAILY OUTPUT	LABOR-HOURS	UNIT	1999 BARE COSTS				TOTAL INCL O&P
						MAT.	LABOR	EQUIP.	TOTAL	
110 3200	52DLH11	E-7	2,000	.040	L.F.	11.05	1.21	.59	12.85	14.95 110
3220	52DLH16		2,000	.040		19.90	1.21	.59	21.70	25
3240	56DLH11		2,000	.040		11.50	1.21	.59	13.30	15.45
3260	56DLH16		2,000	.040		20.50	1.21	.59	22.30	25.50
3280	60DLH12		2,000	.040		12.85	1.21	.59	14.65	16.90
3300	60DLH17		2,000	.040		23	1.21	.59	24.80	28.50
3320	64DLH12		2,200	.036		13.70	1.10	.53	15.33	17.65
3340	64DLH17		2,200	.036		23	1.10	.53	24.63	28
3360	68DLH13		2,200	.036		16.35	1.10	.53	17.98	20.50
3380	68DLH18		2,200	.036		27	1.10	.53	28.63	32
3400	72DLH14		2,200	.036		18.15	1.10	.53	19.78	22.50
3420	72DLH19		2,200	.036		31	1.10	.53	32.63	36.50
4010	SLH series, bolted cross bridging									
4020	Spans to 200', minimum	E-7	16	5	Ton	880	151	73.50	1,104.50	1,325
4040	Average		13	6.154		950	186	90.50	1,226.50	1,475
4060	Maximum		11	7.273		1,200	220	107	1,527	1,800
4200	80SLH15		1,500	.053	L.F.	19	1.61	.78	21.39	24.50
4220	80SLH20		1,500	.053		35.50	1.61	.78	37.89	42.50
4240	88SLH16		1,500	.053		22	1.61	.78	24.39	27.50
4260	88SLH21		1,500	.053		42.50	1.61	.78	44.89	50
4280	96SLH17		1,500	.053		24.50	1.61	.78	26.89	30.50
4300	96SLH22		1,500	.053		48.50	1.61	.78	50.89	57
4320	104SLH18		1,800	.044		28	1.34	.65	29.99	34
4340	104SLH23		1,800	.044		52	1.34	.65	53.99	60
4360	112SLH19		1,800	.044		32	1.34	.65	33.99	38
4380	112SLH24		1,800	.044		62.50	1.34	.65	64.49	71.50
4400	120SLH20		1,800	.044		36.50	1.34	.65	38.49	43.50
4420	120SLH25		1,800	.044		72.50	1.34	.65	74.49	82.50
6000	For welded cross bridging, add						30%			
6100	For L.T.L. lots, add					10%	15%			
6200	For shop prime paint other than mfrs. standard, add					20%				
6300	For bottom chord extensions, add per chord				Ea.	26			26	28.50
7000	Joist girders, minimum	E-5	15	5.333	Ton	770	161	73	1,004	1,225
7020	Average		13	6.154		845	186	84	1,115	1,350
7040	Maximum		11	7.273		970	220	99.50	1,289.50	1,575
8000	Trusses, factory fabricated WT chords, average		11	7.273		1,600	220	99.50	1,919.50	2,250

053 100	Steel Deck	CREW	DAILY OUTPUT	LABOR-HOURS	UNIT	1999 BARE COSTS				TOTAL INCL O&P
						MAT.	LABOR	EQUIP.	TOTAL	
104 0010	**METAL DECKING** Steel decking									104
0200	Cellular units, galvanized, 2" deep, 20-20 gauge, over 15 squares	E-4	1,460	.022	S.F.	2.81	.68	.06	3.55	4.40
0250	18-20 gauge		1,420	.023		3.27	.70	.06	4.03	4.94
0300	18-18 gauge		1,390	.023		3.69	.72	.06	4.47	5.45
0320	16-18 gauge		1,360	.024		4.16	.73	.06	4.95	6
0340	16-16 gauge		1,330	.024		4.57	.75	.06	5.38	6.50
0400	3" deep, galvanized, 20-20 gauge		1,375	.023		2.93	.72	.06	3.71	4.62
0500	18-20 gauge		1,350	.024		3.98	.74	.06	4.78	5.80
0600	18-18 gauge		1,290	.025		4.44	.77	.06	5.27	6.35
0700	16-18 gauge		1,230	.026		4.86	.81	.07	5.74	6.90

5 METALS

053 100 | Steel Deck

		CREW	DAILY OUTPUT	LABOR-HOURS	UNIT	MAT.	LABOR	EQUIP.	TOTAL	TOTAL INCL O&P
						1999 BARE COSTS				
104 0800	16-16 gauge	E-4	1,150	.028	S.F.	5.40	.87	.07	6.34	7.60 **104**
1000	4-1/2" deep, galvanized, 20-18 gauge		1,100	.029		6.15	.90	.07	7.12	8.50
1100	18-18 gauge		1,040	.031		6.80	.96	.08	7.84	9.35
1200	16-18 gauge		980	.033		7.45	1.02	.08	8.55	10.15
1300	16-16 gauge	↓	935	.034	↓	8.05	1.06	.09	9.20	10.90
1500	For acoustical deck, add					15%				
1700	For cells used for ventilation, add					15%				
1900	For multi-story or congested site, add						50%			
2100	Open type, galv., 1-1/2" deep wide rib, 22 gauge, under 50 squares	E-4	4,500	.007	S.F.	.95	.22	.02	1.19	1.47
2200	50-500 squares **CN**		4,900	.007		.78	.20	.02	1	1.25
2400	Over 500 squares		5,100	.006		.72	.20	.02	.94	1.17
2600	20 gauge, under 50 squares		3,865	.008		1.11	.26	.02	1.39	1.71
2650	50-500 squares		4,170	.008		.94	.24	.02	1.20	1.49
2700	Over 500 squares		4,300	.007		.81	.23	.02	1.06	1.34
2900	18 gauge, under 50 squares		3,800	.008		1.44	.26	.02	1.72	2.08
2950	50-500 squares		4,100	.008		1.19	.24	.02	1.45	1.77
3000	Over 500 squares		4,300	.007		1.05	.23	.02	1.30	1.60
3050	16 gauge, under 50 squares		3,700	.009		1.81	.27	.02	2.10	2.50
3060	50-500 squares		4,000	.008		1.62	.25	.02	1.89	2.26
3100	Over 500 squares	↓	4,200	.008		1.36	.24	.02	1.62	1.94
3150	For intermediate rib instead of wide rib, deduct					.01			.01	.02
3160	For narrow rib instead of wide rib, add					.30			.30	.33
3200	3" deep, 22 gauge, under 50 squares	E-4	3,600	.009		1.55	.28	.02	1.85	2.25
3250	50-500 squares		3,800	.008		1.23	.26	.02	1.51	1.85
3260	over 500 squares		4,000	.008		.96	.25	.02	1.23	1.53
3300	20 gauge, under 50 squares		3,400	.009		1.77	.29	.02	2.08	2.51
3350	50-500 squares		3,600	.009		1.35	.28	.02	1.65	2.03
3360	over 500 squares		3,800	.008		1.15	.26	.02	1.43	1.76
3400	18 gauge, under 50 squares		3,200	.010		2.34	.31	.03	2.68	3.17
3450	50-500 squares		3,400	.009		1.78	.29	.02	2.09	2.53
3460	over 500 squares		3,600	.009		1.55	.28	.02	1.85	2.24
3500	16 gauge, under 50 squares		3,000	.011		2.15	.33	.03	2.51	3.01
3550	50-500 squares		3,200	.010		1.89	.31	.03	2.23	2.68
3560	over 500 squares		3,400	.009		1.43	.29	.02	1.74	2.14
3700	4-1/2" deep, long span roof, over 50 squares, 20 gauge		2,700	.012		2.18	.37	.03	2.58	3.11
3800	18 gauge		2,460	.013		3.02	.40	.03	3.45	4.10
3900	16 gauge		2,350	.014		3.76	.42	.03	4.21	4.96
4100	6" deep, long span, 18 gauge		2,000	.016		3.93	.50	.04	4.47	5.30
4200	16 gauge		1,930	.017		4.43	.52	.04	4.99	5.85
4300	14 gauge		1,860	.017		4.68	.54	.04	5.26	6.20
4500	7-1/2" deep, long span, 18 gauge		1,690	.019		4.43	.59	.05	5.07	6
4600	16 gauge		1,590	.020		4.99	.63	.05	5.67	6.70
4700	14 gauge	↓	1,490	.021	↓	5.45	.67	.06	6.18	7.30
4800	For painted instead of galvanized, deduct					2%				
5000	For acoustical perforated, with fiberglass, add				S.F.	1.12			1.12	1.23
5200	Non-cellular composite deck, galv., 2" deep, 22 gauge	E-4	3,860	.008		.96	.26	.02	1.24	1.54
5300	20 gauge		3,600	.009		1.08	.28	.02	1.38	1.73
5400	18 gauge		3,380	.009		1.49	.29	.02	1.80	2.21
5500	16 gauge		3,200	.010		1.75	.31	.03	2.09	2.53
5700	3" deep, galv., 22 gauge		3,200	.010		1.05	.31	.03	1.39	1.76
5800	20 gauge		3,000	.011		1.16	.33	.03	1.52	1.92
5900	18 gauge **CN**		2,850	.011		1.53	.35	.03	1.91	2.35
6000	16 gauge		2,700	.012		2.03	.37	.03	2.43	2.94
6100	Slab form, steel, 28 gauge, 9/16" deep, uncoated		4,000	.008		.46	.25	.02	.73	.98
6200	Galvanized		4,000	.008		.51	.25	.02	.78	1.04
6220	24 gauge, 1" deep, uncoated	↓	3,900	.008	↓	.63	.26	.02	.91	1.18

5

METALS

053 | Metal Decking

053 100 | Steel Deck

		CREW	DAILY OUTPUT	LABOR-HOURS	UNIT	1999 BARE COSTS				TOTAL INCL O&P
						MAT.	LABOR	EQUIP.	TOTAL	
6240	Galvanized	E-4	3,900	.008	S.F.	.68	.26	.02	.96	1.23
6300	24 gauge, 1-5/16" deep, uncoated		3,800	.008		.74	.26	.02	1.02	1.32
6400	Galvanized		3,800	.008		.86	.26	.02	1.14	1.45
6500	22 gauge, 1-5/16" deep, uncoated		3,700	.009		.86	.27	.02	1.15	1.46
6600	Galvanized		3,700	.009		.99	.27	.02	1.28	1.59
6700	22 gauge, 3" deep uncoated		3,600	.009		.82	.28	.02	1.12	1.44
6800	Galvanized		3,600	.009		.95	.28	.02	1.25	1.59
7000	Sheet metal edge closure form, 12" wide with 2 bends									
7100	18 gauge	E-14	360	.022	L.F.	1.76	.72	.23	2.71	3.52
7200	16 gauge	"	360	.022	"	2.09	.72	.23	3.04	3.88
8000	Metal deck and trench, 2" thick, 20 gauge, combination									
8010	60% cellular, 40% non-cellular, inserts and trench	R-4	1,100	.036	S.F.	5.60	1.14	.07	6.81	8.30

054 | Cold Formed Metal Framing

054 100 | Load Bearing Metal Studs

		CREW	DAILY OUTPUT	LABOR-HOURS	UNIT	1999 BARE COSTS				TOTAL INCL O&P
						MAT.	LABOR	EQUIP.	TOTAL	
0010	**BRACING**, shear wall X-bracing, per 10' x 10' bay, one face									
0120	Metal strap, 20 ga x 4" wide	2 Carp	18	.889	Ea.	14.25	24.50		38.75	53.50
0130	6" wide		18	.889		21	24.50		45.50	61
0160	18 ga x 4" wide		16	1		22.50	27.50		50	68
0170	6" wide		16	1		33	27.50		60.50	79.50
0410	Continuous strap bracing, per horizontal row on both faces									
0420	Metal strap, 20 ga x 2" wide, studs 12" O.C.	1 Carp	7	1.143	C.L.F.	43.50	31		74.50	96.50
0430	16" O.C.		8	1		43.50	27.50		71	90.50
0440	24" O.C.		10	.800		43.50	22		65.50	82
0450	18 ga x 2" wide, studs 12" O.C.		6	1.333		60.50	36.50		97	124
0460	16" O.C.		7	1.143		60.50	31		91.50	116
0470	24" O.C.		8	1		60.50	27.50		88	110
0010	**BRIDGING**, solid between studs w/ 1-1/4" leg track, per stud bay									
0200	Studs 12" O.C., 18 ga x 2-1/2" wide	1 Carp	125	.064	Ea.	.56	1.75		2.31	3.36
0210	3-5/8" wide		120	.067		.65	1.82		2.47	3.58
0220	4" wide		120	.067		.67	1.82		2.49	3.60
0230	6" wide		115	.070		.89	1.90		2.79	3.97
0240	8" wide		110	.073		1.14	1.99		3.13	4.37
0300	16 ga x 2-1/2" wide		115	.070		.69	1.90		2.59	3.74
0310	3-5/8" wide		110	.073		.80	1.99		2.79	4
0320	4" wide		110	.073		.84	1.99		2.83	4.04
0330	6" wide		105	.076		1.10	2.08		3.18	4.48
0340	8" wide		100	.080		1.44	2.18		3.62	5
1200	Studs 16" O.C., 18 ga x 2-1/2" wide		125	.064		.71	1.75		2.46	3.53
1210	3-5/8" wide		120	.067		.84	1.82		2.66	3.78
1220	4" wide		120	.067		.86	1.82		2.68	3.81
1230	6" wide		115	.070		1.14	1.90		3.04	4.25
1240	8" wide		110	.073		1.46	1.99		3.45	4.73
1300	16 ga x 2-1/2" wide		115	.070		.88	1.90		2.78	3.96
1310	3-5/8" wide		110	.073		1.03	1.99		3.02	4.25
1320	4" wide		110	.073		1.07	1.99		3.06	4.30
1330	6" wide		105	.076		1.41	2.08		3.49	4.83
1340	8" wide		100	.080		1.84	2.18		4.02	5.45
2200	Studs 24" O.C., 18 ga x 2-1/2" wide		125	.064		1.03	1.75		2.78	3.89

Important: See the Reference Section for critical supporting data - Reference Nos., Crews, & City Cost Indexes

054 | Cold Formed Metal Framing

METALS 5

	054 100	Load Bearing Metal Studs	CREW	DAILY OUTPUT	LABOR-HOURS	UNIT	MAT.	LABOR	EQUIP.	TOTAL	TOTAL INCL O&P	
106	2210	3-5/8" wide	1 Carp	120	.067	Ea.	1.21	1.82		3.03	4.19	106
	2220	4" wide		120	.067		1.25	1.82		3.07	4.23	
	2230	6" wide		115	.070		1.65	1.90		3.55	4.81	
	2240	8" wide		110	.073		2.11	1.99		4.10	5.45	
	2300	16 ga x 2-1/2" wide		115	.070		1.27	1.90		3.17	4.39	
	2310	3-5/8" wide		110	.073		1.49	1.99		3.48	4.75	
	2320	4" wide		110	.073		1.55	1.99		3.54	4.83	
	2330	6" wide		105	.076		2.05	2.08		4.13	5.50	
	2340	8" wide	▼	100	.080	▼	2.66	2.18		4.84	6.35	
	3000	Continuous bridging, per row										
	3100	16 ga x 1-1/2" channel thru studs 12" O.C.	1 Carp	6	1.333	C.L.F.	37	36.50		73.50	98.50	
	3110	16" O.C.		7	1.143		37	31		68	90	
	3120	24" O.C.		8.80	.909		37	25		62	80	
	4100	2" x 2" angle x 18 ga, studs 12" O.C.		7	1.143		66.50	31		97.50	122	
	4110	16" O.C.		9	.889		66.50	24.50		91	111	
	4120	24" O.C.		12	.667		66.50	18.20		84.70	102	
	4200	16 ga, studs 12" O.C.		5	1.600		79	43.50		122.50	156	
	4210	16" O.C.		7	1.143		79	31		110	136	
	4220	24" O.C.	▼	10	.800	▼	79	22		101	122	
110	0010	**FRAMING**, boxed headers/beams										110
	0200	Double, 18 ga x 6" deep	2 Carp	220	.073	L.F.	3.03	1.99		5.02	6.45	
	0210	8" deep		210	.076		3.38	2.08		5.46	7	
	0220	10" deep		200	.080		4.05	2.18		6.23	7.90	
	0230	12" deep		190	.084		5.20	2.30		7.50	9.35	
	0300	16 ga x 8" deep		180	.089		3.87	2.43		6.30	8.10	
	0310	10" deep		170	.094		4.71	2.57		7.28	9.25	
	0320	12" deep		160	.100		4.89	2.73		7.62	9.70	
	0400	14 ga x 10" deep		140	.114		5.50	3.12		8.62	10.95	
	0410	12" deep		130	.123		6.05	3.36		9.41	11.95	
	1210	Triple, 18 ga x 8" deep		170	.094		4.50	2.57		7.07	9	
	1220	10" deep		165	.097		5.35	2.65		8	10.05	
	1230	12" deep		160	.100		7.10	2.73		9.83	12.15	
	1300	16 ga x 8" deep		145	.110		5.25	3.01		8.26	10.50	
	1310	10" deep		140	.114		6.35	3.12		9.47	11.90	
	1320	12" deep		135	.119		6.60	3.24		9.84	12.40	
	1400	14 ga x 10" deep		115	.139		7.55	3.80		11.35	14.30	
	1410	12" deep	▼	110	.145	▼	8.35	3.97		12.32	15.45	
138	0010	**FRAMING**, stud walls w/ top & bottom track, no openings,										138
	0020	headers, beams, bridging or bracing										
	4100	8' high walls, 18 ga x 2-1/2" wide, studs 12" O.C.	2 Carp	54	.296	L.F.	5.20	8.10		13.30	18.50	
	4110	16" O.C.		77	.208		4.18	5.65		9.83	13.50	
	4120	24" O.C.		107	.150		3.14	4.08		7.22	9.85	
	4130	3-5/8" wide, studs 12" O.C.		53	.302		6	8.25		14.25	19.55	
	4140	16" O.C.		76	.211		4.80	5.75		10.55	14.35	
	4150	24" O.C.		105	.152		3.61	4.16		7.77	10.50	
	4160	4" wide, studs 12" O.C.		52	.308		6.20	8.40		14.60	20	
	4170	16" O.C.		74	.216		4.96	5.90		10.86	14.75	
	4180	24" O.C.		103	.155		3.74	4.24		7.98	10.75	
	4190	6" wide, studs 12" O.C.		51	.314		8	8.55		16.55	22	
	4200	16" O.C.		73	.219		6.40	6		12.40	16.45	
	4210	24" O.C.		101	.158		4.84	4.32		9.16	12.10	
	4220	8" wide, studs 12" O.C.		50	.320		10.15	8.75		18.90	25	
	4230	16" O.C.		72	.222		8.15	6.05		14.20	18.55	
	4240	24" O.C.		100	.160		6.15	4.37		10.52	13.65	
	4300	16 ga x 2-1/2" wide, studs 12" O.C.	▼	47	.340	▼	6.15	9.30		15.45	21.50	

163

054 100	Load Bearing Metal Studs	CREW	DAILY OUTPUT	LABOR-HOURS	UNIT	MAT.	LABOR	EQUIP.	TOTAL	TOTAL INCL O&P	
138							1999 BARE COSTS					138
4310	16" O.C.	2 Carp	68	.235	L.F.	4.88	6.40		11.28	15.45	
4320	24" O.C.		94	.170		3.61	4.65		8.26	11.25	
4330	3-5/8" wide, studs 12" O.C.		46	.348		7.05	9.50		16.55	22.50	
4340	16" O.C.		66	.242		5.60	6.60		12.20	16.55	
4350	24" O.C.		92	.174		4.15	4.75		8.90	12	
4360	4" wide, studs 12" O.C.		45	.356		7.25	9.70		16.95	23.50	
4370	16" O.C.		65	.246		5.75	6.70		12.45	16.90	
4380	24" O.C.		90	.178		4.27	4.85		9.12	12.35	
4390	6" wide, studs 12" O.C.		44	.364		9.30	9.95		19.25	26	
4400	16" O.C.		64	.250		7.40	6.80		14.20	18.90	
4410	24" O.C.		88	.182		5.50	4.96		10.46	13.85	
4420	8" wide, studs 12" O.C.		43	.372		12.10	10.15		22.25	29.50	
4430	16" O.C.		63	.254		9.60	6.95		16.55	21.50	
4440	24" O.C.		86	.186		7.15	5.10		12.25	15.85	
5100	10' high walls, 18 ga x 2-1/2" wide, studs 12" O.C.		54	.296		6.25	8.10		14.35	19.65	
5110	16" O.C.		77	.208		4.96	5.65		10.61	14.35	
5120	24" O.C.		107	.150		3.66	4.08		7.74	10.45	
5130	3-5/8" wide, studs 12" O.C.		53	.302		7.15	8.25		15.40	21	
5140	16" O.C.		76	.211		5.70	5.75		11.45	15.30	
5150	24" O.C.		105	.152		4.21	4.16		8.37	11.20	
5160	4" wide, studs 12" O.C.		52	.308		7.40	8.40		15.80	21.50	
5170	16" O.C.		74	.216		5.90	5.90		11.80	15.75	
5180	24" O.C.		103	.155		4.35	4.24		8.59	11.45	
5190	6" wide, studs 12" O.C.		51	.314		9.55	8.55		18.10	24	
5200	16" O.C.		73	.219		7.60	6		13.60	17.75	
5210	24" O.C.		101	.158		5.60	4.32		9.92	13	
5220	8" wide, studs 12" O.C.		50	.320		12.15	8.75		20.90	27	
5230	16" O.C.		72	.222		9.65	6.05		15.70	20	
5240	24" O.C.		100	.160		7.15	4.37		11.52	14.75	
5300	16 ga x 2-1/2" wide, studs 12" O.C.		47	.340		7.45	9.30		16.75	23	
5310	16" O.C.		68	.235		5.85	6.40		12.25	16.50	
5320	24" O.C.		94	.170		4.25	4.65		8.90	11.95	
5330	3-5/8" wide, studs 12" O.C.		46	.348		8.50	9.50		18	24.50	
5340	16" O.C.		66	.242		6.70	6.60		13.30	17.75	
5350	24" O.C.		92	.174		4.88	4.75		9.63	12.80	
5360	4" wide, studs 12" O.C.		45	.356		8.75	9.70		18.45	25	
5370	16" O.C.		65	.246		6.90	6.70		13.60	18.10	
5380	24" O.C.		90	.178		5	4.85		9.85	13.15	
5390	6" wide, studs 12" O.C.		44	.364		11.20	9.95		21.15	28	
5400	16" O.C.		64	.250		8.85	6.80		15.65	20.50	
5410	24" O.C.		88	.182		6.45	4.96		11.41	14.90	
5420	8" wide, studs 12" O.C.		43	.372		14.60	10.15		24.75	32	
5430	16" O.C.		63	.254		11.50	6.95		18.45	23.50	
5440	24" O.C.		86	.186		8.40	5.10		13.50	17.20	
6190	12' high walls, 18 ga x 6" wide, studs 12" O.C.		41	.390		11.10	10.65		21.75	29	
6200	16" O.C.		58	.276		8.75	7.55		16.30	21.50	
6210	24" O.C.		81	.198		6.40	5.40		11.80	15.55	
6220	8" wide, studs 12" O.C.		40	.400		14.15	10.90		25.05	33	
6230	16" O.C.		57	.281		11.15	7.65		18.80	24.50	
6240	24" O.C.		80	.200		8.15	5.45		13.60	17.60	
6390	16 ga x 6" wide, studs 12" O.C.		35	.457		13.15	12.50		25.65	34	
6400	16" O.C.		51	.314		10.25	8.55		18.80	25	
6410	24" O.C.		70	.229		7.40	6.25		13.65	17.95	
6420	8" wide, studs 12" O.C.		34	.471		17.10	12.85		29.95	39	
6430	16" O.C.		50	.320		13.35	8.75		22.10	28.50	
6440	24" O.C.		69	.232		9.60	6.35		15.95	20.50	

Important: See the Reference Section for critical supporting data - Reference Nos., Crews, & City Cost Indexes

			DAILY	LABOR-		1999 BARE COSTS				TOTAL	
	054 100 | Load Bearing Metal Studs	CREW	OUTPUT	HOURS	UNIT	MAT.	LABOR	EQUIP.	TOTAL	INCL O&P	
138	6530	14 ga x 3-5/8" wide, studs 12" O.C.	2 Carp	34	.471	L.F.	13.05	12.85		25.90	34.50	**138**
	6540	16" O.C.		48	.333		10.15	9.10		19.25	25.50	
	6550	24" O.C.		65	.246		7.30	6.70		14	18.55	
	6560	4" wide, studs 12" O.C.		33	.485		13.40	13.25		26.65	36	
	6570	16" O.C.		47	.340		10.45	9.30		19.75	26	
	6580	24" O.C.		64	.250		7.50	6.80		14.30	19	
	6730	12 ga x 3-5/8" wide, studs 12" O.C.		31	.516		19.90	14.10		34	44	
	6740	16" O.C.		43	.372		15.30	10.15		25.45	33	
	6750	24" O.C.		59	.271		10.70	7.40		18.10	23.50	
	6760	4" wide, studs 12" O.C.		30	.533		23	14.55		37.55	48.50	
	6770	16" O.C.		42	.381		17.80	10.40		28.20	36	
	6780	24" O.C.		58	.276		12.40	7.55		19.95	25.50	
	7390	16' high walls, 16 ga x 6" wide, studs 12" O.C.		33	.485		16.95	13.25		30.20	39.50	
	7400	16" O.C.		48	.333		13.15	9.10		22.25	29	
	7410	24" O.C.		67	.239		9.30	6.50		15.80	20.50	
	7420	8" wide, studs 12" O.C.		32	.500		22	13.65		35.65	46	
	7430	16" O.C.		47	.340		17.10	9.30		26.40	33.50	
	7440	24" O.C.		66	.242		12.10	6.60		18.70	23.50	
	7560	14 ga x 4" wide, studs 12" O.C.		31	.516		17.35	14.10		31.45	41	
	7570	16" O.C.		45	.356		13.40	9.70		23.10	30	
	7580	24" O.C.		61	.262		9.45	7.15		16.60	21.50	
	7590	6" wide, studs 12" O.C.		30	.533		22.50	14.55		37.05	47.50	
	7600	16" O.C.		44	.364		17.35	9.95		27.30	34.50	
	7610	24" O.C.		60	.267		12.30	7.30		19.60	25	
	7760	12 ga x 4" wide, studs 12" O.C.		29	.552		30.50	15.05		45.55	57	
	7770	16" O.C.		40	.400		23	10.90		33.90	42.50	
	7780	24" O.C.		55	.291		16	7.95		23.95	30	
	7790	6" wide, studs 12" O.C.		28	.571		33.50	15.60		49.10	61	
	7800	16" O.C.		39	.410		25.50	11.20		36.70	45.50	
	7810	24" O.C.		54	.296		17.70	8.10		25.80	32	
	8590	20' high walls, 14 ga x 6" wide, studs 12" O.C.		29	.552		27.50	15.05		42.55	54	
	8600	16" O.C.		42	.381		21	10.40		31.40	40	
	8610	24" O.C.		57	.281		14.85	7.65		22.50	28.50	
	8620	8" wide, studs 12" O.C.		28	.571		34.50	15.60		50.10	62.50	
	8630	16" O.C.		41	.390		26.50	10.65		37.15	46.50	
	8640	24" O.C.		56	.286		18.75	7.80		26.55	33	
	8790	12 ga x 6" wide, studs 12" O.C.		27	.593		41	16.20		57.20	70.50	
	8800	16" O.C.		37	.432		31.50	11.80		43.30	53	
	8810	24" O.C.		51	.314		21.50	8.55		30.05	37.50	
	8820	8" wide, studs 12" O.C.		26	.615		50.50	16.80		67.30	82	
	8830	16" O.C.		36	.444		38.50	12.15		50.65	61.50	
	8840	24" O.C.		50	.320		26.50	8.75		35.25	43.50	

			DAILY	LABOR-		1999 BARE COSTS				TOTAL	
	055 100 | Metal Stairs	CREW	OUTPUT	HOURS	UNIT	MAT.	LABOR	EQUIP.	TOTAL	INCL O&P	
104	0010	**STAIR** Steel, safety nosing, steel stringers										**104**
	0020	Grating tread and pipe railing, 3'-6" wide	E-4	35	.914	Riser	83.50	28.50	2.35	114.35	146	
	0100	4'-0" wide		30	1.067		92	33	2.74	127.74	165	
	0200	Cement fill metal pan, picket rail, 3'-6" wide		35	.914		71.50	28.50	2.35	102.35	133	

5

METALS

		CREW	DAILY OUTPUT	LABOR-HOURS	UNIT	1999 BARE COSTS				TOTAL INCL O&P
055 100	**Metal Stairs**					MAT.	LABOR	EQUIP.	TOTAL	
104 0300	4'-0" wide	E-4	30	1.067	Riser	79	33	2.74	114.74	151	**104**
0350	Wall rail, both sides, 3'-6" wide **CN**		53	.604		62	18.80	1.55	82.35	104
0400	Cast iron tread and pipe rail, 3'-6" wide		35	.914		183	28.50	2.35	213.85	256
0500	Checkered plate tread, industrial, 3'-6" wide		28	1.143		101	35.50	2.94	139.44	179
0550	Circular, for tanks, 3'-0" wide	↓	33	.970		87.50	30	2.49	119.99	155
0600	For isolated stairs, add						100%			
0800	Custom steel stairs, 3'-6" wide, minimum	E-4	35	.914		103	28.50	2.35	133.85	168
0810	Average		30	1.067		152	33	2.74	187.74	231
0900	Maximum	↓	20	1.600		222	50	4.11	276.11	340
1100	For 4' wide stairs, add					10%	5%			
1300	For 5' wide stairs, add				↓	20%	10%			
1500	Landing, steel pan, conventional	E-4	160	.200	S.F.	35.50	6.20	.51	42.21	51.50
1600	Pre-erected	"	255	.125	"	24	3.90	.32	28.22	33.50
1700	Pre-erected, steel pan tread, 3'-6" wide, 2 line pipe rail	E-2	87	.644	Riser	229	19.15	11.65	259.80	298
1810	Spiral aluminum, 5'-0" diameter, stock units	E-4	45	.711		182	22	1.83	205.83	243
1820	Custom units		45	.711		350	22	1.83	373.83	425
1830	Stock units, 4'-0" diameter, safety treads		50	.640		110	19.90	1.64	131.54	159
1840	Oak treads		50	.640		215	19.90	1.64	236.54	275
1850	5'-0" diameter, safety treads		45	.711		121	22	1.83	144.83	177
1860	Oak treads		45	.711		232	22	1.83	255.83	298
1870	6'-0" diameter, safety treads		40	.800		132	25	2.06	159.06	194
1880	Oak treads		40	.800		254	25	2.06	281.06	325
1900	Spiral, cast iron, 4'-0" diameter, ornamental, minimum		45	.711		166	22	1.83	189.83	225
1920	Maximum		25	1.280		221	40	3.29	264.29	320
2000	Spiral, steel, industrial checkered plate, 4' diameter		45	.711		166	22	1.83	189.83	225
2200	Stock units, 6'-0" diameter		40	.800		199	25	2.06	226.06	267
3900	Industrial ships ladder, 3' W, grating treads, 2 line pipe rail		30	1.067		66	33	2.74	101.74	137
4000	Aluminum	↓	30	1.067	↓	97.50	33	2.74	133.24	171
108 0005	**STAIR TREADS**										**108**
0010	Aluminum grating, 3' long, 1" x 3/16" bars, 6" wide	1 Sswk	24	.333	Ea.	25	10.20		35.20	46
0100	12" wide		22	.364		43.50	11.15		54.65	68.50
0200	1-1/2" x 3/16" bars, 6" wide		24	.333		34.50	10.20		44.70	56.50
0300	12" wide	↓	22	.364		58.50	11.15		69.65	84.50
0400	For abrasive nosings, add					13.50			13.50	14.85
0500	For narrow mesh, add				↓	60%				
0600	Treads, 12" x 3'-6", not incl. stringers. See also div. 033-184									
0700	Cast aluminum, abrasive, 3' long x 12" wide, 5/16" thick	1 Sswk	15	.533	Ea.	99.50	16.30		115.80	140
0800	3/8" thick		15	.533		104	16.30		120.30	144
0900	1/2" thick		15	.533		118	16.30		134.30	160
1000	Cast bronze, abrasive, 3/8" thick		8	1		400	30.50		430.50	495
1100	1/2" thick		8	1		525	30.50		555.50	630
1200	Cast iron, abrasive, 3' long x 12" wide, 3/8" thick		15	.533		74	16.30		90.30	112
1300	1/2" thick	↓	15	.533	↓	85.50	16.30		101.80	124
1400	Fiberglass reinforced plastic with safety nosing,									
1500	1-1/2" thick, 12" wide, 24" long	1 Sswk	22	.364	Ea.	87.50	11.15		98.65	117
1600	30" long		22	.364		91.50	11.15		102.65	122
1700	36" long		22	.364		100	11.15		111.15	131
2000	Steel grating, painted, 3' long, 1-1/4" x 3/16" bars, 6" wide		20	.400		19	12.25		31.25	43.50
2010	12" wide	↓	18	.444		27.50	13.60		41.10	55.50
2100	Add for abrasive nosing, 3' long, painted					12.25			12.25	13.50
2200	Galvanized					20.50			20.50	22.50
2300	Painting 3' treads, in shop, nonstandard paint					.93			.93	1.02
2400	Added coats, standard paint					.48			.48	.53
2500	Expanded steel, 2-1/2" deep, 9" x 3' long, 18 gauge	1 Sswk	20	.400		17.75	12.25		30	42
2600	14 gauge	"	20	.400	↓	22	12.25		34.25	46.50

Important: See the Reference Section for critical supporting data - Reference Nos., Crews, & City Cost Indexes

		055 150 | **Ladders**	CREW	DAILY OUTPUT	LABOR-HOURS	UNIT	1999 BARE COSTS				TOTAL INCL O&P	
							MAT.	LABOR	EQUIP.	TOTAL		
154	0010	**FIRE ESCAPE**										154
	0200	2' wide balcony, 1" x 1/4" bars 1-1/2" O.C.	1 Sswk	5	1.600	L.F.	34	49		83	128	
	0400	1st story cantilever, standard	↓	.09	88.889	Ea.	1,425	2,725		4,150	6,550	
	0700	Platform & stair, 36" x 40"	↓	.17	47.059	Flight	630	1,450		2,080	3,350	
	0900	For 3'-6" wide escapes, add to above					100%	150%				
156	0005	**FIRE ESCAPE STAIRS**										156
	0010	One story, disappearing, stainless steel	2 Sswk	20	.800	V.L.F.	148	24.50		172.50	208	
	0100	Portable ladder				Ea.	173			173	191	
158	0005	**LADDER**										158
	0010	Steel, 20" wide, bolted to concrete, with cage	E-4	50	.640	V.L.F.	53	19.90	1.64	74.54	97	
	0100	Without cage		85	.376		24.50	11.70	.97	37.17	49.50	
	0300	Aluminum, bolted to concrete, with cage		50	.640		100	19.90	1.64	121.54	148	
	0400	Without cage	↓	85	.376		55.50	11.70	.97	68.17	83.50	
	1350	Alternating tread stair, 56/68°, steel, standard paint color	2 Sswk	50	.320		132	9.80		141.80	163	
	1360	Non-standard paint color		50	.320		141	9.80		150.80	173	
	1370	Galvanized steel		50	.320		149	9.80		158.80	182	
	1380	Stainless steel		50	.320		226	9.80		235.80	266	
	1390	68°, aluminum	↓	50	.320	↓	163	9.80		172.80	197	

		055 200 | **Handrails & Railings**										
202	0010	**BUMPER RAILS** For garages, 12 ga. rail, 6" wide, with steel										202
	0020	posts 12'-6" O.C., Minimum	E-4	190	.168	L.F.	8	5.25	.43	13.68	18.90	
	0030	Average		165	.194		10	6.05	.50	16.55	22.50	
	0100	Maximum		140	.229		12	7.10	.59	19.69	27	
	0300	12" channel rail, minimum		160	.200		10	6.20	.51	16.71	23	
	0400	Maximum	↓	120	.267	↓	15	8.30	.69	23.99	32.50	
203	0005	**RAILING, PIPE**										203
	0010	Aluminum, 2 rail, satin finish, 1-1/4" diameter	E-4	160	.200	L.F.	13	6.20	.51	19.71	26.50	
	0030	Clear anodized		160	.200		16.10	6.20	.51	22.81	29.50	
	0040	Dark anodized		160	.200		18.15	6.20	.51	24.86	32	
	0080	1-1/2" diameter, satin finish		160	.200		15.55	6.20	.51	22.26	29	
	0090	Clear anodized		160	.200		17.35	6.20	.51	24.06	31	
	0100	Dark anodized		160	.200		19.25	6.20	.51	25.96	33	
	0140	Aluminum, 3 rail, 1-1/4" diam., satin finish		137	.234		19.90	7.25	.60	27.75	36	
	0150	Clear anodized		137	.234		25	7.25	.60	32.85	41.50	
	0160	Dark anodized		137	.234		27.50	7.25	.60	35.35	44.50	
	0200	1-1/2" diameter, satin finish		137	.234		24	7.25	.60	31.85	40.50	
	0210	Clear anodized		137	.234		27	7.25	.60	34.85	44	
	0220	Dark anodized		137	.234		29.50	7.25	.60	37.35	46.50	
	0500	Steel, 2 rail, on stairs, primed, 1-1/4" diameter		160	.200		9.35	6.20	.51	16.06	22.50	
	0520	1-1/2" diameter		160	.200		10.30	6.20	.51	17.01	23.50	
	0540	Galvanized, 1-1/4" diameter		160	.200		13	6.20	.51	19.71	26.50	
	0560	1-1/2" diameter		160	.200		14.55	6.20	.51	21.26	28	
	0580	Steel, 3 rail, primed, 1-1/4" diameter		137	.234		13.95	7.25	.60	21.80	29.50	
	0600	1-1/2" diameter		137	.234		14.80	7.25	.60	22.65	30.50	
	0620	Galvanized, 1-1/4" diameter		137	.234		19.60	7.25	.60	27.45	35.50	
	0640	1-1/2" diameter **CN**		137	.234		21.50	7.25	.60	29.35	38	
	0700	Stainless steel, 2 rail, 1-1/4" diam. #4 finish		137	.234		32	7.25	.60	39.85	49	
	0720	High polish		137	.234		51.50	7.25	.60	59.35	70.50	
	0740	Mirror polish		137	.234		64.50	7.25	.60	72.35	85	
	0760	Stainless steel, 3 rail, 1-1/2" diam., #4 finish		120	.267		48	8.30	.69	56.99	69	
	0770	High polish		120	.267		79.50	8.30	.69	88.49	103	
	0780	Mirror finish		120	.267		97	8.30	.69	105.99	123	
	0900	Wall rail, alum. pipe, 1-1/4" diam., satin finish	↓	213	.150	↓	7.45	4.67	.39	12.51	17.15	

5

METALS

055 200 | Handrails & Railings

			CREW	DAILY OUTPUT	LABOR-HOURS	UNIT	1999 BARE COSTS				TOTAL INCL O&P	
							MAT.	LABOR	EQUIP.	TOTAL		
203	0905	Clear anodized	E-4	213	.150	L.F.	9.05	4.67	.39	14.11	19	**203**
	0910	Dark anodized		213	.150		11	4.67	.39	16.06	21	
	0915	1-1/2" diameter, satin finish		213	.150		8.25	4.67	.39	13.31	18.05	
	0920	Clear anodized		213	.150		10.35	4.67	.39	15.41	20.50	
	0925	Dark anodized		213	.150		12.80	4.67	.39	17.86	23	
	0930	Steel pipe, 1-1/4" diameter, primed		213	.150		5.70	4.67	.39	10.76	15.25	
	0935	Galvanized		213	.150		8.25	4.67	.39	13.31	18.05	
	0940	1-1/2" diameter		176	.182		5.85	5.65	.47	11.97	17.25	
	0945	Galvanized		213	.150		8.30	4.67	.39	13.36	18.10	
	0955	Stainless steel pipe, 1-1/2" diam., #4 finish		107	.299		25.50	9.30	.77	35.57	46	
	0960	High polish		107	.299		52	9.30	.77	62.07	75	
	0965	Mirror polish		107	.299		61	9.30	.77	71.07	85.50	
206	0005	**RAILINGS, INDUSTRIAL** Welded										**206**
	0010	2 rail, 3'-6" high, 1-1/2" pipe	E-4	255	.125	L.F.	13.70	3.90	.32	17.92	22.50	
	0100	2" angle rail	"	255	.125		12.50	3.90	.32	16.72	21.50	
	0200	For 4" high kick plate, 10 gauge, add					2.86			2.86	3.15	
	0300	1/4" thick, add					3.67			3.67	4.04	
	0500	For curved rails, add					30%	30%				
208	0005	**RAILINGS, ORNAMENTAL**										**208**
	0010	Aluminum, bronze or stainless, minimum	1 Sswk	24	.333	L.F.	97	10.20		107.20	126	
	0100	Maximum		9	.889		545	27		572	650	
	0200	Aluminum ornamental rail, minimum		15	.533		44	16.30		60.30	78.50	
	0300	Maximum		8	1		191	30.50		221.50	266	
	0400	Hand-forged wrought iron, minimum		12	.667		95	20.50		115.50	142	
	0500	Maximum		8	1		405	30.50		435.50	500	
	0550	Steel, minimum		12	.667		43	20.50		63.50	84.50	
	0560	Maximum		8	1		189	30.50		219.50	264	
	0600	Composite metal and wood or glass, minimum		6	1.333		184	41		225	277	
	0700	Maximum		5	1.600		480	49		529	615	

055 300 | Gratings & Floor Plates

			CREW	DAILY OUTPUT	LABOR-HOURS	UNIT	MAT.	LABOR	EQUIP.	TOTAL	TOTAL INCL O&P	
302	0005	**CHECKERED PLATE**										**302**
	0010	1/4" & 3/8", 2000 to 5000 S.F., bolted	E-4	2,900	.011	Lb.	.59	.34	.03	.96	1.31	
	0100	Welded		4,400	.007	"	.55	.23	.02	.80	1.04	
	0300	Pit or trench cover and frame, 1/4" plate, 2' to 3' wide		100	.320	S.F.	15.50	9.95	.82	26.27	36	
	0400	For galvanizing, add				Lb.	.34			.34	.37	
	0500	Platforms, 1/4" plate, no handrails included, rectangular	E-4	4,200	.008		.93	.24	.02	1.19	1.47	
	0600	Circular	"	2,500	.013		1.30	.40	.03	1.73	2.20	
304	0010	**FLOOR GRATING, ALUMINUM**										**304**
	0200	For straight cuts, add				L.F.	2.31			2.31	2.54	
	0300	For curved cuts, add					3.04			3.04	3.34	
	0400	For straight banding, add					2.89			2.89	3.18	
	0500	For curved banding, add					3.63			3.63	3.99	
	0600	For aluminum checkered plate nosings, add					3.38			3.38	3.72	
	0700	For straight toe plate, add					6.25			6.25	6.90	
	0800	For curved toe plate, add					7.70			7.70	8.45	
	1000	For cast aluminum abrasive nosings, add					4.99			4.99	5.50	
	1200	Expanded aluminum, .65# per S.F., on grade	E-4	1,050	.030	S.F.	3.58	.95	.08	4.61	5.75	
	1400	Extruded I bars are 10% less than 3/16" bars										
	1600	Heavy duty, all extruded panels, 3/4" deep, 1.7 # per S.F.	E-4	1,100	.029	S.F.	8.70	.90	.07	9.67	11.35	
	1700	1-1/4" deep, 2.5# per S.F.		1,000	.032		12.80	1	.08	13.88	16	
	1800	1-3/4" deep, 4.8# per S.F.		925	.035		17.35	1.08	.09	18.52	21	
	1900	2-1/4" deep, 6.1# per S.F.		875	.037		19.65	1.14	.09	20.88	23.50	
	2100	For safety serrated surface, add					15%					

5 METALS

Important: See the Reference Section for critical supporting data - Reference Nos., Crews, & City Cost Indexes

055 300 | Gratings & Floor Plates

			CREW	DAILY OUTPUT	LABOR-HOURS	UNIT	MAT.	LABOR	EQUIP.	TOTAL	TOTAL INCL O&P	
304	2400	Close spaced aluminum grating with vinyl tread inserts	E-4	255	.125	S.F.	33.50	3.90	.32	37.72	44.50	304
	2600	For bottom drainage pan, add	"	510	.063	"	13.80	1.95	.16	15.91	18.95	
308	0010	**FLOOR GRATING, FIBERGLASS**										308
	0100	Reinforced polyester, fire retardant, 1" x 4" grid, 1" thick	E-4	510	.063	S.F.	13.75	1.95	.16	15.86	18.90	
	0200	1-1/2" thick		500	.064		16.85	1.99	.16	19	22.50	
	0300	With grit surface, 1" x 4" grid, 1-1/2" thick	↓	500	.064	↓	17.75	1.99	.16	19.90	23.50	
310	0010	**FLOOR GRATING PLANKS**										310
	0020	Aluminum, 9-1/2" wide, 14 ga., 2" rib	E-4	950	.034	L.F.	11.20	1.05	.09	12.34	14.30	
	0200	Galvanized steel, 9-1/2" wide, 14 ga., 2-1/2" rib		950	.034		7	1.05	.09	8.14	9.70	
	0300	4" rib		950	.034		8.25	1.05	.09	9.39	11.05	
	0500	12 gauge, 2-1/2" rib		950	.034		8.60	1.05	.09	9.74	11.50	
	0600	3" rib		950	.034		9.65	1.05	.09	10.79	12.65	
	0800	Stainless steel, type 304, 16 ga., 2" rib		950	.034		18.25	1.05	.09	19.39	22	
	0900	Type 316	↓	950	.034	↓	22	1.05	.09	23.14	26.50	
312	0010	**FLOOR GRATING, STEEL**										312
	0050	Labor for installing, on grade	E-4	845	.038	S.F.		1.18	.10	1.28	2.27	
	0100	Elevated		460	.070	"		2.16	.18	2.34	4.17	
	0300	Platforms, to 12' high, rectangular		3,150	.010	Lb.	1.07	.32	.03	1.42	1.79	
	0400	Circular	↓	2,300	.014	"	1.18	.43	.04	1.65	2.13	
	0800	For straight cuts, add				L.F.	3.74			3.74	4.11	
	0900	For curved cuts, add					4.70			4.70	5.15	
	1000	For straight banding, add					3.59			3.59	3.95	
	1100	For curved banding, add					4.70			4.70	5.15	
	1200	For checkered plate nosings, add					4.04			4.04	4.44	
	1300	For straight toe or kick plate, add					7.25			7.25	8	
	1400	For curved toe or kick plate, add					8.10			8.10	8.90	
	1500	For abrasive nosings, add				↓	5.90			5.90	6.50	
	1510	For stair treads, see division 055-108										
	1600	For safety serrated surface, minimum, add					15%					
	1700	Maximum, add					25%					
	2000	Stainless steel gratings, close spaced, on grade	E-4	340	.094	S.F.	101	2.93	.24	104.17	117	
	2100	Standard grating, 3.3# per S.F.		700	.046		3.29	1.42	.12	4.83	6.35	
	2200	4.5# per S.F.		650	.049		3.72	1.53	.13	5.38	7.05	
	2400	Expanded steel grating, on grade, 3.0# per S.F.		900	.036		2.49	1.11	.09	3.69	4.87	
	2500	3.14# per S.F.		900	.036		2.55	1.11	.09	3.75	4.93	
	2600	4.0# per S.F.		850	.038		3.18	1.17	.10	4.45	5.75	
	2650	4.27# per S.F.		850	.038		3.44	1.17	.10	4.71	6.05	
	2700	5.0# per S.F.		800	.040		4.14	1.24	.10	5.48	6.95	
	2800	6.25# per S.F.		750	.043		4.99	1.33	.11	6.43	8.05	
	2900	7.0# per S.F.	↓	700	.046		5.85	1.42	.12	7.39	9.15	
	3100	For flattened expanded steel grating, add					8%					
	3300	For installation above grade, add				↓		15%				
314	0005	**GRATING FRAME**										314
	0010	Aluminum, for gratings 1" to 1-1/2" deep	1 Sswk	70	.114	L.F.	6.20	3.50		9.70	13.25	
	0100	For each corner, add				Ea.	4.90			4.90	5.40	
370	0010	**TRENCH COVER**										370
	0020	Cast iron grating with bar stops and angle frame, to 18" wide	1 Sswk	20	.400	L.F.	41	12.25		53.25	67.50	
	0100	Frame only (both sides of trench), 1" grating		45	.178		9	5.45		14.45	19.85	
	0150	2" grating	↓	35	.229	↓	14	7		21	28	
	0200	Aluminum, stock units, including frames and										
	0210	3/8" plain cover plate, 4" opening	E-4	205	.156	L.F.	29.50	4.85	.40	34.75	42	
	0300	6" opening		185	.173		36.50	5.40	.44	42.34	50.50	
	0400	10" opening		170	.188		50	5.85	.48	56.33	66.50	
	0500	16" opening	↓	155	.206		69	6.40	.53	75.93	88	
	0700	Add per inch for additional widths to 24"				↓	2.73			2.73	3	

5

METALS

055 300 | Gratings & Floor Plates

			CREW	DAILY OUTPUT	LABOR-HOURS	UNIT	MAT.	LABOR	EQUIP.	TOTAL	TOTAL INCL O&P	
							1999 BARE COSTS					
370	0900	For custom fabrication, add				L.F.	50%					370
	1100	For 1/4" plain cover plate, deduct					12%					
	1500	For cover recessed for tile, 1/4" thick, deduct					12%					
	1600	3/8" thick, add					5%					
	1800	For checkered plate cover, 1/4" thick, deduct					12%					
	1900	3/8" thick, add					2%					
	2100	For slotted or round holes in cover, 1/4" thick, add					3%					
	2200	3/8" thick, add					4%					
	2300	For abrasive cover, add					12%					

055 400 | Castings

			CREW	DAILY OUTPUT	LABOR-HOURS	UNIT	MAT.	LABOR	EQUIP.	TOTAL	TOTAL INCL O&P	
404	0010	**CONSTRUCTION CASTINGS**										404
	0020	Manhole covers and frames see Division 027-152										
	0100	Column bases, cast iron, 16" x 16", approx. 65 lbs.	E-4	46	.696	Ea.	79	21.50	1.79	102.29	128	
	0200	32" x 32", approx. 256 lbs.	"	23	1.391		295	43.50	3.58	342.08	410	
	0400	Cast aluminum for wood columns, 8" x 8"	1 Carp	32	.250		28	6.80		34.80	41.50	
	0500	12" x 12"	"	32	.250		60	6.80		66.80	77	
	0600	Miscellaneous C.I. castings, light sections, less than 150 lbs	E-4	3,200	.010	Lb.	1.21	.31	.03	1.55	1.93	
	1100	Heavy sections, more than 150 lb		4,200	.008		.60	.24	.02	.86	1.11	
	1300	Special low volume items		3,200	.010		2.12	.31	.03	2.46	2.93	
	1500	For ductile iron, add					100%					

055 500 | Metal Specialties

			CREW	DAILY OUTPUT	LABOR-HOURS	UNIT	MAT.	LABOR	EQUIP.	TOTAL	TOTAL INCL O&P	
504	0005	**LAMP POSTS**										504
	0010	Aluminum, 7' high, stock units, post only	1 Carp	16	.500	Ea.	65	13.65		78.65	93	
	0100	Mild steel, plain	"	16	.500	"	39	13.65		52.65	64.50	
508	0005	**WINDOW GUARDS**										508
	0010	Expanded metal, steel angle frame, permanent	E-4	350	.091	S.F.	14.35	2.84	.23	17.42	21	
	0020	Steel bars, 1/2" x 1/2", spaced 5" O.C.	"	290	.110	"	9.90	3.43	.28	13.61	17.50	
	0030	Hinge mounted, add				Opng.	28.50			28.50	31.50	
	0040	Removable type, add				"	18.25			18.25	20	
	0050	For galvanized guards, add				S.F.	35%					
	0070	For pivoted or projected type, add					105%	40%				
	0100	Mild steel, stock units, economy	E-4	405	.079		3.89	2.46	.20	6.55	9	
	0200	Deluxe		405	.079		8	2.46	.20	10.66	13.50	
	0400	Woven wire, stock units, 3/8" channel frame, 3' x 5' opening		40	.800	Opng.	105	25	2.06	132.06	164	
	0500	4' x 6' opening		38	.842		168	26	2.16	196.16	235	
	0800	Basket guards for above, add					144			144	158	
	1000	Swinging guards for above, add					49.50			49.50	54.50	
512	0005	**ALLOY STEEL CHAIN**										512
	0010	Self-colored, cut lengths, w/accessories, 1/4"	E-17	4	4	C.L.F.	330	126	.52	456.52	595	
	0020	3/8"		2	8		415	253	1.04	669.04	920	
	0030	1/2"		1.20	13.333		650	420	1.73	1,071.73	1,475	
	0040	5/8"		.72	22.222		1,350	700	2.89	2,052.89	2,750	
	0050	3/4"		.48	33.333		1,900	1,050	4.33	2,954.33	4,025	
	0060	7/8"		.40	40		2,925	1,275	5.20	4,205.20	5,550	
	0070	1"		.35	45.714		4,325	1,450	5.95	5,780.95	7,400	
	0080	1-1/4"		.24	66.667		7,225	2,100	8.65	9,333.65	11,800	
	0110	Clevis slip hook, 1/4"				Ea.	8.30			8.30	9.15	
	0120	3/8"					12.25			12.25	13.50	
	0130	1/2"					21.50			21.50	23.50	
	0140	5/8"					39.50			39.50	43	
	0150	3/4"					81			81	89	

Important: See the Reference Section for critical supporting data - Reference Nos., Crews, & City Cost Indexes

055 500	Metal Specialties	CREW	DAILY OUTPUT	LABOR- HOURS	UNIT	1999 BARE COSTS				TOTAL INCL O&P	
						MAT.	LABOR	EQUIP.	TOTAL		
512 0160	Eye/sling hook w/ hammerlock coupling, 7/8"				Ea.	185			185	203	**512**
0170	1"					305			305	335	
0180	1-1/4"				▼	495			495	545	

058 100	Exp. Cover Assemblies	CREW	DAILY OUTPUT	LABOR- HOURS	UNIT	1999 BARE COSTS				TOTAL INCL O&P	
						MAT.	LABOR	EQUIP.	TOTAL		
104 0010	**EXPANSION JOINT ASSEMBLIES** Custom units										**104**
0200	Floor cover assemblies, 1" space, aluminum	1 Sswk	38	.211	L.F.	13.20	6.45		19.65	26.50	
0300	Bronze		38	.211		25.50	6.45		31.95	40	
0500	2" space, aluminum		38	.211		15.95	6.45		22.40	29.50	
0600	Bronze		38	.211		27.50	6.45		33.95	42	
0800	Wall and ceiling assemblies, 1" space, aluminum		38	.211		7.90	6.45		14.35	20.50	
0900	Bronze		38	.211		27	6.45		33.45	41.50	
1100	2" space, aluminum		38	.211		13.35	6.45		19.80	26.50	
1200	Bronze		38	.211		24.50	6.45		30.95	39	
1400	Floor to wall assemblies, 1" space, aluminum		38	.211		12	6.45		18.45	25	
1500	Bronze or stainless		38	.211		29	6.45		35.45	44	
1700	Gym floor angle covers, aluminum, 3" x 3" angle		46	.174		10.40	5.30		15.70	21	
1800	3" x 4" angle		46	.174		12.25	5.30		17.55	23.50	
2000	Roof closures, aluminum, flat roof, low profile, 1" space		57	.140		24	4.29		28.29	34.50	
2100	High profile		57	.140		29.50	4.29		33.79	40.50	
2300	Roof to wall, low profile, 1" space		57	.140		13.30	4.29		17.59	22.50	
2400	High profile	▼	57	.140	▼	17.10	4.29		21.39	26.50	

For information about Means Estimating Seminars, see yellow pages 11 and 12 in back of book

Division Notes

	CREW	DAILY OUTPUT	LABOR-HOURS	UNIT	1999 BARE COSTS				TOTAL INCL O&P
					MAT.	LABOR	EQUIP.	TOTAL	

Division 6
Wood & Plastics

Estimating Tips
060 Fasteners & Adhesives

- Common to any wood framed structure are the accessory connector items such as screws, nails, adhesives, hangers, connector plates, straps, angles and holdowns. For typical wood framed buildings, such as residential projects, the aggregate total for these items can be significant, especially in areas where seismic loading is a concern. For floor and wall framing, nail quantities can be figured on a "pounds per thousand board feet basis", with 10 to 25 lbs. per MBF the range. Holdowns, hangers and other connectors should be taken off by the piece.

061 Rough Carpentry

- Lumber is a traded commodity and therefore sensitive to supply and demand in the marketplace. Even in "budgetary" estimating of wood framed projects, it is advisable to call local suppliers for the latest market pricing.
- Common quantity units for wood framed projects are "thousand board feet" (MBF). A board foot is a volume of wood, $1'' \times 1' \times 1'$,

or 144 cubic inches. Board foot quantities are generally calculated using nominal material dimensions—dressed sizes are ignored. Board foot per lineal foot of any stick of lumber can be calculated by dividing the nominal cross sectional area by 12. As an example, 2,000 lineal feet of 2 x 12 equates to 4 MBF by dividing the nominal area, 2 x 12, by 12, which equals 2, and multiplying by 2,000 to give 4,000 board feet. This simple rule applies to all nominal dimensioned lumber.

- Waste is an issue of concern at the quantity takeoff for any area of construction. Framing lumber is sold in even foot lengths, i.e., 10', 12', 14', 16', and depending on spans, wall heights and the grade of lumber, waste is inevitable. A rule of thumb for lumber waste is 5% to 10% depending on material quality and the complexity of the framing.
- Wood in various forms and shapes is used in many projects, even where the main structural framing is steel, concrete or masonry. Plywood as a back-up partition material and 2x boards used as blocking and cant strips around roof edges are two

common examples. The estimator should ensure that the costs of all wood materials are included in the final estimate.

062 Finish Carpentry

- It is necessary to consider the grade of workmanship when estimating labor costs for erecting millwork and interior finish. In practice, there are three grades: premium, custom and economy. The Means daily output for base and case moldings is in the range of 200 to 250 L.F. per carpenter per day. This is appropriate for most average custom grade projects. For premium projects an adjustment to productivity of 25% to 50% should be made depending on the complexity of the job.

Reference Numbers
Reference numbers are shown in bold squares at the beginning of some major classifications. These numbers refer to related items in the Reference Section. The reference information may be an estimating procedure, an alternate pricing method or technical information.

Note: Not all subdivisions listed here necessarily appear in this publication.

060 500	Fasteners & Adhesives	CREW	DAILY OUTPUT	LABOR-HOURS	UNIT	1999 BARE COSTS				TOTAL INCL O&P	
						MAT.	LABOR	EQUIP.	TOTAL		
504 0010	**NAILS** Prices of material only, based on 50# box purchase, copper, plain				Lb.	4.10			4.10	4.51	**504**
0400	Stainless steel, plain					5.40			5.40	5.95	
0500	Box, 3d to 20d, bright					1.13			1.13	1.24	
0520	Galvanized					1.13			1.13	1.24	
0600	Common, 3d to 60d, plain					.77			.77	.85	
0700	Galvanized					.99			.99	1.09	
0800	Aluminum					3.31			3.31	3.64	
1000	Annular or spiral thread, 4d to 60d, plain					.66			.66	.73	
1200	Galvanized					.88			.88	.97	
1400	Drywall nails, plain					.77			.77	.85	
1600	Galvanized					1.10			1.10	1.21	
1800	Finish nails, 4d to 10d, plain					.90			.90	.99	
2000	Galvanized					1.04			1.04	1.14	
2100	Aluminum					4.84			4.84	5.30	
2300	Flooring nails, hardened steel, 2d to 10d, plain					1.22			1.22	1.34	
2400	Galvanized					1.34			1.34	1.47	
2500	Gypsum lath nails, 1-1/8", 13 ga. flathead, blued					1.33			1.33	1.46	
2600	Masonry nails, hardened steel, 3/4" to 3" long, plain					1.61			1.61	1.77	
2700	Galvanized					1.44			1.44	1.58	
2900	Roofing nails, threaded, galvanized					1.21			1.21	1.34	
3100	Aluminum					4.70			4.70	5.15	
3300	Compressed lead head, threaded, galvanized					1.44			1.44	1.58	
3600	Siding nails, plain shank, galvanized					1.33			1.33	1.46	
3800	Aluminum					4			4	4.40	
5000	Add to prices above for cement coating					.07			.07	.08	
5200	Zinc or tin plating					.12			.12	.13	
5500	Vinyl coated sinkers, 8d to 16d					.55			.55	.61	
506 0010	**NAILS** mat. only, for pneumatic tools, framing, per carton of 5000, 2"				Ea.	36.50			36.50	40	**506**
0100	2-3/8"					41.50			41.50	45.50	
0200	Per carton of 4000, 3"					37			37	40.50	
0300	3-1/4"					39			39	43	
0400	Per carton of 5000, 2-3/8", galv.					56.50			56.50	62	
0500	Per carton of 4000, 3", galv.					63.50			63.50	70	
0600	3-1/4", galv.					60			60	66	
0700	Roofing, per carton of 7200, 1"					36			36	39.50	
0800	1-1/4"					36.50			36.50	40.50	
0900	1-1/2"					42			42	46	
1000	1-3/4"					47			47	51.50	
508 0010	**SHEET METAL SCREWS** Steel, standard, #8 x 3/4", plain				C	2.70			2.70	2.97	**508**
0100	Galvanized					3.33			3.33	3.66	
0300	#10 x 1", plain					3.65			3.65	4.02	
0400	Galvanized					4.26			4.26	4.69	
0600	With washers, #14 x 1", plain					10.85			10.85	11.95	
0700	Galvanized					12.05			12.05	13.30	
0900	#14 x 2", plain					15.80			15.80	17.40	
1000	Galvanized					17.80			17.80	19.55	
1500	Self-drilling, with washers, (pinch point) #8 x 3/4", plain					4.68			4.68	5.15	
1600	Galvanized					6.30			6.30	6.90	
1800	#10 x 3/4", plain					6.25			6.25	6.85	
1900	Galvanized					7.15			7.15	7.85	
3000	Stainless steel w/aluminum or neoprene washers, #14 x 1", plain					16.45			16.45	18.10	
3100	#14 x 2", plain					24.50			24.50	27	
512 0010	**TIMBER CONNECTORS** Add up cost of each part for total										**512**
0020	cost of connection										

Important: See the Reference Section for critical supporting data - Reference Nos., Crews, & City Cost Indexes

060 500	Fasteners & Adhesives	CREW	DAILY OUTPUT	LABOR-HOURS	UNIT	1999 BARE COSTS				TOTAL INCL O&P
						MAT.	LABOR	EQUIP.	TOTAL	
512 0100	Connector plates, steel, with bolts, straight	2 Carp	75	.213	Ea.	16	5.80		21.80	27
0110	Tee	"	50	.320		24	8.75		32.75	40
0200	Bolts, machine, sq. hd. with nut & washer, 1/2" diameter, 4" long	1 Carp	140	.057		.68	1.56		2.24	3.19
0300	7-1/2" long		130	.062		.94	1.68		2.62	3.67
0500	3/4" diameter, 7-1/2" long		130	.062		1.62	1.68		3.30	4.42
0600	15" long		95	.084		2.13	2.30		4.43	5.95
0800	Drilling bolt holes in timber, 1/2" diameter		450	.018	Inch		.49		.49	.76
0900	1" diameter		350	.023	"		.62		.62	.98
1100	Framing anchors, 2 or 3 dimensional, 10 gauge, no nails incl.		175	.046	Ea.	.39	1.25		1.64	2.39
1250	Holdowns, 3 gauge base, 10 gauge body		8	1		13.45	27.50		40.95	58
1300	Joist and beam hangers, 18 ga. galv., for 2" x 4" joist		175	.046		.48	1.25		1.73	2.48
1400	2" x 6" to 2" x 10" joist		165	.048		.56	1.32		1.88	2.70
1600	16 ga. galv., 3" x 6" to 3" x 10" joist		160	.050		2.24	1.36		3.60	4.61
1700	3" x 10" to 3" x 14" joist		160	.050		2.60	1.36		3.96	5
1800	4" x 6" to 4" x 10" joist		155	.052		1.96	1.41		3.37	4.38
1900	4" x 10" to 4" x 14" joist		155	.052		2.67	1.41		4.08	5.15
2000	Two-2" x 6" to two-2" x 10" joists		150	.053		2.15	1.46		3.61	4.66
2100	Two-2" x 10" to two-2" x 14" joists		150	.053		2.15	1.46		3.61	4.66
2300	3/16" thick, 6" x 8" joist		145	.055		4.71	1.51		6.22	7.55
2400	6" x 10" joist		140	.057		5.55	1.56		7.11	8.55
2500	6" x 12" joist		135	.059		6.70	1.62		8.32	9.90
2700	1/4" thick, 6" x 14" joist		130	.062		8.30	1.68		9.98	11.75
2800	Joist anchors, 1/4" x 1-1/4" x 18"		140	.057		3.23	1.56		4.79	6
2900	Plywood clips, extruded aluminum H clip, for 3/4" panels					.11			.11	.12
3000	Galvanized 18 ga. back-up clip					.10			.10	.11
3200	Post framing, 16 ga. galv. for 4" x 4" base, 2 piece	1 Carp	130	.062		4.65	1.68		6.33	7.75
3300	Cap		130	.062		2.29	1.68		3.97	5.15
3500	Rafter anchors, 18 ga. galv., 1-1/2" wide, 5-1/4" long		145	.055		.37	1.51		1.88	2.78
3600	10-3/4" long		145	.055		.76	1.51		2.27	3.21
3800	Shear plates, 2-5/8" diameter		120	.067		1.38	1.82		3.20	4.38
3900	4" diameter		115	.070		3.20	1.90		5.10	6.50
4000	Sill anchors, embedded in concrete or block, 18-5/8" long		115	.070		.96	1.90		2.86	4.05
4100	Spike grids, 4" x 4", flat or curved		120	.067		.40	1.82		2.22	3.29
4400	Split rings, 2-1/2" diameter		120	.067		1.15	1.82		2.97	4.13
4500	4" diameter		110	.073		1.77	1.99		3.76	5.05
4700	Strap ties, 16 ga., 1-3/8" wide, 12" long		180	.044		.91	1.21		2.12	2.91
4800	24" long		160	.050		1.42	1.36		2.78	3.71
5000	Toothed rings, 2-5/8" or 4" diameter		90	.089		.97	2.43		3.40	4.89
5200	Truss plates, nailed, 20 gauge, up to 32' span		17	.471	Truss	7	12.85		19.85	27.50
5400	Washers, 2" x 2" x 1/8"				Ea.	.22			.22	.24
5500	3" x 3" x 3/16"				"	.57			.57	.63
6101	Beam hangers, polymer painted									
6102	Bolted, 3 ga., (W x H x L)									
6104	3-1/4" x 9" x 12" top flange	1 Carp	1	8	C	5,450	218		5,668	6,350
6106	5-1/4" x 9" x 12" top flange		1	8		5,675	218		5,893	6,600
6108	5-1/4" x 11" x 11-3/4" top flange		1	8		12,800	218		13,018	14,400
6110	6-7/8" x 9" x 12" top flange		1	8		5,875	218		6,093	6,800
6112	6-7/8" x 11" x 13-1/2" top flange		1	8		13,400	218		13,618	15,100
6114	8-7/8" x 11" x 15-1/2" top flange		1	8		14,400	218		14,618	16,100
6116	Nailed, 3 ga., (W x H x L)									
6118	3-1/4" x 10-1/2" x 10" top flange	1 Carp	1.80	4.444	C	3,775	121		3,896	4,350
6120	3-1/4" x 10-1/2" x 12" top flange		1.80	4.444		4,375	121		4,496	5,000
6122	5-1/4" x 9-1/2" x 10" top flange		1.80	4.444		4,075	121		4,196	4,675
6124	5-1/4" x 9-1/2" x 12" top flange		1.80	4.444		4,600	121		4,721	5,275
6126	5-1/2" x 9-1/2" x 12" top flange		1.80	4.444		4,075	121		4,196	4,675
6128	6-7/8" x 8-1/2" x 12" top flange		1.80	4.444		4,200	121		4,321	4,825

WOOD & PLASTICS 6

For expanded coverage of these items see *Means Interior Cost Data 1999*

060 500	Fasteners & Adhesives	CREW	DAILY OUTPUT	LABOR-HOURS	UNIT	MAT.	LABOR	EQUIP.	TOTAL	TOTAL INCL O&P	
512 6130	7-1/2" x 8-1/2" x 12" top flange	1 Carp	1.80	4.444	C	4,350	121		4,471	4,975	512
6132	8-7/8" x 7-1/2" x 14" top flange	↓	1.80	4.444	↓	4,800	121		4,921	5,475	
6201	Beam and purlin hangers, galvanized, 12 ga.										
6202	Purlin or joist size, 3" x 8"	1 Carp	1.70	4.706	C	690	128		818	960	
6204	3" x 10"		1.70	4.706		775	128		903	1,050	
6206	3" x 12"		1.65	4.848		915	132		1,047	1,200	
6208	3" x 14"		1.65	4.848		1,075	132		1,207	1,375	
6210	3" x 16"		1.65	4.848		1,225	132		1,357	1,550	
6212	4" x 8"		1.65	4.848		690	132		822	970	
6214	4" x 10"		1.65	4.848		805	132		937	1,100	
6216	4" x 12"		1.60	5		865	137		1,002	1,175	
6218	4" x 14"		1.60	5		940	137		1,077	1,250	
6220	4" x 16"		1.60	5		1,075	137		1,212	1,400	
6224	6" x 10"		1.55	5.161		1,175	141		1,316	1,525	
6226	6" x 12"		1.55	5.161		1,250	141		1,391	1,625	
6228	6" x 14"		1.50	5.333		1,350	146		1,496	1,725	
6230	6" x 16"	↓	1.50	5.333	↓	1,550	146		1,696	1,925	
6300	Column bases										
6302	4 x 4, 16 ga.	1 Carp	1.80	4.444	C	960	121		1,081	1,250	
6306	7 ga.		1.80	4.444		1,775	121		1,896	2,150	
6314	6 x 6, 16 ga.		1.75	4.571		1,225	125		1,350	1,550	
6318	7 ga.		1.75	4.571		2,500	125		2,625	2,950	
6326	8 x 8, 7 ga.		1.65	4.848		4,175	132		4,307	4,800	
6330	8 x 10, 7 ga.	↓	1.65	4.848	↓	4,450	132		4,582	5,100	
6590	Joist hangers, heavy duty 12 ga., galvanized										
6592	2" x 4"	1 Carp	1.75	4.571	C	785	125		910	1,050	
6594	2" x 6"		1.65	4.848		835	132		967	1,125	
6596	2" x 8"		1.65	4.848		890	132		1,022	1,200	
6598	2" x 10"		1.65	4.848		950	132		1,082	1,250	
6600	2" x 12"		1.65	4.848		1,075	132		1,207	1,375	
6622	(2) 2" x 6"		1.60	5		1,050	137		1,187	1,375	
6624	(2) 2" x 8"		1.60	5		1,125	137		1,262	1,475	
6626	(2) 2" x 10"		1.55	5.161		1,275	141		1,416	1,625	
6628	(2) 2" x 12"	↓	1.55	5.161	↓	1,550	141		1,691	1,925	
6890	Purlin hangers, painted										
6892	12 ga., 2" x 6"	1 Carp	1.80	4.444	C	955	121		1,076	1,250	
6894	2" x 8"		1.80	4.444		1,000	121		1,121	1,300	
6896	2" x 10"		1.80	4.444		1,025	121		1,146	1,325	
6898	2" x 12"		1.75	4.571		1,125	125		1,250	1,425	
6934	(2) 2" x 6"		1.70	4.706		965	128		1,093	1,250	
6936	(2) 2" x 8"		1.70	4.706		1,075	128		1,203	1,375	
6938	(2) 2" x 10"		1.70	4.706		1,175	128		1,303	1,500	
6940	(2) 2" x 12"	↓	1.65	4.848	↓	1,250	132		1,382	1,575	
516 0010	**WOOD SCREWS** #8, 1" long, steel				C	3.46			3.46	3.81	516
0100	Brass					16.70			16.70	18.35	
0200	#8, 2" long, steel					6.55			6.55	7.20	
0300	Brass					32.50			32.50	35.50	
0400	#10, 1" long, steel					4.22			4.22	4.64	
0500	Brass					22.50			22.50	24.50	
0600	#10, 2" long, steel					7.50			7.50	8.25	
0700	Brass					39.50			39.50	43.50	
0800	#10, 3" long, steel					13.75			13.75	15.10	
1000	#12, 2" long, steel					10.05			10.05	11.05	
1100	Brass					50			50	55	
1500	#12, 3" long, steel				↓	15.90			15.90	17.50	

Important: See the Reference Section for critical supporting data - Reference Nos., Crews, & City Cost Indexes

060 500	Fasteners & Adhesives	CREW	DAILY OUTPUT	LABOR-HOURS	UNIT	1999 BARE COSTS				TOTAL INCL O&P	
						MAT.	LABOR	EQUIP.	TOTAL		
516 2000	#12, 4" long, steel				C	28.50			28.50	31	**516**

061 | Rough Carpentry

061 100	Wood Framing		CREW	DAILY OUTPUT	LABOR-HOURS	UNIT	1999 BARE COSTS				TOTAL INCL O&P	
							MAT.	LABOR	EQUIP.	TOTAL		
102 0010	**BLOCKING**											**102**
2600	Miscellaneous, to wood construction											
2620	2" x 4"		1 Carp	.17	47.059	M.B.F.	540	1,275		1,815	2,625	
2625	Pneumatic nailed			.21	38.095		540	1,050		1,590	2,225	
2660	2" x 8"			.27	29.630		545	810		1,355	1,875	
2665	Pneumatic nailed			.33	24.242		545	660		1,205	1,650	
2720	To steel construction											
2740	2" x 4"		1 Carp	.14	57.143	M.B.F.	540	1,550		2,090	3,050	
2780	2" x 8"		"	.21	38.095	"	545	1,050		1,595	2,225	
104 0010	**BRACING** Let-in, with 1" x 6" boards, studs @ 16" O.C.		1 Carp	1.50	5.333	C.L.F.	52	146		198	287	**104**
0200	Studs @ 24" O.C.		"	2.30	3.478	"	52	95		147	207	
106 0010	**BRIDGING** Wood, for joists 16" O.C., 1" x 3"		1 Carp	1.30	6.154	C.Pr.	34.50	168		202.50	300	**106**
0015	Pneumatic nailed			1.70	4.706		34.50	128		162.50	240	
0100	2" x 3" bridging			1.30	6.154		37	168		205	305	
0105	Pneumatic nailed			1.70	4.706		37	128		165	243	
0300	Steel, galvanized, 18 ga., for 2" x 10" joists at 12" O.C.			1.30	6.154		85	168		253	360	
0400	24" O.C.			1.40	5.714		220	156		376	485	
0600	For 2" x 14" joists at 16" O.C.			1.30	6.154		168	168		336	450	
0700	24" O.C.			1.40	5.714		220	156		376	485	
0900	Compression type, 16" O.C., 2" x 8" joists			2	4		115	109		224	299	
1000	2" x 12" joists			2	4		126	109		235	310	
110 0010	**FRAMING, BEAMS & GIRDERS**	R061 -010	2 Carp	.70	22.857	M.B.F.	540	625		1,165	1,575	**110**
3500	Single, 2" x 6"											
3505	Pneumatic nailed	R061 -030		.81	19.704		540	540		1,080	1,450	
3520	2" x 8"			.86	18.605		545	510		1,055	1,400	
3525	Pneumatic nailed			1	16.048		545	440		985	1,300	
3540	2" x 10"			1	16		605	435		1,040	1,350	
3545	Pneumatic nailed			1.16	13.793		605	375		980	1,250	
3560	2" x 12"			1.10	14.545		645	395		1,040	1,325	
3565	Pneumatic nailed			1.28	12.539		645	340		985	1,250	
3580	2" x 14"			1.17	13.675		820	375		1,195	1,500	
3585	Pneumatic nailed			1.36	11.791		820	320		1,140	1,400	
3600	3" x 8"			1.10	14.545		1,000	395		1,395	1,750	
3620	3" x 10"			1.25	12.800		1,000	350		1,350	1,675	
3640	3" x 12"			1.35	11.852		1,000	325		1,325	1,625	
3660	3" x 14"			1.40	11.429		1,000	310		1,310	1,625	
3680	4" x 8"		F-3	2.66	15.038		1,150	415	167	1,732	2,100	
3700	4" x 10"			3.16	12.658		1,150	350	140	1,640	1,975	
3720	4" x 12"			3.60	11.111		1,150	310	123	1,583	1,900	
3740	4" x 14"			3.96	10.101		1,150	280	112	1,542	1,825	
4000	Double, 2" x 6"		2 Carp	1.25	12.800		540	350		890	1,150	
4005	Pneumatic nailed			1.45	11.034		540	300		840	1,075	
4020	2" x 8"			1.60	10		545	273		818	1,025	

WOOD & PLASTICS 6

061 100 | Wood Framing

		Description	CREW	DAILY OUTPUT	LABOR-HOURS	UNIT	1999 BARE COSTS				TOTAL INCL O&P	
							MAT.	LABOR	EQUIP.	TOTAL		
110	4025	Pneumatic nailed R061 -010	2 Carp	1.86	8.621	M.B.F.	545	235		780	970	**110**
	4040	2" x 10"		1.92	8.333		605	228		833	1,025	
	4045	Pneumatic nailed R061 -030		2.23	7.185		605	196		801	980	
	4060	2" x 12"		2.20	7.273		645	199		844	1,025	
	4065	Pneumatic nailed		2.55	6.275		645	171		816	980	
	4080	2" x 14"		2.45	6.531		820	178		998	1,175	
	4085	Pneumatic nailed		2.84	5.634		820	154		974	1,150	
	5000	Triple, 2" x 6"		1.65	9.697		540	265		805	1,000	
	5005	Pneumatic nailed		1.91	8.377		540	229		769	955	
	5020	2" x 8"		2.10	7.619		545	208		753	925	
	5025	Pneumatic nailed		2.44	6.568		545	179		724	880	
	5040	2" x 10"		2.50	6.400		605	175		780	945	
	5045	Pneumatic nailed		2.90	5.517		605	151		756	905	
	5060	2" x 12"		2.85	5.614		645	153		798	950	
	5065	Pneumatic nailed		3.31	4.840		645	132		777	920	
	5080	2" x 14"		3.15	5.079		820	139		959	1,125	
	5085	Pneumatic nailed		3.35	4.770		820	130		950	1,100	
112	0010	**FRAMING, CEILINGS**										**112**
	6400	Suspended, 2" x 3"	2 Carp	.50	32	M.B.F.	490	875		1,365	1,925	
	6450	2" x 4"		.59	27.119		540	740		1,280	1,775	
	6500	2" x 6"		.80	20		540	545		1,085	1,450	
	6550	2" x 8"		.86	18.605		545	510		1,055	1,400	
114	0010	**FRAMING, JOISTS**										**114**
	2650	Joists, 2" x 4" R061 -010	2 Carp	.83	19.277	M.B.F.	540	525		1,065	1,425	
	2655	Pneumatic nailed R061 -030		.96	16.667		540	455		995	1,300	
	2680	2" x 6"		1.25	12.800		540	350		890	1,150	
	2685	Pneumatic nailed		1.44	11.111		540	305		845	1,075	
	2700	2" x 8"		1.46	10.959		545	299		844	1,075	
	2705	Pneumatic nailed		1.68	9.524		545	260		805	1,000	
	2720	2" x 10"		1.49	10.738		605	293		898	1,125	
	2725	Pneumatic nailed		1.71	9.357		605	255		860	1,075	
	2740	2" x 12"		1.75	9.143		645	250		895	1,100	
	2745	Pneumatic nailed		2.01	7.960		645	217		862	1,050	
	2760	2" x 14"		1.79	8.939		820	244		1,064	1,300	
	2765	Pneumatic nailed		2.06	7.767		820	212		1,032	1,250	
	2780	3" x 6"		1.39	11.511		1,000	315		1,315	1,625	
	2790	3" x 8"		1.90	8.421		1,000	230		1,230	1,475	
	2800	3" x 10"		1.95	8.205		1,000	224		1,224	1,475	
	2820	3" x 12"		1.80	8.889		1,000	243		1,243	1,500	
	2840	4" x 6"		1.60	10		1,150	273		1,423	1,700	
	2850	4" x 8"	F-3	4.15	9.639		1,150	267	107	1,524	1,800	
	2860	4" x 10"	2 Carp	2	8		1,150	218		1,368	1,625	
	2880	4" x 12"		1.80	8.889		1,150	243		1,393	1,650	
	3000	Composite wood joist 9-1/2" deep		.90	17.778	M.L.F.	1,700	485		2,185	2,625	
	3010	11-1/2" deep		.88	18.182		1,825	495		2,320	2,775	
	3020	14" deep		.82	19.512		1,950	535		2,485	3,000	
	3030	16" deep		.78	20.513		2,600	560		3,160	3,725	
	4000	Open web joist 12" deep		.88	18.182		1,875	495		2,370	2,850	
	4010	14" deep		.82	19.512		2,125	535		2,660	3,175	
	4020	16" deep		.78	20.513		2,200	560		2,760	3,300	
	4030	18" deep		.74	21.622		2,300	590		2,890	3,450	
	6000	Composite rim joist, 1-1/4" x 9-1/2"		90	.178		1,925	4.85		1,929.85	2,125	
	6010	1-1/4" x 11-1/2"		.88	18.182		2,050	495		2,545	3,050	
	6020	1-1/4" x 14-1/2"		.82	19.512		2,525	535		3,060	3,625	

Important: See the Reference Section for critical supporting data - Reference Nos., Crews, & City Cost Indexes

		061 100 | Wood Framing		DAILY	LABOR-		1999 BARE COSTS				TOTAL	
			CREW	OUTPUT	HOURS	UNIT	MAT.	LABOR	EQUIP.	TOTAL	INCL O&P	
114	6030	1-1/4" x 16-1/2" R061-010	2 Carp	.78	20.513	M.L.F.	2,625	560		3,185	3,750	114
116	0010	**FRAMING, MISCELLANEOUS**										116
	8500	Firestops, 2" x 4"	2 Carp	.51	31.373	M.B.F.	540	855		1,395	1,950	
	8505	Pneumatic nailed		.62	25.806		540	705		1,245	1,700	
	8520	2" x 6"		.60	26.667		540	730		1,270	1,750	
	8525	Pneumatic nailed		.73	21.858		540	595		1,135	1,525	
	8540	2" x 8"		.60	26.667		545	730		1,275	1,750	
	8560	2" x 12"		.70	22.857		645	625		1,270	1,700	
	8600	Nailers, treated, wood construction, 2" x 4"		.53	30.189		965	825		1,790	2,350	
	8605	Pneumatic nailed		.64	25.157		965	685		1,650	2,125	
	8620	2" x 6"		.75	21.333		965	580		1,545	1,975	
	8625	Pneumatic nailed		.90	17.778		965	485		1,450	1,825	
	8640	2" x 8"		.93	17.204		965	470		1,435	1,800	
	8645	Pneumatic nailed		1.12	14.337		965	390		1,355	1,675	
	8660	Steel construction, 2" x 4"		.50	32		965	875		1,840	2,425	
	8680	2" x 6"		.70	22.857		965	625		1,590	2,025	
	8700	2" x 8"		.87	18.391		965	500		1,465	1,850	
	8760	Rough bucks, treated, for doors or windows, 2" x 6"		.40	40		965	1,100		2,065	2,775	
	8765	Pneumatic nailed		.48	33.333		965	910		1,875	2,475	
	8780	2" x 8"		.51	31.373		965	855		1,820	2,400	
	8785	Pneumatic nailed		.61	26.144		965	715		1,680	2,175	
	8800	Stair stringers, 2" x 10"		.22	72.727		605	1,975		2,580	3,800	
	8820	2" x 12"		.26	61.538		645	1,675		2,320	3,350	
	8840	3" x 10"		.31	51.613		1,000	1,400		2,400	3,350	
	8860	3" x 12"		.38	42.105		1,000	1,150		2,150	2,925	
	8870	Composite LSL, 1-1/4" x 11-1/2"		130	.123	L.F.	2.06	3.36		5.42	7.55	
	8880	1-1/4" x 14-1/2"		130	.123	"	2.52	3.36		5.88	8.05	
118	0010	**FRAMING, COLUMNS**										118
	0400	4" x 4"	2 Carp	.52	30.769	M.B.F.	715	840		1,555	2,100	
	0420	4" x 6"		.55	29.091		1,150	795		1,945	2,525	
	0440	4" x 8"		.59	27.119		1,150	740		1,890	2,450	
	0460	6" x 6"		.65	24.615		1,675	670		2,345	2,900	
	0480	6" x 8"		.70	22.857		1,250	625		1,875	2,350	
	0500	6" x 10"		.75	21.333		1,400	580		1,980	2,475	
120	0010	**FRAMING, ROOFS**										120
	5250	Composite rafter, 9-1/2" deep	2 Carp	575	.028	L.F.	1.93	.76		2.69	3.32	
	5260	11-1/2" deep		575	.028	"	2.06	.76		2.82	3.47	
	6070	Fascia boards, 2" x 8"		.30	53.333	M.B.F.	545	1,450		1,995	2,900	
	6080	2" x 10"		.30	53.333		605	1,450		2,055	2,975	
	7000	Rafters, to 4 in 12 pitch, 2" x 6"		1	16		540	435		975	1,275	
	7060	2" x 8"		1.26	12.698		545	345		890	1,150	
	7300	Hip and valley rafters, 2" x 6"		.76	21.053		540	575		1,115	1,500	
	7360	2" x 8"		.96	16.667		545	455		1,000	1,325	
	7540	Hip and valley jacks, 2" x 6"		.60	26.667		540	730		1,270	1,750	
	7600	2" x 8"		.65	24.615		545	670		1,215	1,650	
	7780	For slopes steeper than 4 in 12, add						30%				
	7790	For dormers or complex roofs, add						50%				
	7800	Rafter tie, 1" x 4", #3	2 Carp	.27	59.259	M.B.F.	1,100	1,625		2,725	3,775	
	7820	Ridge board, #2 or better, 1" x 6" CN		.30	53.333		1,675	1,450		3,125	4,125	
	7840	1" x 8"		.37	43.243		1,450	1,175		2,625	3,425	
	7860	1" x 10"		.42	38.095		1,450	1,050		2,500	3,200	
	7880	2" x 6"		.50	32		540	875		1,415	1,975	
	7900	2" x 8"		.60	26.667		545	730		1,275	1,750	
	7920	2" x 10"		.66	24.242		605	660		1,265	1,725	

WOOD & PLASTICS **6**

For expanded coverage of these items see *Means Interior Cost Data 1999*

WOOD & PLASTICS 6

061 100	Wood Framing		CREW	DAILY OUTPUT	LABOR-HOURS	UNIT	1999 BARE COSTS				TOTAL INCL O&P	
							MAT.	LABOR	EQUIP.	TOTAL		
120	7940	Roof cants, split, 4" x 4"	2 Carp	.86	18.605	M.B.F.	715	510		1,225	1,575	**120**
	7960	6" x 6"		1.80	8.889		1,675	243		1,918	2,225	
	7980	Roof curbs, untreated, 2" x 6"		.52	30.769		540	840		1,380	1,925	
	8000	2" x 12"	↓	.80	20	↓	645	545		1,190	1,575	
122	0010	**FRAMING, SILLS**										**122**
	4482	Ledgers, nailed, 2" x 4"	2 Carp	.50	32	M.B.F.	540	875		1,415	1,975	
	4484	2" x 6"		.60	26.667		540	730		1,270	1,750	
	4486	Bolted, not including bolts, 3" x 8"		.65	24.615		1,000	670		1,670	2,175	
	4488	3" x 12"		.70	22.857		1,000	625		1,625	2,100	
	4490	Mud sills, redwood, construction grade, 2" x 4"		.59	27.119		4,000	740		4,740	5,575	
	4492	2" x 6"		.78	20.513		4,000	560		4,560	5,275	
	4500	Sills, 2" x 4"		.40	40		540	1,100		1,640	2,325	
	4520	2" x 6"		.55	29.091		540	795		1,335	1,850	
	4540	2" x 8"		.67	23.881		545	650		1,195	1,625	
	4600	Treated, 2" x 4"		.36	44.444		965	1,225		2,190	2,950	
	4620	2" x 6"		.50	32		965	875		1,840	2,425	
	4640	2" x 8"		.60	26.667		965	730		1,695	2,200	
	4700	4" x 4"		.60	26.667		1,325	730		2,055	2,625	
	4720	4" x 6"		.70	22.857		1,325	625		1,950	2,450	
	4740	4" x 8"		.80	20		1,325	545		1,870	2,325	
	4760	4" x 10"	↓	.87	18.391	↓	1,325	500		1,825	2,275	
124	0010	**FRAMING, SLEEPERS**										**124**
	0300	On concrete, treated, 1" x 2"	2 Carp	.39	41.026	M.B.F.	1,000	1,125		2,125	2,850	
	0320	1" x 3"		.50	32		1,000	875		1,875	2,475	
	0340	2" x 4"		.99	16.162		965	440		1,405	1,750	
	0360	2" x 6"	↓	1.30	12.308	↓	965	335		1,300	1,575	
126	0010	**FRAMING, SOFFITS & CANOPIES**										**126**
	1300	Canopy or soffit framing, 1" x 4"	2 Carp	.30	53.333	M.B.F.	1,650	1,450		3,100	4,125	
	1340	1" x 8"		.50	32		1,450	875		2,325	2,950	
	1360	2" x 4"		.41	39.024		540	1,075		1,615	2,275	
	1400	2" x 8"		.67	23.881		545	650		1,195	1,625	
	1420	3" x 4"		.50	32		1,000	875		1,875	2,500	
	1460	3" x 8"	↓	.60	26.667	↓	1,000	730		1,730	2,275	
127	0010	**FRAMING, TREATED LUMBER**										**127**
	0020	Water-borne salt, C.C.A., A.C.A., wet, .40 P.C.F. retention										
	0100	2" x 4"				M.B.F.	965			965	1,050	
	0110	2" x 6"					965			965	1,050	
	0120	2" x 8"					965			965	1,050	
	0130	2" x 10"					955			955	1,050	
	0140	2" x 12"					1,100			1,100	1,200	
	0200	4" x 4"					1,325			1,325	1,475	
	0210	4" x 6"					1,325			1,325	1,475	
	0220	4" x 8"				↓	1,325			1,325	1,475	
	0250	Add for .60 P.C.F. retention					40%					
	0260	Add for 2.5 P.C.F. retention					200%					
	0270	Add for K.D.A.T.					20%					
128	0010	**FRAMING, WALLS** R061 -010										**128**
	5860	Headers over openings, 2" x 6"	2 Carp	.36	44.444	M.B.F.	540	1,225		1,765	2,500	
	5865	2" x 6", pneumatic nailed R061 -030		.43	37.209		540	1,025		1,565	2,200	
	5880	2" x 8"		.45	35.556		545	970		1,515	2,125	
	5885	2" x 8", pneumatic nailed		.54	29.630		545	810		1,355	1,875	
	5900	2" x 10"	↓	.53	30.189	↓	605	825		1,430	1,975	

Important: See the Reference Section for critical supporting data - Reference Nos., Crews, & City Cost Indexes

061 100 | Wood Framing

			CREW	DAILY OUTPUT	LABOR-HOURS	UNIT	1999 BARE COSTS				TOTAL INCL O&P	
							MAT.	LABOR	EQUIP.	TOTAL		
128	5905	2" x 10", pneumatic nailed R061 -010	2 Carp	.67	23.881	M.B.F.	605	650		1,255	1,700	**128**
	5920	2" x 12"		.60	26.667		645	730		1,375	1,850	
	5925	2" x 12", pneumatic nailed R061 -030		.72	22.222		645	605		1,250	1,675	
	5940	4" x 12"		.76	21.053		1,150	575		1,725	2,175	
	5945	4" x 12", pneumatic nailed		.92	17.391		1,150	475		1,625	2,025	
	5960	6" x 12"		.84	19.048		1,400	520		1,920	2,375	
	5965	6" x 12", pneumatic nailed		1.01	15.873		1,400	435		1,835	2,225	
	6000	Plates, untreated, 2" x 3"		.43	37.209		490	1,025		1,515	2,150	
	6005	2" x 3", pneumatic nailed		.52	30.769		490	840		1,330	1,875	
	6020	2" x 4" CN		.53	30.189		540	825		1,365	1,900	
	6025	2" x 4", pneumatic nailed		.67	23.881		540	650		1,190	1,625	
	6040	2" x 6"		.75	21.333		540	580		1,120	1,500	
	6045	2" x 6", pneumatic nailed		.90	17.778		540	485		1,025	1,350	
	6120	Studs, 8' high wall, 2" x 3"		.60	26.667		490	730		1,220	1,700	
	6125	2" x 3", pneumatic nailed		.72	22.222		490	605		1,095	1,500	
	6140	2" x 4"		.92	17.391		540	475		1,015	1,350	
	6145	2" x 4", pneumatic nailed		1.10	14.493		540	395		935	1,225	
	6160	2" x 6"		1	16		540	435		975	1,275	
	6165	2" x 6", pneumatic nailed		1.20	13.333		540	365		905	1,175	
	6180	3" x 4"		.80	20		1,000	545		1,545	1,975	
	6185	3" x 4", pneumatic nailed	▼	.96	16.667	▼	1,000	455		1,455	1,850	
	8200	For 12' high walls, deduct						5%				
	8220	For stub wall, 6' high, add						20%				
	8240	3' high, add						40%				
	8250	For second story & above, add						5%				
	8300	For dormer & gable, add	▼					15%				
130	0010	**FURRING** Wood strips, 1" x 2", on walls, on wood	1 Carp	550	.015	L.F.	.18	.40		.58	.82	**130**
	0015	On wood, pneumatic nailed		710	.011		.18	.31		.49	.68	
	0300	On masonry		495	.016		.18	.44		.62	.89	
	0400	On concrete		260	.031		.18	.84		1.02	1.52	
	0600	1" x 3", on walls, on wood		550	.015		.23	.40		.63	.87	
	0605	On wood, pneumatic nailed		710	.011		.23	.31		.54	.73	
	0700	On masonry		495	.016		.23	.44		.67	.94	
	0800	On concrete		260	.031		.23	.84		1.07	1.57	
	0850	On ceilings, on wood		350	.023		.23	.62		.85	1.23	
	0855	On wood, pneumatic nailed		450	.018		.23	.49		.72	1.01	
	0900	On masonry		320	.025		.23	.68		.91	1.32	
	0950	On concrete	▼	210	.038	▼	.23	1.04		1.27	1.89	
132	0010	**GROUNDS** For casework, 1" x 2" wood strips, on wood	1 Carp	330	.024	L.F.	.18	.66		.84	1.24	**132**
	0100	On masonry		285	.028		.18	.77		.95	1.41	
	0200	On concrete		250	.032		.18	.87		1.05	1.57	
	0400	For plaster, 3/4" deep, on wood		450	.018		.18	.49		.67	.96	
	0500	On masonry		225	.036		.18	.97		1.15	1.73	
	0600	On concrete		175	.046		.18	1.25		1.43	2.16	
	0700	On metal lath	▼	200	.040	▼	.18	1.09		1.27	1.92	
134	0010	**INSULATION** See division 072										**134**
136	0010	**LAMINATED** See division 061-804										**136**
138	0010	**PARTITIONS** Wood stud with single bottom plate and										**138**
	0020	double top plate, no waste, std. & better lumber										
	0180	2" x 4" studs, 8' high, studs 12" O.C.	2 Carp	80	.200	L.F.	3.96	5.45		9.41	12.95	
	0185	12" O.C., pneumatic nailed		96	.167		3.96	4.55		8.51	11.50	
	0200	16" O.C.		100	.160		3.24	4.37		7.61	10.40	
	0205	16" O.C., pneumatic nailed	▼	120	.133	▼	3.24	3.64		6.88	9.30	

WOOD & PLASTICS **6**

061 100 | Wood Framing

			CREW	DAILY OUTPUT	LABOR-HOURS	UNIT	MAT.	LABOR	EQUIP.	TOTAL	TOTAL INCL O&P	
138	0300	24" O.C.	2 Carp	125	.128	L.F.	2.52	3.49		6.01	8.25	138
	0305	24" O.C., pneumatic nailed		150	.107		2.52	2.91		5.43	7.35	
	0380	10' high, studs 12" O.C.		80	.200		4.68	5.45		10.13	13.75	
	0385	12" O.C., pneumatic nailed		96	.167		4.68	4.55		9.23	12.30	
	0400	16" O.C.		100	.160		3.78	4.37		8.15	11	
	0405	16" O.C., pneumatic nailed		120	.133		3.78	3.64		7.42	9.90	
	0500	24" O.C.		125	.128		2.88	3.49		6.37	8.65	
	0505	24" O.C., pneumatic nailed		150	.107		2.88	2.91		5.79	7.75	
	0580	12' high, studs 12" O.C.		65	.246		5.40	6.70		12.10	16.50	
	0585	12" O.C., pneumatic nailed		78	.205		5.40	5.60		11	14.75	
	0600	16" O.C.		80	.200		4.32	5.45		9.77	13.35	
	0605	16" O.C., pneumatic nailed		96	.167		4.32	4.55		8.87	11.90	
	0700	24" O.C.		100	.160		3.24	4.37		7.61	10.40	
	0705	24" O.C., pneumatic nailed		120	.133		3.24	3.64		6.88	9.30	
	0780	2" x 6" studs, 8' high, studs 12" O.C.		70	.229		5.95	6.25		12.20	16.30	
	0785	12" O.C., pneumatic nailed		84	.190		5.95	5.20		11.15	14.70	
	0800	16" O.C.		90	.178		4.85	4.85		9.70	13	
	0805	16" O.C., pneumatic nailed		108	.148		4.85	4.04		8.89	11.70	
	0900	24" O.C.		115	.139		3.77	3.80		7.57	10.15	
	0905	24" O.C., pneumatic nailed		138	.116		3.77	3.17		6.94	9.15	
	0980	10' high, studs 12" O.C.		70	.229		7	6.25		13.25	17.50	
	0985	12" O.C., pneumatic nailed		84	.190		7	5.20		12.20	15.90	
	1000	16" O.C.		90	.178		5.65	4.85		10.50	13.90	
	1005	16" O.C., pneumatic nailed		108	.148		5.65	4.04		9.69	12.60	
	1100	24" O.C.		115	.139		4.31	3.80		8.11	10.75	
	1105	24" O.C., pneumatic nailed		138	.116		4.31	3.17		7.48	9.70	
	1180	12' high, studs 12" O.C.		55	.291		8.10	7.95		16.05	21.50	
	1185	12" O.C., pneumatic nailed		66	.242		8.10	6.60		14.70	19.30	
	1200	16" O.C.		70	.229		6.45	6.25		12.70	16.90	
	1205	16" O.C., pneumatic nailed		84	.190		6.45	5.20		11.65	15.30	
	1300	24" O.C.		90	.178		4.85	4.85		9.70	13	
	1305	24" O.C., pneumatic nailed		108	.148		4.85	4.04		8.89	11.70	
	1400	For horizontal blocking, 2" x 4", add		600	.027		.36	.73		1.09	1.55	
	1500	2" x 6", add		600	.027		.54	.73		1.27	1.74	
	1600	For openings, add		250	.064			1.75		1.75	2.75	
	1700	Headers for above openings, material only, add				M.B.F.	545			545	600	
140	0010	**ROUGH HARDWARE** Average % of carpentry material, minimum					.50%					140
	0200	Maximum					1.50%					

061 150 | Sheathing

			CREW	DAILY OUTPUT	LABOR-HOURS	UNIT	MAT.	LABOR	EQUIP.	TOTAL	TOTAL INCL O&P	
154	0010	**SHEATHING** Plywood on roof, CDX										154
	0030	5/16" thick R061 -020	2 Carp	1,600	.010	S.F.	.29	.27		.56	.75	
	0035	Pneumatic nailed R061 -030		1,952	.008		.29	.22		.51	.67	
	0050	3/8" thick		1,525	.010		.32	.29		.61	.80	
	0055	Pneumatic nailed		1,860	.009		.32	.23		.55	.72	
	0100	1/2" thick CN		1,400	.011		.41	.31		.72	.94	
	0105	Pneumatic nailed		1,708	.009		.41	.26		.67	.85	
	0200	5/8" thick		1,300	.012		.49	.34		.83	1.07	
	0205	Pneumatic nailed		1,586	.010		.49	.28		.77	.97	
	0300	3/4" thick		1,200	.013		.58	.36		.94	1.21	
	0305	Pneumatic nailed		1,464	.011		.58	.30		.88	1.11	
	0500	Plywood on walls with exterior CDX, 3/8" thick		1,200	.013		.32	.36		.68	.92	
	0505	Pneumatic nailed		1,488	.011		.32	.29		.61	.81	
	0600	1/2" thick		1,125	.014		.41	.39		.80	1.06	
	0605	Pneumatic nailed		1,395	.011		.41	.31		.72	.94	
	0700	5/8" thick		1,050	.015		.49	.42		.91	1.19	

Important: See the Reference Section for critical supporting data - Reference Nos., Crews, & City Cost Indexes

6
WOOD & PLASTICS

061 150 | Sheathing

		CREW	DAILY OUTPUT	LABOR-HOURS	UNIT	1999 BARE COSTS MAT.	LABOR	EQUIP.	TOTAL	TOTAL INCL O&P	
154	0705	Pneumatic nailed (R061-020)	2 Carp	1,302	.012	S.F.	.49	.34		.83	1.07	154
	0800	3/4" thick		975	.016		.58	.45		1.03	1.34	
	0805	Pneumatic nailed (R061-030)		1,209	.013		.58	.36		.94	1.21	
	1000	For shear wall construction, add						20%				
	1200	For structural 1 exterior plywood, add				S.F.	10%					
	1400	With boards, on roof 1" x 6" boards, laid horizontal	2 Carp	725	.022		.99	.60		1.59	2.04	
	1500	Laid diagonal		650	.025		.99	.67		1.66	2.15	
	1700	1" x 8" boards, laid horizontal		875	.018		1.05	.50		1.55	1.94	
	1800	Laid diagonal		725	.022		1.05	.60		1.65	2.10	
	2000	For steep roofs, add						40%				
	2200	For dormers, hips and valleys, add					5%	50%				
	2400	Boards on walls, 1" x 6" boards, laid regular	2 Carp	650	.025		.99	.67		1.66	2.15	
	2500	Laid diagonal		585	.027		.99	.75		1.74	2.26	
	2700	1" x 8" boards, laid regular		765	.021		1.05	.57		1.62	2.05	
	2800	Laid diagonal		650	.025		1.05	.67		1.72	2.21	
	2850	Gypsum, weatherproof, 1/2" thick		1,125	.014		.21	.39		.60	.84	
	2900	Sealed, 4/10" thick		1,100	.015		.31	.40		.71	.96	
	3000	Wood fiber, regular, no vapor barrier, 1/2" thick		1,200	.013		.38	.36		.74	.99	
	3100	5/8" thick		1,200	.013		.50	.36		.86	1.12	
	3300	No vapor barrier, in colors, 1/2" thick		1,200	.013		.54	.36		.90	1.16	
	3400	5/8" thick		1,200	.013		.67	.36		1.03	1.31	
	3600	With vapor barrier one side, white, 1/2" thick		1,200	.013		.53	.36		.89	1.15	
	3700	Vapor barrier 2 sides, 1/2" thick		1,200	.013		.81	.36		1.17	1.46	
	3800	Asphalt impregnated, 25/32" thick		1,200	.013		.34	.36		.70	.94	
	3850	Intermediate, 1/2" thick		1,200	.013		.28	.36		.64	.88	

061 160 | Subfloor

		CREW	DAILY OUTPUT	LABOR-HOURS	UNIT	1999 BARE COSTS MAT.	LABOR	EQUIP.	TOTAL	TOTAL INCL O&P	
161	0010	**FLOORING, WOOD** See division 095-604										161
164	0010	**SUBFLOOR** Plywood, CDX, 1/2" thick (R061-020)	2 Carp	1,500	.011	SF Flr.	.41	.29		.70	.91	164
	0015	Pneumatic nailed		1,860	.009		.41	.23		.64	.82	
	0100	5/8" thick		1,350	.012		.49	.32		.81	1.05	
	0105	Pneumatic nailed		1,674	.010		.49	.26		.75	.95	
	0200	3/4" thick		1,250	.013		.58	.35		.93	1.19	
	0205	Pneumatic nailed		1,550	.010		.58	.28		.86	1.08	
	0300	1-1/8" thick, 2-4-1 including underlayment		1,050	.015		1.18	.42		1.60	1.95	
	0500	With boards, 1" x 10" S4S, laid regular		1,100	.015		.96	.40		1.36	1.68	
	0600	Laid diagonal		900	.018		.97	.49		1.46	1.83	
	0800	1" x 8" S4S, laid regular		1,000	.016		.86	.44		1.30	1.64	
	0900	Laid diagonal		850	.019		.86	.51		1.37	1.75	
	1100	Wood fiber, T&G, 2' x 8' planks, 1" thick		1,000	.016		1.19	.44		1.63	2	
	1200	1-3/8" thick		900	.018		1.46	.49		1.95	2.37	
168	0010	**UNDERLAYMENT** Plywood, underlayment grade, 3/8" thick (R061-020)	2 Carp	1,500	.011	SF Flr.	.66	.29		.95	1.19	168
	0015	Pneumatic nailed		1,860	.009		.66	.23		.89	1.10	
	0100	1/2" thick		1,450	.011		.75	.30		1.05	1.30	
	0105	Pneumatic nailed		1,798	.009		.75	.24		.99	1.21	
	0200	5/8" thick		1,400	.011		.63	.31		.94	1.18	
	0205	Pneumatic nailed		1,736	.009		.63	.25		.88	1.09	
	0300	3/4" thick		1,300	.012		.86	.34		1.20	1.48	
	0305	Pneumatic nailed		1,612	.010		.86	.27		1.13	1.38	
	0500	Particle board, 3/8" thick		1,500	.011		.38	.29		.67	.88	
	0505	Pneumatic nailed		1,860	.009		.38	.23		.61	.79	
	0600	1/2" thick		1,450	.011		.40	.30		.70	.91	
	0605	Pneumatic nailed		1,798	.009		.40	.24		.64	.82	

WOOD & PLASTICS 6

For expanded coverage of these items see *Means Interior Cost Data 1999*

061 160 | Subfloor

		CREW	DAILY OUTPUT	LABOR-HOURS	UNIT	MAT.	LABOR	EQUIP.	TOTAL	TOTAL INCL O&P	
168	0800	5/8" thick	2 Carp	1,400	.011	SF Flr.	.54	.31		.85	1.08	168
	0805	Pneumatic nailed		1,736	.009		.54	.25		.79	.99	
	0900	3/4" thick		1,300	.012		.60	.34		.94	1.19	
	0905	Pneumatic nailed		1,612	.010		.41	.27		.68	.88	
	1100	Hardboard, underlayment grade, 4' x 4', .215" thick		1,500	.011		.41	.29		.70	.91	

061 200 | Structural Panels

		CREW	DAILY OUTPUT	LABOR-HOURS	UNIT	MAT.	LABOR	EQUIP.	TOTAL	TOTAL INCL O&P	
208	0010	STRESSED SKIN PLYWOOD ROOF PANELS 3/8" group 1 top										208
	0020	skin, 3/8" exterior AD bottom skin										
	0030	1150f stringers, 4' x 8' panels										
	0100	4-1/4" deep	F-3	2,075	.019	SF Roof	3.04	.53	.21	3.78	4.40	
	0200	6-1/8" deep		1,725	.023		3.29	.64	.26	4.19	4.90	
	0300	8-1/8" deep		1,475	.027		3.81	.75	.30	4.86	5.70	
	0500	3/8" top skin, no bottom skin, 5-3/4" deep		1,725	.023		2.88	.64	.26	3.78	4.45	
	0600	7-3/4" deep		1,475	.027		3.29	.75	.30	4.34	5.10	
	0800	For 3-1/2" factory fiberglass insulation, add					.35			.35	.39	
	1000	For 1/2" thick top skin, add					.35			.35	.39	
	1500	Floor panels, substitute 5/8" underlayment as										
	1510	top skin, add to roof panels above				SF Flr.	.41			.41	.45	
	2000	Curved roof panels, 3/8" structural 1 top skin,										
	2010	3/8" exterior AC bottom skin, laminated ribs										
	2200	8' radius, 2-1/4" deep, tie rods not req'd.	F-3	1,150	.035	SF Flr.	5.70	.96	.39	7.05	8.25	
	2400	10' radius, 1-1/2" deep, tie rods are included		950	.042		4.48	1.17	.47	6.12	7.25	
	2600	10' radius, 3-3/8" deep, tie rods not req'd.		1,150	.035		7.80	.96	.39	9.15	10.50	
	2800	12' radius, 2" deep, tie rods are included		950	.042		4.63	1.17	.47	6.27	7.45	
	3000	12' radius, 4-1/2" deep, tie rods not req'd.		1,150	.035		8.10	.96	.39	9.45	10.85	
	4000	Folded plate roofs, structural 1 top skin with intermediate										
	4010	rafters and end chord. Cost of tie rods included										
	4200	Slope 7 in 12, 4' fold, 2" thick, 32' span	F-3	850	.047	SF Flr.	3.71	1.30	.52	5.53	6.70	
	4400	Slope 8-1/2 in 12, 5' fold, 4" thick, 56' span		950	.042		7.40	1.17	.47	9.04	10.50	
	4600	Slope 10 in 12, 8' fold, 4" thick, 52' span		950	.042		4.43	1.17	.47	6.07	7.20	
	4800	Slope 10 in 12, 8' fold, 4" thick, 72' span		1,025	.039		7.80	1.08	.43	9.31	10.70	
	6000	Box beams, structural 1 web										
	6200	24" deep, 2-2" x 4" flanges, 2 webs @ 3/8"	F-3	295	.136	L.F.	8.90	3.76	1.50	14.16	17.30	
	6400	24" deep, 3-2" x 4" flanges, 2 webs @ 1/2"		260	.154		11.05	4.26	1.71	17.02	20.50	
	6600	48" deep, 3-2" x 6" flanges, 2 webs @ 3/4"		140	.286		19.15	7.90	3.17	30.22	37	
	6800	48" deep, 6-2" x 6" flanges, 4 webs @ 3/8",										
	6810	including 2 interior webs	F-3	115	.348	L.F.	39	9.65	3.86	52.51	62.50	
	7000	For exterior AC outer webs, add					1.03			1.03	1.13	
	7200	For medium density overlaid outer webs, add					1.54			1.54	1.69	

061 250 | Wood Decking

		CREW	DAILY OUTPUT	LABOR-HOURS	UNIT	MAT.	LABOR	EQUIP.	TOTAL	TOTAL INCL O&P	
258	0010	ROOF DECKS										258
	0020	For laminated decks, see division 061-808										
	0200	For cementitious decks, see division 035										
	0400	Cedar planks, 3" thick	2 Carp	320	.050	S.F.	5.65	1.36		7.01	8.40	
	0500	4" thick		250	.064		7.65	1.75		9.40	11.15	
	0700	Douglas fir, 3" thick		320	.050		2.12	1.36		3.48	4.48	
	0800	4" thick		250	.064		2.84	1.75		4.59	5.85	
	1000	Hemlock, 3" thick		320	.050		2.12	1.36		3.48	4.48	
	1100	4" thick		250	.064		2.83	1.75		4.58	5.85	
	1300	Western white spruce, 3" thick		320	.050		2.05	1.36		3.41	4.41	
	1400	4" thick		250	.064		2.73	1.75		4.48	5.75	

Important: See the Reference Section for critical supporting data - Reference Nos., Crews, & City Cost Indexes

061 250 | Wood Decking

		CREW	DAILY OUTPUT	LABOR-HOURS	UNIT	1999 BARE COSTS				TOTAL INCL O&P
						MAT.	LABOR	EQUIP.	TOTAL	
259	**0010 DECK, WOOD, PRESSURE TREATED LUMBER**									**259**
0100	Railings and trim , 1″ x 4″	1 Carp	300	.027	L.F.	.50	.73		1.23	1.70
0200	2″ x 4″		300	.027		.50	.73		1.23	1.69
0300	2″ x 6″		300	.027		.88	.73		1.61	2.11
0400	Decking, 1″ x 4″		275	.029	S.F.	1.16	.79		1.95	2.53
0500	2″ x 4″		300	.027		1.49	.73		2.22	2.78
0600	2″ x 6″		320	.025		1.49	.68		2.17	2.71
0650	5/4″ x 6″		320	.025		1.68	.68		2.36	2.91
0700	Redwood decking, 1″ x 4″		275	.029		3.39	.79		4.18	4.98
0800	2″ x 6″		340	.024		5.65	.64		6.29	7.25
0900	5/4″ x 6″		320	.025		4.53	.68		5.21	6.05

061 280 | Mineral Fbr. Cem. Panel

		CREW	DAILY OUTPUT	LABOR-HOURS	UNIT	MAT.	LABOR	EQUIP.	TOTAL	TOTAL INCL O&P
281	**0010 MINERAL FIBER CEMENT PANELS** Including panels, fasteners,									**281**
0100	accessories, trim & sealant									
0130	Architectural, textured finish, 1/8″ thick, minimum	G-3	500	.064	S.F.	5.75	1.70		7.45	9
0140	Maximum		500	.064		7.25	1.70		8.95	10.65
0150	1/4″ thick, minimum		500	.064		6.75	1.70		8.45	10.10
0200	Maximum		500	.064		8.85	1.70		10.55	12.40
0250	3/8″ thick, minimum		500	.064		8.90	1.70		10.60	12.45
0300	Maximum		500	.064		11.75	1.70		13.45	15.60
0350	5/8″ thick, minimum		300	.107		11.75	2.84		14.59	17.40
0400	Maximum		300	.107		17.10	2.84		19.94	23
2000	Flat sheets, 1/8″ thick		1,200	.027		4.05	.71		4.76	5.55
2100	1/4″ thick		1,020	.031		5.80	.83		6.63	7.70
2200	3/8″ thick		885	.036		7.30	.96		8.26	9.55
2300	5/8″ thick		795	.040		9.95	1.07		11.02	12.60
3000	Glasweld, mineral enamel coating, 1/8″ thick		600	.053		3.50	1.42		4.92	6.05
3100	1/4″ thick		322	.099		4.45	2.64		7.09	9.05
4000	Sandwich panel, Glasweld face and back									
4100	1″ thick, perlite core	G-3	322	.099	S.F.	7.25	2.64		9.89	12.15
4200	Polyurethane core		322	.099		7.65	2.64		10.29	12.55
4500	2″ thick, perlite core		322	.099		7.95	2.64		10.59	12.90
4600	Polyurethane core		322	.099		8.35	2.64		10.99	13.35

061 300 | Heavy Timber Constr.

		CREW	DAILY OUTPUT	LABOR-HOURS	UNIT	MAT.	LABOR	EQUIP.	TOTAL	TOTAL INCL O&P
304	**0010 FRAMING, HEAVY** Mill timber, beams, single 6″ x 10″	2 Carp	1.10	14.545	M.B.F.	1,425	395		1,820	2,200	**304**
0100	Single 8″ x 16″		1.20	13.333		1,625	365		1,990	2,375
0200	Built from 2″ lumber, multiple 2″ x 14″		.90	17.778		820	485		1,305	1,675
0210	Built from 3″ lumber, multiple 3″ x 6″		.70	22.857		1,000	625		1,625	2,100
0220	Multiple 3″ x 8″		.80	20		1,000	545		1,545	1,975
0230	Multiple 3″ x 10″		.90	17.778		1,000	485		1,485	1,900
0240	Multiple 3″ x 12″		1	16		1,000	435		1,435	1,800
0250	Built from 4″ lumber, multiple 4″ x 6″		.80	20		1,150	545		1,695	2,125
0260	Multiple 4″ x 8″		.90	17.778		1,150	485		1,635	2,050
0270	Multiple 4″ x 10″		1	16		1,150	435		1,585	1,950
0280	Multiple 4″ x 12″		1.10	14.545		1,150	395		1,545	1,900
0290	Columns, structural grade, 1500f, 4″ x 4″		.60	26.667		1,250	730		1,980	2,525
0300	6″ x 6″		.65	24.615		1,750	670		2,420	2,975
0400	8″ x 8″		.70	22.857		1,875	625		2,500	3,025
0500	10″ x 10″		.75	21.333		1,750	580		2,330	2,850
0600	12″ x 12″		.80	20		1,750	545		2,295	2,775
0800	Floor planks, 2″ thick, T & G, 2″ x 6″		1.05	15.238		700	415		1,115	1,425
0900	2″ x 10″		1.10	14.545		760	395		1,155	1,450

WOOD & PLASTICS **6**

For expanded coverage of these items see *Means Interior Cost Data 1999*

061 300 | Heavy Timber Constr.

			DAILY OUTPUT	LABOR-HOURS	UNIT	1999 BARE COSTS				TOTAL INCL O&P	
						MAT.	LABOR	EQUIP.	TOTAL		
304	1100	3" thick, 3" x 6"	2 Carp	1.05	15.238	M.B.F.	1,150	415		1,565	1,925	304
	1200	3" x 10"		1.10	14.545		1,150	395		1,545	1,900	
	1400	Girders, structural grade, 12" x 12"		.80	20		1,650	545		2,195	2,675	
	1500	10" x 16"	↓	1	16	↓	1,600	435		2,035	2,450	
	2050	Roof planks, see division 061-258										
	2300	Roof purlins, 4" thick, structural grade	2 Carp	1.05	15.238	M.B.F.	1,125	415		1,540	1,900	
	2500	Roof trusses, add timber connectors, division 060-512	"	.45	35.556	"	1,100	970		2,070	2,725	

061 500 | Wood-Metal Systems

			CREW	DAILY OUTPUT	LABOR-HOURS	UNIT	MAT.	LABOR	EQUIP.	TOTAL	INCL O&P	
508	0010	**STRUCTURAL JOISTS** Fabricated "I" joists with wood flanges,										508
	0100	Plywood webs, incl. bridging & blocking, panels 24" O.C.										
	1200	15' to 24' span, 50 psf live load	F-5	2,400	.013	SF Flr.	1.44	.37		1.81	2.16	
	1300	55 psf live load		2,250	.014		1.55	.40		1.95	2.32	
	1400	24' to 30' span, 45 psf live load		2,600	.012		1.80	.34		2.14	2.52	
	1500	55 psf live load	↓	2,400	.013		1.80	.37		2.17	2.56	
	1600	Tubular steel open webs, 45 psf, 24" O.C., 40' span	F-3	6,250	.006		1.80	.18	.07	2.05	2.34	
	1700	55' span		7,750	.005		1.75	.14	.06	1.95	2.21	
	1800	70' span		9,250	.004		2.27	.12	.05	2.44	2.74	
	1900	85 psf live load, 26' span	↓	2,300	.017	↓	2.11	.48	.19	2.78	3.28	

061 800 | Glued-Laminated Const

			CREW	DAILY OUTPUT	LABOR-HOURS	UNIT	MAT.	LABOR	EQUIP.	TOTAL	INCL O&P	
804	0010	**LAMINATED FRAMING** Not including decking										804
	0020	30 lb., short term live load, 15 lb. dead load										
	0200	Straight roof beams, 20' clear span, beams 8' O.C.	F-3	2,560	.016	SF Flr.	1.38	.43	.17	1.98	2.39	
	0300	Beams 16' O.C.		3,200	.013		1.01	.35	.14	1.50	1.80	
	0500	40' clear span, beams 8' O.C.		3,200	.013		2.66	.35	.14	3.15	3.62	
	0600	Beams 16' O.C.	↓	3,840	.010		2.17	.29	.12	2.58	2.97	
	0800	60' clear span, beams 8' O.C.	F-4	2,880	.017		4.56	.45	.28	5.29	6	
	0900	Beams 16' O.C.	"	3,840	.013		3.40	.34	.21	3.95	4.50	
	1100	Tudor arches, 30' to 40' clear span, frames 8' O.C.	F-3	1,680	.024		5.95	.66	.26	6.87	7.85	
	1200	Frames 16' O.C.	"	2,240	.018		4.66	.49	.20	5.35	6.15	
	1400	50' to 60' clear span, frames 8' O.C.	F-4	2,200	.022		6.40	.59	.36	7.35	8.35	
	1500	Frames 16' O.C.		2,640	.018		5.45	.49	.30	6.24	7.10	
	1700	Radial arches, 60' clear span, frames 8' O.C.		1,920	.025		6	.68	.41	7.09	8.10	
	1800	Frames 16' O.C.		2,880	.017		4.61	.45	.28	5.34	6.05	
	2000	100' clear span, frames 8' O.C.		1,600	.030		6.20	.81	.50	7.51	8.65	
	2100	Frames 16' O.C.		2,400	.020		5.45	.54	.33	6.32	7.20	
	2300	120' clear span, frames 8' O.C.		1,440	.033		8.25	.90	.55	9.70	11.10	
	2400	Frames 16' O.C.	↓	1,920	.025		7.55	.68	.41	8.64	9.80	
	2600	Bowstring trusses, 20' O.C., 40' clear span	F-3	2,400	.017		3.72	.46	.18	4.36	5	
	2700	60' clear span	F-4	3,600	.013		3.34	.36	.22	3.92	4.47	
	2800	100' clear span		4,000	.012		4.73	.33	.20	5.26	5.95	
	2900	120' clear span	↓	3,600	.013		5.10	.36	.22	5.68	6.40	
	3000	For less than 1000 B.F., add					20%					
	3050	For over 5000 B.F., deduct					10%					
	3100	For premium appearance, add to S.F. prices					5%					
	3300	For industrial type, deduct					15%					
	3500	For stain and varnish, add					5%					
	3900	For 3/4" laminations, add to straight					25%					
	4100	Add to curved				↓	15%					
	4300	Alternate pricing method: (use nominal footage of										
	4310	components). Straight beams, camber less than 6"	F-3	3.50	11.429	M.B.F.	2,050	315	127	2,492	2,900	
	4400	Columns, including hardware		2	20		2,225	555	222	3,002	3,525	
	4600	Curved members, radius over 32'		2.50	16		2,275	445	177	2,897	3,400	
	4700	Radius 10' to 32'	↓	3	13.333	↓	2,250	370	148	2,768	3,225	

Important: See the Reference Section for critical supporting data - Reference Nos., Crews, & City Cost Indexes

061 | Rough Carpentry

061 800 | Glued-Laminated Const

			CREW	DAILY OUTPUT	LABOR-HOURS	UNIT	MAT.	LABOR	EQUIP.	TOTAL	TOTAL INCL O&P	
804	4900	For complicated shapes, add maximum				M.B.F.	100%					804
	5100	For pressure treating, add to straight					35%					
	5200	Add to curved				▼	45%					
	6000	Laminated veneer members, southern pine or western species										
	6050	1-3/4" wide x 5-1/2" deep	2 Carp	480	.033	L.F.	3.51	.91		4.42	5.30	
	6100	9-1/2" deep		480	.033		3.68	.91		4.59	5.50	
	6150	14" deep		450	.036		5.25	.97		6.22	7.35	
	6200	18" deep	▼	450	.036	▼	6.80	.97		7.77	9.05	
	6300	Parallel strand members, southern pine or western species										
	6350	1-3/4" wide x 9-1/4" deep	2 Carp	480	.033	L.F.	4.21	.91		5.12	6.05	
	6400	11-1/4" deep		450	.036		5.15	.97		6.12	7.25	
	6450	14" deep		400	.040		6.15	1.09		7.24	8.45	
	6500	3-1/2" wide x 9-1/4" deep		480	.033		8.20	.91		9.11	10.45	
	6550	11-1/4" deep		450	.036		10.10	.97		11.07	12.65	
	6600	14" deep		400	.040		12.05	1.09		13.14	14.95	
	6650	7" wide x 9-1/4" deep		450	.036		16.15	.97		17.12	19.30	
	6700	11-1/4" deep		420	.038		19.95	1.04		20.99	23.50	
	6750	14" deep	▼	400	.040	▼	24	1.09		25.09	27.50	
808	0010	**LAMINATED ROOF DECK** Pine or hemlock, 3" thick	2 Carp	425	.038	S.F.	2.84	1.03		3.87	4.74	808
	0100	4" thick		325	.049		3.25	1.34		4.59	5.70	
	0300	Cedar, 3" thick		425	.038		3.26	1.03		4.29	5.20	
	0400	4" thick		325	.049		3.89	1.34		5.23	6.40	
	0600	Fir, 3" thick		425	.038		2.94	1.03		3.97	4.85	
	0700	4" thick	▼	325	.049	▼	3.47	1.34		4.81	5.95	

061 900 | Wood Trusses

			CREW	DAILY OUTPUT	LABOR-HOURS	UNIT	MAT.	LABOR	EQUIP.	TOTAL	TOTAL INCL O&P	
908	0010	**ROOF TRUSSES**										908
	0020	For timber connectors, see div. 060-512	R061 -100									
	0100	Fink (W) or King post type, 2'-0" O.C.										
	0200	Metal plate connected, 4 in 12 slope										
	0210	24' to 29' span	F-3	3,000	.013	SF Flr.	1.30	.37	.15	1.82	2.17	
	0300	30' to 43' span		3,000	.013		1.45	.37	.15	1.97	2.34	
	0400	44' to 60' span	▼	3,000	.013		1.60	.37	.15	2.12	2.50	
	0600	For change in roof pitch, subtract					.05			.05	.06	
	0700	Glued and nailed, add	▼			▼	50%					

062 | Finish Carpentry

062 200 | Millwork Moldings

			CREW	DAILY OUTPUT	LABOR-HOURS	UNIT	MAT.	LABOR	EQUIP.	TOTAL	TOTAL INCL O&P	
208	0010	**MOLDINGS, BASE**										208
	0500	Base, stock pine, 9/16" x 3-1/2"	1 Carp	240	.033	L.F.	1.16	.91		2.07	2.71	
	0550	9/16" x 4-1/2"		200	.040		1.25	1.09		2.34	3.10	
	0561	Base shoe, oak, 3/4" x 1"	▼	240	.033	▼	.76	.91		1.67	2.27	
212	0010	**MOLDINGS, CASINGS**										212
	0090	Apron, stock pine, 5/8" x 2"	1 Carp	250	.032	L.F.	.95	.87		1.82	2.42	
	0110	5/8" x 3-1/2"		220	.036		1.40	.99		2.39	3.10	
	0300	Band, stock pine, 11/16" x 1-1/8"		270	.030		.38	.81		1.19	1.69	
	0350	11/16" x 1-3/4"		250	.032		.60	.87		1.47	2.03	
	0700	Casing, stock pine, 11/16" x 2-1/2"	▼	240	.033	▼	.76	.91		1.67	2.27	

WOOD & PLASTICS | **6**

For expanded coverage of these items see _Means Interior Cost Data 1999_

062 200 | Millwork Moldings

			CREW	DAILY OUTPUT	LABOR-HOURS	UNIT	1999 BARE COSTS				TOTAL INCL O&P	
							MAT.	LABOR	EQUIP.	TOTAL		
212	0750	11/16" x 3-1/2"	1 Carp	215	.037	L.F.	1.46	1.02		2.48	3.21	212
216	0010	**MOLDINGS, CEILINGS**										216
	0600	Bed, stock pine, 9/16" x 1-3/4"	1 Carp	270	.030	L.F.	.60	.81		1.41	1.93	
	0650	9/16" x 2"		240	.033		.72	.91		1.63	2.22	
	1200	Cornice molding, stock pine, 9/16" x 1-3/4"		330	.024		.62	.66		1.28	1.72	
	1300	9/16" x 2-1/4"		300	.027		.83	.73		1.56	2.06	
	2400	Cove scotia, stock pine, 9/16" x 1-3/4"		270	.030		.54	.81		1.35	1.86	
	2500	11/16" x 2-3/4"		255	.031		1.15	.86		2.01	2.62	
	2600	Crown, stock pine, 9/16" x 3-5/8"		250	.032		1.51	.87		2.38	3.03	
	2700	11/16" x 4-5/8"	▼	220	.036	▼	2.17	.99		3.16	3.95	
220	0010	**MOLDINGS, EXTERIOR**										220
	1500	Cornice, boards, pine, 1" x 2"	1 Carp	330	.024	L.F.	.27	.66		.93	1.33	
	1700	1" x 6"		250	.032		.83	.87		1.70	2.28	
	2000	1" x 12"		180	.044		1.27	1.21		2.48	3.31	
	2200	Three piece, built-up, pine, minimum		80	.100		1.47	2.73		4.20	5.90	
	2300	Maximum		65	.123		4.62	3.36		7.98	10.40	
	3000	Corner board, sterling pine, 1" x 4"		200	.040		.46	1.09		1.55	2.23	
	3100	1" x 6"		200	.040		.70	1.09		1.79	2.49	
	3350	Fascia, sterling pine, 1" x 6"		250	.032		.70	.87		1.57	2.14	
	3370	1" x 8"		225	.036		.92	.97		1.89	2.54	
	3400	Trim, exterior, sterling pine, back band		250	.032		.55	.87		1.42	1.98	
	3500	Casing		250	.032		.58	.87		1.45	2.01	
	3600	Crown		250	.032		1.18	.87		2.05	2.67	
	3700	Porch rail with balusters		22	.364		8.70	9.95		18.65	25	
	3800	Screen		395	.020		.31	.55		.86	1.21	
	4100	Verge board, sterling pine, 1" x 4"		200	.040		.32	1.09		1.41	2.07	
	4200	1" x 6"		200	.040		.70	1.09		1.79	2.49	
	4300	2" x 6"		165	.048		1.40	1.32		2.72	3.62	
	4400	2" x 8"	▼	165	.048	▼	1.86	1.32		3.18	4.13	
	4700	For redwood trim, add					200%					
224	0010	**MOLDINGS, TRIM**										224
	0200	Astragal, stock pine, 11/16" x 1-3/4"	1 Carp	255	.031	L.F.	.78	.86		1.64	2.21	
	0250	1-5/16" x 2-3/16"		240	.033		2.60	.91		3.51	4.28	
	0800	Chair rail, stock pine, 5/8" x 2-1/2"		270	.030		.84	.81		1.65	2.20	
	0900	5/8" x 3-1/2"		240	.033		1.26	.91		2.17	2.82	
	1000	Closet pole, stock pine, 1-1/8" diameter		200	.040		.80	1.09		1.89	2.60	
	1100	Fir, 1-5/8" diameter		200	.040		1.19	1.09		2.28	3.03	
	3300	Half round, stock pine, 1/4" x 1/2"		270	.030		.19	.81		1	1.48	
	3350	1/2" x 1"	▼	255	.031	▼	.34	.86		1.20	1.72	
	3400	Handrail, fir, single piece, stock, hardware not included										
	3450	1-1/2" x 1-3/4"	1 Carp	80	.100	L.F.	1.21	2.73		3.94	5.65	
	3470	Pine, 1-1/2" x 1-3/4"		80	.100		1.05	2.73		3.78	5.45	
	3500	1-1/2" x 2-1/2"		76	.105		1.43	2.87		4.30	6.10	
	3600	Lattice, stock pine, 1/4" x 1-1/8"		270	.030		.32	.81		1.13	1.62	
	3700	1/4" x 1-3/4"		250	.032		.30	.87		1.17	1.70	
	3800	Miscellaneous, custom, pine, 1" x 1"		270	.030		.96	.81		1.77	2.33	
	3900	1" x 3"		240	.033		.77	.91		1.68	2.28	
	4100	Birch or oak, nominal 1" x 1"		240	.033		.46	.91		1.37	1.94	
	4200	Nominal 1" x 3"		215	.037		1.57	1.02		2.59	3.33	
	4400	Walnut, nominal 1" x 1"		215	.037		.70	1.02		1.72	2.37	
	4500	Nominal 1" x 3"		200	.040		2.10	1.09		3.19	4.03	
	4700	Teak, nominal 1" x 1"		215	.037		1.05	1.02		2.07	2.75	
	4800	Nominal 1" x 3"		200	.040		3	1.09		4.09	5	
	4900	Quarter round, stock pine, 1/4" x 1/4"	▼	275	.029	▼	.20	.79		.99	1.47	

Important: See the Reference Section for critical supporting data - Reference Nos., Crews, & City Cost Indexes

6 WOOD & PLASTICS

062 | Finish Carpentry

062 200 | Millwork Moldings

			CREW	DAILY OUTPUT	LABOR-HOURS	UNIT	MAT.	LABOR	EQUIP.	TOTAL	TOTAL INCL O&P	
224	4950	3/4" x 3/4"	1 Carp	255	.031	L.F.	.41	.86		1.27	1.80	224
	5600	Wainscot moldings, 1-1/8" x 9/16", 2' high, minimum		76	.105	S.F.	5.60	2.87		8.47	10.65	
	5700	Maximum		65	.123	"	12.35	3.36		15.71	18.90	
228	0010	**MOLDINGS, WINDOW AND DOOR**										228
	2800	Door moldings, stock, decorative, 1-1/8" wide, plain	1 Carp	17	.471	Set	27	12.85		39.85	49.50	
	2900	Detailed		17	.471	"	65	12.85		77.85	91.50	
	3150	Door trim set, 1 head and 2 sides, pine, 2-1/2 wide		5.90	1.356	Opng.	12.45	37		49.45	71.50	
	3170	4-1/2" wide		5.30	1.509	"	22.50	41		63.50	90	
	3200	Glass beads, stock pine, 1/4" x 11/16"		285	.028	L.F.	.29	.77		1.06	1.53	
	3250	3/8" x 1/2"		275	.029		.35	.79		1.14	1.64	
	3270	3/8" x 7/8"		270	.030		.39	.81		1.20	1.70	
	4850	Parting bead, stock pine, 3/8" x 3/4"		275	.029		.28	.79		1.07	1.56	
	4870	1/2" x 3/4"		255	.031		.35	.86		1.21	1.74	
	5000	Stool caps, stock pine, 11/16" x 3-1/2"		200	.040		1.29	1.09		2.38	3.14	
	5100	1-1/16" x 3-1/4"		150	.053		1.93	1.46		3.39	4.41	
	5300	Threshold, oak, 3' long, inside, 5/8" x 3-5/8"		32	.250	Ea.	6.10	6.80		12.90	17.45	
	5400	Outside, 1-1/2" x 7-5/8"		16	.500	"	24.50	13.65		38.15	48.50	
	5900	Window trim sets, including casings, header, stops,										
	5910	stool and apron, 2-1/2" wide, minimum	1 Carp	13	.615	Opng.	15	16.80		31.80	43	
	5950	Average		10	.800		26	22		48	63	
	6000	Maximum		6	1.333		37	36.50		73.50	98	

062 300 | Shelving

			CREW	DAILY OUTPUT	LABOR-HOURS	UNIT	MAT.	LABOR	EQUIP.	TOTAL	TOTAL INCL O&P	
304	0010	**SHELVING** Pine, clear grade, no edge band, 1" x 8"	1 Carp	115	.070	L.F.	1.71	1.90		3.61	4.87	304
	0100	1" x 10"		110	.073		2.31	1.99		4.30	5.65	
	0200	1" x 12"		105	.076		4.18	2.08		6.26	7.85	
	0400	For lumber edge band, by hand, add					1.49			1.49	1.64	
	0420	By machine, add					.95			.95	1.05	
	0600	Plywood, 3/4" thick with lumber edge, 12" wide	1 Carp	75	.107		1.24	2.91		4.15	5.95	
	0700	24" wide		70	.114		2.35	3.12		5.47	7.50	
	0900	Bookcase, clear grade pine, shelves 12" O.C., 8" deep		70	.114	S.F.	4.24	3.12		7.36	9.55	
	1000	12" deep shelves		65	.123	"	4.57	3.36		7.93	10.30	
	1200	Adjustable closet rod and shelf, 12" wide, 3' long		20	.400	Ea.	40.50	10.90		51.40	61.50	
	1300	8' long		15	.533	"	57	14.55		71.55	85.50	
	1500	Prefinished shelves with supports, stock, 8" wide		75	.107	L.F.	3.65	2.91		6.56	8.60	
	1600	10" wide		70	.114	"	4.07	3.12		7.19	9.40	
	1800	Custom, high quality dadoed pine shelving units, minimum				S.F.					28	
	1900	Maximum				"					40	

062 400 | Plastic Laminate

			CREW	DAILY OUTPUT	LABOR-HOURS	UNIT	MAT.	LABOR	EQUIP.	TOTAL	TOTAL INCL O&P	
404	0010	**CONVECTOR COVERS** Laminated plastic on 3/4"										404
	0020	thick particle board, 12" wide, minimum	1 Carp	16	.500	L.F.	20	13.65		33.65	43.50	
	0050	Average		13	.615		25.50	16.80		42.30	54.50	
	0100	Maximum		13	.615		35	16.80		51.80	65	
	0300	Add to above for grille, minimum		150	.053	S.F.	1.60	1.46		3.06	4.05	
	0400	Maximum		75	.107	"	4.50	2.91		7.41	9.55	
408	0010	**COUNTER TOP** Stock, plastic lam., 24" wide w/backsplash, min.	1 Carp	30	.267	L.F.	4.89	7.30		12.19	16.85	408
	0100	Maximum		25	.320		14.25	8.75		23	29.50	
	0300	Custom plastic, 7/8" thick, aluminum molding, no splash		30	.267		15.70	7.30		23	29	
	0400	Cove splash		30	.267		20.50	7.30		27.80	34	
	0600	1-1/4" thick, no splash		28	.286		18.50	7.80		26.30	33	
	0700	Square splash		28	.286		23	7.80		30.80	38	
	0900	Square edge, plastic face, 7/8" thick, no splash		30	.267		19.80	7.30		27.10	33.50	
	1000	With splash		30	.267		26	7.30		33.30	40	

WOOD & PLASTICS 6

062 400 | Plastic Laminate

		CREW	DAILY OUTPUT	LABOR-HOURS	UNIT	1999 BARE COSTS				TOTAL INCL O&P	
						MAT.	LABOR	EQUIP.	TOTAL		
408	1200	For stainless channel edge, 7/8" thick, add				L.F.	2.21			2.21	2.43	408
	1300	1-1/4" thick, add					2.58			2.58	2.84	
	1500	For solid color suede finish, add				↓	1.96			1.96	2.16	
	1700	For end splash, add				Ea.	12.35			12.35	13.60	
	1900	For cut outs, standard, add, minimum	1 Carp	32	.250	↓	2.63	6.80		9.43	13.65	
	2000	Maximum		8	1	↓	3.14	27.50		30.64	46.50	
	2100	Postformed, including backsplash and front edge		30	.267	L.F.	8.55	7.30		15.85	21	
	2110	Mitred, add		12	.667	Ea.		18.20		18.20	28.50	
	2200	Built-in place, 25" wide, plastic laminate		25	.320	L.F.	10.80	8.75		19.55	25.50	
	2300	Ceramic tile mosaic	↓	25	.320		24.50	8.75		33.25	41	
	2500	Marble, stock, with splash, 1/2" thick, minimum	1 Bric	17	.471		30	13		43	53	
	2700	3/4" thick, maximum	"	13	.615		76	17		93	111	
	2900	Maple, solid, laminated, 1-1/2" thick, no splash	1 Carp	28	.286		31	7.80		38.80	46.50	
	3000	With square splash		28	.286	↓	35	7.80		42.80	51	
	3200	Stainless steel		24	.333	S.F.	73	9.10		82.10	95	
	3400	Recessed cutting block with trim, 16" x 20" x 1"		8	1	Ea.	41	27.50		68.50	88.50	
	3600	Table tops, plastic laminate, square edge, 7/8" thick		45	.178	S.F.	7	4.85		11.85	15.35	
	3700	1-1/8" thick	↓	40	.200	"	7.20	5.45		12.65	16.55	

062 500 | Prefin. Wood Paneling

		CREW	DAILY OUTPUT	LABOR-HOURS	UNIT	MAT.	LABOR	EQUIP.	TOTAL	TOTAL INCL O&P	
504	0010	**PANELING, PLYWOOD**										504
	2400	Plywood, prefinished, 1/4" thick, 4' x 8' sheets [R061-020]										
	2410	with vertical grooves. Birch faced, minimum	2 Carp	500	.032	S.F.	.72	.87		1.59	2.16	
	2420	Average		420	.038		1.03	1.04		2.07	2.77	
	2430	Maximum		350	.046		1.49	1.25		2.74	3.60	
	2600	Mahogany, African		400	.040		1.91	1.09		3	3.82	
	2700	Philippine (Lauan)		500	.032		.82	.87		1.69	2.27	
	2900	Oak or Cherry, minimum		500	.032		1.60	.87		2.47	3.13	
	3000	Maximum		400	.040		2.68	1.09		3.77	4.67	
	3200	Rosewood		320	.050		3.81	1.36		5.17	6.35	
	3400	Teak		400	.040		2.68	1.09		3.77	4.67	
	3600	Chestnut		375	.043		3.97	1.16		5.13	6.20	
	3800	Pecan		400	.040		1.70	1.09		2.79	3.59	
	3900	Walnut, minimum		500	.032		2.27	.87		3.14	3.87	
	3950	Maximum		400	.040		4.33	1.09		5.42	6.50	
	4000	Plywood, prefinished, 3/4" thick, stock grades, minimum		320	.050		1.03	1.36		2.39	3.28	
	4100	Maximum		224	.071		4.48	1.95		6.43	8	
	4300	Architectural grade, minimum		224	.071		3.30	1.95		5.25	6.70	
	4400	Maximum		160	.100		5.05	2.73		7.78	9.85	
	4600	Plywood, "A" face, birch, V.C., 1/2" thick, natural		450	.036		1.55	.97		2.52	3.23	
	4700	Select		450	.036		1.70	.97		2.67	3.40	
	4900	Veneer core, 3/4" thick, natural		320	.050		1.65	1.36		3.01	3.97	
	5000	Select		320	.050		1.85	1.36		3.21	4.19	
	5200	Lumber core, 3/4" thick, natural		320	.050		2.47	1.36		3.83	4.87	
	5500	Plywood, knotty pine, 1/4" thick, A2 grade		450	.036		1.34	.97		2.31	3	
	5600	A3 grade		450	.036		1.70	.97		2.67	3.40	
	5800	3/4" thick, veneer core, A2 grade		320	.050		1.75	1.36		3.11	4.08	
	5900	A3 grade		320	.050		1.96	1.36		3.32	4.31	
	6100	Aromatic cedar, 1/4" thick, plywood		400	.040		1.70	1.09		2.79	3.59	
	6200	1/4" thick, particle board	↓	400	.040	↓	.82	1.09		1.91	2.62	

062 550 | Prefin. Hardboard Panel

		CREW	DAILY OUTPUT	LABOR-HOURS	UNIT	MAT.	LABOR	EQUIP.	TOTAL	TOTAL INCL O&P	
554	0010	**PANELING, HARDBOARD**										554
	0050	Not incl. furring or trim, hardboard, tempered, 1/8" thick	2 Carp	500	.032	S.F.	.33	.87		1.20	1.73	
	0100	1/4" thick		500	.032		.37	.87		1.24	1.78	
	0300	Tempered pegboard, 1/8" thick	↓	500	.032	↓	.33	.87		1.20	1.73	

Important: See the Reference Section for critical supporting data - Reference Nos., Crews, & City Cost Indexes

6 WOOD & PLASTICS

062 550		Prefin. Hardboard Panel	CREW	DAILY OUTPUT	LABOR-HOURS	UNIT	1999 BARE COSTS				TOTAL INCL O&P	
							MAT.	LABOR	EQUIP.	TOTAL		
554	0400	1/4" thick	2 Carp	500	.032	S.F.	.41	.87		1.28	1.82	554
	0600	Untempered hardboard, natural finish, 1/8" thick		500	.032		.28	.87		1.15	1.68	
	0700	1/4" thick		500	.032		.31	.87		1.18	1.71	
	0900	Untempered pegboard, 1/8" thick		500	.032		.32	.87		1.19	1.72	
	1000	1/4" thick		500	.032		.35	.87		1.22	1.76	
	1200	Plastic faced hardboard, 1/8" thick		500	.032		.52	.87		1.39	1.94	
	1300	1/4" thick		500	.032		.71	.87		1.58	2.15	
	1500	Plastic faced pegboard, 1/8" thick		500	.032		.50	.87		1.37	1.92	
	1600	1/4" thick		500	.032		.62	.87		1.49	2.05	
	1800	Wood grained, plain or grooved, 1/4" thick, minimum		500	.032		.47	.87		1.34	1.89	
	1900	Maximum		425	.038	↓	.89	1.03		1.92	2.60	
	2100	Moldings for hardboard, wood or aluminum, minimum		500	.032	L.F.	.32	.87		1.19	1.72	
	2200	Maximum	↓	425	.038	"	.90	1.03		1.93	2.61	

062 600		Board Paneling										
604	0010	**PANELING, BOARDS**										604
	6400	Wood board paneling, 3/4" thick, knotty pine	2 Carp	300	.053	S.F.	1.24	1.46		2.70	3.65	
	6500	Rough sawn cedar		300	.053		1.60	1.46		3.06	4.05	
	6700	Redwood, clear, 1" x 4" boards		300	.053		3.75	1.46		5.21	6.40	
	6900	Aromatic cedar, closet lining, boards	↓	275	.058	↓	2.83	1.59		4.42	5.60	

062 700		Misc. Finish Carpentry										
704	0010	**BEAMS, DECORATIVE** Rough sawn cedar, non-load bearing, 4" x 4"	2 Carp	180	.089	L.F.	1.30	2.43		3.73	5.25	704
	0100	4" x 6"		170	.094		2.50	2.57		5.07	6.80	
	0200	4" x 8"		160	.100		3.21	2.73		5.94	7.85	
	0300	4" x 10"		150	.107		4.46	2.91		7.37	9.50	
	0400	4" x 12"		140	.114		5.40	3.12		8.52	10.85	
	0500	8" x 8"	↓	130	.123	↓	7.55	3.36		10.91	13.60	
	1100	Beam connector plates see div. 060-512										
720	0010	**FIREPLACE MANTEL BEAMS** Rough texture wood, 4" x 8"	1 Carp	36	.222	L.F.	4.24	6.05		10.29	14.20	720
	0100	4" x 10"		35	.229	"	5.30	6.25		11.55	15.65	
	0300	Laminated hardwood, 2-1/4" x 10-1/2" wide, 6' long		5	1.600	Ea.	95.50	43.50		139	174	
	0400	8' long		5	1.600	"	133	43.50		176.50	215	
	0600	Brackets for above, rough sawn		12	.667	Pr.	8.75	18.20		26.95	38	
	0700	Laminated	↓	12	.667	"	13.25	18.20		31.45	43	
725	0010	**FIREPLACE MANTELS** 6" molding, 6' x 3'-6" opening, minimum	1 Carp	5	1.600	Opng.	127	43.50		170.50	209	725
	0100	Maximum		5	1.600		154	43.50		197.50	238	
	0300	Prefabricated pine, colonial type, stock, deluxe		2	4		745	109		854	990	
	0400	Economy	↓	3	2.667	↓	251	73		324	390	
730	0010	**GRILLES** and panels, hardwood, sanded										730
	0020	2' x 4' to 4' x 8', custom designs, unfinished, minimum	1 Carp	38	.211	S.F.	12	5.75		17.75	22.50	
	0050	Average		30	.267		26	7.30		33.30	40	
	0100	Maximum		19	.421		40	11.50		51.50	62	
	0300	As above, but prefinished, minimum		38	.211		12	5.75		17.75	22.50	
	0400	Maximum	↓	19	.421		45	11.50		56.50	67.50	
735	0010	**HARDWARE** Finish, see divisions 064-108 & 087										735
	0100	Rough, see division 050										
740	0010	**LOUVERS** Redwood, 2'-0" diameter, full circle	1 Carp	16	.500	Ea.	105	13.65		118.65	138	740
	0100	Half circle		16	.500		104	13.65		117.65	136	
	0200	Octagonal		16	.500		86	13.65		99.65	116	
	0300	Triangular, 5/12 pitch, 5'-0" at base	↓	16	.500	↓	180	13.65		193.65	220	
755	0010	**PARTITIONS** See divisions 092-620, 092-622 and 106-052 to 552										755

WOOD & PLASTICS

6

062 700 | Misc. Finish Carpentry

		CREW	DAILY OUTPUT	LABOR-HOURS	UNIT	1999 BARE COSTS				TOTAL INCL O&P	
						MAT.	LABOR	EQUIP.	TOTAL		
760	0010	**SHUTTERS, EXTERIOR** Aluminum, louvered, 1'-4" wide, 3'-0" long	1 Carp	10	.800	Pr.	29.50	22		51.50	67	**760**
	0400	6'-8" long		9	.889		48	24.50		72.50	90.50	
	1000	Pine, louvered, primed, each 1'-2" wide, 3'-3" long		10	.800	▼	41	22		63	79.50	
	1001	Pine, louvered, primed, each 1'-2" wide, 3'-3" long		20	.400	Ea.	24	10.90		34.90	43.50	
	1100	4'-7" long		10	.800	Pr.	49	22		71	88.50	
	1101	4'-7" long		20	.400	Ea.	34.50	10.90		45.40	55	
	1500	Each 1'-6" wide, 3'-3" long		10	.800	Pr.	45	22		67	84	
	1600	4'-7" long		10	.800	"	61	22		83	102	
	1601	4'-7" long		20	.400	Ea.	38	10.90		48.90	59	
	1620	Hemlock, louvered, 1'-2" wide, 5'-7" long		10	.800	Pr.	59.50	22		81.50	100	
	1630	Each 1'-4" wide, 2'-2" long		10	.800		37	22		59	75.50	
	1670	4'-3" long		10	.800		49	22		71	88	
	1690	5'-11" long		10	.800		60.50	22		82.50	101	
	1700	Door blinds, 6'-9" long, each 1'-3" wide		9	.889		70	24.50		94.50	115	
	1710	1'-6" wide		9	.889		88	24.50		112.50	135	
	1720	Hemlock, solid raised panel, each 1'-4" wide, 3'-3" long		10	.800		60.50	22		82.50	101	
	1740	4'-3" long		10	.800		76.50	22		98.50	119	
	1770	5'-11" long		10	.800		106	22		128	152	
	1800	Door blinds, 6'-9" long, each 1'-3" wide		9	.889		117	24.50		141.50	166	
	1900	1'-6" wide		9	.889		122	24.50		146.50	172	
	2500	Polystyrene, solid raised panel, each 1'-4" wide, 3'-3" long		10	.800		51	22		73	90.50	
	2700	4'-7" long		10	.800		61.50	22		83.50	102	
	4500	Polystyrene, louvered, each 1'-2" wide, 3'-3" long		10	.800		39	22		61	77.50	
	4750	5'-3" long		10	.800		52	22		74	91.50	
	6000	Vinyl, louvered, each 1'-2" x 4'-7" long		10	.800		51	22		73	90.50	
	6200	Each 1'-4" x 6'-8" long	▼	9	.889	▼	77.50	24.50		102	123	
775	0010	**SOFFITS** Wood fiber, no vapor barrier, 15/32" thick	2 Carp	525	.030	S.F.	.73	.83		1.56	2.11	**775**
	0100	5/8" thick		525	.030		.79	.83		1.62	2.18	
	0300	As above, 5/8" thick, with factory finish		525	.030		.81	.83		1.64	2.20	
	0500	Hardboard, 3/8" thick, slotted		525	.030		.98	.83		1.81	2.39	
	1000	Exterior AC plywood, 1/4" thick		420	.038		.50	1.04		1.54	2.19	
	1100	1/2" thick	▼	420	.038	▼	.66	1.04		1.70	2.37	
	1150	For aluminum soffit, see division 076-217										
777	0010	**WOOD SIDING** See division 074-609 & 611										**777**
778	0010	**DOORS AND FRAMES** See division 081 & 082										**778**

063 | Wood Treatment

063 100 | Preservative Treatment

		CREW	DAILY OUTPUT	LABOR-HOURS	UNIT	1999 BARE COSTS				TOTAL INCL O&P	
						MAT.	LABOR	EQUIP.	TOTAL		
102	0011	**LUMBER TREATMENT**										**102**
	0400	Fire retardant, wet				M.B.F.	271			271	298	
	0500	KDAT					255			255	281	
	0700	Salt treated, water borne, .40 lb. retention					122			122	134	
	0800	Oil borne, 8 lb. retention					143			143	157	
	1000	Kiln dried lumber, 1" & 2" thick, softwoods					81.50			81.50	90	
	1100	Hardwoods					87			87	95.50	
	1500	For small size 1" stock, add				▼	11			11	12.10	

6 WOOD & PLASTICS

063 100	Preservative Treatment	CREW	DAILY OUTPUT	LABOR-HOURS	UNIT	1999 BARE COSTS				TOTAL INCL O&P
						MAT.	LABOR	EQUIP.	TOTAL	

102	1700	For full size rough lumber, add					20%					102
104	0010	**PLYWOOD TREATMENT** Fire retardant, 1/4" thick				M.S.F.	204			204	224	104
	0030	3/8" thick					224			224	246	
	0050	1/2" thick					240			240	264	
	0070	5/8" thick					255			255	281	
	0100	3/4" thick					280			280	310	
	0200	For KDAT, add					61			61	67	
	0500	Salt treated water borne, .25 lb., wet, 1/4" thick					112			112	123	
	0530	3/8" thick					117			117	129	
	0550	1/2" thick					122			122	134	
	0570	5/8" thick					133			133	146	
	0600	3/4" thick					138			138	152	
	0800	For KDAT add					61			61	67	
	0900	For .40 lb., per C.F. retention, add					51			51	56	
	1000	For certification stamp, add					30			30	33	

064 100	Custom Casework	CREW	DAILY OUTPUT	LABOR-HOURS	UNIT	1999 BARE COSTS				TOTAL INCL O&P
						MAT.	LABOR	EQUIP.	TOTAL	

102	0010	**CABINETS** Corner china cabinets, stock pine,										102
	0020	80" high, unfinished, minimum	2 Carp	6.60	2.424	Ea.	229	66		295	355	
	0100	Maximum	"	4.40	3.636	"	700	99.50		799.50	925	
	0300	Built-in drawer units, pine, 18" deep, 32" high, unfinished										
	0400	Minimum	2 Carp	53	.302	L.F.	32	8.25		40.25	48	
	0500	Maximum	"	40	.400	"	120	10.90		130.90	149	
	0700	Kitchen base cabinets, hardwood, not incl. counter tops,										
	0710	24" deep, 35" high, prefinished										
	0800	One top drawer, one door below, 12" wide	2 Carp	24.80	.645	Ea.	114	17.60		131.60	153	
	0840	18" wide		23.30	.687		131	18.75		149.75	174	
	0880	24" wide		22.30	.717		140	19.60		159.60	184	
	1000	Four drawers, 12" wide		24.80	.645		113	17.60		130.60	152	
	1040	18" wide		23.30	.687		158	18.75		176.75	204	
	1060	24" wide		22.30	.717		178	19.60		197.60	226	
	1200	Two top drawers, two doors below, 27" wide		22	.727		177	19.85		196.85	226	
	1260	36" wide		20.30	.788		204	21.50		225.50	258	
	1300	48" wide		18.90	.847		242	23		265	305	
	1500	Range or sink base, two doors below, 30" wide		21.40	.748		148	20.50		168.50	195	
	1540	36" wide		20.30	.788		161	21.50		182.50	211	
	1580	48" wide		18.90	.847		191	23		214	247	
	1800	For sink front units, deduct					52			52	57	
	2000	Corner base cabinets, 36" wide, standard	2 Carp	18	.889		163	24.50		187.50	217	
	2100	Lazy Susan with revolving door	"	16.50	.970		203	26.50		229.50	265	
	4000	Kitchen wall cabinets, hardwood, 12" deep with two doors										
	4050	12" high, 30" wide	2 Carp	24.80	.645	Ea.	87	17.60		104.60	124	
	4100	36" wide		24	.667		99.50	18.20		117.70	138	
	4400	15" high, 30" wide		24	.667		93.50	18.20		111.70	132	
	4440	36" wide		22.70	.705		102	19.25		121.25	143	
	4700	24" high, 30" wide		23.30	.687		119	18.75		137.75	161	
	4720	36" wide		22.70	.705		127	19.25		146.25	171	

For expanded coverage of these items see Means Interior Cost Data 1999

WOOD & PLASTICS **6**

064 100	Custom Casework	CREW	DAILY OUTPUT	LABOR-HOURS	UNIT	1999 BARE COSTS				TOTAL INCL O&P	
						MAT.	LABOR	EQUIP.	TOTAL		
102											**102**
5000	30" high, one door, 12" wide	2 Carp	22	.727	Ea.	85	19.85		104.85	125	
5040	18" wide		20.90	.766		90	21		111	132	
5060	24" wide		20.30	.788		110	21.50		131.50	155	
5300	Two doors, 27" wide		19.80	.808		121	22		143	168	
5340	36" wide		18.80	.851		149	23		172	201	
5380	48" wide		18.40	.870		176	23.50		199.50	232	
6000	Corner wall, 30" high, 24" wide		18	.889		141	24.50		165.50	193	
6050	30" wide		17.20	.930		162	25.50		187.50	218	
6100	36" wide		16.50	.970		187	26.50		213.50	248	
6500	Revolving Lazy Susan		15.20	1.053		167	28.50		195.50	229	
7000	Broom cabinet, 84" high, 24" deep, 18" wide		10	1.600		231	43.50		274.50	325	
7500	Oven cabinets, 84" high, 24" deep, 27" wide		8	2		297	54.50		351.50	410	
7750	Valance board trim	↓	396	.040	L.F.	7.55	1.10		8.65	10.05	
9000	For deluxe models of all cabinets, add					40%					
9500	For custom built in place, add					25%	10%				
9550	Rule of thumb, kitchen cabinets not including										
9560	appliances & counter top, minimum	2 Carp	30	.533	L.F.	79	14.55		93.55	110	
9600	Maximum	"	25	.640	"	221	17.45		238.45	271	
9610	For metal cabinets, see division 123-501										
104	**0010** **CASEWORK, FRAMES**										**104**
0050	Base cabinets, counter storage, 36" high, one bay										
0100	18" wide	1 Carp	2.70	2.963	Ea.	89.50	81		170.50	225	
0400	Two bay, 36" wide		2.20	3.636		137	99.50		236.50	305	
1100	Three bay, 54" wide		1.50	5.333		163	146		309	410	
2800	Book cases, one bay, 7' high, 18" wide		2.40	3.333		105	91		196	259	
3500	Two bay, 36" wide		1.60	5		152	137		289	380	
4100	Three bay, 54" wide		1.20	6.667		252	182		434	565	
5100	Coat racks, one bay, 7' high, 24" wide		4.50	1.778		105	48.50		153.50	193	
5300	Two bay, 48" wide		2.75	2.909		147	79.50		226.50	287	
5800	Three bay, 72" wide		2.10	3.810		215	104		319	400	
6100	Wall mounted cabinet, one bay, 24" high, 18" wide		3.60	2.222		58	60.50		118.50	159	
6800	Two bay, 36" wide		2.20	3.636		84	99.50		183.50	249	
7400	Three bay, 54" wide		1.70	4.706		105	128		233	320	
8400	30" high, one bay, 18" wide		3.60	2.222		63	60.50		123.50	165	
9000	Two bay, 36" wide		2.15	3.721		84	102		186	253	
9400	Three bay, 54" wide		1.60	5		105	137		242	330	
9800	Wardrobe, 7' high, single, 24" wide		2.70	2.963		116	81		197	254	
9880	Partition & adjustable shelves, 48" wide		1.70	4.706		147	128		275	365	
9950	Partition, adjustable shelves & drawers, 48" wide	↓	1.40	5.714	↓	221	156		377	490	
106	**0010** **CABINET DOORS**										**106**
2000	Glass panel, hardwood frame										
2200	12" wide, 18" high	1 Carp	34	.235	Ea.	14.55	6.40		20.95	26	
2600	30" high		32	.250		24	6.80		30.80	37.50	
4450	18" wide, 18" high		32	.250		15	6.80		21.80	27.50	
4550	30" high	↓	29	.276	↓	24	7.55		31.55	38.50	
5000	Hardwood, raised panel										
5100	12" wide, 18" high	1 Carp	16	.500	Ea.	20	13.65		33.65	43.50	
5200	30" high		15	.533		32	14.55		46.55	58	
5500	18" wide, 18" high		15	.533		29	14.55		43.55	55	
5600	30" high	↓	14	.571	↓	49	15.60		64.60	78.50	
6000	Plastic laminate on particle board										
6100	12" wide, 18" high	1 Carp	25	.320	Ea.	13	8.75		21.75	28	
6140	30" high		23	.348		21	9.50		30.50	38	
6500	18" wide, 18" high		24	.333		19	9.10		28.10	35.50	
6600	30" high	↓	22	.364	↓	32	9.95		41.95	50.50	

Important: See the Reference Section for critical supporting data - Reference Nos., Crews, & City Cost Indexes

064 100 | Custom Casework

		CREW	DAILY OUTPUT	LABOR-HOURS	UNIT	1999 BARE COSTS				TOTAL INCL O&P	
						MAT.	LABOR	EQUIP.	TOTAL		
106	7000	Plywood, with edge band									**106**
	7010	12" wide, 18" high	1 Carp	27	.296	Ea.	17.15	8.10		25.25	31.50
	7120	30" high		25	.320		29.50	8.75		38.25	46.50
	7650	18" wide, 18" high		26	.308		25	8.40		33.40	40.50
	7750	30" high	↓	24	.333	↓	43	9.10		52.10	62
108	0010	**CABINET HARDWARE**									**108**
	1000	Catches, minimum	1 Carp	235	.034	Ea.	.71	.93		1.64	2.24
	1040	Maximum	"	80	.100	"	4.16	2.73		6.89	8.90
	2000	Door/drawer pulls, handles									
	2200	Handles and pulls, projecting, metal, minimum	1 Carp	160	.050	Ea.	1.51	1.36		2.87	3.81
	2240	Maximum		68	.118		7.20	3.21		10.41	13
	2300	Wood, minimum		160	.050		1.51	1.36		2.87	3.81
	2340	Maximum		68	.118		4.12	3.21		7.33	9.60
	2600	Flush, metal, minimum		160	.050		1.44	1.36		2.80	3.73
	2640	Maximum		68	.118	↓	10.30	3.21		13.51	16.40
	3000	Drawer tracks/glides, minimum		48	.167	Pr.	5.65	4.55		10.20	13.40
	3040	Maximum		24	.333		16.50	9.10		25.60	32.50
	4000	Cabinet hinges, minimum		160	.050		1.40	1.36		2.76	3.69
	4040	Maximum	↓	68	.118	↓	6.20	3.21		9.41	11.85
110	0010	**DRAWERS**									**110**
	0100	Solid hardwood front									
	1000	4" high, 12" wide	1 Carp	17	.471	Ea.	19.10	12.85		31.95	41
	1200	18" wide	"	16	.500	"	24.50	13.65		38.15	48.50
	2800	Plastic laminate on particle board front									
	3000	4" high, 12" wide	1 Carp	17	.471	Ea.	19.55	12.85		32.40	41.50
	3200	18" wide	"	16	.500	"	22.50	13.65		36.15	46.50
	5400	Plywood, flush panel front									
	6000	4" high, 12" wide	1 Carp	17	.471	Ea.	20	12.85		32.85	42
	6200	18" wide	"	16	.500	"	24.50	13.65		38.15	48.50
140	0010	**VANITIES**									**140**
	8000	Vanity bases, 2 doors, 30" high, 21" deep, 24" wide	2 Carp	20	.800	Ea.	123	22		145	170
	8050	30" wide		16	1		129	27.50		156.50	185
	8100	36" wide		13.33	1.200		144	33		177	211
	8150	48" wide	↓	11.43	1.400		182	38		220	260
	9000	For deluxe models of all vanities, add to above					40%				
	9500	For custom built in place, add to above				↓	25%	10%			

064 300 | Stairwork & Handrails

		CREW	DAILY OUTPUT	LABOR-HOURS	UNIT	MAT.	LABOR	EQUIP.	TOTAL	TOTAL INCL O&P	
306	0011	**STAIRS, PREFABRICATED**									**306**
	0100	Box stairs, prefabricated, 3'-0" wide									
	0110	Oak treads, no handrails, 2' high	2 Carp	5	3.200	Flight	216	87.50		303.50	375
	0200	4' high		4	4		430	109		539	645
	0300	6' high		3.50	4.571		600	125		725	855
	0400	8' high		3	5.333		760	146		906	1,075
	0600	With pine treads for carpet, 2' high		5	3.200		86	87.50		173.50	232
	0700	4' high		4	4		159	109		268	345
	0800	6' high		3.50	4.571		233	125		358	450
	0900	8' high	↓	3	5.333		266	146		412	520
	1100	For 4' wide stairs, add				↓	25%				
	1500	Prefabricated stair rail with balusters, 5 risers	2 Carp	15	1.067	Ea.	223	29		252	291
	1700	Basement stairs, prefabricated, soft wood,									
	1710	open risers, 3' wide, 8' high	2 Carp	4	4	Flight	555	109		664	785
	1900	Open stairs, prefabricated prefinished poplar, metal stringers,									
	1910	treads 3'-6" wide, no railings									

WOOD & PLASTICS **6**

		064 300	Stairwork & Handrails	CREW	DAILY OUTPUT	LABOR-HOURS	UNIT	1999 BARE COSTS MAT.	LABOR	EQUIP.	TOTAL	TOTAL INCL O&P	
306	2000		3' high	2 Carp	5	3.200	Flight	224	87.50		311.50	385	**306**
	2100		4' high		4	4		485	109		594	705	
	2200		6' high		3.50	4.571		555	125		680	810	
	2300		8' high		3	5.333		730	146		876	1,025	
	2500		For prefab. 3 piece wood railings & balusters, add for										
	2600		3' high stairs	2 Carp	15	1.067	Ea.	30.50	29		59.50	79.50	
	2700		4' high stairs		14	1.143		49.50	31		80.50	104	
	2800		6' high stairs		13	1.231		61.50	33.50		95	121	
	2900		8' high stairs		12	1.333		93.50	36.50		130	161	
	3100		For 3'-6" x 3'-6" platform, add		4	4		72.50	109		181.50	252	
	3300		Curved stairways, 3'-3" wide, prefabricated, oak, unfinished,										
	3310		incl. curved balustrade system, open one side										
	3400		9' high	2 Carp	.70	22.857	Flight	6,375	625		7,000	8,000	
	3500		10' high		.70	22.857		7,200	625		7,825	8,900	
	3700		Open two sides, 9' high		.50	32		10,000	875		10,875	12,400	
	3800		10' high		.50	32		10,800	875		11,675	13,300	
	4000		Residential, wood, oak treads, prefabricated		1.50	10.667		930	291		1,221	1,475	
	4200		Built in place		.44	36.364		1,325	995		2,320	3,000	
	4400		Spiral, oak, 4'-6" diameter, unfinished, prefabricated,										
	4500		incl. railing, 9' high	2 Carp	1.50	10.667	Flight	4,000	291		4,291	4,850	
308	0010	**STAIR PARTS** Balusters, turned, 30" high, pine, minimum	R064 -100	1 Carp	28	.286	Ea.	3.50	7.80		11.30	16.10	**308**
	0100		Maximum		26	.308		9	8.40		17.40	23	
	0300		30" high birch balusters, minimum		28	.286		6	7.80		13.80	18.85	
	0400		Maximum		26	.308		10	8.40		18.40	24	
	0600		42" high, pine balusters, minimum		27	.296		6	8.10		14.10	19.35	
	0700		Maximum		25	.320		9.50	8.75		18.25	24	
	0900		42" high birch balusters, minimum		27	.296		7.05	8.10		15.15	20.50	
	1000		Maximum		25	.320		15	8.75		23.75	30.50	
	1050		Baluster, stock pine, 1-1/16" x 1-1/16"		240	.033	L.F.	1.97	.91		2.88	3.60	
	1100		1-5/8" x 1-5/8"		220	.036	"	1.82	.99		2.81	3.56	
	1200		Newels, 3-1/4" wide, starting, minimum		7	1.143	Ea.	37	31		68	89.50	
	1300		Maximum		6	1.333		120	36.50		156.50	190	
	1500		Landing, minimum		5	1.600		70	43.50		113.50	146	
	1600		Maximum		4	2		170	54.50		224.50	273	
	1800		Railings, oak, built-up, minimum		60	.133	L.F.	5.30	3.64		8.94	11.60	
	1900		Maximum		55	.145		15	3.97		18.97	23	
	2100		Add for sub rail		110	.073		4	1.99		5.99	7.50	
	2300		Risers, beech, 3/4" x 7-1/2" high		64	.125		5.20	3.41		8.61	11.10	
	2400		Fir, 3/4" x 7-1/2" high		64	.125		1.50	3.41		4.91	7	
	2600		Oak, 3/4" x 7-1/2" high		64	.125		4.75	3.41		8.16	10.55	
	2800		Pine, 3/4" x 7-1/2" high		66	.121		1.50	3.31		4.81	6.85	
	2850		Skirt board, pine, 1" x 10"		55	.145		1.65	3.97		5.62	8.05	
	2900		1" x 12"		52	.154		2	4.20		6.20	8.80	
	3000		Treads, 1-1/16" x 9-1/2" wide, 3' long, oak		18	.444	Ea.	22	12.15		34.15	43	
	3100		4' long, oak		17	.471		29	12.85		41.85	52	
	3300		1-1/16" x 11-1/2" wide, 3' long, oak		18	.444		23.50	12.15		35.65	44.50	
	3400		6' long, oak		14	.571		56	15.60		71.60	86	
	3600		Beech treads, add					40%					
	3800		For mitered return nosings, add				L.F.	8.40			8.40	9.25	
310	0010	**RAILING** Custom design, architectural grade, hardwood, minimum		1 Carp	38	.211	"	12	5.75		17.75	22.50	**310**
	0100		Maximum		30	.267	L.F.	44	7.30		51.30	60	
	0300		Stock interior railing with spindles 6" O.C., 4' long		40	.200		28	5.45		33.45	39.50	
	0400		8' long		48	.167		26	4.55		30.55	35.50	

Important: See the Reference Section for critical supporting data - Reference Nos., Crews, & City Cost Indexes

6 WOOD & PLASTICS

064 400 | Misc. Ornamental Items

		CREW	DAILY OUTPUT	LABOR-HOURS	UNIT	MAT.	LABOR	EQUIP.	TOTAL	TOTAL INCL O&P	
402	**0010**	**COLUMNS** For base plates, see division 055-404										**402**
	0050	Aluminum, round colonial, 6" diameter	2 Carp	80	.200	V.L.F.	14	5.45		19.45	24	
	0100	8" diameter		62.25	.257		17	7		24	30	
	0200	10" diameter		55	.291		22	7.95		29.95	36.50	
	0250	Fir, stock units, hollow round, 6" diameter		80	.200		13	5.45		18.45	23	
	0300	8" diameter		80	.200		14	5.45		19.45	24	
	0350	10" diameter		70	.229		18	6.25		24.25	29.50	
	0400	Solid turned, to 8' high, 3-1/2" diameter		80	.200		7	5.45		12.45	16.30	
	0500	4-1/2" diameter		75	.213		10	5.80		15.80	20	
	0600	5-1/2" diameter		70	.229		13	6.25		19.25	24	
	0800	Square columns, built-up, 5" x 5"		65	.246		11	6.70		17.70	22.50	
	0900	Solid, 3-1/2" x 3-1/2"		130	.123		6	3.36		9.36	11.90	
	1600	Hemlock, tapered, T & G, 12" diam, 10' high		100	.160		30	4.37		34.37	40	
	1700	16' high		65	.246		46	6.70		52.70	61	
	1900	10' high, 14" diameter		100	.160		63	4.37		67.37	76.50	
	2000	18' high		65	.246		60	6.70		66.70	76.50	
	2200	18" diameter, 12' high		65	.246		76	6.70		82.70	94	
	2300	20' high		50	.320		83	8.75		91.75	105	
	2500	20" diameter, 14' high		40	.400		95	10.90		105.90	122	
	2600	20' high		35	.457		110	12.50		122.50	141	
	2800	For flat pilasters, deduct					33%					
	3000	For splitting into halves, add				Ea.	60			60	66	
	4000	Rough sawn cedar posts, 4" x 4"	2 Carp	250	.064	V.L.F.	2.43	1.75		4.18	5.40	
	4100	4" x 6"		235	.068		3.62	1.86		5.48	6.90	
	4200	6" x 6"		220	.073		5.45	1.99		7.44	9.10	
	4300	8" x 8"		200	.080		5.60	2.18		7.78	9.60	
410	**0010**	**MILLWORK, HIGH DENSITY POLYMER**										**410**
	0100	Base, 9/16" x 3-3/16"	1 Carp	230	.035	L.F.	1.20	.95		2.15	2.81	
	0200	Casing, fluted, 5/8" x 3-1/4"		215	.037		1.20	1.02		2.22	2.92	
	0300	Chair rail, 9/16" x 2-1/4"		260	.031		.61	.84		1.45	1.99	
	0400	5/8" x 3-1/8"		230	.035		1.15	.95		2.10	2.76	
	0500	Corner, inside, 1/2" x 1-1/8"		220	.036		.60	.99		1.59	2.22	
	0600	Cove, 13/16" x 3-3/4"		260	.031		1.20	.84		2.04	2.64	
	0700	Crown, 3/4" x 3-13/16"		260	.031		1.20	.84		2.04	2.64	
	0800	Half round, 15/16" x 2"		240	.033		.66	.91		1.57	2.16	

065 100 | Fiberglass

		CREW	DAILY OUTPUT	LABOR-HOURS	UNIT	MAT.	LABOR	EQUIP.	TOTAL	TOTAL INCL O&P	
104	**0010**	**CASTINGS, FIBERGLASS**										**104**
	0100	Angle, 1" x 1" x 1/8" thick	2 Sswk	240	.067	L.F.	1.01	2.04		3.05	4.85	
	0120	3" x 3" x 1/4" thick		200	.080		4.23	2.45		6.68	9.15	
	0140	4" x 4" x 1/4" thick		200	.080		5.45	2.45		7.90	10.50	
	0160	4" x 4" x 3/8" thick		200	.080		8.45	2.45		10.90	13.80	
	0180	6" x 6" x 1/2" thick		160	.100		16.60	3.06		19.66	24	
	1000	Flat sheet, 1/8" thick		140	.114	S.F.	4.32	3.50		7.82	11.15	
	1020	1/4" thick		120	.133		8.30	4.08		12.38	16.65	
	1040	3/8" thick		100	.160		14.20	4.90		19.10	24.50	
	1060	1/2" thick		80	.200		18.25	6.10		24.35	31	

For expanded coverage of these items see _Means Interior Cost Data 1999_

6

WOOD & PLASTICS

065 | Structural Plastics

065 100 | Fiberglass

			CREW	DAILY OUTPUT	LABOR-HOURS	UNIT	1999 BARE COSTS MAT.	LABOR	EQUIP.	TOTAL	TOTAL INCL O&P	
104	2000	Handrail, 42" high, 2" diam. rails pickets 5' O.C.	2 Sswk	32	.500	L.F.	34.50	15.30		49.80	66	104
	3000	Round bar, 1/4" diam.		240	.067		.38	2.04		2.42	4.16	
	3020	1/2" diam.		200	.080		.69	2.45		3.14	5.25	
	3040	3/4" diam.		200	.080		1.66	2.45		4.11	6.30	
	3060	1" diam.		160	.100		2.88	3.06		5.94	8.75	
	3080	1-1/4" diam.		160	.100		3.33	3.06		6.39	9.25	
	3100	1-1/2" diam.		140	.114		3.99	3.50		7.49	10.80	
	3500	Round tube, 1" diam. x 1/8" thick		240	.067		1.57	2.04		3.61	5.45	
	3520	2" diam. x 1/4" thick		200	.080		4.51	2.45		6.96	9.45	
	3540	3" diam. x 1/4" thick		160	.100		8.75	3.06		11.81	15.20	
	4000	Square bar, 1/2" square		240	.067		3.45	2.04		5.49	7.55	
	4020	1" square		200	.080		3.77	2.45		6.22	8.65	
	4040	1-1/2" square		160	.100		6.95	3.06		10.01	13.25	
	4500	Square tube, 1" x 1" x 1/8" thick		240	.067		1.63	2.04		3.67	5.55	
	4520	2" x 2" x 1/8" thick		200	.080		3.08	2.45		5.53	7.85	
	4540	3" x 3" x 1/4" thick		160	.100		8.40	3.06		11.46	14.85	
	5000	Threaded rod, 3/8" diam.		320	.050		3.02	1.53		4.55	6.10	
	5020	1/2" diam.		320	.050		3.35	1.53		4.88	6.50	
	5040	5/8" diam.		280	.057		3.71	1.75		5.46	7.30	
	5060	3/4" diam.		280	.057		4.22	1.75		5.97	7.85	
	6000	Wide flange beam, 4" x 4" x 1/4" thick		120	.133		7.60	4.08		11.68	15.85	
	6020	6" x 6" x 1/4" thick		100	.160		13.75	4.90		18.65	24	
	6040	8" x 8" x 3/8" thick	▼	80	.200	▼	26.50	6.10		32.60	40	
108	0010	**GRATING, FIBERGLASS**										108
	0100	Molded, green (for mod. corrosive environment)										
	0140	1" x 4" mesh, 1" thick	2 Sswk	400	.040	S.F.	10.20	1.22		11.42	13.45	
	0180	1-1/2" square mesh, 1" thick		400	.040		15.25	1.22		16.47	19	
	0220	1-1/4" thick		400	.040		13.50	1.22		14.72	17.10	
	0260	1-1/2" thick		400	.040		16.60	1.22		17.82	20.50	
	0300	2" square mesh, 2" thick	▼	320	.050	▼	23	1.53		24.53	28	
	1000	Orange (for highly corrosive environment)										
	1040	1" x 4" mesh, 1" thick	2 Sswk	400	.040	S.F.	12.30	1.22		13.52	15.80	
	1080	1-1/2" square mesh, 1" thick		400	.040		17.05	1.22		18.27	21	
	1120	1-1/4" thick		400	.040		18.90	1.22		20.12	23	
	1160	1-1/2" thick		400	.040		19.35	1.22		20.57	23.50	
	1200	2" square mesh, 2" thick	▼	320	.050	▼	22.50	1.53		24.03	27.50	
	3000	Pultruded, green (for mod. corrosive environment)										
	3040	1" O.C. bar spacing, 1" thick	2 Sswk	400	.040	S.F.	15.30	1.22		16.52	19.10	
	3080	1-1/2" thick		320	.050		13.75	1.53		15.28	17.90	
	3120	1-1/2" O.C. bar spacing, 1" thick		400	.040		11.85	1.22		13.07	15.30	
	3160	1-1/2" thick	▼	400	.040	▼	16.35	1.22		17.57	20	
	4000	Grating support legs, fixed height, no base				Ea.	43.50			43.50	47.50	
	4040	With base					37.50			37.50	41	
	4080	Adjustable to 60"				▼	63.50			63.50	70	
112	0010	**STAIR TREAD, FIBERGLASS** Isophthalic Resin, 10-1/2" deep										112
	0100	24" wide	2 Sswk	52	.308	Ea.	38.50	9.40		47.90	60	
	0140	30" wide		52	.308		48	9.40		57.40	70.50	
	0180	36" wide		52	.308		58	9.40		67.40	81	
	0220	42" wide	▼	52	.308	▼	67.50	9.40		76.90	91.50	

066 | Plastic Fabrications

066 100	Glass Fiber & Resin	CREW	DAILY OUTPUT	LABOR-HOURS	UNIT	MAT.	LABOR	EQUIP.	TOTAL	TOTAL INCL O&P	
102 0010	**NETTING, FLEXIBLE PLASTIC** 1/8" square mesh	4 Clab	4,000	.008	S.F.	.73	.17		.90	1.07	**102**
0100	1/4" square mesh		4,000	.008		.12	.17		.29	.40	
0120	1/2" square mesh		4,000	.008		.09	.17		.26	.37	
0140	5/8" x 3/4" mesh		4,000	.008		.03	.17		.20	.30	
0160	1-1/4" x 1-1/2" mesh		4,000	.008		.26	.17		.43	.55	
0200	4" square mesh		4,000	.008		.03	.17		.20	.31	
1000	Poly clips				Ea.	.29			.29	.32	

066 500	Solid Surfacing Material	CREW	DAILY OUTPUT	LABOR-HOURS	UNIT	MAT.	LABOR	EQUIP.	TOTAL	TOTAL INCL O&P	
503 0010	**SOLID SURFACE COUNTERTOPS,** Acrylic polymer										**503**
0020	Pricing for orders of 100 L.F. or greater										
0100	25" wide, solid colors	2 Carp	28	.571	L.F.	41	15.60		56.60	69.50	
0200	Patterned colors		28	.571		52	15.60		67.60	81.50	
0300	Premium patterned colors		28	.571		65	15.60		80.60	96	
0400	With silicone attached 4" backsplash, solid colors		27	.593		45	16.20		61.20	75	
0500	Patterned colors		27	.593		57	16.20		73.20	88	
0600	Premium patterned colors		27	.593		71	16.20		87.20	104	
0700	With hard seam attached 4" backsplash, solid colors		23	.696		45	19		64	79.50	
0800	Patterned colors		23	.696		57	19		76	92.50	
0900	Premium patterned colors		23	.696		71	19		90	108	
1000	Pricing for order of 51 - 99 L.F.										
1100	25" wide, solid colors	2 Carp	24	.667	L.F.	47	18.20		65.20	80.50	
1200	Patterned colors		24	.667		60	18.20		78.20	94.50	
1300	Premium patterned colors		24	.667		75	18.20		93.20	111	
1400	With silicone attached 4" backsplash, solid colors		23	.696		52	19		71	87	
1500	Patterned colors		23	.696		65.50	19		84.50	102	
1600	Premium patterned colors		23	.696		81.50	19		100.50	120	
1700	With hard seam attached 4" backsplash, solid colors		20	.800		52	22		74	91.50	
1800	Patterned colors		20	.800		65.50	22		87.50	107	
1900	Premium patterned colors		20	.800		81.50	22		103.50	125	
2000	Pricing for order of 1 - 50 L.F.										
2100	25" wide, solid colors	2 Carp	20	.800	L.F.	55.50	22		77.50	95.50	
2200	Patterned colors		20	.800		70	22		92	112	
2300	Premium patterned colors		20	.800		88	22		110	131	
2400	With silicone attached 4" backsplash, solid colors		19	.842		61	23		84	103	
2500	Patterned colors		19	.842		77	23		100	121	
2600	Premium patterned colors		19	.842		96	23		119	141	
2700	With hard seam attached 4" backsplash, solid colors		4	4		61	109		170	239	
2800	Patterned colors		15	1.067		77	29		106	131	
2900	Premium patterned colors		15	1.067		96	29		125	151	
3000	Sinks, pricing for order of 100 or greater units										
3100	Single bowl, hard seamed, solid colors, 13" x 17"	1 Carp	3	2.667	Ea.	277	73		350	420	
3200	10" x 15"		7	1.143		128	31		159	190	
3300	Cutouts for sinks		8	1			27.50		27.50	43	
3400	Sinks, pricing for order of 51 - 99 units										
3500	Single bowl, hard seamed, solid colors, 13" x 17"	1 Carp	2.55	3.137	Ea.	320	85.50		405.50	485	
3600	10" x 15"		6	1.333		147	36.50		183.50	220	
3700	Cutouts for sinks		7	1.143			31		31	49	
3800	Sinks, pricing for order of 1 - 50 units										
3900	Single bowl, hard seamed, solid colors, 13" x 17"	1 Carp	2	4	Ea.	375	109		484	580	
4000	10" x 15"		4.55	1.758		173	48		221	266	
4100	Cutouts for sinks		5.25	1.524			41.50		41.50	65.50	
4200	Cooktop cutouts, pricing for 100 or greater units		4	2		20	54.50		74.50	108	
4300	51 - 99 units		3.40	2.353		23	64		87	127	
4400	1 - 50 units		3	2.667		27	73		100	145	

6

WOOD & PLASTICS

For expanded coverage of these items see *Means Interior Cost Data 1999*

066 500	Solid Surfacing Material	CREW	DAILY OUTPUT	LABOR-HOURS	UNIT	1999 BARE COSTS				TOTAL INCL O&P
						MAT.	LABOR	EQUIP.	TOTAL	
505 0010	**VANITY TOPS**									505
0015	Solid surface, center bowl, 17" x 19"	1 Carp	12	.667	Ea.	163	18.20		181.20	209
0020	19" x 25"		12	.667		197	18.20		215.20	246
0030	19" x 31"		12	.667		240	18.20		258.20	293
0040	19" x 37"		12	.667		279	18.20		297.20	335
0050	22" x 25"		10	.800		222	22		244	279
0060	22" x 31"		10	.800		259	22		281	320
0070	22" x 37"		10	.800		300	22		322	365
0080	22" x 43"		10	.800		345	22		367	415
0090	22" x 49"		10	.800		380	22		402	455
0110	22" x 55"		8	1		430	27.50		457.50	520
0120	22" x 61"		8	1		495	27.50		522.50	585
0130	22" x 68"		8	1		635	27.50		662.50	740
0140	22" x 73"		8	1		715	27.50		742.50	830
0150	22" x 85"		8	1		825	27.50		852.50	950
0160	Offset bowl, left or right, 22" x 49"		10	.800		455	22		477	535
0170	22" x 55"		8	1		520	27.50		547.50	615
0180	22" x 61"		8	1		590	27.50		617.50	690
0190	22" x 68"		8	1		690	27.50		717.50	805
0200	22" x 73"		8	1		850	27.50		877.50	980
0210	22" x 85"		8	1		995	27.50		1,022.50	1,150
0220	Double bowl, 22" x 61"		8	1		540	27.50		567.50	640
0230	Corner top/bowl, 22" x 49"	▼	8	1	▼	420	27.50		447.50	510
0240	For aggregate colors, add					35%				
0250	For faucets and fittings see 151-141									

For information about Means Estimating Seminars, see yellow pages 11 and 12 in back of book

Important: See the Reference Section for critical supporting data - Reference Nos., Crews, & City Cost Indexes

6 WOOD & PLASTICS

Division 7
Thermal & Moisture Protection

Estimating Tips

071 Waterproofing & Dampproofing

- Be sure of the job specifications before pricing this subdivision. The difference in cost between waterproofing and dampproofing can be great. Waterproofing will hold back standing water. Dampproofing prevents the transmission of water vapor. Also included in this section are vapor retarding membranes.

072 Insulation & Fireproofing

- Insulation and fireproofing products are measured by area, thickness, volume or R value. Specifications may only give what the specific R value should be in a certain situation. The estimator may need to choose the type of insulation to meet that R value.

073 Shingles & Roofing Tiles
074 Preformed Roofing & Siding

- Many roofing and siding products are bought and sold by the square. One square is equal to an area that measures 100 square feet.

This simple change in unit of measure could create a large error if the estimator is not observant. Accessories and fasteners necessary for a complete installation must be figured into any calculations for both material and labor.

075 Membrane Roofing
076 Flashing & Sheet Metal
077 Roof Specialties & Accessories
078 Skylights

- The items in these subdivisions compose a roofing system. No one component completes the installation and all must be estimated. Built-up or single ply membrane roofing systems are made up of many products and installation trades. Wood blocking at roof perimeters or penetrations, parapet coverings, reglets, roof drains, gutters, downspouts, sheet metal flashing, skylights, smoke vents or roof hatches all need to be considered along with the roofing material. Several different installation trades will need to work together on the roofing system. Inherent difficulties in the scheduling and coordination of various trades must be accounted for when estimating labor costs.

079 Joint Sealers

- To complete the weather-tight shell the sealants and caulkings must be estimated. Where different materials meet—at expansion joints, at flashing penetrations, and at hundreds of other locations throughout a construction project—they provide another line of defense against water penetration. Often, an entire system is based on the proper location and placement of caulking or sealants. The detail drawings that are included as part of a set of architectural plans, show typical locations for these materials. When caulking or sealants are shown at typical locations, this means the estimator must include them for all the locations where this detail is applicable. Be careful to keep different types of sealants separate, and remember to consider backer rods and primers if necessary.

Reference Numbers

Reference numbers are shown in bold squares at the beginning of some major classifications. These numbers refer to related items in the Reference Section. The reference information may be an estimating procedure, an alternate pricing method or technical information.

Note: Not all subdivisions listed here necessarily appear in this publication.

071 100	Sheet Waterproofing	CREW	DAILY OUTPUT	LABOR-HOURS	UNIT	1999 BARE COSTS				TOTAL INCL O&P
						MAT.	LABOR	EQUIP.	TOTAL	
102	**0010**	**ELASTOMERIC WATERPROOFING**									**102**
	0050	Acrylic rubber, fluid applied, 20 mils thick	3 Rofc	1,000	.024	S.F.	1.52	.58		2.10	2.67
	0060	50 mil, reinforced, stucco texture	"	600	.040		2.46	.96		3.42	4.37
	0090	EPDM, plain, 45 mils thick	2 Rofc	580	.028		.74	.66		1.40	1.95
	0100	60 mils thick		570	.028		1.08	.68		1.76	2.35
	0300	Nylon reinforced sheets, 45 mils thick		580	.028		1.04	.66		1.70	2.28
	0400	60 mils thick	↓	570	.028	↓	1.44	.68		2.12	2.74
	0600	Vulcanizing splicing tape for above, 2" wide				C.L.F.	33.50			33.50	36.50
	0700	4" wide				"	67			67	73.50
	0900	Adhesive, bonding, 60 SF per gal				Gal.	13.60			13.60	14.95
	1000	Splicing, 75 SF per gal				"	21			21	23
	1200	Neoprene sheets, plain, 45 mils thick	2 Rofc	580	.028	S.F.	1.09	.66		1.75	2.34
	1300	60 mils thick		570	.028		1.82	.68		2.50	3.17
	1500	Nylon reinforced, 45 mils thick		580	.028		1.26	.66		1.92	2.53
	1600	60 mils thick		570	.028		1.28	.68		1.96	2.57
	1800	120 mils thick	↓	500	.032	↓	2.54	.77		3.31	4.12
	1900	Adhesive, splicing, 150 S.F. per gal. per coat				Gal.	18.15			18.15	19.95
	2100	Fiberglass reinforced, fluid applied, 1/8" thick	2 Rofc	500	.032	S.F.	1.50	.77		2.27	2.98
	2200	Polyethylene and rubberized asphalt sheets, 1/8" thick		550	.029		.54	.70		1.24	1.80
	2210	Asphaltic hardboard protection board, 1/8" thick		500	.032		.30	.77		1.07	1.66
	2220	Asphaltic hardboard protection board, 1/4" thick		450	.036		.52	.86		1.38	2.05
	2400	Polyvinyl chloride sheets, plain, 10 mils thick		580	.028		.14	.66		.80	1.29
	2500	20 mils thick		570	.028		.22	.68		.90	1.40
	2700	30 mils thick	↓	560	.029	↓	.30	.69		.99	1.52
	3000	Adhesives, trowel grade, 40-100 SF per gal				Gal.	20			20	22
	3100	Brush grade, 100-250 SF per gal.				"	20			20	22
	3300	Bitumen modified polyurethane, fluid applied, 55 mils thick	2 Rofc	665	.024	S.F.	.65	.58		1.23	1.71
	3600	Vinyl plastic, sprayed on, 25 to 40 mils thick	"	475	.034	"	.98	.81		1.79	2.48
104	**0010**	**MEMBRANE WATERPROOFING** On slabs, 1 ply, felt	G-1	3,000	.019	S.F.	.13	.42	.14	.69	1.02	**104**
	0100	Glass fiber fabric		2,100	.027		.14	.60	.19	.93	1.40
	0300	2 ply, felt		2,500	.022		.25	.51	.16	.92	1.33
	0400	Glass fiber fabric		1,650	.034		.31	.77	.25	1.33	1.94
	0600	3 ply, felt		2,100	.027		.38	.60	.19	1.17	1.67
	0700	Glass fiber fabric	↓	1,550	.036		.41	.82	.26	1.49	2.15
	0900	For installation on walls, add						15%			
	1000	For adhered 1/4" EPS protection board, add	2 Rofc	3,500	.005		.14	.11		.25	.34
	1050	3/8" thick, add		3,500	.005		.16	.11		.27	.37
	1060	1/2" thick, add		3,500	.005	↓	.18	.11		.29	.39
	1070	Fiberglass fabric, black, 20/10 mesh		116	.138	Sq.	9.95	3.32		13.27	16.65
	1080	White, 20/10 mesh		116	.138	"	10.25	3.32		13.57	17
	1100	1/16" urethane, troweled		200	.080	S.F.	.69	1.93		2.62	4.08
	1200	Roller applied, 2 coats	↓	120	.133	"	.60	3.21		3.81	6.20

071 300	Bentonite Waterproofing									
301	**0010**	**BENTONITE**, Panels, 4' x 4', 3/16" thick	1 Rofc	625	.013	S.F.	.70	.31		1.01	1.30	**301**
	0100	Rolls, 3/8" thick, with geotextile fabric both sides	"	550	.015	"	.84	.35		1.19	1.52
	0300	Granular bentonite, 50 lb. bags (.625 C.F.)				Bag	11.65			11.65	12.85
	0400	3/8" thick, troweled on	1 Rofc	475	.017	S.F.	.58	.41		.99	1.34
	0500	Drain board, expanded polystyrene, binder encapsulated, 1-1/2" thick	1 Rohe	1,600	.005		.40	.09		.49	.60
	0510	2" thick		1,600	.005		.50	.09		.59	.71
	0520	3" thick		1,600	.005		.69	.09		.78	.92
	0530	4" thick		1,600	.005		.88	.09		.97	1.13
	0600	With filter fabric, 1-1/2" thick		1,600	.005		.54	.09		.63	.75
	0625	2" thick		1,600	.005		.63	.09		.72	.85
	0650	3" thick		1,600	.005		.82	.09		.91	1.06
	0675	4" thick	↓	1,600	.005	↓	1.01	.09		1.10	1.27

7

THERMAL & MOISTURE PROTECTION

Important: See the Reference Section for critical supporting data - Reference Nos., Crews, & City Cost Indexes

071 | Waterproofing & Dampproofing

071 300 | Bentonite Waterproofing

			CREW	DAILY OUTPUT	LABOR-HOURS	UNIT	1999 BARE COSTS				TOTAL INCL O&P	
							MAT.	LABOR	EQUIP.	TOTAL		
301	0700	Vapor retarder, see polyethelene, 071-922										301

071 450 | Cement. Waterproofing

452	0010	**CEMENTITIOUS WATERPROOFING** One coat cement base										452
	0020	1/8" application, sprayed on	G-2	1,000	.024	S.F.	1.50	.55	.24	2.29	2.76	
	0030	2 coat, cementitious/metallic slurry, troweled, 1/4" thick	1 Cefi	2.48	3.226	C.S.F.	35	84.50		119.50	165	
	0040	3 coat, 3/8" thick		1.84	4.348		52.50	114		166.50	228	
	0050	4 coat, 1/2" thick	↓	1.20	6.667	↓	70	174		244	335	

071 600 | Bitum. Dampproofing

602	0010	**BITUMINOUS ASPHALT COATING** For foundation										602
	0030	Brushed on, below grade, 1 coat	1 Rofc	665	.012	S.F.	.06	.29		.35	.57	
	0100	2 coat		500	.016		.10	.39		.49	.77	
	0300	Sprayed on, below grade, 1 coat, 25.6 S.F./gal.		830	.010		.07	.23		.30	.47	
	0400	2 coat, 20.5 S.F./gal.	↓	500	.016	↓	.14	.39		.53	.82	
	0500	Asphalt coating, with fibers				Gal.	3.40			3.40	3.74	
	0600	Troweled on, asphalt with fibers, 1/16" thick	1 Rofc	500	.016	S.F.	.15	.39		.54	.83	
	0700	1/8" thick		400	.020		.26	.48		.74	1.12	
	1000	1/2" thick	↓	350	.023	↓	.85	.55		1.40	1.89	

071 750 | Water Repellent Coat

754	0010	**RUBBER COATING** Water base liquid, roller applied	2 Rofc	7,000	.002	S.F.	.55	.06		.61	.70	754
	0200	Silicone or stearate, sprayed on CMU, 1 coat	1 Rofc	4,000	.002		.26	.05		.31	.36	
	0300	2 coats	"	3,000	.003	↓	.51	.06		.57	.67	

071 800 | Cementitious Dampprfng

802	0010	**CEMENT PARGING** 2 coats, 1/2" thick, regular P.C.	R071 -010	D-1	250	.064	S.F.	.15	1.57		1.72	2.61	802
	0100	Waterproofed Portland cement		"	250	.064	"	.17	1.57		1.74	2.64	

071 920 | Vapor Retarders

922	0010	**BUILDING PAPER** Aluminum and kraft laminated, foil 1 side	1 Carp	37	.216	Sq.	3.55	5.90		9.45	13.20	922
	0100	Foil 2 sides		37	.216		5.70	5.90		11.60	15.55	
	0300	Asphalt, two ply, 30#, for subfloors		19	.421		10.90	11.50		22.40	30	
	0400	Asphalt felt sheathing paper, 15#	↓	37	.216	↓	2.68	5.90		8.58	12.25	
	0450	Housewrap, exterior, spun bonded polypropylene										
	0470	Small roll	1 Carp	3,800	.002	S.F.	.10	.06		.16	.20	
	0480	Large roll	"	4,000	.002	"	.09	.05		.14	.19	
	0500	Material only, 3' x 111.1' roll				Ea.	33			33	36	
	0520	9' x 111.1' roll				"	93			93	102	
	0600	Polyethylene vapor barrier, standard, .002" thick	1 Carp	37	.216	Sq.	1.15	5.90		7.05	10.55	
	0700	.004" thick		37	.216		2.32	5.90		8.22	11.85	
	0900	.006" thick		37	.216		2.86	5.90		8.76	12.45	
	1200	.010" thick		37	.216		5.45	5.90		11.35	15.30	
	1300	Clear reinforced, fire retardant, .008" thick		37	.216		8.40	5.90		14.30	18.55	
	1350	Cross laminated type, .003" thick		37	.216		6.50	5.90		12.40	16.45	
	1400	.004" thick		37	.216		7.25	5.90		13.15	17.30	
	1500	Red rosin paper, 5 sq rolls, 4 lb per square		37	.216		1.55	5.90		7.45	11	
	1600	5 lbs. per square		37	.216		2	5.90		7.90	11.50	
	1800	Reinf. waterproof, .002" polyethylene backing, 1 side		37	.216		4.92	5.90		10.82	14.70	
	1900	2 sides	↓	37	.216		6.50	5.90		12.40	16.45	
	2100	Roof deck vapor barrier, class 1 metal decks	1 Rofc	37	.216		8.45	5.20		13.65	18.20	
	2200	For all other decks	"	37	.216	↓	5.75	5.20		10.95	15.25	

THERMAL & MOISTURE PROTECTION 7

203

071 920 | Vapor Retarders

			CREW	DAILY OUTPUT	LABOR-HOURS	UNIT	1999 BARE COSTS				TOTAL INCL O&P	
							MAT.	LABOR	EQUIP.	TOTAL		
922	2400	Waterproofed kraft with sisal or fiberglass fibers, minimum	1 Carp	37	.216	Sq.	5.30	5.90		11.20	15.15	922
	2500	Maximum	"	37	.216	↓	13.25	5.90		19.15	24	

072 100 | Building Insulation

			CREW	DAILY OUTPUT	LABOR-HOURS	UNIT	1999 BARE COSTS				TOTAL INCL O&P	
							MAT.	LABOR	EQUIP.	TOTAL		
101	0010	**BLOWN-IN INSULATION** Ceilings, with open access										101
	0020	Cellulose, 3-1/2" thick, R13	G-4	5,000	.005	S.F.	.13	.11	.05	.29	.37	
	0030	5-3/16" thick, R19		3,800	.006		.19	.14	.07	.40	.51	
	0050	6-1/2" thick, R22		3,000	.008		.24	.18	.09	.51	.64	
	0100	8-11/16" thick, R30		2,600	.009		.32	.20	.10	.62	.78	
	0120	10-7/8" thick, R38		1,800	.013		.41	.29	.15	.85	1.07	
	1000	Fiberglass, 5" thick, R11		3,800	.006		.14	.14	.07	.35	.45	
	1050	6" thick, R13		3,000	.008		.15	.18	.09	.42	.54	
	1100	8-1/2" thick, R19		2,200	.011		.21	.24	.12	.57	.74	
	1200	10" thick, R22		1,800	.013		.24	.29	.15	.68	.88	
	1300	12" thick, R26		1,500	.016		.29	.35	.18	.82	1.08	
	2000	Mineral wool, 4" thick, R12		3,500	.007		.16	.15	.08	.39	.50	
	2050	6" thick, R17		2,500	.010		.18	.21	.11	.50	.65	
	2100	9" thick, R23	↓	1,750	.014	↓	.27	.30	.15	.72	.95	
	2500	Wall installation, incl. drilling & patching from outside, two 1"										
	2510	diam. holes @ 16" O.C., top & mid-point of wall, add to above										
	2700	For masonry	G-4	415	.058	S.F.	.06	1.28	.65	1.99	2.79	
	2800	For wood siding	↓	840	.029		.06	.63	.32	1.01	1.41	
	2900	For stucco/plaster	↓	665	.036	↓	.06	.80	.40	1.26	1.78	
105	0010	**INSULATION FASTENERS**										105
	0050	Clips, galv, 1" or 2" insulation				C	16.20			16.20	17.80	
	0100	6" insulation					28.50			28.50	31	
	0300	Screws with plates, galv, for 1" or 2" insulation					14.75			14.75	16.25	
	0500	For 6" insulation				↓	27.50			27.50	30	
108	0010	**MASONRY INSULATION** Vermiculite or perlite, poured										108
	0100	In cores of concrete block, 4" thick wall, .115 CF/SF	D-1	4,800	.003	S.F.	.17	.08		.25	.31	
	0200	6" thick wall, .175 CF/SF		3,000	.005		.25	.13		.38	.48	
	0300	8" thick wall, .258 CF/SF		2,400	.007		.37	.16		.53	.66	
	0400	10" thick wall, .340 CF/SF		1,850	.009		.49	.21		.70	.87	
	0500	12" thick wall, .422 CF/SF	↓	1,200	.013	↓	.61	.33		.94	1.18	
	0550	For sand fill, deduct from above					70%					
	0600	Poured cavity wall, vermiculite or perlite, water repellant	D-1	250	.064	C.F.	1.45	1.57		3.02	4.05	
	0700	Foamed in place, urethane in 2-5/8" cavity	G-2	1,035	.023	S.F.	.38	.53	.23	1.14	1.49	
	0800	For each 1" added thickness, add	"	2,372	.010	"	.12	.23	.10	.45	.60	
109	0600	**PERIMETER INSULATION**, polystyrene, expanded, 1" thick, R4	1 Carp	680	.012	S.F.	.18	.32		.50	.71	109
	0700	2" thick, R8		675	.012	"	.33	.32		.65	.88	
110	0010	**POURED INSULATION** Cellulose fiber, R3.8 per inch		200	.040	C.F.	.45	1.09		1.54	2.22	110
	0040	Ceramic type (perlite), R3.2 per inch		200	.040		1.45	1.09		2.54	3.32	
	0080	Fiberglass wool, R4 per inch		200	.040		.29	1.09		1.38	2.04	
	0100	Mineral wool, R3 per inch		200	.040		.30	1.09		1.39	2.05	
	0300	Polystyrene, R4 per inch		200	.040		1.72	1.09		2.81	3.61	
	0400	Vermiculite or perlite, R2.7 per inch	↓	200	.040	↓	1.45	1.09		2.54	3.32	

7 THERMAL & MOISTURE PROTECTION

Important: See the Reference Section for critical supporting data - Reference Nos., Crews, & City Cost Indexes

			DAILY	LABOR-		1999 BARE COSTS				TOTAL	
	072 100 | Building Insulation	CREW	OUTPUT	HOURS	UNIT	MAT.	LABOR	EQUIP.	TOTAL	INCL O&P	
110	0700	Wood fiber, R3.85 per inch	1 Carp	200	.040	C.F.	.57	1.09		1.66	2.35	**110**
111	0010	**REFLECTIVE INSULATION,** aluminum foil on reinforced scrim	1 Carp	19	.421	C.S.F.	13.90	11.50		25.40	33.50	**111**
	0100	Reinforced with woven polyolefin		19	.421		16.90	11.50		28.40	36.50	
	0500	With single bubble air space, R8.8		15	.533		26	14.55		40.55	51.50	
	0600	With double bubble air space, R9.8	↓	15	.533	↓	28	14.55		42.55	54	
115	0010	**SPRAYED** Fibrous/cementitious, finished wall, 1" thick, R3.7	G-2	2,050	.012	S.F.	.22	.27	.11	.60	.78	**115**
	0100	Attic, 5.2" thick, R19	"	1,550	.015	"	.36	.35	.15	.86	1.12	
	0300	Foam type, incl. preparation										
	0600	3 #/CF, 1" thick, R3.8	G-2	770	.031	S.F.	.42	.71	.31	1.44	1.90	
	0700	2" thick, R7.5	"	475	.051	"	.84	1.16	.50	2.50	3.25	
116	0010	**WALL INSULATION, RIGID**										**116**
	0040	Fiberglass, 1.5#/CF, unfaced, 1" thick, R4.1	1 Carp	1,000	.008	S.F.	.23	.22		.45	.59	
	0060	1-1/2" thick, R6.2		1,000	.008		.30	.22		.52	.67	
	0080	2" thick, R8.3		1,000	.008		.37	.22		.59	.74	
	0120	3" thick, R12.4		800	.010		.44	.27		.71	.91	
	0370	3#/CF, unfaced, 1" thick, R4.3		1,000	.008		.29	.22		.51	.66	
	0390	1-1/2" thick, R6.5		1,000	.008		.56	.22		.78	.96	
	0400	2" thick, R8.7		890	.009		.69	.25		.94	1.15	
	0420	2-1/2" thick, R10.9		800	.010		.86	.27		1.13	1.38	
	0440	3" thick, R13		800	.010		1.02	.27		1.29	1.55	
	0520	Foil faced, 1" thick, R4.3		1,000	.008		.68	.22		.90	1.09	
	0540	1-1/2" thick, R6.5		1,000	.008		.91	.22		1.13	1.34	
	0560	2" thick, R8.7		890	.009		1.13	.25		1.38	1.63	
	0580	2-1/2" thick, R10.9		800	.010		1.34	.27		1.61	1.90	
	0600	3" thick, R13		800	.010		1.46	.27		1.73	2.04	
	0670	6#/CF, unfaced, 1" thick, R4.3		1,000	.008		.65	.22		.87	1.05	
	0690	1-1/2" thick, R6.5		890	.009		1.01	.25		1.26	1.50	
	0700	2" thick, R8.7		800	.010		1.41	.27		1.68	1.98	
	0721	2-1/2" thick, R10.9		800	.010		1.55	.27		1.82	2.13	
	0741	3" thick, R13		730	.011		1.86	.30		2.16	2.52	
	0821	Foil faced, 1" thick, R4.3		1,000	.008		.92	.22		1.14	1.35	
	0840	1-1/2" thick, R6.5		890	.009		1.32	.25		1.57	1.84	
	0850	2" thick, R8.7		800	.010		1.72	.27		1.99	2.32	
	0880	2-1/2" thick, R10.9		800	.010		2.07	.27		2.34	2.71	
	0900	3" thick, R13		730	.011		2.47	.30		2.77	3.19	
	1500	Foamglass, 1-1/2" thick, R3.9		800	.010		1.45	.27		1.72	2.03	
	1550	3" thick, R9	↓	730	.011	↓	2.71	.30		3.01	3.45	
	1600	Isocyanurate, 4' x 8' sheet, foil faced, both sides										
	1610	1/2" thick, R3.9	1 Carp	800	.010	S.F.	.28	.27		.55	.74	
	1620	5/8" thick, R4.5		800	.010		.29	.27		.56	.75	
	1630	3/4" thick, R5.4		800	.010		.30	.27		.57	.76	
	1640	1" thick, R7.2		800	.010		.33	.27		.60	.80	
	1650	1-1/2" thick, R10.8		730	.011		.37	.30		.67	.88	
	1660	2" thick, R14.4 **CN**		730	.011		.47	.30		.77	.98	
	1670	3" thick, R21.6		730	.011		1.10	.30		1.40	1.68	
	1680	4" thick, R28.8		730	.011		1.36	.30		1.66	1.97	
	1700	Perlite, 1" thick, R2.77		800	.010		.25	.27		.52	.71	
	1750	2" thick, R5.55		730	.011		.51	.30		.81	1.03	
	1900	Extruded polystyrene, 25 PSI compressive strength, 1" thick, R5		800	.010		.31	.27		.58	.77	
	1940	2" thick R10		730	.011		.61	.30		.91	1.14	
	1960	3" thick, R15		730	.011		.90	.30		1.20	1.47	
	2100	Expanded polystyrene, 1" thick, R3.85		800	.010		.15	.27		.42	.59	
	2120	2" thick, R7.69		730	.011		.32	.30		.62	.82	
	2140	3" thick, R11.49	↓	730	.011	↓	.52	.30		.82	1.04	

THERMAL & MOISTURE PROTECTION **7**

072 100 | Building Insulation

			CREW	DAILY OUTPUT	LABOR-HOURS	UNIT	MAT.	LABOR	EQUIP.	TOTAL	TOTAL INCL O&P	
118	0010	**WALL OR CEILING INSUL., NON-RIGID**										118
	0040	Fiberglass, kraft faced, batts or blankets										
	0060	3-1/2" thick, R11, 11" wide	1 Carp	1,150	.007	S.F.	.19	.19		.38	.51	
	0080	15" wide CN		1,600	.005		.19	.14		.33	.42	
	0100	23" wide		1,600	.005		.19	.14		.33	.42	
	0140	6" thick, R19, 11" wide		1,000	.008		.28	.22		.50	.65	
	0160	15" wide		1,350	.006		.28	.16		.44	.56	
	0180	23" wide		1,600	.005		.28	.14		.42	.52	
	0200	9" thick, R30, 15" wide		1,150	.007		.50	.19		.69	.85	
	0220	23" wide		1,350	.006		.50	.16		.66	.80	
	0240	12" thick, R38, 15" wide		1,000	.008		.64	.22		.86	1.04	
	0260	23" wide	▼	1,350	.006	▼	.64	.16		.80	.95	
	0400	Fiberglass, foil faced, batts or blankets										
	0420	3-1/2" thick, R11, 15" wide	1 Carp	1,600	.005	S.F.	.29	.14		.43	.53	
	0440	23" wide		1,600	.005		.29	.14		.43	.53	
	0460	6" thick, R19, 15" wide		1,350	.006		.35	.16		.51	.64	
	0480	23" wide		1,600	.005		.35	.14		.49	.60	
	0500	9" thick, R30, 15" wide		1,150	.007		.60	.19		.79	.96	
	0550	23" wide	▼	1,350	.006	▼	.60	.16		.76	.91	
	0800	Fiberglass, unfaced, batts or blankets										
	0820	3-1/2" thick, R11, 15" wide	1 Carp	1,350	.006	S.F.	.17	.16		.33	.44	
	0830	23" wide		1,600	.005		.17	.14		.31	.40	
	0860	6" thick, R19, 15" wide		1,150	.007		.29	.19		.48	.62	
	0880	23" wide		1,350	.006		.29	.16		.45	.57	
	0900	9" thick, R30, 15" wide		1,000	.008		.50	.22		.72	.89	
	0920	23" wide		1,150	.007		.50	.19		.69	.85	
	0940	12" thick, R38, 15" wide		1,000	.008		.64	.22		.86	1.04	
	0960	23" wide	▼	1,150	.007	▼	.64	.19		.83	1	
	1300	Mineral fiber batts, kraft faced										
	1320	3-1/2" thick, R12	1 Carp	1,600	.005	S.F.	.22	.14		.36	.45	
	1340	6" thick, R19		1,600	.005		.33	.14		.47	.57	
	1380	10" thick, R30		1,350	.006	▼	.53	.16		.69	.83	
	1850	Friction fit wire insulation supports, 16" O.C.	▼	960	.008	Ea.	.05	.23		.28	.42	
	1900	For foil backing, add				S.F.	.04			.04	.04	

072 200 | Roof & Deck Insulation

			CREW	DAILY OUTPUT	LABOR-HOURS	UNIT	MAT.	LABOR	EQUIP.	TOTAL	TOTAL INCL O&P	
203	0010	**ROOF DECK INSULATION**										203
	0020	Fiberboard low density, 1/2" thick R1.39	1 Rofc	1,000	.008	S.F.	.17	.19		.36	.52	
	0030	1" thick R2.78		800	.010		.32	.24		.56	.77	
	0080	1 1/2" thick R4.17		800	.010		.48	.24		.72	.95	
	0100	2" thick R5.56		800	.010		.64	.24		.88	1.12	
	0110	Fiberboard high density, 1/2" thick R1.3		1,000	.008		.18	.19		.37	.53	
	0120	1" thick R2.5		800	.010		.34	.24		.58	.79	
	0130	1-1/2" thick R3.8		800	.010		.55	.24		.79	1.03	
	0200	Fiberglass, 3/4" thick R2.78		1,000	.008		.39	.19		.58	.76	
	0400	15/16" thick R3.70		1,000	.008		.51	.19		.70	.89	
	0460	1-1/16" thick R4.17		1,000	.008		.64	.19		.83	1.03	
	0600	1-5/16" thick R5.26		1,000	.008		.88	.19		1.07	1.30	
	0650	2-1/16" thick R8.33		800	.010		.95	.24		1.19	1.47	
	0700	2-7/16" thick R10		800	.010		1.07	.24		1.31	1.60	
	1500	Foamglass, 1-3/4" thick R4.5		800	.010		1.41	.24		1.65	1.97	
	1530	3" thick R7.89		700	.011	▼	2.81	.28		3.09	3.56	
	1600	Tapered for drainage		600	.013	B.F.	.91	.32		1.23	1.55	
	1650	Perlite, 1/2" thick R1.32		1,050	.008	S.F.	.25	.18		.43	.60	
	1655	3/4" thick R2.08		800	.010		.30	.24		.54	.75	
	1660	1" thick R2.78	▼	800	.010	▼	.24	.24		.48	.68	

Important: See the Reference Section for critical supporting data - Reference Nos., Crews, & City Cost Indexes

072 200 | Roof & Deck Insulation

		CREW	DAILY OUTPUT	LABOR-HOURS	UNIT	1999 BARE COSTS				TOTAL INCL O&P	
						MAT.	LABOR	EQUIP.	TOTAL		
203	1670	1-1/2" thick R4.17	1 Rofc	800	.010	S.F.	.36	.24		.60	.82	**203**
	1680	2" thick R5.56		700	.011		.47	.28		.75	.99	
	1685	2-1/2" thick R6.67		700	.011	▼	.71	.28		.99	1.25	
	1690	Tapered for drainage		800	.010	B.F.	.54	.24		.78	1.01	
	1700	Polyisocyanurate, 2#/CF density, 3/4" thick, R5.1		1,500	.005	S.F.	.30	.13		.43	.55	
	1705	1" thick R7.14		1,400	.006		.32	.14		.46	.59	
	1715	1-1/2" thick R10.87		1,250	.006		.36	.15		.51	.67	
	1725	2" thick R14.29		1,100	.007		.45	.18		.63	.79	
	1735	2-1/2" thick R16.67		1,050	.008		.51	.18		.69	.88	
	1745	3" thick R21.74		1,000	.008		.62	.19		.81	1.01	
	1755	3-1/2" thick R25		1,000	.008	▼	.74	.19		.93	1.14	
	1765	Tapered for drainage	▼	1,400	.006	B.F.	.38	.14		.52	.66	
	1900	Extruded Polystyrene										
	1910	15 PSI compressive strength, 1" thick, R5	1 Rofc	1,500	.005	S.F.	.22	.13		.35	.46	
	1920	2" thick, R10		1,250	.006		.35	.15		.50	.65	
	1930	3" thick R15		1,000	.008		.77	.19		.96	1.18	
	1932	4" thick R20		1,000	.008	▼	1.06	.19		1.25	1.50	
	1934	Tapered for drainage		1,500	.005	B.F.	.35	.13		.48	.61	
	1940	25 PSI compressive strength, 1" thick R5		1,500	.005	S.F.	.33	.13		.46	.58	
	1942	2" thick R10		1,250	.006		.66	.15		.81	1	
	1944	3" thick R15		1,000	.008		.99	.19		1.18	1.42	
	1946	4" thick R20		1,000	.008	▼	1.10	.19		1.29	1.54	
	1948	Tapered for drainage		1,500	.005	B.F.	.40	.13		.53	.66	
	1950	40 psi compressive strength, 1" thick R5		1,500	.005	S.F.	.35	.13		.48	.61	
	1952	2" thick R10		1,250	.006		.68	.15		.83	1.02	
	1954	3" thick R15		1,000	.008		1	.19		1.19	1.43	
	1956	4" thick R20		1,000	.008	▼	1.34	.19		1.53	1.80	
	1958	Tapered for drainage		1,400	.006	B.F.	.50	.14		.64	.79	
	1960	60 PSI compressive strength, 1" thick R5		1,450	.006	S.F.	.42	.13		.55	.69	
	1962	2" thick R10		1,200	.007		.75	.16		.91	1.11	
	1964	3" thick R15		975	.008		1.12	.20		1.32	1.57	
	1966	4" thick R20		950	.008	▼	1.55	.20		1.75	2.05	
	1968	Tapered for drainage		1,400	.006	B.F.	.60	.14		.74	.90	
	2010	Expanded polystyrene, 1#/CF density, 3/4" thick R2.89		1,500	.005	S.F.	.18	.13		.31	.42	
	2020	1" thick R3.85		1,500	.005		.18	.13		.31	.42	
	2100	2" thick R7.69		1,250	.006		.33	.15		.48	.64	
	2110	3" thick R11.49		1,250	.006		.50	.15		.65	.82	
	2120	4" thick R15.38		1,200	.007		.56	.16		.72	.90	
	2130	5" thick R19.23		1,150	.007		.70	.17		.87	1.06	
	2140	6" thick R23.26		1,150	.007	▼	.83	.17		1	1.20	
	2150	Tapered for drainage	▼	1,500	.005	B.F.	.34	.13		.47	.59	
	2400	Composites with 2" EPS										
	2410	1" fiberboard	1 Rofc	950	.008	S.F.	.74	.20		.94	1.16	
	2420	7/16" oriented strand board		800	.010		.87	.24		1.11	1.38	
	2430	1/2" plywood		800	.010		.94	.24		1.18	1.45	
	2440	1" perlite	▼	800	.010	▼	.78	.24		1.02	1.28	
	2450	Composites with 1 1/2" polyisocyanurate										
	2460	1" fiberboard	1 Rofc	800	.010	S.F.	.80	.24		1.04	1.30	
	2470	1" perlite		850	.009		.83	.23		1.06	1.30	
	2480	7/16" oriented strand board	▼	800	.010	▼	.95	.24		1.19	1.47	

072 400 | Exterior Insulation

		CREW	DAILY OUTPUT	LABOR-HOURS	UNIT	MAT.	LABOR	EQUIP.	TOTAL	TOTAL INCL O&P	
402	0010	**EXTERIOR INSULATION FINISH SYSTEM**										**402**
	0095	Field applied, 1" EPS insulation	J-1	295	.136	S.F.	1.74	3.26	.14	5.14	7.05	
	0100	With 1/2" cement board sheathing		220	.182		2.77	4.37	.19	7.33	9.95	
	0105	2" EPS insulation	▼	295	.136	▼	1.91	3.26	.14	5.31	7.25	

THERMAL & MOISTURE PROTECTION **7**

072 400 | Exterior Insulation

		CREW	DAILY OUTPUT	LABOR-HOURS	UNIT	MAT.	LABOR	EQUIP.	TOTAL	TOTAL INCL O&P	
402	0110	With 1/2" cement board sheathing	J-1	220	.182	S.F.	2.94	4.37	.19	7.50	10.15	402
	0115	3" EPS insulation		295	.136		2.11	3.26	.14	5.51	7.50	
	0120	With 1/2" cement board sheathing		220	.182		3.14	4.37	.19	7.70	10.35	
	0125	4" EPS insulation		295	.136		2.23	3.26	.14	5.63	7.60	
	0130	With 1/2" cement board sheathing		220	.182		4.28	4.37	.19	8.84	11.60	
	0140	Premium finish add		1,265	.032		.27	.76	.03	1.06	1.51	
	0150	Heavy duty reinforcement add		914	.044		1.65	1.05	.05	2.75	3.48	
	0160	2.5#/S.Y. metal lath substrate add	1 Lath	75	.107	S.Y.	2.12	2.86		4.98	6.60	
	0170	3.4#/S.Y. metal lath substrate add	"	75	.107	"	2.21	2.86		5.07	6.70	
	0180	Color or texture change,	J-1	1,265	.032	S.F.	.72	.76	.03	1.51	2	
	0190	With substrate leveling base coat	1 Plas	530	.015		.72	.39		1.11	1.38	
	0210	With substrate sealing base coat	1 Pord	1,224	.007		.07	.16		.23	.33	
	0220	Prefab. panels, with hat channels 2 1/2" x 3/4", 2" EPS insul.	L-5	1,800	.031		7.70	.96	.31	8.97	10.50	
	0240	3" EPS insulation		1,800	.031		8.55	.96	.31	9.82	11.45	
	0250	4" EPS insulation		1,800	.031		9.55	.96	.31	10.82	12.55	
	0260	1-1/2" profile lightweight metal backing, 1" EPS insulation		1,800	.031		11.75	.96	.31	13.02	15	
	0270	2" EPS insulation		1,800	.031		13.10	.96	.31	14.37	16.50	
	0280	3" EPS insulation		1,800	.031		14.65	.96	.31	15.92	18.15	
	0290	4" EPS insulation		1,800	.031		16.35	.96	.31	17.62	20	
	0310	6"-16 Ga. steel stud back-up, 1" EPS insulation		1,800	.031		13.55	.96	.31	14.82	16.95	
	0320	2" EPS insulation		1,800	.031		15.15	.96	.31	16.42	18.70	
	0330	3" EPS insulation		1,800	.031		16.85	.96	.31	18.12	20.50	
	0340	4" EPS insulation		1,800	.031		18.85	.96	.31	20.12	22.50	
	0350	Premium finish add	J-1	1,265	.032		.27	.76	.03	1.06	1.51	
	0360	Heavy duty reinforcement add	"	914	.044		1.65	1.05	.05	2.75	3.48	
	0370	V groove shape in panel face				L.F.	.52			.52	.57	
	0380	U groove shape in panel face				"	.69			.69	.76	
	0390	Architectural features,										
	0410	Crown moulding 8" high x 4" wide	1 Plas	150	.053	L.F.	1.15	1.37		2.52	3.37	
	0420	Crown moulding 12" high x 6" wide		150	.053		2.15	1.37		3.52	4.47	
	0430	Crown moulding 16" high x 12" wide		150	.053		5.65	1.37		7.02	8.35	
	0440	For higher than one story, add						25%				

072 500 | Fireproofing

		CREW	DAILY OUTPUT	LABOR-HOURS	UNIT	MAT.	LABOR	EQUIP.	TOTAL	TOTAL INCL O&P	
554	0010	**SPRAYED** Mineral fiber or cementitious for fireproofing,										554
	0050	not incl tamping or canvas protection										
	0100	1" thick, on flat plate steel	G-2	3,000	.008	S.F.	.42	.18	.08	.68	.83	
	0200	Flat decking		2,400	.010		.42	.23	.10	.75	.92	
	0400	Beams		1,500	.016		.42	.37	.16	.95	1.20	
	0500	Corrugated or fluted decks		1,250	.019		.63	.44	.19	1.26	1.58	
	0700	Columns, 1-1/8" thick		1,100	.022		.47	.50	.21	1.18	1.53	
	0800	2-3/16" thick		700	.034		.88	.79	.34	2.01	2.55	
	0850	For tamping, add						10%				
	0900	For canvas protection, add	G-2	5,000	.005	S.F.	.06	.11	.05	.22	.29	
	1000	Acoustical sprayed, 1" thick, finished, straight work, minimum		520	.046		.44	1.06	.45	1.95	2.61	
	1100	Maximum		200	.120		.47	2.75	1.18	4.40	6.05	
	1300	Difficult access, minimum		225	.107		.47	2.44	1.05	3.96	5.45	
	1400	Maximum		130	.185		.51	4.23	1.81	6.55	9.10	
	1500	Intumescent epoxy fireproofing on wire mesh, 3/16" thick										
	1550	1 hour rating, exterior use	G-2	136	.176	S.F.	5.40	4.04	1.73	11.17	14.10	
	1600	Magnesium oxychloride, 35# to 40# density, 1/4" thick		3,000	.008		1.14	.18	.08	1.40	1.62	
	1650	1/2" thick		2,000	.012		2.28	.28	.12	2.68	3.07	
	1700	60# to 70# density, 1/4" thick		3,000	.008		1.50	.18	.08	1.76	2.02	
	1750	1/2" thick		2,000	.012		3.03	.28	.12	3.43	3.89	
	2000	Vermiculite cement, troweled or sprayed, 1/4" thick		3,000	.008		1.03	.18	.08	1.29	1.50	
	2050	1/2" thick		2,000	.012		2.04	.28	.12	2.44	2.80	

1999 BARE COSTS

072 700 | Firestopping

		CREW	DAILY OUTPUT	LABOR-HOURS	UNIT	1999 BARE COSTS				TOTAL INCL O&P
						MAT.	LABOR	EQUIP.	TOTAL	
0010	**FIRESTOPPING** R072-030									
0100	Metallic piping, non insulated									
0110	Through walls, 2" diameter	1 Carp	16	.500	Ea.	9.60	13.65		23.25	32
0120	4" diameter		14	.571		14.65	15.60		30.25	40.50
0130	6" diameter		12	.667		19.70	18.20		37.90	50
0140	12" diameter		10	.800		35	22		57	73
0150	Through floors, 2" diameter		32	.250		5.80	6.80		12.60	17.15
0160	4" diameter		28	.286		8.35	7.80		16.15	21.50
0170	6" diameter		24	.333		11	9.10		20.10	26.50
0180	12" diameter		20	.400		18.50	10.90		29.40	37.50
0190	Metallic piping, insulated									
0200	Through walls, 2" diameter	1 Carp	16	.500	Ea.	13.60	13.65		27.25	36.50
0210	4" diameter		14	.571		18.65	15.60		34.25	45
0220	6" diameter		12	.667		23.50	18.20		41.70	54.50
0230	12" diameter		10	.800		39	22		61	77
0240	Through floors, 2" diameter		32	.250		9.80	6.80		16.60	21.50
0250	4" diameter		28	.286		12.35	7.80		20.15	26
0260	6" diameter		24	.333		15	9.10		24.10	31
0270	12" diameter		20	.400		18.50	10.90		29.40	37.50
0280	Non metallic piping, non insulated									
0290	Through walls, 2" diameter	1 Carp	12	.667	Ea.	39.50	18.20		57.70	72
0300	4" diameter		10	.800		49.50	22		71.50	89
0310	6" diameter		8	1		69	27.50		96.50	119
0330	Through floors, 2" diameter		16	.500		31	13.65		44.65	55.50
0340	4" diameter		6	1.333		38.50	36.50		75	99.50
0350	6" diameter		6	1.333		46	36.50		82.50	108
0370	Ductwork, insulated & non insulated, round									
0380	Through walls, 6" diameter	1 Carp	12	.667	Ea.	20	18.20		38.20	50.50
0390	12" diameter		10	.800		40	22		62	78.50
0400	18" diameter		8	1		65	27.50		92.50	115
0410	Through floors, 6" diameter		16	.500		11	13.65		24.65	33.50
0420	12" diameter		14	.571		20	15.60		35.60	46.50
0430	18" diameter		12	.667		35	18.20		53.20	67
0440	Ductwork, insulated & non insulated, rectangular									
0450	With stiffener/closure angle, through walls, 6" x 12"	1 Carp	8	1	Ea.	16.65	27.50		44.15	61.50
0460	12" x 24"		6	1.333		22	36.50		58.50	82
0470	24" x 48"		4	2		63	54.50		117.50	156
0480	With stiffener/closure angle, through floors, 6" x 12"		10	.800		9	22		31	44.50
0490	12" x 24"		8	1		16.20	27.50		43.70	61
0500	24" x 48"		6	1.333		32	36.50		68.50	92.50
0510	Multi trade openings									
0520	Through walls, 6" x 12"	1 Carp	2	4	Ea.	35	109		144	211
0530	12" x 24"	"	1	8		141	218		359	500
0540	24" x 48"	2 Carp	1	16		565	435		1,000	1,300
0550	48" x 96"	"	.75	21.333		2,275	580		2,855	3,425
0560	Through floors, 6" x 12"	1 Carp	2	4		35	109		144	211
0570	12" x 24"	"	1	8		141	218		359	500
0580	24" x 48"	2 Carp	.75	21.333		565	580		1,145	1,525
0590	48" x 96"	"	.50	32		2,275	875		3,150	3,875
0600	Structural penetrations, through walls									
0610	Steel beams, W8 x 10	1 Carp	8	1	Ea.	22	27.50		49.50	67
0620	W12 x 14		6	1.333		35	36.50		71.50	96
0630	W21 x 44		5	1.600		70	43.50		113.50	146
0640	W36 x 135		3	2.667		170	73		243	300
0650	Bar joists, 18" deep		6	1.333		32	36.50		68.50	92.50
0660	24" deep		6	1.333		40	36.50		76.50	102

701

THERMAL & MOISTURE PROTECTION 7

072 700 | Firestopping

			DAILY OUTPUT	LABOR-HOURS	UNIT	1999 BARE COSTS				TOTAL INCL O&P	
		CREW				MAT.	LABOR	EQUIP.	TOTAL		
701	0670	36" deep	1 Carp	5	1.600	Ea.	60	43.50		103.50	135	701
	0680	48" deep	↓	4	2	↓	70	54.50		124.50	163	
	0690	Construction joints, floor slab at exterior wall										
	0700	Precast, brick, block or drywall exterior										
	0710	2" wide joint	1 Carp	125	.064	L.F.	5	1.75		6.75	8.25	
	0720	4" wide joint	"	75	.107	"	10	2.91		12.91	15.60	
	0730	Metal panel, glass or curtain wall exterior										
	0740	2" wide joint	1 Carp	40	.200	L.F.	11.85	5.45		17.30	21.50	
	0750	4" wide joint	"	25	.320	"	16.15	8.75		24.90	31.50	
	0760	Floor slab to drywall partition										
	0770	Flat joint	1 Carp	100	.080	L.F.	4.90	2.18		7.08	8.85	
	0780	Fluted joint	↓	50	.160		10	4.37		14.37	17.85	
	0790	Etched fluted joint	↓	75	.107	↓	6.50	2.91		9.41	11.75	
	0800	Floor slab to concrete/masonry partition										
	0810	Flat joint	1 Carp	75	.107	L.F.	11	2.91		13.91	16.70	
	0820	Fluted joint	"	50	.160	"	13	4.37		17.37	21	
	0830	Concrete/CMU wall joints										
	0840	1" wide	1 Carp	100	.080	L.F.	6	2.18		8.18	10.05	
	0850	2" wide		75	.107		11	2.91		13.91	16.70	
	0860	4" wide	↓	50	.160	↓	21	4.37		25.37	30	
	0870	Concrete/CMU floor joints										
	0880	1" wide	1 Carp	200	.040	L.F.	3	1.09		4.09	5	
	0890	2" wide		150	.053		5.50	1.46		6.96	8.35	
	0900	4" wide	↓	100	.080	↓	10.50	2.18		12.68	15	

Reference box: R072 -030

---

073 100 | Shingles

			DAILY OUTPUT	LABOR-HOURS	UNIT	1999 BARE COSTS				TOTAL INCL O&P	
		CREW				MAT.	LABOR	EQUIP.	TOTAL		
101	0010	**ALUMINUM** Shingles, mill finish, .019" thick	1 Carp	5	1.600	Sq.	150	43.50		193.50	234	101
	0100	.020" thick	"	5	1.600		145	43.50		188.50	228	
	0300	For colors, add				↓	15.15			15.15	16.65	
	0600	Ridge cap, .024" thick	1 Carp	170	.047	L.F.	1.85	1.28		3.13	4.06	
	0700	End wall flashing, .024" thick		170	.047		1.26	1.28		2.54	3.41	
	0900	Valley section, .024" thick		170	.047		2.30	1.28		3.58	4.55	
	1000	Starter strip, .024" thick		400	.020		1.20	.55		1.75	2.18	
	1200	Side wall flashing, .024" thick		170	.047		1.25	1.28		2.53	3.40	
	1500	Gable flashing, .024" thick	↓	400	.020	↓	1.35	.55		1.90	2.35	
103	0010	**FIBER CEMENT** shingles, 16" x 9.35", 500 lb per square	1 Carp	4	2	Sq.	244	54.50		298.50	355	103
	0110	Starters, 16" x 9.35"		3	2.667	C.L.F.	83	73		156	206	
	0120	Hip & ridge, 4.75" x 14"		1	8	"	600	218		818	1,000	
	0200	Shakes, 16" x 9.35", 550 lb per square		2.20	3.636	Sq.	221	99.50		320.50	400	
	0300	Hip & ridge, 4.75 x 14"		1	8	C.L.F.	600	218		818	1,000	
	0400	Hexagonal, 16" x 16"		3	2.667	Sq.	165	73		238	297	
	0500	Square, 16" x 16"	↓	3	2.667	"	148	73		221	278	
104	0010	**ASPHALT SHINGLES**										104
	0100	Standard strip shingles										
	0150	Inorganic, class A, 210-235 lb/sq **CN**	1 Rofc	5.50	1.455	Sq.	26	35		61	89	
	0155	Pneumatic nailed	↓	7	1.143	↓	26	27.50		53.50	76	

Important: See the Reference Section for critical supporting data - Reference Nos., Crews, & City Cost Indexes

073 100 | Shingles

			CREW	DAILY OUTPUT	LABOR-HOURS	UNIT	1999 BARE COSTS MAT.	LABOR	EQUIP.	TOTAL	TOTAL INCL O&P	
104	0200	Organic, class C, 235-240 lb/sq	1 Rofc	5	1.600	Sq.	34.50	38.50		73	105	104
	0205	Pneumatic nailed	↓	6.25	1.280	↓	34.50	31		65.50	91	
	0250	Standard, laminated multi-layered shingles										
	0300	Class A, 240-260 lb/sq	1 Rofc	4.50	1.778	Sq.	34	43		77	112	
	0305	Pneumatic nailed		5.63	1.422		34	34.50		68.50	96.50	
	0350	Class C, 260-300 lb/square, 4 bundles/square		4	2		49.50	48		97.50	138	
	0355	Pneumatic nailed	↓	5	1.600	↓	49.50	38.50		88	121	
	0400	Premium, laminated multi-layered shingles										
	0450	Class A, 260-300 lb, 4 bundles/sq	1 Rofc	3.50	2.286	Sq.	43	55		98	143	
	0455	Pneumatic nailed		4.37	1.831		43	44		87	124	
	0500	Class C, 300-385 lb/square, 5 bundles/square		3	2.667		65.50	64.50		130	183	
	0505	Pneumatic nailed		3.75	2.133		65.50	51.50		117	161	
	0800	#15 felt underlayment		64	.125		2.68	3.01		5.69	8.15	
	0825	#30 felt underlayment		58	.138		5.45	3.32		8.77	11.70	
	0850	Self adhering polyethylene and rubberized asphalt underlayment		22	.364	↓	39.50	8.75		48.25	58.50	
	0900	Ridge shingles		330	.024	L.F.	.72	.58		1.30	1.80	
	0905	Pneumatic nailed	↓	412.50	.019	"	.72	.47		1.19	1.59	
	1000	For steep roofs (7 to 12 pitch or greater), add						50%				
106	0010	SLATE, Buckingham, Virginia, black R073-020										106
	0100	3/16" - 1/4" thick	1 Rots	1.75	4.571	Sq.	540	111		651	785	
	0200	1/4" thick		1.75	4.571		720	111		831	980	
	0900	Pennsylvania black, Bangor, #1 clear		1.75	4.571		435	111		546	670	
	1200	Vermont, unfading, green, mottled green		1.75	4.571		395	111		506	625	
	1300	Semi-weathering green & gray		1.75	4.571		296	111		407	515	
	1400	Purple		1.75	4.571		390	111		501	620	
	1500	Black or gray		1.75	4.571		405	111		516	635	
	1600	Red		1.75	4.571		1,125	111		1,236	1,425	
	1700	Variegated purple	↓	1.75	4.571		375	111		486	600	
107	0010	STEEL Shingles, galvanized, 26 gauge	1 Rots	2.20	3.636	Sq.	149	88		237	315	107
	0200	24 gauge	"	2.20	3.636		157	88		245	325	
	0300	For colored galvanized shingles, add					41			41	45	
	0500	For 1" factory applied polystyrene insulation, add				↓	30			30	33	
108	0010	WOOD 16" No. 1 red cedar shingles, 5" exposure, on roof	1 Carp	2.50	3.200	Sq.	143	87.50		230.50	294	108
	0015	Pneumatic nailed		3.25	2.462		143	67		210	263	
	0200	7-1/2" exposure, on walls		2.05	3.902		95	107		202	273	
	0205	Pneumatic nailed		2.67	2.996		95	82		177	234	
	0300	18" No. 1 red cedar perfections, 5-1/2" exposure, on roof		2.75	2.909		164	79.50		243.50	305	
	0305	Pneumatic nailed		3.57	2.241		164	61		225	278	
	0500	7-1/2" exposure, on walls		2.25	3.556		121	97		218	286	
	0505	Pneumatic nailed		2.92	2.740		121	75		196	251	
	0600	Resquared, and rebutted, 5-1/2" exposure, on roof		3	2.667		199	73		272	335	
	0605	Pneumatic nailed		3.90	2.051		199	56		255	305	
	0900	7-1/2" exposure, on walls		2.45	3.265		146	89		235	300	
	0905	Pneumatic nailed	↓	3.18	2.516		146	68.50		214.50	269	
	1000	Add to above for fire retardant shingles, 16" long					30			30	33	
	1050	18" long				↓	28.50			28.50	31.50	
	1060	Preformed ridge shingles	1 Carp	400	.020	L.F.	1.65	.55		2.20	2.68	
	1100	Hand-split red cedar shakes, 1/2" thick x 24" long, 10" exp. on roof		2.50	3.200	Sq.	138	87.50		225.50	289	
	1105	Pneumatic nailed		3.25	2.462		138	67		205	258	
	1110	3/4" thick x 24" long, 10" exp. on roof		2.25	3.556		138	97		235	305	
	1115	Pneumatic nailed		2.92	2.740		138	75		213	270	
	1200	1/2" thick, 18" long, 8-1/2" exp. on roof		2	4		97.50	109		206.50	279	
	1205	Pneumatic nailed		2.60	3.077		97.50	84		181.50	239	
	1210	3/4" thick x 18" long, 8 1/2" exp. on roof		1.80	4.444		97.50	121		218.50	298	
	1215	Pneumatic nailed		2.34	3.419		97.50	93.50		191	254	
	1255	10" exp. on walls	↓	2	4	↓	110	109		219	293	

073 100 | Shingles

		CREW	DAILY OUTPUT	LABOR-HOURS	UNIT	1999 BARE COSTS				TOTAL INCL O&P	
						MAT.	LABOR	EQUIP.	TOTAL		
108	1260	10" exposure on walls, pneumatic nailed	1 Carp	2.60	3.077	Sq.	110	84		194	253	108
	1700	Add to above for fire retardant shakes, 24" long					30			30	33	
	1800	18" long	↓			↓	30			30	33	
	1810	Ridge shakes	1 Carp	350	.023	L.F.	2.35	.62		2.97	3.56	
	2000	White cedar shingles, 16" long, extras, 5" exposure, on roof		2.40	3.333	Sq.	125	91		216	281	
	2005	Pneumatic nailed		3.12	2.564		125	70		195	248	
	2050	5" exposure on walls		2	4		125	109		234	310	
	2055	Pneumatic nailed		2.60	3.077		125	84		209	270	
	2100	7-1/2" exposure, on walls		2	4		89.50	109		198.50	270	
	2105	Pneumatic nailed		2.60	3.077		89.50	84		173.50	230	
	2150	"B" grade, 5" exposure on walls		2	4		117	109		226	300	
	2155	Pneumatic nailed		2.60	3.077		117	84		201	261	
	2300	For 15# organic felt underlayment on roof, 1 layer, add		64	.125		2.68	3.41		6.09	8.30	
	2400	2 layers, add	↓	32	.250		5.35	6.80		12.15	16.65	
	2600	For steep roofs (7/12 pitch or greater), add to above				↓		50%				
	2700	Panelized systems, No.1 cedar shingles on 5/16" CDX plywood										
	2800	On walls, 8' strips, 7" or 14" exposure	2 Carp	700	.023	S.F.	3.20	.62		3.82	4.50	
	3500	On roofs, 8' strips, 7" or 14" exposure	1 Carp	3	2.667	Sq.	320	73		393	465	
	3505	Pneumatic nailed	"	4	2	"	320	54.50		374.50	435	

073 200 | Roofing Tile

		CREW	DAILY OUTPUT	LABOR-HOURS	UNIT	MAT.	LABOR	EQUIP.	TOTAL	TOTAL INCL O&P	
201	0010	**ALUMINUM** Tiles with accessories, .032" thick, mission tile	1 Carp	2.50	3.200	Sq.	350	87.50		437.50	520	201
	0200	Spanish tiles	"	3	2.667	"	350	73		423	500	
202	0010	**CLAY TILE** ASTM C1167, GR 1, severe weathering, acces. incl.										202
	0200	Lanai tile or Classic tile, 158 pc per sq	1 Rots	1.65	4.848	Sq.	490	117		607	740	
	0300	Americana, 158 pc per sq, most colors		1.65	4.848		490	117		607	740	
	0350	Green, gray or brown		1.65	4.848		490	117		607	740	
	0400	Blue		1.65	4.848		490	117		607	740	
	0600	Spanish tile, 171 pc per sq, red		1.80	4.444		305	108		413	525	
	0800	Blend		1.80	4.444		420	108		528	645	
	0900	Glazed white		1.80	4.444		500	108		608	735	
	1100	Mission tile, 192 pc per sq, machine scored finish, red		1.15	6.957		635	168		803	985	
	1700	French tile, 133 pc per sq, smooth finish, red		1.35	5.926		575	143		718	875	
	1750	Blue or green		1.35	5.926		685	143		828	1,000	
	1800	Norman black 317 pc per sq		1	8		805	194		999	1,225	
	2200	Williamsburg tile, 158 pc per sq, aged cedar		1.35	5.926		490	143		633	785	
	2250	Gray or green		1.35	5.926		490	143		633	785	
	2510	One piece mission tile, natural red, 75 pc per square		1.65	4.848		133	117		250	350	
	2530	Mission Tile, 134 pc per square	↓	1.15	6.957	↓	178	168		346	485	
204	0010	**CONCRETE TILE** Including installation of accessories										204
	0020	Corrugated, 13" x 16-1/2", 90 per sq, 950 lb per sq										
	0050	Earthtone colors, nailed to wood deck	1 Rots	1.35	5.926	Sq.	103	143		246	360	
	0150	Custom blues		1.35	5.926		114	143		257	375	
	0200	Custom greens		1.35	5.926		114	143		257	375	
	0250	Premium colors	↓	1.35	5.926	↓	153	143		296	415	
	0500	Shakes, 13" x 16-1/2", 90 per sq, 950 lb per sq										
	0600	All colors, nailed to wood deck	1 Rots	1.50	5.333	Sq.	185	129		314	425	
	1500	Accessory pieces, ridge & hip, 10" x 16-1/2", 8 lbs. each				Ea.	2.25			2.25	2.48	
	1700	Rake, 6-1/2" x 16-3/4", 9 lbs. each					2.25			2.25	2.48	
	1800	Mansard hip, 10" x 16-1/2", 9.2 lbs. each					2.25			2.25	2.48	
	1900	Hip starter, 10" x 16-1/2", 10.5 lbs. each					9.50			9.50	10.45	
	2000	3 or 4 way apex, 10" each side, 11.5 lbs. each				↓	10.25			10.25	11.30	

7

THERMAL & MOISTURE PROTECTION

Important: See the Reference Section for critical supporting data - Reference Nos., Crews, & City Cost Indexes

074 100 | Preformed Panels

			CREW	DAILY OUTPUT	LABOR-HOURS	UNIT	1999 BARE COSTS				TOTAL INCL O&P	
							MAT.	LABOR	EQUIP.	TOTAL		
101	0010	**ALUMINUM ROOFING** Corrugated or ribbed, .0155" thick, natural	G-3	1,200	.027	S.F.	.61	.71		1.32	1.78	**101**
	0300	Painted		1,200	.027		.87	.71		1.58	2.07	
	0400	Corrugated, .018" thick, on steel frame, natural finish		1,200	.027		.78	.71		1.49	1.97	
	0600	Painted		1,200	.027		.97	.71		1.68	2.18	
	0700	Corrugated, on steel frame, natural, .024" thick		1,200	.027		1.14	.71		1.85	2.36	
	0800	Painted, .024" thick		1,200	.027		1.37	.71		2.08	2.62	
	0900	.032" thick, natural		1,200	.027		1.23	.71		1.94	2.46	
	1200	painted		1,200	.027		1.61	.71		2.32	2.88	
	1300	V-Beam, on steel frame construction, .032" thick, natural		1,200	.027		1.39	.71		2.10	2.64	
	1500	Painted		1,200	.027		1.78	.71		2.49	3.07	
	1600	.040" thick, natural		1,200	.027		1.73	.71		2.44	3.01	
	1800	Painted		1,200	.027		2.15	.71		2.86	3.47	
	1900	.050" thick, natural		1,200	.027		2.10	.71		2.81	3.42	
	2100	Painted		1,200	.027		2.58	.71		3.29	3.94	
	2200	For roofing on wood frame, deduct		4,600	.007	▼	.05	.19		.24	.35	
	2400	Ridge cap, .032" thick, natural	▼	800	.040	L.F.	1.64	1.06		2.70	3.46	
104	0010	**FIBERGLASS** Corrugated panels, roofing, 8 oz per SF	G-3	1,000	.032	S.F.	2.17	.85		3.02	3.72	**104**
	0100	12 oz per SF		1,000	.032	"	2.87	.85		3.72	4.49	
	0300	Corrugated siding, 6 oz per SF		880	.036	S.F.	1.88	.97		2.85	3.58	
	0400	8 oz per SF		880	.036		2.17	.97		3.14	3.90	
	0500	Fire retardant		880	.036		3	.97		3.97	4.81	
	0600	12 oz. siding, textured		880	.036		2.70	.97		3.67	4.48	
	0700	Fire retardant		880	.036		4.02	.97		4.99	5.95	
	0900	Flat panels, 6 oz per SF, clear or colors		880	.036		1.68	.97		2.65	3.36	
	1100	Fire retardant, class A		880	.036		2.91	.97		3.88	4.71	
	1300	8 oz per SF, clear or colors		880	.036		2.17	.97		3.14	3.90	
	1700	Sandwich panels, fiberglass, 1-9/16" thick, panels to 20 SF		180	.178		21	4.73		25.73	30.50	
	1900	As above, but 2-3/4" thick, panels to 100 SF	▼	265	.121	▼	15.60	3.21		18.81	22	
107	0010	**STEEL ROOFING** on steel frame, corrugated or ribbed, 30 ga galv	G-3	1,100	.029	S.F.	.75	.77		1.52	2.04	**107**
	0100	28 ga		1,050	.030		.79	.81		1.60	2.14	
	0300	26 ga		1,000	.032		.86	.85		1.71	2.28	
	0400	24 ga		950	.034		1.02	.90		1.92	2.52	
	0600	Colored, 28 ga		1,050	.030		1.05	.81		1.86	2.42	
	0700	26 ga		1,000	.032		1.12	.85		1.97	2.56	
	0710	Flat profile, 1-3/4" standing seams, 10" wide, standard finish, 26 ga		1,000	.032		2.30	.85		3.15	3.86	
	0715	24 ga		950	.034		2.67	.90		3.57	4.34	
	0720	22 ga		900	.036		3.29	.95		4.24	5.10	
	0725	Zinc aluminum alloy finish, 26 ga		1,000	.032		1.80	.85		2.65	3.31	
	0730	24 ga		950	.034		2.15	.90		3.05	3.77	
	0735	22 ga		900	.036		2.47	.95		3.42	4.20	
	0740	12" wide, standard finish, 26 ga		1,000	.032		2.29	.85		3.14	3.85	
	0745	24 ga		950	.034		2.66	.90		3.56	4.33	
	0750	Zinc aluminum alloy finish, 26 ga		1,000	.032		1.70	.85		2.55	3.20	
	0755	24 ga		950	.034		2.03	.90		2.93	3.63	
	0840	Flat profile, 1" x 3/8" batten, 12" wide, standard finish, 26 ga		1,000	.032		1.79	.85		2.64	3.30	
	0845	24 ga		950	.034		2.10	.90		3	3.71	
	0850	22 ga		900	.036		2.52	.95		3.47	4.25	
	0855	Zinc aluminum alloy finish, 26 ga		1,000	.032		1.51	.85		2.36	2.99	
	0860	24 ga		950	.034		1.68	.90		2.58	3.25	
	0865	22 ga		900	.036		1.95	.95		2.90	3.63	
	0870	16-1/2" wide, standard finish, 24 ga		950	.034		2.13	.90		3.03	3.74	
	0875	22 ga		900	.036		2.39	.95		3.34	4.11	
	0880	Zinc aluminum alloy finish, 24 ga		950	.034		1.60	.90		2.50	3.16	
	0885	22 ga		900	.036		1.78	.95		2.73	3.44	
	0890	Flat profile, 2" x 2" batten, 12" wide, standard finish, 26 ga		1,000	.032		2.05	.85		2.90	3.59	
	0895	24 ga	▼	950	.034	▼	2.45	.90		3.35	4.10	

		074 100	**Preformed Panels**	CREW	DAILY OUTPUT	LABOR-HOURS	UNIT	1999 BARE COSTS MAT.	LABOR	EQUIP.	TOTAL	TOTAL INCL O&P	
107	0900		22 ga	G-3	900	.036	S.F.	3	.95		3.95	4.78	107
	0905		Zinc aluminum alloy finish, 26 ga		1,000	.032		1.69	.85		2.54	3.19	
	0910		24 ga		950	.034		1.83	.90		2.73	3.41	
	0915		22 ga		900	.036		2.25	.95		3.20	3.96	
	0920		16-1/2" wide, standard finish, 24 ga		950	.034		2.27	.90		3.17	3.90	
	0925		22 ga		900	.036		2.64	.95		3.59	4.38	
	0930		Zinc aluminum alloy finish, 24 ga		950	.034		1.80	.90		2.70	3.38	
	0935		22 ga		900	.036	▼	2.05	.95		3	3.74	
	1200		Ridge, galvanized, 10" wide		800	.040	L.F.	1.68	1.06		2.74	3.51	
	1210		20" wide	▼	750	.043	"	2.95	1.13		4.08	5	

		074 200	**Composite Panels**										
201	0010		**EXPOSED AGGREGATE PANELS**										201
	1400		Fiberglass polymer back-up,										
	1500		Small size aggregate	F-3	445	.090	S.F.	4.62	2.49	1	8.11	10.10	
	1600		Medium size aggregate		445	.090		5.20	2.49	1	8.69	10.70	
	1700		Large size aggregate	▼	445	.090	▼	5.35	2.49	1	8.84	10.90	
202	0010		**METAL FACING PANELS**										202
	0400		Textured aluminum, 4' x 8' x 5/16" plywood backing, single face	2 Shee	375	.043	S.F.	2.18	1.35		3.53	4.50	
	0600		Double face		375	.043		3.28	1.35		4.63	5.70	
	0700		4' x 10' x 5/16" plywood backing, single face		375	.043		2.08	1.35		3.43	4.39	
	0900		Double face		375	.043		3.15	1.35		4.50	5.55	
	1000		4' x 12' x 5/15" plywood backing, single face		375	.043		2.99	1.35		4.34	5.40	
	1300		Smooth al, 1/4" panel, fluoropolymer finish, double face		375	.043		5.35	1.35		6.70	8	
	1350		Clear anodized finish, double face		375	.043		5.95	1.35		7.30	8.65	
	1400		Double face textured aluminum, structural panel, 1" EPS insulation	▼	375	.043	▼	3.60	1.35		4.95	6.05	
	1500		Accessories, outside corner	1 Shee	175	.046	L.F.	1.54	1.45		2.99	3.94	
	1600		Inside corner		175	.046		1.04	1.45		2.49	3.39	
	1800		Batten mounting clip		200	.040		.35	1.27		1.62	2.36	
	1900		Low profile batten		480	.017		.72	.53		1.25	1.61	
	2100		High profile batten		480	.017		1.21	.53		1.74	2.15	
	2200		Water table		200	.040		1.28	1.27		2.55	3.38	
	2400		Horizontal joint connector		200	.040		1.05	1.27		2.32	3.12	
	2500		Corner cap		200	.040		1.28	1.27		2.55	3.38	
	2700		H - moulding	▼	480	.017	▼	.68	.53		1.21	1.57	

		074 600	**Cladding/Siding**										
602	0010		**ALUMINUM SIDING** .019" thick, on steel construction, natural	G-3	775	.041	S.F.	.65	1.10		1.75	2.42	602
	0100		Painted		775	.041		.84	1.10		1.94	2.63	
	0400		Farm type, .021" thick on steel frame, natural		775	.041		.75	1.10		1.85	2.54	
	0600		Painted		775	.041		.92	1.10		2.02	2.72	
	0700		Industrial type, corrugated, on steel, .024" thick, mill		775	.041		.95	1.10		2.05	2.76	
	0900		Painted		775	.041		1.15	1.10		2.25	2.98	
	1000		.032" thick, mill		775	.041		1.30	1.10		2.40	3.14	
	1200		Painted		775	.041		1.65	1.10		2.75	3.52	
	1300		V-Beam, on steel frame, .032" thick, mill		775	.041		1.48	1.10		2.58	3.33	
	1500		Painted		775	.041		1.80	1.10		2.90	3.69	
	1600		.040" thick, mill		775	.041		1.80	1.10		2.90	3.69	
	1800		Painted		775	.041		2.15	1.10		3.25	4.07	
	1900		.050" thick, mill		775	.041		2.11	1.10		3.21	4.03	
	2100		Painted		775	.041		2.58	1.10		3.68	4.54	
	2200		Ribbed, 3" profile, on steel frame, .032" thick, natural		775	.041		1.35	1.10		2.45	3.20	
	2400		Painted		775	.041		1.60	1.10		2.70	3.47	
	2500		.040" thick, natural		775	.041		1.60	1.10		2.70	3.47	
	2700		Painted	▼	775	.041	▼	1.87	1.10		2.97	3.77	

THERMAL & MOISTURE PROTECTION — 7

Important: See the Reference Section for critical supporting data - Reference Nos., Crews, & City Cost Indexes

			DAILY	LABOR-		\multicolumn{4}{c}{1999 BARE COSTS}		TOTAL	
074 600	**Cladding/Siding**	CREW	OUTPUT	HOURS	UNIT	MAT.	LABOR	EQUIP.	TOTAL	INCL O&P	
602 2750	.050" thick, natural	G-3	775	.041	S.F.	1.84	1.10		2.94	3.73	**602**
2760	Painted		775	.041		2.10	1.10		3.20	4.02	
3300	For siding on wood frame, deduct from above	↓	2,800	.011	↓	.06	.30		.36	.54	
3400	Screw fasteners, aluminum, self tapping, neoprene washer, 1"				M	123			123	135	
3600	Stitch screws, self tapping, with neoprene washer, 5/8"				"	92.50			92.50	101	
3630	Flashing, sidewall, .032" thick	G-3	800	.040	L.F.	1.75	1.06		2.81	3.59	
3650	End wall, .040" thick		800	.040		2.05	1.06		3.11	3.92	
3670	Closure strips, corrugated, .032" thick		800	.040		.52	1.06		1.58	2.23	
3680	Ribbed, 4" or 8", .032" thick		800	.040		.52	1.06		1.58	2.23	
3690	V-beam, .040" thick	↓	800	.040	↓	.71	1.06		1.77	2.44	
3800	Horizontal, colored clapboard, 8" wide, plain	2 Carp	515	.031	S.F.	1.22	.85		2.07	2.67	
3900	Insulated		515	.031		1.49	.85		2.34	2.97	
4000	Vertical board & batten, colored, non-insulated	↓	515	.031		1.06	.85		1.91	2.50	
4200	For simulated wood design, add				↓	.08			.08	.09	
4300	Corners for above, outside	2 Carp	515	.031	V.L.F.	1.67	.85		2.52	3.17	
4500	Inside corners	"	515	.031	"	.94	.85		1.79	2.36	
4600	Sandwich panels, 1" insulation, single story	G-3	395	.081	S.F.	4.52	2.15		6.67	8.35	
4900	Multi-story	"	345	.093		6.35	2.47		8.82	10.80	
5100	For baked enamel finish 1 side, add				↓	.24			.24	.26	
5200	See also Metal Facing Panels, division 074-202										
606 0010	**STEEL SIDING**, Beveled, vinyl coated, 8" wide, including fasteners	1 Carp	265	.030	S.F.	1.12	.82		1.94	2.53	**606**
0050	10" wide	"	275	.029		1.19	.79		1.98	2.56	
0080	Galv, corrugated or ribbed, on steel frame, 30 gauge	G-3	800	.040		.75	1.06		1.81	2.49	
0100	28 gauge		795	.040		.79	1.07		1.86	2.54	
0300	26 gauge		790	.041		.86	1.08		1.94	2.63	
0400	24 gauge		785	.041		1.03	1.08		2.11	2.82	
0600	22 gauge		770	.042		1.17	1.11		2.28	3.02	
0700	Colored, corrugated/ribbed, on steel frame, 10 yr fnsh, 28 ga.		800	.040		1.08	1.06		2.14	2.85	
0900	26 gauge		795	.040		.95	1.07		2.02	2.72	
1000	24 gauge		790	.041		1.09	1.08		2.17	2.88	
1020	20 gauge		785	.041		1.41	1.08		2.49	3.24	
1200	Factory sandwich panel, 26 ga., 1" insulation, galvanized		380	.084		2.76	2.24		5	6.55	
1300	Colored 1 side		380	.084		3.50	2.24		5.74	7.35	
1500	Galvanized 2 sides		380	.084		4.35	2.24		6.59	8.30	
1600	Colored 2 sides		380	.084		4.50	2.24		6.74	8.45	
1800	Acrylic paint face, regular paint liner	↓	380	.084		3.40	2.24		5.64	7.25	
1900	For 2" thick polystyrene, add					.52			.52	.57	
2000	22 ga, galv, 2" insulation, baked enamel exterior	G-3	360	.089		6.65	2.36		9.01	11	
2100	P.V.F. exterior finish	"	360	.089	↓	7	2.36		9.36	11.40	
607 0010	**VINYL SIDING** Solid PVC panels, 8" to 10" wide, plain	1 Carp	255	.031	S.F.	.61	.86		1.47	2.02	**607**
0100	with 3/8" insulation		255	.031		.75	.86		1.61	2.18	
0200	Soffit and fascia		205	.039	↓	1.33	1.07		2.40	3.14	
0300	Window and door trim moldings		185	.043	L.F.	.31	1.18		1.49	2.20	
0500	Corner posts, outside corner		205	.039		1.04	1.07		2.11	2.82	
0600	Inside corner	↓	205	.039	↓	.58	1.07		1.65	2.32	
609 0010	**WOOD SIDING, BOARDS**										**609**
3200	Wood, cedar bevel, A grade, 1/2" x 6"	1 Carp	250	.032	S.F.	2.02	.87		2.89	3.59	
3300	1/2" x 8"		275	.029		1.66	.79		2.45	3.08	
3500	3/4" x 10", clear grade		300	.027		2.16	.73		2.89	3.52	
3600	"B" grade		300	.027		3.22	.73		3.95	4.69	
3800	Cedar, rough sawn, 1" x 4", A grade, natural		240	.033		3.01	.91		3.92	4.74	
3900	Stained		240	.033		3.39	.91		4.30	5.15	
4100	1" x 12", board & batten, #3 & Btr., natural		260	.031		2.08	.84		2.92	3.61	
4200	Stained		260	.031		2.42	.84		3.26	3.98	
4400	1" x 8" channel siding, #3 & Btr., natural	↓	250	.032	↓	2.02	.87		2.89	3.59	

7

THERMAL & MOISTURE PROTECTION

074 600 | Cladding/Siding

		CREW	DAILY OUTPUT	LABOR-HOURS	UNIT	1999 BARE COSTS				TOTAL INCL O&P	
						MAT.	LABOR	EQUIP.	TOTAL		
609	4500	Stained	1 Carp	250	.032	S.F.	2.30	.87		3.17	3.90	609
	4700	Redwood, clear, beveled, vertical grain, 1/2" x 4"		200	.040		3.22	1.09		4.31	5.25
	4750	1/2" x 6"		225	.036		2.70	.97		3.67	4.50
	4800	1/2" x 8"		250	.032		2.19	.87		3.06	3.78
	5000	3/4" x 10"		300	.027		3.57	.73		4.30	5.10
	5200	Channel siding, 1" x 10", B grade		285	.028		2.30	.77		3.07	3.74
	5250	Redwood, T&G boards, B grade, 1" x 4"	2 Carp	300	.053		2.74	1.46		4.20	5.30
	5270	1" x 8"	"	375	.043		2.36	1.16		3.52	4.43
	5400	White pine, rough sawn, 1" x 8", natural	1 Carp	275	.029		.69	.79		1.48	2.01
	5500	Stained	"	275	.029		1.02	.79		1.81	2.37
611	0010	**WOOD PRODUCT SIDING**										611
	0030	Lap siding, hardboard, 7/16" x 8", primed									
	0050	Wood grain texture finish	2 Carp	650	.025	S.F.	.95	.67		1.62	2.11
	0100	Panels, 7/16" thick, smooth, textured or grooved, primed		700	.023		.81	.62		1.43	1.87
	0200	Stained		700	.023		.88	.62		1.50	1.95
	0700	Particle board, overlaid, 3/8" thick		750	.021		.63	.58		1.21	1.61
	0900	Plywood, medium density overlaid, 3/8" thick		750	.021		1.08	.58		1.66	2.11
	1000	1/2" thick		700	.023		1.26	.62		1.88	2.37
	1100	3/4" thick		650	.025		1.47	.67		2.14	2.68
	1600	Texture 1-11, cedar, 5/8" thick, natural		675	.024		1.50	.65		2.15	2.67
	1700	Factory stained		675	.024		1.74	.65		2.39	2.93
	1900	Texture 1-11, fir, 5/8" thick, natural		675	.024		1.01	.65		1.66	2.13
	2000	Factory stained		675	.024		1.14	.65		1.79	2.27
	2050	Texture 1-11, S.Y.P., 5/8" thick, natural		675	.024		.83	.65		1.48	1.93
	2100	Factory stained		675	.024		.93	.65		1.58	2.04
	2200	Rough sawn cedar, 3/8" thick, natural		675	.024		1.14	.65		1.79	2.27
	2300	Factory stained		675	.024		1.26	.65		1.91	2.41
	2500	Rough sawn fir, 3/8" thick, natural		675	.024		.61	.65		1.26	1.69
	2600	Factory stained		675	.024		.68	.65		1.33	1.77
	2800	Redwood, textured siding, 5/8" thick		675	.024		1.89	.65		2.54	3.10
	3000	Polyvinyl chloride coated, 3/8" thick		750	.021		.93	.58		1.51	1.94
615	0010	**FIBER CEMENT SIDING**										615
	0020	Lap siding, 5/16" thick, 6" wide, smooth texture	2 Carp	415	.039	S.F.	.93	1.05		1.98	2.69
	0025	Woodgrain texture		415	.039		.93	1.05		1.98	2.69
	0030	7-1/2" wide, smooth texture		425	.038		.92	1.03		1.95	2.63
	0035	Woodgrain texture		425	.038		.92	1.03		1.95	2.63
	0040	8" wide, smooth texture		425	.038		.92	1.03		1.95	2.64
	0045	Roughsawn texture		425	.038		.92	1.03		1.95	2.64
	0050	9-1/2" wide, smooth texture		440	.036		.89	.99		1.88	2.54
	0055	Woodgrain texture		440	.036		.89	.99		1.88	2.54
	0060	12" wide, smooth texture		455	.035		.86	.96		1.82	2.46
	0065	Woodgrain texture		455	.035		.86	.96		1.82	2.46
	0070	Panel siding, 5/16" thick, smooth texture		750	.021		.77	.58		1.35	1.76
	0075	Stucco texture		750	.021		.77	.58		1.35	1.76
	0080	Grooved woodgrain texture		750	.021		.77	.58		1.35	1.76
	0085	V - grooved woodgrain texture		750	.021		.77	.58		1.35	1.76
	0090	Wood starter strip		400	.040	L.F.	.25	1.09		1.34	2

Important: See the Reference Section for critical supporting data - Reference Nos., Crews, & City Cost Indexes

075 100 | Built-Up Roofing		CREW	DAILY OUTPUT	LABOR-HOURS	UNIT	1999 BARE COSTS				TOTAL INCL O&P	
						MAT.	LABOR	EQUIP.	TOTAL		
101	**0010** **ASPHALT** Coated felt, #30, 2 sq per roll, not mopped	1 Rofc	58	.138	Sq.	5.45	3.32		8.77	11.70	**101**
	0200 #15, 4 sq per roll, plain or perforated, not mopped		58	.138		2.68	3.32		6	8.65	
	0300 Roll roofing, smooth, #65		15	.533		12.10	12.85		24.95	35.50	
	0500 #90		15	.533		12.65	12.85		25.50	36	
	0520 Mineralized		15	.533		15.50	12.85		28.35	39	
	0540 D.C. (Double coverage), 19" selvage edge	▼	10	.800	▼	30.50	19.30		49.80	66.50	
	0580 Adhesive (lap cement)				Gal.	3.68			3.68	4.05	
	0600 Steep, flat or dead level asphalt, 10 ton lots, bulk				Ton	230			230	253	
	0800 Packaged				"	255			255	281	
102	**0010** **BUILT-UP ROOFING**										**102**
	0120 Asphalt flood coat with gravel/slag surfacing, not including [R075 -020]										
	0140 Insulation, flashing or wood nailers										
	0200 Asphalt base sheet, 3 plies #15 asphalt felt, mopped	G-1	22	2.545	Sq.	35.50	57.50	18.60	111.60	159	
	0350 On nailable decks		21	2.667		39.50	60.50	19.45	119.45	169	
	0500 4 plies #15 asphalt felt, mopped		20	2.800		50	63.50	20.50	134	187	
	0550 On nailable decks		19	2.947		46	67	21.50	134.50	189	
	0700 Coated glass base sheet, 2 plies glass (type IV), mopped		22	2.545		38.50	57.50	18.60	114.60	163	
	0850 3 plies glass, mopped		20	2.800		46	63.50	20.50	130	182	
	0950 On nailable decks		19	2.947		44	67	21.50	132.50	187	
	1100 4 plies glass fiber felt (type IV), mopped		20	2.800		55.50	63.50	20.50	139.50	193	
	1150 On nailable decks		19	2.947		51	67	21.50	139.50	195	
	1200 Coated & saturated base sheet, 3 plies #15 asph. felt, mopped		20	2.800		42	63.50	20.50	126	178	
	1250 On nailable decks		19	2.947		40	67	21.50	128.50	183	
	1300 4 plies #15 asphalt felt, mopped	▼	22	2.545	▼	48	57.50	18.60	124.10	173	
	2000 Asphalt flood coat, smooth surface										
	2200 Asphalt base sheet & 3 plies #15 asphalt felt, mopped	G-1	24	2.333	Sq.	36	53	17.05	106.05	149	
	2400 On nailable decks		23	2.435		34	55	17.75	106.75	152	
	2600 4 plies #15 asphalt felt, mopped		24	2.333		42.50	53	17.05	112.55	156	
	2700 On nailable decks	▼	23	2.435	▼	40	55	17.75	112.75	159	
	2900 Coated glass fiber base sheet, mopped, and 2 plies of										
	2910 glass fiber felt (type IV)	G-1	25	2.240	Sq.	33	51	16.35	100.35	142	
	3100 On nailable decks		24	2.333		31.50	53	17.05	101.55	145	
	3200 3 plies, mopped		23	2.435		40.50	55	17.75	113.25	159	
	3300 On nailable decks		22	2.545		38.50	57.50	18.60	114.60	162	
	3800 4 plies glass fiber felt (type IV), mopped		23	2.435		47.50	55	17.75	120.25	167	
	3900 On nailable decks		22	2.545		45.50	57.50	18.60	121.60	170	
	4000 Coated & saturated base sheet, 3 plies #15 asph. felt, mopped		24	2.333		36.50	53	17.05	106.55	150	
	4200 On nailable decks		23	2.435		34	55	17.75	106.75	152	
	4300 4 plies #15 organic felt, mopped	▼	22	2.545	▼	42.50	57.50	18.60	118.60	167	
	4500 Coal tar pitch with gravel/slag surfacing										
	4600 4 plies #15 tarred felt, mopped	G-1	21	2.667	Sq.	97.50	60.50	19.45	177.45	233	
	4800 3 plies glass fiber felt (type IV), mopped	"	19	2.947	"	80.50	67	21.50	169	227	
	5000 Coated glass fiber base sheet, and 2 plies of										
	5010 glass fiber felt, (type IV), mopped	G-1	19	2.947	Sq.	80	67	21.50	168.50	227	
	5300 On nailable decks		18	3.111		71	70.50	22.50	164	224	
	5600 4 plies glass fiber felt (type IV), mopped		21	2.667		110	60.50	19.45	189.95	247	
	5800 On nailable decks	▼	20	2.800	▼	101	63.50	20.50	185	243	
103	**0010** **CANTS** 4" x 4", treated timber, cut diagonally	1 Rofc	325	.025	L.F.	.80	.59		1.39	1.90	**103**
	0100 Foamglass		325	.025		1.92	.59		2.51	3.13	
	0300 Mineral or fiber, trapezoidal, 1"x 4" x 48"		325	.025		.17	.59		.76	1.21	
	0400 1-1/2" x 5-5/8" x 48"	▼	325	.025	▼	.29	.59		.88	1.34	
104	**0010** **FELT** Glass fibered, #15, no mopping	1 Rofc	58	.138	Sq.	3.70	3.32		7.02	9.75	**104**
	0300 Base sheet, #45, channel vented		58	.138		16.50	3.32		19.82	24	
	0400 #50, coated		58	.138		7.55	3.32		10.87	14	
	0500 Cap, mineral surfaced	▼	58	.138	▼	16.25	3.32		19.57	23.50	

7

THERMAL & MOISTURE PROTECTION

7 THERMAL & MOISTURE PROTECTION

075 100	Built-Up Roofing	CREW	DAILY OUTPUT	LABOR-HOURS	UNIT	1999 BARE COSTS				TOTAL INCL O&P	
						MAT.	LABOR	EQUIP.	TOTAL		
104	0600	Flashing membrane, #65	1 Rofc	16	.500	Sq.	39	12.05		51.05	64	104
	0800	Coal tar fibered, #15, no mopping		58	.138		7.90	3.32		11.22	14.40	
	0900	Asphalt felt, #15, 4 sq per roll, no mopping CN		58	.138		2.68	3.32		6	8.65	
	1100	#30, 2 sq per roll		58	.138		5.45	3.32		8.77	11.70	
	1200	Double coated, #33		58	.138		5.95	3.32		9.27	12.25	
	1400	#40, base sheet		58	.138		6.30	3.32		9.62	12.65	
	1450	Coated and saturated		58	.138		6.50	3.32		9.82	12.85	
	1500	Tarred felt, organic, #15, 4 sq rolls		58	.138		7.80	3.32		11.12	14.30	
	1550	#30, 2 sq roll	↓	58	.138		15.60	3.32		18.92	23	
	1700	Add for mopping above felts, per ply, asphalt, 24 lb per sq	G-1	192	.292		3.06	6.60	2.13	11.79	17.10	
	1800	Coal tar mopping, 30 lb per sq		186	.301		8.05	6.85	2.20	17.10	23	
	1900	Flood coat, with asphalt, 60 lb per sq		60	.933		7.65	21	6.80	35.45	52.50	
	2000	With coal tar, 75 lb per sq	↓	56	1	↓	20	22.50	7.30	49.80	69	
105	0010	**WALKWAY** For built-up roofs, asphalt impregnated, 3' x 6' x 1/2" thk	1 Rofc	400	.020	S.F.	.87	.48		1.35	1.78	105
	0100	3' x 3' x 3/4" thick	"	400	.020		2.07	.48		2.55	3.11	
	0300	Concrete patio blocks, 2" thick, natural	1 Clab	115	.070		1.39	1.49		2.88	3.88	
	0400	Colors	"	115	.070	↓	1.75	1.49		3.24	4.28	

075 150	Cold Applied Roofing	CREW	DAILY OUTPUT	LABOR-HOURS	UNIT	MAT.	LABOR	EQUIP.	TOTAL	TOTAL INCL O&P	
152	0010	**COLD APPLIED** 3-ply system (components listed below)	G-5	50	.800	Sq.		17.70	3.31	21.01	34	152
	0100	Spunbond poly. fabric, 1.35 oz/SY, 36"W, 10.8 Sq/roll				Ea.	123			123	135	
	0200	49" wide, 14.6 Sq./roll					169			169	186	
	0300	2.10 oz./S.Y., 36" wide, 10.8 Sq./roll					185			185	203	
	0400	49" wide, 14.6 Sq./roll				↓	250			250	275	
	0500	Base & finish coat, 3 gal./Sq., 5 gal./can				Gal.	3.25			3.25	3.58	
	0600	Coating, ceramic granules, 1/2 Sq./bag				Ea.	11.50			11.50	12.65	
	0700	Aluminum, 2 gal./Sq.				Gal.	9.25			9.25	10.20	
	0800	Emulsion, fibered or non-fibered, 4 gal./Sq.				"	4.25			4.25	4.68	

075 200	Prepared Roll Roofing	CREW	DAILY OUTPUT	LABOR-HOURS	UNIT	MAT.	LABOR	EQUIP.	TOTAL	TOTAL INCL O&P	
204	0010	**ROLL ROOFING**										204
	0100	Asphalt, mineral surface										
	0200	1 ply #15 organic felt, 1 ply mineral surfaced										
	0300	Selvage roofing, lap 19", nailed & mopped	G-1	27	2.074	Sq.	36.50	47	15.15	98.65	138	
	0400	3 plies glass fiber felt (type IV), 1 ply mineral surfaced										
	0500	Selvage roofing, lapped 19", mopped	G-1	25	2.240	Sq.	54	51	16.35	121.35	165	
	0600	Coated glass fiber base sheet, 2 plies of glass fiber										
	0700	Felt (type IV), 1 ply mineral surfaced selvage										
	0800	Roofing, lapped 19", mopped	G-1	25	2.240	Sq.	57.50	51	16.35	124.85	169	
	0900	On nailable decks	"	24	2.333	"	54.50	53	17.05	124.55	170	
	1000	3 plies glass fiber felt (type III), 1 ply mineral surfaced										
	1100	Selvage roofing, lapped 19", mopped	G-1	25	2.240	Sq.	54	51	16.35	121.35	165	

075 300	Elastomeric Roofing	CREW	DAILY OUTPUT	LABOR-HOURS	UNIT	MAT.	LABOR	EQUIP.	TOTAL	TOTAL INCL O&P	
301	0010	**ELASTOMERIC ROOFING**										301
	0100	For Elastomeric waterproofing, see division 071-102										
	0110	Acrylic rubber, fluid applied, 20 mils thick	G-5	2,000	.020	S.F.	1.80	.44	.08	2.32	2.83	
	0120	50 mils, reinforced		1,200	.033		2.80	.74	.14	3.68	4.50	
	0130	For walking surface, add	↓	900	.044		.85	.98	.18	2.01	2.83	
	0300	Hypalon neoprene, fluid applied, 20 mil thick, not-reinforced	G-1	1,135	.049		2.05	1.12	.36	3.53	4.59	
	0600	Non-woven polyester, reinforced		960	.058		2.07	1.32	.43	3.82	5.05	
	0700	5 coat neoprene deck, 60 mil thick, under 10,000 SF		325	.172		4.56	3.91	1.26	9.73	13.15	
	0900	Over 10,000 SF		625	.090		4.25	2.03	.65	6.93	8.90	
	1300	Vinyl plastic traffic deck, sprayed, 2 to 4 mils thick	↓	625	.090	↓	1.33	2.03	.65	4.01	5.70	

Important: See the Reference Section for critical supporting data - Reference Nos., Crews, & City Cost Indexes

075 300 | Elastomeric Roofing

			DAILY OUTPUT	LABOR-HOURS	UNIT	1999 BARE COSTS				TOTAL INCL O&P	
			CREW			MAT.	LABOR	EQUIP.	TOTAL		
301	1500	Vinyl and neoprene membrane traffic deck	G-1	1,550	.036	S.F.	1.41	.82	.26	2.49	3.25	301
	1600	Polyurethane spray-on with 20 mil silicone rubber coating applied										
	1700	1" thick, R7, minimum	G-2	875	.027	S.F.	1.34	.63	.27	2.24	2.74	
	1800	Maximim		805	.030		1.80	.68	.29	2.77	3.36	
	1900	2" thick, R14, minimum		685	.035		1.90	.80	.34	3.04	3.71	
	2000	Maximum		575	.042		2.41	.96	.41	3.78	4.58	
	2100	3" thick, R21, minimum		500	.048		2.50	1.10	.47	4.07	4.97	
	2200	Maximum		440	.055		3.20	1.25	.53	4.98	6.05	
302	0010	**SINGLE-PLY MEMBRANE**										302
	0800	Chlorosulfonated polyethylene-hypalon (CSPE), 45 mils,										
	0900	0.29 P.S.F., fully adhered	G-5	26	1.538	Sq.	122	34	6.35	162.35	200	
	1100	Loose-laid & ballasted with stone (10 P.S.F.)		51	.784		129	17.35	3.25	149.60	176	
	1200	Partially adhered with fastening strips		35	1.143		124	25.50	4.73	154.23	185	
	1300	Plates with adhesive attachment		35	1.143		122	25.50	4.73	152.23	183	
	3500	Ethylene propylene diene monomer (EPDM), 45 mils, 0.28 P.S.F.										
	3600	Loose-laid & ballasted with stone (10 P.S.F.)	G-5	51	.784	Sq.	62.50	17.35	3.25	83.10	102	
	3700	Partially adhered		35	1.143		53.50	25.50	4.73	83.73	108	
	3800	Fully adhered with adhesive		26	1.538		76	34	6.35	116.35	149	
	4500	60 mils, 0.40 P.S.F.										
	4600	Loose-laid & ballasted with stone (10 P.S.F.)	G-5	51	.784	Sq.	74.50	17.35	3.25	95.10	116	
	4700	Partially adhered		35	1.143		65	25.50	4.73	95.23	120	
	4800	Fully adhered with adhesive		26	1.538		87	34	6.35	127.35	161	
	4810	45 mil, .28 PSF, membrane only CN					34			34	37	
	4820	60 mil, .40 PSF, membrane only					44			44	48.50	
	4850	Seam tape for membrane, 4" x 100' roll				Ea.	67			67	73.50	
	4900	Batten strips, 10' sections					2.45			2.45	2.70	
	4910	Cover tape for batten strips, 6" x 100' roll					126			126	139	
	4930	Plate anchors				M	123			123	135	
	4970	Adhesive for fully adhered systems, 60 S.F./gal.				Gal.	13.60			13.60	14.95	
	7500	Polyisobutylene (PIB), 100 mils, 0.57 P.S.F.										
	7600	Loose-laid & ballasted with stone/gravel (10 P.S.F.)	G-5	51	.784	Sq.	123	17.35	3.25	143.60	169	
	7700	Partially adhered with adhesive		35	1.143		154	25.50	4.73	184.23	218	
	7800	Hot asphalt attachment		35	1.143		147	25.50	4.73	177.23	211	
	7900	Fully adhered with contact cement		26	1.538		159	34	6.35	199.35	241	
	8200	Polyvinyl chloride (PVC), heat welded seams										
	8700	Reinforced, 48 mils, 0.33 P.S.F.										
	8750	Loose-laid & ballasted with stone/gravel (12 P.S.F.)	G-5	51	.784	Sq.	85.50	17.35	3.25	106.10	128	
	8800	Partially adhered with mechanical fasteners		35	1.143		76	25.50	4.73	106.23	132	
	8850	Fully adhered with adhesive		26	1.538		108	34	6.35	148.35	184	
	8860	Reinforced, 60 mils, .40 P.S.F.										
	8870	Loose-laid & ballasted with stone/gravel (12 P.S.F.)	G-5	51	.784	Sq.	89.50	17.35	3.25	110.10	132	
	8880	Partially adhered with mechanical fasteners		35	1.143		79.50	25.50	4.73	109.73	136	
	8890	Fully adhered with adhesive		26	1.538		111	34	6.35	151.35	189	

075 350 | Modified Bit. Roofing

				DAILY OUTPUT	LABOR-HOURS	UNIT	MAT.	LABOR	EQUIP.	TOTAL	TOTAL INCL O&P	
352	0010	**MODIFIED BITUMEN ROOFING**	R075 -030									352
	0020	Base sheet, #15 glass fiber felt, nailed to deck		1 Rofc	58	.138	Sq.	4.38	3.32		7.70	10.50
	0030	Spot mopped to deck		G-1	295	.190		5.25	4.30	1.39	10.94	14.65
	0040	Fully mopped to deck		"	192	.292		6.75	6.60	2.13	15.48	21
	0050	#15 organic felt, nailed to deck		1 Rofc	58	.138		3.37	3.32		6.69	9.40
	0060	Spot mopped to deck		G-1	295	.190		4.21	4.30	1.39	9.90	13.55
	0070	Fully mopped to deck		"	192	.292		5.75	6.60	2.13	14.48	20
	0080	SBS modified, granule surf cap sheet, polyester rein., mopped										
	0600	150 mils		G-1	2,000	.028	S.F.	.38	.63	.20	1.21	1.73
	1100	160 mils			2,000	.028		.44	.63	.20	1.27	1.79

075 350 | Modified Bit. Roofing

			CREW	DAILY OUTPUT	LABOR-HOURS	UNIT	1999 BARE COSTS				TOTAL INCL O&P	
							MAT.	LABOR	EQUIP.	TOTAL		
352	1500	Glass fiber reinforced, mopped, 160 mils	G-1	2,000	.028	S.F.	.36	.63	.20	1.19	1.71	352
	1600	Smooth surface cap sheet, mopped, 145 mils R075 -030		2,100	.027		.36	.60	.19	1.15	1.65	
	1700	Smooth surface flashing, 145 mils		1,260	.044		.36	1.01	.32	1.69	2.49	
	1800	150 mils		1,260	.044		.35	1.01	.32	1.68	2.48	
	1900	Granular surface flashing, 150 mils		1,260	.044		.38	1.01	.32	1.71	2.51	
	2000	160 mils		1,260	.044		.44	1.01	.32	1.77	2.57	
	2100	APP mod., smooth surf. cap sheet, poly. reinf., torched, 160 mils	G-5	2,100	.019		.35	.42	.08	.85	1.20	
	2150	170 mils		2,100	.019		.39	.42	.08	.89	1.24	
	2200	Granule surface cap sheet, poly. reinf., torched, 180 mils		2,000	.020		.38	.44	.08	.90	1.27	
	2250	Smooth surface flashing, torched, 160 mils		1,260	.032		.35	.70	.13	1.18	1.74	
	2300	170 mils		1,260	.032		.39	.70	.13	1.22	1.78	
	2350	Granule surface flashing, torched, 180 mils		1,260	.032		.38	.70	.13	1.21	1.77	
	2400	Fibrated aluminum coating	1 Rofc	3,800	.002		.08	.05		.13	.17	

075 600 | Roof Maint. & Repairs

			CREW	DAILY OUTPUT	LABOR-HOURS	UNIT	MAT.	LABOR	EQUIP.	TOTAL	TOTAL INCL O&P	
604	0010	**ROOF COATINGS** Asphalt CN				Gal.	2.71			2.71	2.98	604
	0200	Asphalt base, fibered aluminum coating					7.65			7.65	8.40	
	0300	Asphalt primer, 5 gallon					3.25			3.25	3.58	
	0600	Coal tar pitch, 200 lb. barrels				Ton	540			540	590	
	0700	Tar roof cement, 5 gal. lots				Gal.	5.55			5.55	6.10	
	0800	Glass fibered roof & patching cement, 5 gallon				"	3.45			3.45	3.80	
	0900	Reinforcing glass membrane, 450 S.F./roll				Ea.	45.50			45.50	50	
	1000	Neoprene roof coating, 5 gal, 2 gal/sq				Gal.	20.50			20.50	22.50	
	1100	Roof patch & flashing cement, 5 gallon					18.95			18.95	21	
	1200	Roof resaturant, glass fibered, 3 gal/sq					7.30			7.30	8.05	
	1300	Mineral rubber, 3 gal/sq					4.71			4.71	5.20	

076 | Flashing & Sheet Metal

076 100 | Sheet Metal Roofing

			CREW	DAILY OUTPUT	LABOR-HOURS	UNIT	1999 BARE COSTS				TOTAL INCL O&P	
							MAT.	LABOR	EQUIP.	TOTAL		
101	0010	**COPPER ROOFING** Batten seam, over 10 sq, 16 oz, 130 lb/sq	1 Shee	1.10	7.273	Sq.	395	231		626	795	101
	0200	18 oz, 145 lb per sq		1	8		440	254		694	880	
	0300	20 oz, 160 lb per sq		1	8		490	254		744	930	
	0400	Standing seam, over 10 squares, 16 oz, 125 lb per sq		1.30	6.154		380	195		575	725	
	0600	18 oz, 140 lb per sq		1.20	6.667		425	212		637	800	
	0700	20 oz, 150 lb per sq		1.10	7.273		460	231		691	865	
	0900	Flat seam, over 10 squares, 16 oz, 115 lb per sq		1.20	6.667		350	212		562	715	
	1000	20 oz, 145 lb per sq		1.10	7.273		440	231		671	845	
	1200	For abnormal conditions or small areas, add					25%	100%				
	1300	For lead-coated copper, add					25%					
102	0010	**LEAD ROOFING** 5 lb. per SF, batten seam	1 Shee	1.20	6.667	Sq.	375	212		587	745	102
	0100	Flat seam	"	1.30	6.154	"	375	195		570	720	
104	0010	**STAINLESS STEEL ROOFING** Type 304, batten seam, 28 gauge	1 Shee	1.20	6.667	Sq.	300	212		512	660	104
	0100	26 gauge	"	1.15	6.957		375	221		596	755	
	0200	For standing seam construction, deduct					2%					
	0500	For flat seam construction, deduct					3%					
	0800	For lead or terne coated stainless, 28 gauge, add					68.50			68.50	75.50	
	0900	For 26 gauge, add					91			91	100	

Important: See the Reference Section for critical supporting data - Reference Nos., Crews, & City Cost Indexes

7 THERMAL & MOISTURE PROTECTION

076 100 | Sheet Metal Roofing

		CREW	DAILY OUTPUT	LABOR-HOURS	UNIT	1999 BARE COSTS				TOTAL INCL O&P	
						MAT.	LABOR	EQUIP.	TOTAL		
105	0010	**ZINC** Copper alloy roofing, batten seam, .020″ thick	1 Shee	1.20	6.667	Sq.	510	212		722	890	105
	0100	.027″ thick		1.15	6.957		615	221		836	1,025	
	0300	.032″ thick		1.10	7.273		695	231		926	1,125	
	0400	.040″ thick	↓	1.05	7.619		820	242		1,062	1,275	
	0600	For standing seam construction, deduct					2%					
	0700	For flat seam construction, deduct				↓	3%					

076 200 | Sheet Mtl Flash & Trim

		CREW	DAILY OUTPUT	LABOR-HOURS	UNIT	MAT.	LABOR	EQUIP.	TOTAL	TOTAL INCL O&P	
201	0010	**DOWNSPOUTS** Aluminum 2″ x 3″, .020″ thick, embossed	1 Shee	190	.042	L.F.	.71	1.34		2.05	2.86	201
	0100	Enameled		190	.042		.68	1.34		2.02	2.83	
	0300	Enameled, .024″ thick, 2″ x 3″		180	.044		1.09	1.41		2.50	3.39	
	0400	3″ x 4″		140	.057		1.41	1.81		3.22	4.37	
	0600	Round, corrugated aluminum, 3″ diameter, .020″ thick		190	.042		.85	1.34		2.19	3.02	
	0700	4″ diameter, .025″ thick		140	.057	↓	1.29	1.81		3.10	4.24	
	0900	Wire strainer, round, 2″ diameter		155	.052	Ea.	1.75	1.64		3.39	4.47	
	1000	4″ diameter		155	.052		1.82	1.64		3.46	4.54	
	1200	Rectangular, perforated, 2″ x 3″		145	.055		2.15	1.75		3.90	5.10	
	1300	3″ x 4″		145	.055	↓	3.10	1.75		4.85	6.15	
	1500	Copper, round, 16 oz., stock, 2″ diameter		190	.042	L.F.	4.88	1.34		6.22	7.45	
	1600	3″ diameter		190	.042		3.84	1.34		5.18	6.30	
	1800	4″ diameter		145	.055		4.66	1.75		6.41	7.85	
	1900	5″ diameter		130	.062		6.70	1.95		8.65	10.45	
	2100	Rectangular, corrugated copper, stock, 2″ x 3″		190	.042		3.25	1.34		4.59	5.65	
	2200	3″ x 4″		145	.055		4.36	1.75		6.11	7.50	
	2400	Rectangular, plain copper, stock, 2″ x 3″		190	.042		4.71	1.34		6.05	7.30	
	2500	3″ x 4″		145	.055	↓	6.20	1.75		7.95	9.50	
	2700	Wire strainers, rectangular, 2″ x 3″		145	.055	Ea.	2.76	1.75		4.51	5.75	
	2800	3″ x 4″		145	.055		4.37	1.75		6.12	7.55	
	3000	Round, 2″ diameter		145	.055		2.59	1.75		4.34	5.55	
	3100	3″ diameter		145	.055		3.62	1.75		5.37	6.70	
	3300	4″ diameter		145	.055		5.60	1.75		7.35	8.85	
	3400	5″ diameter		115	.070	↓	8.10	2.21		10.31	12.35	
	3600	Lead-coated copper, round, stock, 2″ diameter		190	.042	L.F.	4.88	1.34		6.22	7.45	
	3700	3″ diameter		190	.042		4.69	1.34		6.03	7.25	
	3900	4″ diameter		145	.055		5.90	1.75		7.65	9.20	
	4200	6″ diameter, corrugated		105	.076		10.40	2.42		12.82	15.15	
	4300	Rectangular, corrugated, stock, 2″ x 3″		190	.042		4.08	1.34		5.42	6.55	
	4500	Plain, stock, 2″ x 3″		190	.042		6	1.34		7.34	8.70	
	4600	3″ x 4″		145	.055		7.55	1.75		9.30	11	
	4800	Steel, galvanized, round, corrugated, 2″ or 3″ diam, 28 ga		190	.042		.68	1.34		2.02	2.83	
	4900	4″ diameter, 28 gauge		145	.055		.90	1.75		2.65	3.71	
	5100	5″ diameter, 28 gauge		130	.062		1.23	1.95		3.18	4.38	
	5200	26 gauge		130	.062		1.34	1.95		3.29	4.50	
	5400	6″ diameter, 28 gauge		105	.076		1.60	2.42		4.02	5.50	
	5500	26 gauge		105	.076		1.40	2.42		3.82	5.30	
	5700	Rectangular, corrugated, 28 gauge, 2″ x 3″		190	.042		.53	1.34		1.87	2.66	
	5800	3″ x 4″		145	.055		1.48	1.75		3.23	4.35	
	6000	Rectangular, plain, 28 gauge, galvanized, 2″ x 3″		190	.042		.79	1.34		2.13	2.95	
	6100	3″ x 4″		145	.055		1.21	1.75		2.96	4.05	
	6300	Epoxy painted, 24 gauge, corrugated, 2″ x 3″		190	.042		1.01	1.34		2.35	3.19	
	6400	3″ x 4″		145	.055	↓	1.71	1.75		3.46	4.60	
	6600	Wire strainers, rectangular, 2″ x 3″		145	.055	Ea.	1.62	1.75		3.37	4.50	
	6700	3″ x 4″		145	.055		2.61	1.75		4.36	5.60	
	6900	Round strainers, 2″ or 3″ diameter		145	.055		1.20	1.75		2.95	4.04	
	7000	4″ diameter		145	.055		1.42	1.75		3.17	4.28	
	7200	5″ diameter	↓	145	.055	↓	2.20	1.75		3.95	5.15	

THERMAL & MOISTURE PROTECTION **7**

221

076 200 | Sheet Mtl Flash & Trim		CREW	DAILY OUTPUT	LABOR-HOURS	UNIT	1999 BARE COSTS				TOTAL INCL O&P	
						MAT.	LABOR	EQUIP.	TOTAL		
201	**7300**	6" diameter	1 Shee	115	.070	Ea.	2.63	2.21		4.84	6.30	**201**
	7500	Steel pipe, black, extra heavy, 4" diameter		20	.400	L.F.	5.30	12.70		18	25.50	
	7600	6" diameter		18	.444		14.30	14.10		28.40	38	
	7800	Stainless steel tubing, schedule 5, 2" x 3" or 3" diameter		190	.042		12.80	1.34		14.14	16.20	
	7900	3" x 4" or 4" diameter		145	.055		16.30	1.75		18.05	20.50	
	8100	4" x 5" or 5" diameter		135	.059		33.50	1.88		35.38	40	
	8200	Vinyl, rectangular, 2" x 3"		210	.038		.72	1.21		1.93	2.67	
	8300	Round, 2-1/2"	▼	220	.036	▼	.72	1.15		1.87	2.58	
202	**0010**	**DRIP EDGE**, aluminum, .016" thick, 5" wide, mill finish	1 Carp	400	.020	L.F.	.20	.55		.75	1.08	**202**
	0100	White finish		400	.020		.22	.55		.77	1.10	
	0200	8" wide, mill finish		400	.020		.30	.55		.85	1.19	
	0300	Ice belt, 28" wide, mill finish		100	.080		3.42	2.18		5.60	7.20	
	0310	Vented, mill finish		400	.020		1.38	.55		1.93	2.38	
	0320	Painted finish		400	.020		1.50	.55		2.05	2.51	
	0400	Galvanized, 5" wide		400	.020		.22	.55		.77	1.10	
	0500	8" wide, mill finish		400	.020		.33	.55		.88	1.22	
	0510	Rake edge, aluminum, 1-1/2" x 1-1/2"		400	.020		.13	.55		.68	1	
	0520	3-1/2" x 1-1/2"	▼	400	.020	▼	.19	.55		.74	1.07	
203	**0010**	**ELBOWS** Aluminum, 2" x 3", embossed	1 Shee	100	.080	Ea.	.90	2.54		3.44	4.93	**203**
	0100	Enameled		100	.080		1.63	2.54		4.17	5.75	
	0200	3" x 4", .025" thick, embossed		100	.080		3.15	2.54		5.69	7.40	
	0300	Enameled		100	.080		3.15	2.54		5.69	7.40	
	0400	Round corrugated, 3", embossed, .020" thick		100	.080		1.95	2.54		4.49	6.10	
	0500	4", .025" thick		100	.080		2.95	2.54		5.49	7.20	
	0600	Copper, 16 oz. round, 2" diameter		100	.080		11	2.54		13.54	16.05	
	0700	3" diameter		100	.080		5.05	2.54		7.59	9.50	
	0800	4" diameter		100	.080		9.50	2.54		12.04	14.40	
	1000	2" x 3" corrugated		100	.080		5.05	2.54		7.59	9.50	
	1100	3" x 4" corrugated		100	.080		9.75	2.54		12.29	14.65	
	1300	Vinyl, 2-1/2" diameter, 45° or 75°		100	.080		2	2.54		4.54	6.15	
	1400	Tee Y junction	▼	75	.107	▼	8.50	3.39		11.89	14.60	
204	**0010**	**FLASHING** Aluminum, mill finish, .013" thick	1 Shee	145	.055	S.F.	.34	1.75		2.09	3.09	**204**
	0030	.016" thick		145	.055		.50	1.75		2.25	3.27	
	0060	.019" thick		145	.055		.79	1.75		2.54	3.59	
	0100	.032" thick		145	.055		.86	1.75		2.61	3.66	
	0200	.040" thick		145	.055		1.44	1.75		3.19	4.30	
	0300	.050" thick		145	.055	▼	1.80	1.75		3.55	4.70	
	0325	Mill finish 5" x 7" step flashing, .016" thick		1,920	.004	Ea.	.10	.13		.23	.32	
	0350	Mill finish 12" x 12" step flashing, .016" thick	▼	1,600	.005	"	.40	.16		.56	.69	
	0400	Painted finish, add				S.F.	.24			.24	.26	
	0500	Fabric-backed 2 sides, .004" thick	1 Shee	330	.024		.93	.77		1.70	2.22	
	0700	.005" thick		330	.024		1.09	.77		1.86	2.40	
	0750	Mastic-backed, self adhesive		460	.017		2.22	.55		2.77	3.30	
	0800	Mastic-coated 2 sides, .004" thick		330	.024		.93	.77		1.70	2.22	
	1000	.005" thick		330	.024		1.09	.77		1.86	2.40	
	1100	.016" thick	▼	330	.024	▼	1.20	.77		1.97	2.52	
	1300	Asphalt flashing cement, 5 gallon				Gal.	3.34			3.34	3.67	
	1600	Copper, 16 oz, sheets, under 1000 lbs.	1 Shee	115	.070	S.F.	3.05	2.21		5.26	6.80	
	1700	Over 4000 lbs.		155	.052		2.85	1.64		4.49	5.70	
	1900	20 oz sheets, under 1000 lbs.		110	.073		3.82	2.31		6.13	7.80	
	2000	Over 4000 lbs.		145	.055		3.55	1.75		5.30	6.65	
	2200	24 oz sheets, under 1000 lbs.		105	.076		4.60	2.42		7.02	8.80	
	2300	Over 4000 lbs.		135	.059		4.25	1.88		6.13	7.60	
	2500	32 oz sheets, under 1000 lbs.		100	.080		6.10	2.54		8.64	10.65	
	2600	Over 4000 lbs.	▼	130	.062	▼	5.70	1.95		7.65	9.30	

7

THERMAL & MOISTURE PROTECTION

Important: See the Reference Section for critical supporting data - Reference Nos., Crews, & City Cost Indexes

076 200 | Sheet Mtl Flash & Trim		CREW	DAILY OUTPUT	LABOR-HOURS	UNIT	1999 BARE COSTS				TOTAL INCL O&P
						MAT.	LABOR	EQUIP.	TOTAL	
2700	W shape for valleys, 16 oz, 24″ wide	1 Shee	100	.080	L.F.	5.90	2.54		8.44	10.45
2800	Copper, paperbacked 1 side, 2 oz		330	.024	S.F.	.86	.77		1.63	2.15
2900	3 oz		330	.024		1.12	.77		1.89	2.43
3100	Paperbacked 2 sides, 2 oz		330	.024		.91	.77		1.68	2.20
3150	3 oz		330	.024		1.11	.77		1.88	2.42
3200	5 oz		330	.024		1.74	.77		2.51	3.12
3250	7 oz		330	.024		2.72	.77		3.49	4.19
3400	Mastic-backed 2 sides, copper, 2 oz		330	.024		1.02	.77		1.79	2.32
3500	3 oz		330	.024		1.25	.77		2.02	2.58
3700	5 oz		330	.024		1.85	.77		2.62	3.24
3800	Fabric-backed 2 sides, copper, 2 oz		330	.024		1.09	.77		1.86	2.40
4000	3 oz		330	.024		1.35	.77		2.12	2.69
4100	5 oz		330	.024		1.90	.77		2.67	3.29
4300	Copper-clad stainless steel, .015″ thick, under 500 lbs.		115	.070		3.10	2.21		5.31	6.85
4400	Over 2000 lbs.		155	.052		2.99	1.64		4.63	5.85
4600	.018″ thick, under 500 lbs.		100	.080		4.10	2.54		6.64	8.45
4700	Over 2000 lbs.	▼	145	.055	▼	3	1.75		4.75	6
4900	Fabric, asphalt-saturated cotton, specification grade	1 Rofc	35	.229	S.Y.	1.93	5.50		7.43	11.60
5000	Utility grade		35	.229		1.22	5.50		6.72	10.85
5200	Open-mesh fabric, saturated, 40 oz per S.Y.		35	.229		1.35	5.50		6.85	11
5300	Close-mesh fabric, saturated, 17 oz per S.Y.		35	.229		1.42	5.50		6.92	11.05
5500	Fiberglass, resin-coated		35	.229		1.16	5.50		6.66	10.80
5600	Asphalt-coated, 40 oz per S.Y.		35	.229	▼	7.90	5.50		13.40	18.20
5800	Lead, 2.5 lb. per SF, up to 12″ wide		135	.059	S.F.	3.25	1.43		4.68	6.05
5900	Over 12″ wide	▼	135	.059		3.25	1.43		4.68	6.05
6100	Lead-coated copper, fabric-backed, 2 oz	1 Shee	330	.024		1.41	.77		2.18	2.75
6200	5 oz		330	.024		1.70	.77		2.47	3.07
6400	Mastic-backed 2 sides, 2 oz		330	.024		1.10	.77		1.87	2.41
6500	5 oz		330	.024		1.43	.77		2.20	2.77
6700	Paperbacked 1 side, 2 oz		330	.024		.95	.77		1.72	2.25
6800	3 oz		330	.024		1.12	.77		1.89	2.43
7000	Paperbacked 2 sides, 2 oz		330	.024		.98	.77		1.75	2.28
7100	5 oz	▼	330	.024		1.61	.77		2.38	2.97
7300	Polyvinyl chloride, black, .010″ thick	1 Rofc	285	.028		.13	.68		.81	1.30
7400	.020″ thick		285	.028		.19	.68		.87	1.37
7600	.030″ thick		285	.028		.29	.68		.97	1.48
7700	.056″ thick		285	.028		.70	.68		1.38	1.93
7900	Black or white for exposed roofs, .060″ thick	▼	285	.028	▼	1.55	.68		2.23	2.86
8060	PVC tape, 5″ x 45 mils, for joint covers, 100 L.F./roll				Ea.	79.50			79.50	87
8100	Rubber, butyl, 1/32″ thick	1 Rofc	285	.028	S.F.	.70	.68		1.38	1.93
8200	1/16″ thick		285	.028		1.05	.68		1.73	2.31
8300	Neoprene, cured, 1/16″ thick		285	.028		1.48	.68		2.16	2.79
8400	1/8″ thick	▼	285	.028		2.99	.68		3.67	4.45
8500	Shower pan, bituminous membrane, 7 oz	1 Shee	155	.052		1.08	1.64		2.72	3.73
8550	3 ply copper and fabric, 3 oz		155	.052		1.60	1.64		3.24	4.30
8600	7 oz		155	.052		3.30	1.64		4.94	6.15
8650	Copper, 16 oz		100	.080		3.05	2.54		5.59	7.30
8700	Lead on copper and fabric, 5 oz		155	.052		1.70	1.64		3.34	4.41
8800	7 oz		155	.052		2.87	1.64		4.51	5.70
8850	Polyvinyl chloride, .030″ thick		160	.050		.30	1.59		1.89	2.79
8900	Stainless steel sheets, 32 ga, .010″ thick		155	.052		2.16	1.64		3.80	4.92
9000	28 ga, .015″ thick		155	.052		2.55	1.64		4.19	5.35
9100	26 ga, .018″ thick		155	.052		3.16	1.64		4.80	6
9200	24 ga, .025″ thick	▼	155	.052	▼	4.10	1.64		5.74	7.05
9290	For mechanically keyed flashing, add					40%				
9300	Stainless steel, paperbacked 2 sides, .005″ thick	1 Shee	330	.024	S.F.	1.95	.77		2.72	3.34

076 200 | Sheet Mtl Flash & Trim

			DAILY OUTPUT	LABOR-HOURS	UNIT	1999 BARE COSTS				TOTAL INCL O&P	
						MAT.	LABOR	EQUIP.	TOTAL		
204	9400	Terne coated stainless steel, .015" thick, 28 ga	1 Shee	155	.052	S.F.	3.97	1.64		5.61	6.90	204
	9500	.018" thick, 26 ga		155	.052		4.48	1.64		6.12	7.45	
	9600	Zinc and copper alloy (brass), .020" thick		155	.052		3.20	1.64		4.84	6.05	
	9700	.027" thick		155	.052		4.28	1.64		5.92	7.25	
	9800	.032" thick		155	.052		5	1.64		6.64	8.05	
	9900	.040" thick		155	.052		6.10	1.64		7.74	9.25	
205	0010	GUTTERS Aluminum, stock units, 5" box, .027" thick, plain	1 Shee	120	.067	L.F.	.87	2.12		2.99	4.25	205
	0100	Enameled		120	.067		1.07	2.12		3.19	4.47	
	0300	5" box type, .032" thick, plain		120	.067		1.08	2.12		3.20	4.48	
	0400	Enameled		120	.067		1.15	2.12		3.27	4.56	
	0600	5" x 6" combination fascia & gutter, .032" thick, enameled		60	.133		3.45	4.23		7.68	10.35	
	0700	Copper, half round, 16 oz, stock units, 4" wide		120	.067		3.23	2.12		5.35	6.85	
	0900	5" wide		120	.067		3.80	2.12		5.92	7.45	
	1000	6" wide		115	.070		4.25	2.21		6.46	8.10	
	1200	K type, 16 oz, stock, 4" wide		120	.067		3.55	2.12		5.67	7.20	
	1300	5" wide		120	.067		3.65	2.12		5.77	7.30	
	1500	Lead coated copper, half round, stock, 4" wide		120	.067		4.87	2.12		6.99	8.65	
	1600	6" wide		115	.070		6.20	2.21		8.41	10.25	
	1800	K type, stock, 4" wide		120	.067		6.95	2.12		9.07	10.95	
	1900	5" wide		120	.067		6.25	2.12		8.37	10.20	
	2100	Stainless steel, half round or box, stock, 4" wide		120	.067		4.65	2.12		6.77	8.40	
	2200	5" wide		120	.067		5	2.12		7.12	8.80	
	2400	Steel, galv, half round or box, 28 ga, 5" wide, plain		120	.067		.90	2.12		3.02	4.29	
	2500	Enameled		120	.067		.94	2.12		3.06	4.33	
	2700	26 ga, stock, 5" wide		120	.067		.95	2.12		3.07	4.34	
	2800	6" wide		120	.067		1.24	2.12		3.36	4.65	
	3000	Vinyl, O.G., 4" wide	1 Carp	110	.073		.85	1.99		2.84	4.06	
	3100	5" wide		110	.073		1	1.99		2.99	4.22	
	3200	4" half round, stock units		110	.073		.68	1.99		2.67	3.87	
	3250	Joint connectors				Ea.	1.36			1.36	1.50	
	3300	Wood, clear treated cedar, fir or hemlock, 3" x 4"	1 Carp	100	.080	L.F.	6.30	2.18		8.48	10.35	
	3400	4" x 5"	"	100	.080	"	7.30	2.18		9.48	11.45	
206	0010	GUTTER GUARD 6" wide strip, aluminum mesh	1 Carp	500	.016	L.F.	.37	.44		.81	1.10	206
	0100	Vinyl mesh	"	500	.016	"	.22	.44		.66	.93	
207	0010	MANSARD Colored aluminum, with battens, .032" thick										207
	0600	Stock units, straight surfaces	1 Shee	115	.070	S.F.	2.13	2.21		4.34	5.75	
	0700	Concave or convex surfaces		75	.107	"	2.35	3.39		5.74	7.85	
	0800	For framing, to 5' high, add		115	.070	L.F.	2.35	2.21		4.56	6	
	0900	Soffits, to 1' wide		125	.064	S.F.	1.16	2.03		3.19	4.44	
210	0010	REGLET Aluminum, .025" thick, in concrete parapet	1 Carp	225	.036	L.F.	.92	.97		1.89	2.54	210
	0100	Copper, 10 oz.		225	.036	"	1.59	.97		2.56	3.28	
	0300	16 oz.		225	.036	L.F.	2.12	.97		3.09	3.86	
	0400	Galvanized steel, 24 gauge		225	.036		.78	.97		1.75	2.39	
	0600	Stainless steel, .020" thick		225	.036		1.57	.97		2.54	3.26	
	0700	Zinc and copper alloy, 20 oz.		225	.036		1.75	.97		2.72	3.46	
	0900	Counter flashing for above, 12" wide, .032" aluminum	1 Shee	150	.053		1.19	1.69		2.88	3.94	
	1000	Copper, 10 oz.		150	.053		3.32	1.69		5.01	6.30	
	1200	16 oz.		150	.053		3.70	1.69		5.39	6.70	
	1300	Galvanized steel, .020" thick		150	.053		.62	1.69		2.31	3.31	
	1500	Stainless steel, .020" thick		150	.053		2.76	1.69		4.45	5.65	
	1600	Zinc and copper alloy, 20 oz.		150	.053		3.11	1.69		4.80	6.05	
217	0010	SOFFIT Aluminum, residential, stock units, .020" thick	1 Carp	210	.038	S.F.	.99	1.04		2.03	2.73	217
	0100	Baked enamel on steel, 16 or 18 gauge		105	.076		3.74	2.08		5.82	7.40	
	0300	Polyvinyl chloride, white, solid		230	.035		.65	.95		1.60	2.20	
	0400	Perforated		230	.035		.65	.95		1.60	2.20	

Important: See the Reference Section for critical supporting data - Reference Nos., Crews, & City Cost Indexes

THERMAL & MOISTURE PROTECTION 7

076 200	Sheet Mtl Flash & Trim	CREW	DAILY OUTPUT	LABOR-HOURS	UNIT	1999 BARE COSTS				TOTAL INCL O&P	
						MAT.	LABOR	EQUIP.	TOTAL		
217	0500	For colors, add				S.F.	.06			.06	.07	217

077 100	Prefab Roof Specialties	CREW	DAILY OUTPUT	LABOR-HOURS	UNIT	1999 BARE COSTS				TOTAL INCL O&P	
						MAT.	LABOR	EQUIP.	TOTAL		
103	0010	**EXPANSION JOINT**										103
	0300	Butyl or neoprene center with foam insulation, metal flanges										
	0400	Aluminum, .032" thick for openings to 2-1/2"	1 Rofc	165	.048	L.F.	7.40	1.17		8.57	10.15	
	0600	For joint openings to 3-1/2"		165	.048		8.65	1.17		9.82	11.50	
	0610	For joint openings to 5"		165	.048		10.40	1.17		11.57	13.45	
	0620	For joint openings to 8"		165	.048		16.40	1.17		17.57	20	
	0700	Copper, 16 oz. for openings to 2-1/2"		165	.048		10.40	1.17		11.57	13.45	
	0900	For joint openings to 3-1/2"		165	.048		11.95	1.17		13.12	15.15	
	0910	For joint openings to 5"		165	.048		14	1.17		15.17	17.40	
	0920	For joint openings to 8"		165	.048		21	1.17		22.17	25	
	1000	Galvanized steel, 26 ga. for openings to 2-1/2"		165	.048		6.25	1.17		7.42	8.90	
	1200	For joint openings to 3-1/2"		165	.048		7.40	1.17		8.57	10.15	
	1210	For joint openings to 5"		165	.048		9.25	1.17		10.42	12.20	
	1220	For joint openings to 8"		165	.048		15.75	1.17		16.92	19.35	
	1300	Lead-coated copper, 16 oz. for openings to 2-1/2"		165	.048		18.50	1.17		19.67	22.50	
	1500	For joint openings to 3-1/2"		165	.048		22	1.17		23.17	26	
	1600	Stainless steel, .018", for openings to 2-1/2"		165	.048		9.25	1.17		10.42	12.20	
	1800	For joint openings to 3-1/2"		165	.048		10.55	1.17		11.72	13.60	
	1810	For joint openings to 5"		165	.048		13.05	1.17		14.22	16.35	
	1820	For joint openings to 8"		165	.048		19.85	1.17		21.02	24	
	1900	Neoprene, double-seal type with thick center, 4-1/2" wide		125	.064		8.90	1.54		10.44	12.45	
	1950	Polyethylene bellows, with galv steel flat flanges		100	.080		3.45	1.93		5.38	7.10	
	1960	With galvanized angle flanges	▼	100	.080		3.80	1.93		5.73	7.50	
	2000	Roof joint with extruded aluminum cover, 2"	1 Shee	115	.070		24	2.21		26.21	29.50	
	2100	Roof joint, plastic curbs, foam center, standard	1 Rofc	100	.080		8.70	1.93		10.63	12.85	
	2200	Large		100	.080	▼	11.60	1.93		13.53	16.05	
	2300	Transitions, regular, minimum		10	.800	Ea.	75	19.30		94.30	116	
	2350	Maximum		4	2		95.50	48		143.50	188	
	2400	Large, minimum		9	.889		110	21.50		131.50	158	
	2450	Maximum	▼	3	2.667	▼	115	64.50		179.50	238	
	2500	Roof to wall joint with extruded aluminum cover	1 Shee	115	.070	L.F.	20.50	2.21		22.71	26	
	2600	See also divisions 031-132 & 058-104										
	2700	Wall joint, closed cell foam on PVC cover, 9" wide	1 Rofc	125	.064	L.F.	3.10	1.54		4.64	6.05	
	2800	12" wide	"	115	.070	"	3.50	1.68		5.18	6.75	
104	0010	**FASCIA** Aluminum, reverse board and batten,										104
	0100	.032" thick, colored, no furring included	1 Shee	145	.055	S.F.	2.10	1.75		3.85	5.05	
	0300	Steel, galv and enameled, stock, no furring, long panels		145	.055		2.19	1.75		3.94	5.15	
	0600	Short panels	▼	115	.070	▼	3.31	2.21		5.52	7.05	
105	0010	**GRAVEL STOP** Aluminum, .050" thick, 4" face height, mill finish	1 Shee	145	.055	L.F.	2.72	1.75		4.47	5.70	105
	0080	Duranodic finish		145	.055		3.69	1.75		5.44	6.80	
	0100	Painted		145	.055		4.26	1.75		6.01	7.40	
	0300	6" face height		135	.059		3.22	1.88		5.10	6.45	
	0350	Duranodic finish		135	.059		4.35	1.88		6.23	7.70	
	0400	Painted	▼	135	.059	▼	5.05	1.88		6.93	8.45	

077 100 | Prefab Roof Specialties

		CREW	DAILY OUTPUT	LABOR-HOURS	UNIT	MAT.	LABOR	EQUIP.	TOTAL	TOTAL INCL O&P	
105	0600	8" face height	1 Shee	125	.064	L.F.	3.98	2.03		6.01	7.55	105
	0650	Duranodic finish		125	.064		4.98	2.03		7.01	8.65	
	0700	Painted		125	.064		5.05	2.03		7.08	8.75	
	0900	12" face height, .080 thick, 2 piece		100	.080		6.45	2.54		8.99	11.05	
	0950	Duranodic finish		100	.080		6.30	2.54		8.84	10.85	
	1000	Painted		100	.080		7.40	2.54		9.94	12.10	
	1350	Galv steel, 24 ga., 4" leg, plain, with continuous cleat, 4" face		145	.055		1.50	1.75		3.25	4.37	
	1360	6" face height		145	.055		2.27	1.75		4.02	5.20	
	1500	Polyvinyl chloride, 6" face height		135	.059		3.28	1.88		5.16	6.55	
	1600	9" face height		125	.064		3.87	2.03		5.90	7.40	
	1800	Stainless steel, 24 ga., 6" face height		135	.059		7.15	1.88		9.03	10.75	
	1900	12" face height		100	.080		15	2.54		17.54	20.50	
	2100	20 ga., 6" face height		135	.059		8.10	1.88		9.98	11.80	
	2200	12" face height		100	.080		17.10	2.54		19.64	22.50	

077 200 | Roof Accessories

		CREW	DAILY OUTPUT	LABOR-HOURS	UNIT	MAT.	LABOR	EQUIP.	TOTAL	TOTAL INCL O&P	
206	0010	**ROOF HATCHES** With curb, 1" fiberglass insulation, 2'-6" x 3'-0"										206
	0500	Aluminum curb and cover	G-3	10	3.200	Ea.	405	85		490	580	
	0520	Galvanized steel curb and aluminum cover CN		10	3.200		340	85		425	510	
	0540	Galvanized steel curb and cover		10	3.200		300	85		385	465	
	0600	2'-6" x 4'-6", aluminum curb and cover		9	3.556		560	94.50		654.50	765	
	0800	Galvanized steel curb and aluminum cover		9	3.556		475	94.50		569.50	675	
	0900	Galvanized steel curb and cover		9	3.556		460	94.50		554.50	655	
	1200	2'-6" x 8'-0", aluminum curb and cover		6.60	4.848		995	129		1,124	1,300	
	1400	Galvanized steel curb and aluminum cover		6.60	4.848		920	129		1,049	1,200	
	1500	Galvanized steel curb and cover		6.60	4.848		905	129		1,034	1,200	
	1800	For plexiglass panels, 2'-6" x 3'-0", add to above					345			345	380	
207	0010	**SMOKE HATCHES** Unlabeled, not including hand winch operator										207
	0200	For 3'-0" long, add to roof hatches from division 077-206				Ea.	25%	5%				
	0300	For 8'-0" long, add to roof hatches from division 077-206				"	10%	5%				
208	0010	**SMOKE VENT**, insulated, 4' x 4'										208
	0100	Aluminum cover and frame	G-3	13	2.462	Ea.	1,050	65.50		1,115.50	1,250	
	0200	Galvanized steel cover and frame		13	2.462		950	65.50		1,015.50	1,150	
	0300	4' x 8' aluminum cover and frame		8	4		1,425	106		1,531	1,750	
	0400	Galvanized steel cover and frame		8	4		1,250	106		1,356	1,550	
210	0010	**ROOF VENTS** Mushroom for built-up roofs, aluminum	1 Rofc	30	.267	Ea.	24	6.45		30.45	37.50	210
	0100	PVC, 6" high	"	30	.267	"	27.50	6.45		33.95	41.50	
211	0010	**RIDGE VENT**										211
	0100	Aluminum strips, mill finish	1 Rofc	160	.050	L.F.	1.19	1.21		2.40	3.39	
	0150	Painted finish		160	.050	"	2.12	1.21		3.33	4.42	
	0200	Connectors		48	.167	Ea.	1.81	4.02		5.83	8.90	
	0300	End caps		48	.167	"	.76	4.02		4.78	7.75	
	0400	Galvanized strips, with damper and bird screen		160	.050	L.F.	20.50	1.21		21.71	25	
	0430	Molded polyethylene, shingles not included		160	.050	"	2.50	1.21		3.71	4.83	
	0440	End plugs		48	.167	Ea.	.76	4.02		4.78	7.75	
	0450	Flexible roll, shingles not included		160	.050	L.F.	2.24	1.21		3.45	4.54	
212	0010	**VENTS, ONE-WAY** For insul. decks, 1 per M.S.F., plastic, min.	1 Rofc	40	.200	Ea.	12.50	4.82		17.32	22	212
	0100	Maximum		20	.400	"	28.50	9.65		38.15	48	
	0300	Aluminum		30	.267	Ea.	12.50	6.45		18.95	25	
	0800	Polystyrene baffles, 12" wide for 16" O.C. rafter spacing	1 Carp	90	.089		.45	2.43		2.88	4.32	
	0900	For 24" O.C. rafter spacing	"	110	.073		1.05	1.99		3.04	4.27	

Important: See the Reference Section for critical supporting data - Reference Nos., Crews, & City Cost Indexes

7

THERMAL & MOISTURE PROTECTION

077 200	Roof Accessories	CREW	DAILY OUTPUT	LABOR-HOURS	UNIT	1999 BARE COSTS				TOTAL INCL O&P	
						MAT.	LABOR	EQUIP.	TOTAL		
215	0010 **PITCH POCKETS**										**215**
	0100 Adjustable, 4" to 7", welded corners, 4" deep	1 Rofc	48	.167	Ea.	10.50	4.02		14.52	18.45	
	0200 Side extenders, 6"	"	240	.033	"	1.70	.80		2.50	3.25	

078 | Skylights

078 100	Plastic Skylights	CREW	DAILY OUTPUT	LABOR-HOURS	UNIT	1999 BARE COSTS				TOTAL INCL O&P	
						MAT.	LABOR	EQUIP.	TOTAL		
101	0010 **SKYLIGHT** Plastic domes, flush or curb mounted, ten or										**101**
	0100 more units, curb not included, "L" frames										
	0300 Nominal size under 10 S.F., double	G-3	130	.246	S.F.	16.65	6.55		23.20	28.50	
	0400 Single		160	.200		13.25	5.30		18.55	23	
	0600 10 S.F. to 20 S.F., double		315	.102		13.55	2.70		16.25	19.10	
	0700 Single		395	.081		13.50	2.15		15.65	18.20	
	0900 20 S.F. to 30 S.F., double		395	.081		13.90	2.15		16.05	18.65	
	1000 Single		465	.069		11.70	1.83		13.53	15.70	
	1200 30 S.F. to 65 S.F., double		465	.069		14.65	1.83		16.48	18.95	
	1300 Single		610	.052		12.25	1.40		13.65	15.70	
	1500 For insulated 4" curbs, double, add					25%					
	1600 Single, add					30%					
	1800 For integral insulated 9" curbs, double, add					30%					
	1900 Single, add					40%					
	2120 Ventilating insulated plexiglass dome with										
	2130 curb mounting, 36" x 36"	G-3	12	2.667	Ea.	340	71		411	485	
	2150 52" x 52"		12	2.667		510	71		581	675	
	2160 28" x 52"		10	3.200		400	85		485	575	
	2170 36" x 52"		10	3.200		430	85		515	610	
	2180 For electric opening system, add					256			256	281	
	2200 Field fabricated, factory type, aluminum and wire glass	G-3	120	.267	S.F.	13.20	7.10		20.30	25.50	
	2300 Insulated safety glass with aluminum frame		160	.200		78.50	5.30		83.80	95	
	2400 Sandwich panels, fiberglass, for walls, 1-9/16" thick, to 250 SF		200	.160		13.95	4.26		18.21	22	
	2500 250 SF and up		265	.121		12.50	3.21		15.71	18.75	
	2700 As above, but for roofs, 2-3/4" thick, to 250 SF		295	.108		20	2.89		22.89	26.50	
	2800 250 SF and up		330	.097		16.50	2.58		19.08	22	

078 200	Metal Framed Skylights										
202	0010 **SKYROOFS** Translucent panels, 2-3/4" thick, under 5000 SF	G-3	395	.081	SF Hor.	18.40	2.15		20.55	23.50	**202**
	0100 Over 5000 SF		465	.069		16.45	1.83		18.28	21	
	0300 Continuous vaulted, semi-circular, to 8' wide, double glazed		145	.221		42	5.85		47.85	55.50	
	0400 Single glazed		160	.200		27.50	5.30		32.80	39	
	0600 To 20' wide, single glazed		175	.183		30	4.86		34.86	41	
	0700 Over 20' wide, single glazed		200	.160		35.50	4.26		39.76	46	
	0900 Motorized opening type, single glazed, 1/3 opening		145	.221		38	5.85		43.85	50.50	
	1000 Full opening		130	.246		43.50	6.55		50.05	57.50	
	1200 Pyramid type units, self-supporting, to 30' clear opening,										
	1300 square or circular, single glazed, minimum	G-3	200	.160	SF Hor.	20.50	4.26		24.76	29	
	1310 Average		165	.194		28.50	5.15		33.65	39.50	
	1400 Maximum		130	.246		41	6.55		47.55	55	
	1500 Grid type, 4' to 10' modules, single glass glazed, minimum		200	.160		26	4.26		30.26	35.50	
	1550 Maximum		128	.250		43	6.65		49.65	57.50	

THERMAL & MOISTURE PROTECTION   **7**

227

078 200 | Metal Framed Skylights

		CREW	DAILY OUTPUT	LABOR-HOURS	UNIT	1999 BARE COSTS				TOTAL INCL O&P	
						MAT.	LABOR	EQUIP.	TOTAL		
202	1600	Preformed acrylic, minimum	G-3	300	.107	SF Hor.	31	2.84		33.84	39	202
	1650	Maximum	↓	175	.183	↓	44	4.86		48.86	56	
	1800	Dome type units, self-supporting, to 100' clear opening, circular,										
	1900	rise to span ratio = 0.20										
	1920	Minimum	G-3	197	.162	SF Hor.	13	4.32		17.32	21	
	1950	Maximum		113	.283		43.50	7.55		51.05	59.50	
	2100	Rise to span ratio = 0.33, minimum		169	.189		26	5.05		31.05	36.50	
	2150	Maximum		101	.317		51	8.45		59.45	69.50	
	2200	Rise to span ratio = 0.50, minimum		148	.216		39	5.75		44.75	52	
	2250	Maximum		87	.368		57	9.80		66.80	78	
	2400	Ridge units, continuous, to 8' wide, double		130	.246		99.50	6.55		106.05	119	
	2500	Single		200	.160		68	4.26		72.26	81.50	
	2700	Ridge and furrow units, over 4' O.C., double, minimum		200	.160		21.50	4.26		25.76	30	
	2750	Maximum		120	.267		39.50	7.10		46.60	54.50	
	2800	Single, minimum		214	.150		18.20	3.98		22.18	26	
	2850	Maximum		153	.209	↓	38.50	5.55		44.05	51	
	3000	Rolling roof, translucent panels, flat roof, residential, mininum		253	.126	S.F.	16.50	3.36		19.86	23.50	
	3030	Maximum		160	.200		33	5.30		38.30	45	
	3100	Lean-to skyroof, long span, double, minimum		197	.162		22	4.32		26.32	31.50	
	3150	Maximum		101	.317		44	8.45		52.45	61.50	
	3300	Single, minimum		321	.100		16.55	2.65		19.20	22.50	
	3350	Maximum	↓	160	.200	↓	27.50	5.30		32.80	39	

078 400 | Glass Block Skylights

		CREW	DAILY OUTPUT	LABOR-HOURS	UNIT	MAT.	LABOR	EQUIP.	TOTAL	TOTAL INCL O&P	
402	0010	PREFABRICATED Glass block with metal frame, minimum	G-3	265	.121	S.F.	39.50	3.21		42.71	48.50	402
	0100	Maximum	"	160	.200	"	79.50	5.30		84.80	95.50	

079 204 | Sealants & Caulkings

		CREW	DAILY OUTPUT	LABOR-HOURS	UNIT	1999 BARE COSTS				TOTAL INCL O&P	
						MAT.	LABOR	EQUIP.	TOTAL		
204	0010	**CAULKING AND SEALANTS**										204
	0020	Acoustical sealant, elastomeric, cartridges				Ea.	1.95			1.95	2.15	
	0030	Backer rod, polyethylene, 1/4" diameter	1 Bric	4.60	1.739	C.L.F.	1.26	48		49.26	76	
	0050	1/2" diameter		4.60	1.739		2.51	48		50.51	77.50	
	0070	3/4" diameter		4.60	1.739		4.85	48		52.85	80	
	0090	1" diameter	↓	4.60	1.739	↓	8.90	48		56.90	84.50	
	0100	Acrylic latex caulk, white										
	0200	11 fl. oz cartridge				Ea.	1.96			1.96	2.16	
	0500	1/4" x 1/2"	1 Bric	248	.032	L.F.	.16	.89		1.05	1.56	
	0600	1/2" x 1/2"		250	.032		.32	.88		1.20	1.72	
	0800	3/4" x 3/4"		230	.035		.72	.96		1.68	2.28	
	0900	3/4" x 1"		200	.040		.96	1.10		2.06	2.78	
	1000	1" x 1"	↓	180	.044	↓	1.20	1.23		2.43	3.23	
	1400	Butyl based, bulk				Gal.	22			22	24	
	1500	Cartridges				"	26.50			26.50	29.50	
	1700	Bulk, in place 1/4" x 1/2", 154 L.F./gal.	1 Bric	230	.035	L.F.	.14	.96		1.10	1.65	
	1800	1/2" x 1/2", 77 L.F./gal.	"	180	.044	"	.29	1.23		1.52	2.22	
	2000	Latex acrylic based, bulk				Gal.	23			23	25.50	
	2100	Cartridges					26			26	28.50	
	2300	Polysulfide compounds, 1 component, bulk					43.50			43.50	47.50	

Important: See the Reference Section for critical supporting data - Reference Nos., Crews, & City Cost Indexes

204			CREW	DAILY OUTPUT	LABOR-HOURS	UNIT	MAT.	LABOR	EQUIP.	TOTAL	TOTAL INCL O&P	
079 204	**Sealants & Caulkings**						**1999 BARE COSTS**					204
204	2600	1 or 2 component, in place, 1/4" x 1/4", 308 L.F./gal.	1 Bric	145	.055	L.F.	.14	1.52		1.66	2.52	
	2700	1/2" x 1/4", 154 L.F./gal.		135	.059		.28	1.64		1.92	2.85	
	2900	3/4" x 3/8", 68 L.F./gal.		130	.062		.64	1.70		2.34	3.34	
	3000	1" x 1/2", 38 L.F./gal.		130	.062		1.14	1.70		2.84	3.89	
	3200	Polyurethane, 1 or 2 component				Gal.	48.50			48.50	53	
	3500	Bulk, in place, 1/4" x 1/4"	1 Bric	150	.053	L.F.	.16	1.47		1.63	2.46	
	3600	1/2" x 1/4"		145	.055		.31	1.52		1.83	2.72	
	3800	3/4" x 3/8", 68 L.F./gal.		130	.062		.71	1.70		2.41	3.42	
	3900	1" x 1/2"		110	.073		1.26	2.01		3.27	4.50	
	4100	Silicone rubber, bulk				Gal.	34			34	37.50	
	4200	Cartridges				"	40			40	44	
	4400	Neoprene gaskets, closed cell, adhesive, 1/8" x 3/8"	1 Bric	240	.033	L.F.	.20	.92		1.12	1.65	
	4500	1/4" x 3/4"		215	.037		.48	1.03		1.51	2.13	
	4700	1/2" x 1"		200	.040		1.40	1.10		2.50	3.26	
	4800	3/4" x 1-1/2"		165	.048		2.91	1.34		4.25	5.30	
	5500	Resin epoxy coating, 2 component, heavy duty				Gal.	26			26	28.50	
	5800	Tapes, sealant, P.V.C. foam adhesive, 1/16" x 1/4"				C.L.F.	4.60			4.60	5.05	
	5900	1/16" x 1/2"					6.80			6.80	7.50	
	5950	1/16" x 1"					11.30			11.30	12.45	
	6000	1/8" x 1/2"					7.65			7.65	8.40	
	6200	Urethane foam, 2 component, handy pack, 1 C.F.				Ea.	27.50			27.50	30.50	
	6300	50.0 C.F. pack				C.F.	14.05			14.05	15.45	

For information about Means Estimating Seminars, see yellow pages 11 and 12 in back of book

THERMAL & MOISTURE PROTECTION 7

Division Notes

	CREW	DAILY OUTPUT	LABOR-HOURS	UNIT	1999 BARE COSTS				TOTAL INCL O&P
					MAT.	LABOR	EQUIP.	TOTAL	

Division 8
Doors & Windows

Estimating Tips

081 Metal Doors & Frames

- Most metal doors and frames look alike, but there may be significant differences among them. When estimating these items be sure to choose the line item that most closely compares to the specification or door schedule requirements regarding:
  - type of metal
  - metal gauge
  - door core material
  - fire rating
  - finish

082 Wood & Plastic Doors

- Wood and plastic doors vary considerably in price. The primary determinant is the veneer material. Lauan, birch and oak are the most common veneers. Other variables include the following:
  - hollow or solid core
  - fire rating
  - flush or raised panel
  - finish
- If the specifications require compliance with AWI (Architectural Woodwork Institute) standards or acoustical standards, the cost of the door may increase substantially. All wood doors are priced pre-mortised for hinges and predrilled for cylindrical locksets.

083 Special Doors

- There are many varieties of special doors, and they are usually priced per each. Add frames, hardware or operators required for a complete installation.

085 Metal Windows

- Most metal windows are delivered preglazed. However, some metal windows are priced without glass. Refer to 088 Glazing for glass pricing. The grade C indicates commercial grade windows, usually ASTM C-35.

086 Wood & Plastic Windows

- All wood windows are priced preglazed. The two glazing options priced are single pane float glass and insulating glass 1/2″ thick. Add the cost of screens and grills if required.

087 Hardware

- Hardware costs add considerably to the cost of a door. The most efficient method to determine the hardware requirements for a project is to review the door schedule. This schedule, in conjunction with the specifications, is all you should need to take off the door hardware.
- Door hinges are priced by the pair, with most doors requiring 1-1/2 pairs per door. The hinge prices do not include installation labor because it is included in door installation. Hinges are classified according to the frequency of use.

088 Glazing

- Different openings require different types of glass. The three most common types are:
  - float
  - tempered
  - insulating
- Most exterior windows are glazed with insulating glass. Entrance doors and window walls, where the glass is less than 18″ from the floor, are generally glazed with tempered glass. Interior windows and some residential windows are glazed with float glass.

089 Glazed Curtain Walls

- Glazed curtain walls consist of the metal tube framing and the glazing material. The cost data in this subdivision is presented for the metal tube framing alone or the composite wall. If your estimate requires a detailed takeoff of the framing, be sure to add the glazing cost.

Reference Numbers

Reference numbers are shown in bold squares at the beginning of some major classifications. These numbers refer to related items in the Reference Section. The reference information may be an estimating procedure, an alternate pricing method or technical information.

Note: Not all subdivisions listed here necessarily appear in this publication.

081 100 | Steel Doors & Frames

			DAILY	LABOR-		1999 BARE COSTS				TOTAL
		CREW	OUTPUT	HOURS	UNIT	MAT.	LABOR	EQUIP.	TOTAL	INCL O&P
103	0010	**COMMERCIAL STEEL DOORS**			R081 -010						103
	0015	Flush, full panel, hollow core									
	0020	1-3/8" thick, 20 ga., 2'-0"x 6'-8"	2 Carp	20	.800	Ea.	163	22		185	214
	0040	2'-8" x 6'-8"		18	.889		167	24.50		191.50	222
	0060	3'-0" x 6'-8"		17	.941		170	25.50		195.50	228
	0100	3'-0" x 7'-0"		17	.941		178	25.50		203.50	237
	0120	For vision lite, add					51			51	56
	0140	For narrow lite, add					60			60	66
	0160	For bottom louver, add					100			100	110
	0230	For baked enamel finish, add					30%	15%			
	0260	For galvanizing, add					15%				
	0320	Half glass, 20 ga., 2'-0" x 6'-8"	2 Carp	20	.800	Ea.	208	22		230	264
	0340	2'-8" x 6'-8"		18	.889		213	24.50		237.50	273
	0360	3'-0" x 6'-8"		17	.941		218	25.50		243.50	281
	0400	3'-0" x 7'-0"		17	.941		221	25.50		246.50	285
	0500	Hollow core, 1-3/4" thick, full panel, 20 ga., 2'-8" x 6'-8"		18	.889		152	24.50		176.50	205
	0520	3'-0" x 6'-8"		17	.941		173	25.50		198.50	232
	0640	3'-0" x 7'-0"		17	.941		178	25.50		203.50	237
	0680	4'-0" x 7'-0"		15	1.067		225	29		254	294
	0700	4'-0" x 8'-0"		13	1.231		265	33.50		298.50	345
	1000	18 ga., 2'-8" x 6'-8"		17	.941		182	25.50		207.50	241
	1020	3'-0" x 6'-8"		16	1		189	27.50		216.50	251
	1120	3'-0" x 7'-0" **CN**		17	.941		210	25.50		235.50	272
	1180	4'-0" x 7'-0"		14	1.143		230	31		261	300
	1200	4'-0" x 8'-0"		17	.941		272	25.50		297.50	340
	1230	Half glass, 20 ga., 2'-8" x 6'-8"		20	.800		246	22		268	305
	1240	3'-0" x 6'-8"		18	.889		250	24.50		274.50	315
	1260	3'-0" x 7'-0"		18	.889		268	24.50		292.50	335
	1280	4'-0" x 7'-0"		16	1		310	27.50		337.50	390
	1300	4'-0" x 8'-0"		13	1.231		340	33.50		373.50	430
	1320	18 ga., 2'-8" x 6'-8"		18	.889		279	24.50		303.50	345
	1340	3'-0" x 6'-8"		17	.941		274	25.50		299.50	340
	1360	3'-0" x 7'-0"		17	.941		284	25.50		309.50	350
	1380	4'-0" x 7'-0"		15	1.067		400	29		429	485
	1400	4'-0" x 8'-0"		14	1.143		465	31		496	560
	1720	Insulated, 1-3/4" thick, full panel, 18 ga., 3'-0" x 6'-8"		15	1.067		217	29		246	285
	1740	2'-8" x 7'-0"		16	1		214	27.50		241.50	278
	1760	3'-0" x 7'-0"		15	1.067		225	29		254	294
	1800	4'-0" x 8'-0"		13	1.231		315	33.50		348.50	400
	1820	Half glass, 18 ga., 3'-0" x 6'-8"		16	1		280	27.50		307.50	355
	1840	2'-8" x 7'-0"		17	.941		277	25.50		302.50	345
	1860	3'-0" x 7'-0"		16	1		286	27.50		313.50	360
	1900	4'-0" x 8'-0"		14	1.143		420	31		451	515
106	0010	**DOOR FRAMES**									106
	0020	Steel channels with anchors and bar stops									
	0100	6" channel @ 8.2#/L.F., 3' x 7' door, weighs 200#	E-4	13	2.462	Ea.	187	76.50	6.35	269.85	355
	0200	8" channel @ 11.5#/L.F., 6' x 8' door, weighs 300#		9	3.556		272	111	9.15	392.15	510
	0300	8' x 12' door, weighs 450#		6.50	4.923		410	153	12.65	575.65	745
	0400	10" channel @ 15.3#/L.F., 10' x 10' door, weighs 525# **CN**		6	5.333		475	166	13.70	654.70	845
	0500	12' x 12' door, weighs 625#		5.50	5.818		565	181	14.95	760.95	970
	0600	12" channel @ 20.7#/L.F., 12' x 12' door, weighs 825#		4.50	7.111		745	221	18.30	984.30	1,250
	0700	12' x 16' door, weighs 1000#		4	8		905	249	20.50	1,174.50	1,475
	0800	For frames without bar stops, light sections, deduct					15%				
	0900	Heavy sections, deduct					10%				

Important: See the Reference Section for critical supporting data - Reference Nos., Crews, & City Cost Indexes

8

DOORS & WINDOWS

081 100	Steel Doors & Frames		CREW	DAILY OUTPUT	LABOR-HOURS	UNIT	1999 BARE COSTS				TOTAL INCL O&P	
							MAT.	LABOR	EQUIP.	TOTAL		
110	0010	**FIRE DOOR**	R081 -010									**110**
	0015	Steel, flush, "B" label, 90 minute										
	0020	Full panel, 20 ga., 2'-0" x 6'-8"	2 Carp	20	.800	Ea.	208	22		230	264	
	0040	2'-8" x 6'-8"		18	.889		208	24.50		232.50	267	
	0060	3'-0" x 6'-8"		17	.941		210	25.50		235.50	272	
	0080	3'-0" x 7'-0"		17	.941		259	25.50		284.50	325	
	0140	18 ga., 3'-0" x 6'-8"		16	1		230	27.50		257.50	296	
	0160	2'-8" x 7'-0"		17	.941		240	25.50		265.50	305	
	0180	3'-0" x 7'-0"		16	1		236	27.50		263.50	305	
	0200	4'-0" x 7'-0"		15	1.067		300	29		329	375	
	0220	For "A" label, 3 hour, 18 ga., use same price as "B" label										
	0240	For vision lite, add				Ea.	56.50			56.50	62.50	
	0520	Flush, "B" label 90 min., composite, 20 ga., 2'-0" x 6'-8"	2 Carp	18	.889		208	24.50		232.50	267	
	0540	2'-8" x 6'-8"		17	.941		208	25.50		233.50	270	
	0560	3'-0" x 6'-8"		16	1		210	27.50		237.50	274	
	0580	3'-0" x 7'-0"		16	1		234	27.50		261.50	300	
	0640	Flush, "A" label 3 hour, composite, 18 ga., 3'-0" x 6'-8"		15	1.067		250	29		279	320	
	0660	2'-8" x 7'-0"		16	1		255	27.50		282.50	325	
	0680	3'-0" x 7'-0"		15	1.067		260	29		289	330	
	0700	4'-0" x 7'-0"		14	1.143		300	31		331	380	
114	0010	**RESIDENTIAL STEEL DOOR**										**114**
	0020	Prehung, insulated, exterior										
	0030	Embossed, full panel, 2'-8" x 6'-8"	2 Carp	17	.941	Ea.	172	25.50		197.50	230	
	0040	3'-0" x 6'-8"		15	1.067		175	29		204	238	
	0060	3'-0" x 7'-0"		15	1.067		224	29		253	292	
	0070	5'-4" x 6'-8", double		8	2		345	54.50		399.50	465	
	0220	Half glass, 2'-8" x 6'-8"		17	.941		271	25.50		296.50	340	
	0240	3'-0" x 6'-8"		16	1		239	27.50		266.50	305	
	0260	3'-0" x 7'-0"		16	1		292	27.50		319.50	365	
	0270	5'-4" x 6'-8", double		8	2		540	54.50		594.50	680	
	0720	Raised plastic face, full panel, 2'-8" x 6'-8"		16	1		211	27.50		238.50	275	
	0740	3'-0" x 6'-8"		15	1.067		214	29		243	282	
	0760	3'-0" x 7'-0"		15	1.067		227	29		256	296	
	0780	5'-4" x 6'-8", double		8	2		420	54.50		474.50	550	
	0820	Half glass, 2'-8" x 6'-8"		17	.941		250	25.50		275.50	315	
	0840	3'-0" x 6'-8"		16	1		253	27.50		280.50	320	
	0860	3'-0" x 7'-0"		16	1		278	27.50		305.50	350	
	0880	5'-4" x 6'-8", double		8	2		525	54.50		579.50	665	
	1320	Flush face, full panel, 2'-6" x 6'-8"		16	1		190	27.50		217.50	252	
	1340	3'-0" x 6'-8"		15	1.067		193	29		222	258	
	1360	3'-0" x 7'-0"		15	1.067		244	29		273	315	
	1380	5'-4" x 6'-8", double		8	2		380	54.50		434.50	505	
	1420	Half glass, 2'-8" x 6'-8"		17	.941		236	25.50		261.50	300	
	1440	3'-0" x 6'-8"		16	1		238	27.50		265.50	305	
	1460	3'-0" x 7'-0"		16	1		289	27.50		316.50	365	
	1480	5'-4" x 6'-8", double		8	2		470	54.50		524.50	605	
	2300	Interior, residential, closet, bi-fold, 6'-8" x 2'-0" wide		16	1		122	27.50		149.50	177	
	2330	3'-0" wide		16	1		137	27.50		164.50	193	
	2360	4'-0" wide		15	1.067		208	29		237	275	
	2400	5'-0" wide		14	1.143		241	31		272	315	
	2420	6'-0" wide		13	1.231		270	33.50		303.50	350	
118	0010	**STEEL FRAMES, KNOCK DOWN**	R081 -020									**118**
	0020	18 ga., up to 5-3/4 deep										
	0025	6'-8" high, 3'-0" wide, single	2 Carp	16	1	Ea.	86	27.50		113.50	138	
	0040	6'-0" wide, double		14	1.143		96.50	31		127.50	155	

		081 100	Steel Doors & Frames		CREW	DAILY OUTPUT	LABOR-HOURS	UNIT	1999 BARE COSTS				TOTAL INCL O&P	
									MAT.	LABOR	EQUIP.	TOTAL		
118	0100		7'-0" high, 3'-0" wide, single		2 Carp	16	1	Ea.	88.50	27.50		116	140	118
	0140		6'-0" wide, double	R081 -020		14	1.143		93.50	31		124.50	152	
	1000		18 ga., up to 4-7/8" deep, 7'-0" H, 3'-0" W, single	CN		16	1		70.50	27.50		98	121	
	1140		6'-0" wide, double			14	1.143		88.50	31		119.50	146	
	2800		16 ga., up to 3-7/8" deep, 7'-0" high, 3'-0" wide, single			16	1		70	27.50		97.50	120	
	2840		6'-0" wide, double			14	1.143		88	31		119	146	
	3600		5-3/4" deep, 7'-0" high, 4'-0" wide, single			15	1.067		75	29		104	129	
	3640		8'-0" wide, double			12	1.333		109	36.50		145.50	177	
	3700		8'-0" high, 4'-0" wide, single			15	1.067		88.50	29		117.50	144	
	3740		8'-0" wide, double			12	1.333		109	36.50		145.50	177	
	4000		6-3/4" deep, 7'-0" high, 4'-0" wide, single			15	1.067		81.50	29		110.50	136	
	4040		8'-0"			12	1.333		110	36.50		146.50	179	
	4100		8'-0" high, 4'-0" wide, single			15	1.067		93	29		122	148	
	4140		8'-0" wide, double			12	1.333		119	36.50		155.50	189	
	4400		8-3/4" deep, 7'-0" high, 4'-0" wide, single			15	1.067		91	29		120	146	
	4440		8'-0" wide, double			12	1.333		124	36.50		160.50	194	
	4500		8'-0" high, 4'-0" wide, single			15	1.067		105	29		134	161	
	4540		8'-0" wide, double			12	1.333		129	36.50		165.50	200	
	4900		For welded frames, add						29.50			29.50	32.50	
	5400		16 ga., "B" label, up to 5-3/4" deep, 7'-0" high, 4'-0" wide, single		2 Carp	15	1.067		90	29		119	145	
	5440		8'-0" wide, double			12	1.333		116	36.50		152.50	185	
	5800		6-3/4" deep, 7'-0" high, 4'-0" wide, single			15	1.067		93	29		122	148	
	5840		8'-0" wide, double			12	1.333		114	36.50		150.50	183	
	6200		8-3/4" deep, 7'-0" high, 4'-0" wide, single			15	1.067		102	29		131	158	
	6240		8'-0" wide, double			12	1.333		125	36.50		161.50	196	
	6300		For "A" label use same price as "B" label											
	6400		For baked enamel finish, add						30%	15%				
	6500		For galvanizing, add						15%					
	7900		Transom lite frames, fixed, add		2 Carp	155	.103	S.F.	23	2.82		25.82	30	
	8000		Movable, add		"	130	.123	"	27.50	3.36		30.86	36	

		082 050	Wood & Plastic Doors	CREW	DAILY OUTPUT	LABOR-HOURS	UNIT	1999 BARE COSTS				TOTAL INCL O&P	
								MAT.	LABOR	EQUIP.	TOTAL		
054	0010	**WOOD FRAMES**											054
	0400		Exterior frame, incl. ext. trim, pine, 5/4 x 4-9/16" deep	2 Carp	375	.043	L.F.	3.70	1.16		4.86	5.90	
	0420		5-3/16" deep		375	.043		3.95	1.16		5.11	6.20	
	0440		6-9/16" deep		375	.043		5.40	1.16		6.56	7.80	
	0600		Oak, 5/4 x 4-9/16" deep		350	.046		7.95	1.25		9.20	10.70	
	0620		5-3/16" deep		350	.046		8.95	1.25		10.20	11.80	
	0640		6-9/16" deep		350	.046		9.95	1.25		11.20	12.90	
	0800		Walnut, 5/4 x 4-9/16" deep		350	.046		9.35	1.25		10.60	12.25	
	0820		5-3/16" deep		350	.046		13.55	1.25		14.80	16.85	
	0840		6-9/16" deep		350	.046		16	1.25		17.25	19.55	
	1000		Sills, 8/4 x 8" deep, oak, no horns		100	.160		11.30	4.37		15.67	19.25	
	1020		2" horns		100	.160		12.55	4.37		16.92	20.50	
	1040		3" horns		100	.160		14.50	4.37		18.87	23	
	1100		8/4 x 10" deep, oak, no horns		90	.178		15.15	4.85		20	24.50	
	1120		2" horns		90	.178		16.90	4.85		21.75	26.50	
	1140		3" horns		90	.178		18.40	4.85		23.25	28	

Important: See the Reference Section for critical supporting data - Reference Nos., Crews, & City Cost Indexes

8 DOORS & WINDOWS

082 050	Wood & Plastic Doors		CREW	DAILY OUTPUT	LABOR-HOURS	UNIT	1999 BARE COSTS				TOTAL INCL O&P	
							MAT.	LABOR	EQUIP.	TOTAL		
054	2000	Exterior, colonial, frame & trim, 3' opng., in-swing, minimum	2 Carp	22	.727	Ea.	263	19.85		282.85	320	**054**
	2010	Average		21	.762		224	21		245	279	
	2020	Maximum		20	.800		885	22		907	1,000	
	2100	5'-4" opening, in-swing, minimum		17	.941		297	25.50		322.50	365	
	2120	Maximum		15	1.067		885	29		914	1,025	
	2140	Out-swing, minimum		17	.941		305	25.50		330.50	375	
	2160	Maximum		15	1.067		920	29		949	1,050	
	2400	6'-0" opening, in-swing, minimum		16	1		285	27.50		312.50	360	
	2420	Maximum		10	1.600		920	43.50		963.50	1,075	
	2460	Out-swing, minimum		16	1		305	27.50		332.50	380	
	2480	Maximum	↓	10	1.600		1,150	43.50		1,193.50	1,325	
	2600	For two sidelights, add, minimum		30	.533	Opng.	294	14.55		308.55	350	
	2620	Maximum		20	.800	"	940	22		962	1,050	
	2700	Custom birch frame, 3'-0" opening		16	1	Ea.	169	27.50		196.50	228	
	2750	6'-0" opening		16	1		240	27.50		267.50	305	
	2900	Exterior, modern, plain trim, 3' opng., in-swing, minimum		26	.615		24.50	16.80		41.30	53.50	
	2920	Average		24	.667		29.50	18.20		47.70	60.50	
	2940	Maximum	↓	22	.727	↓	36	19.85		55.85	70.50	
	3000	Interior frame, pine, 11/16" x 3-5/8" deep		375	.043	L.F.	3.57	1.16		4.73	5.75	
	3020	4-9/16" deep		375	.043		4.22	1.16		5.38	6.45	
	3200	Oak, 11/16" x 3-5/8" deep		350	.046		4.50	1.25		5.75	6.90	
	3220	4-9/16" deep		350	.046		4.86	1.25		6.11	7.30	
	3240	5-3/16" deep		350	.046		5	1.25		6.25	7.45	
	3400	Walnut, 11/16" x 3-5/8" deep		350	.046		8	1.25		9.25	10.75	
	3420	4-9/16" deep		350	.046		8	1.25		9.25	10.75	
	3440	5-3/16" deep		350	.046	↓	8.30	1.25		9.55	11.10	
	3600	Pocket door frame		16	1	Ea.	67.50	27.50		95	118	
	3800	Threshold, oak, 5/8" x 3-5/8" deep		200	.080	L.F.	2.92	2.18		5.10	6.65	
	3820	4-5/8" deep		190	.084		3.72	2.30		6.02	7.70	
	3840	5-5/8" deep	↓	180	.089	↓	4.47	2.43		6.90	8.75	
	4000	For casing see division 062-212										
062	0010	**WOOD DOOR, ARCHITECTURAL** [R082-120]										**062**
	0015	Flush, int., 1-3/4", 7 ply, hollow core,										
	0020	Lauan face, 2'-0" x 6'-8"	2 Carp	17	.941	Ea.	43	25.50		68.50	87.50	
	0040	2'-6" x 6'-8"		17	.941		51	25.50		76.50	96.50	
	0080	3'-0" x 6'-8"		17	.941		46.50	25.50		72	91.50	
	0100	4'-0" x 6'-8"		16	1		74	27.50		101.50	125	
	0120	Birch face, 2'-0" x 6'-8"		17	.941		46	25.50		71.50	91	
	0140	2'-6" x 6'-8"		17	.941		62	25.50		87.50	109	
	0180	3'-0" x 6'-8"		17	.941		65	25.50		90.50	112	
	0200	4'-0" x 6'-8"		16	1		83	27.50		110.50	135	
	0220	Oak face, 2'-0" x 6'-8"		17	.941		63.50	25.50		89	111	
	0240	2'-6" x 6'-8"		17	.941		67.50	25.50		93	115	
	0280	3'-0" x 6'-8"		17	.941		72.50	25.50		98	120	
	0300	4'-0" x 6'-8"		16	1		92	27.50		119.50	144	
	0320	Walnut face, 2'-0" x 6'-8"		17	.941		128	25.50		153.50	182	
	0340	2'-6" x 6'-8"		17	.941		131	25.50		156.50	185	
	0380	3'-0" x 6'-8"		17	.941		136	25.50		161.50	190	
	0400	4'-0" x 6'-8"	↓	16	1		154	27.50		181.50	212	
	0430	For 7'-0" high, add					13			13	14.30	
	0440	For 8'-0" high, add					18.20			18.20	20	
	0460	For 8'-0" high walnut, add					10.90			10.90	12	
	0480	For prefinishing, clear, add					28.50			28.50	31.50	
	0500	For prefinishing, stain, add					39			39	43	
	1320	M.D. overlay on hardboard, 2'-0" x 6'-8"	2 Carp	17	.941	↓	80.50	25.50		106	129	

8

DOORS & WINDOWS

082 050	Wood & Plastic Doors	CREW	DAILY OUTPUT	LABOR-HOURS	UNIT	MAT.	LABOR	EQUIP.	TOTAL	TOTAL INCL O&P	
062	1340	2'-6" x 6'-8"	2 Carp	17	.941	Ea.	80.50	25.50		106	129	062
	1380	3'-0" x 6'-8"		17	.941		95.50	25.50		121	146	
	1400	4'-0" x 6'-8"		16	1		134	27.50		161.50	190	
	1420	For 7'-0" high, add					7.15			7.15	7.85	
	1440	For 8'-0" high, add					15.80			15.80	17.40	
	1720	H.P. plastic laminate, 2'-0" x 6'-8"	2 Carp	16	1		195	27.50		222.50	258	
	1740	2'-6" x 6'-8"		16	1		195	27.50		222.50	258	
	1780	3'-0" x 6'-8"		15	1.067		222	29		251	290	
	1800	4'-0" x 6'-8"		14	1.143		310	31		341	390	
	1820	For 7'-0" high, add					7.15			7.15	7.85	
	1840	For 8'-0" high, add					18.20			18.20	20	
	2020	5 ply particle core, lauan face, 2'-6" x 6'-8"	2 Carp	15	1.067		65	29		94	118	
	2040	3'-0" x 6'-8"		14	1.143		71	31		102	127	
	2080	3'-0" x 7'-0"		13	1.231		74	33.50		107.50	135	
	2100	4'-0" x 7'-0"		12	1.333		91.50	36.50		128	159	
	2120	Birch face, 2'-6" x 6'-8"		15	1.067		77.50	29		106.50	132	
	2140	3'-0" x 6'-8"		14	1.143		85	31		116	143	
	2180	3'-0" x 7'-0"		13	1.231		87	33.50		120.50	149	
	2200	4'-0" x 7'-0"		12	1.333		106	36.50		142.50	174	
	2220	Oak face, 2'-6" x 6'-8"		15	1.067		85.50	29		114.50	140	
	2240	3'-0" x 6'-8"		14	1.143		94	31		125	153	
	2280	3'-0" x 7'-0"		13	1.231		97	33.50		130.50	159	
	2300	4'-0" x 7'-0"		12	1.333		119	36.50		155.50	188	
	2320	Walnut face, 2'-0" x 6'-8"		15	1.067		95	29		124	150	
	2340	2'-6" x 6'-8"		14	1.143		109	31		140	168	
	2380	3'-0" x 6'-8"		13	1.231		122	33.50		155.50	187	
	2400	4'-0" x 6'-8"		12	1.333		152	36.50		188.50	225	
	2440	For 8'-0" high, add					22			22	24	
	2460	For 8'-0" high walnut, add					12.75			12.75	14.05	
	2480	For solid wood core, add					28.50			28.50	31.50	
	2720	For prefinishing, clear, add					18.60			18.60	20.50	
	2740	For prefinishing, stain, add					39			39	42.50	
	3320	M.D. overlay on hardboard, 2'-6" x 6'-8"	2 Carp	14	1.143		71.50	31		102.50	128	
	3340	3'-0" x 6'-8"		13	1.231		78.50	33.50		112	140	
	3380	3'-0" x 7'-0"		12	1.333		80	36.50		116.50	146	
	3400	4'-0" x 7'-0"		10	1.600		98	43.50		141.50	177	
	3440	For 8'-0" height, add					22			22	24	
	3460	For solid wood core, add					28.50			28.50	31.50	
	3720	H.P. plastic laminate, 2'-6" x 6'-8"	2 Carp	13	1.231		110	33.50		143.50	174	
	3740	3'-0" x 6'-8"		12	1.333		125	36.50		161.50	195	
	3780	3'-0" x 7'-0"		11	1.455		129	39.50		168.50	205	
	3800	4'-0" x 7'-0"		8	2		157	54.50		211.50	259	
	3840	For 8'-0" height, add					22			22	24	
	3860	For solid wood core, add					27			27	29.50	
	4000	Exterior, flush, solid wood stave core, birch, 1-3/4" x 7'-0" x 2'-6"	2 Carp	15	1.067		138	29		167	197	
	4020	2'-8" wide		15	1.067		144	29		173	204	
	4040	3'-0" wide		14	1.143		152	31		183	217	
	4100	Oak faced 1-3/4" x 7'-0" x 2'-6" wide		15	1.067		151	29		180	212	
	4120	2'-8" wide		15	1.067		162	29		191	224	
	4140	3'-0" wide		14	1.143		173	31		204	239	
	4200	Walnut faced, 1-3/4" x 7'-0" x 2'-6" wide		15	1.067		222	29		251	291	
	4220	2'-8" wide		15	1.067		232	29		261	300	
	4240	3'-0" wide		14	1.143		242	31		273	315	
	4300	For 6'-8" high door, deduct from 7'-0" door					18.20			18.20	20	
066	0010	**WOOD DOORS, DECORATOR**										066
	3000	Solid wood, 1-3/4" thick stile and rail										

Important: See the Reference Section for critical supporting data - Reference Nos., Crews, & City Cost Indexes

082 050	Wood & Plastic Doors	CREW	DAILY OUTPUT	LABOR-HOURS	UNIT	1999 BARE COSTS				TOTAL INCL O&P	
						MAT.	LABOR	EQUIP.	TOTAL		
066 3020	Mahogany, 3'-0" x 7'-0", minimum	2 Carp	14	1.143	Ea.	445	31		476	540	**066**
3030	Maximum		10	1.600		650	43.50		693.50	785	
3040	3'-6" x 8'-0", minimum		10	1.600		655	43.50		698.50	790	
3050	Maximum		8	2		755	54.50		809.50	920	
3100	Pine, 3'-0" x 7'-0", minimum		14	1.143		291	31		322	370	
3110	Maximum		10	1.600		525	43.50		568.50	645	
3120	3'-6" x 8'-0", minimum		10	1.600		580	43.50		623.50	705	
3130	Maximum		8	2		1,575	54.50		1,629.50	1,800	
3200	Red oak, 3'-0" x 7'-0", minimum		14	1.143		655	31		686	770	
3210	Maximum		10	1.600		1,175	43.50		1,218.50	1,350	
3220	3'-6" x 8'-0", minimum		10	1.600		830	43.50		873.50	985	
3230	Maximum	↓	8	2	↓	1,375	54.50		1,429.50	1,575	
4000	Hand carved door, mahogany										
4020	3'-0" x 7'-0", minimum	2 Carp	14	1.143	Ea.	650	31		681	765	
4030	Maximum		11	1.455		1,550	39.50		1,589.50	1,800	
4040	3'-6" x 8'-0", minimum		10	1.600		1,050	43.50		1,093.50	1,250	
4050	Maximum		8	2		2,225	54.50		2,279.50	2,525	
4200	Red oak, 3'-0" x 7'-0", minimum		14	1.143		1,275	31		1,306	1,450	
4210	Maximum		11	1.455		3,000	39.50		3,039.50	3,375	
4220	3'-6" x 8'-0", minimum	↓	10	1.600		2,575	43.50		2,618.50	2,925	
4280	For 6'-8" high door, deduct from 7'-0" door					23.50			23.50	25.50	
4400	For custom finish, add					103			103	113	
4600	Side light, mahogany, 7'-0" x 1'-6" wide, minimum	2 Carp	18	.889		258	24.50		282.50	320	
4610	Maximum		14	1.143		700	31		731	820	
4620	8'-0" x 1'-6" wide, minimum		14	1.143		320	31		351	400	
4630	Maximum		10	1.600		810	43.50		853.50	965	
4640	Side light, oak, 7'-0" x 1'-6" wide, minimum		18	.889		345	24.50		369.50	420	
4650	Maximum		14	1.143		805	31		836	935	
4660	8'-0" x 1-6" wide, minimum		14	1.143		420	31		451	510	
4670	Maximum		10	1.600		945	43.50		988.50	1,125	
6520	Interior cafe doors, 2'-6" opening, stock, panel pine		16	1		139	27.50		166.50	196	
6540	3'-0" opening	↓	16	1	↓	145	27.50		172.50	202	
6550	Louvered pine										
6560	2'-6" opening	2 Carp	16	1	Ea.	121	27.50		148.50	176	
8000	3'-0" opening		16	1		129	27.50		156.50	185	
8010	2'-6" opening, hardwood		16	1		173	27.50		200.50	233	
8020	3'-0" opening	↓	16	1	↓	203	27.50		230.50	266	
8800	Pre-hung doors, see division 082-082										
070 0010	**WOOD FIRE DOORS**	R082 -120									**070**
0020	Particle core, 7 face plys, "B" label,										
0040	1 hour, birch face, 1-3/4" x 2'-6" x 6'-8"	2 Carp	14	1.143	Ea.	152	31		183	216	
0080	3'-0" x 6'-8"		13	1.231		157	33.50		190.50	225	
0090	3'-0" x 7'-0"		12	1.333		173	36.50		209.50	248	
0100	4'-0" x 7'-0"		12	1.333		226	36.50		262.50	305	
0140	Oak face, 2'-6" x 6'-8"		14	1.143		171	31		202	238	
0180	3'-0" x 6'-8"		13	1.231		178	33.50		211.50	249	
0190	3'-0" x 7'-0"		12	1.333		185	36.50		221.50	261	
0200	4'-0" x 7'-0"		12	1.333		244	36.50		280.50	325	
0240	Walnut face, 2'-6" x 6'-8"		14	1.143		227	31		258	299	
0280	3'-0" x 6'-8"		13	1.231		243	33.50		276.50	320	
0290	3'-0" x 7'-0"		12	1.333		261	36.50		297.50	345	
0300	4'-0" x 7'-0"		12	1.333		355	36.50		391.50	450	
0440	M.D. overlay on hardboard, 2'-6" x 6'-8"		15	1.067		147	29		176	207	
0480	3'-0" x 6'-8"		14	1.143		149	31		180	213	
0490	3'-0" x 7'-0"		13	1.231		159	33.50		192.50	227	
0500	4'-0" x 7'-0"	↓ ↓	12	1.333	↓	202	36.50		238.50	281	

082 050	Wood & Plastic Doors		CREW	DAILY OUTPUT	LABOR-HOURS	UNIT	1999 BARE COSTS				TOTAL INCL O&P	
							MAT.	LABOR	EQUIP.	TOTAL		
070	0540	H.P. plastic laminate, 2'-6" x 6'-8"	R082 -120	2 Carp	13	1.231	Ea.	233	33.50		266.50	310	070
	0580	3'-0" x 6'-8"			12	1.333		246	36.50		282.50	330
	0590	3'-0" x 7'-0"			11	1.455		253	39.50		292.50	340
	0600	4'-0" x 7'-0"			10	1.600		335	43.50		378.50	435
	0740	90 minutes, birch face, 1-3/4" x 2'-6" x 6'-8"			14	1.143		197	31		228	266
	0780	3'-0" x 6'-8"			13	1.231		204	33.50		237.50	277
	0790	3'-0" x 7'-0"			12	1.333		211	36.50		247.50	290
	0800	4'-0" x 7'-0"			12	1.333		294	36.50		330.50	385
	0840	Oak face, 2'-6" x 6'-8"			14	1.143		187	31		218	255
	0880	3'-0" x 6'-8"			13	1.231		188	33.50		221.50	260
	0890	3'-0" x 7'-0"			12	1.333		204	36.50		240.50	282
	0900	4'-0" x 7'-0"			12	1.333		291	36.50		327.50	380
	0940	Walnut face, 2'-6" x 6'-8"			14	1.143		284	31		315	360
	0980	3'-0" x 6'-8"			13	1.231		310	33.50		343.50	395
	0990	3'-0" x 7'-0"			12	1.333		325	36.50		361.50	415
	1000	4'-0" x 7'-0"			12	1.333		355	36.50		391.50	450
	1140	M.D. overlay on hardboard, 2'-6" x 6'-8"			15	1.067		197	29		226	263
	1180	3'-0" x 6'-8"			14	1.143		204	31		235	273
	1190	3'-0" x 7'-0"			13	1.231		214	33.50		247.50	288
	1200	4'-0" x 7'-0"			12	1.333		297	36.50		333.50	385
	1240	For 8'-0" height, add						52.50			52.50	58
	1260	For 8'-0" height walnut, add						24			24	26.50
	1340	H.P. plastic laminate, 2'-6" x 6'-8"		2 Carp	13	1.231		292	33.50		325.50	375
	1380	3'-0" x 6'-8"			12	1.333		292	36.50		328.50	380
	1390	3'-0" x 7'-0"			11	1.455		292	39.50		331.50	385
	1400	4'-0" x 7'-0"			10	1.600		315	43.50		358.50	420
	2200	Custom architectural "B" label, flush, 1-3/4" thick, birch,										
	2210	Solid core										
	2220	2'-6" x 7'-0"		2 Carp	15	1.067	Ea.	335	29		364	415
	2260	3'-0" x 7'-0"			14	1.143		340	31		371	425
	2300	4'-0" x 7'-0"			13	1.231		470	33.50		503.50	570
	2420	4'-0" x 8'-0"			11	1.455		545	39.50		584.50	665
	2460	For 6'-8" high door, deduct from 7'-0" door						12.50			12.50	13.75
	2480	For oak veneer, add						50%				
	2500	For walnut veneer, add						75%				
074	0010	**WOOD DOORS, PANELED**											074
	0020	Interior, six panel, hollow core, 1-3/8" thick										
	0040	Molded hardboard, 2'-0" x 6'-8"		2 Carp	17	.941	Ea.	40	25.50		65.50	84.50
	0060	2'-6" x 6'-8"			17	.941		43	25.50		68.50	88
	0080	3'-0" x 6'-8"			17	.941		47.50	25.50		73	92.50
	0140	Embossed print, molded hardboard, 2'-0" x 6'-8"			17	.941		43	25.50		68.50	88
	0160	2'-6" x 6'-8"			17	.941		43	25.50		68.50	88
	0180	3'-0" x 6'-8"			17	.941		47.50	25.50		73	92.50
	0540	Six panel, solid, 1-3/8" thick, pine, 2'-0" x 6'-8"			15	1.067		104	29		133	160
	0560	2'-6" x 6'-8"			14	1.143		117	31		148	178
	0580	3'-0" x 6'-8"			13	1.231		134	33.50		167.50	201
	1020	Two panel, bored rail, solid, 1-3/8" thick, pine, 1'-6" x 6'-8"			16	1		190	27.50		217.50	252
	1040	2'-0" x 6'-8"			15	1.067		249	29		278	320
	1060	2'-6" x 6'-8"			14	1.143		285	31		316	365
	1340	Two panel, solid, 1-3/8" thick, fir, 2'-0" x 6'-8"			15	1.067		104	29		133	160
	1360	2'-6" x 6'-8"			14	1.143		117	31		148	178
	1380	3'-0" x 6'-8"			13	1.231		285	33.50		318.50	370
	1740	Five panel, solid, 1-3/8" thick, fir, 2'-0" x 6'-8"			15	1.067		186	29		215	251
	1760	2'-6" x 6'-8"			14	1.143		285	31		316	365
	1780	3'-0" x 6'-8"			13	1.231		285	33.50		318.50	370

Important: See the Reference Section for critical supporting data - Reference Nos., Crews, & City Cost Indexes

082 050	Wood & Plastic Doors	CREW	DAILY OUTPUT	LABOR-HOURS	UNIT	1999 BARE COSTS				TOTAL INCL O&P
						MAT.	LABOR	EQUIP.	TOTAL	
078	0010	**WOOD DOORS, RESIDENTIAL**									078
0200	Exterior, combination storm & screen, pine									
0260	2'-8" wide	2 Carp	10	1.600	Ea.	191	43.50		234.50	279
0280	3'-0" wide		9	1.778		197	48.50		245.50	294
0300	7'-1" x 3'-0" wide		9	1.778		198	48.50		246.50	295
0400	Full lite, 6'-9" x 2'-6" wide		11	1.455		196	39.50		235.50	278
0420	2'-8" wide		10	1.600		201	43.50		244.50	291
0440	3'-0" wide		9	1.778		202	48.50		250.50	299
0500	7'-1" x 3'-0" wide		9	1.778		221	48.50		269.50	320
0700	Dutch door, pine, 1-3/4" x 6'-8" x 2'-8" wide, minimum		12	1.333		525	36.50		561.50	640
0720	Maximum		10	1.600		560	43.50		603.50	685
0800	3'-0" wide, minimum		12	1.333		550	36.50		586.50	665
0820	Maximum		10	1.600		600	43.50		643.50	730
1000	Entrance door, colonial, 1-3/4" x 6'-8" x 2'-8" wide		16	1		289	27.50		316.50	365
1020	6 panel pine, 3'-0" wide		15	1.067		310	29		339	390
1100	8 panel pine, 2'-8" wide		16	1		425	27.50		452.50	515
1120	3'-0" wide	▼	15	1.067		460	29		489	550
1200	For tempered safety glass lites, add					23			23	25.50
1300	Flush, birch, solid core, 1-3/4" x 6'-8" x 2'-8" wide	2 Carp	16	1		66.50	27.50		94	116
1320	3'-0" wide		15	1.067		73	29		102	126
1350	7'-0" x 2'-8" wide		16	1		74	27.50		101.50	125
1360	3'-0" wide	▼	15	1.067		81.50	29		110.50	136
1380	For tempered safety glass lites, add				▼	48			48	53
2700	Interior, closet, bi-fold, w/hardware, no frame or trim incl.									
2720	Flush, birch, 6'-6" or 6'-8" x 2'-6" wide	2 Carp	13	1.231	Ea.	41.50	33.50		75	98.50
2740	3'-0" wide		13	1.231		45	33.50		78.50	103
2760	4'-0" wide		12	1.333		75.50	36.50		112	141
2780	5'-0" wide		11	1.455		82	39.50		121.50	153
2800	6'-0" wide		10	1.600		89	43.50		132.50	167
3000	Raised panel pine, 6'-6" or 6'-8" x 2'-6" wide		13	1.231		114	33.50		147.50	178
3020	3'-0" wide		13	1.231		128	33.50		161.50	193
3040	4'-0" wide		12	1.333		194	36.50		230.50	272
3060	5'-0" wide		11	1.455		225	39.50		264.50	310
3080	6'-0" wide		10	1.600		252	43.50		295.50	345
3200	Louvered, pine 6'-6" or 6'-8" x 2'-6" wide		13	1.231		89.50	33.50		123	151
3220	3'-0" wide		13	1.231		98.50	33.50		132	161
3240	4'-0" wide		12	1.333		154	36.50		190.50	227
3260	5'-0" wide		11	1.455		139	39.50		178.50	216
3280	6'-0" wide	▼	10	1.600	▼	193	43.50		236.50	281
4400	Bi-passing closet, incl. hardware and frame, no trim incl.									
4420	Flush, lauan, 6'-8" x 4'-0" wide	2 Carp	12	1.333	Opng.	137	36.50		173.50	209
4440	5'-0" wide		11	1.455		143	39.50		182.50	220
4460	6'-0" wide		10	1.600		148	43.50		191.50	232
4600	Flush, birch, 6'-8" x 4'-0" wide		12	1.333		158	36.50		194.50	232
4620	5'-0" wide		11	1.455		175	39.50		214.50	256
4640	6'-0" wide		10	1.600		190	43.50		233.50	278
4800	Louvered, pine, 6'-8" x 4'-0" wide		12	1.333		315	36.50		351.50	410
4820	5'-0" wide		11	1.455		345	39.50		384.50	445
4840	6'-0" wide		10	1.600		390	43.50		433.50	495
5000	Paneled, pine, 6'-8" x 4'-0" wide		12	1.333		305	36.50		341.50	395
5020	5'-0" wide		11	1.455		335	39.50		374.50	430
5040	6'-0" wide	▼	10	1.600	▼	375	43.50		418.50	480
6100	Folding accordion, closet, including track and frame									
6120	Vinyl, 2 layer, stock (see also division 106-552)	2 Carp	400	.040	S.F.	2.68	1.09		3.77	4.67
6140	Woven mahogany and vinyl, stock		400	.040		1.37	1.09		2.46	3.23
6160	Wood slats with vinyl overlay, stock	▼	400	.040	▼	9.35	1.09		10.44	12

DOORS & WINDOWS 8

082 050 | Wood & Plastic Doors

		DAILY OUTPUT	LABOR-HOURS	UNIT	1999 BARE COSTS				TOTAL INCL O&P	
		CREW			MAT.	LABOR	EQUIP.	TOTAL		
078	6180	Economy vinyl, stock	2 Carp	400	.040	S.F.	1.44	1.09		2.53	3.30	078
	6200	Rigid PVC	↓	400	.040	↓	4.05	1.09		5.14	6.20
	6220	For custom partition, add					25%	10%			
	7310	Passage doors, flush, no frame included									
	7320	Hardboard, hollow core, 1-3/8" x 6'-8" x 1'-6" wide	2 Carp	18	.889	Ea.	34	24.50		58.50	75.50
	7330	2'-0" wide		18	.889		35.50	24.50		60	77
	7340	2'-6" wide		18	.889		39	24.50		63.50	81
	7350	2'-8" wide		18	.889		40	24.50		64.50	82
	7360	3'-0" wide		17	.941		33	25.50		58.50	77
	7420	Lauan, hollow core, 1-3/8" x 6'-8" x 1'-6" wide		18	.889		19.60	24.50		44.10	59.50
	7440	2'-0" wide		18	.889		22.50	24.50		47	62.50
	7450	2'-4" wide		18	.889		25.50	24.50		50	66
	7460	2'-6" wide		18	.889		25.50	24.50		50	66
	7480	2'-8" wide		18	.889		26.50	24.50		51	67
	7500	3'-0" wide		17	.941		28	25.50		53.50	71.50
	7700	Birch, hollow core, 1-3/8" x 6'-8" x 1'-6" wide		18	.889		27	24.50		51.50	67.50
	7720	2'-0" wide		18	.889		32	24.50		56.50	73.50
	7740	2'-6" wide		18	.889		37.50	24.50		62	79
	7760	2'-8" wide		18	.889		39	24.50		63.50	81
	7780	3'-0" wide **CN**		17	.941		41	25.50		66.50	85.50
	8000	Pine louvered, 1-3/8" x 6'-8" x 1'-6" wide		19	.842		95	23		118	141
	8020	2'-0" wide		18	.889		111	24.50		135.50	160
	8040	2'-6" wide		18	.889		123	24.50		147.50	174
	8060	2'-8" wide		18	.889		134	24.50		158.50	185
	8080	3'-0" wide		17	.941		141	25.50		166.50	197
	8300	Pine paneled, 1-3/8" x 6'-8" x 1'-6" wide		19	.842		95	23		118	141
	8320	2'-0" wide		18	.889		111	24.50		135.50	160
	8340	2'-6" wide		18	.889		123	24.50		147.50	174
	8360	2'-8" wide		18	.889		134	24.50		158.50	185
	8380	3'-0" wide	↓	17	.941	↓	141	25.50		166.50	197
	8550	For over 20 doors, deduct					15%				
082	0010	**PRE-HUNG DOORS**										082
	0300	Exterior, wood, combination storm & screen, 6'-9" x 2'-6" wide	2 Carp	15	1.067	Ea.	200	29		229	266
	0320	2'-8" wide		15	1.067		200	29		229	266
	0340	3'-0" wide	↓	15	1.067		209	29		238	276
	0360	For 7'-0" high door, add				↓	27			27	29.50
	0370	For aluminum storm doors, see division 083-900									
	1600	Entrance door, flush, birch, solid core									
	1620	4-5/8" solid jamb, 1-3/4" x 6'-8" x 2'-8" wide	2 Carp	16	1	Ea.	172	27.50		199.50	232
	1640	3'-0" wide	"	16	1		183	27.50		210.50	245
	1680	For 7'-0" high door, add				↓	17			17	18.70
	2000	Entrance door, colonial, 6 panel pine									
	2020	4-5/8" solid jamb, 1-3/4" x 6'-8" x 2'-8" wide	2 Carp	16	1	Ea.	405	27.50		432.50	490
	2040	3'-0" wide	"	16	1		430	27.50		457.50	515
	2060	For 7'-0" high door, add					17			17	18.70
	2200	For 5-5/8" solid jamb, add				↓	17			17	18.70
	4000	Interior, passage door, 4-5/8" solid jamb									
	4400	Lauan, flush, solid core, 1-3/8" x 6'-8" x 2'-6" wide	2 Carp	20	.800	Ea.	133	22		155	181
	4420	2'-8" wide		20	.800		134	22		156	183
	4440	3'-0" wide		19	.842		157	23		180	208
	4600	Hollow core, 1-3/8" x 6'-8" x 2'-6" wide		20	.800		99	22		121	144
	4620	2'-8" wide		20	.800		99	22		121	144
	4640	3'-0" wide	↓	19	.842		101	23		124	147
	4700	For 7'-0" high door, add					8.70			8.70	9.55
	5000	Birch, flush, solid core, 1-3/8" x 6'-8" x 2'-6" wide	2 Carp	20	.800	↓	138	22		160	187

Important: See the Reference Section for critical supporting data - Reference Nos., Crews, & City Cost Indexes

8

DOORS & WINDOWS

082 050	Wood & Plastic Doors	CREW	DAILY OUTPUT	LABOR-HOURS	UNIT	1999 BARE COSTS				TOTAL INCL O&P
						MAT.	LABOR	EQUIP.	TOTAL	
5020	2'-8" wide	2 Carp	20	.800	Ea.	141	22		163	190
5040	3'-0" wide		19	.842		147	23		170	197
5200	Hollow core, 1-3/8" x 6'-8" x 2'-6" wide		20	.800		110	22		132	156
5220	2'-8" wide		20	.800		132	22		154	181
5240	3'-0" wide	▼	19	.842		115	23		138	163
5280	For 7'-0" high door, add					8.70			8.70	9.55
5500	Hardboard paneled, 1-3/8" x 6'-8" x 2'-6" wide	2 Carp	20	.800		111	22		133	157
5520	2'-8" wide		20	.800		113	22		135	159
5540	3'-0" wide		19	.842		115	23		138	162
6000	Pine paneled, 1-3/8" x 6'-8" x 2'-6" wide		20	.800		190	22		212	244
6020	2'-8" wide		20	.800		200	22		222	254
6040	3'-0" wide	▼	19	.842		207	23		230	264
6500	For 5-5/8" solid jamb, add					19.60			19.60	21.50
6520	For split jamb, deduct				▼	10.90			10.90	11.95

083 050	Access Doors	CREW	DAILY OUTPUT	LABOR-HOURS	UNIT	1999 BARE COSTS				TOTAL INCL O&P
						MAT.	LABOR	EQUIP.	TOTAL	
0010	**ACCESS DOORS**									
1000	Fire rated door with lock									
1100	Metal, 12" x 12"	1 Carp	10	.800	Ea.	105	22		127	151
1150	18" x 18"		9	.889		133	24.50		157.50	184
1200	24" x 24"		9	.889		159	24.50		183.50	213
1250	24" x 36"		8	1		210	27.50		237.50	274
1300	24" x 48"		8	1		256	27.50		283.50	325
1350	36" x 36"		7.50	1.067		310	29		339	390
1400	48" x 48"		7.50	1.067		370	29		399	450
1600	Stainless steel, 12" x 12"		10	.800		169	22		191	221
1650	18" x 18"		9	.889		245	24.50		269.50	305
1700	24" x 24"		9	.889		295	24.50		319.50	365
1750	24" x 36"	▼	8	1	▼	360	27.50		387.50	440
2000	Flush door for finishing									
2100	Metal 8" x 8"	1 Carp	10	.800	Ea.	32	22		54	69.50
2150	12" x 12"	"	10	.800	"	34.50	22		56.50	72
3000	Recessed door for acoustic tile									
3100	Metal, 12" x 12"	1 Carp	4.50	1.778	Ea.	40	48.50		88.50	121
3150	12" x 24"		4.50	1.778		55.50	48.50		104	138
3200	24" x 24"		4	2		71	54.50		125.50	164
3250	24" x 36"	▼	4	2	▼	90.50	54.50		145	186
4000	Recessed door for drywall									
4100	Metal 12" x 12"	1 Carp	6	1.333	Ea.	46	36.50		82.50	108
4150	12" x 24"		5.50	1.455		65	39.50		104.50	134
4200	24" x 36"	▼	5	1.600	▼	99.50	43.50		143	178
6000	Standard door									
6100	Metal, 8" x 8"	1 Carp	10	.800	Ea.	28	22		50	65
6150	12" x 12"		10	.800		30.50	22		52.50	68
6200	18" x 18"		9	.889		38	24.50		62.50	80
6250	24" x 24"		9	.889		50	24.50		74.50	93
6300	24" x 36"		8	1		70	27.50		97.50	120
6350	36" x 36"	▼	8	1		85.50	27.50		113	137

DOORS & WINDOWS   8

8

DOORS & WINDOWS

083 050 | Access Doors

		CREW	DAILY OUTPUT	LABOR-HOURS	UNIT	1999 BARE COSTS MAT.	LABOR	EQUIP.	TOTAL	TOTAL INCL O&P	
054	6500	Stainless steel, 8" x 8"	1 Carp	10	.800	Ea.	53	22		75	93	054
	6550	12" x 12"		10	.800		67.50	22		89.50	109	
	6600	18" x 18"		9	.889		117	24.50		141.50	167	
	6650	24" x 24"		9	.889		153	24.50		177.50	206	

083 100 | Sliding Doors

		CREW	DAILY OUTPUT	LABOR-HOURS	UNIT	MAT.	LABOR	EQUIP.	TOTAL	INCL O&P	
102	0010	**GLASS, SLIDING**										102
	0012	Vinyl clad, 1" insul. glass, 6'-0" x 6'-10" high	2 Carp	4	4	Opng.	680	109		789	920	
	0030	6'-0" x 8'-0" high		4	4	Ea.	1,425	109		1,534	1,750	
	0100	8'-0" x 6'-10" high		4	4	Opng.	1,525	109		1,634	1,850	
	0500	3 leaf, 9'-0" x 6'-10" high		3	5.333		1,375	146		1,521	1,750	
	0600	12'-0" x 6'-10" high		3	5.333		1,725	146		1,871	2,100	
108	0010	**STEEL, SLIDING**										108
	0020	Up to 50' x 18', electric, standard duty, minimum	L-5	360	.156	S.F.	17.65	4.78	1.53	23.96	29.50	
	0100	Maximum		340	.165		29.50	5.05	1.62	36.17	43.50	
	0500	Heavy duty, minimum		297	.189		24	5.80	1.85	31.65	39	
	0600	Maximum		277	.202		63	6.20	1.98	71.18	83	

083 180 | Security Doors

		CREW	DAILY OUTPUT	LABOR-HOURS	UNIT	MAT.	LABOR	EQUIP.	TOTAL	INCL O&P	
182	0010	**SECURITY GATES** See division 106-101										182
184	0010	**VAULT FRONT**										184
	0020	Door and frame, 32" x 78", clear opening										
	0100	1 hour test, 32" door, weighs 750 lbs.	2 Sswk	1.50	10.667	Opng.	2,575	325		2,900	3,450	
	0200	2 hour test, 32" door, weighs 950 lbs.		1.30	12.308		3,075	375		3,450	4,100	
	0250	40" door, weighs 1130 lbs.		1	16		3,225	490		3,715	4,450	
	0300	4 hour test, 32" door, weighs 1025 lbs.		1.20	13.333		2,800	410		3,210	3,825	
	0350	40" door, weighs 1140 lbs.		.90	17.778		3,225	545		3,770	4,550	
	0500	For stainless steel front, including frame, add to above					1,600			1,600	1,750	
	0550	Back, add					1,600			1,600	1,750	
	0600	For time lock, two movement, add	1 Elec	2	4	Ea.	1,275	128		1,403	1,600	
	0650	Three movement, add	"	2	4		1,675	128		1,803	2,050	
	0800	Day gate, painted, steel, 32" wide	2 Sswk	1.50	10.667		1,450	325		1,775	2,200	
	0850	40" wide		1.40	11.429		1,600	350		1,950	2,425	
	0900	Aluminum, 32" wide		1.50	10.667		2,200	325		2,525	3,025	
	0950	40" wide		1.40	11.429		2,450	350		2,800	3,350	
	2050	Security vault door, class I, 3' wide, 3 1/2" thick	E-24	.19	166	Opng.	13,200	5,000	2,850	21,050	26,500	
	2100	Class II, 3' wide, 7" thick		.19	166		15,800	5,000	2,850	23,650	29,300	
	2150	Class III, 9R, 3' wide, 10" thick, minimum		.13	250		21,500	7,525	4,300	33,325	41,600	
	2160	Class V, type 1, 40" door		2.48	12.903	Ea.	4,775	390	222	5,387	6,175	
	2170	Class V, type 2, 40" door		2.48	12.903		4,775	390	222	5,387	6,200	
	2180	Day gate for class V vault	2 Sswk	2	8		935	245		1,180	1,475	

083 200 | Metal-Clad Doors

		CREW	DAILY OUTPUT	LABOR-HOURS	UNIT	MAT.	LABOR	EQUIP.	TOTAL	INCL O&P	
202	0010	**KALAMEIN**										202
	0020	Interior, flush type, 3' x 7'	2 Carp	4.30	3.721	Opng.	165	102		267	340	
204	0010	**TIN CLAD** R083-100										204
	0020	3 ply, 6' x 7', double sliding, manual with hardware	2 Carp	1	16	Opng.	1,300	435		1,735	2,100	
	1000	For electric operator, add	1 Elec	2	4	"	2,300	128		2,428	2,725	

083 250 | Cold Storage Doors

		CREW	DAILY OUTPUT	LABOR-HOURS	UNIT	MAT.	LABOR	EQUIP.	TOTAL	INCL O&P	
251	0010	**COLD STORAGE**										251
	0020	Single, 20 ga. galvanized steel										

Important: See the Reference Section for critical supporting data - Reference Nos., Crews, & City Cost Indexes

083 250 | Cold Storage Doors

		CREW	DAILY OUTPUT	LABOR-HOURS	UNIT	1999 BARE COSTS				TOTAL INCL O&P
						MAT.	LABOR	EQUIP.	TOTAL	
251	0300	Horizontal sliding, 5' x 7', manual operation, 2" thick	2 Carp	2	8	Ea.	2,025	218		2,243	2,575
	0400	4" thick		2	8		2,650	218		2,868	3,275
	0500	6" thick		2	8		2,575	218		2,793	3,200
	0800	5' x 7', power operation, 2" thick		1.90	8.421		4,275	230		4,505	5,075
	0900	4" thick		1.90	8.421		4,350	230		4,580	5,150
	1000	6" thick		1.90	8.421		4,950	230		5,180	5,800
	1300	9' x 10', manual operation, 2" insulation		1.70	9.412		3,450	257		3,707	4,200
	1400	4" insulation		1.70	9.412		3,575	257		3,832	4,325
	1500	6" insulation		1.70	9.412		4,300	257		4,557	5,125
	1800	Power operation, 2" insulation		1.60	10		5,950	273		6,223	6,975
	1900	4" insulation		1.60	10		6,100	273		6,373	7,125
	2000	6" insulation		1.70	9.412		6,900	257		7,157	8,000
	2300	For stainless steel face, add					20%				
	3000	Hinged, lightweight, 3' x 7'-0", galvanized 1 face, 2" thick	2 Carp	2	8	Ea.	985	218		1,203	1,425
	3050	4" thick		1.90	8.421		1,125	230		1,355	1,600
	3300	Aluminum doors, 3' x 7'-0", 4" thick		1.90	8.421		1,025	230		1,255	1,475
	3350	6" thick		1.40	11.429		1,825	310		2,135	2,500
	3600	Stainless steel, 3' x 7'-0", 4" thick		1.90	8.421		1,300	230		1,530	1,800
	3650	6" thick		1.40	11.429		2,200	310		2,510	2,925
	3900	Painted, 3' x 7'-0", 4" thick		1.90	8.421		935	230		1,165	1,375
	3950	6" thick		1.40	11.429		1,800	310		2,110	2,475
	5000	Bi-parting, electric operated									
	5010	6' x 8' opening, galv. faces, 4" thick for cooler	2 Carp	.80	20	Opng.	5,600	545		6,145	7,000
	5050	For freezer, 4" thick		.80	20		6,175	545		6,720	7,625
	5300	For door buck framing and door protection, add		2.50	6.400		460	175		635	780
	6000	Galvanized batten door, galvanized hinges, 4' x 7'		2	8		1,375	218		1,593	1,875
	6050	6' x 8'		1.80	8.889		1,875	243		2,118	2,450
	6500	Fire door, 3 hr., 6' x 8', single slide		.80	20		6,475	545		7,020	7,975
	6550	Double, bi-parting		.70	22.857		8,400	625		9,025	10,200

083 300 | Coiling Doors

		CREW	DAILY OUTPUT	LABOR-HOURS	UNIT	MAT.	LABOR	EQUIP.	TOTAL	TOTAL INCL O&P
302	0010	**COUNTER DOORS**									302
	0020	Manual, incl. frm and hdwe, galv. stl., 4' roll-up, 6' long	2 Carp	2	8	Opng.	605	218		823	1,000
	0300	Galvanized steel, UL label		1.80	8.889		760	243		1,003	1,225
	0600	Stainless steel, 4' high roll-up, 6' long		2	8		1,050	218		1,268	1,500
	0700	10' long		1.80	8.889		1,525	243		1,768	2,050
	2000	Aluminum, 4' high, 4' long		2.20	7.273		695	199		894	1,075
	2020	6' long		2	8		770	218		988	1,200
	2040	8' long		1.90	8.421		885	230		1,115	1,325
	2060	10' long		1.80	8.889		1,050	243		1,293	1,550
	2080	14' long		1.40	11.429		1,550	310		1,860	2,200
	2100	6' high, 4' long		2	8		770	218		988	1,200
	2120	6' long		1.60	10		900	273		1,173	1,425
	2140	10' long		1.40	11.429		1,225	310		1,535	1,850

083 400 | Coiling Grilles

		CREW	DAILY OUTPUT	LABOR-HOURS	UNIT	MAT.	LABOR	EQUIP.	TOTAL	TOTAL INCL O&P
404	0010	**COILING GRILLE**									404
	0015	Aluminum, manual, incl. frame, mill finish									
	0020	Top coiling, 4' high, 4' long	2 Sswk	3.20	5	Opng.	980	153		1,133	1,350
	0030	6' long		3.20	5		1,025	153		1,178	1,400
	0040	8' long		2.40	6.667		1,175	204		1,379	1,650
	0050	12' long		2.40	6.667		1,350	204		1,554	1,875
	0060	16' long		1.60	10		1,700	305		2,005	2,425
	0070	6' high, 4' long		3.20	5		1,050	153		1,203	1,425

DOORS & WINDOWS

8

083 | Special Doors

083 400 | Coiling Grilles

			CREW	DAILY OUTPUT	LABOR-HOURS	UNIT	1999 BARE COSTS MAT.	LABOR	EQUIP.	TOTAL	TOTAL INCL O&P	
404	0080	6' long	2 Sswk	3.20	5	Opng.	1,025	153		1,178	1,400	404
	0090	8' long		2.40	6.667		1,150	204		1,354	1,625	
	0100	12' long		1.60	10		1,425	305		1,730	2,125	
	0110	16' long		1.20	13.333		1,825	410		2,235	2,775	
	0200	Side coiling, 8' high, 12' long		.60	26.667		1,425	815		2,240	3,075	
	0220	18' long		.50	32		2,075	980		3,055	4,075	
	0240	24' long		.40	40		2,850	1,225		4,075	5,375	
	0260	12' high, 12' long		.50	32		2,075	980		3,055	4,075	
	0280	18' long		.40	40		3,075	1,225		4,300	5,625	
	0300	24' long		.28	57.143		4,150	1,750		5,900	7,775	
406	0010	**ROLLING GRILLE SUPPORTS** Overhead framed	E-4	36	.889	L.F.	16.50	27.50	2.28	46.28	71	406

083 550 | Flexible Doors

			CREW	DAILY OUTPUT	LABOR-HOURS	UNIT	MAT.	LABOR	EQUIP.	TOTAL	TOTAL INCL O&P	
551	0010	**FLEXIBLE TRANSPARENT STRIP ENTRANCE**										551
	0100	12" strip width, 2/3 overlap	3 Shee	135	.178	SF Surf	3.55	5.65		9.20	12.65	
	0200	Full overlap		115	.209		4.39	6.65		11.04	15.15	
	0220	8" strip width, 1/2 overlap		140	.171		2.60	5.45		8.05	11.30	
	0240	Full overlap		120	.200		3.45	6.35		9.80	13.65	
	0300	Add for suspension system, header mount				L.F.	6.85			6.85	7.55	
	0400	Wall mount				"	6.95			6.95	7.60	

083 600 | Sectional Overhead Drs

			CREW	DAILY OUTPUT	LABOR-HOURS	UNIT	MAT.	LABOR	EQUIP.	TOTAL	TOTAL INCL O&P	
604	0010	**OVERHEAD, COMMERCIAL** Frames not included										604
	1000	Stock, sectional, heavy duty, wood, 1-3/4" thick, 8' x 8' high	2 Carp	2	8	Ea.	400	218		618	785	
	1100	10' x 10' high		1.80	8.889		610	243		853	1,050	
	1200	12' x 12' high		1.50	10.667		850	291		1,141	1,400	
	1300	Chain hoist, 14' x 14' high		1.30	12.308		1,475	335		1,810	2,150	
	1400	12' x 16' high		1	16		1,300	435		1,735	2,125	
	1500	20' x 8' high		1.30	12.270		1,225	335		1,560	1,875	
	1600	20' x 16' high		.65	24.615		3,075	670		3,745	4,425	
	1800	Center mullion openings, 8' high		4	4		480	109		589	695	
	1900	20' high		2	8		890	218		1,108	1,325	
	2100	For medium duty custom door, deduct					5%	5%				
	2150	For medium duty stock doors, deduct					10%	5%				
	2300	Fiberglass and aluminum, heavy duty, sectional, 12' x 12' high	2 Carp	1.50	10.667	Ea.	1,250	291		1,541	1,825	
	2450	Chain hoist, 20' x 20' high		.50	32		4,275	875		5,150	6,075	
	2600	Steel, 24 ga. sectional, manual, 8' x 8' high		2	8		355	218		573	735	
	2650	10' x 10' high		1.80	8.889		465	243		708	890	
	2700	12' x 12' high		1.50	10.667		625	291		916	1,150	
	2800	Chain hoist, 20' x 14' high		.70	22.857		1,575	625		2,200	2,700	
	2850	For 1-1/4" rigid insulation and 26 ga. galv.										
	2860	back panel, add				S.F.	1.73			1.73	1.90	
	2900	For electric trolley operator, 1/3 H.P., to 12' x 12', add	1 Carp	2	4	Ea.	415	109		524	625	
	2950	Over 12' x 12', 1/2 H.P., add	"	1	8	"	440	218		658	830	
606	0010	**RESIDENTIAL GARAGE DOORS** Including hardware, no frame										606
	0050	Hinged, wood, custom, double door, 9' x 7'	2 Carp	4	4	Ea.	300	109		409	500	
	0070	16' x 7'		3	5.333		510	146		656	795	
	0200	Overhead, sectional, incl. hardware, fiberglass, 9' x 7', standard		5.28	3.030		495	82.50		577.50	675	
	0220	Deluxe		5.28	3.030		580	82.50		662.50	765	
	0300	16' x 7', standard		6	2.667		765	73		838	955	
	0320	Deluxe		6	2.667		955	73		1,028	1,175	
	0500	Hardboard, 9' x 7', standard		8	2		315	54.50		369.50	430	
	0520	Deluxe		8	2		380	54.50		434.50	505	
	0600	16' x 7', standard		6	2.667		555	73		628	725	

8 DOORS & WINDOWS

Important: See the Reference Section for critical supporting data - Reference Nos., Crews, & City Cost Indexes

083 600 | Sectional Overhead Drs

			CREW	DAILY OUTPUT	LABOR-HOURS	UNIT	MAT.	LABOR	EQUIP.	TOTAL	TOTAL INCL O&P	
606	0620	Deluxe	2 Carp	6	2.667	Ea.	650	73		723	825	606
	0700	Metal, 9' x 7', standard		5.28	3.030		225	82.50		307.50	375	
	0720	Deluxe		8	2		495	54.50		549.50	630	
	0800	16' x 7', standard		3	5.333		400	146		546	670	
	0820	Deluxe		6	2.667		680	73		753	865	
	0900	Wood, 9' x 7', standard		8	2		350	54.50		404.50	470	
	0920	Deluxe		8	2		1,000	54.50		1,054.50	1,175	
	1000	16' x 7', standard		6	2.667		705	73		778	890	
	1020	Deluxe		6	2.667		1,475	73		1,548	1,750	
	1800	Door hardware, sectional	1 Carp	4	2		154	54.50		208.50	256	
	1810	Door tracks only		4	2		72	54.50		126.50	166	
	1820	One side only		7	1.143		50	31		81	104	
	3000	Swing-up, including hardware, fiberglass, 9' x 7', standard	2 Carp	8	2		485	54.50		539.50	615	
	3020	Deluxe		8	2		575	54.50		629.50	720	
	3100	16' x 7', standard		6	2.667		610	73		683	785	
	3120	Deluxe		6	2.667		710	73		783	900	
	3200	Hardboard, 9' x 7', standard		8	2		251	54.50		305.50	360	
	3220	Deluxe		8	2		330	54.50		384.50	450	
	3300	16' x 7', standard		6	2.667		350	73		423	500	
	3320	Deluxe		6	2.667		520	73		593	685	
	3400	Metal, 9' x 7', standard		8	2		275	54.50		329.50	390	
	3420	Deluxe		8	2		430	54.50		484.50	560	
	3500	16' x 7', standard		6	2.667		430	73		503	590	
	3520	Deluxe		6	2.667		695	73		768	880	
	3600	Wood, 9' x 7', standard		8	2		300	54.50		354.50	415	
	3620	Deluxe		8	2		480	54.50		534.50	615	
	3700	16' x 7', standard		6	2.667		520	73		593	690	
	3720	Deluxe		6	2.667		735	73		808	925	
	3900	Door hardware only, swing up	1 Carp	4	2		76	54.50		130.50	170	
	3920	One side only		7	1.143		49	31		80	103	
	4000	For electric operator, economy, add		8	1		202	27.50		229.50	265	
	4100	Deluxe, including remote control		8	1		295	27.50		322.50	370	
	4500	For transmitter/receiver control , add to operator				Total	69			69	76	
	4600	Transmitters, additional				"	22.50			22.50	24.50	

083 650 | Multileaf Vert Lift Drs

			CREW	DAILY OUTPUT	LABOR-HOURS	UNIT	MAT.	LABOR	EQUIP.	TOTAL	TOTAL INCL O&P	
652	0010	**TELESCOPING STEEL DOORS** Baked enamel finish										652
	1000	Overhead, .03" thick, electric operated, 10' x 10'	E-3	.80	30	Ea.	8,500	940	103	9,543	11,200	
	2000	20' x 10'		.60	40		11,300	1,250	137	12,687	14,900	
	3000	20' x 16'		.40	60		13,000	1,875	206	15,081	18,000	
654	0010	**VERTICAL LIFT DOORS**										654
	0020	Motorized, 14 ga. stl, incl., frm and ctrl pnl										
	0050	16' x 16' high	L-10	.50	48	Ea.	14,600	1,475	885	16,960	19,700	
	0100	10' x 20' high		1.30	18.462		16,400	570	340	17,310	19,400	
	0120	15' x 20' high		1.30	18.462		20,200	570	340	21,110	23,700	
	0140	20' x 20' high		1	24		24,500	740	445	25,685	28,700	
	0160	25' x 20' high		1	24		27,600	740	445	28,785	32,200	
	0170	32' x 24' high		.75	32		28,600	985	590	30,175	33,900	
	0180	20' x 25' high		1	24		29,100	740	445	30,285	33,800	
	0200	25' x 25' high		.70	34.286		32,900	1,050	635	34,585	38,700	
	0220	25' x 30' high		.70	34.286		35,400	1,050	635	37,085	41,400	
	0240	30' x 30' high		.70	34.286		40,600	1,050	635	42,285	47,100	
	0260	35' x 30' high		.70	34.286		44,800	1,050	635	46,485	52,000	

DOORS & WINDOWS **8**

For expanded coverage of these items see *Means Interior Cost Data 1999*

083 700 | Hangar Doors

		CREW	DAILY OUTPUT	LABOR-HOURS	UNIT	1999 BARE COSTS MAT.	LABOR	EQUIP.	TOTAL	TOTAL INCL O&P	
701	0010 **HANGAR DOOR**										701
	0020 Bi-fold, ovhd., 20 PSF wind load, incl. elec. oper.										
	0100 Without sheeting, 12' high x 40'	2 Sswk	240	.067	S.F.	9.05	2.04		11.09	13.70	
	0200 16' high x 60'		230	.070		16.40	2.13		18.53	22	
	0300 20' high x 80'	▼	220	.073	▼	18.60	2.23		20.83	24.50	

083 720 | Special Purpose Doors

		CREW	DAILY OUTPUT	LABOR-HOURS	UNIT	MAT.	LABOR	EQUIP.	TOTAL	TOTAL INCL O&P	
721	0010 **BULKHEAD CELLAR DOORS**										721
	0020 Steel, not incl. sides, 44" x 62"	1 Carp	5.50	1.455	Ea.	180	39.50		219.50	261	
	0100 52" x 73"		5.10	1.569		200	43		243	288	
	0500 With sides and foundation plates, 57" x 45" x 24"		4.70	1.702		235	46.50		281.50	330	
	0600 42" x 49" x 51"	▼	4.30	1.860	▼	283	51		334	390	
722	0010 **AIR CURTAINS** Incl. motor starters, transformers,										722
	0050 door switches & temperature controls										
	0100 Shipping and receiving doors, unheated, minimal wind stoppage										
	0150 8' high, multiples of 3' wide	2 Shee	3.60	4.444	L.F.	168	141		309	405	
	0160 5'		5	3.200		179	102		281	355	
	0210 10' high, multiples of 4' wide		3.60	4.444		167	141		308	405	
	0250 12' high, 3'-6" wide		3.30	4.848		166	154		320	420	
	0260 12' wide		3.30	4.848		195	154		349	455	
	0350 16' high, 3'-6" wide		3.30	4.848		325	154		479	595	
	0360 12' wide	▼	3.30	4.848	▼	162	154		316	415	
	0500 Maximum wind stoppage										
	0550 10' high, multiples of 4' wide	2 Shee	4.40	3.636	L.F.	124	115		239	315	
	0650 14' high, multiples of 4' wide		3.25	4.923		161	156		317	420	
	0750 20' high, multiples of 8' wide	▼	3	5.333	▼	350	169		519	650	
	1100 Heated, maximum wind stoppage, steam heat										
	1150 10' high, multiples of 4' wide	2 Shee	3	5.333	L.F.	450	169		619	760	
	1250 14' high, multiples of 4' wide		2.80	5.714		550	181		731	885	
	1350 20' high, multiples of 8' wide	▼	2.60	6.154	▼	550	195		745	910	
	1500 Customer entrance doors, unheated, minimal wind stoppage										
	1550 10' high, multiples of 3' wide	2 Shee	3.30	4.848	L.F.	168	154		322	425	
	1560 5'		5	3.200		182	102		284	360	
	1650 Maximum wind stoppage, 12' high, multiples of 4' wide	▼	3.30	4.848		180	154		334	435	
	1700 Heated, minimal wind stoppage, electric heat										
	1750 8' high, multiples of 3' wide	2 Shee	3.10	5.161	L.F.	385	164		549	680	
	1760 Multiples of 5' wide		3.10	5.161		271	164		435	555	
	1850 10' high, multiples of 3' wide		3.10	5.161		440	164		604	740	
	1860 Multiples of 5' wide	▼	3.10	5.161	▼	350	164		514	640	
	1950 Maximum wind stoppage, steam heat										
	1960 12' high, multiples of 4' wide	2 Shee	2.80	5.714	L.F.	280	181		461	590	
	2000 Walk-in coolers and freezers, ambient air, minimal wind stoppage										
	2050 8' high, multiples of 3' wide	2 Shee	3.60	4.444	L.F.	150	141		291	385	
	2060 Multiples of 5' wide		3.60	4.444		150	141		291	385	
	2250 Maximum wind stoppage, 12' high, multiples of 3' wide		3.30	4.848		161	154		315	415	
	2450 Conveyor openings or service windows, unheated, 5' high		4	4		151	127		278	365	
	2460 Heated, electric, 5' high, 2'-6" wide	▼	3.30	4.848		199	154		353	460	
	2700 Entrance with recirculating system, unheated, add				▼	168			168	185	
	2800 Installation of each doorway	2 Shee	.10	160	Total		5,075		5,075	7,900	
723	0010 **FLOOR, COMMERCIAL**										723
	0020 Aluminum tile, steel frame, one leaf, 2' x 2' opng.	2 Sswk	3.50	4.571	Opng.	340	140		480	630	
	0050 3'-6" x 3'-6" opening		3.50	4.571		615	140		755	930	
	0500 Double leaf, 4' x 4' opening	▼	3	5.333	▼	900	163		1,063	1,300	

Important: See the Reference Section for critical supporting data - Reference Nos., Crews, & City Cost Indexes

083 720 | Special Purpose Doors

		CREW	DAILY OUTPUT	LABOR-HOURS	UNIT	1999 BARE COSTS				TOTAL INCL O&P	
						MAT.	LABOR	EQUIP.	TOTAL		
723	0550	5' x 5' opening	2 Sswk	3	5.333	Opng.	1,325	163		1,488	1,750	723
725	0010	**FLOOR, INDUSTRIAL**										725
	0020	Steel 300 psf L.L., single leaf, 2' x 2', 175#	2 Sswk	6	2.667	Opng.	450	81.50		531.50	645	
	0050	3' x 3' opening,300#		5.50	2.909		620	89		709	850	
	0300	Double leaf, 4' x 4' opening,455#		5	3.200		935	98		1,033	1,200	
	0350	5' x 5' opening,645#		4.50	3.556		1,225	109		1,334	1,550	
	1000	Aluminum, 300 psf L.L., single leaf, 2' x 2', 60#		6	2.667		440	81.50		521.50	635	
	1050	3' x 3' opening, 100#		5.50	2.909		690	89		779	925	
	1500	Double leaf, 4' x 4' opening, 160#		5	3.200		1,075	98		1,173	1,375	
	1550	5' x 5' opening, 235#		4.50	3.556		1,450	109		1,559	1,775	
	2000	Aluminum, 150 psf L.L., single leaf, 2' x 2', 60#		6	2.667		405	81.50		486.50	595	
	2050	3' x 3' opening, 95#		5.50	2.909		600	89		689	825	
	2500	Double leaf, 4' x 4' opening, 150#		5	3.200		960	98		1,058	1,225	
	2550	5' x 5' opening, 230#	▼	4.50	3.556	▼	1,300	109		1,409	1,625	
729	0010	**KENNEL DOORS**										729
	0020	2 way, swinging type, 13" x 19" opening	2 Carp	11	1.455	Opng.	195	39.50		234.50	278	
	0100	17" x 29" opening	"	11	1.455	"	195	39.50		234.50	278	
732	0010	**ROLLING SERVICE DOORS** Steel, manual, 20 ga., incl. hardware										732
	0050	8' x 8' high	2 Sswk	1.60	10	Ea.	685	305		990	1,300	
	0100	10' x 10' high		1.40	11.429		900	350		1,250	1,625	
	0200	20' x 10' high		1	16		1,850	490		2,340	2,925	
	0300	12' x 12' high		1.20	13.333		1,175	410		1,585	2,050	
	0400	20' x 12' high		.90	17.778		1,900	545		2,445	3,100	
	0500	14' x 14' high		.80	20		1,550	610		2,160	2,825	
	0600	20' x 16' high		.60	26.667		2,200	815		3,015	3,925	
	0700	10' x 20' high		.50	32		1,325	980		2,305	3,275	
	1000	12' x 12', crank operated, crank on door side		.80	20		1,050	610		1,660	2,275	
	1100	Crank thru wall	▼	.70	22.857		1,175	700		1,875	2,550	
	1300	For vision panel, add				▼	190			190	209	
	1400	For 22 ga., deduct				S.F.	.65			.65	.71	
	1600	3' x 7' pass door within rolling steel door, new construction				Ea.	1,025			1,025	1,125	
	1700	Existing construction	2 Sswk	2	8		1,025	245		1,270	1,575	
	2000	Class A fire doors, manual, 20 ga., 8' x 8' high		1.40	11.429		895	350		1,245	1,625	
	2100	10' x 10' high		1.10	14.545		1,200	445		1,645	2,125	
	2200	20' x 10' high		.80	20		2,375	610		2,985	3,750	
	2300	12' x 12' high		1	16		1,575	490		2,065	2,650	
	2400	20' x 12' high		.80	20		2,750	610		3,360	4,125	
	2500	14' x 14' high		.60	26.667		2,050	815		2,865	3,750	
	2600	20' x 16' high		.50	32		3,425	980		4,405	5,575	
	2700	10' x 20' high	▼	.40	40	▼	2,175	1,225		3,400	4,625	
	3000	For 18 ga. doors, add				S.F.	.68			.68	.75	
	3300	For enamel finish, add				"	.80			.80	.88	
	3600	For safety edge bottom bar, pneumatic, add				L.F.	10.55			10.55	11.60	
	3700	Electric, add					19.85			19.85	22	
	4000	For weatherstripping, extruded rubber, jambs, add					6.80			6.80	7.50	
	4100	Hood, add					4.34			4.34	4.77	
	4200	Sill, add				▼	2.48			2.48	2.73	
	4500	Motor operators, to 14' x 14' opening	2 Sswk	5	3.200	Ea.	645	98		743	890	
	4600	Over 14' x 14', jack shaft type	"	5	3.200		645	98		743	890	
	4700	For fire door, additional fusible link, add				▼	12.40			12.40	13.65	
736	0010	**SHOCK ABSORBING DOORS**										736
	0020	Rigid, no frame, 1-1/2" thick, 5' x 7'	2 Sswk	1.90	8.421	Opng.	1,125	258		1,383	1,700	

DOORS & WINDOWS 8

For expanded coverage of these items see *Means Interior Cost Data 1999*

083 720		Special Purpose Doors	CREW	DAILY OUTPUT	LABOR-HOURS	UNIT	1999 BARE COSTS				TOTAL INCL O&P	
							MAT.	LABOR	EQUIP.	TOTAL		
736	0100	8' x 8'	2 Sswk	1.80	8.889	Opng.	1,575	272		1,847	2,250	736
	0500	Flexible, no frame, insulated, .16" thick, economy, 5' x 7'		2	8		1,375	245		1,620	1,950	
	0600	Deluxe		1.90	8.421		2,075	258		2,333	2,775	
	1000	8' x 8' opening, economy		2	8		2,150	245		2,395	2,800	
	1100	Deluxe	↓	1.90	8.421	↓	2,750	258		3,008	3,500	

083 750		Swing Doors										
752	0010	DOUBLE ACTING										752
	1000	.063" aluminum, 7'-0" high, 4'-0" wide	2 Carp	4.20	3.810	Pr.	345	104		449	545	
	1050	6'-8" wide	"	4	4	"	480	109		589	695	
	2000	Solid core wood, 3/4" thick, metal frame, stainless steel										
	2010	base plate, 7' high opening, 4' wide	2 Carp	4	4	Pr.	705	109		814	945	
	2050	7' wide	"	3.80	4.211	"	1,000	115		1,115	1,275	
754	0010	GLASS DOOR, SWING										754
	0020	Including hardware, 1/2" thick, tempered, 3' x 7' opening	2 Glaz	2	8	Opng.	1,500	213		1,713	1,975	
	0100	6' x 7' opening	"	1.40	11.429	"	3,000	305		3,305	3,775	
756	0010	SWING DOORS										756
	0020	Tubular steel, 7' high, single, 3'-4" opening	2 Sswk	2.50	6.400	Ea.	440	196		636	845	
	0100	Double, 6'-0" opening	"	2	8	Pr.	555	245		800	1,075	

083 800		Sound Retardant Doors										
804	0010	ACOUSTICAL DOORS										804
	0020	Including framed seals, 3' x 7', wood, 27 STC rating	2 Carp	1.50	10.667	Ea.	325	291		616	820	
	0100	Steel, 40 STC rating		1.50	10.667		1,250	291		1,541	1,825	
	0200	45 STC rating		1.50	10.667		1,650	291		1,941	2,275	
	0300	48 STC rating		1.50	10.667		2,075	291		2,366	2,750	
	0400	52 STC rating	↓	1.50	10.667	↓	2,500	291		2,791	3,200	

083 900		Screen & Storm Doors										
904	0010	STORM DOORS & FRAMES Aluminum, residential,										904
	0020	combination storm and screen										
	0400	Clear anodic coating, 6'-8" x 2'-6" wide	2 Carp	15	1.067	Ea.	135	29		164	195	
	0420	2'-8" wide		14	1.143		162	31		193	227	
	0440	3'-0" wide	↓	14	1.143	↓	162	31		193	227	
	0500	For 7' door height, add					5%					
	1000	Mill finish, 6'-8" x 2'-6" wide	2 Carp	15	1.067	Ea.	190	29		219	255	
	1020	2'-8" wide		14	1.143		190	31		221	258	
	1040	3'-0" wide	↓	14	1.143		206	31		237	276	
	1100	For 7'-0" door, add					5%					
	1500	White painted, 6'-8" x 2'-6" wide	2 Carp	15	1.067		166	29		195	229	
	1520	2'-8" wide		14	1.143		177	31		208	244	
	1540	3'-0" wide	↓	14	1.143	↓	189	31		220	257	
	1600	For 7'-0" door, add					5%					
	2000	Wood door & screen, see division 082-078										
	2020											

Important: See the Reference Section for critical supporting data - Reference Nos., Crews, & City Cost Indexes

8 DOORS & WINDOWS

084 100	Aluminum	CREW	DAILY OUTPUT	LABOR-HOURS	UNIT	1999 BARE COSTS				TOTAL INCL O&P
						MAT.	LABOR	EQUIP.	TOTAL	
101	**0010 ALUMINUM FRAMES**									**101**
0020	Entrance, 3' x 7' opening, clear anodized finish	2 Sswk	7	2.286	Opng.	173	70		243	320
0040	Bronze finish		7	2.286		270	70		340	425
0060	Black finish		7	2.286		305	70		375	465
0500	6' x 7' opening, clear finish		6	2.667		270	81.50		351.50	445
0520	Bronze finish		6	2.667		305	81.50		386.50	485
0540	Black finish		6	2.667		360	81.50		441.50	545
1000	With 3' high transoms, 3' x 10' opening, clear finish		6.50	2.462		330	75.50		405.50	505
1050	Bronze finish		6.50	2.462		365	75.50		440.50	540
1100	Black finish		6.50	2.462		430	75.50		505.50	615
1500	With 3' high transoms, 6' x 10' opening, clear finish		5.50	2.909		400	89		489	605
1550	Bronze finish		5.50	2.909		430	89		519	640
1600	Black finish		5.50	2.909		510	89		599	725
103	**0010 ALUMINUM DOORS** Commercial entrance									**103**
0800	Narrow stile, no glazing, standard hardware, pair of 2'-6" x 7'-0"	2 Carp	1.70	9.412	Pr.	625	257		882	1,100
1000	3'-0" x 7'-0", single		3	5.333	Ea.	415	146		561	685
1200	Pair of 3'-0" x 7'-0"		1.70	9.412	Pr.	830	257		1,087	1,325
1500	3'-6" x 7'-0", single		3	5.333	Ea.	445	146		591	720
2000	Medium stile, pair of 2'-6" x 7'-0"		1.70	9.412	Pr.	685	257		942	1,150
2100	3'-0" x 7'-0", single		3	5.333	Ea.	545	146		691	825
2200	Pair of 3'-0" x 7'-0"		1.70	9.412	Pr.	1,025	257		1,282	1,525
2300	3'-6" x 7'-0", single		5.33	3.002	Ea.	640	82		722	835
5000	Flush panel doors, pair of 2'-6" x 7'-0"	2 Sswk	2	8	Pr.	835	245		1,080	1,375
5050	3'-0" x 7'-0", single		2.50	6.400	Ea.	420	196		616	820
5100	Pair of 3'-0" x 7'-0"		2	8	Pr.	865	245		1,110	1,400
5150	3'-6" x 7'-0", single		2.50	6.400	Ea.	515	196		711	925
105	**0010 ALUMINUM DOORS & FRAMES** Entrance, narrow stile, including									**105**
0015	hardware & closer, clear finish, not incl. glass, 2'-6" x 7'-0" opng.	2 Sswk	2	8	Ea.	470	245		715	970
0020	3'-0" x 7'-0" opening		2	8		410	245		655	905
0030	3'-6" x 7'-0" opening		2	8		425	245		670	915
0100	3'-0" x 10'-0" opening, 3' high transom		1.80	8.889		675	272		947	1,250
0200	3'-6" x 10'-0" opening, 3' high transom		1.80	8.889		665	272		937	1,225
0280	5'-0" x 7'-0" opening		2	8		705	245		950	1,225
0300	6'-0" x 7'-0" opening		1.30	12.308	Pr.	685	375		1,060	1,450
0400	6'-0" x 10'-0" opening, 3' high transom		1.10	14.545		620	445		1,065	1,500
0420	7'-0" x 7'-0" opening		1	16		785	490		1,275	1,775
0500	Wide stile, 2'-6" x 7'-0" opening		2	8	Ea.	695	245		940	1,225
0520	3'-0" x 7'-0" opening		2	8		615	245		860	1,125
0540	3'-6" x 7'-0" opening		2	8		645	245		890	1,150
0560	5'-0" x 7'-0" opening		2	8		990	245		1,235	1,550
0580	6'-0" x 7'-0" opening		1.30	12.308	Pr.	950	375		1,325	1,750
0600	7'-0" x 7'-0" opening		1	16	"	1,075	490		1,565	2,075
1100	For full vision doors, with 1/2" glass, add				Leaf	55%				
1200	For non-standard size, add					67%				
1300	Light bronze finish, add					36%				
1400	Dark bronze finish, add					18%				
1500	For black finish, add					36%				
1600	Concealed panic device, add					885			885	975
1700	Electric striker release, add				Opng.	228			228	251
1800	Floor check, add				Leaf	675			675	745
1900	Concealed closer, add				"	450			450	495
2000	Flush 3' x 7' Insulated, 12"x 12" lite, clear finish	2 Sswk	2	8	Ea.	375	245		620	865
109	**0010 BALANCED DOORS**									**109**
0020	Hardware & frame, alum. & glass, 3' x 7', econ.	2 Sswk	.90	17.778	Ea.	730	545		1,275	1,800

8

DOORS & WINDOWS

		084 100	Aluminum	CREW	DAILY OUTPUT	LABOR-HOURS	UNIT	1999 BARE COSTS MAT.	LABOR	EQUIP.	TOTAL	TOTAL INCL O&P	
109	0150		Premium	2 Sswk	.70	22.857	Ea.	775	700		1,475	2,125	109
111	0010	**STOREFRONT SYSTEMS** Aluminum frame, clear 3/8" plate glass,											111
	0020	incl. 3' x 7' door with hardware (400 sq. ft. max. wall)											
	0500	Wall height to 12' high, commercial grade		2 Glaz	150	.107	S.F.	11.35	2.84		14.19	16.75	
	0600	Institutional grade			130	.123		15.10	3.27		18.37	21.50	
	0700	Monumental grade			115	.139		21.50	3.70		25.20	29	
	1000	6' x 7' door with hardware, commercial grade			135	.119		11.60	3.15		14.75	17.55	
	1100	Institutional grade			115	.139		15.85	3.70		19.55	23	
	1200	Monumental grade			100	.160		29.50	4.26		33.76	39	
	1500	For bronze anodized finish, add						15%					
	1600	For black anodized finish, add						30%					
	1700	For stainless steel framing, add to monumental						75%					

084 300 | Stainless Steel

				CREW	DAILY OUTPUT	LABOR-HOURS	UNIT	MAT.	LABOR	EQUIP.	TOTAL	INCL O&P	
301	0010	**STAINLESS STEEL AND GLASS** Entrance unit, narrow stiles											301
	0020	3' x 7' opening, including hardware, minimum		2 Sswk	1.60	10	Opng.	4,175	305		4,480	5,150	
	0050	Average			1.40	11.429		4,525	350		4,875	5,625	
	0100	Maximum			1.20	13.333		4,850	410		5,260	6,075	
	1000	For solid bronze entrance units, statuary finish, add						60%					
	1100	Without statuary finish, add						45%					
	2000	Balanced doors, 3' x 7', economy		2 Sswk	.90	17.778	Ea.	5,675	545		6,220	7,225	
	2100	Premium		"	.70	22.857	"	9,750	700		10,450	12,000	

084 600 | Automatic Doors

				CREW	DAILY OUTPUT	LABOR-HOURS	UNIT	MAT.	LABOR	EQUIP.	TOTAL	INCL O&P	
602	0010	**SLIDING ENTRANCE** 12' x 7'-6" opng., 5' x 7' door, 2 way traf.,											602
	0020	mat activated, panic pushout, incl. operator & hardware,											
	0030	not including glass or glazing		2 Glaz	.70	22.857	Opng.	5,575	610		6,185	7,050	
604	0010	**SLIDING PANELS**											604
	0020	Mall fronts, aluminum & glass, 15' x 9' high		2 Glaz	1.30	12.308	Opng.	2,175	325		2,500	2,900	
	0100	24' x 9' high			.70	22.857		3,175	610		3,785	4,400	
	0200	48' x 9' high, with fixed panels			.90	17.778		5,900	475		6,375	7,200	
	0500	For bronze finish, add						17%					

084 700 | Revolving Doors

				CREW	DAILY OUTPUT	LABOR-HOURS	UNIT	MAT.	LABOR	EQUIP.	TOTAL	INCL O&P	
701	0010	**REVOLVING DOORS** Aluminum, 6'-6" to 7'-0" diameter,											701
	0020	6'-10" to 7' high, stock units, minimum		4 Sswk	.75	42.667	Opng.	14,000	1,300		15,300	17,800	
	0050	Average			.60	53.333		17,200	1,625		18,825	22,000	
	0100	Maximum			.45	71.111		22,700	2,175		24,875	29,000	
	1000	Stainless steel			.30	106		26,200	3,275		29,475	34,800	
	1100	Solid bronze			.15	213		30,500	6,525		37,025	45,600	
	1500	For automatic controls, add		2 Elec	2	8		10,500	255		10,755	12,000	

8 DOORS & WINDOWS

Important: See the Reference Section for critical supporting data - Reference Nos., Crews, & City Cost Indexes

085 100 | Steel Windows

		CREW	DAILY OUTPUT	LABOR-HOURS	UNIT	1999 BARE COSTS MAT.	LABOR	EQUIP.	TOTAL	TOTAL INCL O&P	
102	0010	**STEEL SASH** Custom units, glazing and trim not included R085 -100									102
	0100	Casement, 100% vented	2 Sswk	200	.080	S.F.	33.50	2.45		35.95	41.50
	0200	50% vented		200	.080		28.50	2.45		30.95	36
	0300	Fixed		200	.080		19.20	2.45		21.65	25.50
	1000	Projected, commercial, 40% vented		200	.080		30	2.45		32.45	37.50
	1100	Intermediate, 50% vented		200	.080		33	2.45		35.45	40.50
	1500	Industrial, horizontally pivoted		200	.080		25	2.45		27.45	32
	1600	Fixed		200	.080		21.50	2.45		23.95	28
	2000	Industrial security sash, 50% vented		200	.080		38	2.45		40.45	46
	2100	Fixed		200	.080		32	2.45		34.45	39.50
	2500	Picture window		200	.080		13.75	2.45		16.20	19.60
	3000	Double hung		200	.080		38	2.45		40.45	46
	5000	Mullions for above, open interior face		240	.067	L.F.	6.85	2.04		8.89	11.30
	5100	With interior cover		240	.067	"	11.55	2.04		13.59	16.45
104	0010	**STEEL WINDOWS** Stock, including frame, trim and insulating glass									104
	0020	See also division 131-212-7650 thru 8050									
	1000	Custom units, double hung, 2'-8" x 4'-6" opening R085 -100	2 Sswk	12	1.333	Ea.	480	41		521	605
	1100	2'-4" x 3'-9" opening		12	1.333		395	41		436	510
	1500	Commercial projected, 3'-9" x 5'-5" opening		10	1.600		840	49		889	1,025
	1600	6'-9" x 4'-1" opening		7	2.286		1,100	70		1,170	1,350
	2000	Intermediate projected, 2'-9" x 4'-1" opening		12	1.333		470	41		511	595
	2100	4'-1" x 5'-5" opening		10	1.600		955	49		1,004	1,150

085 200 | Aluminum Windows

		CREW	DAILY OUTPUT	LABOR-HOURS	UNIT	MAT.	LABOR	EQUIP.	TOTAL	TOTAL INCL O&P	
202	0010	**ALUMINUM SASH**									202
	0020	Stock, grade C, glaze & trim not incl., casement	2 Sswk	200	.080	S.F.	26	2.45		28.45	33.50
	0050	Double hung		200	.080		25	2.45		27.45	32
	0100	Fixed casement		200	.080		9.55	2.45		12	15.05
	0150	Picture window		200	.080		11	2.45		13.45	16.60
	0200	Projected window		200	.080		22	2.45		24.45	29
	0250	Single hung		200	.080		11.35	2.45		13.80	17
	0300	Sliding		200	.080		14.70	2.45		17.15	20.50
	1000	Mullions for above, tubular		240	.067	L.F.	3.90	2.04		5.94	8.05
	2000	Custom aluminum sash, grade HC, glazing not included, minimum		200	.080	S.F.	27.50	2.45		29.95	34.50
	2100	Maximum		85	.188	"	35.50	5.75		41.25	49.50
204	0010	**ALUMINUM WINDOWS** Incl. frame and glazing, grade C									204
	0020	See also division 051-230-1600									
	1000	Stock units, casement, 3'-1" x 3'-2" opening	2 Sswk	10	1.600	Ea.	255	49		304	370
	1050	Add for storms					51.50			51.50	56.50
	1600	Projected, with screen, 3'-1" x 3'-2" opening	2 Sswk	10	1.600		182	49		231	291
	1700	Add for storms					48.50			48.50	53.50
	2000	4'-5" x 5'-3" opening	2 Sswk	8	2		257	61		318	395
	2100	Add for storms					67			67	74
	2500	Enamel finish windows, 3'-1" x 3'-2"	2 Sswk	10	1.600		164	49		213	270
	2600	4'-5" x 5'-3"		8	2		245	61		306	380
	3000	Single hung, 2' x 3' opening, enameled, standard glazed		10	1.600		123	49		172	225
	3100	Insulating glass		10	1.600		149	49		198	253
	3300	2'-8" x 6'-8" opening, standard glazed		8	2		261	61		322	400
	3400	Insulating glass		8	2		335	61		396	480
	3700	3'-4" x 5'-0" opening, standard glazed		9	1.778		168	54.50		222.50	285
	3800	Insulating glass		9	1.778		236	54.50		290.50	360
	3890	Awning type, 3' x 3' opening standard glass		14	1.143		300	35		335	395
	3900	Insulating glass		14	1.143		320	35		355	420
	3910	3' x 4' opening, standard glass		10	1.600		350	49		399	475
	3920	Insulating glass		10	1.600		405	49		454	535

For expanded coverage of these items see *Means Interior Cost Data 1999*

8

DOORS & WINDOWS

085 200 | Aluminum Windows

			DAILY OUTPUT	LABOR-HOURS	UNIT	1999 BARE COSTS				TOTAL INCL O&P	
		CREW				MAT.	LABOR	EQUIP.	TOTAL		
204	3930	3' x 5'-4" opening, standard glass	2 Sswk	10	1.600	Ea.	425	49		474	555	204
	3940	Insulating glass		10	1.600		500	49		549	640	
	3950	4' x 5'-4" opening, standard glass		9	1.778		465	54.50		519.50	615	
	3960	Insulating glass		9	1.778		560	54.50		614.50	715	
	4000	Sliding aluminum, 3' x 2' opening, standard glazed		10	1.600		138	49		187	242	
	4100	Insulating glass		10	1.600		153	49		202	258	
	4300	5' x 3' opening, standard glazed		9	1.778		175	54.50		229.50	293	
	4400	Insulating glass		9	1.778		245	54.50		299.50	370	
	4600	8' x 4' opening, standard glazed		6	2.667		252	81.50		333.50	430	
	4700	Insulating glass		6	2.667		405	81.50		486.50	595	
	5000	9' x 5' opening, standard glazed		4	4		380	122		502	645	
	5100	Insulating glass		4	4		610	122		732	895	
	5500	Sliding, with thermal barrier and screen, 6' x 4', 2 track		8	2		520	61		581	680	
	5700	4 track		8	2		635	61		696	810	
	6000	For above units with bronze finish, add					12%					
	6200	For installation in concrete openings, add					5%					

085 500 | Metal Jalousie Windows

			CREW	DAILY OUTPUT	LABOR-HOURS	UNIT	MAT.	LABOR	EQUIP.	TOTAL	TOTAL INCL O&P	
502	0010	**JALOUSIES**										502
	0020	Aluminum incl. glazing & screens, stock, 1'-7" x 3'-2"	2 Sswk	10	1.600	Ea.	111	49		160	212	
	0100	2'-3" x 4'-0"		10	1.600		158	49		207	264	
	0200	3'-1" x 2'-0"		10	1.600		109	49		158	210	
	0300	3'-1" x 5'-3"		10	1.600		226	49		275	340	
	1000	Mullions for above, 2'-0" long		80	.200		8.50	6.10		14.60	20.50	
	1100	5'-3" long		80	.200		14.60	6.10		20.70	27.50	
504	0010	**LOUVERS** See division 041-524, 062-740 & 102-104										504

085 600 | Metal Storm Windows

			CREW	DAILY OUTPUT	LABOR-HOURS	UNIT	MAT.	LABOR	EQUIP.	TOTAL	TOTAL INCL O&P	
601	0010	**STORM WINDOWS** Aluminum, residential										601
	0300	Basement, mill finish, incl. fiberglass screen										
	0320	1'-10" x 1'-0" high	2 Carp	30	.533	Ea.	24.50	14.55		39.05	50	
	0340	2'-9" x 1'-6" high	"	30	.533	"	27	14.55		41.55	52.50	
	1600	Double-hung, combination, storm & screen										
	2000	Average quality, clear anodic coating, 2'-0" x 3'-5" high	2 Carp	30	.533	Ea.	64.50	14.55		79.05	94	
	2020	2'-6" x 5'-0" high		28	.571		81.50	15.60		97.10	114	
	2040	4'-0" x 6'-0" high		25	.640		95.50	17.45		112.95	133	
	2400	White painted, 2'-0" x 3'-5" high		30	.533		63.50	14.55		78.05	93	
	2420	2'-6" x 5'-0" high		28	.571		70	15.60		85.60	102	
	2440	4'-0" x 6'-0" high		25	.640		77	17.45		94.45	112	
	2600	Mill finish, 2'-0" x 3'-5" high		30	.533		58	14.55		72.55	86.50	
	2620	2'-6" x 5'-0" high		28	.571		64.50	15.60		80.10	95.50	
	2640	4'-0" x 6-8" high		25	.640		72.50	17.45		89.95	107	

085 700 | Screens

			CREW	DAILY OUTPUT	LABOR-HOURS	UNIT	MAT.	LABOR	EQUIP.	TOTAL	TOTAL INCL O&P	
701	0010	**SCREENS**										701
	0020	For metal sash, aluminum or bronze mesh, flat screen	2 Sswk	1,200	.013	S.F.	2.91	.41		3.32	3.95	
	0500	Wicket screen, inside window		1,000	.016		4.45	.49		4.94	5.80	
	0800	Security screen, aluminum frame with stainless steel cloth		1,200	.013		15.70	.41		16.11	18	
	0900	Steel grate, painted, on steel frame		1,600	.010		8.45	.31		8.76	9.80	
	1000	For solar louvers, add		160	.100		16.25	3.06		19.31	23.50	
	4000	See also division 055-508										

Important: See the Reference Section for critical supporting data - Reference Nos., Crews, & City Cost Indexes

086 100	Wood Windows	CREW	DAILY OUTPUT	LABOR-HOURS	UNIT	1999 BARE COSTS				TOTAL INCL O&P
						MAT.	LABOR	EQUIP.	TOTAL	
104	0010	**AWNING WINDOW** Including frame, screen, and exterior trim										104

			CREW	DAILY OUTPUT	LABOR-HOURS	UNIT	MAT.	LABOR	EQUIP.	TOTAL	TOTAL INCL O&P
104	0010	**AWNING WINDOW** Including frame, screen, and exterior trim									
	0100	Average quality, builders model, 34" x 22", standard glazed	1 Carp	10	.800	Ea.	154	22		176	204
	0200	Insulating glass		10	.800		221	22		243	278
	0300	40" x 28", standard glazed		9	.889		259	24.50		283.50	320
	0400	Insulating glass		9	.889		370	24.50		394.50	450
	0500	48" x 36", standard glazed		8	1		480	27.50		507.50	575
	0600	Insulating glass		8	1		690	27.50		717.50	805
	1000	34" x 22"		10	.800		194	22		216	248
	1100	40" x 22"		10	.800		214	22		236	270
	1200	36" x 28"		9	.889		228	24.50		252.50	289
	1300	36" x 36"		9	.889		288	24.50		312.50	355
	1400	48" x 28"		8	1		272	27.50		299.50	340
	1500	60" x 36"		8	1		380	27.50		407.50	460
	2000	Metal clad, deluxe, insulating glass, 34" x 22"		10	.800		153	22		175	204
	2100	40" x 22"		10	.800		165	22		187	217
	2200	36" x 25"		9	.889		189	24.50		213.50	245
	2300	40" x 30"		9	.889		229	24.50		253.50	290
	2400	48" x 28"		8	1		221	27.50		248.50	286
	2500	60" x 36"	▼	8	1	▼	253	27.50		280.50	320
108	0010	**BOW-BAY WINDOW** Including frame, screen and exterior trim,										108
	0020	end panels operable									
	1000	Fixed type, builders model, 8' x 5' high, std. glazed, 4 panels	2 Carp	10	1.600	Ea.	840	43.50		883.50	990
	1050	Insulating glass		10	1.600		1,075	43.50		1,118.50	1,250
	1100	10'-0" x 5'-0" high, standard glazed		6	2.667		1,125	73		1,198	1,350
	1200	Insulating glass, 6 panels		6	2.667		1,175	73		1,248	1,425
	1300	Vinyl clad, premium, insulating glass, 6'-0" x 4'-0"		10	1.600		910	43.50		953.50	1,075
	1340	9'-0" x 4'-0"		8	2		1,200	54.50		1,254.50	1,400
	1380	10'-0" x 6'-0"		7	2.286		2,050	62.50		2,112.50	2,350
	1420	12'-0" x 6'-0"		6	2.667		2,600	73		2,673	2,975
	1600	Metal clad, deluxe, insul. glass, 6'-0" x 4'-0" high, 3 panels		10	1.600		760	43.50		803.50	910
	1640	9'-0" x 4'-0" high, 4 panels		8	2		1,075	54.50		1,129.50	1,250
	1680	10'-0" x 5'-0" high, 5 panels		7	2.286		1,475	62.50		1,537.50	1,725
	1720	12'-0" x 6'-0" high, 6 panels		6	2.667		2,050	73		2,123	2,375
	2000	Casement, builders model, bow, 8' x 5' high, std. glazed, 4 panels		10	1.600		1,325	43.50		1,368.50	1,525
	2050	Insulating glass		10	1.600		1,600	43.50		1,643.50	1,825
	2100	12'-0" x 6'-0" high, 6 panels, standard glazed		6	2.667		1,650	73		1,723	1,925
	2200	Insulating glass		6	2.667		1,650	73		1,723	1,950
	2280	6'-0" x 4'-0"		11	1.455		945	39.50		984.50	1,125
	2300	Vinyl clad, premium, insulating glass, 8'-0" x 5'-0"		10	1.600		1,100	43.50		1,143.50	1,300
	2340	10'-0" x 5'-0"		8	2		1,575	54.50		1,629.50	1,800
	2380	10'-0" x 6'-0"		7	2.286		1,650	62.50		1,712.50	1,900
	2420	12'-0" x 6'-0"		6	2.667		1,950	73		2,023	2,275
	2600	Metal clad, deluxe, insul. glass, 8'-0" x 5'-0" high, 4 panels		10	1.600		1,150	43.50		1,193.50	1,350
	2640	10'-0" x 5'-0" high, 5 panels		8	2		1,250	54.50		1,304.50	1,450
	2680	10'-0" x 6'-0" high, 5 panels		7	2.286		1,475	62.50		1,537.50	1,725
	2720	12'-0" x 6'-0" high, 6 panels		6	2.667		2,050	73		2,123	2,375
	3000	Double hung, bldrs. model, bay, 8' x 4' high, std. glazed		10	1.600		920	43.50		963.50	1,100
	3050	Insulating glass		10	1.600		995	43.50		1,038.50	1,175
	3100	9'-0" x 5'-0" high, standard glazed		6	2.667		995	73		1,068	1,225
	3200	Insulating glass		6	2.667		1,050	73		1,123	1,275
	3300	Vinyl clad, premium, insulating glass, 7'-0" x 4'-6"		10	1.600		955	43.50		998.50	1,125
	3340	8'-0" x 4'-6"		8	2		975	54.50		1,029.50	1,150
	3380	8'-0" x 5'-0"		7	2.286		1,025	62.50		1,087.50	1,225
	3420	9'-0" x 5'-0"		6	2.667		1,050	73		1,123	1,275
	3600	Metal clad, deluxe, insul. glass, 7'-0" x 4'-0" high	▼	10	1.600	▼	885	43.50		928.50	1,050

086 100	Wood Windows		CREW	DAILY OUTPUT	LABOR-HOURS	UNIT	1999 BARE COSTS				TOTAL INCL O&P	
							MAT.	LABOR	EQUIP.	TOTAL		
108	3640	8'-0" x 4'-0" high	2 Carp	8	2	Ea.	915	54.50		969.50	1,075	**108**
	3680	8'-0" x 5'-0" high		7	2.286		950	62.50		1,012.50	1,150	
	3720	9'-0" x 5'-0" high		6	2.667		1,000	73		1,073	1,225	
120	0010	**CASEMENT WINDOW** Including frame, screen, and exterior trim										**120**
	0100	Avg. quality, bldrs. model, 2'-0" x 3'-0" H, standard glazed	1 Carp	10	.800	Ea.	158	22		180	208	
	0150	Insulating glass CN		10	.800		190	22		212	244	
	0200	2'-0" x 4'-6" high, standard glazed		9	.889		205	24.50		229.50	263	
	0250	Insulating glass		9	.889		259	24.50		283.50	325	
	0300	2'-3" x 6'-0" high, standard glazed		8	1		220	27.50		247.50	285	
	0350	Insulating glass		8	1		305	27.50		332.50	380	
	0522	Vinyl clad, premium, insulating glass, 2'-0" x 3'-0"		10	.800		192	22		214	246	
	0524	2'-0" x 4'-0"		9	.889		228	24.50		252.50	288	
	0525	2'-0" x 5'-0"		8	1		262	27.50		289.50	330	
	0528	2'-0" x 6'-0"		8	1		310	27.50		337.50	390	
	8000	Solid vinyl, premium, insulating glass, 2'-0" x 3'-0" high		10	.800		161	22		183	212	
	8020	2'-0" x 4'-0" high		9	.889		197	24.50		221.50	255	
	8040	2'-0" x 5'-0" high		8	1		227	27.50		254.50	293	
	8100	Metal clad, deluxe, insulating glass, 2'-0" x 3'-0" high		10	.800		163	22		185	214	
	8120	2'-0" x 4'-0" high		9	.889		196	24.50		220.50	253	
	8140	2'-0" x 5'-0" high		8	1		222	27.50		249.50	288	
	8160	2'-0" x 6'-0" high		8	1		255	27.50		282.50	325	
	8200	For multiple leaf units, deduct for stationary sash										
	8220	2' high				Ea.	16.90			16.90	18.60	
	8240	4'-6" high					19.50			19.50	21.50	
	8260	6' high					26			26	28.50	
	8300	For installation, add per leaf						15%				
124	0010	**DOUBLE HUNG** Including frame, screen, and exterior trim R086 -100										**124**
	0100	Avg. quality, bldrs. model, 2'-0" x 3'-0" high, standard glazed	1 Carp	10	.800	Ea.	105	22		127	151	
	0150	Insulating glass		10	.800		149	22		171	199	
	0200	3'-0" x 4'-0" high, standard glazed		9	.889		138	24.50		162.50	190	
	0250	Insulating glass		9	.889		182	24.50		206.50	238	
	0300	4'-0" x 4'-6" high, standard glazed		8	1		169	27.50		196.50	229	
	0350	Insulating glass		8	1		224	27.50		251.50	289	
	1000	Vinyl clad, premium, insulating glass, 2'-6" x 3'-0"		10	.800		237	22		259	296	
	1100	3'-0" x 3'-6"		10	.800		287	22		309	350	
	1200	3'-0" x 4'-0"		9	.889		310	24.50		334.50	380	
	1300	3'-0" x 4'-6"		9	.889		325	24.50		349.50	395	
	1400	3'-0" x 5'-0"		8	1		335	27.50		362.50	415	
	1500	3'-6" x 6'-0"		8	1		560	27.50		587.50	665	
	1540	Solid vinyl, average quality, insulated glass, 2'-0" x 3'-0"		10	.800		120	22		142	166	
	1542	3'-0" x 4'-0"		9	.889		145	24.50		169.50	197	
	1544	4'-0" x 4'-6"		8	1		174	27.50		201.50	234	
	1560	Premium, insulating glass, 2'-6" x 3'-0"		10	.800		136	22		158	184	
	1562	3'-0" x 3'-6"		9	.889		157	24.50		181.50	211	
	1564	3'-0" x 4'-0"		9	.889		167	24.50		191.50	222	
	1566	3'-0" x 4'-6"		9	.889		172	24.50		196.50	227	
	1568	3'-0" x 5'-0"		8	1		176	27.50		203.50	237	
	1570	3'-6" x 6'-0"		8	1		194	27.50		221.50	256	
	2000	Metal clad, deluxe, insulating glass, 2'-6" x 3'-0" high		10	.800		172	22		194	224	
	2100	3'-0" x 3'-6" high		10	.800		204	22		226	260	
	2200	3'-0" x 4'-0" high		9	.889		217	24.50		241.50	277	
	2300	3'-0" x 4'-6" high		9	.889		235	24.50		259.50	296	
	2400	3'-0" x 5'-0" high		8	1		251	27.50		278.50	320	
	2500	3'-6" x 6'-0" high		8	1		305	27.50		332.50	380	

Important: See the Reference Section for critical supporting data - Reference Nos., Crews, & City Cost Indexes

8

DOORS & WINDOWS

086 100	Wood Windows	CREW	DAILY OUTPUT	LABOR-HOURS	UNIT	1999 BARE COSTS				TOTAL INCL O&P	
						MAT.	LABOR	EQUIP.	TOTAL		
130	**0010**	**PALLADIAN WINDOWS**									**130**
	0020	Aluminum clad, including frame and exterior trim									
	0040	3'-2" x 2'-0" high	2 Carp	11	1.455	Ea.	895	39.50		934.50	1,050
	0060	3'-2" x 4'-10"		11	1.455		1,025	39.50		1,064.50	1,200
	0080	3'-2" x 6'-4"		10	1.600		1,200	43.50		1,243.50	1,400
	0100	4'-0" x 4'-0"		10	1.600		985	43.50		1,028.50	1,150
	0120	4'-0" x 5'-4"	3 Carp	10	2.400		1,150	65.50		1,215.50	1,350
	0140	4'-0" x 6'-0"		9	2.667		1,200	73		1,273	1,450
	0160	4'-0" x 7'-4"		9	2.667		1,300	73		1,373	1,550
	0180	5'-5" x 4'-10"		9	2.667		1,275	73		1,348	1,525
	0200	5'-5" x 6'-10"		9	2.667		1,525	73		1,598	1,800
	0220	5'-5" x 7'-9"		9	2.667		1,650	73		1,723	1,950
	0240	6'-0" x 7'-11"		8	3		2,200	82		2,282	2,550
	0260	8'-0" x 6'-0"		8	3		1,875	82		1,957	2,200
132	**0010**	**PICTURE WINDOW** Including frame and exterior trim									**132**
	0100	Average quality, bldrs. model, 3'-6" x 4'-0" high, standard glazed	2 Carp	12	1.333	Ea.	236	36.50		272.50	320
	0150	Insulating glass		12	1.333		294	36.50		330.50	385
	0200	4'-0" x 4'-6" high, standard glazed		11	1.455		196	39.50		235.50	279
	0250	Insulating glass		11	1.455		200	39.50		239.50	282
	0300	5'-0" x 4'-0" high, standard glazed		11	1.455		198	39.50		237.50	281
	0350	Insulating glass		11	1.455		248	39.50		287.50	335
	0400	6'-0" x 4'-6" high, standard glazed		10	1.600		277	43.50		320.50	375
	0450	Insulating glass		10	1.600		340	43.50		383.50	445
	1000	Vinyl clad, premium, insulating glass, 4'-0" x 4'-0"		12	1.333		415	36.50		451.50	515
	1100	4'-0" x 6'-0"		11	1.455		570	39.50		609.50	690
	1200	5'-0" x 6'-0"		10	1.600		740	43.50		783.50	885
	1300	6'-0" x 6'-0"		10	1.600		755	43.50		798.50	900
	2000	Metal clad, deluxe, insulating glass, 4'-0" x 4'-0" high		12	1.333		255	36.50		291.50	340
	2100	4'-0" x 6'-0" high		11	1.455		375	39.50		414.50	480
	2200	5'-0" x 6'-0" high		10	1.600		415	43.50		458.50	525
	2300	6'-0" x 6'-0" high		10	1.600		475	43.50		518.50	595
140	**0010**	**SLIDING WINDOW** Including frame, screen, and exterior trim									**140**
	0100	Average quality, bldrs. model, 3'-0" x 3'-0" high, standard glazed	1 Carp	10	.800	Ea.	126	22		148	174
	0120	Insulating glass		10	.800		159	22		181	210
	0200	4'-0" x 3'-6" high, standard glazed		9	.889		150	24.50		174.50	203
	0220	Insulating glass		9	.889		188	24.50		212.50	245
	0300	6'-0" x 5'-0" high, standard glazed		8	1		276	27.50		303.50	350
	0320	Insulating glass		8	1		330	27.50		357.50	410
	1000	Vinyl clad, premium, insulating glass, 3'-0" x 3'-0"		10	.800		455	22		477	535
	1050	4'-0" x 3'-6"		9	.889		545	24.50		569.50	635
	1100	5'-0" x 4'-0"		9	.889		630	24.50		654.50	730
	1150	6'-0" x 5'-0"		8	1		780	27.50		807.50	905
	2000	Metal clad, deluxe, insulating glass, 3'-0" x 3'-0" high		10	.800		256	22		278	315
	2050	4'-0" x 3'-6" high		9	.889		315	24.50		339.50	385
	2100	5'-0" x 4'-0" high		9	.889		380	24.50		404.50	455
	2150	6'-0" x 5'-0" high		8	1		460	27.50		487.50	555
142	**0010**	**WEATHERSTRIPPING** See division 087-300									**142**
144	**0010**	**WINDOW GRILLE OR MUNTIN** Snap-in type									**144**
	0020	Colonial or diamond pattern									
	2000	Wood, awning window, glass size 28" x 16" high	1 Carp	30	.267	Ea.	21.50	7.30		28.80	35
	2060	44" x 24" high		32	.250		27.50	6.80		34.30	41.50
	2100	Casement, glass size, 20" x 36" high		30	.267		27	7.30		34.30	41
	2180	20" x 56" high		32	.250		36	6.80		42.80	51

8

DOORS & WINDOWS

		086 100 | Wood Windows	CREW	DAILY OUTPUT	LABOR-HOURS	UNIT	1999 BARE COSTS				TOTAL INCL O&P	
							MAT.	LABOR	EQUIP.	TOTAL		
144	2200	Double hung, glass size, 16" x 24" high	1 Carp	24	.333	Set	57	9.10		66.10	77	144
	2280	32" x 32" high		34	.235	"	102	6.40		108.40	123	
	2500	Picture, glass size, 48" x 48" high		30	.267	Ea.	75.50	7.30		82.80	95	
	2580	60" x 68" high		28	.286	"	118	7.80		125.80	141	
	2600	Sliding, glass size, 14" x 36" high		24	.333	Set	30.50	9.10		39.60	48	
	2680	36" x 36" high		22	.364	"	48	9.95		57.95	68.50	
148	0010	**WOOD SASH** Including glazing but not including trim										148
	0050	Custom, 5'-0" x 4'-0", 1" dbl. glazed, 3/16" thick lites	2 Carp	3.20	5	Ea.	185	137		322	420	
	0100	1/4" thick lites		5	3.200		190	87.50		277.50	345	
	0200	1" thick, triple glazed		5	3.200		435	87.50		522.50	615	
	0300	7'-0" x 4'-6" high, 1" double glazed, 3/16" thick lites		4.30	3.721		445	102		547	645	
	0400	1/4" thick lites		4.30	3.721		500	102		602	710	
	0500	1" thick, triple glazed		4.30	3.721		570	102		672	785	
	0600	8'-6" x 5'-0" high, 1" double glazed, 3/16" thick lites		3.50	4.571		595	125		720	850	
	0700	1/4" thick lites		3.50	4.571		655	125		780	915	
	0800	1" thick, triple glazed		3.50	4.571		660	125		785	920	
	0900	Window frames only, based on perimeter length				L.F.	3.65			3.65	4.02	
	1200	Window sill, stock, per lineal foot					8.35			8.35	9.20	
	1250	Casing, stock					2.90			2.90	3.19	
	3000	Replacement sash, double hung, double glazing, to 12 S.F.	1 Carp	64	.125	S.F.	18.30	3.41		21.71	25.50	
	3100	12 S.F. to 20 S.F.		94	.085		15.30	2.32		17.62	20.50	
	3200	20 S.F. and over		106	.075		13.35	2.06		15.41	17.90	
	3800	Triple glazing for above, add					2.68			2.68	2.95	
	7000	Sash, single lite, 2'-0" x 2'-0" high	1 Carp	20	.400	Ea.	34.50	10.90		45.40	54.50	
	7050	2'-6" x 2'-0" high		19	.421		37	11.50		48.50	59	
	7100	2'-6" x 2'-6" high		18	.444		39	12.15		51.15	62	
	7150	3'-0" x 2'-0" high		17	.471		49.50	12.85		62.35	74.50	
152	0010	**WOOD SCREENS**										152
	0020	Over 3 S.F., 3/4" frames	2 Carp	375	.043	S.F.	3	1.16		4.16	5.15	
	0100	1-1/8" frames	"	375	.043	"	2.95	1.16		4.11	5.10	

		087 100 | Finish Hardware	CREW	DAILY OUTPUT	LABOR-HOURS	UNIT	1999 BARE COSTS				TOTAL INCL O&P	
							MAT.	LABOR	EQUIP.	TOTAL		
101	0010	**AVERAGE** Percentage for hardware, total job cost, minimum									.75%	101
	0050	Maximum									3.50%	
	0500	Total hardware for building, average distribution					85%	15%				
	1000	Door hardware, apartment, interior				Door	104			104	114	
	1500	Hospital bedroom, minimum					236			236	260	
	2000	Maximum					520			520	570	
	2100	Pocket door				Ea.	106			106	117	
	2250	School, single exterior, incl. lever, not incl. panic device				Door	350			350	385	
	2500	Single interior, regular use, no lever included					233			233	256	
	2550	Including handicap lever					315			315	350	
	2600	Heavy use, incl. lever and closer					410			410	450	
	2850	Stairway, single interior					580			580	640	
	3100	Double exterior, with panic device				Pr.	820			820	905	
	3600	Toilet, public, single interior				Door	128			128	141	

8
DOORS & WINDOWS

087 100	Finish Hardware	CREW	DAILY OUTPUT	LABOR-HOURS	UNIT	MAT.	LABOR	EQUIP.	TOTAL	TOTAL INCL O&P	
103 0010	**BOLTS, FLUSH**										103
0020	Standard, concealed	1 Carp	7	1.143	Ea.	17.05	31		48.05	68	
0800	Automatic fire exit	"	5	1.600		265	43.50		308.50	360	
1600	For electric release, add	1 Elec	3	2.667		91	85		176	227	
3000	Barrel, brass, 2" long	1 Carp	40	.200		2.10	5.45		7.55	10.90	
3020	4" long		40	.200		2.05	5.45		7.50	10.85	
3060	6" long	↓	40	.200	↓	4.13	5.45		9.58	13.15	
105 0010	**BUMPER PLATES**										105
0020	1-1/2" x 3/4" U channel	2 Carp	80	.200	L.F.	3.06	5.45		8.51	11.95	
1000	Tear drop, spring-steel, 4" high		15	1.067	Ea.	39.50	29		68.50	89.50	
1100	8" high		15	1.067		79	29		108	133	
1200	10" high	↓	15	1.067	↓	98.50	29		127.50	154	
108 0010	**DEADLOCKS** Mortise, heavy duty, outside key	1 Carp	9	.889	Ea.	109	24.50		133.50	158	108
0020	Double cylinder		9	.889		120	24.50		144.50	171	
0100	Medium duty, outside key		10	.800		84.50	22		106.50	128	
0110	Double cylinder		10	.800		106	22		128	151	
1000	Tubular, standard duty, outside key		10	.800		45.50	22		67.50	84.50	
1010	Double cylinder		10	.800		58.50	22		80.50	99	
1200	Night latch, outside key	↓	10	.800	↓	57.50	22		79.50	97.50	
110 0010	**DOORSTOPS** Holder and bumper, floor or wall	1 Carp	32	.250	Ea.	14	6.80		20.80	26	110
1300	Wall bumper, 4" diameter, with rubber pad, aluminum		32	.250		6.35	6.80		13.15	17.75	
1600	Door bumper, floor type, aluminum		32	.250		3.64	6.80		10.44	14.75	
1900	Plunger type, door mounted	↓	32	.250	↓	22.50	6.80		29.30	35.50	
112 0010	**ENTRANCE LOCKS** Cylinder, grip handle, deadlocking latch	1 Carp	9	.889	Ea.	100	24.50		124.50	148	112
0020	Deadbolt		8	1		121	27.50		148.50	176	
0100	Push and pull plate, dead bolt	↓	8	1		115	27.50		142.50	170	
0900	For handicapped lever, add				↓	126			126	139	
114 0010	**FLOOR CHECKS** For over 3' wide doors, single acting	1 Carp	2.50	3.200	Ea.	385	87.50		472.50	560	114
0500	Double acting	"	2.50	3.200	"	490	87.50		577.50	675	
116 0010	**HINGES** Full mortise, avg. freq., steel base, 4-1/2" x 4-1/2", USP				Pr.	17.55			17.55	19.30	116
0100	5" x 5", USP					28.50			28.50	31	
0200	6" x 6", USP					59.50			59.50	65.50	
0400	Brass base, 4-1/2" x 4-1/2", US10	R087 -100				36			36	40	
0500	5" x 5", US10					51.50			51.50	56.50	
0600	6" x 6", US10					85.50			85.50	94	
0800	Stainless steel base, 4-1/2" x 4-1/2", US32				↓	60			60	66	
0900	For non removable pin, add				Ea.	2.12			2.12	2.33	
0910	For floating pin, driven tips, add					3.86			3.86	4.25	
0930	For hospital type tip on pin, add					11.35			11.35	12.45	
0940	For steeple type tip on pin, add				↓	7.55			7.55	8.35	
0950	Full mortise, high frequency, steel base, 3-1/2" x 3-1/2", US26D				Pr.	25.50			25.50	28	
1000	4-1/2" x 4-1/2", USP					43.50			43.50	48	
1100	5" x 5", USP					49			49	54	
1200	6" x 6", USP					95			95	105	
1400	Brass base, 3-1/2" x 3-1/2", US4					38.50			38.50	42.50	
1430	4-1/2" x 4-1/2", US10					39			39	43	
1500	5" x 5", US10					95.50			95.50	105	
1600	6" x 6", US10					121			121	133	
1800	Stainless steel base, 4-1/2" x 4-1/2", US32				↓	92.50			92.50	102	
1930	For hospital type tip on pin, add				Ea.	14.90			14.90	16.35	
1950	Full mortise, low frequency, steel base, 3-1/2" x 3-1/2", US26D				Pr.	12.10			12.10	13.35	
2000	4-1/2" x 4-1/2", USP					13.35			13.35	14.65	
2100	5" x 5", USP	↓			↓	21			21	23	

DOORS & WINDOWS **8**

For expanded coverage of these items see Means Interior Cost Data 1999

087 100 | Finish Hardware

			CREW	DAILY OUTPUT	LABOR-HOURS	UNIT	1999 BARE COSTS MAT.	LABOR	EQUIP.	TOTAL	TOTAL INCL O&P	
116	2200	6" x 6", USP	R087 -100			Pr.	41.50			41.50	45.50	116
	2300	4-1/2" x 4-1/2", US3					9.65			9.65	10.60	
	2310	5" x 5", US3					29.50			29.50	32.50	
	2400	Brass bass, 4-1/2" x 4-1/2", US10					30			30	33	
	2500	5" x 5", US10					45			45	49.50	
	2800	Stainless steel base, 4-1/2" x 4-1/2", US32					50			50	55	
118	0010	**KICK PLATE** 6" high, for 3' door, stainless steel	1 Carp	15	.533	Ea.	15.95	14.55		30.50	40.50	118
	0500	Bronze		15	.533		19.30	14.55		33.85	44	
	2000	Aluminum, .050, with 3 beveled edges, 10" x 28"		15	.533		14.15	14.55		28.70	38.50	
	2010	10" x 30"		15	.533		15.15	14.55		29.70	39.50	
	2020	10" x 34"		15	.533		17.20	14.55		31.75	42	
	2040	10" x 38"		15	.533		18.95	14.55		33.50	44	
120	0010	**LOCKSET** Standard duty, cylindrical, with sectional trim										120
	0020	Non-keyed, passage	1 Carp	12	.667	Ea.	36	18.20		54.20	68	
	0100	Privacy		12	.667		44	18.20		62.20	77	
	0400	Keyed, single cylinder function **CN**		10	.800		64.50	22		86.50	105	
	0420	Hotel		8	1		87	27.50		114.50	139	
	0500	Lever handled, keyed, single cylinder function		10	.800		87	22		109	131	
	1000	Heavy duty with sectional trim, non-keyed, passages		12	.667		102	18.20		120.20	141	
	1100	Privacy		12	.667		129	18.20		147.20	170	
	1400	Keyed, single cylinder function		10	.800		152	22		174	202	
	1420	Hotel		8	1		226	27.50		253.50	292	
	1600	Communicating		10	.800		179	22		201	232	
	1690	For re-core cylinder, add					25.50			25.50	28	
	1700	Residential, interior door, minimum	1 Carp	16	.500		12.25	13.65		25.90	35	
	1720	Maximum		8	1		32.50	27.50		60	79	
	1800	Exterior, minimum		14	.571		27.50	15.60		43.10	54.50	
	1810	Average		8	1		58	27.50		85.50	107	
	1820	Maximum		8	1		115	27.50		142.50	169	
125	0010	**MORTISE LOCKSET** Comm., wrought knobs & full escutcheon trim										125
	0020	Non-keyed, passage, minimum	1 Carp	9	.889	Ea.	135	24.50		159.50	187	
	0030	Maximum		8	1		218	27.50		245.50	283	
	0040	Privacy, minimum		9	.889		144	24.50		168.50	196	
	0050	Maximum		8	1		236	27.50		263.50	305	
	0100	Keyed, office/entrance/apartment, minimum		8	1		165	27.50		192.50	225	
	0110	Maximum		7	1.143		283	31		314	360	
	0120	Single cylinder, typical, minimum		8	1		142	27.50		169.50	199	
	0130	Maximum		7	1.143		262	31		293	335	
	0200	Hotel, minimum		7	1.143		170	31		201	236	
	0210	Maximum		6	1.333		278	36.50		314.50	365	
	0300	Communication, double cylinder, minimum		8	1		170	27.50		197.50	230	
	0310	Maximum		7	1.143		221	31		252	292	
	1000	Wrought knobs and sectional trim, non-keyed, passage, minimum		10	.800		89.50	22		111.50	133	
	1010	Maximum		9	.889		177	24.50		201.50	233	
	1040	Privacy, minimum		10	.800		105	22		127	150	
	1050	Maximum		9	.889		189	24.50		213.50	246	
	1100	Keyed, entrance, office/apartment, minimum		9	.889		156	24.50		180.50	209	
	1110	Maximum		8	1		225	27.50		252.50	291	
	1120	Single cylinder, typical, minimum		9	.889		150	24.50		174.50	203	
	1130	Maximum		8	1		218	27.50		245.50	283	
	2000	Cast knobs and full escutcheon trim										
	2010	Non-keyed, passage, minimum	1 Carp	9	.889	Ea.	191	24.50		215.50	248	
	2020	Maximum		8	1		310	27.50		337.50	385	
	2040	Privacy, minimum		9	.889		229	24.50		253.50	290	
	2050	Maximum		8	1		335	27.50		362.50	410	

Important: See the Reference Section for critical supporting data - Reference Nos., Crews, & City Cost Indexes

087 100	Finish Hardware	CREW	DAILY OUTPUT	LABOR-HOURS	UNIT	1999 BARE COSTS				TOTAL INCL O&P	
						MAT.	LABOR	EQUIP.	TOTAL		
125											**125**
2120	Keyed, single cylinder, typical, minimum	1 Carp	8	1	Ea.	229	27.50		256.50	295	
2130	Maximum		7	1.143		360	31		391	450	
2200	Hotel, minimum		7	1.143		260	31		291	335	
2210	Maximum		6	1.333		480	36.50		516.50	590	
3000	Cast knob and sectional trim, non-keyed, passage, minimum		10	.800		148	22		170	198	
3010	Maximum		10	.800		295	22		317	360	
3040	Privacy, minimum		10	.800		169	22		191	221	
3050	Maximum		10	.800		295	22		317	360	
3100	Keyed, office/entrance/apartment, minimum		9	.889		191	24.50		215.50	248	
3110	Maximum		9	.889		300	24.50		324.50	370	
3120	Single cylinder, typical, minimum		9	.889		191	24.50		215.50	248	
3130	Maximum		9	.889		370	24.50		394.50	450	
3190	For re-core cylinder, add					26			26	28.50	
3800	Cipher lockset	1 Carp	13	.615		620	16.80		636.80	710	
3900	Keyless, pushbutton type										
4000	Residential/light commercial, deadbolt, standard	1 Carp	9	.889	Ea.	86.50	24.50		111	133	
4010	Heavy duty		9	.889		103	24.50		127.50	151	
4020	Industrial, heavy duty, with deadbolt		9	.889		210	24.50		234.50	269	
4030	Key override		9	.889		233	24.50		257.50	294	
4040	Lever activated handle		9	.889		255	24.50		279.50	320	
4050	Key override		9	.889		283	24.50		307.50	350	
4060	Double sided pushbutton type		8	1		465	27.50		492.50	560	
4070	Key override		8	1		505	27.50		532.50	600	
126	0010	**HASP** Steel, 3" assembly	1 Carp	26	.308	Ea.	2.32	8.40		10.72	15.75	**126**
	0020	4-1/2"		13	.615		2.95	16.80		19.75	30	
	0040	6"		12.50	.640		4.67	17.45		22.12	32.50	
127	0010	**PANIC DEVICE** For rim locks, single door, exit only	1 Carp	6	1.333	Ea.	320	36.50		356.50	410	**127**
	0020	Outside key and pull		5	1.600		370	43.50		413.50	475	
	0200	Bar and vertical rod, exit only		5	1.600		475	43.50		518.50	590	
	0210	Outside key and pull		4	2		565	54.50		619.50	705	
	0400	Bar and concealed rod		4	2		480	54.50		534.50	610	
	0600	Touch bar, exit only		6	1.333		380	36.50		416.50	475	
	0610	Outside key and pull		5	1.600		455	43.50		498.50	570	
	0700	Touch bar and vertical rod, exit only		5	1.600		515	43.50		558.50	640	
	0710	Outside key and pull		4	2		600	54.50		654.50	745	
	1000	Mortise, bar, exit only		4	2		425	54.50		479.50	550	
	1600	Touch bar, exit only		4	2		485	54.50		539.50	620	
	2000	Narrow stile, rim mounted, bar, exit only		6	1.333		510	36.50		546.50	620	
	2010	Outside key and pull		5	1.600		555	43.50		598.50	680	
	2200	Bar and vertical rod, exit only		5	1.600		525	43.50		568.50	650	
	2210	Outside key and pull		4	2		525	54.50		579.50	665	
	2400	Bar and concealed rod, exit only		3	2.667		610	73		683	785	
	3000	Mortise, bar, exit only		4	2		430	54.50		484.50	555	
	3600	Touch bar, exit only		4	2		625	54.50		679.50	770	
129	0010	**PUSH-PULL PLATE**										**129**
	0100	Push plate, .050 thick, 4" x 16", aluminum	1 Carp	12	.667	Ea.	4.85	18.20		23.05	34	
	0500	Bronze		12	.667		10.90	18.20		29.10	40.50	
	1500	Pull handle and push bar, aluminum		11	.727		105	19.85		124.85	146	
	2000	Bronze		10	.800		136	22		158	184	
	3000	Push plate both sides, aluminum		14	.571		12.85	15.60		28.45	38.50	
	3500	Bronze		13	.615		32.50	16.80		49.30	62	
	4000	Door pull, designer style, cast aluminum, minimum		12	.667		59	18.20		77.20	93.50	
	5000	Maximum		8	1		274	27.50		301.50	345	
	6000	Cast bronze, minimum		12	.667		63.50	18.20		81.70	98.50	

087 100 | Finish Hardware

		CREW	DAILY OUTPUT	LABOR-HOURS	UNIT	1999 BARE COSTS				TOTAL INCL O&P	
						MAT.	LABOR	EQUIP.	TOTAL		
129	7000	Maximum	1 Carp	8	1	Ea.	297	27.50		324.50	370	**129**
	8000	Walnut, minimum		12	.667		49	18.20		67.20	82	
	9000	Maximum	↓	8	1	↓	272	27.50		299.50	345	
132	0010	**SPECIAL HINGES**										**132**
	0015	Paumelle, high frequency										
	0020	Steel base, 6" x 4-1/2", US10				Pr.	105			105	115	
	0100	Bronze base, 5" x 4-1/2", US10					141			141	155	
	0200	Paumelle, average frequency, steel base, 4-1/2" x 3-1/2", US10					71			71	78	
	0400	Olive knuckle, low frequency, brass base, 6" x 4-1/2", US10				↓	129			129	142	
	1000	Electric hinge with concealed conductor, average frequency										
	1010	Steel base, 4-1/2" x 4-1/2", US26D				Pr.	235			235	259	
	1100	Bronze base, 4-1/2" x 4-1/2", US26D				"	244			244	268	
	1200	Electric hinge with concealed conductor, high frequency										
	1210	Steel base, 4-1/2" x 4-1/2", US26D				Pr.	175			175	193	
	1600	Double weight, 800 lb., steel base, removable pin, 5" x 6", USP					111			111	122	
	1700	Steel base-welded pin, 5" x 6", USP					123			123	135	
	1800	Triple weight, 2000 lb., steel base, welded pin, 5" x 6", USP					128			128	141	
	2000	Pivot reinf., high frequency, steel base, 7-3/4" door plate, USP					142			142	156	
	2200	Bronze base, 7-3/4" door plate, US10					173			173	190	
	3000	Swing clear, full mortise, full or half surface, high frequency,										
	3010	Steel base, 5" high, USP				Pr.	119			119	131	
	3200	Swing clear, full mortise, average frequency										
	3210	Steel base, 4-1/2" high, USP				Pr.	94.50			94.50	104	
	4000	Wide throw, average frequency, steel base, 4-1/2" x 6", USP					72			72	79.50	
	4200	High frequency, steel base, 4-1/2" x 6", USP				↓	110			110	121	
	4600	Spring hinge, single acting, 6" flange, steel				Ea.	41			41	45.50	
	4700	Brass					72.50			72.50	79.50	
	4900	Double acting, 6" flange, steel					74.50			74.50	82	
	4950	Brass				↓	122			122	134	
	9000	Continuous hinge, steel, full mortise, heavy duty 501-85	2 Carp	64	.250	L.F.	10.30	6.80		17.10	22	

087 200 | Operators

		CREW	DAILY OUTPUT	LABOR-HOURS	UNIT	1999 BARE COSTS				TOTAL INCL O&P	
						MAT.	LABOR	EQUIP.	TOTAL		
202	0010	**AUTOMATIC OPENERS** Swing doors, single	2 Skwk	.80	20	Ea.	2,150	560		2,710	3,250	**202**
	0100	Single operating pair		.50	32	Pr.	4,000	900		4,900	5,825	
	0400	For double simultaneous doors, one way, add		1.20	13.333	↓	335	375		710	960	
	0500	Two way, add		.90	17.778	↓	415	500		915	1,250	
	1000	Sliding doors, 3' wide, including track & hanger, single		.60	26.667	Opng.	3,675	750		4,425	5,225	
	1300	Bi-parting		.50	32		4,425	900		5,325	6,300	
	1450	Activating carpet, single door, one way add		2.20	7.273		680	204		884	1,075	
	1550	Two way, add	↓	1.30	12.308	↓	975	345		1,320	1,625	
	1750	Handicap opener, button operating ♿	1 Carp	1.50	5.333	Ea.	1,250	146		1,396	1,600	
204	0010	**AUTOMATIC OPERATORS** Industrial, sliding doors, to 6' wide	2 Skwk	.60	26.667	Opng.	4,400	750		5,150	6,025	**204**
	0200	To 12' wide	"	.40	40	"	5,375	1,125		6,500	7,700	
	0400	Over 12' wide, add per L.F. of excess ♿				L.F.	590			590	650	
	1000	Swing doors, to 5' wide	2 Skwk	.80	20	Ea.	2,575	560		3,135	3,700	
	1860	Add for controls, wall pushbutton, 3 button		4	4		157	112		269	350	
	1870	Ceiling pull cord	↓	4.30	3.721	↓	135	104		239	315	
206	0010	**DOOR CLOSER** Rack and pinion	1 Carp	6.50	1.231	Ea.	105	33.50		138.50	169	**206**
	0020	Adjustable backcheck, 3 way mount, all sizes, regular arm		6	1.333		111	36.50		147.50	180	
	0040	Hold open arm		6	1.333		121	36.50		157.50	191	
	0100	Fusible link		6.50	1.231		94	33.50		127.50	156	
	0200	Non sized, regular arm		6	1.333		105	36.50		141.50	174	
	0240	Hold open arm	↓	6	1.333	↓	131	36.50		167.50	202	

Important: See the Reference Section for critical supporting data - Reference Nos., Crews, & City Cost Indexes

8 DOORS & WINDOWS

087 200	Operators	CREW	DAILY OUTPUT	LABOR-HOURS	UNIT	1999 BARE COSTS				TOTAL INCL O&P
						MAT.	LABOR	EQUIP.	TOTAL	
206 0400	4 way mount, non sized, regular arm	1 Carp	6	1.333	Ea.	144	36.50		180.50	216
0440	Hold open arm		6	1.333		155	36.50		191.50	229
2000	Backcheck and adjustable power, hinge face mount									
2010	All sizes, regular arm	1 Carp	6.50	1.231	Ea.	133	33.50		166.50	199
2040	Hold open arm		6.50	1.231		143	33.50		176.50	210
2400	Top jamb mount, all sizes, regular arm		6	1.333		133	36.50		169.50	204
2440	Hold open arm		6	1.333		143	36.50		179.50	215
2800	Top face mount, all sizes, regular arm		6.50	1.231		132	33.50		165.50	199
2840	Hold open arm		6.50	1.231		142	33.50		175.50	210
4000	Backcheck, overhead concealed, all sizes, regular arm		5.50	1.455		140	39.50		179.50	217
4040	Concealed arm		5	1.600		150	43.50		193.50	233
4400	Compact overhead, concealed, all sizes, regular arm		5.50	1.455		255	39.50		294.50	345
4440	Concealed arm		5	1.600		266	43.50		309.50	360
4800	Concealed in door, all sizes, regular arm		5.50	1.455		94.50	39.50		134	167
4840	Concealed arm		5	1.600		102	43.50		145.50	181
4900	Floor concealed, all sizes, single acting		2.20	3.636		120	99.50		219.50	288
4940	Double acting		2.20	3.636		155	99.50		254.50	325
5000	For cast aluminum cylinder, deduct					12.70			12.70	14
5040	For delayed action, add					22.50			22.50	24.50
5080	For fusible link arm, add					9.20			9.20	10.15
5120	For shock absorbing arm, add					27.50			27.50	30.50
5160	For spring power adjustment, add					21			21	23.50
6000	Closer-holder, hinge face mount, all sizes, exposed arm	1 Carp	6.50	1.231		97.50	33.50		131	160
7000	Electronic closer-holder, hinge facemount, concealed arm		5	1.600		148	43.50		191.50	232
7400	With built-in detector		5	1.600		445	43.50		488.50	560

087 300	Weatherstripping/Seals									
302 0010	ASTRAGALS One piece overlapping									
0400	Cadmium plated steel, flat, 3/16" x 2"	1 Carp	90	.089	L.F.	2.56	2.43		4.99	6.65
0600	Prime coated steel, flat, 1/8" x 3"		90	.089		3.56	2.43		5.99	7.75
0800	Stainless steel, flat, 3/32" x 1-5/8"		90	.089		13.35	2.43		15.78	18.50
1000	Aluminum, flat, 1/8" x 2"		90	.089		2.57	2.43		5	6.65
1200	Nail on, "T" extrusion		120	.067		.56	1.82		2.38	3.48
1300	Vinyl bulb insert		105	.076		.92	2.08		3	4.28
1600	Screw on, "T" extrusion		90	.089		3.75	2.43		6.18	7.95
1700	Vinyl insert		75	.107		2.52	2.91		5.43	7.35
2000	"L" extrusion, neoprene bulbs		75	.107		1.47	2.91		4.38	6.20
2100	Neoprene sponge insert		75	.107		4.45	2.91		7.36	9.50
2200	Magnetic		75	.107		7.35	2.91		10.26	12.65
2400	Spring hinged security seal, with cam		75	.107		4.71	2.91		7.62	9.80
2600	Spring loaded locking bolt, vinyl insert		45	.178		6.40	4.85		11.25	14.70
2800	Neoprene sponge strip, "Z" shaped, aluminum		60	.133		3.05	3.64		6.69	9.10
2900	Solid neoprene strip, nail on aluminum strip		90	.089		2.55	2.43		4.98	6.60
3000	One piece stile protection									
3020	Neoprene fabric loop, nail on aluminum strips	1 Carp	60	.133	L.F.	.49	3.64		4.13	6.30
3110	Flush mounted aluminum extrusion, 1/2" x 1-1/4"		60	.133		2.44	3.64		6.08	8.45
3140	3/4" x 1-3/8"		60	.133		3.08	3.64		6.72	9.15
3160	1-1/8" x 1-3/4"		60	.133		4.92	3.64		8.56	11.15
3300	Mortise, 9/16" x 3/4"		60	.133		2.61	3.64		6.25	8.60
3320	13/16" x 1-3/8"		60	.133		2.83	3.64		6.47	8.85
3600	Spring bronze strip, nail on type		105	.076		2.23	2.08		4.31	5.70
3620	Screw on, with retainer		75	.107		1.82	2.91		4.73	6.60
3800	Flexible stainless steel housing, pile insert, 1/2" door		105	.076		5.05	2.08		7.13	8.80
3820	3/4" door		105	.076		5.65	2.08		7.73	9.50
4000	Extruded aluminum retainer, flush mount, pile insert		105	.076		1.76	2.08		3.84	5.20

For expanded coverage of these items see Means Interior Cost Data 1999

DOORS & WINDOWS **8**

087 300 | Weatherstripping/Seals

			DAILY	LABOR-			1999 BARE COSTS				TOTAL	
			CREW	OUTPUT	HOURS	UNIT	MAT.	LABOR	EQUIP.	TOTAL	INCL O&P	
302	4080	Mortise, felt insert	1 Carp	90	.089	L.F.	3.14	2.43		5.57	7.25	**302**
	4160	Mortise with spring, pile insert		90	.089		2.46	2.43		4.89	6.55	
	4400	Rigid vinyl retainer, mortise, pile insert		105	.076		1.75	2.08		3.83	5.20	
	4600	Wool pile filler strip, aluminum backing	↓	105	.076	↓	1.76	2.08		3.84	5.20	
	5000	Two piece overlapping astragal, extruded aluminum retainer										
	5010	Pile insert	1 Carp	60	.133	L.F.	2.29	3.64		5.93	8.25	
	5020	Vinyl bulb insert		60	.133		1.48	3.64		5.12	7.40	
	5040	Vinyl flap insert		60	.133		4.58	3.64		8.22	10.80	
	5060	Solid neoprene flap insert		60	.133		4.53	3.64		8.17	10.75	
	5080	Hypalon rubber flap insert		60	.133		4.63	3.64		8.27	10.85	
	5090	Snap on cover, pile insert		60	.133		5.30	3.64		8.94	11.60	
	5400	Magnetic aluminum, surface mounted		60	.133		17.90	3.64		21.54	25.50	
	5500	Interlocking aluminum, 5/8" x 1" neoprene bulb insert		45	.178		2.84	4.85		7.69	10.75	
	5600	Adjustable aluminum, 9/16" x 21/32", pile insert		45	.178		13.50	4.85		18.35	22.50	
	5790	For vinyl bulb, deduct					.37			.37	.41	
	5800	Magnetic, adjustable, 9/16" x 21/32"	1 Carp	45	.178	↓	17.35	4.85		22.20	27	
	6000	Two piece stile protection										
	6010	Cloth backed rubber loop, 1" gap, nail on aluminum strips	1 Carp	45	.178	L.F.	2.91	4.85		7.76	10.85	
	6040	Screw on aluminum strips		45	.178		4.53	4.85		9.38	12.65	
	6100	1-1/2" gap, screw on aluminum extrusion		45	.178		4.08	4.85		8.93	12.15	
	6240	Vinyl fabric loop, slotted aluminum extrusion, 1" gap		45	.178		1.45	4.85		6.30	9.25	
	6300	1-1/4" gap	↓	45	.178		4.30	4.85		9.15	12.40	
304	0010	**THRESHOLD** 3' long door saddles, aluminum	1 Carp	48	.167	L.F.	3.35	4.55		7.90	10.85	**304**
	0100	Aluminum, 8" wide, 1/2" thick		12	.667	Ea.	28.50	18.20		46.70	60	
	0500	Bronze		60	.133	L.F.	27	3.64		30.64	35.50	
	0600	Bronze, panic threshold, 5" wide, 1/2" thick		12	.667	Ea.	56	18.20		74.20	90	
	0700	Rubber, 1/2" thick, 5-1/2" wide		20	.400		28.50	10.90		39.40	48	
	0800	2-3/4" wide	↓	20	.400	↓	13.15	10.90		24.05	31.50	
306	0010	**WEATHERSTRIPPING** Window, double hung, 3' x 5', zinc	1 Carp	7.20	1.111	Opng.	10.60	30.50		41.10	59	**306**
	0100	Bronze		7.20	1.111		19.25	30.50		49.75	68.50	
	0500	As above but heavy duty, zinc		4.60	1.739		12.85	47.50		60.35	88.50	
	0600	Bronze		4.60	1.739		22.50	47.50		70	99	
	1000	Doors, wood frame, interlocking, for 3' x 7' door, zinc		3	2.667		11.55	73		84.55	128	
	1100	Bronze		3	2.667		18.15	73		91.15	135	
	1300	6' x 7' opening, zinc		2	4		12.60	109		121.60	186	
	1400	Bronze	↓	2	4	↓	24	109		133	198	
	1700	Wood frame, spring type, bronze										
	1800	3' x 7' door	1 Carp	7.60	1.053	Opng.	15.30	28.50		43.80	62	
	1900	6' x 7' door	"	7	1.143	"	16.25	31		47.25	67	
	2200	Metal frame, spring type, bronze										
	2300	3' x 7' door	1 Carp	3	2.667	Opng.	25.50	73		98.50	143	
	2400	6' x 7' door	"	2.50	3.200	"	34.50	87.50		122	175	
	2500	For stainless steel, spring type, add					133%					
	2700	Metal frame, extruded sections, 3' x 7' door, aluminum	1 Carp	2	4	Opng.	33.50	109		142.50	209	
	2800	Bronze		2	4		84.50	109		193.50	265	
	3100	6' x 7' door, aluminum		1.20	6.667		42.50	182		224.50	335	
	3200	Bronze	↓	1.20	6.667	↓	100	182		282	395	
	3500	Threshold weatherstripping										
	3650	Door sweep, flush mounted, aluminum	1 Carp	25	.320	Ea.	9.90	8.75		18.65	24.50	
	3700	Vinyl		25	.320		11.75	8.75		20.50	26.50	
	5000	Garage door bottom weatherstrip, 12' aluminum, clear		14	.571		15.90	15.60		31.50	42	
	5010	Bronze		14	.571		60.50	15.60		76.10	91	
	5050	Bottom protection, 12' aluminum, clear		14	.571		19.25	15.60		34.85	45.50	
	5100	Bronze	↓	14	.571	↓	75	15.60		90.60	107	

Important: See the Reference Section for critical supporting data - Reference Nos., Crews, & City Cost Indexes

087 500	Door/Window Acces.	CREW	DAILY OUTPUT	LABOR-HOURS	UNIT	1999 BARE COSTS				TOTAL INCL O&P	
						MAT.	LABOR	EQUIP.	TOTAL		
501	**0010**	**AREA WALL** Galvanized steel, 20 ga., 3'-2" wide, 1' deep	1 Sswk	29	.276	Ea.	12.10	8.45		20.55	29	**501**
	0100	2' deep		23	.348		21.50	10.65		32.15	43	
	0300	16 ga., galv., 3'-2" wide, 1' deep		29	.276		17.05	8.45		25.50	34.50	
	0400	3' deep		23	.348		34.50	10.65		45.15	57.50	
	0600	Welded grating for above, 15 lbs., painted		45	.178		33	5.45		38.45	46	
	0700	Galvanized		45	.178		62.50	5.45		67.95	78.50	
	0900	Translucent plastic cap for above	↓	60	.133	↓	12.50	4.08		16.58	21.50	
504	**0010**	**DETECTION SYSTEMS** See division 168-120										**504**
508	**0010**	**DOOR PROTECTION** Acrylic and neoprene cover for door	1 Carp	100	.080	L.F.	7	2.18		9.18	11.15	**508**
	0020	frame, high impact type										

8

DOORS & WINDOWS

088 100	Glass	CREW	DAILY OUTPUT	LABOR-HOURS	UNIT	1999 BARE COSTS				TOTAL INCL O&P	
						MAT.	LABOR	EQUIP.	TOTAL		
104	**0010**	**ACOUSTICAL GLASS UNITS** 1 lite at 3/8", 1 lite at 3/16", for 1" thick	2 Glaz	100	.160	S.F.	17.40	4.26		21.66	25.50	**104**
	0100	For 4" thick	"	80	.200	"	29.50	5.30		34.80	40.50	
108	**0010**	**BEVELED GLASS** With design patterns, 1/4" thick, 1/2" bevel, minimum	2 Glaz	150	.107	S.F.	29.50	2.84		32.34	37	**108**
	0050	Average		125	.128	↓	70	3.40		73.40	82	
	0100	Maximum	↓	100	.160	↓	123	4.26		127.26	143	
112	**0010**	**CURTAIN WALL** See division 089-200										**112**
116	**0010**	**FACETED** Color tinted glass, 3/4" thick, minimum	2 Glaz	95	.168	S.F.	25	4.48		29.48	34.50	**116**
	0100	Maximum	"	75	.213	"	44	5.65		49.65	57	
118	**0010**	**FLOAT GLASS** 3/16" thick, clear, plain R088-010	2 Glaz	130	.123	S.F.	3.35	3.27		6.62	8.65	**118**
	0200	Tempered, clear		130	.123		4	3.27		7.27	9.40	
	0300	Tinted		130	.123		5	3.27		8.27	10.50	
	0600	1/4" thick, clear, plain **CN**		120	.133		4.22	3.55		7.77	10.05	
	0700	Tinted		120	.133		4	3.55		7.55	9.80	
	0800	Tempered, clear		120	.133		5	3.55		8.55	10.90	
	0900	Tinted		120	.133		6.95	3.55		10.50	13.05	
	1600	3/8" thick, clear, plain		75	.213		6.65	5.65		12.30	15.95	
	1700	Tinted		75	.213		8	5.65		13.65	17.45	
	1800	Tempered, clear		75	.213		10	5.65		15.65	19.65	
	1900	Tinted		75	.213		12.45	5.65		18.10	22.50	
	2200	1/2" thick, clear, plain		55	.291		13.05	7.75		20.80	26	
	2300	Tinted		55	.291		14	7.75		21.75	27	
	2400	Tempered, clear		55	.291		15	7.75		22.75	28.50	
	2500	Tinted		55	.291		18.70	7.75		26.45	32.50	
	2800	5/8" thick, clear, plain		45	.356		14	9.45		23.45	30	
	2900	Tempered, clear		45	.356		16	9.45		25.45	32	
	3200	3/4" thick, clear, plain		35	.457		18	12.15		30.15	38.50	
	3300	Tempered, clear		35	.457		21	12.15		33.15	41.50	
	3600	1" thick, clear, plain	↓	30	.533		30	14.20		44.20	54.50	
	8900	For low emissivity coating for 3/16" and 1/4" only, add to above	↓			↓	100%					
120	**0010**	**FULL VISION** Window system with 3/4" glass mullions, 10' high	H-2	130	.185	S.F.	39.50	4.59		44.09	50.50	**120**
	0100	10' to 20' high, minimum	↓	110	.218	↓	41.50	5.45		46.95	54.50	

For expanded coverage of these items see *Means Interior Cost Data 1999*

8 DOORS & WINDOWS

088 100 | Glass

			CREW	DAILY OUTPUT	LABOR-HOURS	UNIT	MAT.	LABOR	EQUIP.	TOTAL	TOTAL INCL O&P	
							1999 BARE COSTS					
120	0150	Average	H-2	100	.240	S.F.	45	5.95		50.95	58.50	120
	0200	Maximum	↓	80	.300	↓	50.50	7.45		57.95	67	
124	0010	**GLASS BLOCK** See division 042-704 [R042 -700]										124
128	0010	**GLAZING VARIABLES** [R088 -010]										128
	0500	For high rise glazing, from exterior, add per S.F. per story				S.F.					.08	
	0600	For glass replacement, add				"		100%				
	0700	For gasket settings, add				L.F.	3			3	3.30	
	0800	For concrete reglet settings, add				S.F.	20%	25%				
	0900	For sloped glazing, add				"		25%				
	2000	Fabrication, polished edges, 1/4" thick				Inch	.25			.25	.28	
	2100	1/2" thick					.66			.66	.73	
	2500	Mitered edges, 1/4" thick					.66			.66	.73	
	2600	1/2" thick				↓	1.06			1.06	1.17	
132	0010	**INSULATING GLASS** 2 lites 1/8" float, 1/2" thk, under 15 S.F.										132
	0020	Clear	2 Glaz	95	.168	S.F.	5.80	4.48		10.28	13.20	
	0100	Tinted		95	.168		8.60	4.48		13.08	16.30	
	0200	2 lites 3/16" float, for 5/8" thk unit, 15 to 30 S.F., clear [R088 -010]		90	.178		7	4.73		11.73	14.90	
	0300	Tinted		90	.178		7.20	4.73		11.93	15.10	
	0400	1" thk, dbl. glazed, 1/4" float, 30-70 S.F., clear **CN**		75	.213		9.95	5.65		15.60	19.55	
	0500	Tinted		75	.213		12	5.65		17.65	22	
	0600	1" thick double glazed, 1/4" float, 1/4" wire		75	.213		15	5.65		20.65	25	
	0700	1/4" float, 1/4" tempered		75	.213		14.55	5.65		20.20	24.50	
	0800	1/4" wire, 1/4" tempered		75	.213		20	5.65		25.65	30.50	
	0900	Both lites, 1/4" wire		75	.213		19	5.65		24.65	29.50	
	2000	Both lites, light & heat reflective		85	.188		16	5		21	25	
	2500	Heat reflective, film inside, 1" thick unit, clear		85	.188		14	5		19	23	
	2600	Tinted		85	.188		15.10	5		20.10	24	
	3000	Film on weatherside, clear, 1/2" thick unit		95	.168		10	4.48		14.48	17.80	
	3100	5/8" thick unit		90	.178		12.65	4.73		17.38	21	
	3200	1" thick unit	↓	85	.188	↓	13.80	5		18.80	23	
136	0010	**LAMINATED GLASS** Clear float, .03" vinyl, 1/4" thick	2 Glaz	90	.178	S.F.	6.75	4.73		11.48	14.60	136
	0100	3/8" thick		78	.205		10.95	5.45		16.40	20.50	
	0200	.06" vinyl, 1/2" thick		65	.246		12.80	6.55		19.35	24	
	1000	5/8" thick		90	.178		15	4.73		19.73	23.50	
	2000	Bullet-resisting, 1-3/16" thick, to 15 S.F.		16	1		33	26.50		59.50	77	
	2100	Over 15 S.F.		16	1		29.50	26.50		56	72.50	
	2500	2-1/4" thick, to 15 S.F.		12	1.333		48.50	35.50		84	107	
	2600	Over 15 S.F.	↓	12	1.333	↓	43	35.50		78.50	101	
140	0010	**LEAD GLASS** For X-ray, see division 130-911										140
144	0010	**MIRRORS** No frames, wall type, 1/4" plate glass, polished edge										144
	0100	Up to 5 S.F.	2 Glaz	125	.128	S.F.	5.50	3.40		8.90	11.25	
	0200	Over 5 S.F.		160	.100		5.35	2.66		8.01	9.90	
	0500	Door type, 1/4" plate glass, up to 12 S.F.		160	.100		5.70	2.66		8.36	10.35	
	1000	Float glass, up to 10 S.F., 1/8" thick		160	.100		3.36	2.66		6.02	7.75	
	1100	3/16" thick		150	.107		3.91	2.84		6.75	8.60	
	1500	12" x 12" wall tiles, square edge, clear		195	.082		1.43	2.18		3.61	4.90	
	1600	Veined		195	.082		3.65	2.18		5.83	7.35	
	2000	1/4" thick, stock sizes, one way transparent		125	.128		12.45	3.40		15.85	18.90	
	2010	Bathroom, unframed, laminated	↓	160	.100	↓	9.55	2.66		12.21	14.55	
148	0010	**OBSCURE GLASS** 1/8" thick, minimum	2 Glaz	140	.114	S.F.	5.55	3.04		8.59	10.75	148
	0100	Maximum		125	.128		6.65	3.40		10.05	12.50	
	0300	7/32" thick, minimum		120	.133		6.15	3.55		9.70	12.15	
	0400	Maximum	↓	105	.152	↓	7.85	4.05		11.90	14.80	

Important: See the Reference Section for critical supporting data - Reference Nos., Crews, & City Cost Indexes

		088 100	Glass	CREW	DAILY OUTPUT	LABOR-HOURS	UNIT	1999 BARE COSTS				TOTAL INCL O&P	
								MAT.	LABOR	EQUIP.	TOTAL		
152	0010	PATTERNED GLASS Colored, 1/8" thick, minimum		2 Glaz	140	.114	S.F.	6.55	3.04		9.59	11.85	152
	0100	Maximum			125	.128		7.75	3.40		11.15	13.75	
	0300	7/32" thick, minimum			120	.133		7.60	3.55		11.15	13.80	
	0400	Maximum			105	.152		8.30	4.05		12.35	15.30	
160	0010	REFLECTIVE GLASS 1/4" float with fused metallic oxide, tinted		2 Glaz	115	.139	S.F.	8.30	3.70		12	14.80	160
	0500	1/4" float glass with reflective applied coating			115	.139		6.95	3.70		10.65	13.30	
	2000	Solar film on glass, not including glass, minimum			180	.089		3.35	2.36		5.71	7.30	
	2050	Maximum			225	.071		7.75	1.89		9.64	11.45	
164	0010	SANDBLASTED GLASS Float glass, 1/8" thick		2 Glaz	160	.100	S.F.	5.55	2.66		8.21	10.15	164
	0100	3/16" thick			130	.123		6.15	3.27		9.42	11.75	
	0500	Plate glass, 1/4" thick			120	.133		6.40	3.55		9.95	12.40	
	0600	3/8" thick			75	.213		6.95	5.65		12.60	16.30	
168	0010	SHEET GLASS Gray, 1/8" thick		2 Glaz	160	.100	S.F.	2.75	2.66		5.41	7.10	168
	0200	1/4" thick			130	.123	"	3.61	3.27		6.88	8.95	
172	0010	SPANDREL GLASS 1/4" thick, standard colors, up to 1,000 S.F.			110	.145	S.F.	9.45	3.87		13.32	16.30	172
	0200	1,000 to 2,000 S.F.			120	.133	"	8.85	3.55		12.40	15.15	
	0300	For custom colors, add					Total	10%					
	0500	For 3/8" thick, add					S.F.	5.75			5.75	6.35	
	1000	For double coated, 1/4" thick, add						2.15			2.15	2.37	
	1200	For insulation on panels, add						3.55			3.55	3.91	
	2000	Panels, insulated, with aluminum backed fiberglass, 1" thick		2 Glaz	120	.133		8.55	3.55		12.10	14.80	
	2100	2" thick		"	120	.133		9.95	3.55		13.50	16.35	
	2500	With galvanized steel backing, add						2.95			2.95	3.25	
176	0010	WINDOW GLASS Clear float, stops, putty bed, 1/8" thick [R088-100]		2 Glaz	480	.033	S.F.	2.75	.89		3.64	4.38	176
	0500	3/16" thick, clear			480	.033		3.40	.89		4.29	5.10	
	0600	Tinted			480	.033		3.85	.89		4.74	5.60	
	0700	Tempered			480	.033		4.66	.89		5.55	6.50	
180	0010	WINDOW WALL See division 089-200											180
184	0010	WIRE GLASS 1/4" thick, rough obscure (chicken wire)		2 Glaz	135	.119	S.F.	9.40	3.15		12.55	15.15	184
	1000	Polished wire, 1/4" thick, diamond, clear			135	.119		12.15	3.15		15.30	18.15	
	1500	Pinstripe, obscure			135	.119		11.40	3.15		14.55	17.30	

		088 400	Plastic Glazing										
404	0010	PLEXIGLASS ACRYLIC Clear, masked, 1/8" thick, cut sheets		2 Glaz	170	.094	S.F.	2.76	2.50		5.26	6.85	404
	0200	Full sheets			195	.082		1.44	2.18		3.62	4.90	
	0500	1/4" thick, cut sheets			165	.097		4.88	2.58		7.46	9.30	
	0600	Full sheets			185	.086		2.66	2.30		4.96	6.45	
	0900	3/8" thick, cut sheets			155	.103		8.95	2.75		11.70	14.05	
	1000	Full sheets			180	.089		4.82	2.36		7.18	8.90	
	1300	1/2" thick, cut sheets			135	.119		10.30	3.15		13.45	16.15	
	1400	Full sheets			150	.107		10	2.84		12.84	15.30	
	1700	3/4" thick, cut sheets			115	.139		36.50	3.70		40.20	45.50	
	1800	Full sheets			130	.123		21	3.27		24.27	28	
	2100	1" thick, cut sheets			105	.152		41	4.05		45.05	51.50	
	2200	Full sheets			125	.128		25.50	3.40		28.90	33	
	3000	Colored, 1/8" thick, cut sheets			170	.094		8.50	2.50		11	13.15	
	3200	Full sheets			195	.082		5.50	2.18		7.68	9.35	
	3500	1/4" thick, cut sheets			165	.097		9.50	2.58		12.08	14.40	
	3600	Full sheets			185	.086		6.50	2.30		8.80	10.65	
	4000	Mirrors, untinted, cut sheets, 1/8" thick			185	.086		4	2.30		6.30	7.90	
	4200	1/4" thick			180	.089		7.25	2.36		9.61	11.60	
408	0010	POLYCARBONATE Clear, masked, cut sheets, 1/8" thick		2 Glaz	170	.094	S.F.	4.75	2.50		7.25	9	408
	0500	3/16" thick			165	.097		5.75	2.58		8.33	10.30	

DOORS & WINDOWS   8

088 400 | Plastic Glazing

			CREW	DAILY OUTPUT	LABOR-HOURS	UNIT	1999 BARE COSTS				TOTAL INCL O&P	
							MAT.	LABOR	EQUIP.	TOTAL		
408	1000	1/4" thick	2 Glaz	155	.103	S.F.	6.35	2.75		9.10	11.20	**408**
	1500	3/8" thick	↓	150	.107	↓	11.70	2.84		14.54	17.20	
412	0010	**VINYL GLASS** Steel mesh reinforced, stock sizes, .090" thick	2 Glaz	170	.094	S.F.	6.65	2.50		9.15	11.10	**412**
	0500	.120" thick		170	.094		8.40	2.50		10.90	13.05	
	1000	.250" thick	↓	155	.103	↓	11.70	2.75		14.45	17.10	
	1500	For non-standard sizes, add					15%					

089 200 | Glazed Curtain Wall

			CREW	DAILY OUTPUT	LABOR-HOURS	UNIT	1999 BARE COSTS				TOTAL INCL O&P	
							MAT.	LABOR	EQUIP.	TOTAL		
202	0010	**CURTAIN WALLS** Aluminum, stock, including glazing, minimum	H-1	205	.156	S.F.	18.70	4.46		23.16	28	**202**
	0050	Average, single glazed		195	.164		25.50	4.69		30.19	36	
	0150	Average, double glazed		180	.178		35.50	5.10		40.60	47.50	
	0200	Maximum	↓	160	.200	↓	95.50	5.70		101.20	115	
204	0010	**TUBE FRAMING** For window walls and store fronts, aluminum, stock										**204**
	0050	Plain tube frame, mill finish, 1-3/4" x 1-3/4"	2 Glaz	103	.155	L.F.	6	4.13		10.13	12.90	
	0150	1-3/4" x 4"		98	.163		7.50	4.34		11.84	14.85	
	0200	1-3/4" x 4-1/2" **CN**		95	.168		8.35	4.48		12.83	15.95	
	0250	2" x 6"		89	.180		12.60	4.78		17.38	21	
	0350	4" x 4"		87	.184		12.35	4.89		17.24	21	
	0400	4-1/2" x 4-1/2"		85	.188		13.80	5		18.80	23	
	0450	Glass bead		240	.067		1.59	1.77		3.36	4.45	
	1000	Flush tube frame, mill finish, 1/4" glass, 1-3/4" x 4", open header		80	.200		7.40	5.30		12.70	16.20	
	1050	Open sill		82	.195		6.40	5.20		11.60	14.95	
	1100	Closed back header		83	.193		10.35	5.15		15.50	19.15	
	1150	Closed back sill		85	.188		9.80	5		14.80	18.40	
	1200	Vertical mullion, one piece		75	.213		10.95	5.65		16.60	20.50	
	1250	Two piece		73	.219		11.70	5.85		17.55	22	
	1300	90° or 180° vertical corner post		75	.213		18.40	5.65		24.05	29	
	1400	1-3/4" x 4-1/2", open header		80	.200		9	5.30		14.30	18	
	1450	Open sill		82	.195		7.45	5.20		12.65	16.05	
	1500	Closed back header		83	.193		11.40	5.15		16.55	20.50	
	1550	Closed back sill		85	.188		10.60	5		15.60	19.25	
	1600	Vertical mullion, one piece		75	.213		11.80	5.65		17.45	21.50	
	1650	Two piece		73	.219		12.50	5.85		18.35	22.50	
	1700	90° or 180° vertical corner post		75	.213		12.80	5.65		18.45	23	
	2000	Flush tube frame, mill fin. for ins. glass, 2" x 4-1/2", open header		75	.213		10.50	5.65		16.15	20	
	2050	Open sill		77	.208		9.20	5.55		14.75	18.50	
	2100	Closed back header		78	.205		11.45	5.45		16.90	21	
	2150	Closed back sill		80	.200		11.30	5.30		16.60	20.50	
	2200	Vertical mullion, one piece		70	.229		12.70	6.10		18.80	23	
	2250	Two piece		68	.235		13.55	6.25		19.80	24.50	
	2300	90° or 180° vertical corner post		70	.229		12.80	6.10		18.90	23.50	
	5000	Flush tube frame, mill fin., thermal brk., 2-1/4" x 4-1/2", open header		74	.216		11.55	5.75		17.30	21.50	
	5050	Open sill		75	.213		10.10	5.65		15.75	19.75	
	5100	Vertical mullion, one piece		69	.232		13.95	6.15		20.10	25	
	5150	Two piece		67	.239		14.90	6.35		21.25	26	
	5200	90° or 180° vertical corner post		69	.232		13.40	6.15		19.55	24	
	6980	Door stop (snap in)	↓	380	.042	↓	2.16	1.12		3.28	4.09	
	7000	For joints, 90°, clip type, add				Ea.	18.45			18.45	20.50	

8

DOORS & WINDOWS

Important: See the Reference Section for critical supporting data - Reference Nos., Crews, & City Cost Indexes

089 200	Glazed Curtain Wall	CREW	DAILY OUTPUT	LABOR-HOURS	UNIT	1999 BARE COSTS				TOTAL INCL O&P	
						MAT.	LABOR	EQUIP.	TOTAL		
204 7050	Screw spline joint, add				Ea.	13.90			13.90	15.30	**204**
7100	For joint other than 90°, add				↓	29			29	32	
8000	For bronze anodized aluminum, add					15%					
8020	For black finish, add					27%					
8050	For stainless steel materials, add					350%					
8100	For monumental grade, add					50%					
8150	For steel stiffener, add	2 Glaz	200	.080	L.F.	7.10	2.13		9.23	11.05	
8200	For 2 to 5 stories, add per story				Story		5%				
206 0010	**WINDOW WALLS** Aluminum, stock, including glazing, minimum	H-2	160	.150	S.F.	24.50	3.73		28.23	32.50	**206**
0050	Average		140	.171	"	29.50	4.27		33.77	39	
0100	Maximum	↓	110	.218	S.F.	93	5.45		98.45	110	
0500	For translucent sandwich wall systems, see div. 074-104										
0850	Cost of the above walls depends on material,										
0860	finish, repetition, and size of units.										
0870	The larger the opening, the lower the S.F. cost										
1200	Double glazed acoustical window wall for airports,										
1220	including 1" thick glass with 2" x 4-1/2" tube frame	H-2	40	.600	S.F.	60.50	14.95		75.45	89.50	

For information about Means Estimating Seminars, see yellow pages 11 and 12 in back of book

DOORS & WINDOWS

8

Division Notes

		CREW	DAILY OUTPUT	LABOR-HOURS	UNIT	1999 BARE COSTS				TOTAL INCL O&P
						MAT.	LABOR	EQUIP.	TOTAL	

Division 9
Finishes

Estimating Tips

General

- Room Finish Schedule: A complete set of plans should contain a room finish schedule. If one is not available, it would be well worth the time and effort to put one together. A room finish schedule should contain the room number, room name (for clarity), floor materials, base materials, wainscot materials, wainscot height, wall materials (for each wall), ceiling materials and special instructions.

- Surplus Finishes: Review the specifications to determine if there is any requirement to provide certain amounts of extra materials for the owner's maintenance department. In some cases the owner may require a substantial amount of materials, especially when it is a special order item or long lead time item.

092 Lath, Plaster & Gypsum Board

- Lath is estimated by the square yard for both gypsum and metal lath, plus usually 5% allowance for waste. Furring, channels and accessories are measured by the linear foot. An extra foot should be allowed for each accessory miter or stop.

- Plaster is also estimated by the square yard. Deductions for openings vary by preference, from zero deduction to 50% of all openings over 2 feet in width. Some estimators deduct a percentage of the total yardage for openings. The estimator should allow one extra square foot for each linear foot of horizontal interior or exterior angle located below the ceiling level. Also, double the areas of small radius work.

- Each room should be measured, perimeter times maximum wall height. Ceiling areas are equal to length times width.

- Drywall accessories, studs, track, and acoustical caulking are all measured by the linear foot. Drywall taping is figured by the square foot. Gypsum wallboard is estimated by the square foot. No material deductions should be made for door or window openings under 32 S.F. Coreboard can be

obtained in a 1″ thickness for solid wall and shaft work. Additions should be made to price out the inside or outside corners.

- Different types of partition construction should be listed separately on the quantity sheets. There may be walls with studs of various widths, double studded, and similar or dissimilar surface materials. Shaft work is usually different construction from surrounding partitions requiring separate quantities and pricing of the work.

093 Tile
094 Terrazzo

- Tile and terrazzo areas are taken off on a square foot basis. Trim and base materials are measured by the linear foot. Accent tiles are listed per each. Two basic methods of installation are used. Mud set is approximately 30% more expensive than the thin set. In terrazzo work, be sure to include the linear footage of embedded decorative strips, grounds, machine rubbing and power cleanup.

095 Acoustical Treatment & Wood Flooring

- Acoustical systems fall into several categories. The takeoff of these materials is by the square foot of area with a 5% allowance for waste. Do not forget about scaffolding, if applicable, when estimating these systems.

- Wood flooring is available in strip, parquet, or block configuration. The latter two types are set in adhesives with quantities estimated by the square foot. The laying pattern will influence labor costs and material waste. In addition to the material and labor for laying wood floors, the estimator must make allowances for sanding and finishing these areas unless the flooring is prefinished.

096 Flooring & Carpet

- Most of the various types of flooring are all measured on a square foot basis. Base is measured by the linear foot. If adhesive materials are to be quantified, they are estimated at a specified coverage rate by the gallon depending upon the specified type and the manufacturer's recommendations.

- Sheet flooring is measured by the square yard. Roll widths vary, so consideration should be given to use the most economical width, as waste must be figured into the total quantity. Consider also the installation methods available, direct glue down or stretched.

099 Painting & Wall Coverings

- Painting is one area where bids vary to a greater extent than almost any other section of a project. This arises from the many methods of measuring surfaces to be painted. The estimator should check the plans and specifications carefully to be sure of the required number of coats.

- Protection of adjacent surfaces is not included in painting costs. When considering the method of paint application, an important factor is the amount of protection and masking required. These must be estimated separately and may be the determining factor in choosing the method of application.

- Wall coverings are estimated by the square foot. The area to be covered is measured, length by height of wall above baseboards, to calculate the square footage of each wall. This figure is divided by the number of square feet in the single roll which is being used. Deduct, in full, the areas of openings such as doors and windows. Where a pattern match is required allow 25%-30% waste. One gallon of paste should be sufficient to hang 12 single rolls of light to medium weight paper.

Reference Numbers

Reference numbers are shown in bold squares at the beginning of some major classifications. These numbers refer to related items in the Reference Section. The reference information may be an estimating procedure, an alternate pricing method or technical information.

Note: Not all subdivisions listed here necessarily appear in this publication.

091 300 | Suspension Systems

		CREW	DAILY OUTPUT	LABOR-HOURS	UNIT	1999 BARE COSTS				TOTAL INCL O&P
						MAT.	LABOR	EQUIP.	TOTAL	
304	0010	**SUSPENSION SYSTEMS** For boards and tile									304
	0050	Class A suspension system, 15/16" T bar, 2' x 4' grid **CN**	1 Carp	800	.010	S.F.	.31	.27		.58	.77
	0300	2' x 2' grid	"	650	.012		.39	.34		.73	.96
	0350	For 9/16" grid, add					.13			.13	.14
	0360	For fire rated grid, add					.07			.07	.08
	0370	For colored grid, add					.15			.15	.17
	0400	Concealed Z bar suspension system, 12" module	1 Carp	520	.015		.34	.42		.76	1.03
	0600	1-1/2" carrier channels, 4' O.C., add	"	470	.017		.17	.46		.63	.92
	0700	Carrier channels for ceilings with									
	0900	recessed lighting fixtures, add	1 Carp	460	.017	S.F.	.33	.47		.80	1.11
	1040	Hanging wire, 12 ga., 4' long		65	.123	C.S.F.	1.47	3.36		4.83	6.90
	1080	8' long		65	.123	"	2.96	3.36		6.32	8.55
	1200	Seismic bracing, 2-1/2" compression post with four bracing wires		50	.160	Ea.	2.06	4.37		6.43	9.10

092 050 | Furring & Lathing

		CREW	DAILY OUTPUT	LABOR-HOURS	UNIT	1999 BARE COSTS				TOTAL INCL O&P
						MAT.	LABOR	EQUIP.	TOTAL	
052	0010	**ACCESSORIES, PLASTER** Casing bead, expanded flange, galvanized	1 Lath	2.70	2.963	C.L.F.	36	79.50		115.50	159	052
	0100	Zinc alloy	"	2.70	2.963		48	79.50		127.50	172
	0900	Channels, cold rolled, 16 ga., 3/4" deep, galvanized					19.05			19.05	21
	1200	1-1/2" deep, 16 ga., galvanized					26			26	28.50
	1620	Corner bead, expanded bullnose, 3/4" radius, #10, galvanized	1 Lath	2.60	3.077		57	82.50		139.50	187
	1640	Zinc alloy		2.70	2.963		61	79.50		140.50	186
	1650	#1, galvanized		2.55	3.137		35	84		119	165
	1660	Zinc alloy		2.70	2.963		58	79.50		137.50	183
	1670	Expanded wing, 2-3/4" wide, galv. #1		2.65	3.019		50	81		131	176
	1680	Zinc alloy		2.70	2.963		59.50	79.50		139	185
	1700	Inside corner, (corner rite) 3" x 3", painted		2.60	3.077		26.50	82.50		109	153
	1750	Strip-ex, 4" wide, painted		2.55	3.137		17.20	84		101.20	145
	1800	Expansion joint, 3/4" grounds, limited expansion, galv., 1 piece		2.70	2.963		47.50	79.50		127	171
	1900	Zinc alloy		2.70	2.963		68.50	79.50		148	195
	2100	Extreme expansion, galvanized, 2 piece		2.60	3.077		78.50	82.50		161	211
	2300	Zinc alloy		2.70	2.963		144	79.50		223.50	277
	2500	Joist clips for lath, 2-1/2" flange		1.90	4.211	M	73.50	113		186.50	250
	2600	4-1/2" flange		1.80	4.444	"	83.50	119		202.50	270
	2800	Metal base, galvanized and painted, 2-1/2" high		2.40	3.333	C.L.F.	44	89.50		133.50	183
	2900	Stud clips for gypsum lath, field clip		2.35	3.404	M	71	91		162	215
	3100	Resilient		2.30	3.478		106	93		199	257
	3200	Starter/finisher		2.20	3.636		72.50	97.50		170	226
054	0010	**FURRING** Beams & columns, 3/4" galvanized channels,									054
	0030	12" O.C.	1 Lath	155	.052	S.F.	.21	1.38		1.59	2.30
	0050	16" O.C.		170	.047		.17	1.26		1.43	2.08
	0070	24" O.C.		185	.043		.11	1.16		1.27	1.87
	0100	Ceilings, on steel, 3/4" channels, galvanized, 12" O.C.		210	.038		.19	1.02		1.21	1.74
	0300	16" O.C.		290	.028		.17	.74		.91	1.30
	0400	24" O.C.		420	.019		.11	.51		.62	.89
	0600	1-1/2" channels, galvanized, 12" O.C.		190	.042		.26	1.13		1.39	1.98
	0700	16" O.C.		260	.031		.23	.82		1.05	1.50
	0900	24" O.C.		390	.021		.16	.55		.71	.99

Important: See the Reference Section for critical supporting data - Reference Nos., Crews, & City Cost Indexes

			DAILY	LABOR-			1999 BARE COSTS				TOTAL	
	092 050 | Furring & Lathing	CREW	OUTPUT	HOURS	UNIT	MAT.	LABOR	EQUIP.	TOTAL		INCL O&P	
054	1000	Walls, 3/4" channels, galvanized, 12" O.C.	1 Lath	235	.034	S.F.	.19	.91		1.10		1.58	**054**
	1200	16" O.C.		265	.030		.17	.81		.98		1.40	
	1300	24" O.C.		350	.023		.11	.61		.72		1.05	
	1500	1-1/2" channels, galvanized, 12" O.C.		210	.038		.26	1.02		1.28		1.82	
	1600	16" O.C.		240	.033		.23	.89		1.12		1.60	
	1800	24" O.C.		305	.026		.16	.70		.86		1.22	
	8000	Suspended ceilings, including carriers											
	8200	1-1/2" carriers, 24" O.C. with:											
	8300	3/4" channels, 16" O.C.	1 Lath	165	.048	S.F.	.51	1.30		1.81		2.51	
	8320	24" O.C.		200	.040		.45	1.07		1.52		2.11	
	8400	1-1/2" channels, 16" O.C.		155	.052		.57	1.38		1.95		2.70	
	8420	24" O.C.		190	.042		.50	1.13		1.63		2.24	
	8600	2" carriers, 24" O.C. with:											
	8700	3/4" channels, 16" O.C.	1 Lath	155	.052	S.F.	.17	1.38		1.55		2.26	
	8720	24" O.C.		190	.042		.12	1.13		1.25		1.82	
	8800	1-1/2" channels, 16" O.C.		145	.055		.59	1.48		2.07		2.87	
	8820	24" O.C.		180	.044		.52	1.19		1.71		2.35	
056	0010	**GYPSUM LATH** Plain or perforated, nailed, 3/8" thick [R092 -050]	1 Lath	85	.094	S.Y.	3.38	2.52		5.90		7.50	**056**
	0100	1/2" thick, nailed		80	.100		3.96	2.68		6.64		8.40	
	0300	Clipped to steel studs, 3/8" thick		75	.107		3.78	2.86		6.64		8.45	
	0400	1/2" thick		70	.114		3.96	3.06		7.02		8.95	
	0600	Firestop gypsum base, to steel studs, 3/8" thick		70	.114		3.96	3.06		7.02		8.95	
	0700	1/2" thick		65	.123		4.05	3.30		7.35		9.40	
	0900	Foil back, to steel studs, 3/8" thick		75	.107		3.69	2.86		6.55		8.35	
	1000	1/2" thick		70	.114		3.78	3.06		6.84		8.75	
	1500	For ceiling installations, add		216	.037			.99		.99		1.49	
	1600	For columns and beams, add		170	.047			1.26		1.26		1.89	
058	0010	**METAL LATH** Diamond, expanded, 2.5 lb. per S.Y., painted [R092 -060]				S.Y.	1.63			1.63		1.79	**058**
	0100	Galvanized, 2.5 lb. per S.Y.					1.87			1.87		2.06	
	0300	3.4 lb. per S.Y., painted					1.91			1.91		2.10	
	0400	Galvanized					2.10			2.10		2.31	
	0600	For 15# asphalt sheathing paper, add					.24			.24		.27	
	0900	Flat rib, 1/8" high, 2.75 lb., painted					1.54			1.54		1.69	
	1000	Foil backed					2			2		2.20	
	1200	3.4 lb. per S.Y., painted					1.87			1.87		2.05	
	1300	Galvanized					2.75			2.75		3.03	
	1500	For 15# asphalt sheating paper, add					.24			.24		.27	
	1800	High rib, 3/8" high, 3.4 lb. per S.Y., painted					2.89			2.89		3.18	
	1900	Galvanized					3.73			3.73		4.10	
	2400	High rib, 3/4" high, painted, .60 lb. per S.F.				S.F.	.53			.53		.58	
	2500	.75 lb. per S.F.				"	.66			.66		.73	
	2700	Stucco mesh, painted, 1.8 lb.				S.Y.	3.03			3.03		3.33	
	2800	3.6 lb.					3.36			3.36		3.70	
	3000	K-lath, perforated, absorbent paper, regular					2.33			2.33		2.56	
	3100	Heavy duty					2.75			2.75		3.03	
	3300	Waterproof, heavy duty, grade B backing					2.70			2.70		2.97	
	3400	Fire resistant backing					2.98			2.98		3.28	
	3600	2.5 lb. diamond painted, on wood framing, on walls	1 Lath	85	.094		1.63	2.52		4.15		5.55	
	3700	On ceilings		75	.107		1.63	2.86		4.49		6.05	
	3900	3.4 lb. diamond painted, on wood framing, on walls		80	.100		1.87	2.68		4.55		6.05	
	4000	On ceilings		70	.114		1.87	3.06		4.93		6.65	
	4200	3.4 lb. diamond painted, wired to steel framing		75	.107		1.87	2.86		4.73		6.35	
	4300	On ceilings		60	.133		1.87	3.57		5.44		7.40	
	4500	Columns and beams, wired to steel		40	.200		1.87	5.35		7.22		10.10	
	4600	Cornices, wired to steel		35	.229		1.87	6.15		8.02		11.25	

9

FINISHES

			CREW	DAILY OUTPUT	LABOR-HOURS	UNIT	1999 BARE COSTS				TOTAL INCL O&P	
		**092 050	Furring & Lathing**					MAT.	LABOR	EQUIP.	TOTAL		
058	4800	Screwed to steel studs, 2.5 lb.	R092 -060	1 Lath	80	.100	S.Y.	1.63	2.68		4.31	5.80	058
	4900	3.4 lb.			75	.107		1.91	2.86		4.77	6.40	
	5100	Rib lath, painted, wired to steel, on walls, 2.5 lb.			75	.107		1.54	2.86		4.40	5.95	
	5200	3.4 lb.			70	.114		2.89	3.06		5.95	7.75	
	5400	4.0 lb.			65	.123		2.52	3.30		5.82	7.70	
	5500	For self-furring lath, add						.06			.06	.07	
	5700	Suspended ceiling system, incl. 3.4 lb. diamond lath, painted		1 Lath	15	.533		8.65	14.30		22.95	31	
	5800	Galvanized		"	15	.533		8.90	14.30		23.20	31.50	
	6000	Hollow metal stud partitions, 3.4 lb. painted lath both sides											
	6010	Non-load bearing, 25 ga., w/rib lath 2-1/2" studs, 12" O.C.		1 Lath	20.30	.394	S.Y.	7.80	10.55		18.35	24.50	
	6300	16" O.C.			21.10	.379		7.30	10.15		17.45	23	
	6350	24" O.C.			22.70	.352		6.90	9.45		16.35	22	
	6400	3-5/8" studs, 16" O.C.			19.50	.410		7.50	11		18.50	24.50	
	6600	24" O.C.			20.40	.392		7.05	10.50		17.55	23.50	
	6700	4" studs, 16" O.C.			20.40	.392		8.25	10.50		18.75	25	
	6900	24" O.C.			21.60	.370		7.25	9.95		17.20	23	
	7000	6" studs, 16" O.C.			19.50	.410		8.25	11		19.25	25.50	
	7100	24" O.C.			21.10	.379		7.65	10.15		17.80	23.50	
	7200	L.B. partitions, 16 ga., w/rib lath, 2-1/2" studs, 16" O.C.			20	.400		9.20	10.70		19.90	26	
	7300	3-5/8" studs, 16 ga.			19.70	.406		10.55	10.90		21.45	28	
	7500	4" studs, 16 ga.			19.50	.410		11	11		22	28.50	
	7600	6" studs, 16 ga.			18.70	.428		12.65	11.45		24.10	31	

		**092 100	Gypsum Plaster**										
104	0010	**GAUGING PLASTER** 100 lb. bags, less than 1 ton				Bag	14.75			14.75	16.25	104
	0100	Over 1 ton				"	14			14	15.40	
108	0010	**GYPSUM PLASTER** 80# bag, less than 1 ton	R092 -105			Bag	13			13	14.30	108
	0100	Over 1 ton				"	11.90			11.90	13.10	
	0300	2 coats, no lath included, on walls	J-1	105	.381	S.Y.	3.15	9.15	.40	12.70	17.95	
	0400	On ceilings	"	92	.435		3.15	10.45	.46	14.06	20	
	0600	2 coats on and incl. 3/8" gypsum lath on steel, on walls	J-2	97	.495		6.95	12.15	.44	19.54	26.50	
	0700	On ceilings	"	83	.578		6.95	14.20	.51	21.66	29.50	
	0900	3 coats, no lath included, on walls	J-1	87	.460		4.32	11.05	.49	15.86	22	
	1000	On ceilings	"	78	.513		4.32	12.35	.54	17.21	24.50	
	1200	3 coats on and including painted metal lath, on wood studs	J-2	86	.558		6.35	13.70	.49	20.54	28.50	
	1300	On ceilings	"	76.50	.627		6.35	15.40	.55	22.30	31	
	1600	For irregular or curved surfaces, add						30%				
	1800	For columns & beams, add						50%				
112	0010	**KEENES CEMENT** In 100 lb. bags, less than 1 ton				Bag	20.50			20.50	23	112
	0100	Over 1 ton				"	20			20	22	
	0300	Finish only, add to plaster prices, standard	J-1	215	.186	S.Y.	1.93	4.48	.20	6.61	9.20	
	0400	High quality	"	144	.278	"	1.96	6.70	.29	8.95	12.75	
116	0010	**PERLITE OR VERMICULITE PLASTER** 100 lb. bags Under 200 bags				Bag	15.20			15.20	16.70	116
	0100	Over 200 bags				"	14			14	15.40	
	0300	2 coats, no lath included, on walls	R092 -115	J-1	92	.435	S.Y.	3.78	10.45	.46	14.69	20.50	
	0400	On ceilings		"	79	.506		3.78	12.20	.54	16.52	23.50	
	0600	2 coats, on and incl. 3/8" gypsum lath, on metal studs		J-2	84	.571		7.50	14	.50	22	30.50	
	0700	On ceilings		"	70	.686		7.50	16.80	.60	24.90	34.50	
	0900	3 coats, no lath included, on walls		J-1	74	.541		6.10	13	.57	19.67	27.50	
	1000	On ceilings		"	63	.635		6.10	15.30	.67	22.07	31	
	1200	3 coats, on and incl. painted metal lath, on metal studs		J-2	72	.667		9	16.35	.59	25.94	35.50	
	1300	On ceilings		"	61	.787		9	19.30	.69	28.99	40	
	1500	3 coats, on and incl. suspended metal lath ceiling			37	1.297		14.75	32	1.14	47.89	66	
	1700	For irregular or curved surfaces, add to above						30%				
	1800	For columns and beams, add to above						50%				
	1900	For soffits, add to ceiling prices						40%				

9 FINISHES

Important: See the Reference Section for critical supporting data - Reference Nos., Crews, & City Cost Indexes

092 | Lath, Plaster & Gypsum Board

		092 150	Veneer Plaster	CREW	DAILY OUTPUT	LABOR-HOURS	UNIT	1999 BARE COSTS				TOTAL INCL O&P	
								MAT.	LABOR	EQUIP.	TOTAL		
154	0010	THIN COAT Plaster, 1 coat veneer, not incl. lath		J-1	3,600	.011	S.F.	.10	.27	.01	.38	.53	154
	1000	In 50 lb. bags					Bag	7.75			7.75	8.55	

		092 300	Aggregate Coatings	CREW	DAILY OUTPUT	LABOR-HOURS	UNIT	MAT.	LABOR	EQUIP.	TOTAL	TOTAL INCL O&P	
304	0010	STUCCO, 3 coats 1" thick, float finish, with mesh, on wood frame R092 -300		J-2	135	.356	S.Y.	5.40	8.70	.31	14.41	19.60	304
	0100	On masonry construction, no mesh incl.		J-1	200	.200		2.04	4.81	.21	7.06	9.85	
	0300	For trowel finish, add		1 Plas	170	.047			1.21		1.21	1.85	
	0400	For 3/4" thick, on masonry, deduct		J-1	880	.045		.56	1.09	.05	1.70	2.35	
	0600	For coloring and special finish, add, minimum			685	.058		.36	1.40	.06	1.82	2.62	
	0700	Maximum		↓	200	.200		1.26	4.81	.21	6.28	9	
	0900	For soffits, add		J-2	155	.310		1.94	7.60	.27	9.81	14.05	
	1000	Exterior stucco, with bonding agent, 3 coats, on walls, no mesh incl.		J-1	200	.200		3.25	4.81	.21	8.27	11.20	
	1200	Ceilings			180	.222		3.25	5.35	.24	8.84	12.05	
	1300	Beams			80	.500		3.25	12.05	.53	15.83	22.50	
	1500	Columns		↓	100	.400		3.25	9.60	.42	13.27	18.80	
	1600	Mesh, painted, nailed to wood, 1.8 lb.		1 Lath	60	.133		3.03	3.57		6.60	8.70	
	1800	3.6 lb.			55	.145		3.36	3.90		7.26	9.55	
	1900	Wired to steel, painted, 1.8 lb.			53	.151		3.03	4.05		7.08	9.40	
	2100	3.6 lb.		↓	50	.160	↓	3.36	4.29		7.65	10.10	

		092 600	Gypsum Board Systems	CREW	DAILY OUTPUT	LABOR-HOURS	UNIT	MAT.	LABOR	EQUIP.	TOTAL	TOTAL INCL O&P	
602	0010	BLUEBOARD For use with thin coat											602
	0100	plaster application (see division 092-154)											
	1000	3/8" thick, on walls or ceilings, standard, no finish included		2 Carp	1,900	.008	S.F.	.19	.23		.42	.57	
	1100	With thin coat plaster finish			875	.018		.29	.50		.79	1.11	
	1400	On beams, columns, or soffits, standard, no finish included			675	.024		.22	.65		.87	1.26	
	1450	With thin coat plaster finish			475	.034		.33	.92		1.25	1.82	
	3000	1/2" thick, on walls or ceilings, standard, no finish included			1,900	.008		.19	.23		.42	.57	
	3100	With thin coat plaster finish			875	.018		.29	.50		.79	1.11	
	3300	Fire resistant, no finish included			1,900	.008		.19	.23		.42	.57	
	3400	With thin coat plaster finish			875	.018		.29	.50		.79	1.11	
	3450	On beams, columns, or soffits, standard, no finish included			675	.024		.22	.65		.87	1.26	
	3500	With thin coat plaster finish			475	.034		.33	.92		1.25	1.82	
	3700	Fire resistant, no finish included			675	.024		.22	.65		.87	1.26	
	3800	With thin coat plaster finish			475	.034		.33	.92		1.25	1.82	
	5000	5/8" thick, on walls or ceilings, fire resistant, no finish included			1,900	.008		.25	.23		.48	.64	
	5100	With thin coat plaster finish			875	.018		.35	.50		.85	1.18	
	5500	On beams, columns, or soffits, no finish included			675	.024		.29	.65		.94	1.34	
	5600	With thin coat plaster finish			475	.034		.40	.92		1.32	1.89	
	6000	For high ceilings, over 8' high, add			3,060	.005		.10	.14		.24	.33	
	6500	For over 3 stories high, add per story		↓	6,100	.003	↓	.05	.07		.12	.17	
604	0010	CEILINGS Gypsum drywall, fire rated, finished											604
	0100	Screwed to grid, channel or joists, 1/2" thick		2 Carp	765	.021	S.F.	.24	.57		.81	1.16	
	0200	5/8" thick			765	.021		.21	.57		.78	1.13	
	0300	Over 8' high, 1/2" thick			615	.026		.24	.71		.95	1.38	
	0400	5/8" thick		↓	615	.026	↓	.21	.71		.92	1.35	
	0600	Grid suspension system, direct hung											
	0700	1-1/2" C.R.C., with 7/8" hi hat furring channel, 16" O.C.		2 Carp	600	.027	S.F.	.73	.73		1.46	1.95	
	0800	24" O.C.			900	.018		.67	.49		1.16	1.50	
	0900	3-5/8" C.R.C., with 7/8" hi hat furring channel, 16" O.C.			600	.027		.77	.73		1.50	2	
	1000	24" O.C.		↓	900	.018	↓	.68	.49		1.17	1.51	
608	0010	DRYWALL Gypsum plasterboard, nailed or screwed to studs											608
	0100	unless otherwise noted											
	0150	3/8" thick, on walls, standard, no finish included		2 Carp	2,000	.008	S.F.	.16	.22		.38	.52	
	0200	On ceilings, standard, no finish included		↓	1,800	.009	↓	.16	.24		.40	.56	

9

FINISHES

For expanded coverage of these items see *Means Interior Cost Data 1999*

092 600	Gypsum Board Systems	CREW	DAILY OUTPUT	LABOR-HOURS	UNIT	1999 BARE COSTS				TOTAL INCL O&P
						MAT.	LABOR	EQUIP.	TOTAL	
608	0250	On beams, columns, or soffits, no finish included	2 Carp	675	.024	S.F.	.19	.65		.84	1.23	608
	0300	1/2" thick, on walls, standard, no finish included **CN**		2,000	.008		.15	.22		.37	.51
	0350	Taped and finished		965	.017		.24	.45		.69	.98
	0400	Fire resistant, no finish included		2,000	.008		.24	.22		.46	.60
	0450	Taped and finished		965	.017		.33	.45		.78	1.07
	0500	Water resistant, no finish included		2,000	.008		.23	.22		.45	.59
	0550	Taped and finished		965	.017		.32	.45		.77	1.06
	0600	Prefinished, vinyl, clipped to studs		900	.018		.56	.49		1.05	1.38
	1000	On ceilings, standard, no finish included		1,800	.009		.15	.24		.39	.55
	1050	Taped and finished		765	.021		.24	.57		.81	1.17
	1100	Fire resistant, no finish included		1,800	.009		.24	.24		.48	.64
	1150	Taped and finished		765	.021		.33	.57		.90	1.26
	1200	Water resistant, no finish included		1,800	.009		.23	.24		.47	.63
	1250	Taped and finished		765	.021		.32	.57		.89	1.25
	1500	On beams, columns, or soffits, standard, no finish included		675	.024		.18	.65		.83	1.21
	1550	Taped and finished		475	.034		.28	.92		1.20	1.76
	1600	Fire resistant, no finish included		675	.024		.28	.65		.93	1.32
	1650	Taped and finished		475	.034		.38	.92		1.30	1.87
	1700	Water resistant, no finish included		675	.024		.26	.65		.91	1.31
	1750	Taped and finished		475	.034		.37	.92		1.29	1.85
	2000	5/8" thick, on walls, standard, no finish included		2,000	.008		.24	.22		.46	.60
	2050	Taped and finished		965	.017		.33	.45		.78	1.07
	2100	Fire resistant, no finish included		2,000	.008		.21	.22		.43	.57
	2150	Taped and finished		965	.017		.30	.45		.75	1.04
	2200	Water resistant, no finish included		2,000	.008		.29	.22		.51	.66
	2250	Taped and finished		965	.017		.38	.45		.83	1.13
	2300	Prefinished, vinyl, clipped to studs		900	.018		.65	.49		1.14	1.47
	3000	On ceilings, standard, no finish included		1,800	.009		.24	.24		.48	.64
	3050	Taped and finished		765	.021		.33	.57		.90	1.26
	3100	Fire resistant, no finish included		1,800	.009		.21	.24		.45	.61
	3150	Taped and finished		765	.021		.30	.57		.87	1.23
	3200	Water resistant, no finish included		1,800	.009		.29	.24		.53	.70
	3250	Taped and finished		765	.021		.38	.57		.95	1.32
	3500	On beams, columns, or soffits, no finish included		675	.024		.27	.65		.92	1.32
	3550	Taped and finished		475	.034		.38	.92		1.30	1.86
	3600	Fire resistant, no finish included		675	.024		.24	.65		.89	1.29
	3650	Taped and finished		475	.034		.35	.92		1.27	1.83
	3700	Water resistant, no finish included		675	.024		.33	.65		.98	1.39
	3750	Taped and finished		475	.034		.44	.92		1.36	1.93
	4000	Fireproofing, beams or columns, 2 layers, 1/2" thick, incl finish		330	.048		.57	1.32		1.89	2.71
	4050	5/8" thick		300	.053		.60	1.46		2.06	2.95
	4100	3 layers, 1/2" thick		225	.071		.99	1.94		2.93	4.14
	4150	5/8" thick		210	.076		.90	2.08		2.98	4.26
	5050	For 1" thick coreboard on columns	↓	480	.033		.53	.91		1.44	2.01
	5100	For foil-backed board, add					.07			.07	.08
	5200	For high ceilings, over 8' high, add	2 Carp	3,060	.005			.14		.14	.22
	5270	For textured spray, add	2 Lath	1,600	.010	↓	.12	.27		.39	.53
	5300	For over 3 stories high, add per story	2 Carp	6,100	.003	↓		.07		.07	.11
	5350	For finishing corners, inside or outside, add	"	1,100	.015	L.F.	.06	.40		.46	.69
	5500	For acoustical sealant, add per bead	1 Carp	500	.016	"	.03	.44		.47	.72
	5550	Sealant, 1 quart tube				Ea.	4.75			4.75	5.20
	5600	Sound deadening board, 1/4" gypsum	2 Carp	1,800	.009	S.F.	.18	.24		.42	.57
	5650	1/2" wood fiber	"	1,800	.009	"	.26	.24		.50	.67
612	0010	**METAL STUDS, DRYWALL** Partitions, 10' high, with runners R092 -610										612
	0050	See also Studding, division 051-230									

9

FINISHES

274

092 600	Gypsum Board Systems		CREW	DAILY OUTPUT	LABOR-HOURS	UNIT	1999 BARE COSTS				TOTAL INCL O&P	
							MAT.	LABOR	EQUIP.	TOTAL		
612	2000	Non-load bearing, galvanized, 25 ga. 1-5/8" wide, 16" O.C. R092 -610	1 Carp	450	.018	S.F.	.15	.49		.64	.92	**612**
	2100	24" O.C.		520	.015		.12	.42		.54	.79	
	2200	2-1/2" wide, 16" O.C.		440	.018		.17	.50		.67	.96	
	2250	24" O.C.		510	.016		.13	.43		.56	.82	
	2300	3-5/8" wide, 16" O.C.		430	.019		.19	.51		.70	1.01	
	2350	24" O.C.		500	.016		.16	.44		.60	.87	
	2400	4" wide, 16" O.C.		420	.019		.22	.52		.74	1.06	
	2450	24" O.C.		490	.016		.18	.45		.63	.89	
	2500	6" wide, 16" O.C.		410	.020		.27	.53		.80	1.14	
	2550	24" O.C.		480	.017		.22	.45		.67	.96	
	2600	20 ga. studs, 1-5/8" wide, 16" O.C.		450	.018		.41	.49		.90	1.21	
	2650	24" O.C.		520	.015		.33	.42		.75	1.02	
	2700	2-1/2" wide, 16" O.C.		440	.018		.44	.50		.94	1.26	
	2750	24" O.C.		510	.016		.35	.43		.78	1.06	
	2800	3-5/8" wide, 16" O.C. CN		430	.019		.54	.51		1.05	1.39	
	2850	24" O.C.		500	.016		.43	.44		.87	1.17	
	2900	4" wide, 16" O.C.		420	.019		.56	.52		1.08	1.44	
	2950	24" O.C.		490	.016		.45	.45		.90	1.19	
	3000	6" wide, 16" O.C.		410	.020		.72	.53		1.25	1.63	
	3050	24" O.C.	▼	480	.017	▼	.58	.45		1.03	1.35	
	5000	Load bearing studs, see division 054-138	▼									
620	0010	**PARTITION WALL** Stud wall, 8' to 12' high										**620**
	0050	1/2", interior, gypsum drywall, standard, taped both sides										
	0500	Installed on and incl., 2" x 4" wood studs, 16" O.C.	2 Carp	310	.052	S.F.	.89	1.41		2.30	3.20	
	1000	Metal studs, NLB, 25 ga., 16" O.C., 3-5/8" wide		350	.046		.68	1.25		1.93	2.70	
	1200	6" wide		330	.048		.76	1.32		2.08	2.92	
	1400	Water resistant, on 2" x 4" wood studs, 16" O.C.		310	.052		1.05	1.41		2.46	3.37	
	1600	Metal studs, NLB, 25 ga., 16" O.C., 3-5/8" wide		350	.046		.83	1.25		2.08	2.87	
	1800	6" wide		330	.048		.91	1.32		2.23	3.08	
	2000	Fire res., 2 layers, 1-1/2 hr., on 2" x 4" wood studs, 16" O.C.		210	.076		1.55	2.08		3.63	4.97	
	2200	Metal studs, NLB, 25 ga., 16" O.C., 3-5/8" wide		250	.064		1.33	1.75		3.08	4.21	
	2400	6" wide		230	.070		1.41	1.90		3.31	4.54	
	2600	Fire & water res., 2 layers, 1-1/2 hr., 2"x4" studs, 16" O.C.		210	.076		1.55	2.08		3.63	4.97	
	2800	Metal studs, NLB, 25 ga., 16" O.C., 3-5/8" wide		250	.064		1.33	1.75		3.08	4.21	
	3000	6" wide	▼	230	.070	▼	1.41	1.90		3.31	4.54	
	3200	5/8", interior, gypsum drywall, standard, taped both sides										
	3400	Installed on and including 2" x 4" wood studs, 16" O.C.	2 Carp	300	.053	S.F.	1.06	1.46		2.52	3.45	
	3600	24" O.C.		330	.048		.97	1.32		2.29	3.15	
	3800	Metal studs, NLB, 25 ga., 16" O.C., 3-5/8" wide		340	.047		.84	1.28		2.12	2.95	
	4000	6" wide		320	.050		.93	1.36		2.29	3.17	
	4200	24" O.C., 3-5/8" wide		360	.044		.84	1.21		2.05	2.84	
	4400	6" wide		340	.047		.93	1.28		2.21	3.04	
	4800	Water resistant, on 2" x 4" wood studs, 16" O.C.		300	.053		1.17	1.46		2.63	3.57	
	5000	24" O.C.		330	.048		1.08	1.32		2.40	3.26	
	5200	Metal studs, NLB, 25 ga. 16" O.C., 3-5/8" wide		340	.047		.95	1.28		2.23	3.07	
	5400	6" wide		320	.050		1.03	1.36		2.39	3.29	
	5600	24" O.C., 3-5/8" wide		360	.044		.95	1.21		2.16	2.96	
	5800	6" wide		340	.047		1.03	1.28		2.31	3.16	
	6000	Fire res., 2 layers, 2 hr., on 2" x 4" wood studs, 16" O.C.		205	.078		1.33	2.13		3.46	4.82	
	6200	24" O.C.		235	.068		1.33	1.86		3.19	4.39	
	6400	Metal studs, NLB, 25 ga., 16" O.C., 3-5/8" wide		245	.065		1.21	1.78		2.99	4.13	
	6600	6" wide		225	.071		1.29	1.94		3.23	4.47	
	6800	24" O.C., 3-5/8" wide		265	.060		1.21	1.65		2.86	3.92	
	7000	6" wide		245	.065		1.29	1.78		3.07	4.22	
	7200	Fire & water res., 2 layers, 2 hr., 2" x 4" studs, 16" O.C.	▼	205	.078	▼	1.43	2.13		3.56	4.92	

9

FINISHES

For expanded coverage of these items see *Means Interior Cost Data 1999*

9 FINISHES

092 600 | Gypsum Board Systems

		Description	CREW	DAILY OUTPUT	LABOR-HOURS	UNIT	MAT.	LABOR	EQUIP.	TOTAL	TOTAL INCL O&P	
							1999 BARE COSTS					
620	7400	24" O.C.	2 Carp	235	.068	S.F.	1.33	1.86		3.19	4.39	620
	7600	Metal studs, NLB, 25 ga., 16" O.C., 3-5/8" wide		245	.065		1.21	1.78		2.99	4.13	
	7800	6" wide		225	.071		1.29	1.94		3.23	4.47	
	8000	24" O.C., 3-5/8" wide		265	.060		1.21	1.65		2.86	3.92	
	8200	6" wide		245	.065		1.29	1.78		3.07	4.22	
	8600	1/2" blueboard, mesh tape both sides										
	8620	Installed on and including 2" x 4" wood studs, 16" O.C.	2 Carp	300	.053	S.F.	.99	1.46		2.45	3.37	
	8640	Metal studs, NLB, 25 ga., 16" O.C., 3-5/8" wide		340	.047		.77	1.28		2.05	2.87	
	8660	6" wide		320	.050		.85	1.36		2.21	3.09	
	9000	Exterior, 1/2" gypsum sheathing, 1/2" gypsum finished, interior,										
	9100	including foil faced insulation, metal studs, 20 ga.										
	9200	16" O.C., 3-5/8" wide	2 Carp	290	.055	S.F.	1.28	1.51		2.79	3.78	
	9400	6" wide	"	270	.059	"	1.28	1.62		2.90	3.96	
622	0010	**PLASTER PARTITION WALL**										622
	0400	Stud walls, 3.4 lb. metal lath, 3 coat gypsum plaster, 2 sides										
	0600	2" x 4" wood studs, 16" O.C.	J-2	315	.152	S.F.	2.29	3.74	.13	6.16	8.35	
	0700	2-1/2" metal studs, 25 ga., 12" O.C.		325	.148		2.12	3.62	.13	5.87	8	
	0800	3-5/8" metal studs, 25 ga., 16" O.C.		320	.150		2.14	3.68	.13	5.95	8.10	
	0900	Gypsum lath, 2 coat vermiculite plaster, 2 sides										
	1000	2" x 4" wood studs, 16" O.C.	J-2	355	.135	S.F.	2.55	3.32	.12	5.99	8	
	1200	2-1/2" metal studs, 25 ga., 12" O.C.		365	.132		2.26	3.22	.12	5.60	7.55	
	1300	3-5/8" metal studs, 25 ga., 16" O.C.		360	.133		2.33	3.27	.12	5.72	7.70	
624	0010	**SHAFT WALL** Cavity type on 25 ga. J track & C-H studs, 24" O.C.										624
	0030	1" thick coreboard wall liner on shaft side										
	0040	2-hour assembly with double layer										
	0060	5/8" fire rated gypsum board on room side	2 Carp	220	.073	S.F.	1	1.99		2.99	4.22	
	0100	3-hour assembly with triple layer										
	0300	5/8" fire rated gypsum board on room side	2 Carp	180	.089	S.F.	1.24	2.43		3.67	5.20	
	0400	4-hour assembly, 1" coreboard, 5/8" fire rated gypsum board										
	0600	and 3/4" galv. metal furring channels, 24" O.C., with										
	0700	Double layer 5/8" fire rated gypsum board on room side	2 Carp	110	.145	S.F.	1.12	3.97		5.09	7.50	
	0900	For taping & finishing, add per side	1 Carp	1,050	.008	"	.09	.21		.30	.43	
	1000	For insulation, see div. 072										

092 800 | Drywall Accessories

		Description	CREW	DAILY OUTPUT	LABOR-HOURS	UNIT	MAT.	LABOR	EQUIP.	TOTAL	TOTAL INCL O&P	
804	0010	**ACCESSORIES, DRYWALL** Casing bead, galvanized steel [R092-610]	1 Carp	2.90	2.759	C.L.F.	15.90	75.50		91.40	136	804
	0100	Vinyl		3	2.667		24	73		97	142	
	0300	Corner bead, galvanized steel, 1" x 1"		4	2		10.25	54.50		64.75	97.50	
	0400	1-1/4" x 1-1/4"		3.50	2.286		10.40	62.50		72.90	109	
	0600	Vinyl corner bead		4	2		15.75	54.50		70.25	103	
	0700	Door casing, vinyl, for 2" wall systems		2.50	3.200		27.50	87.50		115	168	
	0900	Furring channel, galv. steel, 7/8" deep, standard		2.60	3.077		19.25	84		103.25	153	
	1000	Resilient		2.55	3.137		19.50	85.50		105	157	
	1100	J trim, galvanized steel, 1/2" wide		3	2.667		17.50	73		90.50	134	
	1120	5/8" wide		2.95	2.712		17.75	74		91.75	136	
	1140	L trim, galvanized		3	2.667		17.50	73		90.50	134	
	1150	U trim, galvanized		2.95	2.712		19	74		93	137	
	1160	Screws #6 x 1" A				M	6.85			6.85	7.55	
	1170	#6 x 1-5/8" A				"	10.50			10.50	11.55	
	1200	Studs and runners for partitions, see also div. 051-230										
	1500	Z stud, galvanized steel, 1-1/2" wide	1 Carp	2.60	3.077	C.L.F.	26	84		110	161	
	1600	2" wide	"	2.55	3.137	"	29.50	85.50		115	168	

Important: See the Reference Section for critical supporting data - Reference Nos., Crews, & City Cost Indexes

			093 100	Ceramic Tile	CREW	DAILY OUTPUT	LABOR-HOURS	UNIT	MAT.	LABOR	EQUIP.	TOTAL	TOTAL INCL O&P	
102	0010	**CERAMIC TILE**												102
	0050	Base, using 1' x 4" high pc. with 1" x 1" tiles, mud set			D-7	82	.195	L.F.	3.99	4.69		8.68	11.35	
	0100	Thin set			"	128	.125		3.79	3.01		6.80	8.60	
	0300	For 6" high base, 1" x 1" tile face, add							.63			.63	.69	
	0400	For 2" x 2" tile face, add to above							.33			.33	.36	
	0600	Cove base, 4-1/4" x 4-1/4" high, mud set			D-7	91	.176		3.63	4.23		7.86	10.25	
	0700	Thin set				128	.125		3.63	3.01		6.64	8.45	
	0900	6" x 4-1/4" high, mud set				100	.160		2.58	3.85		6.43	8.55	
	1000	Thin set				137	.117		2.38	2.81		5.19	6.75	
	1200	Sanitary cove base, 6" x 4-1/4" high, mud set				93	.172		3.49	4.14		7.63	9.95	
	1300	Thin set				124	.129		3.29	3.10		6.39	8.20	
	1500	6" x 6" high, mud set				84	.190		4.26	4.58		8.84	11.45	
	1600	Thin set				117	.137		4.06	3.29		7.35	9.35	
	1800	Bathroom accessories, average				82	.195	Ea.	9.50	4.69		14.19	17.40	
	1900	Bathtub, 5', rec. 4-1/4" x 4-1/4" tile wainscot, adhesive set 6' high				2.90	5.517		200	133		333	415	
	2100	7' high wainscot				2.50	6.400		227	154		381	480	
	2200	8' high wainscot				2.20	7.273		253	175		428	535	
	2400	Bullnose trim, 4-1/4" x 4-1/4", mud set				82	.195	L.F.	3.74	4.69		8.43	11.05	
	2500	Thin set				128	.125		3.54	3.01		6.55	8.35	
	2700	6" x 4-1/4" bullnose trim, mud set				84	.190		3.68	4.58		8.26	10.80	
	2800	Thin set				124	.129		3.48	3.10		6.58	8.40	
	3000	Floors, natural clay, random or uniform, thin set, color group 1				183	.087	S.F.	3.55	2.10		5.65	7	
	3100	Color group 2				183	.087		3.82	2.10		5.92	7.30	
	3300	Porcelain type, 1 color, color group 2, 1" x 1"				183	.087		4.11	2.10		6.21	7.65	
	3400	2" x 2" or 2" x 1", thin set				190	.084		4.34	2.03		6.37	7.75	
	3600	For random blend, 2 colors, add							.75			.75	.83	
	3700	4 colors, add							1.08			1.08	1.19	
	3900	For color group 3, add							.44			.44	.48	
	4000	For abrasive non-slip tile, add							.43			.43	.47	
	4300	Specialty tile, 4-1/4" x 4-1/4" x 1/2", decorator finish			D-7	183	.087		6.95	2.10		9.05	10.75	
	4500	Add for epoxy grout, 1/16" joint, 1" x 1" tile				800	.020		.53	.48		1.01	1.29	
	4600	2" x 2" tile				820	.020		.50	.47		.97	1.24	
	4800	Pregrouted sheets, walls, 4-1/4" x 4-1/4", 6" x 4-1/4"												
	4810	and 8-1/2" x 4-1/4", 4 S.F. sheets, silicone grout			D-7	240	.067	S.F.	4.07	1.60		5.67	6.85	
	5100	Floors, unglazed, 2 S.F. sheets,												
	5110	urethane adhesive			D-7	180	.089	S.F.	4.05	2.14		6.19	7.60	
	5400	Walls, interior, thin set, 4-1/4" x 4-1/4" tile **CN**				190	.084		2.21	2.03		4.24	5.40	
	5500	6" x 4-1/4" tile				190	.084		2.25	2.03		4.28	5.45	
	5700	8-1/2" x 4-1/4" tile				190	.084		3.19	2.03		5.22	6.50	
	5800	6" x 6" tile				200	.080		2.56	1.92		4.48	5.65	
	5810	8" x 8" tile				225	.071		2.66	1.71		4.37	5.45	
	5820	12" x 12" tile				300	.053		2.92	1.28		4.20	5.10	
	5830	16" x 16" tile				500	.032		3.17	.77		3.94	4.63	
	6000	Decorated wall tile, 4-1/4" x 4-1/4", minimum				270	.059		2.94	1.43		4.37	5.35	
	6100	Maximum				180	.089		15.25	2.14		17.39	19.95	
	6300	Exterior walls, frostproof, mud set, 4-1/4" x 4-1/4"				102	.157		4.02	3.77		7.79	10	
	6400	1-3/8" x 1-3/8"				93	.172		3.71	4.14		7.85	10.20	
	6600	Crystalline glazed, 4-1/4" x 4-1/4", mud set, plain				100	.160		3.67	3.85		7.52	9.75	
	6700	4-1/4" x 4-1/4", scored tile				100	.160		3.97	3.85		7.82	10.05	
	6900	6" x 6" plain				93	.172		4.43	4.14		8.57	10.95	
	7000	For epoxy grout, 1/16" joints, 4-1/4" tile, add				800	.020		.32	.48		.80	1.06	
	7200	For tile set in dry mortar, add				1,735	.009			.22		.22	.33	
	7300	For tile set in portland cement mortar, add				290	.055			1.33		1.33	1.96	
	9000	Regrout tile 4-1/2 x 4-1/2, or larger, wall			1 Tilf	100	.080		.13	2.13		2.26	3.29	
	9220	Floor			"	125	.064		.14	1.71		1.85	2.67	

FINISHES 9

093 100 | Ceramic Tile

			CREW	DAILY OUTPUT	LABOR-HOURS	UNIT	1999 BARE COSTS				TOTAL INCL O&P	
							MAT.	LABOR	EQUIP.	TOTAL		
104	0010	CERAMIC TILE PANELS Insulated, over 1000 S.F., 1-1/2" thick	D-7	220	.073	S.F.	8.50	1.75		10.25	11.95	104
	0100	2-1/2" thick	"	220	.073	"	9.10	1.75		10.85	12.60	

093 300 | Quarry Tile

			CREW	DAILY OUTPUT	LABOR-HOURS	UNIT	MAT.	LABOR	EQUIP.	TOTAL	TOTAL INCL O&P	
304	0010	QUARRY TILE Base, cove or sanitary, 2" or 5" high, mud set										304
	0100	1/2" thick	D-7	110	.145	L.F.	3.51	3.50		7.01	9	
	0300	Bullnose trim, red, mud set, 6" x 6" x 1/2" thick		120	.133		3.74	3.21		6.95	8.85	
	0400	4" x 4" x 1/2" thick		110	.145		3.75	3.50		7.25	9.30	
	0600	4" x 8" x 1/2" thick, using 8" as edge		130	.123		3.56	2.96		6.52	8.30	
	0700	Floors, mud set, 1,000 S.F. lots, red, 4" x 4" x 1/2" thick		120	.133	S.F.	3.67	3.21		6.88	8.80	
	0900	6" x 6" x 1/2" thick		140	.114		2.65	2.75		5.40	7	
	1000	4" x 8" x 1/2" thick		130	.123		3.67	2.96		6.63	8.40	
	1300	For waxed coating, add					.59			.59	.65	
	1500	For colors other than green, add					.34			.34	.37	
	1600	For abrasive surface, add					.41			.41	.45	
	1800	Brown tile, imported, 6" x 6" x 3/4"	D-7	120	.133		4.35	3.21		7.56	9.55	
	1900	8" x 8" x 1"		110	.145		4.93	3.50		8.43	10.55	
	2100	For thin set mortar application, deduct		700	.023			.55		.55	.81	
	2200	For epoxy grout & mortar, 6" x 6" x 1/2", add		350	.046		1.10	1.10		2.20	2.84	
	2700	Stair tread, 6" x 6" x 3/4", plain		50	.320		3.96	7.70		11.66	15.75	
	2800	Abrasive		47	.340		4.50	8.20		12.70	17.05	
	3000	Wainscot, 6" x 6" x 1/2", thin set, red		105	.152		3.35	3.66		7.01	9.10	
	3100	Colors other than green		105	.152		3.74	3.66		7.40	9.50	
	3300	Window sill, 6" wide, 3/4" thick		90	.178	L.F.	4.32	4.28		8.60	11.05	
	3400	Corners		80	.200	Ea.	4.72	4.81		9.53	12.30	

093 500 | Glass Mosaics

			CREW	DAILY OUTPUT	LABOR-HOURS	UNIT	MAT.	LABOR	EQUIP.	TOTAL	TOTAL INCL O&P	
501	0010	GLASS MOSAICS 3/4" tile on 12" sheets, standard grout										501
	0300	Color group 1 & 2	D-7	73	.219	S.F.	14.65	5.25		19.90	24	
	0350	Color group 3		73	.219		15.45	5.25		20.70	25	
	0400	Color group 4		73	.219		21	5.25		26.25	31	
	0450	Color group 5		73	.219		23.50	5.25		28.75	34	
	0500	Color group 6		73	.219		28.50	5.25		33.75	39.50	
	0600	Color group 7		73	.219		33	5.25		38.25	44.50	
	0700	Color group 8, golds, silvers & specialties		64	.250		48	6		54	62	

093 700 | Metal Tile

			CREW	DAILY OUTPUT	LABOR-HOURS	UNIT	MAT.	LABOR	EQUIP.	TOTAL	TOTAL INCL O&P	
701	0010	METAL TILE 4' x 4' sheet, 24 ga., tile pattern, nailed										701
	0200	Stainless steel	2 Carp	512	.031	S.F.	21.50	.85		22.35	25	
	0400	Aluminized steel	"	512	.031	"	11.45	.85		12.30	13.95	

093 900 | Tile Setting Materials

			CREW	DAILY OUTPUT	LABOR-HOURS	UNIT	MAT.	LABOR	EQUIP.	TOTAL	TOTAL INCL O&P	
910	0010	CEMENTITIOUS BACKERBOARD										910
	0070	Cementitious backerboard, on floor, 3' x 4'x 1/2" sheets	2 Carp	525	.030	S.F.	1.03	.83		1.86	2.45	
	0080	3' x 5' x 1/2" sheets		525	.030		1.03	.83		1.86	2.45	
	0090	3' x 6' x 1/2" sheets		525	.030		1.03	.83		1.86	2.44	
	0100	3' x 4'x 5/8" sheets		525	.030		1.07	.83		1.90	2.48	
	0110	3' x 5' x 5/8" sheets		525	.030		1.05	.83		1.88	2.47	
	0120	3' x 6' x 5/8" sheets		525	.030		1.06	.83		1.89	2.47	
	0150	On wall, 3' x 4'x 1/2" sheets		350	.046		1.03	1.25		2.28	3.10	
	0160	3' x 5' x 1/2" sheets		350	.046		1.03	1.25		2.28	3.10	
	0170	3' x 6' x 1/2" sheets		350	.046		1.03	1.25		2.28	3.09	
	0180	3' x 4'x 5/8" sheets		350	.046		1.07	1.25		2.32	3.13	
	0190	3' x 5' x 5/8" sheets		350	.046		1.05	1.25		2.30	3.12	

9 FINISHES

Important: See the Reference Section for critical supporting data - Reference Nos., Crews, & City Cost Indexes

		093 900 | **Tile Setting Materials**	CREW	DAILY OUTPUT	LABOR-HOURS	UNIT	1999 BARE COSTS				TOTAL INCL O&P	
							MAT.	LABOR	EQUIP.	TOTAL		
910	0200	3' x 6' x 5/8" sheets	2 Carp	350	.046	S.F.	1.06	1.25		2.31	3.12	910
	0250	On counter, 3' x 4'x 1/2" sheets		180	.089		1.03	2.43		3.46	4.96	
	0260	3' x 5' x 1/2" sheets		180	.089		1.03	2.43		3.46	4.96	
	0270	3' x 6' x 1/2" sheets		180	.089		1.03	2.43		3.46	4.95	
	0300	3' x 4'x 5/8" sheets		180	.089		1.07	2.43		3.50	4.99	
	0310	3' x 5' x 5/8" sheets		180	.089		1.05	2.43		3.48	4.98	
	0320	3' x 6' x 5/8" sheets		180	.089		1.06	2.43		3.49	4.98	

9

FINISHES

		094 100 | **Portland Cem. Terrazzo**	CREW	DAILY OUTPUT	LABOR-HOURS	UNIT	1999 BARE COSTS				TOTAL INCL O&P	
							MAT.	LABOR	EQUIP.	TOTAL		
104	0010	**TERRAZZO, CAST IN PLACE** Cove base, 6" high, 16ga. zinc R094-100	1 Mstz	20	.400	L.F.	2.63	10.60		13.23	18.60	104
	0100	Curb, 6" high and 6" wide		6	1.333		4.33	35.50		39.83	57.50	
	0300	Divider strip for floors, 14 ga., 1-1/4" deep, zinc		375	.021		.89	.57		1.46	1.82	
	0400	Brass		375	.021		1.65	.57		2.22	2.66	
	0600	Heavy top strip 1/4" thick, 1-1/4" deep, zinc		300	.027		1.50	.71		2.21	2.70	
	0900	Galv. bottoms, brass		300	.027		2.53	.71		3.24	3.83	
	1200	For thin set floors, 16 ga., 1/2" x 1/2", zinc		350	.023		.78	.61		1.39	1.76	
	1300	Brass		350	.023		1.48	.61		2.09	2.52	
	1500	Floor, bonded to concrete, 1-3/4" thick, gray cement	J-3	130	.123	S.F.	2.31	2.97	.76	6.04	7.75	
	1600	White cement, mud set		130	.123		2.63	2.97	.76	6.36	8.10	
	1800	Not bonded, 3" total thickness, gray cement		115	.139		2.89	3.36	.86	7.11	9.10	
	1900	White cement, mud set		115	.139		3.15	3.36	.86	7.37	9.40	
	2100	For Venetian terrazzo, 1" topping, add					50%	50%				
	2200	For heavy duty abrasive terrazzo, add					50%	50%				
	2400	Bonded conductive floor for hospitals	J-3	90	.178	S.F.	3.15	4.29	1.10	8.54	11	
	2500	Epoxy terrazzo, 1/4" thick, minimum		100	.160		3.57	3.86	.99	8.42	10.70	
	2550	Average		75	.213		3.93	5.15	1.31	10.39	13.35	
	2600	Maximum		60	.267		4.31	6.45	1.64	12.40	16.05	
	2700	Monolithic terrazzo, 1/2" thick										
	2710	10' panels	J-3	125	.128	S.F.	2.10	3.09	.79	5.98	7.75	
	3000	Stairs, cast in place, pan filled treads		30	.533	L.F.	2.12	12.90	3.29	18.31	25	
	3100	Treads and risers		14	1.143	"	4.91	27.50	7.05	39.46	54	
	3300	For stair landings, add to floor prices						50%				
	3400	Stair stringers and fascia	J-3	30	.533	S.F.	3.46	12.90	3.29	19.65	26.50	
	3600	For abrasive metal nosings on stairs, add		150	.107	L.F.	5.90	2.58	.66	9.14	11.05	
	3700	For abrasive surface finish, add		600	.027	S.F.	.72	.64	.16	1.52	1.93	
	3900	For raised abrasive strips, add		150	.107	L.F.	.67	2.58	.66	3.91	5.25	
	4000	Wainscot, bonded, 1-1/2" thick		30	.533	S.F.	2.69	12.90	3.29	18.88	25.50	
	4200	Epoxy terrazzo, 1/4" thick		40	.400	"	3.93	9.65	2.46	16.04	21.50	
108	0010	**TILE OR TERRAZZO BASE** Scratch coat only	1 Mstz	150	.053	S.F.	.30	1.42		1.72	2.42	108
	0500	Scratch and brown coat only	"	75	.107	"	.55	2.83		3.38	4.80	

		094 200 | **Precast Terrazzo**	CREW	DAILY OUTPUT	LABOR-HOURS	UNIT	MAT.	LABOR	EQUIP.	TOTAL	TOTAL INCL O&P	
201	0010	**TERRAZZO, PRECAST** Base, 6" high, straight	1 Mstz	35	.229	L.F.	8.70	6.05		14.75	18.55	201
	0100	Cove		30	.267		10.50	7.10		17.60	22	
	0300	8" high base, straight		30	.267		9.50	7.10		16.60	21	
	0400	Cove		25	.320		11.55	8.50		20.05	25.50	

For expanded coverage of these items see _Means Interior Cost Data 1999_

094 200	Precast Terrazzo	CREW	DAILY OUTPUT	LABOR-HOURS	UNIT	1999 BARE COSTS				TOTAL INCL O&P	
						MAT.	LABOR	EQUIP.	TOTAL		
201	0600	For white cement, add				L.F.	.32			.32	.35	201
	0700	For 16 ga. zinc toe strip, add					.95			.95	1.05	
	0900	Curbs, 4" x 4" high	1 Mstz	19	.421		22	11.20		33.20	40.50	
	1000	8" x 8" high	"	15	.533		29.50	14.15		43.65	53.50	
	1200	Floor tiles, non-slip, 1" thick, 12" x 12"	D-1	29	.552	S.F.	14.85	13.55		28.40	37.50	
	1300	1-1/4" thick, 12" x 12"		29	.552		16.10	13.55		29.65	38.50	
	1500	16" x 16"		23	.696		17.50	17.10		34.60	46	
	1600	1-1/2" thick, 16" x 16"		21	.762		16	18.75		34.75	46.50	
	1800	For Venetian terrazzo, add					4.75			4.75	5.20	
	1900	For white cement, add					.45			.45	.50	
	2400	Stair treads, 1-1/2" thick, non-slip, three line pattern	2 Mstz	70	.229	L.F.	32	6.05		38.05	44	
	2500	Nosing and two lines		70	.229		32	6.05		38.05	44	
	2700	2" thick treads, straight		60	.267		34	7.10		41.10	48	
	2800	Curved		50	.320		44.50	8.50		53	61.50	
	3000	Stair risers, 1" thick, to 6" high, straight sections		60	.267		9.50	7.10		16.60	21	
	3100	Cove		50	.320		11.50	8.50		20	25	
	3300	Curved, 1" thick, to 6" high, vertical		48	.333		20.50	8.85		29.35	35.50	
	3400	Cove		38	.421		27.50	11.20		38.70	46.50	
	3600	Stair tread and riser, single piece, straight, minimum		60	.267		41	7.10		48.10	55.50	
	3700	Maximum		40	.400		52.50	10.60		63.10	73.50	
	3900	Curved tread and riser, minimum		40	.400		57	10.60		67.60	78	
	4000	Maximum		32	.500		72	13.25		85.25	98.50	
	4200	Stair stringers, notched, 1" thick		25	.640		23.50	17		40.50	50.50	
	4300	2" thick		22	.727		27.50	19.30		46.80	59	
	4500	Stair landings, structural, non-slip, 1-1/2" thick		85	.188	S.F.	25.50	5		30.50	35.50	
	4600	3" thick		75	.213		36	5.65		41.65	48	
	4800	Wainscot, 12" x 12" x 1" tiles	1 Mstz	12	.667		8.50	17.70		26.20	35.50	
	4900	16" x 16" x 1-1/2" tiles	"	8	1		11.70	26.50		38.20	52.50	

095 100	Acoustical Ceilings		CREW	DAILY OUTPUT	LABOR-HOURS	UNIT	1999 BARE COSTS				TOTAL INCL O&P	
							MAT.	LABOR	EQUIP.	TOTAL		
102	0010	**CEILING TILE**, Stapled or cemented											102
	0100	12" x 12" or 12" x 24", not including furring											
	0600	Mineral fiber, vinyl coated, 5/8" thick	CN	1 Carp	1,000	.008	S.F.	.56	.22		.78	.96	
	0700	3/4" thick			1,000	.008		1.23	.22		1.45	1.69	
	0900	Fire rated, 3/4" thick, plain faced			1,000	.008		.82	.22		1.04	1.25	
	1000	Plastic coated face			1,000	.008		.96	.22		1.18	1.39	
	1200	Aluminum faced, 5/8" thick, plain			1,000	.008		1.05	.22		1.27	1.50	
	3700	Wall application of above, add			3,100	.003			.07		.07	.11	
	3900	For ceiling primer, add						.12			.12	.13	
	4000	For ceiling cement, add						.31			.31	.34	
104	0010	**SUSPENDED ACOUSTIC CEILING TILES**, Not including											104
	0100	suspension system											
	0300	Fiberglass boards, film faced, 2' x 2' or 2' x 4', 5/8" thick		1 Carp	625	.013	S.F.	.49	.35		.84	1.09	
	0400	3/4" thick			600	.013		1.03	.36		1.39	1.70	
	0500	3" thick, thermal, R11			450	.018		1.21	.49		1.70	2.09	
	0600	Glass cloth faced fiberglass, 3/4" thick			500	.016		1.61	.44		2.05	2.46	
	0700	1" thick			485	.016		1.81	.45		2.26	2.70	
	0820	1-1/2" thick, nubby face			475	.017		2.15	.46		2.61	3.09	

Important: See the Reference Section for critical supporting data - Reference Nos., Crews, & City Cost Indexes

095 100 | Acoustical Ceilings

			CREW	DAILY OUTPUT	LABOR-HOURS	UNIT	1999 BARE COSTS				TOTAL INCL O&P	
							MAT.	LABOR	EQUIP.	TOTAL		
104	1110	Mineral fiber tile, lay-in, 2' x 2' or 2' x 4', 5/8" thick, fine texture	1 Carp	625	.013	S.F.	.40	.35		.75	.99	104
	1115	Rough textured		625	.013		.97	.35		1.32	1.62	
	1125	3/4" thick, fine textured		600	.013		1.09	.36		1.45	1.77	
	1130	Rough textured		600	.013		1.36	.36		1.72	2.07	
	1135	Fissured		600	.013		1.60	.36		1.96	2.33	
	1150	Tegular, 5/8" thick, fine textured		470	.017		.96	.46		1.42	1.79	
	1155	Rough textured		470	.017		1.25	.46		1.71	2.11	
	1165	3/4" thick, fine textured		450	.018		1.36	.49		1.85	2.26	
	1170	Rough textured		450	.018		1.53	.49		2.02	2.44	
	1175	Fissured	▼	450	.018		2.39	.49		2.88	3.39	
	1180	For aluminum face, add					4.36			4.36	4.80	
	1185	For plastic film face, add					.33			.33	.36	
	1190	For fire rating, add					.33			.33	.36	
	1300	Mirror faced panels, 15/16" thick, 2' x 2'	1 Carp	500	.016		5.65	.44		6.09	6.95	
	1900	Eggcrate, acrylic, 1/2" x 1/2" x 1/2" cubes		500	.016		1.32	.44		1.76	2.14	
	2100	Polystyrene eggcrate, 3/8" x 3/8" x 1/2" cubes		510	.016		1.10	.43		1.53	1.88	
	2200	1/2" x 1/2" x 1/2" cubes		500	.016		1.48	.44		1.92	2.32	
	2400	Luminous panels, prismatic, acrylic		400	.020		1.59	.55		2.14	2.61	
	2500	Polystyrene		400	.020		.82	.55		1.37	1.76	
	2700	Flat white acrylic		400	.020		2.79	.55		3.34	3.93	
	2800	Polystyrene		400	.020		1.90	.55		2.45	2.95	
	3000	Drop pan, white, acrylic		400	.020		4.08	.55		4.63	5.35	
	3100	Polystyrene		400	.020		3.41	.55		3.96	4.61	
	3600	Perforated aluminum sheets, .024" thick, corrugated, painted		490	.016		1.62	.45		2.07	2.48	
	3700	Plain		500	.016		1.46	.44		1.90	2.30	
	3720	Mineral fiber, 24" x 24" or 48", reveal edge, painted, 5/8" thick		600	.013		.90	.36		1.26	1.56	
	3740	3/4" thick	▼	575	.014	▼	1.48	.38		1.86	2.23	
106	0010	**SUSPENDED CEILINGS, COMPLETE** Including standard										106
	0100	suspension system but not incl. 1-1/2" carrier channels										
	0600	Fiberglass ceiling board, 2' x 4' x 5/8", plain faced,	1 Carp	500	.016	S.F.	.99	.44		1.43	1.78	
	0700	Offices, 2' x 4' x 3/4"		380	.021		1.10	.57		1.67	2.11	
	0800	Mineral fiber, on 15/16" T bar susp. 2' x 2' x 3/4" lay-in board		345	.023		1.56	.63		2.19	2.72	
	0810	2' x 4' x 5/8" tile		380	.021		1.37	.57		1.94	2.41	
	0820	Tegular, 2' x 2' x 5/8" tile on 9/16" grid		250	.032		1.84	.87		2.71	3.39	
	0830	2' x 4' x 3/4" tile		225	.036		1.92	.97		2.89	3.64	
	0900	Luminous panels, prismatic, acrylic		255	.031		2.47	.86		3.33	4.07	
	1200	Metal pan with acoustic pad, steel		75	.107		3.19	2.91		6.10	8.10	
	1300	Painted aluminum		75	.107		5.65	2.91		8.56	10.85	
	1500	Aluminum, degreased finish		75	.107		3.68	2.91		6.59	8.65	
	1600	Stainless steel		75	.107		6.75	2.91		9.66	12.05	
	1800	Tile, Z bar suspension, 5/8" mineral fiber tile		150	.053		1.43	1.46		2.89	3.86	
	1900	3/4" mineral fiber tile	▼	150	.053	▼	1.52	1.46		2.98	3.96	
	2400	For strip lighting, see division 166-100										
	2500	For rooms under 500 S.F., add				S.F.		25%				

095 250 | Acoustical Space Units

			CREW	DAILY OUTPUT	LABOR-HOURS	UNIT	MAT.	LABOR	EQUIP.	TOTAL	INCL O&P	
254	0010	**SOUND ABSORBING PANELS** Perforated steel facing, painted with										254
	0100	fiberglass or mineral filler, no backs, 2-1/4" thick, modular										
	0200	space units, ceiling or wall hung, white or colored	1 Carp	100	.080	S.F.	7.20	2.18		9.38	11.40	
	0300	Fiberboard sound deadening panels, 1/2" thick	"	600	.013	"	.27	.36		.63	.87	
	0500	Fiberglass panels, 4' x 8' x 1" thick, with										
	0600	glass cloth face for walls, cemented	1 Carp	155	.052	S.F.	4.10	1.41		5.51	6.75	
	0700	1-1/2" thick, dacron covered, inner aluminum frame,										
	0710	wall mounted	1 Carp	300	.027	S.F.	6.95	.73		7.68	8.80	

FINISHES 9

095 | Acoustical Treatment & Wood Flooring

095 250 | Acoustical Space Units

			CREW	DAILY OUTPUT	LABOR-HOURS	UNIT	1999 BARE COSTS				TOTAL INCL O&P	
							MAT.	LABOR	EQUIP.	TOTAL		
254	0900	Mineral fiberboard panels, fabric covered, 30"x 108",										254
	1000	3/4" thick, concealed spline, wall mounted	1 Carp	150	.053	S.F.	5	1.46		6.46	7.80	

095 300 | Acoustical Insulation

			CREW	DAILY OUTPUT	LABOR-HOURS	UNIT	MAT.	LABOR	EQUIP.	TOTAL	TOTAL INCL O&P	
304	0011	**BARRIERS** Plenum										304
	0600	Aluminum foil, fiberglass reinf., parallel with joists	1 Carp	275	.029	S.F.	.73	.79		1.52	2.05	
	0700	Perpendicular to joists		180	.044		.73	1.21		1.94	2.71	
	0900	Aluminum mesh, kraft paperbacked		275	.029		.69	.79		1.48	2.01	
	0970	Fiberglass batts, kraft faced, 3-1/2" thick		1,400	.006		.25	.16		.41	.53	
	0980	6" thick		1,300	.006		.42	.17		.59	.72	
	1000	Sheet lead, 1 lb., 1/64" thick, perpendicular to joists		150	.053		1.55	1.46		3.01	3.99	
	1100	Vinyl foam reinforced, 1/8" thick, 1.0 lb. per S.F.		150	.053		2.70	1.46		4.16	5.25	
308	0010	**SOUND ATTENUATION** Blanket, 1" thick	1 Carp	925	.009	S.F.	.21	.24		.45	.60	308
	0500	1-1/2" thick		920	.009		.26	.24		.50	.65	
	1000	2" thick		915	.009		.32	.24		.56	.74	
	1500	3" thick		910	.009		.44	.24		.68	.86	
	3000	Thermal or acoustical batt above ceiling, 2" thick		900	.009		.42	.24		.66	.84	
	3100	3" thick		900	.009		.63	.24		.87	1.07	
	3200	4" thick		900	.009		.77	.24		1.01	1.23	
	3400	Urethane plastic foam, open cell, on wall, 2" thick	2 Carp	2,050	.008		2.51	.21		2.72	3.10	
	3500	3" thick		1,550	.010		3.33	.28		3.61	4.10	
	3600	4" thick		1,050	.015		4.68	.42		5.10	5.80	
	3700	On ceiling, 2" thick		1,700	.009		2.50	.26		2.76	3.15	
	3800	3" thick		1,300	.012		3.33	.34		3.67	4.19	
	3900	4" thick		900	.018		4.68	.49		5.17	5.90	
	4000	Nylon matting 0.4" thick, with carbon black spinerette										
	4010	plus polyester fabric, on floor	J-4	4,000	.004	S.F.	1.86	.10		1.96	2.19	
	4200	Fiberglass reinf. backer board underlayment, 7/16" thick, on floor	"	800	.020	"	1.56	.48		2.04	2.43	

095 600 | Wood Strip Flooring

			CREW	DAILY OUTPUT	LABOR-HOURS	UNIT	MAT.	LABOR	EQUIP.	TOTAL	TOTAL INCL O&P	
604	0010	**WOOD** Fir, vertical grain, 1" x 4", not incl. finish, B & better	1 Carp	255	.031	S.F.	2.30	.86		3.16	3.88	604
	0100	C grade & better	"	255	.031	"	2.16	.86		3.02	3.73	
	0600	Gym floor, in mastic, over 2 ply felt, #2 & better										
	0700	25/32" thick maple	1 Carp	100	.080	S.F.	3.25	2.18		5.43	7	
	0900	33/32" thick maple		98	.082		3.75	2.23		5.98	7.65	
	1000	For 1/2" corkboard underlayment, add		750	.011		.73	.29		1.02	1.26	
	1300	For #1 grade maple, add					.40			.40	.44	
	1600	Maple flooring, over sleepers, #2 & better										
	1700	25/32" thick	1 Carp	85	.094	S.F.	3.35	2.57		5.92	7.70	
	1900	33/32" thick	"	83	.096		3.90	2.63		6.53	8.45	
	2000	For #1 grade, add					.40			.40	.44	
	2200	For 3/4" subfloor, add	1 Carp	350	.023		.87	.62		1.49	1.94	
	2300	With two 1/2" subfloors, 25/32" thick	"	69	.116		4.20	3.17		7.37	9.60	
	2500	Maple, incl. finish, #2 & btr., 25/32" thick, on rubber										
	2600	Sleepers, with two 1/2" subfloors	1 Carp	76	.105	S.F.	4.50	2.87		7.37	9.45	
	2800	With steel spline, double connection to channels	"	73	.110		4.80	2.99		7.79	10	
	2900	For 33/32" maple, add					.52			.52	.57	
	3100	For #1 grade maple, add					.40			.40	.44	
	3500	For termite proofing all of the above, add					.20			.20	.22	
	3700	Portable hardwood, prefinished panels	1 Carp	83	.096		6	2.63		8.63	10.75	
	3720	Insulated with polystyrene, 1" thick, add		165	.048		.52	1.32		1.84	2.65	
	3750	Running tracks, Sitka spruce surface, 25/32" x 2-1/4"		62	.129		11	3.52		14.52	17.65	
	3770	3/4" plywood surface, finished		100	.080		2.60	2.18		4.78	6.30	
	3800	See also resilient gym floors, division 096-601-4400										
	4000	Maple, strip, 25/32" x 2-1/4", not incl. finish, select	1 Carp	170	.047	S.F.	2.95	1.28		4.23	5.25	
	4100	#2 & better		170	.047		2.65	1.28		3.93	4.94	

Important: See the Reference Section for critical supporting data - Reference Nos., Crews, & City Cost Indexes

095 600 | Wood Strip Flooring

			CREW	DAILY OUTPUT	LABOR-HOURS	UNIT	1999 BARE COSTS				TOTAL INCL O&P	
							MAT.	LABOR	EQUIP.	TOTAL		
604	4300	33/32" x 3-1/4", not incl. finish, #1 grade	1 Carp	170	.047	S.F.	3.25	1.28		4.53	5.60	**604**
	4400	#2 & better	↓	170	.047	↓	2.90	1.28		4.18	5.20	
	4600	Oak, white or red, 25/32" x 2-1/4", not incl. finish										
	4700	#1 common CN	1 Carp	170	.047	S.F.	2.70	1.28		3.98	4.99	
	4900	Select quartered, 2-1/4" wide		170	.047		3.02	1.28		4.30	5.35	
	5000	Clear		170	.047		3.25	1.28		4.53	5.60	
	5200	Parquetry, standard, 5/16" thick, not incl. finish, oak, minimum		160	.050		1.54	1.36		2.90	3.84	
	5300	Maximum		100	.080		5.10	2.18		7.28	9.10	
	5500	Teak, minimum		160	.050		4.15	1.36		5.51	6.70	
	5600	Maximum		100	.080		7.25	2.18		9.43	11.45	
	5650	13/16" thick, select grade oak, minimum		160	.050		8	1.36		9.36	10.95	
	5700	Maximum		100	.080		12.15	2.18		14.33	16.80	
	5800	Custom parquetry, including finish, minimum		100	.080		13.40	2.18		15.58	18.20	
	5900	Maximum		50	.160		17.75	4.37		22.12	26.50	
	6100	Prefinished, white oak, prime grade, 2-1/4" wide		170	.047		5.85	1.28		7.13	8.45	
	6200	3-1/4" wide		185	.043		7.50	1.18		8.68	10.10	
	6400	Ranch plank		145	.055		7.25	1.51		8.76	10.35	
	6500	Hardwood blocks, 9" x 9", 25/32" thick		160	.050		4.85	1.36		6.21	7.50	
	6700	Parquetry, 5/16" thick, oak, minimum		160	.050		3.30	1.36		4.66	5.80	
	6800	Maximum		100	.080		8.05	2.18		10.23	12.35	
	7000	Walnut or teak, parquetry, minimum		160	.050		4.50	1.36		5.86	7.10	
	7100	Maximum	↓	100	.080	↓	7.85	2.18		10.03	12.10	
	7200	Acrylic wood parquet blocks, 12" x 12" x 5/16",										
	7210	irradiated, set in epoxy	1 Carp	160	.050	S.F.	6.55	1.36		7.91	9.35	
	7400	Yellow pine, 3/4" x 3-1/8", T & G, C & better, not incl. finish	"	200	.040		2.15	1.09		3.24	4.09	
	7500	Refinish wood floor, sand, 2 cts poly, wax, soft wood, min.	1 Clab	400	.020		.65	.43		1.08	1.39	
	7600	Hard wood, max		130	.062		.98	1.32		2.30	3.16	
	7800	Sanding and finishing, 2 coats polyurethane	↓	295	.027	↓	.65	.58		1.23	1.63	
	7900	Subfloor and underlayment, see division 061-164 & 168										
	8015	Transition molding, 2 1/4" wide, 5' long	1 Carp	19.20	.417	Ea.	10.15	11.35		21.50	29	
	8300	Floating floor, wood composition strip, complete.	1 Clab	133	.060	S.F.	4.16	1.29		5.45	6.60	
	8310	Floating floor components, T & G wood composite strips					3.87			3.87	4.26	
	8320	Film					.09			.09	.10	
	8330	Foam					.19			.19	.21	
	8340	Adhesive					.06			.06	.07	
	8350	Installation kit				↓	.16			.16	.18	
	8360	Trim, 2" wide x 3' long				L.F.	2.25			2.25	2.48	
	8370	Reducer moulding				"	4.15			4.15	4.57	

095 650 | Wood Block Flooring

			CREW	DAILY OUTPUT	LABOR-HOURS	UNIT	MAT.	LABOR	EQUIP.	TOTAL	TOTAL INCL O&P	
651	0010	**WOOD BLOCK FLOORING** End grain flooring, coated, 2" thick	1 Carp	295	.027	S.F.	2.70	.74		3.44	4.13	**651**
	0400	Natural finish, 1" thick, fir		125	.064		2.80	1.75		4.55	5.85	
	0600	1-1/2" thick, pine		125	.064		2.75	1.75		4.50	5.80	
	0700	2" thick, pine	↓	125	.064	↓	2.72	1.75		4.47	5.75	

095 800 | Wood Comp. Flooring

			CREW	DAILY OUTPUT	LABOR-HOURS	UNIT	MAT.	LABOR	EQUIP.	TOTAL	TOTAL INCL O&P	
801	0010	**WOOD COMPOSITION** Gym floors										**801**
	0100	2-1/4" x 6-7/8" x 3/8", on 2" grout setting bed	D-7	150	.107	S.F.	4.60	2.57		7.17	8.85	
	0200	Thin set, on concrete	"	250	.064		4.20	1.54		5.74	6.90	
	0300	Sanding and finishing, add	1 Carp	200	.040	↓	.62	1.09		1.71	2.40	

FINISHES **9**

096 150 | Marble Flooring

			CREW	DAILY OUTPUT	LABOR-HOURS	UNIT	1999 BARE COSTS				TOTAL INCL O&P	
							MAT.	LABOR	EQUIP.	TOTAL		
151	0010	**MARBLE** Thin gauge tile, 12" x 6", 3/8", White Carara	D-7	60	.267	S.F.	8.60	6.40		15	18.95	151
	0100	Travertine		60	.267		9.45	6.40		15.85	19.90	
	0200	12" x 12" x 3/8", thin set, floors		60	.267		8.60	6.40		15	18.95	
	0300	On walls	↓	52	.308	↓	8.60	7.40		16	20.50	

096 250 | Slate Flooring

			CREW	DAILY OUTPUT	LABOR-HOURS	UNIT	MAT.	LABOR	EQUIP.	TOTAL	INCL O&P	
251	0010	**SLATE TILE** Vermont, 6" x 6" x 1/4" thick, thin set	D-7	180	.089	S.F.	4.14	2.14		6.28	7.70	251
	0200	See also division 025-128										
252	0010	**SLATE & STONE FLOORS** See division 025-166										252
254	0010	**CONCRETE FLOORS** And toppings, see division 033-454										254

096 350 | Brick Flooring

			CREW	DAILY OUTPUT	LABOR-HOURS	UNIT	MAT.	LABOR	EQUIP.	TOTAL	INCL O&P	
354	0010	**FLOORING**										354
	0020	Acid proof shales, red, 8" x 3-3/4" x 1-1/4" thick	D-7	.43	37.209	M	695	895		1,590	2,100	
	0050	2-1/4" thick	D-1	.40	40		755	985		1,740	2,350	
	0200	Acid proof clay brick, 8" x 3-3/4" x 2-1/4" thick	"	.40	40	↓	755	985		1,740	2,350	
	0260	Cast ceramic, pressed, 4" x 8" x 1/2", unglazed	D-7	100	.160	S.F.	4.94	3.85		8.79	11.15	
	0270	Glazed		100	.160		6.60	3.85		10.45	12.95	
	0280	Hand molded flooring, 4" x 8" x 3/4", unglazed		95	.168		6.55	4.05		10.60	13.20	
	0290	Glazed		95	.168		8.20	4.05		12.25	15	
	0300	8" hexagonal, 3/4" thick, unglazed		85	.188		7.15	4.53		11.68	14.60	
	0310	Glazed	↓	85	.188		12.95	4.53		17.48	21	
	0400	Heavy duty industrial, cement mortar bed, 2" thick, not incl. brick	D-1	80	.200		.71	4.92		5.63	8.45	
	0450	Acid proof joints, 1/4" wide	"	65	.246		1.13	6.05		7.18	10.65	
	0500	Pavers, 8" x 4", 1" to 1-1/4" thick, red	D-7	95	.168		2.88	4.05		6.93	9.15	
	0510	Ironspot	"	95	.168		4.07	4.05		8.12	10.50	
	0540	1-3/8" to 1-3/4" thick, red	D-1	95	.168		2.78	4.14		6.92	9.50	
	0560	Ironspot		95	.168		4.02	4.14		8.16	10.85	
	0580	2-1/4" thick, red		90	.178		2.83	4.37		7.20	9.90	
	0590	Ironspot	↓	90	.178	↓	4.38	4.37		8.75	11.60	
	0800	For sidewalks and patios with pavers, see division 025-158										
	0870	For epoxy joints, add	D-1	600	.027	S.F.	2.15	.66		2.81	3.39	
	0880	For Furan underlayment, add	"	600	.027		1.78	.66		2.44	2.98	
	0890	For waxed surface, steam cleaned, add	D-5	1,000	.008	↓	.15	.22		.37	.51	

096 600 | Resilient Flooring

			CREW	DAILY OUTPUT	LABOR-HOURS	UNIT	MAT.	LABOR	EQUIP.	TOTAL	INCL O&P	
601	0010	**RESILIENT FLOORING** R096 -600										601
	0800	Base, cove, rubber or vinyl, .080" thick										
	1100	Standard colors, 2-1/2" high	1 Tilf	315	.025	L.F.	.39	.68		1.07	1.43	
	1150	4" high		315	.025		.49	.68		1.17	1.54	
	1200	6" high		315	.025		.73	.68		1.41	1.80	
	1450	1/8" thick, standard colors, 2-1/2" high		315	.025		.40	.68		1.08	1.44	
	1500	4" high	CN	315	.025	↓	.56	.68		1.24	1.62	
	1600	Corners, 2-1/2" high		315	.025	Ea.	1.02	.68		1.70	2.12	
	1630	4" high		315	.025		1.07	.68		1.75	2.18	
	1660	6" high		315	.025	↓	1.38	.68		2.06	2.52	
	1700	Conductive flooring, rubber tile, 1/8" thick		315	.025	S.F.	2.50	.68		3.18	3.75	
	1800	Homogeneous vinyl tile, 1/8" thick		315	.025		3.42	.68		4.10	4.76	
	2200	Cork tile, standard finish, 1/8" thick		315	.025		2.42	.68		3.10	3.66	
	2250	3/16" thick		315	.025		2.84	.68		3.52	4.12	
	2300	5/16" thick		315	.025		4.26	.68		4.94	5.70	
	2350	1/2" thick		315	.025		4.74	.68		5.42	6.20	
	2500	Urethane finish, 1/8" thick		315	.025		3.65	.68		4.33	5	
	2550	3/16" thick	↓	315	.025	↓	4.81	.68		5.49	6.30	

Important: See the Reference Section for critical supporting data - Reference Nos., Crews, & City Cost Indexes

9 / **FINISHES**

096 600	Resilient Flooring		CREW	DAILY OUTPUT	LABOR-HOURS	UNIT	1999 BARE COSTS				TOTAL INCL O&P
							MAT.	LABOR	EQUIP.	TOTAL	
601 2600	5/16" thick	R096 -600	1 Tilf	315	.025	S.F.	6.60	.68		7.28	8.25 |601
2650	1/2" thick			315	.025		9.20	.68		9.88	11.10
3700	Polyethylene, in rolls, no base incl., landscape surfaces			275	.029		2.19	.78		2.97	3.56
3800	Nylon action surface, 1/8" thick			275	.029		2.35	.78		3.13	3.73
3900	1/4" thick			275	.029		3.39	.78		4.17	4.88
4000	3/8" thick			275	.029		4.25	.78		5.03	5.85
4100	Golf tee surface with foam back			235	.034		4.20	.91		5.11	5.95
4200	Practice putting, knitted nylon surface		↓	235	.034	↓	3.57	.91		4.48	5.25
4400	Polyurethane, thermoset, prefabricated in place, indoor										
4500	3/8" thick for basketball, gyms, etc.		1 Tilf	100	.080	S.F.	3.37	2.13		5.50	6.85
4600	1/2" thick for professional sports			95	.084		3.93	2.24		6.17	7.65
4700	Outdoor, 1/4" thick, smooth, for tennis			100	.080		3.01	2.13		5.14	6.45
4800	Rough, for track, 3/8" thick			95	.084		3.49	2.24		5.73	7.15
5000	Poured in place, indoor, with finish, 1/4" thick			80	.100		2.48	2.66		5.14	6.65
5050	3/8" thick			65	.123		3.01	3.28		6.29	8.15
5100	1/2" thick			50	.160		4	4.26		8.26	10.70
5500	Polyvinyl chloride, sheet goods for gyms, 1/4" thick			80	.100		3.43	2.66		6.09	7.70
5600	3/8" thick			60	.133		3.87	3.55		7.42	9.50
5900	Rubber, sheet goods, 36" wide, 1/8" thick			120	.067		3.01	1.78		4.79	5.95
5950	3/16" thick			100	.080		4.28	2.13		6.41	7.85
6000	1/4" thick			90	.089		4.95	2.37		7.32	8.95
6050	Tile, marbleized colors, 12" x 12", 1/8" thick			400	.020		3.03	.53		3.56	4.12
6100	3/16" thick			400	.020		4.34	.53		4.87	5.55
6300	Special tile, plain colors, 1/8" thick			400	.020		3.83	.53		4.36	5
6350	3/16" thick			400	.020		5.15	.53		5.68	6.45
6410	Raised, radial or square, minimum			400	.020		5	.53		5.53	6.30
6430	Maximum			400	.020		6	.53		6.53	7.40
6450	For golf course, skating rink, etc., 1/4" thick			275	.029		5.90	.78		6.68	7.60
6700	Synthetic turf, 3/8" thick		↓	90	.089	↓	3.16	2.37		5.53	7
6750	Interlocking 2' x 2' squares, 1/2" thick, not										
6810	cemented, for playgrounds, minimum		1 Tilf	210	.038	S.F.	2.61	1.02		3.63	4.37
6850	Maximum			190	.042		6.70	1.12		7.82	9
7000	Vinyl composition tile, 12" x 12", 1/16" thick			500	.016		.68	.43		1.11	1.38
7050	Embossed			500	.016		.86	.43		1.29	1.58
7100	Marbleized			500	.016		.86	.43		1.29	1.58
7150	Solid			500	.016		.97	.43		1.40	1.70
7200	3/32" thick, embossed			500	.016		.87	.43		1.30	1.59
7250	Marbleized			500	.016		.98	.43		1.41	1.71
7300	Solid			500	.016		1.43	.43		1.86	2.20
7350	1/8" thick, marbleized **CN**			500	.016		.98	.43		1.41	1.71
7400	Solid			500	.016		2.01	.43		2.44	2.84
7450	Conductive			500	.016		3.71	.43		4.14	4.71
7500	Vinyl tile, 12" x 12", .050" thick, minimum			500	.016		1.54	.43		1.97	2.32
7550	Maximum			500	.016		3.01	.43		3.44	3.94
7600	1/8" thick, minimum			500	.016		1.94	.43		2.37	2.76
7650	Solid colors			500	.016		4.35	.43		4.78	5.40
7700	Marbleized or Travertine pattern			500	.016		3.11	.43		3.54	4.05
7750	Florentine pattern			500	.016		3.57	.43		4	4.56
7800	Maximum			500	.016		7.35	.43		7.78	8.70
8000	Vinyl sheet goods, backed, .065" thick, minimum			250	.032		1.25	.85		2.10	2.64
8050	Maximum			200	.040		2.30	1.07		3.37	4.11
8100	.080" thick, minimum			230	.035		1.48	.93		2.41	3
8150	Maximum			200	.040		2.61	1.07		3.68	4.45
8200	.125" thick, minimum			230	.035		1.63	.93		2.56	3.16
8250	Maximum		↓	200	.040	↓	3.57	1.07		4.64	5.50
8700	Adhesive cement, 1 gallon does 200 to 300 S.F.					Gal.	13.40			13.40	14.75

FINISHES 9

For expanded coverage of these items see _Means Interior Cost Data 1999_

096 600 | Resilient Flooring

			CREW	DAILY OUTPUT	LABOR-HOURS	UNIT	1999 BARE COSTS				TOTAL INCL O&P	
							MAT.	LABOR	EQUIP.	TOTAL		
601	8800	Asphalt primer, 1 gallon per 300 S.F.				Gal.	8.40			8.40	9.25	**601**
	8900	Emulsion, 1 gallon per 140 S.F.	R096 -600				8.65			8.65	9.55	
	8950	Latex underlayment, liquid, fortified					27.50			27.50	30	

096 780 | Resilient Accessories

			CREW	DAILY OUTPUT	LABOR-HOURS	UNIT	MAT.	LABOR	EQUIP.	TOTAL	TOTAL INCL O&P	
781	0010	**STAIR TREADS AND RISERS** See index for materials other										**781**
	0100	than rubber and vinyl										
	0300	Rubber, molded tread, 12" wide, 5/16" thick, black	1 Tilf	115	.070	L.F.	6.55	1.85		8.40	10	
	0400	Colors		115	.070		6.75	1.85		8.60	10.20	
	0600	1/4" thick, black		115	.070		6.25	1.85		8.10	9.65	
	0700	Colors		115	.070		6.80	1.85		8.65	10.20	
	0900	Grip strip safety tread, colors, 5/16" thick		115	.070		9.85	1.85		11.70	13.60	
	1000	3/16" thick		120	.067		7.45	1.78		9.23	10.80	
	1200	Landings, smooth sheet rubber, 1/8" thick		120	.067	S.F.	2.47	1.78		4.25	5.35	
	1300	3/16" thick		120	.067	"	3.59	1.78		5.37	6.60	
	1500	Nosings, 3" wide, 3/16" thick, black		140	.057	L.F.	2.41	1.52		3.93	4.90	
	1600	Colors		140	.057		2.53	1.52		4.05	5.05	
	1800	Risers, 7" high, 1/8" thick, flat		250	.032		3.26	.85		4.11	4.85	
	1900	Coved		250	.032		2.29	.85		3.14	3.78	
	2100	Vinyl, molded tread, 12" wide, colors, 1/8" thick		115	.070		2.63	1.85		4.48	5.65	
	2200	1/4" thick		115	.070		4.33	1.85		6.18	7.50	
	2300	Landing material, 1/8" thick		200	.040	S.F.	2.73	1.07		3.80	4.58	
	2400	Riser, 7" high, 1/8" thick, coved		175	.046	L.F.	1.61	1.22		2.83	3.57	
	2500	Tread and riser combined, 1/8" thick		80	.100	"	4.06	2.66		6.72	8.40	

096 850 | Sheet Carpet

			CREW	DAILY OUTPUT	LABOR-HOURS	UNIT	MAT.	LABOR	EQUIP.	TOTAL	TOTAL INCL O&P	
852	0010	**CARPET** Commercial grades, direct cement										**852**
	0700	Nylon, level loop, 26 oz., light to medium traffic	1 Tilf	75	.107	S.Y.	12.70	2.84		15.54	18.20	
	0720	28 oz., light to medium traffic		75	.107		13.95	2.84		16.79	19.55	
	0900	32 oz., medium traffic		75	.107		18	2.84		20.84	24	
	1100	40 oz., medium to heavy traffic		75	.107		27	2.84		29.84	33.50	
	2920	Nylon plush, 30 oz., medium traffic		57	.140		13.30	3.74		17.04	20	
	3000	36 oz., medium traffic		75	.107		16.75	2.84		19.59	22.50	
	3100	42 oz., medium to heavy traffic		70	.114		19.30	3.05		22.35	26	
	3200	46 oz., medium to heavy traffic		70	.114		23	3.05		26.05	30	
	3300	54 oz., heavy traffic		70	.114		26	3.05		29.05	33	
	3340	60 oz., heavy traffic		70	.114		37	3.05		40.05	45	
	3665	Olefin, 24 oz., light to medium traffic		75	.107		7.75	2.84		10.59	12.75	
	3670	26 oz., medium traffic		75	.107		8.30	2.84		11.14	13.35	
	3680	28 oz., medium to heavy traffic		75	.107		9.40	2.84		12.24	14.55	
	3700	32 oz., medium to heavy traffic		75	.107		11.55	2.84		14.39	16.95	
	3730	42 oz., heavy traffic		70	.114		17.05	3.05		20.10	23.50	
	4110	Wool, level loop, 40 oz., medium traffic		75	.107		57.50	2.84		60.34	67.50	
	4500	50 oz., medium to heavy traffic		75	.107		59	2.84		61.84	69	
	4700	Patterned, 32 oz., medium to heavy traffic		70	.114		58.50	3.05		61.55	68.50	
	4900	48 oz., heavy traffic		70	.114		59.50	3.05		62.55	70	
	5000	For less than full roll, add					25%					
	5100	For small rooms, less than 12' wide, add						25%				
	5200	For large open areas (no cuts), deduct						25%				
	5600	For bound carpet baseboard, add	1 Tilf	300	.027	L.F.	1	.71		1.71	2.15	
	5610	For stairs, not incl. price of carpet, add	"	30	.267	Riser		7.10		7.10	10.50	
	5620	For borders and patterns, add to labor						18%				
	8950	For tackless, stretched installation, add padding to above										
	9000	Sponge rubber pad, minimum	1 Tilf	150	.053	S.Y.	2.55	1.42		3.97	4.90	

CN

Important: See the Reference Section for critical supporting data - Reference Nos., Crews, & City Cost Indexes

096 850 | Sheet Carpet

			CREW	DAILY OUTPUT	LABOR-HOURS	UNIT	1999 BARE COSTS				TOTAL INCL O&P	
							MAT.	LABOR	EQUIP.	TOTAL		
852	9100	Maximum	1 Tilf	150	.053	S.Y.	6.85	1.42		8.27	9.60	**852**
	9200	Felt pad, minimum		150	.053		3.04	1.42		4.46	5.45	
	9300	Maximum		150	.053		5.10	1.42		6.52	7.70	
	9400	Bonded urethane pad, minimum		150	.053		3.10	1.42		4.52	5.50	
	9500	Maximum		150	.053		5.50	1.42		6.92	8.15	
	9600	Prime urethane pad, minimum		150	.053		1.80	1.42		3.22	4.08	
	9700	Maximum	↓	150	.053		3.30	1.42		4.72	5.75	
	9850	For "branded" fiber, add					25%					
856	0010	**CARPET MAINTENANCE** Steam clean, per cleaning, minimum	1 Clab	3,000	.003	S.F.	.07	.06		.13	.17	**856**
	0500	Maximum	"	2,000	.004	"	.10	.09		.19	.24	

096 900 | Carpet Tile

			CREW	DAILY OUTPUT	LABOR-HOURS	UNIT	1999 BARE COSTS				TOTAL INCL O&P	
							MAT.	LABOR	EQUIP.	TOTAL		
901	0010	**CARPET TILE**										**901**
	0100	Tufted nylon, 18" x 18", hard back, 20 oz.	1 Tilf	150	.053	S.Y.	16.60	1.42		18.02	20.50	
	0110	26 oz.		150	.053		28.50	1.42		29.92	33	
	0200	Cushion back, 20 oz.		150	.053		21	1.42		22.42	25	
	0210	26 oz.	↓	150	.053	↓	32.50	1.42		33.92	38	

097 | Special Flooring & Floor Treatment

097 200 | Epoxy-Marble Flooring

			CREW	DAILY OUTPUT	LABOR-HOURS	UNIT	1999 BARE COSTS				TOTAL INCL O&P	
							MAT.	LABOR	EQUIP.	TOTAL		
201	0010	**COMPOSITION FLOORING** Acrylic, 1/4" thick	C-6	520	.092	S.F.	1.16	2.08	.14	3.38	4.68	**201**
	0100	3/8" thick		450	.107		1.52	2.41	.17	4.10	5.60	
	0300	Cupric oxychloride, on bond coat, minimum		480	.100		2.57	2.26	.16	4.99	6.50	
	0400	Maximum		420	.114		4.30	2.58	.18	7.06	8.95	
	0600	Epoxy, with colored quartz chips, broadcast, minimum		675	.071		2.04	1.60	.11	3.75	4.86	
	0700	Maximum		490	.098		2.47	2.21	.15	4.83	6.35	
	0900	Trowelled, minimum		560	.086		2.63	1.93	.13	4.69	6.05	
	1000	Maximum	↓	480	.100	↓	3.83	2.26	.16	6.25	7.90	
	1200	Heavy duty epoxy topping, 1/4" thick,										
	1300	500 to 1,000 S.F.	C-6	420	.114	S.F.	4.02	2.58	.18	6.78	8.65	
	1500	1,000 to 2,000 S.F.		450	.107		2.94	2.41	.17	5.52	7.15	
	1600	Over 10,000 S.F.	↓	480	.100		2.62	2.26	.16	5.04	6.55	
	1800	Epoxy terrazzo, 1/4" thick, chemical resistant, minimum	J-3	200	.080		4.28	1.93	.49	6.70	8.10	
	1900	Maximum		150	.107		7	2.58	.66	10.24	12.25	
	2100	Conductive, minimum		210	.076		5.40	1.84	.47	7.71	9.20	
	2200	Maximum	↓	160	.100		7.55	2.41	.62	10.58	12.55	
	2400	Mastic, hot laid, 2 coat, 1-1/2" thick, standard, minimum	C-6	690	.070		2.99	1.57	.11	4.67	5.85	
	2500	Maximum		520	.092		3.83	2.08	.14	6.05	7.60	
	2700	Acidproof, minimum		605	.079		3.83	1.79	.12	5.74	7.15	
	2800	Maximum		350	.137		5.30	3.10	.21	8.61	10.90	
	3000	Neoprene, trowelled on, 1/4" thick, minimum		545	.088		2.94	1.99	.14	5.07	6.50	
	3100	Maximum		430	.112		4.04	2.52	.17	6.73	8.55	
	3150	Polyacrylate terrazzo, 1/4" thick, minimum		735	.065		2.38	1.47	.10	3.95	5.05	
	3170	Maximum		480	.100		3.15	2.26	.16	5.57	7.15	
	3200	3/8" thick, minimum		620	.077		2.89	1.75	.12	4.76	6.05	
	3220	Maximum		480	.100		4.61	2.26	.16	7.03	8.75	
	3300	Conductive terrazzo, 1/4" thick, minimum		450	.107		5.70	2.41	.17	8.28	10.25	
	3330	Maximum	↓	305	.157	↓	7.35	3.55	.25	11.15	13.85	

For expanded coverage of these items see Means Interior Cost Data 1999

097 200	Epoxy-Marble Flooring	CREW	DAILY OUTPUT	LABOR-HOURS	UNIT	1999 BARE COSTS				TOTAL INCL O&P	
						MAT.	LABOR	EQUIP.	TOTAL		
201	3350	3/8" thick, minimum	C-6	365	.132	S.F.	7.70	2.97	.21	10.88	13.35	201
	3370	Maximum		255	.188		9.80	4.25	.29	14.34	17.65	
	3450	Granite, conductive, 1/4" thick, minimum		695	.069		5.85	1.56	.11	7.52	8.95	
	3470	Maximum		420	.114		7.75	2.58	.18	10.51	12.75	
	3500	3/8" thick, minimum		695	.069		8.45	1.56	.11	10.12	11.85	
	3520	Maximum		380	.126		10.55	2.85	.20	13.60	16.25	
	3600	Polyester, with colored quartz chips, 1/16" thick, minimum		1,065	.045		2.10	1.02	.07	3.19	3.97	
	3700	Maximum		560	.086		3.18	1.93	.13	5.24	6.65	
	3900	1/8" thick, minimum		810	.059		2.54	1.34	.09	3.97	4.97	
	4000	Maximum		675	.071		3.67	1.60	.11	5.38	6.65	
	4200	Polyester, heavy duty, compared to epoxy, add		2,590	.019		.72	.42	.03	1.17	1.47	
	4300	Polyurethane, with suspended vinyl chips, minimum		1,065	.045		5.70	1.02	.07	6.79	7.95	
	4500	Maximum	▼	860	.056	▼	8.25	1.26	.09	9.60	11.10	

098 150	Glazed Coatings	CREW	DAILY OUTPUT	LABOR-HOURS	UNIT	1999 BARE COSTS				TOTAL INCL O&P	
						MAT.	LABOR	EQUIP.	TOTAL		
150	0010	WALL COATINGS Acrylic glazed coatings, minimum	1 Pord	525	.015	S.F.	.23	.38		.61	.83	150
	0100	Maximum		305	.026		.49	.65		1.14	1.54	
	0300	Epoxy coatings, minimum		525	.015		.30	.38		.68	.91	
	0400	Maximum		170	.047		.93	1.17		2.10	2.81	
	0600	Exposed aggregate, troweled on, 1/16" to 1/4", minimum		235	.034		.47	.85		1.32	1.82	
	0700	Maximum (epoxy or polyacrylate)		130	.062		1	1.53		2.53	3.44	
	0900	1/2" to 5/8" aggregate, minimum		130	.062		.92	1.53		2.45	3.35	
	1000	Maximum		80	.100		1.57	2.49		4.06	5.55	
	1500	Exposed aggregate, sprayed on, 1/8" aggregate, minimum		295	.027		.44	.68		1.12	1.51	
	1600	Maximum		145	.055		.80	1.37		2.17	2.98	
	1800	High build epoxy, 50 mil, minimum		390	.021		.50	.51		1.01	1.33	
	1900	Maximum		95	.084		.87	2.10		2.97	4.17	
	2100	Laminated epoxy with fiberglass, minimum		295	.027		.55	.68		1.23	1.64	
	2200	Maximum		145	.055		.99	1.37		2.36	3.19	
	2400	Sprayed perlite or vermiculite, 1/16" thick, minimum		2,935	.003		.19	.07		.26	.31	
	2500	Maximum		640	.013		.56	.31		.87	1.10	
	2700	Vinyl plastic wall coating, minimum		735	.011		.25	.27		.52	.69	
	2800	Maximum		240	.033		.62	.83		1.45	1.95	
	3000	Urethane on smooth surface, 2 coats, minimum		1,135	.007		.18	.18		.36	.47	
	3100	Maximum		665	.012		.43	.30		.73	.93	
	3600	Ceramic-like glazed coating, cementitious, minimum		440	.018		.37	.45		.82	1.10	
	3700	Maximum		345	.023		.61	.58		1.19	1.55	
	3900	Resin base, minimum		640	.013		.25	.31		.56	.76	
	4000	Maximum	▼	330	.024	▼	.42	.60		1.02	1.38	

9 FINISHES

Important: See the Reference Section for critical supporting data - Reference Nos., Crews, & City Cost Indexes

104		099 100	**Exterior Painting**		DAILY OUTPUT	LABOR-HOURS	UNIT	1999 BARE COSTS				TOTAL INCL O&P	104
				CREW				MAT.	LABOR	EQUIP.	TOTAL		
	0010	**DOORS AND WINDOWS**	R099 -100										
	0100	Door frames & trim, only											
	0110	Brushwork, primer	R099 -220	1 Pord	512	.016	L.F.	.05	.39		.44	.65	
	0120	Finish coat, exterior latex			512	.016		.05	.39		.44	.66	
	0130	Primer & 1 coat, exterior latex			300	.027		.10	.66		.76	1.13	
	0135	2 coats, exterior latex, both sides			15	.533	Ea.	4.62	13.30		17.92	25.50	
	0140	Primer & 2 coats, exterior latex			265	.030	L.F.	.15	.75		.90	1.31	
	0150	Doors, flush, both sides, incl. frame & trim											
	0160	Roll & brush, primer		1 Pord	10	.800	Ea.	3.42	19.90		23.32	34.50	
	0170	Finish coat, exterior latex			10	.800		3.93	19.90		23.83	35	
	0180	Primer & 1 coat, exterior latex			7	1.143		7.35	28.50		35.85	51.50	
	0190	Primer & 2 coats, exterior latex			5	1.600		11.30	40		51.30	73.50	
	0200	Brushwork, stain, sealer & 2 coats polyurethane			4	2		13.10	50		63.10	90.50	
	0210	Doors, French, both sides, 10-15 lite, incl. frame & trim											
	0220	Brushwork, primer		1 Pord	6	1.333	Ea.	1.71	33		34.71	53	
	0230	Finish coat, exterior latex			6	1.333		1.97	33		34.97	53	
	0240	Primer & 1 coat, exterior latex			3	2.667		3.68	66.50		70.18	106	
	0250	Primer & 2 coats, exterior latex			2	4		5.55	99.50		105.05	158	
	0260	Brushwork, stain, sealer & 2 coats polyurethane			2.50	3.200		4.70	79.50		84.20	127	
	0270	Doors, louvered, both sides, incl. frame & trim											
	0280	Brushwork, primer		1 Pord	7	1.143	Ea.	3.42	28.50		31.92	47.50	
	0290	Finish coat, exterior latex			7	1.143		3.93	28.50		32.43	48	
	0300	Primer & 1 coat, exterior latex			4	2		7.35	50		57.35	84	
	0310	Primer & 2 coats, exterior latex			3	2.667		11.05	66.50		77.55	114	
	0320	Brushwork, stain, sealer & 2 coats polyurethane			4.50	1.778		13.10	44.50		57.60	82	
	0330	Doors, panel, both sides, incl. frame & trim											
	0340	Roll & brush, primer		1 Pord	6	1.333	Ea.	3.42	33		36.42	55	
	0350	Finish coat, exterior latex			6	1.333		3.93	33		36.93	55.50	
	0360	Primer & 1 coat, exterior latex			3	2.667		7.35	66.50		73.85	110	
	0370	Primer & 2 coats, exterior latex			2.50	3.200		11.05	79.50		90.55	134	
	0380	Brushwork, stain, sealer & 2 coats polyurethane			3	2.667		13.10	66.50		79.60	116	
	0400	Windows, per ext. side, based on 15 SF											
	0410	1 to 6 lite											
	0420	Brushwork, primer		1 Pord	13	.615	Ea.	.68	15.30		15.98	24	
	0430	Finish coat, exterior latex			13	.615		.78	15.30		16.08	24.50	
	0440	Primer & 1 coat, exterior latex			8	1		1.45	25		26.45	39.50	
	0450	Primer & 2 coats, exterior latex			6	1.333		2.18	33		35.18	53.50	
	0460	Stain, sealer & 1 coat varnish			7	1.143		1.86	28.50		30.36	45.50	
	0470	7 to 10 lite											
	0480	Brushwork, primer		1 Pord	11	.727	Ea.	.68	18.10		18.78	28	
	0490	Finish coat, exterior latex			11	.727		.78	18.10		18.88	28.50	
	0500	Primer & 1 coat, exterior latex			7	1.143		1.45	28.50		29.95	45	
	0510	Primer & 2 coats, exterior latex			5	1.600		2.18	40		42.18	63.50	
	0520	Stain, sealer & 1 coat varnish			6	1.333		1.86	33		34.86	53	
	0530	12 lite											
	0540	Brushwork, primer		1 Pord	10	.800	Ea.	.68	19.90		20.58	31	
	0550	Finish coat, exterior latex			10	.800		.78	19.90		20.68	31.50	
	0560	Primer & 1 coat, exterior latex			6	1.333		1.45	33		34.45	52.50	
	0570	Primer & 2 coats, exterior latex			5	1.600		2.18	40		42.18	63.50	
	0580	Stain, sealer & 1 coat varnish			6	1.333		1.84	33		34.84	53	
	0590	For oil base paint, add						10%					
106	0010	**SIDING** Exterior, Alkyd (oil base)	R099 -100										106
	0500	Spray		2 Pord	4,550	.004	S.F.	.07	.09		.16	.21	
	0800	Paint 2 coats, brushwork	R099 -220		1,300	.012		.10	.31		.41	.58	
	1000	Spray			4,550	.004		.14	.09		.23	.28	

9

FINISHES

099 100	Exterior Painting		CREW	DAILY OUTPUT	LABOR-HOURS	UNIT	1999 BARE COSTS MAT.	LABOR	EQUIP.	TOTAL	TOTAL INCL O&P	
106	1200	Stucco, rough, oil base, paint 2 coats, brushwork	R099-100	2 Pord	1,300	.012	S.F.	.10	.31		.41	.58	**106**
	1400	Roller			1,625	.010		.10	.25		.35	.49	
	1600	Spray	R099-220		2,925	.005		.11	.14		.25	.33	
	1800	Texture 1-11 or clapboard, oil base, primer coat, brushwork			1,300	.012		.08	.31		.39	.55	
	2000	Spray			4,550	.004		.08	.09		.17	.21	
	2400	Paint 2 coats, brushwork			810	.020		.14	.49		.63	.90	
	2600	Spray			2,600	.006		.16	.15		.31	.40	
	3400	Stain 2 coats, brushwork			950	.017		.08	.42		.50	.73	
	4000	Spray			3,050	.005		.09	.13		.22	.30	
	4200	Wood shingles, oil base primer coat, brushwork			1,300	.012		.07	.31		.38	.55	
	4400	Spray			3,900	.004		.07	.10		.17	.23	
	5000	Paint 2 coats, brushwork			810	.020		.12	.49		.61	.88	
	5200	Spray			2,275	.007		.11	.18		.29	.39	
	6500	Stain 2 coats, brushwork			950	.017		.08	.42		.50	.73	
	7000	Spray			2,660	.006		.11	.15		.26	.35	
	8000	For latex paint, deduct						10%					
108	0010	**FENCES**	R099-100										**108**
	0100	Chain link or wire metal, water base											
	0110	Roll & brush, first coat	R099-220	1 Pord	960	.008	S.F.	.05	.21		.26	.38	
	0120	Second coat			1,280	.006	"	.05	.16		.21	.29	
	0130	Spray, first coat			2,275	.004	S.F.	.05	.09		.14	.19	
	0140	Second coat			2,600	.003	"	.05	.08		.13	.18	
	0150	Picket, water base											
	0160	Roll & brush, first coat		1 Pord	865	.009	S.F.	.06	.23		.29	.41	
	0170	Second coat			1,050	.008	S.F.	.06	.19		.25	.35	
	0180	Spray, first coat			2,275	.004		.06	.09		.15	.19	
	0190	Second coat			2,600	.003		.06	.08		.14	.18	
	0200	Stockade, water base											
	0210	Roll & brush, first coat		1 Pord	1,040	.008	S.F.	.06	.19		.25	.35	
	0220	Second coat			1,200	.007	S.F.	.06	.17		.23	.31	
	0230	Spray, first coat			2,275	.004		.06	.09		.15	.19	
	0240	Second coat			2,600	.003		.06	.08		.14	.18	
112	0010	**MISCELLANEOUS,** for painting metals see Div. 051-255	R099-100										**112**
	0100	Railing, ext., decorative wood, incl. cap & baluster											
	0110	newels & spindles @ 12" O.C.	R099-220										
	0120	Brushwork, stain, sand, seal & varnish											
	0130	First coat		1 Pord	90	.089	L.F.	.37	2.21		2.58	3.80	
	0140	Second coat		"	120	.067	"	.37	1.66		2.03	2.95	
	0150	Rough sawn wood, 42" high, 2"x2" verticals, 6" O.C.											
	0160	Brushwork, stain, each coat		1 Pord	90	.089	L.F.	.12	2.21		2.33	3.52	
	0170	Wrought iron, 1" rail, 1/2" sq. verticals											
	0180	Brushwork, zinc chromate, 60" high, bars 6" O.C.											
	0190	Primer		1 Pord	130	.062	L.F.	.24	1.53		1.77	2.60	
	0200	Finish coat			130	.062		.15	1.53		1.68	2.50	
	0210	Additional coat			190	.042		.17	1.05		1.22	1.79	
	0220	Shutters or blinds, single panel, 2'x4', paint all sides											
	0230	Brushwork, primer		1 Pord	20	.400	Ea.	.50	9.95		10.45	15.80	
	0240	Finish coat, exterior latex			20	.400		.41	9.95		10.36	15.70	
	0250	Primer & 1 coat, exterior latex			13	.615		.80	15.30		16.10	24.50	
	0260	Spray, primer			35	.229		.72	5.70		6.42	9.50	
	0270	Finish coat, exterior latex			35	.229		.88	5.70		6.58	9.65	
	0280	Primer & 1 coat, exterior latex			20	.400		.78	9.95		10.73	16.10	
	0290	For louvered shutters, add					S.F.	10%					
	0300	Stair stringers, exterior, metal											
	0310	Roll & brush, zinc chromate, to 14", each coat		1 Pord	320	.025	L.F.	.05	.62		.67	1	
	0320	Rough sawn wood, 4" x 12"											

9 FINISHES

Important: See the Reference Section for critical supporting data - Reference Nos., Crews, & City Cost Indexes

099 100	Exterior Painting		CREW	DAILY OUTPUT	LABOR-HOURS	UNIT	1999 BARE COSTS				TOTAL INCL O&P	
							MAT.	LABOR	EQUIP.	TOTAL		
112	0330	Roll & brush, exterior latex, each coat	R099 -100	1 Pord	215	.037	L.F.	.06	.93		.99	1.49	**112**
	0340	Trellis/lattice, 2"x2" @ 3" O.C. with 2"x8" supports										
	0350	Spray, latex, per side, each coat	R099 -220	1 Pord	475	.017	S.F.	.06	.42		.48	.71
	0450	Decking, Ext., sealer, alkyd, brushwork, sealer coat			1,140	.007		.04	.17		.21	.32
	0460	1st coat			1,140	.007		.05	.17		.22	.32
	0470	2nd coat			1,300	.006		.03	.15		.18	.27
	0500	Paint, alkyd, brushwork, primer coat			1,140	.007		.06	.17		.23	.34
	0510	1st coat			1,140	.007		.06	.17		.23	.34
	0520	2nd coat			1,300	.006	S.F.	.04	.15		.19	.28
	0600	Sand paint, alkyd, brushwork, 1 coat			150	.053	S.F.	.07	1.33		1.40	2.10
116	0010	**SIDING**	R099 -100										**116**
	0100	Aluminum siding										
	0110	Brushwork, primer	R099 -220	2 Pord	2,275	.007	S.F.	.05	.18		.23	.32
	0120	Finish coat, exterior latex			2,275	.007	S.F.	.04	.18		.22	.31
	0130	Primer & 1 coat exterior latex			1,300	.012	S.F.	.09	.31		.40	.57
	0140	Primer & 2 coats exterior latex			975	.016	"	.12	.41		.53	.76
	0150	Mineral Fiber shingles										
	0160	Brushwork, primer		2 Pord	1,495	.011	S.F.	.08	.27		.35	.49
	0170	Finish coat, industrial enamel			1,495	.011	S.F.	.08	.27		.35	.49
	0180	Primer & 1 coat enamel			810	.020	S.F.	.15	.49		.64	.92
	0190	Primer & 2 coats enamel			540	.030		.23	.74		.97	1.38
	0200	Roll, primer			1,625	.010		.09	.25		.34	.47
	0210	Finish coat, industrial enamel			1,625	.010		.08	.25		.33	.47
	0220	Primer & 1 coat enamel			975	.016		.17	.41		.58	.81
	0230	Primer & 2 coats enamel			650	.025		.25	.61		.86	1.21
	0240	Spray, primer			3,900	.004		.07	.10		.17	.23
	0250	Finish coat, industrial enamel			3,900	.004		.07	.10		.17	.24
	0260	Primer & 1 coat enamel			2,275	.007		.13	.18		.31	.42
	0270	Primer & 2 coats enamel			1,625	.010		.20	.25		.45	.60
	0280	Waterproof sealer, first coat			4,485	.004		.06	.09		.15	.21
	0290	Second coat			5,235	.003		.06	.08		.14	.18
	0300	Rough wood incl. shingles, shakes or rough sawn siding										
	0310	Brushwork, primer		2 Pord	1,280	.013	S.F.	.10	.31		.41	.59
	0320	Finish coat, exterior latex			1,280	.013	S.F.	.06	.31		.37	.55
	0330	Primer & 1 coat exterior latex			960	.017	S.F.	.16	.41		.57	.81
	0340	Primer & 2 coats exterior latex			700	.023	S.F.	.23	.57		.80	1.12
	0350	Roll, primer			2,925	.005	S.F.	.13	.14		.27	.36
	0360	Finish coat, exterior latex			2,925	.005		.08	.14		.22	.29
	0370	Primer & 1 coat exterior latex			1,790	.009		.21	.22		.43	.57
	0380	Primer & 2 coats exterior latex			1,300	.012		.28	.31		.59	.78
	0390	Spray, primer			3,900	.004		.11	.10		.21	.28
	0400	Finish coat, exterior latex			3,900	.004		.06	.10		.16	.22
	0410	Primer & 1 coat exterior latex			2,600	.006		.17	.15		.32	.42
	0420	Primer & 2 coats exterior latex			2,080	.008		.23	.19		.42	.54
	0430	Waterproof sealer, first coat			4,485	.004		.11	.09		.20	.26
	0440	Second coat			4,485	.004		.06	.09		.15	.21
	0450	Smooth wood incl. butt, T&G, beveled, drop or B&B siding										
	0460	Brushwork, primer		2 Pord	2,325	.007	S.F.	.07	.17		.24	.34
	0470	Finish coat, exterior latex			1,280	.013	S.F.	.06	.31		.37	.55
	0480	Primer & 1 coat exterior latex			800	.020	S.F.	.13	.50		.63	.91
	0490	Primer & 2 coats exterior latex			630	.025		.20	.63		.83	1.19
	0500	Roll, primer			2,275	.007		.08	.18		.26	.36
	0510	Finish coat, exterior latex			2,275	.007		.07	.18		.25	.34
	0520	Primer & 1 coat exterior latex			1,300	.012		.15	.31		.46	.63
	0530	Primer & 2 coats exterior latex			975	.016		.22	.41		.63	.87
	0540	Spray, primer			4,550	.004		.06	.09		.15	.20

For expanded coverage of these items see *Means Interior Cost Data 1999*

FINISHES **9**

				DAILY	LABOR-		1999 BARE COSTS				TOTAL	
099 100		**Exterior Painting**	CREW	OUTPUT	HOURS	UNIT	MAT.	LABOR	EQUIP.	TOTAL	INCL O&P	
116	0550	Finish coat, exterior latex	2 Pord	4,550	.004	S.F.	.06	.09		.15	.19	**116**
	0560	Primer & 1 coat exterior latex		2,600	.006		.12	.15		.27	.36	
	0570	Primer & 2 coats exterior latex		1,950	.008		.18	.20		.38	.51	
	0580	Waterproof sealer, first coat		5,230	.003		.06	.08		.14	.19	
	0590	Second coat		5,980	.003		.06	.07		.13	.17	
	0600	For oil base paint, add				S.F.	10%					
120	0010	**TRIM**										**120**
	0100	Door frames & trim (see Doors, interior or exterior)										
	0110	Fascia, latex paint, one coat coverage										
	0120	1" x 4", brushwork	1 Pord	640	.013	L.F.	.02	.31		.33	.50	
	0130	Roll		1,280	.006		.02	.16		.18	.26	
	0140	Spray		2,080	.004		.01	.10		.11	.16	
	0150	1" x 6" to 1" x 10", brushwork		640	.013		.06	.31		.37	.54	
	0160	Roll		1,230	.007		.06	.16		.22	.31	
	0170	Spray		2,100	.004		.04	.09		.13	.20	
	0180	1" x 12", brushwork		640	.013		.06	.31		.37	.54	
	0190	Roll		1,050	.008		.06	.19		.25	.35	
	0200	Spray		2,200	.004		.04	.09		.13	.19	
	0210	Gutters & downspouts, metal, zinc chromate paint										
	0220	Brushwork, gutters, 5", first coat	1 Pord	640	.013	L.F.	.05	.31		.36	.54	
	0230	Second coat		960	.008		.05	.21		.26	.37	
	0240	Third coat		1,280	.006		.04	.16		.20	.28	
	0250	Downspouts, 4", first coat		640	.013		.05	.31		.36	.54	
	0260	Second coat		960	.008		.05	.21		.26	.37	
	0270	Third coat		1,280	.006		.04	.16		.20	.28	
	0280	Gutters & downspouts, wood										
	0290	Brushwork, gutters, 5", primer	1 Pord	640	.013	L.F.	.05	.31		.36	.53	
	0300	Finish coat, exterior latex		640	.013		.05	.31		.36	.53	
	0310	Primer & 1 coat exterior latex		400	.020		.10	.50		.60	.87	
	0320	Primer & 2 coats exterior latex		325	.025		.15	.61		.76	1.10	
	0330	Downspouts, 4", primer		640	.013		.05	.31		.36	.53	
	0340	Finish coat, exterior latex		640	.013		.05	.31		.36	.53	
	0350	Primer & 1 coat exterior latex		400	.020		.10	.50		.60	.87	
	0360	Primer & 2 coats exterior latex		325	.025		.07	.61		.68	1.02	
	0370	Molding, exterior, up to 14" wide										
	0380	Brushwork, primer	1 Pord	640	.013	L.F.	.05	.31		.36	.54	
	0390	Finish coat, exterior latex		640	.013		.06	.31		.37	.54	
	0400	Primer & 1 coat exterior latex		400	.020		.12	.50		.62	.89	
	0410	Primer & 2 coats exterior latex		315	.025		.12	.63		.75	1.10	
	0420	Stain & fill		1,050	.008		.05	.19		.24	.34	
	0430	Shellac		1,850	.004		.05	.11		.16	.22	
	0440	Varnish		1,275	.006		.06	.16		.22	.31	
124	0350	**WALLS, MASONRY (CMU)**										**124**
	0360	Concrete masonry units (CMU), smooth surface										
	0370	Brushwork, latex, first coat	1 Pord	640	.013	S.F.	.03	.31		.34	.52	
	0380	Second coat		960	.008		.03	.21		.24	.35	
	0390	Waterproof sealer, first coat		736	.011		.06	.27		.33	.48	
	0400	Second coat		1,104	.007		.04	.18		.22	.32	
	0410	Roll, latex, paint, first coat		1,465	.005		.04	.14		.18	.25	
	0420	Second coat		1,790	.004		.03	.11		.14	.20	
	0430	Waterproof sealer, first coat		1,680	.005		.06	.12		.18	.25	
	0440	Second coat		2,060	.004		.04	.10		.14	.19	
	0450	Spray, latex, paint, first coat		1,950	.004		.03	.10		.13	.19	
	0460	Second coat		2,600	.003		.02	.08		.10	.15	
	0470	Waterproof sealer, first coat		2,245	.004		.04	.09		.13	.18	
	0480	Second coat		2,990	.003		.03	.07		.10	.13	

9

FINISHES

Important: See the Reference Section for critical supporting data - Reference Nos., Crews, & City Cost Indexes

099 100 | Exterior Painting

		CREW	DAILY OUTPUT	LABOR-HOURS	UNIT	1999 BARE COSTS				TOTAL INCL O&P	
						MAT.	LABOR	EQUIP.	TOTAL		
124	0490	Concrete masonry unit (CMU), porous										**124**
	0500	Brushwork, latex, first coat	1 Pord	640	.013	S.F.	.07	.31		.38	.55	
	0510	Second coat		960	.008		.03	.21		.24	.36	
	0520	Waterproof sealer, first coat		736	.011		.05	.27		.32	.46	
	0530	Second coat		1,104	.007		.04	.18		.22	.32	
	0540	Roll latex, first coat		1,465	.005		.05	.14		.19	.26	
	0550	Second coat		1,790	.004		.03	.11		.14	.20	
	0560	Waterproof sealer, first coat		1,400	.006		.06	.14		.20	.29	
	0570	Second coat		2,200	.004		.04	.09		.13	.19	
	0580	Spray latex, first coat		5,600	.001		.03	.04		.07	.08	
	0590	Second coat		6,400	.001		.02	.03		.05	.08	
	0600	Waterproof sealer, first coat		6,000	.001		.06	.03		.09	.12	
	0610	Second coat		6,400	.001		.03	.03		.06	.09	

099 200 | Interior Painting

			CREW	DAILY OUTPUT	LABOR-HOURS	UNIT	MAT.	LABOR	EQUIP.	TOTAL	TOTAL INCL O&P	
204	0010	**CABINETS AND CASEWORK**	R099 -100									**204**
	1000	Primer coat, oil base, brushwork		1 Pord	650	.012	S.F.	.04	.31		.35	.52	
	2000	Paint, oil base, brushwork, 1 coat	R099 -220		650	.012		.05	.31		.36	.53	
	3000	Stain, brushwork, wipe off			650	.012		.04	.31		.35	.51	
	4000	Shellac, 1 coat, brushwork			650	.012		.05	.31		.36	.52	
	4500	Varnish, 3 coats, brushwork, sand after 1st coat			325	.025		.16	.61		.77	1.11	
	5000	For latex paint, deduct						10%					
214	0010	**DOORS & WINDOWS, LATEX**	R099 -100										**214**
	0100	Doors flush, both sides, incl. frame & trim											
	0110	Roll & brush, primer	R099 -220	1 Pord	10	.800	Ea.	2.99	19.90		22.89	34	
	0120	Finish coat, latex			10	.800		3.04	19.90		22.94	34	
	0130	Primer & 1 coat latex			7	1.143		6.05	28.50		34.55	50	
	0140	Primer & 2 coats latex			5	1.600		8.90	40		48.90	71	
	0160	Spray, both sides, primer			20	.400		3.15	9.95		13.10	18.70	
	0170	Finish coat, latex			20	.400		3.19	9.95		13.14	18.75	
	0180	Primer & 1 coat latex			11	.727		6.40	18.10		24.50	34.50	
	0190	Primer & 2 coats latex			8	1		9.40	25		34.40	48.50	
	0200	Doors, French, both sides, 10-15 lite, incl. frame & trim											
	0210	Roll & brush, primer		1 Pord	6	1.333	Ea.	1.50	33		34.50	52.50	
	0220	Finish coat, latex			6	1.333		1.52	33		34.52	52.50	
	0230	Primer & 1 coat latex			3	2.667		3.02	66.50		69.52	105	
	0240	Primer & 2 coats latex			2	4		4.45	99.50		103.95	157	
	0260	Doors, louvered, both sides, incl. frame & trim											
	0270	Roll & brush, primer		1 Pord	7	1.143	Ea.	2.99	28.50		31.49	47	
	0280	Finish coat, latex			7	1.143		3.04	28.50		31.54	47	
	0290	Primer & 1 coat, latex			4	2		5.85	50		55.85	82.50	
	0300	Primer & 2 coats, latex			3	2.667		9.05	66.50		75.55	112	
	0320	Spray, both sides, primer			20	.400		3.15	9.95		13.10	18.70	
	0330	Finish coat, latex			20	.400		3.19	9.95		13.14	18.75	
	0340	Primer & 1 coat, latex			11	.727		6.40	18.10		24.50	34.50	
	0350	Primer & 2 coats, latex			8	1		9.60	25		34.60	48.50	
	0360	Doors, panel, both sides, incl. frame & trim											
	0370	Roll & brush, primer		1 Pord	6	1.333	Ea.	3.15	33		36.15	54.50	
	0380	Finish coat, latex			6	1.333		3.04	33		36.04	54.50	
	0390	Primer & 1 coat, latex			3	2.667		6.05	66.50		72.55	109	
	0400	Primer & 2 coats, latex			2.50	3.200		9.05	79.50		88.55	132	
	0420	Spray, both sides, primer			10	.800		3.15	19.90		23.05	34	
	0430	Finish coat, latex			10	.800		3.19	19.90		23.09	34	
	0440	Primer & 1 coat, latex			5	1.600		6.40	40		46.40	68	

For expanded coverage of these items see Means Interior Cost Data 1999

FINISHES

9

		099 200 | Interior Painting		CREW	DAILY OUTPUT	LABOR-HOURS	UNIT	1999 BARE COSTS				TOTAL INCL O&P	
								MAT.	LABOR	EQUIP.	TOTAL		
214	0450	Primer & 2 coats, latex	R099 -100	1 Pord	4	2	Ea.	9.60	50		59.60	86.50	**214**
	0460	Windows, per interior side, base on 15 SF											
	0470	1 to 6 lite	R099 -220										
	0480	Brushwork, primer		1 Pord	13	.615	Ea.	.59	15.30		15.89	24	
	0490	Finish coat, enamel			13	.615		.60	15.30		15.90	24	
	0500	Primer & 1 coat enamel			8	1		1.19	25		26.19	39.50	
	0510	Primer & 2 coats enamel		↓	6	1.333	↓	1.79	33		34.79	53	
	0530	7 to 10 lite											
	0540	Brushwork, primer		1 Pord	11	.727	Ea.	.59	18.10		18.69	28	
	0550	Finish coat, enamel			11	.727		.60	18.10		18.70	28	
	0560	Primer & 1 coat enamel			7	1.143		1.19	28.50		29.69	45	
	0570	Primer & 2 coats enamel		↓	5	1.600	↓	1.79	40		41.79	63	
	0590	12 lite											
	0600	Brushwork, primer		1 Pord	10	.800	Ea.	.59	19.90		20.49	31	
	0610	Finish coat, enamel			10	.800		.60	19.90		20.50	31	
	0620	Primer & 1 coat enamel			6	1.333		1.19	33		34.19	52.50	
	0630	Primer & 2 coats enamel		↓	5	1.600		1.79	40		41.79	63	
	0650	For oil base paint, add		↓			↓	10%					
216	0010	**DOORS AND WINDOWS, ALKYD (OIL BASE)**	R099 -100										**216**
	0500	Flush door & frame, 3' x 7', oil, primer, brushwork		1 Pord	10	.800	Ea.	1.71	19.90		21.61	32.50	
	1000	Paint, 1 coat	R099 -220		10	.800		1.79	19.90		21.69	32.50	
	1400	Stain, brushwork, wipe off			18	.444		.84	11.05		11.89	17.85	
	1600	Shellac, 1 coat, brushwork			25	.320		.94	7.95		8.89	13.25	
	1800	Varnish, 3 coats, brushwork, sand after 1st coat			9	.889		3.31	22		25.31	37.50	
	2000	Panel door & frame, 3' x 7', oil, primer, brushwork			6	1.333		1.61	33		34.61	53	
	2200	Paint, 1 coat			6	1.333		1.79	33		34.79	53	
	2600	Stain, brushwork, panel door, 3' x 7', not incl. frame			16	.500		.84	12.45		13.29	19.95	
	2800	Shellac, 1 coat, brushwork			22	.364		.94	9.05		9.99	14.90	
	3000	Varnish, 3 coats, brushwork, sand after 1st coat		↓	7.50	1.067	↓	3.31	26.50		29.81	44	
	4400	Windows, including frame and trim, per side											
	4600	Colonial type, 6/6 lites, 2' x 3', oil, primer, brushwork		1 Pord	14	.571	Ea.	.26	14.25		14.51	22.50	
	5800	Paint, 1 coat			14	.571		.28	14.25		14.53	22.50	
	6200	3' x 5' opening, 6/6 lites, primer coat, brushwork			12	.667		.64	16.60		17.24	26	
	6400	Paint, 1 coat			12	.667		.71	16.60		17.31	26.50	
	6800	4' x 8' opening, 6/6 lites, primer coat, brushwork			8	1		1.36	25		26.36	39.50	
	7000	Paint, 1 coat			8	1		1.51	25		26.51	39.50	
	8000	Single lite type, 2' x 3', oil base, primer coat, brushwork			33	.242		.26	6.05		6.31	9.55	
	8200	Paint, 1 coat			33	.242		.28	6.05		6.33	9.55	
	8600	3' x 5' opening, primer coat, brushwork			20	.400		.64	9.95		10.59	15.95	
	8800	Paint, 1 coat			20	.400		.71	9.95		10.66	16.05	
	9200	4' x 8' opening, primer coat, brushwork			14	.571		1.36	14.25		15.61	23.50	
	9400	Paint, 1 coat		↓	14	.571	↓	1.51	14.25		15.76	23.50	
218	0010	**FLOORS**											**218**
	0100	Concrete											
	0110	Brushwork, latex, block filler											
	0120	1st coat		1 Pord	975	.008	S.F.	.10	.20		.30	.42	
	0130	2nd coat			1,150	.007		.07	.17		.24	.34	
	0140	3rd coat		↓	1,300	.006	↓	.05	.15		.20	.29	
	0150	Roll, latex, block filler											
	0160	1st coat		1 Pord	2,600	.003	S.F.	.13	.08		.21	.26	
	0170	2nd coat			3,250	.002		.08	.06		.14	.18	
	0180	3rd coat		↓	3,900	.002	↓	.06	.05		.11	.14	
	0190	Spray, latex, block filler											
	0200	1st coat		1 Pord	2,600	.003	S.F.	.11	.08		.19	.24	
	0210	2nd coat			3,250	.002		.06	.06		.12	.16	
	0220	3rd coat		↓	3,900	.002	↓	.05	.05		.10	.13	

099 200 | Interior Painting

		CREW	DAILY OUTPUT	LABOR-HOURS	UNIT	1999 BARE COSTS				TOTAL INCL O&P
						MAT.	LABOR	EQUIP.	TOTAL	
220	**0010**	**MISCELLANEOUS PAINTING** R099-100								**220**
	2400	Floors, conc./wood, oil base, primer/sealer coat, brushwork	2 Pord	1,950	.008	S.F.	.05	.20		.25	.36
	2450	Roller R099-220		5,200	.003		.05	.08		.13	.18
	2600	Spray		6,000	.003		.05	.07		.12	.16
	2650	Paint 1 coat, brushwork		1,950	.008		.04	.20		.24	.35
	2800	Roller		5,200	.003		.04	.08		.12	.17
	2850	Spray		6,000	.003		.04	.07		.11	.15
	3000	Stain, wood floor, brushwork, 1 coat		4,550	.004		.04	.09		.13	.17
	3200	Roller		5,200	.003		.04	.08		.12	.17
	3250	Spray		6,000	.003		.04	.07		.11	.15
	3400	Varnish, wood floor, brushwork		4,550	.004		.05	.09		.14	.19
	3450	Roller		5,200	.003		.06	.08		.14	.18
	3600	Spray	▼	6,000	.003	▼	.06	.07		.13	.16
	3650	For dust proofing or anti skid, see division 033-454									
	3800	Grilles, per side, oil base, primer coat, brushwork	1 Pord	520	.015	S.F.	.08	.38		.46	.68
	3850	Spray		1,140	.007		.09	.17		.26	.37
	3920	Paint 2 coats, brushwork		325	.025		.18	.61		.79	1.14
	3940	Spray	▼	650	.012	▼	.21	.31		.52	.70
	5000	Pipe, to 4" diameter, primer or sealer coat, oil base, brushwork	2 Pord	1,250	.013	L.F.	.05	.32		.37	.55
	5100	Spray		2,165	.007		.05	.18		.23	.33
	5350	Paint 2 coats, brushwork		775	.021		.09	.51		.60	.89
	5400	Spray		1,240	.013		.10	.32		.42	.60
	6300	To 16" diameter, primer or sealer coat, brushwork		310	.052		.20	1.29		1.49	2.19
	6350	Spray		540	.030		.22	.74		.96	1.38
	6500	Paint 2 coats, brushwork		195	.082		.37	2.04		2.41	3.54
	6550	Spray	▼	310	.052	▼	.42	1.29		1.71	2.43
	7000	Trim, wood, incl. puttying, under 6" wide									
	7200	Primer coat, oil base, brushwork	1 Pord	650	.012	L.F.	.02	.31		.33	.49
	7250	Paint, 1 coat, brushwork		650	.012		.02	.31		.33	.50
	7450	3 coats		325	.025		.07	.61		.68	1.01
	7500	Over 6" wide, primer coat, brushwork		650	.012		.04	.31		.35	.52
	7550	Paint, 1 coat, brushwork		650	.012		.05	.31		.36	.52
	7650	3 coats		325	.025	▼	.14	.61		.75	1.09
	8000	Cornice, simple design, primer coat, oil base, brushwork		650	.012	S.F.	.04	.31		.35	.52
	8250	Paint, 1 coat		650	.012		.05	.31		.36	.52
	8350	Ornate design, primer coat		350	.023		.04	.57		.61	.92
	8400	Paint, 1 coat		350	.023		.05	.57		.62	.92
	8600	Balustrades, primer coat, oil base, brushwork		520	.015		.04	.38		.42	.64
	8650	Paint, 1 coat		520	.015		.05	.38		.43	.64
	8900	Trusses and wood frames, primer coat, oil base, brushwork		800	.010		.04	.25		.29	.43
	8950	Spray		1,200	.007		.04	.17		.21	.30
	9220	Paint 2 coats, brushwork		500	.016		.09	.40		.49	.71
	9240	Spray		600	.013		.10	.33		.43	.62
	9260	Stain, brushwork, wipe off		600	.013		.04	.33		.37	.55
	9280	Varnish, 3 coats, brushwork	▼	275	.029		.16	.72		.88	1.28
	9350	For latex paint, deduct				▼	10%				
224	**0010**	**WALLS AND CEILINGS** R099-100								**224**
	0100	Concrete, dry wall or plaster, oil base, primer or sealer coat									
	0200	Smooth finish, brushwork R099-220	1 Pord	1,150	.007	S.F.	.04	.17		.21	.32
	0240	Roller		2,040	.004		.04	.10		.14	.19
	0300	Sand finish, brushwork		975	.008		.04	.20		.24	.35
	0340	Roller		1,150	.007		.04	.17		.21	.32
	0380	Spray		2,275	.004		.03	.09		.12	.17
	0800	Paint 2 coats, smooth finish, brushwork		680	.012		.07	.29		.36	.53
	0840	Roller		800	.010		.07	.25		.32	.46
	0880	Spray	▼	1,625	.005	▼	.06	.12		.18	.26

For expanded coverage of these items see *Means Interior Cost Data 1999*

9 FINISHES

099 200 | Interior Painting

			CREW	DAILY OUTPUT	LABOR-HOURS	UNIT	1999 BARE COSTS MAT.	LABOR	EQUIP.	TOTAL	TOTAL INCL O&P	
224	0900	Sand finish, brushwork	1 Pord	605	.013	S.F.	.07	.33		.40	.58	224
	0940	Roller		1,020	.008		.07	.20		.27	.38	
	0980	Spray		1,700	.005		.06	.12		.18	.25	
	1200	Paint 3 coats, smooth finish, brushwork		510	.016		.10	.39		.49	.72	
	1240	Roller		650	.012		.11	.31		.42	.59	
	1241	Light clean and paint, roller		375	.021		.14	.53		.67	.97	
	1280	Spray		1,625	.005		.10	.12		.22	.30	
	1600	Glaze coating, 5 coats, spray, clear		900	.009		.60	.22		.82	1	
	1640	Multicolor		900	.009		.77	.22		.99	1.19	
	1700	For latex paint, deduct					10%					
	1800	For ceiling installations, add						25%				
	2000	Masonry or concrete block, oil base, primer or sealer coat										
	2100	Smooth finish, brushwork	1 Pord	1,224	.007	S.F.	.04	.16		.20	.30	
	2180	Spray		2,400	.003		.06	.08		.14	.20	
	2200	Sand finish, brushwork		1,089	.007		.07	.18		.25	.36	
	2280	Spray		2,400	.003		.06	.08		.14	.20	
	2800	Paint 2 coats, smooth finish, brushwork		756	.011		.14	.26		.40	.55	
	2880	Spray		1,360	.006		.13	.15		.28	.36	
	2900	Sand finish, brushwork		672	.012		.14	.30		.44	.60	
	2980	Spray		1,360	.006		.13	.15		.28	.36	
	3600	Glaze coating, 5 coats, spray, clear		900	.009		.60	.22		.82	1	
	3620	Multicolor		900	.009		.77	.22		.99	1.19	
	4000	Block filler, 1 coat, brushwork		425	.019		.11	.47		.58	.85	
	4100	Silicone, water repellent, 2 coats, spray		2,000	.004		.23	.10		.33	.41	
	4120	For latex paint, deduct					10%					
228	0010	**VARNISH** 1 coat + sealer, on wood trim, no sanding included	1 Pord	400	.020	S.F.	.06	.50		.56	.83	228
	0100	Hardwood floors, 2 coats, no sanding included, roller	"	1,890	.004	"	.11	.11		.22	.28	

099 700 | Wallpaper

			CREW	DAILY OUTPUT	LABOR-HOURS	UNIT	1999 BARE COSTS MAT.	LABOR	EQUIP.	TOTAL	TOTAL INCL O&P	
701	0010	**WALL COVERING** Including sizing, add 10%-30% waste at takeoff										701
	0050	Aluminum foil	1 Pape	275	.029	S.F.	.76	.73		1.49	1.96	
	0100	Copper sheets, .025" thick, vinyl backing		240	.033		4.03	.84		4.87	5.70	
	0300	Phenolic backing		240	.033		5.20	.84		6.04	7.05	
	0600	Cork tiles, light or dark, 12" x 12" x 3/16"		240	.033		2.59	.84		3.43	4.13	
	0700	5/16" thick		235	.034		2.69	.85		3.54	4.27	
	0900	1/4" basketweave		240	.033		4.13	.84		4.97	5.80	
	1000	1/2" natural, non-directional pattern		240	.033		5.95	.84		6.79	7.85	
	1100	3/4" natural, non-directional pattern		240	.033		7.80	.84		8.64	9.90	
	1200	Granular surface, 12" x 36", 1/2" thick		385	.021		.89	.52		1.41	1.78	
	1300	1" thick		370	.022		1.16	.54		1.70	2.11	
	1500	Polyurethane coated, 12" x 12" x 3/16" thick		240	.033		2.79	.84		3.63	4.35	
	1600	5/16" thick		235	.034		3.96	.85		4.81	5.65	
	1800	Cork wallpaper, paperbacked, natural		480	.017		1.58	.42		2	2.38	
	1900	Colors		480	.017		1.96	.42		2.38	2.80	
	2100	Flexible wood veneer, 1/32" thick, plain woods		100	.080		1.59	2.01		3.60	4.83	
	2200	Exotic woods		95	.084		2.42	2.11		4.53	5.90	
	2400	Gypsum-based, fabric-backed, fire										
	2500	resistant for masonry walls, minimum, 21 oz./S.Y.	1 Pape	800	.010	S.F.	.63	.25		.88	1.07	
	2700	Maximum, (small quantities)	"	640	.013		.97	.31		1.28	1.55	
	2750	Acrylic, modified, semi-rigid PVC, .028" thick	2 Carp	330	.048		.80	1.32		2.12	2.96	
	2800	.040" thick	"	320	.050		1.05	1.36		2.41	3.30	
	3000	Vinyl wall covering, fabric-backed, lightweight, (12-15 oz./S.Y.)	1 Pape	640	.013		.51	.31		.82	1.04	
	3300	Medium weight, type 2, (20-24 oz./S.Y.)		480	.017		.63	.42		1.05	1.34	
	3400	Heavy weight, type 3, (28 oz./S.Y.)		435	.018		1.15	.46		1.61	1.97	
	3600	Adhesive, 5 gal. lots, (18SY/Gal.)				Gal.	8.05			8.05	8.85	

Important: See the Reference Section for critical supporting data - Reference Nos., Crews, & City Cost Indexes

9 FINISHES

099 700 | Wallpaper

			CREW	DAILY OUTPUT	LABOR-HOURS	UNIT	MAT.	LABOR	EQUIP.	TOTAL	TOTAL INCL O&P	
								1999 BARE COSTS					
701	3700	Wallpaper, average workmanship, solid pattern, low cost paper	1 Pape	640	.013	S.F.	.26	.31		.57	.77	701
	3900	basic patterns (matching required), avg. cost paper		535	.015		.47	.38		.85	1.09	
	4000	Paper at $50 per double roll, quality workmanship		435	.018		.89	.46		1.35	1.69	
	4100	Linen wall covering, paper backed										
	4150	Flame treatment, minimum				S.F.	.58			.58	.64	
	4180	Maximum					1.14			1.14	1.25	
	4200	Grass cloths with lining paper, minimum	1 Pape	400	.020		.57	.50		1.07	1.40	
	4300	Maximum	"	350	.023		1.83	.57		2.40	2.89	

099 860 | Special Wall Surfaces

			CREW	DAILY OUTPUT	LABOR-HOURS	UNIT	MAT.	LABOR	EQUIP.	TOTAL	TOTAL INCL O&P	
861	0010	FIBERGLASS REINFORCED PLASTIC panels, .090" thick, on walls										861
	0020	Adhesive mounted, embossed surface	2 Carp	640	.025	S.F.	1.09	.68		1.77	2.27	
	0030	Smooth surface		640	.025		1.23	.68		1.91	2.42	
	0040	Fire rated, embossed surface		640	.025		1.57	.68		2.25	2.80	
	0050	Nylon rivet mounted, on drywall, embossed surface		480	.033		1.21	.91		2.12	2.76	
	0060	Smooth surface		480	.033		1.35	.91		2.26	2.92	
	0070	Fire rated, embossed surface		480	.033		1.69	.91		2.60	3.29	
	0080	On masonry, embossed surface		320	.050		1.21	1.36		2.57	3.48	
	0090	Smooth surface		320	.050		1.35	1.36		2.71	3.64	
	0100	Fire rated, embossed surface		320	.050		1.69	1.36		3.05	4.01	
	0110	Nylon rivet and adhesive mounted, on drywall, embossed surface		240	.067		1.30	1.82		3.12	4.29	
	0120	Smooth surface		240	.067		1.46	1.82		3.28	4.47	
	0130	Fire rated, embossed surface		240	.067		1.79	1.82		3.61	4.83	
	0140	On masonry, embossed surface		190	.084		1.30	2.30		3.60	5.05	
	0150	Smooth surface		190	.084		1.46	2.30		3.76	5.25	
	0160	Fire rated, embossed surface		190	.084		1.79	2.30		4.09	5.60	
	0170	For moldings add	1 Carp	250	.032	L.F.	.22	.87		1.09	1.61	
	0180	On ceilings, for lay in grid system, embossed surface		400	.020	S.F.	1.09	.55		1.64	2.06	
	0190	Smooth surface		400	.020		1.23	.55		1.78	2.21	
	0200	Fire rated, embossed surface		400	.020		1.57	.55		2.12	2.59	
863	0010	RAISED PANEL SYSTEM, 3/4" MDO										863
	0100	Standard, paint grade	2 Carp	300	.053	S.F.	9.70	1.46		11.16	12.95	
	0110	Oak veneer		300	.053		12.15	1.46		13.61	15.65	
	0120	Maple veneer		300	.053		14.20	1.46		15.66	17.90	
	0130	Cherry veneer		300	.053		17.55	1.46		19.01	21.50	
	0300	Class I fire rated, paint grade		300	.053		17.55	1.46		19.01	21.50	
	0310	Oak veneer		300	.053		19.80	1.46		21.26	24.50	
	0320	Maple veneer		300	.053		21.50	1.46		22.96	26.50	
	0330	Cherry veneer		300	.053		25.50	1.46		26.96	30.50	
	5000	For prefinished paneling, see division 062-504 and 062-554										
865	0010	SLATWALL PANELS AND ACCESSORIES										865
	0100	Slatwall panel, 4' x 8' x 3/4" T, MDF, paint grade	1 Carp	500	.016	S.F.	1.15	.44		1.59	1.96	
	0110	Melamine finish		500	.016		1.65	.44		2.09	2.51	
	0120	High pressure plastic laminate finish		500	.016		1.95	.44		2.39	2.84	
	0130	Aluminum channel inserts, add					1.40			1.40	1.54	
	0200	Accessories, corner forms, 8' L				L.F.	3.10			3.10	3.41	
	0210	T-connector, 8' L					4.50			4.50	4.95	
	0220	J-mold, 8' L					1.15			1.15	1.27	
	0230	Edge cap, 8' L					.75			.75	.83	
	0240	Finish end cap, 8' L					2.75			2.75	3.03	
	0300	Display hook, 4" L				Ea.	.85			.85	.94	
	0310	6" L					.90			.90	.99	
	0320	8" L					1			1	1.10	
	0330	10" L					1.10			1.10	1.21	
	0340	12" L					1.20			1.20	1.32	
	0350	Acrylic, 4" L					.70			.70	.77	

9 FINISHES

099 860 | Special Wall Surfaces

		CREW	DAILY OUTPUT	LABOR-HOURS	UNIT	1999 BARE COSTS				TOTAL INCL O&P
						MAT.	LABOR	EQUIP.	TOTAL	
865	0360	6" L				Ea.	.80			.80	.88	865
	0370	8" L					.85			.85	.94
	0380	10" L					.95			.95	1.05
	0400	Waterfall hanger, metal, 12" - 16"					5.40			5.40	5.95
	0410	Acrylic					7.70			7.70	8.45
	0500	Shelf bracket, metal, 8"					4.20			4.20	4.62
	0510	10"					4.50			4.50	4.95
	0520	12"					4.80			4.80	5.30
	0530	14"					5.40			5.40	5.95
	0540	16"					6			6	6.60
	0550	Acrylic, 8"					2.50			2.50	2.75
	0560	10"					2.80			2.80	3.08
	0570	12"					3.10			3.10	3.41
	0580	14"					3.50			3.50	3.85
	0600	Shelf, acrylic, 12" x 16" x 1/4"					21			21	23
	0610	12" x 24" x 1/4"					35			35	38.50

099 900 | Surface Preparation

		CREW	DAILY OUTPUT	LABOR-HOURS	UNIT	MAT.	LABOR	EQUIP.	TOTAL	TOTAL INCL O&P
902	0010	**REMOVAL** Existing lead paint, by chemicals, per application R020-890									902
	0020	See also, Div 020-800, Haz. Mat'l Abatement									
	0050	Baseboard, to 6" wide	1 Pord	64	.125	L.F.	1.35	3.11		4.46	6.25
	0070	To 12" wide		32	.250	"	2.65	6.20		8.85	12.40
	0200	Balustrades, one side		28	.286	S.F.	3.01	7.10		10.11	14.20
	1400	Cabinets, simple design		32	.250		2.63	6.20		8.83	12.40
	1420	Ornate design		25	.320		3.38	7.95		11.33	15.90
	1600	Cornice, simple design		60	.133		1.41	3.32		4.73	6.65
	1620	Ornate design		20	.400		4.16	9.95		14.11	19.85
	2800	Doors, one side, flush		84	.095		1.02	2.37		3.39	4.75
	2820	Two panel		80	.100		1.06	2.49		3.55	4.98
	2840	Four panel		45	.178		1.87	4.43		6.30	8.80
	2880	For trim, one side, add		64	.125	L.F.	1.35	3.11		4.46	6.25
	3000	Fence, picket, one side		30	.267	S.F.	2.83	6.65		9.48	13.25
	3200	Grilles, one side, simple design		30	.267		2.83	6.65		9.48	13.25
	3220	Ornate design		25	.320		3.38	7.95		11.33	15.90
	4400	Pipes, to 4" diameter		90	.089	L.F.	.96	2.21		3.17	4.45
	4420	To 8" diameter		50	.160		1.68	3.98		5.66	7.95
	4440	To 12" diameter		36	.222		2.34	5.55		7.89	11
	4460	To 16" diameter		20	.400		4.19	9.95		14.14	19.85
	4500	For hangers, add		40	.200	Ea.	2.10	4.98		7.08	9.90
	4800	Siding		90	.089	S.F.	.96	2.21		3.17	4.45
	5000	Trusses, open		55	.145	SF Face	1.53	3.62		5.15	7.25
	6200	Windows, one side only, double hung, 1/1 light, 24" x 48" high		4	2	Ea.	21	50		71	99.50
	6220	30" x 60" high		3	2.667		28	66.50		94.50	133
	6240	36" x 72" high		2.50	3.200		34	79.50		113.50	160
	6280	40" x 80" high		2	4		42.50	99.50		142	199
	6400	Colonial window, 6/6 light, 24" x 48" high		2	4		42.50	99.50		142	199
	6420	30" x 60" high		1.50	5.333		56.50	133		189.50	265
	6440	36" x 72" high		1	8		84.50	199		283.50	400
	6480	40" x 80" high		1	8		84.50	199		283.50	400
	6600	8/8 light, 24" x 48" high		2	4		42.50	99.50		142	199
	6620	40" x 80" high		1	8		84.50	199		283.50	400
	6800	12/12 light, 24" x 48" high		1	8		84.50	199		283.50	400
	6820	40" x 80" high		.75	10.667		113	266		379	530
	6840	Window frame & trim items, included in pricing above									
903	0010	**LEAD PAINT ENCAPSULATION**, water based polymer coating,14 mil DFT									903
	0020	Interior, brushwork, trim, under 6"	1 Pord	240	.033	L.F.	1.99	.83		2.82	3.46

9 FINISHES

099 900 | Surface Preparation

		CREW	DAILY OUTPUT	LABOR-HOURS	UNIT	1999 BARE COSTS MAT.	LABOR	EQUIP.	TOTAL	TOTAL INCL O&P	
903	0030	6" to 12" wide	1 Pord	180	.044	L.F.	2.64	1.11		3.75	4.59	**903**
	0040	Balustrades		300	.027		1.59	.66		2.25	2.77	
	0050	Pipe to 4" diameter		500	.016		.95	.40		1.35	1.66	
	0060	To 8" diameter		375	.021		1.27	.53		1.80	2.21	
	0070	To 12" diameter		250	.032		1.90	.80		2.70	3.31	
	0080	To 16" diameter		170	.047		2.80	1.17		3.97	4.87	
	0090	Cabinets, ornate design		200	.040	S.F.	2.38	1		3.38	4.14	
	0100	Simple design	↓	250	.032	"	1.90	.80		2.70	3.31	
	0110	Doors, 3'x 7', both sides, incl. frame & trim										
	0120	Flush	1 Pord	6	1.333	Ea.	24.50	33		57.50	78	
	0130	French, 10-15 lite		3	2.667		4.88	66.50		71.38	107	
	0140	Panel		4	2		29.50	50		79.50	108	
	0150	Louvered	↓	2.75	2.909	↓	27	72.50		99.50	141	
	0160	Windows, per interior side, per 15 S.F.										
	0170	1 to 6 lite	1 Pord	14	.571	Ea.	16.85	14.25		31.10	40.50	
	0180	7 to 10 lite		7.50	1.067		18.60	26.50		45.10	61	
	0190	12 lite		5.75	1.391		25	34.50		59.50	80.50	
	0200	Radiators		8	1	↓	59.50	25		84.50	104	
	0210	Grilles, vents		275	.029	S.F.	1.73	.72		2.45	3.01	
	0220	Walls, roller, drywall or plaster		1,000	.008		.47	.20		.67	.82	
	0230	With spunbonded reinforcing fabric		720	.011		.53	.28		.81	1	
	0240	Wood		800	.010		.59	.25		.84	1.03	
	0250	Ceilings, roller, drywall or plaster		900	.009		.53	.22		.75	.92	
	0260	Wood		700	.011	↓	.68	.28		.96	1.19	
	0270	Exterior, brushwork, gutters and downspouts		300	.027	L.F.	1.59	.66		2.25	2.77	
	0280	Columns		400	.020	S.F.	1.19	.50		1.69	2.07	
	0290	Spray, siding	↓	600	.013	"	.79	.33		1.12	1.38	
	0300	Miscellaneous										
	0310	Electrical conduit, brushwork, to 2" diameter	1 Pord	500	.016	L.F.	.95	.40		1.35	1.66	
	0320	Brick, block or concrete, spray		500	.016	S.F.	.95	.40		1.35	1.66	
	0330	Steel, flat surfaces and tanks to 12"		500	.016		.95	.40		1.35	1.66	
	0340	Beams, brushwork		400	.020		1.19	.50		1.69	2.07	
	0350	Trusses	↓	400	.020	↓	1.19	.50		1.69	2.07	
904	0010	**SANDING** And puttying interior trim, compared to										**904**
	0100	Painting 1 coat, on quality work				L.F.		100%				
	0300	Medium work						50%				
	0400	Industrial grade				↓		25%				
	0500	Surface protection, placement and removal										
	0510	Basic drop cloths	1 Pord	6,400	.001	S.F.		.03		.03	.05	
	0520	Masking with paper		800	.010		.02	.25		.27	.40	
	0530	Volume cover up (using plastic sheathing, or building paper)	↓	16,000	.001	↓		.01		.01	.02	
910	0010	**SURFACE PREPARATION, EXTERIOR**										**910**
	0015	Doors, per side, not incl. frames or trim										
	0020	Scrape & sand										
	0030	Wood, flush	1 Pord	616	.013	S.F.		.32		.32	.49	
	0040	Wood, detail		496	.016			.40		.40	.61	
	0050	Wood, louvered		280	.029			.71		.71	1.09	
	0060	Wood, overhead	↓	616	.013	↓		.32		.32	.49	
	0070	Wire brush										
	0080	Metal, flush	1 Pord	640	.013	S.F.		.31		.31	.48	
	0090	Metal, detail		520	.015			.38		.38	.59	
	0100	Metal, louvered		360	.022			.55		.55	.85	
	0110	Metal or fibr., overhead		640	.013			.31		.31	.48	
	0120	Metal, roll up		560	.014			.36		.36	.54	
	0130	Metal, bulkhead	↓	640	.013	↓		.31		.31	.48	

9

FINISHES

For expanded coverage of these items see *Means Interior Cost Data 1999*

099 900 | Surface Preparation

		CREW	DAILY OUTPUT	LABOR-HOURS	UNIT	MAT.	1999 BARE COSTS LABOR	EQUIP.	TOTAL	TOTAL INCL O&P	
910	0140	Power wash, based on 2500 lb. operating pressure										910
	0150	Metal, flush	B-9	2,240	.018	S.F.		.39	.08	.47	.70	
	0160	Metal, detail		2,120	.019			.41	.09	.50	.74	
	0170	Metal, louvered		2,000	.020			.44	.09	.53	.79	
	0180	Metal or fibr., overhead		2,400	.017			.36	.08	.44	.65	
	0190	Metal, roll up		2,400	.017			.36	.08	.44	.65	
	0200	Metal, bulkhead		2,200	.018			.40	.08	.48	.72	
	0400	Windows, per side, not incl. trim										
	0410	Scrape & sand										
	0420	Wood, 1-2 lite	1 Pord	320	.025	S.F.		.62		.62	.95	
	0430	Wood, 3-6 lite		280	.029			.71		.71	1.09	
	0440	Wood, 7-10 lite		240	.033			.83		.83	1.27	
	0450	Wood, 12 lite		200	.040			1		1	1.52	
	0460	Wood, Bay / Bow		320	.025			.62		.62	.95	
	0470	Wire brush										
	0480	Metal, 1-2 lite	1 Pord	480	.017	S.F.		.41		.41	.63	
	0490	Metal, 3-6 lite		400	.020			.50		.50	.76	
	0500	Metal, Bay / Bow		480	.017			.41		.41	.63	
	0510	Power wash, based on 2500 lb. operating pressure										
	0520	1-2 lite	B-9	4,400	.009	S.F.		.20	.04	.24	.36	
	0530	3-6 lite		4,320	.009			.20	.04	.24	.37	
	0540	7-10 lite		4,240	.009			.21	.04	.25	.37	
	0550	12 lite		4,160	.010			.21	.04	.25	.38	
	0560	Bay / Bow		4,400	.009			.20	.04	.24	.36	
	0600	Siding, scrape and sand, light=10-30%, med.=30-70%										
	0610	Heavy=70-100%, % of surface to sand										
	0620	For Chemical Washing, see Division 045-108										
	0630	For Steam Cleaning, see Division 042-166										
	0640	For Sand Blasting, see Division 042-150										
	0650	Texture 1-11, light	1 Pord	480	.017	S.F.		.41		.41	.63	
	0660	Med.		440	.018			.45		.45	.69	
	0670	Heavy		360	.022			.55		.55	.85	
	0680	Wood shingles, shakes, light		440	.018			.45		.45	.69	
	0690	Med.		360	.022			.55		.55	.85	
	0700	Heavy		280	.029			.71		.71	1.09	
	0710	Clapboard, light		520	.015			.38		.38	.59	
	0720	Med.		480	.017			.41		.41	.63	
	0730	Heavy		400	.020			.50		.50	.76	
	0740	Wire brush										
	0750	Aluminum, light	1 Pord	600	.013	S.F.		.33		.33	.51	
	0760	Med.		520	.015			.38		.38	.59	
	0770	Heavy		440	.018			.45		.45	.69	
	0780	Pressure wash, based on 2500 lb.. operating pressure										
	0790	Stucco	B-9	3,080	.013	S.F.		.28	.06	.34	.51	
	0800	Aluminum or vinyl		3,200	.013			.27	.06	.33	.49	
	0810	Siding, masonry, brick & block		2,400	.017			.36	.08	.44	.65	
	1300	Miscellaneous, wire brush										
	1310	Metal, pedestrian gate	1 Pord	100	.080	S.F.		1.99		1.99	3.05	
920	0010	**SURFACE PREPARATION, INTERIOR**										920
	0020	Doors										
	0030	Scrape & sand										
	0040	Wood, flush	1 Pord	616	.013	S.F.		.32		.32	.49	
	0050	Wood, detail		496	.016			.40		.40	.61	
	0060	Wood, louvered		280	.029			.71		.71	1.09	
	0070	Wire brush										
	0080	Metal, flush	1 Pord	640	.013	S.F.		.31		.31	.48	

Important: See the Reference Section for critical supporting data - Reference Nos., Crews, & City Cost Indexes

099 900	Surface Preparation	CREW	DAILY OUTPUT	LABOR-HOURS	UNIT	1999 BARE COSTS				TOTAL INCL O&P
						MAT.	LABOR	EQUIP.	TOTAL	
920	0090	Metal, detail	1 Pord	520	.015	S.F.		.38		.38	.59	920
	0100	Metal, louvered	↓	360	.022	↓		.55		.55	.85
	0110	Hand wash									
	0120	Wood, flush	1 Pord	2,160	.004	S.F.		.09		.09	.14
	0130	Wood, detailed		2,000	.004			.10		.10	.15
	0140	Wood, louvered		1,360	.006			.15		.15	.22
	0150	Metal, flush		2,160	.004			.09		.09	.14
	0160	Metal, detail		2,000	.004			.10		.10	.15
	0170	Metal, louvered	↓	1,360	.006	↓		.15		.15	.22
	0400	Windows, per side, not incl. trim									
	0410	Scrape & sand									
	0420	Wood, 1-2 lite	1 Pord	360	.022	S.F.		.55		.55	.85
	0430	Wood, 3-6 lite		320	.025			.62		.62	.95
	0440	Wood, 7-10 lite		280	.029			.71		.71	1.09
	0450	Wood, 12 lite		240	.033			.83		.83	1.27
	0460	Wood, Bay / Bow	↓	360	.022	↓		.55		.55	.85
	0470	Wire brush									
	0480	Metal, 1-2 lite	1 Pord	520	.015	S.F.		.38		.38	.59
	0490	Metal, 3-6 lite		440	.018			.45		.45	.69
	0500	Metal, Bay / Bow	↓	520	.015	↓		.38		.38	.59
	0600	Walls, sanding, light=10-30%									
	0610	Med.=30-70%, heavy=70-100%, % of surface to sand									
	0620	For Chemical Washing, see Division 045-108									
	0630	For Steam Cleaning, see Division 042-166									
	0640	For Sand Blasting, see Division 042-150									
	0650	Walls, sand									
	0660	Drywall, gypsum, plaster, light	1 Pord	3,077	.003	S.F.		.06		.06	.10
	0670	Drywall, gypsum, plaster, med.		2,160	.004			.09		.09	.14
	0680	Drywall, gypsum, plaster, heavy		923	.009			.22		.22	.33
	0690	Wood, T&G, light		2,400	.003			.08		.08	.13
	0700	Wood, T&G, med.		1,600	.005			.12		.12	.19
	0710	Wood, T&G, heavy	↓	800	.010	↓		.25		.25	.38
	0720	Walls, wash									
	0730	Drywall, gypsum, plaster	1 Pord	3,200	.002	S.F.		.06		.06	.10
	0740	Wood, T&G		3,200	.002			.06		.06	.10
	0750	Masonry, brick & block, smooth		2,800	.003			.07		.07	.11
	0760	Masonry, brick & block, coarse	↓	2,000	.004	↓		.10		.10	.15

For information about Means Estimating Seminars, see yellow pages 11 and 12 in back of book

FINISHES

9

Division Notes

	CREW	DAILY OUTPUT	LABOR-HOURS	UNIT	1999 BARE COSTS				TOTAL INCL O&P
					MAT.	LABOR	EQUIP.	TOTAL	

Division 10
Specialties

Estimating Tips

General

- The items in this division are usually priced per square foot or each.
- Many items in Division 10 require some type of support system that is not usually furnished with the item. Examples of these systems include blocking for the attachment of grab bars and support angles for ceiling hung toilet partitions. The required blocking or supports must be added to the estimate in the appropriate division.
- Some items in Division 10, such as lockers, may require assembly before installation. Verify the amount of assembly required. Assembly can often exceed installation time.

101 Visual Display Boards, Compartments & Cubicles

- Toilet partitions are priced by the stall. A stall consists of a side wall, pilaster and door with hardware. Toilet tissue holders and grab bars are extra.

106 Partitions & Storage Shelving

- The required acoustical rating of a folding partition can have a significant impact on costs. Verify the sound transmission coefficient rating of the panel priced to the specification requirements.

Reference Numbers

Reference numbers are shown in bold squares at the beginning of some major classifications. These numbers refer to related items in the Reference Section. The reference information may be an estimating procedure, an alternate pricing method or technical information.

Note: Not all subdivisions listed here necessarily appear in this publication.

10 SPECIALTIES

101 100	Chalkboards	CREW	DAILY OUTPUT	LABOR-HOURS	UNIT	MAT.	LABOR	EQUIP.	TOTAL	TOTAL INCL O&P
104						1999 BARE COSTS				104
0010	**CHALKBOARDS** Porcelain enamel steel									
0100	Freestanding, reversible									
0120	Economy, wood frame, 4' x 6'									
0140	Chalkboard both sides				Ea.	460			460	510
0160	Chalkboard one side, cork other side				"	400			400	440
0200	Standard, lightweight satin finished aluminum, 4' x 6'									
0220	Chalkboard both sides				Ea.	475			475	525
0240	Chalkboard one side, cork other side				"	420			420	460
0300	Deluxe, heavy duty extruded aluminum, 4' x 6'									
0320	Chalkboard both sides				Ea.	1,025			1,025	1,150
0340	Chalkboard one side, cork other side				"	1,200			1,200	1,325
0400	Sliding chalkboards									
0450	Vertical, one sliding board with back panel, wall mounted									
0500	8' x 4'	2 Carp	8	2	Ea.	1,275	54.50		1,329.50	1,475
0520	8' x 8'		7.50	2.133		1,775	58		1,833	2,050
0540	8' x 12'	↓	7	2.286	↓	2,225	62.50		2,287.50	2,525
0600	Two sliding boards, with back panel									
0620	8' x 4'	2 Carp	8	2	Ea.	1,950	54.50		2,004.50	2,225
0640	8' x 8'		7.50	2.133		2,725	58		2,783	3,100
0660	8' x 12'	↓	7	2.286	↓	3,375	62.50		3,437.50	3,800
0700	Horizontal, two track									
0800	4' x 8', 2 sliding panels	2 Carp	8	2	Ea.	1,125	54.50		1,179.50	1,300
0820	4' x 12', 2 sliding panels		7.50	2.133		1,375	58		1,433	1,600
0840	4' x 16', 4 sliding panels	↓	7	2.286	↓	1,850	62.50		1,912.50	2,150
0900	Four track, four sliding panels									
0920	4' x 8'	2 Carp	8	2	Ea.	1,800	54.50		1,854.50	2,075
0940	4' x 12'		7.50	2.133		2,225	58		2,283	2,550
0960	4' x 16'	↓	7	2.286	↓	2,775	62.50		2,837.50	3,150
1200	Vertical, motor operated									
1400	One sliding panel with back panel									
1450	10' x 4'	2 Carp	4	4	Ea.	4,975	109		5,084	5,650
1500	10' x 10'		3.75	4.267		5,750	116		5,866	6,500
1550	10' x 16'	↓	3.50	4.571	↓	6,525	125		6,650	7,400
1700	Two sliding panels with back panel									
1750	10' x 4'	2 Carp	4	4	Ea.	7,975	109		8,084	8,950
1800	10' x 10'		3.75	4.267		9,075	116		9,191	10,200
1850	10' x 16'	↓	3.50	4.571	↓	10,200	125		10,325	11,400
2000	Three sliding panels with back panel									
2100	10' x 4'	2 Carp	4	4	Ea.	10,600	109		10,709	11,800
2150	10' x 10'		3.75	4.267		12,200	116		12,316	13,600
2200	10' x 16'	↓	3.50	4.571	↓	13,800	125		13,925	15,400
2400	For projection screen, glass beaded, add				S.F.	4.50			4.50	4.95
2500	For remote control, 1 panel control, add				Ea.	395			395	435
2600	2 panel control, add				"	695			695	765
2800	For units without back panels, deduct				S.F.	4.91			4.91	5.40
2850	For liquid chalk porcelain panels, add				"	3.64			3.64	4
3000	Swing leaf, any comb. of chalkboard & cork, aluminum frame									
3100	Floor style, 5 panels									
3150	30" x 40" panels				Ea.	1,150			1,150	1,250
3200	48" x 40" panels				"	1,300			1,300	1,450
3300	Wall mounted, 5 panels									
3400	30" x 40" panels	2 Carp	16	1	Ea.	1,225	27.50		1,252.50	1,400
3450	48" x 40" panels	"	16	1	"	1,400	27.50		1,427.50	1,600
3600	Extra panels for swing leaf units									
3700	30" x 40" panels				Ea.	201			201	221
3750	48" x 40" panels				"	237			237	260

Important: See the Reference Section for critical supporting data - Reference Nos., Crews, & City Cost Indexes

101 100 | Chalkboards

		CREW	DAILY OUTPUT	LABOR-HOURS	UNIT	1999 BARE COSTS MAT.	LABOR	EQUIP.	TOTAL	TOTAL INCL O&P	
104	3900	Wall hung										104
	4000	Aluminum frame and chalktrough										
	4200	3' x 4'	2 Carp	16	1	Ea.	128	27.50		155.50	184	
	4300	3' x 5'		15	1.067		186	29		215	251	
	4500	4' x 8'		14	1.143		285	31		316	365	
	4600	4' x 12'		13	1.231		410	33.50		443.50	510	
	4700	Wood frame and chalktrough										
	4800	3' x 4'	2 Carp	16	1	Ea.	130	27.50		157.50	186	
	5000	3' x 5'		15	1.067		184	29		213	248	
	5100	4' x 5'		14	1.143		186	31		217	253	
	5300	4' x 8'		13	1.231		264	33.50		297.50	345	
	5400	Liquid chalk, white porcelain enamel, wall hung										
	5420	Deluxe units, aluminum trim and chalktrough										
	5450	4' x 4'	2 Carp	16	1	Ea.	174	27.50		201.50	235	
	5500	4' x 8'		14	1.143		305	31		336	385	
	5550	4' x 12'		12	1.333		420	36.50		456.50	520	
	5700	Wood trim and chalktrough										
	5900	4' x 4'	2 Carp	16	1	Ea.	360	27.50		387.50	445	
	6000	4' x 6'		15	1.067		445	29		474	535	
	6200	4' x 8'		14	1.143		520	31		551	625	
	6300	Liquid chalk, felt tip markers					1.40			1.40	1.54	
	6500	Erasers					1.30			1.30	1.43	
	6600	Board cleaner, 8 oz. bottle					3.80			3.80	4.18	

101 600 | Toilet Compartments

		CREW	DAILY OUTPUT	LABOR-HOURS	UNIT	1999 BARE COSTS MAT.	LABOR	EQUIP.	TOTAL	TOTAL INCL O&P	
602	0010	**PARTITIONS, TOILET**										602
	0100	Cubicles, ceiling hung, marble	2 Marb	2	8	Ea.	1,300	220		1,520	1,775	
	0200	Painted metal	2 Carp	4	4		385	109		494	590	
	0250	Phenolic		4	4		730	109		839	970	
	0300	Plastic laminate on particle board		4	4		495	109		604	715	
	0500	Stainless steel		4	4		965	109		1,074	1,250	
	0600	For handicap units, incl. 52" grab bars, add					276			276	305	
	0800	Floor & ceiling anchored, marble	2 Marb	2.50	6.400		1,425	176		1,601	1,850	
	1000	Painted metal	2 Carp	5	3.200		355	87.50		442.50	525	
	1050	Phenolic		5	3.200		785	87.50		872.50	1,000	
	1100	Plastic laminate on particle board		5	3.200		470	87.50		557.50	650	
	1300	Stainless steel		5	3.200		1,125	87.50		1,212.50	1,350	
	1400	For handicap units, incl. 52" grab bars, add					250			250	275	
	1600	Floor mounted, marble	2 Marb	3	5.333		840	147		987	1,150	
	1700	Painted metal	2 Carp	7	2.286		355	62.50		417.50	490	
	1750	Phenolic		7	2.286		730	62.50		792.50	900	
	1800	Plastic laminate on particle board		7	2.286		495	62.50		557.50	645	
	2000	Stainless steel		7	2.286		1,075	62.50		1,137.50	1,300	
	2100	For handicap units, incl. 52" grab bars, add					250			250	275	
	2200	For juvenile units, deduct					35			35	38.50	
	2400	Floor mounted, headrail braced, marble	2 Marb	3	5.333		980	147		1,127	1,300	
	2500	Painted metal	2 Carp	6	2.667		355	73		428	505	
	2550	Phenolic		6	2.667		760	73		833	950	
	2600	Plastic laminate on particle board **CN**		6	2.667		515	73		588	680	
	2800	Stainless steel		6	2.667		1,100	73		1,173	1,325	
	2900	For handicap units, incl. 52" grab bars, add					250			250	275	
	3000	Wall hung partitions, painted metal	2 Carp	7	2.286		450	62.50		512.50	595	
	3300	Stainless steel	"	7	2.286		1,075	62.50		1,137.50	1,300	
	3400	For handicap units, incl. 52" grab bars, add					250			250	275	
	4000	Screens, entrance, floor mounted, 58" high, 48" wide										

SPECIALTIES 10

101 600 | Toilet Compartments

		Description	CREW	DAILY OUTPUT	LABOR-HOURS	UNIT	1999 BARE COSTS				TOTAL INCL O&P	
							MAT.	LABOR	EQUIP.	TOTAL		
602	4100	Marble	2 Marb	9	1.778	Ea.	530	49		579	655	602
	4200	Painted metal	2 Carp	15	1.067		166	29		195	228	
	4300	Plastic laminate on particle board		15	1.067		294	29		323	370	
	4500	Stainless steel		15	1.067		585	29		614	690	
	4600	Urinal screen, 18" wide, ceiling braced, marble	D-1	6	2.667		530	65.50		595.50	680	
	4700	Painted metal	2 Carp	8	2		126	54.50		180.50	225	
	4800	Plastic laminate on particle board		8	2		235	54.50		289.50	345	
	5000	Stainless steel		8	2		420	54.50		474.50	545	
	5100	Floor mounted, head rail braced										
	5200	Marble	D-1	6	2.667	Ea.	490	65.50		555.50	640	
	5300	Painted metal	2 Carp	8	2		166	54.50		220.50	268	
	5400	Plastic laminate on particle board		8	2		212	54.50		266.50	320	
	5600	Stainless steel		8	2		500	54.50		554.50	635	
	5700	Pilaster, flush, marble	D-1	9	1.778		600	43.50		643.50	730	
	5800	Painted metal	2 Carp	10	1.600		216	43.50		259.50	305	
	5900	Plastic laminate on particle board		10	1.600		270	43.50		313.50	365	
	6100	Stainless steel		10	1.600		370	43.50		413.50	475	
	6200	Post braced, marble	D-1	9	1.778		590	43.50		633.50	715	
	6300	Painted metal	2 Carp	10	1.600		222	43.50		265.50	315	
	6400	Plastic laminate on particle board		10	1.600		276	43.50		319.50	375	
	6600	Stainless steel		10	1.600		365	43.50		408.50	475	
	6700	Wall hung, bracket supported										
	6800	Painted metal	2 Carp	10	1.600	Ea.	225	43.50		268.50	315	
	6900	Plastic laminate on particle board		10	1.600		108	43.50		151.50	188	
	7100	Stainless steel		10	1.600		335	43.50		378.50	440	
	7400	Flange supported, painted metal		10	1.600		165	43.50		208.50	251	
	7500	Plastic laminate on particle board		10	1.600		160	43.50		203.50	245	
	7700	Stainless steel		10	1.600		390	43.50		433.50	500	
	7800	Wedge type, painted metal		10	1.600		195	43.50		238.50	283	
	8100	Stainless steel		10	1.600		400	43.50		443.50	510	

101 850 | Shower Compartments

		Description	CREW	DAILY OUTPUT	LABOR-HOURS	UNIT	MAT.	LABOR	EQUIP.	TOTAL	TOTAL INCL O&P	
852	0010	**PARTITIONS, SHOWER** Floor mounted, no plumbing										852
	0100	Cabinet, incl. base, no door, painted steel, 1" thick walls	2 Shee	5	3.200	Ea.	585	102		687	805	
	0300	With door, fiberglass		4.50	3.556		480	113		593	700	
	0600	Galvanized and painted steel, 1" thick walls		5	3.200		620	102		722	845	
	0800	Stall, 1" thick wall, no base, enameled steel		5	3.200		595	102		697	815	
	1200	Stainless steel		5	3.200		1,050	102		1,152	1,300	
	1400	For double entry type, no doors, deduct					10%					
	1500	Circular fiberglass, cabinet 36" diameter,	2 Shee	4	4		490	127		617	735	
	1700	One piece, 36" diameter, less door		4	4		420	127		547	660	
	1800	With door		3.50	4.571		680	145		825	970	
	2000	Curved shell shower, no door needed		3	5.333		620	169		789	950	
	2300	For fiberglass seat, add to both above					88			88	97	
	2400	Glass stalls, with doors, no receptors, chrome on brass	2 Shee	3	5.333		1,025	169		1,194	1,400	
	2700	Anodized aluminum	"	4	4		705	127		832	970	
	2900	Marble shower stall, stock design, with shower door	2 Marb	1.20	13.333		1,575	365		1,940	2,300	
	3000	With curtain		1.30	12.308		1,375	340		1,715	2,050	
	3200	Receptors, precast terrazzo, 32" x 32"		14	1.143		295	31.50		326.50	375	
	3300	48" x 34"		9.50	1.684		375	46.50		421.50	480	
	3500	Plastic, simulated terrazzo receptor, 32" x 32"		14	1.143		85.50	31.50		117	143	
	3600	32" x 48"		12	1.333		119	36.50		155.50	188	
	3800	Precast concrete, colors, 32" x 32"		14	1.143		148	31.50		179.50	212	
	3900	48" x 48"		8	2		182	55		237	287	
	4100	Shower doors, economy plastic, 24" wide	1 Shee	9	.889		88	28		116	141	
	4200	Tempered glass door, economy		8	1		170	32		202	237	

Important: See the Reference Section for critical supporting data - Reference Nos., Crews, & City Cost Indexes

101 850	Shower Compartments	CREW	DAILY OUTPUT	LABOR-HOURS	UNIT	1999 BARE COSTS				TOTAL INCL O&P
						MAT.	LABOR	EQUIP.	TOTAL	
852 4400	Folding, tempered glass, aluminum frame	1 Shee	6	1.333	Ea.	207	42.50		249.50	294 **852**
4500	Sliding, tempered glass, 48″ opening		6	1.333		176	42.50		218.50	260
4700	Deluxe, tempered glass, chrome on brass frame, minimum		8	1		215	32		247	287
4800	Maximum		1	8		620	254		874	1,075
4850	On anodized aluminum frame, minimum		2	4		101	127		228	310
4900	Maximum		1	8		340	254		594	770
5100	Shower enclosure, tempered glass, anodized alum. frame									
5120	2 panel & door, corner unit, 32″ x 32″	1 Shee	2	4	Ea.	340	127		467	570
5140	Neo-angle corner unit, 16″ x 24″ x 16″	″	2	4		595	127		722	850
5200	Shower surround, 3 wall, polypropylene, 32″ x 32″	1 Carp	4	2		189	54.50		243.50	293
5220	PVC, 32″ x 32″		4	2		220	54.50		274.50	330
5240	Fiberglass		4	2		264	54.50		318.50	375
5250	2 wall, polypropylene, 32″ x 32″		4	2		182	54.50		236.50	287
5270	PVC		4	2		220	54.50		274.50	330
5290	Fiberglass		4	2		245	54.50		299.50	355
5300	Tub doors, tempered glass & frame, minimum	1 Shee	8	1		145	32		177	209
5400	Maximum		6	1.333		335	42.50		377.50	430
5600	Chrome plated, brass frame, minimum		8	1		189	32		221	257
5700	Maximum		6	1.333		375	42.50		417.50	480
5900	Tub/shower enclosure, temp. glass, alum. frame, minimum		2	4		258	127		385	480
6200	Maximum		1.50	5.333		505	169		674	820
6500	On chrome-plated brass frame, minimum		2	4		360	127		487	590
6600	Maximum		1.50	5.333		735	169		904	1,075
6800	Tub surround, 3 wall, polypropylene	1 Carp	4	2		145	54.50		199.50	245
6900	PVC		4	2		220	54.50		274.50	330
7000	Fiberglass, minimum		4	2		245	54.50		299.50	355
7100	Maximum		3	2.667		420	73		493	580

102 100	Metal Wall Louvers	CREW	DAILY OUTPUT	LABOR-HOURS	UNIT	1999 BARE COSTS				TOTAL INCL O&P
						MAT.	LABOR	EQUIP.	TOTAL	
104 0010	**LOUVERS** Aluminum with screen, residential, 8″ x 8″	1 Carp	38	.211	Ea.	6.95	5.75		12.70	16.65 **104**
0100	12″ x 12″		38	.211		8.75	5.75		14.50	18.65
0200	12″ x 18″		35	.229		12.70	6.25		18.95	24
0250	14″ x 24″		30	.267		16.35	7.30		23.65	29.50
0300	18″ x 24″		27	.296		18.85	8.10		26.95	34
0500	24″ x 30″		24	.333		27.50	9.10		36.60	45
0700	Triangle, adjustable, small		20	.400		23	10.90		33.90	42
0800	Large		15	.533		61.50	14.55		76.05	90.50
1200	Extruded aluminum, see division 157-482									
2100	Midget, aluminum, 3/4″ deep, 1″ diameter	1 Carp	85	.094	Ea.	.56	2.57		3.13	4.66
2150	3″ diameter		60	.133		1.32	3.64		4.96	7.20
2200	4″ diameter		50	.160		1.66	4.37		6.03	8.70
2250	6″ diameter		30	.267		2.25	7.30		9.55	13.95
2300	Ridge vent strip, mill finish	1 Shee	155	.052	L.F.	1.95	1.64		3.59	4.69
2330	Soffit vent, continuous, 3″ wide, aluminum, mill finish	1 Carp	200	.040		.44	1.09		1.53	2.20
2340	Baked enamel finish		200	.040		.52	1.09		1.61	2.29
2400	Under eaves vent, aluminum, mill finish, 16″ x 4″		48	.167	Ea.	1.49	4.55		6.04	8.80
2500	16″ x 8″		48	.167		1.79	4.55		6.34	9.10

SPECIALTIES 10

102 | Louvers, Corner Protection & Access Flooring

102 100 | Metal Wall Louvers

			CREW	DAILY OUTPUT	LABOR-HOURS	UNIT	1999 BARE COSTS				TOTAL INCL O&P	
							MAT.	LABOR	EQUIP.	TOTAL		
104	7000	Vinyl gable vent, 8" x 8"	1 Carp	38	.211	Ea.	9.50	5.75		15.25	19.50	**104**
	7020	12" x 12"		38	.211		17	5.75		22.75	28	
	7080	12" x 18"		35	.229		22	6.25		28.25	34	
	7200	18" x 24"		30	.267		26	7.30		33.30	40	

102 600 | Wall & Corner Guards

			CREW	DAILY OUTPUT	LABOR-HOURS	UNIT	MAT.	LABOR	EQUIP.	TOTAL	TOTAL INCL O&P	
604	0010	**CORNER GUARDS** Steel angle w/anchors, 1" x 1" x 1/4", 1.5#/L.F.	2 Carp	160	.100	L.F.	2.77	2.73		5.50	7.35	**604**
	0100	2" x 2" x 1/4" angles, 3.2#/L.F.		150	.107		6.50	2.91		9.41	11.75	
	0200	3" x 3" x 5/16" angles, 6.1#/L.F.		140	.114		8.45	3.12		11.57	14.20	
	0300	4" x 4" x 5/16" angles, 8.2#/L.F.		120	.133		12.45	3.64		16.09	19.40	
	0350	For angles drilled and anchored to masonry, add					15%	120%				
	0370	Drilled and anchored to concrete, add					20%	170%				
	0400	For galvanized angles, add					35%					
	0450	For stainless steel angles, add				L.F.	100%					
	0500	Steel door track/wheel guards, 4'-0" high	E-4	22	1.455	Ea.	129	45	3.74	177.74	229	
	0800	Pipe bumper for truck doors, 8' long, 6" diameter, filled		20	1.600		179	50	4.11	233.11	293	
	0900	8" diameter		20	1.600		275	50	4.11	329.11	400	
608	0010	**CORNER PROTECTION**										**608**
	0100	Stainless steel, 16 ga., adhesive mount, 3-1/2" leg	1 Sswk	80	.100	L.F.	16.30	3.06		19.36	23.50	
	0200	12 ga. stainless, adhesive mount	"	80	.100		24.50	3.06		27.56	32.50	
	0300	For screw mount, add						10%				
	0500	Vinyl acrylic, adhesive mount, 3" leg	1 Carp	128	.063		5.30	1.71		7.01	8.55	
	0550	1-1/2" leg		160	.050		3.70	1.36		5.06	6.20	
	0600	Screw mounted, 3" leg		80	.100		6.15	2.73		8.88	11.05	
	0650	1-1/2" leg		100	.080		3.60	2.18		5.78	7.40	
	0700	Clear plastic, screw mounted, 2-1/2"		60	.133		2.99	3.64		6.63	9.05	
	1000	Vinyl cover, alum. retainer, surface mount, 3" x 3"		48	.167		6.40	4.55		10.95	14.20	
	1050	2" x 2"		48	.167		4.43	4.55		8.98	12	
	1100	Flush mounted, 3" x 3"		32	.250		12.35	6.80		19.15	24.50	
	1150	2" x 2"		32	.250		9.10	6.80		15.90	21	
612	0010	**WALLGUARD** See division 055										**612**
	0400	Rub rail, vinyl, adhesive mounted	1 Carp	185	.043	L.F.	3.60	1.18		4.78	5.80	
	0500	Neoprene, aluminum backing, 1-1/2" x 2"		110	.073		6	1.99		7.99	9.70	
	1000	Trolley rail, PVC, clipped to wall, 5" high		185	.043		5.05	1.18		6.23	7.40	
	1050	8" high		180	.044		5.90	1.21		7.11	8.40	
	1200	Bed bumper, vinyl acrylic, alum. retainer, 21" long		10	.800	Ea.	33	22		55	71	
	1300	53" long with aligner		9	.889	"	98	24.50		122.50	146	
	1400	Bumper, vinyl cover, alum. retain., cush. mnt., 1-1/2" x 2-3/4"		80	.100	L.F.	9.90	2.73		12.63	15.15	
	1500	2" x 4-1/4"		80	.100		13.20	2.73		15.93	18.85	
	1600	Surface mounted, 1-3/4" x 3-5/8"		80	.100		8.90	2.73		11.63	14.10	
	2000	Crash rail, vinyl cover, alum. retainer, 1" x 4"		110	.073		10.10	1.99		12.09	14.20	
	2100	1" x 8"		90	.089		12.60	2.43		15.03	17.65	
	2150	Vinyl inserts, aluminum plate, 1" x 2-1/2"		110	.073		9.85	1.99		11.84	13.95	
	2200	1" x 5"		90	.089		14.25	2.43		16.68	19.45	
	3000	Handrail/bumper, vinyl cover, alum. retainer										
	3010	Bracket mounted, flat rail, 5-1/2"	1 Carp	80	.100	L.F.	15.20	2.73		17.93	21	
	3100	6-1/2"		80	.100		18.70	2.73		21.43	25	
	3200	Bronze bracket, 1-3/4" diam. rail		80	.100		15.20	2.73		17.93	21	

102 700 | Access Flooring

			CREW	DAILY OUTPUT	LABOR-HOURS	UNIT	MAT.	LABOR	EQUIP.	TOTAL	TOTAL INCL O&P	
705	0010	**PEDESTAL ACCESS FLOORS** Computer room application, metal										**705**
	0020	Particle board or steel panels, no covering, under 6,000 S.F.	2 Carp	400	.040	S.F.	5.65	1.09		6.74	7.90	
	0300	Metal covered, over 6,000 S.F.		450	.036		5.75	.97		6.72	7.90	
	0400	Aluminum, 24" panels		500	.032		20.50	.87		21.37	24	

10 SPECIALTIES

308

Important: See the Reference Section for critical supporting data - Reference Nos., Crews, & City Cost Indexes

102 | Louvers, Corner Protection & Access Flooring

102 700 | Access Flooring

		CREW	DAILY OUTPUT	LABOR-HOURS	UNIT	1999 BARE COSTS				TOTAL INCL O&P	
						MAT.	LABOR	EQUIP.	TOTAL		
705	0600	For carpet covering, add				S.F.	3.92			3.92	4.32	705
	0700	For vinyl floor covering, add					4.18			4.18	4.59	
	0900	For high pressure laminate covering, add					2.59			2.59	2.85	
	0910	For snap on stringer system, add	2 Carp	1,000	.016		1.02	.44		1.46	1.81	
	0950	Office applications, to 8" high, steel panels,										
	0960	no covering, over 6,000 S.F.	2 Carp	500	.032	S.F.	6.65	.87		7.52	8.65	
	1000	Machine cutouts after initial installation	1 Carp	10	.800	Ea.	33	22		55	70.50	
	1050	Pedestals, 6" to 12"	2 Carp	85	.188		4.93	5.15		10.08	13.55	
	1100	Air conditioning grilles, 4" x 12"	1 Carp	17	.471		50	12.85		62.85	75	
	1150	4" x 18"	"	14	.571		66.50	15.60		82.10	97.50	
	1200	Approach ramps, minimum	2 Carp	85	.188	S.F.	15.35	5.15		20.50	25	
	1300	Maximum	"	60	.267	"	25.50	7.30		32.80	39.50	
	1500	Handrail, 2 rail, aluminum	1 Carp	15	.533	L.F.	73	14.55		87.55	103	

103 | Fireplaces, Exterior Specialties & Flagpoles

103 050 | Prefabricated Fireplaces

		CREW	DAILY OUTPUT	LABOR-HOURS	UNIT	1999 BARE COSTS				TOTAL INCL O&P	
						MAT.	LABOR	EQUIP.	TOTAL		
054	0010	**FIREPLACE, PREFABRICATED** Free standing or wall hung										054
	0100	with hood & screen, minimum	1 Carp	1.30	6.154	Ea.	905	168		1,073	1,250	
	0150	Average		1	8		1,075	218		1,293	1,550	
	0200	Maximum		.90	8.889		2,600	243		2,843	3,250	
	0500	Chimney dbl. wall, all stainless, over 8'-6", 7" diam., add		33	.242	V.L.F.	30.50	6.60		37.10	44	
	0600	10" diameter, add		32	.250		43.50	6.80		50.30	59	
	0700	12" diameter, add		31	.258		57	7.05		64.05	73.50	
	0800	14" diameter, add		30	.267		72	7.30		79.30	90.50	
	1000	Simulated brick chimney top, 4' high, 16" x 16"		10	.800	Ea.	155	22		177	206	
	1100	24" x 24"		7	1.143	"	289	31		320	370	
	1500	Simulated logs, gas fired, 40,000 BTU, 2' long, minimum		7	1.143	Set	405	31		436	495	
	1600	Maximum		6	1.333		590	36.50		626.50	710	
	1700	Electric, 1,500 BTU, 1'-6" long, minimum		7	1.143		107	31		138	167	
	1800	11,500 BTU, maximum		6	1.333		232	36.50		268.50	315	
	2000	Fireplace, built-in, 36" hearth, radiant		1.30	6.154	Ea.	465	168		633	775	
	2100	Recirculating, small fan		1	8		665	218		883	1,075	
	2150	Large fan		.90	8.889		1,225	243		1,468	1,725	
	2200	42" hearth, radiant		1.20	6.667		595	182		777	935	
	2300	Recirculating, small fan		.90	8.889		775	243		1,018	1,225	
	2350	Large fan		.80	10		1,475	273		1,748	2,050	
	2400	48" hearth, radiant		1.10	7.273		1,100	199		1,299	1,525	
	2500	Recirculating, small fan		.80	10		1,375	273		1,648	1,925	
	2550	Large fan		.70	11.429		2,125	310		2,435	2,825	
	3000	See through, including doors		.80	10		1,750	273		2,023	2,350	
	3200	Corner (2 wall)	▼	1	8	▼	865	218		1,083	1,300	

103 100 | Fireplace Accessories

		CREW	DAILY OUTPUT	LABOR-HOURS	UNIT	1999 BARE COSTS				TOTAL INCL O&P	
						MAT.	LABOR	EQUIP.	TOTAL		
104	0010	**FIREPLACE ACCESSORIES** Chimney screens, galv., 13" x 13" flue	1 Bric	8	1	Ea.	46	27.50		73.50	94	104
	0050	Galv., 24" x 24" flue		5	1.600		121	44		165	202	
	0200	Stainless steel, 13" x 13" flue		8	1		256	27.50		283.50	325	
	0250	20" x 20" flue	▼	5	1.600	▼	350	44		394	455	

103 100 | Fireplace Accessories

			CREW	DAILY OUTPUT	LABOR-HOURS	UNIT	1999 BARE COSTS				TOTAL INCL O&P	
							MAT.	LABOR	EQUIP.	TOTAL		
104	0400	Cleanout doors and frames, cast iron, 8" x 8"	1 Bric	12	.667	Ea.	17	18.40		35.40	47	104
	0450	12" x 12"		10	.800		32.50	22		54.50	70	
	0500	18" x 24"		8	1		102	27.50		129.50	155	
	0550	Cast iron frame, steel door, 24" x 30"		5	1.600		220	44		264	310	
	0800	Damper, rotary control, steel, 30" opening		6	1.333		61.50	37		98.50	125	
	0850	Cast iron, 30" opening		6	1.333		69	37		106	133	
	0880	36" opening		6	1.333		74.50	37		111.50	139	
	0900	48" opening		6	1.333		102	37		139	170	
	0920	60" opening		6	1.333		242	37		279	325	
	0950	72" opening		5	1.600		294	44		338	395	
	1000	84" opening, special order		5	1.600		615	44		659	750	
	1050	96" opening, special order		4	2		625	55		680	775	
	1200	Steel plate, poker control, 60" opening		8	1		218	27.50		245.50	283	
	1250	84" opening, special opening		5	1.600		400	44		444	510	
	1400	"Universal" type, chain operated, 32" x 20" opening		8	1		149	27.50		176.50	207	
	1450	48" x 24" opening		5	1.600		263	44		307	360	
	1600	Dutch Oven door and frame, cast iron, 12" x 15" opening		13	.615		89	17		106	125	
	1650	Copper plated, 12" x 15" opening		13	.615		171	17		188	216	
	1800	Fireplace forms, no accessories, 32" opening		3	2.667		455	73.50		528.50	615	
	1900	36" opening		2.50	3.200		550	88.50		638.50	740	
	2000	40" opening		2	4		660	110		770	895	
	2100	78" opening		1.50	5.333		940	147		1,087	1,250	
	2400	Squirrel and bird screens, galvanized, 8" x 8" flue		16	.500		43	13.80		56.80	69	
	2450	13" x 13" flue		12	.667		50.50	18.40		68.90	84	

103 200 | Stoves

			CREW	DAILY OUTPUT	LABOR-HOURS	UNIT	MAT.	LABOR	EQUIP.	TOTAL	TOTAL INCL O&P	
201	0010	**WOODBURNING STOVES** Cast iron, minimum	2 Carp	1.30	12.308	Ea.	740	335		1,075	1,350	201
	0020	Average		1	16		1,200	435		1,635	1,975	
	0030	Maximum		.80	20		2,000	545		2,545	3,050	
	0050	For gas log lighter, add					35			35	38.50	

103 460 | Cupolas

			CREW	DAILY OUTPUT	LABOR-HOURS	UNIT	MAT.	LABOR	EQUIP.	TOTAL	TOTAL INCL O&P	
464	0010	**CUPOLA** Stock units, pine, painted, 18" sq., 28" high, alum. roof	1 Carp	4.10	1.951	Ea.	94.50	53.50		148	188	464
	0100	Copper roof		3.80	2.105		134	57.50		191.50	238	
	0300	23" square, 33" high, aluminum roof		3.70	2.162		221	59		280	335	
	0400	Copper roof		3.30	2.424		222	66		288	350	
	0600	30" square, 37" high, aluminum roof		3.70	2.162		335	59		394	465	
	0700	Copper roof		3.30	2.424		345	66		411	485	
	0900	Hexagonal, 31" wide, 46" high, copper roof		4	2		460	54.50		514.50	590	
	1000	36" wide, 50" high, copper roof		3.50	2.286		535	62.50		597.50	690	
	1200	For deluxe stock units, add to above					25%					
	1400	For custom built units, add to above					50%	50%				
	1600	Fiberglass, 5'-0" base, 63" high minimum	F-3	6	6.667		2,225	185	74	2,484	2,825	
	1650	Maximum		4	10		2,575	277	111	2,963	3,400	
	1700	6'-0" base, 63" high, minimum		5	8		3,450	222	88.50	3,760.50	4,250	
	1750	Maximum		3	13.333		3,550	370	148	4,068	4,650	

103 520 | Ground Set Flagpoles

			CREW	DAILY OUTPUT	LABOR-HOURS	UNIT	MAT.	LABOR	EQUIP.	TOTAL	TOTAL INCL O&P	
524	0010	**FLAGPOLE**										524
	0050	Not including base or foundation										
	0100	Aluminum, tapered, ground set 20' high	K-1	2	8	Ea.	590	196	90	876	1,050	
	0200	25' high		1.70	9.412		710	231	106	1,047	1,250	
	0300	30' high		1.50	10.667		920	262	120	1,302	1,525	
	0400	35' high		1.40	11.429		1,275	280	128	1,683	1,975	
	0500	40' high		1.20	13.333		1,675	325	150	2,150	2,525	
	0600	50' high		1	16		2,625	390	180	3,195	3,675	

10 SPECIALTIES

Important: See the Reference Section for critical supporting data - Reference Nos., Crews, & City Cost Indexes

103 520 | Ground Set Flagpoles

		CREW	DAILY OUTPUT	LABOR-HOURS	UNIT	1999 BARE COSTS				TOTAL INCL O&P	
						MAT.	LABOR	EQUIP.	TOTAL		
524	0700	60' high	K-1	.90	17.778	Ea.	4,125	435	199	4,759	5,450	524
	0800	70' high		.80	20		5,625	490	224	6,339	7,200	
	1100	Counterbalanced, internal halyard, 20' high		1.80	8.889		1,875	218	99.50	2,192.50	2,500	
	1200	30' high		1.50	10.667		2,425	262	120	2,807	3,200	
	1300	40' high		1.30	12.308		3,275	300	138	3,713	4,225	
	1400	50' high		1	16		5,775	390	180	6,345	7,150	
	2820	Aluminum, electronically operated, 30' high		1.40	11.429		3,800	280	128	4,208	4,750	
	2840	35' high		1.30	12.308		3,975	300	138	4,413	5,000	
	2860	39' high		1.10	14.545		4,500	355	163	5,018	5,675	
	2880	45' high		1	16		5,625	390	180	6,195	6,975	
	2900	50' high		.90	17.778		6,175	435	199	6,809	7,700	
	3000	Fiberglass, tapered, ground set, 23' high		2	8		835	196	90	1,121	1,325	
	3100	29"-7" high		1.50	10.667		1,225	262	120	1,607	1,850	
	3200	36'-1" high		1.40	11.429		1,475	280	128	1,883	2,200	
	3300	39'-5" high		1.20	13.333		1,575	325	150	2,050	2,425	
	3400	49'-2" high		1	16		4,000	390	180	4,570	5,200	
	3500	59' high	▼	.90	17.778	▼	5,125	435	199	5,759	6,525	
	4300	Concrete, direct imbedded installation										
	4400	Internal halyard, 20' high	K-1	2.50	6.400	Ea.	1,200	157	72	1,429	1,650	
	4500	25' high		2.50	6.400		1,350	157	72	1,579	1,825	
	4600	30' high		2.30	6.957		1,575	171	78	1,824	2,075	
	4700	40' high		2.10	7.619		1,900	187	85.50	2,172.50	2,450	
	4800	50' high		1.90	8.421		3,400	206	94.50	3,700.50	4,175	
	5000	60' high		1.80	8.889		5,300	218	99.50	5,617.50	6,300	
	5100	70' high		1.60	10		6,375	245	112	6,732	7,500	
	5200	80' high		1.40	11.429		8,275	280	128	8,683	9,675	
	5300	90' high		1.20	13.333		10,900	325	150	11,375	12,700	
	5500	100' high	▼	1	16	▼	11,300	390	180	11,870	13,200	
	6400	Wood poles, tapered, clear vertical grain fir with tilting										
	6410	base, not incl. foundation, 4" butt, 25' high	K-1	1.90	8.421	Ea.	465	206	94.50	765.50	940	
	6800	6" butt, 30' high	"	1.30	12.308	"	580	300	138	1,018	1,250	
	7300	Foundations for flagpoles, including										
	7400	excavation and concrete, to 35' high poles	C-1	10	3.200	Ea.	390	82.50		472.50	555	
	7600	40' to 50' high		3.50	9.143		705	236		941	1,150	
	7700	Over 60' high	▼	2	16	▼	790	415		1,205	1,525	

103 540 | Wall-Mounted Flagpoles

		CREW	DAILY OUTPUT	LABOR-HOURS	UNIT	MAT.	LABOR	EQUIP.	TOTAL	TOTAL INCL O&P	
544	0010	**FLAGPOLE** Structure mounted										544
	0100	Fiberglass, vertical wall set, 19'-8" long	K-1	1.50	10.667	Ea.	1,000	262	120	1,382	1,625	
	0200	23' long		1.40	11.429		960	280	128	1,368	1,625	
	0300	26'-3" long		1.30	12.308		1,425	300	138	1,863	2,200	
	0800	19'-8" long outrigger		1.30	12.308		1,125	300	138	1,563	1,875	
	1300	Aluminum, vertical wall set, tapered, with base, 20' high		1.20	13.333		725	325	150	1,200	1,475	
	1400	29'-6" high		1	16		1,550	390	180	2,120	2,525	
	2400	Outrigger poles with base, 12' long		1.30	12.308		1,100	300	138	1,538	1,825	
	2500	14' long	▼	1	16	▼	1,125	390	180	1,695	2,025	

SPECIALTIES 10

104 100	Directories	CREW	DAILY OUTPUT	LABOR-HOURS	UNIT	1999 BARE COSTS				TOTAL INCL O&P
						MAT.	LABOR	EQUIP.	TOTAL	
104	**0010**	**DIRECTORY BOARDS**									**104**
	0050	Plastic, glass covered, 30" x 20"	2 Carp	3	5.333	Ea.	275	146		421	535
	0100	36" x 48"		2	8		540	218		758	940
	0300	Grooved cork, 30" x 20"		3	5.333		283	146		429	540
	0400	36" x 48"		2	8		440	218		658	830
	0600	Black felt, 30" x 20"		3	5.333		290	146		436	550
	0700	36" x 48"		2	8		585	218		803	990
	0900	Outdoor, weatherproof, black plastic, 36" x 24"		2	8		300	218		518	675
	1000	36" x 36"		1.50	10.667		635	291		926	1,150
	1800	Indoor, economy, open face, 18" x 24"		7	2.286		88	62.50		150.50	195
	1900	24" x 36"		7	2.286		120	62.50		182.50	230
	2000	36" x 24"		6	2.667		138	73		211	267
	2100	36" x 48"		6	2.667		212	73		285	350
	2400	Building directory, alum., black felt panels, 1 door, 24" x 18"		4	4		265	109		374	465
	2500	36" x 24"		3.50	4.571		325	125		450	555
	2600	48" x 32"		3	5.333		390	146		536	655
	2700	36" x 48", 2 door		2.50	6.400		635	175		810	970
	2800	36" x 60"		2	8		680	218		898	1,100
	2900	48" x 60"	↓	1	16		730	435		1,165	1,500
	3100	For bronze enamel finish, add					15%				
	3200	For bronze anodized finish, add					25%				
	3400	For illuminated directory, single door unit, add				↓	138			138	151
	3500	For 6" header panel, 6 letters per foot, add				L.F.	22.50			22.50	24.50

104 150	Bulletin Boards	CREW	DAILY OUTPUT	LABOR-HOURS	UNIT	MAT.	LABOR	EQUIP.	TOTAL	TOTAL INCL O&P
151	**0010**	**BULLETIN BOARD** Cork sheets, unbacked, no frame, 1/4" thick	2 Carp	290	.055	S.F.	5.65	1.51		7.16	8.55	**151**
	0100	1/2" thick		290	.055		5.80	1.51		7.31	8.75
	0300	Fabric-face, no frame, on 7/32" cork underlay		290	.055		4.11	1.51		5.62	6.90
	0400	On 1/4" cork on 1/4" hardboard		290	.055		5.50	1.51		7.01	8.40
	0600	With edges wrapped		290	.055		5.85	1.51		7.36	8.80
	0700	On 7/16" fire retardant core		290	.055		4.10	1.51		5.61	6.90
	0900	With edges wrapped	↓	290	.055		4.40	1.51		5.91	7.20
	1000	Designer fabric only, cut to size					1.82			1.82	2
	1200	1/4" vinyl cork, on 1/4" hardboard, no frame	2 Carp	290	.055		4.48	1.51		5.99	7.30
	1300	On 1/4" coreboard		290	.055	↓	6	1.51		7.51	8.95
	2000	For map and display rail, economy, add		385	.042	L.F.	1.44	1.13		2.57	3.36
	2100	Deluxe, add		350	.046	"	2.74	1.25		3.99	4.97
	2120	Prefabricated, 1/4" cork, 3' x 5' with aluminum frame		16	1	Ea.	97.50	27.50		125	150
	2140	Wood frame		16	1		126	27.50		153.50	181
	2160	4' x 4' with aluminum frame		16	1		81	27.50		108.50	132
	2180	Wood frame		16	1		170	27.50		197.50	230
	2200	4' x 8' with aluminum frame		14	1.143		154	31		185	218
	2210	With wood frame		14	1.143		170	31		201	236
	2220	4' x 12' with aluminum frame	↓	12	1.333	↓	224	36.50		260.50	305
	2230	Bulletin board case, single glass door, with lock									
	2240	36" x 24", economy	2 Carp	12	1.333	Ea.	245	36.50		281.50	325
	2250	Deluxe		12	1.333		375	36.50		411.50	475
	2260	42" x 30", economy		12	1.333		275	36.50		311.50	365
	2270	Deluxe	↓	12	1.333	↓	410	36.50		446.50	515
	2300	Glass enclosed cabinets, alum., cork panel, hinged doors									
	2400	3' x 3', 1 door	2 Carp	12	1.333	Ea.	420	36.50		456.50	525
	2500	4' x 4', 2 door		11	1.455		700	39.50		739.50	835
	2600	4' x 7', 3 door		10	1.600		1,150	43.50		1,193.50	1,350
	2800	4' x 10', 4 door	↓	8	2		1,425	54.50		1,479.50	1,650
	2900	For lights, add per door opening	1 Elec	13	.615	↓	138	19.65		157.65	181

Important: See the Reference Section for critical supporting data - Reference Nos., Crews, & City Cost Indexes

10 SPECIALTIES

104 150 | Bulletin Boards

			CREW	DAILY OUTPUT	LABOR-HOURS	UNIT	1999 BARE COSTS				TOTAL INCL O&P	
							MAT.	LABOR	EQUIP.	TOTAL		
151	3100	Horizontal sliding units, 4 doors, 4' x 8' 8' x 4'	2 Carp	9	1.778	Ea.	1,225	48.50		1,273.50	1,425	**151**
	3200	4' x 12'		7	2.286		1,550	62.50		1,612.50	1,800	
	3400	8 doors, 4' x 16'		5	3.200		3,250	87.50		3,337.50	3,700	
	3500	4' x 24'	▼	4	4	▼	4,025	109		4,134	4,600	
155	0010	**CONTROL BOARDS** Magnetic, porcelain finish, 18" x 24", framed	2 Carp	8	2	Ea.	145	54.50		199.50	246	**155**
	0100	24" x 36"		7.50	2.133		205	58		263	320	
	0200	36" x 48"		7	2.286		325	62.50		387.50	460	
	0300	48" x 72"		6	2.667		310	73		383	455	
	0400	48" x 96"	▼	5	3.200	▼	650	87.50		737.50	850	

104 300 | Signs

			CREW	DAILY OUTPUT	LABOR-HOURS	UNIT	MAT.	LABOR	EQUIP.	TOTAL	TOTAL INCL O&P	
304	0010	**SIGNS** Letters, 2" high, 3/8" deep, cast bronze	1 Carp	24	.333	Ea.	15.35	9.10		24.45	31	**304**
	0140	1/2" deep, cast aluminum		18	.444		15.35	12.15		27.50	36	
	0160	Cast bronze		32	.250		21.50	6.80		28.30	35	
	0300	6" high, 5/8" deep, cast aluminum		24	.333		17.15	9.10		26.25	33	
	0400	Cast bronze		24	.333		42.50	9.10		51.60	61	
	0600	8" high, 3/4" deep, cast aluminum		14	.571		27.50	15.60		43.10	55	
	0700	Cast bronze		20	.400		64	10.90		74.90	87.50	
	0900	10" high, 1" deep, cast aluminum		18	.444		37.50	12.15		49.65	60	
	1000	Bronze		18	.444		71	12.15		83.15	97	
	1200	12" high, 1-1/4" deep, cast aluminum		12	.667		45	18.20		63.20	78	
	1500	Cast bronze		18	.444		103	12.15		115.15	132	
	1600	14" high, 2-5/16" deep, cast aluminum		12	.667		57	18.20		75.20	91.50	
	1800	Fabricated stainless steel, 6" high, 2" deep		20	.400		89.50	10.90		100.40	116	
	1900	12" high, 3" deep		18	.444		106	12.15		118.15	136	
	2100	18" high, 3" deep		12	.667		160	18.20		178.20	205	
	2200	24" high, 4" deep		10	.800		223	22		245	281	
	2700	Acrylic, on high density foam,		20	.400		17.80	10.90		28.70	37	
	2800	12" high, 2" deep	▼	18	.444		39.50	12.15		51.65	62.50	
	3900	Plaques, custom, 20" x 30", for up to 450 letters, cast aluminum	2 Carp	4	4		740	109		849	980	
	4000	Cast bronze		4	4		980	109		1,089	1,250	
	4200	30" x 36", up to 900 letters cast aluminum		3	5.333		1,475	146		1,621	1,850	
	4300	Cast bronze		3	5.333		1,925	146		2,071	2,325	
	4500	36" x 48", for up to 1300 letters, cast bronze		2	8		2,900	218		3,118	3,550	
	4800	Signs, reflective alum. street signs, dbl. face, 2-way, w/bracket		30	.533		39	14.55		53.55	66	
	4900	4-way	▼	30	.533		69.50	14.55		84.05	99.50	
	5100	Exit signs, 24 ga. alum., 14" x 12" surface mounted	1 Carp	30	.267		12.70	7.30		20	25.50	
	5200	10" x 7"		20	.400		7.75	10.90		18.65	26	
	5400	Bracket mounted, double face, 12" x 10"	▼	30	.267		22	7.30		29.30	36	
	5500	Sticky back, stock decals, 14" x 10"	1 Clab	50	.160		4.65	3.43		8.08	10.50	
	6000	Interior elec., wall mount, fiberglass panels, 2 lamps, 6"	1 Elec	8	1		47	32		79	99	
	6100	8"	"	8	1		39.50	32		71.50	91	
	6400	Replacement sign faces, 6" or 8"	1 Clab	50	.160	▼	21	3.43		24.43	28.50	

104 560 | Turnstiles

			CREW	DAILY OUTPUT	LABOR-HOURS	UNIT	MAT.	LABOR	EQUIP.	TOTAL	TOTAL INCL O&P	
561	0010	**TURNSTILES** One way, 4 arm, 46" diameter, economy, manual	2 Carp	5	3.200	Ea.	249	87.50		336.50	410	**561**
	0100	Electric		1.20	13.333		835	365		1,200	1,500	
	0300	High security, galv., 5'-5" diameter, 7' high, manual		1	16		2,250	435		2,685	3,150	
	0350	Electric		.60	26.667		2,775	730		3,505	4,225	
	0420	Three arm, 24" opening, light duty, manual		2	8		585	218		803	990	
	0450	Heavy duty		1.50	10.667		600	291		891	1,125	
	0460	Manual, with registering & controls, light duty		2	8		680	218		898	1,100	
	0470	Heavy duty		1.50	10.667		1,200	291		1,491	1,775	
	0480	Electric, heavy duty	▼	1.10	14.545	▼	1,000	395		1,395	1,725	
	0500	For coin or token operating, add					239			239	263	

SPECIALTIES 10

104 | Identifying & Pedestrian Control Devices

104 560	Turnstiles	CREW	DAILY OUTPUT	LABOR-HOURS	UNIT	1999 BARE COSTS				TOTAL INCL O&P
						MAT.	LABOR	EQUIP.	TOTAL	
561	**1200** One way gate with horizontal bars, 5'-5" diameter									**561**
	1300 7' high, recreation or transit type	2 Carp	.80	20	Ea.	2,550	545		3,095	3,650
	1500 For electronic counter, add				"	170			170	187

105 | Lockers, Protective Covers & Postal Specialties

105 050	Metal Lockers	CREW	DAILY OUTPUT	LABOR-HOURS	UNIT	1999 BARE COSTS				TOTAL INCL O&P
						MAT.	LABOR	EQUIP.	TOTAL	
054	**0011** **LOCKERS** Steel, baked enamel R105-050									**054**
	0110 Single tier box locker, 12" x 15" x 72"	1 Shee	8	1	Ea.	128	32		160	190
	0120 18" x 15" x 72"		8	1		148	32		180	213
	0130 12" x 18" x 72"		8	1		132	32		164	196
	0140 18" x 18" x 72"		8	1		153	32		185	218
	0410 Double tier, 12" x 15" x 36"		21	.381		110	12.10		122.10	140
	0420 18" x 15" x 36"		21	.381		116	12.10		128.10	146
	0430 12" x 18" x 36"		21	.381		94	12.10		106.10	122
	0440 18" x 18" x 36"		21	.381		95.50	12.10		107.60	124
	0500 Two person, 18" x 15" x 72"		8	1		173	32		205	240
	0510 18" x 18" x 72"		8	1		201	32		233	271
	0520 Duplex, 15" x 15" x 72"		8	1		195	32		227	265
	0530 15" x 21" x 72"		8	1	▼	222	32		254	294
	0600 5 tier box lockers, minimum		30	.267	Opng.	27	8.45		35.45	43
	0700 Maximum		24	.333		38.50	10.60		49.10	59
	0900 6 tier box lockers, minimum		36	.222		24	7.05		31.05	37.50
	1000 Maximum		30	.267	▼	32.50	8.45		40.95	48.50
	1100 Wire meshed wardrobe, floor. mtd., open front varsity type		7.50	1.067	Ea.	144	34		178	212
	2100 Locker bench, laminated maple, top only		100	.080	L.F.	14	2.54		16.54	19.30
	2200 Pedestals, steel pipe	▼	25	.320	Ea.	25.50	10.15		35.65	44
	2400 16-person locker unit with clothing rack									
	2500 72 wide x 15" deep x 72" high	1 Shee	8	1	Ea.	355	32		387	440
	2550 18" deep	"	8	1	"	365	32		397	450
	3000 Wall mounted lockers, 4 person, with coat bar									
	3100 48" wide x 18" deep x 12" high	1 Shee	8	1	Ea.	187	32		219	256
	3250 Rack w/ 24 wire mesh baskets		1.50	5.333	Set	270	169		439	560
	3260 30 baskets		1.25	6.400		232	203		435	570
	3270 36 baskets		.95	8.421		325	267		592	775
	3280 42 baskets	▼	.80	10	▼	365	320		685	900
	3300 For built-in lock with 2 keys, add				Ea.	4.63			4.63	5.10
	3600 For hanger rods, add				"	1.65			1.65	1.82

105 200	Fire Protect. Specialties	CREW	DAILY OUTPUT	LABOR-HOURS	UNIT	MAT.	LABOR	EQUIP.	TOTAL	TOTAL INCL O&P
220	**0010** **FIRE EQUIPMENT CABINETS** Not equipped, 20 ga. steel box,									**220**
	0040 recessed, D.S. glass in door, box size given									
	1000 Portable extinguisher, single, 8" x 12" x 27", alum. door & frame	Q-12	8	2	Ea.	75.50	58.50		134	172
	1100 Steel door and frame	"	8	2	"	46	58.50		104.50	140
	3000 Hose rack assy., 1-1/2" valve & 100' hose, 24" x 40" x 5-1/2"									
	3100 Aluminum door and frame	Q-12	6	2.667	Ea.	159	78		237	292
	3200 Steel door and frame		6	2.667		101	78		179	229
	3300 Stainless steel door and frame	▼	6	2.667	▼	244	78		322	385

10 SPECIALTIES

314   **Important: See the Reference Section for critical supporting data - Reference Nos., Crews, & City Cost Indexes**

105 200 | Fire Protect. Specialties

			CREW	DAILY OUTPUT	LABOR-HOURS	UNIT	MAT.	LABOR	EQUIP.	TOTAL	TOTAL INCL O&P	
220	4000	Hose rack assy., 2-1/2" x 1-1/2" valve, 100' hose, 24" x 40" x 8"										**220**
	4100	Aluminum door and frame	Q-12	6	2.667	Ea.	160	78		238	294	
	4200	Steel door and frame		6	2.667		108	78		186	237	
	4300	Stainless steel door and frame	↓	6	2.667	↓	221	78		299	360	
	5000	Hose rack assy., 2-1/2" x 1-1/2" valve, 100' hose										
	5100	Aluminum door and frame	Q-12	5	3.200	Ea.	204	93.50		297.50	365	
	5200	Steel door and frame		5	3.200		129	93.50		222.50	284	
	5300	Stainless steel door and frame	↓	5	3.200	↓	285	93.50		378.50	455	
	8000	Valve cabinet for 2-1/2" FD angle valve, 18" x 18" x 8"										
	8100	Aluminum door and frame	Q-12	12	1.333	Ea.	70.50	39		109.50	137	
	8200	Steel door and frame		12	1.333		54	39		93	119	
	8300	Stainless steel door and frame	↓	12	1.333	↓	98.50	39		137.50	168	
225	0010	**FIRE EXTINGUISHERS**										**225**
	0120	CO2, portable with swivel horn, 5 lb.				Ea.	101			101	111	
	0140	With hose and "H" horn, 10 lb.					150			150	165	
	0160	15 lb.					172			172	189	
	0180	20 lb.					207			207	228	
	0360	Wheeled type, cart mounted, 50 lb.				↓	1,200			1,200	1,325	
	1000	Dry chemical, pressurized										
	1040	Standard type, portable, painted, 2-1/2 lb.				Ea.	25			25	27.50	
	1060	5 lb.					40			40	44	
	1080	10 lb.					65			65	71.50	
	1100	20 lb.					90			90	99	
	1120	30 lb.					145			145	160	
	1300	Standard type, wheeled, 150 lb.					1,400			1,400	1,550	
	2000	ABC all purpose type, portable, 2-1/2 lb.					25			25	27.50	
	2060	5 lb.					40			40	44	
	2080	9-1/2 lb.					60			60	66	
	2100	20 lb.					85			85	93.50	
	2300	Wheeled, 45 lb.					650			650	715	
	2360	150 lb.					1,450			1,450	1,600	
	3000	Dry chemical, outside cartridge to -65°F, painted, 9 lb.					200			200	220	
	3060	26 lb.					250			250	275	
	5000	Pressurized water, 2-1/2 gallon, stainless steel					75			75	82.50	
	5060	With anti-freeze					110			110	121	
	6000	Soda & acid, 2-1/2 gallon, stainless steel					75			75	82.50	
	9400	Installation of extinguishers, 12 or more, on wood	1 Carp	30	.267			7.30		7.30	11.45	
	9420	On masonry or concrete	"	15	.533	↓		14.55		14.55	23	

105 380 | Canopies

			CREW	DAILY OUTPUT	LABOR-HOURS	UNIT	MAT.	LABOR	EQUIP.	TOTAL	TOTAL INCL O&P	
384	0010	**CANOPIES** Wall hung, .032", aluminum, prefinished, 8' x 10'	K-2	1.30	18.462	Ea.	1,250	525	138	1,913	2,450	**384**
	0300	8' x 20'		1.10	21.818		2,500	620	163	3,283	4,000	
	0500	10' x 10'		1.30	18.462		1,400	525	138	2,063	2,625	
	0700	10' x 20'		1.10	21.818		2,600	620	163	3,383	4,100	
	1000	12' x 20'		1	24		3,200	680	180	4,060	4,925	
	1360	12' x 30'		.80	30		4,700	850	224	5,774	6,925	
	1700	12' x 40'	↓	.60	40		5,000	1,125	299	6,424	7,825	
	1900	For free standing units, add				↓	20%	10%				
	2300	Aluminum entrance canopies, flat soffit, .032"										
	2500	3'-6" x 4'-0", clear anodized	2 Carp	4	4	Ea.	525	109		634	750	
	2700	Bronze anodized		4	4		925	109		1,034	1,200	
	3000	Polyurethane painted		4	4		750	109		859	995	
	3300	4'-6" x 10'-0", clear anodized		2	8		1,450	218		1,668	1,950	
	3500	Bronze anodized		2	8		1,850	218		2,068	2,375	
	3700	Polyurethane painted	↓	2	8		1,550	218		1,768	2,050	
	4000	Wall downspout, 10 L.F., clear anodized	1 Carp	7	1.143	↓	90	31		121	148	

105 380 | Canopies

			CREW	DAILY OUTPUT	LABOR-HOURS	UNIT	MAT.	LABOR	EQUIP.	TOTAL	TOTAL INCL O&P	
384	4300	Bronze anodized	1 Carp	7	1.143	Ea.	157	31		188	222	384
	4500	Polyurethane painted	↓	7	1.143	↓	135	31		166	198	
	4700	Canvas awnings, including canvas, frame & lettering										
	5000	Minimum	2 Carp	100	.160	S.F.	36	4.37		40.37	46.50	
	5300	Average		90	.178		43.50	4.85		48.35	55.50	
	5500	Maximum		80	.200	↓	88.50	5.45		93.95	106	
	7000	Carport, baked vinyl finish, .032", 20' x 10', no foundations, min.	K-2	4	6	Car	2,250	170	45	2,465	2,825	
	7250	Maximum		2	12	"	5,350	340	90	5,780	6,575	
	7500	Walkway cover, to 12' wide, stl., vinyl finish, .032",no fndtns., min.		250	.096	S.F.	13.20	2.72	.72	16.64	20	
	7750	Maximum	↓	200	.120	"	14.35	3.40	.90	18.65	22.50	

105 510 | Mail Chutes

			CREW	DAILY OUTPUT	LABOR-HOURS	UNIT	MAT.	LABOR	EQUIP.	TOTAL	TOTAL INCL O&P	
511	0010	**MAIL CHUTES** Aluminum & glass, 14-1/4" wide, 4-5/8" deep	2 Shee	4	4	Floor	630	127		757	890	511
	0100	8-5/8" deep		3.80	4.211		700	134		834	980	
	0300	8-3/4" x 3-1/2", aluminum		5	3.200		535	102		637	745	
	0400	Bronze or stainless		4.50	3.556	↓	830	113		943	1,075	
	0600	Lobby collection boxes, aluminum		5	3.200	Ea.	1,725	102		1,827	2,050	
	0700	Bronze or stainless	↓	4.50	3.556	"	2,125	113		2,238	2,525	

105 520 | Mail Boxes

			CREW	DAILY OUTPUT	LABOR-HOURS	UNIT	MAT.	LABOR	EQUIP.	TOTAL	TOTAL INCL O&P	
521	0010	**MAIL BOXES** Horiz., key lock, 5"H x 6"W x 15"D, alum., rear load	1 Carp	34	.235	Ea.	34	6.40		40.40	47.50	521
	0100	Front loading		34	.235		38.50	6.40		44.90	52	
	0200	Double, 5"H x 12"W x 15"D, rear loading		26	.308		60.50	8.40		68.90	79.50	
	0300	Front loading		26	.308		65.50	8.40		73.90	85	
	0500	Quadruple, 10"H x 12"W x 15"D, rear loading		20	.400		109	10.90		119.90	137	
	0600	Front loading		20	.400		120	10.90		130.90	149	
	0800	Vertical, front load, 15"H x 5"W x 6"D, alum., per compartment		34	.235		28.50	6.40		34.90	41.50	
	0900	Bronze, duranodic finish		34	.235		45.50	6.40		51.90	60.50	
	1000	Steel, enameled		34	.235		26	6.40		32.40	38.50	
	1700	Alphabetical directories, 120 names		10	.800		139	22		161	188	
	1800	Letter collection box		6	1.333		570	36.50		606.50	690	
	1900	Letter slot, residential		20	.400		50	10.90		60.90	72	
	2000	Post office type		8	1		189	27.50		216.50	251	
	2200	Post office counter window, with grille		2	4		510	109		619	730	
	2250	Key keeper, single key, aluminum		26	.308		39	8.40		47.40	56	
	2300	Steel, enameled	↓	26	.308	↓	69	8.40		77.40	89	

106 | Partitions & Storage Shelving

106 050 | Wire Mesh Partitions

			CREW	DAILY OUTPUT	LABOR-HOURS	UNIT	MAT.	LABOR	EQUIP.	TOTAL	TOTAL INCL O&P	
052	0010	**PARTITIONS, WOVEN WIRE** For tool or stockroom enclosures										052
	0100	Channel frame, 1-1/2" diamond mesh, 10 ga. wire, painted										
	0300	Wall panels, 4'-0" wide, 7' high	2 Carp	25	.640	Ea.	85.50	17.45		102.95	122	
	0400	8' high		23	.696		95.50	19		114.50	135	
	0600	10' high	↓	18	.889		113	24.50		137.50	162	
	0700	For 5' wide panels, add					5%					
	0900	Ceiling panels, 10' long, 2' wide	2 Carp	25	.640		80.50	17.45		97.95	116	
	1000	4' wide		15	1.067		95.50	29		124.50	151	
	1200	Panel with service window & shelf, 5' wide, 7' high		20	.800		265	22		287	325	
	1300	8' high	↓	15	1.067		269	29		298	340	

10 SPECIALTIES

106 050 | Wire Mesh Partitions

			DAILY OUTPUT	LABOR-HOURS	UNIT	1999 BARE COSTS				TOTAL INCL O&P	
						MAT.	LABOR	EQUIP.	TOTAL		
052	1500	Sliding doors, full height, 3' wide, 7' high	2 Carp	6	2.667	Ea.	255	73		328	395	052
	1600	10' high		5	3.200		292	87.50		379.50	455	
	1800	6' wide sliding door, 7' full height		5	3.200		330	87.50		417.50	495	
	1900	10' high		4	4		380	109		489	590	
	2100	Swinging doors, 3' wide, 7' high, no transom		6	2.667		178	73		251	310	
	2200	7' high, 3' transom		5	3.200		227	87.50		314.50	385	

106 100 | Folding Gates

			CREW	DAILY OUTPUT	LABOR-HOURS	UNIT	MAT.	LABOR	EQUIP.	TOTAL	TOTAL INCL O&P	
101	0010	**SECURITY GATES** For roll up type, see division 083-302										101
	0300	Scissors type folding gate, ptd. steel, single, 6' high, 5-1/2' wide	2 Sswk	4	4	Opng.	112	122		234	345	
	0350	6-1/2' wide		4	4		121	122		243	355	
	0400	7-1/2' wide		4	4		155	122		277	395	
	0600	Double gate, 7-1/2' high, 8' wide		2.50	6.400		170	196		366	545	
	0650	10' wide		2.50	6.400		214	196		410	595	
	0700	12' wide		2	8		265	245		510	740	
	0750	14' wide		2	8		310	245		555	790	
	0900	Door gate, folding steel, 4' wide, 61" high		4	4		52.50	122		174.50	282	
	1000	71" high		4	4		54.50	122		176.50	284	
	1200	81" high		4	4		59.50	122		181.50	289	
	1300	Window gates, 2' to 4' wide, 31" high		4	4		28.50	122		150.50	256	
	1500	55" high		3.75	4.267		40	131		171	283	
	1600	79" high		3.50	4.571		46	140		186	305	

106 150 | Demountable Partitions

			CREW	DAILY OUTPUT	LABOR-HOURS	UNIT	MAT.	LABOR	EQUIP.	TOTAL	TOTAL INCL O&P	
152	0010	**PARTITIONS, MOVABLE OFFICE** Demountable, add for doors										152
	0100	Do not deduct door openings from total L.F.										
	0900	Demountable gypsum system on 2" to 2-1/2"										
	1000	steel studs, 9' high, 3" to 3-3/4" thick										
	1200	Vinyl clad gypsum **CN**	2 Carp	48	.333	L.F.	23	9.10		32.10	39.50	
	1300	Fabric clad gypsum		44	.364		71	9.95		80.95	93.50	
	1500	Steel clad gypsum		40	.400		75.50	10.90		86.40	100	
	1600	1.75 system, aluminum framing, vinyl clad hardboard,										
	1800	paper honeycomb core panel, 1-3/4" to 2-1/2" thick										
	1900	9' high	2 Carp	48	.333	L.F.	48	9.10		57.10	67.50	
	2100	7' high		60	.267		44	7.30		51.30	59.50	
	2200	5' high		80	.200		38	5.45		43.45	50	
	2250	Unitized gypsum system										
	2300	Unitized panel, 9' high, 2" to 2-1/2" thick										
	2350	Vinyl clad gypsum	2 Carp	48	.333	L.F.	64.50	9.10		73.60	85.50	
	2400	Fabric clad gypsum	"	44	.364	"	116	9.95		125.95	144	
	2500	Unitized mineral fiber system										
	2510	Unitized panel, 9' high, 2-1/4" thick, aluminum frame										
	2550	Vinyl clad mineral fiber	2 Carp	48	.333	L.F.	82.50	9.10		91.60	105	
	2600	Fabric clad mineral fiber	"	44	.364	"	105	9.95		114.95	131	
	2800	Movable steel walls, modular system										
	2900	Unitized panels, 9' high, 48" wide										
	3100	Baked enamel, pre-finished	2 Carp	60	.267	L.F.	75.50	7.30		82.80	94.50	
	3200	Fabric clad steel		56	.286	"	115	7.80		122.80	139	
	5310	Trackless wall, cork finish, semi-acoustic, 1-5/8" thick, minimum		325	.049	S.F.	18.35	1.34		19.69	22	
	5320	Maximum		190	.084		22	2.30		24.30	27.50	
	5330	Acoustic, 2" thick, minimum		305	.052		18.95	1.43		20.38	23.50	
	5340	Maximum		225	.071		28	1.94		29.94	34	
	5500	For acoustical partitions, add, minimum					1.33			1.33	1.46	
	5550	Maximum					4.87			4.87	5.35	
	5700	For doors, see div. 081 & 082										
	5800	For door hardware, see div. 087										

		106 150 | **Demountable Partitions**	CREW	DAILY OUTPUT	LABOR-HOURS	UNIT	MAT.	1999 BARE COSTS LABOR	EQUIP.	TOTAL	TOTAL INCL O&P	
152	6100	In-plant modular office system, w/prehung steel door										152
	6200	3" thick honeycomb core panels										
	6250	12' x 12', 2 wall	2 Clab	3.80	4.211	Ea.	3,125	90.50		3,215.50	3,575	
	6300	4 wall		1.90	8.421		4,800	181		4,981	5,575	
	6350	16' x 16', 2 wall		3.60	4.444		4,650	95.50		4,745.50	5,275	
	6400	4 wall		1.80	8.889		6,700	191		6,891	7,650	

		106 300 | **Portable Partitions**	CREW	DAILY OUTPUT	LABOR-HOURS	UNIT	MAT.	LABOR	EQUIP.	TOTAL	TOTAL INCL O&P	
302	0010	**PARTITIONS, HOSPITAL** Curtain track, box channel, ceiling mounted	1 Carp	135	.059	L.F.	3.68	1.62		5.30	6.60	302
	0100	Suspended	"	100	.080	"	5.25	2.18		7.43	9.20	
	0300	Curtains, nylon mesh tops, fire resistant, 11 oz. per lineal yard										
	0310	Polyester oxford cloth, 9' ceiling height	1 Carp	425	.019	L.F.	7.60	.51		8.11	9.15	
	0500	8' ceiling height		425	.019		6.85	.51		7.36	8.35	
	0700	Designer oxford cloth		425	.019		16.65	.51		17.16	19.10	
	0800	I.V. track systems										
	0820	I.V. track, oval	1 Carp	135	.059	L.F.	2.94	1.62		4.56	5.80	
	0830	I.V. trolley		32	.250	Ea.	25	6.80		31.80	38.50	
	0840	I.V. pendent, (tree, 5 hook)		32	.250	"	92.50	6.80		99.30	113	
304	0010	**PARTITIONS, PORTABLE** Divider panels, free standing, fiber core										304
	0020	Fabric face straight										
	0100	3'-0" long, 4'-0" high	2 Carp	100	.160	L.F.	91	4.37		95.37	107	
	0200	5'-0" high		90	.178		92	4.85		96.85	109	
	0500	6'-0" high		75	.213		104	5.80		109.80	123	
	0900	5'-0" long, 4'-0" high		175	.091		68.50	2.50		71	79.50	
	1000	5'-0" high		150	.107		84.50	2.91		87.41	97.50	
	1500	6"0" high		125	.128		88	3.49		91.49	103	
	1600	6'-0" long, 5'-0" high		162	.099		67.50	2.70		70.20	78	
	3100	Curved, 3'-0" long, 5'-0" high		90	.178		85	4.85		89.85	101	
	3150	6'-0" high		75	.213		98.50	5.80		104.30	117	
	3200	Economical panels, fabric face, 4'-0" long, 5'-0" high		132	.121		123	3.31		126.31	140	
	3250	6'-0" high		112	.143		22	3.90		25.90	30	
	3300	5'-0" long, 5'-0" high		150	.107		25	2.91		27.91	32	
	3350	6'-0" high		125	.128		31	3.49		34.49	39.50	
	3380	3'-0" curved, 5'-0" high		90	.178		85	4.85		89.85	101	
	3390	6'-0" high		75	.213		98.50	5.80		104.30	117	
	3450	Acoustical panels, 60 to 90 NRC, 3'-0" long, 5'-0" high		90	.178		77	4.85		81.85	92	
	3550	6'-0" high		75	.213		87.50	5.80		93.30	106	
	3600	5'-0" long, 5'-0" high		150	.107		53.50	2.91		56.41	63.50	
	3650	6'-0" high		125	.128		59.50	3.49		62.99	71	
	3700	6'-0" long, 5'-0" high		162	.099		52	2.70		54.70	61	
	3750	6'-0" high		138	.116		56.50	3.17		59.67	67.50	
	3800	Economy acoustical panels, 40 NRC, 4'-0" long, 5'-0" high		132	.121		30.50	3.31		33.81	39	
	3850	6'-0" high		112	.143		34.50	3.90		38.40	44	
	3900	5'-0" long, 6'-0" high		125	.128		31	3.49		34.49	39.50	
	3950	6'-0" long, 5'-0" high		162	.099		30.50	2.70		33.20	38	
	4000	Metal chalkboard, 6'-6" high, chalkboard, 1 side		125	.128		75.50	3.49		78.99	88.50	
	4100	Metal chalkboard, 2 sides		120	.133		86.50	3.64		90.14	101	
	4300	Tackboard, both sides		123	.130		68.50	3.55		72.05	80.50	

		106 520 | **Panel Partitions**	CREW	DAILY OUTPUT	LABOR-HOURS	UNIT	MAT.	LABOR	EQUIP.	TOTAL	TOTAL INCL O&P	
522	0010	**PARTITIONS, FOLDING LEAF** Acoustic, wood										522
	0100	Vinyl faced, to 18' high, 6 psf, minimum	2 Carp	60	.267	S.F.	34.50	7.30		41.80	49.50	
	0150	Average		45	.356		41	9.70		50.70	60.50	
	0200	Maximum		30	.533		53	14.55		67.55	81.50	
	0400	Formica or hardwood finish, minimum		60	.267		35.50	7.30		42.80	50.50	
	0500	Maximum		30	.533		38	14.55		52.55	64.50	

10 SPECIALTIES

Important: See the Reference Section for critical supporting data - Reference Nos., Crews, & City Cost Indexes

106 520 | Panel Partitions

			CREW	DAILY OUTPUT	LABOR-HOURS	UNIT	MAT.	LABOR	EQUIP.	TOTAL	TOTAL INCL O&P	
							1999 BARE COSTS					
522	0600	Wood, low acoustical type, 4.5 psf, to 14' high	2 Carp	50	.320	S.F.	25.50	8.75		34.25	42.50	522
	1100	Steel, acoustical, 9 to 12 lb. per S.F., vinyl faced, minimum		60	.267		36.50	7.30		43.80	51.50	
	1200	Maximum		30	.533		44.50	14.55		59.05	72	
	1700	Aluminum framed, acoustical, to 12' high, 5.5 psf, minimum		60	.267		24.50	7.30		31.80	38.50	
	1800	Maximum		30	.533		30	14.55		44.55	55.50	
	2000	6.5 lb. per S.F., minimum		60	.267		26	7.30		33.30	40	
	2100	Maximum		30	.533		32	14.55		46.55	58	
524	0010	**PARTITIONS, OPERABLE** Acoustic air wall, 1-5/8" thick, minimum	2 Carp	375	.043	S.F.	21	1.16		22.16	25	524
	0100	Maximum		365	.044		36	1.20		37.20	41.50	
	0300	2-1/4" thick, minimum		360	.044		23.50	1.21		24.71	28	
	0400	Maximum		330	.048		41	1.32		42.32	47.50	
	0600	For track type, add to above				L.F.	76.50			76.50	84.50	
	0700	Overhead track type, acoustical, 3" thick, 11 psf, minimum	2 Carp	350	.046	S.F.	53	1.25		54.25	60	
	0800	Maximum	"	300	.053	"	63.50	1.46		64.96	72.50	

106 550 | Accordion Partitions

			CREW	DAILY OUTPUT	LABOR-HOURS	UNIT	MAT.	LABOR	EQUIP.	TOTAL	TOTAL INCL O&P	
552	0010	**PARTITIONS, FOLDING ACCORDION**										552
	0100	Vinyl covered, over 150 S.F., frame not included										
	0300	Residential, 1.25 lb. per S.F., 8' maximum height	2 Carp	300	.053	S.F.	11.60	1.46		13.06	15.05	
	0400	Commercial, 1.75 lb. per S.F., 8' maximum height		225	.071		11.60	1.94		13.54	15.80	
	0600	2 lb. per S.F., 17' maximum height		150	.107		14.10	2.91		17.01	20	
	0700	Industrial, 4 lb. per S.F., 20' maximum height		75	.213		23.50	5.80		29.30	35	
	0900	Acoustical, 3 lb. per S.F., 17' maximum height		100	.160		17.50	4.37		21.87	26	
	1200	5 lb. per S.F., 20' maximum height		95	.168		26	4.60		30.60	36	
	1300	5.5 lb. per S.F., 17' maximum height		90	.178		29	4.85		33.85	39.50	
	1400	Fire rated, 4.5 psf, 20' maximum height		160	.100		29	2.73		31.73	36.50	
	1500	Vinyl clad wood or steel, electric operation, 5.0 psf		160	.100		29	2.73		31.73	36.50	
	1900	Wood, non-acoustic, birch or mahogany, to 10' high		300	.053		15.55	1.46		17.01	19.40	

106 750 | Storage & Shelving

			CREW	DAILY OUTPUT	LABOR-HOURS	UNIT	MAT.	LABOR	EQUIP.	TOTAL	TOTAL INCL O&P	
752	0010	**PARTS BINS** Metal, gray baked enamel finish										752
	0100	6'-3" high, 3' wide										
	0300	12 bins, 18" wide x 12" high, 12" deep	2 Clab	10	1.600	Ea.	229	34.50		263.50	305	
	0400	24" deep		10	1.600		272	34.50		306.50	355	
	0600	72 bins, 6" wide x 6" high, 12" deep		8	2		440	43		483	555	
	0700	24" deep		8	2		745	43		788	890	
	1000	7'-3" high, 3' wide										
	1200	14 bins, 18" wide x 12" high, 12" deep	2 Clab	10	1.600	Ea.	238	34.50		272.50	315	
	1300	24" deep		10	1.600		325	34.50		359.50	415	
	1500	84 bins, 6" wide x 6" high, 12" deep		8	2		510	43		553	635	
	1600	24" deep		8	2		865	43		908	1,025	
754	0010	**SHELVING** Metal, industrial, cross-braced, 3' wide, 12" deep	1 Sswk	175	.046	SF Shlf	10.25	1.40		11.65	13.85	754
	0100	24" deep		330	.024		6.55	.74		7.29	8.60	
	0300	4' wide, 12" deep		185	.043		9.80	1.32		11.12	13.25	
	0400	24" deep		380	.021		6.55	.64		7.19	8.40	
	1200	Enclosed sides, cross-braced back, 3' wide, 12" deep		175	.046		12.35	1.40		13.75	16.15	
	1300	24" deep		290	.028		7.85	.84		8.69	10.15	
	1500	Fully enclosed, sides and back, 3' wide, 12" deep		150	.053		14.45	1.63		16.08	18.90	
	1600	24" deep		255	.031		9.10	.96		10.06	11.75	
	1800	4' wide, 12" deep		150	.053		13.70	1.63		15.33	18.10	
	1900	24" deep		290	.028		8.80	.84		9.64	11.25	
	2200	Wide span, 1600 lb. capacity per shelf, 6' wide, 24" deep		380	.021		6.30	.64		6.94	8.10	
	2400	36" deep		440	.018		5.55	.56		6.11	7.10	

106 | Partitions & Storage Shelving

106 750	Storage & Shelving		CREW	DAILY OUTPUT	LABOR-HOURS	UNIT	1999 BARE COSTS				TOTAL INCL O&P	
							MAT.	LABOR	EQUIP.	TOTAL		
754	2600	8' wide, 24" deep	1 Sswk	440	.018	SF Shlf	5.75	.56		6.31	7.30	754
	2800	36" deep	↓	520	.015		4.80	.47		5.27	6.10	
	4000	Pallet racks, steel frame 5,000 lb. capacity, 8' long, 36" deep	2 Sswk	450	.036		4.07	1.09		5.16	6.45	
	4200	42" deep		500	.032		3.57	.98		4.55	5.75	
	4400	48" deep	↓	520	.031	↓	3.17	.94		4.11	5.20	

107 | Telephone Specialties

107 550	Telephone Enclosures		CREW	DAILY OUTPUT	LABOR-HOURS	UNIT	1999 BARE COSTS				TOTAL INCL O&P	
							MAT.	LABOR	EQUIP.	TOTAL		
551	0010	TELEPHONE ENCLOSURE										551
	0300	Shelf type, wall hung, minimum	2 Carp	5	3.200	Ea.	705	87.50		792.50	915	
	0400	Maximum		5	3.200		2,425	87.50		2,512.50	2,800	
	0600	Booth type, painted steel, indoor or outdoor, minimum		1.50	10.667		2,975	291		3,266	3,725	
	0700	Maximum (stainless steel)		1.50	10.667		9,850	291		10,141	11,300	
	1300	Outdoor, acoustical, on post		3	5.333		1,250	146		1,396	1,600	
	1400	Phone carousel, pedestal mounted with dividers		.60	26.667		4,800	730		5,530	6,425	
	1900	Outdoor, drive-up type, wall mounted		4	4		775	109		884	1,025	
	2000	Post mounted, stainless steel posts	↓	3	5.333	↓	1,200	146		1,346	1,550	
	2200	Directory shelf, wall mounted, stainless steel										
	2300	3 binders	2 Carp	8	2	Ea.	910	54.50		964.50	1,075	
	2500	4 binders		7	2.286		1,050	62.50		1,112.50	1,250	
	2600	5 binders		6	2.667		1,275	73		1,348	1,525	
	2800	Table type, stainless steel, 4 binders		8	2		1,125	54.50		1,179.50	1,300	
	2900	7 binders	↓	7	2.286	↓	1,375	62.50		1,437.50	1,600	

108 | Toilet & Bath Accessories & Scales

108 200	Bath Accessories		CREW	DAILY OUTPUT	LABOR-HOURS	UNIT	1999 BARE COSTS				TOTAL INCL O&P	
							MAT.	LABOR	EQUIP.	TOTAL		
204	0010	BATH ACCESSORIES										204
	0200	Curtain rod, stainless steel, 5' long, 1" diameter	1 Carp	13	.615	Ea.	19.60	16.80		36.40	48	
	0300	1-1/4" diameter	"	13	.615	"	22.50	16.80		39.30	51.50	
	0500	Dispenser units, combined soap & towel dispensers,										
	0510	mirror and shelf, flush mounted	1 Carp	10	.800	Ea.	258	22		280	320	
	0600	Towel dispenser and waste receptacle,										
	0610	18 gallon capacity	1 Carp	10	.800	Ea.	271	22		293	335	
	0800	Grab bar, straight, 1-1/4" diameter, stainless steel, 18" long		24	.333		34	9.10		43.10	52	
	0900	24" long		23	.348		37	9.50		46.50	56	
	1000	30" long		22	.364		40	9.95		49.95	59.50	
	1100	36" long		20	.400		42.50	10.90		53.40	64	
	1200	1-1/2" diameter, 24" long		23	.348		39.50	9.50		49	58.50	
	1300	36" long		20	.400		46	10.90		56.90	68	
	1500	Tub bar, 1-1/4" diameter, 24" x 36"	↓	14	.571	↓	84.50	15.60		100.10	118	

Important: See the Reference Section for critical supporting data - Reference Nos., Crews, & City Cost Indexes

108 200 | Bath Accessories

		CREW	DAILY OUTPUT	LABOR-HOURS	UNIT	1999 BARE COSTS				TOTAL INCL O&P	
						MAT.	LABOR	EQUIP.	TOTAL		
204	1600	Plus vertical arm	1 Carp	12	.667	Ea.	72	18.20		90.20	108	**204**
	1900	End tub bar, 1″ diameter, 90° angle, 16″ x 32″		12	.667		100	18.20		118.20	139
	2300	Hand dryer, surface mounted, electric, 115 volt, 20 amp		4	2		485	54.50		539.50	620
	2400	230 volt, 10 amp		4	2		485	54.50		539.50	620
	2600	Hat and coat strip, stainless steel, 4 hook, 36″ long		24	.333		44	9.10		53.10	63
	2700	6 hook, 60″ long		20	.400		69	10.90		79.90	93
	3000	Mirror, with stainless steel 3/4″ square frame, 18″ x 24″		20	.400		59.50	10.90		70.40	82
	3100	36″ x 24″		15	.533		108	14.55		122.55	142
	3200	48″ x 24″		10	.800		144	22		166	193
	3300	72″ x 24″		6	1.333		196	36.50		232.50	274
	3500	With 5″ stainless steel shelf, 18″ x 24″		20	.400		88	10.90		98.90	114
	3600	36″ x 24″		15	.533		132	14.55		146.55	168
	3700	48″ x 24″		10	.800		165	22		187	217
	3800	72″ x 24″		6	1.333		295	36.50		331.50	385
	4100	Mop holder strip, stainless steel, 5 holders, 48″ long		20	.400		60	10.90		70.90	83
	4200	Napkin/tampon dispenser, recessed		15	.533		320	14.55		334.55	375
	4300	Robe hook, single, regular		36	.222		10.85	6.05		16.90	21.50
	4400	Heavy duty, concealed mounting		36	.222		11.40	6.05		17.45	22
	4600	Soap dispenser, chrome, surface mounted, liquid		20	.400		39.50	10.90		50.40	60.50
	4700	Powder		20	.400		65.50	10.90		76.40	89
	5000	Recessed stainless steel, liquid		10	.800		65.50	22		87.50	107
	5100	Powder		10	.800		255	22		277	315
	5300	Soap tank, stainless steel, 1 gallon		10	.800		153	22		175	203
	5400	5 gallon		5	1.600		210	43.50		253.50	300
	5600	Shelf, stainless steel, 5″ wide, 18 ga., 24″ long		24	.333		34	9.10		43.10	51.50
	5700	48″ long		16	.500		60.50	13.65		74.15	88
	5800	8″ wide shelf, 18 ga., 24″ long		22	.364		42	9.95		51.95	61.50
	5900	48″ long		14	.571		84	15.60		99.60	117
	6000	Toilet seat cover dispenser, stainless steel, recessed		20	.400		90	10.90		100.90	116
	6050	Surface mounted		15	.533		25	14.55		39.55	50.50
	6100	Toilet tissue dispenser, surface mounted, SS, single roll		30	.267		10.25	7.30		17.55	23
	6200	Double roll		24	.333		15.20	9.10		24.30	31
	6400	Towel bar, stainless steel, 18″ long		23	.348		29.50	9.50		39	47
	6500	30″ long		21	.381		33.50	10.40		43.90	53.50
	6700	Towel dispenser, stainless steel, surface mounted		16	.500		36	13.65		49.65	61
	6800	Flush mounted, recessed		10	.800		134	22		156	183
	7000	Towel holder, hotel type, 2 guest size		20	.400		11.80	10.90		22.70	30
	7200	Towel shelf, stainless steel, 24″ long, 8″ wide		20	.400		45	10.90		55.90	66.50
	7400	Tumbler holder, tumbler only		30	.267		19.40	7.30		26.70	33
	7500	Soap, tumbler & toothbrush		30	.267		19.40	7.30		26.70	33
	7700	Wall urn ash receiver, surface mount, 11″ long		12	.667		98.50	18.20		116.70	137
	7800	7-1/2″, long		18	.444		68.50	12.15		80.65	94.50
	8000	Waste receptacles, stainless steel, with top, 13 gallon		10	.800		180	22		202	232
	8100	36 gallon	▼	8	1	▼	286	27.50		313.50	360
208	0010	**MEDICINE CABINETS** With mirror, st. st. frame, 16″ x 22″, unlighted	1 Carp	14	.571	Ea.	63.50	15.60		79.10	94	**208**
	0100	Wood frame		14	.571		92.50	15.60		108.10	126
	0300	Sliding mirror doors, 20″ x 16″ x 4-3/4″, unlighted		7	1.143		82.50	31		113.50	140
	0400	24″ x 19″ x 8-1/2″, lighted		5	1.600		139	43.50		182.50	222
	0600	Triple door, 30″ x 32″, unlighted, plywood body		7	1.143		205	31		236	275
	0700	Steel body		7	1.143		270	31		301	345
	0900	Oak door, wood body, beveled mirror, single door		7	1.143		120	31		151	181
	1000	Double door		6	1.333		305	36.50		341.50	395
	1200	Hotel cabinets, stainless, with lower shelf, unlighted		10	.800		165	22		187	217
	1300	Lighted	▼	5	1.600	▼	245	43.50		288.50	340

108 800 | Scales	CREW	DAILY OUTPUT	LABOR-HOURS	UNIT	1999 BARE COSTS				TOTAL INCL O&P	
					MAT.	LABOR	EQUIP.	TOTAL		
801	0010	**SCALES** Built-in floor scale, not incl. foundations									801
	0100	Dial type, 5 ton capacity, 8' x 6' platform	3 Carp	.50	48	Ea.	4,350	1,300		5,650	6,825
	0300	9' x 7' platform		.40	60		7,250	1,650		8,900	10,600
	0400	10 ton capacity, steel platform, 8' x 6' platform		.40	60		7,050	1,650		8,700	10,300
	0600	9' x 7' platform		.35	68.571		8,250	1,875		10,125	12,000
	0700	Truck scales, incl. steel weigh bridge,									
	0800	not including foundation, pits									
	0900	Dial type, mech., 24' x 10' platform, 20 ton cap.	3 Carp	.30	80	Ea.	8,300	2,175		10,475	12,600
	1000	30 ton capacity		.20	120		9,375	3,275		12,650	15,500
	1100	50 ton capacity, 50' x 10' platform		.14	171		14,200	4,675		18,875	23,000
	1200	70' x 10' platform		.12	200		17,700	5,450		23,150	28,000
	1400	60 ton capacity, 60' x 10' platform		.13	184		16,200	5,050		21,250	25,700
	1500	70' x 10' platform		.10	240		18,400	6,550		24,950	30,500
	1550	Digital, electronic, 100 ton capacity, steel deck 12' x 10' platform		.20	120		8,650	3,275		11,925	14,700
	1600	40' x 10' platform		.14	171		13,700	4,675		18,375	22,500
	1640	60' x 10' platform		.13	184		23,200	5,050		28,250	33,400
	1680	70' x 10' platform		.12	200		27,300	5,450		32,750	38,700
	2000	For standard automatic printing device, add					1,900			1,900	2,100
	2100	For remote reading electronic system, add					2,000			2,000	2,200
	2300	Concrete foundation pits for above, 8' x 6', 5 C.Y. required	C-1	.50	64		710	1,650		2,360	3,375
	2400	14' x 6' platform, 10 C.Y. required		.35	91.429		1,000	2,375		3,375	4,825
	2600	50' x 10' platform, 30 C.Y. required		.25	128		1,400	3,300		4,700	6,750
	2700	70' x 10' platform, 40 C.Y. required		.15	213		2,350	5,525		7,875	11,300
	2750	Crane scales, dial, 1 ton capacity					805			805	890
	2780	5 ton capacity					960			960	1,050
	2800	Digital, 1 ton capacity					1,800			1,800	1,975
	2850	10 ton capacity					3,425			3,425	3,775
	2900	Low profile electronic warehouse scale,									
	3000	not incl. printer, 4' x 4' platform, 10,000 lb. capacity	2 Carp	.30	53.333	Ea.	2,225	1,450		3,675	4,750
	3300	5' x 7' platform, 10,000 lb. capacity		.25	64		3,400	1,750		5,150	6,500
	3400	20,000 lb. capacity		.20	80		3,950	2,175		6,125	7,775
	3500	For printers, incl. time, date & numbering, add					905			905	995
	3800	Portable, beam type, capacity 1000#, platform 18" x 24"					490			490	540
	3900	Dial type, capacity 2000#, platform 24" x 24"					850			850	935
	4000	Digital type, capacity 1000#, platform 24" x 30"					1,450			1,450	1,600
	4100	Portable contractor truck scales, 50 ton cap., 40' x 10' platform					25,300			25,300	27,800
	4200	60' x 10' platform					28,300			28,300	31,100

109 | Wardrobe & Closet Specialties

109 010 | Coat Racks/Wardrobes	CREW	DAILY OUTPUT	LABOR-HOURS	UNIT	1999 BARE COSTS				TOTAL INCL O&P	
					MAT.	LABOR	EQUIP.	TOTAL		
011	0010	**COAT RACKS & WARDROBES**									011
	0020	Floor model hat & coat racks, 6 hangers									
	0050	Standing, beech wood, 21" x 21" x 72", chrome				Ea.	98.50			98.50	108
	0100	18 gauge tubular steel, 21" x 21" x 69", wood walnut				"	224			224	246
	0500	16 gauge steel frame, 22 gauge steel shelves									
	0650	Single pedestal, 30" x 18" x 63"				Ea.	181			181	199
	0800	Single face rack, 29" x 18-1/2" x 62"					263			263	289
	0900	51" x 18-1/2" x 70"					224			224	246

10 SPECIALTIES

Important: See the Reference Section for critical supporting data - Reference Nos., Crews, & City Cost Indexes

109 010	Coat Racks/Wardrobes		CREW	DAILY OUTPUT	LABOR-HOURS	UNIT	1999 BARE COSTS				TOTAL INCL O&P	
							MAT.	LABOR	EQUIP.	TOTAL		
011	0910	Double face rack, 39" x 26" x 70"				Ea.	395			395	435	**011**
	0920	63" x 26" x 70"				↓	440			440	480	
	0940	For 2" ball casters, add				Set	47.50			47.50	52.50	
	1400	Utility hook strips, 3/8" x 2-1/2" x 18", 6 hooks	1 Carp	48	.167	Ea.	29.50	4.55		34.05	39.50	
	1500	34" long, 12 hooks	"	48	.167	"	47	4.55		51.55	58.50	
	1650	Wall mounted racks, 16 gauge steel frame, 22 gauge steel shelves										
	1850	12" x 15" x 26", 6 hangers	1 Carp	32	.250	Ea.	113	6.80		119.80	135	
	2000	12" x 15" x 50", 12 hangers	"	32	.250	"	171	6.80		177.80	199	
	2150	Wardrobe cabinet, steel, baked enamel finish										
	2300	36" x 21" x 78", incl. top shelf & hanger rod				Ea.	375			375	410	

For information about Means Estimating Seminars, see yellow pages 11 and 12 in back of book

SPECIALTIES **10**

	CREW	DAILY OUTPUT	LABOR-HOURS	UNIT	1999 BARE COSTS				TOTAL INCL O&P
					MAT.	LABOR	EQUIP.	TOTAL	

Division 11
Equipment

Estimating Tips
General
- The items in this division are usually priced per square foot or each. Many of these items are purchased by the owner for installation by the contractor. Check the specifications for responsibilities, and include time for receiving, storage, installation and mechanical and electrical hook-ups in the appropriate divisions.

- Many items in Division 11 require some type of support system that is not usually furnished with the item. Examples of these systems include blocking for the attachment of casework and support angles for ceiling hung projection screens. The required blocking or supports must be added to the estimate in the appropriate division.
- Some items in Division 11 may require assembly or electrical hook-ups. Verify the amount of assembly required or the need for a hard electrical connection and add the appropriate costs.

Reference Numbers
Reference numbers are shown in bold squares at the beginning of some major classifications. These numbers refer to related items in the Reference Section. The reference information may be an estimating procedure, an alternate pricing method or technical information.

Note: Not all subdivisions listed here necessarily appear in this publication.

110 050 | Contractor Equipment

		CREW	DAILY OUTPUT	LABOR-HOURS	UNIT	1999 BARE COSTS				TOTAL INCL O&P	
						MAT.	LABOR	EQUIP.	TOTAL		
051	0010	CONTRACTOR EQUIPMENT See division 016 R016-410									051

110 100 | Maintenance Equipment

		CREW	DAILY OUTPUT	LABOR-HOURS	UNIT	MAT.	LABOR	EQUIP.	TOTAL	TOTAL INCL O&P	
121	0010	**VACUUM CLEANING**									121
	0020	Central, 3 inlet, residential	1 Skwk	.90	8.889	Total	955	249		1,204	1,450
	0200	Commercial		.70	11.429		1,400	320		1,720	2,050
	0400	5 inlet system, residential		.50	16		1,025	450		1,475	1,825
	0600	7 inlet system, commercial		.40	20		1,100	560		1,660	2,075
	0800	9 inlet system		.30	26.667		1,200	750		1,950	2,500
	4010	Rule of thumb: First 1200 S.F., installed									1,044
	4020	For each additional S.F., add				S.F.					.17

110 200 | Security/Vault Equip

		CREW	DAILY OUTPUT	LABOR-HOURS	UNIT	MAT.	LABOR	EQUIP.	TOTAL	TOTAL INCL O&P	
261	0010	**SAFE**									261
	0015	Office, 4 hr. rating, 30" x 18" x 18" inside				Ea.	3,250			3,250	3,575
	0100	60" x 36" x 18"					7,050			7,050	7,750
	0200	1 hr. rating, 30" x 18" x 18"					1,775			1,775	1,975
	0250	40" x 18" x 18"					2,025			2,025	2,225
	0300	60" x 36" x 18", double door					5,000			5,000	5,500
	0400	Data, 4 hr. rating, 23-1/2" x 19-1/2" x 17" inside					3,025			3,025	3,325
	0450	52" x 19" x 17"					7,375			7,375	8,100
	0500	63" x 34" x 16", double door					11,500			11,500	12,600
	0550	52" x 34" x 16", inside					10,800			10,800	11,900
	0600	1 hr. rating, 27" x 19" x 16"					4,975			4,975	5,475
	0700	63" x 34" x 16"					8,875			8,875	9,775
	0750	Diskette, 1 hr., 14" x 12" x 11", inside					3,100			3,100	3,400
	0800	Money, "B" label, 9" x 14" x 14"					445			445	490
	0900	Tool resistive, 24" x 24" x 20"					3,025			3,025	3,325
	1050	Tool and torch resistive, 24" x 24" x 20"					7,675			7,675	8,450
	1150	Jewelers, 23" x 20" x 18"					9,825			9,825	10,800
	1200	63" x 25" x 18"					19,700			19,700	21,600
	1300	For handling into building, add, minimum	A-2	8.50	2.824			61	20	81	118
	1400	Maximum	"	.78	30.769			665	220	885	1,275

110 300 | Teller & Service Equip

		CREW	DAILY OUTPUT	LABOR-HOURS	UNIT	MAT.	LABOR	EQUIP.	TOTAL	TOTAL INCL O&P	
301	0010	**BANK EQUIPMENT**									301
	0020	Alarm system, police	2 Elec	1.60	10	Ea.	3,600	320		3,920	4,425
	0100	With vault alarm	"	.40	40		14,300	1,275		15,575	17,600
	0400	Bullet resistant teller window, 44" x 60"	1 Glaz	.60	13.333		2,725	355		3,080	3,550
	0500	48" x 60"	"	.60	13.333		3,975	355		4,330	4,925
	3000	Counters for banks, frontal only	2 Carp	1	16	Station	1,350	435		1,785	2,150
	3100	Complete with steel undercounter	"	.50	32	"	2,625	875		3,500	4,250
	4600	Door and frame, bullet-resistant, with vision panel, minimum	2 Sswk	1.10	14.545	Ea.	2,200	445		2,645	3,250
	4700	Maximum		1.10	14.545		4,200	445		4,645	5,450
	4800	Drive-up window, drawer & mike, not incl. glass, minimum		1	16		4,200	490		4,690	5,525
	4900	Maximum		.50	32		6,925	980		7,905	9,425
	5000	Night depository, with chest, minimum		1	16		6,175	490		6,665	7,700
	5100	Maximum		.50	32		9,775	980		10,755	12,600
	5200	Package receiver, painted		3.20	5		1,000	153		1,153	1,375
	5300	Stainless steel		3.20	5		1,700	153		1,853	2,150
	5400	Partitions, bullet-resistant, 1-3/16" glass, 8' high	2 Carp	10	1.600	L.F.	145	43.50		188.50	229
	5450	Acrylic	"	10	1.600	"	275	43.50		318.50	375
	5500	Pneumatic tube systems, 2 lane drive-up, complete	L-3	.25	64	Total	18,900	1,900		20,800	23,700
	5550	With T.V. viewer	"	.20	80	"	36,600	2,375		38,975	44,000
	5570	Safety deposit boxes, minimum	1 Sswk	44	.182	Opng.	42	5.55		47.55	56

Important: See the Reference Section for critical supporting data - Reference Nos., Crews, & City Cost Indexes

110 300 | Teller & Service Equip

		DAILY OUTPUT	LABOR-HOURS	UNIT	1999 BARE COSTS				TOTAL INCL O&P	
					MAT.	LABOR	EQUIP.	TOTAL		
301	5580	Maximum, 10" x 15" opening	1 Sswk	19	.421	Opng.	89	12.90		101.90	122	**301**
	5590	Teller locker, average	↓	15	.533		1,150	16.30		1,166.30	1,300	
	5600	Pass thru, bullet-resist. window, painted steel, 24" x 36"	2 Sswk	1.60	10	Ea.	1,400	305		1,705	2,100	
	5700	48" x 48"	↓	1.20	13.333		1,950	410		2,360	2,900	
	5800	72" x 40"	▼	.80	20	▼	2,350	610		2,960	3,700	
	5900	For stainless steel frames, add					20%					
	6000	Surveillance system, 16 mm film camera, complete	2 Elec	1	16	Ea.	3,750	510		4,260	4,900	
	6010	For each additional camera, add					1,575			1,575	1,725	
	6100	Surveillance system, video camera, complete	2 Elec	1	16	▼	11,800	510		12,310	13,800	
	6110	For each additional camera, add					725			725	800	
	6200	Twenty-four hour teller, single unit,										
	6300	automated deposit, cash and memo	L-3	.25	64	Ea.	35,600	1,900		37,500	42,100	
	7000	Vault front, see division 083-184										

110 400 | Ecclesiastical Equip

			CREW	DAILY OUTPUT	LABOR-HOURS	UNIT	MAT.	LABOR	EQUIP.	TOTAL	TOTAL INCL O&P	
401	0010	**CHURCH EQUIPMENT**										**401**
	0020	Altar, wood, custom design, plain	1 Carp	1.40	5.714	Ea.	1,400	156		1,556	1,800	
	0050	Deluxe	"	.20	40		6,850	1,100		7,950	9,250	
	0070	Granite or marble, average	2 Marb	.50	32		5,350	880		6,230	7,250	
	0090	Deluxe	"	.20	80		15,000	2,200		17,200	19,900	
	0100	Arks, prefabricated, plain	2 Carp	.80	20		5,000	545		5,545	6,350	
	0130	Deluxe, maximum	"	.20	80	▼	50,000	2,175		52,175	58,500	
	0150	Baptistry, fiberglass, 3'-6" deep, x 13'-7" long,										
	0160	steps at both ends, incl. plumbing, minimum	L-8	1	20	Ea.	2,000	565		2,565	3,075	
	0200	Maximum	"	.70	28.571		4,000	810		4,810	5,675	
	0250	Add for filter, heater and lights				▼	985			985	1,075	
	0300	Carillon, 4 octave (48 bells), with keyboard				System	500,000			500,000	550,000	
	0320	2 octave (24 bells)					175,000			175,000	192,500	
	0340	3 to 4 bell peal, minimum					45,000			45,000	49,500	
	0360	Maximum				▼	250,000			250,000	275,000	
	0380	Cast bronze bell, average				Ea.	35,000			35,000	38,500	
	0400	Electronic, tape, minimum					6,600			6,600	7,250	
	0410	Solid state with keyboard, maximum					32,000			32,000	35,200	
	0500	Reconciliation room, wood, prefabricated, single, plain	1 Carp	.60	13.333		1,600	365		1,965	2,325	
	0550	Deluxe		.40	20		5,000	545		5,545	6,350	
	0650	Double, plain		.40	20		4,000	545		4,545	5,250	
	0700	Deluxe		.20	40		12,500	1,100		13,600	15,500	
	1000	Lecterns, wood, plain		5	1.600		450	43.50		493.50	565	
	1100	Deluxe		2	4	▼	2,750	109		2,859	3,200	
	1500	Pews, bench type, hardwood, minimum		20	.400	L.F.	50.50	10.90		61.40	72.50	
	1550	Maximum	▼	15	.533		100	14.55		114.55	133	
	1570	For kneeler, add				▼	12.50			12.50	13.75	
	2000	Pulpits, hardwood, prefabricated, plain	1 Carp	2	4	Ea.	1,000	109		1,109	1,275	
	2100	Deluxe		1.60	5	"	6,600	137		6,737	7,475	
	2500	Railing, hardwood, average	▼	25	.320	L.F.	125	8.75		133.75	152	
	2700	Safes, see division 110-261										
	3000	Seating, individual, oak, contour, laminated	1 Carp	21	.381	Person	110	10.40		120.40	137	
	3100	Cushion seat		21	.381		100	10.40		110.40	126	
	3200	Fully upholstered		21	.381		90	10.40		100.40	115	
	3300	Combination, self-rising	▼	21	.381		250	10.40		260.40	291	
	3500	For cherry, add				▼	30%					
	4000	Steeples, translucent fiberglass, 30" square, 15' high	F-3	2	20	Ea.	2,000	555	222	2,777	3,300	
	4150	25' high		1.80	22.222		3,000	615	246	3,861	4,525	
	4350	Opaque fiberglass, 24" square, 14' high		2	20		2,000	555	222	2,777	3,300	
	4500	28' high	▼	1.80	22.222	▼	3,125	615	246	3,986	4,675	

EQUIPMENT 11

110 400 | Ecclesiastical Equip

		CREW	DAILY OUTPUT	LABOR-HOURS	UNIT	MAT.	LABOR	EQUIP.	TOTAL	TOTAL INCL O&P	
401	4600	Aluminum, baked finish, 14' high, 16" square				Ea.	1,550			1,550	1,700	401
	4620	20' high, 3'-6" base					4,125			4,125	4,525	
	4640	35' high, 8' base					15,600			15,600	17,100	
	4660	60' high, 14' base					35,100			35,100	38,600	
	4680	152' high, custom					345,500			345,500	380,000	
	4700	Porcelain enamel steeples, custom, 40' high	F-3	.50	80		10,400	2,225	885	13,510	15,900	
	4800	60' high	"	.30	133		18,000	3,700	1,475	23,175	27,200	
	5000	Wall cross, aluminum, extruded, 2" x 2" section	1 Carp	34	.235	L.F.	28	6.40		34.40	41	
	5150	4" x 4" section		29	.276		40	7.55		47.55	56	
	5300	Bronze, extruded, 1" x 2" section		31	.258		56	7.05		63.05	72.50	
	5350	2-1/2" x 2-1/2" section		34	.235		83	6.40		89.40	102	
	5450	Solid bar stock, 1/2" x 3" section		29	.276		109	7.55		116.55	132	
	5600	Fiberglass, stock		34	.235		22.50	6.40		28.90	35	
	5700	Stainless steel, 4" deep, channel section		29	.276		88	7.55		95.55	109	
	5800	4" deep box section		29	.276		109	7.55		116.55	132	

110 500 | Library Equipment

		CREW	DAILY OUTPUT	LABOR-HOURS	UNIT	MAT.	LABOR	EQUIP.	TOTAL	TOTAL INCL O&P	
501	0010	**LIBRARY EQUIPMENT**										501
	0020	Bookshelf, mtl, 90" high, 10" shelf, dbl face	1 Carp	11.50	.696	L.F.	89.50	19		108.50	128	
	0300	Single face	"	12	.667	"	65	18.20		83.20	100	
	0600	For 8" shelving, subtract from above					10%					
	0700	For 12" shelving, add to above					10%					
	0800	For 42" high with countertop, subtract from above					20%					
	1100	Magazine shelving, 90" high, 12" deep, single face	1 Carp	11.50	.696	L.F.	83	19		102	122	
	1400	Double face		11.50	.696	"	149	19		168	194	
	2500	Carrels, hardwood, 36" x 24", minimum		5	1.600	Ea.	560	43.50		603.50	690	
	2650	Maximum		4	2		735	54.50		789.50	895	
	2700	Card catalog file, 60 trays, complete					3,775			3,775	4,150	
	2720	Alternate method: each tray					63			63	69.50	
	3100	Chairs, wood					105			105	116	
	3500	Charging desk, built-in, with counter, plastic laminated top	1 Carp	7	1.143	L.F.	395	31		426	485	
	3700	Reading table, laminated top, 60" x 36"				Ea.	600			600	660	
	3800	Mobile compacted shelving, hand crank, 9'-0" high										
	3820	Double face, including track, 3' section				Ea.	820			820	900	
	3840	For electrical operation, add					25%					

110 600 | Theater/Stage Equip

		CREW	DAILY OUTPUT	LABOR-HOURS	UNIT	MAT.	LABOR	EQUIP.	TOTAL	TOTAL INCL O&P	
601	0010	**STAGE EQUIPMENT**										601
	0050	Control boards with dimmers and breakers, minimum	1 Elec	1	8	Ea.	2,300	255		2,555	2,900	
	0100	Average		.50	16		8,050	510		8,560	9,625	
	0150	Maximum		.20	40		34,500	1,275		35,775	39,900	
	0500	Curtain track, straight, light duty	2 Carp	20	.800	L.F.	18.25	22		40.25	54.50	
	0600	Heavy duty		18	.889		53.50	24.50		78	97	
	0700	Curved sections		12	1.333		110	36.50		146.50	179	
	1000	Curtains, velour, medium weight		600	.027	S.F.	12.65	.73		13.38	15.05	
	1150	Silica based yarn, fireproof		50	.320		29	8.75		37.75	45.50	
	1500	Flooring, portable oak parquet, 3' x 3' sections					10.05			10.05	11.05	
	1600	Cart to carry 225 S.F. of flooring				Ea.	305			305	335	
	2000	Lights, border, quartz, reflector, vented,										
	2100	colored or white	1 Elec	20	.400	L.F.	138	12.75		150.75	171	
	2500	Spotlight, follow spot, with transformer, 2,100 watt	"	4	2	Ea.	1,325	64		1,389	1,550	
	2600	For no transformer, deduct					475			475	525	
	3000	Stationary spot, fresnel quartz, 6" lens	1 Elec	4	2		115	64		179	223	
	3100	8" lens		4	2		181	64		245	295	
	3500	Ellipsoidal quartz, 1,000W, 6" lens		4	2		239	64		303	360	

11 EQUIPMENT

Important: See the Reference Section for critical supporting data - Reference Nos., Crews, & City Cost Indexes

110 600 | Theater/Stage Equip

		CREW	DAILY OUTPUT	LABOR-HOURS	UNIT	MAT.	LABOR	EQUIP.	TOTAL	TOTAL INCL O&P
601						1999 BARE COSTS				
3600	12" lens	1 Elec	4	2	Ea.	420	64		484	555
4000	Strobe light, 1 to 15 flashes per second, quartz		3	2.667		540	85		625	720
4500	Color wheel, portable, five hole, motorized	↓	4	2		101	64		165	207
5000	Stages, portable with steps, folding legs, stock, 8" high				SF Stg.	7.50			7.50	8.20
5100	16" high					15.50			15.50	17.05
5200	32" high					21			21	23.50
5300	40" high					16.25			16.25	17.90
6000	Telescoping platforms, extruded alum., straight, minimum	4 Carp	157	.204		28	5.55		33.55	39.50
6100	Maximum		77	.416		43	11.35		54.35	65.50
6500	Pie-shaped, minimum		150	.213		31	5.80		36.80	43
6600	Maximum	↓	70	.457		68.50	12.50		81	95
6800	For 3/4" plywood covered deck, deduct					2.73			2.73	3
7000	Band risers, steel frame, plywood deck, minimum	4 Carp	275	.116		22	3.18		25.18	29
7100	Maximum	"	138	.232	↓	41.50	6.35		47.85	55.50
7500	Chairs for above, self-storing, minimum	2 Carp	43	.372	Ea.	80	10.15		90.15	104
7600	Maximum	"	40	.400	"	113	10.90		123.90	141
8000	Rule of thumb: total stage equipment, minimum	4 Carp	100	.320	SF Stg.	72.50	8.75		81.25	93.50
8100	Maximum	"	25	1.280	"	400	35		435	495
604	**MOVIE EQUIPMENT**									
0020	Changeover, minimum				Ea.	350			350	385
0100	Maximum					685			685	750
0400	Film transport, incl. platters and autowind, minimum					5,000			5,000	5,500
0500	Maximum					10,800			10,800	11,900
0800	Lamphouses, incl. rectifiers, xenon, 1,000 watt	1 Elec	2	4		5,875	128		6,003	6,675
0900	1,600 watt		2	4		6,625	128		6,753	7,475
1000	2,000 watt		1.50	5.333		6,825	170		6,995	7,775
1100	4,000 watt	↓	1.50	5.333		9,675	170		9,845	10,900
1400	Lenses, anamorphic, minimum					895			895	980
1500	Maximum					2,100			2,100	2,300
1800	Flat 35 mm, minimum					585			585	645
1900	Maximum					1,125			1,125	1,250
2200	Pedestals, for projectors					560			560	615
2300	Console type					10,000			10,000	11,000
2600	Projector mechanisms, incl. soundhead, 35 mm, minimum					9,925			9,925	10,900
2700	Maximum				↓	13,200			13,200	14,500
3000	Projection screens, rigid, in wall, acrylic, 1/4" thick	2 Glaz	195	.082	S.F.	31.50	2.18		33.68	38.50
3100	1/2" thick	"	130	.123	"	36.50	3.27		39.77	45.50
3300	Electric operated, heavy duty, 400 S.F.	2 Carp	1	16	Ea.	2,225	435		2,660	3,125
3320	Theatre projection screens, matte white, including frames	"	200	.080	S.F.	3.26	2.18		5.44	7.05
3400	Also see division 111-321									
3700	Sound systems, incl. amplifier, mono, minimum	1 Elec	.90	8.889	Ea.	1,825	284		2,109	2,450
3800	Dolby/Super Sound, maximum		.40	20		13,700	640		14,340	16,000
4100	Dual system, 2 channel, front surround, minimum		.70	11.429		3,050	365		3,415	3,925
4200	Dolby/Super Sound, 4 channel, maximum	↓	.40	20		12,600	640		13,240	14,900
4500	Sound heads, 35 mm					3,150			3,150	3,475
4900	Splicer, wet type, minimum					560			560	615
5000	Tape type, maximum					1,000			1,000	1,100
5300	Speakers, recessed behind screen, minimum	1 Elec	2	4		1,325	128		1,453	1,650
5400	Maximum	"	1	8		2,325	255		2,580	2,950
5700	Seating, painted steel, upholstered, minimum	2 Carp	35	.457		92	12.50		104.50	121
5800	Maximum	"	28	.571		237	15.60		252.60	285
6100	Rewind tables, minimum					1,125			1,125	1,225
6200	Maximum				↓	3,550			3,550	3,900
7000	For automation, varying sophistication, minimum	1 Elec	1	8	System	1,300	255		1,555	1,800
7100	Maximum	2 Elec	.30	53.333	"	4,200	1,700		5,900	7,175

EQUIPMENT 11

		111 020 | **Barber Shop Equipment**	CREW	DAILY OUTPUT	LABOR-HOURS	UNIT	|1999 BARE COSTS| MAT.	LABOR	EQUIP.	TOTAL	TOTAL INCL O&P	
021	0010	**BARBER EQUIPMENT**										021
	0020	Chair, hydraulic, movable, minimum	1 Carp	24	.333	Ea.	375	9.10		384.10	425	
	0050	Maximum	"	16	.500		2,700	13.65		2,713.65	3,000	
	0200	Wall hung styling station with mirrors, minimum	L-2	8	2		238	48		286	335	
	0300	Maximum	"	4	4		1,500	96.50		1,596.50	1,800	
	0500	Sink, hair washing basin, rough plumbing not incl.	1 Plum	8	1		365	32.50		397.50	450	
	1000	Sterilizer, liquid solution for tools					125			125	138	
	1100	Total equipment, rule of thumb, per chair, minimum	L-8	1	20		1,475	565		2,040	2,500	
	1150	Maximum	"	1	20		4,000	565		4,565	5,275	

		111 040 | **Cash Register/Checking**										
041	0010	**CHECKOUT COUNTER**										041
	0020	Supermarket conveyor, single belt	2 Clab	10	1.600	Ea.	1,925	34.50		1,959.50	2,175	
	0100	Double belt, power take-away		9	1.778		3,300	38		3,338	3,675	
	0400	Double belt, power take-away, incl. side scanning		7	2.286		4,100	49		4,149	4,575	
	0800	Warehouse or bulk type		6	2.667		4,800	57		4,857	5,375	
	1000	Scanning system, 2 lanes, w/registers, scan gun & memory				System	12,000			12,000	13,200	
	1100	10 lanes, single processor, full scan, with scales				"	114,000			114,000	125,500	
	2000	Register, restaurant, minimum				Ea.	500			500	550	
	2100	Maximum					2,200			2,200	2,425	
	2150	Store, minimum					500			500	550	
	2200	Maximum					2,200			2,200	2,425	

		111 060 | **Display Cases**										
061	0010	**REFRIGERATED FOOD CASES**										061
	0030	Dairy, multi-deck, 12' long	Q-5	3	5.333	Ea.	6,425	157		6,582	7,325	
	0100	For rear sliding doors, add					905			905	1,000	
	0200	Delicatessen case, service deli, 12' long, single deck	Q-5	3.90	4.103		4,350	121		4,471	4,975	
	0300	Multi-deck, 18 S.F. shelf display		3	5.333		6,875	157		7,032	7,825	
	0400	Freezer, self-contained, chest-type, 30 C.F.		3.90	4.103		3,175	121		3,296	3,675	
	0500	Glass door, upright, 78 C.F.		3.30	4.848		6,050	143		6,193	6,875	
	0600	Frozen food, chest type, 12' long		3.30	4.848		4,375	143		4,518	5,025	
	0700	Glass door, reach-in, 5 door		3	5.333		8,400	157		8,557	9,500	
	0800	Island case, 12' long, single deck		3.30	4.848		4,950	143		5,093	5,675	
	0900	Multi-deck		3	5.333		10,400	157		10,557	11,700	
	1000	Meat case, 12' long, single deck		3.30	4.848		3,550	143		3,693	4,125	
	1050	Multi-deck		3.10	5.161		6,175	152		6,327	7,025	
	1100	Produce, 12' long, single deck		3.30	4.848		4,750	143		4,893	5,450	
	1200	Multi-deck		3.10	5.161		5,300	152		5,452	6,050	

		111 100 | **Laundry/Dry Cleaning**										
101	0010	**LAUNDRY EQUIPMENT** Not incl. rough-in										101
	0500	Dryers, gas fired residential, 16 lb. capacity, average	1 Plum	3	2.667	Ea.	505	87		592	690	
	1000	Commercial, 30 lb. capacity, coin operated, single		3	2.667		2,325	87		2,412	2,675	
	1100	Double stacked		2	4		4,975	130		5,105	5,675	
	1500	Industrial, 30 lb. capacity		2	4		1,975	130		2,105	2,375	
	1600	50 lb. capacity		1.70	4.706		2,450	153		2,603	2,925	
	2000	Dry cleaners, electric, 20 lb. capacity	L-1	.20	80		27,900	2,575		30,475	34,600	
	2050	25 lb. capacity		.17	94.118		36,300	3,025		39,325	44,600	
	2100	30 lb. capacity		.15	106		38,300	3,450		41,750	47,300	
	2150	60 lb. capacity		.09	177		60,000	5,725		65,725	74,500	
	3500	Folders, blankets & sheets, minimum	1 Elec	.17	47.059		22,800	1,500		24,300	27,300	
	3700	King size with automatic stacker		.10	80		43,500	2,550		46,050	51,500	

111 100 | Laundry/Dry Cleaning

			CREW	DAILY OUTPUT	LABOR-HOURS	UNIT	1999 BARE COSTS				TOTAL INCL O&P	
							MAT.	LABOR	EQUIP.	TOTAL		
101	3800	For conveyor delivery, add	1 Elec	.45	17.778	Ea.	5,175	565		5,740	6,550	101
	4500	Ironers, institutional, 110", single roll		.20	40		23,800	1,275		25,075	28,100	
	4700	Lint collector, ductwork not included, 8,000 to 10,000 C.F.M.	Q-10	.30	80		6,825	2,375		9,200	11,200	
	4800	Pressers, low capacity air operated	L-6	1.75	6.857		7,300	222		7,522	8,350	
	4820	Hand operated		1.75	6.857		6,300	222		6,522	7,250	
	4830	Extractor		1.75	6.857		5,375	222		5,597	6,225	
	4840	Ironer 48", 240V		3.50	3.429		84,000	111		84,111	92,500	
	4860	Coin dry cleaner 20 lb.		1.75	6.857		21,000	222		21,222	23,400	
	4900	Spreader feeders, 240V, 2 station		.70	17.143		41,000	555		41,555	45,900	
	4920	4 station	↓	.35	34.286		50,500	1,100		51,600	57,000	
	5000	Washers, residential, 4 cycle, average	1 Plum	3	2.667		590	87		677	775	
	5300	Commercial, coin operated, average	"	3	2.667		945	87		1,032	1,175	
	6000	Combination washer/extractor, 20 lb. capacity	L-6	1.50	8		3,375	259		3,634	4,100	
	6100	30 lb. capacity		.80	15		7,000	485		7,485	8,425	
	6200	50 lb. capacity		.68	17.647		7,975	570		8,545	9,625	
	6300	75 lb. capacity		.30	40		15,700	1,300		17,000	19,200	
	6350	125 lb. capacity		.16	75		20,000	2,425		22,425	25,700	
	6380	Washer extractor/dryer, 110 lb., 240V		1	12		6,525	390		6,915	7,750	
	6400	Washer extractor, 135 lb, 240V		1	12		17,000	390		17,390	19,300	
	6450	Pass through		1	12		51,000	390		51,390	56,500	
	6500	200 lb.		1	12		50,000	390		50,390	56,000	
	6550	Pass through		1	12		54,500	390		54,890	60,500	
	6600	Hand operated presser		.70	17.143		4,725	555		5,280	6,000	
	6620	Mushroom press 115V	↓	.70	17.143	↓	5,575	555		6,130	6,975	

111 320 | Projection Screens

			CREW	DAILY OUTPUT	LABOR-HOURS	UNIT	MAT.	LABOR	EQUIP.	TOTAL	TOTAL INCL O&P	
321	0010	**PROJECTION SCREENS** Wall or ceiling hung, matte white										321
	0100	Manually operated, economy	2 Carp	500	.032	S.F.	4.20	.87		5.07	6	
	0300	Intermediate		450	.036	↓	4.90	.97		5.87	6.95	
	0400	Deluxe		400	.040		6.80	1.09		7.89	9.20	
	0600	Electric operated, matte white, 25 S.F., economy		5	3.200	Ea.	615	87.50		702.50	810	
	0700	Deluxe		4	4		1,375	109		1,484	1,675	
	0900	50 S.F., economy		3	5.333		320	146		466	580	
	1000	Deluxe		2	8		1,425	218		1,643	1,925	
	1200	Heavy duty, electric operated, 200 S.F.		1.50	10.667		2,800	291		3,091	3,550	
	1300	400 S.F.	↓	1	16	↓	3,475	435		3,910	4,475	
	1500	Rigid acrylic in wall, for rear projection, 1/4" thick	2 Glaz	30	.533	S.F.	34	14.20		48.20	59	
	1600	1/2" thick (maximum size 10' x 20')	"	25	.640	"	60	17		77	92	

111 400 | Service Station Equip

			CREW	DAILY OUTPUT	LABOR-HOURS	UNIT	MAT.	LABOR	EQUIP.	TOTAL	TOTAL INCL O&P	
401	0010	**AUTOMOTIVE**										401
	0030	Compressors, electric, 1-1/2 H.P., standard controls	L-4	1.50	16	Ea.	320	410		730	995	
	0550	Dual controls		1.50	16		420	410		830	1,100	
	0600	5 H.P., 115/230 volt, standard controls		1	24		1,525	615		2,140	2,675	
	0650	Dual controls	↓	1	24	↓	1,625	615		2,240	2,775	
	1100	Product dispenser with vapor recovery for 6 nozzles, installed, not										
	1110	including piping to storage tanks				Ea.	15,000			15,000	16,500	
	2200	Hoists, single post, 8,000# capacity, swivel arms	L-4	.40	60		3,500	1,550		5,050	6,275	
	2400	Two posts, adjustable frames, 11,000# capacity		.25	96		4,500	2,450		6,950	8,825	
	2500	24,000# capacity		.15	160		6,000	4,100		10,100	13,100	
	2700	7,500# capacity, frame supports		.50	48		5,000	1,225		6,225	7,450	
	2800	Four post, roll on ramp		.50	48		4,500	1,225		5,725	6,900	
	2810	Hydraulic lifts, above ground, 2 post, clear floor, 6000 lb cap		2.67	8.989		4,700	231		4,931	5,550	
	2815	9000 lb capacity		2.29	10.480		11,200	269		11,469	12,700	
	2820	15,000 lb capacity		2	12		25,900	310		26,210	29,000	
	2825	30,000 lb capacity	↓	1.60	15	↓	27,900	385		28,285	31,200	

EQUIPMENT 11

111 400 | Service Station Equip

		CREW	DAILY OUTPUT	LABOR-HOURS	UNIT	1999 BARE COSTS MAT.	LABOR	EQUIP.	TOTAL	TOTAL INCL O&P	
401	2830	4 post, ramp style, 25,000 lb capacity	L-4	2	12	Ea.	10,900	310		11,210	12,500	**401**
	2835	35,000 lb capacity		1	24		51,000	615		51,615	57,000	
	2840	50,000 lb capacity		1	24		57,000	615		57,615	63,500	
	2845	75,000 lb capacity	↓	1	24		66,000	615		66,615	73,500	
	2850	For drive thru tracks, add, minimum					700			700	770	
	2855	Maximum					1,200			1,200	1,325	
	2860	Ramp extensions, 3'(set of 2)					575			575	635	
	2865	Rolling jack platform					2,000			2,000	2,200	
	2870	Elec/hyd jacking beam					5,350			5,350	5,875	
	2880	Scissor lift, portable, 6000 lb capacity				↓	5,250			5,250	5,775	
	3000	Lube equipment, 3 reel type, with pumps, not including piping	L-4	.50	48	Set	6,000	1,225		7,225	8,550	
	4000	Spray painting booth, 26' long, complete	"	.40	60	Ea.	12,000	1,550		13,550	15,600	

111 500 | Parking Control Equip

		CREW	DAILY OUTPUT	LABOR-HOURS	UNIT	1999 BARE COSTS MAT.	LABOR	EQUIP.	TOTAL	TOTAL INCL O&P	
501	0010	**PARKING EQUIPMENT**										**501**
	5000	Barrier gate with programmable controller	2 Elec	3	5.333	Ea.	2,500	170		2,670	3,000	
	5020	Industrial	"	3	5.333		3,500	170		3,670	4,100	
	5100	Card reader	1 Elec	2	4		2,000	128		2,128	2,400	
	5120	Proximity with customer display	2 Elec	1	16		5,000	510		5,510	6,275	
	5200	Cashier booth, average	B-22	1	30		10,000	760	204	10,964	12,400	
	5300	Collector station, pay on foot	2 Elec	.20	80		110,000	2,550		112,550	125,000	
	5320	Credit card only		.50	32		45,000	1,025		46,025	51,000	
	5500	Exit verifier	↓	1	16		22,500	510		23,010	25,600	
	5600	Fee computer	1 Elec	1.50	5.333		13,300	170		13,470	15,000	
	5700	Full sign, 4" letters	"	2	4		900	128		1,028	1,175	
	5800	Inductive loop	2 Elec	4	4		500	128		628	740	
	5900	Ticket spitter with time/date stamp, standard		2	8		6,000	255		6,255	6,975	
	5920	Mag stripe encoding	↓	2	8		16,000	255		16,255	18,000	
	5950	Vehicle detector, microprocessor based	1 Elec	3	2.667		400	85		485	565	
	6000	Parking control software, minimum		.50	16		18,000	510		18,510	20,600	
	6020	Maximum	↓	.20	40	↓	75,000	1,275		76,275	84,500	

111 600 | Loading Dock Equipment

		CREW	DAILY OUTPUT	LABOR-HOURS	UNIT	1999 BARE COSTS MAT.	LABOR	EQUIP.	TOTAL	TOTAL INCL O&P	
601	0010	**LOADING DOCK**										**601**
	0020	Bumpers, rubber blocks 4-1/2" thk, 10" H, 14" long	1 Carp	26	.308	Ea.	37	8.40		45.40	54	
	0200	24" long		22	.364		58	9.95		67.95	79	
	0300	36" long		17	.471		68	12.85		80.85	95	
	0500	12" high, 14" long		25	.320		61	8.75		69.75	81	
	0550	24" long		20	.400		67.50	10.90		78.40	91	
	0600	36" long		15	.533		76	14.55		90.55	107	
	0800	Rubber blocks 6" thick, 10" high, 14" long		22	.364		57.50	9.95		67.45	78.50	
	0850	24" long		18	.444		74	12.15		86.15	101	
	0900	36" long		13	.615		93	16.80		109.80	130	
	0910	20" high, 11" long		13	.615		94.50	16.80		111.30	131	
	0920	Extruded rubber bumpers, T section, 22" x 22" x 3" thick		41	.195		41.50	5.35		46.85	54	
	0940	Molded rubber bumpers, 24" x 12" x 3" thick	↓	20	.400		39	10.90		49.90	60	
	1000	Welded installation of above bumpers	E-14	8	1		2.14	32.50	10.30	44.94	73.50	
	1100	For drilled anchors, add per anchor	1 Carp	36	.222	↓	3.86	6.05		9.91	13.80	
	1300	Steel bumpers, see division 102-604, 608 & 087-508										
	1350	Wood bumpers, see division 111-603										
	2200	Dock boards, heavy duty, 60" x 60", aluminum, 5,000 lb. cap.				Ea.	980			980	1,075	
	2700	9,000 lb. capacity					1,175			1,175	1,275	
	3200	15,000 lb. capacity				↓	1,225			1,225	1,350	
	3600	Door seal for door perimeter, 12" x 12", vinyl covered	1 Carp	26	.308	L.F.	16.30	8.40		24.70	31	
	3900	Folding gates, see division 106-101										

			CREW	DAILY OUTPUT	LABOR-HOURS	UNIT	1999 BARE COSTS				TOTAL INCL O&P	
		**111 600	Loading Dock Equipment**					MAT.	LABOR	EQUIP.	TOTAL		
601	4200	Platform lifter, 6' x 6', portable, 3,000 lb. capacity				Ea.	5,575			5,575	6,125	601
	4250	4,000 lb. capacity					7,825			7,825	8,600	
	4400	Fixed, 6' x 8', 5,000 lb. capacity	E-16	.70	22.857		6,750	720	118	7,588	8,875	
	4500	Levelers, hinged for trucks, 10 ton capacity, 6' x 8'		1.08	14.815		2,900	470	76.50	3,446.50	4,125	
	4650	7' x 8'		1.08	14.815		3,050	470	76.50	3,596.50	4,300	
	4670	Air bag power operated, 10 ton cap., 6'x8'		1.08	14.815		3,325	470	76.50	3,871.50	4,600	
	4680	7' x 8'		1.08	14.815		3,525	470	76.50	4,071.50	4,850	
	4700	Hydraulic, 10 ton capacity, 6' x 8'　　　**CN**		1.08	14.815		5,275	470	76.50	5,821.50	6,750	
	4800	7' x 8'	↓	1.08	14.815		5,675	470	76.50	6,221.50	7,200	
	5000	Lights for loading docks, single arm, 24" long	1 Elec	3.80	2.105		107	67		174	218	
	5700	Double arm, 60" long	"	3.80	2.105		127	67		194	240	
	5800	Loading dock safety restraints, manual style	E-16	1.08	14.815		1,925	470	76.50	2,471.50	3,075	
	5900	Automatic style	"	1.08	14.815		3,425	470	76.50	3,971.50	4,725	
	6200	Shelters, fabric, for truck or train, scissor arms, minimum	1 Carp	1	8		945	218		1,163	1,400	
	6300	Maximum	"	.50	16	↓	1,425	435		1,860	2,225	
603	0010	**DOCK BUMPERS** Bolts not included										603
	0020	2" x 6" to 4" x 8", average	1 Carp	.30	26.667	M.B.F.	885	730		1,615	2,125	
		**111 700	Waste Handling Equip**										
701	0010	**WASTE HANDLING**										701
	0020	Compactors, 115 volt, 250#/hr., chute fed	L-4	1	24	Ea.	8,225	615		8,840	10,000	
	0100	Hand fed		2.40	10		5,675	257		5,932	6,650	
	0300	Multi-bag, 230 volt, 600#/hr, chute fed		1	24		7,025	615		7,640	8,700	
	0400	Hand fed		1	24		6,375	615		6,990	8,000	
	0500	Containerized, hand fed, 2 to 6 C.Y. containers, 250#/hr.		1	24		7,475	615		8,090	9,200	
	0550	For chute fed, add per floor		1	24		860	615		1,475	1,925	
	1000	Heavy duty industrial compactor, 0.5 C.Y. capacity		1	24		5,050	615		5,665	6,525	
	1050	1.0 C.Y. capacity		1	24		7,525	615		8,140	9,250	
	1100	3 C.Y. capacity		.50	48		10,500	1,225		11,725	13,600	
	1150	5.0 C.Y. capacity		.50	48		14,000	1,225		15,225	17,400	
	1200	Combination shredder/compactor (5,000 lbs./hr.)	↓	.50	48		25,200	1,225		26,425	29,700	
	1400	For handling hazardous waste materials, 55 gallon drum packer, std.					11,700			11,700	12,900	
	1410	55 gallon drum packer w/HEPA filter					14,900			14,900	16,400	
	1420	55 gallon drum packer w/charcoal & HEPA filter					19,000			19,000	20,900	
	1430	All of the above made explosion proof, add					9,075			9,075	9,975	
	1500	Crematory, not including building, 1 place	Q-3	.20	160		42,400	4,975		47,375	54,000	
	1750	2 place		.10	320		60,500	9,950		70,450	81,500	
	4400	Incinerator, gas, not incl. chimney, elec. or pipe, 50#/hr., minimum		.80	40		16,200	1,250		17,450	19,700	
	4420	Maximum		.70	45.714		21,100	1,425		22,525	25,400	
	4440	200 lb. per hr., minimum (batch type)		.60	53.333		21,100	1,650		22,750	25,700	
	4460	Maximum (with feeder)		.50	64		41,100	2,000		43,100	48,200	
	4480	400 lb. per hr., minimum (batch type)		.30	106		25,000	3,325		28,325	32,500	
	4500	Maximum (with feeder)		.25	128		47,200	3,975		51,175	58,000	
	4520	800 lb. per hr., with feeder, minimum		.20	160		61,500	4,975		66,475	75,000	
	4540	Maximum		.17	188		84,000	5,850		89,850	101,500	
	4560	1,200 lb. per hr., with feeder, minimum		.15	213		89,000	6,625		95,625	108,000	
	4580	Maximum		.11	290		107,000	9,050		116,050	131,000	
	4600	2,000 lb. per hr., with feeder, minimum		.10	320		155,500	9,950		165,450	186,000	
	4620	Maximum		.05	640		261,500	19,900		281,400	317,500	
	4700	For heat recovery system, add, minimum		.25	128		50,500	3,975		54,475	61,500	
	4710	Add, maximum		.11	290		161,500	9,050		170,550	191,000	
	4720	For automatic ash conveyer, add		.50	64	↓	20,200	2,000		22,200	25,200	
	4750	Large municipal incinerators, incl. stack, minimum		.25	128	Ton/day	12,900	3,975		16,875	20,200	
	4850	Maximum	↓	.10	320	"	34,300	9,950		44,250	53,000	
	5500	Shredder, municipal use, 35 ton per hour				Ea.	192,000			192,000	211,000	

111 | Mercantile, Commercial & Detention Equipment

111 700 | Waste Handling Equip

		CREW	DAILY OUTPUT	LABOR-HOURS	UNIT	MAT.	LABOR	EQUIP.	TOTAL	TOTAL INCL O&P	
701	5600	60 ton per hour				Ea.	409,000			409,000	450,000	701
	5750	Shredder & baler, 50 ton per day					383,500			383,500	422,000	
	5800	Shredder, industrial, minimum					15,100			15,100	16,600	
	5850	Maximum					81,000			81,000	89,000	
	5900	Baler, industrial, minimum					6,050			6,050	6,650	
	5950	Maximum					353,500			353,500	389,000	
	6000	Transfer station compactor, with power unit										
	6050	and pedestal, not including pit, 50 ton per hour				Ea.	121,000			121,000	133,000	

111 900 | Detention Equipment

		CREW	DAILY OUTPUT	LABOR-HOURS	UNIT	MAT.	LABOR	EQUIP.	TOTAL	TOTAL INCL O&P	
901	0010	**DETENTION EQUIPMENT**										901
	0500	Bar front, rolling, 7/8" bars, 4" O.C., 7' high, 5' wide, with hardware	E-4	2	16	Ea.	4,050	500	41	4,591	5,425	
	1000	Doors & frames, 3' x 7', complete, with hardware, single plate		4	8		3,025	249	20.50	3,294.50	3,800	
	1650	Double plate		4	8		3,725	249	20.50	3,994.50	4,575	
	2000	Cells, prefab., 5' to 6' wide, 7' to 8' high, 7' to 8' deep,										
	2010	bar front, cot, not incl. plumbing	E-4	1.50	21.333	Ea.	6,800	665	55	7,520	8,750	
	2500	Cot, bolted, single, painted steel		20	1.600		279	50	4.11	333.11	400	
	2700	Stainless steel		20	1.600		725	50	4.11	779.11	895	
	3000	Toilet apparatus including wash basin, average	L-8	1.50	13.333		1,975	380		2,355	2,775	
	4000	Visitor cubicle, vision panel, no intercom	E-4	2	16		2,300	500	41	2,841	3,475	

114 | Food Service, Residential, Darkroom, Athletic Equipment

114 000 | Food Service Equipment

		CREW	DAILY OUTPUT	LABOR-HOURS	UNIT	MAT.	LABOR	EQUIP.	TOTAL	TOTAL INCL O&P	
002	0010	**RESIDENTIAL APPLIANCES**										002
	0020	Cooking range, 30" free standing, 1 oven, minimum	2 Clab	10	1.600	Ea.	286	34.50		320.50	370	
	0050	Maximum		4	4		1,050	86		1,136	1,275	
	0150	2 oven, minimum		10	1.600		1,125	34.50		1,159.50	1,275	
	0200	Maximum		10	1.600		1,500	34.50		1,534.50	1,700	
	0350	Built-in, 30" wide, 1 oven, minimum	1 Elec	6	1.333		405	42.50		447.50	510	
	0400	Maximum	2 Carp	2	8		1,000	218		1,218	1,450	
	0500	2 oven, conventional, minimum		4	4		880	109		989	1,125	
	0550	1 conventional, 1 microwave, maximum		2	8		1,550	218		1,768	2,050	
	0700	Free-standing, 1 oven, 21" wide range, minimum	2 Clab	10	1.600		277	34.50		311.50	360	
	0750	21" wide, maximum	"	4	4		465	86		551	650	
	0900	Counter top cook tops, 4 burner, standard, minimum	1 Elec	6	1.333		169	42.50		211.50	250	
	0950	Maximum		3	2.667		395	85		480	555	
	1050	As above, but with grille and griddle attachment, minimum		6	1.333		330	42.50		372.50	430	
	1100	Maximum		3	2.667		525	85		610	700	
	1200	Induction cooktop, 30" wide		3	2.667		695	85		780	890	
	1250	Microwave oven, minimum		4	2		92.50	64		156.50	198	
	1300	Maximum		2	4		1,575	128		1,703	1,925	
	1500	Combination range, refrigerator and sink, 30" wide, minimum	L-1	2	8		630	258		888	1,075	
	1550	Maximum		1	16		1,250	515		1,765	2,150	
	1570	60" wide, average		1.40	11.429		2,100	370		2,470	2,850	
	1590	72" wide, average		1.20	13.333		2,350	430		2,780	3,250	
	1600	Office model, 48" wide **CN**		2	8		1,775	258		2,033	2,375	
	1620	Refrigerator and sink only		2.40	6.667		1,775	215		1,990	2,300	
	1640	Combination range, refrigerator, sink, microwave										
	1660	oven and ice maker	L-1	.80	20	Ea.	3,475	645		4,120	4,800	

11 EQUIPMENT

114 000	Food Service Equipment	CREW	DAILY OUTPUT	LABOR-HOURS	UNIT	1999 BARE COSTS				TOTAL INCL O&P
						MAT.	LABOR	EQUIP.	TOTAL	
1750	Compactor, residential size, 4 to 1 compaction, minimum	1 Carp	5	1.600	Ea.	330	43.50		373.50	430
1800	Maximum	"	3	2.667		360	73		433	510
2000	Deep freeze, 15 to 23 C.F., minimum	2 Clab	10	1.600		400	34.50		434.50	490
2050	Maximum		5	3.200		595	68.50		663.50	765
2200	30 C.F., minimum		8	2		730	43		773	875
2250	Maximum	▼	3	5.333		850	114		964	1,125
2450	Dehumidifier, portable, automatic, 15 pint					185			185	204
2550	40 pint					229			229	252
2750	Dishwasher, built-in, 2 cycles, minimum	L-1	4	4		217	129		346	435
2800	Maximum		2	8		445	258		703	875
2950	4 or more cycles, minimum		4	4		234	129		363	450
2960	Average		4	4		310	129		439	540
3000	Maximum	▼	2	8		585	258		843	1,025
3200	Dryer, automatic, minimum	L-2	3	5.333		259	128		387	485
3250	Maximum	"	2	8		530	193		723	890
3300	Garbage disposer, sink type, minimum	L-1	10	1.600		37.50	51.50		89	119
3350	Maximum	"	10	1.600		178	51.50		229.50	274
3550	Heater, electric, built-in, 1250 watt, ceiling type, minimum	1 Elec	4	2		45	64		109	145
3600	Maximum		3	2.667		100	85		185	237
3700	Wall type, minimum		4	2		40	64		104	140
3750	Maximum		3	2.667		90	85		175	226
3900	1500 watt wall type, with blower		4	2		90	64		154	195
3950	3000 watt	▼	3	2.667		130	85		215	270
4150	Hood for range, 2 speed, vented, 30" wide, minimum	L-3	5	3.200		37.50	94.50		132	188
4200	Maximum		3	5.333		278	158		436	550
4300	42" wide, minimum		5	3.200		164	94.50		258.50	325
4330	Custom		5	3.200		206	94.50		300.50	375
4350	Maximum	▼	3	5.333		252	158		410	520
4500	For ventless hood, 2 speed, add					13.40			13.40	14.75
4650	For vented 1 speed, deduct from maximum					26			26	28.50
4850	Humidifier, portable, 8 gallons per day					92.50			92.50	102
5000	15 gallons per day					149			149	164
5200	Icemaker, automatic, 20 lb. per day	1 Plum	7	1.143		505	37.50		542.50	610
5350	51 lb. per day	"	2	4		835	130		965	1,125
5500	Refrigerator, no frost, 10 C.F. to 12 C.F. minimum	2 Clab	10	1.600		415	34.50		449.50	510
5600	Maximum		6	2.667		635	57		692	790
5750	14 C.F. to 16 C.F., minimum		9	1.778		475	38		513	580
5800	Maximum		5	3.200		685	68.50		753.50	865
5950	18 C.F. to 20 C.F., minimum		8	2		525	43		568	650
6000	Maximum		4	4		980	86		1,066	1,200
6150	21 C.F. to 29 C.F., minimum		7	2.286		805	49		854	960
6200	Maximum	▼	3	5.333		2,425	114		2,539	2,825
6400	Sump pump cellar drainer, pedestal, 1/3 H.P., molded PVC base	1 Plum	3	2.667		86.50	87		173.50	227
6450	Solid brass	"	2	4	▼	179	130		309	395
6460	Sump pump, see also division 152-480									
6650	Washing machine, automatic, minimum	1 Plum	3	2.667	Ea.	330	87		417	495
6700	Maximum	"	1	8		865	261		1,126	1,350
6900	Water heater, electric, glass lined, 30 gallon, minimum	L-1	5	3.200		233	103		336	410
6950	Maximum		3	5.333		325	172		497	615
7100	80 gallon, minimum		2	8		450	258		708	885
7150	Maximum	▼	1	16		630	515		1,145	1,475
7180	Water heater, gas, glass lined, 30 gallon, minimum	2 Plum	5	3.200		258	104		362	440
7220	Maximum		3	5.333		360	174		534	660
7260	50 gallon, minimum		2.50	6.400		340	209		549	690
7300	Maximum	▼	1.50	10.667	▼	475	350		825	1,050
7310	Water heater, see also division 153-110									

EQUIPMENT 11

		114 000 | Food Service Equipment	CREW	DAILY OUTPUT	LABOR-HOURS	UNIT	1999 BARE COSTS				TOTAL INCL O&P	
							MAT.	LABOR	EQUIP.	TOTAL		
002	7350	Water softener, automatic, to 30 grains per gallon	2 Plum	5	3.200	Ea.	450	104		554	655	**002**
	7400	To 100 grains per gallon	"	4	4		510	130		640	765	
	7450	Vent kits for dryers	1 Carp	10	.800	▼	9.55	22		31.55	45	
004	0010	**COMMERCIAL KITCHEN EQUIPMENT**										**004**
	0020	Bake oven, gas, one section	Q-1	8	2	Ea.	3,450	58.50		3,508.50	3,900	
	0300	Two sections		7	2.286		6,875	67		6,942	7,675	
	0600	Three sections	▼	6	2.667		9,825	78.50		9,903.50	10,900	
	0900	Electric convection, single deck	L-7	4	7		4,675	184		4,859	5,425	
	1050	Butter pat dispenser	1 Clab	13	.615		395	13.20		408.20	455	
	1100	Bread dispenser, counter top	"	13	.615		455	13.20		468.20	520	
	1300	Broiler, without oven, standard	Q-1	8	2		3,325	58.50		3,383.50	3,750	
	1550	Infra-red	L-7	4	7		4,650	184		4,834	5,375	
	1650	Cabinet, heated, 1 compartment, reach-in	R-18	5.60	4.643		2,050	117		2,167	2,425	
	1655	Pass-thru roll-in		5.60	4.643		2,850	117		2,967	3,325	
	1660	2 compartment, reach-in	▼	4.80	5.417		4,375	136		4,511	5,025	
	1670	Mobile					2,100			2,100	2,300	
	1700	Choppers, 5 pounds	R-18	7	3.714		1,475	93.50		1,568.50	1,775	
	1720	16 pounds		5	5.200		1,850	131		1,981	2,225	
	1740	35 to 40 pounds	▼	4	6.500		3,575	163		3,738	4,175	
	1840	Coffee brewer, 5 burners	1 Plum	3	2.667		895	87		982	1,125	
	1850	Coffee urn, twin 6 gallon urns		2	4		5,675	130		5,805	6,425	
	1860	Single, 3 gallon	▼	3	2.667		4,125	87		4,212	4,675	
	1900	Cup and glass dispenser, drop in	1 Clab	4	2		805	43		848	955	
	1920	Disposable cup, drop in	"	16	.500		226	10.75		236.75	266	
	2350	Cooler, reach-in, beverage, 6' long	Q-1	6	2.667		2,575	78.50		2,653.50	2,950	
	2650	Dish dispenser, drop in, 12"	1 Clab	11	.727		945	15.60		960.60	1,050	
	2660	Mobile	"	10	.800	▼	1,575	17.15		1,592.15	1,775	
	2700	Dishwasher, commercial, rack type										
	2720	10 to 12 racks per hour	Q-1	3.20	5	Ea.	2,350	147		2,497	2,825	
	2750	Semi-automatic 38 to 50 racks per hour	"	1.30	12.308		4,950	360		5,310	6,000	
	2800	Automatic, 190 to 230 racks per hour	Q-2	.70	34.286		10,100	1,050		11,150	12,700	
	2820	235 to 275 racks per hour		.50	48		20,400	1,450		21,850	24,700	
	2840	8,750 to 12,500 dishes per hour	▼	.20	120	▼	38,400	3,650		42,050	47,800	
	2950	Dishwasher hood, canopy type	L-3A	10	1.200	L.F.	3,400	36		3,436	3,800	
	2960	Pant leg type	"	2.50	4.800	Ea.	4,525	145		4,670	5,200	
	2970	Exhaust hood, sst, gutter on all sides, 4' x 4' x 2'	1 Carp	1.80	4.444		2,500	121		2,621	2,950	
	2980	4' x 4' x 7'	"	1.60	5		3,900	137		4,037	4,525	
	3000	Fast food equipment, total package, minimum	6 Skwk	.08	600		103,000	16,800		119,800	140,000	
	3100	Maximum	"	.07	685		140,500	19,200		159,700	185,000	
	3300	Food warmer, counter, 1.2 KW					615			615	675	
	3550	1.6 KW					765			765	840	
	3600	Well, hot food, built-in, rectangular, 12" x 20"	R-30	10	2.600		246	66.50		312.50	370	
	3610	Circular, 7 qt		10	2.600		205	66.50		271.50	330	
	3620	Refrigerated, 2 compartments		10	2.600		1,475	66.50		1,541.50	1,725	
	3630	3 compartments		9	2.889		1,500	74		1,574	1,775	
	3640	4 compartments	▼	8	3.250		2,025	83.50		2,108.50	2,350	
	3800	Food mixers, 20 quarts	L-7	7	4		2,975	105		3,080	3,450	
	3850	40 quarts		5.40	5.185		6,900	136		7,036	7,825	
	3900	60 quarts		5	5.600		9,375	147		9,522	10,500	
	4040	80 quarts		3.90	7.179		10,700	189		10,889	12,100	
	4080	130 quarts		2.20	12.727		14,900	335		15,235	16,900	
	4100	Floor type, 20 quarts		15	1.867		3,900	49		3,949	4,375	
	4120	60 quarts		14	2		8,300	52.50		8,352.50	9,200	
	4140	80 quarts		12	2.333		10,400	61.50		10,461.50	11,500	
	4160	140 quarts	▼	8.60	3.256	▼	18,800	85.50		18,885.50	20,800	

Important: See the Reference Section for critical supporting data - Reference Nos., Crews, & City Cost Indexes

	114 000	Food Service Equipment	CREW	DAILY OUTPUT	LABOR-HOURS	UNIT	1999 BARE COSTS				TOTAL INCL O&P	
							MAT.	LABOR	EQUIP.	TOTAL		
004	4300	Freezers, reach-in, 44 C.F.	Q-1	4	4	Ea.	6,800	117		6,917	7,650	004
	4500	68 C.F.	"	3	5.333		8,300	157		8,457	9,350	
	4600	Freezer, pre-fab, 8' x 8' w/refrigeration	2 Carp	.45	35.556		6,600	970		7,570	8,775	
	4620	8' x 12'		.35	45.714		8,550	1,250		9,800	11,400	
	4640	8' x 16'		.25	64		11,000	1,750		12,750	14,900	
	4660	8' x 20'	↓	.17	94.118		12,300	2,575		14,875	17,600	
	4680	Reach-in, 1 compartment	Q-1	4	4		1,900	117		2,017	2,250	
	4700	2 compartment	"	3	5.333		3,050	157		3,207	3,575	
	4720	Frost cold plate	R-30	9	2.889		9,975	74		10,049	11,100	
	4750	Fryer, with twin baskets, modular model	Q-1	7	2.286		2,925	67		2,992	3,300	
	5000	Floor model on 6" legs	"	5	3.200		3,675	94		3,769	4,200	
	5100	Extra single basket, large					82.50			82.50	91	
	5300	Griddle, 3' long	Q-1	7	2.286		2,350	67		2,417	2,700	
	5550	4' long	"	6	2.667		3,800	78.50		3,878.50	4,300	
	5700	Hot chocolate dispenser	1 Plum	4	2		700	65		765	870	
	5800	Ice cube maker, 50 pounds per day	Q-1	6	2.667		1,525	78.50		1,603.50	1,800	
	5900	250 pounds per day		1.20	13.333		1,825	390		2,215	2,600	
	6050	500 pounds per day		4	4		2,475	117		2,592	2,875	
	6060	With bin		1.20	13.333		3,325	390		3,715	4,275	
	6090	1000 pounds per day, with bin		1	16		7,200	470		7,670	8,625	
	6100	Ice flakers, 300 pounds per day		1.60	10		3,550	294		3,844	4,350	
	6120	600 pounds per day		.95	16.842		3,675	495		4,170	4,800	
	6130	1000 pounds per day		.75	21.333		5,950	625		6,575	7,500	
	6140	2000 pounds per day	↓	.65	24.615		11,200	720		11,920	13,400	
	6160	Ice storage bin, 500 pound capacity	Q-5	1	16		1,425	470		1,895	2,275	
	6180	1000 pound	"	.56	28.571		2,050	840		2,890	3,550	
	6200	Iced tea brewer	1 Plum	3.44	2.326		615	76		691	795	
	6250	Jet spray dispenser	R-18	4.50	5.778		2,050	145		2,195	2,475	
	6300	Juice dispenser, concentrate	"	4.50	5.778		795	145		940	1,100	
	6350	Kettles, steam-jacketed, 20 gallons	L-7	7	4		4,400	105		4,505	5,025	
	6600	60 gallons	"	6	4.667		5,650	123		5,773	6,400	
	6690	Milk dispenser, bulk, 2 flavor	R-30	8	3.250		880	83.50		963.50	1,100	
	6695	3 flavor	"	8	3.250		1,325	83.50		1,408.50	1,575	
	6700	Peelers, small	R-18	8	3.250		1,675	81.50		1,756.50	1,975	
	6720	Large	"	6	4.333	↓	2,825	109		2,934	3,275	
	6750	Pot sink, 3 compartment	1 Plum	7.25	1.103	L.F.	455	36		491	555	
	6760	Pot washer, small		1.60	5	Ea.	11,900	163		12,063	13,300	
	6770	Large		1.20	6.667		29,700	217		29,917	33,000	
	6800	Pulper/extractor, close coupled, 5 HP	↓	1.90	4.211		2,650	137		2,787	3,125	
	6850	Mobile rack w/pan slide					810			810	895	
	6900	Range, restaurant type, 6 burners and 1 standard oven, 36" wide	Q-1	7	2.286		1,975	67		2,042	2,275	
	6950	Convection		7	2.286		2,625	67		2,692	3,000	
	7150	2 standard ovens, 24" griddle, 60" wide		6	2.667		4,100	78.50		4,178.50	4,650	
	7200	1 standard, 1 convection oven		6	2.667		5,275	78.50		5,353.50	5,925	
	7450	Heavy duty, single 34" standard oven, open top		5	3.200		3,575	94		3,669	4,075	
	7500	Convection oven		5	3.200		4,750	94		4,844	5,375	
	7700	Griddle top		6	2.667		4,225	78.50		4,303.50	4,775	
	7750	Convection oven		6	2.667		5,175	78.50		5,253.50	5,825	
	7950	Hood fire protection system, minimum		3	5.333		2,675	157		2,832	3,150	
	8050	Maximum		1	16		18,700	470		19,170	21,300	
	8300	Refrigerators, reach-in type, 44 C.F.		5	3.200		4,525	94		4,619	5,125	
	8310	With glass doors, 68 C.F.	↓	4	4		5,950	117		6,067	6,725	
	8320	Refrigerator, reach-in, 1 compartment	R-18	7.80	3.333		1,875	84		1,959	2,175	
	8330	2 compartment		6.20	4.194		2,375	105		2,480	2,750	
	8340	3 compartment	↓	5.60	4.643		3,375	117		3,492	3,900	
	8350	Pre-fab, with refrigeration, 8' x 8'	2 Carp	.45	35.556	↓	5,225	970		6,195	7,275	

004		**114 000**	**Food Service Equipment**	CREW	DAILY OUTPUT	LABOR-HOURS	UNIT	1999 BARE COSTS				TOTAL INCL O&P	004
								MAT.	LABOR	EQUIP.	TOTAL		
	8360		8' x 12'	2 Carp	.35	45.714	Ea.	6,350	1,250		7,600	8,975	
	8370		8' x 16'		.25	64		7,350	1,750		9,100	10,800	
	8380		8' x 20'	↓	.17	94.118		9,625	2,575		12,200	14,700	
	8390		Pass-thru/roll-in, 1 compartment	R-18	7.80	3.333		2,800	84		2,884	3,200	
	8400		2 compartment		6.24	4.167		3,875	105		3,980	4,425	
	8410		3 compartment	↓	5.60	4.643		5,325	117		5,442	6,025	
	8420		Walk-in, alum, door & floor only, no refrig, 6' x 6' x 7'-6"	2 Carp	1.40	11.429		6,725	310		7,035	7,900	
	8430		10' x 6' x 7'-6"		.55	29.091		9,600	795		10,395	11,900	
	8440		12' x 14' x 7'-6"		.25	64		12,800	1,750		14,550	16,900	
	8450		12' x 20' x 7'-6"	↓	.17	94.118		15,500	2,575		18,075	21,100	
	8460		Refrigerated cabinets, mobile					2,050			2,050	2,250	
	8470		Refrigerator/freezer, reach-in, 1 compartment	R-18	5.60	4.643		4,275	117		4,392	4,875	
	8480		2 compartment		4.80	5.417		5,875	136		6,011	6,650	
	8580		Slicer with table	↓	9	2.889		2,825	72.50		2,897.50	3,200	
	8600		Stainless steel shelving, louvered 4-tier, 20" x 3'	1 Clab	6	1.333		980	28.50		1,008.50	1,125	
	8605		20" x 4'		6	1.333		1,150	28.50		1,178.50	1,325	
	8610		20" x 6'		6	1.333		1,575	28.50		1,603.50	1,800	
	8615		24" x 3'		6	1.333		1,025	28.50		1,053.50	1,175	
	8620		24" x 4'		6	1.333		1,225	28.50		1,253.50	1,375	
	8625		24" x 6'		6	1.333		1,700	28.50		1,728.50	1,900	
	8630		Flat 4-tier, 20" x 3'		6	1.333		900	28.50		928.50	1,025	
	8635		20" x 4'		6	1.333		1,100	28.50		1,128.50	1,250	
	8640		20" x 5'		6	1.333		1,250	28.50		1,278.50	1,425	
	8645		24" x 3'		6	1.333		965	28.50		993.50	1,100	
	8650		24" x 4'		6	1.333		1,200	28.50		1,228.50	1,375	
	8655		24" x 6'		6	1.333		2,300	28.50		2,328.50	2,575	
	8700		Galvanized shelving, louvered 4-tier, 20" x 3'		6	1.333		350	28.50		378.50	430	
	8705		20" x 4'		6	1.333		400	28.50		428.50	485	
	8710		20" x 6'		6	1.333		585	28.50		613.50	690	
	8715		24" x 3'		6	1.333		385	28.50		413.50	470	
	8720		24" x 4'		6	1.333		440	28.50		468.50	525	
	8725		24" x 6'		6	1.333		595	28.50		623.50	700	
	8730		Flat 4-tier, 20" x 3'		6	1.333		355	28.50		383.50	435	
	8735		20" x 4'		6	1.333		410	28.50		438.50	495	
	8740		20" x 6'		6	1.333		550	28.50		578.50	650	
	8745		24" x 3'		6	1.333		380	28.50		408.50	460	
	8750		24" x 4'		6	1.333		440	28.50		468.50	530	
	8755		24" x 6'		6	1.333		595	28.50		623.50	700	
	8760		Stainless steel dunnage rack, 24" x 3'		8	1		525	21.50		546.50	610	
	8765		24" x 4'		8	1		635	21.50		656.50	735	
	8770		Galvanized dunnage rack, 24" x 3'		8	1		124	21.50		145.50	170	
	8775		24" x 4'	↓	8	1	↓	140	21.50		161.50	188	
	8800		Serving counter, straight	1 Carp	40	.200	L.F.	410	5.45		415.45	460	
	8820		Curved section	"	30	.267	"	575	7.30		582.30	640	
	8825		Solid surface, see section 066-503										
	8830		Soft serve ice cream machine, medium	R-18	11	2.364	Ea.	5,900	59.50		5,959.50	6,575	
	8840		Large	"	9	2.889		8,875	72.50		8,947.50	9,850	
	8850		Steamer, electric 27 KW	L-7	7	4		7,200	105		7,305	8,075	
	9100		Electric, 10 KW or gas 100,000 BTU	"	5	5.600		3,200	147		3,347	3,750	
	9150		Toaster, conveyor type, 16-22 slices per minute					925			925	1,025	
	9160		Pop-up, 2 slot					455			455	500	
	9170		Trash compactor, small, up to 125 lb. compacted weight	L-4	4	6		13,500	154		13,654	15,100	
	9175		Large, up to 175 lb. compacted weight	"	3	8		16,500	205		16,705	18,500	
	9180		Tray and silver dispenser, mobile	1 Clab	16	.500	↓	595	10.75		605.75	670	
	9200		For deluxe models of above equipment, add					75%					
	9400		Rule of thumb: Equipment cost based										

Important: See the Reference Section for critical supporting data - Reference Nos., Crews, & City Cost Indexes

114 000 | Food Service Equipment

			CREW	DAILY OUTPUT	LABOR-HOURS	UNIT	1999 BARE COSTS				TOTAL INCL O&P	
							MAT.	LABOR	EQUIP.	TOTAL		
004	9410	on kitchen work area										**004**
	9420	Office buildings, minimum	L-7	77	.364	S.F.	45	9.55		54.55	64.50	
	9450	Maximum		58	.483		76	12.70		88.70	103	
	9550	Public eating facilities, minimum		77	.364		59	9.55		68.55	80	
	9600	Maximum		46	.609		96	16		112	130	
	9750	Hospitals, minimum		58	.483		60.50	12.70		73.20	86.50	
	9800	Maximum		39	.718		101	18.85		119.85	142	
008	0010	**WINE CELLAR**, refrigerated, Redwood interior, carpeted, walk-in type										**008**
	0020	6'-8" high, including racks										
	0200	80 "W x 48"D for 900 bottles	2 Carp	1.50	10.667	Ea.	2,975	291		3,266	3,725	
	0250	80" W x 72" D for 1300 bottles		1.33	12.030		3,900	330		4,230	4,800	
	0300	80" W x 94" D for 1900 bottles		1.17	13.675		4,925	375		5,300	5,975	
	0400	80" W x 124" D for 2500 bottles		1	16		6,000	435		6,435	7,275	
	0600	Portable cabinets, red oak, reach-in temp.& humidity controlled										
	0650	26-5/8"W x 26-1/2"D x 68"H for 235 bottles				Ea.	1,800			1,800	1,975	
	0660	32"W x 21-1/2"D x 73-1/2"H for 144 bottles					2,900			2,900	3,200	
	0670	32"W x 29-1/2"D x 73-1/2"H for 288 bottles					3,100			3,100	3,400	
	0680	39-1/2"W x 29-1/2"D x 86-1/2"H for 440 bottles					3,400			3,400	3,750	
	0690	52-1/2"W x 29-1/2"D x 73-1/2"H for 468 bottles					3,875			3,875	4,275	
	0700	52-1/2"W x 29-1/2"D x 86-1/2"H for 572 bottles					4,000			4,000	4,400	
	0730	Portable, red oak, can be built-in with glass door										
	0750	23-7/8"W x 24"D x 34-1/2"H for 50 bottles				Ea.	895			895	985	

114 580 | Disappearing Stairs

			CREW	DAILY OUTPUT	LABOR-HOURS	UNIT	MAT.	LABOR	EQUIP.	TOTAL	INCL O&P	
581	0010	**DISAPPEARING STAIRWAY** No trim included										**581**
	0100	Custom grade, pine, 8'-6" ceiling, minimum	1 Carp	4	2	Ea.	100	54.50		154.50	196	
	0150	Average		3.50	2.286		105	62.50		167.50	214	
	0200	Maximum		3	2.667		185	73		258	320	
	0500	Heavy duty, pivoted, from 7'7" to 12'10" floor to floor		3	2.667		315	73		388	465	
	0600	16'-0" ceiling		2	4		1,025	109		1,134	1,325	
	0800	Economy folding, pine, 8'-6" ceiling		4	2		83	54.50		137.50	178	
	0900	9'-6" ceiling		4	2		93.50	54.50		148	189	
	1000	Fire escape, galvanized steel, 8'-0" to 10'-4" ceiling	2 Carp	1	16		1,125	435		1,560	1,900	
	1010	10'-6" to 13'-6" ceiling		1	16		1,400	435		1,835	2,225	
	1100	Automatic electric, aluminum, floor to floor height, 8' to 9'		1	16		5,500	435		5,935	6,750	
	1400	11' to 12'		.90	17.778		5,925	485		6,410	7,300	
	1700	14' to 15'		.70	22.857		6,325	625		6,950	7,925	

114 740 | Darkroom Processing

			CREW	DAILY OUTPUT	LABOR-HOURS	UNIT	MAT.	LABOR	EQUIP.	TOTAL	INCL O&P	
741	0010	**DARKROOM EQUIPMENT**										**741**
	0020	Developing sink, 5" deep, 24" x 48"	Q-1	2	8	Ea.	3,350	235		3,585	4,025	
	0050	48" x 52"		1.70	9.412		3,600	276		3,876	4,400	
	0200	10" deep, 24" x 48"		1.70	9.412		4,700	276		4,976	5,575	
	0250	24" x 108"		1.50	10.667		1,700	315		2,015	2,350	
	0500	Dryers, dehumidified filtered air, 36" x 25" x 68" high	L-7	6	4.667		3,475	123		3,598	4,025	
	0550	48" x 25" x 68" high		5	5.600		6,575	147		6,722	7,450	
	2000	Processors, automatic, color print, minimum		4	7		9,500	184		9,684	10,800	
	2050	Maximum		.60	46.667		70,000	1,225		71,225	79,000	
	2300	Black and white print, minimum		2	14		7,200	370		7,570	8,500	
	2350	Maximum		.80	35		47,000	920		47,920	53,000	
	2600	Manual processor, 16" x 20" maximum print size		2	14		7,250	370		7,620	8,550	
	2650	20" x 24" maximum print size		1	28		6,825	735		7,560	8,650	
	3000	Viewing lites, 20" x 24"		6	4.667		300	123		423	520	

EQUIPMENT 11

		CREW	DAILY OUTPUT	LABOR-HOURS	UNIT	1999 BARE COSTS				TOTAL INCL O&P	
						MAT.	LABOR	EQUIP.	TOTAL		
114 740	**Darkroom Processing**										
741	3100	20" x 24" with color correction	L-7	6	4.667	Ea.	425	123		548	660	741
	3500	Washers, round, minimum sheet 11" x 14"	Q-1	2	8		2,175	235		2,410	2,750	
	3550	Maximum sheet 20" x 24"		1	16		2,400	470		2,870	3,350	
	3800	Square, minimum sheet 20" x 24"		1	16		2,150	470		2,620	3,075	
	3900	Maximum sheet 50" x 56"	↓	.80	20	↓	3,350	585		3,935	4,575	
	4500	Combination tank sink, tray sink, washers, with										
	4510	dry side tables, average	Q-1	.45	35.556	Ea.	7,250	1,050		8,300	9,550	
114 760	**Revolving Darkrm Doors**										
768	0010	**DARKROOM DOORS**										768
	0015	Revolving, standard, 2 way, 36" diameter	2 Carp	3.10	5.161	Opng.	1,525	141		1,666	1,900	
	0020	41" diameter		3.10	5.161		1,650	141		1,791	2,050	
	0050	3 way, 51" diameter		1.40	11.429		2,100	310		2,410	2,825	
	1000	4 way, 49" diameter		1.40	11.429		2,475	310		2,785	3,225	
	2000	Hinged safety, 2 way, 41" diameter		2.30	6.957		1,950	190		2,140	2,450	
	2500	3 way, 51" diameter		1.40	11.429		2,500	310		2,810	3,250	
	3000	Pop out safety, 2 way, 41" diameter		3.10	5.161		2,475	141		2,616	2,950	
	4000	3 way, 51" diameter		1.40	11.429		2,500	310		2,810	3,250	
	5000	Wheelchair-type, pop out, 51" diameter		1.40	11.429		2,400	310		2,710	3,150	
	5020	72" diameter	↓	.90	17.778	↓	5,000	485		5,485	6,275	
	9300	For complete dark rooms, see division 130-305										
114 800	**Athletic/Recreational**										
801	0010	**HEALTH CLUB EQUIPMENT**										801
	0020	Abdominal rack, 2 board capacity				Ea.	370			370	410	
	0050	Abdominal board, upholstered					415			415	460	
	0200	Bicycle trainer, minimum					650			650	715	
	0300	Deluxe, electric					3,300			3,300	3,650	
	0400	Bar bell set, chrome plated steel, 25 lbs.					195			195	215	
	0420	100 lbs.					295			295	325	
	0450	200 lbs.				↓	570			570	625	
	0500	Weight plates, cast iron, per lb.				Lb.	4			4	4.40	
	0520	Storage rack, 10 station				Ea.	630			630	695	
	0600	Circuit training apparatus, 12 machines minimum	2 Clab	1.25	12.800	Set	22,000	275		22,275	24,600	
	0700	Average		1	16		27,000	345		27,345	30,200	
	0800	Maximum	↓	.75	21.333		32,000	460		32,460	35,900	
	0820	Dumbbell set, cast iron, with rack and 5 pair				↓	615			615	675	
	0900	Squat racks	2 Clab	5	3.200	Ea.	610	68.50		678.50	780	
	1200	Multi-station gym machine, 5 station					6,350			6,350	6,975	
	1250	9 station					8,900			8,900	9,800	
	1280	Rowing machine, hydraulic					1,175			1,175	1,300	
	1300	Treadmill, manual					1,150			1,150	1,275	
	1320	Motorized					3,450			3,450	3,775	
	1340	Electronic					4,350			4,350	4,775	
	1360	Cardio-testing					6,200			6,200	6,825	
	1400	Treatment/massage tables, minimum					415			415	455	
	1420	Deluxe, with accessories				↓	600			600	660	
	1600	For saunas, see division 130-521										
	1640	For steam baths, see division 130-541										
	1680	For whirlpool baths, see division 152-104										
	1720	For additional exercise equipment, see div. 114-805										
805	0010	**SCHOOL EQUIPMENT**										805
	0200	For exterior equipment, see division 028										
	0201	For chairs & desks, see division 126-230										
	0300	For chalkboards & bulletin boards, see div. 104-151 & 101-104										

11 EQUIPMENT

Important: See the Reference Section for critical supporting data - Reference Nos., Crews, & City Cost Indexes

114 800 | Athletic/Recreational

805			CREW	DAILY OUTPUT	LABOR-HOURS	UNIT	1999 BARE COSTS MAT.	LABOR	EQUIP.	TOTAL	TOTAL INCL O&P	805
	0400	For lockers, see division 105-054										
	1000	Basketball backstops, wall mtd., 6' extended, fixed, minimum	L-2	1	16	Ea.	525	385		910	1,175	
	1100	Maximum		1	16		685	385		1,070	1,350	
	1200	Swing up, minimum		1	16		1,050	385		1,435	1,750	
	1250	Maximum		1	16		1,875	385		2,260	2,675	
	1300	Portable, manual, heavy duty, spring operated		1.90	8.421		8,825	203		9,028	10,000	
	1400	Ceiling suspended, stationary, minimum		.78	20.513		1,300	495		1,795	2,200	
	1450	Fold up, with accessories, maximum		.40	40		4,725	965		5,690	6,725	
	1600	For electrically operated, add	1 Elec	1	8		1,550	255		1,805	2,075	
	2000	Benches, folding, in wall, 14' table, 2 benches	L-4	2	12	Set	500	310		810	1,025	
	3000	Bleachers, telescoping, manual to 15 tier, minimum	F-5	65	.492	Seat	50	13.70		63.70	76.50	
	3100	Maximum		60	.533		75	14.85		89.85	106	
	3300	16 to 20 tier, minimum		60	.533		120	14.85		134.85	156	
	3400	Maximum		55	.582		150	16.15		166.15	191	
	3600	21 to 30 tier, minimum		50	.640		125	17.80		142.80	166	
	3700	Maximum		40	.800		150	22		172	200	
	3900	For integral power operation, add, minimum	2 Elec	300	.053		25	1.70		26.70	30	
	4000	Maximum	"	250	.064		40	2.04		42.04	47	
	4100	Boxing ring, elevated, 22' x 22'	L-4	.10	240	Ea.	8,000	6,150		14,150	18,500	
	4110	For cellular plastic foam padding, add		.10	240		3,825	6,150		9,975	13,900	
	4120	Floor level, including posts and ropes only, 20' x 20'		.80	30		1,175	770		1,945	2,525	
	4130	Canvas, 30' x 30'		5	4.800		1,075	123		1,198	1,400	
	4150	Exercise equipment, bicycle trainer					345			345	380	
	4180	Chinning bar, adjustable, wall mounted	1 Carp	5	1.600		245	43.50		288.50	340	
	4200	Exercise ladder, 16' x 1'-7", suspended	L-2	3	5.333		1,050	128		1,178	1,375	
	4210	High bar, floor plate attached	1 Carp	4	2		910	54.50		964.50	1,075	
	4240	Parallel bars, adjustable		4	2		2,000	54.50		2,054.50	2,275	
	4270	Uneven parallel bars, adjustable		4	2		2,375	54.50		2,429.50	2,700	
	4280	Wall mounted, adjustable	L-2	1.50	10.667	Set	750	257		1,007	1,225	
	4300	Rope, ceiling mounted, 18' long	1 Carp	3.66	2.186	Ea.	147	59.50		206.50	255	
	4330	Side horse, vaulting		5	1.600		1,450	43.50		1,493.50	1,675	
	4360	Treadmill, motorized, deluxe, training type		5	1.600		4,650	43.50		4,693.50	5,200	
	4390	Weight lifting multi-station, minimum	2 Clab	1	16		2,775	345		3,120	3,600	
	4450	Maximum	"	.50	32		12,000	685		12,685	14,300	
	4480	For additional exercise equipment, see div. 114-801										
	4500	Gym divider curtain, mesh top, vinyl bottom, manual	L-4	500	.048	S.F.	6	1.23		7.23	8.55	
	4700	Electric roll up	L-7	400	.070		9	1.84		10.84	12.75	
	5500	Gym mats, 2" thick, naugahyde covered					2.64			2.64	2.90	
	5600	Vinyl/nylon covered					5.05			5.05	5.55	
	5800	Wall pads, 1-1/2" thick	2 Carp	640	.025		5.70	.68		6.38	7.30	
	6000	Wrestling mats, 1" thick, heavy duty					5.20			5.20	5.75	
	7000	Scoreboards, baseball, minimum	R-3	1.30	15.385	Ea.	2,300	485	104	2,889	3,400	
	7200	Maximum		.05	400		16,800	12,600	2,725	32,125	40,500	
	7300	Football, minimum		.86	23.256		5,675	735	158	6,568	7,500	
	7400	Maximum		.20	100		52,500	3,150	680	56,330	63,500	
	7500	Basketball (one side), minimum		2.07	9.662		2,425	305	65.50	2,795.50	3,175	
	7600	Maximum		.30	66.667		16,800	2,100	455	19,355	22,100	
	7700	Hockey-basketball (four sides), minimum		.25	80		8,600	2,525	545	11,670	13,900	
	7800	Maximum		.15	133		26,300	4,200	905	31,405	36,200	

114 880 | Bowling Alleys

881	0010	BOWLING ALLEYS Including alley, pinsetter, scorer,										881
	0020	counters and misc. supplies, minimum	4 Carp	.20	160	Lane	32,200	4,375		36,575	42,400	
	0150	Average		.19	168		34,500	4,600		39,100	45,200	
	0300	Maximum		.18	177		36,700	4,850		41,550	48,000	

EQUIPMENT **11**

114 880	Bowling Alleys	CREW	DAILY OUTPUT	LABOR-HOURS	UNIT	1999 BARE COSTS				TOTAL INCL O&P	
						MAT.	LABOR	EQUIP.	TOTAL		
881 0400	Combo table ball rack, add				Lane	550			550	605	881
0600	For automatic scorer, add, minimum					6,000			6,000	6,600	
0700	Maximum				↓	6,800			6,800	7,475	

114 960 | Shooting Ranges

		CREW	DAILY OUTPUT	LABOR-HOURS	UNIT	MAT.	LABOR	EQUIP.	TOTAL	TOTAL INCL O&P	
961 0010	**SHOOTING RANGE** Incl. bullet traps, target provisions, controls,										961
0100	separators, ceiling system, etc. Not incl. structural shell										
0200	Commercial	L-9	.64	56.250	Point	7,200	1,400		8,600	10,200	
0300	Law enforcement		.28	128		15,300	3,175		18,475	22,000	
0400	National Guard armories		.71	50.704		4,925	1,250		6,175	7,475	
0500	Reserve training centers		.71	50.704		5,025	1,250		6,275	7,575	
0600	Schools and colleges		.32	112		10,800	2,775		13,575	16,500	
0700	Major academies	↓	.19	189		20,100	4,700		24,800	29,800	
0800	For acoustical treatment, add					10%	10%				
0900	For lighting, add					28%	25%				
1000	For plumbing, add					5%	5%				
1100	For ventilating system, add, minimum					40%	40%				
1200	Add, average					25%	25%				
1300	Add, maximum				↓	35%	35%				

11 EQUIPMENT

115 000	Industrial Equipment	CREW	DAILY OUTPUT	LABOR-HOURS	UNIT	1999 BARE COSTS				TOTAL INCL O&P	
						MAT.	LABOR	EQUIP.	TOTAL		
005 0010	**EQUIPMENT INSTALLATION**										005
0020	Industrial equipment, minimum	E-2	12	4.667	Ton		139	84.50	223.50	335	
0200	Maximum	"	2	28	"		835	505	1,340	2,000	

115 010 | Specialized Equipment

011 0010	**VOCATIONAL SHOP EQUIPMENT**										011
0020	Benches, work, wood, average	2 Carp	5	3.200	Ea.	400	87.50		487.50	575	
0100	Metal, average		5	3.200		225	87.50		312.50	385	
0400	Combination belt & disc sander, 6"		4	4		965	109		1,074	1,225	
0700	Drill press, floor mounted, 12", 1/2 H.P.	↓	4	4		400	109		509	610	
0800	Dust collector, not incl. ductwork, 6" diameter	1 Shee	1.10	7.273		2,300	231		2,531	2,900	
1000	Grinders, double wheel, 1/2 H.P.	2 Carp	5	3.200		254	87.50		341.50	415	
1300	Jointer, 4", 3/4 H.P.		4	4		1,050	109		1,159	1,325	
1600	Kilns, 16 C.F., to 2000°		4	4		2,100	109		2,209	2,475	
1900	Lathe, woodworking, 10", 1/2 H.P.		4	4		550	109		659	775	
2200	Planer, 13" x 6"		4	4		1,175	109		1,284	1,450	
2500	Potter's wheel, motorized		4	4		525	109		634	750	
2800	Saws, band, 14", 3/4 H.P.		4	4		580	109		689	805	
3100	Metal cutting band saw, 14"		4	4		690	109		799	930	
3400	Radial arm saw, 10", 2 H.P.		4	4		800	109		909	1,050	
3700	Scroll saw, 24"		4	4		1,350	109		1,459	1,650	
4000	Table saw, 10", 3 H.P.		4	4		1,675	109		1,784	2,000	
4300	Welder AC arc, 30 amp capacity	↓	4	4	↓	425	109		534	640	

Important: See the Reference Section for critical supporting data - Reference Nos., Crews, & City Cost Indexes

116 | Laboratory, Planetarium, Observatory Equipment

116 000	Laboratory Equipment	CREW	DAILY OUTPUT	LABOR-HOURS	UNIT	1999 BARE COSTS				TOTAL INCL O&P
						MAT.	LABOR	EQUIP.	TOTAL	
001	0010 **LABORATORY EQUIPMENT**									001
	0020 Cabinets, base, door units, metal	2 Carp	18	.889	L.F.	115	24.50		139.50	165
	0300 Drawer units		18	.889		228	24.50		252.50	289
	0700 Tall storage cabinets, open, 7' high		20	.800		240	22		262	299
	0900 With glazed doors		20	.800		274	22		296	335
	1300 Wall cabinets, metal, 12-1/2" deep, open		20	.800		69	22		91	111
	1500 With doors		20	.800		144	22		166	193
	1550 Counter tops, not incl. base cabinets, acidproof, minimum		82	.195	S.F.	16	5.35		21.35	26
	1600 Maximum		70	.229		24	6.25		30.25	36.50
	1650 Stainless steel		82	.195		61	5.35		66.35	75.50
	2000 Fume hood, with countertop & base, not including HVAC									
	2020 Simple, minimum	2 Carp	5.40	2.963	L.F.	505	81		586	680
	2050 Complex, including fixtures		2.40	6.667		1,200	182		1,382	1,575
	2100 Special, maximum		1.70	9.412		1,250	257		1,507	1,775
	2200 Ductwork, minimum	2 Shee	1	16	Hood	835	510		1,345	1,700
	2250 Maximum	"	.50	32	"	4,225	1,025		5,250	6,225
	2300 Service fixtures, average				Ea.	43			43	47
	2500 For sink assembly with hot and cold water, add	1 Plum	1.40	5.714		475	186		661	800
	2550 Glassware washer, distilled water rinse, minimum	L-1	1.80	8.889		3,825	287		4,112	4,650
	2600 Maximum	"	1	16		17,100	515		17,615	19,600
	2650 Glove box, fiberglass, bacteriological					11,000			11,000	12,100
	2700 Controlled atmosphere					8,500			8,500	9,350
	2750 Radioisotope					11,000			11,000	12,100
	2770 Carcinogenic					11,000			11,000	12,100
	2800 Sink, one piece plastic, flask wash, hose, free standing	1 Plum	1.60	5		1,250	163		1,413	1,625
	2850 Epoxy resin sink, 25" x 16" x 10"	"	2	4		138	130		268	350
	3000 Utility table, acid resistant top with drawers	2 Carp	30	.533	L.F.	102	14.55		116.55	135
	3800 Titration unit, four 2000 ml reservoirs				Ea.	8,350			8,350	9,200
	4200 Alternate pricing method: as percent of lab furniture									
	4400 Installation, not incl. plumbing & duct work				% Furn.					22%
	4800 Plumbing, final connections, simple system									10%
	5000 Moderately complex system									15%
	5200 Complex system									20%
	5400 Electrical, simple system									10%
	5600 Moderately complex system									20%
	5800 Complex system									35%
	6000 Safety equipment, eye wash, hand held				Ea.	210			210	231
	6200 Deluge shower				"	500			500	550
	6300 Rule of thumb: lab furniture including installation & connection									
	6320 High school				S.F.					27.75
	6340 College									40.75
	6360 Clinical, health care									35
	6380 Industrial									56.50

117 | Medical Equipment

117 000	Medical Equipment	CREW	DAILY OUTPUT	LABOR-HOURS	UNIT	1999 BARE COSTS				TOTAL INCL O&P
						MAT.	LABOR	EQUIP.	TOTAL	
001	0010 **MEDICAL EQUIPMENT**									001
	0015 Autopsy table, standard	1 Plum	1	8	Ea.	6,000	261		6,261	7,000

117 000	Medical Equipment	CREW	DAILY OUTPUT	LABOR-HOURS	UNIT	1999 BARE COSTS				TOTAL INCL O&P
						MAT.	LABOR	EQUIP.	TOTAL	
001 0020	Deluxe	1 Plum	.60	13.333	Ea.	7,500	435		7,935	8,900	001
0300	Blood pressure unit, mercurial, wall					151			151	166
0400	Diagnostic set, wall					440			440	485
0700	Distiller, water, steam heated, 50 gal. capacity	1 Plum	1.40	5.714		12,500	186		12,686	14,100
0750	Exam room furnishings, average per room					4,375			4,375	4,800
1800	Heat therapy unit, humidified, 26" x 78" x 28"				▼	2,300			2,300	2,525
2100	Hubbard tank with accessories, stainless steel,									
2110	125 GPM at 45 psi water pressure				Ea.	17,800			17,800	19,600
2150	For electric overhead hoist, add					1,950			1,950	2,150
2500	Incubators, minimum					2,475			2,475	2,725
2600	Maximum					13,600			13,600	15,000
2900	K-Module for heat therapy, 20 oz. capacity, 75° to 110°F					275			275	305
3200	Mortuary refrigerator, end operated, 2 capacity					9,700			9,700	10,700
3300	6 capacity					18,000			18,000	19,800
3600	Paraffin bath, 126°F, auto controlled					1,050			1,050	1,175
3900	Parallel bars for walking training, 12'-0"					870			870	955
4200	Refrigerator, blood bank, 28.6 C.F. emergency signal					5,425			5,425	5,950
4210	Reach-in, 16.9 C.F.					5,775			5,775	6,375
4400	Scale, physician's, with height rod					238			238	261
4600	Station, dietary, medium, with ice					11,000			11,000	12,100
4700	Medicine					4,975			4,975	5,475
5000	Scrub, surgical, stainless steel, single station, minimum	1 Plum	3	2.667		3,550	87		3,637	4,025
5100	Maximum					5,375			5,375	5,925
5600	Sterilizers, floor loading, 26" x 62" x 42", single door, steam					109,000			109,000	120,000
5650	Double door, steam					140,000			140,000	154,000
5800	General purpose, 20" x 20" x 38", single door					10,500			10,500	11,600
6000	Portable, counter top, steam, minimum					2,625			2,625	2,875
6020	Maximum					4,075			4,075	4,500
6050	Portable, counter top, gas, 17"x15"x32-1/2"				▼	27,000			27,000	29,700
6150	Manual washer/sterilizer, 16"X16"X26"	1 Plum	2	4		37,000	130		37,130	40,900
6200	Steam generators, electric 10 KW to 180 KW, freestanding									
6250	Minimum	1 Elec	3	2.667	Ea.	5,500	85		5,585	6,175
6300	Maximum	"	.70	11.429		19,800	365		20,165	22,300
6500	Surgery table, minor minimum	1 Sswk	.70	11.429		9,525	350		9,875	11,100
6520	Maximum		.50	16		17,900	490		18,390	20,600
6550	Major surgery table, minimum		.50	16		21,600	490		22,090	24,700
6570	Maximum	▼	.50	16		71,500	490		71,990	79,500
6700	Surgical lights, doctor's office, single arm	2 Elec	2	8		995	255		1,250	1,475
6750	Dual arm		1	16		3,800	510		4,310	4,950
6800	Major operating room, dual head, minimum		1	16		12,000	510		12,510	14,000
6850	Maximum	▼	1	16		20,000	510		20,510	22,800
7000	Tables, physical therapy, walk off, electric	2 Carp	3	5.333		2,200	146		2,346	2,650
7150	Standard, vinyl top with base cabinets, minimum		3	5.333		825	146		971	1,125
7200	Maximum	▼	2	8		1,950	218		2,168	2,500
7500	Thermometer, electric, portable					275			275	300
8100	Utensil washer-sanitizer	1 Plum	2	4		7,500	130		7,630	8,450
8200	Bed pan washer-sanitizer	"	2	4		4,900	130		5,030	5,600
8400	Whirlpool bath, mobile, sst, 18" x 24" x 60"		2	4		3,250			3,250	3,575
8450	Fixed, incl. mixing valves	1 Plum	2	4		6,325	130		6,455	7,150
8700	X-ray, mobile, minimum					9,200			9,200	10,100
8750	Maximum					52,000			52,000	57,000
8900	Stationary, minimum					28,800			28,800	31,600
8950	Maximum					149,500			149,500	164,500
9150	Developing processors, minimum					8,000			8,000	8,800
9200	Maximum				▼	21,000			21,000	23,100
9800	For hospital partitions, see division 106-302									

Important: See the Reference Section for critical supporting data - Reference Nos., Crews, & City Cost Indexes

117 000	Medical Equipment	CREW	DAILY OUTPUT	LABOR-HOURS	UNIT	1999 BARE COSTS				TOTAL INCL O&P	
						MAT.	LABOR	EQUIP.	TOTAL		
001	9900	For casework & furniture, see division 123-501 & 126-210										001

117 400	Dental Equipment										
401	0010	**DENTAL EQUIPMENT**										401
	0020	Central suction system, minimum	1 Plum	1.20	6.667	Ea.	2,625	217		2,842	3,225
	0100	Maximum	"	.90	8.889		20,300	290		20,590	22,700
	0300	Air compressor, minimum	1 Skwk	.80	10		2,900	281		3,181	3,625
	0400	Maximum		.50	16		20,000	450		20,450	22,700
	0600	Chair, electric or hydraulic, minimum		.50	16		3,300	450		3,750	4,325
	0700	Maximum	▼	.25	32		7,000	900		7,900	9,125
	0800	Doctor's/assistant's stool, minimum					375			375	415
	0850	Maximum					735			735	810
	1000	Drill console with accessories, minimum	1 Skwk	1.60	5		2,975	140		3,115	3,500
	1100	Maximum		1.60	5		8,500	140		8,640	9,575
	2000	Light, ceiling mounted, minimum		8	1		1,725	28		1,753	1,950
	2100	Maximum	▼	8	1		3,450	28		3,478	3,850
	2200	Unit light, minimum	2 Skwk	5.33	3.002		1,275	84		1,359	1,525
	2210	Maximum		5.33	3.002		4,200	84		4,284	4,750
	2220	Track light, minimum		3.20	5		2,275	140		2,415	2,725
	2230	Maximum	▼	3.20	5		5,000	140		5,140	5,725
	2300	Sterilizers, steam portable, minimum					2,900			2,900	3,200
	2350	Maximum					10,000			10,000	11,000
	2600	Steam, institutional					12,500			12,500	13,700
	2650	Dry heat, electric, portable, 3 trays					975			975	1,075
	2700	Ultra-sonic cleaner, portable, minimum					785			785	865
	2750	Maximum (institutional)					1,750			1,750	1,925
	3000	X-ray unit, wall, minimum	1 Skwk	4	2		3,975	56		4,031	4,475
	3010	Maximum		4	2		7,500	56		7,556	8,350
	3100	Panoramic unit	▼	.60	13.333		18,000	375		18,375	20,400
	3105	Deluxe, minimum	2 Skwk	1.60	10		26,000	281		26,281	29,000
	3110	Maximum	"	1.60	10		60,000	281		60,281	66,500
	3500	Developers, X-ray, average	1 Plum	5.33	1.501		3,500	49		3,549	3,925
	3600	Maximum	"	5.33	1.501	▼	6,500	49		6,549	7,225

For information about Means Estimating Seminars, see yellow pages 11 and 12 in back of book

EQUIPMENT **11**

	CREW	DAILY OUTPUT	LABOR-HOURS	UNIT	1999 BARE COSTS				TOTAL INCL O&P
					MAT.	LABOR	EQUIP.	TOTAL	

Division 12
Furnishings

Estimating Tips

General
- The items in this division are usually priced per square foot or each. Most of these items are purchased by the owner and placed by the supplier. Do not assume the items in Division 12 will be purchased and installed by the supplier. Check the specifications for responsibilities and include receiving, storage, installation and mechanical and electrical hook-ups in the appropriate divisions.

- Some items in this division require some type of support system that is not usually furnished with the item. Examples of these systems include blocking for the attachment of casework and heavy drapery rods. The required blocking must be added to the estimate in the appropriate division.

Reference Numbers
Reference numbers are shown in bold squares at the beginning of some major classifications. These numbers refer to related items in the Reference Section. The reference information may be an estimating procedure, an alternate pricing method or technical information.

Note: Not all subdivisions listed here necessarily appear in this publication.

123 010 | Metal Casework

		CREW	DAILY OUTPUT	LABOR-HOURS	UNIT	1999 BARE COSTS				TOTAL INCL O&P	
						MAT.	LABOR	EQUIP.	TOTAL		
015	0010	**KEY CABINETS**									015
	0020	Wall mounted, 60 key capacity	1 Carp	20	.400	Ea.	166	10.90		176.90	200
	0200	Drawer type, 600 key capacity	1 Clab	15	.533		2,125	11.45		2,136.45	2,350
	0300	2,400 key capacity		20	.400		4,775	8.60		4,783.60	5,275
	0400	Tray type, 20 key capacity		50	.160		96	3.43		99.43	111
	0500	50 key capacity		40	.200		127	4.29		131.29	147

123 500 | Hospital Casework

		CREW	DAILY OUTPUT	LABOR-HOURS	UNIT	MAT.	LABOR	EQUIP.	TOTAL	TOTAL INCL O&P	
501	0010	**CABINETS**									501
	0500	Hospital, base cabinets, laminated plastic	2 Carp	10	1.600	L.F.	174	43.50		217.50	260
	1000	Stainless steel	"	10	1.600		218	43.50		261.50	310
	1200	For all drawers, add					8.85			8.85	9.75
	1300	Cabinet base trim, 4" high, enameled steel	2 Carp	200	.080		26.50	2.18		28.68	32.50
	1400	Stainless steel		200	.080		40	2.18		42.18	47.50
	1450	Counter top, laminated plastic, no backsplash		40	.400		30	10.90		40.90	50
	1650	With backsplash		40	.400		37.50	10.90		48.40	58
	1800	For sink cutout, add		12.20	1.311	Ea.		36		36	56.50
	1900	Stainless steel counter top		40	.400	L.F.	83	10.90		93.90	109
	2000	For drop-in stainless 43" x 21" sink, add				Ea.	425			425	465
	2050	For solid surface countertops see 066-503									
	2100	Nurses station, door type, laminated plastic	2 Carp	10	1.600	L.F.	202	43.50		245.50	291
	2200	Enameled steel		10	1.600		175	43.50		218.50	262
	2300	Stainless steel		10	1.600		233	43.50		276.50	325
	2400	For drawer type, add					82			82	90
	2500	Wall cabinets, laminated plastic	2 Carp	15	1.067		130	29		159	189
	2600	Enameled steel		15	1.067		145	29		174	206
	2700	Stainless steel		15	1.067		226	29		255	295
	2800	For glass doors, add					21			21	23
	3500	Kitchen, base cabinets, metal, minimum	2 Carp	30	.533		42.50	14.55		57.05	70
	3600	Maximum		25	.640		108	17.45		125.45	147
	3700	Wall cabinets, metal, minimum		30	.533		42.50	14.55		57.05	70
	3800	Maximum		25	.640		98	17.45		115.45	136
	5000	School, 24" deep, metal, 84" high units		15	1.067		213	29		242	281
	5150	Counter height units		20	.800		166	22		188	218
	5450	Wood, custom fabricated, 32" high counter		20	.800		138	22		160	187
	5600	Add for counter top		56	.286		16.15	7.80		23.95	30
	5800	84" high wall units		15	1.067		144	29		173	205
	6000	Laminated plastic finish is same price as wood									

123 800 | Display Casework

		CREW	DAILY OUTPUT	LABOR-HOURS	UNIT	MAT.	LABOR	EQUIP.	TOTAL	TOTAL INCL O&P	
801	0010	**DISPLAY CASES** Free standing, all glass										801
	0020	Aluminum frame, 42" high x 36" x 12" deep	2 Carp	8	2	Ea.	445	54.50		499.50	575
	0100	70" high x 48" x 18" deep	"	6	2.667		1,100	73		1,173	1,325
	0500	For wood bases, add					9%				
	0600	For hardwood frames, deduct					8%				
	0700	For bronze, baked enamel finish, add					10%				
	2000	Wall mounted, glass front, aluminum frame									
	2010	Non-illuminated, one section 3' x 4' x 1'-4"	2 Carp	5	3.200	Ea.	930	87.50		1,017.50	1,150
	2100	5' x 4' x 1'-4"		5	3.200		1,125	87.50		1,212.50	1,350
	2200	6' x 4' x 1'-4"		4	4		1,200	109		1,309	1,500
	2500	Two sections, 8' x 4' x 1'-4"		2	8		1,625	218		1,843	2,150
	2600	10' x 4' x 1'-4"		2	8		1,750	218		1,968	2,275
	3000	Three sections, 16' x 4' x 1'-4"		1.50	10.667		2,625	291		2,916	3,325
	3500	For fluorescent lights, add				Section	97.50			97.50	107
	4000	Table exhibit cases, 2' wide, 3' high, 4' long, flat top	2 Carp	5	3.200	Ea.	670	87.50		757.50	875
	4100	3' wide, 3' high, 4' long, sloping top	"	3	5.333	"	1,000	146		1,146	1,325

12 FURNISHINGS

Important: See the Reference Section for critical supporting data - Reference Nos., Crews, & City Cost Indexes

123 | Manufactured Casework

123 900	Residential Casework	CREW	DAILY OUTPUT	LABOR-HOURS	UNIT	1999 BARE COSTS				TOTAL INCL O&P	
						MAT.	LABOR	EQUIP.	TOTAL		
905	0010 **IRONING CENTER**										905
	0020 Including cabinet, board & light, minimum	1 Carp	2	4	Ea.	239	109		348	435	
	0100 Maximum, see also division 111-101	"	1.50	5.333	"	480	146		626	760	

125 | Window Treatment

125 100	Blinds	CREW	DAILY OUTPUT	LABOR-HOURS	UNIT	1999 BARE COSTS				TOTAL INCL O&P	
						MAT.	LABOR	EQUIP.	TOTAL		
103	0010 **BLINDS, INTERIOR**										103
	0020 Horizontal, 1" aluminum slats, solid color, stock	1 Carp	590	.014	S.F.	2.55	.37		2.92	3.38	
	0090 Custom, minimum		590	.014		2.30	.37		2.67	3.11	
	0100 Maximum		440	.018		7.50	.50		8	9.05	
	0250 2" aluminum slats, custom, minimum		590	.014		3.88	.37		4.25	4.84	
	0350 Maximum		440	.018		6.40	.50		6.90	7.80	
	0450 Stock, minimum		590	.014		2.66	.37		3.03	3.51	
	0500 Maximum		440	.018		4.94	.50		5.44	6.25	
	0600 2" steel slats, stock, minimum		590	.014		1.50	.37		1.87	2.23	
	0630 Maximum		440	.018		4.18	.50		4.68	5.40	
	0750 Custom, minimum		590	.014		1.35	.37		1.72	2.07	
	0850 Maximum		400	.020		6.70	.55		7.25	8.20	
	1500 Vertical, 3" to 5" PVC or cloth strips, minimum **CN**		460	.017		6.55	.47		7.02	7.95	
	1600 Maximum		400	.020		8.25	.55		8.80	9.90	
	1800 4" aluminum slats, minimum		460	.017		2.93	.47		3.40	3.97	
	1900 Maximum		400	.020		6.95	.55		7.50	8.45	
	1950 Mylar mirror-finish strips, to 8" wide, minimum		460	.017		11.60	.47		12.07	13.50	
	1970 Maximum		400	.020	↓	17.40	.55		17.95	20	
	3000 Wood folding panels with movable louvers, 7" x 20" each		17	.471	Pr.	36	12.85		48.85	59.50	
	3300 8" x 28" each		17	.471		52	12.85		64.85	77	
	3450 9" x 36" each		17	.471		62	12.85		74.85	88	
	3600 10" x 40" each		17	.471		70	12.85		82.85	97	
	4000 Fixed louver type, stock units, 8" x 20" each		17	.471		54	12.85		66.85	79.50	
	4150 10" x 28" each		17	.471		74	12.85		86.85	102	
	4300 12" x 36" each		17	.471		94	12.85		106.85	123	
	4450 18" x 40" each		17	.471		112	12.85		124.85	143	
	5000 Insert panel type, stock, 7" x 20" each		17	.471		13.90	12.85		26.75	35.50	
	5150 8" x 28" each		17	.471		25.50	12.85		38.35	48	
	5300 9" x 36" each		17	.471		32.50	12.85		45.35	55.50	
	5450 10" x 40" each		17	.471		35	12.85		47.85	58	
	5600 Raised panel type, stock, 10" x 24" each		17	.471		93	12.85		105.85	122	
	5650 12" x 26" each		17	.471		108	12.85		120.85	139	
	5700 14" x 30" each		17	.471		122	12.85		134.85	154	
	5750 16" x 36" each	↓	17	.471		137	12.85		149.85	171	
	6000 For custom built pine, add					22%					
	6500 For custom built hardwood blinds, add				↓	42%					

125 200	Shades	CREW	DAILY OUTPUT	LABOR-HOURS	UNIT	1999 BARE COSTS				TOTAL INCL O&P	
						MAT.	LABOR	EQUIP.	TOTAL		
201	0010 **SHADES**										201
	0020 Basswood, roll-up, stain finish, 3/8" slats	1 Carp	300	.027	S.F.	10.05	.73		10.78	12.20	
	0200 7/8" slats		300	.027		8.50	.73		9.23	10.50	
	0300 Vertical side slide, stain finish, 3/8" slats	↓	300	.027	↓	12.50	.73		13.23	14.90	

FURNISHINGS 12

		CREW	DAILY OUTPUT	LABOR-HOURS	UNIT	1999 BARE COSTS				TOTAL INCL O&P	
125 200	**Shades**					MAT.	LABOR	EQUIP.	TOTAL		
201	0400	7/8" slats	1 Carp	300	.027	S.F.	12.50	.73		13.23	14.90	201
	0500	For fire retardant finishes, add					16%					
	0600	For "B" rated finishes, add					20%					
	0900	Mylar, single layer, non-heat reflective	1 Carp	685	.012		4.99	.32		5.31	6	
	1000	Double layered, heat reflective		685	.012		6.90	.32		7.22	8.10	
	1100	Triple layered, heat reflective	↓	685	.012		6.90	.32		7.22	8.10	
	1200	For metal roller instead of wood, add per				Shade	2.63			2.63	2.89	
	1300	Vinyl coated cotton, standard	1 Carp	685	.012	S.F.	1	.32		1.32	1.60	
	1400	Lightproof decorator shades		685	.012		1.02	.32		1.34	1.63	
	1500	Vinyl, lightweight, 4 gauge		685	.012		.37	.32		.69	.91	
	1600	Heavyweight, 6 gauge		685	.012		1.15	.32		1.47	1.77	
	1700	Vinyl laminated fiberglass, 6 ga., translucent		685	.012		1.08	.32		1.40	1.68	
	1800	Lightproof		685	.012		2.63	.32		2.95	3.39	
	3000	Woven aluminum, 3/8" thick, lightproof and fireproof	↓	350	.023	↓	3.79	.62		4.41	5.15	
125 300	**Drape/Curtain Hardware**										
301	0010	**DRAPERY HARDWARE**										301
	0030	Standard traverse, per foot, minimum	1 Carp	59	.136	L.F.	1.38	3.70		5.08	7.30	
	0100	Maximum		51	.157	"	6.95	4.28		11.23	14.40	
	4000	Traverse rods, adjustable, 28" to 48"		22	.364	Ea.	16.55	9.95		26.50	34	
	4020	48" to 84"		20	.400		22.50	10.90		33.40	42	
	4040	66" to 120"		18	.444		26	12.15		38.15	47.50	
	4060	84" to 156"		16	.500		29.50	13.65		43.15	54	
	4080	100" to 180"		14	.571		29	15.60		44.60	56	
	4090	156" to 228"		13	.615		39.50	16.80		56.30	69.50	
	4100	228" to 312"		13	.615		45.50	16.80		62.30	76.50	
	4200	Double rods, adjustable, 30" to 48"		9	.889		36	24.50		60.50	77.50	
	4220	48" to 86"		9	.889		52	24.50		76.50	95	
	4240	86" to 150"		8	1		60	27.50		87.50	109	
	4260	100" to 180"		7	1.143		52.50	31		83.50	107	
	4300	Tray and curtain rod, adjustable, 30" to 48"		9	.889		23	24.50		47.50	63.50	
	4320	48" to 86"		9	.889		28.50	24.50		53	69	
	4340	86" to 150"		8	1		37.50	27.50		65	84	
	4360	100" to 180"	↓	7	1.143		44.50	31		75.50	98	
	4600	Valance, pinch pleated fabric, 12" deep, up to 54" long, minimum					28			28	31	
	4610	Maximum					70			70	77	
	4620	Up to 77" long, minimum					43			43	47.50	
	4630	Maximum					113			113	124	
	5000	Stationary rods, first 2 feet				↓	6.95			6.95	7.65	
	5020	Each additional foot, add				L.F.	2.79			2.79	3.07	

126 | Furniture & Accessories

		CREW	DAILY OUTPUT	LABOR-HOURS	UNIT	1999 BARE COSTS				TOTAL INCL O&P	
126 100	**Landscape Partitions**					MAT.	LABOR	EQUIP.	TOTAL		
107	0010	**POSTS**										107
	0020	Portable for pedestrian traffic control, standard, minimum				Ea.	87.50			87.50	96	
	0100	Maximum					134			134	147	
	0300	Deluxe posts, minimum					148			148	163	
	0400	Maximum				↓	285			285	315	
	0600	Ropes for above posts, plastic covered, 1-1/2" diameter				L.F.	8.10			8.10	8.95	

Important: See the Reference Section for critical supporting data - Reference Nos., Crews, & City Cost Indexes

126 100	Landscape Partitions	CREW	DAILY OUTPUT	LABOR-HOURS	UNIT	1999 BARE COSTS				TOTAL INCL O&P	
						MAT.	LABOR	EQUIP.	TOTAL		
107	0700 Chain core				L.F.	8.25			8.25	9.05	107

126 200 | Furniture

		CREW	DAILY OUTPUT	LABOR-HOURS	UNIT	MAT.	LABOR	EQUIP.	TOTAL	INCL O&P	
203	0010 **BANK FURNITURE** See division 126-222										203
204	0010 **CHURCH FURNITURE** See division 110-401										204
206	0010 **FURNITURE, DORMITORY**										206
	1000 Chest, four drawer, minimum				Ea.	350			350	385	
	1020 Maximum				"	560			560	615	
	1050 Built-in, minimum	2 Carp	13	1.231	L.F.	105	33.50		138.50	169	
	1150 Maximum		10	1.600		208	43.50		251.50	298	
	1200 Desk top, built-in, laminated plastic, 24" deep, minimum		50	.320		23	8.75		31.75	39	
	1300 Maximum		40	.400		68	10.90		78.90	91.50	
	1450 30" deep, minimum		50	.320		28.50	8.75		37.25	45.50	
	1550 Maximum		40	.400		126	10.90		136.90	155	
	1750 Dressing unit, built-in, minimum		12	1.333		169	36.50		205.50	243	
	1850 Maximum	↓	8	2	↓	500	54.50		554.50	640	
	8000 Rule of thumb: total cost for furniture, minimum				Student					1,900	
	8050 Maximum				"					3,600	
210	0010 **FURNITURE, HOSPITAL**										210
	0020 Beds, manual, minimum				Ea.	790			790	865	
	0100 Maximum					1,375			1,375	1,500	
	0300 Manual and electric beds, minimum					985			985	1,075	
	0400 Maximum					2,125			2,125	2,325	
	0600 All electric hospital beds, minimum					1,300			1,300	1,450	
	0700 Maximum					3,250			3,250	3,575	
	0900 Manual, nursing home beds, minimum					620			620	685	
	1000 Maximum					1,300			1,300	1,425	
	1020 Overbed table, laminated top, minimum					265			265	291	
	1040 Maximum				↓	660			660	725	
	1100 Patient wall systems, not incl. plumbing, minimum				Room	790			790	870	
	1200 Maximum				"	1,450			1,450	1,600	
	2000 Geriatric chairs, minimum				Ea.	255			255	281	
	2020 Maximum				"	595			595	655	
214	0010 **FURNITURE, HOTEL**										214
	0020 Standard quality set, minimum				Room	1,475			1,475	1,625	
	0200 Maximum				"	7,175			7,175	7,875	
218	0010 **FURNITURE, LIBRARY**										218
	0100 Attendant desk, 36" x 62" x 29" high	1 Carp	16	.500	Ea.	2,200	13.65		2,213.65	2,425	
	0200 Book display, "A" frame display, both sides, 42" x 42" x 60" high		16	.500		2,350	13.65		2,363.65	2,600	
	0220 Table with bulletin board, 42" x 24" x 49" high		16	.500		1,200	13.65		1,213.65	1,350	
	0800 Card catalogue, 30 tray unit		16	.500		2,375	13.65		2,388.65	2,650	
	0840 60 tray unit	↓	16	.500		4,075	13.65		4,088.65	4,500	
	0880 72 tray unit	2 Carp	16	1		4,425	27.50		4,452.50	4,900	
	1000 Carrels, single face, initial unit	1 Carp	16	.500		725	13.65		738.65	820	
	1500 Double face, initial unit	2 Carp	16	1		1,100	27.50		1,127.50	1,250	
	4000 Dictionary stand, stationary	1 Carp	16	.500		670	13.65		683.65	760	
	4020 Revolving		16	.500		233	13.65		246.65	278	
	4200 Exhibit case, table style, 60" x 28" x 36"		11	.727		3,325	19.85		3,344.85	3,675	
	7000 Tables, card catalog reference, 24" x 60" x 42"	↓	16	.500	↓	750	13.65		763.65	845	

FURNISHINGS 12

126 200	Furniture	CREW	DAILY OUTPUT	LABOR-HOURS	UNIT	1999 BARE COSTS				TOTAL INCL O&P	
						MAT.	LABOR	EQUIP.	TOTAL		
222	**0010** **FURNITURE, OFFICE**										**222**
0020	Desks, 29" high, double pedestal, 30" x 60", metal, minimum				Ea.	325			325	360	
0030	Maximum					920			920	1,000	
0160	Wood, minimum					405			405	445	
0180	Maximum					1,675			1,675	1,850	
0600	Desks, single pedestal, 30" x 60", metal, minimum					270			270	297	
0620	Maximum					700			700	770	
0640	Wood, minimum					350			350	385	
0650	Maximum					525			525	580	
0670	Executive return, 24" x 42", with box, file, wood, minimum					240			240	264	
0680	Maximum					525			525	580	
0720	Desks, secretarial, 30" x 60", metal, minimum					185			185	204	
0730	Maximum					525			525	580	
0740	Return, 20" x 42", minimum					175			175	193	
0750	Maximum					350			350	385	
0800	Wood, minimum					485			485	535	
0810	Maximum					1,700			1,700	1,875	
0820	Return, 20" x 42", minimum					310			310	340	
0830	Maximum					900			900	990	
0900	Desktop organizer, 72" x 14" x 36" high, wood, minimum					100			100	110	
0920	Maximum					225			225	248	
0940	59" x 12" x 23" high, steel, minimum					130			130	143	
0960	Maximum					200			200	220	
2000	Chairs, office type, executive, minimum					240			240	264	
2150	Maximum					1,525			1,525	1,675	
2200	Management, minimum					200			200	220	
2250	Maximum					1,150			1,150	1,275	
2280	Task, minimum					130			130	143	
2290	Maximum					445			445	490	
2300	Arm kit, minimum					41.50			41.50	45.50	
2320	Maximum				▼	85			85	93.50	
6000	Table, conference										
6050	Boat, 96" x 42", minimum				Ea.	545			545	600	
6150	Maximum					3,075			3,075	3,400	
6720	Rectangle, 96" x 42", minimum					605			605	665	
6740	Maximum				▼	2,550			2,550	2,800	
226	**0010** **FURNITURE, RESTAURANT**										**226**
0020	Bars, built-in, front bar	1 Carp	5	1.600	L.F.	180	43.50		223.50	267	
0200	Back bar	"	5	1.600	"	131	43.50		174.50	213	
0300	Booth seating see division 127-401										
2000	Chair, bentwood side chair, metal, minimum				Ea.	65.50			65.50	72	
2020	Maximum					83.50			83.50	91.50	
2600	Upholstered seat & back, arms, minimum					186			186	204	
2620	Maximum				▼	480			480	530	
230	**0010** **FURNITURE, SCHOOL**										**230**
1000	Chair, molded plastic,										
1100	Integral tablet arm, minimum				Ea.	52.50			52.50	58	
1150	Maximum					204			204	224	
2000	Desk, single pedestal, top book compartment, minimum					48			48	53	
2020	Maximum					52.50			52.50	57.50	
2200	Flip top, minimum					62			62	68	
2220	Maximum				▼	67			67	73.50	
262	**0010** **TABLES, FOLDING** Laminated plastic tops										**262**
1000	Tubular steel legs with glides										
1020	18" x 60", minimum				Ea.	157			157	172	
1040	Maximum				▼	560			560	615	

Important: See the Reference Section for critical supporting data - Reference Nos., Crews, & City Cost Indexes

126 200	Furniture	CREW	DAILY OUTPUT	LABOR-HOURS	UNIT	1999 BARE COSTS				TOTAL INCL O&P	
						MAT.	LABOR	EQUIP.	TOTAL		
262	1840	36" x 96", minimum				Ea.	255			255	280	**262**
	1860	Maximum					1,125			1,125	1,250	
	2000	Round, wood stained, plywood top, 60" diameter, minimum					235			235	259	
	2020	Maximum					1,300			1,300	1,425	

126 500 | Furniture Accessories

			CREW	DAILY OUTPUT	LABOR-HOURS	UNIT	MAT.	LABOR	EQUIP.	TOTAL	TOTAL INCL O&P	
501	0010	**ASH/TRASH RECEIVERS**										**501**
	1000	Ash urn, cylindrical metal										
	1020	8" diameter, 20" high	1 Clab	60	.133	Ea.	106	2.86		108.86	122	
	1060	10" diameter, 26" high	"	60	.133	"	150	2.86		152.86	170	
	2000	Combination ash/trash urn, metal										
	2020	8" diameter, 20" high	1 Clab	60	.133	Ea.	98	2.86		100.86	113	
	2050	10" diameter, 26" high	"	60	.133	"	168	2.86		170.86	190	
	4000	Trash receptacle, metal										
	4020	8" diameter, 15" high	1 Clab	60	.133	Ea.	56	2.86		58.86	66	
	4040	10" diameter, 18" high		60	.133		96	2.86		98.86	110	
	5040	16" x 8" x 14" high	↓	60	.133	↓	28	2.86		30.86	35.50	
	5500	Trash receptacle, plastic, with lid										
	5520	35 gallon	1 Clab	60	.133	Ea.	118	2.86		120.86	134	
	5540	45 gallon	"	60	.133	"	144	2.86		146.86	163	
507	0010	**MULTI-MEDIA BOARDS** See division 101-104										**507**

126 900 | Floor Mats & Frames

			CREW	DAILY OUTPUT	LABOR-HOURS	UNIT	MAT.	LABOR	EQUIP.	TOTAL	TOTAL INCL O&P	
901	0010	**FLOOR MATS**										**901**
	0020	Recessed, in-laid black rubber, 3/8" thick, solid	1 Clab	155	.052	S.F.	19.05	1.11		20.16	22.50	
	0050	Perforated		155	.052		24.50	1.11		25.61	28.50	
	0100	1/2" thick, solid		155	.052		20	1.11		21.11	23.50	
	0150	Perforated		155	.052		22.50	1.11		23.61	26	
	0200	In colors, 3/8" thick, solid		155	.052		21.50	1.11		22.61	25.50	
	0250	Perforated		155	.052		24	1.11		25.11	28	
	0300	1/2" thick, solid		155	.052		30	1.11		31.11	34.50	
	0350	Perforated		155	.052		31	1.11		32.11	35.50	
	0500	Link mats, including nosings, aluminum, 3/8" thick **CN**		155	.052		16.55	1.11		17.66	19.95	
	0550	Black rubber with galvanized tie rods		155	.052		13	1.11		14.11	16.05	
	0600	Steel, galvanized, 3/8" thick		155	.052		6.25	1.11		7.36	8.65	
	0650	Vinyl, in colors	↓	155	.052	↓	16.85	1.11		17.96	20.50	
	0750	Add for nosings, rubber				L.F.	6.05			6.05	6.65	
	0850	Recess frames for above mats, aluminum	1 Carp	100	.080		3.74	2.18		5.92	7.55	
	0870	Bronze	"	100	.080	↓	7.90	2.18		10.08	12.15	
	0900	Skate lock tile, 24" x 24" x 1/2" thick, rubber, black	1 Clab	125	.064	S.F.	12.70	1.37		14.07	16.10	
	0950	Color		125	.064	"	14.60	1.37		15.97	18.25	
	1000	12" x 24" border, black		75	.107	L.F.	21	2.29		23.29	26.50	
	1100	Color		75	.107	"	25.50	2.29		27.79	31.50	
	1150	12" x 12" outside corner, black		100	.080	S.F.	20.50	1.72		22.22	25	
	1200	Color		100	.080		23.50	1.72		25.22	28.50	
	1500	Duckboard, aluminum slats		155	.052		17.45	1.11		18.56	21	
	1700	Hardwood strips on rubber base, to 54" wide		155	.052		12.25	1.11		13.36	15.25	
	1800	Assembled with brass rods and vinyl spacers, to 48" wide		155	.052		16.55	1.11		17.66	19.95	
	1850	Tire fabric, 3/4" thick		155	.052		10.05	1.11		11.16	12.80	
	1900	Vinyl, 36" wide, in colors, hollow top & bottoms		155	.052		4.16	1.11		5.27	6.30	
	1950	Solid top & bottom members	↓	155	.052	↓	7.60	1.11		8.71	10.10	

FURNISHINGS 12

127 100 | Miscellaneous Seating

		CREW	DAILY OUTPUT	LABOR-HOURS	UNIT	1999 BARE COSTS				TOTAL INCL O&P
						MAT.	LABOR	EQUIP.	TOTAL	
101	0010	**SEATING**									101
	0100	Bleachers, See div. 028-708 & 114-805									
	0101	Benches, See 028-704, 105-054 & 114-805									
	0500	Classroom, movable chair & desk type, minimum				Set					65
	0600	Maximum				"					120
	1000	Lecture hall, pedestal type, minimum	2 Carp	22	.727	Ea.	101	19.85		120.85	142
	1200	Maximum		14.50	1.103		355	30		385	445
	2000	Auditorium chair, all veneer construction		22	.727		113	19.85		132.85	155
	2200	Veneer back, padded seat		22	.727		142	19.85		161.85	188
	2350	Fully upholstered, spring seat	↓	22	.727		144	19.85		163.85	190
	2450	For tablet arms, add					37.50			37.50	41
	2500	For fire retardancy, CATB-133, add			↓		15			15	16.50

127 400 | Booths & Tables

		CREW	DAILY OUTPUT	LABOR-HOURS	UNIT	1999 BARE COSTS				TOTAL INCL O&P
						MAT.	LABOR	EQUIP.	TOTAL	
401	0010	**BOOTHS**									401
	1000	Banquet, upholstered seat and back, custom									
	1500	Straight, minimum	2 Carp	40	.400	L.F.	104	10.90		114.90	131
	1520	Maximum		36	.444		201	12.15		213.15	241
	1600	"L" or "U" shape, minimum		35	.457		106	12.50		118.50	136
	1620	Maximum	↓	30	.533	↓	188	14.55		202.55	229
	1800	Upholstered outside finished backs for									
	1810	single booths and custom banquets									
	1820	Minimum	2 Carp	44	.364	L.F.	11.95	9.95		21.90	28.50
	1840	Maximum	"	40	.400	"	36	10.90		46.90	56.50
	3000	Fixed seating, one piece plastic chair and									
	3010	plastic laminate table top									
	3100	Two seat, 24" x 24" table, minimum	F-7	30	1.067	Ea.	415	26		441	500
	3120	Maximum		26	1.231		415	30		445	500
	3200	Four seat, 24" x 48" table, minimum		28	1.143		415	28		443	500
	3220	Maximum	↓	24	1.333	↓	705	32.50		737.50	825
	5000	Mount in floor, wood fiber core with									
	5010	plastic laminate face, single booth									
	5050	24" wide	F-7	30	1.067	Ea.	160	26		186	217
	5100	48" wide	"	28	1.143	"	203	28		231	268

128 150 | Planters

		CREW	DAILY OUTPUT	LABOR-HOURS	UNIT	1999 BARE COSTS				TOTAL INCL O&P
						MAT.	LABOR	EQUIP.	TOTAL	
155	0010	**PLANTERS**									155
	1000	Fiberglass, hanging, 12" diameter, 7" high				Ea.	38			38	41.50
	1500	Rectangular, 48" long, 16" high x 15" wide					310			310	340
	1650	60" long, 30" high, 28" wide					660			660	725
	2000	Round, 12" diameter, 13" high					66			66	72.50
	2050	25" high					78.50			78.50	86.50
	5000	Square, 10" side, 20" high					91			91	100
	5100	14" side, 15" high					92.50			92.50	102
	6000	Metal bowl, 32" diameter, 8" high, minimum					310			310	340
	6050	Maximum				↓	465			465	510
	8750	Wood, fiberglass liner, square									
	8780	14" square, 15" high, minimum				Ea.	171			171	188

12 FURNISHINGS

128 150	Planters	CREW	DAILY OUTPUT	LABOR-HOURS	UNIT	1999 BARE COSTS				TOTAL INCL O&P	
						MAT.	LABOR	EQUIP.	TOTAL		
155	8800	Maximum				Ea.	210			210	232	155
	9400	Plastic cylinder, molded, 10" diameter, 10" high					6.95			6.95	7.65	
	9500	11" diameter, 11" high				↓	22			22	24	

For information about Means Estimating Seminars, see yellow pages 11 and 12 in back of book

FURNISHINGS 12

Division Notes

	CREW	DAILY OUTPUT	LABOR-HOURS	UNIT	1999 BARE COSTS				TOTAL INCL O&P
					MAT.	LABOR	EQUIP.	TOTAL	

Division 13
Special Construction

Estimating Tips

General

- The items and systems in this division are usually estimated, purchased, supplied and installed as a unit by one or more subcontractors. The estimator must ensure that all parties are operating from the same set of specifications and assumptions and that all necessary items are estimated and will be provided. Many times the complex items and systems are covered but the more common ones such as excavation or a crane are overlooked for the very reason that everyone assumes nobody could miss them. The estimator should be the central focus and be able to ensure that all systems are complete.

- Another area where problems can develop in this division is at the interface between systems. The estimator must ensure, for instance, that anchor bolts, nuts and washers are estimated and included for the air-supported structures and pre-engineered buildings to be bolted to their foundations.

Utility supply is a common area where essential items or pieces of equipment can be missed or overlooked due to the fact that each subcontractor may feel it is the others' responsibility. The estimator should also be aware of certain items which may be supplied as part of a package but installed by others, and ensure that the installing contractor's estimate includes the cost of installation. Conversely, the estimator must also ensure that items are not costed by two different subcontractors, resulting in an inflated overall estimate.

131 Pre-Engineered Structures, Aquatic Facilities & Ice Rinks

- The foundations and floor slab, as well as rough mechanical and electrical, should be estimated, as this work is required for the assembly and erection of the structure. Generally, as noted in the book, the pre-engineered building comes as a shell and additional features must be included by the estimator. Here again, the estimator must have a clear understanding of the scope of each portion of the work and all the necessary interfaces.

132 Tanks, Tank Covers, Filtration Equipment

- The prices in this subdivision for above and below ground storage tanks do not include foundations or hold-down slabs. The estimator should refer to Divisions 2 and 3 for foundation system pricing. In addition to the foundations, required tank accessories such as tank gauges, leak detection devices, and additional manholes and piping must be added to the tank prices.

Reference Numbers

Reference numbers are shown in bold squares at the beginning of some major classifications. These numbers refer to related items in the Reference Section. The reference information may be an estimating procedure, an alternate pricing method or technical information.

Note: Not all subdivisions listed here necessarily appear in this publication.

130 100 | Air Supported Structures

		CREW	DAILY OUTPUT	LABOR-HOURS	UNIT	1999 BARE COSTS				TOTAL INCL O&P
						MAT.	LABOR	EQUIP.	TOTAL	
111	0010	**AIR SUPPORTED STRUCTURES**									111
	0020	Site preparation, incl. anchor placement and utilities R130 -110	B-11B	2,000	.008	SF Flr.	.66	.20	.52	1.38	1.61
	0030	For concrete curb, see division 033-130									
	0050	Warehouse, polyester/vinyl fabric, 28 oz., over 10 yr. life, welded									
	0060	Seams, tension cables, primary & auxiliary inflation system,									
	0070	airlock, personnel doors and liner									
	0100	5000 S.F.	4 Clab	5,000	.006	SF Flr.	15.40	.14		15.54	17.15
	0250	12,000 S.F.	"	6,000	.005		10.20	.11		10.31	11.40
	0400	24,000 S.F.	8 Clab	12,000	.005		7.80	.11		7.91	8.80
	0500	50,000 S.F.	"	12,500	.005	↓	6.20	.11		6.31	6.95
	0700	12 oz. reinforced vinyl fabric, 5 yr. life, sewn seams,									
	0710	accordian door, including liner									
	0750	3000 S.F.	4 Clab	3,000	.011	SF Flr.	6.40	.23		6.63	7.40
	0800	12,000 S.F.	"	6,000	.005		5.10	.11		5.21	5.80
	0850	24,000 S.F.	8 Clab	12,000	.005		4.15	.11		4.26	4.75
	0950	Deduct for single layer					.58			.58	.64
	1000	Add for welded seams					.90			.90	.99
	1050	Add for double layer, welded seams included				↓	1.85			1.85	2.04
	1250	Tedlar/vinyl fabric, 28 oz., with liner, over 10 yr. life,									
	1260	incl. overhead and personnel doors									
	1300	3000 S.F.	4 Clab	3,000	.011	SF Flr.	13.50	.23		13.73	15.20
	1450	12,000 S.F.	"	6,000	.005		9.45	.11		9.56	10.60
	1550	24,000 S.F.	8 Clab	12,000	.005		7.40	.11		7.51	8.35
	1700	Deduct for single layer				↓	1.29			1.29	1.42
	2250	Greenhouse/shelter, woven polyethylene with liner, 2 yr. life,									
	2260	sewn seams, including doors									
	2300	3000 S.F.	4 Clab	3,000	.011	SF Flr.	5.90	.23		6.13	6.85
	2350	12,000 S.F.	"	6,000	.005		3.09	.11		3.20	3.58
	2450	24,000 S.F.	8 Clab	12,000	.005		2.69	.11		2.80	3.14
	2550	Deduct for single layer				↓	.56			.56	.62
	2600	Tennis/gymnasium, polyester/vinyl fabric, 28 oz., over 10 yr. life,									
	2610	including thermal liner, heat and lights									
	2650	7200 S.F.	4 Clab	6,000	.005	SF Flr.	13	.11		13.11	14.50
	2750	13,000 S.F.	"	6,500	.005		9.50	.11		9.61	10.60
	2850	Over 24,000 S.F.	8 Clab	12,000	.005		8.50	.11		8.61	9.55
	2860	For low temperature conditions, add				↓	.53			.53	.58
	2870	For average shipping charges, add				Total	2,000			2,000	2,200
	2900	Thermal liner, translucent reinforced vinyl				SF Flr.	.75			.75	.83
	2950	Metalized mylar fabric and mesh, double liner				"	1.30			1.30	1.43
	3050	Stadium/convention center, teflon coated fiberglass, heavy weight,									
	3060	over 20 yr. life, incl. thermal liner and heating system									
	3100	Minimum	9 Clab	26,000	.003	SF Flr.	35	.06		35.06	38.50
	3110	Maximum	"	19,000	.004	"	40	.08		40.08	44
	3400	Doors, air lock, 15' long, 10' x 10'	2 Carp	.80	20	Ea.	12,500	545		13,045	14,700
	3600	15' x 15'	"	.50	32		18,500	875		19,375	21,800
	3700	For each added 5' length, add				↓	2,000			2,000	2,200
	3900	Revolving personnel door, 6' diameter, 6'-6" high	2 Carp	.80	20	↓	8,800	545		9,345	10,500
	4200	Double wall, self supporting, shell only, minimum				SF Flr.					21
	4300	Maximum	↓			"					39
115	0010	**AIR SUPPORTED STORAGE TANK COVERS** Vinyl polyester										115
	0100	scrim, double layer, with hardware, blower, standby & controls									
	0200	Round, 75' diameter	B-2	4,500	.009	S.F.	5.60	.19		5.79	6.45
	0300	100' diameter		5,000	.008		5	.17		5.17	5.80
	0400	150' diameter		5,000	.008		3.55	.17		3.72	4.19
	0500	Rectangular, 20' x 20'	↓	4,500	.009		12	.19		12.19	13.50

Important: See the Reference Section for critical supporting data - Reference Nos., Crews, & City Cost Indexes

130 100 | Air Supported Structures

			DAILY OUTPUT	LABOR-HOURS	UNIT	1999 BARE COSTS				TOTAL INCL O&P	
		CREW				MAT.	LABOR	EQUIP.	TOTAL		
115	0600	30' x 40'	B-2	4,500	.009	S.F.	12	.19		12.19	13.50	115
	0700	50' x 60'	↓	4,500	.009		12	.19		12.19	13.50	
	0800	For single wall construction, deduct, minimum					.27			.27	.30	
	0900	Maximum					1.06			1.06	1.17	
	1000	For maximum resistance to atmosphere or cold, add				↓	.33			.33	.36	
	1100	For average shipping charges, add				Total	1,050			1,050	1,150	

130 250 | Integrated Ceilings

			DAILY OUTPUT	LABOR-HOURS	UNIT	MAT.	LABOR	EQUIP.	TOTAL	TOTAL INCL O&P	
251	0010	**INTEGRATED CEILINGS** Lighting, ventilating & acoustical									251
	0100	Luminaire, incl. connections, 5' x 5' modules, 50% lighted	L-3	90	.178	S.F.	3.53	5.25		8.78	12.05	
	0200	100% lighted	"	50	.320		5.25	9.45		14.70	20.50	
	0400	For ventilating capacity with perforations, add				↓					.27	
	0600	For supply air diffuser, add				Ea.					64	
	0700	Dimensionaire, 2' x 4' board system, see also div. 095-106	L-3	50	.320	L.F.	8.20	9.45		17.65	23.50	
	0900	Tile system	1 Carp	250	.032	S.F.	2.14	.87		3.01	3.72	
	1000	For air bar suspension, add, minimum									.61	
	1100	Maximum				↓					1.05	
	1200	For vaulted coffer, including fixture, stock, add				Ea.	80			80	88	
	1300	Custom, add, minimum									133	
	1400	Average									345	
	1500	Maximum				↓					575	
	1600	Grid sys. & ceil. tile only, flat profile, 5'x5' mod., 50% lghtd.										
	1620	Mineral fiber panels	1 Carp	700	.011	S.F.	3.23	.31		3.54	4.04	
	1630	Glass fiber panels	"	700	.011		3.47	.31		3.78	4.31	
	1640	100% lighted, add					1.68			1.68	1.85	
	1650	For supply air linear diffuser, add					.53			.53	.58	
	1660	For vaulted coffer, 5' x 5' modules, 50% lighted	1 Carp	493	.016	↓	4.12	.44		4.56	5.25	
	1800	Radiant hot water system with finished acoustic ceiling,										
	1810	not including supply piping. Heating only (gross S.F.)				S.F.						
	2100	Elementary schools, minimum									5.30	
	2200	Maximum									6.65	
	2400	High schools and colleges, minimum									4.75	
	2500	Maximum									6.65	
	2700	Libraries, minimum									4.80	
	2800	Maximum									5.75	
	3000	Hospitals, minimum									6.50	
	3100	Maximum									8.15	
	3300	Office buildings, minimum									4.60	
	3400	Maximum									6.15	
	3600	For combined heating and cooling, add, minimum									30%	
	3700	Maximum									40%	
	4000	Radiant electric ceiling board, strapped between joists	1 Elec	250	.032	↓	1.74	1.02		2.76	3.44	
	4100	2' x 4' heating panel for grid system, manila finish		25	.320	Ea.	40.50	10.20		50.70	60	
	4200	Textured epoxy finish		22	.364		41.50	11.60		53.10	63	
	4300	Vinyl finish		19	.421		44	13.45		57.45	68.50	
	4400	Hair cell, ABS plastic finish	↓	13	.615	↓	74	19.65		93.65	111	
	4500	2' x 4' alternate blank panel, for use with above										
	4600	Manila finish	1 Elec	50	.160	Ea.	9.70	5.10		14.80	18.30	
	4700	Textured epoxy finish		45	.178		21.50	5.65		27.15	32	
	4800	Vinyl finish		40	.200		23.50	6.40		29.90	35.50	
	4900	Hair cell, ABS plastic finish	↓	25	.320	↓	64	10.20		74.20	86	

130 300 | Special Purpose Rooms

			DAILY OUTPUT	LABOR-HOURS	UNIT	MAT.	LABOR	EQUIP.	TOTAL	TOTAL INCL O&P	
305	0010	**DARKROOMS** Shell, complete except for door, 64 S.F., 8' high	2 Carp	128	.125	SF Flr.	62	3.41		65.41	73.50	305
	0100	12' high	↓	64	.250	↓	75	6.80		81.80	93.50	

130 300 | Special Purpose Rooms

305			CREW	DAILY OUTPUT	LABOR-HOURS	UNIT	1999 BARE COSTS MAT.	LABOR	EQUIP.	TOTAL	TOTAL INCL O&P	
305	0500	120 S.F. floor, 8' high	2 Carp	120	.133	SF Flr.	46	3.64		49.64	56.50	305
	0600	12' high		60	.267		57	7.30		64.30	74	
	0800	240 S.F. floor, 8' high		120	.133		32	3.64		35.64	41	
	0900	12' high		60	.267		45	7.30		52.30	61	
	1200	Mini-cylindrical, revolving, unlined, 4' diameter		3.50	4.571	Ea.	3,175	125		3,300	3,675	
	1400	5'-6" diameter		2.50	6.400		5,000	175		5,175	5,775	
	1600	Add for lead lining, inner cylinder, 1/32" thick					1,275			1,275	1,425	
	1700	1/16" thick					1,500			1,500	1,650	
	1800	Add for lead lining, inner and outer cylinder, 1/32" thick					2,375			2,375	2,600	
	1900	1/16" thick					2,475			2,475	2,725	
	2000	For darkroom door, see division 114-768										

130 320 | Athletic Rooms

321			CREW	DAILY OUTPUT	LABOR-HOURS	UNIT	MAT.	LABOR	EQUIP.	TOTAL	TOTAL INCL O&P	
321	0010	**SPORT COURT** Floors, No. 2 & better maple, 25/32" thick				SF Flr.					6.30	321
	0100	Walls, laminated plastic bonded to galv. steel studs				SF Wall					8.45	
	0150	Laminated fiberglass surfacing, minimum									2.10	
	0180	Maximum									2.37	
	0300	Squash, regulation court in existing building, minimum				Court					14,500	
	0400	Maximum				"					27,100	
	0450	Rule of thumb for components:										
	0470	Walls	3 Carp	.15	160	Court	10,100	4,375		14,475	18,000	
	0500	Floor	"	.25	96		4,525	2,625		7,150	9,100	
	0550	Lighting	2 Elec	.60	26.667		1,425	850		2,275	2,850	
	0600	Handball, racquetball court in existing building, minimum	C-1	.20	160		1,750	4,125		5,875	8,425	
	0800	Maximum	"	.10	320		23,400	8,275		31,675	38,700	
	0900	Rule of thumb for components: walls	3 Carp	.12	200		9,225	5,450		14,675	18,700	
	1000	Floor		.25	96		4,800	2,625		7,425	9,400	
	1100	Ceiling		.33	72.727		2,000	1,975		3,975	5,325	
	1200	Lighting	2 Elec	.60	26.667		2,475	850		3,325	4,000	

130 340 | Audiometric Rooms

341			CREW	DAILY OUTPUT	LABOR-HOURS	UNIT	MAT.	LABOR	EQUIP.	TOTAL	TOTAL INCL O&P	
341	0010	**AUDIOMETRIC ROOMS** Under 500 S.F. surface	4 Carp	98	.327	SF Surf	39	8.90		47.90	57	341
	0100	Over 500 S.F. surface	"	120	.267	"	37	7.30		44.30	52.50	

130 360 | Clean Rooms

361			CREW	DAILY OUTPUT	LABOR-HOURS	UNIT	MAT.	LABOR	EQUIP.	TOTAL	TOTAL INCL O&P	
361	0010	**CLEAN ROOMS**										361
	1100	Clean room, soft wall, 12' x 12', Class 100	1 Carp	.18	44.444	Ea.	12,200	1,225		13,425	15,300	
	1110	Class 1,000		.18	44.444		9,450	1,225		10,675	12,300	
	1120	Class 10,000		.21	38.095		8,050	1,050		9,100	10,500	
	1130	Class 100,000		.21	38.095		8,050	1,050		9,100	10,500	
	2800	Ceiling grid support, slotted channel struts 4'-0" O.C., ea. way				S.F.					6.50	
	3000	Ceiling panel, vinyl coated foil on mineral substrate										
	3020	Sealed, non-perforated				S.F.					1.40	
	4000	Ceiling panel seal, silicone sealant, 150 L.F./gal.	1 Carp	150	.053	L.F.	.22	1.46		1.68	2.53	
	4100	Two sided adhesive tape	"	240	.033	"	.10	.91		1.01	1.54	
	4200	Clips, one per panel				Ea.	.85			.85	.94	
	6000	HEPA filter, 2'x4', 99.97% eff., 3" dp beveled frame (silicone seal)					260			260	286	
	6040	6" deep skirted frame (channel seal)					300			300	330	
	6100	99.99% efficient, 3" deep beveled frame (silicone seal)					280			280	310	
	6140	6" deep skirted frame (channel seal)					320			320	350	
	6200	99.999% efficient, 3" deep beveled frame (silicone seal)					320			320	350	
	6240	6" deep skirted frame (channel seal)					360			360	395	
	7000	Wall panel systems, including channel strut framing										
	7020	Polyester coated aluminum, particle board				S.F.					20	
	7100	Porcelain coated aluminum, particle board									35	

Important: See the Reference Section for critical supporting data - Reference Nos., Crews, & City Cost Indexes

13 SPECIAL CONSTRUCTION

130 360 | Clean Rooms

		CREW	DAILY OUTPUT	LABOR-HOURS	UNIT	1999 BARE COSTS				TOTAL INCL O&P	
						MAT.	LABOR	EQUIP.	TOTAL		
361	7400	Wall panel support, slotted channel struts, to 12' high				S.F.					18	**361**

130 380 | Cold Storage Rooms

		CREW	DAILY OUTPUT	LABOR-HOURS	UNIT	MAT.	LABOR	EQUIP.	TOTAL	TOTAL INCL O&P	
381	0010	**REFRIGERATION** Curbs, 12" high, 4" thick, concrete	2 Carp	58	.276	L.F.	3.02	7.55		10.57	15.15	**381**
	1000	Doors, see division 083-251										
	2400	Finishes, 2 coat portland cement plaster, 1/2" thick	1 Plas	48	.167	S.F.	.83	4.28		5.11	7.45	
	2500	For galvanized reinforcing mesh, add	1 Lath	335	.024		.56	.64		1.20	1.58	
	2700	3/16" thick latex cement	1 Plas	88	.091		1.42	2.34		3.76	5.15	
	2900	For glass cloth reinforced ceilings, add	"	450	.018		.34	.46		.80	1.07	
	3100	Fiberglass panels, 1/8" thick	1 Carp	149.45	.054		1.88	1.46		3.34	4.37	
	3200	Polystyrene, plastic finish ceiling, 1" thick		274	.029		1.74	.80		2.54	3.16	
	3400	2" thick		274	.029		1.97	.80		2.77	3.42	
	3500	4" thick		219	.037		2.20	1		3.20	3.99	
	3800	Floors, concrete, 4" thick	1 Cefi	93	.086		.82	2.25		3.07	4.25	
	3900	6" thick	"	85	.094		1.24	2.46		3.70	5.05	
	4000	Insulation, 1" to 6" thick, cork				B.F.	.80			.80	.88	
	4100	Urethane					.79			.79	.87	
	4300	Polystyrene, regular					.54			.54	.59	
	4400	Bead board					.41			.41	.45	
	4600	Installation of above, add per layer	2 Carp	657.60	.024	S.F.	.25	.66		.91	1.33	
	4700	Wall and ceiling juncture		298.90	.054	L.F.	1.28	1.46		2.74	3.71	
	4900	Partitions, galvanized sandwich panels, 4" thick, stock		219.20	.073	S.F.	5.40	1.99		7.39	9.10	
	5000	Aluminum or fiberglass		219.20	.073	"	5.95	1.99		7.94	9.65	
	5200	Prefab walk-in, 7'-6" high, aluminum, incl. door & floors,										
	5210	not incl. partitions or refrigeration, 6' x 6' O.D. nominal	2 Carp	54.80	.292	SF Flr.	95	7.95		102.95	118	
	5500	10' x 10' O.D. nominal		82.20	.195		76	5.30		81.30	92	
	5700	12' x 14' O.D. nominal		109.60	.146		68	3.99		71.99	81.50	
	5800	12' x 20' O.D. nominal		109.60	.146		60	3.99		63.99	72.50	
	6100	For 8'-6" high, add					5%					
	6300	Rule of thumb for complete units, w/o doors & refrigeration, cooler	2 Carp	146	.110		86	2.99		88.99	99	
	6400	Freezer		109.60	.146		102	3.99		105.99	118	
	6600	Shelving, plated or galvanized, steel wire type		360	.044	SF Hor.	7.60	1.21		8.81	10.25	
	6700	Slat shelf type		375	.043		9	1.16		10.16	11.75	
	6900	For stainless steel shelving, add					300%					
	7000	Vapor barrier, on wood walls	2 Carp	1,644	.010	S.F.	.11	.27		.38	.54	
	7200	On masonry walls	"	1,315	.012	"	.29	.33		.62	.84	
	7500	For air curtain doors, see division 083-722										

130 520 | Saunas

		CREW	DAILY OUTPUT	LABOR-HOURS	UNIT	MAT.	LABOR	EQUIP.	TOTAL	TOTAL INCL O&P	
521	0010	**SAUNA** Prefabricated, incl. heater & controls, 7' high, 6' x 4', C/C	L-7	2.20	12.727	Ea.	3,075	335		3,410	3,900	**521**
	0050	6' x 4', C/P		2	14		2,525	370		2,895	3,350	
	0400	6' x 5', C/C		2	14		3,200	370		3,570	4,100	
	0450	6' x 5', C/P		2	14		2,875	370		3,245	3,725	
	0600	6' x 6', C/C		1.80	15.556		3,750	410		4,160	4,775	
	0650	6' x 6', C/P		1.80	15.556		3,000	410		3,410	3,950	
	0800	6' x 9', C/C		1.60	17.500		4,675	460		5,135	5,875	
	0850	6' x 9', C/P		1.60	17.500		4,675	460		5,135	5,875	
	1000	8' x 12', C/C		1.10	25.455		6,300	670		6,970	8,000	
	1050	8' x 12', C/P		1.10	25.455		5,275	670		5,945	6,850	
	1200	8' x 8', C/C		1.40	20		4,925	525		5,450	6,250	
	1250	8' x 8', C/P		1.40	20		4,350	525		4,875	5,600	
	1400	8' x 10', C/C		1.20	23.333		5,400	615		6,015	6,875	
	1450	8' x 10', C/P		1.20	23.333		4,725	615		5,340	6,150	
	1600	10' x 12', C/C		1	28		7,100	735		7,835	8,950	
	1650	10' x 12', C/P		1	28		6,375	735		7,110	8,175	

SPECIAL CONSTRUCTION 13

130 520 | Saunas

			CREW	DAILY OUTPUT	LABOR-HOURS	UNIT	1999 BARE COSTS				TOTAL INCL O&P	
							MAT.	LABOR	EQUIP.	TOTAL		
521	1700	Door only, cedar, 2'x6', with tempered insulated glass window	2 Carp	3.40	4.706	Ea.	279	128		407	505	521
	1800	Prehung, incl. jambs, pulls & hardware	"	12	1.333		330	36.50		366.50	420	
	2500	Heaters only (incl. above), wall mounted, to 200 C.F.					460			460	510	
	2750	To 300 C.F.					490			490	535	
	3000	Floor standing, to 720 C.F., 10,000 watts, w/controls	1 Elec	3	2.667		955	85		1,040	1,175	
	3250	To 1,000 C.F., 16,000 watts	"	3	2.667		1,575	85		1,660	1,850	

130 540 | Steam Baths

			CREW	DAILY OUTPUT	LABOR-HOURS	UNIT	MAT.	LABOR	EQUIP.	TOTAL	TOTAL INCL O&P	
541	0010	STEAM BATH Heater, timer & head, single, to 140 C.F.	1 Plum	1.20	6.667	Ea.	720	217		937	1,125	541
	0500	To 300 C.F.		1.10	7.273		785	237		1,022	1,225	
	1000	Commercial size, with blow-down assembly, to 800 C.F.		.90	8.889		2,700	290		2,990	3,425	
	1500	To 2500 C.F.		.80	10		4,250	325		4,575	5,175	
	2000	Multiple, motels, apts., 2 baths, w/ blow-down assm., 500 C.F.	Q-1	1.30	12.308		2,875	360		3,235	3,725	
	2500	4 baths	"	.70	22.857		3,825	670		4,495	5,225	
	2700	Conversion unit for residential tub, including door					2,225			2,225	2,450	

130 560 | Vaults

			CREW	DAILY OUTPUT	LABOR-HOURS	UNIT	MAT.	LABOR	EQUIP.	TOTAL	TOTAL INCL O&P	
564	0010	VAULT FRONT See division 083-184										564

130 810 | Acoustical Enclosures

			CREW	DAILY OUTPUT	LABOR-HOURS	UNIT	MAT.	LABOR	EQUIP.	TOTAL	TOTAL INCL O&P	
811	0010	ACOUSTICAL Enclosure, 4" thick wall and ceiling panels										811
	0020	8# per S.F., up to 12' span	3 Carp	72	.333	SF Surf	25	9.10		34.10	41.50	
	0300	Better quality panels, 10.5# per S.F.		64	.375		28	10.25		38.25	47	
	0400	Reverb-chamber, 4" thick, parallel walls		60	.400		30	10.90		40.90	50	
	0600	Skewed wall, parallel roof, 4" thick panels		55	.436		30	11.90		41.90	52	
	0700	Skewed walls, skewed roof, 4" layers, 4" air space		48	.500		35.50	13.65		49.15	60.50	
	0900	Sound-absorbing panels, pntd mtl, 2'-6" x 8', under 1,000 S.F.		215	.112		9.05	3.05		12.10	14.75	
	1100	Over 1000 S.F.		240	.100		8.70	2.73		11.43	13.85	
	1200	Fabric faced		240	.100		7.25	2.73		9.98	12.30	
	1500	Flexible transparent curtain, clear	3 Shee	215	.112		5.60	3.54		9.14	11.65	
	1600	50% foam		215	.112		7.85	3.54		11.39	14.10	
	1700	75% foam		215	.112		7.85	3.54		11.39	14.10	
	1800	100% foam		215	.112		7.85	3.54		11.39	14.10	
	3100	Audio masking system, including speakers, amplification										
	3110	and signal generator										
	3200	Ceiling mounted, 5,000 S.F.	2 Elec	2,400	.007	S.F.	.96	.21		1.17	1.38	
	3300	10,000 S.F.		2,800	.006		.78	.18		.96	1.13	
	3400	Plenum mounted, 5,000 S.F.		3,800	.004		.81	.13		.94	1.09	
	3500	10,000 S.F.		4,400	.004		.54	.12		.66	.76	
813	0010	ANECHOIC CHAMBERS Standard units, 7' ceiling heights										813
	0100	Area for pricing is net inside dimensions										
	0300	200 cycles per second cutoff, 25 S.F. floor area				SF Flr.					1,375	
	0400	50 S.F.									1,100	
	0600	75 S.F.									1,050	
	0700	100 S.F.									1,025	
	0900	For 150 cycles per second cutoff, add					30%	30%				
	1000	For 100 cycles per second cutoff, add					45%	45%				
815	0010	MUSIC Practice room, modular, perforated steel, under 500 S.F.	2 Carp	70	.229	SF Surf	21.50	6.25		27.75	33.50	815
	0100	Over 500 S.F.	"	80	.200	"	16.30	5.45		21.75	26.50	

130 910 | Radiation Protection

			CREW	DAILY OUTPUT	LABOR-HOURS	UNIT	MAT.	LABOR	EQUIP.	TOTAL	TOTAL INCL O&P	
911	0010	SHIELDING LEAD										911
	0100	Laminated lead in wood doors, 1/16" thick, no hardware				S.F.	27			27	29.50	

13 SPECIAL CONSTRUCTION

130 910 | Radiation Protection

		CREW	DAILY OUTPUT	LABOR-HOURS	UNIT	1999 BARE COSTS				TOTAL INCL O&P
						MAT.	LABOR	EQUIP.	TOTAL	
911	0200	Lead lined door frame, not incl. hardware,									911
	0210	1/16" thick lead, butt prepared for hardware	1 Lath	2.40	3.333	Ea.	330	89.50		419.50	500
	0300	Lead sheets, 1/16" thick	2 Lath	135	.119	S.F.	4.75	3.18		7.93	9.95
	0400	1/8" thick		120	.133		10	3.57		13.57	16.35
	0500	Lead shielding, 1/4" thick		135	.119		10.60	3.18		13.78	16.40
	0550	1/2" thick		120	.133		21.50	3.57		25.07	29
	0600	Lead glass, 1/4" thick, 2.0 mm LE, 12" x 16"	2 Glaz	13	1.231	Ea.	256	32.50		288.50	330
	0700	24" x 36"		8	2		915	53		968	1,075
	0800	36" x 60"		2	8		1,850	213		2,063	2,350
	0850	Window frame with 1/16" lead and voice passage, 36" x 60"		2	8		410	213		623	775
	0870	24" x 36" frame		8	2		325	53		378	440
	0900	Lead gypsum board, 5/8" thick with 1/16" lead		160	.100	S.F.	4.95	2.66		7.61	9.50
	0910	1/8" lead **CN**		140	.114		10.50	3.04		13.54	16.20
	0930	1/32" lead	2 Lath	200	.080		2.90	2.14		5.04	6.40
	0950	Lead headed nails (average 1 lb. per sheet)				Lb.	5.25			5.25	5.80
	1000	Butt joints in 1/8" lead or thicker, 2" batten strip x 7' long	2 Lath	240	.067	Ea.	15	1.79		16.79	19.20
	1200	X-ray protection, average radiography or fluoroscopy									
	1210	room, up to 300 S.F. floor, 1/16" lead, minimum	2 Lath	.25	64	Total	4,600	1,725		6,325	7,625
	1500	Maximum, 7'-0" walls	"	.15	106	"	6,000	2,850		8,850	10,900
	1600	Deep therapy X-ray room, 250 KV capacity,									
	1800	up to 300 S.F. floor, 1/4" lead, minimum	2 Lath	.08	200	Total	15,500	5,350		20,850	25,100
	1900	Maximum, 7'-0" walls	"	.06	266	"	21,000	7,150		28,150	33,800
	2000	X-ray viewing panels, clear lead plastic									
	2010	7 mm thick, 0.3 mm LE, 2.3 lbs/S.F.	H-3	139	.115	S.F.	122	2.73		124.73	138
	2020	12 mm thick, 0.5 mm LE, 3.9 lbs/S.F.		82	.195		165	4.63		169.63	189
	2030	18 mm thick, 0.8mm LE, 5.9 lbs/S.F.		54	.296		179	7.05		186.05	208
	2040	22 mm thick, 1.0 mm LE, 7.2 lbs/S.F.		44	.364		183	8.65		191.65	214
	2050	35 mm thick, 1.5 mm LE, 11.5 lbs/S.F.		28	.571		205	13.55		218.55	247
	2060	46 mm thick, 2.0 mm LE, 15.0 lbs/S.F.		21	.762		270	18.10		288.10	325
	2090	For panels 12 S.F. to 48 S.F., add crating charge				Ea.					50
	4000	X-ray barriers, modular, panels mounted within framework for									
	4002	attaching to floor, wall or ceiling, upper portion is clear lead									
	4005	plastic window panels 48"H, lower portion is opaque leaded									
	4008	steel panels 36"H, structural supports not incl.									
	4010	1-section barrier, 36"W x 84"H overall									
	4020	0.5 mm LE panels	H-3	6.40	2.500	Ea.	2,750	59.50		2,809.50	3,125
	4030	0.8 mm LE panels		6.40	2.500		2,925	59.50		2,984.50	3,325
	4040	1.0 mm LE panels		5.33	3.002		3,025	71		3,096	3,425
	4050	1.5 mm LE panels		5.33	3.002		3,225	71		3,296	3,650
	4060	2-section barrier, 72"W x 84"H overall									
	4070	0.5 mm LE panels	H-3	4	4	Ea.	5,700	95		5,795	6,425
	4080	0.8 mm LE panels		4	4		6,025	95		6,120	6,775
	4090	1.0 mm LE panels		3.56	4.494		6,150	107		6,257	6,950
	5000	1.5 mm LE panels		3.20	5		6,650	119		6,769	7,500
	5010	3-section barrier, 108"W x 84"H overall									
	5020	0.5 mm LE panels	H-3	3.20	5	Ea.	8,150	119		8,269	9,150
	5030	0.8 mm LE panels		3.20	5		8,550	119		8,669	9,575
	5040	1.0 mm LE panels		2.67	5.993		8,750	142		8,892	9,850
	5050	1.5 mm LE panels		2.46	6.504		9,500	154		9,654	10,700
	7000	X-ray barriers, mobile, mounted within framework w/casters on									
	7005	bottom, clear lead plastic window panels on upper portion,									
	7010	opaque on lower, 30"W x 75"H overall, incl. framework									
	7020	24"H upper w/0.5 mm LE, 48"H lower w/0.8 mm LE	1 Carp	16	.500	Ea.	1,875	13.65		1,888.65	2,075
	7030	48"W x 75"H overall, incl. framework									
	7040	36"H upper w/0.5 mm LE, 36"H lower w/0.8 mm LE	1 Carp	16	.500	Ea.	3,300	13.65		3,313.65	3,650
	7050	36"H upper w/1.0 mm LE, 36"H lower w/1.5 mm LE	"	16	.500	"	4,000	13.65		4,013.65	4,425

SPECIAL CONSTRUCTION 13

130 910 | Radiation Protection

		CREW	DAILY OUTPUT	LABOR-HOURS	UNIT	1999 BARE COSTS				TOTAL INCL O&P	
						MAT.	LABOR	EQUIP.	TOTAL		
911	7060	72"W x 75"H overall, incl. framework									**911**
	7070	36"H upper w/0.5 mm LE, 36"H lower w/0.8 mm LE	1 Carp	16	.500	Ea.	3,925	13.65		3,938.65	4,325	
	7080	36"H upper w/1.0 mm LE, 36"H lower w/1.5 mm LE	"	16	.500	"	4,950	13.65		4,963.65	5,475	
912	0010	**SHIELDING, RADIO FREQUENCY**										**912**
	0020	Prefabricated or screen-type copper or steel, minimum	2 Carp	180	.089	SF Surf	24	2.43		26.43	30.50	
	0100	Average	↓	155	.103		26	2.82		28.82	33	
	0150	Maximum	↓	145	.110	↓	31	3.01		34.01	38.50	

131 200 | Pre-Eng. Structures

		CREW	DAILY OUTPUT	LABOR-HOURS	UNIT	1999 BARE COSTS				TOTAL INCL O&P	
						MAT.	LABOR	EQUIP.	TOTAL		
202	0010	**DOMES**										**202**
	0020	Revolv alum, elec drv for astronomy observation, shell only, stk units										
	0600	10'-0" diameter, 800#, dome	2 Carp	.25	64	Ea.	9,400	1,750		11,150	13,100	
	0700	Base		.67	23.881		3,250	650		3,900	4,600	
	0900	18'-0" diameter, 2,500#, dome		.17	94.118		27,500	2,575		30,075	34,400	
	1000	Base		.33	48.485		9,100	1,325		10,425	12,100	
	1200	24'-0" diameter, 4,500#, dome		.08	200		51,000	5,450		56,450	64,500	
	1300	Base	↓	.25	64	↓	16,400	1,750		18,150	20,800	
	1500	Bulk storage, shell only, dual radius hemispher. arch, steel										
	1600	framing, corrugated steel covering, 150' diameter	E-2	550	.102	SF Flr.	27	3.03	1.84	31.87	37	
	1700	400' diameter	"	720	.078		22	2.32	1.41	25.73	29.50	
	1800	Wood framing, wood decking, to 400' diameter	F-4	400	.120	↓	20	3.25	1.98	25.23	29	
	1900	Radial framed wood (2" x 6"), 1/2" thick										
	2000	plywood, asphalt shingles, 50' diameter	F-3	2,000	.020	SF Flr.	25	.55	.22	25.77	28.50	
	2100	60' diameter		1,900	.021		19	.58	.23	19.81	22	
	2200	72' diameter		1,800	.022		17	.62	.25	17.87	19.95	
	2300	116' diameter		1,730	.023		15	.64	.26	15.90	17.80	
	2400	150' diameter	↓	1,500	.027	↓	13	.74	.30	14.04	15.80	
203	0010	**GEODESIC DOME** Shell only, interlocking plywood panels [R131-310]										**203**
	0400	30' diameter	F-5	1.60	20	Ea.	9,950	555		10,505	11,800	
	0500	34' diameter		1.14	28.070		12,200	780		12,980	14,600	
	0600	39' diameter	↓	1	32		14,200	890		15,090	17,000	
	0700	45' diameter	F-3	1.13	35.556		17,100	985	395	18,480	20,800	
	0750	55' diameter		1	40		25,500	1,100	445	27,045	30,200	
	0800	60' diameter		1	40		33,200	1,100	445	34,745	38,700	
	0850	65' diameter	↓	.80	50	↓	39,400	1,375	555	41,330	46,100	
	1100	Aluminum panel, with 6" insulation										
	1200	100' diameter				SF Flr.					30	
	1300	500' diameter				"					25	
	1600	Aluminum framed, plexiglass closure panels										
	1700	40' diameter				SF Flr.					80	
	1800	200' diameter				"					70	
	2100	Aluminum framed, aluminum closure panels										
	2200	40' diameter				SF Flr.					50	
	2300	100' diameter									30	
	2400	200' diameter									25	
	2500	For VRP faced bonded fiberglass insulation, add									10	
	2700	Aluminum framed, fiberglass sandwich panel closure	↓			↓						

131 200 | Pre-Eng. Structures

			CREW	DAILY OUTPUT	LABOR-HOURS	UNIT	1999 BARE COSTS MAT.	LABOR	EQUIP.	TOTAL	TOTAL INCL O&P	
203	2800	6' diameter	R131-310	2 Carp	150	.107	SF Flr.	33	2.91		35.91	41	203
	2900	28' diameter		"	350	.046	"	25	1.25		26.25	29.50	
204	0010	**GARAGE COSTS**	R131-910										204
	0020	Public parking, average					Car					9,850	
	0100	See also division 171-410											
	0300	Residential, prefab shell, stock, wood, single car, minimum		2 Carp	1	16	Total	2,550	435		2,985	3,475	
	0350	Maximum			.67	23.881		5,575	650		6,225	7,150	
	0400	Two car, minimum			.67	23.881		3,650	650		4,300	5,025	
	0450	Maximum			.50	32		7,525	875		8,400	9,650	
205	0010	**SILOS** Concrete stave industrial, not incl. foundations, conical or		D-8	.11	363	Ea.	15,000	9,175		24,175	30,700	205
	0100	sloping bottoms, 12' diameter, 35' high											
	0200	16' diameter, 45' high			.08	500		20,000	12,600		32,600	41,600	
	0400	25' diameter, 75' high			.05	800		49,000	20,200		69,200	85,500	
	0500	Steel, factory fab., 30,000 gallon cap., painted, minimum		L-5	1	56		10,900	1,725	550	13,175	15,700	
	0700	Maximum			.50	112		16,900	3,450	1,100	21,450	26,000	
	0800	Epoxy lined, minimum			1	56		17,500	1,725	550	19,775	23,000	
	1000	Maximum			.50	112		21,800	3,450	1,100	26,350	31,400	
207	0010	**TENSION STRUCTURES** Rigid steel/alum. frame, vyl. coated polyester											207
	0100	fabric shell, 60' clear span, not incl. foundations or floors											
	0200	6,000 S.F.		B-41	1,000	.044	SF Flr.	10.20	.98	.18	11.36	12.95	
	0300	12,000 S.F.			1,100	.040		8.65	.89	.16	9.70	11.10	
	0400	80' clear span, 20,800 S.F.			1,220	.036		9.35	.80	.14	10.29	11.70	
	0410	100' clear span, 10,000 S.F.		L-5	2,175	.026		9.90	.79	.25	10.94	12.60	
	0430	26,000 S.F.			2,300	.024		8.85	.75	.24	9.84	11.35	
	0450	36,000 S.F.			2,500	.022		8.50	.69	.22	9.41	10.80	
	0460	120' clear span, 24,000 S.F.			3,000	.019		11.40	.57	.18	12.15	13.75	
	0470	150' clear span, 30,000 S.F.			6,000	.009		10.45	.29	.09	10.83	12.10	
	0480	200' clear span, 40,000 S.F.		E-6	8,000	.016		14.35	.48	.16	14.99	16.85	
	0500	For roll-up door, 12' x 14', add		L-2	1	16	Ea.	3,600	385		3,985	4,550	
	0600	For personnel doors, add, minimum					SF Flr.	5%					
	0700	Add, maximum						15%					
	0800	For site work, simple foundation, etc., add, minimum									1.25	1.95	
	0900	Add, maximum									2.75	3.05	

131 210 | Pre-Engineered Buildings

			CREW	DAILY OUTPUT	LABOR-HOURS	UNIT	MAT.	LABOR	EQUIP.	TOTAL	TOTAL INCL O&P	
211	0010	**HANGARS** Prefabricated steel T hangars, Galv. steel roof &											211
	0100	walls, incl. electric bi-folding doors, 4 or more units,											
	0110	not including floors or foundations, minimum		E-2	1,275	.044	SF Flr.	8.85	1.31	.79	10.95	12.90	
	0130	Maximum			1,063	.053		9.50	1.57	.95	12.02	14.25	
	0900	With bottom rolling doors, minimum			1,386	.040		8.05	1.20	.73	9.98	11.75	
	1000	Maximum			966	.058		9.50	1.73	1.05	12.28	14.65	
	1200	Alternate pricing method:											
	1300	Galv. roof and walls, electric bi-folding doors, minimum		E-2	1.06	52.830	Plane	10,600	1,575	955	13,130	15,400	
	1500	Maximum			.91	61.538		11,700	1,825	1,100	14,625	17,300	
	1600	With bottom rolling doors, minimum			1.25	44.800		8,400	1,325	810	10,535	12,500	
	1800	Maximum			.97	57.732		10,000	1,725	1,050	12,775	15,200	
	2000	Circular type, prefab., steel frame, plastic skin, electric											
	2010	door, including foundations, 80' diameter,											
	2020	for up to 5 light planes, minimum		E-2	.50	112	Total	81,000	3,325	2,025	86,350	97,000	
	2200	Maximum		"	.25	224	"	91,000	6,675	4,050	101,725	116,500	
212	0010	**PRE-ENGINEERED STEEL BUILDINGS**	R131-210										212
	0100	Clear span rigid frame, 26 ga. colored roofing and siding											
	0200	30' to 40' wide, 10' eave height		E-2	1,800	.031	SF Flr.	4.36	.93	.56	5.85	7.05	
	0300	14' eave height			1,800	.031		4.42	.93	.56	5.91	7.10	

	**131 210	Pre-Engineered Buildings**	CREW	DAILY OUTPUT	LABOR-HOURS	UNIT	1999 BARE COSTS				TOTAL INCL O&P	
						MAT.	LABOR	EQUIP.	TOTAL		
212 0400	16' eave height R131 -210	E-2	1,400	.040	SF Flr.	4.77	1.19	.72	6.68	8.15	212
0500	20' eave height		1,000	.056		5.20	1.67	1.01	7.88	9.75	
0600	24' eave height		1,000	.056		6.30	1.67	1.01	8.98	11	
0700	50' to 100' wide, 10' eave height		1,800	.031		4.05	.93	.56	5.54	6.70	
0800	14' eave height		1,800	.031		4.04	.93	.56	5.53	6.70	
0900	16' eave height		1,400	.040		4.35	1.19	.72	6.26	7.70	
1000	20' eave height		1,000	.056		4.61	1.67	1.01	7.29	9.10	
1100	24' eave height		1,000	.056		5.25	1.67	1.01	7.93	9.85	
1200	Clear span tapered beam frame, 26 ga. colored roofing and siding										
1300	30' wide, 10' eave height	E-2	1,800	.031	SF Flr.	4.77	.93	.56	6.26	7.50	
1400	14' eave height		1,800	.031		5.40	.93	.56	6.89	8.15	
1500	16' eave height		1,400	.040		5.85	1.19	.72	7.76	9.30	
1600	20' eave height		1,000	.056		6.55	1.67	1.01	9.23	11.25	
1700	40' wide, 10' eave height		1,800	.031		4.39	.93	.56	5.88	7.05	
1800	14' eave height		1,800	.031		4.78	.93	.56	6.27	7.50	
1900	16' eave height		1,400	.040		5	1.19	.72	6.91	8.40	
2000	20' eave height		1,000	.056		5.60	1.67	1.01	8.28	10.20	
2100	50' to 80' wide, 10' eave height		1,800	.031		4.42	.93	.56	5.91	7.10	
2200	14' eave height		1,800	.031		4.50	.93	.56	5.99	7.20	
2300	16' eave height		1,400	.040		4.69	1.19	.72	6.60	8.05	
2400	20' eave height		1,000	.056		5.55	1.67	1.01	8.23	10.20	
2500	Single post 2-span frame, 26 ga. colored roofing and siding										
2600	80' wide, 14' eave height	E-2	1,800	.031	SF Flr.	3.71	.93	.56	5.20	6.30	
2700	16' eave height		1,400	.040		3.96	1.19	.72	5.87	7.25	
2800	20' eave height		1,000	.056		4.25	1.67	1.01	6.93	8.70	
2900	24' eave height		1,000	.056		4.65	1.67	1.01	7.33	9.15	
3000	100' wide, 14' eave height		1,800	.031		3.45	.93	.56	4.94	6.05	
3100	16' eave height		1,400	.040		3.83	1.19	.72	5.74	7.10	
3200	20' eave height		1,000	.056		4.23	1.67	1.01	6.91	8.70	
3300	24' eave height		1,000	.056		4.45	1.67	1.01	7.13	8.95	
3400	120' wide, 14' eave height		1,800	.031		3.51	.93	.56	5	6.10	
3500	16' eave height		1,400	.040		3.64	1.19	.72	5.55	6.90	
3600	20' eave height		1,000	.056		3.89	1.67	1.01	6.57	8.30	
3700	24' eave height		1,000	.056		4.32	1.67	1.01	7	8.80	
3800	Double post 3-span frame, 26 ga. colored roofing and siding										
3900	150' wide, 14' eave height	E-2	1,800	.031	SF Flr.	3.28	.93	.56	4.77	5.85	
4000	16' eave height		1,400	.040		3.24	1.19	.72	5.15	6.45	
4100	20' eave height		1,000	.056		3.42	1.67	1.01	6.10	7.80	
4200	24' eave height		1,000	.056		4.33	1.67	1.01	7.01	8.80	
4300	Triple post 4-span frame, 26 ga. colored roofing and siding										
4400	160' wide, 14' eave height	E-2	1,800	.031	SF Flr.	3.02	.93	.56	4.51	5.55	
4500	16' eave height		1,400	.040		3.14	1.19	.72	5.05	6.35	
4600	20' eave height		1,000	.056		3.42	1.67	1.01	6.10	7.80	
4700	24' eave height		1,000	.056		4.16	1.67	1.01	6.84	8.60	
4800	200' wide, 14' eave height		1,800	.031		3.04	.93	.56	4.53	5.60	
4900	16' eave height		1,400	.040		3.21	1.19	.72	5.12	6.40	
5000	20' eave height		1,000	.056		3.42	1.67	1.01	6.10	7.80	
5100	24' eave height		1,000	.056		3.74	1.67	1.01	6.42	8.15	
5200	Accessory items: add to the basic building cost above										
5250	Eave overhang, 2' wide, 26 ga., with soffit	E-2	360	.156	L.F.	8.45	4.63	2.81	15.89	20.50	
5300	4' wide, without soffit		300	.187		11.25	5.55	3.37	20.17	26	
5350	With soffit		250	.224		14.30	6.65	4.05	25	32	
5400	6' wide, without soffit		250	.224		17.30	6.65	4.05	28	35	
5450	With soffit		200	.280		23	8.35	5.05	36.40	45.50	
5500	Entrance canopy, incl. frame, 4' x 4'		25	2.240	Ea.	146	66.50	40.50	253	320	
5550	4' x 8'		19	2.947	"	258	87.50	53.50	399	495	

T3 SPECIAL CONSTRUCTION

Important: See the Reference Section for critical supporting data - Reference Nos., Crews, & City Cost Indexes

				DAILY	LABOR-		1999 BARE COSTS				TOTAL
	131 210	**Pre-Engineered Buildings**	CREW	OUTPUT	HOURS	UNIT	MAT.	LABOR	EQUIP.	TOTAL	INCL O&P
212	5600	End wall roof overhang, 4' wide, without soffit	E-2	850	.066	L.F.	8.35	1.96	1.19	11.50	13.90	**212**
	5650	With soffit	↓	500	.112	"	10.80	3.33	2.02	16.15	20
	5700	Doors, H.M. self-framing, incl. butts, lockset and trim									
	5750	Single leaf, 3070 (3' x 7'), economy	2 Sswk	5	3.200	Opng.	266	98		364	475
	5800	Deluxe		4	4		315	122		437	570
	5825	Glazed		4	4		355	122		477	615
	5850	3670 (3'-6" x 7')		4	4		300	122		422	555
	5900	4070 (4' x 7')		3	5.333		320	163		483	655
	5950	Double leaf, 6070 (6' x 7')		2	8		445	245		690	940
	6000	Glazed		2	8		625	245		870	1,125
	6050	Framing only, for openings, 3' x 7'	E-2	25	2.240		83.50	66.50	40.50	190.50	254
	6100	10' x 10'		21	2.667		275	79.50	48	402.50	495
	6150	For windows below, 2020 (2' x 2')		25	2.240		88.50	66.50	40.50	195.50	259
	6200	4030 (4' x 3')	↓	22	2.545	↓	108	76	46	230	305
	6250	Flashings, 26 ga., corner or eave, painted	2 Sswk	240	.067	L.F.	1.97	2.04		4.01	5.90
	6300	Galvanized		240	.067		1.72	2.04		3.76	5.65
	6350	Rake flashing, painted		240	.067		2.14	2.04		4.18	6.10
	6400	Galvanized		240	.067		1.91	2.04		3.95	5.85
	6450	Ridge flashing, 18" wide, painted		240	.067		2.30	2.04		4.34	6.25
	6500	Galvanized		240	.067		2.05	2.04		4.09	6
	6550	Gutter, eave type, 26 ga., painted		320	.050		2.22	1.53		3.75	5.25
	6600	Galvanized		320	.050		2.06	1.53		3.59	5.05
	6650	Valley type, between buildings, painted		120	.133		3.71	4.08		7.79	11.60
	6700	Galvanized	↓	120	.133	↓	3.40	4.08		7.48	11.25
	6750	Insulation, rated .6 lb density, vinyl faced									
	6800	1-1/2" thick, R5	2 Carp	2,300	.007	S.F.	.17	.19		.36	.49
	6850	3" thick, R10		2,300	.007		.27	.19		.46	.60
	6900	4" thick, R13		2,300	.007		.33	.19		.52	.67
	6950	Foil/scrim/kraft (FSK) faced, 1-1/2" thick, R5		2,300	.007		.20	.19		.39	.52
	7000	2" thick, R6		2,300	.007		.24	.19		.43	.56
	7050	3" thick, R10		2,300	.007		.32	.19		.51	.65
	7100	4" thick, R13		2,300	.007		.39	.19		.58	.73
	7150	Metalized polyester/scrim/kraft (PSK) facing, 1-1/2" thick, R5		2,300	.007		.30	.19		.49	.63
	7200	2" thick, R6		2,300	.007		.35	.19		.54	.69
	7250	3" thick, R11		2,300	.007		.43	.19		.62	.77
	7300	4" thick, R13		2,300	.007		.50	.19		.69	.85
	7350	Vinyl/scrim/foil (VSF), 1-1/2" thick, R5		2,300	.007		.29	.19		.48	.62
	7400	2" thick, R6		2,300	.007		.33	.19		.52	.66
	7450	3" thick, R10		2,300	.007		.42	.19		.61	.76
	7500	4" thick, R13	↓	2,300	.007		.47	.19		.66	.82
	7550	For heavy duty vinyl and foil, add					.07			.07	.08
	7600	For .75 lb. density, add				↓	.12			.12	.13
	7650	Sash, single slide, glazed, with screens, 2020 (2'x 2')	E-1	22	1.091	Opng.	57	33	3.74	93.74	124
	7700	3030 (3' x 3')		14	1.714		129	51.50	5.90	186.40	237
	7750	4030 (4' x 3')		13	1.846		171	55.50	6.35	232.85	292
	7800	6040 (6' x 4')		12	2		345	60.50	6.85	412.35	490
	7850	Double slide sash, 3030 (3' x 3')		14	1.714		135	51.50	5.90	192.40	244
	7900	6040 (6' x 4')		12	2		360	60.50	6.85	427.35	510
	7950	Fixed glass, no screens, 3030 (3' x 3')		14	1.714		124	51.50	5.90	181.40	232
	8000	6040 (6' x 4')		12	2		330	60.50	6.85	397.35	480
	8050	Prefinished storm sash, 3030 (3' x 3')	↓	70	.343	↓	54	10.35	1.18	65.53	78.50
	8100	Siding and roofing, see division 074									
	8200	Skylight, fiberglass panels, to 30 S.F.	E-1	10	2.400	Ea.	86	72.50	8.25	166.75	230
	8250	Larger sizes, add for excess over 30 S.F.	"	300	.080	S.F.	2.87	2.41	.27	5.55	7.65
	8300	Roof vents, circular with damper, birdscreen									
	8350	and operator hardware, painted	↓								

R131
-210

SPECIAL CONSTRUCTION 13

	131 210 | **Pre-Engineered Buildings**	CREW	DAILY OUTPUT	LABOR-HOURS	UNIT	1999 BARE COSTS MAT.	LABOR	EQUIP.	TOTAL	TOTAL INCL O&P	
212	8400	26 ga., 12" diameter	1 Sswk	4	2	Ea.	69	61		130	188	**212**
	8450	20" diameter		3	2.667		142	81.50		223.50	305	
	8500	24 ga., 24" diameter		2	4		281	122		403	535	
	8550	Galvanized		2	4		236	122		358	485	
	8600	Continuous, 26 ga., 10' long, 9" wide	2 Sswk	4	4		207	122		329	450	
	8650	12" wide	"	4	4		248	122		370	495	

R131 -210

	131 220 | **Metal Building Systems**	CREW	DAILY OUTPUT	LABOR-HOURS	UNIT	MAT.	LABOR	EQUIP.	TOTAL	TOTAL INCL O&P	
221	0010	**SHELTERS** Aluminum frame, acrylic glazing, 3' x 9' x 8' high	2 Sswk	1.14	14.035	Ea.	2,550	430		2,980	3,575	**221**
	0100	9' x 12' x 8' high	"	.73	21.918	"	4,250	670		4,920	5,900	

	131 230 | **Greenhouses**	CREW	DAILY OUTPUT	LABOR-HOURS	UNIT	MAT.	LABOR	EQUIP.	TOTAL	TOTAL INCL O&P	
231	0010	**GREENHOUSE** Shell only, stock units, not incl. 2' stub walls,										**231**
	0020	foundation, floors, heat or compartments										
	0300	Residential type, free standing, 8'-6" long x 7'-6" wide	2 Carp	59	.271	SF Flr.	35	7.40		42.40	50	
	0400	10'-6" wide		85	.188		27	5.15		32.15	37.50	
	0600	13'-6" wide		108	.148		24	4.04		28.04	33	
	0700	17'-0" wide		160	.100		27	2.73		29.73	34	
	0900	Lean-to type, 3'-10" wide		34	.471		31	12.85		43.85	54	
	1000	6'-10" wide		58	.276		24	7.55		31.55	38.50	
	1500	Commercial, custom, truss frame, incl. equip., plumbing, elec.,										
	1510	benches and controls, under 2,000 S.F., minimum				SF Flr.					30	
	1550	Maximum									38	
	1700	Over 5,000 S.F., minimum									23	
	1750	Maximum									30	
	2000	Institutional, custom, rigid frame, including compartments and										
	2010	multi-controls, under 500 S.F., minimum				SF Flr.					77	
	2050	Maximum									99	
	2150	Over 2,000 S.F., minimum									39	
	2200	Maximum									65	
	2400	Concealed rigid frame, under 500 S.F., minimum									90	
	2450	Maximum									110	
	2550	Over 2,000 S.F., minimum									68	
	2600	Maximum									79	
	2800	Lean-to type, under 500 S.F., minimum									80	
	2850	Maximum									120	
	3000	Over 2,000 S.F., minimum									44	
	3050	Maximum									73	
	3600	For 1/4" clear plate glass, add				SF Surf	1.40			1.40	1.54	
	3700	For 1/4" tempered glass, add				"	3.15			3.15	3.47	
	3900	For cooling, add, minimum				SF Flr.	2.10			2.10	2.31	
	4000	Maximum					5.20			5.20	5.70	
	4200	For heaters, 13.6 MBH, add					4			4	4.40	
	4300	60 MBH, add					1.50			1.50	1.65	
	4500	For benches, 2' x 3'-6", add				SF Hor.	17.90			17.90	19.70	
	4600	3' x 10', add				S.F.	9.70			9.70	10.65	
	4800	For controls, add, minimum				Total	1,800			1,800	1,975	
	4900	Maximum				"	10,700			10,700	11,800	
	5100	For humidification equipment, add				M.C.F.	4.60			4.60	5.05	
	5200	For vinyl shading, add				S.F.	.97			.97	1.07	
	6000	Geodesic hemisphere, 1/8" plexiglass glazing										
	6050	8' diameter	2 Carp	2	8	Ea.	2,025	218		2,243	2,575	
	6150	24' diameter		.35	45.714		10,400	1,250		11,650	13,400	
	6250	48' diameter		.20	80		28,200	2,175		30,375	34,400	

13 SPECIAL CONSTRUCTION

Important: See the Reference Section for critical supporting data - Reference Nos., Crews, & City Cost Indexes

131 | Pre-Eng. Structures, Aquatic Facilities and Ice Rinks

		131 240	Portable Buildings	CREW	DAILY OUTPUT	LABOR-HOURS	UNIT	MAT.	LABOR	EQUIP.	TOTAL	TOTAL INCL O&P	
									1999 BARE COSTS				
242	0010	**COMFORT STATIONS** Prefab., stock, w/doors, windows & fixt.											242
	0100	Not incl. interior finish or electrical											
	0300	Mobile, on steel frame, minimum					S.F.	37.50			37.50	41	
	0350	Maximum						63.50			63.50	70	
	0400	Permanent, including concrete slab, minimum	B-12J	50	.320		138	8.55	12.75	159.30	179	
	0500	Maximum	"	43	.372		200	9.95	14.80	224.75	251	
	0600	Alternate pricing method, mobile, minimum				Fixture	1,725			1,725	1,900	
	0650	Maximum					2,575			2,575	2,825	
	0700	Permanent, minimum	B-12J	.70	22.857		9,900	610	910	11,420	12,800	
	0750	Maximum	"	.50	32		16,500	855	1,275	18,630	20,900	
244	0010	**GARDEN HOUSE** Prefab wood, no floors or foundations										244
	0100	32 to 200 S.F., minimum	2 Carp	200	.080	SF Flr.	13.45	2.18		15.63	18.25	
	0300	Maximum	"	48	.333	"	24.50	9.10		33.60	41.50	
245	0010	**KIOSKS** Round, 5' diameter, 8' high, 1/4" fiberglass wall				Ea.	5,500			5,500	6,050	245
	0100	1" insulated double wall, fiberglass					6,250			6,250	6,875	
	0500	Rectangular, 5' x 9', 7'-6" high, 1/4" fiberglass wall					8,000			8,000	8,800	
	0600	1" insulated double wall, fiberglass					9,500			9,500	10,500	
247	0010	**PORTABLE BOOTHS** Prefab aluminum with doors, windows, ext. roof										247
	0100	lights wiring & insulation, 15 S.F. building, O.D., painted, minimum				S.F.	255			255	281	
	0300	30 S.F. building, minimum					175			175	193	
	0400	50 S.F. building, minimum					130			130	143	
	0600	80 S.F. building, minimum					105			105	116	
	0700	100 S.F. building, minimum					97			97	107	
	0900	Acoustical booth, 27 Db @ 1,000 Hz, 15 S.F. floor				Ea.	2,875			2,875	3,175	
	1000	7' x 7'-6", including light & ventilation					5,900			5,900	6,500	
	1200	Ticket booth, galv. steel, not incl. foundations., 4' x 4'					4,425			4,425	4,875	
	1300	4' x 6'					5,175			5,175	5,700	

		131 250	Grandstands/Bleachers	CREW	DAILY OUTPUT	LABOR-HOURS	UNIT	MAT.	LABOR	EQUIP.	TOTAL	TOTAL INCL O&P	
251	0010	**GRANDSTANDS** Permanent, municipal, including foundation										251
	0050	Steel understructure w/aluminum closed deck, minimum				Seat					130	
	0100	Maximum									225	
	0300	Steel, minimum									25	
	0400	Maximum									80	
	0600	Aluminum, extruded, stock design, minimum									70	
	0700	Maximum									115	
	0900	Composite, steel, wood and plastic, stock design, minimum									30	
	1000	Maximum									90	

		131 520	Swimming Pools	CREW	DAILY OUTPUT	LABOR-HOURS	UNIT	MAT.	LABOR	EQUIP.	TOTAL	TOTAL INCL O&P	
521	0010	**SWIMMING POOL ENCLOSURE** Translucent, free standing,										521
	0020	not including foundations, heat or light										
	0200	Economy, minimum	2 Carp	200	.080	SF Hor.	10	2.18		12.18	14.45	
	0300	Maximum		100	.160		21	4.37		25.37	30	
	0400	Deluxe, minimum		100	.160		23	4.37		27.37	32.50	
	0600	Maximum		70	.229		250	6.25		256.25	285	
	0700	For motorized roof, 40% opening, solid roof, add					3			3	3.30	
	0800	Skylight type roof, add					4.50			4.50	4.95	
	0900	Air-inflated, including blowers and heaters, minimum									3.50	
	1000	Maximum									6.70	
523	0010	**SWIMMING POOL EQUIPMENT** Diving stand, stainless steel, 3 meter	2 Carp	.40	40	Ea.	4,725	1,100		5,825	6,900	523
	0300	1 meter		2.70	5.926		2,650	162		2,812	3,175	
	0600	Diving boards, 16' long, aluminum		2.70	5.926		1,250	162		1,412	1,625	
	0700	Fiberglass		2.70	5.926		900	162		1,062	1,250	

SPECIAL CONSTRUCTION **13**

131 520	Swimming Pools	CREW	DAILY OUTPUT	LABOR-HOURS	UNIT	1999 BARE COSTS				TOTAL INCL O&P
						MAT.	LABOR	EQUIP.	TOTAL	
523 0900	Filter system, sand or diatomite type, incl. pump, 6,000 gal./hr.	2 Plum	1.80	8.889	Total	1,000	290		1,290	1,550	**523**
1020	Add for chlorination system, 800 S.F. pool		3	5.333	Ea.	270	174		444	560
1040	5,000 S.F. pool		3	5.333	"	560	174		734	880
1100	Gutter system, stainless steel, with grating, stock,									
1110	contains supply and drainage system	E-1	20	1.200	L.F.	160	36	4.12	200.12	244
1120	Integral gutter and 5' high wall system, stainless steel	"	10	2.400	"	240	72.50	8.25	320.75	400
1200	Ladders, heavy duty, stainless steel, 2 tread	2 Carp	7	2.286	Ea.	251	62.50		313.50	375
1500	4 tread		6	2.667		340	73		413	490
1800	Lifeguard chair, stainless steel, fixed		2.70	5.926		1,300	162		1,462	1,675
1900	Portable					2,000			2,000	2,200
2100	Lights, underwater, 12 volt, with transformer, 300 watt	1 Elec	.40	20		150	640		790	1,125
2200	110 volt, 500 watt, standard		.40	20		140	640		780	1,100
2400	Low water cutoff type		.40	20		90	640		730	1,050
2800	Heaters, see division 155-150									
3000	Pool covers, reinforced vinyl, material only				S.F.	.35			.35	.39
3050	Automatic, electric									9.65
3100	Vinyl water tube, material only, minimum					.25			.25	.28
3200	Maximum					.45			.45	.50
3250	Sealed air bubble polyethylene solar blanket					.17			.17	.19
3300	Slides, tubular, fiberglass, aluminum handrails & ladder, 5'-0", straight	2 Carp	1.60	10	Ea.	1,025	273		1,298	1,550
3320	8'-0", curved		3	5.333		9,600	146		9,746	10,800
3400	10'-0", curved		1	16		12,000	435		12,435	13,900
3420	12'-0", straight with platform		1.20	13.333		950	365		1,315	1,625
4500	Hydraulic lift, movable pool bottom, single ram									
4520	Under 1,000 S.F. area	L-9	.03	1,200	Ea.	80,000	29,700		109,700	136,500
4600	Four ram lift, over 1,000 S.F.	"	.02	1,800		96,500	44,600		141,100	179,000
5000	Removable access ramp, stainless steel	2 Clab	2	8		5,800	172		5,972	6,650
5500	Removable stairs, stainless steel, collapsible	"	2	8		3,000	172		3,172	3,575
525 0010	**SWIMMING POOLS** Residential in-ground, vinyl lined, concrete sides										**525**
0020	Sides including equipment, sand bottom	B-52	300	.187	SF Surf	10.40	4.71	1.44	16.55	20.50
0100	Metal or polystyrene sides [R131-520]	B-14	410	.117		8.70	2.66	.53	11.89	14.30
0200	Add for vermiculite bottom					.67			.67	.74
0500	Gunite bottom and sides, white plaster finish									
0600	12' x 30' pool	B-52	145	.386	SF Surf	17.25	9.75	2.98	29.98	37.50
0720	16' x 32' pool		155	.361		15.50	9.10	2.79	27.39	34.50
0750	20' x 40' pool		250	.224		13.85	5.65	1.73	21.23	26
0810	Concrete bottom and sides, tile finish									
0820	12' x 30' pool	B-52	80	.700	SF Surf	17.40	17.65	5.40	40.45	52.50
0830	16' x 32' pool		95	.589		14.40	14.85	4.55	33.80	44.50
0840	20' x 40' pool		130	.431		11.45	10.85	3.33	25.63	33.50
1100	Motel, gunite with plaster finish, incl. medium									
1150	capacity filtration & chlorination	B-52	115	.487	SF Surf	21	12.30	3.76	37.06	47
1200	Municipal, gunite with plaster finish, incl. high									
1250	capacity filtration & chlorination	B-52	100	.560	SF Surf	27.50	14.10	4.32	45.92	57
1350	Add for formed gutters				L.F.	47.50			47.50	52.50
1360	Add for stainless steel gutters				"	141			141	155
1600	For water heating system, see division 155-150									
1700	Filtration and deck equipment only, as % of total				Total				20%	20%
1800	Deck equipment, rule of thumb, 20' x 40' pool				SF Pool					1.30
1900	5000 S.F. pool				"					1.90
3000	Painting pools, preparation + 3 coats, 20' x 40' pool, epoxy	2 Pord	.33	48.485	Total	595	1,200		1,795	2,500
3100	Rubber base paint, 18 gallons	"	.33	48.485		450	1,200		1,650	2,350
3500	42' x 82' pool, 75 gallons, epoxy paint	3 Pord	.14	171		2,500	4,275		6,775	9,275
3600	Rubber base paint	"	.14	171		1,925	4,275		6,200	8,650

13 SPECIAL CONSTRUCTION

131 | Pre-Eng. Structures, Aquatic Facilities and Ice Rinks

131 580	Therapeutic Pools	CREW	DAILY OUTPUT	LABOR-HOURS	UNIT	1999 BARE COSTS				TOTAL INCL O&P	
						MAT.	LABOR	EQUIP.	TOTAL		
584	0010	THERAPEUTIC POOLS See division 152-104										584

131 600	Ice Rinks	CREW	DAILY OUTPUT	LABOR-HOURS	UNIT	1999 BARE COSTS				TOTAL INCL O&P	
						MAT.	LABOR	EQUIP.	TOTAL		
601	0010	ICE SKATING Equipment incl. refrigeration and plumbing, not										601
	0020	including building or slab, 85' x 200' rink										
	0300	55° system, 5 mos., 100 ton				Total					325,000	
	0700	90° system, 12 mos., 135 ton				"					350,000	
	1000	Dasher boards, 1/2" H.D. polyethylene faced steel frame, 3' acrylic										
	1020	screen at sides, 5' acrylic ends, 85' x 200'	F-5	.06	533	Ea.	115,000	14,800		129,800	150,000	
	1100	Fiberglass & aluminum construction, same sides and ends	"	.06	533		125,000	14,800		139,800	161,000	
	1200	Subsoil heating system (recycled from compressor), 85' x 200'	Q-7	.27	118		18,500	3,700		22,200	26,000	
	1300	Subsoil insulation, 2 lb. polystyrene with vapor barrier, 85' x 200'	2 Carp	.14	114	▼	24,000	3,125		27,125	31,300	

132 | Tanks, Tank Covers, Filtration Equipment

132 050	Ground Storage Tanks	CREW	DAILY OUTPUT	LABOR-HOURS	UNIT	1999 BARE COSTS				TOTAL INCL O&P	
						MAT.	LABOR	EQUIP.	TOTAL		
051	0010	TANKS Not incl. pipe or pumps, prestress conc., 250,000 gal.				Ea.					275,000	051
	0100	500,000 gallons									375,000	
	0300	1,000,000 gallons									530,000	
	0400	2,000,000 gallons									800,000	
	0600	4,000,000 gallons									1,270,000	
	0700	6,000,000 gallons									1,730,000	
	0750	8,000,000 gallons									2,200,000	
	0800	10,000,000 gallons									2,650,000	
	0900	Steel, ground level, not incl. foundations, 100,000 gallons									96,000	
	1000	250,000 gallons									135,000	
	1200	500,000 gallons									205,000	
	1250	750,000 gallons									245,000	
	1300	1,000,000 gallons									330,000	
	1500	2,000,000 gallons									605,000	
	1600	4,000,000 gallons									940,000	
	1800	6,000,000 gallons									1,300,000	
	1850	8,000,000 gallons									1,700,000	
	1900	10,000,000 gallons				▼					2,607,000	
	2100	Steel standpipes, 100' to overflow, no foundations										
	2200	500,000 gallons				Ea.					270,000	
	2400	750,000 gallons									330,000	
	2500	1,000,000 gallons									400,000	
	2700	1,500,000 gallons									560,000	
	2800	2,000,000 gallons				▼					690,000	
	3000	Steel, storage, above ground, including supports, coating,										
	3020	fittings, not including mat, pumps or piping										
	3040	Single wall, interior, 275 gallon capacity	Q-5	5	3.200	Ea.	228	94.50		322.50	395	
	3060	550 gallon capacity	"	2.70	5.926		1,150	175		1,325	1,550	
	3080	1,000 gallon capacity	Q-7	2.40	13.333		1,475	415		1,890	2,250	
	3100	1,500 gallon capacity		4.75	6.737		1,750	210		1,960	2,250	
	3120	2,000 gallon capacity		4.60	6.957		1,850	217		2,067	2,350	
	3140	5,000 gallon capacity	▼	3.20	10	▼	4,200	310		4,510	5,100	

132 050	Ground Storage Tanks	CREW	DAILY OUTPUT	LABOR-HOURS	UNIT	1999 BARE COSTS				TOTAL INCL O&P	
						MAT.	LABOR	EQUIP.	TOTAL		
051	3150	10,000 gallon capacity	Q-7	2	16	Ea.	6,900	500		7,400	8,350	**051**
	3160	15,000 gallon capacity		1.70	18.824		9,650	590		10,240	11,500	
	3170	20,000 gallon capacity		1.45	22.069		11,300	690		11,990	13,500	
	3180	25,000 gallon capacity		1.30	24.615		14,100	770		14,870	16,700	
	3190	30,000 gallon capacity		1.10	29.091		17,000	910		17,910	20,100	
	3320	Double wall, 500 gallon capacity	Q-5	2.40	6.667		2,900	197		3,097	3,500	
	3330	2000 gallon capacity	Q-7	4.15	7.711		4,450	241		4,691	5,275	
	3340	4000 gallon capacity		3.60	8.889		7,925	278		8,203	9,150	
	3350	6000 gallon capacity		2.40	13.333		9,425	415		9,840	11,000	
	3360	8000 gallon capacity		2	16		10,800	500		11,300	12,700	
	3370	10000 gallon capacity		1.80	17.778		11,900	555		12,455	13,900	
	3380	15000 gallon capacity		1.50	21.333		17,700	665		18,365	20,500	
	3390	20000 gallon capacity		1.30	24.615		20,200	770		20,970	23,400	
	3400	25000 gallon capacity		1.15	27.826		24,500	870		25,370	28,300	
	3410	30000 gallon capacity		1	32		26,900	1,000		27,900	31,100	
	4000	Fixed roof oil storage tanks, steel, (1 BBL=42 GAL w/ fdn 3'D x 1'W)										
	4200	5,000 barrels				Ea.					142,000	
	4300	25,000 barrels									199,000	
	4500	55,000 barrels									336,000	
	4600	100,000 barrels									540,000	
	4800	150,000 barrels									760,000	
	4900	225,000 barrels									1,145,000	
	5100	Floating roof gasoline tanks, steel, 5,000 barrels									120,000	
	5200	25,000 barrels									268,000	
	5400	55,000 barrels									405,000	
	5500	100,000 barrels									614,000	
	5700	150,000 barrels									830,000	
	5800	225,000 barrels									1,150,000	
	6000	Wood tanks, ground level, 2" cypress, 3,000 gallons	C-1	.19	168		5,000	4,350		9,350	12,400	
	6100	2-1/2" cypress, 10,000 gallons		.12	266		13,500	6,900		20,400	25,700	
	6300	3" redwood or 3" fir, 20,000 gallons		.10	320		20,600	8,275		28,875	35,700	
	6400	30,000 gallons		.08	400		25,500	10,300		35,800	44,400	
	6600	45,000 gallons		.07	457		38,600	11,800		50,400	61,000	
	6700	Larger sizes, minimum				Gal.					.66	
	6900	Maximum				"					.81	
	7000	Vinyl coated fabric pillow tanks, freestanding, 5,000 gallons	4 Clab	4	8	Ea.	1,900	172		2,072	2,375	
	7100	Supporting embankment not included, 25,000 gallons	6 Clab	2	24		6,700	515		7,215	8,175	
	7200	50,000 gallons	8 Clab	1.50	42.667		9,750	915		10,665	12,200	
	7300	100,000 gallons	9 Clab	.90	80		15,100	1,725		16,825	19,300	
	7400	150,000 gallons		.50	144		19,100	3,100		22,200	25,900	
	7500	200,000 gallons		.40	180		22,300	3,850		26,150	30,600	
	7600	250,000 gallons		.30	239		26,800	5,150		31,950	37,600	

132 100 | Elevated Storage Tanks

101	0010	**TANKS** Not incl. pipe or pumps										**101**
	3000	Elevated water tanks, 100' to bottom capacity line, incl painting										
	3010	50,000 gallons				Ea.					163,000	
	3300	100,000 gallons									225,000	
	3400	250,000 gallons									310,000	
	3600	500,000 gallons									500,000	
	3700	750,000 gallons									690,000	
	3900	1,000,000 gallons									797,000	

132 150 | Undergrnd Storage Tanks

151	0010	**TANKS**										**151**
	0210	Fiberglass, underground, single wall, U.L. listed, not including										

132 150	Undergrnd Storage Tanks	CREW	DAILY OUTPUT	LABOR-HOURS	UNIT	MAT.	LABOR	EQUIP.	TOTAL	TOTAL INCL O&P	
						1999 BARE COSTS					
151	0220	manway or hold-down strap										**151**
	0240	2,000 gallon capacity	Q-7	4.57	7.002	Ea.	2,325	219		2,544	2,875
	0250	4,000 gallon capacity		3.55	9.014		3,350	282		3,632	4,125
	0260	6,000 gallon capacity		2.67	11.985		3,750	375		4,125	4,700
	0280	10,000 gallon capacity		2	16		5,100	500		5,600	6,350
	0284	15,000 gallon capacity		1.68	19.048		7,425	595		8,020	9,075
	0290	20,000 gallon capacity	▼	1.45	22.069		10,300	690		10,990	12,400
	0500	For manway, fittings and hold-downs, add				▼	20%	15%			
	1020	Fiberglass, underground, double wall, U.L. listed									
	1030	includes manways, not incl. hold-down straps									
	1040	600 gallon capacity	Q-5	2.42	6.612	Ea.	3,075	195		3,270	3,675
	1050	1,000 gallon capacity	"	2.25	7.111		4,150	210		4,360	4,875
	1060	2,500 gallon capacity	Q-7	4.16	7.692		5,900	240		6,140	6,875
	1070	3,000 gallon capacity		3.90	8.205		6,350	256		6,606	7,375
	1080	4,000 gallon capacity		3.64	8.791		6,925	275		7,200	8,050
	1090	6,000 gallon capacity		2.42	13.223		8,025	415		8,440	9,450
	1100	8,000 gallon capacity **CN**		2.08	15.385		9,675	480		10,155	11,300
	1110	10,000 gallon capacity		1.82	17.582		10,500	550		11,050	12,300
	1120	12,000 gallon capacity	▼	1.70	18.824	▼	12,800	590		13,390	15,000
	2210	Fiberglass, underground, single wall, U.L. listed, including									
	2220	hold-down straps, no manways									
	2240	2,000 gallon capacity	Q-7	3.55	9.014	Ea.	2,475	282		2,757	3,150
	2250	4,000 gallon capacity		2.90	11.034		3,525	345		3,870	4,400
	2260	6,000 gallon capacity		2	16		4,075	500		4,575	5,225
	2280	10,000 gallon capacity		1.60	20		5,425	625		6,050	6,900
	2284	15,000 gallon capacity		1.39	23.022		7,750	720		8,470	9,625
	2290	20,000 gallon capacity	▼	1.14	28.070	▼	10,800	875		11,675	13,100
	3020	Fiberglass, underground, double wall, U.L. listed									
	3030	includes manways and hold-down straps									
	3040	600 gallon capacity	Q-5	1.86	8.602	Ea.	3,250	254		3,504	3,925
	3050	1,000 gallon capacity	"	1.70	9.412		4,300	277		4,577	5,150
	3060	2,500 gallon capacity	Q-7	3.29	9.726		6,050	305		6,355	7,125
	3070	3,000 gallon capacity		3.13	10.224		6,500	320		6,820	7,625
	3080	4,000 gallon capacity		2.93	10.922		7,075	340		7,415	8,325
	3090	6,000 gallon capacity		1.86	17.204		8,350	535		8,885	10,000
	3100	8,000 gallon capacity		1.65	19.394		9,975	605		10,580	11,900
	3110	10,000 gallon capacity		1.48	21.622		10,800	675		11,475	12,900
	3120	12,000 gallon capacity	▼	1.40	22.857	▼	13,100	715		13,815	15,500
	5000	Steel underground, sti-P3, set in place, not incl. hold-down bars.									
	5500	Excavation, pad, pumps and piping not included									
	5510	Single wall, 500 gallon capacity, 7 gauge shell	Q-5	2.70	5.926	Ea.	820	175		995	1,175
	5520	1,000 gallon capacity, 7 gauge shell	"	2.50	6.400		1,625	189		1,814	2,050
	5530	2,000 gallon capacity, 1/4" thick shell	Q-7	4.60	6.957		2,425	217		2,642	3,000
	5535	2,500 gallon capacity, 7 gauge shell	Q-5	3	5.333		2,850	157		3,007	3,375
	5540	5,000 gallon capacity, 1/4" thick shell	Q-7	3.20	10		4,550	310		4,860	5,475
	5580	15,000 gallon capacity, 5/16" thick shell		1.70	18.824		10,200	590		10,790	12,200
	5600	20,000 gallon capacity, 5/16" thick shell		1.50	21.333		13,500	665		14,165	15,900
	5610	25,000 gallon capacity, 3/8" thick shell		1.30	24.615		17,600	770		18,370	20,600
	5620	30,000 gallon capacity, 3/8" thick shell		1.10	29.091		23,800	910		24,710	27,600
	5630	40,000 gallon capacity, 3/8" thick shell		.90	35.556		30,700	1,100		31,800	35,400
	5640	50,000 gallon capacity, 3/8" thick shell	▼	.80	40	▼	38,300	1,250		39,550	44,100
191	0010	**TANK LEAK DETECTION SYSTEMS** Liquid and vapor										**191**
	0100	For hydrocarbons and hazardous liquids/vapors									
	0120	Controller, data acquisition, incl. printer, modem, RS232 port									
	0140	24 channel, for use with all probes				Ea.	2,000			2,000	2,200

SPECIAL CONSTRUCTION **13**

132 | Tanks, Tank Covers, Filtration Equipment

132 150	Undergrnd Storage Tanks	CREW	DAILY OUTPUT	LABOR-HOURS	UNIT	1999 BARE COSTS				TOTAL INCL O&P	
						MAT.	LABOR	EQUIP.	TOTAL		
191	0160	9 channel, for external monitoring				Ea.	1,550			1,550	1,700	**191**
	0200	Probes									
	0210	Well monitoring									
	0220	Liquid phase detection				Ea.	695			695	765
	0230	Hydrocarbon vapor, fixed position					640			640	705
	0240	Hydrocarbon vapor, float mounted					570			570	625
	0250	Both liquid and vapor hydrocarbon					860			860	945
	0300	Secondary containment, liquid phase									
	0310	Pipe trench/manway sump				Ea.	310			310	340
	0320	Double wall pipe and manual sump					300			300	330
	0330	Double wall fiberglass annular space					279			279	305
	0340	Double wall steel tank annular space					279			279	305
	0500	Accessories									
	0510	Modem, non-dedicated phone line				Ea.	250			250	275
	0600	Monitoring, internal									
	0610	Automatic tank gauge, incl. overfill				Ea.	1,000			1,000	1,100
	0620	Product line				"	1,075			1,075	1,175
	0700	Monitoring, special									
	0710	Cathodic protection				Ea.	595			595	655
	0720	Annular space chemical monitor				"	810			810	890

133 | Utility Control Systems

133 300	Power Control Systems	CREW	DAILY OUTPUT	LABOR-HOURS	UNIT	1999 BARE COSTS				TOTAL INCL O&P	
						MAT.	LABOR	EQUIP.	TOTAL		
311	0010	**RADIO TOWERS** Guyed, 50'h, 40 lb. sec., 70MPH basic wind spd.	2 Sswk	1	16	Ea.	1,575	490		2,065	2,625	**311**
	0100	Wind load 90 MPH basic wind speed	"	1	16		1,575	490		2,065	2,625
	0300	190' high, 40 lb. section, wind load 70 MPH basic wind speed	K-2	.33	72.727		4,225	2,050	545	6,820	8,875
	0400	200' high, 70 lb. section, wind load 90 MPH basic wind speed		.33	72.727		8,525	2,050	545	11,120	13,600
	0600	300' high, 70 lb. section, wind load 70 MPH basic wind speed		.20	120		12,100	3,400	900	16,400	20,300
	0700	270' high, 90 lb. section, wind load 90 MPH basic wind speed		.20	120		14,000	3,400	900	18,300	22,400
	0800	400' high, 100 lb. section, wind load 70 MPH basic wind speed		.14	171		20,400	4,850	1,275	26,525	32,300
	0900	Self-supporting, 60' high, wind load 70 MPH basic wind speed		.80	30		3,275	850	224	4,349	5,350
	0910	60' high, wind load 90MPH basic wind speed		.45	53.333		5,875	1,500	400	7,775	9,575
	1000	120' high, wind load 70MPH basic wind speed		.40	60		8,025	1,700	450	10,175	12,300
	1200	190' high, wind load 90 MPH basic wind speed		.20	120		19,500	3,400	900	23,800	28,400
	2000	For states west of Rocky Mountains, add for shipping					10%				

135 | Recording & Transportation Control Instrumentation

135 600	Airport Control	CREW	DAILY OUTPUT	LABOR-HOURS	UNIT	1999 BARE COSTS				TOTAL INCL O&P	
						MAT.	LABOR	EQUIP.	TOTAL		
611	0010	**CONTROL TOWERS**, 12' x 10', incl. instruments, min.				Ea.					450,000	**611**
	0100	Maximum									1,000,000
	0500	With standard 40' tower, average									600,000
	1000	Temporary portable control towers, 8' x 12',									

Important: See the Reference Section for critical supporting data - Reference Nos., Crews, & City Cost Indexes

135 | Recording & Transportation Control Instrumentation

135 600	Airport Control	CREW	DAILY OUTPUT	LABOR-HOURS	UNIT	1999 BARE COSTS				TOTAL INCL O&P	
						MAT.	LABOR	EQUIP.	TOTAL		
611	1010	complete with one position communications, minimum				Ea.					275,000	611
	2000	For fixed facilities, depending on height, minimum									50,000	
	2010	Maximum				↓					100,000	

137 | Security Access and Surveillance

137 100	Security Access	CREW	DAILY OUTPUT	LABOR-HOURS	UNIT	1999 BARE COSTS				TOTAL INCL O&P	
						MAT.	LABOR	EQUIP.	TOTAL		
104	0010	**ACCESS CONTROL**										104
	0020	Card type, 1 time zone, minimum				Ea.	292			292	320	
	0040	Maximum					985			985	1,075	
	0060	3 time zones, minimum					720			720	790	
	0080	Maximum				↓	1,650			1,650	1,825	
	0100	System with printer, and control console, 3 zones				Total	8,075			8,075	8,875	
	0120	6 zones				"	10,600			10,600	11,700	
	0140	For each door, minimum, add				Ea.	1,175			1,175	1,300	
	0160	Maximum, add				"	1,750			1,750	1,925	

For information about Means Estimating Seminars, see yellow pages 11 and 12 in back of book

SPECIAL CONSTRUCTION **13**

375

Division Notes

	CREW	DAILY OUTPUT	LABOR-HOURS	UNIT	1999 BARE COSTS				TOTAL INCL O&P
					MAT.	LABOR	EQUIP.	TOTAL	

Division 14
Conveying Systems

Estimating Tips

General
- Many products in Division 14 will require some type of support or blocking for installation not included with the item itself. Examples are supports for conveyors or tube systems, attachment points for lifts, and footings for hoists or cranes. Add these supports in the appropriate division.

141 Dumbwaiters
142 Elevators
- Dumbwaiters and elevators are estimated and purchased in a method similar to buying a car. The manufacturer has a base unit with standard features. Added to this base unit price will be whatever options the owner or specifications require. Increased load capacity, additional vertical travel, additional stops, higher speed, and cab finish options are items to be considered. When developing an estimate for dumbwaiters and elevators, remember that some items needed by the installers may have to be included as part of the general contract.

Examples are:
- shaftway
- rail support brackets
- machine room
- electrical supply
- sill angles
- electrical connections
- pits
- roof penthouses
- pit ladders

Check the job specifications and drawings before pricing.

143 Escalators & Moving Walks
- Escalators and moving walks are specialty items installed by specialty contractors. There are numerous options associated with these items. For specific options contact a manufacturer or contractor. In a method similar to estimating dumbwaiters and elevators, you should verify the extent of general contract work and add items as necessary.

144 Lifts
145 Material Handling Systems
146 Hoists & Cranes
- Products such as correspondence lifts, conveyors, chutes, pneumatic tube systems, material handling cranes and hoists as well as other items specified in this subdivision may require trained installers. The general contractor might not have any choice as to who will perform the installation or when it will be performed. Long lead times are often required for these products, making early decisions in scheduling necessary.

Reference Numbers
Reference numbers are shown in bold squares at the beginning of some major classifications. These numbers refer to related items in the Reference Section. The reference information may be an estimating procedure, an alternate pricing method or technical information.

Note: Not all subdivisions listed here necessarily appear in this publication.

141 | Dumbwaiters

141 100	Manual Dumbwaiters	CREW	DAILY OUTPUT	LABOR-HOURS	UNIT	1999 BARE COSTS				TOTAL INCL O&P	
						MAT.	LABOR	EQUIP.	TOTAL		
101	0010	**DUMBWAITERS** 2 stop, hand, minimum	2 Elev	.75	21.333	Ea.	2,225	705		2,930	3,525	**101**
	0100	Maximum		.50	32	"	5,050	1,050		6,100	7,150	
	0300	For each additional stop, add	↓	.75	21.333	Stop	800	705		1,505	1,950	

141 200	Electric Dumbwaiters	CREW	DAILY OUTPUT	LABOR-HOURS	UNIT	MAT.	LABOR	EQUIP.	TOTAL	TOTAL INCL O&P	
201	0010	**DUMBWAITERS** 2 stop, electric, minimum	2 Elev	.13	123	Ea.	5,600	4,075		9,675	12,300	**201**
	0100	Maximum		.11	145	"	16,800	4,825		21,625	25,800	
	0600	For each additional stop, add	↓	.54	29.630	Stop	2,475	980		3,455	4,175	

142 | Elevators

142 010	Elevators	CREW	DAILY OUTPUT	LABOR-HOURS	UNIT	1999 BARE COSTS				TOTAL INCL O&P	
						MAT.	LABOR	EQUIP.	TOTAL		
011	0012	**ELEVATORS OR LIFTS**										**011**
	7000	Residential, cab type, 1 floor, 2 stop, minimum	2 Elev	.20	80	Ea.	7,850	2,650		10,500	12,700	
	7100	Maximum		.10	160		13,300	5,300		18,600	22,600	
	7200	2 floor, 3 stop, minimum		.12	133		11,700	4,425		16,125	19,500	
	7300	Maximum		.06	266		19,000	8,850		27,850	34,300	
	7700	Stair climber (chair lift), single seat, minimum		1	16		3,800	530		4,330	4,975	
	7800	Maximum		.20	80		5,225	2,650		7,875	9,775	
	8000	Wheelchair lift, minimum		1	16		5,200	530		5,730	6,525	
	8500	Maximum		.50	32		12,300	1,050		13,350	15,200	
	8700	Stair lift, minimum		1	16		10,300	530		10,830	12,100	
	8900	Maximum	↓	.20	80	↓	16,300	2,650		18,950	21,900	
014	0010	**ELEVATORS**	R142-100									**014**
	0020	For multi-story buildings, housing project, minimum				% total					2.50%	
	0100	Maximum	R142-200								4.50%	
	0300	Office building, minimum									2.50%	
	0400	Maximum	R142-300			↓					10%	
	0410	Holeless hydraulic passenger, 2000 lb capacity, 2 story	2 Elev	.10	160	Ea.	26,800	5,300		32,100	37,400	
	0420	2500 lb capacity	R142-400	.10	160		29,400	5,300		34,700	40,400	
	0425	Electric freight, base unit, 4000 lb, 200 fpm, 4 stop, std. fin.	↓	.05	320		59,000	10,600		69,600	81,000	
	0450	For 5000 lb capacity, add					4,250			4,250	4,675	
	0500	For 6000 lb capacity, add					7,475			7,475	8,225	
	0525	For 7000 lb capacity, add					10,000			10,000	11,000	
	0550	For 8000 lb capacity, add					13,900			13,900	15,300	
	0575	For 10000 lb capacity, add					16,500			16,500	18,100	
	0600	For 12000 lb capacity, add					20,000			20,000	22,000	
	0625	For 16000 lb capacity, add					24,000			24,000	26,400	
	0650	For 20000 lb capacity, add					26,400			26,400	29,000	
	0675	For increased speed, 250 fpm, add					8,475			8,475	9,325	
	0700	300 fpm, geared electric, add					10,600			10,600	11,600	
	0725	350 fpm, geared electric, add					12,500			12,500	13,700	
	0750	400 fpm, geared electric, add					13,800			13,800	15,200	
	0775	500 fpm, gearless electric, add					17,300			17,300	19,000	
	0800	600 fpm, gearless electric, add					19,100			19,100	21,000	
	0825	700 fpm, gearless electric, add					22,300			22,300	24,500	
	0850	800 fpm, gearless electric, add	↓			↓	24,800			24,800	27,300	

Important: See the Reference Section for critical supporting data - Reference Nos., Crews, & City Cost Indexes

142 010	Elevators		CREW	DAILY OUTPUT	LABOR-HOURS	UNIT	1999 BARE COSTS				TOTAL INCL O&P	
							MAT.	LABOR	EQUIP.	TOTAL		
014	0875	For class "B" loading, add	R142 -100				Ea.	1,475			1,475	1,625	014
	0900	For class "C-1" loading, add						3,650			3,650	4,025	
	0925	For class "C-2" loading, add	R142 -200					4,350			4,350	4,800	
	0950	For class "C-3" loading, add						5,900			5,900	6,475	
	0975	For travel over 40 V.L.F., add	R142 -300	2 Elev	7.25	2.207	V.L.F.	90	73		163	210	
	1000	For number of stops over 4, add			.27	59.259	Stop	1,800	1,975		3,775	4,975	
	1025	Hydraulic freight, base unit, 2000 lb, 50 fpm, 2 stop, std. fin.	R142 -400		.10	160	Ea.	30,400	5,300		35,700	41,500	
	1050	For 2500 lb capacity, add						2,050			2,050	2,250	
	1075	For 3000 lb capacity, add						2,975			2,975	3,275	
	1100	For 3500 lb capacity, add						5,125			5,125	5,625	
	1125	For 4000 lb capacity, add						5,475			5,475	6,025	
	1150	For 4500 lb capacity, add						6,475			6,475	7,125	
	1175	For 5000 lb capacity, add						8,825			8,825	9,700	
	1200	For 6000 lb capacity, add						9,075			9,075	10,000	
	1225	For 7000 lb capacity, add						14,400			14,400	15,800	
	1250	For 8000 lb capacity, add						15,600			15,600	17,100	
	1275	For 10000 lb capacity, add						16,400			16,400	18,000	
	1300	For 12000 lb capacity, add						19,500			19,500	21,400	
	1325	For 16000 lb capacity, add						25,400			25,400	27,900	
	1350	For 20000 lb capacity, add						28,200			28,200	31,000	
	1375	For increased speed, 100 fpm, add						635			635	700	
	1400	125 fpm, add						1,175			1,175	1,300	
	1425	150 fpm, add						2,200			2,200	2,425	
	1450	175 fpm, add						3,450			3,450	3,800	
	1475	For class "B" loading, add						1,450			1,450	1,600	
	1500	For class "C-1" loading, add						3,625			3,625	4,000	
	1525	For class "C-2" loading, add						4,350			4,350	4,775	
	1550	For class "C-3" loading, add						5,950			5,950	6,550	
	1575	For travel over 20 V.L.F., add		2 Elev	7.25	2.207	V.L.F.	200	73		273	330	
	1600	For number of stops over 2, add			.27	59.259	Stop	375	1,975		2,350	3,375	
	1625	Electric pass., base unit, 2000 lb, 200 fpm, 4 stop, std. fin.			.05	320	Ea.	60,000	10,600		70,600	82,000	
	1650	For 2500 lb capacity, add						2,425			2,425	2,675	
	1675	For 3000 lb capacity, add						3,625			3,625	4,000	
	1700	For 3500 lb capacity, add						5,175			5,175	5,700	
	1725	For 4000 lb capacity, add						5,250			5,250	5,775	
	1750	For 4500 lb capacity, add						6,900			6,900	7,600	
	1775	For 5000 lb capacity, add						8,750			8,750	9,625	
	1800	For increased speed, 250 fpm, geared electric, add						2,000			2,000	2,200	
	1825	300 fpm, geared electric, add						4,075			4,075	4,475	
	1850	350 fpm, geared electric, add						4,800			4,800	5,275	
	1875	400 fpm, geared electric, add						6,825			6,825	7,525	
	1900	500 fpm, gearless electric, add						31,600			31,600	34,800	
	1925	600 fpm, gearless electric, add						33,400			33,400	36,700	
	1950	700 fpm, gearless electric, add						36,500			36,500	40,200	
	1975	800 fpm, gearless electric, add						40,300			40,300	44,300	
	2000	For travel over 40 V.L.F., add		2 Elev	7.25	2.207	V.L.F.	90	73		163	210	
	2025	For number of stops over 4, add			.27	59.259	Stop	2,125	1,975		4,100	5,325	
	2050	Hyd. pass., base unit, 1500 lb, 100 fpm, 2 stop, std. fin. **CN**			.10	160	Ea.	26,600	5,300		31,900	37,200	
	2075	For 2000 lb capacity, add						475			475	525	
	2100	For 2500 lb capacity, add						1,050			1,050	1,150	
	2125	For 3000 lb capacity, add						2,600			2,600	2,875	
	2150	For 3500 lb capacity, add						4,450			4,450	4,900	
	2175	For 4000 lb capacity, add						5,125			5,125	5,650	
	2200	For 4500 lb capacity, add						6,125			6,125	6,750	
	2225	For 5000 lb capacity, add						8,525			8,525	9,375	
	2250	For increased speed, 125 fpm, add						725			725	800	

CONVEYING SYSTEMS 14

	142 010 | Elevators	CREW	DAILY OUTPUT	LABOR-HOURS	UNIT	1999 BARE COSTS MAT.	LABOR	EQUIP.	TOTAL	TOTAL INCL O&P	
014 2275	150 fpm, add R142-100				Ea.	1,575			1,575	1,725	014
2300	175 fpm, add					2,650			2,650	2,900	
2325	200 fpm, add R142-200					4,275			4,275	4,700	
2350	For travel over 12 V.L.F., add	2 Elev	7.25	2.207	V.L.F.	200	73		273	330	
2375	For number of stops over 2, add R142-300		.27	59.259	Stop	395	1,975		2,370	3,400	
2400	Electric hospital, base unit, 4000 lb, 200 fpm, 4 stop, std fin.		.05	320	Ea.	64,000	10,600		74,600	86,500	
2425	For 4500 lb capacity, add R142-400					4,250			4,250	4,675	
2450	For 5000 lb capacity, add					5,550			5,550	6,125	
2475	For increased speed, 250 fpm, geared electric, add					2,100			2,100	2,300	
2500	300 fpm, geared electric, add					4,000			4,000	4,400	
2525	350 fpm, geared electric, add					4,900			4,900	5,375	
2550	400 fpm, geared electric, add					6,800			6,800	7,500	
2575	500 fpm, gearless electric, add					30,200			30,200	33,300	
2600	600 fpm, gearless electric, add					33,400			33,400	36,700	
2625	700 fpm, gearless electric, add					36,500			36,500	40,100	
2650	800 fpm, gearless electric, add					40,300			40,300	44,300	
2675	For travel over 40 V.L.F., add	2 Elev	7.25	2.207	V.L.F.	90	73		163	210	
2700	For number of stops over 4, add		.27	59.259	Stop	2,525	1,975		4,500	5,750	
2725	Hydraulic hospital, base unit, 4000 lb, 100 fpm, 2 stop, std. fin.		.10	160	Ea.	36,400	5,300		41,700	48,000	
2750	For 4000 lb capacity, add					4,500			4,500	4,950	
2775	For 4500 lb capacity, add					5,125			5,125	5,625	
2800	For 5000 lb capacity, add					7,450			7,450	8,200	
2825	For increased speed, 125 fpm, add					1,225			1,225	1,350	
2850	150 fpm, add					2,050			2,050	2,250	
2875	175 fpm, add					3,225			3,225	3,550	
2900	200 fpm, add					4,700			4,700	5,175	
2925	For travel over 12 V.L.F., add	2 Elev	7.25	2.207	V.L.F.	200	73		273	330	
2950	For number of stops over 2, add	"	.27	59.259	Stop	2,675	1,975		4,650	5,900	
2975	Passenger elevator options										
3000	2 car group automatic controls	2 Elev	.66	24.242	Ea.	2,125	805		2,930	3,575	
3025	3 car group automatic controls		.44	36.364		3,150	1,200		4,350	5,300	
3050	4 car group automatic controls		.33	48.485		5,200	1,600		6,800	8,150	
3075	5 car group automatic controls		.26	61.538		6,875	2,050		8,925	10,700	
3100	6 car group automatic controls		.22	72.727		10,400	2,400		12,800	15,100	
3125	Intercom service		3	5.333		310	177		487	610	
3150	Duplex car selective collective		.66	24.242		2,375	805		3,180	3,825	
3175	Center opening 1 speed doors		2	8		1,200	265		1,465	1,725	
3200	Center opening 2 speed doors		2	8		1,550	265		1,815	2,100	
3225	Rear opening doors (opposite front)		2	8		3,375	265		3,640	4,100	
3250	Side opening 2 speed doors		2	8		5,075	265		5,340	5,975	
3275	Automatic emergency power switching		.66	24.242		850	805		1,655	2,150	
3300	Manual emergency power switching		8	2		256	66.50		322.50	380	
3325	Cab finishes (based on 3500 lb cab size)										
3350	Acrylic panel ceiling				Ea.	355			355	390	
3375	Aluminum eggcrate ceiling					430			430	475	
3400	Stainless steel doors					725			725	800	
3425	Carpet flooring					299			299	330	
3450	Epoxy flooring					249			249	274	
3475	Quarry tile flooring					375			375	410	
3500	Slate flooring					520			520	575	
3525	Textured rubber flooring					100			100	110	
3550	Stainless steel walls					2,225			2,225	2,450	
3575	Stainless steel returns at door					475			475	520	
3625	Hall finishes, stainless steel doors					725			725	800	
3650	Stainless steel frames					725			725	800	
3675	12 month maintenance contract									2,640	

Important: See the Reference Section for critical supporting data – Reference Nos., Crews, & City Cost Indexes

14 CONVEYING SYSTEMS

142 010	Elevators		CREW	DAILY OUTPUT	LABOR-HOURS	UNIT	1999 BARE COSTS				TOTAL INCL O&P
							MAT.	LABOR	EQUIP.	TOTAL	
014	3700	Signal devices, hall lanterns	2 Elev	8	2	Ea.	296	66.50		362.50	425	**014**
	3725	Position indicators, up to 3	R142-100	9.40	1.702		206	56.50		262.50	315
	3750	Position indicators, per each over 3	R142-200	32	.500		57	16.60		73.60	87.50
	3775	High speed heavy duty door opener					1,300			1,300	1,425
	3800	Variable voltage, O.H. gearless machine, min.	R142-300	.16	100		21,000	3,325		24,325	28,100
	3815	Maximum		.07	228		46,300	7,575		53,875	62,500
	3825	Basement installed geared machine	R142-400	.33	48.485		10,900	1,600		12,500	14,400
	3850	Freight elevator options									
	3875	Doors, bi-parting	2 Elev	.66	24.242	Ea.	3,450	805		4,255	5,000
	3900	Power operated door and gate	"	.66	24.242		13,800	805		14,605	16,400
	3925	Finishes, steel plate floor					605			605	665
	3950	14 ga. 1/4" x 4' steel plate walls					1,450			1,450	1,600
	3975	12 month maintenance contract									1,992
	4000	Signal devices, hall lanterns	2 Elev	8	2		296	66.50		362.50	425
	4025	Position indicators, up to 3		9.40	1.702		206	56.50		262.50	315
	4050	Position indicators, per each over 3		32	.500		57	16.60		73.60	87.50
	4075	Variable voltage basement installed geared machine		.66	24.242		12,000	805		12,805	14,400
	4100	Hospital elevator options									
	4125	2 car group automatic controls	2 Elev	.66	24.242	Ea.	2,125	805		2,930	3,575
	4150	3 car group automatic controls		.44	36.364		3,125	1,200		4,325	5,275
	4175	4 car group automatic controls		.33	48.485		5,150	1,600		6,750	8,075
	4200	5 car group automatic controls		.26	61.538		6,825	2,050		8,875	10,600
	4225	6 car group automatic controls		.22	72.727		10,300	2,400		12,700	15,000
	4250	Intercom service		3	5.333		300	177		477	600
	4275	Duplex car selective collective		.66	24.242		2,375	805		3,180	3,825
	4300	Center opening 1 speed doors		2	8		1,200	265		1,465	1,725
	4325	Center opening 2 speed doors		2	8		1,550	265		1,815	2,100
	4350	Rear opening doors (opposite front)		2	8		3,375	265		3,640	4,100
	4375	Side opening 2 speed doors		2	8		5,075	265		5,340	5,975
	4400	Automatic emergency power switching		.66	24.242		850	805		1,655	2,150
	4425	Manual emergency power switching		8	2		256	66.50		322.50	380
	4450	Cab finishes (based on 3500# cab size)									
	4475	Aluminum eggcrate ceiling				Ea.	430			430	475
	4500	Stainless steel doors					725			725	800
	4525	Epoxy flooring					249			249	274
	4550	Quarry tile flooring					375			375	415
	4575	Textured rubber flooring					100			100	110
	4600	Stainless steel walls					2,225			2,225	2,450
	4625	Stainless steel returns at door					475			475	520
	4675	Hall finishes, stainless steel doors					725			725	800
	4700	Stainless steel frames					725			725	800
	4725	12 month maintenance contract									3,985
	4750	Signal devices, hall lanterns	2 Elev	8	2		296	66.50		362.50	425
	4775	Position indicators, up to 3		9.40	1.702		206	56.50		262.50	315
	4800	Position indicators, per each over 3		32	.500		57	16.60		73.60	87.50
	4825	High speed heavy duty door opener					1,300			1,300	1,425
	4850	Variable voltage, O.H. gearless machine, min.	2 Elev	.16	100		21,000	3,325		24,325	28,100
	4865	Maximum		.07	228		46,300	7,575		53,875	62,500
	4875	Basement installed geared machine		.33	48.485		10,900	1,600		12,500	14,400
	5000	Drilling for piston, casing included, 18" diameter	B-48	80	.700	V.L.F.	23.50	16.85	24.50	64.85	79

CONVEYING SYSTEMS **14**

143 100 | Escalators

				DAILY OUTPUT	LABOR-HOURS	UNIT	1999 BARE COSTS				TOTAL INCL O&P	
			CREW				MAT.	LABOR	EQUIP.	TOTAL		
101	0010	**ESCALATORS**	R143 -100									101
	1000	Glass, 32" wide x 10' floor to floor height	M-1	.07	457	Ea.	57,500	14,400	1,225	73,125	86,500	
	1010	48" wide x 10' floor to floor height		.07	457		61,500	14,400	1,225	77,125	91,000	
	1020	32" wide x 15' floor to floor height		.06	533		59,000	16,800	1,425	77,225	92,000	
	1030	48" wide x 15' floor to floor height		.06	533		62,500	16,800	1,425	80,725	95,500	
	1040	32" wide x 20' floor to floor height		.05	653		59,500	20,600	1,725	81,825	98,500	
	1050	48" wide x 20' floor to floor height		.05	653		63,500	20,600	1,725	85,825	103,000	
	1060	32" wide x 25' floor to floor height		.04	800		63,000	25,200	2,125	90,325	110,000	
	1070	48" wide x 25' floor to floor height		.04	800		68,000	25,200	2,125	95,325	115,500	
	1080	Enameled steel, 32" wide x 10' floor to floor height		.07	457		57,500	14,400	1,225	73,125	86,500	
	1090	48" wide x 10' floor to floor height		.07	457		61,500	14,400	1,225	77,125	91,000	
	1110	32" wide x 15' floor to floor height		.06	533		59,000	16,800	1,425	77,225	92,000	
	1120	48" wide x 15' floor to floor height		.06	533		62,500	16,800	1,425	80,725	95,500	
	1130	32" wide x 20' floor to floor height		.05	653		59,500	20,600	1,725	81,825	98,500	
	1140	48" wide x 20' floor to floor height		.05	653		63,500	20,600	1,725	85,825	103,000	
	1150	32" wide x 25' floor to floor height		.04	800		63,000	25,200	2,125	90,325	110,000	
	1160	48" wide x 25' floor to floor height		.04	800		68,000	25,200	2,125	95,325	115,500	
	1170	Stainless steel, 32" wide x 10' floor to floor height		.07	457		62,500	14,400	1,225	78,125	92,000	
	1180	48" wide x 10' floor to floor height		.07	457		66,000	14,400	1,225	81,625	95,500	
	1500	32" wide x 15' floor to floor height		.06	533		65,000	16,800	1,425	83,225	98,500	
	1700	48" wide x 15' floor to floor height		.06	533		68,500	16,800	1,425	86,725	102,500	
	2300	32" wide x 25' floor to floor height		.04	800		69,500	25,200	2,125	96,825	117,000	
	2500	48" wide x 25' floor to floor height		.04	800		74,500	25,200	2,125	101,825	122,500	

143 200 | Moving Walks

			CREW	DAILY OUTPUT	LABOR-HOURS	UNIT	MAT.	LABOR	EQUIP.	TOTAL	TOTAL INCL O&P	
201	0010	**MOVING RAMPS AND WALKS** Walk, 27" tread width, minimum	M-1	6.50	4.923	L.F.	690	155	13.10	858.10	1,000	201
	0100	300' to 500', maximum	R143 -200	4.43	7.223		955	227	19.20	1,201.20	1,425	
	0300	48" tread width walk, minimum		4.43	7.223		1,550	227	19.20	1,796.20	2,075	
	0400	100' to 350', maximum		3.82	8.377		1,825	264	22.50	2,111.50	2,425	
	0600	Ramp, 12° incline, 36" tread width, minimum		5.27	6.072		1,275	191	16.15	1,482.15	1,700	
	0700	70' to 90' maximum		3.82	8.377		1,825	264	22.50	2,111.50	2,425	
	0900	48" tread width, minimum		3.57	8.964		1,850	282	24	2,156	2,500	
	1000	40' to 70', maximum		2.91	10.997		2,350	345	29.50	2,724.50	3,125	

144 | Lifts

144 010 | Lifts

			CREW	DAILY OUTPUT	LABOR-HOURS	UNIT	1999 BARE COSTS				TOTAL INCL O&P	
							MAT.	LABOR	EQUIP.	TOTAL		
011	0010	**CORRESPONDENCE LIFT** 1 floor 2 stop, 25 lb capacity, electric	2 Elev	.20	80	Ea.	4,825	2,650		7,475	9,325	011
	0100	Hand, 5 lb capacity	"	.20	80	"	1,825	2,650		4,475	6,025	
015	0010	**PARCEL LIFT** 20" x 20", 100 lb capacity, electric, per floor	2 Mill	.25	64	Ea.	7,475	1,850		9,325	11,000	015

Important: See the Reference Section for critical supporting data - Reference Nos., Crews, & City Cost Indexes

14 CONVEYING SYSTEMS

145 010	Mat. Handling Systems		CREW	DAILY OUTPUT	LABOR-HOURS	UNIT	1999 BARE COSTS				TOTAL INCL O&P	
							MAT.	LABOR	EQUIP.	TOTAL		
011	0010	**MOTORIZED CAR** Distribution systems, single track,										**011**
	0100	20 lb. per car capacity, material handling										
	0200	Minimum, 50' to 100'	4 Mill	.19	168	Station	23,000	4,850		27,850	32,500	
	0300	Maximum, 100' to 200'	"	.15	213	"	31,000	6,125		37,125	43,300	
	0400	Larger system, incl. hospital transport, track type,										
	0500	fully automated material handling system										
	0600	Minimum, complete system	4 Mill	.05	640	Ea.	102,000	18,400		120,400	139,500	
	0700	Maximum, with several stations	E-6	.05	2,560	"	594,500	77,500	25,000	697,000	819,500	

145 500 | Conveyors

			CREW	DAILY OUTPUT	LABOR-HOURS	UNIT	MAT.	LABOR	EQUIP.	TOTAL	INCL O&P	
501	0010	**MATERIAL HANDLING** Conveyors, gravity type 2" rollers, 3" O.C.										**501**
	0050	10' sections with 2 supports, 600 lb. capacity, 18" wide				Ea.	355			355	390	
	0100	24" wide					410			410	450	
	0150	1400 lb. capacity, 18" wide					325			325	360	
	0200	24" wide					365			365	400	
	0350	Horizontal belt, center drive and takeup, 60 fpm										
	0400	16" belt, 26.5' length	2 Mill	.50	32	Ea.	2,500	920		3,420	4,125	
	0450	24" belt, 41.5' length		.40	40		3,725	1,150		4,875	5,825	
	0500	61.5' length		.30	53.333		4,775	1,525		6,300	7,550	
	0600	Inclined belt, 10' rise with horizontal loader and										
	0620	End idler assembly, 27.5' length, 18" belt	2 Mill	.30	53.333	Ea.	7,300	1,525		8,825	10,300	
	0700	24" belt	"	.15	106	"	8,200	3,075		11,275	13,600	
	3600	Monorail, overhead, manual, channel type										
	3700	125 lb. per L.F.	1 Mill	26	.308	L.F.	8.10	8.85		16.95	22	
	3900	500 lb. per L.F.	"	21	.381	"	7.35	10.95		18.30	24.50	
	4000	Trolleys for above, 2 wheel, 125 lb. capacity				Ea.	45			45	49.50	
	4200	4 wheel, 250 lb. capacity					165			165	181	
	4300	8 wheel, 1,000 lb. capacity					360			360	395	
504	0010	**VERTICAL CONVEYOR** Automatic selective										**504**
	0100	to 10 floors, base price	2 Mill	.13	120	Floor	22,900	3,450		26,350	30,400	
	0200	Add for electrical hook-up and testing	2 Elec	.50	32	"		1,025		1,025	1,525	

145 600 | Chutes

			CREW	DAILY OUTPUT	LABOR-HOURS	UNIT	MAT.	LABOR	EQUIP.	TOTAL	INCL O&P	
601	0010	**CHUTES** Linen or refuse, incl. sprinklers, 12' floor height										**601**
	0050	Aluminized steel, 16 ga., 18" diameter	2 Shee	3.50	4.571	Floor	735	145		880	1,025	
	0100	24" diameter		3.20	5		760	159		919	1,075	
	0200	30" diameter		3	5.333		840	169		1,009	1,200	
	0300	36" diameter		2.80	5.714		985	181		1,166	1,350	
	0400	Galvanized steel, 16 ga., 18" diameter		3.50	4.571		700	145		845	995	
	0500	24" diameter		3.20	5		790	159		949	1,100	
	0600	30" diameter		3	5.333		885	169		1,054	1,250	
	0700	36" diameter		2.80	5.714		1,050	181		1,231	1,425	
	0800	Stainless steel, 18" diameter		3.50	4.571		920	145		1,065	1,225	
	0900	24" diameter		3.20	5		970	159		1,129	1,325	
	1000	30" diameter		3	5.333		1,150	169		1,319	1,550	
	1005	36" diameter		2.80	5.714		1,725	181		1,906	2,175	
	1200	Linen chute bottom collector, aluminized steel		4	4	Ea.	980	127		1,107	1,275	
	1300	Stainless steel		4	4		1,250	127		1,377	1,575	
	1500	Refuse, bottom hopper, aluminized steel, 18" diameter		3	5.333		600	169		769	925	
	1600	24" diameter		3	5.333		665	169		834	1,000	
	1800	36" diameter		3	5.333		895	169		1,064	1,250	
	2900	Package chutes, spiral type, minimum		4.50	3.556	Floor	1,700	113		1,813	2,050	
	3000	Maximum		1.50	10.667	"	4,425	340		4,765	5,400	

145 800 | Tube Systems

			CREW	DAILY OUTPUT	LABOR-HOURS	UNIT	1999 BARE COSTS MAT.	LABOR	EQUIP.	TOTAL	TOTAL INCL O&P	
801	0010	**PNEUMATIC TUBE SYSTEM** Single tube, 2 stations, blower										801
	0020	100' long, stock										
	0100	3" diameter	2 Stpi	.12	133	Total	4,750	4,375		9,125	11,900	
	0300	4" diameter	"	.09	177	"	5,350	5,825		11,175	14,700	
	0400	Twin tube, two stations or more, conventional system										
	0600	2-1/2" round	2 Stpi	62.50	.256	L.F.	9.55	8.40		17.95	23.50	
	0700	3" round		46	.348		11	11.40		22.40	29.50	
	0900	4" round		49.60	.323		12.15	10.55		22.70	29.50	
	1000	4" x 7" oval		37.60	.426		17.75	13.95		31.70	40.50	
	1050	Add for blower		2	8	System	3,750	262		4,012	4,525	
	1110	Plus for each round station, add		7.50	2.133	Ea.	420	70		490	570	
	1150	Plus for each oval station, add		7.50	2.133	"	420	70		490	570	
	1200	Alternate pricing method: base cost, minimum		.75	21.333	Total	4,600	700		5,300	6,100	
	1300	Maximum		.25	64	"	9,125	2,100		11,225	13,200	
	1500	Plus total system length, add, minimum		93.40	.171	L.F.	5.90	5.60		11.50	15	
	1600	Maximum		37.60	.426	"	17.45	13.95		31.40	40	
	1800	Completely automatic system, 4" round, 15 to 50 stations		.29	55.172	Station	14,500	1,800		16,300	18,600	
	2200	51 to 144 stations		.32	50		11,200	1,650		12,850	14,900	
	2400	6" round or 4" x 7" oval, 15 to 50 stations		.24	66.667		18,100	2,175		20,275	23,200	
	2800	51 to 144 stations		.23	69.565		15,200	2,275		17,475	20,200	

146 010 | Hoists & Cranes

			CREW	DAILY OUTPUT	LABOR-HOURS	UNIT	1999 BARE COSTS MAT.	LABOR	EQUIP.	TOTAL	TOTAL INCL O&P	
011	0010	**MATERIAL HANDLING**										011
	1500	Cranes, portable hydraulic, floor type, 2,000 lb capacity				Ea.	1,750			1,750	1,925	
	1600	4,000 lb capacity					3,025			3,025	3,325	
	1800	Movable gantry type, 12' to 15' range, 2,000 lb capacity					2,700			2,700	2,975	
	1900	6,000 lb capacity					3,900			3,900	4,300	
	2100	Hoists, electric overhead, chain, hook hung, 15' lift, 1 ton cap.					1,750			1,750	1,925	
	2200	3 ton capacity					2,175			2,175	2,400	
	2500	5 ton capacity					4,850			4,850	5,325	
	2600	For hand-pushed trolley, add					15%					
	2700	For geared trolley, add					30%					
	2800	For motor trolley, add					75%					
	3000	For lifts over 15', 1 ton, add				L.F.	16.60			16.60	18.25	
	3100	5 ton, add				"	38.50			38.50	42.50	
	3300	Lifts, scissor type, portable, electric, 36" high, 2,000 lb				Ea.	3,150			3,150	3,475	
	3400	48" high, 4,000 lb				"	3,600			3,600	3,950	

146 050 | Rails

			CREW	DAILY OUTPUT	LABOR-HOURS	UNIT	1999 BARE COSTS MAT.	LABOR	EQUIP.	TOTAL	TOTAL INCL O&P	
054	0010	**CRANE RAIL** Box beam bridge, no equipment included	E-4	3,400	.009	Lb.	.87	.29	.02	1.18	1.53	054
	0200	Running track only, 104 lb per yard	"	5,600	.006	"	.46	.18	.01	.65	.86	

146 300 | Overhead Bridge Cranes

			CREW	DAILY OUTPUT	LABOR-HOURS	UNIT	1999 BARE COSTS MAT.	LABOR	EQUIP.	TOTAL	TOTAL INCL O&P	
301	0010	**OVERHEAD BRIDGE CRANES**										301
	0100	1 girder, 20' span, 3 ton	M-3	1	34	Ea.	14,300	975	111	15,386	17,300	
	0125	5 ton		1	34		15,600	975	111	16,686	18,800	
	0150	7.5 ton		1	34		18,600	975	111	19,686	22,100	

Important: See the Reference Section for critical supporting data - Reference Nos., Crews, & City Cost Indexes

146 300	Overhead Bridge Cranes	CREW	DAILY OUTPUT	LABOR-HOURS	UNIT	1999 BARE COSTS				TOTAL INCL O&P	
						MAT.	LABOR	EQUIP.	TOTAL		
301 0175	10 ton	M-3	.80	42.500	Ea.	24,500	1,225	139	25,864	28,900	**301**
0200	15 ton		.80	42.500		31,400	1,225	139	32,764	36,600	
0225	30' span, 3 ton		1	34		14,800	975	111	15,886	17,900	
0250	5 ton		1	34		16,300	975	111	17,386	19,500	
0275	7.5 ton		1	34		19,600	975	111	20,686	23,100	
0300	10 ton		.80	42.500		25,300	1,225	139	26,664	29,900	
0325	15 ton		.80	42.500		32,800	1,225	139	34,164	38,100	
0350	2 girder, 40' span, 3 ton	M-4	.50	72		24,600	2,050	390	27,040	30,700	
0375	5 ton		.50	72		25,700	2,050	390	28,140	31,900	
0400	7.5 ton		.50	72		28,200	2,050	390	30,640	34,600	
0425	10 ton		.40	90		32,900	2,575	490	35,965	40,600	
0450	15 ton		.40	90		44,700	2,575	490	47,765	53,500	
0475	25 ton		.30	119		53,000	3,425	650	57,075	64,500	
0500	50' span, 3 ton		.50	72		28,000	2,050	390	30,440	34,400	
0525	5 ton		.50	72		29,000	2,050	390	31,440	35,500	
0550	7.5 ton		.50	72		31,100	2,050	390	33,540	37,800	
0575	10 ton		.40	90		35,700	2,575	490	38,765	43,700	
0600	15 ton		.40	90		46,800	2,575	490	49,865	56,000	
0625	25 ton		.30	119		55,500	3,425	650	59,575	67,000	

For information about Means Estimating Seminars, see yellow pages 11 and 12 in back of book

CONVEYING SYSTEMS **14**

Division Notes

	CREW	DAILY OUTPUT	LABOR-HOURS	UNIT	1999 BARE COSTS				TOTAL INCL O&P
					MAT.	LABOR	EQUIP.	TOTAL	

Division 15
Mechanical

Estimating Tips

151 Pipe & Fittings
This subdivision is primarily basic pipe and related materials. The pipe may be used by any of the mechanical disciplines, i.e., plumbing, fire protection, heating, and air conditioning.

- The piping section lists the add to labor for elevated pipe installation. These adds apply to all elevated pipe, fittings, valves, insulation, etc., that are placed above 10' high. CAUTION: the correct percentage may vary for the same pipe. For example, the percentage add for the basic pipe installation should be based on the maximum height that the craftsman must install for that particular section. If the pipe is to be located 14' above the floor but it is suspended on threaded rod from beams, the bottom flange of which is 18' high (4' rods), then the height is actually 18' and the add is 20%. The pipe coverer, however, does not have to go above the 14' and so his add should be 10%.

- Most pipe is priced first as straight pipe with a joint (coupling, weld, etc.) every 10' and a hanger usually every 10'. There are exceptions with hanger spacing such as: for cast iron pipe (5') and plastic pipe (3 per 10'). Following each type of pipe there are several lines listing sizes and the amount to be subtracted to delete couplings and hangers. This is for pipe that is to be buried or supported together on trapeze hangers. The reason that the couplings are deleted is that these runs are usually long and frequently longer lengths of pipe are used. By deleting the couplings the estimator is expected to look up and add back the correct reduced number of couplings.

- When preparing an estimate it may be necessary to approximate the fittings. Fittings usually run between 25% and 50% of the cost of the pipe. The lower percentage is for simpler runs, and the higher number is for complex areas like mechanical rooms.

152 Plumbing Fixtures
- Plumbing fixture costs usually require two lines, the fixture itself and its "rough-in, supply and waste".

153 Plumbing Appliances
- Remember that gas and oil fired units need venting.

154 Fire Protection
- Include all valves needed for fire protection system.

155 Heating
- When estimating the cost of an HVAC system, check to see who is responsible for providing and installing the temperature control system. It is possible to overlook controls, assuming that they would be included in the electrical estimate.
- When looking up a boiler be careful on specified capacity. Some manufacturers rate their products on output while others use input.
- Include HVAC insulation for pipe, boiler and duct (wrap and liner).

156 HVAC Piping Specialties
- Be careful when looking up mechanical items to get the correct pressure rating and connection type (thread, weld, flange).

157 Air Conditioning & Ventilation
- Combination heating and cooling units are sized by the air conditioning requirements. (See Reference No. R157-020 for preliminary sizing guide.)
- A ton of air conditioning is nominally 400 CFM.
- Rectangular duct is taken off by the linear foot for each size, but its cost is usually estimated by the pound. Remember that SMACNA standards now base duct on internal pressure.
- Prefabricated duct is estimated and purchased like pipe: straight sections and fittings.
- Note that cranes or other lifting equipment are not included on any lines in Division 15. For example, if a crane is required to lift a heavy piece of pipe into place high above a gym floor, or to put a rooftop unit on the roof of a four-story building, etc., it must be added. Due to the potential for extreme variation—from nothing additional required, to a major crane or helicopter—we feel that including a nominal amount for "lifting contingency" would be useless and detract from the accuracy of the estimate. When using equipment rental from Means do not forget to include the cost of the operator(s).

Reference Numbers
Reference numbers are shown in bold squares at the beginning of some major classifications. These numbers refer to related items in the Reference Section. The reference information may be an estimating procedure, an alternate pricing method or technical information.

Note: Not all subdivisions listed here necessarily appear in this publication.

151 100 | Miscellaneous Fittings

		CREW	DAILY OUTPUT	LABOR-HOURS	UNIT	1999 BARE COSTS				TOTAL INCL O&P
						MAT.	LABOR	EQUIP.	TOTAL	
101	**0010 AVERAGE** Square foot and percent of total										**101**
	0100 job cost for plumbing, see division 171									
105	**0010 BACKFLOW PREVENTER** Includes valves										**105**
	0020 and four test cocks, corrosion resistant, automatic operation									
	4000 Reduced pressure principle									
	4100 Threaded, valves are ball									
	4120 3/4" pipe size	1 Plum	16	.500	Ea.	685	16.30		701.30	780
	4140 1" pipe size		14	.571		730	18.65		748.65	830
	4150 1-1/4" pipe size		12	.667		1,025	21.50		1,046.50	1,150
	4160 1-1/2" pipe size		10	.800		990	26		1,016	1,125
	4180 2" pipe size		7	1.143		1,050	37.50		1,087.50	1,200
	5000 Flanged, bronze, valves are OS&Y									
	5060 2-1/2" pipe size	Q-1	5	3.200	Ea.	2,725	94		2,819	3,150
	5080 3" pipe size		4.50	3.556		3,375	104		3,479	3,875
	5100 4" pipe size		3	5.333		4,800	157		4,957	5,500
	5120 6" pipe size	Q-2	3	8		10,200	243		10,443	11,600
	5600 Flanged, iron, valves are OS&Y									
	5660 2-1/2" pipe size	Q-1	5	3.200	Ea.	1,825	94		1,919	2,150
	5680 3" pipe size		4.50	3.556		1,950	104		2,054	2,275
	5700 4" pipe size		3	5.333		2,575	157		2,732	3,050
	5720 6" pipe size	Q-2	3	8		4,050	243		4,293	4,825
	5740 8" pipe size		2	12		8,175	365		8,540	9,550
	5760 10" pipe size		1	24		11,200	730		11,930	13,400
110	**0010 CLEANOUTS**										**110**
	0060 Floor type									
	0080 Round or square, scoriated nickel bronze top									
	0100 2" pipe size	1 Plum	10	.800	Ea.	84	26		110	132
	0120 3" pipe size		8	1		113	32.50		145.50	175
	0140 4" pipe size		6	1.333		113	43.50		156.50	191
	0980 Round top, recessed for terrazzo									
	1000 2" pipe size	1 Plum	9	.889	Ea.	84	29		113	137
	1080 3" pipe size		6	1.333		113	43.50		156.50	191
	1100 4" pipe size		4	2		113	65		178	224
	1120 5" pipe size	Q-1	6	2.667		137	78.50		215.50	270
115	**0010 CLEANOUT TEE**										**115**
	0100 Cast iron, B&S, with countersunk plug									
	0200 2" pipe size	1 Plum	4	2	Ea.	79.50	65		144.50	187
	0220 3" pipe size		3.60	2.222		87	72.50		159.50	206
	0240 4" pipe size		3.30	2.424		108	79		187	239
	0280 6" pipe size	Q-1	5	3.200		291	94		385	460
	0500 For round smooth access cover, same price									
	4000 Plastic, tees and adapters. Add plugs									
	4010 ABS, DWV									
	4020 Cleanout tee, 1-1/2" pipe size	1 Plum	15	.533	Ea.	5.20	17.40		22.60	32
	4030 2" pipe size	Q-1	27	.593		6.85	17.40		24.25	34
	4040 3" pipe size		21	.762		13.75	22.50		36.25	49
	4050 4" pipe size		16	1		25.50	29.50		55	72.50
	4100 Cleanout plug, 1-1/2" pipe size	1 Plum	32	.250		1	8.15		9.15	13.45
	4110 2" pipe size	Q-1	56	.286		1.26	8.40		9.66	14.10
	4120 3" pipe size		36	.444		2.17	13.05		15.22	22
	4130 4" pipe size		30	.533		4.40	15.65		20.05	28.50
	4180 Cleanout adapter fitting, 1-1/2" pipe size	1 Plum	32	.250		1.69	8.15		9.84	14.20
	4190 2" pipe size	Q-1	56	.286		2.36	8.40		10.76	15.30
	4200 3" pipe size		36	.444		6.65	13.05		19.70	27

Important: See the Reference Section for critical supporting data - Reference Nos., Crews, & City Cost Indexes

			DAILY	LABOR-		1999 BARE COSTS				TOTAL	
151 100	**Miscellaneous Fittings**	CREW	OUTPUT	HOURS	UNIT	MAT.	LABOR	EQUIP.	TOTAL	INCL O&P	
115	4210	4" pipe size	Q-1	30	.533	Ea.	10.95	15.65		26.60	35.50	115
	5000	PVC, DWV										
	5010	Cleanout tee, 1-1/2" pipe size	1 Plum	15	.533	Ea.	1.43	17.40		18.83	28	
	5020	2" pipe size	Q-1	27	.593		2.04	17.40		19.44	28.50	
	5030	3" pipe size		21	.762		5.40	22.50		27.90	40	
	5040	4" pipe size	↓	16	1		10.30	29.50		39.80	56	
	5090	Cleanout plug, 1-1/2" pipe size	1 Plum	32	.250		.74	8.15		8.89	13.15	
	5100	2" pipe size	Q-1	56	.286		.89	8.40		9.29	13.70	
	5110	3" pipe size		36	.444		2.21	13.05		15.26	22	
	5120	4" pipe size		30	.533		3.25	15.65		18.90	27	
	5130	6" pipe size	↓	24	.667		9.90	19.55		29.45	40.50	
	5170	Cleanout adapter fitting, 1-1/2" pipe size	1 Plum	32	.250		.74	8.15		8.89	13.15	
	5180	2" pipe size	Q-1	56	.286		.89	8.40		9.29	13.70	
	5190	3" pipe size		36	.444		2.21	13.05		15.26	22	
	5200	4" pipe size		30	.533		3.25	15.65		18.90	27	
	5210	6" pipe size	↓	24	.667	↓	9.90	19.55		29.45	40.50	
120	0010	**CONNECTORS** Flexible, corrugated, 7/8" O.D., 1/2" I.D.										120
	0050	Gas, seamless brass, steel fittings										
	0200	12" long	1 Plum	36	.222	Ea.	8.50	7.25		15.75	20.50	
	0220	18" long		36	.222		10.55	7.25		17.80	22.50	
	0240	24" long		34	.235		12.50	7.65		20.15	25.50	
	0280	36" long		32	.250		14.90	8.15		23.05	29	
	0340	60" long	↓	30	.267	↓	22.50	8.70		31.20	38	
	2000	Water, copper tubing, dielectric separators										
	2100	12" long	1 Plum	36	.222	Ea.	8.10	7.25		15.35	19.90	
	2260	24" long	"	34	.235	"	12	7.65		19.65	25	
125	0010	**DRAINS**										125
	0140	Cornice, C.I., 45° or 90° outlet										
	0180	1-1/2" & 2" pipe size	Q-1	14	1.143	Ea.	72	33.50		105.50	131	
	0200	3" and 4" pipe size	"	12	1.333		98	39		137	168	
	0260	For galvanized body, add					19.60			19.60	21.50	
	0280	For polished bronze dome, add				↓	16.65			16.65	18.30	
	0400	Deck, auto park, C.I., 13" top										
	0440	3", 4", 5", and 6" pipe size	Q-1	8	2	Ea.	425	58.50		483.50	560	
	0480	For galvanized body, add				"	202			202	223	
	2000	Floor, medium duty, C.I., deep flange, 7" dia top										
	2040	2" and 3" pipe size	Q-1	12	1.333	Ea.	60	39		99	126	
	2080	For galvanized body, add					25			25	27.50	
	2120	For polished bronze top, add				↓	33			33	36.50	
	2400	Heavy duty, with sediment bucket, C.I., 12" dia. loose grate										
	2420	2", 3", 4", 5", and 6" pipe size	Q-1	9	1.778	Ea.	205	52		257	305	
	2460	For polished bronze top, add				"	85			85	93.50	
	2500	Heavy duty, cleanout & trap w/bucket, C.I., 15" top										
	2540	2", 3", and 4" pipe size	Q-1	6	2.667	Ea.	1,475	78.50		1,553.50	1,750	
	2560	For galvanized body, add					440			440	485	
	2580	For polished bronze top, add				↓	465			465	510	
	2780	Shower, with strainer, uniform diam. trap, bronze top										
	2800	1-1/2", 2" and 3" pipe size	Q-1	8	2	Ea.	126	58.50		184.50	228	
	2820	4" pipe size	"	7	2.286	↓	139	67		206	255	
	2840	For galvanized body, add					49.50			49.50	54.50	
	3860	Roof, flat metal deck, C.I. body, 12" C.I. dome										
	3890	3" pipe size	Q-1	14	1.143	Ea.	117	33.50		150.50	180	
	3920	6" pipe size	"	10	1.600	"	85.50	47		132.50	165	
	4620	Main, all aluminum, 12" low profile dome										
	4640	2", 3" and 4" pipe size	Q-1	14	1.143	Ea.	130	33.50		163.50	194	
	4980	Scupper floor, oblique strainer, C.I.										

MECHANICAL **15**

			DAILY	LABOR-			1999 BARE COSTS			TOTAL
151 100	**Miscellaneous Fittings**	CREW	OUTPUT	HOURS	UNIT	MAT.	LABOR	EQUIP.	TOTAL	INCL O&P
125	5000	6" x 7" top, 2", 3" and 4" pipe size	Q-1	16	1	Ea.	84	29.50		113.50	137	125
	5100	8" x 12" top, 5" and 6" pipe size	"	14	1.143		165	33.50		198.50	233
	5160	For galvanized body, add					40%				
	5200	For polished bronze strainer, add					85%				
	5980	Trench, floor, hvy duty, modular, C.I., 12" x 12" top									
	6000	2", 3", 4", 5", & 6" pipe size	Q-1	8	2	Ea.	266	58.50		324.50	380
	6100	For polished bronze top, add				"	130			130	143
	6960	Backwater valve, soil pipe, C.I. body									
	6980	Bronze gate and automatic flapper valves									
	7000	3" and 4" pipe size	Q-1	13	1.231	Ea.	530	36		566	635
	7100	5" and 6" pipe size	"	13	1.231	"	850	36		886	990
	7240	Bronze flapper valve, bolted cover									
	7260	2" pipe size	Q-1	16	1	Ea.	177	29.50		206.50	240
	7300	4" pipe size	"	13	1.231		340	36		376	430
	7340	6" pipe size	Q-2	17	1.412		490	43		533	605
141	0010	**FAUCETS/FITTINGS**									141
	0150	Bath, faucets, diverter spout combination, sweat	1 Plum	8	1	Ea.	66	32.50		98.50	122
	0200	For integral stops, IPS unions, add				"	31			31	34
	0810	Bidet									
	0812	Fitting, over the rim, swivel spray/pop-up drain	1 Plum	8	1	Ea.	115	32.50		147.50	177
	0840	Flush valves, with vacuum breaker									
	0850	Water closet									
	0860	Exposed, rear spud	1 Plum	8	1	Ea.	105	32.50		137.50	166
	0920	Urinal									
	0930	Exposed, stall	1 Plum	8	1	Ea.	96	32.50		128.50	156
	0940	Wall, (washout)		8	1		96	32.50		128.50	156
	0950	Pedestal, top spud		8	1		98	32.50		130.50	158
	0960	Concealed, stall		8	1		110	32.50		142.50	171
	0970	Wall (washout)		8	1		115	32.50		147.50	177
	0971	Automatic flush sensor and operator for									
	0972	urinals or water closets	1 Plum	16	.500	Ea.	315	16.30		331.30	370
	1000	Kitchen sink faucets, top mount, cast spout		10	.800		46	26		72	90
	1100	For spray, add		24	.333		12	10.85		22.85	29.50
	2000	Laundry faucets, shelf type, IPS or copper unions		12	.667		38	21.50		59.50	75
	2100	Lavatory faucet, centerset, without drain		10	.800		33	26		59	76
	2810	Automatic sensor and operator, with faucet head		6.15	1.301		279	42.50		321.50	370
	3000	Service sink faucet, cast spout, pail hook, hose end		14	.571		68.50	18.65		87.15	104
	4000	Shower by-pass valve with union		18	.444		46	14.50		60.50	72.50
	4200	Shower thermostatic mixing valve, concealed		8	1		222	32.50		254.50	294
	5000	Sillcock, compact, brass, IPS or copper to hose		24	.333		4.80	10.85		15.65	22
156	0010	**HYDRANTS**									156
	0050	Wall type, moderate climate, bronze, encased									
	0200	3/4" IPS connection	1 Plum	16	.500	Ea.	226	16.30		242.30	274
	0300	1" IPS connection	"	14	.571		226	18.65		244.65	277
	0500	Anti-siphon type					30.50			30.50	33.50
	1000	Non-freeze, bronze, exposed									
	1100	3/4" IPS connection, 4" to 9" thick wall	1 Plum	14	.571	Ea.	194	18.65		212.65	241
	1120	10" to 14" thick wall		12	.667		207	21.50		228.50	261
	1140	15" to 19" thick wall		12	.667		226	21.50		247.50	282
	1160	20" to 24" thick wall		10	.800		201	26		227	261
	1200	For 1" IPS connection, add					15%	10%			
	1240	For 3/4" adapter type vacuum breaker, add				Ea.	19.25			19.25	21
	1280	For anti-siphon type, add				"	40.50			40.50	44.50
	2000	Non-freeze bronze, encased									

Important: See the Reference Section for critical supporting data - Reference Nos., Crews, & City Cost Indexes

151 100	Miscellaneous Fittings	CREW	DAILY OUTPUT	LABOR-HOURS	UNIT	1999 BARE COSTS				TOTAL INCL O&P	
						MAT.	LABOR	EQUIP.	TOTAL		
156 2100	3/4" IPS connection, 5" to 9" thick wall	1 Plum	14	.571	Ea.	264	18.65		282.65	320	**156**
2140	15" to 19" thick wall	"	12	.667		297	21.50		318.50	360	
2280	Anti-siphon type				↓	124			124	136	
3000	Ground box type, bronze frame, 3/4" IPS connection										
3080	Non-freeze, all bronze, polished face, set flush										
3100	2 feet depth of bury	1 Plum	8	1	Ea.	282	32.50		314.50	360	
3180	6 feet depth of bury		7	1.143		370	37.50		407.50	460	
3220	8 feet depth of bury	↓	5	1.600	↓	415	52		467	535	
3400	For 1" IPS connection, add					15%	10%				
3550	For 2" connection, add					445%	24%				
3600	For tapped drain port in box, add				↓	25.50			25.50	28.50	
5000	Moderate climate, all bronze, polished face										
5020	and scoriated cover, set flush										
5100	3/4" IPS connection	1 Plum	16	.500	Ea.	201	16.30		217.30	246	
5120	1" IPS connection	"	14	.571		200	18.65		218.65	248	
5200	For tapped drain port in box, add				↓	25.50			25.50	28.50	
165 0010	**SHOCK ABSORBERS**										**165**
0490	Copper										
0500	3/4" male I.P.S. For 1 to 11 fixtures	1 Plum	12	.667	Ea.	15.60	21.50		37.10	50	
0600	1" male I.P.S., For 12 to 32 fixtures		8	1		40	32.50		72.50	93.50	
0700	1-1/4" male I.P.S. For 33 to 60 fixtures		8	1		46	32.50		78.50	100	
0800	1-1/2" male I.P.S. For 61 to 113 fixtures		8	1		63	32.50		95.50	119	
0900	2" male I.P.S. For 114 to 154 fixtures		8	1		100	32.50		132.50	160	
1000	2-1/2" male I.P.S. For 155 to 330 fixtures	↓	4	2	↓	220	65		285	340	
170 0010	**SUPPORTS/CARRIERS** For plumbing fixtures										**170**
0500	Drinking fountain, wall mounted										
0600	Plate type with studs, top back plate	1 Plum	7	1.143	Ea.	26.50	37.50		64	85.50	
0700	Top front and back plate		7	1.143		33	37.50		70.50	92.50	
0800	Top & bottom, front & back plates, w/bearing jacks	↓	7	1.143	↓	47.50	37.50		85	109	
3000	Lavatory, concealed arm										
3050	Floor mounted, single										
3100	High back fixture	1 Plum	6	1.333	Ea.	128	43.50		171.50	207	
3200	Flat slab fixture		6	1.333		149	43.50		192.50	230	
3220	Paraplegic		6	1.333	↓	136	43.50		179.50	216	
3250	Floor mounted, back to back										
3300	High back fixtures	1 Plum	5	1.600	Ea.	182	52		234	279	
3400	Flat slab fixtures		5	1.600		185	52		237	282	
3430	Paraplegic	↓	5	1.600		191	52		243	289	
3500	Wall mounted, in stud or masonry										
3600	High back fixture	1 Plum	6	1.333	Ea.	75.50	43.50		119	149	
3700	Flat slab fixture	"	6	1.333	"	96.50	43.50		140	172	
4600	Sink, floor mounted										
4650	Exposed arm system										
4700	Single heavy fixture	1 Plum	5	1.600	Ea.	298	52		350	405	
4750	Single heavy sink with slab		5	1.600		201	52		253	300	
4800	Back to back, standard fixtures		5	1.600		230	52		282	330	
4850	Back to back, heavy fixtures		5	1.600		268	52		320	375	
4900	Back to back, heavy sink with slab	↓	5	1.600	↓	268	52		320	375	
4950	Exposed offset arm system										
5000	Single heavy deep fixture	1 Plum	5	1.600	Ea.	203	52		255	305	
5100	Plate type system										
5200	With bearing jacks, single fixture	1 Plum	5	1.600	Ea.	271	52		323	375	
5300	With exposed arms, single heavy fixture		5	1.600		350	52		402	460	
5400	Wall mounted, exposed arms, single heavy fixture		5	1.600		126	52		178	218	
6000	Urinal, floor mounted, 2" or 3" coupling, blowout type		6	1.333		136	43.50		179.50	215	
6100	With fixture or hanger bolts, blowout or washout	↓	6	1.333	↓	96.50	43.50		140	172	

MECHANICAL 15

151 100	Miscellaneous Fittings		DAILY	LABOR-		1999 BARE COSTS				TOTAL	
		CREW	OUTPUT	HOURS	UNIT	MAT.	LABOR	EQUIP.	TOTAL	INCL O&P	
170	6200	With bearing plate	1 Plum	6	1.333	Ea.	108	43.50		151.50	185	170
	6300	Wall mounted, plate type system	↓	6	1.333	↓	97	43.50		140.50	172	
	6980	Water closet, siphon jet										
	7000	Horizontal, adjustable, caulk										
	7040	Single, 4" pipe size	1 Plum	6	1.333	Ea.	205	43.50		248.50	292	
	7050	4" pipe size, paraplegic		6	1.333		205	43.50		248.50	292	
	7060	5" pipe size		6	1.333		271	43.50		314.50	365	
	7100	Double, 4" pipe size		5	1.600		205	52		257	305	
	7110	4" pipe size, paraplegic		5	1.600		205	52		257	305	
	7120	5" pipe size	↓	5	1.600	↓	271	52		323	375	
	7160	Horizontal, adjustable, extended, caulk										
	7180	Single, 4" pipe size	1 Plum	6	1.333	Ea.	205	43.50		248.50	292	
	7200	5" pipe size		6	1.333		365	43.50		408.50	470	
	7240	Double, 4" pipe size		5	1.600		277	52		329	385	
	7260	5" pipe size	↓	5	1.600	↓	365	52		417	485	
	7400	Vertical, adjustable, caulk or thread										
	7440	Single, 4" pipe size	1 Plum	6	1.333	Ea.	242	43.50		285.50	330	
	7460	5" pipe size		6	1.333		305	43.50		348.50	400	
	7480	6" pipe size		5	1.600		360	52		412	475	
	7520	Double, 4" pipe size		5	1.600		420	52		472	545	
	7540	5" pipe size		5	1.600		485	52		537	615	
	7560	6" pipe size	↓	4	2	↓	535	65		600	690	
	7600	Vertical, adjustable, extended, caulk										
	7620	Single, 4" pipe size	1 Plum	6	1.333	Ea.	242	43.50		285.50	330	
	7640	5" pipe size		6	1.333		305	43.50		348.50	400	
	7680	6" pipe size		5	1.600		360	52		412	475	
	7720	Double, 4" pipe size		5	1.600		420	52		472	545	
	7740	5" pipe size		5	1.600		485	52		537	615	
	7760	6" pipe size	↓	4	2	↓	535	65		600	690	
	7780	Water closet, blow out										
	7800	Vertical offset, caulk or thread										
	7820	Single, 4" pipe size	1 Plum	6	1.333	Ea.	184	43.50		227.50	268	
	7840	Double, 4" pipe size	"	5	1.600	"	315	52		367	425	
	7880	Vertical offset, extended, caulk										
	7900	Single, 4" pipe size	1 Plum	6	1.333	Ea.	231	43.50		274.50	320	
	7920	Double, 4" pipe size	"	5	1.600	"	360	52		412	480	
	7960	Vertical, for floor mounted back-outlet										
	7980	Single, 4" thread, 2" vent	1 Plum	6	1.333	Ea.	187	43.50		230.50	272	
	8000	Double, 4" thread, 2" vent	"	6	1.333	"	335	43.50		378.50	435	
	8040	Vertical, for floor mounted back-outlet, extended										
	8060	Single, 4" caulk, 2" vent	1 Plum	6	1.333	Ea.	179	43.50		222.50	263	
	8080	Double, 4" caulk, 2" vent	"	6	1.333	"	278	43.50		321.50	370	
	8200	Water closet, residential										
	8220	Vertical centerline, floor mount										
	8240	Single, 3" caulk, 2" or 3" vent	1 Plum	6	1.333	Ea.	147	43.50		190.50	228	
	8260	4" caulk, 2" or 4" vent		6	1.333		185	43.50		228.50	270	
	8280	3" copper sweat, 3" vent		6	1.333		147	43.50		190.50	228	
	8300	4" copper sweat, 4" vent	↓	6	1.333	↓	185	43.50		228.50	269	
	8400	Vertical offset, floor mount										
	8420	Single, 3" or 4" caulk, vent	1 Plum	4	2	Ea.	184	65		249	300	
	8440	3" or 4" copper sweat, vent		5	1.600		184	52		236	281	
	8460	Double, 3" or 4" caulk, vent		4	2		315	65		380	445	
	8480	3" or 4" copper sweat, vent	↓	5	1.600	↓	315	52		367	425	
	9000	Water cooler (electric), floor mounted										
	9100	Plate type with bearing plate, single	1 Plum	6	1.333	Ea.	108	43.50		151.50	185	

Important: See the Reference Section for critical supporting data - Reference Nos., Crews, & City Cost Indexes

		151 100	Miscellaneous Fittings	CREW	DAILY OUTPUT	LABOR-HOURS	UNIT	MAT.	LABOR	EQUIP.	TOTAL	TOTAL INCL O&P	
								1999 BARE COSTS					
181	0010		**TRAPS**										181
	0030		Cast iron, service weight										
	0050		Long P trap, 2" pipe size										
	1100		12" long	Q-1	16	1	Ea.	21.50	29.50		51	68	
	1140		18" long	"	16	1	"	24	29.50		53.50	71	
	1180		Running trap, single hub, with vent										
	2080		3" pipe size, 3" vent	Q-1	14	1.143	Ea.	40	33.50		73.50	95	
	2120		4" pipe size, 4" vent	"	13	1.231		54.50	36		90.50	115	
	2300		For double hub, vent, add					10%	20%				
	3000		P trap, B&S, 2" pipe size	Q-1	16	1		14.30	29.50		43.80	60.50	
	3040		3" pipe size	"	14	1.143	↓	21	33.50		54.50	74	
	3350		Deep seal trap, B&S										
	3400		1-1/4" pipe size	Q-1	14	1.143	Ea.	17.10	33.50		50.60	70	
	3410		1-1/2" pipe size		14	1.143		17.10	33.50		50.60	70	
	3420		2" pipe size		14	1.143		16.20	33.50		49.70	69	
	3440		3" pipe size	↓	12	1.333		26.50	39		65.50	88.50	
	3800		Drum trap, 4" x 5", 1-1/2" tapping	Q-2	17	1.412		13.50	43		56.50	80	
	3820		2" tapping	"	17	1.412	↓	13.50	43		56.50	80	
	3840		For galvanized, add					60%					
	4700		Copper, drainage, drum trap										
	4800		3" x 5" solid, 1-1/2" pipe size	1 Plum	16	.500	Ea.	35	16.30		51.30	63	
	4840		3" x 6" swivel, 1-1/2" pipe size	"	16	.500	"	46	16.30		62.30	75	
	5100		P trap, standard pattern										
	5200		1-1/4" pipe size	1 Plum	18	.444	Ea.	25	14.50		39.50	49.50	
	5240		1-1/2" pipe size		17	.471		25	15.35		40.35	51	
	5260		2" pipe size		15	.533		38	17.40		55.40	68.50	
	5280		3" pipe size	↓	11	.727	↓	92.50	23.50		116	138	
	5340		With cleanout and slip joint										
	5360		1-1/4" pipe size	1 Plum	18	.444	Ea.	37	14.50		51.50	62.50	
	5400		1-1/2" pipe size	"	17	.471	"	32	15.35		47.35	58.50	
185	0010		**VACUUM BREAKERS** Hot or cold water										185
	1030		Anti-siphon, brass										
	1040		1/4" size	1 Plum	24	.333	Ea.	17.55	10.85		28.40	36	
	1050		3/8" size		24	.333		17.55	10.85		28.40	36	
	1060		1/2" size		24	.333		20	10.85		30.85	38.50	
	1080		3/4" size		20	.400		24	13.05		37.05	46.50	
	1100		1" size		19	.421		37	13.75		50.75	61.50	
	1120		1-1/4" size		15	.533		51.50	17.40		68.90	83.50	
	1140		1-1/2" size		13	.615		76	20		96	114	
	1160		2" size	↓	11	.727	↓	118	23.50		141.50	166	
	1300		For polished chrome, (1/4" thru 1"), add					50%					
	1900		Vacuum relief, water service, bronze										
	2000		1/2" size	1 Plum	30	.267	Ea.	16.70	8.70		25.40	31.50	
195	0010		**VENT FLASHING**										195
	1000		Aluminum with lead ring										
	1020		1-1/4" pipe	1 Plum	20	.400	Ea.	6.45	13.05		19.50	27	
	1030		1-1/2" pipe		20	.400		6.45	13.05		19.50	27	
	1040		2" pipe		18	.444		6.90	14.50		21.40	29.50	
	1050		3" pipe		17	.471		7.65	15.35		23	32	
	1060		4" pipe	↓	16	.500	↓	9.25	16.30		25.55	34.50	
	1350		Copper with neoprene ring										
	1400		1-1/4" pipe	1 Plum	20	.400	Ea.	14.05	13.05		27.10	35	
	1430		1-1/2" pipe		20	.400		14.05	13.05		27.10	35	
	1440		2" pipe		18	.444		14.85	14.50		29.35	38.50	
	1450		3" pipe	↓	17	.471		17.40	15.35		32.75	42.50	

MECHANICAL 15

For expanded coverage of these items see _Means Mechanical or Plumbing Cost Data 1999_

151 100	Miscellaneous Fittings	CREW	DAILY OUTPUT	LABOR-HOURS	UNIT	1999 BARE COSTS				TOTAL INCL O&P	
						MAT.	LABOR	EQUIP.	TOTAL		
195	1460	4" pipe	1 Plum	16	.500	Ea.	19.20	16.30		35.50	45.50	195

151 200	Pipe, Ident/Test & General										
200	0010	PIPING See also divisions 026 & 027 for site work										200
	1000	Add to labor for elevated installation									
	1080	10' to 15' high						10%			
	1100	15' to 20' high						20%			
	1120	20' to 25' high						25%			
	1140	25' to 30' high						35%			
	1160	30' to 35' high						40%			
	1180	35' to 40' high						50%			
	1200	Over 40' high						55%			

151 250	Brass Pipe										
251	0010	PIPE, BRASS Plain end,										251
	0900	Field threaded, coupling & clevis hanger 10' O.C.									
	0920	Regular weight									
	0980	1/8" diameter	1 Plum	62	.129	L.F.	2.21	4.21		6.42	8.80
	1120	1/2" diameter		48	.167		3.99	5.45		9.44	12.65
	1140	3/4" diameter		46	.174		5.40	5.65		11.05	14.55
	1160	1" diameter		43	.186		7.80	6.05		13.85	17.75
	1180	1-1/4" diameter	Q-1	72	.222		11.65	6.50		18.15	23
	1200	1-1/2" diameter		65	.246		13.80	7.20		21	26
	1220	2" diameter		53	.302		18.75	8.85		27.60	34

151 300	Cast Iron Pipe										
301	0010	PIPE, CAST IRON Soil, on hangers 5' O.C.										301
	0020	Single hub, service wt., lead & oakum joints 10' O.C.									
	2120	2" diameter	Q-1	63	.254	L.F.	4.42	7.45		11.87	16.15
	2140	3" diameter		60	.267		5.95	7.85		13.80	18.40
	2160	4" diameter CN		55	.291		8.25	8.55		16.80	22
	2180	5" diameter	Q-2	76	.316		10.50	9.60		20.10	26
	2200	6" diameter	"	73	.329		12.50	10		22.50	29
	2220	8" diameter	Q-3	59	.542		19.95	16.85		36.80	47.50
	2240	10" diameter		54	.593		33.50	18.45		51.95	65
	2260	12" diameter		48	.667		47	20.50		67.50	83
	2320	For service weight, double hub, add					65%				
	2340	For extra heavy, single hub, add					48%	4%			
	2360	For extra heavy, double hub, add					29%	4%			
	2400	Lead for caulking, (1#/dia in)				Lb.	.98			.98	1.08
	2420	Oakum for caulking, (1/8#/dia in)				"	4.66			4.66	5.15
	4000	No hub, couplings 10' O.C.									
	4100	1-1/2" diameter	Q-1	71	.225	L.F.	5.10	6.60		11.70	15.60
	4120	2" diameter		67	.239		5.20	7		12.20	16.35
	4160	4" diameter		58	.276		8.85	8.10		16.95	22

151 400	Copper Pipe & Tubing										
401	0010	PIPE, COPPER Solder joints										401
	1000	Type K tubing, couplings & clevis hangers 10' O.C.									
	1100	1/4" diameter	1 Plum	84	.095	L.F.	1.19	3.10		4.29	6
	1200	1" diameter		66	.121		2.98	3.95		6.93	9.30
	1260	2" diameter		40	.200		7	6.50		13.50	17.65
	2000	Type L tubing, couplings & hangers 10' O.C.									

Important: See the Reference Section for critical supporting data - Reference Nos., Crews, & City Cost Indexes

15
MECHANICAL

151 400 | Copper Pipe & Tubing

			CREW	DAILY OUTPUT	LABOR-HOURS	UNIT	1999 BARE COSTS				TOTAL INCL O&P	
							MAT.	LABOR	EQUIP.	TOTAL		
401	2100	1/4" diameter	1 Plum	88	.091	L.F.	.87	2.96		3.83	5.45	401
	2120	3/8" diameter		84	.095		1.08	3.10		4.18	5.90	
	2140	1/2" diameter CN		81	.099		1.26	3.22		4.48	6.25	
	2160	5/8" diameter		79	.101		1.72	3.30		5.02	6.90	
	2180	3/4" diameter		76	.105		1.80	3.43		5.23	7.20	
	2200	1" diameter		68	.118		2.46	3.84		6.30	8.50	
	2220	1-1/4" diameter		58	.138		3.11	4.50		7.61	10.20	
	2240	1-1/2" diameter		52	.154		3.94	5		8.94	11.95	
	2260	2" diameter	▼	42	.190		5.95	6.20		12.15	15.95	
	2280	2-1/2" diameter	Q-1	62	.258		8.70	7.55		16.25	21	
	2300	3" diameter		56	.286		11.75	8.40		20.15	25.50	
	2320	3-1/2" diameter		43	.372		15.20	10.90		26.10	33.50	
	2340	4" diameter		39	.410		18.65	12.05		30.70	39	
	2360	5" diameter	▼	34	.471		50	13.80		63.80	76	
	2380	6" diameter	Q-2	40	.600		59.50	18.25		77.75	93	
	2400	8" diameter	"	36	.667		104	20.50		124.50	146	
	2410	For other than full hard temper, add				▼	21%					
	2590	For silver solder, add						15%				
	4000	Type DWV tubing, couplings & hangers 10' O.C.										
	4100	1-1/4" diameter	1 Plum	60	.133	L.F.	2.56	4.35		6.91	9.40	
	4120	1-1/2" diameter		54	.148		3.20	4.83		8.03	10.80	
	4140	2" diameter	▼	44	.182		4.18	5.95		10.13	13.60	
	4160	3" diameter	Q-1	58	.276		7	8.10		15.10	19.95	
	4180	4" diameter		40	.400		12.30	11.75		24.05	31.50	
	4200	5" diameter	▼	36	.444		34.50	13.05		47.55	58	
	4220	6" diameter	Q-2	42	.571	▼	45.50	17.40		62.90	76.50	

151 450 | Corrosion Resistant Pipe

			CREW	DAILY OUTPUT	LABOR-HOURS	UNIT	MAT.	LABOR	EQUIP.	TOTAL	TOTAL INCL O&P	
451	0010	**PIPE, CORROSION RESISTANT** No couplings or hangers										451
	0020	Iron alloy, drain, mechanical joint										
	1000	1-1/2" diameter	Q-1	70	.229	L.F.	22.50	6.70		29.20	35	
	1100	2" diameter		66	.242		26	7.10		33.10	39.50	
	1120	3" diameter		60	.267		36	7.85		43.85	52	
	1140	4" diameter	▼	52	.308	▼	46.50	9.05		55.55	64.50	
	2980	Plastic, epoxy, fiberglass filament wound										
	3000	2" diameter	Q-1	62	.258	L.F.	5.85	7.55		13.40	17.95	
	3100	3" diameter		51	.314		9.45	9.20		18.65	24.50	
	3120	4" diameter		45	.356		11.25	10.45		21.70	28	
	3140	6" diameter	▼	32	.500	▼	16.65	14.70		31.35	40.50	
	3980	Polyester, fiberglass filament wound										
	4000	2" diameter	Q-1	62	.258	L.F.	9.40	7.55		16.95	22	
	4100	3" diameter		51	.314		12.55	9.20		21.75	28	
	4120	4" diameter		45	.356		14.20	10.45		24.65	31.50	
	4140	6" diameter	▼	32	.500	▼	22.50	14.70		37.20	47	
	4980	Polypropylene, acid resistant, fire retardant, schedule 40										
	5000	1-1/2" diameter	Q-1	68	.235	L.F.	1.46	6.90		8.36	12.05	
	5100	2" diameter		62	.258		1.97	7.55		9.52	13.65	
	5120	3" diameter		51	.314		4.14	9.20		13.34	18.50	
	5140	4" diameter	▼	45	.356	▼	5.90	10.45		16.35	22.50	
	5980	Proxylene, fire retardant, Schedule 40										
	6000	1-1/2" diameter	Q-1	68	.235	L.F.	2.74	6.90		9.64	13.45	
	6100	2" diameter		62	.258		3.56	7.55		11.11	15.40	
	6120	3" diameter		51	.314		6.70	9.20		15.90	21.50	
	6140	4" diameter	▼	45	.356	▼	9.40	10.45		19.85	26	

MECHANICAL 15

151 500 | Glass Pipe

		CREW	DAILY OUTPUT	LABOR-HOURS	UNIT	MAT.	LABOR	EQUIP.	TOTAL	TOTAL INCL O&P	
501	0010	**PIPE, GLASS** Borosilicate, couplings & hangers 10' O.C.									**501**
	0020	Drainage									
	1100	1-1/2" diameter	Q-1	52	.308	L.F.	7.90	9.05		16.95	22.50
	1120	2" diameter		44	.364		10.45	10.65		21.10	27.50
	1140	3" diameter		39	.410		13.90	12.05		25.95	33.50
	1160	4" diameter		30	.533		25	15.65		40.65	51
	1180	6" diameter		26	.615		45	18.05		63.05	77

151 550 | Plastic Pipe

		CREW	DAILY OUTPUT	LABOR-HOURS	UNIT	MAT.	LABOR	EQUIP.	TOTAL	TOTAL INCL O&P	
551	0010	**PIPE, PLASTIC**									**551**
	0020	Fiberglass reinforced, couplings 10' O.C., hangers 3 per 10'									
	0200	High strength									
	0240	2" diameter	Q-1	58	.276	L.F.	9.55	8.10		17.65	23
	0260	3" diameter		51	.314		13.55	9.20		22.75	29
	0280	4" diameter		47	.340		16.40	10		26.40	33
	0300	6" diameter		38	.421		24.50	12.35		36.85	46
	0320	8" diameter	Q-2	48	.500		39	15.20		54.20	66
	1800	PVC, couplings 10' O.C., hangers 3 per 10'									
	1820	Schedule 40									
	1860	1/2" diameter	1 Plum	54	.148	L.F.	1.37	4.83		6.20	8.80
	1870	3/4" diameter		51	.157		1.52	5.10		6.62	9.40
	1880	1" diameter		46	.174		1.82	5.65		7.47	10.60
	1890	1-1/4" diameter		42	.190		1.86	6.20		8.06	11.45
	1900	1-1/2" diameter		36	.222		2.06	7.25		9.31	13.25
CN	1910	2" diameter	Q-1	59	.271		2.29	7.95		10.24	14.55
	1920	2-1/2" diameter		56	.286		2.99	8.40		11.39	16
	1930	3" diameter		53	.302		3.51	8.85		12.36	17.30
	1940	4" diameter		48	.333		4.92	9.80		14.72	20.50
	1950	5" diameter		43	.372		6.90	10.90		17.80	24
	1960	6" diameter		39	.410		7.75	12.05		19.80	27
	4100	DWV type, schedule 40, couplings 10' O.C., hangers 3 per 10'									
	4120	ABS									
	4140	1-1/4" diameter	1 Plum	42	.190	L.F.	1.88	6.20		8.08	11.45
	4150	1-1/2" diameter	"	36	.222		1.84	7.25		9.09	13.05
	4160	2" diameter	Q-1	59	.271		1.96	7.95		9.91	14.20
	4400	PVC									
	4410	1-1/4" diameter	1 Plum	42	.190	L.F.	1.88	6.20		8.08	11.45
	4420	1-1/2" diameter	"	36	.222		1.98	7.25		9.23	13.20
	4460	2" diameter	Q-1	59	.271		2.16	7.95		10.11	14.40
	4470	3" diameter		53	.302		3.32	8.85		12.17	17.10
	4480	4" diameter		48	.333		4.64	9.80		14.44	19.95
	4490	6" diameter		39	.410		7.75	12.05		19.80	27
	5360	CPVC, couplings 10' O.C., hangers 3 per 10'									
	5380	Schedule 40									
	5460	1/2" diameter	1 Plum	54	.148	L.F.	2.09	4.83		6.92	9.60
	5470	3/4" diameter		51	.157		2.47	5.10		7.57	10.45
	5480	1" diameter		46	.174		3.18	5.65		8.83	12.10
	5490	1-1/4" diameter		42	.190		3.76	6.20		9.96	13.55
	5500	1-1/2" diameter		36	.222		4.37	7.25		11.62	15.80
	5510	2" diameter	Q-1	59	.271		5.20	7.95		13.15	17.75
	5520	2-1/2" diameter		56	.286		8.15	8.40		16.55	21.50
	5530	3" diameter		53	.302		9.75	8.85		18.60	24

Important: See the Reference Section for critical supporting data - Reference Nos., Crews, & City Cost Indexes

151 600 | Stainless Steel Pipe

		CREW	DAILY OUTPUT	LABOR-HOURS	UNIT	1999 BARE COSTS				TOTAL INCL O&P
						MAT.	LABOR	EQUIP.	TOTAL	
601	0010	**PIPE, STAINLESS STEEL**									601
	3500	Threaded, couplings and hangers 10' O.C.									
	3520	Schedule 40, type 304									
	3540	1/4" diameter	1 Plum	54	.148	L.F.	3.54	4.83		8.37	11.20
	3550	3/8" diameter		53	.151		4.04	4.92		8.96	11.90
	3560	1/2" diameter		52	.154		5.20	5		10.20	13.30
	3580	1" diameter		45	.178		8.05	5.80		13.85	17.65
	3610	2" diameter	Q-1	57	.281		15.45	8.25		23.70	29.50
	3640	4" diameter	Q-2	51	.471		44.50	14.30		58.80	70.50
	3740	For small quantities, add					10%				
	4250	Schedule 40, type 316									
	4290	1/4" diameter	1 Plum	54	.148	L.F.	4.12	4.83		8.95	11.85
	4300	3/8" diameter		53	.151		4.90	4.92		9.82	12.85
	4310	1/2" diameter		52	.154		6.55	5		11.55	14.80
	4320	3/4" diameter		51	.157		7.65	5.10		12.75	16.20
	4330	1" diameter		45	.178		10.25	5.80		16.05	20
	4360	2" diameter	Q-1	57	.281		20	8.25		28.25	34.50
	4390	4" diameter	Q-2	51	.471		59	14.30		73.30	86.50
	4490	For small quantities, add					10%				

151 700 | Steel Pipe

		CREW	DAILY OUTPUT	LABOR-HOURS	UNIT	1999 BARE COSTS				TOTAL INCL O&P
						MAT.	LABOR	EQUIP.	TOTAL	
701	0010	**PIPE, STEEL**									701
	0012	The steel pipe in this section does not include fittings such as ells,tees									
	0014	For fittings either add a % (usually 25 to 35%) or see									
	0015	the Mechanical or Plumbing Cost Data									
	0020	All pipe sizes are to Spec. A-53 unless noted otherwise R151 -050									
	0050	Schedule 40, threaded, with couplings, and clevis type									
	0060	hangers sized for covering, 10' O.C.									
	0540	Black, 1/4" diameter	1 Plum	66	.121	L.F.	1.20	3.95		5.15	7.30
	0550	3/8" diameter		65	.123		1.35	4.01		5.36	7.60
	0560	1/2" diameter		63	.127		1.15	4.14		5.29	7.50
	0570	3/4" diameter		61	.131		1.29	4.28		5.57	7.90
	0580	1" diameter		53	.151		1.64	4.92		6.56	9.25
	0590	1-1/4" diameter	Q-1	89	.180		1.98	5.30		7.28	10.20
	0600	1-1/2" diameter		80	.200		2.26	5.85		8.11	11.40
	0610	2" diameter **CN**		64	.250		3.06	7.35		10.41	14.45
	0620	2-1/2" diameter		50	.320		4.81	9.40		14.21	19.55
	0630	3" diameter		43	.372		6.05	10.90		16.95	23
	0640	3-1/2" diameter		40	.400		7.90	11.75		19.65	26.50
	0650	4" diameter		36	.444		8.75	13.05		21.80	29.50
	0660	5" diameter		26	.615		19.50	18.05		37.55	49
	0670	6" diameter	Q-2	31	.774		22	23.50		45.50	59.50
	1290	Galvanized, 1/4" diameter	1 Plum	66	.121		1.46	3.95		5.41	7.60
	1300	3/8" diameter		65	.123		1.63	4.01		5.64	7.90
	1310	1/2" diameter		63	.127		1.34	4.14		5.48	7.70
	1320	3/4" diameter		61	.131		1.53	4.28		5.81	8.20
	1330	1" diameter		53	.151		1.79	4.92		6.71	9.40
	1340	1-1/4" diameter	Q-1	89	.180		2.20	5.30		7.50	10.40
	1350	1-1/2" diameter		80	.200		2.79	5.85		8.64	11.95
	1360	2" diameter		64	.250		3.36	7.35		10.71	14.80
	1370	2-1/2" diameter		50	.320		5.35	9.40		14.75	20
	1380	3" diameter		43	.372		6.70	10.90		17.60	24
	1390	3-1/2" diameter		40	.400		8.70	11.75		20.45	27.50
	1400	4" diameter		36	.444		9.70	13.05		22.75	30.50
	1410	5" diameter		26	.615		16.10	18.05		34.15	45

MECHANICAL 15

151 700 | Steel Pipe

		CREW	DAILY OUTPUT	LABOR-HOURS	UNIT	1999 BARE COSTS				TOTAL INCL O&P	
						MAT.	LABOR	EQUIP.	TOTAL		
701	1420	6" diameter	R151 -050	Q-2	31	.774	L.F.	22	23.50		45.50	59.50	701
	1430	8" diameter			27	.889		31.50	27		58.50	75.50
	1440	10" diameter			23	1.043		45.50	32		77.50	98
	1450	12" diameter			18	1.333		62.50	40.50		103	131
	2000	Welded, sch. 40, on yoke & roll hangers, sized for covering,										
	2010	10' O.C. (no hangers incl. for 14" diam. and up)										
	2040	Black, 1" diameter		Q-15	93	.172	L.F.	2.12	5.05	.54	7.71	10.55
	2070	2" diameter			61	.262		3	7.70	.82	11.52	15.85
	2090	3" diameter			43	.372		4.84	10.90	1.16	16.90	23
	2110	4" diameter			37	.432		6.65	12.70	1.34	20.69	28
	2120	5" diameter			32	.500		10.30	14.70	1.55	26.55	35
	2130	6" diameter		Q-16	36	.667		16.20	20.50	1.38	38.08	50.50
	2140	8" diameter			29	.828		22	25	1.71	48.71	64
	2150	10" diameter			24	1		41.50	30.50	2.07	74.07	94.50
	2160	12" diameter			19	1.263		51	38.50	2.61	92.11	118

151 800 | Grooved-Joint Pipe

		CREW	DAILY OUTPUT	LABOR-HOURS	UNIT	1999 BARE COSTS				TOTAL INCL O&P	
						MAT.	LABOR	EQUIP.	TOTAL		
801	0010	**PIPE, GROOVED-JOINT STEEL FITTINGS & VALVES**										801
	0020	Pipe includes coupling & clevis type hanger 10' O.C.										
	1000	Schedule 40, black										
	1040	3/4" diameter	1 Plum	71	.113	L.F.	1.76	3.67		5.43	7.50
	1050	1" diameter		63	.127		1.89	4.14		6.03	8.35
	1060	1-1/4" diameter		58	.138		2.35	4.50		6.85	9.40
	1070	1-1/2" diameter		51	.157		2.67	5.10		7.77	10.70
	1080	2" diameter		40	.200		3.16	6.50		9.66	13.40
	1090	2-1/2" diameter	Q-1	57	.281		4.43	8.25		12.68	17.40
	1100	3" diameter		50	.320		5.45	9.40		14.85	20.50
	1110	4" diameter		45	.356		7.75	10.45		18.20	24.50
	1120	5" diameter		37	.432		12	12.70		24.70	32.50
	1130	6" diameter	Q-2	42	.571		15.50	17.40		32.90	43.50
	1800	Galvanized									
	1840	3/4" diameter	1 Plum	71	.113	L.F.	1.99	3.67		5.66	7.75
	1850	1" diameter		63	.127		2.29	4.14		6.43	8.75
	1860	1-1/4" diameter		58	.138		2.60	4.50		7.10	9.65
	1870	1-1/2" diameter		51	.157		2.89	5.10		7.99	10.95
	1880	2" diameter		40	.200		3.45	6.50		9.95	13.70
	1890	2-1/2" diameter	Q-1	57	.281		4.88	8.25		13.13	17.85
	1900	3" diameter		50	.320		6.10	9.40		15.50	21
	1910	4" diameter		45	.356		8.65	10.45		19.10	25.50
	1920	5" diameter		37	.432		15.05	12.70		27.75	36
	1930	6" diameter	Q-2	42	.571		20	17.40		37.40	48.50
	3990	Fittings: cplg. & labor required at joints not incl. in fitting									
	3994	price. Add 1 per joint for installed price.									
	4000	Elbow, 90° or 45°, painted									
	4030	3/4" diameter	1 Plum	50	.160	Ea.	11.45	5.20		16.65	20.50
	4040	1" diameter		50	.160		11.45	5.20		16.65	20.50
	4050	1-1/4" diameter		40	.200		11.45	6.50		17.95	22.50
	4060	1-1/2" diameter		33	.242		11.45	7.90		19.35	24.50
	4070	2" diameter		25	.320		11.45	10.45		21.90	28.50
	4080	2-1/2" diameter	Q-1	40	.400		11.45	11.75		23.20	30.50
	4090	3" diameter		33	.485		20.50	14.25		34.75	44
	4100	4" diameter		25	.640		22	18.80		40.80	53
	4110	5" diameter		20	.800		53.50	23.50		77	94.50
	4120	6" diameter	Q-2	25	.960		63.50	29		92.50	114
	4250	For galvanized elbows, add					26%				

Important: See the Reference Section for critical supporting data - Reference Nos., Crews, & City Cost Indexes

151 800 | Grooved-Joint Pipe

			CREW	DAILY OUTPUT	LABOR-HOURS	UNIT	1999 BARE COSTS MAT.	LABOR	EQUIP.	TOTAL	TOTAL INCL O&P	
801	4690	Tee, painted										**801**
	4700	3/4" diameter	1 Plum	38	.211	Ea.	17.70	6.85		24.55	30	
	4740	1" diameter		33	.242		17.70	7.90		25.60	31.50	
	4750	1-1/4" diameter		27	.296		17.70	9.65		27.35	34	
	4760	1-1/2" diameter		22	.364		17.70	11.85		29.55	37.50	
	4770	2" diameter		17	.471		17.70	15.35		33.05	43	
	4780	2-1/2" diameter	Q-1	27	.593		17.70	17.40		35.10	46	
	4790	3" diameter		22	.727		25	21.50		46.50	60	
	4800	4" diameter		17	.941		38	27.50		65.50	83.50	
	4810	5" diameter		13	1.231		89	36		125	152	
	4820	6" diameter	Q-2	17	1.412		103	43		146	178	
	4900	For galvanized tees, add					24%					
	4906	Couplings, rigid style, painted										
	4908	1" diameter	1 Plum	100	.080	Ea.	8.35	2.61		10.96	13.10	
	4909	1-1/4" diameter		100	.080		8.35	2.61		10.96	13.10	
	4910	1-1/2" diameter		67	.119		8.35	3.89		12.24	15.05	
	4912	2" diameter		50	.160		8.55	5.20		13.75	17.30	
	4914	2-1/2" diameter	Q-1	80	.200		9.85	5.85		15.70	19.70	
	4916	3" diameter		67	.239		11.50	7		18.50	23.50	
	4918	4" diameter		50	.320		16.30	9.40		25.70	32	
	4920	5" diameter		40	.400		21	11.75		32.75	41.50	
	4922	6" diameter	Q-2	50	.480		28.50	14.60		43.10	53	
	4940	Flexible, standard, painted										
	4950	3/4" diameter	1 Plum	100	.080	Ea.	6	2.61		8.61	10.55	
	4960	1" diameter		100	.080		6	2.61		8.61	10.55	
	4970	1-1/4" diameter		80	.100		7.95	3.26		11.21	13.70	
	4980	1-1/2" diameter		67	.119		8.70	3.89		12.59	15.45	
	4990	2" diameter		50	.160		9.20	5.20		14.40	18	
	5000	2-1/2" diameter	Q-1	80	.200		11	5.85		16.85	21	
	5010	3" diameter		67	.239		12.15	7		19.15	24	
	5020	3-1/2" diameter		57	.281		17.75	8.25		26	32	
	5030	4" diameter		50	.320		17.75	9.40		27.15	34	
	5040	5" diameter		40	.400		27	11.75		38.75	48	
	5050	6" diameter	Q-2	50	.480		32	14.60		46.60	57.50	

151 950 | Valves

			CREW	DAILY OUTPUT	LABOR-HOURS	UNIT	MAT.	LABOR	EQUIP.	TOTAL	TOTAL INCL O&P	
951	0010	**VALVES, BRASS**										**951**
	0500	Gas cocks, threaded										
	0530	1/2" size	1 Plum	24	.333	Ea.	7.65	10.85		18.50	25	
	0540	3/4" size		22	.364		10.30	11.85		22.15	29.50	
	0550	1" size		19	.421		12.60	13.75		26.35	35	
	0560	1-1/4" size		15	.533		34.50	17.40		51.90	64.50	
955	0010	**VALVES, BRONZE**										**955**
	1020	Angle, 150 lb., rising stem, threaded										
	1030	1/8" size	1 Plum	24	.333	Ea.	33.50	10.85		44.35	53.50	
	1040	1/4" size		24	.333		33.50	10.85		44.35	53.50	
	1050	3/8" size		24	.333		33.50	10.85		44.35	53.50	
	1060	1/2" size		22	.364		33.50	11.85		45.35	55	
	1070	3/4" size		20	.400		46	13.05		59.05	70.50	
	1080	1" size		19	.421		66.50	13.75		80.25	94	
	1100	1-1/2" size		13	.615		115	20		135	157	
	1110	2" size		11	.727		180	23.50		203.50	234	
	1380	Ball, 150 psi, threaded										
	1400	1/4" size	1 Plum	24	.333	Ea.	6.40	10.85		17.25	23.50	
	1430	3/8" size		24	.333		6.40	10.85		17.25	23.50	
	1450	1/2" size		22	.364		6.40	11.85		18.25	25	

MECHANICAL 15

151 950	Valves	CREW	DAILY OUTPUT	LABOR-HOURS	UNIT	1999 BARE COSTS				TOTAL INCL O&P	
						MAT.	LABOR	EQUIP.	TOTAL		
955	1460	3/4" size	1 Plum	20	.400	Ea.	10.55	13.05		23.60	31.50	955
1470	1" size		19	.421		13.30	13.75		27.05	35.50
1480	1-1/4" size		15	.533		22.50	17.40		39.90	51
1490	1-1/2" size		13	.615		28.50	20		48.50	62
1500	2" size		11	.727		35.50	23.50		59	75.50
1750	Check, swing, class 150, regrinding disc, threaded									
1800	1/8" size	1 Plum	24	.333	Ea.	25	10.85		35.85	44
1830	1/4" size		24	.333		25	10.85		35.85	44
1840	3/8" size		24	.333		25	10.85		35.85	44
1850	1/2" size		24	.333		25	10.85		35.85	44
1860	3/4" size		20	.400		36	13.05		49.05	59.50
1870	1" size		19	.421		53	13.75		66.75	79.50
1880	1-1/4" size		15	.533		74.50	17.40		91.90	109
1890	1-1/2" size		13	.615		88	20		108	127
1900	2" size		11	.727		129	23.50		152.50	178
1910	2-1/2" size	Q-1	15	1.067		170	31.50		201.50	234
2000	For 200 lb, add					5%	10%			
2040	For 300 lb, add					15%	15%			
2850	Gate, N.R.S., soldered, 125 psi									
2900	3/8" size	1 Plum	24	.333	Ea.	15.10	10.85		25.95	33
2920	1/2" size		24	.333		13.30	10.85		24.15	31
2940	3/4" size		20	.400		13.05	13.05		28.15	36.50
2950	1" size		19	.421		21.50	13.75		35.25	45.00
2960	1-1/4" size		15	.533		47	17.40		64.40	78
2970	1-1/2" size		13	.615		36.50	20		56.50	71
2980	2" size		11	.727		52	23.50		75.50	93
2990	2-1/2" size	Q-1	15	1.067		130	31.50		161.50	191
3000	3" size	"	13	1.231		171	36		207	243
3850	Rising stem, soldered, 300 psi									
3950	1" size	1 Plum	19	.421	Ea.	41	13.75		54.75	66.50
3980	2" size	"	11	.727		110	23.50		133.50	157
4000	3" size	Q-1	13	1.231		350	36		386	440
4250	Threaded, class 150									
4310	1/4" size	1 Plum	24	.333	Ea.	19.60	10.85		30.45	38
4320	3/8" size		24	.333		19.60	10.85		30.45	38
4330	1/2" size		24	.333		18.30	10.85		29.15	36.50
4340	3/4" size		20	.400		21.50	13.05		34.55	44
4350	1" size		19	.421		29	13.75		42.75	52.50
4360	1-1/4" size		15	.533		39	17.40		56.40	69.50
4370	1-1/2" size		13	.615		48	20		68	83.50
4380	2" size		11	.727		66	23.50		89.50	109
4390	2-1/2" size	Q-1	15	1.067		153	31.50		184.50	217
4400	3" size	"	13	1.231		216	36		252	293
4500	For 300 psi, threaded, add					100%	15%			
4540	For chain operated type, add					15%				
4850	Globe, class 150, rising stem, threaded									
4920	1/4" size	1 Plum	24	.333	Ea.	20.50	10.85		31.35	39
4940	3/8" size		24	.333		20.50	10.85		31.35	39
4950	1/2" size		24	.333		21.50	10.85		32.35	40.50
4960	3/4" size		20	.400		29.50	13.05		42.55	52.50
4970	1" size		19	.421		46	13.75		59.75	71.50
4980	1-1/4" size		15	.533		72	17.40		89.40	106
4990	1-1/2" size		13	.615		87	20		107	126
5000	2" size		11	.727		131	23.50		154.50	181
5010	2-1/2" size	Q-1	15	1.067		264	31.50		295.50	340
5020	3" size	"	13	1.231		375	36		411	470

Important: See the Reference Section for critical supporting data - Reference Nos., Crews, & City Cost Indexes

	151 950	Valves	CREW	DAILY OUTPUT	LABOR-HOURS	UNIT	1999 BARE COSTS				TOTAL INCL O&P	
							MAT.	LABOR	EQUIP.	TOTAL		
955	5120	For 300 lb threaded, add					50%	15%				**955**
	5600	Relief, pressure & temperature, self-closing, ASME, threaded										
	5640	3/4" size	1 Plum	28	.286	Ea.	61.50	9.30		70.80	81.50	
	5650	1" size		24	.333		93.50	10.85		104.35	119	
	5660	1-1/4" size		20	.400		178	13.05		191.05	216	
	5670	1-1/2" size		18	.444		340	14.50		354.50	395	
	5680	2" size	↓	16	.500	↓	370	16.30		386.30	435	
	5950	Pressure, poppet type, threaded										
	6000	1/2" size	1 Plum	30	.267	Ea.	12.90	8.70		21.60	27.50	
	6040	3/4" size	"	28	.286	"	59.50	9.30		68.80	79.50	
	6400	Pressure, water, ASME, threaded										
	6440	3/4" size	1 Plum	28	.286	Ea.	39.50	9.30		48.80	57.50	
	6450	1" size		24	.333		76.50	10.85		87.35	100	
	6460	1-1/4"size		20	.400		129	13.05		142.05	162	
	6470	1-1/2" size		18	.444		173	14.50		187.50	212	
	6480	2" size		16	.500		250	16.30		266.30	300	
	6490	2-1/2" size	↓	15	.533	↓	810	17.40		827.40	915	
	6900	Reducing, water pressure										
	6920	300 psi to 25-75 psi, threaded or sweat										
	6940	1/2" size	1 Plum	24	.333	Ea.	100	10.85		110.85	126	
	6950	3/4" size		20	.400		100	13.05		113.05	130	
	6960	1" size		19	.421		155	13.75		168.75	191	
	6970	1-1/4" size		15	.533		277	17.40		294.40	330	
	6980	1-1/2" size	↓	13	.615		420	20		440	490	
	8350	Tempering, water, sweat connections										
	8400	1/2" size	1 Plum	24	.333	Ea.	37	10.85		47.85	57.50	
	8440	3/4" size	"	20	.400	"	45.50	13.05		58.55	70	
	8650	Threaded connections										
	8700	1/2" size	1 Plum	24	.333	Ea.	45.50	10.85		56.35	66.50	
	8740	3/4" size		20	.400		177	13.05		190.05	214	
	8750	1" size		19	.421		195	13.75		208.75	235	
	8760	1-1/4" size		15	.533		310	17.40		327.40	365	
	8770	1-1/2" size		13	.615		335	20		355	400	
	8780	2" size	↓	11	.727	↓	505	23.50		528.50	590	
960	0010	**VALVES, IRON BODY**										**960**
	1020	Butterfly, wafer type, gear actuator, 200 lb.										
	1030	2" size	1 Plum	14	.571	Ea.	109	18.65		127.65	147	
	1040	2-1/2" size	Q-1	9	1.778		112	52		164	202	
	1050	3" size		8	2		116	58.50		174.50	217	
	1060	4" size	↓	5	3.200		145	94		239	300	
	1070	5" size	Q-2	5	4.800		175	146		321	415	
	1080	6" size	"	5	4.800	↓	198	146		344	440	
	1650	Gate, 125 lb., N.R.S.										
	2150	Flanged										
	2200	2" size	1 Plum	5	1.600	Ea.	250	52		302	355	
	2240	2-1/2" size	Q-1	5	3.200		256	94		350	425	
	2260	3" size	↓	4.50	3.556		287	104		391	475	
	2280	4" size	↓	3	5.333		410	157		567	690	
	2300	6" size	Q-2	3	8	↓	700	243		943	1,150	
	3550	OS&Y, 125 lb., flanged										
	3600	2" size	1 Plum	5	1.600	Ea.	179	52		231	275	
	3660	3" size	Q-1	4.50	3.556		209	104		313	390	
	3680	4" size	"	3	5.333		222	157		379	480	
	3700	6" size	Q-2	3	8	↓	490	243		733	910	
	3900	For 250 lb, flanged, add					200%	10%				
	5450	Swing check, 125 lb., threaded										

MECHANICAL 15

			DAILY	LABOR-		1999 BARE COSTS				TOTAL	
151 950 | Valves		CREW	OUTPUT	HOURS	UNIT	MAT.	LABOR	EQUIP.	TOTAL	INCL O&P	
960	5500	2" size	1 Plum	11	.727	Ea.	270	23.50		293.50	335	960
	5540	2-1/2" size	Q-1	15	1.067		350	31.50		381.50	435	
	5550	3" size		13	1.231		375	36		411	470	
	5560	4" size	↓	10	1.600	↓	600	47		647	730	
	5950	Flanged										
	6000	2" size	1 Plum	5	1.600	Ea.	126	52		178	218	
	6040	2-1/2" size	Q-1	5	3.200		160	94		254	320	
	6050	3" size		4.50	3.556		239	104		343	420	
	6060	4" size	↓	3	5.333		271	157		428	535	
	6070	6" size	Q-2	3	8	↓	460	243		703	880	
975	0010	**VALVES, PLASTIC**										975
	1100	Angle, PVC, threaded										
	1110	1/4" size	1 Plum	26	.308	Ea.	47.50	10.05		57.55	67	
	1120	1/2" size		26	.308		47.50	10.05		57.55	67	
	1130	3/4" size		25	.320		55.50	10.45		65.95	77	
	1140	1" size	↓	23	.348	↓	66.50	11.35		77.85	90	
	1150	Ball, PVC, socket or threaded, single union										
	1200	1/4" size	1 Plum	26	.308	Ea.	19.75	10.05		29.80	36.50	
	1220	3/8" size		26	.308		19.75	10.05		29.80	36.50	
	1230	1/2" size		26	.308		19.75	10.05		29.80	36.50	
	1240	3/4" size		25	.320		23.50	10.45		33.95	42	
	1250	1" size		23	.348		28.50	11.35		39.85	48	
	1260	1-1/4" size		21	.381		37.50	12.40		49.90	60.50	
	1270	1-1/2" size		20	.400		47	13.05		60.05	72	
	1280	2" size	↓	17	.471		67.50	15.35		82.85	98	
	1360	For PVC, flanged, add				↓	100%	15%				
	1650	CPVC, socket or threaded, single union										
	1700	1/2" size	1 Plum	26	.308	Ea.	36.50	10.05		46.55	55	
	1720	3/4" size		25	.320		43.50	10.45		53.95	63.50	
	1730	1" size		23	.348		51	11.35		62.35	73	
	1750	1-1/4" size		21	.381		88	12.40		100.40	116	
	1760	1-1/2" size	↓	20	.400		88	13.05		101.05	117	
	1840	For CPVC, flanged, add					65%	15%				
	1880	For true union, socket or threaded, add				↓	50%	5%				
	2050	Polypropylene, threaded										
	2100	1/4" size	1 Plum	26	.308	Ea.	30	10.05		40.05	48	
	2120	3/8" size		26	.308		30	10.05		40.05	48	
	2130	1/2" size		26	.308		30	10.05		40.05	48	
	2140	3/4" size		25	.320		37.50	10.45		47.95	57	
	2150	1" size		23	.348		44.50	11.35		55.85	66	
	2160	1-1/4" size		21	.381		64.50	12.40		76.90	90	
	2170	1-1/2" size		20	.400		74	13.05		87.05	101	
	2180	2" size	↓	17	.471	↓	101	15.35		116.35	135	
	4850	Foot valve, PVC, socket or threaded										
	4900	1/2" size	1 Plum	34	.235	Ea.	36.50	7.65		44.15	51.50	
	4930	3/4" size		32	.250		41.50	8.15		49.65	58	
	4940	1" size		28	.286		54	9.30		63.30	73.50	
	4950	1-1/4" size		27	.296		104	9.65		113.65	129	
	4960	1-1/2" size	↓	26	.308	↓	104	10.05		114.05	129	
	6350	Y sediment strainer, PVC, socket or threaded										
	6400	1/2" size	1 Plum	26	.308	Ea.	29.50	10.05		39.55	47.50	
	6440	3/4" size		24	.333		32	10.85		42.85	52	
	6450	1" size		23	.348		39	11.35		50.35	60	
	6460	1-1/4" size		21	.381		64.50	12.40		76.90	90	
	6470	1-1/2" size	↓	20	.400	↓	67.50	13.05		80.55	94.50	

15 MECHANICAL

Important: See the Reference Section for critical supporting data - Reference Nos., Crews, & City Cost Indexes

151 950	Valves	CREW	DAILY OUTPUT	LABOR-HOURS	UNIT	1999 BARE COSTS				TOTAL INCL O&P	
						MAT.	LABOR	EQUIP.	TOTAL		
980	0010	**VALVES, STEEL**									980
	0800	Cast									
	1350	Check valve, swing type, 150 lb., flanged									
	1400	2" size	1 Plum	8	1	Ea.	465	32.50		497.50	565
	1440	2-1/2" size	Q-1	5	3.200		595	94		689	795
	1450	3" size		4.50	3.556		595	104		699	815
	1460	4" size	↓	3	5.333		875	157		1,032	1,200
	1540	For 300 lb., flanged, add					50%	15%			
	1548	For 600 lb., flanged, add				↓	110%	20%			
	1950	Gate valve, 150 lb., flanged									
	2000	2" size	1 Plum	8	1	Ea.	515	32.50		547.50	615
	2040	2-1/2"size	Q-1	5	3.200		725	94		819	940
	2050	3" size		4.50	3.556		725	104		829	960
	2060	4" size	↓	3	5.333		885	157		1,042	1,200
	2070	6" size	Q-2	3	8	↓	1,400	243		1,643	1,925
	3650	Globe valve, 150 lb., flanged									
	3700	2" size	1 Plum	8	1	Ea.	675	32.50		707.50	790
	3740	2-1/2" size	Q-1	5	3.200		855	94		949	1,075
	3750	3" size		4.50	3.556		855	104		959	1,100
	3760	4" size	↓	3	5.333		1,250	157		1,407	1,600
	3770	6" size	Q-2	3	8	↓	1,950	243		2,193	2,525
	5150	Forged									
	5650	Check valve, class 800, horizontal, socket									
	5698	Threaded									
	5700	1/4" size	1 Plum	24	.333	Ea.	41	10.85		51.85	61.50
	5720	3/8" size		24	.333		41	10.85		51.85	61.50
	5730	1/2" size		24	.333		41	10.85		51.85	61.50
	5740	3/4" size		20	.400		41	13.05		54.05	65
	5750	1" size		19	.421		52	13.75		65.75	78
	5760	1-1/4" size	↓	15	.533	↓	99.50	17.40		116.90	136

MECHANICAL 15

152 100	Fixtures	CREW	DAILY OUTPUT	LABOR-HOURS	UNIT	1999 BARE COSTS				TOTAL INCL O&P	
						MAT.	LABOR	EQUIP.	TOTAL		
101	0010	**FIXTURES** Includes trim fittings unless otherwise noted	R151 -420								101
	0080	For rough-in, supply, waste, and vent, see add for each type									
	0120	For electric water coolers, see division 153-105									
	0160	For color, unless otherwise noted, add	↓			Ea.	20%				
104	0010	**BATHS**									104
	0100	Tubs, recessed porcelain enamel on cast iron, with trim									
	0140	42" x 37"	Q-1	5	3.200	Ea.	660	94		754	865
	0180	48" x 42"		4	4		1,050	117		1,167	1,325
	0220	72" x 36"		3	5.333		1,100	157		1,257	1,450
	2000	Enameled formed steel, 4'-6" long		5.80	2.759		245	81		326	395
	2200	5' long		5.50	2.909		229	85.50		314.50	380
	4000	Soaking, acrylic with pop-up drain, 60" x 32" x 21" deep		5.50	2.909		690	85.50		775.50	890
	4100	60" x 48" x 18-1/2" deep	↓	5	3.200	↓	565	94		659	760
	6000	Whirlpool, bath with vented overflow, molded fiberglass									
	6100	66" x 48" x 24"	Q-1	1	16	Ea.	1,950	470		2,420	2,850
	9600	Rough-in, supply, waste and vent, for all above tubs, add	"	2.07	7.729	"	129	227		356	485

152 100	Fixtures		CREW	DAILY OUTPUT	LABOR-HOURS	UNIT	1999 BARE COSTS				TOTAL INCL O&P
							MAT.	LABOR	EQUIP.	TOTAL	
116	0010	**DRINKING FOUNTAIN** For connection to cold water supply										**116**
	1000	Wall mounted, non-recessed									
	1400	Bronze, with no back	1 Plum	4	2	Ea.	930	65		995	1,125
	1800	Cast aluminum, enameled, for correctional institutions		4	2		550	65		615	705
	2000	Fiberglass, 12" back, single bubbler unit		4	2		490	65		555	640
	2040	Dual bubbler		3.20	2.500		710	81.50		791.50	910
	2400	Precast stone, no back		4	2		445	65		510	590
	2700	Stainless steel, single bubbler, no back		4	2		630	65		695	790
	2740	With back		4	2		670	65		735	840
	2780	Dual handle & wheelchair projection type		4	2		465	65		530	610
	2820	Dual level for handicapped type		3.20	2.500		930	81.50		1,011.50	1,150
	3300	Vitreous china									
	3340	7" back	1 Plum	4	2	Ea.	380	65		445	515
	3940	For vandal-resistant bottom plate, add					47.50			47.50	52
	3960	For freeze-proof valve system, add	1 Plum	2	4		370	130		500	610
	3980	For rough-in, supply and waste, add	"	2.21	3.620		53.50	118		171.50	238
	4000	Wall mounted, semi-recessed									
	4200	Poly-marble, single bubbler	1 Plum	4	2	Ea.	595	65		660	755
	4600	Stainless steel, satin finish, single bubbler		4	2		430	65		495	570
	4900	Vitreous china, single bubbler		4	2		430	65		495	575
	5980	For rough-in, supply and waste, add		1.83	4.372		53.50	143		196.50	275
	6000	Wall mounted, fully recessed									
	6400	Poly-marble, single bubbler	1 Plum	4	2	Ea.	680	65		745	845
	6800	Stainless steel, single bubbler		4	2		550	65		615	705
	7560	For freeze-proof valve system, add		2	4		450	130		580	695
	7580	For rough-in, supply and waste, add		1.83	4.372		53.50	143		196.50	275
	7600	Floor mounted, pedestal type									
	7700	Aluminum, architectural style, C.I. base	1 Plum	2	4	Ea.	410	130		540	650
	7780	Wheelchair handicap unit		2	4		975	130		1,105	1,275
	8400	Stainless steel, architectural style		2	4		1,050	130		1,180	1,375
	8600	Enameled iron, heavy duty service, 2 bubblers		2	4		980	130		1,110	1,275
	8660	4 bubblers		2	4		1,050	130		1,180	1,350
	8880	For freeze-proof valve system, add		2	4		300	130		430	530
	8900	For rough-in, supply and waste, add		1.83	4.372		53.50	143		196.50	275
	9100	Deck mounted									
	9500	Stainless steel, circular receptor	1 Plum	4	2	Ea.	272	65		337	400
	9760	White enameled steel, 14" x 9" receptor		4	2		225	65		290	345
	9860	White enameled cast iron, 24" x 16" receptor		3	2.667		214	87		301	370
	9980	For rough-in, supply and waste, add		1.83	4.372		53.50	143		196.50	275
120	0010	**HOT WATER DISPENSERS**										**120**
	0160	Commercial, 100 cup, 11.3 amp	1 Plum	14	.571	Ea.	315	18.65		333.65	375
	3180	Household, 60 cup	"	14	.571	"	168	18.65		186.65	213
124	0010	**INDUSTRIAL SAFETY FIXTURES** Rough-in not included										**124**
	1000	Eye wash fountain									
	1400	Plastic bowl, pedestal mounted	Q-1	4	4	Ea.	170	117		287	365
	1600	Unmounted		4	4		122	117		239	310
	1800	Wall mounted		4	4		131	117		248	320
	2000	Stainless steel, pedestal mounted		4	4		262	117		379	465
	2200	Unmounted		4	4		173	117		290	370
	2400	Wall mounted		4	4		170	117		287	365
	3000	Eye wash, portable, self-contained					365			365	400
	4000	Eye and face wash, combination fountain									
	4200	Stainless steel, pedestal mounted	Q-1	4	4	Ea.	370	117		487	590
	4400	Unmounted		4	4		278	117		395	485

R151 -420

15 MECHANICAL

		152 100	Fixtures	CREW	DAILY OUTPUT	LABOR-HOURS	UNIT	MAT.	LABOR	EQUIP.	TOTAL	TOTAL INCL O&P	
								1999 BARE COSTS					
124	4600		Wall mounted	Q-1	4	4	Ea.	288	117		405	495	124
	5000		Shower, single head, drench, ball valve, pull, freestanding		4	4		236	117		353	440	
	5200		Horizontal or vertical supply		4	4		221	117		338	420	
	6000		Multi-nozzle, eye/face wash combination		4	4		480	117		597	710	
	6400		Multi-nozzle, 12 spray, shower only		4	4		1,025	117		1,142	1,300	
	6600		For freeze-proof, add		6	2.667		415	78.50		493.50	575	
128	0010	**INTERCEPTORS**											128
	0150		Grease, cast iron, 4 GPM, 8 lb. fat capacity	1 Plum	4	2	Ea.	355	65		420	490	
	0200		7 GPM, 14 lb. fat capacity		4	2		490	65		555	640	
	1000		10 GPM, 20 lb. fat capacity		4	2		575	65		640	735	
	1040		15 GPM, 30 lb. fat capacity		4	2		855	65		920	1,050	
	1060		20 GPM, 40 lb. fat capacity		3	2.667		1,050	87		1,137	1,275	
	1120		Fabricated steel, 50 GPM, 100 lb. fat capacity	Q-1	2	8		1,625	235		1,860	2,150	
	1160		100 GPM, 200 lb. fat capacity	"	2	8		4,125	235		4,360	4,875	
	1560		For chemical add-port, add					77.50			77.50	85	
	1580		For seepage pan, add					7%					
	3000		Hair, cast iron, 1-1/4" and 1-1/2" pipe connection	1 Plum	8	1	Ea.	132	32.50		164.50	196	
	3100		For chrome-plated cast iron, add					81.50			81.50	89.50	
	4000		Oil, fabricated steel, 10 GPM, 2" pipe size	1 Plum	4	2		745	65		810	920	
	4100		15 GPM, 2" or 3" pipe size		4	2		1,025	65		1,090	1,225	
	4120		20 GPM, 2" or 3" pipe size		3	2.667		1,225	87		1,312	1,475	
	4220		100 GPM, 3" pipe size	Q-1	2	8		4,125	235		4,360	4,875	
	6000		Solids, precious metals recovery, C.I., 1-1/4" to 2" pipe	1 Plum	4	2		200	65		265	320	
	6100		Dental Lab., large, C.I., 1-1/2" to 2" pipe	"	3	2.667		700	87		787	900	
136	0010	**LAVATORIES** With trim, white unless noted otherwise											136
	0500		Vanity top, porcelain enamel on cast iron										
	0600		20" x 18"	Q-1	6.40	2.500	Ea.	160	73.50		233.50	287	
	0640		33" x 19" oval		6.40	2.500		320	73.50		393.50	465	
	0720		18" round		6.40	2.500		164	73.50		237.50	292	
	0860		For color, add					25%					
	1000		Cultured marble, 19" x 17", single bowl	Q-1	6.40	2.500	Ea.	98.50	73.50		172	219	
	1040		25" x 19", single bowl	"	6.40	2.500	"	126	73.50		199.50	249	
	1580		For color, same price										
	1900		Stainless steel, self-rimming, 25" x 22", single bowl, ledge	Q-1	6.40	2.500	Ea.	149	73.50		222.50	275	
	1960		17" x 22", single bowl		6.40	2.500		145	73.50		218.50	270	
	2600		Steel, enameled, 20" x 17", single bowl		5.80	2.759		83	81		164	215	
	2660		19" round		5.80	2.759		81	81		162	213	
	2900		Vitreous china, 20" x 16", single bowl		5.40	2.963		163	87		250	310	
	2960		20" x 17", single bowl		5.40	2.963		157	87		244	305	
	3020		19" round, single bowl		5.40	2.963		154	87		241	300	
	3200		22" x 13", single bowl		5.40	2.963		172	87		259	320	
	3560		For color, add					50%					
	3580		Rough-in, supply, waste and vent for all above lavatories	Q-1	2.30	6.957	Ea.	74	204		278	390	
	4000		Wall hung										
	4040		Porcelain enamel on cast iron, 16" x 14", single bowl	Q-1	8	2	Ea.	330	58.50		388.50	450	
	4180		20" x 18", single bowl		8	2		185	58.50		243.50	292	
	4240		22" x 19", single bowl		8	2		340	58.50		398.50	465	
	4580		For color, add					30%					
	6000		Vitreous china, 18" x 15", single bowl with backsplash	Q-1	7	2.286	Ea.	178	67		245	298	
	6180		19" x 19", corner style	"	8	2	"	360	58.50		418.50	485	
	6500		For color, add					30%					
	6960		Rough-in, supply, waste and vent for above lavatories	Q-1	1.66	9.639	Ea.	162	283		445	610	
140	0010	**LAUNDRY SINKS** With trim											140
	0020		Porcelain enamel on cast iron, black iron frame										
	0050		24" x 20", single compartment	Q-1	6	2.667	Ea.	335	78.50		413.50	490	
	0100		24" x 23", single compartment	"	6	2.667	"	365	78.50		443.50	520	

R151 -420

MECHANICAL 15

	152 100	Fixtures	CREW	DAILY OUTPUT	LABOR-HOURS	UNIT	MAT.	LABOR	EQUIP.	TOTAL	TOTAL INCL O&P	
140	2000	Molded stone, on wall hanger or legs										**140**
	2020	22" x 23", single compartment	Q-1	6	2.667	Ea.	115	78.50		193.50	245	
	2100	45" x 21", double compartment	"	5	3.200	"	206	94		300	370	
	3000	Plastic, on wall hanger or legs										
	3020	18" x 23", single compartment	Q-1	6.50	2.462	Ea.	80	72.50		152.50	197	
	3300	40" x 24", double compartment		5.50	2.909		184	85.50		269.50	330	
	5000	Stainless steel, counter top, 22" x 17" single compartment		6	2.667		242	78.50		320.50	385	
	5100	22" x 22", single compartment		6	2.667		305	78.50		383.50	455	
	5200	33" x 22", double compartment		5	3.200		281	94		375	450	
	9600	Rough-in, supply, waste and vent, for all laundry sinks		2.14	7.477		85	219		304	430	
148	0010	**SHOWERS**										**148**
	1500	Stall, with door and trim										
	3000	Fiberglass, one piece, with 3 walls, 32" x 32" square	Q-1	2.40	6.667	Ea.	297	196		493	620	
	3100	36" x 36" square		2.40	6.667		340	196		536	670	
	3250	64" x 65-3/4" x 81-1/2" fold. seat, whlchr. ♿		1.80	8.889		2,725	261		2,986	3,400	
	4000	Polypropylene, with molded-stone floor, 30" x 30"		2	8		345	235		580	730	
	4200	Rough-in, supply, waste and vent for above showers		2.05	7.805		80.50	229		309.50	435	
	5000	Built-in, head, arm, 4 GPM valve	1 Plum	4	2		67.50	65		132.50	173	
	5200	Head, arm, by-pass, integral stops, handles	"	3.60	2.222		183	72.50		255.50	310	
	6000	Group, w/pressure balancing valve, rough-in and rigging not included										
	6800	Column, 6 heads, no receptors, less partitions	Q-1	3	5.333	Ea.	4,325	157		4,482	4,975	
	6900	With stainless steel partitions		1	16		10,300	470		10,770	12,000	
	7600	5 heads, no receptors, less partitions		3	5.333		3,525	157		3,682	4,100	
	7700	With stainless steel partitions		1	16		8,525	470		8,995	10,100	
	8000	Wall, 2 heads, no receptors, less partitions		4	4		1,775	117		1,892	2,125	
	8100	With stainless steel partitions		2	8		3,500	235		3,735	4,200	
152	0010	**SINKS** With faucets and drain										**152**
	2000	Kitchen, counter top style, P.E. on C.I., 24" x 21" single bowl	Q-1	5.60	2.857	Ea.	195	84		279	340	
	2100	30" x 21" single bowl		5.60	2.857		360	84		444	520	
	2200	32" x 21" double bowl		4.80	3.333		283	98		381	460	
	2300	42" x 21" double bowl		4.80	3.333		405	98		503	595	
	3000	Stainless steel, self rimming, 19" x 18" single bowl		5.60	2.857		251	84		335	405	
	3100	25" x 22" single bowl		5.60	2.857		276	84		360	430	
	4000	Steel, enameled, with ledge, 24" x 21" single bowl		5.60	2.857		97.50	84		181.50	234	
	4100	32" x 21" double bowl		4.80	3.333		114	98		212	273	
	4960	For color sinks except stainless steel, add					10%					
	4980	For rough-in, supply, waste and vent, counter top sinks	Q-1	2.14	7.477		85	219		304	430	
	5000	Kitchen, raised deck, P.E. on C.I.										
	5100	32" x 21", dual level, double bowl	Q-1	2.60	6.154	Ea.	345	181		526	650	
	5700	For color, add					20%					
	5790	For rough-in, supply, waste & vent, sinks	Q-1	1.85	8.649		85	254		339	480	
	6650	Service, floor, corner, P.E. on C.I., 28" x 28"		4.40	3.636		495	107		602	700	
	6790	For rough-in, supply, waste & vent, floor service sinks		1.64	9.756		207	286		493	660	
	7000	Service, wall, P.E. on C.I., roll rim, 22" x 18"		4	4		370	117		487	585	
	7100	24" x 20"		4	4		400	117		517	620	
	8600	Vitreous china, 22" x 20"		4	4		415	117		532	640	
	8960	For stainless steel rim guard, front or side, add					29			29	32	
	8980	For rough-in, supply, waste & vent, wall service sinks	Q-1	1.30	12.308		320	360		680	900	
168	0010	**URINALS**										**168**
	3000	Wall hung, vitreous china, with hanger & self-closing valve										
	3100	Siphon jet type	Q-1	3	5.333	Ea.	271	157		428	535	
	3120	Blowout type		3	5.333		370	157		527	640	
	3300	Rough-in, supply, waste & vent		2.83	5.654		87.50	166		253.50	350	
	5000	Stall type, vitreous china, includes valve		2.50	6.400		645	188		833	995	
	5100	3" seam cover, add		12	1.333		114	39		153	185	
	5200	6" seam cover, add		12	1.333		158	39		197	233	

R151 -420

Important: See the Reference Section for critical supporting data - Reference Nos., Crews, & City Cost Indexes

			CREW	DAILY OUTPUT	LABOR-HOURS	UNIT	1999 BARE COSTS				TOTAL INCL O&P	
	152 100	**Fixtures**					MAT.	LABOR	EQUIP.	TOTAL		
168	6980	Rough-in, supply, waste and vent	Q-1	1.99	8.040	Ea.	126	236		362	500	168
176	0010	**WASH FOUNTAINS** Rigging not included										176
	1900	Group, foot control										
	2000	Precast terrazzo, circular, 36" diam., 5 or 6 persons	Q-2	3	8	Ea.	2,600	243		2,843	3,250	
	2100	54" diameter for 8 or 10 persons		2.50	9.600		3,250	292		3,542	4,025	
	2400	Semi-circular, 36" diam. for 3 persons		3	8		2,400	243		2,643	3,025	
	2500	54" diam. for 4 or 5 persons		2.50	9.600		2,900	292		3,192	3,625	
	2700	Quarter circle (corner), 54" for 3 persons		3.50	6.857		2,875	209		3,084	3,475	
	3000	Stainless steel, circular, 36" diameter		3.50	6.857		1,425	209		1,634	1,875	
	3100	54" diameter		2.80	8.571		1,650	261		1,911	2,200	
	3400	Semi-circular, 36" diameter		3.50	6.857		1,300	209		1,509	1,750	
	3500	54" diameter		2.80	8.571		2,300	261		2,561	2,925	
	5610	Group, infrared control, barrier free										
	5614	Precast terrazzo										
	5620	Semi-circular 36" diam. for 3 persons	Q-2	3	8	Ea.	3,575	243		3,818	4,300	
	5630	46" diam. for 4 persons		2.80	8.571		3,850	261		4,111	4,650	
	5640	Circular, 54" diam. for 8 persons, button control		2.50	9.600		6,000	292		6,292	7,050	
	5700	Rough-in, supply, waste and vent for above wash fountains	Q-1	1.82	8.791		117	258		375	520	
	6200	Duo for small washrooms, stainless steel		2	8		1,975	235		2,210	2,500	
	6500	Rough-in, supply, waste & vent for duo fountains		2.02	7.921		68	232		300	425	
180	0010	**WATER CLOSETS**	R151 -420									180
	0030	For automatic flush, see 151-141-0972										
	0150	Tank type, vitreous china, incl. seat, supply pipe w/stop										
	0200	Wall hung, one piece	Q-1	5.30	3.019	Ea.	435	88.50		523.50	615	
	0400	Two piece, close coupled		5.30	3.019		380	88.50		468.50	550	
	0960	For rough-in, supply, waste, vent and carrier		2.73	5.861		254	172		426	540	
	1000	Floor mounted, one piece		5.30	3.019		440	88.50		528.50	615	
	1100	Two piece, close coupled, water saver **CN**		5.30	3.019		133	88.50		221.50	280	
	1960	For color, add					30%					
	1980	For rough-in, supply, waste and vent	Q-1	3.05	5.246	Ea.	116	154		270	360	
	3000	Bowl only, with flush valve, seat										
	3100	Wall hung	Q-1	5.80	2.759	Ea.	345	81		426	505	
	3200	For rough-in, supply, waste and vent, single WC		2.56	6.250		265	183		448	570	
	3300	Floor mounted		5.80	2.759		286	81		367	440	
	3350	With wall outlet		5.80	2.759		455	81		536	625	
	3400	For rough-in, supply, waste and vent, single WC		2.84	5.634		128	165		293	390	

152 400 | Pumps

			CREW	DAILY OUTPUT	LABOR-HOURS	UNIT	MAT.	LABOR	EQUIP.	TOTAL	INCL O&P	
401	0010	**PUMPS** For fire pumps see Division 154-145										401
410	0010	**PUMPS, CIRCULATING** Heated or chilled water application										410
	0600	Bronze, sweat connections, 1/40 HP, in line										
	0640	3/4" size	Q-1	16	1	Ea.	110	29.50		139.50	166	
	1000	Flange connection, 3/4" to 1-1/2" size										
	1040	1/12 HP	Q-1	6	2.667	Ea.	300	78.50		378.50	450	
	1060	1/8 HP		6	2.667		520	78.50		598.50	690	
	1100	1/3 HP		6	2.667		570	78.50		648.50	745	
	1140	2" size, 1/6 HP		5	3.200		740	94		834	955	
	1180	2-1/2" size, 1/4 HP		5	3.200		965	94		1,059	1,225	
	2000	Cast iron, flange connection										
	2040	3/4" to 1-1/2" size, in line, 1/12 HP	Q-1	6	2.667	Ea.	197	78.50		275.50	335	
	2100	1/3 HP		6	2.667		365	78.50		443.50	525	
	2140	2" size, 1/6 HP		5	3.200		400	94		494	580	
	2180	2-1/2" size, 1/4 HP		5	3.200		525	94		619	720	

MECHANICAL 15

152 400 | Pumps

		Description	CREW	DAILY OUTPUT	LABOR-HOURS	UNIT	MAT.	LABOR	EQUIP.	TOTAL	TOTAL INCL O&P	
410	2220	3″ size, 1/4 HP	Q-1	4	4	Ea.	535	117		652	765	410
	2600	For non-ferrous impeller, add					3%					
415	0010	**PUMPS, CONDENSATE RETURN SYSTEM**										415
	0200	Simplex, 3/4 H.P. mtr, float switch, controls, 10 Gal. C.I. rcvr, 6-15GPM	Q-1	1	16	Ea.	1,125	470		1,595	1,950	
	1000	Duplex, 2 pumps, 3/4 H.P. motors, float switch,										
	1060	alternator asssembly, 15 Gal. C.I. receiver	Q-1	.50	32	Ea.	3,025	940		3,965	4,750	
430	0010	**PUMPS, GENERAL UTILITY** With motor										430
	0200	Multi-stage, horizontal split, for boiler feed applications										
	0300	Two stage, 3″ discharge x 4″ suction, 75 HP	Q-7	.30	106	Ea.	14,900	3,325		18,225	21,500	
	0340	Four stage, 3″ discharge x 4″ suction, 150 HP	"	.18	177	"	24,500	5,550		30,050	35,400	
	2000	Single stage										
	2060	End suction, 1″D. x 2″S., 3 HP	Q-1	.50	32	Ea.	3,650	940		4,590	5,425	
	2100	1-1/2″D. x 3″S., 10 HP	"	.40	40		3,925	1,175		5,100	6,100	
	2140	2″D. x 3″S., 15 HP	Q-2	.60	40		4,275	1,225		5,500	6,550	
	3000	Double suction, 2″D. x 2-1/2″S., 10 HP	Q-1	.30	53.333		5,050	1,575		6,625	7,925	
	3060	3″D. x 4″S., 15 HP	Q-2	.46	52.174		5,200	1,600		6,800	8,125	
	3180	6″D. x 8″S., 60 HP	Q-3	.30	106		8,475	3,325		11,800	14,400	
	3190	75 HP, to 2500 GPM		.28	114		10,700	3,550		14,250	17,200	
	3220	100 HP, to 3000 GPM		.26	123		14,400	3,825		18,225	21,700	
	3240	150 HP, to 4000 GPM		.24	133		17,700	4,150		21,850	25,800	
440	0010	**PUMPS, GRINDER SYSTEM** Complete, incl. check valve, tank, std.										440
	0020	controls incl. alarm/disconnect panel w/wire. Excavation not included										
	0260	Simplex, 9 GPM at 60 PSIG, 60 gal. tank				Ea.	3,225			3,225	3,550	
	0300	For manway, 26″ I.D., 18″ high, add					610			610	670	
	0340	26″ I.D., 36″ high, add					755			755	830	
	0380	43″ I.D., 4′ high, add					925			925	1,025	
450	0010	**PUMPS, PRESSURE BOOSTER SYSTEM**										450
	0200	Pump system, with diaphragm tank, control, press. switch										
	0300	1 HP pump	Q-1	1.30	12.308	Ea.	2,950	360		3,310	3,800	
	0400	1-1/2 HP pump		1.25	12.800		2,975	375		3,350	3,850	
	0420	2 HP pump		1.20	13.333		3,050	390		3,440	3,950	
	0440	3 HP pump		1.10	14.545		3,100	425		3,525	4,050	
	0460	5 HP pump	Q-2	1.50	16		3,425	485		3,910	4,525	
	0480	7-1/2 HP pump		1.42	16.901		3,825	515		4,340	4,975	
	0500	10 HP pump		1.34	17.910		4,000	545		4,545	5,225	
	1000	Pump/ energy storage system, diaphragm tank, 3 HP pump										
	1100	motor, PRV, switch, gauge, control center, flow switch										
	1200	125 lb. working pressure	Q-2	.70	34.286	Ea.	8,575	1,050		9,625	11,000	
	1300	250 lb. working pressure	"	.64	37.500	"	9,375	1,150		10,525	12,000	
460	0010	**PUMPS, PEDESTAL SUMP** With float control										460
	0400	Molded PVC base, 21 GPM at 15′ head, 1/3 HP	1 Plum	5	1.600	Ea.	86.50	52		138.50	174	
	0800	Iron base, 21 GPM at 15′ head, 1/3 HP		5	1.600		110	52		162	200	
	1200	Solid brass, 21 GPM at 15′ head, 1/3 HP		5	1.600		179	52		231	276	
465	0010	**PUMPS, SEWAGE EJECTOR** With operating and level controls										465
	0100	Simplex system incl. tank, cover, pump 15′ head										
	0500	37 gal PE tank, 12 GPM, 1/2 HP, 2″ discharge	Q-1	3.20	5	Ea.	360	147		507	615	
	0510	3″ discharge		3.10	5.161		385	151		536	655	
	0530	87 GPM, .7 HP, 2″ discharge		3.20	5		560	147		707	835	
	0540	3″ discharge		3.10	5.161		605	151		756	895	
	0600	45 gal. coated stl tank, 12 GPM, 1/2 HP, 2″ discharge		3	5.333		645	157		802	940	
	0610	3″ discharge		2.90	5.517		670	162		832	980	
	0630	87 GPM, .7 HP, 2″ discharge		3	5.333		825	157		982	1,150	
	0640	3″ discharge		2.90	5.517		875	162		1,037	1,200	
	0660	134 GPM, 1 HP, 2″ discharge		2.80	5.714		895	168		1,063	1,250	
	0680	3″ discharge		2.70	5.926		940	174		1,114	1,300	

15 MECHANICAL

Important: See the Reference Section for critical supporting data - Reference Nos., Crews, & City Cost Indexes

152 | Plumbing Fixtures

152 400 | Pumps

		CREW	DAILY OUTPUT	LABOR-HOURS	UNIT	MAT.	LABOR	EQUIP.	TOTAL	TOTAL INCL O&P	
465	0700	70 gal. PE tank, 12 GPM, 1/2 HP, 2" discharge	Q-1	2.60	6.154	Ea.	720	181		901	1,075	**465**
	0710	3" discharge		2.40	6.667		775	196		971	1,150	
	0730	87 GPM, 0.7 HP, 2" discharge		2.50	6.400		935	188		1,123	1,300	
	0740	3" discharge		2.30	6.957		995	204		1,199	1,400	
	0760	134 GPM, 1 HP, 2" discharge		2.20	7.273		1,025	213		1,238	1,450	
	0770	3" discharge	↓	2	8	↓	1,100	235		1,335	1,550	
480	0010	**PUMPS, SUBMERSIBLE** Sump										**480**
	7000	Sump pump, automatic										
	7100	Plastic, 1-1/4" discharge, 1/4 HP	1 Plum	6	1.333	Ea.	117	43.50		160.50	194	
	7140	1/3 HP		5	1.600		142	52		194	235	
	7160	1/2 HP		5	1.600		170	52		222	266	
	7180	1-1/2" discharge, 1/2 HP		4	2		190	65		255	310	
	7500	Cast iron, 1-1/4" discharge, 1/4 HP		6	1.333		138	43.50		181.50	218	
	7540	1/3 HP		6	1.333		155	43.50		198.50	237	
	7560	1/2 HP	↓	5	1.600	↓	195	52		247	294	
490	0010	**PUMPS, WELL** Water system, with pressure control										**490**
	1000	Deep well, jet, 42 gal. galvanized tank										
	1040	3/4 HP	1 Plum	.80	10	Ea.	485	325		810	1,025	
	3000	Shallow well, jet, 30 gal. galvanized tank										
	3040	1/2 HP	1 Plum	2	4	Ea.	550	130		680	805	

153 | Plumbing Appliances

153 100 | Water Appliances

		CREW	DAILY OUTPUT	LABOR-HOURS	UNIT	MAT.	LABOR	EQUIP.	TOTAL	TOTAL INCL O&P	
105	0010	**WATER COOLER**	R151 -430									**105**
	0030	See line 153-105-9800 for rough-in, waste & vent										
	0040	for all water coolers										
	0100	Wall mounted, non-recessed										
	0140	4 GPH	Q-1	4	4	Ea.	420	117		537	645	
	0160	8 GPH, barrier free, sensor operated		4	4		770	117		887	1,025	
	0180	8.2 GPH	↓	4	4		500	117		617	730	
	0600	For hot and cold water, add					128			128	141	
	0640	For stainless steel cabinet, add					66			66	72.50	
	1000	Dual height, 8.2 GPH	Q-1	3.80	4.211		690	124		814	940	
	1040	14.3 GPH	"	3.80	4.211		735	124		859	995	
	1240	For stainless steel cabinet, add					106			106	116	
	2600	Wheelchair type, 8 GPH	Q-1	4	4		1,175	117		1,292	1,475	
	3300	Semi-recessed, 8.1 GPH		4	4		690	117		807	935	
	3320	12 GPH	↓	4	4	↓	730	117		847	985	
	4600	Floor mounted, flush-to-wall										
	4640	4 GPH	1 Plum	3	2.667	Ea.	480	87		567	660	
	4680	8.2 GPH		3	2.667		510	87		597	690	
	4720	14.3 GPH	↓	3	2.667		530	87		617	710	
	4960	For hot and cold water, add					154			154	169	
	4980	For stainless steel cabinet, add					89.50			89.50	98.50	
	5000	Dual height, 8.2 GPH	1 Plum	2	4		745	130		875	1,025	
	5040	14.3 GPH	"	2	4		760	130		890	1,025	
	5120	For stainless steel cabinet, add					131			131	144	

For expanded coverage of these items see *Means Mechanical or Plumbing Cost Data 1999*

MECHANICAL 15

			CREW	DAILY OUTPUT	LABOR-HOURS	UNIT	1999 BARE COSTS				TOTAL INCL O&P	
153 100		**Water Appliances**					MAT.	LABOR	EQUIP.	TOTAL		
105	9800	For supply, waste & vent, all coolers	R151 -430	1 Plum	2.21	3.620	Ea.	53.50	118		171.50	238	105
110	0010	**WATER HEATERS**											110
	0020	For solar, see division 155-471											
	1000	Residential, electric, glass lined tank, 5 yr, 10 gal., single element		1 Plum	2.30	3.478	Ea.	202	113		315	395	
	1040	20 gallon, single element			2.20	3.636		239	119		358	445	
	1060	30 gallon, double element			2.20	3.636		259	119		378	465	
	1080	40 gallon, double element			2	4		280	130		410	510	
	1100	52 gallon, double element			2	4		320	130		450	555	
	1180	120 gallon, double element			1.40	5.714		720	186		906	1,075	
	2000	Gas fired, glass lined tank, 5 yr, vent not incl., 20 gallon			2.10	3.810		280	124		404	500	
	2040	30 gallon			2	4		287	130		417	515	
	2100	75 gallon			1.50	5.333		630	174		804	960	
	2120	100 gallon			1.30	6.154		1,025	201		1,226	1,450	
	3000	Oil fired, glass lined tank, 5 yr, vent not included, 30 gallon			2	4		775	130		905	1,050	
	3040	50 gallon			1.80	4.444		1,025	145		1,170	1,375	
	3060	70 gallon			1.50	5.333		1,450	174		1,624	1,875	
	3080	85 gallon			1.40	5.714		1,975	186		2,161	2,450	
	4000	Commercial, 100° rise. NOTE: for each size tank, a range of											
	4010	heaters between the ones shown are available											
	4020	Electric											
	4100	5 gal., 3 kW, 12 GPH, 208V		1 Plum	2	4	Ea.	1,200	130		1,330	1,500	
	4120	10 gal., 6 kW, 25 GPH, 208V			2	4		1,325	130		1,455	1,650	
	4140	50 gal., 9 kW, 37 GPH, 208V			1.80	4.444		1,800	145		1,945	2,225	
	4160	50 gal., 36 kW, 148 GPH, 208V			1.80	4.444		2,775	145		2,920	3,275	
	4300	200 gal., 15 kW, 61 GPH, 480V		Q-1	1.70	9.412		8,550	276		8,826	9,825	
	4320	200 gal., 120 kW, 490 GPH, 480V			1.70	9.412		11,700	276		11,976	13,200	
	4460	400 gal., 30 kW, 123 GPH, 480V			1	16		11,600	470		12,070	13,500	
	5400	Modulating step control, 2-5 steps		1 Elec	5.30	1.509		1,125	48		1,173	1,300	
	5440	6-10 steps			3.20	2.500		1,450	80		1,530	1,700	
	5460	11-15 steps			2.70	2.963		1,775	94.50		1,869.50	2,100	
	5480	16-20 steps			1.60	5		2,100	160		2,260	2,550	
	6000	Gas fired, flush jacket, std. controls, vent not incl.											
	6040	75 MBH input, 73 GPH		1 Plum	1.40	5.714	Ea.	1,175	186		1,361	1,550	
	6060	98 MBH input, 95 GPH			1.40	5.714		2,000	186		2,186	2,475	
	6080	120 MBH input, 114 GPH CN			1.20	6.667		2,200	217		2,417	2,750	
	6180	200 MBH input, 192 GPH			.60	13.333		3,250	435		3,685	4,225	
	6200	240 MBH input, 230 GPH			.50	16		3,300	520		3,820	4,450	
	6900	For low water cutoff, add			8	1		189	32.50		221.50	258	
	6960	For bronze body hot water circulator, add			4	2		720	65		785	895	
	8000	Oil fired, flush jacket, std. controls, vent not incl.											
	8060	103 MBH gross output, 116 GPH		1 Plum	1.10	7.273	Ea.	1,675	237		1,912	2,200	
	8080	122 MBH gross output, 141 GPH			1	8		1,850	261		2,111	2,425	
	8100	137 MBH gross output, 161 GPH			.80	10		1,900	325		2,225	2,575	
	8160	225 MBH gross output, 256 GPH		Q-1	.80	20		2,675	585		3,260	3,850	
	8180	262 MBH gross output, 315 GPH			.70	22.857		2,700	670		3,370	4,000	
	8280	735 MBH gross output, 880 GPH			.40	40		6,075	1,175		7,250	8,450	
	8300	840 MBH gross output, 1000 GPH			.40	40		7,050	1,175		8,225	9,550	
	8900	For low water cutoff, add		1 Plum	8	1		189	32.50		221.50	258	
	8960	For bronze body hot water circulator, add		"	4	2		720	65		785	895	
150	0010	**WATER TREATMENT, POTABLE**											150
	5800	Softener systems, automatic, intermediate sizes											
	5820	available, may be used in multiples.											
	6000	Hardness capacity between regenerations and flow											
	6100	150,000 grains, 37 GPM cont., 51 GPM peak		Q-1	1.20	13.333	Ea.	2,350	390		2,740	3,200	
	6200	300,000 grains, 81 GPM cont., 113 GPM peak			1	16		3,700	470		4,170	4,775	

Important: See the Reference Section for critical supporting data - Reference Nos., Crews, & City Cost Indexes

T5 MECHANICAL

153 100 | Water Appliances

			CREW	DAILY OUTPUT	LABOR-HOURS	UNIT	1999 BARE COSTS				TOTAL INCL O&P	
							MAT.	LABOR	EQUIP.	TOTAL		
150	6300	750,000 grains, 160 GPM cont., 230 GPM peak	Q-1	.80	20	Ea.	6,750	585		7,335	8,325	150
	6400	900,000 grains, 185 GPM cont., 270 GPM peak	↓	.70	22.857	↓	8,225	670		8,895	10,100	
160	0010	**WATER SUPPLY METERS**										160
	2000	Domestic/commercial, bronze										
	2020	Threaded										
	2060	5/8" diameter, to 20 GPM	1 Plum	16	.500	Ea.	66.50	16.30		82.80	98	
	2080	3/4" diameter, to 30 GPM		14	.571		115	18.65		133.65	155	
	2100	1" diameter, to 50 GPM	↓	12	.667	↓	157	21.50		178.50	205	
	2300	Threaded/flanged										
	2340	1-1/2" diameter, to 100 GPM	1 Plum	8	1	Ea.	485	32.50		517.50	585	
	2360	2" diameter, to 160 GPM	"	6	1.333	"	645	43.50		688.50	775	
	2600	Flanged, compound										
	2640	3" diameter, 320 GPM	Q-1	3	5.333	Ea.	2,900	157		3,057	3,425	
	2660	4" diameter, to 500 GPM		1.50	10.667		4,525	315		4,840	5,450	
	2680	6" diameter, to 1,000 GPM		1	16		6,500	470		6,970	7,875	
	2700	8" diameter, to 1,800 GPM	↓	.80	20	↓	12,800	585		13,385	15,000	

154 100 | Fire Systems

			CREW	DAILY OUTPUT	LABOR-HOURS	UNIT	1999 BARE COSTS				TOTAL INCL O&P	
							MAT.	LABOR	EQUIP.	TOTAL		
101	0010	**AUTOMATIC FIRE SUPPRESSION SYSTEMS**										101
	0040	For detectors and control stations, see division 168-120										
	1000	Dispersion nozzle, CO2, 3" x 5"	1 Plum	18	.444	Ea.	50	14.50		64.50	77	
	1100	FM200, 1-1/2"	"	14	.571		50	18.65		68.65	83	
	2000	Extinguisher, CO2 system, high pressure, 75 lb. cylinder	Q-1	6	2.667	↓	950	78.50		1,028.50	1,175	
	2400	FM200 system, filled, with mounting bracket										
	2540	196 lb. container	Q-1	4	4	Ea.	5,800	117		5,917	6,550	
	3000	Electro/mechanical release	L-1	4	4		115	129		244	320	
	3400	Manual pull station	1 Plum	6	1.333	↓	45	43.50		88.50	116	
	6000	Average FM200 system, minimum				C.F.	1.25			1.25	1.38	
	6020	Maximum				"	2.50			2.50	2.75	
135	0010	**FIRE HOSE AND EQUIPMENT**										135
	0200	Adapters, rough brass, straight hose threads										
	0220	One piece, female to male, rocker lugs										
	0240	1" x 1"				Ea.	17.85			17.85	19.65	
	2200	Hose, less couplings										
	2260	Synthetic jacket, lined, 300 lb. test, 1-1/2" diameter	Q-12	2,600	.006	L.F.	1.33	.18		1.51	1.73	
	2280	2-1/2" diameter		2,200	.007		2.17	.21		2.38	2.71	
	2360	High strength, 500 lb. test, 1-1/2" diameter		2,600	.006		1.59	.18		1.77	2.02	
	2380	2-1/2" diameter	↓	2,200	.007	↓	2.75	.21		2.96	3.35	
	2600	Hose rack, swinging, for 1-1/2" diameter hose,										
	2620	Enameled steel, 50' & 75' lengths of hose	Q-12	20	.800	Ea.	29.50	23.50		53	68	
	2640	100' and 125' lengths of hose	"	20	.800	"	32.50	23.50		56	71	
	3750	Hydrants, wall, w/caps, single, flush, polished brass										
	3800	2-1/2" x 2-1/2"	Q-12	5	3.200	Ea.	91.50	93.50		185	242	
	3840	2-1/2" x 3"	"	5	3.200	"	122	93.50		215.50	276	
	3950	Double, flush, polished brass										
	4000	2-1/2" x 2-1/2" x 4"	Q-12	5	3.200	Ea.	245	93.50		338.50	410	
	4040	2-1/2" x 2-1/2" x 6"	"	4.60	3.478	↓	355	102		457	545	

MECHANICAL 15

For expanded coverage of these items see *Means Mechanical or Plumbing Cost Data 1999*

154 100	Fire Systems	CREW	DAILY OUTPUT	LABOR-HOURS	UNIT	1999 BARE COSTS				TOTAL INCL O&P
						MAT.	LABOR	EQUIP.	TOTAL	
135 4200	For polished chrome, add				Ea.	10%				**135**
4350	Double, projecting, polished brass									
4400	2-1/2" x 2-1/2" x 4"	Q-12	5	3.200	Ea.	113	93.50		206.50	266
4450	2-1/2" x 2-1/2" x 6"	"	4.60	3.478	"	218	102		320	395
4460	Valve control, dbl. flush/projecting hydrant, cap &									
4470	chain, ext. rod & cplg., escutcheon, polished brass	Q-12	8	2	Ea.	133	58.50		191.50	236
5600	Nozzles, brass									
5620	Adjustable fog, 3/4" booster line				Ea.	75.50			75.50	83
5630	1" booster line					78			78	86
5640	1-1/2" leader line					82			82	90.50
5660	2-1/2" direct connection					160			160	176
5780	For chrome plated, add					8%				
5850	Electrical fire, adjustable fog, no shock									
5900	1-1/2"				Ea.	255			255	281
5920	2-1/2"					350			350	385
5980	For polished chrome, add					6%				
6200	Heavy duty, comb. adj. fog and str. stream, with handle									
6210	1" booster line				Ea.	244			244	269
7140	Standpipe connections, wall, w/plugs & chains									
7160	Single, flush, brass, 2-1/2" x 2-1/2"	Q-12	5	3.200	Ea.	65	93.50		158.50	214
7180	2-1/2" x 3"	"	5	3.200	"	71	93.50		164.50	220
7240	For polished chrome, add					15%				
7280	Double, flush, polished brass									
7300	2-1/2" x 2-1/2" x 4"	Q-12	5	3.200	Ea.	219	93.50		312.50	385
7330	2-1/2" x 2-1/2" x 6"	"	4.60	3.478	"	315	102		417	500
7400	For polished chrome, add					15%				
7440	For sill cock combination, add				Ea.	38.50			38.50	42.50
7900	Three way, flush, polished brass									
7920	2-1/2" (3) x 4"	Q-12	4.80	3.333	Ea.	715	97.50		812.50	940
7930	2-1/2" (3) x 6"	"	4.80	3.333		725	97.50		822.50	945
8000	For polished chrome, add					9%				
8020	Three way, projecting, polished brass									
8040	2-1/2"(3) x 4"	Q-12	4.80	3.333	Ea.	475	97.50		572.50	670
145 0010	**FIRE PUMPS** Including controller, fittings and relief valve									**145**
0030	Diesel									
0050	500 GPM, 50 psi, 27 HP, 4" pump	Q-13	.64	50	Ea.	45,700	1,550		47,250	53,000
0200	750 GPM, 50 psi, 44 HP, 5" pump		.60	53.333		48,700	1,650		50,350	56,000
0400	1000 GPM, 100 psi, 89 HP, 4" pump		.56	57.143		52,000	1,775		53,775	60,000
0700	2,000 GPM, 100 psi, 167 HP, 6" pump		.34	94.118		55,500	2,925		58,425	65,500
0950	3500 GPM, 100 psi, 300 HP, 10" pump		.24	133		97,500	4,125		101,625	113,500
3000	Electric									
3100	250 GPM, 55 psi, 15 HP, 3,550 RPM, 2" pump	Q-13	.70	45.714	Ea.	10,700	1,425		12,125	14,000
3200	500 GPM, 50 psi, 27 HP, 1770 RPM, 4" pump		.68	47.059		13,300	1,450		14,750	16,800
3350	750 GPM, 50 psi, 44 HP, 1,770 RPM, 5" pump		.64	50		19,400	1,550		20,950	23,700
3400	750 GPM, 100 psi, 66 HP, 3550 RPM, 4" pump		.58	55.172		17,200	1,700		18,900	21,500
5000	For jockey pump 1", 3 HP, with control, add	Q-12	2	8		2,375	234		2,609	2,975
160 0010	**FIRE VALVES**									**160**
0080	Wheel handle, 300 lb., 1-1/2"	1 Spri	12	.667	Ea.	22.50	21.50		44	57.50
0090	2-1/2"	"	7	1.143	"	55	37		92	117
0100	For polished brass, add					35%				
0110	For polished chrome, add					50%				
170 0010	**SPRINKLER SYSTEM COMPONENTS**									**170**
0600	Accelerator	1 Spri	8	1	Ea.	335	32.50		367.50	420

15 MECHANICAL

			DAILY	LABOR-		1999 BARE COSTS				TOTAL	
154 100		**Fire Systems**	CREW	OUTPUT	HOURS	UNIT	MAT.	LABOR	EQUIP.	TOTAL	INCL O&P
170	2600	Sprinkler heads, not including supply piping										**170**
	2640	Dry, pendent, 1/2" orifice, 3/4" or 1" NPT									
	2700	11" to 12-3/4" length	1 Spri	14	.571	Ea.	33.50	18.55		52.05	65
	2710	13" to 14-3/4" length		13	.615		37.50	20		57.50	72
	2720	15" to 16-3/4" length		13	.615		37.50	20		57.50	72
	2730	17" to 18-3/4" length		13	.615		40.50	20		60.50	75
	3600	Foam-water, pendent or upright, 1/2" NPT	↓	12	.667	↓	32	21.50		53.50	68
	3700	Standard spray, pendent or upright, brass, 135° to 286°F									
	3730	1/2" NPT, 7/16" orifice	1 Spri	16	.500	Ea.	5.30	16.25		21.55	30.50
	3740	1/2" NPT, 1/2" orifice CN	"	16	.500		3.64	16.25		19.89	28.50
	3860	For wax and lead coating, add					4.25			4.25	4.68
	3880	For wax coating, add					1.50			1.50	1.65
	3900	For lead coating, add				↓	3			3	3.30
	3920	For 360°F, same cost									
	3930	For 400°F, add				Ea.	5.50			5.50	6.05
	3940	For 500°F, add				"	5.50			5.50	6.05
	4500	Sidewall, horizontal, brass, 135° to 286°F									
	4520	1/2" NPT, 1/2" orifice	1 Spri	16	.500	Ea.	5.25	16.25		21.50	30.50
	4540	For 360°F, same cost									
	4800	Recessed pendent, brass, 135° to 286°F									
	4820	1/2" NPT, 3/8" orifice	1 Spri	10	.800	Ea.	5.30	26		31.30	45.50
	4830	1/2" NPT, 7/16" orifice		10	.800		5.30	26		31.30	45.50
	4840	1/2" NPT, 1/2" orifice	↓	10	.800	↓	3.64	26		29.64	43.50
	6200	Valves									
	6500	Check, swing, C.I. body, brass fittings, auto. ball drip									
	6520	4" size	Q-12	3	5.333	Ea.	95.50	156		251.50	340
	6800	Check, wafer, butterfly type, C.I. body, bronze fittings									
	6820	4" size	Q-12	4	4	Ea.	152	117		269	345
	7000	Deluge, assembly, incl. trim, pressure									
	7020	operated relief, emergency release, gauges									
	7040	2" size	Q-12	2	8	Ea.	930	234		1,164	1,375
	7060	3" size	"	1.50	10.667	"	1,100	310		1,410	1,675

MECHANICAL 15

			DAILY	LABOR-		1999 BARE COSTS				TOTAL	
155 100		**Boilers**	CREW	OUTPUT	HOURS	UNIT	MAT.	LABOR	EQUIP.	TOTAL	INCL O&P
101	0010	**AVERAGE** Square foot and percent of total										**101**
	0020	job cost, see division 171									
105	0010	**BOILERS, GENERAL** Prices do not include flue piping, elec. wiring,	R155 -050									**105**
	0020	gas or oil piping, boiler base, pad, or tankless unless noted									
	0100	Boiler H.P.: 10 KW = 34 lbs/steam/hr = 33,475 BTU/hr.									
	0120										
	0150	To convert SFR to BTU rating: Hot water, 150 x SFR;									
	0160	Forced hot water, 180 x SFR; steam, 240 x SFR	↓								
110	0010	**BOILERS, ELECTRIC, ASME** Standard controls and trim										**110**
	1000	Steam, 6 KW, 20.5 MBH	Q-19	1.20	20	Ea.	7,900	605		8,505	9,625
	1160	60 KW, 205 MBH		1	24		9,575	725		10,300	11,600
	1220	120 KW, 409 MBH		.75	32		11,000	970		11,970	13,600
	1280	210 KW, 716 MBH	↓	.55	43.636		15,000	1,325		16,325	18,500
	1380	510 KW, 1740 MBH	Q-21	.36	88.889	↓	20,200	2,750		22,950	26,400

155 100 | Boilers

		CREW	DAILY OUTPUT	LABOR-HOURS	UNIT	1999 BARE COSTS				TOTAL INCL O&P	
						MAT.	LABOR	EQUIP.	TOTAL		
110	1480	1,080 KW, 3685 MBH	Q-21	.25	128	Ea.	30,900	3,950		34,850	40,000	**110**
	1600	2,340 KW, 7984 MBH	↓	.16	200		58,500	6,175		64,675	74,000	
	2000	Hot water, 12 KW, 41 MBH	Q-19	1.30	18.462		3,325	560		3,885	4,500	
	2100	60 KW, 205 MBH		1.10	21.818		4,625	660		5,285	6,100	
	2220	240 KW, 820 MBH	↓	.55	43.636		10,400	1,325		11,725	13,500	
	2320	480 KW, 1636 MBH	Q-21	.46	69.565		18,000	2,150		20,150	23,100	
	2500	1,200 KW, 4095 MBH		.34	94.118		33,000	2,900		35,900	40,700	
	2680	2,400 KW, 8191 MBH		.25	128		56,000	3,950		59,950	68,000	
	2820	3,600 KW, 12,283 MBH	↓	.16	200	↓	70,500	6,175		76,675	87,500	
115	0010	**BOILERS, GAS FIRED** Natural or propane, standard controls										**115**
	1000	Cast iron, with insulated jacket										
	2000	Steam, gross output, 81 MBH	Q-7	1.40	22.857	Ea.	1,075	715		1,790	2,250	
	2080	203 MBH		.90	35.556		1,825	1,100		2,925	3,675	
	2180	400 MBH		.60	53.333		3,000	1,675		4,675	5,825	
	2240	765 MBH		.45	71.111		4,825	2,225		7,050	8,675	
	2320	1,875 MBH		.28	114		10,400	3,575		13,975	16,800	
	2440	4,720 MBH		.14	228		19,900	7,150		27,050	32,700	
	2480	6,100 MBH		.12	266		47,600	8,325		55,925	65,000	
	2540	6,970 MBH		.08	400		53,500	12,500		66,000	78,000	
	3000	Hot water, gross output, 80 MBH		1.46	21.918		1,075	685		1,760	2,200	
	3140	320 MBH CN		.80	40		2,500	1,250		3,750	4,650	
	3260	1,088 MBH		.50	64		6,050	2,000		8,050	9,700	
	3380	3,264 MBH		.30	106		15,600	3,325		18,925	22,200	
	3480	6,100 MBH		.12	266		47,600	8,325		55,925	65,000	
	3540	6,970 MBH	↓	.08	400	↓	53,500	12,500		66,000	78,000	
	4000	Steel, insulating jacket										
	4500	Steam, not including burner, gross output										
	5500	1,440 MBH	Q-6	.35	68.571	Ea.	15,000	2,100		17,100	19,700	
	5580	3,065 MBH	"	.18	133		21,400	4,075		25,475	29,800	
	5640	7,200 MBH	Q-7	.18	177		39,300	5,550		44,850	51,500	
	5720	17,990 MBH	"	.10	320	↓	73,500	10,000		83,500	96,000	
	6000	Hot water, including burner & one zone valve, gross output										
	6020	72 MBH	Q-6	2	12	Ea.	1,750	365		2,115	2,475	
	6120	227 MBH		1.30	18.462		4,050	565		4,615	5,300	
	6180	480 MBH		.70	34.286		6,600	1,050		7,650	8,850	
	6240	960 MBH		.45	53.333		11,400	1,625		13,025	15,000	
	6340	3,000 MBH	↓	.15	160	↓	31,600	4,900		36,500	42,100	
	7000	For tankless water heater on smaller gas units, add					10%					
120	0010	**BOILERS, OIL FIRED** Standard controls, flame retention burner										**120**
	1000	Cast iron, with insulated flush jacket										
	2000	Steam, gross output, 109 MBH	Q-7	1.20	26.667	Ea.	1,300	835		2,135	2,700	
	2060	207 MBH		.90	35.556		1,800	1,100		2,900	3,650	
	2180	1,084 MBH		.42	76.190		7,400	2,375		9,775	11,700	
	2280	3,000 MBH		.23	139		13,500	4,350		17,850	21,500	
	2380	5,520 MBH		.13	246		41,000	7,700		48,700	56,500	
	2460	6,970 MBH	↓	.08	400	↓	53,000	12,500		65,500	77,000	
	3000	Hot water, same price as steam										
	4000	For tankless coil in smaller sizes, add					15%					
	5000	Steel, insulated jacket, burner										
	6000	Steam, full water leg construction, gross output										
	6020	144 MBH	Q-6	1.60	15	Ea.	3,200	460		3,660	4,225	
	6120	468 MBH		.80	30		5,900	915		6,815	7,900	
	6180	1,008 MBH		.45	53.333		9,150	1,625		10,775	12,600	
	6280	2,400 MBH	↓	.22	109	↓	17,500	3,325		20,825	24,300	

15 MECHANICAL

Important: See the Reference Section for critical supporting data - Reference Nos., Crews, & City Cost Indexes

155 | Heating

155 100 | Boilers

		CREW	DAILY OUTPUT	LABOR-HOURS	UNIT	MAT.	LABOR	EQUIP.	TOTAL	TOTAL INCL O&P
120	6400 Larger sizes are same as steel, gas fired									
	7000 Hot water, gross output, 155 MBH	Q-6	1.90	12.632	Ea.	3,175	385		3,560	4,075
	7120 420 MBH		.80	30		5,175	915		6,090	7,075
	7200 840 MBH		.50	48		10,200	1,475		11,675	13,400
	7260 1,680 MBH		.33	72.727		18,500	2,225		20,725	23,700
	7320 3,150 MBH		.13	184		23,900	5,650		29,550	34,900
	7340 For tankless coil in steam or hot water, add					7%				
125	0010 **BOILERS, GAS/OIL** Combination with burners and controls									
	1000 Cast iron with insulated jacket									
	2000 Steam, gross output, 720 MBH	Q-7	.40	80	Ea.	6,550	2,500		9,050	11,000
	2080 1,600 MBH		.26	123		10,100	3,850		13,950	16,900
	2140 2,700 MBH		.19	168		14,200	5,250		19,450	23,600
	2280 5,520 MBH		.10	320		49,400	10,000		59,400	69,500
	2340 6,390 MBH		.07	457		53,000	14,300		67,300	79,500
	2380 6,970 MBH		.05	640		56,000	20,000		76,000	92,000
	3000 Hot water, gross output, 584 MBH		.54	59.259		7,350	1,850		9,200	10,900
	3060 1,460 MBH		.45	71.111		17,800	2,225		20,025	23,000
	3160 4,088 MBH		.22	145		49,000	4,550		53,550	61,000
	3300 13,500 MBH, 403.3 BHP		.02	1,600		106,500	50,000		156,500	193,000
	4000 Steel, insulated jacket, skid base, tubeless									
	4500 Steam, 150 psi gross output, 335 MBH, 10 BHP	Q-6	.65	36.923	Ea.	8,875	1,125		10,000	11,500
	4560 670 MBH, 20 BHP		.45	53.333		11,000	1,625		12,625	14,600
	4640 1,339 MBH, 40 BHP		.28	85.714		17,400	2,625		20,025	23,200
	4720 2,511 MBH, 75 BHP		.17	141		23,800	4,325		28,125	32,700
	5000 Hot water, gross output, 525 MBH		.60	40		9,500	1,225		10,725	12,400
	5080 1,050 MBH		.38	63.158		15,300	1,925		17,225	19,800
	5140 2,310 MBH		.20	120		28,500	3,675		32,175	37,000
	5180 3,150 MBH		.09	266		37,900	8,150		46,050	54,000
150	0010 **SWIMMING POOL HEATERS** Not including wiring, external									
	0020 piping, base or pad,									
	0060 Gas fired, input, 115 MBH	Q-6	3	8	Ea.	1,500	245		1,745	2,025
	0100 135 MBH		2	12		1,675	365		2,040	2,400
	0160 155 MBH		1.50	16		1,775	490		2,265	2,700
	0200 190 MBH		1	24		2,300	735		3,035	3,650
	0280 500 MBH		.40	60		5,600	1,825		7,425	8,925
	0400 1,800 MBH		.14	171		15,600	5,250		20,850	25,000
	0440 3,750 MBH		.09	266		31,400	8,150		39,550	46,900
	2000 Electric, 12 KW, 4,800 gallon pool	Q-19	3	8		1,475	242		1,717	2,000
	2020 15 KW, 7,200 gallon pool		2.80	8.571		1,475	260		1,735	2,025
	2040 24 KW, 9,600 gallon pool		2.40	10		1,975	305		2,280	2,625
	2100 55 KW, 24,000 gallon pool		1.20	20		2,825	605		3,430	4,025

155 200 | Boiler Accessories

		CREW	DAILY OUTPUT	LABOR-HOURS	UNIT	MAT.	LABOR	EQUIP.	TOTAL	TOTAL INCL O&P
260	0010 **INDUCED DRAFT FANS**									
	1000 Breeching installation									
	1800 Hot gas, 600°F, variable pitch pulley and motor									
	1860 8" diam. inlet, 1/4 H.P., 1phase, 1120 CFM	Q-9	4	4	Ea.	1,575	114		1,689	1,900
	1900 12" diam. inlet, 3/4 H.P., 3 phase, 2960 CFM		3	5.333		2,150	152		2,302	2,575
	1940 18" diam. inlet, 3 H.P., 3 phase, 9120 CFM		2	8		3,325	229		3,554	4,000
	1980 24" diam. inlet, 7-1/2 H.P., 3 phase, 17,760 CFM		.80	20		5,200	570		5,770	6,625
	2300 For multi-blade damper at fan inlet, add					20%				

MECHANICAL 15

155 400 | Warm Air Systems

		CREW	DAILY OUTPUT	LABOR-HOURS	UNIT	1999 BARE COSTS				TOTAL INCL O&P	
						MAT.	LABOR	EQUIP.	TOTAL		
401	0010	**DUCT FURNACES** Includes burner, controls, stainless steel									**401**
	0020	heat exchanger. Gas fired, electric ignition									
	0030	Indoor installation									
	0100	120 MBH output	Q-5	4	4	Ea.	1,150	118		1,268	1,425
	0130	200 MBH output		2.70	5.926		1,575	175		1,750	2,000
	0140	240 MBH output		2.30	6.957		1,650	205		1,855	2,100
	0180	320 MBH output	↓	1.60	10		2,000	295		2,295	2,650
	0300	For powered venter and adapter, add				↓	250			250	275
	0500	For required flue pipe, see division 155-680									
	1000	Outdoor installation, with vent cap									
	1020	75 MBH output	Q-5	4	4	Ea.	1,500	118		1,618	1,825
	1060	120 MBH output		4	4		1,925	118		2,043	2,300
	1100	187 MBH output		3	5.333		2,750	157		2,907	3,275
	1140	300 MBH output		1.80	8.889		3,600	262		3,862	4,350
	1180	450 MBH output	↓	1.40	11.429		4,250	335		4,585	5,175
	1400	For powered venter, add				↓	10%				
408	0010	**DUCT HEATERS** Electric, 480 V, 3 Ph									**408**
	0020	Finned tubular insert, 500°F									
	0100	8" wide x 6" high, 4.0 kW	Q-20	16	1.250	Ea.	510	36.50		546.50	620
	0120	12" high, 8.0 kW		15	1.333		815	39		854	960
	0140	18" high, 12.0 kW		14	1.429		1,150	42		1,192	1,325
	0160	24" high, 16.0 kW		13	1.538		1,475	45		1,520	1,700
	0180	30" high, 20.0 kW		12	1.667		1,800	48.50		1,848.50	2,050
	0300	12" wide x 6" high, 6.7 kW		15	1.333		525	39		564	635
	0360	24" high, 26.7 kW		12	1.667		1,525	48.50		1,573.50	1,750
	0700	24" wide x 6" high, 17.8 kW		13	1.538		630	45		675	760
	0760	24" high, 71.1 kW	↓	10	2	↓	1,900	58.50		1,958.50	2,200
	8000	To obtain BTU multiply kW by 3413									
420	0010	**FURNACES** Hot air heating, blowers, standard controls									**420**
	0020	not including gas, oil or flue piping									
	1000	Electric, UL listed									
	1020	10.2 MBH	Q-20	5	4	Ea.	315	117		432	525
	1040	17.1 MBH		4.80	4.167		330	122		452	550
	1060	27.3 MBH		4.60	4.348		395	127		522	630
	1100	34.1 MBH	↓	4.40	4.545	↓	405	133		538	650
	3000	Gas, AGA certified, direct drive models									
	3020	45 MBH input	Q-9	4	4	Ea.	435	114		549	660
	3040	60 MBH input		3.80	4.211		580	120		700	825
	3060	75 MBH input		3.60	4.444		615	127		742	875
	3100	100 MBH input	↓	3.20	5	↓	655	143		798	940
	6000	Oil, UL listed, atomizing gun type burner									
	6020	56 MBH output	Q-9	3.60	4.444	Ea.	745	127		872	1,025
	6030	84 MBH output		3.50	4.571		775	131		906	1,050
	6040	95 MBH output		3.40	4.706		790	134		924	1,075
	6060	134 MBH output		3.20	5		1,100	143		1,243	1,425
	6080	151 MBH output		3	5.333		1,225	152		1,377	1,550
440	0010	**HEATING & VENTILATING UNITS** Classroom									**440**
	0020	Includes filter, heating/cooling coils, standard controls									
	0080	750 CFM, 2 tons cooling	Q-6	2	12	Ea.	2,950	365		3,315	3,800
	0120	1250 CFM, 3 tons cooling		1.40	17.143		3,600	525		4,125	4,750
	0140	1500 CFM, 4 tons cooling	↓	.80	30		3,850	915		4,765	5,625
	0500	For electric heat, add					35%				
	1000	For no cooling, deduct				↓	25%	10%			
451	0010	**INFRA-RED UNIT**									**451**
	0020	Gas fired, unvented, electric ignition, 100% shutoff. Piping and									

Important: See the Reference Section for critical supporting data - Reference Nos., Crews, & City Cost Indexes

15 MECHANICAL

155 400 | **Warm Air Systems**			DAILY OUTPUT	LABOR-HOURS	UNIT	1999 BARE COSTS				TOTAL INCL O&P	
			CREW			MAT.	LABOR	EQUIP.	TOTAL		
451	0030	wiring not included									**451**
	0060	Input, 15 MBH	Q-5	7	2.286	Ea.	340	67.50		407.50	470
	0120	45 MBH		5	3.200		380	94.50		474.50	565
	0160	60 MBH		4	4		445	118		563	670
	0240	120 MBH	↓	2	8	↓	805	236		1,041	1,250
461	0010	**MAKE-UP AIR UNIT**									**461**
	0020	Indoor suspension, natural/LP gas, direct fired,									
	0030	standard control. For flue see division 155-680									
	0040	70°F temperature rise, MBH is input									
	0100	2000 CFM, 168 MBH	Q-6	3	8	Ea.	4,800	245		5,045	5,675
	0160	6000 CFM, 502 MBH		1.50	16		6,050	490		6,540	7,425
	0220	12,000 CFM, 1005 MBH	↓	1	24		7,550	735		8,285	9,400
	0300	24,000 CFM, 2007 MBH	Q-7	1	32		9,825	1,000		10,825	12,300
	0400	50,000 CFM, 4180 MBH	"	.80	40		13,500	1,250		14,750	16,700
	0600	For discharge louver assembly, add					10%				
	0700	For filters, add					20%				
	0800	For air shut-off damper section, add					10%				
	0900	For vertical unit, add				↓	10%				
471	0010	**SOLAR ENERGY**	R155 -610								**471**
	0020	System/Package prices, not including connecting									
	0030	pipe, insulation, or special heating/plumbing fixtures									
	0500	Hot water, standard package, low temperature									
	0540	1 collector, circulator, fittings, 65 gal. tank	Q-1	.50	32	Ea.	1,875	940		2,815	3,500
	0580	2 collectors, circulator, fittings, 120 gal. tank		.40	40		2,725	1,175		3,900	4,775
	0620	3 collectors, circulator, fittings, 120 gal. tank	↓	.40	40	↓	3,300	1,175		4,475	5,400
	0700	Medium temperature package									
	0720	1 collector, circulator, fittings, 80 gal. tank	Q-1	.50	32	Ea.	2,025	940		2,965	3,650
	0740	2 collectors, circulator, fittings, 120 gal. tank		.40	40		2,925	1,175		4,100	5,000
	0780	3 collectors, circulator, fittings, 120 gal. tank	↓	.30	53.333	↓	3,775	1,575		5,350	6,525
	0980	For each additional 120 gal. tank, add					720			720	795
	2300	Circulators, air									
	2310	Blowers									
	2400	Reversible fan, 20" diameter, 2 speed	Q-9	18	.889	Ea.	96.50	25.50		122	146
	2520	Space & DHW system, less duct work	"	.50	32		1,325	915		2,240	2,900
	2870	1/12 HP, 30 GPM	Q-1	10	1.600	↓	299	47		346	400
	3000	Collector panels, air with aluminum absorber plate									
	3010	Wall or roof mount									
	3040	Flat black, plastic glazing									
	3080	4' x 8'	Q-9	6	2.667	Ea.	600	76		676	780
	3200	Flush roof mount, 10' to 16' x 22" wide	"	96	.167	L.F.	146	4.76		150.76	168
	3300	Collector panels, liquid with copper absorber plate									
	3330	Alum. frame, 4' x 8', 5/32" single glazing	Q-1	9.50	1.684	Ea.	545	49.50		594.50	675
	3390	Alum. frame, 4' x 10', 5/32" single glazing	"	6	2.667	"	650	78.50		728.50	830
	3440	Flat black									
	3450	Alum. frame, 3' x 8'	Q-1	9	1.778	Ea.	455	52		507	580
	3500	Alum. frame, 4' x 8.5'		5.50	2.909		520	85.50		605.50	700
	3520	Alum. frame, 4' x 10.5'		10	1.600		630	47		677	760
	3540	Alum. frame, 4' x 12.5'		5	3.200		775	94		869	995
	3600	Liquid, full wetted, plastic, alum. frame, 3' x 10'		5	3.200		195	94		289	355
	3650	Collector panel mounting, flat roof or ground rack		7	2.286	↓	64.50	67		131.50	173
	3670	Roof clamps	↓	70	.229	Set	1.61	6.70		8.31	11.90
	3700	Roof strap, teflon	1 Plum	205	.039	L.F.	9.50	1.27		10.77	12.40
	3900	Differential controller with two sensors									
	3930	Thermostat, hard wired	↓ 1 Plum	8	1	Ea.	92	32.50		124.50	151

MECHANICAL **15**

155 400 | Warm Air Systems

		CREW	DAILY OUTPUT	LABOR-HOURS	UNIT	1999 BARE COSTS				TOTAL INCL O&P	
						MAT.	LABOR	EQUIP.	TOTAL		
471	4100	Five station with digital read-out	1 Plum	3	2.667	Ea.	202	87		289	355	**471**
	4300	Heat exchanger	R155 -610									
	4580	Fluid to fluid package includes two circulating pumps										
	4590	expansion tank, check valve, relief valve										
	4600	controller, high temperature cutoff and sensors	Q-1	2.50	6.400	Ea.	695	188		883	1,050	
	4650	Heat transfer fluid										
	4700	Propylene glycol, inhibited anti-freeze	1 Plum	28	.286	Gal.	8.80	9.30		18.10	24	
	8250	Water storage tank with heat exchanger and electric element										
	8280	80 gal. with 2" x 1/2 lb. density insulation	1 Plum	1.60	5	Ea.	360	163		523	640	
	8300	80 gal. with 2" x 2 lb. density insulation		1.60	5		700	163		863	1,025	
	8350	120 gal. with 2" x 1/2 lb. density insulation		1.40	5.714		390	186		576	705	
	8380	120 gal. with 2" x 2 lb. density insulation		1.40	5.714		790	186		976	1,150	
	8400	120 gal. with 2" x 2 lb. density insul., 40 S.F. heat coil		1.40	5.714		910	186		1,096	1,275	
480	0010	**SPACE HEATERS** Cabinet, grilles, fan, controls, burner,										**480**
	0020	thermostat, no piping. For flue see division 155-680										
	1000	Gas fired, floor mounted										
	1100	60 MBH output	Q-5	10	1.600	Ea.	580	47		627	705	
	1140	100 MBH output		8	2		630	59		689	780	
	1180	180 MBH output		6	2.667		860	78.50		938.50	1,075	
	2000	Suspension mounted, propeller fan, 20 MBH output		8.50	1.882		400	55.50		455.50	525	
	2040	60 MBH output		7	2.286		495	67.50		562.50	640	
	2060	80 MBH output		6	2.667		550	78.50		628.50	725	
	2100	130 MBH output		5	3.200		730	94.50		824.50	950	
	2240	320 MBH output		2	8		1,500	236		1,736	2,000	
	2500	For powered venter and adapter, add					259			259	285	
	5000	Wall furnace, 17.5 MBH output	Q-5	6	2.667		460	78.50		538.50	630	
	5020	24 MBH output		5	3.200		485	94.50		579.50	675	
	5040	35 MBH output		4	4		650	118		768	895	
	6000	Oil fired, suspension mounted, 94 MBH output		4	4		1,850	118		1,968	2,200	
	6040	140 MBH output		3	5.333		1,975	157		2,132	2,425	
	6060	184 MBH output		3	5.333		2,125	157		2,282	2,600	

155 600 | Heating System Access.

		CREW	DAILY OUTPUT	LABOR-HOURS	UNIT	MAT.	LABOR	EQUIP.	TOTAL	TOTAL INCL O&P	
601	0010	**HEAT EXCHANGERS**										**601**
	0016	Shell & tube type, 4 pass, 3/4" O.D. copper tubes,										
	0020	C.I. heads, C.I. tube sheet, steel shell										
	0100	Hot water 40°F to 180°F, by steam at 10 PSI										
	0120	8 GPM	Q-5	6	2.667	Ea.	875	78.50		953.50	1,075	
	0140	10 GPM		5	3.200		1,325	94.50		1,419.50	1,600	
	0160	40 GPM		4	4		2,050	118		2,168	2,425	
	0180	64 GPM		2	8		3,125	236		3,361	3,800	
	0200	96 GPM		1	16		4,150	470		4,620	5,275	
	0220	120 GPM	Q-6	1.50	16		5,450	490		5,940	6,725	
	1000	Hot water 40°F to 140°F, by water at 200°F										
	1020	7 GPM	Q-5	6	2.667	Ea.	1,075	78.50		1,153.50	1,300	
	1040	16 GPM		5	3.200		1,525	94.50		1,619.50	1,825	
	1060	34 GPM		4	4		2,325	118		2,443	2,725	
	1100	74 GPM		1.50	10.667		4,125	315		4,440	5,025	
	3000	Plate type,										
	3100	400 GPM	Q-6	.80	30	Ea.	18,400	915		19,315	21,600	
	3120	800 GPM	"	.50	48		32,300	1,475		33,775	37,800	
	3140	1200 GPM	Q-7	.34	94.118		66,500	2,950		69,450	77,500	
	3160	1800 GPM	"	.24	133		99,500	4,175		103,675	116,000	
610	0010	**HEAT TRANSFER PACKAGES** Complete, controls,										**610**
	0020	expansion tank, converter, air separator										

Important: See the Reference Section for critical supporting data - Reference Nos., Crews, & City Cost Indexes

155 600	Heating System Access.	CREW	DAILY OUTPUT	LABOR-HOURS	UNIT	1999 BARE COSTS				TOTAL INCL O&P	
						MAT.	LABOR	EQUIP.	TOTAL		
610	**1000**	Hot water, 180°F enter, 200°F leaving, 15# steam									**610**
	1010	One pump system, 28 GPM	Q-6	.75	32	Ea.	10,300	980		11,280	12,800
	1020	35 GPM		.70	34.286		11,300	1,050		12,350	14,000
	1040	55 GPM		.65	36.923		12,200	1,125		13,325	15,200
	1060	130 GPM		.55	43.636		16,000	1,325		17,325	19,600
	1080	255 GPM		.40	60		21,900	1,825		23,725	26,900
	1100	550 GPM	↓	.30	80	↓	28,600	2,450		31,050	35,200
620	**0010**	**HEAT RECOVERY PACKAGES**									**620**
	0100	Air to air									
	2000	Kitchen exhaust, commercial, heat pipe exchanger									
	2040	Combined supply/exhaust air volume									
	2080	2.5 to 6.0 MCFM	Q-10	2.80	8.571	MCFM	4,400	254		4,654	5,250
	2120	6 to 16 MCFM		5	4.800		2,700	142		2,842	3,200
	2160	16 to 22 MCFM	↓	6	4	↓	2,125	119		2,244	2,500
	7950										
630	**0010**	**HYDRONIC HEATING** Terminal units, not incl. main supply pipe									**630**
	1000	Radiation									
	1100	Panel, baseboard, C.I., including supports, no covers	Q-5	46	.348	L.F.	24	10.25		34.25	42
	1150	Fin tube, wall hung, 14" slope top cover, with damper									
	1200	1-1/4" copper tube, 4-1/4" alum. fin	Q-5	38	.421	L.F.	34.50	12.40		46.90	57
	1250	1-1/4" steel tube, 4-1/4" steel fin	"	36	.444	"	31	13.10		44.10	54
	1500	Note: fin tube may also require corners, caps, etc.									
	1990	Convector unit, floor recessed, flush, with trim									
	2000	for under large glass wall areas, no damper	Q-5	20	.800	L.F.	23	23.50		46.50	61
	2100	For unit with damper, add				"	9.95			9.95	10.95
	3000	Radiators, cast iron									
	3100	Free standing or wall hung, 6 tube, 25" high	Q-5	96	.167	Section	21	4.91		25.91	30.50
	3200	4 tube, 19" high	"	96	.167	"	14.40	4.91		19.31	23.50
	3250	Adj. brackets, 2 per wall radiator up to 30 sections	1 Stpi	32	.250	Ea.	31	8.20		39.20	46.50
	3950	Unit heaters, propeller, 115 V 2 psi steam, 60°F entering air									
	4000	Horizontal, 12 MBH	Q-5	12	1.333	Ea.	330	39.50		369.50	425
	4060	43.9 MBH		8	2		440	59		499	575
	4140	96.8 MBH		6	2.667		650	78.50		728.50	835
	4180	157.6 MBH		4	4		795	118		913	1,050
	4240	286.9 MBH		2	8		1,175	236		1,411	1,650
	4250	326.0 MBH		1.90	8.421		1,375	248		1,623	1,900
	4260	364.0 MBH		1.80	8.889		1,525	262		1,787	2,075
	4270	404.0 MBH	↓	1.60	10		1,925	295		2,220	2,575
	4300	For vertical diffuser, add					136			136	149
	4310	Vertical flow, 40.0 MBH	Q-5	11	1.455		465	43		508	575
	4314	58.5 MBH		8	2		535	59		594	680
	4326	131.0 MBH		4	4		710	118		828	960
	4346	297.0 MBH	↓	1.80	8.889		2,000	262		2,262	2,625
	4354	420 MBH	Q-6	1.80	13.333		2,425	410		2,835	3,275
	4358	500 MBH		1.40	17.143		2,750	525		3,275	3,825
	4362	570 MBH		1.40	17.143		2,850	525		3,375	3,950
	4366	620 MBH		1.30	18.462		3,275	565		3,840	4,450
	4370	960 MBH	↓	1.10	21.818	↓	5,750	665		6,415	7,325
	9500	To convert SFR to BTU rating: Hot water, 150 x SFR									
	9510	Forced hot water, 180 x SFR; steam, 240 x SFR									
640	**0010**	**HUMIDIFIERS**									**640**
	0520	Steam, room or duct, filter, regulators, auto. controls, 220 V									
	0540	11 lb. per hour	Q-5	6	2.667	Ea.	2,000	78.50		2,078.50	2,325
	0560	22 lb. per hour	↓	5	3.200	↓	2,200	94.50		2,294.50	2,575

MECHANICAL 15

155 600 | Heating System Access.

		CREW	DAILY OUTPUT	LABOR-HOURS	UNIT	1999 BARE COSTS				TOTAL INCL O&P	
						MAT.	LABOR	EQUIP.	TOTAL		
640	0580	33 lb. per hour	Q-5	4	4	Ea.	2,250	118		2,368	2,650	**640**
	0600	50 lb. per hour		4	4		2,750	118		2,868	3,200	
	0620	100 lb. per hour	↓	3	5.333	↓	3,300	157		3,457	3,900	
651	0010	**INSULATION**										**651**
	0100	Rule of thumb, as a percentage of total mechanical costs				Job				10%		
	0110	Insulation req'd is based on the surface size/area to be covered										
	1000	Boiler, 1-1/2" calcium silicate, 1/2" cement finish	Q-14	50	.320	S.F.	2.41	8.75		11.16	16.80	
	1020	2" fiberglass	"	80	.200	"	1.42	5.50		6.92	10.40	
	2000	Breeching, 2" calcium silicate with 1/2" cement finish, no lath										
	2020	Rectangular	Q-14	42	.381	S.F.	2.73	10.45		13.18	19.85	
	2040	Round	"	38.70	.413	"	3.06	11.35		14.41	21.50	
	3000	Ductwork										
	3020	Blanket type, fiberglass, flexible										
	3030	Fire resistant liner, black coating one side										
	3050	1/2" thick, 2 lb. density	Q-14	380	.042	S.F.	.28	1.15		1.43	2.17	
	3060	1" thick, 1-1/2 lb. density		350	.046		.37	1.25		1.62	2.43	
	3070	1-1/2" thick, 1-1/2 lb. density		320	.050		.65	1.37		2.02	2.92	
	3080	2" thick, 1-1/2 lb. density	↓	300	.053	↓	.76	1.46		2.22	3.20	
	3140	FRK vapor barrier wrap, .75 lb. density										
	3160	1" thick	Q-14	350	.046	S.F.	.26	1.25		1.51	2.31	
	3170	1-1/2" thick		320	.050		.23	1.37		1.60	2.46	
	3180	2" thick		300	.053		.28	1.46		1.74	2.67	
	3190	3" thick		260	.062		.41	1.69		2.10	3.17	
	3200	4" thick	↓	242	.066	↓	.55	1.81		2.36	3.53	
	3210	Vinyl jacket, same as FRK										
	3280	Unfaced, 1 lb. density										
	3310	1" thick	Q-14	360	.044	S.F.	.23	1.22		1.45	2.22	
	3320	1-1/2" thick		330	.048		.31	1.33		1.64	2.48	
	3330	2" thick	↓	310	.052	↓	.37	1.41		1.78	2.69	
	3490	Board type, fiberglass liner, 3 lb. density										
	3500	Fire resistant, black pigmented, 1 side										
	3520	1" thick	Q-14	150	.107	S.F.	1.10	2.92		4.02	5.95	
	3540	1-1/2" thick		130	.123		1.36	3.37		4.73	6.95	
	3560	2" thick	↓	120	.133	↓	1.64	3.65		5.29	7.70	
	3600	FRK vapor barrier										
	3620	1" thick	Q-14	150	.107	S.F.	1.21	2.92		4.13	6.05	
	3630	1-1/2" thick		130	.123		1.56	3.37		4.93	7.15	
	3640	2" thick	↓	120	.133	↓	1.90	3.65		5.55	8	
	3680	No finish										
	3700	1" thick	Q-14	170	.094	S.F.	.65	2.58		3.23	4.87	
	3710	1-1/2" thick		140	.114		.97	3.13		4.10	6.10	
	3720	2" thick		130	.123		1.30	3.37		4.67	6.90	
	3800	1/2" cement over 1" wire mesh, incl. corner bead		100	.160		1.17	4.38		5.55	8.40	
	3820	Glass cloth, pasted on		170	.094		1.91	2.58		4.49	6.25	
	3900	Weatherproof, non-metallic, 2 lb. per S.F.	↓	100	.160	↓	2.73	4.38		7.11	10.10	
	4000	Pipe covering (price copper tube one size less than IPS)										
	6600	Fiberglass, with all service jacket										
	6840	1" wall, 1/2" iron pipe size	Q-14	240	.067	L.F.	.92	1.83		2.75	3.96	
	6870	1" iron pipe size		220	.073		1.07	1.99		3.06	4.40	
	6900	2" iron pipe size		200	.080		1.43	2.19		3.62	5.10	
	6920	3" iron pipe size		180	.089		1.77	2.44		4.21	5.90	
	6940	4" iron pipe size		150	.107		1.88	2.92		4.80	6.80	
	7320	2" wall, 1/2" iron pipe size		220	.073		2.92	1.99		4.91	6.45	
	7440	6" iron pipe size		100	.160		5.75	4.38		10.13	13.45	
	7460	8" iron pipe size	↓	80	.200	↓	7.15	5.50		12.65	16.70	

Important: See the Reference Section for critical supporting data - Reference Nos., Crews, & City Cost Indexes

155 600	Heating System Access.	CREW	DAILY OUTPUT	LABOR-HOURS	UNIT	MAT.	LABOR	EQUIP.	TOTAL	TOTAL INCL O&P	
651						**1999 BARE COSTS**					**651**
7480	10" iron pipe size	Q-14	70	.229	L.F.	8.55	6.25		14.80	19.50	
7490	12" iron pipe size	↓	65	.246	↓	9.40	6.75		16.15	21.50	
7600	For fiberglass with standard canvas jacket, deduct					5%					
7660	For fittings, add 3 L.F. for each fitting										
7680	plus 4 L.F. for each flange of the fitting										
7700	For equipment or congested areas, add						20%				
7720	Finishes, for .010" aluminum jacket, add	Q-14	200	.080	S.F.	.27	2.19		2.46	3.84	
7740	For .016" aluminum jacket, add		200	.080		.34	2.19		2.53	3.91	
7760	For .010" stainless steel, add	↓	160	.100	↓	1.18	2.74		3.92	5.70	
7780	For single layer of felt, add					10%	10%				
7879	Rubber tubing, flexible closed cell foam										
7880	3/8" wall, 1/4" iron pipe size	1 Asbe	120	.067	L.F.	.36	2.03		2.39	3.68	
7910	1/2" iron pipe size		115	.070		.46	2.12		2.58	3.93	
7920	3/4" iron pipe size		115	.070		.53	2.12		2.65	4	
7930	1" iron pipe size		110	.073		.60	2.21		2.81	4.23	
7950	1-1/2" iron pipe size		110	.073		.88	2.21		3.09	4.54	
8100	1/2" wall, 1/4" iron pipe size		90	.089		.56	2.71		3.27	4.99	
8130	1/2" iron pipe size		89	.090		.68	2.74		3.42	5.15	
8140	3/4" iron pipe size		89	.090		.75	2.74		3.49	5.25	
8150	1" iron pipe size		88	.091		.84	2.77		3.61	5.40	
8170	1-1/2" iron pipe size		87	.092		1.18	2.80		3.98	5.80	
8180	2" iron pipe size		86	.093		1.19	2.83		4.02	5.90	
8200	3" iron pipe size		85	.094		2.65	2.87		5.52	7.55	
8300	3/4" wall, 1/4" iron pipe size		90	.089		.86	2.71		3.57	5.30	
8330	1/2" iron pipe size		89	.090		1.12	2.74		3.86	5.65	
8340	3/4" iron pipe size		89	.090		1.38	2.74		4.12	5.95	
8350	1" iron pipe size		88	.091		1.56	2.77		4.33	6.20	
8370	1-1/2" iron pipe size		87	.092		2.39	2.80		5.19	7.15	
8380	2" iron pipe size		86	.093		2.80	2.83		5.63	7.65	
8400	3" iron pipe size		85	.094		4.26	2.87		7.13	9.30	
8444	1" wall, 1/2" iron pipe size		86	.093		2.18	2.83		5.01	6.95	
8445	3/4" iron pipe size		84	.095		2.64	2.90		5.54	7.60	
8446	1" iron pipe size		84	.095		3.07	2.90		5.97	8.05	
8447	1-1/4" iron pipe size		82	.098		3.47	2.97		6.44	8.60	
8448	1-1/2" iron pipe size		82	.098		4.03	2.97		7	9.25	
8449	2" iron pipe size		80	.100		5.40	3.05		8.45	10.85	
8450	2-1/2" iron pipe size	↓	80	.100	↓	7	3.05		10.05	12.60	
8456	Rubber insulation tape, 1/8" x 2" x 30'				Ea.	13.45			13.45	14.80	
671	0010 **TANKS**										**671**
1505	Fiberglass and steel single / double wall storage, see Div 132-150										
1510	Tank leak detection systems, see Div 132-190										
2000	Steel, liquid expansion, ASME, painted, 15 gallon capacity	Q-5	17	.941	Ea.	325	28		353	395	
2020	24 gallon capacity		14	1.143		325	33.50		358.50	410	
2040	30 gallon capacity		12	1.333		360	39.50		399.50	455	
2060	40 gallon capacity		10	1.600		410	47		457	520	
2080	60 gallon capacity		8	2		490	59		549	630	
2100	80 gallon capacity		7	2.286		505	67.50		572.50	655	
2120	100 gallon capacity		6	2.667		655	78.50		733.50	840	
3000	Steel ASME expansion, rubber diaphragm, 19 gal. cap. accept.		12	1.333		1,250	39.50		1,289.50	1,425	
3020	31 gallon capacity		8	2		1,400	59		1,459	1,625	
3040	61 gallon capacity		6	2.667		1,950	78.50		2,028.50	2,275	
3080	119 gallon capacity		4	4		2,200	118		2,318	2,600	
3100	158 gallon capacity		3.80	4.211		3,075	124		3,199	3,600	
3140	317 gallon capacity		2.80	5.714		4,650	168		4,818	5,375	
3180	528 gallon capacity	↓	2.40	6.667	↓	7,550	197		7,747	8,600	
5950											

MECHANICAL 15

	155 600	Heating System Access.	CREW	DAILY OUTPUT	LABOR-HOURS	UNIT	1999 BARE COSTS				TOTAL INCL O&P	
							MAT.	LABOR	EQUIP.	TOTAL		
680	0010	**VENT CHIMNEY** Prefab metal, U.L. listed										680
	0020	Gas, double wall, galvanized steel										
	0080	3" diameter	Q-9	72	.222	V.L.F.	3.38	6.35		9.73	13.55	
	0100	4" diameter		68	.235		4.14	6.70		10.84	15	
	0120	5" diameter		64	.250		4.91	7.15		12.06	16.50	
	0140	6" diameter		60	.267		5.70	7.60		13.30	18.15	
	0160	7" diameter		56	.286		8.40	8.15		16.55	22	
	0180	8" diameter		52	.308		9.35	8.80		18.15	24	
	0200	10" diameter		48	.333		19.75	9.55		29.30	36.50	
	0220	12" diameter		44	.364		26.50	10.40		36.90	45	
	0260	16" diameter	↓	40	.400		73.50	11.45		84.95	98.50	
	0300	20" diameter	Q-10	36	.667		92	19.75		111.75	132	
	0480	38" diameter	"	24	1		292	29.50		321.50	365	
	7800	All fuel, double wall, stainless steel, 6" diameter	Q-9	60	.267		28	7.60		35.60	42.50	
	7802	7" diameter		56	.286		36	8.15		44.15	52	
	7804	8" diameter		52	.308		42	8.80		50.80	59.50	
	7806	10" diameter		48	.333		61	9.55		70.55	82.50	
	7808	12" diameter		44	.364		82	10.40		92.40	106	
	7810	14" diameter	↓	42	.381	↓	108	10.90		118.90	135	
	9000	High temp. (2000° F), steel jacket, acid resist. refractory lining										
	9010	11 ga. galvanized jacket, U.L. listed										
	9020	Straight section, 48" long, 10" diameter	Q-10	13.30	1.805	Ea.	335	53.50		388.50	450	
	9040	18" diameter		7.40	3.243		510	96		606	710	
	9070	36" diameter	↓	2.70	8.889		1,075	263		1,338	1,600	
	9090	48" diameter	Q-11	2.70	11.852		1,575	360		1,935	2,300	
	9120	Tee section, 10" diameter	Q-10	4.40	5.455		705	162		867	1,025	
	9140	18" diameter		2.40	10		1,000	296		1,296	1,550	
	9170	36" diameter	↓	.80	30		3,300	890		4,190	5,000	
	9190	48" diameter	Q-11	.90	35.556		5,600	1,075		6,675	7,850	
	9220	Cleanout pier section, 10" diameter	Q-10	3.50	6.857		545	203		748	915	
	9240	18" diameter		1.90	12.632		730	375		1,105	1,375	
	9270	36" diameter	↓	.75	32		1,350	950		2,300	2,950	
	9290	48" diameter	Q-11	.75	42.667		2,050	1,300		3,350	4,250	
	9320	For drain, add					53%					
	9330	Elbow, 30° and 45°, 10" diameter	Q-10	6.60	3.636		460	108		568	670	
	9350	18" diameter		3.70	6.486		745	192		937	1,125	
	9380	36" diameter	↓	1.30	18.462		1,850	545		2,395	2,875	
	9400	48" diameter	Q-11	1.35	23.704		3,225	720		3,945	4,675	
	9430	For 60° and 90° elbow, add					112%					
	9440	End cap, 10" diameter	Q-10	26	.923		790	27.50		817.50	910	
	9460	18" diameter		15	1.600		1,125	47.50		1,172.50	1,300	
	9490	36" diameter	↓	5	4.800		1,950	142		2,092	2,375	
	9510	48" diameter	Q-11	5.30	6.038		2,875	183		3,058	3,425	
	9540	Increaser (1 diameter), 10" diameter	Q-10	6.60	3.636		390	108		498	595	
	9560	18" diameter		3.70	6.486		610	192		802	970	
	9590	36" diameter	↓	1.30	18.462		1,200	545		1,745	2,175	
	9592	42" diameter	Q-11	1.60	20		1,325	605		1,930	2,400	
	9594	48" diameter		1.35	23.704		1,575	720		2,295	2,875	
	9596	54" diameter	↓	1.10	29.091	↓	1,925	880		2,805	3,475	
	9600	For expansion joints, add to straight section					7%					
	9610	For 1/4" hot rolled steel jacket, add					157%					
	9620	For 2950°F very high temperature, add					89%					
	9630	26 ga. aluminized jacket, straight section, 48" long										
	9640	10" diameter	Q-10	15.30	1.569	V.L.F.	197	46.50		243.50	288	
	9660	18" diameter		8.50	2.824		300	83.50		383.50	460	
	9690	36" diameter	↓	3.10	7.742	↓	695	229		924	1,125	

15 MECHANICAL

Important: See the Reference Section for critical supporting data - Reference Nos., Crews, & City Cost Indexes

155 600	Heating System Access.	CREW	DAILY OUTPUT	LABOR-HOURS	UNIT	1999 BARE COSTS MAT.	LABOR	EQUIP.	TOTAL	TOTAL INCL O&P	
680 9810	Draw band, galv. stl., 11 gauge, 10" diameter	Q-10	32	.750	Ea.	53	22		75	93	**680**
9820	12" diameter		30	.800		57	23.50		80.50	99.50	
9830	18" diameter		26	.923		71.50	27.50		99	121	
9860	36" diameter	↓	18	1.333		121	39.50		160.50	195	
9880	48" diameter	Q-11	20	1.600	↓	207	48.50		255.50	305	

156 200	Heat/Cool Piping Misc.	CREW	DAILY OUTPUT	LABOR-HOURS	UNIT	1999 BARE COSTS MAT.	LABOR	EQUIP.	TOTAL	TOTAL INCL O&P	
201 0010	**AUTOMATIC AIR VENT**										**201**
0020	Cast iron body, stainless steel internals, float type										
0060	1/2" NPT inlet, 300 psi	1 Stpi	12	.667	Ea.	76	22		98	117	
0220	3/4" NPT inlet, 250 psi	"	10	.800		242	26		268	305	
0340	1-1/2" NPT inlet, 250 psi	Q-5	12	1.333	↓	755	39.50		794.50	890	
205 0010	**AIR CONTROL**										**205**
0030	Air separator, with strainer										
0040	2" diameter	Q-5	6	2.667	Ea.	505	78.50		583.50	675	
0080	2-1/2" diameter		5	3.200		575	94.50		669.50	775	
0100	3" diameter		4	4		875	118		993	1,150	
0120	4" diameter	↓	3	5.333		1,275	157		1,432	1,650	
0130	5" diameter	Q-6	3.60	6.667		1,625	204		1,829	2,075	
0140	6" diameter	"	3.40	7.059	↓	1,925	216		2,141	2,450	
225 0010	**EXPANSION JOINTS**										**225**
0100	Bellows type, neoprene cover, flanged spool										
0120	6" face to face, 1" diameter	1 Stpi	13	.615	Ea.	142	20		162	187	
0140	1-1/4" diameter		11	.727		145	24		169	196	
0160	1-1/2" diameter	↓	10.60	.755		145	24.50		169.50	198	
0180	2" diameter	Q-5	13.30	1.203		147	35.50		182.50	216	
0190	2-1/2" diameter		12.40	1.290		153	38		191	226	
0200	3" diameter		11.40	1.404		174	41.50		215.50	254	
0480	10" face to face, 2" diameter		13	1.231		216	36.50		252.50	293	
0500	2-1/2" diameter		12	1.333		228	39.50		267.50	310	
0520	3" diameter		11	1.455		231	43		274	320	
0540	4" diameter		8	2		258	59		317	375	
0560	5" diameter		7	2.286		315	67.50		382.50	445	
0580	6" diameter	↓	6	2.667	↓	320	78.50		398.50	475	
235 0010	**FLEXIBLE METAL HOSE** Connectors, standard lengths										**235**
0100	Bronze braided, bronze ends										
0120	3/8" diameter x 12"	1 Stpi	26	.308	Ea.	15.50	10.10		25.60	32.50	
0160	3/4" diameter x 12"		20	.400		23	13.10		36.10	45.50	
0180	1" diameter x 18"		19	.421		34	13.80		47.80	58.50	
0200	1-1/2" diameter x 18"		13	.615		59	20		79	95.50	
0220	2" diameter x 18"	↓	11	.727	↓	72.50	24		96.50	116	
240 0010	**HEATING CONTROL VALVES**										**240**
0100	Radiator supply, 1/2" diameter	1 Stpi	24	.333	Ea.	34.50	10.90		45.40	54.50	
0120	3/4" diameter		20	.400		29	13.10		42.10	52	
0140	1" diameter	↓	19	.421	↓	36	13.80		49.80	61	

MECHANICAL 15

For expanded coverage of these items see _Means Mechanical or Plumbing Cost Data 1999_

156 200 | Heat/Cool Piping Misc.

			DAILY OUTPUT	LABOR-HOURS	UNIT	1999 BARE COSTS				TOTAL INCL O&P	
						MAT.	LABOR	EQUIP.	TOTAL		
240	0160	1-1/4" diameter	1 Stpi	15	.533	Ea.	51	17.45		68.45	82.50	240
	0500	For low pressure steam, add					25%					
260	0010	**PRESSURE REGULATOR**										260
	3000	Steam, high capacity, bronze body, stainless steel trim										
	3020	Threaded, 1/2" diameter	1 Stpi	24	.333	Ea.	630	10.90		640.90	705	
	3030	3/4" diameter		24	.333		630	10.90		640.90	705	
	3040	1" diameter		19	.421		700	13.80		713.80	790	
	3060	1-1/4" diameter		15	.533		775	17.45		792.45	880	
	3080	1-1/2" diameter		13	.615		885	20		905	1,000	
	3100	2" diameter	↓	11	.727		1,075	24		1,099	1,225	
	3120	2-1/2" diameter	Q-5	12	1.333		1,350	39.50		1,389.50	1,550	
	3140	3" diameter	"	11	1.455	↓	1,550	43		1,593	1,775	
	3500	Flanged connection, iron body, 125 lb. W.S.P.										
	3520	3" diameter	Q-5	11	1.455	Ea.	1,700	43		1,743	1,950	
	3540	4" diameter	"	5	3.200	"	2,150	94.50		2,244.50	2,500	
270	0010	**STEAM CONDENSATE METER**										270
	0100	500 lb. per hour	1 Stpi	14	.571	Ea.	2,300	18.70		2,318.70	2,550	
	0140	1500 lb. per hour	"	7	1.143	"	2,475	37.50		2,512.50	2,775	
272	0010	**STEAM TRAP**										272
	0030	Cast iron body, threaded										
	0040	Inverted bucket										
	0050	1/2" pipe size	1 Stpi	12	.667	Ea.	97	22		119	139	
	0070	3/4" pipe size		10	.800		166	26		192	223	
	0100	1" pipe size		9	.889		256	29		285	325	
	0120	1-1/4" pipe size	↓	8	1	↓	385	33		418	475	
	1000	Float & thermostatic, 15 psi										
	1010	3/4" pipe size	1 Stpi	16	.500	Ea.	93.50	16.40		109.90	128	
	1020	1" pipe size		15	.533		143	17.45		160.45	185	
	1040	1-1/2" pipe size		9	.889		254	29		283	325	
	1060	2" pipe size	↓	6	1.333	↓	460	43.50		503.50	575	
280	0010	**VENTURI FLOW** Measuring device										280
	0050	1/2" diameter	1 Stpi	24	.333	Ea.	105	10.90		115.90	133	
	0120	1" diameter		19	.421		94	13.80		107.80	124	
	0140	1-1/4" diameter		15	.533		134	17.45		151.45	174	
	0160	1-1/2" diameter		13	.615		140	20		160	185	
	0180	2" diameter	↓	11	.727		166	24		190	219	
	0220	3" diameter	Q-5	14	1.143		335	33.50		368.50	420	
	0240	4" diameter	"	11	1.455		375	43		418	480	
	0280	6" diameter	Q-6	3.50	6.857		660	210		870	1,050	
	0500	For meter, add				↓	1,175			1,175	1,300	

156 600 | Strainers

			DAILY OUTPUT	LABOR-HOURS	UNIT	1999 BARE COSTS				TOTAL INCL O&P	
						MAT.	LABOR	EQUIP.	TOTAL		
608	0010	**STRAINERS, Y TYPE** Bronze body										608
	0050	Screwed, 150 lb., 1/4" pipe size	1 Stpi	24	.333	Ea.	15	10.90		25.90	33	
	0070	3/8" pipe size		24	.333		15	10.90		25.90	33	
	0100	1/2" pipe size		20	.400		15	13.10		28.10	36.50	
	0140	1" pipe size		17	.471		24.50	15.40		39.90	50.50	
	0160	1-1/2" pipe size		14	.571		53	18.70		71.70	87	
	0180	2" pipe size		13	.615		70.50	20		90.50	108	
	0182	3" pipe size	↓	12	.667		269	22		291	330	
	0200	300 lb., 2-1/2" pipe size	Q-5	17	.941		270	28		298	340	
	0220	3" pipe size		16	1		535	29.50		564.50	635	
	0240	4" pipe size	↓	15	1.067	↓	1,225	31.50		1,256.50	1,400	
	0500	For 300 lb rating 1/4" thru 2", add					15%					

Important: See the Reference Section for critical supporting data - Reference Nos., Crews, & City Cost Indexes

15 MECHANICAL

156 600 | Strainers

			DAILY OUTPUT	LABOR-HOURS	UNIT	MAT.	LABOR	EQUIP.	TOTAL	TOTAL INCL O&P	
								1999 BARE COSTS			
			CREW								
608	1000	Flanged, 150 lb., 1-1/2" pipe size	1 Stpi	11	.727	Ea.	320	24		344	385	608
	1020	2" pipe size	"	8	1		430	33		463	525	
	1030	2-1/2" pipe size	Q-5	5	3.200		660	94.50		754.50	875	
	1040	3" pipe size		4.50	3.556		815	105		920	1,050	
	1060	4" pipe size		3	5.333		1,250	157		1,407	1,625	
	1100	6" pipe size	Q-6	3	8		2,375	245		2,620	2,975	
	1106	8" pipe size	"	2.60	9.231		3,675	282		3,957	4,475	
	1500	For 300 lb rating, add					40%					
612	0010	STRAINERS, Y TYPE Iron body										612
	0050	Screwed, 250 lb., 1/4" pipe size	1 Stpi	20	.400	Ea.	8.75	13.10		21.85	29.50	
	0070	3/8" pipe size		20	.400		8.75	13.10		21.85	29.50	
	0100	1/2" pipe size		20	.400		8.75	13.10		21.85	29.50	
	0140	1" pipe size		16	.500		14.10	16.40		30.50	40.50	
	0160	1-1/2" pipe size		12	.667		22	22		44	57	
	0180	2" pipe size		8	1		34	33		67	87	
	0220	3" pipe size	Q-5	11	1.455		198	43		241	283	
	0240	4" pipe size	"	5	3.200		335	94.50		429.50	515	
	0500	For galvanized body, add					50%					
	1000	Flanged, 125 lb., 1-1/2" pipe size	1 Stpi	11	.727	Ea.	113	24		137	161	
	1020	2" pipe size	"	8	1		84.50	33		117.50	143	
	1040	3" pipe size	Q-5	4.50	3.556		107	105		212	277	
	1060	4" pipe size	"	3	5.333		183	157		340	440	
	1080	5" pipe size	Q-6	3.40	7.059		320	216		536	675	
	1100	6" pipe size	"	3	8		375	245		620	780	
	1500	For 250 lb rating, add					20%					
	2000	For galvanized body, add					50%					
	2500	For steel body, add					40%					

157 100 | A.C. & Vent. Units

			CREW	DAILY OUTPUT	LABOR-HOURS	UNIT	MAT.	LABOR	EQUIP.	TOTAL	TOTAL INCL O&P	
101	0010	**AVERAGE** Square foot and percent of total										101
	0100	job cost, see division 171										
105	0010	**AIR CONDITIONING REQUIREMENTS**	R157 -020									105
	0100	For various types of buildings										
110	0010	**ABSORPTION COLD GENERATORS** Water chiller										110
	0020	Steam or hot water, water cooled										
	0050	100 ton	Q-7	.13	240	Ea.	110,500	7,500		118,000	133,000	
	0400	420 ton	"	.10	323	"	184,000	10,100		194,100	217,500	
125	0010	**CENTRAL STATION AIR-HANDLING UNIT** Chilled water										125
	1000	Modular, capacity at 700 FPM face velocity										
	1100	1300 CFM	Q-5	1.20	13.333	Ea.	2,325	395		2,720	3,175	
	1300	3200 CFM	"	.80	20		3,075	590		3,665	4,300	
	1500	8000 CFM	Q-6	.60	40		5,425	1,225		6,650	7,800	
	1700	20,000 CFM		.31	77.419		11,100	2,375		13,475	15,800	
	1900	33,500 CFM		.19	126		19,200	3,850		23,050	27,100	
	2100	63,000 CFM	Q-7	.13	246		37,400	7,700		45,100	52,500	
	2200	For hot water heating coil, add					21%	5%				

MECHANICAL 15

157 100 | A.C. & Vent. Units

			CREW	DAILY OUTPUT	LABOR-HOURS	UNIT	MAT.	LABOR	EQUIP.	TOTAL	TOTAL INCL O&P	
130	0010	**COMPUTER ROOM UNITS**										130
	1000	Air cooled, includes remote condenser but not										
	1020	interconnecting tubing or refrigerant										
	1080	3 ton	Q-5	.50	32	Ea.	6,375	945		7,320	8,450	
	1120	5 ton		.45	35.556		8,575	1,050		9,625	11,000	
	1160	6 ton		.30	53.333		15,800	1,575		17,375	19,700	
	1200	8 ton		.27	59.259		17,800	1,750		19,550	22,300	
	1240	10 ton		.25	64		18,500	1,875		20,375	23,300	
	1280	15 ton		.22	72.727		20,700	2,150		22,850	26,100	
	1290	18 ton	▼	.20	80		23,600	2,350		25,950	29,500	
	1320	20 ton	Q-6	.29	82.759		24,700	2,525		27,225	31,000	
	1360	23 ton	"	.28	85.714	▼	26,000	2,625		28,625	32,600	
150	0010	**FAN COIL AIR CONDITIONING** Cabinet mounted, filters, controls										150
	0100	Chilled water, 1/2 ton cooling	Q-5	8	2	Ea.	730	59		789	895	
	0120	1 ton cooling		6	2.667		830	78.50		908.50	1,025	
	0180	3 ton cooling	▼	4	4		1,675	118		1,793	2,000	
	0200	10 ton cooling	Q-6	2.80	8.571		2,400	262		2,662	3,050	
	0220	15 ton cooling		1.50	16		3,350	490		3,840	4,425	
	0240	20 ton cooling		.80	30		4,300	915		5,215	6,125	
	0260	30 ton cooling	▼	.60	40		6,350	1,225		7,575	8,850	
	0940	Direct expansion, for use w/air cooled condensing, 1.5 ton cooling	Q-5	5	3.200		440	94.50		534.50	630	
	1000	5 ton cooling	"	3	5.333		1,000	157		1,157	1,350	
	1040	10 ton cooling	Q-6	2.60	9.231		2,250	282		2,532	2,900	
	1060	20 ton cooling		.70	34.286		4,175	1,050		5,225	6,175	
	1100	40 ton cooling	▼	.45	53.333	▼	8,775	1,625		10,400	12,100	
	1510	For condensing unit add see division 157-230.										
160	0010	**HEAT PUMPS** (Not including interconnecting tubing)										160
	1000	Air to air, split system, not including curbs, pads, or ductwork										
	1020	2 ton cooling, 8.5 MBH heat @ 0°F	Q-5	1.20	13.333	Ea.	2,100	395		2,495	2,900	
	1060	5 ton cooling, 27 MBH heat @ 0°F		.50	32		4,275	945		5,220	6,125	
	1080	7.5 ton cooling, 33 MBH heat @ 0°F	▼	.30	53.333		6,575	1,575		8,150	9,600	
	1100	10 ton cooling, 50 MBH heat @ 0°F	Q-6	.38	63.158		8,500	1,925		10,425	12,300	
	1120	15 ton cooling, 64 MBH heat @ 0°F	"	.26	92.308		11,700	2,825		14,525	17,200	
	1180	40 ton cooling, 193 MBH heat @ 0°F	Q-7	.12	266	▼	33,900	8,325		42,225	49,800	
	1300	Supplementary electric heat coil, included										
	1500	Single package, not including curbs, pads, or plenums										
	1520	2 ton cooling, 6.5 MBH heat @ 0°F	Q-5	1.50	10.667	Ea.	2,525	315		2,840	3,250	
	1580	4 ton cooling, 13 MBH heat @ 0°F		.96	16.667		3,800	490		4,290	4,925	
	1640	7.5 ton cooling, 35 MBH heat @ 0°F	▼	.40	40		8,225	1,175		9,400	10,800	
	1660	15 ton cooling, 56 MBH heat @ 0°F	Q-6	.30	80	▼	16,100	2,450		18,550	21,400	
	2000	Water source to air, single package										
	2100	1 ton cooling, 13 MBH heat @ 75°F	Q-5	2	8	Ea.	965	236		1,201	1,425	
	2140	2 ton cooling, 19 MBH heat @ 75°F		1.70	9.412		1,225	277		1,502	1,750	
	2220	5 ton cooling, 29 MBH heat @ 75°F	▼	.90	17.778	▼	2,125	525		2,650	3,150	
	3960	For supplementary heat coil, add				▼	10%					
	4000	For increase in capacity thru use										
	4020	of solar collector, size boiler at 60%										
170	0010	**PACKAGED TERMINAL AIR CONDITIONER** Cabinet, wall sleeve,										170
	0100	louver, electric heat, thermostat, manual changeover, 208 V										
	0200	6,000 BTUH cooling, 8800 BTU heat	Q-5	6	2.667	Ea.	890	78.50		968.50	1,100	
	0220	9,000 BTUH cooling, 13,900 BTU heat		5	3.200		915	94.50		1,009.50	1,150	
	0240	12,000 BTUH cooling, 13,900 BTU heat		4	4		945	118		1,063	1,225	
	0260	15,000 BTUH cooling, 13,900 BTU heat	▼	3	5.333		1,025	157		1,182	1,375	
	0500	For hot water coil, increase heat by 10%, add					5%	10%				
	1000	For steam, increase heat output by 30%, add				▼	8%	10%				

15 MECHANICAL

Important: See the Reference Section for critical supporting data - Reference Nos., Crews, & City Cost Indexes

157 100	A.C. & Vent. Units		DAILY	LABOR-		1999 BARE COSTS				TOTAL
		CREW	OUTPUT	HOURS	UNIT	MAT.	LABOR	EQUIP.	TOTAL	INCL O&P
180	**0010**	**ROOF TOP AIR CONDITIONERS** Standard controls, curb, economizer										**180**
	1000	Single zone, electric cool, gas heat									
	1090	2 ton cooling, 55 MBH heating	Q-5	1.40	11.429	Ea.	4,050	335		4,385	4,950
	1100	3 ton cooling, 60 MBH heating		1.30	12.308		4,225	365		4,590	5,175
	1120	4 ton cooling, 95 MBH heating		1.10	14.545		4,500	430		4,930	5,600
	1140	5 ton cooling, 112 MBH heating		.56	28.571		4,675	840		5,515	6,425
	1145	6 ton cooling, 140 MBH heating		.52	30.769		5,350	905		6,255	7,275
	1150	7.5 ton cooling, 170 MBH heating		.40	40		7,325	1,175		8,500	9,825
	1160	10 ton cooling, 200 MBH heating	Q-6	.46	52.174		9,500	1,600		11,100	12,800
	1170	12.5 ton cooling, 230 MBH heating		.38	63.158		10,800	1,925		12,725	14,800
	1190	18 ton cooling, 330 MBH heating		.28	85.714		15,100	2,625		17,725	20,600
	1200	20 ton cooling, 360 MBH heating	Q-7	.32	100		20,600	3,125		23,725	27,300
	1210	25 ton cooling, 450 MBH heating		.27	118		24,600	3,700		28,300	32,700
	1220	30 ton cooling, 540 MBH heating		.22	145		29,100	4,550		33,650	38,900
	1260	50 ton cooling, 810 MBH heating		.13	246		46,000	7,700		53,700	62,000
	1265	60 ton cooling, 900 MBH heating		.11	290		52,000	9,100		61,100	71,500
	1270	80 ton cooling, 1000 MBH heating		.09	355		86,500	11,100		97,600	112,500
	1275	90 ton cooling, 1200 MBH heating		.08	400		102,500	12,500		115,000	131,500
	1280	100 ton cooling, 1350 MBH heating		.07	457		110,000	14,300		124,300	142,500
	2000	Multizone, electric cool, gas heat, economizer									
	2100	15 ton cooling, 360 MBH heating	Q-7	.22	145	Ea.	39,900	4,550		44,450	51,000
	2120	20 ton cooling, 360 MBH heating **CN**		.21	152		45,500	4,750		50,250	57,000
	2200	40 ton cooling, 540 MBH heating		.12	266		66,500	8,325		74,825	86,000
	2210	50 ton cooling, 540 MBH heating		.11	290		78,000	9,100		87,100	100,000
	2220	70 ton cooling, 1500 MBH heating		.09	355		109,000	11,100		120,100	137,000
	2240	80 ton cooling, 1500 MBH heating		.08	400		125,000	12,500		137,500	156,500
	2260	90 ton cooling, 1500 MBH heating		.07	457		140,500	14,300		154,800	176,000
	2280	105 ton cooling, 1500 MBH heating		.06	533		164,000	16,700		180,700	205,000
	2400	For hot water heat coil, deduct					5%				
	2500	For steam heat coil, deduct					2%				
	2600	For electric heat, deduct					3%	5%			
185	**0010**	**SELF-CONTAINED SINGLE PACKAGE**										**185**
	0100	Air cooled, for free blow or duct, including remote condenser									
	0200	3 ton cooling	Q-5	1	16	Ea.	4,300	470		4,770	5,450
	0220	5 ton cooling	Q-6	1.20	20		5,400	610		6,010	6,875
	0240	10 ton cooling	Q-7	1	32		9,225	1,000		10,225	11,600
	0260	20 ton cooling		.90	35.556		15,000	1,100		16,100	18,200
	0280	30 ton cooling		.80	40		18,900	1,250		20,150	22,700
	0340	60 ton cooling	Q-8	.40	80		37,600	2,500	124	40,224	45,300
	0490	For duct mounting no price change									
	0500	For hot water or steam heating coils, add				Ea.	10%	10%			
	1000	Water cooled for free blow or duct, not including tower									
	1010	Constant volume									
	1100	3 ton cooling	Q-6	1	24	Ea.	3,000	735		3,735	4,400
	1120	5 ton cooling	"	1	24		3,575	735		4,310	5,025
	1140	10 ton cooling	Q-7	.90	35.556		7,275	1,100		8,375	9,675
	1160	20 ton cooling		.80	40		13,000	1,250		14,250	16,200
	1180	30 ton cooling		.70	45.714		17,300	1,425		18,725	21,200
190	**0010**	**WATER CHILLERS**, With standard controls										**190**
	0020	Centrifugal liquid chiller, water cooled									
	0030	not including water tower									
	0100	Open drive, 2000 ton	Q-7	.07	477	Ea.	474,000	14,900		488,900	543,500
	0490	Reciprocating, packaged w/integral air cooled condenser, 15 ton cool		.37	86.486		12,300	2,700		15,000	17,600
	0500	20 ton cooling		.34	94.955		15,500	2,975		18,475	21,600

MECHANICAL T5

157 100 | A.C. & Vent. Units

		CREW	DAILY OUTPUT	LABOR-HOURS	UNIT	1999 BARE COSTS				TOTAL INCL O&P	
						MAT.	LABOR	EQUIP.	TOTAL		
190	0520	40 ton cooling	Q-7	.30	108	Ea.	24,900	3,375		28,275	32,500	190
	0600	100 ton cooling	↓	.25	129	↓	53,000	4,025		57,025	64,000	
	0680	Water cooled, single compressor, semi-hermetic, tower not incl.										
	0700	2 to 5 ton cooling	Q-5	.57	28.021	Ea.	5,625	825		6,450	7,425	
	0740	8 ton cooling	"	.31	52.117		9,825	1,525		11,350	13,100	
	0760	10 ton cooling	Q-6	.36	67.039		9,975	2,050		12,025	14,100	
	0800	20 ton cooling	Q-7	.38	83.990		12,800	2,625		15,425	18,000	
	0820	30 ton cooling	"	.33	96.096	↓	15,400	3,000		18,400	21,500	
	0980	Water cooled, multiple compress., semi-hermetic, tower not incl.										
	1000	15 ton cooling	Q-6	.36	65.934	Ea.	16,000	2,025		18,025	20,600	
	1020	20 ton cooling	Q-7	.41	78.049		17,400	2,450		19,850	22,900	
	1060	30 ton cooling		.31	101		19,900	3,175		23,075	26,700	
	1100	50 ton cooling		.28	113		25,100	3,550		28,650	33,000	
	1160	100 ton cooling		.18	179		48,700	5,625		54,325	62,000	
	1180	120 ton cooling		.16	196		52,000	6,125		58,125	66,500	
	1200	140 ton cooling	↓	.16	202	↓	56,500	6,325		62,825	71,500	
	1450	Water cooled, dual compressors, direct drive, tower not incl.										
	1500	80 ton cooling	Q-7	.14	222	Ea.	46,400	6,950		53,350	61,500	
	1520	100 ton cooling		.14	228		54,000	7,150		61,150	70,500	
	1540	120 ton cooling		.14	231		59,500	7,250		66,750	76,500	
	1580	150 ton cooling		.13	240		72,500	7,525		80,025	91,000	
	1620	200 ton cooling		.13	250		88,500	7,800		96,300	109,000	
	1660	250 ton cooling	↓	.12	260	↓	102,000	8,125		110,125	124,500	

157 200 | System Components

		CREW	DAILY OUTPUT	LABOR-HOURS	UNIT	MAT.	LABOR	EQUIP.	TOTAL	INCL O&P	
201	0010	**COILS, FLANGED**										201
	0500	Chilled water cooling, 6 rows, 24" x 48"	Q-5	2	8	Ea.	2,575	236		2,811	3,175	
	1000	Direct expansion cooling, 6 rows, 24" x 48"		2	8		2,775	236		3,011	3,400	
	1500	Hot water heating, 1 row, 24" x 48"		3	5.333		1,025	157		1,182	1,375	
	2000	Steam heating, 1 row, 24" x 48"	↓	3	5.333	↓	1,450	157		1,607	1,850	
225	0010	**CONDENSERS** Ratings are for 30°F TD, R-22										225
	0080	Air cooled, belt drive, propeller fan										
	0240	50 ton	Q-6	.48	50	Ea.	9,275	1,525		10,800	12,500	
	0280	59 ton		.32	75		10,300	2,300		12,600	14,800	
	0320	73 ton		.29	82.759		12,200	2,525		14,725	17,200	
	0360	86 ton		.27	88.889		14,400	2,725		17,125	19,900	
	0380	88 ton	↓	.26	92.308	↓	15,400	2,825		18,225	21,300	
	1550	Air cooled, direct drive, propeller fan										
	1590	1 ton	Q-5	3.80	4.211	Ea.	525	124		649	770	
	1600	1-1/2 ton		3.60	4.444		625	131		756	890	
	1620	2 ton		3.20	5		690	147		837	985	
	1640	5 ton		2	8		1,400	236		1,636	1,900	
	1660	10 ton		1.40	11.429		2,175	335		2,510	2,900	
	1690	16 ton		1.10	14.545		3,050	430		3,480	4,025	
	1720	26 ton	↓	.74	21.622		3,975	635		4,610	5,325	
	1760	41 ton	Q-6	.54	44.444		7,150	1,350		8,500	9,925	
	1800	63 ton	"	.32	75	↓	11,300	2,300		13,600	15,900	
240	0010	**COOLING TOWERS** Packaged units										240
	0070	Galvanized steel										
	0080	Draw thru, single flow										
	0100	Belt drive, 60 tons	Q-6	90	.267	TonAC	74	8.15		82.15	93.50	
	0150	95 tons		100	.240		65	7.35		72.35	82.50	
	0200	110 tons	↓	109	.220	↓	63.50	6.75		70.25	80	

Important: See the Reference Section for critical supporting data - Reference Nos., Crews, & City Cost Indexes

			DAILY	LABOR-		1999 BARE COSTS				TOTAL
157 200	**System Components**	CREW	OUTPUT	HOURS	UNIT	MAT.	LABOR	EQUIP.	TOTAL	INCL O&P
240	0250	125 tons	Q-6	120	.200	TonAC	62	6.10		68.10	77.50	**240**
	1500	Induced air, double flow									
	1900	Gear drive, 150 ton	Q-6	126	.190	TonAC	86	5.80		91.80	103
	2000	300 ton		129	.186		58.50	5.70		64.20	72.50
	2100	600 ton		132	.182		55	5.55		60.55	69
	2150	840 ton		142	.169		44	5.15		49.15	56.50
	2200	Up to 1,000 tons	▼	150	.160	▼	41	4.89		45.89	53
	3000	For higher capacities, use multiples									
	3500	For pumps and piping, add	Q-6	38	.632	TonAC	37	19.30		56.30	69.50
	4000	For absorption systems, add				"	75%	75%			
	4500	For rigging, see division 016-460									
	5000	Fiberglass									
	5010	Draw thru									
	5100	60 tons	Q-6	1.50	16	Ea.	2,775	490		3,265	3,800
	5120	125 tons		.99	24.242		5,675	740		6,415	7,375
	5140	300 tons		.43	55.814		13,300	1,700		15,000	17,200
	5160	600 tons		.22	109		24,200	3,325		27,525	31,700
	5180	1000 tons	▼	.15	160	▼	41,500	4,900		46,400	53,000
	6000	Stainless steel									
	6010	Draw thru									
	6100	60 tons	Q-6	1.50	16	Ea.	7,200	490		7,690	8,675
	6120	110 tons		.99	24.242		11,500	740		12,240	13,800
	6140	300 tons		.43	55.814		31,900	1,700		33,600	37,700
	6160	600 tons		.22	109		49,400	3,325		52,725	59,500
	6180	1000 tons	▼	.15	160	▼	80,500	4,900		85,400	96,000
250	0010	**DUCTWORK**	R157 -050								**250**
	0020	Fabricated rectangular, includes fittings, joints, supports,									
	0030	allowance for flexible connections, no insulation									
	0031	NOTE: Fabrication and installation are combined									
	0040	as LABOR cost.									
	0100	Aluminum, alloy 3003-H14, under 100 lb.	Q-10	75	.320	Lb.	3.50	9.50		13	18.60
	0110	100 to 500 lb.		80	.300		1.44	8.90		10.34	15.40
	0120	500 to 1,000 lb.		95	.253		1.36	7.50		8.86	13.15
	0140	1,000 to 2,000 lb.		120	.200		1.34	5.95		7.29	10.65
	0150	2,000 to 5,000 lb. **CN**		130	.185		1.33	5.45		6.78	9.95
	0160	Over 5,000 lb.		145	.166		1.33	4.90		6.23	9.05
	0500	Galvanized steel, under 200 lb.		235	.102		3.50	3.03		6.53	8.55
	0520	200 to 500 lb.		245	.098		.65	2.90		3.55	5.20
	0540	500 to 1,000 lb.		255	.094		.47	2.79		3.26	4.85
	0560	1,000 to 2,000 lb.		265	.091		.44	2.68		3.12	4.65
	0570	2,000 to 5,000 lb.		275	.087		.40	2.59		2.99	4.46
	0580	Over 5,000 lb. **CN**		285	.084		.40	2.50		2.90	4.32
	1000	Stainless steel, type 304, under 100 lb.		165	.145		3.50	4.31		7.81	10.55
	1020	100 to 500 lb.		175	.137		1.31	4.06		5.37	7.75
	1030	500 to 1,000 lb.		190	.126		1.12	3.74		4.86	7.05
	1040	1,000 to 2,000 lb.		200	.120		1.07	3.56		4.63	6.70
	1050	2,000 to 5,000 lb.		225	.107		1.03	3.16		4.19	6.05
	1060	Over 5,000 lb.	▼	235	.102	▼	1.03	3.03		4.06	5.85
	1100	For medium pressure ductwork, add						15%			
	1200	For high pressure ductwork, add						40%			
	1300	Flexible, coated fiberglass fabric on corr. resist. metal helix									
	1400	pressure to 12" (WG) UL-181									
	1500	Non-insulated, 3" diameter	Q-9	400	.040	L.F.	.95	1.14		2.09	2.83
	1540	5" diameter		320	.050		1.11	1.43		2.54	3.44
	1560	6" diameter	▼	280	.057	▼	1.25	1.63		2.88	3.92

MECHANICAL 15

		157 200	System Components		DAILY OUTPUT	LABOR-HOURS	UNIT	1999 BARE COSTS				TOTAL INCL O&P	
				CREW				MAT.	LABOR	EQUIP.	TOTAL		
250	1580		7" diameter	Q-9	240	.067	L.F.	1.64	1.91		3.55	4.76	250
	1600		8" diameter	R157-050	200	.080		1.69	2.29		3.98	5.40	
	1640		10" diameter		160	.100		2.17	2.86		5.03	6.85	
	1660		12" diameter		120	.133		2.57	3.81		6.38	8.75	
	1900		Insulated, 1" thick with 3/4 lb., PE jacket, 3" diameter		380	.042		1.64	1.20		2.84	3.67	
	1910		4" diameter		340	.047		1.64	1.34		2.98	3.89	
	1920		5" diameter		300	.053		1.92	1.52		3.44	4.48	
	1940		6" diameter		260	.062		2.15	1.76		3.91	5.10	
	1960		7" diameter		220	.073		2.57	2.08		4.65	6.05	
	1980		8" diameter		180	.089		2.63	2.54		5.17	6.85	
	2020		10" diameter		140	.114		3.42	3.27		6.69	8.80	
	2040		12" diameter		100	.160		3.98	4.57		8.55	11.50	
	3490		Rigid fiberglass duct board, foil reinf. kraft facing										
	3500		Rectangular, 1" thick, alum. faced, (FRK), std. weight	Q-10	350	.069	SF Surf	.81	2.03		2.84	4.05	
290	0010	**FANS**		R157-040									290
	0020	Air conditioning and process air handling											
	0030	Axial flow, compact, low sound, 2.5" S.P.											
	0050	3,800 CFM, 5 HP	Q-20	3.40	5.882	Ea.	3,375	172		3,547	4,000		
	0080	6,400 CFM, 5 HP		2.80	7.143		3,775	209		3,984	4,475		
	0100	10,500 CFM, 7-1/2 HP		2.40	8.333		4,700	244		4,944	5,550		
	0120	15,600 CFM, 10 HP		1.60	12.500		5,925	365		6,290	7,100		
	0200	In-line centrifugal, supply/exhaust booster											
	0220	aluminum wheel/hub, disconnect switch, 1/4" S.P.											
	0240	500 CFM, 10" diameter connection	Q-20	3	6.667	Ea.	700	195		895	1,075		
	0260	1,380 CFM, 12" diameter connection		2	10		800	292		1,092	1,325		
	0280	1,520 CFM, 16" diameter connection		2	10		915	292		1,207	1,450		
	0300	2,560 CFM, 18" diameter connection		1	20		1,000	585		1,585	2,000		
	0320	3,480 CFM, 20" diameter connection		.80	25		1,200	730		1,930	2,450		
	0326	5,080 CFM, 20" diameter connection		.75	26.667		1,325	780		2,105	2,675		
	1500	Vaneaxial, low pressure, 2000 CFM, 1/2 HP		3.60	5.556		1,575	162		1,737	2,000		
	1520	4,000 CFM, 1 HP		3.20	6.250		1,625	183		1,808	2,050		
	1540	8,000 CFM, 2 HP		2.80	7.143		1,900	209		2,109	2,425		
	2500	Ceiling fan, right angle, extra quiet, 0.10" S.P.											
	2520	95 CFM	Q-20	20	1	Ea.	138	29		167	197		
	2540	210 CFM		19	1.053		152	31		183	215		
	2560	385 CFM		18	1.111		183	32.50		215.50	251		
	2580	885 CFM		16	1.250		365	36.50		401.50	455		
	2600	1,650 CFM		13	1.538		520	45		565	640		
	2620	2,960 CFM		11	1.818		690	53		743	840		
	2640	For wall or roof cap, add	1 Shee	16	.500		100	15.90		115.90	135		
	2660	For straight thru fan, add					10%						
	2680	For speed control switch, add	1 Elec	16	.500		48	15.95		63.95	77		
	3000	Paddle blade air circulator, 3 speed switch											
	3020	42", 5,000 CFM high, 3000 CFM low	1 Elec	2.40	3.333	Ea.	76.50	106		182.50	244		
	3040	52", 6,500 CFM high, 4000 CFM low	"	2.20	3.636	"	94.50	116		210.50	277		
	3100	For antique white motor, same cost											
	3200	For brass plated motor, same cost											
	3300	For light adaptor kit, add				Ea.	27.50			27.50	30.50		
	3500	Centrifugal, airfoil, motor and drive, complete											
	3520	1000 CFM, 1/2 HP	Q-20	2.50	8	Ea.	870	234		1,104	1,325		
	3540	2,000 CFM, 1 HP		2	10		935	292		1,227	1,475		
	3560	4,000 CFM, 3 HP		1.80	11.111		1,200	325		1,525	1,825		
	3580	8,000 CFM, 7-1/2 HP		1.40	14.286		1,775	420		2,195	2,600		
	3600	12,000 CFM, 10 HP		1	20		2,600	585		3,185	3,750		
	4500	Corrosive fume resistant, plastic											
	4600	roof ventilators, centrifugal, V belt drive, motor											

Important: See the Reference Section for critical supporting data - Reference Nos., Crews, & City Cost Indexes

157 200	System Components		CREW	DAILY OUTPUT	LABOR-HOURS	UNIT	1999 BARE COSTS				TOTAL INCL O&P
							MAT.	LABOR	EQUIP.	TOTAL	
290	4620	1/4" S.P., 250 CFM, 1/4 HP	R157 -040	Q-20	6	3.333	Ea.	2,125	97.50		2,222.50	2,475	290
	4640	895 CFM, 1/3 HP			5	4		2,300	117		2,417	2,725
	4660	1630 CFM, 1/2 HP			4	5		2,725	146		2,871	3,225
	4680	2240 CFM, 1 HP			3	6.667		2,850	195		3,045	3,425
	5000	Utility set, centrifugal, V belt drive, motor										
	5020	1/4" S.P., 1200 CFM, 1/4 HP		Q-20	6	3.333	Ea.	2,925	97.50		3,022.50	3,375
	5040	1520 CFM, 1/3 HP			5	4		2,925	117		3,042	3,400
	5060	1850 CFM, 1/2 HP			4	5		2,950	146		3,096	3,450
	5080	2180 CFM, 3/4 HP			3	6.667		2,950	195		3,145	3,525
	5100	1/2" S.P., 3600 CFM, 1 HP			2	10		3,950	292		4,242	4,800
	5120	4250 CFM, 1-1/2 HP			1.60	12.500		3,975	365		4,340	4,950
	5140	4800 CFM, 2 HP			1.40	14.286		4,000	420		4,420	5,050
	6000	Propeller exhaust, wall shutter, 1/4" S.P.										
	6020	Direct drive, two speed										
	6100	375 CFM, 1/10 HP		Q-20	10	2	Ea.	198	58.50		256.50	305
	6120	730 CFM, 1/7 HP			9	2.222		220	65		285	340
	6140	1000 CFM, 1/8 HP			8	2.500		305	73		378	450
	6200	4720 CFM, 1 HP			5	4		490	117		607	720
	6300	V-belt drive, 3 phase										
	6320	6175 CFM, 3/4 HP		Q-20	5	4	Ea.	410	117		527	630
	6340	7500 CFM, 3/4 HP			5	4		435	117		552	655
	6360	10,100 CFM, 1 HP			4.50	4.444		530	130		660	780
	6380	14,300 CFM, 1-1/2 HP			4	5		620	146		766	910
	6650	Residential, bath exhaust, grille, back draft damper										
	6660	50 CFM		Q-20	24	.833	Ea.	13.95	24.50		38.45	53
	6670	110 CFM			22	.909		49.50	26.50		76	95.50
	6680	Light combination, squirrel cage, 100 watt, 70 CFM			24	.833		61	24.50		85.50	105
	6700	Light/heater combination, ceiling mounted										
	6710	70 CFM, 1450 watt		Q-20	24	.833	Ea.	74	24.50		98.50	119
	6800	Heater combination, recessed, 70 CFM			24	.833		35	24.50		59.50	76
	6820	With 2 infrared bulbs			23	.870		53	25.50		78.50	97.50
	6900	Kitchen exhaust, grille, complete, 160 CFM			22	.909		62	26.50		88.50	109
	6910	180 CFM			20	1		52.50	29		81.50	103
	6920	270 CFM			18	1.111		94.50	32.50		127	154
	6930	350 CFM			16	1.250		74	36.50		110.50	138
	6940	Residential roof jacks and wall caps										
	6944	Wall cap with back draft damper										
	6946	3" & 4" dia. round duct		1 Shee	11	.727	Ea.	12.70	23		35.70	50
	6948	6" dia. round duct		"	11	.727	"	30.50	23		53.50	69.50
	6958	Roof jack with bird screen and back draft damper										
	6960	3" & 4" dia. round duct		1 Shee	11	.727	Ea.	12.20	23		35.20	49.50
	6962	3-1/4" x 10" rectangular duct		"	10	.800	"	22.50	25.50		48	64
	6980	Transition										
	6982	3-1/4" x 10" to 6" dia. round		1 Shee	20	.400	Ea.	13.75	12.70		26.45	35
	7000	Roof exhauster, centrifugal, aluminum housing, 12" galvanized										
	7020	curb, bird screen, back draft damper, 1/4" S.P.										
	7100	Direct drive, 320 CFM, 11" sq. damper		Q-20	7	2.857	Ea.	380	83.50		463.50	550
	7120	600 CFM, 11" sq. damper			6	3.333		390	97.50		487.50	580
	7140	815 CFM, 13" sq. damper			5	4		390	117		507	610
	7160	1450 CFM, 13" sq. damper			4.20	4.762		510	139		649	775
	7180	2050 CFM, 16" sq. damper			4	5		510	146		656	785
	7200	V-belt drive, 1650 CFM, 12" sq. damper			6	3.333		785	97.50		882.50	1,025
	7220	2750 CFM, 21" sq. damper			5	4		885	117		1,002	1,150
	7230	3500 CFM, 21" sq. damper			4.50	4.444		985	130		1,115	1,275
	7240	4910 CFM, 23" sq. damper			4	5		1,200	146		1,346	1,550
	7260	8525 CFM, 28" sq. damper			3	6.667		1,500	195		1,695	1,950

MECHANICAL 15

		157 200 System Components	CREW	DAILY OUTPUT	LABOR-HOURS	UNIT	1999 BARE COSTS				TOTAL INCL O&P	
							MAT.	LABOR	EQUIP.	TOTAL		
290	7280	13,760 CFM, 35" sq. damper — R157 -040	Q-20	2	10	Ea.	2,025	292		2,317	2,675	290
	7300	20,558 CFM, 43" sq. damper	↓	1	20	↓	4,075	585		4,660	5,375	
	7320	For 2 speed winding, add					15%					
	7340	For explosionproof motor, add				Ea.	273			273	300	
	7360	For belt driven, top discharge, add					15%					
	7500	Utility set, steel construction, pedestal, 1/4" S.P.										
	7520	Direct drive, 150 CFM, 1/8 HP	Q-20	6.40	3.125	Ea.	565	91.50		656.50	765	
	7540	485 CFM, 1/6 HP		5.80	3.448		715	101		816	940	
	7560	1950 CFM, 1/2 HP		4.80	4.167		835	122		957	1,100	
	7580	2410 CFM, 3/4 HP		4.40	4.545		1,550	133		1,683	1,900	
	7600	3328 CFM, 1-1/2 HP	↓	3	6.667	↓	1,725	195		1,920	2,200	
	7680	V-belt drive, drive cover, 3 phase										
	7700	800 CFM, 1/4 HP	Q-20	6	3.333	Ea.	695	97.50		792.50	915	
	7720	1,300 CFM, 1/3 HP		5	4		745	117		862	995	
	7740	2,000 CFM, 1 HP		4.60	4.348		765	127		892	1,025	
	7760	2,900 CFM, 3/4 HP	↓	4.20	4.762	↓	1,125	139		1,264	1,475	
	8500	Wall exhausters, centrifugal, auto damper, 1/8" S.P.										
	8520	Direct drive, 610 CFM, 1/20 HP	Q-20	14	1.429	Ea.	160	42		202	241	
	8540	796 CFM, 1/12 HP		13	1.538		165	45		210	252	
	8560	822 CFM, 1/6 HP		12	1.667		261	48.50		309.50	360	
	8580	1,320 CFM, 1/4 HP	↓	12	1.667	↓	263	48.50		311.50	365	
	9500	V-belt drive, 3 phase										
	9520	2,800 CFM, 1/4 HP	Q-20	9	2.222	Ea.	725	65		790	900	
	9540	3,740 CFM, 1/2 HP	"	8	2.500	"	755	73		828	945	

		157 400 Accessories										
401	0010	**AIR FILTERS**										401
	0050	Activated charcoal type, full flow				MCFM	600			600	660	
	2000	Electronic air cleaner, duct mounted										
	2150	400 - 1000 CFM	1 Shee	2.30	3.478	Ea.	495	110		605	715	
	2200	1000 - 1400 CFM		2.20	3.636		675	115		790	920	
	2250	1400 - 2000 CFM	↓	2.10	3.810	↓	745	121		866	1,000	
	2950	Mechanical media filtration units										
	3000	High efficiency type, with frame, non-supported				MCFM	45			45	49.50	
	3100	Supported type					55			55	60.50	
	4000	Medium efficiency, extended surface					5			5	5.50	
	4500	Permanent washable					20			20	22	
	5000	Renewable disposable roll				↓	120			120	132	
	5500	Throwaway glass or paper media type				Ea.	3.89			3.89	4.28	
425	0010	**CONTROL SYSTEMS, PNEUMATIC** (Sub's quote incl. mat. & labor)										425
	0011	and nominal 50 Ft. of tubing.										
	0100	Heating and Ventilating, split system										
	0200	Mixed air control, economizer cycle, panel readout, tubing										
	0220	Up to 10 tons				Ea.					4,603	
	0240	For 10 to 20 tons, add									8%	
	0260	For over 20 tons, add				↓					17%	
	0300	Heating coil, hot water, 3 way valve,										
	0320	Freezestat, limit control on discharge, readout				Ea.					3,292	
	0500	Cooling coil, chilled water, room										
	0520	Thermostat, 3 way valve				Ea.					1,394	
	0600	Cooling tower, fan cycle, damper control,										
	0620	Control system including water readout in/out at panel	Q-19	.67	35.821	Ea.	3,450	1,075		4,525	5,425	
	1000	Unit ventilator, day/night operation,										
	1100	freezestat, ASHRAE, cycle 2				Ea.					3,407	
	2000	Compensated hot water from boiler, valve control,										

Important: See the Reference Section for critical supporting data - Reference Nos., Crews, & City Cost Indexes

15 MECHANICAL

		157 400	Accessories	CREW	DAILY OUTPUT	LABOR-HOURS	UNIT	1999 BARE COSTS				TOTAL INCL O&P	
								MAT.	LABOR	EQUIP.	TOTAL		
425	2100		readout and reset at panel, up to 60 GPM				Ea.					6,087	425
	2120		For 120 GPM, add									7%	
	2140		For 240 GPM, add									12%	
	3000		Boiler room combustion air, damper to 5 SF, controls									2,756	
	3500		Fan coil, heating and cooling valves, 4 pipe control system	Q-19	3	8		780	242		1,022	1,225	
	3600		Heat exchanger system controls	"	.86	27.907		1,675	845		2,520	3,125	
	4000		Pneumatic thermostat, including controlling room radiator valve									572	
	4060		Pump control system	Q-19	3	8	↓	800	242		1,042	1,250	
	4500		Air supply for pneumatic control system										
	4600		Tank mounted duplex compressor, starter, alternator,										
	4620		piping, dryer, PRV station and filter										
	4630		1/2 HP				Ea.					8,790	
	4660		1-1/2 HP									10,680	
	4690		5 HP				↓					23,320	
430	0010	**CONTROL SYSTEMS, ELECTRONIC**											430
	0020		For electronic costs, add to division 157-425				Ea.					15%	
450	0010	**DIFFUSERS** Aluminum, opposed blade damper unless noted											450
	0100		Ceiling, linear, also for sidewall										
	0500		Perforated, 24" x 24" lay-in panel size, 6" x 6"	1 Shee	16	.500	Ea.	79.50	15.90		95.40	112	
	0520		8" x 8"		15	.533		82	16.95		98.95	117	
	0530		9" x 9"		14	.571		84	18.15		102.15	120	
	0540		10" x 10"		14	.571		86	18.15		104.15	123	
	0560		12" x 12"		12	.667		86	21		107	128	
	0590		16" x 16"		11	.727		107	23		130	154	
	0600		18" x 18"		10	.800		114	25.50		139.50	166	
	0610		20" x 20"		10	.800		130	25.50		155.50	183	
	0620		24" x 24"		9	.889		163	28		191	223	
	1000		Rectangular, 1 to 4 way blow, 6" x 6"		16	.500		46	15.90		61.90	75.50	
	1010		8" x 8"		15	.533		54	16.95		70.95	86	
	1014		9" x 9"		15	.533		56	16.95		72.95	88	
	1016		10" x 10"		15	.533		67	16.95		83.95	100	
	1020		12" x 6"		15	.533		57	16.95		73.95	89	
	1040		12" x 9"		14	.571		69.50	18.15		87.65	105	
	1060		12" x 12"		12	.667		82	21		103	123	
	1070		14" x 6"		13	.615		61	19.55		80.55	98	
	1074		14" x 14"		12	.667		101	21		122	144	
	1150		18" x 18"		9	.889		149	28		177	208	
	1160		21" x 21"		8	1		189	32		221	257	
	1170		24" x 12"		10	.800		134	25.50		159.50	188	
	1500		Round, butterfly damper, 6" diameter		18	.444		15	14.10		29.10	38.50	
	1520		8" diameter		16	.500		15.90	15.90		31.80	42	
	1540		10" diameter		14	.571		19.75	18.15		37.90	49.50	
	1560		12" diameter		12	.667		26	21		47	61.50	
	1580		14" diameter		10	.800		32.50	25.50		58	75.50	
	2000		T bar mounting, 24" x 24" lay-in frame, 6" x 6"		16	.500		75.50	15.90		91.40	108	
	2020		9" x 9"		14	.571		84	18.15		102.15	120	
	2040		12" x 12"		12	.667		111	21		132	155	
	2060		15" x 15"		11	.727		143	23		166	193	
	2080		18" x 18"	↓	10	.800	↓	149	25.50		174.50	204	
	6000		For steel diffusers instead of aluminum, deduct					10%					
460	0010	**GRILLES**											460
	0020		Aluminum										
	1000		Air return, 6" x 6"	1 Shee	26	.308	Ea.	12.35	9.75		22.10	29	
	1020		10" x 6"		24	.333	↓	14.95	10.60		25.55	33	
	1080		16" x 8"		22	.364		21.50	11.55		33.05	41.50	
	1100		12" x 12"	↓	22	.364		21.50	11.55		33.05	41.50	

MECHANICAL 15

157 400	Accessories	CREW	DAILY OUTPUT	LABOR-HOURS	UNIT	1999 BARE COSTS				TOTAL INCL O&P	
						MAT.	LABOR	EQUIP.	TOTAL		
460	1120	24" x 12"	1 Shee	18	.444	Ea.	38.50	14.10		52.60	64	**460**
	1220	24" x 18"		16	.500		45.50	15.90		61.40	74.50	
	1280	36" x 24"		14	.571		90.50	18.15		108.65	128	
	3000	Filter grille with filter, 12" x 12"		24	.333		35	10.60		45.60	55	
	3020	18" x 12"		20	.400		51.50	12.70		64.20	76	
	3040	24" x 18"		18	.444		74	14.10		88.10	104	
	3060	24" x 24"	▼	16	.500		97	15.90		112.90	132	
	6000	For steel grilles instead of aluminum in above, deduct				▼	10%					
470	0010	**REGISTERS**										**470**
	0980	Air supply										
	1000	Ceiling/wall, O.B. damper, anodized aluminum										
	1010	One or two way deflection, adj. curved face bars										
	1020	8" x 4"	1 Shee	26	.308	Ea.	17.50	9.75		27.25	34.50	
	1120	12" x 12"		18	.444		32	14.10		46.10	57	
	1240	20" x 6"		18	.444		28.50	14.10		42.60	53.50	
	1340	24" x 8"		13	.615		38.50	19.55		58.05	73	
	1350	24" x 18"	▼	12	.667	▼	72	21		93	112	
	2700	Above registers in steel instead of aluminum, deduct					10%					
	4000	Floor, toe operated damper, enameled steel										
	4020	4" x 6"	1 Shee	32	.250	Ea.	14.45	7.95		22.40	28	
	4100	8" x 10"		22	.364		18.30	11.55		29.85	38	
	4140	10" x 10"		20	.400		22	12.70		34.70	43.50	
	4220	14" x 14"		16	.500		73	15.90		88.90	105	
	4240	14" x 20"	▼	15	.533	▼	97.50	16.95		114.45	134	
	4980	Air return										
	5000	Ceiling or wall, fixed 45° face blades										
	5010	Adjustable O.B. damper, anodized aluminum										
	5020	4" x 8"	1 Shee	26	.308	Ea.	21.50	9.75		31.25	38.50	
	5060	6" x 10"		19	.421		24	13.35		37.35	47.50	
	5280	24" x 24"		11	.727		103	23		126	149	
	5300	24" x 36"	▼	8	1	▼	159	32		191	225	
	6000	For steel construction instead of aluminum, deduct					10%					
480	0010	**DUCT ACCESSORIES**										**480**
	0050	Air extractors, 12" x 4"	1 Shee	24	.333	Ea.	14.45	10.60		25.05	32.50	
	0100	8" x 6"		22	.364		14.45	11.55		26	34	
	0200	20" x 8"		16	.500		32.50	15.90		48.40	60.50	
	0280	24" x 12"		10	.800		44.50	25.50		70	88.50	
	1000	Duct access door, insulated, 6" x 6"		14	.571		13.65	18.15		31.80	43	
	1020	10" x 10"		11	.727		16.25	23		39.25	54	
	1040	12" x 12"		10	.800		17.55	25.50		43.05	59	
	1050	12" x 18"		9	.889		27.50	28		55.50	74	
	1070	18" x 18"		8	1		31	32		63	84	
	1074	24" x 18"		8	1	▼	37	32		69	90.50	
	2000	Fabrics for flexible connections, with metal edge		100	.080	L.F.	1.55	2.54		4.09	5.65	
	2100	Without metal edge		160	.050	"	1.11	1.59		2.70	3.68	
	3000	Fire damper, curtain type, 1-1/2 hr rated, vertical, 6" x 6"		24	.333	Ea.	19.70	10.60		30.30	38	
	3020	8" x 6"		22	.364		19.70	11.55		31.25	39.50	
	3240	16" x 14"		18	.444		31.50	14.10		45.60	56.50	
	3400	24" x 20"		8	1		39.50	32		71.50	93	
	5990	Multi-blade dampers, opposed blade, 8" x 6"		24	.333		18	10.60		28.60	36.50	
	5994	8" x 8"		22	.364		18.80	11.55		30.35	38.50	
	5996	10" x 10"		21	.381		21	12.10		33.10	42.50	
	6000	12" x 12"		21	.381		24.50	12.10		36.60	46	
	6020	12" x 18"		18	.444		32.50	14.10		46.60	57.50	
	6030	14" x 10"		20	.400		23.50	12.70		36.20	45.50	
	6031	14" x 14"	▼	17	.471	▼	29	14.95		43.95	55	

15 MECHANICAL

	157 400	Accessories	CREW	DAILY OUTPUT	LABOR-HOURS	UNIT	1999 BARE COSTS				TOTAL INCL O&P	
							MAT.	LABOR	EQUIP.	TOTAL		
480	6033	16" x 12"	1 Shee	17	.471	Ea.	29	14.95		43.95	55	480
	6035	16" x 16"		16	.500		35.50	15.90		51.40	63.50	
	6037	18" x 14"		16	.500		35.50	15.90		51.40	63.50	
	6038	18" x 18"		15	.533		42	16.95		58.95	72.50	
	6070	20" x 16"		14	.571		43	18.15		61.15	75	
	6072	20" x 20"		13	.615		51	19.55		70.55	86.50	
	6074	22" x 18"		14	.571		51	18.15		69.15	84	
	6076	24" x 16"		11	.727		50	23		73	91	
	6078	24" x 20"		8	1		59.50	32		91.50	115	
	6080	24" x 24"		8	1		69	32		101	126	
	6110	26" x 26"		6	1.333		82.50	42.50		125	156	
	6133	30" x 30"	Q-9	6.60	2.424		109	69.50		178.50	228	
	6135	32" x 32"		6.40	2.500		124	71.50		195.50	247	
	6180	48" x 36"		5.60	2.857		209	81.50		290.50	355	
	7000	Splitter damper assembly, self-locking, 1' rod	1 Shee	24	.333		14.30	10.60		24.90	32	
	7020	3' rod		22	.364		18.50	11.55		30.05	38.50	
	7040	4' rod		20	.400		20.50	12.70		33.20	42	
	7060	6' rod		18	.444		25	14.10		39.10	49.50	
	8000	Multi-blade dampers, parallel blade										
	8100	8" x 8"	1 Shee	24	.333	Ea.	52.50	10.60		63.10	74.50	
	8140	16" x 10"		20	.400		55.50	12.70		68.20	80.50	
	8200	24" x 16"		11	.727		74	23		97	117	
	8260	30" x 18"		7	1.143		86	36.50		122.50	151	
	9000	Silencers, noise control for air flow, duct				MCFM	41.50			41.50	45.50	
	9100	Louvers, galvanized				"	60			60	66	
	9200	Plenums, measured by panel surface				S.F.	8.25			8.25	9.10	
	9300	Vent, air transfer				Ea.	139			139	153	
482	0010	**LOUVERS**										482
	0100	Aluminum, extruded, with screen, mill finish										
	1000	Brick vent, (see also division 041-524)										
	1100	Standard, 4" deep, 8" wide, 5" high	1 Shee	24	.333	Ea.	22	10.60		32.60	40.50	
	1200	Modular, 4" deep, 7-3/4" wide, 5" high		24	.333		23	10.60		33.60	42	
	1300	Speed brick, 4" deep, 11-5/8" wide, 3-7/8" high		24	.333		23	10.60		33.60	42	
	1400	Fuel oil brick, 4" deep, 8" wide, 5" high		24	.333		41	10.60		51.60	61.50	
	2000	Cooling tower and mechanical equip., screens, light weight		40	.200	S.F.	9.25	6.35		15.60	20	
	2020	Standard weight		35	.229		27	7.25		34.25	41	
	2500	Dual combination, automatic, intake or exhaust		20	.400		37	12.70		49.70	60	
	2520	Manual operation		20	.400		27	12.70		39.70	49	
	2540	Electric or pneumatic operation		20	.400		27	12.70		39.70	49	
	2560	Motor, for electric or pneumatic		14	.571	Ea.	325	18.15		343.15	390	
	3000	Fixed blade, continuous line										
	3100	Mullion type, stormproof	1 Shee	28	.286	S.F.	27	9.05		36.05	43.50	
	3200	Stormproof		28	.286		27	9.05		36.05	43.50	
	3300	Vertical line		28	.286		36	9.05		45.05	53.50	
	3500	For damper to use with above, add					50%	30%				
	3520	Motor, for damper, electric or pneumatic	1 Shee	14	.571	Ea.	325	18.15		343.15	390	
	4000	Operating, 45°, manual, electric or pneumatic		24	.333	S.F.	27	10.60		37.60	46	
	4100	Motor, for electric or pneumatic		14	.571	Ea.	325	18.15		343.15	390	
	4200	Penthouse, roof		56	.143	S.F.	15.50	4.54		20.04	24	
	4300	Walls		40	.200		38	6.35		44.35	52	
	5000	Thinline, under 4" thick, fixed blade		40	.200		15	6.35		21.35	26.50	
	5010	Finishes, applied by mfr. at additional cost, available in colors										
	5020	Prime coat only, add				S.F.	1.85			1.85	2.04	
	5040	Baked enamel finish coating, add					3.65			3.65	4.02	
	5060	Anodized finish, add					4.15			4.15	4.57	

MECHANICAL 15

For expanded coverage of these items see *Means Mechanical or Plumbing Cost Data 1999*

157 400 | Accessories

		CREW	DAILY OUTPUT	LABOR-HOURS	UNIT	1999 BARE COSTS				TOTAL INCL O&P	
						MAT.	LABOR	EQUIP.	TOTAL		
482	5080	Duranodic finish, add				S.F.	7.85			7.85	8.65	**482**
	5100	Fluoropolymer finish coating, add					12			12	13.20	
	9980	For small orders (under 10 pieces), add				↓	25%					
490	0010	**VENTILATORS** Base, damper & bird screen, CFM in 5 MPH wind										**490**
	0500	Rotary syphon, galvanized, 6" neck diameter, 185 CFM	Q-9	16	1	Ea.	49	28.50		77.50	98.50	
	0560	12" neck diameter, 310 CFM [R157 -040]		10	1.600		69	45.50		114.50	147	
	0660	24" neck diameter, 1,530 CFM		8	2		281	57		338	400	
	0700	36" neck diameter, 3,800 CFM		6	2.667		765	76		841	965	
	0720	42" neck diameter, 4,500 CFM	↓	4	4	↓	1,250	114		1,364	1,550	
	1280	Spinner ventilators, wind driven, galvanized										
	1300	4" neck diameter, 180 CFM	Q-9	20	.800	Ea.	50	23		73	90.50	
	1340	6" neck diameter, 250 CFM		16	1		56	28.50		84.50	106	
	1400	12" neck diameter, 770 CFM		10	1.600		81	45.50		126.50	160	
	1500	24" neck diameter, 3,100 CFM		8	2		236	57		293	350	
	1540	36" neck diameter, 5,500 CFM	↓	6	2.667	↓	690	76		766	880	
	2000	Stationary, gravity, syphon, galvanized										
	2160	6" neck diameter, 66 CFM	Q-9	16	1	Ea.	43	28.50		71.50	92	
	2240	12" neck diameter, 160 CFM		10	1.600		90.50	45.50		136	171	
	2340	24" neck diameter, 900 CFM		8	2		247	57		304	360	
	2380	36" neck diameter, 2,000 CFM		6	2.667		775	76		851	970	
	4200	Stationary mushroom, aluminum, 16" orifice diameter		10	1.600		246	45.50		291.50	340	
	4220	26" orifice diameter		9	1.778		365	51		416	480	
	4230	30" orifice diameter		8.60	1.860		530	53		583	670	
	4240	38" orifice diameter		8	2		760	57		817	925	
	4250	42" orifice diameter		7.60	2.105		1,000	60		1,060	1,200	
	4260	50" orifice diameter	↓	7	2.286	↓	1,200	65.50		1,265.50	1,425	
	5000	Relief vent										
	5500	Rectangular, aluminum, galvanized curb										
	5510	intake/exhaust, 0.05" SP										
	5600	500 CFM, 12" x 16"	Q-9	8	2	Ea.	545	57		602	690	
	5640	1000 CFM, 12" x 24"		6.60	2.424		645	69.50		714.50	820	
	5680	3000 CFM, 20" x 42"	↓	4	4	↓	1,125	114		1,239	1,400	

157 600 | Miscellaneous

		CREW	DAILY OUTPUT	LABOR-HOURS	UNIT	MAT.	LABOR	EQUIP.	TOTAL	TOTAL INCL O&P	
601	0010	**BALANCING, AIR** (Subcontractor's quote incl. material & labor)										**601**
	0900	Heating and ventilating equipment										
	1000	Centrifugal fans, utility sets				Ea.					237	
	1100	Heating and ventilating unit									355	
	1200	In-line fan									355	
	1300	Propeller and wall fan									67	
	1400	Roof exhaust fan									158	
	2000	Air conditioning equipment, central station									513	
	2100	Built-up low pressure unit									473	
	2200	Built-up high pressure unit									552	
	2500	Multi-zone A.C. and heating unit									355	
	2600	For each zone over one, add									79	
	2700	Package A.C. unit									197	
	2800	Rooftop heating and cooling unit									276	
	3000	Supply, return, exhaust, registers & diffusers, avg. height ceiling									47.30	
	3100	High ceiling									71	
	3200	Floor height				↓					39.45	
604	0010	**BALANCING, WATER** (Subcontractor's quote incl. material & labor)										**604**
	0050	Air cooled condenser				Ea.					139	

			DAILY	LABOR-		1999 BARE COSTS				TOTAL
157 600	**Miscellaneous**	CREW	OUTPUT	HOURS	UNIT	MAT.	LABOR	EQUIP.	TOTAL	INCL O&P
604 0100	Cabinet unit heater				Ea.					47.60	**604**
0200	Chiller									337
0300	Convector									39.70
0500	Cooling tower									258
0600	Fan coil unit, unit ventilator									71.40
0700	Fin tube and radiant panels									79.35
0800	Main and duct re-heat coils									73.40
0810	Heat exchanger									73.40
1000	Pumps									175
1100	Unit heater				↓					55.55
610 0010	**DUST COLLECTION SYSTEMS** Commercial / industrial										**610**
0120	Central vacuum units									
0130	Includes stand, filters and motorized shaker									
0200	500 CFM, 10″ inlet, 2 HP	Q-20	2.40	8.333	Ea.	3,075	244		3,319	3,775
0220	1000 CFM, 10″ inlet, 3 HP		2.20	9.091		3,225	266		3,491	3,950
0240	1500 CFM, 10″ inlet, 5 HP		2	10		3,375	292		3,667	4,175
0260	3000 CFM, 13″ inlet, 10 HP		1.50	13.333		8,625	390		9,015	10,100
0280	5000 CFM, 16″ inlet, 2 @ 10 HP	↓	1	20	↓	9,125	585		9,710	10,900
1000	Vacuum tubing, galvanized									
1100	2-1/8″ OD, 16 ga.	Q-9	440	.036	L.F.	1.71	1.04		2.75	3.49
1110	2-1/2″ OD, 16 ga.		420	.038		1.98	1.09		3.07	3.87
1120	3″ OD, 16 ga.		400	.040		2.51	1.14		3.65	4.54
1130	3-1/2″ OD, 16 ga.		380	.042		3.58	1.20		4.78	5.80
1140	4″ OD, 16 ga.		360	.044		3.65	1.27		4.92	6
1150	5″ OD, 14 ga.		320	.050		6.75	1.43		8.18	9.65
1160	6″ OD, 14 ga.		280	.057		7.60	1.63		9.23	10.90
1170	8″ OD, 14 ga.		200	.080		11.45	2.29		13.74	16.15
1180	10″ OD, 12 ga.		160	.100		11.45	2.86		14.31	17.05
1190	12″ OD, 12 ga.		120	.133		33	3.81		36.81	42
1200	14″ OD, 12 ga.	↓	80	.200	↓	36	5.70		41.70	49
1940	Hose, flexible wire reinforced rubber									
1956	3″ dia.	Q-9	400	.040	L.F.	5.05	1.14		6.19	7.35
1960	4″ dia.		360	.044		6.90	1.27		8.17	9.50
1970	5″ dia.		320	.050		7.80	1.43		9.23	10.75
1980	6″ dia.	↓	280	.057	↓	9.15	1.63		10.78	12.65
2000	90° Elbow, slip fit									
2110	2-1/8″ dia.	Q-9	70	.229	Ea.	7.15	6.55		13.70	18
2120	2-1/2″ dia.		65	.246		9.95	7.05		17	22
2130	3″ dia.		60	.267		13.55	7.60		21.15	27
2140	3-1/2″ dia.		55	.291		16.75	8.30		25.05	31.50
2150	4″ dia.		50	.320		21	9.15		30.15	37
2160	5″ dia.		45	.356		39.50	10.15		49.65	59.50
2170	6″ dia.		40	.400		54.50	11.45		65.95	78
2180	8″ dia.	↓	30	.533	↓	103	15.25		118.25	138
2400	45° Elbow, slip fit									
2410	2-1/8″ dia.	Q-9	70	.229	Ea.	6.25	6.55		12.80	17.05
2420	2-1/2″ dia.		65	.246		9.20	7.05		16.25	21
2430	3″ dia.		60	.267		11.15	7.60		18.75	24
2440	3-1/2″ dia.		55	.291		14	8.30		22.30	28.50
2450	4″ dia.		50	.320		18.15	9.15		27.30	34
2460	5″ dia.		45	.356		30.50	10.15		40.65	49.50
2470	6″ dia.		40	.400		41.50	11.45		52.95	63.50
2480	8″ dia.	↓	35	.457	↓	81	13.05		94.05	110
2800	90° TY, slip fit thru 6″ dia									
2810	2-1/8″ dia.	Q-9	42	.381	Ea.	14	10.90		24.90	32.50
2820	2-1/2″ dia.	↓	39	.410	↓	18.20	11.75		29.95	38

MECHANICAL 15

			DAILY	LABOR-			1999 BARE COSTS			TOTAL	
157 600	**Miscellaneous**	CREW	OUTPUT	HOURS	UNIT	MAT.	LABOR	EQUIP.	TOTAL	INCL O&P	
610	2830	3" dia.	Q-9	36	.444	Ea.	24	12.70		36.70	46	**610**
	2840	3-1/2" dia.		33	.485		30.50	13.85		44.35	55	
	2850	4" dia.		30	.533		44	15.25		59.25	72	
	2860	5" dia.		27	.593		83.50	16.95		100.45	119	
	2870	6" dia.	▼	24	.667		114	19.05		133.05	155	
	2880	8" dia., butt end					242			242	266	
	2890	10" dia., butt end					435			435	480	
	2900	12" dia., butt end					435			435	480	
	2910	14" dia., butt end					795			795	870	
	2920	6 x 4" dia., butt end					69.50			69.50	76.50	
	2930	8" x 4" dia., butt end					132			132	145	
	2940	10" x 4" dia., butt end					172			172	190	
	2950	12" x 4" dia., butt end				▼	198			198	218	
	3100	90° Elbow, butt end segmented										
	3110	8" dia., butt end, segmented				Ea.	156			156	172	
	3120	10" dia., butt end, segmented					257			257	282	
	3130	12" dia., butt end, segmented					360			360	395	
	3140	14" dia., butt end, segmented				▼	405			405	450	
	3200	45° Elbow, butt end segmented										
	3210	8" dia., butt end, segmented				Ea.	133			133	146	
	3220	10" dia., butt end, segmented					186			186	204	
	3230	12" dia., butt end, segmented					190			190	209	
	3240	14" dia., butt end, segmented				▼	234			234	257	
	3400	All butt end fittings require one coupling per joint.										
	3410	Labor for fitting included with couplings.										
	3460	Compression coupling, galvanized, neoprene gasket										
	3470	2-1/8" dia.	Q-9	44	.364	Ea.	9.90	10.40		20.30	27	
	3480	2-1/2" dia.		44	.364		9.90	10.40		20.30	27	
	3490	3" dia.		38	.421		14.30	12.05		26.35	34.50	
	3500	3-1/2" dia.		35	.457		16	13.05		29.05	38	
	3510	4" dia.		33	.485		17.35	13.85		31.20	40.50	
	3520	5" dia.		29	.552		19.80	15.75		35.55	46.50	
	3530	6" dia.		26	.615		23	17.60		40.60	53	
	3540	8" dia.		22	.727		42	21		63	79	
	3550	10" dia.		20	.800		61.50	23		84.50	104	
	3560	12" dia.		18	.889		77	25.50		102.50	124	
	3570	14" dia.	▼	16	1	▼	109	28.50		137.50	165	
	3800	Air gate valves, galvanized										
	3810	2-1/8" dia.	Q-9	30	.533	Ea.	83	15.25		98.25	115	
	3820	2-1/2" dia.		28	.571		87	16.35		103.35	121	
	3830	3" dia.		26	.615		101	17.60		118.60	139	
	3840	4" dia.		23	.696		132	19.90		151.90	177	
	3850	6" dia.	▼	18	.889	▼	182	25.50		207.50	241	

For information about Means Estimating Seminars, see yellow pages 11 and 12 in back of book

Important: See the Reference Section for critical supporting data - Reference Nos., Crews, & City Cost Indexes

Division 16 Electrical

Estimating Tips

160 Raceways

- Conduit should be taken off in three main categories: power distribution, branch power, and branch lighting, so the estimator can concentrate on systems and components, therefore making it easier to ensure all items have been accounted for.
- For cost modifications for elevated conduit installation, add the percentages to labor according to the height of installation and only the quantities exceeding the different height levels, not to the total conduit quantities.

161 Conductors & Grounding

- Remember that aluminum wiring of equal ampacity is larger in diameter than copper and may require larger conduit.
- If more than three wires at a time are being pulled, deduct percentages from the labor hours of that grouping of wires.
- When taking off grounding system, identify separately the type and size of wire and list each unique type of ground connection.

162 Boxes & Wiring Devices

- The estimator should take the weights of materials into consideration when completing a takeoff. Topics to consider include: How will the materials be supported? What methods of support are available? How high will the support structure have to reach? Will the final support structure be able to withstand the total burden? Is the support material included or separate from the fixture, equipment and material specified?

163 Motors, Starters, Boards & Switches

- Supports and concrete pads may be shown on drawings for the larger equipment, or the support system may be just a piece of plywood for the back of a panelboard. In either case, it must be included in the costs.

164 Transformers & Bus Ducts

- Do not overlook the costs for equipment used in the installation. If scaffolding or highlifts are available in the field, contractors may use them in lieu of the proposed ladders and rolling staging.

166 Lighting

- Fixtures should be taken off room by room, using the fixture schedule, specifications, and the ceiling plan. For large concentrations of lighting fixtures in the same area deduct the percentages from labor hours.

168 Special Systems

- When estimating material costs for special systems, it is always prudent to obtain manufacturers' quotations for equipment prices and special installation requirements which will affect the total costs.

Reference Numbers

Reference numbers are shown in bold squares at the beginning of some major classifications. These numbers refer to related items in the Reference Section. The reference information may be an estimating procedure, an alternate pricing method or technical information.

Note: Not all subdivisions listed here necessarily appear in this publication.

160 100 | Cable Trays		CREW	DAILY OUTPUT	LABOR-HOURS	UNIT	1999 BARE COSTS				TOTAL INCL O&P
						MAT.	LABOR	EQUIP.	TOTAL	
105	0010	**CABLE TRAY LADDER TYPE** w/ ftngs & supports, 4" dp., to 15' elev.									105
	0160	Galvanized steel tray								
	0170	4" rung spacing, 6" wide	2 Elec	98	.163	L.F.	8.15	5.20		13.35	16.75
	0200	12" wide		86	.186		9.80	5.95		15.75	19.65
	0400	18" wide		82	.195		11.40	6.20		17.60	22
	0600	24" wide		78	.205		13.10	6.55		19.65	24
	3200	Aluminum tray, 4" deep, 6" rung spacing, 6" wide		134	.119		10.25	3.81		14.06	16.95
	3220	12" wide		124	.129		11.45	4.12		15.57	18.75
	3230	18" wide		114	.140		12.75	4.48		17.23	21
	3240	24" wide		106	.151		14.90	4.82		19.72	23.50
	9980	Allow. for tray ftngs., 5% min.-20% max.									
150	0010	**WIREWAY** to 15' high									150
	0100	Screw cover, with fittings and supports, 2-1/2" x 2-1/2"	1 Elec	45	.178	L.F.	8.30	5.65		13.95	17.60
	0200	4" x 4"	"	40	.200		9.15	6.40		15.55	19.60
	0400	6" x 6"	2 Elec	60	.267		15.50	8.50		24	30
	0600	8" x 8"	"	40	.400		20	12.75		32.75	41
	9980	Allow. for wireway ftngs., 5% min.-20% max.									

160 200 | Conduits

			CREW	DAILY OUTPUT	LABOR-HOURS	UNIT	MAT.	LABOR	EQUIP.	TOTAL	TOTAL INCL O&P
205	0010	**CONDUIT** To 15' high, includes 2 terminations, 2 elbows and									205
	0020	11 beam clamps per 100 L.F. `R160 -205`									
	0300	Aluminum, 1/2" diameter	1 Elec	100	.080	L.F.	1.18	2.55		3.73	5.10
	0500	3/4" diameter		90	.089		1.57	2.84		4.41	5.95
	0700	1" diameter		80	.100		2.13	3.19		5.32	7.10
	1000	1-1/4" diameter		70	.114		2.75	3.65		6.40	8.50
	1030	1-1/2" diameter		65	.123		3.41	3.93		7.34	9.60
	1050	2" diameter		60	.133		4.52	4.25		8.77	11.30
	1070	2-1/2" diameter		50	.160		7.05	5.10		12.15	15.40
	1100	3" diameter	2 Elec	90	.178		9.40	5.65		15.05	18.85
	1130	3-1/2" diameter		80	.200		11.60	6.40		18	22.50
	1140	4" diameter		70	.229		14.05	7.30		21.35	26.50
	1750	Rigid galvanized steel, 1/2" diameter	1 Elec	90	.089		1.46	2.84		4.30	5.85
	1770	3/4" diameter		80	.100		1.72	3.19		4.91	6.65
	1800	1" diameter		65	.123		2.38	3.93		6.31	8.45
	1830	1-1/4" diameter		60	.133		3.13	4.25		7.38	9.80
	1850	1-1/2" diameter		55	.145		3.72	4.64		8.36	11.05
	1870	2" diameter		45	.178		4.96	5.65		10.61	13.95
	1900	2-1/2" diameter		35	.229		8.20	7.30		15.50	19.90
	1930	3" diameter	2 Elec	50	.320		10.40	10.20		20.60	26.50
	1950	3-1/2" diameter		44	.364		12.70	11.60		24.30	31.50
	1970	4" diameter		40	.400		15.25	12.75		28	36
	2500	Steel, intermediate conduit (IMC), 1/2" diameter	1 Elec	100	.080		1.15	2.55		3.70	5.10
	2530	3/4" diameter		90	.089		1.36	2.84		4.20	5.75
	2550	1" diameter		70	.114		1.86	3.65		5.51	7.50
	2570	1-1/4" diameter		65	.123		2.43	3.93		6.36	8.50
	2600	1-1/2" diameter		60	.133		2.79	4.25		7.04	9.40
	2630	2" diameter		50	.160		3.62	5.10		8.72	11.65
	2650	2-1/2" diameter		40	.200		7.05	6.40		13.45	17.30
	2670	3" diameter	2 Elec	60	.267		9.10	8.50		17.60	22.50
	2700	3-1/2" diameter		54	.296		11.20	9.45		20.65	26.50
	2730	4" diameter		50	.320		13	10.20		23.20	29.50
	5000	Electric metallic tubing (EMT), 1/2" diameter	1 Elec	170	.047		.36	1.50		1.86	2.64
	5020	3/4" diameter **CN**		130	.062		.53	1.96		2.49	3.52
	5040	1" diameter		115	.070		.86	2.22		3.08	4.26
	5060	1-1/4" diameter		100	.080		1.31	2.55		3.86	5.25

Important: See the Reference Section for critical supporting data - Reference Nos., Crews, & City Cost Indexes

160 200	Conduits		CREW	DAILY OUTPUT	LABOR-HOURS	UNIT	1999 BARE COSTS				TOTAL INCL O&P	
							MAT.	LABOR	EQUIP.	TOTAL		
205	5080	1-1/2" diameter	1 Elec	90	.089	L.F.	1.64	2.84		4.48	6.05	**205**
	5100	2" diameter		80	.100		2.12	3.19		5.31	7.10	
	5120	2-1/2" diameter		60	.133		5.75	4.25		10	12.65	
	5140	3" diameter	2 Elec	100	.160		6.30	5.10		11.40	14.60	
	5160	3-1/2" diameter		90	.178		8.05	5.65		13.70	17.35	
	5180	4" diameter		80	.200		9.15	6.40		15.55	19.60	
	9900	Add to labor for higher elevated installation										
	9910	15' to 20' high, add						10%				
	9920	20' to 25' high, add						20%				
	9930	25' to 30' high, add						25%				
	9940	30' to 35' high, add						30%				
	9950	35' to 40' high, add						35%				
	9960	Over 40' high, add						40%				
	9980	Allow. for cond. ftngs., 5% min.-20% max.										
230	0010	**CONDUIT IN CONCRETE SLAB** Including terminations,										**230**
	0020	fittings and supports										
	3230	PVC, schedule 40, 1/2" diameter	1 Elec	270	.030	L.F.	.35	.95		1.30	1.79	
	3250	3/4" diameter		230	.035		.42	1.11		1.53	2.12	
	3270	1" diameter		200	.040		.56	1.28		1.84	2.52	
	3300	1-1/4" diameter		170	.047		.77	1.50		2.27	3.09	
	3330	1-1/2" diameter		140	.057		.96	1.82		2.78	3.78	
	3350	2" diameter		120	.067		1.22	2.13		3.35	4.52	
	4350	Rigid galvanized steel, 1/2" diameter		200	.040		1.21	1.28		2.49	3.24	
	4400	3/4" diameter		170	.047		1.47	1.50		2.97	3.86	
	4450	1" diameter		130	.062		2.14	1.96		4.10	5.30	
	4500	1-1/4" diameter		110	.073		2.71	2.32		5.03	6.45	
	4600	1-1/2" diameter		100	.080		3.30	2.55		5.85	7.45	
	4800	2" diameter		90	.089		4.40	2.84		7.24	9.10	
240	0010	**CONDUIT IN TRENCH** Includes terminations and fittings										**240**
	0020	Does not include excavation or backfill, see div. 022-200										
	0200	Rigid galvanized steel, 2" diameter	1 Elec	150	.053	L.F.	4.24	1.70		5.94	7.20	
	0400	2-1/2" diameter	"	100	.080		7.35	2.55		9.90	11.90	
	0600	3" diameter	2 Elec	160	.100		9.55	3.19		12.74	15.25	
	0800	3-1/2" diameter		140	.114		12.15	3.65		15.80	18.80	
	1000	4" diameter		100	.160		13.95	5.10		19.05	23	
	1200	5" diameter		80	.200		29	6.40		35.40	41.50	
	1400	6" diameter		60	.267		41.50	8.50		50	58	
275	0010	**MOTOR CONNECTIONS**										**275**
	0020	Flexible conduit and fittings, 115 volt, 1 phase, up to 1 HP motor	1 Elec	8	1	Ea.	3.95	32		35.95	52	
	0200	25 HP motor		2.70	2.963		14.70	94.50		109.20	157	
	0400	50 HP motor		2.20	3.636		43.50	116		159.50	221	
	0600	100 HP motor		1.50	5.333		109	170		279	375	
290	0010	**WIREMOLD RACEWAY**										**290**
	0090	Surface, metal, straight section										
	0100	No. 500	1 Elec	100	.080	L.F.	.63	2.55		3.18	4.51	
	0110	No. 700		100	.080		.71	2.55		3.26	4.60	
	0400	No. 1500, small pancake		90	.089		1.33	2.84		4.17	5.70	
	0600	No. 2000, base & cover, blank		90	.089		1.29	2.84		4.13	5.65	
	0800	No. 3000, base & cover, blank		75	.107		2.53	3.40		5.93	7.90	
	1000	No. 4000, base & cover, blank		65	.123		4.30	3.93		8.23	10.60	
	1200	No. 6000, base & cover, blank		50	.160		7.05	5.10		12.15	15.40	
	2400	Fittings, elbows, No. 500		40	.200	Ea.	1.14	6.40		7.54	10.80	
	2800	Elbow cover, No. 2000		40	.200		2.30	6.40		8.70	12.10	
	2880	Tee, No. 500		42	.190		2.21	6.10		8.31	11.55	

R160 -205

ELECTRICAL 16

160 200 | Conduits

		CREW	DAILY OUTPUT	LABOR-HOURS	UNIT	1999 BARE COSTS				TOTAL INCL O&P
						MAT.	LABOR	EQUIP.	TOTAL	
290	2900	No. 2000	1 Elec	27	.296	Ea.	7.05	9.45		16.50	22
	3000	Switch box, No. 500		16	.500		8.10	15.95		24.05	33
	3400	Telephone outlet, No. 1500		16	.500		8.60	15.95		24.55	33.50
	3600	Junction box, No. 1500		16	.500		6	15.95		21.95	30.50
	3800	Plugmold wired sections, No. 2000									
	4000	1 circuit, 6 outlets, 3 ft. long	1 Elec	8	1	Ea.	21.50	32		53.50	71
	4100	2 circuits, 8 outlets, 6 ft. long		5.30	1.509		36	48		84	112
	4200	Tele-power poles, aluminum, 4 outlets		3.70	2.162		142	69		211	259

160 500 | Ducts

		CREW	DAILY OUTPUT	LABOR-HOURS	UNIT	1999 BARE COSTS				TOTAL INCL O&P
						MAT.	LABOR	EQUIP.	TOTAL	
540	0010	**TRENCH DUCT** Steel with cover									
	0020	Standard adjustable, depths to 4"									
	0100	Straight, single compartment, 9" wide	2 Elec	40	.400	L.F.	66	12.75		78.75	91.50
	0200	12" wide		32	.500		75	15.95		90.95	107
	0400	18" wide		26	.615		100	19.65		119.65	140
	0600	24" wide		22	.727		128	23		151	176
	0800	30" wide		20	.800		155	25.50		180.50	209
	1000	36" wide		16	1		186	32		218	253
	1200	Horizontal elbow, 9" wide		5.40	2.963	Ea.	245	94.50		339.50	410
	1400	12" wide		4.60	3.478		282	111		393	475
	1600	18" wide		4	4		365	128		493	590
	1800	24" wide		3.20	5		510	160		670	800
	2000	30" wide		2.60	6.154		680	196		876	1,050
	2200	36" wide		2.40	6.667		890	213		1,103	1,300
	2400	Vertical elbow, 9" wide		5.40	2.963		85	94.50		179.50	235
	2600	12" wide		4.60	3.478		93	111		204	268
	2800	18" wide		4	4		107	128		235	310
	3000	24" wide		3.20	5		132	160		292	385
	3200	30" wide		2.60	6.154		147	196		343	455
	3400	36" wide		2.40	6.667		161	213		374	495
	3600	Cross, 9" wide		4	4		405	128		533	635
	3800	12" wide		3.20	5		425	160		585	710
	4000	18" wide		2.60	6.154		510	196		706	855
	4200	24" wide		2.20	7.273		645	232		877	1,050
	4400	30" wide		2	8		840	255		1,095	1,300
	4600	36" wide		1.80	8.889		1,050	284		1,334	1,575
	4800	End closure, 9" wide		14.40	1.111		25	35.50		60.50	80.50
	5000	12" wide		12	1.333		29	42.50		71.50	95.50
	5200	18" wide		10	1.600		44	51		95	125
	5400	24" wide		8	2		58	64		122	160
	5600	30" wide		6.60	2.424		72	77.50		149.50	195
	5800	36" wide		5.80	2.759		86	88		174	227
	6000	Tees, 9" wide		4	4		244	128		372	460
	6200	12" wide		3.60	4.444		284	142		426	520
	6400	18" wide		3.20	5		365	160		525	640
	6600	24" wide		3	5.333		520	170		690	830
	6800	30" wide		2.60	6.154		680	196		876	1,050
	7000	36" wide		2	8		890	255		1,145	1,350
	7200	Riser, and cabinet connector, 9" wide		5.40	2.963		107	94.50		201.50	259
	7400	12" wide		4.60	3.478		124	111		235	300
	7600	18" wide		4	4		153	128		281	360
	7800	24" wide		3.20	5		184	160		344	440
	8000	30" wide		2.60	6.154		213	196		409	530
	8200	36" wide		2	8		248	255		503	655
	8400	Insert assembly, cell to conduit adapter, 1-1/4"	1 Elec	16	.500		42	15.95		57.95	70

16 ELECTRICAL

Important: See the Reference Section for critical supporting data - Reference Nos., Crews, & City Cost Indexes

160 500 | Ducts

		CREW	DAILY OUTPUT	LABOR-HOURS	UNIT	1999 BARE COSTS				TOTAL INCL O&P
						MAT.	LABOR	EQUIP.	TOTAL	
560	0010	**UNDERFLOOR DUCT**									560
	0100	Duct, 1-3/8″ x 3-1/8″ blank, standard	2 Elec	160	.100	L.F.	7.25	3.19		10.44	12.70
	0200	1-3/8″ x 7-1/4″ blank, super duct		120	.133		14.45	4.25		18.70	22.50
	0400	7/8″ or 1-3/8″ insert type, 24″ O.C., 1-3/8″ x 3-1/8″, std.		140	.114		9.70	3.65		13.35	16.10
	0600	1-3/8″ x 7-1/4″, super duct		100	.160		16.90	5.10		22	26.50
	0800	Junction box, single duct, 1 level, 3-1/8″	1 Elec	4	2	Ea.	218	64		282	335
	1000	Junction box, single duct, 1 level, 7-1/4″		2.70	2.963		238	94.50		332.50	405
	1200	1 level, 2 duct, 3-1/8″		3.20	2.500		290	80		370	440
	1400	Junction box, 1 level, 2 duct, 7-1/4″		2.30	3.478		785	111		896	1,025
	1600	Triple duct, 3-1/8″		2.30	3.478		495	111		606	710
	1800	Insert to conduit adapter, 3/4″ & 1″		32	.250		9.95	7.95		17.90	23
	2000	Support, single cell		27	.296		15.80	9.45		25.25	31.50
	2200	Super duct		16	.500		15.80	15.95		31.75	41.50
	2400	Double cell		16	.500		15.80	15.95		31.75	41.50
	2600	Triple cell		11	.727		14.90	23		37.90	51
	2800	Vertical elbow, standard duct		10	.800		47.50	25.50		73	90.50
	3000	Super duct		8	1		47.50	32		79.50	100
	3200	Cabinet connector, standard duct		32	.250		36	7.95		43.95	52
	3400	Super duct		27	.296		36	9.45		45.45	54
	3600	Conduit adapter, 1″ to 1-1/4″		32	.250		36	7.95		43.95	52
	3800	2″ to 1-1/4″		27	.296		36	9.45		45.45	54
	4000	Outlet, low tension (tele, computer, etc.)		8	1		38	32		70	89
	4200	High tension, receptacle (120 volt)		8	1		41.50	32		73.50	93

161 100 | Conductors

		CREW	DAILY OUTPUT	LABOR-HOURS	UNIT	1999 BARE COSTS				TOTAL INCL O&P
						MAT.	LABOR	EQUIP.	TOTAL	
105	0010	**ARMORED CABLE**									105
	0050	600 volt, copper (BX), #14, 2 conductor, solid	1 Elec	2.40	3.333	C.L.F.	45	106		151	209
	0100	3 conductor, solid		2.20	3.636		57	116		173	236
	0150	#12, 2 conductor, solid		2.30	3.478		45.50	111		156.50	216
	0200	3 conductor, solid		2	4		66.50	128		194.50	265
	0250	#10, 2 conductor, solid		2	4		80	128		208	279
	0300	3 conductor, solid		1.60	5		105	160		265	355
	0350	#8, 3 conductor, solid		1.30	6.154		172	196		368	485
	0400	3 conductor with PVC jacket, in cable tray, #6		3.10	2.581		239	82.50		321.50	385
	0450	#4	2 Elec	5.40	2.963		310	94.50		404.50	480
	0500	#2		4.60	3.478		410	111		521	615
	0550	#1		4	4		555	128		683	800
	0600	1/0		3.60	4.444		655	142		797	930
	0650	2/0		3.40	4.706		795	150		945	1,100
	0700	3/0		3.20	5		930	160		1,090	1,275
	0750	4/0		3	5.333		1,075	170		1,245	1,425
	0800	250 kcmil	3 Elec	3.60	6.667		1,200	213		1,413	1,650
	0850	350 kcmil		3.30	7.273		1,600	232		1,832	2,100
	0900	500 kcmil		3	8		2,075	255		2,330	2,675
	1050	5 kV, copper, 3 conductor with PVC jacket,									
	1060	non-shielded, in cable tray, #4	2 Elec	380	.042	L.F.	4.70	1.34		6.04	7.15
	1100	#2		360	.044		6.15	1.42		7.57	8.85

ELECTRICAL 16

161 100 | Conductors

			CREW	DAILY OUTPUT	LABOR-HOURS	UNIT	1999 BARE COSTS				TOTAL INCL O&P	
							MAT.	LABOR	EQUIP.	TOTAL		
105	1200	#1	2 Elec	300	.053	L.F.	7.80	1.70		9.50	11.15	105
	1400	1/0		290	.055		9	1.76		10.76	12.55	
	1600	2/0		260	.062		10.45	1.96		12.41	14.45	
	2000	4/0		240	.067		13.90	2.13		16.03	18.50	
	2100	250 kcmil	3 Elec	330	.073		19	2.32		21.32	24.50	
	2150	350 kcmil		315	.076		23.50	2.43		25.93	29	
	2200	500 kcmil		270	.089		28.50	2.84		31.34	35.50	
	2400	15 KV, copper, 3 conductor with PVC jacket galv steel armored										
	2500	grounded neutral, in cable tray, #2	2 Elec	300	.053	L.F.	9.90	1.70		11.60	13.45	
	2600	#1		280	.057		10.60	1.82		12.42	14.40	
	2800	1/0		260	.062		12.15	1.96		14.11	16.30	
	2900	2/0		220	.073		16	2.32		18.32	21	
	3000	4/0		190	.084		18.20	2.69		20.89	24	
	3100	250 kcmil	3 Elec	270	.089		20.50	2.84		23.34	26.50	
	3150	350 kcmil		240	.100		24	3.19		27.19	31.50	
	3200	500 kcmil		210	.114		32	3.65		35.65	40.50	
	3400	15 kV, copper, 3 conductor with PVC jacket,										
	3450	ungrounded neutral, in cable tray, #2	2 Elec	260	.062	L.F.	10.65	1.96		12.61	14.65	
	3500	#1		230	.070		11.80	2.22		14.02	16.30	
	3600	1/0		200	.080		13.50	2.55		16.05	18.65	
	3700	2/0		190	.084		16.65	2.69		19.34	22.50	
	3800	4/0		160	.100		20	3.19		23.19	27	
	4000	250 kcmil	3 Elec	210	.114		23.50	3.65		27.15	31	
	4050	350 kcmil		195	.123		30.50	3.93		34.43	40	
	4100	500 kcmil		180	.133		37.50	4.25		41.75	47.50	
	9010	600 volt, copper (MC) steel clad, #14, 2 wire	1 Elec	2.40	3.333	C.L.F.	43	106		149	207	
	9020	3 wire		2.20	3.636		55.50	116		171.50	234	
	9040	#12, 2 wire		2.30	3.478		46.50	111		157.50	218	
	9050	3 wire		2	4		68.50	128		196.50	267	
	9070	#10, 2 wire		2	4		80	128		208	279	
	9080	3 wire		1.60	5		120	160		280	370	
	9100	#8, 2 wire, stranded		1.80	4.444		152	142		294	380	
	9110	3 wire, stranded		1.30	6.154		216	196		412	530	
	9200	600 volt, copper (MC) alum. clad, #14, 2 wire		2.65	3.019		43	96.50		139.50	192	
	9210	3 wire		2.45	3.265		55.50	104		159.50	217	
	9220	4 wire		2.20	3.636		77.50	116		193.50	258	
	9230	#12, 2 wire		2.55	3.137		46.50	100		146.50	202	
	9240	3 wire		2.20	3.636		68.50	116		184.50	249	
	9250	4 wire		2	4		89.50	128		217.50	290	
	9260	#10, 2 wire		2.20	3.636		80	116		196	261	
	9270	3 wire		1.80	4.444		120	142		262	345	
	9280	4 wire		1.55	5.161		185	165		350	450	
135	0010	**CONTROL CABLE**										135
	0020	600 volt, copper, #14 THWN wire with PVC jacket, 2 wires	1 Elec	9	.889	C.L.F.	13	28.50		41.50	57	
	0100	4 wires		7	1.143		22.50	36.50		59	79.50	
	0200	6 wires		6	1.333		36	42.50		78.50	103	
	0300	8 wires		5.30	1.509		42	48		90	118	
	0400	10 wires		4.80	1.667		50	53		103	135	
	0500	12 wires		4.30	1.860		58.50	59.50		118	153	
	0600	14 wires		3.80	2.105		68	67		135	175	
	0700	16 wires		3.50	2.286		79	73		152	196	
	0800	18 wires		3.30	2.424		82	77.50		159.50	206	
	0900	20 wires		3	2.667		94	85		179	230	
	1000	22 wires		2.80	2.857		96	91		187	242	
137	0010	**FIBER OPTICS**										137
	0020	Fiber optics cable only. Added costs depend on the type of fiber										

Important: See the Reference Section for critical supporting data - Reference Nos., Crews, & City Cost Indexes

161 100	Conductors	CREW	DAILY OUTPUT	LABOR-HOURS	UNIT	1999 BARE COSTS				TOTAL INCL O&P
						MAT.	LABOR	EQUIP.	TOTAL	
137 0030	special connectors, optical modems, and networking parts.									137
0040	Specialized tools & techniques cause installation costs to vary.									
0070	Cable, minimum, bulk simplex	1 Elec	8	1	C.L.F.	24	32		56	74
0080	Cable, maximum, bulk plenum quad	"	2.29	3.493	"	117	111		228	296
0150	Fiber optic jumper				Ea.	55.50			55.50	61
0200	Fiber optic pigtail					30			30	33
0300	Fiber optic connector	1 Elec	24	.333		16.90	10.65		27.55	34.50
0350	Fiber optic finger splice		32	.250		32	7.95		39.95	47
0400	Transceiver (low cost bi-directional)		8	1		263	32		295	335
0450	Multi-channel rack enclosure (10 modules)		2	4		385	128		513	615
0500	Fiber optic patch panel (12 ports)	▼	6	1.333	▼	178	42.50		220.50	260
140 0010	MINERAL INSULATED CABLE 600 volt									140
0100	1 conductor, #12	1 Elec	1.60	5	C.L.F.	253	160		413	515
0200	#10		1.60	5		264	160		424	530
0400	#8		1.50	5.333		299	170		469	585
0500	#6	▼	1.40	5.714		335	182		517	645
0600	#4	2 Elec	2.40	6.667		465	213		678	830
0800	#2		2.20	7.273		620	232		852	1,025
0900	#1		2.10	7.619		710	243		953	1,150
1000	1/0		2	8		815	255		1,070	1,275
1100	2/0		1.90	8.421		975	269		1,244	1,475
1200	3/0		1.80	8.889		1,150	284		1,434	1,700
1400	4/0	▼	1.60	10		1,325	320		1,645	1,925
1410	250 kcmil	3 Elec	2.40	10		1,500	320		1,820	2,125
1420	350 kcmil		1.95	12.308		1,675	395		2,070	2,425
1430	500 kcmil	▼	1.95	12.308	▼	2,100	395		2,495	2,875
145 0010	NON-METALLIC SHEATHED CABLE 600 volt									145
0100	Copper with ground wire, (Romex)									
0150	#14, 2 conductor	1 Elec	2.70	2.963	C.L.F.	15.40	94.50		109.90	158
0200	3 conductor		2.40	3.333		25.50	106		131.50	187
0250	#12, 2 conductor		2.50	3.200		22.50	102		124.50	178
0300	3 conductor		2.20	3.636		37	116		153	214
0350	#10, 2 conductor		2.20	3.636		37	116		153	214
0400	3 conductor		1.80	4.444		57.50	142		199.50	276
0450	#8, 3 conductor		1.50	5.333		116	170		286	380
0500	#6, 3 conductor	▼	1.40	5.714	▼	179	182		361	470
0550	SE type SER aluminum cable, 3 RHW and									
0600	1 bare neutral, 3 #8 & 1 #8	1 Elec	1.60	5	C.L.F.	96	160		256	345
0650	3 #6 & 1 #6	"	1.40	5.714		109	182		291	390
0700	3 #4 & 1 #6	2 Elec	2.40	6.667		122	213		335	455
0750	3 #2 & 1 #4		2.20	7.273		179	232		411	540
0800	3 #1/0 & 1 #2		2	8		271	255		526	680
0850	3 #2/0 & 1 #1		1.80	8.889		320	284		604	775
0900	3 #4/0 & 1 #2/0	▼	1.60	10	▼	455	320		775	975
150 0010	SHIELDED CABLE Splicing & terminations not included									150
0040	Copper, XLP shielding, 5 kV, #6	2 Elec	4.40	3.636	C.L.F.	85	116		201	267
0050	#4		4.40	3.636		110	116		226	294
0100	#2		4	4		130	128		258	335
0200	#1		4	4		150	128		278	355
0400	1/0		3.80	4.211		165	134		299	385
0600	2/0		3.60	4.444		205	142		347	440
0800	4/0	▼	3.20	5		270	160		430	535
1000	250 kcmil	3 Elec	4.50	5.333		325	170		495	615
1200	350 kcmil	▼	3.90	6.154	▼	420	196		616	755

ELECTRICAL 16

			DAILY OUTPUT	LABOR-HOURS	UNIT	1999 BARE COSTS				TOTAL INCL O&P	
	161 100	**Conductors**	CREW			MAT.	LABOR	EQUIP.	TOTAL		
150	1400	500 kcmil	3 Elec	3.60	6.667	C.L.F.	520	213		733	890	150
	1600	15 KV, ungrounded neutral, #1	2 Elec	4	4		184	128		312	395	
	1800	1/0		3.80	4.211		220	134		354	445	
	2000	2/0		3.60	4.444		250	142		392	485	
	2200	4/0		3.20	5		330	160		490	605	
	2400	250 kcmil	3 Elec	4.50	5.333		365	170		535	655	
	2600	350 kcmil		3.90	6.154		460	196		656	800	
	2800	500 kcmil		3.60	6.667		570	213		783	945	
165	0010	**WIRE** R160 -205										165
	0020	600 volt type THW, copper, solid, #14	1 Elec	13	.615	C.L.F.	4.45	19.65		24.10	34.50	
	0030	#12		11	.727		6.30	23		29.30	41.50	
	0040	#10		10	.800		9.70	25.50		35.20	48.50	
	0050	Stranded, #14		13	.615		5.10	19.65		24.75	35	
	0100	#12 R161 -125		11	.727		7.20	23		30.20	42.50	
	0120	#10		10	.800		11.10	25.50		36.60	50	
	0140	#8		8	1		18.30	32		50.30	67.50	
	0160	#6		6.50	1.231		29.50	39.50		69	91	
	0180	#4	2 Elec	10.60	1.509		48	48		96	125	
	0200	#3		10	1.600		57	51		108	139	
	0220	#2		9	1.778		71	56.50		127.50	163	
	0240	#1		8	2		89.50	64		153.50	194	
	0260	1/0		6.60	2.424		110	77.50		187.50	237	
	0280	2/0		5.80	2.759		137	88		225	282	
	0300	3/0		5	3.200		172	102		274	340	
	0350	4/0		4.40	3.636		214	116		330	410	
	0400	250 kcmil	3 Elec	6	4		258	128		386	475	
	0420	300 kcmil		5.70	4.211		305	134		439	540	
	0450	350 kcmil		5.40	4.444		355	142		497	600	
	0480	400 kcmil		5.10	4.706		410	150		560	675	
	0490	500 kcmil		4.80	5		500	160		660	790	
	0530	Aluminum, stranded, #8	1 Elec	9	.889		12.30	28.50		40.80	56	
	0540	#6	"	8	1		15.70	32		47.70	65	
	0560	#4	2 Elec	13	1.231		19.60	39.50		59.10	80	
	0580	#2		10.60	1.509		26.50	48		74.50	102	
	0600	#1		9	1.778		39	56.50		95.50	128	
	0620	1/0		8	2		46	64		110	146	
	0640	2/0		7.20	2.222		55	71		126	167	
	0680	3/0		6.60	2.424		68	77.50		145.50	191	
	0700	4/0		6.20	2.581		76	82.50		158.50	207	
	0720	250 kcmil	3 Elec	8.70	2.759		90.50	88		178.50	232	
	0740	300 kcmil		8.10	2.963		128	94.50		222.50	282	
	0760	350 kcmil		7.50	3.200		130	102		232	296	
	0780	400 kcmil		6.90	3.478		152	111		263	335	
	0800	500 kcmil		6	4		168	128		296	375	
	0850	600 kcmil		5.70	4.211		212	134		346	435	
	0880	700 kcmil		5.10	4.706		245	150		395	495	
	0900	750 kcmil		4.80	5		248	160		408	510	
	0920	Type THWN-THHN, copper, solid, #14	1 Elec	13	.615		4.45	19.65		24.10	34.50	
	0940	#12		11	.727		6.30	23		29.30	41.50	
	0960	#10		10	.800		9.70	25.50		35.20	48.50	
	1000	Stranded, #14		13	.615		5.10	19.65		24.75	35	
	1200	#12		11	.727		7.20	23		30.20	42.50	
	1250	#10 **CN**		10	.800		11.10	25.50		36.60	50	
	1300	#8		8	1		18.30	32		50.30	67.50	
	1350	#6		6.50	1.231		29.50	39.50		69	91	
	1400	#4	2 Elec	10.60	1.509		48	48		96	125	

Important: See the Reference Section for critical supporting data - Reference Nos., Crews, & City Cost Indexes

161 500 | Terminations

		CREW	DAILY OUTPUT	LABOR-HOURS	UNIT	1999 BARE COSTS				TOTAL INCL O&P
						MAT.	LABOR	EQUIP.	TOTAL	
520	0010	**CABLE TERMINATIONS**									520
	0015	Wire connectors, screw type, #22 to #14	1 Elec	260	.031	Ea.	.05	.98		1.03	1.53
	0020	#18 to #12		240	.033		.06	1.06		1.12	1.66
	0025	#18 to #10		240	.033		.09	1.06		1.15	1.69
	0030	Screw-on connectors, insulated, #18 to #12		240	.033		.08	1.06		1.14	1.68
	0035	#16 to #10		230	.035		.10	1.11		1.21	1.77
	0040	#14 to #8		210	.038		.22	1.22		1.44	2.06
	0045	#12 to #6		180	.044		.35	1.42		1.77	2.51
	0050	Terminal lugs, solderless, #16 to #10		50	.160		.44	5.10		5.54	8.15
	0100	#8 to #4		30	.267		.55	8.50		9.05	13.30
	0150	#2 to #1		22	.364		.74	11.60		12.34	18.15
	0200	1/0 to 2/0		16	.500		2.05	15.95		18	26.50
	0250	3/0		12	.667		2.50	21.50		24	35
	0300	4/0		11	.727		2.50	23		25.50	37.50
	0350	250 kcmil		9	.889		2.50	28.50		31	45.50
	0400	350 kcmil		7	1.143		3.25	36.50		39.75	58
	0450	500 kcmil	▼	6	1.333	▼	6.30	42.50		48.80	70.50
	1600	Crimp 1 hole lugs, copper or aluminum, 600 volt									
	1620	#14	1 Elec	60	.133	Ea.	.26	4.25		4.51	6.65
	1630	#12		50	.160		.36	5.10		5.46	8.05
	1640	#10		45	.178		.36	5.65		6.01	8.90
	1780	#8		36	.222		1.35	7.10		8.45	12.10
	1800	#6		30	.267		2.36	8.50		10.86	15.30
	2000	#4		27	.296		2.66	9.45		12.11	17.10
	2200	#2		24	.333		3.37	10.65		14.02	19.60
	2400	#1		20	.400		3.87	12.75		16.62	23.50
	2600	2/0		15	.533		4.45	17		21.45	30.50
	2800	3/0		12	.667		5.20	21.50		26.70	37.50
	3000	4/0		11	.727		5.65	23		28.65	40.50
	3200	250 kcmil		9	.889		6.85	28.50		35.35	50
	3400	300 kcmil		8	1		7.65	32		39.65	56
	3500	350 kcmil		7	1.143		7.80	36.50		44.30	63
	3600	400 kcmil		6.50	1.231		9.25	39.50		48.75	68.50
	3800	500 kcmil	▼	6	1.333	▼	11.20	42.50		53.70	76

161 800 | Grounding

		CREW	DAILY OUTPUT	LABOR-HOURS	UNIT	MAT.	LABOR	EQUIP.	TOTAL	TOTAL INCL O&P
810	0010	**GROUNDING**									810
	0030	Rod, copper clad, 8' long, 1/2" diameter	1 Elec	5.50	1.455	Ea.	15.65	46.50		62.15	87
	0050	3/4" diameter		5.30	1.509		30	48		78	105
	0080	10' long, 1/2" diameter		4.80	1.667		17.50	53		70.50	99
	0100	3/4" diameter		4.40	1.818		29.50	58		87.50	119
	0130	15' long, 3/4" diameter		4	2	▼	79	64		143	183
	0390	Bare copper wire, #8 stranded		11	.727	C.L.F.	15.80	23		38.80	52
	0400	#6	▼	10	.800		24	25.50		49.50	64
	0600	#2	2 Elec	10	1.600		58.50	51		109.50	141
	0800	3/0		6.60	2.424		155	77.50		232.50	287
	1000	4/0	▼	5.70	2.807		193	89.50		282.50	345
	1200	250 kcmil	3 Elec	7.20	3.333	▼	229	106		335	410
	1800	Water pipe ground clamps, heavy duty									
	2000	Bronze, 1/2" to 1" diameter	1 Elec	8	1	Ea.	11.45	32		43.45	60
	2100	1-1/4" to 2" diameter		8	1		14.90	32		46.90	64
	2200	2-1/2" to 3" diameter		6	1.333		39.50	42.50		82	107
	2800	Brazed connections, #6 wire		12	.667		10.45	21.50		31.95	43.50
	3000	#2 wire		10	.800		14	25.50		39.50	53.50
	3100	3/0 wire		8	1		21	32		53	70.50
	3200	4/0 wire	▼	7	1.143	▼	24	36.50		60.50	81

ELECTRICAL 16

161 800 | Grounding

		CREW	DAILY OUTPUT	LABOR-HOURS	UNIT	1999 BARE COSTS				TOTAL INCL O&P	
						MAT.	LABOR	EQUIP.	TOTAL		
810	3400	250 kcmil wire	1 Elec	5	1.600	Ea.	28	51		79	108	810
	3600	500 kcmil wire	↓	4	2	↓	34.50	64		98.50	134	

162 | Boxes & Wiring Devices

162 100 | Boxes

		CREW	DAILY OUTPUT	LABOR-HOURS	UNIT	1999 BARE COSTS				TOTAL INCL O&P	
						MAT.	LABOR	EQUIP.	TOTAL		
110	0010	**OUTLET BOXES**										**110**
	0020	Pressed steel, octagon, 4"	1 Elec	20	.400	Ea.	1.44	12.75		14.19	20.50	
	0060	Covers, blank		64	.125		.59	3.99		4.58	6.60	
	0100	Extension		40	.200		2.29	6.40		8.69	12.05	
	0150	Square, 4"		20	.400		1.98	12.75		14.73	21.50	
	0200	Extension		40	.200		2.41	6.40		8.81	12.20	
	0250	Covers, blank		64	.125		.68	3.99		4.67	6.70	
	0300	Plaster rings		64	.125		1.09	3.99		5.08	7.15	
	0650	Switchbox		27	.296		2.21	9.45		11.66	16.60	
	1100	Concrete, floor, 1 gang		5.30	1.509		56	48		104	134	
	2000	Poke-thru fitting, fire rated, for 3-3/4" floor		6.80	1.176		90	37.50		127.50	155	
	2040	For 7" floor		6.80	1.176		90	37.50		127.50	155	
	2100	Pedestal, 15 amp, duplex receptacle & blank plate		5.25	1.524		93	48.50		141.50	175	
	2120	Duplex receptacle and telephone plate		5.25	1.524		93.50	48.50		142	176	
	2140	Pedestal, 20 amp, duplex recept. & phone plate		5	1.600		94	51		145	180	
	2200	Abandonment plate	↓	32	.250	↓	27.50	7.95		35.45	42	
130	0010	**PULL BOXES & CABINETS**										**130**
	0100	Sheet metal, pull box, NEMA 1, type SC, 6" W x 6" H x 4" D	1 Elec	8	1	Ea.	8.35	32		40.35	56.50	
	0200	8" W x 8" H x 4" D		8	1		11.45	32		43.45	60	
	0300	10" W x 12" H x 6" D		5.30	1.509		20	48		68	94	
	0400	16" W x 20" H x 8" D		4	2		76	64		140	179	
	0500	20" W x 24" H x 8" D		3.20	2.500		89	80		169	217	
	0600	24" W x 36" H x 8" D		2.70	2.963		125	94.50		219.50	279	
	0650	Hinged cabinets, NEMA 1, 6" W x 6" H x 4" D		8	1		8.50	32		40.50	57	
	0800	12" W x 16" H x 6" D		4.70	1.702		28	54.50		82.50	112	
	1000	20" W x 20" H x 6" D		3.60	2.222		56	71		127	168	
	1200	20" W x 20" H x 8" D		3.20	2.500		113	80		193	243	
	1400	24" W x 36" H x 8" D		2.70	2.963		180	94.50		274.50	340	
	1600	24" W x 42" H x 8" D	↓	2	4	↓	273	128		401	490	
	2100	Pull box, NEMA 3R, type SC, raintight & weatherproof										
	2150	6" L x 6" W x 6" D	1 Elec	10	.800	Ea.	14.30	25.50		39.80	54	
	2200	8" L x 6" W x 6" D		8	1		17.85	32		49.85	67	
	2250	10" L x 6" W x 6" D		7	1.143		23.50	36.50		60	80.50	
	2300	12" L x 12" W x 6" D		5	1.600		33.50	51		84.50	114	
	2350	16" L x 16" W x 6" D		4.50	1.778		67	56.50		123.50	159	
	2400	20" L x 20" W x 6" D		4	2		92	64		156	197	
	2450	24" L x 18" W x 8" D		3	2.667		100	85		185	237	
	2500	24" L x 24" W x 10" D		2.50	3.200		239	102		341	415	
	2550	30" L x 24" W x 12" D		2	4		246	128		374	460	
	2600	36" L x 36" W x 12" D	↓	1.50	5.333	↓	325	170		495	610	
	2800	Cast iron, pull boxes for surface mounting										
	3000	NEMA 4, watertight & dust tight										
	3050	6" L x 6" W x 6" D	1 Elec	4	2	Ea.	134	64		198	243	
	3100	8" L x 6" W x 6" D		3.20	2.500		181	80		261	320	

16 ELECTRICAL

Important: See the Reference Section for critical supporting data - Reference Nos., Crews, & City Cost Indexes

162 100 | Boxes

			DAILY OUTPUT	LABOR-HOURS	UNIT	1999 BARE COSTS				TOTAL INCL O&P	
			CREW			MAT.	LABOR	EQUIP.	TOTAL		
130	3150	10" L x 6" W x 6" D	1 Elec	2.50	3.200	Ea.	225	102		327	400	130
	3200	12" L x 12" W x 6" D		2.30	3.478		380	111		491	585	
	3250	16" L x 16" W x 6" D		1.30	6.154		780	196		976	1,150	
	3300	20" L x 20" W x 6" D		.80	10		1,450	320		1,770	2,075	
	3350	24" L x 18" W x 8" D		.70	11.429		1,525	365		1,890	2,225	
	3400	24" L x 24" W x 10" D		.50	16		2,325	510		2,835	3,350	
	3450	30" L x 24" W x 12" D		.40	20		2,925	640		3,565	4,150	
	3500	36" L x 36" W x 12" D		.20	40		4,850	1,275		6,125	7,225	
	6000	J.I.C. wiring boxes, NEMA 12, dust tight & drip tight										
	6050	6" L x 8" W x 4" D	1 Elec	10	.800	Ea.	33	25.50		58.50	74	
	6100	8" L x 10" W x 4" D		8	1		41	32		73	92.50	
	6150	12" L x 14" W x 6" D		5.30	1.509		67	48		115	146	
	6200	14" L x 16" W x 6" D		4.70	1.702		79.50	54.50		134	169	
	6250	16" L x 20" W x 6" D		4.40	1.818		170	58		228	274	
	6300	24" L x 30" W x 6" D		3.20	2.500		250	80		330	395	
	6350	24" L x 30" W x 8" D		2.90	2.759		267	88		355	425	
	6400	24" L x 36" W x 8" D		2.70	2.963		295	94.50		389.50	465	
	6450	24" L x 42" W x 8" D		2.30	3.478		320	111		431	520	
	6500	24" L x 48" W x 8" D		2	4		350	128		478	575	
	7000	Cabinets, current transformer										
	7050	Single door, 24" H x 24" W x 10" D	1 Elec	1.60	5	Ea.	107	160		267	355	
	7100	30" H x 24" W x 10" D		1.30	6.154		120	196		316	425	
	7150	36" H x 24" W x 10" D		1.10	7.273		134	232		366	490	
	7200	30" H x 30" W x 10" D		1	8		140	255		395	535	
	7250	36" H x 30" W x 10" D		.90	8.889		189	284		473	635	
	7300	36" H x 36" W x 10" D		.80	10		206	320		526	700	
	7500	Double door, 48" H x 36" W x 10" D		.60	13.333		360	425		785	1,025	
	7550	24" H x 24" W x 12" D		1	8		193	255		448	590	

162 300 | Wiring Devices

			CREW	DAILY OUTPUT	LABOR-HOURS	UNIT	MAT.	LABOR	EQUIP.	TOTAL	TOTAL INCL O&P	
310	0010	**LOW VOLTAGE SWITCHING**										310
	3600	Relays, 120 V or 277 V standard	1 Elec	12	.667	Ea.	26	21.50		47.50	60.50	
	3800	Flush switch, standard		40	.200		9.05	6.40		15.45	19.50	
	4000	Interchangeable		40	.200		11.80	6.40		18.20	22.50	
	4100	Surface switch, standard		40	.200		6.60	6.40		13	16.85	
	4200	Transformer 115 V to 25 V		12	.667		93	21.50		114.50	134	
	4400	Master control, 12 circuit, manual		4	2		94	64		158	199	
	4500	25 circuit, motorized		4	2		102	64		166	208	
	4600	Rectifier, silicon		12	.667		30.50	21.50		52	65.50	
	4800	Switchplates, 1 gang, 1, 2 or 3 switch, plastic		80	.100		3	3.19		6.19	8.05	
	5000	Stainless steel		80	.100		8.10	3.19		11.29	13.65	
	5400	2 gang, 3 switch, stainless steel		53	.151		15.65	4.82		20.47	24.50	
	5500	4 switch, plastic		53	.151		6.70	4.82		11.52	14.55	
	5800	3 gang, 9 switch, stainless steel		32	.250		50	7.95		57.95	67	
320	0010	**WIRING DEVICES**										320
	0200	Toggle switch, quiet type, single pole, 15 amp **CN**	1 Elec	40	.200	Ea.	4.25	6.40		10.65	14.25	
	0600	3 way, 15 amp		23	.348		6.50	11.10		17.60	24	
	0900	4 way, 15 amp		15	.533		18.95	17		35.95	46.50	
	1650	Dimmer switch, 120 volt, incandescent, 600 watt, 1 pole		16	.500		10.80	15.95		26.75	36	
	2460	Receptacle, duplex, 120 volt, grounded, 15 amp		40	.200		1.14	6.40		7.54	10.80	
	2470	20 amp		27	.296		5.95	9.45		15.40	20.50	
	2490	Dryer, 30 amp		15	.533		11.70	17		28.70	38.50	
	2500	Range, 50 amp		11	.727		16	23		39	52	
	2600	Wall plates, stainless steel, 1 gang		80	.100		2.05	3.19		5.24	7.05	
	2800	2 gang		53	.151		4.10	4.82		8.92	11.70	
	3200	Lampholder, keyless		26	.308		5.20	9.80		15	20.50	

ELECTRICAL 16

162 300 | Wiring Devices

		CREW	DAILY OUTPUT	LABOR-HOURS	UNIT	1999 BARE COSTS				TOTAL INCL O&P	
						MAT.	LABOR	EQUIP.	TOTAL		
320	3400	Pullchain with receptacle	1 Elec	22	.364	Ea.	8.25	11.60		19.85	26.50	320

163 | Motors, Starters, Boards & Switches

163 100 | Starters & Controls

				CREW	DAILY OUTPUT	LABOR-HOURS	UNIT	1999 BARE COSTS				TOTAL INCL O&P	
								MAT.	LABOR	EQUIP.	TOTAL		
130	0010	**MOTOR STARTERS & CONTROLS**											130
	0050	Magnetic, FVNR, with enclosure and heaters, 480 volt	R163-520										
	0100	5 HP, size 0		1 Elec	2.30	3.478	Ea.	187	111		298	370	
	0200	10 HP, size 1		"	1.60	5		210	160		370	470	
	0300	25 HP, size 2		2 Elec	2.20	7.273		395	232		627	780	
	0400	50 HP, size 3			1.80	8.889		645	284		929	1,125	
	0500	100 HP, size 4			1.20	13.333		1,425	425		1,850	2,200	
	0600	200 HP, size 5			.90	17.778		3,350	565		3,915	4,525	
	0700	Combination, with motor circuit protectors, 5 HP, size 0		1 Elec	1.80	4.444		610	142		752	880	
	0800	10 HP, size 1		"	1.30	6.154		630	196		826	990	
	0900	25 HP, size 2		2 Elec	2	8		885	255		1,140	1,350	
	1000	50 HP, size 3			1.32	12.121		1,275	385		1,660	1,975	
	1200	100 HP, size 4			.80	20		2,775	640		3,415	4,000	
	1400	Combination, with fused switch, 5 HP, size 0		1 Elec	1.80	4.444		465	142		607	720	
	1600	10 HP, size 1		"	1.30	6.154		495	196		691	840	
	1800	25 HP, size 2		2 Elec	2	8		805	255		1,060	1,275	
	2000	50 HP, size 3			1.32	12.121		1,375	385		1,760	2,075	
	2200	100 HP, size 4			.80	20		2,375	640		3,015	3,575	

163 200 | Boards

			CREW	DAILY OUTPUT	LABOR-HOURS	UNIT	1999 BARE COSTS				TOTAL INCL O&P	
							MAT.	LABOR	EQUIP.	TOTAL		
205	0010	**CIRCUIT BREAKERS** (in enclosure)										205
	0100	Enclosed (NEMA 1), 600 volt, 3 pole, 30 amp	1 Elec	3.20	2.500	Ea.	330	80		410	485	
	0200	60 amp		2.80	2.857		330	91		421	500	
	0400	100 amp		2.30	3.478		380	111		491	580	
	0600	225 amp		1.50	5.333		880	170		1,050	1,225	
	0700	400 amp	2 Elec	1.60	10		1,500	320		1,820	2,125	
	0800	600 amp		1.20	13.333		2,175	425		2,600	3,025	
	1000	800 amp		.94	17.021		2,825	545		3,370	3,900	
245	0010	**PANELBOARDS** (Commercial use)										245
	0050	NQOD, w/20 amp 1 pole bolt-on circuit breakers										
	0100	3 wire, 120/240 volts, 100 amp main lugs										
	0150	10 circuits	1 Elec	1	8	Ea.	390	255		645	805	
	0200	14 circuits		.88	9.091		455	290		745	935	
	0250	18 circuits		.75	10.667		500	340		840	1,050	
	0300	20 circuits		.65	12.308		560	395		955	1,200	
	0350	225 amp main lugs, 24 circuits	2 Elec	1.20	13.333		635	425		1,060	1,325	
	0400	30 circuits		.90	17.778		740	565		1,305	1,675	
	0450	36 circuits		.80	20		845	640		1,485	1,875	
	0500	38 circuits		.72	22.222		910	710		1,620	2,050	
	0550	42 circuits		.66	24.242		950	775		1,725	2,200	
	0600	4 wire, 120/208 volts, 100 amp main lugs, 12 circuits	1 Elec	1	8		440	255		695	865	
	0650	16 circuits		.75	10.667		505	340		845	1,075	
	0700	20 circuits		.65	12.308		590	395		985	1,225	
	0750	24 circuits **CN**		.60	13.333		640	425		1,065	1,350	

16 ELECTRICAL

Important: See the Reference Section for critical supporting data - Reference Nos., Crews, & City Cost Indexes

163 200	Boards	CREW	DAILY OUTPUT	LABOR-HOURS	UNIT	1999 BARE COSTS				TOTAL INCL O&P
						MAT.	LABOR	EQUIP.	TOTAL	
245 0800	30 circuits	1 Elec	.53	15.094	Ea.	740	480		1,220	1,525
0850	225 amp main lugs, 32 circuits	2 Elec	.90	17.778		830	565		1,395	1,775
0900	34 circuits		.84	19.048		850	610		1,460	1,850
0950	36 circuits		.80	20		870	640		1,510	1,925
1000	42 circuits		.68	23.529		975	750		1,725	2,200
1200	NEHB,w/20 amp, 1 pole bolt-on circuit breakers									
1250	4 wire, 277/480 volts, 100 amp main lugs, 12 circuits	1 Elec	.88	9.091	Ea.	840	290		1,130	1,350
1300	20 circuits	"	.60	13.333		1,250	425		1,675	2,000
1350	225 amp main lugs, 24 circuits	2 Elec	.90	17.778		1,425	565		1,990	2,425
1400	30 circuits		.80	20		1,725	640		2,365	2,850
1450	36 circuits		.72	22.222		2,000	710		2,710	3,250
1600	NQOD panel, w/20 amp, 1 pole, circuit breakers									
1650	3 wire, 120/240 volt with main circuit breaker									
1700	100 amp main, 12 circuits	1 Elec	.80	10	Ea.	540	320		860	1,075
1750	20 circuits	"	.60	13.333		690	425		1,115	1,400
1800	225 amp main, 30 circuits	2 Elec	.68	23.529		1,325	750		2,075	2,575
1850	42 circuits		.52	30.769		1,525	980		2,505	3,150
1900	400 amp main, 30 circuits		.54	29.630		1,825	945		2,770	3,450
1950	42 circuits		.50	32		2,050	1,025		3,075	3,775
2000	4 wire, 120/208 volts with main circuit breaker									
2050	100 amp main, 24 circuits	1 Elec	.47	17.021	Ea.	805	545		1,350	1,700
2100	30 circuits	"	.40	20		910	640		1,550	1,950
2200	225 amp main, 32 circuits	2 Elec	.72	22.222		1,550	710		2,260	2,750
2250	42 circuits		.56	28.571		1,675	910		2,585	3,225
2300	400 amp main, 42 circuits		.48	33.333		2,275	1,075		3,350	4,100
2350	600 amp main, 42 circuits		.40	40		3,375	1,275		4,650	5,625
2400	NEHB, with 20 amp, 1 pole circuit breaker									
2450	4 wire, 277/480 volts with main circuit breaker									
2500	100 amp main, 24 circuits	1 Elec	.42	19.048	Ea.	1,650	610		2,260	2,725
2550	30 circuits	"	.38	21.053		1,950	670		2,620	3,125
2600	225 amp main, 30 circuits	2 Elec	.72	22.222		2,450	710		3,160	3,750
2650	42 circuits	"	.56	28.571		3,025	910		3,935	4,700
250 0010	**PANELBOARD & LOAD CENTER CIRCUIT BREAKERS**									
0050	Bolt-on, 10,000 amp IC, 120 volt, 1 pole									
0100	15 to 50 amp	1 Elec	10	.800	Ea.	10	25.50		35.50	49
0200	60 amp		8	1		10	32		42	58.50
0300	70 amp		8	1		19	32		51	68.50
0350	240 volt, 2 pole									
0400	15 to 50 amp	1 Elec	8	1	Ea.	22	32		54	71.50
0500	60 amp		7.50	1.067		22	34		56	75
0600	80 to 100 amp		5	1.600		85	51		136	170
0700	3 pole, 15 to 60 amp		6.20	1.290		69.50	41		110.50	138
0800	70 amp		5	1.600		87.50	51		138.50	173
0900	80 to 100 amp		3.60	2.222		99.50	71		170.50	215
1000	22,000 amp I.C., 240 volt, 2 pole, 70 - 225 amp		2.70	2.963		430	94.50		524.50	610
1100	3 pole, 70 - 225 amp		2.30	3.478		475	111		586	690
1200	14,000 amp I.C., 277 volts, 1 pole, 15 - 30 amp		8	1		26.50	32		58.50	76.50
1300	22,000 amp I.C., 480 volts, 2 pole, 70 - 225 amp		2.70	2.963		430	94.50		524.50	610
1400	3 pole, 70 - 225 amp		2.30	3.478		475	111		586	690
260 0010	**SWITCHBOARDS** Incoming main service section									
0100	Aluminum bus bars, not including CT's or PT's									
0200	No main disconnect, includes CT compartment									
0300	120/208 volt, 4 wire, 600 amp	2 Elec	1	16	Ea.	2,650	510		3,160	3,700
0400	800 amp		.88	18.182		2,650	580		3,230	3,800
0500	1000 amp		.80	20		3,175	640		3,815	4,450

ELECTRICAL 16

For expanded coverage of these items see *Means Electrical Cost Data 1999*

163 200 | Boards

		CREW	DAILY OUTPUT	LABOR-HOURS	UNIT	1999 BARE COSTS				TOTAL INCL O&P
						MAT.	LABOR	EQUIP.	TOTAL	
260	0600	1200 amp	2 Elec	.72	22.222	Ea.	3,175	710		3,885	4,550
	0700	1600 amp		.66	24.242		3,175	775		3,950	4,650
	0800	2000 amp		.62	25.806		3,425	825		4,250	5,000
	1000	3000 amp		.56	28.571		4,525	910		5,435	6,350
	2000	Fused switch & CT compartment									
	2100	120/208 volt, 4 wire, 400 amp	2 Elec	1.12	14.286	Ea.	3,450	455		3,905	4,475
	2200	600 amp		.94	17.021		4,125	545		4,670	5,350
	2300	800 amp		.84	19.048		7,175	610		7,785	8,775
	2400	1200 amp		.68	23.529		9,325	750		10,075	11,400
	2900	Pressure switch & CT compartment									
	3000	120/208 volt, 4 wire, 800 amp	2 Elec	.80	20	Ea.	6,450	640		7,090	8,025
	3100	1200 amp		.66	24.242		12,500	775		13,275	14,900
	3200	1600 amp		.62	25.806		13,300	825		14,125	15,800
	3300	2000 amp		.56	28.571		14,100	910		15,010	16,900
	4400	Circuit breaker, molded case & CT compartment									
	4600	3 pole, 4 wire, 600 amp	2 Elec	.94	17.021	Ea.	5,550	545		6,095	6,900
	4800	800 amp		.84	19.048		6,650	610		7,260	8,200
	5000	1200 amp		.68	23.529		9,050	750		9,800	11,100
	5100	Copper bus bars, not incl. CT's or PT's, add, minimum					15%				
264	0010	**SWITCHBOARDS** distribution section									
	0100	Aluminum bus bars, not including breakers									
	0200	120/208 or 277/480 volt, 4 wire, 600 amp	2 Elec	1	16	Ea.	1,150	510		1,660	2,025
	0300	800 amp		.88	18.182		1,450	580		2,030	2,475
	0400	1000 amp		.80	20		1,725	640		2,365	2,850
	0500	1200 amp		.72	22.222		2,375	710		3,085	3,650
	0600	1600 amp		.66	24.242		2,775	775		3,550	4,200
	0700	2000 amp		.62	25.806		3,250	825		4,075	4,800
	0800	2500 amp		.60	26.667		3,675	850		4,525	5,300
	0900	3000 amp		.56	28.571		4,450	910		5,360	6,250
	0950	4000 amp		.52	30.769		6,475	980		7,455	8,600
266	0010	**SWITCHBOARDS** feeder section group mounted devices									
	0030	Circuit breakers									
	0160	FA frame, 15 to 60 amp, 240 volt, 1 pole	1 Elec	8	1	Ea.	63	32		95	117
	0280	FA frame, 70 to 100 amp, 240 volt, 1 pole		7	1.143		83	36.50		119.50	146
	0420	KA frame, 70 to 225 amp		3.20	2.500		660	80		740	845
	0430	LA frame, 125 to 400 amp		2.30	3.478		1,375	111		1,486	1,700
	0460	MA frame, 450 to 600 amp		1.60	5		2,400	160		2,560	2,900
	0470	700 to 800 amp		1.30	6.154		3,150	196		3,346	3,775
	0480	MAL frame, 1000 amp		1	8		3,500	255		3,755	4,225
	0490	PA frame, 1200 amp		.80	10		7,100	320		7,420	8,300
	0500	Branch circuit, fusible switch, 600 volt, double 30/30 amp		4	2		565	64		629	720
	0550	60/60 amp		3.20	2.500		565	80		645	745
	0600	100/100 amp		2.70	2.963		680	94.50		774.50	890
	0650	Single, 30 amp		5.30	1.509		445	48		493	560
	0700	60 amp		4.70	1.702		445	54.50		499.50	570
	0750	100 amp		4	2		445	64		509	585
	0800	200 amp		2.70	2.963		815	94.50		909.50	1,025
	0850	400 amp		2.30	3.478		1,550	111		1,661	1,900
	0900	600 amp		1.80	4.444		1,900	142		2,042	2,300
	0950	800 amp		1.30	6.154		3,475	196		3,671	4,125
	1000	1200 amp		.80	10		3,650	320		3,970	4,475
268	0010	**SWITCHBOARD INSTRUMENTS** 3 phase, 4 wire									
	0100	AC indicating, ammeter & switch	1 Elec	8	1	Ea.	1,200	32		1,232	1,350

Important: See the Reference Section for critical supporting data - Reference Nos., Crews, & City Cost Indexes

163 200 | Boards

			CREW	DAILY OUTPUT	LABOR-HOURS	UNIT	1999 BARE COSTS				TOTAL INCL O&P	
							MAT.	LABOR	EQUIP.	TOTAL		
268	0200	Voltmeter & switch	1 Elec	8	1	Ea.	1,200	32		1,232	1,350	268
	0300	Wattmeter		8	1		2,350	32		2,382	2,650	
	0400	AC recording, ammeter		4	2		4,200	64		4,264	4,725	
	0500	Voltmeter		4	2		4,200	64		4,264	4,725	
	0600	Ground fault protection, zero sequence		2.70	2.963		3,700	94.50		3,794.50	4,225	
	0700	Ground return path		2.70	2.963		3,700	94.50		3,794.50	4,225	
	0800	3 current transformers, 5 to 800 amp		2	4		1,550	128		1,678	1,900	
	0900	1000 to 1500 amp		1.30	6.154		2,225	196		2,421	2,750	
	1200	2000 to 4000 amp		1	8		2,625	255		2,880	3,275	
	1300	Fused potential transformer, maximum 600 volt		8	1		705	32		737	825	

163 300 | Switches

			CREW	DAILY OUTPUT	LABOR-HOURS	UNIT	MAT.	LABOR	EQUIP.	TOTAL	TOTAL INCL O&P	
320	0010	**CONTROL STATIONS**										320
	0050	NEMA 1, heavy duty, stop/start	1 Elec	8	1	Ea.	106	32		138	165	
	0100	Stop/start, pilot light		6.20	1.290		143	41		184	220	
	0200	Hand/off/automatic		6.20	1.290		78.50	41		119.50	148	
	0400	Stop/start/reverse		5.30	1.509		143	48		191	230	
360	0010	**SAFETY SWITCHES**										360
	0100	General duty 240 volt, 3 pole NEMA 1, fusible, 30 amp	1 Elec	3.20	2.500	Ea.	63	80		143	189	
	0200	60 amp		2.30	3.478		107	111		218	284	
	0300	100 amp		1.90	4.211		183	134		317	400	
	0400	200 amp		1.30	6.154		395	196		591	730	
	0500	400 amp	2 Elec	1.80	8.889		995	284		1,279	1,525	
	0600	600 amp	"	1.20	13.333		1,850	425		2,275	2,650	
	2900	Heavy duty, 240 volt, 3 pole NEMA 1 fusible										
	2910	30 amp	1 Elec	3.20	2.500	Ea.	102	80		182	231	
	3000	60 amp		2.30	3.478		172	111		283	355	
	3300	100 amp		1.90	4.211		271	134		405	500	
	3500	200 amp		1.30	6.154		465	196		.661	810	
	3700	400 amp	2 Elec	1.80	8.889		1,075	284		1,359	1,600	
	3900	600 amp	"	1.20	13.333		1,950	425		2,375	2,775	

163 500 | Motors

			CREW	DAILY OUTPUT	LABOR-HOURS	UNIT	MAT.	LABOR	EQUIP.	TOTAL	TOTAL INCL O&P	
520	0010	**MOTORS** 230/460 volts, 60 HZ										520
	0050	Dripproof, Class B, 1.15 service factor										
	0060	1800 RPM, 1/4 HP	1 Elec	5.33	1.501	Ea.	114	48		162	197	
	0070	1/3 HP		5.33	1.501		125	48		173	210	
	0080	1/2 HP		5.33	1.501		147	48		195	234	
	0090	3/4 HP		5.33	1.501		163	48		211	251	
	0100	1 HP		4.50	1.778		172	56.50		228.50	274	
	0250	5 HP		4.50	1.778		259	56.50		315.50	370	
	0350	10 HP		4	2		460	64		524	600	
	0450	20 HP **CN**	2 Elec	5.20	3.077		735	98		833	955	

164 100 | Transformers

			CREW	DAILY OUTPUT	LABOR-HOURS	UNIT	1999 BARE COSTS				TOTAL INCL O&P	
							MAT.	LABOR	EQUIP.	TOTAL		
120	0010	**DRY TYPE TRANSFORMER**										120
	0050	Single phase, 240/480 volt primary, 120/240 volt secondary										

ELECTRICAL 16

164 100 | Transformers

		Description	CREW	DAILY OUTPUT	LABOR-HOURS	UNIT	1999 BARE COSTS				TOTAL INCL O&P	
							MAT.	LABOR	EQUIP.	TOTAL		
120	0100	1 kVA	1 Elec	2	4	Ea.	141	128		269	345	120
	0300	2 kVA		1.60	5		212	160		372	470	
	0500	3 kVA		1.40	5.714		262	182		444	560	
	0700	5 kVA		1.20	6.667		360	213		573	715	
	0900	7.5 kVA		1.10	7.273		500	232		732	895	
	1100	10 kVA		.80	10		640	320		960	1,175	
	1300	15 kVA	2 Elec	1.20	13.333		875	425		1,300	1,600	
	1500	25 kVA	CN	1	16		1,125	510		1,635	2,000	
	1700	37.5 kVA		.80	20		1,450	640		2,090	2,550	
	1900	50 kVA		.70	22.857		1,725	730		2,455	3,000	
	2100	75 kVA		.65	24.615		2,275	785		3,060	3,675	
	2190	480V primary 120/240V secondary, nonvent., 15 kVA		1.20	13.333		815	425		1,240	1,525	
	2200	25 kVA		.90	17.778		1,300	565		1,865	2,275	
	2210	37 kVA		.75	21.333		1,550	680		2,230	2,725	
	2220	50 kVA		.65	24.615		1,850	785		2,635	3,200	
	2300	3 phase, 480 volt primary 120/208 volt secondary										
	2310	Ventilated, 3 kVA	1 Elec	1	8	Ea.	425	255		680	850	
	2700	6 kVA		.80	10		590	320		910	1,125	
	2900	9 kVA		.70	11.429		670	365		1,035	1,275	
	3100	15 kVA	2 Elec	1.10	14.545		875	465		1,340	1,650	
	3300	30 kVA		.90	17.778		1,050	565		1,615	2,000	
	3500	45 kVA		.80	20		1,225	640		1,865	2,300	
	3700	75 kVA		.70	22.857		1,900	730		2,630	3,175	
	3900	112.5 kVA	R-3	.90	22.222		2,525	700	151	3,376	4,000	
	4100	150 kVA		.85	23.529		3,275	745	160	4,180	4,900	
	4300	225 kVA		.65	30.769		4,450	970	209	5,629	6,575	
	4500	300 kVA		.55	36.364		5,625	1,150	247	7,022	8,200	
	4700	500 kVA		.45	44.444		9,325	1,400	300	11,025	12,600	
	4800	750 kVA		.35	57.143		16,300	1,800	390	18,490	21,100	
160	0010	OIL FILLED TRANSFORMER Pad mounted, primary delta or Y,										160
	0050	5 kV or 15 kV, with taps, 277/480 V secondary, 3 phase										
	0100	150 kVA	R-3	.65	30.769	Ea.	6,000	970	209	7,179	8,275	
	0200	300 kVA		.45	44.444		7,500	1,400	300	9,200	10,700	
	0300	500 kVA		.40	50		10,000	1,575	340	11,915	13,800	
	0400	750 kVA		.38	52.632		13,500	1,675	355	15,530	17,800	
	0500	1000 kVA		.26	76.923		16,000	2,425	520	18,945	21,800	
	0600	1500 kVA		.23	86.957		19,000	2,750	590	22,340	25,700	
	0700	2000 kVA		.20	100		24,000	3,150	680	27,830	31,900	
	0800	3750 kVA		.16	125		45,000	3,950	850	49,800	56,500	

164 200 | Bus Ducts/Busways

		Description	CREW	DAILY OUTPUT	LABOR-HOURS	UNIT	1999 BARE COSTS				TOTAL INCL O&P	
							MAT.	LABOR	EQUIP.	TOTAL		
220	0010	COPPER BUS DUCT 10 ft long										220
	0050	3 pole 4 wire, plug-in/indoor, straight section, 225 amp	2 Elec	40	.400	L.F.	136	12.75		148.75	169	
	1000	400 amp		32	.500		136	15.95		151.95	174	
	1500	600 amp		26	.615		136	19.65		155.65	180	
	2400	800 amp		20	.800		162	25.50		187.50	216	
	2450	1000 amp		18	.889		179	28.50		207.50	239	
	2500	1350 amp		16	1		255	32		287	330	
	2510	1600 amp		12	1.333		289	42.50		331.50	385	
	2520	2000 amp		10	1.600		365	51		416	475	
	2550	Feeder, 600 amp		28	.571		119	18.25		137.25	159	
	2600	800 amp		22	.727		145	23		168	194	
	2700	1000 amp		20	.800		162	25.50		187.50	216	
	2800	1350 amp		18	.889		238	28.50		266.50	305	
	2900	1600 amp		14	1.143		272	36.50		308.50	355	

16 ELECTRICAL

Important: See the Reference Section for critical supporting data - Reference Nos., Crews, & City Cost Indexes

164 200	Bus Ducts/Busways	CREW	DAILY OUTPUT	LABOR-HOURS	UNIT	1999 BARE COSTS				TOTAL INCL O&P
						MAT.	LABOR	EQUIP.	TOTAL	
3000	2000 amp	2 Elec	12	1.333	L.F.	350	42.50		392.50	450
3100	Elbows, 225 amp		4	4	Ea.	805	128		933	1,075
3200	400 amp		3.60	4.444		805	142		947	1,100
3300	600 amp		3.20	5		805	160		965	1,125
3400	800 amp		2.80	5.714		875	182		1,057	1,225
3500	1000 amp		2.60	6.154		975	196		1,171	1,375
3600	1350 amp		2.40	6.667		1,150	213		1,363	1,575
3700	1600 amp		2.20	7.273		1,250	232		1,482	1,725
3800	2000 amp		1.80	8.889		1,550	284		1,834	2,125
4000	End box, 225 amp		34	.471		109	15		124	143
4100	400 amp		32	.500		109	15.95		124.95	144
4200	600 amp		28	.571		109	18.25		127.25	148
4300	800 amp		26	.615		109	19.65		128.65	150
4400	1000 amp		24	.667		109	21.50		130.50	152
4500	1350 amp		22	.727		109	23		132	155
4600	1600 amp		20	.800		109	25.50		134.50	158
4700	2000 amp		18	.889		133	28.50		161.50	190
4800	Cable tap box end, 225 amp		3.20	5		780	160		940	1,100
5000	400 amp		2.60	6.154		780	196		976	1,150
5100	600 amp		2.20	7.273		780	232		1,012	1,200
5200	800 amp		2	8		875	255		1,130	1,350
5300	1000 amp		1.60	10		960	320		1,280	1,525
5400	1350 amp		1.40	11.429		1,150	365		1,515	1,825
5500	1600 amp		1.20	13.333		1,300	425		1,725	2,050
5600	2000 amp		1	16		1,475	510		1,985	2,400
5700	Switchboard stub, 225 amp		5.40	2.963		815	94.50		909.50	1,050
5800	400 amp		4.60	3.478		815	111		926	1,075
5900	600 amp		4	4		815	128		943	1,100
6000	800 amp		3.20	5		985	160		1,145	1,325
6100	1000 amp		3	5.333		1,150	170		1,320	1,500
6200	1350 amp		2.60	6.154		1,475	196		1,671	1,925
6300	1600 amp		2.40	6.667		1,650	213		1,863	2,150
6400	2000 amp		2	8		2,025	255		2,280	2,600
6490	Tee fittings, 225 amp		2.40	6.667		1,100	213		1,313	1,550
6500	400 amp		2	8		1,100	255		1,355	1,600
6600	600 amp		1.80	8.889		1,100	284		1,384	1,650
6700	800 amp		1.60	10		1,275	320		1,595	1,875
6800	1350 amp		1.20	13.333		1,850	425		2,275	2,650
7000	1600 amp		1	16		2,100	510		2,610	3,075
7100	2000 amp		.80	20		2,475	640		3,115	3,675
7200	Plug-in switches, 600 volt, 3 pole, 30 amp	1 Elec	4	2		360	64		424	490
7300	60 amp		3.60	2.222		390	71		461	535
7400	100 amp		2.70	2.963		780	94.50		874.50	1,000
7500	200 amp	2 Elec	3.20	5		985	160		1,145	1,325
7600	400 amp		1.40	11.429		2,575	365		2,940	3,375
7700	600 amp		.90	17.778		3,350	565		3,915	4,550
7800	800 amp		.66	24.242		5,800	775		6,575	7,525
7900	1200 amp		.50	32		10,900	1,025		11,925	13,400
8000	Plug-in circuit breakers, molded case, 15 to 50 amp	1 Elec	4.40	1.818		460	58		518	590
8100	70 to 100 amp	"	3.10	2.581		510	82.50		592.50	685
8200	150 to 225 amp	2 Elec	3.40	4.706		1,375	150		1,525	1,750
8300	250 to 400 amp		1.40	11.429		2,425	365		2,790	3,225
8400	500 to 600 amp		1	16		3,275	510		3,785	4,375
8500	700 to 800 amp		.64	25		4,025	800		4,825	5,650
8600	900 to 1000 amp		.56	28.571		5,775	910		6,685	7,725
8700	1200 amp		.44	36.364		6,950	1,150		8,100	9,350

220

		165 100 | **Power Systems**	CREW	DAILY OUTPUT	LABOR-HOURS	UNIT	1999 BARE COSTS				TOTAL INCL O&P
							MAT.	LABOR	EQUIP.	TOTAL	
120	0010	**GENERATOR SET**										120
	0020	Gas or gasoline operated, includes battery,									
	0050	charger, muffler & transfer switch									
	0200	3 phase 4 wire, 277/480 volt, 7.5 kW	R-3	.83	24.096	Ea.	6,000	760	164	6,924	7,925
	0300	10 kW		.71	28.169		8,200	890	191	9,281	10,600
	0400	20 kW		.63	31.746		10,000	1,000	216	11,216	12,700
	0500	35 kW		.55	36.364		12,000	1,150	247	13,397	15,200
	0600	80 kW		.40	50		20,500	1,575	340	22,415	25,400
	0700	100 kW		.33	60.606		22,500	1,925	410	24,835	28,100
	0800	115 kW		.28	71.429		40,500	2,250	485	43,235	48,500
	0900	185 kW		.25	80		61,000	2,525	545	64,070	71,500
	2000	Diesel engine, including battery, charger,									
	2010	muffler, transfer switch & fuel tank, 30 kW	R-3	.55	36.364	Ea.	16,000	1,150	247	17,397	19,600
	2100	50 kW		.42	47.619		19,700	1,500	325	21,525	24,200
	2200	75 kW		.35	57.143		25,700	1,800	390	27,890	31,400
	2300	100 kW		.31	64.516		28,500	2,050	440	30,990	34,900
	2400	125 kW		.29	68.966		30,400	2,175	470	33,045	37,200
	2500	150 kW		.26	76.923		35,000	2,425	520	37,945	42,700
	2600	175 kW		.25	80		36,500	2,525	545	39,570	44,600
	2700	200 kW		.24	83.333		38,000	2,625	565	41,190	46,400
	2800	250 kW		.23	86.957		41,500	2,750	590	44,840	50,500
	2900	300 kW		.22	90.909		49,800	2,875	615	53,290	60,000
	3000	350 kW		.20	100		53,500	3,150	680	57,330	64,500
	3100	400 kW		.19	105		65,500	3,325	715	69,540	78,000
	3200	500 kW		.18	111		76,000	3,500	755	80,255	89,500

165 200 | Capacitors

			CREW	DAILY OUTPUT	LABOR-HOURS	UNIT	MAT.	LABOR	EQUIP.	TOTAL	TOTAL INCL O&P
210	0010	**CAPACITORS** Indoor										210
	0020	240 volts, single & 3 phase, 0.5 kVAR	1 Elec	2.70	2.963	Ea.	262	94.50		356.50	430
	0100	1.0 kVAR		2.70	2.963		315	94.50		409.50	490
	0150	2.5 kVAR		2	4		375	128		503	605
	0200	5.0 kVAR		1.80	4.444		445	142		587	700
	0250	7.5 kVAR		1.60	5		515	160		675	805
	0300	10 kVAR		1.50	5.333		615	170		785	935
	0350	15 kVAR		1.30	6.154		850	196		1,046	1,225
	0400	20 kVAR		1.10	7.273		1,050	232		1,282	1,525
	0450	25 kVAR		1	8		1,250	255		1,505	1,750
	1000	480 volts, single & 3 phase, 1 kVAR		2.70	2.963		246	94.50		340.50	410
	1050	2 kVAR		2.70	2.963		283	94.50		377.50	450
	1100	5 kVAR		2	4		360	128		488	585
	1150	7.5 kVAR		2	4		390	128		518	620
	1200	10 kVAR		2	4		430	128		558	665
	1250	15 kVAR		2	4		535	128		663	775
	1300	20 kVAR		1.60	5		595	160		755	895
	1350	30 kVAR		1.50	5.333		745	170		915	1,075
	1400	40 kVAR		1.20	6.667		945	213		1,158	1,375
	1450	50 kVAR		1.10	7.273		1,075	232		1,307	1,525

16 ELECTRICAL

Important: See the Reference Section for critical supporting data - Reference Nos., Crews, & City Cost Indexes

				DAILY	LABOR-		1999 BARE COSTS				TOTAL	
166 100		**Lighting**	CREW	OUTPUT	HOURS	UNIT	MAT.	LABOR	EQUIP.	TOTAL	INCL O&P	
110	0010	**EXIT AND EMERGENCY LIGHTING**										110
	0080	Exit light ceiling or wall mount, incandescent, single face	1 Elec	8	1	Ea.	35	32		67	86	
	0100	Double face		6.70	1.194		44	38		82	106	
	0200	L.E.D. standard, single face		8	1		59	32		91	113	
	0220	double face		6.70	1.194		62	38		100	125	
	0240	L.E.D. w/ battery unit, single face		4.40	1.818		100	58		158	197	
	0260	double face	▼	4	2	▼	102	64		166	208	
	0300	Emergency light units, battery operated										
	0350	Twin sealed beam light, 25 watt, 6 volt each										
	0500	Lead battery operated	1 Elec	4	2	Ea.	160	64		224	272	
	0700	Nickel cadmium battery operated		4	2		480	64		544	625	
	0900	Self-contained fluorescent lamp pack	▼	10	.800	▼	106	25.50		131.50	155	
115	0010	**EXTERIOR FIXTURES** With lamps										115
	0200	Wall mounted, incandescent, 100 watt	1 Elec	8	1	Ea.	28	32		60	78.50	
	0400	Quartz, 500 watt		5.30	1.509		51	48		99	128	
	0600	Mercury vapor, 100 watt		5.30	1.509		220	48		268	315	
	0800	Wall pack, mercury vapor, 175 watt		4	2		160	64		224	272	
	1000	250 watt		4	2		210	64		274	325	
	1100	Low pressure sodium, 35 watt		4	2		208	64		272	325	
	1150	55 watt	▼	4	2	▼	280	64		344	405	
	1200	Floodlights with ballast and lamp,										
	1400	pole mounted, pole not included										
	1500	Mercury vapor, 250 watt	1 Elec	2.40	3.333	Ea.	200	106		306	380	
	1600	400 watt	2 Elec	4.40	3.636		265	116		381	465	
	1800	1000 watt	"	4	4		365	128		493	590	
	1950	Metal halide, 175 watt	1 Elec	2.70	2.963		285	94.50		379.50	455	
	2000	400 watt	2 Elec	4.40	3.636		345	116		461	555	
	2200	1000 watt	"	4	4		450	128		578	685	
	2340	High pressure sodium, 70 watt	1 Elec	2.70	2.963		190	94.50		284.50	350	
	2400	400 watt	2 Elec	4.40	3.636		335	116		451	545	
	2600	1000 watt	"	4	4		500	128		628	740	
	2650	Roadway area luminaire, low pressure sodium, 135 watt	1 Elec	2	4		535	128		663	780	
	2700	180 watt	"	2	4		565	128		693	810	
	2720	Mercury vapor, 400 watt	2 Elec	4.40	3.636		340	116		456	550	
	2730	1000 watt		4	4		425	128		553	660	
	2750	Metal halide, 400 watt		4.40	3.636		410	116		526	625	
	2760	1000 watt		4	4		490	128		618	730	
	2780	High pressure sodium, 400 watt		4.40	3.636		465	116		581	685	
	2790	1000 watt	▼	4	4	▼	515	128		643	755	
	2800	Light poles, anchor base										
	2820	not including concrete bases										
	2840	Aluminum pole, 8' high	1 Elec	4	2	Ea.	435	64		499	575	
	3000	20' high	R-3	2.90	6.897		595	218	47	860	1,025	
	3200	30' high		2.60	7.692		1,100	243	52	1,395	1,650	
	3400	35' high		2.30	8.696		1,200	275	59	1,534	1,800	
	3600	40' high	▼	2	10		1,375	315	68	1,758	2,050	
	3800	Bracket arms, 1 arm	1 Elec	8	1		75	32		107	130	
	4000	2 arms		8	1		150	32		182	213	
	4200	3 arms		5.30	1.509		225	48		273	320	
	4400	4 arms	▼	5.30	1.509		300	48		348	400	
	4500	Steel pole, galvanized, 8' high		3.80	2.105		410	67		477	550	
	4600	20' high	R-3	2.60	7.692		730	243	52	1,025	1,225	
	4800	30' high		2.30	8.696		860	275	59	1,194	1,425	
	5000	35' high		2.20	9.091		940	287	61.50	1,288.50	1,525	
	5200	40' high	▼	1.70	11.765		1,150	370	80	1,600	1,925	
	5400	Bracket arms, 1 arm	1 Elec	8	1	▼	120	32		152	180	

ELECTRICAL 16

For expanded coverage of these items see *Means Electrical Cost Data 1999*

166 100	Lighting	CREW	DAILY OUTPUT	LABOR-HOURS	UNIT	1999 BARE COSTS				TOTAL INCL O&P
						MAT.	LABOR	EQUIP.	TOTAL	
115	5600	2 arms	1 Elec	8	1	Ea.	185	32		217	252	115

166 100		Lighting	CREW	DAILY OUTPUT	LABOR-HOURS	UNIT	MAT.	LABOR	EQUIP.	TOTAL	TOTAL INCL O&P	
115	5600	2 arms	1 Elec	8	1	Ea.	185	32		217	252	**115**
	5800	3 arms		5.30	1.509		200	48		248	292	
	6000	4 arms		5.30	1.509		280	48		328	380	
130	0010	**INTERIOR LIGHTING FIXTURES** Including lamps, mounting										**130**
	0030	hardware and connections										
	0100	Fluorescent, C.W. lamps, troffer, recess mounted in grid, RS										
	0200	Acrylic lens, 1'W x 4'L, two 40 watt	1 Elec	5.70	1.404	Ea.	46	45		91	118	
	0210	1'W x 4'L, three 40 watt		5.40	1.481		52	47.50		99.50	128	
	0300	2'W x 2'L, two U40 watt		5.70	1.404		50	45		95	122	
	0400	2'W x 4'L, two 40 watt		5.30	1.509		49	48		97	126	
	0500	2'W x 4'L, three 40 watt		5	1.600		53	51		104	135	
	0600	2'W x 4'L, four 40 watt CN		4.70	1.702		56	54.50		110.50	143	
	0700	4'W x 4'L, four 40 watt	2 Elec	6.40	2.500		265	80		345	410	
	0800	4'W x 4'L, six 40 watt		6.20	2.581		275	82.50		357.50	430	
	0900	4'W x 4'L, eight 40 watt		5.80	2.759		287	88		375	445	
	0910	Acrylic lens, 1'W x 4'L, two 32 watt	1 Elec	5.70	1.404		55	45		100	128	
	0930	2'W x 2'L, two U32 watt		5.70	1.404		75	45		120	150	
	0940	2'W x 4'L, two 32 watt		5.30	1.509		67	48		115	146	
	0950	2'W x 4'L, three 32 watt		5	1.600		72	51		123	156	
	0960	2'W x 4'L, four 32 watt		4.70	1.702		75	54.50		129.50	164	
	1000	Surface mounted, RS										
	1030	Acrylic lens with hinged & latched door frame										
	1100	1'W x 4'L, two 40 watt	1 Elec	7	1.143	Ea.	71	36.50		107.50	133	
	1110	1'W x 4'L, three 40 watt		6.70	1.194		82	38		120	147	
	1200	2'W x 2'L, two U40 watt		7	1.143		85	36.50		121.50	148	
	1300	2'W x 4'L, two 40 watt		6.20	1.290		77	41		118	146	
	1400	2'W x 4'L, three 40 watt		5.70	1.404		91	45		136	167	
	1500	2'W x 4'L, four 40 watt		5.30	1.509		91	48		139	172	
	1600	4'W x 4'L, four 40 watt	2 Elec	7.20	2.222		350	71		421	490	
	1700	4'W x 4'L, six 40 watt		6.60	2.424		385	77.50		462.50	540	
	1800	4'W x 4'L, eight 40 watt		6.20	2.581		395	82.50		477.50	560	
	1900	2'W x 8'L, four 40 watt		6.40	2.500		140	80		220	273	
	2000	2'W x 8'L, eight 40 watt		6.20	2.581		168	82.50		250.50	310	
	2100	Strip fixture										
	2130	Surface mounted										
	2200	4' long, one 40 watt RS	1 Elec	8.50	.941	Ea.	26	30		56	73.50	
	2300	4' long, two 40 watt RS		8	1		28	32		60	78.50	
	2400	4' long, one 40 watt, SL		8	1		38	32		70	89.50	
	2500	4' long, two 40 watt, SL		7	1.143		52	36.50		88.50	112	
	2600	8' long, one 75 watt, SL	2 Elec	13.40	1.194		39	38		77	100	
	2700	8' long, two 75 watt, SL	"	12.40	1.290		47	41		88	113	
	2800	4' long, two 60 watt, HO	1 Elec	6.70	1.194		74	38		112	139	
	2900	8' long, two 110 watt, HO	2 Elec	10.60	1.509		80	48		128	160	
	3000	Pendent mounted, industrial, white porcelain enamel										
	3100	4' long, two 40 watt, RS	1 Elec	5.70	1.404	Ea.	43	45		88	115	
	3200	4' long, two 60 watt, HO	"	5	1.600		70	51		121	154	
	3300	8' long, two 75 watt, SL	2 Elec	8.80	1.818		81	58		139	176	
	3400	8' long, two 110 watt, HO	"	8	2		106	64		170	213	
	3470	Troffer, air handling, 2'W x 4'L with four 40 watt, RS	1 Elec	4	2		76	64		140	179	
	3480	2'W x 2'L with two U40 watt RS		5.50	1.455		66	46.50		112.50	142	
	3490	Air connector insulated, 5" diameter		20	.400		52	12.75		64.75	76	
	3500	6" diameter		20	.400		53	12.75		65.75	77.50	
	3580	Mercury vapor, integral ballast, ceiling, recess mounted,										
	3590	prismatic glass lens, floating door										
	3600	2'W x 2'L, 250 watt DX lamp	1 Elec	3.20	2.500	Ea.	235	80		315	380	

16 ELECTRICAL

Important: See the Reference Section for critical supporting data - Reference Nos., Crews, & City Cost Indexes

			DAILY	LABOR-			1999 BARE COSTS				TOTAL	
166 100 | Lighting		**CREW**	**OUTPUT**	**HOURS**	**UNIT**	**MAT.**	**LABOR**	**EQUIP.**	**TOTAL**	**INCL O&P**	
130	3700	2'W x 2'L, 400 watt DX lamp	1 Elec	2.90	2.759	Ea.	255	88		343	415	130
	3800	Surface mtd., prismatic lens, 2'W x 2'L, 250 watt DX lamp		2.70	2.963		240	94.50		334.50	405	
	3900	2'W x 2'L, 400 watt DX lamp	↓	2.40	3.333	↓	270	106		376	455	
	4000	High bay, aluminum reflector										
	4030	Single unit, 400 watt DX lamp	2 Elec	4.60	3.478	Ea.	248	111		359	440	
	4100	Single unit, 1000 watt DX lamp		4	4		390	128		518	620	
	4200	Twin unit, two 400 watt DX lamps	↓	3.20	5	↓	420	160		580	700	
	4450	Incandescent, high hat can, round alzak reflector, prewired										
	4470	100 watt	1 Elec	8	1	Ea.	52	32		84	105	
	4480	150 watt		8	1		76	32		108	131	
	4500	300 watt	↓	6.70	1.194	↓	175	38		213	250	
	4600	Square glass lens with metal trim, prewired										
	4630	100 watt	1 Elec	6.70	1.194	Ea.	36	38		74	96.50	
	4700	200 watt		6.70	1.194		55	38		93	118	
	4800	300 watt		5.70	1.404		85	45		130	161	
	4900	Ceiling/wall, surface mounted, metal cylinder, 75 watt		10	.800		33	25.50		58.50	74.50	
	4920	150 watt	↓	10	.800	↓	51	25.50		76.50	94	
	5200	Ceiling, surface mounted, opal glass drum										
	5300	8", one 60 watt lamp	1 Elec	10	.800	Ea.	32	25.50		57.50	73	
	5400	10", two 60 watt lamps		8	1		36	32		68	87	
	5500	12", four 60 watt lamps		6.70	1.194		50	38		88	112	
	6010	Vapor tight, incandescent, ceiling mounted, 200 watt		6.20	1.290		43	41		84	109	
	6100	Fluorescent, surface mounted, 2 lamps, 4'L, RS, 40 watt		3.20	2.500		86	80		166	214	
	6200	Mercury vapor with ballast, 175 watt	↓	3.20	2.500	↓	320	80		400	470	
	6300	Explosionproof										
	6510	Incandescent, ceiling mounted, 200 watt	1 Elec	4	2	Ea.	630	64		694	790	
	6600	Fluorescent, RS, 4' long, ceiling mounted, two 40 watt		2.70	2.963		1,575	94.50		1,669.50	1,900	
	6700	Mercury vapor with ballast, surface mounted, 175 watt		2.70	2.963		525	94.50		619.50	720	
	6850	Vandalproof, surface mounted, fluorescent, two 40 watt		3.20	2.500		171	80		251	305	
	6860	Incandescent, one 150 watt	↓	8	1	↓	51	32		83	104	
	7500	Ballast replacement, by weight of ballast, to 15' high										
	7520	Indoor fluorescent, less than 2 lbs.	1 Elec	10	.800	Ea.		25.50		25.50	38	
	7540	Two 40W, watt reducer, 2 to 5 lbs.		9.40	.851		18	27		45	60.50	
	7560	Two F96 slimline, over 5 lbs.		8	1		31	32		63	81.50	
	7580	Vaportite ballast, less than 2 lbs.		9.40	.851			27		27	40.50	
	7600	2 lbs. to 5 lbs.		8.90	.899			28.50		28.50	43	
	7620	Over 5 lbs.		7.60	1.053			33.50		33.50	50	
	7630	Electronic ballast for two tubes		8	1		30	32		62	80.50	
	7640	Dimmable ballast one lamp		8	1		43	32		75	94.50	
	7650	Dimmable ballast two-lamp	↓	7.60	1.053	↓	67	33.50		100.50	124	
140	0010	**LAMPS**										140
	0080	Fluorescent, rapid start, cool white, 2' long, 20 watt	1 Elec	1	8	C	261	255		516	665	
	0100	4' long, 40 watt		.90	8.889		225	284		509	675	
	0200	Slimline, 4' long, 40 watt		.90	8.889		845	284		1,129	1,350	
	0400	High output, 4' long, 60 watt		.90	8.889		970	284		1,254	1,500	
	0500	8' long, 110 watt		.80	10		580	320		900	1,100	
	0560	Twin tube compact lamp		.90	8.889		405	284		689	870	
	0570	Double twin tube compact lamp		.80	10		880	320		1,200	1,450	
	0600	Mercury vapor, mogul base, deluxe white, 100 watt		.30	26.667		3,250	850		4,100	4,850	
	0700	250 watt		.30	26.667		4,275	850		5,125	5,975	
	0800	400 watt		.30	26.667		3,425	850		4,275	5,050	
	0900	1000 watt		.20	40		7,850	1,275		9,125	10,500	
	1000	Metal halide, mogul base, 175 watt		.30	26.667		4,100	850		4,950	5,800	
	1200	400 watt		.30	26.667		4,450	850		5,300	6,150	
	1300	1000 watt		.20	40		11,200	1,275		12,475	14,300	
	1350	High pressure sodium, 70 watt	↓	.30	26.667	↓	4,100	850		4,950	5,800	

ELECTRICAL 16

166 100	Lighting	CREW	DAILY OUTPUT	LABOR-HOURS	UNIT	1999 BARE COSTS				TOTAL INCL O&P
						MAT.	LABOR	EQUIP.	TOTAL	
140	**1380** 250 watt	1 Elec	.30	26.667	C	4,700	850		5,550	6,450	**140**
1400	400 watt		.30	26.667		6,650	850		7,500	8,575
1450	1000 watt		.20	40		15,100	1,275		16,375	18,500
3000	Guards, fluorescent lamp, 4' long		1	8		550	255		805	985
3200	8' long		.90	8.889		790	284		1,074	1,300

167 100	Electric Utilities	CREW	DAILY OUTPUT	LABOR-HOURS	UNIT	1999 BARE COSTS				TOTAL INCL O&P
						MAT.	LABOR	EQUIP.	TOTAL	
110	**0010** **ELECTRIC & TELEPHONE SITE WORK** Not including excavation										**110**
0200	backfill and cast in place concrete									
0400	Hand holes, precast concrete, with concrete cover									
0600	2' x 2' x 3' deep	R-3	2.40	8.333	Ea.	335	263	56.50	654.50	820
0800	3' x 3' x 3' deep		1.90	10.526		430	335	71.50	836.50	1,050
1000	4' x 4' x 4' deep		1.40	14.286		545	450	97	1,092	1,375
1200	Manholes, precast with iron racks & pulling irons, C.I. frame									
1400	and cover, 4' x 6' x 7' deep	B-13	2	28	Ea.	1,075	650	275	2,000	2,475
1600	6' x 8' x 7' deep		1.90	29.474		1,325	685	289	2,299	2,850
1800	6' x 10' x 7' deep		1.80	31.111		1,475	725	305	2,505	3,075
2000	Poles, wood, preservative treatment, see also div. 166-115, 20' high	R-3	3.10	6.452		236	204	44	484	615
2400	25' high		2.90	6.897		250	218	47	515	650
2600	30' high		2.60	7.692		270	243	52	565	720
2800	35' high		2.40	8.333		350	263	56.50	669.50	840
3000	40' high		2.30	8.696		430	275	59	764	950
3200	45' high		1.70	11.765		525	370	80	975	1,225
3400	Cross arms with hardware & insulators									
3600	4' long	1 Elec	2.50	3.200	Ea.	110	102		212	274
3800	5' long		2.40	3.333		128	106		234	300
4000	6' long		2.20	3.636		147	116		263	335
4200	Underground duct, banks ready for concrete fill, min. of 7.5"									
4400	between conduits, ctr. to ctr.(for wire & cable see div. 161)									
4580	PVC, type EB, 1 @ 2" diameter	2 Elec	480	.033	L.F.	.56	1.06		1.62	2.21
4600	2 @ 2" diameter		240	.067		1.12	2.13		3.25	4.41
4800	4 @ 2" diameter		120	.133		2.24	4.25		6.49	8.80
5000	2 @ 3" diameter		200	.080		1.54	2.55		4.09	5.50
5200	4 @ 3" diameter		100	.160		3.08	5.10		8.18	11.05
5400	2 @ 4" diameter		160	.100		2.38	3.19		5.57	7.40
5600	4 @ 4" diameter		80	.200		4.77	6.40		11.17	14.80
5800	6 @ 4" diameter		54	.296		7.15	9.45		16.60	22
6200	Rigid galvanized steel, 2 @ 2" diameter		180	.089		8.50	2.84		11.34	13.60
6400	4 @ 2" diameter		90	.178		17	5.65		22.65	27
6800	2 @ 3" diameter		100	.160		18.75	5.10		23.85	28
7000	4 @ 3" diameter		50	.320		37.50	10.20		47.70	56.50
7200	2 @ 4" diameter		70	.229		27	7.30		34.30	40.50
7400	4 @ 4" diameter		34	.471		54	15		69	82
7600	6 @ 4" diameter		22	.727		81	23		104	124

Important: See the Reference Section for critical supporting data - Reference Nos., Crews, & City Cost Indexes

16 ELECTRICAL

168 100	Special Systems		CREW	DAILY OUTPUT	LABOR-HOURS	UNIT	1999 BARE COSTS				TOTAL INCL O&P	
							MAT.	LABOR	EQUIP.	TOTAL		
105	0010	**CLOCKS**										105
	0080	12" diameter, single face	1 Elec	8	1	Ea.	75.50	32		107.50	131	
	0100	Double face	"	6.20	1.290	"	144	41		185	220	
110	0010	**CLOCK SYSTEMS**, not including wires & conduits										110
	0100	Time system components, master controller	1 Elec	.33	24.242	Ea.	1,625	775		2,400	2,950	
	0200	Program bell		8	1		53.50	32		85.50	107	
	0400	Combination clock & speaker		3.20	2.500		189	80		269	325	
	0600	Frequency generator		2	4		6,350	128		6,478	7,175	
	0800	Job time automatic stamp recorder, minimum		4	2		445	64		509	585	
	1000	Maximum	▼	4	2		680	64		744	845	
	1200	Time stamp for correspondence, hand operated					340			340	375	
	1400	Fully automatic					495			495	545	
	1600	Master time clock system, clocks & bells, 20 room	4 Elec	.20	160		4,075	5,100		9,175	12,100	
	1800	50 room	"	.08	400		9,400	12,800		22,200	29,400	
	2000	Time clock, 100 cards in & out, 1 color	1 Elec	3.20	2.500		950	80		1,030	1,175	
	2200	2 colors		3.20	2.500		1,025	80		1,105	1,250	
	2400	With 3 circuit program device, minimum		2	4		375	128		503	605	
	2600	Maximum		2	4		545	128		673	790	
	2800	Metal rack for 25 cards		7	1.143		60	36.50		96.50	121	
	3000	Watchman's tour station		8	1		63	32		95	117	
	3200	Annunciator with zone indication		1	8		230	255		485	635	
	3400	Time clock with tape	▼	1	8	▼	610	255		865	1,050	
112	0010	**CLOCK AND PROGRAM SYSTEMS** For electronic scoreboards										112
	0100	See division 114-805										
115	0010	**DOCTORS IN-OUT REGISTER**										115
	0050	Register, 200 names	4 Elec	.64	50	Ea.	9,475	1,600		11,075	12,800	
	0100	Combination control and recall, 200 names	"	.64	50		12,400	1,600		14,000	16,000	
	0200	Recording register	1 Elec	.50	16		4,825	510		5,335	6,075	
	0300	Transformers	"	4	2		162	64		226	274	
	0400	Pocket pages				▼	835			835	920	
120	0010	**DETECTION SYSTEMS**, not including wires & conduits										120
	0100	Burglar alarm, battery operated, mechanical trigger	1 Elec	4	2	Ea.	249	64		313	370	
	0200	Electrical trigger		4	2		297	64		361	420	
	0400	For outside key control, add		8	1		70.50	32		102.50	125	
	0600	For remote signaling circuitry, add		8	1		112	32		144	171	
	0800	Card reader, flush type, standard		2.70	2.963		835	94.50		929.50	1,050	
	1000	Multi-code		2.70	2.963		1,075	94.50		1,169.50	1,325	
	1200	Door switches, hinge switch		5.30	1.509		52.50	48		100.50	130	
	1400	Magnetic switch		5.30	1.509		62	48		110	140	
	1600	Exit control locks, horn alarm		4	2		310	64		374	435	
	1800	Flashing light alarm		4	2		350	64		414	480	
	2000	Indicating panels, 1 channel	▼	2.70	2.963		330	94.50		424.50	505	
	2200	10 channel	2 Elec	3.20	5		1,125	160		1,285	1,500	
	2400	20 channel		2	8		2,200	255		2,455	2,800	
	2600	40 channel	▼	1.14	14.035		4,000	450		4,450	5,075	
	2800	Ultrasonic motion detector, 12 volt	1 Elec	2.30	3.478		206	111		317	395	
	3000	Infrared photoelectric detector	"	2.30	3.478	▼	170	111		281	355	
	3594	Fire, alarm control panel										
	3600	4 zone	2 Elec	2	8	Ea.	920	255		1,175	1,375	
	3800	8 zone		1	16		1,400	510		1,910	2,325	
	4000	12 zone	▼	.67	23.988		1,825	765		2,590	3,150	
	4020	Alarm device	1 Elec	8	1		122	32		154	182	
	4050	Actuating device		8	1		292	32		324	370	
	4200	Battery and rack	▼	4	2		690	64		754	855	

ELECTRICAL 16

168 100	Special Systems	CREW	DAILY OUTPUT	LABOR-HOURS	UNIT	1999 BARE COSTS				TOTAL INCL O&P	
						MAT.	LABOR	EQUIP.	TOTAL		
120	4400	Automatic charger	1 Elec	8	1	Ea.	445	32		477	540	120
	4600	Signal bell		8	1		49.50	32		81.50	102	
	4800	Trouble buzzer or manual station		8	1		37	32		69	88	
	5000	Detector, rate of rise		8	1		33.50	32		65.50	84.50	
	5200	Smoke detector, ceiling type		6.20	1.290		75	41		116	144	
	5400	Duct type		3.20	2.500		250	80		330	395	
	5600	Strobe and horn		5.30	1.509		95	48		143	177	
	5800	Fire alarm horn		6.70	1.194		36.50	38		74.50	97	
	6000	Door holder, electro-magnetic		4	2		77.50	64		141.50	181	
	6200	Combination holder and closer		3.20	2.500		430	80		510	595	
	6400	Code transmitter		4	2		690	64		754	855	
	6600	Drill switch		8	1		86.50	32		118.50	143	
	6800	Master box		2.70	2.963		3,100	94.50		3,194.50	3,550	
	7000	Break glass station		8	1		50	32		82	103	
	7800	Remote annunciator, 8 zone lamp		1.80	4.444		175	142		317	405	
	8000	12 zone lamp	2 Elec	2.60	6.154		300	196		496	625	
	8200	16 zone lamp	"	2.20	7.273		300	232		532	675	
125	0010	**DOORBELL SYSTEM** Incl. transformer, button & signal										125
	0100	6" bell	1 Elec	4	2	Ea.	55.50	64		119.50	157	
	0200	Buzzer	"	4	2	"	54.50	64		118.50	156	
130	0010	**ELECTRIC HEATING**										130
	0200	Snow melting for paved surface embedded mat heaters & controls	1 Elec	130	.062	S.F.	6.10	1.96		8.06	9.65	
	0400	Cable heating, radiant heat plaster, no controls, in South		130	.062		7.25	1.96		9.21	10.95	
	0600	In North		90	.089		7.25	2.84		10.09	12.25	
	0800	Cable on 1/2" board not incl. controls, tract housing		90	.089		6.10	2.84		8.94	10.95	
	1000	Custom housing		80	.100		7.10	3.19		10.29	12.55	
	1100	Rule of thumb: Baseboard units, including control		4.40	1.818	KW	75	58		133	169	
	1200	Duct heaters, including controls		5.30	1.509	"	61.50	48		109.50	140	
	1300	Baseboard heaters, 2' long, 375 watt		8	1	Ea.	36	32		68	87	
	1400	3' long, 500 watt		8	1		40	32		72	91.50	
	1600	4' long, 750 watt		6.70	1.194		48	38		86	110	
	1800	5' long, 935 watt		5.70	1.404		59	45		104	132	
	2000	6' long, 1125 watt		5	1.600		65	51		116	148	
	2400	8' long, 1500 watt		4	2		81	64		145	185	
	2950	Wall heaters with fan, 120 to 277 volt										
	2970	surface mounted, residential, 750 watt	1 Elec	7	1.143	Ea.	95	36.50		131.50	160	
	3000	1500 watt		5	1.600		118	51		169	207	
	3040	2250 watt		4	2		235	64		299	355	
	3070	4000 watt		3.50	2.286		240	73		313	375	
	3600	Thermostats, integral		16	.500		23	15.95		38.95	49.50	
	3800	Line voltage, 1 pole		8	1		23	32		55	73	
145	0010	**NURSE CALL SYSTEMS**										145
	0100	Single bedside call station	1 Elec	8	1	Ea.	140	32		172	202	
	0200	Ceiling speaker station		8	1		47	32		79	99	
	0400	Emergency call station		8	1		79	32		111	135	
	0600	Pillow speaker		8	1		149	32		181	212	
	0800	Double bedside call station		4	2		245	64		309	365	
	1000	Duty station		4	2		117	64		181	225	
	1200	Standard call button		8	1		56.50	32		88.50	110	
	1400	Lights, corridor, dome or zone indicator		8	1		50	32		82	103	
	1600	Master control station for 20 stations	2 Elec	.65	24.615	Total	2,450	785		3,235	3,875	
150	0010	**PUBLIC ADDRESS SYSTEM**										150
	0100	Conventional, office	1 Elec	5.33	1.501	Speaker	91	48		139	172	

Important: See the Reference Section for critical supporting data - Reference Nos., Crews, & City Cost Indexes

16 ELECTRICAL

168 | Special Systems

168 100 | Special Systems

		CREW	DAILY OUTPUT	LABOR-HOURS	UNIT	MAT.	LABOR	EQUIP.	TOTAL	TOTAL INCL O&P	
150 0200	Industrial	1 Elec	2.70	2.963	Speaker	176	94.50		270.50	335	**150**
0400	Explosionproof system is 3 times cost of central control										
0600	Installation costs run about 120% of material cost										
155 0010	**SOUND SYSTEM** not including rough-in wires, cables & conduits										**155**
0100	Components, outlet, projector	1 Elec	8	1	Ea.	40	32		72	91.50	
0200	Microphone		4	2		44.50	64		108.50	145	
0400	Speakers, ceiling or wall		8	1		77	32		109	132	
0600	Trumpets		4	2		140	64		204	250	
0800	Privacy switch		8	1		55	32		87	108	
1000	Monitor panel		4	2		250	64		314	370	
1200	Antenna, AM/FM		4	2		140	64		204	250	
1400	Volume control		8	1		51	32		83	104	
1600	Amplifier, 250 watts		1	8		1,050	255		1,305	1,525	
1800	Cabinets		1	8		545	255		800	980	
2000	Intercom, 25 station capacity, master station	2 Elec	2	8		1,300	255		1,555	1,825	
2200	Remote station	1 Elec	8	1		105	32		137	164	
2400	Intercom outlets		8	1		61.50	32		93.50	115	
2600	Handset		4	2		190	64		254	305	
2800	Emergency call system, 12 zones, annunciator		1.30	6.154		615	196		811	970	
3000	Bell		5.30	1.509		64	48		112	143	
3200	Light or relay		8	1		32	32		64	82.50	
3400	Transformer		4	2		140	64		204	250	
3600	House telephone, talking station		1.60	5		300	160		460	570	
3800	Press to talk, release to listen		5.30	1.509		70	48		118	149	
4000	System-on button					42			42	46	
4200	Door release	1 Elec	4	2		74	64		138	177	
4400	Combination speaker and microphone		8	1		128	32		160	189	
4600	Termination box		3.20	2.500		40	80		120	163	
4800	Amplifier or power supply		5.30	1.509		460	48		508	575	
5000	Vestibule door unit		16	.500	Name	84.50	15.95		100.45	117	
5200	Strip cabinet		27	.296	Ea.	160	9.45		169.45	190	
5400	Directory		16	.500	"	75	15.95		90.95	107	
160 0010	**T.V. SYSTEMS** not including rough-in wires, cables & conduits										**160**
0100	Master TV antenna system										
0200	VHF reception & distribution, 12 outlets	1 Elec	6	1.333	Outlet	152	42.50		194.50	231	
0400	30 outlets		10	.800		100	25.50		125.50	148	
0600	100 outlets		13	.615		102	19.65		121.65	142	
0800	VHF & UHF reception & distribution, 12 outlets		6	1.333		151	42.50		193.50	230	
1000	30 outlets		10	.800		100	25.50		125.50	148	
1200	100 outlets		13	.615		102	19.65		121.65	142	
1400	School and deluxe systems, 12 outlets		2.40	3.333		199	106		305	380	
1600	30 outlets		4	2		175	64		239	289	
1800	80 outlets		5.30	1.509		168	48		216	257	
2000	Closed circuit, surveillance, one station (camera & monitor)	2 Elec	2.60	6.154	Total	965	196		1,161	1,350	
2200	For additional camera stations, add	1 Elec	2.70	2.963	Ea.	540	94.50		634.50	735	
2400	Industrial quality, one station (camera & monitor)	2 Elec	2.60	6.154	Total	2,000	196		2,196	2,500	
2600	For additional camera stations, add	1 Elec	2.70	2.963	Ea.	1,225	94.50		1,319.50	1,500	
2610	For low light, add		2.70	2.963		985	94.50		1,079.50	1,225	
2620	For very low light, add		2.70	2.963		7,250	94.50		7,344.50	8,125	
2800	For weatherproof camera station, add		1.30	6.154		755	196		951	1,125	
3000	For pan and tilt, add		1.30	6.154		1,950	196		2,146	2,450	
3200	For zoom lens - remote control, add, minimum		2	4		1,800	128		1,928	2,175	
3400	Maximum		2	4		6,575	128		6,703	7,425	
3410	For automatic iris for low light, add		2	4		1,575	128		1,703	1,925	

ELECTRICAL 16

168 100	Special Systems	CREW	DAILY OUTPUT	LABOR-HOURS	UNIT	1999 BARE COSTS				TOTAL INCL O&P
						MAT.	LABOR	EQUIP.	TOTAL	
160 3600	Educational T.V. studio, basic 3 camera system, black & white,									**160**
3800	electrical & electronic equip. only, minimum	4 Elec	.80	40	Total	9,350	1,275		10,625	12,200
4000	Maximum (full console)		.28	114		39,800	3,650		43,450	49,300
4100	As above, but color system, minimum		.28	114		52,500	3,650		56,150	63,500
4120	Maximum		.12	266		229,000	8,500		237,500	264,500
4200	For film chain, black & white, add	1 Elec	1	8	Ea.	10,700	255		10,955	12,200
4250	Color, add		.25	32		13,000	1,025		14,025	15,800
4400	For video tape recorders, add, minimum		1	8		2,250	255		2,505	2,850
4600	Maximum	4 Elec	.40	80		18,700	2,550		21,250	24,400
170 0010	**RESIDENTIAL WIRING**									**170**
0020	20' avg. runs and #14/2 wiring incl. unless otherwise noted									
1000	Service & panel, includes 24' SE-AL cable, service eye, meter,									
1010	Socket, panel board, main bkr., ground rod, 15 or 20 amp									
1020	1-pole circuit breakers, and misc. hardware									
1100	100 amp, with 10 branch breakers	1 Elec	1.19	6.723	Ea.	400	214		614	760
1110	With PVC conduit and wire		.92	8.696		425	277		702	885
1120	With RGS conduit and wire		.73	10.959		540	350		890	1,125
1150	150 amp, with 14 branch breakers		1.03	7.767		620	248		868	1,050
1170	With PVC conduit and wire		.82	9.756		680	310		990	1,225
1180	With RGS conduit and wire		.67	11.940		875	380		1,255	1,525
1200	200 amp, with 18 branch breakers	2 Elec	1.80	8.889		810	284		1,094	1,325
1220	With PVC conduit and wire		1.46	10.959		870	350		1,220	1,475
1230	With RGS conduit and wire		1.24	12.903		1,125	410		1,535	1,875
1800	Lightning surge suppressor for above services, add	1 Elec	32	.250		39.50	7.95		47.45	55.50
2000	Switch devices									
2100	Single pole, 15 amp, Ivory, with a 1-gang box, cover plate,									
2110	Type NM (Romex) cable	1 Elec	17.10	.468	Ea.	6.80	14.90		21.70	30
2120	Type MC (BX) cable		14.30	.559		16.30	17.85		34.15	44.50
2130	EMT & wire		5.71	1.401		17.05	44.50		61.55	86
2150	3-way, #14/3, type NM cable		14.55	.550		10.30	17.55		27.85	37.50
2170	Type MC cable		12.31	.650		20	20.50		40.50	53
2180	EMT & wire		5	1.600		19.40	51		70.40	98
2200	4-way, #14/3, type NM cable		14.55	.550		22.50	17.55		40.05	51
2220	Type MC cable		12.31	.650		32.50	20.50		53	66.50
2230	EMT & wire		5	1.600		31.50	51		82.50	112
2250	S.P., 20 amp, #12/2, type NM cable		13.33	.600		12	19.15		31.15	41.50
2270	Type MC cable		11.43	.700		20	22.50		42.50	55.50
2280	EMT & wire		4.85	1.649		23	52.50		75.50	104
2290	S.P. rotary dimmer, 600W, no wiring		17	.471		15.15	15		30.15	39
2300	S.P. rotary dimmer, 600W, type NM cable		14.55	.550		18.25	17.55		35.80	46
2320	Type MC cable		12.31	.650		27.50	20.50		48	61.50
2330	EMT & wire		5	1.600		29.50	51		80.50	109
2350	3-way rotary dimmer, type NM cable		13.33	.600		21	19.15		40.15	51.50
2370	Type MC cable		11.43	.700		30.50	22.50		53	67
2380	EMT & wire		4.85	1.649		32	52.50		84.50	114
2400	Interval timer wall switch, 20 amp, 1-30 min., #12/2									
2410	Type NM cable	1 Elec	14.55	.550	Ea.	29	17.55		46.55	58
2420	Type MC cable		12.31	.650		34.50	20.50		55	69
2430	EMT & wire		5	1.600		39.50	51		90.50	120
2500	Decorator style									
2510	S.P., 15 amp, type NM cable	1 Elec	17.10	.468	Ea.	9.95	14.90		24.85	33.50
2520	Type MC cable		14.30	.559		19.45	17.85		37.30	48
2530	EMT & wire		5.71	1.401		20	44.50		64.50	89
2550	3-way, #14/3, type NM cable		14.55	.550		13.45	17.55		31	41
2570	Type MC cable		12.31	.650		23.50	20.50		44	56.50

Important: See the Reference Section for critical supporting data - Reference Nos., Crews, & City Cost Indexes

16 ELECTRICAL

168 100	Special Systems	CREW	DAILY OUTPUT	LABOR-HOURS	UNIT	1999 BARE COSTS				TOTAL INCL O&P
						MAT.	LABOR	EQUIP.	TOTAL	
170	2580	EMT & wire	1 Elec	5	1.600	Ea.	22.50	51		73.50	102
	2600	4-way, #14/3, type NM cable		14.55	.550		26	17.55		43.55	54.50
	2620	Type MC cable		12.31	.650		35.50	20.50		56	70
	2630	EMT & wire		5	1.600		35	51		86	115
	2650	S.P., 20 amp, #12/2, type NM cable		13.33	.600		15.15	19.15		34.30	45
	2670	Type MC cable		11.43	.700		23.50	22.50		46	59
	2680	EMT & wire		4.85	1.649		26	52.50		78.50	107
	2700	S.P., slide dimmer, type NM cable		17.10	.468		25.50	14.90		40.40	51
	2720	Type MC cable		14.30	.559		35	17.85		52.85	65
	2730	EMT & wire		5.71	1.401		37	44.50		81.50	108
	2750	S.P., touch dimmer, type NM cable		17.10	.468		25	14.90		39.90	50
	2770	Type MC cable		14.30	.559		34.50	17.85		52.35	64.50
	2780	EMT & wire		5.71	1.401		36.50	44.50		81	107
	2800	3-way touch dimmer, type NM cable		13.33	.600		39	19.15		58.15	71.50
	2820	Type MC cable		11.43	.700		48.50	22.50		71	87
	2830	EMT & wire	▼	4.85	1.649	▼	50	52.50		102.50	134
	3000	Combination devices									
	3100	S.P. switch/15 amp recpt., Ivory, 1-gang box, plate									
	3110	Type NM cable	1 Elec	11.43	.700	Ea.	14.70	22.50		37.20	49.50
	3120	Type MC cable		10	.800		24	25.50		49.50	64.50
	3130	EMT & wire		4.40	1.818		26	58		84	115
	3150	S.P. switch/pilot light, type NM cable		11.43	.700		15.30	22.50		37.80	50.50
	3170	Type MC cable		10	.800		25	25.50		50.50	65
	3180	EMT & wire		4.43	1.806		26.50	57.50		84	115
	3190	2-S.P. switches, 2-#14/2, no wiring		14	.571		5.20	18.25		23.45	33.50
	3200	2-S.P. switches, 2-#14/2, type NM cables		10	.800		17.05	25.50		42.55	57
	3220	Type MC cable		8.89	.900		32.50	28.50		61	78.50
	3230	EMT & wire		4.10	1.951		28	62		90	124
	3250	3-way switch/15 amp recpt., #14/3, type NM cable		10	.800		21	25.50		46.50	61
	3270	Type MC cable		8.89	.900		31	28.50		59.50	77
	3280	EMT & wire		4.10	1.951		30	62		92	126
	3300	2-3 way switches, 2-#14/3, type NM cables		8.89	.900		28	28.50		56.50	74
	3320	Type MC cable		8	1		44	32		76	96
	3330	EMT & wire		4	2		35	64		99	134
	3350	S.P. switch/20 amp recpt., #12/2, type NM cable		10	.800		24.50	25.50		50	65
	3370	Type MC cable		8.89	.900		30	28.50		58.50	76
	3380	EMT & wire	▼	4.10	1.951	▼	35	62		97	132
	3400	Decorator style									
	3410	S.P. switch/15 amp recpt., type NM cable	1 Elec	11.43	.700	Ea.	17.85	22.50		40.35	53
	3420	Type MC cable		10	.800		27.50	25.50		53	68
	3430	EMT & wire		4.40	1.818		29	58		87	119
	3450	S.P. switch/pilot light, type NM cable		11.43	.700		18.45	22.50		40.95	54
	3470	Type MC cable		10	.800		28	25.50		53.50	68.50
	3480	EMT & wire		4.40	1.818		29.50	58		87.50	119
	3500	2-S.P. switches, 2-#14/2, type NM cables		10	.800		20	25.50		45.50	60
	3520	Type MC cable		8.89	.900		35.50	28.50		64	82
	3530	EMT & wire		4.10	1.951		31	62		93	127
	3550	3-way/15 amp recpt., #14/3, type NM cable		10	.800		24	25.50		49.50	64.50
	3570	Type MC cable		8.89	.900		34	28.50		62.50	80.50
	3580	EMT & wire		4.10	1.951		33	62		95	130
	3650	2-3 way switches, 2-#14/3, type NM cables		8.89	.900		31.50	28.50		60	77.50
	3670	Type MC cable		8	1		47.50	32		79.50	99.50
	3680	EMT & wire		4	2		38	64		102	137
	3700	S.P. switch/20 amp recpt., #12/2, type NM cable		10	.800		27.50	25.50		53	68.50
	3720	Type MC cable		8.89	.900		33	28.50		61.50	79.50
	3730	EMT & wire	▼	4.10	1.951	▼	38.50	62		100.50	135

ELECTRICAL 16

168 100	Special Systems	CREW	DAILY OUTPUT	LABOR-HOURS	UNIT	1999 BARE COSTS				TOTAL INCL O&P
						MAT.	LABOR	EQUIP.	TOTAL	
4000	Receptacle devices									
4010	Duplex outlet, 15 amp recpt., Ivory, 1-gang box, plate									
4015	Type NM cable	1 Elec	14.55	.550	Ea.	5.40	17.55		22.95	32
4020	Type MC cable		12.31	.650		14.90	20.50		35.40	47.50
4030	EMT & wire		5.33	1.501		15.65	48		63.65	88.50
4050	With #12/2, type NM cable		12.31	.650		6.85	20.50		27.35	38.50
4070	Type MC cable		10.67	.750		15	24		39	52.50
4080	EMT & wire		4.71	1.699		17.65	54		71.65	100
4100	20 amp recpt., #12/2, type NM cable		12.31	.650		11.70	20.50		32.20	44
4120	Type MC cable		10.67	.750		19.80	24		43.80	58
4130	EMT & wire	↓	4.71	1.699	↓	22.50	54		76.50	106
4140	For GFI see line 4300 below									
4150	Decorator style, 15 amp recpt., type NM cable	1 Elec	14.55	.550	Ea.	8.55	17.55		26.10	35.50
4170	Type MC cable		12.31	.650		18.05	20.50		38.55	51
4180	EMT & wire		5.33	1.501		18.80	48		66.80	92
4200	With #12/2, type NM cable		12.31	.650		10	20.50		30.50	42
4220	Type MC cable		10.67	.750		18.15	24		42.15	56
4230	EMT & wire		4.71	1.699		21	54		75	104
4250	20 amp recpt. #12/2, type NM cable		12.31	.650		14.85	20.50		35.35	47.50
4270	Type MC cable		10.67	.750		23	24		47	61.50
4280	EMT & wire		4.71	1.699		25.50	54		79.50	109
4300	GFI, 15 amp recpt., type NM cable		12.31	.650		33	20.50		53.50	67.50
4320	Type MC cable		10.67	.750		42.50	24		66.50	83
4330	EMT & wire		4.71	1.699		43.50	54		97.50	129
4350	GFI with #12/2, type NM cable		10.67	.750		34.50	24		58.50	74
4370	Type MC cable		9.20	.870		42.50	27.50		70	88.50
4380	EMT & wire		4.21	1.900		45.50	60.50		106	141
4400	20 amp recpt., #12/2 type NM cable		10.67	.750		36	24		60	75.50
4420	Type MC cable		9.20	.870		44	27.50		71.50	90
4430	EMT & wire		4.21	1.900		47	60.50		107.50	142
4500	Weather-proof cover for above receptacles, add	↓	32	.250	↓	4.40	7.95		12.35	16.80
4550	Air conditioner outlet, 20 amp-240 volt recpt.									
4560	30' of #12/2, 2 pole circuit breaker									
4570	Type NM cable	1 Elec	10	.800	Ea.	40	25.50		65.50	82
4580	Type MC cable		9	.889		50	28.50		78.50	98
4590	EMT & wire		4	2		50	64		114	151
4600	Decorator style, type NM cable		10	.800		43.50	25.50		69	86
4620	Type MC cable		9	.889		54	28.50		82.50	102
4630	EMT & wire	↓	4	2	↓	54	64		118	155
4650	Dryer outlet, 30 amp-240 volt recpt., 20' of #10/3									
4660	2 pole circuit breaker									
4670	Type NM cable	1 Elec	6.41	1.248	Ea.	50.50	40		90.50	115
4680	Type MC cable		5.71	1.401		54	44.50		98.50	127
4690	EMT & wire	↓	3.48	2.299	↓	53.50	73.50		127	169
4700	Range outlet, 50 amp-240 volt recpt., 30' of #8/3									
4710	Type NM cable	1 Elec	4.21	1.900	Ea.	80	60.50		140.50	179
4720	Type MC cable		4	2		100	64		164	206
4730	EMT & wire		2.96	2.703		74.50	86		160.50	211
4750	Central vacuum outlet, Type NM cable		6.40	1.250		43	40		83	107
4770	Type MC cable		5.71	1.401		56	44.50		100.50	129
4780	EMT & wire	↓	3.48	2.299	↓	53	73.50		126.50	168
4800	30 amp-110 volt locking recpt., #10/2 circ. bkr.									
4810	Type NM cable	1 Elec	6.20	1.290	Ea.	50	41		91	117
4820	Type MC cable		5.40	1.481		68	47.50		115.50	146
4830	EMT & wire	↓	3.20	2.500	↓	61.50	80		141.50	187
4900	Low voltage outlets									

170

Important: See the Reference Section for critical supporting data - Reference Nos., Crews, & City Cost Indexes

168 100	Special Systems	CREW	DAILY OUTPUT	LABOR-HOURS	UNIT	1999 BARE COSTS				TOTAL INCL O&P
						MAT.	LABOR	EQUIP.	TOTAL	
170 4910	Telephone recpt., 20' of 4/C phone wire	1 Elec	26	.308	Ea.	6.70	9.80		16.50	22	170
4920	TV recpt., 20' of RG59U coax wire, F type connector	"	16	.500	"	11.40	15.95		27.35	36.50
4950	Door bell chime, transformer, 2 buttons, 60' of bellwire									
4970	Economy model	1 Elec	11.50	.696	Ea.	48.50	22		70.50	86.50
4980	Custom model		11.50	.696		83.50	22		105.50	125
4990	Luxury model, 3 buttons	↓	9.50	.842	↓	228	27		255	291
6000	Lighting outlets									
6050	Wire only (for fixture), type NM cable	1 Elec	32	.250	Ea.	3.95	7.95		11.90	16.30
6070	Type MC cable		24	.333		11.05	10.65		21.70	28
6080	EMT & wire		10	.800		10.90	25.50		36.40	50
6100	Box (4"), and wire (for fixture), type NM cable		25	.320		7.85	10.20		18.05	24
6120	Type MC cable		20	.400		14.95	12.75		27.70	35.50
6130	EMT & wire	↓	11	.727	↓	14.80	23		37.80	51
6200	Fixtures (use with lines 6050 or 6100 above)									
6210	Canopy style, economy grade	1 Elec	40	.200	Ea.	25.50	6.40		31.90	37.50
6220	Custom grade		40	.200		46	6.40		52.40	60
6250	Dining room chandelier, economy grade		19	.421		76	13.45		89.45	104
6260	Custom grade		19	.421		225	13.45		238.45	268
6270	Luxury grade		15	.533		495	17		512	570
6310	Kitchen fixture (fluorescent), economy grade		30	.267		51.50	8.50		60	69
6320	Custom grade		25	.320		161	10.20		171.20	192
6350	Outdoor, wall mounted, economy grade		30	.267		27	8.50		35.50	42
6360	Custom grade		30	.267		101	8.50		109.50	124
6370	Luxury grade		25	.320		227	10.20		237.20	265
6410	Outdoor PAR floodlights, 1 lamp, 150 watt		20	.400		20	12.75		32.75	41
6420	2 lamp, 150 watt each		20	.400		33.50	12.75		46.25	56
6430	For infrared security sensor, add		32	.250		87	7.95		94.95	108
6450	Outdoor, quartz-halogen, 300 watt flood		20	.400		38	12.75		50.75	61
6600	Recessed downlight, round, pre-wired, 50 or 75 watt trim		30	.267		35	8.50		43.50	51
6610	With shower light trim		30	.267		42	8.50		50.50	58.50
6620	With wall washer trim		28	.286		52.50	9.10		61.60	71.50
6630	With eye-ball trim	↓	28	.286		49	9.10		58.10	67.50
6640	For direct contact with insulation, add					1.60			1.60	1.76
6700	Porcelain lamp holder	1 Elec	40	.200		3.50	6.40		9.90	13.40
6710	With pull switch		40	.200		3.75	6.40		10.15	13.70
6750	Fluorescent strip, 1-20 watt tube, wrap around diffuser, 24"		24	.333		51.50	10.65		62.15	72.50
6760	1-40 watt tube, 48"		24	.333		65	10.65		75.65	87.50
6770	2-40 watt tubes, 48"		20	.400		78	12.75		90.75	105
6780	With residential ballast		20	.400		89.50	12.75		102.25	118
6800	Bathroom heat lamp, 1-250 watt		28	.286		30.50	9.10		39.60	47
6810	2-250 watt lamps	↓	28	.286	↓	51	9.10		60.10	69.50
6820	For timer switch, see line 2400									
6900	Outdoor post lamp, incl. post, fixture, 35' of #14/2									
6910	Type NMC cable	1 Elec	3.50	2.286	Ea.	166	73		239	292
6920	Photo-eye, add		27	.296		29	9.45		38.45	46
6950	Clock dial time switch, 24 hr., w/enclosure, type NM cable		11.43	.700		51	22.50		73.50	89.50
6970	Type MC cable		11	.727		60	23		83	101
6980	EMT & wire	↓	4.85	1.649	↓	61	52.50		113.50	146
7000	Alarm systems									
7050	Smoke detectors, box, #14/3, type NM cable	1 Elec	14.55	.550	Ea.	28.50	17.55		46.05	57.50
7070	Type MC cable		12.31	.650		36	20.50		56.50	70.50
7080	EMT & wire	↓	5	1.600		35	51		86	115
7090	For relay output to security system, add				↓	11.75			11.75	12.95
8000	Residential equipment									
8050	Disposal hook-up, incl. switch, outlet box, 3' of flex									
8060	20 amp-1 pole circ. bkr., and 25' of #12/2									

ELECTRICAL 16

168 100	Special Systems	CREW	DAILY OUTPUT	LABOR-HOURS	UNIT	1999 BARE COSTS				TOTAL INCL O&P
						MAT.	LABOR	EQUIP.	TOTAL	
8070	Type NM cable	1 Elec	10	.800	Ea.	19.80	25.50		45.30	60
8080	Type MC cable		8	1		29	32		61	79.50
8090	EMT & wire		5	1.600		32	51		83	112
8100	Trash compactor or dishwasher hook-up, incl. outlet box,									
8110	3' of flex, 15 amp-1 pole circ. bkr., and 25' of #14/2									
8120	Type NM cable	1 Elec	10	.800	Ea.	13.75	25.50		39.25	53
8130	Type MC cable		8	1		25	32		57	75
8140	EMT & wire		5	1.600		26	51		77	105
8150	Hot water sink dispensor hook-up, use line 8100									
8200	Vent/exhaust fan hook-up, type NM cable	1 Elec	32	.250	Ea.	3.95	7.95		11.90	16.30
8220	Type MC cable		24	.333		11.05	10.65		21.70	28
8230	EMT & wire		10	.800		10.90	25.50		36.40	50
8250	Bathroom vent fan, 50 CFM (use with above hook-up)									
8260	Economy model	1 Elec	15	.533	Ea.	21.50	17		38.50	49
8270	Low noise model		15	.533		30	17		47	58.50
8280	Custom model		12	.667		111	21.50		132.50	154
8300	Bathroom or kitchen vent fan, 110 CFM									
8310	Economy model	1 Elec	15	.533	Ea.	53	17		70	84
8320	Low noise model	"	15	.533	"	72	17		89	105
8350	Paddle fan, variable speed (w/o lights)									
8360	Economy model (AC motor)	1 Elec	10	.800	Ea.	94.50	25.50		120	142
8370	Custom model (AC motor)		10	.800		165	25.50		190.50	220
8380	Luxury model (DC motor)		8	1		325	32		357	410
8390	Remote speed switch for above, add		12	.667		21	21.50		42.50	55
8500	Whole house exhaust fan, ceiling mount, 36", variable speed									
8510	Remote switch, incl. shutters, 20 amp-1 pole circ. bkr.									
8520	30' of #12/2, type NM cable	1 Elec	4	2	Ea.	680	64		744	840
8530	Type MC cable		3.50	2.286		690	73		763	870
8540	EMT & wire		3	2.667		695	85		780	885
8600	Whirlpool tub hook-up, incl. timer switch, outlet box									
8610	3' of flex, 20 amp-1 pole GFI circ. bkr.									
8620	30' of #12/2, type NM cable	1 Elec	5	1.600	Ea.	75.50	51		126.50	160
8630	Type MC cable		4.20	1.905		81.50	61		142.50	181
8640	EMT & wire		3.40	2.353		84	75		159	205
8650	Hot water heater hook-up, incl. 1-2 pole circ. bkr., box;									
8660	3' of flex, 20' of #10/2, type NM cable	1 Elec	5	1.600	Ea.	20	51		71	98.50
8670	Type MC cable		4.20	1.905		32	61		93	127
8680	EMT & wire		3.40	2.353		28	75		103	143
9000	Heating/air conditioning									
9050	Furnace/boiler hook-up, incl. firestat, local on-off switch									
9060	Emergency switch, and 40' of type NM cable	1 Elec	4	2	Ea.	42	64		106	142
9070	Type MC cable		3.50	2.286		57.50	73		130.50	172
9080	EMT & wire		1.50	5.333		58.50	170		228.50	320
9100	Air conditioner hook-up, incl. local 60 amp disc. switch									
9110	3' sealtite, 40 amp, 2 pole circuit breaker									
9130	40' of #8/2, type NM cable	1 Elec	3.50	2.286	Ea.	133	73		206	255
9140	Type MC cable		3	2.667		171	85		256	315
9150	EMT & wire		1.30	6.154		141	196		337	450
9200	Heat pump hook-up, 1-40 & 1-100 amp 2 pole circ. bkr.									
9210	Local disconnect switch, 3' sealtite									
9220	40' of #8/2 & 30' of #3/2									
9230	Type NM cable	1 Elec	1.30	6.154	Ea.	340	196		536	670
9240	Type MC cable		1.08	7.407		425	236		661	820
9250	EMT & wire		.94	8.511		340	271		611	780
9500	Thermostat hook-up, using low voltage wire									
9520	Heating only	1 Elec	24	.333	Ea.	6.40	10.65		17.05	23

16 ELECTRICAL

Important: See the Reference Section for critical supporting data - Reference Nos., Crews, & City Cost Indexes

168 100	Special Systems	CREW	DAILY OUTPUT	LABOR-HOURS	UNIT	1999 BARE COSTS				TOTAL INCL O&P	
						MAT.	LABOR	EQUIP.	TOTAL		
170	9530	Heating/cooling	1 Elec	20	.400	Ea.	8	12.75		20.75	28	170

For information about Means Estimating Seminars, see yellow pages 11 and 12 in back of book

ELECTRICAL 16

Division Notes

	CREW	DAILY OUTPUT	LABOR-HOURS	UNIT	1999 BARE COSTS				TOTAL INCL O&P
					MAT.	LABOR	EQUIP.	TOTAL	

Division 17
Square Foot & Cubic Foot Costs

Estimating Tips

- The cost figures in Division 17 were derived from approximately 10,200 projects contained in the Means database of completed construction projects, and include the contractor's overhead and profit, but do not generally include architectural fees or land costs. The figures have been adjusted to January of the current year. New projects are added to our files each year, and outdated projects are discarded. For this reason, certain costs may not show a uniform annual progression. In no case are all subdivisions of a project listed.

- These projects were located throughout the U.S. and reflect a tremendous variation in square foot (S.F.) and cubic foot (C.F.) costs. This is due to differences, not only in labor and material costs, but also in individual owners' requirements. For instance, a bank in a large city would have different features than one in a rural area. This is true of all the different types of buildings analyzed. Therefore, caution should be exercised when using Division 17 costs. For example, for court houses, costs in the database are local court house costs and will not apply to the larger, more elaborate federal court houses. As a general rule, the projects in the 1/4 column do not include any site work or equipment, while the projects in the 3/4 column may include both equipment and site work. The median figures do not generally include site work.

- None of the figures "go with" any others. All individual cost items were computed and tabulated separately. Thus the sum of the median figures for Plumbing, HVAC and Electrical will not normally total up to the total Mechanical and Electrical costs arrived at by separate analysis and tabulation of the projects.

- Each building was analyzed as to total and component costs and percentages. The figures were arranged in ascending order with the results tabulated as shown. The 1/4 column shows that 25% of the projects had lower costs, 75% higher. The 3/4 column shows that 75% of the projects had lower costs, 25% had higher. The median column shows that 50% of the projects had lower costs, 50% had higher.

- There are two times when square foot costs are useful. The first is in the conceptual stage when no details are available. Then square foot costs make a useful starting point. The second is after the bids are in and the costs can be worked back into their appropriate units for information purposes. As soon as details become available in the project design, the square foot approach should be discontinued and the project priced as to its particular components. When more precision is required or for estimating the replacement cost of specific buildings, the current edition of *Means Square Foot Costs* should be used.

- In using the figures in Division 17, it is recommended that the median column be used for preliminary figures if no additional information is available. The median figures, when multiplied by the total city construction cost index figures (see City Cost Indexes) and then multiplied by the project size modifier in Reference Number R171-100, should present a fairly accurate base figure, which would then have to be adjusted in view of the estimator's experience, local economic conditions, code requirements and the owner's particular requirements. There is no need to factor the percentage figures, as these should remain constant from city to city. All tabulations mentioning air conditioning had at least partial air conditioning.

- The editors of this book would greatly appreciate receiving cost figures on one or more of your recent projects which would then be included in the averages for next year. All cost figures received will be kept confidential except that they will be averaged with other similar projects to arrive at S.F. and C.F. cost figures for next year's book. See the last page of the book for details and the discount available for submitting one or more of your projects.

171 000 | S.F. & C.F. Costs

			UNIT	UNIT COSTS			% OF TOTAL		
				1/4	MEDIAN	3/4	1/4	MEDIAN	3/4
010	0010	**APARTMENTS** Low Rise (1 to 3 story) R171-100	S.F.	42	52.60	70.35				**010**
	0020	Total project cost	C.F.	3.76	4.99	6.15				
	0100	Site work	S.F.	3.59	5.25	8.30	6.30%	11%	15.70%	
	0500	Masonry		.78	2.04	3.34	1.50%	4%	6.50%	
	1500	Finishes		4.42	5.75	7.45	9%	10.70%	12.90%	
	1800	Equipment		1.37	2.08	3.01	2.70%	4.10%	6.20%	
	2720	Plumbing		3.24	4.20	5.32	6.70%	9%	10.10%	
	2770	Heating, ventilating, air conditioning		2.08	2.57	3.77	4.20%	5.80%	7.70%	
	2900	Electrical		2.42	3.17	4.32	5.20%	6.70%	8.40%	
	3100	Total: Mechanical & Electrical		8.35	10.65	13.40	16%	18.20%	23%	
	9000	Per apartment unit, total cost	Apt.	38,700	59,000	88,100				
	9500	Total: Mechanical & Electrical	"	7,400	11,600	15,200				
020	0010	**APARTMENTS** Mid Rise (4 to 7 story)	S.F.	54.65	66.40	81.25				**020**
	0020	Total project costs	C.F.	4.35	6	8.20				
	0100	Site work	S.F.	2.23	4.41	7.92	5.20%	6.70%	9.10%	
	0500	Masonry		3.70	5.10	7.25	5.90%	7.60%	10.60%	
	1500	Finishes		7	8.95	11.35	10.50%	13.10%	17.70%	
	1800	Equipment		1.85	2.61	3.43	2.80%	3.50%	4.70%	
	2500	Conveying equipment		1.26	1.55	1.90	2%	2.30%	2.70%	
	2720	Plumbing		3.26	5.25	5.80	6.30%	7.40%	10%	
	2900	Electrical		3.84	5.25	6.10	6.70%	7.50%	8.90%	
	3100	Total: Mechanical & Electrical		11.05	15	18.40	18.80%	21.60%	25.10%	
	9000	Per apartment unit, total cost	Apt.	63,000	74,300	123,000				
	9500	Total: Mechanical & Electrical	"	11,900	13,800	17,400				
030	0010	**APARTMENTS** High Rise (8 to 24 story)	S.F.	63.10	76.20	93.10				**030**
	0020	Total project costs	C.F.	5.35	7.50	9.10				
	0100	Site work	S.F.	1.93	3.70	5.20	2.60%	4.80%	6.20%	
	0500	Masonry		3.65	6.65	8.45	4.70%	9.60%	11.10%	
	1500	Finishes		7	8.75	10.35	9.70%	11.80%	13.70%	
	1800	Equipment		2.03	2.50	3.30	2.80%	3.50%	4.30%	
	2500	Conveying equipment		1.44	2.18	3.11	2.20%	2.80%	3.40%	
	2720	Plumbing		4.65	5.50	6.90	7%	9.10%	10.60%	
	2900	Electrical		4.34	5.50	7.40	6.40%	7.70%	8.90%	
	3100	Total: Mechanical & Electrical		13	16.60	19.95	18%	22.30%	24.40%	
	9000	Per apartment unit, total cost	Apt.	65,300	70,900	100,200				
	9500	Total: Mechanical & Electrical	"	14,200	16,200	17,200				
040	0010	**AUDITORIUMS**	S.F.	64.95	87.95	119				**040**
	0020	Total project costs	C.F.	4.11	5.75	8.60				
	2720	Plumbing	S.F.	4.15	5.75	7.25	6.30%	7.20%	9.05%	
	2900	Electrical		5.30	7.30	9.15	6.80%	9%	10.60%	
	3100	Total: Mechanical & Electrical		10.10	13.65	23.70	14.40%	18.50%	23.60%	
050	0010	**AUTOMOTIVE SALES**	S.F.	47.30	54.40	80.15				**050**
	0020	Total project costs	C.F.	3.15	3.84	4.93				
	2720	Plumbing	S.F.	2.37	3.84	4.19	4.70%	6.50%	7%	
	2770	Heating, ventilating, air conditioning		3.42	5.30	5.65	6.40%	10.30%	10.40%	
	2900	Electrical		3.77	5.85	6.70	7.30%	10%	12.40%	
	3100	Total: Mechanical & Electrical		8.50	12.25	15.35	17%	20.40%	26%	
060	0010	**BANKS**	S.F.	94.20	118	149				**060**
	0020	Total project costs	C.F.	6.80	9.25	12.10				
	0100	Site work	S.F.	10.25	17.50	25.65	7.50%	13.90%	17.60%	
	0500	Masonry		4.81	8.20	17.85	2.90%	6.20%	11.40%	
	1500	Finishes		8.10	11.25	14.50	5.50%	7.70%	10.30%	
	1800	Equipment		3.80	7.90	17.60	3.30%	7.70%	12.50%	
	2720	Plumbing		3	4.29	6.25	2.80%	3.90%	4.90%	
	2770	Heating, ventilating, air conditioning		5.80	7.60	10.10	5%	7.20%	8.50%	
	2900	Electrical		9.15	12.05	15.70	8.30%	10.30%	12.20%	
	3100	Total: Mechanical & Electrical		21.65	29.15	35.35	16.60%	20.40%	24.40%	
	3500	See also division 110-301								

171 000 | S.F. & C.F. Costs

			UNIT	UNIT COSTS 1/4	UNIT COSTS MEDIAN	UNIT COSTS 3/4	% OF TOTAL 1/4	% OF TOTAL MEDIAN	% OF TOTAL 3/4	
130	0010	**CHURCHES** R171-100	S.F.	63.25	79.45	103				130
	0020	Total project costs	C.F.	3.97	5	6.60				
	1800	Equipment	S.F.	.77	1.82	4	1%	2.20%	4.70%	
	2720	Plumbing		2.49	3.49	5.15	3.60%	5%	6.30%	
	2770	Heating, ventilating, air conditioning		5.85	7.60	11.20	7.60%	10%	12.20%	
	2900	Electrical		5.25	7.25	9.40	7.30%	8.80%	10.90%	
	3100	Total: Mechanical & Electrical	↓	13.65	20.50	29.80	19.60%	22.90%	25.40%	
	3500	See also division 110-401								
150	0010	**CLUBS, COUNTRY**	S.F.	66.75	79.20	104				150
	0020	Total project costs	C.F.	5.55	6.75	9.35				
	2720	Plumbing	S.F.	4.54	6.15	14.05	6.40%	9.60%	10%	
	2900	Electrical		5.40	7.75	10.20	7%	9.30%	11.40%	
	3100	Total: Mechanical & Electrical	↓	11.70	15.05	27.95	16.10%	22.30%	30.90%	
170	0010	**CLUBS, SOCIAL** Fraternal	S.F.	53.95	77.05	100.90				170
	0020	Total project costs	C.F.	3.25	5.20	6.05				
	2720	Plumbing	S.F.	3.10	4.30	5.20	5.20%	6.80%	8.30%	
	2770	Heating, ventilating, air conditioning		4.98	6	7.75	8.70%	11%	14.40%	
	2900	Electrical		4.36	6.80	8.20	7.30%	9.50%	11.40%	
	3100	Total: Mechanical & Electrical	↓	11.60	15.75	22.50	19.30%	23%	33.10%	
180	0010	**CLUBS, Y.M.C.A.**	S.F.	62.85	79.40	106				180
	0020	Total project costs	C.F.	3.17	5.30	7.90				
	2720	Plumbing	S.F.	4.58	8.65	9.70	6%	9.70%	16%	
	2900	Electrical		5.25	6.90	10.15	6.40%	9.20%	10.80%	
	3100	Total: Mechanical & Electrical	↓	11.55	16.45	24.45	16.20%	21.90%	28.50%	
190	0010	**COLLEGES** Classrooms & Administration	S.F.	78.60	104	137				190
	0020	Total project costs	C.F.	5.70	8	12.95				
	0500	Masonry	S.F.	5	10.05	11.20	5.10%	11.90%	12.10%	
	2720	Plumbing		3.85	7.55	14	4.80%	8.90%	11.80%	
	2900	Electrical		6.45	9.95	12.25	8.40%	10%	12.10%	
	3100	Total: Mechanical & Electrical	↓	13.80	25.25	39.10	14.20%	24.70%	32.50%	
210	0010	**COLLEGES** Science, Engineering, Laboratories	S.F.	125	151	184				210
	0020	Total project costs	C.F.	7.40	10.85	12.30				
	1800	Equipment	S.F.	7.20	16.30	17.85	5.30%	9.70%	15%	
	2900	Electrical		10.70	15.45	23.30	7.10%	9.60%	15.40%	
	3100	Total: Mechanical & Electrical	↓	39.45	47	72.75	28%	31.60%	39.80%	
	3500	See also division 116-001								
230	0010	**COLLEGES** Student Unions	S.F.	82.50	116	136				230
	0020	Total project costs	C.F.	4.60	6.05	7.80				
	3100	Total: Mechanical & Electrical	S.F.	21.70	31	33.60	17.40%	23.90%	28.90%	
250	0010	**COMMUNITY CENTERS**	"	61.65	82.80	110				250
	0020	Total project costs	C.F.	4.45	6.35	8.25				
	1800	Equipment	S.F.	1.71	2.90	4.75	2%	3.30%	6.40%	
	2720	Plumbing		3.31	5.65	8.25	4.90%	7.10%	9.50%	
	2770	Heating, ventilating, air conditioning		5.45	7.85	10.80	7.70%	10.80%	13%	
	2900	Electrical		5.45	7.40	11.20	7.40%	9.40%	11.50%	
	3100	Total: Mechanical & Electrical	↓	20.20	24.15	34.80	25.10%	27.90%	34.40%	
280	0010	**COURT HOUSES**	S.F.	97.50	112	130				280
	0020	Total project costs	C.F.	7.45	9	12.50				
	2720	Plumbing	S.F.	4.68	6.55	9.40	5.90%	7.40%	9.60%	
	2900	Electrical		8.60	10.65	13.75	8.40%	9.80%	11.60%	
	3100	Total: Mechanical & Electrical	↓	19.35	26.95	34.10	20.10%	25.70%	28.70%	
300	0010	**DEPARTMENT STORES**	S.F.	36.35	49.15	62.05				300
	0020	Total project costs	C.F.	1.91	2.52	3.43				
	2720	Plumbing	S.F.	1.18	1.43	2.17	3.10%	4.20%	5.90%	
	2770	Heating, ventilating, air conditioning	↓	3.30	5.10	7.70	8.30%	12.40%	14.80%	

SQUARE FOOT 17

171 000 | S.F. & C.F. Costs

			UNIT	UNIT COSTS			% OF TOTAL		
				1/4	MEDIAN	3/4	1/4	MEDIAN	3/4
300	2900	Electrical	R171 -100	S.F.	4.17	5.70	6.75	9.10%	12.20%	14.90%	**300**
	3100	Total: Mechanical & Electrical	↓		7.35	9.40	15.40	20%	19%	26.50%	
310	0010	**DORMITORIES** Low Rise (1 to 3 story)		S.F.	59.40	84.85	102				**310**
	0020	Total project costs		C.F.	4.12	6.30	9.45				
	2720	Plumbing		S.F.	4.15	5.55	7	8%	9%	9.60%	
	2770	Heating, ventilating, air conditioning			4.38	5.25	7	4.60%	8%	10%	
	2900	Electrical			4.38	6.65	8.40	6.60%	8.90%	9.60%	
	3100	Total: Mechanical & Electrical	↓		23	23.75	24.15	22.50%	24.10%	27.10%	
	9000	Per bed, total cost		Bed	29,200	32,400	69,600				
320	0010	**DORMITORIES** Mid Rise (4 to 8 story)		S.F.	84.65	110	136				**320**
	0020	Total project costs		C.F.	9.35	10.25	12.25				
	2900	Electrical		S.F.	6.25	9.85	12.60	7.40%	8.90%	10.40%	
	3100	Total: Mechanical & Electrical		"	17.70	24.15	31.55	18.10%	22.20%	30.60%	
	9000	Per bed, total cost		Bed	12,000	27,500	56,500				
340	0010	**FACTORIES**		S.F.	32.30	47.65	73.70				**340**
	0020	Total project costs		C.F.	2.06	3.05	5.10				
	0100	Site work		S.F.	3.65	6.65	10.90	5%	10.70%	18.20%	
	2720	Plumbing			1.90	3.21	5.45	4.60%	6.20%	8.70%	
	2770	Heating, ventilating, air conditioning			3.36	4.82	6.50	5.20%	8.50%	11.40%	
	2900	Electrical			3.93	6.25	9.60	8.10%	10.60%	14.20%	
	3100	Total: Mechanical & Electrical	↓		9.45	15.20	23.10	20.40%	28.10%	34.50%	
360	0010	**FIRE STATIONS**		S.F.	61.90	83.15	107				**360**
	0020	Total project costs		C.F.	3.72	5.05	6.73				
	0500	Masonry		S.F.	9.35	16.75	22.15	8.70%	14.10%	19.30%	
	1140	Roofing			2.07	5.60	6.40	1.90%	4.90%	5%	
	1580	Painting			1.61	2.40	2.46	1.40%	2.10%	2.20%	
	1800	Equipment			1.37	1.91	4.58	1.10%	2.50%	4.80%	
	2720	Plumbing			4.03	6	8.75	6.20%	7.70%	9.70%	
	2770	Heating, ventilating, air conditioning			3.47	5.70	8.70	4.90%	7.50%	9.30%	
	2900	Electrical			4.49	7.60	10.60	7%	9.20%	11.30%	
	3100	Total: Mechanical & Electrical	↓		19.55	24.15	28.60	18.80%	23.80%	27.30%	
370	0010	**FRATERNITY HOUSES** and Sorority Houses		S.F.	63.55	81.80	115				**370**
	0020	Total project costs		C.F.	6.10	6.60	8.45				
	2720	Plumbing		S.F.	4.76	5.50	10.05	5.20%	8%	10.90%	
	2900	Electrical			4.19	9.05	11.10	6.60%	9.90%	10.70%	
	3100	Total: Mechanical & Electrical	↓		11.60	16.45	19.80	14.60%	19.90%	20.70%	
380	0010	**FUNERAL HOMES**		S.F.	67.05	91.45	166				**380**
	0020	Total project costs		C.F.	6.85	7.60	14.60				
	2900	Electrical		S.F.	2.96	5.45	6.40	3.60%	4.40%	11%	
	3100	Total: Mechanical & Electrical		"	10.65	15.90	20.60	12.90%	17.80%	18.80%	
390	0010	**GARAGES, COMMERCIAL** (Service)		S.F.	36.60	58.10	79.90				**390**
	0020	Total project costs		C.F.	2.32	3.48	5				
	1800	Equipment		S.F.	2.20	4.81	7.50	3.70%	6.80%	8.30%	
	2720	Plumbing			2.47	3.71	7.40	4.70%	7.40%	11%	
	2730	Heating & ventilating			3.45	4.70	6.50	5.30%	6.90%	9.60%	
	2900	Electrical			3.49	5.35	7.40	7.20%	9%	11.20%	
	3100	Total: Mechanical & Electrical	↓		8.20	15.25	22.50	14.10%	22.50%	26.90%	
400	0010	**GARAGES, MUNICIPAL** (Repair)		S.F.	55.80	74.30	106				**400**
	0020	Total project costs		C.F.	3.48	4.42	7.20				
	0500	Masonry		S.F.	5.25	10.25	15.90	6%	10%	15.50%	
	2720	Plumbing			2.50	4.80	9.05	3.60%	6.70%	8%	
	2730	Heating & ventilating			4.28	6.20	11.95	6.20%	7.40%	13.50%	
	2900	Electrical			4.12	6.20	9.35	6.30%	8%	11.70%	
	3100	Total: Mechanical & Electrical	↓		10.50	18.90	31	15.20%	25.50%	32.70%	

17 SQUARE FOOT

171 000 | S.F. & C.F. Costs

			UNIT	UNIT COSTS			% OF TOTAL			
				1/4	MEDIAN	3/4	1/4	MEDIAN	3/4	
410	0010	**GARAGES, PARKING** R171-100	S.F.	21	30.45	54.45				410
	0020	Total project costs	C.F.	2.04	2.77	4.12				
	2720	Plumbing	S.F.	.56	.95	1.47	2.20%	3.40%	3.90%	
	2900	Electrical		.89	1.34	1.98	4.30%	5.20%	6.30%	
	3100	Total: Mechanical & Electrical	↓	1.18	3.48	4.46	6.50%	9.40%	12.80%	
	3200									
	9000	Per car, total cost	Car	9,100	11,500	14,900				
430	0010	**GYMNASIUMS**	S.F.	60.20	76.85	98.05				430
	0020	Total project costs	C.F.	3.01	3.84	5				
	1800	Equipment	S.F.	1.27	2.59	5.15	2.10%	3.40%	6.70%	
	2720	Plumbing		3.82	4.71	5.75	5.40%	7.30%	7.90%	
	2770	Heating, ventilating, air conditioning		4.11	6.20	12.55	9%	11.10%	22.60%	
	2900	Electrical		4.58	5.75	7.80	6.60%	8.30%	10.70%	
	3100	Total: Mechanical & Electrical	↓	14.15	19.50	26.15	20.60%	26.20%	29.40%	
	3500	See also division 114-801 & 114-805								
460	0010	**HOSPITALS**	S.F.	125	145	215				460
	0020	Total project costs	C.F.	8.80	10.85	15.65				
	1800	Equipment	S.F.	3.02	5.65	9.75	2.50%	3.80%	6%	
	2720	Plumbing		10.35	13.95	18.35	7.80%	9.40%	11.80%	
	2770	Heating, ventilating, air conditioning		14.65	19.45	26.20	8.40%	14.60%	17%	
	2900	Electrical		12.45	16.20	25.50	9.90%	12%	14.50%	
	3100	Total: Mechanical & Electrical	↓	35.50	47.05	77.10	26.60%	33.10%	39%	
	9000	Per bed or person, total cost	Bed	44,700	94,900	137,800				
	9900	See also division 117-001								
480	0010	**HOUSING** For the Elderly	S.F.	57	71.95	88.35				480
	0020	Total project costs	C.F.	4.04	5.65	7.20				
	0100	Site work	S.F.	3.97	6.40	9.05	5.10%	8.20%	12.10%	
	0500	Masonry		1.36	6.10	9.50	2.10%	7.10%	12.20%	
	1800	Equipment		1.37	1.86	3.02	1.90%	3.20%	4.40%	
	2510	Conveying systems		1.39	1.86	2.53	1.80%	2.30%	2.90%	
	2720	Plumbing		4.23	5.65	7.45	8.10%	9.70%	10.90%	
	2730	Heating, ventilating, air conditioning		2.17	3.08	4.60	3.30%	5.60%	7.20%	
	2900	Electrical		4.25	5.70	7.45	7.30%	8.50%	10.50%	
	3100	Total: Mechanical & Electrical	↓	14.60	17.15	22.70	18.10%	22%	29.10%	
	9000	Per rental unit, total cost	Unit	53,000	61,700	68,900				
	9500	Total: Mechanical & Electrical	"	11,700	12,600	15,800				
500	0010	**HOUSING** Public (Low Rise)	S.F.	49.55	66.60	86.60				500
	0020	Total project costs	C.F.	3.80	5.30	6.60				
	0100	Site work	S.F.	6.10	8.80	14.25	8.40%	11.70%	16.50%	
	1800	Equipment		1.30	2.13	3.48	2.30%	3%	5.10%	
	2720	Plumbing		3.25	4.56	5.80	6.80%	9%	11.60%	
	2730	Heating, ventilating, air conditioning		1.74	3.37	3.69	4.20%	6%	6.40%	
	2900	Electrical		2.90	4.32	6	5.10%	6.60%	8.30%	
	3100	Total: Mechanical & Electrical	↓	13.75	17.70	19.85	14.50%	17.60%	26.50%	
	9000	Per apartment, total cost	Apt.	52,600	59,800	75,100				
	9500	Total: Mechanical & Electrical	"	9,600	12,300	14,300				
510	0010	**ICE SKATING RINKS**	S.F.	43.10	74.75	105				510
	0020	Total project costs	C.F.	3.01	3.08	3.55				
	2720	Plumbing	S.F.	1.53	2.87	2.93	3.10%	5.60%	6.70%	
	2900	Electrical		4.38	6.70	7.10	6.70%	15%	15.80%	
	3100	Total: Mechanical & Electrical	↓	7.55	10.90	13.60	9.90%	25.90%	29.80%	
520	0010	**JAILS**	S.F.	126	161	208				520
	0020	Total project costs	C.F.	11.75	14.40	18.40				
	1800	Equipment	S.F.	5.20	14.40	24.45	4%	9.40%	15.20%	
	2720	Plumbing		12.70	16.10	21.25	7%	8.90%	13.80%	
	2770	Heating, ventilating, air conditioning		11.25	15	29	7.50%	9.40%	17.70%	
	2900	Electrical	↓	13.70	17.60	21.80	8.10%	11.40%	12.40%	

SQUARE FOOT 17

171 000 | S.F. & C.F. Costs

			UNIT	UNIT COSTS			% OF TOTAL		
				1/4	MEDIAN	3/4	1/4	MEDIAN	3/4
520	3100	Total: Mechanical & Electrical R171-100	S.F.	35	62.10	73.50	29.20%	31.10%	34.10%	520
530	0010	**LIBRARIES**	S.F.	74.35	94.70	121				530
	0020	Total project costs	C.F.	5.20	6.50	8.45			
	0500	Masonry	S.F.	4.25	9.50	16.35	5.80%	9.50%	11.90%
	1800	Equipment		1.08	2.90	4.50	1.20%	2.80%	4.50%
	2720	Plumbing		3.01	4.23	5.75	3.60%	4.90%	5.70%
	2770	Heating, ventilating, air conditioning		6.45	10.90	14.20	8%	11%	14.60%
	2900	Electrical		7.65	10.10	12.50	8.30%	11%	12.10%
	3100	Total: Mechanical & Electrical		22.05	30.15	37.75	18.90%	25.30%	27.60%
550	0010	**MEDICAL CLINICS**	S.F.	73.15	90.60	114				550
	0020	Total project costs	C.F.	5.45	7.05	9.40			
	1800	Equipment	S.F.	2	4.21	6.55	1.80%	5.20%	7.40%
	2720	Plumbing		4.93	6.95	9.30	6.10%	8.40%	10%
	2770	Heating, ventilating, air conditioning		6.10	7.70	11.30	6.70%	9%	11.30%
	2900	Electrical		6.20	8.85	11.75	8.10%	10%	12.20%
	3100	Total: Mechanical & Electrical		19.55	27.50	38.45	22%	27.60%	34.30%
	3500	See also division 117-001							
570	0010	**MEDICAL OFFICES**	S.F.	68.65	85.10	105				570
	0020	Total project costs	C.F.	5.10	7	9.60			
	1800	Equipment	S.F.	2.30	4.55	6.50	3%	5.80%	7.20%
	2720	Plumbing		3.85	5.95	8.05	5.70%	6.80%	8.60%
	2770	Heating, ventilating, air conditioning		4.66	6.85	8.90	6.20%	8%	9.70%
	2900	Electrical		5.45	7.95	11.05	7.60%	9.80%	11.40%
	3100	Total: Mechanical & Electrical		13.55	19.45	28.90	18.50%	22%	24.90%
590	0010	**MOTELS**	S.F.	44.05	65.20	84.10				590
	0020	Total project costs	C.F.	3.84	5.40	8.85			
	2720	Plumbing	S.F.	4.46	5.70	6.80	9.40%	10.60%	12.50%
	2770	Heating, ventilating, air conditioning		2.72	4.05	7.25	5.60%	5.60%	10%
	2900	Electrical		4.16	5.30	6.90	7.10%	8.20%	10.40%
	3100	Total: Mechanical & Electrical		14.15	17.70	30.35	18.50%	21%	24.40%
	5000								
	9000	Per rental unit, total cost	Unit	22,400	42,600	46,000			
	9500	Total: Mechanical & Electrical	"	4,400	6,600	7,700			
600	0010	**NURSING HOMES**	S.F.	66.15	87.50	107				600
	0020	Total project costs	C.F.	5.30	6.95	9.25			
	1800	Equipment	S.F.	2.22	2.96	4.77	2.40%	3.70%	6%
	2720	Plumbing		6.25	7.95	11	9.40%	10.70%	14.20%
	2770	Heating, ventilating, air conditioning		6.20	8.60	11	9.30%	11.40%	11.80%
	2900	Electrical		6.85	8.55	11.50	9.70%	11%	13%
	3100	Total: Mechanical & Electrical		16.30	22.80	33.40	26%	29.90%	30.50%
	3200								
	9000	Per bed or person, total cost	Bed	28,700	35,200	47,000			
610	0010	**OFFICES** Low Rise (1 to 4 story)	S.F.	55.75	71	94.55				610
	0020	Total project costs	C.F.	4.04	5.65	7.65			
	0100	Site work	S.F.	4.20	7.15	11.10	5.30%	9.70%	14%
	0500	Masonry		1.94	4.53	8.55	2.90%	5.80%	8.70%
	1800	Equipment		.70	1.26	3.47	1.20%	1.50%	4%
	2720	Plumbing		2.12	3.21	4.55	3.70%	4.50%	6.10%
	2770	Heating, ventilating, air conditioning		4.58	6.35	9.40	7.20%	10.50%	11.90%
	2900	Electrical		4.72	6.55	9.15	7.50%	9.60%	11.10%
	3100	Total: Mechanical & Electrical		11.10	15.40	22.55	18%	21.80%	26.50%
620	0010	**OFFICES** Mid Rise (5 to 10 story)	S.F.	61.50	74.55	101				620
	0020	Total project costs	C.F.	4.30	5.45	7.90			
	2720	Plumbing	S.F.	1.86	2.88	4.14	2.80%	3.70%	4.50%
	2770	Heating, ventilating, air conditioning		4.67	6.70	10.65	7.60%	9.40%	11%

See R171-100 and City Cost Indexes in the Reference Section

17 SQUARE FOOT

171 000 | S.F. & C.F. Costs

			UNIT	UNIT COSTS			% OF TOTAL			
				1/4	MEDIAN	3/4	1/4	MEDIAN	3/4	
620	2900	Electrical	S.F. R171-100	4.57	5.85	8.85	6.50%	8.20%	10%	620
	3100	Total: Mechanical & Electrical		11.65	14.90	29.55	17.90%	22.30%	29.90%	
630	0010	**OFFICES** High Rise (11 to 20 story)	S.F.	74.45	95.40	117				630
	0020	Total project costs	C.F.	4.75	6.60	9.50				
	2900	Electrical	S.F.	3.80	5.50	8.30	5.80%	7%	10.50%	
	3100	Total: Mechanical & Electrical	"	13.90	17.80	32.55	15.40%	17%	34.10%	
640	0010	**POLICE STATIONS**	S.F.	89.15	119	150				640
	0020	Total project costs	C.F.	7.20	8.85	12.10				
	0500	Masonry	S.F.	10.60	15.65	18.85	7.80%	10.60%	11.30%	
	1800	Equipment		1.47	6.35	10.20	2.10%	5.20%	9.80%	
	2720	Plumbing		5.07	8.20	11.45	5.60%	6.90%	11.30%	
	2770	Heating, ventilating, air conditioning		7.90	10.50	14.30	7%	10.70%	12%	
	2900	Electrical		9.90	14.80	18.80	9.70%	11.90%	14.90%	
	3100	Total: Mechanical & Electrical		31.50	37.60	52.90	25.10%	31.30%	36.40%	
650	0010	**POST OFFICES**	S.F.	69.45	87.95	110				650
	0020	Total project costs	C.F.	4.24	5.45	6.45				
	2720	Plumbing	S.F.	3.22	4.13	5.05	4.40%	5.60%	5.60%	
	2770	Heating, ventilating, air conditioning		5.04	6.20	6.90	7%	8.40%	10.20%	
	2900	Electrical		5.90	8.20	9.65	7.20%	9.60%	11%	
	3100	Total: Mechanical & Electrical		10.35	20.05	25.30	16.40%	20.40%	22.30%	
660	0010	**POWER PLANTS**	S.F.	528	673	1,208				660
	0020	Total project costs	C.F.	13.65	27.65	63.75				
	2900	Electrical	S.F.	35.10	74.45	111	9.50%	12.80%	18.40%	
	8100	Total: Mechanical & Electrical	"	76.80	166	370	28.80%	32.50%	52.60%	
670	0010	**RELIGIOUS EDUCATION**	S.F.	56.10	72.85	84.95				670
	0020	Total project costs	C.F.	3.21	4.59	5.95				
	2720	Plumbing	S.F.	2.40	3.41	4.81	4.30%	4.90%	8.40%	
	2770	Heating, ventilating, air conditioning		5.65	6.35	7	9.80%	10%	11.80%	
	2900	Electrical		4.58	6.05	8	7.90%	9.10%	10.30%	
	3100	Total: Mechanical & Electrical		11.85	20.55	24.40	17.60%	20.50%	22%	
690	0010	**RESEARCH** Laboratories and facilities	S.F.	86.20	124	180				690
	0020	Total project costs	C.F.	6.45	10.80	14.55				
	1800	Equipment	S.F.	3.83	7.55	18.40	1.60%	5.20%	9.30%	
	2720	Plumbing		8.65	11.35	17.80	7.30%	9.20%	13.30%	
	2770	Heating, ventilating, air conditioning		7.75	26.10	30.90	7.20%	16.50%	17.50%	
	2900	Electrical		9.95	16.65	28.40	9.60%	12%	15.80%	
	3100	Total: Mechanical & Electrical		22.90	43.50	82.90	23.90%	35.60%	41.70%	
700	0010	**RESTAURANTS**	S.F.	83.05	107	139				700
	0020	Total project costs	C.F.	6.95	9.15	12.10				
	1800	Equipment	S.F.	4.97	13.30	20.10	6.10%	13.40%	15.60%	
	2720	Plumbing		6.60	8.25	11.45	6.10%	8.30%	9.10%	
	2770	Heating, ventilating, air conditioning		8.75	11.65	15.20	9.40%	12.20%	13%	
	2900	Electrical		8.85	10.90	14.20	8.50%	10.60%	11.60%	
	3100	Total: Mechanical & Electrical		27.70	28.90	36.55	21.10%	24%	29.50%	
	5000									
	9000	Per seat unit, total cost	Seat	3,075	4,100	5,200				
	9500	Total: Mechanical & Electrical	"	645	918	1,207				
720	0010	**RETAIL STORES**	S.F.	38.75	52.25	68.65				720
	0020	Total project costs	C.F.	2.64	3.77	5.20				
	2720	Plumbing	S.F.	1.42	2.37	4.04	3.20%	4.60%	6.80%	
	2770	Heating, ventilating, air conditioning		3.06	4.19	6.30	6.70%	8.70%	10.10%	
	2900	Electrical		3.50	4.74	6.90	7.30%	9.90%	11.60%	
	3100	Total: Mechanical & Electrical		9.35	12	15.55	17.10%	21.40%	23.80%	
740	0010	**SCHOOLS** Elementary	S.F.	61.70	76.10	91.70				740
	0020	Total project costs	C.F.	4.18	5.35	6.90				
	0500	Masonry	S.F.	5.70	9	12.60	7.10%	10.50%	15.20%	
	1800	Equipment		1.95	3.29	5.80	2.50%	4.20%	7.60%	
	2720	Plumbing		3.57	5.20	6.95	5.70%	7.10%	9.40%	
	2730	Heating, ventilating, air conditioning		5.50	8.75	12.20	8.10%	10.80%	15.20%	

SQUARE FOOT **17**

For expanded coverage of these items see *Means Square Foot Costs 1999*

171 000 | S.F. & C.F. Costs

			UNIT	UNIT COSTS			% OF TOTAL			
				1/4	MEDIAN	3/4	1/4	MEDIAN	3/4	
740	2900	Electrical	S.F.	5.75	7.35	9.40	8.40%	10%	11.60%	740
	3100	Total: Mechanical & Electrical	↓	19.60	22.05	31.10	22.20%	28%	30.40%	
	9000	Per pupil, total cost	Ea.	6,150	10,200	33,200				
	9500	Total: Mechanical & Electrical	"	2,075	2,625	10,500				
760	0010	SCHOOLS Junior High & Middle	S.F.	62.65	77.50	90.80				760
	0020	Total project costs	C.F.	4.12	5.40	6.05				
	0500	Masonry	S.F.	7.05	9.45	11.60	8.80%	11.10%	14%	
	1800	Equipment		2.10	3.38	5.25	2.60%	4.30%	5.90%	
	2720	Plumbing		4.17	4.72	6.15	5.60%	6.90%	8.10%	
	2770	Heating, ventilating, air conditioning		4.80	9.25	12.95	8.70%	12.70%	17.40%	
	2900	Electrical		6.05	7.60	9.15	7.80%	9.40%	10.60%	
	3100	Total: Mechanical & Electrical	↓	17.50	22	29.20	23.30%	25.70%	27.30%	
	9000	Per pupil, total cost	Ea.	7,900	9,000	11,300				
780	0010	SCHOOLS Senior High	S.F.	67.65	77.50	109				780
	0020	Total project costs	C.F.	4.33	5.65	7.95				
	1800	Equipment	S.F.	1.80	4.24	6.25	2.30%	3.70%	5.40%	
	2720	Plumbing		3.50	6.40	10.55	5%	6.90%	12%	
	2770	Heating, ventilating, air conditioning		7.80	8.90	17.10	8.90%	11%	15%	
	2900	Electrical		6.65	8.75	14.55	8.30%	10.10%	12.60%	
	3100	Total: Mechanical & Electrical	↓	18.20	23.95	44.75	19.80%	23.10%	27.80%	
	9000	Per pupil, total cost	Ea.	6,450	11,000	16,400				
800	0010	SCHOOLS Vocational	S.F.	55.25	77.20	98.25				800
	0020	Total project costs	C.F.	3.38	4.90	6.85				
	0500	Masonry	S.F.	3.25	8.05	12.25	4%	10.90%	19.90%	
	1800	Equipment	"	1.73	2.39	5.95	2.90%	3.40%	4.70%	
	2720	Plumbing	S.F.	3.54	5.40	7.75	5.40%	7%	8.50%	
	2770	Heating, ventilating, air conditioning		5.10	9.20	15.40	8.80%	11.90%	14.60%	
	2900	Electrical		5.70	7.90	11.15	8.40%	11.20%	13.80%	
	3100	Total: Mechanical & Electrical	↓	15	20	26.40	21.70%	27.30%	34.70%	
	9000	Per pupil, total cost	Ea.	7,700	20,600	30,800				
830	0010	SPORTS ARENAS	S.F.	48.30	64.70	94.90				830
	0020	Total project costs	C.F.	2.63	4.67	6.05				
	2720	Plumbing	S.F.	2.50	4.17	7.85	4.50%	6.30%	9.40%	
	2770	Heating, ventilating, air conditioning		5.15	7.15	9.30	5.80%	10.20%	13.50%	
	2900	Electrical		4.16	6.55	8.20	7.70%	9.80%	12.30%	
	3100	Total: Mechanical & Electrical	↓	12.25	21.75	28	13.40%	22.50%	30.80%	
850	0010	SUPERMARKETS	S.F.	44.65	52.45	60.80				850
	0020	Total project costs	C.F.	2.49	3.01	4.56				
	2720	Plumbing	S.F.	2.50	3.15	3.65	5.40%	6%	7.50%	
	2770	Heating, ventilating, air conditioning		3.68	4.51	5.35	8.60%	8.60%	9.60%	
	2900	Electrical		5.25	6.45	7.65	10.40%	12.40%	13.60%	
	3100	Total: Mechanical & Electrical	↓	14.60	15.80	21.85	17.40%	20.40%	28.40%	
860	0010	SWIMMING POOLS	S.F.	78.95	121	169				860
	0020	Total project costs	C.F.	5.80	7.25	7.90				
	2720	Plumbing	S.F.	6.70	7.65	10.65	9.70%	12.30%	12.30%	
	2900	Electrical		5.45	8.45	11	7.50%	8%	8%	
	3100	Total: Mechanical & Electrical	↓	11.60	24.45	45.30	17.70%	24.90%	31.60%	
870	0010	TELEPHONE EXCHANGES	S.F.	96.20	141	180				870
	0020	Total project costs	C.F.	5.95	9.10	13.20				
	2720	Plumbing	S.F.	4.06	6.25	9.15	4.50%	5.80%	6.90%	
	2770	Heating, ventilating, air conditioning		8.25	18.55	23.55	11.80%	11.80%	18.40%	
	2900	Electrical		9.80	15.15	27.45	10.90%	14%	17.80%	
	3100	Total: Mechanical & Electrical	↓	21	28.90	71.70	27.30%	33.40%	44.40%	
910	0010	THEATERS	S.F.	60.50	77.60	114				910
	0020	Total project costs	C.F.	2.79	4.13	6.05				
	2720	Plumbing	S.F.	1.88	2.18	6.55	2.90%	4.70%	6.80%	
	2770	Heating, ventilating, air conditioning	↓	5.85	7.10	8.80	8%	12.20%	13.40%	

R171 -100

17 SQUARE FOOT

See R171-100 and City Cost Indexes in the Reference Section

171 000	S.F. & C.F. Costs		UNIT	UNIT COSTS			% OF TOTAL			
				1/4	MEDIAN	3/4	1/4	MEDIAN	3/4	
910	2900	Electrical	R171 -100	S.F.	5.30	7.15	14.55	8%	10%	12.40%	**910**
	3100	Total: Mechanical & Electrical		↓	13.60	20.35	41.75	22.90%	26.60%	27.50%	
940	0010	**TOWN HALLS** City Halls & Municipal Buildings		S.F.	67.50	85.35	111				**940**
	0020	Total project costs		C.F.	5.05	7.05	9.45				
	2720	Plumbing		S.F.	2.40	4.95	8.25	4.20%	6.10%	7.90%	
	2770	Heating, ventilating, air conditioning			5.10	10.10	14.80	7%	9%	13.50%	
	2900	Electrical			6.05	8.15	11.55	8.20%	9.50%	11.70%	
	3100	Total: Mechanical & Electrical		↓	21.40	22.25	28.20	22%	25.10%	30.40%	
970	0010	**WAREHOUSES** And Storage Buildings		S.F.	25.15	35.15	54.15				**970**
	0020	Total project costs		C.F.	1.27	2.06	3.42				
	0100	Site work		S.F.	2.52	5.15	7.80	6%	12.80%	19.80%	
	0500	Masonry			1.57	3.58	7.70	5%	7.70%	13.20%	
	1800	Equipment			.41	.87	4.89	1%	2.40%	6.50%	
	2720	Plumbing			.84	1.51	2.82	2.90%	4.80%	6.60%	
	2730	Heating, ventilating, air conditioning			.96	2.70	3.63	2.40%	5%	8.90%	
	2900	Electrical			1.49	2.81	4.65	5%	7.30%	10%	
	3100	Total: Mechanical & Electrical		↓	4.16	6.10	13.95	12.80%	18.40%	26.20%	
990	0010	**WAREHOUSE & OFFICES** Combination		S.F.	30.30	40.60	53.90				**990**
	0020	Total project costs		C.F.	1.57	2.30	3.40				
	1800	Equipment		S.F.	.51	1.04	1.61	1.20%	2.40%	2.70%	
	2720	Plumbing			1.17	2.02	3.19	3.60%	4.70%	6.30%	
	2770	Heating, ventilating, air conditioning			1.86	2.94	4.13	5%	5.60%	9.60%	
	2900	Electrical			2.05	3.03	4.81	5.70%	7.70%	10%	
	3100	Total: Mechanical & Electrical		↓	4.84	8.05	10.20	13.80%	18.60%	23%	

For information about Means Estimating Seminars, see yellow pages 11 and 12 in back of book

SQUARE FOOT 17

Division Notes

	CREW	DAILY OUTPUT	LABOR-HOURS	UNIT	1999 BARE COSTS				TOTAL INCL O&P
					MAT.	LABOR	EQUIP.	TOTAL	

Reference Section

All the reference information is in one section making it easy to find what you need to know . . . and easy to use the book on a daily basis. This section is visually identified by a vertical gray bar on the edge of pages.

In the reference number information that follows, you'll see the background that relates to the "reference numbers" that appeared in the Unit Price Sections. You'll find reference tables, explanations and estimating information that support how we arrived at the unit price data. Also included are alternate pricing methods, technical data and estimating procedures along with information on design and economy in construction.

Also in this Reference Section, we've included Change Orders, information on pricing changes to contract documents; Crew Listings, a full listing of all the crews, equipment and their costs; Historical Cost Indexes for cost comparisons over time; City Cost Indexes and Location Factors for adjusting costs to the region you are in; and an explanation of all Abbreviations used in the book.

Table of Contents

R010-005 Tips for Accurate Estimating

1. Use pre-printed or columnar forms for orderly sequence of dimensions and locations and for recording telephone quotations.
2. Use only the front side of each paper or form except for certain pre-printed summary forms.
3. Be consistent in listing dimensions: For example, length x width x height. This helps in rechecking to ensure that, the total length of partitions is appropriate for the building area.
4. Use printed (rather than measured) dimensions where given.
5. Add up multiple printed dimensions for a single entry where possible.
6. Measure all other dimensions carefully.
7. Use each set of dimensions to calculate multiple related quantities.
8. Convert foot and inch measurements to decimal feet when listing. Memorize decimal equivalents to .01 parts of a foot (1/8″ equals approximately .01′).
9. Do not "round off" quantities until the final summary.
10. Mark drawings with different colors as items are taken off.
11. Keep similar items together, different items separate.
12. Identify location and drawing numbers to aid in future checking for completeness.
13. Measure or list everything on the drawings or mentioned in the specifications.
14. It may be necessary to list items not called for to make the job complete.
15. Be alert for: Notes on plans such as N.T.S. (not to scale); changes in scale throughout the drawings; reduced size drawings; discrepancies between the specifications and the drawings.
16. Develop a consistent pattern of performing an estimate. For example:
    a. Start the quantity takeoff at the lower floor and move to the next higher floor.
    b. Proceed from the main section of the building to the wings.
    c. Proceed from south to north or vice versa, clockwise or counterclockwise.
    d. Take off floor plan quantities first, elevations next, then detail drawings.
17. List all gross dimensions that can be either used again for different quantities, or used as a rough check of other quantities for verification (exterior perimeter, gross floor area, individual floor areas, etc.).
18. Utilize design symmetry or repetition (repetitive floors, repetitive wings, symmetrical design around a center line, similar room layouts, etc.). Note: Extreme caution is needed here so as not to omit or duplicate an area.
19. Do not convert units until the final total is obtained. For instance, when estimating concrete work, keep all units to the nearest cubic foot, then summarize and convert to cubic yards.
20. When figuring alternatives, it is best to total all items involved in the basic system, then total all items involved in the alternates. Therefore you work with positive numbers in all cases. When adds and deducts are used, it is often confusing whether to add or subtract a portion of an item; especially on a complicated or involved alternate.

R010-010 Architectural Fees

Tabulated below are typical percentage fees by project size, for good professional architectural service. Fees may vary from those listed depending upon degree of design difficulty and economic conditions in any particular area.

Rates can be interpolated horizontally and vertically. Various portions of the same project requiring different rates should be adjusted proportionately.

For alterations, add 50% to the fee for the first $500,000 of project cost and add 25% to the fee for project cost over $500,000.

Architectural fees tabulated below include Engineering Fees.

Building Types	Total Project Size in Thousands of Dollars						
	100	250	500	1,000	5,000	10,000	50,000
Factories, garages, warehouses, repetitive housing	9.0%	8.0%	7.0%	6.2%	5.3%	4.9%	4.5%
Apartments, banks, schools, libraries, offices, municipal buildings	11.7	10.8	8.5	7.3	6.4	6.0	5.6
Churches, hospitals, homes, laboratories, museums, research	14.0	12.8	11.9	10.9	8.5	7.8	7.2
Memorials, monumental work, decorative furnishings	—	16.0	14.5	13.1	10.0	9.0	8.3

R010-020 Construction Time Requirements

Table at left is average construction time in months for different types of building projects. Table at right is the construction time in months for different size projects. Design time runs 25% to 40% of construction time.

Type Building	Construction Time	Project Value	Construction Time
Industrial Buildings	12 Months	Under $1,400,000	10 Months
Commercial Buildings	15 Months	Up to $3,800,000	15 Months
Research & Development	18 Months	Up to $19,000,000	21 Months
Institutional Buildings	20 Months	over $19,000,000	28 Months

R010-030 Engineering Fees

Typical **Structural Engineering Fees** based on type of construction and total project size. These fees are included in Architectural Fees.

Type of Construction	Total Project Size (in thousands of dollars)			
	$500	$500-$1,000	$1,000-$5,000	Over $5000
Industrial buildings, factories & warehouses	Technical payroll times 2.0 to 2.5	1.60%	1.25%	1.00%
Hotels, apartments, offices, dormitories, hospitals, public buildings, food stores		2.00%	1.70%	1.20%
Museums, banks, churches and cathedrals		2.00%	1.75%	1.25%
Thin shells, prestressed concrete, earthquake resistive		2.00%	1.75%	1.50%
Parking ramps, auditoriums, stadiums, convention halls, hangars & boiler houses		2.50%	2.00%	1.75%
Special buildings, major alterations, underpinning & future expansion		Add to above 0.5%	Add to above 0.5%	Add to above 0.5%

For complex reinforced concrete or unusually complicated structures, add 20% to 50%.

Typical **Mechanical and Electrical Engineering Fees** based on the size of the subcontract. These fees are included in Architectural Fees.

Type of Construction	Subcontract Size							
	$25,000	$50,000	$100,000	$225,000	$350,000	$500,000	$750,000	$1,000,000
Simple structures	6.4%	5.7%	4.8%	4.5%	4.4%	4.3%	4.2%	4.1%
Intermediate structures	8.0	7.3	6.5	5.6	5.1	5.0	4.9	4.8
Complex structures	12.0	9.0	9.0	8.0	7.5	7.5	7.0	7.0

For renovations, add 15% to 25% to applicable fee.

R010-040 Builder's Risk Insurance

Builder's Risk Insurance is insurance on a building during construction. Premiums are paid by the owner or the contractor. Blasting, collapse and underground insurance would raise total insurance costs above those listed. Floater policy for materials delivered to the job runs $.75 to $1.25 per $100 value. Contractor equipment insurance runs $.50 to $1.50 per $100 value. Insurance for miscellaneous tools to $1,500 value runs from $3.00 to $7.50 per $100 value.

Tabulated below are New England Builder's Risk insurance rates in dollars per $100 value for $1,000 deductible. For $25,000 deductible, rates can be reduced 13% to 34%. On contracts over $1,000,000, rates may be lower than those tabulated. Policies are written annually for the total completed value in place. For "all risk" insurance (excluding flood, earthquake and certain other perils) add $.025 to total rates below.

Coverage	Frame Construction (Class 1)		Average	Brick Construction (Class 4)		Average	Fire Resistive (Class 6)		Average
	Range			Range			Range		
Fire Insurance	$.300 to	$.420	$.394	$.132 to	$.189	$.174	$.052 to	$.080	$.070
Extended Coverage	.115 to	.150	.144	.080 to	.105	.101	.081 to	.105	.100
Vandalism	.012 to	.016	.015	.008 to	.011	.011	.008 to	.011	.010
Total Annual Rate	$.427 to	$.586	$.553	$.220 to	$.305	$.286	$.141 to	$.196	$.180

R010-050 General Contractor's Overhead

The table below shows a contractor's overhead as a percentage of direct cost in two ways. The figures on the right are for the overhead, markup based on both material and labor. The figures on the left are based on the entire overhead applied only to the labor. This figure would be used if the owner supplied the materials or if a contract is for labor only. Note: Some of these markups are included in the labor rates shown on Reference Table R010-070.

Items of General Contractor's Indirect Costs	% of Direct Costs	
	As a Markup of Labor Only	As a Markup of Both Material and Labor
Field Supervision	6.0%	2.9%
Main Office Expense (see details below)	16.2	7.7
Tools and Minor Equipment	1.0	0.5
Workers' Compensation & Employers' Liability. See R010-060	18.3	8.7
Field Office, Sheds, Photos, Etc.	1.5	0.7
Performance and Payment Bond, 0.7% to 1.5%. See R010-080	2.3	1.1
Unemployment Tax See R010-100 (Combined Federal and State)	7.0	3.3
Social Security and Medicare, See R010-100	7.7	3.7
Sales Tax — add if applicable 42/80 x % as markup of total direct costs including both material and labor. See R010-090		
Sub Total	60.0%	28.5%
*Builder's Risk Insurance ranges from .141% to .586%. See R010-040	0.6	0.3
*Public Liability Insurance	3.2	1.5
Grand Total	63.8%	30.3%

*Paid by Owner or Contractor

Main Office Expense

A General Contractor's main office expense consists of many items not detailed in the front portion of the book. The percentage of main office expense declines with increased annual volume of the contractor. Typical main office expense ranges from 2% to 20% with the median about 7.2% of total volume. This equals about 7.7% of direct costs. The following are approximate percentages of total overhead for different items usually included in a General Contractor's main office overhead. With different accounting procedures, these percentages may vary.

Item	Typical Range			Average
Managers', clerical and estimators' salaries	40 %	to	55 %	48%
Profit sharing, pension and bonus plans	2	to	20	12
Insurance	5	to	8	6
Estimating and project management (not including salaries)	5	to	9	7
Legal, accounting and data processing	0.5	to	5	3
Automobile and light truck expense	2	to	8	5
Depreciation of overhead capital expenditures	2	to	6	4
Maintenance of office equipment	0.1	to	1.5	1
Office rental	3	to	5	4
Utilities including phone and light	1	to	3	2
Miscellaneous	5	to	15	8
Total				100%

R010-060   Workers' Compensation Insurance Rates by Trade

The table below tabulates the national averages for Workers' Compensation insurance rates by trade and type of building. The average "Insurance Rate" is multiplied by the "% of Building Cost" for each trade. This produces the "Workers' Compensation Cost" by % of total labor cost, to be added for each trade by building type to determine the weighted average Workers' Compensation rate for the building types analyzed.

Trade	Insurance Rate (% Labor Cost)			% of Building Cost			Workers' Compensation		
	Range		Average	Office Bldgs.	Schools & Apts.	Mfg.	Office Bldgs.	Schools & Apts.	Mfg.
Excavation, Grading, etc.	4.0 % to	26.6%	11.6%	4.8%	4.9%	4.5%	0.56%	0.57%	0.52%
Piles & Foundations	8.1 to	80.1	30.3	7.1	5.2	8.7	2.15	1.58	2.64
Concrete	7.2 to	39.4	19.1	5.0	14.8	3.7	0.96	2.83	0.71
Masonry	5.3 to	48.6	18.0	6.9	7.5	1.9	1.24	1.35	0.34
Structural Steel	8.1 to	132.9	42.8	10.7	3.9	17.6	4.58	1.67	7.53
Miscellaneous & Ornamental Metals	5.4 to	34.0	14.1	2.8	4.0	3.6	0.39	0.56	0.51
Carpentry & Millwork	7.0 to	49.9	19.9	3.7	4.0	0.5	0.74	0.80	0.10
Metal or Composition Siding	7.3 to	36.7	18.3	2.3	0.3	4.3	0.42	0.05	0.79
Roofing	8.1 to	88.8	34.6	2.3	2.6	3.1	0.80	0.90	1.07
Doors & Hardware	3.6 to	28.8	11.8	0.9	1.4	0.4	0.11	0.17	0.05
Sash & Glazing	5.3 to	30.0	14.7	3.5	4.0	1.0	0.51	0.59	0.15
Lath & Plaster	5.5 to	37.4	15.8	3.3	6.9	0.8	0.52	1.09	0.13
Tile, Marble & Floors	3.3 to	29.5	10.3	2.6	3.0	0.5	0.27	0.31	0.05
Acoustical Ceilings	4.2 to	26.7	12.4	2.4	0.2	0.3	0.30	0.02	0.04
Painting	5.6 to	41.1	15.6	1.5	1.6	1.6	0.23	0.25	0.25
Interior Partitions	7.0 to	49.9	19.9	3.9	4.3	4.4	0.78	0.86	0.88
Miscellaneous Items	2.8 to	112.5	19.3	5.2	3.7	9.7	1.00	0.71	1.87
Elevators	2.0 to	20.0	8.9	2.1	1.1	2.2	0.19	0.10	0.20
Sprinklers	2.7 to	20.6	9.1	0.5	—	2.0	0.05	—	0.18
Plumbing	3.0 to	15.9	9.0	4.9	7.2	5.2	0.44	0.65	0.47
Heat., Vent., Air Conditioning	3.9 to	29.2	12.8	13.5	11.0	12.9	1.73	1.41	1.65
Electrical	3.2 to	11.6	7.0	10.1	8.4	11.1	0.71	0.59	0.78
Total	2.0 % to	132.9%	—	100.0%	100.0%	100.0%	18.68%	17.06%	20.91%
			Overall Weighted Average	18.88%					

Workers' Compensation Insurance Rates by States

The table below lists the weighted average Workers' Compensation base rate for each state with a factor comparing this with the national average of 18.3%.

State	Weighted Average	Factor	State	Weighted Average	Factor	State	Weighted Average	Factor
Alabama	30.4%	166	Kentucky	19.8%	108	North Dakota	16.7%	91
Alaska	12.8	70	Louisiana	27.2	149	Ohio	16.1	88
Arizona	17.6	96	Maine	21.8	119	Oklahoma	21.9	120
Arkansas	13.0	71	Maryland	11.9	65	Oregon	19.2	105
California	18.6	102	Massachusetts	26.5	145	Pennsylvania	22.9	125
Colorado	26.4	144	Michigan	21.7	119	Rhode Island	22.3	122
Connecticut	21.1	115	Minnesota	37.6	205	South Carolina	13.9	76
Delaware	12.3	67	Mississippi	23.8	130	South Dakota	17.9	98
District of Columbia	25.2	138	Missouri	15.8	86	Tennessee	14.4	79
Florida	28.0	153	Montana	37.3	204	Texas	24.7	135
Georgia	24.9	136	Nebraska	15.7	86	Utah	12.9	70
Hawaii	16.7	91	Nevada	17.8	97	Vermont	14.9	81
Idaho	11.8	64	New Hampshire	23.5	128	Virginia	11.8	64
Illinois	27.5	150	New Jersey	11.0	60	Washington	12.0	66
Indiana	7.1	39	New Mexico	23.1	126	West Virginia	14.0	77
Iowa	14.6	80	New York	17.0	93	Wisconsin	13.9	76
Kansas	11.0	60	North Carolina	14.0	77	Wyoming	8.9	49
			Weighted Average for U.S. is	18.9% of payroll = 100%				

Rates in the following table are the base or manual costs per $100 of payroll for Workers' Compensation in each state. Rates are usually applied to straight time wages only and not to premium time wages and bonuses.

The weighted average skilled worker rate for 35 trades is 18.3%. For bidding purposes, apply the full value of Workers' Compensation directly to total labor costs, or if labor is 38%, materials 42% and overhead and profit 20% of total cost, carry 38/80 x 18.3% = 8.7% of cost (before overhead and profit) into overhead. Rates vary not only from state to state but also with the experience rating of the contractor.

Rates are the most current available at the time of publication.

R010-060 Workers' Compensation Insurance Rates by Trade and State (cont.)

State	Carpentry — 3 stories or less	Carpentry — interior cab. work	Carpentry — general	Concrete Work — NOC	Concrete Work — flat (flr, sdwk.)	Electrical Wiring — inside	Excavation — earth NOC	Excavation — rock	Glaziers	Insulation Work	Lathing	Masonry	Painting & Decorating	Pile Driving	Plastering	Plumbing	Roofing	Sheet Metal Work (HVAC)	Steel Erection — door & sash	Steel Erection — inter., ornam.	Steel Erection — structure	Steel Erection — NOC	Tile Work — (interior ceramic)	Waterproofing	Wrecking
	5651	5437	5403	5213	5221	5190	6217	6217	5462	5479	5443	5022	5474	6003	5480	5183	5551	5538	5102	5102	5040	5057	5348	9014	5701
AL	36.41	14.51	34.52	32.46	16.19	9.57	20.98	20.98	24.55	20.09	15.28	29.78	21.80	64.38	33.79	11.96	57.62	25.78	16.41	16.41	54.68	66.67	16.19	6.92	54.68
AK	11.88	6.71	10.26	10.24	6.88	5.18	8.62	8.62	11.65	9.76	7.11	8.12	10.50	38.68	11.01	4.79	14.75	7.23	13.03	13.03	25.07	25.07	7.20	5.45	25.07
AZ	21.46	9.11	26.40	16.82	9.92	8.14	9.97	9.97	14.03	24.32	8.83	17.04	12.65	23.74	13.75	9.23	20.48	11.54	14.38	14.38	50.31	27.82	7.33	6.15	50.31
AR	15.17	10.19	14.14	13.26	7.02	6.32	9.49	9.49	10.53	11.82	9.78	11.16	12.11	19.97	10.44	7.22	24.38	9.62	8.74	8.74	21.42	23.91	7.25	4.95	21.42
CA	28.28	9.07	28.28	13.18	13.18	9.93	7.88	7.88	16.37	23.34	10.90	14.55	19.45	21.18	18.12	11.59	39.79	15.33	13.55	13.55	23.93	21.91	7.74	19.45	21.91
CO	32.22	13.54	20.95	20.96	15.13	9.20	15.99	15.99	15.98	21.31	14.68	31.06	21.85	41.53	37.40	14.50	59.10	12.54	11.83	11.83	73.46	43.49	15.04	12.26	73.46
CT	19.25	13.14	24.07	20.48	14.17	6.68	8.42	8.42	21.25	34.89	13.64	25.45	18.84	24.54	17.60	12.25	33.98	13.93	13.67	13.67	50.86	35.83	13.94	5.02	50.86
DE	13.83	13.83	11.56	9.28	6.29	5.82	8.54	8.54	11.18	11.56	10.51	9.85	13.41	11.64	10.51	5.41	22.89	10.29	10.21	10.21	26.69	10.21	7.54	9.85	25.69
DC	17.51	10.80	18.88	29.27	16.85	8.76	17.96	17.96	28.00	22.14	12.07	29.67	16.85	42.12	17.06	12.83	22.39	15.78	33.66	33.66	66.43	42.52	16.42	4.66	66.43
FL	36.73	23.28	29.09	29.92	16.15	10.51	15.27	15.27	17.81	25.18	24.27	28.72	28.92	49.35	28.79	13.37	53.20	17.12	18.67	18.67	43.71	50.61	12.56	7.89	43.71
GA	31.04	16.02	32.71	20.67	15.12	9.82	19.55	19.55	21.97	19.15	26.66	23.03	21.20	40.88	19.50	11.14	40.72	18.65	13.74	13.74	36.03	53.09	13.08	10.16	36.03
HI	17.38	11.62	28.20	13.74	12.27	7.10	7.73	7.73	20.55	21.19	10.94	17.87	11.33	20.15	15.61	6.06	33.24	8.02	11.11	11.11	31.71	21.70	9.78	11.54	31.71
ID	12.68	6.50	16.29	9.46	9.04	5.37	6.87	6.87	8.58	11.01	8.97	9.38	8.83	18.27	8.94	5.04	21.34	7.76	7.51	7.51	26.68	24.89	5.35	7.44	26.68
IL	18.94	12.54	23.02	36.76	13.56	9.82	10.43	10.43	27.14	25.52	17.50	23.22	14.93	52.40	15.82	14.48	37.32	17.03	25.56	25.56	78.98	70.79	15.71	7.31	78.98
IN	7.26	3.59	6.96	7.15	3.63	3.16	3.96	3.96	5.31	7.74	4.20	5.31	5.56	12.13	5.48	3.02	12.13	4.44	5.63	5.63	21.36	12.17	3.33	4.00	21.36
IA	10.26	6.79	14.54	12.94	6.78	4.20	5.47	5.47	13.54	18.19	5.82	12.97	7.52	19.90	13.44	6.97	21.56	7.44	12.92	12.92	53.87	36.20	7.13	5.05	17.98
KS	9.56	7.37	9.65	11.58	7.05	4.20	4.33	4.33	6.34	14.80	8.60	11.77	8.23	16.03	12.20	5.20	26.06	6.39	5.35	5.35	17.84	32.15	4.14	4.04	17.84
KY	16.47	15.26	24.34	22.12	9.53	8.95	13.67	13.67	15.96	14.68	10.62	22.00	19.83	30.44	13.94	10.38	32.83	15.92	13.21	13.21	45.21	33.73	10.60	8.85	45.21
LA	29.63	22.95	26.01	20.86	14.80	10.48	24.73	24.73	23.88	20.82	14.00	21.27	27.63	54.39	21.42	13.78	49.11	21.35	15.51	15.51	66.41	36.94	12.82	10.26	66.41
ME	13.43	10.42	43.59	24.29	11.12	10.02	14.75	14.75	13.35	15.39	14.03	17.26	16.07	31.44	17.95	11.12	32.07	17.70	15.08	15.08	37.84	61.81	11.66	9.37	37.84
MD	10.55	5.95	10.55	11.35	5.05	5.15	9.25	9.25	13.20	13.05	6.45	11.35	6.75	27.45	6.65	5.55	22.80	7.00	9.15	9.15	26.80	18.50	6.35	3.50	26.80
MA	15.25	11.38	22.82	39.38	16.79	6.14	10.72	10.72	16.78	27.76	14.02	29.03	14.95	29.58	13.75	8.49	67.48	13.20	16.07	16.07	80.76	79.51	16.46	6.62	62.87
MI	15.62	10.60	17.70	31.29	11.92	5.77	12.80	12.80	11.11	25.59	11.14	18.30	17.86	64.20	20.11	8.04	38.89	11.27	14.73	14.73	41.58	39.66	10.84	10.42	41.58
MN	27.66	28.76	49.94	30.20	22.59	8.28	19.73	19.73	30.01	37.20	23.93	29.89	19.91	80.08	23.93	15.88	88.83	13.91	27.21	27.21	132.92	31.83	29.54	8.71	132.92
MS	21.77	16.71	28.09	17.88	11.02	10.84	13.03	13.03	18.18	21.21	14.41	22.67	17.90	77.98	20.06	10.84	36.25	20.16	14.52	14.52	37.42	42.77	12.76	8.68	37.42
MO	18.72	7.52	11.84	13.28	10.66	5.79	11.05	11.05	8.97	19.99	9.67	13.57	11.77	16.88	14.22	7.10	25.74	9.76	13.82	13.82	47.98	31.96	5.83	5.71	47.98
MT	30.63	15.19	47.53	32.78	24.74	11.01	26.63	26.63	25.04	38.54	19.54	48.55	41.06	53.68	27.00	15.03	85.06	16.93	33.95	33.95	86.98	55.25	13.30	12.10	86.98
NE	16.19	9.23	15.94	17.31	12.73	5.43	10.19	10.19	12.36	16.79	9.34	18.07	11.58	22.01	14.62	9.06	33.91	13.86	10.75	10.75	25.03	26.96	7.88	5.47	25.03
NV	15.42	15.42	15.42	12.86	12.86	8.82	10.62	10.62	12.65	17.78	20.58	14.66	27.12	10.39	20.58	11.30	25.86	29.24	12.56	12.56	27.12	27.12	10.61	10.39	27.12
NH	21.36	9.69	18.75	35.09	10.57	6.26	13.35	13.35	12.80	48.52	10.92	30.06	16.38	29.92	13.78	7.97	66.23	15.20	13.84	13.84	50.50	39.47	11.61	7.64	50.50
NJ	10.40	8.85	10.40	10.83	8.37	4.23	7.95	7.95	7.36	12.78	7.88	10.77	11.49	11.32	7.88	5.48	29.41	6.20	8.03	8.03	22.52	12.96	5.74	5.37	29.92
NM	25.35	13.09	20.45	19.66	13.39	6.34	10.95	10.95	14.58	31.49	16.25	26.97	16.87	33.27	15.64	12.10	43.41	16.46	21.26	21.26	67.82	29.91	9.67	10.65	67.82
NY	15.39	6.82	14.50	24.92	15.16	6.89	9.69	9.69	16.92	15.82	13.00	18.61	12.55	27.49	13.03	9.03	32.78	14.13	14.12	14.12	27.11	24.09	10.56	7.48	32.53
NC	14.43	9.91	15.37	16.82	6.37	7.64	7.73	7.73	9.52	12.92	8.00	12.94	10.40	17.95	14.47	6.93	25.39	10.26	13.32	13.32	39.03	16.68	7.68	3.86	39.03
ND	13.89	13.89	13.89	12.17	12.17	6.16	9.91	9.91	9.85	9.38	13.16	12.20	11.42	29.82	13.16	8.05	31.37	8.05	13.89	13.89	29.82	29.82	9.63	31.37	29.82
OH	13.27	13.27	13.27	13.45	13.45	6.02	13.45	13.45	18.28	16.31	16.31	15.62	18.28	13.45	16.31	7.96	28.62	24.12	13.45	13.45	13.45	13.45	13.08	13.45	13.45
OK	26.16	11.79	19.33	15.51	11.50	7.41	18.45	18.45	13.55	18.00	12.05	16.80	17.10	36.26	18.29	8.72	42.26	12.45	11.17	11.17	69.45	46.76	11.42	8.20	69.45
OR	29.51	9.28	20.03	16.50	10.61	6.63	12.94	12.94	14.95	19.68	12.27	12.43	19.83	22.05	13.85	8.38	38.11	11.85	11.01	11.01	61.69	25.54	9.92	12.10	61.69
PA	15.80	15.80	19.85	28.18	11.84	7.80	12.74	12.74	15.69	19.85	18.94	18.96	22.05	32.12	18.94	11.57	43.30	12.58	24.71	24.71	59.05	24.71	11.83	18.96	81.34
RI	21.32	11.02	18.77	22.11	15.55	5.86	12.16	12.16	17.02	20.95	11.89	20.66	19.86	41.76	14.19	7.25	29.48	10.52	13.18	13.18	78.79	49.67	14.34	9.90	78.79
SC	20.63	13.45	18.64	10.77	8.42	7.13	8.76	8.76	14.61	12.03	7.43	10.83	10.94	19.82	17.30	7.52	26.81	11.23	9.33	9.33	17.82	23.27	6.07	4.47	17.82
SD	14.61	9.94	20.92	21.64	9.04	5.72	11.73	11.73	14.77	17.73	10.89	13.30	13.25	32.14	16.96	9.50	29.39	11.35	11.51	11.51	46.57	33.20	9.85	5.59	46.57
TN	15.36	9.54	13.55	13.95	8.57	5.37	9.59	9.59	10.44	18.17	9.13	15.12	12.87	19.05	12.54	6.76	26.46	12.10	9.44	9.44	42.77	16.07	6.12	5.79	42.77
TX	28.29	18.95	28.29	25.59	19.49	11.60	18.16	18.16	14.17	25.84	13.47	23.12	18.29	43.91	20.99	12.88	47.24	22.92	14.29	14.29	50.47	31.01	9.94	11.38	54.81
UT	11.17	11.17	11.17	17.68	7.99	7.05	6.38	6.38	9.55	10.95	12.32	13.75	15.29	17.24	10.60	6.41	26.27	6.30	8.99	8.99	23.51	23.51	6.06	5.83	26.53
VT	9.68	7.02	14.48	29.73	11.12	4.56	9.17	9.17	12.21	11.43	8.08	10.80	7.69	18.34	10.19	7.90	26.72	10.85	10.23	10.23	25.83	48.65	6.67	7.28	25.83
VA	10.97	7.45	8.50	12.32	5.85	3.48	6.44	6.44	13.64	9.75	7.10	9.11	8.54	27.35	7.24	5.56	19.12	7.55	11.52	11.52	21.92	30.41	5.83	2.78	21.92
WA	8.78	8.78	8.78	8.53	8.53	3.68	8.82	8.82	8.71	9.93	8.78	11.54	10.23	20.99	11.78	5.41	23.39	3.94	16.21	16.21	16.21	16.21	9.81	10.23	16.21
WV	15.43	15.43	15.43	20.87	20.87	5.01	9.14	9.14	7.33	7.33	16.88	16.02	16.88	10.56	16.88	7.40	12.04	7.33	14.98	14.98	13.27	14.98	16.02	4.87	13.27
WI	10.47	10.13	19.98	8.91	8.50	5.73	7.00	7.00	9.67	13.88	13.48	15.43	12.10	17.04	11.81	5.95	29.35	8.64	11.52	11.52	36.51	18.72	8.45	3.92	36.51
WY	8.13	8.13	8.13	8.13	8.13	8.13	8.13	8.13	8.13	8.13	8.13	8.13	8.13	8.13	8.13	8.13	8.13	8.13	8.13	8.13	8.13	8.13	8.13	8.13	8.13
AVG.	18.27	11.79	19.92	19.12	11.74	7.04	11.59	11.59	14.71	18.86	12.43	18.01	15.62	30.30	15.84	8.99	34.61	12.77	14.05	14.05	42.77	32.59	10.33	8.46	42.49

486

R010-060 Workers' Compensation (cont.) (Canada in Canadian dollars)

Province		Alberta	British Columbia	Manitoba	Ontario	New Brunswick	Newfndld. & Labrador	Northwest Territories	Nova Scotia	Prince Edward Island	Quebec	Saskat-chewan	Yukon
Carpentry—3 stories or less	Rate	3.10	6.74	9.10	9.63	3.62	5.11	5.25	5.99	5.68	14.46	5.21	2.35
	Code	25401	70600	40102	723	422	403	4-41	4013	401	80110	B12-02	4-042
Carpentry—interior cab. work	Rate	2.34	3.30	9.10	9.63	4.57	5.11	5.25	5.99	5.68	14.46	4.38	2.35
	Code	42133	60412	40102	723	427	403	4-41	4013	401	80110	B11-25	4-042
CARPENTRY—general	Rate	3.10	6.74	9.10	9.63	3.62	5.11	5.25	5.99	5.68	14.46	5.21	2.35
	Code	25401	70600	40102	723	422	403	4-41	4013	401	80110	B12-02	4-042
CONCRETE WORK—NOC	Rate	4.43	6.74	9.16	13.06	3.62	5.11	5.25	7.87	5.68	17.01	8.20	3.25
	Code	42104	70604	40110	745	422	403	4-41	4222	401	80100	B14-04	2-032
CONCRETE WORK—flat (flr. sidewalk)	Rate	4.43	6.74	9.16	13.06	3.62	5.11	5.25	7.87	5.68	17.01	8.20	3.25
	Code	42104	70604	40110	745	422	403	4-41	4222	401	80100	B14-04	2-032
ELECTRICAL Wiring—inside	Rate	2.03	4.18	5.37	4.39	1.36	4.14	3.50	3.30	3.40	7.75	4.38	2.35
	Code	42124	71100	40203	704	426	400	4-46	4261	402	80170	B11-05	4-041
EXCAVATION—earth NOC	Rate	2.65	4.05	9.10	5.03	2.73	5.11	4.50	3.96	5.68	8.23	5.89	3.25
	Code	40604	72607	40706	711	421	403	4-43	4214	401	80030	R11-06	2-016
EXCAVATION—rock	Rate	2.65	4.05	9.10	5.03	2.73	5.11	4.50	3.96	5.68	8.23	5.89	3.25
	Code	40604	72607	40706	711	421	403	4-43	4214	401	80030	R11-06	2-016
GLAZIERS	Rate	2.62	2.47	5.37	9.81	4.57	3.90	5.25	7.75	3.40	15.13	8.20	2.35
	Code	42121	60236	40109	751	423	402	4-41	4233	402	80150	B13-04	4-042
INSULATION WORK	Rate	3.09	6.29	9.10	9.81	4.57	3.90	5.25	7.75	5.68	15.72	5.21	3.25
	Code	42184	70504	40102	751	423	402	4-41	4234	401	80120	B12-07	2-035
LATHING	Rate	6.13	6.29	9.10	9.63	4.57	3.90	5.25	6.14	3.40	15.72	8.20	3.25
	Code	42135	70500	40102	723	427	402	4-41	4271	402	80120	B13-02	2-036
MASONRY	Rate	4.43	6.74	9.10	11.87	4.57	5.11	5.25	7.75	5.68	17.01	8.20	3.25
	Code	42102	70602	40102	741	423	403	4-41	4231	401	80100	B15-01	2-032
PAINTING & DECORATING	Rate	3.65	6.29	9.16	11.60	4.57	3.90	5.25	6.14	3.40	15.72	5.21	3.25
	Code	42111	70501	40105	719	427	402	4-41	4275	402	80120	B12-01	2-036
PILE DRIVING	Rate	4.43	17.50	9.10	8.64	3.62	9.15	4.50	7.87	5.68	8.23	5.21	3.25
	Code	42159	72502	40706	732	422	404	4-43	4221	401	80030	B12-10	2-030
PLASTERING	Rate	6.13	6.29	9.16	11.60	4.57	3.90	5.25	6.14	3.40	15.72	8.20	3.25
	Code	42135	70502	40108	719	427	402	4-41	4271	402	80120	B13-02	2-036
PLUMBING	Rate	2.03	3.89	5.37	4.93	1.95	3.70	3.50	3.70	3.40	8.39	4.38	2.35
	Code	42122	70712	40204	707	424	401	4-46	4241	402	80160	B11-01	4-039
ROOFING	Rate	8.56	6.74	9.16	11.87	7.36	5.11	5.25	7.75	5.68	23.51	8.20	3.25
	Code	42118	70600	40403	728	430	403	4-41	4235	401	80130	B15-02	2-031
SHEET METAL WORK (HVAC)	Rate	2.03	3.89	9.10	4.93	1.95	3.70	3.50	7.75	3.40	8.39	4.38	2.35
	Code	42117	70714	40402	707	424	401	4-46	4236	402	80160	B11-07	4-040
STEEL ERECTION—door & sash	Rate	3.09	17.50	9.10	18.50	3.62	9.15	5.25	7.87	5.68	30.18	8.20	3.25
	Code	42106	72509	40502	748	422	404	4-41	4223	401	80080	B15-03	2-012
STEEL ERECTION—inter., ornam.	Rate	3.09	17.50	9.10	18.50	3.62	5.11	5.25	7.87	5.68	30.18	8.20	3.25
	Code	42106	72509	40502	748	422	403	4-41	4223	401	80080	B15-03	2-012
STEEL ERECTION—structure	Rate	3.09	17.50	9.10	18.50	3.62	9.15	5.25	10.71	5.68	30.18	8.20	3.25
	Code	42106	72509	40502	748	422	404	4-41	4227	401	80080	B15-04	2-012
STEEL ERECTION—NOC	Rate	3.09	17.50	9.10	18.50	3.62	9.15	5.25	10.71	5.68	30.18	8.20	3.25
	Code	42106	72509	40502	748	422	404	4-41	4227	401	80080	B15-03	2-012
TILE WORK—inter. (ceramic)	Rate	3.13	6.29	3.72	11.60	4.57	3.90	5.25	6.14	3.40	15.72	8.20	3.25
	Code	42113	70506	40103	719	427	402	4-41	4276	402	80120	B13-01	2-034
WATERPROOFING	Rate	3.65	6.74	9.10	9.63	4.57	5.11	5.25	7.75	3.40	23.51	4.38	3.25
	Code	42139	70620	40102	723	423	403	4-41	4239	402	80130	B11-17	2-030
WRECKING	Rate	12.86	6.74	9.16	18.50	2.73	5.11	4.50	3.96	5.68	36.09	8.20	3.25
	Code	42108	70600	40106	748	421	403	4-43	4211	401	80220	B14-07	2-030

R010-070   Contractor's Overhead & Profit

Below are the **average** installing contractor's percentage mark-ups applied to base labor rates to arrive at typical billing rates.

Column A: Labor rates are based on union wages averaged for 30 major U.S. cities. Base rates including fringe benefits are listed hourly and daily. These figures are the sum of the wage rate and employer-paid fringe benefits such as vacation pay, employer-paid health and welfare costs, pension costs, plus appropriate training and industry advancement funds costs.

Column B: Workers' Compensation rates are the national average of state rates established for each trade.

Column C: Column C lists average fixed overhead figures for all trades. Included are Federal and State Unemployment costs set at 7.0%; Social Security Taxes (FICA) set at 7.65%; Builder's Risk Insurance costs set at 0.34%; and Public Liability costs set at 1.55%. All the percentages except those for Social Security Taxes vary from state to state as well as from company to company.

Columns D and E: Percentages in Columns D and E are based on the presumption that the installing contractor has annual billing of $1,500,000 and up. Overhead percentages may increase with smaller annual billing. The overhead percentages for any given contractor may vary greatly and depend on a number of factors, such as the contractor's annual volume, engineering and logistical support costs, and staff requirements. The figures for overhead and profit will also vary depending on the type of job, the job location, and the prevailing economic conditions. All factors should be examined very carefully for each job.

Column F: Column F lists the total of Columns B, C, D, and E.

Column G: Column G is Column A (hourly base labor rate) multiplied by the percentage in Column F (O&P percentage).

Column H: Column H is the total of Column A (hourly base labor rate) plus Column G (Total O&P).

Column I: Column I is Column H multiplied by eight hours.

		A		B	C	D	E	F		G	H		I
		Base Rate Incl. Fringes		Workers' Comp. Ins.	Average Fixed Over-head	Over-head	Profit	Total Overhead & Profit			Rate with O & P		
Abbr.	Trade	Hourly	Daily					%	Amount		Hourly		Daily
Skwk	Skilled Workers Average (35 trades)	$28.05	$224.40	18.3%	16.5%	13.0%	10.0%	57.8%	$16.20		$44.25		$354.00
	Helpers Average (5 trades)	20.85	166.80	19.7		11.0		57.2	11.95		32.80		262.40
	Foreman Average, Inside ($.50 over trade)	28.55	228.40	18.3		13.0		57.8	16.50		45.05		360.40
	Foreman Average, Outside ($2.00 over trade)	30.05	240.40	18.3		13.0		57.8	17.35		47.40		379.20
Clab	Common Building Laborers	21.45	171.60	19.9		11.0		57.4	12.30		33.75		270.00
Asbe	Asbestos/Insulation Workers/Pipe Coverers	30.45	243.60	18.9		16.0		61.4	18.70		49.15		393.20
Boil	Boilermakers	32.85	262.80	16.6		16.0		59.1	19.40		52.25		418.00
Bric	Bricklayers	27.60	220.80	18.0		11.0		55.5	15.30		42.90		343.20
Brhe	Bricklayer Helpers	21.60	172.80	18.0		11.0		55.5	12.00		33.60		268.80
Carp	Carpenters	27.30	218.40	19.9		11.0		57.4	15.65		42.95		343.60
Cefi	Cement Finishers	26.15	209.20	11.7		11.0		49.2	12.85		39.00		312.00
Elec	Electricians	31.90	255.20	7.0		16.0		49.5	15.80		47.70		381.60
Elev	Elevator Constructors	33.15	265.20	8.9		16.0		51.4	17.05		50.20		401.60
Eqhv	Equipment Operators, Crane or Shovel	29.35	234.80	11.6		14.0		52.1	15.30		44.65		357.20
Eqmd	Equipment Operators, Medium Equipment	28.40	227.20	11.6		14.0		52.1	14.80		43.20		345.60
Eqlt	Equipment Operators, Light Equipment	27.20	217.60	11.6		14.0		52.1	14.15		41.35		330.80
Eqol	Equipment Operators, Oilers	24.05	192.40	11.6		14.0		52.1	12.55		36.60		292.80
Eqmm	Equipment Operators, Master Mechanics	30.05	240.40	11.6		14.0		52.1	15.65		45.70		365.60
Glaz	Glaziers	26.60	212.80	14.7		11.0		52.2	13.90		40.50		324.00
Lath	Lathers	26.80	214.40	12.4		11.0		49.9	13.35		40.15		321.20
Marb	Marble Setters	27.50	220.00	18.0		11.0		55.5	15.25		42.75		342.00
Mill	Millwrights	28.75	230.00	11.9		11.0		49.4	14.20		42.95		343.60
Mstz	Mosaic and Terrazzo Workers	26.55	212.40	10.3		11.0		47.8	12.70		39.25		314.00
Pord	Painters, Ordinary	24.90	199.20	15.6		11.0		53.1	13.20		38.10		304.80
Psst	Painters, Structural Steel	26.00	208.00	51.6		11.0		89.1	23.15		49.15		393.20
Pape	Paper Hangers	25.10	200.80	15.6		11.0		53.1	13.35		38.45		307.60
Pile	Pile Drivers	27.20	217.60	30.3		16.0		72.8	19.80		47.00		376.00
Plas	Plasterers	25.70	205.60	15.8		11.0		53.3	13.70		39.40		315.20
Plah	Plasterer Helpers	21.60	172.80	15.8		11.0		53.3	11.50		33.10		264.80
Plum	Plumbers	32.60	260.80	9.0		16.0		51.5	16.80		49.40		395.20
Rodm	Rodmen (Reinforcing)	30.40	243.20	32.6		14.0		73.1	22.20		52.60		420.80
Rofc	Roofers, Composition	24.10	192.80	34.6		11.0		72.1	17.40		41.50		332.00
Rots	Roofers, Tile and Slate	24.20	193.60	34.6		11.0		72.1	17.45		41.65		333.20
Rohe	Roofer Helpers (Composition)	18.10	144.80	34.6		11.0		72.1	13.05		31.15		249.20
Shee	Sheet Metal Workers	31.75	254.00	12.8		16.0		55.3	17.55		49.30		394.40
Spri	Sprinkler Installers	32.50	260.00	9.1		16.0		51.6	16.75		49.25		394.00
Stpi	Steamfitters or Pipefitters	32.75	262.00	9.0		16.0		51.5	16.85		49.60		396.80
Ston	Stone Masons	27.70	221.60	18.0		11.0		55.5	15.35		43.05		344.40
Sswk	Structural Steel Workers	30.60	244.80	42.8		14.0		83.3	25.50		56.10		448.80
Tilf	Tile Layers	26.65	213.20	10.3		11.0		47.8	12.75		39.40		315.20
Tilh	Tile Layer Helpers	21.45	171.60	10.3		11.0		47.8	10.25		31.70		253.60
Trlt	Truck Drivers, Light	21.75	174.00	15.6		11.0		53.1	11.55		33.30		266.40
Trhv	Truck Drivers, Heavy	22.10	176.80	15.6		11.0		53.1	11.75		33.85		270.80
Sswl	Welders, Structural Steel	30.60	244.80	42.8		14.0		83.3	25.50		56.10		448.80
Wrck	*Wrecking	21.45	171.60	42.5	▼	11.0	▼	80.0	17.15		38.60		308.80

*Not included in Averages.

R010-080 Performance Bond

This table shows the cost of a Performance Bond for a construction job scheduled to be completed in 12 months. Add 1% of the premium cost per month for jobs requiring more than 12 months to complete. The rates are "standard" rates offered to contractors that the bonding company considers financially sound and capable of doing the work. Preferred rates are offered by some bonding companies based upon financial strength of the contractor. Actual rates vary from contractor to contractor and from bonding company to bonding company. Contractors should prequalify through a bonding agency before submitting a bid on a contract that requires a bond.

Contract Amount	Building Construction Class B Projects			Highways & Bridges					
				Class A New Construction			Class A-1 Highway Resurfacing		
First $ 100,000 bid	$25.00 per M			$15.00 per M			$9.40 per M		
Next 400,000 bid	$ 2,500	plus	$15.00 per M	$ 1,500	plus	$10.00 per M	$ 940	plus	$7.20 per M
Next 2,000,000 bid	8,500	plus	10.00 per M	5,500	plus	7.00 per M	3,820	plus	5.00 per M
Next 2,500,000 bid	28,500	plus	7.50 per M	19,500	plus	5.50 per M	15,820	plus	4.50 per M
Next 2,500,000 bid	47,250	plus	7.00 per M	33,250	plus	5.00 per M	28,320	plus	4.50 per M
Over 7,500,000 bid	64,750	plus	6.00 per M	45,750	plus	4.50 per M	39,570	plus	4.00 per M

R010-090 Sales Tax by State

State sales tax on materials is tabulated below (5 states have no sales tax). Many states allow local jurisdictions, such as a county or city, to levy additional sales tax.

Some projects may be sales tax exempt, particularly those constructed with public funds.

State	Tax (%)	State	Tax (%)	State	Tax (%)	State	Tax (%)
Alabama	4	Illinois	6.25	Montana	0	Rhode Island	7
Alaska	0	Indiana	5	Nebraska	4.5	South Carolina	5
Arizona	5	Iowa	5	Nevada	6.75	South Dakota	4
Arkansas	4.625	Kansas	4.9	New Hampshire	0	Tennessee	6
California	7.25	Kentucky	6	New Jersey	6	Texas	6.25
Colorado	3	Louisiana	4	New Mexico	5	Utah	4.75
Connecticut	6	Maine	6	New York	4	Vermont	5
Delaware	0	Maryland	5	North Carolina	4	Virginia	4.5
District of Columbia	5.75	Massachusetts	5	North Dakota	5	Washington	6.5
Florida	6	Michigan	6	Ohio	5	West Virginia	6
Georgia	4	Minnesota	6.5	Oklahoma	4.5	Wisconsin	5
Hawaii	4	Mississippi	7	Oregon	0	Wyoming	4
Idaho	5	Missouri	4.225	Pennsylvania	6	Average	4.71 %

R010-100 Unemployment Taxes and Social Security Taxes

Mass. State Unemployment tax ranges from 1.8% to 7.7% plus an experience rating assessment the following year, on the first $10,800 of wages. Federal Unemployment tax is 6.2% of the first $7,000 of wages. This is reduced by a credit for payment to the state. The minimum Federal Unemployment tax is .8% after all credits.

Combined rates in Mass. thus vary from 2.6% to 8.5% of the first $10,800 of wages. Combined average U.S. rate is about 7.0% of the first $7,000. Contractors with permanent workers will pay less since the average annual wages for skilled workers is $28.05 x 2,000 hours or about $56,100 per year. The average combined rate for U.S. would thus be 7.0% x $7,000 ÷ $56,100 = 0.9% of total wages for permanent employees.

Rates vary not only from state to state but also with the experience rating of the contractor.

Social Security (FICA) for 1999 is estimated at time of publication to be 7.65% of wages up to $68,400.

R010-110 Overtime

One way to improve the completion date of a project or eliminate negative float from a schedule is to compress activity duration times. This can be achieved by increasing the crew size or working overtime with the proposed crew.

To determine the costs of working overtime to compress activity duration times, consider the following examples. Below is an overtime efficiency and cost chart based on a five, six, or seven day week with an eight through twelve hour day. Payroll percentage increases for time and one half and double time are shown for the various working days.

Days per Week	Hours per Day	Production Efficiency					Payroll Cost Factors	
		1 Week	2 Weeks	3 Weeks	4 Weeks	Average 4 Weeks	@ 1-1/2 Times	@ 2 Times
5	8	100%	100%	100%	100%	100 %	100 %	100 %
	9	100	100	95	90	96.25	105.6	111.1
	10	100	95	90	85	91.25	110.0	120.0
	11	95	90	75	65	81.25	113.6	127.3
	12	90	85	70	60	76.25	116.7	133.3
6	8	100	100	95	90	96.25	108.3	116.7
	9	100	95	90	85	92.50	113.0	125.9
	10	95	90	85	80	87.50	116.7	133.3
	11	95	85	70	65	78.75	119.7	139.4
	12	90	80	65	60	73.75	122.2	144.4
7	8	100	95	85	75	88.75	114.3	128.6
	9	95	90	80	70	83.75	118.3	136.5
	10	90	85	75	65	78.75	121.4	142.9
	11	85	80	65	60	72.50	124.0	148.1
	12	85	75	60	55	68.75	126.2	152.4

R015-010   Heavy Lifting

Hydraulic Climbing Jacks

The use of hydraulic heavy lift systems is an alternative to conventional type crane equipment. The lifting, lowering, pushing, or pulling mechanism is a hydraulic climbing jack moving on a square steel jackrod from 1-5/8" to 4" square, or a steel cable. The jackrod or cable can be vertical or horizontal, stationary or movable, depending on the individual application. When the jackrod is stationary, the climbing jack will climb the rod and push or pull the load along with itself. When the climbing jack is stationary, the jackrod is movable with the load attached to the end and the climbing jack will lift or lower the jackrod with the attached load. The heavy lift system is normally operated by a single control lever located at the hydraulic pump.

The system is flexible in that one or more climbing jacks can be applied wherever a load support point is required, and the rate of lift synchronized.

Economic benefits have been demonstrated on projects such as: erection of ground assembled roofs and floors, complete bridge spans, girders and trusses, towers, chimney liners and steel vessels, storage tanks, and heavy machinery. Other uses are raising and lowering offshore work platforms, caissons, tunnel sections and pipelines.

R015-100   Steel Tubular Scaffolding

On new construction, tubular scaffolding is efficient up to 60' high or five stories. Above this it is usually better to use a hung scaffolding if construction permits. Swing scaffolding operations may interfere with tenants. In this case, the tubular is more practical at all heights.

In repairing or cleaning the front of an existing building the cost of tubular scaffolding per S.F. of building front increases as the height increases above the first tier. The first tier cost is relatively high due to leveling and alignment.

The minimum efficient crew for erection is three workers. For heights over 50', a crew of four is more efficient. Use two or more on top and two at the bottom for handing up or hoisting. Four workers can erect and dismantle about nine frames per hour up to five stories. From five to eight stories they will average six frames per hour. With 7' horizontal spacing

this will run about 400 S.F. and 265 S.F. of wall surface, respectively. Time for placing planks must be added to the above. On heights above 50', five planks can be placed per labor-hour.

The cost per 1,000 S.F. of building front in the table below was developed by pricing the materials required for a typical tubular scaffolding system eleven frames long and two frames high. Planks were figured five wide for standing plus two wide for materials.

Frames are 5' wide and usually spaced 7' O.C. horizontally. Sidewalk frames are 6' wide. Rental rates will be lower for jobs over three months duration.

For jobs under twenty-five frames, add 50% to rental cost. These figures do not include accessories which are listed separately below. Large quantities for long periods can reduce rental rates by 20%.

Item	Unit	Monthly Rent	Per 1,000 S.F. of Building Front	
			No. of Pieces	Rental per Month
5' Wide Standard Frame, 6'-4" High	Ea.	$ 3.75	24	$ 90.00
Leveling Jack & Plate		1.50	24	36.00
Cross Brace		.60	44	26.40
Side Arm Bracket, 21"		1.50	12	18.00
Guardrail Post		1.00	12	12.00
Guardrail, 7' Section		.75	22	16.50
Stairway Section		10.00	2	20.00
Stairway Starter Bar		.10	1	.10
Stairway Inside Handrail		5.00	2	10.00
Stairway Outside Handrail		5.00	2	10.00
Walk-Thru Frame Guardrail	↓	2.00	2	4.00
			Total	$243.00
			Per C.S.F., 1 Use/Mo.	$ 24.30

Scaffolding is often used as falsework over 15' high during construction of cast-in-place concrete beams and slabs. Two foot wide scaffolding is generally used for heavy beam construction. The span between frames depends upon the load to be carried with a maximum span of 5'.

Heavy duty shoring frames with a capacity of 10,000#/leg can be spaced up to 10' O. C. depending upon form support design and loading.

Scaffolding used as horizontal shoring requires less than half the material required with conventional shoring.

On new construction, erection is done by carpenters.

Rolling towers supporting horizontal shores can reduce labor and speed the job. For maintenance work, catwalks with spans up to 70' can be supported by the rolling towers.

R015-200 Pump Staging

Pump staging is generally not available for rent. Purchase prices for individual items are shown in the chart below.

Item	Unit	Purchase, Each	Per 2,400 S.F. of Building Front	
			No. of Pieces	Cost to Buy
Aluminum pole section, 24' long	Ea.	$335.00	6	$2,010.00
Aluminum splice joint, 6' long		68.00	3	204.00
Aluminum foldable brace		48.50	3	145.50
Aluminum pump jack		111.00	3	333.00
Aluminum support for workbench/back safety rail		59.00	3	177.00
Aluminum scaffold plank/workbench, 14" wide x 24' long		545.00	4	2,180.00
Safety net, 22' long		267.00	2	534.00
Aluminum plank end safety rail		183.00	2	366.00
			Total System	$5,949.50
			Per C.S.F., 1 Use	$247.90

This system in place will cover up to 2,400 square feet of wall area. The cost in place will depend on how many uses are realized during the life of the equipment. Several options are given in Division 015-257. The above prices are bare costs.

R016-410 Contractor Equipment

Rental Rates shown in the front of the book pertain to late model high quality machines in excellent working condition, rented from equipment dealers. Rental rates from contractors may be substantially lower than the rental rates from equipment dealers depending upon economic conditions. For older, less productive machines, reduce rates by a maximum of 15%. Any overtime must be added to the base rates. For shift work, rates are lower. Usual rule of thumb is 150% of one shift rate for two shifts; 200% for three shifts.

For periods of less than one week, operated equipment is usually more economical to rent than renting bare equipment and hiring an operator.

Equipment moving and mobilization costs must be added to rental rates where applicable. A large crane, for instance, may take two days to erect and two days to dismantle.

Rental rates vary throughout the country with larger cities generally having lower rates. Lease plans for new equipment are available for periods in excess of six months with a percentage of payments applying toward purchase.

Monthly rental rates vary from 2% to 5% of the cost of the equipment

depending on the anticipated life of the equipment and its wearing parts. Weekly rates are about 1/3 the monthly rates and daily rental rates about 1/3 the weekly rate.

The hourly operating costs for each piece of equipment include costs to the user such as fuel, oil, lubrication, normal expendables for the equipment, and a percentage of mechanic's wages chargeable to maintenance. The hourly operating costs listed do not include the operator's wages.

The daily cost for equipment used in the standard crews is figured by dividing the weekly rate by five, then adding eight times the hourly operating cost to give the total daily equipment cost, not including the operator. This figure is in the right hand column of Division 016 under Crew Equip. Cost./Day

Pile Driving rates shown for pile hammer and extractor do not include leads, crane, boiler or compressor. Vibratory pile driving requires an added field specialist during set-up and pile driving operation for the electric model. The hydraulic model requires a field specialist for set-up only. Up to 125 reuses of sheet piling are possible using vibratory drivers. For normal conditions, crane capacity for hammer type and size are as follows.

Crane Capacity	Hammer Type and Size		
	Air or Steam	Diesel	Vibratory
25 ton	to 8,750 ft.-lb.		70 H.P.
40 ton	15,000 ft.-lb.	to 32,000 ft.-lb.	170 H.P.
60 ton	25,000 ft.-lb.		300 H.P.
100 ton		112,000 ft.-lb.	

Cranes should be specified for the job by size, building and site characteristics, availability, performance characteristics, and duration of time required.

Backhoes & Shovels rent for about the same as equivalent size cranes but maintenance and operating expense is higher. Crane operators rate must

be adjusted for high boom heights. Average adjustments: for 150' boom add $.75 per hour; over 185', add $1.25 per hour; over 210', add $1.75 per hour; over 250', add $2.50 per hour and over 295', add $4.00 per hour.

Tower Cranes

Capacity In Kip-Feet	Typical Jib Length in Feet	Speed at Maximum Reach and Load	Purchase Price (New)		Monthly Rental, to 6 mo.	
			Crane & 80' Mast	Mast Sections	Crane & 80' Mast	Mast Sections
725	100	350 FPM	$245,800	$650 /L.F.	$ 8,170	$17.30 /L.F.
900	100	500	291,800	750	7,010	18.60
*1100	130	1000	404,500	880	10,130	23.50
1450	150	1000	563,200	1,240	14,480	27.70
2150	200	1000	691,200	1,420	18,200	32.10
3000	200	1000	983,000	1,450	23,780	38.10

*Most widely used.

Tower Cranes of the climbing or static type have jibs from 50' to 200' and capacities at maximum reach range from 4,000 to 14,000 pounds. Lifting capacities increase up to maximum load as the hook radius decreases.

Typical rental rates, based on purchase price are about 2% to 3% per month.

Erection and dismantling runs between $12,500 and $80,000. Climbing operation takes ten labor hours per 20' climb. Crane dead time is about five hours per 40' climb. If crane is bolted to side of the building add cost of ties and extra mast sections. Mast sections cost $500 to $1,500 per vertical foot or can be rented at 2% to 3% of purchase price per month. Contractors using climbers claim savings of $1.50 per C.Y. of concrete placed, plus $.15 per S.F. of formwork. Climbing cranes have from 80' to 180' of mast while static cranes have 80' to 800' of mast.

Truck Cranes can be converted to tower cranes by using tower attachments. Mast heights over 400' have been used. See Division 016-460 for rental rates of high boom cranes.

A single 100' high material **Hoist and Tower** can be erected and dismantled for about $15,000; a double 100' high hoist and tower for about $20,000. Erection costs for additional heights are $100 and $125 per vertical foot respectively up to 150' and $100 to $150 per vertical foot over 150' high. A 40' high portable Buck hoist costs about $5,000 to erect and dismantle. Additional heights run $80 per vertical foot to 80' and $100 per vertical foot for the next 100'. Most material hoists do not meet local code requirements for carrying personnel.

A 150' high **Personnel Hoist** requires about 500 to 800 labor hours to erect and dismantle with costs ranging from $12,000 to $28,000. Budget erection cost is $150 per vertical foot for all trades. Local code requirements or labor scarcity requiring overtime can add up to 50% to any of the above erection costs.

Earthmoving Equipment: The selection of earthmoving equipment depends upon the type and quantity of material, moisture content, haul distance, haul road, time available, and equipment available. Short haul cut and fill operations may require dozers only, while another operation may require excavators, a fleet of trucks, and spreading and compaction equipment. Stockpiled material and granular material are easily excavated with front end loaders. Scrapers are most economically used with hauls between 300' and 1-1/2 miles if adequate haul roads can be maintained. Shovels are often used for blasted rock and any material where a vertical face of 8' or more can be excavated. Special conditions may dictate the use of draglines, clamshells, or backhoes. Spreading and compaction equipment must be matched to the soil characteristics, the compaction required and the rate the fill is being supplied.

R020-820   Asbestos Removal Process

Asbestos removal is accomplished by a specialty contractor who understands the federal and state regulations regarding the handling and disposal of the material. The process of asbestos removal is divided into many individual steps. An accurate estimate can be calculated only after all the steps have been priced.

The steps are generally as follows:

1. Obtain an asbestos abatement plan from an industrial hygienist.
2. Monitor the air quality in and around the removal area and along the path of travel between the removal area and transport area. This establishes the background contamination.
3. Construct a two part decontamination chamber at entrance to removal area.
4. Install a HEPA filter to create a negative pressure in the removal area.
5. Install wall, floor and ceiling protection as required by the plan, usually 2 layers of fireproof 6 mil polyethylene.
6. Industrial hygienist visually inspects work area to verify compliance with plan.
7. Provide temporary supports for conduit and piping affected by the removal process.
8. Proceed with asbestos removal and bagging process. Monitor air quality as described in Step #2. Discontinue operations when contaminate levels exceed applicable standards.
9. Document the legal disposal of materials in accordance with EPA standards.
10. Thoroughly clean removal area including all ledges, crevices and surfaces.
11. Post abatement inspection by industrial hygienist to verify plan compliance.
12. Provide a certificate from a licensed industrial hygienist attesting that contaminate levels are within acceptable standards before returning area to regular use.

R020-880   Underground Storage Tank Removal

Underground Storage Tank Removal can be divided into two categories: Non-Leaking and Leaking. Prior to removing an underground storage tank, tests should be made, with the proper authorities present, to determine whether a tank has been leaking or the surrounding soil has been contaminated.

To safely remove Liquid Underground Storage Tanks:

1. Excavate to the top of the tank.
2. Disconnect all piping.
3. Open all tank vents and access ports.
4. Remove all liquids and/or sludge.
5. Purge the tank with an inert gas.
6. Provide access to the inside of the tank and clean out the interior using proper personal protective equipment (PPE).
7. Excavate soil surrounding the tank using proper PPE for on-site personnel.
8. Pull and properly dispose of the tank.
9. Clean up the site of all contaminated material.
10. Install new tanks or close the excavation.

R020-890   Lead Paint Remediation Methods

Lead paint remediation can be accomplished by the following methods.
1. Abrasive blast
2. Chemical stripping
3. Power tool cleaning with vacuum collection system
4. Encapsulation
5. Remove and replace
6. Enclosure

Each of these methods has strengths and weakness depending on the specific circumstances of the project. The following is an overview of each method.

1. **Abrasive blasting** is usually accomplished with sand or recyclable metallic blast. Before work can begin, the area must be contained to ensure the blast material with lead does not escape to the atmosphere. The use of vacuum blast greatly reduces the containment requirements. Lead abatement equipment that may be associated with this work includes a negative air machine. In addition, it is necessary to have an industrial hygienist monitor the project on a continual basis. When the work is complete, the spent blast sand with lead must be disposed of as a hazardous material. If metallic shot was used, the lead is separated from the shot and disposed of as hazardous material. Worker protection includes disposable clothing and respiratory protection.

2. **Chemical stripping** requires strong chemicals be applied to the surface to remove the lead paint. Before the work can begin, the area under/adjacent to the work area must be covered to catch the chemical and removed lead. After the chemical is applied to the painted surface it is usually covered with paper. The chemical is left in place for the specified period, then the paper with lead paint is pulled or scraped off. The process may require several chemical applications. The paper with chemicals and lead paint adhered to it, plus the containment and loose scrapings collected by a HEPA (High Efficiency Particulate Air Filter) vac, must be disposed of as a hazardous material. The chemical stripping process usually requires a neutralizing agent and several wash downs after the paint is removed. Worker protection includes a neoprene or other compatible protective clothing and respiratory protection with face shield. An industrial hygienist is required intermittently during the process.

3. **Power tool cleaning** is accomplished using shrouded needle blasting guns. The shrouding with different end configurations is held up against the surface to be cleaned. The area is blasted with hardened needles and the shroud captures the lead with a HEPA vac and deposits it in a holding tank. An industrial hygienist monitors the project, protective clothing and a respirator is required until air samples prove otherwise. When the work is complete the lead must be disposed of as a hazardous material.

4. **Encapsulation** is a method that leaves the well bonded lead paint in place after the peeling paint has been removed. Before the work can begin, the area under/adjacent to the work must be covered to catch the scrapings. The scraped surface is then washed with a detergent and rinsed. The prepared surface is covered with approximately 10 mils of paint. A reinforcing fabric can also be embedded in the paint covering. The scraped paint and containment must be disposed of as a hazardous material. Workers must wear protective clothing and respirators.

5. **Remove and replace** is an effective way to remove lead paint from windows, gypsum walls and concrete masonry surfaces. The painted materials are removed and new materials are installed. Workers should wear a respirator and tyvek suit. The demolished materials must be disposed of as hazardous waste if it fails the TCLP (Toxicity Characteristic Leachate Process) test.

6. **Enclosure** is the process that permanently seals lead painted materials in place. This process has many applications such as covering lead painted drywall with new dry wall, covering exterior construction with tyvek paper then residing, or covering lead painted structural members with aluminum or plastic. The seams on all enclosing materials must be securely sealed. An industrial hygienist monitors the project, and protective clothing and a respirator is required until air samples prove otherwise.

All the processes require clearance monitoring and wipe testing as required by the hygienist.

R021-440   Wellpoints

A single stage wellpoint system is usually limited to dewatering an average 15' depth below normal ground water level. Multi-stage systems are employed for greater depth with the pumping equipment installed only at the lowest header level. Ejectors, with unlimited lift capacity, can be economical when two or more stages of wellpoints can be replaced or when horizontal clearance is restricted, such as in deep trenches or tunneling projects, and where low water flows are expected. Wellpoints are usually spaced on 2-1/2' to 10' centers along a header pipe. Wellpoint spacing, header size, and pump size are all determined by the expected flow, as dictated by soil conditions.

In almost all soils encountered in wellpoint dewatering, the wellpoints may be jetted into place. Cemented soils and stiff clays may require sand wicks about 12" in diameter around each wellpoint to increase efficiency and eliminate weeping into the excavation. These sand wicks require 1/2 to 3 C.Y. of washed filter sand and are installed by using a 12" diameter steel casing and hole puncher jetted into the ground 2' deeper than the wellpoint. Rock may require predrilled holes.

Labor required for the complete installation and removal of a single stage wellpoint system is in the range of 3/4 to 2 labor-hours per linear foot of header, depending upon jetting conditions, wellpoint spacing, etc.

Continuous pumping is necessary except in some free draining soil where temporary flooding is permissible (as in trenches which are backfilled after each day's work). Good practice requires provision of a stand-by pump during the continuous pumping operation.

Systems for continuous trenching below the water table should be installed three to four times the length of expected daily progress to insure uninterrupted digging, and header pipe size should not be changed during the job.

For pervious free draining soils, deep wells in place of wellpoints may be economical because of lower installation and maintenance costs. Daily production ranges between two to three wells per day, for 25' to 40' depths, to one well per day for depths over 50'.

Detailed analysis and estimating for any dewatering problem is available at no cost from wellpoint manufacturers. Major firms will quote "sufficient equipment" quotes or their affiliates offer lump sum proposals to cover complete dewatering responsibility.

Description for 200' System with 8" Header		Quantities	1st Month		Thereafter	
			Unit	Total	Unit	Total
Equipment & Material	Wellpoints 25' long, 2" diameter @ 5' O.C.	40 Each	$ 28.00	$ 1,120.00	$ 21.00	$ 840.00
	Header pipe, 8" diameter	200 L.F.	5.10	1,020.00	2.55	510.00
	Discharge pipe, 8" diameter	100 L.F.	3.45	345.00	2.24	224.00
	8" valves	3 Each	135.00	405.00	87.75	263.25
	Combination Jetting & Wellpoint pump (standby)	1 Each	2,175.00	2,175.00	1,522.50	1,522.50
	Wellpoint pump, 8" diameter	1 Each	2,150.00	2,150.00	1,505.00	1,505.00
	Transportation to and from site	1 Day	619.05	619.05	—	—
	Fuel 30 days x 60 gal./day	1800 Gallons	1.40	2,520.00	1.40	2,520.00
	Lubricants for 30 days x 16 lbs./day	480 Lbs.	1.00	480.00	1.00	480.00
	Sand for points	40 C.Y.	15.60	624.00	—	—
	Equipment & Materials Subtotal			$11,458.05		$ 7,864.75
Labor	Technician to supervise installation	1 Week	$ 30.05	$ 1,202.00	—	—
	Labor for installation and removal of system	300 Labor-hours	21.45	6,435.00	—	—
	4 Operators straight time 40 hrs./wk. for 4.33 wks.	693 Hrs.	27.20	18,849.60	$ 27.20	$18,849.60
	4 Operators overtime 2 hrs./wk. for 4.33 wks.	35 Hrs.	54.40	1,904.00	54.40	1,904.00
	Labor Subtotal			$28,390.60		$20,753.60
	Total Cost			$39,848.65		$28,618.35
	Monthly Cost/L.F. Header			$ 199.24		$ 143.09

R021-610   Steel Sheet Piling

Limiting weights are 22 to 38#/S.F. of wall surface with 27#/S.F. average for usual types and sizes. (Weights of piles themselves are from 30.7#/L.F. to 57#/L.F. but they are 15″ to 21″ wide.) Lightweight sections 12″ to 28″ wide from 3 ga. to 12 ga. thick are also available for shallow excavations. Piles may be driven two at a time with an impact or vibratory hammer (use vibratory to pull) hung from a crane without leads. A reasonable estimate of the life of steel sheet piling is 10 uses with up to 125 uses possible if a vibratory hammer is used. Used piling costs from 50% to 80% of new piling depending on location and market conditions. Sheet piling and H piles can be rented for about 30% of the delivered mill price for the first month and 5% per month thereafter. Allow 1 labor-hour per pile for cleaning and trimming after driving. These costs increase with depth and hydrostatic head. Vibratory drivers are faster in wet granular soils and are excellent for pile extraction. Pulling difficulty increases with the time in the ground and may cost more than driving. It is often economical to abandon the sheet piling, especially if it can be used as the outer wall form. Allow about 1/3 additional length or more, for toeing into ground. Add bracing, waler and strut costs. Waler costs can equal the cost per ton of sheeting.

Cost of Sheet Piling & Production Rate by Ton & S.F.

Depth of Excavation	15′ Depth		20′ Depth		25′ Depth	
Description of Pile	22 psf = 90.9 S.F./Ton		27 psf = 74 S.F./Ton		38 psf = 52.6 S.F./Ton	
Type of operation: Left in Place or Removed	Drive & Left	Drive & Extract*	Drive & Left	Drive & Extract*	Drive & Left	Drive & Extract*
Labor & Equip. to Drive 1 ton	$ 341.29	$510.98	$ 284.67	$427.00	$ 194.17	$289.81
Piling (75% Salvage)	785.00	196.25	785.00	196.25	785.00	196.25
Cost/Ton in Place	$1,126.29	$707.23	$1,069.67	$623.25	$ 979.17	$486.06
Production Rate, Tons/Day	10.81	7.22	12.96	8.64	19.00	12.73
Cost, S.F. in Place (incl. 33% toe in)	$ 12.39	$ 7.78	$ 14.45	$ 8.42	$ 18.62	$ 9.24
Production Rate, S.F./Day (incl. 33% toe in)	983	656	960	640	1,000	670

Crew B-40	Bare Costs		Inc. Subs O&P		Installation Cost per Ton
	Hr.	Daily	Hr.	Daily	
1 Pile Driver Foreman	$29.20	$ 233.60	$50.45	$ 403.60	for 15′ Deep Excavation plus 33% toe-in
4 Pile Drivers	27.20	870.40	47.00	1,504.00	$3,689.30 per day/10.81 tons = $341.29 per ton
1 Equip. Oper. Oiler	24.05	192.40	36.60	292.80	Installation Cost per S.F.
2 Equip. Oper. (crane)	29.35	469.60	44.65	714.40	$\frac{2000\#}{22\#/S.F.}$ = 90.9 S.F. x 10.81 tons = 983 S.F. per day
1 Crane, 40 Ton		704.30		774.73	
Vibratory Hammer & Gen.		1,219.00		1,340.90	$3,689.30 / 983 S.F. per day = $3.75 per S.F.
64 L.H., Daily Totals		$3,689.30		$5,030.43	

*For driving & extracting, two mobilizations & demobilizations will be necessary

SITE WORK 2

REFERENCE NOS.

R021-620 Wood Sheet Piling

Wood sheet piling may be used for depths to 20' where there is no ground water. If moderate ground water is encountered Tongue & Groove sheeting will help to keep it out. When considerable ground water is present, steel sheeting must be used.

For estimating purposes on trench excavation, sizes are as follows:

Depth	Sheeting	Wales	Braces	B.F. per S.F.
To 8'	3 x 12's	6 x 8's, 2 line	6 x 8's, @ 10'	4.0 @ 8'
8' x 12'	3 x 12's	10 x 10's, 2 line	10 x 10's, @ 9'	5.0 average
12' to 20'	3 x 12's	12 x 12's, 3 line	12 x 12's, @ 8'	7.0 average

Sheeting to be toed in at least 2' depending upon soil conditions. A five person crew with an air compressor and sheeting driver can drive and brace 440 SF/day at 8' deep, 360 SF/day at 12' deep, and 320 SF/day at 16' deep. For normal soils, piling can be pulled in 1/3 the time to install. Pulling difficulty increases with the time in the ground. Production can be increased by high pressure jetting. Figures below assume 50% of lumber is salvaged and includes pulling costs. Some jurisdictions require an equipment operator in addition to Crew B-31.

Sheeting Pulled

Daily Cost Crew B-31	L.H./ Day	Hourly Cost	Daily Cost	8' Depth, 440 S.F./Day		16' Depth, 320 S.F./Day	
				To Drive (1 Day)	To Pull (1/3 Day)	To Drive (1 Day)	To Pull (1/3 Day)
1 Foreman	8	$23.45	$ 187.60	$ 187.60	$ 62.47	$ 187.60	$ 62.47
3 Laborers	24	21.45	514.80	514.80	171.43	514.80	171.43
1 Carpenter	8	27.30	218.40	218.40	72.73	218.40	72.73
1 Air Compressor			116.80	116.80	38.89	116.80	38.89
1 Sheeting Driver			5.95	5.95	1.98	5.95	1.98
2 -50 Ft. Air Hoses, 1-1/2" Diam.			12.80	12.80	4.26	12.80	4.26
Lumber (50% salvage)				1.76 MBF 576.40		2.24 MBF 733.60	
Total			$1,056.35	$1,632.75	$351.76	$1,789.95	$351.76
Total/S.F.				$ 3.71	$ .80	$ 5.59	$ 1.10
Total (Drive and Pull)/S.F.				$ 4.51		$ 6.69	

Sheeting Left in Place

Daily Cost	8' Depth 440 S.F./Day		10' Depth 400 S.F./Day		12' Depth 360 S.F./Day		16' Depth 320 S.F./Day		18' Depth 305 S.F./Day		20' Depth 280 S.F./Day	
Crew B-31		$1,073.30		$1,073.30		$1,073.30		$1,073.30		$1,073.30		$1,073.30
Lumber	1.76 M	1,152.80	1.8 M	1,179.00	1.8 M	1,179.00	2.24 M	1,467.20	2.1 M	1,375.50	1.9 M	1,244.50
Total in Place		$2,226.10		$2,252.30		$2,252.30		$2,540.50		$2,448.80		$2,317.80
Total/S.F.		$ 5.06		$ 5.63		$ 6.26		$ 7.94		$ 8.03		$ 8.28
Total/M.B.F.		$1,264.83		$1,251.28		$1,251.28		$1,134.15		$1,166.10		$1,219.89

R022-220   Compacting Backfill

Compaction of fill in embankments, around structures, in trenches, and under slabs is important to control settlement. Factors affecting compaction are:

1. Soil gradation
2. Moisture content
3. Equipment used
4. Depth of fill per lift
5. Density required

Production Rate:

$$\frac{1.75' \text{ plate width x 50 F.P.M. x 50 min./hr. x .67' lift}}{27 \text{ C.F. per C.Y.}} = 108.5 \text{ C.Y./hr.}$$

Production Rate for 4 Passes:

$$\frac{108.5 \text{ C.Y.}}{4 \text{ passes}} = 27.125 \text{ C.Y./hr. x 8 hrs.} = 217 \text{ C.Y./day}$$

The costs for testing and soil analyses are listed in Division 014-108. Also, see Division 022 for further backfill, borrow, and compaction costs.

Example:

Compact granular fill around a building foundation using a 21″ wide x 24″ vibratory plate in 8″ lifts. Operator moves at 50 FPM working a 50 minute hour to develop 95% Modified Proctor Density with 4 passes.

	Compacting 217 C.Y. with 21″ Wide Vibratory Plate	L.H./ Day	Hourly Cost	Daily Cost	C.Y. Cost
1	Laborer	8	$21.45	$171.60	$ .79
1	Vibratory Plate Compactor			52.00	.24
Total for	217 C.Y./day			$223.60	$1.03

R022-240 Excavating

The selection of equipment used for structural excavation and bulk excavation or for grading is determined by the following factors.

1. Quantity of material.
2. Type of material.
3. Depth or height of cut.
4. Length of haul.
5. Condition of haul road.
6. Accessibility of site.
7. Moisture content and dewatering requirements.
8. Availability of excavating and hauling equipment.

Some additional costs must be allowed for hand trimming the sides and bottom of concrete pours and other excavation below the general excavation.

Number of B.C.Y. per truck = 1.5 C.Y. bucket × 8 passes = 12 loose C.Y.

$$= 12 \times \frac{100}{118} = 10.2 \text{ B.C.Y. per truck}$$

Truck Haul Cycle:

Load truck 8 passes	=	4 minutes
Haul distance 1 mile	=	9 minutes
Dump time	=	2 minutes
Return 1 mile	=	7 minutes
Spot under machine	=	1 minute
		23 minute cycle

When planning excavation and fill, the following should also be considered.

1. Swell factor.
2. Compaction factor.
3. Moisture content.
4. Density requirements.

A typical example for scheduling and estimating the cost of excavation of a 15′ deep basement on a dry site when the material must be hauled off the site, is outlined below.

Assumptions:

1. Swell factor, 18%.
2. No mobilization or demobilization.
3. Allowance included for idle time and moving on job.
4. No dewatering, sheeting, or bracing.
5. No truck spotter or hand trimming.

Fleet Haul Production per day in B.C.Y.

$$4 \text{ trucks} \times \frac{50 \text{ min. hour}}{23 \text{ min. haul cycle}} \times 8 \text{ hrs.} \times 10.2 \text{ B.C.Y.}$$

$$= 4 \times 2.2 \times 8 \times 10.2 = 718 \text{ B.C.Y./day}$$

	Excavating Cost with a 1-1/2 C.Y. Hydraulic Excavator 15′ Deep, 2 Mile Round Trip Haul	L.H./Day	Hourly Cost	Daily Cost	Subtotal	Unit Price
1	Equipment Operator	8	$29.35	$ 234.80		
1	Oiler	8	24.05	192.40		
4	Truck Drivers	32	22.10	707.20	$1,134.40	$1.58
1	Hydraulic Excavator			712.70		
4	Dump Trucks			1,769.00	2,481.70	3.45
Total for 720 B.C.Y.				$3,616.10	$3,616.10	$5.02

Description	1-1/2 C.Y. Hyd. Backhoe 15′ Deep		1-1/2 C.Y. Power Shovel 7′ Bank		1-1/2 C.Y. Dragline 7′ Deep		2-1/2 C.Y. Trackloader Stockpile	
Operator (and Oiler, if required)		$ 427.20		$ 427.20		$ 427.20		$ 427.20
Truck Drivers	3 Ea.	530.40	4 Ea.	707.20	3 Ea.	530.40	4 Ea.	707.20
Equipment Rental		712.70		864.00		892.85		865.10
20 C.Y. Trailer Dump Trucks	3 Ea.	1,326.75	4 Ea.	1,769.00	3 Ea.	1,326.75	4 Ea.	1,769.00
Total Cost per Day		$2,997.05		$3,767.40		$3,177.20		$3,768.50
Daily Production, C.Y. Bank Measure		720.00		960.00		640.00		1000.00
Cost per C.Y.		$ 4.16		$ 3.92		$ 4.96		$ 3.77

Add the mobilization and demobilization costs to the total excavation costs. When equipment is rented for more than three days, there is often no mobilization charge by the equipment dealer. On larger jobs outside of urban areas, scrapers can move earth economically provided a dump site or fill area and adequate haul roads are available. Excavation within sheeting bracing or cofferdam bracing is usually done with a clamshell and production is low, since the clamshell may have to be guided by hand between the bracing. When excavating or filling an area enclosed with a wellpoint system, add 10% to 15% to the cost to allow for restricted access. When estimating earth excavation quantities for structures, allow work space outside the building footprint for construction of the foundation, and a slope of 1:1 unless sheeting is used.

R022-250 Excavating Equipment

The table below lists THEORETICAL hourly production in C.Y./hr. bank measure for some typical excavation equipment. Figures assume 50 minute hours, 83% job efficiency, 100% operator efficiency, 90° swing and properly sized hauling units, which must be modified for adverse digging and loading conditions. Actual production costs in the front of the book average about 50% of the theoretical values listed here.

Equipment	Soil Type	B.C.Y. Weight	% Swell	1 C.Y.	1-1/2 C.Y	2 C.Y.	2-1/2 C.Y.	3 C.Y.	3-1/2 C.Y.	4 C.Y.
Hydraulic Excavator	Moist loam, sandy clay	3400 lb.	40%	85	125	175	220	275	330	380
"Backhoe"	Sand and gravel	3100	18	80	120	160	205	260	310	365
15' Deep Cut	Common earth	2800	30	70	105	150	190	240	280	330
	Clay, hard, dense	3000	33	65	100	130	170	210	255	300
	Moist loam, sandy clay	3400	40	170 (6.0)	245 (7.0)	295 (7.8)	335 (8.4)	385 (8.8)	435 (9.1)	475 (9.4)
Power Shovel	Sand and gravel	3100	18	165 (6.0)	225 (7.0)	275 (7.8)	325 (8.4)	375 (8.8)	420 (9.1)	460 (9.4)
Optimum Cut (Ft.)	Common earth	2800	30	145 (7.8)	200 (9.2)	250 (10.2)	295 (11.2)	335 (12.1)	375 (13.0)	425 (13.8)
	Clay, hard, dense	3000	33	120 (9.0)	175 (10.7)	220 (12.2)	255 (13.3)	300 (14.2)	335 (15.1)	375 (16.0)
	Moist loam, sandy clay	3400	40	130 (6.6)	180 (7.4)	220 (8.0)	250 (8.5)	290 (9.0)	325 (9.5)	385 (10.0)
Drag Line	Sand and gravel	3100	18	130 (6.6)	175 (7.4)	210 (8.0)	245 (8.5)	280 (9.0)	315 (9.5)	375 (10.0)
Optimum Cut (Ft.)	Common earth	2800	30	110 (8.0)	160 (9.0)	190 (9.9)	220 (10.5)	250 (11.0)	280 (11.5)	310 (12.0)
	Clay, hard, dense	3000	33	90 (9.3)	130 (10.7)	160 (11.8)	190 (12.3)	225 (12.8)	250 (13.3)	280 (12.0)

				Wheel Loaders				Track Loaders		
				3 C.Y.	4 C.Y.	6 C.Y.	8 C.Y.	2-1/4 C.Y.	3 C.Y.	4 C.Y.
	Moist loam, sandy clay	3400	40	260	340	510	690	135	180	250
Loading Tractors	Sand and gravel	3100	18	245	320	480	650	130	170	235
	Common earth	2800	30	230	300	460	620	120	155	220
	Clay, hard, dense	3000	33	200	270	415	560	110	145	200
	Rock, well-blasted	4000	50	180	245	380	520	100	130	180

R022-510 Vibroflotation and Vibro Replacement Soil Compaction

Vibroflotation is a proprietary system of compacting sandy soils in place to increase relative density to about 70%. Typical bearing capacities attained will be 6000 psf for saturated sand and 12,000 psf for dry sand. Usual range is 4000 to 8000 psf capacity. Costs in the front of the book are for a vertical foot of compacted cylinder 6' to 10' in diameter.

Vibro replacement is a proprietary system of improving cohesive soils in place to increase bearing capacity. Most silts and clays above or below the water table can be strengthened by installation of stone columns.

The process consists of radial displacement of the soil by vibration. The created hole is then backfilled in stages with coarse granular fill which is thoroughly compacted and displaced into the surrounding soil in the form of a column.

The total project cost would depend on the number and depth of the compacted cylinders. The installing company guarantees relative soil density of the sand cylinders after compaction and the bearing capacity of the soil after the replacement process. Detailed estimating information is available from the installer at no cost.

R023-620 Wood Bearing Piles

Untreated Southern Yellow Pine is most generally used for pile foundations cut off below the low water line. These are driven with the bark on. All piles cut off above the low water line should be treated with a preservative or encased in concrete for the section above water.

Item	Unit Cost	Units	Quantity	Total Cost	Cost/Pile	Cost/L.F.
50' Treated Piles, 13" Butt, 7" Tip	$ 10.91	L.F.	200	$109,100.00	$545.50	$10.91
Installation, Crew B-19	3,203.65	Day	13	41,647.45	208.24	4.16
Mobilization & Demobilization, Crew B-19	3,203.65	Day	3	9,610.95	48.05	.96
Transportation of Equip. One Way	1,090.75	Day	1	1,090.75	5.45	.11
Totals				$161,449.15	$807.24	$16.14

The above figures are based on driving 800 L.F. daily which can be considered average. Time is included for moving rig, cutoff & ordinary delays. A general observation is that the cost of a pile in place complete is about two times the cost of pile only. See also equipment rental division 016-408 and R016-410 for equipment capacities.

R023-710 Pressure Injected Footings

Pressure Injected Footings are end bearing foundation units consisting of expanded bulbous bases formed by high energy blows, and concrete shafts either cased or uncased, to transfer the load.

The bulb must be formed in granular, rock or hardpan soils. Bearing is achieved at a predetermined depth so that the length of the shaft is usually less than for friction piles. High load capacity reduces the number of units and caps required.

Mobilization and demobilization costs below assume maximum distance of 50 miles.

100 Uncased Pressure Injected Footings 25' Long	Quantity Unit	Unit Cost	Total	Cost/Pile	Cost/L.F.
Crew B-44, 8 piles/day	13 Days	$2,528.25	$32,867.25	$328.67	$13.15
Mobilization & Demobilization	2 Days	2,528.25	5,056.50	50.57	2.02
Transportation one way	1.5 Days	1,090.75	1,636.13	16.36	.65
Total Installation			$39,559.88	$395.60	$15.82
Shafts: 100 ea. 17" diam., 25' deep	155 C.Y.	$ 101.40	$15,717.00	$157.17	$ 6.29
Pressure Bulbs: 100 ea. @ .75 C.Y.	75 C.Y.	101.40	7,605.00	76.05	3.04
Waste: 15% of above	35 C.Y.	101.40	3,549.00	35.49	1.42
Total Material Cost			$26,871.00	$268.71	$10.75
Total Cost in Place			$66,430.88	$664.31	$26.57

R023-810 Caissons

The three principal types of caissons are:

(1) **Belled Caissons,** which except for shallow depths and poor soil conditions, are generally recommended. They provide more bearing than shaft area. Because of its conical shape, no horizontal reinforcement of the bell is required.

(2) **Straight Shaft Caissons** are used where relatively light loads are to be supported by caissons that rest on high value bearing strata. While the shaft is larger in diameter than for belled types this is more than offset by the saving in time and labor.

(3) **Keyed Caissons** are used when extremely heavy loads are to be carried. A keyed or socketed caisson transfers its load into rock by a combination of end-bearing and shear reinforcing of the shaft. The most economical shaft often consists of a steel casing, a steel wide flange core and concrete. Allowable compressive stresses of 0.225 f'c for concrete, 16,000 psi for the wide flange core, and 9,000 psi for the steel casing are commonly used. The usual range of shaft diameter is 18" to 84". The number of sizes specified for any one project should be limited due to the problems of casing and auger storage. When hand work is to be performed, shaft diameters should not be less than 32". When inspection of borings is required a minimum shaft diameter of 30" is recommended. Concrete caissons are intended to be poured against earth excavation so permanent forms which add to cost should not be used if the excavation is clean and the earth sufficiently impervious to prevent excessive loss of concrete.

Soil Conditions for Belling		
Good	Requires Handwork	Not Recommended
Clay	Hard Shale	Silt
Sandy Clay	Limestone	Sand
Silty Clay	Sandstone	Gravel
Clayey Silt	Weathered Mica	Igneous Rock
Hard-pan		
Soft Shale		
Decomposed Rock		

R024-520   Single Track R.R. Siding

Description and Bare Costs	Quantity Unit	Unit Cost	Bare Cost/L.F.	Cost with O&P/L.F.
Rail @ 100 lb/yard	.033 Ton	$ 800.00	$26.40	$ 29.04
Spikes, Plates, and Bolts	.200 CWT	50.00	10.00	11.00
6" x 8" x 8'-6" Treated Timber Ties, 22" O.C.	.550 Ea.	26.87	14.78	16.26
Ballast, 1-1/2" Stone @ .67 Ton/L.F.	.667 Ton	13.00	8.67	9.54
Crew B-14 @ 57 L.F./Day	.018 Day	1,310.40	23.59	35.04
Total Cost in Place			$83.44	$100.88

R024-530   Single Track, Steel Ties, Concrete Bed

Description and Bare Costs	Quantity Unit	Unit Cost	Bare Cost/L.F.	Cost with O&P/L.F.
Rail, Prime Grade @ 100 lb/yard	.033 Ton	$ 800.00	$ 26.40	$ 29.04
Fasteners and Plates	.200 CWT	50.00	10.00	11.00
6" WF Beams @ 30" O.C. 6'-6" Long	2.600 L.F.	6.55	17.03	18.73
Concrete, 9' wide, 10" thick	.280 C.Y.	92.00	25.76	28.34
Crew B-14 @ 22 L.F./Day	.045 Day	1,310.40	58.97	87.60
Total Cost in Place			$138.16	$174.71

R025-110 Bituminous Paving

| City | Bituminous Asphalt per Ton* | Pavement (3") 6.13 S.Y./ton | | | | Sidewalks (2") 9.2 S.Y./ton | | | |
| | | Cost per S.Y. | | | Per Ton | Cost per S.Y. | | | Per Ton |
		Material*	Installation	Total	Total	Material*	Installation	Total	Total
Atlanta	$24.65	$4.02	$ .71	$4.73	$28.99	$2.68	$1.12	$3.80	$34.96
Baltimore	30.75	5.02	.77	5.79	35.47	3.34	1.31	4.65	42.78
Boston	35.00	5.71	1.02	6.73	41.27	3.80	2.16	5.96	54.83
Buffalo	32.00	5.22	.99	6.21	38.05	3.48	2.03	5.51	50.69
Chicago	33.00	5.38	1.02	6.40	39.24	3.59	2.16	5.75	52.90
Cincinnati	33.50	5.46	.87	6.33	38.82	3.64	1.71	5.35	49.22
Cleveland	29.80	4.86	.96	5.82	35.70	3.24	2.05	5.29	48.67
Columbus	26.75	4.36	.86	5.22	32.00	2.91	1.65	4.56	41.95
Dallas	23.95	3.91	.70	4.61	28.24	2.60	1.09	3.69	33.95
Denver	24.50	4.00	.74	4.74	29.04	2.66	1.24	3.90	35.88
Detroit	28.25	4.61	.96	5.57	34.15	3.07	2.02	5.09	46.83
Houston	29.55	4.82	.74	5.56	34.10	3.21	1.24	4.45	40.94
Indianapolis	26.50	4.32	.87	5.19	31.83	2.88	1.68	4.56	41.95
Kansas City	23.50	3.83	.89	4.72	28.92	2.55	1.76	4.31	39.65
Los Angeles	30.00	4.89	1.01	5.90	36.20	3.26	2.16	5.42	49.86
Memphis	28.20	4.60	.69	5.29	32.44	3.07	1.09	4.16	38.27
Milwaukee	26.40	4.31	.99	5.30	32.47	2.87	2.05	4.92	45.26
Minneapolis	27.75	4.53	.94	5.47	33.53	3.02	1.95	4.97	45.72
Nashville	27.75	4.53	.73	5.26	32.24	3.02	1.21	4.23	38.92
New Orleans	37.00	6.04	.69	6.73	41.27	4.02	1.09	5.11	47.01
New York City	43.05	7.02	1.24	8.26	50.65	4.68	2.93	7.61	70.01
Philadelphia	26.50	4.32	1.00	5.32	32.58	2.88	2.17	5.05	46.46
Phoenix	23.25	3.79	.77	4.56	27.93	2.53	1.35	3.88	35.70
Pittsburgh	31.30	5.11	.90	6.01	36.86	3.40	1.80	5.20	47.84
St. Louis	24.50	4.00	.97	4.97	30.48	2.66	2.06	4.72	43.42
San Antonio	27.10	4.42	.67	5.09	31.22	2.95	1.00	3.95	36.34
San Diego	27.50	4.49	1.01	5.50	33.74	2.99	2.16	5.15	47.38
San Francisco	32.00	5.22	1.03	6.25	38.30	3.48	2.17	5.65	51.98
Seattle	33.25	5.42	.96	6.38	39.12	3.61	2.01	5.62	51.70
Washington, D.C.	35.00	5.71	.77	6 48	39.73	3.80	1.34	5.14	47.29
Average	$29.40	$4.80	$ .88	$5.68	$34.82	$3.20	$1.73	$4.92	$45.28

Assumed density is 145 lb. per C.F.
*Includes delivery within 20 miles

Table below shows quantities and bare costs for 1000 S.Y. of Bituminous Paving.

| Item | Roads and Parking Areas, 3" Thick (025-104-0460) | | | Sidewalks, 2" Thick (025-128-0010) | | |
	Quantities		Cost	Quantities		Cost
Bituminous asphalt	163 tons @ $29.40	per ton	$4,792.20	109 tons @ $29.40	per ton	$3,204.60
Installation using	Crew B-25B @ $4,331.00	/4900SY/ day x 1000	883.88	Crew B-37 @ $1,244.45	/720 SY/day x 1000	1,728.40
Total per 1000 S.Y.			$5,676.08			$4,933.00
Total per S.Y.			$ 5.68			$ 4.93
Total per Ton			$ 34.82			$ 45.26

R027-110   Concrete Pipe

Prices given are for inside 20 mile delivery zone. Add $1.70 per ton of pipe for each additional 10 miles. Minimum truckload is 10 tons. The non-reinforced pipe listed in the front of the book is designation ASTM C14-59 extra strength. The reinforced pipe listed is ASTM C76-65T class 3, no gaskets. The installation cost given includes shaping bottom of the trench, placing the pipe, and backfilling and tamping to the top of the pipe only.

Site Work   R029   Landscaping

R029-310   Seeding

The type of grass is determined by light, shade and moisture content of soil plus intended use. Fertilizer should be disked 4″ before seeding. For steep slopes disk five tons of mulch and lay two tons of hay or straw on surface per acre after seeding. Surface mulch can be staked, lightly disked or tar emulsion sprayed. Material for mulch can be wood chips, peat moss, partially rotted hay or straw, wood fibers and sprayed emulsions. Hemp seed blankets with fertilizer are also available. For spring seeding, watering is necessary. Late fall seeding may have to be reseeded in the spring. Hydraulic seeding, power mulching, and aerial seeding can be used on large areas.

R029-545   Cost of Trees: Based on Pin Oak (Quercus palustris)

Tree Diameter	Normal Height	Catalog List Price of Tree	Guying Material	Equipment Charge	Installation Labor	Total
2 to 3 inch	14 feet	$ 130	$ 15.00	$ 58.31	$ 73.76	$ 277.07
3 to 4 inch	16 feet	248	18.00	97.18	122.93	486.11
4 to 5 inch	18 feet	390	65.00	116.62	147.52	719.14
6 to 7 inch	22 feet	790	80.00	145.78	184.40	1,200.18
8 to 9 inch	26 feet	1,300	100.00	194.37	245.87	1,840.23

Installation Time & Cost for Planting Trees, Bare Costs													
Ball Size Diam. X Depth	Soil in Ball	Weight of Ball	Hole Diam. Req'd	Hole Excavation	Amount of Soil Displ.	Topsoil Handled	Time Required in Labor-Hours					Cost	
							Dig & Lace	Handle Ball	Dig Hole	Plant & Prune	Water & Guy	Total L.H.	Crew	Total per Tree
Inches	C.F.	Lbs.	Feet	C.F.	C.F.	C.F.								
12 x 12	.70	56.00	2.00	4.00	3.00	11.00	.25	.17	.33	.25	.07	1.10	1 Clab	$ 23.60
18 x 16	2.00	160.00	2.50	8.00	6.00	21.00	.50	.33	.47	.35	.08	1.70	2 Clab	36.47
24 x 18	4.00	320.00	3.00	13.00	9.00	38.00	1.00	.67	1.08	.82	.20	3.80	3 Clab	81.51
30 x 21	7.50	600.00	4.00	27.00	19.50	76.00	.82	.71	.79	1.22	.26	3.80		123.46
36 x 24	12.50	980.00	4.50	38.00	25.50	114.00	1.08	.95	1.11	1.32	.30	4.76		154.65
42 x 27	19.00	1,520.00	5.50	64.00	45.00	185.00	1.90	1.27	1.87	1.43	.34	6.80	B-6	220.93
48 x 30	28.00	2,040.00	6.00	85.00	57.00	254.00	2.41	1.60	2.06	1.55	.39	8.00	@	259.92
54 x 33	38.50	3,060.00	7.00	127.00	88.50	370.00	2.86	1.90	2.39	1.76	.45	9.40	$32.49	305.41
60 x 36	52.00	4,160.00	7.50	159.00	107.00	474.00	3.26	2.17	2.73	2.00	.51	10.70	per	347.64
66 x 39	68.00	5,440.00	8.00	196.00	128.00	596.00	3.61	2.41	3.07	2.26	.58	11.90	Labor-	386.63
72 x 42	87.00	7,160.00	9.00	267.00	180.00	785.00	3.90	2.60	3.71	2.78	.70	13.70	hour	445.11

R031-010   Wall Form Materials

Aluminum Forms

Approximate weight is 3 lbs. per S.F.C.A. Standard widths are available from 4" to 36" with 36" most common. Standard lengths of 2', 4', 6' to 8' are available. Forms are lightweight and fewer ties are needed with the wider widths. The form face is either smooth or textured.

Cost of bare forms per S.F.C.A. with different facing surface is listed below. Typical material cost including usual accessories but not including form ties is $18.05 per S.F.C.A.

Forms may also be rented.

Purchase Cost Per S.F.								
Finish	3' x 8'	2' x 8'	12" x 8'	6" x 8'	3' x 4'	2' x 4'	12" x 4'	6" x 4'
Smooth Aluminum (.096" Face Sheet)	$12.47	$15.40	$20.86	$34.55	$13.36	$18.38	$24.12	$37.69
Textured Brick Aluminum	$14.51	$18.49	$24.95	$39.68	$16.67	$22.02	$28.26	$44.98

Metal Framed Plywood Forms

Manufacturers claim over 75 reuses of plywood and over 300 reuses of steel frames. Sale price for steel framed forms is $8.50 per S.F. for 2' x 8' to $12.15 per S.F. for 2' x 3'. Narrower forms range between $11.75 per S.F. to $21.60 per S.F. for 1' x 3'. Many specials such as corners, fillers, pilasters, etc. are available. Monthly rental is generally about 15% of purchase price for first month and 9% per month thereafter with 90% of rental applied to purchase for the first month and decreasing percentages thereafter. Aluminum framed forms cost 25% to 30% more than steel framed.

Rule of thumb purchase price including corners, specials, etc.; steel framed $12.20 per S.F.; aluminum framed $17.30 per S.F.

Reconditioned steel framed forms are rented for an average of $1.30 per S.F per month. Aluminum framed forms are rented for $2.25 per S.F. for the first month and $1.50 per S.F. per month thereafter.

After the first month, extra days may be prorated from the monthly charge. Rental rates do not include ties, accessories, cleaning, loss of hardware or freight in and out. Approximate weight is 5 lbs. per S.F. for steel; 3 lbs. per S.F. for aluminum.

Forms can be rented with option to buy.

Plywood Forms, Job Fabricated

There are two types of plywood used for concrete forms.

1. Exterior plyform which is completely waterproof. This is face oiled to facilitate stripping. Ten reuses can be expected with this type with 25 reuses possible.
2. An overlaid type consists of a resin fiber fused to exterior plyform. No oiling is required except to facilitate cleaning. This is available in both high density (HDO) and medium density overlaid (MDO). Using HDO, 50 reuses can be expected with 200 possible.

Plyform is available in 5/8" and 3/4" thickness. High density overlaid is available in 3/8", 1/2", 5/8" and 3/4" thickness.

5/8" thick is sufficient for most building forms, while 3/4" is best on heavy construction.

For prices on plywood and framing lumber see R061-010 and R061-020.

Plywood Forms, Modular, Prefabricated

There are many plywood forming systems without frames. Most of these are manufactured from 1-1/8" (HDO) plywood and have some hardware attached. These are used principally for foundation walls 8' or less high. With care and maintenance, 100 reuses can be attained with decreasing quality of surface finish. Sale price of 2' x 8' panels is $6.20 per S.F. with 4" x 8' fillers costing $24.40 per S.F. Typical forms and accessories for 1000 S.F. of form area cost $6,970 not including form ties.

Steel Forms

Approximate weight is 6-1/2 lbs. per S.F.C.A. including accessories. Standard widths are available from 2" to 24", with 24" most common. Standard lengths are from 2' to 8', with 4' the most common. Forms are easily ganged into modular units.

Sale price for typical hand set job is $12.15 per S.F.C.A. including all usual accessories, specials, corners, etc., but not including ties or freight. Cost of bare forms runs from $9.75 for wide forms to over $21.60 per S.F.C.A. for narrow widths and/or short lengths. Forms are usually leased for 15% of the purchase price per month prorated daily over 30 days. Standard 6000 lb. wall ties for 12" walls cost $84 per hundred and 24" long ties cost $121 per hundred.

Rental may be applied to sale price and usually rental forms are bought. With careful handling and cleaning 200 to 400 reuses are possible.

Straight wall gang forms up to 12' x 20' or 8' x 30' can be fabricated. These crane handled forms usually cost from $20.50 to $28.20 per S.F.C.A. or can be leased for 8.7% per month. Straight wall gang forms utilizing 40,000 lb. and 60,000 lb. ties with sizes from 24' x 10' to 24' x 24' cost from $21.50 to $29.75 per S.F. including all accessories and taper ties. Rental rates on these gang forms run from $.30 to $.65 per week per S.F.C.A.

Individual job analysis is available from the manufacturer at no charge.

R031-020   Floor Pans and Domes

For 8' to 15' Ceiling Heights Using Crew C-2 at $26.66 per Labor-Hour	20" Pans (031-150)				19" Domes (031-150)			
	(Line 3500)	1 Use	(Line 3650)	4 Use	(Line 4000)	1 Use	(Line 4150)	4 Use
Rent 20" pans or 19" domes		$ 82.00		$ 25.00		$ 82.00		$ 25.00
Adjustable shores and accessories		46.00		25.25		46.00		25.25
110 S.F. 3/4" plyform at $880 per M.S.F.		96.80		31.46		96.80		31.46
210 B.F. supporting joists, girts and braces at $540 per MBF		113.40		36.86		113.40		36.86
Labor handle, place, strip, oil and clean pans and domes	1.5 L.H.	39.99	1.0 L.H.	26.66	1.8 L.H.	47.99	1.2 L.H.	31.99
Make up, erect; remove wood decking	10 L.H.	266.60	8.5 L.H.	226.61	10.0 L.H.	266.60	8.5 L.H.	226.61
Total per 100 S.F. of floor area		$644.79		$371.84		$652.79		$377.17

The figures above are for closed deck forming. For open deck forming deduct $25 per 100 S.F. for four uses. For pan rental, figure $.82 per S.F. of total area for one use; $.41 for two uses; $.30 for three uses; and $.25 for four uses, which is about the most that can be expected for any one project. The purchase price of fiberglass long pans is $5.75 per S.F.C.A. or an average $11.50 per S.F. of floor area.*

For two-way grid system, 19" steel domes are leased for $.25 to $.75 and 30" fiberglass domes are leased for $.50 to $2.25 per S.F. of floor area.

The purchase price of custom size fiberglass forms runs from $6.00 to $17.00 per S.F.C.A.*

For slab height from 15' to 20' add $55 per 100 S.F. and for 20' to 35' add $75 per 100 S.F., both for four uses.

*It is necessary to divide the purchase cost by the number of expected uses to determine the cost per use.

R031-030   Slipforms

The slipform method of forming may be used for forming circular silo and multi-celled storage bin type structures over 30' high, and building core shear walls over eight stories high. The shear walls, usually enclose elevator shafts, stairwells, mechanical spaces, and toilet rooms. Reuse of the form on duplicate structures will reduce the height necessary and spread the cost of building the form. Slipform systems can be used to cast chimneys, towers, piers, dams, underground shafts or other structures capable of being extruded.

Slipforms are usually 4' high and are raised semi-continuously by jacks climbing on rods which are embedded in the concrete. The jacks are powered by a hydraulic, pneumatic, or electric source and are available in 3, 6, and 22 ton capacities. Interior work decks and exterior scaffolds must be provided for placing inserts, embedded items, reinforcing steel, and concrete. Scaffolds below the form for finishers may be required. The interior work decks are often used as roof slab forms on silos and bin work. Form raising rates will range from 6" to 20" per hour for silos; 6" to 30" per hour for buildings; and 6" to 48" per hour for shaft work.

Reinforcing bars and stressing strands are usually hoisted by crane or gin pole and the concrete material can be hoisted by crane, winch-powered skip, or pumps. The slipform system is operated on a continuous 24 hour day when a monolithic structure is desired. For least cost, the system is operated only during normal working hours.

The cost of the form, including equipment setting, will range from $8.00 to $20 per square foot of actual form built. The cost of work decks and scaffolds will range from $2.50 to $10 per square foot of total deck area. Slipform equipment rental will range from $.20 to $.60 per S.F. of surface formed. Operating costs of slipform systems will range from $.07 to $.40 per S.F. of surface formed.

Placing concrete will range from 0.5 to 1.5 labor-hours per C.Y. Finishing (if required) and curing costs will range from $.06 to $.30 per S.F. of surface. Bucks, blockouts, keyways, weldplates, etc. are extra. Recent in-place total costs have ranged from $275 to $900 per C.Y.

3

CONCRETE

REFERENCE NOS.

R031-040 Forms for Reinforced Concrete

Design Economy

Avoid many sizes in proportioning beams and columns.

From story to story avoid changing column dimensions. Gain strength by adding steel or using a richer mix. If a change in size of column is necessary, vary one dimension only to minimize form alterations. Keep beams and columns the same width.

From floor to floor in a multi-story building vary beam depth, not width, as that will leave slab panel form unchanged. It is cheaper to vary the strength of a beam from floor to floor by means of steel area than by 2" changes in either width or depth.

Cost Factors

Material includes the cost of lumber, cost of rent for metal pans or forms if used, nails, form ties, form oil, bolts and accessories.

Labor includes the cost of carpenters to make up, erect, remove and repair, plus common labor to clean and move. Having carpenters remove forms minimizes repairs.

Improper alignment and condition of forms will increase finishing cost. When forms are heavily oiled, concrete surfaces must be neutralized before finishing. Special curing compounds will cause spillages to spall off in first frost. Gang forming methods will reduce costs on large projects.

Materials Used

Boards are seldom used unless their architectural finish is required. Generally, steel, fiberglass and plywood are used for contact surfaces. Labor on plywood is 10% less than with boards. The plywood is backed up with

2 x 4's at 12" to 32" O.C. Walers are generally 2 - 2 x 4's. Column forms are held together with steel yokes or bands. Shoring is with adjustable shoring or scaffolding for high ceilings.

Reuse

Floor and column forms can be reused four or possibly five times without excessive repair. Remember to allow for 10% waste on each reuse.

When modular sized wall forms are made, up to twenty uses can be expected with exterior plyform.

When forms are reused, the cost to erect, strip, clean and move will not be affected. 10% replacement of lumber should be included and about one hour of carpenter time for repairs on each reuse per 100 S.F.

The reuse cost for certain accessory items normally rented on a monthly basis will be lower than the cost for the first use.

After fifth use, new material required plus time needed for repair prevent form cost from dropping further and it may go up. Much depends on care in stripping, the number of special bays, changes in beam or column sizes and other factors.

Costs for multiple use of formwork may be developed as follows:

2 Uses	3 Uses	4 Uses
$\dfrac{(\text{1st Use + Reuse})}{2} = \text{avg. cost/2 uses}$	$\dfrac{(\text{1st Use + 2 Reuse})}{3} = \text{avg. cost/3 uses}$	$\dfrac{(\text{1st use + 3 Reuse})}{4} = \text{avg. cost/4 uses}$

R031-050 Forms In Place

This section assumes that all cuts are made with power saws, that adjustable shores are employed and that maximum use is made of commercial form ties and accessories. Bare costs are used in the table below.

BEAM AND GIRDER, INTERIOR, 12" Wide (Line 031-138-2000)			First Use			Reuse		
Item	Cost	Unit	Quantity	Material	Installation	Quantity	Material	Installation
5/8" exterior plyform	$760.00	M.S.F.	115 S.F.	$ 87.40		11.5 S.F.	$ 8.75	
Lumber	540.00	M.B.F.	200 B.F.	108.00		20.0 B.F.	10.80	
Accessories, incl. adjustable shores			Allow	24.00		Allow	24.00	
Make up, crew C-2	26.66	L.H.	6.4 L.H.		$170.60	1.0 L.H.		$ 26.65
Erect and strip			8.3 L.H.		221.30	8.3 L.H.		221.30
Clean and move			1.3 L.H.		34.65	1.3 L.H.		34.65
Total per 100 S.F.C.A.			16.0 L.H.	$219.40	$426.55	10.6 L.H.	$43.55	$282.60

For structural steel frame with beams encased, subtract 1.2 labor-hours, and 50 B.F. lumber or about $55.00 per 100 S.F.C.A. for the first use and $32 for each reuse.

BOX CULVERT, 5' to 8' Square or Rectangular (Line 031-146-0010)			First Use			Reuse		
Item	Cost	Unit	Quantity	Material	Installation	Quantity	Material	Installation
3/4" exterior plyform	$880.00	M.S.F.	110 S.F.	$ 96.80		11.0 S.F.	$ 9.70	
Lumber	540.00	M.B.F.	170 B.F.	91.80		17.0 B.F.	9.20	
Accessories			Allow	19.00		Allow	19.00	
Build in place, crew C-1	25.84	L.H.	14.5 L.H.		$374.70	14.5 L.H.		$374.70
Strip and salvage			4.3 L.H.		111.10	4.3 L.H.		111.10
Total per 100 S.F.C.A.			18.8 L.H.	$207.60	$485.80	18.8 L.H.	$37.90	$485.80

R031-050 Forms in Place (cont.)

COLUMNS, 24" x 24" (Line 031-142-6500)			First Use			Reuse		
Item	Cost	Unit	Quantity	Material	Installation	Quantity	Material	Installation
5/8" exterior plyform	$760.00	M.S.F.	120 S.F.	$ 91.20		12.0 S.F.	$ 9.10	
Lumber	540.00	M.B.F.	125 B.F.	67.50		12.5 B.F.	6.75	
Clamps, chamfer strips and accessories			Allow	24.00		Allow	24.00	
Make up, crew C-1	25.84	L.H.	5.8 L.H.		$149.85	1.0 L.H.		$ 25.85
Erect and strip			9.8 L.H.		253.25	9.8 L.H.		253.25
Clean and move			1.2 L.H.		31.00	1.2 L.H.		31.00
Total per 100 S.F.C.A.			16.8 L.H.	$182.70	$434.10	12.0 L.H.	$39.85	$310.10

FLAT SLAB WITH DROP PANELS (Line 031-150-2000)			First Use			Reuse		
Item	Cost	Unit	Quantity	Material	Installation	Quantity	Material	Installation
5/8" exterior plyform	$760.00	M.S.F.	115 S.F.	$ 87.40		11.5 S.F.	$ 8.75	
Lumber	540.00	M.B.F.	210 B.F.	113.40		21.0 B.F.	11.35	
Accessories, incl. adjustable shores			Allow	24.00		Allow	24.00	
Make up, crew C-2	26.66	L.H.	3.5 L.H.		$ 93.30	1.0 L.H.		$ 26.65
Erect and strip			6.0 L.H.		159.95	6.0 L.H.		159.95
Clean and move			1.2 L.H.		32.00	1.2 L.H.		32.00
Total per 100 S.F.C.A.			10.7 L.H.	$224.80	$285.25	8.2 L.H.	$44.10	$218.60

Drop panels included but column caps figure with columns.

FOOTINGS, SPREAD (Line 031-158-5000)			First Use			Reuse		
Item	Cost	Unit	Quantity	Materials	Installation	Quantity	Material	Installation
Lumber	$540.00	M.B.F.	260 B.F.	$140.40		26 B.F.	$14.05	
Accessories			Allow	7.65		Allow	7.65	
Make up, crew C-1	25.84	L.H.	4.7 L.H.		$121.45	1.0 L.H.		$ 25.85
Erect and strip			4.2 L.H.		108.55	4.2 L.H.		108.55
Clean and move			1.6 L.H.		41.35	1.6 L.H.		41.35
Total per 100 S.F.C.A.			10.5 L.H.	$148.05	$271.35	6.8 L.H.	$21.70	$175.75

FOUNDATION WALL, 8' High (Line 031-182-2000)			First Use			Reuse		
Item	Cost	Unit	Quantity	Material	Installation	Quantity	Material	Installation
5/8" exterior plyform	$760.00	M.S.F.	110 S.F.	$ 83.60		11.0 S.F.	$ 8.35	
Lumber	540.00	M.B.F.	140 B.F.	75.60		14.0 B.F.	7.55	
Accessories			Allow	24.00		Allow	24.00	
Make up, crew C-2	26.66	L.H.	5.0 L.H.		$133.30	1.0 L.H.		$ 26.65
Erect and strip			6.5 L.H.		173.30	6.5 L.H.		173.30
Clean and move			1.5 L.H.		40.00	1.5 L.H.		40.00
Total per 100 S.F.C.A.			13.0 L.H.	$183.20	$346.60	9.0 L.H.	$39.90	$239.95

PILE CAPS, Square or Rectangular (Line 031-158-3000)			First Use			Reuse		
Item	Cost	Unit	Quantity	Material	Installation	Quantity	Material	Installation
5/8" exterior plyform	$760.00	M.S.F.	110 S.F.	$ 83.60		11.0 S.F.	$ 8.35	
Lumber	540.00	M.B.F.	160 B.F.	86.40		16.0 B.F.	8.65	
Accessories			Allow	7.28		Allow	7.28	
Make up, crew C-1	25.84	L.H.	4.5 L.H.		$116.30	1.0 L.H.		$ 25.85
Erect and strip			5.0 L.H.		129.20	5.0 L.H.		129.20
Clean and move			1.5 L.H.		38.75	1.5 L.H.		38.75
Total per 100 S.F.C.A.			11.0 L.H.	$177.28	$284.25	7.5 L.H.	$24.28	$193.80

STAIRS, Average Run (Inclined Length x Width) (Line 031-174-0010)			First Use			Reuse		
Item	Cost	Unit	Quantity	Materials	Installation	Quantity	Material	Installation
5/8" exterior plyform	$760.00	M.S.F.	110 S.F.	$ 83.60		11.0 S.F.	$ 8.35	
Lumber	540.00	M.B.F.	425 B.F.	229.50		42.5 B.F.	22.95	
Accessories			Allow	19.00		Allow	19.00	
Build in place, crew C-2	26.66	L.H.	25.0 L.H.		$666.50	25.0 L.H.		$666.50
Strip and salvage			4.0 L.H.		106.65	4.0 L.H.		106.65
Total per 100 S.F.			29.0 L.H.	$332.10	$773.15	29.0 L.H.	$50.30	$773.15

CONCRETE 3

REFERENCE NOS.

R031-060 Formwork Labor Hours

Item	Unit	Hours Required			Total Hours	Multiple Use		
		Fabricate	Erect & Strip	Clean & Move	1 Use	2 Use	3 Use	4 Use
Beam and Girder, interior beams, 12" wide	100 S.F.	6.4	8.3	1.3	16.0	13.3	12.4	12.0
Hung from steel beams		5.8	7.7	1.3	14.8	12.4	11.6	11.2
Beam sides only, 36" high		5.8	7.2	1.3	14.3	11.9	11.1	10.7
Beam bottoms only, 24" wide		6.6	13.0	1.3	20.9	18.1	17.2	16.7
Box out for openings		9.9	10.0	1.1	21.0	16.6	15.1	14.3
Buttress forms, to 8' high		6.0	6.5	1.2	13.7	11.2	10.4	10.0
Centering, steel, 3/4" rib lath			1.0		1.0			
3/8" rib lath or slab form			0.9		0.9			
Chamfer strip or keyway	100 L.F.		1.5		1.5	1.5	1.5	1.5
Columns, fiber tube 8" diameter			20.6		20.6			
12"			21.3		21.3			
16"			22.9		22.9			
20"			23.7		23.7			
24"			24.6		24.6			
30"			25.6		25.6			
Round Steel, 12" diameter			22.0		22.0	22.0	22.0	22.0
16"			25.6		25.6	25.6	25.6	25.6
20"			30.5		30.5	30.5	30.5	30.5
24"			37.7		37.7	37.7	37.7	37.7
Plywood 8" x 8"	100 S.F.	7.0	11.0	1.2	19.2	16.2	15.2	14.7
12" x 12"		6.0	10.5	1.2	17.7	15.2	14.4	14.0
16" x 16"		5.9	10.0	1.2	17.1	14.7	13.8	13.4
24" x 24"		5.8	9.8	1.2	16.8	14.4	13.6	13.2
Steel framed plywood 8" x 8"			10.0	1.0	11.0	11.0	11.0	11.0
12" x 12"			9.3	1.0	10.3	10.3	10.3	10.3
16" x 16"			8.5	1.0	9.5	9.5	9.5	9.5
24" x 24"			7.8	1.0	8.8	8.8	8.8	8.8
Drop head forms, plywood		9.0	12.5	1.5	23.0	19.0	17.7	17.0
Coping forms		8.5	15.0	1.5	25.0	21.3	20.0	19.4
Culvert, box			14.5	4.3	18.8	18.8	18.8	18.8
Curb forms, 6" to 12" high, on grade		5.0	8.5	1.2	14.7	12.7	12.1	11.7
On elevated slabs		6.0	10.8	1.2	18.0	15.5	14.7	14.3
Edge forms to 6" high, on grade	100 L.F.	2.0	3.5	0.6	6.1	5.6	5.4	5.3
7" to 12" high	100 S.F.	2.5	5.0	1.0	8.5	7.8	7.5	7.4
Equipment foundations		10.0	18.0	2.0	30.0	25.5	24.0	23.3
Flat slabs, including drops		3.5	6.0	1.2	10.7	9.5	9.0	8.8
Hung from steel		3.0	5.5	1.2	9.7	8.7	8.4	8.2
Closed deck for domes		3.0	5.8	1.2	10.0	9.0	8.7	8.5
Open deck for pans		2.2	5.3	1.0	8.5	7.9	7.7	7.6
Footings, continuous, 12" high		3.5	3.5	1.5	8.5	7.3	6.8	6.6
Spread, 12" high		4.7	4.2	1.6	10.5	8.7	8.0	7.7
Pile caps, square or rectangular		4.5	5.0	1.5	11.0	9.3	8.7	8.4
Grade beams, 24" deep		2.5	5.3	1.2	9.0	8.3	8.0	7.9
Lintel or Sill forms		8.0	17.0	2.0	27.0	23.5	22.3	21.8
Spandrel beams, 12" wide		9.0	11.2	1.3	21.5	17.5	16.2	15.5
Stairs			25.0	4.0	29.0	29.0	29.0	29.0
Trench forms in floor		4.5	14.0	1.5	20.0	18.3	17.7	17.4
Walls, Plywood, at grade, to 8' high		5.0	6.5	1.5	13.0	11.0	9.7	9.5
8' to 16'		7.5	8.0	1.5	17.0	13.8	12.7	12.1
16' to 20'		9.0	10.0	1.5	20.5	16.5	15.2	14.5
Foundation walls, to 8' high		4.5	6.5	1.0	12.0	10.3	9.7	9.4
8' to 16' high		5.5	7.5	1.0	14.0	11.8	11.0	10.6
Retaining wall to 12' high, battered		6.0	8.5	1.5	16.0	13.5	12.7	12.3
Radial walls to 12' high, smooth		8.0	9.5	2.0	19.5	16.0	14.8	14.3
But in 2' chords		7.0	8.0	1.5	16.5	13.5	12.5	12.0
Prefabricated modular, to 8' high		—	4.3	1.0	5.3	5.3	5.3	5.3
Steel, to 8' high		—	6.8	1.2	8.0	8.0	8.0	8.0
8' to 16' high		—	9.1	1.5	10.6	10.3	10.2	10.2
Steel framed plywood to 8' high		—	6.8	1.2	8.0	7.5	7.3	7.2
8' to 16' high		—	9.3	1.2	10.5	9.5	9.2	9.0

3 CONCRETE

REFERENCE NOS.

R032-010 Reinforcing Steel Weights and Measures

Bar Designation No.**	Nominal Weight Lb./Ft.	U.S. Customary Units Nominal Dimensions*			SI Units Nominal Dimensions*			
		Diameter in.	Cross Sectional Area, in.^{2}	Perimeter in.	Nominal Weight kg/m	Diameter mm	Cross Sectional Area, cm^2	Perimeter mm
3	.376	.375	.11	1.178	.560	9.52	.71	29.9
4	.668	.500	.20	1.571	.994	12.70	1.29	39.9
5	1.043	.625	.31	1.963	1.552	15.88	2.00	49.9
6	1.502	.750	.44	2.356	2.235	19.05	2.84	59.8
7	2.044	.875	.60	2.749	3.042	22.22	3.87	69.8
8	2.670	1.000	.79	3.142	3.973	25.40	5.10	79.8
9	3.400	1.128	1.00	3.544	5.059	28.65	6.45	90.0
10	4.303	1.270	1.27	3.990	6.403	32.26	8.19	101.4
11	5.313	1.410	1.56	4.430	7.906	35.81	10.06	112.5
14	7.650	1.693	2.25	5.320	11.384	43.00	14.52	135.1
18	13.600	2.257	4.00	7.090	20.238	57.33	25.81	180.1

* The nominal dimensions of a deformed bar are equivalent to those of a plain round bar having the same weight per foot as the deformed bar.

** Bar numbers are based on the number of eighths of an inch included in the nominal diameter of the bars.

R032-020 Metric Rebar Specification - ASTM A615-81

Grade 300 (300 MPa* = 43,560 psi; +8.7% vs. Grade 40)				
Grade 400 (400 MPa* = 58,000 psi; −3.4% vs. Grade 60)				
Bar No.	Diameter mm	Area mm^2	Equivalent in.^{2}	Comparison with U.S. Customary Bars
10	11.3	100	.16	Between #3 & #4
15	16.0	200	.31	#5 (.31 in.^{2})
20	19.5	300	.47	#6 (.44 in.^{2})
25	25.2	500	.78	#8 (.79 in.^{2})
30	29.9	700	1.09	#9 (1.00 in.^{2})
35	35.7	1000	1.55	#11 (1.56 in.^{2})
45	43.7	1500	2.33	#14 (2.25 in.^{2})
55	56.4	2500	3.88	#18 (4.00 in.^{2})

* MPa = megapascals

R032-030 Common Stock Styles of Welded Wire Fabric

This table provides some of the basic specifications, sizes, and weights of welded wire fabric used for reinforcing concrete.

	New Designation Spacing — Cross Sectional Area (in.) — (Sq. in. 100)	Old Designation Spacing — Wire Gauge (in.) — (AS & W)		Steel Area per Foot Longitudinal		Transverse		Approximate Weight per 100 S.F.	
				in.	cm	in.	cm	lbs	kg
Rolls	6 x 6 — W1.4 x W1.4	6 x 6 — 10 x 10		.028	.071	.028	.071	21	9.53
	6 x 6 — W2.0 x W2.0	6 x 6 — 8 x 8	1	.040	.102	.040	.102	29	13.15
	6 x 6 — W2.9 x W2.9	6 x 6 — 6 x 6		.058	.147	.058	.147	42	19.05
	6 x 6 — W4.0 x W4.0	6 x 6 — 4 x 4		.080	.203	.080	.203	58	26.91
	4 x 4 — W1.4 x W1.4	4 x 4 — 10 x 10		.042	.107	.042	.107	31	14.06
	4 x 4 — W2.0 x W2.0	4 x 4 — 8 x 8	1	.060	.152	.060	.152	43	19.50
	4 x 4 — W2.9 x W2.9	4 x 4 — 6 x 6		.087	.227	.087	.227	62	28.12
	4 x 4 — W4.0 x W4.0	4 x 4 — 4 x 4		.120	.305	.120	.305	85	38.56
Sheets	6 x 6 — W2.9 x W2.9	6 x 6 — 6 x 6		.058	.147	.058	.147	42	19.05
	6 x 6 — W4.0 x W4.0	6 x 6 — 4 x 4		.080	.203	.080	.203	58	26.31
	6 x 6 — W5.5 x W5.5	6 x 6 — 2 x 2	2	.110	.279	.110	.279	80	36.29
	4 x 4 — W1.4 x W1.4	4 x 4 — 4 x 4		.120	.305	.120	.305	85	38.56

NOTES: 1. Exact W—number size for 8 gauge is W2.1
 2. Exact W—number size for 2 gauge is W5.4

CONCRETE 3

REFERENCE NOS.

R032-040 Weight of Steel Reinforcing Per Square Foot of Wall (PSF)

Reinforced Weights: The table below suggests the weights per square foot for reinforcing steel in walls. Weights are approximate and will be the same for all grades of steel bars. For bars in two directions, add weights for each size and spacing.

C/C Spacing in Inches	Bar Size								
	#3 Wt. (PSF)	#4 Wt. (PSF)	#5 Wt. (PSF)	#6 Wt. (PSF)	#7 Wt. (PSF)	#8 Wt. (PSF)	#9 Wt. (PSF)	#10 Wt. (PSF)	#11 Wt. (PSF)
2"	2.26	4.01	6.26	9.01	12.27				
3"	1.50	2.67	4.17	6.01	8.18	10.68	13.60	17.21	21.25
4"	1.13	2.01	3.13	4.51	6.13	8.10	10.20	12.91	15.94
5"	.90	1.60	2.50	3.60	4.91	6.41	8.16	10.33	12.75
6"	.752	1.34	2.09	3.00	4.09	5.34	6.80	8.61	10.63
8"	.564	1.00	1.57	2.25	3.07	4.01	5.10	6.46	7.97
10"	.451	.802	1.25	1.80	2.45	3.20	4.08	5.16	6.38
12"	.376	.668	1.04	1.50	2.04	2.67	3.40	4.30	5.31
18"	.251	.445	.695	1.00	1.32	1.78	2.27	2.86	3.54
24"	.188	.334	.522	.751	1.02	1.34	1.70	2.15	2.66
30"	.150	.267	.417	.600	.817	1.07	1.36	1.72	2.13
36"	.125	.223	.348	.501	.681	.890	1.13	1.43	1.77
42"	.107	.191	.298	.429	.584	.753	.97	1.17	1.52
48"	.094	.167	.261	.376	.511	.668	.85	1.08	1.33

R032-050 Minimum Wall Reinforcement Weight (PSF)

This table lists the approximate minimum wall reinforcement weights per S.F. according to the specification of .12% of gross area for vertical bars and .20% of gross area for horizontal bars.

Location	Wall Thickness	Bar Size	Horizontal Steel Spacing C/C	Sq. In. Req'd per S.F.	Total Wt. per S.F.	Bar Size	Vertical Steel Spacing C/C	Sq. In. Req'd per S.F.	Total Wt. per S.F.	Horizontal & Vertical Steel Total Weight per S.F.
Both Faces	10"	#4	18"	.24	.89#	#3	18"	.14	.50#	1.39#
	12"	#4	16"	.29	1.00	#3	16"	.17	.60	1.60
	14"	#4	14"	.34	1.14	#3	13"	.20	.69	1.84
	16"	#4	12"	.38	1.34	#3	11"	.23	.82	2.16
	18"	#5	17"	.43	1.47	#4	18"	.26	.89	2.36
One Face	6"	#3	9"	.15	.50	#3	18"	.09	.25	.75
	8"	#4	12"	.19	.67	#3	11"	.12	.41	1.08
	10"	#5	15"	.24	.83	#4	16"	.14	.50	1.34

R032-060 Reinforcing Extras

There are two "base prices" used, the price for bars shipped directly from the mill and the base price for the bars shipped from the steel fabricator's shops. The extras for both methods of pricing are listed below and are listed in dollars per ton for A615-87 steel.

Extras	Item	From Mill	From Fabricator	Extras	Item	From Mill	From Fabricator
Base Price		$284	$475				$500 minimum
Size Extras	#3	$102	$105	Quantity Extras	Under 20 tons	$12	$30
	#4	41	51		20-49 tons	—	20
	#5	30	35		50-99 tons	—	15
	#6	25	25		100-300 tons	—	10
	#7 to 11	20	20		Over 300 tons	—	—
	#14	41	40	Bending	Light #2 thru #11	N/A	85
	#18	41	40		Heavy #4 up		44
Grade Extras	Grade 40	—	—	Detailing	Under 50 tons		46
	Grade 60	$ 13	$ 13		50-100 tons		35
	Grade 75	N/A	N/A		150-500 tons		31
					Over 500 tons		21

"Listing only" runs $4 to $5 per ton.

R032-070   Bend, Place and Tie Reinforcing

See R32-060 for shop bending costs. Field bending costs are usually higher.

Placing and tying by rodmen for footings and slabs runs from nine hrs. per ton for heavy bars to fifteen hrs. per ton for light bars. For beams, columns, and walls, production runs from eight hrs. per ton for heavy bars to twenty hrs. per ton for light bars. Overall average for typical reinforced concrete buildings is about fourteen hrs. per ton. These production figures include placing of accessories and usual inserts. Equipment handling is necessary for the larger size bars so that installation costs for the very heavy bars will not decrease proportionately.

For tie wire, figure 5 lbs. 16 ga. black annealed wire at $.80 per lb. for each ton of bars. Use of commercial accessories improves placing accuracy and costs of $61 per ton of bars is offset by reduced labor time.

Installation costs for splicing reinforcing bars includes allowance for equipment to hold the bars in place while splicing as well as necessary scaffolding for iron workers.

R032-080 Reinforcing Key Prices

Costs given below are for over 50 tons delivered and bent from fabricators in metropolitan area. Over 500 tons, subtract $11 per ton. Figures are for domestic A615, grade 60. Economic conditions can alter the prices substantially.

City	Bare Costs Per Ton					Costs Incl. Subs O & P		
	Material		Installation		Total per Ton	Material	Installation	Total
	50 Tons Fabricated	Allow for Accessories	Unload, Sort and Pile	Place & Tie	in Place	Incl. 10% Mark-Up	Include Mark-up	per Ton
Atlanta	485	$61	$18	$452	$1,016	$601	$ 557	$1,158
Baltimore	500	61	20	504	1,086	617	648	1,265
Boston	480	61	26	640	1,207	595	879	1,474
Buffalo	515	61	24	589	1,188	634	790	1,424
Chicago	475	61	27	663	1,225	590	917	1,507
Cincinnati	425	61	21	515	1,022	535	665	1,200
Cleveland	495	61	23	581	1,161	612	780	1,392
Columbus	490	61	21	518	1,090	606	670	1,276
Dallas	455	61	16	393	925	568	456	1,024
Denver	465	61	17	433	976	579	523	1,102
Detroit	415	61	23	579	1,079	524	775	1,299
Houston	445	61	16	411	934	557	486	1,043
Indianapolis	485	61	21	534	1,102	601	699	1,300
Kansas City	455	61	21	524	1,061	568	680	1,248
Los Angeles	480	61	26	645	1,212	595	885	1,480
Memphis	490	61	17	415	982	606	494	1,100
Milwaukee	500	61	24	588	1,172	617	788	1,405
Minneapolis	470	61	22	557	1,111	584	737	1,321
Nashville	465	61	17	431	975	579	520	1,099
New Orleans	460	61	16	391	927	573	452	1,025
New York	480	61	36	904	1,482	595	1,335	1,930
Philadelphia	485	61	25	635	1,207	601	872	1,473
Phoenix	495	61	19	473	1,048	612	594	1,206
Pittsburgh	520	61	23	583	1,187	639	781	1,420
St. Louis	495	61	22	562	1,140	612	743	1,355
San Antonio	415	61	15	373	864	524	423	947
San Diego	505	61	26	645	1,237	623	885	1,508
San Francisco	520	61	26	649	1,256	639	892	1,531
Seattle	465	61	23	568	1,117	579	754	1,333
Washington, D.C.	490	61	19	482	1,052	606	610	1,216
Average per ton	475	$61	$22	$546	$1,104	$592	$ 710	$1,300

Note: For field fabrication, add $75 to $100 per ton to shop fabrication costs.

Base price of A615, Grade 40 is $284 per ton.

Place and tie costs do not include crane charge.

Bare Material Costs		Total Installed Costs		
Item	Cost per Ton	Description	Bare Costs	Cost Incl. Subs O&P
Base price at mill	$284	Unload, sort & pile, C-5 crew	$ 22	$ 34
Grade 60 extra	13	Place & tie, 14 hours at $39.44 per hour	546	676
Size and length extras	30			
Freight from mill	26	Total Installation Cost	$ 568	$ 710
Warehouse handling & storage	22			
Shearing and shop bending	43	Bare Material Cost	$ 536	
Drafting, detailing and listing	34	Material Cost + 10%		$ 590
Trucking to job site	23	Total Installed Cost	$1,104	$1,300
Reinforcing Delivered	$475			
Accessories Delivered	61			
Total Material Delivered	$536			

Engineering or design fees are not included above.

Material delivered for small jobs can range up to $700 per ton with erection going as high as $725 per ton if job consists of small bars.

Restrictive union practices can increase field labor costs above those listed above.

514

R032-090   Prestressed Concrete, Post-tensioned

In post-tensioned concrete the steel tendons are tensioned after the concrete has reached about 3/4 of its ultimate strength. The cableways are grouted after tensioning to provide bond between the steel and concrete. If bond is to be prevented, the tendons are coated with a corrosion-preventative grease and wrapped with waterproof paper or plastic. Bonded tendons are usually used when ultimate strength (beams & girders) is a controlling factor.

High strength concrete is used to fully utilize the steel, thereby reducing the size and weight of the member. A plasticizing agent may be added to reduce water content. Maximum size aggregate ranges from 1/2" to 1-1/2" depending on the spacing of the tendons.

The types of steel commonly used are bars and strands. Job conditions determine which is best suited. Bars are best for vertical prestresses since they are easy to support. The trend is for steel manufacturers to supply a finished package, cut to length, which reduces field preparation to a minimum.

Bars vary from 3/4" to 1-3/8" diameter. Table below gives time in labor-hours per tendon and the labor cost per pound for placing, tensioning and grouting (if required) a 75' beam. Tendons used in buildings are not usually grouted; tendons for bridges usually are grouted. For strands the table indicates the labor-hours and labor cost per pound for typical prestressed units 100' long. Simple span beams usually require one end stressing regardless of lengths. Continuous beams are usually stressed from two ends. Long slabs are poured from the center outward and stressed in 75' increments after the initial 150' center pour. Prices below do not include subcontractor's overhead and profit. Prices in the front of the book do not include subcontractor's overhead and profit, except in the column headed Total Incl. O & P.

Labor Hours per Tendon and Labor Costs per Pound of Prestress Steel						
Length	100' Beam		75' Beam		100' Slab	
Type Steel	Strand		Bars		Strand	
Diameter	0.5"		3/4"	1-3/8"	0.5"	0.6"
Number	12	24	1	1	1	1
Force in Kips	298	595	42	143	25	35
Preparation & Placing Cables	6.6	10.0	0.7	2.3	0.8	0.8
Stressing Cables	2.4	3.0	0.6	1.3	0.4	0.4
Grouting, if required	3.0	3.5	0.5	1.0		
Total Labor Hours	12.0	16.5	1.8	4.6	1.2	1.2
Prestressing Steel Weights	640#	1280#	115#	380#	53#	74#
Labor cost per lb. Bonded Non-bonded	$.68 to $.88	$.54 to $.65	$.54 to $.65 $.47 to $.56	$.38 to $.45 $.41 to $.49	$.87 to $1.09	$.87 to $1.09

Flat Slab construction — 4000 psi concrete with span to depth ratio between 36 and 44. Two way post-tensioned steel averages 1.0 lb. per S.F. for 24' to 28' bays (usually strand) and additional reinforcing steel averages .5 lb. per S.F.

Pan and Joist construction — 4000 psi concrete with span to depth ratio 28 to 30. Post-tensioned steel averages .8 lb. per S.F. and reinforcing steel about 1.0 lb. per S.F. Placing and stressing averages 40 hours per ton of total material.

Beam construction — 4000 to 5000 psi concrete. Steel weights vary greatly. Tendon material cost runs from $1.00 to $1.85 per lb. for grouted and $.85 to $1.18 per lb. for non-grouted strand. Bar costs are from $.85 to $1.34 per lb.

From the chart above placing and stressing cost for grouted tendons runs from $.38 to $.88 per lb; non-grouted placing costs run from $.41 to $1.09 per lb. Labor cost per pound goes down as the size and length of the tendon increases. The primary economic consideration is the cost per kip for the member.

Post-tensioning becomes feasible for beams and girders over 30' long; for continuous two-way slabs over 20' clear; also in transferring upper building loads over longer spans at lower levels. Post-tension suppliers will provide engineering services at no cost to the user. Substantial economies are possible by using post-tensioned Lift Slabs. See R033-120 for Lift Slabs.

R033-010   Proportionate Quantities

The tables below show both quantities per S.F. of floor areas as well as form and reinforcing quantities per C.Y. Unusual structural requirements would increase the ratios below. High strength reinforcing would reduce the steel weights. Figures are for 3000 psi concrete and 60,000 psi reinforcing unless specified otherwise.

Type of Construction	Live Load	Span	Per S.F. of Floor Area				Per C.Y. of Concrete		
			Concrete	Forms	Reinf.	Pans	Forms	Reinf.	Pans
Flat Plate	50 psf	15 Ft.	.46 C.F.	1.06 S.F.	1.71 lb.		62 S.F.	101 lb.	
		20	.63	1.02	2.40		44	104	
		25	.79	1.02	3.03		35	104	
	100	15	.46	1.04	2.14		61	126	
		20	.71	1.02	2.72		39	104	
		25	.83	1.01	3.47		33	113	
Flat Plate (waffle construction) 20″ domes	50	20	.43	1.00	2.10	.84 S.F.	63	135	53 S.F.
		25	.52	1.00	2.90	.89	52	150	46
		30	.64	1.00	3.70	.87	42	155	37
	100	20	.51	1.00	2.30	.84	53	125	45
		25	.64	1.00	3.20	.83	42	135	35
		30	.76	1.00	4.40	.81	36	160	29
Waffle Construction 30″ domes	50	25	.69	1.06	1.83	.68	42	72	40
		30	.74	1.06	2.39	.69	39	87	39
		35	.86	1.05	2.71	.69	33	85	39
		40	.78	1.00	4.80	.68	35	165	40
Flat Slab (two way with drop panels)	50	20	.62	1.03	2.34		45	102	
		25	.77	1.03	2.99		36	105	
		30	.95	1.03	4.09		29	116	
	100	20	.64	1.03	2.83		43	119	
		25	.79	1.03	3.88		35	133	
		30	.96	1.03	4.66		29	131	
	200	20	.73	1.03	3.03		38	112	
		25	.86	1.03	4.23		32	133	
		30	1.06	1.03	5.30		26	135	
One Way Joists 20″ Pans	50	15	.36	1.04	1.40	.93	78	105	70
		20	.42	1.05	1.80	.94	67	120	60
		25	.47	1.05	2.60	.94	60	150	54
	100	15	.38	1.07	1.90	.93	77	140	66
		20	.44	1.08	2.40	.94	67	150	58
		25	.52	1.07	3.50	.94	55	185	49
One Way Joists 8″ x 16″ filler blocks	50	15	.34	1.06	1.80	.81 Ea.	84	145	64 Ea.
		20	.40	1.08	2.20	.82	73	145	55
		25	.46	1.07	3.20	.83	63	190	49
	100	15	.39	1.07	1.90	.81	74	130	56
		20	.46	1.09	2.80	.82	64	160	48
		25	.53	1.10	3.60	.83	56	190	42
One Way Beam & Slab	50	15	.42	1.30	1.73		84	111	
		20	.51	1.28	2.61		68	138	
		25	.64	1.25	2.78		53	117	
	100	15	.42	1.30	1.90		84	122	
		20	.54	1.35	2.69		68	154	
		25	.69	1.37	3.93		54	145	
	200	15	.44	1.31	2.24		80	137	
		20	.58	1.40	3.30		65	163	
		25	.69	1.42	4.89		53	183	
Two Way Beam & Slab	100	15	.47	1.20	2.26		69	130	
		20	.63	1.29	3.06		55	131	
		25	.83	1.33	3.79		43	123	
	200	15	.49	1.25	2.70		41	149	
		20	.66	1.32	4.04		54	165	
		25	.88	1.32	6.08		41	187	

R033-010   Proportionate Quantities (cont.)

4000 psi Concrete and 60,000 psi Reinforcing—Form and Reinforcing Quantities per C.Y.					
Item	Size	Forms	Reinforcing	Minimum	Maximum
Columns (square tied)	10″ x 10″	130 S.F.C.A.	#5 to #11	220 lbs.	875 lbs.
	12″ x 12″	108	#6 to #14	200	955
	14″ x 14″	92	#7 to #14	190	900
	16″ x 16″	81	#6 to #14	187	1082
	18″ x 18″	72	#6 to #14	170	906
	20″ x 20″	65	#7 to #18	150	1080
	22″ x 22″	59	#8 to #18	153	902
	24″ x 24″	54	#8 to #18	164	884
	26″ x 26″	50	#9 to #18	169	994
	28″ x 28″	46	#9 to #18	147	864
	30″ x 30″	43	#10 to #18	146	983
	32″ x 32″	40	#10 to #18	175	866
	34″ x 34″	38	#10 to #18	157	772
	36″ x 36″	36	#10 to #18	175	852
	38″ x 38″	34	#10 to #18	158	765
	40″ x 40″	32	#10 to #18	143	692

Item	Size	Form	Spiral	Reinforcing	Minimum	Maximum
Columns (spirally reinforced)	12″ diameter	34.5 L.F.	190 lbs.	#4 to #11	165 lbs.	1505 lbs.
		34.5	190	#14 & #18	—	1100
	14″	25	170	#4 to #11	150	970
		25	170	#14 & #18	800	1000
	16″	19	160	#4 to #11	160	950
		19	160	#14 & #18	605	1080
	18″	15	150	#4 to #11	160	915
		15	150	#14 & #18	480	1075
	20″	12	130	#4 to #11	155	865
		12	130	#14 & #18	385	1020
	22″	10	125	#4 to #11	165	775
		10	125	#14 & #18	320	995
	24″	9	120	#4 to #11	195	800
		9	120	#14 & #18	290	1150
	26″	7.3	100	#4 to #11	200	729
		7.3	100	#14 & #18	235	1035
	28″	6.3	95	#4 to #11	175	700
		6.3	95	#14 & #18	200	1075
	30″	5.5	90	#4 to #11	180	670
		5.5	90	#14 & #18	175	1015
	32″	4.8	85	#4 to #11	185	615
		4.8	85	#14 & #18	155	955
	34″	4.3	80	#4 to #11	180	600
		4.3	80	#14 & #18	170	855
	36″	3.8	75	#4 to #11	165	570
		3.8	75	#14 & #18	155	865
	40″	3.0	70	#4 to #11	165	500
		3.0	70	#14 & #18	145	765

CONCRETE 3

REFERENCE NOS.

R033-010 Proportionate Quantities (cont.)

3000 psi Concrete and 60,000 psi Reinforcing—Form and Reinforcing Quantities per C.Y.

Item	Type	Loading	Height	C.Y./L.F.	Forms/C.Y.	Reinf./C.Y.
Retaining Walls	Cantilever	Level Backfill	4 Ft.	.2 C.Y.	49 S.F.	35 lbs.
			8	.5	42	45
			12	.8	35	70
			16	1.1	32	85
			20	1.6	28	105
		Highway Surcharge	4	.3	41	35
			8	.5	36	55
			12	.8	33	90
			16	1.2	30	120
			20	1.7	27	155
		Railroad Surcharge	4	.4	28	45
			8	.8	25	65
			12	1.3	22	90
			16	1.9	20	100
			20	2.6	18	120
	Gravity, with Vertical Face	Level Backfill	4	.4	37	None
			7	.6	27	
			10	1.2	20	
		Sloping Surcharge	4	.3	31	
			7	.8	21	
			10	1.6	15	↓

		Live Load in Kips per Linear Foot							
		Under 1 Kip		2 to 3 Kips		4 to 5 Kips		6 to 7 Kips	
	Span	Forms	Reinf.	Forms	Reinf.	Forms	Reinf.	Forms	Reinf.
Beams	10 Ft.	—	—	90 S.F.	170 #	85 S.F.	175 #	75 S.F.	185 #
	16	130 S.F.	165 #	85	180	75	180	65	225
	20	110	170	75	185	62	200	51	200
	26	90	170	65	215	62	215	—	—
	30	85	175	60	200	—	—	—	—

Item	Size	Type	Forms per C.Y.	Reinforcing per C.Y.
Spread Footings	Under 1 C.Y.	1,000 psf soil	24 S.F.	44 lbs.
		5,000	24	42
		10,000	24	52
	1 C.Y. to 5 C.Y.	1,000	14	49
		5,000	14	50
		10,000	14	50
	Over 5 C.Y.	1,000	9	54
		5,000	9	52
		10,000	9	56
Pile Caps (30 Ton Concrete Piles)	Under 5 C.Y.	shallow caps	20	65
		medium	20	50
		deep	20	40
	5 C.Y. to 10 C.Y.	shallow	14	55
		medium	15	45
		deep	15	40
	10 C.Y. to 20 C.Y.	shallow	11	60
		medium	11	45
		deep	12	35
	Over 20 C.Y.	shallow	9	60
		medium	9	45
		deep	10	40

R033-010 Proportionate Quantities (cont.)

Item	Size	Pile Spacing	50 T Pile	100 T Pile	50 T Pile	100 T Pile
			3000 psi Concrete and 60,000 psi Reinforcing — Form and Reinforcing Quantities per C.Y.			
Pile Caps (Steel H Piles)	Under 5 C.Y.	24" O.C.	24 S.F.	24 S.F.	75 lbs.	90 lbs.
		30"	25	25	80	100
		36"	24	24	80	110
	5 C.Y. to 10 C.Y.	24"	15	15	80	110
		30"	15	15	85	110
		36"	15	15	75	90
	Over 10 C.Y.	24"	13	13	85	90
		30"	11	11	85	95
		36"	10	10	85	90

		8" Thick		10" Thick		12" Thick		15" Thick	
	Height	Forms	Reinf.	Forms	Reinf.	Forms	Reinf.	Forms	Reinf.
Basement Walls	7 Ft.	81 S.F.	44 lbs.	65 S.F.	45 lbs.	54 S.F.	44 lbs.	41 S.F.	43 lbs.
	8		44		45		44		43
	9		46		45		44		43
	10		57		45		44		43
	12		83		50		52		43
	14		116		65		64		51
	16				86		90		65
	18						106		70

R033-020 Materials for One C.Y. of Concrete

This is an approximate method of figuring quantities of cement, sand and coarse aggregate for a field mix with waste allowance included.

With crushed gravel as coarse aggregate, to determine barrels of cement required, divide 10 by total mix; that is, for 1:2:4 mix, 10 divided by 7 = 1-3/7 barrels.

If the coarse aggregate is crushed stone, use 10-1/2 instead of 10 as given for gravel.

To determine tons of sand required, multiply barrels of cement by parts of sand and then by 0.2; that is, for the 1:2:4 mix, as above, 1-3/7 x 2 x .2 = .57 tons.

Tons of crushed gravel are in the same ratio to tons of sand as parts in the mix, or 4/2 x .57 = 1.14 tons.

1 bag cement = 94#	1 C.Y. sand or crushed gravel = 2700#	1 C.Y. crushed stone = 2575#
4 bags = 1 barrel	1 ton sand or crushed gravel = 20 C.F.	1 ton crushed stone = 21 C.F.

Average carload of cement is 692 bags; of sand or gravel is 56 tons.

Do not stack stored cement over 10 bags high.

R033-030 Metric Equivalents of Cement Content for Concrete Mixes

94 Pound Bags per Cubic Yard	Kilograms per Cubic Meter	94 Pound Bags per Cubic Yard	Kilograms per Cubic Meter
1.0	55.77	7.0	390.4
1.5	83.65	7.5	418.3
2.0	111.5	8.0	446.2
2.5	139.4	8.5	474.0
3.0	167.3	9.0	501.9
3.5	195.2	9.5	529.8
4.0	223.1	10.0	557.7
4.5	251.0	10.5	585.6
5.0	278.8	11.0	613.5
5.5	306.7	11.5	641.3
6.0	334.6	12.0	669.2
6.5	362.5	12.5	697.1

a. If you know the cement content in pounds per cubic yard, multiply by .5933 to obtain kilograms per cubic meter.

b. If you know the cement content in 94 pound bags per cubic yard, multiply by 55.77 to obtain kilograms per cubic meter.

CONCRETE 3

REFERENCE NOS.

519

R033-040 Metric Equivalents of Common Concrete Strengths
(to convert other psi values to megapascals, multiply by 0.006895)

U.S. Values psi	SI Value Megapascals	Non-SI Metric Value kgf/cm²*
2000	14	140
2500	17	175
3000	21	210
3500	24	245
4000	28	280
4500	31	315
5000	34	350
6000	41	420
7000	48	490
8000	55	560
9000	62	630
10,000	69	705

* kilograms force per square centimeter

R033-050 Quantities of Cement, Sand and Stone for One C.Y. of Concrete per Various Mixes

This table can be used to determine the quantities of the ingredients for smaller quantities of site mixed concrete.

Concrete (C.Y.)	Mix = 1:1:1-3/4			Mix = 1:2:2.25			Mix = 1:2.25:3			Mix = 1:3:4		
	Cement (sacks)	Sand (C.Y.)	Stone (C.Y.)	Cement (sacks)	Sand (C.Y.)	Stone (C.Y.)	Cement (sacks)	Sand (C.Y.)	Stone (C.Y.)	Cement (sacks)	Sand (C.Y.)	Stone (C.Y.)
1	10	.37	.63	7.75	.56	.65	6.25	.52	.70	5	.56	.74
2	20	.74	1.26	15.50	1.12	1.30	12.50	1.04	1.40	10	1.12	1.48
3	30	1.11	1.89	23.25	1.68	1.95	18.75	1.56	2.10	15	1.68	2.22
4	40	1.48	2.52	31.00	2.24	2.60	25.00	2.08	2.80	20	2.24	2.96
5	50	1.85	3.15	38.75	2.80	3.25	31.25	2.60	3.50	25	2.80	3.70
6	60	2.22	3.78	46.50	3.36	3.90	37.50	3.12	4.20	30	3.36	4.44
7	70	2.59	4.41	54.25	3.92	4.55	43.75	3.64	4.90	35	3.92	5.18
8	80	2.96	5.04	62.00	4.48	5.20	50.00	4.16	5.60	40	4.48	5.92
9	90	3.33	5.67	69.75	5.04	5.85	56.25	4.68	6.30	45	5.04	6.66
10	100	3.70	6.30	77.50	5.60	6.50	62.50	5.20	7.00	50	5.60	7.40
11	110	4.07	6.93	85.25	6.16	7.15	68.75	5.72	7.70	55	6.16	8.14
12	120	4.44	7.56	93.00	6.72	7.80	75.00	6.24	8.40	60	6.72	8.88
13	130	4.82	8.20	100.76	7.28	8.46	81.26	6.76	9.10	65	7.28	9.62
14	140	5.18	8.82	108.50	7.84	9.10	87.50	7.28	9.80	70	7.84	10.36
15	150	5.56	9.46	116.26	8.40	9.76	93.76	7.80	10.50	75	8.40	11.10
16	160	5.92	10.08	124.00	8.96	10.40	100.00	8.32	11.20	80	8.96	11.84
17	170	6.30	10.72	131.76	9.52	11.06	106.26	8.84	11.90	85	9.52	12.58
18	180	6.66	11.34	139.50	10.08	11.70	112.50	9.36	12.60	90	10.08	13.32
19	190	7.04	11.98	147.26	10.64	12.36	118.76	9.84	13.30	95	10.64	14.06
20	200	7.40	12.60	155.00	11.20	13.00	125.00	10.40	14.00	100	11.20	14.80
21	210	7.77	13.23	162.75	11.76	13.65	131.25	10.92	14.70	105	11.76	15.54
22	220	8.14	13.86	170.05	12.32	14.30	137.50	11.44	15.40	110	12.32	16.28
23	230	8.51	14.49	178.25	12.88	14.95	143.75	11.96	16.10	115	12.88	17.02
24	240	8.88	15.12	186.00	13.44	15.60	150.00	12.48	16.80	120	13.44	17.76
25	250	9.25	15.75	193.75	14.00	16.25	156.25	13.00	17.50	125	14.00	18.50
26	260	9.64	16.40	201.52	14.56	16.92	162.52	13.52	18.20	130	14.56	19.24
27	270	10.00	17.00	209.26	15.12	17.56	168.76	14.04	18.90	135	15.02	20.00
28	280	10.36	17.64	217.00	15.68	18.20	175.00	14.56	19.60	140	15.68	20.72
29	290	10.74	18.28	224.76	16.24	18.86	181.26	15.08	20.30	145	16.24	21.46

R033-060 Concrete Material Net Prices

Costs below are C.Y. of concrete delivered; per ton of bulk cement; per bag cement delivered T.L.L.; per ton for stone and sand aggregates loaded at plant (no trucking included) and per 4 C.F. bag for perlite or vermiculite aggregate delivered T.L.L.

City	Ready Mix Concrete Regular Weight 3000 psi	Ready Mix Concrete Regular Weight 5000 psi	Cement T.L. Lots Bulk per Ton	Cement T.L. Lots Bags per Bag	Aggregates per Ton Crushed Stone 1-1/2"	Aggregates per Ton Crushed Stone 3/4"	Aggregates per Ton Sand	Vermiculite or Perlite 4 C.F. Bag
Atlanta	$58.00	$63.00	$72.25	$6.35	$13.70	$14.00	$13.45	$10.35
Baltimore	62.00	67.00	77.00	6.75	9.25	9.80	9.50	8.85
Boston	66.00	72.00	84.00	7.40	11.45	12.65	10.50	9.75
Buffalo	67.00	73.00	89.20	7.65	11.65	12.10	12.30	10.10
Chicago	58.00	64.00	83.35	7.10	9.25	9.55	8.00	8.70
Cincinnati	55.00	60.00	78.35	6.65	8.30	10.30	7.00	10.15
Cleveland	60.00	66.00	84.55	7.20	13.40	13.95	9.90	9.95
Columbus	59.00	63.00	82.05	7.30	12.10	12.35	6.75	10.10
Dallas	65.00	70.00	77.15	6.50	16.10	16.35	7.05	9.80
Denver	59.00	67.00	82.05	8.00	10.10	10.30	8.00	10.20
Detroit	63.00	68.00	80.85	7.10	8.80	9.00	8.70	10.00
Houston	64.00	70.00	84.85	7.25	18.30	18.50	10.50	8.95
Indianapolis	60.00	68.00	79.60	6.85	7.85	8.40	9.90	9.95
Kansas City	56.00	61.00	74.65	6.35	7.80	8.15	7.00	10.20
Los Angeles	54.00	62.00	87.05	7.75	12.95	13.55	12.50	10.40
Memphis	60.00	64.00	77.15	6.65	11.75	12.10	9.35	9.60
Milwaukee	62.00	67.00	87.20	7.45	8.75	9.15	8.55	9.40
Minneapolis	63.00	68.00	84.30	7.20	14.65	14.90	9.85	9.55
Nashville	55.00	60.00	74.65	6.40	9.25	9.45	11.60	9.10
New Orleans	58.00	66.00	78.05	6.60	15.15	15.75	13.05	9.80
New York City	70.00	78.00	88.35	7.60	24.25	25.85	14.50	10.35
Philadelphia	56.00	64.00	80.85	7.05	10.90	11.50	7.50	9.15
Phoenix	65.00	70.00	78.90	7.15	11.20	11.70	11.50	9.70
Pittsburgh	61.00	65.00	75.90	6.55	14.85	15.00	10.50	9.55
St. Louis	53.00	59.00	77.15	6.55	9.70	10.60	9.55	9.80
San Antonio	53.00	57.00	74.65	6.35	7.85	8.05	5.90	8.90
San Diego	57.00	63.00	84.55	7.20	9.20	9.40	8.10	8.95
San Francisco	78.00	83.00	88.00	7.55	18.50	20.55	12.50	9.55
Seattle	60.00	66.00	82.05	7.05	14.80	15.25	11.90	9.95
Washington, D.C.	73.00	83.00	80.85	6.95	13.80	14.80	11.25	10.20
Average	$61.00	$66.90	$81.00	$7.00	$12.20	$12.75	$ 9.90	$ 9.70

R033-070 Ready Mix Material Prices

Table below lists national average prices per C.Y. of concrete. Prices in the key cities for different strengths can be closely estimated by factoring against the 3000 psi or 5000 psi price from R033-060 above or from the City Cost Indexes, cast-in-place concrete material factor.

Strength in psi	Design Mix Nominal Mix	Bags per C.Y.	Heavy Weight Regular	Heavy Weight High Early	Light Weight 110# per C.F.	Light Weight All Light Weight	Admixtures and Special Items Add to Each C.Y. of Concrete for the Following Items:		
2,000	1:3:5	4.5	$57.95	$63.60	$72.45	$86.95	Calcium chloride, 1%	$1.75	perC.Y.
2,500	1:2-1/2:4-1/2	5	59.15	65.40	73.95	88.75	" 2%	3.00	perC.Y.
3,000	1:2:4	5.5	61.00	67.90	76.25	91.50	Water reducing agent	.25	per bag
3,500	1:2:3-1/2	6	62.55	70.05	78.20	93.85	Set retarder	.62	per bag
3,750	1:2:3-1/4	6.3	63.45	71.35	79.30	95.20	High early cement	1.25	per bag
4,000	1:2:3	6.5	64.05	72.20	80.05	96.10	White cement	9.55	per bag
4,500	1:2:2-1/2	7	65.90	74.65	82.40	98.85	Pump aid	2.12	perC.Y.
5,000	1:1-1/2:2-1/2	7.5	65.90	75.30	82.40	98.85	Winter concrete	4.50	perC.Y.
	1:1:2	8	70.50	77.55	—	Perlite	Synthetic Fiber Reinforcing	8.50	perC.Y.
	1:4 topping	7.5	67.00	73.70	—	1:6=			
	1:3 topping	8.5	71.90	79.10	—	$89.10			
	1:2 topping	11	78.25	86.10	—				

R033-080   Field Mix Concrete

Presently most building jobs are built with ready mix concrete except for isolated locations and some larger jobs requiring over 10,000 C.Y. where land is readily available for setting up a temporary batch plant.

The most economical mix is a controlled mix using local aggregate proportioned by trial to give the required strength with the least cost of material.

Costs tabulated below are based on a job that requires 10,000 C.Y. of concrete to be placed within an eight month period.

28 Day Strength	(Line 0020)	3000 psi	(Line 0010)	2250 psi
Mix Proportions	1:2-½:3		1:2-¾:4	
Cement, trucked in bulk @ $4.05 per Cwt.	6-1/4 bags @ 94 lb./bag	$23.79	5 bags @ 94 lb./bag	$19.04
Sand @ $9.75 per ton	.68 tons	6.63	.7 tons	6.83
Stone @ $13.80 per ton, maximum size aggregate 3/4"	.9 tons	12.42	1.02 tons	14.08
Set up plant, foundations, piping, etc.		1.93		1.93
Plant, trucks and equipment rental		7.15		7.15
Labor to mix and transport to forms		12.36		12.36
Supervision		2.68		2.68
Cost per C.Y. FOB Forms		$66.96		$64.04

See R033-060 for material prices in the major cities.

R033-090   Placing Ready Mixed Concrete

For ground pours allow for 5% waste when figuring quantities.

Prices in the front of the book assume normal deliveries. If deliveries are made before 8 A.M. or after 5 P.M. or on Saturday afternoons add $25 per C.Y. Large volume discounts are not included in prices in front of book.

For the lower floors without truck access, concrete may be wheeled in rubber tired buggies, conveyer handled, crane handled or pumped. Pumping is economical if there is top steel. Conveyers are more efficient for thick slabs. Concrete pump with an operator can be rented from $620 per day for up to 25 C.Y. to $1,495 per day for a 400 C.Y. pour. Figures include travel time if done at straight time. Pumping lightweight concrete costs an extra $2.35 per C.Y.

At higher floors the rubber tired buggies may be hoisted by a hoisting tower then wheeled to location. Placement by a conveyer is limited to three floors and is best for high volume pours. Pumped concrete is best when building has no crane access. Concrete may be pumped directly as high as thirty-six stories using special pumping techniques. Normal maximum height is about fifteen stories.

Best pumping aggregate is screened and graded bank gravel rather than crushed stone.

Pumping downward is more difficult than pumping upwards. Horizontal distance from pump to pour may increase preparation time prior to pour. Placing by cranes, either mobile, climbing or tower types continues as the most efficient method for high rise concrete buildings.

	Cost per C.Y. for Wheeled Concrete, Dumped Only (Add to appropriate placing cost)							
	10 C.F. Walking Cart				18 C.F. Riding Cart			
Item	Hourly Cost	Wheeled up to 50 ft.	Wheeled up to 150 ft.	Wheeled up to 250 ft.	Hourly Cost	Wheeled up to 50 ft.	Wheeled up to 150 ft.	Wheeled up to 250 ft.
Laborer	$21.45	$5.36	$ 7.16	$ 9.55	$21.45	$2.15	$2.79	$3.65
.125 Labor foreman	2.93	.73	.98	1.30	2.93	.29	.38	.50
Concrete cart	6.24	1.56	2.08	2.78	10.42	1.04	1.35	1.77
Total Cost/C.Y.		$7.65	$10.22	$13.63		$3.48	$4.52	$5.92
Hourly production		4 C.Y.				10 C.Y.		

R033-100 Average C.Y. of Concrete

Rubbing and floor finish not included — 4 uses of forms assumed, 5 story building.

Item	Description	Strength in psi	Ready Mix	Place	Forms	Reinforcing	Total
			Cost for 1 C.Y. of Concrete				
Beams	10' span	4000	$64.00	$55.06	$280.83	$ 99.75	$499.64
5 kip/L.F.	25'		64.00	36.71	256.41	122.55	479.67
Beam & Slab, 1 way	15'	4000	64.00	16.94	365.40	53.68	500.02
125 psf Sup. L.	25'		64.00	14.68	234.90	67.76	381.34
Beam & Slab, 2 way	15'	4000	64.00	16.94	306.36	57.20	444.50
125 psf Sup. L.	25'		64.00	14.68	190.92	54.12	323.72
Columns,	16" x 16"	4000	64.00	36.71	346.68	310.66	758.05
square tied	24" x 24"		64.00	23.94	230.04	256.76	574.74
Columns, Tied	16" diameter	4000	64.00	36.71	214.89	296.94	612.54
reinforced	24" diameter		64.00	23.94	145.62	294.00	527.56
Flat Plate,	15' span	4000	64.00	20.02	181.78	55.44	321.24
125 psf Sup. L.	25'		64.00	16.94	98.34	49.72	229.00
Flat Slab with drops,	20'	4000	64.00	20.02	138.03	52.36	274.41
125 psf Sup. L.	30'		64.00	16.94	93.09	57.64	231.67
Grade Wall,	8" thick	3000	61.00	12.87	277.02	18.92	369.81
8' high	15" thick		61.00	11.03	140.22	18.49	230.74
Metal Pan Joists,	15' span	4000	64.00	16.94	247.41	20.21	348.56
125 psf Sup. L.	25' span		64.00	14.68	203.99	34.83	317.50
Pile Caps	under 5 C.Y.	3000	61.00	12.87	55.80	21.32	150.99
"	over 10 C.Y.		61.00	5.39	27.90	19.68	113.97
Slab	4" thick	3500	62.50	10.53	15.28	15.39	103.70
on Grade	6" thick		62.50	7.02	9.96	10.26	89.74
Spread	under 1 C.Y.	3000	61.00	21.06	60.72	23.00	165.78
Footings	over 5 C.Y.		61.00	10.53	22.77	27.00	121.30
Strip	9" x 18" plain	3000	61.00	10.53	75.60	—	147.13
Footings	12" x 36" reinforced		61.00	7.47	37.80	20.00	126.27
Waffle 30" Domes	20' span	4000	64.00	16.94	207.09	62.06	350.09
125 psf Sup. L.	30' span		64.00	14.68	164.61	59.16	302.45

*Placement by direct chute assumed. All others, placement by pump assumed.

When form and reinforcing quantity ratios are available compute the C.Y. costs directly from the unit prices in the front of the book. See also R033-010 for relative quantities per C.Y. for additional items. Higher strength concrete would change the ratios below. The tables below show typical examples of how the total costs per C.Y. have been derived. Ready mix concrete prices are U.S. average and equipment handling is assumed.

Beams 5 kip/L.F.	10' Span				25' Span			
	Material		Installation		Material		Installation	
4000 psi concrete	1 C.Y.	$ 64.00	1 C.Y.	$ 55.06	1 C.Y.	$ 64.00	1 C.Y.	$ 36.71
Formwork, 4 uses	69 S.F. @ .88	60.72	69 S.F. @ 3.19	220.11	63 S.F. @ .88	55.44	63 S.F.@ 3.19	200.97
Reinforcing steel	175 # @ .27	47.25	175 # @ .30	52.50	215 # @ .27	58.05	215 # @ .30	64.50
Total per C.Y.		$171.97		$327.67		$177.49		$302.18

Beam & Slab, One Way	125 psf, 15' Span				125 psf, 25' Span			
	Material		Installation		Material		Installation	
4000 psi concrete	1 C.Y.	$ 64.00	1 C.Y.	$ 16.94	1 C.Y.	$ 64.00	1 C.Y.	$ 14.68
Formwork, 4 uses	84 S.F. @ .92	77.28	84 S.F. @ 3.43	288.12	54 S.F. @ .92	49.68	54 S.F. @ 3.43	185.22
Reinforcing steel	122 # @ .27	32.94	122 # @ .17	20.74	154 # @ .27	41.58	154 # @ .17	26.18
Total per C.Y.		$174.22		$325.80		$155.26		$226.08

Beam & Slab, Two Way	125 psf, 15' Span				125 psf, 25' Span			
	Material		Installation		Material		Installation	
4000 psi concrete	1 C.Y.	$ 64.00	1 C.Y.	$ 16.94	1 C.Y.	$ 64.00	1 C.Y.	$ 14.68
Formwork, 4 uses	69 S.F. @ .84	57.96	69 S.F. @ 3.60	248.40	43 S.F. @ .84	36.12	43 S.F. @ 3.60	154.80
Reinforcing steel	130 # @ .27	35.10	130 # @ .17	22.10	123 # @ .27	33.21	123 # @ .17	20.91
Total per C.Y.		$157.06		$287.44		$133.33		$190.39

CONCRETE 3

REFERENCE NOS.

R033-100 Average C.Y. of Concrete (cont.)

Columns, Square Tied	16" Square				24" Square			
	Material		Installation		Material		Installation	
4000 psi concrete	1 C.Y.	$ 64.00	1 C.Y.	$ 36.71	1 C.Y.	$ 64.00	1 C.Y.	$ 23.94
Formwork, 4 uses	81 S.F. @ .76	61.56	81 S.F. @ 3.52	285.12	54 S.F. @ .76	41.04	54 S.F. @ 3.50	189.00
Reinforcing steel, avg.	634 # @ .27	171.18	634 # @ .22	139.48	524 # @ .27	141.48	524 # @ .22	115.28
Total per C.Y.		$296.74		$461.31		$246.52		$328.22

Note: Reinforcing of 16" and 24" square columns can vary from 144 lb. to 972 lb. per C.Y.

Columns, Round Tied Reinforced	16" Diameter				24" Diameter			
	Material		Installation		Material		Installation	
4000 psi concrete	1 C.Y.	$ 64.00	1 C.Y.	$ 36.71	1 C.Y.	$ 64.00	1 C.Y.	$ 23.94
Formwork, fiber forms	19 L.F. @ 5.40	102.60	19 L.F. @ 5.91	112.29	9 L.F. @ 9.82	88.38	9 L.F. @ 6.36	57.24
Reinforcing steel, avg.	606 # @ .27	163.62	606 # @ .22	133.32	600 # @ .27	162.00	600 # @ .22	132.00
Ties	70 # @ .54	37.80	70 # @ .32	22.40	18 # @ .54	9.72	18 # @ .32	5.76
Total per C.Y.		$368.02		$304.72		$324.10		$218.94

Note: Reinforcing of 16" and 24" diameter columns vary from 160 lb. to 1150 lb. per C.Y.

Flat Plate	125 psf, 15' Span				125 psf, 25' Span			
	Material		Installation		Material		Installation	
4000 psi concrete	1 C.Y.	$ 64.00	1 C.Y.	$ 20.02	1 C.Y.	$ 64.00	1 C.Y.	$ 16.94
Formwork, 4 uses	61 S.F. @ .69	42.09	61 S.F. @ 2.29	139.69	33 S.F. @ .69	22.77	33 S.F. @ 2.29	75.57
Reinforcing steel	126 # @ .27	34.02	126 # @ .17	21.42	113 # @ .27	30.51	113 # @ .17	19.21
Total per C.Y.		$140.11		$181.13		$117.28		$111.72

Flat Slab with drops	125 psf, 20' Span				125 psf, 30' Span			
	Material		Installation		Material		Installation	
4000 psi concrete	1 C.Y.	$ 64.00	1 C.Y.	$ 20.02	1 C.Y.	$ 64.00	1 C.Y.	$ 16.94
Formwork, 4 uses	43 S.F. @ .85	36.55	43 S.F. @ 2.36	101.48	29 S.F. @ .85	24.65	29 S.F. @ 2.36	68.44
Reinforcing steel	119 # @ .27	32.13	119 # @ .17	20.23	131 # @ .27	35.37	131 # @ .17	22.27
Total per C.Y.		$132.68		$141.73		$124.02		$107.65

Grade Wall, 8' High	8" Thick				15" Thick			
	Material		Installation		Material		Installation	
3000 psi concrete	1 C.Y.	$ 61.00	1 C.Y.	$ 12.87	1 C.Y.	$ 61.00	1 C.Y.	$ 11.03
Formwork, 4 uses	81 S.F. @ .76	61.56	81 S.F. @ 2.66	215.46	41 S.F. @ .76	31.16	41 S.F. @ 2.66	109.06
Reinforcing steel	44 # @ .27	11.88	44 # @ .16	7.04	43 # @ .27	11.61	43 # @ .16	6.88
Total per C.Y.		$134.44		$235.37		$103.77		$126.97

Metal Pan Joists, 30" One Way	125 psf, 15' Span				125 psf, 25' Span			
	Material		Installation		Material		Installation	
4000 psi concrete	1 C.Y.	$ 64.00	1 C.Y.	$ 16.94	1 C.Y.	$ 64.00	1 C.Y.	$ 14.68
Formwork, 4 uses	49 S.F. @ .71	34.79	49 S.F. @ 2.66	130.34	40 S.F. @ .71	28.40	40 S.F. @ 2.66	106.40
Reinforcing steel	47 # @ .27	12.69	47 # @ .16	7.52	81 # @ .27	21.87	81 # @ .16	12.96
30" metal pans	44 S.F. @ 1.50	66.00	44 S.F. @ .37	16.28	37 S.F. @ 1.50	55.50	37 # @ .37	13.69
Total per C.Y.		$177.48		$171.08		$169.77		$147.73

Pile Caps	Under 5 C.Y.				Over 10 C.Y.			
	Material		Installation		Material		Installation	
3000 psi concrete	1.05 C.Y.	$61.00	1 C.Y.	$12.87	1.05 C.Y.	$61.00	1 C.Y.	$ 5.39
Formwork, 4 uses	20 S.F. @ .63	12.60	20 S.F. @ 2.16	43.20	10 S.F. @ .63	6.30	10 S.F. @ 2.16	21.60
Reinforcing steel	52 # @ .27	14.04	52 # @ .14	7.28	48 # @ .27	12.96	48 # @ .14	6.72
Total per C.Y.		$87.64		$63.35		$80.26		$33.71

Slab on Grade	4" Thick				6" Thick			
	Material		Installation		Material		Installation	
3500 psi concrete	1.05 C.Y.	$62.50	1 C.Y.	$10.53	1.05 C.Y.	$62.50	1 C.Y.	$ 7.02
Formwork, 4 uses	9.2 L.F. @ .28	2.58	9.2 L.F. @ 1.38	12.70	6 L.F. @ .28	1.68	6 L.F. @ 1.38	8.28
W.W. Fabric	81 # @ .08	6.48	81 # @ .11	8.91	54 # @ .08	4.32	54 # @ .11	5.94
Total per C.Y.		$71.56		$32.14		$68.50		$21.24

R033-100 Average C.Y. of Concrete (cont.)

Spread Footings	Under 1 C.Y.				Over 5 C.Y.			
	Material		Installation		Material		Installation	
3000 psi concrete	1.05 C.Y.	$61.00	1.05 C.Y.	$21.06	1.05 C.Y.	$61.00	1 C.Y.	$10.53
Formwork, 4 uses	24 S.F. @ .53	12.72	24 S.F. @ 2.00	48.00	9 S.F. @ .53	4.77	9 S.F. @ 2.00	18.00
Reinforcing steel	46 # @ .27	12.42	46 # @ .23	10.58	54 # @ .27	14.58	54 # @ .23	12.42
Total per C.Y.		$86.14		$79.64		$80.35		$40.95

Strip Footings	9' x 18" Plain				12" x 36" Reinforced			
	Material		Installation		Material		Installation	
3000 psi concrete	1.05 C.Y.	$61.00	1.05 C.Y.	$10.53	1.05 C.Y.	$61.00	1 C.Y.	$ 7.47
Formwork, 4 uses	36 S.F. @ .40	14.40	36 S.F. @ 1.70	61.20	18 S.F. @ .40	7.20	18 S.F. @ 1.70	30.60
Reinforcing steel	—		—		40 # @ .27	10.80	40 # @ .23	9.20
Total per C.Y.		$75.40		$71.73		$79.00		$47.27

Waffle, 30" Domes	125 psf, 20' Span				125 psf, 30' Span			
	Material		Installation		Material		Installation	
4000 psi concrete	1 C.Y.	$ 64.00	1 C.Y.	$ 16.94	1 C.Y.	$ 64.00	1 C.Y.	$ 14.68
Formwork, 4 uses	39 S.F. @ .71	27.69	39 S.F. @ 2.72	106.08	31 S.F. @ .71	22.01	31 S.F. @ 2.72	84.32
Reinforcing steel & mesh	107 # @ .35	37.45	107 # @ .23	24.61	102 # @ .35	35.70	102 # @ .23	23.46
Metal Domes	39 S.F. @ 1.50	58.50	39 S.F. @ .38	14.82	31 S.F. @ 1.50	46.50	31 S.F. @ .38	11.78
Total per C.Y.		$187.64		$162.45		$168.21		$134.24

CONCRETE 3

R033-120   Lift Slabs

The cost advantage of the lift slab method is due to placing all concrete, reinforcing steel, inserts and electrical conduit at ground level and in reduction of formwork. Minimum economical project size is about 30,000 S.F. Slabs may be tilted for parking garage ramps.

It is now used in all types of buildings and has gone up to 22 stories high in apartment buildings. Current trend is to use post-tensioned flat plate slabs with spans from 22' to 35'. Cylindrical void forms are used when deep slabs are required. One pound of prestressing steel is about equal to seven pounds of conventional reinforcing.

To be considered cured for stressing and lifting, a slab must have attained 75% of design strength. Seven days are usually sufficient with four to five days possible if high early strength cement is used. Slabs can be stacked using two coats of a non-bonding agent to insure that slabs do not stick to each other. Lifting is done by companies specializing in this work. Lift rate is 5' to 15' per hour with an average of 10' per hour. Total areas up to 33,000 S.F. have been lifted at one time. 24 to 36 jacking columns are common. Most economical bay sizes are 24' to 28' with four to fourteen stories most efficient. Continuous design reduces reinforcing steel cost. Use of post-tensioned slabs allows larger bay sizes. Supplementary reinforcing, post-tensioned tendons, accessories and field labor cost about $2.47 per pound of tendon.

Table below shows the usual range and average S.F. cost of a typical lift slab project that has been designed to take advantage of lift slab techniques. Figures include Subs Overhead and Profit.

Typical Cost per S.F. for Lift Slabs Project Incl. Subs O & P				
Item	Description	Typical Range	Average	Sub Total
Concrete Slabs	Edge and bulkhead forms	$ .05 to $ .39	$ .22	$4.99
	Reinforcing and post-tensioning steel	1.51 to 2.70	2.11	
	Reinforcing and accessories	.06 to .15	.11	
	Concrete cast at ground level	1.45 to 2.45	1.95	
	Float or trowel finish	.49 to .70	.60	
Lifting Slabs	Separating and curing compound	.03 to .08	.06	2.63
	Lifting collars	.25 to .52	.39	
	Lifting and welding	1.27 to 3.08	2.18	
Columns	Fabricated columns & accessories	.70 to 2.05	1.38	1.78
	Set initial stage	.05 to .22	.14	
	Column splicing	.13 to .38	.26	
Miscellaneous	Grout plates, patching, bolts, etc.	.05 to .23	.14	0.14
Total in place per S.F.		$6.04 to $12.95	$9.54	$9.54

R033-130   Granolithic Finish and Base

Description	Granolithic Topping, Mix 1:1-1-1/2				Granite Base	
	(Line 033-454-0850)	1" Topping	(Line 033-454-0950)	2" Topping	(Line 033-130-0200)	1" x 5" Straight
Cement @ $7.00 per bag	3.60 bags	$ 25.20	7.20 bags	50.40	1.70 bags	$ 11.90
Sand @ $14.35 per C.Y.	3.60 C.F.	1.91	7.20 C.F.	3.83	1.70 C.F.	.90
Peastone @ $20.00 Per C.Y.	5.40 C.F.	4.00	10.80 C.F.	8.00	1.85 C.F.	1.37
Clean, Mix, Place forms, and finish: Crew C-10 @ $24.58 per L.H.	4.07 L.H.	$100.04	4.80 L.H.	$117.98	13.70 L.H.	$336.75
TOTALS	100 S.F.	$131.15	100 S.F.	$180.21	100 L.F.	$350.92

R033-140   Integral Floor Finish

Mix 1:1:2		(Line 033-454-0450)	1/2" Thick	(Line 033-454-0600)	1" Thick
Cement @	$ 7.00 per bag	1.7 bag	$11.90	3.4 bag	$ 23.80
Sand @	$14.35 per C.Y.	0.8 C.F.	.43	1.6 C.F.	.85
Gravel @	$18.25 per C.Y.	3.2 C.F.	2.16	6.4 C.F.	4.33
Mix, place and finish using Crew C-10 @	$24.58 per L.H.	2.52 L.H.	61.94	3.21 L.H.	78.90
Total for 100 S.F.			$76.43		$107.88

3 CONCRETE

REFERENCE NOS.

R034-010 Precast Concrete Wall Panels

Panels are either solid or insulated with plain, colored or textured finishes. Transportation is an important cost factor. Prices below are based on delivery within 50 miles of a plant including fabricators' overhead and profit. Engineering data is available from fabricators to assist with construction details. Usual minimum job size for economical use of panels is about 5000 S.F. Small jobs can double the prices below. For large, highly repetitive jobs, deduct up to 15% from the prices below.

Panel Cost and Maximum Size   Base price for panels based on 50 S.F. or larger panels is as follows:

Thickness	Cost per S.F.	Maximum Size	Thickness	Cost per S.F.	Maximum Size
3″	$ 7.78	50 S.F.	6″	11.80	300 S.F.
4″	8.55	150 S.F.	7″	12.75	300 S.F.
5″	10.25	200 S.F.	8″	13.25	300 S.F.

The above prices are for gray smooth form or board finish one side. Add $1.55 per S.F. for white facing; $8.25 per C.F. for solid white for full thickness. For fluted surface (not broken) add $.46 to $1.03 per S.F. For broken rib finish add $.80 to $1.55 to fluted surface. Add $.35 to $1.30 per S.F. for exposed local aggregate and $1.00 to $3.90 per S.F. for special facing aggregates.

Sandblasting runs $.55 to $1.00 per S.F., and bush hammering runs $1.75 to $3.45 per S.F. Loose hardware (not included above) runs $1.00 to $2.00 per S.F.

2″ thick panels cost about the same as 3″ thick panels and maximum panel size is less. For building panels faced with granite, marble or stone, add the material prices from Division 044 to the plain panel price above. There is a growing trend toward aggregate facings and broken rib finish rather than plain gray concrete panels.

Composite Panels   Add to above panel prices for core insulation, mesh and shear ties per S.F.

Type	1″ Thick	1-1/2″ Thick	2″ Thick
Fiberglass	$1.73	$2.40	$3.10
Polystyrene (E.P.S. Board)	1.10	1.55	2.00

Erection   Table shows cost ranges for erection including Subcontractor's O & P using a six person crew, crane, operator & oiler.

Panel Size L x H	Area per Panel	Low Rise — Hyd. 55 Ton Crane @ $4,455 Daily Cost — Daily Production Range Pieces	Area	Erection Cost Per Piece	Per S.F.	High Rise — Daily Production Range Pieces	Area	55 Ton Crane $4,455 Daily Cost Piece Cost	S.F. Cost	90 Ton Crane $4,696 Daily Cost Piece Cost	S.F. Cost	150 Ton Crane $5,291 Daily Cost Piece Cost	S.F. Cost
4′ x 4′	16 S.F.	10	160 S.F.	$445	$27.85	9	144 S.F.	$495	$30.94	$ 522	$32.61	$ 588	$36.74
		20	320	223	13.90	18	288	247	15.47	261	16.30	294	18.37
4′ x 8′	32	10	320	445	13.90	9	288	495	15.47	522	16.30	588	18.37
		19	608	234	7.35	17	544	262	8.19	276	8.63	311	9.73
8′ x 8′	64	9	576	495	7.75	8	512	557	8.70	587	9.17	661	10.33
		18	1152	247	3.85	16	1024	278	4.35	293	4.59	331	5.17
10′ x 10′	100	8	800	557	5.55	7	700	636	6.36	671	6.71	756	7.56
		15	1500	297	2.95	14	1400	318	3.18	335	3.35	378	3.78
15′ x 10′	150	7	1050	636	4.25	6	900	—	—	783	5.22	882	5.88
		12	1800	371	2.45	11	1650	—	—	427	2.85	481	3.21
20′ x 10′	200	6	1200	742	3.70	5	1000	—	—	939	4.70	1058	5.29
		8	1600	557	2.80	7	1400	—	—	671	3.35	756	3.78
30′ x 10′	300	5	1500	891	2.95	4	1200	—	—	1174	3.91	1323	4.41
		7	2100	636	2.10	7	2100	—	—	671	2.24	756	2.52

Total Cost in Place for Low Rise Construction Including Subcontractor's O & P

Description	Gray 4′ x 8′ x 4″	Gray 20′ x 10′ x 6″	White Face 4′ x 8′ x 4″	White Face 20′ x 10′ x 6″	Exposed Aggregate 4′ x 8′ x 4″	Exposed Aggregate 20′ x 10′ x 6″
Panel, steel form, broomed finish	$ 8.55	$11.80	$10.10	$13.35	$10.55	$13.80
Caulking, grout, etc. (runs about $2.45 per L.F.)	.87	.34	.87	.34	.87	.34
Erect, plumb, align (from above)	13.90	3.70	13.90	3.70	13.90	3.70
Total in place per S.F.	$23.32	$15.84	$24.87	$17.39	$25.32	$17.84

No allowance has been made for supporting steel framework. On one story buildings, panels may rest on grade beams and require only wind bracing and fasteners. On multi-story buildings panels can span from column to column and floor to floor. Plastic designed steel framed structures may have large deflections which slow down erection and raise costs.

Large panels are more economical than small panels on a S.F. basis. When figuring areas include all protrusions, returns, etc. Overhangs can triple erection costs. Panels over 45′ have been produced. Larger flat units should be prestressed. Vacuum lifting of smooth finish panels eliminates inserts and can speed erection.

R034-020 Tilt Up Concrete Panels

The advantage of tilt up construction is in the low cost of forms and the placing of concrete and reinforcing. Panels up to 75' high and 5-1/2" thick have been tilted using strongbacks. Tilt up has been used for one to five story buildings and is well suited for warehouses, stores, offices, schools and residences.

The panels are cast in forms on the floor slab. Most jobs use 5-1/2" thick solid reinforced concrete panels. Sandwich panels with a layer of insulating materials are also used. Where dampness is a factor, lightweight aggregate is used at an added cost of $.50 per square foot. Optimum panel size is 300 to 500 S.F.

Slabs are usually poured with 3000 psi concrete which permits tilting seven days after pouring. Slabs may be stacked on top of each other and are separated from each other by either two coats of bond breaker or a film of polyethylene. Use of high early strength cement allows tilting two days

after a pour. Tilting up is done with a roller outrigger crane with a capacity of at least 1-1/2 times the weight of the panel at the required reach. Exterior precast columns can be set at the same time as the panels; interior precast columns can be set first and the panels clipped directly to them. The use of cast-in-place concrete columns is diminishing due to shrinkage problems. Structural steel columns are sometimes used if crane rails are planned. Panels can be clipped to the columns or lowered between the flanges. Steel channels with anchors may be used as edge forms for the slab. When the panels are lifted the channels form an integral steel column to take structural loads. Roof loads can be carried directly by the panels for wall heights to 14'. For soft ground requiring mats for cranes, add 100% to costs below.

Below are typical costs per S.F., for panels of 300 to 500 S.F. 20' high, not including contractor's overhead and profit.

Item	5-1/2" Thick			7-1/2" Thick		
	Material	Installation	Total	Material	Installation	Total
Prepare pouring surface	$ .02	$ .08	$ .10	$ .02	$ .08	$ .10
Erect, strip side forms	.08	.34	.42	.13	.46	.59
Place concrete, 3000 psi	1.06	.13	1.19	1.44	.46	1.90
Steel trowel finish & curing	.02	.41	.43	.02	.41	.43
Reinforcing, inserts & misc. items	1.04	.57	1.61	1.44	.77	2.21
Panel erection and aligning	—	.72	.72	—	.72	.72
Total per S.F. of Wall	$ 2.22	$ 2.25	$ 4.47	$ 3.05	$ 2.90	$ 5.95
Site precast concrete columns, add	1.11	1.24	2.35	1.07	1.24	2.31
Total per S.F. of Wall	$ 3.33	$ 3.49	$ 6.82	$ 4.12	$ 4.14	$ 8.26
Panels only, per C.Y.	$119.88	$121.50	$241.38	$131.76	$125.28	$257.04

Requirements of local building codes may be a limiting factor and should be checked. Building floor slabs should be poured first and should be a minimum of 5" thick with 100% compaction of soil or 6" thick with less than 100% compaction.

Setting times as fast as nine minutes per panel have been observed, but a safer expectation would be four panels per hour with a crane and a four man setting crew. If crane erects from inside building, some provision must be made to get crane out after walls are erected. Good yarding procedure is important to minimize delays. Equalizing three point lifting beams and self-releasing pick-up hooks speed erection. If panels must be carried to their final location, setting time per panel will be increased and erection costs may fall in the erection cost range of architectural precast wall panels. Placing panels into slots formed in continuous footers will speed erection.

Reinforcing should be with #5 bars with vertical bars on the bottom. If surface is to be sandblasted, stainless steel chairs should be used to prevent rust staining.

Use of a broom finish is popular since the unavoidable surface blemishes are concealed. Many West Coast jobs have used exposed aggregate panels which cost an additional $.50 to $3.60 per S.F. depending on the aggregate and construction method employed. Eastern prices for exposed aggregate tend to be considerably higher. Panel connections run $1.60 to $6.00 per L.F. Precast columns run from three to five times the C.Y. price of the panels only.

R034-030   Prestressed Precast Concrete Structural Units

See also R032-090 for post-tensioned prestressed concrete.

Type	Location	Depth	Span in Ft.		Live Load Lb. per S.F.	Cost per S.F. Incl. Subs O & P	
						Delivered	Erected
Double Tee	Floor	28" to 34"	60 to 80		50 to 80	$ 5.85 to 8.15	$ 7.00 to 9.80
	Roof	12" to 24"	30 to 50		40	$ 3.90 to 6.20	$ 4.70 to 7.45
	Wall	Width 8'	Up to 55' high		Wind	$4.10 to $ 8.25	$4.90 to $ 9.90
Multiple Tee	Roof	8" to 12"	15 to 40		40	$ 4.10	$ 4.90
	Floor	8" to 12"	15 to 30		100	$ 4.30	$ 4.95
Plank or	Roof or Floor		Roof	Floor	40 for Roof		
		4"	13	12		$ 3.75	$ 4.70
		6"	22	18		4.00	5.20
		8"	26	25		4.25	5.30
		10"	33	29	100 for	4.75	5.95
		12"	42	32	Floor	5.00	6.25
Single Tee	Roof	28"	40			$ 6.75 *	$ 7.75 *
		32"	80			8.00 *	9.20 *
		36"	100		40	10.70 *	12.30 *
		48"	120			11.25 *	12.95 *
AASHO Girder	Bridges	Type 4	100		Highway	$107/L.F.	$161/L.F.
		5	110			142	178
		6	125			178	214
Box Beam	Bridges	15"	40		Highway	$115/L.F.	$173/L.F.
		27"	to			153	191
		33"	100			176	211

*Costs are for 10' wide members; for 8' wide members add $.55 per S.F.

The costs above are based on a project of 10,000 to 20,000 S.F. with a haul distance of 25 to 50 miles.

The majority of precast projects today utilize double tees rather than single tees because of speed and ease of installation. As a result casting beds at manufacturing plants are normally formed for double tees. Single tee projects will therefore require an initial set up charge of approximately $7,000 to be spread over the individual single tee costs. The prices above for single tees includes this cost based on a project size of 15,000 square feet.

For floors, a 2" to 3" topping is field cast over the shapes. For roofs, insulating concrete or rigid insulation is placed over the shapes. Topping is not included in the above costs.

Hauling costs (incl. above) run from $.50 per S.F. for short haul to $.90 for 50 mile haul. Member lengths up to 40' are standard haul, 40' to 60' require special permits and lengths over 60' must be escorted. Over width and/or over length can add up to 100% on hauling costs.

Multiple tee erection runs between $.55 to $1.50 per S.F. and single tee erection runs between $.70 to $1.70 per S.F. Large heavy members may require two cranes for lifting which would increase these erection costs by about 45%. An eight man erection crew and crane will run about $3,000 to $5,500 per day depending on location and size of crane. The crew can install 12 to 20 double tees, or 45 to 70 quad tees or planks per day.

The cost of supporting beams must be added to the above costs. Simple support beams run from $42 to $110 per L.F. delivered. Inverted tee beams run from $72 to $260 per L.F. delivered. Standard sized columns run $40 to $160 per L.F. delivered.

Grouting of connections must be added to the above. Typical costs are about $20 per connection but can go as high as $50 per connection.

Grouting planks run between $.40 to $.60 per S.F.

Single story buildings, including double tee roof members, supporting columns and girders, but no foundations, cost from $8.75 to $30.00 per S.F. of floor area. Parking garages run from $11.00 to $22.00 per S.F. above the foundations. Optimum parking garage design runs .02 C.Y. per S.F with overall costs for concrete in place running between $540 to $750 per C.Y.

Several system buildings utilizing precast members are available. Heights can go to 22 stories for apartment buildings with costs ranging from $11.00 to $40.00 per S.F. depending on the system, its components, its location and the degree of interior finish supplied. Optimum design ratio is 3 S.F. of surface to 1 S.F. of floor area.

R035-010 Lightweight Concrete

Vermiculite or Perlite come in bags of 4 C.F. under various trade names. Weight is about 8 lbs. per C.F. For insulating roof fill use 1:6 mix. For structural deck use 1:4 mix over gypsum boards, steeltex, steel centering, etc. supported by closely spaced joists or bulb trees. For structural slabs use 1:3:2 vermiculite sand concrete over steeltex, metal lath, steel centering, etc. on joists spaced 2'-0" O.C. for maximum L.L. of 80#/S.F. Use same mix for slab base fill over steel flooring or regular reinforced concrete slab when tile, terrazzo or other finish is to be laid over.

For slabs on grade use 1:3:2 mix when tile, etc. finish is to be laid over. If radiant heating units are installed use a 1:6 mix for a base. After coils are in place, cover with a regular granolithic finish (mix 1:3:2) to a minimum depth of 1-1/2" over top of units.

Reinforce all slabs with 6 x 6 or 10 x 10 welded wire mesh.

Vermiculite concrete can be purchased ready mixed but the following breakdown is included for field mix. Prices given below are for a one story building and assume 50 C.Y. or more. For less than 50 C.Y. add 10%. For over one story add $3.35 per C.Y. Screed finish cost is included below. Ready mix 1:6 costs $86.20 per C.Y. delivered in truckload lots.

See R033-070 for prices of ready mix lightweight concrete.

Quantities per C.Y., Field Mix			(Line 0110) 1:6 Mix Insulating Roof Fill		1:3:2 Mix Lightweight Structural Concrete	
Portland cement @	$ 7.00	per bag	5.0 bags	$ 35.00	6.2 bags	$ 43.40
Vermiculite or Perlite @	$ 9.70	per bag	7.5 bags	72.75		
treated type @	$11.30	per bag			4.7 bags	53.11
Sand @	$14.35	per C.Y.			12.5 C.F.	6.64
Plant and Water				6.37		6.37
Labor, machine mix, hoist and place, Crew C-8 @	$35.60	per L.H.	1.12 L.H.	39.87	.91 L.H.	32.40
Total in place, per C.Y.				$153.99		$141.92

R041-100   Cement Mortar (material only)

Type N - 1:1:6 mix by volume. Use everywhere above grade except as noted below.

- 1:3 mix using conventional masonry cement which saves handling two separate bagged materials.

Type M - 1:1/4:3 mix by volume, or 1 part cement, 1/4 (10% by wt.) lime, 3 parts sand. Use for heavy loads and where earthquakes or hurricanes may occur. Also for reinforced brick, sewers, manholes and everywhere below grade.

Cost and Mix Proportions of Various Types of Mortar

Components	Type Mortar and Mix Proportions by Volume										
	M		S		N		O		K	PM	PL
	1:1:6	1:1/4:3	1/2:1:4	1:1/2:4	1:3	1:1:6	1:3	1:2:9	1:3:12	1:1:6	1:1/2:4
Portland cement @ $7.00 per bag	$ 7.00	$ 7.00	$ 3.50	$ 7.00	—	$ 7.00	—	$ 7.00	$ 7.00	$ 7.00	$ 7.00
Masonry cement @ $5.80 per bag	5.80	—	5.80	—	5.80	—	5.80	—	—	5.80	—
Lime @ $5.45 per 50 lb. bag	—	1.36	—	2.73	—	5.45	—	10.90	16.35	—	2.73
Masonry sand @ $21.85 per C.Y.*	4.86	2.43	3.24	3.24	2.43	4.86	2.43	7.28	9.71	4.86	3.24
Mixing machine incl. fuel**	1.36	.68	.91	.91	.68	1.36	.68	2.04	2.72	1.36	.91
Total for Materials	$19.02	$11.47	$13.45	$13.88	$8.91	$18.67	$8.91	$27.22	$35.78	$19.02	$13.88
Total C.F.	6	3	4	4	3	6	3	9	12	6	4
Approximate Cost per C.F.	$ 3.17	$ 3.82	$ 3.36	$ 3.47	$2.97	$ 3.11	$2.97	$ 3.02	$ 2.98	$ 3.17	$ 3.47

*Includes 10 mile haul
**Based on a daily rental, 10 C.F., 25 H.P. mixer, mix 200 C.F./Day

Mix Proportions by Volume, Compressive Strength and Cost of Mortar

Where Used	Allowable Proportions by Volume					Compressive Strength @ 28 days	Cost per Cubic Foot
	Mortar Type	Portland Cement	Masonry Cement	Hydrated Lime	Masonry Sand		
Plain Masonry		1	1	—	6		$3.17
	M	1	—	1/4	3	2500 psi	3.82
		1/2	1	—	4		3.36
	S	1	—	1/4 to 1/2	4	1800 psi	3.47
		—	1	—	3		2.97
	N	1	—	1/2 to 1-1/4	6	750 psi	3.11
		—	1	—	3		2.97
	O	1	—	1-1/4 to 2-1/2	9	350 psi	3.02
	K	1	—	2-1/2 to 4	12	75 psi	2.98
Reinforced Masonry	PM	1	1	—	6	2500 psi	3.17
	PL	1	—	1/4 to 1/2	4	2500 psi	3.47

Note: The total aggregate should be between 2.25 to 3 times the sum of the cement and lime used.

The labor cost to mix the mortar is included in the labor cost on brickwork.

Machine mixing is usually specified on jobs of any size. There is a large price saving over hand mixing and mortar is more uniform.

There are two types of mortar color used. Prices in Section 041-012 are for the inert additive type with about 100 lbs. per M brick as the typical quantity required. These colors are also available in smaller batch size bags (1 lb. to 15 lb.) which can be placed directly into the mixer without measuring. The other type is premixed and replaces the masonry cement with ranges in price from $5 to $15 per 70 lb. bag. Dark green color has the highest cost.

R041-200   Miscellaneous Mortar (material only)

Quantities	Glass Block Mortar		Gypsum Cement Mortar	
White Portland cement at $17.50 per bag	7 bags	$122.50		
Gypsum cement at $11.25 per 80 lb. bag			11.25 bags	$126.56
Lime at $5.45 per 50 lb. bag	280 lbs.	30.52		
Sand at $21.85 per C.Y.*	1 C.Y.	21.85	1 C.Y.	21.85
Mixing machine and fuel		6.12		6.12
Total per C.Y.		$180.99		$154.53
Approximate Total per C.F.		$ 6.70		$ 5.72

* Includes 10 mile haul

R041-500   Masonry Reinforcing

Horizontal joint reinforcing helps prevent wall cracks where wall movement may occur and in many locations is required by code. Horizontal joint reinforcing is generally not considered to be structural reinforcing and an unreinforced wall may still contain joint reinforcing.

Reinforcing strips come in 10' and 12' lengths and in truss and ladder shapes, with and without drips. Field labor runs between 2.7 to 5.3 hours per 1000 L.F. for wall thicknesses up to 12".

The wire meets ASTM A82 for cold drawn steel wire and the typical size is 9 ga. sides and ties with 3/16" diameter also available. Typical finish is mill galvanized with zinc coating at .10 oz. per S.F. Class I (.40 oz. per S.F.) and Class III (.80 oz per S.F.) are also available, as is hot dipped galvanizing at 1.50 oz. per S.F.

R042-050   Brick Chimneys

Quantities	16" x 16"		20" x 20"		20" x 24"		20" x 32"	
Brick at $300 per M	28 brick	$ 8.40	37 brick	$11.10	42 brick	$12.60	51 brick	$15.30
Type M mortar at $3.82 per C.F.	.5 C.F.	1.91	.6 C.F.	2.29	1.0 C.F.	3.82	1.3 C.F.	4.97
Flue tile (per foot)	8" x 8"	3.35	12" x 12"	5.70	2 @ 8" x 12"	9.60	2 @ 12" x 12"	11.40
Install tile & brick, crew D-1	.055 day	21.65	.073 day	28.73	.083 day	32.67	.10 day	39.36
Total per V.L.F.		$35.31		$47.82		$58.69		$71.03

Material costs include 3% waste for brick and 25% waste for mortar.

Labor costs are bare costs and do not include contractor's O&P.

Labor for chimney brick using D-1 crew is 31 hours per thousand brick or about $750 per thousand brick. An 8" x 12" flue takes 33 brick and two 8" x 8" flues take 37 brick.

R042-055   Industrial Chimneys

Foundation requirements in C.Y. of concrete for various sized chimneys.

Size Chimney	2 Ton Soil	3 Ton Soil	Size Chimney	2 Ton Soil	3 Ton Soil	Size Chimney	2 Ton Soil	3 Ton Soil
75' x 3'-0"	13 C.Y.	11 C.Y.	160' x 6'-6"	86 C.Y.	76 C.Y.	300' x 10'-0"	325 C.Y.	245 C.Y.
85' x 5'-6"	19	16	175' x 7'-0"	108	95	350' x 12'-0"	422	320
100' x 5'-0"	24	20	200' x 6'-0"	125	105	400' x 14'-0"	520	400
125' x 5'-6"	43	36	250' x 8'-0"	230	175	500' x 18'-0"	725	575

R042-100   Economy in Bricklaying

Have adequate supervision. Be sure bricklayers are always supplied with materials so there is no waiting. Place best bricklayers at corners and openings.

Use only screened sand for mortar. Otherwise, labor time will be wasted picking out pebbles. Use seamless metal tubs for mortar as they do not leak or catch the trowel. Locate stack and mortar for easy wheeling.

Have brick delivered for stacking. This makes for faster handling, reduces chipping and breakage, and requires less storage space. Many dealers will deliver select common in 2' x 3' x 4' pallets or face brick packaged. This affords quick handling with a crane or forklift and easy tonging in units of ten, which reduces waste.

Use wider bricks for one wythe wall construction. Keep scaffolding away from wall to allow mortar to fall clear and not stain wall.

On large jobs develop specialized crews for each type of masonry unit.

Consider designing for prefabricated panel construction on high rise projects.

Avoid excessive corners or openings. Each opening adds about 50% to labor cost for area of opening.

Bolting stone panels and using window frames as stops reduces labor costs and speeds up erection.

R042-120 Common and Face Brick Prices

Prices are based on truckload lot purchases for Common Building Brick and Facing Brick. Prices are per M, (thousand), brick.

City	Material			Installation				Total			
	Brick per M Delivered		Mortar 3/8" Joint	Common in 8" Wall		Face Brick, 4" Veneer		Common in 8" Wall		Face Brick, 4" Veneer	
	Common	Face		Bare Costs	Incl. O & P	Bare Costs	Incl. O & P	Bare Costs	Incl. O & P	Bare Costs	Incl. O & P
Atlanta	$190	$235	$38.90	$363	$ 564	$ 435	$ 686	$ 597	$ 822	$ 709	$ 988
Baltimore	235	305	for 8" Wall	444	690	532	840	725	999	879	1,221
Boston	305	430	and	768	1,195	922	1,455	1,121	1,583	1,397	1,977
Buffalo	260	325	$32.05	671	1,043	805	1,270	978	1,381	1,172	1,674
Chicago	260	350	for 4" Wall	709	1,103	851	1,343	1,016	1,441	1,244	1,775
Cincinnati	230	305		536	833	643	1,014	811	1,136	989	1,395
Cleveland	235	325		646	1,005	775	1,224	927	1,314	1,142	1,627
Columbus	235	300		537	836	645	1,018	818	1,145	986	1,393
Dallas	200	350		362	562	434	685	607	832	827	1,117
Denver	210	285		429	667	515	812	684	947	840	1,170
Detroit	235	335		679	1,055	814	1,285	959	1,364	1,191	1,700
Houston	250	370		406	632	487	769	703	958	901	1,224
Indianapolis	255	280		554	862	665	1,050	856	1,194	986	1,402
Kansas City	260	285		545	848	654	1,032	852	1,185	980	1,391
Los Angeles	250	275		687	1,068	824	1,300	983	1,394	1,139	1,647
Memphis	190	220		391	607	469	740	625	865	727	1,024
Milwaukee	280	360		655	1,018	786	1,240	982	1,378	1,189	1,683
Minneapolis	255	375		620	965	744	1,175	922	1,296	1,163	1,635
Nashville	185	215		382	593	458	723	611	846	711	1,002
New Orleans	210	285		337	524	405	639	592	805	730	997
New York City	300	380		918	1,428	1,102	1,739	1,266	1,811	1,525	2,205
Philadelphia	245	340		713	1,109	856	1,351	1,004	1,429	1,238	1,771
Phoenix	340	375		416	646	499	787	805	1,074	917	1,247
Pittsburgh	210	290		596	927	715	1,128	851	1,207	1,046	1,492
St. Louis	235	240		641	997	769	1,214	922	1,306	1,048	1,521
San Antonio	260	410		343	533	411	649	649	870	865	1,148
San Diego	285	460		602	936	722	1,139	934	1,301	1,228	1,696
San Francisco	350	480		783	1,218	940	1,483	1,183	1,657	1,467	2,062
Seattle	340	425		619	962	742	1,171	1,008	1,390	1,212	1,688
Washington, D.C.	210	275		454	705	544	859	709	986	860	1,206
Average	$250	$330	▼	$560	$ 870	$ 670	$1,061	$ 860	$1,200	$1,040	$1,470

Common building brick manufactured according to ASTM C62 and facing brick manufactured according to ASTM C216 are the two standard bricks available for general building use.

Building brick is made in three grades; SW, where high resistance to damage caused by cyclic freezing is required; MW, where moderate resistance to cyclic freezing is needed; and NW, where little resistance to cyclic freezing is needed. Facing brick is made in only the two grades SW and MW. Additionally, facing brick is available in three types; FBS, for general use; FBX, for general use where a higher degree of precision and lower permissible variation in size than FBS is needed; and FBA, for general use to produce characteristic architectural effects resulting from non-uniformity in size and texture of the units.

In figuring above installation costs, a D-8 Crew (with a daily output of 1.5 M) was used for the 4" veneer. A D-8 Crew (with a daily output of 1.8 M) was used for the 8" solid wall.

In figuring the total cost including overhead and profit, an allowance of 10% was added to the sum of the cost of the brick and mortar. Also, 3% breakage was included for both the bare costs and the costs with overhead and profit. If bricks are delivered palletized with 280 to 300 per pallet, or packaged, allow only 1-1/2% for breakage. Then add $10 per M to the cost of brick and deduct two hours helper time. The net result is a savings of $30 to $40 per M in place. Packaged or palletized delivery is practical when

a job is big enough to have a crane or other equipment available to handle a package of brick. This is so on all industrial work but not always true on small commercial buildings.

The prices above are for bricks used in commercial, apartment house or industrial construction. If it is possible to obtain the price of the actual brick to be used, it should be done and substituted in the table. The use of buff and gray face is increasing, and there is a continuing trend to the Norman, Roman, Jumbo and SCR brick.

See R042-500 for brick quantities per S.F. and mortar quantities per M brick. (Average prices for the various sizes are listed in Division 4)

Common red clay brick for backup is not used that often. Concrete block is the most usual backup material with occasional use of sand lime or cement brick. Sand lime cost about $15 per M less than red clay and cement brick are about $5 per M less than red clay. These figures may be substituted in the common brick breakdown for the cost of these items in place, as labor is about the same. Building brick is commonly used in solid walls for strength and as a fire stop.

Brick panels built on the ground and then crane erected to the upper floors have proven to be economical. This allows the work to be done under cover and without scaffolding.

MASONRY 4

REFERENCE NOS.

R042-180 Brick in Place

Table below is for common bond with 3/8″ concave joints and
includes 3% waste for brick and 25% waste for mortar.
Crew costs are bare costs.

Item	8″ Common Brick Wall 8″ x 2-2/3″ x 4″		Select Common Face 8″ x 2-2/3″ x 4″		Red Face Brick 8″ x 2-2/3″ x 4″	
1030 brick delivered	$250 per M	$257.50	$300 per M	$ 309.00	$330 per M	$ 339.90
Type N mortar @ $2.96 per C.F.	12.5 C.F.	38.88	10.3 C.F.	32.03	10.3 C.F.	32.03
Installation using indicated crew	Crew D-8 @ .556 Days	560.45	Crew D-8 @ .667 Days	672.34	Crew D-8 @ .667 Days	672.34
Total per M in place		$856.83		$1,013.37		$1,044.27
Total per S.F. of wall	13.5 bricks/S.F.	$ 11.57	6.75 bricks/S.F.	$ 6.84	6.75 bricks/S.F.	$ 7.05

R042-185 Reinforced Brick in Walls

Table below is for common bond with 3/8″ concave joints and includes 3%
waste. Standard 4″ x 2-2/3″ x 8″ (nominal) bricks.

Item	8″ to 9″ Thick Wall		16″ to 17″ Thick Wall	
1030 brick (select common)	$300 per M	$314.15	$305 per M	$309.00
Type PL mortar @ $3.47 per C.F.	12.5 C.F.	43.38	13.9 C.F.	48.23
Reinforcing bars delivered	50 lb. @ .27 per lb.	13.50	50 lb. @ .26 per lb.	13.50
Installation incl. reinforcing using indicated crew	Crew D-8 @ .517 Days	508.93	Crew D-8 @ .513 Days	517.10
Total per M in place		$879.96		$887.83
Total per S.F. of wall	8″ wall, 13.5 brick/S.F.	$ 11.88	16″ wall, 27.0 brick/S.F.	$ 23.97

R042-200 Concrete Block

Concrete masonry units, 8″ high x 16″ long, sand aggregate, 3/8 joints for partitions,
tooled joint one side, not including Subs O&P, 113 blocks per 100 S.F.

City	Material Per Block, Delivered 4″ Thick	8″ Thick	113 Block, Delivered 4″ Thick	8″ Thick	Mortar 3/8″ Joint	Installation 4″ Thick	8″ Thick	Total Per 100 S.F. 4″ Thick	8″ Thick
Atlanta	$.70	$1.00	$79.10	$113.00	$10.89 for 4″ block	$152.05	$174.23	$242.04	$309.02
Baltimore	.63	.83	71.19	93.79	$21.79 for 8″ block	186.08	213.24	268.16	328.82
Boston	.59	.85	66.67	96.05		322.19	369.21	399.75	487.05
Buffalo	.69	1.15	77.97	129.95		281.35	322.41	370.21	474.15
Chicago	.59	.90	66.67	101.70		297.53	340.95	375.09	464.44
Cincinnati	.58	.85	65.54	96.05		224.61	257.39	301.04	375.23
Cleveland	.66	.92	74.58	103.96		271.03	310.57	356.50	436.32
Columbus	.59	.82	66.67	92.66		225.41	258.31	302.97	372.76
Dallas	.71	1.02	80.23	115.26		151.67	173.81	242.79	310.86
Denver	.76	1.04	85.88	117.52		179.82	206.06	276.59	345.37
Detroit	.73	1.05	82.49	118.65		284.58	326.10	377.96	466.54
Houston	.69	.96	77.97	108.48		170.33	195.19	259.19	325.46
Indianapolis	.70	1.00	79.10	113.00		232.48	266.40	322.47	401.19
Kansas City	.79	1.20	89.27	135.60		228.66	262.02	328.82	419.41
Los Angeles	.62	.92	70.06	103.96		287.95	329.97	368.90	455.72
Memphis	.65	.90	73.45	101.70		163.79	187.69	248.13	311.18
Milwaukee	.73	1.09	82.49	123.17		274.66	314.74	368.04	459.70
Minneapolis	.72	1.00	81.36	113.00		260.16	298.12	352.41	432.91
Nashville	.76	1.02	85.88	115.26		160.06	183.42	256.83	320.47
New Orleans	.79	1.15	89.27	129.95		141.46	162.10	241.62	313.84
New York City	.75	1.00	84.75	113.00		385.14	441.34	480.78	576.13
Philadelphia	.68	.79	76.84	89.27		299.13	342.79	386.86	453.85
Phoenix	.60	.85	67.80	96.05		174.34	199.78	253.03	317.62
Pittsburgh	.70	.89	79.10	100.57		249.91	286.37	339.90	408.73
St. Louis	.75	1.10	84.75	124.30		268.81	308.03	364.45	454.12
San Antonio	.75	.95	84.75	107.35		143.68	164.64	239.32	293.78
San Diego	.61	.85	68.93	96.05		252.35	289.17	332.17	407.01
San Francisco	.85	1.45	96.05	163.85		328.55	376.49	435.49	562.13
Seattle	.85	1.25	96.05	141.25		259.45	297.31	366.39	460.35
Washington D.C.	.65	1.05	73.45	118.65		190.20	217.96	274.54	358.40
Average	$.70	$1.00	$78.61	$112.44		$234.91	$269.19	$324.41	$403.42

Cost for 100 S.F. of 8″ x 16″ Concrete Block Partitions to Four Stories High, Tooled Joint One Side

8″ x 16″ Sand Aggregate	4″ Thick Block		8″ Thick Block		12″ Thick Block	
113 block delivered	$.70 Ea.	$ 79.10	$1.00 Ea.	$113.00	$1.55 Ea.	$175.15
Type N mortar at $3.11 per C.F.	3.5 C.F.	10.89	7.0 C.F.	21.77	10.4 C.F.	32.34
Installation Crew	D-8 .233 days	234.86	D-8 .267 days	269.14	D-9 .294 days	347.16
Total per 100 S.F.		$324.85		$403.91		$554.65
Add for filling cores solid	6.7 C.F.	$108.75	25.8 C.F.	$218.26	42.2 C.F.	$287.97

Cost for 100 S.F. of 8″ x 16″ Concrete Block Backup with Trowel Cut Joints

8″ x 16″ Sand Aggregate	4″ Thick Block		8″ Thick Block		12″ Thick Block	
113 block delivered	$.70 Ea.	$ 79.10	$1.00 Ea.	$113.00	$1.55 Ea.	$175.15
Type N mortar at $3.11 per C.F.	5.8 C.F.	18.04	9.3 C.F.	28.92	12.7 C.F.	39.50
Installation crew	D-8 .225 days	226.80	D-8 .250 days	252.00	D-9 .270 days	318.82
Total per 100 S.F.		$323.94		$393.92		$533.47

Special block: corner, jamb and head block are same price as ordinary block of same size. Tabulated on the next page are national average prices per block. Labor on specials is about the same as equal sized regular block. Bond beam and 16″ high lintel blocks cost 30% more than regular units of equal size. Lintel blocks are 8″ long and 8″ or 16″ high. Costs in individual cities may be factored from the table above.

Use of motorized mortar spreader box will speed construction of continuous walls. Hollow non-load bearing units are made according to ASTM C129 and hollow load bearing units according to ASTM C90.

R042-220 Interlocking Grout Block Walls

8″ x 16″ Sand Aggregate	8″ Thick Block		12″ Thick Block		16″ Thick Block	
Cost per 100 S.F. of 8″ x 16″ Interlocking Grout Block, "Open End" Type						
113 block delivered	$1.50 ea.	$169.50	$2.30 ea.	$259.90	$3.55 ea.	$ 401.15
Type M mortar @ $3.82 per C.F.	7.0 C.F.	26.74	10.4 C.F.	39.73	13.8 C.F.	52.72
Minimum reinf. req.@ $ .27 per lb.						
8″ block #4 @ 24″ horiz.	33.4 lb.	9.02				
32″ vert.	40.0 lb.	10.80				
12″ & 16″ block #5 @ 8″ horiz.			99.7 lb.	26.92	99.7 lb.	26.92
16″ vert.			50.2 lb.	13.55	50.2 lb.	13.55
Type PM mortar @ $3.17 C.F.	25.8 C.F.	81.79	42.2 C.F.	133.77	56.1 C.F.	177.84
Installation Crew D-4 @ $893.35 per day	.408 days	364.49	.455 days	406.47	.541 days	483.30
Total per 100 S.F.		$662.34		$880.34		$1,155.48
1 S.F.		6.62		8.80		11.55

R042-500 Brick, Block & Mortar Quantities

Type Brick	Nominal Size (incl. mortar) L H W	Modular Coursing	Number of Brick per S.F.	3/8" Joint	1/2" Joint	Bond Type	Description	Factor
	Running Bond Number of Brick per S.F. of Wall - Single Wythe with 3/8" Joints			**C.F. of Mortar per M** Bricks, Waste Included			**For Other Bonds Standard Size** Add to S.F. Quantities in Table to Left	
Standard	8 x 2-2/3 x 4	3C=8"	6.75	10.3	12.9	Common	full header every fifth course	+20%
Economy	8 x 4 x 4	1C=4"	4.50	11.4	14.6		full header every sixth course	+16.7%
Engineer	8 x 3-1/5 x 4	5C=16"	5.63	10.6	13.6	English	full header every second course	+50%
Fire	9 x 2-1/2 x 4-1/2	2C=5"	6.40	550 # Fireclay	—	Flemish	alternate headers every course	+33.3%
Jumbo	12 x 4 x 6 or 8	1C=4"	3.00	23.8	30.8		every sixth course	+5.6%
Norman	12 x 2-2/3 x 4	3C=8"	4.50	14.0	17.9	Header = W x H exposed		+100%
Norwegian	12 x 3-1/5 x 4	5C=16"	3.75	14.6	18.6	Rowlock = H x W exposed		+100%
Roman	12 x 2 x 4	2C=4"	6.00	13.4	17.0	Rowlock stretcher = L x W exposed		+33.3%
SCR	12 x 2-2/3 x 6	3C=8"	4.50	21.8	28.0	Soldier = H x L exposed		—
Utility	12 x 4 x 4	1C=4"	3.00	15.4	19.6	Sailor = W x L exposed		-33.3%

Concrete Blocks Nominal Size	Approximate Weight per S.F. Standard	Lightweight	Blocks per 100 S.F.	Mortar per M block Partitions	Back up
2" x 8" x 16"	20 PSF	15 PSF	113	16 C.F.	36 C.F.
4"	30	20		31	51
6"	42	30		46	66
8"	55	38		62	82
10"	70	47		77	97
12"	85	55		92	112

R042-550 Brick Veneer in Place

Table below is for running bond with 3/8" concave joints and includes 3% waste for brick and 25% waste for mortar.

Item	Buff Face Brick 4" x 2-2/3"x 8"		Red Norman Face Brick 4" x 2-2/3" x 12"		Buff Roman Face Brick 4" x 2" x 12"	
1030 brick delivered	$380 per M	$ 391.40	$670 per M	$ 690.10	$690 per M	$ 710.70
Type N mortar @ $3.11 per C.F.	10.3 C.F.	32.03	14.0 C.F.	43.54	13.4 C.F.	41.67
Installation using indicated crew	Crew D-8 @ .667 days	672.34	Crew D-8 @ .690 days	695.52	Crew D-8 @ .667 days	672.34
Total per M in place		$1,095.77		$1,429.16		$1,424.71
Total per S.F. of wall	6.75 brick/S.F.	$ 7.40	4.50 brick/S.F.	$ 6.43	6.00 brick/S.F.	$ 8.55

R042-700 Glass Block

Cost of Blocks Each, Zone 1 Contractor Prices

Nominal Size (Including Mortar)	Truckload or Carload				Less Than Truckload or Carload			
	Regular	Thinline	Essex	Solar Reflective	Regular	Thinline	Essex	Solar Reflective
6" x 6"	$ 3.20	$ 2.90	—	$ 7.00	$ 3.50	$ 3.50	—	$ 8.00
8" x 8"	3.85	2.95	$ 4.70	9.00	4.20	4.00	$ 4.85	10.00
Solid 8" x 8"	21.60	—	—	—	22.35	—	—	—
4" x 8"	2.70	2.70	—	—	2.90	3.00	—	—
12" x 12"	10.35	—	—	—	11.25	—	—	—

Size	Per 100 S.F.		Per 1000 Block				
	No. of Block	Mortar ¼" Joint	Asphalt Emulsion	Caulk	Expansion Joint	Panel Anchors	Wall Mesh
6" x 6"	410 ea.	5.0 C.F.	.17 gal.	1.5 gal.	80 L.F.	20 ea.	500 L.F.
8" x 8"	230	3.6	.33	2.8	140	36	670
12" x 12"	102	2.3	.67	6.0	312	80	1000
Approximate quantity per 100 S.F.			.07 gal.	0.6 gal.	32 L.F.	9 ea.	51, 68, 102 L.F.

Accessories required for installation — Wall ties: run galvanized double steel mesh full length of joint at $.30 per L.F. — fiberglass expansion joints at sides and top at $.35 per L.F. — silicone caulking one gallon at $50.00 does 95 L.F. — asphalt emulsion one gallon does 600 L.F. at $4.50 per gallon in one gallon lots. If blocks are not set in wall chase use 2'-0" long wall anchors at $1.50 each at 2'-0" O.C.

Quantities	Cost per 100 S.F. (Truckload or Carload Price)					
	6" x 6" Block		8" x 8" Block		12" x 12" Block	
Block, delivered	410 @ $3.50	$1,435.00	230 @ $4.50	$1,035.00	102 @ $12.00	$1,224.00
Mortar @ $6.40 per C.F.	5.0 C.F.	32.00	3.6 C.F.	23.04	2.3 C.F.	14.72
Ties, expansion joints, etc.		92.82		92.82		102.00
Crew D-8@ $1,008.00 per day	.690 days	695.52	.465 days	468.72	.417 days	420.34
Total per 100 S.F.		$2,255.34		$1,619.58		$1,761.06

R044-500 Ashlar Veneer

Quantities			Low Price Stone		High Price Stone	
4" Stone, random ashlar,	2.5	tons average	$200 per ton	$ 500.00	$350 per ton	$ 875.00
Type N mortar 1:1:6 @	$3.11	C.F.	20 C.F.	62.20	20 C.F.	62.20
50 - 1" x 1/4" x 6" galv. stone anchors @	$1.10	each		55.00		55.00
Crew D-8 @	$1,008.00 per day		.714 days	719.71	.833 days	839.66
Total per 100 S.F.				$1,336.91		$1,831.86

Stone coverage varies from 30 S.F. to 50 S.F. per ton. Mortar quantities include 1" backing bed.

R045-100 Cleaning Face Brick

On smooth brick a person can clean 70 S.F. an hour; on rough brick 50 S.F. per hour. Use one gallon muriatic acid to 20 gallons of water for 1000 S.F. Do not use acid solution until wall is at least seven days old, but a mild soap solution may be used after two days. Commercial cleaners cost from $9 to $12 per gallon.

Time has been allowed for clean-up in brick prices.

R050-310   Coating Structural Steel

On field-welded jobs, the shop-applied primer coat is necessarily omitted. All painting must be done in the field and usually consists of red oxide rust inhibitive paint or an aluminum paint (see Division 099-208 for paint material costs). The table below shows paint coverage and daily production for field painting.

See Division 051-255 for hot-dipped galvanizing and for field-applied cold galvanizing and other paints and protective coatings.

See division 051-255 for steel surface preparation treatments such as wire brushing, pressure washing and sand blasting.

Type Construction	Surface Area per Ton	Coat	One Gallon Covers		In 8 Hrs. Person Covers		Average per Ton Spray	
			Brush	Spray	Brush	Spray	Gallons	Labor-hours
Light Structural	300 S.F. to 500 S.F.	1st	500 S.F.	455 S.F.	640 S.F.	2000 S.F.	0.9 gals.	1.6 L.H.
		2nd	450	410	800	2400	1.0	1.3
		3rd	450	410	960	3200	1.0	1.0
Medium	150 S.F. to 300 S.F.	All	400	365	1600	3200	0.6	0.6
Heavy Structural	50 S.F. to 150 S.F.	1st	400	365	1920	4000	0.2	0.2
		2nd	400	365	2000	4000	0.2	0.2
		3rd	400	365	2000	4000	0.2	0.2
Weighted Average	225 S.F.	All	400	365	1350	3000	0.6	0.6

R050-510   High Strength Bolts

Common bolts (A307) are usually used in secondary connections (see Division 050-508).

High strength bolts (A325 and A490) are usually specified for primary connections such as column splices, beam and girder connections to columns, column bracing, connections for supports of operating equipment or of other live loads which produce impact or reversal of stress, and in structures carrying cranes of over 5-ton capacity.

Allow 20 field bolts per ton of steel for a 6 story office building, apartment house or light industrial building. For 6 to 12 stories allow 25 bolts per ton, and above 12 stories, 30 bolts per ton. On power stations, 20 to 25 bolts per ton are needed.

R050-520   Welded Structural Steel

Usual weight reductions with welded design run 10% to 20% compared with bolted or riveted connections. This amounts to about the same total cost compared with bolted structures since field welding is more expensive than bolts. For normal spans of 18' to 24' figure 6 to 7 connections per ton.

Trusses — For welded trusses add 4% to weight of main members for connections. Up to 15% less steel can be expected in a welded truss compared to one that is shop bolted. Cost of erection is the same whether shop bolted or welded.

General — Typical electrodes for structural steel welding are E6010, E6011, E60T and E70T. Typical buildings vary between 2# to 8# of weld rod per

ton of steel. Buildings utilizing continuous design require about three times as much welding as conventional welded structures. In estimating field erection by welding, it is best to use the average linear feet of weld per ton to arrive at the welding cost per ton. The type, size and position of the weld will have a direct bearing on the cost per linear foot. A typical field welder will deposit 1.8# to 2# of weld rod per hour manually. Using semiautomatic methods can increase production by as much as 50% to 75%. Below is the cost per hour for manual welding.

Welded Structural Steel in Field, per Hour					
Item			No Operating Engr.	1/2 Operating Engr.	1 Operating Engr.
2 lb. weld rod	$1.30	per lb.	$ 2.60	$ 2.60	$ 2.60
Equipment (for welding only)	$205	/40 hrs.	5.13	5.13	5.13
Operating cost @	$5.17	per hour	5.17	5.17	5.17
Welder Foreman 1 hr. @	$32.60	per hour	32.60	32.60	32.60
Operating engineer @	$27.20	per hour	—	13.60	27.20
Total per Welder Hour (Bare Costs)			$45.50	$59.10	$72.70

The welding costs per ton of structural steel will vary from $45 to $290 per ton depending on Union requirements, crew size, design and inspection required.

R051-210   Structural Steel

1 to 2 Story Building	Bare Costs for 100 Ton Job with High Strength Bolts					Including Subs O&P		
City	Fabricated and Delivered	Unloading and Sorting	Erection Equipment	Field Erection Labor	Total per Ton in Place	Material	Installation	Total
Atlanta	$1,115	$29	$ 57	$183	$1,384	$1,227	$418	$1,645
Baltimore	1,065	34	73	217	1,389	1,172	499	1,671
Boston	1,225	45	91	288	1,649	1,348	657	2,005
Buffalo	1,300	40	73	253	1,666	1,430	571	2,001
Chicago	1,150	47	92	297	1,586	1,265	677	1,942
Cincinnati	1,090	35	62	224	1,411	1,199	502	1,701
Cleveland	1,100	41	61	261	1,463	1,210	574	1,784
Columbus	1,095	34	57	218	1,404	1,205	486	1,691
Dallas	1,200	23	65	148	1,436	1,320	355	1,675
Denver	1,260	26	62	166	1,514	1,386	388	1,774
Detroit	1,240	44	77	276	1,637	1,364	621	1,985
Houston	1,160	24	67	154	1,405	1,276	370	1,646
Indianapolis	1,300	37	56	232	1,625	1,430	512	1,942
Kansas City	1,300	35	73	223	1,631	1,430	512	1,942
Los Angeles	1,400	45	71	282	1,798	1,540	626	2,166
Memphis	1,140	26	66	162	1,394	1,254	387	1,641
Milwaukee	1,200	40	64	251	1,555	1,320	558	1,878
Minneapolis	1,115	38	97	240	1,490	1,227	570	1,797
Nashville	1,265	26	65	165	1,521	1,392	389	1,781
New Orleans	1,250	23	56	147	1,476	1,375	346	1,721
New York City	1,370	71	108	448	1,997	1,507	991	2,498
Philadelphia	1,195	48	87	306	1,636	1,315	689	2,004
Phoenix	1,110	31	56	199	1,396	1,221	447	1,668
Pittsburgh	1,200	40	91	252	1,583	1,320	587	1,907
St. Louis	1,275	37	89	237	1,638	1,403	555	1,958
San Antonio	1,120	22	53	139	1,334	1,232	327	1,559
San Diego	1,350	45	71	282	1,748	1,485	626	2,111
San Francisco	1,315	45	78	282	1,720	1,447	634	2,081
Seattle	1,165	38	65	239	1,507	1,282	536	1,818
Washington, DC	1,110	33	79	207	1,429	1,221	487	1,708
U.S. Average	$1,200	$37	$ 72	$232	$1,541	$1,320	$530	$1,850

Adjustments to the Base Price Above	Bare Costs	Incl. Subs O&P
For 3 to 6 story building, add per ton	$ 40	$ 50
For 7 to 15 stories, add per ton	70	100
For over 15 stories, add per ton	125	150
For field welded connections, add per ton	155	225
For multi-story masonry wall bearing construction, add to erection costs	30%	30%

General Average per Ton in Place for A36 Steel (High Strength Steel Base Price May Be Substituted)				
Item	Description	Bare Costs		Incl. Subs O&P
		Itemized	Summary	
Material	Base price	$ 490		
	Extras and delivery to shop	110		
	Drafting	90	$1,200	$1,320
	Shop fabrication & warehouse rehandling	415		
	Shop coat paint	55		
	Trucking to job site	40		
Installation	Unload and shake out, 1.2 hours @ $30.60	37		
	Erect and plumb, 5.6 hours @ $30.60	171	$ 341	$ 530
	Field bolt, 2.0 hours @ $30.60	61		
	Crane & minor erection equip. (incl. oper. & oiler) 20 tons/day	72		
	Total per ton in place	$1,541	$1,541	$1,850

R051-220   Steel Estimating Quantities

One estimate on erection is that a crane can handle 35 to 60 pieces per day. Say the average is 45. With usual sizes of beams, girders, and columns, this would amount to about 20 tons per day. Allowing a bare cost of $1,440 per day for the crane including crew equipment cost per day plus operator and oiler, this would amount amount to $72 per ton for the crane charge. The type of connection greatly affects the speed of erection. Moment connections for continuous design slow down production and increase erection costs.

Short open web bar joists can be set at the rate of 75 to 80 per day, with 50 per day being the average for setting long span joists.

After main members are calculated, add the following for usual allowances: base plates 2% to 3%; column splices 4% to 5%; and miscellaneous details 4% to 5%, for a total of 10% to 13% in addition to main members.

The ratio of column to beam tonnage varies depending on type of steels used, typical spans, story heights and live loads.

It is more economical to keep the column size constant and to vary the strength of the column by using high strength steels. This also saves floor space. Buildings have recently gone as high as ten stories with 8" high strength columns. For light columns under W8X31 lb. sections, concrete filled steel columns are economical.

High strength steels may be used in columns and beams to save floor space and to meet head room requirements. High strength steels in some sizes sometimes require long lead times.

Round, square and rectangular columns, both plain and concrete filled, are readily available and save floor area, but are higher in cost per pound than rolled columns. For high unbraced columns, tube columns may be less expensive.

Below are average minimum figures for the weights of the structural steel frame for different types of buildings using A36 steel, rolled shapes and simple joints. For economy in domes, rise to span ratio = .13. Open web joist framing systems will reduce weights by 10% to 40%. Composite design can reduce steel weight by up to 25% but additional concrete floor slab thickness may be required. Continuous design can reduce the weights up to 20%. There are many building codes with different live load requirements and different structural requirements, such as hurricane and earthquake loadings which can alter the figures.

*See R131-310 for Domes and Thin Shell Structures.

Structural Steel Weights per S.F. of Floor Area									
Type of Building	No. of Stories	Avg. Spans	L.L. #/S.F.	Lbs. Per S.F.	Type of Building	No. of Stories	Avg. Spans	L.L. #/S.F.	Lbs. Per S.F.
Steel Frame Mfg.	1	20'x20'	40	8	Apartments	2-8	20'x20'	40	8
		30'x30'		13		9-25			14
		40'x40'		18	Office	to 10	Various	80	10
Parking garage	4	Various	80	8.5		20			18
Domes (Schwedler)*	1	200'	30	10		30			26
		300'		15		over 50			35

R051-225   Common Structural Steel Specifications

ASTM A36 is the all-purpose carbon grade steel widely used in building and bridge construction.

The other high-strength steels listed below may each have certain advantages over ASTM A36 stuctural carbon steel, depending on the application. They have proven to be economical choices where, due to lighter members, the reduction of dead load and the associated savings in shipping cost can be significant.

ASTM A588 atmospheric weathering, high-strength low-alloy steels can be used in the bare (uncoated) condition, where exposure to normal atmosphere causes a tightly adherant oxide to form on the surface protecting the steel from further oxidation. ASTM A242 corrosion-resistant, high-strength low-alloy steels have enhanced atmospheric corrosion resistance of at least two times that of carbon structural steels with copper, or four times that of carbon structural steels without copper. The reduction or elimination of maintenance resulting from the use of these steels often offsets their higher initial cost.

Steel Type	ASTM Designation	Minimum Yield Stress in KSI	Shapes Available
Carbon	A36	36	All structural shape groups, and plates & bars up thru 8" thick
	A529	42	Structural shape group 1, and plates & bars up thru 1/2" thick
High-Strength Low-Alloy Manganese-Vanadium	A441	40	Plates & bars over 4" up thru 8" thick
		42	Structural shape groups 4 & 5, and plates & bars over 1-1/2" up thru 4" thick
		46	Structural shape group 3, and plates & bars over 3/4" up thru 1-1/2" thick
		50	Structural shape groups 1 & 2, and plates & bars up thru 3/4" thick
High-Strength Low-Alloy Columbium-Vanadium	A572	42	All structural shape groups, and plates & bars up thru 6" thick
		50	All structural shape groups, and plates & bars up thru 4" thick
		60	Structural shape groups 1 & 2, and plates & bars up thru 1-1/4" thick
		65	Structural shape group 1, and plates & bars up thru 1-1/4" thick
Corrosion-Resistant High-Strength Low-Alloy	A242	42	Structural shape groups 4 & 5, and plates & bars over 1-1/2" up thru 4" thick
		46	Structural shape group 3, and plates & bars over 3/4" up thru 1-1/2" thick
		50	Structural shape groups 1 & 2, and plates & bars up thru 3/4" thick
Weathering High-Strength Low-Alloy	A588	42	Plates & bars over 5" up thru 8" thick
		46	Plates & bars over 4" up thru 5" thick
		50	All structural shape groups, and plates & bars up thru 4" thick
Quenched and Tempered Low-Alloy	A852	70	Plates & bars up thru 4" thick
Quenched and Tempered Alloy	A514	90	Plates & bars over 2-1/2" up thru 6" thick
		100	Plates & bars up thru 2-1/2" thick

R051-230   High Strength Steels

The mill price of these steels may be higher than A36 carbon steel but their proper use can achieve overall savings thru total reduced weights. For columns with L/r over 100, A36 steel is best; under 100, high strength steels are economical. For heavy columns high strength steels are economical when cover plates are eliminated. There is no economy using high strength steels for clip angles or supports or for beams where deflection governs. Thinner members are more economical than thick.

Below is a table of the high strength steels available and their base price with specification extras. A36 is included for comparison purposes. See also R051-210 for typical in place costs of A36 steel. The per ton erection and fabricating costs of the high strength steels will be higher than for A36 since the same number of pieces, but less weight, will be installed. See R051-240 for extras for Jumbo 14" A36 columns.

							Special Column Sections W 14 x 455 Thru 805		
		High Strength Low Alloy Steels							
Type of Steel	Minimum Yield Stress in KSI	Base	Size Extra		Spec. Extra	Base + Size & Spec. Extra		Size Extra	Spec. Extra	Base + Size Spec. Extra
A36 Carbon Structural	36	$490	$0	— $140	—	$490	— $630	$120	—	$610
A529 Carbon Structural	42				—	490	— 630		—	610
A441 Manganese-Vanadium	40 to 50				—	490	— 630		—	610
A572 Columian-Vanadium	42				—	490	— 630		—	610
	50				$40	530	— 670		$45	655
	60				35	525	— 665		40	650
	65				40	530	— 670		50	660
A242 Corrosion-Resistant	42 to 50				—	490	— 630		—	610
A588 Weathering	42 to 50				60	550	— 690		60	670

Table title: Mill Base Prices and Range of Size and Specification Extras per Ton for Structural Shapes

R051-235   Common Steel Sections

The upper portion of this table shows the name, shape, common designation and basic characteristics of commonly used steel sections. The lower portion explains how to read the designations used for the above illustrated common sections.

Shape & Designation	Name & Characteristics	Shape & Designation	Name & Characteristics
W	W Shape / Parallel flange surfaces	MC	Miscellaneous Channel / Infrequently rolled by some producers
S	American Standard Beam (I Beam) / Sloped inner flange	L	Angle / Equal or unequal legs, constant thickness
M	Miscellaneous Beams / Cannot be classified as W, HP or S; infrequently rolled by some producers	T	Structural Tee / Cut from W, M or S on center of web
C	American Standard Channel / Sloped inner flange	HP	Bearing Pile / Parallel flanges and equal flange and web thickness

Common drawing designations follow:

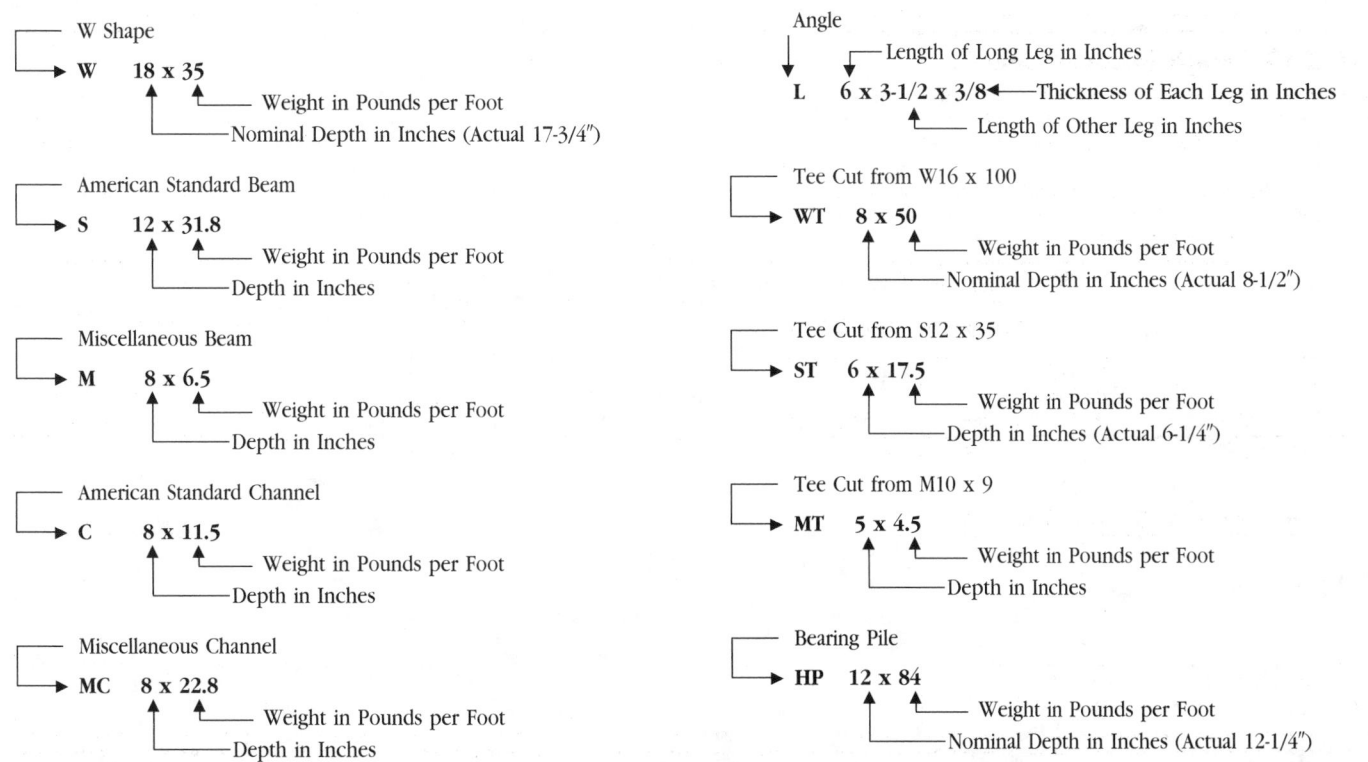

W Shape
W 18 x 35
— Weight in Pounds per Foot
— Nominal Depth in Inches (Actual 17-3/4")

American Standard Beam
S 12 x 31.8
— Weight in Pounds per Foot
— Depth in Inches

Miscellaneous Beam
M 8 x 6.5
— Weight in Pounds per Foot
— Depth in Inches

American Standard Channel
C 8 x 11.5
— Weight in Pounds per Foot
— Depth in Inches

Miscellaneous Channel
MC 8 x 22.8
— Weight in Pounds per Foot
— Depth in Inches

Angle
L 6 x 3-1/2 x 3/8
— Length of Long Leg in Inches
— Thickness of Each Leg in Inches
— Length of Other Leg in Inches

Tee Cut from W16 x 100
WT 8 x 50
— Weight in Pounds per Foot
— Nominal Depth in Inches (Actual 8-1/2")

Tee Cut from S12 x 35
ST 6 x 17.5
— Weight in Pounds per Foot
— Depth in Inches (Actual 6-1/4")

Tee Cut from M10 x 9
MT 5 x 4.5
— Weight in Pounds per Foot
— Depth in Inches

Bearing Pile
HP 12 x 84
— Weight in Pounds per Foot
— Nominal Depth in Inches (Actual 12-1/4")

R051-240 Structural Steel Extras
Principal Extras in Dollars Per Ton

Item quantity — using 5 tons per size as base price. Under 5 tons to 3 tons inclusive add $6 per ton. Under 3 tons to 2 tons inclusive add $11 per ton. Under 2 tons to 1 ton inclusive add $16 per ton. Under 1 ton to 1/2 ton add $55 per ton. Under 1/2 ton add $100 per ton.

Size Extras:

W Shapes: W36 x 393-135, W33 x 354-118, W30 x 326-90, W27 x 217-84, W24 x 176; W21 x 166, W18 x 311-130, W14 x 426-145, W12 x 336-136, add $20; W24 x 162-55, W21 x 147-44, W18 x 119-35, W16 x 100-26, W14 x 132-22, W12 x 120-14, W10 x 112-12, W8 x 67-10, W6 x 25-9, W5 x 19-16, and W4 x 13, no add; W14 x 808-455 add $124. For specification extras see R051-230.

Miscellaneous Shapes: M14 x 18 to M5 x 18.9 no add. Standard Beams: S24 x 121-80, S20 x 96-66 add $140; S18 x 70 & 54.7, S15 x 50 & 42.9, S12 x 50-31.8, S10 x 35 & 25.4, S8 x 23 & 18.4, S7 x 20 & 15.3 and S6 x 17.25 & 12.5 add $80.

Standard Channels: C15 x 50-33.9, C12 x 30-20.7, C10 x 30-15.3, C9 x 20-13.4, C8 x 18.75-11.5, C7 x 14.75-9.8 add $60.

Miscellaneous Channels: MC18 x 58 to MC13 x 31.8 add $80; MC12 x 50 to MC6 x 18 add $80. Car Building Sections: add $120. Bulb Angles: add $140. Angles Equal & Unequal Leg: add $80.

Cambering: Channels, standard beams, tees & wide flange shapes to 300 lbs./ft. add $60.

Galvanizing under 1 ton, $600; over 20 tons, $500 per ton. For color coating of galvanizing add 40% to prices.

Government Specifications: MIL-S-20166 B, class U, type 1, Grade M-medium add $20; Grade HT-high tensile add $15. American Association of State Highway & Transportation Officials (AASHTO): Without Charpy Impacts, Grade M 188 no add; Grade M 222 add $65; M 223, Gr. 50 add $15.

Cut Lengths: 10' to 20' add $20; 20' to 25' add $20; 25' to 65' no add; over 65' to 90' no add; over 90' add $5.

Milling One or Two Ends: 10 lbs./ft. to 100 lbs./ft., over 10' add $70; over 100 lbs./ft., over 10' add $60.

Handling & Loading: Under 5 tons add $10. Banding or Wiring: add $10. Special loading: add $10. Special marking: add $10.

Special Testing: Bend Tests, $10, Charpy Impact Testing: one set of tests — impact strength values only, $5; one set of tests — impact strength values, lateral expansion, percent shear fracture, $7.

Special Straightening: For tolerances not more restrictive than 50% of standard camber or sweep, $7.

Splitting Beams to Produce Tees: over 15 lbs./ft.-30 lbs./ft., $60, over 30 lbs./ft.-150 lbs./ft. $40, over 150 lbs./ft., $35.

Chemistry: Ladle analysis limits for carbon, manganese, silicon & copper tests run from $10 to $30 per ton.

R051-245 Installation Time for Structural Steel Building Components

The following tables show the expected average installation times for various structural steel shapes. Table A presents installation times for columns, Table B for beams, Table C for light framing and bolts, and Table D for structural steel for various project types.

Table A		
Description	**Labor-Hours**	**Unit**
Columns		
Steel, Concrete Filled		
3-1/2" Diameter	.933	Ea.
6-5/8" Diameter	1.120	Ea.
Steel Pipe		
3" Diameter	.933	Ea.
8" Diameter	1.120	Ea.
12" Diameter	1.244	Ea.
Structural Tubing		
4" x 4"	.966	Ea.
8" x 8"	1.120	Ea.
12" x 8"	1.167	Ea.
W Shape 2 Tier		
W8 x 31	.052	L.F.
W8 x 67	.057	L.F.
W10 x 45	.054	L.F.
W10 x 112	.058	L.F.
W12 x 50	.054	L.F.
W12 x 190	.061	L.F.
W14 x 74	.057	L.F.
W14 x 176	.061	L.F.

Table B				
Description	**Labor-Hours**	**Unit**	**Labor-Hours**	**Unit**
Beams, W Shape				
W6 x 9	.949	Ea.	.093	L.F.
W10 x 22	1.037	Ea.	.085	L.F.
W12 x 26	1.037	Ea.	.064	L.F.
W14 x 34	1.333	Ea.	.069	L.F.
W16 x 31	1.333	Ea.	.062	L.F.
W18 x 50	2.162	Ea.	.088	L.F.
W21 x 62	2.222	Ea.	.077	L.F.
W24 x 76	2.353	Ea.	.072	L.F.
W27 x 94	2.581	Ea.	.067	L.F.
W30 x 108	2.857	Ea.	.067	L.F.
W33 x 130	3.200	Ea.	.071	L.F.
W36 x 300	3.810	Ea.	.077	L.F.

Table C		
Description	**Labor-Hours**	**Unit**
Light Framing		
Angles 4" and Larger	.011	lbs.
Less than 4"	.018	lbs.
Channels 8" and Larger	.009	lbs.
Less than 8"	.016	lbs.
Cross Bracing Angles	.009	lbs.
Rods	.034	lbs.
Hanging Lintels	.028	lbs.
High Strength Bolts in Place		
3/4" Bolts	.097	Ea.
7/8" Bolts	.100	Ea.

Table D				
Description	**Labor-Hours**	**Unit**	**Labor-Hours**	**Unit**
Apartments, Nursing Homes, etc.				
1-2 Stories	4.211	Piece	7.767	Ton
3-6 Stories	4.444	Piece	7.921	Ton
7-15 Stories	4.923	Piece	9.014	Ton
Over 15 Stories	5.333	Piece	9.209	Ton
Offices, Hospitals, etc.				
1-2 Stories	4.211	Piece	7.767	Ton
3-6 Stories	4.741	Piece	8.889	Ton
7-15 Stories	4.923	Piece	9.014	Ton
Over 15 Stories	5.120	Piece	9.209	Ton
Industrial Buildings				
1 Story	3.478	Piece	6.202	Ton

METALS **5**

R051-250   Subpurlins

Bulb tee and truss tee subpurlins are structural members designed to support and reinforce a variety of roof deck systems such as precast cement fiber roof deck tiles, monolithic roof deck systems, and gypsum or lightweight concrete over formboard. Other uses include interstitial service ceiling systems, wall panel systems, and joist anchoring in bond beams.

The table below shows bare costs for material only for subpurlins spaced at 32-5/8" O.C., purchased in lots of 6,000 to 16,000 L.F. (see Division 051-210 for pricing on a square foot basis). Maximum span is based on a 3-span condition with a total allowable vertical load of 55 psf.

	Bulb Tees, Painted			Truss Tees, Painted								
Type	Wt. per L.F.	Cost per L.F.	Max. Span	Size	Wt. per L.F.	Cost per L.F.	Max. Span	Size	Wt. per L.F.	Cost per L.F.	Max. Span
112	1.48 #	$ .85	5'-6"	2"	1.10 #	$ .74	5'-9"	2-1/2"	1.39 #	$ .84	7'-0"
158	1.68	.86	6'-4"		1.27	.79	6'-0"		1.85	1.15	8'-9"
168	1.87	.91	7'-8"		1.33	.80	6'-0"	3"	1.14	.74	7'-3"
178	2.15	1.00	8'-9"		1.78	1.08	7'-6"		1.88	1.15	9'-0"
218	3.19	1.26	10'-2"	2-1/2"	1.12	.74	6'-9"	3-1/2"	1.17	.75	7'-9"
228	3.87	1.60	12'-0"		1.34	.80	6'-9"		1.90	1.17	10'-9"

R055-810 Dimensions and Weights of Sheet Steel

Gauge No.	Approximate Thickness				Weight		
	Inches (in fractions)	Inches (in decimal parts)		Millimeters	per S.F. in Ounces	per S.F. in Lbs.	per Square Meter in Kg.
	Wrought Iron	Wrought Iron	Steel	Steel			
0000000	1/2"	.5	.4782	12.146	320	20.000	97.650
000000	15/32"	.46875	.4484	11.389	300	18.750	91.550
00000	7/16"	.4375	.4185	10.630	280	17.500	85.440
0000	13/32"	.40625	.3886	9.870	260	16.250	79.330
000	3/8"	.375	.3587	9.111	240	15.000	73.240
00	11/32"	.34375	.3288	8.352	220	13.750	67.130
0	5/16"	.3125	.2989	7.592	200	12.500	61.030
1	9/32"	.28125	.2690	6.833	180	11.250	54.930
2	17/64"	.265625	.2541	6.454	170	10.625	51.880
3	1/4"	.25	.2391	6.073	160	10.000	48.820
4	15/64"	.234375	.2242	5.695	150	9.375	45.770
5	7/32"	.21875	.2092	5.314	140	8.750	42.720
6	13/64"	.203125	.1943	4.935	130	8.125	39.670
7	3/16"	.1875	.1793	4.554	120	7.500	36.320
8	11/64"	.171875	.1644	4.176	110	6.875	33.570
9	5/32"	.15625	.1495	3.797	100	6.250	30.520
10	9/64"	.140625	.1345	3.416	90	5.625	27.460
11	1/8"	.125	.1196	3.038	80	5.000	24.410
12	7/64"	.109375	.1046	2.657	70	4.375	21.360
13	3/32"	.09375	.0897	2.278	60	3.750	18.310
14	5/64"	.078125	.0747	1.897	50	3.125	15.260
15	9/128"	.0713125	.0673	1.709	45	2.813	13.730
16	1/16"	.0625	.0598	1.519	40	2.500	12.210
17	9/160"	.05625	.0538	1.367	36	2.250	10.990
18	1/20"	.05	.0478	1.214	32	2.000	9.765
19	7/160"	.04375	.0418	1.062	28	1.750	8.544
20	3/80"	.0375	.0359	.912	24	1.500	7.324
21	11/320"	.034375	.0329	.836	22	1.375	6.713
22	1/32"	.03125	.0299	.759	20	1.250	6.103
23	9/320"	.028125	.0269	.683	18	1.125	5.490
24	1/40"	.025	.0239	.607	16	1.000	4.882
25	7/320"	.021875	.0209	.531	14	.875	4.272
26	3/160"	.01875	.0179	.455	12	.750	3.662
27	11/640"	.0171875	.0164	.417	11	.688	3.357
28	1/64"	.015625	.0149	.378	10	.625	3.052

METALS 5

REFERENCE NOS.

R061-010 Thirty City Lumber Prices

Prices for boards are for #2 or better or sterling, whichever is in best supply. Dimension lumber is "Standard or Better" either Southern Yellow Pine (S.Y.P.), Spruce-Pine-Fir (S.P.F.), Hem-Fir (H.F.) or Douglas Fir (D.F.). The species of lumber used in a geographic area is listed by city. Plyform is 3/4" BB oil sealed fir or S.Y.P. whichever prevails locally, 3/4" CDX is S.Y.P. or Fir.

These are prices at the time of publication and should be checked against the current market price. Relative differences between cities will stay approximately constant.

City	Species	Contractor Purchases per M.B.F. S4S								Contractor Purchases per M.S.F.	
		Dimensions						Boards		3/4" Ext. Plyform	3/4" Thick CDX T&G
		2"x4"	2"x6"	2"x8"	2"x10"	2"x12"	4"x4"	1"x6"	1"x12"		
Atlanta	S.P.F.	$549	$525	$586	$683	$832	$ 687	$ 832	$1,417	$ 750	$600
Baltimore	S.P.F.	630	558	526	566	619	786	1,300	1,800	922	765
Boston	S.P.F.	455	415	420	515	569	1,203	1,120	1,360	860	610
Buffalo	H.F.	535	576	515	531	556	855	1,400	1,660	796	662
Chicago	S.P.F.	486	468	501	561	603	575	1,061	1,350	665	578
Cincinnati	S.P.F.	599	683	691	795	883	863	1,200	1,460	936	686
Cleveland	S.P.F.	612	574	591	744	812	656	1,082	2,099	859	625
Columbus	S.P.F.	515	657	633	745	826	620	1,560	2,140	989	579
Dallas	S.P.F.	575	489	525	640	550	557	800	1,131	797	589
Denver	H.F.	574	574	588	639	599	799	848	1,243	997	672
Detroit	S.P.F.	645	690	600	859	895	495	1,240	2,050	860	593
Houston	S.Y.P.	480	488	488	494	583	669	777	940	821	525
Indianapolis	S.Y.F.	510	495	620	620	670	740	665	890	785	590
Kansas City	D.F.	570	550	540	570	600	720	1,050	1,300	800	600
Los Angeles	D.F.	500	510	515	510	530	580	1,063	1,231	950	565
Memphis	S.P.F.	635	595	630	695	700	890	999	1,445	795	780
Milwaukee	S.P.F.	640	612	612	743	806	702	850	1,280	847	688
Minneapolis	S.P.F.	509	496	488	563	686	562	1,316	1,549	948	562
Nashville	S.P.F.	527	540	523	567	564	720	1,100	1,500	758	563
New Orleans	S.Y.P.	525	525	510	510	519	610	650	1,700	725	535
New York City	S.P.F.	490	475	550	555	525	870	880	1,210	887	594
Philadelphia	H.F.	555	540	540	565	615	670	1,325	1,600	1,204	650
Phoenix	D.F.	559	599	650	699	699	945	1,295	1,599	1,090	690
Pittsburgh	H.F.	554	591	648	681	714	666	1,260	1,590	1,030	624
St. Louis	S.P.F.	444	428	375	365	460	715	1,000	719	516	516
San Antonio	S.Y.P.	440	443	443	475	535	514	677	1,345	825	583
San Diego	D.F.	475	475	500	515	535	640	1,164	1,295	1,070	675
San Francisco	D.F.	490	490	471	484	484	700	1,100	1,490	1,159	644
Seattle	S.P.F.	577	567	599	693	720	681	720	1,188	811	591
Washington, DC	S.P.F.	516	499	533	627	651	757	900	1,330	938	612
Average		$539	$538	$547	$607	$645	$ 715	$1,041	$1,430	$ 880	$618

To convert square feet of surface to board feet, 4% waste included.

S4S Size	Multiply S.F. by	T & G Size	Multiply S.F. by	Flooring Size	Multiply S.F. by
				25/32" x 2-1/4"	1.37
1 x 4	1.18	1 x 4	1.27	25/32" x 3-1/4"	1.29
1 x 6	1.13	1 x 6	1.18	15/32" x 1-1/2"	1.54
1 x 8	1.11	1 x 8	1.14	1" x 3"	1.28
1 x 10	1.09	2 x 6	2.36	1" x 4"	1.24

R061-020   Plywood

There are two types of plywood used in construction: interior, which is moisture resistant but not waterproofed, and exterior, which is waterproofed.

The grade of the exterior surface of the plywood sheets is designated by the first letter: A, for smooth surface with patches allowed; B, for solid surface with patches and plugs allowed; C, which may be surface plugged or may have knot holes up to 1″ wide; and D, which is used only for interior type plywood and may have knot holes up to 2-1/2″ wide. "Structural Grade" is specifically designed for engineered applications such as box beams. All CC & DD grades have roof and floor spans marked on them.

Underlayment grade plywood runs from 1/4″ to 1-1/4″ thick. Thicknesses 5/8″ and over have optional tongue and groove joints which eliminates the need for blocking the edges. Underlayment 19/32″ and over may be referred to as Sturd-i-Floor.

The price of plywood can fluctuate widely due to geographic and economic conditions. When one or two local prices are known, the relative prices for other types and sizes may be found by direct factoring of the prices in the table below.

Typical uses for various plywood grades are as follows:

AA-AD Interior — cupboards, shelving, paneling, furniture

BB Plyform — concrete form plywood

CDX — wall and roof sheathing

Structural — box beams, girders, stressed skin panels

AA-AC Exterior — fences, signs, siding, soffits, etc.

Underlayment — base for resilient floor coverings

Overlaid HDO — high density for concrete forms & highway signs

Overlaid MDO — medium density for painting, siding, soffits & signs

303 Siding — exterior siding, textured, striated, embossed, etc.

Grade	National Average Price in Lots of 10 MSF, per MSF-January 1999						
	Type	4'x8'	Type		4'x8'	4'x10'	
Sanded Grade	1/4″ Interior AD	$ 449	1/4″ Exterior AC		$ 496	$ 520	
	3/8″	500	3/8″		556	572	
	1/2″	598	1/2″		662	704	
	5/8″	708	5/8″		774	864	
	3/4″	748	3/4″		840	978	
	1″	870	1″		1,090	1,127	
	1-1/4″	1,043	Exterior AA, add		150	155	
	Interior AA, add	150	Exterior AB, add		145	140	
Unsanded Grade 4' x 8' Sheets	5/16″ CDX	$ 318	CD Structural 1	Underlayment			
	3/8″	320	5/16″, 4'x8' sheets $378	3/8″, 4'x8' sheets		$ 453	
	1/2″	410	3/8″ 475	1/2″		526	
	5/8″	493	1/2″ 545	5/8″T&G		627	
	3/4″	582	5/8″ 648	3/4″T&G		810	
	3/4″ T&G	618	3/4″ 724	1-1/8″ 2-4-1 T&G		1,269	
Form Plywood	5/8″ Exterior, oiled BB, plyform	$ 865	5/8″ HDO (overlay 2 sides)			$2,148	
	3/4″ Exterior, oiled BB, plyform	975	3/4″ HDO (overlay 2 sides)			2,214	
Overlaid 4'x8' Sheets	Overlay 2 Sides MDO		Overlay 1 Side MDO				
	3/8″ thick	$1,025	3/8″ thick			$ 918	
	1/2″	1,199	1/2″			1,058	
	5/8″	1,299	5/8″			1,188	
	3/4″	1,399	3/4″			1,393	
303 Siding	Fir, rough sawn, natural finish, 3/8″ thick	$ 682	Texture 1-11	5/8″ thick, Fir		$1,037	
	Redwood	1,890		Redwood		1,890	
	Cedar	1,890		Cedar		1,537	
	Southern Yellow Pine	693		Southern Yellow		832	
Waferboard/O.S.B.	1/4″ sheathing	$ 150	19/32″ T&G			$ 270	
	7/16″ sheathing	174	23/32″ T&G			298	

For 2 MSF to 10 MSF, add 10%. For less than 2 MSF, add 15%.

R061-030 Lumber Product Material Prices

The price of forest products fluctuates widely from location to location and from season to season depending upon economic conditions. The table below indicates National Average material prices in effect Jan. 1, 1999. The table shows relative differences between various sizes, grades and species. These percentage differentials remain fairly constant even though lumber prices in general may change significantly during the year.

Availability of certain items depends upon geographic location and must be checked prior to firm price bidding.

	National Average Contractor Price, Quantity Purchase						Heavy Timbers, Fir		
	Dimension Lumber, S4S, #2 & Better, KD								
	Species	2"x4"	2"x6"	2"x8"	2"x10"	2"x12"			
Framing Lumber per MBF	Douglas Fir	$ 530	$ 537	$ 544	$ 566	$ 578	3" x 4" thru 3" x 12"		$1,012
	Spruce	507	492	515	560	589	4" x 4" thru 4" x 12"		1,153
	Southern Yellow Pine	487	485	512	523	568	6" x 6" thru 6" x 12"		1,739
	Hem-Fir	555	570	573	604	621	8" x 8" thru 8" x 12"		1,868
							10" x 10" and 10" x 12"		1,739

	S4S "D" Quality or Clear, KD						S4S # 2 & Better or Sterling, KD				
	Species	1"x4"	1"x6"	1"x8"	1"x10"	1"x12"	Species	1"x4"	1"x6"	1"x8"	1"x10"	1"x12"
Boards per MBF *See also Cedar Siding	Sugar Pine	$ 980	$1,400	$1,400	$1,505	$2,226	Sugar Pine	$630	$658	$686	$875	$1,022
	Idaho Pine	1,288	1,631	1,645	1,764	2,240	Idaho Pine	826	826	812	826	945
	Engleman Spruce	1,288	1,778	1,778	1,848	2,198	Engleman Spruce	546	588	560	805	889
	So. Yellow Pine	1,190	1,351	1,358	1,400	1,526	So. Yellow Pine	546	574	700	840	973
	Ponderosa Pine	1,064	1,456	1,526	1,631	2,198	Ponderosa Pine	560	651	609	749	861
	Redwood, CVG	3,940	3,800	3,750	3,770	5,940						

Flooring per MSF	1" x 4" Vertical grain, Fir "B" & better	$2,300	2-1/4" x 25/32" Maple, select	$2,970	
	2-1/4" x 25/32", Oak, clear	3,200	#2 & better	2,700	
	Select	3,050	2-1/4" x 33/32" Maple, #2 & better	3,132	
	#1 common	2,700	3-1/4" x 33/32" Maple, #2 & better	2,916	
	Oak, prefinished, standard & better	3,750	Parquet, unfinished, 5/16", minimum	1,539	
	Standard	3,200	Maximum	4,800	
Siding per MBF	Clapboard, Cedar, beveled		*Rough sawn, Cedar, T&G, "A" grade 1" x 4"	$2,434	
	1/2" x 6" thru 1/2" x 8", clear	$1,404	1" x 6"	3,201	
	"A" grade	1,377	"STK" grade, 1" x 6"	1,614	
	"B" grade	972	Board, "STK" grade, 1" x 8"	1,561	
	3/4" x 10" "clear"	2,700	Board & Batten 1" x 12"	1,561	
	"A" grade	2,430	Cedar channel siding 1" x 8", #3 & better	$1,513	
	Redwood, beveled		Factory stained	1,693	
	1/2" x 6" thru 1/2" x 8", vertical grain, clear	1,907	White Pine siding, T&G, rough sawn	$ 507	
	3/4" x 10" vertical grain, clear	2,889	Factory stained	556	
Shingles per CSF	Red Cedar		White Cedar shingles		
	5X—16" long #1 regular	$ 157	16" long, extra grade (East Coast)	$ 144	
	#2	177	Clear, 1st grade (East Coast)	120	
	18" long perfections #1	202			
	#2	133	Fire retardant Red Cedar		
	Resquared & Rebutted #1	118	5X — 16" long	$ 257	
			18" long perfections	275	
	Handsplit shakes, resawn		Handsplit & resawn		
	24" long, 1/2" to 3/4"	182	24" long, 1/2" to 3/4"	275	
	18" long, 1/2" to 3/4"	162	18" long, 1/2" to 3/4"	290	

6 WOOD & PLASTICS

REFERENCE NOS.

R061-100 Wood Roof Trusses

Loading figures represent live load. An additional load of 10 psf on the top chord and 10 psf on the bottom chord is included in the truss design. Spacing is 24″ O.C.

Span in Feet	Cost per Truss for Different Live Loads and Roof Pitches					
	Flat	4 in 12 Pitch		5 in 12 Pitch		8 in 12 Pitch
	40 psf	30 psf	40 psf	30 psf	40 psf	30 psf
20	$ 70.00	$39.00	$41.00	$42.00	$46.00	$ 54.00
22	77.00	43.00	45.00	46.00	50.00	58.00
24	84.00	47.00	49.00	50.00	54.00	62.00
26	91.00	50.00	52.00	53.00	57.00	65.00
28	98.00	52.00	54.00	55.00	59.00	67.00
30	105.00	63.00	65.00	66.00	70.00	78.00
32	112.00	65.00	67.00	68.00	72.00	80.00
34	119.00	76.00	78.00	79.00	83.00	91.00
36	126.00	80.00	82.00	83.00	87.00	95.00
38	133.00	83.00	85.00	86.00	90.00	98.00
40	140.00	90.00	92.00	93.00	97.00	105.00

R064-100 Wood Stair, Residential

Item				Quantity	Unit Cost	Bare Costs	Costs Incl. Subs O & P
One Flight with 8′-6″ Story Height, 3′-6″ Wide Oak Treads Open One Side, Built in Place							
Treads 10-1/2″ x 1-1/16″ thick				11 Ea.	$ 26.00	$ 286.00	$ 314.60
Landing tread nosing, rabbeted				1 Ea.	7.80	7.80	8.60
Risers 3/4″ thick				12 Ea.	11.50	138.00	151.80
Single end starting step	(range	$175 to	$210)	1 Ea.	193.00	193.00	212.40
Balusters	(range	$4.50 to	$25.00)	22 Ea.	7.75	170.50	187.60
Newels, starting & landing	(range	$40 to	$120)	2 Ea.	73.00	146.00	160.60
Rail starter	(range	$39 to	$190)	1 Ea.	67.00	67.00	73.80
Handrail	(range	$4.50 to	$10.30)	26 L.F.	7.00	182.00	200.20
Cove trim				50 L.F.	.55	27.50	30.20
Rough stringers three - 2 x 12's, 14′ long				84 B.F.	.65	54.60	60.00
Carpenter's installation: Bare Cost				36 Hrs.	$ 27.30	$ 982.80	
Incl. Subs O & P					$ 42.95		$1,546.20
				Total per Flight		$2,255.20	$2,946.00

Add for rail return on second floor and for varnishing or other finish. Adjoining walls or landings must be figured separately.

WOOD & PLASTICS 6

REFERENCE NOS.

R071-010   1/2" Pargeting (rough dampproofing plaster)

1:2-1/2 Mix, 4.5 C.F. Covers 100 S.F., 2 Coats, Waste Included		Regular Portland Cement		Waterproofed Portland Cement	
1.7 lbs. integral waterproofing admixture				$.67 per lb.	$ 1.14
1.7 bags Portland cement		$7.00 per bag	$ 11.90	$7.00 per bag	11.90
4.25 C.F. sand @	$21.85 per C.Y.		3.44		3.44
Labor mix, apply, crew D-1 @	$393.60 per day	.40 days	157.44	.40 days	157.44
Total Bare Cost per 100 S.F.			$172.78		$173.92

R072-030   Firestopping

Firestopping is the sealing of structural, mechanical, electrical and other penetrations through fire-rated assemblies. The basic components of firestop systems are safing insulation and firestop sealant on both sides of wall penetrations and the top side of floor penetrations.

Pipe penetrations are assumed to be through concrete, grout, or joint compound and can be sleeved or unsleeved. Costs for the penetrations and sleeves are not included. An annular space of 1" is assumed. Escutcheons are not included.

Metallic pipe is assumed to be copper, aluminum, cast iron or similar metallic material. Insulated metallic pipe is assumed to be covered with a thermal insulating jacket of varying thickness and materials.

Non-metallic pipe is assumed to be PVC, CPVC, FR Polypropylene or similar plastic piping material. Intumescent firestop sealant or wrap strips are included. Collars on both sides of wall penetrations and a sheet metal plate on the underside of floor penetrations are included.

Ductwork is assumed to be sheet metal, stainless steel or similar metallic material. Duct penetrations are assumed to be through concrete, grout or joint compound. Costs for penetrations and sleeves are not included. An annular space of 1/2" is assumed.

Multi-trade openings include costs for sheet metal forms, firestop mortar, wrap strips, collars and sealants as necessary.

Structural penetrations joints are assumed to be 1/2" or less. CMU walls are assumed to be within 1-1/2" of metal deck. Drywall walls are assumed to be tight to the underside of metal decking.

Metal panel, glass or curtain wall systems include a spandrel area of 5' filled with mineral wool foil-faced insulation. Fasteners and stiffeners are included.

R073-020   Roof Slate

16", 18" and 20" are standard lengths and slate usually comes in random widths. For standard 3/16" thickness use 1-1/2" copper nails. Allow for 3% breakage.

Quantities per Square	Unfading Vermont Colored	Weathering Sea Green	Buckingham, Virginia Black
Slate delivered (incl. punching)	$395.00	$295.50	$550.00
# 30 Felt, 2.5 lbs. copper nails	17.87	17.87	17.87
Slate roofer 4.6 hrs. @ $24.20 per hr.	111.32	111.32	111.32
Total Bare Cost per Square	$524.19	$424.69	$679.19

R075-020 Built-Up Roofing

Asphalt is available in kegs of 100 lbs. each; coal tar pitch in 560 lb. kegs. Prepared roofing felts are available in a wide range of sizes, weights & characteristics. However, the most commonly used are #15 (432 S.F. per roll, 13 lbs. per square) and #30 (216 S.F. per roll, 27 lbs. per square).

Inter-ply bitumen varies from 24 lbs. per sq. (asphalt) to 30 lbs. per sq. (coal tar) per ply, ± 25%. Flood coat bitumen also varies from 60 lbs. per sq. (asphalt) to 75 lbs. per sq. (coal tar), ± 25%. Expendable equipment (mops, brooms, screeds, etc.) runs about 16% of the bitumen cost. For new, inexperienced crews this factor may be much higher.

Rigid insulation board is typically applied in two layers. The first is mechanically attached to nailable decks or spot or solid mopped to non-nailable decks; the second layer is then spot or solid mopped to the first layer. Membrane application follows the insulation, except in protected membrane roofs, where the membrane goes down first and the insulation on top, followed with ballast (stone or concrete pavers). Insulation and related labor costs are NOT included in Div. 075-102; they can be found in Div. 072-203.

Prices shown are bare costs only.

4-Ply Felt on Insulated Deck	Asphalt/Glass Fiber Felt		Coal Tar/Organic Felt	
4 ply #15 felt (1/4 roll per ply)	$3.70 per ply	$ 14.80	$7.89 per ply	$ 31.56
156 lbs. hot asphalt/195 lbs. hot tar	$255 per ton	19.89	$538 per ton	52.46
500 lbs. roofing stone (1/4″ to 1/2″)	$21.00 per ton	5.25	$21.00 per ton	5.25
Installation, crew G-1, @ $1,678.50 per day	.050 days	83.93	.048 days	80.57
Total Bare Cost per Square in-place		$123.87		$169.84

R075-030 Modified Bitumen Roofing

The cost of modified bitumen roofing is highly dependent on the type of installation that is planned. Installation is based on the type of modifier used in the bitumen. The two most popular modifiers are atactic polypropylene (APP) and styrene butadiene styrene (SBS). The modifiers are added to heated bitumen during the manufacturing process to change its characteristics. A polyethylene, polyester or fiberglass reinforcing sheet is then sandwiched between layers of this bitumen. When completed, the result is a pre-assembled, built-up roof that has increased elasticity and weatherablility. Some manufacturers include a surfacing material such as ceramic or mineral granules, metal particles or sand.

The preferred method of adhering SBS-modified bitumen roofing to the substrate is with hot-mopped asphalt (much the same as built-up roofing). This installation method requires a tar kettle/pot to heat the asphalt, as well as the labor, tools and equipment necessary to distribute and spread the hot asphalt.

The alternative method for applying APP and SBS modified bitumen is as follows. A skilled installer uses a torch to melt a small pool of bitumen off the membrane. This pool must form across the entire roll for proper adhesion. The installer must unroll the roofing at a pace slow enough to melt the bitumen, but fast enough to prevent damage to the rest of the membrane.

Modified bitumen roofing provides the advantages of both built-up and single-ply roofing. Labor costs are reduced over those of built-up roofing because only a single ply is necessary. The elasticity of single-ply roofing is attained with the reinforcing sheet and polymer modifiers. Modifieds have some self-healing characteristics and because of their multi-layer construction, they offer the reliability and safety of built-up roofing.

THERMAL & MOISTURE PROTECTION 7

REFERENCE NOS.

R081-010 Hollow Metal Doors

Table below lists material prices only, not including hardware or labor.

Door Thickness and Size		Full Flush Doors, 18 Ga.				Flush Fire Doors			
		Hollow Metal		Composite Core		Hollow Metal "B"		Hollow Metal Class "A"	
		Plain	Glazed	Plain	Glazed	20 Ga.	18 Ga.	16 Ga.	18 Ga.
1-3/4"	3'-0" x 6'-8"	$195	$245	$218	$270	$185	$215	$260	$227
	3'-0" x 7'-0"	205	255	225	275	195	221	275	232
	3'-6" x 7'-0"	230	280	265	315	230	270	305	284
	3'-0" x 8'-0"	260	315	285	340	240	280	322	294
	4'-0" x 8'-0"	300	350	320	385	305	342	355	347
1-3/8"	2'-0" x 6'-8"	145	190	—	—	165	—	—	—
	2'-6" x 6'-8"	155	205	—	—	180	—	—	—
	3'-0" x 7'-0"	165	215	—	—	190	—	—	—

*Indicates 20 gauge doors.

R081-020 Hollow Metal Frames

Table below lists material prices only.

Frame for Opening Size	16 Ga. Frames		16 Ga. UL Frames		16 Ga. Drywall Frames		16 Ga. UL Drywall Frames	
	6-3/4" Deep	8-3/4" Deep	6-3/4" Deep	8-3/4" Deep	7-1/8" Deep	8-1/4" Deep	7-1/8" Deep	8-1/4" Deep
2'-0" x 6'-8"	$ 74	$78	$ 86	$ 91	$ 80	$ 88	$ 98	$104
2'-6" x 6'-8"	74	78	86	91	80	88	98	104
3'-0" x 7'-0"	80	80	88	94	84	90	101	107
3'-6" x 7'-0"	85	83	91	96	86	92	102	111
4'-0" x 7'-0"	90	92	102	107	94	102	111	117
6'-0" x 7'-0"	90	92	102	107	94	102	111	117
8'-0" x 8'-0"	107	97	119	112	111	103	129	139

For welded frames, add $26.00.
For galvanized frames, add $22.00.

R082-120   Wood Doors

Table below lists price per door only, not including frame, hardware or labor. For pre-hung exterior door units up to 3' x 7', add $162 per door for wood frame and hardware for types not listed under pre-hung. Pricing is for ten or more doors. Doors are factory trimmed for butts and locksets.

Door Thickness and Size		Flush Type Doors (Interior)					Architectural (1-3/4" Ext., 1-3/8" Int.)					
		Hollow Core		Solid Particle Core			Pine		Fir		Pre-hung	
		Lauan	Birch	Lauan	Birch	Oak	Panel	Glazed	Panel	Glazed	Pine Panel	Flush Birch S. C.
1-3/4"	2'-6" x 6'-8"	$47	$48	$70	$78 *	$ 88	$330	$365	$200	$215	$455	$210
	3'-0" x 6'-8"	48	55	75	82 *	91	345	380	205	221	475	225
	3'-0" x 7'-0"	54	60	85	89 *	99	385	410	225	234	480	235
	3'-6" x 7'-0"	69	69	90	93 *	108	—	—	—	—	—	—
	4'-0" x 7'-0"	72	73	93	96 *	115	—	—	—	—	—	—
1-3/8"	2'-0" x 6'-8"	30	44	68	74 *	84	104	—	121	—	177	160
	2'-6" x 6'-8"	33	42	66	80 *	88	125	159	135	—	260	165
	3'-0" x 6'-8"	37	47	74	87 *	94	148	175	155	—	285	170
	3'-0" x 7'-0"	46	55	86	92 *	98	190	—	175	—	—	—

*Add to the above for the following birch face door types:

Solid wood stave core, add $31   1 hour label door, add $72   8' high door, add 31%
3/4 hour label door, add $61   1-1/2 hour label door, add $82

R083-100   Tin Clad Fire Door

6' x 7' Opening, A Label	Double Sliding Door		Double Swing Door	
Doors, 3 Ply		$ 510.00		$ 510.00
Frame, 9" channel (15 lb. per L.F.), installed	325 lb.	295.75	325 lb.	295.75
Track, hangers, hardware	13.75 L.F.	693.00	Straps and hinges	326.00
2 Carpenters @ $27.30 per hour each	16 hrs. total	436.80	16 hrs. total	436.80
Complete in place		$1,935.55		$1,568.55

R085-100   Steel Sash

Ironworker crew will erect 25 S.F. or 1.3 sash unit per hour, whichever is less.

Mechanic will point 30 L.F. per hour.

Painter will paint 90 S.F. per coat per hour.

Glazier production depends on light size.

Allow 1 lb. special steel sash putty per 16" x 20" light.

R086-100   D. H. Wood Window (ready hung)

Ponderosa pine and vinyl clad sash, exterior primed with insulating glass.

Description	2'-0" x 3'-0"				3'-0" x 4'-0"			
	Wood		Vinyl Clad		Wood		Vinyl Clad	
Window and frame, 2 lights		$100.00		$185.00		$135.00		$245.00
Removable grilles		12.20		13.50		11.44		19.50
Aluminum storm/screen		56.00		63.00		75.00		85.00
Interior trim set		15.00		15.00		19.00		19.00
Carpenter @ $27.30 per hr.	1.7 hr.	46.41	1.7 hr.	46.41	2 hr.	54.60	2 hr.	54.60
Complete in place		$229.61		$322.91		$295.04		$423.10

Mullions for above windows are about $28 each. Aluminum clad
double-hung windows run about $27.50 per S.F. of glass.

R087-100   Hinges

All closer equipped doors should have ball bearing hinges. Lead lined or extremely heavy doors require special strength hinges. Usually 1-1/2 pair of hinges are used per door up to 7'-6" high openings. Table below shows typical hinge requirements.

Use Frequency	Type Hinge Required	Type of Opening	Type of Structure
High	Heavy weight	Entrances	Banks, Office buildings, Schools, Stores & Theaters
	ball bearing	Toilet Rooms	Office buildings and Schools
Average	Standard	Entrances	Dwellings
	weight	Corridors	Office buildings and Schools
	ball bearing	Toilet Rooms	Stores
Low	Plain bearing	Interior	Dwellings

Door Thickness	Weight of Doors in Pounds per Square Foot				
	White Pine	Oak	Hollow Core	Solid Core	Hollow Metal
1-3/8"	3 psf	6 psf	1-1/2 psf	3-1/2 — 4 psf	6-1/2 psf
1-3/4"	3-1/2	7	2-1/2	4-1/2 — 5-1/4	6-1/2
2-1/4"	4-1/2	9	—	5-1/2 — 6-3/4	6-1/2

R088-010   Glazing Labor

Glass sizes are estimated by the "united inch" (height + width). Table below shows the number of lights glazed in an eight hour period by the crew size indicated, for glass up to 1/4" thick. Square or nearly square lights are more economical on a S.F. basis. Long slender lights will have a high S.F.

installation cost. For insulated glass reduce production by 33%. For 1/2" float glass reduce production by 50%. Production time for glazing with two glaziers per day averages: 1/4" float glass 120 S.F.; 1/2" float glass 55 S.F.; 1/2" insulated glass 95 S.F.; 3/4" insulated glass 75 S.F.

Glazing Method	United Inches per Light							
	40"	60"	80"	100"	135"	165"	200"	240"
Number of Men in Crew	1	1	1	1	2	3	3	4
Industrial sash, putty	60	45	24	15	18	—	—	—
With stops, putty bed	50	36	21	12	16	8	4	3
Wood stops, rubber	40	27	15	9	11	6	3	2
Metal stops, rubber	30	24	14	9	9	6	3	2
Structural glass	10	7	4	3	—	—	—	—
Corrugated glass	12	9	7	4	4	4	3	—
Storefronts	16	15	13	11	7	6	4	4
Skylights, putty glass	60	36	21	12	16	—	—	—
Thiokol set	15	15	11	9	9	6	3	2
Vinyl set, snap on	18	18	13	12	12	7	5	4
Maximum area per light	2.8 S.F.	6.3 S.F.	11.1 S.F.	17.4 S.F.	31.6 S.F.	47 S.F.	69 S.F.	100 S.F.
Daily Bare Crew Cost	$212.80	$212.80	$212.80	$212.80	$425.60	$638.40	$638.40	$851.20

R088-100   Commercial Window Glass

Description		1/2" Insulated Glass		3/8" Float Glass	
Glass per S.F.		$ 5.94	R Value	$ 6.47	R Value
Allow for breakage 10%		.59	1.68	.65	.93
Glazing compound 1/2 lb. per S.F. @	$1.10 per lb.	.55	Average	.55	Average
Labor, 2 glaziers @	$26.60 per hour each (75 S.F./day)	5.67		5.67	
Total in place, per S.F.		$12.75		$13.34	

DOORS & WINDOWS 8

REFERENCE NOS.

R092-050   Gypsum Lath

Item	Nailed to Wood Studs		Clipped to Steel Studs	
	Quantities and Bare Cost	Incl. O & P	Quantities and Bare Cost	Incl. O & P
105 S.Y. 3/8" gypsum lath	$3.30 per S.Y. $346.50	$381.15	$3.30 per S.Y. $346.50	$381.15
Fasteners, nails & clips	6 lb. @ $1.55/lb. 9.30	10.23	600 @ $.07 ea. 42.00	46.20
Lather @ $26.80 and $40.15 per hr.	9.4 hours 251.92	377.41	10.7 hours 286.76	429.61
Total per 100 S.Y. in place	$607.72	$768.79	$675.26	$856.96

Regular lath comes in 16" x 32" and 16" x 48" sheets. Firestop gypsum base
comes in 4' x 8' sheets. For nailing use 1-1/8" No. 13 ga. flathead blued nails.

R092-060   Metal Lath

Painted Diamond Expanded	Nailed to Wood Studs		Screwed to Steel Studs	
	Quantity and Bare Cost	Incl. O & P	Quantity and Bare Cost	Incl. O & P
105 S.Y. 3.4 lb. lath	$2.35 per S.Y. $246.75	$271.43	$2.35 per S.Y. $246.75	$271.43
Fasteners, corner beads, etc.	12.00	13.20	12.30	13.53
Lather @ $26.80 and $40.15 per hr.	10.0 hrs. 268.00	401.50	10.7 hrs. 286.76	429.61
Total per 100 S.Y. in place	$526.75	$686.13	$545.81	$714.57

Material prices of other types of lath in Div. 092 may be substituted in the table above to arrive at their costs in place. For nailing on studs use a 1" roofing nail with 7/16" head. For nailing on ceiling use 1-1/2" No. 11 ga. barbed roofers nail with 7/16" head for holding power.

R092-105   Gypsum Plaster

Quantities for 100 S.Y.	2 Coat, 5/8" Thick		3 Coat, 3/4" Thick		
	Base	Finish	Scratch	Brown	Finish
	1:3 Mix	2:1 Mix	1:2 Mix	1:3 Mix	2:1 Mix
Gypsum plaster	1300 lb.		1350 lb.	650 lb.	
Sand	1.75 C.Y.		1.85 C.Y.	1.35 C.Y.	
Finish hydrated lime		340 lb.			340 lb.
Gauging plaster		170 lb.			170 lb.

Total, in Place for 100 S.Y. on Walls	2 Coat, 5/8" Thick			3 Coat, 3/4" Thick		
	Quantities	Bare Cost	Incl. O & P	Quantities	Bare Cost	Incl. O & P
Gypsum plaster @ $ 13.00 per 80 lb. bag	1300 lb.	$ 211.25	$ 232.38	2000 lb.	$ 325.00	$ 357.50
Finish hydrated lime @ $ 5.45 per 50 lb. bag	340 lb.	37.06	40.77	340 lb.	37.06	40.77
Gauging plaster @ $ 16.40 per 100 lb. bag	170 lb.	27.88	30.67	170 lb.	27.88	30.67
Sand @ $ 21.85 per C.Y.	1.7 C.Y.	37.15	40.87	2.0 C.Y.	43.70	48.07
J-1 crew @ $1,004.85 & $1,521.90 per day	.87 days	874.22	1,324.05	1.05 days	1,055.09	1,598.00
Cleaning, staging, handling, patching	.09 days	87.77	134.04	.10 days	97.52	148.93
Total per 100 S.Y. in place		$1,275.33	$1,802.78		$1,586.25	$2,223.94

R092-115   Vermiculite or Perlite Plaster

Proportions: Over lath, scratch coat and brown coat, 100# gypsum plaster to 2 C.F. aggregate; over masonry, 100# gypsum plaster to 3 C.F. aggregate.

Quantities for 100 S.Y.	2 Coat, 5/8" Thick			3 Coat, 3/4" Thick		
	Quantities	Bare Cost	Incl. O & P	Quantities	Bare Cost	Incl. O & P
Gypsum plaster @ $ 13.00 per 80 lb. bag	1250 lb.	$ 203.13	$ 223.44	2250 lb.	$ 365.63	$ 402.19
Vermiculite or perlite @ $ 13.60 per bag	7.8 bags	106.08	116.69	11.3 bags	153.68	169.05
Finish hydrated lime @ $ 5.45 per 50 lb. bag	340 lb.	37.06	40.77	340 lb.	37.06	40.77
Gauging plaster @ $ 16.40 per 100 lb. bag	170 lb.	27.88	30.67	170 lb.	27.88	30.67
J-1 crew @ $1,004.85 & $1,521.90 per day	1.00 days	1004.85	1,521.90	1.25 days	1,256.06	1,902.38
Cleaning, staging, handling, patching	.09 day	87.77	134.04	.10 days	97.52	148.93
Total per 100 S.Y. in place		$1,466.77	$2,067.51		$1,937.83	$2,693.99

R092-300   Stucco

Quantities for 100 S.Y. 3 Coats, 1″ Thick	On Wood Frame			On Masonry		
	Quantities	Bare Cost	Incl. O & P	Quantities	Bare Cost	Incl. O & P
Portland cement @ $ 7.00 per bag	29 bags	$ 203.00	$ 223.30	21 bags	$147.00	$ 161.70
Sand @ $21.85 per C.Y.	2.6 C.Y.	56.81	62.49	2 C.Y.	43.70	48.07
Hydrated lime @ $ 5.45 per 50 lb. bag	180 lb.	19.62	21.58	120 lb.	13.08	14.39
Painted stucco mesh @ $ 3.65 per S.Y., 3.6#	105 S.Y.	383.25	421.58	—	—	—
Furring nails and jute fiber		18.40	20.24	—	—	—
Mix and install with crew indicated	J-2 @ .74 days	902.25	1,363.89	J-1 @ .5 days	502.43	760.95
Cleaning, staging, handling, patching	.10 days	117.68	180.07	.10 days	96.24	147.95
Total per 100 S.Y. in place		$1,701.01	$2,293.15		$802.45	$1,133.06

R092-610   Studs, Joists and Track

Material prices per 1000 L.F., for galvanized studs, joists and tracks. Panhead, framing screws, 7/16″ long are $9.30 per thousand; 1-5/8″ long are $14.50 per thousand.

Non-load bearing, 20 ga. stud and track are primarily used for curtain wall. (C.W.)

Size	Non-Load Bearing				Load Bearing 1-5/8″ Flange—Light Gauge Structural					
	25 Ga.		20 Ga. (C.W.)		18 Ga.		16 Ga.		14 Ga.	
	Stud	Track	Stud	Track	Stud	Track	Stud	Track	Stud	Track
1-5/8″	$140	$137	$240	$230						
2-1/2″	165	162	280	270	$500	$450	$ 610	$ 560	$ 760	$ 710
3-5/8″	195	192	325	315	560	510	690	640	850	800
4″	220	217	360	350	590	540	730	680	900	850
6″	310	305	460	448	750	700	920	870	1,130	980
8″					895	845	1,125	1,075	1,400	1,350

Size	Non-Load Bearing				Load Bearing—Extra Wide Flange — 2″ Flange							
	25 Ga.		22 Ga.		18 Ga.		16 Ga.		14 Ga.		12 Ga.	
	C-H Stud	J-Track	C-H Stud	J-Track	Joist	Track	Joist	Track	Joist	Track	Joist	Track
2-1/2″	$480	$380	$780	$510	—	$620	—	$ 880	—	$1,025	—	—
4″	576	540	990	750	$ 800	790	$1,020	990	$1,170	1,110	$1,900	$1,860
6″					840	810	1,050	1,000	1,280	1,200	2,000	1,910
8″					1,020	980	1,240	1,170	1,550	1,500	2,440	2,200
10″							1,520	1,420	1,880	1,780	2,900	2,700
12″							1,390	1,290	2,140	2,040	3,320	3,120

R094-100 Terrazzo Floor

5000 S.F. 5/8" Terrazzo Topping Quantities for 100 S.F.	Bonded to Concrete 1-1/8" Bed, 1:4 Mix			Not Bonded 2-1/8" Bed and 1/4" Sand		
	Quantities	Bare Cost	Incl. O & P	Quantities	Bare Cost	Incl. O & P
Portland cement @ $7.00 per bag	6 bags	$ 42.00	$ 46.20	8 bags	$ 56.00	$ 61.60
Sand @ $21.85 per C.Y.	10 C.F.	8.09	8.90	20 C.F.	16.19	17.81
Divider strips for 4' square panels, zinc, 14 ga.	55 L.F.	49.50	54.45	55 L.F.	49.50	54.45
Terrazzo fill, domestic aggregates	600 lb.	174.00	191.40	600 lb.	174.00	191.40
15 lb. tarred felt	—	—	—		2.00	2.20
Mesh 2 x 2 #14 galvanized	—	—	—		13.60	14.96
Crew J-3 @ $484.90 & $679.55 per day	.77 days	373.37	523.25	.87 days	421.86	591.21
Total per 100 S.F. in place	Bonded	$646.96	$824.20	Not Bonded	$733.15	$933.63

2' x 2' panels require 1 L.F. divider strip per S.F.; 6' x 6' panels need .33 L.F. per S.F.

R096-600 Resilient Flooring and Base

Description	12" x 12" x 3/32" V.C. Tile			4" x 1/8" Vinyl Base		
	Quantities	Bare Cost	Incl. O & P	Quantities	Bare Cost	Incl. O & P
Vinyl composition, tile, standard line	100 S.F.	$ 94.00	$103.40	100 L.F.	$ 56.00	$ 61.60
Vinyl cement @ $12.23 per gallon	0.8 gallon	9.78	10.76	0.5 gallon	6.12	6.73
Tile layer @ $26.65 and $39.40 per hour	1.6 hrs.	42.64	63.04	2.55 hrs.	67.96	100.47
Total per 100 S.F. in place		$146.42	$177.20	100 L.F.	$130.08	$168.80

9 FINISHES

REFERENCE NOS.

560

R099-100   Paint

Material prices per gallon in 5 gallon lots, up to 25 gallons. For 100 gallons, deduct 10%.

Exterior, Alkyd (oil base)	
Flat	$21.00
Gloss	19.50
Primer	19.25

Exterior, Latex (water base)	
Acrylic stain	16.75
Gloss enamel	22.00
Flat	17.00
Primer	18.00
Semi-gloss	21.00

Interior, Alkyd (oil base)	
Enamel undercoater	18.00
Flat	15.75
Gloss	21.50
Primer sealer	17.00
Semi-gloss	20.00

Interior, Latex (water base)	
Enamel undercoater	15.50
Flat	15.20
Floor and deck	16.00
Gloss	21.00
Primer sealer	15.25
Semi-gloss	16.65

Masonry, Exterior	
Alkali resistant primer	$19.00
Block filler, epoxy	20.00
Block filler, latex	9.75
Latex, flat or semi-gloss	16.75

Masonry, Interior	
Alkali resistant primer	15.50
Block filler, epoxy	23.00
Block filler, latex	10.00
Floor, alkyd	17.00
Floor, latex	19.50
Latex, flat acrylic	11.00
Latex, flat emulsion	10.75
Latex, sealer	12.00
Latex, semi-gloss	18.50

Varnish and Stain	
Alkyd clear	21.00
Polyurethane, clear	19.50
Primer sealer	14.00
Semi-transparent stain	16.00
Solid color stain	15.50

Metal Coatings	
Galvanized	24.25
High heat	32.00
Machinery enamel, alkyd	20.50
Normal heat	22.00

Rust inhibitor ferrous metal	$21.50
Zinc chromate	19.50

Heavy Duty Coatings	
Acrylic urethane	50.50
Chlorinated rubber	29.50
Coal tar epoxy	22.00
Metal pretreatment (polyvinyl butyral)	23.00
Polyamide epoxy finish	31.00
Polyamide epoxy primer	31.00
Silicone alkyd	31.50
2 component solvent based acrylic epoxy	32.50
2 component solvent based polyester epoxy	40.75
Vinyl	26.50
Zinc rich primer	47.50

Special Coatings/Miscellaneous	
Aluminum	22.50
Dry fall out, flat	12.00
Fire retardant, intumescent	36.00
Linseed oil	10.00
Shellac	18.00
Swimming pool, epoxy or urethane base	35.75
Swimming pool, rubber base	24.75
Texture paint	11.00
Turpentine	10.50
Water repellent 5% silicone	17.50

R099-220   Painting

Item	Coat	One Gallon Covers			In 8 Hours a Laborer Covers			Labor-Hours per 100 S.F.		
		Brush	Roller	Spray	Brush	Roller	Spray	Brush	Roller	Spray
Paint wood siding	prime	250 S.F.	225 S.F.	290 S.F.	1150 S.F.	1300 S.F.	2275 S.F.	.695	.615	.351
	others	270	250	290	1300	1625	2600	.615	.492	.307
Paint exterior trim	prime	400	—	—	650	—	—	1.230	—	—
	1st	475	—	—	800	—	—	1.000	—	—
	2nd	520	—	—	975	—	—	.820	—	—
Paint shingle siding	prime	270	255	300	650	975	1950	1.230	.820	.410
	others	360	340	380	800	1150	2275	1.000	.695	.351
Stain shingle siding	1st	180	170	200	750	1125	2250	1.068	.711	.355
	2nd	270	250	290	900	1325	2600	.888	.603	.307
Paint brick masonry	prime	180	135	160	750	800	1800	1.066	1.000	.444
	1st	270	225	290	815	975	2275	.981	.820	.351
	2nd	340	305	360	815	1150	2925	.981	.695	.273
Paint interior plaster or drywall	prime	400	380	495	1150	2000	3250	.695	.400	.246
	others	450	425	495	1300	2300	4000	.615	.347	.200
Paint interior doors and windows	prime	400	—	—	650	—	—	1.230	—	—
	1st	425	—	—	800	—	—	1.000	—	—
	2nd	450	—	—	975	—	—	.820	—	—

R099-700 Wall Covering

Quantities for 100 S.F.	Medium Price Paper			Expensive Paper		
	Quantities	Bare Cost	Incl. O & P	Quantities	Bare Cost	Incl. O & P
Paper @ $25.30 and $52.00 per double roll	1.6 dbl. rolls	$40.48	$ 44.53	1.6 dbl. rolls	$ 83.20	$ 91.52
Wall sizing @ $15.00 per gallon	.25 gallon	3.75	4.13	.25 gallon	3.75	4.13
Vinyl wall paste @ $ 7.80 per gallon	0.4 gallon	3.12	3.43	0.4 gallon	3.12	3.43
Apply sizing @ $24.55 and $37.60 per hour	0.3 hour	7.37	11.28	0.3 hour	7.37	11.28
Apply paper @ $24.55 and $37.60 per hour	1.2 hours	29.46	45.12	1.5 hours	36.83	56.40
Total including waste allowance per 100 S.F.		$84.18	$108.49		$134.27	$166.76

This is equivalent to about $67.80 and $104.23 per double roll
complete in place. Most wallpapers now come in double rolls
only. To remove old paper allow 1.3 hours per 100 S.F.

R105-050 Steel Lockers

Price per Opening, Material Only, Based on 100 Openings, Not Including Locks						
Single Tier	1-Wide	3-Wide	Double Tier	1-Wide	3-Wide
12" x 12" x 60"	$121.50	$101.50	12" x 12" x 30"	$72.00	$61.50
12" x 15" x 60"	128.00	106.00	12" x 12" x 36"	76.00	63.50
12" x 18" x 60"	136.00	111.50	12" x 15" x 36"	78.00	66.50
12" x 12" x 72"	134.00	113.00	12" x 18" x 36"	84.00	69.00
12" x 15" x 72"	141.00	117.00	15" x 15" x 36"	84.00	74.00
12" x 18" x 72"	147.00	121.00	Multiple Tier	1-Wide	3-Wide
15" x 18" x 72"	159.00	135.50	5-High 12" x 12" x 12"	$33.00	$28.00
18" x 18" x 72"	164.50	140.50	12" x 15" x 12"	34.50	29.00
18" x 21" x 72"	175.00	147.50	15" x 15" x 12"	38.00	33.00
			6-High 12" x 12" x 12"	31.00	27.00
			12" x 15" x 12"	34.00	28.00

R130-110 Air Supported Structures

Air supported structures are made from fabrics that can be classified into two groups: temporary and permanent. Temporary fabrics include nylon, woven polyethylene, vinyl film, and vinyl coated dacron. These have lifespans that range from five to fifteen plus years, and costs that vary accordingly from $7 to $15 per S.F. floor area. This is a package cost for a fabric shell, tension cables, primary and back-up inflation systems and doors. The lower cost structures are used for construction shelters, bulk storage and pond covers. The more expensive are used for recreational structures and warehouses.

Permanent fabrics are teflon coated fiberglass. The package cost runs $35 to $40 per S.F. floor area. The life of this structure is twenty plus years. The high cost limits its application to architectural designed structures which call for a clear span covered area, such as stadiums and convention centers. Both temporary and permanent structures are available in translucent fabrics which eliminates the need for daytime lighting.

Areas to be covered vary from 10,000 S.F. to any area up to 1000 foot wide by any lenght. Height restrictions range from a maximum of 1/2 of width

to a minimum of 1/6 of the width. Erection of even the largest of the temporary structures requires no more than a week ($.07 to $.25 per S.F. floor area).

Centrifugal fans provide the inflation necessary to support the structure during application of live loads. Airlocks are usually used at large entrances to prevent loss of static pressure. Some manufacturers employ propeller fans which generate sufficient airflow (30,000 CFM) to eliminate the need for airlocks. These fans may also be automatically controlled to resist high wind conditions, regulate humidity (air changes), and provide cooling and heat.

Insulation can be provided with the addition of a second or even third interior liner, creating a dead air space with an "R" value of four to nine. Some structures allow for the liner to be collapsed into the outer shell to enable the internal heat to melt accumulated snow. For cooling or air conditioning, the exterior face of the liner can be aluminized to reflect the sun's heat. Costs for the liner run about $.75 to $1.60 per S.F. floor area.

13

SPECIAL CONTRUCTION

REFERENCE NOS.

R131-210 Pre-engineered Steel Buildings

These buildings are manufactured by many companies and normally erected by franchised dealers throughout the U.S. The four basic types are: Rigid Frames, Truss type, Post and Beam and the Sloped Beam type. Most popular roof slope is low pitch of 1″ in 12″. The minimum economical area of these buildings is about 3000 S.F. of floor area. Bay sizes are usually 20′ to 24′ but can go as high as 30′ with heavier girts and purlins. Eave heights are usually 12′ to 24′ with 18′ to 20′ most typical.

Material prices shown in Division 131-212 are bare costs for the building shell only and do not include floors, foundations, interior finishes or utilities.

Costs assume at least three bays of 24′ each, a 1″ in 12″ roof slope, and they are based on 30 psf roof load and 20 psf wind load and no unusual requirements. Costs include the structural frame, 26 ga. colored steel roofing, 26 ga. colored steel siding, fasteners, closures, trim and flashing but no allowance for doors, windows, skylights, gutters or downspouts. Very large projects would generally cost less for materials than the prices shown.

Conditions at the site, weather, shape and size of the building, and labor availability will affect the erection cost of the building.

R131-310 Dome Structures

Steel — The four types are Lamella, Schwedler, Arch and Geodesic. For maximum economy, rise should be about 15 to 20% of diameter. Most common diameters are in the 200′ to 300′ range. Lamella domes weigh about 5 P.S.F. of floor area less than Schwedler domes. Schwedler dome weight in lbs. per S.F. approaches 0.046 times the diameter. Domes below 125′ diameter weigh .07 times diameter and the cost per ton of steel is higher. Recent costs per ton erected range from $2,800 to $5,600 per ton of steel. See R051-220 for estimating weights.

Wood — Small domes are of sawn lumber, larger ones are laminated. In larger sizes, triaxial and triangular cost about the same; radial domes cost more. Radial domes are economical in the 60′ to 70′ diameter range. Most economical range of all types is 80′ to 200′ diameters. Diameters can run over 400′. Costs above foundations run $23 per S.F. for under 50′ diameter. At 77′ diameter, radial cost $19 and triaxial cost $21 per S.F. Cost is $10.25 for 120′; $14.50 for 200′; $18.75 for 300′ and $24.00 for 400′ diameter. Prices include 2″ decking and tension tie ring in place.

Plywood — See division 061-208 for stressed skin and folded plates. Stock prefab geodesic domes are available with diameters from 24′ to 60′. Shells cost between $6.80 to $18.00 per S.F. of floor with erection running between $.48 and $2.20 per S.F. Canopies without sides run about $1.00 per S.F. less.

Fiberglass — Aluminum framed translucent sandwich panels with spans from 5′ to 45′ are commercially available.

Aluminum — Stressed skin aluminum panels form geodesic domes with spans ranging from 82′ to 232′. Total in-place cost for the entire range is $35.00 per S.F. of floor. For vinyl faced fiberglass insulation with an R rating of 20, add $10.00 per S.F. of floor. An aluminum space truss, triangulated or nontriangulated, with aluminum or clear acrylic closure panels can be used for clear spans of 40′ to 415′. The aluminum panel costs range from $22 to $24 per S.F. floor and the acrylic panel from $70 to $75 per S.F. floor total in-place costs.

R131-520 Swimming Pools

Pool prices given per square foot of surface area include pool structure, filter and chlorination equipment where required, pumps, related piping, diving boards, ladders, maintenance kit, skimmer and vacuum system. Decks and electrical service to equipment are not included.

Residential in-ground pool construction can be divided into two categories: vinyl lined and gunite. Vinyl lined pool walls are constructed of different materials including wood, concrete, plastic or metal. The bottom is often graded with sand over which the vinyl liner is installed. Costs are generally in the $15 to $22 per S.F. range. Vermiculite or soil cement bottoms may be substituted for an added cost.

Gunite pool construction is used both in residential and municipal installations. These structures are steel reinforced for strength and finished

with a white cement limestone plaster. Residential costs run from $25 to $36 per S.F. surface. Municipal costs vary from $50 to $70 because plumbing codes require more expensive materials, chlorination equipment and higher filtration rates.

Municipal pools greater than 1,800 S.F. require gutter systems to control waves. This gutter may be formed into the concrete wall. Often a vinyl, stainless steel gutter or gutter and wall system is specified, which will raise the pool cost an additional $51 to $150 per L.F. of gutter installed and up to $290 per L.F. if a gutter and wall system is installed.

Competition pools usually require tile bottoms and sides with contrasting lane striping. Add $12 per S.F. of wall or bottom to be tiled.

R131-910 Public Garage Costs

Cars	Per Car	Levels	Per Car	Levels	Per Car	Levels	Per S.F.
200 to 400	$ 9,500	3	$9,950	6	$ 9,750	Surface	$26.25
400 to 600	$ 9,775	4	$9,950	7	$10,100	2	$26.85
600 to 800	$10,200	5	$9,750	8 to 12	$10,575	3 or 4	$32.85
800 to 1200	$10,575					5 or 6	$32.40
Average	$10,025	Average per Car (310 to 320 S.F. per car)			$10,025	7 or 8	$34.05

The above costs refer to open wall construction. For enclosed structures, costs must be added for sprinkler and ventilation systems.

R142-100 Freight Elevators

Capacities run from 2,000 lbs. to over 100,000 lbs. with 3,000 lbs. to 10,000 lbs. most common. Travel speeds are generally lower and control less intricate than on passenger elevators. Unit prices in division 142-010 are for hydraulic and geared elevators.

R142-200 Elevator Selective Costs See R142-400 for cost development.

	Passenger		Freight		Hospital	
A. Base Unit	Hydraulic	Electric	Hydraulic	Electric	Hydraulic	Electric
Capacity	1,500 Lb.	2,000 Lb.	2,000 Lb.	4,000 Lb.	4,000 lb.	4,000 lb.
Speed	100 F.P.M.	200 F.P.M.	50 F.P.M.	200 F.P.M.	100 F.P.M.	200 F.P.M.
#Stops/Travel Ft.	2/12	4/40	2/20	4/40	2/20	4/40
Push Button Oper.	Yes	Yes	Yes	Yes	Yes	Yes
Telephone Box & Wire	"	"	"	"	"	"
Emergency Lighting	"	"	No	No	"	"
Cab	Plastic Lam. Walls	Plastic Lam. Walls	Painted Steel	Painted Steel	Plastic Lam. Walls	Plastic Lam. Walls
Cove Lighting	Yes	Yes	No	No	Yes	Yes
Floor	V.C.T.	V.C.T.	Wood w/Safety Treads	Wood w/Safety Treads	V.C.T.	V.C.T.
Doors, & Speedside Slide	Yes	Yes	Yes	Yes	Yes	Yes
Gates, Manual	No	No	No	No	No	No
Signals, Lighted Buttons	Car and Hall	Car and Hall	Car and Hall	Car and Hall	Car and Hall	Car and Hall
O.H. Geared Machine	N.A.	Yes	N.A.	Yes	N.A.	Yes
Variable Voltage Contr.	"	"	N.A.	"	"	"
Emergency Alarm	Yes	"	Yes	"	Yes	"
Class "A" Loading	N.A.	N.A.	"	"	N.A.	N.A.
Base Cost	$37,250	$82,150	$41,500	$81,070	$48,005	$86,500
B. Capacity Adjustment						
2,000 Lb.	$ 525					
2,500	1,150	$ 2,675	$ 2,242			
3,000	2,865	4,000	3,271			
3,500	4,890	5,700	5,628			
4,000	5,640	5,780	6,022			
4,500	6,750	7,600	7,115		$ 5,628	$ 4,685
5,000	9,380	9,620	9,700	$ 4,682	8,198	6,118
6,000			9,990	8,216		
7,000			15,786	11,000		
8,000			17,120	15,333		
10,000			18,008	18,106		
12,000			21,435	22,000		
16,000			27,950	26,422		
20,000			31,016	29,008		
C. Travel Over Base	$ 331 V.L.F.	$ 207 V.L.F.	$ 331 V.L.F.	$ 207 V.L.F.	$ 331 V.L.F.	$ 207 V.L.F.
D. Additional Stops	$ 3,410 Ea.	$ 5,319 Ea.	$ 3,385 Ea.	$ 4,963 Ea.	$ 5,904 Ea.	$ 5,752 Ea.
E. Speed Adjustment						
100 F.P.M.			$ 700			
125	$ 800		1,296		$ 1,341	
150	1,725		2,418		2,259	
175	2,910		3,792		3,546	
200	4,695				5,174	
Geared 250		$ 2,200		$ 9,318		$ 2,366
4 Flrs. 300		4,485		11,625		4,410
Min. 350		5,270		13,720		5,380
400		7,518		15,210		7,493
Gearless 500		34,780		19,050		33,255
4 Flrs. 600		36,729		21,050		36,730
Min. 700		40,202		24,522		40,098
800		44,300		27,258		44,305
1,000		Spec. Applic.		Spec. Applic.		Spec. Applic.
1,200		"		"		"
F. Other Than Class						
"A" Loading						
"B"			$ 1,600	$ 1,625		
"C-1"			3,994	4,018		
"C-2"			4,782	4,796		
"C-3"			6,542	6,585		

Conveying Systems   R142 | Elevators

R142-200   Elevator Selective Costs (cont.)   See R142-400 for cost development.

	Passenger	Freight	Hospital
G. Options			
1. Controls			
Automatic, 2 car group	$3,564		$3,557
3 car group	5,304		5,269
4 car group	8,156		8,087
5 car group	10,663		10,588
6 car group	15,097		14,990
Emergency, fireman service	Yes		Yes
Intercom service	600		600
Selective collective, single car	—		—
Duplex car	3,816		3,816
2. Doors			
Center opening, 1 speed	$1,720		$1,720
2 speed	2,100		2,100
Rear opening-opposite front	4,012		4,012
Side opening, 2 speed	5,980		5,980
Freight, bi-parting	—	$ 5,000	—
Power operated door and gate	—	16,500	—
3. Emergency power switching, automatic	$2,150		$2,150
Manual	382		382
4. Finishes based on 3500# cab			
Ceilings, acrylic panel	$390		
Aluminum egg crate	475		$475
Doors, stainless steel	800 Ea.		800 Ea.
Floors, carpet, class "A" — (Finish flooring by others)	329		—
Epoxy	274		274
Quarry tile	410		412
Slate	573		—
Steel plate	—	$ 665	
Textured rubber	110		110
Walls, plastic laminate	Std.		Std.
Stainless steel	2,445		2,445
Return at door	520		520
Steel plate, 1/4" x 4' high, 14 ga. above	—	$ 1,601	—
Entrance, doors, baked enamel	Std.		Std.
Stainless steel	800 Ea.		800 Ea.
Frames, baked enamel	Std.		Std.
Stainless steel	800 Ea.		800 Ea.
5. Maintenance contract - 12 months	$2,840	$ 2,122	$4,125
6. Signal devices			
Hall lantern, each	$426	$ 426	$426
Position indicator, car or lobby	312	312	312
Add for over three each, per floor	88	88	88
7. Specialties			
High speed, heavy duty door opener	$1,420		$1,420
Variable voltage, O.H. gearless machine	28,100-62,400		28,100-62,400
Basement installed geared machine	$14,400	$14,450	$14,400

R142-300   Passenger Elevators

Electric elevators are used generally but hydraulic elevators can be used for lifts up to 70' and where large capacities are required. Hydraulic speeds are limited to 200 F.P.M. but cars are self leveling at the stops. On low rises, hydraulic installation runs about 15% less than standard electric types but on higher rises this installation cost advantage is reduced. Maintenance of hydraulic elevators is about the same as electric type but underground portion is not included in the maintenance contract.

In electric elevators there are several control systems available, the choice of which will be based upon elevator use, size, speed and cost criteria. The two types of drives are geared for low speeds and gearless for 450 F.P.M. and over.

The tables on the preceding pages illustrate typical installed costs of the various types of elevators available.

R142-400   Elevator Cost Development

Requirement: One passenger elevator, five story hydraulic, 2,500 lb. capacity, 12' floor to floor, speed 150 F.P.M., emergency power switching and maintenance contract.

Description	Total Cost
A. Base Elevator, (Hydraulic Passenger)	$37,250
B. Capacity Adjustment (2,500 lb.)	1,150
C. Excess Travel Over Base (4 x 12') = 48' — (12' for Base Unit) = 36' x $331 per V.L.F.	11,916
D. Stops Over Base 5 — (2 for base unit) = 3 x $3,410 Ea.	10,230
E. Speed Adjustment (150 F.P.M.)	1,725
F. Options:	
1. Intercom Service	600
2. Emergency power switching, automatic	2,150
3. Stainless steel entrance doors 5 x $800 Ea.	4,000
4. Maintenance Contract (12 months)	2,840
5. Position indicators 2 x $312	624
Total Cost	$72,485

R143-100   Escalators

Moving stairs can be used for buildings where 600 or more people are to be carried to the second floor or beyond. Freight cannot be carried on escalators and at least one elevator must be available for this function. Carrying capacity is 5,000 to 8,000 people per hour. Power requirement is 2 to 3 KW per hour and incline angle is 30°.

Table below indicates approximate installed costs per unit, with O & P. Installation outside major cities would increase the costs below, due to mechanics additional travel expense.

Vertical Rise	Stainless Steel Balustrade		Enameled Steel or Glass Balustrade	
	32" Wide	48" Wide	32" Wide	48" Wide
10'	$ 91,993	$ 95,528	$ 86,384	$ 90,911
15'	96,939	100,700	91,811	95,612
20'	105,625	109,606	98,664	103,091
25'	117,203	122,383	109,822	115,273

R143-200   Moving Ramps and Walks

These are a specialized form of conveyor 3' to 6' wide with capacities of 3,600 to 18,000 persons per hour. Maximum speed is 140 F.P.M. and normal incline is 0° to 15°.

Local codes will determine the maximum angle. A 48" wide, 150' long indoor moving ramp costs between $290,000 to $330,000 per unit installed, including O & P. Outdoor units would require additional weather protection. Approximate cost add is $250 per foot.

CONVEYING SYSTEMS 14

REFERENCE NOS.

R151-050 Pipe Material Considerations

1. Malleable fittings should be used for gas service.
2. Malleable fittings are used where there are stresses/strains due to expansion and vibration.
3. Cast fittings may be broken as an aid to disassembling of heating lines frozen by long use, temperature and minerals.
4. Cast iron pipe is extensively used for underground and submerged service.
5. Type M (light wall) copper tubing is available in hard temper only and is used for nonpressure and less severe applications than K and L.
6. Type L (medium wall) copper tubing, available hard or soft for interior service.
7. Type K (heavy wall) copper tubing, available in hard or soft temper for use where conditions are severe. For underground and interior service.
8. Hard drawn tubing requires fewer hangers or supports but should not be bent. Silver brazed fittings are recommended, however soft solder is normally used.
9. Type DMV (very light wall) copper tubing designed for drainage, waste and vent plus other non-critical pressure services.

Domestic/Imported Pipe and Fittings Cost

The prices shown in this publication for steel/cast iron pipe and steel, cast iron, malleable iron fittings are based on domestic production sold at the normal trade discounts. The above listed items of foreign manufacture may be available at prices of 1/3 to 1/2 those shown. Some imported items after minor machining or finishing operations are being sold as domestic to further complicate the system.

Caution: Most pipe prices in this book also include a coupling and pipe hangers which for the larger sizes can add significantly to the per foot cost and should be taken into account when comparing "book cost" with quoted supplier's cost.

R151-420 Plumbing Fixture Installation Time

Item	Rough-In	Set	Total Hours	Item	Rough-In	Set	Total Hours
Bathtub	5	5	10	Shower head only	2	1	3
Bathtub and shower, cast iron	6	6	12	Shower drain	3	1	4
Fire hose reel and cabinet	4	2	6	Shower stall, slate		15	15
Floor drain to 4 inch diameter	3	1	4	Slop sink	5	3	8
Grease trap, single, cast iron	5	3	8	Test 6 fixtures			14
Kitchen gas range		4	4	Urinal, wall	6	2	8
Kitchen sink, single	4	4	8	Urinal, pedestal or floor	6	4	10
Kitchen sink, double	6	6	12	Water closet and tank	4	3	7
Laundry tubs	4	2	6	Water closet and tank, wall hung	5	3	8
Lavatory wall hung	5	3	8	Water heater, 45 gals. gas, automatic	5	2	7
Lavatory pedestal	5	3	8	Water heaters, 65 gals. gas, automatic	5	2	7
Shower and stall	6	4	10	Water heaters, electric, plumbing only	4	2	6

Fixture prices in front of book are based on the cost per fixture set in place. The rough-in cost, which must be added for each fixture, includes carrier, if required, some supply, waste and vent pipe connecting fittings and stops. The lengths of rough-in pipe are nominal runs which would connect to the larger runs and stacks. The supply runs and DWV runs and stacks must be accounted for in separate entries. In the eastern half of the United States it is common for the plumber to carry these to a point 5' outside the building.

R151-430 Water Cooler Application

Type of Service	Requirement
Office, School or Hospital	12 persons per gallon per hour
Office, Lobby or Department Store	4 or 5 gallons per hour per fountain
Light manufacturing	7 persons per gallon per hour
Heavy manufacturing	5 persons per gallon per hour
Hot heavy manufacturing	4 persons per gallon per hour
Hotel	.08 gallons per hour per room
Theatre	1 gallon per hour per 100 seats

Mechanical · R155 · Heating

R155-035 Heating (42° degrees latitude)

$$\text{Approximate S.F. radiation} = \frac{\text{S.F. sash}}{2} + \frac{\text{S.F. wall + roof — sash}}{20} + \frac{\text{C.F. building}}{200}$$

R155-050 Factor for Determining Heat Loss for Various Types of Buildings

General: While the most accurate estimates of heating requirements would naturally be based on detailed information about the building being considered, it is possible to arrive at a reasonable approximation using the following procedure:

1. Calculate the cubic volume of the room or building.
2. Select the appropriate factor from Table 1 below. Note that the factors apply only to inside temperatures listed in the first column and to 0°F outside temperature.
3. If the building has bad north and west exposures, multiply the heat loss factor by 1.1.
4. If the outside design temperature is other than 0°F, multiply the factor from Table 1 by the factor from Table 2.
5. Multiply the cubic volume by the factor selected from Table 1. This will give the estimated BTUH heat loss which must be made up to maintain inside temperature.

Table 1 — Building Type	Conditions	Qualifications	Loss Factor*
Factories & Industrial Plants General Office Areas at 70°F	One Story	Skylight in Roof	6.2
		No Skylight in Roof	5.7
	Multiple Story	Two Story	4.6
		Three Story	4.3
		Four Story	4.1
		Five Story	3.9
		Six Story	3.6
	All Walls Exposed	Flat Roof	6.9
		Heated Space Above	5.2
	One Long Warm Common Wall	Flat Roof	6.3
		Heated Space Above	4.7
	Warm Common Walls on Both Long Sides	Flat Roof	5.8
		Heated Space Above	4.1
Warehouses at 60°F	All Walls Exposed	Skylights in Roof	5.5
		No Skylight in Roof	5.1
		Heated Space Above	4.0
	One Long Warm Common Wall	Skylight in Roof	5.0
		No Skylight in Roof	4.9
		Heated Space Above	3.4
	Warm Common Walls on Both Long Sides	Skylight in Roof	4.7
		No Skylight in Roof	4.4
		Heated Space Above	3.0

*Note: This table tends to be conservative particularly for new buildings designed for minimum energy consumption.

Table 2 — Outside Design Temperature Correction Factor (for Degrees Fahrenheit)									
Outside Design Temperature	50	40	30	20	10	0	-10	-20	-30
Correction Factor	.29	.43	.57	.72	.86	1.00	1.14	1.28	1.43

R155-610 Solar Heating (Space and Hot Water)

Collectors should face as close to due South as possible, however, variations of up to 20 degrees on either side of true South are acceptable. Local climate and collector type may influence the choice between east or west deviations. Obviously they should be located so they are not shaded from the sun's rays. Incline collectors at a slope of latitude minus 5 degrees for domestic hot water and latitude plus 15 degrees for space heating.

Flat plate collectors consist of a number of components as follows: Insulation to reduce heat loss through the bottom and sides of the collector. The enclosure which contains all the components in this assembly is usually weatherproof and prevents dust, wind and water from coming in contact with the absorber plate. The cover plate usually consists of one or more layers of a variety of glass or plastic and reduces the reradiation by creating an air space which traps the heat between the cover and the absorber plates.

The absorber plate must have a good thermal bond with the fluid passages. The absorber plate is usually metallic and treated with a surface coating which improves absorptivity. Black or dark paints or selective coatings are used for this purpose, and the design of this passage and plate combination helps determine a solar system's effectiveness.

Heat transfer fluid passage tubes are attached above and below or integral with the absorber plate for the purpose of transferring thermal energy from the absorber plate to a heat transfer medium. The heat exchanger is a device for transferring thermal energy from one fluid to another.

Piping and storage tanks should be well insulated to minimize heat losses.

Size domestic water heating storage tanks to hold 20 gallons of water per user, minimum, plus 10 gallons per dishwasher or washing machine. For domestic water heating an optimum collector size is approximately 3/4 square foot of area per gallon of water storage. For space heating of residences and small commercial applications the collector is commonly sized between 30% and 50% of the internal floor area. For space heating of large commercial applications, collector areas less than 30% of the internal floor area can still provide significant heat reductions.

A supplementary heat source is recommended for Northern states for December through February.

The solar energy transmission per square foot of collector surface varies greatly with the material used. Initial cost, heat transmittance and useful life are obviously interrelated.

MECHANICAL 15

REFERENCE NOS.

R155-720 Demolition (Selective vs. Removal for Replacement)

Demolition can be divided into two basic categories.

One type of demolition involves the removal of material with no concern for its replacement. The labor-hours to estimate this work are found in Div. 020-700 under "Selective Demolition". It is selective in that individual items or all the material installed as a system or trade grouping such as plumbing or heating systems are removed. This may be accomplished by the easiest way possible, such as sawing, torch cutting, or sledge hammer as well as simple unbolting.

The second type of demolition is the removal of some item for repair or replacement. This removal may involve careful draining, opening of unions, disconnecting and tagging of electrical connections, capping of pipes/ducts to prevent entry of debris or leakage of the material contained as well as transport of the item away from its in-place location to a truck/dumpster. An approximation of the time required to accomplish this type of demolition is to use half of the time indicated as necessary to install a new unit. For example; installation of a new pump might be listed as requiring 6 labor-hours so if we had to estimate the removal of the old pump we would allow an additional 3 hours for a total of 9 hours. That is, the complete replacement of a defective pump with a new pump would be estimated to take 9 labor-hours.

R157-020 Air Conditioning Requirements

BTU's per hour per S.F. of floor area and S.F. per ton of air conditioning.

Type of Building	BTU per S.F.	S.F. per Ton	Type of Building	BTU per S.F.	S.F. per Ton	Type of Building	BTU per S.F.	S.F. per Ton
Apartments, Individual	26	450	Dormitory, Rooms	40	300	Libraries	50	240
Corridors	22	550	Corridors	30	400	Low Rise Office, Exterior	38	320
Auditoriums & Theaters	40	300/18*	Dress Shops	43	280	Interior	33	360
Banks	50	240	Drug Stores	80	150	Medical Centers	28	425
Barber Shops	48	250	Factories	40	300	Motels	28	425
Bars & Taverns	133	90	High Rise Office—Ext. Rms.	46	263	Office (small suite)	43	280
Beauty Parlors	66	180	Interior Rooms	37	325	Post Office, Individual Office	42	285
Bowling Alleys	68	175	Hospitals, Core	43	280	Central Area	46	260
Churches	36	330/20*	Perimeter	46	260	Residences	20	600
Cocktail Lounges	68	175	Hotel, Guest Rooms	44	275	Restaurants	60	200
Computer Rooms	141	85	Corridors	30	400	Schools & Colleges	46	260
Dental Offices	52	230	Public Spaces	55	220	Shoe Stores	55	220
Dept. Stores, Basement	34	350	Industrial Plants, Offices	38	320	Shop'g. Ctrs., Supermarkets	34	350
Main Floor	40	300	General Offices	34	350	Retail Stores	48	250
Upper Floor	30	400	Plant Areas	40	300	Specialty	60	200

*Persons per ton
12,000 BTU = 1 ton of air conditioning

R157-040 Recommended Ventilation Air Changes

Table below lists range of time in minutes per change for various types of facilities.

Assembly Halls	2-10	Dance Halls	2-10	Laundries	1-3
Auditoriums	2-10	Dining Rooms	3-10	Markets	2-10
Bakeries	2-3	Dry Cleaners	1-5	Offices	2-10
Banks	3-10	Factories	2-5	Pool Rooms	2-5
Bars	2-5	Garages	2-10	Recreation Rooms	2-10
Beauty Parlors	2-5	Generator Rooms	2-5	Sales Rooms	2-10
Boiler Rooms	1-5	Gymnasiums	2-10	Theaters	2-8
Bowling Alleys	2-10	Kitchens-Hospitals	2-5	Toilets	2-5
Churches	5-10	Kitchens-Restaurant	1-3	Transformer Rooms	1-5

CFM air required for changes = Volume of room in cubic feet ÷ Minutes per change.

15 MECHANICAL REFERENCE NOS.

R157-050 Ductwork

Duct Weight in Pounds per L.F., Straight Runs

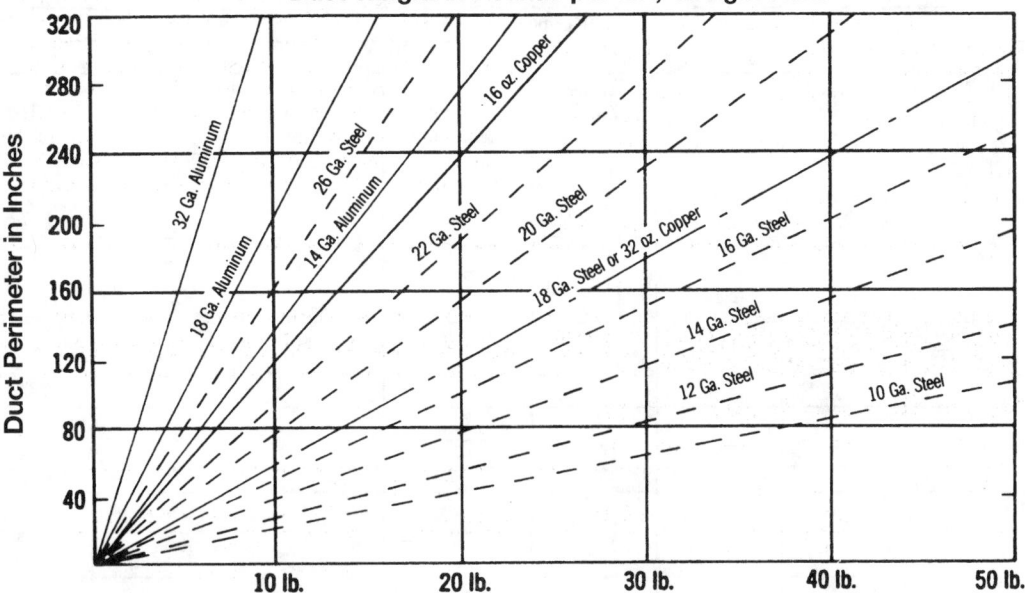

Add to the above for fittings; 90° elbow is 3 L.F.; 45° elbow is 2.5 L.F.; offset is 4 L.F.; transition offset is 6 L.F.; square-to-round transition is 4 L.F.; 90° reducing elbow is 5 L.F. For bracing and waste, add 20% to aluminum and copper, 15% to steel.

R160-205 Conductors in Conduit

Table below lists maximum number of conductors for various sized conduit using THW, TW or THWN insulations.

Copper Wire Size	Conduit Size																											
	1/2"			3/4"			1"			1-1/4"			1-1/2"			2"			2-1/2"			3"		3-1/2"		4"	
	TW	THW	THWN	TW	THW	THWN	TW	THW	THWN	TW	THW	THWN	TW	THW	THWN	TW	THW	THWN	TW	THW	THWN	THW	THWN	THW	THWN	THW	THWN
#14	9	6	13	15	10	24	25	16	39	44	29	69	60	40	94	99	65	154	142	93		143		192			
#12	7	4	10	12	8	18	19	13	29	35	24	51	47	32	70	78	53	114	111	76	164	117		157			
#10	5	4	6	9	6	11	15	11	18	26	19	32	36	26	44	60	43	73	85	61	104	95	160	127		163	
#8	2	1	3	4	3	5	7	5	9	12	10	16	17	13	22	28	22	36	40	32	51	49	79	66	106	85	136
#6		1	1		2	4		4	6		7	11		10	15		16	26		23	37	36	57	48	76	62	98
#4		1	1		1	2		3	4		5	7		7	9		12	16		17	22	27	35	36	47	47	60
#3		1	1		1	1		2	3		4	6		6	8		10	13		15	19	23	29	31	39	40	51
#2		1	1		1	1		2	3		4	5		5	7		9	11		13	16	20	25	27	33	34	43
#1					1	1		1	1		3	3		4	5		6	8		9	12	14	18	19	25	25	32
1/0					1	1		1	1		2	3		3	4		5	7		8	10	12	15	16	21	21	27
2/0					1	1		1	1		1	2		3	3		5	6		7	8	10	13	14	17	18	22
3/0					1	1		1	1		1	1		2	3		4	5		6	7	9	11	12	14	15	18
4/0						1		1	1		1	1		1	2		3	4		5	6	7	9	10	12	13	15
250 kcmil								1	1		1	1		1	1		2	3		4	4	6	7	8	10	10	12
300								1	1		1	1		1	1		2	3		3	4	5	6	7	8	9	11
350									1		1	1		1	1		1	2		3	3	4	5	6	7	8	9
400											1	1		1	1		1	1		2	3	4	5	5	6	7	8
500											1	1		1	1		1	1		1	2	3	4	4	5	6	7
600												1		1	1		1	1		1	1	3	3	4	4	5	5
700														1	1		1	1		1	1	2	3	3	4	4	5
750														1	1		1	1		1	1	2	2	3	3	4	4

ELECTRICAL 16

REFERENCE NOS.

R161-125 Minimum Copper and Aluminum Wire Size Allowed for Various Types of Insulation Various Types of Insulation

Minimum Wire Sizes

Amperes	Copper THW THWN or XHHW	Copper THHN XHHW *	Aluminum THW XHHW	Aluminum THHN XHHW *	Amperes	Copper THW THWN or XHHW	Copper THHN XHHW *	Aluminum THW XHHW	Aluminum THHN XHHW *
15A	#14	#14	#12	#12	195		2/0		
20	#12	#12			200	3/0			
25			#10	#10	205			250kcmil	4/0
30	#10	#10			225		3/0		
40		#8	#8		230	4/0		300kcmil	250kcmil
45				#8	250			350kcmil	
50	#8		#6		255	250kcmil			300kcmil
55		#8			260		4/0		
60				#6	270			400kcmil	
65	#6		#4		280				350kcmil
75		#6	#3	#4	285	300kcmil			
85	#4			#3	290		250kcmil		
90			#2		305				400kcmil
95		#4			310	350kcmil		500kcmil	
100	#3		#1	#2	320		300kcmil		
110		#3			335	400kcmil			
115	#2			#1	340			600kcmil	
120			1/0		350		350kcmil		500kcmil
130	#1	#2			375			700kcmil	
135			2/0	1/0	380	500kcmil	400kcmil		
150	1/0	#1		2/0	385			750kcmil	600kcmil
155			3/0		420	600kcmil			700kcmil
170		1/0			430		500kcmil		
175	2/0			3/0	435				750kcmil
180			4/0		475		600kcmil		

*Dry Locations Only

Notes:

1. Size #14 to 4/0 is in AWG units (American Wire Gauge).
2. Size 250 to 750 is in kcmil units (Thousand Circular Mils).
3. Use next higher ampere value if exact value is not listed in table.
4. For loads that operate continuously increase ampere value by 25% to obtain proper wire size.
5. Refer to Table R161-120 for the maximum circuit length for the various size wires.
6. Table R161-125 has been written for estimating purpose only, based on ambient temperature of 30° C (86° F); for ambient temperature other than 30° C (86° F), ampacity correction factors will be applied.

R163-520 Motors

230/460 Volt A.C., three phase 60 cycle ball bearing squirrel cage
induction motors, NEMA design B Standard Line, including
installation and Subs O & P. (No conduit, wire, or terminations)

Horsepower	Synchronous Speed in RPM	Drip Proof, Class B Insulation 1.15 Service Factor			Totally Enclosed, Class B Insulation, 1.0 or 1.15 Service Factor		
		Motor Only	With Manual Starter	With Magnetic Starter	Motor Only	With Manual Starter	With Magnetic Starter
1	1,200	$ 302	$544	$ 579	$ 297	$539	$ 574
	1,800	274	519	554	287	529	564
2	1,200	390	634	761	349	594	721
	1,800	320	564	691	316	559	686
	3,600	287	529	661	297	539	671
3	1,200	495	739	866	415	659	786
	1,800	337	579	711	318	564	691
	3,600	308	554	681	317	559	691
5	1,200	575	926	1,045	540	891	1,010
	1,800	370	721	840	351	701	820
	3,600	369	721	840	367	716	840
7.5	1,800	511	862	1,290	476	827	1,255
10		601	952	1,380	571	922	1,350
15		794	—	1,565	754	—	1,540
20		957	—	2,095	937	—	2,070
25		1,108	—	2,255	1,138	—	2,280
30		1,259	—	2,385	1,409	—	2,535
40		1,816	—	4,025	1,641	—	3,850
50		2,064	—	4,275	1,864	—	4,075
60		2,698	—	7,225	2,873	—	7,400
75		3,045	—	7,575	3,445	—	7,975
100		4,025	—	8,550	4,500	—	9,025
125		4,720	—	16,000	5,895	—	17,200
150		6,210	—	17,500	7,135	—	18,400
200		8,090	—	19,325	9,315	—	20,625

ELECTRICAL 16

REFERENCE NOS.

R171-100 Square Foot Project Size Modifier

One factor that affects the S.F. cost of a particular building is the size. In general, for buildings built to the same specifications in the same locality, the larger building will have the lower S.F. cost. This is due mainly to the decreasing contribution of the exterior walls plus the economy of scale usually achievable in larger buildings. The Area Conversion Scale shown below will give a factor to convert costs for the typical size building to an adjusted cost for the particular project.

The Square Foot Base Size lists the median costs, most typical project size in our accumulated data and the range in size of the projects.

The Size Factor for your project is determined by dividing your project area in S.F. by the typical project size for the particular Building Type. With this factor, enter the Area Conversion Scale at the appropriate Size Factor and determine the appropriate cost multiplier for your building size.

Example: Determine the cost per S.F. for a 100,000 S.F. Mid-rise apartment building.

$$\frac{\text{Proposed building area} = 100,000 \text{ S.F.}}{\text{Typical size from below} = 50,000 \text{ S.F.}} = 2.00$$

Enter Area Conversion scale at 2.0, intersect curve, read horizontally the appropriate cost multiplier of .94. Size adjusted cost becomes .94 x $66.40 = $62.40 based on national average costs.

Note: For Size Factors less than .50, the Cost Multiplier is 1.1
For Size Factors greater than 3.5, the Cost Multiplier is .90

Square Foot Base Size							
Building Type	**Median Cost per S.F.**	**Typical Size Gross S.F.**	**Typical Range Gross S.F.**	**Building Type**	**Median Cost per S.F.**	**Typical Size Gross S.F.**	**Typical Range Gross S.F.**
Apartments, Low Rise	$ 52.60	21,000	9,700.00 - 37,200	Jails	$161.00	13,700	7,500 - 28,000
Apartments, Mid Rise	66.40	50,000	32,000 - 100,000	Libraries	94.70	12,000	7,000 - 31,000
Apartments, High Rise	76.20	310,000	100,000 - 650,000	Medical Clinics	90.60	7,200	4,200 - 15,700
Auditoriums	87.95	25,000	7,600 - 39,000	Medical Offices	85.10	6,000	4,000 - 15,000
Auto Sales	54.40	20,000	10,800 - 28,600	Motels	65.20	27,000	15,800 - 51,000
Banks	118.00	4,200	2,500 - 7,500	Nursing Homes	87.50	23,000	15,000 - 37,000
Churches	79.45	9,000	5,300 - 13,200	Offices, Low Rise	71.00	8,600	4,700 - 19,000
Clubs, Country	79.20	6,500	4,500 - 15,000	Offices, Mid Rise	74.55	52,000	31,300 - 83,100
Clubs, Social	77.05	10,000	6,000 - 13,500	Offices, High Rise	94.50	260,000	151,000 - 468,000
Clubs, YMCA	79.40	28,300	12,800 - 39,400	Police Stations	119.00	10,500	4,000 - 19,000
Colleges (Class)	104.00	50,000	23,500 - 98,500	Post Offices	87.95	12,400	6,800 - 30,000
Colleges (Science Lab)	151.00	45,600	16,600 - 80,000	Power Plants	673.00	7,500	1,000 - 20,000
College (Student Union)	116.00	33,400	16,000 - 85,000	Religious Education	72.85	9,000	6,000 - 12,000
Community Center	82.80	9,400	5,300 - 16,700	Research	124.00	19,000	6,300 - 45,000
Court Houses	112.00	32,400	17,800 - 106,000	Restaurants	107.00	4,400	2,800 - 6,000
Dept. Stores	49.15	90,000	44,000 - 122,000	Retail Stores	52.25	7,200	4,000 - 17,600
Dormitories, Low Rise	84.85	24,500	13,400 - 40,000	Schools, Elementary	76.10	41,000	24,500 - 55,000
Dormitories, Mid Rise	110.00	55,600	36,100 - 90,000	Schools, Jr. High	77.50	92,000	52,000 - 119,000
Factories	47.65	26,400	12,900 - 50,000	Schools, Sr. High	77.50	101,000	50,500 - 175,000
Fire Stations	83.15	5,800	4,000 - 8,700	Schools, Vocational	77.20	37,000	20,500 - 82,000
Fraternity Houses	81.80	12,500	8,200 - 14,800	Sports Arenas	64.70	15,000	5,000 - 40,000
Funeral Homes	91.45	7,800	4,500 - 11,000	Supermarkets	52.45	20,000	12,000 - 30,000
Garages, Commercial	58.10	9,300	5,000 - 13,600	Swimming Pools	121.00	13,000	7,800 - 22,000
Garages, Municipal	74.30	8,300	4,500 - 12,600	Telephone Exchange	141.00	4,500	1,200 - 10,600
Garages, Parking	30.45	163,000	76,400 - 225,300	Theaters	77.60	10,500	8,800 - 17,500
Gymnasiums	76.85	19,200	11,600 - 41,000	Town Halls	85.35	10,800	4,800 - 23,400
Hospitals	145.00	55,000	27,200 - 125,000	Warehouses	35.15	25,000	8,000 - 72,000
House (Elderly)	71.95	37,000	21,000 - 66,000	Warehouse & Office	40.60	25,000	8,000 - 72,000
Housing (Public)	66.60	36,000	14,400 - 74,400				
Ice Rinks	74.75	29,000	27,200 - 33,600				

Change Orders

Change Order Considerations

A Change Order is a written document, usually prepared by the design professional, and signed by the owner, the architect/engineer and the contractor. A change order states the agreement of the parties to: an addition, deletion, or revision in the work; an adjustment in the contract sum, if any; or an adjustment in the contract time, if any. Change orders, or "extras" in the construction process occur after execution of the construction contract and impact architects/engineers, contractors and owners.

Change orders that are properly recognized and managed can ensure orderly, professional and profitable progress for all who are involved in the project. There are many causes for change orders and change order requests. In all cases, change orders or change order requests should be addressed promptly and in a precise and prescribed manner. The following paragraphs include information regarding change order pricing and procedures.

The Causes of Change Orders

Reasons for issuing change orders include:

- Unforeseen field conditions that require a change in the work
- Correction of design discrepancies, errors or omissions in the contract documents
- Owner-requested changes, either by design criteria, scope of work, or project objectives
- Completion date changes for reasons unrelated to the construction process
- Changes in building code interpretations, or other public authority requirements that require a change in the work
- Changes in availability of existing or new materials and products

Procedures

Properly written contract documents must include the correct change order procedures for all parties—owners, design professionals and contractors—to follow in order to avoid costly delays and litigation.

Being "in the right" is not always a sufficient or acceptable defense. The contract provisions requiring notification and documentation must be adhered to within a defined or reasonable time frame.

The appropriate method of handling change orders is by a written proposal and acceptance by all parties involved. Prior to starting work on a project, all parties should identify their authorized agents who may sign and accept change orders, as well as any limits placed on their authority.

Time may be a critical factor when the need for a change arises. For such cases, the contractor might be directed to proceed on a "time and materials" basis, rather than wait for all paperwork to be processed—a delay that could impede progress. In this situation, the contractor must still follow the prescribed change order procedures, including but not limited to, notification and documentation.

All forms used for change orders should be dated and signed by the proper authority. Lack of documentation can be very costly, especially if legal judgments are to be made and if certain field personnel are no longer available. For time and material change orders, the contractor should keep accurate daily records of all labor and material allocated to the change. Forms that can be used to document change order work are available in *Means Forms for Building Construction Professionals.*

Owners or awarding authorities who do considerable and continual building construction (such as the federal government) realize the inevitability of change orders for numerous reasons, both predictable and unpredictable. As a result, the federal government, the American Institute of Architects (AIA), the Engineers Joint Contract Documents Committee (EJCDC) and other contractor, legal and technical organizations have developed standards and procedures to be followed by all parties to achieve contract continuance and timely completion, while being financially fair to all concerned.

In addition to the change order standards put forth by industry associations, there are also many books available on the subject.

Pricing Change Orders

When pricing change orders, regardless of their cause, the most significant factor is *when* the change occurs. The need for a change may be perceived in the field or requested by the architect/engineer *before* any of the actual installation has begun, or may evolve or appear *during* construction when the item of work in question is partially installed. In the latter cases, the original sequence of construction is disrupted, along with all contiguous and supporting systems. Change orders cause the greatest impact when they occur *after* the installation has been completed and must be uncovered, or even replaced. Post-completion changes may be caused by necessary design changes, product failure, or changes in the owner's requirements that are not discovered until the building or the systems begin to function.

Specified procedures of notification and record keeping must be adhered to and enforced regardless of the stage of construction: *before, during,* or *after* installation. Some bidding documents anticipate change orders by requiring that unit prices including overhead and profit percentages—for additional as well as deductible changes—be listed. Generally these unit prices do not fully take into account the ripple effect, or impact on other trades, and should be used for general guidance only.

When pricing change orders, it is important to classify the time frame in which the change occurs. There are two basic time frames for change orders: *pre-installation change orders*, which occur before the start of construction, and *post-installation change orders*, which involve reworking after the original installation. Change orders that occur between these stages may be priced according to the extent of work completed using a combination of techniques developed for pricing *pre-* and *post-installation* changes.

The following factors are the basis for a check list to use when preparing a change order estimate.

REFERENCE

Factors To Consider When Pricing Change Orders

As an estimator begins to prepare a change order, the following questions should be reviewed to determine their impact on the final price.

General

- Is the change order work *pre-installation* or *post-installation*?

Change order work costs vary according to how much of the installation has been completed. Once workers have the project scoped in their mind, even though they have not started, it can be difficult to refocus.

Consequently they may spend more than the normal amount of time understanding the change. Also, modifications to work in place such as trimming or refitting usually take more time than was initially estimated. The greater the amount of work in place, the more reluctant workers are to change it. Psychologically they may resent the change and as a result the rework takes longer than normal. Post-installation change order estimates must include demolition of existing work as required to accomplish the change. If the work is performed at a later time, additional obstacles such as building finishes may be present which must be

protected. Regardless of whether the change occurs pre-installation or post-installation, attempt to isolate the identifiable factors and price them separately. For example, add shipping costs that may be required pre-installation or any demolition required post-installation. Then analyze the potential impact on productivity of psychological and/or learning curve factors and adjust the output rates accordingly. One approach is to break down the typical workday into segments and quantify the impact on each segment. The following chart may be useful as a guide:

Task	Activities (Productivity) Expressed as Percentages of a Workday		
	Means Mechanical Cost Data (for New Construction)	Pre-Installation Change Orders	Post-Installation Change Orders
1. Study plans	3%	6%	6%
2. Material procurement	3%	3%	3%
3. Receiving and storing	3%	3%	3%
4. Mobilization	5%	5%	5%
5. Site movement	5%	5%	8%
6. Layout and marking	8%	10%	12%
7. Actual installation	64%	59%	54%
8. Clean-up	3%	3%	3%
9. Breaks—non-productive	6%	6%	6%
Total	100%	100%	100%

Change Order Installation Efficiency

The labor-hours expressed (for new construction) are based on average installation time, using an efficiency level of approximately 60-65%. For change order situations, adjustments to this efficiency level should reflect the daily labor-hour allocation for that particular occurrence.

If any of the specific percentages expressed in the above chart do not apply to a particular project situation, then those percentage points should be reallocated to the appropriate task(s). Example: Using data for new construction, assume there is no new material being utilized. The percentages for Tasks 2 and 3 would therefore be reallocated to other tasks. If the time required for Tasks 2 and 3 can now be applied to installation, we can add the time allocated for *Material Procurement* and *Receiving and Storing* to the *Actual Installation* time for new construction, thereby increasing the Actual Installation percentage.

This chart shows that, due to reduced productivity, labor costs will be higher than those for new construction by 5% to 15% for pre-installation change orders and by 15% to 25% for post-installation change orders. Each job and change order is unique and must be examined individually. Many factors, covered elsewhere in this section, can each have a significant impact on productivity and change order costs. All such factors should be considered in every case.

- Will the change substantially delay the original completion date?

A significant change in the project may cause the original completion date to be extended. The extended schedule may subject the contractor to new wage rates dictated by relevant labor contracts. Project supervision and other project overhead must also be extended beyond the original completion date. The schedule extension may also put installation into a new weather season. For example, underground piping scheduled for October installation was delayed until January. As a result, frost penetrated the trench area, thereby changing the degree of difficulty of the task. Changes and delays may have a ripple effect throughout the project. This effect must be analyzed and negotiated with the owner.

- What is the net effect of a deduct change order?

In most cases, change orders resulting in a deduction or credit reflect only bare costs. The contractor may retain the overhead and profit based on the original bid.

Materials

- Will you have to pay more or less for the new material, required by the change order, than you paid for the original purchase?

The same material prices or discounts will usually apply to materials purchased for change orders as new construction. In some instances, however, the contractor may forfeit the advantages of competitive pricing for change orders. Consider the following example:

A contractor purchased over $20,000 worth of fan coil units for an installation, and obtained the maximum discount. Some time later it was determined the project required an additional matching unit. The contractor has to purchase this unit from the original supplier to ensure a match. The supplier at this time may not discount the unit because of the small quantity, and the fact that he is no longer in a competitive situation. The impact of quantity on purchase can add between 0% and 25% to material prices and/or subcontractor quotes.

- If materials have been ordered or delivered to the job site, will they be subject to a cancellation charge or restocking fee?

Check with the supplier to determine if ordered materials are subject to a cancellation charge. Delivered materials not used as result of a change order may be subject to a restocking fee if returned to the supplier. Common restocking charges run between 20% and 40%. Also, delivery charges to return the goods to the supplier must be added.

Labor

- How efficient is the existing crew at the actual installation?

Is the same crew that performed the initial work going to do the change order? Possibly the change consists of the installation of a unit identical to one already installed; therefore the change should take less time. Be sure to consider this potential productivity increase and modify the productivity rates accordingly.

- If the crew size is increased, what impact will that have on supervision requirements?

Under most bargaining agreements or management practices, there is a point at which a working foreman is replaced by a nonworking foreman. This replacement increases project overhead by adding a nonproductive worker. If additional workers are added to accelerate the project or to perform changes while maintaining the schedule, be sure to add additional supervision time if warranted. Calculate the hours involved and the additional cost directly if possible.

- What are the other impacts of increased crew size?

The larger the crew, the greater the potential for productivity to decrease. Some of the factors that cause this productivity loss are: overcrowding (producing restrictive conditions in the working space), and possibly a shortage of any special tools and equipment required. Such factors affect not only the crew working on the elements directly involved in the change order, but other crews whose movement may also be hampered.

As the crew increases, check its basic composition for changes by the addition or deletion of apprentices or nonworking foreman and quantify the potential effects of equipment shortages or other logistical factors.

- As new crews, unfamiliar with the project, are brought onto the site, how long will it take them to become oriented to the project requirements?

The orientation time for a new crew to become 100% effective varies with the site and type of project. Orientation is easiest at a new construction site, and most difficult at existing, very restrictive renovation sites. The type of work also affects orientation time. When all elements of the work are exposed, such as concrete or masonry work, orientation is decreased. When the work is concealed or less visible, such as existing electrical systems, orientation takes longer. Usually orientation can be accomplished in one day or less. Costs for added orientation should be itemized and added to the total estimated cost.

- How much actual production can be gained by working overtime?

Short term overtime can be used effectively to accomplish more work in a day. However, as overtime is scheduled to run beyond several weeks, studies have shown marked decreases in output. The following chart shows the effect of long term overtime on worker efficiency. If the anticipated change requires extended overtime to keep the job on schedule, these factors can be used as a guide to predict the impact on time and cost. Add project overhead, particularly supervision, that may also be incurred.

Days per Week	Hours per Day	Production Efficiency					Payroll Cost Factors	
		1 Week	2 Weeks	3 Weeks	4 Weeks	Average 4 Weeks	@ 1-1/2 Times	@ 2 Times
5	8	100%	100%	100%	100%	100%	100%	100%
	9	100	100	95	90	96.25	105.6	111.1
	10	100	95	90	85	91.25	110.0	120.0
	11	95	90	75	65	81.25	113.6	127.3
	12	90	85	70	60	76.25	116.7	133.3
6	8	100	100	95	90	96.25	108.3	116.7
	9	100	95	90	85	92.50	113.0	125.9
	10	95	90	85	80	87.50	116.7	133.3
	11	95	85	70	65	78.75	119.7	139.4
	12	90	80	65	60	73.75	122.2	144.4
7	8	100	95	85	75	88.75	114.3	128.6
	9	95	90	80	70	83.75	118.3	136.5
	10	90	85	75	65	78.75	121.4	142.9
	11	85	80	65	60	72.50	124.0	148.1
	12	85	75	60	55	68.75	126.2	152.4

Effects of Overtime

Caution: Under many labor agreements, Sundays and holidays are paid at a higher premium than the normal overtime rate.

The use of long-term overtime is counterproductive on almost any construction job; that is, the longer the period of overtime, the lower the actual production rate. Numerous studies have been conducted, and while they have resulted in slightly different numbers, all reach the same conclusion. The figure above tabulates the effects of overtime work on efficiency.

As illustrated, there can be a difference between the *actual* payroll cost per hour and the *effective* cost per hour for overtime work. This is due to the reduced production efficiency with the increase in weekly hours beyond 40. This difference between actual and effective cost results from overtime work over a prolonged period. Short-term overtime work does not result in as great a reduction in efficiency, and in such cases, effective cost may not vary significantly from the actual payroll cost. As the total hours per week are increased on a regular basis, more time is lost because of fatigue, lowered morale, and an increased accident rate.

As an example, assume a project where workers are working 6 days a week, 10 hours per day. From the figure above (based on productivity studies), the average effective productive hours over a four-week period are:

$$0.875 \times 60 = 52.5$$

Depending upon the locale and day of week, overtime hours may be paid at time and a half or double time. For time and a half, the overall (average) *actual* payroll cost (including regular and overtime hours) is determined as follows:

$$\frac{40 \text{ reg. hrs.} + (20 \text{ overtime hrs.} \times 1.5)}{60 \text{ hrs.}} = 1.167$$

Based on 60 hours, the payroll cost per hour will be 116.7% of the normal rate at 40 hours per week. However, because the effective production (efficiency) for 60 hours is reduced to the equivalent of 52.5 hours, the effective cost of overtime is calculated as follows:

For time and a half:

$$\frac{40 \text{ reg. hrs.} + (20 \text{ overtime hrs.} \times 1.5)}{52.5 \text{ hrs.}} = 1.33$$

Installed cost will be 133% of the normal rate (for labor).

Thus, when figuring overtime, the actual cost per unit of work will be higher than the apparent overtime payroll dollar increase, due to the reduced productivity of the longer workweek. These efficiency calculations are true only for those cost factors determined by hours worked. Costs that are applied weekly or monthly, such as equipment rentals, will not be similarly affected.

Equipment

- What equipment is required to complete the change order?

Change orders may require extending the rental period of equipment already on the job site, or the addition of special equipment brought in to accomplish the change work. In either case, the additional rental charges and operator labor charges must be added.

Summary

The preceding considerations and others you deem appropriate should be analyzed and applied to a change order estimate. The impact of each should be quantified and listed on the estimate to form an audit trail.

Change orders that are properly identified, documented, and managed help to ensure the orderly, professional and profitable progress of the work. They also minimize potential claims or disputes at the end of the project.

REFERENCE

Crew A-1

Crew A-1	Hr.	Daily	Hr.	Daily	Bare Costs	Incl. O&P
1 Building Laborer	$21.45	$171.60	$33.75	$270.00	$21.45	$33.75
1 Gas Eng. Power Tool		68.65		75.50	8.58	9.44
8 L.H., Daily Totals		$240.25		$345.50	$30.03	$43.19

Crew A-1A

Crew A-1A	Hr.	Daily	Hr.	Daily	Bare Costs	Incl. O&P
1 Skilled Worker	$28.05	$224.40	$44.25	$354.00	$28.05	$44.25
1 Shot Blaster, 20"		131.90		145.10	16.49	18.14
8 L.H., Daily Totals		$356.30		$499.10	$44.54	$62.39

Crew A-2

Crew A-2	Hr.	Daily	Hr.	Daily	Bare Costs	Incl. O&P
2 Laborers	$21.45	$343.20	$33.75	$540.00	$21.55	$33.60
1 Truck Driver (light)	21.75	174.00	33.30	266.40		
1 Light Truck, 1.5 Ton		171.95		189.15	7.16	7.88
24 L.H., Daily Totals		$689.15		$995.55	$28.71	$41.48

Crew A-2A

Crew A-2A	Hr.	Daily	Hr.	Daily	Bare Costs	Incl. O&P
2 Laborers	$21.45	$343.20	$33.75	$540.00	$21.55	$33.60
1 Truck Driver (light)	21.75	174.00	33.30	266.40		
1 Light Truck, 1.5 ton		171.95		189.15		
1 Concrete Saw		110.65		121.70	11.78	12.95
24 L.H., Daily Totals		$799.80		$1117.25	$33.33	$46.55

Crew A-3

Crew A-3	Hr.	Daily	Hr.	Daily	Bare Costs	Incl. O&P
1 Truck Driver (heavy)	$22.10	$176.80	$33.85	$270.80	$22.10	$33.85
1 Dump Truck, 12 Ton		364.30		400.75	45.54	50.09
8 L.H., Daily Totals		$541.10		$671.55	$67.64	$83.94

Crew A-3A

Crew A-3A	Hr.	Daily	Hr.	Daily	Bare Costs	Incl. O&P
1 Truck Driver (light)	$21.75	$174.00	$33.30	$266.40	$21.75	$33.30
1 Pickup truck (4x4)		145.50		160.05	18.19	20.01
8 L.H., Daily Totals		$319.50		$426.45	$39.94	$53.31

Crew A-3B

Crew A-3B	Hr.	Daily	Hr.	Daily	Bare Costs	Incl. O&P
1 Equip. Oper. (medium)	$28.40	$227.20	$43.20	$345.60	$25.25	$38.53
1 Truck Driver (heavy)	22.10	176.80	33.85	270.80		
1 Dump Truck, 16 Ton		442.25		486.50	27.64	30.40
16 L.H., Daily Totals		$846.25		$1102.90	$52.89	$68.93

Crew A-3C

Crew A-3C	Hr.	Daily	Hr.	Daily	Bare Costs	Incl. O&P
1 Equip. Oper. (light)	$27.20	$217.60	$41.35	$330.80	$27.20	$41.35
1 Wheeled Skid Steer Loader		230.25		253.30	28.78	31.66
8 L.H., Daily Totals		$447.85		$584.10	$55.98	$73.01

Crew A-4

Crew A-4	Hr.	Daily	Hr.	Daily	Bare Costs	Incl. O&P
2 Carpenters	$27.30	$436.80	$42.95	$687.20	$26.50	$41.33
1 Painter, Ordinary	24.90	199.20	38.10	304.80		
24 L.H., Daily Totals		$636.00		$992.00	$26.50	$41.33

Crew A-5

Crew A-5	Hr.	Daily	Hr.	Daily	Bare Costs	Incl. O&P
2 Laborers	$21.45	$343.20	$33.75	$540.00	$21.48	$33.70
.25 Truck Driver (light)	21.75	43.50	33.30	66.60		
.25 Light Truck, 1.5 Ton		42.99		47.30	2.39	2.63
18 L.H., Daily Totals		$429.69		$653.90	$23.87	$36.33

Crew A-6

Crew A-6	Hr.	Daily	Hr.	Daily	Bare Costs	Incl. O&P
1 Chief Of Party	$27.20	$217.60	$41.35	$330.80	$25.63	$38.97
1 Instrument Man	24.05	192.40	36.60	292.80		
16 L.H., Daily Totals		$410.00		$623.60	$25.63	$38.97

Crew A-7

Crew A-7	Hr.	Daily	Hr.	Daily	Bare Costs	Incl. O&P
1 Chief Of Party	$27.20	$217.60	$41.35	$330.80	$24.03	$36.92
1 Instrument Man	24.05	192.40	36.60	292.80		
1 Rodman/Chainman	20.85	166.80	32.80	262.40		
24 L.H., Daily Totals		$576.80		$886.00	$24.03	$36.92

Crew A-8

Crew A-8	Hr.	Daily	Hr.	Daily	Bare Costs	Incl. O&P
1 Chief Of Party	$27.20	$217.60	$41.35	$330.80	$23.24	$35.89
1 Instrument Man	24.05	192.40	36.60	292.80		
2 Rodmen/Chainmen	20.85	333.60	32.80	524.80		
32 L.H., Daily Totals		$743.60		$1148.40	$23.24	$35.89

Crew A-9

Crew A-9	Hr.	Daily	Hr.	Daily	Bare Costs	Incl. O&P
1 Asbestos Foreman	$30.95	$247.60	$49.95	$399.60	$30.51	$49.25
7 Asbestos Workers	30.45	1705.20	49.15	2752.40		
64 L.H., Daily Totals		$1952.80		$3152.00	$30.51	$49.25

Crew A-10

Crew A-10	Hr.	Daily	Hr.	Daily	Bare Costs	Incl. O&P
1 Asbestos Foreman	$30.95	$247.60	$49.95	$399.60	$30.51	$49.25
7 Asbestos Workers	30.45	1705.20	49.15	2752.40		
64 L.H., Daily Totals		$1952.80		$3152.00	$30.51	$49.25

Crew A-10A

Crew A-10A	Hr.	Daily	Hr.	Daily	Bare Costs	Incl. O&P
1 Asbestos Foreman	$30.95	$247.60	$49.95	$399.60	$30.62	$49.42
2 Asbestos Workers	30.45	487.20	49.15	786.40		
24 L.H., Daily Totals		$734.80		$1186.00	$30.62	$49.42

Crew A-10B

Crew A-10B	Hr.	Daily	Hr.	Daily	Bare Costs	Incl. O&P
1 Asbestos Foreman	$30.95	$247.60	$49.95	$399.60	$30.58	$49.35
3 Asbestos Workers	30.45	730.80	49.15	1179.60		
32 L.H., Daily Totals		$978.40		$1579.20	$30.58	$49.35

Crew A-10C

Crew A-10C	Hr.	Daily	Hr.	Daily	Bare Costs	Incl. O&P
3 Asbestos Workers	$30.45	$730.80	$49.15	$1179.60	$30.45	$49.15
1 Flatbed truck		171.95		189.15	7.16	7.88
24 L.H., Daily Totals		$902.75		$1368.75	$37.61	$57.03

Crew A-10D

Crew A-10D	Hr.	Daily	Hr.	Daily	Bare Costs	Incl. O&P
2 Asbestos Workers	$30.45	$487.20	$49.15	$786.40	$28.58	$44.89
1 Equip. Oper. (crane)	29.35	234.80	44.65	357.20		
1 Equip. Oper. Oiler	24.05	192.40	36.60	292.80		
1 Hydraulic crane, 33 ton		694.95		764.45	21.72	23.89
32 L.H., Daily Totals		$1609.35		$2200.85	$50.30	$68.78

Crew A-11

Crew A-11	Hr.	Daily	Hr.	Daily	Bare Costs	Incl. O&P
1 Asbestos Foreman	$30.95	$247.60	$49.95	$399.60	$30.51	$49.25
7 Asbestos Workers	30.45	1705.20	49.15	2752.40		
2 Chipping Hammers		25.90		28.50	.40	.45
64 L.H., Daily Totals		$1978.70		$3180.50	$30.91	$49.70

Crew A-12

Crew A-12	Hr.	Daily	Hr.	Daily	Bare Costs	Incl. O&P
1 Asbestos Foreman	$30.95	$247.60	$49.95	$399.60	$30.51	$49.25
7 Asbestos Workers	30.45	1705.20	49.15	2752.40		
1 Large Prod. Vac. Loader		524.90		577.40	8.20	9.02
64 L.H., Daily Totals		$2477.70		$3729.40	$38.71	$58.27

Crew A-13

Crew A-13	Hr.	Daily	Hr.	Daily	Bare Costs	Incl. O&P
1 Equip. Oper. (light)	$27.20	$217.60	$41.35	$330.80	$27.20	$41.35
1 Large Prod. Vac. Loader		524.90		577.40	65.61	72.17
8 L.H., Daily Totals		$742.50		$908.20	$92.81	$113.52

Crews

Crew B-1

Crew No.	Bare Costs Hr.	Daily	Incl. Subs O&P Hr.	Daily	Cost Per L.H. Bare Costs	Incl. O&P
1 Labor Foreman (outside)	$23.45	$187.60	$36.90	$295.20	$22.12	$34.80
2 Laborers	21.45	343.20	33.75	540.00		
24 L.H., Daily Totals		$530.80		$835.20	$22.12	$34.80

Crew B-2

Crew No.	Hr.	Daily	Hr.	Daily	Bare Costs	Incl. O&P
1 Labor Foreman (outside)	$23.45	$187.60	$36.90	$295.20	$21.85	$34.38
4 Laborers	21.45	686.40	33.75	1080.00		
40 L.H., Daily Totals		$874.00		$1375.20	$21.85	$34.38

Crew B-3

Crew No.	Hr.	Daily	Hr.	Daily	Bare Costs	Incl. O&P
1 Labor Foreman (outside)	$23.45	$187.60	$36.90	$295.20	$23.16	$35.88
2 Laborers	21.45	343.20	33.75	540.00		
1 Equip. Oper. (med.)	28.40	227.20	43.20	345.60		
2 Truck Drivers (heavy)	22.10	353.60	33.85	541.60		
1 F.E. Loader, T.M., 2.5 C.Y.		865.10		951.60		
2 Dump Trucks, 16 Ton		884.50		972.95	36.45	40.10
48 L.H., Daily Totals		$2861.20		$3646.95	$59.61	$75.98

Crew B-3A

Crew No.	Hr.	Daily	Hr.	Daily	Bare Costs	Incl. O&P
4 Laborers	$21.45	$686.40	$33.75	$1080.00	$22.84	$35.64
1 Equip. Oper. (med.)	28.40	227.20	43.20	345.60		
1 Hyd. Excavator, 1.5 C.Y.		712.70		783.95	17.82	19.60
40 L.H., Daily Totals		$1626.30		$2209.55	$40.66	$55.24

Crew B-3B

Crew No.	Hr.	Daily	Hr.	Daily	Bare Costs	Incl. O&P
2 Laborers	$21.45	$343.20	$33.75	$540.00	$23.35	$36.14
1 Equip. Oper. (med.)	28.40	227.20	43.20	345.60		
1 Truck Drivers (heavy)	22.10	176.80	33.85	270.80		
1 Backhoe Loader, 80 H.P.		290.10		319.10		
1 Dump Trucks, 16 Ton		442.25		486.50	22.89	25.17
32 L.H., Daily Totals		$1479.55		$1962.00	$46.24	$61.31

Crew B-3C

Crew No.	Hr.	Daily	Hr.	Daily	Bare Costs	Incl. O&P
3 Laborers	$21.45	$514.80	$33.75	$810.00	$23.19	$36.11
1 Equip. Oper. (med.)	28.40	227.20	43.20	345.60		
1 F.E. Loader, 3.75 C.Y.		576.90		634.60	18.03	19.83
32 L.H., Daily Totals		$1318.90		$1790.20	$41.22	$55.94

Crew B-4

Crew No.	Hr.	Daily	Hr.	Daily	Bare Costs	Incl. O&P
1 Labor Foreman (outside)	$23.45	$187.60	$36.90	$295.20	$21.89	$34.29
4 Laborers	21.45	686.40	33.75	1080.00		
1 Truck Driver (heavy)	22.10	176.80	33.85	270.80		
1 Tractor, 4 x 2, 195 H.P.		324.05		356.45		
1 Platform Trailer		150.70		165.75	9.89	10.88
48 L.H., Daily Totals		$1525.55		$2168.20	$31.78	$45.17

Crew B-5

Crew No.	Hr.	Daily	Hr.	Daily	Bare Costs	Incl. O&P
1 Labor Foreman (outside)	$23.45	$187.60	$36.90	$295.20	$23.72	$36.90
4 Laborers	21.45	686.40	33.75	1080.00		
2 Equip. Oper. (med.)	28.40	454.40	43.20	691.20		
1 Air Compr., 250 C.F.M.		124.90		137.40		
2 Air Tools & Accessories		39.00		42.90		
2-50 Ft. Air Hoses, 1.5" Dia.		16.40		18.05		
1 F.E. Loader, T.M., 2.5 C.Y.		865.10		951.60	18.67	20.53
56 L.H., Daily Totals		$2373.80		$3216.35	$42.39	$57.43

Crew B-5A

Crew No.	Hr.	Daily	Hr.	Daily	Bare Costs	Incl. O&P
1 Foreman	$23.45	$187.60	$36.90	$295.20	$23.36	$36.24
6 Laborers	21.45	1029.60	33.75	1620.00		
2 Equip. Oper. (med.)	28.40	454.40	43.20	691.20		
1 Equip. Oper. (light)	27.20	217.60	41.35	330.80		
2 Truck Drivers (heavy)	22.10	353.60	33.85	541.60		
1 Air Compr. 365 C.F.M.		174.30		191.75		
2 Pavement Breakers		39.00		42.90		
8 Air Hoses w/Coup.,1"		41.60		45.75		
2 Dump Trucks, 12 Ton		728.60		801.45	10.24	11.27
96 L.H., Daily Totals		$3226.30		$4560.65	$33.60	$47.51

Crew B-5B

Crew No.	Hr.	Daily	Hr.	Daily	Bare Costs	Incl. O&P
1 Powderman	$28.05	$224.40	$44.25	$354.00	$25.19	$38.70
2 Equip. Oper. (med.)	28.40	454.40	43.20	691.20		
3 Truck Drivers (heavy)	22.10	530.40	33.85	812.40		
1 F.E. Ldr. 2-1/2 CY		450.50		495.55		
3 Dump Trucks, 16 Ton		1326.75		1459.45		
1 Air Compr. 365 C.F.M.		174.30		191.75	40.66	44.72
48 L.H., Daily Totals		$3160.75		$4004.35	$65.85	$83.42

Crew B-5C

Crew No.	Hr.	Daily	Hr.	Daily	Bare Costs	Incl. O&P
3 Laborers	$21.45	$514.80	$33.75	$810.00	$23.79	$36.67
1 Equip. Oper. (medium)	28.40	227.20	43.20	345.60		
2 Truck Drivers (heavy)	22.10	353.60	33.85	541.60		
1 Equip. Oper. (crane)	29.35	234.80	44.65	357.20		
1 Equip. Oper. Oiler	24.05	192.40	36.60	292.80		
2 Dump Truck, 16 Ton		884.50		972.95		
1 F.E. Lder, 3.75 C.Y.		1142.00		1256.20		
1 Hyd. Crane, 25 Ton		547.90		602.70	40.23	44.25
64 L.H., Daily Totals		$4097.20		$5179.05	$64.02	$80.92

Crew B-6

Crew No.	Hr.	Daily	Hr.	Daily	Bare Costs	Incl. O&P
2 Laborers	$21.45	$343.20	$33.75	$540.00	$23.37	$36.28
1 Equip. Oper. (light)	27.20	217.60	41.35	330.80		
1 Backhoe Loader, 48 H.P.		218.80		240.70	9.12	10.03
24 L.H., Daily Totals		$779.60		$1111.50	$32.49	$46.31

Crew B-6A

Crew No.	Hr.	Daily	Hr.	Daily	Bare Costs	Incl. O&P
.5 Labor Foreman (outside)	$23.45	$93.80	$36.90	$147.60	$24.63	$38.16
1 Laborers	21.45	171.60	33.75	270.00		
1 Equip. Oper. (med.)	28.40	227.20	43.20	345.60		
1 Vacuum Trk.,5000 Gal		320.15		352.15	16.01	17.61
20 L.H., Daily Totals		$812.75		$1115.35	$40.64	$55.77

Crew B-7

Crew No.	Hr.	Daily	Hr.	Daily	Bare Costs	Incl. O&P
1 Labor Foreman (outside)	$23.45	$187.60	$36.90	$295.20	$22.94	$35.85
4 Laborers	21.45	686.40	33.75	1080.00		
1 Equip. Oper. (med.)	28.40	227.20	43.20	345.60		
1 Chipping Machine		214.10		235.50		
1 F.E. Loader, T.M., 2.5 C.Y.		865.10		951.60		
2 Chain Saws, 36"		106.10		116.70	24.69	27.16
48 L.H., Daily Totals		$2286.50		$3024.60	$47.63	$63.01

Crew B-7A

Crew No.	Hr.	Daily	Hr.	Daily	Bare Costs	Incl. O&P
2 Laborers	$21.45	$343.20	$33.75	$540.00	$23.37	$36.28
1 Equip. Oper. (light)	27.20	217.60	41.35	330.80		
1 Rake w/Tractor		218.90		240.80		
2 Chain Saws, 18"		54.60		60.05	11.40	12.54
24 L.H., Daily Totals		$834.30		$1171.65	$34.77	$48.82

CREWS

Crews

Crew No.	Bare Costs Hr.	Daily	Incl. Subs O&P Hr.	Daily	Cost Per Labor-Hour Bare Costs	Incl. O&P
Crew B-8						
1 Labor Foreman (outside)	$23.45	$187.60	$36.90	$295.20	$23.93	$36.89
2 Laborers	21.45	343.20	33.75	540.00		
2 Equip. Oper. (med.)	28.40	454.40	43.20	691.20		
1 Equip. Oper. Oiler	24.05	192.40	36.60	292.80		
2 Truck Drivers (heavy)	22.10	353.60	33.85	541.60		
1 Hyd. Crane, 25 Ton		549.35		604.30		
1 F.E. Loader, T.M., 2.5 C.Y.		865.10		951.60		
2 Dump Trucks, 16 Ton		884.50		972.95	35.92	39.51
64 L.H., Daily Totals		$3830.15		$4889.65	$59.85	$76.40
Crew B-9						
1 Labor Foreman (outside)	$23.45	$187.60	$36.90	$295.20	$21.85	$34.38
4 Laborers	21.45	686.40	33.75	1080.00		
1 Air Compr., 250 C.F.M.		124.90		137.40		
2 Air Tools & Accessories		39.00		42.90		
2-50 Ft. Air Hoses, 1.5" Dia.		16.40		18.05	4.51	4.96
40 L.H., Daily Totals		$1054.30		$1573.55	$26.36	$39.34
Crew B-9A						
2 Laborers	$21.45	$343.20	$33.75	$540.00	$21.67	$33.78
1 Truck Driver (heavy)	22.10	176.80	33.85	270.80		
1 Water Tanker		205.10		225.60		
1 Tractor		324.05		356.45		
2-50 Ft. Disch. Hoses		10.80		11.90	22.50	24.75
24 L.H., Daily Totals		$1059.95		$1404.75	$44.17	$58.53
Crew B-9B						
2 Laborers	$21.45	$343.20	$33.75	$540.00	$21.67	$33.78
1 Truck Driver (heavy)	22.10	176.80	33.85	270.80		
2-50 Ft. Disch. Hoses		10.80		11.90		
1 Water Tanker		205.10		225.60		
1 Tractor		324.05		356.45		
1 Pressure Washer		55.40		60.95	24.81	27.29
24 L.H., Daily Totals		$1115.35		$1465.70	$46.48	$61.07
Crew B-9C						
1 Labor Foreman (outside)	$23.45	$187.60	$36.90	$295.20	$21.85	$34.38
4 Laborers	21.45	686.40	33.75	1080.00		
1 Air Compr., 250 C.F.M.		124.90		137.40		
2-50 Ft. Air Hoses, 1.5" Dia.		16.40		18.05		
2 Breaker, Pavement, 60 lb.		39.00		42.90	4.51	4.96
40 L.H., Daily Totals		$1054.30		$1573.55	$26.36	$39.34
Crew B-10						
1 Equip. Oper. (med.)	$28.40	$227.20	$43.20	$345.60	$26.08	$40.05
.5 Laborer	21.45	85.80	33.75	135.00		
12 L.H., Daily Totals		$313.00		$480.60	$26.08	$40.05
Crew B-10A						
1 Equip. Oper. (med.)	$28.40	$227.20	$43.20	$345.60	$26.08	$40.05
.5 Laborer	21.45	85.80	33.75	135.00		
1 Roll. Compact., 2K Lbs.		93.05		102.35	7.75	8.53
12 L.H., Daily Totals		$406.05		$582.95	$33.83	$48.58
Crew B-10B						
1 Equip. Oper. (med.)	$28.40	$227.20	$43.20	$345.60	$26.08	$40.05
.5 Laborer	21.45	85.80	33.75	135.00		
1 Dozer, 200 H.P.		838.95		922.85	69.91	76.90
12 L.H., Daily Totals		$1151.95		$1403.45	$95.99	$116.95
Crew B-10C						
1 Equip. Oper. (med.)	$28.40	$227.20	$43.20	$345.60	$26.08	$40.05
.5 Laborer	21.45	85.80	33.75	135.00		
1 Dozer, 200 H.P.		838.95		922.85		
1 Vibratory Roller, Towed		112.90		124.20	79.32	87.25
12 L.H., Daily Totals		$1264.85		$1527.65	$105.40	$127.30
Crew B-10D						
1 Equip. Oper. (med.)	$28.40	$227.20	$43.20	$345.60	$26.08	$40.05
.5 Laborer	21.45	85.80	33.75	135.00		
1 Dozer, 200 H.P		838.95		922.85		
1 Sheepsft. Roller, Towed		124.90		137.40	80.32	88.35
12 L.H., Daily Totals		$1276.85		$1540.85	$106.40	$128.40
Crew B-10E						
1 Equip. Oper. (med.)	$28.40	$227.20	$43.20	$345.60	$26.08	$40.05
.5 Laborer	21.45	85.80	33.75	135.00		
1 Tandem Roller, 5 Ton		152.85		168.15	12.74	14.01
12 L.H., Daily Totals		$465.85		$648.75	$38.82	$54.06
Crew B-10F						
1 Equip. Oper. (med.)	$28.40	$227.20	$43.20	$345.60	$26.08	$40.05
.5 Laborer	21.45	85.80	33.75	135.00		
1 Tandem Roller, 10 Ton		245.10		269.60	20.43	22.47
12 L.H., Daily Totals		$558.10		$750.20	$46.51	$62.52
Crew B-10G						
1 Equip. Oper. (med.)	$28.40	$227.20	$43.20	$345.60	$26.08	$40.05
.5 Laborer	21.45	85.80	33.75	135.00		
1 Sheepsft. Roll., 130 H.P.		558.70		614.55	46.56	51.21
12 L.H., Daily Totals		$871.70		$1095.15	$72.64	$91.26
Crew B-10H						
1 Equip. Oper. (med.)	$28.40	$227.20	$43.20	$345.60	$26.08	$40.05
.5 Laborer	21.45	85.80	33.75	135.00		
1 Diaphr. Water Pump, 2"		29.15		32.05		
1-20 Ft. Suction Hose, 2"		6.50		7.15		
2-50 Ft. Disch. Hoses, 2"		8.80		9.70	3.70	4.07
12 L.H., Daily Totals		$357.45		$529.50	$29.78	$44.12
Crew B-10I						
1 Equip. Oper. (med.)	$28.40	$227.20	$43.20	$345.60	$26.08	$40.05
.5 Laborer	21.45	85.80	33.75	135.00		
1 Diaphr. Water Pump, 4"		64.40		70.85		
1-20 Ft. Suction Hose, 4"		12.50		13.75		
2-50 Ft. Disch. Hoses, 4"		17.00		18.70	7.83	8.61
12 L.H., Daily Totals		$406.90		$583.90	$33.91	$48.66
Crew B-10J						
1 Equip. Oper. (med.)	$28.40	$227.20	$43.20	$345.60	$26.08	$40.05
.5 Laborer	21.45	85.80	33.75	135.00		
1 Centr. Water Pump, 3"		37.90		41.70		
1-20 Ft. Suction Hose, 3"		9.50		10.45		
2-50 Ft. Disch. Hoses, 3"		10.80		11.90	4.85	5.34
12 L.H., Daily Totals		$371.20		$544.65	$30.93	$45.39

Crew No.	Bare Costs		Incl. Subs O & P		Cost Per Labor-Hour	
Crew B-10K	Hr.	Daily	Hr.	Daily	Bare Costs	Incl. O&P
1 Equip. Oper. (med.)	$28.40	$227.20	$43.20	$345.60	$26.08	$40.05
.5 Laborer	21.45	85.80	33.75	135.00		
1 Centr. Water Pump, 6"		159.45		175.40		
1-20 Ft. Suction Hose, 6"		21.50		23.65		
2-50 Ft. Disch. Hoses, 6"		39.00		42.90	18.33	20.16
12 L.H., Daily Totals		$532.95		$722.55	$44.41	$60.21

Crew No.	Bare Costs		Incl. Subs O & P		Cost Per Labor-Hour	
Crew B-10L	Hr.	Daily	Hr.	Daily	Bare Costs	Incl. O&P
1 Equip. Oper. (med.)	$28.40	$227.20	$43.20	$345.60	$26.08	$40.05
.5 Laborer	21.45	85.80	33.75	135.00		
1 Dozer, 75 H.P.		307.75		338.50	25.65	28.21
12 L.H., Daily Totals		$620.75		$819.10	$51.73	$68.26

Crew No.	Bare Costs		Incl. Subs O & P		Cost Per Labor-Hour	
Crew B-10M	Hr.	Daily	Hr.	Daily	Bare Costs	Incl. O&P
1 Equip. Oper. (med.)	$28.40	$227.20	$43.20	$345.60	$26.08	$40.05
.5 Laborer	21.45	85.80	33.75	135.00		
1 Dozer, 300 H.P.		1131.00		1244.10	94.25	103.68
12 L.H., Daily Totals		$1444.00		$1724.70	$120.33	$143.73

Crew No.	Bare Costs		Incl. Subs O & P		Cost Per Labor-Hour	
Crew B-10N	Hr.	Daily	Hr.	Daily	Bare Costs	Incl. O&P
1 Equip. Oper. (med.)	$28.40	$227.20	$43.20	$345.60	$26.08	$40.05
.5 Laborer	21.45	85.80	33.75	135.00		
1 F.E. Loader, T.M., 1.5 C.Y.		350.10		385.10	29.18	32.09
12 L.H., Daily Totals		$663.10		$865.70	$55.26	$72.14

Crew No.	Bare Costs		Incl. Subs O & P		Cost Per Labor-Hour	
Crew B-10O	Hr.	Daily	Hr.	Daily	Bare Costs	Incl. O&P
1 Equip. Oper. (med.)	$28.40	$227.20	$43.20	$345.60	$26.08	$40.05
.5 Laborer	21.45	85.80	33.75	135.00		
1 F.E. Loader, T.M., 2.25 C.Y.		510.40		561.45	42.53	46.79
12 L.H., Daily Totals		$823.40		$1042.05	$68.61	$86.84

Crew No.	Bare Costs		Incl. Subs O & P		Cost Per Labor-Hour	
Crew B-10P	Hr.	Daily	Hr.	Daily	Bare Costs	Incl. O&P
1 Equip. Oper. (med.)	$28.40	$227.20	$43.20	$345.60	$26.08	$40.05
.5 Laborer	21.45	85.80	33.75	135.00		
1 F.E. Loader, T.M., 2.5 C.Y.		865.10		951.60	72.09	79.30
12 L.H., Daily Totals		$1178.10		$1432.20	$98.17	$119.35

Crew No.	Bare Costs		Incl. Subs O & P		Cost Per Labor-Hour	
Crew B-10Q	Hr.	Daily	Hr.	Daily	Bare Costs	Incl. O&P
1 Equip. Oper. (med.)	$28.40	$227.20	$43.20	$345.60	$26.08	$40.05
.5 Laborer	21.45	85.80	33.75	135.00		
1 F.E. Loader, T.M., 5 C.Y.		1142.00		1256.20	95.17	104.68
12 L.H., Daily Totals		$1455.00		$1736.80	$121.25	$144.73

Crew No.	Bare Costs		Incl. Subs O & P		Cost Per Labor-Hour	
Crew B-10R	Hr.	Daily	Hr.	Daily	Bare Costs	Incl. O&P
1 Equip. Oper. (med.)	$28.40	$227.20	$43.20	$345.60	$26.08	$40.05
.5 Laborer	21.45	85.80	33.75	135.00		
1 F.E. Loader, W.M., 1 C.Y.		244.65		269.10	20.39	22.43
12 L.H., Daily Totals		$557.65		$749.70	$46.47	$62.48

Crew No.	Bare Costs		Incl. Subs O & P		Cost Per Labor-Hour	
Crew B-10S	Hr.	Daily	Hr.	Daily	Bare Costs	Incl. O&P
1 Equip. Oper. (med.)	$28.40	$227.20	$43.20	$345.60	$26.08	$40.05
.5 Laborer	21.45	85.80	33.75	135.00		
1 F.E. Loader, W.M., 1.5 C.Y.		320.10		352.10	26.68	29.34
12 L.H., Daily Totals		$633.10		$832.70	$52.76	$69.39

Crew No.	Bare Costs		Incl. Subs O & P		Cost Per Labor-Hour	
Crew B-10T	Hr.	Daily	Hr.	Daily	Bare Costs	Incl. O&P
1 Equip. Oper. (med.)	$28.40	$227.20	$43.20	$345.60	$26.08	$40.05
.5 Laborer	21.45	85.80	33.75	135.00		
1 F.E. Ldr, W.M.,2.5CY		450.50		495.55	37.54	41.30
12 L.H., Daily Totals		$763.50		$976.15	$63.62	$81.35

Crew No.	Bare Costs		Incl. Subs O & P		Cost Per Labor-Hour	
Crew B-10U	Hr.	Daily	Hr.	Daily	Bare Costs	Incl. O&P
1 Equip. Oper. (med.)	$28.40	$227.20	$43.20	$345.60	$26.08	$40.05
.5 Laborer	21.45	85.80	33.75	135.00		
1 F.E. Loader, W.M., 5.5 C.Y.		926.50		1019.15	77.21	84.93
12 L.H., Daily Totals		$1239.50		$1499.75	$103.29	$124.98

Crew No.	Bare Costs		Incl. Subs O & P		Cost Per Labor-Hour	
Crew B-10V	Hr.	Daily	Hr.	Daily	Bare Costs	Incl. O&P
1 Equip. Oper. (med.)	$28.40	$227.20	$43.20	$345.60	$26.08	$40.05
.5 Laborer	21.45	85.80	33.75	135.00		
1 Dozer, 700 H.P.		2765.00		3041.50	230.42	253.46
12 L.H., Daily Totals		$3078.00		$3522.10	$256.50	$293.51

Crew No.	Bare Costs		Incl. Subs O & P		Cost Per Labor-Hour	
Crew B-10W	Hr.	Daily	Hr.	Daily	Bare Costs	Incl. O&P
1 Equip. Oper. (med.)	$28.40	$227.20	$43.20	$345.60	$26.08	$40.05
.5 Laborer	21.45	85.80	33.75	135.00		
1 Dozer, 105 H.P.		417.45		459.20	34.79	38.27
12 L.H., Daily Totals		$730.45		$939.80	$60.87	$78.32

Crew No.	Bare Costs		Incl. Subs O & P		Cost Per Labor-Hour	
Crew B-10X	Hr.	Daily	Hr.	Daily	Bare Costs	Incl. O&P
1 Equip. Oper. (med.)	$28.40	$227.20	$43.20	$345.60	$26.08	$40.05
.5 Laborer	21.45	85.80	33.75	135.00		
1 Dozer, 410 H.P.		1380.00		1518.00	115.00	126.50
12 L.H., Daily Totals		$1693.00		$1998.60	$141.08	$166.55

Crew No.	Bare Costs		Incl. Subs O & P		Cost Per Labor-Hour	
Crew B-10Y	Hr.	Daily	Hr.	Daily	Bare Costs	Incl. O&P
1 Equip. Oper. (med.)	$28.40	$227.20	$43.20	$345.60	$26.08	$40.05
.5 Laborer	21.45	85.80	33.75	135.00		
1 Vibratory Drum Roller		351.85		387.05	29.32	32.25
12 L.H., Daily Totals		$664.85		$867.65	$55.40	$72.30

Crew No.	Bare Costs		Incl. Subs O & P		Cost Per Labor-Hour	
Crew B-11	Hr.	Daily	Hr.	Daily	Bare Costs	Incl. O&P
1 Equipment Oper. (med.)	$28.40	$227.20	$43.20	$345.60	$24.93	$38.48
1 Laborer	21.45	171.60	33.75	270.00		
16 L.H., Daily Totals		$398.80		$615.60	$24.93	$38.48

Crew No.	Bare Costs		Incl. Subs O & P		Cost Per Labor-Hour	
Crew B-11A	Hr.	Daily	Hr.	Daily	Bare Costs	Incl. O&P
1 Equipment Oper. (med.)	$28.40	$227.20	$43.20	$345.60	$24.93	$38.48
1 Laborer	21.45	171.60	33.75	270.00		
1 Dozer, 200 H.P.		838.95		922.85	52.43	57.68
16 L.H., Daily Totals		$1237.75		$1538.45	$77.36	$96.16

Crew No.	Bare Costs		Incl. Subs O & P		Cost Per Labor-Hour	
Crew B-11B	Hr.	Daily	Hr.	Daily	Bare Costs	Incl. O&P
1 Equipment Oper. (med.)	$28.40	$227.20	$43.20	$345.60	$24.93	$38.48
1 Laborer	21.45	171.60	33.75	270.00		
1 Dozer, 200 H.P.		838.95		922.85		
1 Air Powered Tamper		14.80		16.30		
1 Air Compr. 365 C.F.M.		174.30		191.75		
2-50 Ft. Air Hoses, 1.5" Dia.		16.40		18.05	65.28	71.81
16 L.H., Daily Totals		$1443.25		$1764.55	$90.21	$110.29

Crew No.	Bare Costs		Incl. Subs O & P		Cost Per Labor-Hour	
Crew B-11C	Hr.	Daily	Hr.	Daily	Bare Costs	Incl. O&P
1 Equipment Oper. (med.)	$28.40	$227.20	$43.20	$345.60	$24.93	$38.48
1 Laborer	21.45	171.60	33.75	270.00		
1 Backhoe Loader, 48 H.P.		218.80		240.70	13.68	15.04
16 L.H., Daily Totals		$617.60		$856.30	$38.61	$53.52

Crew No.	Bare Costs		Incl. Subs O & P		Cost Per Labor-Hour	
Crew B-11K	Hr.	Daily	Hr.	Daily	Bare Costs	Incl. O&P
1 Equipment Oper. (med.)	$28.40	$227.20	$43.20	$345.60	$24.93	$38.48
1 Laborer	21.45	171.60	33.75	270.00		
1 Trencher, 8' D., 16" W.		535.90		589.50	33.49	36.84
16 L.H., Daily Totals		$934.70		$1205.10	$58.42	$75.32

Crews

Crew No.	Bare Costs Hr.	Daily	Incl. Subs O & P Hr.	Daily	Cost Per Labor-Hour Bare Costs	Incl. O&P
Crew B-11L	Hr.	Daily	Hr.	Daily	Bare Costs	Incl. O&P
1 Equipment Oper. (med.)	$28.40	$227.20	$43.20	$345.60	$24.93	$38.48
1 Laborer	21.45	171.60	33.75	270.00		
1 Grader, 30,000 Lbs.		575.55		633.10	35.97	39.57
16 L.H., Daily Totals		$974.35		$1248.70	$60.90	$78.05
Crew B-11M	Hr.	Daily	Hr.	Daily	Bare Costs	Incl. O&P
1 Equipment Oper. (med.)	$28.40	$227.20	$43.20	$345.60	$24.93	$38.48
1 Laborer	21.45	171.60	33.75	270.00		
1 Backhoe Loader, 80 H.P.		290.10		319.10	18.13	19.94
16 L.H., Daily Totals		$688.90		$934.70	$43.06	$58.42
Crew B-11N	Hr.	Daily	Hr.	Daily	Bare Costs	Incl. O&P
1 Labor Foreman	$23.45	$187.60	$36.90	$295.20	$23.65	$36.27
2 Equipment Operators (med.)	28.40	454.40	43.20	691.20		
6 Truck Drivers (hvy.)	22.10	1060.80	33.85	1624.80		
1 F.E. Loader, 5.5 C.Y.		926.50		1019.15		
1 Dozer, 400 H.P.		1380.00		1518.00		
6 Off Hwy. Tks. 50 Ton		8400.00		9240.00	148.70	163.57
72 L.H., Daily Totals		$12409.30		$14388.35	$172.35	$199.84
Crew B-11Q	Hr.	Daily	Hr.	Daily	Bare Costs	Incl. O&P
1 Equipment Operator (med.)	$28.40	$227.20	$43.20	$345.60	$26.08	$40.05
.5 Laborer	21.45	85.80	33.75	135.00		
1 Dozer, 140 H.P.		506.00		556.60	42.17	46.38
12 L.H., Daily Totals		$819.00		$1037.20	$68.25	$86.43
Crew B-11R	Hr.	Daily	Hr.	Daily	Bare Costs	Incl. O&P
1 Equipment Operator (med.)	$28.40	$227.20	$43.20	$345.60	$26.08	$40.05
.5 Laborer	21.45	85.80	33.75	135.00		
1 Dozer, 215 H.P.		838.95		922.85	69.91	76.90
12 L.H., Daily Totals		$1151.95		$1403.45	$95.99	$116.95
Crew B-11S	Hr.	Daily	Hr.	Daily	Bare Costs	Incl. O&P
1 Equipment Operator	$28.40	$227.20	$43.20	$345.60	$26.08	$40.05
.5 Laborer	21.45	85.80	33.75	135.00		
1 Dozer, 285 H.P.		1131.00		1244.10	94.25	103.68
12 L.H., Daily Totals		$1444.00		$1724.70	$120.33	$143.73
Crew B-11T	Hr.	Daily	Hr.	Daily	Bare Costs	Incl. O&P
1 Equipment Operator (med.)	$28.40	$227.20	$43.20	$345.60	$26.08	$40.05
.5 Laborer	21.45	85.80	33.75	135.00		
1 Dozer, 370 H.P.		1380.00		1518.00	115.00	126.50
12 L.H., Daily Totals		$1693.00		$1998.60	$141.08	$166.55
Crew B-11U	Hr.	Daily	Hr.	Daily	Bare Costs	Incl. O&P
1 Equipment Operator (med.)	$28.40	$227.20	$43.20	$345.60	$26.08	$40.05
.5 Laborer	21.45	85.80	33.75	135.00		
1 Dozer, 520 H.P.		1720.00		1892.00	143.33	157.67
12 L.H., Daily Totals		$2033.00		$2372.60	$169.41	$197.72
Crew B-11V	Hr.	Daily	Hr.	Daily	Bare Costs	Incl. O&P
3 Laborer	$21.45	$514.80	$33.75	$810.00	$21.45	$33.75
1 Roll. Compact., 2K Lbs.		93.05		102.35	3.88	4.26
24 L.H., Daily Totals		$607.85		$912.35	$25.33	$38.01
Crew B-12	Hr.	Daily	Hr.	Daily	Bare Costs	Incl. O&P
1 Equip. Oper. (crane)	$29.35	$234.80	$44.65	$357.20	$26.70	$40.63
1 Equip. Oper. Oiler	24.05	192.40	36.60	292.80		
16 L.H., Daily Totals		$427.20		$650.00	$26.70	$40.63

Crew No.	Bare Costs Hr.	Daily	Incl. Subs O & P Hr.	Daily	Cost Per Labor-Hour Bare Costs	Incl. O&P
Crew B-12A	Hr.	Daily	Hr.	Daily	Bare Costs	Incl. O&P
1 Equip. Oper. (crane)	$29.35	$234.80	$44.65	$357.20	$26.70	$40.63
1 Equip. Oper. Oiler	24.05	192.40	36.60	292.80		
1 Hyd. Excavator, 1 C.Y.		547.60		602.35	34.23	37.65
16 L.H., Daily Totals		$974.80		$1252.35	$60.93	$78.28
Crew B-12B	Hr.	Daily	Hr.	Daily	Bare Costs	Incl. O&P
1 Equip. Oper. (crane)	$29.35	$234.80	$44.65	$357.20	$26.70	$40.63
1 Equip. Oper. Oiler	24.05	192.40	36.60	292.80		
1 Hyd. Excavator, 1.5 C.Y.		712.70		783.95	44.54	49.00
16 L.H., Daily Totals		$1139.90		$1433.95	$71.24	$89.63
Crew B-12C	Hr.	Daily	Hr.	Daily	Bare Costs	Incl. O&P
1 Equip. Oper. (crane)	$29.35	$234.80	$44.65	$357.20	$26.70	$40.63
1 Equip. Oper. Oiler	24.05	192.40	36.60	292.80		
1 Hyd. Excavator, 2 C.Y.		1023.00		1125.30	63.94	70.33
16 L.H., Daily Totals		$1450.20		$1775.30	$90.64	$110.96
Crew B-12D	Hr.	Daily	Hr.	Daily	Bare Costs	Incl. O&P
1 Equip. Oper. (crane)	$29.35	$234.80	$44.65	$357.20	$26.70	$40.63
1 Equip. Oper. Oiler	24.05	192.40	36.60	292.80		
1 Hyd. Excavator, 3.5 C.Y.		2156.00		2371.60	134.75	148.23
16 L.H., Daily Totals		$2583.20		$3021.60	$161.45	$188.86
Crew B-12E	Hr.	Daily	Hr.	Daily	Bare Costs	Incl. O&P
1 Equip. Oper. (crane)	$29.35	$234.80	$44.65	$357.20	$26.70	$40.63
1 Equip. Oper. Oiler	24.05	192.40	36.60	292.80		
1 Hyd. Excavator, .5 C.Y.		336.45		370.10	21.03	23.13
16 L.H., Daily Totals		$763.65		$1020.10	$47.73	$63.76
Crew B-12F	Hr.	Daily	Hr.	Daily	Bare Costs	Incl. O&P
1 Equip. Oper. (crane)	$29.35	$234.80	$44.65	$357.20	$26.70	$40.63
1 Equip. Oper. Oiler	24.05	192.40	36.60	292.80		
1 Hyd. Excavator, .75 C.Y.		449.10		494.00	28.07	30.88
16 L.H., Daily Totals		$876.30		$1144.00	$54.77	$71.51
Crew B-12G	Hr.	Daily	Hr.	Daily	Bare Costs	Incl. O&P
1 Equip. Oper. (crane)	$29.35	$234.80	$44.65	$357.20	$26.70	$40.63
1 Equip. Oper. Oiler	24.05	192.40	36.60	292.80		
1 Power Shovel, .5 C.Y.		472.00		519.20		
1 Clamshell Bucket, .5 C.Y		49.90		54.90	32.62	35.88
16 L.H., Daily Totals		$949.10		$1224.10	$59.32	$76.51
Crew B-12H	Hr.	Daily	Hr.	Daily	Bare Costs	Incl. O&P
1 Equip. Oper. (crane)	$29.35	$234.80	$44.65	$357.20	$26.70	$40.63
1 Equip. Oper. Oiler	24.05	192.40	36.60	292.80		
1 Power Shovel, 1 C.Y.		506.80		557.50		
1 Clamshell Bucket, 1 C.Y.		67.10		73.80	35.87	39.46
16 L.H., Daily Totals		$1001.10		$1281.30	$62.57	$80.09
Crew B-12I	Hr.	Daily	Hr.	Daily	Bare Costs	Incl. O&P
1 Equip. Oper. (crane)	$29.35	$234.80	$44.65	$357.20	$26.70	$40.63
1 Equip. Oper. Oiler	24.05	192.40	36.60	292.80		
1 Power Shovel, .75 C.Y.		491.60		540.75		
1 Dragline Bucket, .75 C.Y.		33.55		36.90	32.82	36.10
16 L.H., Daily Totals		$952.35		$1227.65	$59.52	$76.73

CREWS

583

Crew No.	Bare Costs		Incl. Subs O & P		Cost Per Labor-Hour	
	Hr.	Daily	Hr.	Daily	Bare Costs	Incl. O&P
Crew B-12J					Bare Costs	Incl. O&P
1 Equip. Oper. (crane)	$29.35	$234.80	$44.65	$357.20	$26.70	$40.63
1 Equip. Oper. Oiler	24.05	192.40	36.60	292.80		
1 Gradall, 3 Ton, .5 C.Y.		636.75		700.40	39.80	43.78
16 L.H., Daily Totals		$1063.95		$1350.40	$66.50	$84.41
Crew B-12K	Hr.	Daily	Hr.	Daily	Bare Costs	Incl. O&P
1 Equip. Oper. (crane)	$29.35	$234.80	$44.65	$357.20	$26.70	$40.63
1 Equip. Oper. Oiler	24.05	192.40	36.60	292.80		
1 Gradall, 3 Ton, 1 C.Y.		805.75		886.35	50.36	55.40
16 L.H., Daily Totals		$1232.95		$1536.35	$77.06	$96.03
Crew B-12L	Hr.	Daily	Hr.	Daily	Bare Costs	Incl. O&P
1 Equip. Oper. (crane)	$29.35	$234.80	$44.65	$357.20	$26.70	$40.63
1 Equip. Oper. Oiler	24.05	192.40	36.60	292.80		
1 Power Shovel, .5 C.Y.		472.00		519.20		
1 F.E. Attachment, .5 C.Y.		55.05		60.55	32.94	36.23
16 L.H., Daily Totals		$954.25		$1229.75	$59.64	$76.86
Crew B-12M	Hr.	Daily	Hr.	Daily	Bare Costs	Incl. O&P
1 Equip. Oper. (crane)	$29.35	$234.80	$44.65	$357.20	$26.70	$40.63
1 Equip. Oper. Oiler	24.05	192.40	36.60	292.80		
1 Power Shovel, .75 C.Y		491.60		540.75		
1 F.E. Attachment, .75 C.Y.		102.50		112.75	37.13	40.84
16 L.H., Daily Totals		$1021.30		$1303.50	$63.83	$81.47
Crew B-12N	Hr.	Daily	Hr.	Daily	Bare Costs	Incl. O&P
1 Equip. Oper. (crane)	$29.35	$234.80	$44.65	$357.20	$26.70	$40.63
1 Equip. Oper. Oiler	24.05	192.40	36.60	292.80		
1 Power Shovel, 1 C.Y.		506.80		557.50		
1 F.E. Attachment, 1 C.Y.		131.00		144.10	39.86	43.85
16 L.H., Daily Totals		$1065.00		$1351.60	$66.56	$84.48
Crew B-12O	Hr.	Daily	Hr.	Daily	Bare Costs	Incl. O&P
1 Equip. Oper. (crane)	$29.35	$234.80	$44.65	$357.20	$26.70	$40.63
1 Equip. Oper. Oiler	24.05	192.40	36.60	292.80		
1 Power Shovel, 1.5 C.Y.		704.30		774.75		
1 F.E. Attachment, 1.5 C.Y.		159.70		175.65	54.00	59.40
16 L.H., Daily Totals		$1291.20		$1600.40	$80.70	$100.03
Crew B-12P	Hr.	Daily	Hr.	Daily	Bare Costs	Incl. O&P
1 Equip. Oper. (crane)	$29.35	$234.80	$44.65	$357.20	$26.70	$40.63
1 Equip. Oper. Oiler	24.05	192.40	36.60	292.80		
1 Crawler Crane, 40 Ton		704.30		774.75		
1 Dragline Bucket, 1.5 C.Y.		48.15		52.95	47.03	51.73
16 L.H., Daily Totals		$1179.65		$1477.70	$73.73	$92.36
Crew B-12Q	Hr.	Daily	Hr.	Daily	Bare Costs	Incl. O&P
1 Equip. Oper. (crane)	$29.35	$234.80	$44.65	$357.20	$26.70	$40.63
1 Equip. Oper. Oiler	24.05	192.40	36.60	292.80		
1 Hyd. Excavator, 5/8 C.Y.		394.25		433.70	24.64	27.10
16 L.H., Daily Totals		$821.45		$1083.70	$51.34	$67.73
Crew B-12R	Hr.	Daily	Hr.	Daily	Bare Costs	Incl. O&P
1 Equip. Oper. (crane)	$29.35	$234.80	$44.65	$357.20	$26.70	$40.63
1 Equip. Oper. Oiler	24.05	192.40	36.60	292.80		
1 Hyd. Excavator, 1.5 C.Y.		712.70		783.95	44.54	49.00
16 L.H., Daily Totals		$1139.90		$1433.95	$71.24	$89.63

Crew No.	Bare Costs		Incl. Subs O & P		Cost Per Labor-Hour	
	Hr.	Daily	Hr.	Daily	Bare Costs	Incl. O&P
Crew B-12S					Bare Costs	Incl. O&P
1 Equip. Oper. (crane)	$29.35	$234.80	$44.65	$357.20	$26.70	$40.63
1 Equip. Oper. Oiler	24.05	192.40	36.60	292.80		
1 Hyd. Excavator, 2.5 C.Y.		1695.00		1864.50	105.94	116.53
16 L.H., Daily Totals		$2122.20		$2514.50	$132.64	$157.16
Crew B-12T	Hr.	Daily	Hr.	Daily	Bare Costs	Incl. O&P
1 Equip. Oper. (crane)	$29.35	$234.80	$44.65	$357.20	$26.70	$40.63
1 Equip. Oper. Oiler	24.05	192.40	36.60	292.80		
1 Crawler Crane, 75 Ton		929.30		1022.25		
1 F.E. Attachment, 3 C.Y.		300.15		330.15	76.84	84.52
16 L.H., Daily Totals		$1656.65		$2002.40	$103.54	$125.15
Crew B-12V	Hr.	Daily	Hr.	Daily	Bare Costs	Incl. O&P
1 Equip. Oper. (crane)	$29.35	$234.80	$44.65	$357.20	$26.70	$40.63
1 Equip. Oper. Oiler	24.05	192.40	36.60	292.80		
1 Crawler Crane, 75 Ton		929.30		1022.25		
1 Dragline Bucket, 3 C.Y.		82.70		90.95	63.25	69.58
16 L.H., Daily Totals		$1439.20		$1763.20	$89.95	$110.21
Crew B-13	Hr.	Daily	Hr.	Daily	Bare Costs	Incl. O&P
1 Labor Foreman (outside)	$23.45	$187.60	$36.90	$295.20	$23.24	$36.16
4 Laborers	21.45	686.40	33.75	1080.00		
1 Equip. Oper. (crane)	29.35	234.80	44.65	357.20		
1 Equip. Oper. Oiler	24.05	192.40	36.60	292.80		
1 Hyd. Crane, 25 Ton		549.35		604.30	9.81	10.79
56 L.H., Daily Totals		$1850.55		$2629.50	$33.05	$46.95
Crew B-13A	Hr.	Daily	Hr.	Daily	Bare Costs	Incl. O&P
1 Foreman	$23.45	$187.60	$36.90	$295.20	$23.91	$36.93
2 Laborers	21.45	343.20	33.75	540.00		
2 Equipment Operator	28.40	454.40	43.20	691.20		
2 Truck Drivers (heavy)	22.10	353.60	33.85	541.60		
1 Crane, 75 Ton		929.30		1022.25		
1 F.E. Lder, 3.75 C.Y.		1142.00		1256.20		
2 Dump Trucks, 12 Ton		728.60		801.45	50.00	55.00
56 L.H., Daily Totals		$4138.70		$5147.90	$73.91	$91.93
Crew B-13B	Hr.	Daily	Hr.	Daily	Bare Costs	Incl. O&P
1 Labor Foreman (outside)	$23.45	$187.60	$36.90	$295.20	$23.24	$36.16
4 Laborers	21.45	686.40	33.75	1080.00		
1 Equip. Oper. (crane)	29.35	234.80	44.65	357.20		
1 Equip. Oper. Oiler	24.05	192.40	36.60	292.80		
1 Hyd. Crane, 55 Ton		793.10		872.40	14.16	15.58
56 L.H., Daily Totals		$2094.30		$2897.60	$37.40	$51.74
Crew B-13C	Hr.	Daily	Hr.	Daily	Bare Costs	Incl. O&P
1 Labor Foreman (outside)	$23.45	$187.60	$36.90	$295.20	$23.24	$36.16
4 Laborers	21.45	686.40	33.75	1080.00		
1 Equip. Oper. (crane)	29.35	234.80	44.65	357.20		
1 Equip. Oper. Oiler	24.05	192.40	36.60	292.80		
1 Crawler Crane, 100 Ton		1131.00		1244.10	20.20	22.22
56 L.H., Daily Totals		$2432.20		$3269.30	$43.44	$58.38
Crew B-14	Hr.	Daily	Hr.	Daily	Bare Costs	Incl. O&P
1 Labor Foreman (outside)	$23.45	$187.60	$36.90	$295.20	$22.74	$35.54
4 Laborers	21.45	686.40	33.75	1080.00		
1 Equip. Oper. (light)	27.20	217.60	41.35	330.80		
1 Backhoe Loader, 48 H.P.		218.80		240.70	4.56	5.01
48 L.H., Daily Totals		$1310.40		$1946.70	$27.30	$40.55

CREWS

Crew B-15

Crew No.	Bare Costs Hr.	Daily	Incl. Subs O & P Hr.	Daily	Cost Per Labor-Hour Bare Costs	Incl. O&P
1 Equipment Oper. (med)	$28.40	$227.20	$43.20	$345.60	$23.81	$36.51
.5 Laborer	21.45	85.80	33.75	135.00		
2 Truck Drivers (heavy)	22.10	353.60	33.85	541.60		
2 Dump Trucks, 16 Ton		884.50		972.95		
1 Dozer, 200 H.P.		838.95		922.85	61.55	67.71
28 L.H., Daily Totals		$2390.05		$2918.00	$85.36	$104.22

Crew B-16

Crew No.	Bare Costs Hr.	Daily	Incl. Subs O & P Hr.	Daily	Cost Per Labor-Hour Bare Costs	Incl. O&P
1 Labor Foreman (outside)	$23.45	$187.60	$36.90	$295.20	$22.11	$34.56
2 Laborers	21.45	343.20	33.75	540.00		
1 Truck Driver (heavy)	22.10	176.80	33.85	270.80		
1 Dump Truck, 16 Ton		442.25		486.50	13.82	15.20
32 L.H., Daily Totals		$1149.85		$1592.50	$35.93	$49.76

Crew B-17

Crew No.	Bare Costs Hr.	Daily	Incl. Subs O & P Hr.	Daily	Cost Per Labor-Hour Bare Costs	Incl. O&P
2 Laborers	$21.45	$343.20	$33.75	$540.00	$23.05	$35.67
1 Equip. Oper. (light)	27.20	217.60	41.35	330.80		
1 Truck Driver (heavy)	22.10	176.80	33.85	270.80		
1 Backhoe Loader, 48 H.P.		218.80		240.70		
1 Dump Truck, 12 Ton		364.30		400.75	18.22	20.04
32 L.H., Daily Totals		$1320.70		$1783.05	$41.27	$55.71

Crew B-18

Crew No.	Bare Costs Hr.	Daily	Incl. Subs O & P Hr.	Daily	Cost Per Labor-Hour Bare Costs	Incl. O&P
1 Labor Foreman (outside)	$23.45	$187.60	$36.90	$295.20	$22.12	$34.80
2 Laborers	21.45	343.20	33.75	540.00		
1 Vibrating Compactor		52.00		57.20	2.17	2.38
24 L.H., Daily Totals		$582.80		$892.40	$24.29	$37.18

Crew B-19

Crew No.	Bare Costs Hr.	Daily	Incl. Subs O & P Hr.	Daily	Cost Per Labor-Hour Bare Costs	Incl. O&P
1 Pile Driver Foreman	$29.20	$233.60	$50.45	$403.60	$27.59	$45.54
4 Pile Drivers	27.20	870.40	47.00	1504.00		
2 Equip. Oper. (crane)	29.35	469.60	44.65	714.40		
1 Equip. Oper. Oiler	24.05	192.40	36.60	292.80		
1 Crane, 40 Ton & Access.		704.30		774.75		
60 L.F. Leads, 15K Ft. Lbs.		105.00		115.50		
1 Hammer, 15K Ft. Lbs.		321.75		353.95		
1 Air Compr., 600 C.F.M.		272.90		300.20		
2-50 Ft. Air Hoses, 3" Dia.		33.70		37.05	22.46	24.71
64 L.H., Daily Totals		$3203.65		$4496.25	$50.05	$70.25

Crew B-19A

Crew No.	Bare Costs Hr.	Daily	Incl. Subs O & P Hr.	Daily	Cost Per Labor-Hour Bare Costs	Incl. O&P
1 Pile Driver Foreman	$29.20	$233.60	$50.45	$403.60	$27.59	$45.54
4 Pile Drivers	27.20	870.40	47.00	1504.00		
2 Equip. Oper. (crane)	29.35	469.60	44.65	714.40		
1 Equip. Oper. Oiler	24.05	192.40	36.60	292.80		
1 Crawler Crane, 75 Ton		929.30		1022.25		
60 Leads, 25K ft.lbs		126.00		138.60		
1 Air Compressor, 750 CFM		288.55		317.40		
4-50 Ft. Air Hose, 3" Dia.		67.40		74.15	22.05	24.26
64 L.H., Daily Totals		$3177.25		$4467.20	$49.64	$69.80

Crew B-20

Crew No.	Bare Costs Hr.	Daily	Incl. Subs O & P Hr.	Daily	Cost Per Labor-Hour Bare Costs	Incl. O&P
1 Labor Foreman (out)	$23.45	$187.60	$36.90	$295.20	$24.32	$38.30
1 Skilled Worker	28.05	224.40	44.25	354.00		
1 Laborer	21.45	171.60	33.75	270.00		
24 L.H., Daily Totals		$583.60		$919.20	$24.32	$38.30

Crew B-20A

Crew No.	Bare Costs Hr.	Daily	Incl. Subs O & P Hr.	Daily	Cost Per Labor-Hour Bare Costs	Incl. O&P
1 Labor Foreman	$23.45	$187.60	$36.90	$295.20	$25.90	$39.90
1 Laborer	21.45	171.60	33.75	270.00		
1 Plumber	32.60	260.80	49.40	395.20		
1 Plumber Apprentice	26.10	208.80	39.55	316.40		
32 L.H., Daily Totals		$828.80		$1276.80	$25.90	$39.90

Crew B-21

Crew No.	Bare Costs Hr.	Daily	Incl. Subs O & P Hr.	Daily	Cost Per Labor-Hour Bare Costs	Incl. O&P
1 Labor Foreman (out)	$23.45	$187.60	$36.90	$295.20	$25.04	$39.21
1 Skilled Worker	28.05	224.40	44.25	354.00		
1 Laborer	21.45	171.60	33.75	270.00		
.5 Equip. Oper. (crane)	29.35	117.40	44.65	178.60		
.5 S.P. Crane, 5 Ton		135.73		149.30	4.85	5.33
28 L.H., Daily Totals		$836.73		$1247.10	$29.89	$44.54

Crew B-21A

Crew No.	Bare Costs Hr.	Daily	Incl. Subs O & P Hr.	Daily	Cost Per Labor-Hour Bare Costs	Incl. O&P
1 Labor Foreman	$23.45	$187.60	$36.90	$295.20	$26.59	$40.85
1 Laborer	21.45	171.60	33.75	270.00		
1 Plumber	32.60	260.80	49.40	395.20		
1 Plumber Apprentice	26.10	208.80	39.55	316.40		
1 Equip. Oper. (crane)	29.35	234.80	44.65	357.20		
1 S.P. Crane, 12 Ton		391.35		430.50	9.78	10.76
40 L.H., Daily Totals		$1454.95		$2064.50	$36.37	$51.61

Crew B-22

Crew No.	Bare Costs Hr.	Daily	Incl. Subs O & P Hr.	Daily	Cost Per Labor-Hour Bare Costs	Incl. O&P
1 Labor Foreman (out)	$23.45	$187.60	$36.90	$295.20	$25.32	$39.57
1 Skilled Worker	28.05	224.40	44.25	354.00		
1 Laborer	21.45	171.60	33.75	270.00		
.75 Equip. Oper. (crane)	29.35	176.10	44.65	267.90		
.75 S.P. Crane, 5 Ton		203.59		223.95	6.79	7.46
30 L.H., Daily Totals		$963.29		$1411.05	$32.11	$47.03

Crew B-22A

Crew No.	Bare Costs Hr.	Daily	Incl. Subs O & P Hr.	Daily	Cost Per Labor-Hour Bare Costs	Incl. O&P
1 Labor Foreman (out)	$23.45	$187.60	$36.90	$295.20	$24.51	$38.34
1 Skilled Worker	28.05	224.40	44.25	354.00		
2 Laborers	21.45	343.20	33.75	540.00		
.75 Equipment Oper. (crane)	29.35	176.10	44.65	267.90		
.75 Crane, 5 Ton		203.59		223.95		
1 Generator, 5 KW		45.90		50.50		
1 Butt Fusion Machine		.45		.50	6.58	7.24
38 L.H., Daily Totals		$1181.24		$1732.05	$31.09	$45.58

Crew B-22B

Crew No.	Bare Costs Hr.	Daily	Incl. Subs O & P Hr.	Daily	Cost Per Labor-Hour Bare Costs	Incl. O&P
1 Skilled Worker	$28.05	$224.40	$44.25	$354.00	$24.75	$39.00
1 Laborer	21.45	171.60	33.75	270.00		
1 Electro Fusion Machine		.45		.50	.03	.03
16 L.H., Daily Totals		$396.45		$624.50	$24.78	$39.03

Crew B-23

Crew No.	Bare Costs Hr.	Daily	Incl. Subs O & P Hr.	Daily	Cost Per Labor-Hour Bare Costs	Incl. O&P
1 Labor Foreman (outside)	$23.45	$187.60	$36.90	$295.20	$21.85	$34.38
4 Laborers	21.45	686.40	33.75	1080.00		
1 Drill Rig, Wells		1776.00		1953.60		
1 Light Truck, 3 Ton		179.45		197.40	48.89	53.77
40 L.H., Daily Totals		$2829.45		$3526.20	$70.74	$88.15

Crew B-23A

Crew No.	Bare Costs Hr.	Daily	Incl. Subs O & P Hr.	Daily	Cost Per Labor-Hour Bare Costs	Incl. O&P
1 Labor Foreman (outside)	$23.45	$187.60	$36.90	$295.20	$24.43	$37.95
1 Laborers	21.45	171.60	33.75	270.00		
1 Equip. Operator, medium	28.40	227.20	43.20	345.60		
1 Drill Rig, Wells		1776.00		1953.60		
1 Pickup Truck, 3/4 Ton		128.95		141.85	79.37	87.31
24 L.H., Daily Totals		$2491.35		$3006.25	$103.80	$125.26

Crew B-23B

Crew No.	Bare Costs Hr.	Daily	Incl. Subs O & P Hr.	Daily	Cost Per Labor-Hour Bare Costs	Incl. O&P
1 Labor Foreman (outside)	$23.45	$187.60	$36.90	$295.20	$24.43	$37.95
1 Laborers	21.45	171.60	33.75	270.00		
1 Equip. Operator, medium	28.40	227.20	43.20	345.60		
1 Drill Rig, Wells		1776.00		1953.60		
1 Pickup Truck, 3/4 Ton		128.95		141.85		
1 Pump, Cntfgl, 6"		159.45		175.40	86.02	94.62
24 L.H., Daily Totals		$2650.80		$3181.65	$110.45	$132.57

Crew B-24

Crew No.	Bare Costs Hr.	Daily	Incl. Subs O & P Hr.	Daily	Cost Per Labor-Hour Bare Costs	Incl. O&P
1 Cement Finisher	$26.15	$209.20	$39.00	$312.00	$24.97	$38.57
1 Laborer	21.45	171.60	33.75	270.00		
1 Carpenter	27.30	218.40	42.95	343.60		
24 L.H., Daily Totals		$599.20		$925.60	$24.97	$38.57

Crew B-25

Crew No.	Bare Costs Hr.	Daily	Incl. Subs O & P Hr.	Daily	Cost Per Labor-Hour Bare Costs	Incl. O&P
1 Labor Foreman	$23.45	$187.60	$36.90	$295.20	$23.53	$36.61
7 Laborers	21.45	1201.20	33.75	1890.00		
3 Equip. Oper. (med.)	28.40	681.60	43.20	1036.80		
1 Asphalt Paver, 130 H.P		1299.00		1428.90		
1 Tandem Roller, 10 Ton		245.10		269.60		
1 Roller, Pneumatic Wheel		244.20		268.60	20.32	22.35
88 L.H., Daily Totals		$3858.70		$5189.10	$43.85	$58.96

Crew B-25B

Crew No.	Bare Costs Hr.	Daily	Incl. Subs O & P Hr.	Daily	Cost Per Labor-Hour Bare Costs	Incl. O&P
1 Labor Foreman	$23.45	$187.60	$36.90	$295.20	$23.93	$37.16
7 Laborers	21.45	1201.20	33.75	1890.00		
4 Equip. Oper. (medium)	28.40	908.80	43.20	1382.40		
1 Asphalt Paver, 130 H.P.		1299.00		1428.90		
2 Rollers, Steel Wheel		490.20		539.20		
1 Roller, Pneumatic Wheel		244.20		268.60	21.18	23.30
96 L.H., Daily Totals		$4331.00		$5804.30	$45.11	$60.46

Crew B-25C

Crew No.	Bare Costs Hr.	Daily	Incl. Subs O & P Hr.	Daily	Cost Per Labor-Hour Bare Costs	Incl. O&P
1 Labor Foreman	$23.45	$187.60	$36.90	$295.20	$24.10	$37.42
3 Laborers	21.45	514.80	33.75	810.00		
2 Equip. Oper. (medium)	28.40	454.40	43.20	691.20		
1 Asphalt Paver, 130 H.P.		1299.00		1428.90		
1 Rollers, Steel Wheel		245.10		269.60	32.17	35.39
48 L.H., Daily Totals		$2700.90		$3494.90	$56.27	$72.81

Crew B-26

Crew No.	Bare Costs Hr.	Daily	Incl. Subs O & P Hr.	Daily	Cost Per Labor-Hour Bare Costs	Incl. O&P
1 Labor Foreman (outside)	$23.45	$187.60	$36.90	$295.20	$24.14	$37.95
6 Laborers	21.45	1029.60	33.75	1620.00		
2 Equip. Oper. (med.)	28.40	454.40	43.20	691.20		
1 Rodman (reinf.)	30.40	243.20	52.60	420.80		
1 Cement Finisher	26.15	209.20	39.00	312.00		
1 Grader, 30,000 Lbs.		575.55		633.10		
1 Paving Mach. & Equip.		1321.00		1453.10	21.55	23.71
88 L.H., Daily Totals		$4020.55		$5425.40	$45.69	$61.66

Crew B-27

Crew No.	Bare Costs Hr.	Daily	Incl. Subs O & P Hr.	Daily	Cost Per Labor-Hour Bare Costs	Incl. O&P
1 Labor Foreman (outside)	$23.45	$187.60	$36.90	$295.20	$21.95	$34.54
3 Laborers	21.45	514.80	33.75	810.00		
1 Berm Machine		70.30		77.35	2.20	2.42
32 L.H., Daily Totals		$772.70		$1182.55	$24.15	$36.96

Crew B-28

Crew No.	Bare Costs Hr.	Daily	Incl. Subs O & P Hr.	Daily	Cost Per Labor-Hour Bare Costs	Incl. O&P
2 Carpenters	$27.30	$436.80	$42.95	$687.20	$25.35	$39.88
1 Laborer	21.45	171.60	33.75	270.00		
24 L.H., Daily Totals		$608.40		$957.20	$25.35	$39.88

Crew B-29

Crew No.	Bare Costs Hr.	Daily	Incl. Subs O & P Hr.	Daily	Cost Per Labor-Hour Bare Costs	Incl. O&P
1 Labor Foreman (outside)	$23.45	$187.60	$36.90	$295.20	$23.24	$36.16
4 Laborers	21.45	686.40	33.75	1080.00		
1 Equip. Oper. (crane)	29.35	234.80	44.65	357.20		
1 Equip. Oper. Oiler	24.05	192.40	36.60	292.80		
1 Gradall, 3 Ton, 1/2 C.Y.		636.75		700.40	11.37	12.51
56 L.H., Daily Totals		$1937.95		$2725.60	$34.61	$48.67

Crew B-30

Crew No.	Bare Costs Hr.	Daily	Incl. Subs O & P Hr.	Daily	Cost Per Labor-Hour Bare Costs	Incl. O&P
1 Equip. Oper. (med.)	$28.40	$227.20	$43.20	$345.60	$24.20	$36.97
2 Truck Drivers (heavy)	22.10	353.60	33.85	541.60		
1 Hyd. Excavator, 1.5 C.Y.		712.70		783.95		
2 Dump Trucks, 16 Ton		884.50		972.95	66.55	73.21
24 L.H., Daily Totals		$2178.00		$2644.10	$90.75	$110.18

Crew B-31

Crew No.	Bare Costs Hr.	Daily	Incl. Subs O & P Hr.	Daily	Cost Per Labor-Hour Bare Costs	Incl. O&P
1 Labor Foreman (outside)	$23.45	$187.60	$36.90	$295.20	$23.02	$36.22
3 Laborers	21.45	514.80	33.75	810.00		
1 Carpenter	27.30	218.40	42.95	343.60		
1 Air Compr., 250 C.F.M.		124.90		137.40		
1 Sheeting Driver		11.20		12.30		
2-50 Ft. Air Hoses, 1.5" Dia.		16.40		18.05	3.81	4.19
40 L.H., Daily Totals		$1073.30		$1616.55	$26.83	$40.41

Crew B-32

Crew No.	Bare Costs Hr.	Daily	Incl. Subs O & P Hr.	Daily	Cost Per Labor-Hour Bare Costs	Incl. O&P
1 Laborer	$21.45	$171.60	$33.75	$270.00	$26.66	$40.84
3 Equip. Oper. (med.)	28.40	681.60	43.20	1036.80		
1 Grader, 30,000 Lbs.		575.55		633.10		
1 Tandem Roller, 10 Ton		245.10		269.60		
1 Dozer, 200 H.P.		838.95		922.85	51.86	57.05
32 L.H., Daily Totals		$2512.80		$3132.35	$78.52	$97.89

Crew B-32A

Crew No.	Bare Costs Hr.	Daily	Incl. Subs O & P Hr.	Daily	Cost Per Labor-Hour Bare Costs	Incl. O&P
1 Laborer	$21.45	$171.60	$33.75	$270.00	$26.08	$40.05
2 Equip. Oper. (medium)	28.40	454.40	43.20	691.20		
1 Grader, 30,000 Lbs.		575.55		633.10		
1 Roller, Vibratory, 29,000 Lbs.		421.75		463.95	41.55	45.71
24 L.H., Daily Totals		$1623.30		$2058.25	$67.63	$85.76

Crew B-32B

Crew No.	Bare Costs Hr.	Daily	Incl. Subs O & P Hr.	Daily	Cost Per Labor-Hour Bare Costs	Incl. O&P
1 Laborer	$21.45	$171.60	$33.75	$270.00	$26.08	$40.05
2 Equip. Oper. (medium)	28.40	454.40	43.20	691.20		
1 Dozer, 200 H.P.		838.95		922.85		
1 Roller, Vibratory, 29,000 Lbs.		421.75		463.95	52.53	57.78
24 L.H., Daily Totals		$1886.70		$2348.00	$78.61	$97.83

Crew B-32C

Crew No.	Bare Costs Hr.	Daily	Incl. Subs O & P Hr.	Daily	Cost Per Labor-Hour Bare Costs	Incl. O&P
1 Labor Foreman	$23.45	$187.60	$36.90	$295.20	$25.26	$39.00
2 Laborers	21.45	343.20	33.75	540.00		
3 Equip. Oper. (medium)	28.40	681.60	43.20	1036.80		
1 Grader, 30,000 Lbs.		575.55		633.10		
1 Roller, Steel Wheel		245.10		269.60		
1 Dozer, 200 H.P.		838.95		922.85	34.58	38.03
48 L.H., Daily Totals		$2872.00		$3697.55	$59.84	$77.03

Crew B-33

Crew No.	Bare Costs Hr.	Daily	Incl. Subs O & P Hr.	Daily	Cost Per Labor-Hour Bare Costs	Incl. O&P
1 Equip. Oper. (med.)	$28.40	$227.20	$43.20	$345.60	$26.41	$40.50
.5 Laborer	21.45	85.80	33.75	135.00		
.25 Equip. Oper. (med.)	28.40	56.80	43.20	86.40		
14 L.H., Daily Totals		$369.80		$567.00	$26.41	$40.50

Crew No.	Bare Costs		Incl. Subs O & P		Cost Per Labor-Hour	
Crew B-33A	Hr.	Daily	Hr.	Daily	Bare Costs	Incl. O&P
1 Equip. Oper. (med.)	$28.40	$227.20	$43.20	$345.60	$26.41	$40.50
.5 Laborer	21.45	85.80	33.75	135.00		
.25 Equip. Oper. (med.)	28.40	56.80	43.20	86.40		
1 Scraper, Towed, 7 C.Y.		77.50		85.25		
1 Dozer, 300 H.P.		1131.00		1244.10		
.25 Dozer, 300 H.P.		282.75		311.00	106.52	117.17
14 L.H., Daily Totals		$1861.05		$2207.35	$132.93	$157.67

Crew No.	Bare Costs		Incl. Subs O & P		Cost Per Labor-Hour	
Crew B-33B	Hr.	Daily	Hr.	Daily	Bare Costs	Incl. O&P
1 Equip. Oper. (med.)	$28.40	$227.20	$43.20	$345.60	$26.41	$40.50
.5 Laborer	21.45	85.80	33.75	135.00		
.25 Equip. Oper. (med.)	28.40	56.80	43.20	86.40		
1 Scraper, Towed, 10 C.Y.		202.45		222.70		
1 Dozer, 300 H.P.		1131.00		1244.10		
.25 Dozer, 300 H.P.		282.75		311.00	115.44	126.99
14 L.H., Daily Totals		$1986.00		$2344.80	$141.85	$167.49

Crew No.	Bare Costs		Incl. Subs O & P		Cost Per Labor-Hour	
Crew B-33C	Hr.	Daily	Hr.	Daily	Bare Costs	Incl. O&P
1 Equip. Oper. (med.)	$28.40	$227.20	$43.20	$345.60	$26.41	$40.50
.5 Laborer	21.45	85.80	33.75	135.00		
.25 Equip. Oper. (med.)	28.40	56.80	43.20	86.40		
1 Scraper, Towed, 12 C.Y.		202.45		222.70		
1 Dozer, 300 H.P.		1131.00		1244.10		
.25 Dozer, 300 H.P.		282.75		311.00	115.44	126.99
14 L.H., Daily Totals		$1986.00		$2344.80	$141.85	$167.49

Crew No.	Bare Costs		Incl. Subs O & P		Cost Per Labor-Hour	
Crew B-33D	Hr.	Daily	Hr.	Daily	Bare Costs	Incl. O&P
1 Equip. Oper. (med.)	$28.40	$227.20	$43.20	$345.60	$26.41	$40.50
.5 Laborer	21.45	85.80	33.75	135.00		
.25 Equip. Oper. (med.)	28.40	56.80	43.20	86.40		
1 S.P. Scraper, 14 C.Y.		1617.00		1778.70		
.25 Dozer, 300 H.P.		282.75		311.00	135.70	149.27
14 L.H., Daily Totals		$2269.55		$2656.70	$162.11	$189.77

Crew No.	Bare Costs		Incl. Subs O & P		Cost Per Labor-Hour	
Crew B-33E	Hr.	Daily	Hr.	Daily	Bare Costs	Incl. O&P
1 Equip. Oper. (med.)	$28.40	$227.20	$43.20	$345.60	$26.41	$40.50
.5 Laborer	21.45	85.80	33.75	135.00		
.25 Equip. Oper. (med.)	28.40	56.80	43.20	86.40		
1 S.P. Scraper, 24 C.Y.		1912.00		2103.20		
.25 Dozer, 300 H.P.		282.75		311.00	156.77	172.44
14 L.H., Daily Totals		$2564.55		$2981.20	$183.18	$212.94

Crew No.	Bare Costs		Incl. Subs O & P		Cost Per Labor-Hour	
Crew B-33F	Hr.	Daily	Hr.	Daily	Bare Costs	Incl. O&P
1 Equip. Oper. (med.)	$28.40	$227.20	$43.20	$345.60	$26.41	$40.50
.5 Laborer	21.45	85.80	33.75	135.00		
.25 Equip. Oper. (med.)	28.40	56.80	43.20	86.40		
1 Elev. Scraper, 11 C.Y.		688.55		757.40		
.25 Dozer, 300 H.P.		282.75		311.00	69.38	76.32
14 L.H., Daily Totals		$1341.10		$1635.40	$95.79	$116.82

Crew No.	Bare Costs		Incl. Subs O & P		Cost Per Labor-Hour	
Crew B-33G	Hr.	Daily	Hr.	Daily	Bare Costs	Incl. O&P
1 Equip. Oper. (med.)	$28.40	$227.20	$43.20	$345.60	$26.41	$40.50
.5 Laborer	21.45	85.80	33.75	135.00		
.25 Equip. Oper. (med.)	28.40	56.80	43.20	86.40		
1 Elev. Scraper, 20 C.Y.		984.55		1083.00		
.25 Dozer, 300 H.P.		282.75		311.00	90.52	99.57
14 L.H., Daily Totals		$1637.10		$1961.00	$116.93	$140.07

Crew No.	Bare Costs		Incl. Subs O & P		Cost Per Labor-Hour	
Crew B-33H	Hr.	Daily	Hr.	Daily	Bare Costs	Incl. O&P
.25 Laborer	$21.45	$42.90	$33.75	$67.50	$27.20	$41.57
1 Equipment Operator (med.)	28.40	227.20	43.20	345.60		
.2 Equipment Operator (med.)	28.40	45.44	43.20	69.12		
1 Scraper, 32-44 C.Y.		2328.00		2560.80		
.2 Dozer, 298 kw		276.00		303.60	224.48	246.93
11 L.H., Daily Totals		$2919.54		$3346.62	$251.68	$288.50

Crew No.	Bare Costs		Incl. Subs O & P		Cost Per Labor-Hour	
Crew B-33J	Hr.	Daily	Hr.	Daily	Bare Costs	Incl. O&P
1 Equipment Operator (med.)	$28.40	$227.20	$43.20	$345.60	$28.40	$43.20
1 Scraper 17 C.Y.		1617.00		1778.70	202.13	222.34
8 L.H., Daily Totals		$1844.20		$2124.30	$230.53	$265.54

Crew No.	Bare Costs		Incl. Subs O & P		Cost Per Labor-Hour	
Crew B-34A	Hr.	Daily	Hr.	Daily	Bare Costs	Incl. O&P
1 Truck Driver (heavy)	$22.10	$176.80	$33.85	$270.80	$22.10	$33.85
1 Dump Truck, 12 Ton		364.30		400.75	45.54	50.09
8 L.H., Daily Totals		$541.10		$671.55	$67.64	$83.94

Crew No.	Bare Costs		Incl. Subs O & P		Cost Per Labor-Hour	
Crew B-34B	Hr.	Daily	Hr.	Daily	Bare Costs	Incl. O&P
1 Truck Driver (heavy)	$22.10	$176.80	$33.85	$270.80	$22.10	$33.85
1 Dump Truck, 16 Ton		442.25		486.50	55.28	60.81
8 L.H., Daily Totals		$619.05		$757.30	$77.38	$94.66

Crew No.	Bare Costs		Incl. Subs O & P		Cost Per Labor-Hour	
Crew B-34C	Hr.	Daily	Hr.	Daily	Bare Costs	Incl. O&P
1 Truck Driver (heavy)	$22.10	$176.80	$33.85	$270.80	$22.10	$33.85
1 Truck Tractor, 40 Ton		426.30		468.95		
1 Dump Trailer, 16.5 C.Y.		132.60		145.85	69.86	76.85
8 L.H., Daily Totals		$735.70		$885.60	$91.96	$110.70

Crew No.	Bare Costs		Incl. Subs O & P		Cost Per Labor-Hour	
Crew B-34D	Hr.	Daily	Hr.	Daily	Bare Costs	Incl. O&P
1 Truck Driver (heavy)	$22.10	$176.80	$33.85	$270.80	$22.10	$33.85
1 Truck Tractor, 40 Ton		426.30		468.95		
1 Dump Trailer, 20 C.Y.		135.10		148.60	70.18	77.19
8 L.H., Daily Totals		$738.20		$888.35	$92.28	$111.04

Crew No.	Bare Costs		Incl. Subs O & P		Cost Per Labor-Hour	
Crew B-34E	Hr.	Daily	Hr.	Daily	Bare Costs	Incl. O&P
1 Truck Driver (heavy)	$22.10	$176.80	$33.85	$270.80	$22.10	$33.85
1 Truck, Off Hwy., 25 Ton		699.00		768.90	87.38	96.11
8 L.H., Daily Totals		$875.80		$1039.70	$109.48	$129.96

Crew No.	Bare Costs		Incl. Subs O & P		Cost Per Labor-Hour	
Crew B-34F	Hr.	Daily	Hr.	Daily	Bare Costs	Incl. O&P
1 Truck Driver (heavy)	$22.10	$176.80	$33.85	$270.80	$22.10	$33.85
1 Truck, Off Hwy., 22 C.Y.		1038.00		1141.80	129.75	142.73
8 L.H., Daily Totals		$1214.80		$1412.60	$151.85	$176.58

Crew No.	Bare Costs		Incl. Subs O & P		Cost Per Labor-Hour	
Crew B-34G	Hr.	Daily	Hr.	Daily	Bare Costs	Incl. O&P
1 Truck Driver (heavy)	$22.10	$176.80	$33.85	$270.80	$22.10	$33.85
1 Truck, Off Hwy., 34 C.Y.		1400.00		1540.00	175.00	192.50
8 L.H., Daily Totals		$1576.80		$1810.80	$197.10	$226.35

Crew No.	Bare Costs		Incl. Subs O & P		Cost Per Labor-Hour	
Crew B-34H	Hr.	Daily	Hr.	Daily	Bare Costs	Incl. O&P
1 Truck Driver (heavy)	$22.10	$176.80	$33.85	$270.80	$22.10	$33.85
1 Truck, Off Hwy., 42 C.Y.		1598.00		1757.80	199.75	219.73
8 L.H., Daily Totals		$1774.80		$2028.60	$221.85	$253.58

Crew No.	Bare Costs		Incl. Subs O & P		Cost Per Labor-Hour	
Crew B-34J	Hr.	Daily	Hr.	Daily	Bare Costs	Incl. O&P
1 Truck Driver (heavy)	$22.10	$176.80	$33.85	$270.80	$22.10	$33.85
1 Truck, Off Hwy., 60 C.Y.		2238.00		2461.80	279.75	307.73
8 L.H., Daily Totals		$2414.80		$2732.60	$301.85	$341.58

Crew B-34K	Bare Costs Hr.	Daily	Incl. Subs O&P Hr.	Daily	Bare Costs	Incl. O&P
1 Truck Driver (heavy)	$22.10	$176.80	$33.85	$270.80	$22.10	$33.85
1 Truck Tractor, 240 H.P.		530.20		583.20		
1 Low Bed Trailer		383.75		422.15	114.24	125.67
8 L.H., Daily Totals		$1090.75		$1276.15	$136.34	$159.52

Crew B-35	Hr.	Daily	Hr.	Daily	Bare Costs	Incl. O&P
1 Laborer Foreman (out)	$23.45	$187.60	$36.90	$295.20	$26.49	$40.92
1 Skilled Worker	28.05	224.40	44.25	354.00		
1 Welder (plumber)	32.60	260.80	49.40	395.20		
1 Laborer	21.45	171.60	33.75	270.00		
1 Equip. Oper. (crane)	29.35	234.80	44.65	357.20		
1 Equip. Oper. Oiler	24.05	192.40	36.60	292.80		
1 Electric Welding Mach.		49.70		54.65		
1 Hyd. Excavator, .75 C.Y.		449.10		494.00	10.39	11.43
48 L.H., Daily Totals		$1770.40		$2513.05	$36.88	$52.35

Crew B-35A	Hr.	Daily	Hr.	Daily	Bare Costs	Incl. O&P
1 Laborer Foreman (out)	$23.45	$187.60	$36.90	$295.20	$25.77	$39.90
2 Laborers	21.45	343.20	33.75	540.00		
1 Skilled Worker	28.05	224.40	44.25	354.00		
1 Welder (plumber)	32.60	260.80	49.40	395.20		
1 Equip. Oper. (crane)	29.35	234.80	44.65	357.20		
1 Equip. Oper. Oiler	24.05	192.40	36.60	292.80		
1 Welder, 300 amp		82.35		90.60		
1 Crane, 75 Ton		929.30		1022.25	18.07	19.87
56 L.H., Daily Totals		$2454.85		$3347.25	$43.84	$59.77

Crew B-36	Hr.	Daily	Hr.	Daily	Bare Costs	Incl. O&P
1 Labor Foreman (outside)	$23.45	$187.60	$36.90	$295.20	$24.63	$38.16
2 Laborers	21.45	343.20	33.75	540.00		
2 Equip. Oper. (med.)	28.40	454.40	43.20	691.20		
1 Dozer, 200 H.P.		838.95		922.85		
1 Aggregate Spreader		71.05		78.15		
1 Tandem Roller, 10 Ton		245.10		269.60	28.88	31.77
40 L.H., Daily Totals		$2140.30		$2797.00	$53.51	$69.93

Crew B-36A	Hr.	Daily	Hr.	Daily	Bare Costs	Incl. O&P
1 Labor Foreman	$23.45	$187.60	$36.90	$295.20	$25.71	$39.60
2 Laborers	21.45	343.20	33.75	540.00		
4 Equip. Oper. (med.)	28.40	908.80	43.20	1382.40		
1 Dozer, 200 H.P.		838.95		922.85		
1 Aggregate Spreader		71.05		78.15		
1 Roller, Steel Wheel		245.10		269.60		
1 Roller, Pneumatic Wheel		244.20		268.60	24.99	27.49
56 L.H., Daily Totals		$2838.90		$3756.80	$50.70	$67.09

Crew B-36B	Hr.	Daily	Hr.	Daily	Bare Costs	Incl. O&P
1 Labor Foreman	$23.45	$187.60	$36.90	$295.20	$25.26	$38.88
2 Laborers	21.45	343.20	33.75	540.00		
4 Equip. Oper. (medium)	28.40	908.80	43.20	1382.40		
1 Truck Driver, Heavy	22.10	176.80	33.85	270.80		
1 Grader, 30,000 Lbs.		575.55		633.10		
1 F.E. Loader, crl, 1.5 C.Y.		399.70		439.65		
1 Dozer, 300 H.P.		1131.00		1244.10		
1 Roller, vibratory		421.75		463.95		
1 Truck, Tractor, 240 H.P.		530.20		583.20		
1 Water Tanker, 5000 Gal.		205.10		225.60	50.99	56.09
64 L.H., Daily Totals		$4879.70		$6078.00	$76.25	$94.97

Crew B-37	Hr.	Daily	Hr.	Daily	Bare Costs	Incl. O&P
1 Labor Foreman (outside)	$23.45	$187.60	$36.90	$295.20	$22.74	$35.54
4 Laborers	21.45	686.40	33.75	1080.00		
1 Equip. Oper. (light)	27.20	217.60	41.35	330.80		
1 Tandem Roller, 5 Ton		152.85		168.15	3.18	3.50
48 L.H., Daily Totals		$1244.45		$1874.15	$25.92	$39.04

Crew B-38	Hr.	Daily	Hr.	Daily	Bare Costs	Incl. O&P
1 Labor Foreman (outside)	$23.45	$187.60	$36.90	$295.20	$24.39	$37.79
2 Laborers	21.45	343.20	33.75	540.00		
1 Equip. Oper. (light)	27.20	217.60	41.35	330.80		
1 Equip. Oper. (medium)	28.40	227.20	43.20	345.60		
1 Backhoe Loader, 48 H.P.		218.80		240.70		
1 Demol.Hammer,(1200 l		180.15		198.15		
1 F.E. Loader (170 H.P.)		576.90		634.60		
1 Pavt. Rem. Bucket		44.45		48.90	25.51	28.06
40 L.H., Daily Totals		$1995.90		$2633.95	$49.90	$65.85

Crew B-39	Hr.	Daily	Hr.	Daily	Bare Costs	Incl. O&P
1 Labor Foreman (outside)	$23.45	$187.60	$36.90	$295.20	$22.74	$35.54
4 Laborers	21.45	686.40	33.75	1080.00		
1 Equip. Oper. (light)	27.20	217.60	41.35	330.80		
1 Air Compr., 250 C.F.M.		124.90		137.40		
2 Air Tools & Accessories		39.00		42.90		
2-50 Ft. Air Hoses, 1.5" Dia.		16.40		18.05	3.76	4.13
48 L.H., Daily Totals		$1271.90		$1904.35	$26.50	$39.67

Crew B-40	Hr.	Daily	Hr.	Daily	Bare Costs	Incl. O&P
1 Pile Driver Foreman	$29.20	$233.60	$50.45	$403.60	$27.59	$45.54
4 Pile Drivers	27.20	870.40	47.00	1504.00		
2 Equip. Oper. (crane)	29.35	469.60	44.65	714.40		
1 Equip. Oper. Oiler	24.05	192.40	36.60	292.80		
1 Crane, 40 Ton		704.30		774.75		
1 Vibratory Hammer & Gen.		1219.00		1340.90	30.05	33.06
64 L.H., Daily Totals		$3689.30		$5030.45	$57.64	$78.60

Crew B-41	Hr.	Daily	Hr.	Daily	Bare Costs	Incl. O&P
1 Labor Foreman (outside)	$23.45	$187.60	$36.90	$295.20	$22.29	$34.95
4 Laborers	21.45	686.40	33.75	1080.00		
.25 Equip. Oper. (crane)	29.35	58.70	44.65	89.30		
.25 Equip. Oper. Oiler	24.05	48.10	36.60	73.20		
.25 Crawler Crane, 40 Ton		176.07		193.70	4.00	4.40
44 L.H., Daily Totals		$1156.87		$1731.40	$26.29	$39.35

Crew B-42	Hr.	Daily	Hr.	Daily	Bare Costs	Incl. O&P
1 Labor Foreman (outside)	$23.45	$187.60	$36.90	$295.20	$24.16	$38.66
4 Laborers	21.45	686.40	33.75	1080.00		
1 Equip. Oper. (crane)	29.35	234.80	44.65	357.20		
1 Equip. Oper. Oiler	24.05	192.40	36.60	292.80		
1 Welder	30.60	244.80	56.10	448.80		
1 Hyd. Crane, 25 Ton		549.35		604.30		
1 Gas Welding Machine		82.35		90.60		
1 Horz. Boring Csg. Mch.		490.70		539.75	17.54	19.29
64 L.H., Daily Totals		$2668.40		$3708.65	$41.70	$57.95

Crew B-43	Hr.	Daily	Hr.	Daily	Bare Costs	Incl. O&P
1 Labor Foreman (outside)	$23.45	$187.60	$36.90	$295.20	$23.53	$36.57
3 Laborers	21.45	514.80	33.75	810.00		
1 Equip. Oper. (crane)	29.35	234.80	44.65	357.20		
1 Equip. Oper. Oiler	24.05	192.40	36.60	292.80		
1 Drill Rig & Augers		1776.00		1953.60	37.00	40.70
48 L.H., Daily Totals		$2905.60		$3708.80	$60.53	$77.27

Crew No.	Bare Costs		Incl. Subs O & P		Cost Per Labor-Hour	
	Hr.	Daily	Hr.	Daily	Bare Costs	Incl. O&P
Crew B-44						
1 Pile Driver Foreman	$29.20	$233.60	$50.45	$403.60	$27.27	$45.19
4 Pile Drivers	27.20	870.40	47.00	1504.00		
2 Equip. Oper. (crane)	29.35	469.60	44.65	714.40		
1 Laborer	21.45	171.60	33.75	270.00		
1 Crane, 40 Ton, & Access.		704.30		774.75		
45 L.F. Leads, 15K Ft. Lbs.		78.75		86.65	12.24	13.46
64 L.H., Daily Totals		$2528.25		$3753.40	$39.51	$58.65

Crew B-45	Hr.	Daily	Hr.	Daily	Bare Costs	Incl. O&P
1 Equip. Oper. (med.)	$28.40	$227.20	$43.20	$345.60	$25.25	$38.53
1 Truck Driver (heavy)	22.10	176.80	33.85	270.80		
1 Dist. Tank Truck, 3K Gal.		364.95		401.45		
1 Tractor, 4 x 2, 250 H.P.		380.00		418.00	46.56	51.22
16 L.H., Daily Totals		$1148.95		$1435.85	$71.81	$89.75

Crew B-46	Hr.	Daily	Hr.	Daily	Bare Costs	Incl. O&P
1 Pile Driver Foreman	$29.20	$233.60	$50.45	$403.60	$24.66	$40.95
2 Pile Drivers	27.20	435.20	47.00	752.00		
3 Laborers	21.45	514.80	33.75	810.00		
1 Chain Saw, 36" Long		53.05		58.35	1.11	1.22
48 L.H., Daily Totals		$1236.65		$2023.95	$25.77	$42.17

Crew B-47	Hr.	Daily	Hr.	Daily	Bare Costs	Incl. O&P
1 Blast Foreman	$23.45	$187.60	$36.90	$295.20	$24.03	$37.33
1 Driller	21.45	171.60	33.75	270.00		
1 Equip. Oper. (light)	27.20	217.60	41.35	330.80		
1 Crawler Type Drill, 4"		311.55		342.70		
1 Air Compr., 600 C.F.M.		272.90		300.20		
2-50 Ft. Air Hoses, 3" Dia.		33.70		37.05	25.76	28.33
24 L.H., Daily Totals		$1194.95		$1575.95	$49.79	$65.66

Crew B-47A	Hr.	Daily	Hr.	Daily	Bare Costs	Incl. O&P
1 Drilling Foreman	$23.45	$187.60	$36.90	$295.20	$25.62	$39.38
1 Equip. Oper. (heavy)	29.35	234.80	44.65	357.20		
1 Oiler	24.05	192.40	36.60	292.80		
1 Quarry Drill		482.05		530.25	20.09	22.09
24 L.H., Daily Totals		$1096.85		$1475.45	$45.71	$61.47

Crew B-47C	Hr.	Daily	Hr.	Daily	Bare Costs	Incl. O&P
1 Laborer	$21.45	$171.60	$33.75	$270.00	$24.33	$37.55
1 Equip. Oper. (light)	27.20	217.60	41.35	330.80		
1 Air Compressor, 750 CFM		288.55		317.40		
2-50' Air Hose, 3"		33.70		37.05		
1 Air Track Drill, 4"		311.55		342.70	39.61	43.57
16 L.H., Daily Totals		$1023.00		$1297.95	$63.94	$81.12

Crew B-47E	Hr.	Daily	Hr.	Daily	Bare Costs	Incl. O&P
1 Laborer Forman	$23.45	$187.60	$36.90	$295.20	$21.95	$34.54
3 Laborers	21.45	514.80	33.75	810.00		
1 Truck, Flatbed, 3 ton		179.45		197.40	5.61	6.17
32 L.H., Daily Totals		$881.85		$1302.60	$27.56	$40.71

Crew B-48	Hr.	Daily	Hr.	Daily	Bare Costs	Incl. O&P
1 Labor Foreman (outside)	$23.45	$187.60	$36.90	$295.20	$24.06	$37.25
3 Laborers	21.45	514.80	33.75	810.00		
1 Equip. Oper. (crane)	29.35	234.80	44.65	357.20		
1 Equip. Oper. Oiler	24.05	192.40	36.60	292.80		
1 Equip. Oper. (light)	27.20	217.60	41.35	330.80		
1 Centr. Water Pump, 6"		159.45		175.40		
1-20 Ft. Suction Hose, 6"		21.50		23.65		
1-50 Ft. Disch. Hose, 6"		19.50		21.45		
1 Drill Rig & Augers		1776.00		1953.60	35.29	38.82
56 L.H., Daily Totals		$3323.65		$4260.10	$59.35	$76.07

Crew B-49	Hr.	Daily	Hr.	Daily	Bare Costs	Incl. O&P
1 Labor Foreman (outside)	$23.45	$187.60	$36.90	$295.20	$25.11	$39.64
3 Laborers	21.45	514.80	33.75	810.00		
2 Equip. Oper. (crane)	29.35	469.60	44.65	714.40		
2 Equip. Oper. Oilers	24.05	384.80	36.60	585.60		
1 Equip. Oper. (light)	27.20	217.60	41.35	330.80		
2 Pile Drivers	27.20	435.20	47.00	752.00		
1 Hyd. Crane, 25 Ton		549.35		604.30		
1 Centr. Water Pump, 6"		159.45		175.40		
1-20 Ft. Suction Hose, 6"		21.50		23.65		
1-50 Ft. Disch. Hose, 6"		19.50		21.45		
1 Drill Rig & Augers		1776.00		1953.60	28.70	31.57
88 L.H., Daily Totals		$4735.40		$6266.40	$53.81	$71.21

Crew B-50	Hr.	Daily	Hr.	Daily	Bare Costs	Incl. O&P
2 Pile Driver Foremen	$29.20	$467.20	$50.45	$807.20	$26.34	$43.58
6 Pile Drivers	27.20	1305.60	47.00	2256.00		
2 Equip. Oper. (crane)	29.35	469.60	44.65	714.40		
1 Equip. Oper. Oiler	24.05	192.40	36.60	292.80		
3 Laborers	21.45	514.80	33.75	810.00		
1 Crane, 40 Ton		704.30		774.75		
60 L.F. Leads, 15K Ft. Lbs.		105.00		115.50		
1 Hammer, 15K Ft. Lbs.		321.75		353.95		
1 Air Compr., 600 C.F.M.		272.90		300.20		
2-50 Ft. Air Hoses, 3" Dia.		33.70		37.05		
1 Chain Saw, 36" Long		53.05		58.35	13.31	14.64
112 L.H., Daily Totals		$4440.30		$6520.20	$39.65	$58.22

Crew B-51	Hr.	Daily	Hr.	Daily	Bare Costs	Incl. O&P
1 Labor Foreman (outside)	$23.45	$187.60	$36.90	$295.20	$21.83	$34.20
4 Laborers	21.45	686.40	33.75	1080.00		
1 Truck Driver (light)	21.75	174.00	33.30	266.40		
1 Light Truck, 1.5 Ton		171.95		189.15	3.58	3.94
48 L.H., Daily Totals		$1219.95		$1830.75	$25.41	$38.14

Crew B-52	Hr.	Daily	Hr.	Daily	Bare Costs	Incl. O&P
1 Carpenter Foreman	$29.30	$234.40	$46.10	$368.80	$25.21	$39.60
1 Carpenter	27.30	218.40	42.95	343.60		
3 Laborers	21.45	514.80	33.75	810.00		
1 Cement Finisher	26.15	209.20	39.00	312.00		
.5 Rodman (reinf.)	30.40	121.60	52.60	210.40		
.5 Equip. Oper. (med.)	28.40	113.60	43.20	172.80		
.5 F.E. Ldr., T.M., 2.5 C.Y.		432.55		475.80	7.72	8.50
56 L.H., Daily Totals		$1844.55		$2693.40	$32.93	$48.10

Crew B-53	Hr.	Daily	Hr.	Daily	Bare Costs	Incl. O&P
1 Equip. Oper. (light)	$27.20	$217.60	$41.35	$330.80	$27.20	$41.35
1 Trencher, Chain, 12 H.P.		97.30		107.05	12.16	13.38
8 L.H., Daily Totals		$314.90		$437.85	$39.36	$54.73

CREWS

Crew B-54

Crew No.	Bare Costs Hr.	Bare Costs Daily	Incl. Subs O & P Hr.	Incl. Subs O & P Daily	Cost Per Labor-Hour Bare Costs	Cost Per Labor-Hour Incl. O&P
1 Equip. Oper. (light)	$27.20	$217.60	$41.35	$330.80	$27.20	$41.35
1 Trencher, Chain, 40 H.P.		210.90		232.00	26.36	29.00
8 L.H., Daily Totals		$428.50		$562.80	$53.56	$70.35

Crew B-54A

Crew No.	Bare Costs Hr.	Bare Costs Daily	Incl. Subs O & P Hr.	Incl. Subs O & P Daily	Cost Per Labor-Hour Bare Costs	Cost Per Labor-Hour Incl. O&P
.17 Labor Forman	$23.45	$31.89	$36.90	$50.18	$27.68	$42.28
1 Equipment Operator (med.)	28.40	227.20	43.20	345.60		
1 Wheel Trencher, 67HP		399.35		439.30	42.67	46.93
9.36 L.H., Daily Totals		$658.44		$835.08	$70.35	$89.21

Crew B-54B

Crew No.	Bare Costs Hr.	Bare Costs Daily	Incl. Subs O & P Hr.	Incl. Subs O & P Daily	Cost Per Labor-Hour Bare Costs	Cost Per Labor-Hour Incl. O&P
.25 Labor Foreman	$23.45	$46.90	$36.90	$73.80	$27.41	$41.94
1 Equipment Operator (med.)	28.40	227.20	43.20	345.60		
1 Wheel Trencher,150HP		565.20		621.70	56.52	62.17
10 L.H., Daily Totals		$839.30		$1041.10	$83.93	$104.11

Crew B-55

Crew No.	Bare Costs Hr.	Bare Costs Daily	Incl. Subs O & P Hr.	Incl. Subs O & P Daily	Cost Per Labor-Hour Bare Costs	Cost Per Labor-Hour Incl. O&P
2 Laborers	$21.45	$343.20	$33.75	$540.00	$21.55	$33.60
1 Truck Driver (light)	21.75	174.00	33.30	266.40		
1 Auger, 4" to 36" Dia		459.30		505.25		
1 Flatbed 3 Ton Truck		179.45		197.40	26.61	29.28
24 L.H., Daily Totals		$1155.95		$1509.05	$48.16	$62.88

Crew B-56

Crew No.	Bare Costs Hr.	Bare Costs Daily	Incl. Subs O & P Hr.	Incl. Subs O & P Daily	Cost Per Labor-Hour Bare Costs	Cost Per Labor-Hour Incl. O&P
1 Laborer	$21.45	$171.60	$33.75	$270.00	$24.33	$37.55
1 Equip. Oper. (light)	27.20	217.60	41.35	330.80		
1 Crawler Type Drill, 4"		311.55		342.70		
1 Air Compr., 600 C.F.M.		272.90		300.20		
1-50 Ft. Air Hose, 3" Dia.		16.85		18.55	37.58	41.34
16 L.H., Daily Totals		$990.50		$1262.25	$61.91	$78.89

Crew B-57

Crew No.	Bare Costs Hr.	Bare Costs Daily	Incl. Subs O & P Hr.	Incl. Subs O & P Daily	Cost Per Labor-Hour Bare Costs	Cost Per Labor-Hour Incl. O&P
1 Labor Foreman (outside)	$23.45	$187.60	$36.90	$295.20	$24.49	$37.83
2 Laborers	21.45	343.20	33.75	540.00		
1 Equip. Oper. (crane)	29.35	234.80	44.65	357.20		
1 Equip. Oper. (light)	27.20	217.60	41.35	330.80		
1 Equip. Oper. Oiler	24.05	192.40	36.60	292.80		
1 Power Shovel, 1 C.Y.		506.80		557.50		
1 Clamshell Bucket, 1 C.Y.		67.10		73.80		
1 Centr. Water Pump, 6"		159.45		175.40		
1-20 Ft. Suction Hose, 6"		21.50		23.65		
20-50 Ft. Disch. Hoses, 6"		390.00		429.00	23.85	26.24
48 L.H., Daily Totals		$2320.45		$3075.35	$48.34	$64.07

Crew B-58

Crew No.	Bare Costs Hr.	Bare Costs Daily	Incl. Subs O & P Hr.	Incl. Subs O & P Daily	Cost Per Labor-Hour Bare Costs	Cost Per Labor-Hour Incl. O&P
2 Laborers	$21.45	$343.20	$33.75	$540.00	$23.37	$36.28
1 Equip. Oper. (light)	27.20	217.60	41.35	330.80		
1 Backhoe Loader, 48 H.P.		218.80		240.70		
1 Small Helicopter		3510.00		3861.00	155.37	170.90
24 L.H., Daily Totals		$4289.60		$4972.50	$178.74	$207.18

Crew B-59

Crew No.	Bare Costs Hr.	Bare Costs Daily	Incl. Subs O & P Hr.	Incl. Subs O & P Daily	Cost Per Labor-Hour Bare Costs	Cost Per Labor-Hour Incl. O&P
1 Truck Driver (heavy)	$22.10	$176.80	$33.85	$270.80	$22.10	$33.85
1 Truck, 30 Ton		324.05		356.45		
1 Water tank, 5000 Gal.		205.10		225.60	66.14	72.76
8 L.H., Daily Totals		$705.95		$852.85	$88.24	$106.61

Crew B-59A

Crew No.	Bare Costs Hr.	Bare Costs Daily	Incl. Subs O & P Hr.	Incl. Subs O & P Daily	Cost Per Labor-Hour Bare Costs	Cost Per Labor-Hour Incl. O&P
2 Laborers	$21.45	$343.20	$33.75	$540.00	$21.67	$33.78
1 Truck Driver (heavy)	22.10	176.80	33.85	270.80		
1 Water tank, 5K w/pum		205.10		225.60		
1 Truck, 30 Ton		324.05		356.45	22.05	24.25
24 L.H., Daily Totals		$1049.15		$1392.85	$43.72	$58.03

Crew B-60

Crew No.	Bare Costs Hr.	Bare Costs Daily	Incl. Subs O & P Hr.	Incl. Subs O & P Daily	Cost Per Labor-Hour Bare Costs	Cost Per Labor-Hour Incl. O&P
1 Labor Foreman (outside)	$23.45	$187.60	$36.90	$295.20	$24.88	$38.34
2 Laborers	21.45	343.20	33.75	540.00		
1 Equip. Oper. (crane)	29.35	234.80	44.65	357.20		
2 Equip. Oper. (light)	27.20	435.20	41.35	661.60		
1 Equip. Oper. Oiler	24.05	192.40	36.60	292.80		
1 Crawler Crane, 40 Ton		704.30		774.75		
45 L.F. Leads, 15K Ft. Lbs.		78.75		86.65		
1 Backhoe Loader, 48 H.P.		218.80		240.70	17.89	19.68
56 L.H., Daily Totals		$2395.05		$3248.90	$42.77	$58.02

Crew B-61

Crew No.	Bare Costs Hr.	Bare Costs Daily	Incl. Subs O & P Hr.	Incl. Subs O & P Daily	Cost Per Labor-Hour Bare Costs	Cost Per Labor-Hour Incl. O&P
1 Labor Foreman (outside)	$23.45	$187.60	$36.90	$295.20	$23.00	$35.90
3 Laborers	21.45	514.80	33.75	810.00		
1 Equip. Oper. (light)	27.20	217.60	41.35	330.80		
1 Cement Mixer, 2 C.Y.		259.45		285.40		
1 Air Compr., 160 C.F.M.		97.40		107.15	8.92	9.81
40 L.H., Daily Totals		$1276.85		$1828.55	$31.92	$45.71

Crew B-62

Crew No.	Bare Costs Hr.	Bare Costs Daily	Incl. Subs O & P Hr.	Incl. Subs O & P Daily	Cost Per Labor-Hour Bare Costs	Cost Per Labor-Hour Incl. O&P
2 Laborers	$21.45	$343.20	$33.75	$540.00	$23.37	$36.28
1 Equip. Oper. (light)	27.20	217.60	41.35	330.80		
1 Loader, Skid Steer		111.25		122.40	4.64	5.10
24 L.H., Daily Totals		$672.05		$993.20	$28.01	$41.38

Crew B-63

Crew No.	Bare Costs Hr.	Bare Costs Daily	Incl. Subs O & P Hr.	Incl. Subs O & P Daily	Cost Per Labor-Hour Bare Costs	Cost Per Labor-Hour Incl. O&P
4 Laborers	$21.45	$686.40	$33.75	$1080.00	$22.60	$35.27
1 Equip. Oper. (light)	27.20	217.60	41.35	330.80		
1 Loader, Skid Steer		111.25		122.40	2.78	3.06
40 L.H., Daily Totals		$1015.25		$1533.20	$25.38	$38.33

Crew B-64

Crew No.	Bare Costs Hr.	Bare Costs Daily	Incl. Subs O & P Hr.	Incl. Subs O & P Daily	Cost Per Labor-Hour Bare Costs	Cost Per Labor-Hour Incl. O&P
1 Laborer	$21.45	$171.60	$33.75	$270.00	$21.60	$33.53
1 Truck Driver (light)	21.75	174.00	33.30	266.40		
1 Power Mulcher (small)		114.95		126.45		
1 Light Truck, 1.5 Ton		171.95		189.15	17.93	19.72
16 L.H., Daily Totals		$632.50		$852.00	$39.53	$53.25

Crew B-65

Crew No.	Bare Costs Hr.	Bare Costs Daily	Incl. Subs O & P Hr.	Incl. Subs O & P Daily	Cost Per Labor-Hour Bare Costs	Cost Per Labor-Hour Incl. O&P
1 Laborer	$21.45	$171.60	$33.75	$270.00	$21.60	$33.53
1 Truck Driver (light)	21.75	174.00	33.30	266.40		
1 Power Mulcher (large)		282.35		310.60		
1 Light Truck, 1.5 Ton		171.95		189.15	28.39	31.23
16 L.H., Daily Totals		$799.90		$1036.15	$49.99	$64.76

Crew B-66

Crew No.	Bare Costs Hr.	Bare Costs Daily	Incl. Subs O & P Hr.	Incl. Subs O & P Daily	Cost Per Labor-Hour Bare Costs	Cost Per Labor-Hour Incl. O&P
1 Equip. Oper. (light)	$27.20	$217.60	$41.35	$330.80	$27.20	$41.35
1 Backhoe Ldr. w/Attchmt.		200.50		220.55	25.06	27.57
8 L.H., Daily Totals		$418.10		$551.35	$52.26	$68.92

Crew B-67

Crew No.	Bare Costs		Incl. Subs O & P		Cost Per Labor-Hour	
Crew B-67	Hr.	Daily	Hr.	Daily	Bare Costs	Incl. O&P
1 Millwright	$28.75	$230.00	$42.95	$343.60	$27.98	$42.15
1 Equip. Oper. (light)	27.20	217.60	41.35	330.80		
1 Forklift		171.10		188.20	10.69	11.76
16 L.H., Daily Totals		$618.70		$862.60	$38.67	$53.91

Crew B-68

Crew No.	Bare Costs		Incl. Subs O & P		Cost Per Labor-Hour	
Crew B-68	Hr.	Daily	Hr.	Daily	Bare Costs	Incl. O&P
2 Millwrights	$28.75	$460.00	$42.95	$687.20	$28.23	$42.42
1 Equip. Oper. (light)	27.20	217.60	41.35	330.80		
1 Forklift		171.10		188.20	7.13	7.84
24 L.H., Daily Totals		$848.70		$1206.20	$35.36	$50.26

Crew B-69

Crew No.	Bare Costs		Incl. Subs O & P		Cost Per Labor-Hour	
Crew B-69	Hr.	Daily	Hr.	Daily	Bare Costs	Incl. O&P
1 Labor Foreman (outside)	$23.45	$187.60	$36.90	$295.20	$23.53	$36.57
3 Laborers	21.45	514.80	33.75	810.00		
1 Equip Oper. (crane)	29.35	234.80	44.65	357.20		
1 Equip Oper. Oiler	24.05	192.40	36.60	292.80		
1 Truck Crane, 80 Ton		1098.00		1207.80	22.88	25.16
48 L.H., Daily Totals		$2227.60		$2963.00	$46.41	$61.73

Crew B-69A

Crew No.	Bare Costs		Incl. Subs O & P		Cost Per Labor-Hour	
Crew B-69A	Hr.	Daily	Hr.	Daily	Bare Costs	Incl. O&P
1 Labor Foreman	$23.45	$187.60	$36.90	$295.20	$23.73	$36.73
3 Laborers	21.45	514.80	33.75	810.00		
1 Equip. Oper. (medium)	28.40	227.20	43.20	345.60		
1 Concrete Finisher	26.15	209.20	39.00	312.00		
1 Curb Paver		453.90		499.30	9.46	10.40
48 L.H., Daily Totals		$1592.70		$2262.10	$33.19	$47.13

Crew B-69B

Crew No.	Bare Costs		Incl. Subs O & P		Cost Per Labor-Hour	
Crew B-69B	Hr.	Daily	Hr.	Daily	Bare Costs	Incl. O&P
1 Labor Foreman	$23.45	$187.60	$36.90	$295.20	$23.73	$36.73
3 Laborers	21.45	514.80	33.75	810.00		
1 Equip. Oper. (medium)	28.40	227.20	43.20	345.60		
1 Cement Finisher	26.15	209.20	39.00	312.00		
1 Curb/Gutter Paver		905.75		996.35	18.87	20.76
48 L.H., Daily Totals		$2044.55		$2759.15	$42.60	$57.49

Crew B-70

Crew No.	Bare Costs		Incl. Subs O & P		Cost Per Labor-Hour	
Crew B-70	Hr.	Daily	Hr.	Daily	Bare Costs	Incl. O&P
1 Labor Foreman (outside)	$23.45	$187.60	$36.90	$295.20	$24.71	$38.25
3 Laborers	21.45	514.80	33.75	810.00		
3 Equip. Oper. (med.)	28.40	681.60	43.20	1036.80		
1 Motor Grader, 30,000 Lb.		575.55		633.10		
1 Grader Attach., Ripper		62.90		69.20		
1 Road Sweeper, S.P.		194.30		213.75		
1 F.E. Loader, 1-3/4 C.Y.		320.10		352.10	20.59	22.65
56 L.H., Daily Totals		$2536.85		$3410.15	$45.30	$60.90

Crew B-71

Crew No.	Bare Costs		Incl. Subs O & P		Cost Per Labor-Hour	
Crew B-71	Hr.	Daily	Hr.	Daily	Bare Costs	Incl. O&P
1 Labor Foreman (outside)	$23.45	$187.60	$36.90	$295.20	$24.71	$38.25
3 Laborers	21.45	514.80	33.75	810.00		
3 Equip. Oper. (med.)	28.40	681.60	43.20	1036.80		
1 Pvmt. Profiler, 450 H.P.		3431.00		3774.10		
1 Road Sweeper, S.P.		194.30		213.75		
1 F.E. Loader, 1-3/4 C.Y.		320.10		352.10	70.45	77.50
56 L.H., Daily Totals		$5329.40		$6481.95	$95.16	$115.75

Crew B-72

Crew No.	Bare Costs		Incl. Subs O & P		Cost Per Labor-Hour	
Crew B-72	Hr.	Daily	Hr.	Daily	Bare Costs	Incl. O&P
1 Labor Foreman (outside)	$23.45	$187.60	$36.90	$295.20	$25.18	$38.87
3 Laborers	21.45	514.80	33.75	810.00		
4 Equip. Oper. (med.)	28.40	908.80	43.20	1382.40		
1 Pvmt. Profiler, 450 H.P.		3431.00		3774.10		
1 Hammermill, 250 H.P.		1176.00		1293.60		
1 Windrow Loader		919.85		1011.85		
1 Mix Paver 165 H.P.		1435.00		1578.50		
1 Roller, Pneu. Tire, 12 T.		244.20		268.60	112.59	123.85
64 L.H., Daily Totals		$8817.25		$10414.25	$137.77	$162.72

Crew B-73

Crew No.	Bare Costs		Incl. Subs O & P		Cost Per Labor-Hour	
Crew B-73	Hr.	Daily	Hr.	Daily	Bare Costs	Incl. O&P
1 Labor Foreman (outside)	$23.45	$187.60	$36.90	$295.20	$26.04	$40.05
2 Laborers	21.45	343.20	33.75	540.00		
5 Equip. Oper. (med.)	28.40	1136.00	43.20	1728.00		
1 Road Mixer, 310 H.P.		993.15		1092.45		
1 Roller, Tandem, 12 Ton		245.10		269.60		
1 Hammermill, 250 H.P.		1176.00		1293.60		
1 Motor Grader, 30,000 Lb.		575.55		633.10		
.5 F.E. Loader, 1-3/4 C.Y.		160.05		176.05		
.5 Truck, 30 Ton		162.02		178.25		
.5 Water Tank 5000 Gal.		102.55		112.80	53.35	58.69
64 L.H., Daily Totals		$5081.22		$6319.05	$79.39	$98.74

Crew B-74

Crew No.	Bare Costs		Incl. Subs O & P		Cost Per Labor-Hour	
Crew B-74	Hr.	Daily	Hr.	Daily	Bare Costs	Incl. O&P
1 Labor Foreman (outside)	$23.45	$187.60	$36.90	$295.20	$25.34	$38.89
1 Laborer	21.45	171.60	33.75	270.00		
4 Equip. Oper. (med.)	28.40	908.80	43.20	1382.40		
2 Truck Drivers (heavy)	22.10	353.60	33.85	541.60		
1 Motor Grader, 30,000 Lb.		575.55		633.10		
1 Grader Attach., Ripper		62.90		69.20		
2 Stabilizers, 310 H.P.		1394.10		1533.50		
1 Flatbed Truck, 3 Ton		179.45		197.40		
1 Chem. Spreader, Towed		99.30		109.25		
1 Vibr. Roller, 29,000 Lb.		421.75		463.95		
1 Water Tank 5000 Gal.		205.10		225.60		
1 Truck, 30 Ton		324.05		356.45	50.97	56.07
64 L.H., Daily Totals		$4883.80		$6077.65	$76.31	$94.96

Crew B-75

Crew No.	Bare Costs		Incl. Subs O & P		Cost Per Labor-Hour	
Crew B-75	Hr.	Daily	Hr.	Daily	Bare Costs	Incl. O&P
1 Labor Foreman (outside)	$23.45	$187.60	$36.90	$295.20	$25.80	$39.61
1 Laborer	21.45	171.60	33.75	270.00		
4 Equip. Oper. (med.)	28.40	908.80	43.20	1382.40		
1 Truck Driver (heavy)	22.10	176.80	33.85	270.80		
1 Motor Grader, 30,000 Lb.		575.55		633.10		
1 Grader Attach., Ripper		62.90		69.20		
2 Stabilizers, 310 H.P.		1394.10		1533.50		
1 Dist. Truck, 3000 Gal.		364.95		401.45		
1 Vibr. Roller, 29,000 Lb.		421.75		463.95	50.34	55.38
56 L.H., Daily Totals		$4264.05		$5319.60	$76.14	$94.99

Crew B-76

Crew No.	Bare Costs		Incl. Subs O & P		Cost Per Labor-Hour	
Crew B-76	Hr.	Daily	Hr.	Daily	Bare Costs	Incl. O&P
1 Dock Builder Foreman	$29.20	$233.60	$50.45	$403.60	$27.55	$45.71
5 Dock Builders	27.20	1088.00	47.00	1880.00		
2 Equip. Oper. (crane)	29.35	469.60	44.65	714.40		
1 Equip. Oper. Oiler	24.05	192.40	36.60	292.80		
1 Crawler Crane, 50 Ton		844.70		929.15		
1 Barge, 400 Ton		452.65		497.90		
1 Hammer, 15K Ft. Lbs.		321.75		353.95		
60 L.F. Leads, 15K Ft. Lbs.		105.00		115.50		
1 Air Compr., 600 C.F.M.		272.90		300.20		
2-50 Ft. Air Hoses, 3" Dia.		33.70		37.05	28.20	31.02
72 L.H., Daily Totals		$4014.30		$5524.55	$55.75	$76.73

Crew No.	Bare Costs		Incl. Subs O & P		Cost Per Labor-Hour	
Crew B-77	Hr.	Daily	Hr.	Daily	Bare Costs	Incl. O&P
1 Labor Foreman	$23.45	$187.60	$36.90	$295.20	$21.91	$34.29
3 Laborers	21.45	514.80	33.75	810.00		
1 Truck Driver (light)	21.75	174.00	33.30	266.40		
1 Crack Cleaner, 25 H.P.		76.90		84.60		
1 Crack Filler, Trailer Mtd.		144.35		158.80		
1 Flatbed Truck, 3 Ton		179.45		197.40	10.02	11.02
40 L.H., Daily Totals		$1277.10		$1812.40	$31.93	$45.31
Crew B-78	Hr.	Daily	Hr.	Daily	Bare Costs	Incl. O&P
1 Labor Foreman	$23.45	$187.60	$36.90	$295.20	$21.83	$34.20
4 Laborers	21.45	686.40	33.75	1080.00		
1 Truck Driver (light)	21.75	174.00	33.30	266.40		
1 Paint Striper, S.P.		203.75		224.15		
1 Flatbed Truck, 3 Ton		179.45		197.40		
1 Pickup Truck, 3/4 Ton		128.95		141.85	10.67	11.74
48 L.H., Daily Totals		$1560.15		$2205.00	$32.50	$45.94
Crew B-79	Hr.	Daily	Hr.	Daily	Bare Costs	Incl. O&P
1 Labor Foreman	$23.45	$187.60	$36.90	$295.20	$21.91	$34.29
3 Laborers	21.45	514.80	33.75	810.00		
1 Truck Driver (light)	21.75	174.00	33.30	266.40		
1 Thermo. Striper, T.M.		243.70		268.05		
1 Flatbed Truck, 3 Ton		179.45		197.40		
2 Pickup Truck, 3/4 Ton		257.90		283.70	17.03	18.73
40 L.H., Daily Totals		$1557.45		$2120.75	$38.94	$53.02
Crew B-80	Hr.	Daily	Hr.	Daily	Bare Costs	Incl. O&P
1 Labor Foreman	$23.45	$187.60	$36.90	$295.20	$23.46	$36.33
1 Laborer	21.45	171.60	33.75	270.00		
1 Truck Driver (light)	21.75	174.00	33.30	266.40		
1 Equip. Oper. (light)	27.20	217.60	41.35	330.80		
1 Flatbed Truck, 3 Ton		179.45		197.40		
1 Fence Post Auger, TM		360.00		396.00	16.86	18.54
32 L.H., Daily Totals		$1290.25		$1755.80	$40.32	$54.87
Crew B-80A	Hr.	Daily	Hr.	Daily	Bare Costs	Incl. O&P
3 Laborer	$21.45	$514.80	$33.75	$810.00	$21.45	$33.75
1 Flatbed Truck, 3 Ton		179.45		197.40	7.48	8.22
24 L.H., Daily Totals		$694.25		$1007.40	$28.93	$41.97
Crew B-80B	Hr.	Daily	Hr.	Daily	Bare Costs	Incl. O&P
3 Laborer	$21.45	$514.80	$33.75	$810.00	$22.89	$35.65
1 Equip. Oper. (light)	27.20	217.60	41.35	330.80		
1 Crane, flatbed mnt		282.80		311.10	8.84	9.72
32 L.H., Daily Totals		$1015.20		$1451.90	$31.73	$45.37
Crew B-81	Hr.	Daily	Hr.	Daily	Bare Costs	Incl. O&P
1 Laborer	$21.45	$171.60	$33.75	$270.00	$23.98	$36.93
1 Equip. Oper. (med.)	28.40	227.20	43.20	345.60		
1 Truck Driver (heavy)	22.10	176.80	33.85	270.80		
1 Hydromulcher, T.M.		281.75		309.95		
1 Tractor Truck, 4x2		324.05		356.45	25.24	27.77
24 L.H., Daily Totals		$1181.40		$1552.80	$49.22	$64.70
Crew B-82	Hr.	Daily	Hr.	Daily	Bare Costs	Incl. O&P
1 Laborer	$21.45	$171.60	$33.75	$270.00	$24.33	$37.55
1 Equip. Oper. (light)	27.20	217.60	41.35	330.80		
1 Horiz. Borer, 6 H.P.		47.80		52.60	2.99	3.29
16 L.H., Daily Totals		$437.00		$653.40	$27.32	$40.84

Crew No.	Bare Costs		Incl. Subs O & P		Cost Per Labor-Hour	
Crew B-83	Hr.	Daily	Hr.	Daily	Bare Costs	Incl. O&P
1 Tugboat Captain	$28.40	$227.20	$43.20	$345.60	$24.93	$38.48
1 Tugboat Hand	21.45	171.60	33.75	270.00		
1 Tugboat, 250 H.P.		438.50		482.35	27.41	30.15
16 L.H., Daily Totals		$837.30		$1097.95	$52.34	$68.63
Crew B-84	Hr.	Daily	Hr.	Daily	Bare Costs	Incl. O&P
1 Equip. Oper. (med.)	$28.40	$227.20	$43.20	$345.60	$28.40	$43.20
1 Rotary Mower/Tractor		208.50		229.35	26.06	28.67
8 L.H., Daily Totals		$435.70		$574.95	$54.46	$71.87
Crew B-85	Hr.	Daily	Hr.	Daily	Bare Costs	Incl. O&P
3 Laborers	$21.45	$514.80	$33.75	$810.00	$22.97	$35.66
1 Equip. Oper. (med.)	28.40	227.20	43.20	345.60		
1 Truck Driver (heavy)	22.10	176.80	33.85	270.80		
1 Aerial Lift Truck		569.50		626.45		
1 Brush Chipper, 130 H.P.		214.10		235.50		
1 Pruning Saw, Rotary		24.15		26.55	20.19	22.21
40 L.H., Daily Totals		$1726.55		$2314.90	$43.16	$57.87
Crew B-86	Hr.	Daily	Hr.	Daily	Bare Costs	Incl. O&P
1 Equip. Oper. (med.)	$28.40	$227.20	$43.20	$345.60	$28.40	$43.20
1 Stump Chipper, S.P.		162.40		178.65	20.30	22.33
8 L.H., Daily Totals		$389.60		$524.25	$48.70	$65.53
Crew B-86A	Hr.	Daily	Hr.	Daily	Bare Costs	Incl. O&P
1 Equip. Oper. (medium)	$28.40	$227.20	$43.20	$345.60	$28.40	$43.20
1 Grader, 30,000 Lbs.		575.55		633.10	71.94	79.14
8 L.H., Daily Totals		$802.75		$978.70	$100.34	$122.34
Crew B-86B	Hr.	Daily	Hr.	Daily	Bare Costs	Incl. O&P
1 Equip. Oper. (medium)	$28.40	$227.20	$43.20	$345.60	$28.40	$43.20
1 Dozer, 200 H.P.		838.95		922.85	104.87	115.36
8 L.H., Daily Totals		$1066.15		$1268.45	$133.27	$158.56
Crew B-87	Hr.	Daily	Hr.	Daily	Bare Costs	Incl. O&P
1 Laborer	$21.45	$171.60	$33.75	$270.00	$27.01	$41.31
4 Equip. Oper. (med.)	28.40	908.80	43.20	1382.40		
2 Feller Bunchers, 50 H.P.		779.80		857.80		
1 Log Chipper, 22" Tree		2042.00		2246.20		
1 Dozer, 105 H.P.		307.75		338.50		
1 Chainsaw, Gas, 36" Long		53.05		58.35	79.57	87.52
40 L.H., Daily Totals		$4263.00		$5153.25	$106.58	$128.83
Crew B-88	Hr.	Daily	Hr.	Daily	Bare Costs	Incl. O&P
1 Laborer	$21.45	$171.60	$33.75	$270.00	$27.41	$41.85
6 Equip. Oper. (med.)	28.40	1363.20	43.20	2073.60		
2 Feller Bunchers, 50 H.P.		779.80		857.80		
1 Log Chipper, 22" Tree		2042.00		2246.20		
2 Log Skidders, 50 H.P.		744.30		818.75		
1 Dozer, 105 H.P.		307.75		338.50		
1 Chainsaw, Gas, 36" Long		53.05		58.35	70.12	77.14
56 L.H., Daily Totals		$5461.70		$6663.20	$97.53	$118.99
Crew B-89	Hr.	Daily	Hr.	Daily	Bare Costs	Incl. O&P
1 Equip. Oper. (light)	$27.20	$217.60	$41.35	$330.80	$24.48	$37.33
1 Truck Driver (light)	21.75	174.00	33.30	266.40		
1 Truck, Stake Body, 3 Ton		179.45		197.40		
1 Concrete Saw		110.65		121.70		
1 Water Tank, 65 Gal.		12.00		13.20	18.88	20.77
16 L.H., Daily Totals		$693.70		$929.50	$43.36	$58.10

Crews

Crew B-89A

	Bare Hr.	Bare Daily	Incl. Hr.	Incl. Daily	Bare Costs	Incl. O&P
1 Skilled Worker	$28.05	$224.40	$44.25	$354.00	$24.75	$39.00
1 Laborer	21.45	171.60	33.75	270.00		
1 Core Drill (large)		61.35		67.50	3.83	4.22
16 L.H., Daily Totals		$457.35		$691.50	$28.58	$43.22

Crew B-89B

	Bare Hr.	Bare Daily	Incl. Hr.	Incl. Daily	Bare Costs	Incl. O&P
1 Equip. Oper. (light)	$27.20	$217.60	$41.35	$330.80	$24.48	$37.33
1 Truck Driver, Light	21.75	174.00	33.30	266.40		
1 Wall Saw, Hydraulic, 10 H.P.		133.05		146.35		
1 Generator, Diesel, 100 KW		203.70		224.05		
1 Water Tank, 65 Gal.		12.00		13.20		
1 Flatbed Truck, 3 Ton		179.45		197.40	33.01	36.31
16 L.H., Daily Totals		$919.80		$1178.20	$57.49	$73.64

Crew B-90

	Bare Hr.	Bare Daily	Incl. Hr.	Incl. Daily	Bare Costs	Incl. O&P
1 Labor Foreman (outside)	$23.45	$187.60	$36.90	$295.20	$23.30	$36.07
3 Laborers	21.45	514.80	33.75	810.00		
2 Equip. Oper. (light)	27.20	435.20	41.35	661.60		
2 Truck Drivers (heavy)	22.10	353.60	33.85	541.60		
1 Road Mixer, 310 H.P.		993.15		1092.45		
1 Dist. Truck, 2000 Gal.		340.30		374.35	20.84	22.92
64 L.H., Daily Totals		$2824.65		$3775.20	$44.14	$58.99

Crew B-90A

	Bare Hr.	Bare Daily	Incl. Hr.	Incl. Daily	Bare Costs	Incl. O&P
1 Labor Foreman	$23.45	$187.60	$36.90	$295.20	$25.71	$39.60
2 Laborers	21.45	343.20	33.75	540.00		
4 Equip. Oper. (medium)	28.40	908.80	43.20	1382.40		
2 Graders, 30,000 Lbs.		1151.10		1266.20		
1 Roller, Steel Wheel		245.10		269.60		
1 Roller, Pneumatic Wheel		244.20		268.60	29.29	32.22
56 L.H., Daily Totals		$3080.00		$4022.00	$55.00	$71.82

Crew B-90B

	Bare Hr.	Bare Daily	Incl. Hr.	Incl. Daily	Bare Costs	Incl. O&P
1 Labor Foreman	$23.45	$187.60	$36.90	$295.20	$25.26	$39.00
2 Laborers	21.45	343.20	33.75	540.00		
3 Equip. Oper. (medium)	28.40	681.60	43.20	1036.80		
1 Roller, Steel Wheel		245.10		269.60		
1 Roller, Pneumatic Wheel		244.20		268.60		
1 Road Mixer, 310 H.P.		993.15		1092.45	30.88	33.97
48 L.H., Daily Totals		$2694.85		$3502.65	$56.14	$72.97

Crew B-91

	Bare Hr.	Bare Daily	Incl. Hr.	Incl. Daily	Bare Costs	Incl. O&P
1 Labor Foreman (outside)	$23.45	$187.60	$36.90	$295.20	$25.26	$38.88
2 Laborers	21.45	343.20	33.75	540.00		
4 Equip. Oper. (med.)	28.40	908.80	43.20	1382.40		
1 Truck Driver (heavy)	22.10	176.80	33.85	270.80		
1 Dist. Truck, 3000 Gal.		364.95		401.45		
1 Aggreg. Spreader, S.P.		635.50		699.05		
1 Roller, Pneu. Tire, 12 Ton		244.20		268.60		
1 Roller, Steel, 10 Ton		245.10		269.60	23.28	25.61
64 L.H., Daily Totals		$3106.15		$4127.10	$48.54	$64.49

Crew B-92

	Bare Hr.	Bare Daily	Incl. Hr.	Incl. Daily	Bare Costs	Incl. O&P
1 Labor Foreman (outside)	$23.45	$187.60	$36.90	$295.20	$21.95	$34.54
3 Laborers	21.45	514.80	33.75	810.00		
1 Crack Cleaner, 25 H.P.		76.90		84.60		
1 Air Compressor		76.30		83.95		
1 Tar Kettle, T.M.		22.95		25.25		
1 Flatbed Truck, 3 Ton		179.45		197.40	11.11	12.22
32 L.H., Daily Totals		$1058.00		$1496.40	$33.06	$46.76

Crew B-93

	Bare Hr.	Bare Daily	Incl. Hr.	Incl. Daily	Bare Costs	Incl. O&P
1 Equip. Oper. (med.)	$28.40	$227.20	$43.20	$345.60	$28.40	$43.20
1 Feller Buncher, 50 H.P.		389.90		428.90	48.74	53.61
8 L.H., Daily Totals		$617.10		$774.50	$77.14	$96.81

Crew B-94A

	Bare Hr.	Bare Daily	Incl. Hr.	Incl. Daily	Bare Costs	Incl. O&P
1 Laborer	$21.45	$171.60	$33.75	$270.00	$21.45	$33.75
1 Diaph. Water Pump, 2"		29.15		32.05		
1-20 Ft. Suction Hose, 2"		6.50		7.15		
2-50 Ft. Disch. Hoses, 2"		8.80		9.70	5.56	6.11
8 L.H., Daily Totals		$216.05		$318.90	$27.01	$39.86

Crew B-94B

	Bare Hr.	Bare Daily	Incl. Hr.	Incl. Daily	Bare Costs	Incl. O&P
1 Laborer	$21.45	$171.60	$33.75	$270.00	$21.45	$33.75
1 Diaph. Water Pump, 4"		64.40		70.85		
1-20 Ft. Suction Hose, 4"		12.50		13.75		
2-50 Ft. Disch. Hoses, 4"		17.00		18.70	11.74	12.91
8 L.H., Daily Totals		$265.50		$373.30	$33.19	$46.66

Crew B-94C

	Bare Hr.	Bare Daily	Incl. Hr.	Incl. Daily	Bare Costs	Incl. O&P
1 Laborer	$21.45	$171.60	$33.75	$270.00	$21.45	$33.75
1 Centr. Water Pump, 3"		37.90		41.70		
1-20 Ft. Suction Hose, 3"		9.50		10.45		
2-50 Ft. Disch. Hoses, 3"		10.80		11.90	7.28	8.00
8 L.H., Daily Totals		$229.80		$334.05	$28.73	$41.75

Crew B-94D

	Bare Hr.	Bare Daily	Incl. Hr.	Incl. Daily	Bare Costs	Incl. O&P
1 Laborer	$21.45	$171.60	$33.75	$270.00	$21.45	$33.75
1 Centr. Water Pump, 6"		159.45		175.40		
1-20 Ft. Suction Hose, 6"		21.50		23.65		
2-50 Ft. Disch. Hoses, 6"		39.00		42.90	27.49	30.24
8 L.H., Daily Totals		$391.55		$511.95	$48.94	$63.99

Crew B-95

	Bare Hr.	Bare Daily	Incl. Hr.	Incl. Daily	Bare Costs	Incl. O&P
1 Equip. Oper. (crane)	$29.35	$234.80	$44.65	$357.20	$25.40	$39.20
1 Laborer	21.45	171.60	33.75	270.00		
16 L.H., Daily Totals		$406.40		$627.20	$25.40	$39.20

Crew B-95A

	Bare Hr.	Bare Daily	Incl. Hr.	Incl. Daily	Bare Costs	Incl. O&P
1 Equip. Oper. (crane)	$29.35	$234.80	$44.65	$357.20	$25.40	$39.20
1 Laborer	21.45	171.60	33.75	270.00		
1 Hyd. Excavator, 5/8 C.Y.		394.25		433.70	24.64	27.10
16 L.H., Daily Totals		$800.65		$1060.90	$50.04	$66.30

Crew B-95B

	Bare Hr.	Bare Daily	Incl. Hr.	Incl. Daily	Bare Costs	Incl. O&P
1 Equip. Oper. (crane)	$29.35	$234.80	$44.65	$357.20	$25.40	$39.20
1 Laborer	21.45	171.60	33.75	270.00		
1 Hyd. Excavator, 1.5 C.Y.		712.70		783.95	44.54	49.00
16 L.H., Daily Totals		$1119.10		$1411.15	$69.94	$88.20

Crew B-95C

	Bare Hr.	Bare Daily	Incl. Hr.	Incl. Daily	Bare Costs	Incl. O&P
1 Equip. Oper. (crane)	$29.35	$234.80	$44.65	$357.20	$25.40	$39.20
1 Laborer	21.45	171.60	33.75	270.00		
1 Hyd. Excavator, 2.5 C.Y.		1695.00		1864.50	105.94	116.53
16 L.H., Daily Totals		$2101.40		$2491.70	$131.34	$155.73

Crew C-1

	Bare Hr.	Bare Daily	Incl. Hr.	Incl. Daily	Bare Costs	Incl. O&P
3 Carpenters	$27.30	$655.20	$42.95	$1030.80	$25.84	$40.65
1 Laborer	21.45	171.60	33.75	270.00		
32 L.H., Daily Totals		$826.80		$1300.80	$25.84	$40.65

Crew No.	Bare Costs Hr.	Daily	Incl. Subs O & P Hr.	Daily	Cost Per Labor-Hour Bare Costs	Incl. O&P
Crew C-2	Hr.	Daily	Hr.	Daily	Bare Costs	Incl. O&P
1 Carpenter Foreman (out)	$29.30	$234.40	$46.10	$368.80	$26.66	$41.94
4 Carpenters	27.30	873.60	42.95	1374.40		
1 Laborer	21.45	171.60	33.75	270.00		
48 L.H., Daily Totals		$1279.60		$2013.20	$26.66	$41.94

Crew C-2A	Hr.	Daily	Hr.	Daily	Bare Costs	Incl. O&P
1 Carpenter Foreman	$29.30	$234.40	$46.10	$368.80	$26.47	$41.28
3 Carpenters	27.30	655.20	42.95	1030.80		
1 Cement Finisher	26.15	209.20	39.00	312.00		
1 Laborer	21.45	171.60	33.75	270.00		
48 L.H., Daily Totals		$1270.40		$1981.60	$26.47	$41.28

Crew C-3	Hr.	Daily	Hr.	Daily	Bare Costs	Incl. O&P
1 Rodman Foreman	$32.40	$259.20	$56.10	$448.80	$28.01	$46.92
4 Rodmen (reinf.)	30.40	972.80	52.60	1683.20		
1 Equip. Oper. (light)	27.20	217.60	41.35	330.80		
2 Laborers	21.45	343.20	33.75	540.00		
3 Stressing Equipment		40.50		44.55		
.5 Grouting Equipment		117.63		129.40	2.47	2.72
64 L.H., Daily Totals		$1950.93		$3176.75	$30.48	$49.64

Crew C-4	Hr.	Daily	Hr.	Daily	Bare Costs	Incl. O&P
1 Rodman Foreman	$32.40	$259.20	$56.10	$448.80	$30.90	$53.48
3 Rodmen (reinf.)	30.40	729.60	52.60	1262.40		
3 Stressing Equipment		40.50		44.55	1.27	1.39
32 L.H., Daily Totals		$1029.30		$1755.75	$32.17	$54.87

Crew C-5	Hr.	Daily	Hr.	Daily	Bare Costs	Incl. O&P
1 Rodman Foreman	$32.40	$259.20	$56.10	$448.80	$29.63	$49.68
4 Rodmen (reinf.)	30.40	972.80	52.60	1683.20		
1 Equip. Oper. (crane)	29.35	234.80	44.65	357.20		
1 Equip. Oper. Oiler	24.05	192.40	36.60	292.80		
1 Hyd. Crane, 25 Ton		549.35		604.30	9.81	10.79
56 L.H., Daily Totals		$2208.55		$3386.30	$39.44	$60.47

Crew C-6	Hr.	Daily	Hr.	Daily	Bare Costs	Incl. O&P
1 Labor Foreman (outside)	$23.45	$187.60	$36.90	$295.20	$22.57	$35.15
4 Laborers	21.45	686.40	33.75	1080.00		
1 Cement Finisher	26.15	209.20	39.00	312.00		
2 Gas Engine Vibrators		74.90		82.40	1.56	1.72
48 L.H., Daily Totals		$1158.10		$1769.60	$24.13	$36.87

Crew C-7	Hr.	Daily	Hr.	Daily	Bare Costs	Incl. O&P
1 Labor Foreman (outside)	$23.45	$187.60	$36.90	$295.20	$23.26	$36.05
5 Laborers	21.45	858.00	33.75	1350.00		
1 Cement Finisher	26.15	209.20	39.00	312.00		
1 Equip. Oper. (med.)	28.40	227.20	43.20	345.60		
1 Equip. Oper. (oiler)	24.05	192.40	36.60	292.80		
2 Gas Engine Vibrators		74.90		82.40		
1 Concrete Bucket, 1 C.Y.		24.45		26.90		
1 Hyd. Crane, 55 Ton		793.10		872.40	12.40	13.63
72 L.H., Daily Totals		$2566.85		$3577.30	$35.66	$49.68

Crew C-7A	Hr.	Daily	Hr.	Daily	Bare Costs	Incl. O&P
1 Labor Foreman (outside)	$23.45	$187.60	$36.90	$295.20	$21.86	$34.17
5 Laborers	21.45	858.00	33.75	1350.00		
2 Truck Driver (Heavy)	22.10	353.60	33.85	541.60		
2 Conc. Transit Mixers		1822.40		2004.65	28.48	31.32
64 L.H., Daily Totals		$3221.60		$4191.45	$50.34	$65.49

Crew C-7B	Hr.	Daily	Hr.	Daily	Bare Costs	Incl. O&P
1 Labor Foreman (outside)	$23.45	$187.60	$36.90	$295.20	$23.01	$35.86
5 Laborers	21.45	858.00	33.75	1350.00		
1 Equipment Operator (heavy)	29.35	234.80	44.65	357.20		
1 Equipment Oiler	24.05	192.40	36.60	292.80		
1 Conc. Bucket, 2 C.Y.		32.10		35.30		
1 Truck Crane, 165 Ton		1941.00		2135.10	30.83	33.91
64 L.H., Daily Totals		$3445.90		$4465.60	$53.84	$69.77

Crew C-7C	Hr.	Daily	Hr.	Daily	Bare Costs	Incl. O&P
1 Labor Foreman (outside)	$23.45	$187.60	$36.90	$295.20	$23.44	$36.51
5 Laborers	21.45	858.00	33.75	1350.00		
2 Equipment Operator (Medium)	28.40	454.40	43.20	691.20		
2 Wheel Loader, 4 C.Y.		1153.80		1269.20	18.03	19.83
64 L.H., Daily Totals		$2653.80		$3605.60	$41.47	$56.34

Crew C-7D	Hr.	Daily	Hr.	Daily	Bare Costs	Incl. O&P
1 Labor Foreman	$23.45	$187.60	$36.90	$295.20	$22.73	$35.55
5 Laborers	21.45	858.00	33.75	1350.00		
1 Equipment Operator (med.)	28.40	227.20	43.20	345.60		
1 Concrete Conveyer		180.90		199.00	3.23	3.55
56 L.H., Daily Totals		$1453.70		$2189.80	$25.96	$39.10

Crew C-8	Hr.	Daily	Hr.	Daily	Bare Costs	Incl. O&P
1 Labor Foreman (outside)	$23.45	$187.60	$36.90	$295.20	$24.07	$37.05
3 Laborers	21.45	514.80	33.75	810.00		
2 Cement Finishers	26.15	418.40	39.00	624.00		
1 Equip. Oper. (med.)	28.40	227.20	43.20	345.60		
1 Concrete Pump (small)		645.60		710.15	11.53	12.68
56 L.H., Daily Totals		$1993.60		$2784.95	$35.60	$49.73

Crew C-8A	Hr.	Daily	Hr.	Daily	Bare Costs	Incl. O&P
1 Labor Foreman (outside)	$23.45	$187.60	$36.90	$295.20	$23.35	$36.03
3 Laborers	21.45	514.80	33.75	810.00		
2 Cement Finishers	26.15	418.40	39.00	624.00		
48 L.H., Daily Totals		$1120.80		$1729.20	$23.35	$36.03

Crew C-8B	Hr.	Daily	Hr.	Daily	Bare Costs	Incl. O&P
1 Labor Foreman (outside)	$23.45	$187.60	$36.90	$295.20	$23.24	$36.27
3 Laborers	21.45	514.80	33.75	810.00		
1 Equipment Operator	28.40	227.20	43.20	345.60		
1 Vibrating Screed		47.40		52.15		
1 Vibratory Roller		421.75		463.95		
1 Dozer, 200 HP		838.95		922.85	32.70	35.97
40 L.H., Daily Totals		$2237.70		$2889.75	$55.94	$72.24

Crew C-8C	Hr.	Daily	Hr.	Daily	Bare Costs	Incl. O&P
1 Labor Forman	$23.45	$187.60	$36.90	$295.20	$23.73	$36.73
3 Laborers	21.45	514.80	33.75	810.00		
1 Cement Finisher	26.15	209.20	39.00	312.00		
1 Equipment Operator (med.)	28.40	227.20	43.20	345.60		
1 Shotcrete Rig, 12 CY/hr		376.00		413.60	7.83	8.62
48 L.H., Daily Totals		$1514.80		$2176.40	$31.56	$45.35

Crew C-8D	Hr.	Daily	Hr.	Daily	Bare Costs	Incl. O&P
1 Labor Foreman	$23.45	$187.60	$36.90	$295.20	$24.56	$37.75
1 Laborers	21.45	171.60	33.75	270.00		
1 Cement Finisher	26.15	209.20	39.00	312.00		
1 Equipment Operator (light)	27.20	217.60	41.35	330.80		
1 Compressor, 250 CFM		124.90		137.40		
2 Hoses, 1", 50'		10.40		11.45	4.23	4.65
32 L.H., Daily Totals		$921.30		$1356.85	$28.79	$42.40

Crew C-8E

Crew No.	Bare Costs		Incl. Subs O & P		Cost Per Labor-Hour	
Crew C-8E	Hr.	Daily	Hr.	Daily	Bare Costs	Incl. O&P
1 Labor Foreman	$23.45	$187.60	$36.90	$295.20	$24.56	$37.75
1 Laborers	21.45	171.60	33.75	270.00		
1 Cement Finisher	26.15	209.20	39.00	312.00		
1 Equipment Operator (light)	27.20	217.60	41.35	330.80		
1 Compressor, 250 CFM		124.90		137.40		
2 Hoses, 1", 50'		10.40		11.45		
1 Concrete Pump (small)		645.60		710.15	24.40	26.84
32 L.H., Daily Totals		$1566.90		$2067.00	$48.96	$64.59

Crew C-10

Crew No.	Hr.	Daily	Hr.	Daily	Bare Costs	Incl. O&P
1 Laborer	$21.45	$171.60	$33.75	$270.00	$24.58	$37.25
2 Cement Finishers	26.15	418.40	39.00	624.00		
24 L.H., Daily Totals		$590.00		$894.00	$24.58	$37.25

Crew C-11

Crew No.	Hr.	Daily	Hr.	Daily	Bare Costs	Incl. O&P
1 Struc. Steel Foreman	$32.60	$260.80	$59.75	$478.00	$29.96	$53.07
6 Struc. Steel Workers	30.60	1468.80	56.10	2692.80		
1 Equip. Oper. (crane)	29.35	234.80	44.65	357.20		
1 Equip. Oper. Oiler	24.05	192.40	36.60	292.80		
1 Truck Crane, 150 Ton		1553.00		1708.30	21.57	23.73
72 L.H., Daily Totals		$3709.80		$5529.10	$51.53	$76.80

Crew C-12

Crew No.	Hr.	Daily	Hr.	Daily	Bare Costs	Incl. O&P
1 Carpenter Foreman (out)	$29.30	$234.40	$46.10	$368.80	$27.00	$42.23
3 Carpenters	27.30	655.20	42.95	1030.80		
1 Laborer	21.45	171.60	33.75	270.00		
1 Equip. Oper. (crane)	29.35	234.80	44.65	357.20		
1 Hyd. Crane, 12 Ton		443.45		487.80	9.24	10.16
48 L.H., Daily Totals		$1739.45		$2514.60	$36.24	$52.39

Crew C-13

Crew No.	Hr.	Daily	Hr.	Daily	Bare Costs	Incl. O&P
1 Struc. Steel Worker	$30.60	$244.80	$56.10	$448.80	$29.50	$51.72
1 Welder	30.60	244.80	56.10	448.80		
1 Carpenter	27.30	218.40	42.95	343.60		
1 Gas Welding Machine		82.35		90.60	3.43	3.77
24 L.H., Daily Totals		$790.35		$1331.80	$32.93	$55.49

Crew C-14

Crew No.	Hr.	Daily	Hr.	Daily	Bare Costs	Incl. O&P
1 Carpenter Foreman (out)	$29.30	$234.40	$46.10	$368.80	$26.61	$42.53
5 Carpenters	27.30	1092.00	42.95	1718.00		
4 Laborers	21.45	686.40	33.75	1080.00		
4 Rodmen (reinf.)	30.40	972.80	52.60	1683.20		
2 Cement Finishers	26.15	418.40	39.00	624.00		
1 Equip. Oper. (crane)	29.35	234.80	44.65	357.20		
1 Equip. Oper. Oiler	24.05	192.40	36.60	292.80		
1 Crane, 80 Ton, & Tools		1098.00		1207.80	7.63	8.39
144 L.H., Daily Totals		$4929.20		$7331.80	$34.24	$50.92

Crew C-14A

Crew No.	Hr.	Daily	Hr.	Daily	Bare Costs	Incl. O&P
1 Carpenter Foreman (out)	$29.30	$234.40	$46.10	$368.80	$27.41	$43.74
16 Carpenters	27.30	3494.40	42.95	5497.60		
4 Rodmen (reinf.)	30.40	972.80	52.60	1683.20		
2 Laborers	21.45	343.20	33.75	540.00		
1 Cement Finisher	26.15	209.20	39.00	312.00		
1 Equip. Oper. (med)	28.40	227.20	43.20	345.60		
1 Gas Engine Vibrator		37.45		41.20		
1 Concrete Pump (small)		645.60		710.15	3.42	3.76
200 L.H., Daily Totals		$6164.25		$9498.55	$30.83	$47.50

Crew C-14B

Crew No.	Hr.	Daily	Hr.	Daily	Bare Costs	Incl. O&P
1 Carpenter Foreman (out)	$29.30	$234.40	$46.10	$368.80	$27.36	$43.55
16 Carpenters	27.30	3494.40	42.95	5497.60		
4 Rodmen (reinf.)	30.40	972.80	52.60	1683.20		
2 Laborers	21.45	343.20	33.75	540.00		
2 Cement Finishers	26.15	418.40	39.00	624.00		
1 Equip. Oper. (med)	28.40	227.20	43.20	345.60		
1 Gas Engine Vibrator		37.45		41.20		
1 Concrete Pump (small)		645.60		710.15	3.28	3.61
208 L.H., Daily Totals		$6373.45		$9810.55	$30.64	$47.16

Crew C-14C

Crew No.	Hr.	Daily	Hr.	Daily	Bare Costs	Incl. O&P
1 Carpenter Foreman (out)	$29.30	$234.40	$46.10	$368.80	$26.13	$41.64
6 Carpenters	27.30	1310.40	42.95	2061.60		
2 Rodmen (reinf)	30.40	486.40	52.60	841.60		
4 Laborers	21.45	686.40	33.75	1080.00		
1 Cement Finisher	26.15	209.20	39.00	312.00		
1 Gas Engine Vibrator		37.45		41.20	.33	.37
112 L.H., Daily Totals		$2964.25		$4705.20	$26.46	$42.01

Crew C-14D

Crew No.	Hr.	Daily	Hr.	Daily	Bare Costs	Incl. O&P
1 Carpenter Foreman (out)	$29.30	$234.40	$46.10	$368.80	$27.16	$42.96
18 Carpenters	27.30	3931.20	42.95	6184.80		
2 Rodmen (reinf.)	30.40	486.40	52.60	841.60		
2 Laborers	21.45	343.20	33.75	540.00		
1 Cement Finisher	26.15	209.20	39.00	312.00		
1 Equip. Oper. (med.)	28.40	227.20	43.20	345.60		
1 Gas Engine Vibrator		37.45		41.20		
1 Concrete Pump (small)		645.60		710.15	3.42	3.76
200 L.H., Daily Totals		$6114.65		$9344.15	$30.58	$46.72

Crew C-14E

Crew No.	Hr.	Daily	Hr.	Daily	Bare Costs	Incl. O&P
1 Carpenter Foreman (out)	$29.30	$234.40	$46.10	$368.80	$26.91	$43.88
2 Carpenters	27.30	436.80	42.95	687.20		
4 Rodmen (reinf.)	30.40	972.80	52.60	1683.20		
3 Laborers	21.45	514.80	33.75	810.00		
1 Cement Finisher	26.15	209.20	39.00	312.00		
1 Gas Engine Vibrator		37.45		41.20	.43	.47
88 L.H., Daily Totals		$2405.45		$3902.40	$27.34	$44.35

Crew C-14F

Crew No.	Hr.	Daily	Hr.	Daily	Bare Costs	Incl. O&P
1 Laborer Foreman (out)	$23.45	$187.60	$36.90	$295.20	$24.81	$37.60
2 Laborers	21.45	343.20	33.75	540.00		
6 Cement Finishers	26.15	1255.20	39.00	1872.00		
1 Gas Engine Vibrator		37.45		41.20	.52	.57
72 L.H., Daily Totals		$1823.45		$2748.40	$25.33	$38.17

Crew C-14G

Crew No.	Hr.	Daily	Hr.	Daily	Bare Costs	Incl. O&P
1 Laborer Foreman	$23.45	$187.60	$36.90	$295.20	$24.42	$37.20
2 Laborers	21.45	343.20	33.75	540.00		
4 Cement Finishers	26.15	836.80	39.00	1248.00		
1 Gas Engine Vibrator		37.45		41.20	.67	.74
56 L.H., Daily Totals		$1405.05		$2124.40	$25.09	$37.94

Crew C-14H

Crew No.	Hr.	Daily	Hr.	Daily	Bare Costs	Incl. O&P
1 Carpenter Foreman (out)	$29.30	$234.40	$46.10	$368.80	$26.98	$42.89
2 Carpenters	27.30	436.80	42.95	687.20		
1 Rodman (reinf.)	30.40	243.20	52.60	420.80		
1 Laborer	21.45	171.60	33.75	270.00		
1 Cement Finisher	26.15	209.20	39.00	312.00		
1 Gas Engine Vibrator		37.45		41.20	.78	.86
48 L.H., Daily Totals		$1332.65		$2100.00	$27.76	$43.75

Crew No.	Bare Costs Hr.	Daily	Incl. Subs O & P Hr.	Daily	Cost Per Labor-Hour Bare Costs	Incl. O&P
Crew C-15	**Hr.**	**Daily**	**Hr.**	**Daily**	**Bare Costs**	**Incl. O&P**
1 Carpenter Foreman (out)	$29.30	$234.40	$46.10	$368.80	$25.66	$40.43
2 Carpenters	27.30	436.80	42.95	687.20		
3 Laborers	21.45	514.80	33.75	810.00		
2 Cement Finishers	26.15	418.40	39.00	624.00		
1 Rodman (reinf.)	30.40	243.20	52.60	420.80		
72 L.H., Daily Totals		$1847.60		$2910.80	$25.66	$40.43
Crew C-16	**Hr.**	**Daily**	**Hr.**	**Daily**	**Bare Costs**	**Incl. O&P**
1 Labor Foreman (outside)	$23.45	$187.60	$36.90	$295.20	$25.48	$40.51
3 Laborers	21.45	514.80	33.75	810.00		
2 Cement Finishers	26.15	418.40	39.00	624.00		
1 Equip. Oper. (med.)	28.40	227.20	43.20	345.60		
2 Rodmen (reinf.)	30.40	486.40	52.60	841.60		
1 Concrete Pump (small)		645.60		710.15	8.97	9.86
72 L.H., Daily Totals		$2480.00		$3626.55	$34.45	$50.37
Crew C-17	**Hr.**	**Daily**	**Hr.**	**Daily**	**Bare Costs**	**Incl. O&P**
2 Skilled Worker Foremen	$30.05	$480.80	$47.40	$758.40	$28.45	$44.88
8 Skilled Workers	28.05	1795.20	44.25	2832.00		
80 L.H., Daily Totals		$2276.00		$3590.40	$28.45	$44.88
Crew C-17A	**Hr.**	**Daily**	**Hr.**	**Daily**	**Bare Costs**	**Incl. O&P**
2 Skilled Worker Foremen	$30.05	$480.80	$47.40	$758.40	$28.46	$44.88
8 Skilled Workers	28.05	1795.20	44.25	2832.00		
.125 Equip. Oper. (crane)	29.35	29.35	44.65	44.65		
.125 Crane, 80 Ton, & Tools		137.25		151.00	1.69	1.86
81 L.H., Daily Totals		$2442.60		$3786.05	$30.15	$46.74
Crew C-17B	**Hr.**	**Daily**	**Hr.**	**Daily**	**Bare Costs**	**Incl. O&P**
2 Skilled Worker Foremen	$30.05	$480.80	$47.40	$758.40	$28.47	$44.87
8 Skilled Workers	28.05	1795.20	44.25	2832.00		
.25 Equip. Oper. (crane)	29.35	58.70	44.65	89.30		
.25 Crane, 80 Ton, & Tools		274.50		301.95		
.25 Walk Behind Power Tools		12.24		13.45	3.50	3.85
82 L.H., Daily Totals		$2621.44		$3995.10	$31.97	$48.72
Crew C-17C	**Hr.**	**Daily**	**Hr.**	**Daily**	**Bare Costs**	**Incl. O&P**
2 Skilled Worker Foremen	$30.05	$480.80	$47.40	$758.40	$28.48	$44.87
8 Skilled Workers	28.05	1795.20	44.25	2832.00		
.375 Equip. Oper. (crane)	29.35	88.05	44.65	133.95		
.375 Crane, 80 Ton & Tools		411.75		452.95	4.96	5.46
83 L.H., Daily Totals		$2775.80		$4177.30	$33.44	$50.33
Crew C-17D	**Hr.**	**Daily**	**Hr.**	**Daily**	**Bare Costs**	**Incl. O&P**
2 Skilled Worker Foremen	$30.05	$480.80	$47.40	$758.40	$28.49	$44.87
8 Skilled Workers	28.05	1795.20	44.25	2832.00		
.5 Equip. Oper. (crane)	29.35	117.40	44.65	178.60		
.5 Crane, 80 Ton & Tools		549.00		603.90	6.54	7.19
84 L.H., Daily Totals		$2942.40		$4372.90	$35.03	$52.06
Crew C-17E	**Hr.**	**Daily**	**Hr.**	**Daily**	**Bare Costs**	**Incl. O&P**
2 Skilled Worker Foremen	$30.05	$480.80	$47.40	$758.40	$28.45	$44.88
8 Skilled Workers	28.05	1795.20	44.25	2832.00		
1 Hyd. Jack with Rods		60.50		66.55	.76	.83
80 L.H., Daily Totals		$2336.50		$3656.95	$29.21	$45.71

Crew No.	Bare Costs Hr.	Daily	Incl. Subs O & P Hr.	Daily	Cost Per Labor-Hour Bare Costs	Incl. O&P
Crew C-18	**Hr.**	**Daily**	**Hr.**	**Daily**	**Bare Costs**	**Incl. O&P**
.125 Labor Foreman (out)	$23.45	$23.45	$36.90	$36.90	$21.67	$34.10
1 Laborer	21.45	171.60	33.75	270.00		
1 Concrete Cart, 10 C.F.		49.90		54.90	5.54	6.10
9 L.H., Daily Totals		$244.95		$361.80	$27.21	$40.20
Crew C-19	**Hr.**	**Daily**	**Hr.**	**Daily**	**Bare Costs**	**Incl. O&P**
.125 Labor Foreman (out)	$23.45	$23.45	$36.90	$36.90	$21.67	$34.10
1 Laborer	21.45	171.60	33.75	270.00		
1 Concrete Cart, 18 C.F.		83.35		91.70	9.26	10.19
9 L.H., Daily Totals		$278.40		$398.60	$30.93	$44.29
Crew C-20	**Hr.**	**Daily**	**Hr.**	**Daily**	**Bare Costs**	**Incl. O&P**
1 Labor Foreman (outside)	$23.45	$187.60	$36.90	$295.20	$23.16	$35.98
5 Laborers	21.45	858.00	33.75	1350.00		
1 Cement Finisher	26.15	209.20	39.00	312.00		
1 Equip. Oper. (med.)	28.40	227.20	43.20	345.60		
2 Gas Engine Vibrators		74.90		82.40		
1 Concrete Pump (small)		645.60		710.15	11.26	12.38
64 L.H., Daily Totals		$2202.50		$3095.35	$34.42	$48.36
Crew C-21	**Hr.**	**Daily**	**Hr.**	**Daily**	**Bare Costs**	**Incl. O&P**
1 Labor Foreman (outside)	$23.45	$187.60	$36.90	$295.20	$23.16	$35.98
5 Laborers	21.45	858.00	33.75	1350.00		
1 Cement Finisher	26.15	209.20	39.00	312.00		
1 Equip. Oper. (med.)	28.40	227.20	43.20	345.60		
2 Gas Engine Vibrators		74.90		82.40		
1 Concrete Conveyer		180.90		199.00	4.00	4.40
64 L.H., Daily Totals		$1737.80		$2584.20	$27.16	$40.38
Crew C-22	**Hr.**	**Daily**	**Hr.**	**Daily**	**Bare Costs**	**Incl. O&P**
1 Rodman Foreman	$32.40	$259.20	$56.10	$448.80	$30.60	$52.70
4 Rodmen (reinf.)	30.40	972.80	52.60	1683.20		
.125 Equip. Oper. (crane)	29.35	29.35	44.65	44.65		
.125 Equip. Oper. Oiler	24.05	24.05	36.60	36.60		
.125 Hyd. Crane, 25 Ton		68.67		75.55	1.63	1.80
42 L.H., Daily Totals		$1354.07		$2288.80	$32.23	$54.50
Crew C-23	**Hr.**	**Daily**	**Hr.**	**Daily**	**Bare Costs**	**Incl. O&P**
2 Skilled Worker Foremen	$30.05	$480.80	$47.40	$758.40	$28.18	$44.16
6 Skilled Workers	28.05	1346.40	44.25	2124.00		
1 Equip. Oper. (crane)	29.35	234.80	44.65	357.20		
1 Equip. Oper. Oiler	24.05	192.40	36.60	292.80		
1 Crane, 90 Ton		1012.00		1113.20	12.65	13.92
80 L.H., Daily Totals		$3266.40		$4645.60	$40.83	$58.08
Crew C-24	**Hr.**	**Daily**	**Hr.**	**Daily**	**Bare Costs**	**Incl. O&P**
2 Skilled Worker Foremen	$30.05	$480.80	$47.40	$758.40	$28.18	$44.16
6 Skilled Workers	28.05	1346.40	44.25	2124.00		
1 Equip. Oper. (crane)	29.35	234.80	44.65	357.20		
1 Equip. Oper. Oiler	24.05	192.40	36.60	292.80		
1 Truck Crane, 150 Ton		1553.00		1708.30	19.41	21.35
80 L.H., Daily Totals		$3807.40		$5240.70	$47.59	$65.51
Crew C-25	**Hr.**	**Daily**	**Hr.**	**Daily**	**Bare Costs**	**Incl. O&P**
2 Rodmen (reinf.)	$30.40	$486.40	$52.60	$841.60	$24.25	$41.88
2 Rodmen Helpers	18.10	289.60	31.15	498.40		
32 L.H., Daily Totals		$776.00		$1340.00	$24.25	$41.88

Crew No.	Bare Costs Hr.	Daily	Incl. Subs O & P Hr.	Daily	Cost Per Labor-Hour Bare Costs	Incl. O&P
Crew D-1	Hr.	Daily	Hr.	Daily	Bare Costs	Incl. O&P
1 Bricklayer	$27.60	$220.80	$42.90	$343.20	$24.60	$38.25
1 Bricklayer Helper	21.60	172.80	33.60	268.80		
16 L.H., Daily Totals		$393.60		$612.00	$24.60	$38.25
Crew D-2	Hr.	Daily	Hr.	Daily	Bare Costs	Incl. O&P
3 Bricklayers	$27.60	$662.40	$42.90	$1029.60	$25.39	$39.52
2 Bricklayer Helpers	21.60	345.60	33.60	537.60		
.5 Carpenter	27.30	109.20	42.95	171.80		
44 L.H., Daily Totals		$1117.20		$1739.00	$25.39	$39.52
Crew D-3	Hr.	Daily	Hr.	Daily	Bare Costs	Incl. O&P
3 Bricklayers	$27.60	$662.40	$42.90	$1029.60	$25.30	$39.36
2 Bricklayer Helpers	21.60	345.60	33.60	537.60		
.25 Carpenter	27.30	54.60	42.95	85.90		
42 L.H., Daily Totals		$1062.60		$1653.10	$25.30	$39.36
Crew D-4	Hr.	Daily	Hr.	Daily	Bare Costs	Incl. O&P
1 Bricklayer	$27.60	$220.80	$42.90	$343.20	$24.50	$37.86
2 Bricklayer Helpers	21.60	345.60	33.60	537.60		
1 Equip. Oper. (light)	27.20	217.60	41.35	330.80		
1 Grout Pump, 50 C.F./hr		67.35		74.10		
1 Hoses & Hopper		31.50		34.65		
1 Accessories		10.50		11.55	3.42	3.76
32 L.H., Daily Totals		$893.35		$1331.90	$27.92	$41.62
Crew D-5	Hr.	Daily	Hr.	Daily	Bare Costs	Incl. O&P
1 Bricklayer	$27.60	$220.80	$42.90	$343.20	$27.60	$42.90
8 L.H., Daily Totals		$220.80		$343.20	$27.60	$42.90
Crew D-6	Hr.	Daily	Hr.	Daily	Bare Costs	Incl. O&P
3 Bricklayers	$27.60	$662.40	$42.90	$1029.60	$24.71	$38.44
3 Bricklayer Helpers	21.60	518.40	33.60	806.40		
.25 Carpenter	27.30	54.60	42.95	85.90		
50 L.H., Daily Totals		$1235.40		$1921.90	$24.71	$38.44
Crew D-7	Hr.	Daily	Hr.	Daily	Bare Costs	Incl. O&P
1 Tile Layer	$26.65	$213.20	$39.40	$315.20	$24.05	$35.55
1 Tile Layer Helper	21.45	171.60	31.70	253.60		
16 L.H., Daily Totals		$384.80		$568.80	$24.05	$35.55
Crew D-8	Hr.	Daily	Hr.	Daily	Bare Costs	Incl. O&P
3 Bricklayers	$27.60	$662.40	$42.90	$1029.60	$25.20	$39.18
2 Bricklayer Helpers	21.60	345.60	33.60	537.60		
40 L.H., Daily Totals		$1008.00		$1567.20	$25.20	$39.18
Crew D-9	Hr.	Daily	Hr.	Daily	Bare Costs	Incl. O&P
3 Bricklayers	$27.60	$662.40	$42.90	$1029.60	$24.60	$38.25
3 Bricklayer Helpers	21.60	518.40	33.60	806.40		
48 L.H., Daily Totals		$1180.80		$1836.00	$24.60	$38.25
Crew D-10	Hr.	Daily	Hr.	Daily	Bare Costs	Incl. O&P
1 Bricklayer Foreman	$29.60	$236.80	$46.05	$368.40	$25.95	$40.16
1 Bricklayer	27.60	220.80	42.90	343.20		
2 Bricklayer Helpers	21.60	345.60	33.60	537.60		
1 Equip. Oper. (crane)	29.35	234.80	44.65	357.20		
1 Truck Crane, 12.5 Ton		391.35		430.50	9.78	10.76
40 L.H., Daily Totals		$1429.35		$2036.90	$35.73	$50.92

Crew No.	Bare Costs Hr.	Daily	Incl. Subs O & P Hr.	Daily	Cost Per Labor-Hour Bare Costs	Incl. O&P
Crew D-11	Hr.	Daily	Hr.	Daily	Bare Costs	Incl. O&P
1 Bricklayer Foreman	$29.60	$236.80	$46.05	$368.40	$26.27	$40.85
1 Bricklayer	27.60	220.80	42.90	343.20		
1 Bricklayer Helper	21.60	172.80	33.60	268.80		
24 L.H., Daily Totals		$630.40		$980.40	$26.27	$40.85
Crew D-12	Hr.	Daily	Hr.	Daily	Bare Costs	Incl. O&P
1 Bricklayer Foreman	$29.60	$236.80	$46.05	$368.40	$25.10	$39.04
1 Bricklayer	27.60	220.80	42.90	343.20		
2 Bricklayer Helpers	21.60	345.60	33.60	537.60		
32 L.H., Daily Totals		$803.20		$1249.20	$25.10	$39.04
Crew D-13	Hr.	Daily	Hr.	Daily	Bare Costs	Incl. O&P
1 Bricklayer Foreman	$29.60	$236.80	$46.05	$368.40	$26.18	$40.63
1 Bricklayer	27.60	220.80	42.90	343.20		
2 Bricklayer Helpers	21.60	345.60	33.60	537.60		
1 Carpenter	27.30	218.40	42.95	343.60		
1 Equip. Oper. (crane)	29.35	234.80	44.65	357.20		
1 Truck Crane, 12.5 Ton		391.35		430.50	8.15	8.97
48 L.H., Daily Totals		$1647.75		$2380.50	$34.33	$49.60
Crew E-1	Hr.	Daily	Hr.	Daily	Bare Costs	Incl. O&P
1 Welder Foreman	$32.60	$260.80	$59.75	$478.00	$30.13	$52.40
1 Welder	30.60	244.80	56.10	448.80		
1 Equip. Oper. (light)	27.20	217.60	41.35	330.80		
1 Gas Welding Machine		82.35		90.60	3.43	3.77
24 L.H., Daily Totals		$805.55		$1348.20	$33.56	$56.17
Crew E-2	Hr.	Daily	Hr.	Daily	Bare Costs	Incl. O&P
1 Struc. Steel Foreman	$32.60	$260.80	$59.75	$478.00	$29.77	$52.20
4 Struc. Steel Workers	30.60	979.20	56.10	1795.20		
1 Equip. Oper. (crane)	29.35	234.80	44.65	357.20		
1 Equip. Oper. Oiler	24.05	192.40	36.60	292.80		
1 Crane, 90 Ton		1012.00		1113.20	18.07	19.88
56 L.H., Daily Totals		$2679.20		$4036.40	$47.84	$72.08
Crew E-3	Hr.	Daily	Hr.	Daily	Bare Costs	Incl. O&P
1 Struc. Steel Foreman	$32.60	$260.80	$59.75	$478.00	$31.27	$57.32
1 Struc. Steel Worker	30.60	244.80	56.10	448.80		
1 Welder	30.60	244.80	56.10	448.80		
1 Gas Welding Machine		82.35		90.60	3.43	3.77
24 L.H., Daily Totals		$832.75		$1466.20	$34.70	$61.09
Crew E-4	Hr.	Daily	Hr.	Daily	Bare Costs	Incl. O&P
1 Struc. Steel Foreman	$32.60	$260.80	$59.75	$478.00	$31.10	$57.01
3 Struc. Steel Workers	30.60	734.40	56.10	1346.40		
1 Gas Welding Machine		82.35		90.60	2.57	2.83
32 L.H., Daily Totals		$1077.55		$1915.00	$33.67	$59.84
Crew E-5	Hr.	Daily	Hr.	Daily	Bare Costs	Incl. O&P
2 Struc. Steel Foremen	$32.60	$521.60	$59.75	$956.00	$30.22	$53.73
5 Struc. Steel Workers	30.60	1224.00	56.10	2244.00		
1 Equip. Oper. (crane)	29.35	234.80	44.65	357.20		
1 Welder	30.60	244.80	56.10	448.80		
1 Equip. Oper. Oiler	24.05	192.40	36.60	292.80		
1 Crane, 90 Ton		1012.00		1113.20		
1 Gas Welding Machine		82.35		90.60	13.68	15.05
80 L.H., Daily Totals		$3511.95		$5502.60	$43.90	$68.78

Crew E-6	Hr.	Daily	Hr.	Daily	Bare Costs	Incl. O&P
3 Struc. Steel Foremen	$32.60	$782.40	$59.75	$1434.00	$30.27	$53.93
9 Struc. Steel Workers	30.60	2203.20	56.10	4039.20		
1 Equip. Oper. (crane)	29.35	234.80	44.65	357.20		
1 Welder	30.60	244.80	56.10	448.80		
1 Equip. Oper. Oiler	24.05	192.40	36.60	292.80		
1 Equip. Oper. (light)	27.20	217.60	41.35	330.80		
1 Crane, 90 Ton		1012.00		1113.20		
1 Gas Welding Machine		82.35		90.60		
1 Air Compr., 160 C.F.M.		97.40		107.15		
2 Impact Wrenches		57.60		63.35	9.76	10.74
128 L.H., Daily Totals		$5124.55		$8277.10	$40.03	$64.67

Crew E-7	Hr.	Daily	Hr.	Daily	Bare Costs	Incl. O&P
1 Struc. Steel Foreman	$32.60	$260.80	$59.75	$478.00	$30.22	$53.73
4 Struc. Steel Workers	30.60	979.20	56.10	1795.20		
1 Equip. Oper. (crane)	29.35	234.80	44.65	357.20		
1 Equip. Oper. Oiler	24.05	192.40	36.60	292.80		
1 Welder Foreman	32.60	260.80	59.75	478.00		
2 Welders	30.60	489.60	56.10	897.60		
1 Crane, 90 Ton		1012.00		1113.20		
2 Gas Welding Machines		164.70		181.15	14.71	16.18
80 L.H., Daily Totals		$3594.30		$5593.15	$44.93	$69.91

Crew E-8	Hr.	Daily	Hr.	Daily	Bare Costs	Incl. O&P
1 Struc. Steel Foreman	$32.60	$260.80	$59.75	$478.00	$30.05	$53.15
4 Struc. Steel Workers	30.60	979.20	56.10	1795.20		
1 Welder Foreman	32.60	260.80	59.75	478.00		
4 Welders	30.60	979.20	56.10	1795.20		
1 Equip. Oper. (crane)	29.35	234.80	44.65	357.20		
1 Equip. Oper. Oiler	24.05	192.40	36.60	292.80		
1 Equip. Oper. (light)	27.20	217.60	41.35	330.80		
1 Crane, 90 Ton		1012.00		1113.20		
4 Gas Welding Machines		329.40		362.35	12.90	14.19
104 L.H., Daily Totals		$4466.20		$7002.75	$42.95	$67.34

Crew E-9	Hr.	Daily	Hr.	Daily	Bare Costs	Incl. O&P
2 Struc. Steel Foremen	$32.60	$521.60	$59.75	$956.00	$30.27	$53.93
5 Struc. Steel Workers	30.60	1224.00	56.10	2244.00		
1 Welder Foreman	32.60	260.80	59.75	478.00		
5 Welders	30.60	1224.00	56.10	2244.00		
1 Equip. Oper. (crane)	29.35	234.80	44.65	357.20		
1 Equip. Oper. Oiler	24.05	192.40	36.60	292.80		
1 Equip. Oper. (light)	27.20	217.60	41.35	330.80		
1 Crane, 90 Ton		1012.00		1113.20		
5 Gas Welding Machines		411.75		452.95	11.12	12.24
128 L.H., Daily Totals		$5298.95		$8468.95	$41.39	$66.17

Crew E-10	Hr.	Daily	Hr.	Daily	Bare Costs	Incl. O&P
1 Welder Foreman	$32.60	$260.80	$59.75	$478.00	$31.60	$57.92
1 Welder	30.60	244.80	56.10	448.80		
4 Gas Welding Machines		329.40		362.35		
1 Truck, 3 Ton		179.45		197.40	31.80	34.98
16 L.H., Daily Totals		$1014.45		$1486.55	$63.40	$92.90

Crew E-11	Hr.	Daily	Hr.	Daily	Bare Costs	Incl. O&P
2 Painters, Struc. Steel	$26.00	$416.00	$49.15	$786.40	$25.16	$43.35
1 Building Laborer	21.45	171.60	33.75	270.00		
1 Equip. Oper. (light)	27.20	217.60	41.35	330.80		
1 Air Compressor 250 C.F.M.		124.90		137.40		
1 Sand Blaster		31.50		34.65		
1 Sand Blasting Accessories		10.50		11.55	5.22	5.74
32 L.H., Daily Totals		$972.10		$1570.80	$30.38	$49.09

Crew E-12	Hr.	Daily	Hr.	Daily	Bare Costs	Incl. O&P
1 Welder Foreman	$32.60	$260.80	$59.75	$478.00	$29.90	$50.55
1 Equip. Oper. (light)	27.20	217.60	41.35	330.80		
1 Gas Welding Machine		82.35		90.60	5.15	5.66
16 L.H., Daily Totals		$560.75		$899.40	$35.05	$56.21

Crew E-13	Hr.	Daily	Hr.	Daily	Bare Costs	Incl. O&P
1 Welder Foreman	$32.60	$260.80	$59.75	$478.00	$30.80	$53.62
.5 Equip. Oper. (light)	27.20	108.80	41.35	165.40		
1 Gas Welding Machine		82.35		90.60	6.86	7.55
12 L.H., Daily Totals		$451.95		$734.00	$37.66	$61.17

Crew E-14	Hr.	Daily	Hr.	Daily	Bare Costs	Incl. O&P
1 Welder Foreman	$32.60	$260.80	$59.75	$478.00	$32.60	$59.75
1 Gas Welding Machine		82.35		90.60	10.29	11.32
8 L.H., Daily Totals		$343.15		$568.60	$42.89	$71.07

Crew E-16	Hr.	Daily	Hr.	Daily	Bare Costs	Incl. O&P
1 Welder Foreman	$32.60	$260.80	$59.75	$478.00	$31.60	$57.92
1 Welder	30.60	244.80	56.10	448.80		
1 Gas Welding Machine		82.35		90.60	5.15	5.66
16 L.H., Daily Totals		$587.95		$1017.40	$36.75	$63.58

Crew E-17	Hr.	Daily	Hr.	Daily	Bare Costs	Incl. O&P
1 Structural Steel Foreman	$32.60	$260.80	$59.75	$478.00	$31.60	$57.92
1 Structural Steel Worker	30.60	244.80	56.10	448.80		
1 Power Tool		2.00		2.20	.13	.14
16 L.H., Daily Totals		$507.60		$929.00	$31.73	$58.06

Crew E-18	Hr.	Daily	Hr.	Daily	Bare Costs	Incl. O&P
1 Structural Steel Foreman	$32.60	$260.80	$59.75	$478.00	$30.56	$54.25
3 Structural Steel Workers	30.60	734.40	56.10	1346.40		
1 Equipment Operator (med.)	28.40	227.20	43.20	345.60		
1 Crane, 20 Ton		512.55		563.80	12.81	14.10
40 L.H., Daily Totals		$1734.95		$2733.80	$43.37	$68.35

Crew E-19	Hr.	Daily	Hr.	Daily	Bare Costs	Incl. O&P
1 Structural Steel Worker	$30.60	$244.80	$56.10	$448.80	$30.13	$52.40
1 Structural Steel Foreman	32.60	260.80	59.75	478.00		
1 Equip. Oper. (light)	27.20	217.60	41.35	330.80		
1 Power Tool		2.00		2.20		
1 Crane, 20 ton		512.55		563.80	21.44	23.58
24 L.H., Daily Totals		$1237.75		$1823.60	$51.57	$75.98

Crew E-20	Hr.	Daily	Hr.	Daily	Bare Costs	Incl. O&P
1 Structural Steel Foreman	$32.60	$260.80	$59.75	$478.00	$29.88	$52.69
5 Structural Steel Workers	30.60	1224.00	56.10	2244.00		
1 Equip. Oper. (crane)	29.35	234.80	44.65	357.20		
1 Oiler	24.05	192.40	36.60	292.80		
1 Power Tool		2.00		2.20		
1 Crane, 40 ton		694.70		764.15	10.89	11.97
64 L.H., Daily Totals		$2608.70		$4138.35	$40.77	$64.66

Crew E-22	Hr.	Daily	Hr.	Daily	Bare Costs	Incl. O&P
1 Skilled Worker Foreman	$30.05	$240.40	$47.40	$379.20	$28.72	$45.30
2 Skilled Worker	28.05	448.80	44.25	708.00		
24 L.H., Daily Totals		$689.20		$1087.20	$28.72	$45.30

Crew E-24

Crew No.	Bare Costs Hr.	Daily	Incl. Subs O & P Hr.	Daily	Cost Per Labor-Hour Bare Costs	Incl. O&P
3 Structural Steel Worker	$30.60	$734.40	$56.10	$1346.40	$30.05	$52.88
1 Equipment Operator (medium)	28.40	227.20	43.20	345.60		
1-25 ton crane		549.35		604.30	17.17	18.88
32 L.H., Daily Totals		$1510.95		$2296.30	$47.22	$71.76

Crew F-2A

Crew No.	Hr.	Daily	Hr.	Daily	Bare Costs	Incl. O&P
2 Carpenters	$27.30	$436.80	$42.95	$687.20	$27.30	$42.95
16 L.H., Daily Totals		$436.80		$687.20	$27.30	$42.95

Crew F-3

Crew No.	Hr.	Daily	Hr.	Daily	Bare Costs	Incl. O&P
4 Carpenters	$27.30	$873.60	$42.95	$1374.40	$27.71	$43.29
1 Equip. Oper. (crane)	29.35	234.80	44.65	357.20		
1 Hyd. Crane, 12 Ton		443.45		487.80	11.09	12.19
40 L.H., Daily Totals		$1551.85		$2219.40	$38.80	$55.48

Crew F-4

Crew No.	Hr.	Daily	Hr.	Daily	Bare Costs	Incl. O&P
4 Carpenters	$27.30	$873.60	$42.95	$1374.40	$27.10	$42.17
1 Equip. Oper. (crane)	29.35	234.80	44.65	357.20		
1 Equip. Oper. Oiler	24.05	192.40	36.60	292.80		
1 Hyd. Crane, 55 Ton		793.10		872.40	16.52	18.18
48 L.H., Daily Totals		$2093.90		$2896.80	$43.62	$60.35

Crew F-5

Crew No.	Hr.	Daily	Hr.	Daily	Bare Costs	Incl. O&P
1 Carpenter Foreman	$29.30	$234.40	$46.10	$368.80	$27.80	$43.74
3 Carpenters	27.30	655.20	42.95	1030.80		
32 L.H., Daily Totals		$889.60		$1399.60	$27.80	$43.74

Crew F-6

Crew No.	Hr.	Daily	Hr.	Daily	Bare Costs	Incl. O&P
2 Carpenters	$27.30	$436.80	$42.95	$687.20	$25.37	$39.61
2 Building Laborers	21.45	343.20	33.75	540.00		
1 Equip. Oper. (crane)	29.35	234.80	44.65	357.20		
1 Hyd. Crane, 12 Ton		443.45		487.80	11.09	12.19
40 L.H., Daily Totals		$1458.25		$2072.20	$36.46	$51.80

Crew F-7

Crew No.	Hr.	Daily	Hr.	Daily	Bare Costs	Incl. O&P
2 Carpenters	$27.30	$436.80	$42.95	$687.20	$24.38	$38.35
2 Building Laborers	21.45	343.20	33.75	540.00		
32 L.H., Daily Totals		$780.00		$1227.20	$24.38	$38.35

Crew G-1

Crew No.	Hr.	Daily	Hr.	Daily	Bare Costs	Incl. O&P
1 Roofer Foreman	$26.10	$208.80	$44.90	$359.20	$22.67	$39.03
4 Roofers, Composition	24.10	771.20	41.50	1328.00		
2 Roofer Helpers	18.10	289.60	31.15	498.40		
1 Application Equipment		165.70		182.25		
1 Tar Kettle/Pot		45.90		50.50		
1 Crew Truck		197.30		217.05	7.30	8.03
56 L.H., Daily Totals		$1678.50		$2635.40	$29.97	$47.06

Crew G-2

Crew No.	Hr.	Daily	Hr.	Daily	Bare Costs	Incl. O&P
1 Plasterer	$25.70	$205.60	$39.40	$315.20	$22.92	$35.42
1 Plasterer Helper	21.60	172.80	33.10	264.80		
1 Building Laborer	21.45	171.60	33.75	270.00		
1 Grouting Equipment		235.25		258.75	9.80	10.78
24 L.H., Daily Totals		$785.25		$1108.75	$32.72	$46.20

Crew G-3

Crew No.	Hr.	Daily	Hr.	Daily	Bare Costs	Incl. O&P
2 Sheet Metal Workers	$31.75	$508.00	$49.30	$788.80	$26.60	$41.53
2 Building Laborers	21.45	343.20	33.75	540.00		
32 L.H., Daily Totals		$851.20		$1328.80	$26.60	$41.53

Crew G-4

Crew No.	Hr.	Daily	Hr.	Daily	Bare Costs	Incl. O&P
1 Labor Foreman (outside)	$23.45	$187.60	$36.90	$295.20	$22.12	$34.80
2 Building Laborers	21.45	343.20	33.75	540.00		
1 Light Truck, 1.5 Ton		171.95		189.15		
1 Air Compr., 160 C.F.M.		97.40		107.15	11.22	12.35
24 L.H., Daily Totals		$800.15		$1131.50	$33.34	$47.15

Crew G-5

Crew No.	Hr.	Daily	Hr.	Daily	Bare Costs	Incl. O&P
1 Roofer Foreman	$26.10	$208.80	$44.90	$359.20	$22.10	$38.04
2 Roofers, Composition	24.10	385.60	41.50	664.00		
2 Roofer Helpers	18.10	289.60	31.15	498.40		
1 Application Equipment		165.70		182.25	4.14	4.56
40 L.H., Daily Totals		$1049.70		$1703.85	$26.24	$42.60

Crew G-6A

Crew No.	Hr.	Daily	Hr.	Daily	Bare Costs	Incl. O&P
2 Roofers Composition	$24.10	$385.60	$41.50	$664.00	$24.10	$41.50
1 Small Compressor		20.95		23.05		
2 Pneumatic Nailers		39.90		43.90	3.80	4.18
16 L.H., Daily Totals		$446.45		$730.95	$27.90	$45.68

Crew G-7

Crew No.	Hr.	Daily	Hr.	Daily	Bare Costs	Incl. O&P
1 Carpenter	$27.30	$218.40	$42.95	$343.60	$27.30	$42.95
1 Small Compressor		20.95		23.05		
1 Pneumatic Nailer		19.95		21.95	5.11	5.62
8 L.H., Daily Totals		$259.30		$388.60	$32.41	$48.57

Crew H-1

Crew No.	Hr.	Daily	Hr.	Daily	Bare Costs	Incl. O&P
2 Glaziers	$26.60	$425.60	$40.50	$648.00	$28.60	$48.30
2 Struc. Steel Workers	30.60	489.60	56.10	897.60		
32 L.H., Daily Totals		$915.20		$1545.60	$28.60	$48.30

Crew H-2

Crew No.	Hr.	Daily	Hr.	Daily	Bare Costs	Incl. O&P
2 Glaziers	$26.60	$425.60	$40.50	$648.00	$24.88	$38.25
1 Building Laborer	21.45	171.60	33.75	270.00		
24 L.H., Daily Totals		$597.20		$918.00	$24.88	$38.25

Crew H-3

Crew No.	Hr.	Daily	Hr.	Daily	Bare Costs	Incl. O&P
1 Glazier	$26.60	$212.80	$40.50	$324.00	$23.73	$36.65
1 Helper	20.85	166.80	32.80	262.40		
16 L.H., Daily Totals		$379.60		$586.40	$23.73	$36.65

Crew J-1

Crew No.	Hr.	Daily	Hr.	Daily	Bare Costs	Incl. O&P
3 Plasterers	$25.70	$616.80	$39.40	$945.60	$24.06	$36.88
2 Plasterer Helpers	21.60	345.60	33.10	529.60		
1 Mixing Machine, 6 C.F.		42.45		46.70	1.06	1.17
40 L.H., Daily Totals		$1004.85		$1521.90	$25.12	$38.05

Crew J-2

Crew No.	Hr.	Daily	Hr.	Daily	Bare Costs	Incl. O&P
3 Plasterers	$25.70	$616.80	$39.40	$945.60	$24.52	$37.42
2 Plasterer Helpers	21.60	345.60	33.10	529.60		
1 Lather	26.80	214.40	40.15	321.20		
1 Mixing Machine, 6 C.F.		42.45		46.70	.88	.97
48 L.H., Daily Totals		$1219.25		$1843.10	$25.40	$38.39

Crew J-3

Crew No.	Hr.	Daily	Hr.	Daily	Bare Costs	Incl. O&P
1 Terrazzo Worker	$26.55	$212.40	$39.25	$314.00	$24.15	$35.70
1 Terrazzo Helper	21.75	174.00	32.15	257.20		
1 Terrazzo Grinder, Electric		45.70		50.25		
1 Terrazzo Mixer		52.80		58.10	6.16	6.77
16 L.H., Daily Totals		$484.90		$679.55	$30.31	$42.47

Crew No.	Bare Costs Hr.	Daily	Incl. Subs O & P Hr.	Daily	Cost Per Labor-Hour Bare Costs	Incl. O&P
Crew J-4	Hr.	Daily	Hr.	Daily	Bare Costs	Incl. O&P
1 Tile Layer	$26.65	$213.20	$39.40	$315.20	$24.05	$35.55
1 Tile Layer Helper	21.45	171.60	31.70	253.60		
16 L.H., Daily Totals		$384.80		$568.80	$24.05	$35.55
Crew K-1	Hr.	Daily	Hr.	Daily	Bare Costs	Incl. O&P
1 Carpenter	$27.30	$218.40	$42.95	$343.60	$24.52	$38.13
1 Truck Driver (light)	21.75	174.00	33.30	266.40		
1 Truck w/Power Equip.		179.45		197.40	11.22	12.34
16 L.H., Daily Totals		$571.85		$807.40	$35.74	$50.47
Crew K-2	Hr.	Daily	Hr.	Daily	Bare Costs	Incl. O&P
1 Struc. Steel Foreman	$32.60	$260.80	$59.75	$478.00	$28.32	$49.72
1 Struc. Steel Worker	30.60	244.80	56.10	448.80		
1 Truck Driver (light)	21.75	174.00	33.30	266.40		
1 Truck w/Power Equip.		179.45		197.40	7.48	8.22
24 L.H., Daily Totals		$859.05		$1390.60	$35.80	$57.94
Crew L-1	Hr.	Daily	Hr.	Daily	Bare Costs	Incl. O&P
1 Electrician	$31.90	$255.20	$47.70	$381.60	$32.25	$48.55
1 Plumber	32.60	260.80	49.40	395.20		
16 L.H., Daily Totals		$516.00		$776.80	$32.25	$48.55
Crew L-2	Hr.	Daily	Hr.	Daily	Bare Costs	Incl. O&P
1 Carpenter	$27.30	$218.40	$42.95	$343.60	$24.08	$37.88
1 Carpenter Helper	20.85	166.80	32.80	262.40		
16 L.H., Daily Totals		$385.20		$606.00	$24.08	$37.88
Crew L-3	Hr.	Daily	Hr.	Daily	Bare Costs	Incl. O&P
1 Carpenter	$27.30	$218.40	$42.95	$343.60	$29.56	$45.73
.5 Electrician	31.90	127.60	47.70	190.80		
.5 Sheet Metal Worker	31.75	127.00	49.30	197.20		
16 L.H., Daily Totals		$473.00		$731.60	$29.56	$45.73
Crew L-3A	Hr.	Daily	Hr.	Daily	Bare Costs	Incl. O&P
1 Carpenter Foreman (outside)	$29.30	$234.40	$46.10	$368.80	$30.12	$47.17
.5 Sheet Metal Worker	31.75	127.00	49.30	197.20		
12 L.H., Daily Totals		$361.40		$566.00	$30.12	$47.17
Crew L-4	Hr.	Daily	Hr.	Daily	Bare Costs	Incl. O&P
2 Skilled Workers	$28.05	$448.80	$44.25	$708.00	$25.65	$40.43
1 Helper	20.85	166.80	32.80	262.40		
24 L.H., Daily Totals		$615.60		$970.40	$25.65	$40.43
Crew L-5	Hr.	Daily	Hr.	Daily	Bare Costs	Incl. O&P
1 Struc. Steel Foreman	$32.60	$260.80	$59.75	$478.00	$30.71	$54.99
5 Struc. Steel Workers	30.60	1224.00	56.10	2244.00		
1 Equip. Oper. (crane)	29.35	234.80	44.65	357.20		
1 Hyd. Crane, 25 Ton		549.35		604.30	9.81	10.79
56 L.H., Daily Totals		$2268.95		$3683.50	$40.52	$65.78
Crew L-5A	Hr.	Daily	Hr.	Daily	Bare Costs	Incl. O&P
1 Structural Steel Foreman	$32.60	$260.80	$59.75	$478.00	$30.79	$54.15
2 Structural Steel Worker	30.60	489.60	56.10	897.60		
1 Equip. Oper. (crane)	29.35	234.80	44.65	357.20		
1 Crane, SP, 25 Ton		547.90		602.70	17.12	18.83
32 L.H., Daily Totals		$1533.10		$2335.50	$47.91	$72.98

Crew No.	Bare Costs Hr.	Daily	Incl. Subs O & P Hr.	Daily	Cost Per Labor-Hour Bare Costs	Incl. O&P
Crew L-6	Hr.	Daily	Hr.	Daily	Bare Costs	Incl. O&P
1 Plumber	$32.60	$260.80	$49.40	$395.20	$32.37	$48.83
.5 Electrician	31.90	127.60	47.70	190.80		
12 L.H., Daily Totals		$388.40		$586.00	$32.37	$48.83
Crew L-7	Hr.	Daily	Hr.	Daily	Bare Costs	Incl. O&P
2 Carpenters	$27.30	$436.80	$42.95	$687.20	$26.29	$41.00
1 Building Laborer	21.45	171.60	33.75	270.00		
.5 Electrician	31.90	127.60	47.70	190.80		
28 L.H., Daily Totals		$736.00		$1148.00	$26.29	$41.00
Crew L-8	Hr.	Daily	Hr.	Daily	Bare Costs	Incl. O&P
2 Carpenters	$27.30	$436.80	$42.95	$687.20	$28.36	$44.24
.5 Plumber	32.50	130.40	49.40	197.60		
20 L.H., Daily Totals		$567.20		$884.80	$28.36	$44.24
Crew L-9	Hr.	Daily	Hr.	Daily	Bare Costs	Incl. O&P
1 Labor Foreman (inside)	$21.95	$175.60	$34.55	$276.40	$24.76	$40.44
2 Building Laborers	21.45	343.20	33.75	540.00		
1 Struc. Steel Worker	30.60	244.80	56.10	448.80		
.5 Electrician	31.90	127.60	47.70	190.80		
36 L.H., Daily Totals		$891.20		$1456.00	$24.76	$40.44
Crew L-10	Hr.	Daily	Hr.	Daily	Bare Costs	Incl. O&P
1 Structural Steel Foreman	$32.60	$260.80	$59.75	$478.00	$30.85	$53.50
1 Structural Steel Worker	30.60	244.80	56.10	448.80		
1 Equip. Oper. (crane)	29.35	234.80	44.65	357.20		
1 Hyd. Crane, 12 Ton		443.45		487.80	18.48	20.32
24 L.H., Daily Totals		$1183.85		$1771.80	$49.33	$73.82
Crew M-1	Hr.	Daily	Hr.	Daily	Bare Costs	Incl. O&P
3 Elevator Constructors	$33.15	$795.60	$50.20	$1204.80	$31.49	$47.68
1 Elevator Apprentice	26.50	212.00	40.10	320.80		
5 Hand Tools		85.00		93.50	2.66	2.92
32 L.H., Daily Totals		$1092.60		$1619.10	$34.15	$50.60
Crew Q-1	Hr.	Daily	Hr.	Daily	Bare Costs	Incl. O&P
1 Plumber	$32.60	$260.80	$49.40	$395.20	$29.35	$44.48
1 Plumber Apprentice	26.10	208.80	39.55	316.40		
16 L.H., Daily Totals		$469.60		$711.60	$29.35	$44.48
Crew Q-1C	Hr.	Daily	Hr.	Daily	Bare Costs	Incl. O&P
1 Plumber	$32.60	$260.80	$49.40	$395.20	$29.03	$44.05
1 Plumber Apprentice	26.10	208.80	39.55	316.40		
1 Equip. Oper. (medium)	28.40	227.20	43.20	345.60		
1 Trencher, Chain		535.90		589.50	22.33	24.56
24 L.H., Daily Totals		$1232.70		$1646.70	$51.36	$68.61
Crew Q-2	Hr.	Daily	Hr.	Daily	Bare Costs	Incl. O&P
2 Plumbers	$32.60	$521.60	$49.40	$790.40	$30.43	$46.12
1 Plumber Apprentice	26.10	208.80	39.55	316.40		
24 L.H., Daily Totals		$730.40		$1106.80	$30.43	$46.12
Crew Q-3	Hr.	Daily	Hr.	Daily	Bare Costs	Incl. O&P
1 Plumber Foreman (ins)	$33.10	$264.80	$50.15	$401.20	$31.10	$47.13
2 Plumbers	32.60	521.60	49.40	790.40		
1 Plumber Apprentice	26.10	208.80	39.55	316.40		
32 L.H., Daily Totals		$995.20		$1508.00	$31.10	$47.13

Crew No.	Bare Costs Hr.	Bare Costs Daily	Incl. Subs O & P Hr.	Incl. Subs O & P Daily	Cost Per Labor-Hour Bare Costs	Cost Per Labor-Hour Incl. O&P
Crew Q-4	Hr.	Daily	Hr.	Daily	Bare Costs	Incl. O&P
1 Plumber Foreman (ins)	$33.10	$264.80	$50.15	$401.20	$31.10	$47.13
1 Plumber	32.60	260.80	49.40	395.20		
1 Welder (plumber)	32.60	260.80	49.40	395.20		
1 Plumber Apprentice	26.10	208.80	39.55	316.40		
1 Electric Welding Mach.		49.70		54.65	1.55	1.71
32 L.H., Daily Totals		$1044.90		$1562.65	$32.65	$48.84
Crew Q-5	Hr.	Daily	Hr.	Daily	Bare Costs	Incl. O&P
1 Steamfitter	$32.75	$262.00	$49.60	$396.80	$29.48	$44.65
1 Steamfitter Apprentice	26.20	209.60	39.70	317.60		
16 L.H., Daily Totals		$471.60		$714.40	$29.48	$44.65
Crew Q-6	Hr.	Daily	Hr.	Daily	Bare Costs	Incl. O&P
2 Steamfitters	$32.75	$524.00	$49.60	$793.60	$30.57	$46.30
1 Steamfitter Apprentice	26.20	209.60	39.70	317.60		
24 L.H., Daily Totals		$733.60		$1111.20	$30.57	$46.30
Crew Q-7	Hr.	Daily	Hr.	Daily	Bare Costs	Incl. O&P
1 Steamfitter Foreman (ins)	$33.25	$266.00	$50.35	$402.80	$31.24	$47.31
2 Steamfitters	32.75	524.00	49.60	793.60		
1 Steamfitter Apprentice	26.20	209.60	39.70	317.60		
32 L.H., Daily Totals		$999.60		$1514.00	$31.24	$47.31
Crew Q-8	Hr.	Daily	Hr.	Daily	Bare Costs	Incl. O&P
1 Steamfitter Foreman (ins)	$33.25	$266.00	$50.35	$402.80	$31.24	$47.31
1 Steamfitter	32.75	262.00	49.60	396.80		
1 Welder (steamfitter)	32.75	262.00	49.60	396.80		
1 Steamfitter Apprentice	26.20	209.60	39.70	317.60		
1 Electric Welding Mach.		49.70		54.65	1.55	1.71
32 L.H., Daily Totals		$1049.30		$1568.65	$32.79	$49.02
Crew Q-9	Hr.	Daily	Hr.	Daily	Bare Costs	Incl. O&P
1 Sheet Metal Worker	$31.75	$254.00	$49.30	$394.40	$28.58	$44.38
1 Sheet Metal Apprentice	25.40	203.20	39.45	315.60		
16 L.H., Daily Totals		$457.20		$710.00	$28.58	$44.38
Crew Q-10	Hr.	Daily	Hr.	Daily	Bare Costs	Incl. O&P
2 Sheet Metal Workers	$31.75	$508.00	$49.30	$788.80	$29.63	$46.02
1 Sheet Metal Apprentice	25.40	203.20	39.45	315.60		
24 L.H., Daily Totals		$711.20		$1104.40	$29.63	$46.02
Crew Q-11	Hr.	Daily	Hr.	Daily	Bare Costs	Incl. O&P
1 Sheet Metal Foreman (ins)	$32.25	$258.00	$50.10	$400.80	$30.29	$47.04
2 Sheet Metal Workers	31.75	508.00	49.30	788.80		
1 Sheet Metal Apprentice	25.40	203.20	39.45	315.60		
32 L.H., Daily Totals		$969.20		$1505.20	$30.29	$47.04
Crew Q-12	Hr.	Daily	Hr.	Daily	Bare Costs	Incl. O&P
1 Sprinkler Installer	$32.50	$260.00	$49.25	$394.00	$29.25	$44.33
1 Sprinkler Apprentice	26.00	208.00	39.40	315.20		
16 L.H., Daily Totals		$468.00		$709.20	$29.25	$44.33
Crew Q-13	Hr.	Daily	Hr.	Daily	Bare Costs	Incl. O&P
1 Sprinkler Foreman (ins)	$33.00	$264.00	$50.05	$400.40	$31.00	$46.99
2 Sprinkler Installers	32.50	520.00	49.25	788.00		
1 Sprinkler Apprentice	26.00	208.00	39.40	315.20		
32 L.H., Daily Totals		$992.00		$1503.60	$31.00	$46.99

Crew No.	Bare Costs Hr.	Bare Costs Daily	Incl. Subs O & P Hr.	Incl. Subs O & P Daily	Cost Per Labor-Hour Bare Costs	Cost Per Labor-Hour Incl. O&P
Crew Q-14	Hr.	Daily	Hr.	Daily	Bare Costs	Incl. O&P
1 Asbestos Worker	$30.45	$243.60	$49.15	$393.20	$27.40	$44.23
1 Asbestos Apprentice	24.35	194.80	39.30	314.40		
16 L.H., Daily Totals		$438.40		$707.60	$27.40	$44.23
Crew Q-15	Hr.	Daily	Hr.	Daily	Bare Costs	Incl. O&P
1 Plumber	$32.60	$260.80	$49.40	$395.20	$29.35	$44.48
1 Plumber Apprentice	26.10	208.80	39.55	316.40		
1 Electric Welding Mach.		49.70		54.65	3.11	3.42
16 L.H., Daily Totals		$519.30		$766.25	$32.46	$47.90
Crew Q-16	Hr.	Daily	Hr.	Daily	Bare Costs	Incl. O&P
2 Plumbers	$32.60	$521.60	$49.40	$790.40	$30.43	$46.12
1 Plumber Apprentice	26.10	208.80	39.55	316.40		
1 Electric Welding Mach.		49.70		54.65	2.07	2.28
24 L.H., Daily Totals		$780.10		$1161.45	$32.50	$48.40
Crew Q-17	Hr.	Daily	Hr.	Daily	Bare Costs	Incl. O&P
1 Steamfitter	$32.75	$262.00	$49.60	$396.80	$29.48	$44.65
1 Steamfitter Apprentice	26.20	209.60	39.70	317.60		
1 Electric Welding Mach.		49.70		54.65	3.11	3.42
16 L.H., Daily Totals		$521.30		$769.05	$32.59	$48.07
Crew Q-17A	Hr.	Daily	Hr.	Daily	Bare Costs	Incl. O&P
1 Steamfitter	$32.75	$262.00	$49.60	$396.80	$29.43	$44.65
1 Steamfitter Apprentice	26.20	209.60	39.70	317.60		
1 Equip. Oper. (crane)	29.35	234.80	44.65	357.20		
1 Truck Crane, 12 Ton		443.45		487.80		
1 Electric Welding Mach.		49.70		54.65	20.55	22.60
24 L.H., Daily Totals		$1199.55		$1614.05	$49.98	$67.25
Crew Q-18	Hr.	Daily	Hr.	Daily	Bare Costs	Incl. O&P
2 Steamfitters	$32.75	$524.00	$49.60	$793.60	$30.57	$46.30
1 Steamfitter Apprentice	26.20	209.60	39.70	317.60		
1 Electric Welding Mach.		49.70		54.65	2.07	2.28
24 L.H., Daily Totals		$783.30		$1165.85	$32.64	$48.58
Crew Q-19	Hr.	Daily	Hr.	Daily	Bare Costs	Incl. O&P
1 Steamfitter	$32.75	$262.00	$49.60	$396.80	$30.28	$45.67
1 Steamfitter Apprentice	26.20	209.60	39.70	317.60		
1 Electrician	31.90	255.20	47.70	381.60		
24 L.H., Daily Totals		$726.80		$1096.00	$30.28	$45.67
Crew Q-20	Hr.	Daily	Hr.	Daily	Bare Costs	Incl. O&P
1 Sheet Metal Worker	$31.75	$254.00	$49.30	$394.40	$29.24	$45.04
1 Sheet Metal Apprentice	25.40	203.20	39.45	315.60		
.5 Electrician	31.90	127.60	47.70	190.80		
20 L.H., Daily Totals		$584.80		$900.80	$29.24	$45.04
Crew Q-21	Hr.	Daily	Hr.	Daily	Bare Costs	Incl. O&P
2 Steamfitters	$32.75	$524.00	$49.60	$793.60	$30.90	$46.65
1 Steamfitter Apprentice	26.20	209.60	39.70	317.60		
1 Electrician	31.90	255.20	47.70	381.60		
32 L.H., Daily Totals		$988.80		$1492.80	$30.90	$46.65
Crew Q-22	Hr.	Daily	Hr.	Daily	Bare Costs	Incl. O&P
1 Plumber	$32.60	$260.80	$49.40	$395.20	$29.35	$44.48
1 Plumber Apprentice	26.10	208.80	39.55	316.40		
1 Truck Crane, 12 Ton		443.45		487.80	27.72	30.49
16 L.H., Daily Totals		$913.05		$1199.40	$57.07	$74.97

CREWS

Crew Q-22A

Crew No.	Bare Costs Hr.	Bare Costs Daily	Incl. Subs O & P Hr.	Incl. Subs O & P Daily	Cost Per Labor-Hour Bare Costs	Cost Per Labor-Hour Incl. O&P
1 Plumber	$32.60	$260.80	$49.40	$395.20	$27.38	$41.84
1 Plumber Apprentice	26.10	208.80	39.55	316.40		
1 Laborer	21.45	171.60	33.75	270.00		
1 Equip. Oper. (crane)	29.35	234.80	44.65	357.20		
1 Truck Crane, 12 Ton		443.45		487.80	13.86	15.24
32 L.H., Daily Totals		$1319.45		$1826.60	$41.24	$57.08

Crew Q-23

Crew No.	Bare Costs Hr.	Bare Costs Daily	Incl. Subs O & P Hr.	Incl. Subs O & P Daily	Cost Per Labor-Hour Bare Costs	Cost Per Labor-Hour Incl. O&P
1 Plumber Foreman	$34.60	$276.80	$52.40	$419.20	$31.87	$48.33
1 Plumber	32.60	260.80	49.40	395.20		
1 Equip. Oper. (medium)	28.40	227.20	43.20	345.60		
1 Power Tools		2.00		2.20		
1 Crane, 20 Ton		512.55		563.80	21.44	23.58
24 L.H., Daily Totals		$1279.35		$1726.00	$53.31	$71.91

Crew R-1

Crew No.	Bare Costs Hr.	Bare Costs Daily	Incl. Subs O & P Hr.	Incl. Subs O & P Daily	Cost Per Labor-Hour Bare Costs	Cost Per Labor-Hour Incl. O&P
1 Electrician Foreman	$32.40	$259.20	$48.45	$387.60	$28.30	$42.86
3 Electricians	31.90	765.60	47.70	1144.80		
2 Helpers	20.85	333.60	32.80	524.80		
48 L.H., Daily Totals		$1358.40		$2057.20	$28.30	$42.86

Crew R-1A

Crew No.	Bare Costs Hr.	Bare Costs Daily	Incl. Subs O & P Hr.	Incl. Subs O & P Daily	Cost Per Labor-Hour Bare Costs	Cost Per Labor-Hour Incl. O&P
1 Electrician	$31.90	$255.20	$47.70	$381.60	$26.38	$40.25
1 Helper	20.85	166.80	32.80	262.40		
16 L.H., Daily Totals		$422.00		$644.00	$26.38	$40.25

Crew R-2

Crew No.	Bare Costs Hr.	Bare Costs Daily	Incl. Subs O & P Hr.	Incl. Subs O & P Daily	Cost Per Labor-Hour Bare Costs	Cost Per Labor-Hour Incl. O&P
1 Electrician Foreman	$32.40	$259.20	$48.45	$387.60	$28.45	$43.11
3 Electricians	31.90	765.60	47.70	1144.80		
2 Helpers	20.85	333.60	32.80	524.80		
1 Equip. Oper. (crane)	29.35	234.80	44.65	357.20		
1 S.P. Crane, 5 Ton		271.45		298.60	4.85	5.33
56 L.H., Daily Totals		$1864.65		$2713.00	$33.30	$48.44

Crew R-3

Crew No.	Bare Costs Hr.	Bare Costs Daily	Incl. Subs O & P Hr.	Incl. Subs O & P Daily	Cost Per Labor-Hour Bare Costs	Cost Per Labor-Hour Incl. O&P
1 Electrician Foreman	$32.40	$259.20	$48.45	$387.60	$31.59	$47.39
1 Electrician	31.90	255.20	47.70	381.60		
.5 Equip. Oper. (crane)	29.35	117.40	44.65	178.60		
.5 S.P. Crane, 5 Ton		135.73		149.30	6.79	7.46
20 L.H., Daily Totals		$767.53		$1097.10	$38.38	$54.85

Crew R-4

Crew No.	Bare Costs Hr.	Bare Costs Daily	Incl. Subs O & P Hr.	Incl. Subs O & P Daily	Cost Per Labor-Hour Bare Costs	Cost Per Labor-Hour Incl. O&P
1 Struc. Steel Foreman	$32.60	$260.80	$59.75	$478.00	$31.26	$55.15
3 Struc. Steel Workers	30.60	734.40	56.10	1346.40		
1 Electrician	31.90	255.20	47.70	381.60		
1 Gas Welding Machine		82.35		90.60	2.06	2.26
40 L.H., Daily Totals		$1332.75		$2296.60	$33.32	$57.41

Crew R-5

Crew No.	Bare Costs Hr.	Bare Costs Daily	Incl. Subs O & P Hr.	Incl. Subs O & P Daily	Cost Per Labor-Hour Bare Costs	Cost Per Labor-Hour Incl. O&P
1 Electrician Foreman	$32.40	$259.20	$48.45	$387.60	$27.93	$42.35
4 Electrician Linemen	31.90	1020.80	47.70	1526.40		
2 Electrician Operators	31.90	510.40	47.70	763.20		
4 Electrician Groundmen	20.85	667.20	32.80	1049.60		
1 Crew Truck		197.30		217.05		
1 Tool Van		212.10		233.30		
1 Pickup Truck, 3/4 Ton		128.95		141.85		
.2 Crane, 55 Ton		158.62		174.50		
.2 Crane, 12 Ton		88.69		97.55		
.2 Auger, Truck Mtd.		355.20		390.70		
1 Tractor w/Winch		296.95		326.65	16.34	17.97
88 L.H., Daily Totals		$3895.41		$5308.40	$44.27	$60.32

Crew R-6

Crew No.	Bare Costs Hr.	Bare Costs Daily	Incl. Subs O & P Hr.	Incl. Subs O & P Daily	Cost Per Labor-Hour Bare Costs	Cost Per Labor-Hour Incl. O&P
1 Electrician Foreman	$32.40	$259.20	$48.45	$387.60	$27.93	$42.35
4 Electrician Linemen	31.90	1020.80	47.70	1526.40		
2 Electrician Operators	31.90	510.40	47.70	763.20		
4 Electrician Groundmen	20.85	667.20	32.80	1049.60		
1 Crew Truck		197.30		217.05		
1 Tool Van		212.10		233.30		
1 Pickup Truck, 3/4 Ton		128.95		141.85		
.2 Crane, 55 Ton		158.62		174.50		
.2 Crane, 12 Ton		88.69		97.55		
.2 Auger, Truck Mtd.		355.20		390.70		
1 Tractor w/Winch		296.95		326.65		
3 Cable Trailers		419.70		461.65		
.5 Tensioning Rig		135.95		149.55		
.5 Cable Pulling Rig		813.50		894.85	31.90	35.09
88 L.H., Daily Totals		$5264.56		$6814.45	$59.83	$77.44

Crew R-7

Crew No.	Bare Costs Hr.	Bare Costs Daily	Incl. Subs O & P Hr.	Incl. Subs O & P Daily	Cost Per Labor-Hour Bare Costs	Cost Per Labor-Hour Incl. O&P
1 Electrician Foreman	$32.40	$259.20	$48.45	$387.60	$22.78	$35.41
5 Electrician Groundmen	20.85	834.00	32.80	1312.00		
1 Crew Truck		197.30		217.05	4.11	4.52
48 L.H., Daily Totals		$1290.50		$1916.65	$26.89	$39.93

Crew R-8

Crew No.	Bare Costs Hr.	Bare Costs Daily	Incl. Subs O & P Hr.	Incl. Subs O & P Daily	Cost Per Labor-Hour Bare Costs	Cost Per Labor-Hour Incl. O&P
1 Electrician Foreman	$32.40	$259.20	$48.45	$387.60	$28.30	$42.86
3 Electrician Linemen	31.90	765.60	47.70	1144.80		
2 Electrician Groundmen	20.85	333.60	32.80	524.80		
1 Pickup Truck, 3/4 Ton		128.95		141.85		
1 Crew Truck		197.30		217.05	6.80	7.48
48 L.H., Daily Totals		$1684.65		$2416.10	$35.10	$50.34

Crew R-9

Crew No.	Bare Costs Hr.	Bare Costs Daily	Incl. Subs O & P Hr.	Incl. Subs O & P Daily	Cost Per Labor-Hour Bare Costs	Cost Per Labor-Hour Incl. O&P
1 Electrician Foreman	$32.40	$259.20	$48.45	$387.60	$26.44	$40.34
1 Electrician Lineman	31.90	255.20	47.70	381.60		
2 Electrician Operators	31.90	510.40	47.70	763.20		
4 Electrician Groundmen	20.85	667.20	32.80	1049.60		
1 Pickup Truck, 3/4 Ton		128.95		141.85		
1 Crew Truck		197.30		217.05	5.10	5.61
64 L.H., Daily Totals		$2018.25		$2940.90	$31.54	$45.95

Crew R-10

Crew No.	Bare Costs Hr.	Bare Costs Daily	Incl. Subs O & P Hr.	Incl. Subs O & P Daily	Cost Per Labor-Hour Bare Costs	Cost Per Labor-Hour Incl. O&P
1 Electrician Foreman	$32.40	$259.20	$48.45	$387.60	$30.14	$45.34
4 Electrician Linemen	31.90	1020.80	47.70	1526.40		
1 Electrician Groundman	20.85	166.80	32.80	262.40		
1 Crew Truck		197.30		217.05		
3 Tram Cars		528.75		581.65	15.13	16.64
48 L.H., Daily Totals		$2172.85		$2975.10	$45.27	$61.98

Crew R-11

Crew No.	Bare Costs Hr.	Bare Costs Daily	Incl. Subs O & P Hr.	Incl. Subs O & P Daily	Cost Per Labor-Hour Bare Costs	Cost Per Labor-Hour Incl. O&P
1 Electrician Foreman	$32.40	$259.20	$48.45	$387.60	$28.90	$43.69
4 Electricians	31.90	1020.80	47.70	1526.40		
1 Helper	20.85	166.80	32.80	262.40		
1 Common Laborer	21.45	171.60	33.75	270.00		
1 Crew Truck		197.30		217.05		
1 Crane, 12 Ton		443.45		487.80	11.44	12.59
56 L.H., Daily Totals		$2259.15		$3151.25	$40.34	$56.28

Crews

Crew No.	Bare Costs		Incl. Subs O & P		Cost Per Labor-Hour	
Crew R-12	Hr.	Daily	Hr.	Daily	Bare Costs	Incl. O&P
1 Carpenter Foreman	$27.80	$222.40	$43.75	$350.00	$25.62	$40.90
4 Carpenters	27.30	873.60	42.95	1374.40		
4 Common Laborers	21.45	686.40	33.75	1080.00		
1 Equip. Oper. (med.)	28.40	227.20	43.20	345.60		
1 Steel Worker	30.60	244.80	56.10	448.80		
1 Dozer, 200 H.P.		838.95		922.85		
1 Pickup Truck, 3/4 Ton		128.95		141.85	11.00	12.10
88 L.H., Daily Totals		$3222.30		$4663.50	$36.62	$53.00
Crew R-13	Hr.	Daily	Hr.	Daily	Bare Costs	Incl. O&P
1 Electrician Foreman	$32.40	$259.20	$48.45	$387.60	$30.25	$45.47
3 Electricians	31.90	765.60	47.70	1144.80		
1 Equipment Operator	29.35	234.80	44.65	357.20		
1 Equipment Oiler	24.05	192.40	36.60	292.80		
.25-1 Hyd Crane, 33 Ton		173.74		191.10	3.62	3.98
48 L.H., Daily Totals		$1625.74		$2373.50	$33.87	$49.45
Crew R-15	Hr.	Daily	Hr.	Daily	Bare Costs	Incl. O&P
1 Electrician Foreman	$32.40	$259.20	$48.45	$387.60	$31.20	$46.77
4 Electricians	31.90	1020.80	47.70	1526.40		
1 Equipment Operator	27.20	217.60	41.35	330.80		
1 Aerial Lift Truck		281.35		309.50	5.86	6.45
48 L.H., Daily Totals		$1778.95		$2554.30	$37.06	$53.22
Crew R-18	Hr.	Daily	Hr.	Daily	Bare Costs	Incl. O&P
.25 Electrician Foreman	$32.40	$64.80	$48.45	$96.90	$25.14	$38.59
1 Electrician	31.90	255.20	47.70	381.60		
2 Helpers	20.85	333.60	32.80	524.80		
26 L.H., Daily Totals		$653.60		$1003.30	$25.14	$38.59
Crew R-19	Hr.	Daily	Hr.	Daily	Bare Costs	Incl. O&P
.5 Electrician Foreman	$32.40	$129.60	$48.45	$193.80	$32.00	$47.85
2 Electricians	31.90	510.40	47.70	763.20		
20 L.H., Daily Totals		$640.00		$957.00	$32.00	$47.85
Crew R-21	Hr.	Daily	Hr.	Daily	Bare Costs	Incl. O&P
1 Electrician Foreman	$32.40	$259.20	$48.45	$387.60	$31.94	$47.77
3 Electricians	31.90	765.60	47.70	1144.80		
.1 Equip. Oper. (med.)	28.40	22.72	43.20	34.56		
.1 Hyd. Crane 25 Ton		54.79		60.25	1.67	1.84
32. L.H., Daily Totals		$1102.31		$1627.21	$33.61	$49.61
Crew R-22	Hr.	Daily	Hr.	Daily	Bare Costs	Incl. O&P
.66 Electrician Foreman	$32.40	$171.07	$48.45	$255.82	$27.23	$41.41
2 Helpers	20.85	333.60	32.80	524.80		
2 Electricians	31.90	510.40	47.70	763.20		
37.28 L.H., Daily Totals		$1015.07		$1543.82	$27.23	$41.41

Historical Cost Indexes

The table below lists both the Means Historical Cost Index based on Jan. 1, 1993 = 100 as well as the computed value of an index based on Jan. 1, 1999 costs. Since the Jan. 1, 1999 figure is estimated, space is left to write in the actual index figures as they become available through either the quarterly "Means Construction Cost Indexes" or as printed in the "Engineering News-Record." To compute the actual index based on Jan. 1, 1999 = 100, divide the Historical Cost Index for a particular year by the actual Jan. 1, 1999 Construction Cost Index. Space has been left to advance the index figures as the year progresses.

Year	Historical Cost Index Jan. 1, 1993 = 100		Current Index Based on Jan. 1, 1999 = 100		Year	Historical Cost Index Jan. 1, 1993 = 100	Current Index Based on Jan. 1, 1999 = 100		Year	Historical Cost Index Jan. 1, 1993 = 100	Current Index Based on Jan. 1, 1999 = 100	
	Est.	Actual	Est.	Actual		Actual	Est.	Actual		Actual	Est.	Actual
Oct 1999					July 1984	82.0	70.4		July 1966	22.7	19.5	
July 1999					1983	80.2	68.9		1965	21.7	18.6	
April 1999					1982	76.1	65.4		1964	21.2	18.2	
Jan 1999	116.4		100.0	100.0	1981	70.0	60.1		1963	20.7	17.8	
July 1998		115.1	98.9		1980	62.9	54.0		1962	20.2	17.4	
1997		112.8	96.9		1979	57.8	49.7		1961	19.8	17.0	
1996		110.2	94.7		1978	53.5	46.0		1960	19.7	16.9	
1995		107.6	92.4		1977	49.5	42.5		1959	19.3	16.6	
1994		104.4	89.7		1976	46.9	40.3		1958	18.8	16.2	
1993		101.7	87.4		1975	44.8	38.5		1957	18.4	15.8	
1992		99.4	85.4		1974	41.4	35.6		1956	17.6	15.1	
1991		96.8	83.2		1973	37.7	32.4		1955	16.6	14.3	
1990		94.3	81.0		1972	34.8	29.9		1954	16.0	13.7	
1989		92.1	79.2		1971	32.1	27.6		1953	15.8	13.6	
1988		89.9	77.2		1970	28.7	24.7		1952	15.4	13.2	
1987		87.7	75.3		1969	26.9	23.1		1951	15.0	12.9	
1986		84.2	72.4		1968	24.9	21.4		1950	13.7	11.8	
1985		82.6	71.0		1967	23.5	20.2		1949	13.3	11.4	

Adjustments to Costs

The Historical Cost Index can be used to convert National Average building costs at a particular time to the approximate building costs for some other time.

Example:

Estimate and compare construction costs for different years in the same city.

To estimate the National Average construction cost of a building in 1970, knowing that it cost $900,000 in 1999:

INDEX in 1970 = 28.7

INDEX in 1999 = 116.4

Time Adjustment using the Historical Cost Indexes:

$$\frac{\text{Index for Year A}}{\text{Index for Year B}} \times \text{Cost in Year B} = \text{Cost in Year A}$$

$$\frac{\text{INDEX 1970}}{\text{INDEX 1999}} \times \text{Cost 1999} = \text{Cost 1970}$$

$$\frac{28.7}{116.4} \times \$900,000 = .247 \times \$900,000 = \$222,300$$

The construction cost of the building in 1970 is $222,300.

Note: The City Cost Indexes for Canada can be used to convert U.S. National averages to local costs in Canadian dollars.

How to Use the City Cost Indexes

What you should know before you begin

Means City Cost Indexes (CCI) are an extremely useful tool to use when you want to compare costs from city to city and region to region.

This publication contains average construction cost indexes for 689 U.S. and Canadian cities covering over 930 three-digit zip code locations, as listed directly under each city.

Keep in mind that a City Cost Index number is a *percentage ratio* of a specific city's cost to the national average cost of the same item at a stated time period.

In other words, these index figures represent relative construction *factors* (or, if you prefer, multipliers) for Material and Installation costs, as well as the weighted average for Total In Place costs for each CSI MasterFormat division. Installation costs include both labor and equipment rental costs.

The 30 City Average Index is the average of 30 major U.S. cities and serves as a National Average.

Index figures for both material and installation are based on the 30 major city average of 100 and represent the cost relationship as of July 1, 1998. The index for each division is computed from representative material and labor quantities for that division. The weighted average for each city is a weighted total of the components listed above it, but does not include relative productivity between trades or cities.

As changes occur in local material prices, labor rates and equipment rental rates, the impact of these changes should be accurately measured by the change in the City Cost Index for each particular city (as compared to the 30 City Average).

> *Therefore, if you know (or have estimated) building costs in one city today, you can easily convert those costs to expected building costs in another city.*

> *In addition, by using the Historical Cost Index, you can easily convert National Average building costs at a particular time to the approximate building costs for some other time. The City Cost Indexes can then be applied to calculate the costs for a particular city.*

Quick Calculations

Location Adjustment Using the City Cost Indexes:

$$\frac{\text{Index for City A}}{\text{Index for City B}} \times \text{Cost in City B} = \text{Cost in City A}$$

Time Adjustment for the National Average Using the Historical Cost Index:

$$\frac{\text{Index for Year A}}{\text{Index for Year B}} \times \text{Cost in Year B} = \text{Cost in Year A}$$

Adjustment from the National Average:

$$\frac{\text{Index for City A}}{100} \times \text{National Average Cost} = \text{Cost in City A}$$

Since each of the other R.S. Means publications contains many different items, any *one* item multiplied by the particular city index may give incorrect results. However, the larger the number of items compiled, the closer the results should be to actual costs for that particular city.

The City Cost Indexes for Canadian cities are calculated using Canadian material and equipment prices and labor rates, in Canadian dollars. Therefore, indexes for Canadian cities can be used to convert U.S. National Average prices to local costs in Canadian dollars.

How to use this section

1. Compare costs from city to city.

In using the Means Indexes, remember that an index number is not a fixed number but a *ratio:* It's a percentage ratio of a building component's cost at any stated time to the National Average cost of that same component at the same time period. Put in the form of an equation:

$$\frac{\text{Specific City Cost}}{\text{National Average Cost}} \times 100 = \text{City Index Number}$$

Therefore, when making cost comparisons between cities, do not subtract one city's index number from the index number of another city and read the result as a percentage difference. Instead, divide one city's index number by that of the other city. The resulting number may then be used as a multiplier to calculate cost differences from city to city.

The formula used to find cost differences between cities for the purpose of comparison is as follows:

$$\frac{\text{City A Index}}{\text{City B Index}} \times \text{City B Cost (Known)} = \text{City A Cost (Unknown)}$$

In addition, you can use *Means CCI* to calculate and compare costs division by division between cities using the same basic formula. (Just be sure that you're comparing similar divisions.)

2. Compare a specific city's construction costs with the National Average.

When you're studying construction location feasibility, it's advisable to compare a prospective project's cost index with an index of the National Average cost.

For example, divide the weighted average index of construction costs of a specific city by that of the 30 City Average, which = 100.

$$\frac{\text{City Index}}{100} = \% \text{ of National Average}$$

As a result, you get a ratio that indicates the relative cost of construction in that city in comparison with the National Average.

3. Covert U.S. National Average to actual costs in Canadian City.

$$\frac{\text{Index for Canadian City}}{100} \times \text{National Average Cost} = \text{Cost in Canadian City in \$ CAN}$$

4. Adjust construction cost data based on a National Average.

When you use a source of construction cost data which is based on a National Average (such as *Means cost data publications*), it is necessary to adjust those costs to a specific location.

$$\frac{\text{City Index}}{100} \quad \text{x} \quad \frac{\text{"Book" Cost Based on}}{\text{National Average Costs}} \quad = \quad \frac{\text{City Cost}}{\text{(Unknown)}}$$

5. When applying the City Cost Indexes to demolition projects, use the appropriate division installation index. For example, for removal of existing doors and windows, use the Division 8 index.

What you might like to know about how we developed the Indexes

To create a reliable index, R.S. Means researched the building type most often constructed in the United States and Canada. Because it was concluded that no one type of building completely represented the building construction industry, nine different types of buildings were combined to create a composite model.

The exact material, labor and equipment quantities are based on detailed analysis of these nine building types, then each quantity is weighted in proportion to expected usage. These various material items, labor hours, and equipment rental rates are thus combined to form a composite building representing as closely as possible the actual usage of materials, labor and equipment used in the North American Building Construction Industry.

The following structures were chosen to make up that composite model:

1. Factory, 1 story
2. Office, 2–4 story
3. Store, Retail
4. Town Hall, 2–3 story
5. High School, 2–3 story
6. Hospital, 4–8 story
7. Garage, Parking
8. Apartment, 1–3 story
9. Hotel/Motel, 2–3 story

For the purposes of ensuring the timeliness of the data, the components of the index for the composite model have been streamlined. They currently consist of:

- specific quantities of 66 commonly used construction materials;
- specific labor-hours for 21 building construction trades; and
- specific days of equipment rental for 6 types of construction equipment (normally used to install the 66 material items by the 21 trades.)

A sophisticated computer program handles the updating of all costs for each city on a quarterly basis. Material and equipment price quotations are gathered quarterly from 689 cities in the United States and Canada. These prices and the latest negotiated labor wage rates for 21 different building trades are used to compile the quarterly update of the City Cost Index.

The 30 major U.S. cities used to calculate the National Average are:

Atlanta, GA
Baltimore, MD
Boston, MA
Buffalo, NY
Chicago, IL
Cincinnati, OH
Cleveland, OH
Columbus, OH
Dallas, TX
Denver, CO
Detroit, MI
Houston, TX
Indianapolis, IN
Kansas City, MO
Los Angeles, CA

Memphis, TN
Milwaukee, WI
Minneapolis, MN
Nashville, TN
New Orleans, LA
New York, NY
Philadelphia, PA
Phoenix, AZ
Pittsburgh, PA
St. Louis, MO
San Antonio, TX
San Diego, CA
San Francisco, CA
Seattle, WA
Washington, DC

F.Y.I.: The CSI MasterFormat Divisions

1. General Requirements
2. Site Work
3. Concrete
4. Masonry
5. Metals
6. Wood & Plastics
7. Thermal & Moisture Protection
8. Doors & Windows
9. Finishes
10. Specialties
11. Equipment
12. Furnishings
13. Special Construction
14. Conveying Systems
15. Mechanical
16. Electrical

The information presented in the CCI is organized according to the Construction Specifications Institute (CSI) MasterFormat.

What the CCI does not indicate

The weighted average for each city is a total of the components listed above weighted to reflect typical usage, but it does *not* include the productivity variations between trades or cities.

In addition, the CCI does not take into consideration factors such as the following:

- managerial efficiency
- competitive conditions
- automation
- restrictive union practices
- unique local requirements
- regional variations due to specific building codes

City Cost Indexes

UNITED STATES / ALABAMA

DIVISION		30 CITY AVERAGE MAT.	INST.	TOTAL	ANNISTON 362 MAT.	INST.	TOTAL	BIRMINGHAM 350-352 MAT.	INST.	TOTAL	BUTLER 369 MAT.	INST.	TOTAL	DECATUR 356 MAT.	INST.	TOTAL	DOTHAN 363 MAT.	INST.	TOTAL
2	SITE WORK	100.0	100.0	100.0	96.6	91.8	92.9	89.7	93.7	92.8	111.6	86.8	92.5	89.1	93.1	92.2	108.8	86.8	91.9
031	CONCRETE FORMWORK	100.0	100.0	100.0	96.9	43.5	51.1	95.4	78.5	81.0	83.1	58.2	61.8	97.0	62.0	67.0	96.2	58.2	63.7
032	CONCRETE REINFORCEMENT	100.0	100.0	100.0	92.7	68.6	79.1	92.7	82.1	86.7	103.5	68.6	83.8	92.7	65.7	77.5	103.5	68.6	83.8
033	CAST IN PLACE CONCRETE	100.0	100.0	100.0	94.0	47.0	74.3	96.2	69.2	84.9	91.7	58.4	77.7	93.5	67.5	82.6	91.7	58.4	77.7
3	CONCRETE	100.0	100.0	100.0	92.7	51.6	72.0	89.6	77.1	83.3	93.6	61.9	77.6	88.4	66.2	77.2	93.3	61.9	77.4
4	MASONRY	100.0	100.0	100.0	83.0	42.7	58.0	84.1	74.0	77.8	86.2	45.0	60.6	84.2	63.9	71.6	87.2	45.0	61.0
5	METALS	100.0	100.0	100.0	96.1	86.9	92.8	98.8	93.6	96.9	95.0	85.9	91.7	98.4	87.3	94.4	95.2	85.9	91.8
6	WOOD & PLASTICS	100.0	100.0	100.0	87.7	44.2	65.2	95.4	79.7	87.3	81.6	61.1	71.0	94.4	61.9	77.6	95.8	61.1	77.9
7	THERMAL & MOISTURE PROTECTION	100.0	100.0	100.0	97.4	44.8	73.0	97.6	68.8	84.2	97.5	53.9	77.3	97.3	60.7	80.4	97.6	53.9	77.4
8	DOORS & WINDOWS	100.0	100.0	100.0	93.3	49.0	82.6	97.0	76.2	91.9	93.3	60.3	85.3	97.0	58.7	87.7	93.4	60.3	85.4
092	LATH, PLASTER & GYPSUM BOARD	100.0	100.0	100.0	106.2	42.9	65.4	113.3	79.6	91.6	105.2	60.4	76.3	109.8	61.2	78.5	110.4	60.4	78.1
095	ACOUSTICAL TREATMENT & WOOD FLOORING	100.0	100.0	100.0	96.0	42.9	61.6	100.2	79.6	86.8	96.0	60.4	72.9	100.2	61.2	74.9	96.0	60.4	72.9
096	FLOORING & CARPET	100.0	100.0	100.0	96.0	43.2	83.2	99.0	60.9	89.7	102.1	34.1	85.6	99.0	63.5	90.3	109.7	34.1	91.3
099	PAINTING & WALL COVERINGS	100.0	100.0	100.0	90.8	37.8	59.7	90.8	57.5	71.2	90.8	64.1	75.1	90.8	59.9	72.7	90.8	64.1	75.1
9	FINISHES	100.0	100.0	100.0	96.9	42.9	69.2	99.2	73.3	85.9	100.4	54.4	76.8	98.7	62.0	79.9	103.4	54.4	78.3
10-14	TOTAL DIV. 10-14	100.0	100.0	100.0	100.0	62.7	92.1	100.0	83.2	96.4	100.0	66.7	93.0	100.0	69.5	93.6	100.0	66.7	93.0
15	MECHANICAL	100.0	100.0	100.0	97.3	47.1	74.9	100.0	70.7	86.9	98.1	50.8	77.0	100.0	61.7	82.9	98.1	50.8	77.0
16	ELECTRICAL	100.0	100.0	100.0	93.2	34.1	53.4	95.5	69.4	77.9	96.4	48.1	63.8	95.9	63.3	73.9	94.5	48.1	63.2
1-16	WEIGHTED AVERAGE	100.0	100.0	100.0	95.1	53.1	74.8	96.5	77.0	87.1	96.3	59.5	78.6	96.3	68.4	82.8	96.6	59.5	78.7

ALABAMA

DIVISION		EVERGREEN 364 MAT.	INST.	TOTAL	GADSDEN 359 MAT.	INST.	TOTAL	HUNTSVILLE 357-358 MAT.	INST.	TOTAL	JASPER 355 MAT.	INST.	TOTAL	MOBILE 365-366 MAT.	INST.	TOTAL	MONTGOMERY 360-361 MAT.	INST.	TOTAL
2	SITE WORK	112.1	87.1	92.9	95.3	92.9	93.5	88.8	93.1	92.1	94.8	92.6	93.1	100.3	87.3	90.3	100.8	87.1	90.3
031	CONCRETE FORMWORK	78.3	60.4	62.9	92.1	55.8	61.0	97.0	61.5	66.6	99.5	45.6	53.3	97.0	65.4	69.9	95.1	58.7	63.9
032	CONCRETE REINFORCEMENT	104.0	68.6	84.1	101.5	81.5	90.2	92.7	70.7	80.3	92.7	68.8	79.2	95.7	64.2	77.9	95.7	81.1	87.5
033	CAST IN PLACE CONCRETE	91.7	61.5	79.0	93.5	61.8	80.2	91.0	65.0	80.1	103.6	54.9	83.2	96.1	65.5	83.3	97.8	58.7	81.4
3	CONCRETE	93.6	63.9	78.6	92.7	64.4	78.5	87.2	66.0	76.5	96.2	55.2	75.5	90.1	66.7	78.3	90.5	64.6	77.6
4	MASONRY	86.2	50.7	64.2	82.2	53.4	64.3	84.0	60.8	69.6	79.8	46.6	59.2	84.7	64.3	72.1	85.1	44.7	60.0
5	METALS	95.1	85.9	91.8	96.1	92.4	94.8	98.4	88.9	95.0	96.0	86.2	92.5	97.0	86.4	93.2	97.1	91.6	95.1
6	WOOD & PLASTICS	76.4	61.1	68.5	82.3	55.7	68.6	94.4	61.1	77.2	90.3	45.3	67.0	94.4	66.7	80.1	92.0	61.1	76.0
7	THERMAL & MOISTURE PROTECTION	97.5	56.3	78.4	97.4	59.1	79.6	97.3	62.3	81.0	97.4	49.9	75.4	97.2	64.2	81.9	97.1	56.7	78.4
8	DOORS & WINDOWS	93.4	60.3	85.3	93.3	57.5	84.6	97.0	59.4	87.9	93.3	52.6	83.4	97.0	62.9	88.7	97.0	63.1	88.8
092	LATH, PLASTER & GYPSUM BOARD	103.0	60.4	75.5	104.1	54.8	72.3	109.8	60.4	77.9	106.4	44.0	66.2	109.8	66.2	81.6	109.8	60.4	77.9
095	ACOUSTICAL TREATMENT & WOOD FLOORING	96.0	60.4	72.9	96.0	54.8	69.3	100.2	60.4	74.4	94.7	44.0	61.9	100.2	66.2	78.1	100.2	60.4	74.4
096	FLOORING & CARPET	99.5	34.1	83.6	94.0	54.5	84.4	99.0	52.7	87.7	97.0	41.6	83.5	107.3	66.1	97.2	107.3	34.1	89.5
099	PAINTING & WALL COVERINGS	90.8	64.1	75.1	90.8	52.4	68.3	90.8	56.3	70.6	90.8	46.4	64.7	90.8	65.2	75.8	90.8	64.1	75.1
9	FINISHES	99.3	55.8	77.0	95.9	54.5	74.7	98.7	58.9	78.3	96.9	44.8	70.2	102.5	65.7	83.6	102.6	54.3	77.9
10-14	TOTAL DIV. 10-14	100.0	69.1	93.5	100.0	75.2	94.8	100.0	77.9	95.3	100.0	63.8	92.4	100.0	77.8	95.3	100.0	75.1	94.7
15	MECHANICAL	98.1	53.9	78.4	106.5	54.0	83.1	100.0	56.6	80.6	106.5	40.4	77.0	100.0	60.2	82.2	100.0	50.9	78.1
16	ELECTRICAL	92.2	48.1	62.5	94.9	69.4	77.7	95.9	70.8	79.0	95.2	34.6	54.3	95.9	65.8	75.6	95.5	48.1	63.5
1-16	WEIGHTED AVERAGE	95.9	61.3	79.2	97.0	66.2	82.2	96.1	68.4	82.7	97.5	53.3	76.2	97.0	68.9	83.4	97.1	60.8	79.6

ALABAMA / ALASKA

DIVISION		PHENIX CITY 368 MAT.	INST.	TOTAL	SELMA 367 MAT.	INST.	TOTAL	TUSCALOOSA 354 MAT.	INST.	TOTAL	ANCHORAGE 995-996 MAT.	INST.	TOTAL	FAIRBANKS 997 MAT.	INST.	TOTAL	JUNEAU 998 MAT.	INST.	TOTAL
2	SITE WORK	116.5	87.2	93.9	108.5	86.8	91.8	89.2	92.6	91.9	127.1	136.7	134.5	113.1	136.7	131.3	124.2	136.7	133.9
031	CONCRETE FORMWORK	88.0	51.6	56.8	84.7	58.2	62.0	97.0	48.1	55.1	129.9	120.0	121.4	131.4	126.1	126.9	131.2	120.0	121.6
032	CONCRETE REINFORCEMENT	102.8	79.9	89.9	103.5	68.6	83.8	92.7	81.5	86.4	143.8	115.1	127.7	120.9	115.2	117.7	107.0	115.1	111.6
033	CAST IN PLACE CONCRETE	91.7	59.4	78.2	91.6	58.4	77.7	94.8	56.0	78.5	178.7	119.1	153.8	151.3	119.7	138.0	179.4	119.1	154.2
3	CONCRETE	96.1	61.5	78.6	92.3	61.9	76.9	89.0	59.1	73.9	152.2	117.8	134.9	130.2	120.7	125.4	147.8	117.9	132.7
4	MASONRY	86.1	51.0	64.3	91.2	45.0	62.5	84.3	47.2	61.2	174.8	124.1	143.3	167.6	124.1	140.6	155.1	124.1	135.8
5	METALS	95.0	91.4	93.7	95.0	85.9	91.7	97.5	92.3	95.6	129.8	102.6	120.0	129.9	102.9	120.2	130.2	102.6	120.2
6	WOOD & PLASTICS	87.0	50.9	68.4	83.4	61.1	71.9	94.4	48.4	70.7	123.4	118.5	120.8	123.7	126.4	125.1	123.4	118.5	120.8
7	THERMAL & MOISTURE PROTECTION	97.9	57.5	79.2	97.4	53.9	77.2	97.3	55.2	77.8	198.8	112.6	158.8	195.0	114.5	157.7	195.5	112.6	157.1
8	DOORS & WINDOWS	93.3	57.2	84.6	93.3	60.3	85.3	97.0	59.3	87.9	132.1	112.3	127.3	129.1	116.6	126.1	129.1	112.3	125.0
092	LATH, PLASTER & GYPSUM BOARD	107.3	49.8	70.2	105.7	60.4	76.4	109.8	47.3	69.5	126.5	119.0	121.7	126.5	127.2	127.0	126.5	119.0	121.7
095	ACOUSTICAL TREATMENT & WOOD FLOORING	96.0	49.8	66.1	96.0	60.4	72.9	100.2	47.3	65.9	133.0	119.0	123.9	133.0	127.2	129.3	133.0	119.0	123.9
096	FLOORING & CARPET	104.9	34.1	87.7	103.0	34.1	86.3	99.0	50.1	87.1	130.2	130.9	130.4	130.4	130.9	130.5	130.2	130.9	130.4
099	PAINTING & WALL COVERINGS	90.8	64.1	75.1	90.8	64.1	75.1	90.8	55.2	69.9	122.9	117.8	119.9	122.9	122.6	122.7	122.9	117.8	119.9
9	FINISHES	101.9	48.7	74.6	100.5	54.4	76.9	98.7	48.4	72.9	143.9	122.2	132.8	141.9	127.4	134.5	142.6	122.2	132.1
10-14	TOTAL DIV. 10-14	100.0	74.3	94.6	100.0	66.7	93.0	100.0	71.9	94.1	100.0	116.7	103.5	100.0	117.7	103.7	100.0	116.7	103.5
15	MECHANICAL	98.1	51.6	77.4	98.1	50.8	77.0	100.0	44.7	75.3	106.9	109.9	108.2	106.9	118.3	112.0	106.9	111.9	109.1
16	ELECTRICAL	95.6	48.1	63.6	94.1	48.1	63.1	95.8	69.4	78.0	158.6	118.6	131.6	161.4	118.6	132.5	161.4	118.6	132.5
1-16	WEIGHTED AVERAGE	96.9	60.1	79.2	96.2	59.5	78.5	96.2	62.2	79.8	133.1	117.7	125.7	129.3	120.7	125.1	131.2	118.0	124.8

ALASKA / ARIZONA

DIVISION		KETCHIKAN 999 MAT.	INST.	TOTAL	CHAMBERS 865 MAT.	INST.	TOTAL	FLAGSTAFF 860 MAT.	INST.	TOTAL	GLOBE 855 MAT.	INST.	TOTAL	KINGMAN 864 MAT.	INST.	TOTAL	MESA/TEMPE 852 MAT.	INST.	TOTAL
2	SITE WORK	176.2	136.7	145.8	69.1	100.1	93.0	89.4	100.4	97.8	86.5	100.2	97.1	69.0	100.1	93.0	77.4	99.8	94.7
031	CONCRETE FORMWORK	114.3	120.0	119.2	96.0	73.5	76.7	106.6	79.6	83.5	92.3	73.5	76.2	93.1	73.6	76.4	99.6	71.9	75.8
032	CONCRETE REINFORCEMENT	111.5	115.1	113.6	105.2	73.2	87.2	104.4	82.9	92.3	106.2	79.2	91.0	106.0	79.3	91.0	107.0	79.4	91.4
033	CAST IN PLACE CONCRETE	289.3	119.0	218.0	95.6	77.9	87.9	95.7	80.4	89.3	95.0	77.4	87.6	95.3	78.0	88.0	95.7	71.5	85.6
3	CONCRETE	223.3	117.8	170.1	100.9	74.8	87.7	122.1	80.2	100.9	108.3	75.6	91.8	100.5	75.8	88.1	100.5	73.1	86.7
4	MASONRY	178.2	124.0	144.5	102.1	70.1	82.2	101.6	71.2	82.7	110.9	70.0	85.5	102.1	71.1	82.9	111.0	58.2	78.2
5	METALS	129.0	102.4	119.4	98.2	71.0	88.4	98.6	76.1	90.5	100.4	71.7	90.1	98.8	71.1	88.8	100.6	76.1	91.8
6	WOOD & PLASTICS	113.4	118.5	116.0	103.1	72.9	87.5	110.8	80.4	95.1	98.0	73.0	85.1	99.5	72.9	85.8	102.3	78.0	89.8
7	THERMAL & MOISTURE PROTECTION	199.1	115.1	160.1	108.1	75.1	92.8	110.3	78.2	95.4	106.5	74.1	91.5	108.1	75.4	92.9	105.6	70.4	89.3
8	DOORS & WINDOWS	129.1	112.3	125.0	102.5	72.5	95.2	102.6	79.4	97.0	101.9	72.6	94.8	102.7	72.5	95.4	102.0	77.5	96.0
092	LATH, PLASTER & GYPSUM BOARD	123.3	119.0	120.5	92.4	72.1	79.3	95.4	79.8	85.3	94.9	72.1	80.2	90.6	72.1	78.7	96.5	77.3	84.1
095	ACOUSTICAL TREATMENT & WOOD FLOORING	133.0	119.0	123.9	96.9	72.1	80.9	98.3	79.8	86.3	102.7	72.1	82.9	98.3	72.1	81.4	102.7	77.3	86.3
096	FLOORING & CARPET	126.6	130.9	127.6	97.2	65.5	89.5	99.7	65.8	91.5	102.9	65.5	93.8	96.3	74.4	91.0	104.3	74.4	97.0
099	PAINTING & WALL COVERINGS	122.9	112.1	116.6	103.7	63.2	79.9	103.7	63.2	79.9	115.0	63.2	84.6	103.7	63.2	79.9	115.0	67.9	87.4
9	FINISHES	143.7	121.5	132.3	95.1	70.9	82.7	98.2	75.3	86.4	100.1	70.9	85.1	94.8	72.4	83.3	100.0	72.0	85.6
10-14	TOTAL DIV. 10-14	100.0	116.7	103.5	100.0	83.0	96.4	100.0	84.0	96.6	100.0	83.1	96.4	100.0	83.0	96.4	100.0	77.6	95.3
15	MECHANICAL	96.9	111.9	103.6	98.8	76.5	88.9	99.9	83.2	92.4	96.5	73.7	86.3	98.8	75.2	88.3	99.8	71.7	87.3
16	ELECTRICAL	161.5	118.6	132.6	103.4	61.1	74.9	101.8	63.5	75.9	106.4	61.1	75.8	103.4	61.1	74.9	101.0	42.9	61.8
1-16	WEIGHTED AVERAGE	140.6	118.0	129.7	99.1	74.2	87.1	102.8	78.2	91.0	101.4	73.9	88.1	99.1	74.4	87.2	100.6	69.5	85.6

ARIZONA / ARKANSAS

DIVISION		PHOENIX 850,853 MAT.	INST.	TOTAL	PRESCOTT 863 MAT.	INST.	TOTAL	SHOW LOW 859 MAT.	INST.	TOTAL	TUCSON 856-857 MAT.	INST.	TOTAL	BATESVILLE 725 MAT.	INST.	TOTAL	CAMDEN 717 MAT.	INST.	TOTAL
2	SITE WORK	77.1	100.7	95.2	75.6	100.1	94.5	88.5	100.2	97.5	73.6	100.4	94.2	73.2	82.4	80.2	74.1	81.9	80.1
031	CONCRETE FORMWORK	100.9	79.4	82.5	101.1	73.5	77.4	101.8	73.5	77.6	100.4	79.1	82.2	75.4	54.3	57.3	74.3	38.2	43.3
032	CONCRETE REINFORCEMENT	105.1	83.2	92.8	104.4	73.2	86.8	107.0	73.2	87.9	104.0	82.9	92.1	95.5	55.6	73.0	96.9	30.6	59.5
033	CAST IN PLACE CONCRETE	95.8	79.9	89.2	95.6	77.9	88.2	95.0	77.4	87.6	88.7	79.8	85.0	79.5	53.6	68.7	81.5	39.4	63.8
3	CONCRETE	100.1	80.0	90.0	106.7	74.7	90.6	110.7	74.6	92.5	96.5	79.8	88.0	78.2	55.0	66.5	79.9	38.4	59.0
4	MASONRY	97.6	75.6	84.0	102.1	70.1	82.2	110.9	70.0	85.5	97.8	71.1	81.2	94.3	53.2	68.8	105.6	46.1	68.6
5	METALS	102.0	78.0	93.3	98.6	70.9	88.6	100.2	71.7	89.9	101.3	76.7	92.4	95.5	65.2	84.6	95.2	55.4	80.9
6	WOOD & PLASTICS	103.6	79.9	91.4	105.0	72.9	88.4	108.3	73.0	90.1	103.7	79.9	91.4	77.2	56.0	66.2	76.9	39.5	57.6
7	THERMAL & MOISTURE PROTECTION	105.6	78.3	92.9	108.7	75.1	93.1	106.8	74.7	91.9	106.3	74.3	91.5	94.5	54.3	75.8	94.2	41.0	69.6
8	DOORS & WINDOWS	101.1	79.1	95.8	102.6	72.5	95.3	100.9	72.6	94.0	99.7	79.1	94.7	96.9	51.1	85.8	93.1	33.8	78.7
092	LATH, PLASTER & GYPSUM BOARD	97.0	79.3	85.6	93.2	72.1	79.6	98.3	72.1	81.4	97.3	79.3	85.7	88.0	55.6	67.1	88.0	38.6	56.1
095	ACOUSTICAL TREATMENT & WOOD FLOORING	102.7	79.3	87.5	98.3	72.1	81.4	101.4	72.1	82.4	104.1	79.3	88.0	97.6	55.6	70.4	97.6	38.6	59.4
096	FLOORING & CARPET	104.6	74.6	97.3	98.0	65.5	90.1	106.1	65.5	96.2	103.7	65.8	94.5	110.4	61.9	98.6	110.3	20.5	88.4
099	PAINTING & WALL COVERINGS	115.0	64.3	85.3	103.7	63.2	79.9	115.0	63.2	84.6	112.3	63.2	83.5	97.1	54.6	72.2	97.1	42.7	65.2
9	FINISHES	100.1	76.9	88.2	96.2	70.9	83.2	101.4	70.9	85.8	99.9	75.0	87.1	97.3	56.1	76.2	97.3	36.1	65.9
10-14	TOTAL DIV. 10-14	100.0	84.0	96.6	100.0	83.0	96.4	100.0	83.1	96.4	100.0	84.1	96.6	100.0	60.2	91.6	100.0	55.1	90.5
15	MECHANICAL	99.8	83.2	92.4	99.9	73.1	87.9	96.5	75.2	87.0	99.7	77.7	89.9	96.6	52.6	77.0	96.6	40.5	71.6
16	ELECTRICAL	110.5	63.5	78.8	101.3	61.1	74.2	101.6	61.1	74.3	104.6	67.3	79.4	100.5	60.6	73.6	94.7	50.5	64.9
1-16	WEIGHTED AVERAGE	100.6	79.0	90.2	100.3	73.6	87.4	101.5	74.0	88.2	99.4	77.6	88.9	93.9	59.1	77.1	93.8	47.5	71.5

ARKANSAS

DIVISION		FAYETTEVILLE 727 MAT.	INST.	TOTAL	FORT SMITH 729 MAT.	INST.	TOTAL	HARRISON 726 MAT.	INST.	TOTAL	HOT SPRINGS 719 MAT.	INST.	TOTAL	JONESBORO 724 MAT.	INST.	TOTAL	LITTLE ROCK 720-722 MAT.	INST.	TOTAL
2	SITE WORK	72.8	81.6	79.5	76.1	82.4	80.9	78.0	82.4	81.4	77.6	81.7	80.7	101.1	96.9	97.9	76.1	82.4	80.9
031	CONCRETE FORMWORK	70.0	31.5	37.1	98.8	50.0	57.0	81.2	54.3	58.2	71.1	35.7	40.8	82.9	58.1	61.6	92.9	54.8	60.2
032	CONCRETE REINFORCEMENT	95.6	26.6	59.5	96.6	71.4	82.4	90.5	55.6	72.8	94.9	29.1	57.9	92.5	56.4	72.2	96.9	61.4	76.9
033	CAST IN PLACE CONCRETE	79.5	41.3	63.5	82.3	53.8	70.4	88.3	53.6	73.8	83.5	38.2	64.5	86.7	66.3	78.2	83.3	53.8	70.9
3	CONCRETE	77.8	35.4	56.4	82.6	56.1	69.2	85.3	55.0	70.0	83.0	36.6	59.6	83.2	62.3	72.7	82.7	56.4	69.4
4	MASONRY	85.2	35.4	54.2	94.6	58.1	71.9	94.5	53.2	68.8	78.8	35.6	52.0	87.1	58.8	69.5	92.5	58.1	71.1
5	METALS	95.5	52.4	80.0	97.4	71.2	88.0	96.5	65.2	85.2	95.2	53.9	80.3	91.7	82.5	88.4	97.0	67.8	86.5
6	WOOD & PLASTICS	71.7	30.9	50.6	99.4	49.5	73.6	83.8	56.0	69.4	73.3	35.8	53.9	80.7	59.5	69.8	95.6	56.0	75.2
7	THERMAL & MOISTURE PROTECTION	95.5	42.5	70.9	96.1	55.1	77.1	94.9	54.3	76.1	94.5	42.0	70.2	105.7	61.1	85.0	94.7	55.8	76.6
8	DOORS & WINDOWS	96.9	28.3	80.3	97.6	50.8	86.3	97.6	51.1	86.4	93.1	33.9	78.7	98.9	56.9	88.7	97.6	53.0	86.8
092	LATH, PLASTER & GYPSUM BOARD	85.4	29.7	49.5	94.8	48.9	65.2	91.6	55.6	68.4	87.0	34.7	53.3	99.8	58.8	73.3	94.8	55.6	69.5
095	ACOUSTICAL TREATMENT & WOOD FLOORING	97.6	29.7	53.6	101.7	48.9	67.5	101.7	55.6	71.8	97.6	34.7	56.9	102.3	58.8	74.1	101.7	55.6	71.8
096	FLOORING & CARPET	106.8	61.9	95.9	119.3	70.4	107.4	113.5	61.9	100.9	108.5	61.9	97.2	79.2	56.0	73.6	120.7	70.4	108.5
099	PAINTING & WALL COVERINGS	97.1	31.0	58.3	97.1	71.4	82.0	97.1	54.6	72.2	97.1	45.6	66.9	85.1	64.2	72.9	97.1	54.6	72.2
9	FINISHES	95.8	37.0	65.7	102.1	56.0	78.4	99.8	56.1	77.4	96.9	41.6	68.5	92.1	58.4	74.9	102.5	57.8	79.6
10-14	TOTAL DIV. 10-14	100.0	53.9	90.3	100.0	67.8	93.2	100.0	60.2	91.6	100.0	54.7	90.4	100.0	69.2	93.5	100.0	68.7	93.4
15	MECHANICAL	96.7	35.4	69.3	100.0	53.4	79.2	96.6	52.6	77.0	96.6	41.9	72.2	100.1	53.6	79.4	100.0	58.0	81.3
16	ELECTRICAL	90.4	27.2	47.8	96.0	61.7	72.9	98.1	60.6	72.8	98.2	49.4	65.3	101.2	60.7	73.9	97.0	65.4	75.7
1-16	WEIGHTED AVERAGE	92.4	40.8	67.5	96.1	60.7	79.0	95.3	59.1	77.8	93.0	46.9	70.7	95.2	64.8	80.5	95.9	62.3	79.7

ARKANSAS / CALIFORNIA

DIVISION		PINE BLUFF 716 MAT.	INST.	TOTAL	RUSSELLVILLE 728 MAT.	INST.	TOTAL	TEXARKANA 718 MAT.	INST.	TOTAL	WEST MEMPHIS 723 MAT.	INST.	TOTAL	ALHAMBRA 917-918 MAT.	INST.	TOTAL	ANAHEIM 928 MAT.	INST.	TOTAL
2	SITE WORK	79.1	82.4	81.6	74.6	82.3	80.6	95.7	83.0	86.0	109.2	96.9	99.8	86.2	109.3	104.0	88.7	111.5	106.2
031	CONCRETE FORMWORK	74.8	54.7	57.6	76.6	49.3	53.2	83.9	48.1	53.2	86.7	58.1	62.2	109.3	118.7	117.3	103.3	120.1	117.7
032	CONCRETE REINFORCEMENT	96.8	61.4	76.9	96.2	57.8	74.5	96.2	54.6	72.8	92.5	56.4	72.2	113.6	115.2	114.5	108.5	115.2	112.3
033	CAST IN PLACE CONCRETE	83.5	53.8	71.1	83.4	53.5	70.9	83.4	51.7	70.1	89.7	66.3	79.9	88.8	115.0	99.8	97.2	121.2	107.2
3	CONCRETE	84.0	56.4	70.0	81.4	53.2	67.2	79.6	51.5	65.4	88.7	62.3	75.4	112.6	115.5	114.1	114.2	118.6	116.4
4	MASONRY	108.7	58.1	77.3	94.7	53.2	68.9	94.7	41.3	61.5	77.6	58.8	65.9	129.3	120.3	123.7	87.4	122.9	109.4
5	METALS	96.1	67.8	85.9	95.5	65.7	84.8	88.5	64.4	79.8	91.0	82.5	87.9	91.2	97.2	93.4	110.5	101.4	107.2
6	WOOD & PLASTICS	72.7	56.0	64.1	79.0	49.5	63.8	83.5	51.6	67.0	88.5	59.5	73.5	101.9	117.1	109.8	99.6	117.5	108.9
7	THERMAL & MOISTURE PROTECTION	94.6	55.8	76.6	95.8	53.6	76.3	95.6	50.0	74.4	106.3	61.1	85.4	104.4	115.5	109.5	120.9	120.7	120.8
8	DOORS & WINDOWS	93.1	53.0	83.3	96.9	48.1	85.1	98.1	48.7	86.1	98.9	56.9	88.7	98.6	113.4	102.2	105.7	114.2	107.7
092	LATH, PLASTER & GYPSUM BOARD	86.5	55.6	66.5	88.0	48.9	62.8	90.1	51.1	64.9	102.4	58.8	74.3	90.8	118.1	108.4	91.1	118.2	108.6
095	ACOUSTICAL TREATMENT & WOOD FLOORING	97.6	55.6	70.4	97.6	48.9	66.0	104.4	51.1	69.9	102.3	58.8	74.1	115.3	118.1	117.1	122.0	118.2	119.5
096	FLOORING & CARPET	108.3	70.4	99.1	110.0	61.9	98.3	111.3	53.4	97.2	82.3	56.0	75.9	119.7	115.7	118.7	127.1	116.8	124.6
099	PAINTING & WALL COVERINGS	97.1	54.6	72.2	97.1	54.6	72.2	97.1	45.0	66.5	85.1	64.2	72.9	119.4	114.9	116.7	116.7	118.8	117.9
9	FINISHES	96.8	57.8	76.8	97.4	52.3	74.2	99.9	48.9	73.8	94.0	58.4	75.8	112.8	117.7	115.3	116.4	119.1	117.8
10 - 14	TOTAL DIV. 10 - 14	100.0	68.7	93.4	100.0	59.3	91.4	100.0	49.3	89.3	100.0	69.2	93.5	100.0	110.8	102.3	100.0	117.1	103.6
15	MECHANICAL	100.0	56.4	80.5	96.7	46.0	74.1	100.0	47.4	76.5	96.8	53.6	77.6	96.8	116.5	105.5	100.2	118.7	108.4
16	ELECTRICAL	94.9	65.4	75.0	96.0	60.6	72.1	98.1	50.5	66.0	103.9	60.7	74.7	116.0	118.2	117.5	93.4	109.6	104.3
1 - 16	WEIGHTED AVERAGE	95.4	62.0	79.3	94.1	56.9	76.1	94.7	53.4	74.8	95.2	64.8	80.5	103.2	114.4	108.6	104.9	115.1	109.8

CALIFORNIA

DIVISION		BAKERSFIELD 932-933 MAT.	INST.	TOTAL	BERKELEY 947 MAT.	INST.	TOTAL	EUREKA 955 MAT.	INST.	TOTAL	FRESNO 936-938 MAT.	INST.	TOTAL	INGLEWOOD 903-905 MAT.	INST.	TOTAL	LONG BEACH 906-908 MAT.	INST.	TOTAL
2	SITE WORK	94.4	108.3	105.1	119.4	105.3	108.6	105.0	107.1	106.6	95.8	108.3	105.5	81.0	107.3	101.2	88.1	107.3	102.9
031	CONCRETE FORMWORK	94.0	119.2	115.6	120.0	138.1	135.5	111.6	123.2	121.6	100.0	123.7	120.3	106.4	118.5	116.8	98.9	118.5	115.7
032	CONCRETE REINFORCEMENT	108.1	114.7	111.9	97.3	115.8	107.7	104.8	115.9	111.0	108.6	115.0	112.2	111.9	115.2	113.7	111.0	115.2	113.3
033	CAST IN PLACE CONCRETE	92.8	119.7	104.1	122.2	121.3	121.8	100.7	117.1	107.6	101.6	117.7	108.3	83.3	115.3	96.7	94.8	115.3	103.4
3	CONCRETE	111.4	117.5	114.5	122.9	126.7	124.8	126.6	118.5	122.6	116.2	118.9	117.5	105.6	115.7	110.7	116.6	115.7	116.1
4	MASONRY	106.4	115.3	111.9	132.0	127.4	129.2	112.2	132.7	124.9	109.0	116.8	113.8	85.8	120.4	107.3	95.4	120.4	110.9
5	METALS	105.0	99.6	103.1	107.3	106.1	106.8	110.1	101.2	106.9	106.9	100.8	104.7	102.9	98.4	101.3	102.8	98.4	101.2
6	WOOD & PLASTICS	87.9	117.6	103.3	124.6	140.1	132.6	114.6	123.5	119.2	101.1	123.5	112.7	103.9	117.0	110.7	95.8	117.0	106.7
7	THERMAL & MOISTURE PROTECTION	104.3	110.1	107.0	111.7	132.4	121.3	111.4	115.5	113.3	101.0	112.2	106.2	113.3	116.9	115.0	113.6	116.9	115.1
8	DOORS & WINDOWS	104.5	113.8	106.8	105.2	130.1	111.2	105.5	114.0	107.6	104.5	116.6	107.4	96.2	113.4	100.4	96.2	113.4	100.3
092	LATH, PLASTER & GYPSUM BOARD	91.1	118.2	108.6	92.3	141.1	123.8	96.4	124.2	114.3	90.0	124.2	112.1	86.7	118.1	106.9	84.1	118.1	106.0
095	ACOUSTICAL TREATMENT & WOOD FLOORING	122.0	118.2	119.5	115.5	141.1	132.1	128.9	124.2	125.9	122.0	124.2	123.4	112.6	118.1	116.1	112.6	118.1	116.1
096	FLOORING & CARPET	122.7	77.5	111.7	118.6	116.8	118.1	127.4	132.5	128.6	123.4	116.8	121.8	118.8	115.7	118.0	115.9	115.7	115.8
099	PAINTING & WALL COVERINGS	119.9	110.8	114.5	121.5	127.9	125.3	119.3	74.4	92.9	119.3	99.4	107.6	116.6	114.9	115.6	116.6	114.9	115.6
9	FINISHES	118.0	111.5	114.7	117.5	134.0	126.0	122.2	120.4	121.3	118.1	120.0	119.1	111.5	117.6	114.6	110.5	117.6	114.1
10 - 14	TOTAL DIV. 10 - 14	100.0	117.4	103.7	100.0	139.2	108.3	100.0	135.4	107.5	100.0	135.6	107.5	100.0	110.3	102.2	100.0	110.3	102.2
15	MECHANICAL	100.2	102.2	101.1	96.8	137.4	114.9	96.9	101.0	98.7	100.2	112.6	105.7	96.8	116.4	105.6	96.8	116.4	105.6
16	ELECTRICAL	93.8	95.1	94.7	112.7	135.8	128.3	103.5	106.0	105.2	92.6	102.3	99.2	112.1	118.2	116.2	111.6	118.2	116.1
1 - 16	WEIGHTED AVERAGE	104.3	107.0	105.6	109.5	127.9	118.4	108.6	112.1	110.3	105.3	112.3	108.7	101.2	114.3	107.5	103.0	114.3	108.5

CALIFORNIA

DIVISION		LOS ANGELES 900-902 MAT.	INST.	TOTAL	MARYSVILLE 959 MAT.	INST.	TOTAL	MODESTO 953 MAT.	INST.	TOTAL	MOJAVE 935 MAT.	INST.	TOTAL	OAKLAND 946 MAT.	INST.	TOTAL	OXNARD 930 MAT.	INST.	TOTAL
2	SITE WORK	87.0	110.6	105.1	100.7	107.3	105.8	94.0	107.6	104.5	89.2	108.2	103.8	127.5	105.1	110.3	95.4	105.9	103.5
031	CONCRETE FORMWORK	106.4	119.7	117.8	97.5	123.6	119.9	95.4	123.7	119.6	105.7	119.0	117.1	107.7	138.5	134.0	101.1	120.1	117.4
032	CONCRETE REINFORCEMENT	111.9	115.2	113.8	104.8	115.0	110.6	108.6	115.2	112.3	110.1	114.8	112.7	99.5	115.6	108.6	108.1	114.8	111.9
033	CAST IN PLACE CONCRETE	90.7	118.4	102.3	112.5	117.2	114.5	100.7	117.6	107.8	84.7	120.4	99.6	118.9	121.4	120.0	98.5	120.2	107.6
3	CONCRETE	113.1	117.3	115.2	126.2	118.6	122.4	115.4	118.9	117.1	104.8	117.6	111.3	125.3	126.9	126.1	114.7	118.1	116.4
4	MASONRY	100.6	122.9	114.4	113.0	115.8	114.8	111.3	120.1	116.8	106.7	120.4	115.2	139.6	124.9	130.5	112.0	114.4	113.5
5	METALS	109.3	99.5	105.8	109.4	100.4	106.1	106.5	101.2	104.6	105.9	98.8	103.4	102.5	105.4	103.5	104.5	100.4	103.0
6	WOOD & PLASTICS	99.4	117.1	108.5	99.1	123.5	111.7	92.5	123.5	108.5	104.1	115.4	109.9	107.8	140.7	124.8	97.6	117.6	107.9
7	THERMAL & MOISTURE PROTECTION	113.8	119.4	116.3	110.6	115.4	112.9	110.0	112.7	111.3	104.8	111.9	108.1	111.8	131.2	120.8	109.9	115.5	112.5
8	DOORS & WINDOWS	100.5	114.0	103.7	104.7	117.5	107.8	103.7	115.0	106.4	100.0	112.6	103.1	105.3	128.5	110.9	103.2	114.3	105.9
092	LATH, PLASTER & GYPSUM BOARD	87.0	118.2	107.1	90.8	124.2	112.4	92.2	124.2	112.8	99.2	115.8	110.0	86.8	141.7	122.3	91.1	118.2	108.6
095	ACOUSTICAL TREATMENT & WOOD FLOORING	120.8	118.2	119.1	127.5	124.2	125.4	122.0	124.2	123.4	127.5	115.8	120.0	116.9	141.7	133.0	122.0	118.2	119.5
096	FLOORING & CARPET	117.2	116.8	117.1	122.1	116.8	120.8	122.7	116.8	121.3	129.7	76.7	116.8	113.0	116.8	113.9	122.7	116.8	121.3
099	PAINTING & WALL COVERINGS	116.6	120.0	118.6	119.9	104.4	110.6	119.3	98.7	107.2	119.3	101.7	109.0	121.5	123.9	123.0	119.3	114.3	116.4
9	FINISHES	112.8	118.9	115.9	119.1	120.2	119.7	118.0	119.3	118.7	122.1	110.0	115.9	115.7	134.0	125.1	117.8	118.7	118.3
10 - 14	TOTAL DIV. 10 - 14	100.0	115.9	103.4	100.0	135.4	107.5	100.0	135.6	107.5	100.0	114.1	103.0	100.0	139.5	108.3	100.0	117.5	103.7
15	MECHANICAL	100.1	118.6	108.3	96.9	116.0	105.4	100.2	112.6	105.7	96.9	111.4	103.4	100.1	137.5	116.8	100.2	118.7	108.5
16	ELECTRICAL	106.8	121.4	116.6	97.9	107.2	104.2	101.7	107.1	105.3	92.1	95.1	94.1	111.4	129.6	123.7	97.6	111.8	107.1
1 - 16	WEIGHTED AVERAGE	105.0	116.5	110.6	107.3	113.7	110.4	106.0	113.3	109.5	102.8	108.9	105.8	109.9	126.5	117.9	105.4	113.8	109.4

CALIFORNIA

DIVISION		PALM SPRINGS 922			PALO ALTO 943			PASADENA 910 - 912			REDDING 960			RICHMOND 948			RIVERSIDE 925		
		MAT.	INST.	TOTAL	MAT.	INST.	TOTAL	MAT.	INST.	TOTAL	MAT.	INST.	TOTAL	MAT.	INST.	TOTAL	MAT.	INST.	TOTAL
2	SITE WORK	81.3	108.7	102.4	114.6	105.2	107.4	83.3	109.3	103.3	100.4	107.3	105.7	126.9	105.2	110.2	87.6	109.1	104.1
031	CONCRETE FORMWORK	95.3	118.9	115.5	101.4	132.3	127.8	94.2	118.7	115.2	100.7	123.2	120.0	124.2	138.4	136.4	104.4	119.9	117.7
032	CONCRETE REINFORCEMENT	110.3	115.0	112.9	97.3	116.2	107.9	114.6	115.2	114.9	104.8	114.8	110.4	97.3	115.6	107.6	107.1	114.9	111.0
033	CAST IN PLACE CONCRETE	88.6	119.4	101.5	103.3	118.0	109.5	84.3	115.0	97.2	111.5	117.1	113.9	118.9	121.5	120.0	96.3	121.1	106.7
3	CONCRETE	106.0	117.3	111.7	110.8	123.1	117.0	107.1	115.5	111.3	125.0	118.4	121.7	126.3	126.9	126.6	113.7	118.4	116.1
4	MASONRY	84.4	115.9	103.9	109.3	135.5	125.5	109.7	120.3	116.3	113.0	112.2	112.5	131.8	124.9	127.5	85.2	115.9	104.2
5	METALS	110.0	100.0	106.4	100.4	106.4	102.6	91.3	97.2	93.4	108.9	99.8	105.6	100.5	105.5	102.3	110.5	100.7	107.0
6	WOOD & PLASTICS	94.0	117.4	106.1	104.7	132.5	119.0	85.5	117.1	101.8	98.7	123.5	111.5	129.7	140.7	135.4	99.6	117.5	108.9
7	THERMAL & MOISTURE PROTECTION	119.6	116.9	118.3	111.3	132.0	120.9	103.9	115.5	109.2	110.6	113.8	112.1	112.2	130.2	120.5	120.0	116.5	118.4
8	DOORS & WINDOWS	102.5	114.2	105.3	105.3	123.2	109.6	98.6	113.5	102.2	106.0	115.0	108.2	105.3	127.8	110.7	105.7	114.2	107.7
092	LATH, PLASTER & GYPSUM BOARD	88.7	118.1	107.6	85.7	133.3	116.4	85.5	118.1	106.5	91.1	124.2	112.5	94.2	141.7	124.9	90.8	118.2	108.4
095	ACOUSTICAL TREATMENT & WOOD FLOORING	120.7	118.1	119.0	116.9	141.7	133.0	115.3	118.1	117.1	128.9	124.2	125.9	116.9	141.7	133.0	120.7	118.2	119.0
096	FLOORING & CARPET	124.9	111.0	121.5	111.9	116.8	113.1	114.1	115.7	114.5	121.9	116.8	120.7	120.4	116.8	119.5	127.0	116.8	124.5
099	PAINTING & WALL COVERINGS	116.7	118.8	117.9	121.5	126.8	124.6	119.4	114.9	116.7	119.3	104.4	110.6	121.5	123.9	123.0	116.7	118.8	117.9
9	FINISHES	114.7	117.5	116.1	114.6	129.4	122.2	110.1	117.7	114.0	119.3	120.2	119.8	119.1	134.0	126.8	116.0	119.1	117.6
10 - 14	TOTAL DIV. 10 - 14	100.0	116.7	103.5	100.0	138.2	108.1	100.0	110.8	102.3	100.0	135.3	107.4	100.0	139.4	108.3	100.0	117.1	103.6
15	MECHANICAL	96.9	116.6	105.7	96.8	145.4	118.5	96.8	116.5	105.5	100.2	112.1	105.5	96.8	137.4	114.9	100.1	118.6	108.4
16	ELECTRICAL	99.3	102.3	101.4	111.2	129.7	123.7	111.4	118.2	116.0	102.7	86.5	91.8	112.6	122.4	119.3	93.6	102.3	99.5
1 - 16	WEIGHTED AVERAGE	102.5	112.0	107.1	105.0	127.8	116.0	100.6	114.4	107.2	108.4	108.9	108.6	109.3	125.3	117.0	104.6	112.8	108.6

CALIFORNIA

DIVISION		SACRAMENTO 942,956 - 958			SALINAS 939			SAN BERNARDINO 923 - 924			SAN DIEGO 919 - 921			SAN FRANCISCO 940 - 941			SAN JOSE 951		
		MAT.	INST.	TOTAL	MAT.	INST.	TOTAL	MAT.	INST.	TOTAL	MAT.	INST.	TOTAL	MAT.	INST.	TOTAL	MAT.	INST.	TOTAL
2	SITE WORK	93.2	111.9	107.6	111.4	108.3	109.0	67.1	108.7	99.1	87.3	102.6	99.1	128.3	112.0	115.8	130.8	101.6	108.3
031	CONCRETE FORMWORK	106.1	123.9	121.3	102.3	128.0	124.3	105.5	118.7	116.8	103.9	119.4	117.2	107.9	139.9	135.3	103.4	138.9	133.8
032	CONCRETE REINFORCEMENT	100.6	115.1	108.7	108.6	115.4	112.5	107.1	114.8	111.5	110.9	114.7	113.1	113.3	116.3	115.0	105.1	116.2	111.4
033	CAST IN PLACE CONCRETE	103.2	117.6	109.2	97.3	117.8	105.9	66.4	118.3	88.2	92.4	109.1	99.4	118.8	123.4	120.7	117.1	121.9	119.1
3	CONCRETE	117.7	118.9	118.3	128.8	120.9	124.8	83.1	116.8	100.1	113.6	114.0	113.8	127.1	128.4	127.8	123.5	127.5	125.5
4	MASONRY	112.1	116.1	114.6	106.4	122.1	116.1	92.5	113.6	105.6	102.3	114.9	110.1	139.9	135.4	137.1	147.6	136.1	140.5
5	METALS	98.6	100.2	99.2	109.5	101.8	106.8	110.2	99.7	106.4	108.5	99.9	105.4	107.1	108.5	107.6	111.0	108.5	110.1
6	WOOD & PLASTICS	103.0	123.6	113.6	103.6	128.6	116.5	104.8	117.4	111.3	99.1	117.5	108.6	107.8	141.0	124.9	105.0	140.6	123.4
7	THERMAL & MOISTURE PROTECTION	121.0	115.1	118.3	106.3	121.5	113.4	118.5	115.4	117.1	113.5	109.2	111.5	111.8	135.3	122.7	106.6	134.7	119.6
8	DOORS & WINDOWS	116.6	115.0	116.2	104.6	121.1	108.6	102.5	113.5	105.2	104.4	114.3	106.8	107.3	131.3	113.1	95.3	129.5	103.6
092	LATH, PLASTER & GYPSUM BOARD	85.2	124.2	110.4	92.3	129.5	116.3	92.7	118.1	109.1	90.0	118.1	108.1	89.0	141.7	123.0	91.6	141.7	124.0
095	ACOUSTICAL TREATMENT & WOOD FLOORING	125.1	124.2	124.5	127.5	129.5	128.8	122.0	118.1	119.5	123.6	118.1	120.0	126.5	141.7	136.4	113.8	141.7	131.9
096	FLOORING & CARPET	115.4	116.8	115.7	123.6	116.8	121.9	129.1	111.2	124.8	123.5	116.1	121.7	113.0	116.8	113.9	119.7	116.8	119.0
099	PAINTING & WALL COVERINGS	119.7	106.5	111.9	119.3	127.4	124.1	116.7	113.6	114.9	116.7	116.5	116.5	121.5	140.2	132.5	121.3	127.4	124.9
9	FINISHES	116.2	120.6	118.5	120.7	126.8	123.8	115.9	117.0	116.4	115.9	118.8	117.4	117.8	136.2	127.2	117.0	134.4	125.9
10 - 14	TOTAL DIV. 10 - 14	100.0	135.7	107.5	100.0	136.2	107.6	100.0	111.6	102.4	100.0	117.3	103.7	100.0	140.2	108.5	100.0	139.3	108.3
15	MECHANICAL	100.0	116.9	107.6	96.9	112.5	103.9	96.9	116.2	105.5	100.1	116.7	107.5	100.1	167.5	130.2	100.2	145.9	120.6
16	ELECTRICAL	102.1	108.2	106.2	92.5	115.5	108.0	99.3	105.1	103.2	93.8	94.3	94.1	107.3	149.4	135.7	108.0	131.8	124.1
1 - 16	WEIGHTED AVERAGE	106.9	114.5	110.5	107.3	116.7	111.8	100.1	111.8	105.7	104.9	109.5	107.1	111.0	138.2	124.2	110.1	129.7	119.6

CALIFORNIA

DIVISION		SAN LUIS OBISPO 934			SAN MATEO 944			SAN RAFAEL 949			SANTA ANA 926 - 927			SANTA BARBARA 931			SANTA CRUZ 950		
		MAT.	INST.	TOTAL	MAT.	INST.	TOTAL	MAT.	INST.	TOTAL	MAT.	INST.	TOTAL	MAT.	INST.	TOTAL	MAT.	INST.	TOTAL
2	SITE WORK	102.5	107.9	106.7	123.7	105.2	109.5	105.9	111.9	110.5	79.6	108.7	102.0	95.8	108.3	105.4	131.6	101.5	108.4
031	CONCRETE FORMWORK	108.6	118.6	117.2	110.1	138.5	134.4	118.2	138.1	135.3	106.2	118.8	117.0	101.6	119.9	117.3	103.4	128.4	124.9
032	CONCRETE REINFORCEMENT	110.1	114.7	112.7	97.3	116.1	107.9	98.0	116.1	108.2	110.9	114.8	113.1	108.1	114.9	111.9	122.4	115.5	118.5
033	CAST IN PLACE CONCRETE	104.6	117.1	109.9	115.0	121.5	117.7	133.9	120.1	128.1	84.9	118.3	98.9	98.1	119.9	107.3	116.2	120.5	118.0
3	CONCRETE	124.5	116.3	120.4	121.8	127.0	124.4	149.4	126.1	137.7	103.4	116.9	110.2	114.5	117.9	116.2	125.4	122.3	123.8
4	MASONRY	108.2	111.9	110.5	131.5	135.3	133.8	107.8	135.2	124.8	81.1	114.0	101.5	106.8	115.8	112.4	147.9	122.6	132.2
5	METALS	106.6	99.1	103.9	100.3	106.3	102.5	101.7	101.9	101.8	110.2	99.8	106.5	104.3	100.3	102.9	112.3	106.8	110.3
6	WOOD & PLASTICS	106.8	117.6	112.3	114.2	140.7	127.9	116.6	140.4	128.9	107.0	117.4	112.4	97.6	117.6	107.9	105.0	128.8	117.3
7	THERMAL & MOISTURE PROTECTION	106.1	110.7	108.2	111.8	132.4	121.4	128.2	131.6	129.8	120.1	116.2	118.3	105.3	113.0	108.9	106.6	125.8	115.5
8	DOORS & WINDOWS	102.3	112.5	104.8	105.2	128.9	111.0	115.4	128.7	118.7	101.6	113.5	104.5	104.5	114.3	106.9	97.4	121.1	103.1
092	LATH, PLASTER & GYPSUM BOARD	100.3	118.1	111.8	88.9	141.7	123.0	90.5	141.7	123.6	93.7	118.1	109.4	91.1	118.2	108.6	93.8	129.5	116.8
095	ACOUSTICAL TREATMENT & WOOD FLOORING	127.5	118.1	121.4	116.9	141.7	133.0	125.1	141.7	135.9	122.0	118.1	119.5	122.0	118.2	119.5	123.4	129.5	127.4
096	FLOORING & CARPET	130.9	104.8	124.6	115.1	116.8	115.5	124.8	116.8	122.9	129.9	114.9	126.3	122.7	104.8	118.4	119.7	116.8	119.0
099	PAINTING & WALL COVERINGS	119.3	114.3	116.4	121.5	126.8	124.6	117.0	122.0	119.9	116.7	114.9	115.6	119.3	114.3	116.4	121.3	127.4	124.9
9	FINISHES	123.5	116.0	119.7	116.5	134.4	125.7	118.3	133.6	126.1	117.3	117.8	117.5	118.1	116.6	117.3	119.1	127.0	123.1
10 - 14	TOTAL DIV. 10 - 14	100.0	111.9	102.5	100.0	139.5	108.3	100.0	138.5	108.1	100.0	111.6	102.4	100.0	117.5	103.7	100.0	136.9	107.8
15	MECHANICAL	96.9	116.7	105.7	96.8	129.7	111.5	96.8	147.5	119.4	96.9	116.7	105.7	100.2	118.7	108.4	100.2	112.7	105.8
16	ELECTRICAL	92.1	106.8	102.0	111.2	134.7	127.0	104.9	106.0	105.7	99.3	106.9	104.5	90.4	106.4	101.2	105.9	124.7	118.6
1 - 16	WEIGHTED AVERAGE	106.2	111.5	108.8	108.0	127.1	117.2	111.2	125.7	118.2	102.3	112.3	107.2	104.6	112.9	108.6	110.8	118.4	114.5

DIVISION		CALIFORNIA															COLORADO		
		SANTA ROSA			STOCKTON			SUSANVILLE			VALLEJO			VAN NUYS			ALAMOSA		
		954			952			961			945			913 - 916			811		
		MAT.	INST.	TOTAL	MAT.	INST.	TOTAL	MAT.	INST.	TOTAL	MAT.	INST.	TOTAL	MAT.	INST.	TOTAL	MAT.	INST.	TOTAL
2	SITE WORK	94.1	107.8	104.7	93.7	107.6	104.4	108.1	107.3	107.5	89.2	112.1	106.8	99.1	109.3	107.0	146.2	94.7	106.6
031	CONCRETE FORMWORK	96.9	138.0	132.1	95.4	123.6	119.6	99.3	123.1	119.7	108.2	137.8	133.5	103.0	118.7	116.4	98.2	59.7	65.3
032	CONCRETE REINFORCEMENT	105.8	116.2	111.7	108.6	115.2	112.3	104.8	115.0	110.5	99.2	116.1	108.8	114.6	115.2	114.9	110.6	66.5	85.8
033	CAST IN PLACE CONCRETE	110.5	119.6	114.3	98.1	117.6	106.3	101.6	117.0	108.0	106.6	119.4	112.0	88.9	115.0	99.8	106.6	60.6	87.3
3	CONCRETE	126.2	126.0	126.1	114.1	118.8	116.5	129.2	118.3	123.7	119.3	125.8	122.6	125.7	115.6	120.6	120.3	62.1	90.9
4	MASONRY	111.8	133.5	125.3	111.6	120.1	116.9	112.1	115.8	114.4	79.9	133.5	113.2	129.4	120.3	123.7	136.3	67.3	93.4
5	METALS	110.7	104.1	108.3	106.5	101.1	104.6	109.6	99.9	106.1	101.9	102.9	102.2	90.5	97.3	92.9	100.8	77.9	92.5
6	WOOD & PLASTICS	95.6	140.3	118.7	92.5	123.5	108.5	100.9	123.5	112.6	102.1	140.4	121.9	94.9	117.1	106.4	100.8	57.3	78.4
7	THERMAL & MOISTURE PROTECTION	120.7	130.4	125.2	110.0	111.0	110.5	111.5	114.8	113.0	126.3	131.3	128.6	105.3	115.5	110.0	104.9	75.6	91.4
8	DOORS & WINDOWS	103.3	130.9	110.0	103.7	115.0	106.4	105.9	115.0	108.1	117.9	131.0	121.1	98.4	113.5	102.1	94.0	57.1	85.0
092	LATH, PLASTER & GYPSUM BOARD	89.5	141.7	123.2	92.2	124.2	112.8	91.3	124.2	112.5	85.7	141.7	121.9	87.8	118.1	107.4	88.6	55.6	67.3
095	ACOUSTICAL TREATMENT & WOOD FLOORING	122.0	141.7	134.8	122.0	124.2	123.4	127.5	124.2	125.4	125.1	141.7	135.9	114.0	118.1	116.6	94.2	55.6	69.2
096	FLOORING & CARPET	125.5	116.8	123.4	122.7	116.8	121.3	122.7	116.8	121.3	117.4	116.8	119.0	117.4	115.7	117.0	117.9	87.8	110.6
099	PAINTING & WALL COVERINGS	116.7	125.1	121.7	119.3	103.4	110.0	119.3	85.0	99.2	117.7	120.1	119.1	119.4	114.9	116.7	127.5	38.0	74.9
9	FINISHES	116.0	133.5	125.0	118.0	119.9	119.0	120.0	117.8	118.9	114.7	133.0	124.1	112.2	117.7	115.0	106.3	61.9	83.5
10 - 14	TOTAL DIV. 10 - 14	100.0	137.5	107.9	100.0	135.6	107.5	100.0	135.3	107.4	100.0	137.8	108.0	100.0	110.8	102.3	100.0	73.1	94.3
15	MECHANICAL	96.9	166.3	127.9	100.2	112.6	105.7	97.1	108.7	102.3	100.0	136.6	116.4	96.8	115.0	104.9	96.5	67.2	83.4
16	ELECTRICAL	99.8	106.1	104.0	101.7	106.7	105.1	103.3	86.5	92.0	98.1	127.1	117.6	111.4	118.2	116.0	96.6	71.8	79.9
1 - 16	WEIGHTED AVERAGE	107.2	129.0	117.7	105.9	113.2	109.4	108.6	108.3	108.5	105.5	127.0	115.9	104.5	114.1	109.2	104.9	70.3	88.2

DIVISION		COLORADO																	
		BOULDER			COLORADO SPRINGS			DENVER			DURANGO			FORT COLLINS			FORT MORGAN		
		803			808 - 809			800 - 802			813			805			807		
		MAT.	INST.	TOTAL	MAT.	INST.	TOTAL	MAT.	INST.	TOTAL	MAT.	INST.	TOTAL	MAT.	INST.	TOTAL	MAT.	INST.	TOTAL
2	SITE WORK	119.4	101.4	105.5	120.7	99.2	104.1	119.9	108.6	111.2	138.2	94.7	104.8	132.6	102.5	109.4	120.9	102.4	106.6
031	CONCRETE FORMWORK	109.1	58.8	66.0	92.4	76.4	78.7	102.9	78.1	81.7	113.0	58.6	66.4	102.3	71.7	76.1	110.8	71.3	76.9
032	CONCRETE REINFORCEMENT	101.2	60.7	78.4	97.5	79.7	87.5	97.0	80.2	87.5	110.6	69.8	87.8	101.6	71.0	84.4	102.2	71.0	84.6
033	CAST IN PLACE CONCRETE	105.2	71.7	91.2	109.4	81.4	97.7	102.3	81.0	93.4	122.6	59.9	96.3	118.8	75.3	100.6	103.2	70.2	89.4
3	CONCRETE	110.1	64.2	86.9	102.5	79.2	90.7	107.0	79.8	93.3	123.5	62.1	92.5	121.2	73.3	97.0	108.7	71.3	89.8
4	MASONRY	98.8	50.6	68.9	105.1	71.3	84.1	104.1	79.5	88.8	123.0	67.3	88.4	115.4	70.8	87.7	114.4	69.9	86.7
5	METALS	98.5	68.4	87.7	100.2	85.3	94.9	103.5	86.0	97.2	101.5	80.2	93.8	99.5	81.1	92.9	98.3	80.0	91.7
6	WOOD & PLASTICS	112.6	60.0	85.5	90.9	77.7	84.1	101.9	79.3	90.2	115.8	56.4	85.2	101.6	72.3	86.5	114.3	72.3	92.6
7	THERMAL & MOISTURE PROTECTION	105.2	67.4	87.6	105.5	80.8	94.1	104.9	83.1	94.8	105.0	69.8	88.7	105.5	76.2	91.9	105.1	75.9	91.6
8	DOORS & WINDOWS	93.9	52.5	83.8	96.3	79.4	92.2	95.5	80.9	92.0	101.2	58.5	90.9	93.8	71.0	88.3	93.8	70.4	88.1
092	LATH, PLASTER & GYPSUM BOARD	103.0	58.9	74.5	91.0	76.9	81.9	100.5	78.8	86.5	94.9	54.7	69.0	99.3	71.5	81.4	103.5	71.5	82.9
095	ACOUSTICAL TREATMENT & WOOD FLOORING	89.7	58.9	69.8	96.6	76.9	83.8	95.2	78.8	84.6	94.2	54.7	68.6	89.7	71.5	77.9	89.7	71.5	77.9
096	FLOORING & CARPET	116.5	87.8	109.5	111.5	71.2	101.7	112.5	91.6	107.4	122.8	87.8	114.3	112.3	87.8	106.4	117.1	87.8	109.9
099	PAINTING & WALL COVERINGS	116.0	40.3	71.6	116.1	65.2	86.2	116.0	78.1	93.8	127.5	38.0	74.9	116.0	64.7	85.9	116.0	70.5	89.3
9	FINISHES	104.2	62.0	82.6	102.4	73.9	87.8	103.6	80.4	91.7	108.0	61.4	84.1	102.9	73.9	88.1	104.4	74.4	89.0
10 - 14	TOTAL DIV. 10 - 14	100.0	69.4	93.5	100.0	84.6	96.7	100.0	84.2	96.7	100.0	72.9	94.3	100.0	74.2	94.6	100.0	73.9	94.5
15	MECHANICAL	96.6	70.1	84.8	100.0	73.9	88.3	99.9	81.6	91.7	96.5	48.1	74.9	99.9	76.6	89.5	96.6	76.0	87.4
16	ELECTRICAL	89.6	75.9	80.4	94.7	83.6	87.2	96.7	85.2	89.0	95.6	67.8	76.9	89.6	75.9	80.4	89.9	78.4	82.1
1 - 16	WEIGHTED AVERAGE	100.0	69.5	85.3	100.7	80.3	90.8	101.9	84.8	93.6	105.6	66.0	86.5	103.2	77.9	91.0	100.7	77.7	89.6

DIVISION		COLORADO																	
		GLENWOOD SPRINGS			GOLDEN			GRAND JUNCTION			GREELEY			MONTROSE			PUEBLO		
		816			804			815			806			814			810		
		MAT.	INST.	TOTAL	MAT.	INST.	TOTAL	MAT.	INST.	TOTAL	MAT.	INST.	TOTAL	MAT.	INST.	TOTAL	MAT.	INST.	TOTAL
2	SITE WORK	156.7	102.0	114.6	137.8	101.6	110.0	136.5	101.1	109.3	116.7	101.4	104.9	149.1	97.5	109.4	127.5	95.7	103.1
031	CONCRETE FORMWORK	95.3	69.1	72.8	92.6	71.2	74.3	110.3	54.1	62.2	99.2	58.7	64.6	94.7	54.7	60.4	104.4	76.9	80.8
032	CONCRETE REINFORCEMENT	108.4	69.2	86.3	102.2	62.7	79.9	109.8	76.9	91.3	101.2	60.7	78.4	108.2	77.0	90.6	102.4	79.6	89.6
033	CAST IN PLACE CONCRETE	106.5	60.4	87.2	103.3	74.7	91.3	118.0	52.9	90.7	99.3	71.7	87.7	106.5	53.9	84.5	105.8	81.8	95.8
3	CONCRETE	125.6	66.7	95.9	120.0	71.5	95.5	119.5	58.8	88.9	104.4	64.1	84.1	116.1	59.4	87.5	108.2	79.5	93.7
4	MASONRY	110.2	63.1	81.0	118.3	60.0	82.0	145.5	60.0	92.4	109.3	50.6	72.8	115.7	60.1	81.2	106.5	69.0	83.2
5	METALS	101.2	80.0	93.6	98.4	80.9	92.1	101.8	79.2	93.7	99.5	68.4	88.3	100.2	79.8	92.9	103.4	86.4	97.3
6	WOOD & PLASTICS	95.0	69.6	81.9	94.7	72.3	83.1	109.4	56.0	81.8	98.3	60.0	78.5	94.3	56.2	74.6	103.6	77.9	90.4
7	THERMAL & MOISTURE PROTECTION	104.7	73.9	90.4	106.0	72.2	90.4	103.6	63.4	85.0	104.7	67.4	87.4	104.9	66.4	87.1	103.0	79.3	92.0
8	DOORS & WINDOWS	100.4	71.7	93.4	93.9	74.3	89.1	101.1	58.3	90.7	93.8	52.5	83.8	101.0	58.4	90.7	95.5	79.5	91.6
092	LATH, PLASTER & GYPSUM BOARD	100.6	68.8	80.1	96.6	71.5	80.4	106.4	54.7	73.0	98.2	58.9	72.9	83.8	54.7	65.0	91.2	76.9	82.0
095	ACOUSTICAL TREATMENT & WOOD FLOORING	92.8	68.8	77.2	89.7	71.5	77.9	92.8	54.7	68.1	89.7	58.9	69.8	94.2	54.7	68.6	101.0	76.9	85.4
096	FLOORING & CARPET	117.1	80.8	108.3	109.9	87.8	104.5	122.2	87.8	113.8	111.0	87.8	105.4	120.2	72.6	108.6	118.8	91.6	112.2
099	PAINTING & WALL COVERINGS	127.4	71.0	94.3	116.0	72.4	90.4	127.4	38.0	74.9	116.0	40.3	71.6	127.5	38.0	74.9	127.5	60.4	88.1
9	FINISHES	107.7	71.6	89.2	102.5	74.6	88.2	108.5	58.8	83.0	101.6	62.0	81.3	106.4	55.8	80.5	106.6	77.7	91.8
10 - 14	TOTAL DIV. 10 - 14	100.0	73.4	94.4	100.0	73.9	94.5	100.0	67.9	93.2	100.0	69.4	93.5	100.0	68.5	93.4	100.0	85.2	96.9
15	MECHANICAL	96.5	75.9	87.3	96.6	76.3	87.5	99.8	42.2	74.1	99.9	68.0	85.7	96.5	43.0	72.6	99.8	78.5	90.3
16	ELECTRICAL	91.7	75.9	81.1	89.9	78.4	82.1	95.2	67.8	76.7	89.6	75.9	80.4	95.2	67.8	76.7	96.6	71.8	79.9
1 - 16	WEIGHTED AVERAGE	105.0	75.6	90.8	102.4	76.9	90.1	106.9	63.7	86.0	100.2	69.1	85.2	103.9	63.3	84.3	102.6	79.2	91.3

COST INDEXES

DIVISION		COLORADO SALIDA 812			CONNECTICUT BRIDGEPORT 066			BRISTOL 060			HARTFORD 061			MERIDEN 064			NEW BRITAIN 060		
		MAT.	INST.	TOTAL	MAT.	INST.	TOTAL	MAT.	INST.	TOTAL	MAT.	INST.	TOTAL	MAT.	INST.	TOTAL	MAT.	INST.	TOTAL
2	SITE WORK	137.9	98.3	107.4	113.0	99.6	102.7	112.1	99.6	102.5	112.3	99.6	102.5	111.4	100.4	103.0	112.3	99.6	102.5
031	CONCRETE FORMWORK	105.9	59.3	66.0	97.6	94.3	94.8	97.6	104.5	103.5	96.7	104.5	103.4	94.0	104.6	103.1	98.0	104.5	103.6
032	CONCRETE REINFORCEMENT	107.2	66.6	84.3	116.7	124.1	120.9	116.7	124.0	120.8	116.7	124.0	120.9	116.7	124.1	120.9	116.7	124.0	120.8
033	CAST IN PLACE CONCRETE	122.1	59.7	95.9	112.2	112.6	112.4	105.1	112.6	108.3	104.9	112.6	108.1	101.2	112.6	106.0	106.9	112.6	109.3
3	CONCRETE	117.8	61.6	89.4	117.4	105.8	111.6	114.0	110.3	112.1	113.8	110.3	112.0	111.8	110.3	111.0	114.9	110.3	112.5
4	MASONRY	146.7	69.6	98.8	106.2	111.7	109.6	95.9	111.7	105.7	95.8	111.7	105.7	95.6	111.7	105.6	96.0	111.7	105.7
5	METALS	99.2	77.6	91.4	101.9	103.8	102.6	101.9	103.8	102.6	102.3	103.8	102.8	101.7	103.9	102.5	98.5	103.8	100.4
6	WOOD & PLASTICS	107.8	57.1	81.6	100.2	88.4	94.1	100.2	102.3	101.3	100.2	102.3	101.3	100.2	102.3	101.3	100.2	102.3	101.3
7	THERMAL & MOISTURE PROTECTION	103.7	76.2	91.0	101.2	107.8	104.3	101.4	105.7	103.4	100.2	105.7	102.8	101.4	105.7	103.4	101.4	105.7	103.4
8	DOORS & WINDOWS	94.0	57.0	85.1	109.1	103.7	107.8	109.1	106.2	108.4	109.1	106.2	108.4	111.7	111.3	111.6	109.1	106.2	108.4
092	LATH, PLASTER & GYPSUM BOARD	88.1	55.6	67.1	98.0	87.0	90.9	98.0	101.4	100.2	98.0	101.4	100.2	100.8	101.4	101.2	98.0	101.4	100.2
095	ACOUSTICAL TREATMENT & WOOD FLOORING	94.2	55.6	69.2	101.7	87.0	92.2	101.7	101.4	101.5	101.7	101.4	101.5	114.0	101.4	105.8	101.7	101.4	101.5
096	FLOORING & CARPET	125.6	72.6	112.7	96.9	113.7	101.0	96.9	113.7	101.0	96.9	113.7	101.0	96.9	113.7	101.0	96.9	113.7	101.0
099	PAINTING & WALL COVERINGS	127.5	38.0	74.9	89.3	103.1	97.4	89.3	113.3	103.4	89.3	113.3	103.4	89.3	115.1	102.3	89.3	113.3	103.4
9	FINISHES	107.8	58.7	82.6	98.5	97.5	98.0	98.5	106.8	102.8	98.5	106.8	102.7	101.2	106.6	104.0	98.5	106.8	102.8
10 - 14	TOTAL DIV. 10 - 14	100.0	72.5	94.2	100.0	110.8	102.3	100.0	112.6	102.7	100.0	112.6	102.7	100.0	112.6	102.7	100.0	112.6	102.7
15	MECHANICAL	96.5	67.1	83.4	100.0	102.2	101.0	100.0	104.0	101.8	100.0	104.1	101.8	96.7	104.1	100.0	100.0	104.1	101.8
16	ELECTRICAL	95.5	71.8	79.5	104.7	98.1	100.2	104.7	94.6	97.9	103.9	101.2	102.1	104.7	94.6	97.9	104.7	99.5	101.2
1 - 16	WEIGHTED AVERAGE	104.8	70.4	88.2	104.4	102.5	103.5	103.4	104.2	103.8	103.3	105.3	104.3	102.9	104.5	103.7	103.0	105.0	104.0

DIVISION		CONNECTICUT NEW HAVEN 065			NEW LONDON 063			NORWALK 068			STAMFORD 069			WATERBURY 067			WILLIMANTIC 062		
		MAT.	INST.	TOTAL	MAT.	INST.	TOTAL	MAT.	INST.	TOTAL	MAT.	INST.	TOTAL	MAT.	INST.	TOTAL	MAT.	INST.	TOTAL
2	SITE WORK	112.0	100.4	103.1	104.4	100.4	101.3	112.8	99.6	102.6	113.5	99.6	102.8	112.5	99.6	102.6	112.8	99.4	102.5
031	CONCRETE FORMWORK	97.4	104.7	103.6	94.6	104.5	103.1	97.6	94.4	94.9	97.6	94.6	95.0	97.6	104.7	103.7	98.3	104.0	103.1
032	CONCRETE REINFORCEMENT	116.7	124.1	120.9	91.6	124.0	109.9	116.7	124.4	121.0	116.7	124.4	121.0	116.7	124.1	120.9	116.7	121.6	119.4
033	CAST IN PLACE CONCRETE	108.7	112.6	110.4	92.7	112.6	101.0	110.5	115.4	112.5	112.3	115.4	113.6	112.3	112.7	112.4	104.9	112.0	107.9
3	CONCRETE	116.7	110.4	113.5	100.8	110.3	105.6	116.6	106.9	111.7	117.4	107.0	112.2	117.4	110.4	113.9	113.9	109.4	111.6
4	MASONRY	96.1	111.7	105.8	94.6	111.7	105.2	96.3	116.9	109.1	96.4	116.9	109.1	96.4	111.7	105.9	95.8	110.8	105.1
5	METALS	98.5	103.9	100.5	98.7	103.8	100.5	101.9	104.4	102.8	101.9	104.5	102.8	101.9	103.9	102.6	102.2	102.6	102.3
6	WOOD & PLASTICS	100.2	102.3	101.3	100.2	102.3	101.3	100.2	88.4	94.1	100.2	88.4	94.1	100.2	102.3	101.3	100.2	102.3	101.3
7	THERMAL & MOISTURE PROTECTION	101.5	108.6	104.8	101.4	105.7	103.4	101.4	109.5	105.1	101.4	109.5	105.1	101.4	107.9	104.4	101.7	103.4	102.5
8	DOORS & WINDOWS	109.1	111.3	109.6	112.5	105.5	110.8	109.1	103.7	107.8	109.1	103.7	107.8	109.1	111.3	109.6	112.5	105.8	110.9
092	LATH, PLASTER & GYPSUM BOARD	98.0	101.4	100.2	98.0	101.4	100.2	98.0	87.0	90.9	98.0	87.0	90.9	98.0	101.4	100.2	98.0	101.4	100.2
095	ACOUSTICAL TREATMENT & WOOD FLOORING	101.7	101.4	101.5	101.7	101.4	101.5	101.7	87.0	92.2	101.7	87.0	92.2	101.7	101.4	101.5	101.7	101.4	101.5
096	FLOORING & CARPET	96.9	113.7	101.0	96.9	113.7	101.0	96.9	119.6	102.4	96.9	119.6	102.4	96.9	113.7	101.0	96.9	113.5	100.9
099	PAINTING & WALL COVERINGS	89.3	106.8	99.6	89.3	113.3	103.4	89.3	110.9	102.0	89.3	110.9	102.0	89.3	115.1	102.3	89.3	101.4	96.4
9	FINISHES	98.5	106.0	102.4	98.1	106.8	102.5	98.5	99.6	99.1	98.5	99.6	99.1	98.4	106.6	102.6	98.6	105.2	102.0
10 - 14	TOTAL DIV. 10 - 14	100.0	112.6	102.7	100.0	112.6	102.7	100.0	110.7	102.3	100.0	110.8	102.3	100.0	112.6	102.7	100.0	110.8	102.3
15	MECHANICAL	100.0	104.1	101.8	96.7	104.1	100.0	100.0	102.2	101.0	100.0	102.3	101.0	100.0	104.1	101.8	100.0	102.1	100.9
16	ELECTRICAL	104.7	99.5	101.1	99.4	103.8	102.3	104.7	93.7	97.2	104.7	114.0	111.0	103.8	98.1	100.0	104.7	99.5	101.1
1 - 16	WEIGHTED AVERAGE	103.2	105.3	104.2	100.2	105.8	102.9	103.7	102.7	103.2	103.9	106.1	105.0	103.7	105.0	104.4	103.8	104.0	103.9

DIVISION		D.C. WASHINGTON 200 - 205			DELAWARE DOVER 199			NEWARK 197			WILMINGTON 198			FLORIDA DAYTONA BEACH 321			FORT LAUDERDALE 333		
		MAT.	INST.	TOTAL	MAT.	INST.	TOTAL	MAT.	INST.	TOTAL	MAT.	INST.	TOTAL	MAT.	INST.	TOTAL	MAT.	INST.	TOTAL
2	SITE WORK	105.3	92.0	95.1	110.8	114.2	113.4	110.8	114.2	113.4	94.2	114.4	109.8	123.7	87.6	95.9	108.7	74.0	82.0
031	CONCRETE FORMWORK	97.2	81.3	83.6	102.6	92.9	94.3	102.6	92.9	94.3	102.3	92.9	94.2	97.5	69.7	73.7	94.6	66.2	70.3
032	CONCRETE REINFORCEMENT	102.0	98.0	99.7	95.4	90.3	92.5	96.3	90.3	92.9	96.3	90.3	92.9	95.7	79.8	86.7	95.7	84.3	89.3
033	CAST IN PLACE CONCRETE	117.0	86.7	104.3	85.5	90.1	87.4	85.5	90.1	87.4	80.3	90.1	84.4	93.3	73.6	85.0	97.8	72.7	87.3
3	CONCRETE	108.3	87.9	98.0	100.6	92.8	96.7	100.8	92.8	96.8	98.2	92.8	95.5	88.8	74.3	81.5	90.8	73.3	82.0
4	MASONRY	93.4	82.9	86.9	101.8	88.2	93.3	101.8	88.2	93.3	105.8	88.2	94.8	85.1	70.5	76.1	85.3	67.4	74.2
5	METALS	97.6	113.1	103.2	97.1	113.3	103.0	97.1	113.3	103.0	96.8	113.3	102.8	99.3	93.1	97.1	98.7	95.6	97.6
6	WOOD & PLASTICS	96.3	81.2	88.5	106.2	93.1	99.4	106.2	93.1	99.4	106.2	93.1	99.4	95.1	70.0	82.2	89.7	67.6	78.3
7	THERMAL & MOISTURE PROTECTION	95.1	84.0	90.0	99.6	101.4	100.4	99.8	101.4	100.5	99.3	101.4	100.3	97.4	72.2	85.7	97.4	70.4	84.9
8	DOORS & WINDOWS	100.4	90.1	97.9	99.4	97.0	98.9	99.4	97.0	98.9	99.2	97.0	98.6	99.3	67.2	91.5	97.0	67.1	89.7
092	LATH, PLASTER & GYPSUM BOARD	100.7	80.4	87.6	103.4	92.6	96.4	103.4	92.6	96.4	103.1	92.6	96.3	109.8	69.6	83.8	109.3	67.1	82.1
095	ACOUSTICAL TREATMENT & WOOD FLOORING	97.1	80.4	86.3	100.2	92.6	95.3	100.2	92.6	95.3	98.8	92.6	94.8	100.2	69.6	80.3	100.2	67.1	78.7
096	FLOORING & CARPET	92.4	91.2	92.1	77.2	91.4	80.7	77.2	91.4	80.7	77.0	91.4	80.5	112.8	72.0	102.8	112.8	74.9	103.6
099	PAINTING & WALL COVERINGS	101.0	96.9	98.6	87.4	95.2	92.0	87.4	95.2	92.0	87.4	95.2	92.0	106.6	91.5	97.7	103.1	63.3	79.7
9	FINISHES	94.0	84.3	89.1	93.6	92.6	93.1	93.7	92.6	93.1	93.3	92.6	93.0	107.3	72.4	89.4	105.2	61.5	85.7
10 - 14	TOTAL DIV. 10 - 14	100.0	94.5	98.8	100.0	103.1	100.6	100.0	103.1	100.6	100.0	103.1	100.6	100.0	79.5	95.7	100.0	80.0	95.8
15	MECHANICAL	100.0	90.6	95.9	100.2	89.6	95.5	100.2	89.6	95.5	100.2	89.6	95.5	100.0	74.2	88.5	100.0	69.6	86.4
16	ELECTRICAL	99.7	98.1	98.6	100.9	97.4	98.6	100.9	97.4	98.6	100.7	97.4	98.5	95.9	70.7	78.9	95.9	84.6	88.2
1 - 16	WEIGHTED AVERAGE	99.6	92.0	96.0	99.5	97.2	98.4	99.5	97.2	98.4	98.8	97.2	98.0	98.7	75.9	87.7	97.9	75.1	86.9

COST INDEXES

FLORIDA

DIVISION		FORT MYERS 339			GAINESVILLE 326			JACKSONVILLE 320,322			LAKELAND 338			MELBOURNE 329			MIAMI 330 - 332,340		
		MAT.	INST.	TOTAL	MAT.	INST.	TOTAL	MAT.	INST.	TOTAL	MAT.	INST.	TOTAL	MAT.	INST.	TOTAL	MAT.	INST.	TOTAL
2	SITE WORK	121.1	87.2	95.0	135.2	86.9	98.0	123.8	88.0	96.2	123.2	86.9	95.3	132.0	87.5	97.8	108.0	73.9	81.8
031	CONCRETE FORMWORK	87.1	58.7	62.8	87.0	61.8	65.4	97.2	62.5	67.5	82.1	58.8	62.1	91.0	69.3	72.4	94.4	66.2	70.2
032	CONCRETE REINFORCEMENT	96.7	73.2	83.5	101.6	64.6	80.8	95.7	65.3	78.6	99.1	73.7	84.8	96.7	79.8	87.2	95.7	84.2	89.3
033	CAST IN PLACE CONCRETE	102.1	65.1	86.6	107.1	55.9	85.7	94.2	61.1	80.3	104.3	66.3	88.4	108.3	73.1	93.5	95.3	71.2	85.2
3	CONCRETE	91.8	65.4	78.5	99.2	61.9	80.4	89.2	64.1	76.6	93.3	65.8	79.5	98.2	73.9	86.0	89.6	72.8	81.1
4	MASONRY	79.1	53.6	63.3	96.5	50.3	67.8	84.3	60.2	69.3	90.1	62.6	73.0	79.8	69.6	73.5	81.1	67.3	72.5
5	METALS	100.8	89.9	96.9	97.0	85.9	93.0	98.9	87.3	94.7	100.5	89.7	96.6	107.2	93.1	102.1	99.2	95.2	97.8
6	WOOD & PLASTICS	86.1	58.8	72.0	87.5	63.2	75.0	95.1	63.2	78.7	80.5	58.8	69.3	88.1	70.0	78.7	89.7	67.6	78.3
7	THERMAL & MOISTURE PROTECTION	97.1	58.7	79.3	97.7	63.5	81.8	97.7	67.1	83.5	97.1	60.9	80.3	97.6	71.8	85.6	100.5	71.6	87.1
8	DOORS & WINDOWS	98.0	59.3	88.7	97.5	58.7	88.1	99.3	59.5	89.6	98.0	56.6	87.9	98.5	67.2	90.9	97.0	67.8	89.9
092	LATH, PLASTER & GYPSUM BOARD	105.7	58.0	74.9	105.9	62.6	77.9	109.8	62.6	79.3	103.3	58.0	74.1	106.2	69.6	82.6	109.3	67.1	82.1
095	ACOUSTICAL TREATMENT & WOOD FLOORING	96.0	58.0	71.4	94.7	62.6	73.9	100.2	62.6	75.8	94.7	58.0	70.9	96.0	69.6	78.9	100.2	67.1	78.7
096	FLOORING & CARPET	108.6	51.4	94.7	109.3	44.1	93.4	112.8	61.5	100.3	105.9	65.2	96.0	109.6	72.0	100.4	121.5	74.8	110.2
099	PAINTING & WALL COVERINGS	106.6	59.0	78.7	106.6	56.8	77.4	106.6	61.0	79.9	106.6	59.0	78.7	106.6	90.4	97.1	103.1	63.3	79.7
9	FINISHES	104.4	56.9	80.1	105.3	57.8	81.0	107.3	61.9	84.0	103.0	59.7	80.8	105.4	72.0	88.3	108.1	67.1	87.1
10 - 14	TOTAL DIV. 10 - 14	100.0	73.9	94.5	100.0	77.0	95.2	100.0	73.7	94.5	100.0	73.9	94.5	100.0	79.1	95.6	100.0	80.0	95.8
15	MECHANICAL	98.2	60.3	81.3	99.7	64.9	84.2	100.0	62.6	83.3	98.2	63.5	82.7	100.0	73.7	88.3	100.0	68.9	86.1
16	ELECTRICAL	99.5	51.9	67.4	96.4	53.8	67.6	95.4	68.8	77.5	96.3	60.2	72.0	95.6	67.4	76.6	96.4	81.5	86.3
1 - 16	WEIGHTED AVERAGE	98.2	64.2	81.8	100.0	64.6	82.9	98.6	68.6	84.1	98.5	67.3	83.5	100.6	75.1	88.3	98.0	74.3	86.6

FLORIDA

DIVISION		ORLANDO 327 - 328,347			PANAMA CITY 324			PENSACOLA 325			SARASOTA 342			ST. PETERSBURG 337			TALLAHASSEE 323		
		MAT.	INST.	TOTAL	MAT.	INST.	TOTAL	MAT.	INST.	TOTAL	MAT.	INST.	TOTAL	MAT.	INST.	TOTAL	MAT.	INST.	TOTAL
2	SITE WORK	124.3	87.2	97.6	139.5	85.0	97.6	136.9	87.3	98.7	125.3	87.0	95.8	124.6	87.0	95.7	125.2	86.7	95.5
031	CONCRETE FORMWORK	97.3	64.9	69.5	96.1	34.4	43.2	85.3	63.2	66.3	97.2	58.7	64.2	94.5	58.7	63.8	97.3	48.1	55.2
032	CONCRETE REINFORCEMENT	95.7	79.7	86.7	99.9	64.0	79.7	102.4	64.3	80.9	95.7	73.6	83.3	99.1	73.6	84.8	95.7	64.6	78.2
033	CAST IN PLACE CONCRETE	101.4	73.4	89.7	98.9	41.1	74.7	98.9	65.2	84.8	106.7	66.2	89.8	105.4	66.2	89.0	97.6	55.4	79.9
3	CONCRETE	92.8	72.1	82.3	97.1	44.1	70.4	95.9	65.7	80.6	95.4	65.8	80.4	95.0	65.8	80.3	90.9	55.8	73.2
4	MASONRY	82.2	70.5	75.0	88.7	35.0	55.4	86.4	63.2	72.0	86.1	62.6	71.5	124.4	62.6	86.0	87.1	49.1	63.5
5	METALS	108.4	92.7	102.8	97.6	72.9	88.7	97.5	86.9	93.7	102.3	89.6	97.7	101.3	89.6	97.1	99.2	85.6	94.3
6	WOOD & PLASTICS	95.1	64.0	79.1	93.7	34.6	63.2	81.9	64.1	72.7	95.1	58.8	76.4	91.9	58.8	74.8	95.1	46.5	70.1
7	THERMAL & MOISTURE PROTECTION	97.7	71.6	85.6	97.9	36.6	69.5	97.6	63.5	81.8	97.4	60.9	80.4	97.3	59.8	79.9	97.8	55.6	78.3
8	DOORS & WINDOWS	99.3	63.1	90.5	97.0	32.4	81.4	97.0	61.0	88.3	99.3	55.9	88.7	98.0	56.4	87.9	99.3	49.7	87.2
092	LATH, PLASTER & GYPSUM BOARD	109.8	63.4	79.9	108.0	33.1	59.1	103.8	63.5	77.8	109.8	58.0	76.4	107.5	58.0	75.6	109.8	45.4	68.2
095	ACOUSTICAL TREATMENT & WOOD FLOORING	100.2	63.4	76.4	94.7	33.1	54.8	94.7	63.5	74.5	100.2	58.0	72.9	94.7	58.0	70.9	100.2	45.4	64.7
096	FLOORING & CARPET	112.8	72.0	102.8	112.1	23.6	90.6	106.6	65.4	96.6	112.8	64.9	101.1	111.3	65.2	100.1	112.8	47.8	97.0
099	PAINTING & WALL COVERINGS	106.6	70.1	85.2	106.6	31.2	62.4	106.6	70.9	85.7	106.6	59.0	78.7	106.6	59.0	78.7	106.6	50.2	73.5
9	FINISHES	107.3	66.5	86.3	106.9	31.5	68.2	104.3	64.4	83.9	107.3	59.7	82.9	105.4	59.7	82.0	107.4	47.5	76.7
10 - 14	TOTAL DIV. 10 - 14	100.0	78.7	95.5	100.0	59.3	91.4	100.0	66.3	92.9	100.0	73.8	94.5	100.0	69.5	93.6	100.0	71.3	94.0
15	MECHANICAL	100.0	65.4	84.5	100.0	32.0	69.7	100.0	63.6	83.8	99.9	61.0	82.5	100.0	63.5	83.7	100.0	50.6	78.0
16	ELECTRICAL	96.5	57.2	70.0	94.1	42.8	59.5	99.5	64.1	75.6	95.9	47.0	62.9	96.3	61.8	73.1	96.5	53.0	67.1
1 - 16	WEIGHTED AVERAGE	100.4	70.6	86.1	99.6	45.6	73.5	99.2	68.5	84.4	100.0	64.6	82.9	101.5	67.5	85.1	99.1	58.6	79.6

FLORIDA / GEORGIA

DIVISION		TAMPA 335 - 336,346			WEST PALM BEACH 334,349			ALBANY 317			ATHENS 306			ATLANTA 300 - 303,399			AUGUSTA 308 - 309		
		MAT.	INST.	TOTAL	MAT.	INST.	TOTAL	MAT.	INST.	TOTAL	MAT.	INST.	TOTAL	MAT.	INST.	TOTAL	MAT.	INST.	TOTAL
2	SITE WORK	124.9	87.0	95.8	105.4	74.0	81.2	108.9	75.9	83.5	119.2	91.4	97.8	113.8	93.3	98.0	109.6	91.8	95.9
031	CONCRETE FORMWORK	98.9	58.8	64.6	97.2	65.5	70.1	96.9	52.3	58.7	89.4	46.4	52.5	98.3	75.5	78.8	94.9	63.3	67.8
032	CONCRETE REINFORCEMENT	95.7	73.7	83.3	98.3	65.3	79.7	95.7	88.4	91.6	105.9	84.8	94.0	102.0	89.7	95.1	107.8	73.3	88.3
033	CAST IN PLACE CONCRETE	103.2	66.3	87.7	93.0	67.7	82.4	99.6	49.2	78.5	103.8	51.4	81.9	103.8	71.6	90.3	98.1	57.7	81.2
3	CONCRETE	93.8	65.8	79.7	88.1	67.8	77.9	91.8	59.8	75.6	100.8	56.2	78.3	98.4	76.9	87.6	94.6	63.6	79.0
4	MASONRY	86.0	62.6	71.4	85.0	59.6	69.2	87.2	39.9	57.8	74.0	56.8	63.3	89.5	68.5	76.4	89.6	50.4	65.2
5	METALS	102.2	89.7	97.7	97.5	88.4	94.2	96.6	92.3	95.1	91.9	76.0	86.2	93.6	80.0	88.7	92.3	71.9	84.9
6	WOOD & PLASTICS	96.9	58.8	77.2	94.4	67.6	80.6	94.4	52.9	73.0	93.4	45.4	68.6	98.6	77.7	87.8	95.0	66.3	80.2
7	THERMAL & MOISTURE PROTECTION	97.6	60.9	80.6	97.1	67.2	83.2	97.5	57.8	79.1	94.4	54.2	75.7	94.5	74.4	85.2	93.9	60.1	78.2
8	DOORS & WINDOWS	99.3	55.7	88.7	96.2	62.9	88.1	97.0	57.0	87.3	92.3	52.7	82.7	95.8	75.5	90.9	92.4	61.2	84.8
092	LATH, PLASTER & GYPSUM BOARD	109.8	58.0	76.4	109.8	67.1	82.3	109.8	52.0	72.5	117.9	44.2	70.3	119.5	77.4	92.4	118.4	65.7	84.4
095	ACOUSTICAL TREATMENT & WOOD FLOORING	100.2	58.0	72.9	96.0	67.1	77.3	100.2	52.0	68.9	106.1	44.2	66.0	106.1	77.4	87.5	106.1	65.7	79.9
096	FLOORING & CARPET	112.8	65.2	101.2	115.4	62.6	102.5	112.8	39.9	95.0	84.3	71.9	81.3	85.9	76.1	83.6	84.8	50.8	76.6
099	PAINTING & WALL COVERINGS	106.6	59.0	78.7	103.1	58.5	76.9	103.1	50.3	72.1	93.9	41.9	63.4	93.9	78.2	84.7	93.9	47.8	66.8
9	FINISHES	107.3	59.7	82.9	105.2	64.0	84.1	105.3	48.8	76.4	93.5	49.8	71.1	93.8	76.0	84.7	93.1	59.5	75.9
10 - 14	TOTAL DIV. 10 - 14	100.0	73.9	94.5	100.0	79.9	95.8	100.0	74.8	94.7	100.0	63.1	92.2	100.0	81.7	96.1	100.0	74.0	94.5
15	MECHANICAL	100.0	63.5	83.7	98.2	69.0	85.2	100.0	56.1	80.4	96.8	54.4	77.9	100.1	77.9	90.2	100.1	53.3	79.2
16	ELECTRICAL	95.4	61.8	72.8	95.9	66.8	76.3	90.0	65.9	73.7	94.7	81.4	85.7	93.6	84.2	87.3	95.7	58.3	70.5
1 - 16	WEIGHTED AVERAGE	99.8	67.6	84.2	96.8	69.2	83.5	97.4	61.6	80.1	95.1	64.5	80.3	96.9	79.3	88.4	95.7	62.7	79.8

COST INDEXES

GEORGIA

DIVISION	COLUMBUS 318-319			DALTON 307			GAINESVILLE 305			MACON 310-312			SAVANNAH 313-314			STATESBORO 304		
	MAT.	INST.	TOTAL	MAT.	INST.	TOTAL	MAT.	INST.	TOTAL	MAT.	INST.	TOTAL	MAT.	INST.	TOTAL	MAT.	INST.	TOTAL
2 SITE WORK	108.9	76.0	83.6	120.3	94.1	100.1	118.0	89.7	96.2	109.7	94.1	97.7	109.5	77.8	85.1	120.3	73.1	84.0
031 CONCRETE FORMWORK	98.5	66.9	71.5	78.5	24.5	32.2	93.9	44.2	51.3	95.9	67.8	71.8	96.7	62.4	67.3	73.0	27.4	34.0
032 CONCRETE REINFORCEMENT	95.7	88.4	91.6	104.0	66.5	82.9	104.9	83.0	92.5	98.1	88.7	92.8	101.4	73.6	85.7	105.8	24.1	59.8
033 CAST IN PLACE CONCRETE	99.2	49.8	78.5	100.8	31.3	71.7	109.1	35.8	78.4	97.9	53.7	79.4	96.1	57.4	79.9	103.6	31.8	73.5
3 CONCRETE	91.7	66.4	78.9	99.5	36.6	67.8	102.6	49.4	75.8	91.2	68.1	79.6	90.8	64.3	77.4	99.9	31.2	65.2
4 MASONRY	87.2	40.3	58.1	76.6	22.8	43.2	85.6	28.3	50.0	102.5	48.0	68.7	91.0	60.1	71.8	75.9	27.7	46.0
5 METALS	97.0	92.6	95.4	94.3	66.3	84.2	91.8	70.3	84.1	91.9	93.4	92.5	97.3	88.3	94.1	97.2	70.5	87.6
6 WOOD & PLASTICS	96.4	72.9	84.3	76.1	24.1	49.3	98.3	45.4	71.0	99.5	71.6	85.1	94.5	62.5	77.9	68.2	26.9	46.9
7 THERMAL & MOISTURE PROTECTION	97.2	60.0	79.9	99.6	27.3	66.1	94.3	42.2	70.1	96.0	65.3	81.7	97.6	60.9	80.6	95.1	28.5	64.2
8 DOORS & WINDOWS	97.0	68.1	90.0	93.2	20.5	75.5	92.3	45.4	80.9	95.4	68.4	88.9	97.0	58.5	87.6	94.0	24.0	77.0
092 LATH, PLASTER & GYPSUM BOARD	109.8	72.6	85.8	99.8	22.2	49.7	119.5	44.2	70.9	116.0	71.2	87.1	109.8	61.8	78.8	102.9	25.1	52.7
095 ACOUSTICAL TREATMENT & WOOD FLOORING	100.2	72.6	82.3	119.1	22.2	56.3	106.1	44.2	66.0	92.5	71.2	78.7	100.2	61.8	75.3	115.5	25.1	56.9
096 FLOORING & CARPET	112.8	40.5	95.2	85.1	17.2	68.6	85.9	17.2	69.2	88.0	46.9	78.0	112.8	59.9	99.9	99.4	59.9	89.8
099 PAINTING & WALL COVERINGS	103.1	48.2	70.9	84.2	17.8	45.2	93.9	41.9	63.4	104.8	58.9	77.9	103.1	59.8	77.7	92.4	17.8	48.6
9 FINISHES	105.2	60.4	82.2	100.3	22.1	60.2	94.0	38.6	65.6	91.3	63.1	76.8	105.5	61.6	83.0	103.7	32.4	67.2
10-14 TOTAL DIV. 10-14	100.0	77.3	95.2	100.0	23.3	83.8	100.0	63.1	92.2	100.0	78.9	95.6	100.0	75.1	94.8	100.0	58.0	91.1
15 MECHANICAL	100.0	45.9	75.8	96.9	19.4	62.4	96.8	22.9	63.9	100.0	51.7	78.4	100.0	55.7	80.2	96.1	24.4	64.1
16 ELECTRICAL	91.3	46.9	61.4	102.3	20.6	47.2	94.7	81.3	85.7	89.4	67.8	74.8	92.7	65.4	74.3	94.9	25.9	48.4
1-16 WEIGHTED AVERAGE	97.5	59.7	79.2	96.8	34.8	66.9	96.0	52.2	74.9	95.7	67.6	82.2	97.8	65.6	82.3	96.8	36.9	67.9

GEORGIA / HAWAII / IDAHO

DIVISION	VALDOSTA 316			WAYCROSS 315			HILO 967			HONOLULU 968			STATES & POSS., GUAM 969			BOISE 836-837		
	MAT.	INST.	TOTAL	MAT.	INST.	TOTAL	MAT.	INST.	TOTAL	MAT.	INST.	TOTAL	MAT.	INST.	TOTAL	MAT.	INST.	TOTAL
2 SITE WORK	120.2	76.3	86.4	116.8	75.4	84.9	117.9	109.3	111.3	119.3	109.3	111.6	117.9	102.0	105.7	84.4	102.5	98.3
031 CONCRETE FORMWORK	81.8	53.6	57.6	81.9	35.1	41.8	104.4	152.3	145.4	104.9	152.3	145.5	104.7	39.7	49.0	99.1	88.5	90.0
032 CONCRETE REINFORCEMENT	101.6	53.3	74.3	101.6	80.5	89.7	109.1	125.8	118.5	108.1	125.8	118.1	108.1	28.0	63.2	101.1	80.1	89.3
033 CAST IN PLACE CONCRETE	97.5	58.1	81.0	110.1	37.5	79.7	187.6	132.4	164.5	180.8	132.4	160.5	189.4	49.0	130.6	100.0	93.1	97.1
3 CONCRETE	95.4	56.9	76.0	99.8	46.7	73.0	158.6	138.6	148.5	155.2	138.6	146.8	159.3	42.0	100.2	103.8	88.3	96.0
4 MASONRY	93.9	52.9	68.4	94.8	35.5	57.9	130.9	139.0	136.0	131.0	139.0	136.0	130.0	36.4	71.9	136.3	79.5	101.1
5 METALS	96.8	81.0	91.1	95.8	87.9	92.9	117.0	109.7	114.4	116.2	109.7	113.9	115.9	57.3	94.8	113.5	80.1	101.4
6 WOOD & PLASTICS	78.0	51.6	64.3	80.8	33.8	56.5	102.0	155.6	129.7	101.9	155.6	129.6	101.9	39.8	69.8	97.1	88.4	92.6
7 THERMAL & MOISTURE PROTECTION	97.2	62.7	81.2	97.0	40.7	70.9	110.0	135.0	121.6	111.9	135.0	122.6	110.3	44.0	79.6	96.2	83.6	90.3
8 DOORS & WINDOWS	92.7	47.8	81.8	92.7	43.1	80.7	103.4	145.9	113.7	108.8	145.9	117.8	103.2	36.1	86.9	95.2	80.7	91.7
092 LATH, PLASTER & GYPSUM BOARD	102.7	50.6	69.1	103.8	32.2	57.6	91.4	157.3	133.9	108.5	157.3	140.0	91.1	37.8	56.7	89.6	87.8	88.5
095 ACOUSTICAL TREATMENT & WOOD FLOORING	94.7	50.6	66.1	94.7	32.2	54.2	123.4	157.3	145.4	122.0	157.3	144.9	122.0	37.8	67.4	101.0	87.8	92.5
096 FLOORING & CARPET	104.9	47.8	91.0	106.4	39.9	90.2	125.2	135.6	127.7	123.7	135.6	126.6	123.7	37.5	102.7	104.6	69.4	96.0
099 PAINTING & WALL COVERINGS	103.1	43.6	68.2	103.1	41.9	67.2	120.0	153.2	139.5	119.3	153.2	139.2	119.3	43.8	75.0	113.2	63.8	84.2
9 FINISHES	101.4	50.6	75.5	101.7	35.7	67.8	120.4	150.7	135.9	121.8	150.7	136.6	119.1	39.6	78.3	99.5	82.7	90.9
10-14 TOTAL DIV. 10-14	100.0	73.6	94.4	100.0	60.0	91.6	100.0	132.1	106.8	100.0	132.1	106.8	100.0	85.4	96.9	100.0	85.9	97.0
15 MECHANICAL	100.0	48.7	77.1	97.7	31.7	68.3	100.2	126.5	111.9	100.2	126.5	111.9	100.2	34.6	71.0	99.8	85.7	93.5
16 ELECTRICAL	88.2	38.8	54.9	95.0	70.7	78.6	109.3	128.7	122.4	112.4	128.7	123.4	109.6	36.7	60.4	82.5	79.9	80.8
1-16 WEIGHTED AVERAGE	97.3	55.6	77.1	97.7	52.0	75.6	115.4	130.7	122.8	116.0	130.7	123.1	115.2	47.2	82.4	101.9	85.1	93.8

IDAHO / ILLINOIS

DIVISION	COEUR D'ALENE 838			IDAHO FALLS 834			LEWISTON 835			POCATELLO 832			TWIN FALLS 833			BLOOMINGTON 617		
	MAT.	INST.	TOTAL	MAT.	INST.	TOTAL	MAT.	INST.	TOTAL	MAT.	INST.	TOTAL	MAT.	INST.	TOTAL	MAT.	INST.	TOTAL
2 SITE WORK	83.8	94.5	92.0	82.4	101.2	96.9	89.0	95.6	94.0	86.8	102.5	98.9	94.4	100.6	99.2	89.5	95.9	94.4
031 CONCRETE FORMWORK	98.9	48.6	55.8	93.5	45.2	52.1	109.6	81.7	85.7	99.0	88.3	89.8	102.5	44.1	52.5	80.1	100.2	97.3
032 CONCRETE REINFORCEMENT	114.1	96.4	104.1	108.9	68.8	86.3	114.1	99.5	105.8	101.4	80.0	89.3	110.2	61.5	82.8	99.5	123.6	113.1
033 CAST IN PLACE CONCRETE	104.3	80.6	94.4	91.8	65.6	80.8	108.4	95.3	102.9	99.2	93.0	96.6	101.8	52.5	81.1	96.3	101.5	98.5
3 CONCRETE	111.5	69.6	90.4	96.8	57.8	77.1	115.7	89.7	102.6	103.5	88.1	95.7	112.9	51.3	81.8	90.4	105.5	98.0
4 MASONRY	136.3	68.3	94.1	130.5	40.6	74.6	136.4	95.1	110.7	130.7	77.3	97.6	134.3	49.5	81.6	108.7	108.4	108.5
5 METALS	99.1	86.2	94.4	113.4	73.9	99.2	97.2	89.0	94.3	113.8	79.7	101.5	113.7	70.2	98.0	94.0	116.5	102.1
6 WOOD & PLASTICS	94.0	43.1	67.7	87.6	42.8	64.5	101.9	76.3	88.7	97.1	88.4	92.6	100.7	44.2	71.6	82.8	96.0	89.6
7 THERMAL & MOISTURE PROTECTION	165.4	72.9	122.5	95.8	58.9	78.7	165.1	87.1	128.9	96.4	79.7	88.6	97.4	56.3	78.3	95.3	102.8	98.8
8 DOORS & WINDOWS	116.0	55.7	101.4	98.5	49.4	86.6	116.0	78.6	106.9	95.2	76.3	90.6	98.5	44.8	85.5	92.8	102.5	95.2
092 LATH, PLASTER & GYPSUM BOARD	147.9	41.3	79.1	85.2	40.8	56.5	150.6	75.6	102.2	89.6	87.8	88.5	89.7	42.2	59.1	103.5	95.5	98.3
095 ACOUSTICAL TREATMENT & WOOD FLOORING	149.4	41.3	79.3	95.6	40.8	60.1	149.4	75.6	101.6	101.0	87.8	92.5	96.9	42.2	61.5	88.9	95.5	93.1
096 FLOORING & CARPET	143.2	73.9	126.3	101.5	69.4	93.7	146.2	96.5	134.1	104.9	69.4	96.3	106.2	69.4	97.2	93.6	97.8	94.6
099 PAINTING & WALL COVERINGS	141.6	79.9	105.4	113.2	52.8	77.8	141.6	79.9	105.4	113.2	66.9	86.0	113.2	39.4	69.9	86.8	90.2	88.8
9 FINISHES	169.1	54.8	110.5	96.4	49.4	72.3	170.5	83.5	125.9	99.6	83.1	91.1	99.8	47.9	73.2	90.9	97.8	94.4
10-14 TOTAL DIV. 10-14	100.0	76.0	94.9	100.0	63.0	92.2	100.0	99.8	100.0	100.0	85.9	97.0	100.0	61.7	91.9	100.0	105.0	101.0
15 MECHANICAL	111.4	59.4	88.3	102.2	63.0	84.7	100.6	94.4	97.9	99.8	85.7	93.5	99.8	45.9	75.7	96.9	109.3	102.5
16 ELECTRICAL	87.8	48.4	61.3	86.0	79.0	81.3	84.7	90.3	88.5	85.6	79.0	81.2	88.1	40.7	56.2	96.1	96.5	96.3
1-16 WEIGHTED AVERAGE	115.5	65.7	91.4	101.4	65.2	83.9	113.5	90.6	102.4	101.9	84.4	93.5	104.1	54.7	80.3	95.0	103.8	99.3

City Cost Indexes

ILLINOIS

DIVISION		CARBONDALE 629 MAT.	INST.	TOTAL	CENTRALIA 628 MAT.	INST.	TOTAL	CHAMPAIGN 618-619 MAT.	INST.	TOTAL	CHICAGO 606 MAT.	INST.	TOTAL	DECATUR 625 MAT.	INST.	TOTAL	EAST ST. LOUIS 620-622 MAT.	INST.	TOTAL
2	SITE WORK	99.4	96.6	97.2	99.8	97.2	97.8	99.1	96.8	97.3	84.2	94.0	91.7	82.6	96.7	93.5	101.6	97.2	98.2
031	CONCRETE FORMWORK	92.5	98.8	97.9	95.7	102.9	101.9	89.2	96.9	95.8	105.1	126.4	123.3	100.9	94.4	95.4	92.8	107.8	105.7
032	CONCRETE REINFORCEMENT	99.7	107.9	104.3	99.7	128.5	116.0	99.5	121.2	111.7	95.8	168.0	136.5	93.6	120.0	108.5	99.1	128.6	115.7
033	CAST IN PLACE CONCRETE	90.1	96.5	92.8	90.6	96.1	92.9	111.8	106.2	109.4	103.5	131.0	115.0	98.4	95.4	97.2	92.2	110.4	99.8
3	CONCRETE	82.8	100.9	91.9	83.4	106.5	95.0	102.3	105.2	103.8	97.3	134.9	116.3	92.1	100.1	96.2	84.2	113.6	99.0
4	MASONRY	75.7	101.6	91.8	75.7	112.2	98.4	130.7	102.9	113.4	95.6	130.3	117.2	71.9	104.7	92.3	75.8	113.0	98.9
5	METALS	93.5	119.2	102.8	93.5	129.1	106.4	94.1	114.0	101.3	97.1	128.9	108.6	97.2	113.1	102.9	94.6	129.3	107.1
6	WOOD & PLASTICS	99.8	98.2	99.0	103.2	100.9	102.0	92.2	93.5	92.9	101.8	124.8	113.7	103.0	90.4	96.5	96.1	108.0	102.2
7	THERMAL & MOISTURE PROTECTION	91.5	93.3	92.4	91.6	101.4	96.1	96.1	103.4	99.5	100.6	126.4	112.6	96.5	95.6	96.1	91.5	105.3	97.9
8	DOORS & WINDOWS	88.8	101.4	91.8	88.8	114.1	94.9	93.5	100.7	95.3	105.0	136.4	112.6	99.7	98.8	99.5	88.9	114.4	95.1
092	LATH, PLASTER & GYPSUM BOARD	108.8	97.7	101.6	109.8	100.5	103.8	106.1	92.9	97.6	97.4	125.1	115.3	111.1	89.7	97.3	107.7	107.8	107.8
095	ACOUSTICAL TREATMENT & WOOD FLOORING	88.9	97.7	94.6	88.9	100.5	96.4	88.9	92.9	91.5	91.4	125.1	113.2	94.3	89.7	91.3	88.9	107.8	101.2
096	FLOORING & CARPET	111.4	93.6	107.0	112.7	81.8	105.2	97.0	102.8	98.4	84.2	123.8	93.8	100.8	103.9	101.5	110.0	66.1	99.3
099	PAINTING & WALL COVERINGS	94.7	76.4	84.0	94.7	96.4	95.7	86.8	105.7	97.9	77.2	128.3	107.2	86.8	100.8	95.0	94.7	99.4	97.4
9	FINISHES	95.4	94.8	95.1	96.0	98.2	97.1	92.8	98.4	95.7	85.4	125.6	106.0	95.4	96.6	96.0	94.9	98.7	96.9
10-14	TOTAL DIV. 10-14	100.0	109.2	101.9	100.0	110.8	102.3	100.0	102.5	100.5	100.0	128.4	106.0	100.0	102.7	100.6	100.0	111.3	102.4
15	MECHANICAL	96.9	94.0	95.6	96.9	94.9	96.0	96.9	102.5	99.4	100.2	122.4	110.1	100.2	99.3	99.8	100.1	100.5	100.3
16	ELECTRICAL	95.5	103.2	100.7	97.8	103.2	101.5	101.5	91.4	94.7	101.8	129.9	120.7	100.7	90.9	94.1	97.1	102.6	100.8
1-16	WEIGHTED AVERAGE	92.6	100.5	96.4	92.9	104.6	98.6	98.7	100.9	99.8	97.8	125.0	110.9	96.1	99.1	97.5	93.7	106.9	100.1

ILLINOIS

DIVISION		EFFINGHAM 624 MAT.	INST.	TOTAL	GALESBURG 614 MAT.	INST.	TOTAL	JOLIET 604 MAT.	INST.	TOTAL	KANKAKEE 609 MAT.	INST.	TOTAL	LA SALLE 613 MAT.	INST.	TOTAL	NORTH SUBURBAN 600-603 MAT.	INST.	TOTAL
2	SITE WORK	90.7	96.5	95.2	91.9	95.9	95.0	84.9	93.3	91.4	79.5	92.7	89.6	91.4	96.1	95.0	84.2	93.1	91.0
031	CONCRETE FORMWORK	102.5	94.7	95.8	88.9	99.5	98.0	106.2	125.6	122.8	91.6	113.8	110.6	108.4	98.2	99.6	105.1	120.1	117.9
032	CONCRETE REINFORCEMENT	99.5	121.0	111.6	97.2	122.9	111.7	95.8	155.8	129.6	100.1	141.1	123.2	97.9	133.9	118.2	95.8	166.7	135.8
033	CAST IN PLACE CONCRETE	98.2	83.4	92.0	99.2	94.3	97.2	103.4	118.9	109.9	96.4	114.5	104.0	99.1	104.3	101.3	103.5	125.1	112.5
3	CONCRETE	93.3	96.3	94.8	92.7	102.6	97.7	97.3	127.9	112.7	91.5	118.5	105.1	94.2	107.5	100.9	97.3	129.7	113.6
4	MASONRY	79.6	94.3	88.8	108.8	109.3	109.1	97.2	126.7	115.5	95.2	112.5	105.9	108.8	110.0	109.6	95.6	126.5	114.8
5	METALS	94.0	113.3	101.0	94.0	115.8	101.9	95.4	121.0	104.6	94.7	113.8	101.6	94.1	120.9	103.8	96.4	124.3	106.5
6	WOOD & PLASTICS	105.9	93.5	99.5	92.0	95.6	93.8	103.7	125.0	114.7	93.7	114.1	104.2	112.2	93.8	102.7	101.8	117.3	109.8
7	THERMAL & MOISTURE PROTECTION	95.9	94.5	95.3	95.5	101.2	98.2	100.4	123.5	111.1	99.3	114.5	106.4	95.9	102.6	99.0	100.6	121.5	110.3
8	DOORS & WINDOWS	93.4	104.9	96.2	92.8	100.8	94.7	104.9	133.5	111.8	95.8	123.0	102.4	92.8	100.5	94.7	105.0	131.1	111.3
092	LATH, PLASTER & GYPSUM BOARD	110.3	92.9	99.1	106.1	95.0	98.9	95.2	125.3	114.6	92.6	114.1	106.4	112.5	93.2	100.0	97.4	117.4	110.3
095	ACOUSTICAL TREATMENT & WOOD FLOORING	88.9	92.9	91.5	88.9	95.0	92.8	91.4	125.3	113.4	91.4	114.1	106.1	88.9	93.2	91.7	91.4	117.4	108.3
096	FLOORING & CARPET	102.0	102.8	102.2	96.9	105.7	99.1	83.7	115.1	91.3	81.1	118.4	90.2	104.1	90.0	100.7	84.2	115.1	91.7
099	PAINTING & WALL COVERINGS	86.8	91.5	89.6	86.8	97.9	93.4	74.8	110.4	95.7	74.8	98.3	88.6	86.8	89.3	88.3	77.2	118.1	101.2
9	FINISHES	94.6	95.6	95.1	92.4	100.2	96.4	84.9	122.1	103.9	83.4	112.3	98.2	95.6	94.6	95.1	85.4	118.2	102.3
10-14	TOTAL DIV. 10-14	100.0	97.9	99.5	100.0	105.1	101.1	100.0	128.1	105.9	100.0	123.4	104.9	100.0	104.3	100.9	100.0	127.2	105.7
15	MECHANICAL	97.0	100.0	98.4	96.9	105.9	100.9	100.4	109.5	104.4	97.1	105.8	101.0	96.9	104.8	100.5	100.2	115.8	107.2
16	ELECTRICAL	97.2	103.2	101.2	97.6	81.6	86.8	101.4	100.0	100.4	93.3	94.6	94.2	92.9	94.6	94.1	100.5	124.5	116.7
1-16	WEIGHTED AVERAGE	94.6	99.8	97.1	95.7	100.5	98.0	97.6	114.9	105.9	93.8	107.9	100.6	96.1	103.0	99.4	97.6	119.9	108.4

ILLINOIS

DIVISION		PEORIA 615-616 MAT.	INST.	TOTAL	QUINCY 623 MAT.	INST.	TOTAL	ROCKFORD 610-611 MAT.	INST.	TOTAL	ROCK ISLAND 612 MAT.	INST.	TOTAL	SOUTH SUBURBAN 605 MAT.	INST.	TOTAL	SPRINGFIELD 626-627 MAT.	INST.	TOTAL
2	SITE WORK	90.9	95.8	94.7	89.6	96.6	95.0	91.1	96.2	95.1	90.0	95.4	94.1	84.2	93.0	91.0	89.5	96.7	95.0
031	CONCRETE FORMWORK	98.8	100.7	100.4	99.5	96.7	97.1	103.8	111.2	110.1	91.4	99.3	98.2	105.1	120.0	117.8	103.2	97.7	98.5
032	CONCRETE REINFORCEMENT	93.6	123.6	110.5	97.9	113.1	106.4	93.6	151.2	126.1	97.2	122.7	111.6	95.8	166.7	135.8	93.6	120.1	108.5
033	CAST IN PLACE CONCRETE	96.2	108.5	101.4	98.3	95.4	97.1	98.4	106.7	101.9	97.1	98.9	97.8	103.5	124.9	112.4	91.5	99.8	95.0
3	CONCRETE	90.9	108.1	99.6	92.5	99.8	96.2	92.4	117.3	104.9	91.4	104.0	97.7	97.3	129.5	113.5	88.9	103.1	96.1
4	MASONRY	115.1	110.7	112.4	98.3	95.1	96.3	86.3	112.6	102.6	108.7	101.1	104.0	95.6	126.5	114.8	70.6	106.9	93.2
5	METALS	97.2	116.6	104.2	94.1	110.5	100.1	97.2	127.9	108.2	94.1	114.1	101.3	96.4	123.9	106.3	97.2	113.5	103.1
6	WOOD & PLASTICS	103.0	95.5	99.2	102.8	96.6	99.6	103.0	109.1	106.1	94.7	98.2	96.5	101.8	117.3	109.8	102.8	93.5	98.0
7	THERMAL & MOISTURE PROTECTION	96.2	104.0	99.9	95.9	91.7	94.0	96.1	113.3	104.1	95.5	99.3	97.3	100.6	119.4	109.3	96.0	99.4	97.6
8	DOORS & WINDOWS	99.7	103.9	100.7	94.3	100.4	95.8	99.7	119.0	104.4	92.8	102.2	95.1	105.0	131.1	111.3	99.7	100.4	99.9
092	LATH, PLASTER & GYPSUM BOARD	111.1	94.9	100.7	109.8	96.1	100.9	111.1	109.0	109.7	107.2	97.7	101.0	97.4	117.4	110.3	111.1	92.9	99.3
095	ACOUSTICAL TREATMENT & WOOD FLOORING	94.3	94.9	94.7	88.9	96.1	93.5	94.3	109.0	103.8	88.9	97.7	94.6	91.4	117.4	108.3	94.3	92.9	93.4
096	FLOORING & CARPET	100.8	105.7	102.0	100.8	97.6	100.0	100.8	102.8	101.3	97.8	91.4	96.3	84.2	115.1	91.7	101.0	104.0	101.8
099	PAINTING & WALL COVERINGS	86.8	99.8	94.5	86.8	60.7	71.5	86.8	109.2	100.0	86.8	97.9	93.4	77.2	118.1	101.2	86.8	100.5	94.8
9	FINISHES	95.3	100.7	98.1	94.1	92.8	93.4	95.3	108.8	102.2	92.8	97.5	95.2	85.4	118.2	102.3	95.4	98.9	97.2
10-14	TOTAL DIV. 10-14	100.0	105.7	101.2	100.0	98.2	99.6	100.0	116.3	103.4	100.0	101.7	100.4	100.0	127.2	105.7	100.0	103.9	100.8
15	MECHANICAL	100.2	105.4	102.5	96.9	101.8	99.1	100.2	108.8	104.0	96.9	103.2	99.7	100.2	115.8	107.2	100.2	100.2	100.2
16	ELECTRICAL	99.6	88.8	92.3	93.0	79.4	83.8	99.6	112.1	108.1	85.0	92.9	90.3	100.5	124.5	116.7	100.7	93.8	96.0
1-16	WEIGHTED AVERAGE	98.5	102.8	100.6	95.2	95.9	95.6	97.1	112.0	104.3	94.7	100.7	97.6	97.6	119.8	108.3	95.8	100.9	98.3

INDIANA

DIVISION		ANDERSON 460			BLOOMINGTON 474			COLUMBUS 472			EVANSVILLE 476 - 477			FORT WAYNE 467 - 468			GARY 463 - 464		
		MAT.	INST.	TOTAL	MAT.	INST.	TOTAL	MAT.	INST.	TOTAL	MAT.	INST.	TOTAL	MAT.	INST.	TOTAL	MAT.	INST.	TOTAL
2	SITE WORK	83.1	94.2	91.7	77.5	95.2	91.1	74.9	95.1	90.4	83.1	130.1	119.2	83.2	94.4	91.8	83.7	95.7	92.9
031	CONCRETE FORMWORK	89.2	84.1	84.8	103.4	79.5	82.9	88.0	81.1	82.0	90.9	85.4	86.2	88.8	79.5	80.8	89.7	102.2	100.4
032	CONCRETE REINFORCEMENT	95.8	84.7	89.5	89.4	84.9	86.9	89.7	85.0	87.0	97.9	81.5	88.7	95.8	84.0	89.2	95.8	94.9	95.3
033	CAST IN PLACE CONCRETE	106.5	85.4	97.7	104.7	82.4	95.4	104.2	81.6	94.7	100.0	91.7	96.5	113.1	87.2	102.2	111.2	102.6	107.6
3	CONCRETE	98.3	85.1	91.6	102.0	81.1	91.4	100.5	81.5	90.9	104.3	86.9	95.5	101.5	83.6	92.5	100.6	101.0	100.8
4	MASONRY	92.2	82.7	86.3	91.5	81.8	85.5	91.5	81.7	85.4	87.6	87.3	87.4	90.2	87.4	88.5	93.6	96.8	95.6
5	METALS	99.2	90.7	96.1	96.1	74.7	88.4	95.8	74.8	88.2	89.9	86.3	88.6	99.2	90.5	96.0	99.2	98.0	98.8
6	WOOD & PLASTICS	98.6	84.2	91.2	115.6	78.2	96.3	103.5	79.9	91.3	91.3	84.6	87.8	98.2	77.8	87.7	97.6	103.5	100.7
7	THERMAL & MOISTURE PROTECTION	98.5	82.2	90.9	93.1	85.4	89.5	92.5	85.7	89.3	96.5	89.6	93.3	97.9	82.2	90.6	97.8	98.9	98.3
8	DOORS & WINDOWS	102.8	84.4	98.4	105.4	80.6	99.4	100.8	82.2	96.3	98.2	82.5	94.4	102.8	78.7	97.0	102.8	97.3	101.5
092	LATH, PLASTER & GYPSUM BOARD	111.3	84.3	93.8	109.4	78.2	89.3	104.7	80.1	88.8	104.1	82.9	90.4	106.0	77.6	87.7	107.0	104.2	105.2
095	ACOUSTICAL TREATMENT & WOOD FLOORING	89.9	84.3	86.3	82.9	78.2	79.9	82.9	80.1	81.1	94.3	82.9	86.9	89.9	77.6	81.9	89.9	104.2	99.1
096	FLOORING & CARPET	90.2	81.6	88.1	103.5	71.0	95.6	98.4	79.4	93.7	97.0	85.5	94.2	90.2	87.0	89.4	90.2	103.1	93.3
099	PAINTING & WALL COVERINGS	89.7	75.7	81.5	90.9	90.6	90.7	90.9	90.6	90.7	97.0	90.2	93.0	89.7	77.0	82.2	89.7	91.4	90.7
9	FINISHES	92.9	82.9	87.8	97.1	79.0	87.8	94.8	81.5	88.0	96.4	85.9	91.0	92.0	80.3	86.0	92.3	101.6	97.0
10 - 14	TOTAL DIV. 10 - 14	100.0	89.3	97.7	100.0	73.3	94.4	100.0	88.2	97.5	100.0	91.9	98.3	100.0	91.2	98.1	100.0	101.6	100.3
15	MECHANICAL	99.8	83.0	92.3	99.8	85.5	93.4	96.6	87.2	92.4	100.0	86.7	94.1	99.8	86.3	93.8	99.8	92.2	96.4
16	ELECTRICAL	83.2	95.3	91.3	101.4	91.5	94.8	100.0	95.2	96.8	96.2	89.3	91.5	83.6	91.0	88.6	90.4	94.4	93.1
1 - 16	WEIGHTED AVERAGE	97.0	87.3	92.3	98.9	84.2	91.8	96.8	85.9	91.6	96.7	91.5	94.2	97.2	86.9	92.2	97.8	96.9	97.4

INDIANA

DIVISION		INDIANAPOLIS 461 - 462			KOKOMO 469			LAFAYETTE 479			LAWRENCEBURG 470			MUNCIE 473			NEW ALBANY 471		
		MAT.	INST.	TOTAL	MAT.	INST.	TOTAL	MAT.	INST.	TOTAL	MAT.	INST.	TOTAL	MAT.	INST.	TOTAL	MAT.	INST.	TOTAL
2	SITE WORK	83.4	98.2	94.8	80.2	93.8	90.7	75.6	94.9	90.4	73.0	115.3	105.6	76.9	95.4	91.2	71.6	99.7	93.2
031	CONCRETE FORMWORK	89.6	89.5	89.5	94.8	72.1	75.4	84.9	81.7	82.1	84.3	75.4	76.6	88.3	83.7	84.3	83.7	73.4	74.9
032	CONCRETE REINFORCEMENT	95.2	87.4	90.8	86.7	78.5	82.1	89.4	78.4	83.2	88.8	76.3	81.7	98.8	84.6	90.8	90.0	77.1	82.7
033	CAST IN PLACE CONCRETE	102.5	89.6	97.1	105.4	84.0	96.4	104.8	83.5	95.9	98.0	76.1	88.8	110.0	80.8	97.8	101.1	76.0	90.6
3	CONCRETE	100.5	88.6	94.5	95.7	78.2	86.9	100.6	81.2	90.8	94.0	76.4	85.1	100.6	83.3	91.9	99.2	75.4	87.2
4	MASONRY	99.6	88.2	92.5	92.0	82.6	86.2	98.6	80.4	87.3	77.3	74.3	75.4	93.5	82.6	86.8	84.0	77.8	80.1
5	METALS	101.0	79.5	93.2	95.6	87.6	92.7	94.2	72.1	86.2	90.9	85.2	88.8	97.7	89.9	94.9	93.0	81.1	88.7
6	WOOD & PLASTICS	96.3	89.9	93.0	109.2	69.6	88.8	100.0	81.8	90.6	89.9	74.6	82.0	100.3	84.0	91.9	92.4	73.5	82.6
7	THERMAL & MOISTURE PROTECTION	97.2	89.8	93.7	98.2	78.7	89.2	92.4	83.2	88.1	92.3	78.4	85.9	94.8	82.2	88.9	87.6	75.2	81.8
8	DOORS & WINDOWS	109.0	88.4	104.0	96.8	74.0	91.3	99.4	80.6	94.8	100.4	72.8	93.7	100.1	83.4	96.0	97.8	74.0	92.1
092	LATH, PLASTER & GYPSUM BOARD	107.0	89.8	95.9	113.7	69.1	85.0	102.1	82.0	89.2	94.5	74.4	81.5	102.9	84.3	90.9	102.1	72.9	83.2
095	ACOUSTICAL TREATMENT & WOOD FLOORING	88.5	89.8	89.4	84.4	69.1	74.5	78.8	82.0	80.9	94.3	74.4	81.4	84.3	84.3	84.3	90.2	72.9	79.0
096	FLOORING & CARPET	90.0	96.7	91.6	94.3	71.2	88.7	97.1	86.7	94.5	72.0	71.0	71.8	96.4	81.6	92.8	95.1	59.1	86.3
099	PAINTING & WALL COVERINGS	89.7	90.6	90.2	89.7	73.3	80.0	90.9	81.5	85.4	92.9	78.7	84.6	90.9	75.7	82.0	97.0	66.6	79.1
9	FINISHES	92.3	91.2	91.8	93.4	71.4	82.1	93.3	82.7	87.8	87.0	74.8	80.8	93.9	82.7	88.2	94.9	70.2	82.2
10 - 14	TOTAL DIV. 10 - 14	100.0	91.2	98.2	100.0	72.1	94.1	100.0	87.8	97.4	100.0	68.6	93.4	100.0	88.6	97.6	100.0	68.5	93.4
15	MECHANICAL	99.8	90.3	95.6	96.4	82.2	90.1	96.6	82.4	90.3	97.2	79.6	89.4	99.7	82.9	92.2	96.7	75.1	87.1
16	ELECTRICAL	101.3	95.2	97.2	90.5	86.7	88.0	99.2	82.9	88.2	93.1	81.7	85.4	89.9	80.9	83.8	93.6	79.3	83.9
1 - 16	WEIGHTED AVERAGE	99.9	90.5	95.4	95.3	82.0	88.9	96.6	82.6	89.8	93.1	81.9	87.7	97.1	84.6	91.1	94.7	78.3	86.8

INDIANA — IOWA

DIVISION		SOUTH BEND 465 - 466			TERRE HAUTE 478			WASHINGTON 475			BURLINGTON 526			CARROLL 514			CEDAR RAPIDS 522 - 524		
		MAT.	INST.	TOTAL	MAT.	INST.	TOTAL	MAT.	INST.	TOTAL	MAT.	INST.	TOTAL	MAT.	INST.	TOTAL	MAT.	INST.	TOTAL
2	SITE WORK	83.2	94.4	91.8	84.2	129.9	119.4	85.0	129.7	119.4	80.7	94.7	91.5	72.7	94.2	89.2	81.9	93.0	90.4
031	CONCRETE FORMWORK	94.4	77.1	79.6	92.5	81.0	82.7	90.0	80.3	81.7	89.9	44.0	50.6	74.3	63.3	64.9	101.7	75.6	79.4
032	CONCRETE REINFORCEMENT	95.8	71.3	82.0	97.9	78.4	86.9	90.5	78.3	83.6	96.1	77.6	85.7	96.8	67.6	80.4	96.8	79.7	87.1
033	CAST IN PLACE CONCRETE	103.6	84.2	95.5	96.8	88.2	93.2	105.4	91.0	99.4	104.7	52.0	82.6	104.7	71.0	90.6	105.0	76.7	93.1
3	CONCRETE	93.0	80.2	86.6	103.9	83.1	93.4	109.4	83.7	96.5	95.8	54.5	75.0	94.2	67.5	80.8	96.4	77.2	86.7
4	MASONRY	87.5	83.1	84.7	94.5	85.8	89.1	87.8	82.8	84.7	97.3	54.5	70.7	101.3	61.3	76.5	105.1	73.1	85.2
5	METALS	99.2	101.0	99.8	90.6	83.6	88.1	83.1	83.4	83.2	85.1	82.9	84.3	85.6	77.6	82.7	87.9	83.6	86.3
6	WOOD & PLASTICS	98.9	76.1	87.1	94.0	79.7	86.6	94.2	78.8	86.3	98.0	42.3	69.2	79.6	66.2	72.7	106.3	74.7	90.0
7	THERMAL & MOISTURE PROTECTION	97.5	84.9	91.7	96.6	84.5	91.0	96.8	85.3	91.4	96.1	64.5	81.4	96.4	67.8	83.1	97.3	78.4	88.5
8	DOORS & WINDOWS	95.7	76.4	91.0	98.8	80.3	94.3	95.7	83.0	92.6	97.3	53.4	86.6	103.0	60.4	92.7	104.0	76.4	97.3
092	LATH, PLASTER & GYPSUM BOARD	110.8	75.8	88.2	104.1	77.9	87.2	102.3	77.0	86.0	94.0	40.6	59.6	87.6	65.3	73.2	98.1	74.4	82.8
095	ACOUSTICAL TREATMENT & WOOD FLOORING	89.9	75.8	80.8	94.3	77.9	83.7	82.0	77.0	78.7	106.5	40.6	63.8	106.5	65.3	79.8	110.6	74.4	87.1
096	FLOORING & CARPET	90.2	66.3	84.4	97.0	83.8	93.8	98.2	78.5	93.4	112.8	61.1	100.2	105.9	53.7	93.2	130.7	72.0	116.6
099	PAINTING & WALL COVERINGS	89.7	75.0	81.0	97.0	87.2	91.3	97.0	84.5	89.7	111.1	45.4	72.5	111.1	49.2	74.8	115.8	74.0	91.3
9	FINISHES	92.8	74.3	83.3	96.4	82.1	89.0	94.3	80.0	87.0	105.7	46.2	75.2	101.8	60.0	80.4	113.3	74.1	93.2
10 - 14	TOTAL DIV. 10 - 14	100.0	82.2	96.2	100.0	93.0	98.5	100.0	90.7	98.0	100.0	67.0	93.0	100.0	63.1	92.2	100.0	85.1	96.9
15	MECHANICAL	99.8	77.1	89.7	100.0	83.6	92.7	96.7	83.5	90.8	96.8	47.2	74.7	96.6	67.3	83.5	100.2	78.8	90.7
16	ELECTRICAL	96.4	79.6	85.0	93.3	84.9	87.6	94.2	86.0	88.7	97.3	65.5	75.9	98.4	53.9	68.3	93.6	80.5	84.7
1 - 16	WEIGHTED AVERAGE	96.2	82.4	89.5	97.1	88.5	93.0	95.0	88.2	91.8	95.8	61.0	79.0	95.7	67.0	81.9	98.9	79.6	89.6

IOWA

DIVISION		COUNCIL BLUFFS 515			CRESTON 508			DAVENPORT 527 - 528			DECORAH 521			DES MOINES 500 - 503,509			DUBUQUE 520		
		MAT.	INST.	TOTAL	MAT.	INST.	TOTAL	MAT.	INST.	TOTAL	MAT.	INST.	TOTAL	MAT.	INST.	TOTAL	MAT.	INST.	TOTAL
2	SITE WORK	85.5	90.8	89.6	72.7	95.3	90.1	80.5	97.5	93.6	79.5	94.7	91.2	74.8	98.3	92.9	80.1	90.6	88.2
031	CONCRETE FORMWORK	76.6	67.2	68.5	77.3	68.9	70.1	101.3	88.8	90.6	87.8	59.9	63.9	103.5	75.5	79.5	78.6	63.5	65.7
032	CONCRETE REINFORCEMENT	98.8	75.6	85.7	96.2	71.8	82.5	96.8	90.4	93.2	96.1	67.9	80.2	96.8	78.2	86.3	95.4	81.9	87.8
033	CAST IN PLACE CONCRETE	109.3	69.0	92.4	104.4	74.0	91.7	101.2	89.8	96.4	101.9	67.0	87.3	105.3	80.3	94.8	102.8	69.2	88.7
3	CONCRETE	97.9	70.1	83.9	94.2	71.7	82.9	94.5	89.7	92.1	93.8	64.7	79.1	96.7	78.1	87.3	92.7	69.7	81.1
4	MASONRY	106.2	58.6	76.6	101.3	66.5	79.7	105.4	85.9	93.2	119.7	57.0	80.8	99.7	82.9	89.3	106.0	69.4	83.3
5	METALS	92.9	80.7	88.5	86.7	78.1	83.6	87.9	92.8	89.7	86.0	77.5	82.9	87.8	84.8	86.7	86.4	83.7	85.4
6	WOOD & PLASTICS	78.0	67.9	72.8	80.3	70.4	75.2	106.3	88.1	96.9	94.1	61.4	77.2	108.1	73.5	90.2	80.5	61.2	70.5
7	THERMAL & MOISTURE PROTECTION	96.4	63.8	81.3	97.5	70.4	85.0	96.7	87.6	92.5	96.4	56.8	78.0	97.4	77.2	88.1	96.8	68.0	83.4
8	DOORS & WINDOWS	103.0	69.3	94.8	117.2	68.4	105.3	104.0	87.5	100.0	101.1	63.0	91.9	104.0	75.3	97.1	103.0	69.8	95.0
092	LATH, PLASTER & GYPSUM BOARD	87.6	67.4	74.6	87.6	69.7	76.1	98.1	87.9	91.5	92.9	64.0	71.9	94.3	72.8	80.5	88.1	60.5	70.3
095	ACOUSTICAL TREATMENT & WOOD FLOORING	106.5	67.4	81.2	106.5	69.7	82.6	110.6	87.9	95.9	106.5	60.4	76.6	109.2	72.8	85.6	106.5	60.5	76.6
096	FLOORING & CARPET	104.6	58.8	93.5	106.2	53.7	93.5	115.7	84.6	108.1	112.3	76.2	103.6	115.5	53.7	100.4	119.9	50.1	102.9
099	PAINTING & WALL COVERINGS	106.3	60.7	79.6	111.1	60.7	81.5	111.1	88.1	97.6	111.1	52.1	76.5	111.1	75.4	90.1	114.5	60.7	82.9
9	FINISHES	101.8	64.7	82.8	101.8	65.1	83.0	108.0	87.9	97.7	105.4	62.3	83.3	106.4	70.8	88.2	107.4	59.8	83.0
10 - 14	TOTAL DIV. 10 - 14	100.0	80.5	95.9	100.0	65.9	92.8	100.0	89.9	97.9	100.0	71.0	93.9	100.0	86.7	97.2	100.0	81.8	96.2
15	MECHANICAL	100.2	72.2	87.7	96.9	70.1	84.9	100.2	86.9	94.3	96.8	58.7	79.8	100.2	86.2	94.0	99.8	72.4	87.6
16	ELECTRICAL	101.7	81.2	87.9	90.3	70.3	76.8	91.0	88.1	89.1	93.6	53.6	66.6	94.3	85.9	88.6	99.4	75.0	83.0
1 - 16	WEIGHTED AVERAGE	99.0	73.7	86.8	97.1	72.5	85.2	97.9	89.2	93.7	96.9	64.7	81.4	97.8	83.2	90.8	97.5	73.4	85.9

IOWA

DIVISION		FORT DODGE 505			MASON CITY 504			OTTUMWA 525			SHENANDOAH 516			SIBLEY 512			SIOUX CITY 510 - 511		
		MAT.	INST.	TOTAL	MAT.	INST.	TOTAL	MAT.	INST.	TOTAL	MAT.	INST.	TOTAL	MAT.	INST.	TOTAL	MAT.	INST.	TOTAL
2	SITE WORK	77.1	93.8	90.0	77.1	94.7	90.7	80.7	90.1	87.9	84.3	90.0	88.7	87.7	94.2	92.7	88.6	95.1	93.6
031	CONCRETE FORMWORK	77.9	33.8	40.1	82.2	60.1	63.3	84.7	66.3	68.9	73.7	57.8	60.1	75.6	53.7	56.9	101.7	56.8	63.2
032	CONCRETE REINFORCEMENT	96.2	67.1	79.8	96.1	80.4	87.2	96.1	78.0	85.9	98.8	55.5	74.4	98.8	54.4	73.8	96.8	68.7	81.0
033	CAST IN PLACE CONCRETE	98.0	44.3	75.5	98.0	67.1	85.0	105.4	68.5	90.0	105.5	61.1	86.9	103.5	61.0	85.7	104.1	59.0	85.2
3	CONCRETE	89.9	45.3	67.4	90.2	67.2	78.6	95.3	70.0	82.5	95.1	59.7	77.3	94.2	57.4	75.7	96.0	60.8	78.2
4	MASONRY	98.8	36.0	59.8	113.4	57.0	78.4	103.0	64.5	79.1	106.0	46.4	69.0	122.6	50.7	77.9	99.3	61.6	75.9
5	METALS	86.6	75.6	82.7	85.7	82.6	84.6	85.6	83.4	84.8	91.2	77.5	86.2	85.5	71.1	80.3	87.9	78.5	84.5
6	WOOD & PLASTICS	81.1	31.2	55.4	88.3	61.4	74.5	91.2	67.6	79.0	80.7	62.0	71.0	81.8	54.6	67.7	106.3	57.2	81.0
7	THERMAL & MOISTURE PROTECTION	96.6	40.8	70.8	95.9	60.6	79.5	97.0	68.6	83.8	95.5	54.8	76.6	96.0	57.1	77.9	96.7	59.4	79.4
8	DOORS & WINDOWS	108.9	38.9	92.0	98.3	66.2	90.5	103.0	72.1	95.5	91.5	60.8	84.1	98.7	53.5	87.7	104.0	61.9	93.8
092	LATH, PLASTER & GYPSUM BOARD	88.1	29.2	50.1	90.8	60.4	71.2	91.8	67.1	75.9	88.7	61.3	71.0	88.7	53.3	65.9	98.1	56.1	71.0
095	ACOUSTICAL TREATMENT & WOOD FLOORING	106.5	29.2	56.4	106.5	60.4	76.6	106.5	67.1	81.0	106.5	61.3	77.2	106.5	53.3	72.0	110.6	56.1	75.3
096	FLOORING & CARPET	107.9	76.2	100.2	110.4	76.2	102.1	123.6	70.7	110.7	105.5	55.3	93.3	107.0	54.9	94.4	116.3	73.9	106.0
099	PAINTING & WALL COVERINGS	111.1	22.3	59.0	111.1	47.6	73.8	114.5	71.9	89.5	106.3	33.7	63.7	111.1	49.2	74.8	112.4	59.0	81.1
9	FINISHES	103.2	39.4	70.5	104.4	61.8	82.5	109.2	67.5	87.8	102.1	55.1	78.0	104.0	53.5	78.1	109.1	60.0	83.9
10 - 14	TOTAL DIV. 10 - 14	100.0	64.1	92.4	100.0	71.0	93.9	100.0	72.0	94.1	100.0	72.2	94.1	100.0	53.9	90.3	100.0	78.5	95.5
15	MECHANICAL	97.0	68.0	84.1	97.0	59.0	80.0	96.8	77.9	88.4	96.7	42.0	72.3	96.7	53.0	77.2	99.8	53.3	79.0
16	ELECTRICAL	101.0	70.8	80.6	99.5	53.6	68.5	97.0	65.6	75.8	93.6	42.5	59.1	93.6	53.9	66.8	93.6	74.1	80.4
1 - 16	WEIGHTED AVERAGE	96.5	59.9	78.8	96.0	65.7	81.4	97.0	73.3	85.6	95.6	56.6	76.8	96.7	59.5	78.8	98.2	67.0	83.1

DIVISION		IOWA SPENCER 513			WATERLOO 506 - 507			KANSAS BELLEVILLE 669			COLBY 677			DODGE CITY 678			EMPORIA 668		
		MAT.	INST.	TOTAL	MAT.	INST.	TOTAL	MAT.	INST.	TOTAL	MAT.	INST.	TOTAL	MAT.	INST.	TOTAL	MAT.	INST.	TOTAL
2	SITE WORK	87.7	94.2	92.7	81.0	94.9	91.7	107.5	95.2	98.0	104.8	95.2	97.4	106.9	95.2	97.9	97.5	92.6	93.8
031	CONCRETE FORMWORK	86.2	53.7	58.4	102.3	60.1	66.2	93.6	53.8	59.5	102.6	53.8	60.8	96.1	53.8	59.9	81.2	48.9	53.6
032	CONCRETE REINFORCEMENT	98.8	54.4	73.8	96.8	82.0	88.4	101.3	81.9	90.4	103.8	81.9	91.5	101.3	81.9	90.4	99.9	71.9	84.1
033	CAST IN PLACE CONCRETE	103.5	61.0	85.7	105.0	55.9	84.4	116.8	69.1	96.8	112.3	69.1	94.2	114.3	69.1	95.4	112.9	52.2	87.4
3	CONCRETE	95.0	57.4	76.0	96.4	63.6	79.9	109.2	65.9	87.4	106.8	65.9	86.2	107.7	65.9	86.6	102.0	56.0	78.8
4	MASONRY	122.6	50.7	77.9	99.4	64.1	77.4	97.6	54.2	70.6	99.5	54.2	71.3	104.7	54.2	73.3	103.7	53.6	72.6
5	METALS	86.6	71.0	81.0	87.9	83.9	86.5	93.0	89.0	91.6	93.4	89.0	91.8	94.4	89.0	92.4	92.6	84.5	89.7
6	WOOD & PLASTICS	91.0	54.6	72.2	107.1	61.4	83.5	94.6	53.9	73.6	103.7	53.9	78.0	93.9	53.9	73.2	82.8	49.4	65.5
7	THERMAL & MOISTURE PROTECTION	97.3	56.1	78.2	96.4	59.5	79.3	97.4	60.5	80.3	97.4	60.5	80.3	97.3	60.5	80.2	95.5	51.4	75.1
8	DOORS & WINDOWS	113.4	53.5	98.9	99.3	69.4	92.0	99.6	57.7	89.4	99.7	57.7	89.5	99.6	57.7	89.5	97.1	52.2	86.2
092	LATH, PLASTER & GYPSUM BOARD	91.8	53.3	67.0	98.1	60.4	73.8	109.0	52.0	72.2	112.1	52.0	73.3	108.4	52.0	72.0	104.2	47.3	67.5
095	ACOUSTICAL TREATMENT & WOOD FLOORING	106.5	53.3	72.0	110.6	60.4	78.1	87.5	52.0	64.5	87.5	52.0	64.5	87.5	52.0	64.5	87.5	47.3	61.4
096	FLOORING & CARPET	110.2	54.9	96.8	117.5	76.2	107.4	100.3	62.7	91.1	104.0	62.7	94.0	100.1	62.7	91.0	94.7	58.2	85.8
099	PAINTING & WALL COVERINGS	111.1	49.2	74.8	112.4	44.1	72.3	86.8	61.8	72.2	86.8	61.8	72.2	86.8	61.8	72.2	86.8	61.8	72.2
9	FINISHES	105.4	53.5	78.8	108.7	61.3	84.4	94.5	55.8	74.7	96.1	55.8	75.4	94.3	55.8	74.6	91.3	51.5	70.9
10 - 14	TOTAL DIV. 10 - 14	100.0	53.9	90.3	100.0	79.1	95.6	100.0	64.6	92.5	100.0	64.6	92.5	100.0	64.6	92.5	100.0	58.8	91.3
15	MECHANICAL	96.7	53.0	77.2	100.2	58.9	81.8	96.8	60.1	80.5	96.9	60.1	80.5	100.2	60.1	82.4	96.9	63.1	81.8
16	ELECTRICAL	96.2	53.9	67.6	93.6	59.4	70.5	100.7	64.8	76.5	100.7	64.8	76.5	95.4	64.8	74.8	96.0	72.0	79.8
1 - 16	WEIGHTED AVERAGE	99.2	59.5	80.0	97.5	67.3	82.9	98.7	66.7	83.3	98.8	66.7	83.3	99.4	66.7	83.6	96.7	65.3	81.5

COST INDEXES

617

KANSAS

DIVISION		FORT SCOTT 667			HAYS 676			HUTCHINSON 675			INDEPENDENCE 673			KANSAS CITY 660-662			LIBERAL 679		
		MAT.	INST.	TOTAL	MAT.	INST.	TOTAL	MAT.	INST.	TOTAL	MAT.	INST.	TOTAL	MAT.	INST.	TOTAL	MAT.	INST.	TOTAL
2	SITE WORK	94.5	94.3	94.3	110.5	95.2	98.7	86.9	95.5	93.5	108.3	95.6	98.5	86.6	91.8	90.6	111.0	95.5	99.1
031	CONCRETE FORMWORK	104.3	71.0	75.8	99.1	53.8	60.3	85.4	37.1	44.1	114.7	37.6	48.7	103.0	79.5	82.9	93.6	36.8	44.9
032	CONCRETE REINFORCEMENT	99.3	78.2	87.4	101.3	81.9	90.4	101.3	68.1	82.6	100.7	68.1	82.3	96.3	88.7	92.0	102.7	68.1	83.2
033	CAST IN PLACE CONCRETE	104.8	74.1	91.9	88.5	69.1	80.4	82.0	50.7	68.9	114.8	51.5	88.3	89.6	87.0	88.5	88.5	50.6	72.6
3	CONCRETE	98.1	74.4	86.2	97.7	65.9	81.7	82.1	49.7	65.7	109.3	50.2	79.5	90.1	84.5	87.3	99.1	49.5	74.1
4	MASONRY	103.2	62.1	77.7	106.0	54.2	73.8	99.7	52.2	70.2	96.1	53.6	69.7	107.3	83.3	92.4	102.7	52.2	71.3
5	METALS	92.5	88.3	91.0	93.0	89.0	91.6	92.8	84.4	89.8	92.5	84.3	89.6	99.4	92.8	97.0	93.2	84.0	89.9
6	WOOD & PLASTICS	107.2	73.4	89.7	100.1	53.9	76.2	86.2	35.0	59.7	116.3	35.0	74.3	102.8	78.4	90.2	94.6	35.0	63.8
7	THERMAL & MOISTURE PROTECTION	96.6	71.6	85.0	97.7	60.5	80.5	96.1	50.9	75.1	97.5	50.4	75.6	95.0	84.7	90.2	97.8	46.7	74.1
8	DOORS & WINDOWS	97.1	72.8	91.2	99.6	57.7	89.4	99.6	44.4	86.2	97.1	44.4	84.3	98.4	76.2	93.0	99.7	44.4	86.3
092	LATH, PLASTER & GYPSUM BOARD	110.6	72.0	85.7	110.6	52.0	72.8	105.8	32.4	58.4	116.4	32.4	62.2	104.1	77.2	86.8	109.0	32.4	59.6
095	ACOUSTICAL TREATMENT & WOOD FLOORING	87.5	72.0	77.5	87.5	52.0	64.5	87.5	32.4	51.8	87.5	32.4	51.8	94.3	77.2	83.3	87.5	32.4	51.8
096	FLOORING & CARPET	111.0	63.1	99.4	102.6	62.7	92.9	96.9	62.7	88.6	109.1	62.7	97.8	90.0	66.9	84.4	100.3	62.7	91.1
099	PAINTING & WALL COVERINGS	89.6	61.8	73.3	86.8	61.8	72.2	86.8	61.8	72.2	86.8	61.8	72.2	94.7	81.3	86.8	86.8	61.8	72.2
9	FINISHES	97.8	68.8	82.9	95.8	55.8	75.3	91.9	43.3	67.0	98.4	43.7	70.3	93.4	77.0	85.0	95.0	43.3	68.5
10-14	TOTAL DIV. 10-14	100.0	76.8	95.1	100.0	64.6	92.5	100.0	59.8	91.5	100.0	60.4	91.6	100.0	89.0	97.7	100.0	59.7	91.5
15	MECHANICAL	96.9	57.8	79.5	96.9	60.1	80.5	96.9	49.1	75.6	96.9	47.2	74.8	100.0	83.8	92.8	96.9	41.5	72.2
16	ELECTRICAL	94.8	72.0	79.4	98.9	64.8	75.9	90.4	65.0	73.2	94.1	72.0	79.2	104.5	91.0	95.4	95.4	51.3	65.6
1-16	WEIGHTED AVERAGE	97.0	72.2	85.0	97.9	66.7	82.9	93.8	59.3	77.1	98.5	60.3	80.1	98.1	85.6	92.1	97.6	55.4	77.2

KANSAS / KENTUCKY

DIVISION		SALINA 674			TOPEKA 664-666			WICHITA 670-672			ASHLAND 411-412			BOWLING GREEN 421-422			CAMPTON 413-414		
		MAT.	INST.	TOTAL	MAT.	INST.	TOTAL	MAT.	INST.	TOTAL	MAT.	INST.	TOTAL	MAT.	INST.	TOTAL	MAT.	INST.	TOTAL
2	SITE WORK	96.5	95.1	95.4	89.0	92.5	91.7	90.5	95.3	94.2	99.0	85.2	88.4	71.0	100.3	93.6	78.2	100.8	95.6
031	CONCRETE FORMWORK	91.1	56.2	61.2	101.8	54.1	60.9	98.4	57.0	63.0	78.8	102.4	99.0	82.0	81.0	81.2	81.3	41.6	47.3
032	CONCRETE REINFORCEMENT	100.7	93.7	96.8	93.6	97.5	95.8	93.6	94.3	94.0	91.2	103.6	98.2	88.7	71.5	79.0	89.6	77.5	82.8
033	CAST IN PLACE CONCRETE	99.2	58.3	82.0	93.4	63.2	80.8	87.7	65.5	78.4	92.0	104.7	97.3	91.3	78.8	86.0	101.9	48.0	79.3
3	CONCRETE	95.5	65.3	80.3	89.8	66.8	78.2	86.7	68.3	77.4	94.4	104.2	99.3	93.1	78.6	85.8	97.0	51.7	74.2
4	MASONRY	119.8	52.0	77.7	100.0	65.1	78.3	95.0	64.2	75.9	92.8	107.1	101.7	94.2	80.2	85.5	93.2	38.5	59.2
5	METALS	94.2	92.2	93.5	97.2	95.0	96.4	97.2	93.6	95.9	92.1	113.7	99.9	93.6	79.4	88.5	93.1	77.7	87.6
6	WOOD & PLASTICS	88.8	58.0	72.9	98.8	51.3	74.3	96.2	55.2	75.1	74.3	100.6	87.8	86.9	80.6	83.7	85.0	41.0	62.3
7	THERMAL & MOISTURE PROTECTION	96.7	65.4	82.2	96.7	72.5	85.5	96.3	71.3	84.7	88.3	98.9	93.3	87.4	78.0	83.0	96.9	50.4	75.4
8	DOORS & WINDOWS	99.6	60.3	90.1	99.7	62.2	90.6	99.7	61.6	90.5	96.0	97.7	96.5	97.9	73.0	91.9	99.4	48.5	87.1
092	LATH, PLASTER & GYPSUM BOARD	106.9	56.2	74.2	111.1	49.3	71.2	111.1	53.3	73.8	69.2	100.5	89.4	100.0	80.3	87.3	100.0	37.9	59.9
095	ACOUSTICAL TREATMENT & WOOD FLOORING	87.5	56.2	67.2	94.3	49.3	65.2	94.3	53.3	67.8	82.7	100.5	94.2	90.2	80.3	83.8	90.2	37.9	56.3
096	FLOORING & CARPET	98.0	45.8	85.3	101.7	58.2	91.1	100.8	72.2	93.9	76.9	102.5	83.1	93.0	77.3	89.2	94.3	53.7	84.4
099	PAINTING & WALL COVERINGS	86.8	48.7	64.5	86.8	67.7	75.6	86.8	57.5	69.6	99.6	100.4	100.1	97.0	82.4	88.4	97.0	40.9	64.1
9	FINISHES	92.9	53.0	72.4	95.6	54.8	74.7	95.4	59.1	76.8	81.5	102.4	92.2	93.8	80.6	87.0	94.4	43.1	68.1
10-14	TOTAL DIV. 10-14	100.0	71.5	94.0	100.0	79.7	95.7	100.0	73.2	94.3	100.0	102.4	100.5	100.0	74.9	94.7	100.0	71.7	94.0
15	MECHANICAL	100.2	45.2	75.7	100.2	69.6	86.6	100.2	67.3	85.5	96.6	97.5	97.0	100.0	84.4	93.1	96.8	44.0	73.3
16	ELECTRICAL	94.9	71.1	78.8	101.6	72.0	81.6	99.3	71.1	80.3	89.5	94.5	92.8	93.6	89.0	90.5	89.3	35.5	53.0
1-16	WEIGHTED AVERAGE	98.2	65.0	82.2	97.6	71.9	85.2	96.8	71.9	84.8	93.1	99.8	96.4	95.2	83.8	89.7	95.2	52.8	74.7

KENTUCKY

DIVISION		CORBIN 407-409			COVINGTON 410			ELIZABETHTOWN 427			FRANKFORT 406			HAZARD 417-418			HENDERSON 424		
		MAT.	INST.	TOTAL	MAT.	INST.	TOTAL	MAT.	INST.	TOTAL	MAT.	INST.	TOTAL	MAT.	INST.	TOTAL	MAT.	INST.	TOTAL
2	SITE WORK	75.5	100.8	94.9	74.7	116.3	106.7	65.9	100.2	92.3	76.3	102.8	96.7	76.2	100.8	95.1	72.7	130.2	116.9
031	CONCRETE FORMWORK	77.5	41.6	46.8	76.6	81.9	81.1	72.5	80.5	79.4	92.9	73.1	76.0	77.7	41.6	46.8	85.2	84.7	84.8
032	CONCRETE REINFORCEMENT	89.1	77.5	82.6	88.3	96.3	92.8	89.1	96.4	93.2	90.4	90.7	90.6	90.0	77.5	82.9	88.8	96.6	93.2
033	CAST IN PLACE CONCRETE	96.4	48.0	76.1	97.6	85.3	92.4	82.5	77.2	80.3	96.4	81.6	90.2	98.0	48.0	77.0	80.7	96.3	87.2
3	CONCRETE	92.3	51.7	71.8	95.2	86.3	90.8	85.1	82.5	83.8	93.6	79.7	86.6	93.7	51.7	72.5	89.7	91.0	90.3
4	MASONRY	92.6	38.5	59.0	108.8	90.4	97.4	79.2	78.3	78.7	92.8	69.9	78.6	90.8	38.5	58.3	98.8	96.8	97.6
5	METALS	93.0	77.7	87.5	90.9	96.7	93.0	92.9	89.5	91.7	93.0	88.3	91.3	93.2	77.7	87.6	83.1	92.4	86.5
6	WOOD & PLASTICS	81.2	41.0	60.4	81.5	79.5	80.5	80.6	80.6	80.6	94.0	71.7	82.5	81.2	41.0	60.4	89.0	81.1	84.9
7	THERMAL & MOISTURE PROTECTION	96.6	50.4	75.2	92.5	91.0	91.8	86.9	81.5	84.4	96.8	77.2	87.7	96.8	50.4	75.3	96.1	92.5	94.4
8	DOORS & WINDOWS	98.7	48.5	86.5	101.5	83.4	97.1	97.9	86.5	95.1	98.0	77.8	93.1	99.8	48.5	87.4	96.2	86.6	93.8
092	LATH, PLASTER & GYPSUM BOARD	98.9	37.9	59.6	91.8	79.4	83.8	98.4	80.3	86.7	103.2	69.7	81.6	98.9	37.9	59.6	100.2	79.4	86.8
095	ACOUSTICAL TREATMENT & WOOD FLOORING	90.2	37.9	56.3	94.3	79.4	84.7	90.2	80.3	83.8	90.2	69.7	76.9	90.2	37.9	56.3	82.0	79.4	80.3
096	FLOORING & CARPET	92.6	53.7	83.2	69.5	93.9	75.4	90.3	76.0	86.8	97.9	65.7	90.0	92.6	53.7	83.2	96.1	91.5	95.0
099	PAINTING & WALL COVERINGS	97.0	40.9	64.1	92.9	88.3	90.2	97.0	78.3	86.0	97.0	61.8	76.3	97.0	40.9	64.1	97.0	99.0	98.2
9	FINISHES	93.5	43.1	67.6	86.0	84.2	85.1	92.4	79.0	85.6	95.8	69.8	82.4	93.5	43.1	67.7	92.8	87.1	89.9
10-14	TOTAL DIV. 10-14	100.0	71.7	94.0	100.0	91.4	98.2	100.0	92.2	98.3	100.0	87.4	97.3	100.0	71.7	94.0	100.0	96.9	99.3
15	MECHANICAL	96.8	44.0	73.3	97.1	92.2	94.9	96.7	83.1	90.6	100.1	77.9	90.2	96.8	44.0	73.3	96.7	85.8	91.8
16	ELECTRICAL	89.7	35.5	53.2	95.7	86.3	89.3	88.9	89.0	89.0	93.8	78.3	83.4	89.3	35.5	53.0	92.7	86.2	88.3
1-16	WEIGHTED AVERAGE	94.3	52.8	74.3	95.1	91.6	93.4	91.8	85.7	88.9	95.8	80.1	88.2	94.5	52.8	74.4	92.6	93.3	92.9

KENTUCKY

DIVISION		LEXINGTON 403 - 405			LOUISVILLE 400 - 402			OWENSBORO 423			PADUCAH 420			PIKEVILLE 415 - 416			SOMERSET 425 - 426		
		MAT.	INST.	TOTAL	MAT.	INST.	TOTAL	MAT.	INST.	TOTAL	MAT.	INST.	TOTAL	MAT.	INST.	TOTAL	MAT.	INST.	TOTAL
2	SITE WORK	77.0	102.7	96.8	68.0	100.2	92.8	82.2	130.0	119.0	76.2	130.0	117.6	110.2	82.7	89.0	68.6	100.8	93.3
031	CONCRETE FORMWORK	92.9	71.0	74.2	91.7	80.5	82.1	86.4	79.2	80.2	80.2	82.3	82.0	90.2	68.9	72.0	78.2	41.6	46.9
032	CONCRETE REINFORCEMENT	97.9	95.8	96.7	97.9	96.4	97.0	88.8	95.3	92.5	89.4	85.1	87.0	91.7	67.8	78.2	89.1	77.5	82.6
033	CAST IN PLACE CONCRETE	98.7	78.8	90.3	95.4	77.2	87.8	94.2	91.2	92.9	85.9	90.6	87.9	101.1	69.3	87.8	80.7	48.0	67.0
3	CONCRETE	95.7	78.8	87.2	94.0	82.5	88.2	101.8	86.6	94.1	94.0	85.8	89.9	107.6	70.6	89.0	80.2	51.7	65.8
4	MASONRY	91.5	60.0	71.9	91.9	78.3	83.4	91.6	65.0	75.1	94.4	80.4	85.7	93.0	61.9	73.6	85.2	38.5	56.2
5	METALS	95.1	89.7	93.2	95.1	89.5	93.1	85.3	90.3	87.1	82.9	86.8	84.3	92.1	94.8	93.1	93.0	77.7	87.5
6	WOOD & PLASTICS	94.0	71.7	82.5	97.3	80.6	88.7	86.8	79.2	82.9	83.9	81.2	82.5	86.1	69.4	77.5	82.0	41.0	60.8
7	THERMAL & MOISTURE PROTECTION	97.0	70.2	84.6	87.5	81.6	84.8	96.5	79.2	88.5	96.1	82.0	89.6	89.2	63.4	77.3	96.1	50.4	74.9
8	DOORS & WINDOWS	98.8	81.7	94.6	98.8	86.5	95.8	96.2	85.3	93.5	95.3	83.8	92.5	96.8	67.6	89.7	98.7	48.5	86.5
092	LATH, PLASTER & GYPSUM BOARD	104.1	69.7	81.9	107.6	80.3	89.9	99.7	77.4	85.3	98.6	79.4	86.2	72.9	68.3	69.9	99.5	37.9	59.7
095	ACOUSTICAL TREATMENT & WOOD FLOORING	94.3	69.7	78.4	94.3	80.3	85.2	82.0	77.4	79.0	82.0	79.4	80.3	82.7	68.3	73.4	90.2	37.9	56.3
096	FLOORING & CARPET	97.9	53.7	87.1	96.3	76.0	91.4	95.2	77.3	90.8	94.0	77.3	90.0	81.5	63.9	77.2	92.9	53.7	83.4
099	PAINTING & WALL COVERINGS	97.0	72.7	82.7	97.0	78.3	86.0	97.0	102.4	100.1	97.0	75.8	84.5	99.6	45.2	67.7	97.0	40.9	64.1
9	FINISHES	96.7	67.4	81.7	96.4	79.0	87.5	92.9	81.2	86.9	92.1	80.4	86.1	84.1	65.4	74.5	93.2	43.1	67.5
10 - 14	TOTAL DIV. 10 - 14	100.0	92.0	98.3	100.0	92.2	98.3	100.0	95.0	98.9	100.0	88.3	97.5	100.0	73.6	94.5	100.0	71.7	94.0
15	MECHANICAL	100.0	60.3	82.3	100.0	83.1	92.5	100.0	66.3	85.0	96.7	85.5	91.7	96.6	66.3	83.1	96.8	44.0	73.3
16	ELECTRICAL	94.4	65.2	74.7	94.4	89.0	90.7	92.8	83.8	86.7	96.8	88.2	91.0	94.5	63.4	73.5	90.0	35.5	53.2
1 - 16	WEIGHTED AVERAGE	96.6	73.4	85.4	95.9	85.7	90.9	95.0	84.1	89.8	93.0	89.3	91.3	95.9	70.3	83.6	92.2	52.8	73.2

LOUISIANA

DIVISION		ALEXANDRIA 713 - 714			BATON ROUGE 707 - 708			HAMMOND 704			LAFAYETTE 705			LAKE CHARLES 706			MONROE 712		
		MAT.	INST.	TOTAL	MAT.	INST.	TOTAL	MAT.	INST.	TOTAL	MAT.	INST.	TOTAL	MAT.	INST.	TOTAL	MAT.	INST.	TOTAL
2	SITE WORK	103.7	83.8	88.4	112.4	86.1	92.2	113.2	87.7	93.6	113.5	87.5	93.5	114.1	85.8	92.4	103.7	83.6	88.3
031	CONCRETE FORMWORK	78.2	53.4	57.0	96.8	59.2	64.6	70.7	65.5	66.3	96.2	49.4	56.1	96.9	60.2	65.4	77.4	54.3	57.6
032	CONCRETE REINFORCEMENT	98.1	66.0	80.0	100.4	61.8	78.7	98.9	67.2	81.0	100.4	61.2	78.3	100.4	61.7	78.6	97.1	65.9	79.5
033	CAST IN PLACE CONCRETE	95.1	50.4	76.4	91.3	60.9	78.5	92.3	59.6	78.6	91.7	60.0	78.5	96.6	62.6	82.4	95.1	55.6	78.6
3	CONCRETE	88.1	55.8	71.8	92.0	61.1	76.4	90.5	64.5	77.4	92.2	56.5	74.2	94.6	62.1	78.2	87.9	57.9	72.8
4	MASONRY	110.6	53.9	75.4	98.2	64.4	77.2	99.5	61.0	75.6	99.5	52.0	69.9	97.3	64.8	77.1	106.1	51.8	72.4
5	METALS	87.4	75.6	83.2	98.0	74.1	89.4	98.2	77.1	90.6	97.5	73.5	88.9	97.5	74.2	89.1	87.4	74.9	82.9
6	WOOD & PLASTICS	77.3	54.2	65.4	102.6	59.8	80.5	78.6	66.9	72.6	102.6	47.5	74.1	100.4	60.7	79.9	76.3	54.8	65.2
7	THERMAL & MOISTURE PROTECTION	96.0	57.6	78.2	97.3	65.1	82.4	99.4	66.0	83.9	100.2	59.7	81.4	99.9	65.0	83.7	96.0	59.2	78.9
8	DOORS & WINDOWS	99.8	58.5	89.8	103.5	59.9	92.9	99.0	70.8	92.2	103.5	53.3	91.3	103.5	62.3	93.5	99.8	64.0	91.1
092	LATH, PLASTER & GYPSUM BOARD	87.6	53.8	65.8	102.3	59.4	74.6	94.7	66.7	76.6	102.3	46.7	66.4	102.3	60.3	75.2	87.1	54.4	66.0
095	ACOUSTICAL TREATMENT & WOOD FLOORING	100.3	53.8	70.2	97.1	59.4	72.7	98.5	66.7	77.9	97.1	46.7	64.5	97.1	60.3	73.3	100.3	54.4	70.5
096	FLOORING & CARPET	108.5	61.6	97.1	110.8	74.2	101.9	99.4	74.2	93.2	111.1	55.5	97.6	111.1	64.8	99.8	108.2	53.5	94.9
099	PAINTING & WALL COVERINGS	97.1	51.2	70.2	96.8	58.8	74.5	96.8	53.6	71.5	96.8	58.8	74.5	96.8	54.1	71.8	97.1	44.9	66.5
9	FINISHES	98.3	54.5	75.8	102.8	61.9	81.9	98.3	65.8	81.7	102.9	50.5	76.0	102.9	60.4	81.1	98.1	52.6	74.8
10 - 14	TOTAL DIV. 10 - 14	100.0	68.7	93.4	100.0	71.0	93.9	100.0	72.9	94.3	100.0	68.9	93.4	100.0	71.2	93.9	100.0	69.1	93.5
15	MECHANICAL	100.0	47.7	76.7	100.0	54.4	79.6	96.7	60.7	80.6	100.0	59.6	82.0	100.0	65.5	84.6	100.0	54.6	79.7
16	ELECTRICAL	95.9	57.1	69.7	99.0	64.7	75.9	97.5	63.7	74.7	99.6	77.2	84.5	99.0	64.7	75.9	99.0	61.3	73.6
1 - 16	WEIGHTED AVERAGE	96.4	59.2	78.5	99.6	64.8	82.8	97.5	67.5	83.0	99.7	64.2	82.6	99.9	67.0	84.0	96.4	61.3	79.4

LOUISIANA / MAINE

DIVISION		NEW ORLEANS 700 - 701			SHREVEPORT 710 - 711			THIBODAUX 703			AUGUSTA 043			BANGOR 044			BATH 045		
		MAT.	INST.	TOTAL	MAT.	INST.	TOTAL	MAT.	INST.	TOTAL	MAT.	INST.	TOTAL	MAT.	INST.	TOTAL	MAT.	INST.	TOTAL
2	SITE WORK	113.5	88.2	94.1	102.4	83.5	87.8	115.8	87.7	94.1	92.3	95.7	94.9	92.2	98.1	96.7	90.8	96.1	94.8
031	CONCRETE FORMWORK	96.0	67.5	71.6	100.2	57.5	63.7	85.8	66.8	69.5	96.1	40.8	48.7	89.7	85.1	85.8	80.7	45.8	50.8
032	CONCRETE REINFORCEMENT	100.4	67.3	81.7	96.6	66.0	79.4	98.9	67.2	81.1	95.0	42.7	65.5	95.0	106.7	101.6	94.0	99.2	96.9
033	CAST IN PLACE CONCRETE	95.7	64.7	82.7	93.7	60.0	79.6	99.0	64.1	84.4	87.1	49.2	71.2	87.1	76.0	82.4	87.1	52.8	72.7
3	CONCRETE	94.1	67.1	80.5	88.3	60.9	74.5	95.9	66.6	81.1	107.4	45.7	76.3	106.3	85.4	95.8	106.1	57.8	81.8
4	MASONRY	99.6	62.4	76.5	99.4	50.4	68.9	121.7	61.7	84.4	92.6	46.0	63.7	109.5	67.6	83.5	116.8	43.9	71.6
5	METALS	100.4	77.6	92.2	88.2	74.6	83.3	98.8	76.8	90.9	98.1	75.0	89.8	97.9	82.0	92.2	96.2	58.9	82.7
6	WOOD & PLASTICS	96.1	68.9	82.1	101.8	59.6	80.0	88.4	68.3	78.0	99.0	39.1	68.1	91.8	84.9	88.2	86.0	43.7	64.2
7	THERMAL & MOISTURE PROTECTION	100.8	67.6	85.4	95.3	59.3	78.6	101.0	67.1	85.3	101.7	43.5	74.7	101.5	65.3	84.7	101.4	52.1	78.6
8	DOORS & WINDOWS	102.1	71.0	94.6	97.6	61.6	88.9	104.4	71.5	96.5	106.0	47.7	91.9	105.9	76.1	98.7	105.9	40.5	90.1
092	LATH, PLASTER & GYPSUM BOARD	101.2	68.8	80.3	94.8	59.4	71.9	98.9	68.2	79.1	96.6	36.1	57.6	93.2	83.4	86.9	91.0	40.9	58.7
095	ACOUSTICAL TREATMENT & WOOD FLOORING	97.1	68.8	78.8	101.7	59.4	74.3	98.5	68.2	78.8	93.5	36.1	56.3	92.1	83.4	86.5	92.1	58.9	58.9
096	FLOORING & CARPET	111.6	63.6	99.9	119.1	56.2	103.8	107.3	63.0	96.6	96.9	53.8	86.4	94.7	53.8	84.8	92.8	53.8	83.3
099	PAINTING & WALL COVERINGS	98.8	68.9	81.3	97.1	51.2	70.2	98.8	65.3	79.2	89.3	29.5	54.2	89.3	40.8	60.9	89.3	30.1	54.6
9	FINISHES	103.1	66.9	84.5	102.9	56.4	79.1	101.8	66.0	83.4	95.6	41.0	67.6	94.1	74.4	84.0	93.2	44.4	68.2
10 - 14	TOTAL DIV. 10 - 14	100.0	73.3	94.4	100.0	70.8	93.8	100.0	73.1	94.3	100.0	67.9	93.2	100.0	90.9	98.1	100.0	69.2	93.5
15	MECHANICAL	100.0	64.3	84.1	100.0	57.8	81.2	96.7	63.7	82.0	100.0	32.4	69.8	100.0	79.9	91.0	96.7	65.1	82.6
16	ELECTRICAL	99.3	67.0	77.6	96.9	66.0	76.1	95.4	67.0	76.3	104.6	83.8	90.6	101.3	88.5	92.6	98.0	38.8	58.0
1 - 16	WEIGHTED AVERAGE	100.2	69.5	85.4	96.5	63.4	80.5	100.5	69.0	85.3	100.6	57.2	79.6	100.9	82.0	91.8	99.8	56.3	78.8

		MAINE																
		HOULTON			KITTERY			LEWISTON			MACHIAS			PORTLAND			ROCKLAND		
	DIVISION	047			039			042			046			040 - 041			048		
		MAT.	INST.	TOTAL	MAT.	INST.	TOTAL	MAT.	INST.	TOTAL	MAT.	INST.	TOTAL	MAT.	INST.	TOTAL	MAT.	INST.	TOTAL
2	SITE WORK	92.3	96.2	95.3	86.8	95.4	93.4	90.4	98.1	96.3	91.8	96.1	95.1	89.8	98.1	96.2	88.2	95.4	93.7
031	CONCRETE FORMWORK	90.4	80.6	82.0	81.3	53.0	57.1	96.3	84.9	86.5	86.9	63.4	66.8	95.5	84.9	86.4	88.1	63.7	67.2
032	CONCRETE REINFORCEMENT	95.0	102.0	98.9	93.3	55.2	71.8	116.7	106.6	111.0	95.0	84.5	89.1	116.7	106.6	111.0	95.0	84.5	89.1
033	CAST IN PLACE CONCRETE	87.1	57.3	74.6	87.0	58.3	75.0	97.0	75.9	88.1	87.1	58.8	75.2	90.2	75.9	84.2	88.8	59.2	76.4
3	CONCRETE	107.2	75.0	91.0	100.8	56.3	78.4	109.9	85.2	97.4	106.7	65.7	86.0	106.5	85.2	95.8	103.2	66.0	84.4
4	MASONRY	92.6	52.6	67.8	105.8	49.6	70.9	93.7	67.6	77.5	92.6	52.8	67.9	91.6	67.6	76.7	86.3	52.9	65.6
5	METALS	96.4	58.2	82.6	96.1	76.2	88.9	101.3	81.8	94.3	96.4	70.6	87.1	101.3	81.8	94.3	96.3	71.2	87.2
6	WOOD & PLASTICS	96.6	88.5	92.4	86.6	52.6	69.1	99.0	84.9	91.7	92.8	63.7	77.8	99.0	84.9	91.7	94.1	63.7	78.4
7	THERMAL & MOISTURE PROTECTION	101.6	55.3	80.1	101.0	51.8	78.2	101.4	65.3	84.6	101.6	50.8	78.0	101.3	65.3	84.6	101.3	55.9	80.2
8	DOORS & WINDOWS	106.0	64.4	95.9	105.9	57.1	94.1	109.1	76.1	101.1	106.0	57.3	94.2	109.1	76.1	101.1	105.9	57.3	94.1
092	LATH, PLASTER & GYPSUM BOARD	95.6	87.1	90.1	91.0	50.0	64.6	98.5	83.4	88.8	94.0	61.5	73.0	98.5	83.4	88.8	94.2	61.5	73.1
095	ACOUSTICAL TREATMENT & WOOD FLOORING	93.5	87.1	89.4	92.1	50.0	64.8	101.7	83.4	89.8	93.5	61.5	72.8	101.7	83.4	89.8	92.1	61.5	72.3
096	FLOORING & CARPET	96.1	53.8	86.5	93.0	53.8	83.5	96.9	53.8	86.4	95.0	26.4	78.3	96.9	53.8	86.4	95.3	53.8	85.3
099	PAINTING & WALL COVERINGS	89.3	40.8	60.9	89.3	41.5	61.2	89.3	40.8	60.9	89.3	30.9	55.1	89.3	40.8	60.9	89.3	40.8	60.9
9	FINISHES	95.2	73.0	83.8	93.0	51.4	71.6	97.1	74.4	85.5	94.6	53.5	73.5	97.1	74.4	85.5	94.2	59.4	76.3
10 - 14	TOTAL DIV. 10 - 14	100.0	77.2	95.2	100.0	71.3	93.9	100.0	90.8	98.0	100.0	75.5	94.8	100.0	90.8	98.0	100.0	75.5	94.8
15	MECHANICAL	96.7	51.2	76.4	95.9	56.5	78.3	100.0	79.9	91.0	96.7	53.9	77.6	100.0	79.9	91.0	96.7	54.0	77.6
16	ELECTRICAL	104.7	35.5	58.0	98.0	59.5	72.0	104.7	57.2	72.7	104.7	88.4	93.7	104.9	57.2	72.7	104.5	88.4	93.6
1 - 16	WEIGHTED AVERAGE	99.5	61.3	81.1	98.2	61.6	80.6	101.9	76.8	89.8	99.3	67.5	84.0	101.4	76.8	89.5	98.4	68.4	83.9

		MAINE			MARYLAND														
		WATERVILLE			ANNAPOLIS			BALTIMORE			COLLEGE PARK			CUMBERLAND			EASTON		
	DIVISION	049			214			210 - 212			207 - 208			215			216		
		MAT.	INST.	TOTAL	MAT.	INST.	TOTAL	MAT.	INST.	TOTAL	MAT.	INST.	TOTAL	MAT.	INST.	TOTAL	MAT.	INST.	TOTAL
2	SITE WORK	92.5	95.7	95.0	95.7	88.6	90.3	95.7	93.2	93.8	103.4	91.3	94.1	88.8	88.5	88.6	95.0	86.4	88.4
031	CONCRETE FORMWORK	80.0	40.8	46.4	101.5	68.8	73.5	101.5	76.4	80.0	75.2	71.5	72.0	86.4	78.9	80.0	83.1	35.7	42.5
032	CONCRETE REINFORCEMENT	95.0	42.7	65.5	96.9	85.8	90.7	96.9	85.8	90.7	102.0	74.4	86.4	85.7	74.9	79.6	84.9	25.6	51.5
033	CAST IN PLACE CONCRETE	87.1	49.2	71.2	106.4	80.5	95.5	106.4	81.5	95.9	118.0	77.0	100.9	95.0	80.8	89.1	105.5	48.9	81.8
3	CONCRETE	107.9	45.7	76.5	102.8	77.6	90.1	102.8	81.2	91.9	107.5	75.6	91.4	91.4	80.0	85.7	99.2	40.8	69.7
4	MASONRY	103.8	46.0	67.9	90.5	73.7	80.0	90.5	73.7	80.1	106.5	71.6	84.8	87.6	75.6	80.2	102.7	31.2	58.3
5	METALS	96.4	75.0	88.7	94.0	99.3	95.9	94.9	100.0	96.7	92.6	98.5	94.7	94.0	94.1	94.0	94.0	68.6	84.8
6	WOOD & PLASTICS	85.1	39.1	61.4	101.9	67.6	84.2	101.9	77.7	89.4	76.6	71.7	74.1	89.2	78.9	83.9	86.0	36.9	60.6
7	THERMAL & MOISTURE PROTECTION	101.5	43.5	74.6	94.7	77.3	86.6	94.7	78.6	87.2	94.6	76.2	86.1	94.2	81.7	88.4	94.3	39.4	68.9
8	DOORS & WINDOWS	106.0	47.7	91.8	92.9	79.0	89.5	92.9	82.6	90.4	97.1	73.9	91.5	94.0	84.4	91.7	92.1	30.3	77.2
092	LATH, PLASTER & GYPSUM BOARD	90.8	36.1	55.5	101.8	67.9	79.6	101.8	77.6	86.2	92.8	70.5	78.4	98.1	79.0	85.8	97.0	35.6	57.4
095	ACOUSTICAL TREATMENT & WOOD FLOORING	93.5	36.1	56.3	93.3	67.3	76.5	93.3	77.6	83.1	90.2	70.5	77.5	93.3	79.0	84.0	93.3	35.6	55.9
096	FLOORING & CARPET	92.6	53.8	83.2	90.0	84.0	88.5	90.0	84.0	88.5	84.5	82.7	84.1	85.9	76.4	83.6	84.9	20.5	69.3
099	PAINTING & WALL COVERINGS	89.3	29.5	54.2	100.0	80.8	88.7	100.0	80.8	88.7	101.0	83.8	90.9	100.0	57.3	74.9	100.0	29.5	58.6
9	FINISHES	93.5	41.0	66.6	90.8	72.1	81.2	90.8	78.1	84.3	89.1	74.7	81.7	88.7	76.0	82.2	88.5	33.5	60.3
10 - 14	TOTAL DIV. 10 - 14	100.0	67.9	93.2	100.0	81.4	96.1	100.0	83.2	96.5	100.0	80.6	95.9	100.0	85.3	96.9	100.0	62.6	92.1
15	MECHANICAL	96.7	32.4	68.0	100.0	81.2	91.6	100.0	81.2	91.6	96.8	78.7	88.7	96.8	72.1	85.7	96.8	26.0	65.2
16	ELECTRICAL	104.6	83.8	90.6	100.6	90.7	93.9	102.0	97.4	98.9	99.2	87.9	91.6	98.4	81.1	86.7	97.5	37.7	57.1
1 - 16	WEIGHTED AVERAGE	99.9	57.2	79.3	96.9	82.5	90.0	97.1	85.7	91.6	97.5	81.4	89.8	94.0	80.2	87.4	95.6	42.8	70.1

		MARYLAND															MASSACHUSETTS		
		ELKTON			HAGERSTOWN			SALISBURY			SILVER SPRING			WALDORF			BOSTON		
	DIVISION	219			217			218			209			206			020 - 022, 024		
		MAT.	INST.	TOTAL	MAT.	INST.	TOTAL	MAT.	INST.	TOTAL	MAT.	INST.	TOTAL	MAT.	INST.	TOTAL	MAT.	INST.	TOTAL
2	SITE WORK	83.6	87.8	86.9	87.0	88.8	88.4	95.0	86.7	88.6	92.0	83.6	85.5	98.4	83.7	87.1	98.0	104.8	103.2
031	CONCRETE FORMWORK	91.5	80.2	81.8	88.3	77.6	79.1	102.4	40.2	49.1	85.1	70.2	72.4	94.4	67.4	71.3	102.3	134.7	130.1
032	CONCRETE REINFORCEMENT	84.9	77.6	80.8	85.7	74.9	79.6	84.9	77.3	80.6	100.9	76.5	87.1	101.5	76.4	87.4	115.4	169.2	145.7
033	CAST IN PLACE CONCRETE	85.4	60.7	75.1	90.4	67.3	80.7	105.5	49.4	82.0	120.9	77.3	102.6	135.4	72.4	109.0	112.3	142.1	124.8
3	CONCRETE	84.1	73.6	78.8	87.9	74.8	81.3	100.6	51.7	75.9	106.0	75.3	90.5	117.1	72.2	94.5	121.0	142.4	131.7
4	MASONRY	88.8	53.1	66.6	94.7	75.6	82.9	101.5	42.5	64.8	104.5	72.1	84.4	88.7	65.3	74.2	112.5	146.7	133.8
5	METALS	94.0	83.5	90.2	93.9	93.8	93.9	94.0	72.9	86.4	98.4	93.3	96.6	98.5	91.2	95.8	104.2	126.5	112.2
6	WOOD & PLASTICS	95.0	89.0	91.9	87.9	77.5	82.5	106.8	39.5	72.1	84.4	70.5	77.2	94.0	70.5	81.8	103.4	133.8	119.1
7	THERMAL & MOISTURE PROTECTION	93.9	62.4	79.3	93.9	78.2	86.6	94.7	62.4	79.7	94.8	81.9	88.8	95.4	70.3	83.8	101.9	141.0	120.1
8	DOORS & WINDOWS	92.2	73.5	87.6	92.1	78.2	88.7	92.4	34.1	78.3	90.2	73.2	86.1	90.9	68.7	85.5	102.9	138.5	111.6
092	LATH, PLASTER & GYPSUM BOARD	99.7	89.4	93.0	97.6	77.6	84.7	103.4	38.3	61.4	98.1	70.5	80.3	101.2	70.5	81.4	98.5	133.9	121.4
095	ACOUSTICAL TREATMENT & WOOD FLOORING	93.3	89.4	90.8	93.3	77.6	83.1	93.3	38.3	57.7	97.1	70.5	79.9	97.1	70.5	79.9	101.7	133.9	122.6
096	FLOORING & CARPET	87.7	78.4	85.4	85.5	76.4	83.3	91.5	78.4	88.4	90.3	82.7	88.4	94.4	78.4	90.5	96.7	153.0	110.4
099	PAINTING & WALL COVERINGS	100.0	89.6	93.9	100.0	50.0	70.7	100.0	30.1	59.0	106.4	83.8	93.1	106.4	86.1	94.5	90.1	143.3	121.3
9	FINISHES	89.3	82.4	85.8	88.4	74.3	81.2	91.5	45.3	67.8	89.2	73.8	81.3	91.2	71.5	81.1	96.8	139.1	118.5
10 - 14	TOTAL DIV. 10 - 14	100.0	73.6	94.4	100.0	85.1	96.9	100.0	63.3	92.2	100.0	77.3	95.2	100.0	69.6	93.6	100.0	129.5	106.2
15	MECHANICAL	96.8	75.8	87.4	100.0	82.6	92.3	96.8	52.9	77.2	96.8	78.4	88.6	96.8	74.8	87.0	100.0	123.5	110.4
16	ELECTRICAL	100.6	84.5	89.8	98.1	81.1	86.6	95.4	80.3	85.2	96.5	87.9	90.7	92.4	87.9	89.4	99.7	128.2	118.9
1 - 16	WEIGHTED AVERAGE	93.2	77.2	85.4	94.4	81.0	87.9	96.2	60.2	78.8	96.9	80.1	88.8	97.7	77.2	87.8	103.8	130.6	116.7

City Cost Indexes

MASSACHUSETTS

	DIVISION	BROCKTON 023			BUZZARDS BAY 025			FALL RIVER 027			FITCHBURG 014			FRAMINGHAM 017			GREENFIELD 013		
		MAT.	INST.	TOTAL	MAT.	INST.	TOTAL	MAT.	INST.	TOTAL	MAT.	INST.	TOTAL	MAT.	INST.	TOTAL	MAT.	INST.	TOTAL
2	SITE WORK	96.0	100.9	99.8	87.0	100.8	97.6	94.9	101.1	99.6	88.2	100.8	97.9	85.4	100.2	96.8	92.3	99.6	97.9
031	CONCRETE FORMWORK	102.0	119.6	117.1	94.9	117.1	113.9	102.0	117.7	115.4	88.2	112.9	109.4	96.4	115.3	112.6	85.7	105.5	102.7
032	CONCRETE REINFORCEMENT	116.7	168.9	146.1	93.6	147.7	124.1	116.7	151.6	136.4	93.6	157.4	129.5	93.5	168.6	135.8	97.5	136.9	119.7
033	CAST IN PLACE CONCRETE	106.9	136.4	119.2	88.8	137.4	109.1	93.4	138.2	118.0	88.8	133.3	107.4	93.1	125.0	106.3	95.7	111.4	102.3
3	CONCRETE	117.9	133.5	125.8	99.4	128.8	114.3	116.2	130.1	123.2	95.0	127.2	111.3	100.8	127.6	114.3	101.4	112.5	107.0
4	MASONRY	107.6	141.1	128.4	98.6	142.4	125.8	107.3	142.4	129.1	95.0	137.9	121.7	101.1	138.0	124.0	99.4	111.1	106.6
5	METALS	101.2	123.1	109.1	95.8	115.4	102.9	101.2	117.5	107.1	96.0	115.1	102.8	95.7	121.4	105.0	98.2	103.5	100.1
6	WOOD & PLASTICS	102.6	115.4	109.2	99.1	112.6	106.1	102.6	113.0	107.9	93.9	107.4	100.9	101.6	110.1	106.0	91.7	104.0	98.1
7	THERMAL & MOISTURE PROTECTION	101.7	134.7	117.0	101.0	127.0	113.0	101.6	130.0	114.8	101.0	124.4	111.9	101.2	129.6	114.4	101.0	109.3	104.8
8	DOORS & WINDOWS	102.9	126.2	108.5	98.4	120.2	103.7	102.9	121.3	107.3	108.0	119.8	110.8	98.8	123.9	104.9	107.8	109.6	108.2
092	LATH, PLASTER & GYPSUM BOARD	98.5	114.8	109.1	94.7	112.0	105.9	98.5	112.0	107.2	93.2	106.6	101.8	95.8	109.8	104.8	94.8	103.1	100.2
095	ACOUSTICAL TREATMENT & WOOD FLOORING	101.7	114.8	110.2	92.1	112.0	105.0	101.7	112.0	108.3	92.1	106.6	101.5	92.1	109.8	103.5	101.7	103.1	102.6
096	FLOORING & CARPET	97.1	153.0	110.7	95.9	153.0	109.8	96.9	153.0	110.5	94.7	153.0	108.9	96.2	153.0	110.0	94.0	111.6	98.3
099	PAINTING & WALL COVERINGS	89.4	132.9	114.9	89.4	132.9	114.9	89.4	132.9	114.9	89.3	132.9	114.9	90.1	132.9	115.2	89.3	96.0	93.2
9	FINISHES	97.0	126.8	112.2	93.8	125.1	109.8	96.9	125.4	111.5	93.2	122.1	108.0	93.9	123.6	109.1	95.3	105.4	100.5
10 - 14	TOTAL DIV. 10 - 14	100.0	126.4	105.6	99.8	126.0	105.4	100.0	127.1	105.7	100.0	111.1	102.3	100.0	124.8	105.2	100.0	107.4	101.6
15	MECHANICAL	100.0	103.2	101.4	96.7	106.0	100.8	100.0	106.1	102.7	96.9	99.6	98.2	97.7	109.5	103.0	96.9	96.6	96.8
16	ELECTRICAL	99.4	96.7	97.5	94.6	96.7	96.0	98.7	96.7	97.3	105.4	97.3	99.9	99.7	124.4	116.3	105.4	82.0	89.7
1 - 16	WEIGHTED AVERAGE	102.7	116.6	109.4	96.9	115.2	105.7	102.3	115.9	108.9	98.1	112.6	105.1	97.8	120.4	108.7	99.7	100.9	100.3

MASSACHUSETTS

	DIVISION	HYANNIS 026			LAWRENCE 019			LOWELL 018			NEW BEDFORD 027			PITTSFIELD 012			SPRINGFIELD 010 - 011		
		MAT.	INST.	TOTAL	MAT.	INST.	TOTAL	MAT.	INST.	TOTAL	MAT.	INST.	TOTAL	MAT.	INST.	TOTAL	MAT.	INST.	TOTAL
2	SITE WORK	92.5	100.8	98.8	96.7	100.9	100.0	95.8	100.9	99.7	94.6	101.1	99.6	97.0	99.6	99.0	96.1	99.8	99.0
031	CONCRETE FORMWORK	91.4	117.1	113.4	101.8	115.9	113.8	98.1	116.0	113.4	102.0	117.7	115.4	98.0	104.2	103.3	98.2	105.7	104.7
032	CONCRETE REINFORCEMENT	93.6	147.7	124.1	115.7	141.5	130.3	116.7	138.5	129.0	116.7	151.6	136.4	96.8	126.6	113.6	116.7	136.9	128.1
033	CAST IN PLACE CONCRETE	97.5	137.4	114.2	107.2	139.5	120.7	98.0	139.6	115.4	100.4	138.2	116.2	106.9	104.4	105.8	100.4	111.6	105.1
3	CONCRETE	107.1	128.8	118.1	117.9	127.9	122.9	107.7	127.2	117.6	114.7	130.1	122.5	109.5	107.5	108.5	108.9	112.6	110.8
4	MASONRY	106.5	142.4	128.8	107.1	141.1	128.2	93.8	137.9	121.2	107.1	142.4	129.0	94.5	101.9	99.1	94.0	111.3	104.8
5	METALS	97.7	115.4	104.1	98.7	113.5	104.0	98.7	109.2	102.5	101.2	117.5	107.1	98.5	99.2	98.8	101.1	103.6	102.0
6	WOOD & PLASTICS	91.5	112.6	102.4	102.6	110.4	106.6	101.3	110.4	106.0	102.6	113.0	107.9	101.3	104.2	102.8	101.3	104.0	102.7
7	THERMAL & MOISTURE PROTECTION	101.2	127.0	113.1	101.7	134.6	117.0	101.4	133.0	116.1	101.7	130.0	114.8	101.5	106.1	103.6	101.4	109.7	105.3
8	DOORS & WINDOWS	99.1	120.2	104.2	102.9	117.5	106.4	109.1	116.8	111.0	102.9	121.3	107.3	109.1	107.3	108.7	109.1	109.6	109.2
092	LATH, PLASTER & GYPSUM BOARD	92.1	112.0	104.9	98.5	109.8	105.8	98.5	109.8	105.8	98.5	112.0	107.2	98.5	103.3	101.6	98.5	103.1	101.5
095	ACOUSTICAL TREATMENT & WOOD FLOORING	92.1	112.0	105.0	101.7	109.8	106.9	101.7	109.8	106.9	101.7	112.0	108.3	101.7	103.3	102.7	101.7	103.1	102.6
096	FLOORING & CARPET	93.6	153.0	108.1	96.9	153.0	110.5	96.9	153.0	110.5	96.9	153.0	110.5	97.1	101.4	98.2	96.7	111.6	100.3
099	PAINTING & WALL COVERINGS	89.4	131.1	113.9	89.4	132.9	114.9	89.3	132.9	114.9	89.4	132.9	114.9	89.3	96.0	93.2	90.7	96.0	93.8
9	FINISHES	92.9	124.9	109.3	96.9	123.9	110.7	96.9	123.9	110.7	96.8	125.4	111.5	97.0	102.7	99.9	96.9	105.5	101.3
10 - 14	TOTAL DIV. 10 - 14	100.0	126.0	105.5	100.0	125.8	105.4	100.0	125.8	105.4	100.0	127.1	105.7	100.0	108.2	101.7	100.0	109.5	102.0
15	MECHANICAL	100.0	106.0	102.6	100.0	104.0	101.8	100.0	118.0	108.0	100.0	106.1	102.7	100.0	90.1	95.5	100.0	96.7	98.5
16	ELECTRICAL	94.6	96.7	96.0	103.8	107.7	106.4	104.7	107.6	106.7	99.7	96.7	97.7	104.7	84.3	90.9	104.7	90.3	95.0
1 - 16	WEIGHTED AVERAGE	99.3	115.2	107.0	102.6	116.2	109.2	101.4	118.0	109.4	102.2	115.9	108.8	101.7	97.6	99.7	101.9	102.4	102.2

	DIVISION	MASSACHUSETTS WORCESTER 015 - 016			MICHIGAN ANN ARBOR 481			BATTLE CREEK 490			BAY CITY 487			DEARBORN 481			DETROIT 482		
		MAT.	INST.	TOTAL	MAT.	INST.	TOTAL	MAT.	INST.	TOTAL	MAT.	INST.	TOTAL	MAT.	INST.	TOTAL	MAT.	INST.	TOTAL
2	SITE WORK	95.8	100.8	99.7	81.3	94.3	91.3	83.2	88.2	87.1	70.1	93.7	88.2	81.1	94.5	91.4	100.4	96.3	97.2
031	CONCRETE FORMWORK	98.7	113.5	111.4	95.7	103.6	102.4	94.9	87.0	88.2	95.7	98.2	97.9	95.6	116.1	113.1	97.6	116.2	113.5
032	CONCRETE REINFORCEMENT	116.7	157.5	139.7	96.1	118.2	108.5	97.9	86.6	91.5	96.1	117.1	107.9	96.1	118.8	108.9	95.4	118.7	108.6
033	CAST IN PLACE CONCRETE	98.0	133.3	112.8	92.4	116.1	102.3	98.5	106.4	101.8	88.3	95.5	91.3	90.2	119.5	102.5	99.1	119.6	107.7
3	CONCRETE	107.8	127.5	117.7	86.9	111.6	99.4	95.8	92.9	94.4	85.0	101.9	93.5	85.9	118.3	102.2	90.3	116.7	103.6
4	MASONRY	93.8	140.2	122.6	102.6	108.3	106.1	98.5	77.4	85.4	102.3	82.9	90.3	102.4	119.7	113.2	101.3	119.7	112.7
5	METALS	101.1	115.1	106.2	97.7	125.7	107.8	96.3	85.1	92.3	98.3	122.5	107.0	97.8	126.9	108.3	98.3	102.5	99.8
6	WOOD & PLASTICS	101.8	108.3	105.2	91.5	100.9	96.4	94.0	86.4	90.1	91.5	100.9	96.4	91.5	115.8	104.1	94.0	115.8	105.3
7	THERMAL & MOISTURE PROTECTION	101.4	125.1	112.4	93.1	110.2	101.0	92.5	85.0	89.0	91.5	90.0	90.8	92.3	120.2	105.3	90.7	120.2	104.4
8	DOORS & WINDOWS	109.1	122.1	112.2	96.5	105.1	98.6	93.6	80.7	90.4	96.5	98.7	97.1	96.5	112.0	100.3	96.0	113.3	100.2
092	LATH, PLASTER & GYPSUM BOARD	98.5	107.6	104.4	115.9	99.1	105.1	104.1	79.7	88.4	115.9	99.1	105.1	115.9	114.4	115.0	115.9	114.4	115.0
095	ACOUSTICAL TREATMENT & WOOD FLOORING	101.7	107.6	105.5	100.3	99.1	99.5	94.3	79.7	84.9	100.3	99.1	99.5	100.3	114.4	109.5	100.3	114.4	109.5
096	FLOORING & CARPET	96.9	143.5	108.2	95.5	110.7	99.2	96.5	79.9	92.5	95.2	72.8	89.8	95.0	115.7	100.0	95.2	115.7	100.1
099	PAINTING & WALL COVERINGS	89.3	117.1	105.6	96.3	112.1	105.6	97.0	77.3	85.4	96.3	78.9	86.1	96.3	114.9	107.2	98.4	114.9	108.1
9	FINISHES	96.9	118.9	108.2	100.2	105.4	102.9	96.2	84.1	90.0	99.9	92.1	95.9	100.1	116.0	108.3	101.0	116.0	108.7
10 - 14	TOTAL DIV. 10 - 14	100.0	113.2	102.8	100.0	108.1	101.7	100.0	108.0	101.7	100.1	103.3	100.8	100.0	111.2	102.4	100.0	111.2	102.4
15	MECHANICAL	100.0	99.6	99.8	100.0	95.4	98.0	100.1	86.4	94.0	100.0	84.3	93.0	100.0	108.9	104.0	100.0	111.9	105.3
16	ELECTRICAL	104.7	97.3	99.7	94.1	107.9	103.4	93.8	94.7	94.4	93.5	84.2	87.3	94.1	110.6	105.2	95.3	110.5	105.6
1 - 16	WEIGHTED AVERAGE	101.8	112.6	107.0	96.6	105.9	101.1	96.5	87.8	92.3	96.0	93.3	94.7	96.4	113.1	104.5	97.6	111.4	104.3

MICHIGAN

DIVISION		FLINT 484 - 485			GAYLORD 497			GRAND RAPIDS 493,495			IRON MOUNTAIN 498 - 499			JACKSON 492			KALAMAZOO 491		
		MAT.	INST.	TOTAL	MAT.	INST.	TOTAL	MAT.	INST.	TOTAL	MAT.	INST.	TOTAL	MAT.	INST.	TOTAL	MAT.	INST.	TOTAL
2	SITE WORK	68.6	93.8	88.0	78.8	86.4	84.7	82.9	88.1	86.9	86.5	95.1	93.1	100.9	87.0	90.2	83.5	88.2	87.1
031	CONCRETE FORMWORK	96.8	99.9	99.4	89.4	79.0	80.5	96.0	82.7	84.6	83.8	84.3	84.3	86.3	90.2	89.7	94.9	86.7	87.9
032	CONCRETE REINFORCEMENT	96.1	117.8	108.3	91.3	115.4	104.8	97.9	85.7	91.0	90.8	97.2	94.4	88.6	117.2	104.7	97.9	86.9	91.7
033	CAST IN PLACE CONCRETE	93.0	96.8	94.6	98.3	93.4	96.2	98.8	92.5	96.2	116.4	88.8	104.8	98.2	98.7	98.4	100.5	101.4	100.9
3	CONCRETE	87.3	103.2	95.3	92.0	92.1	92.1	96.0	86.1	91.0	102.0	88.5	95.2	87.0	99.4	93.2	98.8	91.1	94.9
4	MASONRY	102.6	97.8	99.6	111.4	79.0	91.3	96.2	62.3	75.1	96.1	87.5	90.7	89.6	83.2	85.6	98.7	85.0	90.2
5	METALS	97.8	123.8	107.1	97.8	115.0	104.0	96.0	82.8	91.2	96.6	91.0	94.6	97.9	120.9	106.1	96.3	85.3	92.3
6	WOOD & PLASTICS	91.5	100.9	96.4	87.8	76.6	82.0	95.6	83.9	89.5	87.5	83.6	85.5	86.3	89.4	87.9	94.0	86.4	90.1
7	THERMAL & MOISTURE PROTECTION	91.5	94.9	93.1	91.5	75.9	84.3	92.8	68.6	81.6	94.1	81.4	88.2	90.8	94.2	92.4	92.5	86.8	89.9
8	DOORS & WINDOWS	96.5	101.4	97.7	95.2	78.5	91.2	93.5	75.0	89.0	100.1	75.2	94.1	94.2	95.1	94.4	93.6	77.5	89.7
092	LATH, PLASTER & GYPSUM BOARD	115.9	99.1	105.1	107.5	73.6	85.6	104.1	77.2	86.7	69.4	83.6	78.5	106.2	86.9	93.8	104.1	79.7	88.4
095	ACOUSTICAL TREATMENT & WOOD FLOORING	100.3	99.1	99.5	91.6	73.6	80.0	94.3	77.2	83.2	91.3	83.6	86.3	93.0	86.9	89.0	94.3	79.7	84.9
096	FLOORING & CARPET	95.2	91.5	94.3	90.0	76.8	86.8	96.5	54.0	86.2	109.0	90.2	104.4	88.2	80.0	86.2	96.5	62.6	88.3
099	PAINTING & WALL COVERINGS	96.3	71.2	81.6	92.9	70.9	80.0	97.0	54.2	71.9	115.0	59.2	82.2	92.9	106.8	101.0	97.0	79.2	86.5
9	FINISHES	99.6	95.3	97.4	96.8	76.9	86.6	96.2	74.1	84.9	98.2	82.5	90.1	97.3	89.5	93.3	96.2	80.8	88.3
10 - 14	TOTAL DIV. 10 - 14	100.0	104.7	101.0	100.0	97.2	99.4	100.0	106.3	101.3	100.0	92.0	98.3	100.0	102.7	100.6	100.0	111.3	102.4
15	MECHANICAL	100.0	90.3	95.7	96.9	83.7	91.0	100.0	66.8	85.2	96.9	88.4	93.1	96.9	89.5	93.6	100.0	83.4	92.6
16	ELECTRICAL	94.1	101.5	99.1	89.8	78.4	82.1	93.9	78.3	83.4	100.1	79.7	86.4	96.3	107.8	104.1	93.5	84.5	87.5
1 - 16	WEIGHTED AVERAGE	96.2	99.6	97.8	96.0	85.6	91.0	96.4	76.8	86.9	97.9	86.4	92.4	95.2	96.5	95.9	96.8	85.7	91.5

MICHIGAN / MINNESOTA

DIVISION		LANSING 488 - 489			MUSKEGAN 494			ROYAL OAK 480,483			SAGINAW 486			TRAVERSE CITY 496			BEMIDJI 566		
		MAT.	INST.	TOTAL	MAT.	INST.	TOTAL	MAT.	INST.	TOTAL	MAT.	INST.	TOTAL	MAT.	INST.	TOTAL	MAT.	INST.	TOTAL
2	SITE WORK	89.4	93.7	92.7	83.4	88.5	87.3	87.2	93.5	92.0	70.9	93.6	88.4	72.4	95.1	89.9	78.3	96.5	92.3
031	CONCRETE FORMWORK	99.4	92.4	93.4	96.3	86.2	87.6	88.1	104.1	101.8	95.7	97.9	97.6	83.5	75.8	76.9	78.1	94.5	92.2
032	CONCRETE REINFORCEMENT	96.1	117.5	108.2	97.8	86.5	91.4	87.1	113.2	101.8	96.1	117.1	107.9	92.2	81.6	86.2	101.7	99.1	100.2
033	CAST IN PLACE CONCRETE	92.4	94.6	93.3	98.5	105.1	101.3	80.8	111.6	93.7	91.2	95.3	92.9	90.8	81.8	87.0	99.5	95.5	97.8
3	CONCRETE	87.2	99.1	93.2	95.9	92.1	94.0	75.2	107.4	91.4	86.4	101.7	94.1	83.6	79.5	81.5	91.3	96.9	94.1
4	MASONRY	95.6	91.9	93.3	98.5	81.8	88.2	95.7	111.6	105.6	104.1	82.9	90.9	94.2	79.3	85.0	100.3	93.1	95.8
5	METALS	97.0	123.1	106.4	95.4	84.5	91.5	97.5	95.6	96.8	97.8	122.2	106.6	96.6	86.0	92.8	92.5	118.5	101.9
6	WOOD & PLASTICS	95.3	91.9	93.5	95.6	83.2	89.2	87.2	102.3	95.0	91.5	100.9	96.4	87.1	74.0	80.3	73.9	96.4	85.6
7	THERMAL & MOISTURE PROTECTION	92.7	90.6	91.7	92.5	81.0	87.2	90.0	112.9	100.6	92.0	90.0	91.0	93.0	73.4	83.9	97.5	90.7	94.4
8	DOORS & WINDOWS	96.5	96.4	96.5	94.4	74.7	89.6	96.7	102.9	98.2	96.5	95.6	96.3	100.1	69.5	92.7	104.3	109.7	105.7
092	LATH, PLASTER & GYPSUM BOARD	119.4	89.8	100.3	105.4	76.4	86.7	113.5	100.4	105.1	115.9	99.1	105.1	69.4	73.6	72.1	109.9	96.9	101.5
095	ACOUSTICAL TREATMENT & WOOD FLOORING	100.3	89.8	93.5	99.8	76.4	84.7	99.0	100.4	99.9	100.3	99.1	99.5	91.3	73.6	79.8	139.9	96.9	112.0
096	FLOORING & CARPET	95.5	80.0	91.7	96.5	78.0	92.0	92.5	105.7	95.7	95.5	72.8	89.9	108.9	76.8	101.1	109.1	124.9	112.9
099	PAINTING & WALL COVERINGS	96.3	81.9	87.8	97.0	66.3	79.0	98.4	106.8	103.3	96.3	78.0	85.6	115.0	70.9	89.1	106.3	85.5	94.1
9	FINISHES	101.2	88.4	94.6	97.4	82.0	89.5	98.9	104.5	101.7	100.0	91.3	95.5	97.4	75.1	86.0	112.6	99.5	105.9
10 - 14	TOTAL DIV. 10 - 14	100.0	102.9	100.6	100.0	109.4	102.0	100.1	104.0	100.9	100.8	103.3	101.3	100.0	78.7	95.5	100.0	94.2	98.8
15	MECHANICAL	100.0	90.3	95.7	100.0	89.4	95.3	97.0	100.3	98.5	100.0	84.0	92.9	96.9	74.2	86.8	97.2	89.7	93.8
16	ELECTRICAL	92.8	91.5	91.9	93.8	74.0	80.5	97.4	99.7	99.0	96.3	84.2	88.1	92.3	77.4	82.3	100.0	92.5	95.0
1 - 16	WEIGHTED AVERAGE	96.4	95.5	96.0	96.6	84.6	90.8	94.3	102.2	98.1	96.5	92.9	94.8	94.5	79.1	87.1	98.1	97.0	97.5

MINNESOTA

DIVISION		BRAINERD 564			DETROIT LAKES 565			DULUTH 556 - 558			MANKATO 560			MINNEAPOLIS 553 - 554			ROCHESTER 559		
		MAT.	INST.	TOTAL	MAT.	INST.	TOTAL	MAT.	INST.	TOTAL	MAT.	INST.	TOTAL	MAT.	INST.	TOTAL	MAT.	INST.	TOTAL
2	SITE WORK	78.7	102.9	97.3	76.7	96.7	92.1	79.3	103.6	98.0	75.9	102.8	96.6	78.9	109.8	102.7	78.5	103.2	97.5
031	CONCRETE FORMWORK	78.4	92.1	90.1	74.3	81.4	80.4	97.7	115.9	113.3	89.8	99.3	97.9	98.4	132.4	127.5	98.2	105.8	104.7
032	CONCRETE REINFORCEMENT	100.6	99.1	99.8	101.7	99.0	100.2	90.7	99.5	95.7	98.4	105.8	102.6	90.0	106.5	99.6	90.7	105.8	99.2
033	CAST IN PLACE CONCRETE	108.1	103.1	106.0	96.5	100.1	98.0	103.9	110.2	106.5	99.6	96.0	98.1	102.4	120.4	109.9	100.4	102.2	101.2
3	CONCRETE	95.1	98.4	96.8	88.9	92.7	90.8	94.1	111.4	102.8	90.8	100.3	95.6	94.9	123.4	109.3	92.4	105.4	99.0
4	MASONRY	122.6	102.1	109.9	122.1	101.9	109.5	107.1	115.2	112.1	108.9	108.2	108.4	107.2	128.6	120.5	106.9	106.1	106.4
5	METALS	93.3	118.3	102.3	92.5	117.9	101.7	93.2	119.7	102.8	93.5	120.4	103.2	95.3	125.2	106.1	93.1	123.0	103.9
6	WOOD & PLASTICS	87.4	88.3	87.9	69.9	74.9	72.5	102.6	115.6	109.3	99.5	96.6	98.0	103.1	132.0	118.0	103.1	104.8	104.0
7	THERMAL & MOISTURE PROTECTION	96.4	106.5	101.1	97.3	95.6	96.5	97.3	121.1	108.4	97.0	102.9	99.8	97.2	134.2	114.4	97.1	107.0	101.7
8	DOORS & WINDOWS	90.3	105.4	94.0	104.3	98.1	102.8	100.9	119.4	105.4	95.6	118.8	101.2	102.0	138.4	110.8	100.9	117.0	104.8
092	LATH, PLASTER & GYPSUM BOARD	83.4	88.8	86.9	108.3	74.7	86.6	93.2	116.9	108.5	87.7	97.3	93.9	93.2	133.7	119.4	93.2	105.9	101.4
095	ACOUSTICAL TREATMENT & WOOD FLOORING	68.0	88.8	81.5	139.9	74.7	97.6	87.2	116.9	106.5	68.0	97.3	87.0	87.2	133.7	117.3	87.2	105.9	99.3
096	FLOORING & CARPET	107.7	124.9	111.9	107.5	124.9	111.7	117.3	99.5	113.0	109.9	124.9	113.5	114.4	124.9	116.9	117.1	98.0	112.4
099	PAINTING & WALL COVERINGS	101.7	85.5	92.2	106.3	81.2	91.6	106.2	107.9	107.2	110.9	79.6	92.5	113.5	118.0	116.2	109.0	99.8	103.6
9	FINISHES	94.4	96.9	95.7	111.7	88.6	99.9	101.0	112.8	107.0	96.2	101.8	99.0	100.5	130.2	115.7	101.0	104.2	102.7
10 - 14	TOTAL DIV. 10 - 14	100.0	91.7	98.2	100.0	94.8	98.9	99.2	105.1	100.5	100.0	100.6	100.1	100.0	111.2	102.4	100.0	101.6	100.3
15	MECHANICAL	96.4	89.3	93.2	97.2	94.1	95.8	99.9	106.4	102.8	96.4	93.2	95.0	99.8	115.5	106.8	99.8	103.7	101.5
16	ELECTRICAL	96.2	101.9	100.1	100.0	61.5	74.0	96.9	101.9	100.3	106.4	90.4	95.6	98.0	109.2	105.6	96.9	90.4	92.5
1 - 16	WEIGHTED AVERAGE	96.0	99.9	97.9	98.8	91.0	95.0	97.9	109.9	103.7	96.4	101.1	98.7	98.5	120.4	109.0	97.7	104.3	100.9

MINNESOTA / MISSISSIPPI

DIVISION		SAINT PAUL 550-551			ST. CLOUD 563			THIEF RIVER FALLS 567			WILLMAR 562			WINDOM 561			BILOXI 395		
		MAT.	INST.	TOTAL	MAT.	INST.	TOTAL	MAT.	INST.	TOTAL	MAT.	INST.	TOTAL	MAT.	INST.	TOTAL	MAT.	INST.	TOTAL
2	SITE WORK	81.1	104.1	98.8	74.3	104.1	97.2	76.3	96.4	91.8	74.1	100.4	94.4	68.9	99.8	92.6	110.9	87.4	92.8
031	CONCRETE FORMWORK	88.3	125.3	120.0	79.2	97.4	94.8	77.1	93.4	91.0	74.7	70.9	71.4	81.5	70.6	72.1	100.3	55.8	62.2
032	CONCRETE REINFORCEMENT	87.5	106.4	98.1	100.0	105.4	103.0	103.4	95.5	99.0	99.0	96.8	97.8	99.0	96.5	97.6	95.7	69.5	80.9
033	CAST IN PLACE CONCRETE	103.3	117.2	109.1	95.4	114.7	103.5	95.5	72.2	85.7	96.9	62.5	82.5	83.8	54.5	71.5	108.3	56.4	86.6
3	CONCRETE	96.2	119.3	107.8	87.1	106.0	96.6	88.6	87.4	88.0	86.7	74.3	80.4	78.0	71.5	74.8	96.3	60.3	78.1
4	MASONRY	117.9	128.6	124.6	108.2	107.2	107.6	100.1	93.1	95.7	112.2	53.2	75.6	121.6	48.3	76.0	88.2	51.0	65.1
5	METALS	91.9	124.7	103.7	94.4	122.8	104.6	92.5	113.2	100.0	93.3	106.0	97.9	93.6	110.1	99.6	98.3	86.0	93.9
6	WOOD & PLASTICS	92.2	122.9	108.0	83.2	88.3	85.8	72.7	96.4	85.0	82.8	76.1	79.3	89.6	76.1	82.6	99.5	56.9	77.5
7	THERMAL & MOISTURE PROTECTION	97.0	132.2	113.3	96.6	123.1	108.9	97.3	80.6	89.6	96.4	61.1	80.0	96.5	57.8	78.6	97.4	57.6	78.9
8	DOORS & WINDOWS	98.0	133.5	106.6	95.6	108.1	98.6	104.3	106.9	105.0	92.9	86.0	91.3	97.6	88.9	95.5	97.0	57.5	87.4
092	LATH, PLASTER & GYPSUM BOARD	88.9	124.5	111.9	81.8	88.8	86.3	109.4	96.9	101.3	81.8	76.2	78.2	84.5	76.2	79.1	110.9	56.0	75.5
095	ACOUSTICAL TREATMENT & WOOD FLOORING	84.4	124.5	110.4	68.0	88.8	81.5	139.9	96.9	112.0	68.0	76.2	73.3	68.0	76.2	73.3	100.2	56.0	71.6
096	FLOORING & CARPET	109.9	124.9	113.6	103.8	120.5	107.9	108.6	120.5	111.5	105.8	59.3	94.5	108.4	52.6	94.8	112.8	52.5	98.1
099	PAINTING & WALL COVERINGS	113.5	113.1	113.3	110.9	81.2	93.4	106.3	81.2	91.6	106.3	75.0	87.9	106.3	79.6	90.7	103.1	46.2	69.7
9	FINISHES	98.2	124.3	111.6	93.3	98.5	96.0	112.2	97.9	104.9	93.7	69.3	81.1	94.5	68.6	81.2	105.4	54.1	79.1
10-14	TOTAL DIV. 10-14	100.0	109.5	102.0	100.0	103.3	100.7	100.0	94.0	98.7	100.0	78.3	95.4	100.0	78.3	95.4	100.0	73.1	94.3
15	MECHANICAL	99.8	112.2	105.3	99.7	104.1	101.7	97.2	80.8	89.9	96.4	64.9	82.3	96.4	60.1	80.2	100.0	53.3	79.2
16	ELECTRICAL	96.0	95.5	95.7	96.2	87.2	90.2	93.6	61.5	71.9	96.2	84.0	88.0	106.4	84.0	91.3	95.9	61.0	72.3
1-16	WEIGHTED AVERAGE	97.8	115.2	106.2	95.5	103.4	99.3	97.2	87.8	92.6	94.5	77.2	86.1	95.3	75.8	85.9	98.8	62.6	81.3

MISSISSIPPI

DIVISION		CLARKSDALE 386			COLUMBUS 397			GREENVILLE 387			GREENWOOD 389			JACKSON 390-392			LAUREL 394		
		MAT.	INST.	TOTAL	MAT.	INST.	TOTAL	MAT.	INST.	TOTAL	MAT.	INST.	TOTAL	MAT.	INST.	TOTAL	MAT.	INST.	TOTAL
2	SITE WORK	108.3	86.5	91.5	110.2	86.9	92.3	115.4	87.2	93.7	112.2	86.2	92.2	107.1	87.2	91.8	117.0	85.4	92.7
031	CONCRETE FORMWORK	80.1	24.9	32.8	77.5	40.5	45.8	77.5	46.5	51.0	92.6	27.3	36.7	94.4	53.2	59.1	77.4	31.8	38.3
032	CONCRETE REINFORCEMENT	103.0	22.6	57.7	102.4	41.0	67.8	103.7	57.7	77.8	103.0	61.6	79.7	95.7	62.6	77.1	103.0	40.3	67.7
033	CAST IN PLACE CONCRETE	105.7	33.2	75.3	110.5	46.3	83.6	108.8	49.4	84.0	113.6	33.0	79.9	106.3	52.5	83.8	108.0	37.1	78.3
3	CONCRETE	93.9	30.0	61.7	97.1	44.8	70.8	98.5	51.6	74.9	100.3	38.2	69.0	94.9	56.5	75.5	98.7	37.6	67.9
4	MASONRY	88.1	27.2	50.3	107.8	32.1	60.8	128.3	48.4	78.7	88.7	25.4	49.4	91.2	48.4	64.6	105.5	32.9	60.4
5	METALS	95.8	67.0	85.4	95.7	74.2	87.9	96.6	81.5	91.2	95.8	79.4	89.9	98.3	83.4	92.9	95.5	72.3	87.1
6	WOOD & PLASTICS	78.9	23.4	50.2	76.0	42.3	58.6	74.0	46.1	59.6	92.6	26.4	58.4	92.6	54.8	73.1	77.0	33.3	54.4
7	THERMAL & MOISTURE PROTECTION	96.9	35.5	68.4	97.0	35.4	68.5	97.3	52.3	76.4	97.4	29.7	66.0	97.1	54.0	77.1	97.2	33.7	67.7
8	DOORS & WINDOWS	96.4	27.8	79.8	96.4	37.3	82.1	96.4	47.9	84.7	96.4	38.4	82.4	97.4	53.8	86.8	93.2	32.2	78.4
092	LATH, PLASTER & GYPSUM BOARD	104.3	21.5	50.9	103.3	41.0	63.1	102.7	44.9	65.4	109.6	24.6	54.7	110.9	53.9	74.1	102.2	31.7	56.7
095	ACOUSTICAL TREATMENT & WOOD FLOORING	94.7	21.5	47.2	94.7	41.0	59.9	94.7	44.9	62.4	94.7	24.6	49.2	100.2	53.9	70.2	94.7	31.7	53.9
096	FLOORING & CARPET	106.0	30.5	87.6	104.6	26.3	85.6	103.6	44.8	89.3	112.8	27.5	92.0	112.8	51.9	98.0	102.8	24.8	83.8
099	PAINTING & WALL COVERINGS	103.1	19.4	54.0	103.1	28.6	59.3	103.1	45.5	69.3	103.1	26.5	58.1	103.1	45.5	69.3	103.1	28.6	59.3
9	FINISHES	101.1	25.1	62.2	100.7	37.3	68.2	100.6	45.9	72.6	104.3	27.2	64.8	105.4	52.2	78.1	100.4	31.0	64.8
10-14	TOTAL DIV. 10-14	100.0	57.8	91.1	100.0	61.6	91.9	100.0	70.8	93.8	100.0	58.3	91.2	100.0	71.9	94.1	100.0	36.8	86.7
15	MECHANICAL	98.7	25.4	66.0	98.1	34.6	69.8	100.0	46.5	76.1	98.7	26.7	66.6	100.0	46.0	75.9	98.2	29.5	67.5
16	ELECTRICAL	94.8	21.3	45.2	91.8	36.8	54.7	94.8	50.8	65.1	94.8	19.6	44.1	96.4	50.8	65.6	94.3	26.9	48.8
1-16	WEIGHTED AVERAGE	96.9	36.6	67.8	98.0	46.2	73.0	100.2	56.1	78.9	98.3	39.2	69.8	98.7	58.0	79.0	98.0	40.7	70.3

MISSISSIPPI / MISSOURI

DIVISION		MCCOMB 396			MERIDIAN 393			TUPELO 388			BOWLING GREEN 633			CAPE GIRARDEAU 637			CHILLICOTHE 646		
		MAT.	INST.	TOTAL	MAT.	INST.	TOTAL	MAT.	INST.	TOTAL	MAT.	INST.	TOTAL	MAT.	INST.	TOTAL	MAT.	INST.	TOTAL
2	SITE WORK	101.4	86.3	89.8	105.9	87.2	91.5	106.3	86.5	91.0	97.7	94.3	95.1	98.5	94.2	95.2	98.4	90.1	92.0
031	CONCRETE FORMWORK	77.4	30.4	37.1	75.8	45.1	49.5	76.3	32.2	38.6	92.3	88.9	89.4	85.1	87.4	87.0	87.0	69.5	72.0
032	CONCRETE REINFORCEMENT	103.7	19.2	56.1	102.4	62.1	79.7	100.9	62.5	79.3	98.0	95.9	96.8	99.2	95.8	97.3	97.3	78.5	86.7
033	CAST IN PLACE CONCRETE	95.9	35.0	70.4	102.7	52.5	81.6	105.7	48.1	81.6	91.3	95.5	93.0	90.3	94.4	92.0	96.2	79.8	89.3
3	CONCRETE	87.0	32.4	59.5	91.5	52.9	72.0	93.3	45.8	69.3	85.9	94.2	90.1	84.8	93.1	89.0	96.3	76.0	86.1
4	MASONRY	109.9	27.3	58.5	87.8	39.4	57.7	115.9	41.9	69.9	120.2	88.3	100.4	114.4	83.7	95.3	108.0	74.0	86.9
5	METALS	95.9	65.4	84.9	96.6	82.4	91.5	95.7	81.5	90.6	95.8	118.3	103.9	96.3	118.3	104.2	93.3	93.2	93.2
6	WOOD & PLASTICS	76.0	31.0	52.7	72.0	44.2	57.7	74.7	31.0	52.1	95.1	90.1	92.5	84.5	89.0	86.8	90.8	69.1	79.6
7	THERMAL & MOISTURE PROTECTION	96.6	32.8	67.0	96.7	50.6	75.4	96.9	44.5	72.6	92.5	90.3	91.5	92.2	86.8	89.7	97.7	76.2	87.7
8	DOORS & WINDOWS	96.4	31.0	80.6	96.4	47.0	84.4	96.4	36.4	81.8	98.2	90.5	96.3	98.2	89.9	96.2	93.2	67.1	86.8
092	LATH, PLASTER & GYPSUM BOARD	103.3	29.3	55.5	101.7	43.0	63.8	102.7	29.3	55.3	103.2	89.3	94.2	99.5	88.2	92.2	109.5	67.7	82.5
095	ACOUSTICAL TREATMENT & WOOD FLOORING	94.7	29.3	52.3	94.7	43.0	61.2	94.7	29.3	52.3	88.7	89.3	89.1	88.7	88.2	88.4	92.3	67.7	76.3
096	FLOORING & CARPET	104.6	30.5	86.6	102.7	36.4	86.6	104.0	33.6	86.9	105.1	95.3	102.7	101.1	95.3	99.7	92.0	68.4	86.3
099	PAINTING & WALL COVERINGS	103.1	27.0	58.4	103.1	44.9	68.9	103.1	45.5	69.3	94.7	101.6	98.7	94.7	94.3	94.5	90.9	71.2	79.4
9	FINISHES	100.2	30.5	64.5	99.6	42.8	70.5	100.3	33.6	66.1	94.3	91.2	92.7	92.5	89.2	90.8	95.0	69.6	82.0
10-14	TOTAL DIV. 10-14	100.0	58.9	91.3	100.0	70.6	93.8	100.0	59.7	91.5	100.0	86.0	97.0	100.0	85.1	96.9	100.0	78.7	95.5
15	MECHANICAL	98.1	23.6	64.9	100.0	46.1	76.0	98.7	33.2	69.5	96.8	97.7	97.2	100.0	96.6	98.5	96.9	58.1	79.6
16	ELECTRICAL	89.5	24.4	45.6	94.3	54.5	67.4	94.3	33.6	53.4	94.4	83.5	87.0	94.4	109.9	104.9	92.2	55.9	67.7
1-16	WEIGHTED AVERAGE	96.4	37.7	68.1	96.7	55.6	76.8	98.1	46.7	73.3	96.5	93.9	95.2	96.5	97.1	96.8	96.2	70.9	84.0

COST INDEXES

City Cost Indexes

MISSOURI

DIVISION	COLUMBIA 652 MAT.	INST.	TOTAL	FLAT RIVER 636 MAT.	INST.	TOTAL	HANNIBAL 634 MAT.	INST.	TOTAL	HARRISONVILLE 647 MAT.	INST.	TOTAL	JEFFERSON CITY 650-651 MAT.	INST.	TOTAL	JOPLIN 648 MAT.	INST.	TOTAL
2 SITE WORK	99.5	95.5	96.5	100.8	94.4	95.9	94.6	94.3	94.4	89.9	92.0	91.5	101.7	95.6	97.0	98.5	96.7	97.1
031 CONCRETE FORMWORK	85.7	66.4	69.2	101.3	88.7	90.5	89.9	88.0	88.3	83.2	84.8	84.5	104.6	73.2	77.7	105.7	62.7	68.9
032 CONCRETE REINFORCEMENT	97.4	97.3	97.3	99.2	101.5	100.5	97.5	96.0	96.6	97.0	88.6	92.3	96.2	103.5	100.3	100.5	78.2	87.9
033 CAST IN PLACE CONCRETE	88.1	76.1	83.1	94.2	96.2	95.0	86.3	95.5	90.2	98.5	87.9	94.0	92.2	79.7	87.0	104.2	77.3	92.9
3 CONCRETE	80.7	78.0	79.3	88.8	95.4	92.1	82.2	93.8	88.0	93.2	87.5	90.3	84.7	83.4	84.0	97.1	72.4	84.6
4 MASONRY	133.3	70.0	93.9	117.2	87.0	98.4	112.1	90.0	98.3	102.8	82.8	90.4	94.4	70.0	79.2	99.0	65.7	78.3
5 METALS	94.6	117.4	102.8	95.7	120.6	104.7	95.8	118.5	104.0	93.3	101.9	96.4	93.7	120.3	103.3	97.3	96.7	97.1
6 WOOD & PLASTICS	88.3	61.8	74.6	105.1	89.0	96.8	92.5	89.0	90.7	85.9	84.3	85.1	109.1	70.3	89.0	104.6	59.8	81.5
7 THERMAL & MOISTURE PROTECTION	91.6	75.3	84.0	92.7	90.0	91.5	92.2	87.9	90.2	96.6	86.8	92.0	91.9	76.6	84.8	98.3	70.0	85.2
8 DOORS & WINDOWS	93.9	83.6	91.4	98.2	91.5	96.5	98.2	89.9	96.2	93.5	84.6	91.4	90.5	89.9	90.4	94.4	73.9	89.4
092 LATH, PLASTER & GYPSUM BOARD	105.0	60.1	76.1	106.9	88.2	94.8	102.1	88.2	93.1	100.9	83.4	89.6	111.9	68.9	84.1	114.4	58.1	78.0
095 ACOUSTICAL TREATMENT & WOOD FLOORING	88.9	60.1	70.2	88.7	88.2	88.4	88.7	88.2	88.4	92.3	83.4	86.5	88.9	68.9	75.9	95.0	58.1	71.1
096 FLOORING & CARPET	107.1	68.0	97.6	108.8	95.3	105.5	104.2	95.3	102.0	88.2	88.6	88.3	114.8	68.0	103.4	115.0	64.1	102.7
099 PAINTING & WALL COVERINGS	94.7	63.8	76.6	94.7	94.3	94.5	94.7	101.6	98.7	95.3	88.6	91.3	94.7	63.8	76.6	90.2	55.7	70.0
9 FINISHES	93.5	64.4	78.6	96.1	90.0	93.0	93.7	90.6	92.1	92.5	85.8	89.0	97.0	69.3	82.8	101.8	61.1	80.9
10-14 TOTAL DIV. 10-14	100.0	91.9	98.3	100.0	86.4	97.1	100.0	85.8	97.0	100.0	90.2	97.9	100.0	93.0	98.5	100.0	84.3	96.7
15 MECHANICAL	100.1	74.5	88.7	96.8	98.4	97.5	96.8	97.7	97.2	96.8	86.4	92.1	100.1	90.9	96.0	100.2	61.7	83.0
16 ELECTRICAL	94.0	83.5	86.9	101.3	109.9	107.1	92.5	83.4	86.4	103.5	92.3	96.0	97.1	83.5	87.9	89.9	73.3	78.7
1-16 WEIGHTED AVERAGE	96.4	81.5	89.2	97.5	98.5	98.0	95.3	93.8	94.6	95.8	89.1	92.6	95.2	86.7	91.1	98.0	73.3	86.1

MISSOURI

DIVISION	KANSAS CITY 640-641 MAT.	INST.	TOTAL	KIRKSVILLE 635 MAT.	INST.	TOTAL	POPLAR BLUFF 639 MAT.	INST.	TOTAL	ROLLA 654-655 MAT.	INST.	TOTAL	SEDALIA 653 MAT.	INST.	TOTAL	SIKESTON 638 MAT.	INST.	TOTAL
2 SITE WORK	91.0	94.9	94.0	90.1	90.8	90.7	75.6	94.7	90.3	99.4	95.4	96.3	88.6	91.8	91.1	79.2	94.7	91.2
031 CONCRETE FORMWORK	105.3	94.4	96.0	79.5	74.8	75.5	79.3	74.8	75.5	93.4	54.0	59.6	90.3	66.5	69.9	80.7	75.2	76.0
032 CONCRETE REINFORCEMENT	95.4	98.3	97.0	98.0	78.6	87.1	101.0	78.9	88.6	97.9	71.8	83.2	96.3	78.1	86.1	100.3	78.9	88.3
033 CAST IN PLACE CONCRETE	96.8	98.2	97.4	94.2	82.4	89.3	72.4	78.1	74.8	90.1	59.8	77.4	94.2	78.8	87.7	77.3	80.4	78.6
3 CONCRETE	93.3	97.2	95.2	96.1	79.4	87.7	75.1	78.0	76.6	82.7	61.9	72.2	93.0	74.4	83.6	78.8	79.0	78.9
4 MASONRY	105.9	92.5	97.5	123.2	68.1	88.9	114.3	75.4	90.1	110.0	60.7	79.4	116.2	62.4	82.8	114.8	76.2	90.8
5 METALS	100.9	107.6	103.3	93.8	97.6	95.2	96.3	98.7	97.1	94.0	102.4	97.0	92.0	96.2	93.5	96.2	98.7	97.1
6 WOOD & PLASTICS	104.9	93.3	98.9	79.0	74.8	76.9	77.9	74.8	76.3	99.6	51.4	74.7	93.0	62.7	77.4	79.2	74.8	76.9
7 THERMAL & MOISTURE PROTECTION	97.1	96.5	96.8	96.3	79.7	88.6	95.9	80.6	88.8	91.9	59.2	76.7	95.2	78.1	87.3	96.1	81.0	89.1
8 DOORS & WINDOWS	98.5	96.2	97.9	102.3	78.2	96.5	103.3	71.0	95.5	93.9	55.3	84.6	97.6	69.3	90.7	103.3	71.0	95.5
092 LATH, PLASTER & GYPSUM BOARD	108.6	92.7	98.3	96.7	73.5	81.7	97.3	73.5	81.9	108.8	49.4	70.5	99.4	61.0	74.6	97.9	73.5	82.1
095 ACOUSTICAL TREATMENT & WOOD FLOORING	100.5	92.7	95.4	86.0	73.5	77.9	88.7	73.5	78.8	88.9	49.4	63.3	87.5	61.0	70.4	88.7	73.5	78.8
096 FLOORING & CARPET	94.5	92.6	94.0	80.9	70.1	78.3	85.6	64.1	87.9	111.3	59.8	98.8	86.6	88.6	87.1	96.2	64.1	88.4
099 PAINTING & WALL COVERINGS	95.3	98.0	96.8	90.3	55.7	70.0	89.6	66.2	75.9	94.7	43.7	64.8	94.7	88.6	91.1	89.6	66.2	75.9
9 FINISHES	97.1	94.1	95.6	89.6	71.1	80.1	91.9	71.1	81.2	95.4	52.4	73.4	90.5	72.0	81.0	92.3	71.3	81.5
10-14 TOTAL DIV. 10-14	100.0	96.9	99.3	100.0	79.3	95.6	100.0	79.4	95.6	100.0	76.8	95.1	100.0	85.5	96.9	100.0	79.7	95.7
15 MECHANICAL	100.0	98.0	99.1	96.8	85.9	92.0	96.8	95.1	96.1	96.8	92.2	94.8	96.8	83.6	90.9	96.8	95.6	96.3
16 ELECTRICAL	104.1	101.5	102.4	93.2	83.4	86.6	93.3	109.9	104.5	91.5	37.7	55.2	92.8	95.6	94.7	91.9	109.9	104.0
1-16 WEIGHTED AVERAGE	99.2	97.9	98.6	97.2	82.0	89.8	94.5	88.9	91.8	94.8	69.5	82.6	95.7	82.1	89.1	95.0	89.2	92.2

MISSOURI / MONTANA

DIVISION	SPRINGFIELD 656-658 MAT.	INST.	TOTAL	ST. JOSEPH 644-645 MAT.	INST.	TOTAL	ST. LOUIS 630-631 MAT.	INST.	TOTAL	BILLINGS 590-591 MAT.	INST.	TOTAL	BUTTE 597 MAT.	INST.	TOTAL	GREAT FALLS 594 MAT.	INST.	TOTAL
2 SITE WORK	91.3	95.3	94.4	92.4	92.0	92.1	97.9	97.8	97.8	79.9	97.5	93.4	85.7	96.2	93.8	89.1	97.0	95.2
031 CONCRETE FORMWORK	104.2	61.0	67.2	105.2	77.6	81.6	103.4	107.5	106.9	94.9	90.2	90.9	80.5	87.9	86.8	101.6	90.5	92.1
032 CONCRETE REINFORCEMENT	93.6	97.4	95.7	98.0	88.4	90.9	90.8	105.1	98.8	96.8	97.9	97.4	105.0	97.7	100.9	96.8	97.9	97.4
033 CAST IN PLACE CONCRETE	99.6	68.1	86.4	96.8	93.7	95.5	90.3	111.4	99.1	115.9	84.8	102.9	119.3	80.9	103.2	126.1	74.4	104.4
3 CONCRETE	93.0	71.9	82.3	93.1	86.4	89.7	85.1	109.6	97.4	101.2	90.0	95.6	101.1	87.6	94.3	106.7	86.5	96.5
4 MASONRY	89.3	71.1	78.0	105.0	77.3	87.8	98.7	108.1	104.5	119.3	97.2	105.6	117.7	95.7	104.0	121.6	101.6	109.2
5 METALS	98.0	102.4	99.6	99.5	101.3	100.1	98.8	124.1	107.9	96.2	96.9	96.5	95.6	96.0	95.7	96.6	96.6	96.6
6 WOOD & PLASTICS	103.0	58.1	79.8	105.4	75.9	90.2	104.2	105.8	105.0	96.9	90.9	93.8	82.7	89.7	86.3	106.1	90.6	98.1
7 THERMAL & MOISTURE PROTECTION	95.9	70.4	84.1	97.8	82.4	90.6	92.3	102.6	97.0	97.2	90.1	93.9	96.5	84.7	91.0	97.4	88.9	93.5
8 DOORS & WINDOWS	99.7	70.1	92.5	99.3	84.3	95.6	95.0	110.5	98.8	103.8	87.4	99.8	100.3	86.6	97.0	104.0	87.3	100.0
092 LATH, PLASTER & GYPSUM BOARD	111.1	56.3	75.7	115.3	74.7	89.1	107.3	105.5	106.1	98.1	91.2	93.6	89.0	89.9	89.6	98.1	90.9	93.4
095 ACOUSTICAL TREATMENT & WOOD FLOORING	94.3	56.3	69.7	99.1	74.7	83.3	92.8	105.5	101.0	110.6	91.2	98.0	107.8	89.9	96.2	110.6	90.9	97.8
096 FLOORING & CARPET	109.6	64.1	98.6	105.8	78.6	92.4	108.5	96.8	105.7	114.0	66.7	102.5	105.7	53.1	92.9	114.0	72.1	103.8
099 PAINTING & WALL COVERINGS	89.6	73.9	80.4	90.9	83.2	86.4	94.7	107.5	102.2	106.3	81.9	92.0	106.3	63.2	81.0	106.3	71.2	85.7
9 FINISHES	98.5	61.9	79.8	98.2	77.7	87.7	96.8	105.1	101.0	107.0	84.9	95.7	102.5	78.8	90.3	107.1	85.0	95.8
10-14 TOTAL DIV. 10-14	100.0	87.2	97.3	100.0	91.7	98.3	100.0	104.2	100.9	100.0	90.4	98.0	100.0	89.5	97.8	100.0	91.3	98.2
15 MECHANICAL	100.2	70.0	86.8	100.2	79.0	90.8	100.0	108.3	103.7	100.2	90.3	95.8	100.2	83.0	92.5	100.2	82.9	92.5
16 ELECTRICAL	100.4	74.0	82.6	103.5	78.6	86.7	98.0	113.5	108.5	92.4	86.9	88.7	103.7	66.6	78.7	92.4	82.3	85.6
1-16 WEIGHTED AVERAGE	97.8	75.9	87.2	99.2	83.5	91.6	96.6	109.1	102.7	100.6	90.9	95.9	100.3	84.5	92.7	101.8	88.6	95.5

COST INDEXES

624

City Cost Indexes

MONTANA

DIVISION		HAVRE 595 MAT.	INST.	TOTAL	HELENA 596 MAT.	INST.	TOTAL	KALISPELL 599 MAT.	INST.	TOTAL	MILES CITY 593 MAT.	INST.	TOTAL	MISSOULA 598 MAT.	INST.	TOTAL	WOLF POINT 592 MAT.	INST.	TOTAL
2	SITE WORK	92.7	96.9	96.0	90.7	96.9	95.5	77.4	96.2	91.8	83.1	96.9	93.7	70.5	96.2	90.3	98.5	96.9	97.3
031	CONCRETE FORMWORK	67.7	92.6	89.1	101.7	90.5	92.1	81.2	87.3	86.5	90.2	92.8	92.4	85.1	87.5	87.1	80.9	90.4	89.0
032	CONCRETE REINFORCEMENT	105.9	92.7	98.5	100.2	97.7	98.8	107.9	102.7	105.0	105.4	99.8	102.2	106.8	102.7	104.5	106.9	87.6	96.0
033	CAST IN PLACE CONCRETE	128.4	83.5	109.6	128.4	83.2	109.5	103.6	85.5	96.0	113.4	83.1	100.7	87.9	85.5	86.9	127.0	77.0	106.1
3	CONCRETE	108.2	89.7	98.9	108.3	89.5	98.8	90.8	89.6	90.2	98.1	90.8	94.4	80.3	90.0	85.2	111.7	85.6	98.5
4	MASONRY	118.7	94.0	103.4	118.6	98.3	106.0	116.8	87.9	98.9	124.7	100.0	109.4	144.5	87.9	109.3	125.7	94.0	106.0
5	METALS	95.1	95.5	95.3	95.8	96.1	95.9	95.0	93.5	94.5	94.5	96.3	95.1	95.7	99.1	96.9	94.5	95.3	94.8
6	WOOD & PLASTICS	72.6	93.4	83.4	106.3	90.6	98.2	87.9	90.6	89.3	96.2	93.4	94.8	87.9	90.6	89.3	85.8	90.6	88.3
7	THERMAL & MOISTURE PROTECTION	96.8	87.3	92.4	97.4	88.8	93.4	96.1	91.3	93.9	96.5	94.8	95.7	95.7	92.4	94.1	97.5	85.3	91.8
8	DOORS & WINDOWS	100.4	88.7	97.6	103.5	87.5	99.6	100.4	86.5	97.0	99.8	88.7	97.2	100.4	90.3	97.9	99.9	87.2	96.8
092	LATH, PLASTER & GYPSUM BOARD	85.8	93.8	91.0	97.5	90.9	93.2	91.1	90.9	90.9	97.1	93.8	95.0	91.1	90.9	90.9	92.4	90.9	91.4
095	ACOUSTICAL TREATMENT & WOOD FLOORING	107.8	93.8	98.7	107.8	90.9	96.8	107.8	90.9	96.8	106.5	93.8	98.2	107.8	90.9	96.8	106.5	90.9	96.4
096	FLOORING & CARPET	102.1	58.5	91.5	114.0	69.8	103.2	107.5	83.0	101.6	113.7	65.6	102.0	107.5	83.0	101.6	109.5	58.5	97.1
099	PAINTING & WALL COVERINGS	106.3	70.3	85.2	106.3	79.0	90.3	106.3	71.2	85.7	106.3	66.4	82.9	106.3	71.2	85.7	106.3	66.4	82.9
9	FINISHES	101.3	84.0	92.4	106.6	85.5	95.8	103.0	85.0	93.8	105.8	84.9	95.1	102.6	85.0	93.6	104.6	81.9	93.0
10 - 14	TOTAL DIV. 10 - 14	100.0	81.6	96.1	100.0	80.9	96.0	100.0	77.2	95.2	100.0	81.4	96.1	100.0	77.2	95.2	100.0	81.3	96.0
15	MECHANICAL	96.9	83.4	90.9	100.2	81.3	91.8	96.9	80.7	89.7	96.9	87.6	92.8	100.2	81.6	91.9	96.9	83.4	90.9
16	ELECTRICAL	92.4	81.8	85.2	92.4	82.1	85.4	98.8	77.9	84.7	92.4	85.3	87.6	100.4	77.9	85.2	92.4	81.8	85.2
1 - 16	WEIGHTED AVERAGE	99.6	87.9	94.0	101.7	88.1	95.1	97.8	85.9	92.0	99.0	90.4	94.8	98.7	86.8	93.0	100.9	87.0	94.2

NEBRASKA

DIVISION		ALLIANCE 693 MAT.	INST.	TOTAL	COLUMBUS 686 MAT.	INST.	TOTAL	GRAND ISLAND 688 MAT.	INST.	TOTAL	HASTINGS 689 MAT.	INST.	TOTAL	LINCOLN 683 - 685 MAT.	INST.	TOTAL	MCCOOK 690 MAT.	INST.	TOTAL
2	SITE WORK	88.5	98.0	95.8	89.4	91.6	91.1	93.6	92.5	92.8	92.9	92.3	92.4	84.5	92.5	90.6	93.7	91.6	92.1
031	CONCRETE FORMWORK	87.2	32.9	40.7	96.8	30.7	40.1	100.2	58.5	64.5	100.9	57.6	63.8	105.8	51.1	59.0	92.9	32.7	41.3
032	CONCRETE REINFORCEMENT	110.0	56.2	79.7	102.5	62.6	80.0	102.0	70.5	84.2	102.0	62.5	79.7	93.6	70.3	80.5	103.8	62.4	80.4
033	CAST IN PLACE CONCRETE	106.0	44.9	80.4	108.6	56.7	86.9	115.0	65.1	94.1	115.0	62.6	93.1	103.9	67.9	88.8	115.0	44.5	85.5
3	CONCRETE	109.7	42.7	75.9	98.0	47.9	72.8	102.6	64.5	83.4	102.6	61.7	82.0	95.2	62.3	78.6	102.3	44.6	73.2
4	MASONRY	110.2	37.4	65.0	116.6	38.7	68.2	107.5	54.0	74.3	116.2	50.7	75.5	97.8	51.2	68.6	107.0	37.3	63.7
5	METALS	103.2	65.1	89.5	92.8	81.9	88.9	94.5	86.3	91.5	92.8	82.3	89.0	97.5	86.0	93.4	93.0	81.1	88.7
6	WOOD & PLASTICS	87.2	31.6	58.5	100.0	28.2	63.0	99.1	56.6	77.2	104.1	56.6	79.6	104.8	46.8	74.9	95.9	31.7	62.8
7	THERMAL & MOISTURE PROTECTION	102.1	43.0	74.7	96.0	39.4	69.8	96.1	62.1	80.4	96.2	57.2	78.1	96.6	56.7	78.1	96.1	40.3	70.2
8	DOORS & WINDOWS	93.5	35.8	79.5	93.2	42.2	80.9	93.2	57.2	84.5	93.2	57.6	84.6	98.9	54.1	88.0	93.3	37.3	79.7
092	LATH, PLASTER & GYPSUM BOARD	101.1	29.1	54.6	107.9	25.5	54.7	107.4	54.8	73.4	109.0	54.8	74.0	111.1	44.6	68.2	106.9	29.1	56.6
095	ACOUSTICAL TREATMENT & WOOD FLOORING	108.9	29.1	57.2	87.5	25.5	47.3	87.5	54.8	66.3	87.5	54.8	66.3	94.3	44.6	62.1	87.5	29.1	49.6
096	FLOORING & CARPET	99.1	36.5	83.9	99.1	46.4	86.3	98.9	44.5	85.7	100.6	46.4	87.4	100.8	46.4	87.6	97.7	36.5	82.9
099	PAINTING & WALL COVERINGS	132.8	27.5	71.0	86.8	63.0	72.9	86.8	52.1	66.5	86.8	63.0	72.9	86.8	52.6	66.8	86.8	27.5	52.0
9	FINISHES	97.7	32.3	64.2	93.4	35.9	63.9	93.5	54.5	73.5	94.2	56.0	74.6	95.7	49.2	71.8	93.0	32.2	61.8
10 - 14	TOTAL DIV. 10 - 14	100.0	63.7	92.3	100.0	61.0	91.8	100.0	78.7	95.5	100.0	70.6	93.8	100.0	77.4	95.2	100.0	61.5	91.9
15	MECHANICAL	95.7	34.0	68.2	96.9	31.0	67.5	100.2	75.6	89.2	96.9	48.9	75.5	100.2	75.6	89.2	96.9	31.5	67.8
16	ELECTRICAL	94.6	23.8	46.9	93.4	56.6	68.6	91.2	56.6	67.9	90.2	56.6	67.5	100.7	56.6	70.9	97.4	23.9	47.8
1 - 16	WEIGHTED AVERAGE	99.4	43.9	72.6	96.5	50.9	74.5	97.5	67.9	83.2	97.0	61.5	79.9	97.9	66.2	82.6	96.8	44.4	71.5

NEBRASKA / NEVADA

DIVISION		NORFOLK 687 MAT.	INST.	TOTAL	NORTH PLATTE 691 MAT.	INST.	TOTAL	OMAHA 680 - 681 MAT.	INST.	TOTAL	VALENTINE 692 MAT.	INST.	TOTAL	CARSON CITY 897 MAT.	INST.	TOTAL	ELKO 898 MAT.	INST.	TOTAL
2	SITE WORK	76.2	90.8	87.5	94.4	92.3	92.8	77.0	91.0	87.8	79.9	94.4	91.1	65.0	103.1	94.3	60.2	102.6	92.8
031	CONCRETE FORMWORK	79.7	57.8	60.9	100.0	57.6	63.7	100.6	68.1	72.8	81.0	29.3	36.7	98.7	92.7	93.6	113.3	90.4	93.7
032	CONCRETE REINFORCEMENT	102.6	63.3	80.5	103.2	62.5	80.3	97.9	71.3	83.0	103.8	50.6	73.8	109.3	115.4	112.8	108.0	78.1	91.2
033	CAST IN PLACE CONCRETE	109.2	69.2	92.4	115.0	62.6	93.1	107.7	71.7	92.7	101.1	52.7	80.8	116.3	96.3	107.9	104.4	91.4	99.0
3	CONCRETE	96.2	63.5	79.8	102.7	61.7	82.1	96.5	70.4	83.3	98.2	43.0	70.4	112.9	97.9	105.3	104.4	88.3	96.3
4	MASONRY	122.1	65.9	87.2	92.3	50.7	66.4	103.4	70.4	82.9	105.1	38.6	63.8	133.4	82.8	102.0	132.5	77.6	98.4
5	METALS	96.8	72.9	88.2	94.7	82.3	90.2	100.1	76.4	91.6	110.0	65.7	94.1	104.3	97.3	101.8	103.8	82.1	96.0
6	WOOD & PLASTICS	82.3	55.8	68.6	98.9	56.6	77.1	101.2	68.9	84.5	82.9	27.4	54.2	96.2	91.7	93.9	113.8	91.7	102.4
7	THERMAL & MOISTURE PROTECTION	93.8	64.6	80.2	96.1	57.2	78.0	91.7	69.5	81.4	94.4	38.5	68.5	103.2	90.9	97.5	103.2	87.5	95.9
8	DOORS & WINDOWS	95.8	57.3	86.5	92.5	57.6	84.0	102.8	66.0	93.9	95.9	34.2	81.0	94.8	100.1	96.1	97.5	87.0	94.9
092	LATH, PLASTER & GYPSUM BOARD	109.4	54.8	74.2	107.7	54.8	73.6	117.8	68.3	85.8	115.2	25.5	57.3	88.7	91.3	90.4	95.6	91.3	92.8
095	ACOUSTICAL TREATMENT & WOOD FLOORING	121.9	54.8	78.4	88.9	54.8	66.8	132.8	68.3	91.0	131.0	25.5	62.6	96.9	91.3	93.3	96.9	91.3	93.3
096	FLOORING & CARPET	123.8	55.7	107.3	98.8	46.4	86.1	129.2	55.7	111.3	126.8	55.7	109.5	106.1	82.7	100.4	112.0	82.7	104.9
099	PAINTING & WALL COVERINGS	138.7	63.0	94.3	86.8	63.0	72.9	138.7	63.0	94.3	138.7	42.3	82.1	113.2	99.3	105.0	113.2	101.5	106.3
9	FINISHES	115.7	57.3	85.8	93.7	56.0	74.4	120.7	65.0	92.2	119.6	34.8	76.1	98.3	90.7	94.4	100.9	90.4	95.5
10 - 14	TOTAL DIV. 10 - 14	100.0	76.3	95.0	100.0	70.6	93.8	100.0	78.3	95.4	100.0	51.5	89.8	100.0	125.8	105.4	100.0	96.6	99.3
15	MECHANICAL	96.6	65.6	82.8	100.2	48.9	77.3	99.9	73.4	88.1	96.4	30.3	66.9	99.8	96.9	98.5	97.9	97.5	97.7
16	ELECTRICAL	89.4	71.2	77.1	94.5	56.6	68.9	88.5	82.1	84.2	88.3	56.5	66.9	93.9	89.3	90.8	93.0	89.3	90.5
1 - 16	WEIGHTED AVERAGE	98.8	68.3	84.1	96.9	61.5	79.8	100.5	74.9	88.2	100.6	48.2	75.3	101.8	95.0	98.5	100.8	90.5	95.8

City Cost Indexes

DIVISION		ELY 893 MAT.	INST.	TOTAL	LAS VEGAS 889-891 MAT.	INST.	TOTAL	RENO 894-895 MAT.	INST.	TOTAL	CHARLESTON 036 MAT.	INST.	TOTAL	CLAREMONT 037 MAT.	INST.	TOTAL	CONCORD 032-033 MAT.	INST.	TOTAL
2	SITE WORK	65.7	102.6	94.1	65.3	104.3	95.3	65.4	103.1	94.4	90.1	95.8	94.5	84.1	95.8	93.1	96.8	97.0	96.9
031	CONCRETE FORMWORK	101.8	90.7	92.3	94.7	105.2	103.7	99.1	92.7	93.6	78.5	48.2	52.5	86.7	48.2	53.7	83.9	75.5	76.7
032	CONCRETE REINFORCEMENT	106.8	78.2	90.7	101.0	118.9	111.1	101.0	118.5	110.8	93.3	70.6	80.5	93.3	70.6	80.5	93.3	93.8	93.6
033	CAST IN PLACE CONCRETE	112.1	94.3	104.7	110.7	109.0	110.0	118.5	96.3	109.2	104.1	64.2	87.4	95.7	64.2	82.5	101.2	92.3	97.5
3	CONCRETE	113.3	89.4	101.3	108.7	108.5	108.6	112.9	98.5	105.6	111.3	58.7	84.8	102.1	58.7	80.2	107.9	84.8	96.2
4	MASONRY	137.6	77.6	100.4	127.3	97.6	108.9	133.5	82.8	102.0	88.6	55.6	68.1	88.7	55.6	68.1	93.7	102.7	99.3
5	METALS	103.9	82.3	96.1	105.0	100.8	103.5	104.9	98.4	102.5	95.8	73.8	87.8	95.9	73.8	87.8	96.0	87.2	92.8
6	WOOD & PLASTICS	100.3	91.7	95.9	90.8	103.1	97.2	96.6	91.7	94.1	83.8	47.5	65.1	92.4	47.5	69.2	85.6	70.1	77.6
7	THERMAL & MOISTURE PROTECTION	103.9	89.8	97.3	102.8	101.3	102.1	103.2	90.9	97.5	100.6	57.6	80.7	100.5	57.6	80.6	101.1	102.7	101.9
8	DOORS & WINDOWS	97.4	87.0	94.9	95.2	106.9	98.0	95.2	99.5	96.2	105.8	50.7	92.5	107.1	50.7	93.5	107.1	76.4	99.7
092	LATH, PLASTER & GYPSUM BOARD	91.9	91.3	91.5	88.6	103.1	97.9	89.6	91.3	90.7	89.4	44.8	60.6	92.6	44.8	61.8	90.0	68.1	75.9
095	ACOUSTICAL TREATMENT & WOOD FLOORING	96.9	91.3	93.3	102.6	103.1	102.9	101.0	91.3	94.7	92.1	44.8	61.4	92.1	44.8	61.4	92.1	68.1	76.5
096	FLOORING & CARPET	109.6	92.5	105.4	106.1	94.0	103.1	106.1	82.7	100.4	91.6	50.2	81.5	94.2	50.2	83.5	92.1	106.6	95.7
099	PAINTING & WALL COVERINGS	113.2	101.5	106.3	113.2	129.7	122.9	113.2	99.3	105.0	89.3	40.3	60.5	89.3	40.3	60.5	89.3	84.6	86.6
9	FINISHES	100.0	92.4	96.1	99.3	105.2	102.3	99.2	90.7	94.8	91.6	46.8	68.6	92.5	46.8	69.1	92.7	81.2	86.8
10-14	TOTAL DIV. 10-14	100.0	96.6	99.3	100.0	114.7	103.1	100.0	125.8	105.4	100.0	62.1	92.0	100.0	62.1	92.0	100.0	99.4	99.9
15	MECHANICAL	97.9	97.5	97.7	99.8	109.5	104.1	99.8	97.0	98.5	96.7	50.2	75.9	96.7	50.2	75.9	100.0	99.8	99.9
16	ELECTRICAL	93.0	89.3	90.5	96.0	111.1	106.2	93.9	89.3	90.8	100.6	46.7	64.3	100.6	46.7	64.3	104.7	77.9	86.6
1-16	WEIGHTED AVERAGE	102.1	91.0	96.7	101.3	106.4	103.8	102.0	95.1	98.7	98.7	58.1	79.1	97.8	58.1	78.7	100.2	86.5	93.6

NEW HAMPSHIRE / NEW JERSEY

DIVISION		KEENE 034 MAT.	INST.	TOTAL	LITTLETON 035 MAT.	INST.	TOTAL	MANCHESTER 031 MAT.	INST.	TOTAL	NASHUA 030 MAT.	INST.	TOTAL	PORTSMOUTH 038 MAT.	INST.	TOTAL	ATLANTIC CITY 082,084 MAT.	INST.	TOTAL
2	SITE WORK	97.4	95.8	96.2	84.0	95.4	92.7	95.9	97.0	96.7	98.1	97.0	97.2	92.0	96.5	95.4	100.9	100.4	100.5
031	CONCRETE FORMWORK	84.3	49.5	54.5	97.9	55.5	61.6	97.7	75.5	78.7	98.2	75.5	78.8	83.9	72.3	74.0	109.2	123.7	121.6
032	CONCRETE REINFORCEMENT	93.3	84.1	88.1	94.2	49.0	68.7	116.7	93.8	103.8	116.7	93.8	103.8	93.3	93.7	93.5	89.0	108.3	99.8
033	CAST IN PLACE CONCRETE	104.7	64.6	87.9	93.9	57.1	78.5	101.6	92.3	97.7	99.0	92.3	96.2	94.0	88.2	91.6	88.8	126.1	104.4
3	CONCRETE	111.1	62.0	86.3	101.6	55.4	78.3	112.2	84.8	98.4	111.0	84.8	97.8	101.6	81.8	91.6	104.4	120.2	112.4
4	MASONRY	94.2	55.6	70.2	101.1	53.5	71.5	94.9	102.7	99.8	95.3	102.7	99.9	89.7	95.8	93.5	97.8	124.8	114.6
5	METALS	95.7	79.9	90.0	96.1	64.7	84.8	96.1	87.2	96.1	101.1	87.2	96.1	97.6	85.7	93.3	95.6	97.4	96.3
6	WOOD & PLASTICS	90.5	47.5	68.3	102.7	54.5	77.8	101.3	70.1	85.2	101.3	70.1	85.2	85.6	70.1	77.6	113.8	123.7	119.0
7	THERMAL & MOISTURE PROTECTION	101.1	59.9	82.0	100.7	63.6	83.5	101.3	102.7	101.9	101.6	102.7	102.1	101.2	98.5	99.9	101.4	117.8	109.0
8	DOORS & WINDOWS	104.2	62.1	94.0	108.3	50.7	94.2	109.1	76.4	101.2	109.1	76.4	101.2	110.1	69.0	100.2	106.7	117.2	109.3
092	LATH, PLASTER & GYPSUM BOARD	92.1	44.8	61.6	104.3	52.0	70.6	98.5	68.1	78.9	98.5	68.1	78.9	90.0	68.1	75.9	101.1	123.5	115.6
095	ACOUSTICAL TREATMENT & WOOD FLOORING	92.1	44.8	61.4	92.1	52.0	66.1	101.7	68.1	79.9	101.7	68.1	79.9	92.1	68.1	76.5	92.1	123.5	112.4
096	FLOORING & CARPET	93.6	82.3	90.9	103.0	50.2	90.2	97.1	106.6	99.4	96.9	106.6	99.2	92.1	106.6	95.7	100.7	130.6	107.9
099	PAINTING & WALL COVERINGS	89.3	40.3	60.5	89.3	78.0	82.7	89.3	84.6	86.6	89.3	84.6	86.6	89.3	48.8	65.5	89.3	138.0	117.9
9	FINISHES	93.6	53.4	73.0	96.9	56.7	76.3	97.1	81.2	89.0	97.3	81.2	89.1	92.4	75.4	83.7	97.4	127.2	112.7
10-14	TOTAL DIV. 10-14	100.0	88.7	97.6	100.0	65.5	92.7	100.0	99.4	99.9	100.0	99.4	99.9	100.0	96.6	99.3	100.0	120.4	104.3
15	MECHANICAL	96.7	53.7	77.5	96.7	63.4	81.8	100.0	83.5	92.6	100.0	83.5	92.6	100.0	79.8	91.0	99.8	117.0	107.4
16	ELECTRICAL	100.6	46.7	64.3	102.5	74.1	83.3	104.9	77.9	86.7	104.7	77.9	86.6	101.4	79.1	86.4	96.3	110.8	106.1
1-16	WEIGHTED AVERAGE	99.3	61.7	81.2	99.3	65.1	82.8	102.4	86.5	94.7	102.4	86.5	94.7	99.4	83.6	91.8	100.1	115.1	107.3

NEW JERSEY

DIVISION		CAMDEN 081 MAT.	INST.	TOTAL	DOVER 078 MAT.	INST.	TOTAL	ELIZABETH 072 MAT.	INST.	TOTAL	HACKENSACK 076 MAT.	INST.	TOTAL	JERSEY CITY 073 MAT.	INST.	TOTAL	LONG BRANCH 077 MAT.	INST.	TOTAL
2	SITE WORK	101.4	100.9	101.0	113.8	101.6	104.4	117.4	101.3	105.0	114.1	101.1	104.1	101.4	101.4	101.4	106.6	101.5	102.7
031	CONCRETE FORMWORK	98.2	123.8	120.1	93.8	121.0	117.1	109.8	117.8	116.7	93.2	119.6	115.9	98.2	124.5	120.7	98.4	123.9	120.2
032	CONCRETE REINFORCEMENT	116.7	112.9	114.6	89.8	122.4	108.2	89.8	122.2	108.1	89.8	122.3	108.1	116.7	126.3	122.1	89.8	126.1	110.3
033	CAST IN PLACE CONCRETE	86.1	120.3	100.4	110.4	128.8	118.1	94.9	123.7	106.9	107.9	126.7	115.8	86.1	127.3	103.4	95.7	127.0	108.8
3	CONCRETE	104.7	119.2	112.0	112.7	123.0	117.9	108.3	119.7	114.1	110.5	121.6	116.1	104.7	124.5	114.7	110.3	124.1	117.3
4	MASONRY	87.3	124.6	110.5	91.6	125.1	112.4	108.8	117.3	114.1	96.1	122.4	112.5	87.3	127.5	112.3	100.4	123.9	115.0
5	METALS	101.0	99.3	100.4	95.6	107.5	99.9	97.4	106.6	100.7	95.4	106.8	99.5	101.0	106.5	103.0	95.6	106.0	99.3
6	WOOD & PLASTICS	101.3	123.7	112.9	100.0	120.1	110.4	117.9	120.1	119.1	100.0	120.0	110.3	101.3	123.5	112.7	102.5	123.7	113.4
7	THERMAL & MOISTURE PROTECTION	101.1	117.0	108.4	101.7	116.6	108.6	102.0	113.4	107.3	101.3	119.4	109.7	101.1	114.9	107.5	101.3	118.6	109.3
8	DOORS & WINDOWS	109.1	117.0	111.0	112.5	120.5	114.5	110.4	120.5	112.8	109.5	119.9	112.0	109.1	123.7	112.6	105.1	123.7	109.6
092	LATH, PLASTER & GYPSUM BOARD	98.5	123.5	114.6	95.3	119.8	111.1	102.2	119.8	113.5	95.3	119.6	111.0	98.5	123.3	114.5	96.9	123.5	114.0
095	ACOUSTICAL TREATMENT & WOOD FLOORING	101.7	123.5	115.8	92.1	119.8	110.0	92.1	119.8	110.0	92.1	119.6	109.9	101.7	123.3	115.7	92.1	123.5	112.4
096	FLOORING & CARPET	96.9	130.3	105.0	95.9	125.4	103.1	101.2	70.2	93.6	95.9	70.2	89.7	96.9	134.4	106.0	97.2	127.2	104.5
099	PAINTING & WALL COVERINGS	89.3	135.6	116.5	89.2	121.0	108.5	89.2	121.1	108.5	89.2	99.6	95.3	89.3	122.1	108.6	89.3	82.4	85.2
9	FINISHES	97.6	126.8	112.6	95.2	121.8	108.9	98.3	108.7	103.6	95.1	107.2	101.3	97.6	126.1	112.2	96.1	119.8	108.3
10-14	TOTAL DIV. 10-14	100.0	120.4	104.3	100.0	112.9	100.6	100.0	97.6	99.5	100.0	99.4	99.9	100.0	104.3	100.9	100.0	103.5	100.7
15	MECHANICAL	100.0	116.7	107.8	99.7	122.6	109.9	100.0	110.0	104.4	99.7	121.7	109.5	100.0	124.7	111.0	99.7	124.7	110.8
16	ELECTRICAL	104.7	116.4	112.6	98.4	113.5	108.6	99.5	124.7	116.5	98.4	128.0	118.4	106.3	136.5	126.7	97.9	122.3	114.4
1-16	WEIGHTED AVERAGE	101.3	116.1	108.4	101.6	117.0	109.0	102.8	113.4	107.9	101.2	116.9	108.8	101.4	122.2	111.4	100.8	118.8	109.5

COST INDEXES

City Cost Indexes

NEW JERSEY

	DIVISION	NEWARK 070-071 MAT.	INST.	TOTAL	NEW BRUNSWICK 088-089 MAT.	INST.	TOTAL	PATERSON 074-075 MAT.	INST.	TOTAL	POINT PLEASANT 087 MAT.	INST.	TOTAL	SUMMIT 079 MAT.	INST.	TOTAL	TRENTON 085-086 MAT.	INST.	TOTAL
2	SITE WORK	121.4	101.3	105.9	114.8	101.5	104.6	115.7	101.3	104.6	116.2	100.9	104.4	114.9	100.6	103.9	102.5	101.5	101.8
031	CONCRETE FORMWORK	96.5	124.0	120.1	101.0	123.9	120.6	97.0	121.5	118.0	94.3	122.7	118.7	98.7	117.7	114.9	97.3	129.2	124.6
032	CONCRETE REINFORCEMENT	116.7	126.3	122.1	89.8	126.0	110.2	116.7	126.3	122.1	89.8	122.1	108.0	89.8	122.2	108.1	116.7	121.1	119.2
033	CAST IN PLACE CONCRETE	97.0	129.8	110.7	109.7	127.0	116.9	109.7	129.7	118.1	109.7	124.5	115.9	91.7	123.6	105.1	96.9	127.5	109.7
3	CONCRETE	109.9	125.3	117.7	124.0	124.1	124.0	116.2	124.2	120.2	123.5	121.9	122.7	104.4	119.6	112.1	109.9	125.6	117.8
4	MASONRY	97.0	126.3	115.2	94.3	122.1	111.6	92.0	126.3	113.3	87.8	121.3	108.6	94.2	117.3	108.5	88.3	124.6	110.9
5	METALS	101.0	109.8	104.2	95.2	105.9	99.0	101.0	109.7	104.2	95.4	103.3	98.2	96.0	106.5	99.8	101.8	104.0	102.6
6	WOOD & PLASTICS	103.1	123.3	113.5	106.2	123.7	115.2	103.1	120.0	111.8	98.3	123.2	111.2	104.4	120.1	112.5	101.3	130.0	116.1
7	THERMAL & MOISTURE PROTECTION	101.4	123.8	111.7	101.4	118.2	109.2	101.8	120.9	110.7	101.5	113.0	106.8	102.2	113.4	107.4	100.0	121.1	109.8
8	DOORS & WINDOWS	115.4	123.5	117.3	100.9	124.7	106.7	115.4	121.2	116.8	103.1	122.1	107.7	118.1	120.5	118.7	109.1	126.4	113.3
092	LATH, PLASTER & GYPSUM BOARD	98.5	123.0	114.3	98.4	123.5	114.6	98.5	119.6	112.1	95.3	123.0	113.2	96.9	119.8	111.6	98.5	130.0	118.8
095	ACOUSTICAL TREATMENT & WOOD FLOORING	101.7	123.0	115.5	92.1	123.5	112.4	101.7	119.6	113.3	92.1	123.0	112.1	92.1	119.8	110.0	101.7	130.0	120.0
096	FLOORING & CARPET	97.1	134.4	106.2	98.3	91.1	96.6	96.9	134.4	106.0	95.9	127.2	103.5	97.2	70.2	90.7	97.1	130.4	105.2
099	PAINTING & WALL COVERINGS	89.2	122.1	108.5	89.3	122.1	108.6	89.2	122.1	108.5	89.3	81.2	84.6	89.2	122.1	108.5	89.3	122.1	108.6
9	FINISHES	98.1	125.7	112.2	97.3	118.1	108.0	97.7	123.8	111.1	96.1	119.2	108.0	96.1	108.7	102.6	97.7	128.9	113.7
10-14	TOTAL DIV. 10-14	100.0	103.7	100.8	100.0	103.5	100.7	100.0	103.3	100.7	100.0	103.1	100.7	100.0	97.5	99.5	100.0	121.1	104.5
15	MECHANICAL	100.0	122.6	110.1	99.6	120.5	108.9	100.0	123.2	110.3	99.6	119.6	108.6	99.7	110.0	104.3	100.0	122.8	110.2
16	ELECTRICAL	106.1	131.4	123.2	97.5	131.8	120.6	106.3	128.2	121.1	96.3	118.5	111.3	99.5	117.0	111.3	104.9	129.2	121.7
1-16	WEIGHTED AVERAGE	103.9	121.4	112.4	101.9	119.2	110.2	104.2	120.4	112.0	101.5	116.0	108.5	101.7	112.0	106.7	102.1	121.5	111.5

NEW JERSEY / NEW MEXICO

	DIVISION	VINELAND 080,083 MAT.	INST.	TOTAL	ALBUQUERQUE 870-872 MAT.	INST.	TOTAL	CARRIZOZO 883 MAT.	INST.	TOTAL	CLOVIS 881 MAT.	INST.	TOTAL	FARMINGTON 874 MAT.	INST.	TOTAL	GALLUP 873 MAT.	INST.	TOTAL
2	SITE WORK	105.6	100.9	102.0	86.2	111.0	105.3	109.1	110.8	110.4	94.8	110.8	107.1	91.6	110.9	106.4	101.8	110.9	108.8
031	CONCRETE FORMWORK	91.0	123.4	118.8	97.0	72.1	75.7	94.8	72.0	75.2	94.8	71.9	75.2	97.0	72.0	75.6	95.5	72.4	75.7
032	CONCRETE REINFORCEMENT	89.0	108.2	99.8	101.7	69.7	83.7	109.1	67.5	85.6	110.2	63.4	83.8	111.0	67.6	86.5	106.3	67.6	84.5
033	CAST IN PLACE CONCRETE	95.7	127.6	109.1	108.9	79.3	96.5	100.9	79.0	91.8	100.8	79.0	91.7	109.5	79.1	96.8	103.0	79.6	93.2
3	CONCRETE	109.7	120.7	115.2	108.1	75.3	91.6	123.6	74.7	98.9	111.1	73.9	92.3	111.9	74.7	93.2	118.7	75.0	96.7
4	MASONRY	84.1	124.8	109.4	110.0	73.5	87.3	114.6	69.7	86.7	114.6	71.3	87.7	124.2	71.6	91.5	114.7	72.7	88.6
5	METALS	95.2	97.2	95.9	106.0	91.4	100.7	104.4	90.3	99.3	104.1	87.7	98.2	105.2	89.6	99.6	104.8	88.7	99.0
6	WOOD & PLASTICS	94.7	123.7	109.7	97.1	73.6	85.0	97.1	73.6	85.0	97.1	73.6	85.0	97.1	73.6	85.0	97.1	73.6	85.0
7	THERMAL & MOISTURE PROTECTION	100.9	118.0	108.9	102.2	79.7	91.7	104.6	78.5	92.5	103.2	78.8	91.9	103.0	78.9	91.8	104.3	79.4	92.7
8	DOORS & WINDOWS	102.5	115.8	105.7	95.1	72.4	89.6	94.2	72.4	88.9	94.4	70.5	88.6	98.1	71.8	91.7	98.1	71.1	91.6
092	LATH, PLASTER & GYPSUM BOARD	93.7	123.5	112.9	89.6	72.2	78.4	88.1	72.2	77.8	88.1	72.2	77.8	88.1	72.2	77.8	88.1	72.2	77.8
095	ACOUSTICAL TREATMENT & WOOD FLOORING	92.1	123.5	112.4	101.0	72.2	82.4	94.2	72.2	80.0	94.2	72.2	80.0	94.2	72.2	80.0	94.2	72.2	80.0
096	FLOORING & CARPET	94.9	130.6	103.6	106.1	77.7	99.2	106.1	70.5	97.4	106.1	70.5	97.4	106.1	77.7	99.2	106.1	77.7	99.2
099	PAINTING & WALL COVERINGS	89.3	81.2	84.6	113.2	66.1	85.5	113.2	72.5	89.3	113.2	72.5	89.3	113.2	72.5	89.3	113.2	72.5	89.3
9	FINISHES	94.9	120.6	108.1	99.7	72.5	85.8	100.0	71.8	85.5	98.8	71.8	85.0	98.5	73.3	85.6	99.5	73.5	86.2
10-14	TOTAL DIV. 10-14	100.0	120.4	104.3	100.0	78.3	95.4	100.0	78.3	95.4	100.0	78.3	95.4	100.0	78.3	95.4	100.0	78.7	95.5
15	MECHANICAL	99.8	114.2	106.2	99.8	77.1	89.7	97.5	75.5	87.7	97.5	75.0	87.5	99.8	75.1	88.8	97.7	75.6	87.9
16	ELECTRICAL	96.3	109.7	105.4	88.3	79.9	82.7	88.3	78.6	81.8	84.7	78.6	80.6	85.6	78.6	80.9	84.5	78.6	80.5
1-16	WEIGHTED AVERAGE	99.1	113.6	106.1	100.6	81.1	91.1	102.5	79.9	91.6	100.2	79.5	90.2	101.8	80.1	91.3	102.0	80.3	91.5

NEW MEXICO

	DIVISION	LAS CRUCES 880 MAT.	INST.	TOTAL	LAS VEGAS 877 MAT.	INST.	TOTAL	ROSWELL 882 MAT.	INST.	TOTAL	SANTA FE 875 MAT.	INST.	TOTAL	SOCORRO 878 MAT.	INST.	TOTAL	TRUTH/CONSEQUENCES 879 MAT.	INST.	TOTAL
2	SITE WORK	99.5	86.2	89.3	90.6	110.9	106.2	96.7	110.8	107.6	85.5	111.0	105.1	87.7	110.6	105.3	111.9	86.0	92.0
031	CONCRETE FORMWORK	94.5	69.8	73.3	94.8	72.0	75.3	97.0	68.9	72.9	97.0	72.1	75.7	94.8	69.0	72.7	90.6	67.2	70.5
032	CONCRETE REINFORCEMENT	105.1	62.8	81.2	108.1	69.6	86.4	110.2	63.4	83.8	109.1	69.6	86.8	110.2	67.6	86.2	103.9	63.0	80.9
033	CAST IN PLACE CONCRETE	95.3	68.4	84.0	106.5	79.1	95.0	100.9	77.6	91.1	102.9	79.3	93.0	104.4	77.6	93.2	114.4	66.9	94.5
3	CONCRETE	88.3	68.7	78.4	108.8	75.1	91.8	112.4	72.0	92.0	106.2	75.3	90.6	107.3	72.9	90.0	100.9	67.1	83.8
4	MASONRY	110.4	63.3	81.1	115.0	71.6	88.0	125.6	69.1	90.5	114.7	71.4	87.8	114.9	71.1	87.7	112.3	66.8	84.1
5	METALS	99.4	78.1	91.7	104.2	90.6	99.3	105.1	87.5	98.8	105.1	91.2	100.1	104.4	89.5	99.1	98.1	78.2	90.9
6	WOOD & PLASTICS	90.5	72.0	80.9	97.1	73.6	85.0	97.1	69.9	83.1	97.1	73.6	85.0	97.1	69.9	83.1	90.5	68.3	79.0
7	THERMAL & MOISTURE PROTECTION	87.8	70.4	79.7	102.6	78.9	91.6	103.3	77.4	91.3	102.4	79.1	91.6	102.5	77.9	91.1	88.6	70.4	80.2
8	DOORS & WINDOWS	87.1	69.6	82.9	94.4	72.4	89.0	94.1	68.5	87.9	94.3	72.4	89.0	94.2	69.8	88.3	87.1	67.6	82.4
092	LATH, PLASTER & GYPSUM BOARD	89.9	72.2	78.5	88.1	72.2	77.8	88.1	68.4	75.4	88.1	72.2	77.8	88.1	68.4	75.4	89.9	68.4	76.0
095	ACOUSTICAL TREATMENT & WOOD FLOORING	97.8	72.2	81.2	94.2	72.2	80.0	94.2	68.4	77.5	94.2	72.2	80.0	94.2	68.4	77.5	97.8	68.4	78.7
096	FLOORING & CARPET	138.6	52.6	117.7	106.1	77.7	99.2	106.1	70.5	97.4	106.1	77.7	99.2	106.1	77.7	99.2	138.6	70.5	122.1
099	PAINTING & WALL COVERINGS	106.2	48.5	72.3	113.2	66.5	85.8	113.2	72.5	89.3	113.2	65.0	84.9	113.2	66.5	85.8	106.2	72.5	86.4
9	FINISHES	116.6	64.2	89.7	98.4	72.6	85.2	98.9	69.5	83.8	98.3	72.4	85.0	98.3	70.3	83.9	117.1	68.3	92.1
10-14	TOTAL DIV. 10-14	100.0	73.6	94.4	100.0	78.3	95.4	100.0	77.6	95.3	100.0	78.3	95.4	100.0	77.6	95.3	100.0	72.9	94.3
15	MECHANICAL	100.0	76.5	89.5	97.7	75.5	87.8	99.8	75.2	88.8	99.8	77.1	89.7	99.7	74.7	87.4	99.2	74.3	88.1
16	ELECTRICAL	86.8	65.2	72.2	88.3	78.6	81.8	86.8	78.6	81.3	88.3	77.2	80.8	85.2	82.9	83.6	91.2	82.8	85.6
1-16	WEIGHTED AVERAGE	97.8	71.4	85.1	100.1	80.2	90.5	101.8	78.6	90.6	100.2	80.3	90.6	99.6	79.8	90.1	99.8	74.4	87.6

COST INDEXES

627

DIVISION		NEW MEXICO TUCUMCARI 884			NEW YORK ALBANY 120-122			BINGHAMTON 137-139			BRONX 104			BROOKLYN 112			BUFFALO 140-142		
		MAT.	INST.	TOTAL	MAT.	INST.	TOTAL	MAT.	INST.	TOTAL	MAT.	INST.	TOTAL	MAT.	INST.	TOTAL	MAT.	INST.	TOTAL
2	SITE WORK	94.4	110.8	107.0	74.8	108.9	101.0	99.1	91.3	93.1	132.8	127.3	128.5	127.4	131.3	130.4	104.3	93.7	96.1
031	CONCRETE FORMWORK	94.8	68.9	72.6	99.9	91.0	92.3	106.2	79.5	83.3	104.9	148.7	142.4	114.7	148.4	143.6	103.5	114.1	112.6
032	CONCRETE REINFORCEMENT	108.1	63.4	83.0	99.0	88.9	93.3	98.0	91.9	94.5	94.9	153.7	128.0	99.2	153.6	129.9	101.4	105.7	103.8
033	CAST IN PLACE CONCRETE	100.8	77.6	91.1	82.4	101.4	90.4	104.3	91.4	98.9	110.0	145.7	124.9	106.1	144.5	122.2	105.1	119.2	111.0
3	CONCRETE	110.5	72.1	91.1	100.8	95.5	98.1	102.2	88.5	95.3	114.8	147.4	131.2	116.0	146.9	131.5	103.7	113.4	108.6
4	MASONRY	129.8	70.7	93.1	93.0	97.9	96.0	108.8	84.9	94.0	100.2	154.2	133.7	119.9	154.1	141.2	106.3	117.9	113.5
5	METALS	104.1	87.6	98.2	99.9	109.8	103.5	95.1	124.2	105.6	96.5	130.6	108.8	106.5	130.7	115.2	100.0	96.6	98.8
6	WOOD & PLASTICS	97.1	69.9	83.1	101.6	89.1	95.1	111.8	76.2	93.4	107.8	149.4	129.2	116.7	149.2	133.5	103.6	114.2	109.1
7	THERMAL & MOISTURE PROTECTION	103.1	77.8	91.4	92.7	95.6	94.1	102.3	88.0	95.7	108.2	141.7	123.8	108.0	141.2	123.4	105.2	107.6	106.3
8	DOORS & WINDOWS	94.1	68.5	87.9	99.7	84.2	95.9	94.4	78.7	90.6	96.9	150.6	109.9	93.6	150.5	107.4	94.1	106.9	97.2
092	LATH, PLASTER & GYPSUM BOARD	88.1	68.4	75.4	100.4	88.4	92.7	107.4	74.7	86.3	99.1	150.7	132.4	102.3	150.7	133.5	96.1	114.1	107.7
095	ACOUSTICAL TREATMENT & WOOD FLOORING	94.2	68.4	77.5	97.1	88.4	91.5	97.1	74.7	82.6	85.1	150.7	127.6	80.6	150.7	126.0	92.8	114.1	106.6
096	FLOORING & CARPET	106.1	70.5	97.4	81.6	88.8	83.4	91.2	84.7	89.6	90.8	147.1	104.5	97.6	147.1	109.6	90.8	118.4	97.5
099	PAINTING & WALL COVERINGS	113.2	72.5	89.3	78.9	83.4	81.5	79.6	70.1	74.0	95.3	133.7	117.8	101.9	131.4	119.2	75.6	115.9	99.2
9	FINISHES	98.8	69.5	83.8	94.6	89.4	91.9	94.7	78.6	86.5	102.7	147.1	125.5	108.0	146.7	127.8	93.0	115.9	104.8
10-14	TOTAL DIV. 10-14	100.0	77.6	95.3	100.2	95.0	99.1	100.0	91.6	98.2	100.0	128.9	106.1	100.0	128.3	106.0	100.0	106.2	101.3
15	MECHANICAL	97.5	75.2	87.6	100.3	90.8	96.1	100.6	85.5	93.8	100.3	146.9	121.1	100.1	143.1	119.3	100.0	97.8	99.1
16	ELECTRICAL	88.3	78.6	81.8	103.2	92.9	96.2	104.1	78.0	86.5	98.5	157.9	138.5	103.6	150.8	135.4	102.0	101.4	101.6
1-16	WEIGHTED AVERAGE	101.2	78.8	90.4	98.5	95.8	97.2	99.4	87.8	93.8	102.4	145.6	123.2	105.6	144.0	124.1	99.8	105.0	102.3

DIVISION		NEW YORK ELMIRA 148-149			FAR ROCKAWAY 116			FLUSHING 113			GLENS FALLS 128			HICKSVILLE 115,117,118			JAMAICA 114		
		MAT.	INST.	TOTAL	MAT.	INST.	TOTAL	MAT.	INST.	TOTAL	MAT.	INST.	TOTAL	MAT.	INST.	TOTAL	MAT.	INST.	TOTAL
2	SITE WORK	99.6	90.5	92.6	132.5	131.3	131.5	132.5	131.3	131.5	64.9	108.5	98.5	119.5	131.6	128.8	125.6	131.3	130.0
031	CONCRETE FORMWORK	79.1	78.3	78.4	91.4	148.5	140.3	96.9	148.6	141.1	78.1	82.2	81.6	90.1	146.9	138.7	96.9	148.6	141.2
032	CONCRETE REINFORCEMENT	102.6	89.3	95.1	99.2	153.6	129.9	101.0	153.7	130.7	98.7	83.6	90.2	99.2	165.1	136.3	99.2	153.7	129.9
033	CAST IN PLACE CONCRETE	98.3	91.8	95.6	114.8	144.5	127.3	114.8	144.5	127.3	79.2	98.4	87.3	97.5	143.1	116.6	106.1	144.5	122.2
3	CONCRETE	96.7	87.7	92.2	122.0	147.0	134.6	122.6	147.0	134.9	91.2	89.6	90.4	106.3	147.8	127.2	114.6	147.0	131.0
4	MASONRY	102.8	83.1	90.6	122.6	154.1	142.2	116.6	154.1	139.9	93.7	91.4	92.3	111.4	150.3	135.6	121.0	154.1	141.6
5	METALS	96.0	124.3	106.2	106.4	131.8	115.5	106.4	131.9	115.6	96.0	107.3	100.1	109.2	133.8	118.1	106.4	131.9	115.6
6	WOOD & PLASTICS	81.1	76.0	78.5	95.3	149.2	123.1	101.4	149.2	126.1	82.5	80.5	81.5	90.0	149.2	120.5	101.4	149.2	126.1
7	THERMAL & MOISTURE PROTECTION	105.8	86.2	96.7	107.7	140.9	123.1	107.8	140.9	123.2	91.4	90.3	90.9	107.2	137.5	121.2	107.6	141.2	123.2
8	DOORS & WINDOWS	96.9	76.9	92.1	91.9	151.2	106.3	91.9	151.2	106.3	93.1	79.1	89.7	91.9	151.2	106.3	91.9	151.2	106.3
092	LATH, PLASTER & GYPSUM BOARD	93.1	74.7	81.2	95.4	150.7	131.1	97.5	150.7	131.8	90.1	79.5	83.3	93.8	150.7	130.5	97.5	150.7	131.8
095	ACOUSTICAL TREATMENT & WOOD FLOORING	90.1	74.7	80.1	80.6	150.7	126.0	80.6	150.7	126.0	88.9	79.5	82.8	80.6	150.7	126.0	80.6	150.7	126.0
096	FLOORING & CARPET	81.4	84.2	82.1	92.1	147.1	105.5	93.6	147.1	106.6	73.6	88.8	77.3	90.8	111.4	95.8	93.6	147.1	106.6
099	PAINTING & WALL COVERINGS	73.5	60.8	66.0	101.9	133.7	120.6	101.9	133.7	120.6	78.9	75.8	77.1	101.9	137.4	122.8	101.9	133.7	120.6
9	FINISHES	88.7	76.6	82.5	105.7	146.9	126.9	106.5	146.9	127.2	88.2	82.2	85.1	104.2	138.9	122.0	106.1	146.9	127.0
10-14	TOTAL DIV. 10-14	100.0	90.1	97.9	100.0	128.3	106.0	100.0	128.3	106.0	100.1	81.0	96.1	100.0	126.6	105.6	100.0	128.3	106.0
15	MECHANICAL	97.0	83.9	91.2	96.5	143.2	117.3	96.5	147.3	119.2	97.0	78.9	88.9	99.9	141.8	118.6	96.5	147.3	119.2
16	ELECTRICAL	98.2	90.7	93.2	117.1	157.9	144.6	117.1	157.9	144.6	95.8	86.6	89.6	103.6	149.8	134.8	101.8	157.9	139.6
1-16	WEIGHTED AVERAGE	96.8	88.8	92.9	106.1	145.3	125.0	106.0	146.1	125.4	93.5	89.2	91.4	103.2	142.6	122.2	103.9	146.1	124.3

DIVISION		NEW YORK JAMESTOWN 147			KINGSTON 124			LONG ISLAND CITY 111			MONTICELLO 127			MOUNT VERNON 105			NEW ROCHELLE 108		
		MAT.	INST.	TOTAL	MAT.	INST.	TOTAL	MAT.	INST.	TOTAL	MAT.	INST.	TOTAL	MAT.	INST.	TOTAL	MAT.	INST.	TOTAL
2	SITE WORK	101.1	89.8	92.4	127.7	130.4	129.8	129.2	131.3	130.8	122.4	130.4	128.6	143.9	127.5	131.3	142.1	127.5	130.9
031	CONCRETE FORMWORK	79.2	75.6	76.2	81.7	108.6	104.8	106.6	148.6	142.6	90.8	108.7	106.1	87.8	132.5	126.1	114.2	132.5	129.9
032	CONCRETE REINFORCEMENT	102.7	89.3	95.2	99.0	126.7	114.6	99.2	153.7	129.9	98.3	126.7	114.3	94.0	151.4	126.3	94.1	151.4	126.4
033	CAST IN PLACE CONCRETE	101.9	94.1	98.6	107.0	115.7	110.6	106.9	144.5	124.2	100.2	115.7	106.7	122.9	134.4	127.7	122.9	134.4	127.7
3	CONCRETE	99.6	84.6	92.1	115.3	114.3	114.8	118.5	147.0	132.9	109.8	114.3	112.1	126.6	136.1	131.4	126.6	136.2	131.4
4	MASONRY	110.5	90.8	98.2	110.0	130.3	122.6	117.0	154.1	140.1	102.4	130.3	119.7	106.3	135.6	124.5	106.3	135.6	124.5
5	METALS	95.9	83.5	91.4	106.8	114.1	109.4	106.3	131.9	115.6	106.3	114.1	109.1	96.2	129.8	108.3	96.5	131.7	109.2
6	WOOD & PLASTICS	80.2	67.2	73.5	84.1	104.8	94.8	108.2	149.2	129.3	95.4	104.8	100.2	93.3	130.5	112.5	122.7	130.5	126.8
7	THERMAL & MOISTURE PROTECTION	104.9	83.2	94.9	109.4	119.9	114.2	107.8	141.2	123.3	109.0	119.9	114.0	109.7	134.1	121.0	110.1	134.1	121.2
8	DOORS & WINDOWS	96.8	75.1	91.5	97.3	116.8	102.0	91.9	151.2	106.3	91.9	116.8	97.9	96.9	139.8	107.3	96.9	141.1	107.6
092	LATH, PLASTER & GYPSUM BOARD	87.9	65.6	73.5	91.4	104.9	100.1	99.6	150.7	132.6	95.1	104.9	101.4	93.5	131.3	117.9	104.1	131.3	121.6
095	ACOUSTICAL TREATMENT & WOOD FLOORING	87.3	65.6	73.2	79.3	104.9	95.9	80.6	150.7	126.0	79.3	104.9	95.9	83.7	131.3	114.5	83.7	131.3	114.5
096	FLOORING & CARPET	84.3	102.0	88.6	85.1	117.0	92.9	95.4	147.1	108.0	87.9	117.0	94.9	83.2	147.1	98.8	89.8	147.1	103.7
099	PAINTING & WALL COVERINGS	75.6	67.5	70.8	99.2	96.8	97.8	101.9	133.7	120.6	99.2	96.8	97.8	92.6	121.0	109.3	92.6	121.0	109.3
9	FINISHES	88.6	78.7	83.6	102.2	107.9	105.1	107.1	146.9	127.5	103.2	107.9	105.6	99.9	133.5	117.1	103.2	133.5	118.7
10-14	TOTAL DIV. 10-14	100.0	92.6	98.4	100.1	106.0	101.3	100.0	128.3	106.0	100.1	106.0	101.3	100.0	124.3	105.1	100.0	124.3	105.1
15	MECHANICAL	96.8	78.6	88.7	97.0	102.5	99.4	100.1	147.3	121.1	97.0	105.4	100.7	97.1	137.1	115.0	97.1	137.2	115.0
16	ELECTRICAL	97.4	84.0	88.4	100.1	100.2	100.2	102.7	157.9	139.9	100.1	94.9	96.6	96.1	128.1	117.7	96.1	128.1	117.7
1-16	WEIGHTED AVERAGE	97.4	83.2	90.6	103.6	112.0	107.6	105.3	146.1	125.0	101.9	111.7	106.6	103.1	132.9	117.5	103.8	133.1	117.9

COST INDEXES

City Cost Indexes

NEW YORK

DIVISION		NEW YORK 100-102 MAT.	INST.	TOTAL	NIAGARA FALLS 143 MAT.	INST.	TOTAL	PLATTSBURGH 129 MAT.	INST.	TOTAL	POUGHKEEPSIE 125-126 MAT.	INST.	TOTAL	QUEENS 110 MAT.	INST.	TOTAL	RIVERHEAD 119 MAT.	INST.	TOTAL
2	SITE WORK	143.5	128.8	132.2	103.5	91.7	94.4	90.3	104.1	100.9	123.6	130.2	128.6	123.2	131.3	129.4	119.7	131.7	129.0
031	CONCRETE FORMWORK	110.5	171.7	162.9	79.1	114.8	109.7	85.3	82.5	82.9	81.7	140.0	131.6	90.4	148.7	140.3	95.0	146.9	139.4
032	CONCRETE REINFORCEMENT	100.4	185.1	148.1	101.4	102.0	101.7	102.8	86.9	93.9	99.0	120.8	111.3	101.0	167.4	138.4	101.1	165.1	137.2
033	CAST IN PLACE CONCRETE	123.8	161.6	139.6	105.5	116.6	110.1	96.9	100.7	98.5	103.6	112.6	107.4	100.9	144.6	119.2	95.7	144.9	116.3
3	CONCRETE	127.3	168.8	148.2	102.2	111.9	107.1	106.1	89.3	97.6	112.2	125.9	119.1	109.8	149.4	129.7	105.8	148.4	127.0
4	MASONRY	113.3	166.9	146.6	120.1	116.9	118.1	91.6	91.4	91.4	104.2	124.0	116.5	111.6	154.1	138.0	118.8	153.6	140.4
5	METALS	109.3	143.0	121.5	96.0	90.8	94.2	103.0	82.5	95.6	106.9	113.0	109.1	106.3	132.0	115.6	109.3	133.8	118.1
6	WOOD & PLASTICS	113.4	173.6	144.5	80.1	114.2	97.7	90.4	80.5	85.3	84.1	149.2	117.7	90.2	149.2	120.6	95.2	147.5	122.2
7	THERMAL & MOISTURE PROTECTION	108.7	158.7	131.9	105.1	105.1	105.1	99.7	89.3	94.9	109.3	122.0	115.2	107.3	141.2	123.0	107.2	138.6	121.8
8	DOORS & WINDOWS	100.5	171.5	117.7	96.8	104.6	98.7	103.1	79.9	97.5	97.3	139.2	107.5	91.9	151.2	106.2	91.9	150.3	106.1
092	LATH, PLASTER & GYPSUM BOARD	105.3	175.8	150.8	87.9	114.1	104.8	110.5	78.8	90.0	91.4	150.7	129.7	93.8	150.7	130.5	95.4	149.0	130.0
095	ACOUSTICAL TREATMENT & WOOD FLOORING	102.9	175.8	150.1	87.3	114.1	104.7	98.1	78.8	85.6	79.3	150.7	125.6	80.6	150.7	126.0	80.6	149.0	124.9
096	FLOORING & CARPET	92.0	161.0	108.8	84.2	112.3	91.0	95.4	86.6	93.2	85.1	138.8	98.1	90.8	147.1	104.5	92.1	111.4	96.8
099	PAINTING & WALL COVERINGS	95.3	151.3	128.2	75.6	99.3	89.5	97.2	78.1	86.0	99.2	96.8	97.8	101.9	133.7	120.6	101.9	137.4	122.8
9	FINISHES	108.0	167.6	138.5	88.7	113.2	101.3	96.7	82.2	89.3	102.0	136.9	119.9	104.4	146.9	126.2	104.8	138.7	122.2
10-14	TOTAL DIV. 10-14	100.0	151.0	110.8	100.0	101.4	100.3	100.1	74.1	94.6	100.1	114.9	103.2	100.0	128.3	106.0	100.0	127.7	105.9
15	MECHANICAL	100.2	164.2	128.8	96.8	101.5	98.9	96.2	77.5	87.8	97.0	102.8	99.5	100.1	147.3	121.1	99.9	143.5	119.3
16	ELECTRICAL	111.8	168.9	150.3	95.1	96.6	96.1	95.3	79.2	84.4	100.1	99.7	99.8	103.6	157.9	140.2	104.7	149.8	135.1
1-16	WEIGHTED AVERAGE	108.9	160.6	133.9	98.1	103.3	100.6	99.2	84.7	92.2	102.8	117.9	110.1	103.3	146.4	124.1	103.7	143.3	122.8

NEW YORK

DIVISION		ROCHESTER 144-146 MAT.	INST.	TOTAL	SCHENECTADY 123 MAT.	INST.	TOTAL	STATEN ISLAND 103 MAT.	INST.	TOTAL	SUFFERN 109 MAT.	INST.	TOTAL	SYRACUSE 130-132 MAT.	INST.	TOTAL	UTICA 133-135 MAT.	INST.	TOTAL
2	SITE WORK	78.6	108.6	101.7	74.8	108.9	101.0	149.2	127.3	132.3	138.1	126.6	129.2	97.7	108.8	106.2	72.8	108.0	99.8
031	CONCRETE FORMWORK	103.1	99.6	100.1	105.6	91.1	93.2	90.7	148.7	140.4	99.7	113.7	111.7	104.9	89.0	91.3	106.1	78.7	82.7
032	CONCRETE REINFORCEMENT	102.2	90.0	95.4	97.7	88.9	92.8	94.9	153.7	128.0	94.1	126.8	112.6	99.0	92.3	95.2	99.0	81.2	89.0
033	CAST IN PLACE CONCRETE	100.0	105.4	102.3	95.7	101.4	98.1	122.9	145.7	132.5	119.2	125.0	121.6	95.7	97.5	96.5	88.4	92.8	90.3
3	CONCRETE	110.2	100.0	105.5	107.6	95.6	101.5	129.1	147.4	138.3	122.3	119.7	121.0	105.7	93.9	99.7	104.2	85.7	94.9
4	MASONRY	102.6	101.3	101.8	93.7	97.9	96.3	115.6	154.2	139.6	107.3	132.7	123.1	99.4	94.8	96.6	92.4	88.3	89.9
5	METALS	103.2	110.6	105.9	99.9	109.8	103.5	96.5	131.7	109.2	96.5	114.5	103.0	99.8	110.6	103.7	98.0	106.0	100.9
6	WOOD & PLASTICS	99.9	100.1	100.0	108.6	89.1	98.6	91.8	149.4	121.5	106.7	109.5	108.2	108.6	87.0	97.5	108.6	76.4	92.0
7	THERMAL & MOISTURE PROTECTION	101.8	101.3	101.6	92.9	95.6	94.2	108.6	141.7	123.9	109.7	127.4	117.9	100.8	95.8	98.5	92.8	92.6	92.7
8	DOORS & WINDOWS	99.8	93.7	98.3	99.7	84.2	95.9	96.9	151.3	110.1	96.9	120.2	102.6	97.4	83.6	94.1	99.7	72.7	93.1
092	LATH, PLASTER & GYPSUM BOARD	97.8	99.8	99.1	100.4	88.4	92.7	93.3	150.7	130.3	98.2	109.6	105.6	100.4	86.3	91.3	100.4	75.3	84.2
095	ACOUSTICAL TREATMENT & WOOD FLOORING	95.6	99.8	98.3	97.1	88.4	91.5	85.1	150.7	127.6	83.7	109.6	100.5	97.1	86.3	90.1	97.1	75.3	83.0
096	FLOORING & CARPET	79.8	106.1	86.2	81.6	88.8	83.4	87.1	147.1	101.7	86.2	77.5	84.1	83.5	87.5	84.4	81.4	82.9	81.8
099	PAINTING & WALL COVERINGS	72.8	104.6	91.5	78.9	83.4	81.5	95.3	133.7	117.8	92.6	96.8	95.1	84.1	89.0	87.0	78.9	83.9	81.8
9	FINISHES	94.0	101.8	98.0	94.3	89.4	91.8	101.9	147.1	125.1	101.1	103.4	102.3	95.8	88.2	91.9	94.2	79.5	86.7
10-14	TOTAL DIV. 10-14	100.0	96.3	99.2	100.0	95.0	98.9	100.0	128.9	106.1	100.0	120.2	104.3	100.0	93.2	98.6	100.0	90.8	98.1
15	MECHANICAL	100.0	93.6	97.2	100.3	93.1	97.1	100.3	143.1	119.4	97.1	119.0	106.9	100.3	93.4	97.2	100.3	84.3	93.2
16	ELECTRICAL	107.1	98.6	101.4	103.6	97.4	99.4	98.5	157.9	138.5	109.3	117.4	114.8	103.6	92.0	95.8	103.6	82.1	89.1
1-16	WEIGHTED AVERAGE	101.2	100.5	100.9	99.4	97.0	98.2	105.1	145.0	124.4	103.8	118.8	111.0	100.3	95.5	97.9	98.6	88.1	93.5

NEW YORK / NORTH CAROLINA

DIVISION		WATERTOWN 136 MAT.	INST.	TOTAL	WHITE PLAINS 106 MAT.	INST.	TOTAL	YONKERS 107 MAT.	INST.	TOTAL	ASHEVILLE 287-288 MAT.	INST.	TOTAL	CHARLOTTE 281-282 MAT.	INST.	TOTAL	DURHAM 277 MAT.	INST.	TOTAL
2	SITE WORK	81.6	110.0	103.4	132.2	127.5	128.6	143.2	127.1	130.8	103.4	73.9	80.7	103.8	73.9	80.8	103.5	84.2	88.6
031	CONCRETE FORMWORK	84.8	85.8	85.6	111.3	132.2	129.2	110.3	142.4	137.8	93.2	48.1	54.6	102.1	48.2	55.9	95.4	48.3	55.0
032	CONCRETE REINFORCEMENT	99.5	71.9	84.0	94.1	151.3	126.4	97.8	109.9	104.6	93.8	45.8	66.8	94.2	45.9	67.0	94.2	52.2	70.6
033	CAST IN PLACE CONCRETE	103.0	97.8	100.8	109.1	134.3	119.6	122.1	134.6	127.3	101.0	55.2	81.8	103.5	55.2	83.3	101.2	55.3	82.0
3	CONCRETE	117.6	88.8	103.1	114.4	136.0	125.3	126.4	133.9	130.1	99.4	52.4	75.7	100.8	52.5	76.4	99.1	53.7	76.2
4	MASONRY	93.5	95.2	94.5	106.5	135.6	124.6	112.9	135.9	127.2	79.4	41.8	56.0	82.9	41.8	57.4	81.5	41.8	56.8
5	METALS	98.0	102.3	99.6	99.4	131.4	110.9	106.4	137.6	117.7	93.7	83.7	90.1	95.0	83.8	91.0	95.1	86.5	92.0
6	WOOD & PLASTICS	85.0	84.4	84.7	115.6	130.5	123.3	114.0	143.6	129.3	95.0	48.9	71.2	106.0	48.9	76.6	97.6	48.9	72.5
7	THERMAL & MOISTURE PROTECTION	93.1	92.2	92.7	109.6	131.7	119.9	110.0	140.7	124.2	95.5	48.5	73.7	95.5	49.7	74.3	96.1	49.2	74.3
8	DOORS & WINDOWS	99.7	75.5	93.8	96.9	141.1	107.7	100.6	155.9	114.0	92.9	47.6	81.9	96.7	47.6	84.8	96.7	49.9	85.4
092	LATH, PLASTER & GYPSUM BOARD	92.5	83.5	86.7	101.4	131.3	120.7	105.0	144.8	130.7	99.7	47.1	65.7	106.4	47.1	68.1	106.4	47.1	68.1
095	ACOUSTICAL TREATMENT & WOOD FLOORING	97.1	83.5	88.3	83.7	131.3	114.5	101.5	144.8	129.5	87.3	47.1	61.2	91.4	47.1	62.7	91.4	47.1	62.7
096	FLOORING & CARPET	74.3	82.9	76.4	88.2	147.1	102.5	87.8	150.1	103.0	89.9	51.0	80.4	93.1	51.0	82.8	93.3	51.0	83.0
099	PAINTING & WALL COVERINGS	78.9	75.1	76.7	92.6	121.0	109.3	92.6	121.0	109.3	105.8	47.8	71.8	105.8	47.8	71.8	105.8	47.8	71.8
9	FINISHES	91.5	83.7	87.5	101.6	133.5	118.0	106.1	141.9	124.4	91.2	48.4	69.2	93.8	48.4	70.6	94.0	48.4	70.6
10-14	TOTAL DIV. 10-14	100.0	91.1	98.1	100.0	124.3	105.1	100.0	142.0	108.9	100.0	71.7	94.0	100.0	71.7	94.0	100.0	71.7	94.0
15	MECHANICAL	100.3	87.3	94.5	100.6	123.2	110.7	100.6	125.4	111.6	100.1	53.3	79.3	100.1	53.4	79.3	100.1	53.5	79.3
16	ELECTRICAL	103.6	79.3	87.2	96.1	131.5	119.9	109.3	143.2	132.2	100.7	51.5	67.5	100.2	46.2	63.8	98.6	49.3	65.3
1-16	WEIGHTED AVERAGE	100.0	89.9	95.1	103.0	130.9	116.5	108.0	136.0	121.5	96.1	56.1	76.8	97.4	55.3	77.1	97.0	57.4	77.9

NORTH CAROLINA

DIVISION		ELIZABETH CITY 279			FAYETTEVILLE 283			GASTONIA 280			GREENSBORO 270,272-274			HICKORY 286			KINSTON 285		
		MAT.	INST.	TOTAL	MAT.	INST.	TOTAL	MAT.	INST.	TOTAL	MAT.	INST.	TOTAL	MAT.	INST.	TOTAL	MAT.	INST.	TOTAL
2	SITE WORK	108.1	85.1	90.4	101.6	84.2	88.2	103.2	73.9	80.7	103.4	84.2	88.6	102.9	83.0	87.6	101.8	83.8	87.9
031	CONCRETE FORMWORK	72.9	31.0	37.0	90.3	48.3	54.3	102.3	48.1	55.9	95.5	48.3	55.0	83.1	30.0	37.6	78.9	30.2	37.2
032	CONCRETE REINFORCEMENT	93.3	55.3	71.9	93.3	52.2	70.2	94.2	45.8	67.0	94.2	52.2	70.6	93.8	24.9	55.0	93.3	25.0	54.8
033	CAST IN PLACE CONCRETE	101.3	39.1	75.3	97.7	55.3	79.9	98.7	55.2	80.5	100.3	55.3	81.5	100.9	42.1	76.3	97.4	43.3	74.8
3	CONCRETE	98.7	41.1	69.7	96.5	53.7	75.0	98.4	52.4	75.2	98.7	53.7	76.0	98.6	36.2	67.1	95.5	36.8	65.9
4	MASONRY	94.5	33.7	56.7	85.5	41.8	58.4	82.5	41.8	57.2	81.3	41.7	56.7	69.0	33.5	46.9	75.3	33.9	49.6
5	METALS	94.3	85.9	91.3	94.0	86.5	91.3	95.8	83.7	91.5	95.8	86.5	92.5	92.9	76.2	86.9	93.5	78.0	87.9
6	WOOD & PLASTICS	76.7	31.5	53.4	91.4	48.9	69.5	106.0	48.9	76.6	97.6	48.9	72.5	87.4	30.6	58.1	82.4	30.6	55.7
7	THERMAL & MOISTURE PROTECTION	95.3	31.8	65.9	95.7	49.2	74.1	95.8	49.7	74.4	96.1	49.2	74.3	95.8	35.0	67.6	95.5	35.7	67.8
8	DOORS & WINDOWS	93.8	35.3	79.6	93.0	49.9	82.5	96.7	47.6	84.8	96.7	49.9	85.4	92.9	30.3	77.7	93.0	30.8	77.9
092	LATH, PLASTER & GYPSUM BOARD	95.8	28.1	52.2	100.0	47.1	65.8	106.4	47.1	68.1	106.4	47.1	68.1	99.7	28.1	53.5	97.3	28.1	52.7
095	ACOUSTICAL TREATMENT & WOOD FLOORING	91.4	28.1	50.4	88.7	47.1	61.7	91.4	47.1	62.7	91.4	47.1	62.7	87.3	28.1	49.0	88.7	28.1	49.5
096	FLOORING & CARPET	84.0	20.0	68.4	90.0	51.0	80.5	93.3	35.0	79.1	93.3	51.0	83.0	89.8	18.2	72.4	86.7	27.3	72.2
099	PAINTING & WALL COVERINGS	105.8	16.2	53.3	105.8	47.8	71.8	105.8	47.8	71.8	105.8	47.8	71.8	105.8	19.1	54.9	105.8	21.0	56.1
9	FINISHES	89.7	26.6	57.4	91.5	48.4	69.4	93.9	45.6	69.2	94.0	48.4	70.6	91.3	26.0	57.8	90.1	28.4	58.4
10-14	TOTAL DIV. 10-14	100.0	70.3	93.7	100.0	71.7	94.0	100.0	71.7	94.0	100.0	71.7	94.0	100.0	66.5	92.9	100.0	66.5	92.9
15	MECHANICAL	96.7	34.5	69.0	100.1	53.5	79.3	100.1	54.0	79.3	100.1	53.5	79.3	96.8	23.7	64.2	96.8	24.7	64.7
16	ELECTRICAL	101.5	38.0	58.6	93.8	49.3	63.8	99.7	46.2	63.6	99.7	49.3	65.7	96.5	37.9	57.0	96.2	38.0	56.9
1-16	WEIGHTED AVERAGE	96.1	45.6	71.7	95.6	57.4	77.1	97.2	55.0	76.8	97.1	57.3	77.9	94.2	41.4	68.7	94.0	42.3	69.0

NORTH CAROLINA / NORTH DAKOTA

DIVISION		MURPHY 289			RALEIGH 275-276			ROCKY MOUNT 278			WILMINGTON 284			WINSTON-SALEM 271			BISMARCK 585		
		MAT.	INST.	TOTAL	MAT.	INST.	TOTAL	MAT.	INST.	TOTAL	MAT.	INST.	TOTAL	MAT.	INST.	TOTAL	MAT.	INST.	TOTAL
2	SITE WORK	105.1	72.7	80.2	104.6	84.2	88.9	106.4	83.0	88.4	104.5	73.9	81.0	103.8	84.2	88.7	82.2	95.3	92.3
031	CONCRETE FORMWORK	99.2	30.1	40.0	98.7	48.2	55.5	82.9	35.9	42.6	95.0	48.3	55.0	97.0	48.2	55.2	93.6	60.5	65.3
032	CONCRETE REINFORCEMENT	93.4	24.8	54.7	94.2	52.2	70.6	93.3	24.9	54.8	94.5	52.2	70.7	94.2	45.9	67.0	105.0	72.8	86.8
033	CAST IN PLACE CONCRETE	104.4	37.3	76.3	107.1	55.3	85.4	99.1	36.3	72.8	100.5	55.3	81.6	103.6	55.3	83.3	104.6	64.7	87.9
3	CONCRETE	102.5	34.5	68.2	102.3	53.7	77.8	99.9	36.7	68.0	99.2	53.7	76.3	100.4	52.5	76.2	96.7	65.0	80.7
4	MASONRY	71.0	30.4	45.8	88.8	41.8	59.6	75.5	35.9	50.9	68.9	41.8	52.1	81.5	41.8	56.8	105.9	61.3	78.2
5	METALS	92.7	75.8	86.6	95.1	86.5	92.0	93.8	76.0	87.4	94.1	86.5	91.3	95.0	83.8	91.0	93.5	76.0	87.2
6	WOOD & PLASTICS	107.0	30.6	67.6	101.6	48.9	74.4	86.2	37.4	61.0	97.0	48.9	72.2	97.6	48.9	72.5	86.8	58.2	72.0
7	THERMAL & MOISTURE PROTECTION	95.8	37.4	68.7	95.9	48.6	74.0	96.0	34.0	67.2	95.5	49.2	74.0	96.1	49.2	74.3	97.9	61.5	81.0
8	DOORS & WINDOWS	92.8	35.1	78.8	93.7	49.9	83.1	93.0	33.0	78.5	93.0	49.9	82.6	96.7	48.1	84.9	104.7	55.4	92.7
092	LATH, PLASTER & GYPSUM BOARD	105.5	28.1	55.6	109.9	47.1	69.4	98.4	35.2	57.6	100.5	47.1	66.0	106.4	47.1	68.1	115.5	57.4	78.0
095	ACOUSTICAL TREATMENT & WOOD FLOORING	87.3	28.1	49.0	91.4	47.1	62.7	88.7	35.2	54.0	88.7	47.1	61.7	91.4	47.1	62.7	141.3	57.4	86.9
096	FLOORING & CARPET	93.7	29.1	78.0	93.3	51.0	83.0	88.1	19.3	71.3	90.5	51.0	80.9	93.3	51.0	83.0	114.5	59.0	101.0
099	PAINTING & WALL COVERINGS	105.8	21.7	56.5	105.8	47.8	71.8	105.8	36.3	65.0	105.8	47.8	71.8	105.8	47.8	71.8	106.3	49.1	72.7
9	FINISHES	93.3	28.2	59.9	94.5	48.4	70.9	91.0	33.3	61.4	91.7	48.4	69.5	94.0	48.4	70.6	115.5	58.7	86.4
10-14	TOTAL DIV. 10-14	100.0	66.6	93.0	100.0	71.7	94.0	100.0	68.4	93.3	100.0	71.7	94.0	100.0	71.7	94.0	100.0	64.5	92.5
15	MECHANICAL	96.8	23.8	64.3	100.1	53.5	79.3	96.8	23.8	64.3	100.1	53.5	79.3	100.1	53.4	79.3	100.5	67.9	85.9
16	ELECTRICAL	102.3	37.9	58.9	99.3	49.3	65.6	105.0	37.9	59.8	100.5	49.3	65.9	99.7	49.3	65.7	94.3	65.5	74.9
1-16	WEIGHTED AVERAGE	95.6	40.4	69.0	97.6	57.3	78.2	95.5	42.8	70.1	95.7	56.3	76.7	97.2	56.9	77.8	100.1	68.0	84.6

NORTH DAKOTA

DIVISION		DEVILS LAKE 583			DICKINSON 586			FARGO 580-581			GRAND FORKS 582			JAMESTOWN 584			MINOT 587		
		MAT.	INST.	TOTAL	MAT.	INST.	TOTAL	MAT.	INST.	TOTAL	MAT.	INST.	TOTAL	MAT.	INST.	TOTAL	MAT.	INST.	TOTAL
2	SITE WORK	87.0	95.1	93.2	93.6	95.1	94.8	82.2	95.3	92.3	89.8	95.1	93.9	86.1	95.1	93.0	87.6	95.3	93.5
031	CONCRETE FORMWORK	99.8	55.3	61.7	81.4	55.3	59.0	94.3	61.0	65.8	90.0	55.3	60.3	83.4	55.3	59.4	84.3	60.4	63.8
032	CONCRETE REINFORCEMENT	105.0	65.9	82.9	106.1	65.9	83.4	96.8	72.7	83.2	103.4	65.9	82.2	105.6	69.1	85.0	107.0	72.8	87.7
033	CAST IN PLACE CONCRETE	115.6	65.4	94.6	104.6	65.4	88.2	107.1	67.5	90.5	104.6	65.4	88.2	114.2	65.4	93.8	104.6	62.9	87.1
3	CONCRETE	105.2	61.7	83.2	103.6	61.7	82.5	96.9	66.2	81.4	101.1	61.7	81.2	103.0	62.3	82.5	99.7	64.3	81.8
4	MASONRY	111.3	56.8	77.5	112.1	56.8	77.7	106.0	60.8	77.9	105.9	56.8	75.4	121.3	56.9	81.3	105.4	61.4	78.1
5	METALS	93.6	73.3	86.3	93.5	73.3	86.2	93.4	75.8	87.1	93.5	73.3	86.2	93.5	74.5	86.7	93.8	75.9	87.4
6	WOOD & PLASTICS	97.1	53.5	74.6	77.1	53.5	64.9	86.8	59.1	72.5	82.7	53.5	67.6	79.3	53.5	66.0	76.7	58.2	67.2
7	THERMAL & MOISTURE PROTECTION	98.5	59.4	80.4	99.0	59.4	80.6	98.4	61.5	81.3	98.6	59.4	80.4	98.2	59.5	80.2	98.3	61.4	81.2
8	DOORS & WINDOWS	104.7	51.2	91.7	104.6	51.2	91.7	104.6	55.8	92.8	104.7	51.2	91.7	104.7	51.9	91.9	104.8	55.4	92.8
092	LATH, PLASTER & GYPSUM BOARD	119.7	52.6	76.4	111.3	52.6	73.4	115.5	58.3	78.6	113.4	52.6	74.1	112.3	52.6	73.8	111.3	57.4	76.5
095	ACOUSTICAL TREATMENT & WOOD FLOORING	141.3	52.6	83.8	141.3	52.6	83.8	141.3	58.3	87.5	141.3	52.6	83.8	141.3	52.6	83.8	141.3	57.4	86.9
096	FLOORING & CARPET	118.7	58.1	104.0	110.4	58.1	97.7	114.3	58.1	100.6	112.8	58.1	99.5	111.4	58.1	98.4	110.2	59.2	97.8
099	PAINTING & WALL COVERINGS	106.3	37.5	65.9	106.3	37.5	65.9	106.3	45.1	70.4	106.3	37.5	65.9	106.3	37.5	65.9	106.3	41.1	68.0
9	FINISHES	117.8	53.4	84.8	114.5	53.4	83.2	115.4	58.5	86.3	115.2	53.4	83.5	114.3	53.4	83.1	113.9	57.8	85.1
10-14	TOTAL DIV. 10-14	100.0	62.3	92.0	100.0	62.3	92.0	100.0	64.6	92.5	100.0	62.3	92.0	100.0	62.3	92.0	100.0	64.5	92.5
15	MECHANICAL	97.2	59.0	80.2	97.2	59.0	80.2	100.5	61.2	82.0	100.5	59.0	82.0	97.2	59.0	80.2	100.5	61.1	82.9
16	ELECTRICAL	93.6	62.1	72.3	107.0	62.1	76.7	94.2	55.2	67.9	98.9	62.1	74.1	93.6	62.1	72.3	103.9	65.5	78.0
1-16	WEIGHTED AVERAGE	101.1	63.7	83.0	101.6	63.7	83.3	100.1	65.1	83.2	101.1	63.7	83.0	100.8	63.9	83.0	101.1	66.5	84.4

COST INDEXES

NORTH DAKOTA / OHIO

		WILLISTON 588			AKRON 442 - 443			ATHENS 457			CANTON 446 - 447			CHILLICOTHE 456			CINCINNATI 451 - 452		
	DIVISION	MAT.	INST.	TOTAL	MAT.	INST.	TOTAL	MAT.	INST.	TOTAL	MAT.	INST.	TOTAL	MAT.	INST.	TOTAL	MAT.	INST.	TOTAL
2	SITE WORK	88.4	95.1	93.5	104.2	108.7	107.7	93.3	96.1	95.5	104.4	107.7	106.9	82.3	112.2	105.3	78.4	111.4	103.8
031	CONCRETE FORMWORK	88.7	55.3	60.1	102.0	100.4	100.6	86.8	74.7	76.4	102.0	89.0	90.9	89.6	94.9	94.2	96.8	84.9	86.6
032	CONCRETE REINFORCEMENT	108.0	65.9	84.3	96.0	96.6	96.4	92.8	73.2	81.7	96.0	81.3	87.7	90.5	83.2	86.4	95.9	82.3	88.3
033	CAST IN PLACE CONCRETE	104.6	65.4	88.2	101.1	108.8	104.3	106.7	86.6	98.3	102.0	105.9	103.6	96.8	98.7	97.6	89.2	88.6	88.9
3	CONCRETE	101.0	61.7	81.2	97.0	101.4	99.2	100.3	78.2	89.2	97.4	92.6	95.0	94.9	93.9	94.4	90.1	85.9	88.0
4	MASONRY	100.9	56.8	73.5	94.9	101.1	98.7	76.1	78.4	77.5	95.8	92.2	93.6	85.3	97.8	93.1	84.8	91.0	88.6
5	METALS	93.8	73.3	86.4	91.0	81.7	87.6	102.8	70.0	91.0	91.0	75.1	85.2	92.4	86.7	90.3	94.9	87.1	92.1
6	WOOD & PLASTICS	85.1	53.5	68.8	97.4	99.5	98.5	83.9	73.5	78.5	97.4	88.2	92.6	94.4	93.3	93.9	97.3	82.6	89.7
7	THERMAL & MOISTURE PROTECTION	98.6	59.4	80.4	108.8	100.9	105.1	108.8	81.6	96.2	109.3	96.7	103.5	96.1	94.2	95.2	93.5	92.2	92.9
8	DOORS & WINDOWS	104.7	51.2	91.8	107.1	98.7	105.0	96.9	70.5	90.5	101.4	82.0	96.7	92.1	86.0	90.6	97.8	80.6	93.6
092	LATH, PLASTER & GYPSUM BOARD	115.0	52.6	74.7	103.1	98.7	100.2	103.9	72.2	83.5	103.1	87.0	92.7	102.4	93.6	96.7	103.8	82.5	90.0
095	ACOUSTICAL TREATMENT & WOOD FLOORING	141.3	52.6	83.8	97.3	98.7	98.2	102.0	72.2	82.7	97.3	87.0	90.6	93.1	93.6	93.4	94.5	82.5	86.7
096	FLOORING & CARPET	113.9	58.1	100.3	113.4	108.5	112.2	104.1	87.5	100.1	113.6	86.5	107.0	84.9	90.1	86.2	85.7	96.1	88.2
099	PAINTING & WALL COVERINGS	106.3	37.5	65.9	103.4	117.7	111.8	104.5	64.7	81.1	103.4	90.4	95.8	100.4	95.0	97.2	100.4	94.6	97.0
9	FINISHES	115.7	53.4	83.8	104.7	103.9	104.3	94.7	75.7	84.9	104.8	88.2	96.3	91.1	94.2	92.7	91.6	87.8	89.6
10 - 14	TOTAL DIV. 10 - 14	100.0	62.3	92.0	100.0	97.8	99.5	100.0	79.3	95.6	100.0	94.4	98.8	100.0	92.8	98.5	100.0	87.0	97.3
15	MECHANICAL	97.2	59.0	80.2	100.0	102.6	101.2	96.6	75.5	87.2	100.0	88.2	94.8	97.4	97.0	97.2	99.9	91.8	96.3
16	ELECTRICAL	99.6	62.1	74.3	98.7	98.2	98.4	99.8	72.9	81.7	97.2	95.9	96.3	97.1	87.5	90.6	94.9	83.9	87.5
1 - 16	WEIGHTED AVERAGE	100.3	63.7	82.6	99.6	100.1	99.8	97.4	77.3	87.7	98.9	91.3	95.3	94.1	94.7	94.4	94.8	90.0	92.5

OHIO

		CLEVELAND 441			COLUMBUS 430 - 432			DAYTON 453 - 454			HAMILTON 450			LIMA 458			LORAIN 440		
	DIVISION	MAT.	INST.	TOTAL	MAT.	INST.	TOTAL	MAT.	INST.	TOTAL	MAT.	INST.	TOTAL	MAT.	INST.	TOTAL	MAT.	INST.	TOTAL
2	SITE WORK	104.0	108.0	107.1	79.8	99.2	94.7	78.0	111.1	103.4	78.3	111.3	103.7	87.8	96.6	94.6	103.6	107.8	106.9
031	CONCRETE FORMWORK	102.1	107.1	106.4	95.9	88.8	89.8	96.7	86.7	88.1	96.7	84.7	86.4	86.8	88.5	88.3	102.1	89.8	91.6
032	CONCRETE REINFORCEMENT	96.4	97.0	96.7	105.6	83.4	93.1	95.9	80.2	87.0	95.9	82.3	88.3	92.8	80.1	85.6	96.0	96.2	96.1
033	CAST IN PLACE CONCRETE	99.1	116.7	106.5	95.8	95.8	95.8	81.3	91.2	85.5	88.9	88.3	88.7	98.1	98.7	98.3	96.2	100.1	97.8
3	CONCRETE	96.1	107.3	101.7	94.8	89.7	92.3	86.2	86.5	86.3	89.9	85.7	87.8	93.6	90.1	91.8	94.6	93.7	94.1
4	MASONRY	99.8	111.3	106.9	88.5	92.3	90.9	84.2	90.7	88.2	84.7	90.6	88.4	103.7	87.7	93.8	91.6	91.2	91.4
5	METALS	92.3	84.4	89.5	96.2	80.0	90.3	94.2	76.1	87.7	94.2	87.0	91.6	102.9	79.5	94.4	91.5	80.8	87.7
6	WOOD & PLASTICS	96.7	104.8	100.9	95.3	88.0	91.6	99.2	86.3	92.6	97.3	82.6	89.7	83.8	88.1	86.0	97.4	86.7	91.9
7	THERMAL & MOISTURE PROTECTION	107.5	112.8	109.9	104.0	96.4	100.5	98.2	90.1	94.4	96.2	92.3	94.4	110.9	95.2	103.6	109.2	100.3	105.1
8	DOORS & WINDOWS	95.6	102.6	97.3	99.9	83.7	96.0	98.2	80.9	94.0	97.8	80.6	93.6	96.9	76.7	92.0	101.4	92.8	99.3
092	LATH, PLASTER & GYPSUM BOARD	102.2	104.1	103.4	103.0	87.3	92.9	103.8	86.3	92.5	103.8	82.5	90.0	103.9	87.3	93.2	103.1	85.4	91.7
095	ACOUSTICAL TREATMENT & WOOD FLOORING	95.9	104.1	101.2	97.4	87.3	90.9	94.5	86.3	89.2	94.5	82.5	86.7	100.5	87.3	92.0	97.3	85.4	89.6
096	FLOORING & CARPET	113.2	116.5	114.0	96.8	93.5	96.0	88.1	88.6	88.2	85.7	96.0	88.2	103.4	93.2	101.0	113.6	94.0	108.9
099	PAINTING & WALL COVERINGS	103.4	117.7	111.8	98.6	96.2	97.2	100.4	90.0	94.3	100.4	90.7	94.7	104.5	83.4	92.1	103.4	117.7	111.8
9	FINISHES	104.3	110.0	107.2	95.6	90.3	92.9	92.4	87.4	89.8	91.6	87.2	89.3	93.8	88.9	91.3	104.8	93.0	98.7
10 - 14	TOTAL DIV. 10 - 14	100.0	104.9	101.0	100.0	91.3	98.2	100.0	86.7	97.2	100.0	86.8	97.2	100.0	89.3	97.7	100.0	99.4	99.9
15	MECHANICAL	100.0	106.5	102.9	100.0	93.6	97.2	100.7	87.0	94.6	100.6	91.6	96.6	96.6	86.6	92.2	100.0	92.7	96.8
16	ELECTRICAL	97.8	106.9	103.9	103.5	88.0	93.1	93.5	88.9	90.4	93.6	86.0	88.5	100.3	89.3	92.9	97.2	96.7	96.9
1 - 16	WEIGHTED AVERAGE	98.4	105.6	101.9	97.5	90.5	94.1	94.6	88.9	91.8	94.9	90.2	92.6	97.9	88.2	93.2	98.4	94.1	96.3

OHIO

		MANSFIELD 448 - 449			MARION 433			SPRINGFIELD 455			STEUBENVILLE 439			TOLEDO 434 - 436			YOUNGSTOWN 444 - 445		
	DIVISION	MAT.	INST.	TOTAL	MAT.	INST.	TOTAL	MAT.	INST.	TOTAL	MAT.	INST.	TOTAL	MAT.	INST.	TOTAL	MAT.	INST.	TOTAL
2	SITE WORK	99.5	107.6	105.7	76.4	99.8	94.4	78.5	110.8	103.4	110.5	108.5	109.0	79.6	100.1	95.3	104.1	108.8	107.7
031	CONCRETE FORMWORK	83.7	88.6	87.9	87.4	85.9	86.1	96.7	87.1	88.4	87.2	93.1	92.3	95.9	99.6	99.1	102.0	91.5	93.0
032	CONCRETE REINFORCEMENT	87.4	81.2	83.9	97.3	80.8	88.0	95.9	80.1	87.0	95.0	90.0	92.2	105.6	87.9	95.6	96.0	90.5	92.9
033	CAST IN PLACE CONCRETE	93.6	98.4	95.6	87.5	93.8	90.1	85.2	90.8	87.6	95.1	101.5	97.7	95.8	106.0	100.1	100.1	107.7	103.3
3	CONCRETE	88.9	89.7	89.3	86.7	87.2	87.0	88.1	86.5	87.3	90.7	94.4	92.6	94.8	99.0	97.0	96.5	96.0	96.2
4	MASONRY	94.2	94.5	94.4	92.0	89.3	90.3	84.5	90.0	87.9	79.7	96.1	89.9	98.2	98.5	98.4	95.0	97.9	96.8
5	METALS	91.5	74.5	85.4	96.9	76.8	89.7	94.2	75.9	87.6	91.7	78.5	87.0	96.0	86.1	92.4	91.0	79.6	86.9
6	WOOD & PLASTICS	82.6	86.7	84.7	90.7	84.4	87.4	99.2	87.3	93.1	86.7	91.2	89.0	95.3	100.9	98.2	97.4	89.2	93.2
7	THERMAL & MOISTURE PROTECTION	108.4	96.3	102.8	103.3	92.7	98.4	98.2	90.0	94.4	116.5	96.6	107.3	106.0	101.3	103.8	109.4	98.9	104.6
8	DOORS & WINDOWS	100.7	77.2	95.0	96.6	74.2	91.2	98.2	78.8	93.5	97.3	93.5	96.4	99.9	94.5	98.6	101.4	89.6	98.5
092	LATH, PLASTER & GYPSUM BOARD	98.4	85.4	90.0	102.1	83.5	90.1	103.8	87.3	93.2	103.7	89.8	94.7	103.0	100.6	101.4	103.1	88.0	93.3
095	ACOUSTICAL TREATMENT & WOOD FLOORING	100.0	85.4	90.6	100.2	83.5	89.4	94.5	87.3	89.9	92.7	89.8	90.8	95.9	100.6	98.9	97.3	88.0	91.3
096	FLOORING & CARPET	107.9	96.4	105.1	95.0	96.4	95.4	88.1	88.6	88.2	124.0	100.2	118.2	95.9	78.3	91.6	113.6	99.1	110.1
099	PAINTING & WALL COVERINGS	103.4	73.7	86.0	98.6	73.7	84.0	100.4	90.0	94.3	114.9	90.9	100.8	98.6	104.4	102.0	103.4	101.1	102.1
9	FINISHES	102.5	88.3	95.2	95.1	86.4	90.7	92.4	87.8	90.1	112.2	93.9	102.8	95.0	96.1	95.6	104.8	93.5	99.0
10 - 14	TOTAL DIV. 10 - 14	100.0	95.1	99.0	100.0	89.0	97.7	100.0	86.6	97.2	100.0	96.7	99.3	100.0	100.0	100.0	100.0	97.9	99.6
15	MECHANICAL	96.8	91.9	94.6	96.8	92.4	94.8	100.7	86.6	94.4	97.2	98.1	97.6	100.0	99.8	99.9	100.0	90.6	95.8
16	ELECTRICAL	93.3	89.2	90.5	94.8	86.7	89.4	93.5	84.7	87.5	86.2	100.8	96.1	104.1	100.9	102.0	97.2	88.3	91.2
1 - 16	WEIGHTED AVERAGE	96.3	90.5	93.5	94.8	88.2	91.6	94.8	88.0	91.5	96.5	96.3	96.4	98.0	97.9	98.0	98.8	93.1	96.1

DIVISION		OHIO			OKLAHOMA														
		ZANESVILLE			ARDMORE			CLINTON			DURANT			ENID			GUYMON		
		437 - 438			734			736			747			737			739		
		MAT.	INST.	TOTAL	MAT.	INST.	TOTAL	MAT.	INST.	TOTAL	MAT.	INST.	TOTAL	MAT.	INST.	TOTAL	MAT.	INST.	TOTAL
2	SITE WORK	79.0	99.7	94.9	106.7	88.5	92.7	108.0	86.9	91.8	100.6	87.0	90.2	109.4	87.3	92.4	113.5	86.3	92.6
031	CONCRETE FORMWORK	83.4	86.7	86.2	87.8	58.1	62.3	86.2	49.6	54.9	76.7	58.0	60.7	94.3	48.8	55.3	96.1	30.5	39.9
032	CONCRETE REINFORCEMENT	96.8	86.5	91.0	96.5	79.2	86.8	97.1	79.2	87.0	97.4	71.3	82.7	96.4	84.2	89.5	97.1	42.5	66.3
033	CAST IN PLACE CONCRETE	92.2	93.0	92.6	97.0	56.3	80.0	93.8	56.3	78.1	90.7	55.9	76.1	93.8	60.1	79.7	93.8	36.7	69.9
3	CONCRETE	90.1	88.3	89.2	89.7	61.2	75.3	89.1	57.5	73.2	84.5	59.5	73.1	90.0	59.4	74.5	92.1	35.3	63.5
4	MASONRY	87.1	87.3	87.2	91.6	58.6	71.1	116.1	58.6	80.4	90.1	59.5	71.1	97.7	59.5	74.0	95.4	30.0	54.8
5	METALS	96.9	78.8	90.4	92.9	64.8	82.8	93.0	64.8	82.8	92.9	61.7	81.7	94.1	67.0	84.3	93.4	44.9	75.9
6	WOOD & PLASTICS	86.5	86.0	86.3	91.4	59.2	74.8	89.9	47.8	68.1	79.5	59.2	69.0	95.0	46.4	69.9	100.8	30.2	64.4
7	THERMAL & MOISTURE PROTECTION	103.4	91.3	97.8	96.2	64.1	81.3	96.3	62.9	80.8	95.7	63.4	80.7	96.4	63.3	81.1	96.8	36.5	68.8
8	DOORS & WINDOWS	96.6	82.9	93.2	96.2	65.2	88.7	96.2	59.0	87.2	96.2	62.9	88.1	96.2	59.3	87.3	96.3	33.0	81.0
092	LATH, PLASTER & GYPSUM BOARD	100.5	85.3	90.7	90.8	59.1	70.4	89.8	47.3	62.3	86.6	59.1	68.9	91.3	45.8	62.0	93.2	29.1	51.9
095	ACOUSTICAL TREATMENT & WOOD FLOORING	100.2	85.3	90.5	93.5	59.1	71.2	93.5	47.3	63.5	93.5	59.1	71.2	93.5	45.8	62.6	94.8	29.1	52.2
096	FLOORING & CARPET	93.2	92.2	93.0	115.7	69.1	104.3	114.0	65.9	102.3	109.5	70.1	100.0	116.3	65.9	104.0	118.8	39.0	99.4
099	PAINTING & WALL COVERINGS	98.6	62.4	77.4	97.1	44.3	66.1	97.1	65.2	78.4	97.1	65.2	78.4	97.1	65.2	78.4	97.1	24.5	54.5
9	FINISHES	94.5	84.9	89.6	100.2	58.1	78.6	99.6	53.2	75.8	97.3	60.7	78.6	100.7	52.3	75.9	102.3	31.4	65.9
10 - 14	TOTAL DIV. 10 - 14	100.0	90.9	98.1	100.0	67.4	93.1	100.0	65.9	92.8	100.0	67.4	93.1	100.0	68.7	93.4	100.0	62.9	92.2
15	MECHANICAL	96.8	86.4	92.1	96.7	64.7	82.4	96.5	64.7	82.3	96.5	64.6	82.3	100.0	64.8	84.3	96.5	35.2	69.2
16	ELECTRICAL	93.7	82.0	85.8	93.7	66.2	75.2	95.4	66.2	75.7	98.1	65.6	76.2	95.4	66.2	75.7	98.1	26.3	49.7
1 - 16	WEIGHTED AVERAGE	94.8	86.6	90.8	95.6	65.5	81.1	96.9	63.9	81.0	94.6	65.1	80.3	97.2	64.4	81.4	97.0	39.4	69.2

DIVISION		OKLAHOMA																	
		LAWTON			MCALESTER			MIAMI			MUSKOGEE			OKLAHOMA CITY			PONCA CITY		
		735			745			743			744			730 - 731			746		
		MAT.	INST.	TOTAL	MAT.	INST.	TOTAL	MAT.	INST.	TOTAL	MAT.	INST.	TOTAL	MAT.	INST.	TOTAL	MAT.	INST.	TOTAL
2	SITE WORK	104.9	88.9	92.6	92.7	86.8	88.2	95.4	84.4	86.9	95.5	83.6	86.4	105.1	89.5	93.1	101.7	87.3	90.6
031	CONCRETE FORMWORK	98.7	67.2	71.7	74.0	53.1	56.1	91.1	58.3	63.0	100.5	44.4	52.5	100.2	57.4	63.6	85.1	59.0	62.7
032	CONCRETE REINFORCEMENT	96.6	84.2	89.6	97.1	61.9	77.3	95.6	79.3	86.4	96.4	47.6	68.9	96.6	84.2	89.6	96.4	79.2	86.7
033	CAST IN PLACE CONCRETE	90.7	60.1	77.9	79.5	59.4	71.1	83.4	58.2	72.8	84.4	45.8	68.2	90.7	58.7	77.3	93.2	49.1	74.7
3	CONCRETE	86.7	67.4	77.0	76.0	56.7	66.3	80.5	63.1	71.7	82.2	46.6	64.3	86.8	62.7	74.7	86.7	59.2	72.9
4	MASONRY	95.2	59.5	73.0	107.0	53.9	74.0	93.6	60.4	73.0	104.9	54.2	73.4	97.1	61.3	74.9	83.2	59.7	68.6
5	METALS	96.2	67.0	85.7	92.9	55.3	79.4	92.8	82.4	89.1	94.0	65.4	83.7	96.4	67.1	85.8	92.8	66.7	83.4
6	WOOD & PLASTICS	99.4	71.4	85.0	76.3	53.4	64.5	95.1	58.9	76.5	101.8	45.8	72.9	101.8	57.1	78.7	88.9	60.1	74.1
7	THERMAL & MOISTURE PROTECTION	96.2	65.8	82.1	95.2	61.2	79.5	95.8	64.2	81.2	96.0	52.2	75.7	95.4	65.2	81.4	96.0	62.9	80.6
8	DOORS & WINDOWS	97.6	72.8	91.6	96.2	57.0	86.7	96.2	64.9	88.6	96.2	45.1	83.8	97.6	65.1	89.7	96.2	67.0	89.1
092	LATH, PLASTER & GYPSUM BOARD	94.8	71.7	79.9	85.5	53.1	64.6	91.9	58.6	70.4	94.0	45.1	62.5	94.8	56.9	70.4	89.8	60.0	70.6
095	ACOUSTICAL TREATMENT & WOOD FLOORING	101.7	71.7	82.2	93.5	53.1	67.3	93.5	58.6	70.9	93.5	45.1	62.1	101.7	56.9	72.7	93.5	60.0	71.8
096	FLOORING & CARPET	119.3	65.9	106.3	108.2	65.9	97.9	117.3	71.2	106.1	120.3	53.1	104.0	119.3	65.9	106.3	113.5	65.9	101.9
099	PAINTING & WALL COVERINGS	97.1	44.3	66.1	97.1	60.8	75.8	97.1	56.6	73.4	97.1	44.2	66.0	97.1	65.2	78.4	97.1	65.2	78.4
9	FINISHES	103.3	64.5	83.4	96.3	55.9	75.6	100.1	60.0	79.5	101.4	46.3	73.2	103.3	59.1	80.6	99.2	60.7	79.4
10 - 14	TOTAL DIV. 10 - 14	100.0	72.0	94.1	100.0	69.6	93.6	100.0	68.4	93.3	100.0	41.5	87.7	100.0	70.9	93.9	100.0	67.9	93.2
15	MECHANICAL	100.0	64.8	84.3	96.5	49.5	75.6	96.7	67.5	83.7	100.0	37.5	72.1	100.0	65.8	84.7	96.5	64.9	82.4
16	ELECTRICAL	98.1	56.1	69.8	95.5	64.0	74.3	97.9	65.3	75.9	94.8	45.1	61.3	96.9	66.2	76.2	94.8	68.7	77.2
1 - 16	WEIGHTED AVERAGE	97.5	66.4	82.5	93.9	59.5	77.3	94.6	67.8	81.6	96.3	50.7	74.3	97.6	66.7	82.7	94.5	66.3	80.9

DIVISION		OKLAHOMA												OREGON					
		POTEAU			SHAWNEE			TULSA			WOODWARD			BEND			EUGENE		
		749			748			740 - 741			738			977			974		
		MAT.	INST.	TOTAL	MAT.	INST.	TOTAL	MAT.	INST.	TOTAL	MAT.	INST.	TOTAL	MAT.	INST.	TOTAL	MAT.	INST.	TOTAL
2	SITE WORK	75.4	83.0	81.3	104.7	87.2	91.3	102.2	84.5	88.5	108.7	86.9	91.9	107.3	106.6	106.7	95.1	106.8	104.1
031	CONCRETE FORMWORK	82.0	58.9	62.2	76.5	56.7	59.5	100.1	58.6	64.5	86.4	49.5	54.8	102.4	102.6	102.6	101.2	107.2	106.3
032	CONCRETE REINFORCEMENT	97.5	79.3	87.3	96.4	68.2	80.5	96.6	84.1	89.6	96.4	79.3	86.7	100.9	97.9	99.2	111.0	100.3	105.0
033	CAST IN PLACE CONCRETE	83.4	57.7	72.6	96.1	55.2	79.0	92.0	58.3	77.9	93.8	56.3	78.1	99.3	102.0	100.4	96.2	102.8	99.0
3	CONCRETE	82.0	63.2	72.5	88.1	58.2	73.0	87.4	64.1	75.7	89.4	57.5	73.3	125.6	101.3	113.3	113.9	104.1	109.0
4	MASONRY	94.7	59.6	72.9	108.3	53.8	74.4	94.3	59.3	72.6	87.8	59.5	70.2	113.8	99.2	104.7	111.8	104.2	107.1
5	METALS	92.9	82.3	89.1	92.8	59.9	81.0	96.8	83.3	91.9	93.0	64.9	82.9	94.5	98.1	95.8	95.5	100.5	97.3
6	WOOD & PLASTICS	85.0	60.3	72.3	79.3	58.9	68.7	100.3	58.9	78.9	90.1	47.6	68.2	102.5	102.9	102.7	97.6	106.5	102.2
7	THERMAL & MOISTURE PROTECTION	95.9	63.6	80.9	95.9	61.6	80.0	96.0	64.1	81.2	96.3	63.1	80.9	110.8	97.2	104.5	109.7	96.8	103.7
8	DOORS & WINDOWS	96.2	66.8	89.1	96.2	58.5	87.1	97.6	65.4	89.8	96.2	58.9	87.2	101.1	104.9	102.0	101.4	108.4	103.1
092	LATH, PLASTER & GYPSUM BOARD	88.7	60.0	70.2	86.6	58.8	68.6	94.8	58.6	71.4	90.1	47.1	62.3	92.4	103.0	99.2	91.1	106.6	101.1
095	ACOUSTICAL TREATMENT & WOOD FLOORING	93.5	60.0	71.8	93.5	58.8	71.0	101.7	58.6	73.8	94.8	47.1	63.9	120.7	103.0	109.2	122.0	106.6	112.1
096	FLOORING & CARPET	112.7	70.1	102.3	109.4	52.4	95.6	119.1	68.6	106.8	114.1	69.1	103.1	123.0	100.4	117.5	121.2	100.4	116.2
099	PAINTING & WALL COVERINGS	97.1	63.2	77.2	97.1	54.9	72.3	97.1	65.2	78.4	97.1	65.2	78.4	125.4	73.3	94.9	125.4	86.7	102.7
9	FINISHES	97.5	61.1	78.9	97.5	55.6	76.0	102.8	60.7	81.2	100.0	53.7	76.3	119.2	99.1	108.9	117.8	103.9	110.7
10 - 14	TOTAL DIV. 10 - 14	100.0	68.0	93.2	100.0	66.6	93.0	100.0	71.7	94.0	100.0	65.9	92.8	100.0	104.6	101.0	100.0	109.9	102.1
15	MECHANICAL	96.7	65.2	82.6	96.5	58.4	79.5	100.0	66.2	84.9	96.5	64.7	82.3	96.9	93.0	95.2	100.2	95.8	98.3
16	ELECTRICAL	95.0	65.3	74.9	98.3	65.6	76.2	98.1	58.8	71.6	97.9	66.2	76.5	104.5	99.5	101.1	102.4	99.2	100.2
1 - 16	WEIGHTED AVERAGE	93.7	67.3	81.0	96.1	62.1	79.7	97.5	66.8	82.7	95.7	64.0	80.4	105.3	99.3	102.4	104.0	101.8	102.9

COST INDEXES

OREGON

DIVISION		KLAMATH FALLS 976			MEDFORD 975			PENDLETON 978			PORTLAND 970 - 972			SALEM 973			VALE 979		
		MAT.	INST.	TOTAL	MAT.	INST.	TOTAL	MAT.	INST.	TOTAL	MAT.	INST.	TOTAL	MAT.	INST.	TOTAL	MAT.	INST.	TOTAL
2	SITE WORK	112.2	106.5	107.8	104.0	106.5	105.9	97.9	98.9	98.7	96.1	106.8	104.4	95.3	106.8	104.2	84.7	98.9	95.6
031	CONCRETE FORMWORK	93.2	102.4	101.1	95.4	102.4	101.4	101.7	100.3	100.5	102.8	107.6	106.9	102.7	107.3	106.7	110.2	102.3	103.4
032	CONCRETE REINFORCEMENT	100.9	97.9	99.2	102.5	97.9	99.9	99.6	98.1	98.8	106.1	100.6	103.0	112.2	100.5	105.6	97.2	98.0	97.7
033	CAST IN PLACE CONCRETE	99.3	101.9	100.4	99.3	101.9	100.4	100.0	103.8	101.6	103.2	105.0	103.9	99.7	104.9	101.9	78.8	103.6	89.1
3	CONCRETE	129.6	101.1	115.2	121.7	101.2	111.3	101.7	101.0	101.4	116.8	105.1	110.9	115.9	104.9	110.4	84.6	101.8	93.3
4	MASONRY	129.6	99.2	110.7	109.0	99.2	102.9	117.9	102.6	108.4	113.5	107.5	109.8	114.0	107.5	110.0	116.2	102.6	107.8
5	METALS	94.5	97.8	95.7	95.0	98.9	96.4	107.9	100.0	105.1	94.1	101.1	96.6	96.1	100.9	97.8	107.6	99.5	104.7
6	WOOD & PLASTICS	92.6	102.9	97.9	91.2	102.9	97.3	96.1	99.6	97.9	99.6	106.5	103.2	99.6	106.5	103.2	105.6	103.1	104.3
7	THERMAL & MOISTURE PROTECTION	111.0	94.3	103.3	110.5	94.3	103.0	97.9	94.4	96.3	109.5	101.2	105.7	109.5	101.0	105.6	97.2	94.7	96.0
8	DOORS & WINDOWS	101.1	104.9	102.0	104.2	105.7	104.6	97.8	103.9	99.2	98.8	108.4	101.1	100.5	108.4	102.4	97.7	102.7	98.9
092	LATH, PLASTER & GYPSUM BOARD	88.7	103.0	97.9	88.1	103.0	97.7	74.4	99.3	90.5	90.0	106.6	100.8	91.9	106.6	101.4	78.1	103.0	94.1
095	ACOUSTICAL TREATMENT & WOOD FLOORING	120.7	103.0	109.2	120.7	103.0	109.2	76.9	99.3	91.4	122.0	106.6	112.1	130.3	106.6	115.0	76.9	103.0	93.8
096	FLOORING & CARPET	119.5	100.4	114.8	119.0	100.4	114.5	83.7	100.4	87.8	121.2	104.1	117.1	121.2	100.4	116.2	86.2	100.4	89.7
099	PAINTING & WALL COVERINGS	125.4	73.3	94.9	125.4	73.3	94.9	103.0	90.0	95.4	125.4	86.7	102.7	125.4	86.7	102.7	103.0	73.3	85.6
9	FINISHES	117.9	99.1	108.3	117.0	99.1	107.8	79.4	99.1	89.5	117.7	104.7	111.0	119.4	103.9	111.5	80.0	99.2	89.8
10 - 14	TOTAL DIV. 10 - 14	100.0	104.5	101.0	100.0	104.5	101.0	100.0	104.6	101.0	100.0	109.9	102.1	100.0	109.9	102.1	100.0	105.0	101.0
15	MECHANICAL	96.9	93.0	95.2	100.2	93.0	97.0	94.2	100.7	97.1	100.2	104.7	102.2	100.2	95.8	98.3	94.2	83.4	89.4
16	ELECTRICAL	102.4	91.8	95.2	107.9	91.8	97.0	90.6	96.8	94.8	102.3	108.1	106.2	102.0	99.2	100.1	90.6	78.6	82.5
1 - 16	WEIGHTED AVERAGE	106.4	97.9	102.3	105.6	98.0	101.9	97.8	99.9	98.8	104.0	105.7	104.8	104.5	102.3	103.5	95.4	93.6	94.5

PENNSYLVANIA

DIVISION		ALLENTOWN 181			ALTOONA 166			BEDFORD 155			BRADFORD 167			BUTLER 160			CHAMBERSBURG 172		
		MAT.	INST.	TOTAL	MAT.	INST.	TOTAL	MAT.	INST.	TOTAL	MAT.	INST.	TOTAL	MAT.	INST.	TOTAL	MAT.	INST.	TOTAL
2	SITE WORK	96.7	108.8	106.0	101.5	108.9	107.2	98.3	107.5	105.4	96.3	108.9	106.0	92.1	109.4	105.4	90.8	107.2	103.4
031	CONCRETE FORMWORK	104.8	103.1	103.3	82.7	89.7	88.7	80.8	89.5	88.3	83.3	90.2	89.2	81.4	103.0	99.9	82.7	88.7	87.9
032	CONCRETE REINFORCEMENT	99.0	108.2	104.2	96.0	94.9	95.4	101.7	83.4	91.4	98.0	94.6	96.0	98.0	94.6	96.0	96.4	89.7	92.7
033	CAST IN PLACE CONCRETE	87.6	99.7	92.7	97.7	91.1	94.9	106.0	90.9	99.6	93.2	89.3	91.6	86.2	102.6	93.1	94.7	93.0	94.0
3	CONCRETE	100.5	104.2	102.3	94.4	92.8	93.6	99.5	90.4	94.9	101.8	92.4	97.0	85.3	103.6	94.5	108.3	92.0	100.1
4	MASONRY	97.5	96.6	96.9	99.3	87.3	91.8	99.3	87.2	91.8	94.9	84.0	88.1	101.3	95.6	97.8	97.9	92.7	94.6
5	METALS	97.9	125.5	107.8	91.7	118.6	101.4	97.9	112.7	103.3	95.9	117.8	103.8	91.4	125.2	103.6	95.6	115.6	102.8
6	WOOD & PLASTICS	108.6	104.0	106.2	81.3	89.2	85.4	84.3	89.3	86.9	88.2	89.3	88.7	83.5	102.7	93.4	92.7	89.3	91.0
7	THERMAL & MOISTURE PROTECTION	100.8	113.1	106.5	99.9	95.8	98.0	96.5	96.1	96.3	100.6	93.4	97.3	99.5	102.4	100.9	99.5	87.9	94.1
8	DOORS & WINDOWS	97.4	105.6	99.4	91.8	99.9	93.8	93.5	96.4	94.2	97.9	96.1	97.5	91.8	110.1	96.2	94.2	89.4	93.0
092	LATH, PLASTER & GYPSUM BOARD	100.4	103.7	102.6	83.7	88.5	86.8	87.2	88.6	88.1	91.6	88.6	89.7	84.3	102.5	96.0	95.9	88.6	91.2
095	ACOUSTICAL TREATMENT & WOOD FLOORING	97.1	103.7	101.4	91.6	88.5	89.6	90.1	88.6	89.1	88.9	88.6	88.7	91.6	102.5	98.6	88.9	88.6	88.7
096	FLOORING & CARPET	83.5	92.1	85.6	78.2	73.1	77.0	83.6	74.1	81.3	77.4	81.9	78.5	79.2	99.1	84.0	81.3	73.9	79.5
099	PAINTING & WALL COVERINGS	84.1	98.9	92.8	80.0	103.1	93.5	87.3	103.1	96.5	84.1	67.9	74.6	80.0	105.2	94.8	84.1	82.0	82.8
9	FINISHES	95.8	100.5	98.2	91.1	87.3	89.2	93.9	87.4	90.5	91.1	85.7	88.3	91.0	102.3	96.8	92.2	85.0	88.5
10 - 14	TOTAL DIV. 10 - 14	100.0	103.2	100.7	100.0	100.9	100.2	100.0	101.0	100.2	100.0	102.0	100.4	100.0	105.3	101.1	100.0	95.6	99.1
15	MECHANICAL	100.3	101.0	100.6	99.9	85.9	93.7	96.6	85.8	91.8	97.0	87.2	92.6	96.6	94.7	95.7	97.0	86.5	92.3
16	ELECTRICAL	103.6	90.6	94.8	88.7	104.0	99.0	91.4	104.0	99.9	92.8	104.0	100.4	89.7	87.2	88.0	90.6	78.8	82.6
1 - 16	WEIGHTED AVERAGE	99.2	102.9	101.0	95.1	96.7	95.9	96.4	95.5	96.0	96.7	96.1	96.4	93.2	101.1	97.0	97.0	91.5	94.4

PENNSYLVANIA

DIVISION		DOYLESTOWN 189			DUBOIS 158			ERIE 164 - 165			GREENSBURG 156			HARRISBURG 170 - 171			HAZLETON 182		
		MAT.	INST.	TOTAL	MAT.	INST.	TOTAL	MAT.	INST.	TOTAL	MAT.	INST.	TOTAL	MAT.	INST.	TOTAL	MAT.	INST.	TOTAL
2	SITE WORK	112.9	90.5	95.7	103.7	107.7	106.7	97.7	109.1	106.5	94.1	108.3	105.0	84.7	107.4	102.2	89.6	108.3	104.0
031	CONCRETE FORMWORK	77.1	120.0	113.8	79.9	89.8	88.3	103.1	95.0	96.2	89.8	103.0	101.1	95.7	90.6	91.3	75.2	93.2	90.6
032	CONCRETE REINFORCEMENT	95.6	99.6	97.9	101.0	94.9	97.6	98.0	90.4	93.7	101.0	99.1	100.0	96.1	91.3	94.7	96.1	99.6	98.1
033	CAST IN PLACE CONCRETE	82.8	106.2	92.6	102.2	93.1	98.4	96.1	97.3	96.6	98.4	102.5	100.1	86.0	93.2	89.0	82.8	90.6	86.1
3	CONCRETE	95.1	111.5	103.4	100.4	93.5	96.9	93.5	96.4	95.0	94.5	103.6	99.1	98.8	93.2	96.0	91.7	95.0	93.3
4	MASONRY	100.3	116.1	110.1	99.5	87.3	91.9	91.0	95.3	93.7	111.3	96.9	102.3	94.8	92.4	93.3	108.2	98.2	102.0
5	METALS	94.5	117.4	102.7	97.9	118.0	105.2	91.8	117.0	100.9	97.8	124.4	107.4	99.7	117.4	106.1	97.7	120.5	105.9
6	WOOD & PLASTICS	82.0	121.4	102.4	83.2	89.3	86.4	104.3	94.0	99.0	94.2	102.8	98.6	99.7	90.8	95.1	79.9	93.7	87.0
7	THERMAL & MOISTURE PROTECTION	99.1	121.1	109.3	96.8	97.6	97.2	100.0	97.2	98.7	96.4	102.5	99.2	103.8	94.1	99.3	100.0	105.0	102.3
8	DOORS & WINDOWS	99.1	127.0	105.9	93.5	99.9	95.0	92.0	91.8	91.9	93.4	110.1	97.5	97.4	99.5	97.9	98.2	103.8	99.5
092	LATH, PLASTER & GYPSUM BOARD	83.5	121.7	108.2	86.7	88.6	87.9	93.4	93.4	93.4	90.1	102.5	98.1	100.4	90.1	93.8	91.1	93.1	92.4
095	ACOUSTICAL TREATMENT & WOOD FLOORING	93.0	121.7	111.6	90.1	88.6	89.1	97.1	93.4	94.7	88.7	102.5	97.6	97.1	90.1	92.6	94.2	94.1	94.1
096	FLOORING & CARPET	68.7	122.8	81.9	83.1	76.8	81.6	84.1	82.3	83.6	87.6	99.1	90.4	83.7	88.0	84.8	74.8	87.8	78.0
099	PAINTING & WALL COVERINGS	83.4	111.0	99.6	87.3	80.1	83.0	86.1	84.1	84.9	87.3	103.1	96.5	84.1	82.0	82.8	84.1	92.3	88.9
9	FINISHES	84.8	119.8	102.7	94.0	85.2	89.5	95.7	91.2	93.4	95.0	102.0	98.6	94.7	88.9	91.8	91.1	91.8	91.4
10 - 14	TOTAL DIV. 10 - 14	100.0	109.2	101.9	100.0	101.0	100.2	100.0	103.8	100.8	100.0	105.5	101.2	100.0	95.7	99.1	100.0	96.5	99.3
15	MECHANICAL	96.6	116.2	105.3	96.6	85.9	91.8	99.9	93.7	97.1	96.6	93.1	95.0	100.3	91.1	96.2	97.0	85.6	91.9
16	ELECTRICAL	92.7	111.7	105.5	92.3	104.0	100.2	91.1	84.3	86.5	92.3	104.0	100.2	102.6	80.8	87.9	94.2	78.3	83.5
1 - 16	WEIGHTED AVERAGE	95.7	113.2	104.2	96.8	96.4	96.6	95.4	96.3	95.8	96.7	103.5	99.9	98.6	94.2	96.5	96.3	94.7	95.5

COST INDEXES

PENNSYLVANIA

DIVISION		INDIANA 157			JOHNSTOWN 159			KITTANNING 162			LANCASTER 175 - 176			LEHIGH VALLEY 180			MONTROSE 188		
		MAT.	INST.	TOTAL	MAT.	INST.	TOTAL	MAT.	INST.	TOTAL	MAT.	INST.	TOTAL	MAT.	INST.	TOTAL	MAT.	INST.	TOTAL
2	SITE WORK	92.2	108.3	104.6	98.9	107.9	105.8	94.9	109.5	106.1	81.9	107.5	101.6	94.1	108.8	105.4	92.8	108.6	104.9
031	CONCRETE FORMWORK	81.7	92.9	91.3	79.9	90.3	88.8	81.4	103.2	100.1	85.5	90.7	90.0	91.9	103.0	101.4	76.0	95.0	92.3
032	CONCRETE REINFORCEMENT	100.3	99.1	99.6	101.7	98.9	100.1	96.6	99.2	98.1	96.1	91.2	93.4	96.1	104.3	100.7	100.7	99.5	100.0
033	CAST IN PLACE CONCRETE	96.5	102.5	99.0	106.9	91.5	100.5	89.6	102.7	95.1	80.6	93.7	86.1	89.6	99.7	93.8	87.9	94.7	90.8
3	CONCRETE	92.1	99.1	95.7	100.1	94.1	97.1	88.1	103.7	96.0	95.0	93.5	94.2	98.9	103.4	101.2	96.9	97.2	97.0
4	MASONRY	96.4	96.9	96.7	97.2	87.9	91.4	104.5	96.9	99.8	103.6	91.9	96.3	97.6	96.4	96.9	97.5	99.3	98.6
5	METALS	97.9	124.5	107.5	97.9	122.7	106.8	91.5	125.4	103.7	95.6	117.0	103.3	97.6	123.9	107.1	95.7	120.5	104.6
6	WOOD & PLASTICS	85.4	89.3	87.4	83.2	89.3	86.4	83.5	102.7	93.4	95.7	90.8	93.2	98.0	103.6	100.9	81.2	93.7	87.6
7	THERMAL & MOISTURE PROTECTION	96.2	100.2	98.0	96.5	97.3	96.9	99.7	102.7	101.1	99.0	94.2	96.8	100.6	109.6	104.7	100.0	106.4	103.0
8	DOORS & WINDOWS	93.5	102.9	95.7	93.4	102.9	95.7	91.8	110.1	96.2	94.2	93.3	94.0	98.2	110.4	101.1	94.4	103.8	96.7
092	LATH, PLASTER & GYPSUM BOARD	87.7	88.6	88.3	86.4	88.5	87.8	84.3	102.5	96.0	96.9	90.1	92.5	96.4	103.4	100.9	90.1	93.1	92.1
095	ACOUSTICAL TREATMENT & WOOD FLOORING	90.1	88.6	89.1	88.7	88.5	88.6	91.6	102.5	98.6	88.9	90.1	89.7	95.7	103.4	100.7	88.9	93.1	91.6
096	FLOORING & CARPET	84.0	99.1	87.7	83.1	90.9	85.0	79.2	99.1	84.0	82.3	88.0	83.7	80.3	122.8	90.6	75.2	92.7	79.5
099	PAINTING & WALL COVERINGS	87.3	103.1	96.5	87.3	103.1	96.5	80.0	103.1	93.5	84.1	91.9	88.7	84.1	98.9	92.8	84.1	90.2	87.7
9	FINISHES	93.7	94.2	94.0	93.4	91.1	92.2	91.2	102.0	96.7	92.2	89.6	90.9	93.9	105.7	99.9	90.9	93.1	91.9
10 - 14	TOTAL DIV. 10 - 14	100.0	103.7	100.8	100.0	101.3	100.3	100.0	105.4	101.1	100.0	96.2	99.2	100.0	100.8	100.2	100.0	99.8	100.0
15	MECHANICAL	96.6	88.5	93.0	96.6	86.4	92.1	96.6	100.2	98.2	97.0	91.7	94.7	97.0	102.2	99.3	97.0	89.2	93.5
16	ELECTRICAL	92.3	104.0	100.2	92.3	104.0	100.2	88.7	104.0	99.0	93.0	75.3	81.0	94.2	102.3	99.7	92.8	88.5	89.9
1 - 16	WEIGHTED AVERAGE	95.3	100.5	97.8	96.4	97.9	97.1	93.7	105.1	99.2	95.6	93.1	94.4	97.2	105.4	101.2	95.5	97.9	96.6

PENNSYLVANIA

DIVISION		NEW CASTLE 161			NORRISTOWN 194			OIL CITY 163			PHILADELPHIA 190 - 191			PITTSBURGH 150 - 152			POTTSVILLE 179		
		MAT.	INST.	TOTAL	MAT.	INST.	TOTAL	MAT.	INST.	TOTAL	MAT.	INST.	TOTAL	MAT.	INST.	TOTAL	MAT.	INST.	TOTAL
2	SITE WORK	92.6	109.5	105.6	104.4	97.3	98.9	90.9	108.7	104.6	111.1	96.4	99.8	97.4	111.2	108.0	85.2	107.4	102.2
031	CONCRETE FORMWORK	81.4	102.7	99.6	78.7	121.4	115.3	81.4	88.2	87.2	106.2	130.3	126.8	103.9	103.6	103.7	74.1	90.5	88.1
032	CONCRETE REINFORCEMENT	95.4	98.9	97.4	94.9	111.0	104.0	96.6	98.6	97.7	97.9	120.5	110.6	102.2	105.5	104.1	95.2	95.8	95.6
033	CAST IN PLACE CONCRETE	87.1	95.7	90.7	80.8	118.1	96.4	84.5	93.5	88.3	99.0	124.2	109.6	102.2	102.8	102.4	95.7	97.7	90.8
3	CONCRETE	85.9	100.9	93.5	94.2	118.2	106.3	84.1	93.6	88.9	107.8	125.9	117.0	98.5	105.0	101.8	98.9	95.6	97.3
4	MASONRY	98.5	97.7	98.0	107.9	117.7	114.0	100.3	88.1	92.7	94.5	127.5	115.0	93.8	106.2	101.5	97.3	91.8	93.9
5	METALS	91.5	122.6	102.7	95.3	118.2	103.6	91.5	119.8	101.7	98.9	121.1	106.9	98.7	126.1	108.5	95.8	119.5	104.3
6	WOOD & PLASTICS	83.5	102.7	93.4	85.1	121.3	103.8	83.5	87.3	85.4	110.1	130.1	120.5	105.9	102.8	104.3	83.2	90.8	87.1
7	THERMAL & MOISTURE PROTECTION	99.6	102.7	101.0	98.8	122.8	109.9	99.5	93.7	96.8	99.9	131.4	114.5	96.7	105.3	100.7	99.0	104.1	101.3
8	DOORS & WINDOWS	91.8	104.7	94.9	93.1	125.6	100.9	91.8	91.3	91.7	99.2	134.0	107.6	93.6	110.7	97.8	94.2	100.9	95.8
092	LATH, PLASTER & GYPSUM BOARD	84.3	102.5	96.0	87.0	121.7	109.4	84.3	86.5	85.7	94.4	130.9	117.9	94.1	102.5	99.5	92.2	90.1	90.8
095	ACOUSTICAL TREATMENT & WOOD FLOORING	91.6	102.5	98.6	96.0	121.7	112.7	91.6	86.5	88.3	96.0	130.9	118.6	90.1	102.5	98.1	88.9	90.1	89.7
096	FLOORING & CARPET	79.2	92.0	82.3	71.6	122.4	84.0	79.2	76.8	78.6	78.9	134.8	92.5	92.2	109.9	96.5	78.1	97.9	82.9
099	PAINTING & WALL COVERINGS	80.0	100.2	91.8	89.4	124.5	110.0	80.0	100.6	92.1	89.4	136.7	117.2	87.3	114.1	103.0	84.1	75.6	79.1
9	FINISHES	91.0	100.3	95.8	88.6	121.9	105.7	90.9	86.7	88.8	92.2	131.9	112.5	97.5	105.4	101.6	90.4	89.9	90.1
10 - 14	TOTAL DIV. 10 - 14	100.0	105.4	101.1	100.0	120.4	104.3	100.0	100.9	100.2	100.0	127.6	105.8	100.0	105.5	101.2	100.0	95.4	99.0
15	MECHANICAL	96.6	94.4	95.6	96.7	118.1	106.2	96.6	91.1	94.1	100.0	125.3	111.3	99.9	103.5	101.5	97.0	85.6	91.9
16	ELECTRICAL	89.7	95.5	93.6	90.9	111.7	104.9	92.3	87.1	88.8	98.2	132.5	121.4	96.1	104.0	101.4	89.9	83.8	85.8
1 - 16	WEIGHTED AVERAGE	93.1	101.6	97.2	95.5	115.9	105.4	93.1	94.7	93.9	99.8	124.9	111.9	97.8	107.5	102.5	95.3	94.5	94.9

PENNSYLVANIA

DIVISION		READING 195 - 196			SCRANTON 184 - 185			STATE COLLEGE 168			STROUDSBURG 183			SUNBURY 178			UNIONTOWN 154		
		MAT.	INST.	TOTAL	MAT.	INST.	TOTAL	MAT.	INST.	TOTAL	MAT.	INST.	TOTAL	MAT.	INST.	TOTAL	MAT.	INST.	TOTAL
2	SITE WORK	108.4	114.5	113.1	97.2	108.8	106.2	87.4	107.7	103.0	91.5	108.3	104.4	96.1	108.4	105.6	92.8	108.4	104.8
031	CONCRETE FORMWORK	105.7	91.5	93.5	104.9	95.2	96.6	81.5	89.6	88.5	84.1	88.7	88.1	88.4	85.4	85.8	72.0	103.0	98.6
032	CONCRETE REINFORCEMENT	96.3	95.7	95.9	99.0	99.8	99.5	97.2	95.2	96.1	99.3	103.8	101.8	98.0	91.3	94.2	101.0	99.1	100.0
033	CAST IN PLACE CONCRETE	72.4	93.4	81.2	91.6	95.1	93.1	88.4	91.1	89.5	86.2	92.5	88.9	93.8	91.9	93.0	96.5	102.5	99.0
3	CONCRETE	94.6	94.6	94.6	102.4	97.6	100.0	102.3	92.8	97.5	95.8	94.6	95.2	102.7	90.5	96.5	91.5	103.5	97.6
4	MASONRY	96.8	93.7	94.8	97.8	105.6	102.6	102.3	92.3	96.1	94.6	91.2	92.5	94.9	90.4	92.1	111.2	96.9	102.3
5	METALS	96.0	119.9	104.6	99.7	122.0	107.8	95.4	119.0	103.9	97.7	121.4	106.2	95.5	117.2	103.3	97.6	124.6	107.4
6	WOOD & PLASTICS	110.1	89.4	99.4	108.6	93.7	100.9	91.4	89.2	90.3	89.5	87.2	88.3	94.6	85.0	89.6	74.6	102.8	89.2
7	THERMAL & MOISTURE PROTECTION	99.9	109.8	104.5	100.7	108.0	104.1	99.2	92.9	96.3	100.3	87.7	94.4	100.5	101.7	101.1	96.0	102.5	99.0
8	DOORS & WINDOWS	99.2	93.9	97.9	97.4	103.8	99.0	94.1	99.9	95.5	98.2	90.4	96.3	94.3	96.4	94.8	93.4	110.1	97.4
092	LATH, PLASTER & GYPSUM BOARD	103.1	88.8	93.9	100.4	93.1	95.7	95.0	88.5	90.8	93.7	86.4	89.0	93.5	84.1	87.4	83.7	102.5	95.8
095	ACOUSTICAL TREATMENT & WOOD FLOORING	98.8	88.8	92.3	97.1	93.1	94.5	87.5	88.5	88.2	95.7	86.4	89.7	87.5	84.1	85.3	88.7	102.5	97.6
096	FLOORING & CARPET	77.2	90.8	80.5	83.5	92.7	85.7	80.9	79.6	80.6	77.7	82.2	78.8	79.2	68.0	76.5	79.7	102.5	85.3
099	PAINTING & WALL COVERINGS	87.4	92.7	90.5	84.1	90.2	87.7	84.1	103.1	95.2	84.1	98.9	92.8	84.1	92.9	89.3	87.3	96.7	92.8
9	FINISHES	93.3	90.8	92.0	95.8	93.8	94.8	91.5	88.7	90.1	92.5	88.3	90.4	91.7	82.6	87.0	91.5	102.1	97.0
10 - 14	TOTAL DIV. 10 - 14	100.0	100.6	100.1	100.0	99.8	100.0	100.0	95.8	99.1	100.0	98.2	99.6	100.0	94.1	98.7	100.0	105.5	101.2
15	MECHANICAL	100.2	100.3	100.2	100.3	89.2	95.3	97.0	91.3	94.5	97.0	89.3	93.6	97.0	89.8	93.8	96.6	88.6	93.0
16	ELECTRICAL	100.9	83.8	89.3	103.6	88.6	93.4	91.4	104.0	99.9	94.2	84.5	87.6	90.4	80.9	84.0	87.4	104.0	98.6
1 - 16	WEIGHTED AVERAGE	98.2	98.2	98.2	99.8	98.7	99.3	96.3	98.1	97.2	96.4	94.3	95.4	96.3	92.7	94.5	95.3	102.6	98.8

PENNSYLVANIA

DIVISION		WASHINGTON 153			WELLSBORO 169			WESTCHESTER 193			WILKES-BARRE 186-187			WILLIAMSPORT 177			YORK 173-174		
		MAT.	INST.	TOTAL	MAT.	INST.	TOTAL	MAT.	INST.	TOTAL	MAT.	INST.	TOTAL	MAT.	INST.	TOTAL	MAT.	INST.	TOTAL
2	SITE WORK	92.8	108.4	104.8	100.7	108.4	106.6	110.9	95.0	98.7	89.3	108.8	104.3	86.4	107.7	102.8	85.5	107.5	102.4
031	CONCRETE FORMWORK	81.9	103.3	100.3	82.4	85.5	85.0	88.4	119.9	115.4	87.2	94.2	93.2	83.7	84.6	84.5	82.5	89.8	88.8
032	CONCRETE REINFORCEMENT	101.0	99.5	100.2	97.2	101.8	99.8	94.0	102.5	98.8	98.0	102.0	100.2	97.2	91.2	93.8	98.0	91.3	94.2
033	CAST IN PLACE CONCRETE	96.5	102.6	99.1	92.6	90.6	91.7	89.4	116.0	100.5	82.8	94.2	87.6	79.4	83.4	81.1	86.2	94.0	89.5
3	CONCRETE	92.3	103.7	98.0	104.7	92.1	98.3	103.0	114.7	108.9	92.8	97.2	95.0	89.3	87.2	88.3	100.5	93.2	96.8
4	MASONRY	97.6	102.9	100.9	102.3	89.7	94.5	102.7	117.4	111.8	109.2	100.9	104.1	87.8	89.7	89.0	95.8	91.9	93.4
5	METALS	97.6	124.1	107.2	95.9	122.2	105.3	95.3	104.5	98.6	95.7	122.4	105.3	95.6	115.9	102.9	97.0	117.5	104.4
6	WOOD & PLASTICS	85.5	102.8	94.4	87.4	85.0	86.1	95.7	121.3	108.9	93.3	93.7	93.5	89.5	85.0	87.1	88.4	88.8	88.6
7	THERMAL & MOISTURE PROTECTION	96.2	104.0	99.8	100.9	101.8	101.3	99.3	121.2	109.4	100.0	94.7	97.5	99.9	89.8	95.2	99.1	106.2	102.4
8	DOORS & WINDOWS	93.4	109.3	97.2	97.9	99.8	98.4	93.1	107.3	96.5	94.4	98.4	95.3	94.4	84.8	92.0	94.2	98.5	95.2
092	LATH, PLASTER & GYPSUM BOARD	87.4	102.5	97.1	91.3	84.1	86.7	90.2	121.7	110.5	93.2	93.1	93.2	92.2	84.1	87.0	94.3	88.1	90.3
095	ACOUSTICAL TREATMENT & WOOD FLOORING	88.7	102.5	97.6	87.5	84.1	85.3	96.0	121.7	112.7	88.9	93.1	91.6	88.9	84.1	85.8	88.9	88.1	88.4
096	FLOORING & CARPET	84.1	99.1	87.7	77.1	81.4	78.2	74.7	122.8	86.4	78.9	81.7	79.6	77.7	79.6	78.2	79.8	88.0	81.8
099	PAINTING & WALL COVERINGS	87.3	96.7	92.8	84.1	92.3	88.9	89.4	124.5	110.0	84.1	92.3	88.9	84.1	92.3	88.9	84.1	82.0	82.8
9	FINISHES	93.4	102.0	97.8	91.1	84.6	87.8	90.4	121.6	106.4	91.4	91.8	91.6	90.7	84.4	87.5	91.2	88.3	89.7
10-14	TOTAL DIV. 10-14	100.0	105.5	101.2	100.0	94.0	98.7	100.0	120.3	104.3	100.0	99.1	99.8	100.0	94.0	98.7	100.0	96.0	99.2
15	MECHANICAL	96.6	100.2	98.2	97.0	90.0	93.9	96.7	117.7	106.1	97.0	88.4	93.2	97.0	89.7	93.8	100.3	91.8	96.5
16	ELECTRICAL	91.3	104.0	99.9	92.8	72.7	79.3	90.8	93.9	92.9	94.2	84.0	87.3	91.1	71.7	78.0	93.0	80.8	84.8
1-16	WEIGHTED AVERAGE	95.2	105.4	100.2	97.6	92.3	95.0	96.8	110.1	103.2	95.9	96.5	96.2	93.9	89.8	92.0	96.8	94.4	95.6

RHODE ISLAND / SOUTH CAROLINA

DIVISION		NEWPORT 028			PROVIDENCE 029			AIKEN 298			BEAUFORT 299			CHARLESTON 294			COLUMBIA 290-292		
		MAT.	INST.	TOTAL	MAT.	INST.	TOTAL	MAT.	INST.	TOTAL	MAT.	INST.	TOTAL	MAT.	INST.	TOTAL	MAT.	INST.	TOTAL
2	SITE WORK	91.0	99.9	97.8	90.4	99.9	97.7	115.6	82.3	90.0	110.8	82.5	89.0	94.3	83.2	85.8	94.0	83.2	85.7
031	CONCRETE FORMWORK	101.9	109.9	108.7	101.0	109.9	108.6	95.0	35.9	44.4	92.8	36.4	44.5	95.4	46.7	53.7	101.1	49.4	56.8
032	CONCRETE REINFORCEMENT	116.7	126.3	122.1	116.7	126.3	122.1	95.4	14.3	49.7	94.5	37.1	62.2	94.2	48.3	68.8	94.2	48.0	68.2
033	CAST IN PLACE CONCRETE	87.0	117.8	99.9	91.9	116.8	102.3	76.1	37.6	60.0	76.1	48.6	64.6	86.0	58.3	74.4	84.3	53.9	71.5
3	CONCRETE	108.2	115.2	111.7	110.5	114.9	112.7	101.6	34.8	68.0	98.7	43.1	70.7	91.7	53.1	72.2	91.3	52.7	71.8
4	MASONRY	100.6	117.8	111.3	103.1	117.8	112.2	72.5	23.9	42.3	87.0	32.4	53.1	88.3	45.5	61.7	87.3	39.8	57.8
5	METALS	101.1	111.7	104.9	101.1	111.7	104.9	92.9	65.0	82.8	92.9	73.8	86.0	95.0	80.4	89.7	95.0	79.7	89.5
6	WOOD & PLASTICS	102.5	107.2	104.9	102.5	107.2	104.9	101.3	36.9	68.0	99.1	36.9	67.0	97.6	46.4	71.2	104.8	50.5	76.8
7	THERMAL & MOISTURE PROTECTION	101.6	110.0	105.5	100.4	109.9	104.8	97.1	31.7	66.7	96.7	40.6	70.7	95.6	52.4	75.6	95.6	50.2	74.6
8	DOORS & WINDOWS	102.9	115.4	105.9	102.9	115.4	105.9	92.9	31.7	78.1	92.9	36.9	79.3	96.7	44.4	84.0	96.7	46.6	84.6
092	LATH, PLASTER & GYPSUM BOARD	98.5	106.5	103.6	98.5	106.5	103.6	101.8	34.6	58.4	106.0	34.6	59.9	106.4	45.5	66.5	106.4	48.7	69.2
095	ACOUSTICAL TREATMENT & WOOD FLOORING	101.7	106.5	104.8	101.7	106.5	104.8	87.3	34.6	53.2	87.3	34.6	53.2	91.4	44.5	61.0	91.4	48.7	63.7
096	FLOORING & CARPET	96.9	124.2	103.5	96.9	124.2	103.5	92.4	21.5	75.2	93.8	39.3	80.6	93.3	50.2	82.8	93.1	50.2	82.7
099	PAINTING & WALL COVERINGS	89.4	120.5	107.7	89.4	120.5	107.7	105.8	32.7	62.9	105.8	32.7	62.9	105.8	48.0	71.9	105.8	48.0	71.9
9	FINISHES	96.9	113.6	105.5	96.8	113.7	105.5	94.2	33.4	63.0	94.9	36.5	65.0	94.2	47.2	70.1	94.1	49.5	71.2
10-14	TOTAL DIV. 10-14	99.8	109.6	101.9	99.0	109.6	101.2	100.0	67.6	93.2	100.0	51.0	89.7	100.0	70.6	93.8	100.0	71.2	93.9
15	MECHANICAL	99.8	104.0	101.7	100.0	104.0	101.8	96.8	39.6	71.3	96.9	31.6	67.8	100.1	55.6	80.3	100.1	46.8	76.3
16	ELECTRICAL	99.7	100.1	100.0	99.2	100.1	99.8	96.6	20.5	45.3	102.5	34.1	56.3	99.6	42.7	61.2	100.8	45.0	63.2
1-16	WEIGHTED AVERAGE	100.9	108.4	104.5	101.2	108.3	104.6	95.6	40.4	68.9	96.4	44.2	71.2	96.3	55.9	76.8	96.3	54.3	76.1

SOUTH CAROLINA / SOUTH DAKOTA

DIVISION		FLORENCE 295			GREENVILLE 296			ROCK HILL 297			SPARTANBURG 293			ABERDEEN 574			MITCHELL 573		
		MAT.	INST.	TOTAL	MAT.	INST.	TOTAL	MAT.	INST.	TOTAL	MAT.	INST.	TOTAL	MAT.	INST.	TOTAL	MAT.	INST.	TOTAL
2	SITE WORK	104.4	83.2	88.1	99.6	82.8	86.7	97.7	82.5	86.0	99.4	82.8	86.6	78.8	94.5	90.9	76.5	94.5	90.3
031	CONCRETE FORMWORK	78.5	49.4	53.6	95.3	49.5	56.1	89.2	35.0	42.8	99.8	49.5	56.7	94.0	48.1	54.7	88.6	48.1	53.9
032	CONCRETE REINFORCEMENT	93.9	48.3	68.2	93.8	48.4	68.2	94.7	23.0	54.3	93.8	48.4	68.2	103.6	65.8	82.3	102.9	65.8	82.0
033	CAST IN PLACE CONCRETE	76.1	58.2	68.6	76.0	58.2	68.6	76.0	42.6	62.1	76.0	58.2	68.6	99.7	59.6	82.9	96.8	59.6	81.3
3	CONCRETE	93.0	54.2	73.5	91.7	54.3	72.9	89.8	37.8	63.6	92.1	54.3	73.0	94.9	56.6	75.5	92.3	56.6	74.3
4	MASONRY	72.5	45.5	55.7	70.6	45.5	55.0	94.9	29.8	54.5	72.5	45.5	55.7	105.9	63.3	79.4	93.7	63.3	74.8
5	METALS	93.9	80.1	88.9	93.8	80.2	88.9	92.9	68.5	84.1	93.8	80.2	88.9	104.7	74.9	94.0	103.5	74.9	93.1
6	WOOD & PLASTICS	78.4	50.5	64.0	97.2	50.5	73.1	94.9	35.6	64.3	102.2	50.5	75.5	96.8	45.7	70.4	95.5	45.7	69.8
7	THERMAL & MOISTURE PROTECTION	95.9	52.8	75.9	95.9	52.8	75.9	95.7	35.3	67.7	96.0	52.8	76.0	96.2	57.0	78.0	96.0	57.0	77.9
8	DOORS & WINDOWS	92.9	46.6	81.7	92.9	46.6	81.7	92.9	33.0	78.4	92.9	46.6	81.7	99.5	49.4	87.3	97.8	49.4	86.0
092	LATH, PLASTER & GYPSUM BOARD	95.2	48.7	65.2	100.7	48.7	67.1	100.2	33.3	57.0	102.3	48.7	67.7	99.8	44.5	64.1	98.9	44.5	63.8
095	ACOUSTICAL TREATMENT & WOOD FLOORING	88.7	48.7	62.8	87.3	48.7	62.3	87.3	33.3	52.3	87.3	48.7	62.3	108.7	44.5	67.1	107.3	44.5	66.6
096	FLOORING & CARPET	84.5	50.2	76.2	90.9	51.2	81.2	90.2	38.9	77.8	92.6	51.2	82.6	114.9	74.3	105.0	114.4	74.3	104.6
099	PAINTING & WALL COVERINGS	105.8	48.0	71.9	105.8	48.0	71.9	105.8	32.7	62.9	105.8	48.0	71.9	106.3	42.9	69.1	106.3	42.9	69.1
9	FINISHES	90.1	49.5	69.3	92.4	49.7	70.5	92.0	35.7	63.1	93.2	49.7	70.9	107.0	52.1	78.8	106.4	52.1	78.5
10-14	TOTAL DIV. 10-14	100.0	71.1	93.9	100.0	71.2	93.9	100.0	67.4	93.1	100.0	71.2	93.9	100.0	63.8	92.4	100.0	63.8	92.4
15	MECHANICAL	100.1	46.8	76.3	100.1	47.0	76.4	96.8	23.5	64.1	100.1	47.0	76.4	100.0	44.6	75.3	96.8	44.6	73.5
16	ELECTRICAL	96.5	25.7	48.8	100.0	40.9	60.2	100.0	25.3	49.6	100.0	40.9	60.2	99.7	51.9	67.4	97.3	51.9	66.6
1-16	WEIGHTED AVERAGE	94.4	52.0	74.0	94.7	54.6	75.3	94.8	39.7	68.2	95.0	54.6	75.5	100.3	58.9	80.3	97.9	58.9	79.1

COST INDEXES

SOUTH DAKOTA / TENNESSEE

DIVISION		MOBRIDGE 576			PIERRE 575			RAPID CITY 577			SIOUX FALLS 570 - 571			WATERTOWN 572			CHATTANOOGA 373 - 374		
		MAT.	INST.	TOTAL	MAT.	INST.	TOTAL	MAT.	INST.	TOTAL	MAT.	INST.	TOTAL	MAT.	INST.	TOTAL	MAT.	INST.	TOTAL
2	SITE WORK	76.4	94.5	90.3	77.2	94.5	90.5	77.5	94.2	90.3	78.4	96.4	92.3	76.3	94.5	90.3	107.5	98.8	100.8
031	CONCRETE FORMWORK	77.4	48.1	52.3	92.1	50.4	56.4	105.5	45.9	54.4	92.2	48.2	54.5	72.3	48.1	51.5	96.6	56.2	62.0
032	CONCRETE REINFORCEMENT	105.6	65.8	83.2	103.1	66.5	82.5	96.8	66.7	79.8	96.8	67.1	80.0	100.2	65.8	80.8	94.2	52.6	70.8
033	CAST IN PLACE CONCRETE	96.8	59.6	81.3	96.8	55.0	79.3	96.1	51.7	77.5	100.2	59.7	83.2	96.8	56.6	81.3	104.2	57.4	84.6
3	CONCRETE	91.8	56.6	74.0	92.5	56.1	74.2	92.4	53.0	72.5	93.3	56.9	75.3	90.7	56.6	73.5	92.0	57.9	74.8
4	MASONRY	104.4	63.3	78.9	104.3	53.4	72.7	104.7	50.3	70.9	101.6	63.3	77.8	126.8	63.3	87.4	95.4	56.1	71.0
5	METALS	103.9	74.9	93.4	104.6	74.3	93.7	106.7	74.6	95.2	107.2	75.5	95.8	103.5	74.9	93.1	96.8	85.1	92.6
6	WOOD & PLASTICS	82.4	45.7	63.5	94.6	49.4	71.3	106.1	46.2	75.2	94.6	45.7	69.3	77.4	45.7	61.0	95.0	57.4	75.6
7	THERMAL & MOISTURE PROTECTION	96.0	57.0	77.9	96.4	54.6	77.0	96.9	52.8	76.4	96.3	58.4	78.7	95.7	57.0	77.7	103.2	57.9	82.2
8	DOORS & WINDOWS	101.0	49.4	88.5	103.0	51.7	90.6	104.0	50.0	90.9	104.0	49.7	90.9	97.8	49.4	86.0	99.9	56.1	89.3
092	LATH, PLASTER & GYPSUM BOARD	93.4	44.5	61.8	98.4	48.3	66.1	99.6	45.0	64.4	99.6	44.5	64.0	91.0	44.5	61.0	91.5	56.6	69.0
095	ACOUSTICAL TREATMENT & WOOD FLOORING	108.7	44.5	67.1	107.3	48.3	69.1	112.8	45.0	68.9	112.8	44.5	68.5	107.3	44.5	66.6	97.3	56.6	70.9
096	FLOORING & CARPET	108.8	74.3	100.5	114.0	53.0	99.2	114.0	62.3	101.4	114.0	74.3	104.4	106.9	74.3	99.0	96.2	62.7	88.0
099	PAINTING & WALL COVERINGS	106.3	42.9	69.1	106.3	42.9	69.1	106.3	42.9	69.1	106.3	42.9	69.1	106.3	42.9	69.1	99.9	53.1	72.5
9	FINISHES	104.1	52.1	77.4	106.2	49.9	77.3	107.5	48.4	77.2	107.4	52.1	79.1	102.8	52.1	76.8	97.2	57.1	76.7
10 - 14	TOTAL DIV. 10 - 14	100.0	63.8	92.4	100.0	64.3	92.5	100.0	61.4	91.9	100.0	63.9	92.4	100.0	63.8	92.4	100.0	70.2	93.7
15	MECHANICAL	96.8	44.6	73.5	100.0	44.5	75.3	100.0	41.2	73.8	100.0	44.6	75.3	96.8	44.6	73.5	100.1	54.1	79.6
16	ELECTRICAL	99.7	51.9	67.4	93.6	51.9	65.4	94.2	51.9	65.7	93.2	63.5	73.2	96.0	51.9	66.2	102.4	69.2	80.0
1 - 16	WEIGHTED AVERAGE	98.7	58.9	79.5	99.8	57.6	79.4	100.5	55.9	79.0	100.4	61.2	81.4	98.8	58.9	79.5	98.5	65.6	82.6

TENNESSEE

DIVISION		COLUMBIA 384			COOKEVILLE 385			JACKSON 383			JOHNSON CITY 376			KNOXVILLE 377 - 379			MCKENZIE 382		
		MAT.	INST.	TOTAL	MAT.	INST.	TOTAL	MAT.	INST.	TOTAL	MAT.	INST.	TOTAL	MAT.	INST.	TOTAL	MAT.	INST.	TOTAL
2	SITE WORK	99.6	84.8	88.2	106.8	85.3	90.3	108.0	95.9	98.7	118.0	87.4	94.5	94.6	87.4	89.1	106.0	85.5	90.2
031	CONCRETE FORMWORK	74.6	54.2	57.2	74.8	31.1	37.4	86.8	40.5	47.2	79.8	58.5	61.5	95.5	58.5	63.8	85.5	30.7	38.6
032	CONCRETE REINFORCEMENT	99.3	54.5	74.0	99.3	24.8	57.3	99.6	51.7	72.6	98.3	52.8	72.7	94.2	52.8	70.9	100.2	24.5	57.6
033	CAST IN PLACE CONCRETE	94.3	54.8	79.3	106.8	38.5	78.2	104.2	40.4	77.5	83.9	64.2	75.6	97.9	58.7	81.5	104.5	39.1	77.1
3	CONCRETE	91.7	57.7	74.5	101.4	35.2	68.0	93.9	44.9	69.2	95.3	61.2	78.1	89.1	59.4	74.1	100.5	35.1	67.5
4	MASONRY	105.8	50.8	71.6	100.4	28.1	55.5	106.2	28.2	57.7	102.7	55.8	73.6	72.7	55.8	62.2	103.4	46.1	67.8
5	METALS	93.3	84.2	90.0	93.4	70.1	85.0	95.3	81.0	90.2	94.5	84.7	90.9	97.0	84.8	92.6	93.4	71.5	85.5
6	WOOD & PLASTICS	68.1	60.3	64.1	68.4	32.1	49.7	82.8	43.0	62.3	70.7	60.9	65.6	87.6	60.9	73.8	80.2	31.2	54.9
7	THERMAL & MOISTURE PROTECTION	94.2	53.5	75.3	94.6	35.6	67.3	101.7	41.2	73.7	97.4	58.8	79.5	95.2	58.2	78.0	94.8	38.4	68.6
8	DOORS & WINDOWS	91.9	55.8	83.1	91.9	29.6	76.8	100.3	46.7	87.3	96.2	59.8	87.4	93.1	59.8	85.0	91.9	29.5	76.8
092	LATH, PLASTER & GYPSUM BOARD	100.5	59.5	74.1	100.5	30.5	55.3	98.9	41.8	62.0	104.0	60.2	75.7	111.4	60.2	78.3	104.7	29.5	56.2
095	ACOUSTICAL TREATMENT & WOOD FLOORING	91.9	59.5	70.9	91.9	30.5	52.1	100.8	41.8	62.5	85.8	60.2	69.2	92.7	60.2	71.6	91.9	29.5	51.5
096	FLOORING & CARPET	87.0	32.5	73.8	87.2	30.3	73.3	88.3	30.3	74.2	90.0	64.4	83.8	95.3	64.4	87.8	90.6	62.4	83.8
099	PAINTING & WALL COVERINGS	91.3	31.3	56.1	91.3	29.1	54.8	92.6	34.7	68.5	97.7	68.0	80.2	97.7	68.0	80.2	91.3	33.2	57.2
9	FINISHES	96.7	49.0	72.2	97.2	30.9	64.2	96.3	38.7	66.8	97.8	60.9	78.9	92.3	60.9	76.2	98.8	37.1	67.2
10 - 14	TOTAL DIV. 10 - 14	100.0	65.4	92.7	100.0	30.2	85.3	100.0	63.1	92.2	100.0	69.4	93.5	100.0	69.5	93.6	100.0	28.7	85.0
15	MECHANICAL	99.4	49.4	77.1	99.4	28.9	68.0	100.1	47.6	76.7	99.9	57.3	80.9	99.9	57.3	80.9	99.4	26.4	66.8
16	ELECTRICAL	90.5	36.2	53.9	93.2	27.7	49.0	98.5	39.6	58.8	90.0	42.3	57.8	99.3	65.0	76.2	92.9	23.0	45.7
1 - 16	WEIGHTED AVERAGE	95.6	56.0	76.5	97.0	39.5	69.2	98.5	51.1	75.6	97.4	61.7	80.2	94.6	65.2	80.4	97.3	40.8	70.0

TENNESSEE / TEXAS

DIVISION		MEMPHIS 375,380 - 381			NASHVILLE 370 - 372			ABILENE 795 - 796			AMARILLO 790 - 791			AUSTIN 786 - 787			BEAUMONT 776 - 777		
		MAT.	INST.	TOTAL	MAT.	INST.	TOTAL	MAT.	INST.	TOTAL	MAT.	INST.	TOTAL	MAT.	INST.	TOTAL	MAT.	INST.	TOTAL
2	SITE WORK	103.0	93.5	95.7	96.3	101.3	100.2	102.6	83.5	87.9	102.6	84.6	88.7	91.1	85.9	87.1	98.5	81.5	85.4
031	CONCRETE FORMWORK	96.5	63.8	68.5	96.3	68.6	72.6	96.6	55.6	61.4	100.3	62.4	67.9	99.3	67.3	71.9	105.8	73.5	78.2
032	CONCRETE REINFORCEMENT	93.5	63.7	76.7	94.4	64.4	77.5	96.6	58.5	75.1	96.6	58.3	75.0	94.5	66.2	78.5	94.6	60.8	75.5
033	CAST IN PLACE CONCRETE	97.2	65.7	85.7	91.3	68.8	82.5	95.4	51.6	77.0	98.9	61.1	83.1	84.8	63.8	76.0	86.3	69.9	79.4
3	CONCRETE	88.7	67.7	78.1	86.7	69.6	78.1	88.8	55.6	72.1	90.8	61.9	76.2	79.2	66.4	72.8	85.9	70.3	78.0
4	MASONRY	82.5	67.5	73.2	84.2	65.7	72.7	97.4	56.0	71.7	101.1	54.2	71.9	97.5	61.2	74.9	100.4	73.4	83.6
5	METALS	96.0	93.7	95.2	99.4	92.4	96.9	97.4	71.2	88.0	97.4	71.0	87.9	98.0	75.3	89.8	97.2	74.2	88.9
6	WOOD & PLASTICS	93.9	65.1	79.0	90.4	70.1	79.9	95.9	57.1	75.9	99.4	65.1	81.7	97.0	70.4	83.2	106.1	76.3	90.7
7	THERMAL & MOISTURE PROTECTION	100.5	68.4	85.6	97.3	67.0	83.3	96.1	61.6	80.1	98.1	57.0	79.0	94.6	65.7	81.2	96.3	72.1	85.1
8	DOORS & WINDOWS	99.8	66.7	91.8	93.4	68.8	87.4	92.9	58.2	84.5	92.9	59.6	84.8	95.8	69.3	89.4	97.0	69.4	90.3
092	LATH, PLASTER & GYPSUM BOARD	101.7	64.2	77.5	100.9	69.6	80.7	94.8	56.8	70.3	94.8	65.0	75.6	96.1	70.3	79.5	93.8	76.6	82.7
095	ACOUSTICAL TREATMENT & WOOD FLOORING	95.9	64.2	75.4	92.7	69.6	77.7	101.7	56.8	72.6	101.7	65.0	77.9	91.1	70.3	77.6	107.7	76.6	87.6
096	FLOORING & CARPET	90.9	57.2	82.7	98.3	70.2	91.5	119.3	62.4	105.5	119.1	55.8	103.7	98.3	64.2	90.0	118.3	80.3	109.0
099	PAINTING & WALL COVERINGS	92.7	60.9	78.8	102.1	64.0	79.8	95.9	71.5	81.6	95.9	51.1	69.6	100.0	57.3	75.0	89.7	68.9	77.5
9	FINISHES	94.9	62.8	78.4	101.1	68.6	84.4	102.8	58.7	80.2	102.8	60.2	81.0	94.4	65.8	79.7	99.0	74.8	86.5
10 - 14	TOTAL DIV. 10 - 14	100.0	75.9	94.9	100.0	73.4	94.4	100.0	66.7	93.0	100.0	65.9	92.8	100.0	69.5	93.6	100.0	75.2	94.8
15	MECHANICAL	100.0	67.5	85.5	100.0	67.1	85.3	100.0	47.6	76.6	100.0	57.5	81.0	100.0	63.2	83.6	100.0	66.2	84.9
16	ELECTRICAL	97.6	79.8	85.6	101.1	61.8	74.6	98.1	51.1	66.4	99.0	62.0	74.1	97.5	69.9	78.9	95.1	77.7	83.4
1 - 16	WEIGHTED AVERAGE	96.4	74.2	85.7	96.6	72.6	85.0	97.4	58.5	78.6	98.0	63.2	81.2	95.3	68.8	82.5	97.0	73.3	85.4

TEXAS

DIVISION		BROWNWOOD 768 MAT.	INST.	TOTAL	BRYAN 778 MAT.	INST.	TOTAL	CHILDRESS 792 MAT.	INST.	TOTAL	CORPUS CHRISTI 783-784 MAT.	INST.	TOTAL	DALLAS 752-753 MAT.	INST.	TOTAL	DEL RIO 788 MAT.	INST.	TOTAL
2	SITE WORK	109.9	82.7	89.0	88.6	81.8	83.4	115.5	83.4	90.8	126.7	81.0	91.5	126.8	83.2	93.2	112.3	84.5	90.9
031	CONCRETE FORMWORK	89.9	40.2	47.3	76.3	54.4	57.6	90.4	61.3	65.5	103.0	53.5	60.6	97.1	68.9	72.9	87.7	34.8	42.4
032	CONCRETE REINFORCEMENT	98.6	35.6	63.1	98.8	66.4	80.6	97.4	56.6	74.4	92.7	55.9	72.0	96.1	65.3	78.7	94.5	24.7	55.2
033	CAST IN PLACE CONCRETE	102.8	47.6	79.7	69.1	73.1	70.8	97.8	59.7	81.9	100.5	58.7	83.0	100.6	68.3	87.0	108.1	40.5	79.8
3	CONCRETE	95.3	43.1	69.0	71.5	64.2	67.8	96.0	60.6	78.1	88.8	57.8	73.1	90.0	69.6	79.7	102.7	36.2	69.2
4	MASONRY	121.6	47.1	75.3	136.2	66.9	93.2	100.9	53.1	71.2	88.1	53.8	66.8	91.9	65.4	75.5	103.2	39.5	63.7
5	METALS	95.2	58.9	82.2	97.8	76.9	90.2	95.1	69.9	86.0	97.5	86.3	93.5	98.9	90.5	95.9	96.7	53.3	81.1
6	WOOD & PLASTICS	93.2	39.1	65.3	73.0	50.5	61.4	94.1	65.1	79.1	106.3	54.3	79.5	101.7	70.9	85.8	85.4	36.3	60.1
7	THERMAL & MOISTURE PROTECTION	96.2	49.6	74.6	90.7	68.4	80.3	96.8	55.9	77.8	97.2	59.0	79.5	93.1	70.2	82.5	94.7	39.4	69.1
8	DOORS & WINDOWS	89.0	39.8	77.1	99.3	59.6	89.7	90.1	60.1	82.8	103.0	52.7	90.8	106.1	67.4	96.7	96.7	30.7	80.7
092	LATH, PLASTER & GYPSUM BOARD	93.3	38.1	57.7	84.3	50.1	62.2	92.8	65.0	74.9	97.3	53.6	69.1	97.6	70.9	80.4	92.8	35.2	55.6
095	ACOUSTICAL TREATMENT & WOOD FLOORING	97.6	38.1	59.1	103.6	50.1	68.9	97.6	65.0	76.5	96.6	53.6	68.7	101.2	70.9	81.6	88.4	35.2	53.9
096	FLOORING & CARPET	117.9	53.3	102.2	89.4	69.7	84.6	116.8	55.8	102.0	110.8	60.7	98.6	109.4	67.9	99.3	95.3	25.6	78.4
099	PAINTING & WALL COVERINGS	95.9	41.1	63.7	85.8	74.8	79.3	95.9	54.2	71.4	115.1	51.2	77.6	102.6	66.9	81.6	100.0	36.1	62.5
9	FINISHES	101.8	42.4	71.4	87.0	58.1	72.2	101.9	60.0	80.4	101.1	54.3	77.1	102.4	68.8	85.2	93.6	34.0	63.0
10-14	TOTAL DIV. 10-14	100.0	59.4	91.4	100.0	67.6	93.2	100.0	65.0	92.6	100.0	70.6	93.8	100.0	71.3	93.9	100.0	55.4	90.6
15	MECHANICAL	96.7	40.7	71.7	96.7	73.6	86.4	96.7	55.8	78.5	100.0	48.0	76.8	100.0	68.6	86.0	96.6	25.7	65.0
16	ELECTRICAL	91.5	38.9	56.0	91.1	75.3	80.5	98.1	53.1	67.8	94.3	64.0	73.9	98.9	73.8	82.0	97.5	27.7	50.4
1-16	WEIGHTED AVERAGE	97.5	48.1	73.7	94.5	70.2	82.7	97.3	60.8	79.7	98.4	61.2	80.5	99.6	72.9	86.7	98.0	39.6	69.8

TEXAS

DIVISION		DENTON 762 MAT.	INST.	TOTAL	EASTLAND 764 MAT.	INST.	TOTAL	EL PASO 798-799,885 MAT.	INST.	TOTAL	FORT WORTH 760-761 MAT.	INST.	TOTAL	GALVESTON 775 MAT.	INST.	TOTAL	GIDDINGS 789 MAT.	INST.	TOTAL
2	SITE WORK	110.3	76.4	84.2	113.5	82.8	89.9	103.1	83.6	88.1	102.4	83.8	88.1	129.6	79.3	90.9	98.1	85.1	88.1
031	CONCRETE FORMWORK	102.6	53.0	60.1	92.6	47.5	54.0	97.3	42.5	50.4	99.4	68.5	73.0	86.3	77.9	79.1	86.1	47.7	53.2
032	CONCRETE REINFORCEMENT	98.2	54.9	73.8	99.3	58.7	76.4	96.6	58.3	75.0	96.6	65.0	78.8	98.6	67.8	81.2	97.4	59.7	76.2
033	CAST IN PLACE CONCRETE	79.5	63.4	72.8	108.7	48.5	83.5	97.7	46.5	76.2	93.7	62.5	80.6	91.8	74.3	84.5	91.7	49.2	73.9
3	CONCRETE	75.9	59.1	67.5	99.9	50.7	75.1	90.4	48.1	68.9	88.2	66.3	77.2	87.1	76.2	81.6	84.6	51.2	67.8
4	MASONRY	128.7	53.9	82.2	96.6	47.3	66.0	96.2	53.9	69.9	93.0	61.3	73.3	95.7	67.0	77.9	110.1	47.2	71.0
5	METALS	95.3	88.2	92.7	94.8	62.9	83.3	97.3	69.8	87.4	97.2	74.3	89.0	99.2	94.1	97.4	96.3	63.5	84.4
6	WOOD & PLASTICS	109.0	54.3	80.8	101.3	49.2	74.4	97.4	41.1	68.3	103.9	70.8	86.8	86.9	82.1	84.4	85.3	49.3	66.7
7	THERMAL & MOISTURE PROTECTION	95.4	59.1	78.5	96.8	50.0	75.1	95.7	57.9	78.2	96.6	65.2	82.0	89.1	71.7	81.0	95.2	50.4	74.4
8	DOORS & WINDOWS	106.0	57.2	94.2	78.1	44.9	70.0	92.9	45.8	81.5	87.6	67.3	82.7	104.9	77.0	98.2	95.9	44.9	83.6
092	LATH, PLASTER & GYPSUM BOARD	98.5	53.7	69.6	93.3	48.6	64.5	94.8	40.2	59.6	94.8	70.9	79.4	89.7	82.5	85.0	91.4	48.6	63.8
095	ACOUSTICAL TREATMENT & WOOD FLOORING	108.5	53.7	73.0	97.6	48.6	65.8	101.7	40.2	61.9	101.7	70.9	81.7	106.3	82.5	90.8	87.0	48.6	62.1
096	FLOORING & CARPET	109.4	62.2	97.9	153.5	38.4	125.5	119.3	62.5	105.5	154.8	62.2	132.3	105.2	69.7	96.6	94.8	38.4	81.1
099	PAINTING & WALL COVERINGS	106.6	41.6	68.4	97.1	42.9	65.3	95.9	40.0	64.8	97.1	72.8	82.8	98.8	64.6	78.7	100.0	42.9	66.5
9	FINISHES	98.5	53.2	75.3	113.9	45.8	79.0	102.8	46.2	73.8	114.7	68.2	90.9	97.3	75.3	86.1	92.3	45.7	68.4
10-14	TOTAL DIV. 10-14	100.0	62.0	92.0	100.0	60.3	91.6	100.0	64.3	92.5	100.0	71.0	93.9	100.0	76.1	95.0	100.0	58.3	91.2
15	MECHANICAL	96.7	53.8	77.5	96.7	41.1	71.9	100.0	46.8	76.3	100.0	67.1	85.3	96.6	71.7	85.5	96.7	41.1	71.9
16	ELECTRICAL	94.0	63.1	73.2	91.5	48.5	62.5	96.9	58.8	71.2	96.9	67.9	77.3	94.0	73.6	80.2	92.3	48.5	62.8
1-16	WEIGHTED AVERAGE	97.5	61.9	80.3	96.8	51.9	75.2	97.3	56.0	77.4	97.7	69.1	83.9	97.6	75.9	87.1	95.2	52.2	74.4

TEXAS

DIVISION		GREENVILLE 754 MAT.	INST.	TOTAL	HOUSTON 770-772 MAT.	INST.	TOTAL	HUNTSVILLE 773 MAT.	INST.	TOTAL	LAREDO 780 MAT.	INST.	TOTAL	LONGVIEW 756 MAT.	INST.	TOTAL	LUBBOCK 793-794 MAT.	INST.	TOTAL
2	SITE WORK	118.3	78.7	87.8	126.4	79.3	90.1	108.7	80.2	86.8	91.3	85.6	86.9	110.2	86.6	92.0	132.8	81.7	93.5
031	CONCRETE FORMWORK	82.5	39.8	45.9	92.5	79.7	81.6	86.2	49.7	54.9	90.6	53.7	59.0	75.4	41.1	46.0	98.7	57.7	63.5
032	CONCRETE REINFORCEMENT	98.9	35.7	63.3	97.2	68.0	80.7	100.0	40.5	66.5	94.5	56.3	73.0	98.1	32.6	61.2	95.8	58.2	74.6
033	CAST IN PLACE CONCRETE	91.5	47.4	73.0	88.9	76.8	83.9	95.0	58.0	79.5	77.4	63.2	71.5	106.5	43.8	80.2	95.6	61.1	81.1
3	CONCRETE	83.0	44.0	63.3	85.2	77.9	81.5	92.3	51.9	71.9	79.0	58.4	68.6	96.3	41.7	68.8	87.6	60.9	74.2
4	MASONRY	132.1	45.6	78.4	95.5	74.7	82.6	135.9	47.4	80.9	97.3	60.2	74.3	129.7	42.1	75.3	97.1	50.0	67.8
5	METALS	96.8	74.6	88.8	101.4	94.5	98.9	97.7	61.8	84.7	98.8	70.5	88.6	89.4	58.0	78.1	101.5	87.4	96.4
6	WOOD & PLASTICS	89.3	39.1	63.4	89.9	82.1	85.9	84.3	52.5	67.9	85.1	54.2	69.1	80.9	42.4	61.0	100.2	60.0	79.5
7	THERMAL & MOISTURE PROTECTION	92.7	50.9	73.3	89.0	75.0	82.5	91.9	51.8	73.3	93.5	62.2	79.0	92.9	43.7	70.1	89.7	60.9	76.3
8	DOORS & WINDOWS	107.0	39.8	90.7	105.2	77.0	98.4	99.3	44.8	86.0	96.4	53.4	86.0	96.0	38.5	82.1	105.3	57.4	93.7
092	LATH, PLASTER & GYPSUM BOARD	93.3	38.1	57.7	91.7	82.5	85.7	88.1	52.1	64.8	93.4	53.6	67.7	90.3	41.7	58.9	95.4	59.6	72.3
095	ACOUSTICAL TREATMENT & WOOD FLOORING	98.5	38.1	59.4	110.4	82.5	92.3	103.6	52.1	70.2	91.1	53.6	66.8	94.3	41.7	60.3	104.4	59.6	75.4
096	FLOORING & CARPET	104.6	62.2	94.3	106.3	69.7	97.4	93.1	38.4	79.8	95.2	60.7	86.8	110.9	54.8	97.3	110.3	52.7	96.3
099	PAINTING & WALL COVERINGS	102.6	35.3	63.1	98.8	74.8	84.7	85.8	44.0	61.3	100.0	56.7	74.6	93.3	47.1	66.2	107.3	45.8	71.2
9	FINISHES	99.3	43.2	70.5	105.5	77.7	91.2	89.6	47.5	68.0	93.2	54.9	73.5	104.6	44.4	73.7	104.3	55.4	79.2
10-14	TOTAL DIV. 10-14	100.0	58.8	91.3	100.0	78.0	95.4	100.0	58.0	91.1	100.0	68.9	93.4	100.0	32.3	85.7	100.0	67.5	93.1
15	MECHANICAL	96.7	39.9	71.4	99.9	76.1	89.3	96.7	38.7	70.8	99.9	56.5	80.5	96.6	39.3	71.0	99.9	54.9	79.8
16	ELECTRICAL	94.0	39.0	56.9	96.9	75.4	82.4	91.1	47.6	61.7	97.5	68.7	78.0	96.3	60.0	71.8	96.5	57.9	70.5
1-16	WEIGHTED AVERAGE	98.9	49.1	74.9	99.4	78.4	89.3	97.9	51.3	75.4	95.2	63.5	79.9	98.4	50.4	75.2	100.0	62.1	81.7

COST INDEXES

637

TEXAS

DIVISION		LUFKIN 759 MAT.	INST.	TOTAL	MCALLEN 785 MAT.	INST.	TOTAL	MCKINNEY 750 MAT.	INST.	TOTAL	MIDLAND 797 MAT.	INST.	TOTAL	ODESSA 797 MAT.	INST.	TOTAL	PALESTINE 758 MAT.	INST.	TOTAL
2	SITE WORK	103.4	86.7	90.5	131.6	81.0	92.7	113.2	79.6	87.4	137.1	81.7	94.5	102.6	84.0	88.3	112.7	86.1	92.3
031	CONCRETE FORMWORK	78.9	58.1	61.1	97.7	50.3	57.1	81.0	61.0	63.9	100.6	55.4	61.9	96.6	55.2	61.2	66.9	47.5	50.3
032	CONCRETE REINFORCEMENT	99.7	45.2	69.0	94.1	55.9	72.5	98.9	59.2	76.5	101.6	58.2	77.1	98.8	58.1	75.9	97.3	59.8	76.2
033	CAST IN PLACE CONCRETE	95.3	51.3	76.9	109.2	56.4	87.1	85.9	65.5	77.3	101.5	57.4	83.0	95.4	56.0	78.9	87.1	47.9	70.6
3	CONCRETE	89.1	54.4	71.6	94.1	55.6	74.7	78.9	64.0	71.4	92.6	58.6	75.5	89.1	57.0	72.9	88.6	50.6	69.5
4	MASONRY	104.0	55.6	73.9	101.4	53.8	71.8	140.4	58.8	89.7	112.8	50.1	73.8	97.4	50.0	68.0	97.7	49.4	67.7
5	METALS	96.1	69.9	86.6	96.4	86.0	92.6	96.7	87.6	93.4	99.7	87.1	95.1	97.6	70.5	87.8	95.9	62.4	83.8
6	WOOD & PLASTICS	89.1	61.2	74.7	103.9	50.5	76.4	87.7	62.3	74.6	106.4	57.0	80.9	95.9	56.9	75.8	76.0	49.0	62.1
7	THERMAL & MOISTURE PROTECTION	92.6	56.6	75.9	97.3	55.4	77.8	92.5	65.0	79.8	90.0	56.1	74.3	96.1	55.0	77.1	93.1	53.3	74.7
8	DOORS & WINDOWS	77.8	56.1	72.5	101.5	51.3	89.4	107.0	61.5	96.0	104.3	55.3	92.4	92.9	55.2	83.8	77.7	45.7	70.0
092	LATH, PLASTER & GYPSUM BOARD	91.1	61.1	71.7	96.2	49.7	66.2	92.8	62.1	73.0	96.6	56.5	70.7	94.8	56.5	70.1	86.9	48.6	62.2
095	ACOUSTICAL TREATMENT & WOOD FLOORING	88.9	61.1	70.9	93.9	49.7	65.2	98.5	62.1	74.9	100.3	56.5	71.9	101.7	56.5	72.4	88.9	48.6	62.8
096	FLOORING & CARPET	149.1	51.1	125.3	110.2	59.0	97.7	104.1	62.2	93.9	112.0	52.1	97.4	119.3	52.1	102.9	139.3	38.4	114.7
099	PAINTING & WALL COVERINGS	93.3	53.1	69.7	115.1	40.5	71.4	102.6	66.2	81.2	107.3	45.2	70.9	95.9	45.8	66.5	93.3	42.9	63.7
9	FINISHES	115.9	56.5	85.5	100.4	50.6	74.9	98.8	61.8	79.8	104.6	53.6	78.4	102.8	53.5	77.6	112.7	45.3	78.1
10 - 14	TOTAL DIV. 10 - 14	100.0	67.7	93.2	100.0	68.8	93.4	100.0	64.6	92.5	100.0	67.8	93.2	100.0	67.4	93.1	100.0	59.9	91.5
15	MECHANICAL	96.6	46.0	74.0	96.7	44.4	73.4	96.7	61.0	80.8	96.6	51.2	76.4	100.0	51.2	78.2	96.6	41.1	71.8
16	ELECTRICAL	98.5	66.5	76.9	93.1	49.2	63.5	94.0	67.5	76.1	96.5	58.0	70.6	98.1	58.0	71.0	91.9	55.0	67.0
1 - 16	WEIGHTED AVERAGE	96.3	60.4	79.0	98.6	57.1	78.6	98.7	66.9	83.4	100.5	60.6	81.3	97.4	59.1	78.9	95.2	53.5	75.1

TEXAS

DIVISION		SAN ANGELO 769 MAT.	INST.	TOTAL	SAN ANTONIO 781 - 782 MAT.	INST.	TOTAL	TEMPLE 765 MAT.	INST.	TOTAL	TEXARKANA 755 MAT.	INST.	TOTAL	TYLER 757 MAT.	INST.	TOTAL	VICTORIA 779 MAT.	INST.	TOTAL
2	SITE WORK	105.6	83.8	88.9	91.1	89.6	89.9	92.3	83.5	85.5	96.6	86.6	88.9	109.6	86.6	91.9	135.6	76.1	89.8
031	CONCRETE FORMWORK	90.3	52.2	57.7	90.6	69.3	72.3	96.2	50.4	56.9	89.8	48.7	54.6	81.4	47.0	51.9	87.1	58.5	62.6
032	CONCRETE REINFORCEMENT	97.8	58.0	75.4	94.5	61.7	76.0	99.0	60.1	77.1	97.2	64.5	78.7	98.1	64.8	79.3	93.5	45.3	66.3
033	CAST IN PLACE CONCRETE	97.0	54.3	79.1	76.0	66.6	72.0	79.5	62.6	72.4	87.7	55.3	74.1	104.6	53.7	83.3	103.1	53.0	82.1
3	CONCRETE	90.9	55.0	72.8	78.3	67.4	72.8	78.5	57.5	67.9	83.5	55.0	69.2	95.9	53.8	74.7	93.7	56.3	74.8
4	MASONRY	118.7	47.5	74.5	97.1	64.6	76.9	128.2	59.5	85.5	143.9	47.7	84.1	134.7	52.3	83.5	112.0	55.7	77.1
5	METALS	95.4	69.2	86.0	99.4	73.7	90.2	94.7	71.2	86.2	89.3	72.2	83.1	95.8	72.9	87.5	97.5	87.3	93.8
6	WOOD & PLASTICS	93.5	54.3	73.3	85.1	72.2	78.4	105.3	49.2	76.3	96.8	48.4	71.8	91.6	45.6	67.8	90.5	61.4	75.5
7	THERMAL & MOISTURE PROTECTION	96.0	53.9	76.5	93.5	67.9	81.6	95.6	59.6	78.9	92.5	56.0	75.6	93.0	58.3	76.9	93.4	57.7	76.9
8	DOORS & WINDOWS	89.1	53.7	80.5	96.4	67.7	89.5	76.0	49.9	69.6	96.0	53.5	85.7	77.7	51.9	71.5	104.7	58.5	93.5
092	LATH, PLASTER & GYPSUM BOARD	93.3	53.9	67.9	93.4	72.1	79.7	94.9	48.6	65.0	94.7	47.9	64.5	91.6	45.0	61.5	89.4	61.1	71.1
095	ACOUSTICAL TREATMENT & WOOD FLOORING	97.6	53.9	69.3	91.1	72.1	78.8	97.6	48.6	65.8	93.0	47.9	63.8	88.9	45.0	60.4	111.8	61.1	78.9
096	FLOORING & CARPET	118.1	46.5	100.7	95.2	61.5	87.0	156.5	63.9	133.9	119.9	49.5	102.8	151.0	56.7	128.1	104.7	51.1	91.6
099	PAINTING & WALL COVERINGS	95.9	47.3	67.4	100.0	56.7	74.6	97.1	47.2	67.9	93.3	55.3	71.0	93.3	55.3	71.0	99.0	53.1	72.0
9	FINISHES	101.7	50.5	75.5	93.2	66.6	79.5	114.0	51.7	82.1	107.2	48.9	77.3	116.9	48.8	82.0	96.2	56.8	76.0
10 - 14	TOTAL DIV. 10 - 14	100.0	65.1	92.6	100.0	73.3	94.4	100.0	60.3	91.6	100.0	65.1	92.6	100.0	64.7	92.6	100.0	64.2	92.4
15	MECHANICAL	96.7	46.6	74.3	99.9	73.5	88.1	96.7	48.9	75.4	96.6	53.4	77.3	96.6	65.1	82.6	96.7	46.0	74.1
16	ELECTRICAL	98.1	46.2	63.1	97.5	68.7	78.0	93.6	58.8	70.1	98.3	63.4	74.8	96.3	60.2	71.9	101.2	66.6	77.9
1 - 16	WEIGHTED AVERAGE	97.2	55.1	76.9	95.2	71.5	83.7	95.3	59.2	77.9	97.8	59.6	79.4	98.9	61.6	80.9	99.7	61.3	81.2

TEXAS / UTAH

DIVISION		WACO 766 - 767 MAT.	INST.	TOTAL	WAXAHACHIE 751 MAT.	INST.	TOTAL	WHARTON 774 MAT.	INST.	TOTAL	WICHITA FALLS 763 MAT.	INST.	TOTAL	LOGAN 843 MAT.	INST.	TOTAL	OGDEN 842,844 MAT.	INST.	TOTAL
2	SITE WORK	101.4	83.8	87.8	116.4	79.5	88.0	144.1	77.6	93.0	102.1	83.5	87.8	99.2	100.5	100.2	85.7	100.5	97.1
031	CONCRETE FORMWORK	98.8	54.6	60.9	81.1	61.8	64.6	79.3	52.8	56.6	98.8	57.9	63.8	101.7	65.6	70.7	101.7	65.6	70.7
032	CONCRETE REINFORCEMENT	96.6	66.0	79.3	98.9	60.8	77.4	98.0	63.4	78.5	96.6	57.9	74.8	101.4	72.5	85.2	101.4	72.5	85.2
033	CAST IN PLACE CONCRETE	86.1	64.7	77.1	90.6	52.9	74.8	105.9	63.5	88.1	92.0	59.3	78.3	92.0	71.7	83.5	93.4	71.7	84.3
3	CONCRETE	84.4	61.1	72.7	82.1	60.3	71.1	97.1	60.6	78.7	87.3	59.2	73.2	112.0	69.5	90.6	100.8	69.5	85.0
4	MASONRY	94.9	62.8	75.0	132.5	60.3	87.7	96.8	48.7	66.9	95.3	58.6	72.5	120.4	63.9	85.3	113.3	63.9	82.6
5	METALS	97.4	74.2	89.0	96.8	87.7	93.5	99.2	88.7	95.4	97.3	71.7	88.1	105.7	77.2	95.4	106.1	77.2	95.7
6	WOOD & PLASTICS	102.6	52.8	76.9	87.7	65.1	76.0	79.3	55.5	67.0	102.6	59.8	80.5	91.3	65.7	78.1	91.3	65.7	78.1
7	THERMAL & MOISTURE PROTECTION	96.8	61.8	80.6	92.7	61.3	78.1	89.4	62.4	76.9	96.8	62.1	80.7	102.9	71.0	88.1	101.5	71.0	87.4
8	DOORS & WINDOWS	87.6	53.5	79.3	107.0	63.1	96.4	104.9	56.7	93.2	87.6	59.6	80.8	90.0	64.0	83.7	90.0	64.0	83.7
092	LATH, PLASTER & GYPSUM BOARD	94.8	52.3	67.4	92.8	65.0	74.9	87.1	55.0	66.4	94.8	59.6	72.1	89.6	64.4	73.3	89.6	64.4	73.3
095	ACOUSTICAL TREATMENT & WOOD FLOORING	101.7	52.3	69.7	98.5	65.0	76.8	106.3	55.0	73.1	101.7	59.6	74.4	101.0	64.4	77.3	101.0	64.4	77.3
096	FLOORING & CARPET	155.0	48.4	129.1	104.1	51.1	91.2	102.5	38.4	86.9	155.5	66.6	133.9	106.1	56.8	94.1	106.1	56.8	94.1
099	PAINTING & WALL COVERINGS	97.1	47.2	67.9	102.6	53.1	73.6	98.8	44.0	66.7	99.1	74.9	84.9	113.2	60.4	82.2	113.2	60.4	82.2
9	FINISHES	114.7	51.9	82.5	99.0	59.3	78.7	96.8	49.5	72.5	115.0	61.5	87.6	100.9	62.7	81.3	99.8	62.7	80.8
10 - 14	TOTAL DIV. 10 - 14	100.0	68.6	93.4	100.0	65.2	92.7	100.0	61.7	91.9	100.0	64.3	92.5	100.0	80.2	95.8	100.0	80.2	95.8
15	MECHANICAL	100.0	59.9	82.1	96.7	57.5	79.2	96.6	37.9	70.4	100.0	56.8	80.7	99.8	72.7	87.7	99.8	72.7	87.7
16	ELECTRICAL	98.1	66.1	76.5	93.9	77.1	82.6	100.5	47.6	64.9	102.9	62.6	75.8	93.9	70.4	78.0	93.9	70.4	78.0
1 - 16	WEIGHTED AVERAGE	97.4	64.1	81.3	98.8	67.2	83.5	99.6	55.8	78.5	98.2	63.4	81.4	101.8	72.8	87.8	99.6	72.8	86.7

UTAH / VERMONT

DIVISION		PRICE 845 MAT.	INST.	TOTAL	PROVO 846-847 MAT.	INST.	TOTAL	SALT LAKE CITY 840-841 MAT.	INST.	TOTAL	BELLOWS FALLS 051 MAT.	INST.	TOTAL	BENNINGTON 052 MAT.	INST.	TOTAL	BRATTLEBORO 053 MAT.	INST.	TOTAL
2	SITE WORK	96.9	97.6	97.4	95.2	98.5	97.7	85.3	100.4	96.9	86.8	94.3	92.6	86.2	93.9	92.2	87.7	94.3	92.8
031	CONCRETE FORMWORK	103.9	41.2	50.2	103.5	65.6	71.1	100.3	65.6	70.6	94.4	34.6	43.2	91.3	31.2	39.8	94.8	34.6	43.2
032	CONCRETE REINFORCEMENT	108.8	72.4	88.3	109.7	72.6	88.8	99.3	72.6	84.3	93.3	74.2	82.5	93.3	24.2	54.3	92.4	74.2	82.2
033	CAST IN PLACE CONCRETE	92.1	55.8	76.9	92.1	71.8	83.6	101.0	71.8	88.7	100.9	50.8	79.9	100.9	41.4	76.0	104.1	50.8	81.8
3	CONCRETE	113.2	53.3	83.0	111.6	69.5	90.4	104.2	69.5	86.7	107.3	48.4	77.6	107.1	34.5	70.5	110.0	48.4	79.0
4	MASONRY	125.3	53.7	80.8	125.3	63.9	87.1	132.2	63.9	89.7	93.9	38.9	59.7	101.6	38.4	62.4	101.1	38.9	62.4
5	METALS	102.8	76.5	93.4	103.1	77.3	93.8	104.9	77.3	95.0	96.0	65.9	85.2	96.0	50.2	79.5	96.0	65.9	85.2
6	WOOD & PLASTICS	94.7	38.8	65.8	93.3	65.7	79.0	91.5	65.7	78.2	101.3	32.5	65.8	97.8	29.8	62.7	101.7	32.5	66.0
7	THERMAL & MOISTURE PROTECTION	105.1	64.1	86.1	105.1	71.0	89.3	103.9	71.0	88.7	100.7	54.3	79.2	100.6	52.7	78.4	100.8	54.3	79.2
8	DOORS & WINDOWS	94.4	49.5	83.5	94.4	64.0	87.0	90.0	64.0	83.7	105.9	39.7	89.8	105.9	25.1	86.3	105.9	39.7	89.8
092	LATH, PLASTER & GYPSUM BOARD	89.4	36.6	55.4	88.9	64.4	73.1	89.6	64.4	73.3	96.3	29.3	53.1	94.7	26.5	50.7	96.3	29.3	53.1
095	ACOUSTICAL TREATMENT & WOOD FLOORING	95.6	36.6	57.4	95.6	64.4	75.4	101.0	64.4	77.3	92.1	29.3	51.4	92.1	26.5	49.6	92.1	29.3	51.4
096	FLOORING & CARPET	107.2	56.8	94.9	106.7	56.8	94.6	105.6	56.8	93.7	96.9	48.0	85.0	95.8	48.0	84.2	97.0	48.0	85.1
099	PAINTING & WALL COVERINGS	113.2	60.4	82.2	113.2	62.9	83.7	113.2	62.9	83.7	89.3	20.0	48.6	89.3	18.6	47.8	89.3	20.0	48.6
9	FINISHES	100.6	44.6	71.9	100.3	63.0	81.2	100.1	63.0	81.1	94.0	34.3	63.4	93.4	32.5	62.2	94.1	34.3	63.5
10-14	TOTAL DIV. 10 - 14	100.0	72.5	94.2	100.0	80.2	95.8	100.0	80.2	95.8	100.0	58.6	91.3	100.0	58.1	91.2	100.0	58.6	91.3
15	MECHANICAL	97.9	52.1	77.5	99.8	72.7	87.7	99.9	72.7	87.8	96.7	41.3	72.0	96.7	20.3	62.6	96.7	41.3	72.0
16	ELECTRICAL	101.7	58.2	72.4	94.4	77.4	82.9	94.8	77.4	83.1	104.7	23.1	49.6	104.7	26.7	52.1	104.7	23.1	49.6
1-16	WEIGHTED AVERAGE	102.5	59.9	81.9	102.1	73.8	88.4	101.0	74.0	88.0	99.3	46.4	73.7	99.5	38.7	70.2	100.0	46.4	74.1

VERMONT

DIVISION		BURLINGTON 054 MAT.	INST.	TOTAL	GUILDHALL 059 MAT.	INST.	TOTAL	MONTPELIER 056 MAT.	INST.	TOTAL	RUTLAND 057 MAT.	INST.	TOTAL	ST. JOHNSBURY 058 MAT.	INST.	TOTAL	WHITE RIVER JCT. 050 MAT.	INST.	TOTAL
2	SITE WORK	89.2	96.3	94.7	85.8	94.7	92.6	88.4	96.1	94.3	89.2	96.3	94.7	85.9	94.7	92.7	89.9	94.2	93.2
031	CONCRETE FORMWORK	86.6	60.5	64.2	90.8	37.2	44.9	87.8	60.5	64.4	98.6	60.5	66.0	87.8	37.4	44.6	87.8	31.3	39.4
032	CONCRETE REINFORCEMENT	116.7	75.4	93.4	94.2	48.2	68.2	92.9	75.4	83.0	116.7	75.4	93.4	92.4	74.5	82.3	93.3	35.7	60.8
033	CAST IN PLACE CONCRETE	103.7	65.8	87.9	97.9	58.7	81.5	104.1	65.8	88.1	98.9	65.9	85.0	97.9	58.7	81.5	104.1	56.3	84.1
3	CONCRETE	112.5	65.3	88.7	104.5	47.6	75.8	109.6	65.2	87.2	111.0	65.3	87.9	104.1	52.1	77.8	112.0	41.8	76.6
4	MASONRY	94.9	68.7	78.6	101.4	45.7	66.8	83.0	68.7	74.1	83.0	68.7	74.1	128.2	45.7	77.0	113.3	38.4	66.8
5	METALS	101.8	70.5	90.5	96.0	59.0	82.7	96.0	70.4	86.8	101.1	70.5	90.0	96.0	59.1	82.7	96.0	53.9	80.8
6	WOOD & PLASTICS	86.9	59.6	72.8	95.4	35.6	64.5	90.3	59.6	74.4	101.8	59.6	80.0	90.3	35.6	62.1	93.9	29.5	60.7
7	THERMAL & MOISTURE PROTECTION	101.0	65.2	84.4	100.4	46.9	75.6	100.5	65.2	84.1	100.9	65.2	84.3	100.3	47.0	75.6	100.8	40.5	72.8
8	DOORS & WINDOWS	109.1	54.4	95.8	105.9	36.0	88.9	105.9	54.4	93.4	109.1	54.4	95.8	105.9	36.0	88.9	105.9	28.2	87.0
092	LATH, PLASTER & GYPSUM BOARD	98.5	57.3	71.9	100.6	32.5	56.7	105.3	57.3	74.3	98.5	57.3	71.9	105.3	32.5	58.4	93.2	26.2	50.0
095	ACOUSTICAL TREATMENT & WOOD FLOORING	101.7	57.3	72.9	92.1	32.5	53.5	92.1	57.3	69.5	101.7	57.3	72.9	92.1	32.5	53.5	92.1	26.2	49.4
096	FLOORING & CARPET	96.9	79.2	92.6	100.1	48.0	87.5	103.8	79.2	97.8	96.9	79.2	92.6	103.8	48.0	90.2	94.7	48.0	83.3
099	PAINTING & WALL COVERINGS	89.3	49.0	65.7	89.3	32.7	56.1	89.3	49.0	65.7	89.3	49.0	65.7	89.3	32.7	56.1	89.3	26.1	52.2
9	FINISHES	96.3	62.7	79.1	95.6	37.9	66.0	97.6	62.7	79.7	96.2	62.7	79.0	97.4	37.9	66.9	93.1	33.1	62.4
10-14	TOTAL DIV. 10 - 14	100.0	67.0	93.0	100.0	59.5	91.5	100.0	67.0	93.0	100.0	67.0	93.0	100.0	59.5	91.5	100.0	58.1	91.2
15	MECHANICAL	100.0	63.7	83.8	96.7	46.6	74.3	96.7	63.7	82.0	100.0	63.7	83.8	96.7	46.6	74.3	96.7	32.3	68.0
16	ELECTRICAL	105.9	50.6	68.6	104.7	33.5	56.6	104.7	50.6	68.2	104.7	50.6	68.2	104.7	33.5	56.6	104.7	36.4	58.6
1-16	WEIGHTED AVERAGE	102.2	65.7	84.6	99.4	49.2	75.2	99.2	65.7	83.0	101.3	65.7	84.1	100.9	49.8	76.2	100.8	43.8	73.3

VIRGINIA

DIVISION		ALEXANDRIA 223 MAT.	INST.	TOTAL	ARLINGTON 222 MAT.	INST.	TOTAL	BRISTOL 242 MAT.	INST.	TOTAL	CHARLOTTESVILLE 229 MAT.	INST.	TOTAL	CULPEPER 227 MAT.	INST.	TOTAL	FAIRFAX 220-221 MAT.	INST.	TOTAL
2	SITE WORK	116.1	85.0	92.2	126.0	84.7	94.2	108.1	82.8	88.6	112.7	85.6	91.9	114.6	84.2	91.2	125.2	84.7	94.0
031	CONCRETE FORMWORK	91.8	74.3	76.8	88.6	73.0	75.2	79.2	40.0	45.6	76.6	48.9	52.9	73.3	67.6	68.4	76.9	72.6	73.3
032	CONCRETE REINFORCEMENT	83.8	89.5	87.0	94.5	81.4	87.1	94.5	69.6	80.5	93.9	76.3	84.0	94.5	74.1	83.0	94.5	81.4	87.1
033	CAST IN PLACE CONCRETE	104.5	82.0	95.1	101.7	78.1	91.8	101.2	49.2	79.4	105.3	62.5	87.4	104.1	50.4	81.6	101.7	77.7	91.7
3	CONCRETE	99.1	81.1	90.0	106.5	77.6	91.9	102.3	50.9	76.4	103.0	60.7	81.7	99.6	64.3	81.8	105.7	77.3	91.4
4	MASONRY	85.8	72.0	77.2	96.5	69.4	79.7	85.0	35.5	54.3	109.0	53.7	74.7	98.7	43.8	64.6	96.5	68.8	79.3
5	METALS	95.1	99.1	96.5	94.0	94.9	94.3	93.2	85.6	90.5	93.3	92.7	93.1	93.3	91.0	92.5	93.3	94.8	93.8
6	WOOD & PLASTICS	96.3	74.4	85.0	91.6	74.4	82.7	82.2	40.2	60.5	79.6	48.5	63.5	78.0	74.4	76.1	82.2	74.4	78.2
7	THERMAL & MOISTURE PROTECTION	95.8	80.4	88.7	96.5	76.9	87.4	96.0	45.3	72.5	95.5	55.2	76.8	95.7	55.6	77.1	96.4	76.6	87.2
8	DOORS & WINDOWS	96.7	77.2	92.0	94.7	74.6	89.9	97.9	46.2	85.4	96.0	52.2	85.4	96.4	73.2	90.7	94.7	74.6	89.9
092	LATH, PLASTER & GYPSUM BOARD	106.4	73.4	85.1	101.6	73.4	83.4	98.1	38.0	59.3	96.5	45.6	63.7	97.3	73.4	81.9	98.4	73.4	82.2
095	ACOUSTICAL TREATMENT & WOOD FLOORING	91.4	73.4	79.7	88.7	73.4	78.8	87.3	38.0	55.4	87.3	45.6	60.3	88.7	73.4	78.8	88.7	73.4	78.8
096	FLOORING & CARPET	93.3	82.9	90.8	91.8	72.3	87.1	87.8	35.0	74.9	86.3	39.4	74.9	86.3	82.9	85.5	88.1	72.3	84.3
099	PAINTING & WALL COVERINGS	117.1	81.7	96.3	117.1	98.0	105.9	105.8	40.8	67.6	105.8	67.1	83.1	117.1	48.7	77.0	117.1	85.2	98.4
9	FINISHES	94.7	76.3	85.3	93.9	75.9	84.7	90.6	38.5	63.9	89.9	48.1	68.5	90.8	69.4	79.8	92.3	74.3	83.0
10-14	TOTAL DIV. 10 - 14	100.0	87.5	97.4	100.0	80.0	95.8	100.0	67.5	93.1	100.0	74.2	94.6	100.0	73.0	94.3	100.0	79.7	95.7
15	MECHANICAL	100.1	84.2	93.0	100.1	77.9	90.2	96.8	51.0	76.4	96.8	67.5	83.8	96.8	48.7	75.4	96.8	76.7	87.9
16	ELECTRICAL	99.8	93.3	95.4	96.2	86.8	89.9	99.7	38.9	58.7	99.6	72.6	81.4	103.9	62.6	76.1	101.6	83.9	89.7
1-16	WEIGHTED AVERAGE	97.7	84.2	91.2	98.7	80.4	89.9	96.4	52.3	75.1	97.6	67.0	82.8	97.1	64.6	81.4	97.9	79.4	88.9

COST INDEXES

VIRGINIA

DIVISION		FARMVILLE 239			FREDERICKSBURG 224 - 225			GRUNDY 246			HARRISONBURG 228			LYNCHBURG 245			NEWPORT NEWS 236		
		MAT.	INST.	TOTAL	MAT.	INST.	TOTAL	MAT.	INST.	TOTAL	MAT.	INST.	TOTAL	MAT.	INST.	TOTAL	MAT.	INST.	TOTAL
2	SITE WORK	109.0	85.2	90.7	114.0	84.2	91.1	106.0	82.2	87.7	122.3	84.1	92.9	106.8	82.9	88.4	107.3	86.1	91.0
031	CONCRETE FORMWORK	93.3	33.7	42.2	77.2	55.3	58.4	83.4	30.2	37.8	71.9	32.7	38.3	78.9	44.1	49.1	95.4	56.2	61.8
032	CONCRETE REINFORCEMENT	94.5	59.9	75.0	95.2	80.5	86.9	93.2	60.1	74.6	94.5	73.9	82.9	93.9	69.9	80.4	94.2	74.5	83.1
033	CAST IN PLACE CONCRETE	105.3	55.0	84.2	103.2	56.6	83.7	101.2	43.7	71.1	101.7	68.9	87.9	101.2	55.5	82.1	102.3	62.1	85.4
3	CONCRETE	102.4	47.8	74.9	99.2	62.2	80.5	101.1	42.6	71.6	103.0	55.5	79.0	100.9	54.9	77.7	99.7	63.4	81.4
4	MASONRY	94.1	41.3	61.3	97.1	45.1	64.8	88.9	36.7	56.5	96.2	37.6	59.8	102.5	42.7	65.3	89.8	53.9	67.5
5	METALS	93.3	73.1	86.0	93.3	91.6	92.7	93.2	75.6	86.9	93.3	90.7	92.3	93.2	86.3	90.7	95.1	91.7	93.9
6	WOOD & PLASTICS	99.1	32.2	64.6	82.2	57.0	69.2	86.6	29.6	57.2	76.4	28.4	51.6	82.2	43.8	62.4	97.6	57.1	76.7
7	THERMAL & MOISTURE PROTECTION	95.7	41.3	70.5	95.7	54.8	76.8	95.9	39.0	69.5	96.1	61.2	80.0	95.7	48.1	73.7	95.6	54.2	76.4
8	DOORS & WINDOWS	96.3	32.7	80.9	96.0	63.8	88.2	97.9	31.8	81.9	96.3	48.3	84.7	96.3	48.2	84.7	96.7	57.7	87.2
092	LATH, PLASTER & GYPSUM BOARD	105.5	28.9	56.1	98.4	55.4	70.7	99.1	27.2	52.7	96.5	25.9	50.9	98.1	41.8	61.8	106.4	54.6	73.0
095	ACOUSTICAL TREATMENT & WOOD FLOORING	87.3	28.9	49.5	88.7	55.4	67.1	87.3	27.2	48.3	87.3	25.9	47.5	87.3	41.8	57.8	91.4	54.6	67.5
096	FLOORING & CARPET	93.3	67.5	87.0	88.1	82.9	86.9	89.4	30.9	75.2	85.8	82.9	85.1	87.8	45.9	77.6	93.3	48.9	82.5
099	PAINTING & WALL COVERINGS	105.8	43.1	69.0	117.1	48.7	77.0	105.8	40.8	67.6	117.1	27.3	64.4	105.8	40.8	67.6	105.8	52.2	74.4
9	FINISHES	93.2	40.7	66.3	91.5	59.3	75.0	91.2	31.2	60.4	90.7	39.6	64.5	90.5	43.5	66.4	94.0	54.2	73.6
10 - 14	TOTAL DIV. 10 - 14	100.0	60.4	91.7	100.0	70.9	93.9	100.0	57.5	91.0	100.0	65.9	92.8	100.0	69.1	93.5	100.0	75.5	94.8
15	MECHANICAL	96.9	33.9	68.8	96.8	68.5	84.2	96.8	34.1	68.9	96.8	41.5	72.1	96.8	66.8	83.4	100.1	60.5	82.5
16	ELECTRICAL	95.0	57.5	69.7	96.6	103.1	101.0	99.7	26.8	50.5	100.0	41.8	60.8	101.5	38.1	58.7	99.7	59.9	72.9
1 - 16	WEIGHTED AVERAGE	96.9	50.7	74.6	96.5	73.1	85.2	96.5	43.1	70.8	97.3	52.9	75.8	97.1	57.3	77.9	97.7	64.9	81.9

VIRGINIA

DIVISION		NORFOLK 233 - 235			PETERSBURG 238			PORTSMOUTH 237			PULASKI 243			RICHMOND 230 - 232			ROANOKE 240 - 241		
		MAT.	INST.	TOTAL	MAT.	INST.	TOTAL	MAT.	INST.	TOTAL	MAT.	INST.	TOTAL	MAT.	INST.	TOTAL	MAT.	INST.	TOTAL
2	SITE WORK	106.4	87.3	91.7	111.0	86.8	92.4	105.6	85.9	90.4	105.4	82.2	87.6	108.0	86.8	91.7	104.2	82.9	87.8
031	CONCRETE FORMWORK	100.7	56.2	62.6	83.4	62.5	65.5	81.3	56.2	59.8	83.4	30.2	37.8	96.4	62.5	67.3	95.2	44.0	51.3
032	CONCRETE REINFORCEMENT	94.2	74.5	83.1	93.9	76.5	84.1	93.2	74.6	83.0	93.2	60.1	74.6	94.2	76.5	84.3	94.2	69.8	80.5
033	CAST IN PLACE CONCRETE	105.3	62.1	87.2	105.3	63.8	87.9	101.3	64.3	85.8	101.2	43.7	77.1	108.8	63.8	90.0	106.1	55.4	84.9
3	CONCRETE	101.5	63.4	82.3	103.0	67.1	84.9	98.1	64.2	81.0	101.1	42.6	71.6	102.9	67.1	84.9	101.5	54.8	78.0
4	MASONRY	97.5	53.9	70.4	101.8	58.3	74.8	94.5	55.0	69.9	82.2	36.7	54.0	86.2	58.3	68.9	89.3	42.5	60.2
5	METALS	94.3	91.8	93.4	93.3	93.6	93.4	94.3	91.4	93.2	93.3	75.6	86.9	95.1	93.6	94.5	94.9	86.2	91.7
6	WOOD & PLASTICS	104.4	57.1	80.0	87.2	65.0	75.7	81.4	57.1	68.9	86.6	29.6	57.2	99.0	65.0	81.4	97.6	43.8	69.8
7	THERMAL & MOISTURE PROTECTION	95.5	54.2	76.3	95.6	58.3	78.3	95.5	54.5	76.5	95.9	39.0	69.5	95.2	58.3	78.1	95.6	49.7	74.3
8	DOORS & WINDOWS	96.7	59.8	87.8	96.0	61.1	87.6	96.8	59.8	87.8	97.9	31.8	81.9	96.7	61.1	88.1	96.7	48.2	84.9
092	LATH, PLASTER & GYPSUM BOARD	106.4	54.6	73.0	99.5	62.7	75.7	96.9	54.6	69.6	99.1	27.2	52.7	106.4	62.7	78.2	106.4	41.8	64.7
095	ACOUSTICAL TREATMENT & WOOD FLOORING	91.4	54.6	67.5	88.7	62.7	71.8	91.4	54.6	67.5	87.3	27.2	48.3	91.4	62.7	72.8	91.4	41.8	59.3
096	FLOORING & CARPET	93.1	48.9	82.3	89.1	72.7	85.1	85.7	48.9	76.7	89.4	30.9	75.2	93.1	72.7	88.1	93.3	43.9	81.3
099	PAINTING & WALL COVERINGS	105.8	54.5	75.7	105.8	67.1	83.1	105.8	45.8	70.6	105.8	40.8	67.6	105.8	67.1	83.1	105.8	40.8	67.6
9	FINISHES	94.0	54.4	73.7	91.4	65.0	77.9	90.3	53.4	71.4	91.2	31.2	60.4	93.9	65.0	79.1	93.9	43.1	67.9
10 - 14	TOTAL DIV. 10 - 14	100.0	75.5	94.8	100.0	77.1	95.2	100.0	75.5	94.8	100.0	57.5	91.0	100.0	77.1	95.2	100.0	69.1	93.5
15	MECHANICAL	100.1	58.3	81.5	96.9	68.6	84.3	100.1	61.2	82.8	96.8	34.1	68.9	100.1	68.6	86.1	100.1	50.1	77.8
16	ELECTRICAL	99.7	59.7	72.8	100.0	72.6	81.5	96.8	63.4	74.3	99.7	26.8	50.5	100.8	72.6	81.8	99.7	44.5	62.5
1 - 16	WEIGHTED AVERAGE	98.3	64.7	82.0	97.4	71.4	84.9	96.8	65.8	81.8	96.1	43.1	70.6	98.0	71.4	85.1	97.8	55.2	77.2

VIRGINIA / WASHINGTON

DIVISION		STAUNTON 244			WINCHESTER 226			CLARKSTON 994			EVERETT 982			OLYMPIA 985			RICHLAND 993		
		MAT.	INST.	TOTAL	MAT.	INST.	TOTAL	MAT.	INST.	TOTAL	MAT.	INST.	TOTAL	MAT.	INST.	TOTAL	MAT.	INST.	TOTAL
2	SITE WORK	109.5	85.2	90.8	121.0	83.7	92.3	92.6	90.7	91.2	90.5	118.9	112.3	93.7	119.1	113.2	93.6	91.3	91.9
031	CONCRETE FORMWORK	82.8	31.1	38.5	75.4	46.3	50.5	107.2	83.1	86.6	106.8	95.4	97.1	89.7	100.4	98.9	110.7	88.3	91.5
032	CONCRETE REINFORCEMENT	93.9	54.9	74.6	93.8	64.8	74.6	107.8	88.4	96.8	109.4	91.2	99.1	112.4	93.4	101.7	102.7	92.9	97.2
033	CAST IN PLACE CONCRETE	105.3	45.9	80.4	101.7	69.9	88.3	106.2	89.1	99.1	92.9	105.2	98.0	97.4	106.3	101.1	106.5	91.3	100.2
3	CONCRETE	102.1	44.0	72.8	102.6	61.7	82.0	110.3	86.0	98.1	106.5	97.5	102.0	109.9	100.5	105.1	110.2	89.9	100.0
4	MASONRY	100.5	34.6	59.6	93.2	41.8	61.2	111.6	90.3	98.4	135.3	101.6	114.4	128.4	102.6	112.4	111.5	93.8	100.5
5	METALS	93.3	82.8	89.5	93.5	90.1	92.3	94.4	83.6	90.5	102.8	85.9	96.7	102.4	87.5	97.0	93.4	86.1	90.8
6	WOOD & PLASTICS	86.6	29.9	57.3	80.0	44.7	61.8	101.5	82.1	91.5	109.4	93.3	101.1	90.2	98.9	94.7	101.7	86.7	94.0
7	THERMAL & MOISTURE PROTECTION	95.5	41.8	70.6	96.3	64.5	81.6	167.4	84.0	128.7	104.7	96.9	101.1	104.3	97.1	101.0	169.0	88.4	131.6
8	DOORS & WINDOWS	96.3	35.9	81.7	98.0	57.1	88.1	116.4	78.2	107.1	103.2	90.8	100.2	103.1	95.1	101.2	116.4	83.2	108.3
092	LATH, PLASTER & GYPSUM BOARD	99.1	26.5	52.3	97.9	42.7	62.2	119.0	81.5	94.8	99.6	92.6	95.1	95.8	98.5	97.5	119.0	86.2	97.9
095	ACOUSTICAL TREATMENT & WOOD FLOORING	87.3	26.5	47.9	88.7	42.7	58.9	123.4	81.5	96.3	114.5	92.6	100.3	113.2	98.5	103.6	123.4	86.2	99.3
096	FLOORING & CARPET	88.9	26.9	73.8	87.3	82.9	86.3	116.5	77.0	106.9	128.8	100.6	122.0	121.3	104.3	117.2	116.8	56.5	102.2
099	PAINTING & WALL COVERINGS	105.8	24.9	58.3	117.1	34.4	68.5	122.9	71.8	93.0	119.8	82.2	97.7	119.8	91.1	103.0	122.9	80.9	98.3
9	FINISHES	91.0	29.8	59.6	91.6	50.7	70.6	133.0	80.5	106.1	121.4	94.6	107.7	118.3	100.0	108.9	133.1	80.8	106.3
10 - 14	TOTAL DIV. 10 - 14	100.0	59.4	91.4	100.0	69.3	93.5	100.0	85.1	96.9	100.0	101.3	100.3	100.0	105.0	101.1	100.0	98.3	99.6
15	MECHANICAL	96.8	38.3	70.7	96.8	72.0	85.8	97.3	85.1	91.9	100.0	99.6	99.8	100.1	102.9	101.4	100.6	99.1	100.0
16	ELECTRICAL	97.8	27.3	50.3	97.3	43.5	61.0	99.8	94.1	95.9	103.2	95.1	97.7	102.9	96.4	98.5	102.4	94.1	96.8
1 - 16	WEIGHTED AVERAGE	97.0	45.1	71.9	97.2	62.1	80.3	107.7	86.7	97.6	106.0	98.4	102.3	105.5	100.9	103.3	108.5	91.3	100.2

WASHINGTON

DIVISION		SEATTLE 980 - 981,987			SPOKANE 990 - 992			TACOMA 983 - 984			VANCOUVER 986			WENATCHEE 988			YAKIMA 989		
		MAT.	INST.	TOTAL	MAT.	INST.	TOTAL	MAT.	INST.	TOTAL	MAT.	INST.	TOTAL	MAT.	INST.	TOTAL	MAT.	INST.	TOTAL
2	SITE WORK	93.9	116.8	111.5	94.3	91.3	92.0	93.3	119.1	113.2	106.2	104.0	104.5	105.4	118.2	115.3	96.5	118.3	113.3
031	CONCRETE FORMWORK	89.7	100.8	99.2	119.3	88.4	92.9	89.7	100.6	99.0	94.2	97.3	96.9	92.8	85.1	86.2	92.2	96.4	95.8
032	CONCRETE REINFORCEMENT	103.3	93.6	97.8	103.4	93.0	97.5	103.3	93.5	97.8	106.3	93.1	98.9	107.6	88.5	96.8	105.9	92.4	98.3
033	CAST IN PLACE CONCRETE	97.4	106.5	101.2	110.4	91.4	102.4	95.4	106.4	100.0	106.6	100.9	104.2	97.5	91.5	95.0	101.9	90.6	97.2
3	CONCRETE	108.7	100.8	104.7	112.7	90.0	101.3	107.7	100.6	104.1	120.3	97.4	108.7	119.7	87.7	103.6	113.4	93.0	103.1
4	MASONRY	128.3	104.0	113.2	113.4	93.8	101.2	128.1	102.6	112.3	130.3	100.8	112.0	130.9	91.5	106.4	119.6	79.1	94.4
5	METALS	104.2	89.8	99.0	93.2	86.3	90.7	103.8	87.7	98.0	102.3	88.9	97.5	102.2	83.9	95.6	102.7	83.9	95.9
6	WOOD & PLASTICS	91.3	98.9	95.2	112.6	86.7	99.2	90.2	98.9	94.7	89.2	96.7	93.0	97.1	83.2	89.9	93.0	98.9	96.0
7	THERMAL & MOISTURE PROTECTION	104.5	100.8	102.8	166.8	88.8	130.7	104.3	100.0	102.3	107.8	89.2	99.2	105.3	85.9	96.3	104.5	84.7	95.3
8	DOORS & WINDOWS	103.3	95.1	101.4	116.2	82.2	107.9	103.8	95.1	101.7	100.3	88.1	97.4	103.4	78.6	97.4	103.4	86.2	99.2
092	LATH, PLASTER & GYPSUM BOARD	93.6	98.5	96.7	122.7	86.2	99.2	96.8	98.5	97.9	98.3	96.6	97.2	98.5	82.2	88.0	96.9	98.5	97.9
095	ACOUSTICAL TREATMENT & WOOD FLOORING	118.6	98.5	105.6	123.4	86.2	99.3	117.3	98.5	105.1	125.7	96.6	106.9	113.2	82.2	93.1	113.2	98.5	103.6
096	FLOORING & CARPET	120.6	100.6	115.8	120.0	85.0	111.5	121.3	104.3	117.2	127.2	77.9	115.2	124.6	85.0	114.9	122.6	80.7	112.4
099	PAINTING & WALL COVERINGS	119.8	91.1	103.0	122.9	84.7	100.5	119.8	91.1	103.0	128.8	79.7	100.0	119.8	80.9	97.0	119.8	84.7	99.2
9	FINISHES	118.8	99.3	108.8	134.7	87.1	110.3	119.2	100.0	109.4	119.4	91.8	105.3	120.5	84.1	101.8	119.0	92.5	105.4
10 - 14	TOTAL DIV. 10 - 14	100.0	105.0	101.1	100.0	98.3	99.6	100.0	105.0	101.1	100.0	90.1	97.9	100.0	97.8	99.5	100.0	101.8	100.4
15	MECHANICAL	100.0	111.8	105.3	100.6	93.2	97.3	100.1	102.9	101.4	100.3	97.3	99.0	96.8	90.3	93.9	100.1	97.6	99.0
16	ELECTRICAL	102.9	100.1	101.0	93.2	88.9	90.3	102.9	96.4	98.5	115.5	105.1	108.5	104.2	89.4	94.2	107.6	94.1	98.5
1 - 16	WEIGHTED AVERAGE	105.7	103.4	104.6	108.4	90.1	99.5	105.6	101.0	103.4	108.0	97.4	102.9	106.8	90.9	99.1	106.0	94.2	100.3

WEST VIRGINIA

DIVISION		BECKLEY 258 - 259			BLUEFIELD 247 - 248			BUCKHANNON 262			CHARLESTON 250 - 253			CLARKSBURG 263 - 264			GASSAWAY 266		
		MAT.	INST.	TOTAL	MAT.	INST.	TOTAL	MAT.	INST.	TOTAL	MAT.	INST.	TOTAL	MAT.	INST.	TOTAL	MAT.	INST.	TOTAL
2	SITE WORK	100.1	85.9	89.2	99.9	85.9	89.1	105.8	85.7	90.4	102.6	86.6	90.3	106.4	85.7	90.5	103.4	86.3	90.2
031	CONCRETE FORMWORK	77.5	89.5	87.8	79.6	66.3	68.2	78.6	88.7	87.3	104.1	92.3	94.0	75.2	88.9	86.9	77.5	87.7	86.3
032	CONCRETE REINFORCEMENT	92.9	67.4	78.5	92.9	67.3	78.5	93.4	85.3	88.8	94.2	78.5	85.4	93.4	85.2	88.8	93.4	82.7	87.4
033	CAST IN PLACE CONCRETE	99.0	98.1	98.6	99.0	97.8	98.5	98.7	93.6	96.6	100.1	100.8	100.4	108.1	93.7	102.1	103.3	103.8	103.5
3	CONCRETE	96.6	89.5	93.0	96.7	79.0	87.8	100.0	90.8	95.3	99.2	93.5	96.4	103.7	90.9	97.3	100.1	93.4	96.7
4	MASONRY	88.1	89.6	89.0	83.9	89.6	87.4	94.3	86.7	89.6	88.0	89.9	89.2	95.5	86.7	90.0	96.7	87.4	90.9
5	METALS	93.2	98.5	95.2	93.3	93.2	93.3	93.5	101.1	96.2	95.1	99.1	96.5	93.5	102.0	96.5	93.2	100.6	95.9
6	WOOD & PLASTICS	81.7	89.9	85.9	83.9	59.6	71.4	82.9	88.8	85.9	107.8	92.7	100.0	79.0	88.8	84.0	81.9	87.7	84.9
7	THERMAL & MOISTURE PROTECTION	95.6	87.2	91.7	95.7	80.2	88.5	96.0	84.6	90.7	95.6	87.8	92.0	95.9	84.6	90.6	95.6	85.9	91.1
8	DOORS & WINDOWS	97.4	81.8	93.6	98.2	62.4	89.5	98.2	82.4	94.4	97.9	82.9	94.3	98.2	89.9	96.2	96.2	80.1	92.3
092	LATH, PLASTER & GYPSUM BOARD	95.4	89.4	91.5	96.0	58.1	71.5	96.3	88.2	91.1	106.4	92.2	97.3	95.2	88.2	90.7	95.8	87.1	90.1
095	ACOUSTICAL TREATMENT & WOOD FLOORING	87.3	89.4	88.6	87.3	58.1	68.4	88.7	88.2	88.4	91.4	92.2	92.0	88.7	88.2	88.4	88.7	87.1	87.6
096	FLOORING & CARPET	85.0	93.2	87.0	85.7	86.0	85.7	85.3	93.1	87.2	93.1	96.6	93.9	84.0	93.1	86.2	85.1	94.4	87.3
099	PAINTING & WALL COVERINGS	105.8	58.7	78.2	105.8	58.7	78.2	105.8	83.1	92.5	105.8	87.0	94.8	105.8	83.1	92.5	105.8	87.0	94.8
9	FINISHES	89.0	87.1	88.0	89.2	68.2	78.5	89.8	89.2	89.5	94.0	93.0	93.5	89.2	89.2	89.2	89.4	89.0	89.2
10 - 14	TOTAL DIV. 10 - 14	100.0	89.5	97.8	100.0	85.4	96.9	100.0	93.3	98.6	100.0	98.2	99.6	100.0	93.3	98.6	100.0	92.7	98.5
15	MECHANICAL	96.8	80.5	89.5	96.7	72.7	86.0	96.7	90.5	94.0	100.1	82.5	92.3	96.7	86.9	92.4	96.7	88.6	93.1
16	ELECTRICAL	93.5	87.0	89.1	97.9	58.4	71.2	100.2	81.3	87.4	99.7	83.9	89.1	100.2	81.3	87.4	100.2	86.1	90.7
1 - 16	WEIGHTED AVERAGE	94.9	87.2	91.2	95.2	75.4	85.6	96.6	88.6	92.7	97.7	88.7	93.4	97.0	88.3	92.8	96.3	89.3	92.9

WEST VIRGINIA

DIVISION		HUNTINGTON 255 - 257			LEWISBURG 249			MARTINSBURG 254			MORGANTOWN 265			PARKERSBURG 261			PETERSBURG 268		
		MAT.	INST.	TOTAL	MAT.	INST.	TOTAL	MAT.	INST.	TOTAL	MAT.	INST.	TOTAL	MAT.	INST.	TOTAL	MAT.	INST.	TOTAL
2	SITE WORK	104.4	86.5	90.6	114.9	85.9	92.6	103.6	84.0	88.6	100.8	86.1	89.5	108.5	86.6	91.7	100.1	86.1	89.3
031	CONCRETE FORMWORK	96.1	94.5	94.7	75.7	89.2	87.3	77.9	42.1	47.2	75.9	89.2	87.3	84.2	90.7	89.8	80.3	83.4	83.0
032	CONCRETE REINFORCEMENT	94.2	89.7	91.7	93.4	67.4	78.7	92.9	66.4	78.0	93.4	90.6	91.8	92.9	85.4	88.7	92.9	70.8	80.5
033	CAST IN PLACE CONCRETE	108.0	103.1	106.9	99.1	98.0	98.6	103.7	51.6	81.9	98.7	93.8	96.6	101.9	91.9	97.1	98.7	92.7	96.2
3	CONCRETE	102.5	98.1	100.3	106.2	89.1	97.6	100.2	52.4	76.1	96.3	92.0	94.2	102.0	91.1	96.5	96.6	85.4	91.0
4	MASONRY	89.4	100.0	96.0	88.4	89.6	89.1	91.3	55.2	68.9	115.2	86.7	97.5	75.1	91.8	85.5	90.8	84.6	87.0
5	METALS	95.1	103.8	98.2	93.4	93.8	93.5	93.7	92.9	93.4	93.6	103.3	97.1	93.9	101.4	96.6	93.7	95.7	94.4
6	WOOD & PLASTICS	97.6	93.7	95.6	79.4	89.8	84.8	81.7	38.9	59.6	79.4	88.8	84.3	84.2	89.6	87.0	84.2	82.1	83.1
7	THERMAL & MOISTURE PROTECTION	95.8	91.8	93.9	96.7	87.6	92.5	95.9	63.8	81.0	95.7	84.6	90.5	95.9	86.2	91.4	95.8	79.5	88.2
8	DOORS & WINDOWS	96.7	86.0	94.1	98.2	78.7	93.5	99.5	51.5	87.9	99.5	90.6	97.4	97.3	81.9	93.5	99.5	78.2	94.4
092	LATH, PLASTER & GYPSUM BOARD	106.4	93.3	98.0	95.2	89.3	91.4	95.8	36.7	57.6	95.2	88.2	90.7	98.9	89.1	92.6	96.3	81.3	86.6
095	ACOUSTICAL TREATMENT & WOOD FLOORING	91.4	93.3	92.6	88.7	89.3	89.1	88.7	36.7	55.0	88.7	88.2	88.4	88.7	89.1	89.0	88.7	81.3	83.9
096	FLOORING & CARPET	93.1	110.5	97.3	84.1	93.2	86.3	85.0	86.0	85.2	84.1	93.1	86.3	88.5	94.6	90.0	85.8	89.4	86.7
099	PAINTING & WALL COVERINGS	105.8	90.3	96.7	105.8	58.7	78.2	105.8	69.3	84.4	105.8	83.1	92.5	105.8	100.4	102.6	105.8	59.5	78.6
9	FINISHES	93.8	97.4	95.7	89.9	87.1	88.5	89.4	52.2	70.3	88.9	89.2	89.1	91.2	92.9	92.1	89.6	81.9	85.6
10 - 14	TOTAL DIV. 10 - 14	100.0	99.3	99.8	100.0	89.4	97.8	100.0	74.0	94.5	100.0	92.3	98.4	100.0	98.6	99.7	100.0	92.4	98.4
15	MECHANICAL	100.1	89.9	95.6	96.7	80.4	89.4	96.8	52.0	76.8	96.7	91.8	94.5	100.1	82.1	92.1	96.7	93.4	95.2
16	ELECTRICAL	99.7	84.2	89.2	93.5	58.4	69.8	103.3	39.3	60.1	100.6	81.3	87.5	100.5	93.5	95.8	105.9	76.1	85.8
1 - 16	WEIGHTED AVERAGE	97.9	93.0	95.5	96.7	81.8	89.5	96.7	58.0	78.0	97.2	89.5	93.5	96.7	90.2	93.6	96.4	85.6	91.2

COST INDEXES

641

WEST VIRGINIA / WISCONSIN

DIVISION		ROMNEY 267			WHEELING 260			BELOIT 535			EAU CLAIRE 547			GREEN BAY 541 - 543			KENOSHA 531		
		MAT.	INST.	TOTAL	MAT.	INST.	TOTAL	MAT.	INST.	TOTAL	MAT.	INST.	TOTAL	MAT.	INST.	TOTAL	MAT.	INST.	TOTAL
2	SITE WORK	102.9	86.0	89.9	109.2	86.1	91.4	82.0	100.6	96.3	79.5	101.0	96.0	81.5	96.9	93.3	86.2	98.5	95.7
031	CONCRETE FORMWORK	74.8	83.2	82.0	86.4	89.1	88.7	97.5	94.6	95.0	96.5	95.6	95.8	117.2	95.5	98.6	111.9	101.1	102.6
032	CONCRETE REINFORCEMENT	93.4	71.1	80.8	92.2	90.9	91.5	102.8	105.7	104.5	100.2	95.3	97.5	96.8	94.6	95.6	109.8	100.0	104.3
033	CAST IN PLACE CONCRETE	103.3	89.8	97.7	100.9	99.8	100.5	105.9	88.7	98.7	97.5	94.2	96.1	100.8	98.2	99.7	115.2	98.9	108.4
3	CONCRETE	99.9	84.4	92.1	102.1	94.1	98.1	97.8	94.7	96.3	93.7	95.1	94.4	95.5	96.3	95.9	104.4	100.1	102.2
4	MASONRY	88.7	81.9	84.5	97.5	92.7	94.5	101.3	88.2	93.2	92.0	100.2	97.1	112.3	99.2	104.2	96.9	105.7	102.4
5	METALS	93.7	96.1	94.6	94.0	103.8	97.5	94.8	99.2	96.4	92.0	94.8	93.0	94.5	94.7	94.6	96.0	99.1	97.1
6	WOOD & PLASTICS	78.3	82.1	80.2	86.6	87.7	87.1	109.1	95.6	102.1	106.3	95.3	100.6	125.8	95.3	110.0	121.5	99.6	110.2
7	THERMAL & MOISTURE PROTECTION	95.8	78.8	87.9	96.0	88.1	92.4	96.3	92.9	94.7	95.7	89.4	92.8	98.4	90.9	95.0	96.7	94.2	95.5
8	DOORS & WINDOWS	99.5	74.0	93.3	98.1	90.0	96.2	106.6	98.2	104.6	105.3	91.9	102.0	103.9	90.1	100.5	101.8	101.5	101.7
092	LATH, PLASTER & GYPSUM BOARD	94.7	81.3	86.0	99.5	87.1	91.5	107.4	96.0	100.0	108.4	95.7	100.2	93.2	95.7	94.8	95.3	100.1	98.4
095	ACOUSTICAL TREATMENT & WOOD FLOORING	88.7	81.3	83.9	88.7	87.1	87.6	95.6	96.0	95.8	106.5	95.7	99.5	101.4	95.7	97.7	86.5	100.1	95.3
096	FLOORING & CARPET	83.8	82.7	83.6	89.4	100.6	92.2	92.9	91.9	92.6	95.6	75.1	90.6	115.0	102.3	111.9	110.1	95.6	106.5
099	PAINTING & WALL COVERINGS	105.8	59.5	78.6	105.8	86.5	94.5	92.8	76.3	83.1	95.4	84.0	88.7	106.3	77.6	89.5	104.3	94.8	98.8
9	FINISHES	88.9	80.5	84.6	91.6	90.7	91.1	99.0	92.6	95.7	100.7	90.6	95.5	105.2	95.2	100.0	102.0	99.7	100.8
10 - 14	TOTAL DIV. 10 - 14	100.0	92.4	98.4	100.0	100.4	100.1	100.0	85.7	97.0	100.0	99.1	99.8	100.0	98.0	99.6	100.0	99.3	99.9
15	MECHANICAL	96.7	93.3	95.2	100.1	91.2	96.1	100.0	93.3	97.0	100.0	89.6	95.4	100.5	91.3	96.4	100.3	97.0	98.8
16	ELECTRICAL	104.7	76.1	85.4	95.7	81.3	86.0	91.5	100.4	97.5	102.2	88.7	93.1	93.5	90.0	91.2	93.3	102.4	99.5
1 - 16	WEIGHTED AVERAGE	96.5	84.9	90.9	97.8	90.7	94.4	98.5	95.4	97.0	97.8	93.3	95.6	99.6	94.0	96.9	99.5	100.0	99.7

WISCONSIN

DIVISION		LA CROSSE 546			LANCASTER 538			MADISON 537			MILWAUKEE 530,532			NEW RICHMOND 540			OSHKOSH 549		
		MAT.	INST.	TOTAL	MAT.	INST.	TOTAL	MAT.	INST.	TOTAL	MAT.	INST.	TOTAL	MAT.	INST.	TOTAL	MAT.	INST.	TOTAL
2	SITE WORK	74.0	100.9	94.7	81.4	100.1	95.8	81.4	101.7	97.0	82.4	90.7	88.8	76.5	101.7	95.9	74.8	97.3	92.1
031	CONCRETE FORMWORK	78.7	95.8	93.3	92.5	98.9	98.0	98.2	106.6	105.4	101.3	110.1	108.9	84.4	84.7	84.6	84.3	84.9	84.8
032	CONCRETE REINFORCEMENT	98.8	91.7	94.8	111.3	76.7	91.8	102.8	91.9	96.7	102.8	100.4	101.5	97.2	94.8	95.8	97.5	90.3	93.4
033	CAST IN PLACE CONCRETE	87.7	93.9	90.3	105.3	88.4	98.2	103.9	101.4	102.8	103.9	105.6	104.6	101.6	88.4	96.0	93.6	86.7	90.7
3	CONCRETE	84.9	94.4	89.7	98.1	90.6	94.3	96.9	101.7	99.3	97.4	105.7	101.6	91.6	88.1	89.8	86.7	86.8	86.8
4	MASONRY	91.3	100.6	97.1	101.3	85.5	91.5	101.2	110.7	107.1	101.2	115.1	109.9	115.8	86.6	97.7	104.5	85.1	92.5
5	METALS	91.9	93.3	92.4	92.4	79.1	87.6	96.3	94.1	95.5	97.8	90.2	95.1	93.9	91.3	92.9	92.1	91.8	92.0
6	WOOD & PLASTICS	86.1	95.3	90.8	108.0	106.3	107.1	109.3	106.3	107.7	113.0	109.1	111.0	93.9	84.1	88.8	93.1	84.3	88.5
7	THERMAL & MOISTURE PROTECTION	94.9	88.8	92.1	95.9	77.1	87.2	95.8	93.9	94.9	95.1	105.7	100.0	96.7	92.8	94.9	97.1	83.3	90.7
8	DOORS & WINDOWS	105.3	87.3	100.9	101.7	87.3	98.2	106.6	101.0	105.3	106.5	106.7	106.5	92.5	83.0	90.2	98.8	84.7	95.4
092	LATH, PLASTER & GYPSUM BOARD	101.1	95.7	97.6	105.8	107.0	106.6	107.4	107.0	107.1	107.4	109.7	108.9	85.6	84.4	84.8	82.8	84.3	83.8
095	ACOUSTICAL TREATMENT & WOOD FLOORING	102.3	95.7	98.0	88.7	107.0	100.6	95.6	107.0	103.0	95.6	109.7	104.7	68.0	84.4	78.6	97.3	84.3	88.9
096	FLOORING & CARPET	88.5	91.7	89.2	92.3	80.4	89.4	92.9	63.1	85.6	95.7	108.5	98.8	107.8	68.8	98.3	105.8	85.6	100.9
099	PAINTING & WALL COVERINGS	95.4	84.0	88.7	92.8	69.8	79.3	92.8	89.9	91.1	95.5	106.7	102.1	110.9	79.9	92.7	106.3	67.9	83.8
9	FINISHES	96.3	94.1	95.1	97.3	94.4	95.8	99.0	96.9	97.9	100.1	109.7	105.0	95.2	80.9	87.9	99.6	83.0	91.1
10 - 14	TOTAL DIV. 10 - 14	100.0	99.3	99.8	100.0	73.5	94.4	100.0	96.5	99.3	100.0	102.5	100.5	100.0	84.9	96.8	100.0	88.8	97.6
15	MECHANICAL	100.0	91.8	96.4	96.7	67.6	83.7	100.0	100.3	100.1	100.0	101.1	100.5	96.4	84.3	91.0	97.2	83.8	91.2
16	ELECTRICAL	102.6	88.7	93.2	91.5	76.0	81.0	92.2	99.7	97.3	93.0	106.1	101.8	99.9	88.7	92.3	100.5	82.1	88.1
1 - 16	WEIGHTED AVERAGE	95.9	93.8	94.8	96.7	82.9	90.0	98.6	100.4	99.5	99.2	103.3	101.2	95.9	87.9	92.1	95.8	86.2	91.2

WISCONSIN / WYOMING

DIVISION		PORTAGE 539			RACINE 534			RHINELANDER 545			SUPERIOR 548			WAUSAU 544			CASPER 826		
		MAT.	INST.	TOTAL	MAT.	INST.	TOTAL	MAT.	INST.	TOTAL	MAT.	INST.	TOTAL	MAT.	INST.	TOTAL	MAT.	INST.	TOTAL
2	SITE WORK	73.0	100.4	94.1	81.8	102.3	97.5	85.2	96.2	93.7	73.7	102.7	96.0	71.3	97.0	91.0	86.6	100.3	97.1
031	CONCRETE FORMWORK	82.6	101.7	98.9	98.0	101.1	100.6	81.0	85.5	84.8	82.4	86.0	85.5	83.6	94.7	93.1	99.5	53.7	60.3
032	CONCRETE REINFORCEMENT	112.0	76.9	92.3	102.8	100.0	101.2	98.8	91.9	94.9	98.4	89.4	93.3	98.8	91.1	94.5	108.2	55.7	78.6
033	CAST IN PLACE CONCRETE	90.3	89.5	90.0	103.9	98.5	101.6	106.1	86.2	97.7	95.7	95.7	95.7	87.2	89.5	88.2	103.6	68.3	88.8
3	CONCRETE	86.4	92.5	89.5	96.9	99.9	98.4	98.2	86.9	92.5	86.4	90.3	88.3	82.2	92.2	87.3	106.6	59.9	83.0
4	MASONRY	100.2	89.5	93.5	101.2	105.7	104.0	122.0	84.9	99.0	115.1	93.4	101.6	104.1	89.9	95.2	107.3	49.7	71.5
5	METALS	93.0	85.1	90.2	96.3	99.1	97.3	92.2	86.7	90.2	94.8	92.2	93.8	92.1	92.1	92.1	105.4	68.2	92.0
6	WOOD & PLASTICS	96.5	106.3	101.5	109.5	99.6	104.4	89.3	85.8	87.5	91.6	84.1	87.7	92.2	95.3	93.8	97.5	52.4	74.2
7	THERMAL & MOISTURE PROTECTION	95.2	85.6	90.7	96.3	93.6	95.0	98.0	78.0	88.7	96.3	91.9	94.3	96.8	81.0	89.5	103.0	65.7	85.7
8	DOORS & WINDOWS	101.9	95.3	100.3	106.6	101.5	105.4	98.8	83.6	95.1	92.4	83.2	90.2	99.0	90.9	97.0	94.0	53.4	84.2
092	LATH, PLASTER & GYPSUM BOARD	101.9	107.0	105.2	107.4	100.1	102.7	81.7	85.9	84.4	85.3	84.4	84.7	82.2	95.7	90.9	88.4	50.7	64.0
095	ACOUSTICAL TREATMENT & WOOD FLOORING	90.1	107.0	101.0	95.6	100.1	98.5	97.3	85.9	89.9	69.4	84.4	79.1	97.3	95.7	96.2	95.6	50.7	66.5
096	FLOORING & CARPET	88.6	91.3	89.3	92.9	95.6	93.5	104.8	85.8	100.2	109.1	97.8	106.4	105.6	86.4	100.9	106.1	53.5	93.3
099	PAINTING & WALL COVERINGS	92.8	80.2	85.4	92.7	95.1	94.1	106.3	80.0	90.8	101.7	100.9	101.2	106.3	80.0	90.8	113.2	70.1	87.9
9	FINISHES	95.4	98.8	97.2	99.0	99.7	99.4	99.8	85.4	92.4	95.1	89.5	92.2	99.3	92.2	95.6	98.8	54.7	76.2
10 - 14	TOTAL DIV. 10 - 14	100.0	89.2	97.7	100.0	99.3	99.9	100.0	83.1	96.4	100.0	91.5	98.2	100.0	98.0	99.6	100.0	78.4	95.5
15	MECHANICAL	96.7	86.2	92.0	100.0	97.0	98.6	97.2	82.2	90.5	96.4	92.3	94.6	97.2	85.0	91.7	99.8	67.1	85.2
16	ELECTRICAL	96.7	87.8	90.7	91.1	103.0	99.1	99.5	81.0	87.0	107.5	92.1	97.1	102.3	81.0	87.9	93.9	65.7	74.8
1 - 16	WEIGHTED AVERAGE	95.1	91.1	93.2	98.6	100.4	99.5	98.4	85.1	92.0	95.8	92.3	94.1	95.3	89.0	92.2	100.4	65.7	83.7

COST INDEXES

City Cost Indexes

WYOMING

DIVISION		CHEYENNE 820			NEWCASTLE 827			RAWLINS 823			RIVERTON 825			ROCK SPRINGS 829-831			SHERIDAN 828		
		MAT.	INST.	TOTAL	MAT.	INST.	TOTAL	MAT.	INST.	TOTAL	MAT.	INST.	TOTAL	MAT.	INST.	TOTAL	MAT.	INST.	TOTAL
2	SITE WORK	86.4	100.3	97.1	86.9	99.8	96.8	104.6	99.8	100.9	97.3	99.9	99.3	89.4	99.8	97.4	94.2	100.3	98.9
031	CONCRETE FORMWORK	99.5	53.8	60.3	91.9	38.0	45.7	98.5	38.0	46.7	90.1	42.0	48.9	105.4	38.0	47.7	90.7	53.6	58.9
032	CONCRETE REINFORCEMENT	101.4	55.8	75.7	109.6	52.1	77.2	109.3	52.1	77.1	110.2	52.2	77.5	110.2	52.1	77.5	109.5	55.8	79.6
033	CAST IN PLACE CONCRETE	103.6	68.3	88.8	104.1	51.0	81.8	104.2	51.0	81.9	104.1	52.4	82.5	104.1	51.0	81.8	104.1	68.2	89.1
3	CONCRETE	105.7	60.0	82.6	106.4	46.4	76.1	121.1	46.4	83.5	115.3	48.6	81.7	107.5	46.5	76.7	112.9	59.9	86.2
4	MASONRY	106.4	52.5	72.9	107.4	40.6	65.9	107.4	40.6	65.9	107.4	43.9	68.0	172.5	40.6	90.6	107.4	52.7	73.4
5	METALS	106.7	68.4	92.9	105.4	65.8	91.1	105.4	65.8	91.1	105.5	66.1	91.3	105.5	67.0	91.6	105.5	68.2	92.1
6	WOOD & PLASTICS	96.5	52.4	73.7	90.1	37.0	62.7	97.4	37.0	66.2	88.3	40.9	63.8	104.4	37.0	69.7	88.7	52.4	70.0
7	THERMAL & MOISTURE PROTECTION	102.9	66.4	86.0	103.4	49.7	78.5	105.4	49.7	79.6	104.6	53.1	80.7	103.6	49.7	78.6	104.3	66.4	86.7
8	DOORS & WINDOWS	94.7	53.4	84.7	100.7	44.3	87.0	100.3	44.3	86.8	100.5	46.4	87.4	101.0	45.1	87.5	100.9	54.9	89.7
092	LATH, PLASTER & GYPSUM BOARD	89.6	50.7	64.5	85.7	34.8	52.9	88.4	34.8	53.8	85.2	38.8	55.3	91.6	34.8	55.0	85.2	50.7	62.9
095	ACOUSTICAL TREATMENT & WOOD FLOORING	102.6	50.7	68.9	95.6	34.8	56.2	95.6	34.8	56.2	95.6	38.8	58.8	95.6	34.8	56.2	95.6	50.7	66.5
096	FLOORING & CARPET	106.1	71.3	97.6	103.3	52.5	91.0	106.0	52.5	93.0	102.7	52.5	90.5	108.3	52.5	94.7	102.9	59.0	92.2
099	PAINTING & WALL COVERINGS	113.0	70.1	87.9	113.2	42.6	71.8	113.2	42.6	71.8	113.2	39.0	69.7	113.2	42.6	71.8	113.2	70.1	87.9
9	FINISHES	100.2	58.4	78.8	97.5	40.3	68.1	100.1	40.3	69.4	98.1	42.7	69.7	99.9	40.3	69.3	97.9	55.8	76.3
10 - 14	TOTAL DIV. 10 - 14	100.0	78.5	95.5	100.0	64.6	92.5	100.0	64.6	92.5	100.0	66.1	92.8	100.0	64.6	92.5	100.0	78.4	95.4
15	MECHANICAL	99.8	56.5	80.5	97.9	39.8	72.0	97.9	39.8	72.0	97.9	50.8	76.9	99.8	39.8	73.1	97.9	54.4	78.5
16	ELECTRICAL	93.8	68.5	76.7	95.4	59.0	70.8	95.4	59.0	70.8	95.4	59.0	70.8	92.8	59.0	70.0	95.4	59.0	70.9
1 - 16	WEIGHTED AVERAGE	100.6	64.9	83.4	100.6	53.4	77.9	103.3	53.4	79.2	102.1	56.8	80.2	105.0	53.6	80.2	101.7	62.7	82.9

WYOMING / CANADA

DIVISION		WHEATLAND 822			WORLAND 824			YELLOWSTONE NAT'L PA 821			BARRIE, ONTARIO			BRANTFORD, ONTARIO			CALGARY, ALBERTA		
		MAT.	INST.	TOTAL	MAT.	INST.	TOTAL	MAT.	INST.	TOTAL	MAT.	INST.	TOTAL	MAT.	INST.	TOTAL	MAT.	INST.	TOTAL
2	SITE WORK	93.0	100.0	98.4	88.2	99.8	97.1	88.3	100.2	97.4	114.0	97.9	101.6	113.1	98.4	101.8	105.1	97.9	99.6
031	CONCRETE FORMWORK	94.4	47.4	54.1	94.8	38.0	46.2	93.5	44.0	51.1	119.8	104.9	107.0	120.2	114.0	114.9	119.0	86.6	91.2
032	CONCRETE REINFORCEMENT	109.6	52.1	77.2	110.2	52.1	77.5	112.2	52.2	78.4	179.5	97.3	133.2	174.2	95.7	130.0	160.8	68.4	108.7
033	CAST IN PLACE CONCRETE	113.5	52.7	88.1	104.1	51.0	81.8	104.1	67.2	88.6	174.7	95.9	141.7	170.4	118.7	148.7	144.3	87.9	120.7
3	CONCRETE	114.4	51.1	82.5	106.7	46.5	76.3	106.9	54.6	80.5	154.4	99.7	126.9	151.7	111.7	131.5	137.1	84.0	110.3
4	MASONRY	108.1	43.7	68.1	107.4	40.6	65.9	107.4	48.1	70.6	155.1	113.6	129.3	150.1	119.0	130.8	146.6	87.9	110.1
5	METALS	105.2	65.9	91.1	105.5	67.0	91.6	105.5	66.3	91.4	106.1	89.9	100.3	100.8	97.9	99.7	100.8	85.8	95.4
6	WOOD & PLASTICS	93.1	48.0	69.8	93.2	37.0	64.2	93.2	41.0	66.3	120.3	104.1	111.9	118.9	113.0	115.9	116.7	85.9	100.8
7	THERMAL & MOISTURE PROTECTION	103.8	52.1	79.8	103.4	49.7	78.5	102.9	57.4	81.8	105.1	101.8	103.6	104.7	108.4	106.4	104.5	86.5	96.2
8	DOORS & WINDOWS	99.3	50.2	87.4	100.9	45.1	87.3	94.2	46.4	82.6	91.2	95.4	92.2	90.5	107.1	94.5	91.3	80.3	88.6
092	LATH, PLASTER & GYPSUM BOARD	86.8	46.2	60.6	86.8	34.8	53.3	87.1	38.9	56.0	169.0	104.3	127.2	144.6	113.5	124.6	149.4	85.5	108.2
095	ACOUSTICAL TREATMENT & WOOD FLOORING	95.6	46.2	63.5	95.6	34.8	56.2	96.9	38.9	59.3	101.7	104.3	103.4	101.7	113.5	109.4	101.7	85.5	91.2
096	FLOORING & CARPET	104.6	52.5	91.9	104.6	52.5	91.9	104.6	64.8	94.9	132.3	113.1	127.6	132.2	113.1	127.5	132.2	87.6	121.4
099	PAINTING & WALL COVERINGS	113.2	42.6	71.8	113.2	42.6	71.8	113.2	42.6	71.8	110.6	103.2	106.3	109.9	113.3	111.9	109.8	83.5	94.4
9	FINISHES	98.4	47.4	72.2	98.0	40.3	68.4	98.3	46.4	71.7	123.2	106.9	114.9	118.7	114.4	116.5	119.9	86.9	102.8
10 - 14	TOTAL DIV. 10 - 14	100.0	72.4	94.2	100.0	64.6	92.5	100.0	73.1	94.3	99.4	90.0	97.4	99.4	93.0	98.0	100.0	91.0	98.1
15	MECHANICAL	97.9	41.5	72.8	97.9	39.8	72.0	97.9	50.2	76.7	103.1	102.8	103.0	103.0	106.5	104.6	103.0	78.6	92.1
16	ELECTRICAL	95.4	59.0	70.8	95.4	59.0	70.8	93.8	59.0	70.3	127.9	107.1	113.8	126.9	111.7	116.7	113.8	87.5	96.0
1 - 16	WEIGHTED AVERAGE	101.7	56.2	79.8	100.8	53.6	78.0	100.0	58.6	80.0	115.3	102.3	109.1	113.2	108.4	110.9	110.3	85.9	98.5

CANADA

DIVISION		CHARLOTTETOWN, PEI			CHICOUTIMI, QUEBEC			CORNWALL, ONTARIO			EDMONTON, ALBERTA			HALIFAX, NOVA SCOTIA			HAMILTON, ONTARIO		
		MAT.	INST.	TOTAL	MAT.	INST.	TOTAL	MAT.	INST.	TOTAL	MAT.	INST.	TOTAL	MAT.	INST.	TOTAL	MAT.	INST.	TOTAL
2	SITE WORK	104.9	95.4	97.6	90.0	98.2	96.3	110.9	97.9	100.9	106.4	97.9	99.9	96.6	97.2	97.0	111.4	98.1	101.1
031	CONCRETE FORMWORK	79.9	66.0	68.0	124.8	99.3	102.9	116.7	105.2	106.8	116.2	86.6	90.8	77.6	80.6	80.2	119.0	118.4	118.4
032	CONCRETE REINFORCEMENT	147.4	59.7	98.0	111.3	92.7	100.8	174.2	95.4	129.8	160.8	68.4	108.7	147.4	60.6	98.5	174.2	97.4	130.9
033	CAST IN PLACE CONCRETE	167.9	66.4	125.4	131.4	107.2	121.2	153.3	108.2	134.4	144.3	87.9	120.7	178.4	79.2	136.8	157.2	111.2	138.0
3	CONCRETE	154.4	65.6	109.6	124.6	100.6	112.5	143.0	104.1	123.4	136.9	84.0	110.2	150.3	76.9	113.3	145.1	107.6	126.2
4	MASONRY	146.3	72.8	100.6	142.9	103.0	118.1	148.8	109.1	124.1	146.6	87.9	110.1	147.1	83.1	107.3	149.8	113.1	127.0
5	METALS	100.8	73.6	91.0	99.2	93.2	97.0	100.8	96.2	99.1	100.8	85.8	95.4	99.9	81.6	93.3	101.0	96.9	99.5
6	WOOD & PLASTICS	80.6	65.5	72.8	132.7	99.4	115.5	116.7	105.1	110.7	112.2	85.9	98.6	80.2	79.8	80.0	117.6	109.4	113.4
7	THERMAL & MOISTURE PROTECTION	105.2	69.1	88.5	104.8	102.7	103.8	104.5	101.1	102.9	104.5	86.5	96.1	103.3	80.8	92.8	104.5	104.3	104.4
8	DOORS & WINDOWS	93.6	57.7	84.9	90.9	84.0	89.2	91.3	99.6	93.3	91.3	80.5	88.7	81.9	74.5	80.1	91.3	103.2	94.2
092	LATH, PLASTER & GYPSUM BOARD	186.1	64.5	107.6	214.9	99.5	140.4	259.2	105.3	159.9	146.3	85.5	107.1	185.6	79.2	116.9	208.3	109.8	144.7
095	ACOUSTICAL TREATMENT & WOOD FLOORING	101.7	64.5	77.6	100.3	99.5	99.8	101.7	105.3	104.0	101.7	85.5	91.2	101.7	79.2	87.1	101.7	109.8	106.9
096	FLOORING & CARPET	110.8	71.8	101.3	132.1	115.2	128.1	132.2	111.5	127.2	132.2	87.6	121.4	110.7	77.1	102.5	132.2	113.1	127.5
099	PAINTING & WALL COVERINGS	109.9	49.2	74.3	109.9	106.6	107.9	109.9	105.5	107.3	109.9	83.5	94.4	109.9	73.8	88.7	109.9	113.3	111.9
9	FINISHES	117.3	65.6	90.8	127.1	103.5	115.0	133.8	107.0	120.1	119.3	86.9	102.7	116.5	79.3	97.4	127.1	111.4	119.0
10 - 14	TOTAL DIV. 10 - 14	99.4	76.1	94.4	99.4	100.0	99.5	99.4	89.4	97.3	100.0	90.7	98.0	99.4	78.6	95.0	100.0	117.3	103.6
15	MECHANICAL	100.7	65.2	84.9	100.7	93.4	97.5	103.0	103.6	103.3	103.0	78.6	92.1	100.7	86.3	94.3	103.0	108.2	105.3
16	ELECTRICAL	123.6	68.0	86.1	123.6	95.2	104.5	123.6	107.8	113.0	113.8	87.5	96.0	123.6	83.8	96.8	126.9	110.9	116.1
1 - 16	WEIGHTED AVERAGE	112.2	70.4	92.0	109.0	97.4	103.4	113.5	103.5	108.7	110.2	85.9	98.5	109.8	83.3	97.0	113.4	107.3	110.5

COST INDEXES

643

CANADA

DIVISION		KINGSTON, ONTARIO			KITCHENER, ONTARIO			LONDON, ONTARIO			MONCTON, NEW BRUNSWICK			MONTREAL, QUEBEC			NORTH BAY, ONTARIO		
		MAT.	INST.	TOTAL	MAT.	INST.	TOTAL	MAT.	INST.	TOTAL	MAT.	INST.	TOTAL	MAT.	INST.	TOTAL	MAT.	INST.	TOTAL
2	SITE WORK	110.9	97.9	100.9	97.7	97.7	97.7	111.4	97.8	100.9	93.4	95.9	95.4	91.6	98.2	96.7	113.1	97.5	101.1
031	CONCRETE FORMWORK	116.7	105.2	106.9	106.5	100.6	101.5	118.3	104.3	106.3	86.6	72.3	74.3	129.0	99.3	103.6	120.2	102.3	104.8
032	CONCRETE REINFORCEMENT	174.2	95.4	129.8	110.0	97.2	102.8	122.4	95.5	107.2	147.4	72.2	105.0	156.2	92.7	120.4	174.2	94.8	129.5
033	CAST IN PLACE CONCRETE	153.3	108.2	134.4	140.1	89.8	119.0	157.2	108.3	136.7	144.0	67.2	111.8	140.1	107.2	126.3	170.4	94.0	138.4
3	CONCRETE	143.0	104.1	123.4	120.9	95.9	108.3	138.2	103.8	120.9	136.4	70.9	103.4	135.2	100.6	117.7	151.7	97.8	124.5
4	MASONRY	148.8	109.2	124.2	146.7	108.0	122.6	149.1	109.6	124.6	144.5	74.7	101.2	145.0	103.0	118.9	150.1	104.4	121.7
5	METALS	100.8	96.1	99.1	100.0	93.3	97.6	99.2	96.4	98.2	100.5	77.9	92.3	101.1	93.3	98.3	100.8	95.9	99.0
6	WOOD & PLASTICS	116.7	105.2	110.8	110.0	99.5	104.6	117.6	103.0	110.1	90.0	72.6	81.0	132.7	99.4	115.5	118.9	103.4	110.9
7	THERMAL & MOISTURE PROTECTION	104.5	100.6	102.7	103.1	99.0	101.2	104.5	102.0	103.4	103.9	71.3	88.8	104.8	102.7	103.8	104.7	98.4	101.8
8	DOORS & WINDOWS	91.3	97.3	92.8	82.8	93.9	85.5	92.1	98.6	93.6	85.3	64.3	80.2	91.3	85.9	90.0	90.5	96.7	92.0
092	LATH, PLASTER & GYPSUM BOARD	259.2	105.5	160.0	150.3	99.6	117.5	208.3	103.2	140.5	191.7	71.7	114.3	149.1	99.5	117.1	144.6	103.6	118.1
095	ACOUSTICAL TREATMENT & WOOD FLOORING	101.7	105.5	104.1	101.7	99.6	100.3	101.7	103.2	102.7	103.1	71.7	82.8	101.7	99.5	100.2	101.7	103.6	102.9
096	FLOORING & CARPET	132.2	111.5	127.2	128.0	113.1	124.4	132.2	113.1	127.5	114.5	53.9	99.8	132.2	115.2	128.1	132.2	111.5	127.2
099	PAINTING & WALL COVERINGS	109.9	107.3	108.4	109.9	103.2	105.9	109.9	113.3	111.9	109.9	59.1	80.1	109.9	106.6	107.9	109.9	104.7	106.8
9	FINISHES	133.8	107.3	120.2	117.4	103.6	110.3	127.1	107.2	116.9	119.0	67.5	92.6	118.0	103.5	110.9	118.7	105.0	111.7
10 - 14	TOTAL DIV. 10 - 14	99.4	89.5	97.3	99.4	114.4	102.5	100.0	115.6	103.3	99.4	76.8	94.6	100.0	100.0	100.0	99.4	87.6	96.9
15	MECHANICAL	103.0	103.8	103.4	100.7	103.4	101.9	103.0	103.8	103.4	100.7	71.7	87.7	103.0	93.4	98.8	103.0	101.3	102.3
16	ELECTRICAL	123.6	106.5	112.1	126.9	106.6	113.2	126.9	106.8	113.3	123.6	81.6	95.3	123.6	95.2	104.5	126.9	107.7	114.0
1 - 16	WEIGHTED AVERAGE	113.5	103.2	108.5	107.1	101.6	104.5	112.4	104.0	108.3	108.8	75.8	92.9	110.5	97.5	104.2	113.2	101.2	107.4

CANADA

DIVISION		OSHAWA, ONTARIO			OTTAWA, ONTARIO			OWEN SOUND, ONTARIO			PETERBOROUGH, ONTARIO			QUEBEC, QUEBEC			REGINA, SASKATCHEWAN		
		MAT.	INST.	TOTAL	MAT.	INST.	TOTAL	MAT.	INST.	TOTAL	MAT.	INST.	TOTAL	MAT.	INST.	TOTAL	MAT.	INST.	TOTAL
2	SITE WORK	113.0	97.9	101.4	110.9	97.9	100.9	114.0	97.6	101.4	113.1	97.9	101.4	91.2	98.2	96.6	108.2	96.2	99.0
031	CONCRETE FORMWORK	120.2	105.5	107.6	116.7	105.4	107.0	119.8	100.3	103.1	120.2	103.4	105.8	129.0	99.4	103.7	91.6	70.7	73.7
032	CONCRETE REINFORCEMENT	174.2	98.1	131.3	174.2	95.4	129.8	179.5	97.3	133.2	174.2	95.5	129.8	147.4	92.7	116.6	113.5	83.2	96.4
033	CAST IN PLACE CONCRETE	169.7	96.1	138.9	153.3	108.3	134.4	174.7	89.6	139.1	170.4	96.0	139.2	137.5	107.2	124.8	152.7	79.9	122.2
3	CONCRETE	151.3	100.5	125.7	143.0	104.3	123.5	154.4	95.8	124.9	151.7	99.1	125.2	132.7	100.7	116.6	126.3	76.7	101.3
4	MASONRY	150.0	112.2	126.5	149.5	109.1	124.4	155.1	110.8	127.6	150.1	112.2	126.5	144.0	103.0	118.6	146.4	78.0	103.9
5	METALS	100.8	96.7	99.3	101.0	96.4	99.3	106.1	93.2	101.5	100.8	96.4	99.2	101.0	93.3	98.3	97.7	83.6	92.6
6	WOOD & PLASTICS	118.9	104.1	111.2	116.7	105.1	110.7	120.3	99.5	109.6	118.9	101.8	110.1	132.7	99.4	115.5	92.2	68.9	80.2
7	THERMAL & MOISTURE PROTECTION	104.6	101.4	103.1	104.5	101.1	102.9	105.1	99.3	102.4	104.7	102.0	103.5	104.8	102.7	103.8	103.8	76.5	91.1
8	DOORS & WINDOWS	90.5	99.9	92.8	91.3	99.6	93.3	91.2	93.8	91.8	90.5	98.7	92.5	91.3	91.1	91.3	86.2	68.2	81.9
092	LATH, PLASTER & GYPSUM BOARD	155.0	104.3	122.3	259.2	105.3	159.9	169.0	99.6	124.2	144.6	102.0	117.1	215.2	99.5	140.5	124.6	68.0	88.1
095	ACOUSTICAL TREATMENT & WOOD FLOORING	101.7	104.3	103.4	101.7	105.3	104.0	101.7	99.6	100.3	101.7	102.0	101.9	101.7	99.5	100.2	99.0	68.0	78.9
096	FLOORING & CARPET	132.2	116.0	128.3	132.2	111.5	127.2	132.3	113.1	127.6	132.2	111.5	127.2	132.2	115.2	128.1	120.2	75.2	109.3
099	PAINTING & WALL COVERINGS	109.9	123.6	117.9	109.9	105.5	107.3	110.6	103.2	106.3	109.9	107.3	108.4	109.9	106.6	107.9	109.9	78.0	91.1
9	FINISHES	120.0	109.9	114.8	133.8	107.0	120.1	123.2	103.6	113.1	118.7	105.8	112.1	127.4	103.5	115.2	111.9	71.9	91.4
10 - 14	TOTAL DIV. 10 - 14	99.4	116.2	102.9	100.0	113.1	102.8	99.4	88.3	97.0	99.4	89.7	97.3	100.0	100.0	100.0	99.4	76.6	94.6
15	MECHANICAL	103.0	105.6	104.2	103.0	104.1	103.5	103.1	101.4	102.4	103.1	104.1	103.5	103.0	93.5	98.8	100.7	79.3	91.1
16	ELECTRICAL	126.9	107.5	113.8	123.6	107.8	113.0	128.8	106.3	113.7	126.9	107.5	113.8	123.6	95.3	104.5	128.8	83.6	98.3
1 - 16	WEIGHTED AVERAGE	113.3	104.7	109.2	113.6	104.2	109.1	115.4	100.8	108.4	113.2	103.2	108.4	111.0	97.7	104.6	107.5	80.0	94.2

CANADA

DIVISION		SAINT JOHN, N B			SARNIA, ONTARIO			SASKATOON, SASKATCHEWAN			ST CATHARINES, ONTARIO			ST JOHNS, NEWFOUNDLAND			SUDBURY, ONTARIO		
		MAT.	INST.	TOTAL	MAT.	INST.	TOTAL	MAT.	INST.	TOTAL	MAT.	INST.	TOTAL	MAT.	INST.	TOTAL	MAT.	INST.	TOTAL
2	SITE WORK	94.0	96.0	95.5	111.4	97.9	101.0	105.2	96.3	98.4	98.9	97.6	97.9	112.7	95.8	99.7	98.9	97.6	97.9
031	CONCRETE FORMWORK	113.9	72.5	78.5	118.3	112.0	112.9	91.3	70.6	73.5	103.6	102.9	103.0	90.4	69.4	72.4	98.6	102.5	101.9
032	CONCRETE REINFORCEMENT	147.4	72.3	105.1	122.4	100.7	110.2	118.5	83.1	98.6	111.0	97.3	103.3	147.4	61.3	98.9	112.0	94.9	102.3
033	CAST IN PLACE CONCRETE	144.8	67.2	112.3	157.2	110.0	137.4	152.7	79.9	122.2	140.1	94.8	121.1	179.6	75.8	136.1	140.1	94.1	120.8
3	CONCRETE	139.1	71.1	104.8	138.2	108.7	123.3	126.9	76.6	101.6	120.8	98.6	109.6	164.0	70.9	117.0	120.6	98.0	109.2
4	MASONRY	144.6	76.5	102.3	149.1	115.4	128.1	146.5	78.0	103.9	146.7	104.6	120.5	147.2	73.9	101.7	146.8	104.4	120.4
5	METALS	100.1	78.4	92.2	98.8	98.1	98.5	97.8	83.4	92.6	97.2	93.5	95.9	102.2	78.4	93.6	97.4	96.1	97.0
6	WOOD & PLASTICS	121.4	72.6	96.2	117.6	111.8	114.6	92.2	68.9	80.2	106.8	104.9	105.8	91.6	68.4	79.6	101.0	103.4	102.2
7	THERMAL & MOISTURE PROTECTION	104.5	72.5	89.7	104.5	107.6	105.9	103.6	75.3	90.5	103.1	100.3	101.8	106.0	70.8	89.7	103.1	98.4	100.9
8	DOORS & WINDOWS	85.2	63.0	79.8	92.1	104.2	95.0	84.8	68.2	80.8	82.4	98.7	86.3	97.6	65.4	89.8	83.3	96.7	86.6
092	LATH, PLASTER & GYPSUM BOARD	208.3	71.7	120.4	208.3	112.3	146.3	124.6	68.0	88.1	147.5	105.1	120.1	192.6	67.4	111.8	144.4	103.6	118.1
095	ACOUSTICAL TREATMENT & WOOD FLOORING	101.7	71.7	82.3	101.7	112.3	108.5	99.0	68.0	78.9	99.0	105.1	102.9	104.4	67.4	80.4	99.0	103.6	102.0
096	FLOORING & CARPET	127.5	53.9	109.6	132.2	117.8	128.7	120.2	75.2	109.3	126.5	113.1	123.2	115.3	64.7	103.0	123.6	111.5	120.7
099	PAINTING & WALL COVERINGS	109.9	75.4	89.6	109.9	117.8	114.5	109.9	66.5	84.4	109.9	113.3	111.9	109.9	70.0	86.5	109.9	104.7	106.8
9	FINISHES	125.3	69.3	96.6	127.1	114.5	120.6	111.9	70.6	90.7	116.0	106.7	111.2	120.4	68.5	93.8	114.6	105.0	109.7
10 - 14	TOTAL DIV. 10 - 14	99.4	76.9	94.6	99.4	91.8	97.8	99.4	76.6	94.6	99.4	87.1	96.8	99.4	76.8	94.6	99.4	113.8	102.4
15	MECHANICAL	100.7	81.0	91.9	103.0	111.6	106.9	100.7	79.3	91.1	100.7	101.3	101.0	100.7	72.6	88.2	100.7	98.0	99.5
16	ELECTRICAL	123.1	101.7	108.6	126.9	116.2	119.7	128.8	83.6	98.3	126.9	107.9	114.1	123.6	75.3	91.0	126.9	107.7	114.0
1 - 16	WEIGHTED AVERAGE	110.1	81.4	96.2	112.3	109.2	110.8	107.3	79.8	94.0	106.5	101.5	104.1	114.7	75.0	95.6	106.4	101.2	103.9

COST INDEXES

DIVISION		CANADA																		
		THUNDER BAY, ONTARIO			TORONTO, ONTARIO			VANCOUVER, B C			VICTORIA, B C			WINDSOR, ONTARIO			WINNIPEG, MANITOBA		
		MAT.	INST.	TOTAL	MAT.	INST.	TOTAL	MAT.	INST.	TOTAL	MAT.	INST.	TOTAL	MAT.	INST.	TOTAL	MAT.	INST.	TOTAL
2	SITE WORK	104.9	97.7	99.4	113.1	98.4	101.8	100.6	99.4	99.7	119.5	99.1	103.8	94.4	98.0	97.1	109.4	96.8	99.7
031	CONCRETE FORMWORK	114.4	104.2	105.7	120.2	115.1	115.8	110.5	106.1	106.8	96.4	105.3	104.0	114.4	105.0	106.3	119.1	85.8	90.6
032	CONCRETE REINFORCEMENT	99.2	96.5	97.7	173.3	102.0	133.1	160.8	108.2	131.2	116.0	108.1	111.5	108.7	95.6	101.3	160.8	69.5	109.4
033	CAST IN PLACE CONCRETE	155.3	108.4	135.7	170.5	119.6	149.2	125.5	109.1	118.6	164.8	108.8	141.3	140.1	109.8	127.4	151.2	85.0	123.5
3	CONCRETE	130.9	103.9	117.3	151.6	113.6	132.5	133.3	107.2	120.1	160.7	106.7	133.4	121.3	104.6	112.9	133.6	82.9	108.0
4	MASONRY	147.7	105.8	121.6	150.9	123.0	133.6	145.2	110.6	123.7	147.3	110.6	124.5	146.7	113.7	126.2	147.1	82.5	106.9
5	METALS	96.5	95.4	96.1	101.0	100.0	100.6	101.2	100.1	100.8	99.8	99.2	99.6	96.7	96.8	96.7	100.8	85.4	95.3
6	WOOD & PLASTICS	118.9	105.3	111.9	118.9	113.0	115.9	103.6	103.8	103.7	94.0	103.8	99.1	118.9	103.5	110.9	116.6	87.1	101.4
7	THERMAL & MOISTURE PROTECTION	103.6	99.5	101.7	104.8	112.5	108.4	104.3	105.6	104.9	105.1	97.9	101.7	103.2	104.1	103.6	104.5	86.1	96.0
8	DOORS & WINDOWS	81.4	98.7	85.6	90.5	108.0	94.8	91.3	102.9	94.1	87.1	102.7	90.9	81.1	98.5	85.3	91.3	77.9	88.0
092	LATH, PLASTER & GYPSUM BOARD	154.1	105.6	122.8	144.6	113.5	124.6	152.0	104.0	121.0	146.7	104.0	119.1	154.1	103.7	121.6	138.5	86.8	105.1
095	ACOUSTICAL TREATMENT & WOOD FLOORING	97.6	105.6	102.8	101.7	113.5	109.4	101.7	104.0	103.2	101.7	104.0	103.2	97.6	103.7	101.5	101.7	86.8	92.0
096	FLOORING & CARPET	132.2	65.1	115.9	132.2	119.9	129.2	132.2	112.5	127.4	126.7	65.3	111.8	132.2	113.9	127.8	132.2	82.9	120.2
099	PAINTING & WALL COVERINGS	109.9	107.7	108.6	112.9	123.6	119.2	109.8	120.5	116.1	109.9	96.5	102.0	109.9	107.2	108.3	109.9	72.2	87.8
9	FINISHES	118.8	98.7	108.5	118.9	117.3	118.1	120.1	108.3	114.0	118.8	97.3	107.8	118.5	107.3	112.7	118.1	84.2	100.7
10 - 14	TOTAL DIV. 10 - 14	99.4	88.5	97.1	100.0	119.7	104.2	100.0	112.1	102.6	99.4	112.1	102.1	99.4	90.1	97.4	100.0	80.1	95.8
15	MECHANICAL	100.7	102.4	101.4	103.0	114.4	108.1	103.0	105.9	104.3	100.7	97.4	99.2	100.7	105.7	102.9	103.0	87.8	96.2
16	ELECTRICAL	126.9	105.8	112.7	127.0	111.7	116.7	125.6	106.7	112.9	126.5	106.8	113.2	126.9	108.2	114.3	128.8	93.1	104.7
1 - 16	WEIGHTED AVERAGE	108.1	101.4	104.9	113.4	111.9	112.7	110.5	105.7	108.2	113.0	102.4	107.9	106.6	104.6	105.6	111.0	87.1	99.5

Location Factors

Costs shown in *Means cost data publications* are based on National Averages for materials and installation. To adjust these costs to a specific location, simply multiply the base cost by the factor and divide by 100 for that city. The data is arranged alphabetically by state and postal zip code numbers. For a city not listed, use the factor for a nearby city with similar economic characteristics.

STATE/ZIP	CITY	MAT.	INST.	TOTAL
ALABAMA				
350-352	Birmingham	96.5	77.0	87.1
354	Tuscaloosa	96.2	62.2	79.8
355	Jasper	97.5	53.3	76.2
356	Decatur	96.3	68.4	82.8
357-358	Huntsville	96.1	68.4	82.7
359	Gadsden	97.0	66.2	82.2
360-361	Montgomery	97.1	60.8	79.6
362	Anniston	95.1	53.1	74.8
363	Dothan	96.6	59.5	78.7
364	Evergreen	95.9	61.3	79.2
365-366	Mobile	97.0	68.9	83.4
367	Selma	96.2	59.5	78.5
368	Phenix City	96.9	60.1	79.2
369	Butler	96.3	59.5	78.6
ALASKA				
995-996	Anchorage	133.1	117.7	125.7
997	Fairbanks	129.3	120.7	125.1
998	Juneau	131.2	118.0	124.8
999	Ketchikan	140.6	118.0	129.7
ARIZONA				
850,853	Phoenix	100.6	79.0	90.2
852	Mesa/Tempe	100.6	69.5	85.6
855	Globe	101.4	73.9	88.1
856-857	Tucson	99.4	77.6	88.9
859	Show Low	101.5	74.0	88.2
860	Flagstaff	102.8	78.2	91.0
863	Prescott	100.3	73.6	87.4
864	Kingman	99.1	74.4	87.2
865	Chambers	99.1	74.2	87.1
ARKANSAS				
716	Pine Bluff	95.4	62.0	79.3
717	Camden	93.8	47.5	71.5
718	Texarkana	94.7	53.4	74.8
719	Hot Springs	93.0	46.9	70.7
720-722	Little Rock	95.9	62.3	79.7
723	West Memphis	95.2	64.8	80.5
724	Jonesboro	95.2	64.8	80.5
725	Batesville	93.9	59.1	77.1
726	Harrison	95.3	59.1	77.8
727	Fayetteville	92.4	40.8	67.5
728	Russellville	94.1	56.9	76.1
729	Fort Smith	96.1	60.7	79.0
CALIFORNIA				
900-902	Los Angeles	105.0	116.5	110.6
903-905	Inglewood	101.2	114.3	107.5
906-908	Long Beach	103.0	114.3	108.5
910-912	Pasadena	100.6	114.4	107.2
913-916	Van Nuys	104.5	114.1	109.2
917-918	Alhambra	103.2	114.4	108.6
919-921	San Diego	104.9	109.5	107.1
922	Palm Springs	102.5	112.0	107.1
923-924	San Bernardino	100.1	111.8	105.7
925	Riverside	104.6	112.8	108.6
926-927	Santa Ana	102.3	112.3	107.2
928	Anaheim	104.9	115.1	109.8
930	Oxnard	105.4	113.8	109.4
931	Santa Barbara	104.6	112.9	108.6
932-933	Bakersfield	104.3	107.0	105.6
934	San Luis Obispo	106.2	111.5	108.8
935	Mojave	102.8	108.9	105.8
936-938	Fresno	105.3	112.3	108.7
939	Salinas	107.3	116.7	111.8
940-941	San Francisco	111.0	138.2	124.2
942,956-958	Sacramento	106.9	114.5	110.5
943	Palo Alto	105.0	127.8	116.0
944	San Mateo	108.0	127.1	117.2
945	Vallejo	105.5	127.0	115.9
946	Oakland	109.9	126.5	117.9
947	Berkeley	109.5	127.9	118.4
948	Richmond	109.3	125.3	117.0
949	San Rafael	111.2	125.7	118.2
950	Santa Cruz	110.8	118.4	114.5

STATE/ZIP	CITY	MAT.	INST.	TOTAL
CALIFORNIA (CONT'D)				
951	San Jose	110.1	129.7	119.6
952	Stockton	105.9	113.2	109.4
953	Modesto	106.0	113.3	109.5
954	Santa Rosa	107.2	129.0	117.7
955	Eureka	108.6	112.1	110.3
959	Marysville	107.3	113.7	110.4
960	Redding	108.4	108.9	108.6
961	Susanville	108.6	108.3	108.5
COLORADO				
800-802	Denver	101.9	84.8	93.6
803	Boulder	100.0	69.5	85.3
804	Golden	102.4	76.9	90.1
805	Fort Collins	103.2	77.9	91.0
806	Greeley	100.2	69.1	85.2
807	Fort Morgan	100.7	77.7	89.6
808-809	Colorado Springs	100.7	80.3	90.8
810	Pueblo	102.6	79.2	91.3
811	Alamosa	104.9	70.3	88.2
812	Salida	104.8	70.4	88.2
813	Durango	105.6	66.0	86.5
814	Montrose	103.9	63.3	84.3
815	Grand Junction	106.9	63.7	86.0
816	Glenwood Springs	105.0	75.6	90.8
CONNECTICUT				
060	New Britain	103.0	105.0	104.0
061	Hartford	103.3	105.3	104.3
062	Willimantic	103.8	104.0	103.9
063	New London	100.2	105.8	102.9
064	Meriden	102.9	104.5	103.7
065	New Haven	103.2	105.3	104.2
066	Bridgeport	104.4	102.5	103.5
067	Waterbury	103.7	105.0	104.4
068	Norwalk	103.7	102.7	103.2
069	Stamford	103.9	106.1	105.0
D.C.				
200-205	Washington	99.6	92.0	96.0
DELAWARE				
197	Newark	99.5	97.2	98.4
198	Wilmington	98.8	97.2	98.0
199	Dover	99.5	97.2	98.4
FLORIDA				
320,322	Jacksonville	98.6	68.6	84.1
321	Daytona Beach	98.7	75.9	87.7
323	Tallahassee	99.1	58.6	79.6
324	Panama City	99.6	45.6	73.5
325	Pensacola	99.2	68.5	84.4
326	Gainesville	100.0	64.6	82.9
327-328,347	Orlando	100.4	70.6	86.1
329	Melbourne	100.6	75.1	88.3
330-332,340	Miami	98.0	74.3	86.6
333	Fort Lauderdale	97.9	75.1	86.9
334,349	West Palm Beach	96.8	69.2	83.5
335-336,346	Tampa	99.8	67.6	84.2
337	St. Petersburg	101.5	67.5	85.1
338	Lakeland	98.5	67.3	83.5
339	Fort Myers	98.2	64.2	81.8
342	Sarasota	100.0	64.6	82.9
GEORGIA				
300-303,399	Atlanta	96.9	79.3	88.4
304	Statesboro	96.8	36.9	67.9
305	Gainesville	96.0	52.2	74.9
306	Athens	95.1	64.5	80.3
307	Dalton	96.8	34.8	66.9
308-309	Augusta	95.7	62.7	79.8
310-312	Macon	95.7	67.6	82.2
313-314	Savannah	97.8	65.6	82.3
315	Waycross	97.7	52.0	75.6
316	Valdosta	97.3	55.6	77.1
317	Albany	97.4	61.6	80.1
318-319	Columbus	97.5	59.7	79.2

STATE/ZIP	CITY	MAT.	INST.	TOTAL
HAWAII				
967	Hilo	115.4	130.7	122.8
968	Honolulu	116.0	130.7	123.1
STATES & POSS.				
969	Guam	115.2	47.2	82.4
IDAHO				
832	Pocatello	101.9	84.4	93.5
833	Twin Falls	104.1	54.7	80.3
834	Idaho Falls	101.4	65.2	83.9
835	Lewiston	113.5	90.6	102.4
836-837	Boise	101.9	85.1	93.8
838	Coeur d'Alene	115.5	65.7	91.4
ILLINOIS				
600-603	North Suburban	97.6	119.9	108.4
604	Joliet	97.6	114.9	105.9
605	South Suburban	97.6	119.8	108.3
606	Chicago	97.8	125.0	110.9
609	Kankakee	93.8	107.9	100.6
610-611	Rockford	97.1	112.0	104.3
612	Rock Island	94.7	100.7	97.6
613	La Salle	96.1	103.0	99.4
614	Galesburg	95.7	100.5	98.0
615-616	Peoria	98.5	102.8	100.6
617	Bloomington	95.0	103.8	99.3
618-619	Champaign	98.7	100.9	99.8
620-622	East St. Louis	93.7	106.9	100.1
623	Quincy	95.2	95.9	95.6
624	Effingham	94.6	99.8	97.1
625	Decatur	96.1	99.1	97.5
626-627	Springfield	95.8	100.9	98.3
628	Centralia	92.9	104.6	98.6
629	Carbondale	92.6	100.5	96.4
INDIANA				
460	Anderson	97.0	87.3	92.3
461-462	Indianapolis	99.9	90.5	95.4
463-464	Gary	97.8	96.9	97.4
465-466	South Bend	96.2	82.4	89.5
467-468	Fort Wayne	97.2	86.9	92.2
469	Kokomo	95.3	82.0	88.9
470	Lawrenceburg	93.1	81.9	87.7
471	New Albany	94.7	78.3	86.8
472	Columbus	96.8	85.9	91.6
473	Muncie	97.1	84.6	91.1
474	Bloomington	98.9	84.2	91.8
475	Washington	95.0	88.2	91.8
476-477	Evansville	96.7	91.5	94.2
478	Terre Haute	97.1	88.5	93.0
479	Lafayette	96.6	82.6	89.8
IOWA				
500-503,509	Des Moines	97.8	83.2	90.8
504	Mason City	96.0	65.7	81.4
505	Fort Dodge	96.5	59.9	78.8
506-507	Waterloo	97.5	67.3	82.9
508	Creston	97.1	72.5	85.2
510-511	Sioux City	98.2	67.0	83.1
512	Sibley	96.7	59.5	78.8
513	Spencer	99.2	59.5	80.0
514	Carroll	95.7	67.0	81.9
515	Council Bluffs	99.0	73.7	86.8
516	Shenandoah	95.6	56.6	76.8
520	Dubuque	97.5	73.4	85.9
521	Decorah	96.9	64.7	81.4
522-524	Cedar Rapids	98.9	79.6	89.6
525	Ottumwa	97.0	73.3	85.6
526	Burlington	95.8	61.0	79.0
527-528	Davenport	97.9	89.2	93.7
KANSAS				
660-662	Kansas City	98.1	85.6	92.1
664-666	Topeka	97.6	71.9	85.2
667	Fort Scott	97.0	72.2	85.0
668	Emporia	96.7	65.3	81.5
669	Belleville	98.7	66.7	83.3
670-672	Wichita	96.8	71.9	84.8
673	Independence	98.5	60.3	80.1
674	Salina	98.2	65.0	82.2
675	Hutchinson	93.8	59.3	77.1
676	Hays	97.9	66.7	82.9
677	Colby	98.8	66.7	83.3

STATE/ZIP	CITY	MAT.	INST.	TOTAL
KANSAS (CONT'D)				
678	Dodge City	99.4	66.7	83.6
679	Liberal	97.6	55.4	77.2
KENTUCKY				
400-402	Louisville	95.9	85.7	90.9
403-405	Lexington	96.6	73.4	85.4
406	Frankfort	95.8	80.1	88.2
407-409	Corbin	94.3	52.8	74.3
410	Covington	95.1	91.6	93.4
411-412	Ashland	93.1	99.8	96.4
413-414	Campton	95.2	52.8	74.7
415-416	Pikeville	95.9	70.3	83.6
417-418	Hazard	94.5	52.8	74.4
420	Paducah	93.0	89.3	91.3
421-422	Bowling Green	95.2	83.8	89.7
423	Owensboro	95.0	84.1	89.8
424	Henderson	92.6	93.3	92.9
425-426	Somerset	92.2	52.8	73.2
427	Elizabethtown	91.8	85.7	88.9
LOUISIANA				
700-701	New Orleans	100.2	69.5	85.4
703	Thibodaux	100.5	69.0	85.3
704	Hammond	97.5	67.5	83.0
705	Lafayette	99.7	64.2	82.6
706	Lake Charles	99.9	67.0	84.0
707-708	Baton Rouge	99.6	64.8	82.8
710-711	Shreveport	96.5	63.4	80.5
712	Monroe	96.4	61.3	79.4
713-714	Alexandria	96.4	59.2	78.5
MAINE				
039	Kittery	98.2	61.6	80.6
040-041	Portland	101.4	76.8	89.5
042	Lewiston	101.9	76.8	89.8
043	Augusta	100.6	57.2	79.6
044	Bangor	100.9	82.0	91.8
045	Bath	99.8	56.3	78.8
046	Machias	99.3	67.5	84.0
047	Houlton	99.5	61.3	81.1
048	Rockland	98.4	68.4	83.9
049	Waterville	99.9	57.2	79.3
MARYLAND				
206	Waldorf	97.7	77.2	87.8
207-208	College Park	97.5	81.4	89.8
209	Silver Spring	96.9	80.1	88.8
210-212	Baltimore	97.1	85.7	91.6
214	Annapolis	96.9	82.5	90.0
215	Cumberland	94.0	80.2	87.4
216	Easton	95.6	42.8	70.1
217	Hagerstown	94.4	81.0	87.9
218	Salisbury	96.2	60.2	78.8
219	Elkton	93.2	77.2	85.4
MASSACHUSETTS				
010-011	Springfield	101.9	102.4	102.2
012	Pittsfield	101.7	97.6	99.7
013	Greenfield	99.7	100.9	100.3
014	Fitchburg	98.1	112.6	105.1
015-016	Worcester	101.8	112.6	107.0
017	Framingham	97.8	120.4	108.7
018	Lowell	101.4	118.0	109.4
019	Lawrence	102.6	116.2	109.2
020-022, 024	Boston	103.8	130.6	116.7
023	Brockton	102.7	116.6	109.4
025	Buzzards Bay	96.9	115.2	105.7
026	Hyannis	99.3	115.2	107.0
027	New Bedford	102.2	115.9	108.8
MICHIGAN				
480,483	Royal Oak	94.3	102.2	98.1
481	Ann Arbor	96.6	105.9	101.1
482	Detroit	97.6	111.4	104.3
484-485	Flint	96.2	99.6	97.8
486	Saginaw	96.5	92.9	94.8
487	Bay City	96.0	93.3	94.7
488-489	Lansing	96.4	95.5	96.0
490	Battle Creek	96.5	87.8	92.3
491	Kalamazoo	96.8	85.7	91.5
492	Jackson	95.2	96.5	95.9
493,495	Grand Rapids	96.4	76.8	86.9
494	Muskegan	96.6	84.6	90.8

STATE/ZIP	CITY	MAT.	INST.	TOTAL
MICHIGAN (CONT'D)				
496	Traverse City	94.5	79.1	87.1
497	Gaylord	96.0	85.6	91.0
498-499	Iron mountain	97.9	86.4	92.4
MINNESOTA				
550-551	Saint Paul	97.8	115.2	106.2
553-554	Minneapolis	98.5	120.4	109.0
556-558	Duluth	97.9	109.9	103.7
559	Rochester	97.7	104.3	100.9
560	Mankato	96.4	101.1	98.7
561	Windom	95.3	75.8	85.9
562	Willmar	94.5	77.2	86.1
563	St. Cloud	95.5	103.4	99.3
564	Brainerd	96.0	99.9	97.9
565	Detroit Lakes	98.8	91.0	95.0
566	Bemidji	98.1	97.0	97.5
567	Thief River Falls	97.2	87.8	92.6
MISSISSIPPI				
386	Clarksdale	96.9	36.6	67.8
387	Greenville	100.2	56.1	78.9
388	Tupelo	98.1	46.7	73.3
389	Greenwood	98.3	39.2	69.8
390-392	Jackson	98.7	58.0	79.0
393	Meridian	96.7	55.6	76.8
394	Laurel	98.0	40.7	70.3
395	Biloxi	98.8	62.6	81.3
396	Mccomb	96.4	37.7	68.1
397	Columbus	98.0	46.2	73.0
MISSOURI				
630-631	St. Louis	96.6	109.1	102.7
633	Bowling Green	96.5	93.9	95.2
634	Hannibal	95.3	93.8	94.6
635	Kirksville	97.2	82.0	89.8
636	Flat River	97.5	98.5	98.0
637	Cape Girardeau	96.5	97.1	96.8
638	Sikeston	95.0	89.2	92.2
639	Poplar Bluff	94.5	88.9	91.8
640-641	Kansas City	99.2	97.9	98.6
644-645	St. Joseph	99.2	83.5	91.6
646	Chillicothe	96.2	70.9	84.0
647	Harrisonville	95.8	89.1	92.6
648	Joplin	98.0	73.3	86.1
650-651	Jefferson City	95.2	86.7	91.1
652	Columbia	96.4	81.5	89.2
653	Sedalia	95.7	82.1	89.1
654-655	Rolla	94.8	69.5	82.6
656-658	Springfield	97.8	75.9	87.2
MONTANA				
590-591	Billings	100.6	90.9	95.9
592	Wolf Point	100.9	87.0	94.2
593	Miles City	99.0	90.4	94.8
594	Great Falls	101.8	88.6	95.5
595	Havre	99.6	87.9	94.0
596	Helena	101.7	88.1	95.1
597	Butte	100.3	84.5	92.7
598	Missoula	98.7	86.8	93.0
599	Kalispell	97.8	85.9	92.0
NEBRASKA				
680-681	Omaha	100.5	74.9	88.2
683-685	Lincoln	97.9	66.2	82.6
686	Columbus	96.5	50.9	74.5
687	Norfolk	98.8	68.3	84.1
688	Grand Island	97.5	67.9	83.2
689	Hastings	97.0	61.5	79.9
690	Mccook	96.8	44.4	71.5
691	North Platte	96.9	61.5	79.8
692	Valentine	100.6	48.2	75.3
693	Alliance	99.4	43.9	72.6
NEVADA				
889-891	Las Vegas	101.3	106.4	103.8
893	Ely	102.1	91.0	96.7
894-895	Reno	102.0	95.1	98.7
897	Carson City	101.8	95.0	98.5
898	Elko	100.8	90.5	95.8
NEW HAMPSHIRE				
030	Nashua	102.4	86.5	94.7
031	Manchester	102.4	86.5	94.7

STATE/ZIP	CITY	MAT.	INST.	TOTAL
NEW HAMPSHIRE (CONT'D)				
032-033	Concord	100.2	86.5	93.6
034	Keene	99.3	61.7	81.2
035	Littleton	99.3	65.1	82.8
036	Charleston	98.7	58.1	79.1
037	Claremont	97.8	58.1	78.7
038	Portsmouth	99.4	83.6	91.8
NEW JERSEY				
070-071	Newark	103.9	121.4	112.4
072	Elizabeth	102.8	113.4	107.9
073	Jersey City	101.4	122.2	111.4
074-075	Paterson	104.2	120.4	112.0
076	Hackensack	101.2	116.9	108.8
077	Long Branch	100.8	118.8	109.5
078	Dover	101.6	117.0	109.0
079	Summit	101.7	112.0	106.7
080,083	Vineland	99.1	113.6	106.1
081	Camden	101.3	116.1	108.4
082,084	Atlantic City	100.1	115.1	107.3
085-086	Trenton	102.1	121.5	111.5
087	Point Pleasant	101.5	116.0	108.5
088-089	New Brunswick	101.9	119.2	110.2
NEW MEXICO				
870-872	Albuquerque	100.6	81.1	91.1
873	Gallup	102.0	80.3	91.5
874	Farmington	101.8	80.1	91.3
875	Santa Fe	100.2	80.3	90.6
877	Las Vegas	100.1	80.2	90.5
878	Socorro	99.6	79.8	90.1
879	Truth/Consequences	99.8	74.4	87.6
880	Las Cruces	97.8	71.4	85.1
881	Clovis	100.2	79.5	90.2
882	Roswell	101.8	78.6	90.6
883	Carrizozo	102.5	79.9	91.6
884	Tucumcari	101.2	78.8	90.4
NEW YORK				
100-102	New York	108.9	160.6	133.9
103	Staten Island	105.1	145.0	124.4
104	Bronx	102.4	145.6	123.2
105	Mount Vernon	103.1	132.9	117.5
106	White Plains	103.0	130.9	116.5
107	Yonkers	108.0	136.0	121.5
108	New Rochelle	103.8	133.1	117.9
109	Suffern	103.8	118.8	111.0
110	Queens	103.3	146.4	124.1
111	Long Island City	105.3	146.1	125.0
112	Brooklyn	105.6	144.0	124.1
113	Flushing	106.0	146.1	125.4
114	Jamaica	103.9	146.1	124.3
115,117,118	Hicksville	103.2	142.6	122.2
116	Far Rockaway	106.1	145.3	125.0
119	Riverhead	103.7	143.3	122.8
120-122	Albany	98.5	95.8	97.2
123	Schenectady	99.4	97.0	98.2
124	Kingston	103.6	112.0	107.6
125-126	Poughkeepsie	102.8	117.9	110.1
127	Monticello	101.9	111.7	106.6
128	Glens Falls	93.5	89.2	91.4
129	Plattsburgh	99.2	84.7	92.2
130-132	Syracuse	100.3	95.5	97.9
133-135	Utica	98.6	88.1	93.5
136	Watertown	100.0	89.9	95.1
137-139	Binghamton	99.4	87.8	93.8
140-142	Buffalo	99.8	105.0	102.3
143	Niagara Falls	98.1	103.3	100.6
144-146	Rochester	101.2	100.5	100.9
147	Jamestown	97.4	83.2	90.6
148-149	Elmira	96.8	88.8	92.9
NORTH CAROLINA				
270,272-274	Greensboro	97.1	57.3	77.9
271	Winston-Salem	97.2	56.9	77.8
275-276	Raleigh	97.6	57.3	78.2
277	Durham	97.0	57.4	77.9
278	Rocky Mount	95.5	42.8	70.1
279	Elizabeth City	96.1	45.6	71.7
280	Gastonia	97.2	55.0	76.8
281-282	Charlotte	97.4	55.3	77.1
283	Fayetteville	95.6	57.4	77.1
284	Wilmington	95.7	56.3	76.7
285	Kinston	94.0	42.3	69.0

STATE/ZIP	CITY	MAT.	INST.	TOTAL
NORTH CAROLINA (CONT'D)				
286	Hickory	94.2	41.4	68.7
287-288	Asheville	96.1	56.1	76.8
289	Murphy	95.6	40.4	69.0
NORTH DAKOTA				
580-581	Fargo	100.1	65.1	83.2
582	Grand Forks	101.1	63.7	83.0
583	Devils Lake	101.1	63.7	83.0
584	Jamestown	100.8	63.9	83.0
585	Bismarck	100.1	68.0	84.6
586	Dickinson	101.6	63.7	83.3
587	Minot	101.1	66.5	84.4
588	Williston	100.3	63.7	82.6
OHIO				
430-432	Columbus	97.5	90.5	94.1
433	Marion	94.8	88.2	91.6
434-436	Toledo	98.0	97.9	98.0
437-438	Zanesville	94.8	86.6	90.8
439	Steubenville	96.5	96.3	96.4
440	Lorain	98.4	94.1	96.3
441	Cleveland	98.4	105.6	101.9
442-443	Akron	99.6	100.1	99.8
444-445	Youngstown	98.8	93.1	96.1
446-447	Canton	98.9	91.3	95.3
448-449	Mansfield	96.3	90.5	93.5
450	Hamilton	94.9	90.2	92.6
451-452	Cincinnati	94.8	90.0	92.5
453-454	Dayton	94.6	88.9	91.8
455	Springfield	94.8	88.0	91.5
456	Chillicothe	94.1	94.7	94.4
457	Athens	97.4	77.3	87.7
458	Lima	97.9	88.2	93.2
OKLAHOMA				
730-731	Oklahoma City	97.6	66.7	82.7
734	Ardmore	95.6	65.5	81.1
735	Lawton	97.5	66.4	82.5
736	Clinton	96.9	63.9	81.0
737	Enid	97.2	64.4	81.4
738	Woodward	95.7	64.0	80.4
739	Guymon	97.0	39.4	69.2
740-741	Tulsa	97.5	66.8	82.7
743	Miami	94.6	67.8	81.6
744	Muskogee	96.3	50.7	74.3
745	Mcalester	93.9	59.5	77.3
746	Ponca City	94.5	66.3	80.9
747	Durant	94.6	65.1	80.3
748	Shawnee	96.1	62.1	79.7
749	Poteau	93.7	67.3	81.0
OREGON				
970-972	Portland	104.0	105.7	104.8
973	Salem	104.5	102.3	103.5
974	Eugene	104.0	101.8	102.9
975	Medford	105.6	98.0	101.9
976	Klamath Falls	106.4	97.9	102.3
977	Bend	105.3	99.3	102.4
978	Pendleton	97.8	99.9	98.8
979	Vale	95.4	93.6	94.5
PENNSYLVANIA				
150-152	Pittsburgh	97.8	107.5	102.5
153	Washington	95.2	105.4	100.2
154	Uniontown	95.3	102.6	98.8
155	Bedford	96.4	95.5	96.0
156	Greensburg	96.7	103.5	99.9
157	Indiana	95.3	100.5	97.8
158	Dubois	96.8	96.4	96.6
159	Johnstown	96.4	97.9	97.1
160	Butler	93.2	101.1	97.0
161	New Castle	93.1	101.6	97.2
162	Kittanning	93.7	105.1	99.2
163	Oil City	93.1	94.7	93.9
164-165	Erie	95.4	96.3	95.8
166	Altoona	95.1	96.7	95.9
167	Bradford	96.7	96.1	96.4
168	State College	96.3	98.1	97.2
169	Wellsboro	97.6	92.3	95.0
170-171	Harrisburg	98.6	94.2	96.5
172	Chambersburg	97.0	91.5	94.4
173-174	York	96.8	94.4	95.6
175-176	Lancaster	95.6	93.1	94.4

STATE/ZIP	CITY	MAT.	INST.	TOTAL
PENNSYLVANIA (CONT'D)				
177	Williamsport	93.9	89.8	92.0
178	Sunbury	96.3	92.7	94.5
179	Pottsville	95.3	94.5	94.9
180	Lehigh Valley	97.2	105.4	101.2
181	Allentown	99.2	102.9	101.0
182	Hazleton	96.3	94.7	95.5
183	Stroudsburg	96.4	94.3	95.4
184-185	Scranton	99.8	98.7	99.3
186-187	Wilkes-Barre	95.9	96.5	96.2
188	Montrose	95.5	97.9	96.6
189	Doylestown	95.7	113.2	104.2
190-191	Philadelphia	99.8	124.9	111.9
193	Westchester	96.8	110.1	103.2
194	Norristown	95.5	115.9	105.4
195-196	Reading	98.2	98.2	98.2
RHODE ISLAND				
028	Newport	100.9	108.4	104.5
029	Providence	101.2	108.3	104.6
SOUTH CAROLINA				
290-292	Columbia	96.3	54.3	76.1
293	Spartanburg	95.0	54.6	75.5
294	Charleston	96.3	55.9	76.8
295	Florence	94.4	52.0	74.0
296	Greenville	94.7	54.6	75.3
297	Rock Hill	94.8	39.7	68.2
298	Aiken	95.6	40.4	68.9
299	Beaufort	96.4	44.2	71.2
SOUTH DAKOTA				
570-571	Sioux Falls	100.4	61.2	81.4
572	Watertown	98.8	58.9	79.5
573	Mitchell	97.9	58.9	79.1
574	Aberdeen	100.3	58.9	80.3
575	Pierre	99.8	57.6	79.4
576	Mobridge	98.7	58.9	79.5
577	Rapid City	100.5	55.9	79.0
TENNESSEE				
370-372	Nashville	96.6	72.6	85.0
373-374	Chattanooga	98.5	65.6	82.6
375,380-381	Memphis	96.4	74.2	85.7
376	Johnson City	97.4	61.7	80.2
377-379	Knoxville	94.6	65.2	80.4
382	Mckenzie	97.3	40.8	70.0
383	Jackson	98.5	51.1	75.6
384	Columbia	95.6	56.0	76.5
385	Cookeville	97.0	39.5	69.2
TEXAS				
750	Mckinney	98.7	66.9	83.4
751	Waxahackie	98.8	67.2	83.5
752-753	Dallas	99.6	72.9	86.7
754	Greenville	98.9	49.1	74.9
755	Texarkana	97.8	59.6	79.4
756	Longview	98.4	50.4	75.2
757	Tyler	98.9	61.6	80.9
758	Palestine	95.2	53.5	75.1
759	Lufkin	96.3	60.4	79.0
760-761	Fort Worth	97.7	69.1	83.9
762	Denton	97.5	61.9	80.3
763	Wichita Falls	98.2	63.4	81.4
764	Eastland	96.8	51.9	75.2
765	Temple	95.3	59.2	77.9
766-767	Waco	97.4	64.1	81.3
768	Brownwood	97.5	48.1	73.7
769	San Angelo	97.2	55.1	76.9
770-772	Houston	99.4	78.4	89.3
773	Huntsville	97.9	51.3	75.4
774	Wharton	99.6	55.8	78.5
775	Galveston	97.6	75.9	87.1
776-777	Beaumont	97.0	73.3	85.5
778	Bryan	94.5	70.2	82.7
779	Victoria	99.7	61.3	81.2
780	Laredo	95.2	63.5	79.9
781-782	San Antonio	95.2	71.5	83.7
783-784	Corpus Christi	98.4	61.2	80.5
785	Mc Allen	98.6	57.1	78.6
786-787	Austin	95.3	68.8	82.5
788	Del Rio	98.0	39.6	69.8
789	Giddings	95.2	52.2	74.4
790-791	Amarillo	98.0	63.2	81.2

STATE/ZIP	CITY	MAT.	INST.	TOTAL
TEXAS (CONT'D)				
792	Childress	97.3	60.8	79.7
793-794	Lubbock	100.0	62.1	81.7
795-796	Abilene	97.4	58.5	78.6
797	Midland	100.5	60.6	81.3
798-799,885	El Paso	97.3	56.0	77.4
UTAH				
840-841	Salt Lake City	101.0	74.0	88.0
842,844	Ogden	99.6	72.8	86.7
843	Logan	101.8	72.8	87.8
845	Price	102.5	59.9	81.9
846-847	Provo	102.1	73.8	88.4
VERMONT				
050	White River Jct.	100.8	43.8	73.3
051	Bellows Falls	99.3	46.4	73.7
052	Bennington	99.5	38.7	70.2
053	Brattleboro	100.0	46.4	74.1
054	Burlington	102.2	65.7	84.6
056	Montpelier	99.2	65.7	83.0
057	Rutland	101.3	65.7	84.1
058	St. Johnsbury	100.9	49.8	76.2
059	Guildhall	99.4	49.2	75.2
VIRGINIA				
220-221	Fairfax	97.9	79.4	88.9
222	Arlington	98.7	80.4	89.9
223	Alexandria	97.7	84.2	91.2
224-225	Fredericksburg	96.5	73.1	85.2
226	Winchester	97.2	62.1	80.3
227	Culpeper	97.1	64.6	81.4
228	Harrisonburg	97.3	52.9	75.8
229	Charlottesville	97.6	67.0	82.8
230-232	Richmond	98.0	71.4	85.1
233-235	Norfolk	98.3	64.7	82.0
236	Newport News	97.7	64.9	81.9
237	Portsmouth	96.8	65.8	81.8
238	Petersburg	97.4	71.4	84.9
239	Farmville	96.9	50.7	74.6
240-241	Roanoke	97.8	55.2	77.2
242	Bristol	96.4	52.3	75.1
243	Pulaski	96.1	43.1	70.6
244	Staunton	97.0	45.1	71.9
245	Lynchburg	97.1	57.3	77.9
246	Grundy	96.5	43.1	70.8
WASHINGTON				
980-981,987	Seattle	105.7	103.4	104.6
982	Everett	106.0	98.4	102.3
983-984	Tacoma	105.6	101.0	103.4
985	Olympia	105.5	100.9	103.3
986	Vancouver	108.0	97.4	102.9
988	Wenatchee	106.8	90.9	99.1
989	Yakima	106.0	94.2	100.3
990-992	Spokane	108.4	90.1	99.5
993	Richland	108.5	91.3	100.2
994	Clarkston	107.7	86.7	97.6
WEST VIRGINIA				
247-248	Bluefield	95.2	75.4	85.6
249	Lewisburg	96.7	81.8	89.5
250-253	Charleston	97.7	88.7	93.4
254	Martinsburg	96.7	58.0	78.0
255-257	Huntington	97.9	93.0	95.5
258-259	Beckley	94.9	87.2	91.2
260	Wheeling	97.8	90.7	94.4
261	Parkersburg	96.7	90.2	93.6
262	Buckhannon	96.6	88.6	92.7
263-264	Clarksburg	97.0	88.3	92.8
265	Morgantown	97.2	89.5	93.5
266	Gassaway	96.3	89.3	92.9
267	Romney	96.5	84.9	90.9
268	Petersburg	96.4	85.6	91.2
WISCONSIN				
530,532	Milwaukee	99.2	103.3	101.2
531	Kenosha	99.5	100.0	99.7
534	Racine	98.6	100.4	99.5
535	Beloit	98.5	95.4	97.0
537	Madison	98.6	100.4	99.5
538	Lancaster	96.7	82.9	90.0
539	Portage	95.1	91.1	93.2
540	New Richmond	95.9	87.9	92.1

STATE/ZIP	CITY	MAT.	INST.	TOTAL
WISCONSIN (CONT'D)				
541-543	Green Bay	99.6	94.0	96.9
544	Wausau	95.3	89.0	92.2
545	Rhinelander	98.4	85.1	92.0
546	La Crosse	95.9	93.8	94.8
547	Eau Claire	97.8	93.3	95.6
548	Superior	95.8	92.3	94.1
549	Oshkosh	95.8	86.2	91.2
WYOMING				
820	Cheyenne	100.6	64.9	83.4
821	Yellowstone Nat'l Park	100.0	58.6	80.0
822	Wheatland	101.7	56.2	79.8
823	Rawlins	103.3	53.4	79.2
824	Worland	100.8	53.6	78.0
825	Riverton	102.1	56.8	80.2
826	Casper	100.4	65.7	83.7
827	Newcastle	100.6	53.4	77.9
828	Sheridan	101.7	62.7	82.9
829-831	Rock Springs	105.0	53.6	80.2
CANADIAN FACTORS (reflect Canadian currency)				
ALBERTA				
	Calgary	110.3	85.9	98.5
	Edmonton	110.2	85.9	98.5
BRITISH COLUMBIA				
	Vancouver	110.5	105.7	108.2
	Victoria	113.0	102.4	107.9
MANITOBA				
	Winnipeg	111.0	87.1	99.5
NEW BRUNSWICK				
	Moncton	108.8	75.8	92.9
	Saint John	110.1	81.4	96.2
NEWFOUNDLAND				
	St. John's	114.7	75.0	95.6
NOVA SCOTIA				
	Halifax	109.8	83.3	97.0
ONTARIO				
	Barrie	115.3	102.3	109.1
	Brantford	113.2	108.4	110.9
	Cornwall	113.5	103.5	108.7
	Hamilton	113.4	107.3	110.5
	Kingston	113.5	103.2	108.5
	Kitchener	107.1	101.6	104.5
	London	112.4	104.0	108.3
	North Bay	113.2	101.2	107.4
	Oshawa	113.3	104.7	109.2
	Ottawa	113.6	104.2	109.1
	Owen Sound	115.4	100.8	108.4
	Peterborough	113.2	103.2	108.4
	Sarnia	112.3	109.2	110.8
	St. Catharines	106.5	101.5	104.1
	Sudbury	106.4	101.2	103.9
	Thunder Bay	108.1	101.4	104.9
	Toronto	113.4	111.9	112.7
	Windsor	106.6	104.6	105.6
PRINCE EDWARD ISLAND				
	Charlottetown	112.2	70.4	92.0
QUEBEC				
	Chicoutimi	109.0	97.4	103.4
	Montreal	110.5	97.5	104.2
	Quebec	111.0	97.7	104.6
SASKATCHEWAN				
	Regina	107.5	80.0	94.2
	Saskatoon	107.3	79.8	94.0

Abbreviations

A	Area Square Feet; Ampere	Cab.	Cabinet	d.f.u.	Drainage Fixture Units
ABS	Acrylonitrile Butadiene Stryrene; Asbestos Bonded Steel	Cair.	Air Tool Laborer	D.H.	Double Hung
A.C.	Alternating Current; Air-Conditioning; Asbestos Cement; Plywood Grade A & C	Calc	Calculated	DHW	Domestic Hot Water
		Cap.	Capacity	Diag.	Diagonal
		Carp.	Carpenter	Diam.	Diameter
		C.B.	Circuit Breaker	Distrib.	Distribution
A.C.I.	American Concrete Institute	C.C.A.	Chromate Copper Arsenate	Dk.	Deck
AD	Plywood, Grade A & D	C.C.F.	Hundred Cubic Feet	D.L.	Dead Load; Diesel
Addit.	Additional	cd	Candela	DLH	Deep Long Span Bar Joist
Adj.	Adjustable	cd/sf	Candela per Square Foot	Do.	Ditto
af	Audio-frequency	CD	Grade of Plywood Face & Back	Dp.	Depth
A.G.A.	American Gas Association	CDX	Plywood, Grade C & D, exterior glue	D.P.S.T.	Double Pole, Single Throw
Agg.	Aggregate			Dr.	Driver
A.H.	Ampere Hours	Cefi.	Cement Finisher	Drink.	Drinking
A hr.	Ampere-hour	Cem.	Cement	D.S.	Double Strength
A.H.U.	Air Handling Unit	CF	Hundred Feet	D.S.A.	Double Strength A Grade
A.I.A.	American Institute of Architects	C.F.	Cubic Feet	D.S.B.	Double Strength B Grade
AIC	Ampere Interrupting Capacity	CFM	Cubic Feet per Minute	Dty.	Duty
Allow.	Allowance	c.g.	Center of Gravity	DWV	Drain Waste Vent
alt.	Altitude	CHW	Chilled Water; Commercial Hot Water	DX	Deluxe White, Direct Expansion
Alum.	Aluminum			dyn	Dyne
a.m.	Ante Meridiem	C.I.	Cast Iron	e	Eccentricity
Amp.	Ampere	C.I.P.	Cast in Place	E	Equipment Only; East
Anod.	Anodized	Circ.	Circuit	Ea.	Each
Approx.	Approximate	C.L.	Carload Lot	E.B.	Encased Burial
Apt.	Apartment	Clab.	Common Laborer	Econ.	Economy
Asb.	Asbestos	C.L.F.	Hundred Linear Feet	EDP	Electronic Data Processing
A.S.B.C.	American Standard Building Code	CLF	Current Limiting Fuse	EIFS	Exterior Insulation Finish System
Asbe.	Asbestos Worker	CLP	Cross Linked Polyethylene	E.D.R.	Equiv. Direct Radiation
A.S.H.R.A.E.	American Society of Heating, Refrig. & AC Engineers	cm	Centimeter	Eq.	Equation
		CMP	Corr. Metal Pipe	Elec.	Electrician; Electrical
A.S.M.E.	American Society of Mechanical Engineers	C.M.U.	Concrete Masonry Unit	Elev.	Elevator; Elevating
		CN	Change Notice	EMT	Electrical Metallic Conduit; Thin Wall Conduit
A.S.T.M.	American Society for Testing and Materials	Col.	Column		
		CO₂	Carbon Dioxide	Eng.	Engine, Engineered
Attchmt.	Attachment	Comb.	Combination	EPDM	Ethylene Propylene Diene Monomer
Avg.	Average	Compr.	Compressor		
A.W.G.	American Wire Gauge	Conc.	Concrete	EPS	Expanded Polystyrene
AWWA	American Water Works Assoc.	Cont.	Continuous; Continued	Eqhv.	Equip. Oper., Heavy
Bbl.	Barrel	Corr.	Corrugated	Eqlt.	Equip. Oper., Light
B. & B.	Grade B and Better; Balled & Burlapped	Cos	Cosine	Eqmd.	Equip. Oper., Medium
		Cot	Cotangent	Eqmm.	Equip. Oper., Master Mechanic
B. & S.	Bell and Spigot	Cov.	Cover	Eqol.	Equip. Oper., Oilers
B. & W.	Black and White	C/P	Cedar on Paneling	Equip.	Equipment
b.c.c.	Body-centered Cubic	CPA	Control Point Adjustment	ERW	Electric Resistance Welded
B.C.Y.	Bank Cubic Yards	Cplg.	Coupling	E.S.	Energy Saver
BE	Bevel End	C.P.M.	Critical Path Method	Est.	Estimated
B.F.	Board Feet	CPVC	Chlorinated Polyvinyl Chloride	esu	Electrostatic Units
Bg. cem.	Bag of Cement	C.Pr.	Hundred Pair	E.W.	Each Way
BHP	Boiler Horsepower; Brake Horsepower	CRC	Cold Rolled Channel	EWT	Entering Water Temperature
		Creos.	Creosote	Excav.	Excavation
B.I.	Black Iron	Crpt.	Carpet & Linoleum Layer	Exp.	Expansion, Exposure
Bit.; Bitum.	Bituminous	CRT	Cathode-ray Tube	Ext.	Exterior
Bk.	Backed	CS	Carbon Steel, Constant Shear Bar Joist	Extru.	Extrusion
Bkrs.	Breakers			f.	Fiber stress
Bldg.	Building	Csc	Cosecant	F	Fahrenheit; Female; Fill
Blk.	Block	C.S.F.	Hundred Square Feet	Fab.	Fabricated
Bm.	Beam	CSI	Construction Specifications Institute	FBGS	Fiberglass
Boil.	Boilermaker			F.C.	Footcandles
B.P.M.	Blows per Minute	C.T.	Current Transformer	f.c.c.	Face-centered Cubic
BR	Bedroom	CTS	Copper Tube Size	f'c.	Compressive Stress in Concrete; Extreme Compressive Stress
Brg.	Bearing	Cu	Copper, Cubic		
Brhe.	Bricklayer Helper	Cu. Ft.	Cubic Foot	F.E.	Front End
Bric.	Bricklayer	cw	Continuous Wave	FEP	Fluorinated Ethylene Propylene (Teflon)
Brk.	Brick	C.W.	Cool White; Cold Water		
Brng.	Bearing	Cwt.	100 Pounds	F.G.	Flat Grain
Brs.	Brass	C.W.X.	Cool White Deluxe	F.H.A.	Federal Housing Administration
Brz.	Bronze	C.Y.	Cubic Yard (27 cubic feet)	Fig.	Figure
Bsn.	Basin	C.Y./Hr.	Cubic Yard per Hour	Fin.	Finished
Btr.	Better	Cyl.	Cylinder	Fixt.	Fixture
BTU	British Thermal Unit	d	Penny (nail size)	Fl. Oz.	Fluid Ounces
BTUH	BTU per Hour	D	Deep; Depth; Discharge	Flr.	Floor
B.U.R.	Built-up Roofing	Dis.;Disch.	Discharge	F.M.	Frequency Modulation; Factory Mutual
BX	Interlocked Armored Cable	Db.	Decibel		
c	Conductivity, Copper Sweat	Dbl.	Double	Fmg.	Framing
C	Hundred; Centigrade	DC	Direct Current	Fndtn.	Foundation
C/C	Center to Center, Cedar on Cedar	DDC	Direct Digital Control	Fori.	Foreman, Inside
		Demob.	Demobilization	Foro.	Foreman, Outside

651

Abbreviations

Fount.	Fountain	J.I.C.	Joint Industrial Council	MD	Medium Duty
FPM	Feet per Minute	K	Thousand; Thousand Pounds;	M.D.O.	Medium Density Overlaid
FPT	Female Pipe Thread		Heavy Wall Copper Tubing	Med.	Medium
Fr.	Frame	K.A.H.	Thousand Amp. Hours	MF	Thousand Feet
F.R.	Fire Rating	KCMIL	Thousand Circular Mils	M.F.B.M.	Thousand Feet Board Measure
FRK	Foil Reinforced Kraft	KD	Knock Down	Mfg.	Manufacturing
FRP	Fiberglass Reinforced Plastic	K.D.A.T.	Kiln Dried After Treatment	Mfrs.	Manufacturers
FS	Forged Steel	kg	Kilogram	mg	Milligram
FSC	Cast Body; Cast Switch Box	kG	Kilogauss	MGD	Million Gallons per Day
Ft.	Foot; Feet	kgf	Kilogram Force	MGPH	Thousand Gallons per Hour
Ftng.	Fitting	kHz	Kilohertz	MH, M.H.	Manhole; Metal Halide; Man-Hour
Ftg.	Footing	Kip.	1000 Pounds	MHz	Megahertz
Ft. Lb.	Foot Pound	KJ	Kiljoule	Mi.	Mile
Furn.	Furniture	K.L.	Effective Length Factor	MI	Malleable Iron; Mineral Insulated
FVNR	Full Voltage Non-Reversing	K.L.F.	Kips per Linear Foot	mm	Millimeter
FXM	Female by Male	Km	Kilometer	Mill.	Millwright
Fy.	Minimum Yield Stress of Steel	K.S.F.	Kips per Square Foot	Min., min.	Minimum, minute
g	Gram	K.S.I.	Kips per Square Inch	Misc.	Miscellaneous
G	Gauss	kV	Kilovolt	ml	Milliliter
Ga.	Gauge	kVA	Kilovolt Ampere	M.L.F.	Thousand Linear Feet
Gal.	Gallon	K.V.A.R.	Kilovar (Reactance)	Mo.	Month
Gal./Min.	Gallon per Minute	KW	Kilowatt	Mobil.	Mobilization
Galv.	Galvanized	KWh	Kilowatt-hour	Mog.	Mogul Base
Gen.	General	L	Labor Only; Length; Long;	MPH	Miles per Hour
G.F.I.	Ground Fault Interrupter		Medium Wall Copper Tubing	MPT	Male Pipe Thread
Glaz.	Glazier	Lab.	Labor	MRT	Mile Round Trip
GPD	Gallons per Day	lat	Latitude	ms	Millisecond
GPH	Gallons per Hour	Lath.	Lather	M.S.F.	Thousand Square Feet
GPM	Gallons per Minute	Lav.	Lavatory	Mstz.	Mosaic & Terrazzo Worker
GR	Grade	lb.; #	Pound	M.S.Y.	Thousand Square Yards
Gran.	Granular	L.B.	Load Bearing; L Conduit Body	Mtd.	Mounted
Grnd.	Ground	L. & E.	Labor & Equipment	Mthe.	Mosaic & Terrazzo Helper
H	High; High Strength Bar Joist;	lb./hr.	Pounds per Hour	Mtng.	Mounting
	Henry	lb./L.F.	Pounds per Linear Foot	Mult.	Multi; Multiply
H.C.	High Capacity	lbf/sq.in.	Pound-force per Square Inch	M.V.A.	Million Volt Amperes
H.D.	Heavy Duty; High Density	L.C.L.	Less than Carload Lot	M.V.A.R.	Million Volt Amperes Reactance
H.D.O.	High Density Overlaid	Ld.	Load	MV	Megavolt
Hdr.	Header	LE	Lead Equivalent	MW	Megawatt
Hdwe.	Hardware	LED	Light Emitting Diode	MXM	Male by Male
Help.	Helpers Average	L.F.	Linear Foot	MYD	Thousand Yards
HEPA	High Efficiency Particulate Air	Lg.	Long; Length; Large	N	Natural; North
	Filter	L & H	Light and Heat	nA	Nanoampere
Hg	Mercury	LH	Long Span Bar Joist, Labor Hours	NA	Not Available; Not Applicable
HIC	High Interrupting Capacity	L.L.	Live Load	N.B.C.	National Building Code
HM	Hollow Metal	L.L.D.	Lamp Lumen Depreciation	NC	Normally Closed
H.O.	High Output	L-O-L	Lateralolet	N.E.M.A.	National Electrical Manufacturers
Horiz.	Horizontal	lm	Lumen		Assoc.
H.P.	Horsepower; High Pressure	lm/sf	Lumen per Square Foot	NEHB	Bolted Circuit Breaker to 600V.
H.P.F.	High Power Factor	lm/W	Lumen per Watt	N.L.B.	Non-Load-Bearing
Hr.	Hour	L.O.A.	Length Over All	NM	Non-Metallic Cable
Hrs./Day	Hours per Day	log	Logarithm	nm	Nanometer
HSC	High Short Circuit	L.P.	Liquefied Petroleum; Low Pressure	No.	Number
Ht.	Height	L.P.F.	Low Power Factor	NO	Normally Open
Htg.	Heating	LR	Long Radius	N.O.C.	Not Otherwise Classified
Htrs.	Heaters	L.S.	Lump Sum	Nose.	Nosing
HVAC	Heating, Ventilation & Air-	Lt.	Light	N.P.T.	National Pipe Thread
	Conditioning	Lt. Ga.	Light Gauge	NQOD	Combination Plug-on/Bolt on
Hvy.	Heavy	L.T.L.	Less than Truckload Lot		Circuit Breaker to 240V.
HW	Hot Water	Lt. Wt.	Lightweight	N.R.C.	Noise Reduction Coefficient
Hyd.;Hydr.	Hydraulic	L.V.	Low Voltage	N.R.S.	Non Rising Stem
Hz.	Hertz (cycles)	M	Thousand; Material; Male;	ns	Nanosecond
I.	Moment of Inertia		Light Wall Copper Tubing	nW	Nanowatt
I.C.	Interrupting Capacity	m/hr; M.H.	Man-hour	OB	Opposing Blade
ID	Inside Diameter	mA	Milliampere	OC	On Center
I.D.	Inside Dimension; Identification	Mach.	Machine	OD	Outside Diameter
I.F.	Inside Frosted	Mag. Str.	Magnetic Starter	O.D.	Outside Dimension
I.M.C.	Intermediate Metal Conduit	Maint.	Maintenance	ODS	Overhead Distribution System
In.	Inch	Marb.	Marble Setter	O.G.	Ogee
Incan.	Incandescent	Mat; Mat'l.	Material	O.H.	Overhead
Incl.	Included; Including	Max.	Maximum	O & P	Overhead and Profit
Int.	Interior	MBF	Thousand Board Feet	Oper.	Operator
Inst.	Installation	MBH	Thousand BTU's per hr.	Opng.	Opening
Insul.	Insulation/Insulated	MC	Metal Clad Cable	Orna.	Ornamental
I.P.	Iron Pipe	M.C.F.	Thousand Cubic Feet	OSB	Oriented Strand Board
I.P.S.	Iron Pipe Size	M.C.F.M.	Thousand Cubic Feet per Minute	O. S. & Y.	Outside Screw and Yoke
I.P.T.	Iron Pipe Threaded	M.C.M.	Thousand Circular Mils	Ovhd.	Overhead
I.W.	Indirect Waste	M.C.P.	Motor Circuit Protector	OWG	Oil, Water or Gas
J	Joule			Oz.	Ounce

652

P.	Pole; Applied Load; Projection	SCFM	Standard Cubic Feet per Minute	T.L.	Truckload
p.	Page	Scaf.	Scaffold	T.M.	Track Mounted
Pape.	Paperhanger	Sch.; Sched.	Schedule	Tot.	Total
P.A.P.R.	Powered Air Purifying Respirator	S.C.R.	Modular Brick	T-O-L	Threadolet
PAR	Weatherproof Reflector	S.D.	Sound Deadening	T.S.	Trigger Start
Pc., Pcs.	Piece, Pieces	S.D.R.	Standard Dimension Ratio	Tr.	Trade
P.C.	Portland Cement; Power Connector	S.E.	Surfaced Edge	Transf.	Transformer
P.C.F.	Pounds per Cubic Foot	Sel.	Select	Trhv.	Truck Driver, Heavy
P.C.M.	Phase Contract Microscopy	S.E.R.;	Service Entrance Cable	Trlr	Trailer
P.E.	Professional Engineer;	S.E.U.	Service Entrance Cable	Trlt.	Truck Driver, Light
	Porcelain Enamel;	S.F.	Square Foot	TV	Television
	Polyethylene; Plain End	S.F.C.A.	Square Foot Contact Area	T.W.	Thermoplastic Water Resistant
Perf.	Perforated	S.F.G.	Square Foot of Ground		Wire
Ph.	Phase	S.F. Hor.	Square Foot Horizontal	UCI	Uniform Construction Index
P.I.	Pressure Injected	S.F.R.	Square Feet of Radiation	UF	Underground Feeder
Pile.	Pile Driver	S.F. Shlf.	Square Foot of Shelf	UGND	Underground Feeder
Pkg.	Package	S4S	Surface 4 Sides	U.H.F.	Ultra High Frequency
Pl.	Plate	Shee.	Sheet Metal Worker	U.L.	Underwriters Laboratory
Plah.	Plasterer Helper	Sin.	Sine	Unfin.	Unfinished
Plas.	Plasterer	Skwk.	Skilled Worker	URD	Underground Residential
Pluh.	Plumbers Helper	SL	Saran Lined		Distribution
Plum.	Plumber	S.L.	Slimline	US	United States
Ply.	Plywood	Sldr.	Solder	USP	United States Primed
p.m.	Post Meridiem	SLH	Super Long Span Bar Joist	UTP	Unshielded Twisted Pair
Pntd.	Painted	S.N.	Solid Neutral	V	Volt
Pord.	Painter, Ordinary	S-O-L	Socketolet	V.A.	Volt Amperes
pp	Pages	S.P.	Static Pressure; Single Pole; Self-	V.C.T.	Vinyl Composition Tile
PP; PPL	Polypropylene		Propelled	VAV	Variable Air Volume
P.P.M.	Parts per Million	Spri.	Sprinkler Installer	VC	Veneer Core
Pr.	Pair	Sq.	Square; 100 Square Feet	Vent.	Ventilation
P.E.S.B.	Pre-engineered Steel Building	S.P.D.T.	Single Pole, Double Throw	Vert.	Vertical
Prefab.	Prefabricated	SPF	Spruce Pine Fir	V.F.	Vinyl Faced
Prefin.	Prefinished	S.P.S.T.	Single Pole, Single Throw	V.G.	Vertical Grain
Prop.	Propelled	SPT	Standard Pipe Thread	V.H.F.	Very High Frequency
PSF; psf	Pounds per Square Foot	Sq. Hd.	Square Head	VHO	Very High Output
PSI; psi	Pounds per Square Inch	Sq. In.	Square Inch	Vib.	Vibrating
PSIG	Pounds per Square Inch Gauge	S.S.	Single Strength; Stainless Steel	V.L.F.	Vertical Linear Foot
PSP	Plastic Sewer Pipe	S.S.B.	Single Strength B Grade	Vol.	Volume
Pspr.	Painter, Spray	sst	Stainless Steel	VRP	Vinyl Reinforced Polyester
Psst.	Painter, Structural Steel	Sswk.	Structural Steel Worker	W	Wire; Watt; Wide; West
P.T.	Potential Transformer	Sswl.	Structural Steel Welder	w/	With
P. & T.	Pressure & Temperature	St.; Stl.	Steel	W.C.	Water Column; Water Closet
Ptd.	Painted	S.T.C.	Sound Transmission Coefficient	W.F.	Wide Flange
Ptns.	Partitions	Std.	Standard	W.G.	Water Gauge
Pu	Ultimate Load	STK	Select Tight Knot	Wldg.	Welding
PVC	Polyvinyl Chloride	STP	Standard Temperature & Pressure	W. Mile	Wire Mile
Pvmt.	Pavement	Stpi.	Steamfitter, Pipefitter	W-O-L	Weldolet
Pwr.	Power	Str.	Strength; Starter; Straight	W.R.	Water Resistant
Q	Quantity Heat Flow	Strd.	Stranded	Wrck.	Wrecker
Quan.; Qty.	Quantity	Struct.	Structural	W.S.P.	Water, Steam, Petroleum
Q.C.	Quick Coupling	Sty.	Story	WT., Wt.	Weight
r	Radius of Gyration	Subj.	Subject	WWF	Welded Wire Fabric
R	Resistance	Subs.	Subcontractors	XFER	Transfer
R.C.P.	Reinforced Concrete Pipe	Surf.	Surface	XFMR	Transformer
Rect.	Rectangle	Sw.	Switch	XHD	Extra Heavy Duty
Reg.	Regular	Swbd.	Switchboard	XHHW; XLPE	Cross-Linked Polyethylene Wire
Reinf.	Reinforced	S.Y.	Square Yard		Insulation
Req'd.	Required	Syn.	Synthetic	XLP	Cross-linked Polyethylene
Res.	Resistant	S.Y.P.	Southern Yellow Pine	Y	Wye
Resi.	Residential	Sys.	System	yd	Yard
Rgh.	Rough	t.	Thickness	yr	Year
RGS	Rigid Galvanized Steel	T	Temperature; Ton	Δ	Delta
R.H.W.	Rubber, Heat & Water Resistant;	Tan	Tangent	%	Percent
	Residential Hot Water	T.C.	Terra Cotta	~	Approximately
rms	Root Mean Square	T & C	Threaded and Coupled	Ø	Phase
Rnd.	Round	T.D.	Temperature Difference	@	At
Rodm.	Rodman	T.E.M.	Transmission Electron Microscopy	#	Pound; Number
Rofc.	Roofer, Composition	TFE	Tetrafluoroethylene (Teflon)	<	Less Than
Rofp.	Roofer, Precast	T. & G.	Tongue & Groove;	>	Greater Than
Rohe.	Roofer Helpers (Composition)		Tar & Gravel		
Rots.	Roofer, Tile & Slate	Th.; Thk.	Thick		
R.O.W.	Right of Way	Thn.	Thin		
RPM	Revolutions per Minute	Thrded	Threaded		
R.R.	Direct Burial Feeder Conduit	Tilf.	Tile Layer, Floor		
R.S.	Rapid Start	Tilh.	Tile Layer, Helper		
Rsr	Riser	THHN	Nylon Jacketed Wire		
RT	Round Trip	THW.	Insulated Strand Wire		
S.	Suction; Single Entrance; South	THWN;	Nylon Jacketed Wire		

Index

INDEX

Index

Index

Division Notes

	CREW	DAILY OUTPUT	LABOR-HOURS	UNIT	1999 BARE COSTS				TOTAL INCL O&P
					MAT.	LABOR	EQUIP.	TOTAL	

Division Notes

	CREW	DAILY OUTPUT	LABOR-HOURS	UNIT	1999 BARE COSTS				TOTAL INCL O&P
					MAT.	LABOR	EQUIP.	TOTAL	

		CREW	DAILY OUTPUT	LABOR-HOURS	UNIT	1999 BARE COSTS				TOTAL INCL O&P
						MAT.	LABOR	EQUIP.	TOTAL	

Division Notes

	CREW	DAILY OUTPUT	LABOR-HOURS	UNIT	1999 BARE COSTS				TOTAL INCL O&P
					MAT.	LABOR	EQUIP.	TOTAL	

Division Notes

	CREW	DAILY OUTPUT	LABOR-HOURS	UNIT	1999 BARE COSTS				TOTAL INCL O&P
					MAT.	LABOR	EQUIP.	TOTAL	

Division Notes

	CREW	DAILY OUTPUT	LABOR-HOURS	UNIT	1999 BARE COSTS				TOTAL INCL O&P
					MAT.	LABOR	EQUIP.	TOTAL	

Division Notes

	CREW	DAILY OUTPUT	LABOR-HOURS	UNIT	1999 BARE COSTS				TOTAL INCL O&P
					MAT.	LABOR	EQUIP.	TOTAL	

690

691

R.S. Means Company, Inc., a CMD Group company, the leading provider of construction cost data in North America, supplies comprehensive construction cost guides, related technical publications and educational services.

CMD Group, a leading worldwide provider of proprietary construction information, is comprised of three synergistic product groups crafted to be the complete resource for reliable, timely and actionable construction market data. In North America, CMD Group encompasses: Architects' First Source, an innovative product selection and specification solution in print and on the Internet; Construction Market Data (CMD), the source for construction project information for sales leads, market construction activity information as well as early planning reports for the design community; Associated Construction Publications, with 14 magazines, one of the largest editorial networks dedicated to U.S. highway and heavy construction coverage; Manufacturer's Survey Associates (MSA), the leading estimating and quantity survey firm in the U.S.; R.S. Means, the authority on construction cost data in North America; CMD Canada, the leading supplier of project information, industry news and forecasting data products for the Canadian construction industry and BIMSA/Mexico, the dominant distributor of information on building projects and construction throughout Mexico. Worldwide, CMD Group includes Byggfakta Scandinavia, providing construction market data to Denmark, Estonia, Finland, Norway and Sweden; and Cordell Building Information Services, the market leader for construction cost information in Australia.

Architects' First Source for Products, available in print, on the Means CostWorks CD, and on the Internet, is a comprehensive, product information source. Through an alliance with The Construction Specifications Institute (CSI) and Thomas Register, Architects' First Source also produces CSI's SPEC-DATA® and MANU-SPEC® as well as CADBlocks℠ which deliver manufacturers' technical product data, proprietary specifications, and dimensionally accurate drawings. Together, these products offer commercial building product information for each member of the building team at each stage of the construction process.

Construction Market Data provides complete, accurate and timely project information through all stages of construction. Construction Market Data supplies industry data through productive leads, project reports, contact lists, market penetration analysis and sales evaluation reports. Any of these products can pinpoint a county, look at a state, or cover the country. Data is delivered via paper, e-mail or the Internet.

Construction Market Data Canada serves the Canadian construction market with reliable and comprehensive information services that cover all facets of construction. Core services include: Buildcore Product Source, a preliminary product selection tool available in print and on the Internet; national construction project lead services such as customized CMD Key products, weekly CMD Bulletins and CMD Online; CanaData, statistical and forecasting information; Daily Commercial News, a construction newspaper reporting on news and projects in Ontario; and Journal of Commerce, reporting on products and news in British Columbia and Alberta.

Manufacturers' Survey Associates is a quantity survey and specification service whose experienced estimating staff examines project documents throughout the bidding process and distributes edited information to its clients. Material estimates, edited plans and specifications and distribution of addenda form the heart of the Manufacturers' Survey Associates product line which is available in print and on CD-ROM. In addition, the Market Intelligence Report is a quarterly compilation of actual project specification data that lets building product manufacturers know precisely when and where their products are specified.

Associated Construction Publications are a group of 14 regional magazines focused on highway and heavy construction and cost data. With one of the world's largest editorial staffs dedicated to construction coverage, the magazine group focuses on local and regional news of projects in all phases of construction, precise project volumes, plus material and labor costs. The magazines have a cumulative ABC and BPA audited circulation of 112,000, an estimated pass-along readership of one-half million, and more than 25,000 pages of annual advertising.

Clark Reports is the premier provider of industrial construction project data, noted for providing earlier, more complete project information in the planning and pre-planning stages for all segments of industrial and institutional construction. Comprehensive reporting and tailored information distinguish the Clark Reports products.

Byggfakta Scandinavia AB, founded in 1936, is the parent company for the leaders of customized construction market data for Denmark, Estonia, Finland, Norway and Sweden. Each company fully covers the local construction market and provides information across several platforms including subscription, ad-hoc basis, electronically and on paper.

Cordell Building Information Services, with its complete range of project and cost and estimating services, is Australia's specialist in the construction information industry. Cordell provides in-depth and historical information on all aspects of construction projects and estimation, including several customized reports, construction and sales leads, and detailed cost information among others.

For more information, please visit our website at www.cmdg.com.

CMD Group Corporate Offices
30 Technology Parkway South #100
Norcross, GA 30092-2912
(770) 417-4000
(770) 417-4002 (fax)
www.cmdg.com

Means Project Cost Report

By filling out and returning the Project Description, you can receive a discount of $20.00 off any one of the Means products advertised in the following pages. The cost information required includes all items marked (✔) except those where no costs occurred. The sum of all major items should equal the Total Project Cost.

$20.00 Discount per product for each report you submit.

DISCOUNT PRODUCTS AVAILABLE—FOR U.S. CUSTOMERS ONLY—STRICTLY CONFIDENTIAL

Project Description (No remodeling projects, please.)

✔ Type Building _____

✔ Location _____

  Capacity _____

✔ Frame _____

✔ Exterior _____

✔ Basement: full ☐ partial ☐ none ☐ crawl ☐

✔ Height in Stories _____

✔ Total Floor Area _____

  Ground Floor Area _____

✔ Volume in C.F. _____

  % Air Conditioned _____ Tons _____

  Comments _____

Owner _____

Architect _____

General Contractor _____

✔ Bid Date _____

  Typical Bay Size _____

✔ Labor Force: _____ % Union   _____ % Non-Union

✔ Project Description (Circle one number in each line)

  1. Economy 2. Average   3. Custom 4. Luxury

  1. Square   2. Rectangular 3. Irregular 4. Very Irregular

			Total Project Cost			$
A	✔		General Conditions			$
B	✔		Site Work			$
	BS		Site Clearing & Improvement			
	BE		Excavation	(	C.Y.)	
	BF		Caissons & Piling	(	L.F.)	
	BU		Site Utilities			
	BP		Roads & Walks Exterior Paving	(	S.Y.)	
C	✔		Concrete			$
	C		Cast in Place	(	C.Y.)	
	CP		Precast	(	S.F.)	
D	✔		Masonry			$
	DB		Brick	(	M)	
	DC		Block	(	M)	
	DT		Tile	(	S.F.)	
	DS		Stone	(	S.F.)	
E	✔		Metals			$
	ES		Structural Steel	(	Tons)	
	EM		Misc. & Ornamental Metals			
F	✔		Wood & Plastics			$
	FR		Rough Carpentry	(	MBF)	
	FF		Finish Carpentry			
	FM		Architectural Millwork			
G	✔		Thermal & Moisture Protection			$
	GW		Waterproofing-Dampproofing	(	S.F.)	
	GN		Insulation	(	S.F.)	
	GR		Roofing & Flashing	(	S.F.)	
	GM		Metal Siding/Curtain Wall	(	S.F.)	
H	✔		Doors and Windows			$
	HD		Doors	(	Ea.)	
	HW		Windows	(	S.F.)	
	HH		Finish Hardware			
	HG		Glass & Glazing	(	S.F.)	
	HS		Storefronts	(	S.F.)	

J	✔		Finishes			$
	JL		Lath & Plaster	(	S.Y.)	
	JD		Drywall	(	S.F.)	
	JM		Tile & Marble	(	S.F.)	
	JT		Terrazzo	(	S.F.)	
	JA		Acoustical Treatment	(	S.F.)	
	JC		Carpet	(	S.Y.)	
	JF		Hard Surface Flooring	(	S.F.)	
	JP		Painting & Wall Covering	(	S.F.)	
K	✔		Specialties			$
	KB		Bathroom Partitions & Access.	(	S.F.)	
	KF		Other Partitions	(	S.F.)	
	KL		Lockers	(	Ea.)	
L	✔		Equipment			$
	LK		Kitchen			
	LS		School			
	LO		Other			
M	✔		Furnishings			$
	MW		Window Treatment			
	MS		Seating	(	Ea.)	
N	✔		Special Construction			$
	NA		Acoustical	(	S.F.)	
	NB		Prefab. Bldgs.	(	S.F.)	
	NO		Other			
P	✔		Conveying Systems			$
	PE		Elevators	(	Ea.)	
	PS		Escalators	(	Ea.)	
	PM		Material Handling			
Q	✔		Mechanical			$
	QP		Plumbing (No. of fixtures)			
	QS		Fire Protection (Sprinklers)			
	QF		Fire Protection (Hose Standpipes)			
	QB		Heating, Ventilating & A.C.			
	QH		Heating & Ventilating (BTU Output)			
	QA		Air Conditioning	(	Tons)	
R	✔		Electrical			$
	RL		Lighting	(	S.F.)	
	RP		Power Service			
	RD		Power Distribution			
	RA		Alarms			
	RG		Special Systems			
S	✔		Mech./Elec. Combined			$

Product Name _____

Product Number _____

Your Name _____

Title _____

Company _____

  ☐ Company

  ☐ Home   Street Address _____

City, State, Zip _____

☐ Please send _____ forms.

Please specify the Means product you wish to receive. Complete the address information as requested and return this form with your check (product cost less $20.00) to address below.

R.S. Means Company, Inc.,
Square Foot Costs Department
100 Construction Plaza, P.O. Box 800
Kingston, MA 02364-9988

R.S. Means Company, Inc. . .
a tradition of excellence in Construction Cost Information and Services since 1942.

Table of Contents

Book Selection Guide

The following table provides definitive information on the content of each cost data publication. The number of lines of data provided in each unit price or assemblies division, as well as the number of reference tables and crews is listed for each book. The presence of other elements such as an historical cost index, city cost indexes, square foot models or cross-referenced index is also indicated. You can use the table to help select the Means' book that has the quantity and type of information you most need in your work.

Unit Cost Divisions	Building Construction Costs	Mechanical	Electrical	Repair & Remodel.	Square Foot	Site Work Landsc.	Assemblies	Interior	Concrete Masonry	Open Shop	Heavy Construc.	Resi-dential	Light Commercial	Facil. Construc.	Plumbing	Western Construction Costs
1	1099	577	578	770		1025		488	995	1097	1074	425	623	1519	652	1090
2	3393	1440	467	2292		8799		1150	1645	3339	5267	1152	1217	5043	1764	3364
3	1507	90	68	760		1297		191	1913	1483	1470	261	206	1347	43	1487
4	874	27	0	669		767		640	1191	856	670	318	403	1141	0	852
5	1578	179	174	672		823		686	697	1546	1112	480	497	1641	97	1558
6	1398	96	83	1303		476		1316	328	1367	593	1520	1459	1395	59	1752
7	1430	141	90	1368		489		611	455	1428	353	814	1075	1422	151	1429
8	1833	60	0	1857		339		1754	727	1813	9	1096	1146	1967	0	1834
9	1679	50	0	1522		186		1775	322	1627	114	1318	1420	1873	50	1668
10	973	57	31	540		205		813	194	973	0	245	447	971	249	973
11	1036	275	206	602		69		901	34	1019	16	115	235	1033	245	1018
12	355	0	0	48		227		1536	30	346	0	73	69	1538	0	346
13	903	413	200	155		462		356	105	879	323	120	1331	965	240	928
14	365	43	0	262		36		330	0	365	40	9	18	363	17	363
15	2719	15388	703	2284		1962		1772	80	2638	2247	986	1612	12986	11640	2595
16	1550	801	10627	1149		1098		1350	62	1565	1034	707	1305	10207	708	1492
17	459	369	459	1		0		0	0	460	0	0	0	459	369	459
Totals	23151	20006	13686	16254		18260		15669	8778	22801	14322	9639	13063	45870	16284	23208

Assembly Divisions	Building Construction Costs	Mechanical	Electrical	Repair & Remodel.	Square Foot	Site Work Landsc.	Assemblies	Interior	Concrete Masonry	Open Shop	Heavy Construc.	Resi-dential	Light Commercial	Facil. Construc.	Plumbing	Western Construction Costs
1		0	0	202	131	738	775	0	689		752	844	131	89	0	
2		0	0	40	33	35	48	0	49		0	589	32	0	0	
3		0	0	443	1114	0	3064	228	1245		0	1546	806	174	0	
4		0	0	714	1353	0	3172	255	1273		0	2046	1171	26	0	
5		0	0	288	227	0	459	0	0		0	1082	223	25	0	
6		0	0	1006	845	0	1295	1723	152		0	1082	738	330	0	
7		0	0	42	89	0	179	159	0		0	723	33	71	0	
8		2257	149	998	1691	0	2720	926	0		0	1694	1191	1114	2084	
9		0	1346	355	357	0	1283	306	0		0	241	358	319	0	
10		0	0	0	0	0	0	0	0		0	0	0	0	0	
11		0	0	365	465	0	724	153	0		0	0	466	201	0	
12		501	160	539	84	2392	699	0	698		541	0	84	119	850	
Totals		2758	1655	4992	6389	3165	14418	3750	4106		1293	9847	5233	2468	2934	

Reference Section	Building Construction Costs	Mechanical	Electrical	Repair & Remodel.	Square Foot	Site Work Landsc.	Assemblies	Interior	Concrete Masonry	Open Shop	Heavy Construc.	Resi-dential	Light Commercial	Facil. Construc.	Plumbing	Western Construction Costs
Tables	150	46	84	66	4	84	223	59	83	146	55	49	70	86	49	148
Models					102								32	43		
Crews	408	408	408	389		408		408	408	391	408	391	391	389	408	408
City Cost Indexes	yes	yes	yes	yes	yes	yes	yes	yes	yes	yes	yes	yes	yes	yes	yes	yes
Historical Cost Indexes	yes	yes	yes	yes	yes	yes	yes	yes	yes	yes	yes	no	yes	yes	yes	yes
Index	yes	yes	yes	yes	no	yes	yes	yes	yes	yes	yes	yes	yes	yes	yes	yes

Means Building Construction Cost Data 1999

Available in Both Softbound and Looseleaf Editions

The "Bible" of the industry comes in the standard softcover edition or the looseleaf edition.

Many customers enjoy the convenience and flexibility of the looseleaf binder, which increases the usefulness of *Means Building Construction Cost Data 1999* by making it easy to add and remove pages. You can insert your own cost information pages, so everything is in one place. Copying pages for faxing is easier also. Whichever edition you prefer, softbound or the convenient looseleaf edition, you'll get the *Design/Build Intelligence* newsletter at no extra cost.

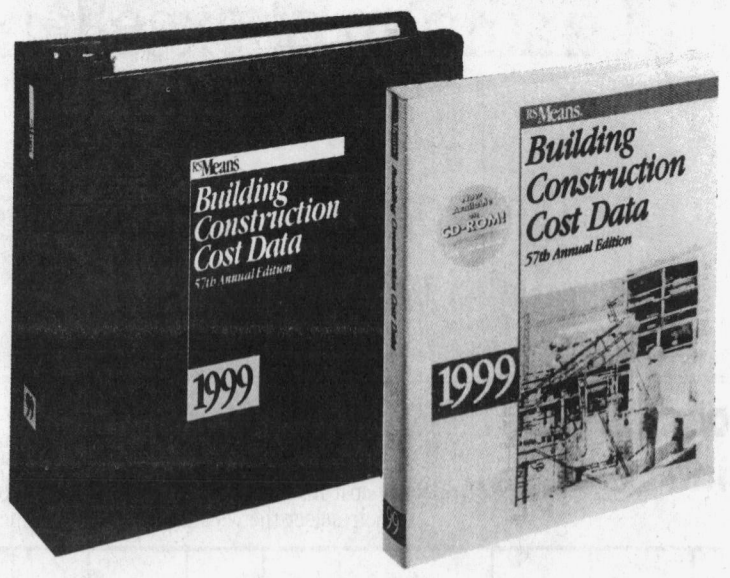

$84.95 per copy, Softbound
Catalog No. 60019

$109.95 per copy, Looseleaf
Catalog No. 61019

Means Building Construction Cost Data 1999

Offers you unchallenged unit price reliability in an easy-to-use arrangement. Whether used for complete, finished estimates or for periodic checks, it supplies more cost facts better and faster than any comparable source. Over 21,000 unit prices for 1999. The City Cost Indexes now cover over 930 areas, for indexing to any project location in North America. Order and get the *Design/Build Intelligence* newsletter sent to you FREE. You'll have year-long access to the Means Estimating **Hotline** FREE with your subscription. Expert assistance when using Means data is just a phone call away.

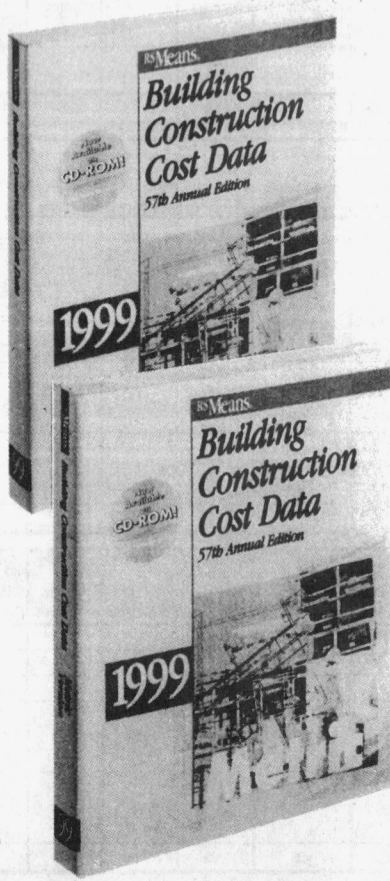

$84.95 per copy
Over 650 pages, illustrated, available Oct. 1998
Catalog No. 60019

Means Building Construction Cost Data 1999

Metric Version

The Federal Government has stated that all federal construction projects must now use metric documentation. The *Metric Version* of *Means Building Construction Cost Data 1999* is presented in metric measurements covering all construction areas. Don't miss out on these billion dollar opportunities. Make the switch to metric today.

$89.95 per copy
Over 650 pages, illustrated, available Nov. 1998
Catalog No. 63019

Annual Cost Guides

Means Mechanical Cost Data 1999

· HVAC · Controls

Total unit and systems price guidance for mechanical construction . . . materials, parts, fittings, and complete labor cost information. Includes prices for piping, heating, air conditioning, ventilation, and all related construction.

Plus new 1999 unit costs for:

· Over 3000 installed HVAC/controls assemblies
· "On Site" Location Factors for close to 1,000 cities and towns in the U.S. and Canada
· Crews, labor and equipment

$87.95 per copy
Over 600 pages, illustrated, available Oct. 1998
Catalog No. 60029

Means Plumbing Cost Data 1999

Comprehensive unit prices and assemblies for plumbing, irrigation systems, commercial and residential fire protection, point-of-use water heaters, and the latest approved materials. This publication and its companion, *Means Mechanical Cost Data*, provide full-range cost estimating coverage for all the mechanical trades.

$87.95 per copy
Over 500 pages, illustrated, available Oct. 1998
Catalog No. 60219

Means Electrical Cost Data 1999

Pricing information for every part of electrical cost planning: More than 17,000 unit and systems costs with design tables; clear specifications and drawings; engineering guides and illustrated estimating procedures; complete labor-hour and materials costs for better scheduling and procurement; the latest electrical products and construction methods.

· A Variety of Special Electrical Systems including Cathodic Protection
· Costs for maintenance, demolition, HVAC/ mechanical, specialties, equipment, and more

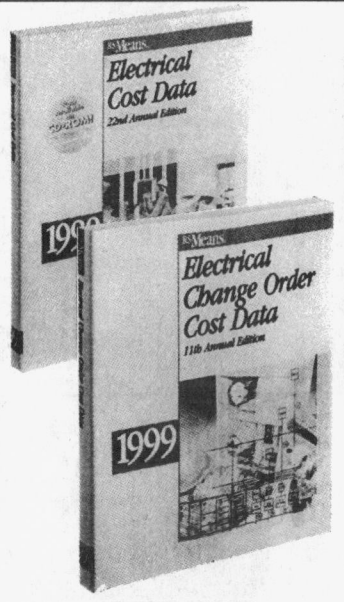

$87.95 per copy
Over 450 pages, illustrated, available Oct. 1998
Catalog No. 60039

Means Electrical Change Order Cost Data 1999

You are provided with electrical unit prices exclusively for pricing change orders—based on the recent, direct experience of contractors and suppliers. Analyze and check your own change order estimates against the experience others have had doing the same work. It also covers productivity analysis and change order cost justifications. With useful information for calculating the effects of change orders and dealing with their administration.

$89.95 per copy
Over 440 pages, available Oct. 1998
Catalog No. 60239

Means Facilities Maintenance & Repair Cost Data 1999

Published in a looseleaf format, *Means Facilities Maintenance & Repair Cost Data* gives you a complete system to manage and plan your facility repair and maintenance costs and budget efficiently. Guidelines for auditing a facility and developing an annual maintenance plan. Budgeting is included, along with reference tables on cost and management and information on frequency and productivity of maintenance operations.

The only nationally recognized source of maintenance and repair costs. Developed in cooperation with the Army Corps of Engineers.

$199.95 per copy
Over 600 pages, illustrated, available Dec. 1998
Catalog No. 60309

Means Square Foot Costs 1999

It's Accurate and Easy To Use!

· **Updated 1999 price information**, based on nationwide figures from suppliers, estimators, labor experts and contractors.

· "How-to-Use" Sections, with **better, clearer examples** of commercial, residential, industrial, and institutional structures.

· More realistic graphics, offering true-to-life illustrations of building projects.

· More extensive information on using square foot cost data, including **sample estimates** and **alternate pricing methods.**

$99.95 per copy
Over 460 pages, illustrated, available Nov. 1998
Catalog No. 60059

Annual Cost Guides

For more information
visit Means Web Site
at http://www.rsmeans.com

Means Repair & Remodeling Cost Data 1999

Commercial/Residential

You can use this valuable tool to estimate commercial and residential renovation and remodeling.

Includes: New 1999 costs for hundreds of unique methods, materials and conditions that only come up in repair and remodeling. PLUS:

- 1999 unit costs for over 15,000 construction components
- 1999 installed costs for over 4,000 assemblies
- 1999 costs for 300+ construction crews
- Over 930 "On Site" localization factors for the U.S. and Canada.

$79.95 per copy
Over 600 pages, illustrated, available Oct. 1998
Catalog No. 60049

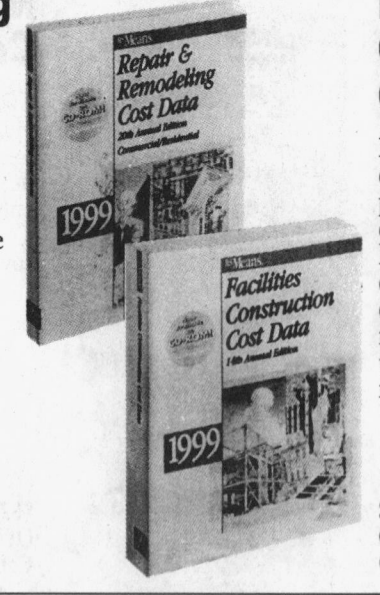

Means Facilities Construction Cost Data 1999

For the maintenance and construction of commercial, industrial, municipal, and institutional properties. Costs are shown for new and remodeling construction and are broken down into materials, labor, equipment, overhead, and profit. Special emphasis is given to sections on mechanical, electrical, furnishings, site work, building maintenance, finish work, and demolition. More than 40,000 unit costs plus assemblies and reference sections are included.

$209.95 per copy
Over 1100 pages, illustrated, available Nov. 1998
Catalog No. 60209

Means Residential Cost Data 1999

Now contains square foot costs for 30 basic home models with the look of today—plus hundreds of custom additions and modifications you can quote right off the page. With costs for the 100 residential systems you're most likely to use in the year ahead. Complete with blank estimating forms, sample estimates and step-by-step instructions.

$74.95 per copy
Over 550 pages, illustrated, available Dec. 1998
Catalog No. 60179

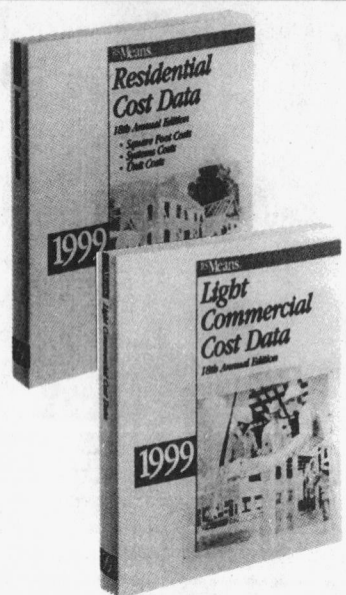

Means Light Commercial Cost Data 1999

Specifically addresses the light commercial market, which is an increasingly specialized niche in the industry. Aids you, the owner/designer/contractor, in preparing all types of estimates, from budgets to detailed bids. Includes new advances in methods and materials. Assemblies section allows you to evaluate alternatives in the early stages of design/planning.

Over 10,000 unit costs for 1999 ensure you have the prices you need... when you need them.

$76.95 per copy
Over 600 pages, illustrated, available Nov. 1998
Catalog No. 60189

Means Assemblies Cost Data 1999

Means Assemblies Cost Data 1999 takes the guesswork out of preliminary or conceptual estimates. Now you don't have to try to calculate the assembled cost by working up individual components costs. We've done all the work for you.

Presents detailed illustrations, descriptions, specifications and costs for every conceivable building assembly—240 types in all—arranged in the easy-to-use UniFormat system. Each illustrated "assembled" cost includes a complete grouping of materials and associated installation costs including the installing contractor's overhead and profit.

$139.95 per copy
Over 570 pages, illustrated, available Oct. 1998
Catalog No. 60069

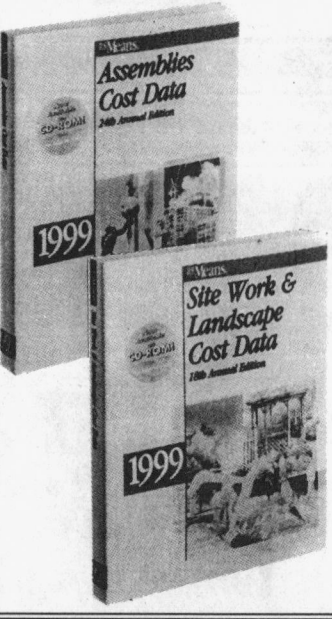

Means Site Work & Landscape Cost Data 1999

Means Site Work & Landscape Cost Data 1999 is organized to assist you in all your estimating needs. Hundreds of fact-filled pages help you make accurate cost estimates efficiently.

New for 1999!

- New or expanded demolition features— ceilings, doors, electrical, flooring, HVAC, millwork, plumbing, roofing, walls and windows
- State-of-the-art segmental retaining walls
- Flywheel trenching costs and details
- Expanded wells section
- Thousands of landscape materials, flowers, shrubs and trees

$89.95 per copy
Over 550 pages, illustrated, available Nov. 1998
Catalog No. 60289

Annual Cost Guides

Means Open Shop Building Construction Cost Data 1999

The latest costs for accurate budgeting and estimating of new commercial and residential construction... renovation work... change orders... cost engineering. *Means Open Shop BCCD* will assist you to...
· Develop benchmark prices for change orders
· Plug gaps in preliminary estimates, budgets
· Estimate complex projects
· Substantiate invoices on contracts
· Price ADA-related renovations

$89.95 per copy
Over 650 pages, illustrated, available Nov. 1998
Catalog No. 60159

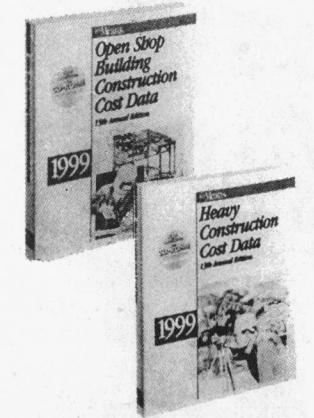

Means Heavy Construction Cost Data 1999

A comprehensive guide to heavy construction costs. Includes costs for highly specialized projects such as tunnels, dams, highways, airports, and waterways. Information on different labor rates, equipment, and material costs is included. Has unit price costs, systems costs, and numerous reference tables for costs and design. Valuable not only to contractors and civil engineers, but also to government agencies and city/town engineers.

$89.95 per copy
Over 450 pages, illustrated, available Dec. 1998
Catalog No. 60169

Means Building Construction Cost Data 1999

Western Edition

This regional edition provides more precise cost information for western North America. Labor rates are based on union rates from 13 western states and western Canada. Included are western practices and materials not found in our national edition: tilt-up concrete walls, glu-lam structural systems, specialized timber construction, seismic restraints, landscape and irrigation systems.

$89.95 per copy
Over 600 pages, illustrated, available Nov. 1998
Catalog No. 60229

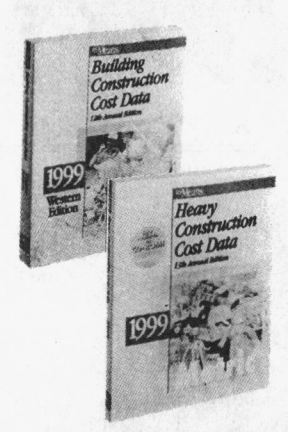

Means Heavy Construction Cost Data 1999

Metric Version

Make sure you have the Means industry standard metric costs for the federal, state, municipal and private marketplace. With thousands of up-to-date metric unit prices in tables by CSI standard divisions. Supplies you with assemblies costs using the metric standard for reliable cost projections in the design stage of your project. Helps you determine sizes, material amounts, and has tips for handling metric estimates.

$89.95 per copy
Over 450 pages, illustrated, available Dec. 1998
Catalog No. 63169

Means Construction Cost Indexes 1999

Who knows what 1999 holds? What materials and labor costs will change unexpectedly? By how much?
· Breakdowns for 305 major cities.
· National averages for 30 key cities.
· Expanded five major city indexes.
· Historical construction cost indexes.

$198.00 per year/$49.50 individual quarters
Catalog No. 60149

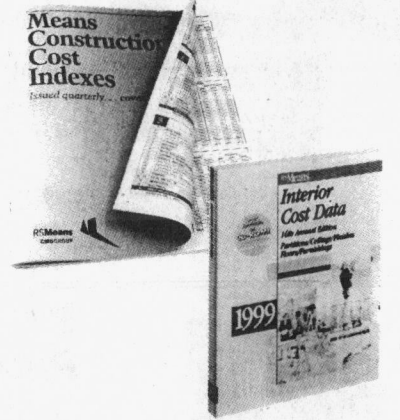

Means Interior Cost Data 1999

Provides you with prices and guidance needed to make accurate interior work estimates. Contains costs on materials, equipment, hardware, custom installations, furnishings, labor costs . . . every cost factor for new and remodel commercial and industrial interior construction, plus more than 50 reference tables. For contractors, facility managers, owners.

Newly expanded information on office furnishings.

$89.95 per copy
Over 550 pages, illustrated, available Nov. 1998
Catalog No. 60099

Means Concrete & Masonry Cost Data 1999

Provides you with cost facts for virtually all concrete/masonry estimating needs, from complicated form work to various sizes and face finishes of brick and block, all in great detail. The comprehensive unit cost section contains more than 15,000 selected entries. The assemblies cost section is illustrated with isometric drawings. A detailed reference section supplements the cost data.

$77.95 per copy
Over 520 pages, illustrated, available Dec. 1998
Catalog No. 60119

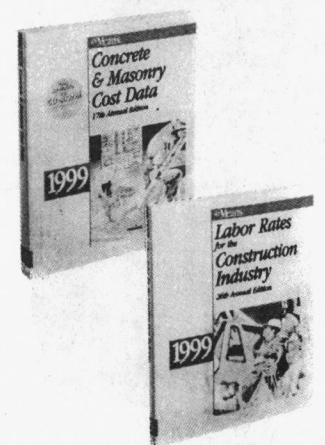

Means Labor Rates for the Construction Industry 1999

Complete information for estimating labor costs, making comparisons and negotiating wage rates by trade for over 300 cities (United States and Canada), 46 construction trades listed by local union number in each city, historical wage rates included for comparison. No similar book is available through the trade.

Each city chart now lists the county and is alphabetically arranged with handy visual flip tabs for quick reference.

$189.95 per copy
Over 330 pages, available Dec. 1998
Catalog No. 60129

Reference Books

For more information
visit Means Web Site
at http://www.rsmeans.com

Total Productive Facilities Management

By Richard W. Sievert, Jr.

A New Operational Standard for Facilities Management.

The TPFM program incorporates the best of today's cost and project management, quality, and value engineering principles. The book includes:

- Realistic ways to collect and use benchmarking data.
- Value Engineering to find economical answers to the organization's needs.
- Selecting the best scheduling, control, and contracting methods for construction projects.

$79.95 per copy
Over 270 pages, illustrated, Hardcover
Catalog No. 67321

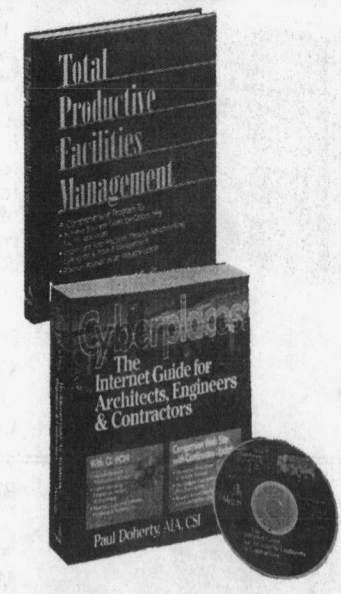

Cyberplaces: The Internet Guide for Architects, Engineers & Contractors

By Paul Doherty

Internet applications for business and project management. Includes book, CD-ROM and Web site.

The CD-ROM offers built-in links to Web sites, tours of captured sites, FREE browser software and a document workshop, plus a test for Continuing Education Credits. **The Web Site** keeps the book current with updates on new technologies, interactive workshops, links to new tools and sites, and reports from top firms.

$59.95 per copy
Over 700 pages, illustrated, Softcover
Catalog No. 67317

Value Engineering: Practical Applications

...For Design, Construction, Maintenance & Operations

By Alphonse Dell'Isola, P.E., leading authority on VE in construction

A tool for immediate application—for engineers, architects, facility managers, owners and contractors. Includes: Making the Case for VE—The Management Briefing, Integrating VE into Planning and Budgeting, Conducting Life Cycle Costing, Integrating VE into the Design Process, and Using VE Methodology in Design Review and Consultant Selection, Case Studies (corporate, commercial, hospital, industrial and civil), an expertly organized VE Workbook, and a Life Cycle cost program on disk.

$79.95 per copy
Over 450 pages, illustrated, Hardcover
Catalog No. 67319

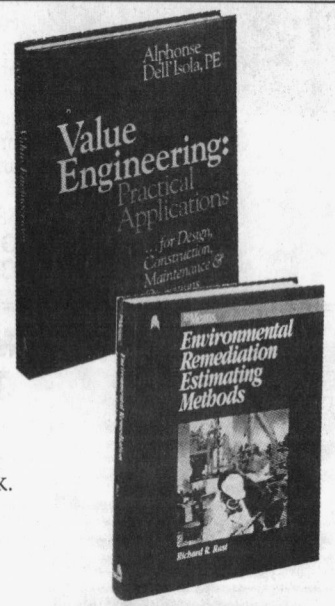

Means Environmental Remediation Estimating Methods

By Richard R. Rast

The first-ever guide to estimating any size environmental remediation project... anywhere in the country.

Field-tested guidelines for estimating 50 standard remediation technologies. This resource will help you: prepare preliminary budgets, develop detailed estimates, compare costs, select solutions, estimate liability, review quotes, and negotiate settlements.

A valuable support tool for *Environmental Restoration* cost data books.

$99.95 per copy
Over 600 pages, illustrated, Hardcover
Catalog No. 64777

The ADA in Practice

(Revised, expanded edition of the award-winning* New ADA: Compliance & Costs)

By Deborah S. Kearney, PhD

Helps you meet and budget for the requirements of the Americans with Disabilities Act. Shows you how to do the job right, by understanding what the law requires, allows, and enforces. Includes an objective "authoritative buyers guide" for 70 ADA-compliant products with specs and purchasing information. With sample evaluation forms and illustrations of ADA-compliant products.

*Winner of the "Distinguished Author of the Year" award from the International Facility Managers Association.

$72.95 per copy
Over 600 pages, illustrated, Softcover
Catalog No. 67147A

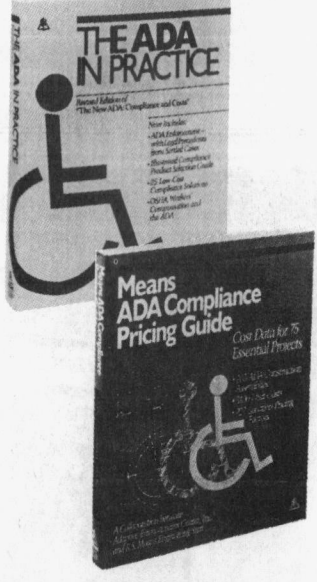

Means ADA Compliance Pricing Guide

Gives you detailed cost estimates for budgeting modification projects, including estimates for each of 260 alternates. You get the 75 most commonly needed modifications for ADA compliance, each an assembly estimate with detailed cost breakdown including materials, labor hours and contractor's overhead. 3,000 additional ADA compliance-related unit cost line items allow you to tailor the estimates to your own site conditions and requirements. Costs easily adjusted for 900 cities and towns.

$72.95 per copy
Over 350 pages, illustrated, Softcover
Catalog No. 67310

6

Cost Planning & Estimating for Facilities Maintenance

In this unique book, a team of facilities management authorities shares their expertise at:
· Evaluating and budgeting maintenance operations
· Maintaining & repairing key building components
· Applying *Means Facilities Maintenance & Repair Cost Data* to your estimating

With the special maintenance requirements of the 10 major building types.

$82.95 per copy
Over 475 pages, Hardcover
Catalog No. 67314

Facilities Planning & Relocation

By David D. Owen

An A–Z working guide—complete with the checklists, schematic diagrams and questionnaires to ensure the success of every office relocation. Complete with a step-by-step manual, 100-page technical reference section, over 50 reproducible forms in hard copy and on computer diskettes.
Winner of the International Facility Managers Assoc. "Distinguished Author of the Year" award.

$109.95 per copy, Textbook-384 pages,
Forms Binder-146 pages, illustrated, Hardcover
Catalog No. 67301

Means Facilities Maintenance Standards

Unique features of this one-of-a-kind working guide for facilities maintenance

A working encyclopedia that points the way to solutions to every kind of maintenance and repair dilemma. With a labor-hours section to provide productivity figures for over 180 maintenance tasks. Included are ready-to-use forms, checklists, worksheets and comparisons, as well as analysis of materials systems and remedies for deterioration and wear.

$159.95 per copy, 600 pages, 205 tables,
checklists and diagrams, Hardcover
Catalog No. 67246

HVAC: Design Criteria, Options, Selection
Expanded Second Edition

By William H. Rowe III, AIA, PE

Includes Indoor Air Quality, CFC Removal, Energy Efficient Systems and Special Systems by Building Type. Helps you solve a wide range of HVAC system design and selection problems effectively and economically. Gives you clear explanations of the latest ASHRAE standards.

$84.95 per copy
Over 600 pages, illustrated, Hardcover
Catalog No. 67306

The Facilities Manager's Reference

By Harvey H. Kaiser, PhD

The tasks and tools the facility manager needs to accomplish the organization's objectives, and develop individual and staff skills. Includes Facilities and Property Management, Administrative Control, Planning and Operations, Support Services, and a complete building audit with forms and instructions, widely used by facilities managers nationwide.

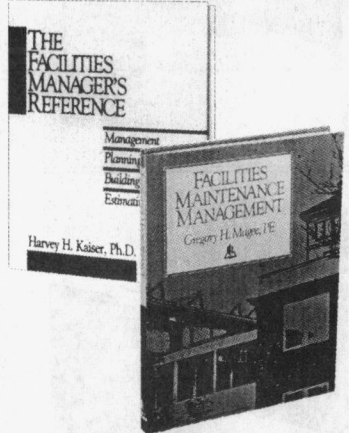

$86.95 per copy, over 250 pages,
with prototype forms and graphics, Hardcover
Catalog No. 67264

Facilities Maintenance Management

By Gregory H. Magee, PE

Now you can get successful management methods and techniques for all aspects of facilities maintenance. This comprehensive reference explains and demonstrates successful management techniques for all aspects of maintenance, repair and improvements for buildings, machinery, equipment and grounds. Plus, guidance for outsourcing and managing internal staffs.

$86.95 per copy
Over 280 pages with illustrations, Hardcover
Catalog No. 67249

Understanding Building Automation Systems

· Direct Digital Control · Energy Management
· Security/Access Control · Life Safety
   · Lighting

By Reinhold A. Carlson, PE & Robert Di Giandomenico

The authors, leading authorities on the design and installation of these systems, describe the major building systems in both an overview with estimating and selection criteria, and in system configuration-level detail.

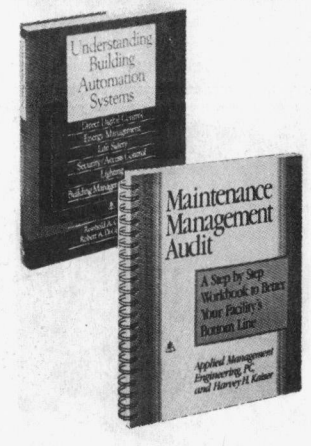

$79.95 per copy
Over 200 pages, illustrated, Hardcover
Catalog No. 67284

Maintenance Management Audit

At a time when many companies and institutions are reorganizing or downsizing, this annual audit program is essential for every organization in need of a proper assessment of its maintenance operation. The forms presented in this easy-to-use workbook allow managers to identify and correct problems, enhance productivity, and impact the bottom line. Includes a set of electronic forms on disk.

Now $32.48 per copy, limited quantity
125 pages, spiral bound, illustrated, Hardcover
Catalog No. 67299

Reference Books

Basics For Builders: Plan Reading & Material Takeoff

By Wayne J. DelPico

For Residential and Light Commercial Construction

A valuable tool for understanding plans and specs, and accurately calculating material quantities. Step-by-step instructions and takeoff procedures based on a full set of working drawings.

$35.95 per copy
Over 420 pages, Softcover
Catalog No. 67307

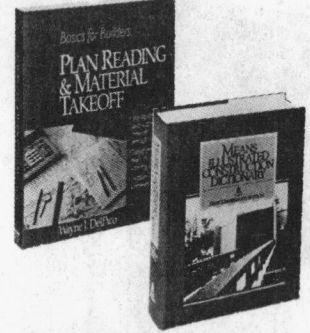

Means Illustrated Construction Dictionary Unabridged Edition

Written in contractor's language, the information is adaptable for report writing, specifications or just intelligent discussion. Contains over 17,000 construction terms, words, phrases, acronyms and abbreviations, slang, regional terminology, and hundreds of illustrations.

$99.95 per copy
Over 700 pages, Hardcover
Catalog No. 67292

Superintending for Contractors:

How to Bring Jobs in On-time, On-Budget
By Paul J. Cook

This book examines the complex role of the superintendent/field project manager, and provides guidelines for the efficient organization of this job. Includes administration of contracts, change orders, purchase orders, and more.

$35.95 per copy
Over 220 pages, illustrated, Softcover
Catalog No. 67233

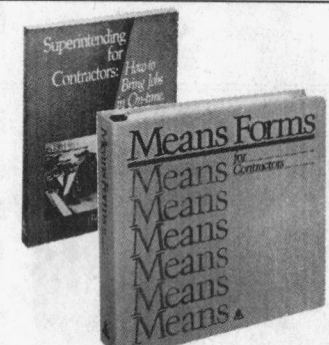

Means Forms for Contractors

For general and specialty contractors

Means editors have created and collected the most needed forms as requested by contractors of various-sized firms and specialties. This book covers all project phases. Includes sample correspondence and personnel administration.
Full-size forms on durable paper for photocopying or reprinting.

$79.95 per copy
Over 400 pages, three-ring binder, more than 80 forms
Catalog No. 67288

Bidding for Contractors:

How to Make Bids that Make Money
By Paul J. Cook

The author shares the benefit of his more than 30 years of experience in construction project management, providing contractors with the tools they need to develop competitive bids.

Now $17.98 per copy, limited quantity
Over 225 pages with graphics, Softcover
Catalog No. 67180

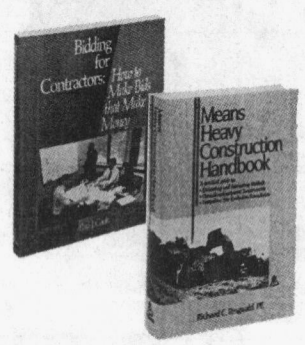

Means Heavy Construction Handbook

Informed guidance for planning, estimating and performing today's heavy construction projects. Provides expert advice on every aspect of heavy construction work including hazardous waste remediation and estimating. To assist planning, estimating, performing, or overseeing work.

$74.95 per copy
Over 430 pages, illustrated, Hardcover
Catalog No. 67148

Estimating for Contractors:

How to Make Estimates that Win Jobs
By Paul J. Cook

Estimating for Contractors is a reference that will be used over and over, whether to check a specific estimating procedure, or to take a complete course in estimating.

$35.95 per copy
Over 225 pages, illustrated, Softcover
Catalog No. 67160

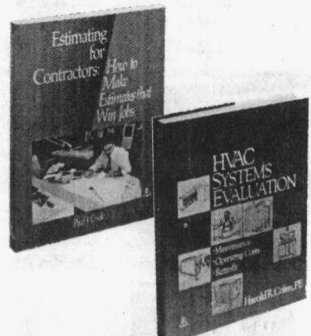

HVAC Systems Evaluation

By Harold R. Colen, PE

You get direct comparisons of how each type of system works, with the relative costs of installation, operation, maintenance and applications by type of building. With requirements for hooking up electrical power to HVAC components. Contains experienced advice for repairing operational problems in existing HVAC systems, ductwork, fans, cooling coils, and much more!

$84.95 per copy
Over 500 pages, illustrated, Hardcover
Catalog No. 67281

Quantity Takeoff for Contractors:

How to Get Accurate Material Counts
By Paul J. Cook

Contractors who are new to material takeoffs or want to be sure they are using the best techniques will find helpful information in this book, organized by CSI MasterFormat division.

Now $17.48 per copy, limited quantity
Over 250 pages, illustrated, Softcover
Catalog No. 67262

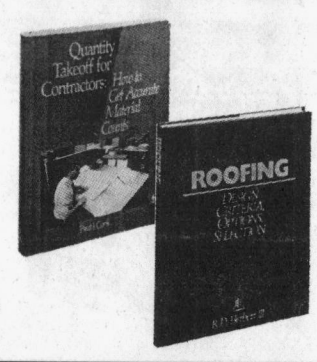

Roofing: Design Criteria, Options, Selection

By R.D. Herbert, III

This book is required reading for those who specify, install or have to maintain roofing systems. It covers all types of roofing technology and systems. You'll get the facts needed to intelligently evaluate and select both traditional and new roofing systems.

Now $31.48 per copy
Over 225 pages with illustrations, Hardcover
Catalog No. 67253

Reference Books

Means Estimating Handbook

This comprehensive reference covers a full spectrum of
technical data for estimating, with information on
sizing, productivity, equipment requirements, codes,
design standards and engineering factors.
Means Estimating Handbook will help you: evaluate
architectural plans and specifications, prepare accurate
quantity takeoffs, prepare estimates from conceptual to
detailed, and evaluate change orders.

$99.95 per copy
Over 900 pages, Hardcover
Catalog No. 67276

Means Repair and Remodeling Estimating Third Edition

By Edward B. Wetherill & R.S. Means

Focuses on the unique problems of estimating
renovations of existing structures. It helps you
determine the true costs of remodeling through careful
evaluation of architectural details and a site visit.
New section on disaster restoration costs.

$69.95 per copy
Over 450 pages, illustrated, Hardcover
Catalog No. 67265A

Successful Estimating Methods:
From Concept to Bid

By John D. Bledsoe, PhD, PE

A highly practical, all-in-one guide to the tips and
practices of today's successful estimator. Presents
techniques for all types of estimates, *and* advanced
topics such as life cycle cost analysis, value engineering,
and automated estimating.
Estimate spreadsheets available at Means Web site.

$64.95 per copy
Over 300 pages, illustrated, Hardcover
Catalog No. 67287

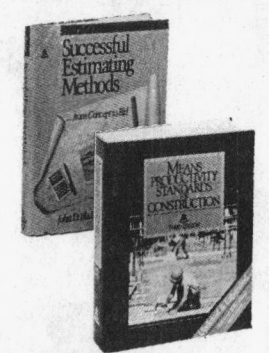

Means Productivity Standards for Construction

Expanded Edition *(Formerly* Man-Hour Standards)

Here is the working encyclopedia of labor productivity
information for construction professionals, with labor
requirements for thousands of construction functions in
CSI MasterFormat.
Completely updated, with over 3,000 new work items.

$159.95 per copy
Over 800 pages, Hardcover
Catalog No. 67236A

Means Electrical Estimating
Methods Second Edition

Expanded version includes sample estimates and cost
information in keeping with the latest version of the
CSI MasterFormat. Contains new coverage of Fiber
Optic and Uninterruptible Power Supply electrical
systems, broken down by components and explained
in detail. A practical companion to *Means Electrical
Cost Data.*

$62.95 per copy
Over 325 pages, Hardcover
Catalog No. 67230A

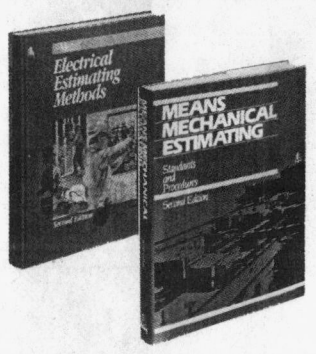

Means Mechanical Estimating
Second Edition

This guide assists you in making a review of plans,
specs and bid packages with suggestions for takeoff
procedures, listings, substitutions and pre-bid
scheduling. Includes suggestions for budgeting labor and
equipment usage. Compares materials and construction
methods to allow you to select the best options for your
job.

$64.95 per copy
Over 350 pages, illustrated, Hardcover
Catalog No. 67294

Means Scheduling Manual
Third Edition

By F. William Horsley

Fast, convenient expertise for keeping your scheduling
skills right in step with today's cost-conscious times.
Covers bar charts, PERT, precedence and CPM
scheduling methods. Now updated to include computer
applications.

$62.95 per copy
Over 200 pages, spiral-bound, Softcover
Catalog No. 67291

Means Graphic Construction Standards

Means Graphic Construction Standards bridges the gap
between design and actual construction methods.
With illustrations of unit assemblies, systems and
components, you can see quickly which construction
methods work best to meet design, budget and time
objectives.

$124.95 per copy
Over 540 pages, illustrated, Hardcover
Catalog No. 67210

Means Square Foot Estimating
Methods Second Edition

By Billy J. Cox and F. William Horsley

Proven techniques for conceptual and design-stage cost
planning. Steps you through the square foot cost
process, demonstrating faster, better ways to relate the
design to the budget. Now updated to the latest version
of UniFormat.

$69.95 per copy
Over 300 pages, illustrated, Hardcover
Catalog No. 67145A

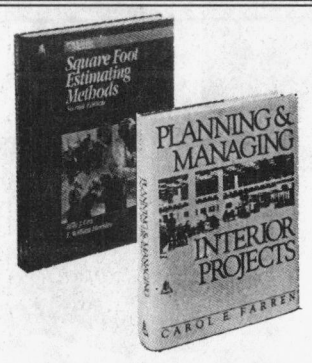

Planning and Managing Interior Projects

By Carol E. Farren

Expert, up-to-date guidance for managing interior
installation projects. Includes: project phases; winning
client support; space planning and design; budgeting,
bidding and purchasing, and more.
For interior designers, architects, facilities professionals.

$76.95 per copy
Over 330 pages, illustrated, Hardcover
Catalog No. 67245

Reference Books

For more information
visit Means Web Site
at http://www.rsmeans.com

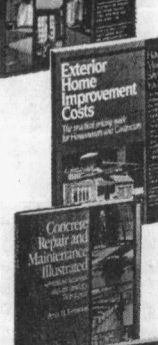

Unit Price Estimating Methods
2nd Edition
$59.95 per copy
Catalog No. 67303

Legal Reference for Design & Construction
By Charles R. Heuer, Esq., AIA
Now $54.98 per copy, limited quantity
Catalog No. 67266

Plumbing Estimating
By Joseph J. Galeno & Sheldon T. Greene
$59.95 per copy
Catalog No. 67283

Structural Steel Estimating
By S. Paul Bunea, PhD
$79.95 per copy
Catalog No. 67241

Business Management for Contractors
How to Make Profits in Today's Market
By Paul J. Cook
Now $17.98 per copy, limited quantity
Catalog No. 67250

Understanding Legal Aspects of Design/Build
By Timothy R. Twomey, Esq., AIA
$79.95 per copy
Catalog No. 67259

Construction Paperwork
An Efficient Management System
By J. Edward Grimes
$52.95 per copy
Catalog No. 67268

Contractor's Business Handbook
By Michael S. Milliner
Now $21.48 per copy, limited quantity
Catalog No. 67255

Basics for Builders: How to Survive and Prosper in Construction
By Thomas N. Frisby
$34.95 per copy
Catalog No. 67273

Successful Interior Projects Through Effective Contract Documents
By Joel Downey & Patricia K. Gilbert
Now $34.98 per copy
Catalog No. 67313

The Building Professional's Guide to Contract Documents
By Waller S. Poage, AIA, CSI, CCS
$64.95 per copy
Catalog No. 67261

Illustrated Construction Dictionary, Condensed
$59.95 per copy
Catalog No. 67282

Managing Construction Purchasing
By John G. McConville, CCC, CPE
Now $31.48 per copy, limited quantity
Catalog No. 67302

Hazardous Material & Hazardous Waste
By Francis J. Hopcroft, PE, David L. Vitale, M. Ed., & Donald L. Anglehart, Esq.
Now $44.98 per copy, limited quantity
Catalog No. 67258

Fundamentals of the Construction Process
By Kweku K. Bentil, AIC
Now $34.98 per copy, limited quantity
Catalog No. 67260

Construction Delays
By Theodore J. Trauner, Jr., PE, PP
$69.95 per copy
Catalog No. 67278

Basics for Builders: Framing & Rough Carpentry
By Scot Simpson
$24.95 per copy
Catalog No. 67298

Interior Home Improvement Costs   NEW!
6th Edition
$19.95 per copy
Catalog No. 67308B

Exterior Home Improvement Costs   NEW!
6th Edition
$19.95 per copy
Catalog No. 67309B

Concrete Repair and Maintenance Illustrated
By Peter H. Emmons
$64.95 per copy
Catalog No. 67146

How to Estimate with Metric Units
Now $24.98 per copy, limited quantity
Catalog No. 67304

Seminars

How to Develop Facility Assessment Programs

This two-day program concentrates on the management process required for planning, conducting, and documenting the physical condition and functional adequacy of buildings and other facilities. Regular facilities condition inspections are one of the facility department's most important duties. However, gathering reliable data hinges on the design of the overall program used to identify and gauge deferred maintenance requirements. Knowing where to look . . . and reporting results effectively are the keys. This seminar is designed to give the facility executive essential steps for conducting facilities inspection programs.

Inspection Program Requirements • Where is the deficiency? • What is the nature of the problem? • How can it be remedied? • How much will it cost in labor, equipment and materials? • When should it be accomplished? • Who is best suited to do the work?

Note: Because of its management focus, this course will not address trade practices and procedures.

Mechanical and Electrical Estimating

This seminar is tailored to fit the needs of those seeking to develop or improve their skills and to have a better understanding of how mechanical and electrical estimates are prepared during the conceptual, planning, budgeting and bidding stages. Learn how to avoid costly omissions and overlaps between these two interrelated specialties by preparing complete and thorough cost estimates for both trades. Featured are order of magnitude, assemblies, and unit price estimating. In combination with the use of **Means Mechanical Cost Data**, **Means Plumbing Cost Data** and **Means Electrical Cost Data**, this seminar will ensure more accurate and complete Mechanical/Electrical estimates for both unit price and preliminary estimating procedures.

Square Foot Cost Estimating

Learn how to make better preliminary estimates with a limited amount of budget and design information. You will benefit from examples of a wide range of systems estimates with specifications limited to building use requirements, budget, building codes, and type of building. And yet, with minimal information, you will obtain a remarkable degree of accuracy.

Workshop sessions will provide you with model square foot estimating problems and other skill-building exercises. The exclusive Means building assemblies square foot cost approach shows how to make very reliable estimates using "bare bones" budget and design information.

Facilities Maintenance and Repair Estimating

With our Facilities Maintenance and Repair Estimating seminar, you'll learn how to plan, budget, and estimate the cost of ongoing and preventive maintenance and repair for all your buildings and grounds. Based on R.S. Means' groundbreaking cost estimating book, this two-day seminar will show you how to decide to either contract out or retain maintenance and repair work in-house. In addition, you'll learn how to prepare budgets and schedules that help cut down on unplanned and costly emergency repair projects. Facilities Maintenance and Repair Estimating crystallizes what facilities professionals have learned over the years, but never had time to organize or document. This program covers a variety of maintenance and repair projects, from underground storage tank removal, roof repair and maintenance, exterior wall renovations, and energy source conversions, to service upgrades and estimating energy-saving alternatives.

Repair and Remodeling Estimating

Repair and remodeling work is becoming increasingly competitive as more professionals enter the market. Recycling existing buildings can pose difficult estimating problems. Labor costs, energy use concerns, building codes, and the limitations of working with an existing structure place enormous importance on the development of accurate estimates. Using the exclusive techniques associated with Means' widely acclaimed **Repair & Remodeling Cost Data**, this seminar sorts out and discusses solutions to the problems of building alteration estimating. Attendees will receive two intensive days of eye-opening methods for handling virtually every kind of repair and remodeling situation . . . from demolition and removal to final restoration.

Unit Price Estimating

This seminar shows how today's advanced estimating techniques and cost information sources can be used to develop more reliable unit price estimates for projects of any size. It demonstrates how to organize data, use plans efficiently, and avoid embarrassing errors by using better methods of checking.

You'll get down-to-earth help and easy-to-apply guidance for:
• making maximum use of construction cost information sources
• organizing estimating procedures in order to save time and reduce mistakes
• sorting out and identifying unusual job requirements to improve estimating accuracy.

Scheduling and Project Management

This seminar helps you successfully establish project priorities, develop realistic schedules, and apply today's advanced management techniques to your construction projects. Hands-on exercises familiarize participants with network approaches such as the Critical Path Method. Special emphasis is placed on cost control, including use of computer-based systems. Through this seminar you'll perfect your scheduling and management skills, ensuring completion of your projects *on time* and *within budget*. Includes hands-on application of **Means Scheduling Manual** and **Means Building Construction Cost Data.**

Managing Facilities Construction and Maintenance

In you're involved in new facility construction, renovation or maintenance projects and are concerned about getting quality work done on time and on or below budget, in Means' seminar **Managing Facilities Construction and Maintenance** you'll learn management techniques needed to effectively plan, organize, control and get the most out of your limited facilities resources.

Learn how to develop budgets, reduce expenditures and check productivity. With the knowledge gained in this course, you'll be better prepared to successfully sell accurate project budgets, timing and manpower needs to senior management . . . plus understand how to evaluate the impact of today's facility decisions on tomorrow's budget.

Call 1-800-448-8182 for more information

Seminars

For more information
visit Means Web Site
at http://www.rsmeans.com

1999 Means Seminar Schedule

Location	Dates
Las Vegas, NV	March 15 - 18
Washington, DC	April 12 - 15
New Orleans, LA	April TBA
Denver, CO	May 17 - 20
San Francisco, CA	June 14 - 17
Hyannis, MA	September TBA
Washington, DC	September TBA
San Diego, CA	October TBA
Orlando, FL	November TBA

Registration Information

Register Early... Save up to $150! Register 30 days before the start date of a seminar and save $150 off your total fee. *Note: This discount can be applied only once per order. It cannot be applied to team discount registrations or any other special offer.*

How to Register Register by phone today! Means toll-free number for making reservations is: **1-800-448-8182, ext. 701.**

Individual Seminar Registration Fee $875 To register by mail, complete the registration form and return with your full fee to: Seminar Division, R.S. Means Company, Inc., 63 Smiths Lane, Kingston, MA 02364.

Federal Government Pricing All Federal Government employees save 25% off regular seminar price. Other promotional discounts cannot be combined with Federal Government discount.

Team Discount Program Two to four seminar registrations: $760 per person—Five or more seminar registrations: $710 per person—Ten or more seminar registrations: Call for pricing.

Consecutive Seminar Offer One individual signing up for two separate courses at the same location during the designated time period pays only $1,400. You get the second course for only $525 (**a 40% discount**). Payment must be received at least ten days prior to seminar dates to confirm attendance.

Refunds Cancellations will be accepted up to ten days prior to the seminar start. There are no refunds for cancellations postmarked later than ten working days prior to the first day of the seminar. A $150 processing fee will be charged for all cancellations. Written notice or telegram is required for all cancellations. Substitutions can be made at any time before the session starts. **No-shows are subject to the full seminar fee.**

AACE Approved Courses The R.S. Means Construction Estimating and Management Seminars described and offered to you here have each been approved for 14 hours (1.4 recertification credits) of credit by the AACE International Certification Board toward meeting the continuing education requirements for re-certification as a Certified Cost Engineer/Certified Cost Consultant.

AIA Continuing Education R.S. Means is registered with the AIA Continuing Education System (AIA/CES) and is committed to developing quality learning activities in accordance with the CES criteria. R.S. Means seminars meet the AIA/CES criteria for Quality Level 2. AIA members will receive (28) learning units (LUs) for each two day R.S. Means Course.

Daily Course Schedule The first day of each seminar session begins at 8:30 A.M. and ends at 4:30 P.M. The second day is 8:00 A.M.–4:00 P.M. Participants are urged to bring a hand-held calculator since many actual problems will be worked out in each session.

Continental Breakfast Your registration includes the cost of a continental breakfast, a morning coffee break, and an afternoon cola break. These informal segments will allow you to discuss topics of mutual interest with other members of the seminar. (You are free to make your own lunch and dinner arrangements.)

Hotel/Transportation Arrangements R.S. Means has arranged to hold a block of rooms at each hotel hosting a seminar. To take advantage of special group rates when making your reservation be sure to mention that you are attending the Means Seminar. You are of course free to stay at the lodging place of your choice. (**Hotel reservations and transportation arrangements should be made directly by seminar attendees.**)

Important Class sizes are limited, so please register as soon as possible.

Registration Form

Call 1-800-448-8182, ext. 701 to register or FAX 1-781-585-7466

Please register the following people for the Means Construction Seminars as shown here. Full payment or deposit is enclosed, and we understand that we must make our own hotel reservations if overnight stays are necessary.

☐ Full payment of $ _____ enclosed.

Name of Registrant(s)
(To appear on certificate of completion)

_____

_____

P.O. #: _____
GOVERNMENT AGENCIES MUST SUPPLY PURCHASE ORDER NUMBER

Firm Name _____

Address _____

City/State/Zip _____

Telephone No. _____ Fax No. _____

E-Mail Address _____

Charge our registration(s) to: ☐ MasterCard ☐ VISA ☐ American Express ☐ Discover

Account No. _____ Exp. Date _____

Cardholder's Signature _____

Seminar Name	City	Dates

Please mail check to: R.S. MEANS COMPANY, INC., 63 Smiths Lane, P.O. Box 800, Kingston, MA 02364 USA

12

Consulting Services Group

Proven Solutions for Managing the Costs of Construction and Facility Operations

Business and government leaders go to great lengths today to control costs and increase return on their construction-related activities. But they seldom have an opportunity to achieve optimum success. No matter what the activity . . . planning and design, new construction, facilities management, or introducing new building technologies . . . it comes down to the same problem: To control costs one must accurately predict them.

R.S. Means Consulting Services Group has a proven record of success helping businesses meet this challenge while greatly increasing opportunity to maximize return on construction investments. With its extensive, highly specialized experience understanding both construction costs and relational database technology, Means Consulting Services Group offers an unparalleled opportunity for businesses to realize dramatic improvement in their construction/facilities cost control programs.

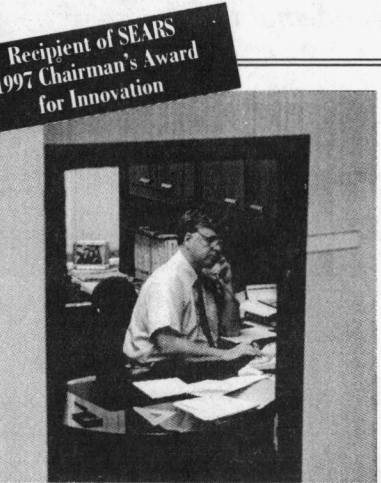

Recipient of SEARS 1997 Chairman's Award for Innovation

Research and Custom Database Development

- **Custom Database Development**—Means expertise in construction cost engineering and database management can be put to work creating customized cost databases and applications.
- **Data Licensing & Integration**—To enhance applications dealing with construction, any segment of Means vast database can be licensed for use, and harnessed in a format compatible with a previously developed proprietary system.
- **Cost Modeling**—Pre-built custom cost models provide organizations that expend countless hours estimating repetitive work with a systematic time-saving estimating solution.
- **Database Auditing & Maintenance**—For clients with in-house data, Means can help organize it, and by linking it with Means database, fill in any gaps that exist and provide necessary updates to maintain current and relevant proprietary cost data.

Estimating and Cost Analysis

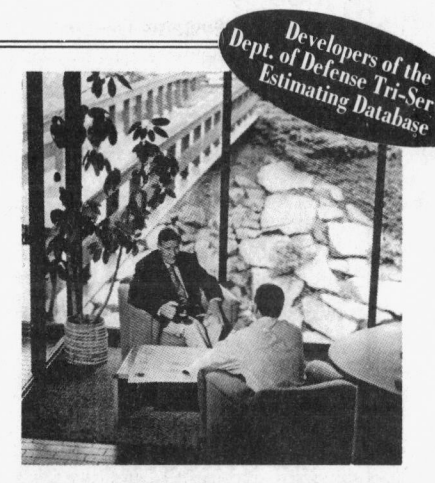

Developers of the Dept. of Defense Tri-Services Estimating Database

- **Estimating Service**—Means expertise is available to perform construction cost estimates, as well as to develop baseline schedules and establish management plans for projects of all sizes and types. Conceptual, budget and detailed estimates are available.
- **Benchmarking**—Means can run baseline estimates on existing project estimates. Gauging estimating accuracy and identifying inefficiencies improves the success ratio, precision and productivity of estimates.
- **Project Feasibility Studies**—The Consulting Services Group can assist in the review and clarification of the most sound and practical construction approach in terms of time, cost and use.
- **Litigation Support**—Means is available to provide opinions of value and to supply expert interpretations or testimony in the resolution of construction cost claims, litigation and mediation.

Staff Assessments, Training and Development

- **Core Curriculum**—Means educational programs, delivered on-site, are designed to sharpen professional skills and to maximize effective use of cost estimating and management tools. On-site training cuts down on travel expenses and time away from the office. The broad curriculum covers such topics as repair, new construction, and conceptual estimating; scheduling and project management; facilities management; metrication; and delivery order contracting (DOC) methods.
- **Custom Curriculum**—Means can custom-tailor courses to meet the specific needs and requirements of clients. The goal is to simultaneously boost skills and broaden cost estimating and management knowledge while focusing on applications that bring immediate benefits to unique operations, challenges, or markets.
- **Staff Assessments and Development Programs**—In addition to custom curricula, Means can work with a client's Human Resources Department or with individual operating units to create programs consistent with long-term employee development objectives.

For more information and a copy of our capabilities brochure, please call 1-800-448-8182 and ask for the Consulting Services Group, or reach us at http://www.rsmeans.com

MeansData™

CONSTRUCTION COSTS FOR SOFTWARE APPLICATIONS
Your construction estimating software is only as good as your cost data.

Software Integration

A proven construction cost database is a mandatory part of any estimating package. We have linked MeansData™ directly into the industry's leading software applications. The following list of software providers can offer you MeansData™ as an added feature for their estimating systems. Visit them on-line at *http://www.rsmeans.com/demo/* for more information and free demos. Or call their numbers listed below.

ACT
Applied Computer Technologies
Facility Management Software
1-910-897-1418

ASSETWORKS, Inc.
Facility Management Software
1-800-659-9001

BSD
Building Systems Design
1-800-875-0047

CDCI
Construction Data Controls, Inc.
1-800-285-3929

CMS
Computerized Micro Solutions
1-800-255-7407

CONAC GROUP
1-604-273-3463

CONSTRUCTIVE COMPUTING, Inc.
1-800-456-2113

ESTIMATING SYSTEMS, Inc.
1-800-967-8572

G2, Inc.
1-800-657-6312

GEAC COMMERCIAL SYSTEMS, Inc.
1-800-851-1115

GRANTLUN CORPORATION
1-602-897-7750

IQ BENECO
1-801-565-1122

MC²
Management Computer Controls
1-800-225-5622

PRISM COMPUTER CORPORATION
Facility Management Software
1-800-774-7622

PYXIS TECHNOLOGIES
1-914-276-3369

SANDERS SOFTWARE, Inc.
1-404-934-8423

STN, Inc.
Workline Maintenance Systems
1-800-321-1969

TIMBERLINE SOFTWARE CORP.
1-800-628-6583

TMA SYSTEMS, Inc.
Facility Management Software
1-918-494-2890

US COST, Inc.
1-800-955-1385

VERTIGRAPH, Inc.
1-800-989-4243

WINESTIMATOR, Inc.
1-800-950-2374

DemoSource™
One-stop shopping for the latest cost estimating software for just $19.95. This evaluation tool includes product literature and demo diskettes for ten or more estimating systems, all of which link to MeansData™. **Call 1-800-334-3509 to order.**

FOR MORE INFORMATION ON ELECTRONIC PRODUCTS CALL
1-800-448-8182 OR FAX 1-800-632-6732.

MeansData™ is a registered trademark of R.S. Means Co., Inc., *A Construction Market Data Group* Company.

14